Principles of
BIOCHEMISTRY

Principles of

BIOCHEMISTRY

FOURTH EDITION

ABRAHAM WHITE, Ph.D. *Dan Danciger Professor of Biochemistry*
Albert Einstein College of Medicine
Yeshiva University

PHILIP HANDLER, Ph.D. *James B. Duke Professor of Biochemistry*
School of Medicine
Duke University

EMIL L. SMITH, Ph.D. *Professor of Biological Chemistry*
UCLA School of Medicine
University of California, Los Angeles

The Blakiston Division
McGRAW-HILL BOOK COMPANY
New York Sydney Toronto London

PRINCIPLES OF BIOCHEMISTRY

Library of Congress Catalog Card Number: 67-26890

69756

1234567890–HDBP–743210698

TO OUR WIVES

Preface

Biochemistry continues to advance at an ever-increasing rate, and the pace and volume of its literature necessitated not only revision of all chapters, but a significant rewriting of many chapters and some degree of recasting of the organization of the previous edition of "Principles of Biochemistry."

In preparing this new edition we have again sought to provide a text for medical students, a useful volume for undergraduate and graduate students of biochemistry, and a review for graduate practicing physicians who wish to remain au courant with developments in biochemistry. As in previous editions, some of the subject matter is treated more extensively than may be necessary for the student in an initial course in Biochemistry. A continuing attempt has been made, in the preparation of this edition, to present separately basic, major principles of a particular subject area and the more detailed and reference material. This has been achieved by separation of large subject areas into several chapters. Moreover, we have again so organized the content of certain larger areas of biochemical knowledge as to permit ready selection by the teacher, as well as by the student, of that information deemed most pertinent to the needs of medical, graduate, or undergraduate students. For example, Chapters 15 and 16 may suffice as an introduction to biological oxidations, or Chapter 23 for an understanding of nitrogen and amino acid metabolism. In the same vein, Chapter 20 on photosynthesis may interest only some medical students, but will be of concern to other students.

Many areas of biochemistry have now achieved scientific maturity in the sense that relatively complete descriptive information appears to be available of the phenomena involved, as, for example, the metabolic pathways of amino acids. This has resulted in a lesser requirement to describe earlier experiments that support currently accepted conclusions. On the other hand, many areas of biochemistry continue to develop at a rapid rate and biochemical theory and practice are being extended into new fields. This new edition has attempted to take cognizance of all these factors.

A totally new introductory chapter, entitled "The Scope of Biochemistry," enumerates and summarizes the major problems to which biochemistry has and currently is addressing itself. The rapidly developing knowledge of lipid chemistry and metabolism has necessitated a complete rewriting of the chapters devoted to those subjects, including a new classification of lipids. The fine and topographical structure of proteins and nucleic acids has been given more extended treatment in the revision of chapters dealing with protein and nucleic acid structure, with particular emphasis on sequence analysis and conformational properties of these macromolecules. This information has been utilized in the more precise description now possible of the mechanism of action of enzymes. More detailed understanding of enzymic reaction mechanisms has led, in turn, to more frequent utilization of modern organic chemical notation in the treatment of these problems.

Throughout this new edition, major emphasis has been placed on one of the most rewarding areas of expanding biochemical sophistication, namely, control and regulatory mechanisms, with particular emphasis on factors that modulate the activity of enzymes. This will be evident in the chapters on the mechanism of enzymic action, the introduction to metabolism, and throughout all chapters concerned with metabolism and endocrine function.

Numerous significant contributions of biochemistry to understanding of genetics at the level of the genome necessitated extensive revisions of the three chapters concerned with the genetic aspects of metabolism. Particular emphasis has been placed upon the relation of the sequence of DNA to protein sequence and structure. Knowledge of the "genetic" code, now practically complete, is presented as are factors involved in the control of protein synthesis and variations in protein structure, including their significance for evolution.

A totally new chapter on blood has been prepared, encompassing several previous chapters dealing with constituents and functions of this body fluid. Included is a considerably expanded treatment of the growing knowledge of immunoglobulins, with emphasis on their structure and the mechanism and control of their synthesis.

Progress in knowledge of photosynthesis justifies a separate chapter devoted to this subject area. Similarly, spectacular advances in knowledge of the biochemistry of cell walls has resulted in a new chapter entitled "Cell Walls of Plants and Microorganisms." The information in this chapter is of significance for medicine because of its relationship to the mechanism of action of certain antibiotics while it is also of great interest to students concerned with problems of comparative biochemistry.

The method used in the preparation of this edition was the same as that for its predecessors. Preparation of the first draft of each of the chapters was assigned to one author. Each first draft was then subjected to the detailed suggestions and criticisms of each of the other authors who also rewrote paragraphs, sections, or, on occasion, even the whole chapter. In addition, the authors convened at intervals, for meetings of several days' duration. These meetings afforded opportunities for criticism (sometimes painful), suggestions, deletions (always welcome), additions (always resisted), and appraisal of the extent to which this edition is in keeping with our general goals. In effect, therefore, each chapter was written by all three authors, who, accordingly, jointly share responsibility for the entire work.

Grateful acknowledgement is made to the numerous colleagues and friends who contributed information and criticism to the preparation of this new edition. Particular thanks are due the teachers and students who have used the book and, on the basis of their experiences, have provided valuable suggestions for its improvement, as well as directing our attention to occasional errors!

Abraham White
Philip Handler
Emil L. Smith

Contents

Part Three **Metabolism**

Part Four **Body Fluids and Specialized Tissues**

Part Five **Biochemistry of the Endocrine Glands**

1. The Scope of Biochemistry

In the early dawn of language, the word "life" was employed to characterize the condition of objects as diverse as grass, trees, insects, worms, birds, fish, and man. Each undergoes a life cycle, reproduces its own kind, and responds in a variety of ways to external stimuli. Over the course of a few millennia, "living" forms have been classified, at first in terms of characteristics visible to the unaided eye, *i.e.*, their gross comparative anatomy, and later with the aid of the light microscope. In the early nineteenth century, Schleiden and Schwann recognized that all these forms were constructed of unit cells of rather similar dimensions and general appearance. This relatively primitive body of information, together with increased understanding of the fossil record, sufficed to permit formulation by Darwin of the most sweeping and compelling biological generalization of all, the concept of historic and continuing biological evolution.

Progress in the physical sciences, in turn, led to increasingly sophisticated questioning by students of biology. Identification of the major atmospheric gases was soon followed by demonstration of the use of oxygen and production of carbon dioxide by animals and the photosynthetic reversal of this relationship in green plants. The general statements of the laws of conservation of energy and matter applicable to the physical world were shown by Lavoisier and Laplace in 1785 to be equally valid in the biological system which they could examine experimentally. Isolation of increasing numbers of purified materials from living forms and recognition of the fact that all contained carbon gave birth to "organic" chemistry. This actually fortified vitalistic thinking until Wöhler synthesized urea in 1828, thereby demonstrating that carbon compounds need not necessarily be formed in living organisms. Formulation of the general principles of catalysis by Berzelius rapidly led to recognition that the ptyalin of saliva, pepsin of gastric juice, and amylase of sprouted malt were biological catalysts. Yeast was believed to be an inert catalyst at that time, and hence studies of the chemistry of fermentation contributed to the decline of vitalistic thinking. Indeed, it is ironic that chemical synthesis of ethanol, by Hennell, had preceded the synthesis of urea, but did not serve as an equivalent philosophical milestone because the living nature of yeast failed to gain acceptance until the work of Pasteur.

Until the major laws of physics and chemistry governing the inanimate universe had been elucidated, it was not possible to formulate the important, penetrating questions concerning the nature of life. These questions, which we shall consider shortly, were not given overt expression until the first quarter of the twentieth century. Meanwhile, inorganic, organic, and physical chemistry flourished, the laws of thermodynamics were enunciated by Willard Gibbs and others, and it became possible to contemplate whether living systems also obey the laws of physics and chemistry. The doctrine of evolution gained acceptance, the principles of genetic inheritance were formulated by Gregor Mendel, and the list of compounds obtained from living organisms grew ever larger. The conducting role of the nervous system was described and the role of glycogen as a storage form of glucose in liver and muscle was demonstrated by Claude Bernard, who also recognized the constancy of the *milieu interieur* (Chap. 33). The germ theory of disease was established and systematic microbiology introduced.

At the turn of the century, Emil Fischer established the structures of many carbohydrates, learned to separate amino acids from hydrolysates of proteins, and initiated much of contemporary biochemical thought by recognizing the optical configuration of carbohydrates and amino acids and by demonstrating the specificity of enzymic action. In postulating the "lock-and-key" concept of enzymic action (Chap. 12), Fischer began the study of the relation of the topography of macromolecules to the phenomena of life. With these studies and the exploitation by Harden and Young (Chap. 18) of the accidental observation by the Buchner brothers that a cell-free extract of yeast could ferment glucose, with the production of alcohol, modern biochemistry began. The term "biochemistry" was introduced by Carl Neuberg in 1903.

In the next half century the pace of biochemical research quickened; since then information and understanding have been increasing exponentially. Intellectual curiosity and philosophical questions have largely shaped the general course of these biochemical investigations. In large measure, however, the quickening tempo of this effort reflects not so much a fundamental human drive for self-understanding as the belief that the knowledge gained will improve agricultural practice and with it animal and human nutrition, as well as assist in the alleviation of human disease. In considerable measure, these goals have been realized.

Research in biochemistry has been addressed to a series of major questions, each of which continues to command attention. These will be briefly considered below.

1. *Of what chemical compounds are living things composed?* Relatively few biochemists, today, devote themselves primarily to such efforts. Obviously, a catalogue of such compounds is a *sine qua non* for the understanding of life in chemical terms. New compounds are, however, continually being recognized, usually in the course of investigations directed toward unraveling metabolic reaction sequences which commence and terminate with well-recognized chemical entities. These compounds will be encountered throughout this book. The ubiquitous distribution of these compounds results in a degree of similarity in the *qualitative* composition of most living organisms, with differences among these forms, as well as among their own tissues and organs, being primarily of a *quantitative* nature. These quantitative differences are paralleled generally by differences in functions and relative rates at which similar functions, processes, or reactions may proceed.

2. *What are the structures of the macromolecules characteristic of living organisms?* The pioneers of biochemistry recognized the presence in nature of substances which they named proteins, nucleic acids, polysaccharides, complex lipids. Procedures were later developed to isolate and purify such materials. New physicochemi-

cal methods revealed molecular weights of from 10,000 to more than 100,000,000 for individual substances. For years, the seemingly herculean task of establishing the complete structures of such molecules appeared to be experimentally unapproachable. However, improved analytical techniques, degradative procedures which take advantage of the known specificities of hydrolytic enzymes, a variety of new physical instruments including the ultracentrifuge, electrophoresis apparatus, recording spectrophotometers and spectropolarimeters, and x-ray crystallographic analysis have revealed the general structures of these molecular species. Detailed three-dimensional models of a few smaller proteins and nucleic acids are available, and there is a rapidly growing understanding of the forces which maintain these molecules, which are primarily long, thin fiber-like structures, folded upon themselves into highly specific compact structures. The biological functioning of these molecules is entirely dependent upon these three-dimensional structures.

Understanding of the structures of these large molecules is rapidly expanding, thereby providing the basis for a more penetrating insight into the mode of operation of enzymes, the structural basis for genetic phenomena, and the fine structure of living cells. Indeed, this is the major theme of this book.

3. *How do enzymes accomplish their catalytic tasks?* In the nineteenth century, degradation of proteins, starch, and fats to their smaller constituents in the digestive tract was recognized as being due to enzymic activity. That fermentation is also the result of such catalysis was shown by the Buchners. Twenty years earlier, Kühne had coined the name "enzyme" (Gk., in yeast) to designate the unorganized "ferments," as distinguished from bacteria, which were also called ferments. The studies of Fischer on the specificity of enzymes were followed by the formulation by Michaelis and others of the elementary rules describing enzymic catalysis and by Sumner's isolation in 1926 of a crystalline enzyme, urease, with the demonstration that it is a protein. Since then, hundreds of enzymes, each more or less specific for one chemical reaction, have been isolated in pure form; each has proved to be a distinct, unique protein, and many of them have been crystallized.

How these proteins function as catalysts is one of the central problems in biochemistry—and one of the oldest. The question was first raised in the year 1800, when the Academy of the First French Republic offered a prize of one kilo of gold for a satisfactory answer to the question, What is the difference between "ferments" and the materials which they are fermenting? The prize was never awarded. In all probability, those who posed this question would have been pleased to award their prize a century later to Emil Fischer, who, however, was aware of the superficiality of his understanding. In the time since, the phenomena operative in enzymic catalysis and their bases in protein structure have been revealed in considerable detail. This fascinating aspect of science, in many respects the heart of biochemistry, is the subject of Part Two of this book.

4. *What substances are required to satisfy the nutritional requirements of man and other organisms?* The small catalogue of these substances, now perhaps complete for man and other mammals, is presented in Part Six. This knowledge is adequate to manage the nutritional affairs of mankind; the inadequate nutrition of half of humanity reflects failures of production and inequities of distribution, not lack of necessary fundamental information.

In the course of studies of the nutrition of bacteria, powerful experimental tools were forged which have influenced all aspects of biochemical research. The ability to estimate bacterial growth quantitatively has been utilized as the basis of sensitive analytical procedures. The fact that a given compound is an essential nutrient because the organism cannot accomplish its synthesis, yet requires that compound for further metabolic transformations, has been exploited in the elucidation both of metabolic pathways and of genetic mechanisms.

5. *By what chemical processes are the materials of the diet transformed into the compounds characteristic of the cells of a given species?* Study of the manifold individual events of metabolism was at the center of biochemical interest until comparatively recently and continues to warrant attention. This is the subject of Part Three of this book. The nature of this research endeavor is readily apparent. Rather large quantities of a small group of organic compounds are ingested daily; concomitantly, CO_2 and H_2O are excreted in the expired air, while urea and a few other compounds appear in the excreted urine. In the growing child, a large collection of compounds is accumulated which are rather different in composition from that of the ingested mixture. It is also evident that, despite the through passage of about a pound of mixed solids per day, an adult remains constant in weight and body composition. Since, in attempting to understand these processes, one cannot readily sample the reaction mixture, how, then, can one ascertain the reactions to which ingested foodstuffs are subjected? This problem was even more vexing for plants since they "ingest" only water, minerals, and CO_2 while growing as complete plants. This experimental impasse was broken by the availability of radioactive isotopes, particularly ^{14}C, and the apparatus with which to measure their abundance (Chap. 13). When this tool was added to increased skill in the handling of tissue preparations in vitro and to the powerful separatory capabilities of column and paper chromatography (Chaps. 2 and 5), a large, intricate network of metabolic sequences and pathways was quickly exposed. This task continues, although the outlines of the major processes appear to have been revealed.

6. *How is the potential energy available from the oxidation of foodstuffs utilized to drive the manifold energy-requiring processes of the living cell?* Among such processes, we need note only the synthesis of hundreds of new molecular species, accumulation of mineral ions and organic compounds against concentration gradients, and the performance of mechanical work. The impossibility of utilizing thermal energy to accomplish useful work at constant temperature makes untenable a simple analogy between food-burning animals and fuel-burning heat engines. Understanding of the solution to this problem, coupling of the oxidation of carbohydrates and fats to the synthesis of one compound, adenosine triphosphate (Chaps. 15 to 17), with subsequent utilization of the energy of this compound for virtually all endergonic processes, is cardinal to the understanding of living cells.

The corollary problem, elucidation of the mechanism by which light energy is harnessed to achieve fixation of CO_2 into carbohydrate, has been a challenging major question in its own right (Chap. 20). Understanding of the primary photochemical events and the subsequent reactions which lead to carbohydrate accumulation has been gained only in the last few years.

7. *What is the structure of a living cell and how is it organized to conduct*

its characteristic chemical functions? The general topography of cells was revealed by light microscopy—an outer membrane, an inner nucleus, and numerous lesser bodies visible in the compound light microscope. The advent of the electron microscope has permitted much more detailed stereoscopic visualization of the finer structure: a network of microcanals, the endoplasmic reticulum, which lead from within the nucleus through the cytoplasm and occasionally to the cell exterior; large complex bodies, the mitochondria; numerous smaller dense bodies frequently attached to the reticulum; an unusual organization of fibers in the spindle apparatus of a dividing cell; and the double-layered structure of the cell membrane. Concomitantly, newly devised techniques made possible the isolation of concentrated preparations of each of these substructures, free of the remainder of the cell. These techniques have revealed the partition of functions within the cell: the nucleus as the site of genetic control and cellular duplication; ribosomes as loci for protein synthesis; mitochondria as units in which oxidative metabolism generates adenosine triphosphate; the lipid layer of the endoplasmic reticulum as the site of the metabolism of certain nonpolar molecules such as steroids; the cell membrane as the site of vectorially organized mechanisms for controlling the general electrolyte composition of the cytoplasm and bringing required nutrients into the cell proper (Chap. 34); and the cytoplasm, a solution of hundreds of individual enzymes which direct the multitudinous reactions by which nutrients are converted into cell constituents. It is the sum of all these chemical activities which constitutes the "life" of the cell.

8. *By what means do cells divide to yield identical daughter cells? What is the chemistry of inheritance? What is a "gene" and how does it function?* No chapter in the history of science has unfolded with such great rapidity or engendered such widespread interest as the answers to these questions. Few hold deeper or more significant implications for the future of man. This is the subject of Chaps. 28, 29, and 30.

It will be evident from the brief foregoing discussion that it is the structure of its complement of proteins which determines the form, shape, and organization of a cell, and it is the structures of those proteins which are enzymes that make possible the chemical reactions which comprise the life of the cell. It follows that the genetic "instructions" to a cell must be the encyclopedia of information required to achieve the precise synthesis of the ensemble of proteins characteristic of that cell. This information is encoded in the structure of the very large molecules of deoxyribonucleic acid. Cell duplication requires perfect reproduction of these molecules with subsequent equal distribution of the information between the cells. Utilization of this information requires its transmittal from nucleus through cytoplasm to the ribosomal protein factories. Changes in the chemical structure of deoxyribonucleic acid become evident as mutations and are transmitted to subsequent generations. How these processes operate has been disclosed largely by studies of their occurrence in a nonpathogenic enterobacterium, *Escherichia coli,* and by study of the phenomena involved in the duplication of bacteriophages, bacterial viruses, each of which is a limited bit of genetic information wrapped in a specific protein coat but capable of self-duplication only by utilizing the synthetic apparatus of a living cell. The totality of this information has made intelligible the laws of genetics, the nature and basis of hereditary diseases, and the biochemical operation of the process of evolution.

Had evolution not been deduced earlier on other grounds, it would surely have become obvious to the biochemist. Whereas the unaided eye reveals the diversity of life, the answers to each of the foregoing questions are essentially identical for *all* living forms. The impressiveness of this oneness of the cardinal aspects of all forms of life is matched only by the remarkable manner in which subtle variations on these themes have given rise to the rich variety and abundance of living forms as well as the over-all balanced activity of the total biosphere.

9. *Since the life of a cell is the totality of thousands of different chemical reactions, each catalyzed by a specific enzyme, how are these synchronized into a harmonious whole?* Clearly it is advantageous to the cell to match the pace of energy-yielding reactions to the requirement for that energy, and to provide the requisite monomeric units (amino acids, nucleotides, sugars) at a rate commensurate with the demands for polymer synthesis (proteins, nucleic acids, polysaccharides). Investigation of the mechanisms by which such regulation occurs constitutes one of the latest chapters in biochemical research. Although details are scanty, some of the outlines are clear; this problem is discussed in many places in this book. Included are arrangements analogous to both the negative and positive feedback systems of electronic engineering; these are intrinsic in the structures of some enzymes which participate in synthetic processes and help to assure a steady flow, but not a surplus, of necessary synthetic intermediates. In other instances, regulation involves repression or derepression of the synthesis of the enzymes which participate in synthetic processes.

In the multicellular, multiple-organed vertebrate not only must the diverse aspects of the metabolism of each cell be synchronized, but the activities of the various organs, muscle, liver, brain, etc., must also operate in harmony. From this it follows that information concerning the metabolic state of muscle, for example, must be transmitted to the liver as required. In large part, this is the role of the endocrine system. Endocrine glands, responding to changes in the chemical composition of the blood, which in turn reflect changes in some tissue or organ, release hormones which are carried in the circulation to the cells of target organs and there modulate specific cellular metabolic activities. This is a major theme of Part Five of this book.

10. *How do the specialized cells of animal tissues and organs make their unique contributions to the total animal economy?* Osteoblasts make bone, muscle cells contract, nerve cells conduct, kidney cells make urine, endocrine cells make hormones, all by mechanisms specific for these cell types. Perhaps because the systems involved are more readily accessible, biochemists have been more successful in learning the generalities of how all cells live than in ascertaining the details of this group of related problems. Nevertheless, partial answers are available; these are presented in Part Four and elsewhere. Much of biochemical effort in the next decade will be directed to expansion of such understanding.

11. *How does an animal regulate the volume and composition of the fluids which constitute the environment of its cells and of the blood which interconnects them?* This area of inquiry has proved to be extremely rewarding. The large body of informa-

tion which has been gathered has contributed significantly to the management of diverse acute disorders of man and to the achievements of modern surgery. The physiological mechanisms involved are extraordinarily sensitive and, as in critically engineered systems, frequently redundant. They have reached a high degree of perfection in man—and thus permit him to range the earth from the equator to the poles, from ocean depths to mountain peaks, and to survive despite enormous variations in the composition and quantity of the food and drink he ingests. These regulatory mechanisms are also discussed in Part Four.

Central to much of this aspect of life are the erythrocyte, a unique cell which has lost its nucleus and, hence, is doomed to a limited lifetime (Chap. 31), and the chemistry of hemoglobin, probably the most thoroughly studied of all proteins and our most detailed source of insight into the correlation of protein structure with physiological function (Chaps. 8 and 32).

The answers presently available to all the questions above have largely been obtained in the last two decades. As we have noted, the rate of progress has, in large measure, been determined by the rate at which new and more powerful tools have become available. A single example will suffice to illustrate this fact. The most important biological concept established in recent years, that deoxyribonucleic acid encodes instructions for the precise ordering of the amino acids in the primary strand of a protein, was most clearly indicated by the demonstration of a single amino acid substitution in the β-chain of hemoglobin (Chap. 28) from individuals with sickle cell anemia. However, this study could not have been undertaken until electrophoretic analysis had shown a difference in the net charge of hemoglobin from normal humans and those with this disease. Furthermore, the precise amino acid substitution could not have been sought until Sanger had provided the analytical techniques for deciphering the amino acid sequences of proteins which he developed to study the structure of insulin. Moreover, his techniques could not have been attempted until others had developed general procedures for column and paper chromatography (Chaps. 2 and 5), etc. The subsequent achievement, the revelation of the detailed three-dimensional structure of several proteins, necessitated a technology for the preparation of large protein crystals to which heavy metals are specifically bonded and which yet are crystallographically isomorphous (Chap. 7). Analysis of the massive body of x-ray data was impossible prior to the development of high-speed digital computers. The expanding knowledge of the primary structure of proteins dates to the commercial availability of the chromatographic amino acid analyzer. Regardless of the specific problem under study, current research in biochemistry almost invariably requires a complement of high-speed refrigerated centrifuges, recording spectrophotometers, scintillation counters, etc. Accordingly, it may be anticipated that the next burst of research progress will follow the next significant addition to the armamentarium of techniques and instruments.

It is evident that biochemistry is not an isolated discipline but has become the very language of biology, basic to the understanding of phenomena in the biological and medical sciences. Thus, drugs, whether antibiotics or tranquilizers, must exert their influence on the chemical structures and metabolic events of the cells affected by these drugs. Most diseases are derangements of cellular structure or function and many are now understood in these terms; a number of these will be considered in the pages which follow.

Since the time of Aristotle, students of biology have sought to correlate structure and function. This endeavor continues to the present time; the search for the explanation of structure and function, and their interrelationships, in terms of the structure and interactions of molecules, is the main theme of biochemistry.

PART ONE THE MAJOR CONSTITUENTS OF CELLS

2. The Carbohydrates. I

Monosaccharides

CLASSIFICATION

The carbohydrates are a large group of compounds which are polyhydroxyaldehydes or polyhydroxyketones, and their derivatives. These compounds may be classified as

Monosaccharides
Derived monosaccharides
Oligosaccharides (di- and trisaccharides, etc.)
High molecular weight polysaccharides

In general, carbohydrates are white solids, sparingly soluble in organic liquids but, except for certain polysaccharides, soluble in water.

Monosaccharides. Although formaldehyde and hydroxyacetaldehyde (glycolaldehyde) conform to the empirical formula of the carbohydrates, CH_2O, the smallest molecules generally termed carbohydrates are glyceraldehyde and dihydroxyacetone. These two compounds are the only possible trioses, the 3-carbon sugars.

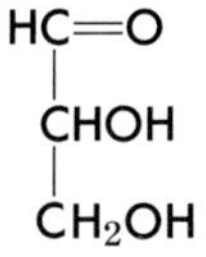

Glyceraldehyde

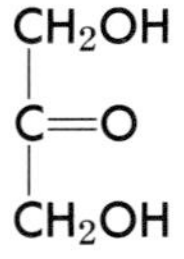

Dihydroxyacetone

Consideration of these formulas reveals several characteristics common to the entire group:

(1) The carbon skeleton is usually unbranched. (2) Each carbon atom except one bears a hydroxyl group. (3) One carbon atom bears a carbonyl oxygen which may reside on a terminal carbon atom, giving an aldehyde, or on a centrally placed carbon atom, giving a ketone.

One of the methods of naming sugars relates to the last point. Thus glyceraldehyde may be termed an *aldo*triose, and dihydroxyacetone is then called a *keto*triose. Among the common ketoses, or ketonic monosaccharides, the carbonyl oxygen is found on the C-2 carbon atom (adjacent to the uppermost one).

Monosaccharides containing four carbon atoms are called tetroses, those containing five carbon atoms are termed pentoses, whereas the hexoses contain six and the heptoses seven carbon atoms. Generic names for the ketoses are formed by insertion of "ul"; thus; pentulose, hexulose, heptulose.

Derived Monosaccharides. In this group is included a variety of compounds structurally very similar to the monosaccharides but deviating in one or another regard from the aldoses and ketoses just described. Three main varieties of carboxylic acids are found among oxidation products of the simple sugars. There are those in which the aldehydic group is oxidized to the carboxyl level, *e.g.*, gluconic acid. There are acids in which the primary hydroxyl group remote from the aldehyde is oxidized to the carboxyl level, *e.g.*, glucuronic acid. Finally there are dicarboxylic acids, oxidized at both ends, such as saccharic acid.

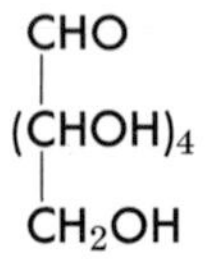

Aldohexose
(glucose)

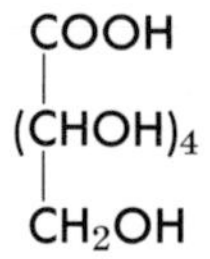

Hexonic acid
(gluconic acid)

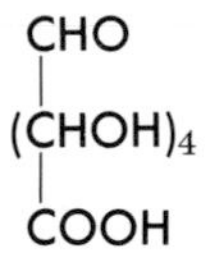

Uronic acid
(glucuronic acid)

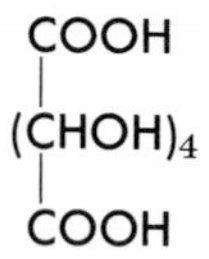

Dicarboxylic acid
(saccharic acid)

Upon reduction of aldoses or ketoses, polyhydric alcohols are obtained, *e.g.*, sorbitol. Related to such sugar alcohols are carbocyclic alcohols, such as inositol.

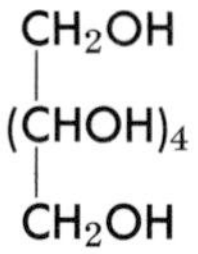

Hexitol (sorbitol)

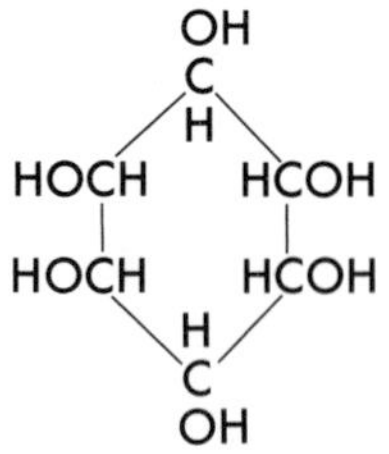

Hexahydroxycyclohexane (inositol)

The carbocyclic alcohols are isomeric with the true monosaccharides, sharing with them the empirical formula $(CH_2O)_n$.

Replacement of a hydroxyl group by an amino group yields an amino sugar; replacement of a hydroxyl group by hydrogen yields a deoxysugar.

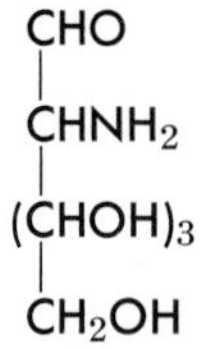

Hexosamine (glucosamine)

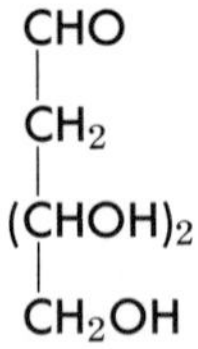

Deoxypentose (deoxyribose)

More than one hydroxyl group may be replaced, as in diamino sugars, dideoxysugars, and mono- and diamino-, di- and trideoxysugars.

Monosaccharides are also known with an empirical formula $C_n(H_2O)_{n-1}$, and these are called anhydrosugars.

MOLECULAR STRUCTURE AND OPTICAL ROTATION

The development of carbohydrate chemistry is so closely associated with polarimetry that a brief discussion of the principles involved is appropriate at this point. Biot, commencing about 1815, in addition to observing the optical rotation of the plane of polarized light passing through certain solids, liquids, and gases, studied the rotatory power of similar materials in solution, noted the dependence of observed rotation upon concentration of solute and length of the path of light through the solution, and formulated the definition of specific rotation, $[\alpha]$,

$$[\alpha] = \frac{100 \times A}{c \times l}$$

where A is the observed rotation (plus or minus) in degrees, c is the concentration in grams per 100 ml. of solution, and l is the length, in decimeters, of the optical path through the solution. The specific rotation is thus the actual rotation imparted to a beam of plane-polarized light passing through 1 decimeter of a hypothetical solution of 100 g. of optically active solute in 100 ml. of solution. The specific rotation is a complex function of the wavelength of light and the nature of the solvent used. Monochromatic light, most frequently the D line of the sodium spectrum, is employed and specified. The temperature must also be specified, as rotatory power is temperature-dependent. These quantities are indicated as follows:

$$[\alpha]_D^{20} = \text{rotation value (solvent used); } e.g., +23^\circ \ (CHCl_3)$$

Since the dependence of observed rotation upon concentration is not always strictly linear, it is customary to specify the concentration of solute at which the rotation is determined.

The variation in optical rotation as a function of the wavelength of monochromatic light is termed *optical rotatory dispersion.* Since rotatory power is a consequence of molecular asymmetry, optical rotatory measurements are particularly useful for studies of helical structures because helices are inherently asymmetric. Thus optical rotatory dispersion studies have contributed to understanding of the secondary structures of proteins (page 155).

Molecular Asymmetry. Pasteur recognized that for a compound to possess optical activity it must be asymmetric. This asymmetry may be restricted to the crystal structure as is the situation with crystalline quartz, but more frequently it resides in the molecule proper. Pasteur's study of the tartaric acids culminated in the definitions of "meso" acid and "racemic" acid,

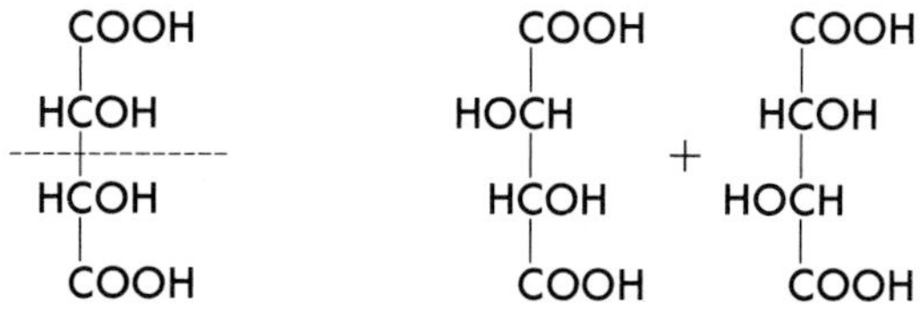

"Meso" acid **"Racemic" acid**

of which the former, optically inert and unresolvable, *i.e.*, a single substance, has a plane of symmetry indicated by the broken line, while the latter, devoid of such symmetry, proved to be a mixture of two optically active antipodes, resolvable into *d* (*dextro-*) and *l* (*levo-*) components.

The independent and simultaneously announced conclusions of van't Hoff and LeBel, in 1874, placed stereochemistry on a sound theoretical basis. It was pointed out that the geometrically most probable orientation of the bonds of tetravalent carbon was to the apexes of an equilateral tetrahedron and that this assumption would account for all the then known phenomena of molecular asymmetry. This hypothesis has proved to be one of the most generally applicable in all of chemistry and, in conjunction with subsidiary hypotheses relating to freedom of rotation about the axes of valence bonds, has accounted for virtually all types of stereochemical relationships. Placing the two central carbon atoms of tartaric acid at the centers of tetrahedrons, the formulas become as shown.

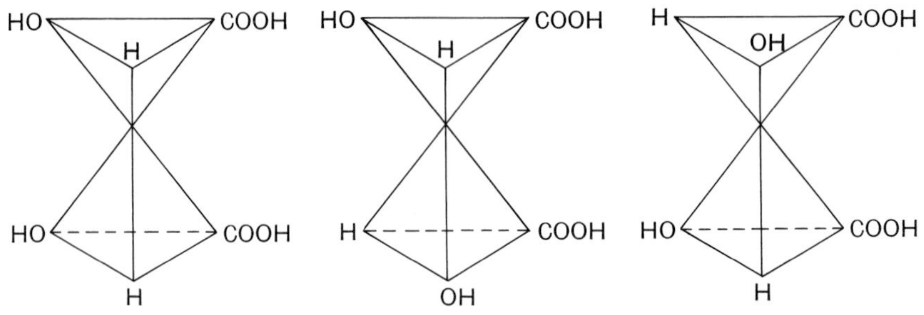

"Meso" acid "Racemic" acid

The simplest type of molecular asymmetry that can arise from this hypothesis results from the occurrence of an atom asymmetrically located in the molecule, and, in the case of carbon, any carbon atom which bears four different substituents on its four valences becomes an asymmetric center. Carbon atoms bearing double or triple bonds are at once excluded from this category. Every compound the formula of which has one asymmetrically situated carbon atom is either optically active, *i.e.*, rotates the plane of polarized light, or is resolvable into optical antipodes, a pair of substances each of which is optically active. Whereas the general chemical properties of such antipodes are very similar (cf. Resolution of *dl* Mixtures and Racemates, below), they will differ from each other in that one will be *dextro*-rotatory, the other *levo*rotatory, while the numerical value of $[\alpha]$, disregarding sign, will be the same for both compounds. The steric formulas of these two compounds will meet the criterion of being mirror images of each other but not superimposable,

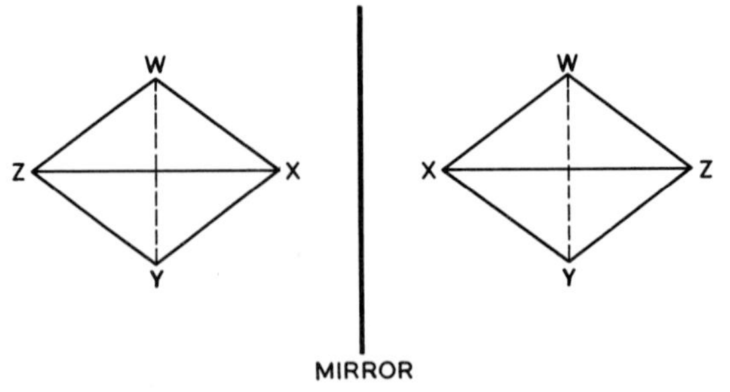

in the same relationship as that existing between the right and left hands. Because of the difficulties involved in drawing numerous tetrahedrons, the convention adopted in stereochemistry is to represent these antipodes in plane projection as

Here an added restriction arises from the two-dimensional representation, *viz.*, the test for superimposability must be conducted without removing the formula from the plane of the paper.

Returning to the tartaric acids, it will be seen that the *d* and *l* components of *racemic* acid fulfill the requirements of being nonsuperimposable mirror images. The *meso* acid, however, is readily superimposed on its mirror image by rotation in the plane of the paper.

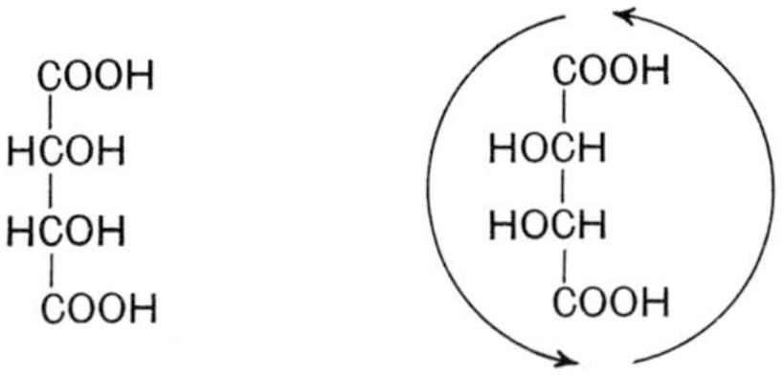

"Meso" acid

Consequently, the *meso* acid is not resolvable and is optically inert, despite the fact that it contains two atomic centers of asymmetry. This situation is termed "internal compensation" and may be pictured as resulting from opposing rotatory powers of opposite sign at the two ends of the molecule. The general conclusion is that, regardless of asymmetrically substituted atoms, optical activity will be absent if the molecule exhibits any plane of symmetry.

Asymmetric Behavior of Certain Symmetrical Compounds. There are certain molecules which, although possessed of a plane of symmetry, behave under certain circumstances as though they were asymmetric. These molecules all contain one carbon atom which bears *a pair of like and a pair of unlike substituents.*

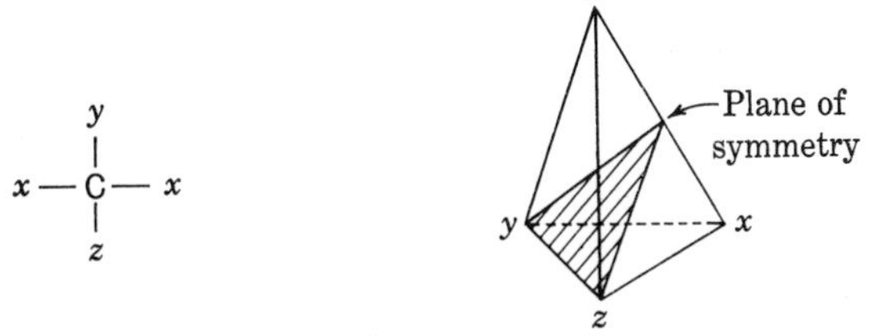

The tetrahedral representation of such a molecule clearly has a plane of symmetry, yet experimentally it can readily be shown that the two *x* substituents do not react equally when presented with an asymmetric reagent. The reason for this resides in an extension of the principles already outlined. Whereas the two *x* substituents are symmetrically disposed, and the two half-molecules generated by cleavage of the molecule along the plane of symmetry are mirror images, these two half-molecules are not superimposable one upon the other. Thus these two *half*-molecules may be said to be antipodal, just as the whole molecules of *d*- and *l*-tartaric acid are antipodal. Precisely as the two tartaric acids, *d* and *l*, will not be distinguished by symmetrical reagents but will react differentially with an asymmetric reagent, so will the two *x* groups in the molecule C(*xxyz*) react at different rates if the reagent is itself unsymmetrical. The carbon atom in C(*xxyz*) is referred to as a *meso* atom. This type of *meso* symmetry is of especial importance in biochemistry because virtually all reactions involve enzymes, and these are highly asymmetrical; examples will be encountered later (cf. Chaps. 12 and 16).

Number of Optical Isomers. The principles outlined permit computation of the expected number of optical isomers of any given structure. A molecule having one center of asymmetry will exist in two configurations corresponding to the *d* and *l* antipodes. Barring *meso* forms and steric incompatibilities, with two centers of asymmetry one finds four possible forms, or two antipodal pairs, *a* and *b*, and *c* and *d*.

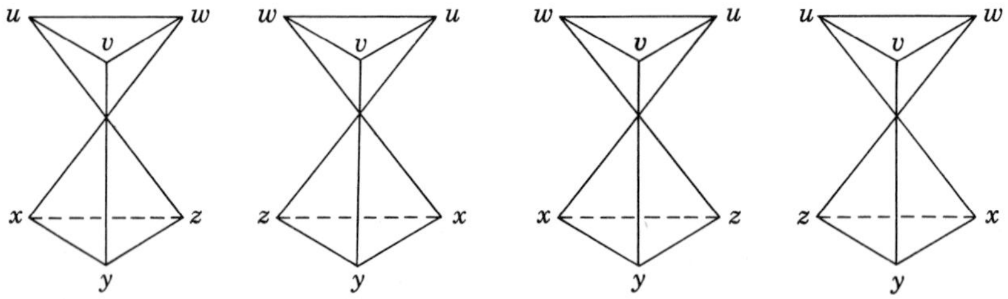

a *b* *c* *d*

All other steric modifications that can be drawn will be found to be superimposable on, hence identical with, one of these. The expected relationship, again barring situations of internal compensation and restricted rotation of the molecule about some bond or bonds, is that for *n* centers of asymmetry, 2^n optically active isomers may be anticipated.

It should also be pointed out that, whereas optical antipodes *a* and *b* will be identical in most chemical and physical properties, *a* will not resemble *c* as closely as two isomers usually resemble each other. The *dl* mixture of *a* + *b* will in general be readily separable from the *dl* mixture of *c* + *d*.

Resolution of *dl* Mixtures and Racemates. When a living cell, in the course of biosynthesis, establishes a new center of asymmetry in a molecule, it virtually invariably produces an optically active product, with complete exclusion of its antipode. Products isolated from living sources, if they contain an asymmetric carbon atom, are optically active and usually pure from the stereochemical point of view. *Meso* forms, however, will be optically inactive. Not only do living cells, or enzymic systems derived therefrom, apparently synthesize only one of two antipodes, but if confronted with a synthetic racemic mixture they will utilize the two antipodes of a given compound at differing rates or in different ways, often completely consuming the one and leaving the other unattacked.

When the chemist, starting with optically inactive materials, synthesizes a compound containing an asymmetrically substituted carbon atom, he almost invariably obtains a product in which the *d* and *l* antipodes are equally represented. Thus reduction of 2-butanone yields an optically inactive mixture of products, since the probability of formation is the same for each antipode. Such *dl* mixtures

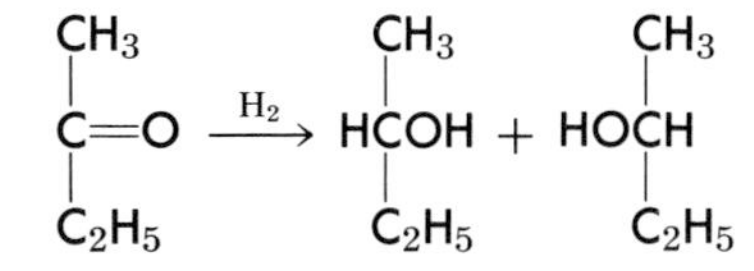

2-Butanone **dl-2-Butanol**

are called racemic compounds. A frequently encountered problem involves the separation, or "resolution," of such materials into pure *d* and *l* components.

The crystals of *d*- and *l*-tartrates (sodium ammonium salts) separate from solution, under certain conditions, in crystalline forms which are themselves of two varieties, one being the nonsuperimposable mirror image of the other. This difference in crystalline forms permitted Pasteur to accomplish the tedious manual resolution of the mixture of crystals. A more generally useful method of resolution involves preparation of diastereoisomers, *i.e.*, derivatives made with one antipodal form of an optically active reagent. Alkaloids such as *l*-brucine or terpene acids such as *d*-camphoric acid are often employed in this manner. If equal quantities of *d*- and *l*-tartaric acids are treated with *l*-brucine, two salts will be produced, *l*-brucine *d*-tartrate and *l*-brucine *l*-tartrate. These two compounds are now no longer antipodes, since the antipode of *l*-brucine *d*-tartrate will necessarily be *d*-brucine *l*-tartrate. They will consequently have different physical properties and can be separated by fractional crystallization or some other physical means.

Another general method of resolution depends upon the basic asymmetry of living systems and their enzymes. Thus in many cases a *dl* mixture can be offered as substrate to a microorganism, a tissue, or an enzyme, and the one antipode will be quantitatively destroyed while the other remains. A modification of this method involves preparation of a derivative of the *dl* mixture, which is then treated with an enzyme that splits one antipodal derivative but not the other, permitting isolation of the latter. These biochemical resolution methods serve to emphasize the fundamental asymmetry of enzyme-catalyzed reactions, attributable to the asymmetry of the enzymes themselves.

Designation of Configuration. The employment of *d* and *l* or (+) and (−) to designate the sign of rotation of plane-polarized light by a given substance is useful to indicate which of a pair of antipodes is being discussed. It gives, however, no information as to the *configuration* of the several substituents about the center or centers of asymmetry in the molecule. Thus naturally occurring glucose is dextrorotatory, $[\alpha]_D^{20} = +52.7°$, while fructose is levorotatory, $[\alpha]_D^{20} = -92.4°$; yet these two hexoses prove to be configurationally intimately related.

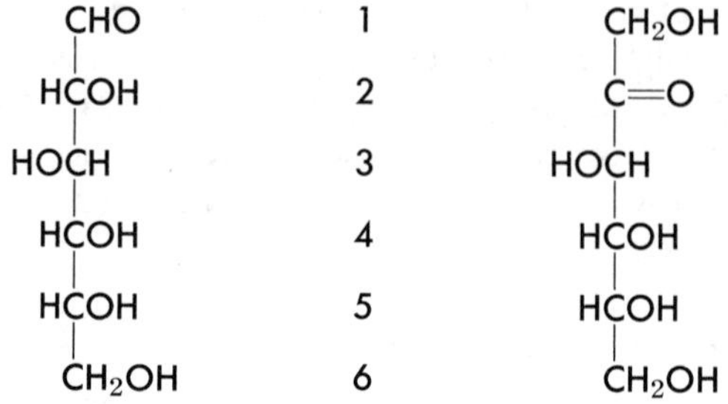

d(+)-Glucose   *l*(−)-Fructose

It will be observed that the configurations about carbon atoms 3, 4, and 5 are identical for the two substances. A convention of nomenclature has been devised based upon configurational rather than optical properties. The actual sign of rota-

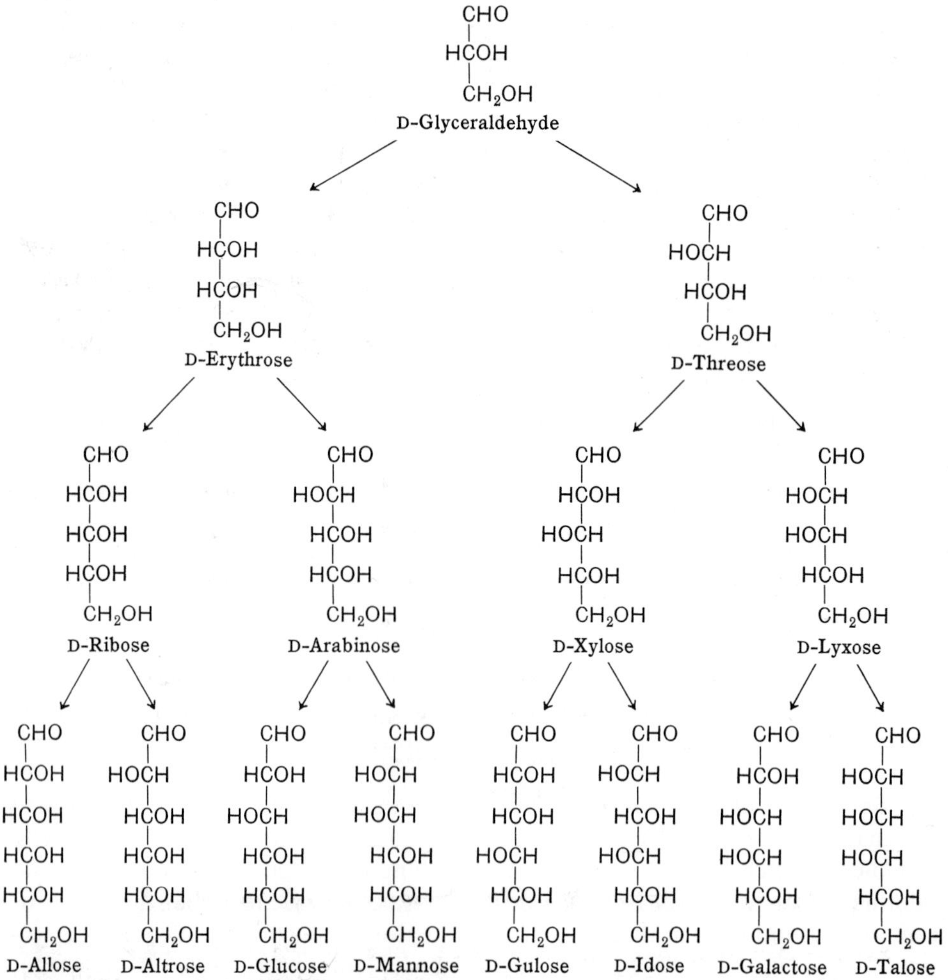

Fig. 2.1. Relationships of the D-aldoses. The formulas of the L-aldoses are in each case the mirror images of those structures given in the figure.

tion may still be indicated by the italic letters *d* and *l* or by (+) or (−), but the configuration is shown by the prefixed symbols in roman small capitals D and L. For the common sugars, the rule relates to that center of asymmetry most remote from the aldehydic end of the molecule; in hexoses this is carbon atom 5. Arbitrary configurations, now known to be correct in the absolute sense, have been assigned to the two glyceraldehydes,

D-Glyceraldehyde **L-Glyceraldehyde**

and all sugars terminating in these configurations are said to belong accordingly to the D or L configurational series. Thus both glucose and fructose are of the D series, and if the sign of rotation is to be included in the name, it is usually indicated as follows: D (+)-glucose, D (−)-fructose. The consistent utilization of these conventions permits the reasonable arrangement of sugars, as will be seen from the accompanying charts (Figs. 2.1 and 2.2) of the D-aldoses and D-ketoses.

It should be pointed out that these conventions can be employed only after configurational relationships have been established by unequivocal chemical means. Optically active materials of undetermined configuration must still be designated with respect to the sign of rotation. It should also be mentioned that some confusion may occasionally arise in nomenclature. Thus the products of oxidation of D-glucose and L-gulose to the dicarboxylic acid level are identical and may be given either of two names with equal justification.

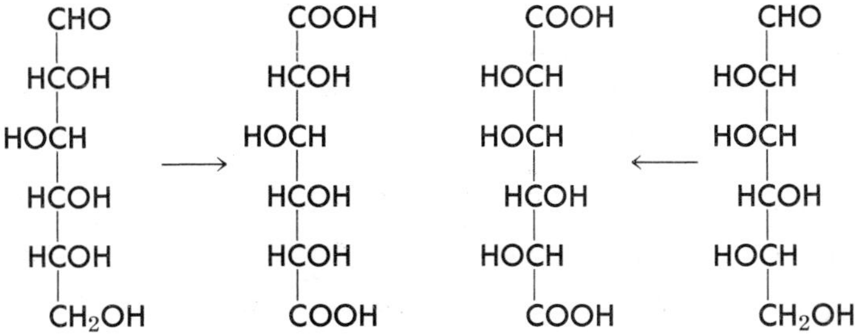

D-Glucose **D-Glucosaccharic acid** **L-Gulosaccharic acid** **L-Gulose**

Identical Products

SPATIAL CONFIGURATION OF MONOSACCHARIDES

Formula of D-Glucose, Linear Form. There are four asymmetric centers in the glucose molecule, from which it could be calculated that D-glucose was one of 2^4 or 16 stereoisomers, and further stereoisomers were excluded. However, it soon

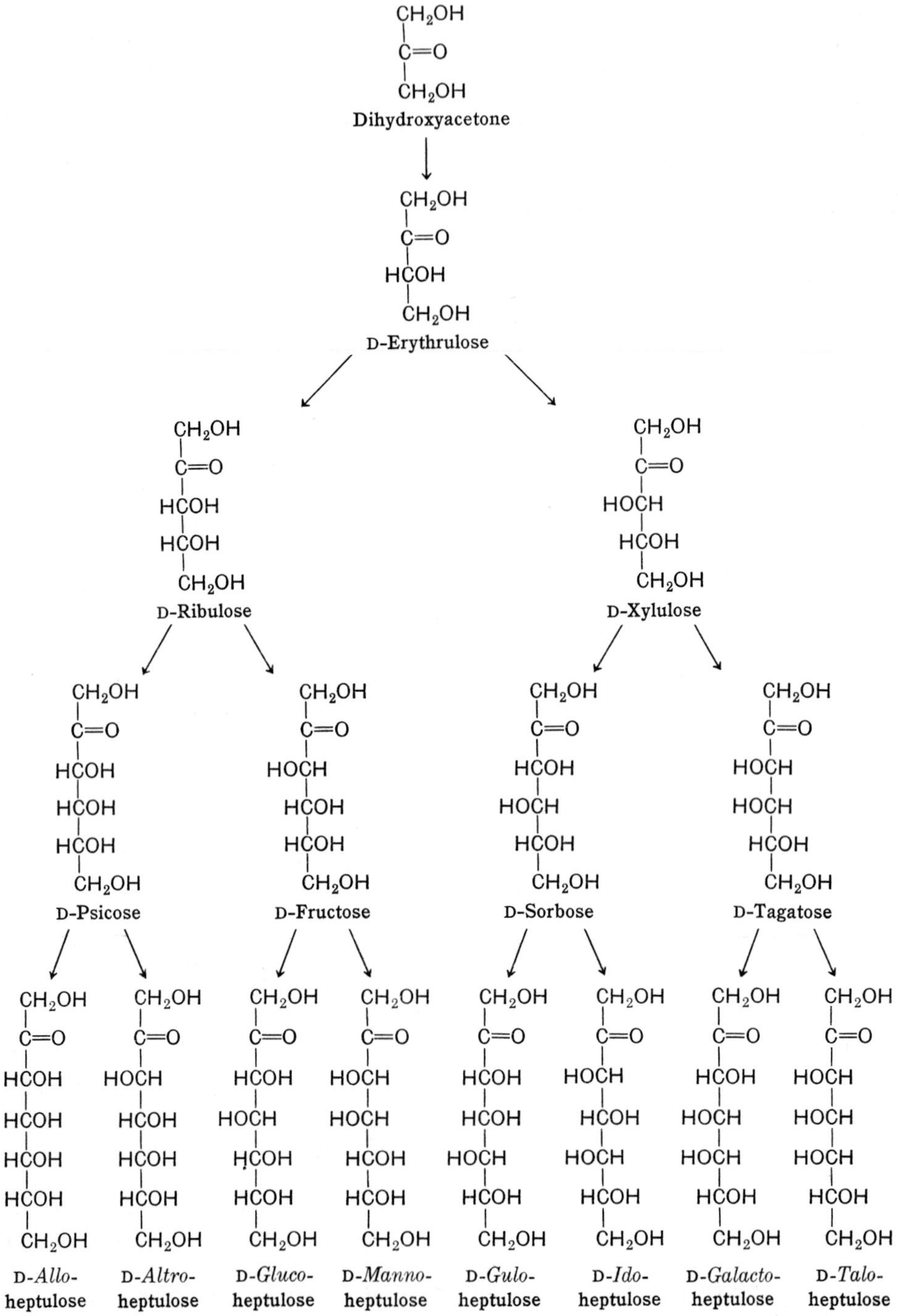

Fig. 2.2. Relationships of the D-ketoses. The formulas of the L-ketoses are in each case the mirror images of those structures given in the figure.

became apparent that even this formulation failed to describe certain chemical properties of glucose.

Cyclic Structure of Glucose. The earliest indication that D-glucose existed in more than one configurational modification came from the finding that the specific rotation of a freshly prepared glucose solution changed under observation in the polarimeter. Ultimately two differing species of D-glucose could be prepared which in aqueous solution were found to exhibit widely differing optical activities. For the one, α-D-glucose, $[\alpha]_D^{20} = +112.2°$; for the other, β-D-glucose, $[\alpha]_D^{20} = +18.7°$. If these solutions were subjected to continued polarimetric observation, the specific rotations of each would approach the value of +52.7°, after which no further change would occur. This finding clearly meant there were two distinct stereoisomeric modifications of glucose which were interconverted in aqueous solution to yield an equilibrium mixture represented by the final rotation. This phenomenon, termed *mutarotation,* is not peculiar to glucose but is a general phenomenon of sugars and has been observed for a variety of pentoses and hexoses, as well as certain disaccharides.

The demonstrated existence of two isomers of D-glucose requires that there be an additional asymmetric center in the formula. It was established that carbon atom 1 is the locus of this asymmetry, and formulas for α- and β-D-glucose have been assigned.

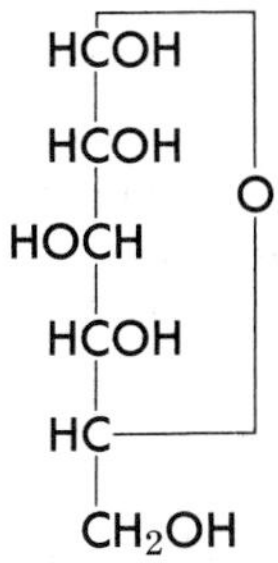

α-D-Glucose
$[\alpha]_D^{20} = +112.2°$

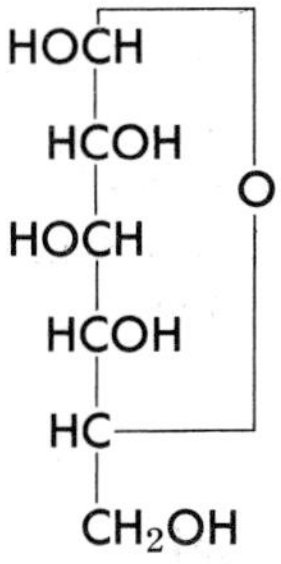

β-D-Glucose
$[\alpha]_D^{20} = +18.7°$

In this formulation there are five centers of asymmetry, hence $2^5 = 32$ possible isomers, permitting an α and β modification of each aldohexose. The α designation is used to indicate that, in plane projection, the hydroxyl group on carbon atom 1 is on the same side of the structure as the ring oxygen; the β modification refers to the form in which the C-1 hydroxyl group is on the side of the structure opposite to the ring oxygen. Among sugars of the D configuration, the α isomer always has an optical rotation higher than that of the β isomer.

This formulation also accounts for the fact that when glucose is treated with methanol and mineral acid, two distinct methyl glucosides are produced. The formulas of these may be written:

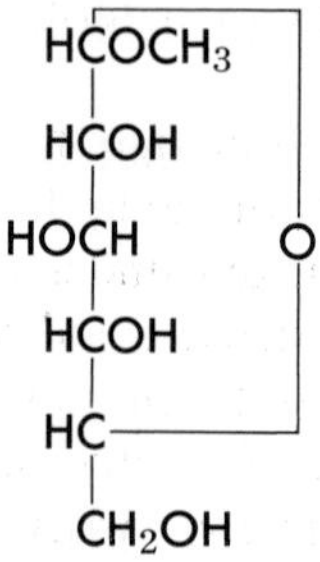

Methyl α-D-glucoside
$[\alpha]_D^{20} = +158.9°$

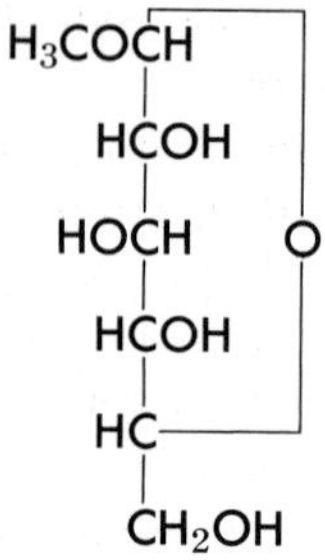

Methyl β-D-glucoside
$[\alpha]_D^{20} = -34.2°$

The oxygen bridge between the first and fifth carbon atoms is a six-membered ring which resembles that of pyran,

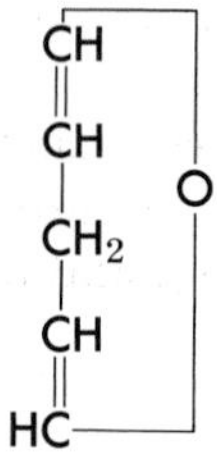

Pyran

and sugars containing such a ring are termed *pyranoses.* Thus the above methyl α-D-glucoside is correctly named methyl α-D-glucopyranoside. The ring structures of the aldopentoses and ketohexoses, as they commonly occur in oligosaccharides, are predominantly five-membered; five-membered rings have also been prepared among certain aldohexoses (galactose, mannose).

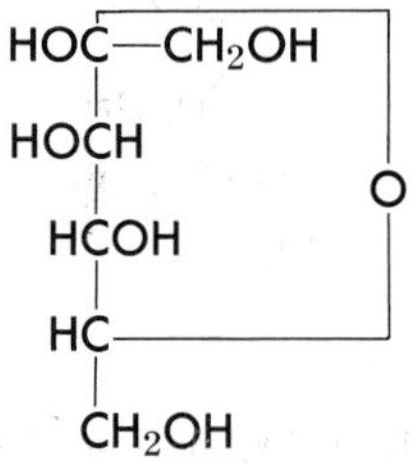

β-D-Fructofuranose

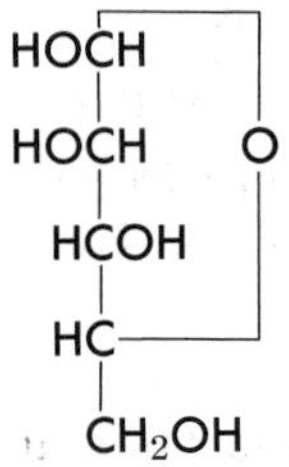

β-D-Arabinofuranose

This nomenclature is based on the similarity of these rings to that of furan,

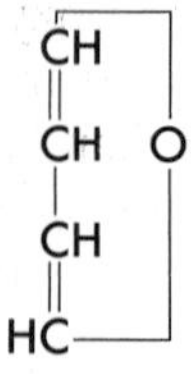

Furan

and sugars containing this type of ring are termed *furanoses.* It is generally supposed that mutarotation, which involves mutual interconversion of α- and β-pyranoses and α- and β-furanoses, occurs through the intermediate formation of the open-chain aldehyde or its hydrate.

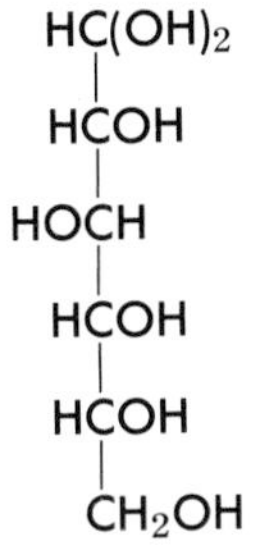

D-Glucose monohydrate

In aqueous solution the open-chain aldehyde is a very minor constituent of the equilibrium mixture. For glucose it has been estimated to represent 0.024 per cent of the total.

Hexagonal or Pentagonal Formulation. The structures of pyranoses and furanoses are more realistically depicted as more or less regular hexagons and pentagons. The plane of the ring is shown as perpendicular to the plane of the paper, and this is emphasized by shading those bonds which are indicated to be nearer to the reader.

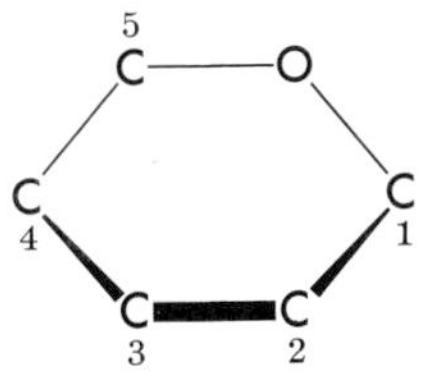

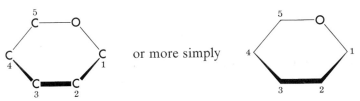

or more simply

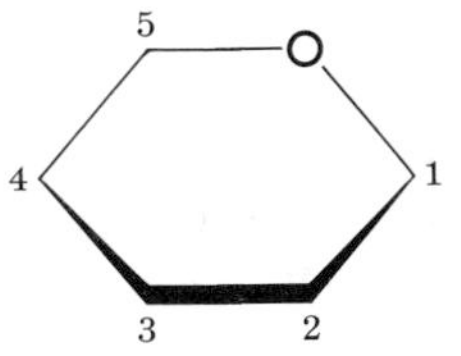

In this convention the two forms of D-glucopyranose are

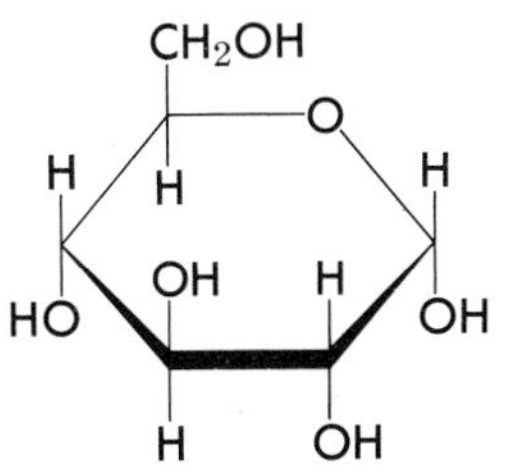

α-D-Glucopyranose

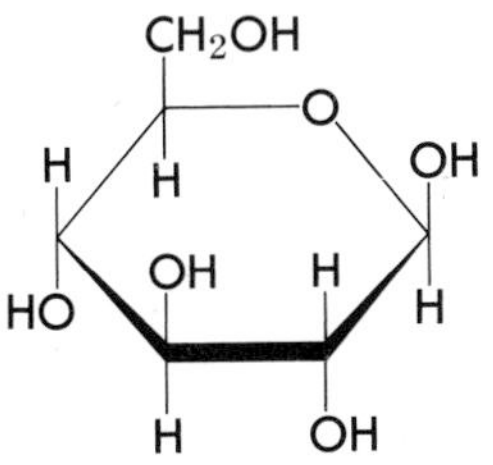

β-D-Glucopyranose

Similarly, the furanoses are represented as follows:

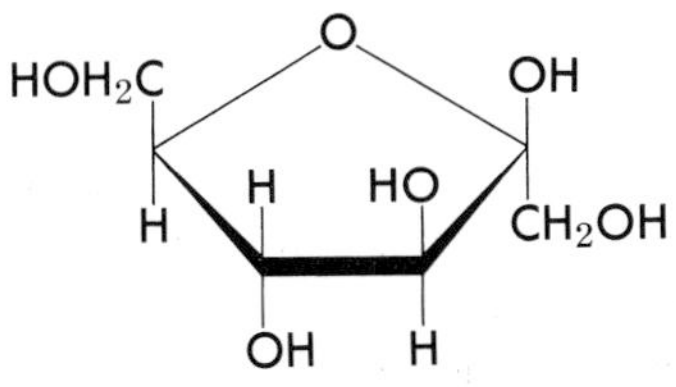

β-D-Fructofuranose

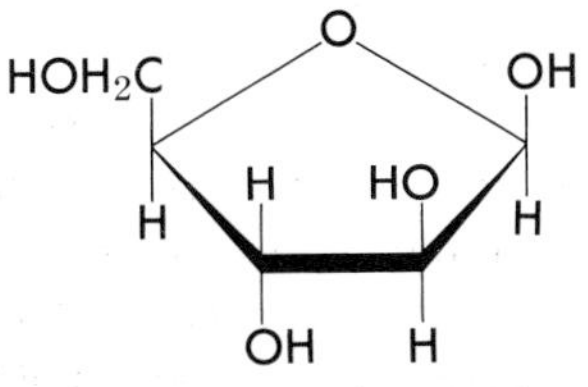

β-D-Arabinofuranose

The transition from one convention to the other relates the left and right sides of the carbon chain structure to the upper and lower aspects, respectively, of the plane of the ring formulation. Exceptions do occur, as will be seen by comparing the arrangement about carbon-5 of glucose in the two conventions. In the linear formulation the hydrogen at this position is to the left of the carbon, while in the hexagonal formulation it is below the plane of the ring. The reader is referred to texts on carbohydrate chemistry for a detailed discussion of this transition.

Conformational Representation. Whereas the six atoms in the benzene ring lie in one plane, the normal valence angle of the carbon atom precludes a stable planar arrangement for the six carbon atoms of cyclohexane. From atomic model studies, two types of arrangement in space, *conformations,* are possible, the "chair" and "boat" forms.

Chair Boat

Cyclohexane

On energetic grounds, the chair form is quite rigid and by far the more stable. Considerable force is needed to distort the chair form or to convert it to another form. The boat form is, however, completely flexible and has an infinite number of possible actual conformations in solution. The twelve hydrogen atoms of cyclohexane now fall into two classes, those on C—H bonds parallel to the axis of symmetry of the ring, *axial,* and those which are parallel to the nonadjacent side of the ring, or *equatorial.* These relationships are shown in the following conformational formulas.

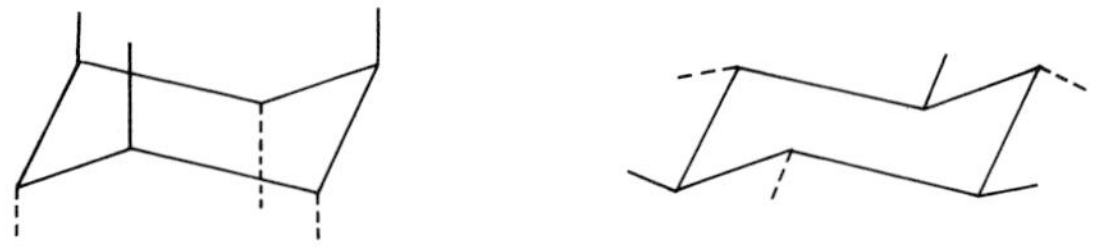

Axial bonds (*a*) **Equatorial bonds (*e*)**

It will be noted from the above, as well as from the previously indicated chair form of cyclohexane, that at each carbon atom one hydrogen is up or directed upwards, and the other is down or directed downwards. The six hydrogens that are up are termed β-hydrogens and are represented by solid lines; those that are down are α-hydrogens and are represented as dotted lines.

The conformational method of representation is far more than an exercise in projectional geometry. It reveals properties of the molecule that are not at once apparent from other representations of structure. A generalization of importance is that a substituent, particularly a large substituent, is at a lower energy state in the equatorial than in the axial location. Therefore, for example, in the equilibrium between the conformational isomers of methylcyclohexane,

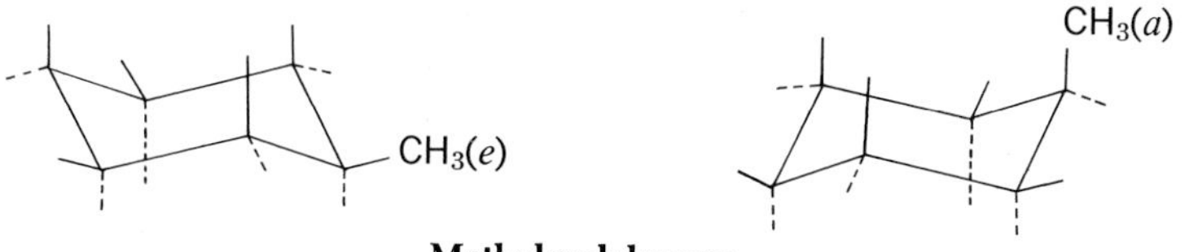

Methylcyclohexane

the equatorial form preponderates. As might be expected, equatorial hydroxyl groups are more readily esterified than axial hydroxyls, and there are many known instances of differences in chemical reactivity of groups, depending upon whether they are axially or equatorially disposed.

Conformational formulas are useful in the field of sugar chemistry and will again be encountered in the consideration of steroid chemistry. Conformational formulation of aldohexoses is particularly meaningful in interpreting reactivity of hydroxyl groups.

The conformational representation of glucose, in this convention, is:

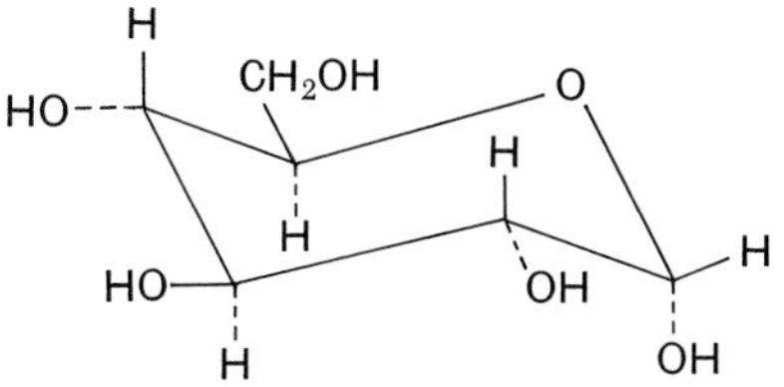

α-D-Glucose

Hemiacetal and Acetal Bonds. A pair of stereoisomers related to each other as are α- and β-D-glucose are said to be *anomers,* and since their sole configurational difference resides in steric arrangement about carbon atom 1, the "carbonyl" carbon of the linear formula, this atom has been termed the *anomeric carbon atom.* Considerable interest attaches to the arrangement of substituents about this reactive center.

When an aldehyde reacts with two equivalents of alcohol, the product is an acetal.

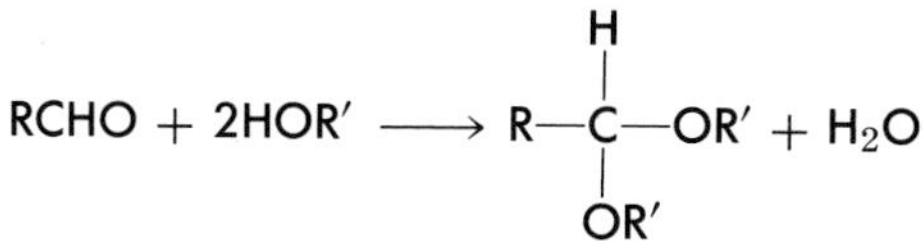

Acetal

If the quantity of alcohol is restricted, the expected product is a hemiacetal,

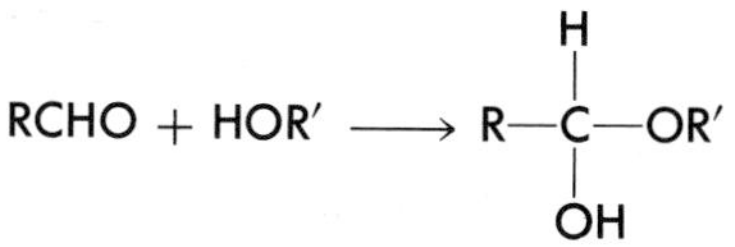

Hemiacetal

and it will be noted that the formerly aldehydic carbon atom now bears four different substituents, —H, —OH, —OR′, and —R. Every carbon atom similarly located is a hemiacetal carbon atom, and this is precisely the arrangement of substituents about the anomeric carbon atom in position 1 of glucopyranose. A distinctive feature of the hemiacetal carbon in monosaccharides is that the —OR′ and R substituents stem from the same molecule, making intramolecular hemiacetals. In this these compounds bear an analogy to the lactones, which are intramolecular esters. Fructofuranose, in which the anomeric carbon atom 2 bears R, R′, OR″, and OH as substituents, has a structure analogous to that of the reaction product of a ketone and alcohol,

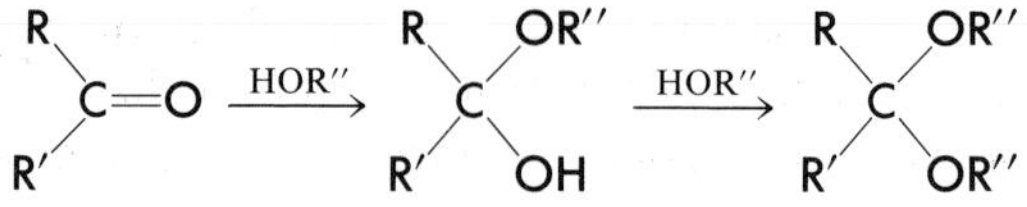

Hemiketal **Ketal**

and may therefore be designated as a hemiketal.

Glycosidic Bond. The residual hydroxyl group of a hemiacetal can react with an alcohol to yield an acetal. This is what occurs when methanol reacts with *glucose,*

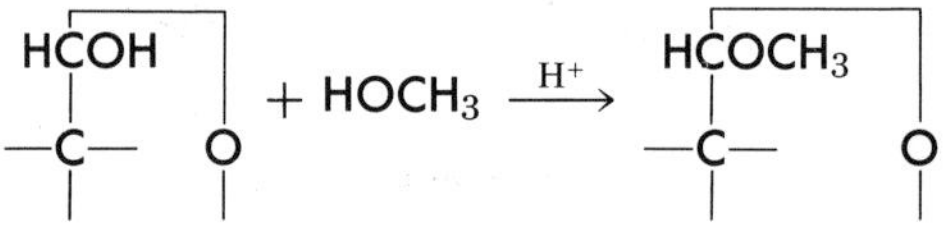

and the product, a mixed acetal, wherein one alcoholic contribution comes from outside while the other is intramolecular, is called a *glucoside.* This is one of a group of compounds known as *glycosides,* formed when the hydroxyl group on the anomeric carbon atom of a monosaccharide has reacted with an alcohol to form an acetal. Since this reaction may occur with either the α- or β-stereoisomer, there exist the corresponding groups of α- and β-glycosides. The members of these groups are named in accordance with the monosaccharides from which they are derived, *e.g.,* α- and β-glucosides, α- and β-galactosides, etc. It should be noted particularly that in reactivity the hemiacetal hydroxyl is quite different from an alcoholic hydroxyl group, and likewise the glycoside oxygen bridge, though bearing some superficial resemblance to an ether bridge, has none of the chemical stability generally found in aliphatic ethers.

Hexoses. Whereas only glucose and fructose occur abundantly in nature as free monosaccharides, several other simple sugars are repeatedly encountered, either as units of disaccharides and polysaccharides or in other types of linkage. The hexoses which commonly are present in these combined forms are the aldohexoses, *viz.*, glucose, galactose, and mannose, and the 2-ketohexose, fructose. Each of these sugars undergoes mutarotation, and each exists in an α and β modification. The formulas at the top of page 25 show the α forms.

Glucose and mannose are epimers (*differ only in the configuration of a single carbon atom*) with respect to carbon atom 2. Glucose and galactose represent an epimeric pair with respect to carbon-4. Fructose differs from the others in that its

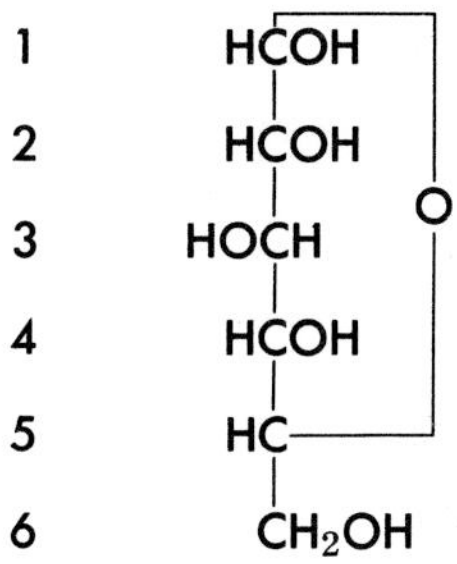

α-D-Glucopyranose

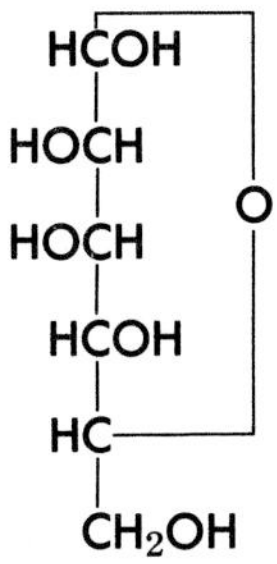

α-D-Mannopyranose

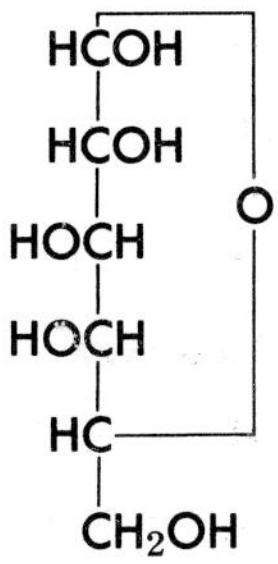

α-D-Galactopyranose

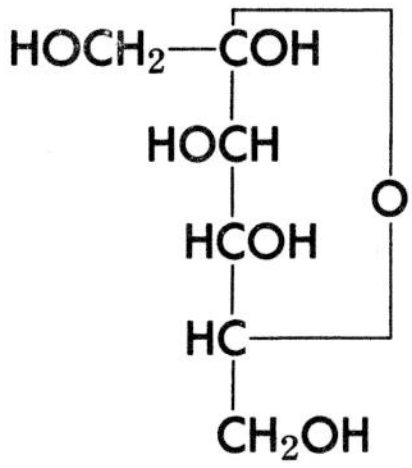

α-D-Fructofuranose

anomeric carbon atom is 2 rather than 1. Fructose is identical with glucose and mannose with respect to configuration about atoms 3, 4, and 5.

The cyclic formulations for mannose, galactose, and fructose are shown:

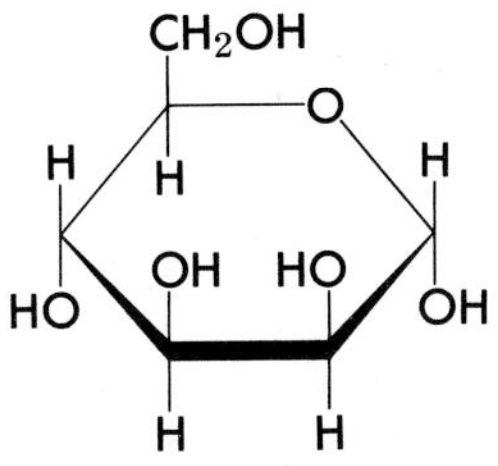

α-D-Mannopyranose

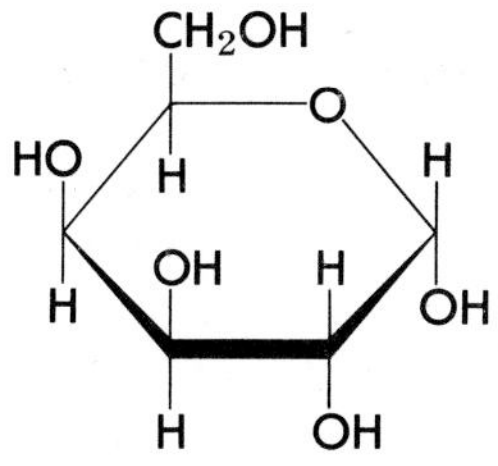

α-D-Galactopyranose

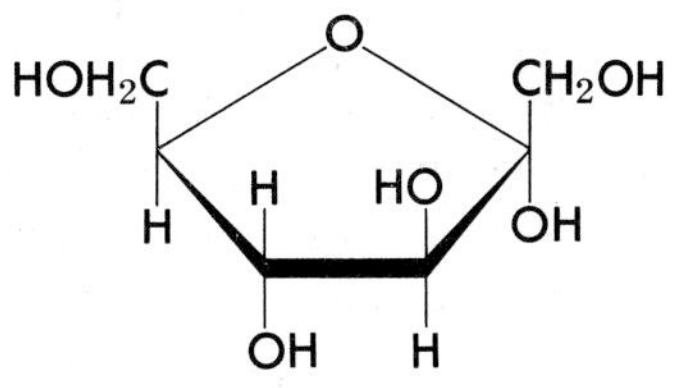

α-D-Fructofuranose

Only one other simple, unsubstituted hexose occurs in natural products; this is the ketohexose, sorbose.

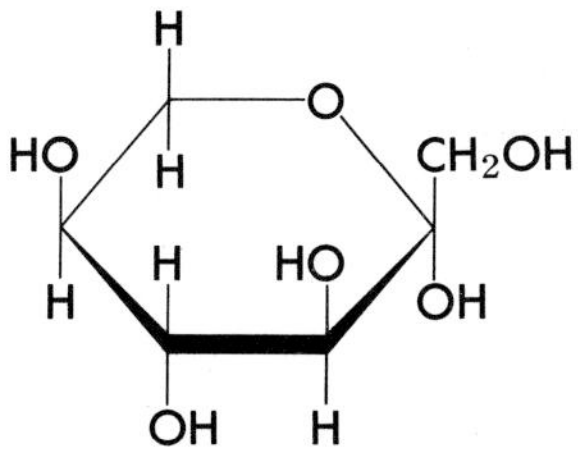

α-L-Sorbopyranose

Pentoses. Among the naturally occurring pentoses are the aldoses, L-arabinose, D-ribose, and D-xylose, and the ketopentose, L-xylulose.

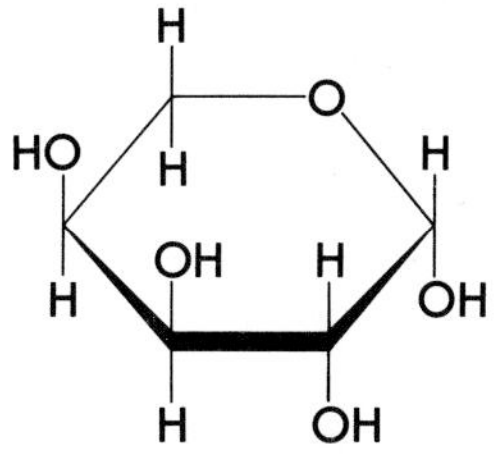

α-L-Arabinopyranose

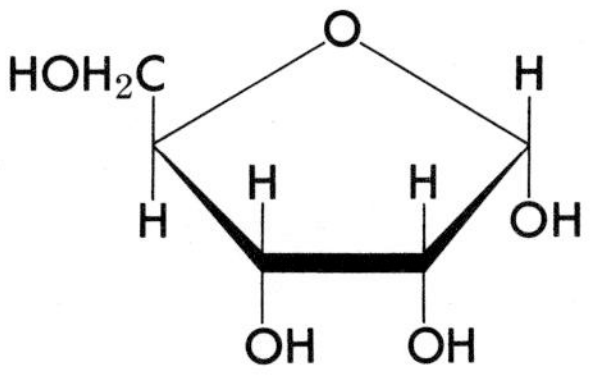

α-D-Ribofuranose

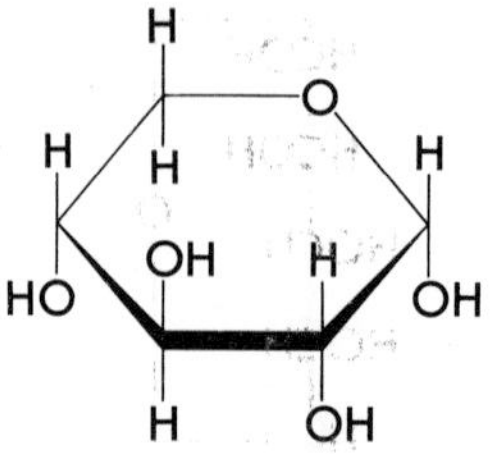

α-D-Xylopyranose

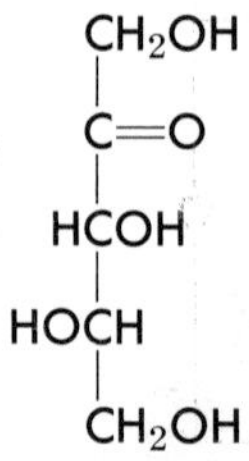

L-Xylulose

Whereas arabinose is found chiefly in plant products, ribose occurs as the characteristic sugar in certain nucelic acids (Chap. 9) of both plants and animals, and xylulose is an abnormal constituent of urine (Chap. 35). Xylose is of significance in structural components of animal tissues (Chap. 38) as well as being widely distributed in plants.

SOME REACTIONS OF MONOSACCHARIDES

Reaction of the Anomeric Carbon. Mention has been made (page 24) of glycosides in which the alcoholic hydroxyl of one monosaccharide portion has lost the elements of water in reaction with the hemiacetal hydroxyl of an adjacent monosaccharide. This type of linkage, the fundamental linkage of most di- and trisaccharides and all polysaccharides, is shown in maltose (4-O-α-D-glucopyranosyl-D-glucose).

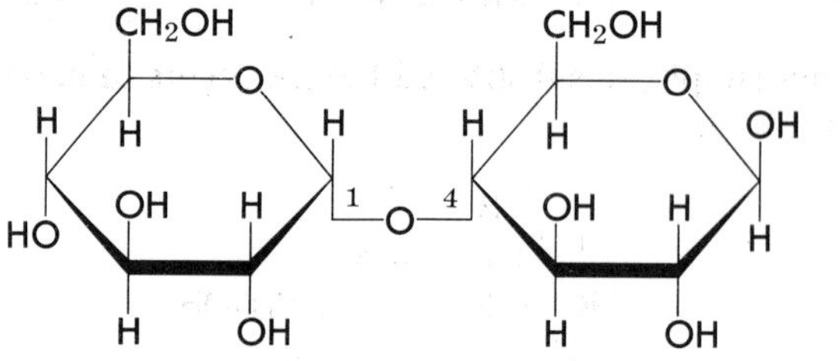

β-Maltose

Here the acetal linkage is between C-1 of one glucose moiety and C-4 of the other.

Phosphoric Acid Esters. Among the many known sugar esters, those of phosphoric acid occupy a unique position in biochemistry. Phosphoric acid esters are encountered with trioses, tetroses, pentoses, hexoses, and heptoses, as well as with derived sugars such as sugar acids and sugar alcohols. Indeed, phosphorylation of all sugars is the initial step in their metabolism. Thus, glucose is converted to glucose 6-phosphate. In many living systems, this compound may be further transformed into α-glucose 1-phosphate. Glucose 6-phosphate may also be converted to fructose 6-phosphate, which upon further phosphorylation yields fructose 1,6-diphosphate. The formulas of these esters are given at the top of page 27.

Methylation of Hydroxyl Groups. In addition to esterification, the hydroxyl groups may undergo etherification. Methylation of the *hemiacetal* hydroxyl on the anomeric carbon atom proceeds readily with methanol and acid catalyst and yields a glycoside, which is an acetal, not an ether. Methylation of the *alcoholic* hydroxyl groups requires much more vigorous conditions, *e.g.*, dimethyl sulfate plus alkali or methyl iodide plus silver oxide. Repeated treatment is often required to obtain

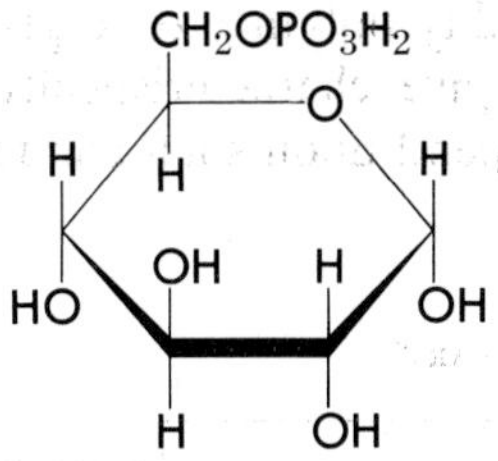

α-D-Glucose 6-phosphate

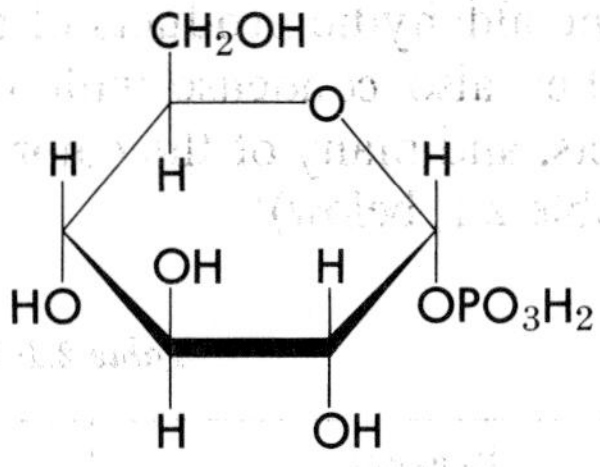

α-D-Glucose 1-phosphate

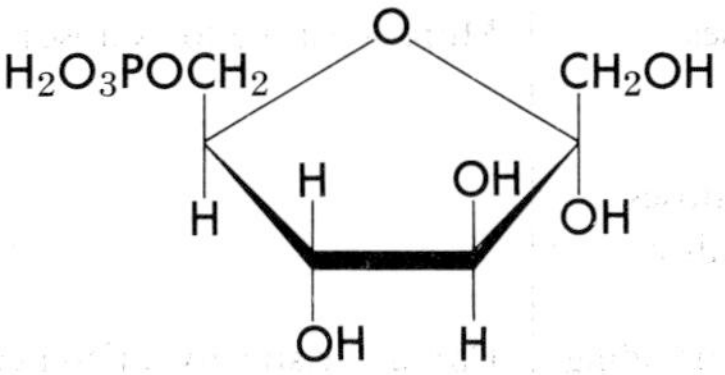

α-D-Fructose 6-phosphate

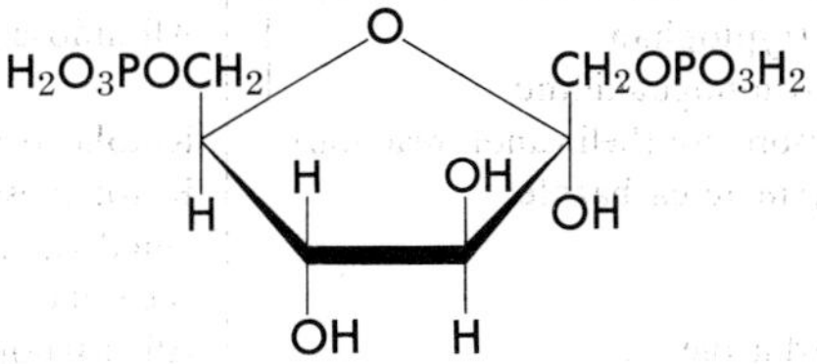

α-D-Fructose 1,6-diphosphate

etherification of all the hydroxyl groups. The reaction has been extensively employed to ascertain which hydroxyl groups in a sugar are free and available for reaction. Thus methyl glucopyranoside is methylated in positions 2, 3, 4, and 6, but not in position 5, which is involved in the hemiacetal link. Similarly with disaccharides and with polysaccharides, exhaustive methylation followed by hydrolysis—the ether link is very resistant to hydrolysis—yields methylated monosaccharides, the study of which may provide much structural information.

Reactions in Acid Solution. In strong mineral acid the characteristic reaction of sugars is dehydration. From pentoses, the product obtained is furfural.

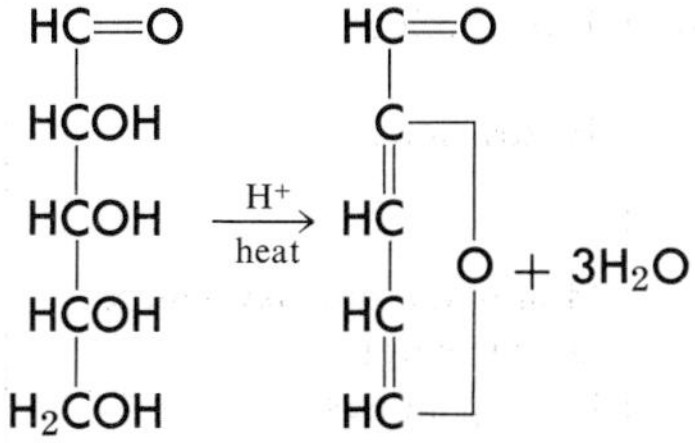

Furfural

With hexoses, the analogous reaction leads to formation of 5-hydroxymethylfurfural, which on further heating is transformed to levulinic acid.

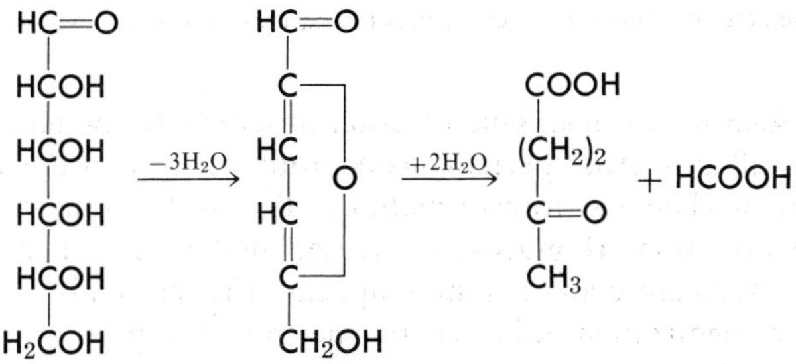

5-Hydroxymethylfurfural　Levulinic acid　Formic acid

The aldehydic products of these reactions readily polymerize to give brown tars. They also condense with various phenols to give characteristically colored products, and many of the color tests for sugars depend upon such condensations (see Table 2.1, below).

Table 2.1: COLOR REACTIONS OF SUGARS*

Reagent	Sugar type	Comment
α-Naphthol (Molisch reaction) Tryptophan Aminoguanidine	All aldoses and ketoses	More sensitive for ketoses
Resorcinol (Seliwanoff reaction)	Ketohexoses	
Cysteine-carbazole	Ketohexoses, ketopentoses, methylpentoses, dihydroxyacetone	
Carbazole	All carbohydrates, including uronic acids, deoxypentoses	Characteristically different colors with different sugars
Cysteine-H_2SO_4	Many sugars, including polysaccharides; generally used for hexoses	Different colors obtained with different sugars
Anthrone	Many sugars, including polysaccharides; generally used for hexoses	Different colors obtained with different sugars
Orcinol	Pentoses, heptuloses, uronic acids	Colors due to other sugars may be corrected for by independent methods. Uronic acids decarboxylate to pentose and then give reaction
Naphthoresorcinol	Uronic acids	
Acetylacetone-*p*-dimethylaminobenzaldehyde	Hexosamines	
Nitrite-indole	Hexosamines	Amino sugars give no color without prior deamination by nitrite
Diphenylamine	Mono- and dideoxypentoses	
Tryptophan-perchloric acid	Deoxypentoses	
Indole-HCl	Deoxypentoses	
Leucofuchsin (Feulgen reaction)	Deoxypentoses	
Thiobarbituric acid	Sialic acids	

* The reagents commonly employed and the carbohydrates for which each is most useful are given. All reactions shown occur in strongly acidic solution. None is entirely specific; except for the tests for hexosamines, all the other reactions are given by polysaccharides as well as monosaccharides.

Reaction in Alkaline Solution. One reaction observed at low temperatures with glucose in dilute alkali merits special consideration because of the analogy that it bears to certain biochemical transformations. This is the Lobry de Bruyn–van Ekenstein transformation. If glucose is treated under these conditions, as the glucose disappears, fructose and mannose appear. The mechanism of this reaction probably involves enolization, followed by the loss of a proton, resulting in the intermediate formation of an unsaturated enolate ion.

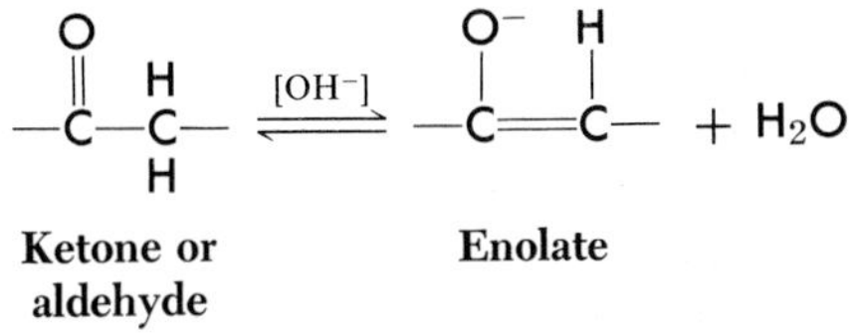

Enolization is a general property of aldehydes or ketones, the α-carbon of which bears a hydrogen atom.

This relationship among glucose, fructose, and mannose is one of the consequences of the identical steric configurations of these compounds about carbon atoms 3, 4, and 5.

Reduction of Monosaccharides. When a simple sugar is treated with H_2 gas under pressure in the presence of a metal catalyst, or with an active metal, such as Ca, in water, the carbonyl group is reduced to an alcoholic hydroxyl group, yielding a polyhydric alcohol. D-Glucose, under these circumstances, yields sorbitol (D-glucitol), which, it will be noted, is also a product of the reduction of L-sorbose.

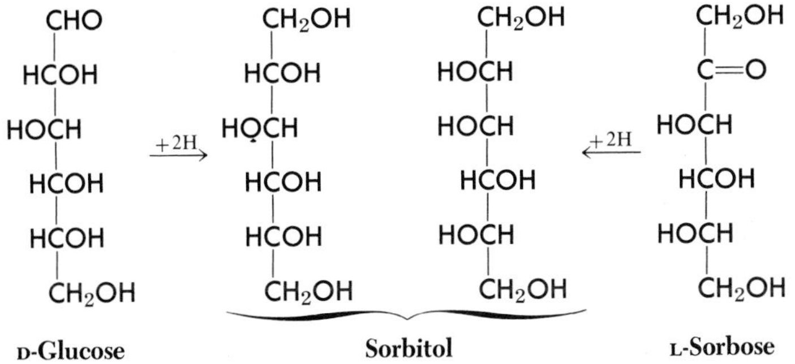

The reduction products of mannose and galactose are termed, respectively, mannitol and dulcitol.

Oxidation of Monosaccharides. Many oxidizing agents attack aldoses. Among these perhaps the simplest to picture is alkaline hypohalite. Sodium hypoiodite oxidizes aldoses to give the sodium salts of the aldonic acids.

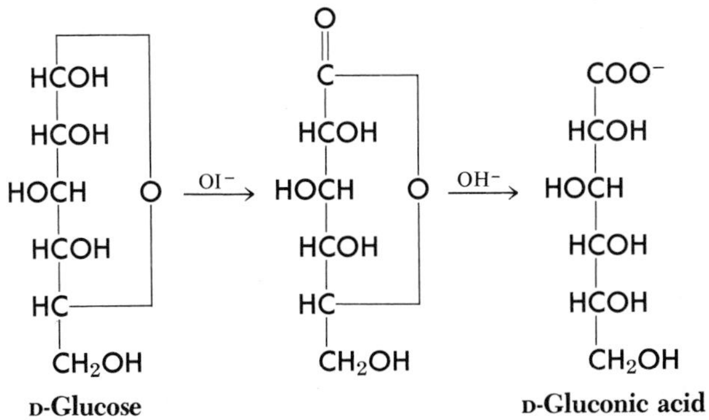

This oxidation is specific for the aldoses and is the basis for an analytical method that distinguishes them from ketoses. In this method the unspent NaOI is reconverted into I_2 by acidification, and the I_2 may then be titrated with sodium thiosulfate. The specific nature of this reaction and the excellent yields obtained make it a valuable preparative method as well.

In alkaline solution monosaccharides are susceptible to oxidation by a variety of agents, including cupric ion, silver ion, and ferricyanide ion. Although these reagents are widely employed in analytical methods for sugars, mixtures of products are generally obtained, rendering them of little value in preparative chemistry.

Identification of Individual Sugars. For the qualitative analysis of a sugar or sugars present in a mixture the biochemist employs several approaches. Various more or less specific color tests for certain classes of sugars have been devised (Table 2.1), and the dietary habits of various microorganisms have been used to advantage. In addition to the native fermentative abilities of microorganisms, various organisms artificially adapted to the consumption of specific carbohydrates have been used for analytical purposes. Chromatographic methods are particularly useful for the identification of carbohydrates (pages 35 *ff.*).

SUGAR ACIDS

Mention has been made of the sugar acids, of which the most important are the following:

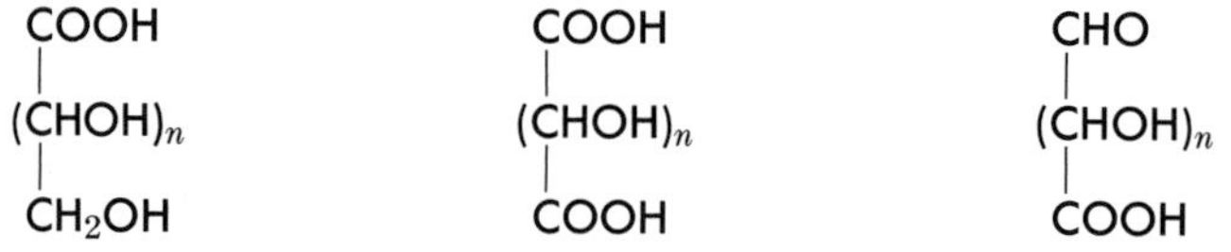

Aldonic **Aldaric** **Uronic**

These compounds, like other γ- and δ-hydroxy acids, tend to form inner esters or lactones, generally with the establishment of a five- or, more rarely, a six-membered ring.

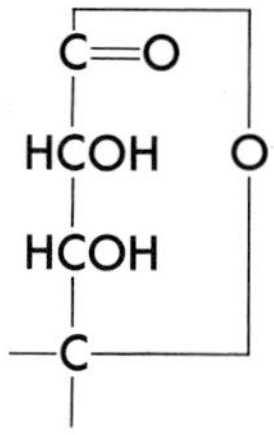

γ-Lactone

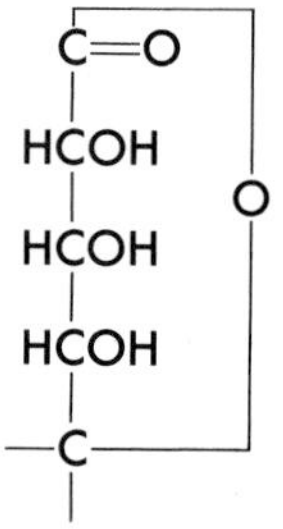

δ-Lactone

These compounds are strong acids, and their salts are soluble in water and yield neutral solutions. Gluconic acid, nontoxic and well metabolized, is often employed for the introduction into the body of a cation such as Ca^{++}.

δ-Gluconolactone may be formed in an aerobic oxidation catalyzed by an enzyme termed *glucose oxidase.* This enzyme, available from the mold *Penicillium notatum,* catalyzes the following reaction.

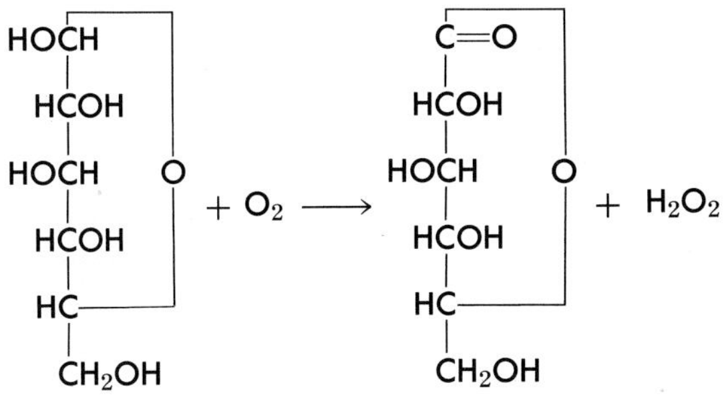

β-D-Glucose δ-Gluconolactone

Glucose oxidase exhibits marked specificity for β-D-glucose and has been applied to the quantitative determination of glucose, since the hydrogen peroxide which is stochiometrically generated may be measured quantitatively.

Glucuronic acid,

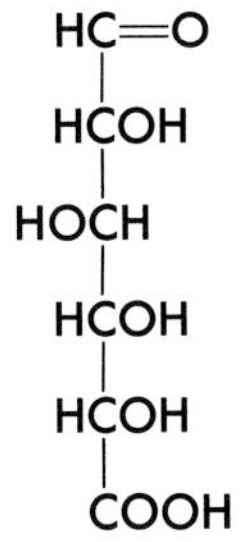

D-Glucuronic acid

occurs in human urine where it is bound in glycosidic linkage to various hydroxylated compounds, *e.g.*, phenols, steroids, etc. The enhanced solubility in water incident to glucosiduronic acid formation from these alcohols may well render them more readily disposable by the body. Glucuronic acid may also form esters, as with bilirubin, a bile pigment (Chap. 31). Glucuronic acid as a component of many polysaccharides is considered in the following chapter as well as in Chap. 38.

A special case of a sugar acid of great biological importance, widely distributed in animal and vegetable nature, is vitamin C, ascorbic acid (L-xyloascorbic acid—vitamin C). From its formula,

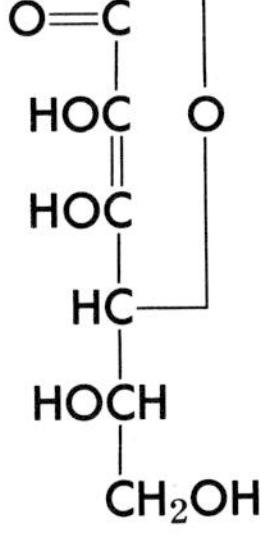

Ascorbic acid

it will be recognized as a γ-lactone of a hexonic acid which differs from other examples previously discussed in that it contains a double bond between carbons-2 and -3, an enediol linkage. By virtue of this arrangement it is a very unstable compound, one of the most readily oxidizable substances isolated from natural sources. Ascorbic acid is readily oxidized by oxygen, in alkaline solution in the presence of traces of metal ions, to dehydroascorbic acid.

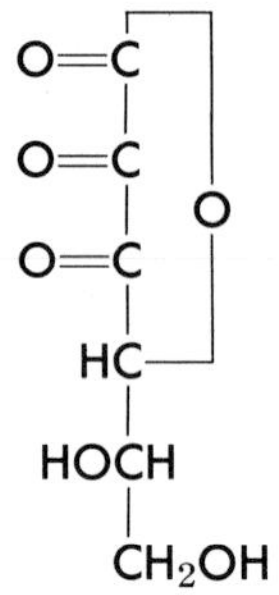

Dehydroascorbic acid

This is a reversible reaction. Further oxidation of dehydroascorbic acid is accompanied by cleavage of the carbon skeleton between carbons-2 and -3.

SUGAR ALCOHOLS OR POLYOLS

Whereas the linear alcohols of the types of sorbitol, dulcitol, and mannitol are of little biochemical interest, the 3-carbon member of the series, glycerol, is of enormous importance. Its chemistry will be considered with the chemistry of lipids in view of the fact that its fatty acid esters belong to this class of substances. Its intensely sweet taste serves as a reminder that structurally it belongs to the same class of compounds as sorbitol.

The carbocyclic polyols of interest in biochemistry are hexitols (page 29). Nine stereoisomeric modifications are possible, and these are indicated as having the ring at right angles to the plane of the paper. Seven of these are internally compensated or *meso* forms, while two are a pair of optically active antipodes. The best-known compound of the group is called *myo*-inositol.

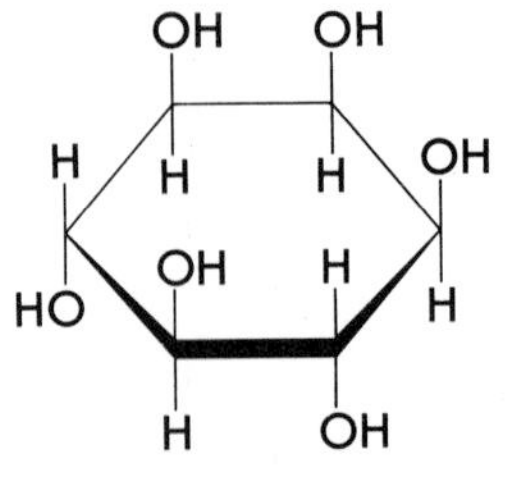

***myo*-Inositol**

Myo-inositol is widely distributed among microorganisms, higher plants, and animals. In plants it is found phosphorylated as phytic acid, the hexaphosphate, or as the mixed magnesium calcium salt of phytic acid, phytin. It also occurs in lower states of phosphorylation in plants as well as in animal tissues and as free inositol in muscle, heart, lung, liver, etc. It is a constituent of certain phospholipids called phosphoinositides (page 72). Biochemical interest in *myo*-inositol also relates to its nutritional essentiality, under certain circumstances, and to its role in lipid as well as in carbohydrate metabolism. *d*-Inositol, *l*-inositol, and scyllitol are stereoisomers of *myo*-inositol of more limited biological distribution.

AMINO SUGARS

In these compounds, the hydroxyl group is replaced by an amino group, —NH_2 on carbon atom 2 of aldohexoses. Two representatives of this class are D-glucosamine and D-galactosamine, which are always found in nature as the N-acetylated compounds.

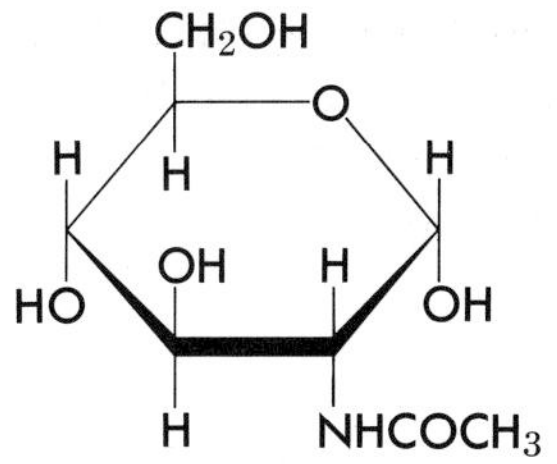

N-Acetyl-D-glucosamine

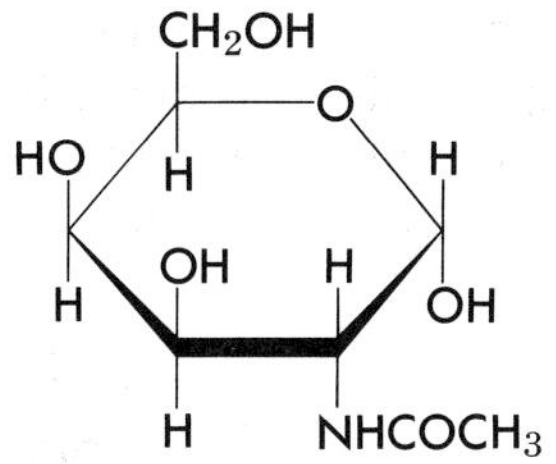

N-Acetyl-D-galactosamine

Glucosamine is the product of hydrolysis of chitin, the major polysaccharide of the shells of insects and crustaceans, and occurs in various mammalian polysaccharides (Chaps. 3 and 38) and in certain proteins (Chap. 6). Galactosamine is found in the characteristic polysaccharide of cartilage, chondroitin sulfate, and in a number of glycosphingolipids (page 74).

Additional amino sugars have been described, particularly in microorganisms, and include the 2-amino derivatives of D-gulose and D-talose (Fig. 2.1, page 16).

DEOXYSUGARS

Several examples of aldohexoses are known in which the terminal —CH_2OH group is replaced by —CH_3; L-rhamnose (6-deoxy-L-mannose) and L-fucose (6-deoxy-L-galactose) are examples.

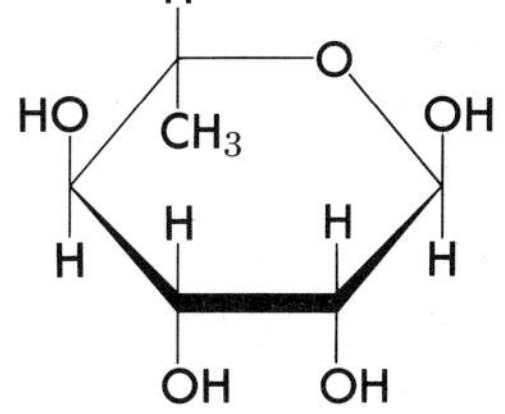

β-L-Rhamnose

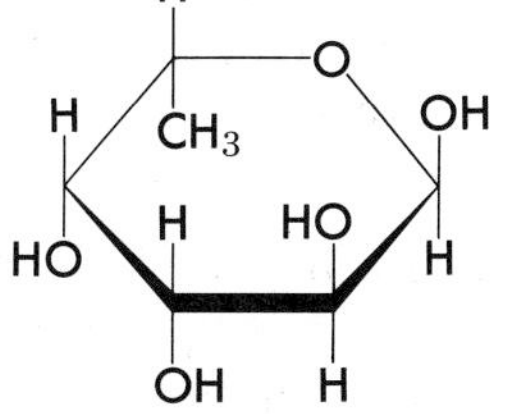

β-L-Fucose

Of great interest in biochemistry is 2-deoxyribose,

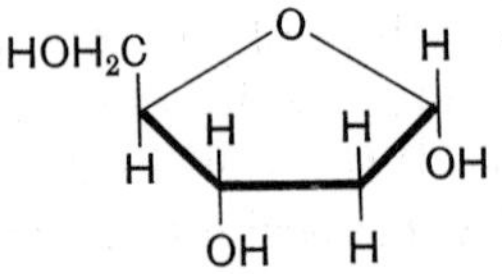

2-Deoxy-D-ribose

which shares with ribose the sugar function in the group of compounds that includes nucleosides, nucleotides, and nucleic acids. The nucleic acids are generally classified according to whether they contain ribose or deoxyribose as their constituent sugar (Chap. 9).

Reference has been made previously to diamino-, dideoxy-, and trideoxy sugars, characteristic particularly of bacteria (page 10).

Sialic Acids. Sialic acids are a group of naturally occurring N- and O-acyl derivatives of a 9-carbon, 3-deoxy-5-amino sugar acid called neuraminic acid. The sialic acids are ubiquitously distributed in bacteria and in animal tissues, having been identified as constituents of lipids, of polysaccharides, and of mucoproteins. Sialic acid from human plasma has the following structure.

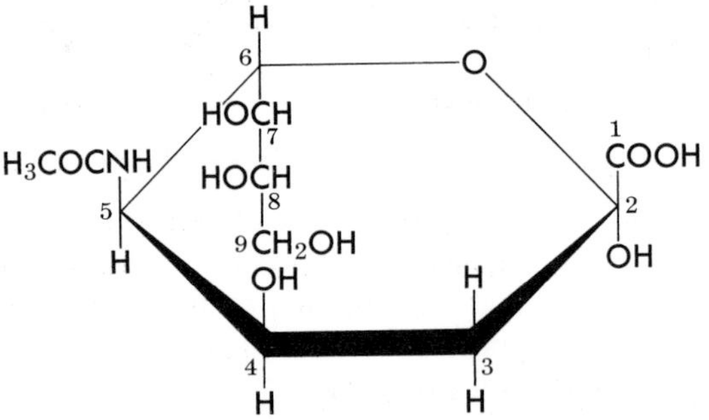

N-Acetylneuraminic acid

This substance may be envisaged as an aldol condensation product of N-acetylmannosamine and pyruvic acid, and certain bacterial enzymes can effect hydrolysis to these products.

Sialic acids present in bovine and sheep submaxillary gland mucin are diacetyl derivatives, with the additional acetyl group present as 7-O-acetyl. An 8-O-methyl derivative of N-acetylneuraminic acid has also been described. In sialic acids found in pig submaxillary mucin, and pig, horse, and ox erythrocytes, the N-acetyl group is replaced by a glycolyl, $-\overset{\overset{\displaystyle O}{\|}}{C}-CH_2OH$, grouping. In mucoproteins containing sialic acid, the latter is linked in the molecule via a glycosidic bond at carbon-2.

Muramic Acid. The mucopeptides of bacterial cell walls ("mureins," Chap. 41) are characterized by yet another complex sugar derivative which is also an N-acetylamino sugar acid. Like sialic acid, N-acetylmuramic acid may be regarded as derived from a 3-carbon acid, in this case D(-)-lactic acid and N-acetylglucosamine. The link between the two portions is an ethereal one, as shown in the formula.

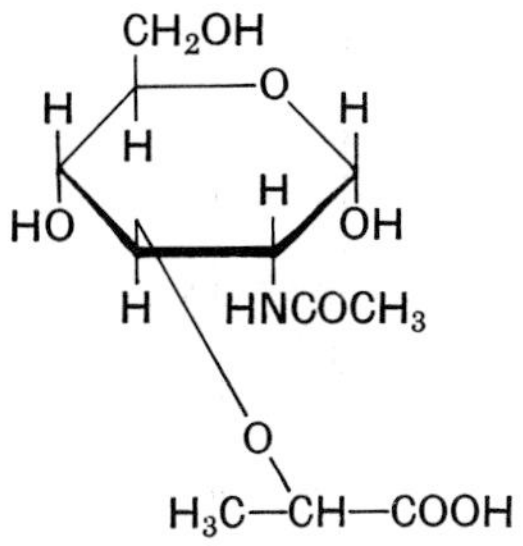

N-Acetylmuramic acid

SPECIAL METHODS OF SEPARATION, PURIFICATION, AND CHARACTERIZATION OF COMPOUNDS OF BIOCHEMICAL INTEREST

The methods to be described are not peculiar to biochemistry. They have, however, been extraordinarily useful to biochemists because of the small quantities of material that can be handled and the mild conditions that may be used. Often simple in application yet very effective in separations achieved, these methods have largely replaced earlier techniques of fractional crystallization and fractional distillation.

Fundamentally, all the methods to be discussed entail the same general principle. The material to be purified or the mixture to be resolved is repeatedly distributed between two phases. One phase may be stationary, while the other flows past it, which is the situation in chromatography. Alternatively, both phases may move past each other in opposite directions, as occurs in countercurrent distribution. Both phases may be liquid, which is the case in countercurrent distribution and in partition chromatography. One phase may be solid, as in sorption chromatography or in ion exchange chromatography. In gas phase chromatography, as the name implies, the mobile phase is indeed gaseous. One or more of these methods has been successfully applied to each class of organic compound of concern to the biochemist. Thus, mixtures of sugars are often separated by partition chromatography on paper, as are amino acids, peptides, etc. Steroid mixtures are separable in most cases by sorption chromatography; gas-phase separations have superseded all other methods for the resolution of fatty acid mixtures. A general survey of some of these methods is introduced at this point even though their application is in no sense peculiar to carbohydrate chemistry. Because they are related to one another, both in theory and in operation, they are here treated together.

Chromatographic procedures are used to detect mixtures of substances and to separate the components of mixtures. The technique of chromatography depends on the use of solid adsorbents which have specific affinities for the adsorbed substances. It is customary to use the adsorbent in a column and allow the solution to percolate slowly down the column (Fig. 2.3, on left). The solutes may be completely adsorbed at the top of the column (Fig. 2.3, middle), in which case an eluting solvent is added to the column; this liberates each substance according to its adsorbing affinity. When the substances are colored, zones or bands may be observed

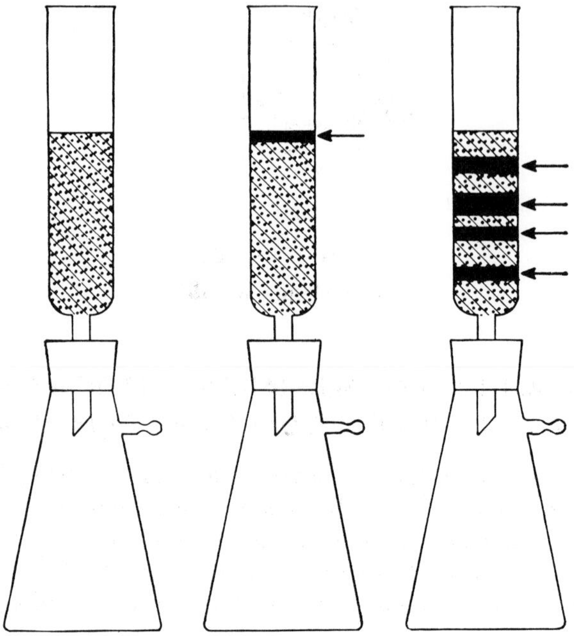

Fig. 2.3. A diagrammatic representation of the simplest type of chromatogram. A column of a suitable adsorbent is prepared (*left*). A solution of a mixture of four substances is poured on the column and is allowed to move slowly either by gravity or with a slight vacuum applied. The mixture is adsorbed as a narrow band at the top of the column (*middle*). When an eluting solvent is poured on the column, the substances are separated and, because they are colored, show four discrete zones (*right*).

on the column (Fig. 2.3, on right). Tswett, the Russian botanist who is generally credited with the discovery of this technique, showed that extracts of a green leaf contain two green pigments (chlorophylls a and b) and a number of yellow pigments (carotenoids).

Although the procedure is called chromatography since the behavior of colored substances is readily observed, the method can be applied to colorless substances also. However, some procedure is necessary for detecting colorless substances, and this may be done by illuminating with ultraviolet light and observing fluorescent substances or by using indicators or chemical procedures which give a color reaction. In most cases, it is more convenient to collect fractions of the eluting solvent as they emerge at the exit of the column. These fractions may then be tested by physical or chemical procedures or by biological assays if specifically active substances are present.

Homogeneous substances will give a single zone on the column or a single continuous fraction in the eluate. When suitable solvents are used, mixtures will give multiple zones or multiple eluted fractions, each of which is separated by a portion of the solvent. When uncharged substances, such as the chlorophylls, are separated by simple adsorption, the process is called *adsorption chromatography*. Ionic substances, *e.g.*, amino acids, can be readily separated by using polar particles as the adsorbing medium. This process is *ion exchange chromatography* (see Chap. 5).

The method of countercurrent distribution depends on the repetitive distribution of a solute between two immiscible solvents in a series of vessels in which the two solvent phases are in contact. The procedure may be carried out in a consecutive number of separatory funnels in which the solute in solvent *A* is shaken with solvent *B* until equilibrium is reached, the two phases are separated, and each is then equilibrated with fresh solvent. The process may be repeated as many times as desired. Automatic systems for the multiple extractions have been devised by Craig and his collaborators. These investigators have shown that at the end of the experiment, a homogeneous substance will be distributed in the various units (each separatory funnel represents a unit) in accordance with a constant termed the *distribution coefficient* or sometimes the *partition coefficient.* This constant represents essentially the ratio of the molecular concentration of the substance in the two solvents, which in turn is dependent upon the solubility coefficients. The separation of a mixture of organic acids is shown in Fig. 2.4. A mixture will give a skewed curve or several distinct curves, depending on the solvents and the number of units used. The method of countercurrent distribution has been successfully used in the analysis and separation of many types of substances.

Partition chromatography applies the principle of countercurrent distribution to column chromatography. The procedure depends on using two immiscible solvent systems with a solid, supporting medium, such as silica gel or starch, in a column. One solvent phase, usually the aqueous one, is more strongly adsorbed to the solid phase. The other solvent, containing a mixture of substances, is allowed to percolate down the column. At each particle, a redistribution of solute occurs

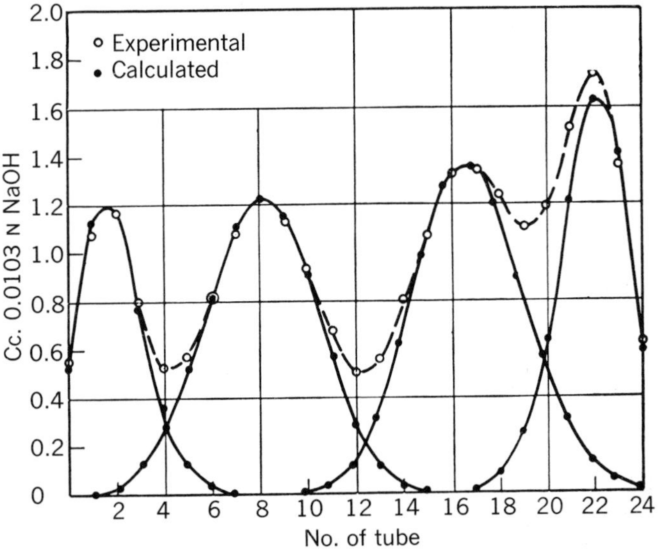

Fig. 2.4. Separation by countercurrent distribution of a mixture of four organic acids. The maxima, from left to right, are acetic, propionic, butyric, and valeric acids, respectively. The acid content of each tube was determined by titration with sodium hydroxide. (*From L. C. Craig and D. Craig, in "Technique of Organic Chemistry," vol.* 3, *p. 171, Interscience Publishers, Inc., New York,* 1950.)

between the mobile and stationary phases, just as it does in a separatory funnel. The process is very efficient since relatively short columns act in the same manner as many separatory funnels. The process, as developed by Martin and Synge, can be employed for the separation of many types of substances.

One of the most ingenious methods of utilizing partition chromatography is the employment of filter paper as the supporting medium; hence the term, *paper chromatography.* In this method, a drop of solution containing a mixture of compounds is placed on one end or one corner of a sheet of filter paper. The filter paper is put into a sealed glass jar or cylinder which contains a small amount of an organic solvent saturated with water. The end of the paper nearest the mixture is inserted into the solvent-water mixture in the bottom of the container, with the paper being placed to hang freely without touching the sides of the vessel. Thus, the solvent will ascend into the paper, and this procedure is termed "ascending paper chromatography." Alternatively, the same end of the filter paper may be put into the solvent mixture contained in a narrow trough mounted near the top of the vessel. In this case, the solvent will descend into the paper; this procedure is termed "descending paper chromatography." After some time, the sheet of paper is removed and sprayed with a chemical developing agent, such as an alkaline silver reagent, which, by formation of a colored spot, serves to localize compounds on the paper. The paper may then be dried, and measurement made of the distance of migration of each compound from the point at which the mixture was placed on the filter paper. Since the relative rate of migration of each substance is a characteristic constant for each solvent system used, the identity of compounds on the filter paper may be established by measurement of the distance of migration with time. This migration rate is expressed as the R_F value, which is defined as the ratio of the distance moved by a particular solute to that moved by the solvent front. Thus, when a substance moves with the solvent front, the R_F is 1; when a substance moves half the distance of the front, the R_F is 0.5. Figure 2.5 illustrates schematically the separation of a mixture of six carbohydrates by filter paper chromatography.

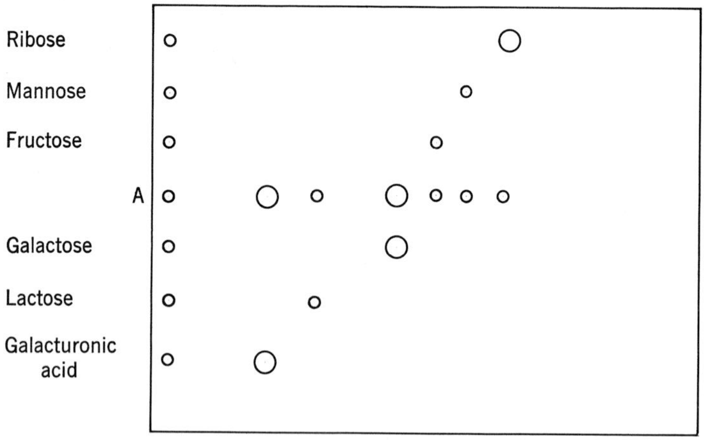

Fig. 2.5. A paper chromatogram of a mixture containing six sugars. The mixture was introduced in the circle at the left marked *A*, and the individual sugars were placed in the other circles at the left as indicated. The chromatogram was run in *s*-collidine. (*From S. M. Partridge, Biochem. J.,* **42,** 238, 1948.)

A variation of the above-described filter paper chromatography has been termed "two-dimensional chromatography." After migration of the substances in one direction is completed, the sheet of filter paper may be turned 90° and placed in a different solvent mixture to obtain further resolution. This type of procedure gives excellent separations; it is illustrated later in Chap. 5 (Fig. 5.4, page 114).

Thin layer chromatography combines many of the advantages of paper chromatography with those of chromatography on columns. Thin films of uniform thickness are prepared of adsorbents such as silica or alumina powder. From this point on, these plates are treated much as described above for paper sheets, with an almost limitless range of available solvents. Because of the inorganic nature of the supporting medium, concentrated sulfuric acid spray followed by heating may be used to detect substances on the chromatogram, in an entirely nonspecific way, by charring and rendering visible any spots of organic material.

Another important technique involves use of a granular dextran (page 50) preparation. Each granule behaves as though it were a small sac of dialyzing membrane. The two phases in a column packed with the dextran in water are (1) the water outside the granules, and (2) the water inside the granules. Small solute molecules become distributed in both phases; large molecules are excluded from the inside phase. If a solution containing a mixture of large and small molecules is placed at the top of such a column, followed by a slow stream of solvent, the large molecules will appear at the effluent end of the column ahead of the small molecules. Columns of this sort conveniently separate large from small molecules regardless of their chemical natures. The cutoff between what is large and what is small is determined by the arbitrary choice of the packing material, granules of differing porosities being available. The process has been likened to a *molecular sieve* and has been termed *gel filtration chromatography.*

Gas-liquid chromatography is a modification applicable to compounds and mixtures of compounds that exert significant vapor pressures at temperatures below those of excessive pyrrolysis. All compounds, including carbohydrates, amino acids, and lipids that can be converted to stable, volatile derivatives, can be separated by vapor phase chromatography. Since the method has been particularly applicable to separation of lipids, it will be considered later (Chap. 4). *Ion exchange chromatography,* particularly useful in the separation of amino acids, will be considered in Chap. 5.

REFERENCES

See list following Chap. 3.

3. The Carbohydrates. II

Oligosaccharides and Polysaccharides

Many compounds on hydrolysis yield monosaccharides either exclusively or together with other products. Some of the properties of the more important compounds of this type will be reviewed.

OLIGOSACCHARIDES

Oligosaccharides yield monosaccharides on hydrolysis and contain, per molecule, two to ten monosaccharide residues (*oligo-,* few). The distinction between oligo- and high molecular weight polysaccharides is arbitrary since properties of higher members of the class of oligosaccharides merge with those of the lower members of the polysaccharide class.

A large number of oligosaccharides have been described; many of these do not occur as such in nature but have been prepared by partial hydrolysis of higher oligosaccharides and polysaccharides. Most of the naturally occurring representatives occur in plant rather than in animal sources. The characteristics of oligosaccharides will be discussed with reference to the three most important disaccharides, sucrose, maltose, and lactose, and other representatives will be referred to only incidentally.

The methods employed in elucidating the structure of sucrose and other oligosaccharides may be mentioned briefly. The foremost of these involves exhaustive methylation, usually with dimethyl sulfate and alkali, followed by hydrolysis and identification of the methylated monosaccharides produced. Under these circumstances methoxyl groups (CH_3—O—) will be found to have replaced all hydroxyl groups except those participating in glycosidic or hemiacetal links, and from the distribution of these methoxyl groups the position of the glycosidic bond can be inferred. Additional information may also be secured from studies of periodate oxidation (page 44).

The determination of configuration of the glycosidic bond, whether α or β, often rests on the use of enzymes with known specificities. Thus there are *glucosidases,* enzymes which promote the hydrolysis specifically of glucosides, whose activity is restricted toward either α- or β-glucosides. Yeast maltase, for instance, attacks only α-glucosides, while almond emulsin is specific for β-glucosides.

SUCROSE

Sucrose, α-D-glucopyranosyl-β-D-fructofuranoside,

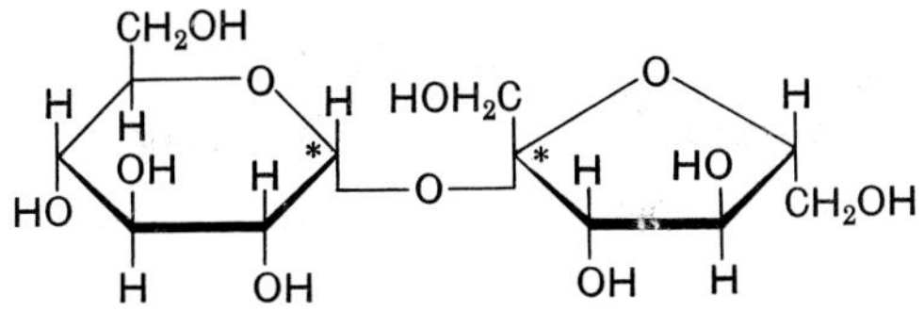

Sucrose

is the common sugar of commerce and the kitchen. It is derived commercially from either cane or beet but occurs in varying amounts in a variety of fruits, seeds, leaves, flowers, and roots. Together with its products of hydrolysis, glucose and fructose, it is the major carbohydrate of maple sugar.

In contrast to most other simple carbohydrates, the oxygen bridge between the two monosaccharide moieties in sucrose is between the anomeric carbon atoms (marked in the above formula with asterisks) of the glucose and fructose portions. As a consequence, sucrose lacks those properties of sugars which depend upon the presence of an active carbonyl group. Thus sucrose does not exhibit mutarotation or react with specific reagents for carbonyl groups, such as phenylhydrazine. Sucrose is not a reducing sugar and hence is not oxidized by alkaline solutions of Cu^{++} ions.

On hydrolysis, sucrose yields an equimolar mixture of glucose and fructose. This mixture will readily reduce Cu^{++} ions to Cu^{+} and will also react with phenylhydrazine to yield a single product, glucose phenylosazone.

All glycosides, including oligo- and polysaccharides, can be hydrolyzed with varying degrees of difficulty. The hydrolysis of sucrose, which proceeds more easily than that of most other disaccharides, has been called the *inversion of sucrose* because during the reaction, which may be followed polarimetrically, the sign of rotation changes from positive (*dextro-*) to negative (*levo-*). This reaction is catalyzed by enzymes called *invertases* and also by H^{+} ions. In fact, in aqueous solution the rate of sucrose inversion is proportional to the activity of H^{+}, other things being equal, and this relationship provided an early method for the estimation of hydrogen ion concentration. The equimolar mixture of glucose and fructose, called "invert sugar," is sweeter than sucrose to the taste. Such a mixture is present in honey.

MALTOSE

This disaccharide is the most common example of a sugar which on hydrolysis yields two identical fragments. It is 4-O-α-D-glucopyranosyl-D-glucopyranose. In contrast to sucrose, maltose possesses an unattached anomeric carbon atom, a hemiacetal grouping, and is consequently a reducing sugar, capable of reaction with carbonyl reagents and exhibiting mutarotation. In the formula (page 42), this anomeric carbon atom is shown in the β configuration, hence the designation β-maltose. Although the anomeric carbon may exist in either the α or β configuration, the

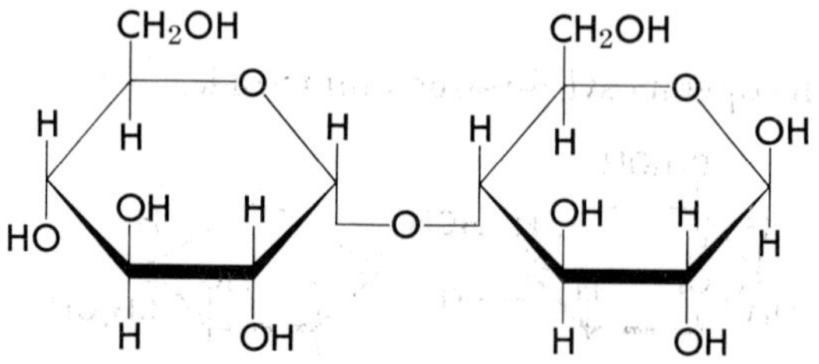

β-Maltose

glycosidic linkage in maltose must be in the α configuration. This linkage joins carbon-1 of one glucose molecule and carbon-4 of the second unit and is abbreviated as an α-1,4 linkage. Occasionally in the literature the convention α-1,4′ is used to indicate that reference is made to C-1 and C-4 of different carbon chains. Also, the mode of linkage is often designated by use of an arrow, *e.g.*, α-1 → 4.

Although maltose does not occur abundantly, as such, in nature, its occurrence has occasionally been reported. This is not surprising as it is the major product of enzymic hydrolysis of starch (page 47). Low molecular weight oligosaccharides comprised of maltose residues have been isolated from mammalian liver and probably represent intermediates in the degradation of glycogen (see below).

Maltose, with its α-1,4 glucosidic bond, is but one of several known glucosyl-glucose disaccharides. *Cellobiose,* in which the bond is β-1,4, is the repeating unit in the polysaccharide, cellulose; maltose is the analogous unit in the starch, amylose. In *gentiobiose* the link is β-1,6, whereas in *isomaltose* it is α-1,6. In *trehalose* the bridge between the two glucose moieties connects the two anomeric carbon atoms and this sugar, consequently, resembles sucrose in having no reducing activity. Trehalose is the principal sugar of insect hemolymph.

LACTOSE

This disaccharide, apparently solely of mammalian origin, is found in milk to the extent of about 5 per cent. It is 4-O-β-D-galactopyranosyl-D-glucopyranose.

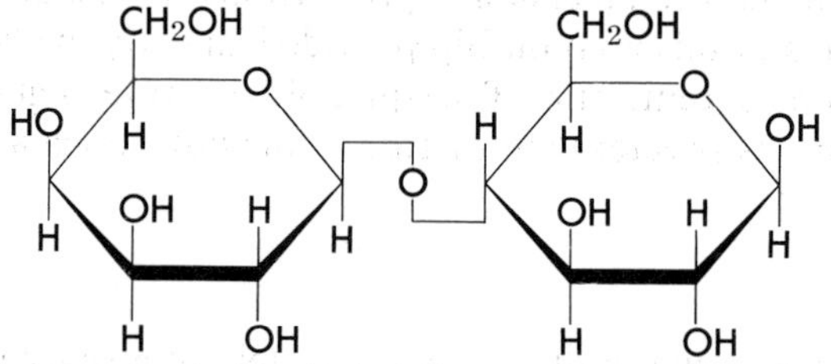

β-Lactose

Upon hydrolysis it yields an equimolar mixture of galactose and glucose. Lactose is a reducing sugar and reacts with carbonyl reagents.

A minor quantity of lactose is also present in milk linked as an α-glycoside to the anomeric (C-2 of N-acetylneuraminic acid (page 34) via the 3-hydroxyl of the galactose moiety of lactose. The compound has been termed neuramin lactose; it has also been isolated from human urine.

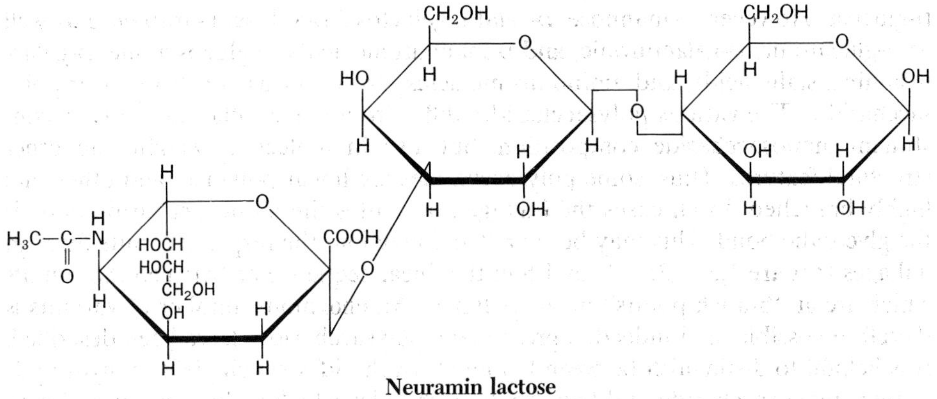

Neuramin lactose

GLYCOSIDES CONTAINING NONCARBOHYDRATE RESIDUES

In addition to glycosides of the oligosaccharide type, there are in nature many examples of compounds containing a sugar or a uronic acid linked in glycosidic bond to some noncarbohydrate residue. Most of these glycosides occur in plants. Upon hydrolysis, one generally obtains a sugar or mixture of sugars and a nonsugar portion, termed the *aglycon.* Although the glycoside is a larger molecule than the aglycon derived from it, the glycoside is usually more soluble in water than the aglycon because of the hydrophilic hydroxyl groups of the sugar.

Among the interesting aglycons of nature are a wide variety of phenol derivatives. The glycosides of various flavones and anthocyanins occur as pigments in flowers, and rutin from buckwheat is such a compound. The poison phlorhizin, found in the roots of many fruit trees, is the glycoside of a closely related polyphenol, phloretin, and has had wide experimental use because of its depressant effect upon the renal threshold for glucose. Indoxyl, from which the dye indigo is prepared, also is found in nature as a glycoside.

In oil of bitter almond is found the glycoside amygdalin, which on hydrolysis yields two moles of glucose and one mole each of benzaldehyde and hydrogen cyanide. Noteworthy also is the fact that β-glucosidase, an enzyme that catalyzes the hydrolysis of amygdalin, is present in emulsin, a preparation derived from almonds, among other sources. The study of this enzyme-catalyzed reaction, dating back to 1830, is one of the earliest in the history of enzyme chemistry.

A number of phenanthrene derivatives are conjugated to sugars in nature. Of particular importance in medicine are the drugs of the digitalis group, the aglycons of which are closely related to the steroids (Chap. 4). The carbohydrate portions of these glycosides are often di- and trisaccharides containing 6-deoxy- and 2,6-dideoxyaldohexoses.

POLYSACCHARIDES

The overwhelming bulk of all carbohydrates of nature exists in the form of molecules of high molecular weight, which on hydrolysis yield exclusively or chiefly monosaccharides or products related to monosaccharides, most frequently

D-glucose. However, D-mannose, D- and L-galactose, D-xylose, L-arabinose as well as D-glucuronic, D-galacturonic, and D-mannuronic acids, D-glucosamine, D-galactosamine, sialic acids, and amino uronic acids also occur as constituents of polysaccharides. The various polysaccharides differ from one another not only in constituent monosaccharide composition, but also in molecular weight and other structural features. Thus, some polysaccharides are linear polymers and others are highly branched. In all cases the linkage that unites the monosaccharide units is the glycosidic bond. This may be α or β and may join the respective units through linkages that are 1,2; 1,3; 1,4; or 1,6 in the linear sequence or between those units which are at "branch points" in the polymer. An enormous number of variants is therefore possible, and indeed, a great many polysaccharides have been described. It is helpful to distinguish between homopolysaccharides which yield on hydrolysis a single monosaccharide and heteropolysaccharides which yield a mixture of constituent building units.

Methods of investigation of polysaccharides include a variety of techniques for estimation of molecular weight. Among these are osmotic pressure measurements, ultracentrifugation studies, observations of viscosity and of light scattering, and end group analysis (see below). Most if not all polysaccharides studied are *polydisperse, i.e.*, even though meeting other criteria of purity, each sample consists of molecules of various molecular weights. For example, a sample of "starch" may contain molecules which have between perhaps 100 and 10,000 glucose units per molecule, yet all are properly called "starch."

Analysis of the products of complete and partial, often enzymic, hydrolysis may reveal the sequence of occurrence of various monosaccharide units in the chain. Study of the mixture obtained when an exhaustively methylated polysaccharide is hydrolyzed is also often revealing. If, for example, an unbranched polyglucoside (1,4) is studied in this way, the chain length may be estimated. One end of such a chain will be occupied by a glucose residue unsubstituted at carbon-4. If the chain has 10 glucose residues, one should find 2,3,6-tri-O-methylglucose and 2,3,4,6-tetra-O-methylglucose in a ratio of 9:1. Wherever a branching of the polysaccharide chain occurs, there must be a glucose residue with an additional substituted hydroxyl group and this should yield a di-O-methylglucose residue. This method of analysis thus gives information as to size, number of end groups, number of branchings, and positions of attachment of glucosidic bonds.

Periodic acid is another useful reagent in studies of polysaccharide structure. Oxidation of sugars with this reagent under controlled conditions is coupled with quantitative measurements of periodic acid consumed and formic acid produced. Glycols are cleaved by periodic acid with formation of the dialdehyde and consumption of a mole of periodate. If three or more contiguous hydroxyl groups are present, the middle carbon atom is liberated as formic acid, the dialdehyde is formed, and two moles of periodate are used. Thus the reaction of methyl α-D-glucose (α-methylglucoside) with periodic acid is as shown on page 45.

Periodic acid treatment of a polysaccharide consisting of repeating units of hexose linked linearly in 1,4-glycoside linkages would yield one mole of formic acid from one end, that at which the hexose is linked glycosidically. Two additional moles of formic acid would derive from the reducing end of the polysaccharide chain,

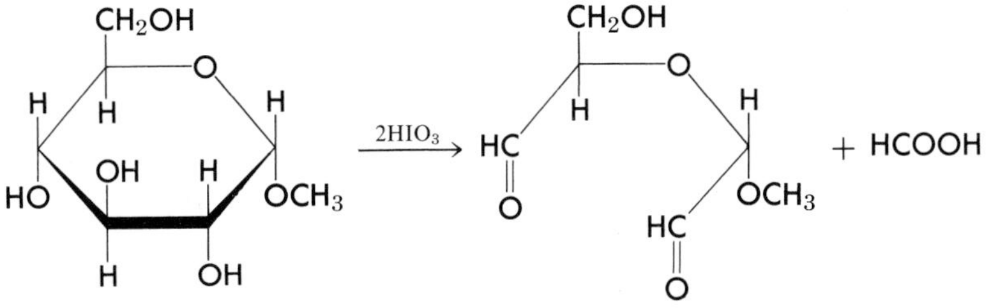

one from the usual *cis*-glycol oxidation, the other from the free aldehydic carbon atom. The number of end groups in a polysaccharide, determined by this procedure, is an index of the degree of linearity and branching of the polysaccharide. If a yield of three moles of formic acid per mole of polysaccharide is exceeded, then the latter must contain branched chains. The relation of periodate consumed to formate produced gives an insight into the degree of branching.

Many of the larger molecular weight polysaccharides are strongly antigenic, *i.e.*, they lead to production of antibodies when injected into suitable animals (Chap. 31). Such an antibody, which reacts specifically with the antigen that elicited its production, also often cross-reacts with polysaccharides of similar structure. If the structure of the initial antigen is known, *i.e.*, the monosaccharide components and the linkages between them, the structures of cross-reacting polysaccharides may be inferred. By the use of a spectrum of antibodies of known specificity and measurement of the cross reaction to an unknown polysaccharide, important information regarding the structure of the latter has been gained in a number of cases.

An excellent approach to studies of polysaccharide structure involves the successive use of selective enzymes that have been characterized with regard to the type of glycosidic bonds they will rupture (page 47). This is considered below.

HOMOPOLYSACCHARIDES

CELLULOSE

Cellulose is unquestionably the most abundant organic compound in the world, comprising 50 per cent or more of all the carbon in vegetation. Though predominantly a plant polysaccharide, it has been found in certain tunicates. The purest source is cotton, which is at least 90 per cent pure cellulose.

On complete hydrolysis, cellulose yields glucose virtually quantitatively; partial hydrolysis yields the disaccharide cellobiose. Hydrolysis of fully methylated cellulose gives 2,3,6-tri-O-methylglucose almost quantitatively, indicating the absence of branching. From these and other considerations, the structure of the repeating unit of the cellulose chain joined by β-1,4 linkages is that shown on page 46.

Cellulose is insoluble in water but will dissolve in ammoniacal solutions of cupric salts. Its molecular weight has been estimated on different preparations to be between 50,000 and 400,000, corresponding roughly to 300 to 2,500 glucose residues per molecule, but these estimates may be far too low because of degradation in the preparation of the sample.

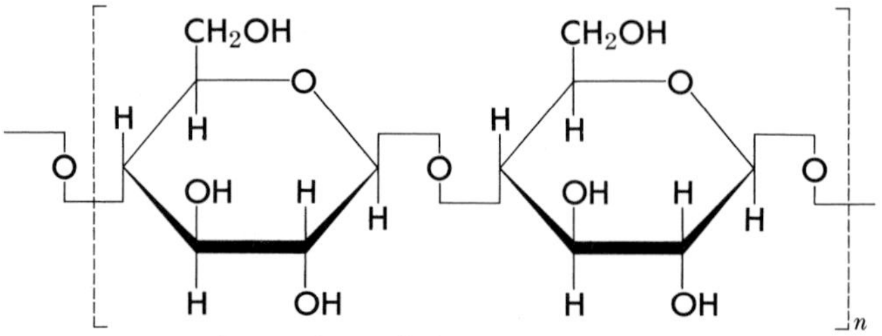

Repeating cellobiose unit of cellulose

STARCHES

Whereas cellulose, in which glucose units are joined in β-1,4 linkages, is the characteristic structural polysaccharide of plant cells, in starches, which serve as nutritional reservoirs in plants, the corresponding link is α-1,4. The repeating disaccharide unit therefore is maltose rather than cellobiose.

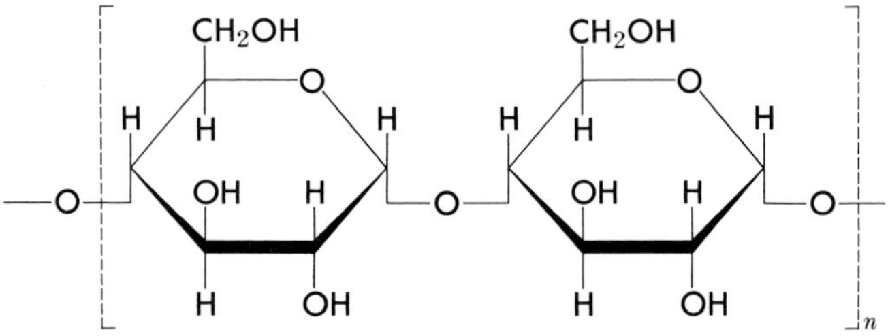

Repeating maltose unit of starch

Native starches are a mixture of two types of compounds which are separable from each other. *Amylose* is the component that is believed to be a long, unbranched chain, similar in this regard to cellulose. *Amylopectin* has been shown on the basis of methylation studies to be a branched-chain polysaccharide, one terminal glucose occurring for every 24 to 30 glucose residues. The glucose residue which is situated at each point of branching is substituted not only on carbon-4 but also on carbon-6. Isolation of the α-1,6 disaccharide, isomaltose, from the products of incomplete hydrolysis of amylopectin proves the constitution of the branch points, which may be represented as shown:

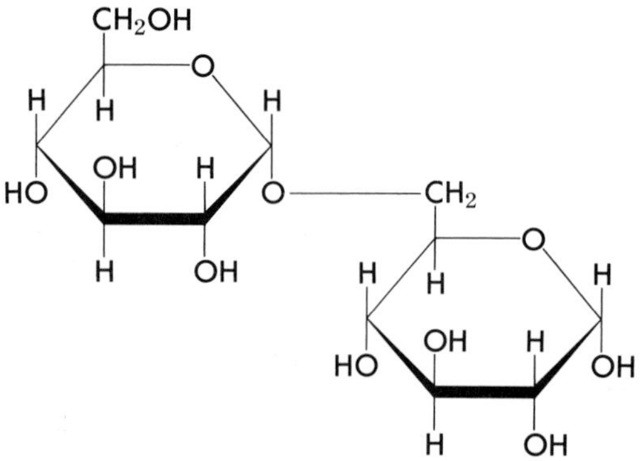

α-Isomaltose

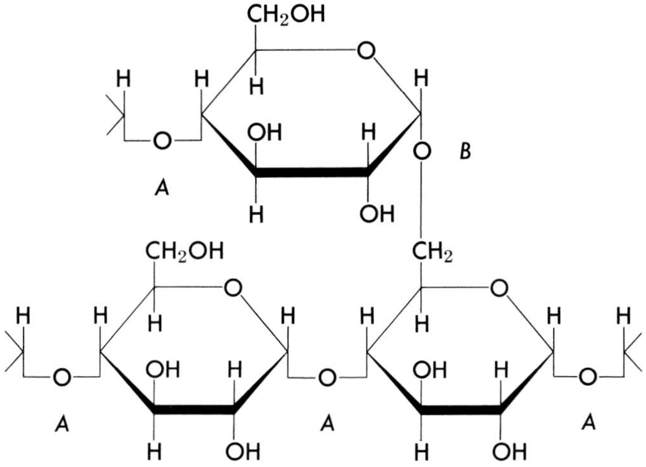

Branch point of amylopectin and glycogen

A = α-1,4-glucosidic bond
B = α-1,6-glucosidic bond

The degree of branching is greater in glycogen than in amylopectin. Indeed, amylopectin may not be rebranched, as is glycogen, but may consist of a single main chain with associated branches.

Molecular weight determinations on various starch preparations have given results ranging from 50,000 up to several millions. In view of the likelihood of some degradation during isolation, the higher values probably represent more truly the sizes of native starch molecules.

The differences between the unbranched amylose and the branched amylopectin are most clearly understood from a consideration of the products of their enzymic hydrolyses. *Amylases,* the enzymes that catalyze the hydrolysis of starches, have been divided into two classes, formerly known as α- and β-amylases. This relatively general terminology has been replaced by more precise terms that indicate the nature of the action of these enzymes. The term α-amylase now connotes an *α-1,4-glucan 4-glucanohydrolase* and the term β-amylase an *α-1,4-glucan maltohydrolase.* The former enzyme, present for example in pancreatic juice and in saliva, splits the α-1,4-glucosidic bonds of a polysaccharide molecule in an apparently random fashion, while the latter enzyme, such as that of barley malt, attacks polysaccharide chains, effecting successive removals of maltose units from the nonreducing ends. Neither type of enzyme exhibits activity against α-1,6 bonds or against β-1,4 bonds. However, the α-1,6 bonds of amylopectin are cleaved by an enzyme, *pullulanase,* obtained from *Aerobacter aerogenes.*

When amylose, which contains exclusively α-1,4-glucosidic bonds, is attacked by α-1,4-glucan 4-glucanohydrolase, random cleavage of these bonds gives rise to a mixture of glucose and maltose (Fig. 3.1), whereas attack by α-1,4-glucan maltohydrolase gives pure maltose in almost quantitative yield (Fig. 3.2). Amylopectin, when treated with α-1,4-glucan 4-glucanohydrolase, undergoes random rupture of α-1,4 bonds and yields, as ultimate products, a mixture of branched and unbranched oligosaccharides in which α-1,6 bonds are abundantly present (Fig. 3.3).

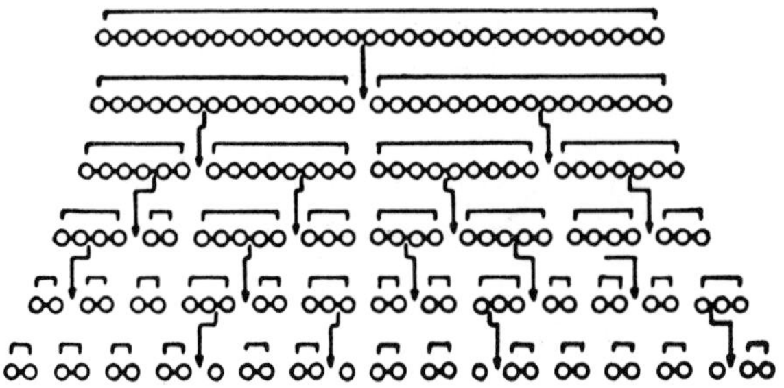

Fig. 3.1. Action of α-1,4-glucan 4-glucanohydrolase on amylose. (o) glucose unit; (-o-o-) α-1,4-glucosidic bond; (↓) action of enzyme. (*According to P. Bernfeld, Advances in Enzymol.,* **12,** 379, 1951.)

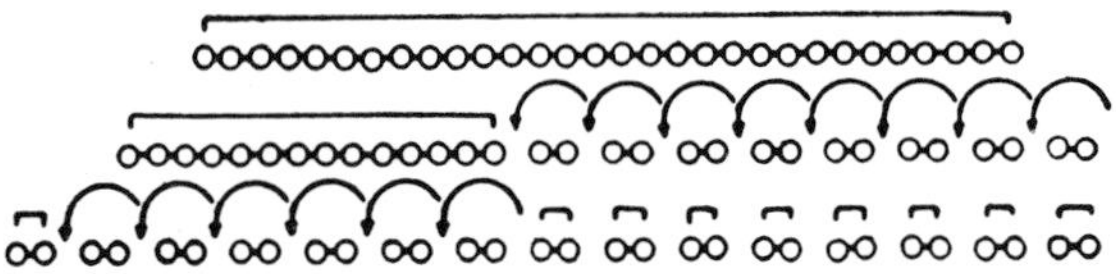

Fig. 3.2. Action of α-1,4-glucan maltohydrolase on amylose (-o-o-) α-1,4-glucosidic linkage: (↶) enzyme action. Maltose yield, 100 per cent. (*According to P. Bernfeld, Advances in Enzymol.,* **12,** 379, 1951.)

When amylopectin is hydrolyzed by the action of α-1,4-glucan maltohydrolase, successive maltose units are liberated, commencing at the nonreducing ends of the polysaccharide molecule. This process continues until a branch point is approached. Since the enzyme has no capacity to hydrolyze the α-1,6 bond that it here encounters, all reaction stops. The polysaccharide fragment that remains after such incomplete hydrolysis is called a *dextrin,* and specifically, since this dextrin is the limit of attack of α-1,4-glucan maltohydrolase upon amylopectin, it is called a *limit dextrin* (Fig. 3.4).

Starch in the native state occurs as microscopically visible granules which can be stained blue-black by iodine. On grinding with water, starch becomes dispersible, and "solutions" that are grossly clear can be prepared by this means. Some degree of degradation of the molecule may accompany such treatment. Starch is readily hydrolyzed by dilute mineral acid, with ultimate production of glucose in quantitative yield. The course of hydrolysis may be followed by the gradual change in color produced by iodine (blue-black → purple → red → none) and by the increase in reducing-sugar concentration.

GLYCOGEN

The counterpart of starch in the animal is glycogen, which occurs in significant amount in liver and muscle and is particularly abundant in molluscs. A glycogen-like polysaccharide has also been described in corn.

Glycogen is a branched-chain polysaccharide, resembling amylopectin rather than amylose. Glycogen has variously been found to have 8 to 12 glucose residues per nonreducing end group, and molecular weights ranging from 270,000 to 100,000,000 have been reported. Even in single preparations, a wide distribution of molecular sizes is encountered.

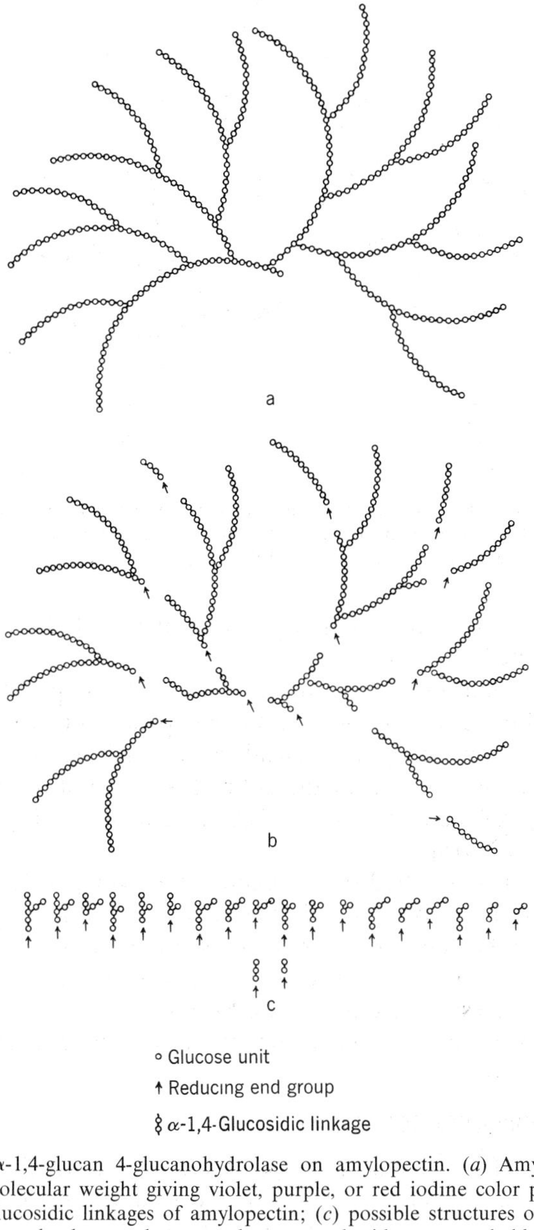

Fig. 3.3. Action of α-1,4-glucan 4-glucanohydrolase on amylopectin. (*a*) Amylopectin model; (*b*) dextrins of medium molecular weight giving violet, purple, or red iodine color produced by splitting of 4 per cent of the glucosidic linkages of amylopectin; (*c*) possible structures of limit dextrins from amylopectin breakdown; the hepta-, hexa-, and pentasaccharides are probably split, more or less rapidly, into lower-molecular oligosaccharides. (*According to P. Bernfeld, Advances in Enzymol.*, **12**, 379, 1951.)

Glycogen occurs in animal cells in particles much smaller than starch granules. It is readily dispersed in water to form opalescent "solutions" which give a violet-red color with iodine. Glycogen is relatively stable in hot alkali and is precipitated from aqueous solution by addition of ethyl alcohol. These properties were used ad-

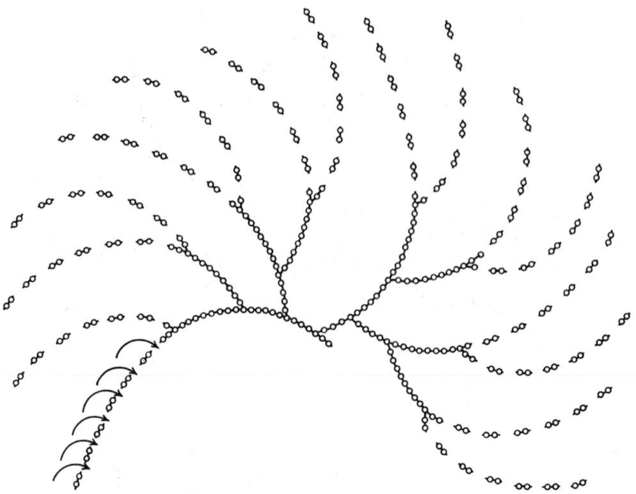

Fig. 3.4. Action of α-1,4-glucan maltohydrolase on amylopectin. Conversion limit, in this case, 61 per cent maltose. At the center, high molecular weight limit dextrin. (⌒↘) enzyme action. (*According to P. Bernfeld, Advances in Enzymol.,* **12**, 379, 1951.)

vantageously in the original studies of glycogen by Claude Bernard and by Pflüger in the middle of the last century.

α-1,4-Glucan maltohydrolase attacks glycogen incompletely, yielding, as with amylopectin, maltose and limit dextrin. On complete acid hydrolysis, purified glycogen gives glucose in nearly theoretical yield. Analysis of tissues for glycogen usually depends upon solution of the entire tissue in hot alkali, precipitation of glycogen with alcohol, acid hydrolysis of this precipitate, and quantitative determination of the glucose formed.

Glycogen is also attached by the bacterial enzyme, *pullulanase,* which splits the α-1,6 bonds of this polysaccharide as well as those of amylopectin.

Knowledge of the details of the structures of the starches and glycogen depend largely upon experiments in which various enzymes of known specificity were employed to degrade these macromolecules. Further discussion of this evidence will be deferred (Chap. 19) until the characteristics of the enzymes employed have been presented (Chap. 10).

FRUCTOSE HOMOPOLYSACCHARIDES

An example of a fructose homopolysaccharide is inulin, derived from the Jerusalem artichoke and certain other plants. It is apparently composed exclusively of fructose residues in furanose form linked to each other in linear fashion by a β-2,1 bond.

OTHER HOMOPOLYSACCHARIDES

In addition to cellulose, starches, and glycogen, other polysaccharides of glucose include the dextrans of yeast and bacteria, in which most of the glucose residues are bound in α-1,6 linkages. Almost all preparations of dextrans exhibit some branching. The linkage at the branch point (1,2, 1,3, 1,4) and the distance between branch points is characteristic of the strain and species of the particular organism from which the dextran was obtained.

A very highly branched galactose polysaccharide accompanies glycogen in certain species of snails. Mannose polysaccharides are found widely distributed in ivory nuts, orchid tubers, pine trees, yeasts, molds, bacteria, and elsewhere. Various pentosans, polymers of L-arabinose or of D-xylose, occur in woods, nuts, and other vegetable products.

Of the uronic acid polysaccharides, galacturonic acid polymers, partially esterified with methyl alcohol, comprise the pectins of fruits and berries; polymers of D-mannuronic acid have been described in seaweed.

Chitin in the shells of crustaceans and insects is a linear polymer of N-acetyl-D-glucosamine. Thus, it is an analogue of cellulose (page 45) in which the hydroxyl group of glucose at C-2 is replaced by an N-acetylamino group. Except for cellulose, chitin is probably the most abundant polysaccharide of nature. Colominic acid, a homopolysaccharide that is a polymer of N-acetylneuraminic acid (page 34), has been prepared from *Escherichia coli* and is also found in other organisms.

HETEROPOLYSACCHARIDES

Polysaccharides which yield on hydrolysis mixtures of monosaccharides and derived products are numerous in both plants and animals. The simplest heteropolysaccharides are those constructed by repetitive use of a mixed disaccharide. The immunochemically specific polysaccharide of type III pneumococcus is an example of such a heteropolysaccharide since, on hydrolysis, it yields equimolar amounts of glucose and glucuronic acid. It has the following structure.

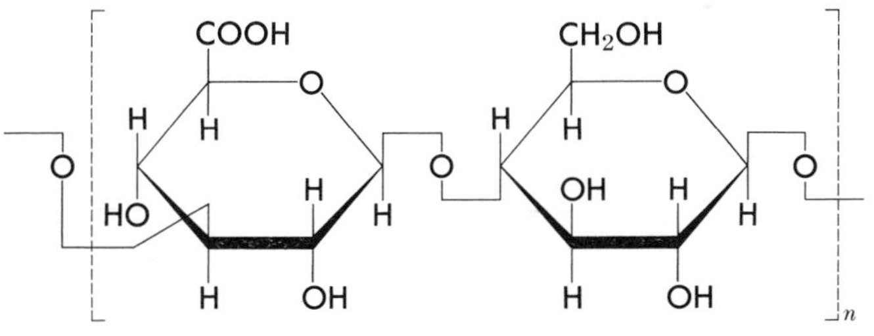

Repeating aldobiuronic acid unit of pneumococcus type III

Other heteropolysaccharides of a similar structure will be discussed later.

Acid Mucopolysaccharides. Widely distributed throughout the animal body in connective tissue are large quantities of a group of related heteropolysaccharides, each of which contains an N-acetylated hexosamine in its characteristic repeating disaccharide unit. Table 3.1 lists the more common mucopolysaccharides, their repeating disaccharide units, and the linkages between the individual hexoses. It will be seen that the repeating structure of each disaccharide involves alternate 1,4 and 1,3 linkages. These polysaccharides can be separated in the laboratory by fractional precipitation by addition of alcohol to a solution of the mucopolysaccharides containing calcium ions.

The most abundant member of this group is probably *hyaluronic acid.*

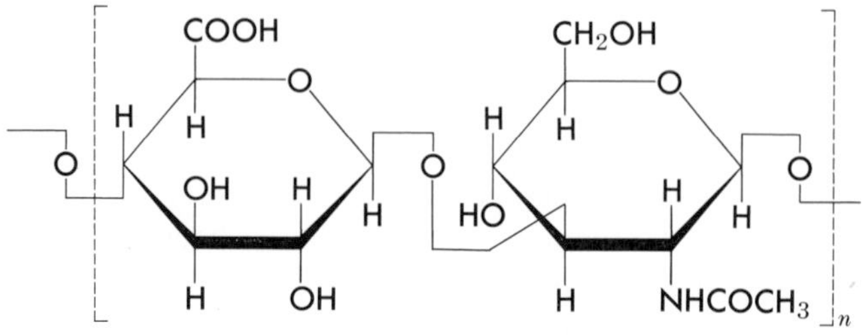

Repeating unit of hyaluronic acid

Solutions of this polysaccharide are characterized by their very high viscosity. Wharton's jelly from umbilical cord is usually employed as the starting material for preparation of hyaluronic acid. However, this polymer is ubiquitous in connective tissue throughout the body. Most of the procedures for its isolation result in partial degradation of the molecule. Thus, the viscosity of both vitreous and synovial fluids considerably exceeds that of a reconstructed solution of the isolated hyaluronic acid. Further, the isolated polymer yields a flocculent precipitate upon acidification, whereas acid produces a *mucin clot* when added to hyaluronate in the native state, because of the presence of small amounts of associated protein.

Chondroitin, which is of limited distribution, is a polymer of β-D-glucuronido, 1,3-N-acetyl-D-galactosamine joined in repeating β-1,4 linkages. The repeating unit of this polysaccharide differs from hyaluronic acid only in that it contains galactosamine rather than glucosamine. It may be regarded, however, as the parent material for two more widely distributed polysaccharides, *chondroitin sulfate A* and *chondroitin sulfate C,* which differ only in the position in which sulfate is esterified to the galactosamine moiety.

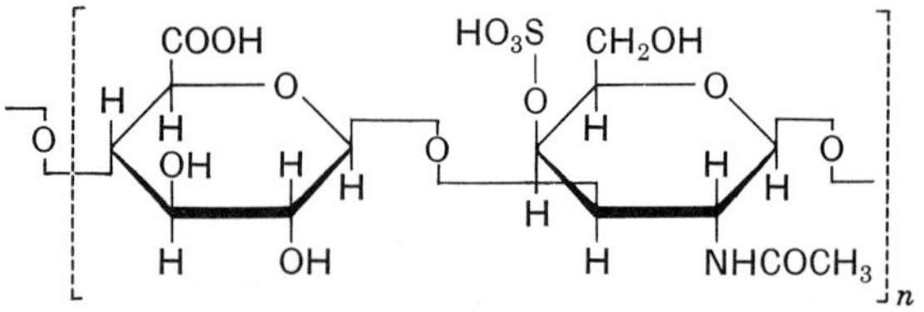

Repeating unit of chondroitin sulfate A

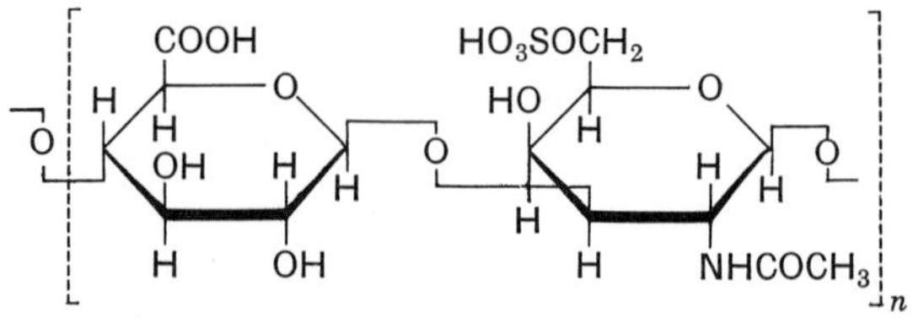

Repeating unit of chondroitin sulfate C

As shown in Table 3.1, the polymer that was long termed *chondroitin sulfate B* yields upon hydrolysis not D-glucuronic acid but L-iduronic, the two differing in configuration only at C-5. Hence, the conventional designation is a misnomer

Table 3.1: SOME MUCOPOLYSACCHARIDES AND THEIR REPEATING DISACCHARIDE UNITS

Polysaccharide	Monosaccharides in the disaccharide units	Linkages*
Hyaluronic acid	D-Glucuronic acid; N-acetyl-D-glucosamine	β-1 → 3; β-1 → 4
Chondroitin	D-Glucuronic acid; N-acetyl-D-galactosamine	β-1 → 3; β-1 → 4
Chondroitin sulfate A (Chondroitin 4-sulfate)	D-Glucuronic acid; N-acetyl-D-galactosamine 4-sulfate	β-1 → 3; β-1 → 4
Chondroitin sulfate C (Chondroitin 6-sulfate)	D-Glucuronic acid; N-acetyl-D-galactosamine 6-sulfate	β-1 → 3; β-1 → 4
Dermatan sulfate (Chondroitin sulfate B)	L-Iduronic acid; N-acetyl-D-galactosamine 4-sulfate	α-1 → 3; β-1 → 4
Keratosulfate	D-Galactose; N-acetyl-D-glucosamine 6-sulfate	β-1 → 4; β-1 → 3

* The linkages are given in the same order as the names of the monosaccharide units, *e.g.*, for chondroitin the linkages, as given, indicate that the linkage of glucuronic acid to acetylgalactosamine is β-1 → 3, and that of acetylgalactosamine to the next glucuronic acid is β-1 → 4, etc.

and current literature refers to this compound as *dermatan sulfate;* it was formerly also called β-heparin. *Keratosulfate* differs from the other members of this group in that the uronic acid component is replaced by D-galactose. Although this polymer is also constructed of alternating β-1,4 and β-1,3 linkages, in this instance the hexosaminidic linkage is β-1,3.

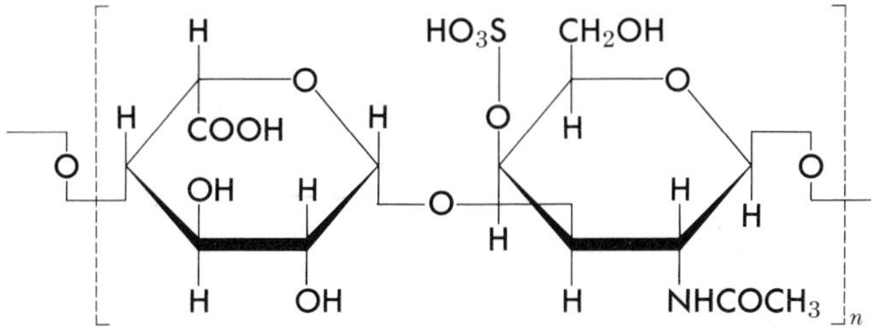

Repeating unit of dermatan sulfate
(chondroitin sulfate B)

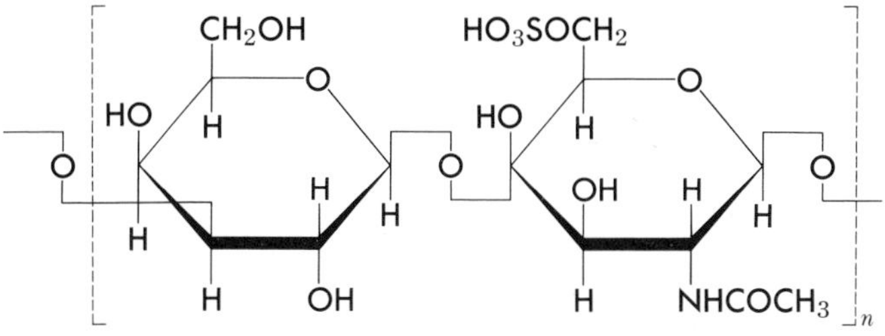

Repeating unit of keratosulfate

Related to the sulfated mucopolysaccharides is *heparin,* a polysaccharide early recognized because of its anticoagulant properties (Chap. 31) and present in liver, lung, the walls of large arteries, and, indeed, wherever mast cells are found. Heparin is a polymer in which the repeating unit consists of D-glucuronic acid usually with an O-sulfate group at C-2, and D-glucosamine N-sulfate with an additional O-sulfate group at C-6. Both the linkages of the polymer are alternating α-1,4; it is believed that seven of each eight of the glucuronic acid residues in heparin carry an O-sulfate group.

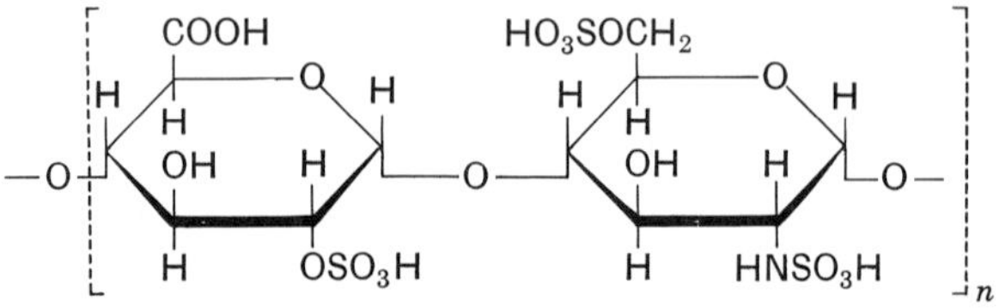

Postulated repeating unit of heparin

Heparan sulfate, also termed *heparitin sulfate,* is a monosulfuric acid ester of an acetylated (N-acetyl) heparin. It is probable that heparan sulfate, as isolated from animal tissues (liver), is a mixture of mucopolysaccharides with varying degrees of sulfation or acetylation of the amino groups.

The acid mucopolysaccharides are present in their ubiquitous sites primarily as mucoproteins, *i.e.,* combinations of the mucopolysaccharides with proteins (page 117). In no instance has the structure of the specific protein been established, but the carbohydrate moiety of the mucopolysaccharide is probably linked with its protein component through a glycosidic bond to hydroxyl groups of serine residues and to γ-carboxyl groups of glutamic acid residues of the protein (page 117 and Chap. 38).

The heteropolysaccharides which have been discussed above are characterized by the presence of two different sugars in each monomer. More complex heteropolysaccharides occur which contain more than two carbohydrates. In plants, the structures of such heteropolysaccharides remain to be established. The vegetable "gums" contain as many as four different types of monosaccharides in a single preparation. Most common in these structures are D-glucuronic acid, D-mannose, D-xylose, and L-rhamnose, but many others are also found in gums from various sources. Agar, derived from certain seaweeds, yields on hydrolysis D- and L-galactose in a ratio of 9:1 and is sulfated in varying degree.

In animal and microbial tissue, a wide variety of heteropolysaccharides have been described which on hydrolysis yield diverse carbohydrates. As is the case for the acid mucopolysaccharides, which occur linked to proteins (see above), these more complex polysaccharides are also bound to proteins, as well as to lipids. Indeed, a large number of proteins and lipids can be classed as glycoproteins (page 118) and glycolipids (gangliosides, page 74) because of the presence of an oligosaccharide moiety. The cell membranes of animals are characterized by the presence of an oligosaccharide which on hydrolysis yields fucose, together with varying amounts of glucosamine, galactosamine, galactose, and occasionally sialic acid (Chaps. 31 and 38). The same sugars are found in the mucoproteins derived from epithelial mucous secretions, *viz.,* those of the respiratory and digestive tracts. These mucoproteins, in contrast to those of connective tissue in which polysaccharides are linked to proteins, have numerous small oligosaccharides linked covalently to a single protein molecule. An example is the "mucin" or mucoprotein of saliva, which contains, per protein molecule, numerous molecules of the disaccharide 6-(N-acetyl-α-neuraminyl)-N-acetyl-D-galactosamine.

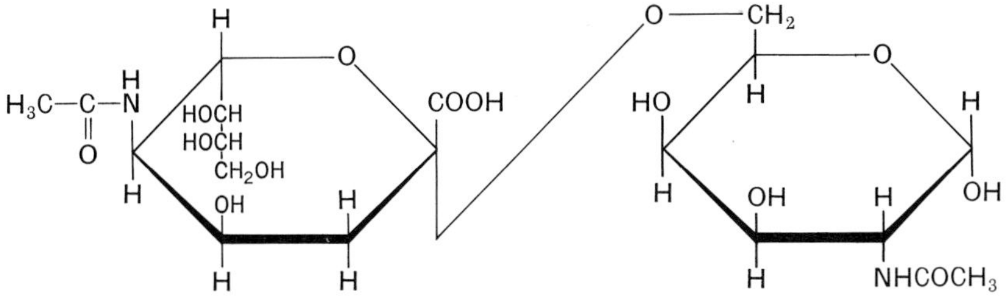

6-(N-Acetyl-α-neuraminyl)-N-acetyl-D-galactosamine

Another group of heteropolysaccharides is found in the mucoproteins of serum among the α-globulins (Chap. 31). On hydrolysis, these yield galactose, mannose, a hexosamine, fucose, and a sialic acid. In all these instances, the sialic acid is found exclusively at the nonreducing end of the polysaccharide, since *neuraminidase,* a bacterial enzyme specific for neuraminic acid glycosides, liberates all the sialic acid of the polymer and is without influence on the remainder of the molecule.

The immunologically specific substances with blood group activity are also heteropolysaccharides; these will be considered in Chap. 31. The polysaccharides of bacterial cell walls will be discussed in Chap. 41.

REFERENCES

Books

Block, R. J., Durrum, E. L., and Zweig, G., "A Manual of Paper Chromatography and Paper Electrophoresis," 2d ed., Academic Press, Inc., New York, 1958.

Brimacombe, J. S., and Webber, J. M., "Mucopolysaccharides," American Elsevier Publishing Company, New York, 1964.

Clark, F., and Grant, J. K., eds., "Biochemistry of Mucopolysaccharides of Connective Tissue," Cambridge University Press, New York, 1961.

Davidson, E. A., "Carbohydrate Chemistry," Holt, Rinehart and Winston, New York, 1967.

Florkin, M., and Stotz, E., "Comprehensive Biochemistry," Carbohydrates, sec. II, vol. 5, American Elsevier Publishing Company, New York, 1963.

Gottschalk, A., "The Chemistry and Biology of Sialic Acids," Cambridge University Press, New York, 1960.

Heftmann, E., ed., "Chromatography," 2d ed., Reinhold Publishing Company, New York, 1967.

Heidelberger, M., and Plescia, O. J., eds., "Symposium on Immunochemical Approaches to Problems in Microbiology," Rutgers University Press, New Brunswick, N.J., 1961.

Jeanloz, R. W., and Balazs, E. A., eds., "The Amino Sugars," vol. IIA, Academic Press, Inc., New York, 1965.

Lederer, E., and Lederer, M., "Chromatography: A Review of Principles and Applications," 2d ed., American Elsevier Publishing Company, New York, 1957.

Pigman, W. W., ed., "The Carbohydrates: Chemistry, Biochemistry and Physiology," Academic Press, Inc., New York, 1957.

Stacey, M., and Barker, S. A., "Polysaccharides of Microorganisms," Oxford University Press, Fair Lawn, N.J., 1960.

Stacey, M., and Barker, S. A., "Carbohydrates of Living Tissues," D. Van Nostrand Company, Inc., Princeton, N.J., 1962.

Whistler, R. L., and Wolfrom, M. L., eds., "Methods in Carbohydrate Chemistry," vol. I, Analysis and Preparation of Sugars; vol. IV, Starch, vol. V, General Polysaccharides, Academic Press, Inc., New York, 1962, 1964, 1965.

Review Articles

Bell, D. J., Natural Monosaccharides and Oligosaccharides: Their Structure and Occurrence, in M. Florkin and H. S. Mason, eds., "Comparative Biochemistry," vol. III, part A, pp. 288–354, Academic Press, Inc., New York, 1962.

Brimacombe, J. S., and Stacey, M., Mucopolysaccharides in Disease, *Advances in Clin. Chem.,* **7,** 199–234, 1964.

Hirst, E. L., The Structure of Polysaccharides, in D. J. Bell and J. K. Grant, eds., "The Structure and Biosynthesis of Macromolecules," pp. 45–62, Cambridge University Press, New York, 1962.

Jeanloz, R. W., Recent Developments in the Biochemistry of the Amino Sugars, *Advances in Enzymol.,* **25,** 433–456, 1963.

Muir, H., Chemistry and Metabolism of Connective Tissue Glycosaminoglycans (Mucopolysaccharides), *Intern. Rev. Connective Tissue Research,* **2,** 101–154, 1964.

Salton, M. J. R., Chemistry and Function of Amino Sugars and Derivatives, *Ann. Rev. Biochem.,* **34,** 143–174, 1965.

Sharon, N., Polysaccharides, *Ann. Rev. Biochem.,* **35,** 485–520, 1966.

Many special phases of carbohydrate chemistry have been reviewed in *Advances in Carbohydrate Chemistry,* Academic Press, Inc., New York, 1945-current.

4. The Lipids

If animal or vegetable tissues are extracted with one or more of the so-called "fat solvents," *e.g.*, ethanol, ether, chloroform, benzene, petroleum ether, etc., a portion of the material may dissolve; the name "lipid" is given to the components of this soluble fraction. The lipid fraction is an operational rather than a structural classification and is defined, simply, as that fraction of any biological material which is extractable by nonpolar solvents. It includes a heterogeneous group of structural types, of which the more prominent members may be grouped as follows, on the basis of chemical composition.

I. Fatty acids
II. Lipids containing glycerol
    A. Neutral fats
        1. Mono-, di-, and triglycerides
        2. Glyceryl ethers
        3. Glycosyl glycerides
    B. Phospholipids
        1. Phosphatides
        2. Phosphoglycerides and phosphoinositides
III. Lipids not containing glycerol
    A. Sphingolipids
        1. Ceramides
        2. Sphingomyelins
        3. Glycosphingolipids
    B. Aliphatic alcohols and waxes
    C. Terpenes
    D. Steroids
IV. Lipids combined with other classes of compounds
    A. Lipoproteins
    B. Proteolipids
    C. Phosphatidopeptides
    D. Lipoamino acids
    E. Lipopolysaccharides

THE FATTY ACIDS

In order to appreciate the physical and chemical properties of many of the classes of lipids, it is necessary to consider some of the properties and reactions of the fatty acids. Whereas fatty acids are rarely found free in nature, they are found in abundance in either *ester* or *amide* linkage in several of the classes of compounds listed above. The fatty acids encountered in nature have, with only occasional exceptions, the following properties.

1. They are, for the most part, monocarboxylic acids with hydrocarbon residues which are both acyclic and unbranched.

2. The number of carbon atoms in the molecule is in most cases even although odd-numbered carbon atom fatty acids are also found in nature.

3. They may be saturated or may contain one or more double bonds.

Saturated Fatty Acids. The common names and formulas of some of the saturated fatty acids are given in Table 4.1. Certain properties of these substances may

Table 4.1: SOME NATURALLY OCCURRING SATURATED FATTY ACIDS

Molecular formula	Common name	Systematic name	Structural formula
$C_2H_4O_2$	Acetic	. . . .	CH_3COOH
$C_3H_6O_2$	Propionic	. . . .	CH_3CH_2COOH
$C_4H_8O_2$	*n*-Butyric	. . . .	$CH_3(CH_2)_2COOH$
$C_6H_{12}O_2$	Caproic	*n*-Hexanoic	$CH_3(CH_2)_4COOH$
$C_8H_{16}O_2$	Caprylic	*n*-Octanoic	$CH_3(CH_2)_6COOH$
$C_9H_{18}O_2$	Pelargonic	*n*-Nonanoic	$CH_3(CH_2)_7COOH$
$C_{10}H_{20}O_2$	Capric	*n*-Decanoic	$CH_3(CH_2)_8COOH$
$C_{12}H_{24}O_2$	Lauric	*n*-Dodecanoic	$CH_3(CH_2)_{10}COOH$
$C_{14}H_{28}O_2$	Myristic	*n*-Tetradecanoic	$CH_3(CH_2)_{12}COOH$
$C_{16}H_{32}O_2$	Palmitic*	*n*-Hexadecanoic	$CH_3(CH_2)_{14}COOH$
$C_{18}H_{36}O_2$	Stearic*	*n*-Octadecanoic	$CH_3(CH_2)_{16}COOH$
$C_{20}H_{40}O_2$	Arachidic	*n*-Eicosanoic	$CH_3(CH_2)_{18}COOH$
$C_{22}H_{44}O_2$	Behenic	*n*-Docosanoic	$CH_3(CH_2)_{20}COOH$
$C_{24}H_{48}O_2$	Lignoceric	*n*-Tetracosanoic	$CH_3(CH_2)_{22}COOH$
$C_{26}H_{52}O_2$	Cerotic	*n*-Hexacosanoic	$CH_3(CH_2)_{24}COOH$
$C_{28}H_{56}O_2$	Montanic	*n*-Octacosanoic	$CH_3(CH_2)_{26}COOH$

* These are the most abundant saturated fatty acids encountered in animal lipids.

be regarded as the sum of contributions by the polar hydrophilic carboxyl group and by the hydrophobic paraffin residue. Thus, acetic and propionic acids are miscible with water. Butyric acid has a limited solubility in water, 5.6 per cent; caproic acid has a solubility of 0.4 per cent; higher members of the series are virtually insoluble in water although readily soluble in nonpolar solvents. As the paraffin residue becomes longer, it dominates in establishing physical properties. Boiling points and melting points of fatty acids rise with increasing chain length. Even-numbered–carbon atom saturated fatty acids of less than 10 carbon atoms are liquids at room temperature; longer-chained members are solids.

The fatty acids dissociate in aqueous solution,

$$RCOOH \rightleftharpoons RCOO^- + H^+$$

where the dissociation constant, $K = \dfrac{[H^+][RCOO^-]}{[RCOOH]}$

and $pK = -\log K$. Except for the first member of the series, formic acid ($pK =$

3.75), all the saturated fatty acids resemble acetic acid in their dissociation constants ($pK = 4.76$).

The mixture of fatty acids obtained upon hydrolysis of lipids derived from various sources generally contains both saturated and unsaturated fatty acids. In typical animal lipids, the most abundant saturated fatty acid is usually palmitic (C_{16}), with stearic (C_{18}) second in amount. Shorter-chain fatty acids (C_{14} and C_{12}) do occur in small quantity, as do longer-chain members (up to C_{28}). Fatty acids of 10 carbon atoms or less are rarely present in animal lipids. An exception is milk fat, which contains appreciable concentrations of lower molecular weight fatty acids (Chap. 34).

Unsaturated Fatty Acids. The names and structures of certain of the more common unsaturated fatty acids are given in Table 4.2. Unsaturation markedly alters certain properties of a fatty acid. In general, the melting point is greatly lowered and solubility in nonpolar solvents is enhanced (Table 4.3). All the common unsaturated fatty acids of nature are liquids at room temperature.

The double bond in the singly unsaturated fatty acids of animal lipids is generally in the 9,10 position. Thus, the two most abundant singly unsaturated fatty acids of animal lipids are oleic acid and palmitoleic acid.

$$CH_3—(CH_2)_7—CH{=}CH—(CH_2)_7—COOH$$

Oleic acid

$$CH_3—(CH_2)_5—CH{=}CH—(CH_2)_7—COOH$$

Palmitoleic acid

Oleic acid is the most widely distributed and most abundant fatty acid in nature.

The introduction of a double bond gives rise to the possibility of *cis-trans* isomerism. The isomeric forms of 9-octadecenoic acid may be written as follows:

$$\begin{array}{l} H—C—(CH_2)_7—CH_3 \\ \quad\ \| \\ H—C—(CH_2)_7—COOH \end{array} \qquad \begin{array}{r} CH_3—(CH_2)_7—C—H \\ \| \quad\ \\ H—C—(CH_2)_7—COOH \end{array}$$

Oleic acid (***cis***) **Elaidic acid** (***trans***)

In this and in certain other known cases in which ethylenic double bonds occur in the fatty acid series, it is the *cis* configuration which is found in nature. However, among other classes of compounds this rule does not necessarily apply. In sphingosine (page 72) the configuration is *trans*, as it is in the abundant ethylene *di*carboxylic acid of nature, fumaric acid.

Fatty acids containing more than one double bond are also commonly found. When two double bonds occur in a carbon skeleton in the relationship

$$—CH{=}CH—CH{=}CH—$$

they are said to be in conjugation. Such conjugated double bonds are usually more readily oxidized than isolated double bonds or double bonds related to each other in any other fashion. Conjugated double bonds are found in the fatty acid series of the vegetable world (eleostearic acid, Table 4.2), among the so-called "drying oils" of the paint industry. Linolenic acid (Table 4.2), in which the double bonds are not conjugated, is the principal constituent of linseed oil, the chief drying oil of indus-

Table 4.2: SOME NATURALLY OCCURRING UNSATURATED FATTY ACIDS

Molecular formula	Common name	Systematic name	Structural formula
$C_{16}H_{30}O_2$	Palmitoleic*	9-Hexadecenoic	$CH_3(CH_2)_5CH{=}CH(CH_2)_7COOH$
$C_{18}H_{34}O_2$	Oleic*	*cis*-9-Octadecenoic	$CH_3(CH_2)_7CH{=}CH(CH_2)_7COOH$
$C_{18}H_{34}O_2$	Elaidic	*trans*-9-Octadecenoic	$CH_3(CH_2)_7CH{=}CH(CH_2)_7COOH$
$C_{18}H_{34}O_2$	Vaccenic	11-Octadecenoic	$CH_3(CH_2)_5CH{=}CH(CH_2)_9COOH$
$C_{18}H_{32}O_2$	Linoleic*	*cis, cis*-9, 12-Octadecadienoic	$CH_3(CH_2)_4CH{=}CHCH_2CH{=}CH(CH_2)_7COOH$
$C_{18}H_{30}O_2$	Linolenic*	9, 12, 15-Octadecatrienoic	$CH_3CH_2CH{=}CHCH_2CH{=}CHCH_2CH{=}CH(CH_2)_7COOH$
$C_{18}H_{30}O_2$	γ-Linolenic	6, 9, 12-Octadecatrienoic	$CH_3(CH_2)_4CH{=}CHCH_2CH{=}CHCH_2CH{=}CH(CH_2)_4COOH$
$C_{18}H_{30}O_2$	Eleostearic	9, 11, 13-Octadecatrienoic	$CH_3(CH_2)_3CH{=}CH{-}CH{=}CH{-}CH{=}CH(CH_2)_7COOH$
$C_{20}H_{32}O_2$	Arachidonic	5, 8, 11, 14-Eicosatetraenoic	$CH_3(CH_2)_4CH{=}CHCH_2CH{=}CHCH_2CH{=}CHCH_2CH{=}CH(CH_2)_3COOH$
$C_{24}H_{46}O_2$	Nervonic	*cis*-15-Tetracosenic	$CH_3(CH_2)_7CH{=}CH(CH_2)_{13}COOH$

* These are the most abundant unsaturated fatty acids in animal lipids.

Table 4.3: Melting Points of the Common 18-carbon Fatty Acids

	Double bonds	M.p., °C.	Solubility in cold ethanol
Stearic acid	0	70°	2.5%
Oleic acid	1	14°	Infinitely soluble
Linoleic acid	2	5°	Infinitely soluble
Linolenic acid	3	−11°	Infinitely soluble

try. Multiple unsaturation as it occurs in the fatty acids of animal lipids is in general not conjugated but rather of the divinylmethane type:

$$-CH{=}CH-CH_2-CH{=}CH-$$

The fatty acids most frequently encountered in mammalian biochemistry exhibiting multiple unsaturation are linoleic acid, containing two double bonds,

$$CH_3(CH_2)_4-CH{=}CH-CH_2-CH{=}CH-(CH_2)_7-COOH$$

Linoleic acid

linolenic acid, with three double bonds,

$$CH_3-CH_2-CH{=}CH-CH_2-CH{=}CH-CH_2-CH{=}CH-(CH_2)_7-COOH$$

Linolenic acid

and arachidonic acid, with four double bonds,

$$CH_3(CH_2)_4CH{=}CHCH_2CH{=}CHCH_2CH{=}CHCH_2CH{=}CH(CH_2)_3COOH$$

Arachidonic acid

Although double bonds rarely occur between the first (carboxyl) and ninth carbon atoms, bonds of this type are present in γ-linolenic and arachidonic acids (Table 4.2); successive double bonds occur one carbon atom remote from conjugation.

Carbocyclic and Hydroxy Fatty Acids. Fatty acids carrying various substituents are occasionally encountered. Tuberculostearic acid, derived from the lipid of the tubercle bacillus, is one of the relatively few branched-chain odd-numbered–carbon atom fatty acids of nature, being 10-methylstearic acid. Occasionally, alicyclic substituents are encountered, *e.g.*, in chaulmoogra oil. Chaulmoogric acid, derived from this source, has the formula

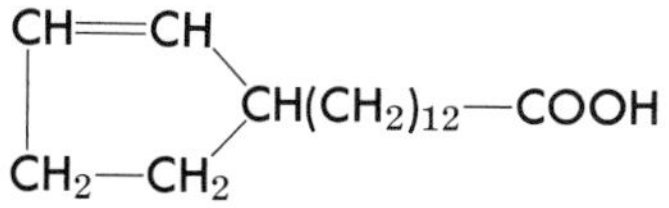

Chaulmoogric acid

An interesting type of bacterial fatty acid containing a cyclopropane ring has been described.

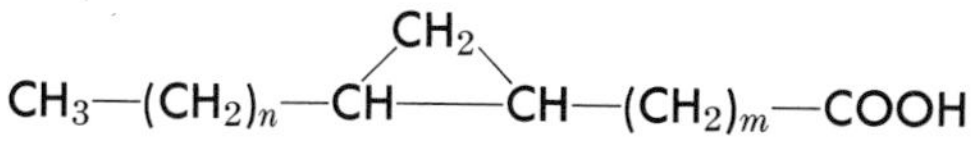

$(n + m = 14)$

Hydroxy fatty acids include cerebronic acid (2-hydroxylignoceric acid),

$$CH_3—(CH_2)_{21}—CH(OH)—COOH$$

Cerebronic acid

present particularly in cerebrosides (page 73) of the nervous system, and the characteristic unsaturated hydroxy acid of castor oil, ricinoleic acid (12-hydroxyoleic acid),

$$CH_3—(CH_2)_5—CH(OH)—CH_2—CH{=}CH—(CH_2)_7—COOH$$

Ricinoleic acid

SOAPS AND DETERGENCY

The generic name "soap" has been given to any salt of long-chain fatty acids. The sodium and potassium soaps of commerce are more or less soluble in water, whereas the salts of fatty acids with other metal ions are quite insoluble. The potassium soaps are more soluble in water than are the sodium soaps, and, in accord with principles already mentioned, the soaps of unsaturated fatty acids are more soluble in water than are those of saturated fatty acids. Thus potassium linoleate gives a clear solution in water, whereas sodium stearate is sparingly soluble, and its solution is grossly opalescent, indicating the presence of micelles of macromolecular dimensions.

All detergents contain both a hydrophobic hydrocarbon structure and a hydrophilic group which may be anionic (as in soaps), cationic, or neutral. The ability of detergents to form stable emulsions and their cleansing action depend on trapping lipid-soluble materials in the interior of the hydrophobic portions of the micelles.

Necessary Conditions for Detergency. Incident to hydrolysis, aqueous solutions of soaps are somewhat alkaline.

$$R—COO^- + Na^+ + H_2O \longrightarrow R—COOH + Na^+ + OH^-$$

If acid is added to such a solution, with increasing H^+ ion concentration the fatty acid is formed and precipitates

$$R—COO^- + H^+ \longrightarrow R—COOH\downarrow$$

from solution, and "soapiness," as well as detergency, disappears. Thus the common soaps are not effective detergents in acid solutions, and if a detergent is required under these circumstances, use is made of synthetic alkyl sulfonates, $R—SO_3^- + Na^+$, which, as salts of far stronger acids, retain their ionic nature even at high H^+ concentrations.

The fatty acid salts of metals other than the alkali metals are practically insoluble in water. Thus calcium, magnesium, and iron soaps are completely devoid of detergent action. This is the chemical basis for "hardness" of water encountered where these ions are present in excessive amounts.

REACTIONS OF FATTY ACIDS

Esterification. The most prominent reaction of the carboxyl group is esterification, in which a molecule of acid and of alcohol react reversibly to yield one molecule of ester and water.

$$R{-}\overset{\overset{\displaystyle O}{\|}}{C}{-}OH + HOR' \rightleftharpoons R{-}\overset{\overset{\displaystyle O}{\|}}{C}{-}OR' + H_2O$$

The reaction of esterification is extremely slow if uncatalyzed but may be accelerated either with heat or by addition of hydrogen ion, or both.

Reactions of the Double Bond. Reactions of double bonds of fatty acids include addition of hydrogen or halogen and oxidation by various reagents. Unsaturated fatty acids or their esters may readily be hydrogenated by gaseous hydrogen in the presence of catalysts, *e.g.*, finely divided platinum, palladium, or active nickel. If reduction is carried to completion, such unsaturated acids as linoleic and linolenic acids are quantitatively transformed into stearic acid.

$$\underset{\textbf{Linoleic acid}}{CH_3(CH_2)_4CH{=}CHCH_2CH{=}CH(CH_2)_7COOH} \xrightarrow[\text{Pt or Ni}]{H_2} \underset{\textbf{Stearic acid}}{CH_3(CH_2)_{16}COOH}$$

Halogens, such as Br_2 and IBr, add readily to double bonds of fatty acids and their esters, the reaction proceeding spontaneously in suitable solvent and to completion in most cases. This reaction is the basis of the "iodine-number determination" (see below).

Whereas the saturated fatty acids are relatively insusceptible to oxidation, unsaturated fatty acids can be oxidized. This occurs slowly and spontaneously in the presence of air and contributes to the processes termed "rancidification." The reaction is believed to involve attack at the double bonds by peroxide radicals, with formation of unstable hydroperoxides which decompose to keto and hydroxy keto acids. Oxidation of double bonds proceeds much more rapidly in the presence of ozone, O_3; an unstable ozonide is believed to form initially, with subsequent cleavage by water under reducing conditions to give rise to two aldehydic groups:

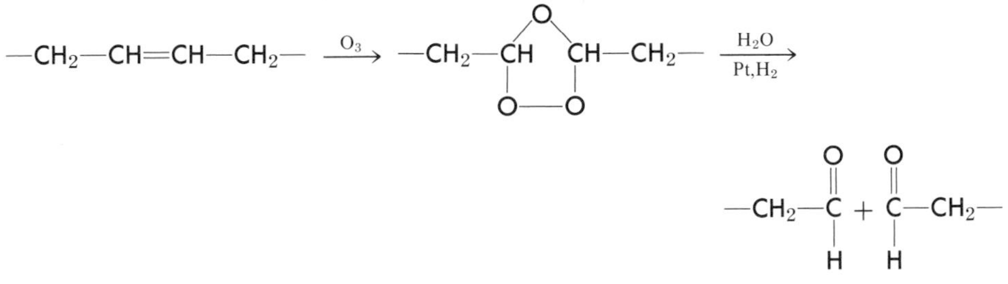

This reaction has been employed to establish the position of double bonds in fatty acid chains; the identification of the fragments derived from ozonolysis, followed by hydrolysis under oxidizing conditions, also permits inference of structure. Potassium permanganate oxidation has also been employed for similar purposes. Under mild conditions, glycols are formed at the sites of double bonds.

$$CH_3(CH_2)_7CH{=}CH(CH_2)_7COOH \xrightarrow{KMnO_4} CH_3(CH_2)_7CHOH{-}CHOH(CH_2)_7COOH$$

Oleic acid **9,10-Dihydroxystearic acid**

Under vigorous conditions the same reagent cleaves the molecule at the double bond and oxidizes the termini to the carboxyl level.

$$CH_3(CH_2)_7CH{=}CH(CH_2)_7COOH$$

Oleic acid

$$\downarrow KMnO_4$$

$$CH_3(CH_2)_7COOH + HOOC(CH_2)_7COOH$$

Pelargonic acid **Azelaic acid**

CHARACTERIZATION OF FATTY ACID MIXTURES

Fatty acids of natural origin are usually obtained from their esters by alkaline hydrolysis. The fatty acids, which separate from the aqueous phase on addition of strong acids, may be collected by filtration or by extraction with ether or petroleum ether. These procedures yield complex mixtures of fatty acids. Accordingly, a description of fatty acid mixtures has been frequently restricted to certain general characteristics indicating average chain length, average degree of unsaturation, etc., rather than to identification and relative quantities of individual fatty acids in the mixture. Among the commonly employed characterizations of such a mixture are the *iodine number,* the *titration equivalent weight,* and the *acetyl number.*

Iodine-number Determination. The determination of the iodine number depends on the fact that most double bonds react quantitatively with Br_2 or IBr at room temperature in acetic acid or methanol solution. By conventional iodometric procedures the quantity of free halogen that remains after a weighed amount of fat or fatty acid has reacted with a measured excess of halogen can be determined. From this the iodine number, defined as the number of grams of iodine equivalent to halogen reacting with 100 g. of lipid, is calculated. The iodine number of saturated fatty acids is zero; of oleic acid, 90; of linoleic acid, 181; of linolenic acid, 274.

Titration Equivalent Weight. The average molecular weight of a fatty acid mixture is usually determined by titration of a weighed sample in aqueous alcohol with standard alkali. Titration is carried to a somewhat basic end point (phenolphthalein) with exclusion of CO_2, and the mean molecular weight is calculated.

$$\text{Mean mol. wt.} = \text{mg. of fatty acid/mEq. of alkali}$$

Acetyl Number. The treatment of a fat or fatty acid mixture with acetic anhydride results in acetylation of all alcoholic hydroxyl groups. The *acetyl number,* which is a measure of the number of such hydroxyl groups, is defined as the number of milligrams of KOH required to neutralize the acetic acid contained in 1 g. of acetylated fat.

Separation and Isolation of Fatty Acids. Quantitative separation of a fatty acid mixture may be achieved by one of several procedures. Classical methods involve separation of unsaturated and saturated fatty acids by crystallization of the latter from cold acetone or precipitation of their lead salts from ethanol. Unsaturated fatty acids may be further separated by fractional crystallization of their bromo derivatives; saturated fatty acids, by meticulous fractional distillation of their methyl esters. More recently, unsaturated fatty acids have been separated as their mercuriacetate adducts.

Separation of fatty acids may be accomplished by chromatographic procedures, *viz.*, thin-layer (page 39), paper (page 38), and, particularly, gas-liquid chromatography. Because of the relative insolubility of fatty acids in polar solvents, their separation on paper entails certain special features, *e.g.*, the use of silicic acid–impregnated and glass fiber paper.

In gas chromatography, a glass or metal tube (column), 1 to 2 m. long and 0.2 to 2 cm. in internal diameter, is filled with a finely divided inert solid, such as diatomaceous earth or ground firebrick impregnated with a nonvolatile liquid (lubricating greases, silicone oils, or polyesters of high molecular weight alcohols and dibasic acids). A mixture of methyl esters of fatty acids is flash-evaporated at one end of the column, the entire length of which is maintained at an elevated temperature (170 to 225°C.). The volatilized esters are swept through the column by a stream of an inert gas, *e.g.*, argon, helium, or nitrogen flowing at a constant rate. Each component of the ester mixture moves on the column at a different rate determined by its ratio of partition between the gas phase and the nonvolatile liquid (stationary) phase. The presence of the individual esters in the gas emerging from the column is detected usually by physical or chemical means. The data are automatically recorded on a chart as a series of peaks. The area under each peak is proportional to the concentration of the particular component in the mixture. Identification of components in a chromatogram is achieved by the use of standards. The resolution of a mixture of the methyl esters of fatty acids by this procedure is illustrated in Fig. 4.1. Members in a homologous group, *e.g.*, straight-chain

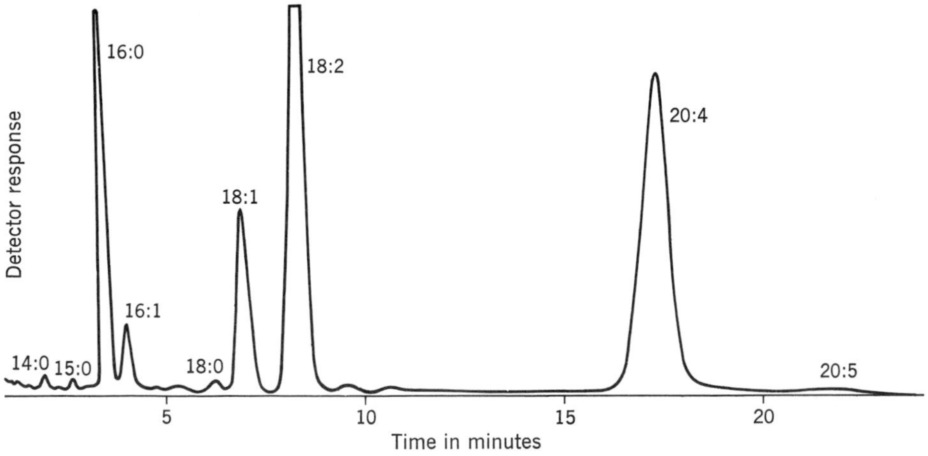

Fig. 4.1. Gas-liquid chromatographic analysis of the methyl esters of the fatty acids present in normal rat plasma as cholesterol esters. The numbers at the top or at one side of the peaks, to the left and right of the colon, indicate, respectively, the number of carbon atoms and double bonds present in each fatty acid. The areas under each peak reflect the relative quantities of each fatty acid present. (*Courtesy, Dr. L. I. Gidez.*)

saturated fatty acid methyl esters, may be identified from a plot of the time necessary to elute a component from the column (retention time) versus the number of carbon atoms in the component. This technique permits quantitative resolution of microgram quantities of mixtures of fatty acids.

Gas-liquid chromatography has also been applied to separation of suitable volatile derivatives of fatty aldehydes, carbohydrates, and amino acids, as well as of steroids (page 84).

LIPIDS CONTAINING GLYCEROL

NEUTRAL FATS

Mono-, Di-, and Triglycerides. The neutral fats comprise by far the most abundant group of lipids in nature. These are esters of fatty acids with the trihydroxyl alcohol glycerol, $CH_2OHCHOHCH_2OH$. One, two, or three of the hydroxyl groups of glycerol may be esterified with a fatty acid; hence the designations mono-, di-, and triglycerides. Quantitatively, the triglycerides represent the most significant group of neutral fats, although mono- and diglycerides are present in nature and are of importance in lipid metabolism.

The generic formula for a triglyceride is

$$\begin{array}{r}
R—COO—CH_2 \\
| \\
R'—COO—CH \\
| \\
R''—COO—CH_2
\end{array}$$

Neutral fat

in which RCOOH, R′COOH, and R″COOH represent three molecules of either the same or different fatty acids. The system for naming the neutral fats is based upon the names of the constituent fatty acids. Thus tristearin contains three stearic acid residues per molecule, and oleodistearin contains one residue of oleic acid and two of stearic acid.

Physical and Chemical Properties. The general physical properties of fats are reminiscent of those of fatty acids. As a class, the fats are insoluble in water, soluble in nonpolar solvents. The solubility is greater, and the melting point lower, the richer the fat is in short-chain and unsaturated fatty acid residues. Saturation as well as increasing chain length tend to result in elevation of melting point such that, whereas tristearin is a solid at room temperature (melting point, 71°C.), triolein (melting point, -17°C.) and tributyrin (melting point, < -75°C.) are liquids.

Table 4.4 contains examples of the fatty acid composition of neutral fats from various sources. Most samples of animal fat contain predominantly esters of palmitic, stearic, palmitoleic, oleic, and linoleic acids in various proportions. Fat from diverse portions of the same organism may differ widely in composition. Thus, subcutaneous fat of man has an average iodine number of 65, whereas fat from the liver, which is richer in unsaturated acids, has an average iodine number of 135. Butter or milk fat is a notable exception in having fatty acids of shorter chain length in higher abundance than elsewhere. Subcutaneous fat of different mammalian species varies considerably in degree of unsaturation, and this is

Table 4.4: APPROXIMATE COMPOSITION, IN MOLAR PERCENTAGE, OF THE FATTY ACID MIXTURES OBTAINED FROM TRIGLYCERIDES FROM VARIOUS SOURCES

Acid	Depot fat				Liver fat, cow	Milk fat, cow
	Human	Cow	Pig	Sheep		
Butyric	..	..	..	..	...	9
Caproic	..	..	..	..	...	3
Caprylic	..	..	..	..	...	2
Capric	..	..	..	..	...	4
Lauric	..	..	..	..	...	3
Myristic	3	7	1	2	3	11
Palmitic	23	29	28	25	35	23
Stearic	6	21	10	26	5	9
Palmitoleic	5	..	..	..	10	4
Oleic	50	41	58	42	36	26
Linoleic	10	2	3	5	8	3
C_{20-22} unsaturated	..	..	..	..	3*	

* Average number of double bonds per molecule = 3.0.

SOURCE: Data adapted from T. P. Hilditch, "The Chemical Constitution of Natural Fats," John Wiley & Sons, Inc., New York, 1940, and from the literature.

reflected in the fact that, whereas the melting point of beef tallow is high, pork lard melts at a considerably lower temperature.

Vegetable fats, or oils, also exhibit diversity in their fatty acid composition in regard to both chain length and degree of unsaturation of the constituent fatty acids; many are liquids at room temperature.

Hydrolysis and Saponification. The most important chemical reaction of the neutral fats is their hydrolysis to yield three molecules of fatty acid and one of glycerol.

$$
\begin{array}{l}
R\text{—COO—}CH_2 \\
\qquad\qquad\;| \\
R'\text{—COO—}CH \\
\qquad\qquad\;| \\
R''\text{—COO—}CH_2
\end{array}
+ 3H_2O \underset{}{\overset{(H^+)}{\rightleftharpoons}}
\begin{array}{c}
R\text{—COOH} \\ + \\ R'\text{—COOH} \\ + \\ R''\text{—COOH}
\end{array}
+
\begin{array}{l}
HOCH_2 \\ \quad | \\ HOCH \\ \quad | \\ HOCH_2
\end{array}
$$

Neutral fat **Fatty acids** **Glycerol**

This reaction proceeds slowly in boiling water. It is reversible, its reversal being the reaction of esterification, and the attainment of equilibrium will be accelerated by the same catalysts, *e.g.*, H^+, employed in esterification. The enzymes that catalyze this reaction in animals and plants are *esterases*, or more specifically, *lipases*.

Neutral fats are readily hydrolyzed by alkali, a reaction termed *saponification*. The carboxylate ions formed, in the presence of a cation, become soaps.

$$
\begin{array}{l}
R\text{—COOCH}_2 \\
\qquad\quad\;| \\
R'\text{—COOCH} \\
\qquad\quad\;| \\
R''\text{—COOCH}_2
\end{array}
+ 3OH^- \longrightarrow
\begin{array}{c}
R\text{—COO}^- \\ + \\ R'\text{—COO}^- \\ + \\ R''\text{—COO}^-
\end{array}
+
\begin{array}{l}
HOCH_2 \\ \quad | \\ HOCH \\ \quad | \\ HOCH_2
\end{array}
$$

Neutral fat **Glycerol**

The reaction is irreversible; the carboxylate ions do not recombine with the hydroxyl groups of glycerol. The progress of the reaction is further favored by the fact that the initial portion of soap that is formed itself acts as a detergent. When the reaction is completed, the products, soap and glycerol, are all soluble in water and insoluble in nonpolar solvents such as ether. This fact gives rise to an operational separation frequently employed in lipid chemistry. The *saponifiable fraction* is defined as that portion of total lipid which, after treatment with hot alkali, is soluble in water and insoluble in ether. Neutral fats are thus said to be saponifiable. Examples of nonsaponifiable materials will be encountered later in this chapter. Although originally signifying "soapmaking," the term saponification has been extended to include alkaline hydrolyses of all esters.

Glycerol. Glycerol is a sweet, viscous liquid, miscible with water and with ethanol, but insoluble in ether. It forms esters with inorganic, as well as organic, acids. Thus, α-glycerolphosphate is an important intermediate in carbohydrate (Chap. 18) and lipid (Chap. 22) metabolism.

CH_2OH
|
$CHOH$
|
$CH_2OPO_3H_2$

α-Glycerolphosphate

Glyceryl Ethers. In glyceryl ethers an α-hydroxyl group of the glycerol is in ethereal linkage with an aliphatic alcohol. The latter may be saturated or unsaturated. Three such compounds have been isolated from animal tissues and from shark oil.

Chimyl alcohol	Batyl alcohol	Selachyl alcohol
$CH_2O(CH_2)_{15}CH_3$	$CH_2O(CH_2)_{17}CH_3$	$CH_2O(CH_2)_8CH{=}CH(CH_2)_7CH_3$
$CHOH$	$CHOH$	$CHOH$
CH_2OH	CH_2OH	CH_2OH

Batyl alcohol is an ether of glycerol and octadecyl alcohol (C_{18} corresponding to stearic acid), while selachyl alcohol contains an unsaturated oleyl alcohol residue (corresponding to oleic acid). These alcohols occur naturally in a form in which the hydroxyl groups are acylated.

Another class of glyceryl ethers consists of derivatives of α-phosphatidic acid (page 69) and will be considered below with other phosphatidic acid–containing lipids.

Glycosyl Glycerides. Glycosyl glycerides of plants include mono- and digalactosyl diglycerides. An α-D-galactosyl (1 → 6) β-D-galactosyl diglyceride has been described in chloroplast lipids. The mono- and digalactosyl diglycerides of green leaves contain a high proportion of linolenic acid.

The glycosyl glycerides include a sulfolipid, a sulfonic acid derivative of 6-deoxyglucosyl diglyceride. The sugar, 6-deoxyglucose, is *quinovose.* This sulfolipid has been found widely distributed in plants, localized mainly in the chloroplasts, and is also present in the chromatophores of photosynthetic bacteria.

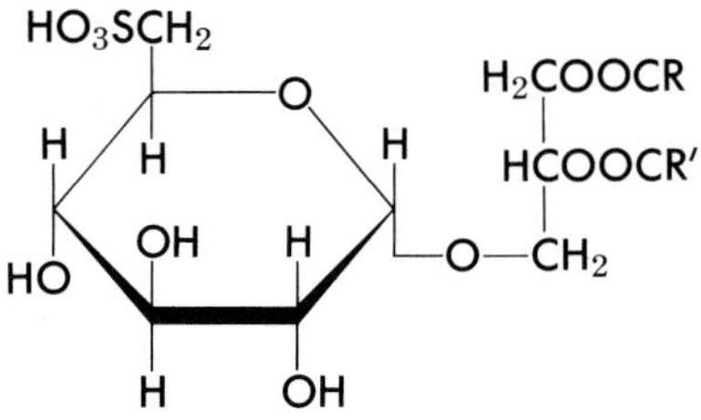

6-Sulfo-6-deoxy-α-glucosyl diglyceride
(6-sulfoquinovosyl diglyceride)

PHOSPHOLIPIDS

The next large class of glycerol-containing lipids to be considered is the phospholipids. This class comprises compounds that are derivatives of glycerol phosphate, and may be further subdivided on the basis of the nature of the linkage in the hydrocarbon chains and of polar groups present in the molecule. In addition, further characterization may be based on the presence or absence of a nitrogen-containing base as a portion of the polar group.

Phosphatides. The larger groups of naturally occurring phospholipids are termed phosphatides. These are derivatives of glycerol phosphate and most frequently contain a nitrogenous base. The compounds of this group may be regarded as derivatives of L-α-phosphatidic acid.

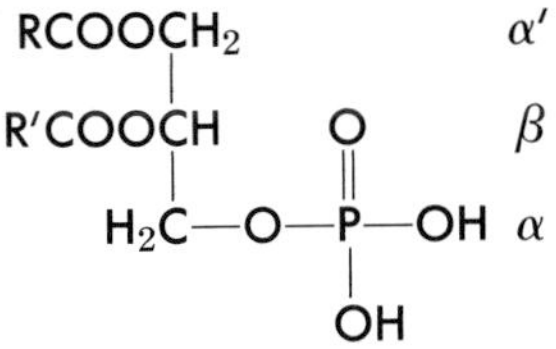

L-α-Phosphatidic acid

This compound is designated as the L form because of its stereochemical relationship to L-glyceryl phosphate. Phosphatidic acid yields, on total hydrolysis, one mole each of glycerol and of phosphoric acid and two moles of fatty acid. The fact that the fatty acids obtained often contain about one-half a double bond per mole indicates that, in phosphatidic acids derived from some natural phosphatides, one molecule of each pair of fatty acids is saturated and the other unsaturated. It will be noted that the phosphoric acid is bound to glycerol in ester linkage at the terminal α-hydroxyl group.

In the phosphatides, the phosphoric acid is also bound in ester linkage to one of the following nitrogenous compounds:

$HO—CH_2—CH_2—\underset{+}{N}\equiv(CH_3)_3$

Choline

$HO—CH_2—CH_2—NH_2$

Ethanolamine

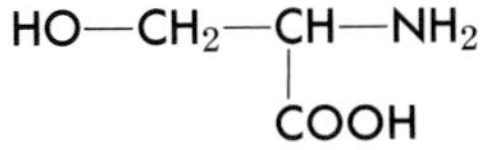

L-Serine

The structural names of the three corresponding phosphatides are phosphatidyl choline, phosphatidyl ethanolamine, and phosphatidyl serine, respectively.

The *phosphatidyl cholines,* also termed lecithins, have the type formula:

$$RCOOCH_2$$
$$R'COOCH$$
$$H_2C-O-\overset{\overset{O}{\|}}{\underset{\underset{O^-}{|}}{P}}-OCH_2CH_2\overset{+}{N}\equiv(CH_3)_3$$

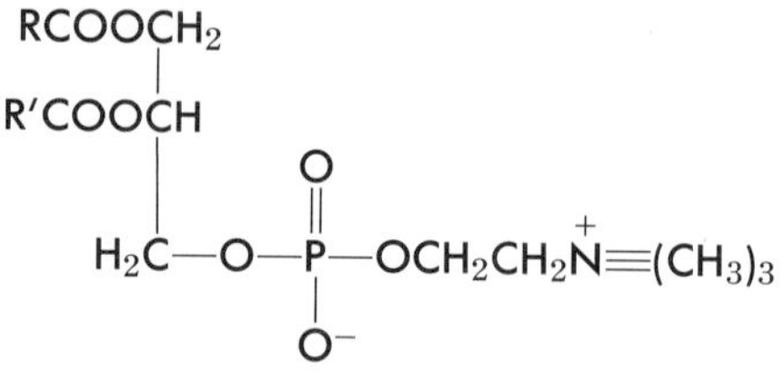

L-α-Lecithin

On complete hydrolysis, choline, phosphoric acid, glycerol, and two molecules of fatty acid are obtained. The formula is written in dipolar form (page 101), which is intended to convey the fact that choline is a strong base, that phosphoric acid is a moderately strong acid, and that over a wide pH range, including physiological pH, both groups exist predominantly in the ionic form. Partial hydrolysis of lecithin with removal only of one fatty acid yields *lysolecithin.* Hydrolysis of the α-ester is catalyzed by an enzyme in brain tissue, while β-ester hydrolysis is catalyzed by enzymes found in snake venoms, microorganisms, and in animal tissues (Chap. 22).

The basicity of choline is characteristic of all quaternary ammonium compounds, *i.e.,* compounds in which N bears four organic substituents. Free choline base,

$$(CH_3)_3\equiv\overset{+}{N}-CH_2-CH_2-OH + OH^-$$
Choline

exists in aqueous solution practically completely dissociated, much like KOH, and in striking contrast to NH_4OH. Salts of choline with strong acids, such as choline chloride,

$$(CH_3)_3\equiv\overset{+}{N}-CH_2-CH_2-OH + Cl^-$$
Choline chloride

react as neutral substances in aqueous solution.

The *phosphatidyl ethanolamines* and *phosphatidyl serines* have the following type structures:

$$RCOOCH_2$$
$$R'COOCH$$
$$H_2C-O-\overset{\overset{O}{\|}}{\underset{\underset{O^-}{|}}{P}}-OCH_2CH_2\overset{+}{N}H_3$$

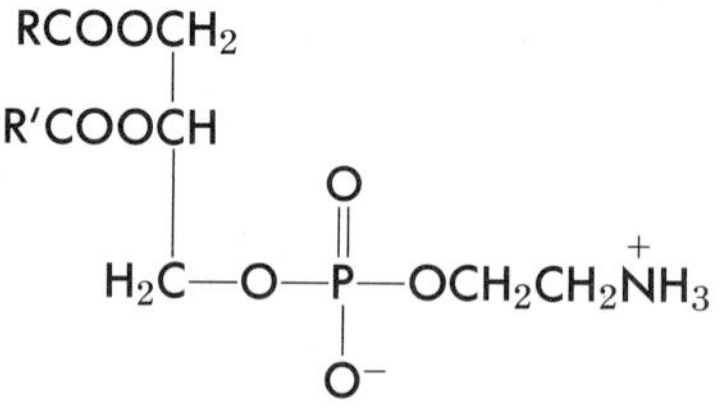

Phosphatidyl ethanolamine

$$RCOOCH_2$$
$$R'COOCH$$
$$H_2C-O-\overset{\overset{O}{\|}}{\underset{\underset{O^-}{|}}{P}}-OCH_2\overset{\underset{\underset{COO^-}{|}}{}}{C}H\overset{+}{N}H_3$$

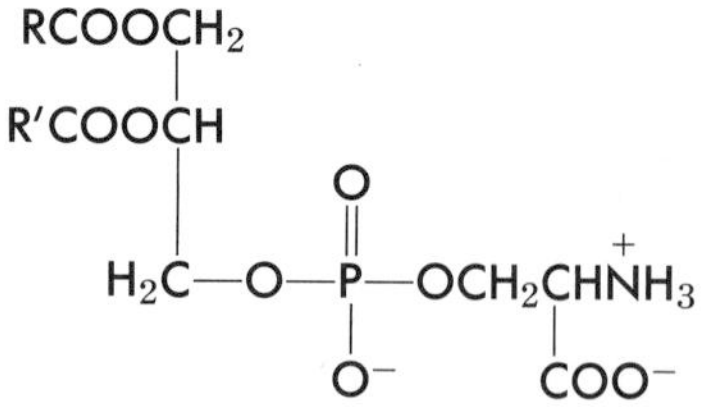

Phosphatidyl serine

Like the phosphatidyl cholines, these compounds are ionic at physiological pH, but since the primary amino group is much less basic than the quaternary ammonium group, the phosphatidyl ethanolamines and phosphatidyl serines are more acidic than the choline-containing phosphatides. Phosphatidyl serine, with its added carboxyl group, is the most strongly acidic member.

Plasmalogens. The plasmalogens are a class of phosphatides in which the fatty acid at the α' position is replaced by an α,β-unsaturated ether. Thus, the ethanolamine-containing plasmalogens have the following generic formula:

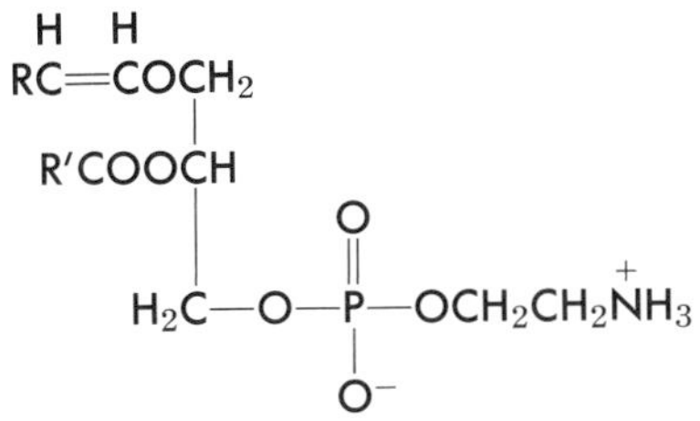

Phosphatidal ethanolamine

The α,β-unsaturated ether linkage of plasmalogens is relatively alkali stable; on mild acid hydrolysis, the α,β-unsaturated ether gives rise to an aldehyde. The three principal classes of plasmalogens are phosphatidal ethanolamine, phosphatidal choline, and phosphatidal serine.

Plasmalogens occurring in nature are of the L configuration. They are present in highest concentrations in brain and heart, and apparently are not found in significant quantity in nonanimal tissues. A few exceptions are selected bacterial species. The aldehydogenic chains may be of lengths varying from C_{12} to C_{18}, including odd-numbered, branched, and unsaturated carbon chains. Nineteen different aldehydes have been obtained from the lipids of human red blood cells. Pure phosphatidal choline of beef heart showed 98 per cent of the fatty acyl chain to be unsaturated, whereas 94 per cent of the aldehydogenic chains were saturated. The α,β-unsaturated ether linkage has the *cis* configuration. Plasmalogens without a nitrogenous base, *i.e.*, α,β-unsaturated ethers of phosphatidic acid, have also been reported in animal tissues, *e.g.*, liver. Neutral lipid plasmalogens occur in low concentrations in structures analogous to triglycerides in which two chains are bound as acyl ester and one as unsaturated ether.

Ether Phosphatides. Glyceryl ether derivatives of phosphatidyl choline are relatively widely distributed in animal tissues; the ether linkage is at the α' position in place of the acyl fatty residue in other phosphatides. Although generally comprising less than a few per cent of the total phospholipids, these glyceryl ether derivatives occur more abundantly in erythrocytes and bone marrow. In terrestrial slugs and various molluscs, up to 25 per cent of the total phospholipid consists of ether derivatives.

Phosphoglycerides and Phosphoinositides. Two groups of nitrogen-free derivatives of glycerol phosphate are present in nature, the phosphoglycerides and the phosphoinositides.

Phosphoglycerides. The phosphoglycerides, present in animal and, particularly, plant tissues, are composed of one mole of glycerol and two moles of L-α-phosphatidic acid. The name diphosphatidyl glycerol describes these compounds, with the following generic formula:

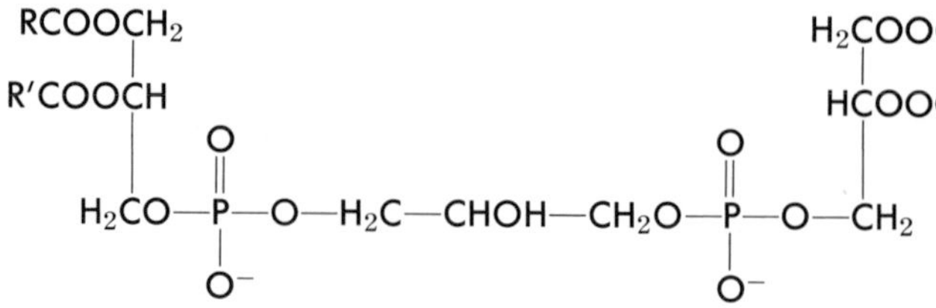

A diphosphatidyl glycerol

The single important example of a diphosphatidyl glycerol reported in animal tissue is cardiolipin, initially isolated from heart tissue. This is the only phosphatide with known immunological properties; it is utilized in the serological diagnosis of syphilis. Similar compounds are also present in plants; however, the phosphatides in plants contain one mole of phosphatidic acid and are, therefore, monophosphatidyl glycerols. Members of the group differ from one another in the nature of the fatty acids yielded on hydrolysis.

Phosphoinositides. A second group of nitrogen-free derivatives of α-phosphatidic acid contains inositol. On hydrolysis, phosphoinositides usually yield one mole of glycerol, one mole of the hexahydroxy alcohol, *myo*-inositol (page 32), two moles of fatty acid, and one to three moles of phosphate. The 4-diphospho- and 4,5-triphosphoinositides are present in brain and represent more than half the total phosphoinositides. A triphosphoinositide, 1-phosphatidyl-*myo*-inositol-4,5-diphosphate, has the generic structure below.

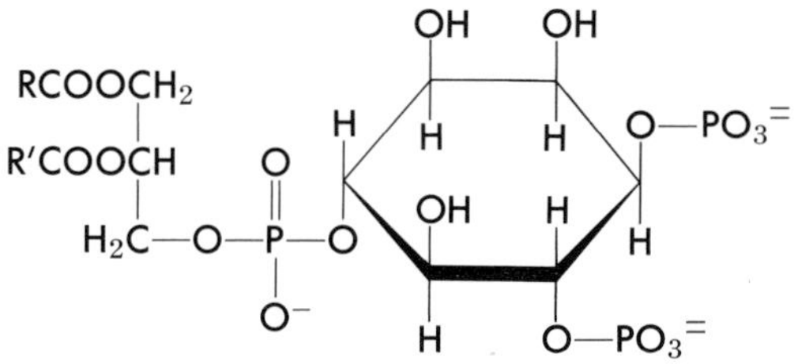

A phosphoinositide

LIPIDS NOT CONTAINING GLYCEROL

SPHINGOLIPIDS

A large number of lipids may be grouped together on the basis of the presence in their structures of a long-chain, aliphatic base. Initially, these compounds were all thought to be derivatives of the base sphingosine or a closely related structure, dihydrosphingosine.

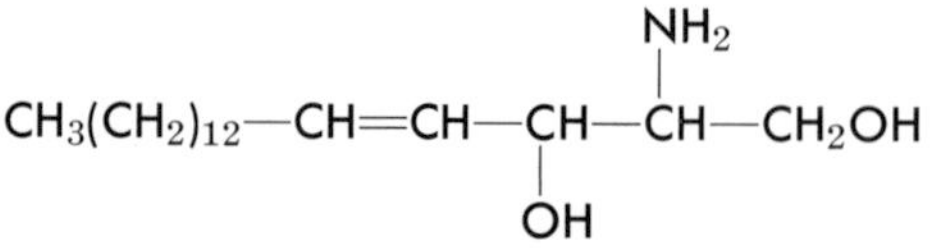

Sphingosine

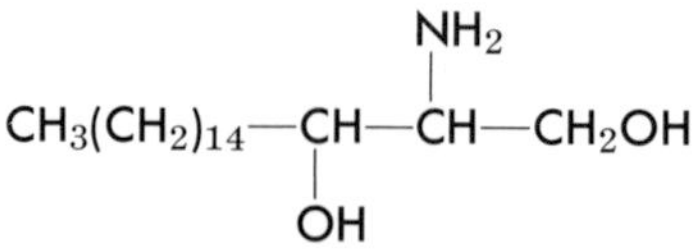

Dihydrosphingosine

Lipids containing these bases are termed sphingolipids. Although the above C_{18} sphingosines are most abundant in sphingolipids, other homologous C_{16}, C_{17}, C_{19}, and C_{20} sphingosines also are found among the naturally occurring sphingolipids.

Ceramides. The term *ceramide* is used to designate the N-acyl fatty acid derivatives of a sphingosine.

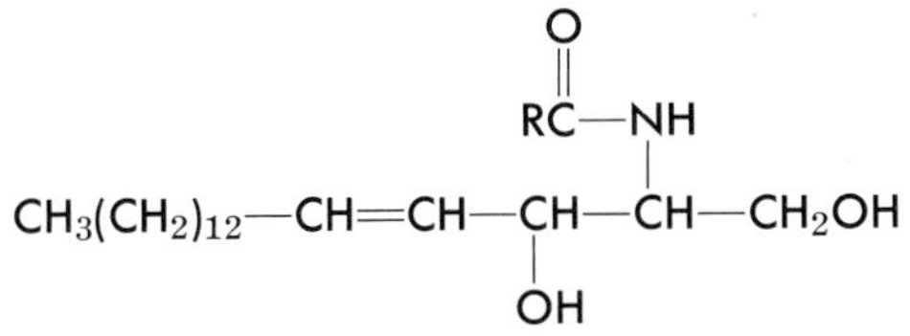

A ceramide

Ceramides are widely distributed in plant and animal tissues, but quantitatively they are minor lipid constituents. The fatty acid residue, present in amide linkage, is of the C_{16}, C_{18}, C_{22}, or C_{24} series; members of the group of ceramide derivatives differ from one another in the variety of their constituent fatty acids.

Sphingomyelins. The sphingomyelins are the only group of phosphorus-containing sphingolipids. Chemically, the sphingomyelins may be regarded as phosphoryl choline derivatives of a ceramide.

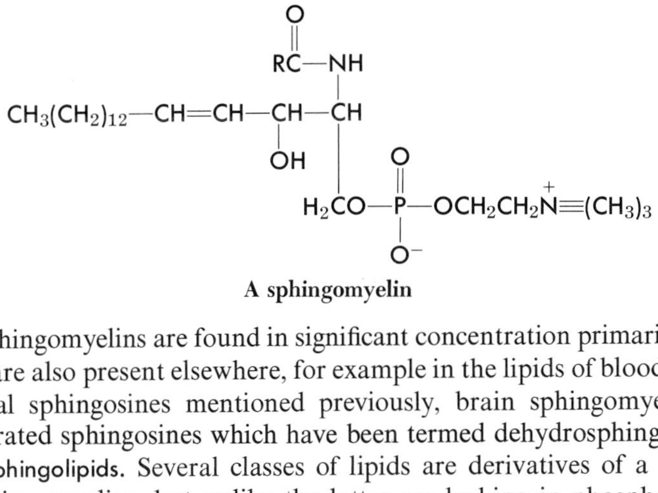

A sphingomyelin

The sphingomyelins are found in significant concentration primarily in nervous tissue, but are also present elsewhere, for example in the lipids of blood. In addition to the usual sphingosines mentioned previously, brain sphingomyelin contains polyunsaturated sphingosines which have been termed dehydrosphingosines.

Glycosphingolipids. Several classes of lipids are derivatives of a ceramide, as are the sphingomyelins, but unlike the latter are lacking in phosphorus and the additional nitrogenous base; they do, however, contain one or more moles of carbohydrate. Glycosphingolipids accumulate in abnormally large amount in certain diseases characterized by disturbances of lipid metabolism (Chap. 22). Three classes of glycosphingolipids are the cerebrosides, the gangliosides, and the ceramide oligosaccharides.

Cerebrosides. The cerebrosides are ceramide monosaccharides occurring most abundantly in the myelin sheath of nerves. On hydrolysis, cerebrosides yield one molecule each of a sphingosine, a fatty acid, and a hexose, most frequently D-galactose, less commonly, D-glucose. For this reason members of this group are sometimes referred to as *galactolipids* or *glucolipids.* The terms *galactocerebrosides* and *glucocerebrosides* are also encountered. The hexose in cerebrosides is present in β-glycosidic linkage. The generic formula is

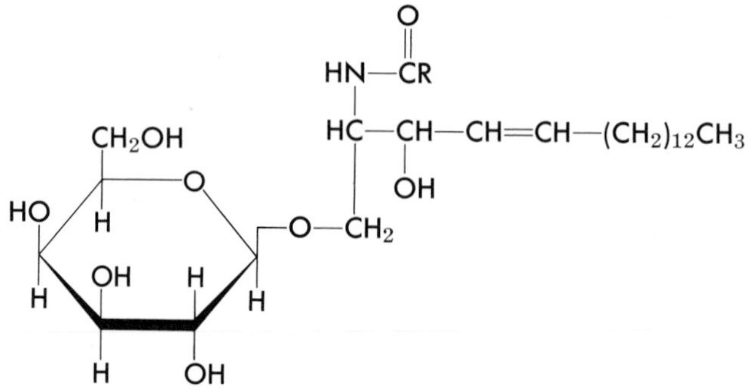

A cerebroside

The variety of fatty acids in the ceramide portion of the cerebrosides has been mentioned previously. In the cerebroside kerasin, the saturated C_{24} acid, lignoceric, is found, while its 2-hydroxy derivative, cerebronic acid, is found in phrenosin. Other 2-hydroxy acids are found among the C_{16} and C_{18} fatty acid constituents of cerebrosides. A sulfate ester analogue of phrenosin is abundant in the white matter of brain in the galactocerebrosides. The sulfate is present in ester linkage at C-3 of the galactose portion of the molecule. Similar sulfur-containing lipids have been reported in other tissues. Members of the group of cerebroside sulfuric esters have been designated as *sulfatides.*

Gangliosides. This class of glycosphingolipids is found in significant concentrations in nerve tissue and in other selected tissues, notably spleen. The structure of gangliosides is related to that of cerebrosides in that gangliosides contain a ceramide linked to carbohydrate. However, in addition to the hexose (galactose or glucose) found in cerebrosides, gangliosides generally contain several additional moles of carbohydrate. These include at least one mole of N-acetylgalactosamine or N-acetylglucosamine, and at least one mole of N-acetylneuraminic acid (a sialic acid, page 34). In gangliosides from horse erythrocytes and spleen, the sialic acid is N-glycolylneuraminic acid (page 34).

Ceramide Oligosaccharides. Glycosphingolipids have been described in which the number of moles of carbohydrate attached to a ceramide may vary from one to several, joined in glycosidic linkage; hence the terms, *ceramide monosaccharide, ceramide disaccharide,* etc. One of these, termed cytolipin H, is ceramide galactosylglucose (ceramide lactoside), which displays immunological activity under certain conditions. Another, cytolipin K, isolated initially from kidney, is probably identical with globoside, the most abundant glycosphingolipid in human erythrocyte stroma. This compound is N-acetylgalactosaminyl (1 $\rightarrow$ 6) galactosyl (1 $\rightarrow$ 4) galactosyl (1 $\rightarrow$ 4) glucosyl ceramide.

A ceramide pentasaccharide isolated from a human adenocarcinoma has one residue of fucose (page 33). Immunological activity has been found to be a general property of glycosphingolipids. In contrast, only one phospholipid, cardiolipin (page 72), has this property.

ALIPHATIC ALCOHOLS AND WAXES

Significant quantities of aliphatic alcohols have been recovered from certain lipid sources. Thus in the feces there are detectable amounts of cetyl alcohol, the primary alcohol corresponding to palmitic acid.

$$CH_3(CH_2)_{14}CH_2OH$$

Cetyl alcohol

In various highly specialized lipids, such alcohols occur as esters of fatty acids. The "head oil" of the sperm whale is largely cetyl palmitate, known as spermaceti, and beeswax is rich in myricyl palmitate.

$$CH_3(CH_2)_{14}{-}\underset{\underset{O}{\|}}{C}{-}OCH_2(CH_2)_{14}CH_3$$

Cetyl palmitate

$$CH_3(CH_2)_{14}{-}\underset{\underset{O}{\|}}{C}{-}OCH_2(CH_2)_{28}CH_3$$

Myricyl palmitate

The generic name "wax" is given to a naturally occurring fatty acid ester of any alcohol other than glycerol. It may be pointed out that the alcoholic components discussed here are part of the *nonsaponifiable fraction* (page 68) of a lipid hydrolysate.

TERPENES

Throughout the biological world there are numerous compounds whose carbon skeletons suggest a structural relationship to isoprene, 2-methyl butadiene.

$$CH_2{=}\overset{\overset{CH_3}{|}}{C}{-}CH{=}CH_2$$

Isoprene

Many of these compounds contain multiples of five carbon atoms so related to each other as to make possible dissection of their structures into isoprene-like fragments. This class of compounds, called *terpenes* (from turpentine), includes essential oils such as citral, pinene, geraniol, camphor, and menthane; the resin acids and rubber; a variety of plant pigments, including carotenes, lycopene, and others; and vitamin A and squalene of animals. Examples of open-chain terpenes are phytol, the alcoholic fragment obtained on hydrolysis of chlorophyll,

$$CH_3{-}\overset{\overset{CH_3}{|}}{CH}{-}CH_2{-}CH_2{-}CH_2{-}\overset{\overset{CH_3}{|}}{CH}{-}CH_2{-}CH_2{-}CH_2{-}\overset{\overset{CH_3}{|}}{CH}{-}CH_2{-}CH_2{-}CH_2{-}\overset{\overset{CH_3}{|}}{C}{=}CH{-}CH_2OH$$

Phytol

and squalene, a hydrocarbon found in shark-liver oil and an intermediate in the biosynthesis of cholesterol.

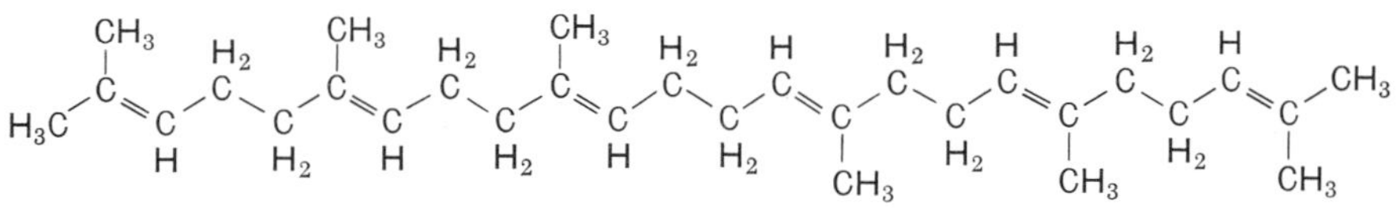

Squalene

Of particular interest are the carotenes, pigments occurring in plants. The formulas of α-, β-, and γ-carotene are shown.

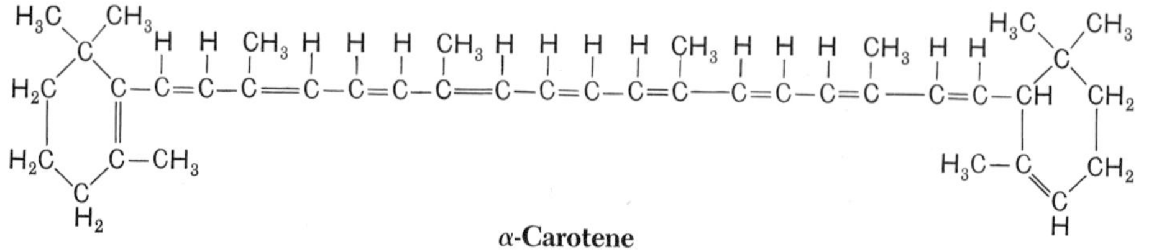

α-Carotene

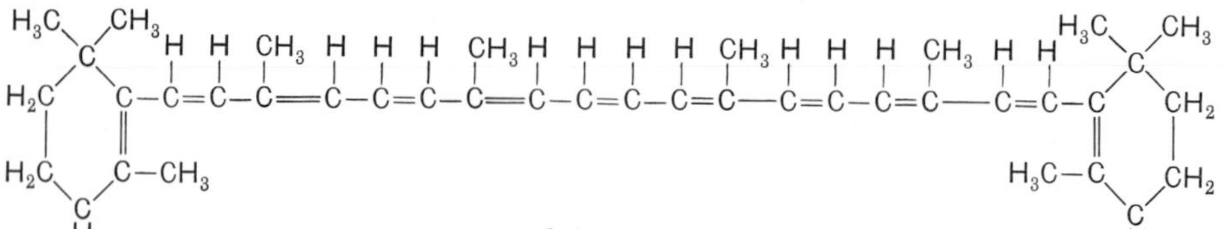

β-Carotene

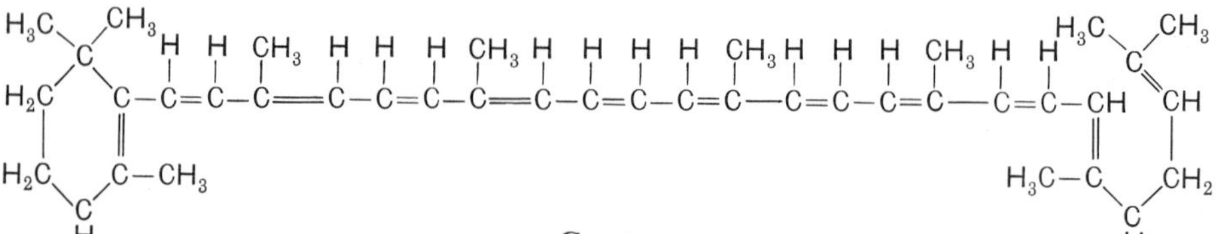

γ-Carotene

β-Carotene, as can be seen, is symmetrical with respect to the arrangement of the two termini of the chain. Interest in these compounds relates to the fact that the carotenes are precursors of the group of vitamins A (Chap. 49); an example is retinol (vitamin A_1).

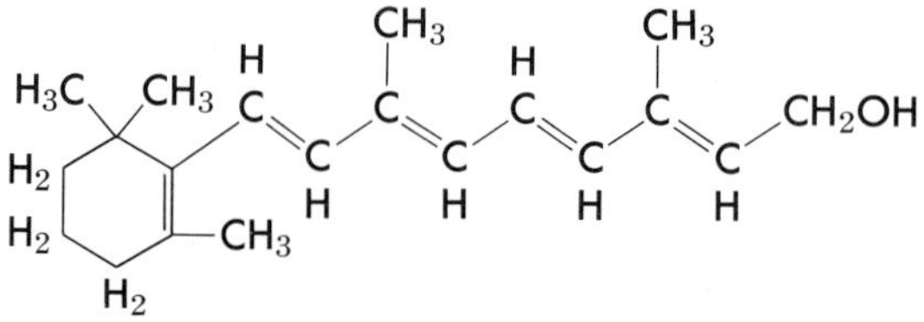

Retinol (vitamin A_1)

Because of the many double bonds in the carotenoid side chain, *cis-trans* isomerism is possible. The major proportion of naturally occurring carotenoids have their double bonds in the all-*trans* configuration, as shown above for vitamin A_1. Small quantities of carotenoids containing one or more *cis* double bonds occur in nature, and isomerization can take place in animal tissues. A *cis* form of vitamin A is present in rat liver, and a carotenoid containing a *cis* double bond is present in the retina of the eye (Chap. 40).

Certain other accessory dietary factors are related chemically to the terpenes. The compounds that have been grouped as vitamins E (Chap. 50) are lipid-soluble aromatic compounds bearing side chains of the type seen among terpenes. The antihemorrhagic factor, vitamin K (Chap. 50), is a substituted 1,4-naphthoquinone bearing a phytyl side chain, as is coenzyme Q, which functions as an electron carrier (Chap. 16).

STEROIDS

This large group of compounds bears some structural resemblance to the terpenes. The diversity of their physiological activity makes it one of the most studied classes of biological compounds. The members of this group may be considered as derivatives of a fused, reduced ring system, perhydrocyclopentanophenanthrene, comprising three fused cyclohexane rings (A, B, and C) in the nonlinear or phenanthrene arrangement,

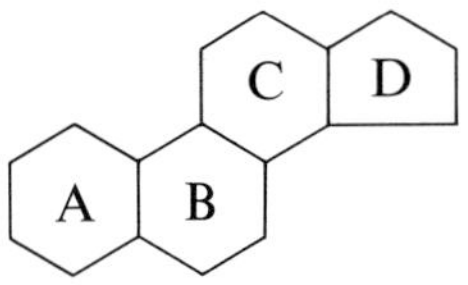

Perhydrocyclopentanophenanthrene

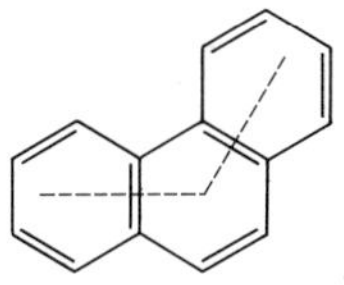

Phenanthrene

and a terminal cyclopentane ring (D). Certain general characteristics of the group may be considered with reference to a typical member, cholestanol; the conventional numbering of the carbon atoms is included in the formula.

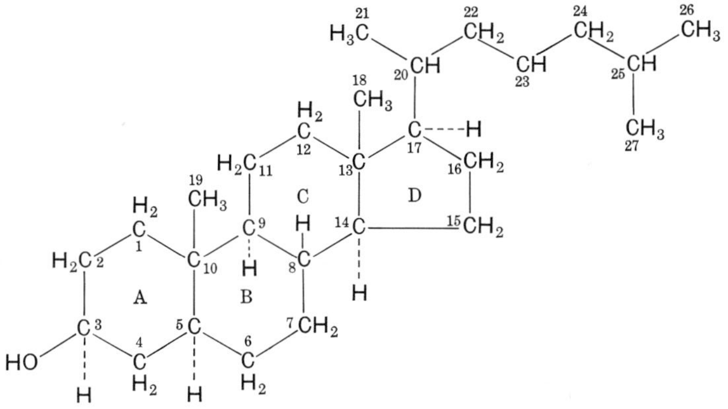

Cholestanol

The shorthand of steroid chemistry transforms this formula into the following:

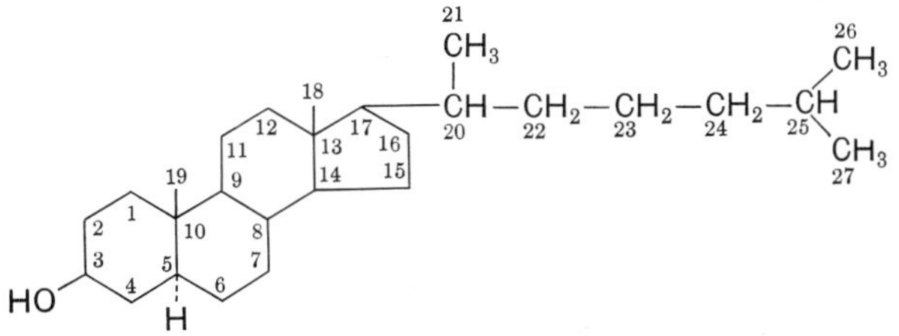

Note that each ring is completely saturated. In this shorthand, whenever double bonds occur they are specifically indicated.

General Considerations. Certain facts of general applicability may be pointed out in regard to this formula. There is an oxygenated substituent on carbon atom 3, a characteristic shared by almost all naturally occurring steroids. There are "angular" methyl groups, numbered 19 and 18, on carbon atoms 10 and 13. This also is a general characteristic. However, as will be seen later, in the estrogens, ring A is aromatic, and under these circumstances carbon atom 10 cannot bear a methyl group. There may be an aliphatic substituent on carbon atom 17. This substituent serves as a convenient basis for classification of steroids, and in the present discussion steroids will be grouped according to the number of carbon atoms in this side chain. It contains 8, 9, or 10 carbon atoms in the sterols (total carbon atoms—27, 28, or 29, respectively), 5 carbon atoms in the bile acids (total of 24 carbon atoms), 2 in the adrenal cortical steroids and in progesterone (total of 21 carbon atoms), and none in the naturally occurring estrogens or androgens (total of either 18 or 19 carbon atoms).

Conformational and Steric Considerations. The conformational formula of cholestanol shows that the molecule is rigid. The small letters in the formula, *a* and *e*, refer to axial and equatorial substitutions, respectively. Not all hydrogen atoms have been represented; some are indicated by solid or dashed lines. The chair conformation of cyclohexane itself is more stable than the boat conformation. In cholestanol rings B and C are locked rigidly in the chair conformation by the *trans*

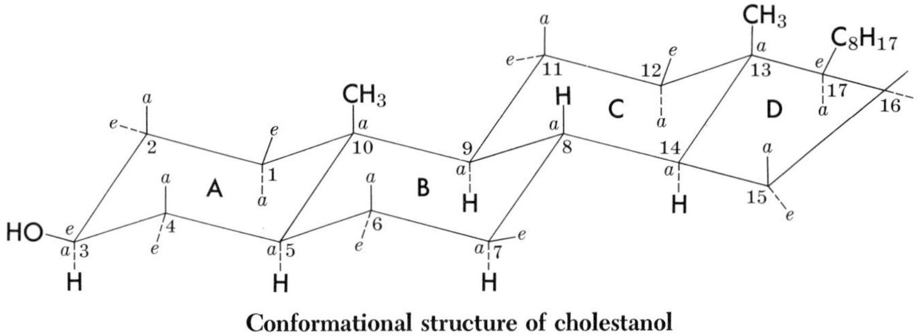

Conformational structure of cholestanol

fusions to rings A and D. Ring A is free to assume the boat form, but the instability associated with the boat form of cyclohexane itself would be enhanced by a strong interaction between methyl and hydroxyl groups at the bow (C-10) and stern (C-3) positions.

In cholestanol there are nine centers of asymmetry in the molecule; these are at carbon atoms 3, 5, 8, 9, 10, 13, 14, 17, and 20. In the structural representation of the steroid (see formulas above), substituents that are β-oriented (to the front) are indicated with a solid bond; those which are α-oriented (to the back) are represented with a dashed line. Thus the angular methyl groups at carbon atoms 10 and 13 are β-oriented, as is the 3-hydroxyl, the 8-hydrogen, and the side chain at carbon atom 17.

Two positions of asymmetry warrant further comment. The hydrogen atom on carbon-5 may be either on the same side of the plane of the molecule as the methyl group on carbon-10, or on the opposite side. In the former case, rings A and B will be *cis* to each other, and the molecule is said to belong to the *normal* configuration. If the hydrogen and methyl are on opposite sides, rings A and B are *trans* to each other, and the molecule is of the *allo* configuration.

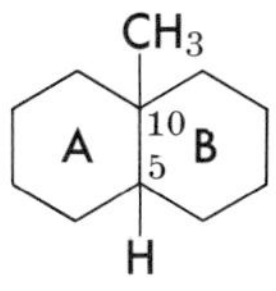

A/B *cis* normal

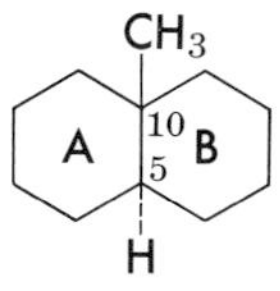

A/B *trans* allo

The hydroxyl group on carbon-3 may similarly be α- or β-oriented.

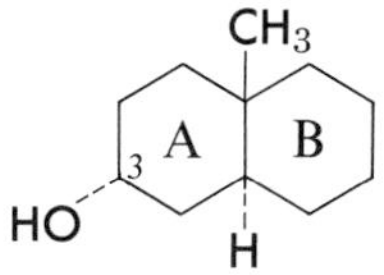

α, allo

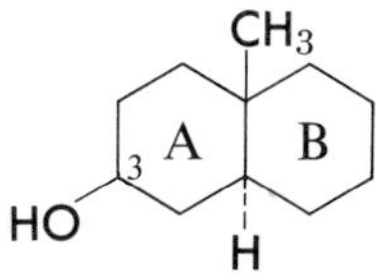

β, allo

Steroids with the 3β-hydroxy structures are precipitated by digitonin, itself the glycoside of a steroid. Many analytical methods for cholesterol depend upon an initial digitonin precipitation. This reagent is useful also in the determination of configuration of other steroids by virtue of this specific reactivity.

Sterols. Steroids with 8 to 10 carbon atoms in the side chain at position 17 and an alcoholic hydroxyl group at position 3 are classed as sterols; the most abundant representative in animal tissues is cholesterol.

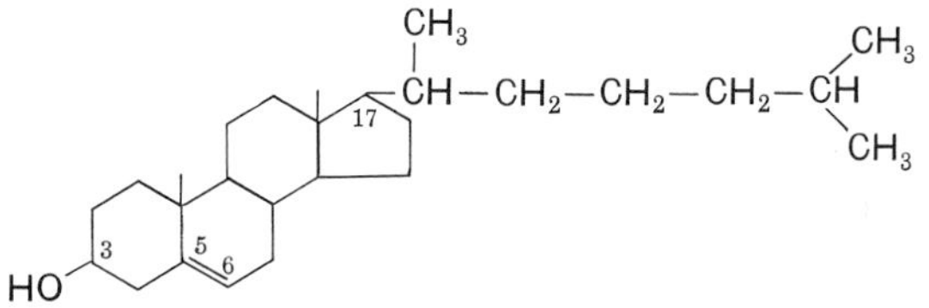

Cholesterol

In addition to a hydroxyl group at position 3, there is a double bond at the 5,6 position. Cholesterol occurs in almost all samples of animal lipid as well as in blood and bile. In the blood about two-thirds of the cholesterol is esterified, chiefly to unsaturated fatty acids, the remainder occurring as the free alcohol (Chap. 22). Reduction of the double bond gives rise to two products, both of which are naturally occurring, *viz.*, coprosterol (β, normal) and β-cholestanol (β, allo). The former is the major fecal sterol; the latter occurs as a minor constituent of the sterols of blood and other tissues. 7-Dehydrocholesterol arises from cholesterol on oxidation and possesses a conjugated pair of double bonds; this sterol is present in skin (Chaps. 22 and 50).

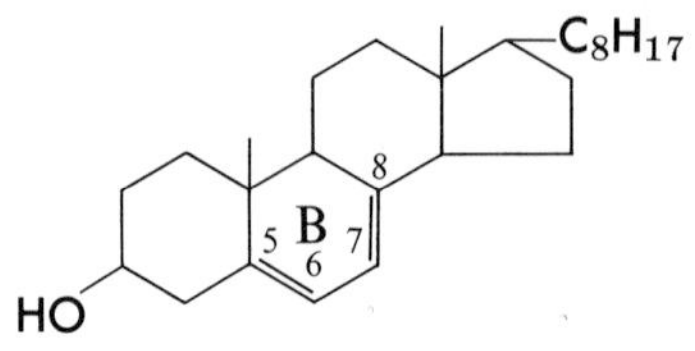

7-Dehydrocholesterol

A similar arrangement of double bonds is in the yeast sterol, ergosterol,

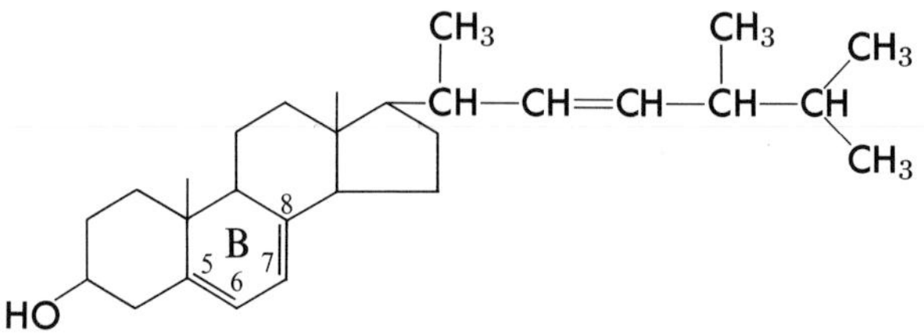

Ergosterol

which, in common with certain other plant sterols, has more than eight carbon atoms in the side chain and, in addition, a double bond in the side chain.

As a result of the conjugated unsaturation of the B ring, ultraviolet irradiation of the above two compounds yields products that result from rupture of the B ring; these possess vitamin D activity (Chap. 50). If one starts with 7-dehydrocholesterol, vitamin D_3 is obtained; from ergosterol, the product is vitamin D_2.

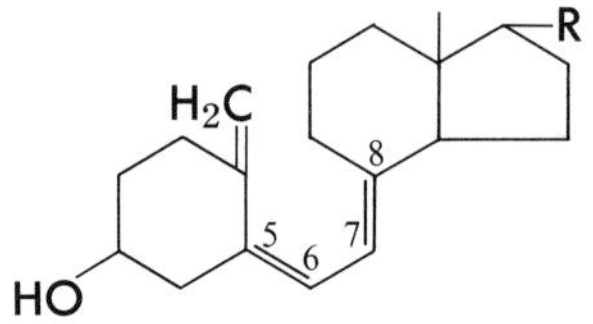

Vitamin D

$R = C_8H_{17}$ in vitamin D_3.
$R = C_9H_{17}$ in vitamin D_2.

Another type of important sterol is represented by lanosterol, first identified in wool lipid.

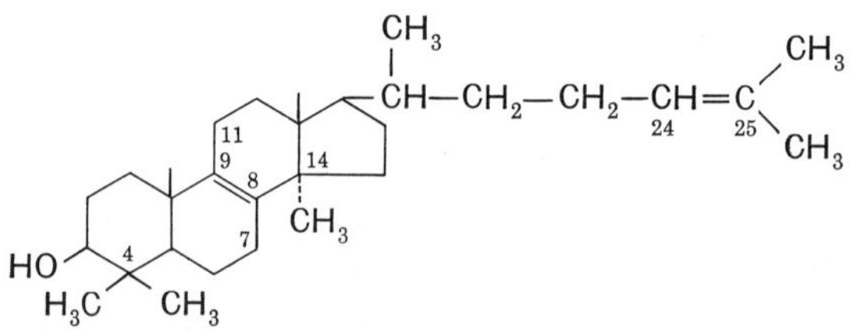

Lanosterol

The distinguishing structural features of this sterol are the *gem* (twin) dimethyl substitution on carbon-4 and the angular methyl group on carbon-14. There is a double bond in the side chain at position 24,25, and another in the nucleus at position 8,9. Lanosterol is an intermediate in cholesterol biosynthesis (Chap. 22).

Bile Acids—C_{24} Steroids. In bile acids, the side chain at carbon-17 is five carbon atoms in length, terminating in a carboxyl group. Four such acids have been isolated from human bile.

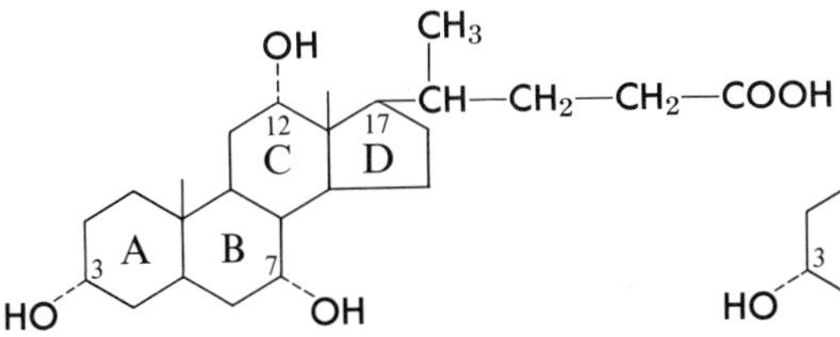

Cholic acid

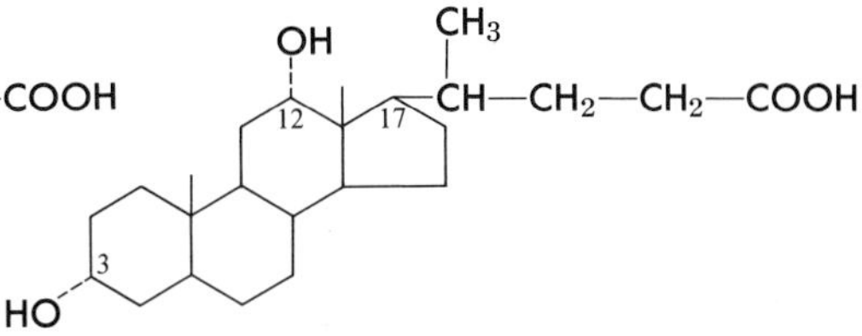

Deoxycholic acid

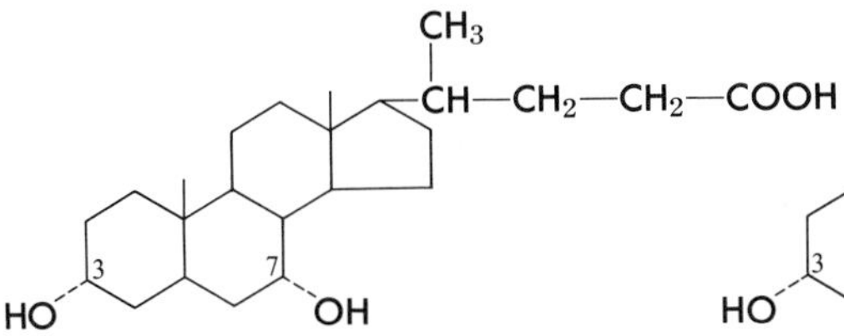

Chenodeoxycholic acid

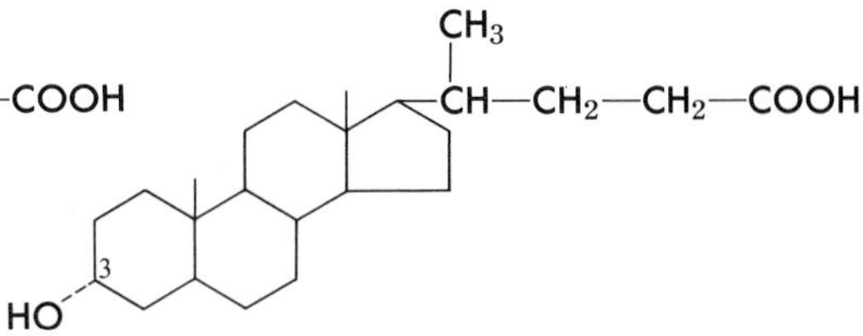

Lithocholic acid

All the hydroxyl groups are of the α configuration; these compounds are therefore not precipitable by digitonin. The A and B rings are of the *cis* or normal configuration. Cholic acid (3α, 7α, 12α-trihydroxycholanic acid) is by far the most abundant bile acid in human bile, but considerable species variation has been found.

In bile these acids are coupled in amide linkage to the amino acids glycine and taurine as glycocholic and taurocholic acids.

$$C_{23}H_{26}(OH)_3\overset{\overset{\large O}{\|}}{C}-\overset{H}{N}-CH_2-COOH$$

Glycocholic acid
(cholylglycine)

$$C_{23}H_{26}(OH)_3\overset{\overset{\large O}{\|}}{C}-\overset{H}{N}-CH_2-CH_2-SO_3H$$

Taurocholic acid
(cholyltaurine)

The salts of these conjugated acids are water soluble and are powerful detergents.

Progesterone and the Adrenal Cortical Steroids—C_{21} Steroids. Since the steroids of endocrinological significance will be discussed in detail later in Part Six, only certain chemical features will be introduced at this point. Among the steroids with two carbon atoms in the side chain at position 17 is progesterone, produced in the corpus luteum.

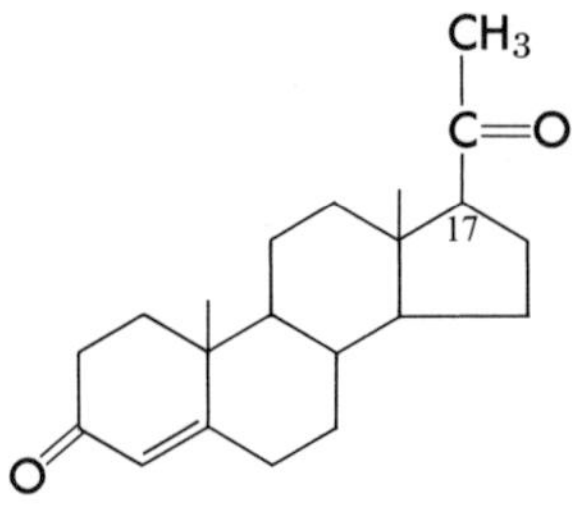

Progesterone

Numerous steroids have been isolated from the adrenal cortex. For purposes of nomenclature, the parent compound is corticosterone; among the various physiologically active modifications of this molecule found in the adrenal cortex is 17-hydroxycorticosterone, termed cortisol.

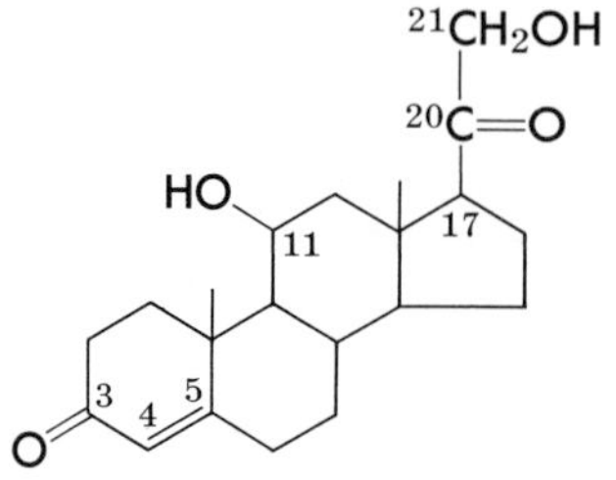

Corticosterone

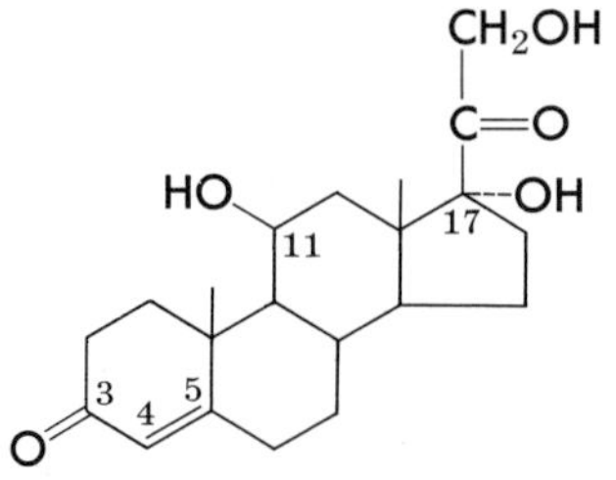

Cortisol
(17-hydroxycorticosterone)

The important structural characteristics of these compounds include the system of conjugated double bonds at carbon atoms 3, 4, and 5, the so-called α, β-unsaturated ketone; the hydroxyl substituent at position 11; and the state of oxidation of the side-chain carbon atoms 20 and 21.

Androgens and Estrogens—C_{19} and C_{18} Steroids. The *androgens,* or male sex hormones, are C_{19} steroids and belong to the class of steroids devoid of a carbon side chain at position 17. An example is testosterone, which is synthesized in the testis.

The *estrogens* differ from all the foregoing steroids in that ring A is aromatic. Consequently, there is no available valence for a methyl group at position 10 and the hydroxyl group at position 3 is phenolic rather than alcoholic in nature. Because of this latter fact, the estrogens behave as weak acids and are extractable from benzene solution with aqueous alkali. The estrogen secreted by the ovary is estradiol-17β.

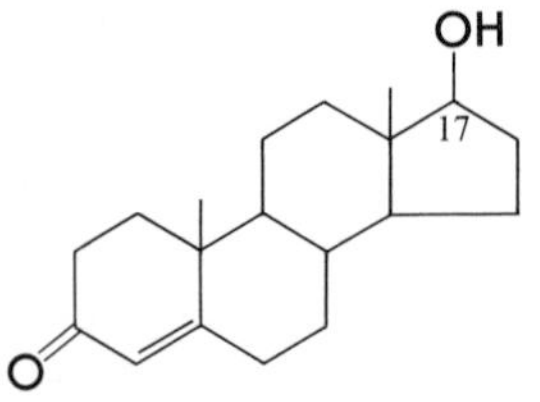

Testosterone

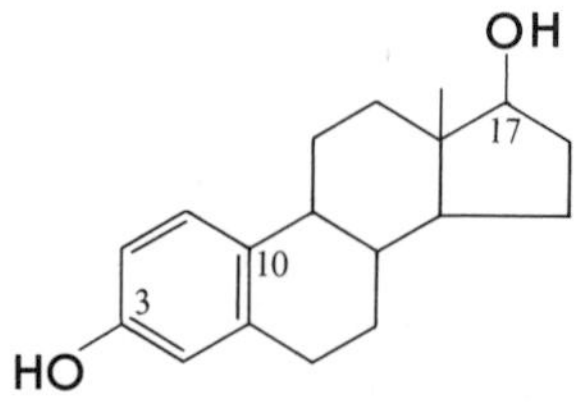

Estradiol-17β

Miscellaneous Steroids. In addition to the several types already discussed, the steroids include many other compounds of interest. Among them are the cardiac-stimulating glycosides, drugs derived from the foxglove, squill, and other plants, which on hydrolysis yield a sugar moiety and a steroid aglycon, *e.g.*, digitoxigenin.

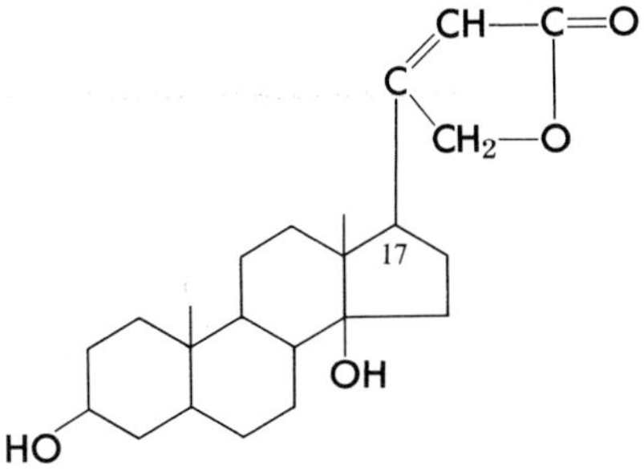

Digitoxigenin

The unusual structural characteristic of this group of compounds is the γ-lactone ring at position 17. The aglycon sapogenins of vegetable origin, as well as certain toad poisons, are also steroids. The compound of sapogenin and sugar is termed a *saponin.* All saponins are surface-active and powerful lytic agents.

An insect steroid hormone that influences molting and metamorphosis has been named ecdysone.

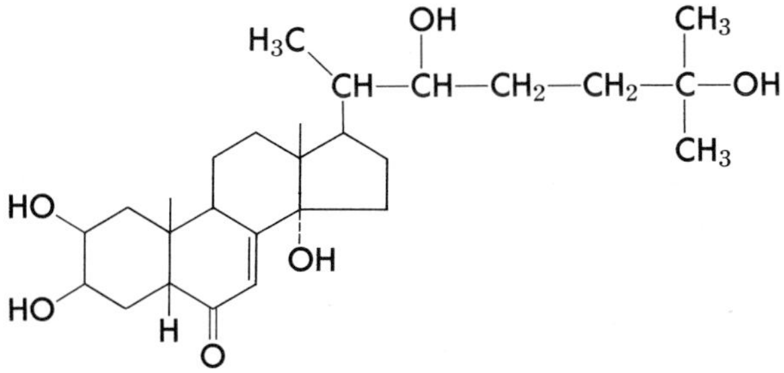

Ecdysone

Aromatic compounds similar to steroids are often carcinogenic. A potent carcinogen is methylcholanthrene.

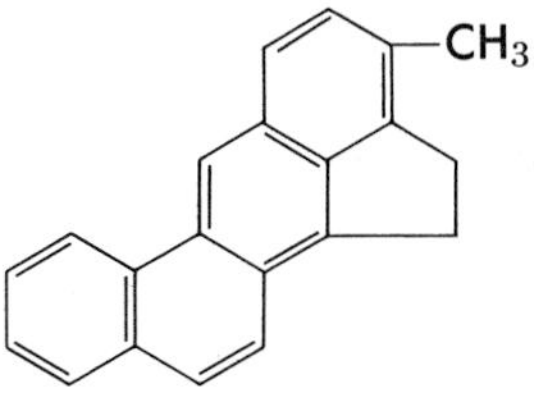

Methylcholanthrene

SPECIAL METHODS OF SEPARATION AND IDENTIFICATION OF STEROIDS

Steroids as a class may be separated from other lipids by virtue of their being nonsaponifiable. The bile acids are, of course, soluble in aqueous alkali, as are the estrogens. However, the estrogens, like other phenols, although extractable from organic solvents by aqueous NaOH or Na_2CO_3, are not extracted by weak bases such as aqueous $NaHCO_3$, whereas the fatty carboxylic acids are. Ketonic steroids may be extracted by treatment with the hydrazide of betaine chloride, Girard's reagent, yielding water-soluble salts of the resultant substituted hydrazones.

$$\underset{/}{\overset{\backslash}{C}}=O + H_2N-NH-\overset{\overset{O}{\|}}{C}-CH_2-\overset{+}{N}(CH_3)_3Cl^- \longrightarrow \underset{/}{\overset{\backslash}{C}}=N-NH-\overset{\overset{O}{\|}}{C}-CH_2-\overset{+}{N}(CH_3)_3Cl^-$$

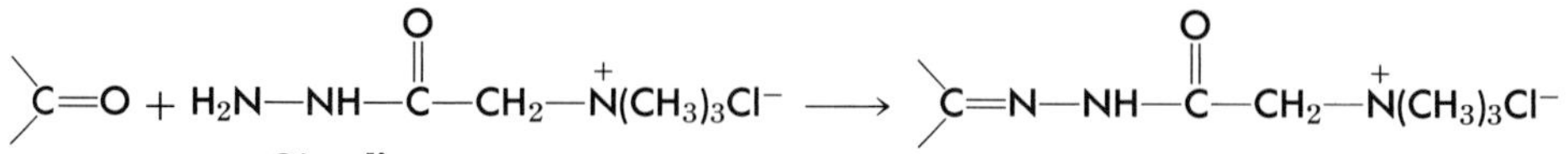

Girard's reagent

Alcoholic steroids may be recovered as their hemisuccinates after treatment with succinic anhydride, thus yielding products soluble in aqueous alkali.

$$H-\overset{|}{\underset{|}{C}}-OH + \overset{\overset{O}{\|}}{C}-CH_2-CH_2-\overset{\overset{O}{\|}}{\underset{\underset{O}{|}}{C}} \longrightarrow H-\overset{|}{\underset{|}{C}}-O-\overset{\overset{O}{\|}}{C}-CH_2-CH_2-COOH$$

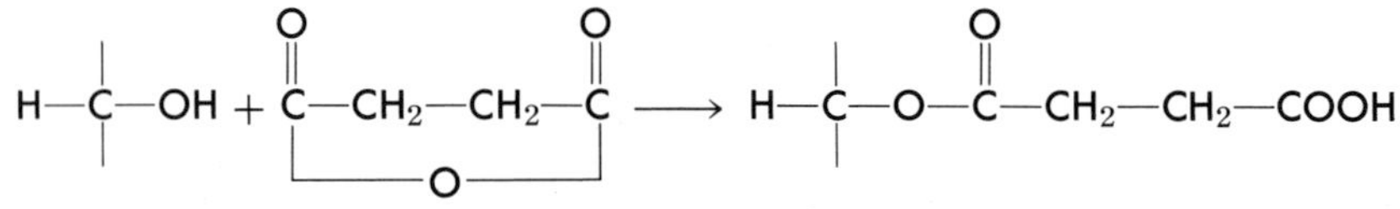

Further separation of mixtures may be achieved by chromatographic and counter-current distribution procedures (Chap. 2). Paper, column, and gas-liquid chromatography are very useful for the separation and identification of steroids. Particularly applicable has been thin-layer chromatography (Chap. 2). The advantages of this procedure lie in the fact that a number of steroid mixtures may be resolved simultaneously, side by side, on the layered adsorbent. Also, after localization of substances on the plate with "developing" reagents (page 38), selected "strips" or areas of the plate's surface may be scraped off and the compounds isolated by extraction of the powder with suitable solvents, etc. Thin-layer chromatography has also been useful for separation of classes of lipids, as well as separations among members of lipid groups other than steroids.

In addition to the conventional criteria employed in the identification of organic compounds, *e.g.*, melting point, optical activity, etc., the steroid chemist utilizes optical rotatory dispersion, proton magnetic resonance spectroscopy and mass spectrometry, and has made special use of absorption spectroscopy, both in identification and as an aid in proof of structure. Although steroids do not absorb light in the visible portion of the spectrum, they do absorb in the ultraviolet and infrared portions. Absorption in the infrared portion of the spectrum is related to the vibrational energies of adjacent pairs of atoms, and the frequencies absorbed by a particular molecule therefore depend upon the occurrence in the substance of such atom pairs. Thus carbonyl groups absorb light maximally at a wavelength of about 5.8 μ†; hydroxyl groups, at 2.8 μ. The absorption due to a complex molecule in this portion of the spectrum is made up of numerous fine peaks, many of which can be ascribed to specific atomic groupings. In the ultraviolet, broader bands of absorption are encountered for certain unsaturated steroids, corresponding to specific electronic arrangements in the molecule. Thus many α,β-unsaturated ketones,

† A micron (μ) $= 10^{-6}$ m. A millimicron (mμ) $= 10^{-3}$ $\mu = 10^{-9}$ m. An angstrom (Å.) $= 1 \times 10^{-8}$ cm. $= 10^{-10}$ m. $= 10^{-4}$ $\mu = 10^{-1}$ mμ. The octave of visible light covers approximately 4,000 to 8,000 Å., 400 to 800 mμ, or 0.4 to 0.8 μ in wavelengths.

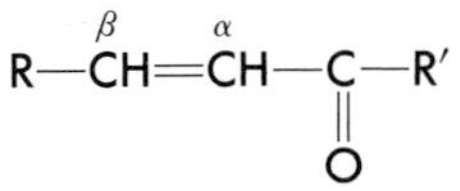

exhibit an absorption maximum in the region 230 to 260 mμ. Combined spectroscopic studies permit the unequivocal identification of steroids for which characteristic absorption spectra have been previously established.

LIPIDS COMBINED WITH OTHER CLASSES OF COMPOUNDS

Lipids are found in nature combined with other classes of compounds; thus the descriptive terms lipoproteins, proteolipids, phosphatidopeptides, lipoamino acids, and lipopolysaccharides (page 57). Examples of some of these will be encountered in subsequent pages of this book.

REFERENCES

Books

Ansell, G. B., and Hawthorne, J. N., "Phospholipids: Chemistry, Metabolism and Function," American Elsevier Publishing Company, New York, 1964.

Burchfield, H. P., and Storrs, E. E., "Biochemical Applications of Gas Chromatography," Academic Press, Inc., New York, 1962.

Cook, R. P., ed., "Cholesterol: Chemistry, Biochemistry, and Pathology," Academic Press, Inc., New York, 1958.

Deuel, H. J., Jr., "The Lipids: Their Chemistry and Biochemistry," vols. I, II, and III, Interscience Publishers, Inc., New York, 1951, 1955, 1957.

Fieser, L. F., and Fieser, M., "Steroids," Reinhold Publishing Corporation, New York, 1959.

Florkin, M., and Mason, H. S., eds., "Comparative Biochemistry," vol. III, part A, chaps. 1 through 5, and 10; vol. IV, part B, chap. 14, Academic Press, Inc., New York, 1962.

Gunstone, F. D., "An Introduction to the Chemistry of Fats and Fatty Acids," John Wiley & Sons, Inc., New York, 1958.

Hanahan, D. J., Gurd, F. R. N., and Zabin, I., "Lipid Chemistry," John Wiley & Sons, Inc., New York, 1960.

Heftmann, E., and Mosettig, E., "Biochemistry of Steroids," Reinhold Publishing Corporation, New York, 1960.

Hilditch, T. P., and Williams, P. N., "The Chemical Constitution of Natural Fats," 4th ed., Chapman & Hall, Ltd., London, 1964.

Klyne, W., "The Chemistry of the Steroids," corr. 1st ed., John Wiley & Sons, Inc., New York, 1961.

Kritchevsky, D., "Cholesterol," John Wiley & Sons, Inc., New York, 1958.

Lovern, J. A., "The Chemistry of Lipids of Biochemical Significance," 2d ed., Methuen & Co., Ltd., London, 1957.

Pecsok, R. L., ed., "Principles and Practice of Gas Chromatography," John Wiley & Sons, Inc., New York, 1959.

Ralston, A. W., "Fatty Acids and Their Derivatives," John Wiley & Sons, Inc., New York, 1948.

Shoppee, C. W., "Chemistry of the Steroids," 2d ed., Academic Press, Inc., New York, 1964.

Review Articles

Asselineau, J., and Lederer, E., Chemistry of Lipids, *Ann. Rev. Biochem.,* **30,** 71–92, 1961.

Carter, H. E., Johnson, P., and Weber, E. J., Glycolipids, *Ann. Rev. Biochem.,* **34,** 109–142, 1965.

Celmer, W. D., and Carter, H. E., Chemistry of Phosphatides and Cerebrosides, *Physiol. Revs.,* **32,** 167–196, 1952.

Hanahan, D. J., and Thompson, G. A., Jr., Complex Lipids, *Ann. Rev. Biochem.,* **32,** 215–240, 1963.

Hawthorne, J. N., The Inositol Phospholipids, *J. Lipid Research,* **1,** 255–280, 1960.

Hawthorne, J. N., and Kemp, P., The Brain Phosphoinositides, *Advances in Lipid Research,* **2,** 127–166, 1964.

Kates, M., Bacterial Lipids, *Advances in Lipid Research,* **2,** 17–90, 1964.

Rapport, M. M., and Norton, W. T., Chemistry of the Lipids, *Ann. Rev. Biochem.,* **31,** 103–138, 1962.

5. The Proteins. I

Amino Acids and Peptides

The name protein (Gk., *protos,* or first) was suggested by Berzelius to Mulder and given by the latter in 1838 to the complex organic nitrogenous substances found in the cells of all animals and plants. Proteins occupy a central position in the architecture and functioning of living matter. They are intimately connected with all phases of chemical and physical activity that constitute the life of the cell since all enzymes, the biological catalysts, are proteins (Chap. 10). Some proteins serve as structural elements of the body, for example, as hair, wool, and collagen, an important constituent of connective tissue; other proteins may be hormones (Part Five), or oxygen carriers (Chap. 32). Still other proteins participate in muscular contraction (Chap. 36), and some are associated with the genes, the hereditary factors (Chap. 28). The antibodies that are concerned with immunological defense mechanisms are proteins (Chap. 31). It is apparent that there is hardly an important physiological function in which proteins do not participate. Indeed, proteins are quantitatively the main material of animal tissue for they may constitute as much as three-fourths of the dry substance.

In essence, the objective of protein chemistry is to explain the physiological functions of these complex molecules in terms of their structure. The experimental approach consists largely in examination of the parts of the molecules, the probable arrangement of these parts in individual proteins, and the chemical and physical behavior of the intact proteins. These tasks are formidable and present considerable technical difficulties because of the great diversity and complexity of proteins.

Before discussing in detail the properties of proteins, it is useful to consider briefly some of the main characteristics of these complex molecules.

1. The proteins are large molecules with molecular weights ranging from approximately 5,000 to many millions. Because of their large size proteins exhibit many unusual properties; special methods have been devised to study these and other macromolecules.

2. Twenty amino acids are commonly found in proteins; they are linked together by peptide bonds which couple the α-carboxyl group of one amino acid residue to the α-amino group of another residue, as shown below.

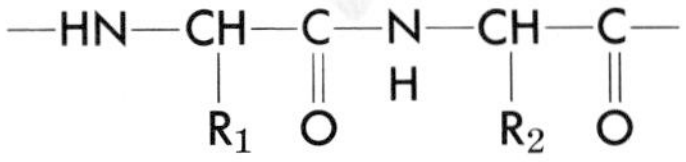

A protein molecule consists of one or more polypeptide chains, each chain containing from approximately 20 to several hundred amino acid residues.

3. In the soluble, globular proteins, the peptide chains are specifically folded in complex structures which are held together by secondary, noncovalent forces, *e.g.*, hydrogen bonds, hydrophobic contacts, and electrostatic interactions.

4. Because the structures of proteins are dependent on weak secondary forces, these molecules are generally labile and readily modified in solution when subjected to alterations in pH, radiation, high temperatures, detergents, organic solvents, etc. Aside from the information that such changes may give concerning proteins, their lability poses technical difficulties and limits the tools and conditions which may be used for investigation.

5. Proteins are extremely reactive and highly specific in behavior. This is because of the presence of large numbers of different kinds of side chains of the amino acid residues and their specific arrangement in these macromolecules. These side chains include, among others, both cationic and anionic groups, aromatic and aliphatic hydroxyls, amides, thiols, complex heterocycles, as well as strongly hydrophobic aliphatic and aromatic side chains which contribute not only to the specific interactions of proteins but also to their structural stability.

In this chapter we shall consider first the fundamental units of the proteins, *viz.*, the amino acids, as well as some of their properties, and some attributes of simple peptides. In the following chapters, present knowledge of the properties of the proteins and the methods employed for their study will be presented.

AMINO ACIDS

The biological unity of living organisms is clearly manifested by the fact that the same 20 amino acids have been found in the proteins of all presently existing species of microorganisms, plants and animals. A few additional amino acids are of limited distribution and occur in only one or a few kinds of proteins.

The amino acids that have been isolated from protein hydrolysates are, with two exceptions, primary α-amino acids, *i.e.*, the carboxyl and amino groups are attached to the same carbon atom.

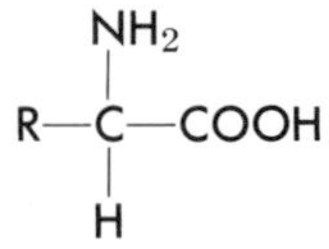

The various α-amino acids possess different R groups attached to the α-carbon. The usual classification of amino acids depends on the number of acidic and basic groups that are present. Thus, the neutral amino acids contain one amino and one carboxyl group. The acidic amino acids have an excess of carboxyl over amino groups. The basic amino acids possess an excess of basic groups. Two amino acids are imino acids rather than primary α-amino acids.

NEUTRAL AMINO ACIDS

Aliphatic Amino Acids. *Glycine* is the simplest of the amino acids and one of the first to be isolated from proteins. It has a characteristic sweet taste.

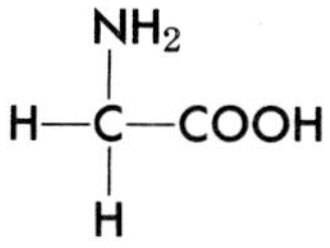

Alanine

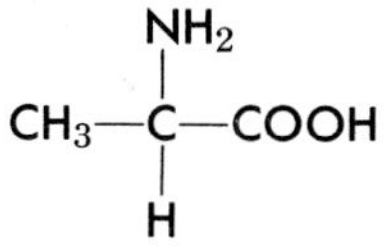

Valine

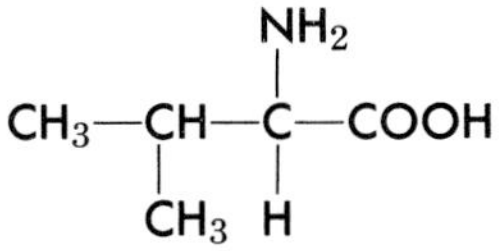

Leucine is one of the few amino acids that are sparingly soluble in water.

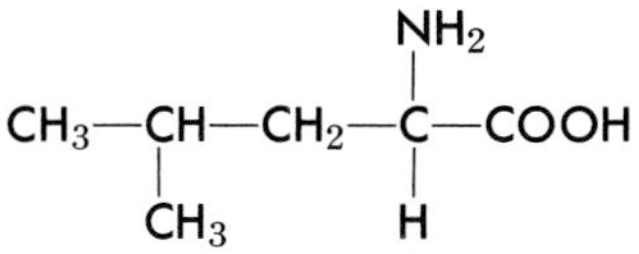

Isoleucine

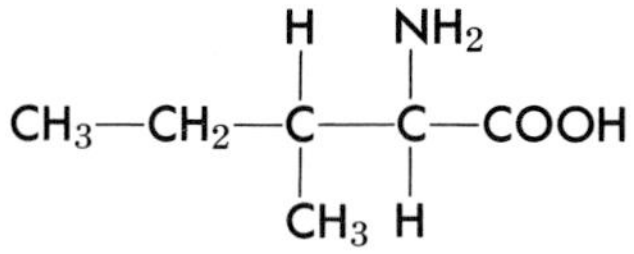

Serine was first obtained from the silk protein, sericin.

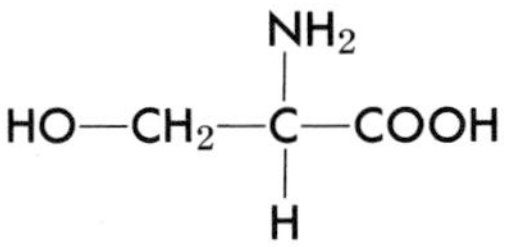

Threonine was one of the last of the common amino acids to be discovered (Table 5.1, page 94).

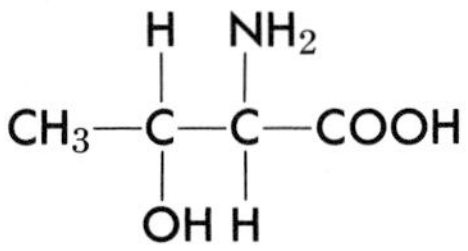

Aromatic Amino Acids. The presence of these amino acids in proteins is responsible for their ultraviolet absorption maxima at about 275 to 280 mμ. The absorption spectra of these amino acids are shown in Fig. 5.1.

Phenylalanine

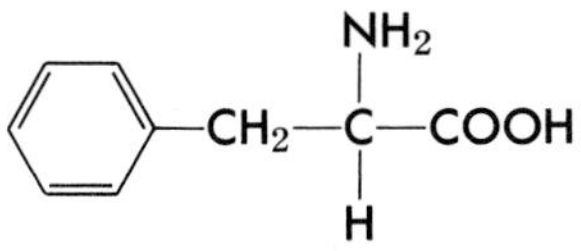

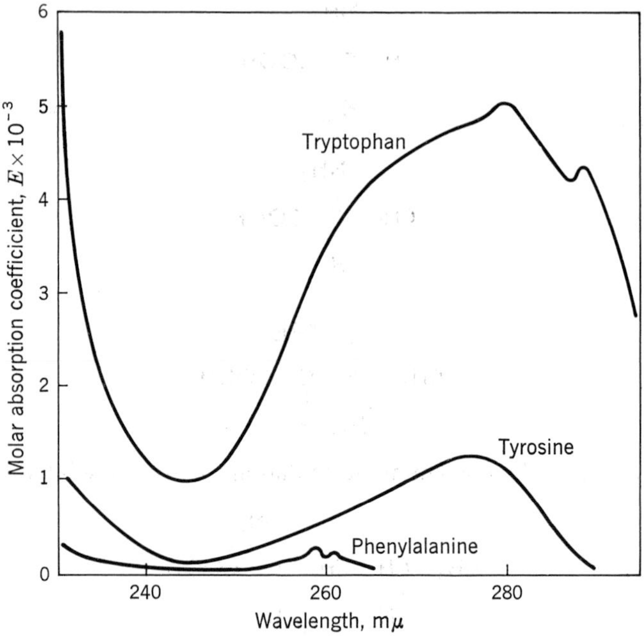

Fig. 5.1. The absorption spectra of tryptophan, tyrosine, and phenylalanine at pH 8. The spectrum for tyrosine undergoes a shift in position at alkaline pH values; this is completed above pH 12. The contribution of phenylalanine to protein absorption at 275 to 280 mμ is negligible. (*After J. S. Fruton and S. Simmonds, "General Biochemistry," 2d ed., p. 73, John Wiley & Sons, Inc., New York,* 1958.)

Tyrosine is sparingly soluble in water. Proteins containing tyrosine all give characteristic reactions for the phenolic group present in this amino acid.

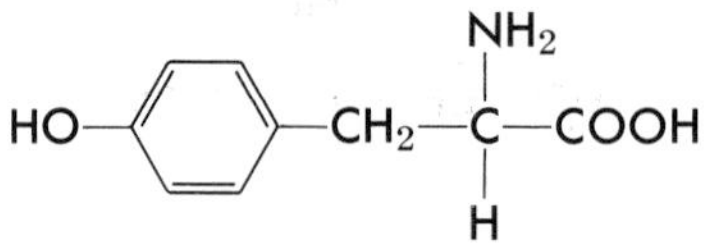

Tryptophan

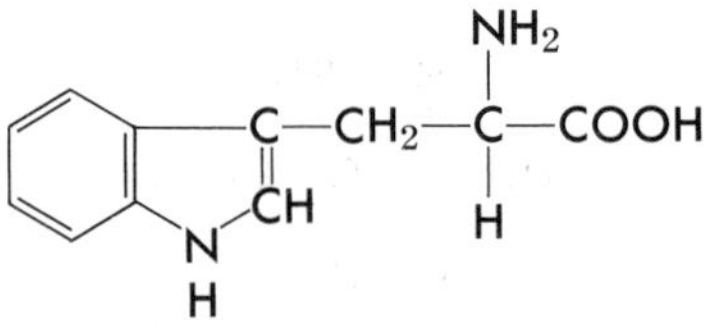

This heterocyclic amino acid is usually included with the aromatic compounds. It is destroyed during acidic hydrolysis of proteins but may be isolated after enzymic hydrolysis. Proteins that contain tryptophan react with reagents for the heterocyclic indole group.

Sulfur-containing Amino Acids. *Cystine* is neutral since it contains two amino and two carboxyl groups; it is sparingly soluble in water. It was discovered in urinary calculi by Wollaston in 1810, but not until 1899 was it obtained from proteins. Cystine and its reduction product, the sulfhydryl-containing *cysteine* (or half-cystine), are together counted as one of the 20 amino acids.

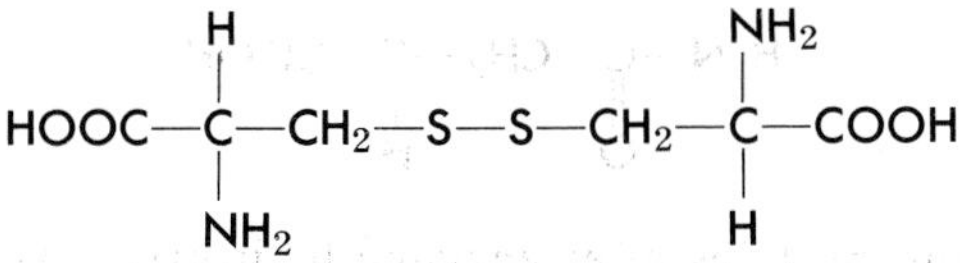

Cysteine is present in many proteins, and these give a positive test for free sulfhydryl groups.

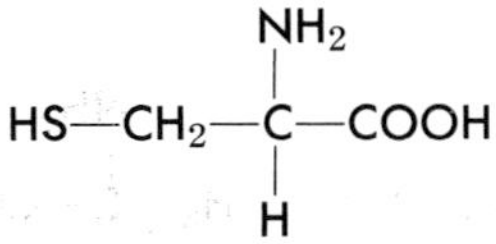

Methionine contains sulfur in thioether linkage.

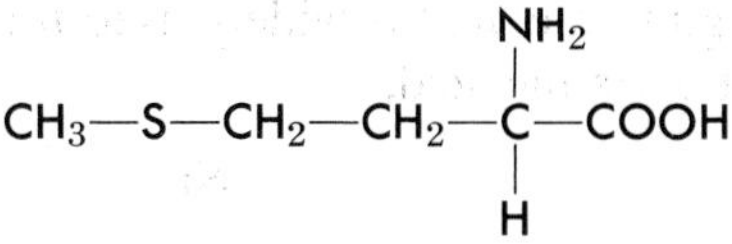

Imino Acids. There are two amino acids that contain an imino nitrogen rather than a primary amino group. The nitrogen is present in a pyrrolidine ring. Unlike other amino acids, proline is very soluble in alcohol and hydroxyproline is somewhat soluble.

Proline is present in all proteins that have been studied.

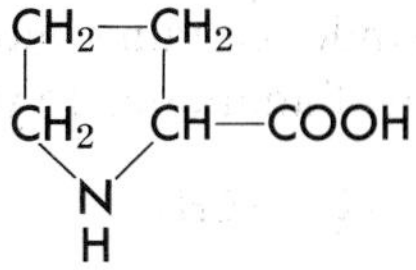

4-*Hydroxyproline* (4-hydroxypyrrolidine-2-carboxylic acid) and the isomeric 3-hydroxyproline have been obtained only from collagen and related proteins.

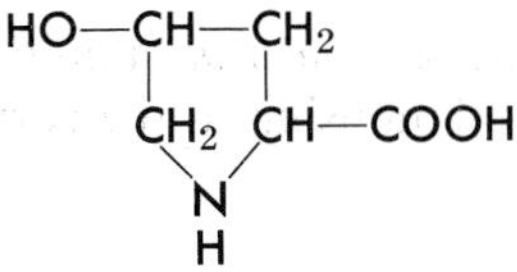

DICARBOXYLIC AMINO ACIDS AND THEIR AMIDES

Aspartic acid

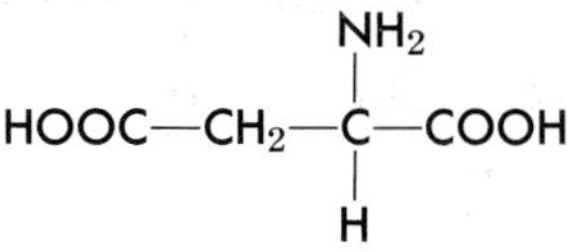

The β-amide of aspartic acid, *asparagine,*

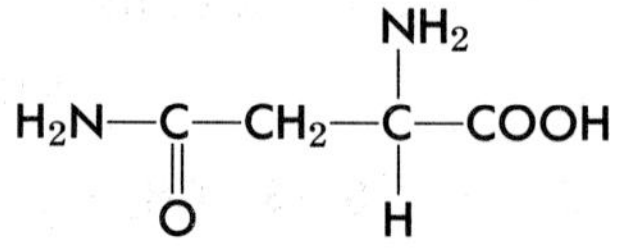

has been isolated from proteins after enzymic hydrolysis. Ammonia is produced from this substance during acidic or alkaline hydrolysis of proteins, giving rise to aspartic acid. Asparagine has long been known as a constituent of plant tissues.

Glutamic acid

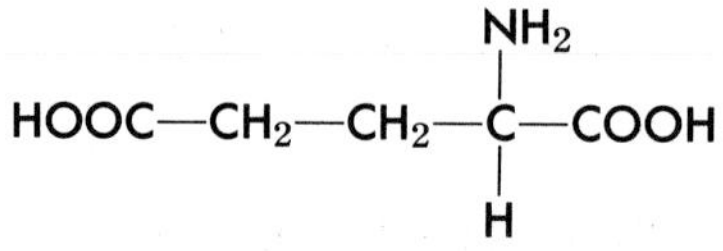

The monosodium salt of glutamic acid is widely used for artificial meat flavoring. *Glutamine,* the γ-amide of glutamic acid,

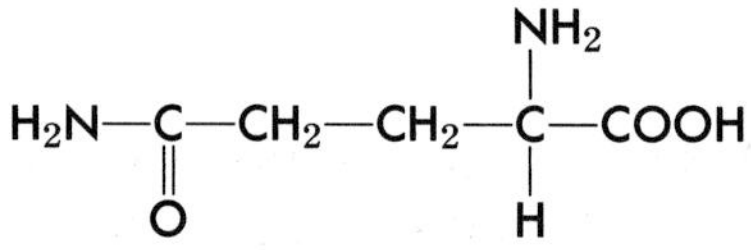

has been isolated from proteins after enzymic hydrolysis. Like asparagine it yields ammonia on acidic or alkaline hydrolysis. The two amides are responsible for the ammonia produced by chemical hydrolysis of proteins. Free glutamine is found in many animal and plant tissues. Under certain conditions, glutamine readily undergoes ring closure, with elimination of ammonia, to yield pyrrolidone carboxylic acid.

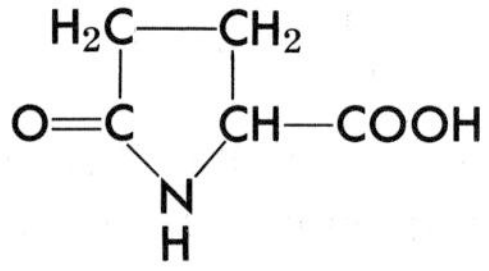

Asparagine and glutamine are *neutral* amino acids. They are listed here for convenience to show their relationship to the corresponding dicarboxylic acids.

BASIC AMINO ACIDS

Histidine

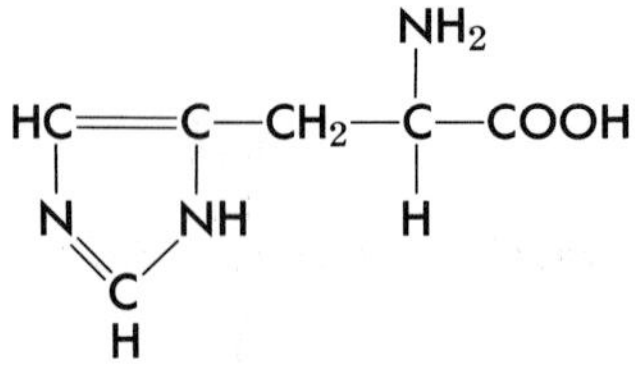

The imidazole group is basic (see below).

Arginine

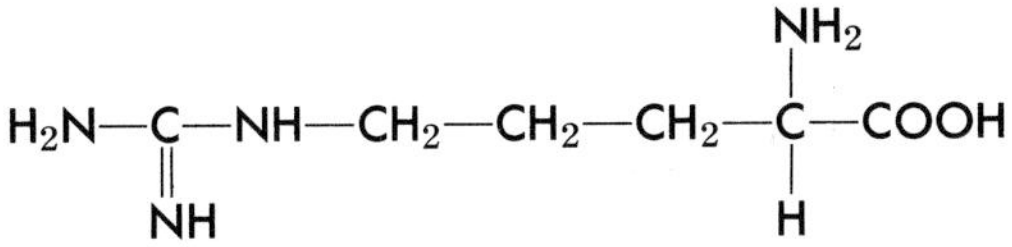

The guanido group is basic.

Lysine is generally abundant in animal proteins but present in limited amounts in plant proteins.

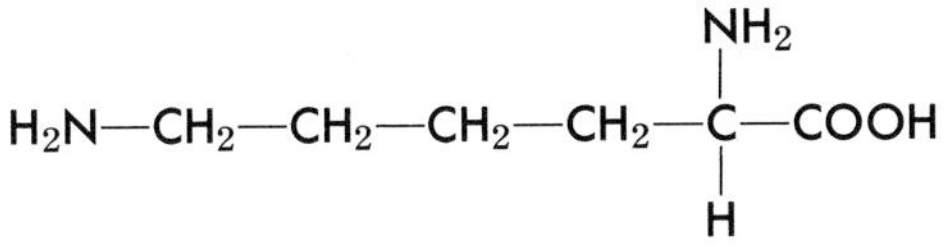

Hydroxylysine has been found, thus far, only in collagen and gelatin.

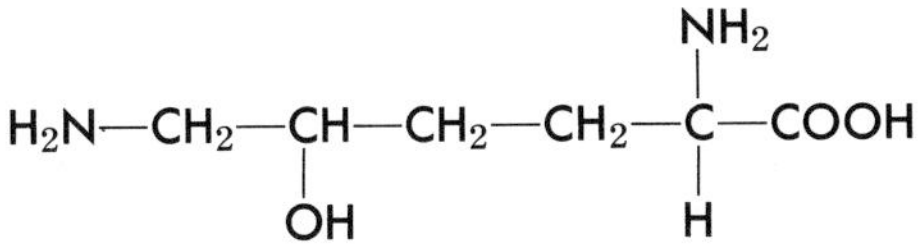

All the above amino acids have been demonstrated to occur in proteins. The hydroxyprolines and hydroxylysine are of limited distribution in proteins; they have been identified with certainty thus far only in collagen and similar structural proteins. In the flagellar proteins of some bacteria and in some plant histones (page 117), ϵ-N-methyllysine is present. Complex derivatives of lysine, desmosine and isodesmosine, are present in the fibrous protein elastin (Chap. 38).

In Table 5.1 (page 94) are given the dates and the discoverers of the amino acids in proteins.

SOME OTHER AMINO ACIDS

Many amino acids other than those commonly found in proteins occur in nature, and some play important roles in metabolism; several of these are listed below.

β-Alanine is found in the important vitamin, pantothenic acid (Chap. 49), and in the naturally occurring peptides, carnosine and anserine (page 112).

$$H_2N{-}CH_2{-}CH_2{-}COOH$$

Ornithine possesses one less carbon atom than its homologue, lysine.

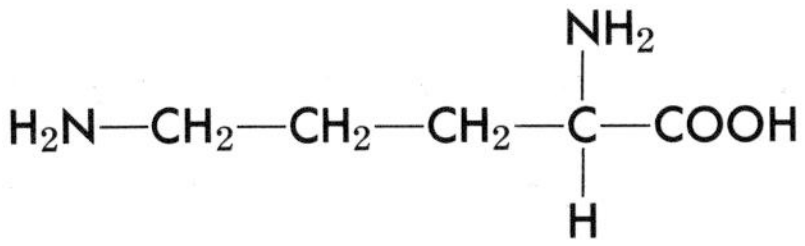

Ornithine occurs in the urine of some birds as dibenzoylornithine, or ornithuric acid.

3,5,3′-*Triiodothyronine* is a constituent of the thyroid gland; its name is derived

Table 5.1: THE DISCOVERY OF THE AMINO ACIDS IN PROTEINS

Amino acid	Discoverer	Date
Leucine	Proust	1819
Glycine	Braconnot	1820
Tyrosine	Liebig; Bopp	1846; 1849
Serine	Cramer	1865
Glutamic acid	Ritthausen	1866
Aspartic acid	Ritthausen	1868
Alanine	Schützenberger and Bourgeois; Weyl	1875; 1888
Phenylalanine	Schulze and Barbieri	1879
Lysine	Drechsel	1889
Arginine	Hedin	1895
Histidine	Kossel; Hedin	1896
Cystine	Mörner	1899
Valine	Fischer	1901
Proline	Fischer	1901
Tryptophan	Hopkins and Cole	1901
Hydroxyproline	Fischer	1902
Isoleucine	Ehrlich	1904
Methionine	Mueller	1922
Threonine	Meyer and Rose	1936
Hydroxylysine	Schryver and associates; Van Slyke and associates	1925; 1938
3-Hydroxyproline	Ogle and associates; Irreverre and associates	1962

from the fact that the thyroxine nucleus (see formula below), devoid of its iodine atoms, is termed *thyronine.*

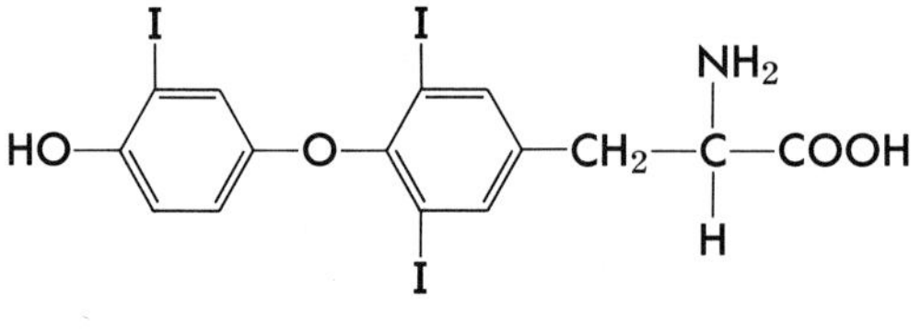

Thyroxine (3,5,3′,5′-tetraiodothyronine)

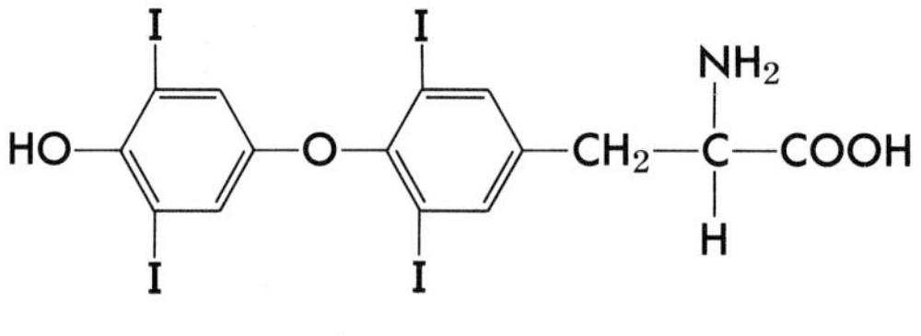

The iodinated compounds above are found in thyroglobulin (Chap. 43).

OPTICAL ACTIVITY OF AMINO ACIDS

With the exception of glycine, all α-amino acids are optically active since the α-carbon atom is a center of asymmetry, as indicated in the formula on page 88. The amino acids of proteins all possess the same configuration as L-alanine, which has been shown to be related to L-glyceraldehyde. The symbols, D or L, here as in the sugar series (page 17), do not refer to the sign of rotation but indicate configurational relationships of similar compounds.

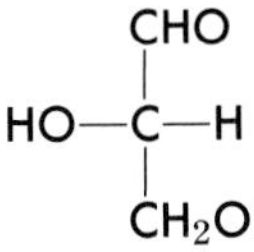

L-Glyceraldehyde

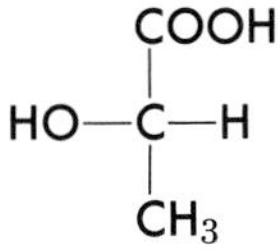

L-Lactic acid

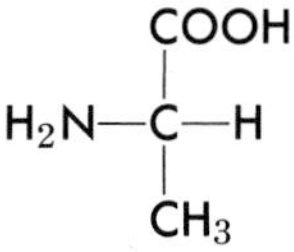

L-Alanine

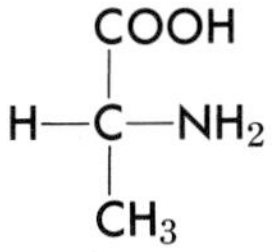

D-Alanine

L-Alanine and L-lactic acid are actually dextrorotatory when examined in the polarimeter, but the present system of nomenclature indicates their structural relationship to L-glyceraldehyde. Only L-amino acids have been obtained from animal and plant proteins under conditions that do not produce racemization. Among the common amino acids, threonine, hydroxylysine, cystine, isoleucine, and the two hydroxyprolines possess two optically active centers. Hence the synthetic compounds are mixtures of four diastereoisomers. Two are designated as L- and D-, respectively. The two additional diastereoisomeric forms are described as L-allo and D-allo; thus the terms allothreonine, alloisoleucine, etc.

The rotation of an amino acid is strongly influenced by the acidity of the solution, and the configuration is therefore described by the chemical relationships rather than by the actual sign of rotation. In general, all amino acids of the L-configuration show positive shifts in the sign of the rotation as the acidity is increased, and amino acids of the D-configuration give the opposite change. This empirical rule is useful in determining the spatial structures of newly isolated amino acids.

Certain D-amino acids have been obtained from microorganisms. D-Glutamic acid is a constituent of the capsular material of *Bacillus anthracis* and some related organisms. D-Glutamic acid and D-alanine are abundant in material comprising the cell walls of many bacteria (Chap. 41). Some D-amino acids have also been obtained from the hydrolytic products of polypeptide antibiotics, such as gramicidin and bacitracin.

SOME PHYSICOCHEMICAL PRINCIPLES

Before proceeding with a discussion of the properties and behavior of the amino acids in solution, it is necessary to review some aspects of the theory of weak electrolytes.

The simplest definition of an acid according to the Brønsted theory is that it is a substance, charged or uncharged, that liberates hydrogen ions (H^+) or protons in solution. A base is a substance that can bind hydrogen ions or protons and remove them from solution. It should be emphasized that in this terminology, ammonia, NH_3, acetate ion, CH_3COO^-, and sulfate ion, $SO_4^=$, are bases, whereas ammonium ion, NH_4^+, acetic acid, CH_3COOH, and bisulfate ion, HSO_4^-, are acids.

It is necessary to distinguish between strong and weak electrolytes. Strong electrolytes are essentially completely ionized in aqueous solution; these include almost all neutral salts, *e.g.*, NaCl, Na_2SO_4, KBr, etc., as well as strong acids, such as hydrochloric and nitric, and strong alkalies, such as NaOH, KOH, etc. Weak electrolytes are only partially ionized in aqueous solution and yield a mixture of the undissociated compound and ions. Many acids (HA) are weak electrolytes, and partially dissociate to produce hydrogen ions and the generalized anion A^-. It follows that a measurement of the hydrogen ion concentration, $[H^+]$, of a solution of a weak acid will not give a measure of the total concentration of the acid since there is a reservoir of the substance in the form of undissociated HA. However, if the free $[H^+]$ is titrated, additional HA will dissociate and titration will eventually give an estimate of the total acid since both the free H^+ and that derivable from the undissociated molecule will be titrated. Hence we distinguish between titratable acidity and actual acidity (hydrogen ion concentration) in dealing with weak acids. With strong acids at concentration levels at which they are completely dissociated, the titratable and actual acidity are the same.

A weak acid of the type HA dissociates reversibly in water to produce hydrogen ion H^+ and A^-.

$$HA \rightleftharpoons H^+ + A^- \tag{1}$$

Applying the mass law equation and expressing concentration in brackets, there is obtained the general equation

$$\frac{[H^+][A^-]}{[HA]} = K_a \tag{2}$$

where K_a is the ionization or dissociation constant of the acid. The equation indicates that K is a measure of the strength of the acid. The higher the value of K, the greater the number of hydrogen ions liberated per mole of acid in solution and hence the stronger the acid. Consequently, different acids may be compared in terms of their K values. It should be noted that the discussion here is of concentrations of H^+, A^-, and HA. More correctly, the equation applies rigidly only to *activity* values, and, throughout the discussion it must be understood that, in referring to concentrations, this is done only for convenience. In fact, most measurements of H^+ are measurements of activity rather than concentration, whether these are done by electrometric methods or by indicators. Equilibrium equations, such as (2), apply precisely only to activity values. The deviations from ideal behavior are due to interionic and other intermolecular forces. At low concentrations where such forces are negligible, differences between concentration and activity values are minimal.

The dissociation of BOH is represented in the same manner.

$$\frac{[B^+][OH^-]}{[BOH]} = K_b \tag{3}$$

Water, the usual solvent, dissociates into hydrogen and hydroxyl ions; the equation for the dissociation of water is

$$\frac{[H^+][OH^-]}{[H_2O]} = K \tag{4}$$

Inasmuch as the amount of undissociated water is large and remains essentially constant, this equation is usually simplified to

$$[H^+][OH^-] = K_w \tag{5}$$

where K_w, the ion product of water, is approximately 10^{-14} at 25°C.

In aqueous solution, equation (5) must be satisfied. Alteration in $[H^+]$ or $[OH^-]$ must always result in immediate compensatory change in the concentration of the other ion. Hence the ion product of water is of fundamental importance in considering the behavior of acids and bases in solution.

In biological work, it is customary to arrange the mass law equations into a common form that is convenient for both acids and bases. Equation (2) may be put into the form

$$[H^+] = \frac{K_a[HA]}{[A^-]} \tag{6}$$

and, taking logarithms of both sides of the equation,

$$\log [H^+] = \log K_a + \log [HA] - \log [A^-] \tag{7}$$

Now, multiplying by -1,

$$-\log [H^+] = -\log K_a - \log [HA] + \log [A^-]$$

The term $-\log [H^+]$ was defined by Sørensen as the pH and $-\log K_a$ as pK_a. In this manner there is obtained the important relationship

$$pH = pK_a + \log \frac{[A^-]}{[HA]} \tag{8}$$

It is important to note that equation (8) is applicable to conjugate acids like NH_4^+ and its base NH_3, as well as to substances ordinarily recognized as acids.

Equation (5) may be transformed to

$$\log [H^+] + \log [OH^-] = \log K_w \tag{9}$$

and by multiplying by -1

$$-\log [H^+] - \log [OH^-] = -\log K_w$$

and by substituting pH for $-\log [H^+]$, pOH for $-\log [OH^-]$, etc., there results

$$pH + pOH = pK_w \tag{10}$$

From these definitions, pH can be used to express either the acidity or the alkalinity of a solution. Since $pK_w = 14$, it is possible to obtain readily the pH of a solution from pOH. In practice, pOH is rarely used, and all acidic or basic solutions are expressed on the pH scale. Since pH is a logarithmic term, it must be handled in the same manner as other logarithmic quantities. For a neutral solution, pH = 7. Since $pK_w = 14$, pOH = 7, or $[H^+] = 10^{-7}\ M = [OH^-]$. It is obviously more convenient to employ the pH scale in place of awkward exponents. This should be apparent since, for a neutral solution where pH = 7.0, the $[H^+] =$ 0.0000001, or $10^{-7}\ M$. For a solution at pH 6.0, the $[H^+] = 0.000001$ or $10^{-6}\ M$. The difference between the quantities involved is more easily appreciated and stated as the pH values of 7 and 6 than by giving the numerical values. The pH range ordinarily encountered encompasses 14 units, from solutions that contain 1.0N acid (pH = 0), through the neutral range to solutions of 1.0N alkali (pH = 14). This is particularly convenient in handling [H^+] changes graphically.

To obtain the pH of a solution from the hydrogen ion concentration, it is merely necessary to take the logarithm of the [H^+] and change the sign. For example, to calculate the pH of a solution that contains $2 \times 10^{-5}M$ hydrogen ion, take the logarithm of 2, which is 0.30, and add −5.00, which gives −4.70. The pH is 4.70. It must be kept in mind that a change of one pH unit is a tenfold change in hydrogen ion concentration since decimal logarithms are used for the pH scale.

The pH values for a number of materials are given in Table 5.2 in order to illustrate the applicability to common materials as well as to biologically important media.

Table 5.2: APPROXIMATE pH VALUES OF COMMON FLUIDS, BIOLOGICAL FLUIDS, AND FOODS

	Material	pH		Material	pH
Acids	Acetic, 0.1N	2.9	Foods	Beers	4–5
	Boric, 0.1N	5.2		Cider	2.8–3.3
	Citric, 0.1N	2.2		Eggs	7.6–8.0
	Hydrochloric, 0.1N	1.1		Grapefruit	3.0–3.3
	Sulfuric, 0.1N	1.2		Lemons	2.2–2.4
Bases	Ammonia, 0.1N	11.1		Limes	1.8–2.0
	Potassium hydroxide, 0.1N	13.0		Oranges	3.0–4.0
	Sodium bicarbonate, 0.1N	8.4		Pickles	3.0–3.6
	Sodium carbonate, 0.1N	11.6		Potatoes	5.6–6.0
	Sodium hydroxide, 0.1N	13.0		Sauerkraut	3.4–3.6
Biological fluids	Blood plasma	7.3–7.5		Soft drinks	2.0–4.0
	Gastric juice	1.2–3.0		Vinegar	2.4–3.4
	Milk, cow's	6.3–6.7		Wines	2.0–4.0
	Saliva	6.5–7.5			
	Urine	5–8			

Buffer Action. Buffered solutions resist the changes in [H^+] which would otherwise result from the addition of an acid or a base. In general, buffer action is exhibited by ions of weak acids or bases. Strong acids and bases are almost completely dissociated in water and have no reservoir of undissociated acid or base. Weak electrolytes that exhibit buffer action have extremely important practical properties and have great value for living systems since most cells can survive within only fairly narrow pH limits.

Practically, the best buffer action is exhibited by a mixture of a weak acid HA and its salt BA. Salts may be regarded as completely dissociated; hence BA exists in solution as B^+ and A^-. Let it be assumed that the weak acid HA, which has a pK_a of 5, contributes only 10^{-5} mole of H^+ and A^- for each mole of acid and thus, as a first approximation, may be regarded as entirely undissociated. The contribution of the ion A^- by the acid is negligible compared with that of the salt BA, and the salt concentration may be used for [A^-]. Equation (8) may thus be expressed as

$$\text{pH} = \text{p}K_a + \log\frac{[\text{salt}]}{[\text{acid}]} \tag{11}$$

where [salt] = [A^-] and [acid] = [HA]. Equation (11) is generally known as the Henderson-Hasselbalch equation.

If the weak acid HA is present at $0.1M$ concentration and the salt BA is likewise present at $0.1M$, then

$$\mathrm{pH} = 5 + \log\frac{0.1}{0.1}$$

and the pH of the solution is 5. If 0.01 mole of BOH is added, there is now formed an additional 0.01 mole of BA and 0.01 mole of water; the latter may be neglected since its contribution to the total water is negligible. The pH of the solution may be again calculated from the increased salt concentration and decreased acid concentration.

$$\mathrm{pH} = 5 + \log\frac{0.11}{0.09}$$

The pH is now 5.09, an increase of only 0.09 pH unit. A similar addition of 0.01 mole of a strong acid would decrease the pH by 0.09 pH unit, giving the solution a final pH of 4.91.

It is important to contrast this behavior with that of a strong acid like hydrochloric at the same pH, *viz.*, 5. The concentration of HCl would be $0.00001M$, and addition of 0.01 mole of base would bring the solution almost to pH 12, a change of 7 pH units, since there is no reservoir of undissociated acid. Even at pH 2 ($0.01M$ HCl), addition of 0.01 mole of base brings the solution to neutrality, a change of 5 pH units.

It must be emphasized that in the regions where weak acids and bases exert their buffer action, from about pH 2 to pH 12, this buffering capacity represents an important means of maintaining constant pH. In the region near neutrality, where most living cells function, certain buffers, *viz.*, those formed from carbonic and phosphoric acids, are of paramount importance.

Equations (8) and (11) have been derived in such a way that they apply to all weak electrolytes and hence describe the titration of all weak acids and bases. The titration curve of a monobasic weak acid is shown in Fig. 5.2. The pH at the mid-

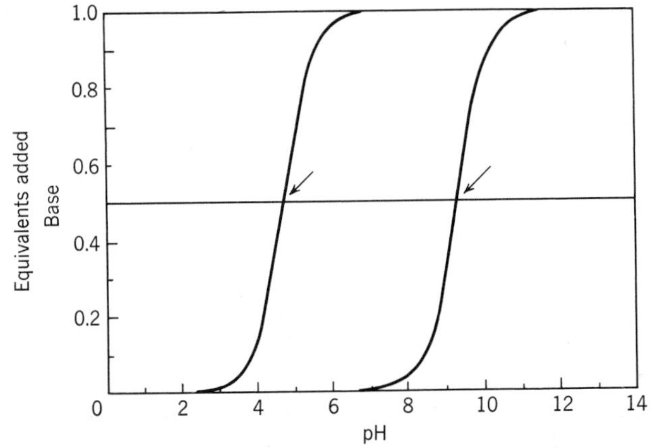

Fig. 5.2. The calculated titration curves for one equivalent of weak acid whose pK_a is 4.7 (acetic acid) and for ammonium ion whose pK_a is 9.3. The pH found when half an equivalent of base is added is the pK value. The shape of this curve is identical for all univalent acids or bases; the position of the curve is determined by the pK of the substance being titrated.

point of the titration curve, where 50 per cent of the acid or base has been used up, gives the value of pK since [salt]/[acid] is 1 and log [salt]/[acid] is zero. Buffering efficiency is at its greatest at the pH corresponding to the pK value and diminishes at more acid or more alkaline values. Thus, buffering efficiency is determined largely by the *ratio* of salt to acid. When the ratio is 1, the pH is at the pK value of the system and maximal buffering is found. This corresponds to the change in slope of the titration curve. The inflection point is the pK. It should be reemphasized that the salt of a weak acid alone, or a weak acid by itself, is not a complete buffer system.

For polybasic acids, each group has its characteristic pK value which may be described by equation (8). Table 5.3 gives the pK_a values for a number of compounds that are either useful or biologically important in buffers. The dissociation of a base represented by K_b in equation (3) is seldom used in biological work. It is more convenient to express all dissociation constants as pK_a values on the same scale: pK_a = pK_w − pK_b, where pK_b = −log K_b.

Table 5.3: **pK_a Values of Some Useful and Biologically Important Compounds**

Compound	pK_a	*Compound*	pK_a
Phosphoric acid (pK_1)	2.0	Citric acid (pK_3)	6.4
Citric acid (pK_1)	3.1	Phosphoric acid (pK_2)	6.7
Formic acid	3.8	Imidazole	7.0
Lactic acid	3.9	Diethylbarbituric acid	8.0
Benzoic acid	4.2	Tris(hydroxymethyl)aminomethane	8.1
Acetic acid	4.7	Boric acid	9.2
Citric acid (pK_2)	4.7	Ammonium ion	9.3
Pyridinium ion	5.3	Ethylammonium ion	9.8
Cacodylic acid	6.2	Triethylammonium ion	10.8
Maleic acid	6.2	Carbonic acid (pK_2)	10.4
Carbonic acid (pK_1)	6.3	Phosphoric acid (pK_3)	12.4

Indicators. Many substances that are themselves weak electrolytes exist in characteristically colored or colorless forms in different pH regions. These are useful as indicators of the pH of a solution. In general, the indicator is a weak electrolyte and may be written as the acid HIn, which can exist in two chromogenic forms:

$$\text{HIn (color A)} \rightleftharpoons \text{In}^- \text{ (color B)} + \text{H}^+$$

From the Henderson-Hasselbalch equation

$$\text{pH} = \text{p}K + \log\frac{[\text{In}^-]}{[\text{HIn}]} \qquad (12)$$

At any pH, [In$^-$]/[HIn] will give color B/color A. The mixture of color A and color B permits a visual comparison with color standards and hence a rapid determination of pH. Some commonly used indicators are listed in Table 5.4.

The applicability of indicator theory may be illustrated with bromocresol green as an example. In solutions more acid than about pH 3.9 this dye is yellow. In solutions more alkaline than approximately pH 5.3 it is blue. These are the two forms of this substance, each of which predominates at the extremes of its titration

Table 5.4: USEFUL pH RANGE OF SOME INDICATORS

Indicator	pK_a	Range of pH	Acid color	Alkaline color
Thymol blue (pK_1)	1.5	1.2–2.8	Red	Yellow
Bromophenol blue	4.0	3.0–4.6	Yellow	Blue
Bromocresol green	4.7	3.8–5.4	Yellow	Blue
Methyl red	5.1	4.4–6.0	Red	Yellow
Chlorophenol red	6.0	4.8–6.4	Yellow	Red
Bromothymol blue	7.0	6.0–7.6	Yellow	Blue
Phenol red	7.9	6.8–8.4	Yellow	Red
Thymol blue (pK_2)	8.9	8.0–9.6	Yellow	Blue
Phenolphthalein	9.7	8.4–10.5	Colorless	Red

SOURCE: Adapted from W. M. Clark, "The Determination of Hydrogen Ions," 3d ed., The Williams & Wilkins Company, Baltimore, 1928.

range. In solutions of pH 4.7, which corresponds to the pK value, the indicator is present as equal quantities of the acid and of its salt. Hence, there are equal quantities of yellow and blue forms, and the solution is green. Changes in pH away from the pK value will alter the ratio of the two forms in accord with the Henderson-Hasselbalch equation. Thus, starting from the acid side the color of bromocresol green will change from yellow to yellow-green, then to green, to blue-green, and, finally, to blue. These characteristic color forms can be used as standards for the rapid determination of the pH of unknowns.

AMINO ACIDS AS ELECTROLYTES

Amino acids are ampholytes, *i.e.*, they behave as acids and as bases since they each contain at least one carboxyl group and one amino group. Each group has its characteristic pK value, and by convention these are designated pK_1, pK_2, etc., starting with the group titrated at the most acid region.

A monoamino monocarboxylic acid, like glycine, exists as *dipolar ions,* also termed *zwitterions,* in which both the acidic and basic groups are ionized.

$$^{+}H_3N—CH_2—COO^{-}$$

However, the molecule is electrically neutral, since the number of positive charges is equal to the number of negative charges. In this condition, the molecule is termed isoelectric. The pH at which a dipolar ion does not migrate in an electrical field is called the *isoelectric point.* In water and in the absence of other solutes, this pH is also the *isoionic point, i.e.*, the pH at which the number of cations is equal to the number of anions. In salt solutions or in solutions containing ions other than those derived from the ampholyte, some of the ionizable groups of the ampholyte may be electrically neutralized by the other ions present. Under these circumstances, there will be differences between the value for the isoelectric point and that for the isoionic point. This is of significance for proteins particularly.

Addition of hydrogen ions to the isoelectric molecule, depicted above, produces a change in charge, since ionization of the carboxylate group is repressed and the molecule acquires a net positive charge.

$$^{+}H_3N—CH_2—COO^- + H^+ \rightleftharpoons {}^{+}H_3N—CH_2—COOH$$

The dipolar ion has accepted a proton, and the predominant form is now a positively charged molecule.

Correspondingly, addition of base to the dipolar ion removes a proton from the ammonium group, leaving the molecule with a net negative charge.

$$^{+}H_3N—CH_2—COO^- + OH^- \rightleftharpoons H_2N—CH_2—COO^- + H_2O$$

It is easier to regard the isoelectric amino acids as salts that are fully ionized and internally neutralized by their own amino and carboxyl groups than to consider them as acids in the conventional sense. Indeed, their properties conform more to those of salts than to non-ionic organic compounds, as shown by the following.

I. In general, the isoelectric amino acids are soluble in water and other polar solvents, and insoluble in nonpolar solvents such as ether, chloroform, benzene, etc. Non-ionic compounds of similar structure are usually soluble in nonpolar solvents and sparingly soluble in water.

2. The α-amino acids melt with decomposition above 200°C., and many do so above 300°C. The great majority of structurally similar non-ionic organic compounds have much lower melting points. Amino acids are intermediate between non-ionic compounds and inorganic salts, which have very high melting points.

3. Formaldehyde combines with uncharged amino groups to form weakly basic derivatives. In the titration of amino acids with base in the presence of formaldehyde, only that pK is altered which may be ascribed to the amino group on the basis of dipolar ion structure. In the reaction of amino acids with formalde-

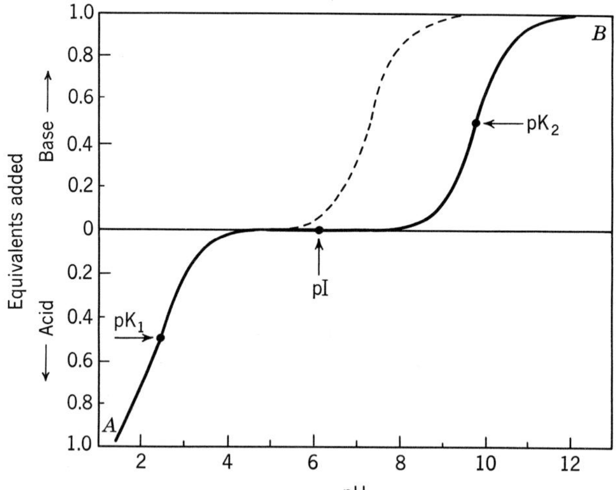

Fig. 5.3. Diagrammatic representation of the titration curve of a monoamino monocarboxylic acid such as glycine with pK_1 equal to 2.3 and pK_2 equal to 9.6. In the acid solution A, the substance present is glycine hydrochloride; in B, it is sodium glycinate. The dashed curve is obtained in the presence of formaldehyde. The isoelectric point, pI (page 105), is 6.

hyde in neutral or slightly basic solutions, one or two molecules of formaldehyde combine with the uncharged amino group to form a mono- or dimethylol derivative.

$$R-\underset{\underset{NH_2}{|}}{CH}-COO^- + CH_2O \rightleftharpoons R\underset{\underset{HN-CH_2OH}{|}}{CH}-COO^- + CH_2O \rightleftharpoons R-\underset{\underset{HOH_2C-N-CH_2OH}{|}}{CH}-COO^-$$

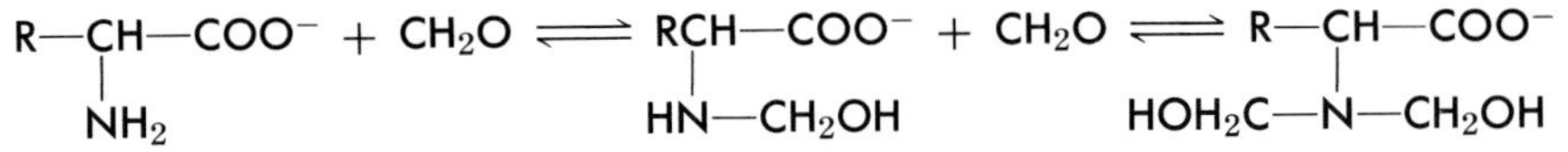

The titration curve of a typical monoamino monocarboxylic acid is shown in Fig. 5.3. It is apparent that there are two titratable groups and that the more alkaline one is due to titration of the amino group as judged by the effect of formaldehyde in shifting the curve. The pK values for the titratable groups of amino acids are given in Table 5.5. The table includes values for the sulfhydryl group of cys-

Table 5.5: APPARENT pK_a AND pI VALUES FOR AMINO ACIDS AT 25°C.

Amino acid	pK_1(COOH)	pK_2	pK_3	pI*
Alanine	2.34	9.69		6.00
Arginine	2.17	9.04 (NH$_3^+$)	12.48 (guanidinium)	10.76
Asparagine	2.02	8.80		5.41
Aspartic acid	1.88	3.65 (COOH)	9.60 (NH$_3^+$)	2.77
Cysteine (30°)	1.96	8.18 (SH)	10.28 (NH$_3^+$)	5.07
Cystine (30°)	<1.0	1.7 (COOH)	7.48 and 9.02 (NH$_3^+$)	4.60
Glutamic acid	2.19	4.25 (COOH)	9.67 (NH$_3^+$)	3.22
Glutamine	2.17	9.13		5.65
Glycine	2.34	9.60		5.97
Histidine	1.82	6.00 (imidazolium)	9.17 (NH$_3^+$)	7.59
Hydroxyproline	1.92	9.73		5.83
Isoleucine	2.36	9.68		6.02
Leucine	2.36	9.60		5.98
Lysine	2.18	8.95 (α-NH$_3^+$)	10.53 (ε-NH$_3^+$)	9.74
Phenylalanine	1.83	9.13		5.48
Proline	1.99	10.60		6.30
Serine	2.21	9.15		5.68
Tryptophan	2.38	9.39		5.89
Tyrosine	2.20	9.11 (NH$_3^+$)	10.07 (OH)	5.66
Valine	2.32	9.62		5.96

*pI = pH at the isoelectric point (page 105).

SOURCE: From E. J. Cohn and J. T. Edsall. "Proteins, Amino Acids and Peptides," Reinhold Publishing Corporation, New York, 1943.

teine and the phenolic group of tyrosine and for groups more obviously acidic or basic. Amino acids, like other weak electrolytes, exert their maximal buffering effect at their pK and not at their pI values. As polyvalent electrolytes, therefore, each has at least two pH regions in which it can serve as a buffer. The pK values for the SH and NH$_3^+$ of cysteine overlap; values given are for the predominant form at the pK indicated.

It is of interest to examine the titration of some of the more complex amino acids. Glutamic acid in the isoelectric state has the structure shown in (II).

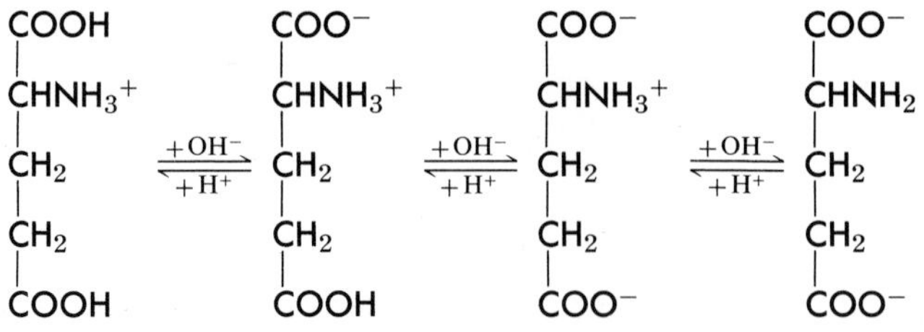

(I) (II) (III) (IV)

Addition of one mole of acid produces (I). Addition of two moles of alkali to (II) gives rise successively to (III) and (IV). Aspartic acid behaves in the same manner as glutamic acid.

Formula (III) below represents the isoelectric state of lysine.

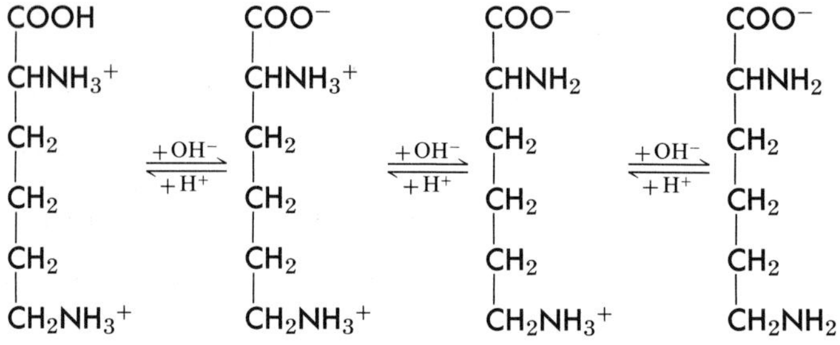

(I) (II) (III) (IV)

(II) is a monoacid salt, such as lysine monohydrochloride; (I) is a diacid salt, such as lysine dihydrochloride; (IV) is a basic salt, such as sodium lysinate. In contrast to lysine, glutamic acid can form only a monohydrochloride but can yield mono- and disodium salts.

The various forms of histidine are shown below; (III) is the isoelectric form.

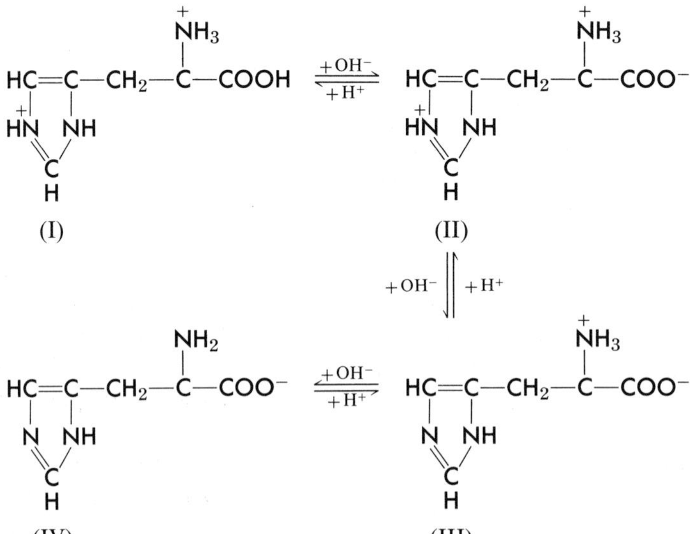

(I) (II) (IV) (III)

Note that the imidazolium group has a *lower* pK_a value than the ammonium group (Table 5.5).

In the case of arginine, the isoelectric form is (III). Observe that the guanidinium group has a very high pK_a value, 12.48 (Table 5.5).

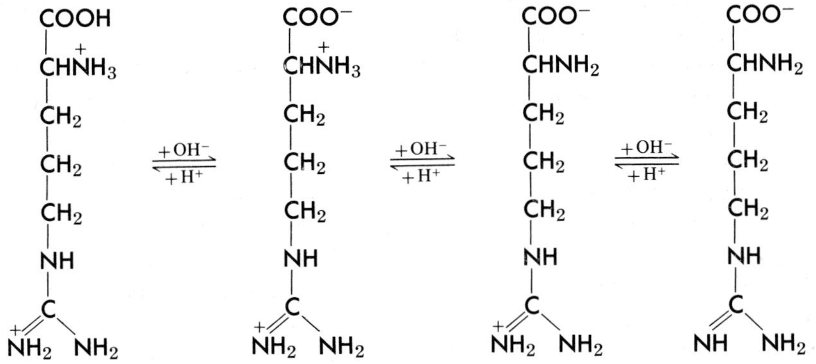

(I) (II) (III) (IV)

Calculation of Isoelectric Point. The isoelectric point (page 101) of an ampholyte is determined by the magnitude of the dissociation constants, K_1 and K_2. The condition of the isoelectric point is that the net charge be equal to zero, or, expressed in other terms, that the number of cations A^+ be equal to the number of anions A^-. If A is the isoelectric molecule, then, at the isoelectric point, its dissociations as acid and as base are equal or

$$A + H^+ \rightleftharpoons A^+ \quad \text{and} \quad A \rightleftharpoons A^- + H^+$$

From the mass law,

$$K_1 = \frac{[H^+][A]}{[A^+]} \quad \text{and} \quad K_2 = \frac{[H^+][A^-]}{[A]} \tag{13}$$

At the isoelectric point $[A^+] = [A^-]$, or

$$[A^+] = \frac{[H^+][A]}{K_1} = [A^-] = \frac{K_2[A]}{[H^+]} \tag{14}$$

which yields

$$[H_I{}^+]^2 = K_1K_2 \quad \text{or} \quad pH_I = \frac{pK_1 + pK_2}{2} \tag{15}$$

where pH_I is the pH at the isoelectric point, or pI. The calculated pI values for the amino acids are given in Table 5.5. For those amino acids which contain more than two ionizable groups, the pI value may be calculated from the two pK values whose groups are titrated to either side of the isoelectric form of the amino acid.

CHEMICAL PROPERTIES OF THE AMINO ACIDS

The chemical reactions of the amino acids are relatively numerous because of the different reactive groups that are present in the same molecule. Aliphatic monoamino monocarboxylic acids give all the reactions expected for carboxyl and amino groups. The other amino acids give these same reactions and, in addition, those reactions characteristic of additional groups that may be present. For example, cysteine gives the reactions characteristic of the sulfhydryl (SH) group, tyrosine the reactions of a phenolic group, etc. Many of these reactions are also given by proteins because they contain these amino acids.

Reaction with Ninhydrin. Amino acids react with ninhydrin (triketohydrindene hydrate) to yield CO_2, ammonia, and an aldehyde containing one less carbon than the amino acid. The reaction also yields a blue or purple color, which is useful for the colorimetric quantitative estimation of amino acids. However, the colorimetric reaction is not specific for amino acids since color with ninhydrin is produced by ammonia and by many amino compounds, including peptides and proteins, in circumstances in which CO_2 is not liberated in the reaction. Therefore, measurement of CO_2 production with ninhydrin is specific for the presence of a free carboxyl group adjacent to the amino group, *i.e.*, free α-amino carboxylic acid functions. Proline and hydroxyprolines give yellow products with ninhydrin.

The reaction of ninhydrin with α-amino acids is as follows:

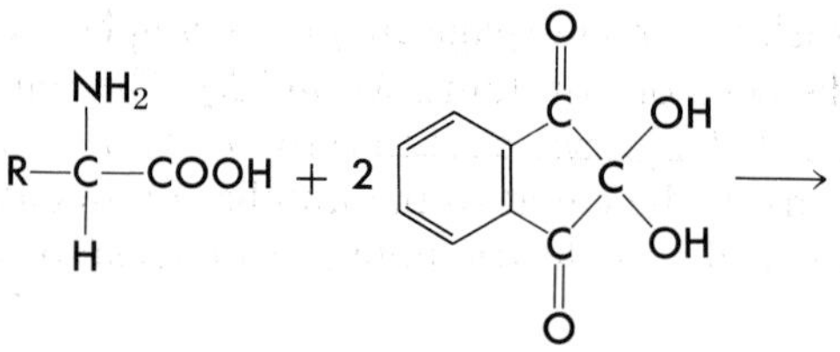

α-Amino acid **Ninhydrin** (triketohydrindene hydrate)

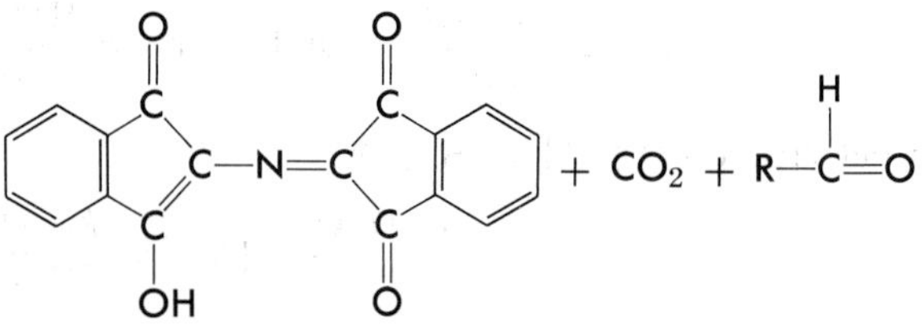

Diketohydrindylidene-diketohydrindamine + Carbon dioxide + Aldehyde

Reactions of Specific Amino Acids. Some of the color reactions given by particular amino acids are in common use as analytical tools. These reactions are also given by most proteins and thus aid in their determination or identification. Several of these reagents are used in sprays for the detection of specific amino acids on paper chromatograms (page 113).

Millon Reaction. A red color is obtained when phenolic compounds are heated with $Hg(NO_3)_2$ in nitric acid containing a trace of nitrous acid. Proteins containing tyrosine give this reaction.

Sakaguchi Reaction. Guanidines in alkaline solution give a red color with the reagent, which contains α-naphthol and sodium hypochlorite. The reaction is given by arginine and by proteins that contain this amino acid.

Nitroprusside Test. Cysteine and proteins that have free sulfhydryl groups give a red color with sodium nitroprusside [$Na_2(NO)Fe(CN)_5$—$2H_2O$] in dilute ammoniacal solution.

Aldehyde Reaction. Indole derivatives give strongly colored products with a number of aromatic aldehydes. With *p*-dimethylaminobenzaldehyde in sulfuric acid, a red-violet color is obtained with tryptophan (Ehrlich reaction). This test can be used for the quantitative estimation of tryptophan in proteins.

Folin's Reaction. In alkaline solution, amino acids give a deep red color with sodium 1,2-naphthoquinone-4-sulfonate. This method is used for rapid quantitative estimation of amino acids; it has been applied, for example, to blood.

Pauly Reaction for Histidine and Tyrosine. Histidine and tyrosine couple with diazotized sulfanilic acid in alkaline solution, giving a red color.

PEPTIDES

Of the greatest importance in understanding the structure of proteins is the fact that amino acids can be linked together in amide bonds through the carboxyl group of one and the amino group of another to form a substituted amide bond, which is termed a peptide bond or linkage. The characteristic peptide bond structure is enclosed in the dotted area:

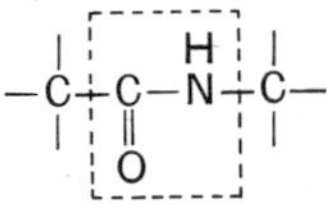

Peptide nomenclature lists the amino acid residues as they occur in the chain starting from the *free* amino group, which is conventionally shown at the left hand part of the structure. The ending *-yl* is used for each of the residues except the terminal one, which bears the unsubstituted carboxyl group. Peptides containing two amino acid residues are dipeptides, those with three amino acid residues, tripeptides, etc. In general, peptides with more than two residues are termed polypeptides. The tetrapeptide alanylglycyltyrosylglutamic acid has the structure indicated.

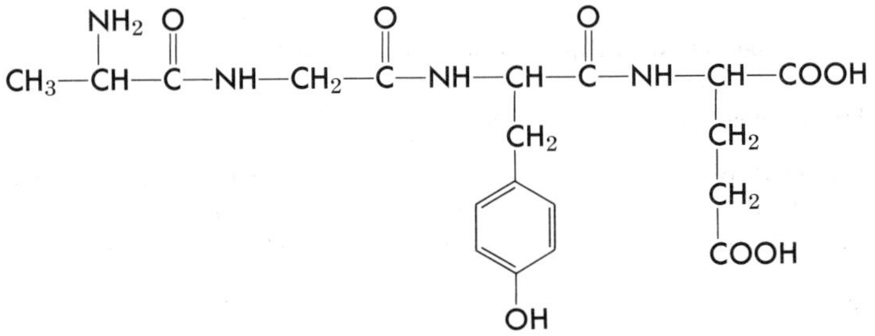

Alanylglycyltyrosylglutamic acid

The names of the amino acid residues and their structures are given in Table 5.6. The distinction between glutamyl and glutaminyl, as well as aspartyl and

Table 5.6: Amino Acid Residues in Proteins*

Residue name	*Abbreviation*	*Residue name*	*Abbreviation*
Alanyl	Ala	Isoleucyl	Ile
Arginyl	Arg	Leucyl	Leu
Asparaginyl	Asn or $AspNH_2$	Lysyl	Lys
Aspartyl	Asp	Methionyl	Met
Cysteinyl	CySH	Phenylalanyl	Phe
Cystyl	CyS-SCy	Prolyl	Pro
Glutaminyl	Gln or $GluNH_2$	Pyrrolidone carboxyl	Pyr
Glutamyl	Glu	Seryl	Ser
Glycyl	Gly	Threonyl	Thr
Histidyl	His	Tryptophanyl	Trp
Hydroxylysyl	Hyl	Tyrosyl	Tyr
Hydroxyprolyl	3 or 4 Hyp	Valyl	Val

* For other amino acids, see *J. Biol. Chem.*, **241**, 2491, 1966.

asparaginyl residues, should be noted. It also should be emphasized that glutamyl derivatives may be of the γ type as well as the α variety.

Peptide Synthesis. Synthesis of complex peptides involves technically difficult problems in organic chemistry. First, in order to couple amino acid residues in specific peptide linkage, the amino and carboxyl groups which are not to be linked must be blocked so as to be unreactive. Further, all other reactive side chains of the amino acids must be blocked or the coupling procedure must be performed by a method which does not cause these groups to react. Some of the protective groups used in peptide synthesis are listed in Table 5.7. Second, the coupling must be done by a method which does not cause racemization or chemical alteration of side chains. Most of the coupling reactions involve activation of the α-carboxyl groups, usually by forming some type of mixed acid anhydride or reactive ester; some of these are given in Table 5.8. Third, the protecting group(s) must be removed quantitatively by mild methods which do not cause rearrangements or racemization, or produce cleavage of the peptide bonds.

Only a single, simple example of the synthesis of a dipeptide, glycyl-L-proline, will be given here. The most important protective group, the carbobenzoxy residue, was devised by Bergmann and Zervas in 1932. The amino group is protected by the

Table 5.7: Some Protective Groups Used in Peptide Synthesis

Protecting group	Blocked group	Method of removal
Carbobenzoxy (benzyloxycarbonyl)	Amino	H_2 and Pd black; HBr in acetic acid
Triphenylmethyl (trityl)	Amino	H_2 and Pd
Tertiary butyloxycarbonyl	Amino	Trifluoroacetic acid
Phthaloyl	Amino	Hydrazine
Trifluoroacetyl	Amino	Weak alkali
Benzyl	Sulfhydryl	Na metal in liquid ammonia
p-Toluene sulfonyl (tosyl)	Amino	Na metal in liquid ammonia
Ethyl or methyl ester	Carboxyl	Weak alkali
Benzyl ester	Carboxyl	Weak alkali; H_2 and Pd
Tertiary butyl ester	Carboxyl	Trifluoroacetic acid; HBr in acetic acid
Acetyl	Hydroxyl	Weak alkali

Table 5.8: SOME COUPLING REAGENTS IN PEPTIDE SYNTHESIS

Coupling reagent	Structure	Method of preparation
Acid chloride	$R-\overset{\overset{O}{\|}}{C}-Cl$	PCl_5 (anhydrous)
Acid azide	$R-\overset{\overset{O}{\|}}{C}-N_3$	From acyl hydrazide with HNO_2
Dicyclohexyl carbodiimide	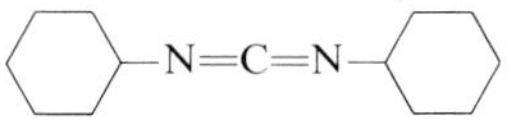	Addition of reagent to compounds to be coupled in organic solvent
p-Nitrophenyl ester	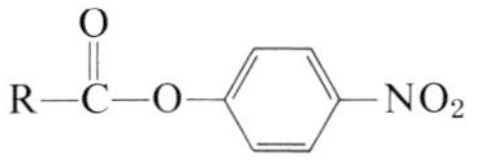	Via acid chloride, mixed anhydride, or the carbodiimide
Mixed organic anhydride, *e.g.*, from isobutylchlorocarbonate	$R-\overset{\overset{O}{\|}}{C}-O-\overset{\overset{O}{\|}}{C}-O-C_4H_9$	$C_4H_9-O-\overset{\overset{O}{\|}}{C}-Cl$ (isobutylchloroformate)

carbobenzoxy residue, which is easily removed later in the form of volatile products by catalytic hydrogenation. This method is suitable for synthesizing optically active di- and polypeptides, and many have been prepared.

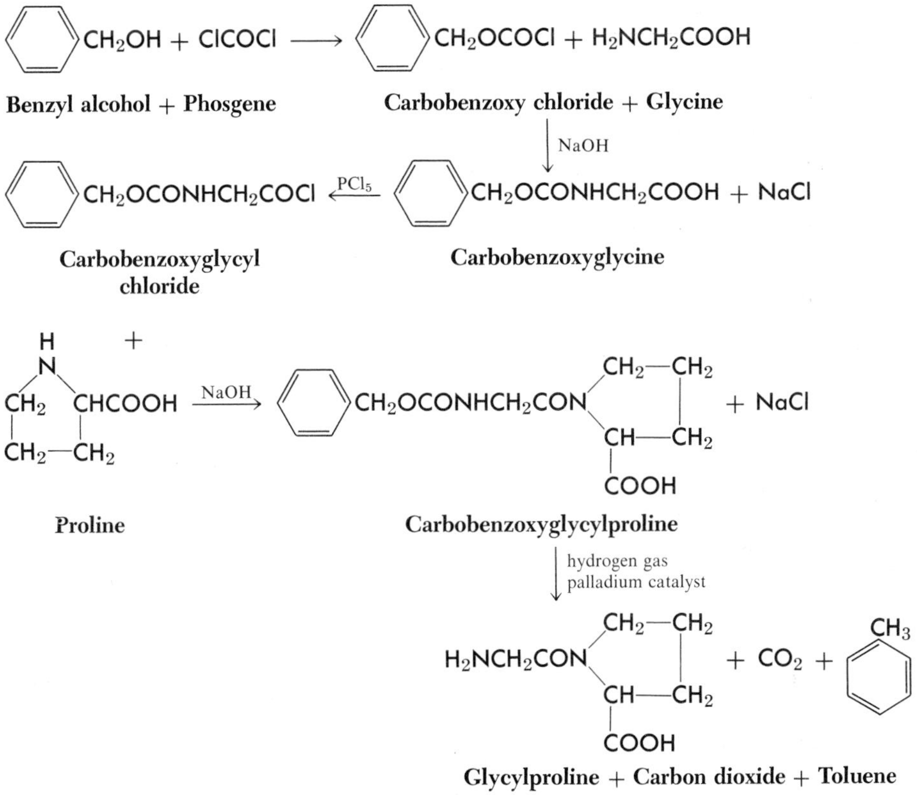

Longer peptides may be synthesized by preparing the several component smaller peptides and, at the last stages, linking them together and removing the protective groups. Preferably, synthesis can also be performed in stepwise fashion, adding one amino acid residue at a time, starting with the residue which ultimately bears the free α-carboxyl group. In this procedure the blocking group on the protected dipeptide ester is removed and coupling is repeated by adding another protected amino acid residue by linking its carboxyl group to the free amino group of the dipeptide ester. In the example below, Z represents the carbobenzoxy group.

$$\mathrm{Z{-}NH{-}CHR{-}\overset{\overset{\displaystyle O}{\|}}{C}{-}X + NH_2{-}CHR'{-}\overset{\overset{\displaystyle O}{\|}}{C}{-}NH{-}CHR''{-}\overset{\overset{\displaystyle O}{\|}}{C}{-}OCH_3} \longrightarrow$$

$$\mathrm{Z{-}NH{-}CHR{-}\overset{\overset{\displaystyle O}{\|}}{C}{-}NH{-}CHR'{-}\overset{\overset{\displaystyle O}{\|}}{C}{-}NH{-}CHR''{-}\overset{\overset{\displaystyle O}{\|}}{C}{-}OCH_3}$$

Z is then removed and the procedure is repeated until the peptide of desired structure is prepared. A novel variation of this procedure has been to link in ester linkage the carboxyl group of an amino acid to an insoluble resin and then couple with a protected amino acid, yielding a protected dipeptide still coupled to the resin. The protecting group may be removed and the procedure repeated many times until the desired product is achieved. The peptide is then liberated from the resin. The advantage is that at each chemical step, impurities and by-products are readily removed by simply washing the resin with appropriate solvents.

With present methods of peptide synthesis, many naturally occurring polypeptides and the small protein hormone, insulin, have been synthesized. Thus, peptide synthesis has served to prove the structures which have been deduced by degradative methods. Synthesis has also been used to prepare analogues of the natural compounds to investigate the role of each side chain and residue in the biological activity as well as to obtain compounds with more desirable properties for therapeutic purposes. Synthetic peptides and their derivatives have been of great value in the study of the specificity and mechanism of action of proteolytic enzymes which hydrolyze peptide bonds (Chap. 12).

THE PEPTIDE LINKAGE IN PROTEINS

In 1902 Hofmeister and Fischer proposed independently that the main type of linkage between the amino acids in proteins is the peptide bond. This hypothesis provided the foundation from which all advances in the chemistry of these complex substances have been made. A brief summary of some of the evidence that indicates the presence of peptide linkages in proteins also indicates some of the properties and methods used for the study of proteins.

1. Intact proteins show little free amino nitrogen, but large amounts are formed after hydrolysis. This is consistent with the view that the majority of the amino groups are linked in peptide bonds.

2. Acid or enzymic hydrolysis of proteins yields progressively increasing quantities of amino and carboxyl groups in equal amounts. This indicates that equivalent amounts of acidic and basic groups are simultaneously liberated.

3. Proteolytic enzymes, which are very specific in hydrolyzing proteins, hydrolyze peptide bonds in synthetic peptides and in peptide derivatives of known structure. It is evident that these enzymes split the same or similar bonds in proteins and in synthetic substrates. The action of proteolytic enzymes on proteins yields a mixture of peptides of varying size and free amino acids.

4. Several antibiotics isolated from microorganisms have proved to be complex peptides. Gramicidin S is a complex cyclodecapeptide

-L-valyl-L-ornithyl-L-leucyl-D-phenylalanyl-L-prolyl-

in which the above unit occurs twice in a closed peptide chain. The presence of the unnatural D-phenylalanine should be noted. Many antibiotics contain D amino acids and a variety of L or D amino acids which have not been encountered in proteins.

5. Several polypeptide hormones have been isolated in pure form and have been synthesized from the constituent amino acids (Part Five).

6. Biuret is obtained by heating urea to about 180°C.

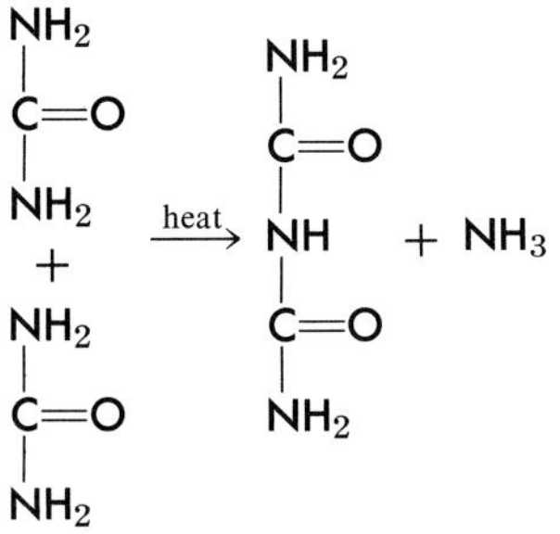

2 Urea ⟶ biuret + ammonia

In strongly alkaline solution, biuret gives a violet color with copper sulfate. The biuret reaction is given by all compounds with two amide or peptide bonds linked directly or through an intermediate carbon atom. Tripeptides and proteins possess the following structure and give the color reaction.

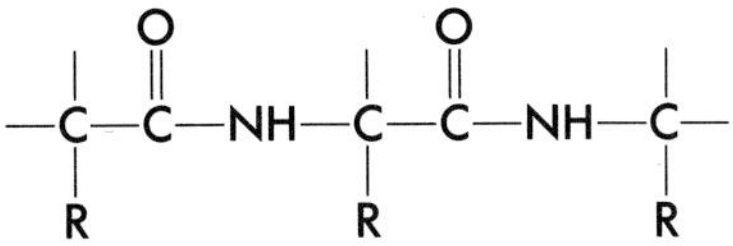

Dipeptides and amino acids (with the exception of histidine, serine, and threonine) do not give this reaction. The biuret reaction is extensively used for detection of proteins and as the basis for a quantitative colorimetric method for protein estimation.

Further evidence concerning the occurrence of peptide bonds in proteins will become apparent later. It should not be assumed that peptide linkages represent the sole method of holding the protein molecule in its specific conformation. There are many properties of proteins which show that protein molecules are far more complex than simple long polypeptides.

Of interest in relation to the above considerations is the occurrence in animal tissues of peptides. Carnosine, β-alanyl-L-histidine, and anserine, which is β-alanyl-1-methyl-L-histidine, are found free in muscle tissue of vertebrates.

$$\underset{\text{HN}\qquad\text{N}}{\overset{\text{COOH}}{\text{H}_2\text{NCH}_2\text{CH}_2\text{CO—NHCHCH}_2\text{C}=\!=\!\text{CH}}}$$

$$\underset{\text{H}_3\text{CN}\qquad\text{N}}{\overset{\text{COOH}}{\text{H}_2\text{NCH}_2\text{CH}_2\text{CO—NHCHCH}_2\text{C}=\!=\!\text{CH}}}$$

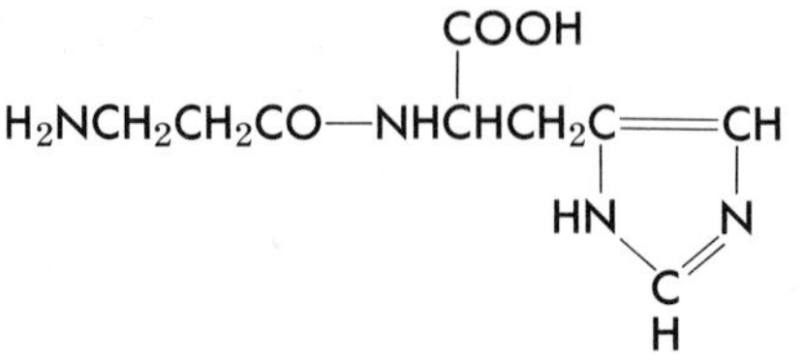

Carnosine (β-alanyl-L-histidine)   **Anserine** (β-alanyl-1-methyl-L-histidine)

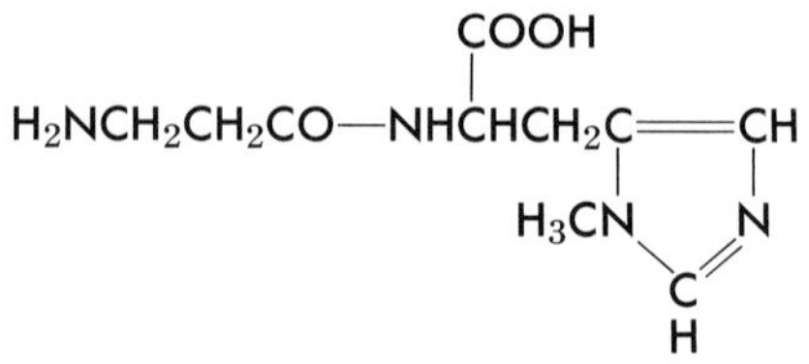

Glutathione, a tripeptide, γ-L-glutamyl-L-cysteinylglycine, was isolated by Hopkins from yeast, muscle, and liver tissue and is widely distributed in nature.

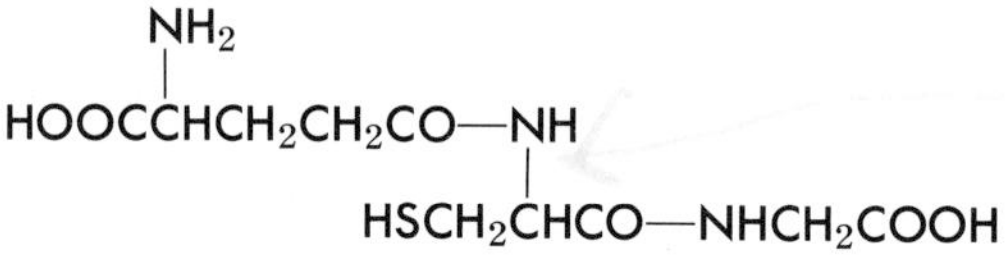

$$\overset{\text{NH}_2}{\text{HOOCCHCH}_2\text{CH}_2\text{CO—NH}}$$
$$\text{HSCH}_2\text{CHCO—NHCH}_2\text{COOH}$$
Glutathione

Carnosine and anserine contain β-alanine, an amino acid that has not been found in proteins, and anserine contains methylhistidine, which has not been identified in proteins, although it has been isolated from the urine of mammals. Glutathione differs from peptides commonly found in proteins in that the γ-carboxyl of glutamic acid is linked in peptide bond; however, the second bond is an α-peptide bond.

ESTIMATION AND SEPARATION OF AMINO ACIDS

Amino acid analysis is of prime importance in determining the composition and structure of proteins and polypeptides. These determinations are generally made after acidic hydrolysis. In addition, estimation of free amino acids in urine, tissue extracts, and body fluids is important in various types of physiological and clinical studies.

Amino Acid Analysis. As in all biological studies, the trend has been to develop rapid quantitative methods that are applicable to micro quantities of material. Some of these methods have been discussed and will be mentioned only briefly.

Gravimetric Methods. The general procedure of isolating and weighing substances finds rare application in quantitative amino acid analysis because of the general lack of reagents that will quantitatively precipitate a specific amino acid from a mixture. However, an example of a quantitative isolation technique is the precipitation of arginine as an insoluble flavianate. Such procedures are also useful for isolation of samples of optically active amino acids from protein hydrolysates.

Colorimetric Methods. Various amino acids give specific color reactions; several of these have been mentioned previously. Such methods are particularly applicable to amino acids which possess characteristic functional groups.

Microbiological Methods. Many microorganisms, such as bacteria and fungi, require complex mixtures of specific amino acids, vitamins, etc., for growth. If the nutritional requirements for growth are known, a medium may be prepared that lacks one essential nutrient. From the growth rate of the organism when some material, *e.g.*, a protein hydrolysate or tissue extract, is added, the amount of the limiting nutrient in the added material may be estimated. Many of the lactic acid bacteria are particularly suitable since determination of lactic acid liberated into the medium is readily made and serves as an index of bacterial growth. Suitable strains of bacteria have been found for the estimation of all the amino acids found in proteins.

Enzymic Methods. Specific enzymes may be used for estimation of certain amino acids and indeed for many other substances also. For example, the enzyme arginase acts only on arginine, yielding ornithine and urea. The urea is easily estimated by colorimetric methods and serves as an index of preexisting arginine.

$$H_2N\underset{\underset{NH}{\|}}{C}NHCH_2CH_2CH_2CHNH_2COOH + H_2O \xrightarrow{\text{arginase}}$$

$$H_2NCH_2CH_2CH_2CHNH_2COOH + H_2NCONH_2$$

Arginine + water ⟶ ornithine + urea

The urea may be treated with the specific enzyme urease, yielding carbon dioxide and ammonia, as an application of another enzymic method.

$$H_2NCONH_2 + H_2O \xrightarrow{\text{urease}} 2NH_3 + CO_2$$

The ammonia or CO_2 can be estimated by a variety of methods. Other enzymes have also been applied to the estimation of amino acids, *e.g.*, specific bacterial decarboxylases that liberate CO_2; this gas is conveniently measured manometrically.

Chromatographic Methods. Some applications of chromatography to analysis have already been discussed (page 35). Partition chromatography on paper is widely employed for detection of amino acids, and various methods are used for quantitative estimation. A ninhydrin spray is used for amino acids in general, and special reagents may be applied for the detection of particular functional groups, *e.g.*, the Pauly reagent (for histidine and tyrosine), the Ehrlich reagent (for tryptophan), the Sakaguchi reagent (for arginine), etc. (page 106).

A representative two-dimensional separation of amino acids is shown in Fig. 5.4. It should be noted that in the solvent system for dimension 1, the greatest movement is found for those amino acids which possess large nonpolar side chains since the stationary phase is the water phase of the solvent mixture. Hence those amino acids with the higher affinity for water, *i.e.*, the more polar compounds including those with additional ionic groups, have the lower R_F values.

Ion exchange chromatography is in wide use for the separation of amino acids; it employs synthetic resins, some being anionic and others cationic substances. Sulfonated polystyrene resin bears many $—SO_3H$ groups. When a column of this resin is treated with a buffer at pH 3 containing Na^+ ions, the resin is converted to the form $—SO_3Na$. Cationic amino acids introduced at the top of such a column will displace Na^+ and will be retarded or strongly adsorbed, the degree of retardation depending on the basicity of the individual amino acid. Thus, aspartic and glutamic acids emerge before most of the neutral amino acids, and well ahead of the basic amino acids. Even the neutral amino acids can be separated readily from

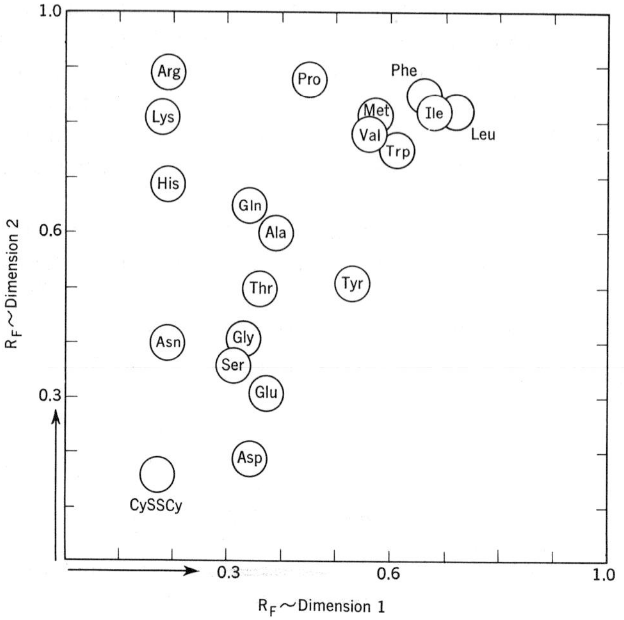

Fig. 5.4. A schematic representation of a two-dimensional paper chromatogram. The solvent for dimension 1 is *n*-butanol-acetic acid-water (250:60:250 vol. per vol.) and for dimension 2, phenol-water-ammonia (120:20:0.3 per cent). Each solvent front moves with an R_F equal to 1 in each dimension. The abbreviations used for amino acid radicals are given in Table 5.6 (page 108).

one another since, despite their similar pK values, their side chains possess different affinities for the resin.

Figure 5.5 illustrates the separation achieved with a synthetic amino acid mixture. In this case the effluent from the column was mixed automatically with a ninhydrin reagent, the color developed by passage through a hot water bath, and the absorbancy recorded as the effluent passed through a light beam directed on a photocell. The change in photocurrent with volume of effluent was plotted automatically by a recorder. Integration of the area under each peak permits quantitative estimations. Identification of each amino acid is readily made, since, under controlled conditions, the position of emergence is constant for each amino acid. These principles have been utilized for construction of amino acid analyzers.

Various types of resins and buffer systems are in use for separation of amino acids, peptides, and proteins. The quantitative analysis of protein hydrolysates is discussed later (page 139).

Electrophoretic Methods. Inasmuch as amino acids are electrolytes, they migrate in an electric field at appropriate pH values. For example, at pH 5, glutamic and aspartic acids possess a net negative charge and migrate toward the anode; histidine, arginine, and lysine migrate toward the cathode; the "neutral" amino acids will remain stationary. Such electrophoretic methods may be applied on a single sheet or strip of paper to effect a group separation. At other pH values, *e.g.*, pH 1.9, even the neutral amino acids may be separated from one another to a considerable degree.

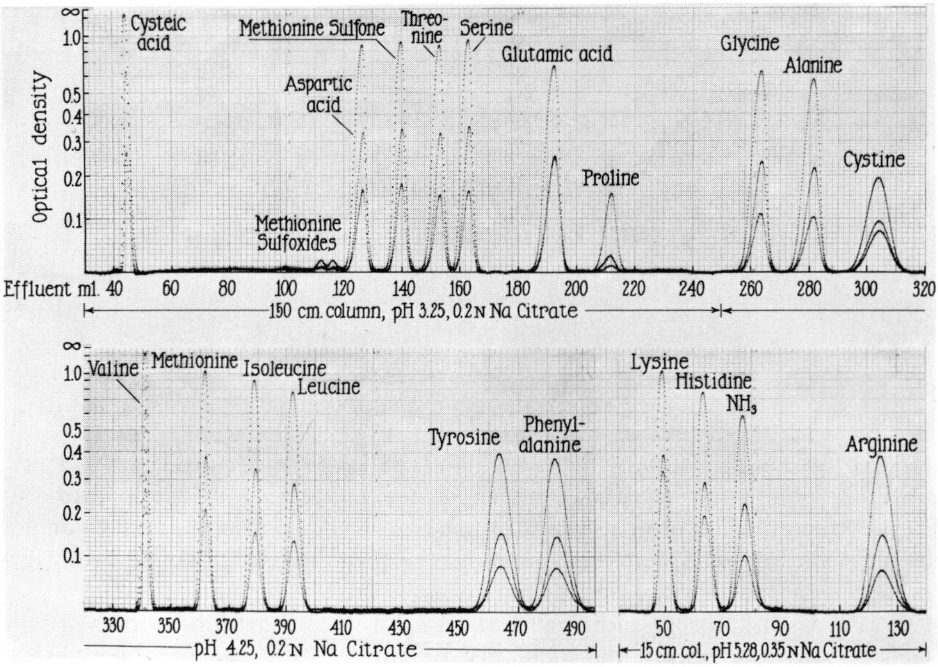

Fig. 5.5. Automatically recorded chromatographic analysis of a synthetic mixture of amino acids on a sulfonated polystyrene resin. (*From D. H. Spackman, W. H. Stein, and S. Moore, Anal. Chem.,* **30,** 1190, 1958.)

A widely used technique involves a two-dimensional separation on paper, employing electrophoresis in one dimension and partition chromatography in the other. Such a system has a high resolving power and is used not only for separation and identification of amino acids but also for peptides. Such methods are useful for critical estimation of purity since a pure substance will yield only a single spot when the entire sheet is sprayed with a general reagent, such as a ninhydrin solution.

REFERENCES

See list following Chap. 7.

6. The Proteins. II

Physical and Chemical Properties. Purity and Purification

CLASSIFICATION OF PROTEINS

No system of classifying the many thousands of proteins satisfactorily depicts their differences and similarities. Simpler compounds are generally grouped according to structure. The proteins are all similar in structure insofar as they contain amino acids, but since there is information regarding the structure of relatively few proteins, other methods of classification must be used.

The classification given below is based upon three general properties of proteins, *viz.*, shape, solubility, and chemical composition. The classes of proteins listed, as well as the subgroups in each class, are designated by general names given to a group of proteins of similar physical and chemical properties. However, within each subgroup differences exist in the amino acid composition and, more significantly, in the order of arrangement of the individual amino acids in the polypeptide chains (page 151). Thus, serum albumin defines a protein present in the blood serum of vertebrates. However, the serum albumin of the horse differs from the serum albumins of man, sheep, cow, dog, etc. Thus, most proteins are species-specific although very similar substances are generally found in related species.

Fibrous Proteins. These are insoluble animal proteins, highly resistant to digestion by proteolytic enzymes. They exist mainly as elongated molecules which may be constituted of several coiled peptide chains which are tightly linked. Because of their insolubility, true physical molecular weights cannot be determined. They include the proteins of silk, wool, skin, hair, horn, nails, hoofs, quills, connective tissue, and bone. This heterogeneous group may be subdivided into several distinct types of proteins. Collagens and elastins are of mesenchymal origin; keratins are derived from ectoderm.

1. *Collagens* are the major proteins of connective tissue (Chap. 38). They are insoluble in water and resistant to animal digestive enzymes but are altered to easily digestible, soluble *gelatins* by boiling in water, dilute acids, or alkalies. Approximately 30 per cent of the total protein in the mammalian body is collagen. The collagens appear to be unique in their high content of hydroxyprolines and in containing hydroxylysine. They are poor in sulfur, since cysteine and cystine are absent, and contain no tryptophan.

2. *Elastins* are present in tendon, arteries, and other elastic tissues (Chap. 38). Although similar to collagens in many respects, they cannot be converted to gelatins.

3. *Keratins* are the proteins of hair, wool, quills, hoofs, nails, etc. These proteins generally contain large amounts of sulfur as cystine. Human hair has about 14 per cent cystine.

Globular Proteins. These proteins are soluble in water or in aqueous media containing salts, acids, bases, or ethanol. In solution such molecules generally are spheroids or ellipsoids. This group includes all the enzymes, oxygen-carrying proteins, protein hormones, as well as others. As will become apparent in later chapters, the most significant description of these proteins is based on their functional properties, *e.g.*, enzymes are most readily recognized by their specific catalytic behavior.

A number of descriptive names that serve to categorize certain proteins are based on their solubility. Since these terms are in common use, they will be mentioned briefly. It should be understood that many of these names connote categories that are not mutually exclusive and that a single protein may be described in several ways.

Albumins are readily soluble in water and coagulable by heat. This is a large group, of which egg albumin and serum albumin are typical examples.

Globulins are insoluble or sparingly soluble in water, but their solubility is greatly increased by the addition of neutral salts such as sodium chloride. These proteins are coagulable by heat. Many globulins are easily prepared from animal or plant tissues since they are readily extracted by salt solutions (5 to 10 per cent NaCl) and are precipitated from the saline solution by dilution with water. Examples are serum globulins, globulins of muscle and other tissues, and globulins of plant seeds.

Histones are basic proteins which are soluble in water and which yield, on hydrolysis, large amounts of arginine or lysine. Histones can be extracted in large amounts from certain glandular tissues, such as thymus and pancreas. Histones are combined with nucleic acids (Chap. 9) within cells.

Protamines are strongly basic proteins of relatively low molecular weight. They are associated with nucleic acids and are obtained in large quantity from ripe sperm cells of fish. They contain no sulfur and have a high nitrogen content (25 to 30 per cent) because of the presence of large quantities of arginine. Tyrosine and tryptophan are absent. Typical protamines are salmine from salmon sperm, clupeine from herring, and sturine from sturgeon.

Conjugated Proteins. Conjugated proteins are combined with characteristic non-amino acid substances. The term *prosthetic group* is generally used to designate the non-amino acid moiety (see also page 214).

Nucleoproteins are combinations of proteins with nucleic acids. The latter will be discussed in Chap. 9.

The protein components of *mucoids or mucoproteins* are combined with large amounts (more than 4 per cent) of carbohydrate, measured as hexosamine. The carbohydrate portions of these conjugated proteins are complex polysaccharides containing N-acetylhexosamine in combination with uronic acids or other sugars. Many contain a sialic acid (page 34). The carbohydrate of some of the mucoproteins is covalently linked by ester linkages to serine residues or to the γ-carboxyl group of glutamic acid residues (Chap. 38).

Well-defined soluble mucoproteins have been obtained from egg white (ovomucoid α), serum (Chap. 31), and human pregnancy urine; the product from the last-named source has gonadotropic activity (Chap. 47). Soluble mucoproteins are not readily denatured by heat, nor are they easily precipitated by common protein precipitants like trichloroacetic or picric acids.

The term *glycoprotein* is frequently restricted to those proteins which contain small amounts of carbohydrate, less than 4 per cent hexosamine. This group includes many common albumins and globulins, *e.g.*, egg albumin, some serum albumins, and certain serum globulins. In some glycoproteins (ovalbumin, γ-globulin) the carbohydrate is covalently linked to the protein by an amide bond from an asparagine residue to carbon-1 of N-acetylglucosamine (Chap. 31). The distinction between glyco- and mucoproteins is arbitrary, and intermediate types exist.

Lipoproteins are water-soluble proteins conjugated with lipid, *e.g.*, lecithin, cholesterol, etc., such as the several lipoproteins of serum (Chap. 31). The *proteolipids* are distinguished from the *lipoproteins* by their solubility in nonpolar solvents and insolubility in water. Brain and nerve tissue are rich sources of proteolipids and lipoproteins (Chap. 37).

Other types of conjugated proteins contain a variety of non-amino acid groups linked to the protein. These include hemoproteins (Chap. 8), flavoproteins, metalloproteins, phosphoproteins, etc.

PROPERTIES OF PROTEINS

Amphoteric Behavior. Like amino acids, proteins are ampholytes, *i.e.*, they act as both acids and bases. Since proteins are electrolytes, they migrate in an electric field and the direction of migration will be determined by the net charge of the molecule. The net charge is influenced by pH, and for each protein there is a pH value at which it will not move in an electric field; this pH value is the isoelectric point (pI). At pH values acid to the isoelectric point, the protein will have a net positive charge and as a cation will migrate to the negative pole (cathode). Correspondingly, at pH values alkaline to the isoelectric point the protein will possess a net negative charge and, as an anion, will migrate to the positive pole (anode). The isoelectric point of a given protein is a constant and aids in characterization of these substances. The isoelectric points for several proteins are given later (Table 6.3, page 126).

The behavior of proteins is strongly influenced by pH. Hence it is desirable to ascertain the amount and kind of charge possessed by a protein molecule and what determines the net charge at a particular pH value. Like amino acids, proteins are dipolar ions (zwitterions) at the isoelectric point, *i.e.*, the sum of the positive charges is equal to the sum of the negative charges and the net charge is zero. The total charge on the protein molecule, the sum of positive and negative charges together, may be at a maximum at the isoelectric point. For simplicity, the isoelectric protein molecule may be depicted as follows:

$$(H_3\overset{+}{N})_m—R—(COO^-)_n$$

The actual numbers of positive and negative charges can be measured by titration of the protein. The number of ionized groups in simple polyvalent acids like phosphoric or tartaric acids is determined by the number of equivalents of hydrogen ions that can be derived from each mole of acid. The same procedure may be employed for proteins. In titrating from the isoelectric point, the total base consumption is determined by the number of acidic groups, and the total acid consumption, by the number of basic groups. Table 6.1 gives these values for several proteins, together with molecular weights of these proteins. Proteins possess a large number of ionic groups, *i.e.*, these molecules are highly polyvalent, and there are large differences among proteins.

Table 6.1: NUMBER OF TITRATABLE GROUPS OF SOME PROTEINS

Protein	Titratable basic groups	Titratable acidic groups	Molecular weight
Ovalbumin	91	82	40,000
Insulin	101	130	5,700
Zein	18–21	30	40,000
Casein	76–90	....	....
Edestin	127–134	....	310,000
Horse hemoglobin	148	113	68,000
Human serum albumin	144	135	69,000
Bovine serum albumin	147	134	69,000

Note: Data are given as equivalents of acid or base reacting with 100,000 g. of protein in order to compare the different proteins.

The ionic groups of a protein are contributed by the side chains of the polyvalent amino acids. For example, a lysine residue in the interior of a peptide chain is combined in peptide linkage through its α-amino group and its α-carboxyl group, but the ϵ-amino group is free, as indicated by the fact that the ϵ-amino group of lysine in proteins may be deaminated with nitrous acid, and when proteins are dinitrophenylated, ϵ-dinitrophenyl-lysine can be isolated after hydrolysis. The guanido group of arginine and the imidazole group of histidine are also free in proteins. Hence, the total acid-binding capacity should be equal to the sum of the number of arginine, histidine, and lysine residues in the protein.

Similarly, the anionic groups of proteins are the γ- and β-carboxylic groups contributed by glutamic and aspartic acids. However, some of these residues are present as glutamine and asparagine, the amides of these acids. The amide content of proteins is estimated by determining the amount of ammonia liberated by mild acid hydrolysis. The number of free carboxyl groups must then be equivalent to the sum of the aspartic and glutamic acids less the number of carboxyl groups combined with ammonia in amide linkage.

In general, the agreement between the computed and observed acid and base consumption is satisfactory. For some proteins, the observed acid and base consumption is significantly greater than the values calculated from the molecular weights, assuming a single free amino group and a single free carboxyl group at the two terminal ends of the polypeptide chains. To account for these additional ionic groups, it is necessary to assume the existence in these proteins of more than one peptide chain.

Titration Curves. As would be expected from the large number and diversity of ionic groups, the titration curves of proteins are complex. Table 6.2 gives the characteristic p*K* values for the major ionic groups that can be titrated in proteins. These are the values that are found in proteins, and comparison with the data in Table 5.5 shows that the characteristic p*K* values are shifted somewhat from the constants found for the free amino acids. Sulfhydryl and phenolic hydroxyl groups are included in Table 6.2. However, since free sulfhydryl groups are usually present in only small amounts, their ionic contribution is usually neglected. The phenolic

Table 6.2: CHARACTERISTIC pK VALUES FOR ACIDIC AND BASIC GROUPS IN PROTEINS

Group	pK_a *at* 25°C
α-Carboxyl (terminal)	3.0–3.2
β-Carboxyl (aspartic)	3.0–4.7
γ-Carboxyl (glutamic)	Approx. 4.4
Imidazolium (histidine)	5.6–7.0
α-Amino (terminal)	7.6–8.4
ε-Amino (lysine)	9.4–10.6
Guanidinium (arginine)	11.6–12.6
Phenolic hydroxyl (tyrosine)	9.8–10.4
Sulfhydryl (cysteine)	Approx. 8–9

SOURCE: From E. J. Cohn and J. T. Edsall, "Proteins, Amino Acids, and Peptides as Ions and Dipolar Ions," Reinhold Publishing Corporation, New York, 1942.

groups of tyrosine contribute significantly in proteins of high tyrosine content, such as insulin. The presence in most proteins of nearly all the ionic groups listed in the table endows proteins with effective buffering capacity over the greater portion of the pH range. However, the only group which has significant buffering capacity at pH values near neutrality is the imidazolium group of histidine, since the number of terminal α-amino groups is usually small. Thus, histidine groups are of great physiological importance as buffer groups of proteins in tissues. The titration curve of horse hemoglobin, shown in Fig. 6.1, illustrates the complex titration curve characteristic of a protein.

Ion Binding of Proteins. As ampholytes, proteins can form salts of both types; *i.e.*, protein anions can bind with cations, and protein cations with anions. Indeed, a mixture of different proteins at a given pH will include both anions and cations if their isoelectric points fall on opposite sides of the pH value, and salts of protein-protein combinations will be formed. This will occur in tissues since both basic and acidic proteins are present. Specific combinations of small ions with proteins also play an important role in tissues and body fluids.

Many ions form insoluble salts with proteins and serve as excellent precipitating agents for proteins. Acids such as phosphotungstic, trichloroacetic, picric, sulfosalicylic, perchloric, etc., are used for deproteinizing solutions, since the anions of these acids will form insoluble salts with proteins when the latter are in the form of cations (acid side of their isoelectric points). A standard method for determining the amount of protein nitrogen in a solution is to estimate the total nitrogen by the

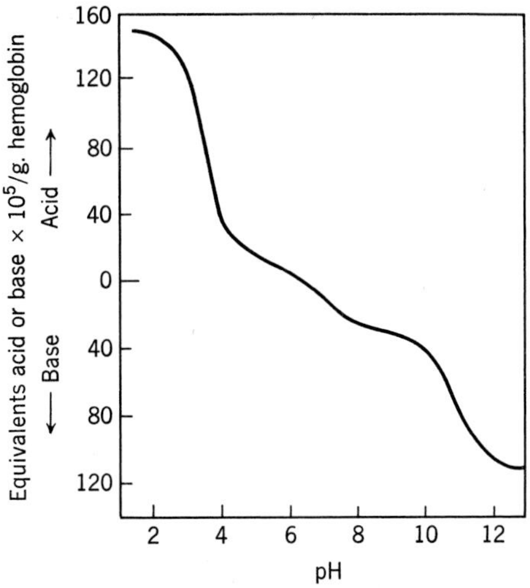

Fig. 6.1. The titration curve of crystalline horse hemoglobin. (*After E. J. Cohn, A. A. Green, and M. H. Blanchard, J. Am. Chem. Soc.*, **59**, 509, 1937.)

Kjeldahl method on an aliquot of the solution and to measure the nitrogen on an aliquot of the filtrate obtained after precipitating the proteins with one of the above acids, such as trichloroacetic. The difference between total nitrogen and nonprotein nitrogen gives the protein nitrogen. An average figure for the nitrogen content of proteins is 16 per cent. Hence multiplying the value for protein nitrogen by the factor 6.25 gives the amount of protein. This approximation is useful in estimating the protein content of heterogeneous material, such as foodstuffs and tissues.

Heavy metal ions are used for precipitating proteins on the alkaline side of their isoelectric points, the proteins behaving as anions. Ions of mercury, copper, silver, zinc, barium, etc., are frequently employed for this purpose. Many acid dyes find practical use for coloring the insoluble proteins, wool and silk.

Many of the ion combinations with proteins involve interactions other than simple salt formation. Ions such as Cu^{++}, Ni^{++}, etc., can form coordination complexes with imidazoles, substituted amines, and ammonia as in the familiar deeply blue cupric ammonia complex, $Cu(NH_3)_4^{++}$. Such complexes are also formed with peptides and proteins since free amino, imidazole, and other groups are present.

Electrophoresis. It has already been mentioned that proteins migrate in an electric field except at the pH of the isoelectric point. This was discovered by W. B. Hardy in 1899 and has been widely employed in the study of proteins because of its usefulness in ascertaining the properties of protein mixtures. The titration curve of a protein may be regarded as a measure of the charged state of the molecule at any pH value. The rate of migration of a molecule in an electric field is largely dependent on the net charge. Hence, the rate of electrophoretic migration will depend on pH in the same manner as does the degree of ionization. Figure 6.2 illustrates the electrophoretic migration and the titration curve of crystalline egg albumin. Although the shape and size of the molecule influence the absolute rate of migration, the major factor is the net charge.

Since proteins differ markedly in their isoelectric points and in their titration

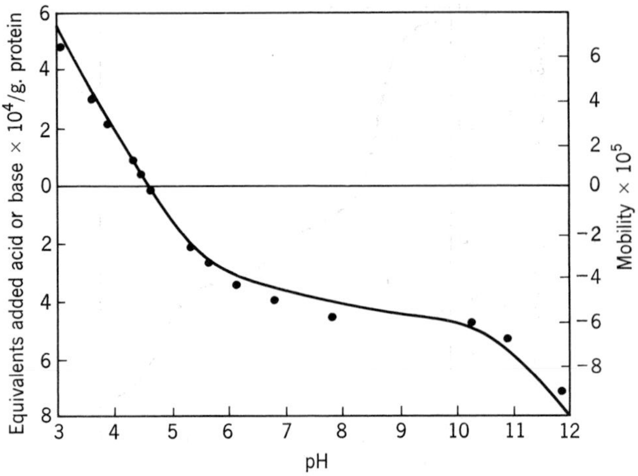

Fig. 6.2. The electrophoretic mobility in square centimeters per volt per second (shown as dots) and the titration curve of crystalline egg albumin. (*Mobility data from L. C. Longsworth, Ann. N.Y. Acad. Sci.*, **41**, 275, 1941; *titration data from R. K. Cannan, A. Kibrick, and A. H. Palmer, Ann. N.Y. Acad. Sci.*, **41**, 243, 1941.)

curves, they differ in electrophoretic mobility at any given pH value. Electrophoretic analysis is used to determine the purity of individual proteins and for quantitative analysis of complex mixtures.

In principle the technique of electrophoresis is simple. A protein solution containing a buffer at a definite pH is placed at the bottom of a U tube. Immediately above the protein solution, a layer of buffer solution is introduced without disturbing the boundary created between protein solution and buffer. Electrodes are inserted in the buffer solution. The whole U tube is immersed in a bath maintained at a constant temperature near 0°C. in order to minimize convection currents that may arise from the heat generated, and also to prevent heat coagulation of temperature-sensitive proteins. Modern electrophoretic technique, largely due to Tiselius, employs U tubes with optically flat surfaces and constructed so that the solutions may be introduced and equilibrated to temperature without boundary disturbance. The rate of migration of the protein in the electric field is measured by observing the movement of the boundary as a function of time. Migration of colored proteins, such as hemoglobin, is readily observed. Since most proteins are colorless, optical systems have been devised that permit visualization and measurement of variations in refractive index in the electrophoresis cell. These differences in refractive index of the various portions of the protein solution indicate alterations in concentration, *i.e.*, refractive index gradients provide a measure of the distribution and quantity of the solute, in this case, protein.

Figure 6.3 shows the electrophoretic patterns of two highly purified proteins (*B* and *C*). The pattern obtained with a sample of normal human plasma (*A*) is also given in order to illustrate the result obtained with a complex mixture of dissolved proteins. With pure substances, only a single peak will be obtained which will be symmetrical and have the form of a probability curve. Multiple peaks and asymmetrical curves indicate mixtures. With mixtures it is possible to assess the

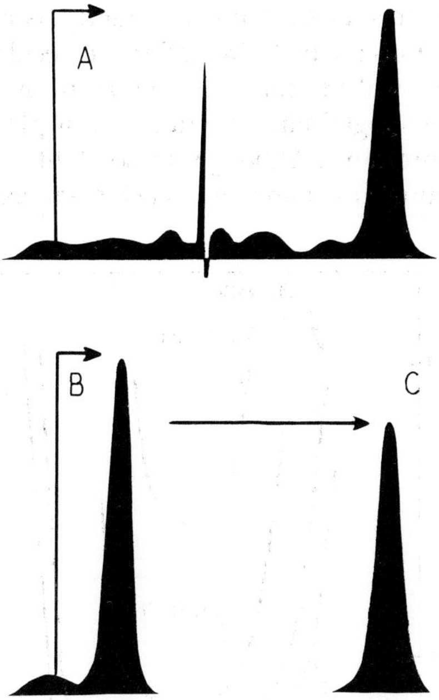

Fig. 6.3. The electrophoretic patterns of *A*, the complex mixture of proteins found in human plasma; *B*, the single peak found with crystalline carboxypeptidase; and *C*, the homogeneous peak found with crystalline bovine serum albumin. The vertical line indicates the starting point of the boundary. The peak at the initial position is due to the presence of the buffer salt.

proportion of the different components by measuring the fraction of the total area attributable to each component. Applications of electrophoretic analysis will be given later (Chaps. 31 and 35).

In addition to the applicability to charged compounds in solution, electrophoresis may also be performed by using a porous inert medium such as starch, silica gel, or moistened filter paper. The mixture of substances in solution is usually applied as a discrete spot or zone, and complete separation of components can be obtained. This process is known as zone electrophoresis and is useful both for analysis and for isolation of materials.

The method of immunoelectrophoresis is described in Chap. 31.

Solubility. Each homogeneous protein has a definite and characteristic solubility in a solution of fixed salt concentration and pH. Therefore, under definite conditions the amount of protein that may be dissolved to form a saturated solution at equilibrium is independent of the amount of excess solid phase suspended in the fluid. This provides a sensitive test for the purity of a protein. Thus, the hemoglobins prepared from several species are pure by this test.

The solubility of proteins is markedly influenced by pH, as might be expected from their amphoteric behavior; solubility is at a minimum at the isoelectric point, and increases with increasing acidity or alkalinity. The explanation is as follows. In the isoelectric condition, electrostatic repulsive forces between solute molecules are at a minimum and crystal-lattice forces in the solid state will be at a maximum.

When the ampholyte molecules exist predominantly as either anions or cations, repulsive forces between ions are high since all the molecules possess excess charges of the same sign and they will be more soluble than in the isoelectric state.

The solubility of β-lactoglobulin as a function of pH is shown in Fig. 6.4 for four different salt concentrations. Although the solubility is markedly enhanced by increased salt concentration, it is minimal in each case near the isoelectric point at pH 5.2 to 5.4.

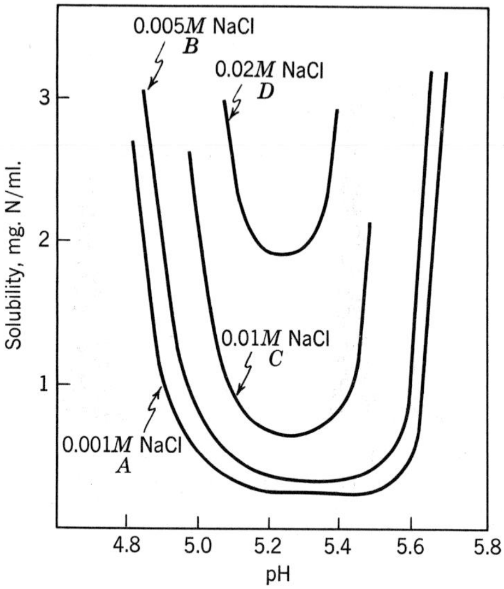

Fig. 6.4. The solubility of β-lactoglobulin as a function of pH at four different values of ionic strength (μ). Curve A, $\mu = 0.001$; curve B, $\mu = 0.005$; curve C, $\mu = 0.01$; curve D, $\mu = 0.02$.

It is apparent from previous comments (pages 116*ff.*) that various kinds of protein differ enormously in their solubility. Certain factors greatly influence protein solubility. These can be utilized for group separations, and, frequently, for purification of individual proteins. The major influences on solubility are considered below.

1. Globulins are sparingly soluble in water, and their solubility is markedly increased by neutral salts. Globulins can often be precipitated by dilution of a saline tissue extract with distilled water or by removal of salts by *dialysis* against water of a protein solution contained in a bag made from a *semipermeable membrane,* such as collodion or cellophane. Semipermeable membranes permit free passage of water and small ions or molecules but not of the larger proteins.

The effect of neutral salts in increasing the solubility of globulins is called the "salting-in" effect. This is shown in Fig. 6.4 for β-lactoglobulin. The logarithm of the solubility is frequently a linear function of a term called the ionic strength, which is readily calculated from the molar concentrations of the ions and their charge, using the expression

$$\mu = \tfrac{1}{2}\Sigma mZ^2$$

where μ is the ionic strength, m the molarity, and Z the charge of the ion. The summation sign Σ denotes that the mZ^2 terms are added for each of the ions. For example, for a $0.1M$ solution of NaCl, the ionic strength $\mu = ½(0.1 \times 1^2 + 0.1 \times 1^2) = 0.1$, or, for a salt of univalent ions, the ionic strength is equal to the molarity. For a $0.1M$ solution of Na_2SO_4, $\mu = ½(0.2 \times 1^2 + 0.1 \times 2^2) = 0.3$.

The solubility of albumins in water is scarcely affected by addition of low concentrations of neutral salts. However, in mixtures of a nonpolar solvent and water, solubility of albumins is increased by neutral salts.

The explanation of the "salting-in" phenomenon is as follows. Solubility of any substance depends on the relative affinity of solute molecules for each other (crystal-lattice formation) and for the solvent molecules. Any factor that decreases interaction of solute molecules will tend to increase solubility. The small ions of neutral salts will interact with the ionic groups of the protein molecules, diminishing protein molecule interactions and, therefore, increasing solubility.

2. Proteins are precipitated from aqueous solution by high concentrations of neutral salts. This is the "salting-out" phenomenon. Di- and trivalent ions are more effective than univalent ions (Fig. 6.5). Commonly used salts are ammonium sulfate, sodium sulfate, magnesium salts, and phosphates. The most effective region of salting out is at the isoelectric point of the protein. The solubility of carboxyhemoglobin (carbon monoxide hemoglobin) as a function of ionic strength is shown in Fig. 6.5. Salting out occurs not only with proteins but with gases, uncharged molecules, and electrolytes.

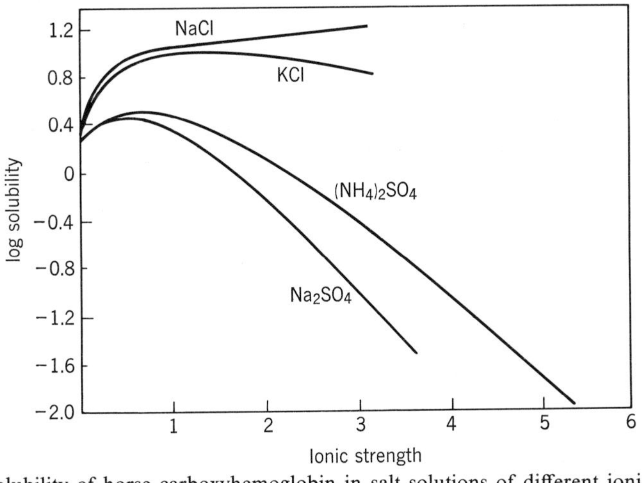

Fig. 6.5. The solubility of horse carboxyhemoglobin in salt solutions of different ionic strength. Both "salting in" and "salting out" are observed with this protein. (*From E. J. Cohn and J. T. Edsall, "Proteins, Amino Acids, and Peptides as Ions and Dipolar Ions," Reinhold Publishing Corporation, New York,* 1942.)

The mechanism of salting out is complex. Hofmeister suggested that it was due to "dehydration" of the protein by the added salt. According to Debye, the salt ions attract around themselves the polarizable water molecules, making less water available for the proteins since, at high salt concentrations, the number of charged groups contributed by the salts is enormous compared with those of the proteins. Since solubility of proteins in water depends on clustering of water molecules around the hydrophilic ionic groups, removal of water molecules to other ions will decrease protein solubility. Qualitatively, the effect of salts on the solubility of proteins and of amino acids is very similar.

3. Proteins are precipitated from aqueous solutions by nonpolar solvents that are miscible with water. Commonly used solvents are methanol, ethanol, and acetone. Protein solubilities in these solvents are markedly affected by neutral salts. Most satisfactory results are obtained by working at low temperatures at which proteins are most stable.

4. The effect of pH on protein solubility has already been noted, and this effect, *i.e.*, minimal solubility at or near the isoelectric point, applies regardless of the precipitating agent used: neutral salts, nonpolar solvents, etc. Some proteins, such as casein of milk, are readily precipitated by adjusting the solution to the pH of the isoelectric point; hence the procedure is described as "isoelectric precipitation." The isoelectric points of some proteins are listed in Table 6.3; they vary from below pH 1.0 for pepsin to pH 10.6 for cytochrome c.

Table 6.3: MOLECULAR WEIGHTS AND ISOELECTRIC POINTS OF SOME PROTEINS

Protein	$M_{S,D}$	$M_{osmotic}$	pI
Cytochrome c	13,000	.....	10.6
Ribonuclease	14,000	15,000	7.8
Myoglobin, horse	17,000	.....	7.0
Growth hormone (somatotropin), human	21,500	.....	6.9
Carboxypeptidase	34,000	.....	6.0
Pepsin	35,500	36,000	Less than 1.0
Ovalbumin, hen	40,000	40,000–46,000	4.6
Hemoglobin, horse	65,000	67,000	6.9
Serum albumin, human	66,500	69,000	4.8
Diphtheria toxin	74,000		
Serum γ-globulins	160,000	177,000	6.4–7.2
Catalase	250,000	.....	5.6
Fibrinogen	330,000	.....	5.5
Urease	480,000	.....	5.1
Thyroglobulin	660,000	.....	4.6
Hemocyanin, octopus	2,800,000		

Note: $M_{S,D}$ = molecular weight calculated from sedimentation and diffusion measurements (page 129); $M_{osmotic}$, from osmotic pressure measurements (page 127).

Molecular Weight. Most of the usual methods for determining the size of small molecules, such as alterations in freezing point, boiling point, vapor pressure, etc., are not applicable to proteins because of their high molecular weight and instability.

From Composition. The minimal molecular weight (M_{min}) of any substance may be computed from the content of any element present in small amount by making the assumption that the molecule contains only one atom of such an element. This computation has been applied to proteins for elements other than the abundant C, H, N, and O. The usual formula may be applied.

$$M_{min} = \frac{\text{atomic weight} \times 100}{\text{per cent of constituent}}$$

Mammalian hemoglobin contains 0.34 per cent iron (atomic weight = 55.8), and the computed M_{min} is 16,700. Actually the true molecular weight (M) of hemoglobin, found by other methods, is about 65,000, which indicates that a molecule of hemoglobin contains four atoms of iron.

In the same manner, minimal molecular weights may be computed from the content of amino acids present in small amounts. Bovine serum albumin has only 0.58 per cent tryptophan, which yields $M_{min} = 35{,}200$. M determined by other methods is almost twice this value, *viz.*, 69,000. Analytical data may be helpful in determining minimal values but do not indicate true molecular weights. In fact, most amino acids are present in such large amounts that minimal values are useless even for purposes of a rough estimate.

By Osmotic Pressure. For reasonable estimates of molecular weight, physical methods are most useful, and one of the first to be applied was that of osmotic pressure determinations. The general formula of the gas law is applicable to osmotic pressure,

$$\pi V = \frac{g}{M} RT$$

where π is the osmotic pressure in atmospheres (1 atmosphere = 760 mm. Hg), V the volume of solution in liters, g the weight of solute in grams, M the molecular weight, R the gas constant, and T the absolute temperature (t°C. + 273.1). R is known (0.082 liter-atmosphere per mole per degree); the other quantities can be measured experimentally, and M then computed. Many such determinations have been made for homogeneous proteins. Table 6.3 gives M values obtained for some proteins from osmotic pressure measurements.

The measurements are made by putting a solution of the protein in a rigid bag of a semipermeable membrane, such as collodion. The bag is placed in a buffer solution, and after equilibrium has been achieved, the increased pressure in the bag is measured by the height of the capillary above the solution (Fig. 6.6). There are difficulties in making such measurements, and these may be mentioned briefly.

1. Equilibrium may be achieved only slowly, and readings must be made for days or weeks until no further change in pressure is recorded. Because of the time required, the measurements are generally made at a low, constant temperature to avoid bacterial growth and protein decomposition. However, osmometers have been devised which attain equilibrium in only a few hours; these are to be preferred.

2. The gas law equation given above is valid only at low concentrations of solute since osmotic pressure generally increases with protein concentration much more rapidly than is predicted by the simple equation. Measurements are usually made at several protein concentrations, and π is obtained by extrapolation to zero protein concentration.

3. The osmotic pressure due to the protein must be obtained at the isoelectric point since higher values of π are found for the protein salts. The bound ions increase the osmotic pressure because of the Donnan equilibrium effect (see below).

4. The total osmotic pressure due to dissolved protein is small, and there are difficulties in obtaining precise readings. The equation also shows that the larger the protein, the smaller is the pressure for a given quantity of protein. In practice, the

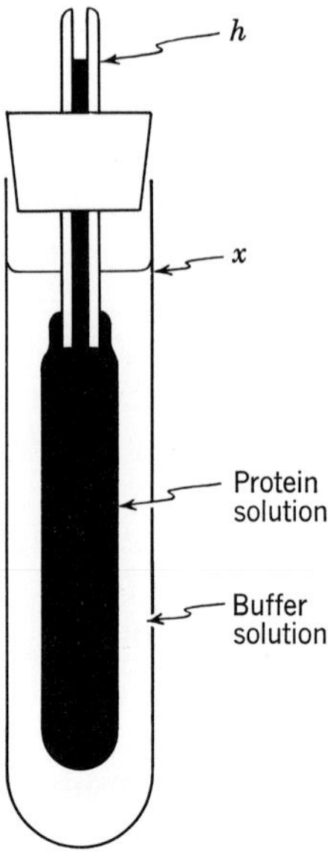

Fig. 6.6. A simple osmometer shown diagrammatically. The measured osmotic pressure is the difference in height between the level of the buffer solution (x) and the protein solution in the capillary (h).

method is not applicable to proteins with molecular weights greater than a few hundred thousand.

5. The observed osmotic pressure is recorded for all the proteins in the solution and therefore gives an average molecular weight. There is no way of determining from such measurements whether or not the system contains a mixture of molecules of different sizes.

Gibbs-Donnan Equilibrium. The measurement of osmotic pressure of a protein solution requires the use of a semipermeable membrane that prevents the migration of the large particles. For a system at equilibrium which contains protein ions bearing a net charge (such as P^-) and diffusible ions, we have the situation

$$Na_a^+, Cl_a^- \,|\, Na_b^+, Cl_b^-, P_b^-$$

where the subscripts a and b refer to the two sides of the membrane. In solution b, $[Na_b^+] = [Cl_b^-] + [P_b^-]$, and in solution a, $[Na_a^+] = [Cl_a^-]$ since electric neutrality must be maintained on each side of the membrane. The values for $[Na_a^+]$ and $[Na_b^+]$ must be different because P_b^- is present on only one side of the membrane. The concentrations of the diffusible ions bear the following relationship to one another, specifically for the example given, and generally for ions X^+ and Y^- in a system at equilibrium:

$$\frac{[Na_a^+]}{[Na_b^+]} = \frac{[Cl_b^-]}{[Cl_a^-]} = \frac{[X_a^+]}{[X_b^+]} = \frac{[Y_b^-]}{[Y_a^-]} \tag{1}$$

This relationship may be derived from the general equilibrium expression where the change in free energy ΔF is equal to zero and is given by the thermodynamic equation

$$\Delta F = 0 = RT \ln \frac{[\mathrm{Na}_a{}^+]}{[\mathrm{Na}_b{}^+]} + RT \ln \frac{[\mathrm{Cl}_a{}^-]}{[\mathrm{Cl}_b{}^-]}$$

which may be solved to give

$$[\mathrm{Na}_a{}^+][\mathrm{Cl}_a{}^-] = [\mathrm{Na}_b{}^+][\mathrm{Cl}_b{}^-]$$

which is the same as equation (1), above. Similarly,

$$[\mathrm{X}_a{}^+][\mathrm{Y}_a{}^-] = [\mathrm{X}_b{}^+][\mathrm{Y}_b{}^-]$$

From these relationships, it is apparent that, at the isoelectric point of the protein at which its net charge is zero, the distribution of electrolytes will not be influenced by the amount of protein in solution and the osmotic pressure will be due only to the protein. Since the Gibbs-Donnan equilibrium effect leads only to small differences in diffusible electrolyte concentrations, it can be made negligible at pH values removed from the isoelectric points of proteins by adding high concentrations of neutral salts.

By Sedimentation. The rate at which a particle in solution is driven down a centrifuge tube under the action of centrifugal force depends (1) on the force applied, (2) on the size, shape, and density of the particle, and (3) on the density and viscosity of the solvent. Large molecules can be sedimented at high centrifugal forces whereas small molecules cannot be sedimented at forces presently attainable in the laboratory. Modern ultracentrifuges attain speeds as high as 75,000 r.p.m. and forces in excess of 400,000 times gravity. Several types of instruments have been devised for obtaining strong centrifugal fields. The first of these, devised by T. Svedberg and his collaborators, is a turbine driven by jets of oil. Others employ air jets or high-speed electrical motors. In all these instruments, the rotors are operated in high vacuum to minimize air resistance and consequent heating due to friction. These ultracentrifuges are equipped with suitable optical systems for recording the positions of the boundaries of the proteins during the sedimentation runs. The optical system generally used is similar to that employed in electrophoresis studies (page 122).

When the solution contains particles which are all of the same size and shape, they will move down the centrifuge tube at the same rate, giving a sharp boundary between the solute in solution and the clear solvent. A number of boundaries will be observed when the solution contains a mixture of substances of different particle size. It is thus possible to determine homogeneity with respect to particle size and, at the same time, to estimate molecular weight.

Particles moving at a uniform speed in a circular path are accelerated toward the center of the path. The acceleration is $\omega^2 x$, where ω is the velocity of rotation in radians per sec. and x is the radius in cm. The product of the mass and the acceleration gives the force F, where

$$F = V(\rho_2 - \rho_1)\omega^2 x = V\rho_2\omega^2 x - V\rho_1\omega^2 x$$

V is the volume of the particle and ρ_2 and ρ_1 are the densities of the particles and the solvent, respectively.

For one mole of solute where M is the molecular weight of the solute, and where $\bar{V}$, the partial specific volume, equals $1/\rho_2$,

$$F = M\omega^2 x - M\bar{V}\rho_1\omega^2 x = M\omega^2 x(1 - \bar{V}\rho_1)$$

The force resisting sedimentation F is $f(dx/dt)$, where dx/dt is the rate of sedimentation and f is the frictional coefficient per mole.

$$f\frac{dx}{dt} = M\omega^2 x(1 - \bar{V}\rho_1)$$

Solving for M gives

$$M = \frac{f}{(1 - \bar{V}\rho_1)} \frac{1}{\omega^2 x} \frac{dx}{dt}$$

where $(dx/dt)/\omega^2 x$ is the rate of sedimentation per unit field of force and is termed s, the sedimentation coefficient. The equation becomes the following:

$$M = \frac{fs}{(1 - \bar{V}\rho_1)}$$

The diffusion coefficient (D) is equal to RT/f, where R is the gas constant, T the absolute temperature, and f the frictional coefficient, as before. Hence, the fundamental equation derived by Svedberg for the molecular weight is

$$M = \frac{RTs}{D(1 - \bar{V}\rho)}$$

The sedimentation constant s has the dimensions of time per unit of force and usually lies between 1×10^{-13} and 200×10^{-13} sec. The factor 1×10^{-13} is called the Svedberg unit (S). D is in cm.^{2} per sec.

The diffusion constant can be measured by observing the spread of an initially sharp boundary, between the protein solution and a solvent, as the protein diffuses into the solvent layer.

Rate of sedimentation depends on the difference in density of solute particles and the solvent. If the particles are of lower density than the solvent, they will rise to the top (as in a cream separator), and flotation will occur. For most proteins $\bar{V}$ has the value of 0.70 to 0.75. The exact value depends on the amino acid composition and can be calculated, when this is accurately known, or it can be determined experimentally.

Highly purified proteins show homogeneous boundaries in the ultracentrifuge; representative sedimentation curves obtained by a refractive-index method are shown in Fig. 6.7. The ultracentrifuge is a useful tool in studying the stability of proteins since under many conditions, *e.g.*, extremes of pH, temperature, etc., aggregation or dissociation may occur, and this can be detected by changes in sedimentation constants. Molecular weights of some proteins obtained by the sedimentation method are given in Table 6.3. Present methods estimate the molecular weight to within 5 per cent of the true value.

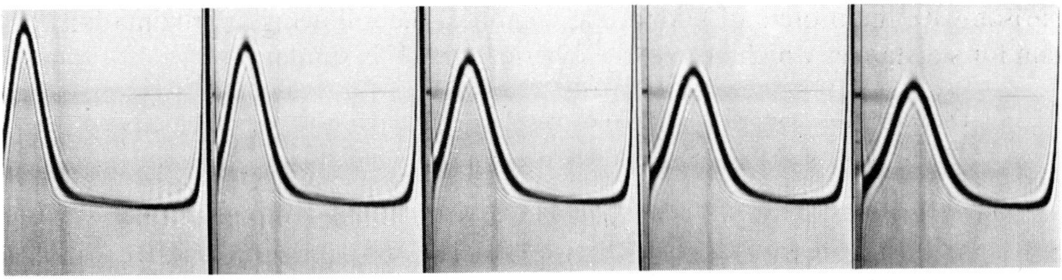

Fig. 6.7. Sedimentation of crystalline ribonuclease at 59,780 r.p.m. in the analytical ultracentrifuge. The direction of sedimentation is from left to right. The first picture was taken 1 hr. after attaining speed; subsequent exposures were made at 16-min. intervals. It will be noted that as the boundary moves down the cell, the height of the boundary decreases and its width increases. This is due to normal diffusion. The boundary at the bottom of the cell represents sedimented protein. When a preparation contains a mixture of proteins or other solutes of different sizes, multiple peaks are found.

One of the most remarkable facts of protein chemistry is that these large and complex molecules have definite and characteristic molecular weights, in contrast to polysaccharides, which are generally polydisperse.

By Sedimentation Equilibrium. Another technique for determining molecular weight in the ultracentrifuge utilizes a low centrifugal force. The ultracentrifuge is operated until an equilibrium distribution of the protein under study is achieved throughout the length of the cell. At equilibrium, there is no net migration of protein since its movement to the bottom of the cell due to the centrifugal force is exactly balanced by its movement to the top of the cell by diffusion. This obviates the necessity of independent measurements of diffusion constants. The equilibrium distribution is a characteristic for each protein and depends upon the molecular weight of the protein. The equation used in the calculation of molecular weight, M_E, by the method of sedimentation equilibrium is

$$M_E = \frac{2\,RT \ln (c_2/c_1)}{[\omega^2(1 - \bar{V}\rho)(x_2^2 - x_1^2)]}$$

where R, T, ω, $\bar{V}$, and ρ are as defined previously (page 129) and c_1 and c_2 the concentrations of a protein at distances x_1 and x_2 from the axis of rotation. This method has the advantage that M_E can be determined with relatively small quantities of protein, provided that the sample is substantially free of contaminating proteins.

By Sedimentation in Sucrose Gradients. It is possible to determine sedimentation constants and hence to estimate *approximate molecular weights* with impure solutions if a suitable technique is available for determining relative concentrations, *e.g.*, enzymic activities. The sample to be studied is layered on a linear sucrose gradient and centrifuged at high speed in a swinging-bucket rotor. Generally, a protein of known s value is added as a standard or "marker." Substances of different sedimentation properties will separate from one another as bands in the gradient. At the end of the run a small hole is punched in the bottom of the centrifuge tube and emerging fractions are collected and analyzed. If samples are collected at different times of centrifugation, a plot of distance of peak activity from the meniscus versus time should be linear. For a specific time of centrifugation in comparison with the protein of known s as standard, the following approximations obtain for substances which are nearly spherical and $\bar{V}$ is similar.

$$\frac{\text{Distance from meniscus of unknown}}{\text{Distance from meniscus of standard}} = \frac{s_{20,w} \text{ of unknown}}{s_{20,w} \text{ of standard}} = \left(\frac{M_{\text{unknown}}}{M_{\text{standard}}}\right)^{2/3}$$

By Filtration on Molecular Sieves. Cross-linked dextran gels (page 39) or polyacrylamide gels which possess pores of small but finite size permit the penetration of molecules up to a certain size limit but exclude larger molecules. The gels are used in columns similar to those employed for chromatography. This technique, termed "gel filtration," permits evaluation of purity by molecular size and allows separation of substances of different molecular weights. When a column has been calibrated with substances of known molecular weights, the position of emergence can be used to approximate the molecular weight. For a homogeneous substance a single symmetrical peak is obtained. When a mixture is present, the region of emergence of a specific protein can be determined if the protein has a readily estimated activity, *e.g.*, enzymic, hormonal, etc.

Figure 6.8 shows a plot of elution volume V_e against molecular weight on a log scale for a number of proteins. A single smooth curve fits all the data except for some proteins with high content of carbohydrate, *e.g.*, ovomucoid, fetuin, γ-globulins, fibrinogen.

Cross-linked dextrans and polyacrylamides of differing and relatively uniform pore size are available and permit study by this method of molecules of molecular weight from a few hundred to several hundred thousand.

Shape of Protein Molecules. From the sedimentation constant s and the diffusion constant D a frictional ratio f/f_o can be calculated from the formula

$$\frac{f}{f_o} = 10^{-8}\left(\frac{1 - \bar{V}\rho}{D^2 s \bar{V}}\right)^{1/3}$$

where the various terms have the same meaning as above (page 130). The molar frictional coefficient for a compact spherical and unhydrated particle is f_o. When f/f_o is equal to 1.0, the molecule is essentially an unhydrated sphere. When f/f_o is greater than 1.0, the molecules may be asymmetrical or hydrated or both. In some instances, hydration may be estimated, and many globular proteins appear to be spherical or nearly spherical molecules. These proteins include ribonuclease, insulin, chymotrypsinogen, pepsin, and others. However, other proteins are highly asymmetrical and exist in solution as pronounced ellipsoids or rods. Fibrinogen, the plasma protein that is the precursor of fibrin of blood clots, and myosin, the main protein of muscle fibers, are examples of long ellipsoids. It is estimated that human fibrinogen has a cross-sectional diameter of 38 Å. and a length of 700 Å. Solutions of such molecules show pronounced light scattering, and the shape of molecules, as well as molecular weight, may be estimated from light-scattering measurements.

The *viscosity* of a solution depends on the molecular weight and shape of the solute molecules at a given solute concentration. Highly asymmetrical molecules show a high intrinsic viscosity as compared with spherical molecules of the same

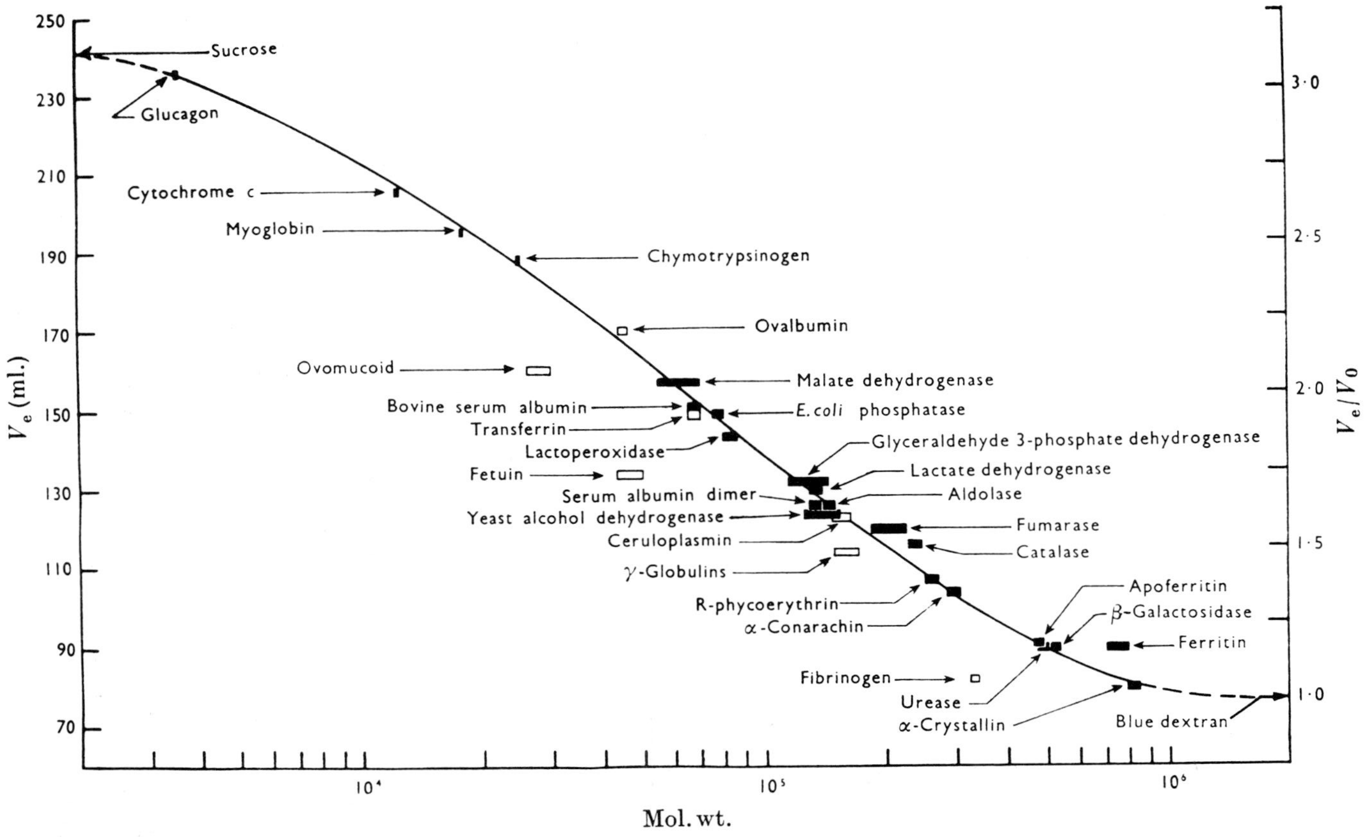

Fig. 6.8. Relationship of elution volume V_e to molecular weight for proteins on a column of cross-linked dextran (Sephadex G-200) (2.5 × 50 cm.) at pH 7.5. Open bars represent glycoproteins. V_0 represents the void volume. (*After P. Andrews, Biochem. J.*, **96**, 595, 1965.)

molecular weight. When the latter is known, the shape can be estimated from the variation of viscosity with solute concentration.

X-ray analysis has shown that insoluble fibrous proteins such as wool and silk exist as long fibers with extremely small cross sections. Thus proteins can range in shape from simple crystalloidal spherical structure to shapes which resemble those of synthetic, linear high polymers.

Criteria of Purity. The ordinary criteria of purity of small molecules are not readily applicable to proteins. Such standards as sharpness of melting point cannot be used since dry proteins decompose on heating. Crystallization of proteins leads to purification in many cases but by itself is not a sufficient criterion of purity, since proteins readily form mixed crystals and repeated crystallization by the same method retains the mixture. The retention of constant composition and physical constants on recrystallization is useful, but if mixed crystals are formed, these criteria are useless. For these reasons special techniques have been devised for assessing the purity of proteins and other large molecules. Many of these techniques are applicable as well to small molecules, and thus have assumed great importance in the isolation and purification of natural products. Some of these procedures have already been mentioned, but it is useful to summarize them here. There are really no tests for purity, only methods for the detection of impurities or inhomogeneity. It is assumed that a substance is pure only after all possible tests have failed to reveal inhomogeneity. The history of protein chemistry is replete with instances in which homogeneity has been thought to be established and the assumption has later been shown to be incorrect when new criteria have been applied.

Constant Solubility. Constant solubility, which is independent of the amount of solid phase present, is an important criterion of purity. Since this depends on the applicability of the phase rule, it may be employed for all substances. In practice, solubility tests should be made at a variety of pH and ionic strength conditions.

Homogeneity in Size as Determined in the Ultracentrifuge. The lower limit in size for this method, with presently available ultracentrifuges, is probably about 4,000 molecular weight (Fig. 6.7).

Homogeneity in Electrical Charge. As determined by electrophoretic mobility in different buffers at several pH values (Fig. 6.3), this criterion can be applied to small electrolytes as well as to proteins.

Homogeneity by Chromatography. Many proteins have been chromatographed on columns of ion exchange resins or other absorbents (page 113). Such methods have not only provided a critical evaluation of homogeneity but have permitted resolution of complex mixtures of proteins. In general basic proteins can be successfully chromatographed on carboxylic acid resins, *e.g.*, carboxymethyl cellulose or carboxylated polystyrene resins. Acidic proteins have been chromatographed on basic ion exchangers, such as diethylaminoethyl cellulose (DEAE cellulose).

Partition chromatography and countercurrent distribution have been successfully employed with relatively few proteins because of the limited stability and solubility of most proteins in nonpolar solvents. Nevertheless, such methods have been important for the study of insulins of various species, for the separation of polypeptides obtained from natural sources, and for the resolution of peptides obtained as degradation products of large proteins.

Chromatography on columns of molecular sieves is routinely used for assessing purity of proteins and polypeptides (page 132).

Analytical Criteria. All chemical substances may be tested for purity by analytical criteria, *i.e.*, the atomic ratios should be constant after recrystallization from different solvents, and the elementary composition calculated from analytical data must yield whole numbers of atoms. For proteins, the amounts of the common elements are very similar and the number of atoms is so large that these criteria are seldom useful. For the elements present in lesser amounts, such as sulfur, or those found in only a few proteins, such as metals, phosphorus, iodine, etc., the analytical data may be useful. More fruitful is the consideration of amino acid content. Residues of amino acids must be present as an integral number of units per molecule of a pure protein. When only a few residues of a particular amino acid are found in a protein, valuable information as to homogeneity may be obtained.

Specific analysis may be used to detect other macromolecular substances that may be present. Tests for sugars can be used to estimate the content of various polysaccharides; however, sugars may be covalently bound to protein in glyco- and mucoproteins (page 118). Measurement of the absorption ratio at 280 mμ and 260 mμ is used to detect the presence of nucleic acids, which have a high absorption maximum at 260 mμ (page 195) in contrast to proteins, which possess a weak maximum near 280 mμ because of the presence of aromatic amino acid residues (page 89). The ratio of absorption of 280 to 260 mμ is near 1.75 for proteins which contain only amino acid residues. This value may be as low as 0.8 for muco- and glycoproteins, and near 0.5 for nucleic acids.

Specific Functional Properties. Many proteins have specific functional properties, *i.e.*, they act as hormones, enzymes, oxygen carriers, etc. Tests for functional purity are extremely sensitive since minute quantities of these active substances can be detected by their physiological effects, catalytic behavior, etc. If a preparation has a number of catalytic or biological effects, and a single protein is responsible for these effects, then repeated fractionation or partial inactivation with respect to one type of activity should not alter the *relative* ratios to other effects. For example, if an enzyme preparation attacks *A* ten times more rapidly than it does *B* and reprecipitation alters the ratio from 10:1 to 100:1, it is evident that different enzymes must be responsible for the two catalytic effects and the preparation must be inhomogeneous.

The criterion of functional homogeneity is one of the most valuable in the study of the proteins because of the sensitivity of the tests. Frequently, an impurity present in as low a concentration as one part in many thousands or millions can be readily detected by a test of functional properties; this is a much greater sensitivity than can be obtained by most physical or chemical analytical methods.

Antigenic Properties. Most proteins and some polysaccharides and lipids are antigenic, *i.e.*, they stimulate the formation of specific antibodies when injected into a suitable test animal, *e.g.*, rabbit. Injection of a homogeneous substance will result in the formation of a single type of specific antibody, whereas a mixture of substances that are antigenic will produce several types of antibody. This biological method for ascertaining the homogeneity of protein preparations may be as sensitive as tests for enzymic or hormonal homogeneity, provided the impurities are antigenic.

The amount of a specific protein in solution can be estimated quantitatively by adding the specific antiserum (Chap. 31), separating the resulting precipitate, and determining the amount of precipitate by a variety of methods.

PURIFICATION OF PROTEINS

In order to follow the isolation of a single protein, some quantitative method of protein estimation is essential. Chemical methods include the determination of protein nitrogen (page 120), use of the biuret reagent (page 111), the ninhydrin reagent (page 106), the phenol reagent (page 106), and the Lowry method, which combines the use of the phenol and biuret methods. Physical methods include measurement of refractive index, estimation of dry weight of salt-free solutions, and measurement of light absorption at 280 mμ (page 89). Estimation of specific activity is used to follow the purification of enzymes (Part Two) or protein hormones (Part Five).

Procedures for Purification. A first and major requirement is that the protein be obtained in solution since many proteins are present in cellular structures which are particulate in character and proteins may be associated with materials, such as lipids, which render them insoluble in water. Tissues must be ground to rupture the cells and to permit access of the solvent to the entire mass of tissue. Proteins may frequently be dissolved by extraction with water or with salt solutions of low ionic strength. They may be solubilized by use of weak detergent solutions, by treatment with nonpolar solvents, such as acetone and ether, to remove lipids, or by treatment with butanol, to disrupt cellular structures. These procedures also frequently denature and coagulate unwanted proteins. Similarly, brief heat treatment at temperatures from 40 to 60°C. is used to denature labile proteins, provided the desired protein is stable to such treatment.

Many of the procedures listed below have been discussed and can be mentioned very briefly. The order in which they are listed is arbitrary, and in specific instances, some methods are more useful than others; the choices depend on the biological source and concentration of the protein.

1. *By differential solubility.* Factors that influence solubility—pH, ionic strength, and nonpolar solvent concentration (alcohol, acetone, etc.)—have already been discussed (pages 123*ff.*).

2. *By specific precipitation.* Nucleic acids may be removed by specific precipitation with basic substances (*e.g.*, with the small protein protamine, which is readily removed by dialysis) or by precipitation with a divalent cation, such as Mn^{++}. Many proteins can be specifically precipitated with other cations, Ag^{+}, Hg^{++}, Ba^{++}, Zn^{++}, etc. (page 121).

3. *By column chromatography.* This procedure (page 134) is applicable with anionic or cationic adsorbents, with inert media using partition methods with

different solvents, with molecular sieves, with neutral adsorbents such as alumina, hydroxyapatite, etc.

4. *By preparative electrophoresis.* In order to perform electrophoretic separation on a reasonable scale, methods have been devised that permit use of inert supporting media such as starch or powdered cellulose. The high voltage is applied through a conducting buffer solution across the length of a vertical column or across a horizontal trough. An alternate method involves continuous flow of a solution through a high voltage field on a nonconducting, almost horizontal surface or on a vertical sheet of filter paper.

5. *By preparative ultracentrifugation.* With preparative ultracentrifuges, the proteins of higher molecular weight can be sedimented from solution, leaving smaller molecules behind in solution. The centrifugal force and the time of sedimentation will determine the partition that is effected.

6. *By enzymic digestion.* Some native proteins are more resistant to the action of proteinases than others. The digestion products, peptides and amino acids, are readily separated from resistant proteins. Digestion of other types of macromolecules may be accomplished with specific enzymes: nucleic acids by digestion with nucleases (page 189), glycogen and starch with amylases (page 47), hyaluronic acid with hyaluronidases (Chap. 38), etc.

Purification of proteins or of other natural products is an empirical procedure, and the various methods are largely "cut and try." In general, procedures 1 and 2 are most suitable for the first attempts since they are adaptable for handling large amounts of material. The methods of chromatography, electrophoresis, and ultracentrifugation are usually applied after considerable purification has been achieved.

At each stage of purification, the preparation must be monitored for specific activity or by physical examination, *e.g.*, electrophoresis or ultracentrifugation. These last procedures may suggest the type of method that may effect further purification. The task is completed, *i.e.*, a pure protein is in hand, when all the criteria of purity, discussed above, have been satisfied to the necessary extent.

REFERENCES

See list following Chap. 7.

7. The Proteins. III

Amino Acid Sequences and Conformation

The complex and specific chemical and biological properties of individual proteins should find explanation in terms of an exact description of their structure in the native state. In this chapter we shall consider primarily the methods being used to attack this problem, as well as some of the findings that pertain to proteins in general. First, the protein must satisfy the criteria of purity described in the preceding chapter before structural studies can be undertaken. Second, the amino acid sequences obtained pertain only to the specific protein from a single species; homologous proteins may differ in structure if obtained from another tissue or another species. This problem is presented in Chap. 30. Specific properties of certain enzymes, protein hormones, antibodies, and other biologically active proteins are also described in later chapters.

For convenience, the structure of proteins may be considered under separate categories. The main mode of linkage is the peptide bond which couples the α-carboxyl group of one amino acid residue to the α-amino group of another residue. Individual proteins may consist of one or more peptide chains. The complete amino acid sequences of several proteins have been established. This is sometimes called the *primary structure.*

If peptide bonds were the only structural linkage in proteins, these molecules would behave as randomly coiled long-chain polypeptides. However, the properties of native, globular proteins indicate an ordered structure in which the peptide chains are folded in a regular manner. Much of the folding is the result of linking carbonyl and amide groups of the peptide chain backbone by hydrogen bonds.

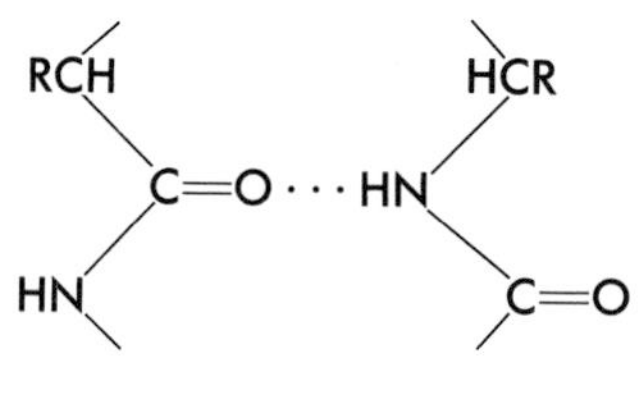

Such folding, produced or maintained by hydrogen bonding, is frequently called the *secondary structure* of the protein. In many proteins the hydrogen bonding produces a regular coiled arrangement, called the α-helix (page 153).

If a globular protein consisted solely of a single helix, these molecules would be elongated structures with a large axial ratio (length to cross section); however, physical measurements reveal that many proteins are spherical or nearly so (page

132). Hence the helical arrangement must be periodically interrupted, permitting numerous bonds with additional folding. This spatial conformation in three dimensions must be maintained by covalent or other bonds. The covalent disulfide bond involving two half-cystine residues is one way in which this is achieved. However, this is not the only type of bonding. Indeed, most of the large globular proteins are completely lacking in disulfide bonds, yet they are stable molecules in solution. These additional types of bonds include hydrogen bonds, salt bonds, and hydrophobic or nonpolar bonds. This has been called the *tertiary structure.*

The term *quaternary structure* has been used for proteins which contain more than a single peptide chain to indicate the spatial relationships among the separate chains or *subunits* (see below).

PRIMARY STRUCTURE OF PROTEINS

AMINO ACID COMPOSITION

As a preliminary to more detailed analysis of the amino acid sequence of a protein, it is essential to establish its amino acid composition. Various methods of analysis have been described (pages 112*ff.*); however, for the analysis of protein hydrolysates, the main quantitative method presently used involves ion exchange chromatography (page 113). For analysis of most of the amino acids the protein is usually hydrolyzed with 6*N* HCl at 110°C. for 20 hr. or longer in an anaerobic atmosphere. The time is usually varied, both to estimate the rate of destruction of labile amino acids, *e.g.*, serine, threonine, and tyrosine, and to ensure complete hydrolysis of more stable peptide bonds, particularly those involving isoleucine and valine residues. Acidic hydrolysis destroys tryptophan, which can be estimated by colorimetric or other methods. Glutamine and asparagine are converted to the respective dicarboxylic acids; however, the amides can be estimated after complete hydrolysis with proteolytic enzymes.

The data in Table 7.1 indicate that representative proteins differ greatly in composition. Insulin, a hormone, lacks tryptophan and methionine. The fibrous protein, collagen, contains hydroxylysine and hydroxyproline. Myoglobin does not contain cysteine or cystine. The data in Table 7.1 are given as grams of amino acid per 100 g. of original protein and add up to approximately 118 g. of amino acids per 100 g. of protein, because of the addition of one molecule of water for each peptide bond hydrolyzed. If the recoveries of the amino acids are given as *residue weights* in grams rather than weights of free amino acids, the total recovery should be 100 per cent for a protein lacking non-amino acid constituents. For example, the hydrolysis of insulin yields 8.6 g. of free phenylalanine per 100 g. of protein (Table 7.1). If calculated as residue weight recovery, $8.6 \times 147/165 = 7.7$ g. per 100 g. of protein, since 165 is the molecular weight of phenylalanine and 147 is its residue weight in polypeptides and proteins.

For an understanding of protein structure, another method of reporting amino acid analyses of proteins is more instructive. When the molecular weight of the protein is known, the numbers of amino acid residues per molecule of protein may be calculated and should give whole numbers, since a fraction of a residue cannot be present. Table 7.2 gives the analysis calculated in this manner for insulin.

Table 7.1: AMINO ACID CONTENT OF PROTEINS, IN PER CENT

Constituent	Insulin (bovine)	Ribonuclease (bovine)	Cytochrome c (equine)	Hemoglobin (human)	Myoglobin (human)	Collagen (bovine)
Alanine	4.6	7.7	3.5	9.0	5.7	7.9
Amide NH_3	1.7	2.1	1.1	0.9	1.1	0.7
Arginine	3.1	4.9	2.7	3.3	2.7	8.0
Aspartic acid	6.7	15.0	7.6	9.6	9.2	5.8
Cysteine	0	0	1.7	1.0	0	0
Cystine	12.2	7.0	0	0	0	0
Glutamic acid	17.9	12.4	13.0	6.6	17.3	9.9
Glycine	5.2	1.6	5.6	4.2	6.3	20.3
Histidine	5.4	4.2	3.4	8.8	8.2	0.7
Hydroxyproline	0	0	0	0	0	10.9
Hydroxylysine	0	0	0	0	0	1.0
Isoleucine	2.3	2.7	5.4	0	5.0	1.3
Leucine	13.5	2.0	5.6	14.0	12.2	3.0
Lysine	2.6	10.5	19.7	9.6	16.1	3.4
Methionine	0	4.0	2.1	1.2	2.5	0.9
Phenylalanine	8.6	3.5	4.5	7.3	6.2	2.0
Proline	2.1	3.9	3.3	4.8	4.0	13.6
Serine	5.3	11.4	0	4.4	4.6	3.5
Threonine	2.0	8.9	8.4	5.2	2.9	1.8
Tryptophan	0	0	1.5	1.9	3.6	0
Tyrosine	12.6	7.6	4.9	2.9	2.4	0.8
Valine	9.7	7.5	2.4	10.3	5.3	2.0

AMINO ACID SEQUENCE OF PROTEINS

Knowledge of the amino acid composition of proteins represents the first step in attempts to understand the structure of these substances. The next is to obtain information concerning the sequence of amino acids in the peptide chain(s). This presents many difficulties because of the possibilities of isomerism arising from the kinds of amino acid, the amount of each, and their positions relative to one another. As an illustration of the last type of isomerism, consider the dipeptides containing glycine and leucine in which there are two possible isomers, glycylleucine and

Table 7.2: AMINO ACID RESIDUES OF BOVINE INSULIN

Amino acid	*Residues*	*Amino acid*	*Residues*
Alanine	3	Leucine	6
Arginine	1	Lysine	1
Aspartic acid	0	Methionine	0
Asparagine	3	Phenylalanine	3
Half-cystine	6	Proline	1
Glutamic acid	4	Serine	3
Glutamine	3	Threonine	1
Glycine	4	Tryptophan	0
Histidine	2	Tyrosine	4
Isoleucine	1	Valine	5
		Total	51

leucylglycine. For a tripeptide with the same two amino acids and tyrosine, there are six possible isomers. In a polypeptide containing 20 different amino acids in which each residue occurs only once, there is the possibility of factorial 20 (20!), or approximately 2×10^{18}, different compounds. Each of these contains the identical amino acids in the same proportions. Such a polypeptide of 20 amino acids would have a molecular weight of only 2,000.

When the proteins with their large molecular weights are contemplated, the magnitude of the problem becomes apparent. Synge has calculated that for a hypothetical protein with a molecular weight of 34,000, which contains only 12 different amino acids with 288 residues, there are 10^{300} isomers possible. If one molecule of each of these existed on earth, the total mass would be 10^{280} g. Fortunately, the total mass of the earth is only 10^{27} g., so the existence of all possible isomers need not be considered.

Protein Subunits. For many proteins, the minimal molecular weight calculated from amino acid analyses may be much smaller than the molecular weight obtained by physical determinations (page 127). In these instances the proteins may consist of a number of subunits, which may or may not be identical. When the subunits are held together by noncovalent forces, dissociation may be achieved by treatment with detergents, 8*M* urea, or 6*M* guanidine salts, usually in the presence of a mercaptide, *e.g.*, mercaptoethanol, in order to reduce disulfide bonds and to prevent oxidation of thiol groups to disulfides which would result in polymer formation.

Many large proteins may be dissociated by such methods, *e.g.*, hemoglobin, aldolase, muscle phosphorylase, and glutamic acid dehydrogenase. In general, the smaller proteins of molecular weight up to 30,000 are likely to consist of a single physical unit that cannot be dissociated, whereas many of the larger ones consist of subunits. Before attempting to determine the amino acid sequence of a protein, it is essential to determine the minimal physical unit as well as the minimal chemical unit. The latter may be estimated from the amino acid composition, from end group determinations (see below), and from peptide maps (page 148).

The minimal molecular weight calculated for insulin from the amino acid composition in Table 7.2 is near 5,700. This value is in agreement with the smallest physical molecular weight estimate obtained in the presence of concentrated solutions of urea or guanidine salts.

End Group Analysis. If a polypeptide or a protein consists of a single peptide chain, there should be only a single residue which bears a free α-amino group (amino-terminal residue) and a single residue with a free α-carboxyl group (carboxyl-terminal residue), as in the following example.

$$H_2NCHRCONH \cdots CONHCHRCOOH$$

Thus, quantitative determination of the number of end groups (amino- or carboxyl-terminal) permits an evaluation of the number of peptide chains per molecule of protein. End group analysis of this type is thus essential prior to undertaking a study of the sequence. Furthermore, quantitative end group methods provide an additional assessment of the purity of the protein, since there must be a simple integral relationship between the number of end groups per molecule of protein.

Amino End Group Methods. A technique of great value consists in the use of a reagent that binds with amino groups of proteins and is not removed by subsequent hydrolysis. Sanger found that 1-fluoro-2,4-dinitrobenzene (FDNB) is a suitable reagent for estimating and identifying amino end groups. The reagent reacts with α- and ε-amino groups to give yellow dinitrophenyl (DNP) derivatives and with other amino acid side chains (imidazole, phenol, thiol) to yield colorless DNP compounds. The reaction of FDNB with an amino acid is illustrated below.

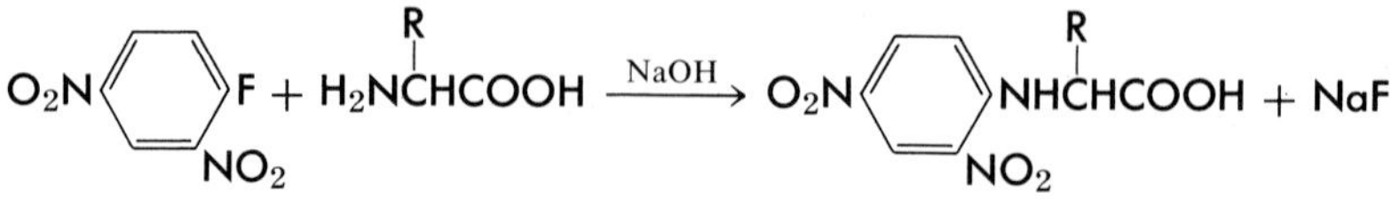

With a protein, the reagent is allowed to react under mildly alkaline conditions, the excess reagent is then removed, and the protein is hydrolyzed with 6*N* HCl. The yellow DNP-amino acids are extracted into a nonpolar solvent such as ether or chloroform and separated by chromatographic procedures. Inasmuch as the ε-DNP lysine remains in the aqueous phase after extracting the α-DNP amino acids, a quantitative estimation of the ε-amino groups may also be made.

Some proteins have more than one free α-amino group per molecule, indicating that multiple peptide chains are present. Insulin has one N-terminal phenyl-

Table 7.3: AMINO-TERMINAL GROUPS, DISULFIDES, AND NUMBER OF CHAINS OF SOME PROTEINS

Protein	Source	Mol. wt.	Amino-terminal groups	No. of S—S bonds	No. of chains
Insulin*	Bovine, porcine, etc., pancreas	5,700	1 Phe, 1 Gly	3	2
Ribonuclease T1*	*Aspergillus oryzae*	11,100	1 Ala	2	1
Cytochrome c*	Vertebrates, heart	12,400	Acetyl-Gly	0	1
Ribonuclease*	Bovine pancreas	13,700	1 Lys	4	1
Lysozyme*	Chicken egg white	14,400	1 Lys	4	1
Myoglobin*	Human muscle	17,000	1 Gly	0	1
Myoglobin*	Whale muscle	17,000	1 Val	0	1
Papain*	Papaya	21,000	1 Ile	3	1
Trypsinogen*,†	Bovine pancreas	24,400	1 Val	6	1
Trypsin*,†	Bovine pancreas	24,000	1 Ile	6	1
Chymotrypsinogen*,†	Bovine pancreas	24,700	1 Half-Cys	5	1
α-Chymotrypsin*,†	Bovine pancreas	24,300	1 Half-Cys, 1 Ala, 1 Ile	5	3
Subtilisin*	*Bacillus subtilis*	27,500	1 Ala	0	1
Serum albumin	Human serum	66,500	1 Asp	17	1
Enolase	Rabbit muscle	67,000	2 Acetyl-Ala	0	2
Hemoglobin*	Human erythrocytes	65,000	4 Val	0	4
Glyceraldehyde 3-phosphate dehydrogenase	Rabbit muscle	140,000	4 Val	0	4
Lactic acid dehydrogenase	Bovine muscle	140,000	?	0	4
Alcohol dehydrogenase	Yeast	140,000	?	0	4
Alcohol dehydrogenase	Equine liver	75,000	....	0	2
γ-Globulin	Human serum	160,000	(See Chap. 31)	17	4

* Amino acid sequences of these proteins have been determined.

† The relationship of trypsinogen and trypsin and of chymotrypsinogen and chymotrypsin is discussed in Chap. 12.

alanine and one N-terminal glycine residue and therefore has two different peptide chains. Human hemoglobin has four polypeptide chains. Table 7.3 records some of the results that have been obtained by the DNP method of determining amino-terminal residues.

Another chemical method for end group determinations is the cyanate method. Treatment of peptides with cyanate in slightly alkaline solution produces the carbamyl derivative.

$$NCO^- + {}^+H_3N-CHR_1-\overset{\overset{\displaystyle O}{\|}}{C}-\overset{\displaystyle H}{N}-CHR_2 \cdots \longrightarrow$$

$$H_2N-\overset{\overset{\displaystyle O}{\|}}{C}-\overset{\displaystyle H}{N}-CHR_1-\overset{\overset{\displaystyle O}{\|}}{C}-\overset{\displaystyle H}{N}-CHR_2 \cdots$$

$$\Big\downarrow \begin{array}{l}\text{acid}\\\text{100°C}\end{array}$$

$$+ \ {}^+H_3N-CHR_2 \cdots$$

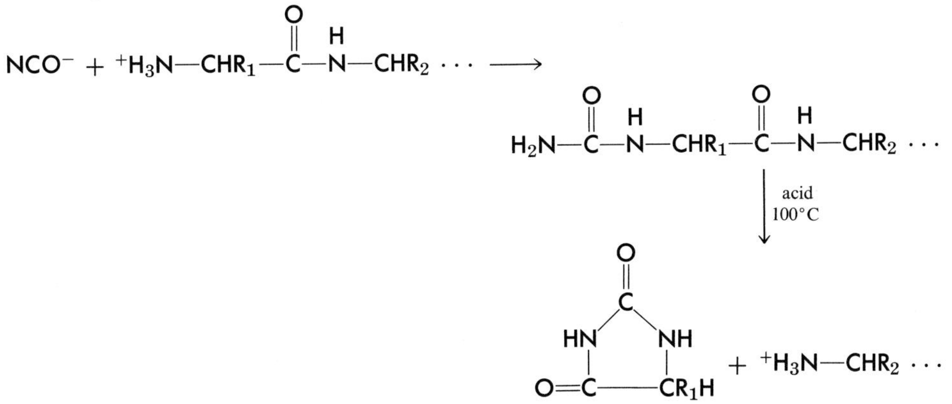

Heating the carbamyl derivative in weak acid liberates the NH_2-terminal residue as the hydantoin derivative which may be readily isolated and identified as the free amino acid after acid hydrolysis.

Another type of approach to amino end group determination involves the use of a reagent that can be applied to the liberation of a derivative of the amino-terminal residue without hydrolysis of the remainder of the peptide chain. Edman's reagent, phenylisothiocyanate, has been successfully used in this way. The procedure is shown in the following reactions.

$$C_6H_5-N{=}C{=}S + H_2N\overset{R'}{C}H\overset{\overset{O}{\|}}{C}-\overset{H}{N}\overset{R''}{C}H\overset{\overset{O}{\|}}{C}-\overset{H}{N}\overset{R'''}{C}H\overset{\overset{O}{\|}}{C}\cdots$$

Phenylisothiocyanate

$$\Big\downarrow \begin{array}{l}\text{Step I}\\\text{weak alkali}\end{array}$$

$$C_6H_5-\overset{H}{N}-\overset{\overset{S}{\|}}{C}-\overset{H}{N}-\overset{R'}{C}H-\overset{\overset{O}{\|}}{C}-\overset{H}{N}\overset{R''}{C}H\overset{\overset{O}{\|}}{C}-\overset{H}{N}\overset{R'''}{C}H\overset{\overset{O}{\|}}{C}\cdots$$

Phenylthiocarbamyl (PTC) peptide (or protein)

$$\Big\downarrow \begin{array}{l}\text{Step II}\\\text{weak acid}\end{array}$$

$$+ \ H_2N\overset{R''}{C}H\overset{\overset{O}{\|}}{C}-\overset{H}{N}\overset{R'''}{C}H\overset{\overset{O}{\|}}{C}\cdots$$

Phenylthiohydantoin (PTH) amino acid

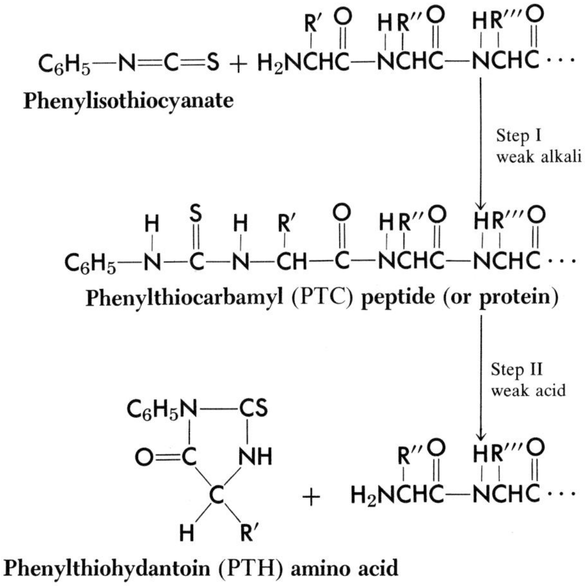

In step I, the PTC peptide (or protein) is prepared. This derivative is then treated (step II) with weak acid at room temperature, thus liberating the *phenylthiohydantoin* (PTH) of the terminal amino acid and a peptide one residue shorter than the original. The PTH amino acid may be identified by its chromatographic behavior. The procedure may then be repeated with the shorter peptide. It is frequently possible to repeat the procedure five or more times for the identification of the amino-terminal peptide sequence.

In addition to chemical methods for determining amino end groups, an enzymic method is also used. The enzyme *leucine aminopeptidase* requires a terminal free α-amino group for its action. Therefore, in a peptide whose sequence is $H_2N—A\cdot B\cdot C\cdot D\cdot \cdots$ where A, B, C, and D are different residues, the action of the aminopeptidase will liberate sequentially the free amino acids. At any given time the amount of A liberated will be greater than that of B, and the amount of B greater than that of C, etc. Therefore, the sequence ABC can be deduced from rate measurements of amino acid liberation. The aminopeptidase acts most rapidly on terminal leucine residues but also liberates all others found in proteins; it does not act at X-Pro bonds (Table 7.4).

Stepwise degradation of a protein by the Edman method or the use of the aminopeptidase (or a carboxypeptidase, see below) can yield meaningful results only when applied to a single peptide chain, or to a protein which has several identical chains.

For many proteins the number of free α-amino groups is in accord with the number of peptide chains. Nevertheless, there are proteins which do not possess free α-amino groups, *e.g.*, cytochrome c (page 178), ovalbumin, the protein of tobacco mosaic virus, and others. The N-terminal α-amino group is N-acetylated in most of these proteins. In a few instances pyrrolidone carboxylic acid (page 92) has been identified as a terminal residue. This residue possesses no free amino group.

Table 7.4: SPECIFICITY OF SOME PROTEOLYTIC ENZYMES

Proteolytic enzyme	Source	Major sites of action*	Other sites of action
Trypsin	Pancreas	Arg, Lys	
Chymotrypsin	Pancreas	Trp, Phe, Tyr	Leu, Met, $AspNH_2$, His
Pepsin	Gastric mucosa	Trp, Phe, Tyr, Met, Leu	Various; acidic, etc.
Carboxypeptidase A	Pancreas	C-terminal bond of Tyr, Trp, Phe, etc.	Does not act at Arg, Lys, Pro
Carboxypeptidase B	Pancreas	Arg, Lys	None
Leucine aminopeptidase	Kidney, intestinal mucosa, etc.	N-terminal bond of various residues	No action at X-Pro bond†
Papain	Papaya	Arg, Lys, Gly, etc.	Wide specificity; does not act at acidic residues
Subtilisin	*Bacillus subtilis*	Aromatic and aliphatic residues	Various
Elastase	Pancreas	Neutral residues	

* Except for the carboxypeptidases, the sites of action refer to the residues bearing the carbonyl group of the peptide bond, *e.g.*, trypsin catalyzed hydrolysis of arginyl and lysyl bonds.

† X = any other amino acid residue.

For other proteins which do not possess free α-amino groups, the nature of the blocking groups has not yet been determined.

Carboxyl End Group Methods. A chemical method for the identification of residues bearing α-carboxyl groups depends on treatment of the protein (or peptide) with hydrazine under anhydrous conditions at 100°C. Cleavage of the peptide bonds by *hydrazinolysis* converts all amino acid residues to amino acid hydrazides except the carboxyl-terminal residue, which remains as the free amino acid and can be isolated and identified chromatographically.

$$\text{Protein} + n(\text{H}_2\text{N—NH}_2) \longrightarrow n(\text{H}_2\text{NCHRCONHNH}_2) + \text{amino acid}$$

An enzymic method involves the use of pancreatic *carboxypeptidases*. *Carboxypeptidase A* liberates only the residue from a protein or peptide that bears a free α-carboxyl group. As in the case of the analogous aminopeptidase (page 144), information concerning the carboxyl-terminal sequence can be obtained by the rate of liberation of successive residues; *e.g.*, in a sequence $H_2N—A{\cdot}B{\cdot}C{\cdot}D{\cdot}\cdots W{\cdot}X{\cdot}Y{\cdot}Z—COOH$, after a given time of hydrolysis, the amount of *Z* liberated will be greater than that of *Y*, etc. Carboxypeptidase A has little or no action on carboxyl-terminal proline, arginine, or lysine residues. *Carboxypeptidase B*, a distinct enzyme, liberates carboxyl-terminal arginine or lysine residues (Table 7.4).

DETERMINATION OF AMINO ACID SEQUENCE

Let us assume that we wish to determine the sequence of a protein consisting of a single peptide chain, whose molecular weight and amino acid composition are known, and whose sequence is as follows:

$$H_2N—A{\cdot}B{\cdot}C{\cdot}D{\cdot}E{\cdot}F{\cdot}\cdots T{\cdot}U{\cdot}V{\cdot}X{\cdot}Y{\cdot}Z—COOH$$

By the use of end group methods, the amino- and carboxyl-terminal residues, *A* and *Z*, respectively, have already been identified. Indeed, perhaps even the amino-terminal sequence, $A{\cdot}B{\cdot}C$, has been ascertained by the use of the stepwise Edman procedure and the sequence $X{\cdot}Y{\cdot}Z$ has been identified by the use of carboxypeptidases. The problem is now to elucidate the remainder of the sequence of the protein. The general method for accomplishing this is as follows: (1) partial hydrolysis of the protein, (2) isolation in pure form of the resulting peptides, (3) determination of the sequences of the smaller peptides, and finally, (4) the deduction of the complete sequence from the sequences of peptides of overlapping structures.

Partial Hydrolysis. Two general methods of partial hydrolysis of polypeptides and proteins are in use: (1) partial acidic hydrolysis, and (2) hydrolysis catalyzed by proteolytic enzymes.

Partial Acidic Hydrolysis. A partial cleavage of a polypeptide or protein can be obtained by limiting the time of hydrolysis, lowering the temperature, working with dilute acid or 12*N* HCl, the actions of which are slower than that of 6*N* HCl. These procedures yield a mixture of free amino acids and a large variety of small peptides. The technique is useful with smaller peptides but is seldom employed with larger peptides or proteins because of several inherent disadvantages as compared to the hydrolysis with enzymes (see below). These limitations are as follows:

(1) the yield of peptides is low, (2) hydrolysis is essentially random (although aspartyl bonds are most susceptible), resulting in a mixture of many components that are difficult to isolate in pure form, and (3) larger peptides are seldom obtained in significant yield. To illustrate the problems, we may use as an example a tetrapeptide with the sequence $A \cdot B \cdot C \cdot D$. A partial acidic hydrolysate will yield the four amino acids as well as three dipeptides and two tripeptides, or nine components in all, excluding the original peptide. In contrast, it is frequently possible to hydrolyze a tetrapeptide with an enzyme, obtaining a theoretical yield of the dipeptides $A \cdot B$ and $C \cdot D$. Thus, only two components need be separated.

Action of Proteolytic Enzymes. These enzymes catalyze only a limited cleavage, producing relatively large fragments in good yield. Since each peptide must be isolated in pure form, usually by chromatographic and ionophoretic methods, the smaller the number of fragments, the simpler is the problem of purification.

Each proteolytic enzyme will hydrolyze only certain types of peptide bonds. For example, crystalline trypsin catalyzes the hydrolysis of peptide bonds in which the carboxyl group of lysine or arginine residues participates. Thus, from a knowledge of the contents of these two amino acids in a given protein or polypeptide, the number of bonds susceptible to trypsin may be calculated. Each peptide resulting from tryptic action, with the possible exception of the original carboxyl-terminal one, should terminate in an arginine or lysine residue. For example, a protein containing 5 arginine and 10 lysine residues should yield, after tryptic hydrolysis, 16 peptides. If one of these lacks a basic residue, it will represent the carboxyl-terminal sequence of the protein.

Other proteinases, *e.g.*, chymotrypsin, pepsin, papain, etc., can hydrolyze proteins or peptides at other loci in the peptide chain. The specificity of certain proteolytic enzymes, which are available in pure form and which are particularly useful for structural work on proteins, is indicated in Table 7.4.

The action of certain proteolytic enzymes on the oxidized B chain of bovine insulin is shown in Fig. 7.1. The sites of action of the three enzymes are limited and relatively specific when compared with the random hydrolysis effected by acid.

For the complete sequence determination of a protein, at least two different forms of enzymic hydrolysis must be used in order to deduce the structure by the method of overlapping sequences. A hypothetical case is illustrated below in which the points of hydrolysis by trypsin are shown by the arrows, yielding peptides I through VI.

$$H_2N\text{—}A\cdot B\cdot C\cdot D\cdot E\cdot F\cdot G\cdot \cdots\cdots\cdots\cdots\cdots \; T\cdot U\cdot V\cdot W\cdot X\cdot Y\cdot Z\text{—}COOH$$

↑ ↑ ↑ ↑ ↑

I II III IV V VI

Peptides I and VI can be tentatively assigned positions if the amino- and carboxyl-terminal sequences are known. There is no way of positioning peptides II, III, IV, and V. If, however, hydrolysis of the protein is accomplished by chymotrypsin, a different series of peptides will be obtained since the action of this enzyme is mainly at aromatic amino acid residues (Table 7.4). Suppose a peptide is obtained having the sequence $C \cdot D \cdot E \cdot F \cdot G \cdot$, etc. It will overlap with the sequences of peptide I and

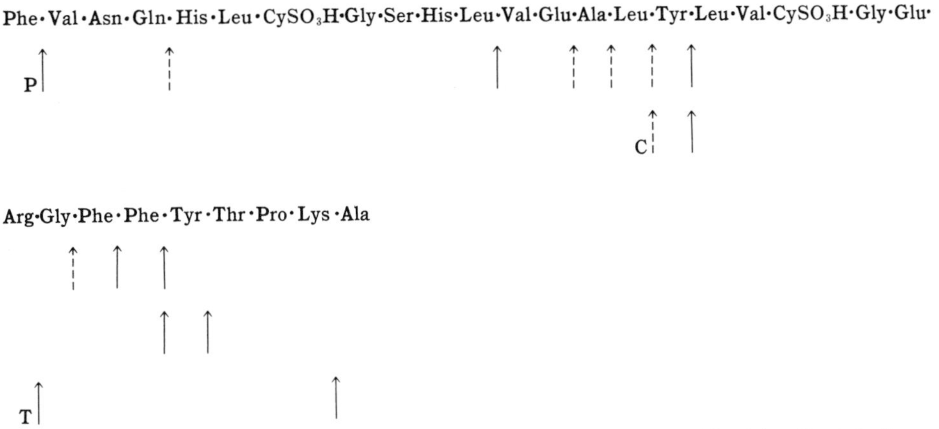

Fig. 7.1. Action of some crystalline proteolytic enzymes on B chain of oxidized insulin; ↑ indicates major sites of action of enzymes, ⇡ indicates other bonds split by enzymes. The oxidation procedure has altered the cystine (CyS-SCy) residues to cysteic acid ($CySO_3H$) residues. Bonds on the line indicated *P* are hydrolyzed by pepsin; *C*, by chymotrypsin; and *T*, by trypsin. (*After F. Sanger and H. Tuppy, Biochem. J.*, **49**, 481, 1951.)

peptide II, thus establishing this overlap. In similar fashion, the entire sequence can be deduced.

For small individual peptides, the sequence can be determined with the enzymic amino- and carboxyl-terminal methods, and with the stepwise phenylthiohydantoin method. For larger peptides, it may be necessary to hydrolyze first with other proteolytic enzymes. Thus a peptide is obtained from a tryptic digest, with the sequence $A\cdot B\cdot C\cdot$Tyr$\cdot E\cdot F\cdot G\cdot H\cdot$Lys$\cdot$, where A,B,C,E,F,G, and H are aliphatic residues of the type of glycine, alanine, serine, etc. Chymotrypsin can be used for hydrolysis at the tyrosine residue, liberating $A\cdot B\cdot C\cdot$Tyr$\cdot$ and $E\cdot F\cdot G\cdot H\cdot$Lys$\cdot$. Since the positioning of the two peptides relative to each other is already established, the one containing tyrosine being amino-terminal and the one containing lysine being carboxyl-terminal, further sequence work is relatively straightforward.

The Structure of Insulin. The work of Sanger and his associates on insulin was an important achievement because it was the first instance in which the structure of a protein was completely established. We shall consider briefly how this was accomplished since it illustrates the use of some of the methods described above as well as others.

Bovine insulin contains 51 amino acid residues (Table 7.2, page 140) and yields one amino-terminal glycine residue and one amino-terminal phenylalanine residue (Table 7.3, page 142). Therefore, insulin has at least two peptide chains which must be held together by other linkages; these are the S—S bonds of cystine. Oxidation of insulin with performic acid cleaved the S—S linkages and yielded a cysteic acid residue, $HO_3S—CH_2—CH(NH—)CO—$, in place of each half-cystine residue. This procedure produced two peptides which could be separated electrophoretically or chromatographically. One was a basic peptide (B chain) possessing amino-terminal phenylalanine, and proved, on analysis, to contain 30 residues; the other, an acidic peptide (A chain) with amino-terminal glycine, contained 21 residues. Each chain was then studied separately.

From partial acidic hydrolysates of the DNP (page 142) derivative of the A peptide, Sanger isolated DNP glycine, DNP glycylisoleucine, DNP glycylisoleucylvaline, and DNP glycylisoleucylvalylglutamic acid. From this evidence, the A peptide chain has the initial sequence, glycylisoleucylvalylglutamyl ···. For the B chain, the initial sequence, derived by the same method, is phenylalanylvalylaspartylglutamyl ···. Thus, this approach succeeded in identifying a portion of the peptide structure of each chain of insulin.

With each chain, partial hydrolysates were then prepared, by both acidic and enzymic hydrolysis. The smaller peptides were separated by chromatographic procedures on columns and on paper, and by electrophoresis, and the structures were identified with the aid of the DNP method. Overlapping sequences were fitted together in the manner of a jigsaw puzzle.

As an example of how overlapping sequences were fitted together, the following di-, tri-, and tetrapeptides from partial acidic hydrolysates permitted deduction of a nonapeptide sequence in the A chain of insulin, with the information that only two tyrosine residues are present in this chain.

Ser·Leu   Glu·Leu   Asp·Tyr
Leu·Tyr   Leu·Glu   Tyr·$CySO_3H$
Tyr·Glu   Glu·Asp
Ser·Leu·Tyr   Leu·Glu·Asp
Leu·Tyr·Glu   Glu·Asp·Tyr
Glu·Leu·Glu
Ser·Leu·Tyr·Glu   Glu·Asp·Tyr·$CySO_3H$

Sequence: Ser·Leu·Tyr·Glu·Leu·Glu·Asp·Tyr·$CySO_3H$

From such studies, the linear sequence of each oxidized chain, A and B, was established. However, the oxidized A chain contained four cysteic acid residues, and the B chain two cysteic acid residues, all present in native insulin in the form of half-cystine residues, cysteine being absent from this protein.

The arrangement of the two chains was established by digesting intact insulin with chymotrypsin and by partial acidic hydrolysis. Since the sequences of both chains were already known, it was necessary only to isolate peptides containing cystine which bridged the various half-cystine residues. It should be emphasized that methods must be used that preclude disulfide interchange, as in the following reversible reaction, resulting in a completely random distribution of disulfides.

$$R_1—S—S—R_2 + R_3—S—S—R_4 \xrightarrow[\text{alkaline pH}]{\text{strong acid or}} R_1—S—S—R_3 + R_2—S—S—R_4 + \text{etc.}$$

Since this occurs in strong acid as well as in neutral or alkaline solutions, the reaction must be inhibited to permit unequivocal identification of disulfide bridges. The complete structure of bovine insulin is shown in Fig. 7.2.

Peptide Mapping. For relatively small proteins consisting of a single peptide chain, the methods described above can be used directly for determining the amino acid sequence. For larger proteins when it is not certain how many peptide chains

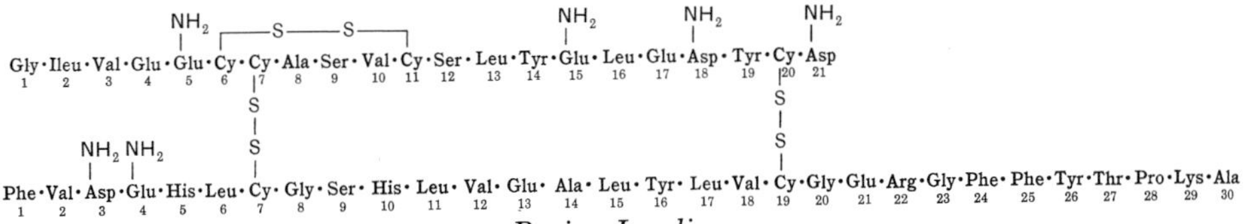

Bovine Insulin

Fig. 7.2. The structure of bovine insulin. The abbreviations used for the amino acid residues are given in Table 5.6. The NH_2 groups indicate the β- and γ-carboxamide groups of asparagine and glutamine, respectively. The presence of the internal disulfide bridge in the glycyl (A) chain of insulin between residues 6 and 11 is noteworthy. It should be noted that the standard convention is followed in which the amino-terminal residues are at the left of each chain.

are present or when several subunits are known to be present, as judged by dissociation with urea, detergents, etc., and it is unknown whether the subunits differ in sequence or are identical, it is useful to obtain peptide maps or "fingerprints" after enzymic hydrolysis. For example, a protein of 50,000 molecular weight, containing 8 arginine and 10 lysine residues, can be dissociated to 25,000 by use of 8*M* urea or 6*M* guanidine salts. Since trypsin catalyzes hydrolysis only at lysine and arginine residues, 10 tryptic peptides may be expected if the protein contains 2 identical subunits; however, if the subunits are different, 19 tryptic peptides might be expected. An answer to this problem may be obtained by complete hydrolysis of the denatured protein with trypsin followed by the preparation of two-dimensional peptide maps on paper—either by two different chromatographic procedures in the two directions or by electrophoresis in one direction and by chromatography in the other (page 113*ff.*). After the paper is dried and sprayed with ninhydrin, it is possible to count the number of distinct spots and estimate the number of distinct tryptic peptides. By spraying with the Sakaguchi reagent (page 116), the arginine-containing peptides may be recognized. The same procedures can be used after chymotryptic hydrolysis to estimate the numbers of peptides containing residues which yield distinctive color reactions, such as tryptophan, tyrosine, histidine, etc.

Peptide mapping has been widely employed in comparing proteins expected to be identical in all but one or a few peptides. Many abnormal hemoglobins (Chap. 30) differ from normal hemoglobin in only a single residue. The peptide containing the amino acid substitution will usually manifest a different electrophoretic or chromatographic mobility from the corresponding peptide of the normal protein. Similarly, the peptide maps of the same protein from different species or varieties may be compared to ascertain whether the proteins differ and to isolate the peptides which show altered behavior on electrophoresis or chromatography. The study of genetic mutants which possess altered amino acid sequences in a given protein has been greatly facilitated by use of these methods (Chap. 30).

Comments on Sequence Determinations. From the foregoing it is evident that many methods are presently available for sequence determinations. Difficulties arise frequently because of two methodological problems, separation of peptides and problems of hydrolysis. Peptides of moderate size are readily purified by available electrophoretic and chromatographic methods, the latter usually involving column procedures with ion exchange resins and partition systems on paper. Larger and hydrophobic peptides are often soluble only in acidic solvents or in strong solutions of urea or guanidine salts. For such peptides gel filtration on resins of different retention characteristics (page 132) has been of great value.

The problem of separation of peptides is directly related to the problem of hydrolysis of large proteins—the less specific the method, the greater the number of peptides to be resolved. Hence, proteolytic cleavage by relatively specific enzymes, *e.g.*, trypsin and chymotrypsin, is desirable. Less specific enzymes (pepsin, papain, and subtilisin) are usually employed only on peptides of modest size.

It would be helpful to have other specific enzymes as well as chemical methods which cleave specifically at a single type of residue. Various methods have been devised both to limit and to extend the action of enzymes. For example, acylation of ϵ-amino groups of lysine residues limits tryptic action to hydrolysis at arginine residues, whereas transformation of cysteine residues to aminoethylcysteine residues by treatment with ethyleneimine permits tryptic action at the carbonyl bonds of such residues.

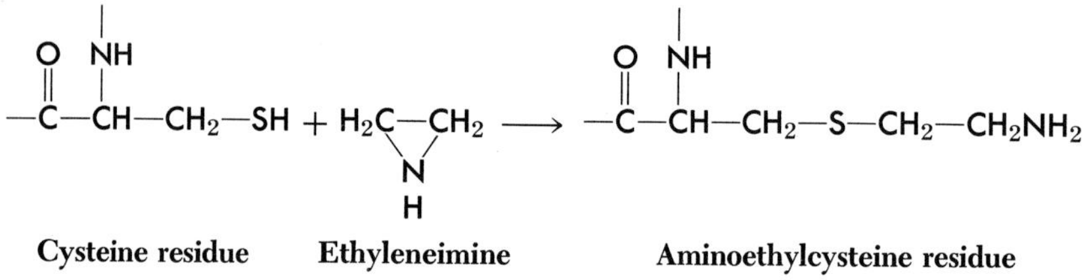

A specific chemical method of value for cleavage at methionine residues has been the use of cyanogen bromide (CNBr) in acid solution, *e.g.*, 70 per cent formic acid.

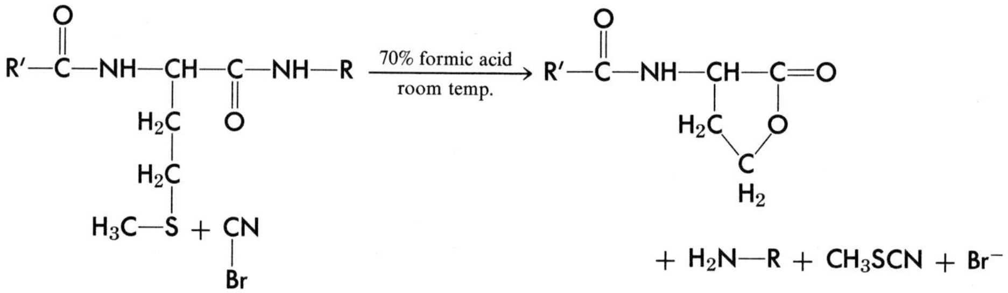

The carboxyl-terminal residue of the peptide now contains homoserine lactone. Hydrolysis with 6*N* HCl under the usual conditions (page 139) liberates both homoserine and its lactone.

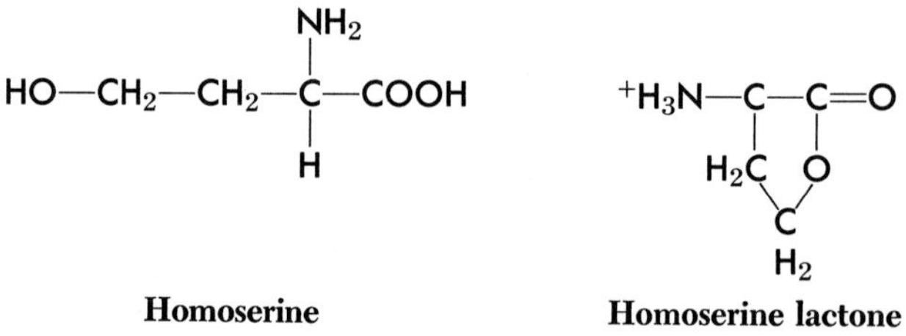

Since most proteins contain relatively few methionine residues, the method offers an approach to obtaining large peptides.

In proteins containing cysteine or cystine residues, treatment of these residues is usually required before the primary sequence can be studied. If cystine residues are present, random disulfide interchange can produce a confusing mixture of artifactual peptides. Cysteine residues will oxidize randomly to form mixtures of disulfide-containing peptides. To prevent such reactions, oxidation by performic acid, as used for insulin, is satisfactory only when tryptophan is absent, since this amino acid is destroyed. Although methionine is converted to the stable methionine sulfone, this does not interfere with sequence work. It is more generally desirable to alkylate cysteine residues, either originally present or produced by reduction of disulfide bonds, by treatment with iodoacetate in the presence of 8*M* urea to form carboxymethylcysteine residues. Reaction of cysteine residues with ethyleneimine to form aminoethylcysteine residues has been noted above.

Some Comments on Primary Structure. Since the elucidation of the complete structure of insulin in 1955, the structures of a number of proteins have been established, *e.g.*, cytochrome c (page 178), hemoglobin (page 173), ribonuclease (page 257), myoglobin, lysozyme, and others (Table 7.3). Certain features of protein structure that have emerged from these studies should be emphasized, although some of these points have been mentioned earlier.

1. There is no apparent regularity or periodicity in the sequences established thus far. Each homogeneous protein is relatively unique and possesses a specific amino acid sequence; however, homologous proteins of different species do resemble one another (Chap. 30), and related proteins of common genetic origin and similar function possess similarities of sequence, *e.g.*, the pancreatic proteinases (Chaps. 12 and 30).

2. With respect to the presence and arrangement of disulfide bridges, all possible types of proteins have been found. Insulin contains an intrachain disulfide bridge (A chain) as well as two interchain disulfide bridges. There are other proteins with intrachain disulfide bridges (ribonuclease, page 151) and proteins lacking disulfide bridges—hemoglobin (page 173) and cytochrome c (page 178). In general, most proteins containing more than one peptide chain appear to lack disulfide bonds. Insulin, γ-globulin, β-lactoglobulin, thyroglobulin, and chymotrypsin are among the exceptions to this rule (see Table 7.3).

3. The distribution of glutamine and asparagine is independent of their relationship to glutamic and aspartic acids, respectively, and they must be regarded as individual amino acids.

4. Only amino acids of the L configuration, with the exception of the optically inactive glycine, have been found.

5. No covalent bonds between amino acids have been detected in globular proteins other than the α-peptide bond and the disulfide bridge. Thus, despite the wide range of composition, size, function, and other properties, the fundamental structure of the globular proteins follows the same relatively simple pattern. In fibrous proteins various types of covalent cross-linking have been found (Chap. 38).

THE CONFORMATION OF PROTEINS

The Hydrogen Bond. In 1936 Mirsky and Pauling suggested that a major factor in maintaining the folded structure of the peptide chain is the presence of hydrogen bonds. The formation of a hydrogen bond is due to the tendency of a hydrogen atom to share the electrons of an oxygen atom. For example, the carbonyl oxygen of one peptide bond shares its electrons with the hydrogen atom of another peptide bond.

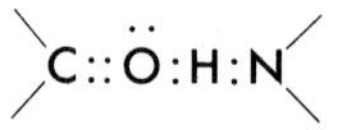

Although individual hydrogen bonds of this type are relatively weak as compared with covalent bonds, they will reinforce each other if a molecule contains many such bonds. Direct evidence for the presence of hydrogen bonds in proteins has come from the use of infrared spectroscopy; strong absorption bands are found at the wavelengths characteristic of such hydrogen bonds.

Another type of evidence that indicates the existence of hydrogen bonds in proteins has been provided by Linderstrøm-Lang and his associates. When a substance is dissolved in a medium containing deuterium oxide (heavy water), an exchange takes place between dissociable hydrogens of the compound and the deuterium of the medium. Ionizable hydrogens and those present in peptide bonds (—CONH—), hydroxyl groups, and amide groups (—$CONH_2$) should be readily exchangeable, whereas carbon-bound hydrogens are not. In the small peptide, leucylglycylglycylglycine, there are six exchangeable hydrogen atoms, *viz.*, three ammonium and three peptide hydrogens, which equilibrate rapidly with hydrogen from the medium. Similarly, in the oxidized A chain of insulin, the 41 exchangeable hydrogens equilibrate very rapidly. Native insulin contains 91 potentially exchangeable hydrogen atoms. However, only about two-thirds of these atoms exchange very rapidly; the others exchange more slowly, the rate of exchange being increased by raising the temperature from 0 to 38°C. or by adding urea or guanidine at 0°. These results are consistent with the view that the action of urea causes an unfolding of the structure of proteins and that hydrogen bonding plays a major role in maintaining this structure.

Pauling and Corey have proposed several structures for polypeptide chains in which maximal stability can be attained by extensive hydrogen bond formation when the known properties of peptide structures are taken into consideration. Since the C—N peptide bond is a partial double bond, the peptide bond group is planar and able to exist in *cis* and *trans* forms. The *trans* configuration (shown below) is preferred in most of the suggested structures of proteins since there is less steric hindrance from large R groups than in the *cis* form.

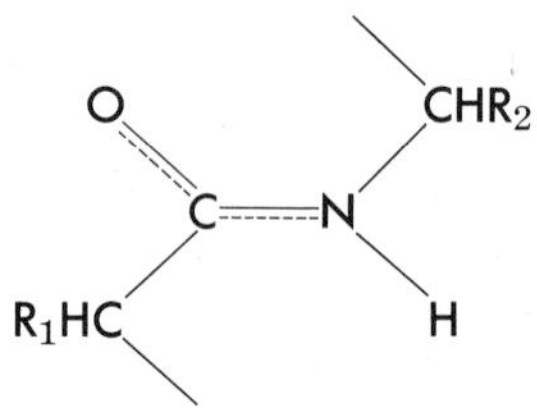

The most important of the regular structures proposed for proteins is the α-helix, in which 3.7 amino acid residues are present in each complete coil and the shape is maintained by hydrogen bonding between the CO and NH groups of adjacent coils. The right-handed helix is favored for L-amino acids and the left-handed helix for D-amino acids (Fig. 7.3). Direct evidence for the existence of the α-helix in proteins has come from x-ray diffraction studies, discussed below (page 157). Some keratins show x-ray patterns characteristic of the α-helix. Many homopolymers of α-amino acids readily form an α-helical structure.

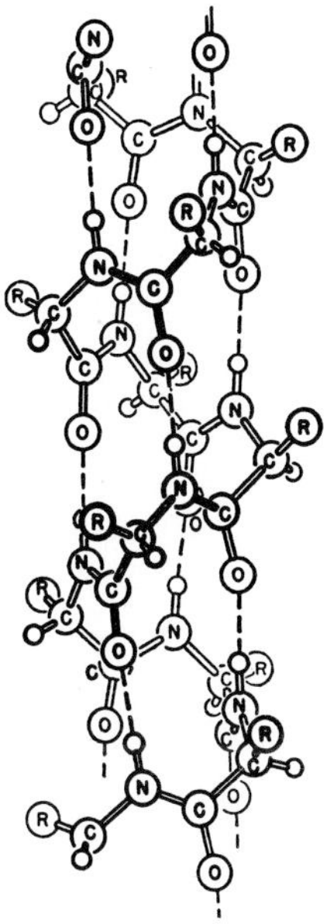

Fig. 7.3. (*a*) The α-helix with 3.7 amino acid residues per turn. The NH and CO of each peptide bond are linked by hydrogen bonds. (*From L. Pauling, R. B. Corey, and H. R. Branson, Proc. Nat. Acad. Sci.,* **37**, 205, 1951.)

Pauling and Corey postulated another structure, the so-called β structure or pleated sheet, on the basis that large numbers of hydrogen bonds would provide stabilization. As shown in Fig. 7.4 the pleated sheet structures can be of the parallel or antiparallel type, depending on the directions of the peptide chains. In a multichain fibrous protein, such as certain types of silk fibroin, the chains are oriented with close packing in an antiparallel structure. X-ray analysis of the structure of lysozyme (page 269) indicates that a portion of the structure is in the β form.

The Ionic Bond. The interaction of two ions leads to attraction or repulsion, depending on the charge of the ions. For a positive group, such as a guanidinium ion of arginine residues, and a negative group, such as a carboxylate ion, the attraction can become strong if these groups can come into juxtaposition. Similarly, the repulsive forces can be high for ionic groups possessing a similar charge.

Ionic bonds of the type between COO^- and ^+H_3NR are presumably stable only in the absence of water, inasmuch as ions are strongly hydrated in the presence of water. Ionic bonds are protected from water in hydrophobic portions of protein

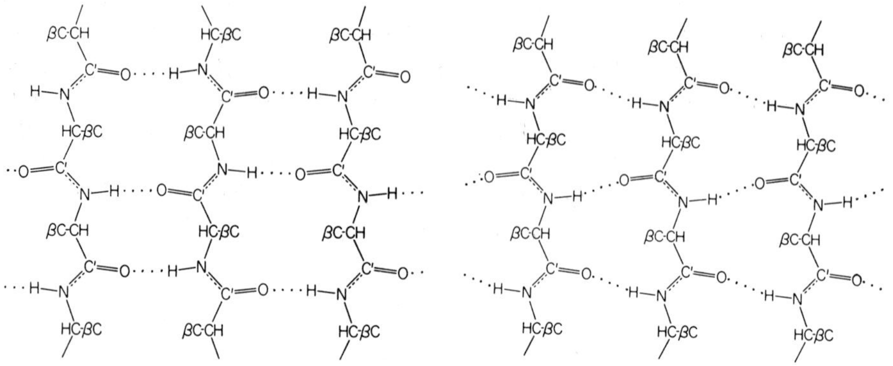

Fig. 7.4. Diagrammatic representations of pleated sheet or β structures. On the left is shown the antiparallel-chain structure and on the right, the parallel-chain structure. (*From L. Pauling and R. B. Corey, Proc. Nat. Acad. Sci.*, **37**, 729–740, 1951.)

molecules (see below) and contribute to the over-all stabilization of the native protein structure.

The Hydrophobic Bond. The side chains or R groups of alanine, valine, leucine, isoleucine, phenylalanine, tyrosine, methionine, and tryptophan are essentially hydrophobic, *i.e.*, they have little or no attraction for water molecules, as compared with the strong hydrogen bonding between water molecules. Such R groups can closely approach each other, with exclusion of water, to form links of different parts of a peptide chain or to bind different chains. When large regions of peptide chains involve only hydrophobic residues, the exclusion of water will result in a very tightly bonded structure. The forces involved are similar to those involved in the coalescence of oil droplets suspended in water and to the tendency of detergents to form micelles (page 62) in which the nonpolar side chains are in the interior, whereas the ionic or polar groups are on the exterior and in contact with water molecules. Thus, folding of the protein tends to produce a globular structure in which most of the nonpolar groups are in the interior and the ionic groups are on the surface of the molecule. Such a structure has been aptly described as an "oil drop with a polar coat."

The formation of hydrophobic bonds in proteins also serves to bring together groups that can form hydrogen bonds or ionic bonds in the absence of water. Thus, there is a cooperative aspect to the proper folding of the peptide chain in the formation of the native globular protein. Each of the types of bonding aids in the formation of the others, with the hydrophobic bonds presumably playing a major part in this phenomenon. Hydrophobic bonds also play an important role in other protein interactions, *e.g.*, the formation of enzyme-substrate complexes (Chap. 12) and antibody-antigen interactions (Chap. 31).

Absorption Spectra of Proteins. In general, proteins absorb ultraviolet light in three rather distinct regions: (1) above 250 mμ with a peak near 280 mμ, where absorption is due entirely to the presence of aromatic residues (page 89); (2) between 210 and 250 mμ, due to multiple factors, *e.g.*, absorption of aromatic and other residues, and absorption due to some types of hydrogen bonds and other interactions concerned in conformation and helix content; (3) below 210 mμ, due to absorption by peptide bonds and many conformation factors. Obviously, the last two regions are extremely complex.

Since the absorption of tyrosine residues shifts with pH because of the ionization of the phenolic groups, whereas that of tryptophan does not, spectrophotometric methods are used in estimating the amounts of these residues in proteins. In many proteins only a portion of the tyrosine residues ionizes normally, but following denaturation complete ionization of all tyrosine residues is usually observed (page 162). Thus, changes in absorption are useful in evaluating alterations in the environment of aromatic residues and in conformation of the protein. Alterations in the state of the protein and in its chromophores are frequently studied by recording difference spectra, *i.e.*, direct plotting of the difference in absorption between the protein under two conditions, such as pH, presence of urea, etc. Changes in the difference spectrum are usually about ten times greater near 230 mμ than at 280 mμ when the native and the denatured protein are compared.

Optical Rotation of Proteins. All protein solutions change the direction of the polarization of polarized light that passes through them. This is partly because of the presence of optically active amino acid residues present in these molecules. From our knowledge of small molecules, it is evident that the sign and magnitude of the optical rotation depend not only on the atoms and groups of atoms bonded to asymmetric carbon atoms but also on the relationships of the asymmetric carbons to one another. Thus changes in the structure or conformation of the protein molecules may be reflected in the specific rotation (page 11).

The specific rotations of proteins are always negative and for globular proteins are usually in the range of -30 to $-60°$. For example, the specific rotation $[\alpha]_D$ for ovalbumin is near $-30°$ over the pH range 3.5 to 11, although the net charge per molecule changes from approximately $+20$ to -30. Clearly, the ionic state of the molecule has little or no effect on the rotation. However, at higher or lower pH values the rotation becomes more negative, *e.g.*, at pH 13, $[\alpha]_D$ is near $-60°$. Similarly, in 8*M* urea solution, $[\alpha]_D = -88°$. Treatment of proteins at high temperatures produces similar increases in negative rotation. Inasmuch as these large increases in negative rotation are not accompanied by changes in primary structure, *i.e.*, rupture of peptide bonds, they must reflect drastic alterations in the conformation of the molecule, dependent upon those properties that are ascribed to secondary or tertiary structure. Such changes are generally described under the term "denaturation."

The term "configuration" usually refers to the D and L forms in which four different groups are attached to a carbon atom. "Conformation" is used in protein chemistry to designate the over-all structure of a molecule in which asymmetry may be produced by a spiral arrangement (helix) or other special folding (see also page 152).

The nature of the changes responsible for the increased levorotation of proteins upon denaturation has been actively investigated. When model synthetic polypeptides in their α-helical form are transformed into a random coil arrangement they become more levorotatory.

Optical rotatory dispersion, i.e., the variation of optical rotation with the wavelength of monochromatic light, has provided additional information concerning the structure of proteins. For synthetic polypeptides, such as poly-L-glutamic acid, the dispersion curves differ for the random-coil form at pH 7 and the helical form at pH 4.3 (Fig. 7.5). Differences in dispersion curves are also found for native and denatured proteins. Various theoretical and empirical formulas have been applied to such data in attempts to calculate the amount of α-helix in proteins. For the present, it may be stated that such methods do not give absolute values; however, they do show that globular proteins vary in helical content, some containing perhaps as low as 0 to 10 per cent and others perhaps as high as 70 to 80 per cent of the amino acid residues in this form. Such methods are useful in following changes in conformation, but absolute determinations of conformation and helical content can be made at present only by the method of x-ray diffraction.

Circular dichroism is a measure of the *difference in absorption* of left and right

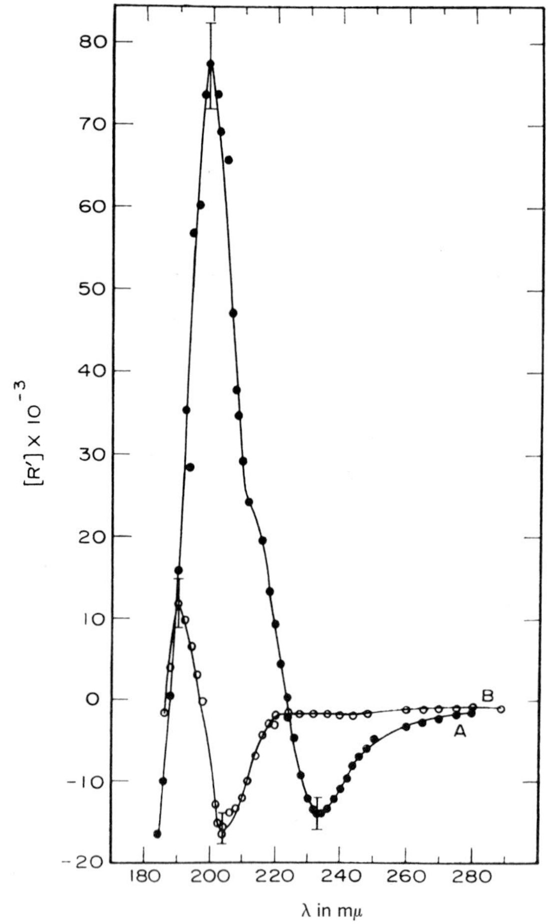

Fig. 7.5. The ultraviolet optical rotatory dispersion of the helical form of poly-α, L-glutamic acid in water at pH 4.3 (curve *A*), and of the random-coil form of the sodium salt at pH 7.1 (curve *B*). The ordinate is the optical rotation per residue. (*From E. R. Blout, I. Schmier, and N. S. Simmons, J. Am. Chem. Soc.,* 84, 3193, 1962.)

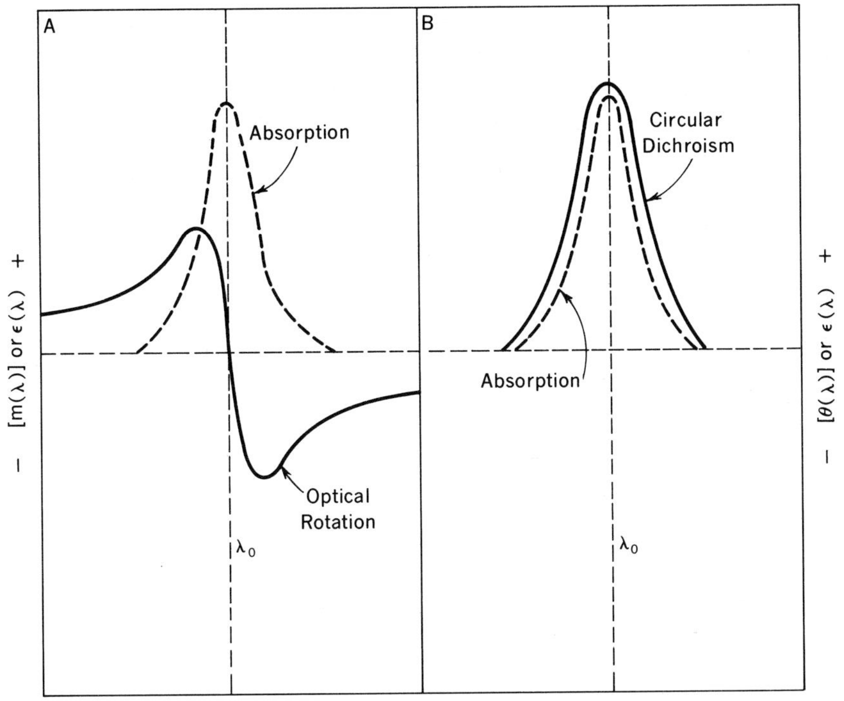

Fig. 7.6. Theoretical relationships for (*A*) an optically active absorption band for light absorption or optical rotation, and for (*B*) circular dichroism. The ordinates are as follows: $m(\lambda)$, optical rotation per mole at wavelength λ; $\epsilon(\lambda)$, extinction coefficient of absorption band; $\theta(\lambda)$, ellipticity or difference in absorption for left and right circularly polarized light. (*From S. Beychok, Science,* **154**, 1288, 1966.)

circularly polarized light. In the region of an optically active absorption band, the optical rotatory dispersion and circular dichroism have the characteristic features shown in Fig. 7.6. The two phenomena illustrated, *i.e.*, unequal velocity of transmission (optical rotation) and unequal absorption (circular dichroism), have been named the Cotton effect, after their discoverer. Such Cotton effects, in simple proteins, are associated with the polypeptide absorption bands in the 190 to 240 mμ region, as well as in the region 280 to 290 mμ dominated by the absorption of the aromatic side chains of tryptophan and tyrosine. The shape and magnitude of rotatory dispersion curves and circular dichroism spectra are very sensitive to changes in conformation of proteins and in the environment of the side-chain chromophores. This is illustrated in Fig. 7.7 for the binding by the enzyme lysozyme with an inhibitor which is known from x-ray studies to interact with tryptophan residues. In conjugated proteins additional Cotton effects associated with the absorption bands of the nonprotein moieties are usually observed; these have been termed *extrinsic* Cotton effects.

X-ray Analysis of Protein Structure. In crystals there is a regular three-dimensional lattice of unit cells, each unit cell having the same relationship to its neighbors and the contents of each cell being the same. The unit cell may be composed of one or a few atoms or molecules. In simple molecules composed of only a few atoms, much of the chemical behavior can be explained without taking into account exact

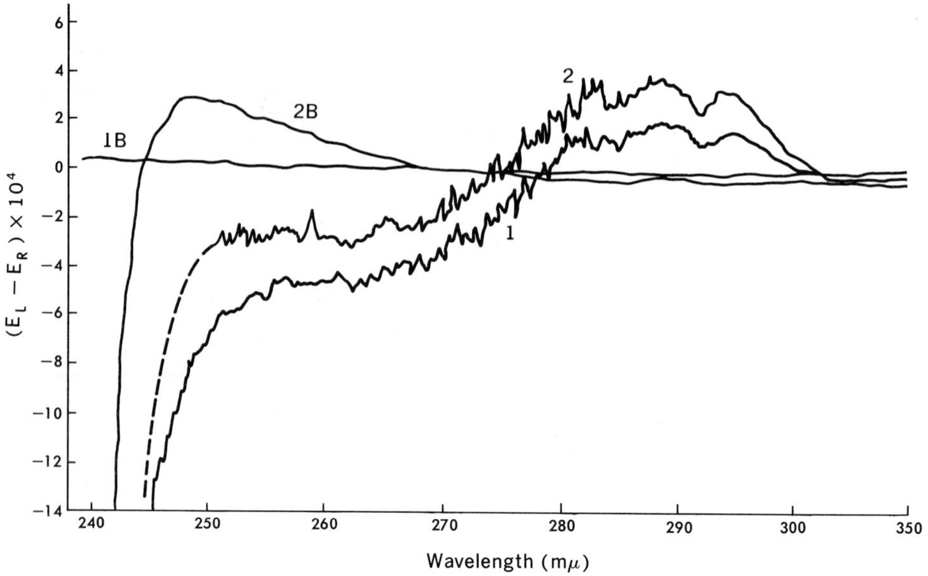

Fig. 7.7. Changes in circular dichroism shown by lysozyme on binding with N-acetyl-D-glucosamine (NAG). Curve 1, lysozyme in 0.05*M* phosphate buffer at pH 7; curve 2, lysozyme with 0.25*M* NAG; curve 1*B*, phosphate buffer alone; curve 2*B*, 0.25 *M* NAG in phosphate buffer. $E_L - E_R$ is the ellipticity or difference in absorption for left (*L*) and right (*R*) circularly polarized light. (*From A. N. Glazer and N. S. Simmons, J. Am. Chem. Soc.,* 88, 2335, 1966.)

dimensions and geometrical relationships. For complex molecules, particularly those as large and diverse as proteins, such relationships become of critical importance. At present, the only method available to obtain this information is the technique of x-ray analysis.

X-rays have very short wavelengths of the order of interatomic distances, *e.g.*, the x-ray beam of 1.542 Å., produced by electron bombardment of copper, is used for much of the work on protein structure. When x-rays strike an atom, they are diffracted (reflected) in proportion to the number of extranuclear electrons in the atom. Thus, heavier atoms, those of higher atomic number, produce more diffraction than lighter ones. The crystal may be regarded as a three-dimensional pattern of electron density, which has high values near the centers of atoms and low or zero values in between.

Figure 7.8 shows a typical roentgenogram of a single protein crystal. The spots form a regular two-dimensional lattice and, clearly, there is a marked symmetry in the pattern of spots in the four quarters of the photograph. Such a photograph has been obtained by mounting a small crystal in a known orientation in the path of a fine beam of monochromatic x-rays. The x-rays, scattered by the crystal, impinge on a photographic plate mounted behind the crystal. The picture in Fig. 7.8 is a two-dimensional lattice, since the photograph has been taken in a single plane, whereas the crystal is three-dimensional. From a series of electron-density photographs in different planes, the three dimensions of simple molecules can be constructed.

Such x-ray diffraction pictures permit measurement of the intensities (ampli-

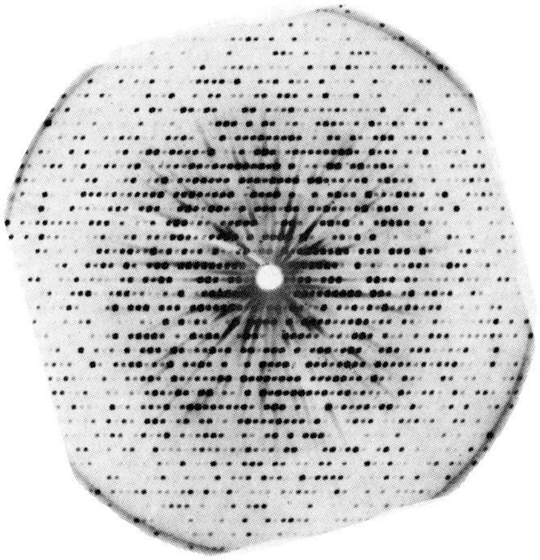

Fig. 7.8. A roentgenogram of crystalline whale myoglobin. The spots form a regular two-dimensional lattice in symmetry. Only a portion of the complete pattern is shown. (*Courtesy of Dr. John C. Kendrew.*)

tudes) but not of the phases of the diffracted x-rays. There are several methods of attempting to determine the phases. The most important in the study of protein crystals is the technique of *isomorphous replacement.* This requires obtaining protein crystals that are identical except for the introduction of a heavy atom in the structure, for instance, replacing a hydrogen by a mercury atom. The diffraction patterns of the two crystals are similar except for the small changes in the intensities of the diffracted rays that are caused by the different scattering power of the two atoms. These changes permit determination of the location of the heavy atom and then the calculation of approximate phase angles. This technique, introduced by Perutz for the study of hemoglobin, has been successfully applied in studies of protein structure.

By measurement of a series of plates, such as shown in Fig. 7.8, with respect to both distance and intensity of the spots, it is possible to prepare by calculation electron-density maps which, at a relatively gross level, indicate the general conformation of a molecule and, at a highly refined level, may permit the assignment in space of each of the heavier atoms of a protein molecule. The first successful study of this type was on the structure of whale myoglobin by Kendrew and his associates.

Figure 7.9 shows the structure of a model of myoglobin. The tubes, shown at the left, are hollow, and the walls follow a spiral path which has the exact dimensions of the Pauling-Corey α-helix. The helical regions are shown at the right as double lines; the nonhelical corners or bends, as single lines. At presently available levels of resolution (2 Å.), it is possible to identify certain groups which are large and characteristic, such as the indole group and the imidazole group.

The information gleaned from the study of the myoglobin molecule permitted, for the first time, direct verification of many of the major deductions concerning the general structure of globular proteins, as well as a precise picture of myoglobin itself. The right-handed α-helix is a major feature of the secondary structure involving about 75 per cent of the 153 residues in myoglobin; the remaining third of the

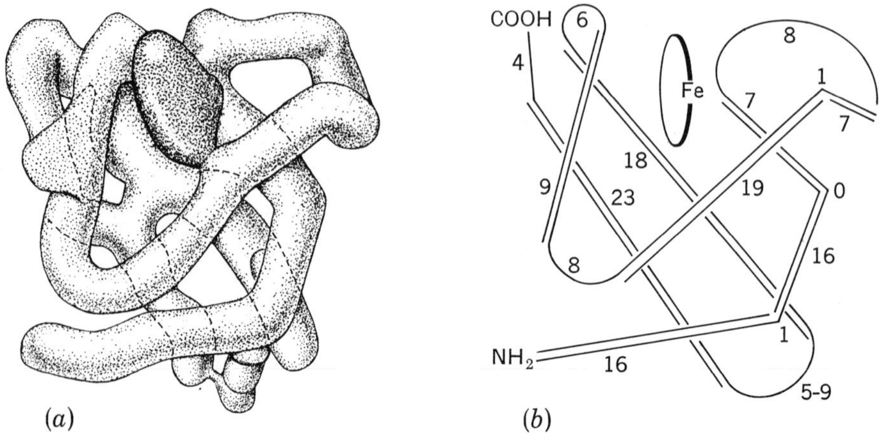

Fig. 7.9. The structure of whale myoglobin as deduced from x-ray crystallography. (*a*) The general conformation of the molecule. The plane of the heme group is not correct in this diagram. (*b*) A schematic representation of the molecule showing the numbers of amino acids in each of the α-helical segments (double lines) and the corners (single lines). (*Courtesy of Dr. John C. Kendrew.*)

residues are distributed over nonhelical regions with no regular pattern apparent. Proline residues do not fit an α-helix because the δ-carbon atom of the pyrrolidine ring occupies the position of the peptide (amide) hydrogen which for other amino acids functions to bond with the carbonyl in the next coil of the helix. The four proline residues of whale myoglobin are present in nonhelical regions; however, there are several such regions without proline. These nonhelical regions are stabilized by various types of interactions: hydrogen bonding involving the hydroxyl group of threonine, hydrophobic bonding between two parallel aromatic rings, etc. Myoglobin is devoid of cysteine and cystine; the disulfide bonds of cystine contribute to the over-all conformation and stability of proteins containing such bonds.

X-ray analysis of egg-white lysozyme by Phillips and his coworkers has revealed the structure of this molecule, which is made up of a single peptide chain of 129 residues with four disulfide bridges (see Fig. 12.11, page 270). The content of α-helix is difficult to assign because all these regions appear in a somewhat distorted form; nevertheless, the helical content is less than half the amount in myoglobin. These regions include residues 5 to 15, 24 to 34, and 88 to 96, and (much modified) residues 80 to 85 and 119 to 122. In one part of the molecule there is a distinct "hairpin" structure containing a β structure or "antiparallel" pleated sheet between residues 41 to 45 and 50 to 54, with residues 46 to 49 folded into the bend of the hairpin. The remaining sections of the peptide chain are folded irregularly.

Of major importance is the finding that lysozyme is folded in such a manner that a sizable cleft runs along one part of the molecule. Competitive inhibitors, and presumably the substrate, fit into this cleft, as fortunately revealed by isomorphous crystals of inhibitor-enzyme complexes. From x-ray analysis it has been possible to ascertain the nature of the binding sites of inhibitors to the enzyme and to postulate a reasonable mechanism of action of the enzyme. These findings are discussed in Chap. 12.

Table 7.5 lists some types of secondary bonding revealed in myoglobin and lysozyme by x-ray analysis. Relatively few such interactions are found among polar residues. In myoglobin, for example, most of the ionic residues (Arg, His, Lys, Glu, Asp) are on the surface of the molecule and such ionic groups are free and surrounded by water; nevertheless, the few such interactions which are present in the interior of the protein play an important role in maintaining its structural integrity. This is particularly the case with peptide carbonyl or NH groups which are not bonded in an α-helical segment. Furthermore, cross-hydrogen bonding from peptide C=O and NH groups between segments of a single chain, as in myoglobin, or between different chains, as in collagen (Chap. 38), aids in stabilizing globular as well as fibrous proteins.

Table 7.5: SOME TYPES OF SECONDARY BONDING IN PROTEINS

Residue	Partner	Type of bonding
Lys	Glu	Ionic
Arg	Asp	Ionic
Glu	Trp (indole N)	Hydrogen
Asp	Arg, His	Ionic
Gln, Asn	Peptide carbonyl	Hydrogen from amide
Ser, Thr (OH)	Peptide carbonyl; amide carbonyl	Hydrogen
His	Chain carbonyl	Hydrogen
Tyr	Chain carbonyl	Hydrogen

The foregoing discussion deals with types of interactions which are relatively simple in that ionic and hydrogen bonding are relatively specific. In contrast, nonpolar contacts are exceedingly numerous and difficult to describe simply; *e.g.*, in myoglobin, Kendrew has estimated that there are about 90 nonpolar contacts between the heme group and the neighboring atoms.

Denaturation and Enzymic Activity. The native state of a globular protein may be defined as the highly ordered conformation in which the biological activity of the protein is manifested. Denaturation then represents the loss of native conformation and hence represents alteration of the uniquely ordered structure into randomly arranged peptide chains. This assumes that, in general, there is a single, native conformation of maximal stability. For globular proteins this is obviously the situation in the solid crystalline state, since the x-ray patterns are regular and all molecules possess the same form. Loss of specific biological activity, particularly enzymic activity, is the most common and, generally, the most sensitive indication of an alteration in conformation sufficiently gross to warrant being described as "denaturation."

Most proteins are denatured by heating, usually over a fairly sharp transition region (page 234), by high concentrations of urea (6 to 8*M*), guanidine salts (4 to 6*M*), by anionic and cationic detergents, by extremes of pH, and by many organic solvents miscible with water. Shaking and foaming protein solutions produce surface films of denatured proteins; a familiar example is the beating of egg white in the formation of meringues. Chemical modification of amino acid side chains or cleavage of peptide bonds does not necessarily accompany denaturation, although some chemical agents may do both. Ultraviolet radiation or other types of radiation may produce denaturation, but this may be secondary to photooxidation of imidazole-, indole-, phenol-, or sulfur-containing side chains.

There are many methods of assessing denaturation, other than loss of enzymic or other biological activity. Denaturation may be manifested by loss of crystallizability, a marked increase in viscosity, a decrease in solubility, or changes in sedimentation and diffusion coefficients. As already noted, there are usually marked changes in exchangeability of hydrogen atoms (page 152), and in optical rotatory dispersion, absorption spectrum, and circular dichroism (page 154). Many native, globular proteins are resistant to digestion by proteolytic enzymes, such as trypsin and chymotrypsin, whereas denatured proteins are rapidly hydrolyzed by these enzymes.

As already noted (page 141), large proteins containing subunits which are held together by noncovalent forces are dissociated by denaturation with 8*M* urea or 6*M* guanidine salts; the individual subunits will be in randomly coiled forms.

Changes in the reactivity of amino acid side chains in proteins may occur on denaturation; these alterations assist in assigning the role of such groups in maintaining native structure. An example is the role of phenolic side chains of tyrosine residues. The titration of phenolic groups may be measured spectrophotometrically, since there is a large shift in the ultraviolet absorption accompanying the change from the undissociated group to the phenolate ion. Ionization of phenolic groups should be complete near pH 12.0; however, in many proteins only a portion of such groups is instantaneously titrated. Heating, treatment with 8*M* urea, or bringing the solution to pH 13.0 or 14.0 may be required to titrate all phenolic groups. Such groups are unavailable for titration in the native protein because they reside in a hydrophobic environment or because they are hydrogen-bonded, or due to both factors.

Similarly, a portion of other groups—imidazole, thiol, carboxyl, ϵ-amino, etc. —in certain native proteins is unavailable for titration. Evidently, the over-all conformation of a native protein, as indicated by the direct evidence for myoglobin and lysozyme and by much indirect evidence for other proteins, involves all the secondary forces discussed above—hydrophobic, ionic, and hydrogen bonding. This is also indicated by the types of agents that produce denaturation. Hydrogen-bonding agents in high concentration—urea, guanidine salts, formamide, etc.—complete with the hydrogen bonds in the native protein and produce an unfolding. High concentrations of H^+ ion (pH 1 to 2) or OH^- ion (pH 12 to 14) disrupt ionic and hydrogen bonds. Agents, such as trichloroacetate ion, which introduce a hydrophobic group and which bind strongly with cationic groups, also cause denaturation. Hydrophobic bonds are broken by nonpolar solvents and by detergents, agents which are water soluble but which possess large hydrophobic groups, *e.g.*, dodecyl sulfonate, bile salts (pages 62 and 81), etc.

All enzymically active proteins lose their catalytic activity on denaturation. This in itself suggests that the activity of such molecules is determined by the juxtaposition or interaction of amino acid side chains present in the native conformation and destroyed by denaturation. Denaturation is in many cases a reversible phenomenon, indicating that the native form is, under proper conditions, a highly stable one, produced by specific interactions. Indeed, present views and understanding of protein biosynthesis (Chap. 29) indicate that protein structure is determined genetically by specification of the amino acid sequence in the synthesis of the linear polypeptide chain or chains and that the specific conformation is attained spontaneously. Another way of stating this is that under specific conditions of pH, ionic strength, temperature, etc., the native conformation is the most stable and hence the most probable one.

The enzyme ribonuclease, whose sequence is shown in Fig. 12.4 (page 257), contains four disulfide bridges. Reduction of these bonds, in the presence of 8*M* urea, produces a total loss of activity and changes in all the physical properties of the protein. After the urea has been removed by dialysis or other means, the enzyme remains inactive; however, reoxidation in air at neutral pH reforms the original disulfide bonds, resulting in essentially complete regeneration of enzymic activity and restoration of all the original physical and chemical properties. Clearly, the over-all conformation is produced spontaneously, the correct disulfide bond formation occurring because the half-cystine residues are in proper juxtaposition within the single-peptide chain of the molecule.

Hemoglobin lacks disulfide bonds and contains four peptide chains, two α and two β (pages 172*ff.*), each chain associated with a heme group (Chap. 8). At acid pH values in acetone, the heme is separated from the denatured globin. Nevertheless, at neutral pH values, native hemoglobin will be reformed from heme and globin. Thus, native conformation is reattained even in this complex molecule with four peptide chains, each with its heme group in the proper position. It can only be concluded that the secondary and tertiary structures are determined by the specific amino acid sequences of the linear polypeptide chains.

Other large proteins, including many enzymes, are dissociated into inactive subunits by various agents. Thus, although the active protein unit may be very large, of the order of several hundred thousand to millions in molecular weight, the primary peptide chains are relatively small. The forces involved in forming the active proteins by an aggregation of smaller units are the same as those involved in folding the individual peptide chain into its proper conformation.

Primary Structure and Biological Activity. Because proteolytic enzymes can cleave peptide bonds in a specific manner, they are used as reagents to degrade biologically active polypeptides and proteins in attempts to ascertain the parts of the molecule that are essential for activity. Among the first successful studies of this type was the demonstration that certain proteinases could hydrolyze specific antibodies to smaller molecules without loss of their antibody properties (Chap. 31).

By the same approach, studies have been made on a number of hormones. Carboxypeptidase can remove C-terminal alanine from the B chain of insulin without impairing the hormonal activity of this substance. Thus, the biological activity must reside elsewhere in the insulin molecule. A further example is the action of pepsin on adrenocorticotropic hormone (ACTH; see Chap. 47); a portion of the carboxyl-terminal end of the molecule can be removed without destroying the hormonal activity. In contrast to this, removal of a few residues from the amino-terminal end of the molecule by leucine aminopeptidase destroys the hormonal activity. Other examples of such studies are mentioned elsewhere in this book.

Amino Acid Substitution. Comparative sequence studies of homologous proteins from different species and from genetic variants within a species have shown that considerable substitution of residues can occur without significant effect on the biological activity of proteins. In such instances it has been possible to deduce which regions of the sequence are critical for conformation and activity and which residues are apparently unessential. Amino acid substitutions have also proved to be useful in supplementing knowledge of the relationships of amino acid side chains and their roles in protein structure. These problems are considered for cytochrome c in Chap. 8 and for other proteins in Chap. 30.

REFERENCES

Books

Alexander, P., and Lundgren, H. P., eds., "A Laboratory Manual of Analytical Methods in Protein Chemistry Including Polypeptides," vol. 4, Pergamon, New York, 1966.

Anfinsen, C. B., Jr., Anson, M. L., Edsall, J. T., and Richards, F. M., eds., "Advances in Protein Chemistry," Academic Press, Inc., New York. (An annual series beginning with vol. I in 1944.)

Bailey, J. L., "Techniques of Protein Chemistry," 2d ed., American Elsevier Publishing Company, New York, 1967.

Bodanszky, M., and Ondetti, M. A., "Peptide Synthesis," Interscience Publishers, Inc., New York, 1966.

Cohn, E. J., and Edsall, J. T., "Proteins, Amino Acids, and Peptides as Ions and Dipolar Ions," Reinhold Publishing Corporation, New York, 1942.

Edsall, J. T., and Wyman, J., "Biophysical Chemistry," vol. I, "Thermodynamics, Electrostatics, and the Biological Significance of the Properties of Matter," Academic Press, Inc., New York, 1958.

Florkin, M., and Stotz, E. H., eds., "Comprehensive Biochemistry," vols. 7 and 8, American Elsevier Publishing Company, New York, 1963.

Gordon, A. H., and Eastoe, J. E., "Practical Chromatographic Techniques," D. Van Nostrand Company, Inc., Princeton, N.J., 1964.

Gottschalk, A., ed., "Glycoproteins, Their Composition, Structure and Function," American Elsevier Publishing Company, New York, 1966.

Greenstein, J. P., and Winitz, M., "Chemistry of the Amino Acids," 3 vols., John Wiley & Sons, Inc., New York, 1961.

Heftmann, E., ed., "Chromatography," 2d ed., Reinhold Publishing Corporation, New York, 1967.

Hirs, C. H. W., ed., "Enzyme Structure," vol. XI, in S. P. Colowick and N. O. Kaplan, eds., "Methods in Enzymology," Academic Press, Inc., New York, 1967.

James, A. T., and Morris, L. J., ed., "New Biochemical Separations," D. Van Nostrand Company, Inc., New York, 1964.

Martin, R. B., "Introduction to Biophysical Chemistry," McGraw-Hill Book Company, New York, 1964.

Meister, A., "Biochemistry of the Amino Acids," 2d ed., 2 vols., Academic Press, Inc., New York, 1965.

Neuberger, A., ed., "Symposium on Protein Structure," Methuen & Co., Ltd., London, 1958.

Neurath, H., ed., "The Proteins: Composition, Structure and Function," 2d ed., vols. I–V, Academic Press, Inc., New York, 1963–1968.

Schachman, H. K., "Ultracentrifugation in Biochemistry," Academic Press, Inc., New York, 1959.

Scheraga, H. A., "Protein Structure," Academic Press, Inc., New York, 1963.

Schröder, E., and Lübke, K., "The Peptides," 2 vols., Academic Press, Inc., New York, 1965–1966.

Svedberg, T., and Pedersen, K. O., "The Ultracentrifuge," Oxford University Press, Fair Lawn, N.J., 1940.

Tanford, C., "Physical Chemistry of Macromolecules," John Wiley & Sons, Inc., New York, 1961.

Williams, J. W., ed., "Ultracentrifugation Analysis," Academic Press, Inc., New York, 1963.

Review Articles

Beychok, S., Circular Dichroism of Biological Macromolecules, *Science,* **154,** 1288–1299, 1966.

Holmes, K. C., and Blow, D. M., The Use of X-ray Diffraction in the Study of Protein and Nucleic Acid Structure, *Methods of Biochemical Analysis,* **13,** 113–239, 1965.

Kendrew, J. C., The Three-dimensional Structure of a Protein Molecule, *Scientific American,* **205,** (Dec.) 96–110, 1961.

Kent, P. W., Structure and Function of Glycoproteins, in P. N. Campbell and G. D. Greville, eds., "Essays in Biochemistry," vol. 3, pp. 105–151, Academic Press, Inc., New York, 1967.

Phillips, D. C., The Three-dimensional Structure of an Enzyme Molecule, *Scientific American,* **215,** (Nov.) 78–90, 1966.

Tschiersch, B., and Mother, K., Amino Acids: Structure and Distribution, in M. Florkin and H. S. Mason, eds., "Comparative Biochemistry," vol. V, part C, pp. 1–90, Academic Press, Inc., New York, 1963.

Vickery, H. B., and Schmidt, C. L. A., The History of the Discovery of the Amino Acids, *Chem. Revs.,* **9,** 169–318, 1931.

8. The Proteins. IV

Hemoproteins and Porphyrins

The hemoproteins include the hemoglobins concerned in oxygen transport, the cytochromes, which are electron-transfer agents, and several enzymes such as catalase and peroxidase. In each instance the colorless protein is linked to an iron-porphyrin compound. Porphyrins are also found linked with other metals; the most important example is chlorophyll, a porphyrin derivative containing magnesium (Chap. 20). The distribution of the hemoproteins embraces almost all living matter. The ability to synthesize porphyrins is possessed by almost all species; only a few species require exogenous porphyrins as growth factors.

The important place that porphyrins occupy in living organisms is also indicated by the findings that porphyrins derived from chlorophyll occur in petroleum, coals, oil shales, and asphalts. Porphyrins have also been found in fossilized excrements (coproliths) of crocodiles. Thus the geologist has provided evidence that essential biological compounds that functioned millions of years ago did not differ from those which function today.

THE PORPHYRINS

The naturally occurring porphyrins are derivatives of the fundamental substance *porphin,* which contains four pyrrole-like rings linked by four CH groups or methene bridges in a ring system.

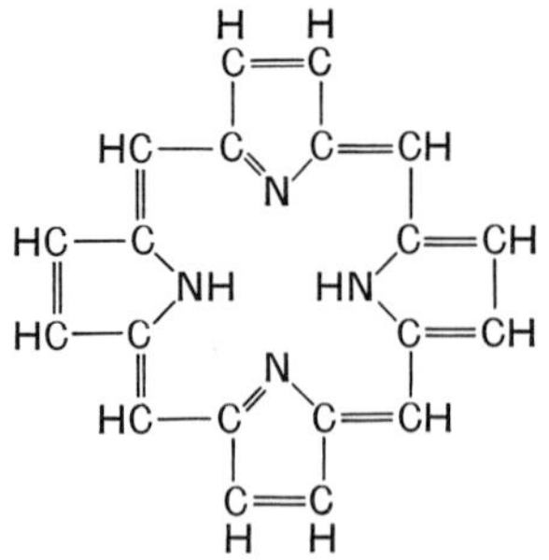

Porphin ($C_{20}H_{14}N_4$)

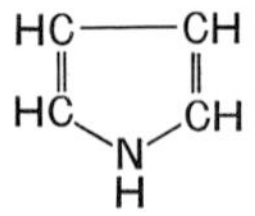

Pyrrole

The porphyrin structure contains a central 16-membered ring formed from 12 carbon and 4 nitrogen atoms contributed by 4 pyrrole rings. The positions assigned to the alternating double and single bonds of porphin are arbitrary because of resonance. Porphin was first synthesized by Hans Fischer and Gleim in 1935; the compound is not known to occur in nature.

It is convenient to use the formula shown below, in which the carbon atoms of the pyrrole rings are represented as the corners of pentagons, with the nitrogen shown and the hydrogen atoms omitted.

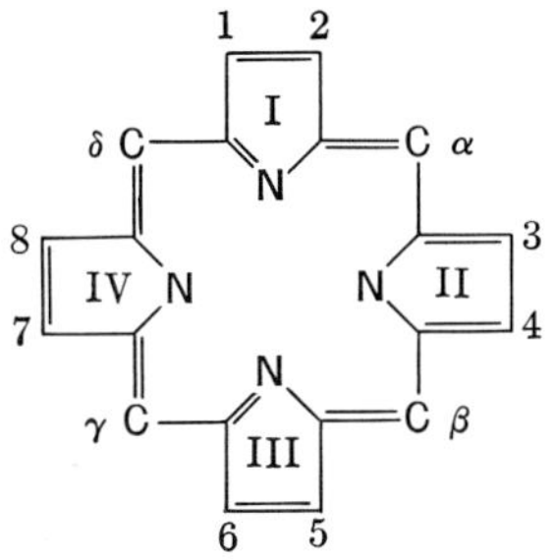

The porphyrins that are found in nature are all compounds in which side chains are substituted for the eight hydrogen atoms numbered in the pyrrole rings in porphin (formula above). For convenience in showing these substitutions, the simplified shorthand formula of Fischer is used for representing the porphyrin nucleus, with all the bridges and rings omitted:

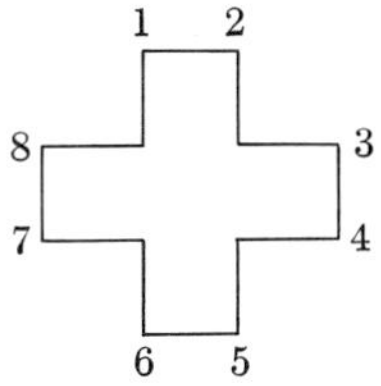

The names of some of the important porphyrins are given in Table 8.1 with the types and positions of the side chains indicated.

Each formula listed in Table 8.1 actually represents a large group of substances, since considerable isomerism is possible. The simplest examples are etio-,

Table 8.1: STRUCTURES OF PORPHYRIN TYPES I AND III

Porphyrin	Formula	Type I	Type III
Etioporphyrin	$C_{32}H_{38}N_4$	1,3,5,7-M,2,4,6,8-E	1,3,5,8-M,2,4,6,7-E
Mesoporphyrin	$C_{34}H_{38}O_4N_4$	1,3,5,7-M,2,4-E,6,8-P	1,3,5,8-M,2,4-E,6,7-P
Protoporphyrin	$C_{34}H_{34}O_4N_4$	1,3,5,7-M,2,4-V,6,8-P	1,3,5,8-M,2,4-V,6,7-P
Coproporphyrin	$C_{36}H_{38}O_8N_4$	1,3,5,7-M,2,4,6,8-P	1,3,5,8-M,2,4,6,7-P
Uroporphyrin	$C_{40}H_{38}O_{16}N_4$	1,3,5,7-A,2,4,6,8-P	1,3,5,8-A,2,4,6,7-P
Deuteroporphyrin	$C_{30}H_{30}O_4N_4$	1,3,5,7-M,2,4-H,6,8-P	1,3,5,8-M,2,4-H,6,7-P
Hematoporphyrin	$C_{34}H_{38}O_6N_4$	1,3,5,7-M,2,4-EOH,6,8-P	1,3,5,8-M,2,4-EOH,6,7-P

Note: Symbols used for the side chains are as follows: M = methyl, E = ethyl, P = propionic acid, V = vinyl, EOH = hydroxyethyl, A = acetic acid, H = hydrogen. Numbers indicate the positions in the porphyrin nucleus at which these side chains are substituted. Except for the uroporphyrins, all the above types of porphyrins are soluble in ether.

copro-, and uroporphyrins, in which only two different kinds of groupings are found. For these porphyrins, four isomers are possible in each case. However, only two of these occur in nature, and these have been designated type I and type III by Fischer. The structures of the isomeric etioporphyrins are indicated in the formulas below, where M and E represent methyl and ethyl.

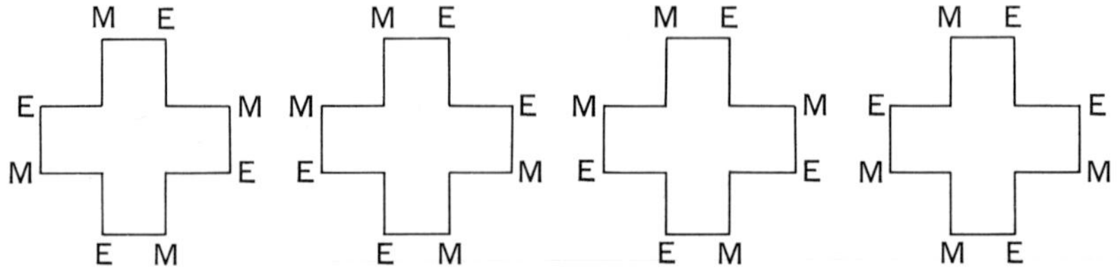

Isomeric etioporphyrins

Type I represents alternating substitution, and type III, unsymmetrical substitution. Porphyrins derived from type III are the most important in nature.

When three types of side chain are present, as in protoporphyrin, the number of possible isomers is 15. Naturally occurring protoporphyrin and certain other compounds derived from it are all related to Fischer's type III etioporphyrin since the methyl groups occupy the same positions. A protoporphyrin is shown in the formula below, where M is methyl ($—CH_3$), V is vinyl ($—CH{=}CH_2$), and P is propionic acid ($—CH_2—CH_2—COOH$).

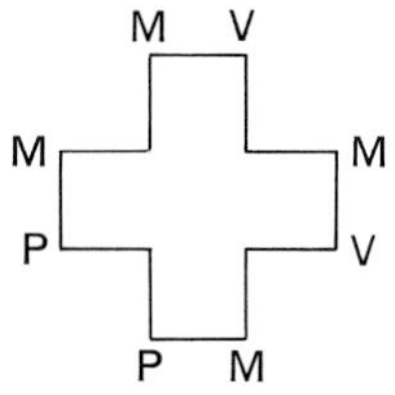

Protoporphyrin type III (No. IX)

This protoporphyrin is type III or, more commonly, No. IX, since it was the ninth in the series of isomers listed by Fischer. Etio-, meso-, and deuteroporphyrin (Table 8.1) are not known to occur in nature.

Properties of Porphyrins. The porphyrins are all weak bases because of the tertiary nitrogens in the two *pyrrolene* nuclei in each porphyrin.

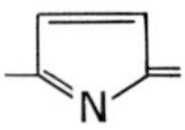

Pyrrolene nucleus

With the exception of the etioporphyrins, porphyrins are also acids, because of carboxyl groups in the side chains, and thus exhibit weakly amphoteric character. Porphyrins are easily precipitated from aqueous solutions in the isoelectric regions from about pH 3 to 4.5. The purification of the porphyrins is based largely on partition in nonpolar solvents in the presence of acids.

The most striking physical property of the porphyrins is their color; they have sharply banded absorption spectra in the visible region. In nonpolar solvents, they have four main bands in the visible and a strong band, the Soret band, in the near ultraviolet near 400 mμ. For protoporphyrin in ether–acetic acid, the main bands are at 632.5, 576, 537, 502, and 395 mμ. Solutions of porphyrins in organic solvents or mineral acids exhibit a strong red fluorescence when illuminated by ultraviolet light. The intense fluorescence is frequently used for detection and estimation of small amounts of free porphyrins. The strong absorption and fluorescence in visible light depends on the resonating character of the conjugated double bonds in the porphyrins. When the double bonds of the methene bridges are reduced by addition of hydrogen atoms, the porphyrins are converted to colorless porphyrinogens, which contain pyrrole nuclei linked by CH_2 groups.

Metal-Porphyrins. Porphyrins possess the ability to combine with many kinds of metals; the most important biological metalloporphyrins are those containing iron and magnesium. Other derivatives that can readily be formed include those with zinc, nickel, cobalt, copper, and silver. Although some complexes such as those with magnesium are labile in dilute acetic acid, stronger acids are required to remove iron or copper.

The metal compounds of porphyrins are not salts, as shown by their solubility in nonpolar solvents and by the fact that complexing occurs with both porphyrin esters and etioporphyrin, which do not contain carboxyl groups. The metal replaces the dissociable hydrogen atoms of two pyrrole rings and is simultaneously bound by coordinate valences to the tertiary nitrogen atoms of the other two pyrrole rings. Since resonance occurs, we do not distinguish among these valences but indicate that the central metal is bound equally to all four nitrogen atoms of the pyrrole rings, which lie in a plane.

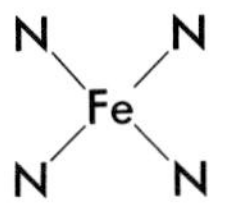

For simplicity, only the central portion of the iron-porphyrin complex is shown.

This discussion will be restricted mainly to the iron compounds containing protoporphyrin since they have the greatest biological interest. Because many important iron-porphyrin compounds were described before their chemistry was clearly understood, the terminology has undergone revisions as our knowledge has developed. The general types of compounds will be presented, designated in the terminology in most common use.

The iron compounds can exist in two forms depending on the valence of the iron, and these are distinguished by the prefixes *ferro-* for divalent iron and *ferri-* for trivalent iron. Ferroprotoporphyrin, or *heme,* contains divalent iron; the group possesses no net charge and may be written as 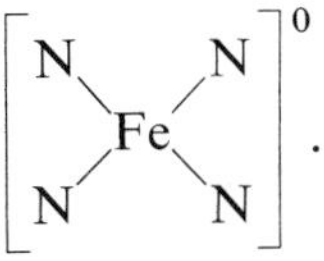 . Heme and similar chelates with other divalent metal ions exist in square-planar forms. Ferriprotoporphyrin, or *hemin,* has a net positive charge and adds an extra ligand. It is usually obtained as the chloride; the resulting pentacoordinate complex is essentially square-pyramidal with the extra ligand perpendicular to the porphyrin plane. The chloride or other halide anions are bound coordinately, not electrostatically, to the iron. Free heme is unstable and rapidly oxidized to hemin. Crystalline hemin is easily obtained by heating a hemoglobin solution with acetic acid containing a little sodium chloride, a procedure frequently employed in legal medicine for the detection of blood.

In the presence of excess alkali, hemin gives rise to a divalent anion of ferriprotoporphyrin hydroxide in which two carboxyl groups of the propionic acid side chains are ionized and the iron is bound to the hydroxyl group of the base and to a mole of water in octahedral form.

$$\left[H_2O-Fe \begin{array}{c} COO \\ OH \\ COO \end{array} \right]^=$$

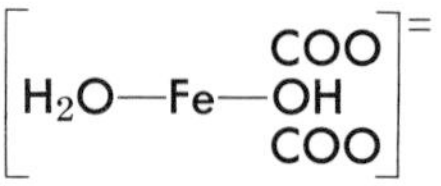

If hemin is dissolved in excess alkali and then titrated with acid, a compound precipitates when two moles of acid have been added. The resulting neutral compound is known as *hematin, hydroxyhemin,* or ferriprotoporphyrin hydroxide. For simplicity the nitrogens are omitted in this and in the above formula.

$$\left[Fe \begin{array}{c} COOH \\ OH \\ COOH \end{array} \right]^0$$

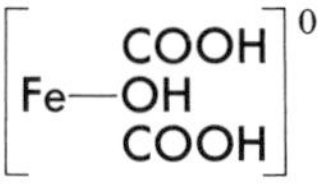

A copper-uroporphyrin III complex, *turacin,* is found as an ornamental red pigment in the feathers of certain South African birds of the genus *Turaco.*

Nitrogenous Compounds of Iron-Porphyrins. One mole of a ferro- or ferriporphyrin can combine with two moles of a nitrogenous base to form a complex compound. In the older terminology these compounds were known as *hemochromogens* (ferro) and *parahematins* (ferri), but they are currently referred to as base ferro- (or ferri-) porphyrins. For example, the ferro compound with pyridine is pyridine ferroporphyrin. The nitrogen of the base is linked to the central iron atom, and hexacoordinate structures are formed.

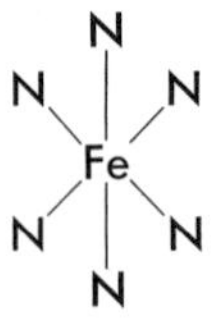

The atoms of N shown at the angles represent the porphyrin nitrogens, which lie in a plane; the vertical N atoms represent the nitrogens of the base, one of which lies above and the other below this plane. This type of complex is analogous to ferrocyanide ion, which is similarly pictured at the top of page 171.

Ferroprotoporphyrin combines with other bases, *e.g.,* piperidine, nicotine, cyanide, and histidine. However, similar compounds are not formed with arginine or lysine. Denatured globin also combines with ferroporphyrin, giving globin

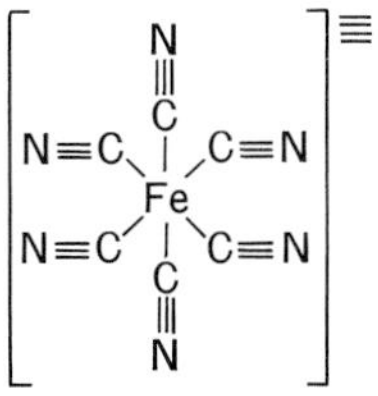

ferroporphyrin. The base-ferroporphyrin compounds react readily with atmospheric oxygen, converting the divalent iron to the trivalent state. The cyanide ferroporphyrin and the pyridine ferroporphyrin complexes are shown as examples.

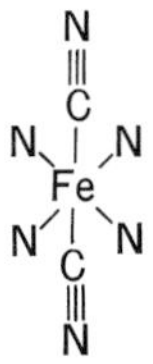

Cyanide ferroporphyrin

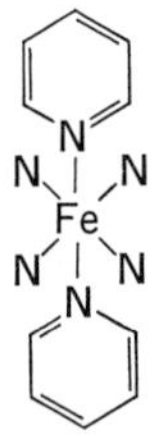

Pyridine ferroporphyrin

Like the uncombined porphyrins, the metal derivatives as well as the nitrogenous compounds of the metal-porphyrins possess characteristic absorption spectra which may be utilized for identification and for quantitative estimation. Formation of pyridine ferroporphyrin has been extensively used for recognition and measurement of ferroporphyrin.

All the nitrogenous ligands with ferroporphyrins form octahedral complexes in which the addition of the first acceptor ligand strongly increases the affinity for the second.

HEMOGLOBIN

Among all the nitrogenous derivatives of ferroprotoporphyrin, hemoglobin is unique in its ability to bind molecular oxygen in a loose and easily reversible combination. In this combination, the iron remains in the ferrous state. The symbol Hb is used for unoxygenated or reduced hemoglobin and HbO_2 for oxyhemoglobin.

The hemoglobins are conjugated proteins composed of colorless basic proteins, the globins, and ferroprotoporphyrin, or heme. The hemoglobins of different species differ quantitatively in such properties as crystal form, solubility, amino acid content, affinity for oxygen, and absorption spectra. These differences are due entirely to the protein moiety since the same ferroprotoporphyrin is present in all vertebrate and in many invertebrate hemoglobins.

The two component portions of hemoglobin can be separated by treatment with acids or bases. In the presence of hydrochloric acid and acetone, the globin is precipitated, and the ferroprotoporphyrin remaining in solution is rapidly oxidized.

$$\text{Hemoglobin} + \text{HCl} \longrightarrow \text{globin} \cdot \text{HCl} + \text{ferroprotoporphyrin}$$

$$\underset{\text{(heme)}}{2\ \text{Ferroprotoporphyrin}} + 2\text{HCl} + \tfrac{1}{2}\text{O}_2 \longrightarrow \underset{\text{(hemin)}}{2\ \text{ferriprotoporphyrin chloride}} + \text{H}_2\text{O}$$

In the presence of a strong reducing agent, such as $Na_2S_2O_4$, the ferroprotoporphyrin is not oxidized. At pH values near neutrality, globin recombines with ferroprotoporphyrin to give hemoglobin, or with ferriprotoporphyrin to yield methemoglobin (ferrihemoglobin).

$$\text{Globin} + \text{ferroprotoporphyrin} \longrightarrow \text{hemoglobin}$$
$$\text{Globin} + \text{ferriprotoporphyrin} \longrightarrow \text{methemoglobin}$$

Hemoglobin treated with oxidizing agents is converted to methemoglobin, which can be cleaved by acids or bases to yield globin and ferriprotoporphyrin.

Globin can also combine with other ferroporphyrins to give synthetic hemoglobins which bind oxygen reversibly. Such artificial hemoglobins have been prepared with ferromesoporphyrin, ferrohematoporphyrin, and others. From these findings it is evident that the vinyl groups present in protoporphyrin, but absent in the other compounds, are not essential for the combination with globin.

STRUCTURE OF HEMOGLOBIN

All mammalian hemoglobins have a molecular weight of approximately 65,000 and are essentially tetramers, consisting of four peptide chains, to each of which is bound a heme group. Normal human beings are capable, genetically, of synthesizing and incorporating into hemoglobin four distinct but related polypeptide chains, designated as α, β, γ, and δ, respectively. With but few exceptions, hemoglobin molecules are constructed by combining two α chains with two β, γ, or δ chains. Normal human adult hemoglobin, called Hb A, contains two α and two β chains; in short form this is designated as Hb A = $\alpha_2{}^A\beta_2{}^A$, to indicate that each chain is from Hb A and that there are two chains of each type. Correspondingly, normal human fetal hemoglobin is Hb F = $\alpha_2{}^A\gamma_2{}^F$.

The individual chains can be separated by removing the heme at an acid pH value, followed by ion exchange chromatography on a carboxylate resin, by countercurrent distribution or by electrophoresis. The complete amino acid sequences of the α, β, and γ chains are shown in Fig. 8.1 with the structures aligned to permit comparison of the sequences. The α chain contains 141 residues, and the β and γ chains contain 146 residues; nevertheless, it is evident that there is close similarity among them.

By themselves the sequences in the three chains yield little information concerning the functional properties of hemoglobin; however, when they are considered in conjunction with x-ray crystallographic studies on the conformation of the chains, a wealth of information emerges. Perutz and associates have demonstrated that each of the four chains in Hb A possesses an over-all conformation similar to that of myoglobin (page 159), and from the positions of the residues in myoglobin and the convolutions of the chains in hemoglobin, they assigned the positions of the residues in the helical and nonhelical regions of the α and β chains of the latter protein.

Figure 8.2 shows in a plane some of these relationships for the β chain only. The helical portions are labeled *A*, *B*, *C*, etc. Since the β and γ chains are of equal length and very similar in amino acid sequence, the relationships presumably apply to the γ chain also. The abbreviations for the residues that differ in the β

10 20

α Val·Leu·Ser·Pro·Ala·Asp·Lys·Thr·Asg·Val·Lys·Ala·Ala·Try·Gly·Lys·Val·Gly·Ala·His·Ala·Gly·Glu·Tyr·

β Val·His·Leu·Thr·Pro·Glu·Glu·Lys·Ser·Ala·Val·Thr·Ala·Leu·Try·Gly·Lys·Val·Asg·Val· Asp·Glu·Val·

γ Gly·His·Phe·Thr·Glu·Glu·Asp·Lys·Ala·Thr·Ileu·Thr·Ser·Leu·Try·Gly·Lys·Val·Asg·Val· Glu·Asp·Ala·

10 20

30 40

α Gly·Ala·Glu·Ala·Leu·Glu·Arg·Met·Phe·Leu·Ser·Phe·Pro·Thr·Thr· Lys·Thr·Tyr·Phe·Pro·His·Phe·Asp·Leu·

β Gly·Gly·Glu·Ala·Leu·Gly·Arg·Leu·Leu·Val·Val·Tyr·Pro·Try·Thr·Glm·Arg·Phe·Phe·Glu·Ser·Phe·Gly·Asp·Leu·

γ Gly·Gly·Glu·Thr·Leu·Gly·Arg·Leu·Leu·Val·Val·Tyr·Pro·Try·Thr·Glm·Arg·Phe·Phe·Asp·Ser·Phe·Gly·Asg·Leu·

30 40

50 60

α Ser·His·Gly·Ser·Ala· Glm·Val·Lys·Gly·His·Gly·Lys·Lys·Val·Ala·Asp·Ala·Leu·Thr·Asg·

β Ser·Thr·Pro·Asp·Ala·Val·Met·Gly·Asg·Pro·Lys·Val·Lys·Ala·His·Gly·Lys·Lys·Val·Leu·Gly·Ala·Phe·Ser·Asp·

γ Ser·Ser·Ala·Ser·Ala·Ileu·Met·Gly·Asg·Pro·Lys·Val·Lys·Ala·His·Gly·Lys·Lys·Val·Leu·Thr·Ser·Leu·Gly·Asp·

50 60 ↑ 70

70 80 90

α Ala·Val·Ala·His·Val·Asp·Asp·Met·Pro·Asg·Ala·Leu·Ser·Ala·Leu·Ser·Asp·Leu·His·Ala·His·Lys·Leu·Arg·Val·

β Gly·Leu·Ala·His·Leu·Asp·Asp·Leu·Lys·Gly·Thr·Phe·Ala·Thr·Leu·Ser·Glm·Leu·His·Cys·Asp·Lys·Leu·His·Val·

γ Ala·Ileu·Lys·His·Leu·Asp·Asp·Leu·Lys·Gly·Thr·Phe·Ala·Glm·Leu·Ser·Glu·Leu·His·Cys·Asp·Lys·Leu·His·Val·

80 90 ↑

100 110

α Asp·Pro·Val·Asg·Phe·Lys·Leu·Leu·Ser·His·Cys·Leu·Leu·Val·Thr·Leu·Ala·Ala·His·Leu·Pro·Ala·Glu·Phe·Thr·

β Asp·Pro·Glm·Asp·Phe·Arg·Leu·Leu·Gly·Asg·Val·Leu·Val·Cys·Val·Leu·Ala·His·His·Phe·Gly·Lys·Glu·Phe·Thr·

γ Asp·Pro·Glu·Asg·Phe·Lys·Leu·Leu·Gly·Asg·Val·Leu·Val·Thr·Val·Leu·Ala·Ileu·His·Phe·Gly·Lys·Glu·Phe·Thr·

100 110 120

120 130 141

α Pro·Ala·Val·His·Ala·Ser·Leu·Asp·Lys·Phe·Leu·Ala·Ser·Val·Ser·Thr·Val·Leu·Thr·Ser·Lys·Tyr·Arg

β Pro·Pro·Val·Glm·Ala·Ala·Tyr·Glm·Lys·Val·Val·Ala·Gly·Val·Ala·Asp·Ala·Leu·Ala·His·Lys·Tyr·His

γ Pro·Glu·Val·Glm·Ala·Ser·Try·Glm·Lys·Met·Val·Thr·Gly·Val·Ala·Ser·Ala·Leu·Ser·Ser·Arg·Tyr·His

130 140 146

Fig. 8.1. Amino acid sequences in α, β, and γ chains derived from human hemoglobins; α and β from adult hemoglobin, and α and γ from fetal hemoglobin. Gaps are left in depicting the sequences only in order to emphasize sequence similarities in homologous portions of the three chains. In these sequences, *Ileu* = isoleucine, *Try* = tryptophan, *Asg* = asparagine and *Glm* = glutamine. The numbers at the top indicate every tenth residue position in the α chain; at the bottom the numbers refer to residues in the β and γ chains.

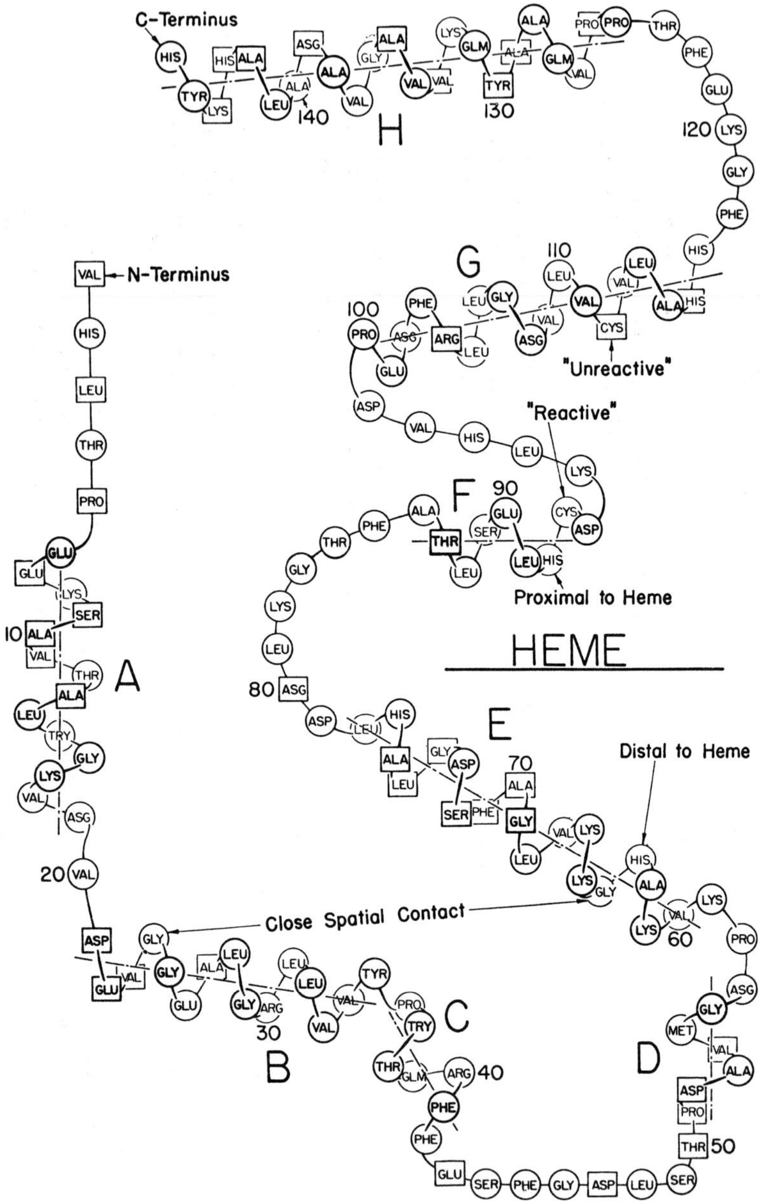

Fig. 8.2. The conformation of the β chain of human hemoglobin shown in a plane. Regions of α-helix are designated *A*, *B*, *C*, *D*, etc., with the boldface residues facing the same plane. Every tenth residue is numbered from the amino- or N-terminus. The heme residues are between regions *E* and *F*. The cysteine residue at position 93 is "reactive" with thiol reagents. In this figure, TRY = tryptophan, ASG = asparagine, and GLM = glutamine. (*From W. A. Schroeder, Ann. Rev. Biochem.*, **32**, 301, 1963.)

and γ chains are shown in rectangles. The over-all structure of the α chain is also given by the structure in the figure, with a few differences. The interhelical region *AB* should be lengthened by two residues, and *CD* should be shortened by one; the helix *D* (residues 54 to 58) is absent since five residues in the β and γ chains have no counterpart in the α chain. Note that these differences between the α and β chains are not in the neighborhood of the heme groups.

The heme group lies within a crevice. The hydrophobic vinyl groups of the porphyrin are surrounded by hydrophobic amino acid side chains. The two propionate side chains of each heme lie in juxtaposition to positively charged nitrogens of a lysine residue and an arginine residue. The iron component of the heme is closely coordinated to an imidazole side chain of histidine (residues 92 in the β and the γ, 87 in the α chains), as had been predicted from the nature of the oxygen-binding curves (Chap. 32). The iron is somewhat more removed from the histidine residue at position 58 in the α chain and position 63 in the β and γ chains. A molecule of water may normally lie between the iron atom and this histidine in methemoglobin. *In Hb this space seems to be vacant with no ligand at the oxygen-binding site.* Thus, the lipophilic oxygen molecule can readily penetrate into the hydrophobic, low dielectric environment of the heme.

In the complete structure of hemoglobin, the four chains mesh together with little free space in the interior. The forces linking the four chains involve only secondary forces, *i.e.*, hydrogen bonding, salt links, and hydrophobic bonds. The hydrophobic interior of the molecule has low dielectric properties, whereas the exterior hydrophilic residues give the molecule its high solubility and charge characteristics.

The unique feature of hemoglobin is not its ability to bind oxygen per se. Ferroporphyrin and its other nitrogenous derivatives (base ferroporphyrins) can also bind oxygen, but, in these instances, the iron is rapidly oxidized to the ferric state. What is unique about hemoglobin is the formation of a stable oxygen complex in which the iron remains in the ferrous form because the heme moiety lies within a cover of hydrophobic groups of the globin. This is also suggested by the behavior of a model system. The complex of phenylethylimidazole and the diethyl ester of heme readily combines with O_2. In water this oxygenated iron is rapidly oxidized to the ferric state; if embedded in a film of polystyrene, the iron is readily oxygenated and deoxygenated with no change in valence.

The structure of oxyhemoglobin may be represented as an octahedral complex with the four porphyrin nitrogens at the angles.

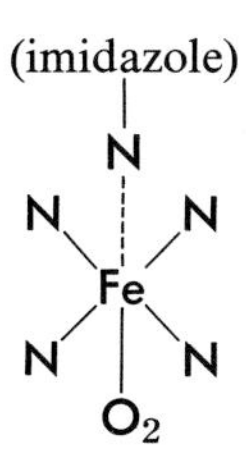

Molecules that contain unpaired electrons are magnetic; those in which all the electrons are paired are diamagnetic—*i.e.*, there is no magnetic moment. Both Hb and O_2 possess magnetic moments. When these two molecules combine, the resulting HbO_2 is diamagnetic because all the unpaired electrons in both molecules have become paired. These magnetic changes probably represent the best available evidence that oxygen combines with the iron of the hemoglobin molecule. These changes are also in accord with the observed stoichiometry of the reaction, *i.e.*, each atom of iron can bind one molecule of oxygen.

Hemoglobin and oxyhemoglobin have distinctive absorption spectra, as shown in Fig. 8.3. Quantitative estimations of Hb and HbO_2 are frequently made by spectrophotometric methods. Hemoglobin has a single band at 552.5 mμ, whereas HbO_2 shows bands at 541.5 and 576 mμ in this part of the spectrum. The strong bands near 400 mμ of all hemoproteins have been termed Soret bands after their discoverer. The absorption near 275 mμ is due to the presence of aromatic amino acids in the protein.

Carboxyhemoglobin (Carbon Monoxide Hemoglobin). Hemoglobin and the nitrogenous base derivatives of ferroprotoporphyrin bind carbon monoxide (CO). Carboxyhemoglobin and other iron compounds with carbon monoxide are sensitive to light, and a photochemical dissociation results. The compound of Hb and NO is extremely stable and dissociates very slowly.

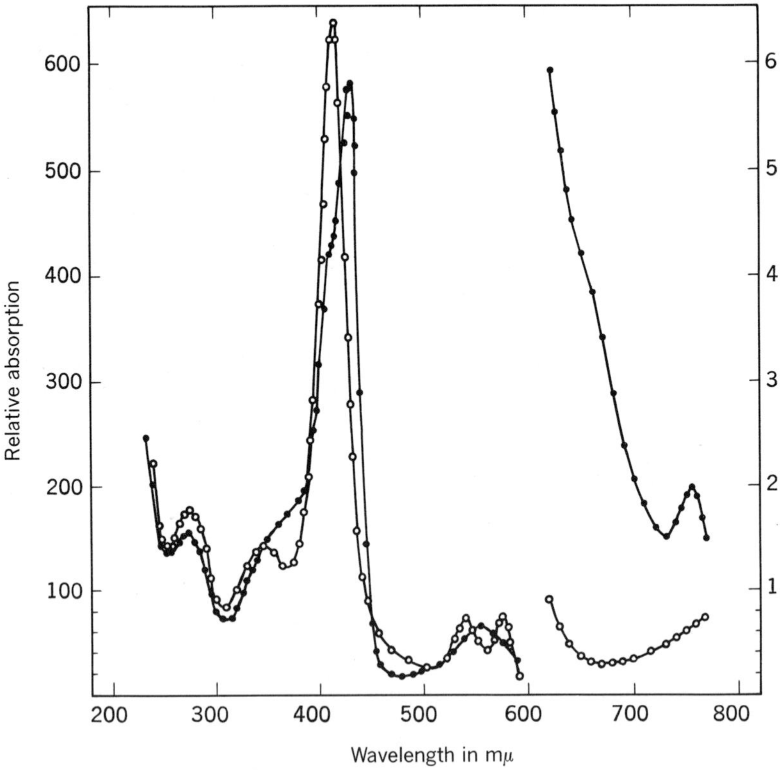

Fig. 8.3. Absorption spectra of human hemoglobin (•) and oxyhemoglobin (○) given on the same relative basis. The scale on the right is multiplied 100 times in order to show the absorption curves above 600 mμ. (*After A. E. Sidwell, Jr., R. H. Munch, E. S. G. Barron, and T. R. Hogness, J. Biol. Chem.,* **123**, 335, 1938.)

Methemoglobin. As already mentioned, when Hb is oxidized, methemoglobin (metHb), which contains ferric iron, is formed. MetHb cannot bind oxygen or carbon monoxide. Being positively charged, however, because of the additional charge on the ferric iron, it combines with hydroxide ion in alkaline solution, or with chloride and other anions in acidic solutions (cf. hemin). The brown color of metHb solutions changes on addition of ions, indicating formation of new compounds. The cyanide metHb complex is bright red. Spectroscopically distinctive derivatives are also formed with azide, sulfide, cyanate, nitric oxide, hydrogen peroxide, fluoride, and other anions.

The oxidation of hemoglobin to methemoglobin is readily accomplished by a large number of oxidizing agents such as peroxides, ferricyanide, quinones, etc. Treatment of oxyhemoglobin with ferricyanide or other oxidizing agents causes complete liberation of the bound oxygen and conversion to methemoglobin. In suitable gasometric apparatus, this provides a rapid method for the quantitative estimation of oxyhemoglobin. Fully oxygenated hemoglobin is more resistant to oxidation by oxidizing agents than reduced hemoglobin.

OTHER IRON-PORPHYRIN PROTEINS

Cytochromes. In 1885, MacMunn observed spectroscopically that many tissues contain pigments that have absorption characteristics similar to those of hematin (ferriprotoporphyrin) and its derivatives; he named these pigments histohematins. In 1925, Keilin confirmed MacMunn's observations and pointed out the biological importance of these pigments. Keilin renamed them cytochromes (cell pigments) to avoid the earlier confusion of nomenclature. One of the cytochromes, cytochrome c, has been obtained in crystalline form; its chemistry will be considered here. The other cytochromes, as well as their important biological role, will be discussed later (Chap. 15).

Cytochrome c has been isolated in pure form from many aerobic species—animals, plants, and fungi—and the primary structures of many of these cytochromes c have been elucidated. The cytochromes c are basic proteins, relatively stable to heat and acids, and consisting of a single peptide chain with 104 or more residues (molecular weight $\approx$ 12,400 to 13,000). In contrast to hemoglobin, acid-acetone does not separate the iron-porphyrin from the protein. The iron-containing group is derived from protoporphyrin, but the two vinyl side chains are reduced and linked by their α-carbon atoms in thioether bonds to the sulfur atoms of two cysteine residues, as shown below, diagrammatically.

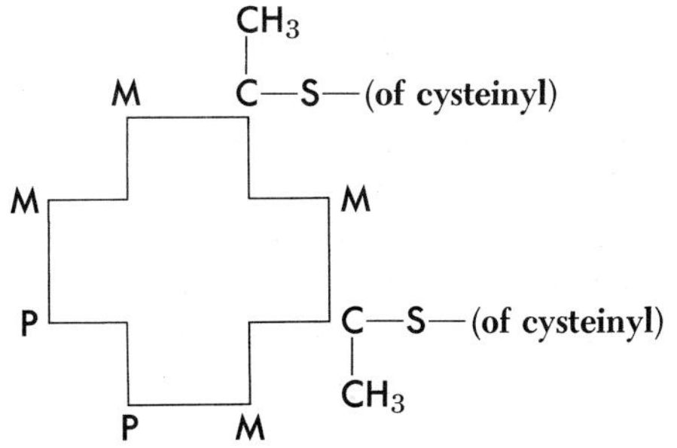

Acetyl-Gly-Asp-Val-Glu-Lys-Gly-Lys-Lys-Ile-Phe-Ile-Met-Lys-CyS-Ser-Gln-CyS-His-Thr-Val-Glu-
10 Heme 20

Lys-Gly-Gly-Lys-His-Lys-Thr-Gly-Pro-Asn-Leu-His-Gly-Leu-Phe-Gly-Arg-Lys-Thr-Gly-Gln-Ala-
30 40

Pro-Gly-Tyr-Ser-Tyr-Thr-Ala-Ala-Asn-Lys-Asn-Lys-Gly-Ile-Ile-Trp-Gly-Glu-Asp-Thr-Leu-Met-Glu-
50 60

Tyr-Leu-Glu-Asn-Pro-Lys-Lys-Tyr-Ile-Pro-Gly-Thr-Lys-Met-Ile-Phe-Val-Gly-Ile-Lys-Lys-Lys-Glu-
70 80

Glu-Arg-Ala-Asp-Leu-Ile-Ala-Tyr-Leu-Lys-Lys-Ala-Thr-Asn-GluCOOH
90 100 104

Fig. 8.4. The amino acid sequence of human heart cytochrome c. (*From H. Matsubara and E. L. Smith, J. Biol. Chem.*, **237**, PC 3575, 1962.) The residues that are underlined are those which have been found to be constant in all species studied to the present, including fungi, plants, and animals.

The complete sequence of human heart cytochrome c is shown in Fig. 8.4. As in all other known vertebrate cytochromes c, N-acetylglycine is present at the amino-terminus, whereas in the cytochromes c of invertebrates, plants, and fungi, four to eight additional residues are present at the N-terminus. Only 35 residues have been found to be constant in the various cytochromes studied to the present. This has provided evidence concerning the residues which are essential for the function and conformation of the protein.

Ferrocytochrome c is diamagnetic (no unpaired electrons) and, unlike hemoglobin, does not combine in the neutral pH range with O_2 and CO. Present evidence suggests that in ferrocytochrome c, in addition to the four ligands with the pyrrole nitrogen atoms, additional stable ligands are formed with the constant imidazole group at residue 18 and a nonnitrogenous group. The nature of the sixth ligand is still unknown, but it may involve the sulfur atom of residue 80, the only constant methionine, or a phenolic group of a tyrosine residue. Note the remarkable constancy of residues 70 to 80, inclusive. Since this is the only invariant long sequence, presumably it plays a vital role in the conformation, perhaps because the sixth ligand is present in this sequence.

X-ray analysis of the structure of the oxidized form of crystalline horse cytochrome c indicates that the heme resides in a hydrophobic pocket and is almost completely covered by protein. Only one edge of the porphyrin is exposed.

Cytochrome c is reversibly denatured by alcohols, heat, etc., indicating that the primary structure determines the over-all conformation (page 163). Present evidence suggests that the hydrophobic groups of the protein, which are concentrated in clusters, play a major role in determining the conformation, although in different species there is considerable substitution of one aliphatic residue for another, *e.g.*, Ile, Val, Leu, Met, and Thr are replaced by one another in various species. Note that Thr functions in some sites of the protein as a replacement for an aliphatic residue, whereas at other sites (residues 40 and 49) in cytochrome c and in other proteins, it can substitute for Ser. Presumably, in the latter situation, it is the β-hydroxyl group which is important. Similarly, there is considerable replacement by one polar (hydrophilic) residue for another. As already noted (page 154), since the polar residues are mainly on the surface of the molecule and the nonpolar residues are in the interior, considerable substitution is possible as long as the proper folding of the molecule is undisturbed. The more limited sequence information available for other proteins, which has been obtained from several species, is in accord with this principle. Evidence that probably all the lysine residues are on the surface of the molecule is indicated by the finding that the ϵ-amino groups can be guanidinated without disturbing the properties of the cytochrome c.

It is assumed that all cytochromes c of this type have essentially the same conformation inasmuch as all will cross-react with any cytochrome oxidase (Chap. 16), and since they all possess similar physical properties, *e.g.*, oxidation-reduction potential, spectra, etc. (Chap. 16). Thus, preliminary judgments concerning conformation can be made on the basis of the distribution of the proline residues in all species, since proline interrupts an α-helix and a stable α-helix is unlikely when it is shorter than six residues (two turns). Proline residues are located at residues 3, 25, 30, 44, 71, 76, 83, and 88 in various species, but only those at 30, 71, and 76 have been found in all species (Fig. 8.4). The presence of two proline residues in the constant sequence at positions 71 and 76 suggests nonhelical structure in this region.

Cytochrome c can be reduced to liberate the modified ferroporphyrin and the colorless protein containing cysteine residues at 14 and 17 (Fig. 8.4). When the reduced protein is isolated, it will recombine spontaneously with a porphyrinogen, a precursor of protoporphyrin (Chap. 31). In the presence of oxygen, the thioether bridges are formed and oxidation of the porphyrinogen occurs, resulting in synthesis of the porphyrin-protein. When Fe^{++} is added, native cytochrome is regenerated.

The role of cytochrome c in aerobic metabolism is discussed later (Chaps. 16 and 17). Genetic and evolutionary implications of the changes in cytochrome c sequences are considered in Chap. 30.

Peroxidases. These enzymes (Chap. 17), in conjunction with hydrogen peroxide, catalyze the oxidation of certain organic compounds. Crystalline horseradish peroxidase is reversibly dissociated by acetone and hydrochloric acid to give an inert colorless protein and ferriprotoporphyrin. In the intact enzyme, the iron is ferric and combined with one hydroxyl group. There also seems to be one carboxyl group combined with the iron; imidazole groups do not appear to play a role in this enzyme.

Peroxidase forms spectroscopically well-defined derivatives with various anions that can displace hydroxyl ion; these include cyanide, azide, fluoride, and the substrate, hydrogen peroxide. In its ability to bind such compounds, peroxidase resembles methemoglobin; however, methemoglobin possesses only weak peroxidatic activity. Reduction of peroxidase gives ferrous iron, which can combine with carbon monoxide. The ferrous compound does not bind oxygen. Further discussion of the enzymic behavior of peroxidase is given in Chap. 17.

Crystalline peroxidases have also been obtained from leukocytes (myeloperoxidase or verdoperoxidase) and from milk (lactoperoxidase). Both these enzymes contain greenish ferriporphyrins of unknown constitution. Neither can be split by acid-acetone, as can hemoglobin and horseradish peroxidase, or by silver salts, as can cytochrome c.

Catalases. These enzymes are present in all aerobic organisms and catalyze the decomposition of hydrogen peroxide and other reactions (Chap. 17). A number of catalases from different sources have been prepared in crystalline form. They contain ferriprotoporphyrin, which is easily removed by acid-acetone; recombination has not been achieved. The enzyme gives characteristic derivatives with cyanide, fluoride, azide, and other anions and with its substrate, hydrogen peroxide. The iron is not readily reduced to the divalent state.

Further discussion of certain of the enzymes which are hemoproteins is presented in Chap. 17.

REFERENCES

Books

Falk, J. E., "Porphyrins and Metalloporphyrins," American Elsevier Publishing Company, New York, 1964.

Falk, J. E., Lemberg, R., and Morton, R. K., eds., "Hematin Enzymes," Pergamon Press, New York, 1961.

Fischer, H., and Orth, H., "Die Chemie des Pyrrols," Bd. I, Akademische Verlagsgesellschaft, M.b.H., Leipzig, 1934; Bd. II, 1 Hälfte, 1937.

Fischer, H., and Stern, A., "Die Chemie des Pyrrols," Bd. II, 2 Hälfte, Akademische Verlagsgesellschaft, M.b.H., Leipzig, 1940.

Ingram, V. M., "Hemoglobin and Its Abnormalities," Charles C Thomas, Publisher, Springfield, Ill., 1961.

Keilin, D., "The History of Cell Respiration and Cytochrome," Cambridge University Press, New York, 1966.

Lehmann, H., and Huntsman, R. G., "Man's Haemoglobins," North Holland Publishing Company, Amsterdam, 1966.

Lemberg, R., and Legge, J. W., "Hematin Compounds and Bile Pigments, Their Constitution, Metabolism and Function," Interscience Publishers, Inc., New York, 1949.

Review Articles

Braunitzer, G., Hilse, K., Rudloff, V., and Hilschmann, N., The Hemoglobins, *Advances in Protein Chem.,* **19,** 1–71, 1964.

Corwin, A. H., The Chemistry of the Porphyrins, in H. Gilman, ed., "Organic Chemistry," vol. II, pp. 1259–1292, John Wiley & Sons, Inc., New York, 1943.

Margoliash, E., and Smith, E. L., Structural and Functional Aspects of Cytochrome c in Relation to Evolution, in V. Bryson and H. J. Vogel, eds., "Evolving Genes and Proteins," Academic Press, Inc., New York, 1965.

Rossi Fanelli, A., Antonini, E., and Caputo, A., Hemoglobin and Myoglobin, *Advances in Protein Chem.,* **19,** 73–222, 1964.

Schroeder, W. A., and Jones, R. T., Some Aspects of the Chemistry and Function of Human and Animal Hemoglobins, *Fortschr. Chem. org. Naturstoffe,* **23,** 113–194, 1965.

9. Nucleic Acids and Nucleoproteins

This chapter presents the chemistry of the nucleic acids, the hereditary determinants of living organisms. The biological significance of nucleic acids is discussed in Chaps. 28, 29, and 30. Although the name *nucleic acids* suggests location in the nuclei of cells, certain nucleic acids are also present in cytoplasm. Nucleic acids may naturally occur bound to proteins; hence the term nucleoprotein.

COMPONENTS OF NUCLEIC ACIDS

Elementary Composition. Nucleic acids contain carbon, hydrogen, oxygen, nitrogen, and, most strikingly, phosphorus. In almost all nucleic acids, there are approximately 15 to 16 per cent nitrogen and 9 to 10 per cent phosphorus.

Hydrolysis Products. Complete acid hydrolysis of a nucleic acid yields a mixture of basic substances called *purines* and *pyrimidines,* a sugar (ribose or deoxyribose), and phosphoric acid. After partial hydrolysis of nucleic acids, *nucleotides* and *nucleosides* can be isolated. Each nucleoside consists of a base and a sugar component; the nucleotides yield on hydrolysis the same components and, in addition, phosphoric acid. The successive degradation of the nucleoproteins is indicated in Fig. 9.1. Each of the components of nucleoproteins will be discussed separately.

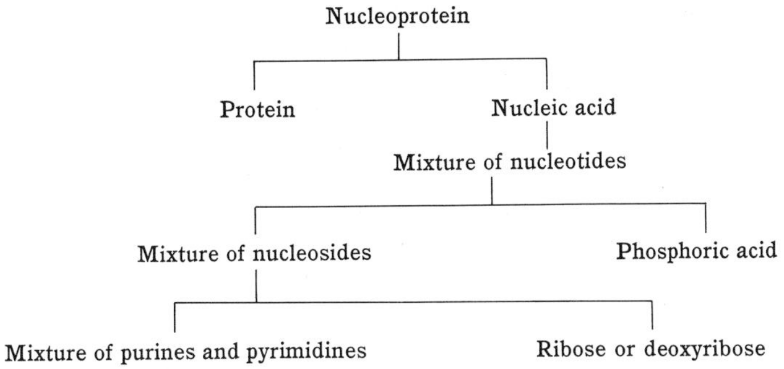

Fig. 9.1. Products arising in the successive hydrolytic degradation of nucleoproteins.

Pyrimidines. Substances of this type isolated from hydrolysates of nucleic acids are all derivatives of the parent heterocyclic compound *pyrimidine;* its structure and the convention of numbering the positions in the ring are indicated below.

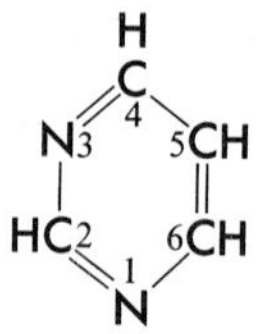

Pyrimidine

The prevalent pyrimidines found in nucleic acids are uracil, thymine, and cytosine. 5-Methylcytosine, 5-hydroxymethylcytosine, and other pyrimidine derivatives are also found in some types of nucleic acids (page 202).

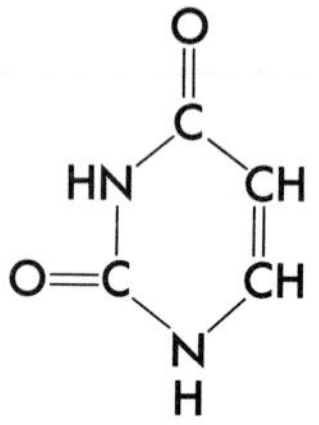

Uracil
(2,4-dioxypyrimidine)

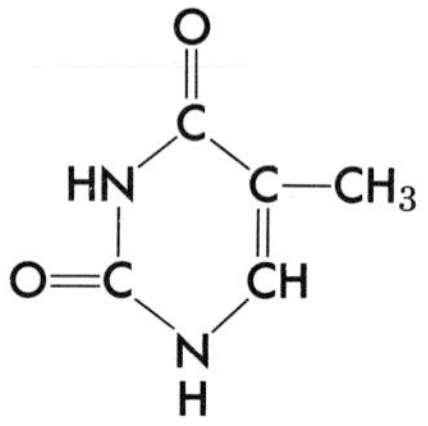

Thymine
(5-methyl-2,4-dioxypyrimidine)

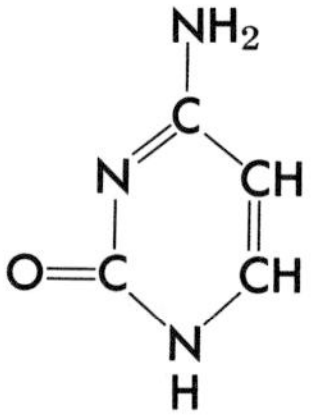

Cytosine
(2-oxy-4-aminopyrimidine)

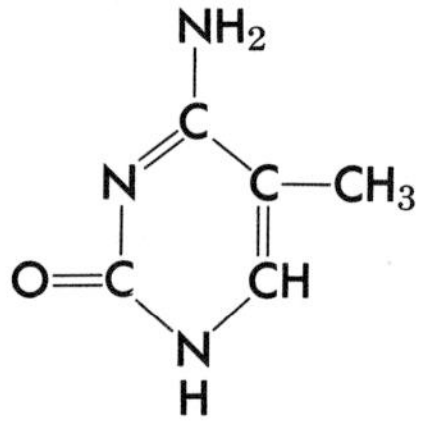

5-Methylcytosine
(5-methyl-2-oxy-4-aminopyrimidine)

The biological importance of pyrimidines is not restricted to nucleic acids. Several pyrimidine nucleotides play important roles in carbohydrate and lipid metabolism (Part Three). Vitamin B_1 (thiamine) is a pyrimidine derivative (Chap. 16). In recent years, certain synthetic pyrimidines have found important biological applications. Alloxan (2,4,5,6-tetraoxypyrimidine) produces experimental diabetes in animals (Chap. 19); this has proved fruitful in experimental investigations of this disease. Thiouracil and related compounds are employed for the treatment of hyperthyroidism (Chap. 43).

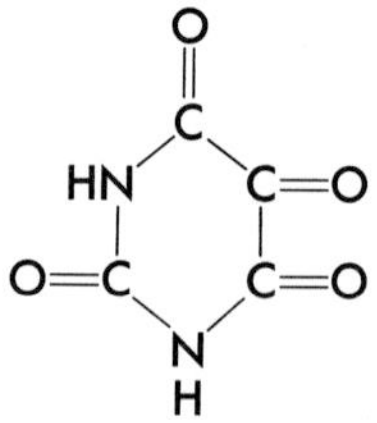

Alloxan

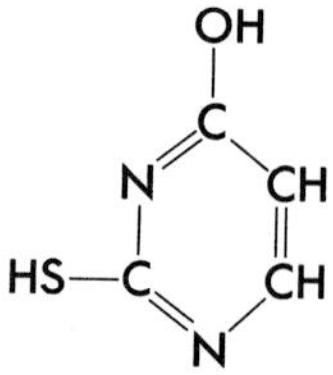

Thiouracil

The pyrimidines all show lactam-lactim tautomerism and may be written in either form illustrated for uracil. At neutral and acid pH values, the lactam forms predominate.

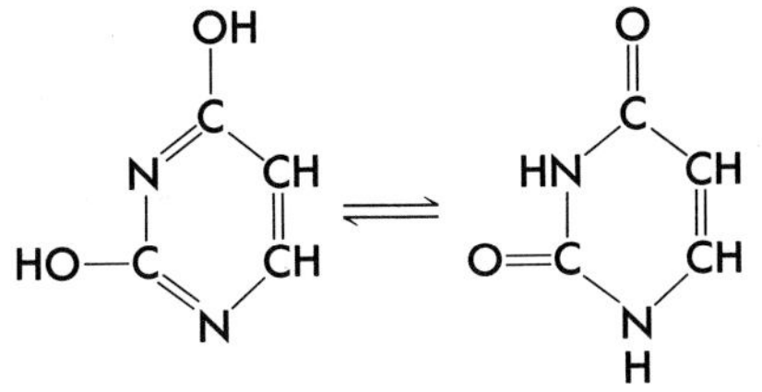

Lactim form of uracil — Lactam form of uracil

Purines. The parent substance, purine, contains the six-membered pyrimidine ring fused to the five-membered imidazole ring.

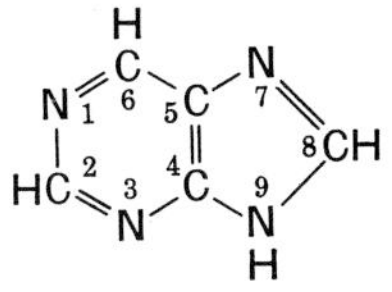

Purine

Adenine and guanine are the major purines of nucleic acids.

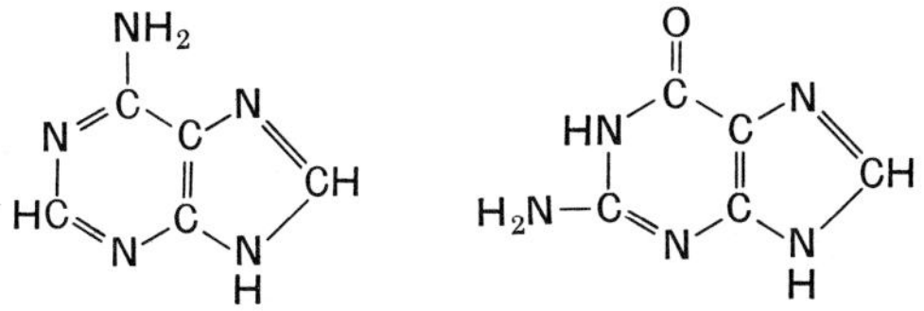

Adenine (6-aminopurine) — **Guanine** (2-amino-6-oxypurine)

Of more limited distribution are hypoxanthine and various methylated and other derivatives of adenine and guanine (page 202). Other important purines are xanthine and uric acid (page 184).

Hypoxanthine (6-oxypurine) — **Xanthine** (2,6-dioxypurine)

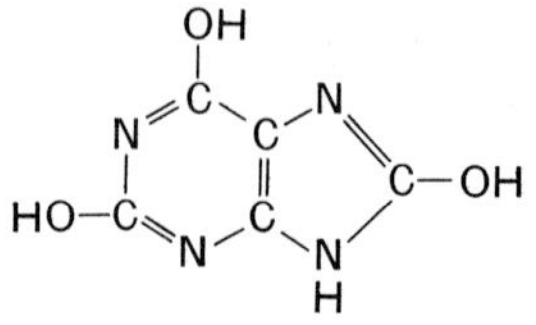

Uric acid (lactim form)

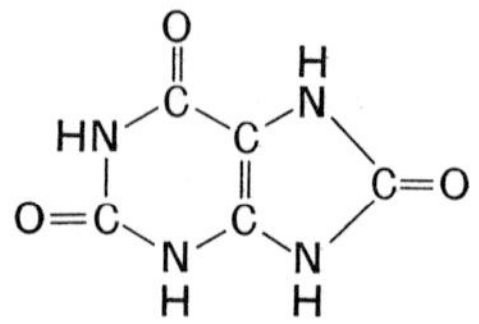

Uric acid (lactam form)
(2,6,8-trioxypurine)

Like the pyrimidines, purines show lactam-lactim tautomerism as illustrated above for uric acid. The lactim form of uric acid possesses weakly acidic properties ($pK' = 5.4, 10.3$) and forms salts such as mono- and disodium or potassium urates. Uric acid and its salts are sparingly soluble in water. This is reflected in the gradual precipitation of urates from urine that is allowed to stand (Chap. 35).

Caffeine (1,3,7-trimethylxanthine) is found in coffee, tea, and other plants; theobromine (3,7-dimethylxanthine) occurs in tea, cocoa, and chocolate. Other purines are also found in plants, and some of these, like caffeine and theobromine, have important pharmacological actions.

Sugars of the Nucleic Acids. The nucleic acids are classified according to the sugar they contain. One type of nucleic acid yields the pentose, D-ribose; hence the name ribose nucleic acid, or ribonucleic acid, is applied to this class of substances. The abbreviated form RNA is commonly used. In nucleic acids, ribose occurs in the furanose form (page 20).

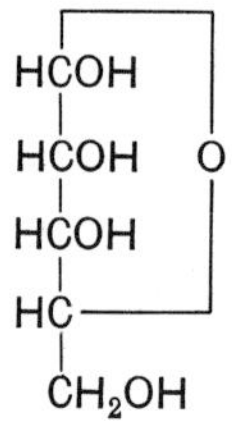

or

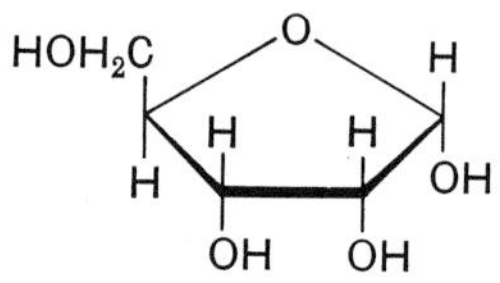

D-Ribose (α-D-ribofuranose)

The sugar in the other type of nucleic acid is D-2-deoxyribose; hence the name deoxyribose nucleic acid, or deoxyribonucleic acid. The abbreviated form is DNA.

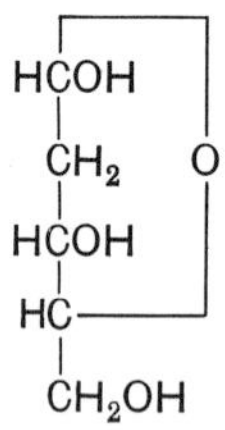

or

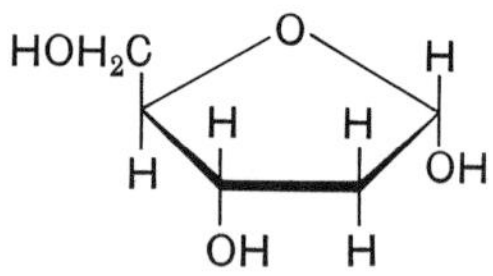

D-2-Deoxyribose (α-D-2-deoxyribofuranose)

NUCLEOSIDES

Partial hydrolysis of nucleic acids yields compounds in which ribose or deoxyribose is conjugated to a purine or pyrimidine base. Adenine linked to ribose is called adenosine, the guanine nucleoside is called guanosine, and, correspondingly, the pyrimidine nucleosides are cytidine and uridine. The analogous nucleosides formed with deoxyribose are called deoxyribonucleosides; thus, adenine deoxyribonucleoside or deoxyadenosine, deoxycytidine, etc. Note that the deoxyribonucleoside of thymine is called thymidine, not deoxythymidine, since this pyrimidine is primarily found in DNA. Thymine also occurs in one type of RNA, termed transfer RNA (page 201). In this instance, the name thymine ribonucleoside has been used. For the deoxyribonucleotide containing thymine as the base (see below), the name thymidylic acid generally suffices.

The purine nucleosides obtained from RNA have a β-glycosidic linkage from carbon-1 of the sugar to the nitrogen in position 9, as shown for adenosine. The pyrimidine nucleosides are N-1 glycosides, as shown for cytidine.

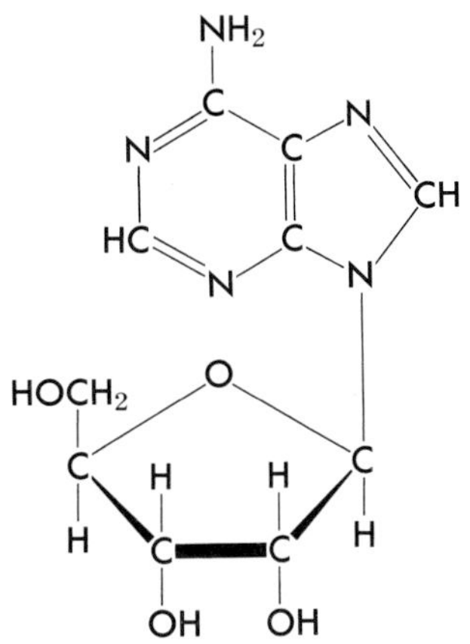

Adenosine
(9-β-D-ribofuranosyladenine)

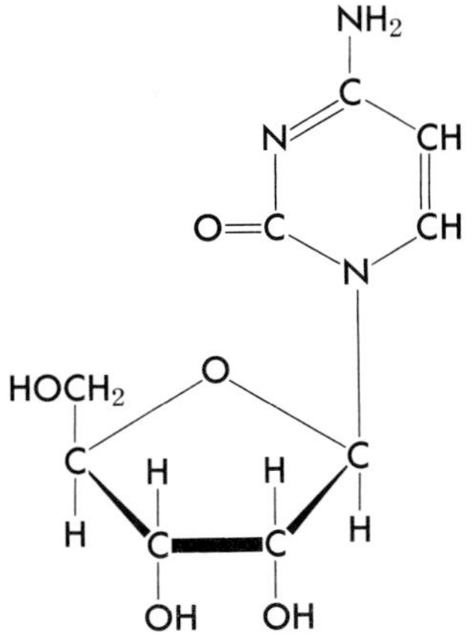

Cytidine
(1-β-D-ribofuranosylcytosine)

As in the case of O-glycosides, the nucleosides are stable in alkali. Purine nucleosides are readily hydrolyzed by acid whereas pyrimidine nucleosides hydrolyze only after relatively prolonged treatment with concentrated acid.

The nucleosides containing deoxyribose possess the same types of glycosidic linkages and are the 9-β-D-2′-deoxyribofuranosides of guanine and adenine and the 1-β-D-2′-deoxyribofuranosides of cytosine and thymine. In the nucleosides and their derivatives, the primed numbers refer to the positions in the sugar moiety.

From the mushroom *Agaricus nebularis,* a toxic, free nucleoside, named *nebularine,* has been identified as 9-β-D-ribofuranosylpurine. This is the first reported biological occurrence of the purine nucleus substituted only in position 9; however, the free base, purine, has not been found in natural sources. Other naturally occurring unusual nucleosides are antibiotics; some of these inhibit protein synthesis (Chap. 29).

NUCLEOTIDES

These are phosphoric esters of the nucleosides and are strong acids. They are called adenylic acid, guanylic acid, thymidylic acid, cytidylic acid, and uridylic acid. The phosphate is always esterified to the sugar moiety.

For the nucleotides containing deoxyribose, phosphorylation of the sugar is possible only at C-3′ and C-5′, since C-1′ and C-4′ are involved in the furanose ring and C-2′ does not bear a hydroxyl group. Both types of phosphate substitution have been found after appropriate conditions of hydrolysis of DNA. The structures of the two deoxyadenylic acids are shown.

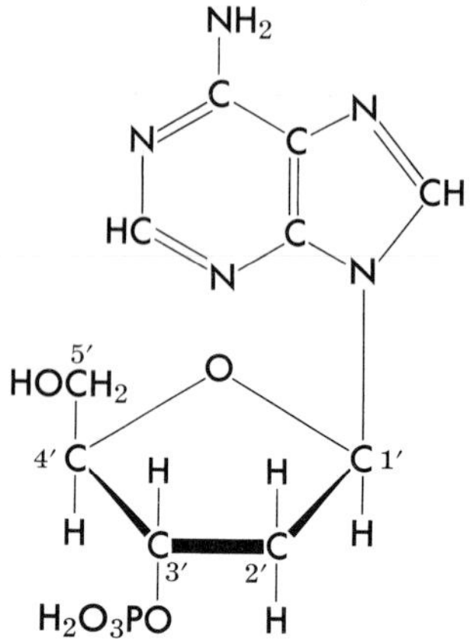

Deoxy-3′-adenylic acid
(deoxyadenosine 3′-phosphate)

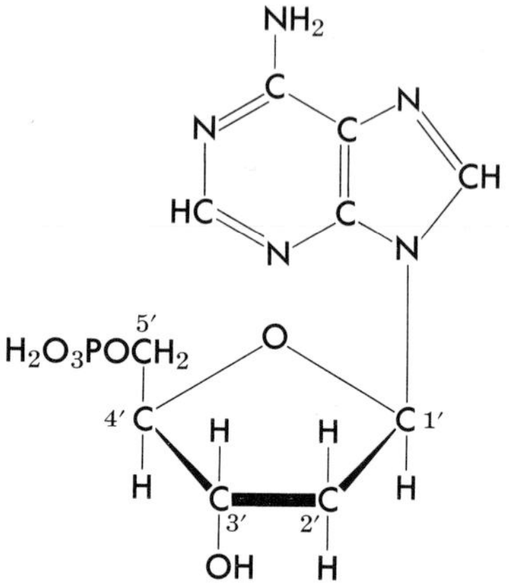

Deoxy-5′-adenylic acid
(deoxyadenosine 5′-phosphate)

Deoxyguanylic, thymidylic, and deoxycytidylic acids esterified at positions 3′ and 5′ are also found in hydrolysates of DNA.

In RNA, only positions 1′ and 4′ are unavailable for esterification, which leaves the possibility of substitution at C-2′, C-3′, and C-5′. Hydrolysis by alkali gives rise to isomeric nucleotides esterified at C-2′ or C-3′. The guanylic, cytidylic, and uridylic acids derived from alkaline hydrolysis of RNA are also mixtures of nucleoside 2′- and 3′-phosphates. RNA, degraded by digestion with pancreatic *ribonuclease,* yields the nucleoside 3′-phosphates. Hydrolysis with snake venom *phosphodiesterase* yields the nucleoside 5′-phosphates. The significance of these different isomeric nucleotides will be considered later.

In muscle and other tissues, there is a free form of adenylic acid which contains ribose, adenosine 5′-phosphate; the latter is also found in compounds that are further phosphorylated, *e.g.*, adenosine diphosphate (ADP) and adenosine triphosphate (ATP). The di- and triphosphates do not occur in nucleic acids. Similar 5′-nucleoside di- and triphosphates of other purines and pyrimidines also occur in tissues, as both the ribose and deoxyribose derivatives. Some of these di- and triphosphates are listed in Table 9.1 with the commonly used abbreviations for their designation. The metabolic roles of these compounds will be discussed in Part Three.

The hydrolytic deamination product of adenylic acid is also found in muscle tissue and in some nucleic acids. This is inosinic acid, or 9-β-5′-phospho-D-ribosylhypoxanthine (page 187).

Although the first nucleotides studied were those obtained by partial hydroly-

Table 9.1: Nomenclature and Abbreviations for Nucleotides and Related Compounds

Nucleotide	Abbreviations*	
Adenosine monophosphate (adenylic acid)	AMP	Ado-5′-P
Adenosine diphosphate	ADP	Ado-5′-PP
Adenosine triphosphate	ATP	Ado-5′-PPP
Guanosine monophosphate (guanylic acid)	GMP	Guo-5′-P
Guanosine diphosphate	GDP	Guo-5′-PP
Guanosine triphosphate	GTP	Guo-5′-PPP
Cytidine monophosphate (cytidylic acid)	CMP	Cyd-5′-P
Cytidine diphosphate	CDP	Cyd-5′-PP
Cytidine triphosphate	CTP	Cyd-5′-PPP
Uridine monophosphate (uridylic acid)	UMP	Urd-5′-P
Uridine diphosphate	UDP	Urd-5′-PP
Uridine triphosphate	UTP	Urd-5′-PPP
Thymidine monophosphate (thymidylic acid)	TMP	dThd-5′-P
Thymidine diphosphate	TDP	dThd-5′-PP
Thymidine triphosphate	TTP	dThd-5′-PPP
Thymidine ribonucleoside 5′-phosphate	TRP	Thd-5′-P
Deoxyadenosine monophosphate	dAMP	dAdo-5′-P
Deoxyadenosine diphosphate	dADP	dAdo-5′-PP
Deoxyadenosine triphosphate	dATP	dAdo-5′-PPP
Deoxyguanosine monophosphate	dGMP	dGuo-5′-P
etc.		

Note: The above nomenclature refers only to 5′ substituents. Other monophosphates are 3′-AMP (Ado-3′-P), 2′-AMP, 3′-GMP, etc. Note that, in the case of the thymidine derivatives, the commonly occurring deoxyribose derivatives are TMP, TDP, and TTP; the ribose derivatives must be specified.

* In each case the first abbreviation is more generally used. For a more complete list of abbreviations, see J. Biol. Chem., **241**, 527, 1966.

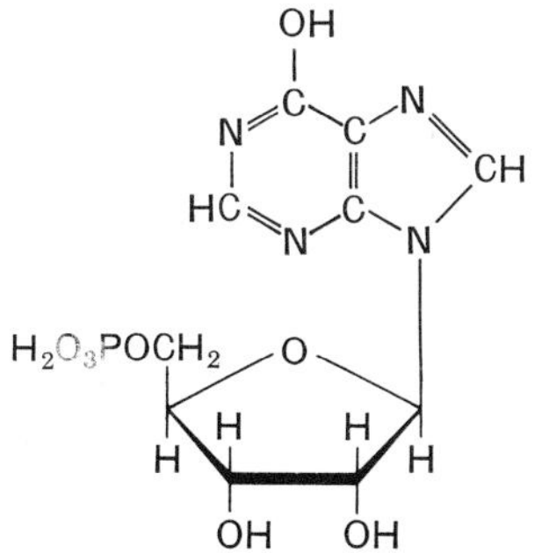

Inosinic acid (lactim form)

sis of nucleic acids, there are other important nucleotides which are not found in nucleic acids and which contain substances other than the usual purines and pyrimidines. The more general definition describes a nucleotide as a compound containing a phosphorylated sugar N-glycosidically linked to a base. Some nucleotides contain a vitamin as a component of their structure and function as coenzymes (Table 10.1, page 215).

METHODS USED FOR THE STUDY OF NUCLEIC ACIDS AND THEIR COMPONENTS

Isolation of Nucleic Acids. Nucleic acids, present in tissues as nucleoproteins, may be prepared either by direct extraction or by first extracting the nucleoproteins, generally with 1*M* salt solution. The soluble nucleoprotein can be split by careful addition of weak acid or alkali, or by saturating the solution with NaCl. Following scission of the nucleoprotein, the nucleic acid may be precipitated by slow addition of alcohol. Saturation with NaCl precipitates the protein. The latter may also be first degraded with proteolytic enzymes, particularly trypsin, and the nucleic acid subsequently isolated. Proteins may be removed from nucleic acids by a phenol extraction procedure. Extraction of tissues with hot trichloroacetic acid also removes nucleic acids.

Shaking a solution containing proteins and nucleic acids with chloroform and octanol produces a readily separable gel of denatured protein at the chloroform-water interface. However, some procedures that denature proteins may also alter nucleic acids.

Identification of DNA and RNA. Recognition of the type of nucleic acid is aided by identification of the sugar. Most of the currently used methods are colorimetric and can be employed for quantitative determinations of the sugars themselves, or in suitable modifications for estimation of nucleic acids, nucleotides, and other derivatives.

Some of the methods for pentose estimation depend on the liberation of furfural after heating with HCl (page 27). The furfural gives a red color with aniline acetate or a yellow color with *p*-bromophenylhydrazine. Ribose gives a distinctive color reaction with orcinol under suitable conditions (Table 2.1, page 28).

When DNA is heated with diphenylamine in acid solution (Dische reaction), a blue color is obtained. In the Feulgen reaction (page 28), deoxyribose, or DNA after partial acid hydrolysis, yields a blue-violet color. Other color reactions are also used for estimation of deoxyribose or DNA (page 28).

Absorption Spectra. The presence of the conjugated ring systems of the purines and pyrimidines in nucleic acid results in marked absorption in the ultraviolet region of the spectrum, with absorption maxima near 260 mμ. Since proteins have a much weaker absorption in this region, about 1 or 2 per cent as much, the spectral properties of the nucleic acids have been useful in locating and estimating these substances in cells and tissues. The photography of cells by using ultraviolet light depends largely on the strong absorption of the nucleic acids and permits, for example, studies of chromosomal behavior in living cells without the necessity of staining. This technique, usually in conjunction with the use of highly purified enzymes that digest protein, has permitted more precise location of nucleic acids. Also, nucleic acids may be removed by specific enzymic digestion, further establishing the intracellular localization of these substances.

Separation and Estimation of Nucleotides. The diverse methods of chromatography have been applied for separation of the purine and pyrimidine bases, as well as the nucleosides and the nucleotides, from one another. Electrophoresis is also useful for separation of nucleotides. It is also possible by these techniques to separate ribose-containing compounds from deoxyribose-containing ones. Separation of nucleotides by these methods depends on the presence of ionizable groups in the purines and pyrimidines: the enolic hydroxyl groups of uracil, cytosine, thymine, and guanine with pK' values in the range of 9 to 12.5, and the amino groups of adenine, guanine, and cytosine with pK' values between 2 and 4.5. Furthermore, the nucleotides all possess the two acidic groups of the substituted phosphoric acid, with a pK' value near 1 for the primary phosphate ionization and pK' near 6 for the secondary dissociation.

When these methods are combined with measurements of ultraviolet absorption spectra, the partial and complete hydrolysis products of nucleic acids can be separated from one another, identified, and estimated with microgram quantities of nucleic acids.

STRUCTURE OF DEOXYRIBONUCLEIC ACIDS

Internucleotide Linkages. The fundamental units of the nucleic acids are the nucleotides. The deoxyribonucleic acids are polynucleotides in which the phosphate residues of each nucleotide act as bridges in forming diester linkages between the deoxyribose moieties. For DNA the internucleotide bonds are between C-3′ and C-5′. Evidence for this has come from the study of partial hydrolysates of DNA; di- and trinucleotides have been isolated in which the nucleotides are linked only by 3′,5′-phosphodiester bonds. Thus DNA is a long-chain polymer in which the internucleotide linkages are of the diester type between C-3′ and C-5′, in the manner illustrated in Fig. 9.2, on the following page, for a portion of a DNA chain.

To show the DNA structure schematically, the type of diagram given at the right in Fig. 9.2 is used, in which the horizontal line represents the carbon chain of the sugar with the base attached at C-1′. The diagonal line indicates the C-3′ phosphate linkage near the middle of the horizontal line; that at the end of the horizontal line denotes the C-5′ phosphate linkage. This diagrammatic notation is used for both DNA and RNA.

For long sequences of polynucleotides, another shorthand system is also used. The letters *A*, *G*, *C*, *U*, and *T* represent the nucleosides, as in Table 9.1. The phosphate group is shown as *p*; when placed to the right of the nucleoside symbol, esterification is at C-3′; when placed to the left, esterification is at C-5′. Thus ApUp is a dinucleotide with a monoester at C-3′ of a uridine and a phosphodiester bond between C-5′ of U and C-3′ of A. Unless it is evident that deoxynucleotides are under discussion, it is useful to specify this, as in d-ApTpGpTp, etc., or as d-ATGT, etc., indicating that all the nucleosides contain deoxyribose.

Action of Nucleases and Other Enzymes. Certain enzymes that hydrolyze DNA are valuable tools in structural studies of these molecules. Pancreatic *deoxyribonuclease* (DNase I) degrades DNA mainly to a mixture of oligonucleotides. The few mononucleotides found are 5′-phosphates (Table 9.2). When the action of the DNase is followed by the action of snake venom *diesterase,* essentially a quantitative yield of 5′-nucleotides is obtained.

In contrast, DNase II from spleen or from a species of *Micrococcus* (*Staphylococcus*) (Table 9.2) hydrolyzes DNA to yield oligonucleotides and 3′-mononucleo-

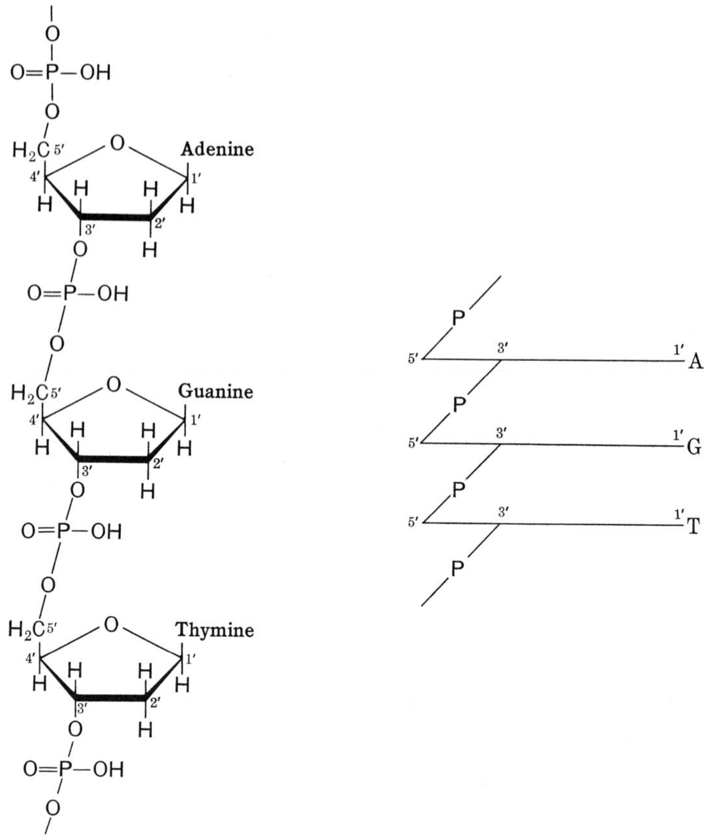

Fig. 9.2. Representation of a portion of a DNA chain, showing the position of the internucleotide linkage between C-3′ and C-5′. Schematic representation is given at the right.

tides. When the action of this enzyme is followed by that of spleen diesterase (Table 9.2), essentially a quantitative yield of 3′-mononucleotides is obtained. These actions are shown in Fig. 9.3 for a section of DNA.

Composition of DNA. Table 9.3 gives the base compositions of DNA from a number of sources. These analyses, as well as many others, show regularities of considerable importance, as first noted by Chargaff and his coworkers. In all cases,

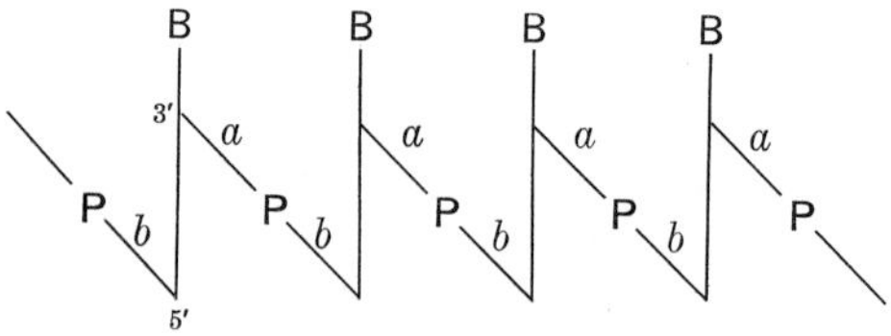

Fig. 9.3. Representation of hydrolysis of a section of DNA; B represents base. The points marked *a* represent the sites of hydrolysis by pancreatic deoxyribonuclease DNase I and venom diesterase. The points marked *b* are the sites of action of deoxyribonuclease DNase II from spleen or from *Micrococcus* followed by the action of a spleen diesterase. Hydrolysis at *a* yields 5′-nucleotides; at *b*, 3′-nucleotides. An endonuclease (I) from *Escherichia coli* also acts at *a* with liberation of the 3′-hydroxyl groups, whereas a phosphatase from the same organism hydrolyzes the 5′-phosphates.

Table 9.2: SPECIFICITY OF SOME NUCLEASES

Enzyme	Source	Substrate	Products
Deoxyribonuclease I . .	Pancreas	DNA	Oligonucleotides; 5′-mononucleotides
Deoxyribonuclease II . .	Spleen, thymus, *Staphylococcus* (*Micrococcus*), other bacteria	DNA	Oligonucleotides; 3′-mononucleotides
Endonuclease I	*Escherichia coli*	Single-stranded DNA; requires 3′-OH end*	Oligonucleotides; 5′-mononucleotides
Ribonuclease	Pancreas	RNA; at pyrimidine nucleotides	Oligonucleotides; Py-3′-P†
Ribonuclease	*Bacillus subtilis*	RNA; most rapid at purine nucleotides	Oligonucleotides; Pu-3′-P†
Ribonuclease T1	*Aspergillus oryzae* (Taka-diastase)	RNA; at 3′-guanylate or 3′-inosinate residues	Oligonucleotides; 3′-GMP
Ribonuclease T2	*A. oryzae* (Taka-diastase)	RNA; at 3′-adenylate residues	Oligonucleotides; 3′-AMP
Diesterase (exonuclease)	Snake venom	DNA*, RNA, or oligonucleotides; requires a free 3′-OH group	5′-Mononucleotides
Diesterase (exonuclease)	Spleen	DNA, RNA, or oligonucleotides; requires a free 5′-OH group	3′-Mononucleotides

* Attacks only single-stranded form.

† Py-3′-P = pyrimidine 3′-phosphate; Pu-3′-P = purine 3′-phosphate.

Table 9.3: DNA COMPOSITION OF VARIOUS SPECIES

Species	Base proportions, moles %				$\frac{A+T}{G+C}$	A/T	G/C	Pu/Py
	G	A	C	T				
Sarcina lutea	37.1	13.4	37.1	12.4	0.35	1.08	1.00	1.02
Alcaligenes faecalis	33.9	16.5	32.8	16.8	0.50	0.98	1.03	1.02
Brucella abortus	20.0	21.0	28.9	21.1	0.73	1.00	1.00	1.00
Escherichia coli K12	24.9	26.0	25.2	23.9	1.00	1.09	0.99	1.08
Salmonella paratyphi A	24.9	24.8	25.0	25.3	1.00	0.98	1.00	0.99
Wheat germ	22.7	27.3	22.8*	27.1	1.19	1.01	1.00	1.00
Bovine thymus	21.5	28.2	22.5*	27.8	1.27	1.01	0.96	0.99
Staphylococcus aureus	21.0	30.8	19.0	29.2	1.50	1.05	1.11	1.07
Human thymus	19.9	30.9	19.8	29.4	1.52	1.05	1.01	1.03
Human liver	19.5	30.3	19.9	30.3	1.54	1.00	0.98	0.99
Saccharomyces cerevisiae . .	18.3	31.7	17.4	32.6	1.80	0.97	1.05	1.00
Pasteurella tularensis	17.6	32.4	17.1	32.9	1.88	0.98	1.03	1.00
Clostridium perfringens	14.0	36.9	12.8	36.3	2.70	1.02	1.09	1.04

Note: G = guanine; A = adenine; C = cytosine; T = thymine; Pu/Py = purine/pyrimidine.

* Cytosine + methylcytosine.

SOURCE: Compiled from the work of several investigators.

the amount of purines (Pu) is equal to the amount of pyrimidines (Py), *i.e.*, Pu/Py is equal to 1, as is also the ratio of adenine (A) to thymine (T). Similarly, the ratio of guanine (G) to cytosine (C) (plus methylcytosine where it occurs) is equal to 1. Although the analyses are reported for the purines and pyrimidines per se, these are also the ratios for the corresponding nucleotides, since for each mole of base, purine or pyrimidine, there is one mole each of deoxyribose and phosphate.

Such analyses have been performed on the DNA of a large number of species, and the same regularities have been found. Nevertheless, each species shows a characteristic composition that is unaffected by age, conditions of growth, various environmental factors, etc. Indeed, in higher organisms, samples of DNA from different organs or tissues of the same species are similar in composition, as indicated by the data in Table 9.3 for human liver and thymus. The characteristic composition of DNA from a given source can be indicated by the ratio of (A + T)/(G + C). In bacteria a wide range of compositions is encountered, some being high in A + T, others in G + C. In higher organisms the range is more limited; in most animals the ratio (A + T)/(G + C) is found to be from 1.3 to 2.2, and in higher plants from 1.1 to 1.7.

Only the two purines, adenine and guanine, and the two pyrimidines, cytosine and thymine, have been found in the DNA of a wide variety of microorganisms: bacteria, actinomycetes, fungi, algae, and protozoa. The presence of methylcytosine is a characteristic feature of the DNA of certain higher plants and animals, the DNA of plants being richer in this pyrimidine than the DNA of animals. Wheat germ, the richest source yet found, contains 6 moles of methylcytosine per 100 moles of bases. In all cases, however, it replaces an equivalent amount of cytosine. The base content of viruses is discussed below (page 205).

The analytical data presented in Table 9.3 were obtained on the total DNA of the various microorganisms or tissues. The DNA obtained from many microorganisms yields a single sharp band on density gradient centrifugation (see below). The DNA of mammalian tissues is, however, inhomogeneous. Chargaff and coworkers were able to separate calf thymus into fractions with (A + T)/(G + C) ratios ranging from 1.0 to 1.8. Nevertheless, for each fraction, A/T = G/C = 1. Differential extraction procedures and chromatographic methods have been used in efforts to obtain further fractionation. As yet, no sample of DNA has been obtained from animal tissues that satisfies the criteria of molecular homogeneity.

Molecular Weight. As they occur naturally, DNA molecules are among the largest known. It is difficult to isolate DNA without fragmentation (page 200). Intracellularly, the entire DNA of bacteria such as *E. coli* is a single circular molecule with a molecular weight of 2×10^9. Viral DNAs range in size from about 1 to 350×10^6 (*e.g.*, Table 9.6, page 205). Such values should be related to the residue weight of a single nucleotide, 300 to 350; thus, there are approximately 3000 nucleotides per million molecular weight of DNA. DNA from mammalian cells consists of fragmented mixtures (page 200).

To describe the mass of such large molecules as DNA or RNA and particles containing these nucleic acids combined with protein, it is useful to refer to the mass of a single molecule. The mass of a single hydrogen atom is defined as one dalton.

Hence, the mass of the DNA of *E. coli* is 2×10^9 daltons; the mass of bacteriophage T7 is 38×10^6 daltons (Table 9.6, page 205).

In view of the wide variations in composition of DNA from different species (Table 9.3), individual nucleic acids found in nature must differ greatly in nucleotide sequence. Considering the possible arrangements of the four kinds of nucleotides in a chain of more than 100,000 nucleotides, the potential isomerism can reach the same astronomical figures already noted for possible variations in protein structure due to different sequences of amino acids (page 141).

Double Helical Structure of DNA. Application of x-ray diffraction analysis to the problem of the native structure of DNA has proved exceedingly fruitful. From the data obtained by Wilkins and coworkers, Watson and Crick proposed that DNA is composed of two chains of polynucleotides which exist in a double helical structure. The main chain of each strand consists of deoxyribose residues joined by 3′,5′ phosphodiester bridges. The two chains are held together in part by hydrogen bonding, each amino group being joined to a keto group, *i.e.*, adenine to thymine, thymine to adenine, guanine to cytosine, etc. In each instance the pairing involves a pyrimidine and a purine base. The types of hydrogen bonding are shown in Fig. 9.4 on the following page.

The two chains of the helix are coiled to permit the proper hydrogen bonding. Moreover, the chains are not identical but are *complementary* in terms of the appropriate base pairing, A to T and C to G, as shown for a hypothetical fragment of two chains of DNA, where the letters represent the deoxyribonucleotides.

$$\longrightarrow$$

———— A G T C A A G T G G C C ————

———— T C A G T T C A C C G G ————

$$\longleftarrow$$

The chains do not run in the same direction with respect to internucleotide linkages, *i.e.*, the chains are *antiparallel.* If, for example, the upper chain is linked 5′-3′ with respect to AG, GT, TC, etc., the lower chain is linked 3′-5′ with respect to TC, CA, AG, etc. Both chains follow right-handed helices, each coiled around the same axis. A schematic diagram of the coiling of the two chains is shown in Fig. 9.5.

In addition to hydrogen bonds, hydrophobic forces between the stacked purines and pyrimidines contribute to maintenance of the rigid, two-stranded structure. Thus, reagents such as formamide and urea, which increase the solubility of the aromatic groups in the surrounding aqueous medium, also tend to denature DNA (see page 195*ff.*).

The suggested base pairing is in accord with a content of G = C and A = T, as established for a large number of DNA samples. The cross section of the helix ascertained from the x-ray diffraction measurements is consonant only with purine-pyrimidine hydrogen bonding, the planes of the bases being perpendicular to the fiber axis. The dimensions of the helix are such that there is insufficient space to permit purine-purine pairing, and would allow too much space for hydrogen bonds to form between two pyrimidine residues on the individual strands.

Since Watson and Crick proposed the above structure in 1953, a great variety

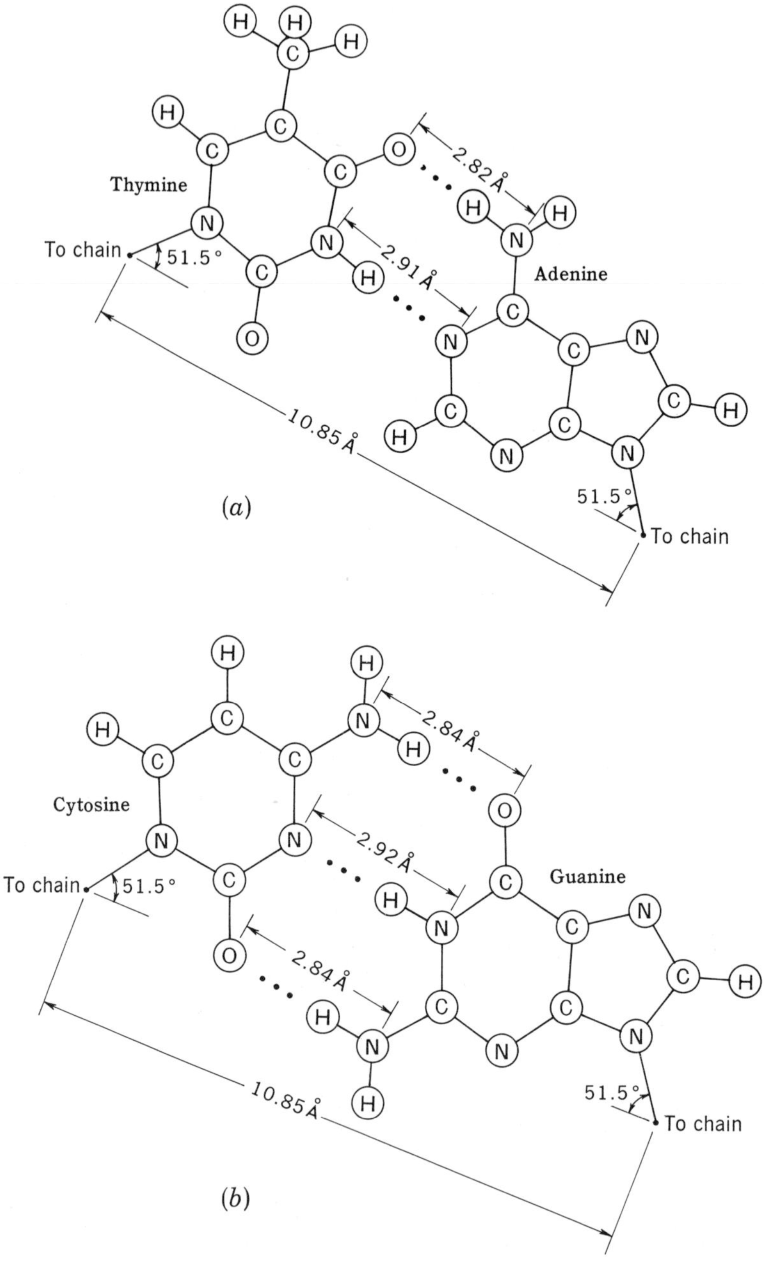

Fig. 9.4. Dimensions and hydrogen bonding of (*a*) thymine to adenine and (*b*) cytosine to guanine. (*From M. H. F. Wilkins and S. Arnott, J. Molecular Biol.*, **11**, 391, 1965.)

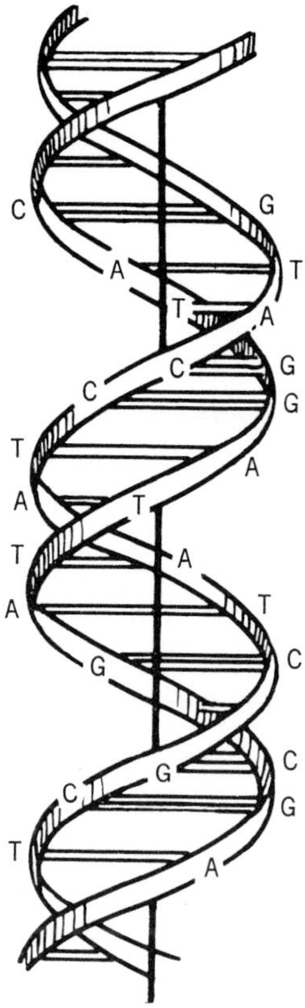

Fig. 9.5. A schematic representation of the double helix of DNA. The two ribbons represent the phosphate-sugar chains, and the horizontal rods represent the bonding between the pairs of bases. The vertical line indicates the fiber axis.

of supporting evidence has been obtained by the study of DNA and of model polynucleotides. The suggested structure depends on the hydrogen-bonding properties of the bases. As in the case of proteins (Chap. 7), individual hydrogen bonds of DNA are weak; it is the large number of such bonds which confers stability on the structure.

The effect of a variety of agents and conditions, *e.g.*, acid, alkali, heat, low ionic strength, on DNA structure can be explained on the basis that the DNA, initially in a firm, helical, two-stranded, native structure, can be converted to a "denatured" state which appears to be a single-stranded, flexible structure. The change from a native to a denatured form is usually very abrupt and is accelerated by reagents, *e.g.*, urea and formamide, which enhance the aqueous solubility of the purine and pyrimidine groups.

Several methods are used to assess the transition from the native to a denatured state, as well as to determine other properties of DNA. These methods are:

1. *Ultraviolet absorption*. All nucleic acids show a strong absorption in the ultraviolet with a maximum near 260 mμ. When native DNA is altered, there is a marked "*hyperchromic* effect," or increase in absorption. This change reflects a de-

crease in hydrogen bonding and is observed not only with DNA but with other nucleic acids, and with many synthetic polynucleotides which also possess a hydrogen bonded structure.

2. *Optical rotation.* Native DNA shows a strong positive rotation which is markedly decreased by procedures that produce denaturation. This change is analogous to the changes in rotation observed with the denaturation of proteins (page 155).

3. *Viscosity.* Solutions of native DNA possess a high viscosity because of the relatively rigid double helical structure and long, rodlike character of the DNA. Disruption of the hydrogen bonds produces a marked decrease in viscosity.

Effect of Temperature. Heating of a sample of DNA in a given ionic environment produces an increase in ultraviolet absorption and a decrease in optical rotation and viscosity at a critical temperature. The entire process is completed over a relatively narrow temperature range (Fig. 9.6).

The temperature midpoint (T_m) is analogous to the melting point of a crystal. DNA preparations from diverse sources possess different T_m values which

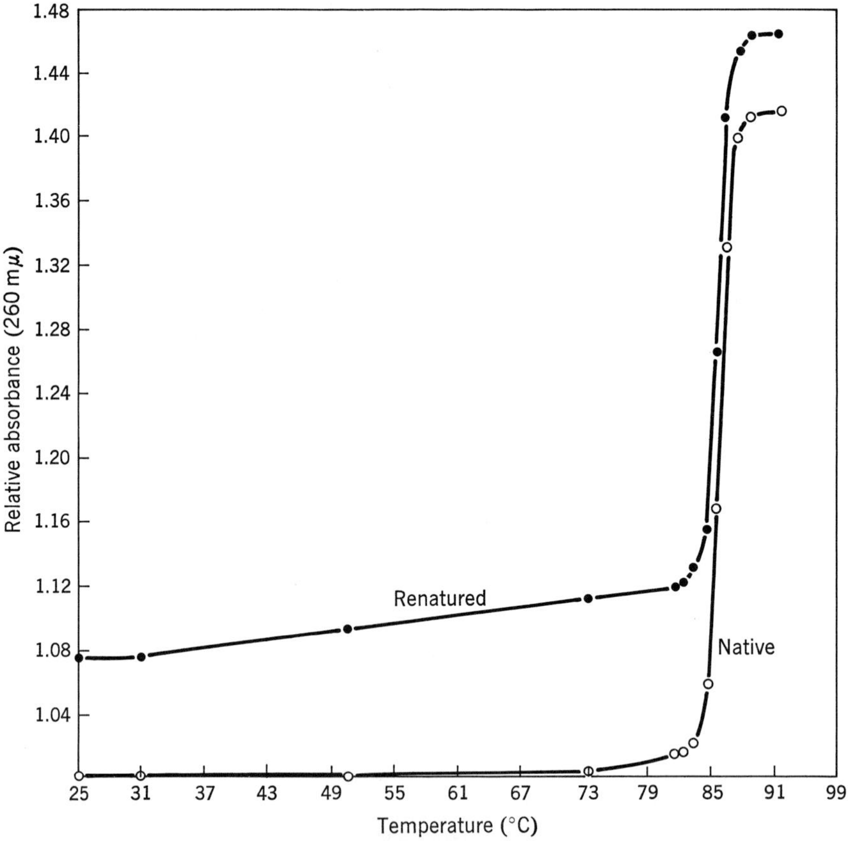

Fig. 9.6. Effect of temperature on the relative absorbance of native and renatured DNA. (*From P. Doty, in D. J. Bell and J. K. Grant, eds., "The Structure and Biosynthesis of Macromolecules," Biochemical Society Symposia; No.* 21, *p.* 8, *Cambridge University Press, New York,* 1962.)

depend on the absolute amounts of guanine (G) + cytosine (C) and adenine (A) + thymine (T). The higher the content of G-C, the higher the transition temperature between the native, two-stranded helix and the single-stranded form (Fig. 9.7). It will be noted (Fig. 9.4) that the G-C pair can form a triply hydrogen bonded structure whereas the A-T pair can form only a doubly hydrogen bonded structure. Indeed, T_m determinations, using careful calibration with DNA preparations of known composition, permit estimations of the G + C and A + T content of an unknown DNA. Such studies must be performed at fixed ionic strength and pH, since these have a marked effect on the stability of DNA. T_m can be lowered by the addition of urea, an agent known to disrupt hydrogen bonds. In $8M$ urea, T_m is decreased by nearly 20°C. DNA in 95 per cent formamide is completely separated into single strands at room temperature.

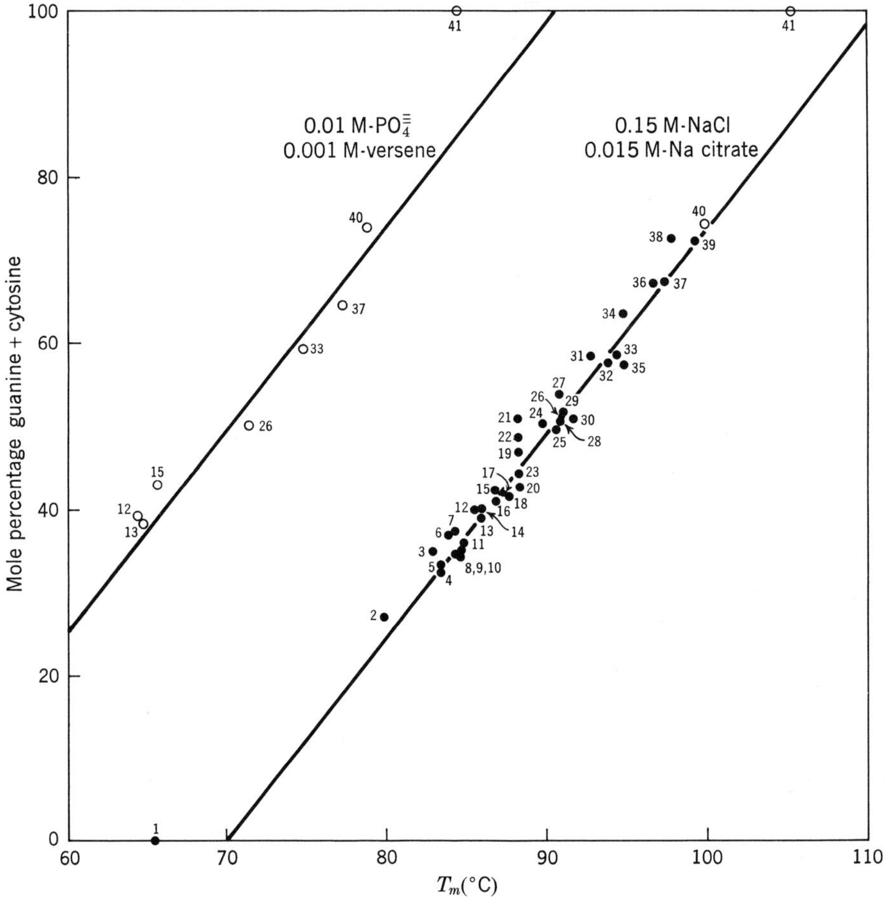

Fig. 9.7. Dependence of the temperature midpoint T_m on the content of guanine + cytosine. Numbers 2 through 40 represent points for samples of DNA from various sources. Number 1 is the value for dAT (page 199); No. 41, for d-poly G + d-poly C (page 199). (*From P. Doty, in D. J. Bell and J. K. Grant, eds., "The Structure and Biosynthesis of Macromolecules," Biochemical Society Symposia, No. 21, p. 8, Cambridge University Press, New York,* 1962.)

Complete rupture of the two-stranded helix by heating is not a readily reversible process. However, if a solution of denatured DNA, prepared by heating, is returned slowly to room temperature, some renatured DNA is obtained. Maximal reversibility (50 to 60 per cent) is usually attained by "annealing" the denatured DNA, *i.e.*, holding the solution at a temperature about 25°C. below T_m and above a concentration of 0.4*M* Na^+ for several hours. Snake venom diesterase (Table 9.2) does not attack native, two-stranded DNA but does hydrolyze the denatured, single-stranded form. When a preparation of partially renatured DNA is treated with the diesterase, the digestion products are readily removed from the renatured material; the latter exhibits all the properties—optical rotation, absorption, and viscosity—of the native form.

Effect of pH. Disruption of the double-stranded structure of DNA also occurs at acid and alkaline pH values at which ionic changes of the substituents on the purine and pyrimidine bases can occur. Near pH 12, ionization of enolic hydroxyl groups occurs, preventing the keto-amino group hydrogen bonding. Similarly, in acid solutions near pH 2 to 3, at which amino groups bind protons, the helix is disrupted. Treatment of DNA at these extreme pH values produces single-stranded molecules of half the molecular weight of native DNA.

Density of DNA. When a concentrated solution of cesium chloride is centrifuged in the analytical ultracentrifuge at high speeds until equilibrium is attained, the opposing processes of sedimentation and diffusion (page 131) produce a stable concentration gradient of the CsCl, *i.e.*, there is a continuous increase in density along the direction of centrifugal force. If the CsCl solution contains a small amount of DNA, at equilibrium the molecules of DNA will be collected in bands at those zones of the centrifuge cell at which their density and the density of the medium are exactly equal. The position of the DNA in the cell can be established by ultraviolet absorption photography. Since the gradient of solution density can be precisely estimated throughout the cell, the density of the DNA sample can be established. The technique is termed *density gradient centrifugation.*

Samples of *bacterial* DNA from different species show very narrow density bands, the density of a DNA preparation depending on the ratio of G-C to A-T pairs within the sample. The triply hydrogen bonded G-C pairs produce a relatively more compact, higher-density structure than do the doubly bonded A-T pairs. Thus, these findings are in accord with the Watson-Crick formulation of DNA structure inasmuch as the density depends on the relative amounts of the predicted types of hydrogen bonding between the base pairs.

Synthetic Polynucleotides. Considerable information concerning hydrogen bonding and base pairing has been obtained by studying polynucleotides synthesized enzymically with use of polynucleotide phosphorylase (Chap. 29) for preparation of RNA-like polymers, and of DNA polymerase (Chap. 28) for preparation of DNA-like polymers.

When polyadenylic acid (poly A) and polyuridylic acid (poly U) are mixed under appropriate conditions, a two-stranded helix is formed, in which, presumably, the hydrogen bonding is between A and U pairs. Heating such an equimolar mixture of the two homopolymers produces a considerable hyperchromic effect and other changes suggestive of the breakdown of a two-stranded helix. Similarly, d-poly G and d-poly C form a strongly hydrogen bonded structure with a very high T_m (Fig. 9.7). There is a strong hyperchromic effect on heating or on titration to pH 12. These effects are readily reversible, presumably because of the ease of base pairing, there being an exact fit of G to C along the entire length of the polymers.

A synthetic copolymer containing dAT has a structure in which each chain possesses a strictly alternating base sequence ··· ATATAT ···. Such a copolymer also forms a two-stranded structure resembling that of DNA. The T_m is low (Fig. 9.7), as expected for A-T hydrogen bonding. Changes in optical density and viscosity are produced by heating or alkaline titration, and these changes are readily reversible after quick cooling.

In summary, most of the physical properties of double-stranded, native DNA are directly related to either the molecular weight or the (A + T)/(G + C) ratio. Intracellularly, the circular DNA of most bacteria and of bacteriophages (page 206) behaves as a single chromosome which is also circular in form. In organisms containing more than one chromosome, the DNA is inhomogeneous, which may indicate that each chromosome is a distinct giant molecule of DNA, varying in size and in base composition.

Finally, although little sequence information is available as yet for DNA, large variations have been demonstrated by several chemical methods. Treatment of DNA below pH 3 and at temperatures above 60°C. cleaves N-glycosidic bonds to purine bases, producing *apurinic acids* which show that in many regions purines are clustered in long stretches. Correspondingly, treatment of DNA with hydrazine produces *apyrimidinic acids,* showing many regions of clustering of pyrimidine nucleotides. Such clustering of purines and pyrimidines, as well as the individual bases, is expected from the relationships of base sequences and protein synthesis (Chap. 29).

DEOXYRIBONUCLEOPROTEINS

DNA is found in the nuclei of cells as parts of the chromosome structure and has been identified in such cytoplasmic organelles as mitochondria and chloroplasts. Basic proteins, either histones or, in sperm cells, protamines, are associated with DNA, undoubtedly as a result of ionic linkages between anionic phosphate groups of DNA and cationic groups of the basic amino acids of these proteins. For each internucleotide phosphate in DNA (as well as RNA), there is a single anionic group, $pK_a' \approx 1$. Hence, samples of nucleic acids at neutral pH values are present as salts containing Na^+, Mg^{++}, etc., or intracellularly either with di- and polyamines such as cadaverine, putrescine, spermine, and spermidine (Chap. 25) or one of the histones or protamines. Bacterial DNA is not associated with protein.

The deoxyribonucleoproteins are viscous materials which are insoluble in $0.15M$ sodium chloride (physiological concentration) but soluble in $1M$ salt solutions. These solubility characteristics are employed in the preparation of nucleoproteins (page 188), which, when precipitated, are so viscous and sticky that they may be collected by winding on a stirring rod. Nucleoproteins are of high molecular weight, with estimated values ranging from 10 million to several hundred million.

A problem in establishing precisely the molecular weights of the nucleic acids arises from the difficulty in obtaining from natural sources chemically pure and undegraded preparations. Even the mild shearing forces of stirring or pipetting solutions of such large molecules cause fragmentation into smaller units. Thus the reported molecular or particle weights of the nucleic acids must be viewed with reservation.

The deoxyribonucleoproteins are highly elongated and threadlike in shape. Approximately 25 to 50 per cent of the dry weight of the molecule is nucleic acid. The amino acid composition of the nucleoproteins is distinctive. The *histones* of plant and animal tissues are relatively small proteins that are rich in arginine or lysine or both, and deficient in tryptophan. Histones are usually obtained by extracting nuclei or nucleoproteins with dilute acids ($0.2M$ HCl) and are then precipitated from solution with alkali at about pH 10, or by salting-out techniques. Further fractionation has been obtained by ion exchange chromatography and other methods. The histones have been separated into at least six discrete fractions that are similar from both thymus and wheat germ.

Protamines have been found only in the ripe sperm of certain families of fish and have never been obtained from somatic cell nuclei. These very basic proteins have usually been prepared by extraction with dilute mineral acids followed by precipitation with ethanol, or by dialysis of the nucleoprotamine against $1M$ HCl; the protamines pass through the membrane. The protamines are relatively small proteins, lacking many amino acids but extremely rich in arginine. In nucleoprotamines the ratio of arginine to phosphate is 1, producing complete neutralization of the charges. Clupein from herring sperm has been fractionated into three distinct polypeptides, each containing 30 or 31 amino acid residues of distinctive but similar sequence.

The biological role of the deoxyribonucleoproteins will be considered in Chaps. 28, 29, and 30.

STRUCTURE OF RIBONUCLEIC ACIDS

There are several distinct types of RNA, distinguishable by characteristic composition, size, and functional properties and by their location within the cell. Some general aspects of the structure of RNA will be given first; this will be followed by a brief discussion of the properties of the presently recognized types of RNA. The special functions of these types will be considered later (Chap. 29).

Internucleotide Linkage. In RNA, the hydroxyl groups at C-2′, C-3′, and C-5′ are available for esterification. However, present evidence indicates that, as in DNA, the internucleotide linkages are between C-3′ and C-5′.

When RNA is treated with a phosphodiesterase from snake venom, the main products are nucleoside 5′-phosphates. Crystalline ribonuclease (RNase) of bovine pancreas hydrolyzes nucleotide linkages at a point distal to the phosphate of a pyrimidine nucleotide that is esterified at the 3′ position. The points of cleavage of this enzyme are shown by the vertical dotted lines marked a.

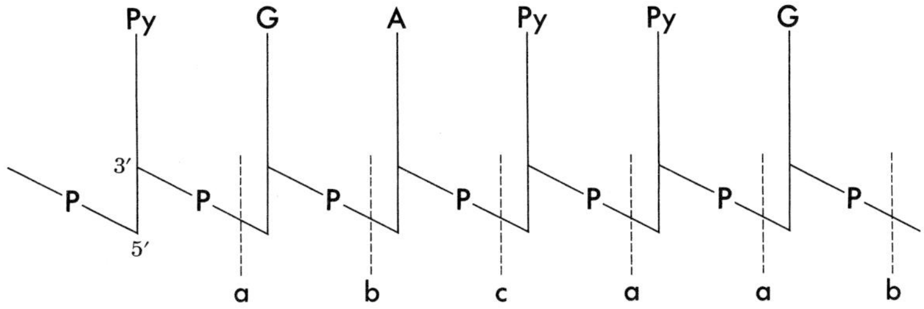

In this schematic structure, *Py* represents a pyrimidine. The action of the enzyme can be followed by the decrease in viscosity of the RNA preparation, appearance of dialyzable products, production of titratable acid groups, etc. The final products of pancreatic RNase action are pyrimidine-containing nucleoside 3′-phosphates and oligonucleotides terminating in a Py-3′-P, *i.e.*, ··· GpApGpPyp. The sites of action of RNase T1 and T2 (Table 9.2) are indicated at points b and c, respectively. RNase of *B. subtilis* acts at both b and c.

Unlike DNA, RNA is hydrolyzed by weak alkali (pH 9 at 100°C.). This treatment leads to the intermediate formation of a phosphate triester followed by hydrolytic scission of one bond to leave 2′,3′ cyclic phosphates, of the type shown below, for each of the purine and pyrimidine mononucleotides present in RNA.

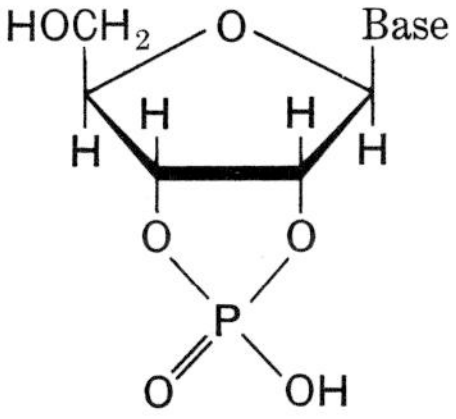

A 2′,3′ cyclic monophosphate nucleotide

The action of stronger alkali on the cyclic compounds produces a random hydrolytic cleavage giving a mixture of the isomeric 2′ and 3′ mononucleotides.

Thus the lability of RNA in alkali is due to formation of labile compounds in which phosphate is triply esterified. The possibility of formation of such bonds exists only with RNA, since in DNA there is no hydroxyl group at C-2′; this explains the relative stability of DNA in weak alkali. Pancreatic RNase can hydrolyze the cyclic 2′,3′-phosphates of cytidine and uridine to yield the 3′-phosphates. The structure and mechanism of action of this enzyme are presented later (page 255*ff.*).

A cyclic nucleotide, adenosine 3′,5′-cyclic monophosphate, cyclic AMP, has proved to be of considerable importance in metabolism (Part Three).

Transfer RNA. Transfer RNA (tRNA) remains in solution after centrifuging a broken cell suspension at 100,000 × gravity for several hours. Under this centrifugal force, cellular debris, nuclei, mitochondria, and microsomes are sedimented (Chap. 15). tRNA comprises approximately 10 to 20 per cent of the cellular RNA and is composed of relatively small molecules with molecular weights of approximately 30,000. tRNA functions as a mediator in peptide bond synthesis, at least one specific tRNA serving for each amino acid. Thus, there are at least 20 different kinds of tRNA; this role of tRNA is presented in Chap. 29.

Alkaline digestion of tRNA releases the end group nucleoside and a mixture of nucleotides, thus providing a specific end group method. The major end group of tRNA is adenosine (over 90 per cent of the total), and it is released in a ratio of one residue per 70 to 80 nucleotides.

Although tRNA is composed largely of the four main types of ribonucleotides —adenylic, guanylic, cytidylic, and uridylic acids—there are, in addition, smaller quantities of other nucleotides. These include the nucleotides of pseudouridine (see below), various methylated adenines and guanines, methylated pyrimidines, such as thymine and 5-methylcytosine, and others. Not all these are present in any one source of tRNA, but pseudouridine is the most abundant and universally distributed. The relative proportions of some of the additional components of the total tRNA of rat liver are given in Table 9.4. A list of some known minor components of tRNA and their abbreviations are given in Table 9.5; the structures of some of these are given in Fig. 9.8.

Table 9.4: RELATIVE AMOUNTS OF SOME MINOR COMPONENTS IN TOTAL TRANSFER RNA OF RAT LIVER

Component	*Transfer RNA**
Pseudouridine	25
5-Methylcytosine	10
6-Methylaminopurine	8.1
6,6-Dimethylaminopurine	0.1
1-Methylguanine	3.3
2-Methylamino-6-hydroxypurine	2.3
2,2-Dimethylamino-6-hydroxypurine	3.0

* Values are in moles per 100 moles of uridine.

SOURCE: From D. B. Dunn, Biochim. et Biophys. Acta, **34**, 286, 1959.

Table 9.5: SOME ADDITIONAL COMPONENTS IN TRANSFER RNA FROM VARIOUS SOURCES

*Component**	*Abbreviation*	*Component**	*Abbreviation*
Pseudouridine	ψU or ψ†	1-Methyladenine	1-MeA
5-Methylcytosine	5-MeC	Inosine	I
6-Methylaminopurine or 6-methyladenine	6-MeA	1-Methylinosine	1-MeI
		Thymine	T
2-Methyladenine	2-MeA	2′-O-Methylguanine	2′-OMeG
6,6-Dimethylaminopurine or dimethyladenine	DiMeA	2′-O-Methyladenine	2′-OMeA
		2′-O-Methylcytidine	2′-OMeC
1-Methylguanine	1-MeG	2′-O-Methylpseudouridine	2′-OMeψ
2-Methylamino-6-hydroxypurine or 2-methylguanine	2-MeG	7-Methylguanine	7-MeG
		N^6-2-Isopentenyladenosine (N^6-3-Methyl-2-butenyladenosine)	6-IPA
2,2-Dimethylamino-6-hydroxypurine or 2-dimethylguanine	2-DiMeG	4-Thiouracil	4-ThioU
Dihydrouracil	DiHU		

* In some cases the name of the nucleoside is given; in others, that of the purine or pyrimidine. In all cases it is the corresponding ribonucleotide which is present in tRNA.

† ψ = pseudo.

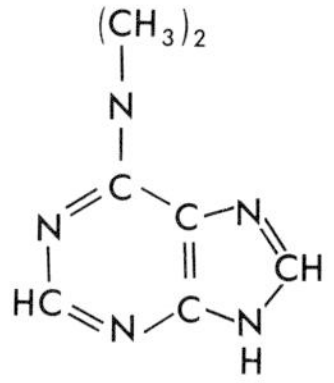

6,6-Dimethylaminopurine (6-Dimethyladenine)

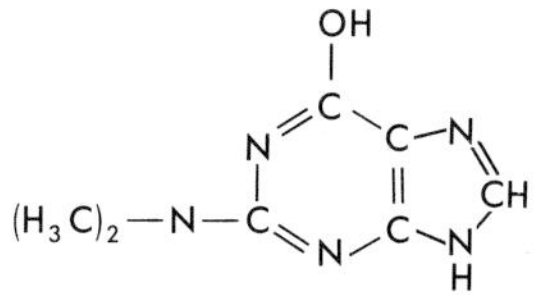

2,2-Dimethylamino-6-hydroxypurine (2-Dimethylguanine)

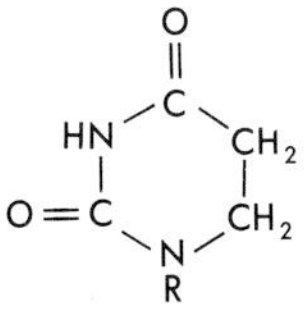

Dihydrouridine (lactam form)

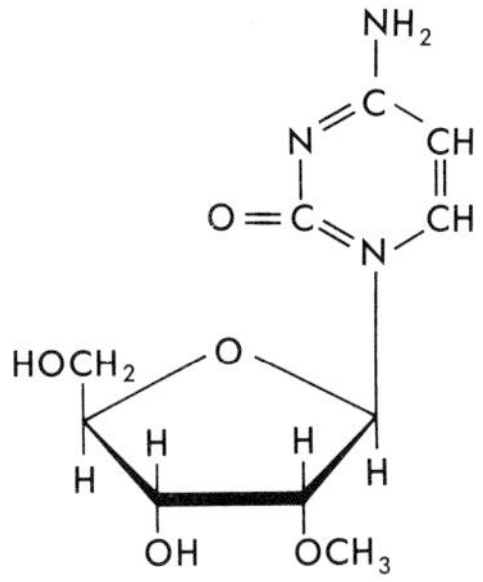

2′-O-Methylcytidine

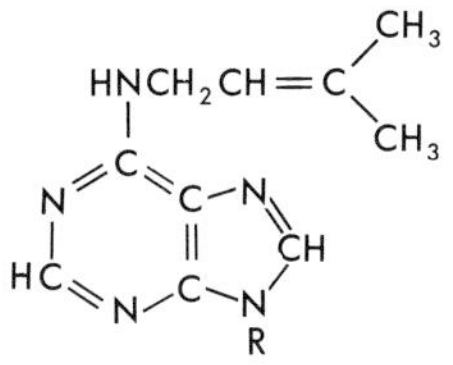

N^6-2-Isopentenyladenosine

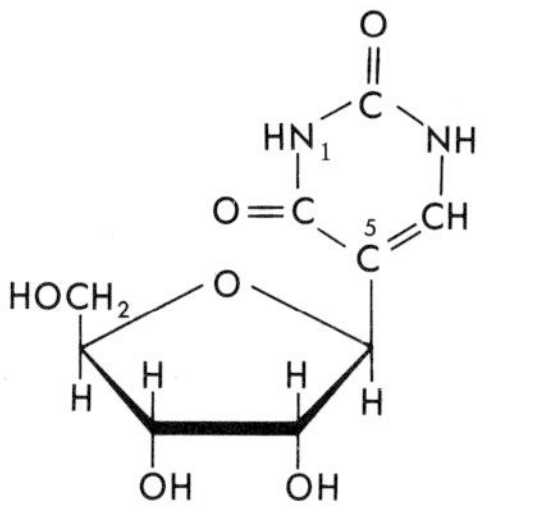

5-Ribosyluracil (pseudouridine)

Fig. 9.8. Structures of some bases or nucleosides from tRNA. *R* represents ribose.

Pseudouridine is of special interest in that the usual N-glycosidic bond is absent; the ribose is directly linked at C-1′ to the 5 position of uracil by a carbon-to-carbon bond.

The possible functional significance of the presence of pseudouridine and the other unusual bases is discussed later (Chap. 29).

The structures of several tRNA species have been determined, the first of these being that of an alanyl tRNA of yeast by Holley and coworkers in 1965. The methods for structure determinations may be briefly summarized as follows. The principle is essentially the same as that indicated for protein sequence determinations (Chap. 7). The single-chain, linear tRNA molecules were hydrolyzed to small fragments by pancreatic RNase and RNase T1, used successively; each component was isolated and its sequence determined. Larger fragments were obtained by limiting the action of the RNase T1 by incubation at 0°C. and in the presence of high Mg^{++}; such large fragments were separated by chromatography on DEAE-cellulose (page 134) in 7*M* urea and degraded to determine the sequence.

The linear sequence of alanyl tRNA is shown in Fig. 9.9. The 77 residues represent a molecular weight of 26,600 as the sodium salt. The usual conventions are followed in that the 5′-phosphate is shown at the left and the 3′-hydroxyl on the right.

Ribosomal RNA. Ribosomes (Chap. 29) contain a large portion of the RNA of a cell, representing as much as 80 per cent of the total in some bacteria. This type of RNA is strongly associated with protein; RNA (of the ribosomal particles of mammalian cells) comprises about 40 to 50 per cent of the dry weight of these

```
                                 Di          Di              Di
                                 |           |               |
               Me                H           H               Me                        Me
               |                 |           |               |                         |
pG-G-G-C-G-U-G-U-G-G-C-G-C-G-U-A-G-U-C-G-G-U-A-G-C-G-C-G-C-U-G-C-C-C-U-U-I-G-C-I-
 1                 10                    20                    30

ψ-G-G-G-A-G-A-G-U*-C-U-C-C-G-G-T-ψ-C-G-A-U-U-C-C-G-G-A-C-U-C-G-U-C-C-A-C-C-AOH
40                   50                    60                    70            77
```

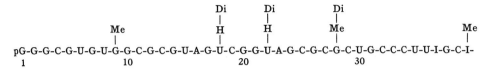

Fig. 9.9. Nucleotide sequence of a yeast alanyl tRNA. Abbreviations are given in Tables 9.1 and 9.5. U* represents a mixture of U and DiHU.

particles. The properties of these particles are best considered in connection with their biological role (Chap. 29).

Ribosomal RNA preparations from various sources, *e.g.*, rat liver or *Escherichia coli,* yield similar values for nucleotide content. Guanylic acid is invariably most abundant; uridylic and cytidylic acids are present in approximately equal amounts and are least abundant. Pseudouridine is present only in trace amounts, and methylated bases are present in small quantities.

Messenger, Template, or "Information" RNA. The properties of this type of RNA (mRNA) and its biological role are presented later (Chaps. 29 and 30).

VIRUSES AS NUCLEOPROTEINS

In 1935, W. M. Stanley isolated the virus that causes tobacco mosaic disease, as a highly purified crystalline protein, and a year later it was shown by Bawden and Pirie that this substance was a nucleoprotein of the ribose type. Since that time, many additional plant viruses have been isolated as crystalline substances, and they are all ribonucleoproteins. All the evidence indicates that the properties of the isolated nucleoproteins are consonant with the properties of the viruses themselves. Procedures that influence the chemical properties of the nucleoprotein, such as treatment by chemical or physical methods, digestion by enzymes, etc., also result in a loss of viral activity.

Table 9.6 gives the nucleic acid content and particle weight (molecular weight) of certain plant and animal viruses; wide variation is evident. All the RNA viruses are single-stranded and contain only the usual four bases (A, G, U, C).

Viruses can reproduce inside living cells but are incapable of doing so independently; this aids in distinguishing them as a group from bacteria and other small free-living organisms. What is so striking is that ability to reproduce, a characteristic once regarded as reserved for "living cells," has been found to be possessed by nucleoproteins, albeit in a host cell.

Preparations of tobacco mosaic virus have been treated by procedures that have permitted isolation of the specific RNA essentially free of protein. Isolated RNA of this type, although very labile, has proved to be infective in suitable host plants, leading to the formation of new tobacco mosaic virus. This evidence indicates that it is the RNA alone which possesses biological activity.

Bacteriophages may be regarded as viruses of bacteria in as much as they are infectious and can reproduce in specific host cells. The types of phage that have

Table 9.6: COMPOSITION AND SIZE OF SOME VIRUSES

Virus (and host)	Nucleic acid	No. of chains	Percentage of nucleic acid	Particle weight × 10^6	
				Virus	Nucleic acid
Tobacco mosaic	RNA	1	6	40	2.2
Bushy stunt (tomato)	RNA	1	15	10.6	1.65
Ring spot (tomato)	RNA	1	44	1.5	0.66
Poliomyelitis (man)	RNA	1	22–30	6.7	2
Encephalitis (equine)	RNA	1	4.4	50	2.2
Influenza, type A (man)	RNA	1	0.8	280	2.2
Bacteriophages T2, T4, T6 (*Escherichia coli*)	DNA	2	61	220	130
Bacteriophage T7 (*E. coli*)	DNA	2	41	38	25
Bacteriophage φX174 (*E. coli*)	DNA	1*	26	6.2	1.6
Bacteriophages MS2, R17 (*E. coli*)	RNA	1	32	3.6	1.1
Adenovirus (man)	DNA	2	5	200	10
Polyoma (vertebrates)	DNA	2	13.4	21	3.4

* φX174 is also obtained in double-helical replicative form (Chap. 29) with a molecular weight of 3.4 × 10^6.

been obtained in highly purified form contain either DNA or RNA (Table 9.6). For certain strains of phage it appears that the protein aids in penetrating the host bacterial cell but only the nucleic acid actually enters the cell. After a brief period of time, the host cell breaks open (undergoes lysis) and many new phage particles are liberated. This evidence indicates that only the specific nucleic acid is concerned in the self-duplicating properties of the phage.

Certain bacteriophages contain 5-hydroxymethylcytosine in place of cytosine in their DNA. The hydroxymethyl group is linked glycosidically to glucose.

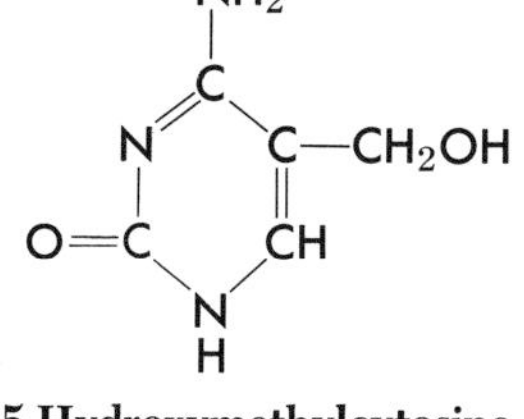

5-Hydroxymethylcytosine
(5-hydroxymethyl-2-oxy-4-aminopyrimidine)

Table 9.7 gives the base content of *E. coli* DNA in comparison with that of three different T-even bacteriophages that can infect these cells. The glucosides may be of either the α or β configuration or the α-glucosyl-β-glucoside structure. In these DNA molecules, A = T and G = HMC. Other bacteriophages of *E. coli,* T1, T3, T5, T7, lack hydroxymethylcytosine; their base content reflects conventional base pairing of DNA, *i.e.*, A = T and G = C.

The DNA of a bacteriophage of *Bacillus subtilis* contains guanine, adenine, cytosine, and, in place of thymine, 5-hydroxymethyluracil.

Table 9.7: BASE CONTENT OF DNA FROM *Escherichia coli* AND CERTAIN BACTERIOPHAGES

Source of DNA	Percentage of total bases							
	A	T	G	C	HMC	HMC-α-G	HMC-β-G	HMC-α-diG
E. coli	25	25	24	26	0	0	0	0
T2	32	32	18	0	4	12	0	1
T4	32	33	18	0	0	12	5	0
T6	32	33	18	0	4	1	0	12

Note: A, T, G, C = the usual four bases of DNA; HMC = 5-hydroxymethylcytosine; HMC-α-G = the α-glucoside; HMC-β-G = the β-glucoside; and HMC-α-diG = the α-glucosyl-β-glucoside.

SOURCE: D. S. Hogness, in "The Molecular Control of Cellular Activity," J. M. Allen, ed., p. 206, McGraw-Hill Book Company, New York, 1962.

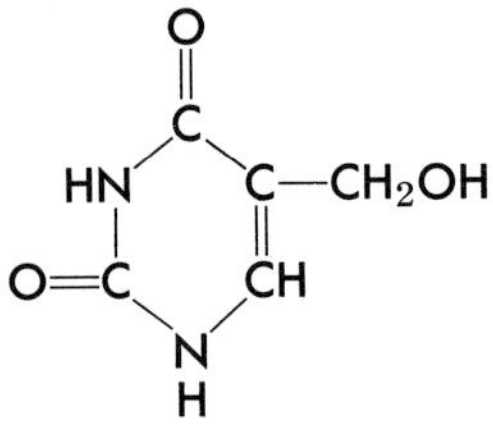

5-Hydroxymethyluracil
(5-hydroxymethyl-2,4-dioxypyrimidine)

Animal viruses contain either DNA or RNA (Table 9.6). Among the DNA group are those causing such familiar diseases as chickenpox, mumps, cowpox (vaccinia), psittacosis, rabies, and polyoma (which produces a malignant tumor in rodents). RNA viruses include the agents producing influenza, poliomyelitis, and encephalitis, among other diseases. Some of these larger viruses contain, in addition to the nucleic acid and protein, lipids and polysaccharides.

For many of the viruses listed in Table 9.6 and others mentioned above, the entire DNA is present in a single molecule, which in some cases has been shown to be circular, *e.g.*, polyoma virus, phage ϕX174, and *E. coli* phage λ. It is likely that the nucleic acids of other viruses exist in circular form during replication in the host cell.

REFERENCES

Books

Bonner, J., and Ts'o, P., eds., "The Nucleohistones," Holden-Day, San Francisco, 1964.

Cantoni, G. L., and Davies, D. R., eds., "Procedures in Nucleic Acid Research," Harper & Row, Publishers, Incorporated, New York, 1966.

Chargaff, E., and Davidson, J. N., eds., "The Nucleic Acids. Chemistry and Biology," vols. I and II, 1955; vol. III, 1960, Academic Press, Inc., New York.

Davidson, J. N., "The Biochemistry of the Nucleic Acids," 5th ed., John Wiley & Sons, Inc., New York, 1965.

Davidson, J. N., and Cohn, W. E., eds., "Progress in Nucleic Acid Research and Molecular

Biology," Academic Press, Inc., New York. (A series of volumes published at intervals since 1963.)

Florkin, M., and Stotz, E. H., eds., "Comprehensive Biochemistry," vol. 8 (Nucleic Acids), American Elsevier Publishing Company, New York, 1963.

Michelson, A. M., "The Chemistry of Nucleosides and Nucleotides," Academic Press, Inc., New York, 1963.

Steiner, R. F., and Beers, R. J., Jr., "Polynucleotides. Natural and Synthetic Nucleic Acids," American Elsevier Publishing Company, New York, 1961.

Review Articles

Burton, K., Sequence Determination in Nucleic Acids, in P. M. Campbell and G. D. Greville, eds., "Essays in Biochemistry," vol. 1, pp. 58–89, Academic Press, Inc., New York, 1965.

Doty, P., Inside Nucleic Acids, *Harvey Lectures,* **55,** 103–140, 1959–1960.

Doty, P., The Relationship of the Interaction of Polynucleotides to the Secondary Structure of Nucleic Acids, in D. J. Bell and J. K. Grant, eds., "The Structure and Biosynthesis of Macromolecules," Biochemical Society Symposia, no. 21, pp. 8–28, Cambridge University Press, New York, 1962.

Felsenfeld, G., and Miles, H. T., The Physical and Chemical Properties of Nucleic Acids, *Ann. Rev. Biochem.,* **36,** 407–448, 1967.

Holley, R. W., Apgar, J., Everett, G. A., Madison, J. T., Marquisee, M., Merrill, S. H., Penswick, J. R., and Zamir, A., Structure of a Ribonucleic Acid, *Science,* **147,** 1462–1465, 1965; also *Science,* **153,** 531–534, 1966.

Josse, J., and Eigner, J., Physical Properties of Deoxyribonucleic Acid, *Ann. Rev. Biochem.,* **35,** 789–834, 1966.

Lehman, I. R., Linn, S., and Richardson, C. C., Nucleases as Reagents in the Study of Nucleic Acid Structure, *Federation Proc.,* **24,** 1466–1472, 1965.

Various authors on Transfer RNA: Chemistry, in *Cold Spring Harbor Symp. Quant. Biol.,* **31,** 109–478, 1966.

Many special phases of nucleic acid chemistry are reviewed in *Progress in Nucleic Acid Research and Molecular Biology,* Academic Press, Inc., New York, 1963—current.

PART TWO CATALYSIS

10. Enzymes. I

Nature and Classification

Biochemical reactions occur with great rapidity through the mediation of natural catalysts called *enzymes.* The high degree of specificity and the great efficiency of enzymes direct transformations of organic compounds through defined reaction sequences. Enzymes are universally present in living organisms, and the occurrence of metabolic reactions common to all cells reflects the specificity of the responsible enzymes.

Enzymic reactions were used by man long before written history. The discovery of fermentation to produce wines was attributed by the Greeks to Bacchus. The making of cheese, the leavening of bread, and the manufacture of vinegar are enzymic processes which stem from antiquity. These practical aspects of enzyme chemistry have occupied an important place in the history of enzymology. The breweries and their research laboratories have provided much information concerning the processes by which living cells utilize sugar, for the fundamental reactions of fermentation in a yeast culture are much the same as in the tissues of mammals. Indeed, the name enzyme, coined by W. Kühne, means "in yeast," but the word is now used to connote a biological catalyst, regardless of origin.

The recognition that living cells are responsible for alcoholic and other types of fermentation was one of the great achievements of the nineteenth century. The work of Louis Pasteur and others destroyed the ancient beliefs that fermentation and putrefaction could spontaneously generate life and established that these chemical processes were caused by microscopic living organisms. However, Pasteur concluded that these processes could be performed only by intact living cells. Enzymology received a great impetus when the Buchner brothers showed that yeast cells, ground with sand and squeezed under high pressures, gave a cell-free juice capable of fermenting sugar with the production of alcohol and carbon dioxide. It is now apparent that yeast juice contains a complex mixture of enzymes required to effect these transformations and that enzymes can function extra- as well as intracellularly.

A substantial part of the study of the chemistry of living cells is today devoted to the enzymes, for it is now understood that all physiological functions, *e.g.*, muscular contraction, nerve conduction, excretion by the kidney, etc., are inextricably linked to the activity of enzymes. A complex process, such as muscular contraction, which requires the utilization of energy, may be dissected into a series of enzyme-catalyzed reactions. Many of these reactions have now been studied in vitro as isolated systems with pure, crystalline enzymes.

Even relatively simple reactions that occur in living cells may be catalyzed by enzymes when the noncatalyzed reaction is too slow for physiological needs. An example is the reversible combination of carbon dioxide and water to form carbonic acid, which in blood is catalyzed by the specific enzyme, carbonic anhydrase. The noncatalyzed reaction would not permit CO_2 interchange between the blood and the tissues, and between the blood and the lungs, at rates sufficient for physiological requirements.

NATURE OF CATALYSIS

Enzymes are catalysts peculiar to living matter, but catalysis itself is a familiar chemical phenomenon. A *catalyst* is defined as a substance that accelerates a chemical reaction but is not consumed in the over-all process. The use of platinum to catalyze the union of the elements of water is a familiar example as is the hydrolysis of sucrose catalyzed by acid. The important characteristic of such reactions is that the amount of catalyst bears no stoichiometric relationship to the quantity of substance altered. The efficiency of a catalyst can be expressed as the moles of substrate transformed per mole of catalyst in unit time. The catalytic efficiency of enzymes is extremely high; this can be seen from the fact that pure enzymes may catalyze the transformation of as many as 10,000 to 1,000,000 moles of substrate per minute per mole of enzyme.

Another important aspect of catalysis is the directed nature of the reaction. It is a common experience in organic chemistry to perform reactions at a high temperature or pressure and find that the desired compound is obtained among a mixture of other products. In general, catalyzed reactions give a more uniform reaction, *i.e.*, the yield of products is high, and this is true of enzymic reactions.

Thermodynamic Principles. In order to discuss the nature of enzymic catalysis, it is necessary first to introduce some thermodynamic principles. Whereas chemistry generally is concerned with molecular transformations, the occurrence and extent of such phenomena are governed by the flow of energy. This is the subject matter of the science of thermodynamics, the major concepts of which can be stated in two unifying principles, the first and second laws of thermodynamics. These laws permit one to understand the direction of chemical events, *i.e.*, whether a reversible reaction will proceed from left to right or from right to left as it is written, whether the progress of such reactions will permit the accomplishment of useful work or whether, in order for the reaction to proceed, energy must be delivered from an external source. The principles of thermodynamics are stated in terms of parameters which were originally invented so as to permit description of energy transformation in physical and chemical systems: enthalpy, entropy, and free energy. Of these, the last has proved to be most useful in understanding biochemical events.

The first law of thermodynamics is essentially the law of conservation of energy. Even in dealing with a finite system of molecules within a container, it is generally impossible to ascertain the magnitude of U, the total energy of that system. But if, to the system, energy is added as heat Q, then

$$Q = \Delta U + W \qquad (1)$$

where ΔU is the change in total energy, and W is the total work, if any, that has been accomplished.

In many instances the addition of energy as heat Q to the system may result in a change of volume, the pressure remaining constant. This change, $P\Delta V$, is in effect a form of work and, hence, is a component of the term W in equation (1). However, this is rarely a useful form of work and hence it has been found convenient to combine this component of W with the change in U, thereby defining a new term, H, *enthalpy* or heat content. The change in enthalpy in any process, at constant pressure, is:

$$\Delta H = \Delta U + P\Delta V \tag{2}$$

The first law can then be restated as

$$Q = \Delta H + W' \tag{3}$$

where W' is, therefore, *useful* work accomplished by input of the quantity of heat Q.

The first law, stated above, constitutes an adequate description of an ideal, reversible system, *i.e.*, one in which the energy utilized to alter the system is released as an exactly equal amount of energy, available to perform yet other work, when the system reverts to its original state. In fact, however, such instances of perfect reversibility do not occur. Some fraction of the increase in enthalpy resulting from adding energy to the system is not available to do useful work when the reverse process is allowed to proceed. Thus, it is common experience that most physical and chemical processes occur spontaneously in only one direction, *e.g.*, water runs downhill, protons and hydroxyl ions react, giving off heat. However, heating water does not drive it uphill, nor does it result in a net redissociation of water to a mixture of protons and hydroxyl ions. All such processes can be described in terms of the concept of equilibrium. Spontaneous changes tend toward the equilibrium state, not away from it; this is a manner of stating the second law of thermodynamics. The simplest description of the second law of thermodynamics is in terms of another thermodynamic quantity, the *entropy, S*, that fraction of the enthalpy which may not be utilized for the performance of useful work since, in most cases, it has increased the random motions of the molecules in the system. Hence, more generally, entropy is a measure of randomness or disorder. The product $T \times S$, in which T is the absolute temperature, represents energy that is wasted in the form of random molecular motions. In terms of S, the second law of thermodynamics states that given the opportunity, any system will undergo spontaneous change in that direction which results in an increase in entropy. Equilibrium is attained when entropy has reached a maximum; no further change may occur spontaneously unless additional energy is supplied from outside the system.

Let us now consider the consequence of adding heat to a system. Since heat represents the kinetic energy of random molecular motion, the addition of heat increases the entropy. If the system is at equilibrium, it follows that

$$Q = T\Delta S \tag{4}$$

If the system is not at equilibrium, however, change in the system may spontaneously increase the entropy even without addition of heat. Thus, in general for systems not at equilibrium,

$$T\Delta S > Q \tag{5}$$

If we combine equations (3) and (4), *i.e.*, combine the first and second laws, for a system at equilibrium we find that

$$\Delta H = T\Delta S - W' \tag{6}$$

However, in biochemistry we are rarely interested in the equilibrium state. Rather, interest is in reactions proceeding, as they must, in the direction which approaches toward equilibrium and at a single temperature. For such systems, equation (5) modifies the statement of equation (6) so that

$$\Delta H < T\Delta S - W' \tag{7}$$

This permits introduction of the thermodynamic parameter of greatest general utility in biochemistry, the quantity called "free energy," F. It is defined as

$$F = H - TS \tag{8}$$

In general, the change in free energy, ΔF, is the energy that becomes available to be utilized for the accomplishment of work, if there are appropriate means, as a system proceeds toward equilibrium. For a process at a single temperature,

$$\Delta F = \Delta H - T\Delta S \tag{9}$$

This is the form in which the laws of thermodynamics are most readily expressed for description of biochemical systems. For a system that is not at equilibrium,

$$\Delta F = -W' \tag{10}$$

Hence, systems not at equilibrium proceed spontaneously only in the direction of *negative* free energy change. When equilibrium has been attained, no further change in free energy can occur spontaneously. Once the system is at equilibrium, the *available* free energy content is zero. Conversely, a system already at equilibrium can be brought to a state remote from equilibrium only if, by some means, free energy can be made available to it. Utilization of free energy in this manner constitutes the performance of work.

Chemical Equilibria. The phenomenon of diffusion demonstrates that the free energy of a solute in solution increases with its concentration. Solutes in a concentrated solution placed in contact with a dilute solution diffuse into the latter until a uniform concentration is achieved. Since this occurs spontaneously, the free energy change for dilution by such diffusion must be negative. The variation of free energy with concentration is logarithmic,

$$F = F^\circ + RT \ln [C] \tag{11}$$

where R is the gas constant, T is absolute temperature, $[C]$ is the molar concentration of solute, and F° is the "standard free energy," *i.e.*, the free energy at a concentration of one mole per liter. Our interest, however, is not in such absolute values, but in the changes associated with chemical reactions.

For any chemical reaction,

$$A + B \rightleftharpoons C + D$$

and

$$\Delta F = \Delta F^\circ + RT \ln \frac{[C][D]}{[A][B]} \qquad (12)$$

where $\Delta F^\circ = F^\circ_C + F^\circ_D - F^\circ_A - F^\circ_B$. Equation (12) is applicable under all conditions. At equilibrium, however, ΔF must be zero. Hence, if $\frac{[C][D]}{[A][B]} = K$, where K is the equilibrium constant, then

$$\Delta F^\circ = -RT \ln K \qquad (13)$$

or at 37°,

$$\Delta F^\circ = -1{,}420 \log_{10} K \qquad (14)$$

Thus, the standard free energy of a chemical reaction, *i.e.*, the free energy made available by reaction of a mole of each reactant to form a mole of each product under standard conditions, can be calculated from measurement of the equilibrium constant. If ΔF° is negative, the process may proceed spontaneously; if it is positive, the reaction can be made to proceed only if, by some means, external free energy is made available. Values for ΔF° are expressed as calories per mole.

Rates of Chemical Reactions. For the same chemical reaction, $A + B \rightleftharpoons C + D$, proceeding by the simplest reaction mechanisms, the velocity of the forward reaction v_1 is proportional to the concentration of A and B, or

$$v_1 = k_1[A][B] \qquad (15)$$

For the reverse reaction,

$$v_2 = k_2[C][D] \qquad (16)$$

where k_1 and k_2 are the individual velocity constants.

At equilibrium, $v_1 = v_2$; hence

$$k_1[A][B] = k_2[C][D]$$

or

$$K = \frac{k_1}{k_2} = \frac{[C][D]}{[A][B]} \qquad (17)$$

Thus K, the thermodynamic equilibrium constant, is also the relationship between the velocity constants. As we have seen, the actual net direction in which reaction will proceed is determined by the initial concentration of each reactant and the value of K; spontaneous reaction always proceeds in the direction of negative free energy change, *i.e.*, toward equilibrium.

If the initial system is not at equilibrium, reaction proceeds at a rate determined by the velocity constants, k_1 and k_2. Frequently, however, in common experience, a system of components remote from equilibrium appears to be in a *metastable* state. It should proceed by increase in entropy and/or decrease in free energy to the equilibrium state, yet fails to do so. This is obviously true of flammable organic compounds exposed to air; a spark may be required for ignition. Or a boulder, lying in a trough on a hillside, will roll downhill only if lifted over the

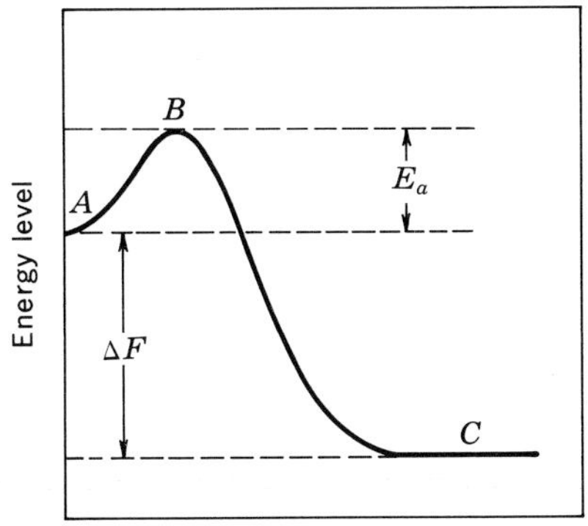

Fig. 10.1. The chemical compound A is in a metastable state. Energy of activation E_a is required for transition from A to B before the reaction liberates energy. In the change from B to C, E_a is recovered, and energy is made available according to the difference in state ΔF between A and C. At C, the new compound has reached an equilibrium state.

barrier that holds it in check. Many chemical systems behave similarly. For reaction to proceed, energy must first be delivered that increases the incidence of excited or reactive molecules in the system. It is precisely for this reason that the chemist must frequently heat a reaction mixture in order to initiate the reaction. As shown in Fig. 10.1, as reaction then proceeds, the energy of activation E_a reemerges and the total energy change resulting from the reaction is the calculated ΔH and E_a. Further consideration of energy of activation is given later (page 232).

Living organisms usually do not influence the rate of metabolic reactions by invoking changes in temperature, nor can most survive at high temperatures. Hence a catalyzed reaction is necessary to make the process go fast enough at the temperature of the organism. Moreover, if important biological reactions proceeded without catalysts, no control could be exercised over their rates. If the organism loses control of the rates of important reactions, the maintenance of normal structure and function becomes impossible and the organism dies.

NATURE OF ENZYMES

Knowledge of the role of enzymes as catalysts has grown with knowledge of catalysis in general. Berzelius, who was one of the first to define and recognize the nature of catalysis, proposed in 1837 that "ferments" were catalysts produced by living cells. Nevertheless, little was known concerning the chemical nature of enzymes until the beginning of this century, when there was a growing conviction that enzymes were probably protein in nature. The announcement by J. B. Sumner in 1926 of the isolation of the enzyme urease as a crystalline protein was greeted with some skepticism, but in the next few years Northrop and Kunitz reported the isolation of crystalline pepsin, trypsin, and chymotrypsin. Since that time, several hundred enzymes have been obtained in highly purified form and more than 100 of these in the crystalline state; all have proved to be proteins.

Until 1926, much of the effort in studying enzymes was concerned with the nature of the process, *i.e.*, enzymes were characterized by the chemical reactions that they catalyzed. With the availability of crystalline enzymes, they have also been investigated from the viewpoint of protein chemistry. Indeed, a great deal of our present knowledge of the chemistry of proteins has come from study of crystalline enzymes and from attempts to understand the nature and mode of action of these catalysts. For the chemical and physical properties of enzymes, we may refer to the general chemistry of the proteins already presented. Many enzyme proteins have been cited as examples in the discussion of the properties of proteins.

The largest part of the dry weight of tissues is protein. Indeed, it has been suggested that most of the proteins of actively functioning tissues such as liver and kidney, with the exception of some structural elements like collagen and elastin, are really enzymes. Regardless of the exact fraction of total body protein that possesses enzymic functions, it is clear that there must be many hundreds of different protein enzymes in the tissues in order to account for the myriad of known metabolic reactions.

Nomenclature. Enzymes are usually named in terms of the reactions that are catalyzed. A customary practice is to add the suffix *-ase* to a major part of the name of the substrate. The enzyme that attacks urea is urease, arginine is acted upon by arginase, tyrosine by tyrosinase, uric acid by uricase, etc. In addition, an older nomenclature has persisted, and we find such names as pepsin, trypsin, rennin, etc. Enzymes may also be classified by groups that catalyze similar chemical reactions, such as proteinases, lipases, oxidases, etc. An International Commission on Enzymes has devised a complete but rather complex system of nomenclature for enzymes. However, since this has not yet gained wide acceptance or general usage, it is not employed in this book.

Many enzymes possess chemical groups that are non-amino acid in nature. These conjugated proteins (*holoenzymes*) may also be classified chemically in terms of these special groupings or prosthetic groups (page 117). Thus, an enzyme may frequently be dissociated into a protein component, termed the *apoenzyme,* and a nonprotein organic *prosthetic* portion. As an example, reddish-brown catalase (Chap. 8) dissociates in acid to yield a colorless protein and ferriprotoporphrin. In other instances, an enzyme may contain only amino acids and a metal. For example, there are enzymes which are copper proteins; ascorbic acid oxidase is one of these. In this enzyme, the copper is tightly bound and is not separated readily from the protein. Many other enzymes require the addition of metal ions in order to activate the enzyme. In some instances these metal ions, frequently called *activators,* function in combination with the protein; in others, the metal ion forms a compound with the substrate, and it is the metal-substrate complex that reacts with the enzyme. Arginase, certain phosphatases, and some peptidases are examples of enzymes that require certain metal ions for activity. The ions of Ca, Co, Cu, Mg, Mn, Mo, Na, K, and Zn are known to participate in enzymic reactions.

Many enzymes require certain organic substances as cofactors in order to function. The *cofactors,* or *coenzymes,* generally act as acceptors or donors of a functional group or of atoms that are removed from or contributed to the substrate. Since these organic cofactors are frequently readily dissociable from the enzyme protein, they may properly be regarded as cosubstrates. The type of

Table 10.1: **Some Organic Coenzymes and Prosthetic Groups of Enzymes**

Coenzyme or prosthetic group	Structure given on page	Enzymic and other functions	Essential nutritional factor or vitamin*
Diphosphopyridine nucleotide (DPN)	322	Hydrogen carrier.	Nicotinic acid
Triphosphopyridine nucleotide (TPN)	356	Hydrogen carrier	Nicotinic acid
Adenosine triphosphate (ATP)	315	Transphosphorylation	None
Pyridoxal phosphate	542	Transaminases, amino acid decarboxylases, racemases, etc.	Pyridoxine
Thiamine pyrophosphate	328	Oxidative decarboxylation, active aldehyde carrier	Thiamine
Flavin mononucleotide (FMN)	362	Hydrogen carrier	Riboflavin
Flavin adenine dinucleotide (FAD)	324	Hydrogen carrier	Riboflavin
Coenzyme A (CoA)	326	Acetyl or other acyl group transfer; fatty acid synthesis and oxidation	Pantothenic acid
Iron-protoporphyrin	168	In catalase, peroxidase, cytochromes, hemoglobin	None
6,8-Dithio-*n*-octanoic acid (lipoic acid)	328	Oxidative decarboxylation; as hydrogen and acyl acceptor	Required by some microorganisms
Tetrahydrofolic acid	550	One carbon transfer	Folic acid
Biotin	1033	CO_2 transfer	Biotin
Cobamide	1039	Group transfer	Cobalamine

* See also Chaps. 49 and 50.

coenzyme concerned in the enzymic process aids in classification. Some of these organic cofactors and prosthetic groups are listed in Table 10.1. Several of these cofactors are derived from components that cannot be synthesized by mammals and are essential nutritive factors, or vitamins. Other vitamins may also function in a similar manner, but their specific role in enzymic reactions has yet to be determined.

Specificity. Urease is a specific enzyme; its primary substrate is urea. In contradistinction, other enzymes can attack many related substrates of similar structure, *e.g.*, some esterases can act upon the esters of different fatty acids with a variety of alcohols. Nevertheless, the esterases are specific in their esterase action; they do not catalyze other hydrolytic reactions, nor do they function as oxidases, decarboxylases, etc. Specificity is evident in the type of reaction that is catalyzed. Almost all enzymes show a high degree of spatial specificity. Arginase acts only on L-arginine; it does not attack D-arginine. D-Amino acid oxidase has no action on L-amino acids, whereas D-amino acids are rapidly oxidized to the corresponding keto acids. The specificity of enzymes is one of their most fundamental and important properties and will be discussed in some detail later (Chap. 12).

The specificity of enzymes is of practical utility in preparative chemistry, particularly in the resolution of racemic compounds. Since D-amino acid oxidase acts only on D-amino acids, the action of the enzyme on a DL-amino acid mixture yields the L-amino acid and the keto acid. In view of the solubility of the keto acid in organic solvents, the two compounds are readily separated from each other. The spatial specificity of hydrolytic enzymes may also be used to advantage when both isomers are desired. For example, enzymes in extracts of kidney or pancreas will cleave only the natural or L form of chloroacetyl-DL-phenylalanine, resulting in a mixture containing the readily separable water-soluble L-phenylalanine and the chloroform-soluble chloroacetyl-D-phenylalanine; D-phenylalanine is obtained by acid hydrolysis of the latter compound. As long ago as 1858 Pasteur (who discovered the optical activity of the natural amino acids) showed that a green mold would ferment dextrorotatory tartaric acid but not levorotatory tartaric acid. Since that time, molds, bacteria, and yeasts as well as highly purified enzymes have been utilized for the destruction of one or the other isomer of various amino acids, sugars, and other compounds.

Enzymes are widely used as chemical reagents for analytical determinations. Some examples of enzymic reactions have been mentioned as used for the estimation of amino acids (page 113). It is obviously a great advantage to use rapid methods of analysis, *e.g.*, those involving the pH meter, colorimeter, spectrophotometer, or change in gas volume. The following examples illustrate some of these methods. Esterases liberate an acid that will lower the pH of a weakly buffered solution. By adding sufficient alkali to maintain constant pH, the extent of reaction can be determined from the alkali consumption. Appearance of amino groups as a result of proteolysis can be measured by the ninhydrin reaction (page 106), with quantitative measurements made in a colorimeter. A change in ultraviolet light absorption accompanies enzymic alteration in the saturation of organic compounds; this can be quantitatively evaluated in the spectrophotometer.

Purified enzymes are used to convert substances whose quantitation is difficult to other products that are easily measured quantitatively by reading available apparatus. It is frequently possible to use several purified enzymes that can catalyze the alteration of a substance by several steps to a product that can be measured conveniently. It is only because of the great specificity manifested by enzymes that such methods are possible. Many compounds present in only minute quantities can be quantitatively measured by such methods.

Other uses of enzymes are in the degradation of large molecules in order to study their structures and their constituent parts. Examples have already been cited that illustrate the use of enzymes for analysis of the structure of polysaccharides, proteins, and nucleic acids. Isolation of L-tryptophan, glutamine, and asparagine from proteins can be accomplished only after enzymic digestion of proteins. Other examples of such procedures will be described later. The coming together of classical chemical procedures and the enzymic methods of the biochemist is permitting rapid progress in the study of complex molecules generally.

CLASSIFICATION OF ENZYMES

Enzymes may be described in chemical terms, as conjugated proteins, and by the reaction catalyzed. The usual schemes of classification are grossly inadequate to deal with the many types of reactions, and most of these catalyzed reactions are understood best in relation to metabolic processes. Here it will suffice to indicate, with a few illustrations, some of the main types of reactions catalyzed by enzymes. Most enzymic reactions have their counterparts in familiar types of chemical reactions, but many novel reactions have been discovered first in biological systems. Many of the more specialized types of reactions are best presented later in the chapters that deal with details of processes of metabolism.

HYDROLYTIC ENZYMES (3)

Hydrolytic enzymes act by catalyzing introduction of the elements of water at a specific bond of the substrate. These reactions are frequently reversible, and the classification of these enzymes as hydrolytic rather than synthetic is arbitrary, based on the more easily measurable phenomenon and on the fact that in aqueous solution, equilibrium favors a predominance of hydrolytic products.

Esterases. These are enzymes that catalyze hydrolysis of ester linkages. More specific classification depends on both the type of acid and the type of alcohol comprising the ester. Some of these enzymes effect hydrolysis of a variety of compounds of similar structure; others are highly specific.

Simple esterases such as liver esterase catalyze reversibly the scission and synthesis of esters of lower alcohols and fatty acids. An example is the reversible hydrolysis of ethyl butyrate to give ethanol and butyric acid.

$$CH_3CH_2CH_2COOC_2H_5 + H_2O \rightleftharpoons CH_3CH_2CH_2COOH + C_2H_5OH$$

The true lipases hydrolyze fats into long-chain fatty acids and glycerol. Examples are *pancreatic lipase* and many plant lipases. The action of the simple esterases and lipases is reversible, and synthesis or hydrolysis proceeds to equilibrium in solution.

Phosphatases hydrolyze esters of phosphoric acid. This is a large and complex group of enzymes, some of which appear to be highly specific. The *monoesterases* hydrolyze monophosphoric esters according to the following general reaction.

$$R—O—\overset{\overset{\displaystyle O}{\|}}{\underset{\underset{\displaystyle OH}{|}}{P}}—OH + H_2O \longrightarrow ROH + P_i$$

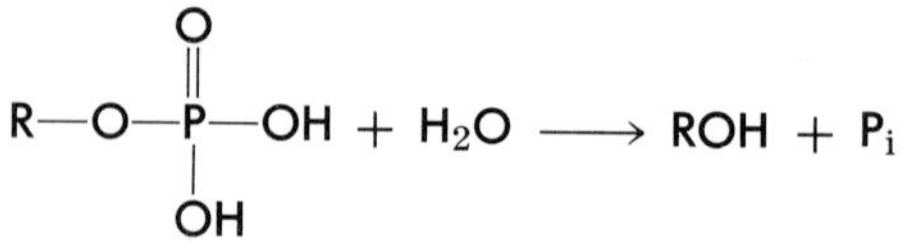

The symbol P_i, employed in the equation above, will be encountered frequently in this textbook and in the biochemical literature. It is an abbreviation for inorganic orthophosphate and implies the mixture of $H_2PO_4^-$ and $HPO_4^=$ which would exist at the specific pH of the reaction medium, calculated from the Henderson-Hasselbalch equation (page 98). Similarly, the abbreviation PP_i will indicate inorganic pyrophosphate, which at pH 7.4 is largely

$$HO—\overset{\overset{\displaystyle O}{\|}}{\underset{\underset{\displaystyle O^-}{|}}{P}}—O—\overset{\overset{\displaystyle O}{\|}}{\underset{\underset{\displaystyle O^-}{|}}{P}}—O^-$$

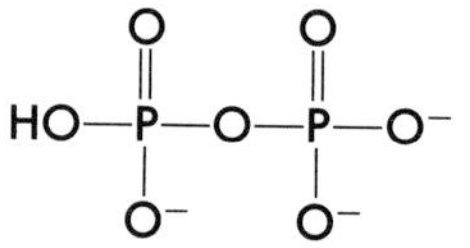

Pyrophosphatases are enzymes that split pyrophosphate linkages. A crystalline pyrophosphatase from baker's yeast splits inorganic pyrophosphate to orthophosphate.

Nucleases cause a depolymerization of nucleic acids, liberating oligo- or mononucleotides. These enzymes are *diesterases* of a special type which split the linkages that bind the individual nucleotides.

$$R—O—\overset{\overset{\displaystyle O}{\|}}{\underset{\underset{\displaystyle OH}{|}}{P}}—O—R' + H_2O \longrightarrow R—O—\overset{\overset{\displaystyle O}{\|}}{\underset{\underset{\displaystyle OH}{|}}{P}}—OH + R'OH$$

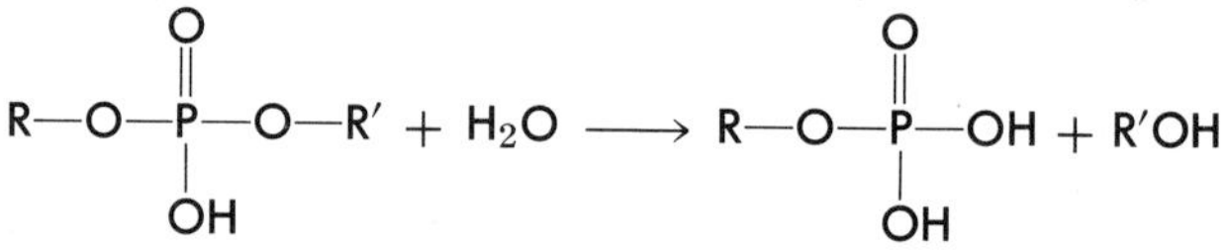

Two specific nucleases have been obtained in crystalline form from mammalian pancreas. *Ribonuclease* hydrolyzes ribonucleic acids, and the site of action appears to be at diester linkages involving pyrimidine nucleotides (page 200). *Deoxyribonuclease* acts in the presence of Mg^{++} ions, causing the depolymerization of deoxyribonucleic acids.

Carbohydrases. These enzymes hydrolyze the glycosidic linkages of simple glycosides, oligosaccharides, and polysaccharides.

Glycosidases hydrolyze simple glycosides and oligosaccharides. An example is yeast *invertase,* which hydrolyzes sucrose to glucose and fructose.

Polysaccharidases act on the complex polysaccharides. α-1,4-Glucan maltohydrolases (*β-amylases*) hydrolyze starch and glycogen to maltose and to residual polysaccharides. Their action has been described (Chap. 3). The polysaccharidases of wheat, barley, soybeans, and other plants are of this type.

α-1,4-Glucan 4-glucanohydrolases (*α-amylases*) can hydrolyze glycogen and starch and the residual polysaccharides of starch (amylodextrins) to give glucose, maltose, and products that no longer give a color with iodine (page 48). These enzymes are widely distributed in plants. Animal amylases of this type have been obtained in crystalline form from pancreas and saliva and are also found in blood and urine.

Proteases (Proteolytic Enzymes). These are enzymes which attack the peptide bonds of proteins and peptides. It is customary to distinguish between the proteinases (endopeptidases) and the peptidases (exopeptidases).

The *peptidases* (*exopeptidases*) act on peptide bonds adjacent to a free amino or carboxyl group. Among the principal types of peptidases are the following.

Carboxypeptidases require the presence of a free carboxyl group in the substrate and split the peptide bond adjacent to this group, liberating a free amino acid. Examples are the carboxypeptidase of mammalian pancreas and carboxypeptidases of kidney, spleen, etc.

$$R{-}{-}{-}\underset{\underset{O}{\|}}{C}{-}\underset{\underset{H}{|}}{N}{-}CHR'{-}COOH + H_2O \longrightarrow R{-}{-}{-}COOH + H_2NCHR'COOH$$

Aminopeptidases act on the peptide bond adjacent to the essential free amino group of simple peptides. An example is an *aminotripeptidase* found in many animal tissues which splits tripeptides such as L-alanylglycylglycine to L-alanine and glycylglycine.

Dipeptidases specifically act only on certain dipeptides; an example is *glycylglycine dipeptidase,* which requires Co^{++} or Mn^{++} for its action.

Proteinases (*endopeptidases*) act on the interior peptide bonds of proteins; they can, however, also split peptide bonds in suitable simple peptides and their derivatives. Examples are *pepsin, trypsin,* and *chymotrypsin* from animals (Chap. 23).

Cathepsins are intracellular proteinases found in most animal tissues. The richest sources are liver, kidney, and spleen. Many of these enzymes are active only in the presence of certain reducing substances such as cysteine, glutathione, HCN, H_2S, and ascorbic acid.

Plant Proteinases. Papain is obtained from the unripe fruit of the papaya, or papaw tree. Similar enzymes are *bromelin,* found in pineapples, and *ficin,* in the milky sap of the fig tree. Ficin has long been used for the clotting of milk in areas of the Near East. Some of the plant proteinases are used commercially for tenderizing meat. Some proteinases of plants and microorganisms have been named from the species of origin, *e.g.*, papain from papaya, subtilisin from *Bacillus subtilis,* etc.

PHOSPHORYLASES

Polysaccharide phosphorylases catalyze reversibly the phosphorolytic cleavage of the α-glucosidic 1,4 linkages of glycogen and starch to α-glucose 1-phosphate. This degradation is not a hydrolysis such as that carried out by the amylases since it involves the elements not of water but of phosphoric acid. The reaction may be written

$$\textbf{Glycogen} + n \textbf{ phosphate} \rightleftharpoons n \textbf{ glucose 1-phosphate}$$

Polynucleotide phosphorylases catalyze the reversible formation of polynucleotides, as in the following example.

$$\textbf{Polynucleotide} + n \textbf{ phosphate} \rightleftharpoons n \textbf{ nucleoside diphosphates}$$

OXIDATION-REDUCTION ENZYMES

Enzymes concerned with *oxidation-reduction* processes play an extremely important role in metabolism and will be considered in detail elsewhere (Chaps. 16 and 17). Here it is desirable to indicate a few of the main types since the study of some of these enzymes has had a major influence on knowledge of enzymes in general.

Dehydrogenases. Oxidation of organic compounds is generally a dehydrogenation process, and there are many *dehydrogenases,* enzymes catalyzing dehydrogenation, which are highly specific. The over-all process for dehydrogenation may be represented as follows.

$$XH_2 + A \longrightarrow X + AH_2$$

An example is the action of *ethanol dehydrogenase,* in which diphosphopyridine nucleotide (DPN, page 322) is the hydrogen acceptor.

$$\text{Ethanol} + \text{DPN}^+ \rightleftharpoons \text{acetaldehyde} + \text{DPNH} + \text{H}^+$$

Oxidases. There are a number of *aerobic oxidases* which can utilize oxygen directly.

Cytochrome oxidase, an iron-porphyrin enzyme, is one of the most important of this group (Chap. 16). The iron of the prosthetic group undergoes reversible oxidation-reduction from the ferrous to the ferric state, accepting electrons which are transferred subsequently to oxygen.

Flavin Enzymes. These enzymes have prosthetic groups that are mono- or dinucleotides containing riboflavin (page 324). This moiety functions in accepting hydrogen atoms and some are subsequently oxidized by molecular oxygen (Chap. 16). An example is *xanthine oxidase,* found in liver and milk; the enzyme contains flavin adenine dinucleotide, iron, and molybdenum, and catalyzes the oxidation of hypoxanthine to xanthine and the latter to uric acid. This enzyme also catalyzes oxidation of a variety of aldehydes.

Copper Enzymes. The *phenol oxidases* are copper-containing proteins that catalyze the oxidation of phenol derivatives to quinones. Examples are the *polyphenol oxidases* of mushrooms and potatoes. *Tyrosinase,* widely distributed in animals and plants, catalyzes the oxidation of tyrosine to the orthoquinone; subsequent oxidation steps, some of which are spontaneous, lead to the formation of the dark pigment *melanin* (Chap. 25).

TRANSFERRING ENZYMES (TRANSFERASES)

There are many types of enzyme which catalyze the transfer of a group from one substance to another. The *transaminases* (Chap. 23) catalyze transfer of an amino group of an amino acid to an α-keto acid. Such a reaction is shown below for *glutamic-aspartic transaminase,* also termed glutamic-oxaloacetic transaminase. Transfer reactions also involve phosphate groups, methyl groups, amide groups, etc. Such reactions are also catalyzed by many hydrolytic enzymes such as the *carbohydrases, phosphatases, esterases,* and *proteinases.*

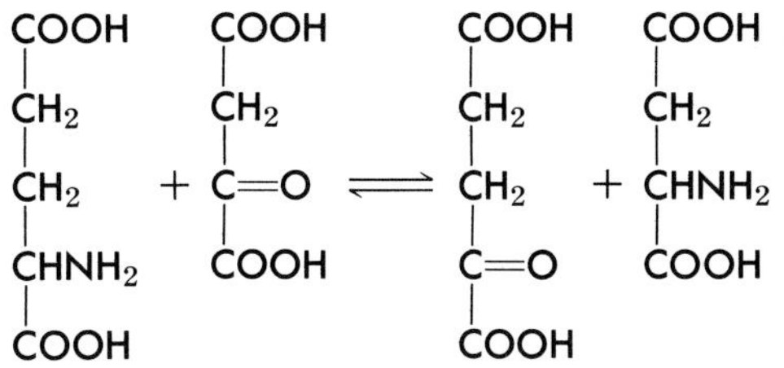

$$\underset{\text{Glutamic acid}}{\begin{array}{c} COOH \\ | \\ CH_2 \\ | \\ CH_2 \\ | \\ CHNH_2 \\ | \\ COOH \end{array}} + \underset{\text{oxaloacetic acid}}{\begin{array}{c} COOH \\ | \\ CH_2 \\ | \\ C{=}O \\ | \\ COOH \end{array}} \rightleftharpoons \underset{\text{α-ketoglutaric acid}}{\begin{array}{c} COOH \\ | \\ CH_2 \\ | \\ CH_2 \\ | \\ C{=}O \\ | \\ COOH \end{array}} + \underset{\text{aspartic acid}}{\begin{array}{c} COOH \\ | \\ CH_2 \\ | \\ CHNH_2 \\ | \\ COOH \end{array}}$$

Glutamic acid + oxaloacetic acid $\rightleftharpoons$ α-ketoglutaric acid + aspartic acid

DECARBOXYLASES

The decarboxylases remove CO_2 from carboxylic acids.

Amino acid decarboxylases of microorganisms are very widespread. These enzymes are responsible for the formation of amines. An example is lysine decarboxylase.

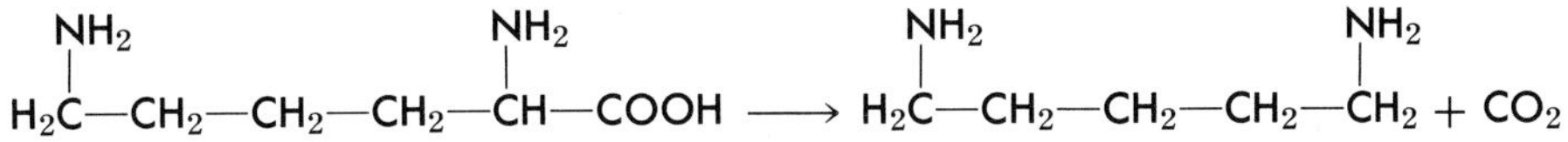

$$\underset{\text{Lysine}}{H_2\overset{\overset{\displaystyle NH_2}{|}}{C}-CH_2-CH_2-CH_2-\overset{\overset{\displaystyle NH_2}{|}}{C}H-COOH} \longrightarrow \underset{\text{Cadaverine}}{H_2\overset{\overset{\displaystyle NH_2}{|}}{C}-CH_2-CH_2-CH_2-\overset{\overset{\displaystyle NH_2}{|}}{C}H_2} + CO_2$$

Keto acid decarboxylases are important in the liberation of CO_2. An example is the catalyzed decarboxylation of oxaloacetate to pyruvate and CO_2.

HYDRASES

These enzymes catalyze addition to or removal of water from their specific substrates. *Fumarase* catalyzes the interconversion of malic and fumaric acids.

$$HOOC-CH_2-CHOH-COOH \rightleftharpoons HOOC-CH{=}CH-COOH + H_2O$$

ISOMERASES

The term *isomerase* denotes those enzymes which catalyze an intramolecular rearrangement, *e.g.*, the interconversion of aldose and ketose sugars. For example, *phosphohexose isomerase* catalyzes the following interconversion.

Glucose 6-phosphate $\rightleftharpoons$ fructose 6-phosphate

In this group of enzymes may also be included the *epimerases, e.g., uridine diphosphate galactose 4-epimerase,* which catalyzes a Walden inversion in which the configuration about carbon-4 of the galactosyl residue is transformed to that of a glucosyl residue (page 426):

Uridine diphosphate galactose $\rightleftharpoons$ uridine diphosphate glucose

Uridine diphosphate glucose plays an important role in several aspects of carbohydrate metabolism (Chap. 19).

REFERENCES

Books

Boyer, P. D., Lardy, H., and Myrbäck, K., eds., "The Enzymes," 2d ed., 8 vols., Academic Press, Inc., New York, 1958–1963.

Colowick, S. P., and Kaplan, N. O., eds., "Methods in Enzymology," 11 vols., Academic Press, Inc., New York, 1954–1968. Additional volumes are being issued periodically.

Dixon, M., and Webb, E. C., "Enzymes," 2d ed., Academic Press, Inc., New York, 1963.

"Enzyme Nomenclature," Recommendations (1964) of the International Union of Biochemistry on the Nomenclature and Classification of Enzymes, Together with Their Units and the Symbols of Enzyme Kinetics, American Elsevier Publishing Company, New York, 1965.

Gutfreund, H., "An Introduction to the Study of Enzymes," Blackwell Scientific Publications, Ltd., Oxford, 1965.

See also list following Chap. 12.

11. Enzymes. II

Kinetics. Inhibition. Metabolic Inhibitors. Control of Enzymic Activity

There are three general ways in which the problem of mechanism of enzyme action is being attacked. One is by a study of the chemical nature of the enzyme itself; this is essentially a special problem in protein chemistry. The second method is to study the mechanism of action of simpler catalysts on the same reaction; the investigation of such model reactions has been exceedingly fruitful. The third approach is to determine how various factors influence the rate of enzymic reactions. We shall first consider this last type of study.

The view that has dominated all attempts to explain mechanisms of enzymic catalysis is that the enzyme forms an intermediate complex with the substrate or substrates. Before proceeding to a more detailed examination of the evidence in support of this concept, it is desirable to summarize briefly some of the early experiments that led to this view.

1. In 1880, Wurtz noted that after addition of the soluble proteinase, *papain,* to the insoluble protein, fibrin, repeated washing of the fibrin did not stop the proteolysis. He concluded that the papain had formed a compound with the fibrin.

2. O'Sullivan and Tompson, in 1890, observed that the enzyme, *invertase,* could withstand higher temperatures in the presence of the substrate, sucrose, than in its absence. To explain this observation they suggested that the enzyme invertase had combined with its substrate, sucrose.

3. The experiments of Emil Fischer in the 1890s indicated that *glycosidases* were highly stereospecific with respect to their substrates. He suggested a similar specificity in the structure of the enzymes since these results could be explained only if enzyme and substrate reacted. His remarks have been widely quoted.

> "Inasmuch as the enzymes are in all probability proteins, . . . it is probable that their molecules also have an asymmetrical structure, and one whose asymmetry is, on the whole, comparable to that of the hexoses. Only if enzyme and fermentable substance have a similar geometrical shape can the two molecules approach each other close enough for the production of a chemical reaction. Metaphorically, we may say that enzyme and glucoside must fit into each other like lock and key."

4. Many investigators observed that rates of substrate conversion did not follow the rate laws of a simple reaction. A. Brown (1892) and V. Henri (1903) both

suggested that product formation depended on the rate of decomposition of an enzyme-substrate complex. It remained for Michaelis and Menten in 1913 to supply an acceptable mathematical formulation in support of this view.

KINETICS OF ENZYMIC REACTIONS

For the biochemist, the problem of rate studies is of paramount importance. It is insufficient to know that a synthetic reaction takes place in living tissue; the reaction must occur rapidly enough to supply the needs of the organism for the products of the reaction. Toxic metabolic products must be eliminated at sufficient velocity to prevent their accumulation to injurious levels. When a needed substance is not supplied or a product is not removed at sufficient speed, derangements may result that are metabolic disorders or disease processes. Correspondingly, metabolic derangements also occur when a physiological process is uncontrolled and the rate is much greater than normal. Thus, there is a dual purpose in considering rates of enzymic processes, *viz.*, to understand the normal and abnormal metabolism of the organism as a whole, and to attempt to elucidate the intimate nature of the enzymic process itself.

The rates of chemical reactions are generally estimated as the change in concentration c of substrate or product per unit of time t. Customary units are moles per liter and seconds. Thus, if c is the initial concentration which decreases with time, the reaction velocity is $-dc/dt$. This rate may depend upon the instantaneous value of c in various ways. It may be independent of c:

$$-\frac{dc}{dt} = k_0 = k_0c^0 \qquad \text{(zero-order reaction)}$$

It may be proportional to c:

$$-\frac{dc}{dt} = k_1c = k_1c^1 \qquad \text{(first-order reaction)}$$

Or it may be proportional to the second, or rarely a higher power of c:

$$-\frac{dc}{dt} = k_2(c \times c) = k_2c^2 \qquad \text{(second-order reaction)}$$

In each case, k is known as the *reaction velocity constant* or *rate constant.* The dimensions of k depend on the order of the reaction and are given by $k = c^{1-n}t^{-1}$, where n is the order of the reaction. Thus k_0 has dimensions of (ct^{-1}); k_1 of (t^{-1}); k_2 of $(c^{-1}t^{-1})$. From this it follows that velocity constants for reactions of different orders may not be added or subtracted. Also, only in first-order reactions is the velocity constant devoid of the dimension of concentration. Thus only for first-order reactions is the half-time, *i.e.*, the time required to halve the initial concentration, a constant at all concentrations.

For enzymic reactions in which the molecular weight of the enzyme is known, rate constants are given in terms of moles of enzyme. When the molecular weight of the enzyme is unknown or the enzyme preparation is impure, the velocity is frequently expressed per milligram of protein per milliliter or in rate per milligram of protein nitrogen per milliliter. By international convention, one unit of any enzyme is that amount which will catalyze the transformation of 1 micromole of substrate per minute under specified conditions.

For all enzymic processes, other conditions being constant, the rate of the reaction depends on the concentration of the enzyme and of its substrate. When the enzyme concentration is constant, the initial velocity of the reaction increases, as the substrate concentration is increased, in a hyperbolic manner toward a maximum velocity. Note that *initial* velocity is specified since substrate will be consumed as the reaction proceeds. In practice, this means that sensitive methods must be used so that the velocity can be estimated accurately before the substrate concentration changes appreciably.

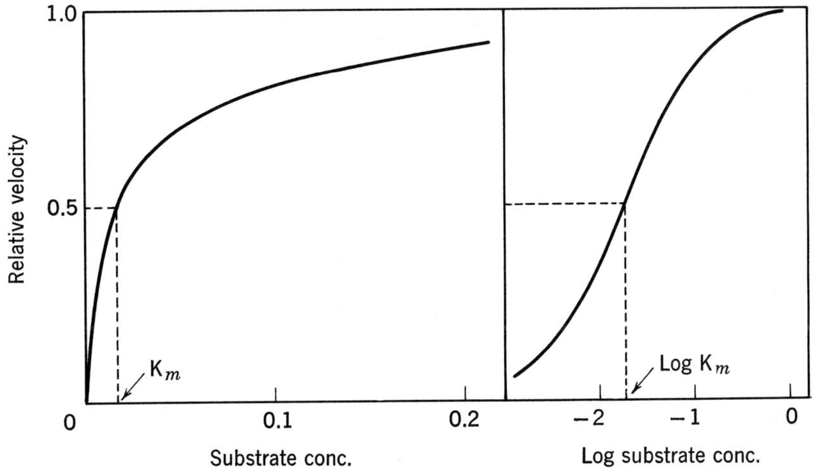

Fig. 11.1. Relative initial velocity as a function of substrate concentration (*left*) and as a function of the logarithm of the substrate concentration (*right*) for the action of yeast invertase on sucrose. The value of the substrate concentration for half the maximal velocity is 0.017*M*, which is equal to K_m; for the semilogarithmic plot, at half-maximal velocity, log substrate concentration is −1.77, which is equal to log K_m.

Figure 11.1 shows the rate of enzymic hydrolysis of sucrose to glucose and fructose, as a function of the sucrose concentration. Observe that the rate does not increase linearly with increase in substrate concentration. Instead, the rate approaches a limiting velocity at high substrate concentrations. This indicates that all of the enzyme has been "saturated" with substrate. In other words, the enzyme must have a finite number of sites for substrate and when these are all occupied, no further augmentation of rate is possible. Here we shall use the explanation of these findings offered in 1913 by Michaelis and Menten.

The Michaelis-Menten Hypothesis. The most important feature of this theory is the assumption that an intermediate enzyme-substrate complex is formed. It is further assumed that the rate of conversion of the substrate to the products of the reaction is determined by the rate of conversion of the enzyme-substrate complex to reaction products and the enzyme. The following scheme may be written to illustrate this concept:

$$\text{Enzyme (E)} + \text{substrate (S)} \rightleftharpoons \text{enzyme-substrate complex (ES)} \\ \downarrow \\ \text{Enzyme} + \text{products (P)}$$

Thus, the rate of product formation depends on the concentration of ES. If the rate of formation of P depended directly on [S], then at constant [E] a linear relationship could be expected between velocity and the concentration of S. Since this is not obtained, Michaelis and Menten proposed the following explanation; this is given in a somewhat simplified form, it being assumed that only a single substrate and a single product are formed. It is also assumed that the process proceeds essentially to completion and that the concentration of the substrate is much greater than that of the enzyme in the system.

$$E + S \underset{k_{-1}}{\overset{k_1}{\rightleftharpoons}} ES \xrightarrow{k_2} P + E \tag{1}$$

where k_1, k_{-1}, and k_2 are the respective velocity constants of the three assumed processes. For the rate of formation of ES, v_f, we may write

$$v_f = k_1([E] - [ES])[S] \tag{2}$$

where [E] – [ES] is the concentration of uncombined enzyme. This states that the rate of formation of ES, which is v_f, is proportional to the concentration of uncombined enzyme and substrate. The rate of disappearance of ES, v_d, is then

$$v_d = k_{-1}[ES] + k_2[ES] \tag{3}$$

since ES can disappear to give the initial reactants (k_{-1}) or by the formation of products (k_2) [see equation (1)].

When the rates of formation and disappearance of ES are equal, *i.e.*, when v_f [equation (2)] equals v_d [equation (3)], then equation (4) describes the steady state.

$$k_1([E] - [ES])[S] = k_{-1}[ES] + k_2[ES] \tag{4}$$

The terms may be rearranged to give

$$\frac{[S]([E] - [ES])}{[ES]} = \frac{k_{-1} + k_2}{k_1} = K_m \tag{5}$$

The term containing the three velocity constants is K_m, *the Michaelis-Menten constant,* and is a particularly useful parameter, characteristic of each enzyme.

The relationships among the substrate concentration, the enzyme concentration and the velocity of the enzyme-catalyzed reaction can be developed in the following manner. From equation (5), by rearrangement to solve for [ES], the steady state concentration of the enzyme-substrate complex is

$$[ES] = \frac{[E][S]}{K_m + [S]} \tag{6}$$

Because it is not generally feasible to measure the concentration of ES, in order to ascertain K_m, it is necessary to derive an expression that relates K_m to the readily measured parameters [E] and [S]. Advantage is taken of the relationship,

$$V = k_2[\text{ES}] \tag{7}$$

where V is the observed *initial* velocity. When the substrate concentration is made so high in relation to the enzyme concentration that essentially all the enzyme is present as ES, then the velocity of the reaction is maximal, and this velocity, V_{max}, has the value

$$V_{max} = k_2[\text{E}] \tag{8}$$

By substituting for ES in equation (7) its value in equation (6), and dividing equation (7) by equation (8), there is obtained the desired expression:

$$V = \frac{V_{max}[\text{S}]}{K_m + [\text{S}]} \quad \text{or} \quad K_m = [\text{S}]\left(\frac{V_{max}}{V} - 1\right) \quad \text{or} \quad V = \frac{V_{max}}{1 + K_m/[\text{S}]} \tag{9}$$

This is the Michaelis-Menten equation.

For experimental determination of K_m, the velocity of the reaction (relative activity of the enzyme) is measured as a function of substrate concentration. These experimentally determined values may be plotted against one another as indicated in Fig. 11.1.

When $V = \frac{1}{2}V_{max}$, it will be seen from equation (9) that K_m is numerically equal to the substrate concentration, or K_m *is equal to the concentration (expressed in moles per liter) of the substrate which gives half the numerical maximal velocity,* V_{max}. Thus, from the data of Fig. 11.1 it is possible to ascertain K_m.

The K_m shown in Fig. 11.1 is indicated to be 0.017*M*. For each enzyme-substrate system, K_m has a characteristic value which is independent of the enzyme concentration but may be dependent on pH, temperature, or other extrinsic factors. If the same enzyme can attack several substrates, the K_m values frequently give a useful comparison of the affinity for different substrates. Such information has been of value in assessing the specificity and binding groups in various enzymes (Chap. 12). In addition K_m values for the several enzymes in a metabolic series of consecutive reactions can occasionally indicate the rate-limiting step in the pathway.

There are many alternative methods of determining K_m. One of the most commonly used depends on rearrangement of equation (9) to give the following form:

$$\frac{[\text{S}]}{V} = \frac{[\text{S}]}{V_{max}} + \frac{K_m}{V_{max}} \tag{10}$$

A plot of $[\text{S}]/V$ vs. [S] gives a straight line. The intercept of the line on the $[\text{S}]/V$ axis is K_m/V_{max} and the slope is $1/V_{max}$ (Fig. 11.2*A*). Thus, K_m can be calculated from the slope and the intercept. When a pure enzyme is used at known concentration, since $V_{max} = k_2[\text{E}]$ [equation (8)], k_2 may be calculated from the slope and the molecular weight of the enzyme. Measurements at high values of [S] which approach V_{max} are unnecessary since V_{max} can be evaluated from the slope.

Two additional methods of plotting kinetic data are also shown in Fig. 11.2. The equation for the linear form of plot *B* (Fig. 11.2) is given later [equation (27), page 237].

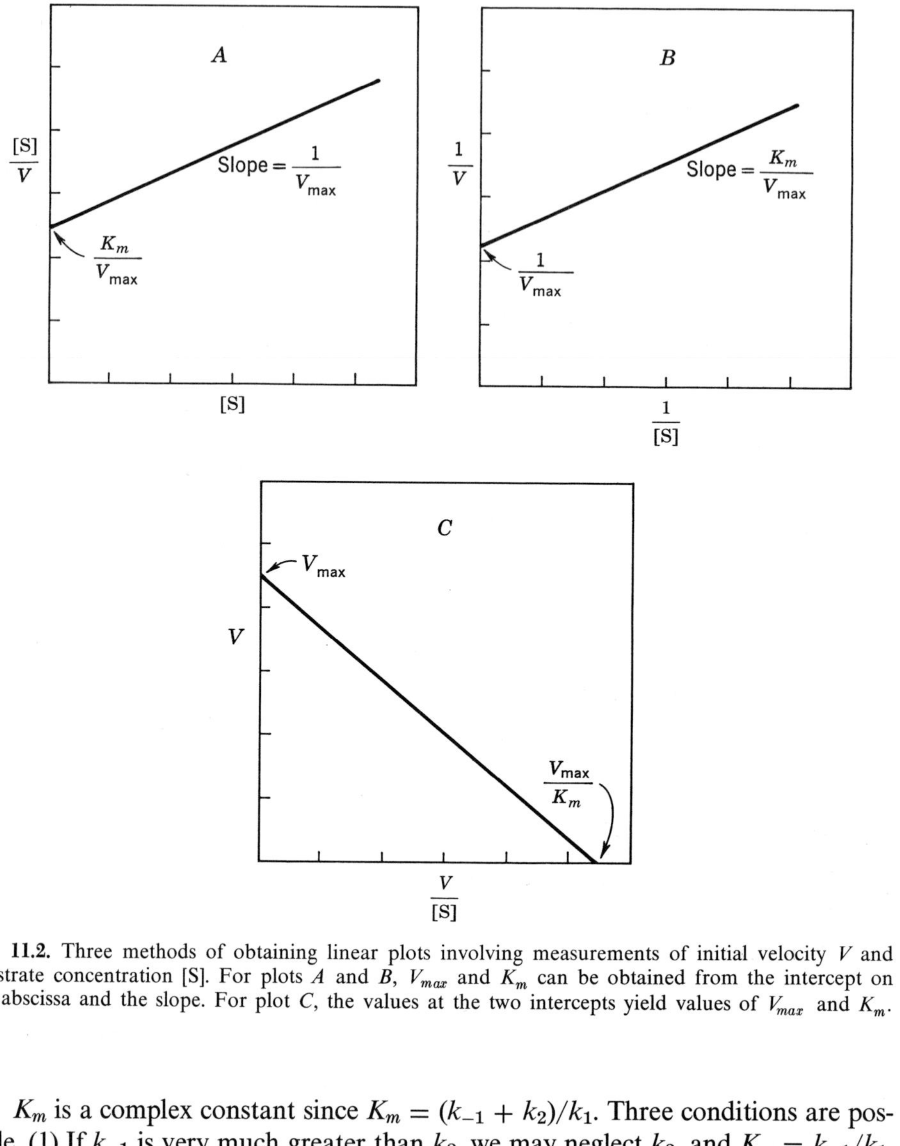

Fig. 11.2. Three methods of obtaining linear plots involving measurements of initial velocity V and substrate concentration [S]. For plots A and B, V_{max} and K_m can be obtained from the intercept on the abscissa and the slope. For plot C, the values at the two intercepts yield values of V_{max} and K_m.

K_m is a complex constant since $K_m = (k_{-1} + k_2)/k_1$. Three conditions are possible. (1) If k_{-1} is very much greater than k_2, we may neglect k_2, and $K_m = k_{-1}/k_1$. K_m is then the thermodynamic equilibrium constant for the reversible formation of ES. (2) If k_2 is very much greater than k_{-1}, then $K_m = k_2/k_1$, and K_m is a steady state constant containing two independent velocity constants. (3) When k_{-1} and k_2 are of the same order of magnitude, all three reaction constants determine the value of K_m. All three conditions have been observed to obtain, depending on the particular enzyme-substrate system that is studied.

It may be noted that the condition represented by equation (1) is seldom achieved since many reactions are readily reversible. More generally,

$$\mathrm{E} + \mathrm{S} \underset{k_{-1}}{\overset{k_1}{\rightleftharpoons}} \mathrm{ES} \underset{k_{-2}}{\overset{k_2}{\rightleftharpoons}} \mathrm{P} + \mathrm{E} \qquad (11)$$

Equation (9) is still applicable inasmuch as [P] is zero when the *initial velocity* is measured. However, for equation (11) the equilibrium constant may be shown to be related to the four velocity constants and to the K_m and V_{max} values

$$K_{eq} = \frac{[\mathrm{P}]_{eq}}{[\mathrm{S}]_{eq}} = \frac{k_1 k_2}{k_{-1} k_{-2}} = \frac{V_{max}^{\mathrm{S}} K_m^{\mathrm{P}}}{V_{max}^{\mathrm{P}} K_m^{\mathrm{S}}} \tag{12}$$

where the superscript *S* pertains to the substrate S and *P* to the product P. K_m and V_{max}^{P} are measured for the reverse reaction in the same manner as already described above for the substrate S. Measurements for the reversible system of equation (11) offer a useful check on the kinetic constants, particularly since K_{eq} can be determined from concentrations alone.

Order of Enzymic Reactions. It has already been indicated that the rate of an enzymic reaction will depend on the substrate concentration [equation (9)]. The course of the reaction with time will be determined by the region of the Michaelis-Menten curve represented by the initial substrate concentration. It is therefore of interest to examine the relationships indicated in equation (9) for circumstances of high and low initial substrate concentration, since these are the conditions which obtain at the beginning and at the end, respectively, of an enzyme-catalyzed reaction. When [S] is much greater than K_m, *i.e.*, at constant enzyme and very high substrate concentrations, on the plateau portion of the curve of Fig. 11.3,

$$-\frac{d[\mathrm{S}]}{dt} = V = V_{max} = k_2[\mathrm{E}] \tag{13}$$

The rate of the reaction is constant for a given enzyme concentration, E. Under these conditions, the enzyme is saturated with substrate and the reaction is proceeding at maximal velocity. In other words, a further increase in [S] will not alter the velocity.

With a fall of the substrate concentration to values much less than K_m, at the bottom portion of the curve in Fig. 11.3,

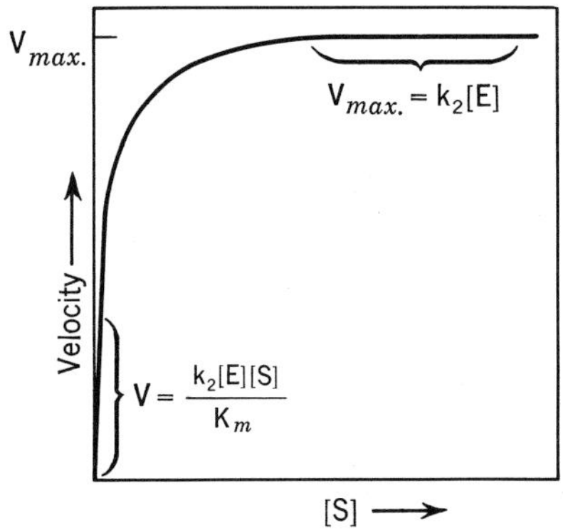

Fig. 11.3. A theoretical plot of velocity vs. substrate concentration. When $V = V_{max}$, the rate is independent of substrate concentration (zero-order reaction). When the substrate concentration is small compared with K_m, the rate is proportional to the substrate concentration and a first-order expression is obtained.

$$-\frac{d[S]}{dt} = V = \frac{V_{max}[S]}{K_m} = \frac{k_2[E][S]}{K_m} \tag{14}$$

That is, the velocity of the reaction is directly proportional to the substrate concentration for a given enzyme concentration, E.

The reaction in which the rate is independent of the substrate concentration is a zero-order reaction with respect to substrate. In terms of the *total amount* of substrate transformed in a given time, zero-order reactions characteristically show a linear relationship between the amount of substrate altered and the time. This follows from the fact that at constant [E], integration of equation (13) yields.

$$S' = k_2[E]t \tag{15}$$

where S' is the *amount* of substrate, in moles per liter, transformed to products in time t, measured from $t = 0$.

The dependence of the reaction velocity on the substrate concentration, at constant enzyme concentration, is a first-order relationship as in equation (14). Integration of equation (14) gives

$$k[E] = \frac{1}{t}\ln\frac{[S_0]}{[S]} \tag{16}$$

where [S] is substrate concentration at time t, starting with a substrate concentration $[S_0]$, and where $k = k_2/K_m$. Equation (16) describes enzymic reactions when the initial substrate concentration is much smaller than the value of K_m.

At intermediate substrate concentrations, when [S] is approximately equal to K_m, the expression becomes more complicated since both first-order and zero-order terms are present.

The Enzyme-Substrate Complex. The assumptions involved in the Michaelis-Menten kinetic scheme were accepted long before any direct chemical proof was available. Many of the reasons for this will be discussed later, in the sections on the inhibition of enzymes and enzymic specificity.

In 1937 Keilin and Mann found direct evidence that peroxidase forms a definite chemical compound with its substrate, hydrogen peroxide. In common with methemoglobin, catalase, and the cytochromes, which are also iron-protoporphyrin proteins, peroxidase in solution is reddish brown and shows characteristic absorption bands when examined spectroscopically. When hydrogen peroxide is added to the enzyme, there is an immediate shift of the absorption bands, indicating the formation of a new chemical compound.

$$\text{Peroxidase} + H_2O_2 \underset{k_{-1}}{\overset{k_1}{\rightleftharpoons}} [\text{peroxidase} \cdot H_2O_2] \tag{17}$$

In the presence of a suitable hydrogen donor, such as an oxidizable dye or ascorbic acid, which may be designated simply as H_2A, a further reaction occurs:

$$[\text{Peroxidase} \cdot H_2O_2] + H_2A \xrightarrow{k_2} \text{peroxidase} + 2H_2O + A \tag{18}$$

That reactions (17) and (18) can be observed directly through the spectroscope constitutes convincing proof of the correctness of the assumption that an ES compound is formed. Although there is evidence which indicates that the product in equation (17) represents a change in valence of the peroxidase iron, the over-all kinetic picture remains the same, the nature of the complex, [peroxidase · H_2O_2], being unspecified.

Chance has utilized the spectroscopic changes to measure the kinetics of reactions (17) and (18). For these rapid reactions it was necessary to devise special equipment. The equipment permits rapid mixing of the solution of the enzyme and of the substrate as they are forced in a continuous flow past a photocell, with the continuous formation of the enzyme-substrate complex occurring in front of the photocell. The latter measures the change in the amount of light transmitted through the reaction vessel, at a wavelength at which a large change in light absorption occurs. The current from the photocell is conducted to a sensitive amplifier, and the recordings are made photographically.

At 25°C. with leukomalachite green, an oxidizable dye, as the hydrogen donor H_2A, k_1 is 1.2×10^7 liter per mole per second, k_{-1} is 0.2 per second, and k_2 is 5.2 per second. K_m measured from the effect of substrate concentration on the initial velocity by the method of Michaelis is 0.41×10^{-6}. It is apparent that k_2 is twenty-six times larger than k_{-1}; hence $K_m = k_2/k_1 = 0.44 \times 10^{-6}$, which is in excellent agreement with the direct measurement of 0.41×10^{-6}. $K_{eq} = k_{-1}/k_1 = 2 \times 10^{-8}$, which is much smaller than K_m. For peroxidase, K_m is not an equilibrium constant, and this is true also for catalase, where the reaction constants k_1, k_{-1}, and k_2 can be estimated directly by the same technique. Direct kinetic measurement of the formation and decomposition of the enzyme-substrate complex, in the case of some dehydrogenases which utilize pyridine nucleotides (page 359), has also been achieved by measurements of changes in absorbancy or of alterations in fluorescence spectra, which occur when the reduced coenzyme is bound to the apoenzyme (page 358). These and many other experiments have provided definitive evidence of the correctness of the Michaelis-Menten concept of an enzyme-substrate complex and the consequent kinetic theory of enzymic action.

For catalase and peroxidase, $K_m = k_2/k_1$, and this is apparently true for other enzymes, as shown by indirect methods. However, the situation where $K_m = k_{-1}/k_1$ appears to be valid for yeast invertase, chymotrypsin, and for other enzymes.

Effect of Temperature. The rate of most chemical reactions depends strongly on temperature, and reactions catalyzed by enzymes are no exception to this rule. The general formulation of the effect of temperature on reaction rate was given by Arrhenius. He postulated that not all the molecules in a system can react; only those molecules which have sufficient energy of activation are capable of reacting.

For an equilibrium between the inactive and active molecules, we may write

$$\textbf{Inactive} \underset{k_{-1}}{\overset{k_1}{\rightleftharpoons}} \textbf{active} \tag{19}$$

and the constant K for this process may be treated as an equilibrium constant.

The effect of temperature on an equilibrium constant for a chemical reaction is given by the van't Hoff equation,

$$2.3 \log K = C - \frac{\Delta H}{RT} \tag{20}$$

where ΔH is the heat of the reaction in calories per mole, R is the gas constant equal to 1.98 cal. per mole per degree, and T is the absolute temperature. C is an integration constant. It follows from this equation that graphs of log K plotted against the reciprocal of the absolute temperature ($1/T$) should give a straight line. The slope of the line is $\Delta H/2.3R$.

Arrhenius assumed that the energy of activation E_a could be obtained for rate processes in the same manner as ΔH for equilibrium processes. Since an increase in temperature will increase the rate of a chemical reaction, there must occur an increase in the rate of formation of active molecules in the equilibrium depicted in equation (19). Thus the effect of temperature is on K, the equilibrium constant, as described in equation (20), where ΔH for this reaction is the heat of activation.

The Arrhenius equation relating a velocity constant k to absolute temperature is

$$2.3 \log k = B - \frac{E_a}{RT} \tag{21}$$

where B is a constant which is a qualitative expression of the frequency of collisions and of the requirement for specific orientation between the colliding molecules. For two temperatures

$$2.3 \log \frac{k'}{k''} = \frac{E_a}{R}\left(\frac{1}{T''} - \frac{1}{T'}\right) \tag{22}$$

where E_a is the energy of activation. As a rule, this equation describes the data for ordinary chemical reactions in a satisfactory manner. Plots of log k vs. $1/T$ may not give straight lines, suggesting that for these reactions the equation is not an adequate description of what is taking place. In general, increased temperatures favor the formation of active molecules, *i.e.*, those with sufficient energy of activation to react, and this process is satisfactorily described by the Arrhenius equation. "Active" molecules are thought to be those molecules having sufficient energy of activation (page 213) to permit them to form an activated complex from which state a spontaneous conversion to the reaction products or to the original reactants may occur.

The effect of temperature on *enzymic* reactions usually is twofold: (1) an increase in rate with temperature until a maximal rate is achieved and (2) a region at high temperatures in which the rate decreases with increase in temperature. The decreased rate at high temperatures is due to thermal inactivation of the enzyme itself, a phenomenon discussed below. In the temperature region in which the enzyme itself is not destroyed by heat, the Arrhenius equation describes the data. Figure 11.4 shows the effect of temperature on the velocity constant k_2 for the hydrolysis of two substrates by carboxypeptidase.

Energy of Activation. The energy of activation is a measure of the energy needed for the conversion of molecules to the reactive state (page 213). A catalyst lowers the activation energy necessary for a reaction that can proceed spontaneously without a catalyst. Table 11.1 shows the activation energies for a number of processes. The decomposition of hydrogen peroxide requires 18,000 cal. per mole; this is lowered to 11,700 when colloidal platinum is the catalyst, and is much

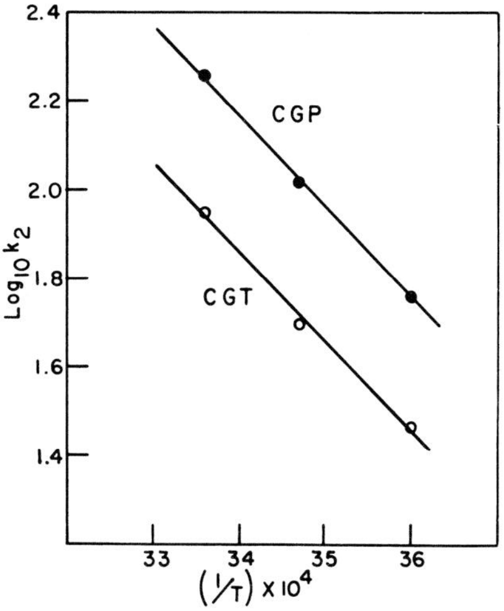

Fig. 11.4. Effect of temperature on the rate of hydrolysis k_2 of carbobenzoxyglycylphenylalanine (CGP) and carbobenzoxyglycyltryptophan (CGT) by crystalline carboxypeptidase. The data are plotted as $\log_{10} k_2$ vs. the reciprocal of the absolute temperature. The apparent activation energies E_a are 9900 cal. per mole for CGT and 9600 cal. per mole for CGP. (*Data from R. Lumry, E. L. Smith, and R. R. Glantz, J. Am. Chem. Soc.,* **73,** 4330, 1951.)

lower for the enzymic reaction. It is obvious that catalase is far more efficient than the inorganic catalyst of this reaction. In fact, catalase is so efficient that only a small activation energy is required in the process. This is consistent with the knowledge that the decomposition of hydrogen peroxide by catalase proceeds at one of the highest rates of any known enzymic reaction.

Inspection of the other data in Table 11.1 shows that the same relationship holds, *viz.*, that a catalyst lowers activation energy and that an enzyme decreases E_a more than an inorganic catalyst does. The effectiveness of enzymes as catalysts is indicated by the high reaction velocities at physiological temperatures. In other

Table 11.1: Energy of Activation for Enzymic and Nonenzymic Catalyses

Process	Catalyst	E_a, cal./mole
Decomposition of hydrogen peroxide	None	18,000
	Colloidal platinum	11,700
	Catalase	<2,000
Hydrolysis of ethyl butyrate	Hydrogen ion	16,800
	Hydroxyl ion	10,200
	Pancreatic lipase	4,500
Hydrolysis of casein	Hydrogen ion	20,600
	Trypsin	12,000
Hydrolysis of sucrose	Hydrogen ion	25,600
	Yeast invertase	8,000–10,000
Hydrolysis of β-methylglucoside	Hydrogen ion	32,600
	β-Glucosidase	12,200

Source: These data have been compiled from the work of several investigators.

words, the decreased E_a permits rapid reaction rates at temperatures much lower than for the uncatalyzed reaction. This may be readily shown by comparing the relative values of the velocity constants for the same reaction at a given temperature (37°C.) when the activation energies are different. The data given in Table 11.1 for the hydrolysis of sucrose by yeast invertase and by hydrogen ion may be taken as an example. Equation (21) may be written for the enzyme (e) as

$$\log k_e = \frac{B_e}{2.3} - \frac{8{,}000}{2.3RT} \tag{23}$$

and for the hydrogen ion (h) as

$$\log k_h = \frac{B_h}{2.3} - \frac{25{,}600}{2.3RT} \tag{24}$$

Let us assume that the values for B_e and B_h are approximately the same. In order to calculate the relative rates of the enzyme-catalyzed and of the hydrogen ion-catalyzed reactions, *i.e.*, the ratio of the two rate constants, equation (24) is subtracted from equation (23), with the values for $R(1.98)$ and $T(37 + 273 = 310)$ substituted. This gives

$$\log \frac{k_e}{k_h} = \frac{25{,}600 - 8{,}000}{2.3 \times 1.98 \times 310} = \frac{17{,}600}{1{,}415} = 12.4 \tag{25}$$

and

$$\frac{k_e}{k_h} = 2.5 \times 10^{12} \tag{26}$$

In other words, the rate constant for the enzymic reaction may be expected to be approximately a trillion times greater than for the hydrogen ion-catalyzed reaction at the same temperature.

Inactivation of Enzymes. Like other proteins, enzymes are denatured at elevated temperatures. The thermal inactivation of an enzyme is readily measured since it can be estimated by the loss of catalytic ability. For ordinary chemical reactions, including catalytic ones, E_a values range from a few thousand up to approximately 40,000 cal. per mole, with the majority in the neighborhood of 15,000 to 25,000 cal. per mole. In marked contrast, the apparent E_a values for the inactivation of enzymes and the denaturation of proteins are all extremely high, from about 40,000 up to 100,000 cal. per mole and, in some cases, even higher. The similarity between enzymes and other proteins in this respect was used as an indication of the protein nature of enzymes many years before any enzyme had been obtained in a highly purified state.

The high E_a values for enzyme inactivation depend greatly on the conditions of measurement, such as pH, presence of substrate, ionic strength, etc. Nevertheless, the high values are almost unique among chemical reactions and point to the complexity of denaturation. Protein denaturation involves rupture of many weak bonds (such as hydrogen bonds, hydrophobic bonds, etc., see Chap. 7). Enzyme inactivation, or protein denaturation in general, at high temperatures is usually a first-order process, *i.e.*, the rate of inactivation is dependent on the first power of the enzyme concentration.

Effect of pH. The pH has a marked influence on the rate of enzymic reactions. Characteristically, each enzyme has a pH value at which the rate is optimal, and

Table 11.2: OPTIMAL pH VALUES FOR SOME HYDROLYTIC ENZYMES

Enzyme	Substrate	Optimal pH
Pepsin	Egg albumin	1.5
Pepsin	Casein	1.8
Pepsin	Hemoglobin	2.2
Pepsin	Carbobenzoxyglutamyltyrosine	4.0
α-Glucosidase	α-Methylglucoside	5.4
α-Glucosidase	Maltose	7.0
Urease	Urea	6.4–6.9
Trypsin	Proteins	7.8
Pancreatic α-amylase	Starch	6.7–7.2
Malt β-amylase	Starch	4.5
Carboxypeptidase	Various substrates	7.5
Plasma alkaline phosphatase	β-Glycerophosphate	9–10
Plasma acid phosphatase	β-Glycerophosphate	4.5–5.0
Arginase	Arginine	9.5–9.9

on each side of this optimum the rate is lower. Table 11.2 gives the optimal pH values for some representative enzymes. It is evident that there is a great spread from the optimal action of pepsin at acid pH values to that of alkaline phosphatase at very alkaline values. Practically, it is necessary in all enzymic studies to control the pH by the addition of suitable buffers, although the type of buffer may influence the optimal pH.

The influence of pH on enzymic reactions is not completely understood; several different types of effect may be involved. Enzymes, like other proteins, are ampholytes and possess many ionic groups. If the enzymic function depends on certain special groupings, these may have to be present in some instances in the un-ionized state and, in others, as ions. In some cases, it has been possible to identify the groups on the enzyme responsible for catalytic effects by comparison of the effect of pH on enzymic activity and known pK' values of titratable groups in the protein (Table 6.2). To cite a few examples, in trypsin, chymotrypsin, and subtilisin, histidine residues are catalytically active (page 254), whereas in papain, a thiol group is involved (see Chap. 12). The requirement of ionic groups for the functioning of enzymes is also suggested by the fact that the ionic strength of the solution affects the rate of some enzymic reactions, notably for carboxypeptidase and urease. This type of effect is usual for nonenzymic ionic catalysis.

In many cases, the substrates of enzymes are themselves electrolytes, and the reaction may depend on a particular ionic or non-ionic form of the substrate. Such an influence is undoubtedly present in the action of pepsin on different proteins, as shown in Table 11.2.

The pH may influence the rate of enzymic action indirectly insofar as many enzymes, like proteins in general, are stable only within a relatively limited pH range, most often near neutrality. Nevertheless, there are many exceptions, and pepsin exhibits the most unusual properties since it is stable at acid pH values and is rapidly inactivated in neutral and alkaline solutions.

Many enzymes are conjugated proteins in which the nonprotein portion is loosely bound to the protein. Since both moieties are essential for activity, conditions that influence the conjugation will destroy the enzymic activity, although this dissociation is frequently reversible, *e.g.*, peroxidase is split into its two components, both inactive at acid pH values; readjustment of the solution to pH 7 restores the activity.

INHIBITION OF ENZYMES

The concept of enzymic action that has developed is that an enzyme reacts with its substrate to form an intermediate complex. Although the entire enzyme molecule may be necessary for its catalytic behavior, there must be a small but definite locus on the surface of the enzyme where the substrate can combine. In those instances in which the enzyme is a conjugated protein, the non-amino acid portion, or prosthetic group, is one place where the substrate may combine or react. Even for those enzymes which are simple proteins, there must be an "active center" or "active site." That this must be so is evident from size considerations alone. The smallest enzymes are proteins of about 12,000 molecular weight, and many enzymes are ten or more times larger than this. In contrast, most substrates are small molecules with molecular weights that are but a tiny fraction of those of the enzymes. The combination of catalase (molecular weight of 250,000) with hydrogen peroxide (molecular weight of 34) illustrates this point.

Perhaps the strongest evidence for the localization of active centers comes from studies with specific inhibitors of enzymes. By specific inhibitors are meant compounds that can combine with an enzyme in such a manner as to prevent the normal substrate-enzyme combination and the catalytic reaction. It is necessary to distinguish these inhibitors from agents that denature proteins.

Competitive Inhibition. The simple scheme for the formation of the enzyme-substrate complex may again be stated as

$$E + S \rightleftharpoons ES \longrightarrow E + P$$

The formation of an enzyme-inhibitor complex may be given as

$$E + I \rightleftharpoons EI$$

where I is the inhibitor and EI is the enzyme-inhibitor complex.

If formation of EI is reversible and there is continuing competition between the substrate and the inhibitor for the same locus on the enzyme, the situation is designated as "competitive inhibition." The actual rate of the catalyzed reaction is then strictly dependent on the relative concentrations of substrate and inhibitor.

This type of inhibition is shown by succinic dehydrogenase, which catalyzes the following reaction in the presence of a suitable hydrogen acceptor (A):

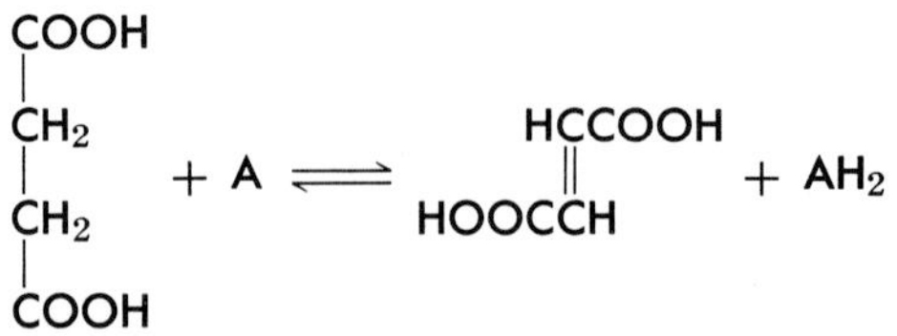

Succinic acid + acceptor ⇌ fumaric acid + reduced acceptor

Many compounds which are structurally similar to succinic acid but which are not dehydrogenated can combine with the enzyme and inhibit catalysis of the normal reaction by blocking the active centers. Several compounds that inhibit succinic acid dehydrogenase are shown below.

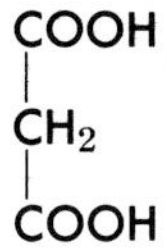
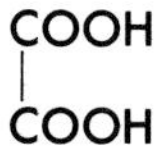
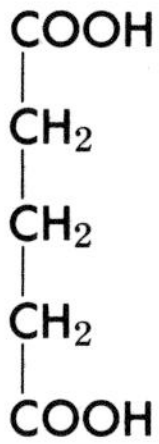
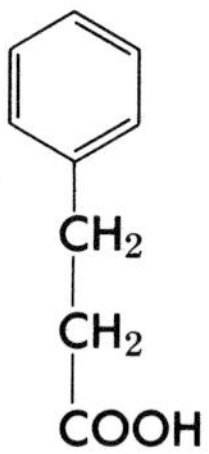

Malonic acid **Oxalic acid** **Glutaric acid** **Phenylpropionic acid**

The most potent of these inhibitors is malonic acid. When the concentration ratio of inhibitor to substrate is 1:50, the enzyme is inhibited 50 per cent. Increasing the concentration of substrate at constant inhibitor concentration decreases the amount of inhibition, and, conversely, decreasing the substrate concentration increases the inhibition. If succinic and malonic acids were bound at different sites on the enzyme, it would be difficult to explain why they should compete with one another. Since they do compete, it must be concluded that they combine with the enzyme at the same locus. The phenomenon of competitive inhibition gives further proof for the concept that the combination of substrate and enzyme is highly specific and that this interaction occurs at a specific place in the structure of the enzyme. Competitive inhibitors have been found for many enzymes, and other examples will be cited later.

Competitive inhibition can be recognized by the effect of inhibitor concentration on the relationship between V and [S]. The linear plot, shown in Fig. 11.2*B*, is given by equation (27).

$$\frac{1}{V} = \frac{[\mathrm{S}] + K_m}{V_{max}[\mathrm{S}]} = \frac{K_m}{V_{max}} \cdot \frac{1}{[\mathrm{S}]} + \frac{1}{V_{max}} \tag{27}$$

A plot of $1/V$ vs. $1/[\mathrm{S}]$ gives a straight line in which the ordinate intercept equals $1/V_{max}$ and the slope is K_m/V_{max}.

For the action of a competitive inhibitor, an equation may be derived incorporating the inhibitor concentration [I] and the dissociation constant of the enzyme-inhibitor complex, K_i, in a manner similar to that used in obtaining equation (9). The following equation gives the relationship among these terms:

$$V = \frac{V_{max}[\mathrm{S}]}{K_m\left(1 + \frac{[\mathrm{I}]}{K_i}\right) + [\mathrm{S}]} \tag{28}$$

Rearrangement of equation (28) in the same way used to obtain equation (27) yields

$$\frac{1}{V} = \frac{K_m}{V_{max}}\left(1 + \frac{[\mathrm{I}]}{K_i}\right)\frac{1}{[\mathrm{S}]} + \frac{1}{V_{max}} \tag{29}$$

When a plot of $1/V$ against $1/[S]$ is made, it is found, for competitive inhibition, that the ordinate intercept, $1/V_{max}$, is the same as in the uninhibited reaction, but that the slope, which is now $\left(\frac{K_m}{V_{max}}\right)\left(1 + \frac{[I]}{K_i}\right)$, is increased by the factor $\left(1 + \frac{[I]}{K_i}\right)$. This is shown graphically in Fig. 11.5. Thus, by using a sufficiently high substrate concentration, the effect of the competitive inhibitor can be overcome and V_{max} can be reached. For noncompetitive inhibition (discussed below) both the slope and the intercept are altered, which readily permits distinguishing between these two types of inhibition, as also shown in Fig. 11.5.

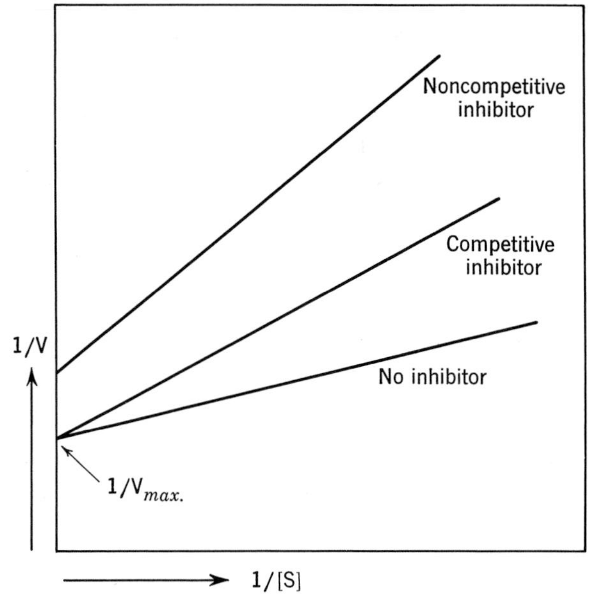

Fig. 11.5. A plot of 1/V vs. 1/[S] for an enzymic reaction with no inhibitor, a competitive inhibitor, and a noncompetitive inhibitor. For competitive inhibition, the slope is changed and the intercept remains the same; for noncompetitive inhibition, both the slope and the intercept are increased.

Noncompetitive Inhibition. In this instance, there is no relationship between the degree of inhibition and the concentration of substrate. Inhibition depends only on the concentration of the inhibitor. In contrast to the competitive type, it may be assumed that the formation of EI occurs at a locus on the enzyme that is not attacked by the substrate.

$$E + I \rightleftharpoons EI$$

and

$$ES + I \rightleftharpoons ESI$$

where EI and ESI are inactive.

Examples of noncompetitive inhibition are the inhibition of many enzymes by heavy metal ions such as Ag^{+}, Hg^{++}, Pb^{++}, etc. Urease is extremely sensitive to traces of these ions.

Many instances of noncompetitive inhibition are actually examples of combination of inhibitor and enzyme to form chemical derivatives. The inhibition may be a result either of irreversible combination of the inhibitor with the same site on the enzyme at which the substrate would combine, or of combination of the inhibitor with some portion of the enzyme molecule not immediately involved in enzyme-substrate combination but which may result in a conformational change so that catalytic activity is lost.

When irreversible combination occurs, it cannot always be easily determined whether the active locus is involved or not. However, such inhibitions give valuable information concerning the chemical nature of the enzyme. Sulfhydryl groups are present in some enzymes, and these groups are, in many instances, essential for the enzymic activity. A widely used agent for the detection of sulfhydryl groups is iodoacetamide, which reacts chemically in the manner illustrated.

$$RSH + ICH_2CONH_2 \longrightarrow RS\text{—}CH_2CONH_2 + HI$$

Various organic mercury derivatives such as *p*-mercuribenzoate, certain arsenicals, and N-ethylmaleimide also react with SH groups.

Those enzymes that contain essential metal ions are inhibited by compounds that form complexes with the metal. Cyanide and hydrogen sulfide inhibit iron-containing enzymes such as peroxidase and catalase. Fluoride and oxalate inhibit enzymes that need magnesium, calcium, or similar ions. Table 11.3 lists some of the reagents commonly employed as inhibitors in enzymic studies. The effects of other inhibitors will be encountered throughout this text.

Certain enzyme inhibitors already mentioned, such as cyanide, carbon monoxide, oxalic acid, etc., are poisonous for living organisms. Since their toxicity

Table 11.3: Some Commonly Used Enzyme Inhibitors

Inhibitor	*Enzyme group that combines with inhibitor*
Cyanide	Fe^{+++}, Cu, Zn, certain other metals
Sulfide	Various metals
Fluoride	Mg, Ca, other metals
Oxalate	Ca, Mg
Carbon monoxide	Fe^{++}, Cu
Diethyldithiocarbamate	Cu
Pyrophosphate	Mg, Mn, Zn, other metals
α,α'-Dipyridyl	Fe
Azide	Fe^{+++}-protoporphyrin enzymes
o-Phenanthroline	Fe, Co, Zn, other metals
Ethylenediaminetetraacetate	Various metal ions
Cysteine and other sulfhydryl compounds	Fe, Cu, and other metals; may also produce reduction
Heavy metals (Ag^+, Hg^{++}, Pb^{++}, etc.)	Sulfhydryl; may also cause nonspecific protein precipitation
Iodoacetate	Sulfhydryl, imidazole, carboxyl, thiol ether
p-Mercuribenzoate; other mercurials	Sulfhydryl
Various arsenicals	Sulfhydryl
N-Ethylmaleimide	Sulfhydryl
Diisopropylphosphofluoridate	Serine hydroxyl

is due to their ability to inhibit enzymic reactions, a substantial part of toxicology must be understood in terms of the specific inhibition of enzymes that are vital to the organism.

METABOLIC INHIBITORS—ANTIMETABOLITES

Utilization of the knowledge that an enzyme may be inhibited by compounds possessing a structure related to the natural substrate is not limited to applications involving individually known enzymic reactions. Indeed, the pharmacologist has long recognized that substances having related structures may compete with physiologically active substances. Although knowledge of the enzymes concerned is, in most cases, lacking, there is evidence suggesting that many pharmacologically active substances alter metabolic processes by virtue of their action on enzymes or enzymic systems.

All higher animals and many microorganisms lack the ability to synthesize certain organic compounds that are essential for survival. Requirements vary for different species and may include certain amino acids, fatty acids, the heterogeneous group of substances known as vitamins, certain purines and pyrimidines, etc. The essential substances needed for survival or for growth are used by the organism for incorporation into proteins, nucleic acids, etc., or they are transformed into other substances that are required by the cells. The utilization of the required *metabolites* is dependent on enzymes, and these enzymes may be inhibited by substances whose structures are related to those of the metabolites. Although, in most cases, the nature of the essential enzymic step is unknown, the term *antimetabolites* is used to describe these inhibitors.

Antimetabolites may also block the utilization of metabolites that are synthesized by the organism itself. This provides an important tool in studying metabolic pathways since, in many instances, the inhibition of growth can be overcome by the addition of the necessary metabolite. Thus, study of antimetabolites has a twofold interest, *i.e.*, the discovery of essential metabolites and growth factors necessary for different species and the search for potent inhibitors of the growth of pathogenic microorganisms and neoplasms. The antimetabolites that have been obtained from living organisms, such as bacteria, fungi, actinomycetes, etc., are commonly called antibiotics.

Although synthetic antibacterial agents such as arsphenamine and the sulfonamides were discovered earlier by empirical search, the rational investigation of antimetabolites began with the observation of D. D. Woods in 1940 that the inhibition of bacterial growth by sulfanilamide is competitively overcome by *p*-aminobenzoic acid, whose role as a growth factor was first recognized in this way. The structural similarity of the two substances is obvious, and the phenomenon is akin to that of competitive inhibition of an isolated enzyme.

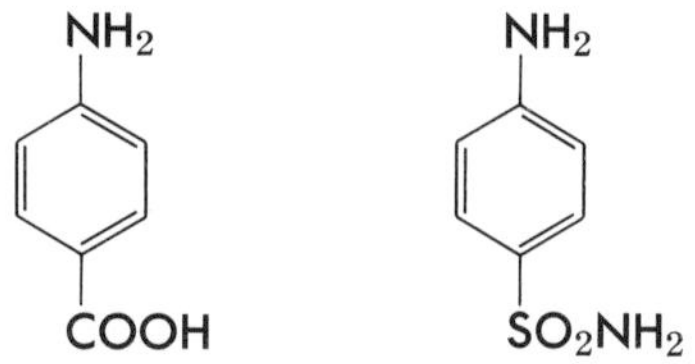

***p*-Aminobenzoic acid** **Sulfanilamide**

Indeed, *p*-aminobenzoic acid will competitively overcome the inhibition of all sulfonamides of the structure $NH_2—C_6H_4—SO_2NHR$, *e.g.*, sulfaguanidine, sulfathiazole, sulfapyridine, and sulfadiazine.

Those organisms which require *p*-aminobenzoic acid (PABA) for growth utilize it for synthesis of folic acid (Chap. 49).

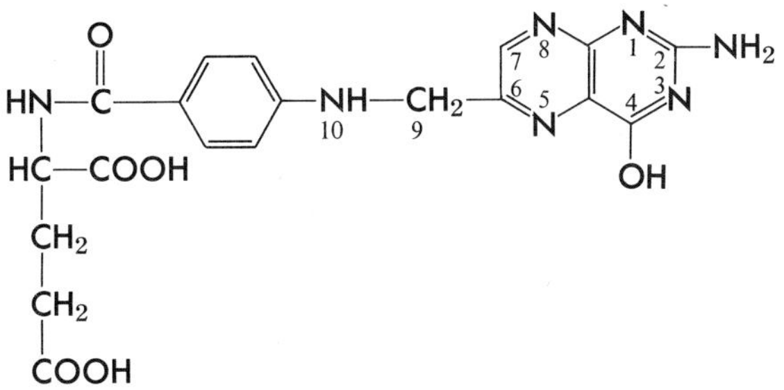

Folic acid (pteroylglutamic acid)

Growth of organisms that require PABA is inhibited by sulfonamides, and this inhibition can be reversed by PABA. Those organisms which need folic acid for growth and cannot utilize PABA are not inhibited by sulfonamides. Thus the sulfonamides inhibit the enzymic step or steps involved in the synthesis of folic acid from *p*-aminobenzoic acid and other metabolic precursors. This simplified picture illustrates how the sulfonamides have been useful in elucidating aspects of the role of PABA in bacterial metabolism. The effective utilization of the sulfonamides in combating bacterial infections in man probably depends on the fact that man requires folic acid and cannot synthesize it from PABA. Thus, the sulfonamides block a metabolic reaction essential for certain bacteria without influencing the metabolism of the host who does not make his folic acid from PABA.

Many thousands of compounds have been synthesized in attempts to discover new antimetabolites. Some of the considerations involved in the search for effective inhibitors may be illustrated by listing a few representative synthetic antimetabolites used in the study of metabolically important compounds.

Metabolite *Antimetabolite*

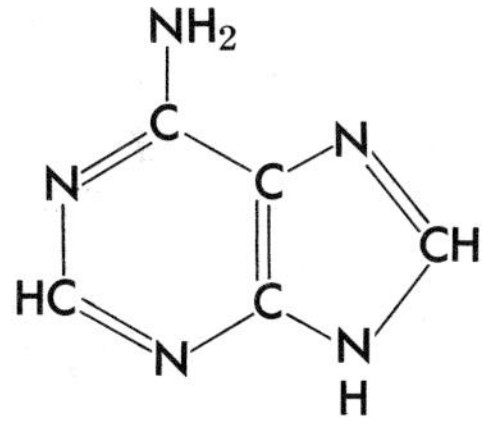

Adenine

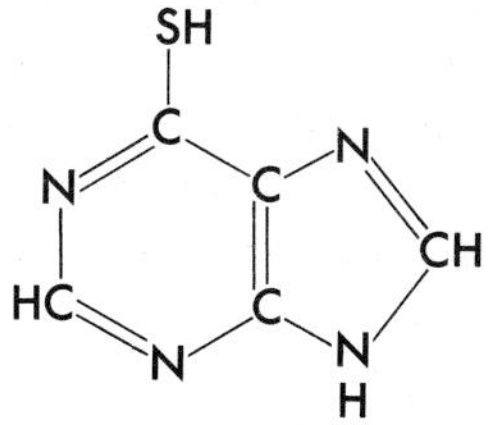

6-Mercaptopurine

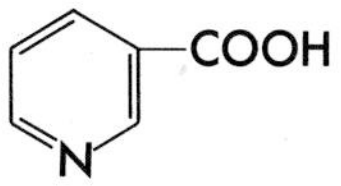

Nicotinic acid

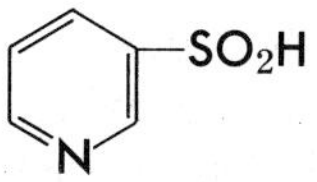

Pyridine-3-sulfonic acid

Certain of the folic acid antagonists have been used experimentally and, to some extent, clinically in the treatment of leukemia and other neoplastic diseases. Thus 4-aminopteroylglutamic acid (aminopterin) inhibits growth of certain types of tumors. This has stimulated a wide search for other potential inhibitors of neoplasms; many of these studies are based upon the antimetabolite concept.

CONTROL OF ENZYMIC ACTIVITY

The rates at which compounds are synthesized or degraded are usually regulated by the metabolic needs of the cell. The various types of control mechanisms that operate at different levels in the cell or the whole organism are discussed in Chap. 13. Here we shall present control mechanisms which are manifested by individual enzymes.

The general characteristics of "regulatable enzymes" were first described in connection with studies of synthetic pathways in bacteria. Essential metabolites, such as amino acids or nucleotides, are synthesized in the bacterial cell in a series of discrete steps, each catalyzed by a different enzyme. One of the first examples discovered was the synthesis of L-isoleucine from L-threonine (Chap. 24).

$$\text{L-Threonine} \longrightarrow \alpha\text{-ketobutyrate} \longrightarrow \longrightarrow \longrightarrow \longrightarrow \text{L-isoleucine}$$

The phenomenon of regulation is illustrated by the fact that the *first step* in the pathway, catalyzed by *threonine deaminase,* is specifically inhibited by L-isoleucine, the end product of the entire pathway. Other examples of such inhibitions in amino acid biosynthesis are listed in Table 23.1.

The properties of enzymes subject to "feedback" regulation can be summarized as follows:

1. The inhibition is not of the simple competitive type, and the feedback inhibitor may bear little or no structural resemblance to the substrate of the enzyme. Moreover, the site of binding of the inhibitor on the enzyme is distinct from the substrate binding site. Since the inhibitor is not a steric analogue of the substrate, the phenomenon was termed by Monod and Jacob "allosteric inhibition" and the inhibitor site on the enzyme, the "allosteric site." More generally, these have been called "regulatory inhibition" and the "regulatory site."

2. The kinetics are anomalous. For enzymes that follow Michaelis-Menten kinetics, plots of velocity vs. substrate concentration show a simple hyperbolic function (Fig. 11.1, left). For regulatable enzymes, sigmoidal curves are usually obtained, indicating kinetics of a second- or higher-order relationship between velocity and substrate concentration, or between velocity and inhibitor concentration.

3. Regulatable enzymes are subject not only to inhibitory regulation but also to activation, *i.e.*, "effectors" (= modifiers, = modulators, = determinants) can be positive or negative in action.

4. Enzymes responsive to effectors have been found to consist of multiple subunits and to contain more than one substrate binding site per molecule of enzyme.

5. The interaction of the effector (or substrate) with the enzyme produces an alteration of the conformation or of the subunit relationships of the protein. Frequently, reagents which alter conformation without loss of enzymic activity may alter or abolish the action of the effectors.

Before further discussion of the general characteristics of such regulatable enzymes, the behavior of myoglobin and hemoglobin will be considered as models, although these oxygen-binding proteins are not enzymes. For myoglobin, a plot of per cent saturation (formation of MbO_2) as a function of O_2 tension follows a rectangular hyperbola (Fig. 32.1), as expected from the simple mass law relationship, where $Mb + O_2 \rightleftharpoons MbO_2$ and $y = Kx/(1 + Kx)$; y is the fraction of myoglobin saturated with oxygen, x is the partial pressure of oxygen, and K is the equilibrium constant. In contrast, for hemoglobin, a similar plot yields a sigmoidal curve (Fig. 32.2) which obeys the relationship known as the Hill equation

$$y = Kx^n/(1 + Kx^n) \tag{30}$$

where x and y have the same meaning as before, but the exponent n gives a measure of the sigmoidicity of the curve. For hemoglobin, under usual conditions, $n = 2.5$.

Myoglobin contains one heme per molecule, whereas mammalian hemoglobins contain four heme groups per molecule (Chap. 8). The sigmoidal relationship for hemoglobin indicates cooperative binding in that combination of one or more equivalents of oxygen with the heme groups *increases* the affinity of the remaining groups for oxygen. X-ray diffraction analysis has shown a conformational difference between unoxygenated and fully oxygenated hemoglobin. In unoxygenated Hb, the heme groups have a low affinity for oxygen. When one or more of the heme groups are bound to oxygen, the conformational relationship among the four subunits alters and the affinity of the unfilled ligand positions for oxygen is greatly increased (Chap. 32). Note that the conformation of the individual α and β subunits of hemoglobin remains largely unchanged. Further, for HbH(β_4) (Chap. 32) $n = 1$, and there is no change in conformational relationship among the subunits.

The behavior of regulatable enzymes is exactly analogous to that of the oxygen-binding proteins. For enzymes obeying Michaelis kinetics, there is a single substrate binding site or a number of sites whose affinity for substrate is unaffected by partial saturation of neighboring sites on the same molecule. For regulatable enzymes, when the velocity is proportional to fractional saturation of substrate sites, equation (9), page 227, may be put in the form

$$K = [S]^n\left(\frac{V_{max} - 1}{V}\right) \tag{31}$$

By rearranging the equation and converting to the logarithmic form, we obtain

$$\log[V/(V_{max} - V)] = n \log[S] - \log K \tag{32}$$

where the terms have the same meaning as before and n is the number of substrate binding sites. When $n = 1$, equation (31) is identical to equation (9). In the ideal case, a plot of $\log[V/(V_{max} - V)]$ vs. $\log[S]$ is linear, with a slope of n. For many enzymes, however, the slope is nonlinear and reflects both n and a changing strength of interaction among the sites. Useful values of n and K are obtained at 50 per cent maximal velocity, *i.e.*, when $V/(V_{max} - V) = 1$. The value of K then reflects the affinity of the enzyme for substrate and the value of n reflects the degree of interaction or cooperativeness among the binding sites. The limiting condition is reflected when $n = 1$ and all interaction is eliminated; then $K = K_m$, as in equation (9).

Modifiers can alter the value of K or the value of n or both. Indeed, in a few instances the value of V_{max} may be influenced as well. Positive modifiers increase the apparent affinity of the enzyme for the substrate; negative modifiers decrease the affinity for substrate.

An enzyme subject to such regulatory influences and for which greatest structural information is presently available is *aspartate transcarbamylase,* which catalyzes the initial step in the formation of CTP, cytidine triphosphate (Chap. 27).

$$H_2N—\underset{\underset{O}{\|}}{C}—OPO_3H_2 + H_2N—\underset{\underset{COOH}{|}}{CH}—CH_2—COOH \xrightarrow{\underset{CTP}{\uparrow}}$$

Carbamyl phosphate   L-Aspartic acid

$$H_2N—\underset{\underset{O}{\|}}{C}—\underset{\underset{H}{}}{N}—\underset{\underset{COOH}{|}}{CH}—CH_2—COOH + P_i$$

Carbamylaspartic acid

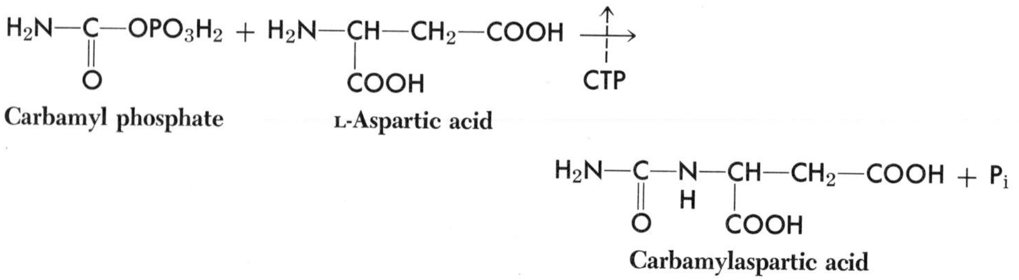

Pardee and Gerhart discovered that CTP acts as a negative modifier, thus functioning as a negative feedback inhibitor, *i.e.,* when the supply of CTP is high, the formation of carbamylaspartate is inhibited; conversely, when the concentration of CTP is low, the rate of formation of carbamylaspartate is limited by the supply of substrate and the concentration of enzyme.

Gerhart and Schachman have demonstrated that the enzyme, aspartate transcarbamylase, molecular weight = 300,000, dissociates on treatment with mercurials into two types of subunits which are readily separable. One type of subunit possesses the entire catalytic activity and is insensitive to CTP; this *catalytic subunit* has a molecular weight of 48,000 and there are four of these per molecule of native enzyme, in accord with the finding that each molecule of native enzyme can bind four molecules of succinate, an unreactive analogue of the substrate aspartate. There are four *regulatory subunits,* each of 28,000 molecular weight, per molecule of enzyme, and each molecule of native enzyme can bind four molecules of CTP. Re-formation of the native enzyme requires removal of the mercurial and the presence of both types of subunits; this occurs optimally at pH 8.5 and 37°C.

When the enzyme is dissociated by treatment with mercurials, the catalytic activity is enhanced approximately fourfold, and this is also the activity of the isolated subunits. Under these conditions, V as a function of [S] exhibits a hyperbolic function (Fig. 11.6; curve $\underline{A}$) whereas in the intact native enzyme, a sigmoidal relationship is found and the entire curve is shifted to the right (curve $\underline{B}$). In the presence of CTP, the curve is shifted further on the abscissa (curve $\underline{C}$). In the presence of the positive modifier, adenosine triphosphate (ATP), CTP is displaced and the curve is shifted to the left. Thus ATP and CTP compete for the same regulatory site.

Of the total protein, the catalytic subunits represent 63 per cent and the regulatory subunits 37 per cent of the structure. The regulatory subunits of the protein serve as a means by which small compounds, CTP or ATP, can effect rapid, reversible control of enzymic activity. As yet, it is unknown whether other regulatable en-

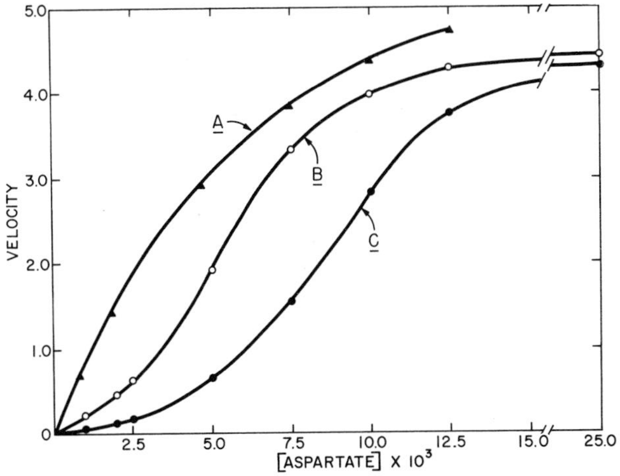

Fig. 11.6. Velocity of aspartate transcarbamylase as a function of substrate concentration. Curve A illustrates the behavior of the enzyme in the presence of a mercurial. In this condition, the enzyme exhibits normal Michaelis kinetics and $n = 1$. The enzyme has been dissociated into separate catalytic and regulatory subunits. Curve B shows the behavior of the native enzyme; it is evident that $n > 1$. Curve C shows the effect of $2.0 \times 10^{-4} M$ CTP. The curve is shifted on the abscissa; K is increased, reflecting a decreased affinity of the enzyme for substrate, *i.e.*, a higher concentration of aspartate is required to attain the same velocity.

zymes possess their regulatory sites on the same subunits as the catalytic subunits or on different ones. Both types are certainly possible, as indicated by the example of hemoglobin, in which each of the four chains can bind oxygen but in which a conformational change is effected by the partial binding of oxygen. The sigmoidal curve for aspartate transcarbamylase in the absence of CTP shows that n is greater than 1; the negative modifier increases the value of K (higher concentrations of S being required for saturation of the enzyme).

Although the regulatory and catalytic sites may or may not be on the same subunits, kinetic evidence indicates that these sites must involve distinctive portions of the polypeptide chains, since inhibition is not competitive. Furthermore, studies of the genetics of many regulatable enzymes in bacteria have shown that mutations (amino acid substitutions) can abolish the regulatable properties of the enzymes without affecting the catalytic activity.

Regulation of the rate of aspartate transcarbamylase activity is accomplished by binding of CTP to the regulating subunit which, in nature, is an integral part of the working enzyme. Many examples of essentially similar phenomena have been observed. However, modification of enzymic activity also may occur by other means. For example, glycogen synthetase and glycogen phosphorylase are subject to phosphorylation and dephosphorylation by specific enzymes (Chap. 19); the properties of the phosphorylated and dephosphorylated forms are strikingly different and appropriate to their respective metabolic situations. The glutamine synthetase (Chap. 23) of *E. coli B*, which consists of 12 identical subunits, is substantially modified by the action of a specific enzyme which transfers AMP, from ATP, into an unidentified covalent linkage on each subunit. The adenylated enzyme then exhibits specificity for Mn^{++} instead of Mg^{++}, is relatively more sensitive to a variety of feedback inhibitors and, in general, is a less active form of the enzyme. Finally, we may note the presence in animal tissues of an enzyme which is specific for transfer of galactose from UDP-galactose to N-acetylglucosamine into glycosidic linkage for formation of N-acetyllactosoamine [reaction (1), below] (page 432). In the presence of a protein previously identified only as the α-lactalbumin of milk (page 825), the specificity of the enzyme is altered and it becomes, specifically, the enzyme which transfers a galactose unit from UDP-galactose to glucose with formation of lactose [reaction (2), below] (page 432). Undoubtedly, many additional types of enzymic modification remain to be discovered and elucidated.

(1) **UDP-D-galactose + N-acetyl-D-glucosamine $\longrightarrow$ N-acetyllactosamine + UDP**

(2) **UDP-D-galactose + D-glucose $\xrightarrow{\alpha\text{-lactalbumin}}$ lactose + UDP**

Further examples of regulatable enzymes and their roles in controlling metabolic rates will be encountered in various chapters in later sections of this book. As yet, little information is available concerning the mechanisms by which modifiers exert their controls. In the following chapter, there is discussed the types of binding involved in determining the specificity and mechanism of action at the catalytic sites of various enzymes. For the present, it may be assumed that the specificity of modifier binding to enzymes involves the same phenomena as substrate binding, but the catalytic events that occur incident to substrate binding are lacking. Instead, modifier binding produces only conformational changes which may be manifested in changes in n or K, or more rarely, in V_{max} as well.

REFERENCES

See list following Chap. 12.

12. Enzymes. III

Specificity of Enzymes. Nature of Active Enzymic Sites. Mechanism of Enzymic Action

The two most remarkable properties of enzymes are their specificity and their catalytic efficiency, and it is in these properties that enzymes differ most strikingly from simple catalysts. Consideration has already been given to the efficiency of enzymes in reducing the activation energies of chemical reactions. At constant temperature, it is not possible to lower the activation energy for a specific, defined chemical reaction. If activation energy for an over-all process has been lowered, as is true for all catalyzed processes, it follows that the over-all reaction must be accomplished by a mechanism, or via a series of intermediates, different from that of the spontaneous, uncatalyzed event. This chapter will examine the mechanisms of enzymic catalysis in order to provide an understanding of the manner in which the enzymic surface modifies the requirement for the needed energy of activation. The subjects of enzymic specificity and mechanism of action are too extensive to allow full discussion here. Certain examples have been selected to illustrate the more general considerations and the methods employed. Further examples are mentioned in succeeding chapters in Part Three.

SPECIFICITY OF ENZYMES

Evidence has been presented that enzymes combine with their substrates. This is indicated by consideration of kinetics, by the phenomenon of competitive inhibition, and by the instances in which direct spectroscopic evidence of the existence of the intermediary compound has been obtained. As previously noted, studies of enzymic specificity have contributed most strikingly to this concept. Experimentally, enzymic specificity is tested by utilizing a number of compounds in which the structure is varied systematically. When the enzyme has *absolute specificity* with respect to a single substrate, limited information can be gained, and some enzymes show this type of specificity. An example, already mentioned, is the reversible reaction of succinic acid dehydrogenase with succinate or fumarate.

When the enzyme reacts with a variety of compounds, it is said to show *relative specificity*. An example is D-amino acid oxidase, which reacts with many D-amino acids but at different rates (Table 12.1).

Table 12.1: RATE OF OXIDATION OF D-AMINO ACIDS AND OTHER SUBSTRATES BY D-AMINO ACID OXIDASE

Substrate	*Oxygen uptake**	*Substrate*	*Oxygen uptake**
D-Tyrosine	190	D-Histidine	6.2
D-Proline	148	D-Threonine	2.1
D-Methionine	80	D-Cystine	1.9
D-Alanine	64	D-Aspartic acid	1.4
D-Serine	42	D-Lysine	0.6
D-Tryptophan	37	D-Glutamic acid	0
D-Valine	35	L-Amino acids	0
D-Phenylalanine	26	D-Peptides	0
D-Isoleucine	22	*N*-Acetylalanine	0
D-Leucine	14	β-Pyridyl-4-alanine	95

* Crude extract of acetone-dried sheep kidney cortex at pH 8.3 used as source of enzyme. Rate of oxygen consumption is calculated from uptake in 10 min., and results are given as microliter per milligram of enzyme preparation per hour.

SOURCE: After H. A. Krebs, chap. 58, in J. B. Sumner and K. Myrbäck, eds., "The Enzymes," vol. II, part I, Academic Press, Inc., New York, 1951.

The general equation for the action of the oxidase is the following.

$$\underset{\displaystyle\text{HNR}'}{\text{R—}\overset{}{\text{CH}}\text{—COOH}} + \tfrac{1}{2}O_2 \longrightarrow \text{R—}\underset{\displaystyle\text{O}}{\overset{}{\underset{\|}{\text{C}}}}\text{—COOH} + R'NH_2$$

R′ is an alkyl or substituted alkyl group or an H atom. R is the side chain of the amino acid. Peptides containing D-amino acids or N-acyl-α-amino acids are not attacked. Dehydrogenation can occur only if there is at least one H atom on the α-carbon and one on the N of the amino group. The enzyme can catalyze oxidation of compounds with large variations in R, many of which do not occur biologically, *e.g.*, β-pyridyl-4-alanine. Compounds with ionic side chains, *e.g.*, glutamic acid, lysine, histidine, etc., are poorly attacked.

Understanding of enzymic specificity must be sought in the nature of the binding forces involved in the formation of the enzyme-substrate compound, and in the nature of the groups in the enzyme responsible for the catalytic action. These groups must be the side chains of the amino acids that are linked in the peptide chains and the organic cofactors or metal ions present in many enzymes.

The first clear formulation that interaction involves more than one group of the substrate, and hence of the enzyme also, emerged from studies of the hydrolysis of simple peptides by certain exopeptidases. In 1926, von Euler and Josephson observed that benzoylglycylglycine was resistant to the action of enzyme preparations that rapidly hydrolyzed glycylglycine. They concluded that a free amino group was required for the action of the enzyme (now called *glycylglycine dipeptidase*) and proposed the "diaffinity theory," which postulates that the enzyme combines with the substrate at the amino group and at the peptide linkage to be hydrolyzed.

The peptide, L-leucylglycine, is rapidly hydrolyzed by *leucine aminopeptidase* of mammalian tissues, yet the antipodal compound, D-leucylglycine, is completely resistant. Bergmann and his coworkers noted in 1935 that in order to explain such *spatial specificity,* it was necessary to postulate a minimum of three points of attachment of the substrate to the enzyme. If only two points of attachment were necessary, the stereoisomeric peptide, D-leucylglycine, could combine with the enzyme just as easily as the L compound. This "polyaffinity" concept has played an important role in our understanding of enzymic action, for the spatial specificity of enzymes can be explained only in terms of multipoint attachment.

Leucine aminopeptidase hydrolyzes di- and polypeptides at the bond adjacent to the free amino group and also acts on L-amino acid amides. The rate of hydrolysis of amides, $R—CH(NH_2)CONH_2$, depends on the nature and size of the substituent R, increasing from $H < CH_3 < C_2H_5 < C_3H_7 < C_4H_9$. The presence of polar or *hydrophilic* groups in R, such as COO^-, NH_3^+, $CONH_2$, CH_2OH, etc., decreases the rate. It has been concluded that nonpolar *hydrophobic* forces (page 154) are involved in the interaction between the R group of the substrate and similar hydrophobic groups in the protein. This view is supported by observations that aliphatic alcohols are inhibitors of this enzyme, presumably because their R groups compete for the protein-binding sites. Such hydrophobic binding is also apparently involved in the action of D-amino acid oxidase.

Similar considerations apply to other proteolytic enzymes, *e.g.*, chymotrypsin and carboxypeptidase, which preferentially hydrolyze peptide bonds involving aromatic or large aliphatic amino acid side chains (Table 7.4, page 144), as well as to various lipases whose affinity for the aliphatic hydrocarbon chains of the substrates is very great, and to a large number of other enzymes that act on various types of compounds which may entirely lack polar or ionic groups or may contain them only in limited number. In all such instances, the complementary shapes of enzyme and substrate must permit a precise fit and the major binding forces are of the hydrophobic type. This appears to be the case for all enzymes that interact with neutral fats, fatty acids, sterols, steroids, terpenes, carotenoids, etc. However, the examples of the aminopeptidase and D-amino acid oxidase, cited above, show that hydrophobic interactions are not limited to nonpolar substances.

Enzymes that react specifically with individual amino acids, purines, nucleotides, etc., must "recognize" the geometry of the side chain, bind with this grouping, and exclude substrates that do not bind with sufficient energy of interaction. The data presented in Table 12.1 for the action of D-amino acid oxidase indicate not only that hydrophobic forces are involved but that the shape of the R group is also an important factor (cf. leucine and isoleucine).

Water-soluble substances, such as the hexoses, with a multiplicity of hydrophilic groups, do not possess groupings that would readily interact with an enzyme by hydrophobic forces. With such compounds, the enzyme-substrate fit must involve complementary structures and binding must be achieved primarily by the reactive carbonyl group and the hydroxyl groups of the substrate. These hydroxyl groups can form hydrogen bonds with suitable oxygen atoms, *e.g.*, those in peptide carbonyl groups or others.

Ionic forces also play a major role in enzyme-substrate interactions. This is probably so for *succinic acid dehydrogenase* (page 236), which catalyzes reversibly the dehydrogenation of succinate or hydrogenation of fumarate. These substrates must possess free carboxylate ions. Moreover, all the competitive inhibitors of this enzyme (page 257) possess at least one carboxyl group. Presumably, enzymic specificity involves reaction with the two carboxylate ions in succinate or fumarate, or with those of the potent inhibitor malonate, for which the binding constant with the enzyme is greater than that of either substrate. Although no information is available concerning the groups in the enzyme that react with the carboxylate ions, presumably they are cationic side chains of amino acid residues in the protein, *e.g.*, guanidinium, ammonium, or imidazolium.

Similar considerations apply to the proteolytic enzyme *trypsin,* which acts only at peptide, amide, or ester bonds involving arginine or lysine residues (Table 7.4, page 144). α-Benzoyl-L-argininamide or α-benzoyl-L-lysinamide serve as suitable model substrates. The specificity for reaction of the cationic group of the substrate, presumably with an anionic group of the enzyme, is so exact that deamidation will not occur with compounds in which the cationic group is displaced by one CH_2 group from the sensitive amide bond. Neither the longer chain of α-benzoyl-L-homoargininamide nor the shorter chain of α-benzoyl-L-ornithinamide provides a suitable structure for reaction with trypsin.

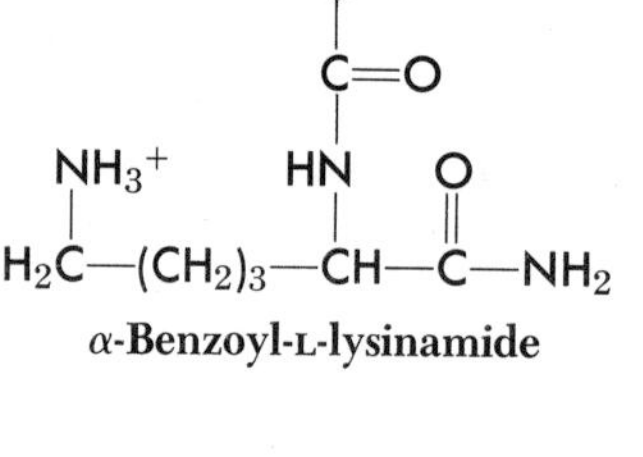

α-Benzoyl-L-lysinamide

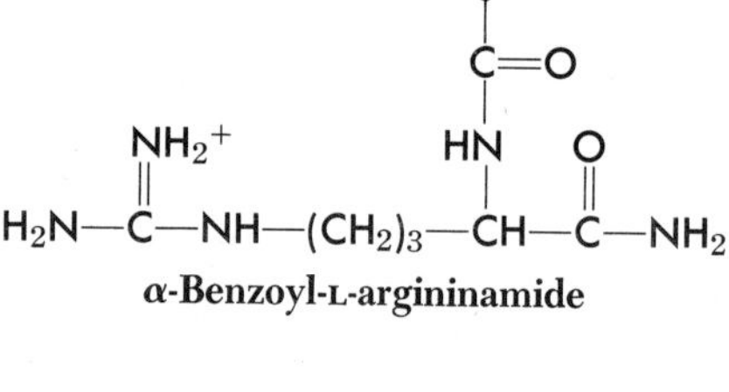

α-Benzoyl-L-argininamide

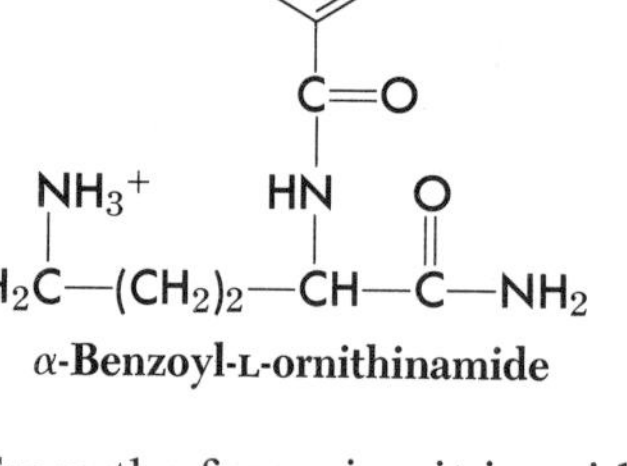

α-Benzoyl-L-ornithinamide

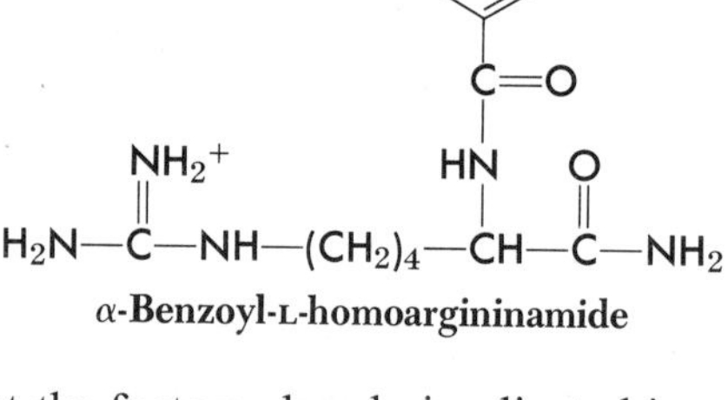

α-Benzoyl-L-homoargininamide

From the foregoing it is evident that the factors already implicated in maintaining protein structure itself, *viz.*, hydrophobic forces, ionic interaction, and hydrogen bonding, are involved in enzyme-substrate interactions. A side chain of an amino acid residue in an enzyme can serve either function and in a similar manner.

Conformation and Enzymic Activity. Although catalytic activity is dependent on the appropriate conformation of each enzyme, it must not be assumed that this conformation is absolute. Indeed, the active site, in the sense of both binding forces and participating groups, may not preexist in the form required until the approach of the substrate. Koshland has suggested that the surface conformation may then be altered, resulting in an "induced fit." Evidence suggestive of a change in enzymic conformation as a consequence of substrate binding has been noted in several instances. Examples include failure of the enzyme to combine with its antibody in the presence of substrates—*adenylic acid deaminase* (page 627) and *creatine kinase* (page 268); stabilization against heat denaturation by substrates (many enzymes); changes in optical rotation—*creatine kinase, triose phosphate dehydrogenase* (page 398); reversal by substrate of inactivation at low temperature—*Neurospora glutamic acid dehydrogenase* (page 536); changes in sedimentation behavior—*acetyl CoA carboxylase* (page 486) and D-*amino acid oxidase* (page 248); and dissociation into discrete subunits—*glutamic acid dehydrogenase.*

Similarly, for several enzymes that catalyze reaction between two substrates, kinetic data suggest a compulsory order of binding. Thus, *lactic acid dehydrogenase,* which catalyzes reduction of pyruvic acid by reduced DPN (page 402), appears to offer no binding site for pyruvate until the nucleotide is fixed in place. X-ray crystallographic data indicate a change in conformation when hemoglobin is oxygenated; similarly, oxidized cytochrome c (page 370) differs in conformation from the reduced form. In sum, these diverse observations indicate that although, generally speaking, the "lock-and-key" theory remains valid, to some extent the "lock" tailors itself to fit the "key" as the latter approaches and is bound.

Enzymic Recognition of *meso*-Carbon Atoms. The multipoint attachment of substrate and enzyme affords a means of discriminating among the four substituent groups about a given carbon atom. Historically, this recognition derives from the observation that the two $—CH_2COOH$ groups of citric acid behave, biologically, in nonidentical manner (page 333). Since the citric acid molecule possesses a plane of symmetry and, hence, exhibits no optical rotation, it was assumed that the two $—CH_2COOH$ groups were intrinsically identical and should not be differentiated in chemical reactions. The dilemma was initially solved by Ogston, who indicated that, if the molecule were to be affixed to an enzymic surface at three points ("three-point attachment"), there could be only one manner of binding, resulting in discrimination between the two identical groups.

A stereochemical solution of the problem was provided by Carter and Schwartz, who showed that in a molecule in which there is a carbon atom that can be represented as C_{aabd}, the two a groups are not geometrically equivalent but bear a mirror-image relationship to each other, as shown below.

In citric acid these groups are, respectively, $a_1 = —CH_2COOH$, $a_2 = —CH_2COOH$, $b = —OH$, $d = —COOH$. The behavioral asymmetry of citric acid in biological systems, therefore, is not the consequence of attachment to an enzymic surface; rather, the asymmetric enzymic surface distinguishes between the two intrinsically, geometrically nonequivalent $—CH_2COOH$ groups. The carbon atom with its valence bonds linked to groups *a,a,b,d* has been termed a "*meso*-carbon" (page 14).

Meso-carbon atoms occur in many substances of biological interest. Some of these are shown in bold type in the structures on the following page.

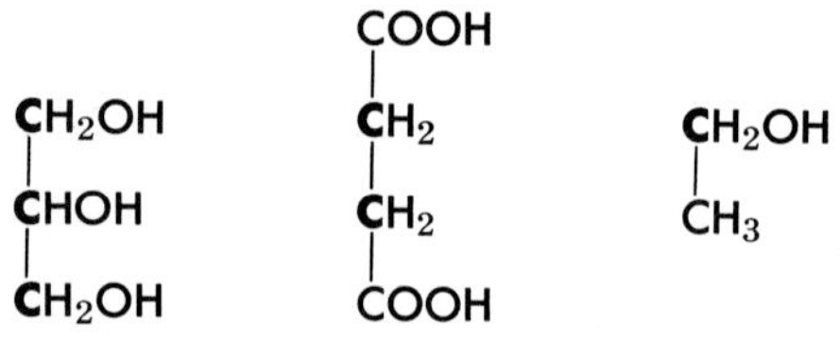

Glycerol **Succinic acid** **Ethanol**

As in the case of citric acid, enzymes that catalyze reactions involving compounds bearing *meso*-carbon atoms invariably discriminate between the two nonequivalent like groups. For example, phosphorylation of glycerol, catalyzed by *glycerokinase,* results exclusively in the formation of L-α-glycerophosphate rather than the DL mixture that would otherwise result.

NATURE OF ENZYMIC ACTIVE SITES

Up to this point we have been concerned chiefly with the factors involved in the formation of the enzyme-substrate complex. Some success has attended efforts to recognize and identify the residues in a given enzyme which are present at that locus at which substrate binding occurs and which may also participate in the catalytic reaction mechanism. Examples of studies on a few enzymes will serve to illustrate the approaches that have been employed to attack these problems.

Chymotrypsin. The specificity of *chymotrypsin* has previously been noted (Table 7.4, page 144), its most sensitive substrates being those which contain aromatic or large aliphatic side chains. Clearly, hydrophobic forces are implicated in the R group interactions. In addition, chymotrypsin exhibits strong esterase action with substrates analogous to peptide derivatives, *e.g.*, acetyl-L-tyrosine ethyl ester as well as acetyl-L-tyrosinamide.

Chymotrypsin and other animal proteinases that are active in the gastrointestinal tract, *e.g.*, *trypsin, pepsin, elastase,* etc., do not contain a prosthetic group or require a cofactor for their action. Nevertheless, chymotrypsin is synthesized in the pancreas as its *proenzyme* or *zymogen, chymotrypsinogen.* Inactive chymotrypsinogen consists of a single peptide chain cross-linked by five disulfide bridges. Hydrolysis by trypsin of a single peptide bond converts chymotrypsinogen to the active *π-chymotrypsin.* This is followed by liberation of the dipeptide, Ser-Arg, yielding *δ-chymotrypsin.* Another dipeptide, Thr-AspNH_2, is then released, yielding the more stable *α-chymotrypsin.* The process is shown schematically in Fig. 12.1, together with the alternate route of α-chymotrypsin formation.

Several features of the activation process should be noted: (1) there are 4 arginine and 11 lysine residues in chymotrypsinogen, yet only a single bond involving arginine is selectively attacked by trypsin; (2) the single peptide chain of the inactive zymogen is converted to the three-chain active chymotrypsin; (3) as a result of the removal of one or two dipeptides, there is a conformational change in the structure, as shown by changes in optical rotatory properties, *i.e.*, circular dichroism and optical rotatory dispersion. This suggests that the conformation of the active enzyme produces a juxtaposition of amino acid side chains that are remote from one another in the inactive zymogen. The novel interaction of groups in

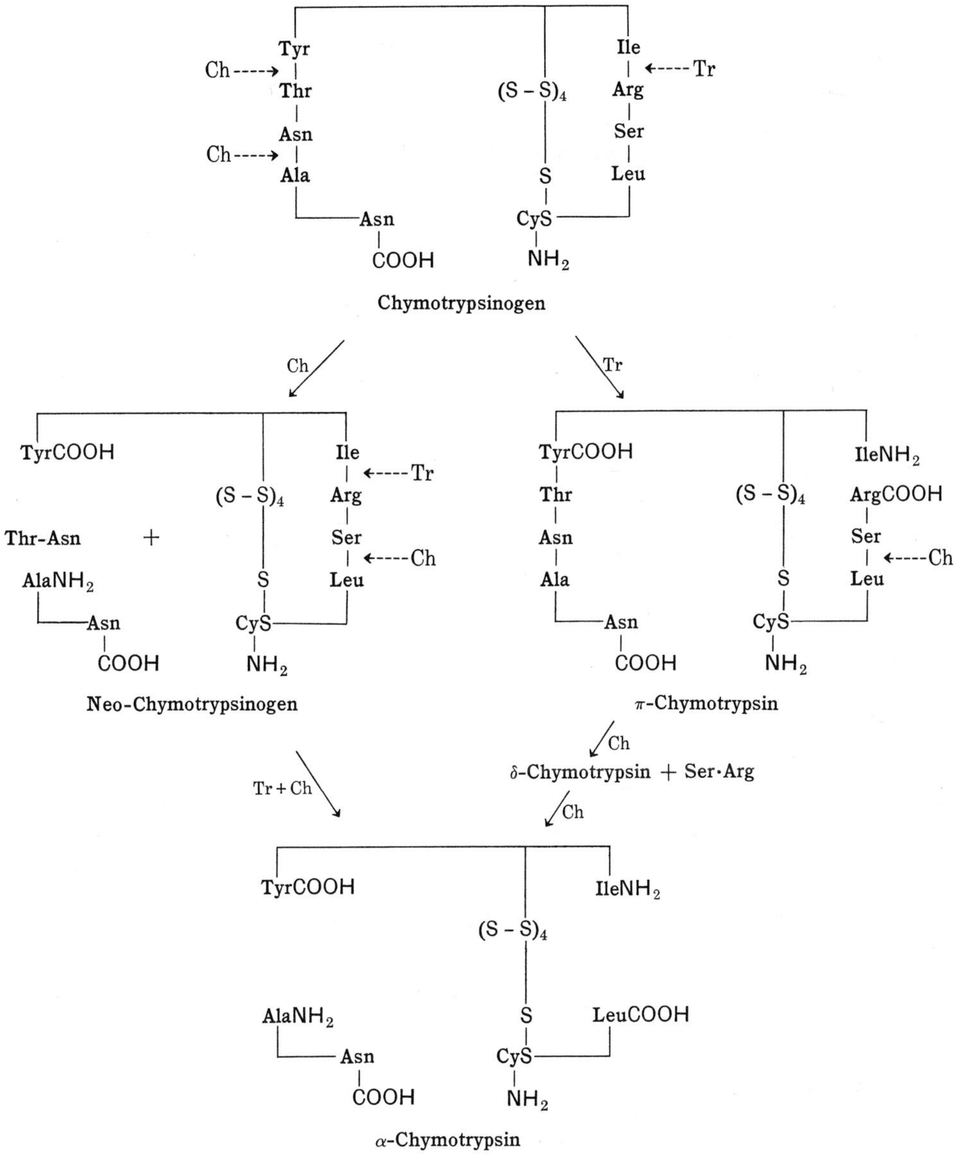

Fig. 12.1. Scheme for activation of chymotrypsinogen A. Hydrolysis of the Arg-Ile bond produces π-chymotrypsin. This is followed by liberation of the dipeptide, Ser-Arg, producing δ-chymotrypsin, which is then converted to α-chymotrypsin by liberation of Thr-Asn. Alternatively, inactive neo-chymotrypsinogen is formed as an intermediate; tryptic action on this protein yields active α-chymotrypsin. Chymotrypsinogen possesses a single peptide chain, whereas α-chymotrypsin contains three chains held together by five disulfide bridges, arbitrarily shown in the diagrams. *Tr* = trypsin; *Ch* = chymotrypsin. (*From the work of Neurath and of Desnuelle and their associates.*)

the enzyme results in a high degree of reactivity of certain groups, as shown below.

Chymotrypsin is irreversibly inactivated by a stoichiometric reaction with certain reactive phosphorus derivatives, *e.g.*, diisopropylphosphofluoridate (DIPF), which forms a diisopropylphosphoryl derivative with a single, specific serine residue of the enzyme.

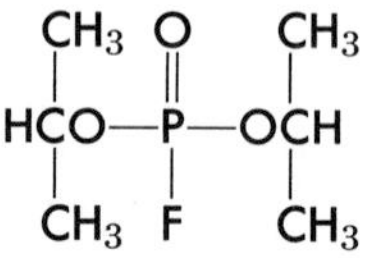

Diisopropylphosphofluoridate

Neither chymotrypsinogen nor denatured chymotrypsin reacts with DIPF and although α-chymotrypsin contains 28 serine residues, only one combines with the reagent. Clearly, the formation of inactive protein by such a specific combination with an inhibitor indicates that only a small well-defined part of chymotrypsin is responsible for its activity and that the reactive serine probably forms a part of the "active site" of the enzyme. Table 12.2 presents the amino acid sequence near the

Table 12.2: SEQUENCES ADJACENT TO THE REACTIVE SERINE IN ENZYMES

Enzyme	*Sequence*
Chymotrypsin	Gly·Asp·SerP*·Gly·
Trypsin	Gly·Asp·SerP·*Gly·
Elastase	Gly·Asp·SerP*·Gly·
Subtilisin	Thr·SerP*·Met·Ala·
Aliesterase	Gly·Glu·SerP*·Ala·
Pseudocholinesterase	Gly·Glu·SerP*·Ala·
Phosphoglucomutase	Thr·Ala·SerP·His·Asp·
Glycogen synthetase	Glu·Ile·SerP·Val·Arg·
Phosphorylase	Glu·Ile·SerP·Val·Arg·

* The reactive serine substituted by DIPF or similar reagents. Phosphoglucomutase does not react with DIPF but does accept or transfer phosphate at a seryl hydroxyl group as part of its mode of action (page 424). Glycogen synthetase and phosphorylase are not affected by DIPF but are normally interconverted to relatively inactive and active forms by phosphorylation of the seryl hydroxyl group (page 436*ff.*).

reactive serine of chymotrypsin and other enzymes that are inhibited by DIPF or similar reagents. Furthermore, the reactivity of the serine residues of the enzymes listed is indicated by the findings that free serine does not react with DIPF or similar reagents, nor do small serine peptides.

Much indirect and direct evidence also implicates a histidine residue in the catalytic activity of chymotrypsin. The effect of pH on the value of the kinetic constant k_2 (page 226) suggests the participation of a grouping with a pK value similar to that of the imidazole of histidine, the active form being the uncharged N of this ring structure. Moreover, treatment of chymotrypsin with L-1-(*p*-toluenesulfonyl) amido-2-phenylethylchloromethyl ketone (TPCK) results in a parallel loss of activity and of histidine residue 57.

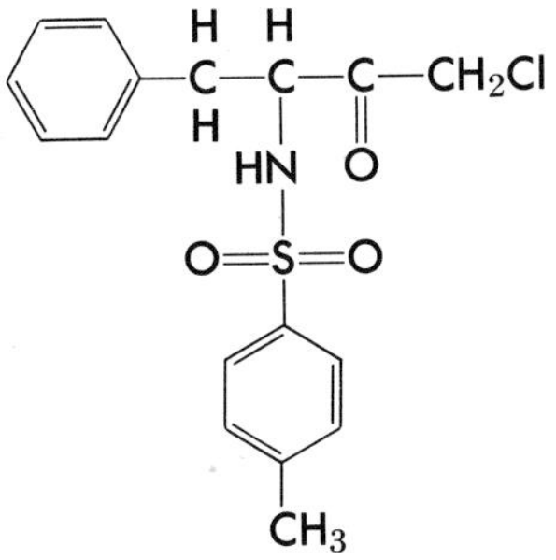

L-1-(*p*-Toluenesulfonyl)amido-2-phenylethylchloromethyl ketone

TPCK reacts with chymotrypsin because of a resemblance to substrates, such as an acylphenylalanine amide or ester, in which the $—NH_2$ or $—OCH_3$ is present in place of the nonhydrolyzable $—CH_2Cl$. Similar results have been obtained for trypsin with the analogous lysine derivative TLCK, or L-5-amino-1-(*p*-toluenesulfonyl) amidopentylchloromethyl ketone.

The reactive serine and the two histidine residues of the protein are remote from one another in the linear, primary amino acid sequence; however, the two histidine residues approach each other as a result of linking of adjacent half-cystine residues in a disulfide bridge. As with other enzymes, denaturation, *i.e.*, unfolding of the protein, results in complete inactivation of chymotrypsin. Again, this implies that the active form of the enzyme depends on the native conformation, which brings together the specific side chains of the amino acid residues essential for determining the specificity and catalytic reactivity of the protein.

As indicated in Table 12.2, there is a large class of enzymes—esterases and proteinases (which are also esterases)—that possess a reactive serine residue. These enzymes must possess certain structural and catalytic features in common; however, in view of their great differences in specificity, other parts of their "active sites" must differ profoundly.

Elucidation of the complete amino acid sequences of chymotrypsin and trypsin (Fig. 12.2), and of partial sequences of elastase and chymotrypsin B, has shown that all these proteolytic enzymes, synthesized in the mammalian pancreas, are structurally related to one another, as indicated by large sections of similar sequences. Indeed, the similarities of sequences at the "reactive" serine (Fig. 12.2) are a reflection of common genetic origin (Chap. 30) rather than of function, the special reactivity being conferred by proximity or interaction with side chains other than the adjacent ones. This is indicated also by the finding that the *subtilisins,* proteolytic enzymes of *Bacillus subtilis* of different genetic origin, possess entirely different amino acid sequences not only near the "reactive" serine but throughout the entire proteins (Fig. 12.3), despite many similarities in specificity and reaction mechanism. The sequences of the two subtilisins shown in Fig. 12.3 differ considerably yet exhibit very similar specificity and mechanism of action. This indicates that changes in sequence that do not influence active sites or conformation can be rather extensive.

Ribonuclease. The action of pancreatic *ribonuclease* on RNA has already been presented (pages 200*ff.*). Ribonuclease (bovine) contains 124 amino acid residues;

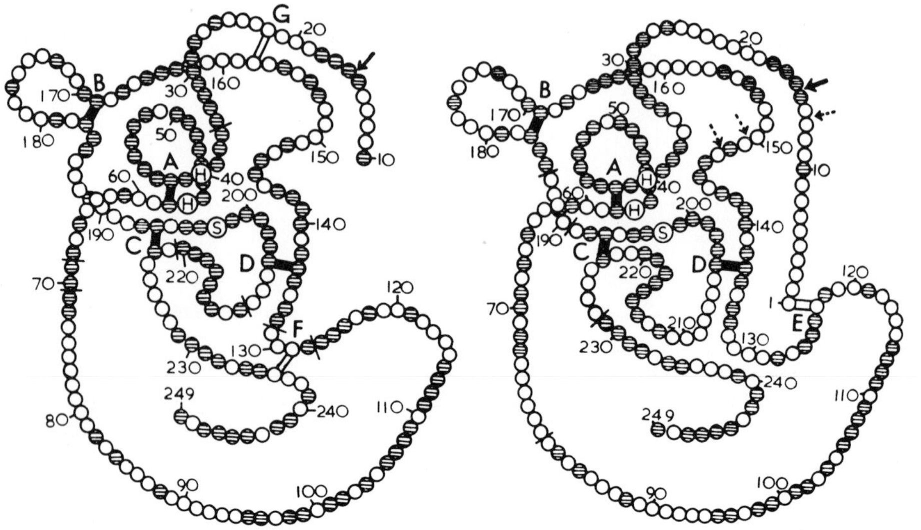

Fig. 12.2. Comparison of the structures of trypsinogen and chymotrypsinogen A. The solid arrows indicate the points of hydrolysis which result in activation of trypsinogen to trypsin and chymotrypsinogen to chymotrypsin; broken arrows indicate additional points of hydrolysis to form α-chymotrypsin. The shaded circles indicate amino acid residues that are chemically similar or identical in the two proteinases. The disulfide bridges are lettered *A* to *G*. *H* indicates histidine residues and *S* the serine residue of the active site that is reactive with DIPF. Deletions are indicated by lines between the circles. (*From B. S. Hartley, J. R. Brown, D. L. Kauffman, and L. B. Smillie, Nature,* **207**, 1157, 1965.)

its composition is given in Table 7.1 (page 140). The complete sequence of the single peptide chain and the positions of the disulfide bridges are shown in Fig. 12.4.

Under conditions of limited digestion by *subtilisin,* ribonuclease (RNase) is hydrolyzed at the peptide bond between residues 20 and 21. The two fragments, S-peptide and S-protein, may be separated at acid pH, but neither is active. These fragments recombine at neutral pH, and enzymic activity is restored. Although the peptide bond is not re-formed, the 20-residue S-peptide binds strongly to the 104-residue S-protein by secondary valence forces. A synthetic peptide containing the amino terminal 13 residues of the S-peptide will regenerate approximately 70 per cent of the enzymic activity with the S-protein, thus demonstrating that residues 14 to 20 play no essential *catalytic* role. However, the undecapeptide, lacking histidine and methionine (residues 12 and 13, respectively) does not restore activity.

Treatment of ribonuclease with iodoacetate at pH 5.5 produces two inactive monocarboxymethylhistidine derivatives of the enzyme. One of these is 1-carboxymethylhistidine ribonuclease substituted at position 119; the other is the 3-carboxymethylhistidine derivative substituted at position 12 in the peptide sequence (Fig. 12.4); these are formed in a ratio of 8 : 1. The structures of these isomeric derivatives are shown on page 258.

Inasmuch as both derivatives of RNase are inactive, both histidine residues are essential for enzymic activity. Moreover, alkylation of one histidine residue prevents

Thr Ile Pro Leu Asp Lys Val Gln Ala Phe Lys Ala
NH_2-Ala-Gln-Ser-Val-Pro-Tyr-Gly-Val-Ser-Gln-Ile-Lys-Ala-Pro-Ala-Leu-His-Ser-Gln-Gly-Tyr-Thr-Gly-Ser-Asn-Val-Lys-Val-Ala-Val-
10 20 30

Leu Thr Gln Ala Asn Val Phe Ala Gly Ala — Tyr Asn Thr
Ile-Asp-Ser-Gly-Ile-Asp-Ser-Ser-His-Pro-Asp-Leu-Lys-Val-Ala-Gly-Gly-Ala-Ser-Met-Val-Pro-Ser-Glu-Thr-Pro-Asn-Phe-Gln-Asp-
40 50 60

Gly Gly Asp Thr Thr Val Ser
Asp-Asn-Ser-His-Gly-Thr-His-Val-Ala-Gly-Thr-Val-Ala-Ala-Leu-Asn-Asn-Ser-Ile-Gly-Val-Leu-Gly-Val-Ala-Pro-Ser-Ser-Ala-Leu-
70 80 90

Asn Ser Ser Ser Gly Val Ser Thr Thr Gly
Tyr-Ala-Val-Lys-Val-Leu-Gly-Asp-Ala-Gly-Ser-Gly-Gln-Tyr-Ser-Trp-Ile-Ile-Asn-Gly-Ile-Glu-Trp-Ala-Ile-Ala-Asn-Asn-Met-Asp-
100 110 120

Ala Thr Met Gln Asn Tyr Arg
Val-Ile-Asn-Met-Ser-Leu-Gly-Gly-Pro-Ser-Gly-Ser-Ala-Ala-Leu-Lys-Ala-Ala-Val-Asp-Lys-Ala-Val-Ala-Ser-Gly-Val-Val-Val-Val-
130 140 150

Ser Asn Ser Thr Asn Ile Ala Asp
Ala-Ala-Ala-Gly-Asn-Glu-Gly-Ser-Thr-Gly-Ser-Ser-Ser-Thr-Val-Gly-Tyr-Pro-Gly-Lys-Tyr-Pro-Ser-Val-Ile-Ala-Val-Gly-Ala-Val-
160 170 180

Asn Ser Asn Ala Glu Ala Gly Val Tyr Tyr
Asp-Ser-Ser-Asn-Gln-Arg-Ala-Ser-Phe-Ser-Ser-Val-Gly-Pro-Glu-Leu-Asp-Val-Met-Ala-Pro-Gly-Val-Ser-Ile-Gln-Ser-Thr-Leu-Pro-
190 200 210

Thr Thr Ala Thr Leu
Gly-Asn-Lys-Tyr-Gly-Ala-Tyr-Asn-Gly-Thr-Ser*-Met-Ala-Ser-Pro-His-Val-Ala-Gly-Ala-Ala-Ala-Leu-Ile-Leu-Ser-Lys-His-Pro-Asn-
220 230 240

Leu Ser Ala Ser Asn Arg Ser Ser Ala Tyr Ser
Trp-Thr-Asn-Thr-Gln-Val-Arg-Ser-Ser-Leu-Gln-Asn-Thr-Thr-Thr-Lys-Leu-Gly-Asp-Ser-Phe-Tyr-Tyr-Gly-Lys-Gly-Leu-Ile-Asn-Val-
250 260 270

Glu
Gln-Ala-Ala-Ala-GlnCOOH
275

Fig. 12.3. Amino acid sequences of two types of subtilisin. The continuous sequence is the subtilisin from *Bacillus subtilis* BPN′ strain; the residues shown above this sequence are the substitutions found in the Carlsberg strain. The bar (—) at residue 56 indicates a deletion. The serine (*) at residue 221 reacts with DIPF; the DIP subtilisins are completely inactivated. Note that the subtilisins lack cysteine and cystine residues and show no significant resemblance to the sequences of the animal proteinases, chymotrypsin and trypsin. (*From E. L. Smith, F. S. Markland, C. B. Kasper, R. J. DeLange, M. Landon, and W. H. Evans, J. Biol. Chem.,* **241**, 5974, 1966.) The amide distribution has been corrected from the aforementioned publication.

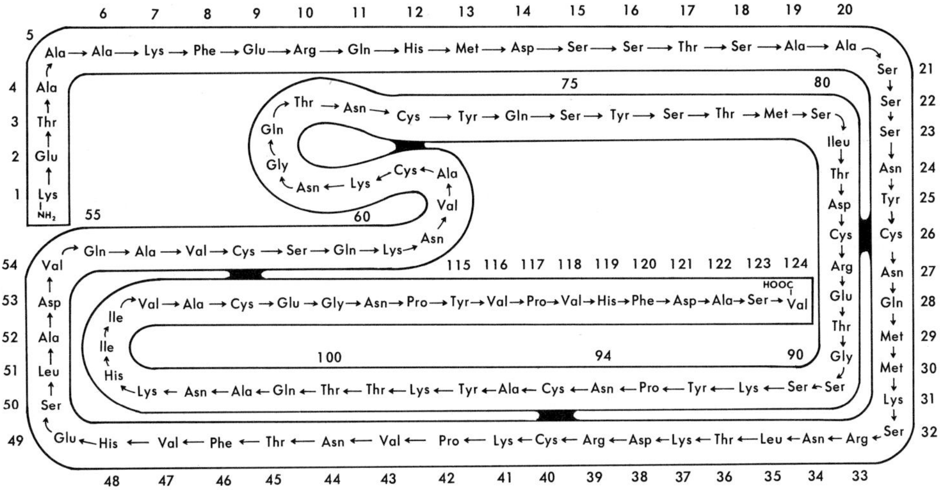

Fig. 12.4. The amino acid sequence of bovine pancreatic ribonuclease. (*From D. G. Smyth, W. H. Stein, and S. Moore, J. Biol. Chem.,* **238**, 227, 1963.)

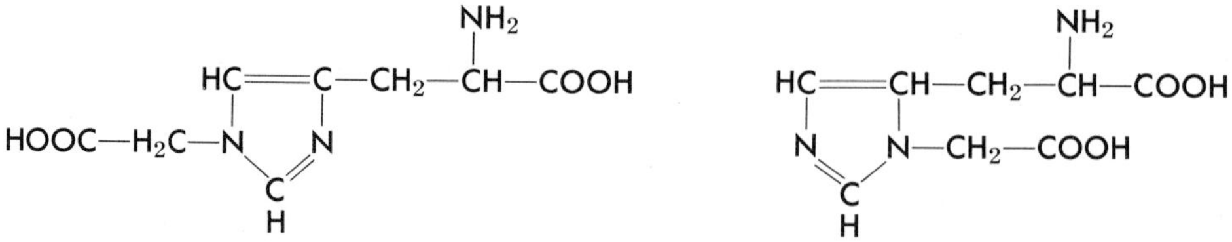

3-Carboxymethylhistidine 1-Carboxymethylhistidine

alkylation of the other. Aggregates of RNase (dimers, trimers, etc.) are formed by lyophilization from 50 per cent acetic acid. When such aggregates are formed with a mixture of the two inactive RNase derivatives, the resulting hybrids manifest about half the activity of an equivalent weight of native enzyme treated in the same fashion. These results indicate that histidine-12 and histidine-119 are both at the active site which can be formed from parts of two different molecules. This activity disappears when the aggregates are broken up by heating at 67°C. for 10 min., conditions which do not denature untreated RNase.

These findings indicate once again that the enzymic activity is a consequence of specific folding or conformation of the peptide chain, *viz.*, active sites are formed by the juxtaposition of amino acid side chains that may be remote from one another in the sequence; however, the folding is a consequence of linear amino acid sequence (Chaps. 7 and 28). Moreover, as in the case of the reactive serine of chymotrypsin, trypsin, and other enzymes described above, the histidine residues of RNase are uniquely reactive with iodoacetate or bromoacetate at pH 5.5. Histidine itself, the other two histidine residues of RNase, and histidine residues of most other proteins do not react with these reagents at this pH value. Obviously, it is the special conformation which confers this unique reactivity.

X-ray analysis has revealed much of the three-dimensional structure of ribonuclease. The portion of the structure shown in Fig. 12.5 indicates that histidines 12 and 119 are near each other, as expected, and that the substrate or an inhibitor fits into a cleft in the molecule. Other features of the structure are noted in the legend of Fig. 12.5.

MECHANISM OF ENZYMIC ACTION

A major goal of biochemistry is to provide a detailed account of the mechanisms by which enzymes catalyze specific reactions. It must be admitted, at the outset, that this goal has not yet been realized completely in a single instance. The simplest concept of the action of an enzyme is that its surface serves to gather together, in favorable proximity, the various reactants involved, *i.e.*, the substrates and cofactors. In effect, this would accomplish a marked local increase in concentration of reactants, as achieved, for example, by platinum black for H_2 and O_2. This could also assure appropriate orientation of the reacting groups in space so that collision must occur and in a manner leading to a specific reaction. If this concept were adequate, only the specific binding capacity of the protein would be involved; however, this concept appears to offer only a partial solution to the remarkable catalytic efficiency of enzymes generally. Thus, many enzymes increase

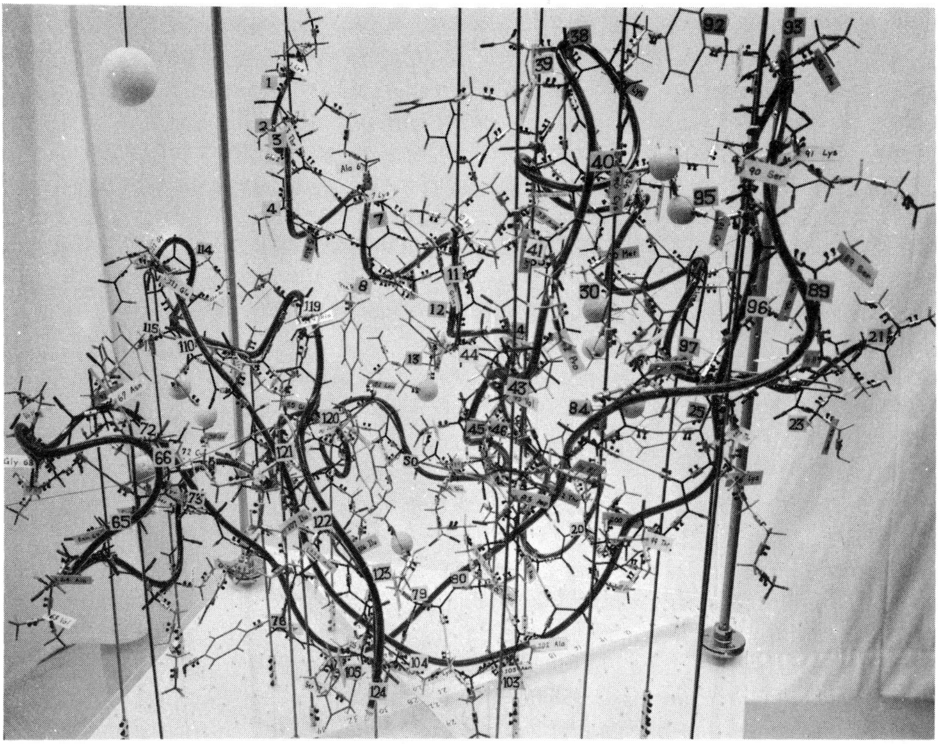

Fig. 12.5. Model of part of the pancreatic ribonuclease-S molecule based on x-ray diffraction analysis at 3.5 Å. resolution. The continuous strand of tubing follows the peptide chain. α-Helix constitutes about 15 per cent of the molecule: residues 2 to 13 (partly visible), residues 26 to 33, and residues 50 to 58 (in the back of the molecule). There is a section of antiparallel β structures comprising residues 71 to 92 and 94 to 110. The pairs of balls represent sulfur atoms in disulfide bridges, *e.g.*, between 40 and 95 at the upper right. The single balls represent sulfur atoms of methionine residues. The imidazole rings of histidines 12 and 119 are, as predicted from the chemical data, near one another in a groove. The competitive inhibitor, 5-iodouridine-2′(3′)-phosphate, is bound in this groove. (*From H. W. Wyckoff, K. D. Hardman, N. M. Allewell, T. Inagami, L. N. Johnson, and F. M. Richards, J. Biol. Chem.*, **242**, 3984, 1967.)

by a factor of 10^6 to 10^{10} the rate at which the same reaction would occur were all the reactants present in $1M$ aqueous solution in the absence of the enzyme. Indeed, this is even true of hydrolytic reactions; enzymically catalyzed hydrolyses proceed about 10^3 times as fast as do the same reactions even when catalyzed by strong acid or base.

Activation by Strain or Bond Distortion. A relatively old hypothesis to account for enzymic catalysis suggested that specific attachment to the enzymic surface induces "strain" or "deformation" in the bonds that are to be broken. Although the concept of strain was not fully defined, the idea is not without merit, if it be interpreted to mean the intermediate formation of a compound or complex in which the susceptible bond is inherently less stable than in the original reactants. An example of strain in this sense is afforded by the fact that the base-catalyzed hydrolysis of ethylene phosphate occurs 10^7 times more rapidly than does that of dimethylphosphate.

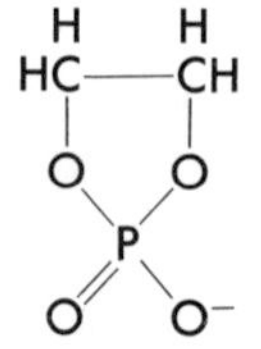

Ethylene phosphate

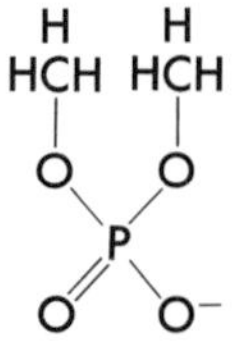

Dimethylphosphate

The strain hypothesis has seemed an attractive explanation for enzymic catalysis. An actual example appears to be reflected in the behavior of horse liver esterase. For the hydrolysis of a series of esters of *m*-hydroxybenzoic acid,

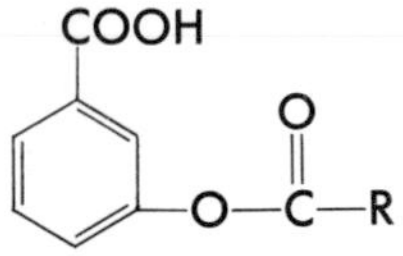

Hofstee found that K_m is almost independent of the chain length of R, whereas V_{max} increases by several orders of magnitude as the chain length is increased. Since the bond to be hydrolyzed is the same in each instance, it can be inferred that the increased binding energy of the longer-chain esters reduced the activation energy for the reaction, *i.e.*, the energy of the tighter bonding of the hydrocarbon moiety is offset by the strain energy induced in the acyl portion of the molecule.

Functional Groups at the Catalytic Site. The most likely explanation of enzymic catalytic efficiency that has gained general acceptance is the presence at the "active site" of amino acid residues or other groupings, such as metal ions, which can serve as acids or bases or as nucleophilic (electron-donating) or electrophilic (electron-attracting) agents and, thus, actually participate in the reaction mechanism. In the case of enzymes containing metal ions, the metal ion itself can bind with groupings in the substrate and act as a strain-producing agent by forming a chelated intermediary compound. At the same time, the metal ion, because of its positive charge, is a strong electrophilic agent which can act as an effective participant in the reaction. As already noted (page 214), many enzymes contain bound metal ions which do not alter their valence state during the enzymic reaction, *e.g.*, Mg^{++}, Mn^{++}, Zn^{++}, and which probably function as electrophilic groups. This has been suggested for certain hydrolytic enzymes, notably the peptidases.

There are many enzymes that lack metal ions or other non-amino acid groups. In these instances, the groupings at the "active sites" have been sought by other methods. As described previously, it is possible to demonstrate by various procedures the essentiality of an unaltered form of several amino acid residues for the functioning of a given enzyme. Methionine residues can be oxidized to the corresponding sulfoxide, $CH_3\underset{\underset{O}{\|}}{S}—CH_2—$, histidine residues are destroyed by photooxidation, sulfhydryl groups react with heavy metals or with iodoacetate. Enz— below connotes the remainder of the enzyme.

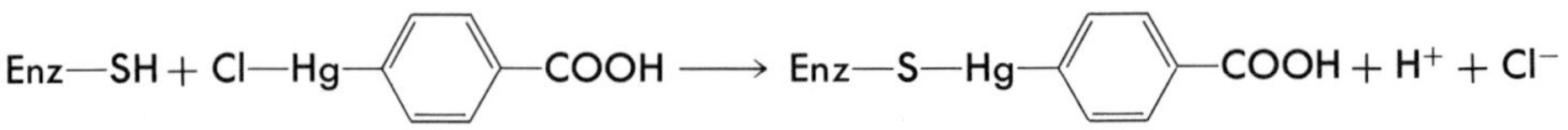

$$Enz—SH + ICH_2COO^- \longrightarrow Enz—S—CH_2—COO^- + H^+ + I^-$$

Histidine residues also react with iodoacetate, as in the case of ribonuclease (page 256). Lysine residues can react with fluorodinitrobenzene (page 142) or with various acylating agents and can be converted to homoarginine by reaction with isothiourea.

$$Enz—(CH_2)_4—NH_2 + HS—\overset{\overset{\displaystyle NH}{\|}}{C}—NH_2 \longrightarrow Enz—(CH_2)_4—\underset{\underset{\displaystyle H}{|}}{N}—\overset{\overset{\displaystyle }{}}{C}—NH_2 + H_2S$$

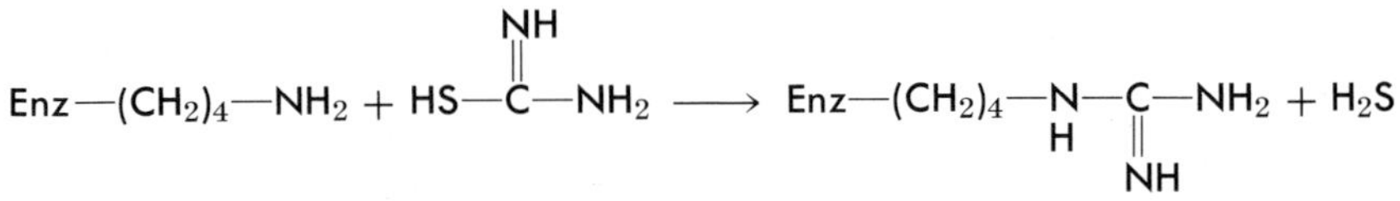

It is essential that in each case the claimed formation of a derivative of an amino acid side chain be demonstrated by isolation of the compound. The fact that iodoacetate not only can inactivate many thiol enzymes, *e.g.*, papain, various dehydrogenases, etc., but will also inactivate pancreatic RNase by formation of carboxymethylhistidine illustrates this point. Further, RNase T1 (page 201) is also inactivated by iodoacetate but with this enzyme, there is a stoichiometric reaction with a single γ-carboxyl group of a glutamic acid residue to form a glycolic ester.

$$—\overset{\overset{\displaystyle O}{\|}}{C}—\underset{\underset{\displaystyle NH}{|}}{\overset{\overset{\displaystyle H}{|}}{C}}—CH_2—CH_2—\overset{\overset{\displaystyle O}{\|}}{C}—O—CH_2—COOH$$

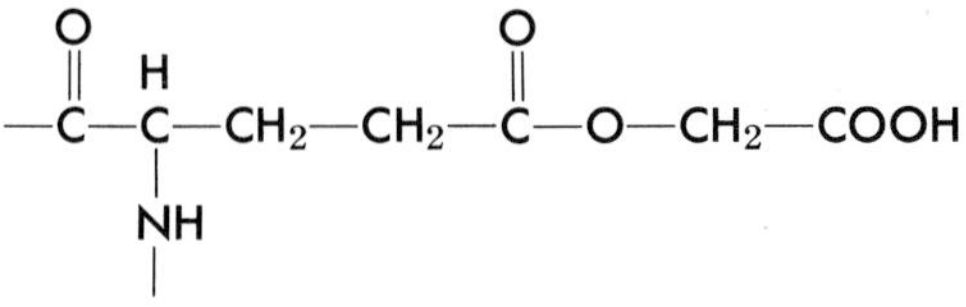

In some proteins, iodoacetate reacts at acid pH values with methionine residues to form the carboxymethyl sulfonium derivatives. Thus, one reagent, iodoacetate, can inhibit different enzymes by formation of distinct derivatives with specific cysteine, histidine, glutamic acid, and, perhaps, methionine and lysine residues.

Another useful procedure is examination of the titration curve of the enzyme in the region of its optimal pH. The presence of titratable amino acid residues in this region suggests that these groups may participate in the enzymic reaction mechanism and are required in either the protonated or the unprotonated form. Since most enzymes exhibit pH optima in the region 6.5 to 7.5, and since the imidazole group of histidine is the only amino acid group normally titratable in this range, there is now a long list of enzymes for which it has been suggested that histidine has a specific function. Only in a few instances has direct proof been furnished that histidine is the responsible residue, *e.g.*, chymotrypsin, trypsin, pancreatic RNase.

The kind of evidence described above does not establish the participation of these various amino acid residues in the catalytic event. Titration of histidine, binding of a sulfhydryl group, or oxidation of methionine could have altered the conformation of the protein, thereby influencing the catalytic site. In each instance it is necessary to demonstrate that the amino acid residues in question are indeed at the active site. This information can be provided only through knowledge of the complete, three-dimensional structure of the enzyme as it might be reconstructed from x-ray crystallographic analysis, as in the case of lysozyme (page 269). Moreover, demonstration of the presence of a specific residue at the active site does not necessarily indicate that it participates as a *reactant* in the catalytic process.

More acceptable as evidence is a demonstration that a specific residue can react with substrate, or a suitable model compound, to form a stable, covalently bonded derivative. Evidence of this type is available in several instances. As noted previously, several hydrolytic enzymes containing individual serine residues that react with DIPF also have reactive histidine residues. For chymotrypsin and trypsin, this has been shown by formation of a histidine derivative when the enzyme was exposed to a substance of a character similar to that of a model synthetic substrate. Hypotheses regarding the mechanism of action of these enzymes, therefore, are based on the nucleophilic character of the seroxide group, —OC—CHNHR—CH_2O^-, and the ease with which the imidazolium group can serve as proton donor. A serine residue has also been implicated in the functioning of *phosphoglucomutase* (page 424); the phosphate ester of one serine residue can be isolated after treatment of the enzyme with the substrate, glucose 1,6-diphosphate. It is of interest that there is a histidine residue immediately adjacent to this serine in the primary structure of the enzyme. However, evidence is lacking that this serine residue is phosphorylated and dephosphorylated at a rate commensurate with the over-all catalysis.

The sulfhydryl groups of cysteine residues have been implicated as participating in the catalytic mechanisms of a variety of enzymes. Thus, proteolysis by *papain* and *ficin* entails intermediate formation of an enzyme-bound thioester with the carboxyl group of the ruptured peptide bond.

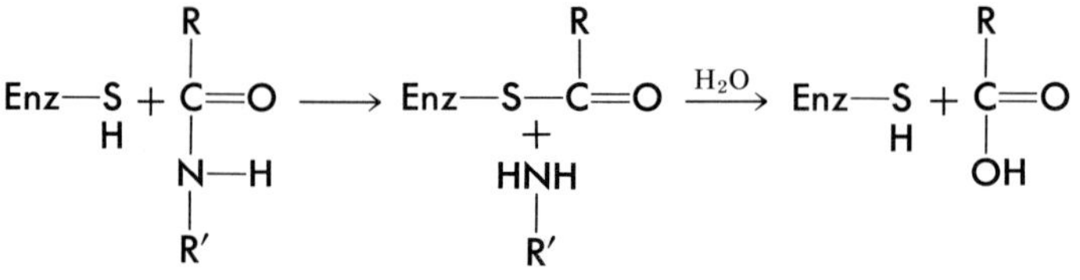

(a) (b) (c)

This has been demonstrated with substrates of the type, R—C(=S)—NHR′, which form dithio acyl intermediates, Enz—S—C(R)=S, and possess a distinctive spectral absorption not shown by the native enzyme. The rates of formation and decomposition of the dithio compound are consistent with the catalytic rate of formation of products. The thiol group of papain is in the sequence, -Ser-Gly-CySH-Trp-.

Enzyme sulfhydryl groups are thought to participate in electron transfer in several oxidative enzymes such as *dehydrolipoyl transacetylase* (page 328) and by addition to the aldehyde function of 3-phosphoglyceraldehyde so that the first oxidation product is the 3-phosphoglyceryl thioester of the enzyme (page 399), similar in general structure to that shown in (*b*) above.

In yet another instance, phosphate has been shown to be covalently bound to the imidazole nitrogen of a histidine residue of a protein during the operation of the α-ketoglutarate dehydrogenase complex (page 331).

Another form of covalent bonding of substrates is apparent in the formation of Schiff bases (aldimines) between carbonyl compounds and the ε-amino group of lysine in some enzymes. Thus when dihydroxyacetone phosphate is incubated with *aldolase* (page 397) in the absence of 3-phosphoglyceraldehyde (its normal partner in the reaction catalyzed by this enzyme), the following reaction occurs.

$$\text{Enz—(CH}_2)_4\text{—NH}_2 + \overset{\displaystyle \text{CH}_2\text{OH}}{\underset{\displaystyle \text{CH}_2\text{—O—PO}_3\text{H}_2}{\text{O}=\text{C}}} \rightleftharpoons \text{Enz—(CH}_2)_4\text{—N}=\overset{\displaystyle \text{CH}_2\text{OH}}{\underset{\displaystyle \text{CH}_2\text{O—PO}_3\text{H}_2}{\text{C}}}$$

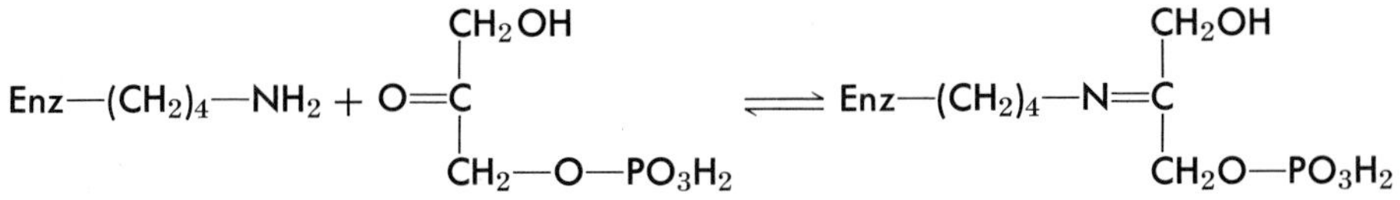

This was demonstrated by reducing the aldimine double bond with borohydride and then hydrolyzing the product completely with acid. The hydrolysate contained lysine with its ϵ-amino group linked as a secondary amine to dihydroxyacetone phosphate. Similar results have been obtained with *transaldolase* (page 398). Formation of this Schiff base can account for the observed labilization of a hydrogen on the adjacent carbon.

$$-\text{N}=\overset{\displaystyle \text{CH}_2\text{OH}}{\underset{\displaystyle \text{CH}_2\text{OPO}_3\text{H}_2}{\text{C}}} \rightleftharpoons \underset{\displaystyle \text{H}}{-\text{N}}-\overset{\displaystyle \text{HCOH}}{\underset{\displaystyle \text{CH}_2\text{OPO}_3\text{H}_2}{\text{C}}}$$

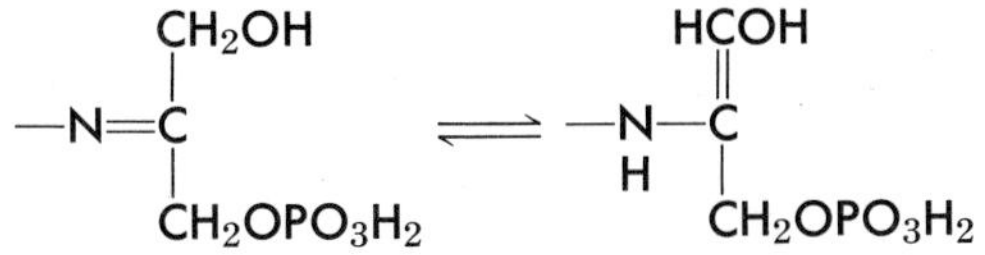

Of interest in this regard is the manner in which pyridoxal phosphate functions in transaminases. This mechanism is discussed in detail elsewhere (page 542). It need only be noted here that in transamination the aldehyde function of pyridoxal phosphate initially participates in formation of a Schiff base with a lysine ϵ-amino group of the apoenzyme. This also occurs in muscle phosphorylase. This is not merely a means of binding the pyridoxal phosphate to the enzyme; this is accomplished elsewhere on the pyridoxal phosphate molecule. The initial binding serves rather to accelerate the enzymic reaction since *transaldimination* is much more rapid than is initial formation of the Schiff base. Thus the initial step in this reaction may be visualized as in Fig. 12.6.

Studies such as those cited above have led to the postulation of reaction mechanisms for a variety of enzymes. The total evidence available for chymotrypsin and ribonuclease is greater than that for other enzymes; the most generally accepted hypotheses are depicted in Figs. 12.7 and 12.8. The mechanism shown for ribonuclease rests, among other observations, on the isolation of cyclic phosphate esters from the reaction mixture and the fact that hydrolysis of cytidine 2',3'-cyclic phosphate by ribonuclease is exceedingly rapid.

In general, two types of phenomena, *transfer* and *exchange,* have suggested formation in the catalytic process of an intermediate in which a group from one of the reactants is covalently linked to a site on the enzyme, as in the above-cited formation of a thiol ester of papain. Many hydrolytic enzymes catalyze not only the hydrolysis for which each is regarded as specific, but group transfer reactions as well. This was originally observed as the process of *transpeptidation,* as shown below.

1. Hydrolysis:

$$\underset{\displaystyle }{R-\overset{\displaystyle \text{O}}{\overset{\|}{\text{C}}}-\text{NH}-\overset{\displaystyle \text{R}'}{\underset{\displaystyle }{\text{CH}}}-\text{CO}-\text{NH}-\text{Y}} \xrightarrow{\text{HOH}} R-\overset{\displaystyle \text{O}}{\overset{\|}{\text{C}}}-\text{NH}-\overset{\displaystyle \text{R}'}{\text{CH}}-\text{COOH} + \text{H}_2\text{N}-\text{Y}$$

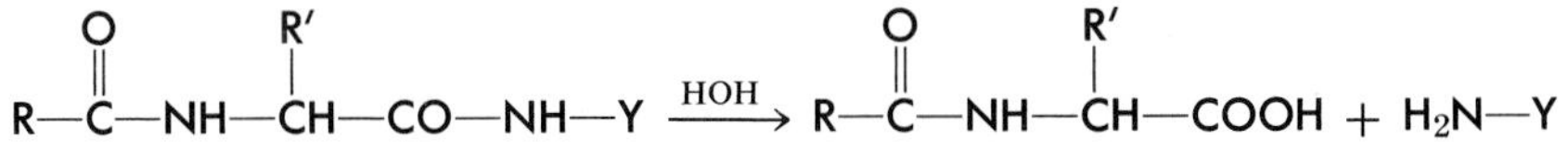

2. Transpeptidation:

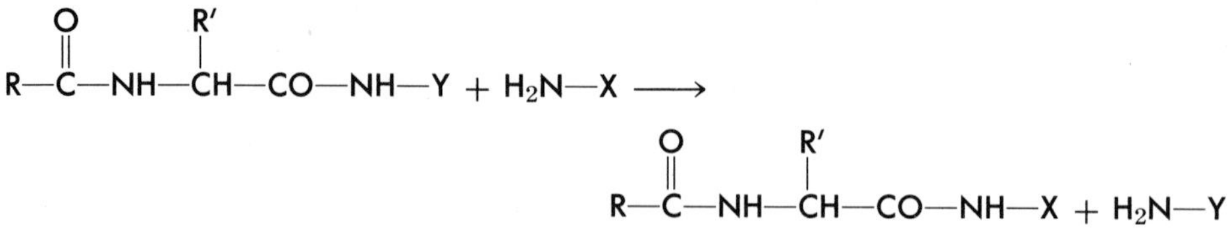

Such proteolytic enzymes as trypsin, chymotrypsin, papain, and subtilisin all catalyze transamidation and transesterification as well. Similar reactions have been observed with amylases, in which, instead of hydrolysis, a new glycoside results when the medium contains an appropriate alcohol, which displaces the group leaving. Esterases normally catalyze hydrolysis of esters, but in the presence of an alcohol in the medium can catalyze transesterification.

$$\mathrm{R{-}CO{-}OA + BOH \rightleftharpoons R{-}CO{-}OB + AOH}$$

On the basis of such transfers, it has been suggested that the actual mechanism might be as follows.

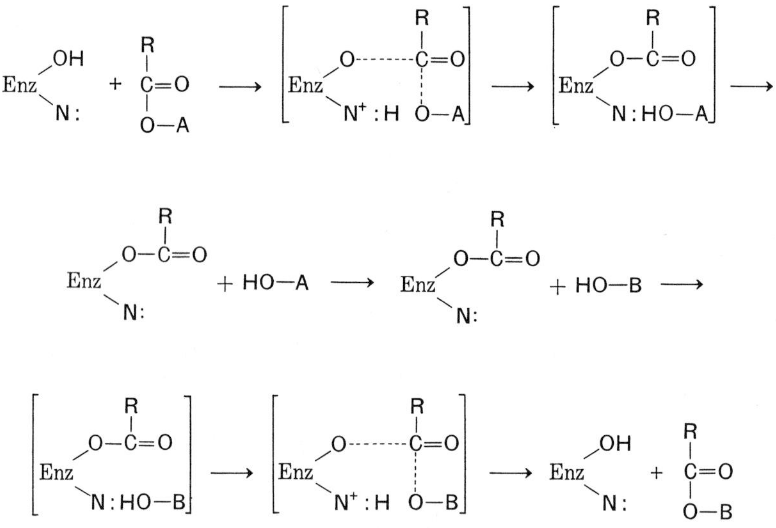

The —OH and N= groups could be those of serine and histidine, respectively. In effect, an acyl-enzyme is formed, as in the afore-mentioned case of a thiol ester with papain (page 262).

A similar mechanism has been invoked to explain the frequently observed exchange reactions. Thus, for an esterase such as that above, the use of ^{14}C-labeled compounds (Chap. 13) has demonstrated that the enzyme catalyzes the following exchange.

$$\mathrm{CH_3COO{-}CH_3 + {}^{14}CH_3OH \rightleftharpoons CH_3COO{-}{}^{14}CH_3 + CH_3OH}$$

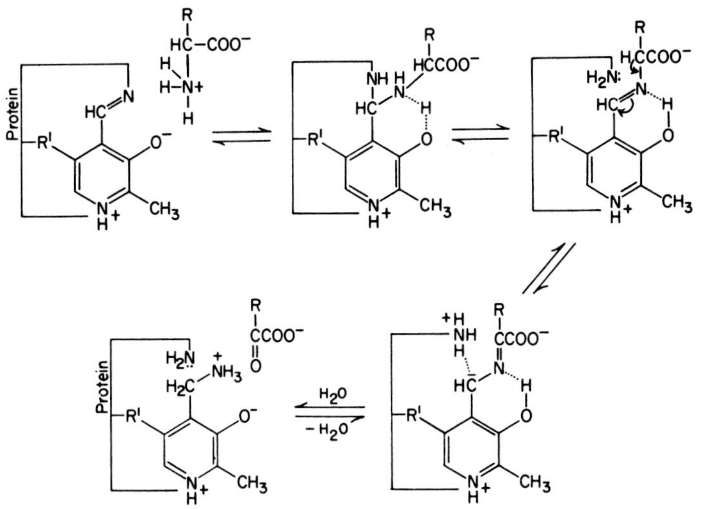

Fig. 12.6. Speculative scheme for catalysis of a half-reaction of enzymic transamination illustrating how the nonhydrogen-bonded aldimine structure facilitates interaction with the amino acid substrate, and how the liberated ε-amino group of lysine might act as a general acid-base catalyst for the electron shifts initiated by pyridoxal. (*From E. E. Snell, Brookhaven Symposia in Biology, No.* 15, *Brookhaven National Laboratory, Upton, N.Y., p.* 39, 1962.)

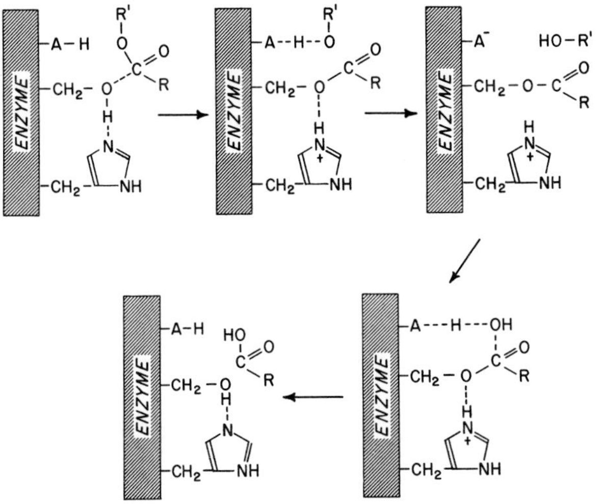

Fig. 12.7. A suggested mechanism for hydrolysis of an ester catalyzed by chymotrypsin. As shown, catalysis requires concerted operation of serine residue 195, a histidine imidazole group (residue 57), and an unidentified acid function (—A—H), possibly an imidazolium group. (*Modified from F. H. Westheimer, Advances in Enzymol.,* **24**, 464, 1962.)

On the other hand, if the same enzyme is incubated with unlabeled ester and labeled acid $^{14}CH_3{}^{14}COOH$, no exchange into the ester occurs, in accord with the proposed mechanism. This is a general phenomenon; enzymes that catalyze such exchanges, *viz.*,

$$A—B + B^* \rightleftharpoons A—B^* + B$$

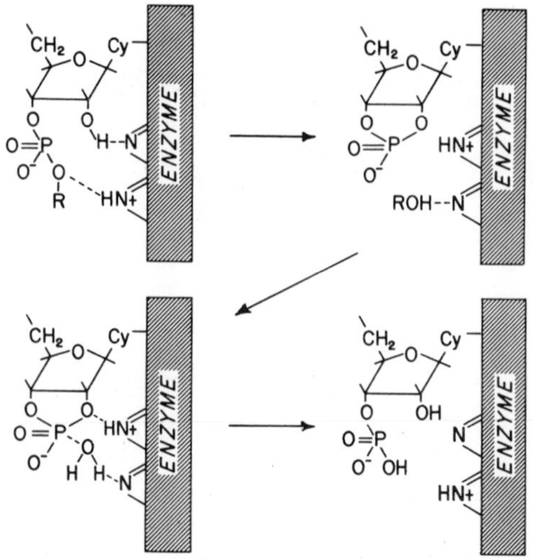

Fig. 12.8. Generalized representation of cooperative function of histidine residues 12 and 119 in catalyzing hydrolysis of an internucleotide linkage of ribonucleic acid by ribonuclease. *Cy* indicates the cytosine residue. Note the postulated formation of the cyclic phosphate nucleotide, on the enzymic surface, as an intermediate.

fail to catalyze exchange with the other component of the system (A); *i.e.*, A*—B is not formed when enzyme is incubated with A—B and A*. From this it follows that there must be formation of an intermediate in which A is covalently bonded to the enzyme in a manner that retains the bond energy of A—B.

$$\textbf{Enz—H} + \textbf{A—B} \rightleftharpoons \textbf{Enz—A} + \textbf{BH}$$

It is this reaction which is observed as an "exchange." Since it occurs reversibly at a velocity equal to that of normal hydrolytic catalysis, and Enz—A retains much of the bond energy of A—B, ΔF for this process must be small. Hydrolysis, then, represents the subsequent hydrolytic rupture of Enz—A to Enz—H + AOH, and it is this step which must proceed with a favorable negative change in free energy. Thus for many of these enzymes (chymotrypsin, papain, etc.), the reaction sequence has been demonstrated to proceed in three steps:

$$\text{E} + \text{S} \underset{k_{-1}}{\overset{k_1}{\rightleftharpoons}} \text{ES} \underset{k_{-2}}{\overset{k_2}{\rightleftharpoons}} \underset{\substack{+\\ \text{P}_1}}{\text{acyl—E}} + \text{H}_2\text{O} \xrightarrow{k_3} \text{E} + \text{P}_2$$

The first step represents formation of the Michaelis complex; the second step (k_2) is a transfer reaction forming the acyl enzyme, liberating the first product (P_1), and finally, the hydrolytic reaction (k_3) yields the second product (P_2).

It should be understood that observation of an exchange process such as that discussed above, of itself, does not constitute proof of participation of the enzyme in the manner indicated. It is conceivable that on an enzymic surface, binding of an ester might occur in a manner which makes it susceptible to direct attack by the exchanging group, *i.e.*, the alcohol in the series above. Additional evidence of a more direct nature is required before these exchange studies may be accepted as an indication of participation of the enzyme as a reactant.

Covalent bonding of intermediates to group(s) on the enzyme is not limited to hydrolytic enzymes. Indeed, the first convincing evidence for this was provided by *sucrose phosphorylase* (page 431). This enzyme catalyzes the following reaction.

(*a*) **Sucrose + P_i $\rightleftharpoons$ α-D-glucose 1-phosphate + fructose**

The enzyme also catalyzes each of the following reactions.

(*b*) **Glucose 1-phosphate + $^{32}P_i$ $\rightleftharpoons$ glucose-1-$^{32}PO_4$ + P_i**
(*c*) **Sucrose + sorbose $\rightleftharpoons$ glucose-sorboside + fructose**

These observations are best reconciled by the following reaction mechanism.

(*d*) **Enz—OH + glucose—O—R $\rightleftharpoons$ Enz—O—glucose + HOR**
(*e*) **Enz—O—glucose + HO—Y $\rightleftharpoons$ Enz—OH + glucose—O—Y**

The nature of the group on the enzyme which functions as described is not known; however, by using ^{14}C-labeled glucose in the sucrose, direct evidence has been obtained that a glucosyl-enzyme is formed and that the fructose moiety is not bound to the enzyme. This scheme is also supported by the fact that in both sucrose and the glucose 1-phosphate which is formed in reaction (*a*), the glucose is in the α configuration. Were the phosphorolysis of sucrose a normal metathetical reaction, simply occurring on the enzymic surface, it would proceed as a "back-sided attack" and sucrose, an α-glucoside, should yield β-glucose 1-phosphate.

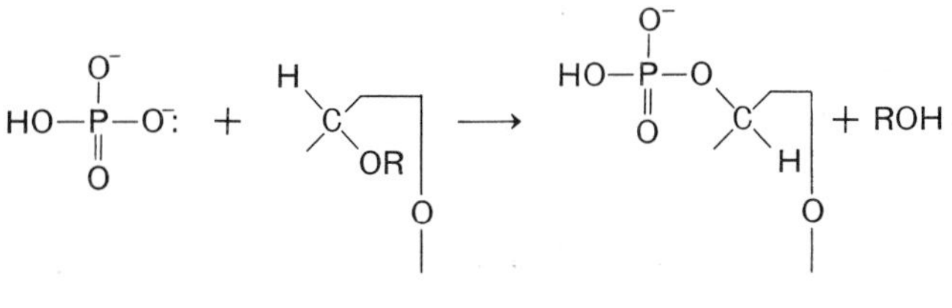

Since the actual product is the α form, the mechanism must consist of two consecutive double displacements, the first of which involves a group on the enzyme, represented below as an oxygen function for convenience since the actual group is unknown.

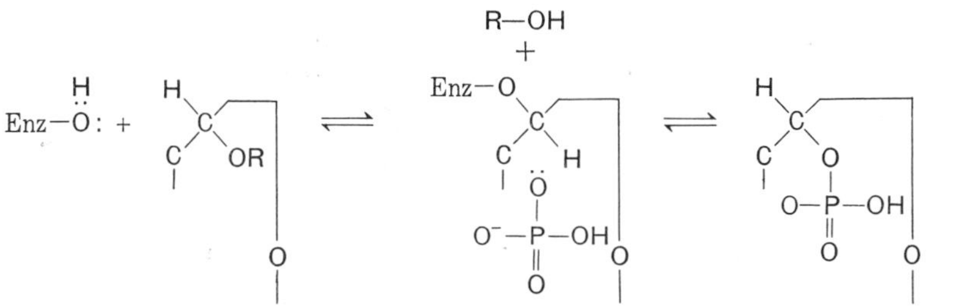

Enzyme + sucrose **β-Glucosyl enzyme + attacking P_i** **α-Glucose 1-phosphate**

The combination of exchange data and the observed stereochemical events led Rose to postulate the scheme shown in Fig. 12.9 for the mechanism of action of *phosphoglucose isomerase* (page 395). In this and many other reactions in carbo-

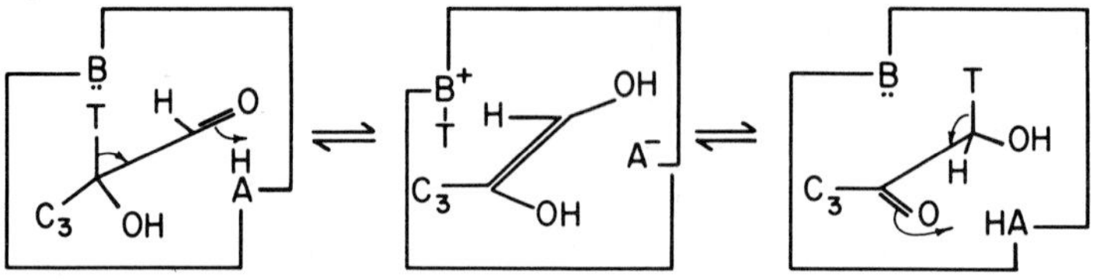

Fig. 12.9. Generalized mechanism by which an enzyme may catalyze formation of a *cis*-enediol from an aldose or ketose. *B* and *HA* are not specifically defined but represent a base and an acid, respectively, on the enzymic surface. This scheme indicates the manner in which a base, fixed in position with relation to the substrate, assures stereospecific migration of a proton, indicated in the figure as *T* (tritium) for labeling purposes.

hydrate metabolism, the hydrogen atom α to a carbonyl group is labilized and removed. In this instance, it was shown by the use of tritium labeling that the hydrogen atom is removed in a specific stereochemical manner and is not diluted by the protons of the water. In the conversion of glucose 6-phosphate to fructose 6-phosphate, a reaction postulated to involve intermediate formation of a *cis*-enediol, the tritium label behaves as shown below (T = tritium, ^3H).

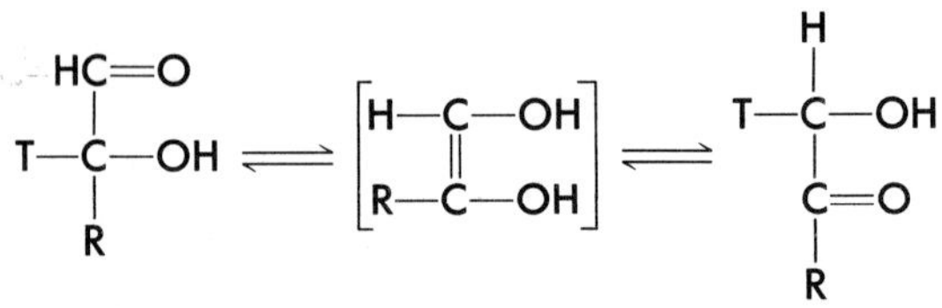

Glucose 6-phosphate | **Postulated *cis*-enediol** | **Fructose 6-phosphate**

Although the actual enzymic groups were not identified, the absolute steric specificity and the failure of the migrating proton to dilute with the medium require postulation of an arrangement involving a nucleophilic agent and an acid, such as that depicted in Fig. 12.9, above.

Although group transfer reactions, such as transpeptidation and transglycosylation discussed above, proceed by transient formation of an intermediate in which the group to be transferred is covalently bound to the enzyme, many transfer reactions seem to occur as metathetical reactions between the substrates at an appropriate locus on the enzymic surface. Thus, *kinases,* each of which catalyzes transfer of a phosphate group from ATP to a specific acceptor, do not appear to form intermediate phosphate-enzyme compounds. *Creatine kinase* (page 853) which catalyzes the reaction

$$\textbf{ATP + creatine} \rightleftharpoons \textbf{ADP + creatine phosphate}$$

will not catalyze exchange of creatine-^{14}C with creatine phosphate or of $AD^{32}P$ with ATP, as might be expected if an intermediate phosphate-enzyme compound were formed. Nevertheless, catalysis of the reaction entails more than merely binding both substrates in such a manner that they can react. A series of observations strongly suggests the concerted participation of both a histidine residue and a sulfhydryl group of the enzyme, as well as a Mg^{++} ion, as depicted in Fig. 12.10.

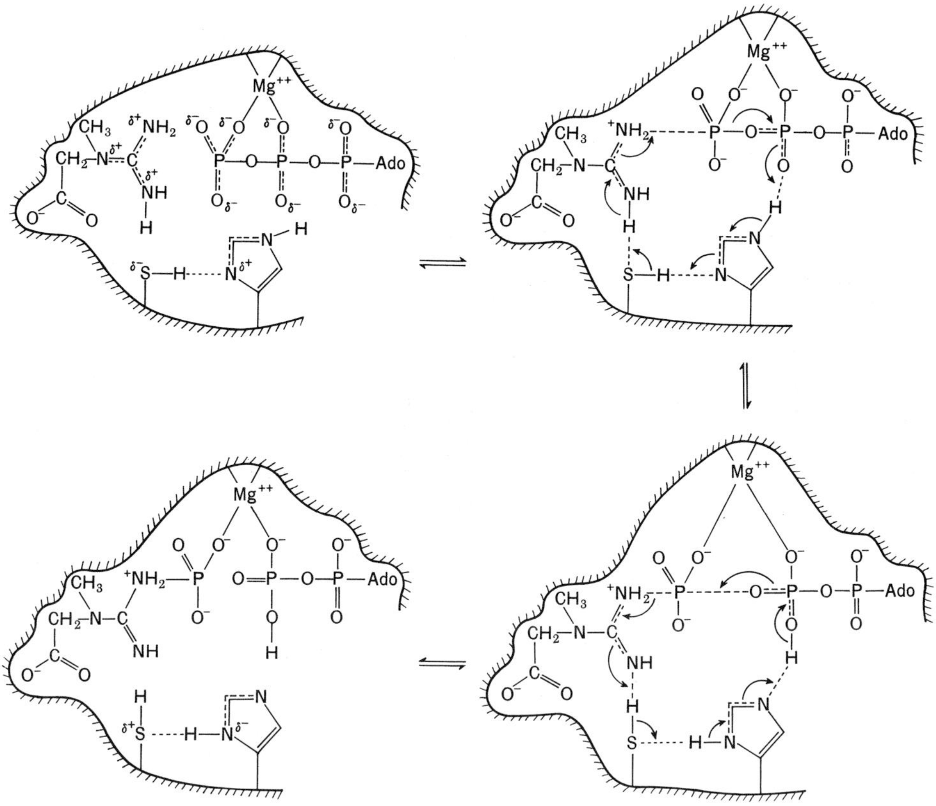

Fig. 12.10. Schematic representation of the transfer of a phosphate group from ATP to creatine, catalyzed by muscle creatine kinase. *Ado* represents the adenosine portion of ATP. The upper left diagram shows the disposition of all groups at the instant of binding of creatine and ATP. The lower left diagram represents completion of the reaction just prior to departure of creatine phosphate and ADP from the enzymic surface.

Lysozyme. The elucidation of the complete, three-dimensional structure of egg white lysozyme by Phillips and his coworkers (page 160) has provided a detailed insight into both the nature of enzyme-substrate interaction and the mechanism of action of this enzyme. The x-ray diffraction analysis has revealed that there is a cleft in the lysozyme molecule, and that the polysaccharide fits into this cleft. The structure was derived mainly from study of the complex with an inhibitor, a trisaccharide containing three N-acetylglucosamine units, although the most sensitive substrates contain alternating units of N-acetylglucosamine and N-acetylmuramic acid (NAM) (Chap. 41).

A complex network of hydrogen bonds and nonpolar interactions obtains between the inhibitor (or substrate) and the enzyme. Only a few of the interactions are indicated in the schematic representation of Fig. 12.11 for a hexasaccharide

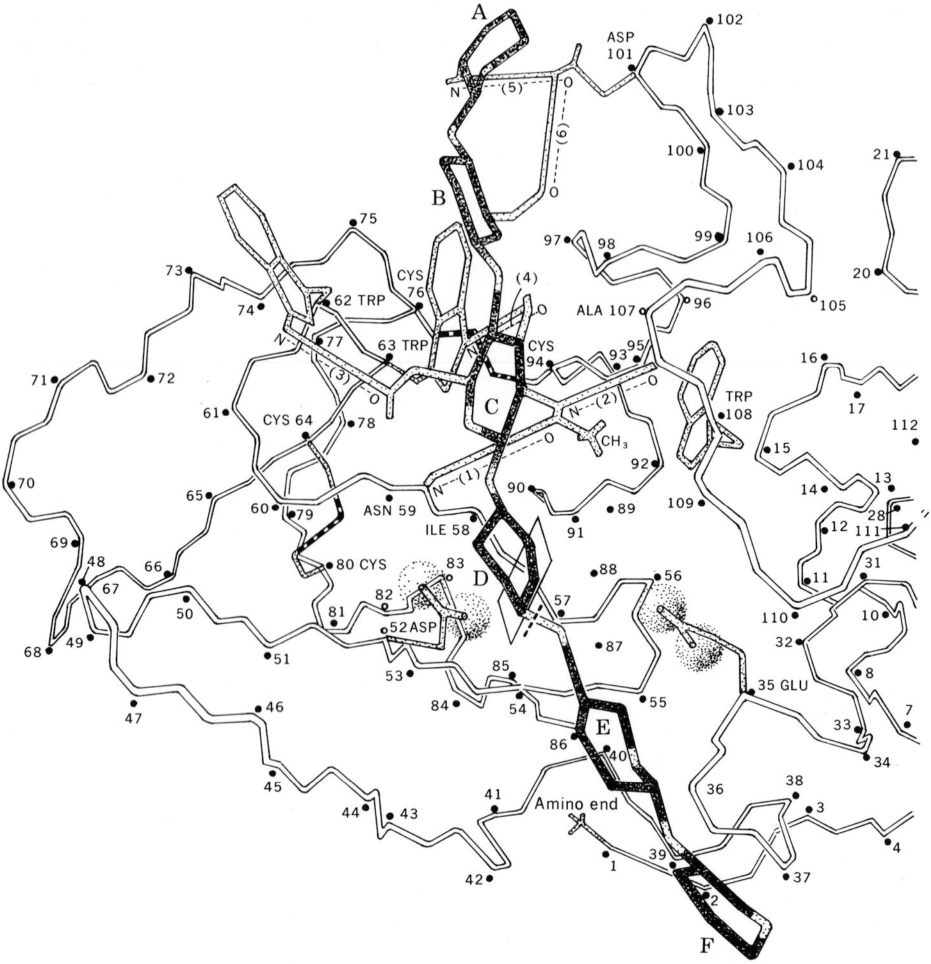

Fig. 12.11. A portion of the model of egg white lysozyme is shown with the main interactions with a hexasaccharide; the six rings of the hexasaccharides are indicated as *A* to *F*. The saccharide is in a deep crevice of the enzyme molecule. (*Adapted from D. C. Phillips, Scientific American,* **215** (Nov.), 78, 1966.)

substrate. The folding of the enzyme has permitted interaction of the side chains of sugar unit C to the peptide nitrogen of residue 59, to the peptide carbonyl of residue 107, and to the side chain of the tryptophan residue 108. Two oxygen atoms of the sugar unit also interact with the side chains of residues 62 and 63, both tryptophans. These are only some of the interactions of sugar unit C! Hence a complete description will not be attempted here. The most important points to be noted are the following: (1) The total number of interactions between enzyme and substrate is, indeed, extremely large, with a bewildering variety of interactions in which participate not only many amino acid side chains but also peptide bond carbonyl and amide groups; these contacts include many types of polar and nonpolar groups. A detailed view of the correctness of the "polyaffinity" of enzyme-substrate interaction is finally at hand. (2) When tri-N-acetylglucosamine is bound to the enzyme, there is a shift in parts of the enzyme structure. The movements are small, of the order of only 0.75 Å., but this is sufficient to offer support for the "induced-fit" theory. (3) Sugar unit D is forced out of its normally stable "chair" conformation into a conformation in which carbon atoms 1, 2, and 5 and oxygen atom 5 are all in a plane. Thus, a "deformation" or "strain" is induced in the structure, producing a weakening of the glycosidic bond between D and E. (4) The reactive groups near this bond are the carboxyl side chains of the aspartic acid residue 52 and glutamic acid residue 35. Residue 52 is in a highly polar environment and is probably ionic under most conditions; in contrast, residue 35 is in a nonpolar area and is likely to be un-ionized. The concerted action of these two groups brings about the cleavage of the susceptible bond between the glycosidic oxygen and carbon atom 1 of unit D.

The course of enzymic events is postulated to occur in the following steps, shown in Fig. 12.12: (1) Lysozyme becomes bound to six monosaccharide units (or residues) of the polysaccharide substrate with deformation at unit D; (2) Residue 35 (Glu) transfers its proton to the glycosidic oxygen, thus producing a cleavage between the oxygen and carbon atom 1 of D, and hence a carbonium ion (C^+) is created when the oxygen has been separated from this carbon atom; (3) The carbonium ion is stabilized temporarily by the interaction with the COO^- of the aspartic acid residue 52 until reaction occurs with an OH^- ion from the solvent, thereby completing the reaction.

Although the above description of the mechanism is still tentative, the picture is in accord with many of the general deductions of the mechanism of enzymic action that were made on the basis of indirect evidence. The chemistry involved is fully consonant with our knowledge of such reaction mechanisms.

In lysozyme, the prime reacting groups are two carboxyls, one acting as a proton donor and the other as an acceptor of a cation. It is an obvious extrapolation to imagine similar functioning in other enzymes, not only of two carboxyls, one of which is protonated, but also of two imidazoles, or of phenols, etc., or of instances in which two different side chains can serve these functions. X-ray diffraction studies are in progress on chymotrypsin, ribonuclease, papain, subtilisin, carboxypeptidase, and a variety of other enzymes. High resolution at 2 to 2.5 Å. is required to provide atomic resolution that will picture clearly the enzyme-substrate or enzyme-inhibitor complexes. From such studies, a deeper insight can be expected into the mechanism of action of these and other enzymes.

Thus, a substantial body of both direct and circumstantial evidence indicates that enzymes do not merely offer convenient surfaces on which defined reactions may occur; functional groups on enzymic surfaces also directly participate in the reaction mechanism. This may take the form of acid or base catalysis, or nucleophilic or electrophilic attack. The substrate may be altered through transient complexes or defined transition states or may become covalently linked to the enzyme during the reaction. These suggestions have indicated that the proposed mechanisms might be examined experimentally by use of model systems, *e.g.*, study of the possible catalytic influence of histidine or histidyl-seryl peptides on the hydrolysis of esters or amides. Perhaps the most striking conclusion from such studies is that in these model systems, catalysis is most effective if the presumed catalytic group is

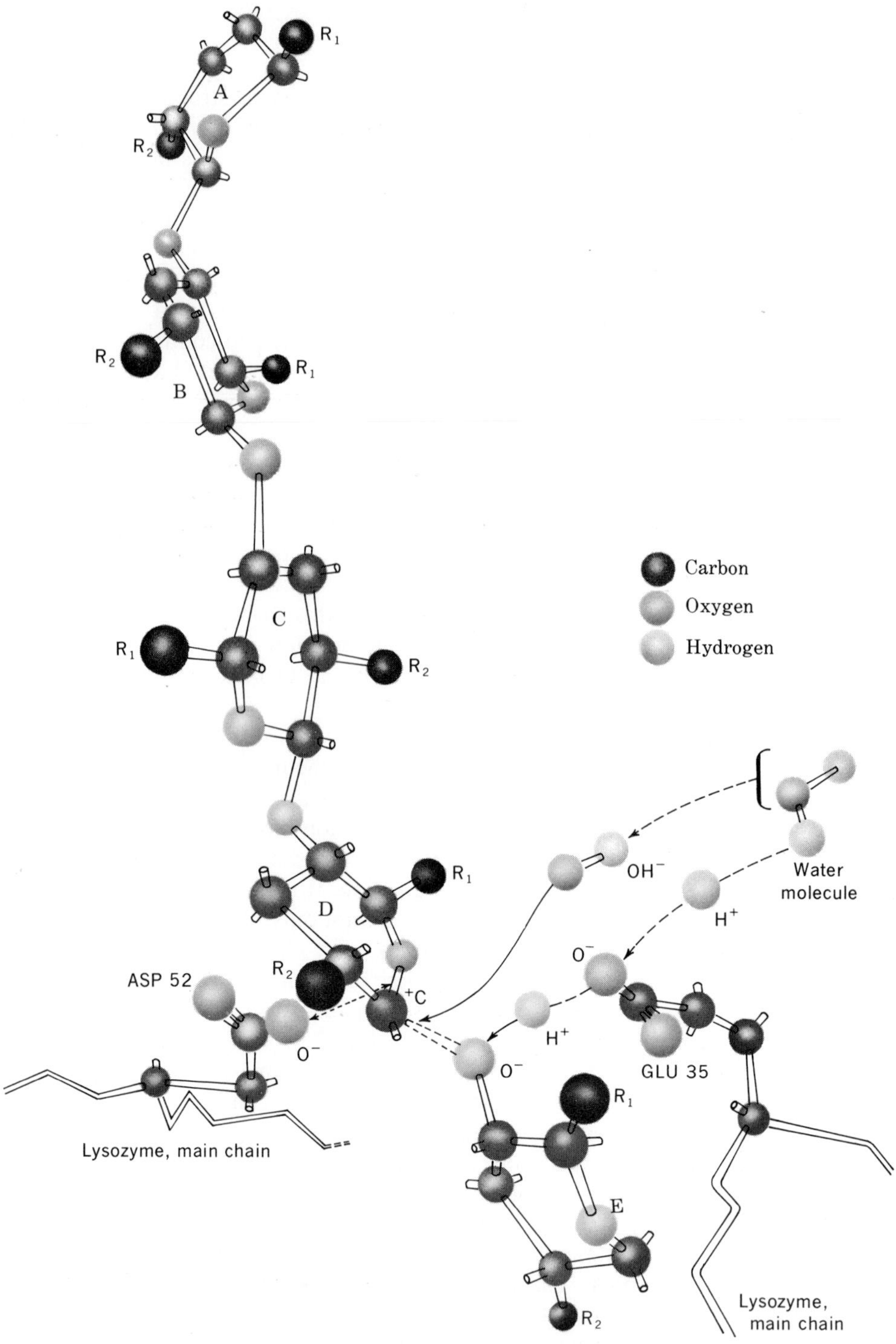

Fig. 12.12. The mechanism of hydrolysis of a polysaccharide (Chap. 4) by lysozyme. The participating groups of the enzyme are the —COOH of glutamyl 35 and the $—COO^-$ of aspartyl 52. The rings of the sugars are designated *A* to *E*. The sugar side chains are $R_1 = CH_2OH$ and $R_2 = —NH—\underset{\underset{O}{\|}}{C}—CH_3$.

(*Adapted from D. C. Phillips, Scientific American,* **215** (Nov.), 78, 1966.)

appropriately located within the very molecule to be hydrolyzed. Thus, regardless of the absolute concentration of the molecule, the catalytic activity of a histidine residue within a synthetic peptide is equivalent to that which might be exhibited by a 5 or 10*M* solution of histidine. In a sense, this is precisely what is accomplished by enzymes; the forces and groups that bind the substrate ensure almost instantaneous reaction with the functional catalytic groups. Although studies of the role and nature of these functional groups have been conducted only for a few years, it is abundantly apparent that these groups do, indeed, exist and that when combined with those aspects of enzymic structure which confer binding specificity, they account for both the nature of the reaction that is catalyzed and the remarkable catalytic efficiency of enzymes.

REFERENCES

Books

Bray, H. G., and White, K., "Kinetics and Thermodynamics in Biochemistry," 2d ed., Academic Press, Inc., New York, 1967.

Brookhaven Symposia in Biology, "Enzyme Models and Enzyme Structure," No. 15, Brookhaven National Laboratory, Upton, N.Y., 1962.

Bruice, T. C., and Benkovic, S., "Bioorganic Mechanisms," vols. I, II, W. A. Benjamin, Inc., New York, 1966.

Florkin, M., and Stotz, E. H., eds., "Comprehensive Biochemistry," sec. III (vols. 12 to 16), Biochemical Reaction Mechanisms, Elsevier Publishing Company, Amsterdam, distributed by American Elsevier Publishing Company, New York, 1964–1966.

Goodwin, T. W., Harris, J. I., and Hartley, B. S., eds., "Structure and Activity of Enzymes," Academic Press, Inc., New York, 1964.

Ingraham, L. L., "Biochemical Mechanisms," John Wiley & Sons, Inc., New York, 1962.

Klotz, I. M., "Energy Changes in Biochemical Reactions," Academic Press, Inc., New York, 1967.

Kosower, E. M., "Molecular Biochemistry," McGraw-Hill Book Company, New York, 1962.

McElroy, W. D., and Glass, B., eds., "The Mechanism of Enzyme Action," The Johns Hopkins Press, Baltimore, 1954.

Nord, F. F., ed., "Advances in Enzymology," Interscience Publishers, Inc., New York. (This is an annual series since 1941 and contains numerous review articles.)

Waley, S. G., "Mechanisms of Organic and Enzymic Reactions," Oxford University Press, London, 1962.

Webb, J. L., "Enzyme and Metabolic Inhibitors," vols. I to III, Academic Press, Inc., New York, 1963–1966.

Review Articles

Atkinson, D. E., Regulation of Enzyme Action, *Ann. Rev. Biochem.,* **35,** 85–124, 1966.

Bender, M. L., and Breslow, R., Mechanisms of Organic Reactions, in M. Florkin and E. H. Stotz, eds., "Comprehensive Biochemistry," vol. II, pp. 1–218, Elsevier Publishing Company, Amsterdam, distributed by American Elsevier Publishing Company, New York, 1962.

Bender, M. L., and Kezdy, F. J., Mechanism of Action of Proteolytic Enzymes, *Ann. Rev. Biochem.,* **34,** 49–76, 1965.

Chance, B., Enzyme-substrate Compounds, *Advances in Enzymol.,* **12,** 153–190, 1951.

Eigen, M., and Hammes, G., Elementary Steps in Enzyme Reactions, *Advances in Enzymol.,* **25,** 1–38, 1963.

Gutfreund, H., and Knowles, J. R., The Foundations of Enzyme Action, in P. N. Campbell and G. D. Greville, eds., "Essays in Biochemistry," vol. 3, pp. 25–72, Academic Press, Inc., New York, 1967.

Jencks, W. P., Mechanism of Enzyme Action, *Ann. Rev. Biochem.,* **32,** 639–676, 1963.

Monod, J., Wyman, J., and Changeux, J. P., On the Nature of Allosteric Transitions: A Plausible Model, *J. Molecular Biol.,* **12,** 88–118, 1965.

Neurath, H., The Activation of Zymogens, *Advances in Protein Chem.,* **12,** 319–386, 1957.

Phillips, D. C., The Three-dimensional Structure of an Enzyme Molecule, *Scientific American,* **215** (Nov.), 78–90, 1966.

Rose, I. A., Mechanism of Enzyme Action, *Ann. Rev. Biochem.,* **35,** 23–56, 1966.

Smith, E. L., and Hill, R. L., Leucine Aminopeptidase, in P. D. Boyer, H. Lardy, and K. Myrbäck, eds., "The Enzymes," 2d ed., vol. IV, pp. 37–62, Academic Press, Inc., New York, 1960.

Westheimer, F. H., Mechanisms Related to Enzyme Catalysis, *Advances in Enzymol.,* **24,** 441–482, 1962.

13. Introduction to Metabolism

Regulatory Mechanisms. Experimental Approaches to the Study of Metabolism

Living organisms, which constitute open systems in continuous exchange with the external environment, engage in continuing chemical change that comprises their *metabolism.* The magnitude of this metabolism and of the tasks involved may be appreciated from an over-all consideration of metabolic phenomena in microorganisms and in man.

A culture of bacteria, *e.g.*, *Escherichia coli,* can double in number every 20 min. in a simple medium containing only glucose and inorganic salts. The components of the medium are depleted, but little is added to the medium by the growing culture. Each cell then contains a normal complement of approximately 400 molecules of each of about 2,500 different proteins, 10 to 300 million molecules of each of about 1,000 kinds of organic compounds, and a variety of nucleic acids. It is evident that these cells have engaged in a remarkable biosynthetic performance, the net synthesis of each of the cell constituents having proceeded at the rate required to ensure harmonious growth of the culture with no significant over- or underproduction of any component.

No less impressive are the metabolic activities of a human adult who maintains constant weight for perhaps 40 years, during which he processes a total of about 6 tons of solid food and 10,000 gal. of water, yet remains constant in both weight and body composition. Again it will be evident that remarkable control is exercised over the multitudinous metabolic processes and the energy-yielding reactions which make them possible.

These metabolic accomplishments of both simpler and more complex organisms are achieved even in the face of disturbing conditions in the environment. The fact that an organism can maintain a normal, constant internal state, termed *homeostasis,* despite the numerous complex metabolic reactions which it performs and the continuing alterations in its environment, is due to the sensitivity of specific regulatory mechanisms. Indeed, the failure of one or more of such mechanisms to function normally leads to the metabolic aberrations underlying various diseases.

An understanding of metabolism requires knowledge of the chemistry of the participating molecules (metabolites), the reactions to which they are subjected, the enzymes participating in these reactions, and the regulatory mechanisms which determine the rates of the sequential reactions by which any given metabolite (A) is converted to a second (B). Such a series of steps comprises a metabolic pathway, and the operation of the manifold *metabolic pathways* and their integrative functioning constitute *metabolism.*

It is the purpose of this chapter to indicate some general aspects of metabolic processes which regulate metabolism, and the major experimental tools available for the study of metabolic processes. Some chemical species important in metabolism and aspects of the mechanism of enzymic action have been presented in previous chapters. Only in the last few years has knowledge of metabolism, the structure of the enzymes involved, and the mechanism of enzymic action become sufficiently detailed to permit examination of the devices which regulate metabolism.

REGULATION OF METABOLISM

If metabolism is to proceed normally, the rate of each individual reaction must be controlled in a manner that permits it to proceed at a rate commensurate with the demands of the cell, *e.g.*, for the required building blocks for synthesis of proteins and nucleic acids, for structural components of walls and membranes, for cell division, or simply to provide energy required for endergonic processes. Hence, the metabolic apparatus is controlled in a manner that balances supply and demand under diverse circumstances.

A prime function of control mechanisms is to ensure that energy is not wasted and to permit synthesis only of enough of a particular metabolite to meet the immediate needs of the cell. Whereas mammals and plants can store reserve energy in the form of carbohydrate or lipid, microorganisms do not, and their immediate energy needs must be met from their environment or at the cost of intracellular constituents. No organism has a significant reserve store of amino acids, although seeds of plants and eggs of many animals can store proteins which, on hydrolysis, provide amino acids for developing embryos.

Regulatory phenomena in the organism occur at the cellular, organ, and organism levels. These regulatory processes may involve communication between anatomical compartments, *e.g.*, between various organs via the nervous or the circulatory systems, between extra- and intracellular spaces, or between the cytoplasm and some intracellular structure such as a mitochondrion or the nucleus. Alternatively, regulation may be the simple consequence of the chemical environment of enzymic systems.

The integrative metabolic regulation of the diverse organs and tissues in multicellular organisms is achieved by way of two channels of communication, the nervous and the circulatory systems. Physical connecting pathways for integrative functioning are provided by anatomically distinguishable tracts of the nervous system. The circulatory fluids transport substances between and among the various organs and tissues of the organism. Such substances may be nutrients, products of metabolism, waste products to be eliminated, or "humoral" agents which effectively alter rates of specific metabolic processes on reaching their target organs. These agents are the *hormones;* their collaborative, integrative functioning with the nervous system in the mammalian organism is considered in Part Five of this book. These regulatory mechanisms are of prime significance at the level of inter-organ functioning.

Control mechanisms which operate at the enzymic level may be evident in two aspects of enzymic functioning: (1) the inherent properties of the enzyme, and (2) the metabolic pathway whose rate is determined by the enzyme. The functioning of the enzymic system as a control site may in turn be affected by (1) the quantity of the enzyme present, as well as the capacity of the cell to synthesize new required enzyme in response to the presence of substrate; (2) the influence of products of a series of enzymic reactions on the rate of sequential reactions in a metabolic pathway; and (3) allosteric effects in the enzymic system. This last phenomenon and its significance for rates of enzymic reactions have been considered in Chap. 11.

General Aspects of Metabolic Reaction Sequences

1. If a metabolic pathway is to provide a substance for use by the cell, the pathway must be essentially irreversible, *viz.*, it must proceed with a substantial release of free energy. Although there usually are freely reversible reactions in a pathway, it is the essentially irreversible steps which render the entire process unidirectional.

2. There are numerous instances in which, physiologically, two metabolites are interconvertible. In almost all instances in which such interconversions are physiologically meaningful, *i.e.*, if in metabolism there is a requirement for conversion of A to B and at other times of B to A, this is made possible by reaction sequences which are totally or partially independent (Fig. 13.1) and each of which, overall, proceeds with a large, negative ΔF. If these two reactions are also independently controlled, each can be utilized for differing metabolic purposes.

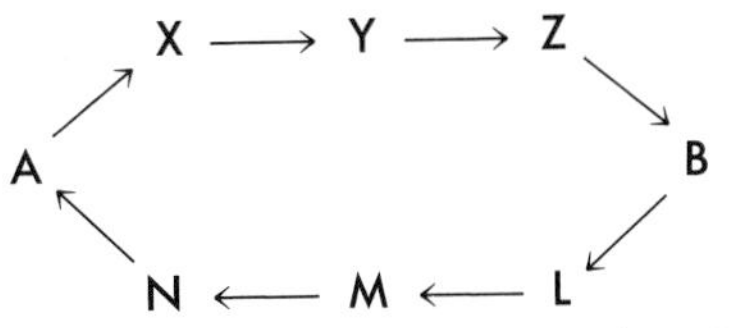

Fig. 13.1. Independent reaction sequences for conversion of A to B and B to A.

3. The synthetic sequences which lead to formation of required end products, such as amino acids, purine and pyrimidine nucleosides and nucleotides, and steroids, begin with the simpler cell materials that arise from the metabolism of the ubiquitous carbohydrates or fatty acids.

4. In each metabolic pathway there occurs a "committed step," the reaction which produces the first metabolite which has no role in metabolism other than to serve as an intermediate in the biosynthesis of the end product of the sequence of reactions. This may be identified as the reaction step $c \rightarrow l$ in the sequence shown in Fig. 13.2. In most instances, the committed step proceeds with a large loss

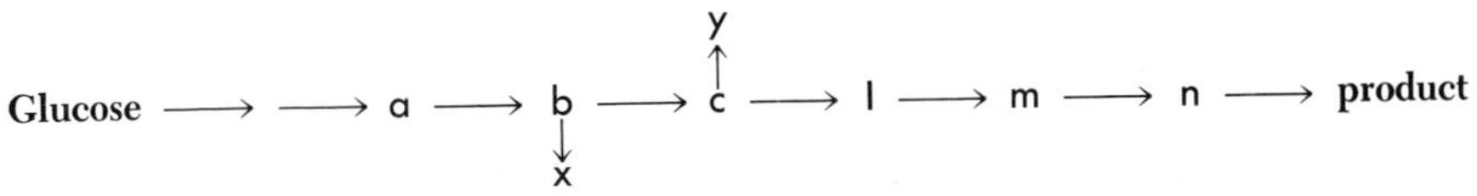

Fig. 13.2. Formulation of a synthetic pathway from glucose to an end product. a, b, and c may be intermediates in glucose metabolism common to many metabolic pathways. The reaction $c \rightarrow l$ is the committed step since l, m, and n are intermediates with no metabolic role other than to serve as precursors of the final product of the pathway.

in free energy so that the reaction is essentially irreversible. It is apparent that a metabolic control intended to restrict the formation of end product would function most satisfactorily were it to govern the committed step. Inhibition later in the sequence would result in accumulation of intermediates such as *l*, *m*, or *n* which have no alternative roles in metabolism and, hence, might be deleterious to the cell were they to increase significantly in concentration. Further, biosynthesis of unneeded compounds would result in wasteful use of energy.

Control of the committed step in metabolic reaction sequences has proved to be an almost invariant attribute of metabolism; control of intermediary steps is exercised less frequently. Although instances of all four types of regulation of enzymic activity to be cited below have been observed in mammalian systems, much of the available information has been derived from the study of bacteria. The latter lend themselves readily to investigations of control mechanisms and participate in diverse reaction sequences which do not occur in mammals, such as synthesis of all the α-amino acids present in proteins and the wide variety of sugars found in bacterial cell walls. Moreover, the metabolism of bacteria is characterized by the prominence of biosynthetic activity associated with their relatively brief generation times. Studies of mammalian metabolism, however, have revealed increasing numbers of examples of control mechanisms, and it may be anticipated that all metabolic reaction sequences are governed in this manner. It also seems likely that many instances of metabolic disease will prove to be aberrations in the operation of the fine controls of metabolism. Thus, whereas genetic lack of an enzyme renders impossible the process for which that enzyme is responsible, and is tolerable only for relatively unessential metabolic phenomena, complete failure of any of the cardinal processes of metabolism would be incompatible with life. However, a minor defect in the regulatory process governing a central metabolic process could be manifested as disease. Undoubtedly, many disorders of metabolism resulting from changes in the rate controlling aspects of the structure of enzymes will be identified in the future.

Mechanisms Utilized in the Regulation of Metabolism. In a general way, five types of control devices are of significance in effecting controls in single cells. Each of these mechanisms may also be observed in mammalian cells. However, in a multicellular animal, one group of cells, *e.g.*, the liver, may produce a compound which is then utilized elsewhere, *e.g.*, in skeletal muscle. If supply is to be regulated by demand, the producing cells require information concerning the metabolic state of the consumers. This is provided by hormones which can influence the rates of diverse metabolic processes by affecting the rate of operation of one of the several general mechanisms considered below.

1. *Regulation of the rate of entry of nutrients into cells.* A most important set of controls are those which govern the passage of nutrients across cell membranes. In most organisms, the cell membranes utilize two processes which regulate transfer of substances from extra- to intracellular compartments. One is a "*selective permeability*" which may be based upon the properties of the nutrient, including size, solubility, charge, etc. The second is "*active transport,*" in which energy is expended to transport a nutrient against a concentration gradient so that it accumulates either intra- or extracellularly. The entrance of individual sugars into some cells may be made possible by the presence of a specific protein in the cell membrane

that facilitates passage of the sugar across the membrane in the direction of the concentration gradient. These proteins have been termed *permeases;* the best known is the galactose permease of *E. coli.* A formally similar, facilitated transport system for glucose exists in the membranes of all mammalian cells and indeed is a major site of insulin action (page 448). Diverse amino acids, sulfate, and phosphate are brought into the cell by active transport systems. Many foreign compounds are restrained from entry by the barrier imposed by the cell membrane. Maintenance of the intracellular [K^+] appears to be accomplished by extrusion of sodium from the cell rather than by selective entry of potassium. It is apparent that if the activity of these cell transport systems is in some manner correlated with cellular metabolic demands, the cell is in a general way assured of a constant, normal composition. Details of such mechanisms, however, are unknown. An example of the response of transport mechanisms to physiological circumstances is seen with *E. coli.* When grown in a mixture of glucose and lactose, the biosynthesis of galactose permease by *E. coli* remains repressed and the organism preferentially utilizes glucose. At a higher level, the hypothalamic control of human appetite, influenced significantly by the level of blood glucose, may serve a similar function.

2. *Repression of enzyme synthesis by an end product of a metabolic sequence* (page 277) *is a coarse control of metabolism* (Chap. 29). If, for example, the synthesis of the group of enzymes required for histidine formation is repressed by the presence of histidine in the medium of a bacterial culture, newly formed cells will contain decreasing quantities of these enzymes but the enzymes initially present at the time of histidine addition are destroyed slowly, if at all. Similarly in mammalian systems, in the presence of a repressor, several days may be required for the enzymic level to diminish significantly as a result of its own continuing degradation.

An interesting problem arises in the case of branched metabolic pathways which result in synthesis of more than one desired product, *e.g.*,

$$\mathrm{A} \longrightarrow \mathrm{B} \longrightarrow \underset{\searrow \mathrm{D}}{\mathrm{C}} \longrightarrow \mathrm{E} \longrightarrow \mathrm{F} \longrightarrow \mathrm{G}$$

where *D, E,* and *G* might all be useful end products. In bacteria, various solutions to this problem have been observed. Thus there may be several, genetically independent enzymes for the step $B \rightarrow C$ which can be repressed by sufficient concentrations of *D, E,* and *G* respectively. Again there may be one enzyme for this step repressible by one of the products and one which is not repressible by any.

3. *Induction of formation of an enzyme by a nutrient* which is its substrate may be extremely rapid in bacteria, occurring within minutes of exposure of the cell to the appropriate inducer. An increase in the cell complement of a specific enzyme represents a relatively coarse metabolic control in multicellular organisms. In known instances, increase in cellular enzyme concentration in mammals occurs over several hours or days in response to the inducer.

4. *Inhibition of an enzyme* already present can serve as the fine control mechanism for regulating metabolic events since this permits instantaneous response to changing intracellular environment. Feedback inhibition of the enzyme responsible for the committed step in a given pathway, by the product of the pathway, is the most common form of control of biosynthetic processes. The advantages to the cell

of such self-regulation will be evident. The phenomenon of allosteric modification of enzymic function has been considered earlier (page 242*ff.*). Of particular significance is the typical sigmoid curve relating velocity and either substrate or modifier concentration (Fig. 11.6, page 245), since at low concentration little effect will be evident whereas in the intermediate concentration range, below saturation, relatively small changes in concentration result in large changes in velocity. Hence, allosteric control can serve efficiently as a basis of a regulatory mechanism for reaction rates.

5. *Stimulation by a metabolite of enzymes which function in the pathways which achieve its utilization* is a type of metabolic control which has been observed in mammalian as well as bacterial cells.

Examples of these control devices in the regulation of metabolism will be encountered in the succeeding chapters of this book.

EXPERIMENTAL APPROACHES TO THE STUDY OF METABOLISM

Many techniques have been developed to ascertain what chemical reactions are occurring in living systems, to measure the effects of assorted variables upon these reactions, and to determine the rates at which these reactions occur in health and in disease. In all experimental science, in contrast to purely observational science, it is necessary to apply some disturbing condition to the system under observation and to measure the effect of that disturbance. In biochemistry, this means that one must experimentally modify or alter the organism or tissue being studied. Present understanding of metabolism has been obtained from experiments in which the applied disturbance has ranged from the addition of a minute amount of an isotopically labeled nutrient to the diet of an intact animal to experiments in which the animal was sacrificed, an organ removed, an enzyme isolated from the organ, and the kinetics of the reaction catalyzed by the enzyme studied.

LEVELS OF ORGANIZATION

In *the intact organism* the constituent substrates and enzymes bear definite geometrical and chemical relationships to each other. The rates at which materials are delivered to and removed from a given tissue are affected or modified by the activities of other tissues. Many processes are regulated by products discharged into the blood stream at remote points. The various membranes of the body serve to compartmentalize materials and to limit the rates at which substrates enter or leave certain compartments. Hence, one may anticipate that the rate at which the normal liver *in situ* performs various reactions may differ from the rate at which the same organ, isolated and perfused, sliced or minced, conducts these processes. The intact animal offers many experimental difficulties, chiefly because many processes are occurring simultaneously. There are, however, several techniques for exploring what is happening in the intact animal. These include administration of diets deficient in one or another normal nutrient, addition by dietary or parenteral route of either an abnormal material or an excess of a normal material, or introduction into the animal of some metabolite labeled in such a fashion that the subsequent distribution of the label can be studied. A variety of surgical procedures have also been employed which give the experimenter access to some previously inaccessible tissue or body fluid. These include the establishment of fistulas from various segments of the gastrointestinal tract, the biliary tract, or the lymphatic system, as well as procedures that permit sampling of blood at various points in the circulatory system. Study of the effects of partial or complete excision of one or another organ is also useful, although here the magnitude of the derangement is somewhat greater.

The next level of organization is the perfused, isolated organ or extremity. In such studies, one is apt to lose certain of the regulatory mechanisms, *e.g.*, hormonal and/or nervous control, which operate upon the organ or the extremity in its normal locus. However, such experiments have been widely employed to determine the organ in which a particular reaction occurs and the products that may be formed after a given precursor is added to the perfusing fluid.

The next level of organization is the sliced organ. Liver, kidney, brain, and other tissues, cut into slices approximately 50 μ in thickness, offer to the bathing fluid a surface sufficient to permit adequate exchange of nutrients and waste products so that viability is maintained for several hours. A fraction of cell membranes is incised in the slicing operation, but most cellular constituents remain within cells. Although reaction rates within the cell of a sliced liver, shaken in a crudely simulated extracellular fluid, may deviate from normal rates, this technique has proved extremely useful because of the simplicity of subsequent experimental operations. There is complete control, not only of the organ and the previous nutritional status of the animal, but also of the composition of the bath fluid and of the gas phase with which it is equilibrated. By judicious addition of precursors to the bath fluid, detailed pathways of metabolism may be delineated.

A yet lower level of organization remains after mincing the tissue mechanically. After rupture of the cell membranes, many relationships which obtained between parts in the normal cell no longer exist. This procedure has proved particularly useful in studies designed to determine the location of metabolic processes within the cell as it permits separation of various particulate structures of the cell—nuclei, membranes, mitochondria, and microsomes (page 285)—by differential centrifugation. By addition of suitable substrates, the presence or absence of a given enzymic activity in these cellular fractions may be ascertained.

The lowest level of organization is the enzyme in solution. A purified enzyme is, in general, a prerequisite to complete understanding of the reaction which it catalyzes. It permits studies of the kinetic and regulatory properties of the enzyme. The thermodynamic characteristics of the reaction and the equilibrium constants can be established, and the effects of inhibitors and activators can be ascertained.

It is by synthesis of information obtained with these and other techniques that existing understanding of metabolism has been obtained. The ultimate objective of this field of research is to obtain as complete an understanding as possible of the individual metabolic reactions, their rates, the factors controlling these rates, and those deviations from these rates which are termed disease.

TECHNIQUES USED IN STUDIES OF METABOLISM

Balance Studies. Determination of the amount of a substance ingested and the quantity of the same compound or of its metabolic products excreted has long been used to study metabolic processes in vivo. This permits construction of a balance sheet between intake and output and makes possible inferences regarding the level and nature of the metabolic activity involving one or another dietary constituent.

The balance technique has been particularly informative, albeit limited, in studies of a variety of substances, *e.g.*, proteins, minerals, etc.

Laboratory Animals. A large variety of animal species is employed in studies of metabolism. The nature of the problem may determine which particular species is most useful. Thus valuable nutritional data have been obtained with mice, rats, dogs, guinea pigs, chickens, and monkeys. Studies requiring large numbers of animals have generally employed mice and rats, which have the advantages that they breed well under laboratory conditions, are omnivorous, and have a rapid rate of growth and development; furthermore their relatively short life spans, *viz.*, under three years, permit study of several generations of animals. Inbreeding is rapidly possible in these species; hence, pure strains may be established, and the significance of genetic factors examined. Growth studies with the rat have provided much of the data on which the science of nutrition is based (Part Six). However, data obtained with one species occasionally differ from those obtained with another species in similar experiments. Studies on monkeys and other primates are often conducted to secure data most likely to contribute to understanding of human metabolism.

Naturally Occurring and Induced Metabolic Alterations. Surgical extirpation of or severe damage to an organ *in situ* is one of the oldest approaches to studies of metabolism. For example, the role of the pancreas in the etiology of diabetes mellitus was discovered in studies of the surgically depancreatized dog. Similarly, knowledge of the metabolic role of the liver has derived from observation of hepatectomized dogs and of human beings with various liver disorders.

Several metabolic disorders were termed by Garrod "inborn errors of metabolism" because each is present throughout life and is hereditary. The inborn errors listed by Garrod in 1908 were cystinuria, alkaptonuria, congenital porphyria, pentosuria, congenital steatorrhea, and albinism. Many additional metabolic alterations of man have been described which have a genetic basis (Chap. 30). Individuals with such conditions have provided experimental approach to problems of metabolism, particularly the elucidation of metabolic pathways. The hereditary nature of these errors of metabolism indicates that they arise as a consequence of the development of mutant strains. Thus the human mutants resemble microorganisms in which existence of a mutation can be discerned as a consequence of the organism's inability to conduct a particular metabolic transformation or group of transformations (Chap. 28). Such bacterial mutants have been singularly useful in elucidating biosynthetic pathways.

Most abnormalities in metabolism, whether induced by abnormal diet, by drug, by poison, or by disease, are reflections of changes in the specific rates of one or more body processes. These changes in rates may in turn result in changes in body composition, but it is only from a study of rates that changes in composition can be fully understood. For example, the concentration of glucose of the blood in the normal animal is relatively constant. It is, however, subject to a rapid turnover, and the constancy of its concentration is the result of a complex series of mechanisms which serve to balance the rate of glucose production against the rate of glucose utilization. Under various abnormal circumstances, the concentration of glucose in the blood is found to rise, and indeed such a rise is a cardinal manifestation of diabetes mellitus. Clearly this rise in concentration might result from an abnormal decrease in the rate of glucose utilization or from an increase in the rate of glucose production. By appropriate application of the isotopic technique (page 287), problems of this kind may be analyzed in terms of individual rates and a better insight gained into the nature of the pathological processes.

Experimental Fistulas. Application of surgical techniques can provide experimental animals with suitable fistulas which permit sampling of blood or lymph entering and leaving particular organs, as well as making accessible the secretions of the various segments of the gastrointestinal and biliary tracts. The use of fistulas affords an in vivo approach to the study of reactions conducted by a particular organ or tissue.

Catheterization of Blood Vessels. The contribution of a particular organ to metabolism can be evaluated by comparison of the chemical composition of its arterial supply and venous drainage. Intravascular catheterization also permits withdrawal of blood samples and adding specific substances to the arterial flow. This is the in vivo counterpart of the isolated organ perfusion approach described below and complements the fistula technique.

Organ Perfusion. By perfusion of an isolated organ in vitro it is possible to introduce a known substance and, by analysis of the fluid emerging from the organ, learn the metabolic fate in that organ of the perfused compound. In practice, the organ is placed in a closed system in which a suitable oxygenated fluid circulates under positive pressure. The circulating medium may be defibrinated blood or whole blood containing a suitable anticoagulant (Chap. 31) or erythrocytes suspended in solutions of physiologically isotonic mixtures of salts designed to simulate normal plasma in pH, ionic strength, and ionic composition. This approach, which has been extensively used with liver, heart, kidney, and small intestine, has contributed significantly to knowledge of the metabolic roles of these organs. Inferences drawn from the data are obviously tempered by the artificial conditions under which the tissue is operating.

Histochemistry. The procedures of histochemistry are designed to map the cellular distribution of compounds, enzymes, and metabolic systems. This is achieved by treating a suitably prepared tissue section with a solution containing a reagent which reacts specifically with some tissue component. The colored reaction product remains at its site of formation, thus providing a direct visualization of the location of the substance in question. In this manner it has been possible to determine the distribution of various important metabolites and enzymes and to study variations in the relative quantity and distribution of these substances in diverse normal and altered metabolic states. The techniques of histochemistry also have been applied to separated cellular components, *e.g.*, nuclei and mitochondria, obtained by differential centrifugation (page 285).

Metabolism of Tissue Slices, Minces, and Broken Cell Preparations. In 1912, Warburg initiated studies of the metabolism of tissue slices, using the manometric technique. The apparatus, invented 10 years earlier by Barcroft and Haldane, measures quantitatively changes in gas volume or pressure. The Warburg procedure permits study of the metabolism of small segments of a particular tissue or organ by determining manometrically the rate of oxygen utilization and carbon dioxide production (the respiratory quotient, page 293). In addition, it is possible to examine the rate of utilization of a foodstuff or substrate added to the medium in which the tissue has been placed, and to determine the nature and amount of metabolites produced during the experimental period. This technique has been extended to minces of tissue and to broken cell or cell-free preparations of tissues (see below), as well as to tissue slices.

Biochemical Cytology. The diverse chemical mechanisms upon which cellular functions are based are not randomly dispersed within the cytological units, *e.g.*, nucleus, cytoplasm, etc., of cells, but are specifically localized. This intracellular organization of structural and chemical units permits sequences and cycles of reactions to occur coupled to one another in a manner that could not be achieved if participants and catalysts were randomly distributed throughout the cell. Biochemical cytology has contributed explanations for the special biological functions of discrete structural units. Knowledge has derived from use of light, phase contrast, and electron microscopy, from application of histochemical techniques and from disruption and fractionation of cells, with physical and chemical examination of the components obtained.

The structure of the cytoplasmic membrane is of great importance inasmuch as it is the barrier between the external and internal environments of the cell. The active intervention of the cytoplasmic membrane in the movement of substances in and out of the cell is strikingly exemplified by the capacity of the membrane of mammalian cells, for example, to distinguish between sodium and potassium ions. The membrane facilitates potassium entry into the cell and utilizes metabolically derived energy to discharge sodium ions into the surrounding medium. Furthermore, the plasma membrane engages in "active transport" processes (page 780*ff.*) specific for amino acids, sugars, and other small molecules. In bacteria, the plasma membrane is also the site of oxidative metabolism.

The variety of functions of the plasma membrane suggests that there is a diversity of membrane composition not only among cell types but also in different areas of the same membrane, *i.e.*, a mosaic of membrane constituents which contributes to the selective specificity of the membrane as a whole, as well as making possible diverse chemical specificity in different areas of the same membrane. Analysis of the membrane reveals the presence of proteins, lipids, and carbohydrates. The lipids include phospholipids, glycosphingolipids, and cholesterol; sialic acids are present among the carbohydrate components. Combined x-ray diffraction and electron microscopic examination of membranes of specialized cells, such as those of nerve fibers, has suggested that their structure is basically a bimolecular leaflet of lipid molecules covered by two protein layers in which many types of proteins may be distributed. The total thickness of the membrane has been estimated to be approximately 75Å. It seems likely that, in many instances, the proteins of the inner and outer layers differ. This would facilitate the establishment of chemical gradients across the membrane. In addition, many cells have a mucoprotein covering which varies in thickness among cell types and probably modifies permeability as well as playing a role in pinocytosis.

Many cell types have a system of membrane-bound channels which may extend from the nuclear membrane to the cellular membrane, thus permeating large regions of cytoplasm. This complex system of membranes, the *endoplasmic reticulum,* consists of lipoprotein membranes in the form of tubules, vesicles, or connected vesicles (*cisternae*). Two types of endoplasmic reticulum have been described, depending on the number of attached particles of ribonucleoprotein, the *ribosomes*

(Chap. 29). The *smooth-surfaced reticulum* has relatively few or no particles; the *rough-surfaced reticulum* which is lined with ribosomes was originally termed the *ergastoplasm.* The endoplasmic reticulum functions in a number of important processes, *e.g.*, the synthesis of steroids (Chap. 22), and as a precursor for other membrane systems. Ribosomes attached to the endoplasmic reticulum function in protein synthesis (Chap. 29).

The largest structure within cells in general is the *nucleus,* visible under the light microscope without staining because its density and refractility differ from those of the surrounding cytoplasm. Within the nucleus are chromatin granules, dense aggregates of DNA, and lesser quantities of RNA. The nucleus usually contains a distinct body, the *nucleolus,* which is rich in RNA and devoid of DNA. In the electron microscope, the nucleus is seen to have two membranes. The inner one may be considered the true nuclear membrane; the intermembrane space appears frequently to be continuous with the channels of the endoplasmic reticulum.

Within the cytoplasm are particles or organelles of varying sizes; the largest and most dense of these are the *mitochondria,* which by electron microscopy appear as minute spheres, rods, or filaments enclosed by a double membrane. A number of highly important integrated enzymic systems, *e.g.*, the oxidative and respiratory enzymic systems (Chap. 16), are concentrated in cell mitochondria. The inner membrane of the mitochondrion is folded into villi or crests (*cristae*). A correlation appears to exist between the complexity of mitochondrial structure, *i.e.*, the number of cristae, and the level of oxidative activity of the cell, *e.g.*, kidney cells compared to fibroblasts. Within differentiated cells, mitochondria are frequently aggregated within that area in which the demand for energy is maximal, *e.g.*, in muscle as rings around the contractile myofibrils; in nerves at the synapse; in sperm at the neck of the cell at the point at which the head joins the whiplash tail; and in cells of the renal tubule, in the folds of the absorbing cell surface.

A *lysosome* is a cytoplasmic body surrounded by a single membrane; these are present in many cell types but are especially abundant in liver and kidney. Lysosomes contain a number of hydrolytic enzymes and have been postulated to be the site of intracellular digestion accompanying pinocytosis and phagocytosis, as well as contributing to post-mortem autolysis.

The *Golgi apparatus* is a cytoplasmic organelle revealed in the electron microscope as an irregularly arranged, interlacing network of fibrils and of small and large vesicles. This structure appears to play an important but unknown role in the secretory activities of cells; secretory products may be seen aggregated as granules within Golgi vesicles for transport to the cell surface. Other inclusions visible in the phase contrast or electron microscope may be present in cytoplasm. For example, *pigment granules* are found in specialized cells.

Centrifugal Separation of Subcellular Particles. In characterizing chemically and functionally diverse submicroscopic cellular particles, it is useful to separate them by physical means. Cell membranes may be disrupted by subjecting a suspension of cells in a suitable isotonic medium, *e.g.*, 0.25*M* sucrose, to ultrasonic vibration, or by utilization of a mechanical blendor, such as a Waring blendor or a hand- or motor-driven pestle rotating in a mortar in which the tissue, in suitable medium, is subjected to a shearing force that disrupts cells and yields broken cell preparations, loosely termed "homogenates." These have few unbroken cells and are suspensions of nuclei, mitochondria, microsomes, other cellular organelles, and disrupted cellular membranes, as well as the soluble phase of the cytoplasm. Differential centrifugation of the broken cell suspension at low temperatures separates the individual particulate fractions of the suspension. Table 13.1 indicates the usual fractions obtained by this technique, as well as the distribution of several important enzymic systems and reaction pathways in these fractions.

Table 13.1: Distribution of Typical Cellular Components Achieved by Fractional Centrifugation of a Broken Cell Preparation of Liver

Fraction	Centrifugal field for separation, × gravity	Time of centrifugation, min.	Some enzymic or other activities
Cellular debris; nuclei; membranes	1,000–6,000	10	Nucleic acid synthesis (nuclei)
Mitochondria; lysosomes; microbodies	10,000–15,000	30	Electron transport; oxidative phosphorylation; citric acid cycle; fatty acid oxidation; amino acid oxidation; urea synthesis; hydrolases; fatty acid elongation
Microsomes; ribosomes; reticulum; Golgi apparatus	100,000	60	Protein synthesis; hydroxylating systems; cytochrome b_5; hydrolases; glucuronyl transferases; mucopolysaccharide, phospholipid, triglyceride, and steroid synthesis; steroid reductases; phosphatases
"Soluble fraction," or supernate	. . .	. . .	Glycolytic system; hexose monophosphate pathway; glycogen synthesis; glycogenolysis; fatty acid synthesis; purine and pyrimidine catabolism; peptidases; amino acyl synthetases; transaminases

It should be strongly emphasized that the fractionation procedure and the description of the fractions obtained are often arbitrary and have operational rather than anatomical significance. For example, the terms "particle" or "particulate fraction" merely indicate a water-insoluble complex that is sedimented in a suitable gravitational field. A continuum is often found between particles and soluble proteins; indeed, the same complex may behave as a particle under one set of conditions and as soluble proteins under another. Moreover, the particulate state need not be an expression of molecular size; it may be a reflection of a preponderance of water-repelling groups in the molecular unit. "Microsomes" are particles which may be sedimented at 100,000 × gravity and which include both ribosomes, which exist as such in the cell, and fragments of the endoplasmic reticulum as well as some of the lysosomes and perhaps of the Golgi apparatus. The heterogeneity of the microsomes is further indicated by studies utilizing density gradient centrifugation (page 198); this technique permits recognition of microsomes of varying

RNA content. The mitochondrial fraction may also contain lysosomes. These findings emphasize the arbitrary nature of the fractionation procedure.

Despite the above limitations, fractionation of broken cell preparations has proved to be an invaluable and powerful technique. As centrifugal procedures become more refined, and as the electron microscope is used to define more rigorously the composition of fractions obtained by centrifugation, knowledge of the intracellular localization of enzymic reactions becomes increasingly precise.

Studies of Tissue Culture. Cells and tissues can be grown in suitable vessels containing accurately defined media. This capability is, in large measure, due to the isolation and identification of essential growth factors, *e.g.*, vitamins and amino acids required for cellular proliferation. This technique therefore provides opportunity to examine biochemical processes in *growing* cell populations and in successive generations of cells over periods of time longer than those afforded by the Warburg technique (page 283) for resting cells, slices, and broken cell preparations. A number of different types of cells, both normal and malignant, of several mammalian species, including the human, have been grown in vitro by tissue culture techniques. Tissue culture techniques have also provided another approach to the study of virus propagation in cells, permitting examination of chemical alterations induced in the host cells in vitro by the viral agent as well as insight into the neoplastic changes induced by certain viruses. The significance of data obtained from tissue culture studies for phenomena in vivo may well be limited, and caution must be exercised in their interpretation. Perhaps the most serious difficulty encountered in such studies is the tendency of animal cells to dedifferentiate during multiplication; a culture derived from liver cells cannot be considered identical to liver cells in vivo. Care must also be exercised against contamination of the cultures by microorganisms.

Studies With Microorganisms. Application of biochemical knowledge and techniques to the discipline of microbiology has provided understanding of bacterial metabolism. In turn, use of microorganisms has furnished the biochemist with simple living forms for laboratory studies. These lower forms have relatively simple requirements for growth and development, they reproduce rapidly, and some of them can be grown in large quantities. Thus microorganisms are a ready source for preparation of individual enzymes as well as broken cell preparations for study of metabolic transformations. Not infrequently microorganisms can be utilized to perform chemical transformations that the organic chemist finds difficult to duplicate, as well as to produce compounds less readily synthesized in the laboratory. In addition, mutant strains of microorganisms can easily be obtained experimentally. Numerous contributions of studies with microorganisms to our understanding of biochemical genetics are considered in Chaps. 28, 29, and 30. Studies with mutant strains of microorganisms have also discerned diverse, specific metabolic "blocks" which have been of aid in delineating normal metabolic pathways and control mechanisms.

Application of Isotopic Tracers. The availability of isotopes of the common elements permits labeling of molecules in order to follow or trace the metabolic pathways of specific portions of their structures. In addition, the isotopic labeling technique is very useful for tagging specific or active sites of large molecules, *e.g.*, enzymes. The labeling technique is particularly valuable for metabolic studies of normal body constituents which, when administered, are otherwise impossible to trace because they are indistinguishable from the same molecules already present in the body. The availability of isotopes of the common elements permits such studies since, for most purposes, the chemical differences between isotopes of any given element are negligible, although distinct isotope effects, *e.g.*, differential rates of reaction between normal and tritiated compounds, have been observed.

The isotopes that have been most frequently used in biochemical studies are those of the common elements of organic compounds, *viz.*, hydrogen, nitrogen, carbon, sulfur, phosphorus, and oxygen. In addition, isotopes of iodine, sodium, potassium, iron, and calcium have been utilized. Except for the stable isotopes of hydrogen (^{2}H), nitrogen (^{15}N), and oxygen (^{18}O), those unstable isotopes which emit radiation as they decay (radioactive isotopes) are most generally employed. The quantitative detection of radioactivity is technically simpler than is the estimation of mass of a stable isotope (see below). Also, the availability of radioactive isotopes with relatively long half-lives of decay makes possible their use in long-term experiments. The sensitivity of methods for measuring radioactivity when combined with paper and thin layer chromatography has provided a powerful tool for the detection and identification of compounds in experimental work.

Measurement of Isotope Abundance. The primary instrument for the determination of abundance of a stable isotope is the mass spectrometer, which is used for analyzing mixtures of stable isotopes of carbon, oxygen, or nitrogen, as well as hydrogen. In the case of hydrogen, where the mass ratio of deuterium (^{2}H) to hydrogen (^{1}H) is relatively large, other methods have also been employed, *e.g.*, measurements of density or refractive index of water containing the hydrogen under study. Since no useful radioactive isotopes of oxygen or nitrogen exist, the mass spectrometer is an essential tool for studies of the metabolic roles of these elements.

Radioactive isotopes are measured by detecting the radiations they emit in either a Geiger-Müller or a proportional counter or a liquid scintillation counter.

Biological Applications of Isotopes. In the application of the isotopic technique to biological problems, the first consideration is the form in which the isotope is administered. In many experiments, the isotopic atom may be a constituent of a simple molecule:

$$^{14}CO_2,\ ^{2}H_2O,\ ^{3}H_2O,\ ^{24}NaCl,\ NaH_2{}^{32}PO_4,\ K^{131}I$$

In other experiments, organic compounds of greater or lesser complexity labeled in one or more constituent atoms must first be prepared. Often this is done by application of classical methods of organic syntheses; many isotopically labeled compounds useful in biological research are now available commercially. In some cases the biosynthetic route is either the best or the only available method. Thus isotopic serum albumin may be prepared by inclusion of a labeled amino acid in the diet of an animal and subsequent isolation of albumin from the animal's serum. Similarly, ^{14}C may be incorporated into glucose by allowing photosynthesis (Chap. 20) to proceed in an atmosphere of $^{14}CO_2$, with subsequent hydrolysis of the starch that accumulates in the leaves. Not infrequently, it is desirable to include more than one isotopic label in the same substance. Thus for study of the fate of both the carbon skeleton and the nitrogen atom of the amino acid glycine, one may synthesize glycine-^{14}C and glycine-^{15}N separately and by simply mixing the two products obtain what is, in effect, a doubly labeled material. Compounds with more than one isotopic label may also be synthesized by methods that permit incorporation of more than one isotope into a single molecule.

Certain types of problem lend themselves particularly to the isotopic approach. The simplest of these is the analysis of a mixture by the so-called *isotope dilution technique.* Consider a mixture in which material A occurs in unknown abundance and for which a satisfactory analytical method is not available. A sample of material A containing isotope is prepared, and a known weight of pure isotopic A is added to the mixture. Material A is now isolated from the mixture without regard to yield but is carefully purified. The isotope concentration in this product will be determined by the quantity of nonisotopic A initially present in the mixture, the quantity of isotopic A added, and the isotope concentration of the latter material. These relationships are expressed mathematically as follows:

Let a = grams of isotopic material A added
Let a_0 = grams of nonisotopic material A initially present
b = isotope concentration in material added
c = isotope concentration in material isolated

Then
$$a \times b = c(a + a_0)$$
$$a_0 = a\left(\frac{b}{c} - 1\right)$$

A second application of isotopes relates to the problem of *anatomical distribution.* The isotopic material may be administered and, from subsequent analyses for the appropriate isotope in various tissues or products derived from them, the distribution of the isotopic atom may be ascertained. In the case of radioactive isotopes an additional tool is available, *viz.*, radioautography. In this procedure photographic film is applied to a cut section of a tissue, and, after adequate exposure and development, those portions of the film which were close to radioactive areas in the tissue will be found to have darkened. With isotopes of satisfactory radiation characteristics this technique can be refined to permit resolution at a histological level.

Isotopes have been widely used in the study of the *precursor-product relationship,* that is, to determine whether compound A is converted into B in the organism. Isotopic compound A is administered; compound B is isolated, carefully purified, and then analyzed for isotope. Application of this technique has been particularly useful in the demonstration in the animal of reactions that had previously been susceptible to study only in a simpler system. By degradation of the product and determination of the distribution of isotope among its atoms, one may often procure information as to the mechanism of the transformation.

The isotopic technique has been extremely important in the *analysis of rates of processes,* particularly in the intact animal. The quantity of any tissue constituent may be reasonably constant in the adult animal in balance, but this constancy may result from a balance between rates of synthesis and degradation. Prior to the advent of the isotopic technique, no satisfactory method was available for measurement of these rates. With isotopes, two approaches have been used in studies of this kind. The body store of a given compound may be labeled in a preliminary period by administration of the labeled material, and the subsequent disappearance of isotope followed. Alternatively, an isotopically labeled precursor of the material may be administered and a study made of the appearance of isotope in the product. From the rates of change in isotope concentration, the rates of synthesis and destruction may be calculated. From studies of this type, the concept of continuous *turnover* (synthesis, degradation, and replacement) of certain body constituents, even at constant composition, has been evolved. This concept of a dynamic steady state was dramatically emphasized in the work of Schoenheimer and his collaborators.

REFERENCES

Books

Bourne, G. H., "Division of Labor in Cells," Academic Press, Inc., New York, 1962.

Brachet, J., "Biochemical Cytology," Academic Press, Inc., New York, 1957.

Calvin, M., Heidelberger, C., Reid, J. C., Tolbert, B. M., and Yankwich, P. E., "Isotopic Carbon," John Wiley & Sons, Inc., New York, 1949.

Damm, H. C., and Besch, P. K., eds., "The Handbook of Biochemistry and Biophysics," The World Publishing Company, Cleveland, 1966.

Dixon, M., "Manometric Methods," Cambridge University Press, New York, 1951.

Extermann, R. C., ed., "Radioisotopes in Scientific Research," vol. III, Pergamon Press, New York, 1958.

Garrod, A. E., "Inborn Errors of Metabolism," reprinted with supplement by H. Harris, Oxford University Press, Fair Lawn, N.J., 1963.

Kamen, M. D., "Radioactive Tracers in Biology: An Introduction to Tracer Methodology," 3d ed., Academic Press, Inc., New York, 1957.

Roodyn, D. B., "Enzyme Cytology," Academic Press, Inc., New York, 1967.

Umbreit, W. W., Burris, R. H., and Stauffer, J. F., "Manometric Techniques and Related Methods for the Study of Tissue Metabolism," 4th ed., Burgess Publishing Company, Minneapolis, 1964.

Review Articles

Anderson, N. G., Techniques for Mass Isolation of Cellular Components, in G. Oster and A. W. Pollister, eds., "Physical Techniques in Biological Research," vol. III, pp. 229–352, Academic Press, Inc., New York, 1956.

Brachet, J., The Living Cell, *Scientific American,* **205** (Sept.), 50–61, 1961.

De Duve, C., The Separation and Characterization of Subcellular Particles, *Harvey Lectures,* **59,** 49–87, 1963–1964.

De Duve, C., and Wattiaux, R., Functions of Lysosomes, *Ann. Rev. Physiol.,* **28,** 435–492, 1966.

Eagle, H., Metabolic Studies with Normal and Malignant Cells in Culture, *Harvey Lectures,* **54,** 156–175, 1958–1959.

Finean, J. B., The Molecular Organization of Cell Membranes, *Prog. Biophys. Chem.,* **16,** 143–170, 1966.

Hogeboom, G. H., Fractionation of Cell Components of Animal Tissues, in S. P. Colowick and N. O. Kaplan, eds., "Methods in Enzymology," vol. I, pp. 16–19, Academic Press, Inc., New York, 1955.

Holter, H., How Things Get into Cells, *Scientific American,* **205** (Sept.), 167–183, 1961.

Porter, K. R., The Endoplasmic Reticulum, in T. W. Goodwin and O. Lindberg, eds., "Biological Structure and Function," vol. I, pp. 127–155, Academic Press, Inc., New York, 1961.

Puck, T. T., Quantitative Growth of Mammalian Cells, *Harvey Lectures,* **55,** 1–12, 1959–1960.

14. General Metabolism

Energy Considerations

Transformation of energy necessarily accompanies the variety of chemical reactions that make possible the characteristic properties, *e.g.*, movement, respiration, reproduction, growth, and response to stimuli, which distinguish living cells from nonliving structures. The total metabolism, which is manifest as the energy released by all chemical transformations in the animal and must derive ultimately from the oxidation of foodstuffs, may appear as heat or as external mechanical work. Even during muscular activity, the major portion of the energy appears as heat because of the relative inefficiency of the muscles as mechanical devices. During rest, practically all this energy appears as heat.

As will be developed in detail in subsequent chapters, biological oxidations normally proceed only at the rate at which the free energy (ΔF) liberated is required for the performance of useful work. The latter takes many forms. Living cells are effective transducers of chemical potential energy into other forms of energy, *viz.*, chemical, mechanical, electrical, and osmotic, and in some organisms, even into electromagnetic energy (light). Thus, the free energy derived from the oxidation of glucose can be utilized for the synthesis of proteins, fatty acids, nucleic acids, or steroids; for the contraction of muscles, conduction of the nervous impulse or, in the electric eel, generation of an electrical charge; for secretion of hypertonic urine or maintenance of a large concentration gradient for Na^+ and K^+ within and outside of cells; and in the firefly, for production of light. Of the total energy, ΔH, released during glucose oxidation, only the free-energy component, ΔF, can be utilized in processes indicated above, whereas the entropic component, $T\Delta S$, must appear as heat. However, the transfer of energy occurs in quanta, and if a chemical reaction which yields, for example, -8000 cal./mole is coupled to a process which requires $+5000$ cal./mole (as the synthesis of the peptide bond), the difference, 3000 cal./mole, cannot be saved for future use but must immediately appear as heat. In the subsequent fate of the molecule which had been synthesized, *e.g.*, hydrolysis of that peptide bond, again only the free energy thus released may be utilized for performance of some task if the means be available. The entropy must again appear as heat, and if no mode is available to take advantage of the free energy change, it, too, will appear as heat. If all such processes are summed, each 24 hr., in a 70-kg. adult, about 2000 Cal. will have been generated and released. Since neither his weight, structure, nor composition will have changed significantly over this period, all this energy will have appeared as heat, regardless of any intermediary transformations, except for that work which was done upon the environment, such as lifting weights, etc. Since the energy loss, as heat, is also irretrievably dissipated to the environment, there is engendered a requirement for daily provision of new, external sources of energy, *viz.*, foodstuffs which can be oxidized. The total process, which incidentally provides the heat necessary for the maintenance of body temperature in an external environment cooler than 37°C., is not, as might appear, wasteful. It is the sum of these activities, made possible by transient use of the free energy of oxidation, that makes possible the highly ordered structures and vital activities of the living organism.

CALORIC VALUES OF FOODSTUFFS

The Calorie referred to in metabolic studies is the kilo-calorie (1000 calories), the amount of heat required to raise the temperature of 1,000 g. of water from 15 to 16°C. The energy that may be derived from oxidation of compounds, including those of food, can be measured in a bomb calorimeter. The over-all energy release accompanying a chemical reaction, ΔH (page 210), is independent of the reaction mechanism. Much evidence demonstrates that ΔH for the reaction

$$\text{Glucose} + 6O_2 \longrightarrow 6CO_2 + 6H_2O$$

is identical whether it occurs in a bomb calorimeter or in man.

The first measurements of the heat that can be produced by oxidation of foodstuffs outside the body were conducted by Rubner, who compared the values obtained in the bomb calorimeter with those given by direct measurement of heat production by a dog placed in a calorimeter and then fed a known quantity of carbohydrate, protein, or lipid. For carbohydrate and lipid, the values were similar whether the foodstuff was burned inside or outside the body. However, the in vivo value for protein (4.1 Cal. per g.) was less than that obtained in the bomb calorimeter (5.3 Cal. per g.). This discrepancy is due to the fact that under physiological conditions, the nitrogen of proteins is not oxidized but is excreted mainly as urea.

Since the carbohydrates, lipids, and proteins of the natural foodstuffs include mono- and polysaccharides, short- and long-chain fatty acids, saturated and unsaturated fatty acids, etc., the caloric values of individual members of each of these major classes are variable. Thus, glucose yields 3.75 Cal. per g., whereas glycogen gives 4.3 Cal. per g. Again, animal proteins appear to yield higher values than do plant proteins, and most animal lipids liberate 9.5 Cal. per g., although butter and lard give 9.2 Cal. per g. Therefore, the caloric values given for the three classes of foodstuff represent averages. The average caloric values of the three major foodstuffs are given in Table 14.1. If additional allowance is made for the possibility of incomplete digestion and/or absorption, the values can be rounded off to 4, 9, and 4 for carbohydrate, lipid, and protein, respectively.

DIRECT CALORIMETRY

Heat production of a human subject can be measured in a calorimeter by placing the subject in an insulated chamber lined with a large network of constantan

Table 14.1: AVERAGE METABOLIC ENERGY DERIVED FROM THE THREE MAJOR FOODSTUFFS

Foodstuff	*Cal./g.*
Carbohydrate	4.1
Lipid	9.3
Protein	4.1

thermocouples. Measurements of this type yield values of approximately 1500 to 1800 Cal. per day for the average total heat production of an adult, postabsorptive (after absorption of food from the intestine has ceased), resting male subject. This technique permits the study of effects of activity or exercise, food consumption, various occupations, and environmental temperature on heat production, inasmuch as the large size of the chamber permits a moderate degree of normal activities.

THE RESPIRATORY QUOTIENT

Metabolic heat production is the consequence of the oxidation of foodstuffs by atmospheric oxygen with production of carbon dioxide. The magnitude and nature of this gaseous exchange vary with the type of foodstuff, or the mixture of foodstuffs, undergoing oxidation. The theoretical relationship between oxygen consumption and carbon dioxide production can be calculated from the stoichiometry of the equations for the oxidation of carbohydrate, lipid, and protein, respectively.

The complete oxidation of glucose may be represented as

$$C_6H_{12}O_6 + 6O_2 \longrightarrow 6CO_2 + 6H_2O$$

The molar ratio of carbon dioxide produced to oxygen utilized is one. This ratio is defined as the respiratory quotient, frequently abbreviated as R.Q. For all carbohydrates its value is 1.

The equation for the oxidation of a typical triglyceride, tripalmitin, is

$$C_{51}H_{98}O_6 + 72.5O_2 \longrightarrow 51CO_2 + 49H_2O$$

The calculated value for the R.Q. is $51/72.5 = 0.703$. The fact that the value is less than 1 is a reflection of the highly reduced nature of fatty acids as compared to carbohydrates. Therefore, more oxygen must be consumed in the oxidation of lipid, per mole of CO_2 produced, than is the case with carbohydrate. It is evident that the R.Q. for oxidation of tripalmitin will differ somewhat from that for other triglycerides, *e.g.*, triolein (0.713), tristearin (0.699). However, the average R.Q. for either mixed dietary or body lipids is 0.71. Therefore, during fasting, when energy production is derived almost entirely from reserve calories in the form of depot lipids, the R.Q. approaches 0.71. Conversely, when a marked degree of conversion of carbohydrate to lipid occurs as in animals force-fed carbohydrate (fattening of geese), the R.Q. may even exceed 1.0. This may be seen from the following equation:

$$\underset{\textbf{Glucose}}{4C_6H_{12}O_6} + O_2 \longrightarrow \underset{\textbf{Palmitic acid}}{C_{16}H_{32}O_2} + 8CO_2 + 8H_2O$$

The R.Q. for this synthesis of fatty acid from carbohydrate (lipogenesis, see Chap. 21) is 8.0, and values approaching 2.0 have been obtained experimentally in young hogs fed a mixture of starch and glucose. Under normal circumstances, however, the R.Q. seldom exceeds 1.0 and the extent to which the R.Q. approaches 1 at any time during the metabolism of mammals reflects the degree to which carbohydrate oxidation predominates in the metabolic mixture.

Calculation of a theoretical R.Q. for oxidation of protein is complicated inasmuch as oxidation of proteins is not carried completely to carbon dioxide and water and the nitrogenous end products of protein oxidation are for the most part excreted in the urine. However, calculations by Loewy in 1910 provided a basis for establishing the R.Q. for proteins. Analyses showed that 100 g. of meat protein contained 52.4 g. of carbon, 7.3 g. of hydrogen, 22.7 g. of oxygen, 16.7 g. of nitrogen, and 1.0 g. of sulfur. Of this quantity of ingested meat protein, it was estimated that urine and feces contained 10.9 g. of carbon, 2.9 g. of hydrogen, 15.0 g. of oxygen, 16.7 g. of nitrogen, and 1.0 g. of sulfur. The urinary and fecal excretion of these elements, subtracted from the quantities contained in 100 g. of meat protein, left 41.5 g. of carbon, 4.4 g. of hydrogen, and 7.7 g. of oxygen for complete oxidation to carbon dioxide and water. Calculation reveals that this would require 138.2 g. of oxygen from extraneous sources, with production of 152.2 g. of carbon dioxide. One gram of oxygen at 0°C. and 760 mm. Hg (standard conditions of temperature and pressure) occupies 0.6997 liter, whereas 1 g. of carbon dioxide, under standard conditions, occupies 0.5089 liter. Therefore, $138.2 \times 0.6997 = 96.68$ liters of oxygen would be required to produce $152.2 \times 0.5089 = 77.45$ liters of carbon dioxide from the complete oxidation of that portion of those elements, derived from 100 g. of protein, which are not excreted in the urine or feces but are completely oxidized. Thus, the R.Q. for protein would be 77.45/96.68, or 0.801.

INDIRECT CALORIMETRY

Since the total energy released as heat during any period of time results from the oxidation of a mixture of protein, carbohydrate, and lipid, estimation of the actual amounts oxidized of each of these major components permits calculation of the necessarily associated heat production while obviating the need for the complex apparatus required for direct calorimetry. This approach is termed *indirect calorimetry*. In practice, protein oxidation is estimated from urinary nitrogen excretion while the extent of oxidation of lipid and carbohydrate is calculated from measurement of the total oxygen consumption and R.Q., after applying corrections for the oxygen utilized and the CO_2 produced by the oxidation of the protein metabolized during the same period.

The heat production associated with a given amount of oxygen consumed and carbon dioxide produced varies with the type of food being oxidized. For example, a liter of oxygen consumed during combustion of carbohydrate results in a greater liberation of heat (5.0 to 5.4 Cal.) than when lipid (4.5 to 4.7 Cal.) or protein (4.3 to 4.7 Cal.) is oxidized.

It is rare, however, that a single foodstuff is being oxidized. Hence, the nature of the mixture of foodstuffs burned, as indicated by the R.Q., must be considered and can be calculated as follows. The amount of protein catabolized is obtained from measurement of urinary nitrogen during the period under study. Each gram of urinary nitrogen is equivalent to 6.25 g. of protein (100/16 = 6.25, 16 being the average percentage of nitrogen in proteins) and represents the production of 4.76 liters of carbon dioxide and the consumption of 5.94 liters of oxygen. Subtraction of the volumes of oxygen and carbon dioxide exchanged in the catabolism of protein from the total oxygen consumption and carbon dioxide production yields a ratio for these two gases which is termed the *nonprotein respiratory quotient.* Table 14.2 gives the percentages of lipid and carbohydrate undergoing combustion, per liter of oxygen consumed, as calculated for nonprotein respiratory quotients ranging from 0.707 to 1.0, and the relation of these quotients to heat production.

Table 14.2: RELATION OF NONPROTEIN RESPIRATORY QUOTIENTS TO CALORIES PER LITER OF OXYGEN USED AND TO PERCENTAGES OF CARBOHYDRATE AND LIPID METABOLIZED

Nonprotein respiratory quotient	Cal./liter O_2	Calories derived from	
		Carbohydrate, per cent	Lipid, per cent
0.707	4.686	0	100
0.72	4.702	4.8	95.2
0.74	4.727	12.0	88.0
0.76	4.751	19.2	80.8
0.78	4.776	26.3	73.7
0.80	4.801	33.4	66.6
0.82	4.825	40.3	59.7
0.84	4.850	47.2	52.8
0.86	4.875	54.1	45.9
0.88	4.899	60.8	39.2
0.90	4.924	67.5	32.5
0.92	4.948	74.1	25.9
0.94	4.973	80.7	19.3
0.96	4.998	87.2	12.8
0.98	5.022	93.6	6.4
1.00	5.047	100	0

Therefore, from knowledge of nitrogen excretion, oxygen consumption, and carbon dioxide production during a given period, the quantity of protein, carbohydrate, and lipid catabolized can be estimated, and the heat production calculated.

In clinical practice, the heat production is determined under resting, postabsorptive conditions (see below, Basal Metabolism). In these circumstances, urinary nitrogen is not measured, the R.Q. of the postabsorptive state is assumed to be 0.82, and heat production can be calculated directly from the oxygen consumption on the basis that, at this R.Q., 1 liter of oxygen is equivalent to 4.825 Cal. (Table 14.2). Only minor errors result from these assumptions.

BASAL METABOLISM

It is not possible to assess the relative significance at a specified time in metabolism of the energy relationships among foodstuffs, heat production, metabolic energy, and stored energy. However, the significance of food and of heat produced as a result of work can be delimited. This is accomplished by measurement of energy exchange in a postabsorptive period and in the resting state, thus minimizing energy utilization due to work on the environment. Under these controlled conditions, heat production becomes the major means of energy loss from the body, and stored energy can be the only source of this heat. Since energy cannot be created or destroyed, the decrease in stored energy becomes equal to the heat loss, and measurements of the latter afford an estimate of the total metabolism. Since this measurement is done under resting, postabsorptive conditions, it estimates the *basal metabolism.* The basal metabolism reflects the energy requirements for maintenance and conduct of those cellular and tissue processes which are fundamental to the continuing activities of the organism, *e.g.*, the metabolic activity of muscle, brain, renal, liver, and other cells, plus the heat released as a result of the mechanical work represented by contraction of the muscles involved in respiration, circulation, and peristalsis. The total energy requirement of these processes, the basal metabolism, comprises approximately 50 per cent of the total energy expenditure required for the diverse activities of a normal 24-hr. day. The *basal metabolic rate,* abbreviated B.M.R., is not the minimal metabolism necessary for mere maintenance of life, since there are circumstances, *e.g.*, during sleep, when the metabolic rate may be lower than the basal rate.

The apparatus commonly employed for measurements of basal metabolism by indirect calorimetry is the Benedict-Roth apparatus. From the volume of oxygen consumed by the subject, corrected to dry oxygen at standard temperature and pressure, it is possible to calculate heat production by use of the value 4.825, the caloric equivalent of 1 liter of oxygen for an R.Q. of 0.82 (Table 14.2).

The basal metabolism is usually given in terms of *Calories per hour.* In practice the basal metabolic rate is determined over a 10- to 15-min. period and expressed per square meter of body surface or as a percentage above or below certain standard values (see below).

Factors Affecting the Basal Metabolism. The basal metabolism is quite constant in a given individual and in similar individuals of the same species. Many factors affect basal metabolism, *e.g.*, body size, age, sex, climatic conditions, diet, physical training, drugs, etc. The basal metabolism may deviate from normal values in a variety of pathological states, and in certain of these its measurement provides a useful diagnostic tool.

Influence of Body Size on Basal Metabolism. There are four important factors in heat loss from the organism: (1) the temperature difference between the environment and the organism, (2) the nature of the surface that radiates the heat, (3) the area of that surface, and (4) the thermal conductance of the environment. Under the conditions of determination of the basal metabolism, surface area is the most important of these factors, and, under similar physiological conditions, the basal metabolism of various mammals is roughly proportional to the surface area. The relation of heat production to body weight and to surface area is seen in Table 14.3. It is evident from the table that, although the *heat production per kilogram* may vary widely among various species and is *inversely* related to *body weight,* the *heat production per square meter of body surface is essentially constant.*

Table 14.3: RELATION OF DAILY BASAL HEAT PRODUCTION TO BODY WEIGHT AND SURFACE AREA

	Body weight, kg.	Metabolism per kg. of body weight per day, Cal.	Metabolism per m.² of body surface per day, Cal.
Horse	441.0	11.3	948
Pig	128.0	19.1	1078
Man	64.3	32.1	1042
Dog	15.2	51.5	1039
Goose	3.5	66.7	969
Fowl	2.0	71.0	943
Mouse	0.018	212.0	1188

SOURCE: After G. Lusk, "The Elements of the Science of Nutrition," 4th ed, W. B. Saunders Company, Philadelphia, 1928.

Thus, knowledge of the surface area of the subject was required in establishing the standards of basal metabolism, since heat production in the postabsorptive, resting state is a function of the total surface area.

Determination of the total surface area in man was conducted initially by E. F. Du Bois and D. Du Bois, who made measurements by means of flexible but inelastic paper molds. The laborious nature of this method led to a formula for surface area that was accurate but fairly simple and based on the height and weight of the subject.

$$A = W^{0.425} \times H^{0.725} \times C$$

A is the surface area in square centimeters; W, the weight in kilograms; H, the height in centimeters; and C, a constant with a value of 71.84. The formula can also be expressed as follows:

$$\log A = 0.425 \log W + 0.725 \log H + 1.8564$$

In order to avoid this calculation, nomograms have been constructed so that approximate surface area can be ascertained, the height and weight of the subject being known.

A somewhat simpler expression of metabolic rate, independent of surface area, has been derived from a relatively consistent general relation of the metabolic rate and the body mass of homeotherms. This is expressed by the approximation $B = 3W^{3/4}$, where B is the hourly metabolic rate in Calories and W is the body weight in kilograms.

Influence of Age and Sex on Basal Metabolism. Both age and sex affect the basal metabolism, the values being higher in childhood than in adult life, and uniformly higher in the male as compared with the female in the same age group. The heat production of premature babies is extremely low, and the basal metabolism of the newborn is significantly lower (25 Cal.) than that of infants a few weeks of age (55 Cal.). Heat production gradually declines with advancing age from 45 to 50 Cal. per m.² per hr. in the ten-year-old male to approximately 35 Cal. at the age of sixty-five. Table 14.4 shows some values of basal heat production in relation to age and sex.

Table 14.4: BASAL HEAT PRODUCTION IN RELATION TO AGE AND SEX

Age, yr.	Average Cal./hr./m.^{2} body surface		Age, yr.	Average Cal./hr./m.^{2} body surface	
	Males	Females		Males	Females
5	53.0	51.6	20–24	41.0	36.9
6	52.7	50.7	25–29	40.3	36.6
7	52.0	49.3			
8	51.2	48.1	30–34	39.8	36.2
9	50.4	46.9	35–39	39.2	35.8
10	49.5	45.8	40–44	38.3	35.3
11	48.6	44.6	45–49	37.8	35.0
12	47.8	43.4			
13	47.1	42.0	50–54	37.2	34.5
14	46.2	41.0	55–59	36.6	34.1
15	45.3	39.6			
16	44.7	38.5	60–64	36.0	33.8
17	43.7	37.4	65–69	35.5	33.4
18	42.9	37.3	70–74	34.8	32.8
19	42.1	37.2	75–79	34.2	32.3

Other Factors Affecting the Basal Metabolism. The environmental temperature influences the basal metabolic rate. Individuals living in a tropical climate have a lower rate than do individuals in temperate or colder climates. A minimal basal metabolic rate is observed at normal environmental temperature, *i.e.*, 20 to 25°C. Environmental temperatures above 30°C. cause a slight rise in the metabolic rate and in body temperature. When the temperature falls below 15°C., muscular tone increases, shivering may ensue, and heat production increases. However, elevation of the basal metabolic rate in colder climates is apparently independent of the increased heat production caused by shivering and is related to an augmentation of the basal oxygen consumption.

Muscular training, as in athletes, may be reflected in a slightly elevated basal metabolic rate.

Certain hormones, e.g., epinephrine and thyroxine, cause an increase in the basal metabolic rate.

Pathological Alterations in Basal Metabolic Rate. As suggested above, determination of the basal metabolic rate may be of value as an aid in diagnosis of disease and in following the response to therapeutic measures. In *hyperthyroidism,* the basal metabolic rate may be increased 50 to 75 per cent above normal standards; in *hypothyroidism,* the rate may be as much as 40 per cent lower than normal (Chap. 43). *Fever* increases heat production approximately 13 per cent of the basal metabolic rate for each degree centigrade increase in body temperature; this is probably the most frequently encountered reason for increased basal metabolic rate. The augmenting effects of temperature increments on rates of metabolic processes, *e.g.*, enzymic reactions, have been described (page 231). Table 14.5 lists a few of the circumstances in which the basal metabolic rate may be above or below normal.

Table 14.5: Some Circumstances in Which Basal Metabolic Rate May Vary

Below normal	*Above normal*
During sleep	Athletic training
Malnutrition	Latter half of pregnancy
Hypothyroidism	Hyperthyroidism
Hypophyseal insufficiency	Fever
Addison's disease	Cardiorenal disease with dyspnea
Drug administration, *e.g.*, anesthetics	Leukemia
Elevated environmental temperature, *e.g.*, in tropical climates	Polycythemia
	Drug administration, *e.g.*, epinephrine, caffeine, thyroid, or thyroxine

FACTORS AFFECTING THE TOTAL METABOLISM

It has been indicated previously that conditions for determining the basal metabolism are defined so that the variables of food and voluntary muscular activity are eliminated. The extent to which food and muscular work may alter metabolism from the basal level will be considered briefly.

Effect of Food. Caloric restriction may be accompanied by a considerable decline in total metabolism. Conversely, ingestion of food is followed by an increase in heat production above normal basal level. This occurs immediately after eating and can be related to the digestion and absorption of foodstuffs. Additional elevation of metabolism may then result as a consequence of subsequent metabolic transformations of absorbed products.

Of the three major foodstuffs, the ingestion of protein causes the greatest elevation in the total metabolism. The stimulating effect of food on the heat production of the organism is called the *specific dynamic action,* which is the extra heat produced by the organism, over and above the basal heat production, as a result of food ingestion. In the case of protein, the specific dynamic effect amounts to approximately 30 per cent, for carbohydrate 6 per cent, and for lipid 4 per cent, respectively, of the energy value of the food ingested. Thus, ingestion of 25 g. of protein, equivalent to 100 Cal., leads to 30 Cal. of extra heat production over the basal metabolic rate. These calories are wasted as heat, and therefore only 70 Cal. of potentially useful energy can be derived from the 25 g. of protein. Thus it is essential, in calculating the caloric value or equivalent of a diet, to make provision for the calories dissipated as heat as a result of the specific dynamic effect.

The specific dynamic effect of foodstuffs is a consequence of the extra energy released incident to metabolism of the food. Reactions in the metabolism of foodstuffs in the liver are apparently responsible in large part for the specific dynamic action of these substances.

Effect of Muscular Work. Muscles utilize the energy of oxidation for the performance of mechanical work with an efficiency of approximately 30 per cent. Moreover, when engaged in hard work, the total energy required for performing this work may be many times that reflected by the basal metabolic rate. Table 14.6 indicates the influence of muscular activity on the degree of caloric expenditure.

Effect of Mental Effort. Mental effort, in contrast to muscular effort, produces little increase in metabolism. The mental effort associated with the preparation for

Table 14.6: APPROXIMATE CALORIC EXPENDITURE OF MAN AS AFFECTED BY MUSCULAR ACTIVITY

Type of muscular activity	*Cal./hr.*
Sleeping	65
Awake, lying still	77
Awake, sitting up, at rest	100
Reading aloud	105
Standing relaxed	105
Standing at attention	115
Light to extreme muscular exercise	170–600

examinations or solution of a mathematical problem led to an increase of only 3 or 4 per cent in the metabolism.

Intense emotion may elevate the metabolism 5 to 10 per cent above the basal level. The metabolism may fall *below* the basal level during *sleep,* since the muscles are more completely relaxed than when the individual is awake, provided the sleep is undisturbed.

DAILY CALORIC REQUIREMENT

It is apparent from previous considerations that the total caloric requirement of man varies considerably with age, sex, diet, and daily activity and is not the same at all periods of life of a given individual. Therefore, it is not possible to establish standard caloric requirements; instead the range of values under various circumstances is indicated. To prevent loss or wasting of body tissues, the caloric intake of the food ingested must be at least equivalent to the total heat production during the same period. This heat production is equal to the basal metabolic rate plus the energy expended in performing the day's activities, including work. Table 14.7 indicates the average caloric expenditure of normal individuals in various types of activity.

Table 14.7: APPROXIMATE CALORIC EXPENDITURE IN VARIOUS OCCUPATIONS

Activity or occupation	*Cal./hr.*
Seamstress	110
Typist	140
Housemaid	150
Bookbinder	170
Shoemaker	180
Carpenter, metalworker, industrial painter	240
Stonemason	400
Lumberman, in cold environment	500

Inasmuch as the first law of thermodynamics is obeyed by living organisms, the balance that obtains between caloric intake and energy expenditure is the prime factor, under normal circumstances, that determines whether weight gain or weight loss occurs over a period of time. Weight gain in relation to lipid metabolism is discussed in Chap. 21.

REFERENCES

Books

Brody, S., "Bioenergetics and Growth," Reinhold Publishing Corporation, New York, 1945.

Consolazio, C. F., Johnson, R. E., and Pecora, L. J., "Physiological Measurements of Metabolic Functions in Man," McGraw-Hill Book Company, New York, 1963.

Kleiber, M., "The Fire of Life: An Introduction to Animal Energetics," John Wiley & Sons, Inc., New York, 1961.

Review Articles

Hardy, J. D., Physiology of Temperature Regulation, *Physiol. Revs.,* **41,** 521–606, 1961.

Keys, A., Energy Requirements of Adults, in "Handbook of Nutrition," American Medical Association, pp. 259–274, McGraw-Hill Book Company, New York, 1951.

Keys, A., Undernutrition, in G. G. Duncan, ed., "Diseases of Metabolism," 4th ed., pp. 501–528, W. B. Saunders Company, Philadelphia, 1959.

Kleiber, M., Body Size and Metabolic Rate, *Physiol. Revs.,* **27,** 511–541, 1947.

Kleiber, M., and Rogers, T. A., Energy Metabolism, *Ann. Rev. Physiol.,* **23,** 15–36, 1961.

Sadhu, D. P., The Specific Dynamic Action of Nutrients, with Special Reference to the Effects of Vitamins and Hormones, *Missouri Univ. Agr. Expt. Sta. Res. Bull.* 408, pp. 3–64, 1947.

Strang, J. M., Obesity, in G. G. Duncan, ed., "Diseases of Metabolism," 4th ed., pp. 529–633, W. B. Saunders Company, Philadelphia, 1959.

Symposium on Energy Balance, *Am. J. Clin. Nutr.,* **8,** 527–774, 1960.

15. Biological Oxidations. I

Oxidation-Reduction Reactions. High Energy Phosphate Compounds

Oxidation in living cells serves two chief functions: (1) to provide energy for endergonic cellular processes, and (2) to transform dietary materials into cellular constituents. Many of the problems involved in biological oxidations may be introduced by considering the oxidation of glucose, which may be represented as follows.

$$C_6H_{12}O_6 + 6O_2 \longrightarrow 6CO_2 + 6H_2O + \text{energy}$$

The complete oxidation of a mole of glucose according to the above equation would result in a free-energy release, under standard conditions, of 686,000 cal. Were this energy released, in the manner of the heat released in a bunsen flame, intracellular oxidation of a relatively small amount of glucose would suffice to disrupt cellular structure. Moreover, the equation above provides no information concerning the means by which the cell may utilize for its functions the energy arising from oxidation. In most man-made machines the energy derived from the oxidation of fuel is released as heat which is used to expand a gas in a manner that permits useful work to be accomplished. Within cells, however, a large fraction of the free energy available from oxidations is retained as chemical energy rather than released as an equivalent amount of heat. Indeed, since the temperature of mammalian cells is relatively constant, heat, *of itself,* could not even in theory be employed for useful work (page 210). Work can be derived from heat only in a device that operates with a temperature differential, *e.g.*, a steam engine. However, mammalian cells which operate isothermally nevertheless employ the energy derived from oxidations to perform useful work. This work, therefore, is not accomplished by using the *heat* released by oxidation of glucose or other foodstuffs.

In the course of many biological oxidations, the free energy released is used to synthesize a compound whose chemical energy is, in turn, available for doing work. Further, whether the material being oxidized is carbohydrate, lipid, or amino acid, the free energy available from each oxidative step is immediately employed, with but few exceptions, to synthesize one compound, *adenosine triphosphate* (ATP) (page 315); the energy of ATP is then available for the manifold endergonic processes of the body. The standard free energy required to synthesize ATP from adenosine diphosphate and inorganic phosphate is approximately +7000 cal. per mole. For the synthesis of ATP at 37°C., pH 7.4, and at prevailing biological concentrations of phosphate and adenosine diphosphate, ΔF is of the order of +8,000 to 10,000 cal. per mole. To obtain maximal benefit from the oxidation of a given metabolite, its complete oxidation should, therefore, be subdivided into the largest possible number of steps, each of which will yield approximately 10,000 cal. per mole, *i.e.*, sufficient energy for formation of a mole of ATP. Thus, the living organism differs from man-made, fuel-consuming engines in the important fact that a portion of the loss of energy as heat is eliminated; the metabolic apparatus operates with incomplete heat loss, and much of the free energy of oxidation is trapped directly as another form of potential chemical energy.

It is implicit in these concepts that, unless released as heat, energy *may* be transferred repeatedly under isothermal conditions. Thus, energy supplied to the organism as glucose may next appear locked into a molecule of ATP, may then appear in a newly synthesized protein molecule, and may finally emerge as heat when the protein undergoes hydrolysis to component amino acids. Despite the numerous steps in such transformations, which are here oversimplified, the overall energy change is identical with that which may be observed, as heat production, in a bomb calorimeter as described in the previous chapter. To the extent that each of the consecutive transfers of energy in a given series of reactions involves the production of heat and of unavailable energy as entropy, metabolic transformations are inefficient, insofar as the energy economy of the organism is concerned. However, it is this very "inefficiency" which gives *direction* to metabolic events. For example, in the reaction series,

$$A \rightleftharpoons B \rightleftharpoons C \rightleftharpoons D \rightleftharpoons E$$

if ΔF for each step were zero, *viz.*, if the equilibrium constant for each reaction were 1, A would be transformed into an equal mixture of A, B, C, D, and E. If, however, ΔF for each step were large and negative, particularly that for $D \rightleftharpoons E$, then A would be converted almost entirely to E.

At this point one may ask why an animal requires a source of energy, apart from that required to do work on the environment such as exercise, etc. This question is particularly pointed in respect to the adult who remains at constant weight and fixed body composition in an isothermal environment. In general, *a major use of the energy derived from biological oxidations is to maintain the body in a state remote from equilibrium.* Thus, cells contain large quantities of polysaccharides, proteins, lipids, and nucleic acids in the presence of relatively small concentrations of their constituents, *i.e.*, glucose, amino acids, etc. Yet, at equilibrium, in the presence of appropriate enzymes, quite the opposite situation would prevail. Again, the ionic composition of the solution bathing body cells is remarkably different from that within the cell, despite permeability of the cell membrane to the ions on both sides. Free energy from the oxidation of glucose is employed to synthesize polysaccharides, proteins, etc., at a rate equal to that at which they are degraded as they tend toward equilibrium. Similarly, energy is employed to expel various ions from within the cell in opposition to the tendency to attain equilibrium across the cell membrane. As large molecules are hydrolyzed, or as ions return to the cell, energy is lost as a consequence of both entropy change and lack of means of utilizing the resultant free-energy release. Over a period of time, therefore, since the rates in both directions are equal, all the energy supplied appears ultimately as heat, but the disequilibrium has been maintained. Thus, the "order" of the foodstuffs is altered, through oxidation, to maintain the high degree of "order" of the cell. The sum of such processes in the organism may be presumed to comprise a major fraction of the basal metabolic rate (page 296). If the supply of food or oxygen ceases, the tendency toward equilibrium is not counterbalanced, and the expected equilibria are attained.

It is the purpose of these chapters to describe biological oxidations and particularly the means by which the oxidation of foodstuffs by molecular oxygen may occur at 37°C. without disrupting cellular structure while the free energy of oxidation is made available for the endergonic processes of living cells.

OXIDATION-REDUCTION REACTIONS

The term oxidation is restricted to those reactions between molecules which involve electron transfer. Net loss of an electron is oxidation; acquisition of an electron is reduction.

$$Fe^{++} \rightleftharpoons Fe^{+++} + e$$
$$Co^{++} \rightleftharpoons Co^{+++} + e$$
$$Na \rightleftharpoons Na^{+} + e$$

The above equations describe oxidation of metals at an electrode. In solution, oxidation of a metal ion occurs by transfer of an electron directly to an acceptor ion or molecule. Many factors affect the ease of electron transfer to or from metal ions. In general, the process is facilitated (1) by bridging groups that can combine with both reacting species, *i.e.*,

$$A + X^{-}B^{+} \longrightarrow AX^{-}B^{+} \longrightarrow A^{+}X^{-}B \longrightarrow A^{+}X^{-} + B$$

or (2) by coordination of the metal ion to ligands, particularly to ligand groups with relatively large, highly resonant structures.

That loss of an electron is an "oxidation" is readily understood for most inorganic substances but may not be quite so obvious for the reaction $H_2 + \frac{1}{2}O_2 \rightleftharpoons H_2O$, since the H—O bond is covalent and a pair of electrons is shared between the oxygen and hydrogen atoms. Even here, however, since the pair of electrons is somewhat closer to the oxygen nucleus than it is to the hydrogen nucleus, it can be stated that the hydrogen is oxidized since it has partially lost an electron while the oxygen has partially gained electrons. The polarity of water and the electrolysis of water by an electric current are further evidence that the reaction between H_2 and O_2 is primarily an electron transfer from one atom to another.

A more complex problem arises in the case of organic compounds for which oxidation is generally synonymous with dehydrogenation. Consider the oxidation of hydroquinone, which may be represented as:

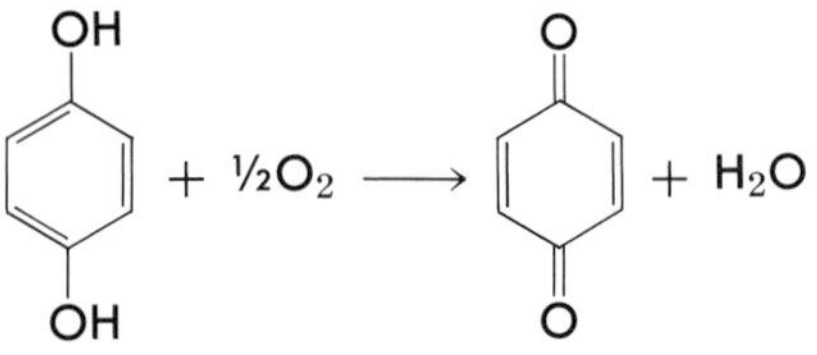

The reaction may also be regarded as proceeding in steps, the first being an acidic dissociation, and the second, withdrawal of electrons.

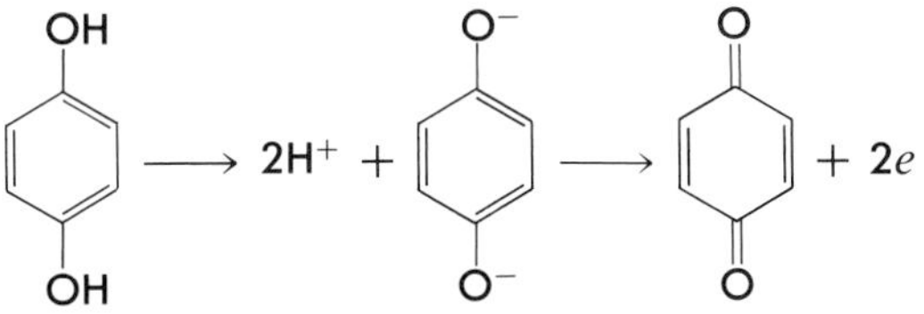

If the oxidant were ferric ion, for example, the over-all reaction would be

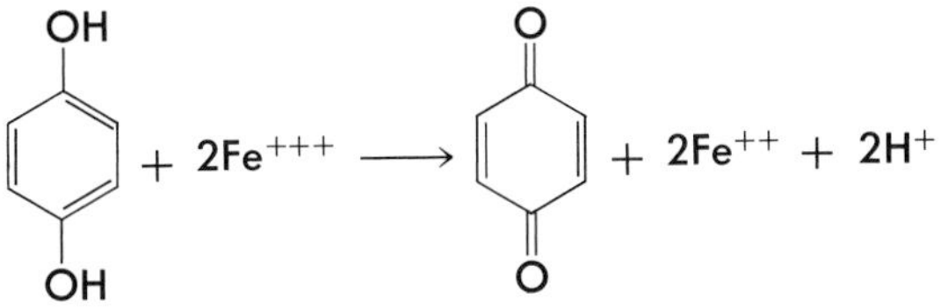

In the example given, protons and electrons depart independently from the molecule being oxidized. In theory, the departing group may be an intact hydrogen atom H·, whereas in many biological systems, oxidation frequently occurs by transfer of a hydride ion, $H:^-$, to the acceptor molecule.

One-electron Transfer: Free Radicals. In the example above, two hydrogen atoms and two electrons were removed from the hydroquinone molecule. However, the ferric ion, Fe^{+++}, like several biological oxidants, can accept but one electron. Since the likelihood of a ternary collision, involving simultaneously two 1-electron acceptor ions and one 2-electron donor, is remote, it appears that the mechanism shown above is inadequate and that electron transfer can occur in successive 1-electron steps, even from organic molecules. Such a process may again be illustrated by the oxidation of hydroquinone.

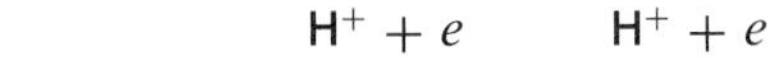

Hydroquinone Semiquinone Quinone

Since the intermediate *semiquinone* is formed by loss of a proton and an electron, it is a free radical, *i.e.*, distributed through the molecule is an odd or unpaired electron. Such free radicals vary widely in their stability. Some, *e.g.*, $CH_3\cdot$, exist for no more than millimicroseconds; others, with large resonant structures, and particularly those with complex conjugated double-bond systems, are relatively stable and, hence, readily detected. The latter group is of interest and significance in biological oxidations.

The odd electron does not confer a negative charge upon the semiquinone molecule; rather, it represents an unfilled valence bond. However, it does confer other distinctive properties. The magnetic moments of the two members of an electron pair are aligned in opposite directions so they cancel. Since the magnetic moment of an odd electron is unopposed, the molecule has a net magnetic moment equal to that of one unpaired electron, *viz.*, one Bohr magneton, and is said to be paramagnetic (page 175). Molecules with no magnetic moment are termed diamagnetic. Hence, these can be distinguished by devices which measure magnetic susceptibility.

A more sensitive and useful technique for detection and estimation of such free radicals is *electron spin resonance spectrometry*. When a population of free-radical molecules is in zero magnetic field, the spins and magnetic moments of their unpaired electrons are randomly oriented and all have equal energy. When placed in a magnetic field, however, the unpaired electrons must align themselves with their magnetic moments either parallel or antiparallel to the applied field. Those aligned with the field have energies $\frac{1}{2}g\beta H$ less than the zero-field value, and those arranged antiparallel have energies $\frac{1}{2}g\beta H$ greater than the zero-field value. Thus, the difference in energy between these two groups of unpaired electrons is $g\beta H$, where β is a constant, the Bohr magneton, H is the magnetic field strength in gauss, and g, the "spectroscopic splitting factor," is a function characteristic of the unpaired electrons of a particular organic free radical or paramagnetic metal. The precise value of g is determined by electron spin resonance spectrometry in the following manner. If electromagnetic radiation is applied to a solution of free radicals in a magnetic field, some of the antiparallel (lower-energy) electrons absorb the incident radiant energy, reverse their magnetic moments, and, hence, are at the higher-energy level. In a given magnetic field, H, this is possible only at that frequency, υ, at which the energy of the absorbed quanta, $h\upsilon$, exactly equals the difference in energy levels between the two classes of unpaired electrons, *viz.*, when $h\upsilon = g\beta H$. Thus, the value of the "spectroscopic splitting factor" becomes

$$g = \frac{h\upsilon}{\beta H}$$

In principle, electron resonance absorption can occur at any frequency of applied radiation if the strength of the magnetic field is adjusted to satisfy the equation above. In practice, it is convenient to use a generator of fixed frequency output, vary the magnetic field strength, and measure the absorption of the radiant energy by the sample.

For an isolated electron, g is 2.0023. Since organic free radicals yield g values very close to 2.00, the remainder of these molecules exerts relatively little influence on the unpaired electron.

Organic free radicals are generally intensely colored, *i.e.*, they absorb light energy in the visible portion of the spectrum. This has permitted their detection and estimation by spectrophotometry, although this procedure is not so sensitive as electron spin resonance spectrometry.

The occurrence of oxidation-reduction reactions demonstrates that molecules and atoms vary in their affinity for electrons. A familiar expression of this varying affinity is to be found in the electromotive series of the elements. It will be recalled that metals high in the series will displace any metal below from its salts. Thus for

$$2Na + Cu^{++} \longrightarrow 2Na^{+} + Cu$$

the Na is oxidized and Cu^{++} is reduced. Further, the more remote two metals are from each other in the series, the greater the relative oxidizing and reducing powers. A similar series can also be compiled for organic substances, expressing their relative affinities for electrons. A few systems of interest are presented in Table 15.1.

Table 15.1: ELECTRODE POTENTIALS OF SOME REDUCTION-OXIDATION SYSTEMS

System	E_0', volts*	pH
$H_2O/\frac{1}{2}O_2$	0.82	7.0
NO_2^-/NO_3^-	0.42	7.0
Phenylalanine/dihydroxyphenylalanine	0.37	7.0
Ferrocyanide/ferricyanide	0.36	7.0
$H_2O_2/\frac{1}{2}O_2 + H_2O$	0.30	7.0
Cytochrome a Fe^{++}/Fe^{+++}	0.29	7.0
Cytochrome c Fe^{++}/Fe^{+++}	0.22	7.0
2,6-Dichlorophenolindophenol red/ox†	0.22	7.0
Butyryl CoA/crotonyl CoA	0.19	7.0
Hemoglobin/methemoglobin	0.17	7.0
Cytochrome b_2 Fe^{++}/Fe^{+++}	0.12	7.4
Ubiquinone red/ox	0.10	7.4
Ascorbic acid/dehydroascorbic acid	0.08	6.4
Cytochrome b Fe^{++}/Fe^{+++}	0.07	7.4
Succinic acid/fumaric acid	0.03	7.0
Methylene blue red/ox	0.01	7.0
"Old" yellow enzyme $FMNH_2/FMN$	−0.12	7.0
Alanine/pyruvic acid + NH_4^+	−0.13	7.0
Glutamic acid/α-ketoglutaric acid + NH_4^+	−0.14	7.0
Malic acid/oxaloacetic acid	−0.17	7.0
Lactic acid/pyruvic acid	−0.19	7.0
Ethanol/acetaldehyde	−0.20	7.0
Glutathione red/ox	−0.23	7.0
β-Hydroxybutyric acid/acetoacetic acid	−0.27	7.0
3-Phosphoglyceraldehyde + P_i‡/1,3-diphosphoglyceric acid	−0.29	7.0
Lipoic acid red/ox	−0.29	7.0
DPNH + H^+/DPN^+	−0.32	7.0
Malic acid/pyruvic acid + CO_2	−0.33	7.0
Xanthine/uric acid	−0.36	7.0
Ferredoxin red/ox (algal)	−0.41	7.5
Acetaldehyde + CoA/acetyl CoA	−0.41	7.0
Ferredoxin red/ox (*Clostridium pasteurianum*)	−0.42	7.5
$\frac{1}{2}H_2/H^+$	−0.42	7.0
Ferredoxin red/ox (spinach)	−0.43	7.5
Acetaldehyde/acetic acid	−0.60	7.0
α-Ketoglutaric acid/succinic acid + CO_2	−0.67	7.0
Pyruvic acid/acetic acid + CO_2	−0.70	7.0

* The values shown for E_0' are the potentials that would be exhibited by a potentiometer interposed between a standard hydrogen electrode and an inert electrode in a solution containing equimolar amounts, at the pH specified, of the oxidized and reduced member of each pair, were the latter electroactive. Circuit closure is effected by a salt bridge between the hydrogen electrode and the solution under study.

† red/ox = reduced form/oxidized form.

‡P_i = inorganic orthophosphate (see page 218).

Clearly, oxygen can oxidize ethyl alcohol; yet at room temperature this reaction does not occur readily. Again, many substances should be capable of reducing acetaldehyde to ethanol, whereas relatively few such reactions do occur. Rather, in order to oxidize alcohol, vigorous conditions may be required, *e.g.*, chromic acid at elevated temperature, and to reduce acetaldehyde, a strong reducing agent such as sodium amalgam may be needed. A pair of substances that should interact, yet do not, is termed a *sluggish system.* This failure is due to a relatively high activation energy, E_a (page 232), for the reacting molecules, so that the expected reaction proceeds too slowly to be observed. In contrast are compounds that are readily oxidized and reduced in reversible manner. A useful example of such systems is the following reaction.

$$\text{Methylene blue} + 2e + 2H^+ \rightleftharpoons \text{leukomethylene blue}$$

A compound is said to be *autoxidizable* if its reduced form is readily oxidized by atmospheric oxygen in the absence of a catalyst. Michaelis suggested that solutions of compounds, which are readily oxidized and reduced in reversible systems, contain both oxidized and reduced forms and also have significant amounts of intermediary free radical, whereas sluggish systems such as ethanol/acetaldehyde contain relatively little of the free radical.

Quantitative Aspects of Oxidation-Reduction Reactions. When a metal strip, *e.g.*, zinc, here termed an electrode, is placed in an aqueous solution, some of the zinc atoms in the metal surface give up electrons and form zinc ions (Zn^{++}) which enter the solution at the metal surface. The electrons remain on the metal and give it a negative charge, thus building up an electrical double layer. The potential difference across this layer is the *electrode potential.* The process is reversible and, after a short time, zinc ions in solution begin to recombine with electrons on the surface of the metal to form zinc atoms. At equilibrium, the rate at which ions from the solution combine with electrons on the metal surface to form atoms is equal to the rate at which zinc atoms lose electrons to form ions in the solution, and the electrode exhibits its reversible potential.

The ionization of molecular hydrogen is also an oxidation.

$$\tfrac{1}{2}H_2 \rightleftharpoons H^+ + e$$

For this reaction the dissociation constant may be formulated as

$$K = \frac{[H^+][e]}{[H_2]^{1/2}}$$

where $[e]$ represents the *electron pressure* at the place where ionization occurs. This electron pressure must be closely related to what is measured as an "electrode potential."

Inasmuch as hydrogen is a gas, it cannot be used directly as an electrode. However, an inert metal, *e.g.*, platinum, covered with finely divided platinum (platinum black) adsorbs hydrogen; reversible dissociation of this hydrogen occurs on the surface, and the arrangement then functions as a hydrogen electrode. The potential of this electrode, from the formulation above, is a function of the ratio

$[H^+]/[H_2]^{1/2}$, *i.e.*, of the oxidized and reduced forms of hydrogen. The potential can be determined by comparison with a second electrode and measurement of the potential difference between the two electrodes. If the potential of one electrode is arbitrarily designated as zero, all other electrode potentials can be referred to it and their potentials expressed numerically in volts. This reference electrode is the *hydrogen electrode* whose potential is taken as zero for a solution containing hydrogen ions at unit activity in equilibrium with hydrogen gas at 1 atmosphere.

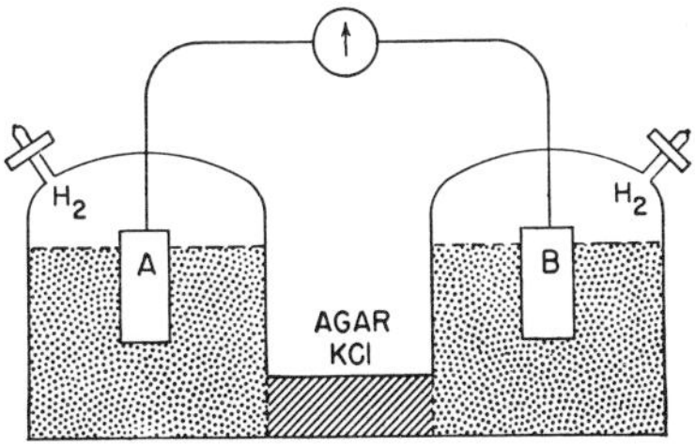

Fig. 15.1. An electrolytic cell consisting of two hydrogen electrodes.

Consider a cell consisting of two hydrogen electrodes, as depicted in Fig. 15.1, in which the two solutions are initially of equal $[H^+]$, but the hydrogen pressure in half-cell B is less than 1 atmosphere. If now the two electrodes are connected by a conducting wire, electrons will move through the wire from electrode A to electrode B. The following reactions occur.

$$\text{At A:}\quad \tfrac{1}{2}H_2 \longrightarrow H^+ + e$$
$$\text{At B:}\quad H^+ + e \longrightarrow \tfrac{1}{2}H_2$$

Effectively, the net result is transfer of hydrogen ions from B to A; current flows until, in both vessels, the *ratio* $[H^+]/[H_2]^{1/2}$ is identical. To maintain electrical neutrality, anions, *e.g.*, Cl^-, must flow through the salt bridge from B to A. If a potentiometer is interposed between B and A, the tendency of electrons to flow from A to B may be measured. This is the potential difference, E, which is expressed in volts. In this instance electrode A would be negative with respect to B in the exterior circuit since electrons are entering B and leaving A.

Ions in dilute solution obey the gas laws, and from these laws a quantitative expression can be derived for the potential difference between the electrodes A and B.

$$-E = \frac{RT}{F} \ln [H_2]_B^{1/2} \quad \text{or} \quad -E = \frac{RT}{2F} \ln [H_2]_B \qquad (1)$$

where R is the molar gas constant, 8.315 joules per degree, T is the absolute temperature, and F is the faraday, 96,500 coulombs. At 30°C., substituting in (1) and converting to $\log_{10}$,

$$\frac{8.315 \times 303 \times 2.303}{2 \times 96{,}500} = 0.03$$

hence

$$-E = 0.03 \log [H_2]_B \quad \text{volts} \qquad (2)$$

This expression is valid when A is the normal hydrogen electrode and the $[H^+]$ in B is unity ($1N$). If the H_2 pressure in B had been 1 atmosphere but the $[H^+]$ less than $1N$, electrons would have flowed from electrode B to electrode A, the electrode reactions would have been reversed, and the electrode potential at B would be expressed

$$E = \frac{RT}{\mathsf{F}} \ln [H^+]_B$$

Substituting as above and converting to $\log_{10}$,

$$E = 0.06 \log [H^+]_B$$

or

$$-E = 0.06 \text{ pH}_B \quad (3)$$

By combining both expressions, a statement is obtained for the potential difference between the normal hydrogen electrode and a hydrogen electrode in any other dilute aqueous solution.

$$E = \frac{RT}{\mathsf{F}} \ln \frac{[H^+]}{[H_2]^{1/2}} \quad (4)$$

$$E = \frac{RT}{\mathsf{F}} \ln [H^+] - \frac{RT}{2\mathsf{F}} \ln [H_2] \quad (5)$$

At 30°,

$$E = 0.03 \log \frac{1}{[H_2]} - 0.06 \text{ pH} \quad (6)$$

If $[H_2]$ is maintained at some known pressure, one can perform potentiometric pH measurements. If the pH is known, the hydrogen pressure can be determined.

Since the oxidation of an organic compound occurs by dehydrogenation, a solution of such a compound may be thought of as exerting an infinitesimal hydrogen pressure. Various autoxidizable dyes can be reduced by gaseous hydrogen in the presence of a suitable catalyst, *e.g.*, platinum black; with the reduced form of methylene blue (MbH_2) as an example, the reaction is

$$MbH_2 \rightleftharpoons Mb + H_2$$

A solution of the reduced dye, by this formulation, exerts a small but real hydrogen pressure which can be measured with an inert electrode or used to reduce some other autoxidizable substance. For this reaction

$$K = \frac{[Mb][H_2]}{[MbH_2]}$$

and

$$[H_2] = \frac{K[MbH_2]}{[Mb]}$$

The hydrogen pressure in equilibrium with this system thus depends on the *ratio* of reduced to oxidized form of the dye; the potential of an inert electrode in this system is that of a hydrogen electrode at this hydrogen pressure and varies with pH as does any hydrogen electrode.

We may, therefore, substitute for $[H_2]$ in equation (5),

$$[H_2] = K\frac{[\text{reductant}]}{[\text{oxidant}]}$$

so that

$$E = \frac{RT}{\mathsf{F}} \ln [H^+] - \frac{RT}{2\mathsf{F}} \ln K\frac{[\text{reductant}]}{[\text{oxidant}]} \tag{7}$$

or

$$E = \frac{RT}{\mathsf{F}} \ln [H^+] - \frac{RT}{2\mathsf{F}} \ln K + \frac{RT}{2\mathsf{F}} \ln \frac{[\text{oxidant}]}{[\text{reductant}]} \tag{8}$$

For any system $(RT/2\mathsf{F}) \ln K$ must be a constant, and if measurements are made in well-buffered media so that $(RT/\mathsf{F}) \ln [H^+]$ is constant, these may be combined as a new constant E_0'. Since for reactions in general one or more electrons may be involved per molecule, the term n is introduced for the number of electrons transferred per mole. The new expression may be formulated as

$$E = E_0' + \frac{RT}{n\mathsf{F}} \ln \frac{[\text{oxidant}]}{[\text{reductant}]} \tag{9}$$

At 30° this simplifies to

$$E = E_0' + \frac{0.06}{n} \log \frac{[\text{oxidant}]}{[\text{reductant}]} \tag{10}$$

It may be seen that E_0' is the potential of the half-reduced system at some stipulated pH and temperature since, with a ratio value of 1, the second term is then zero.

The *redox* (oxidation-reduction) *potential*, E, of a given solution in expression (10) is, then, the "electron pressure" that this solution exerts on an inert electrode and can be measured with appropriate apparatus and a potentiometer. Returning to the simplest example, the dissociation of hydrogen to protons and electrons, it is apparent that such a solution must also exert a "proton pressure." Protons do not readily "flow" over wires, and no simple *instrumental* arrangement permits a *direct* measurement of proton pressure. The electrometric measurement of pH described above is, rather, a measurement of the effects of variation in proton pressure on electron pressure, and it is the latter which is measured, thereby permitting calculation of the pH. Thus, the pH of a solution represents its proton pressure, exactly as E measures the electron pressure. Just as a system composed of an acid and its salt resists changes in pH, or buffers a given solution, a system composed of oxidant and reductant also resists changes in redox potential E, or *poises* the solution.

Table 15.1 includes E_0' values for some systems of biological interest. The word "system" is used here in the sense of a mixture of the oxidized and reduced forms of a given substance, *e.g.*, quinone-hydroquinone, pyruvate-lactate, etc. The values shown are for electrode potentials. It may be remarked that many handbooks and reference works use a different convention for designating the electron pressure of an oxidation-reduction system, *viz.*, the "electrode E.M.F." (electromotive force). The latter is particularly convenient in considerations of the behavior of metallic electrodes. Since electrode potential = −electrode E.M.F., the convention with respect to sign, + or −, is reversed in such tables.

From equation (6) it will be seen that of two systems at the same pH, the one with the more negative potential is the stronger reducing agent inasmuch as it is in equilibrium with a greater hydrogen pressure.

The oxidized member of a system is inherently capable of being reduced by the reduced member of a system with a more negative E'_0. According to equation (6), a solution which is 0.03 volt more negative than another has ten times the hydrogen pressure of the second. When two solutions of different potentials are mixed, reaction may proceed until equilibrium is attained; the final solution has a potential between that of the two original solutions. Thus, were a solution containing A and its reduced form, AH_2, mixed with a solution of B and BH_2, reaction would proceed until both systems attained the same potential. At that point,

$$E'_{0(B)} + \frac{RT}{n\mathsf{F}} \ln \frac{[B]}{[BH_2]} = E'_{0(A)} + \frac{RT}{n\mathsf{F}} \ln \frac{[A]}{[AH_2]}$$

and

$$E'_{0(B)} - E'_{0(A)} = \frac{RT}{n\mathsf{F}} \ln \frac{[A]}{[AH_2]} - \frac{RT}{n\mathsf{F}} \ln \frac{[B]}{[BH_2]}$$

from which

$$\Delta E'_0 = \frac{RT}{n\mathsf{F}} \ln \frac{[A][BH_2]}{[B][AH_2]} \tag{11}$$

Hence the difference in potential between the original solutions determines the relative amount of each reactant at equilibrium or, stated in another way, the extent to which reaction proceeds. The course of such a reaction may be followed by titration, analogous to acidimetric procedures. Instead of an indicator whose color changes with acidic dissociation, *e.g.*, phenolphthalein, a small amount of dye whose color is dependent on its state of oxidation may be used, *e.g.*, methylene blue. The same potentiometer and electrode may be employed for either type of titration. For acidimetry, potential changes are followed as the $[H^+]$ changes; during oxidation-reduction reactions, the potential changes with the "hydrogen pressure" or "electron pressure." Typical electrometric titration curves are shown in Fig. 15.2. Note the resemblance to the curves obtained by plotting pH changes in acidimetric titrations, as expected from the similarity in form of expression (10) and the Henderson-Hasselbalch equation (page 98).

Energy Relations in Oxidative Reactions. Since interest in oxidative reactions derives in large measure from the fact that they yield energy, let us consider some quantitative aspects of oxidative changes in relation to energy production.

If the reduced form of one system is mixed with the oxidized form of another, reaction proceeds according to the equation

$$AH_2 + B \rightleftharpoons A + BH_2$$

For this reaction, the standard free-energy change, in calories per mole, may be calculated from equilibrium data in the usual manner. K is the equilibrium constant.

$$K = \frac{[A][BH_2]}{[AH_2][B]} \tag{12}$$

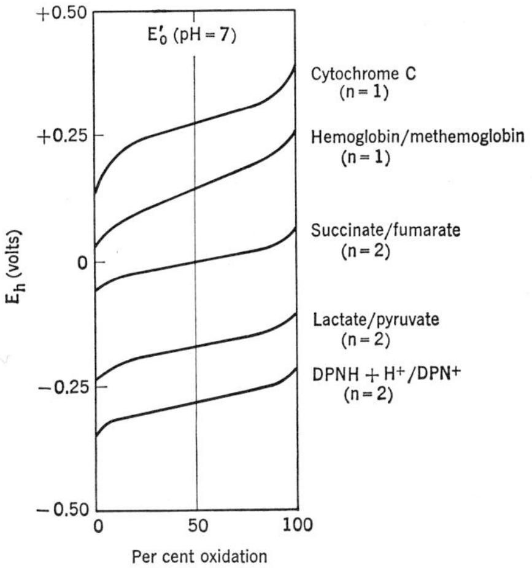

Fig. 15.2. Potentiometric titration curves. E_h = observed potential difference (in volts); n = number of electrons transferred per mole oxidized.

Actual determination of K, therefore, depends on the availability of adequate analytical methods for the various components. When the difference in potential, $\Delta E'_0$, between the two reacting systems is large, equilibrium may lie so far in one direction that accurate determination of the final concentration of AH_2 and B may be impossible. The free-energy change associated with the reaction may, however, by calculated from knowledge of the potentials of the two reacting systems. By substitution of the expression for K of equation (12) in equation (11) one obtains:

$$\Delta E'_0 = \frac{RT}{n\mathsf{F}} \ln K \qquad (13)$$

or

$$n\mathsf{F}\,\Delta E'_0 = RT \ln K$$

Since

$$-\Delta F^\circ = RT \ln K \qquad \text{(page 212)}$$

then

$$-\Delta F^\circ = n\mathsf{F}\,\Delta E'_0 \qquad (14)$$

where ΔF° is the standard free energy of the reaction, n is the number of electrons (or hydrogens) involved, F is the faraday (96,500 coulombs), and $\Delta E'_0$ is the difference between the E'_0 values of the two systems. The units of $\mathsf{F}\,\Delta E'_0$ are coulomb-volts or joules, which are readily converted to the usual units of free energy since 4.18 joules equal 1 g.-cal. The value obtained for ΔF° is that for the oxidation of one mole of reductant. Consider as an example the oxidation of malic acid to oxaloacetic acid by cytochrome c under circumstances such that equimolar concentrations always exist of each of the reactants of the two systems. Since the E'_0 values are −0.17 volt and 0.2 volt, respectively (Table 15.1), then

$$\Delta F^\circ = -n\mathsf{F}\,\Delta E'_0 = \frac{-2 \times 96{,}500 \times [0.2 - (-0.17)]}{4.18} = -18{,}246 \text{ cal.}$$

The oxidation of one mole of malic acid by cytochrome c, under these circumstances, results in the release of 18,246 cal., which could then be available under physiological circumstances for doing useful work. Were the malic acid oxidized by molecular oxygen, 45,715 cal. would be released, since E_0' for the reduction of oxygen is +0.82 volt.

HIGH-ENERGY PHOSPHATE COMPOUNDS

Earlier in this chapter it was stated that within cells a mechanism exists by means of which the free energy available from oxidative reactions may be utilized to drive endergonic processes. This is accomplished largely by trapping this energy through the formation of a special class of compounds, most of which are anhydrides of phosphoric acid. Present concepts of the biological significance of these compounds are due largely to Meyerhof, Lipmann, and Kalckar.

The free-energy change associated with hydrolysis of a given compound may be calculated from the equilibrium constant in the usual manner.

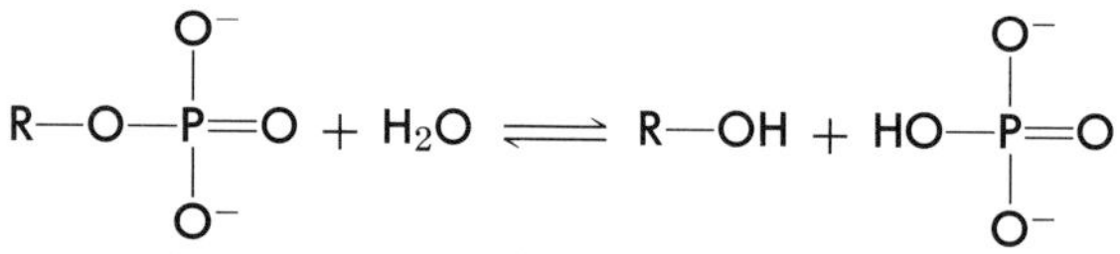

$$K = \frac{[ROH][HPO_4^{=}]}{[ROPO_3^{=}]}$$

The free-energy changes for the diverse phosphate derivatives found in biological systems range from about −2,000 to −13,000 cal. per mole. The smaller values, −2,000 to −5,000 cal. per mole are observed with simple esters, of which α-glycerophosphate and glucose 6-phosphate are typical. These are stable and, in the laboratory, can be hydrolyzed only by prolonged digestion with hot acid.

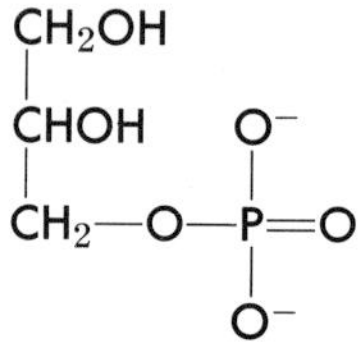

α-Glycerophosphate

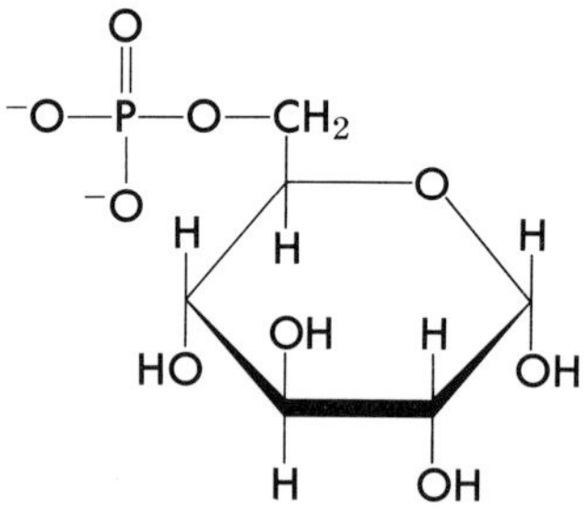

Glucose 6-phosphate

Numerous esters of this type occur in intermediary metabolism. A smaller group consists of those organic phosphates whose hydrolysis occurs with a $\Delta F°$ between −5,000 and −13,000 cal. per mole. These are relatively unstable and may be quantitatively hydrolyzed by $1N$ H^+ in a few minutes. Examples of such high-energy phosphate-containing compounds are the following:

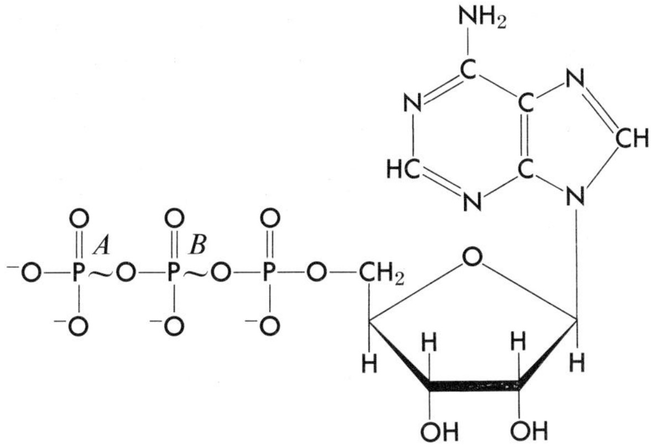

Adenosine triphosphate (ATP)

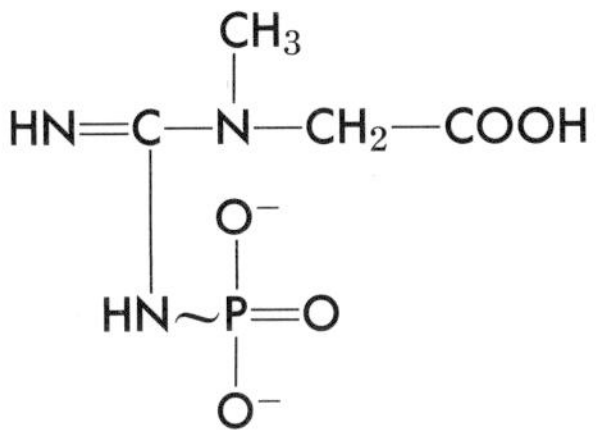

Creatine phosphate

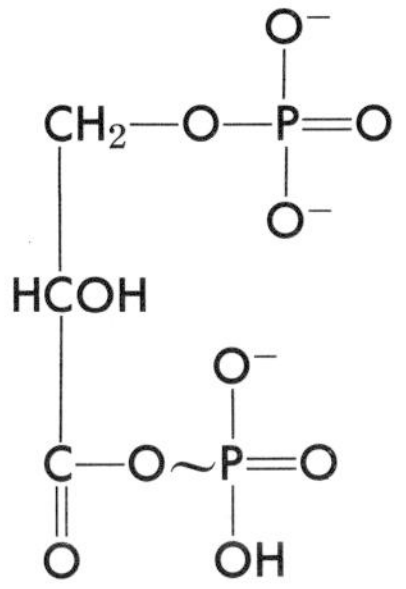

1,3-Diphosphoglyceric acid

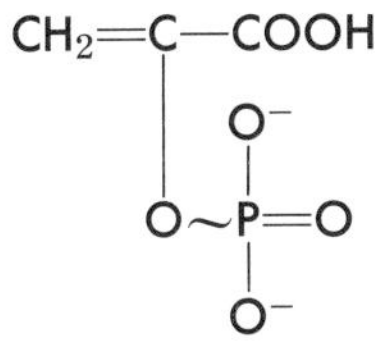

Phosphoenolpyruvic acid

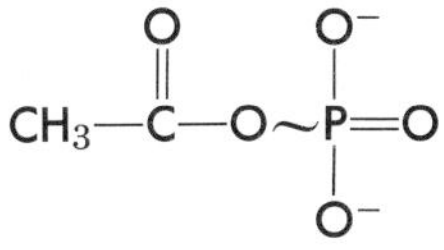

Acetyl phosphate

Three classes of high-energy phosphate compounds are known: acid anhydrides, phosphate esters of enols, and derivatives of phosphamic acid R—NH—PO_3H. It will be recalled that acid anhydrides, *e.g.*, P_2O_5, acetyl chloride, etc., are unstable in water and hydrolyze rapidly, with liberation of considerable heat. The free energy of hydrolysis, ΔF°, for phosphoenolpyruvate (13,000 cal. per mole at 25°C.) is among the highest of all known naturally occurring high-energy phosphate compounds. By convention, the ~ denotes the bond whose hydrolysis is accompanied by the release of a large amount of free energy. The relatively high potential energy that is made available on hydrolysis is a property of the structure of the phosphate compound as a whole and does not merely reside in the P—O bond which is ruptured by hydrolysis.

Several factors contribute in varying degree to the large release of free energy associated with hydrolysis of this group of compounds. In many, such as the anhydrides, the molecule is inherently unstable, because of the electron-withdrawing properties of the phosphoryl group which make less stable the electrophilic carbonyl carbon atom of the acyl group. A second contribution arises from the fact that the resonance energy of the hydrolysis products may substantially exceed that of the "high-energy compound." This is readily seen from the fact that the possible number of resonant forms of creatine + phosphate considerably exceeds that of creatine phosphate; similarly the number of resonant forms of acetate ion + P_i exceeds the number possible in acetyl phosphate. In general, the greater the number of possible resonant forms, the greater the stability of the system. In these instances, this reflects the fact, for example, that the π electrons of the oxygen linking the acetyl and the phosphate of acetyl phosphate cannot simultaneously satisfy the demand for electrons of the carbonyl and phosphoryl functions.

$$H_3C-\overset{\overset{\textstyle O}{\|}}{C}-O-\overset{\overset{\textstyle O}{\|}}{\underset{\underset{\textstyle O^-}{|}}{P}}-O^-$$

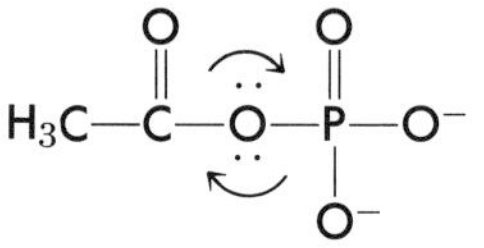

These phenomena are most evident in phosphoenolpyruvate. Finally, hydrolysis of many members of this group, at pH 7, results in an increase in charge. This is seen by comparison of the hydrolysis of glycerol and acetyl phosphates.

(a)
$$\begin{array}{l} CH_2OH \\ | \\ CHOH \quad\quad O^- \\ | \quad\quad\quad\; | \\ CH_2-O-P=O \\ \quad\quad\quad\; | \\ \quad\quad\quad\; O^- \end{array} + H_2O \rightleftharpoons \begin{array}{l} CH_2OH \\ | \\ CHOH \\ | \\ CH_2OH \end{array} + \begin{array}{l} O^- \\ | \\ HO-P=O \\ | \\ O^- \end{array}$$

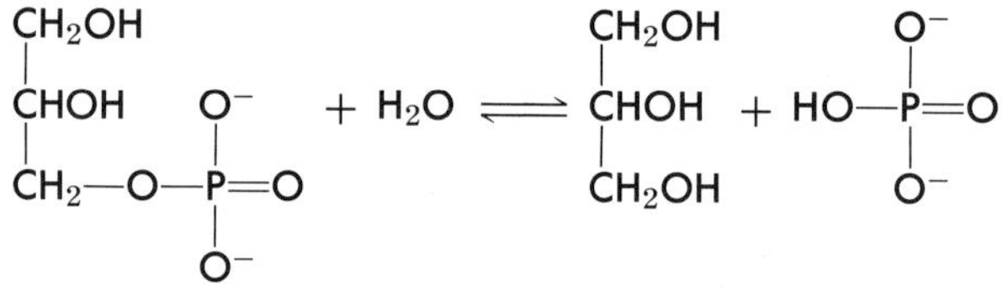

(b)
$$CH_3-\overset{\overset{\textstyle O}{\|}}{C}-O\sim\overset{\overset{\textstyle O^-}{|}}{\underset{\underset{\textstyle O^-}{|}}{P}}=O + H_2O \rightleftharpoons CH_3-\overset{\overset{\textstyle O}{\|}}{\underset{\underset{\textstyle O^-}{|}}{C}} + HO-\overset{\overset{\textstyle O^-}{|}}{\underset{\underset{\textstyle O^-}{|}}{P}}=O + H^+$$

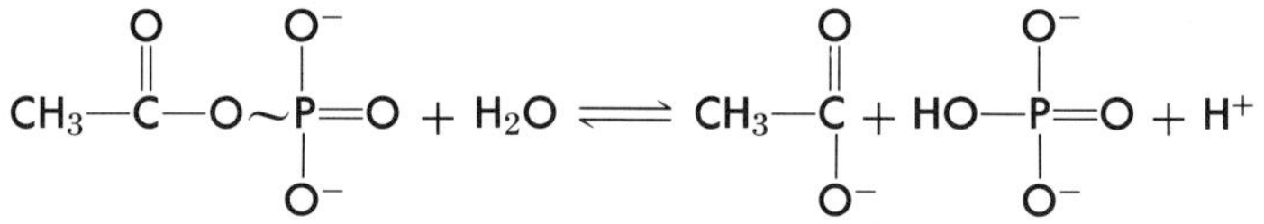

It will be seen that hydrolysis of a mole of acetyl phosphate results in liberation of a proton. Removal of the latter by the buffered medium makes a large contribution to the total change in free energy, driving the reaction, as represented above, to the right.

Other instances of high-energy compounds, in the sense that the free-energy change accompanying their hydrolysis is in the range $-6,000$ to $-11,000$ cal. per mole, are found in living systems. Particularly noteworthy are acyl thioesters, *i.e.*, compounds of the general structure R—CO—S—R' such as the fatty acyl esters of coenzyme A (page 480). The energy-rich nature of these compounds largely results from the restrictions on resonance resulting from low interaction between the π electrons of the sulfur bridge and the carbonyl function. This is exaggerated in acetoacetyl thioesters by the electrophilic terminal acyl group. Other high-energy classes include amino acyl esters of the ribose moiety of nucleic acids (page 318), as well as sulfonium compounds of the general structure

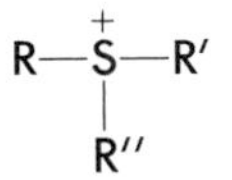

Table 15.2 presents the free-energy changes associated with hydrolysis of diverse compounds of biological interest.

Table 15.2: FREE ENERGY OF HYDROLYSIS OF SOME COMPOUNDS OF BIOLOGICAL INTEREST

Compound	$-\Delta F°$ *at pH 7 cal./mole*	*Compound*	$-\Delta F°$ *at pH 7 cal./mole*
Acetyladenylate	13,300	ATP, terminal bond (A)	7,400
Phosphoenolpyruvate	13,000	Sucrose	6,600
Acetyl phosphate	10,500	Phosphodiesters	*ca.* 6,000
Acetoacetyl CoA	10,500	Glucose 1-phosphate	5,000
S-Adenosylmethionine	10,000	Ethyl acetate	4,900
Phosphocreatine	9,000	Alanylglycine	4,000
Ethyl glycinate	8,500	Glutamine	3,400
Acetyl CoA	7,700	α-Glycerol phosphate	3,000
Uridine diphosphate glucose	7,600	Lactose	3,000
ATP, inner bond (B)	7,500	Peptide bond (internal in large polypeptide only)	500

Lipmann suggested that the unique merit of phosphate esters, aside from requirements of enzymic specificity, resides in the fact that phosphate confers kinetic stability on thermodynamically labile molecules. Thus, while $\Delta F°$ for hydrolysis of acetic anhydride, acetyl phosphate, and inorganic pyrophosphate are all of the same magnitude, they are stable in water for a few seconds, several hours, and years, respectively. It is the fact that ATP is stable in water which permits utilization of the energy of oxidation to drive endergonic reactions, since its simple hydrolysis would be wasteful and pointless in the biological economy.

It should be noted that the various values for free-energy changes cited above have been for $\Delta F°$, the free-energy change when each of the reactants is in the standard equimolar state. Since the actual free-energy change is defined by

$$-\Delta F = RT \ln K - RT \ln \frac{[c][d]}{[a][b]}$$

then, if the reactants in a biological system are maintained in a steady state in which [c][d] is not equal to [a][b], the true value for ΔF may be significantly greater or less than $\Delta F°$.

Phosphate Transfer. The participation of phosphate in metabolic processes was established by Harden and Young (page 393) in their classical studies of fermentation by yeast extracts. Many studies of metabolism in contracting muscle demonstrated that the energy for contraction, under anaerobic conditions, is in some manner supplied by the glycolytic process that converts glycogen to lactic acid. This process is described in detail in Chap. 18. Lundsgaard, however, observed that contraction also proceeded in muscles poisoned with iodoacetate, an inhibitor which prevents glycolysis. Further, these muscles continued to contract until their supply of creatine phosphate was exhausted by conversion to creatine and P_i. Lohmann then demonstrated that muscle does not possess an enzyme system for catalysis of this reaction unless adenine nucleotides are present; he formulated the process as follows:

(*a*) **Creatine ~ phosphate + ADP ⟶ creatine + ATP**		$\Delta F^\circ = -1500$ **cal.**
(*b*) **ATP ⟶ ADP + phosphate**		$\Delta F^\circ = -7500$ **cal.**
Net: Creatine ~ phosphate ⟶ creatine + phosphate		$\Delta F^\circ = -9000$ **cal.**

Thus, the reaction in which ATP is "hydrolyzed" seemed to be the immediate source of energy for the contractile process. This concept has since been expanded: *ATP is the immediate source of energy for most endergonic biological systems.*

The mechanism by which an exergonic reaction can drive an endergonic process may now be described. Consider the synthesis of an ester and the simultaneous apparent hydrolysis of ATP to AMP + PP_i.

(*a*) **RCOOH + HO—R′ ⇌ RCOOR′ +** H_2O	$\Delta F^\circ = +4000$ **cal.**
(*b*) **ATP +** H_2O **⇌ AMP +** PP_i	$\Delta F^\circ = -7500$ **cal.**
Net: RCOOH + HO—R′ + ATP ⟶ RCOOR′ + AMP + PP_i	$\Delta F^\circ = -3500$ **cal.**

The energy released in reaction (*b*) can be utilized to drive (*a*) *only* if they are coupled by way of a common intermediate, as in the following hypothetical case:

(*c*) **RCOOH + ATP ⇌ RCOO—AMP +** PP_i	$\Delta F^\circ = +2500$ **cal.**
(*d*) **RCOO—AMP + HO—R′ ⇌ RCOOR′ + AMP**	$\Delta F^\circ = -6000$ **cal.**
Net: RCOOH + HO—R′ + ATP ⟶ RCOOR′ + AMP + PP_i	$\Delta F^\circ = -3500$ **cal.**

The intermediate, RCOO—AMP, is an anhydride of the carboxylic acid and the phosphate of adenylic acid.

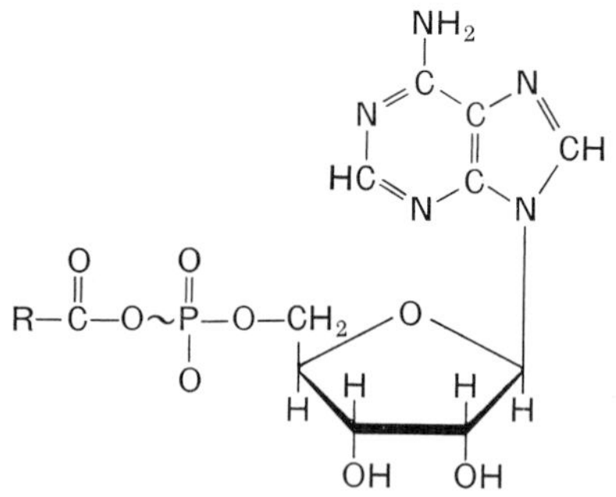

An acyl adenylate (RCOO—AMP)

There are numerous instances of such reactions in metabolism (*e.g.*, pages 326 and 479).

In the formation of an acyl adenylate, it is the bond between the pyrophosphate and adenylate moieties of ATP which is ruptured, so that pyrophosphate, rather than orthophosphate, appears as the final product. Hydrolytic cleavage of the two anhydride bonds of ATP proceeds with essentially identical changes in free energy. From the relatively small change in ΔF, it will be recognized that synthesis of the organic ester by this mechanism can proceed but is not strongly favored. However, synthesis is assured by an additional consideration. Pyrophosphate appears to have relatively few metabolic fates; most prominent among these is hydrolysis to orthophosphate catalyzed by *pyrophosphatases* present in all cells. The $\Delta F°$ for pyrophosphate hydrolysis is -7000 cal.; therefore, the immediate hydrolysis of the pyrophosphate formed in the reaction sequence required for ester synthesis effectively renders the latter irreversible. Instances of processes in which pyrophosphate hydrolysis renders irreversible an otherwise reversible process are nucleotide (page 626), polynucleotide (page 667), peptide bond (page 662) syntheses, and fatty acid activation (page 480).

Enzymes that catalyze transfer of phosphate from ATP to an acceptor are designated *kinases* and may be considered in two categories. Since $\Delta F°$ for the hydrolysis of each of the various high-energy compounds is of the same magnitude, kinases catalyzing transfer among these compounds may operate readily in both directions and actually do so in metabolism, *e.g.*,

$$\text{ADP} + \text{creatine phosphate} \rightleftharpoons \text{ATP} + \text{creatine} \qquad \Delta F° = -1500 \text{ cal.}, K = 10$$

In contrast, phosphate transfer with formation of low-energy compounds may be expected to proceed significantly in the forward direction only.

$$\text{ATP} + \text{glucose} \longrightarrow \text{ADP} + \text{glucose 6-phosphate} \qquad \Delta F° = -4500 \text{ cal.}, K = 4000$$

From these facts a basic concept emerges: to supply energy for endergonic biological processes, the respiratory process must, in some manner, be coupled with the synthesis of energy-rich phosphate compounds, specifically ATP. The latter is the "unit of currency" in metabolic energy transformations. The equation describing the over-all oxidation of glucose may now be stated as

$$C_6H_{12}O_6 + x\text{P}_i + x\text{ADP} \longrightarrow 6CO_2 + 6H_2O + x\text{ATP} + \text{unavailable energy}$$

The fraction of the total ΔF for the oxidation of glucose, or any other metabolic fuel, which is employed for ATP synthesis represents the true efficiency of the cellular respiration process, insofar as its objective is to supply energy for endergonic processes. $\Delta F°$ for ATP synthesis is about $+7000$ cal. per mole. Under physiological conditions, the true ΔF is about $+8{,}000$ to $+12{,}000$ cal. per mole. Since $\Delta F°$ for complete oxidation of one mole of glucose is $-686{,}000$ cal., this could *potentially* provide energy for the synthesis of 50 to 85 moles of ATP from ADP and P_i. However, simultaneous collisions between many molecules are unlikely, and it is apparent that if the combustion of a molecule of glucose is to result in the formation of a large number of ATP molecules, the total oxidation process must be accomplished by the summation of many individual oxidations. The manner in which this is achieved by coupling ATP synthesis with oxidations occurring in steps is considered in the following chapter.

REFERENCES

Books

Bray, H. G., and White, K., "Kinetics and Thermodynamics in Biochemistry," 2d ed., Academic Press, Inc., New York, 1966.

Edsall, J. T., and Wyman, J., "Biophysical Chemistry," vol. 1, chap. 4, Academic Press, Inc., New York, 1958.

Howett, L. F., "Oxidation-Reduction Potentials in Bacteriology and Biochemistry," 6th ed., E. and S. Livingston, Ltd., Edinburgh, 1950.

Ingram, D. J. E., "Free Radicals as Studied by Electron Spin Resonance," Academic Press, Inc., New York, 1958.

Kaplan, N. O., and Kennedy, E. P., eds., "Current Aspects of Biochemical Energetics," Academic Press, Inc., New York, 1966.

Klotz, I. M., "Chemical Thermodynamics," 2d ed., W. A. Benjamin, Inc., New York, 1964.

Klotz, I. M., "Energy Changes in Biochemical Reactions," Academic Press, Inc., New York, 1967.

Krebs, H. A., and Kornberg, H. L., "Energy Transformation in Living Matter," Springer-Verlag OHG, Berlin, 1957.

Lehninger, A. L., "Bioenergetics," W. A. Benjamin, Inc., New York, 1965.

Racker, E., "Mechanisms in Bioenergetics," Academic Press, Inc., New York, 1965.

Singer, T. P., ed., "Biological Oxidations," John Wiley & Sons, Inc., New York, 1967.

Review Articles

Axelrod, B., Enzymatic Phosphate Transfer, *Advances in Enzymol.,* **17,** 159–188, 1956.

Bock, R. M., Adenine Nucleotides and Properties of Pyrophosphate Compounds, in P. D. Boyer, H. Lardy, and K. Myrbäck, eds., "The Enzymes," vol. II, pp. 3–38, Academic Press, Inc., New York, 1960.

Ennor, A. H., and Morrison, J. F., Biochemistry of the Phosphagens and Related Guanidines, *Physiol. Revs.,* **38,** 631–674, 1958.

George, P., and Griffith, S. J., Electron Transfer and Enzyme Catalysis, in P. D. Boyer, H. Lardy, and K. Myrbäck, eds., "The Enzymes," vol. I, pp. 347–390, Academic Press, Inc., New York, 1959.

George, P., and Rutman, R. J., The 'High Energy Phosphate Bond' Concept, in J. A. V. Butler and B. Katz, eds., "Progress in Biophysics and Biophysical Chemistry," vol. X, pp. 1–53, Pergamon Press, New York, 1960.

Huennekens, F. M., and Whiteley, H. R., Phosphoric Acid Anhydrides and Other Energy-rich Compounds, in M. Florkin and H. S. Mason, eds., "Comparative Biochemistry," vol. I, pp. 107–180, Academic Press, Inc., New York, 1960.

Ingraham, L. L., and Pardee, A. B., Free Energy and Entropy in Metabolism, in D. M. Greenberg, ed., "Metabolic Pathways," vol. I, pp. 2–46, 3d ed., Academic Press, Inc., New York, 1967.

Leach, S. J., The Mechanism of Enzymic Oxidoreduction, *Advances in Enzymol.,* **15,** 1–48, 1954.

Lipmann, F., Biosynthetic Mechanisms, *Harvey Lectures,* **44,** 99–123, 1948–1949.

Michaelis, L., Fundamentals of Oxidation and Reduction, in D. E. Green, ed., "Currents in Biochemical Research," pp. 207–227, Interscience Publishers, Inc., New York, 1946.

Pullman, B., and Pullman, A., Electronic Structure of Energy-rich Phosphates, *Radiation Research,* Suppl. **2,** pp. 160–181, 1960.

16. Biological Oxidations. II

Citric Acid Cycle. The Mitochondrion. Electron Transport. Oxidative Phosphorylation

The energy requirements of most living cells are met by the oxidation of organic compounds by molecular oxygen with liberation of free energy. These processes underlie the respiration of cells and are accomplished by a highly ordered array of enzymes, organized in *mitochondria.* The mitochondria receive from the cytoplasm oxidizable substrates, such as pyruvic or fatty acids. An appropriate group of enzymes catalyzes a series of consecutive dehydrogenations and transformations of these substrates, resulting in their complete oxidation to CO_2 and H_2O. The electrons removed from the substrates during these oxidations flow through an organized arrangement of electron carriers, from that of lowest to that of highest potential, and thence to oxygen. In the course of this electron flow, much of the free energy thus made available appears as newly synthesized ATP, the common form of energy utilizable in the endergonic processes of living cells. In this chapter we shall briefly describe the enzymes and coenzymes required for dehydrogenation of the organic substrates, and then consider the major reaction sequence in which these dehydrogenations occur (the citric acid cycle), the organized array of electron carriers which convey electrons from the dehydrogenated substrates to O_2, and the means by which the free energy of this process is harnessed to the synthesis of ATP.

OXIDATIVE ENZYMES, COENZYMES, AND RESPIRATORY CARRIERS

Biological oxidations are catalyzed by enzymes which function in conjunction with a group of coenzymes and electron carriers. These substances, their properties, and their mechanism of action are presented, in some detail, in the following chapter. For appreciation of the subject matter of the present chapter, the following summary may suffice.

Each oxidative enzyme consists of a specific protein which functions in conjunction with a coenzyme or prosthetic group. The protein moiety confers substrate specificity on the system, activates both substrate and prosthetic group, and, frequently, alters the redox potential of the latter, which invariably participates in the reaction. Although there is a great variety of oxidative enzymes, there are only a few coenzymes; each enzyme is specific not only for its substrate but for its coenzyme as well. Only rarely does an oxidative enzyme catalyze a direct reaction in which the metabolite (MH_2) reacts with O_2, as would appear from the over-all reaction at the top of the following page.

$$MH_2 + \frac{1}{2}O_2 \longrightarrow M + H_2O$$

Rather, a transfer of electrons occurs from substrate to coenzyme,

$$MH_2 + Co \longrightarrow M + CoH_2$$

with subsequent oxidation of the reduced coenzyme in an independent process.

Coenzymes of Electron Transport. The coenzyme most frequently employed as acceptor of electrons from the substrate is *diphosphopyridine nucleotide* (DPN^+).

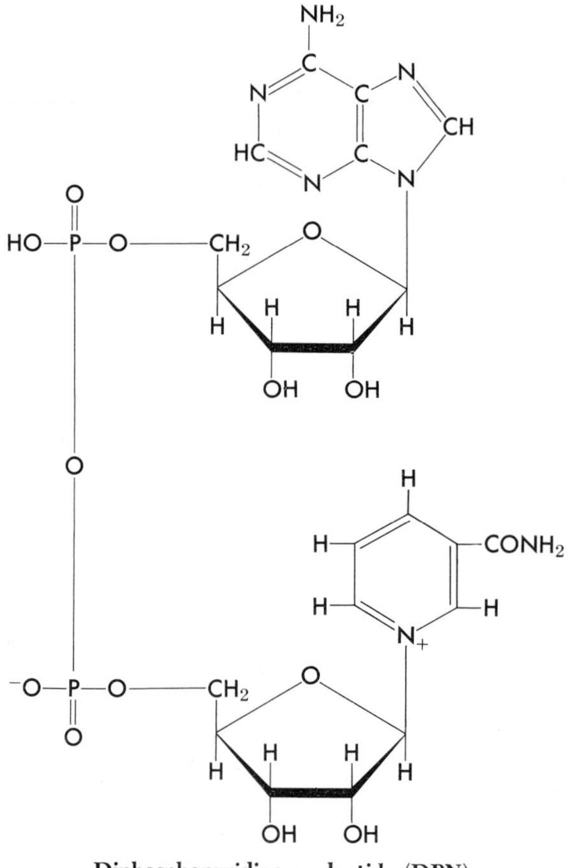

Diphosphopyridine nucleotide (DPN)

When bound to the appropriate site on a dehydrogenase protein, a hydride ion is transferred from the substrate to the nicotinamide moiety and a proton is liberated into the medium.

$$MH_2 + DPN^+ \longrightarrow M + DPNH + H^+$$

Details of this process are described in the following chapter. Less frequently, the coenzyme may be a phosphorylated derivative of DPN^+, *triphosphopyridine nucleotide* (TPN^+) (page 356), which functions in the same manner. In either case, the reduced coenzyme is not metabolized further on the same enzymic surface. Both the oxidized substrate and the reduced pyridine nucleotide (DPNH or TPNH) probably then dissociate from the dehydrogenase. DPNH and TPNH are not autoxidizable in the presence of O_2; reoxidation to DPN^+ or TPN^+ requires the participation of a second group of enzymes, to be described below.

This pair of pyridine nucleotides has been abbreviated as DPN^+ and TPN^+, respectively, since their discovery. Because DPN^+ is not a nucleotide of "diphosphopyridine," because this name does not indicate the presence either of nicotinamide or of adenine, and because this nomenclature is not in keeping with that used for many related compounds, an International Commission, in 1961, suggested that these two coenzymes be designated as *nicotinamide adenine dinucleotide* (NAD) and *nicotinamide adenine dinucleotide phosphate* (NADP). Their reduced forms, in this convention, are denoted as $NADH_2$ and $NADPH_2$, respectively. Both groups of terms are in current use. Until it is determined which terminology will achieve general acceptance, it has seemed appropriate to continue the use of DPN and TPN in this edition.

A second, large group of oxidative enzymes employs as cofactor either of two derivatives of the vitamin riboflavin (Chap. 49). These are flavin mononucleotide (FMN) and flavin adenine dinucleotide (FAD), respectively. In contrast to the ready dissociation of pyridine nucleotides from dehydrogenases, the two flavin nucleotides are invariably tightly bound, the combination being termed a *flavoprotein.* Many flavoproteins function by transfer of electrons, effectively as hydrogen, from an organic substrate to the riboflavin component of a flavin coenzyme. Of particular interest is a mitochondrial *DPNH dehydrogenase,* which catalyzes reduction of its flavin coenzyme by reduced pyridine nucleotide, as shown on page 324. The fate of the reduced flavoprotein will be considered later in this chapter. It may be noted that the riboflavin moiety in the structure on page 324 is a derivative of D-ribitol whereas the adenosine is a β-ribofuranoside derivative.

Of significance in respiration is a group of electron carriers which, properly speaking, are not enzymes. One of these is the hydrophobic compound, *ubiquinone* or coenzyme Q (page 340).

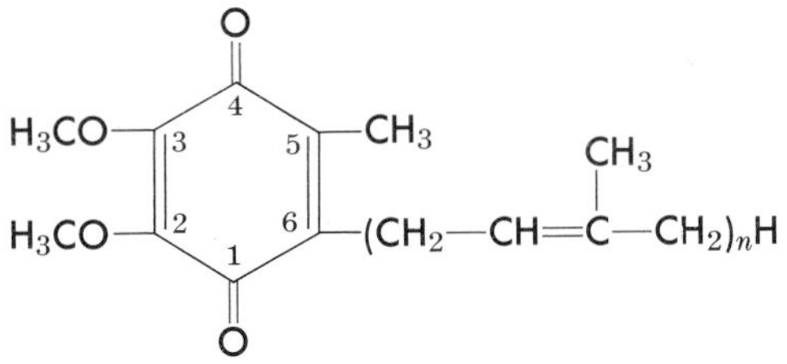

Ubiquinone

In ubiquinone obtained from diverse biological sources, n in the above formula varies from 6 in some yeasts to 10 in mammalian liver. Ubiquinone forms a relatively stable semiquinone on partial reduction and hence could participate in both 1- and 2-electron transfers.

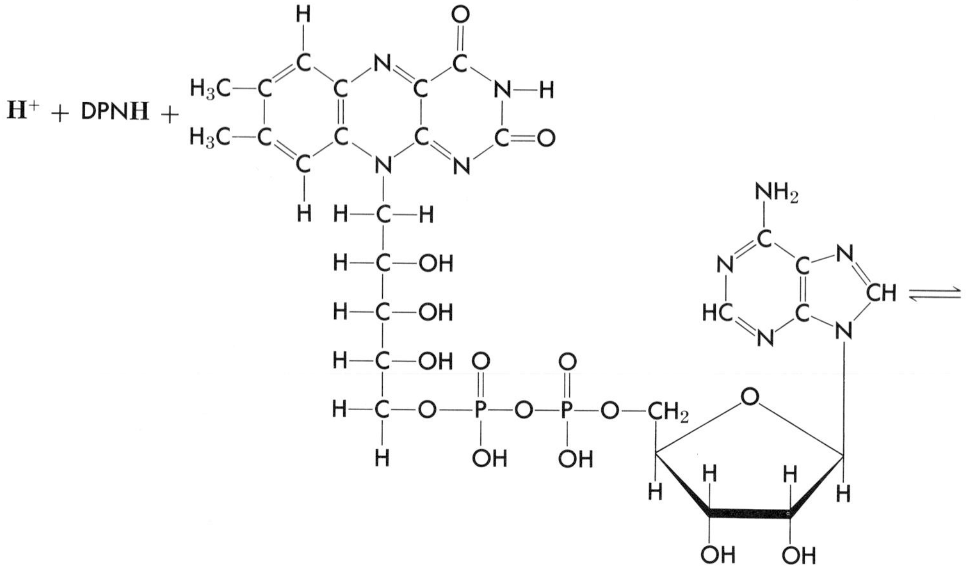

Flavin adenine dinucleotide (FAD)

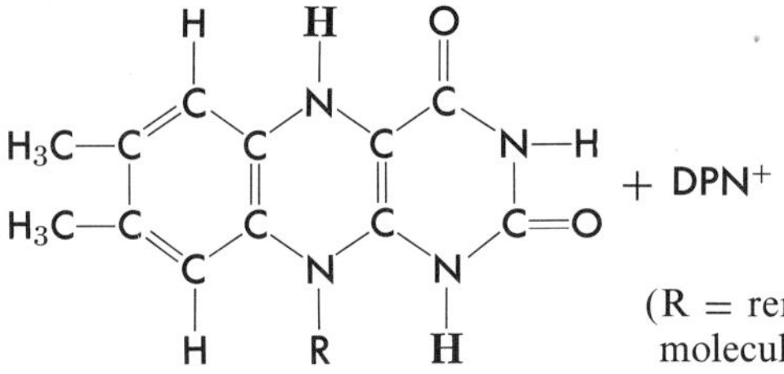

(R = remainder of molecule, above)

FADH$_2$

A molecular arrangement that functions in electron transport in mitochondria, microsomes, and chloroplasts, but which is poorly understood, is nonheme iron. This may be present in simple iron-proteins, or, in the case of several flavoproteins, bound to protein by linkages that are unknown.

All the other known electron carriers in mammalian cells are *cytochromes.* These are hemoproteins (page 177) differing from one another in their proteins, in the nature of the side chains of the porphyrin, and in the mode of attachment of the heme-like group to the protein. The cytochromes are most conveniently identified by the absorption spectra of their reduced forms. Of the group, cytochrome b is the member of lowest potential, which increases through c_1 and c to cytochromes a and a_3. The latter pair constitute *cytochrome oxidase,* a hemoprotein which, like the other members, undergoes alternate reduction and reoxidation of the iron constituent during normal function. Uniquely, the reduced form of cytochrome oxidase can reduce molecular oxygen. Indeed, it is the only component of mitochondria that can readily do so, and reoxidation of this cytochrome is the predominant mechanism for reduction of oxygen in mammalian tissues.

CITRIC ACID CYCLE—GENERAL CONSIDERATIONS

Although many organic compounds are oxidized by respiring tissues, one group of reactions may be set apart as the major reaction sequence which provides electrons to the transport system that accomplishes reduction of oxygen while generating ATP. This series of reactions is known as the *citric acid cycle,* the *tricarboxylic acid cycle,* or the *Krebs cycle.*

Figure 16.1 presents the major intermediates of the citric acid cycle. The reaction sequence may be pictured as commencing (reaction 1) by condensation of acetyl CoA with oxaloacetic acid to yield the 6-carbon tricarboxylic acid, citric acid. By successive loss and recapture of water, this compound is rearranged into isocitric acid, which is oxidized and decarboxylated to α-ketoglutaric acid. Succinic acid is formed from this compound by oxidative decarboxylation, and then, by an oxidation, a hydration, and a second oxidation, a molecule of oxaloacetic acid is again formed.

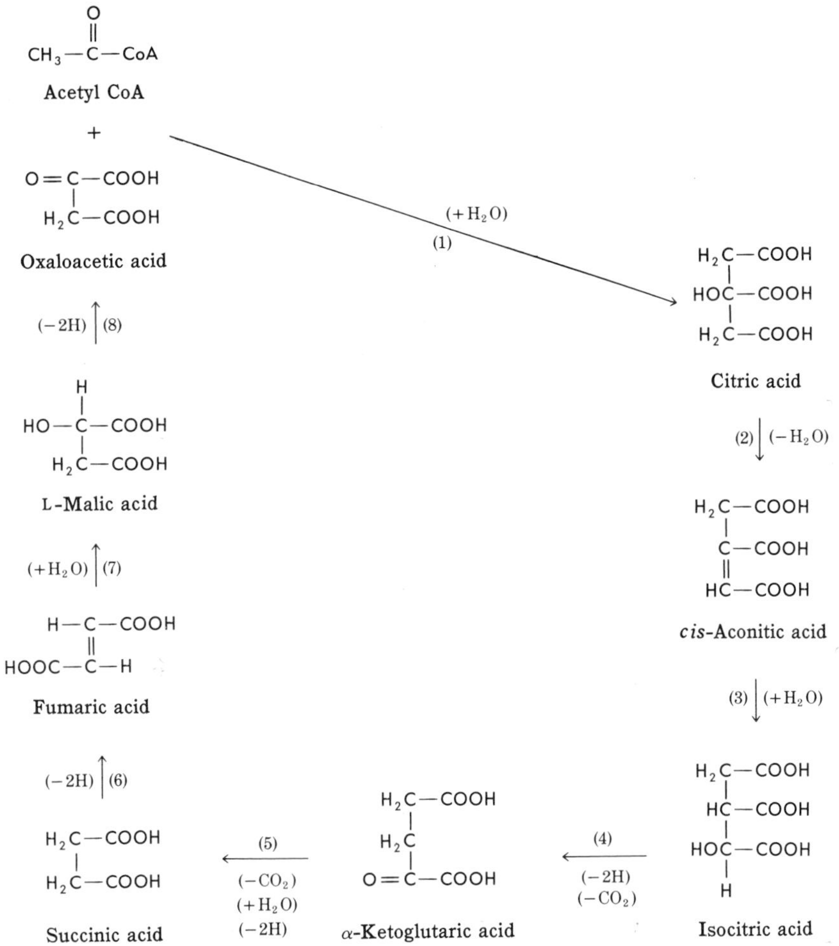

Fig. 16.1. Major intermediates of the citric acid cycle. The numbers in parentheses are for reference to identify these reactions in the text.

For each revolution of the cycle, one mole of acetyl CoA is consumed and two moles of CO_2 are evolved. Oxaloacetic acid, which is utilized in the initial condensation, is regenerated, permitting the process to operate in a continuous manner as long as acetyl CoA continues to enter the cycle and hydrogen atoms and CO_2 are removed. Four of the individual reactions are dehydrogenations, three of which result in the formation of reduced pyridine nucleotides. It is the oxidation of DPNH via the electron transport chain, to be considered later (page 338*ff.*), which delivers most of the energy available from respiration (Chap. 14). The net accomplishment of the citric acid cycle, per revolution, may be represented as follows.

$$CH_3COOH \longrightarrow 2CO_2 + 8e + 4H^+ + \text{energy}$$

The electrons are delivered to O_2 via the electron transport chain. Some aspects of the individual reactions of the cycle may now be considered.

CITRIC ACID CYCLE—INDIVIDUAL REACTIONS

Acetyl Coenzyme A. Much of the succeeding chapters dealing with the metabolism of carbohydrates, lipids, and amino acids will relate to their transformation into products that can enter the tricarboxylic acid cycle. The most important of these products, quantitatively, is "active acetyl," the acetyl thioester of coenzyme A. This coenzyme is composed of adenosine 3′-phosphate-5′-pyrophosphate, bound in ester linkage to the vitamin pantothenic acid, which, in turn, is attached in amide linkage to β-mercaptoethylamine. The acetyl group considered here is linked to the sulfur of coenzyme A as a thioester. In this text, coenzyme A is frequently abbreviated as either CoA or CoA—SH. The —SH shown in the latter abbreviation refers to the sulfhydryl group of coenzyme A and is not intended to represent an additional sulfhydryl group.

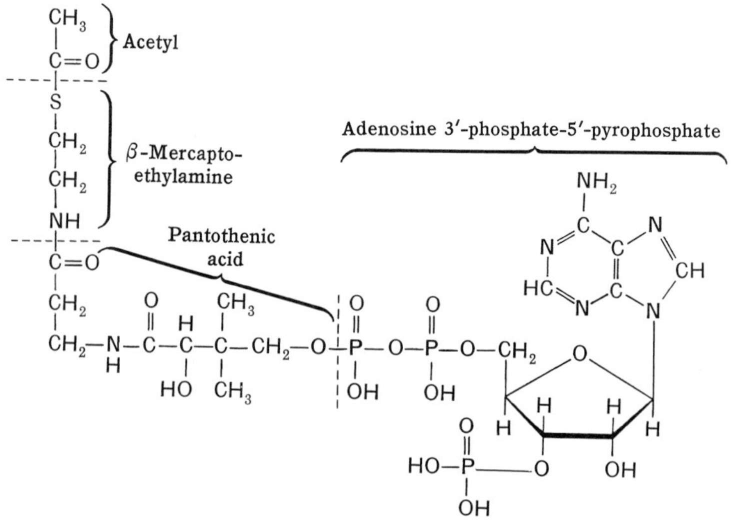

Acetyl coenzyme A (acetyl CoA)

Acetate in the reactive form of acetyl CoA is utilized in a variety of biological processes. As will be seen subsequently, acetyl CoA is the precursor for biosynthesis of fatty acids and sterols, can give rise to acetoacetic acid, and is the biological acetylating agent in the synthesis of such compounds as acetylcholine, acetylsulfanilamide, N-acetyl sugars, and the acetylated N-termini of peptide chains.

Acetyl CoA may be synthesized by organisms from acetic acid itself. In view of the relatively small quantities of acetic acid normally available to mammalian tissues, this may be, quantitatively, a relatively unimportant process. However, by virtue of this activation of acetate, biochemists have utilized this reaction to study the fates of acetyl CoA. The activation requires adenosine triphosphate (ATP) and yields adenylic acid (AMP) and inorganic pyrophosphate (PP_i). Although catalyzed by a single enzyme, the reaction appears to proceed in two steps.

(*a*)	$\text{ATP} + \text{acetate} \rightleftharpoons \text{acetyl-AMP} + PP_i$
(*b*)	$\text{Acetyl-AMP} + \text{CoA} \rightleftharpoons \text{AMP} + \text{acetyl CoA}$
Sum:	$\text{ATP} + \text{acetate} + \text{CoA} \rightleftharpoons \text{AMP} + PP_i + \text{acetyl CoA}$

Acetyladenylate is the mixed anhydride of the carboxyl of acetic acid and the phosphate of adenylic acid. Its structure is that shown on page 318 for the general form of an acyl adenylate. No *free* acetyladenylate (acetyl-AMP) appears during the reaction. This is an invariant property of all systems in which an acyl adenylate participates as an intermediate. It is bound tightly to the locus on the enzyme where it is formed, and the subsequent reaction occurs with this enzyme-bound intermediate. This reaction sequence seems likely since reaction (*b*) is catalyzed by the enzyme with CoA and synthetic AMP-acetate as substrates and since the enzyme catalyzes exchange of $^{32}PP_i$ with ATP, *i.e.*, reversal of reaction (*a*), but only in the presence of acetate. Similar enzyme-bound acyl adenylates are formed as intermediates in the activation of long-chain fatty acids (Chap. 21) and of amino acids (Chap. 29). Hydrolysis of acetyl CoA

$$CH_3CO\text{—}SCoA + HOH \rightleftharpoons CH_3COO^- + HS\text{—}CoA + H^+$$

proceeds with a free-energy change, $\Delta F°$, of approximately $-10{,}000$ cal. per mole. This is the energy which permits the functioning of acetyl CoA in biological acetylations and which drives the formation of citric acid.

Origin of Acetyl Coenzyme A. As seen in Fig. 16.1, the source of acetyl CoA is essentially irrelevant to the cyclic process here considered, but because acetyl CoA is the fuel that is oxidized by this process, it is appropriate to examine its origin. The two major sources derive from the metabolism of glucose (Chap. 18) and fatty acids (Chap. 21). The process of glycolysis (Chap. 18) consists of a series of transformations by which each glucose molecule is transformed, in the cell cytoplasm, into two molecules of pyruvic acid. The latter compound may then enter a mitochondrion, where it is oxidized according to the following over-all equation.

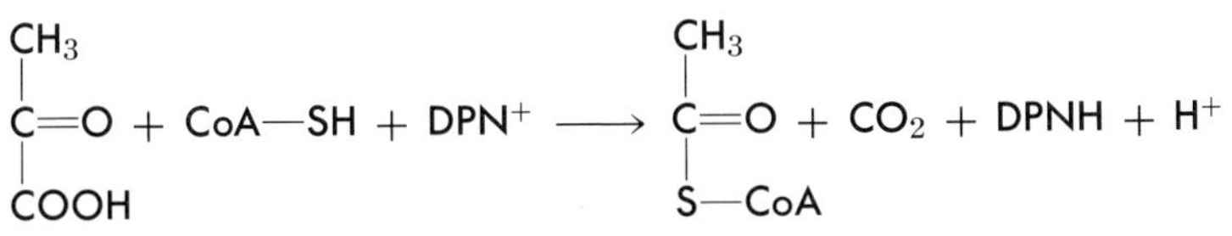

This process requires the cooperation of three enzymes and four cofactors present in a tightly knit complex which constitutes *pyruvic acid dehydrogenase*. The entire process proceeds with a favorable free-energy change; $\Delta F°$ for formation of acetyl CoA and DPNH in the manner shown is about -8000 cal.

The initial event is a reaction catalyzed by *pyruvic acid decarboxylase* between pyruvic acid and thiamine pyrophosphate (ThPP), the pyrophosphate ester of vitamin B_1 (Chap. 49), resulting in evolution of CO_2 and formation of a compound that serves as "active acetaldehyde," α-hydroxyethyl thiamine pyrophosphate.

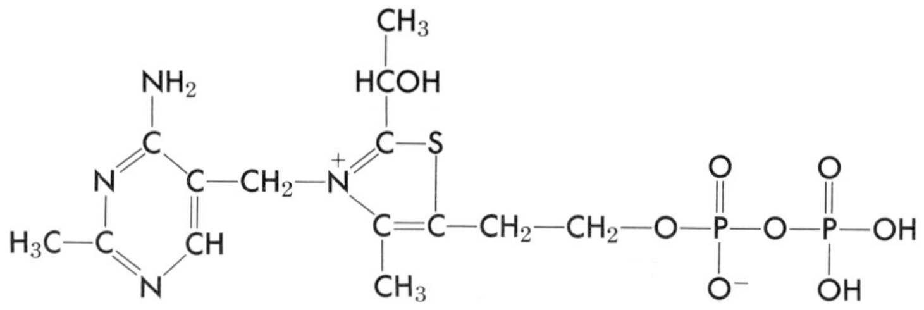

α-Hydroxyethyl thiamine pyrophosphate

The "active acetaldehyde" is transferred to lipoic acid, linked to the ε-amino group of a lysine residue of the enzyme, *dihydrolipoyl transacetylase*.

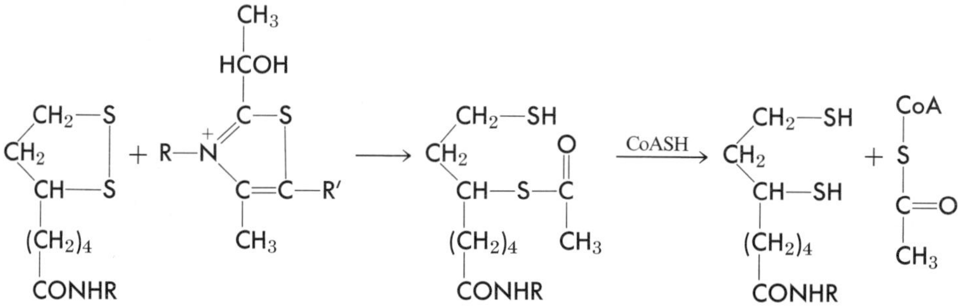

Lipoamide **α-Hydroxyethyl thiamine pyrophosphate** **Acetyl-lipoamide** **Dihydrolipoamide** **Acetyl CoA**

Effectively, the aldehyde is oxidized to acetic acid while the lipoamide disulfide is reduced to a disulfhydryl compound. The acetyllipoamide is next attacked by the sulfhydryl group of coenzyme A; the acetyl group is transferred to the thiol group of coenzyme A, liberating the disulfhydryl form of lipoamide.

At this stage, with respect to the pyruvic acid, the reaction is complete; it has been oxidized to acetyl CoA + CO_2. The thiamine pyrophosphate has also completed its catalytic cycle but the lipoamide, originally a disulfide, is in the disulfhydryl form. It is reoxidized by a flavoprotein, *dihydrolipoyl dehydrogenase,* which catalyzes the following reaction.

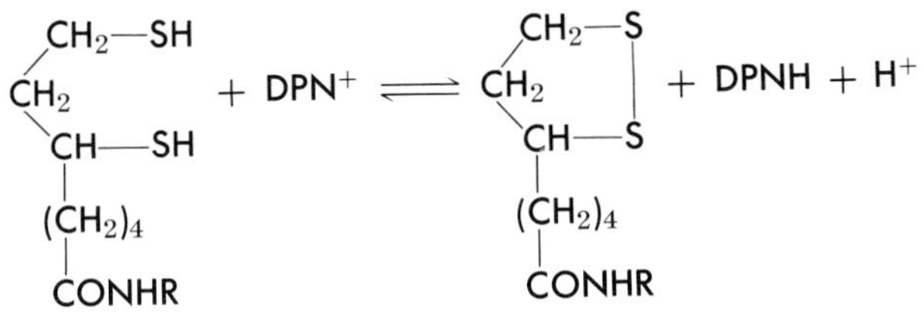

$$\begin{array}{c} \mathrm{CH_2-SH} \\ \mathrm{CH_2} \\ \mathrm{CH-SH} \\ | \\ (\mathrm{CH_2})_4 \\ | \\ \mathrm{CONHR} \end{array} + \mathrm{DPN^+} \rightleftharpoons \begin{array}{c} \mathrm{CH_2-S} \\ \mathrm{CH_2} \quad | \\ \mathrm{CH-S} \\ | \\ (\mathrm{CH_2})_4 \\ | \\ \mathrm{CONHR} \end{array} + \mathrm{DPNH} + \mathrm{H^+}$$

The entire reaction sequence occurs on the surface of a multienzyme complex. As isolated from *Escherichia coli*, the complex has a molecular weight of about 4.5×10^6 and is composed of about 12 molecules of the decarboxylase (molecular weight = 183,000), 6 molecules of the dihydrolipoyl dehydrogenase (molecular weight = 112,000), and 24 molecules of the transacetylase (molecular weight = 70,000). Since there are two lipoyl residues per 70,000 g., the last may actually be 48 molecules of a unit of molecular weight = 35,000. Evidence suggests that a similar or identical complex must exist in the living cell. Figure 16.2 schematically depicts a proposed mechanism for the functioning of the complex.

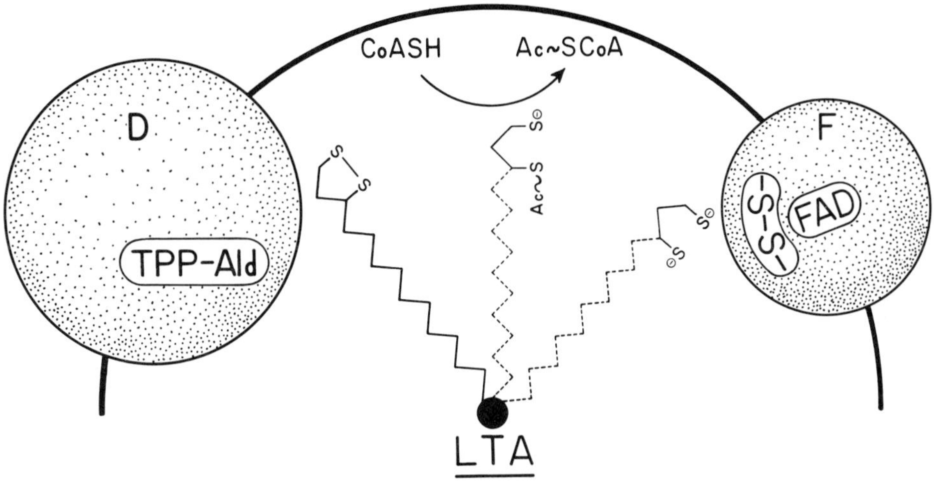

Fig. 16.2. A schematic representation of the possible rotation of a lipoyllysyl moiety between α-hydroxy-ethylthiamine pyrophosphate (TPP-Ald) bound to pyruvic acid decarboxylase (D), the site for acetyl transfer to CoA, and the reactive disulfide of the flavoprotein, dihydrolipoyl dehydrogenase (F). The lipoyllysyl moiety is an integral part of dihydrolipoyl transacetylase (LTA). The net charge on the lipoyl moiety during its cycle of transformations may be 0, minus 1, or minus 2. This change in net charge may provide the driving force for displacement of the lipoyl moiety from one site to the next within the complex. (*Courtesy of Dr. L. J. Reed.*)

Acetyl CoA + Oxaloacetic Acid (Reaction 1, Fig. 16.1). The crystalline "condensing enzyme," *citrate synthetase*, catalyzing the formation of citric acid, was isolated from pig heart by Ochoa and Stern. The equilibrium for the reaction

$$\mathrm{H_2O} + \underset{\substack{\| \\ \mathrm{O}}}{\mathrm{CH_3C}}-\mathrm{S-CoA} + \begin{array}{c} \mathrm{O=C-COOH} \\ | \\ \mathrm{H_2C-COOH} \end{array} \rightleftharpoons \begin{array}{c} \mathrm{H_2C-COOH} \\ | \\ \mathrm{HOC-COOH} \\ | \\ \mathrm{H_2C-COOH} \end{array} + \mathrm{HS-CoA}$$

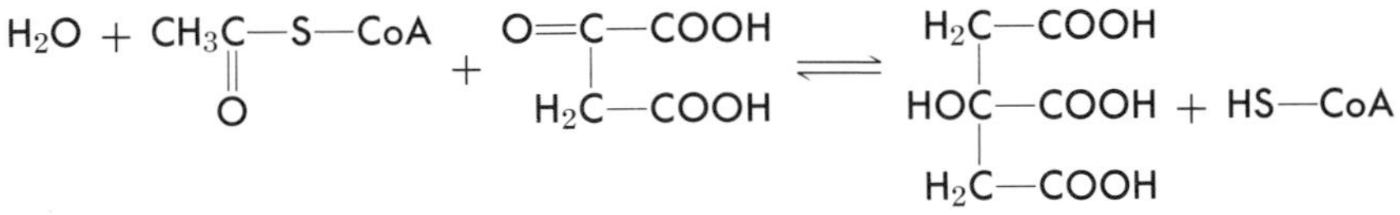

Acetyl CoA   Oxaloacetic acid   Citric acid   CoA

lies far to the right; under physiological conditions, $K = 5 \times 10^5$, $\Delta F^\circ = -7800$ cal.

In forming citric acid, acetyl CoA behaves as though the methyl rather than the carboxyl carbon of acetic acid is being activated. This is true of a number of reactions of acetyl CoA and is in contrast to such transacylating fates of acetyl CoA as the formation of acetylcholine, acetylsulfanilamide, and acetyl phosphate.

However, the mechanism of this condensation has not been established. In the absence of oxaloacetate, the enzyme fails to exchange hydrogen on the "activated methyl" group with protons from the medium. Intermediate formation of citryl CoA has been suggested but not demonstrated. Particularly noteworthy is the fact that the enzyme is inhibited by physiological levels of ATP, which serves as a negative modifier, markedly increasing K_m for acetyl CoA. Thus, when the requirement for oxidizing acetyl CoA in order to provide ATP is minimal, the metabolism of acetyl CoA is shunted into other channels such as acetoacetate and fatty acid formation.

Formation of α-Ketoglutaric Acid (Reaction 4, Fig. 16.1). Isocitric acid (Reactions 2 and 3, Fig. 16.1; see pages 333–334) is oxidized by transfer of electrons to a pyridine nucleotide. Animal tissues contain two *isocitric acid dehydrogenases,* specific, respectively, for transfer to TPN^+ and DPN^+. The Mg^{++}-dependent, DPN-specific enzyme is present exclusively in mitochondria; the enzyme in cytoplasm is Mn^{++}-dependent and TPN-specific. The mechanism of action of the latter has been shown to involve intermediary formation of enzyme-bound oxalosuccinate.

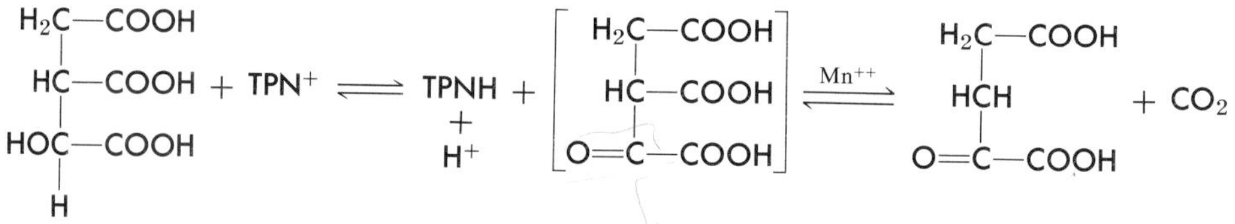

Isocitric acid Oxalosuccinic acid α-Ketoglutaric acid

The TPN-specific isocitric acid dehydrogenases are typical of a group of β-hydroxy acid dehydrogenases. In no instance does the β-keto acid formed in the initial dehydrogenation actually leave the enzymic surface; the second step, decarboxylation, follows immediately but is slow and rate-limiting in the absence of Mn^{++}. If, however, free β-keto acid (oxalosuccinic acid in this instance) is added to the enzyme, decarboxylation occurs readily if Mn^{++} is present. Decarboxylation is catalyzed by formation of an unstable, enzyme-bound chelate of Mn^{++} and the β-keto acid. Similar enzymes include "malic enzyme" (page 408) and 6-phosphogluconic acid dehydrogenase (page 416). The equilibrium position of the net reaction favors α-ketoglutaric acid formation; under physiological conditions, $\Delta F^\circ \approx -5000$ cal.

There is no evidence for oxalosuccinate as an intermediate in the process catalyzed by the mitochondrial enzyme. The velocity of the catalysis in this case is markedly dependent on the concentrations of substrate, DPN^+, Mg^{++}, and a positive modifier, AMP. Each enzyme molecule appears to possess two catalytic sites, at each of which may be bound one isocitrate, one Mg^{++} and one DPN^+. In addition there are two regulatable AMP-binding sites and two regulatable isocitrate-binding sites. Binding of AMP appears to induce a conformational change resulting in an enhanced affinity for isocitrate at all isocitrate-binding sites as well as for DPN^+ and Mg^{++} at the catalytic sites. Isocitrate binding at the catalytic sites is cooperative and enhances the ease of binding at all unfilled sites. The net result, then, is to facilitate reduction of DPN^+ by isocitrate under conditions of low supply of ATP.

α-Ketoglutaric acid represents a point of convergence in the metabolic pathways of carbohydrates, lipids, and certain amino acids. Glutamic acid can by transamination or oxidation yield α-ketoglutaric acid; this appears to be a significant source of the latter acid in mammalian metabolism (Chap. 23). Moreover, all amino acids that can yield glutamic acid in their metabolism are also potential precursors of α-ketoglutaric acid; these include ornithine, proline, glutamine, and histidine (Chap. 26). Conversely, glutamic acid and its derivatives are formed from α-ketoglutaric acid, which is produced by the citric acid cycle.

α-Ketoglutaric Acid $\longrightarrow$ Succinic Acid (Reaction 5, Fig. 16.1). The oxidation of α-ketoglutaric acid is analogous to that of pyruvic acid (page 327). The process, for which $\Delta F^\circ = -8000$ cal. per mole, is catalyzed by a multienzyme complex, molecular weight $\approx 2 \times 10^6$, which seems to be an octomer of a fundamental unit consisting of at least one molecule of each of the three participating enzymes.

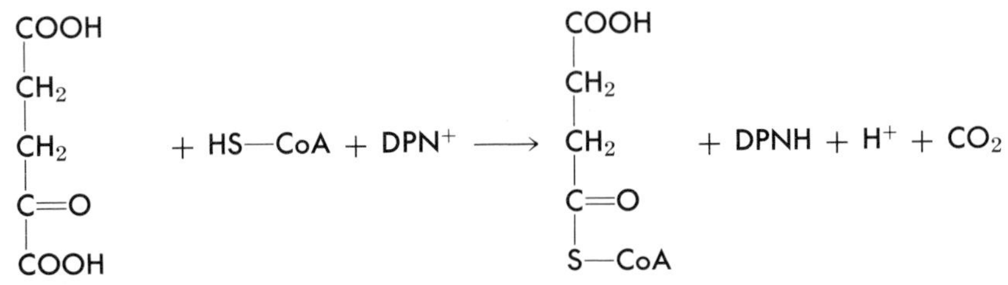

α-Ketoglutaric acid **Succinyl CoA**

In the initial event, α-ketoglutaric acid reacts, on the enzymic surface, with thiamine pyrophosphate, to form "active succinic semialdehyde," *α-hydroxy-γ-carboxypropyl thiamine pyrophosphate* (see page 332). The 4-carbon chain is then transferred to enzyme-bound lipoic acid, forming a succinyl lipoyl enzyme. The succinyl group is then transferred to coenzyme A and the dihydrolipoamide is reoxidized by DPN^+, as in the case of pyruvic acid oxidation. The thioester energy of succinyl CoA may be utilized for acylation reactions, and, by condensation with glycine, to initiate porphyrin synthesis (page 585). However, most of the succinyl CoA formed by operation of the citric acid cycle is converted to succinate in a reaction in which the energy of the thioester is conserved by formation of GTP.

$$\textbf{Succinyl CoA} + \textbf{GDP} + \textbf{P}_i \longrightarrow \textbf{succinate} + \textbf{GTP} + \textbf{CoA}$$

GDP and GTP are guanosine di- and triphosphate, respectively. Little is known of

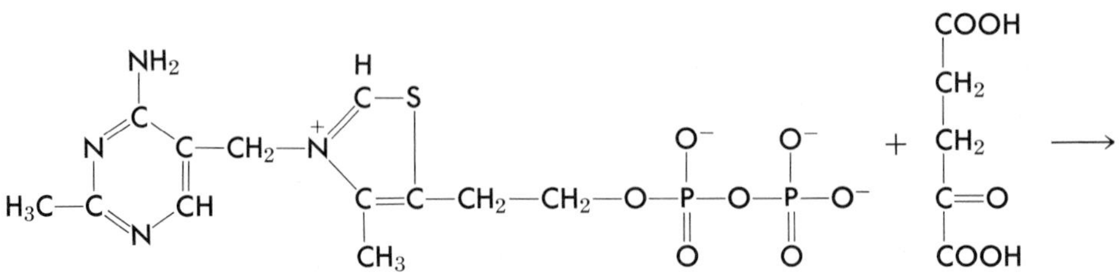

Thiamine pyrophosphate + α-Ketoglutaric acid ⟶

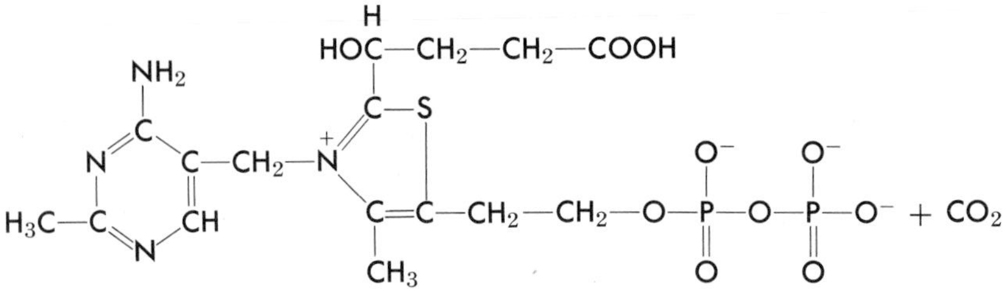

α-Hydroxy-γ-carboxypropyl thiamine pyrophosphate

the mechanism of this complex reaction, which is catalyzed by *succinic acid thiokinase.* A nucleoside diphosphokinase catalyzes phosphate transfer from GTP to ADP; during this process a histidine residue on the enzyme accepts and transfers the phosphate group.

$$\text{GTP} + \text{ADP} \rightleftharpoons \text{GDP} + \text{ATP}$$

The GTP may also be used directly for diverse metabolic processes, e.g., the activation of fatty acids (page 480).

Dehydrogenation of Succinic Acid (Reaction 6, Fig. 16.1). The oxidation of succinic acid to fumaric acid, catalyzed by *succinic acid dehydrogenase,* is the only dehydrogenation in the citric acid cycle in which pyridine nucleotides do not participate.

$$\text{FAD} + \text{HOOC—CH}_2\text{—CH}_2\text{—COOH} \rightleftharpoons \begin{array}{c} \text{H—C—COOH} \\ \| \\ \text{HOOC—C—H} \end{array} + \text{FADH}_2$$

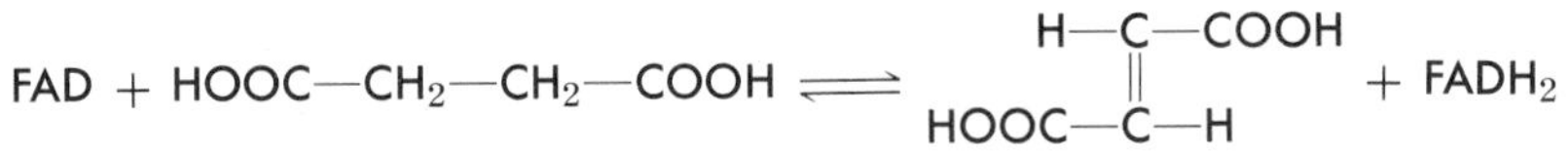

Succinic acid   Fumaric acid

Succinic acid dehydrogenase obtained from heart muscle contains four atoms of iron and one mole of flavin per mole of protein, molecular weight 200,000. The flavin appears to be FAD, but, unlike other flavoproteins, the flavin component cannot be removed except by extensive proteolysis. The reaction is specific for the *trans* form; maleic acid, the *cis*-isomer, is not produced by this enzyme. Malonate is a specific competitive inhibitor; when present in sufficiently high concentration, malonate can be employed to interrupt the citric acid cycle at this point, with resultant accumulation of succinate.

Malic Acid Formation (Reaction 7, Fig. 16.1). The reversible hydration of fumaric acid to yield L-malic acid is catalyzed by *fumarase*, a tetramer of four identical polypeptide chains.

$$\begin{array}{ccc}
& & H_2C\!\!-\!\!COOH \\
H\!\!-\!\!C\!\!-\!\!COOH & & | \\
\| & + \; H_2O \rightleftharpoons & HOC\!\!-\!\!COOH \\
HOOC\!\!-\!\!C\!\!-\!\!H & & | \\
& & H
\end{array}$$

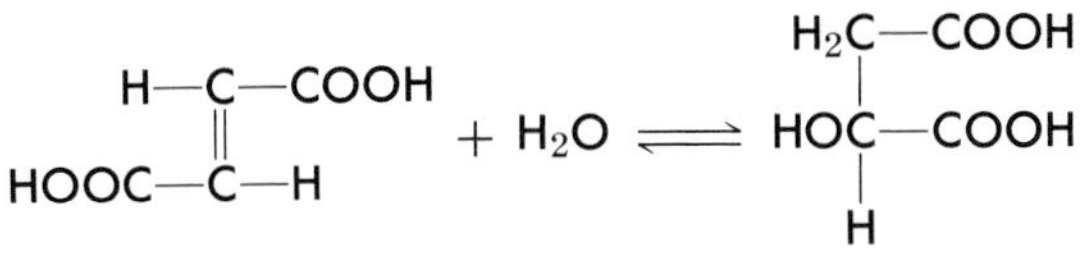

Fumaric acid   L-Malic acid

For the reaction, as written, $K = 4$; thus it is freely reversible. The absolute steric specificity, *viz.*, for the *trans*-unsaturated acid and the L-hydroxy acid, is noteworthy.

Regeneration of Oxaloacetic Acid (Reaction 8, Fig. 16.1). Malic acid is oxidized in the presence of malic acid dehydrogenase and DPN^+ to yield oxaloacetic acid.

$$\begin{array}{ccc}
H_2C\!\!-\!\!COOH & & H_2C\!\!-\!\!COOH \\
| & & | \\
HOC\!\!-\!\!COOH + DPN^+ \rightleftharpoons & & O\!\!=\!\!C\!\!-\!\!COOH \quad + \; DPNH + H^+ \\
| & & \\
H & &
\end{array}$$

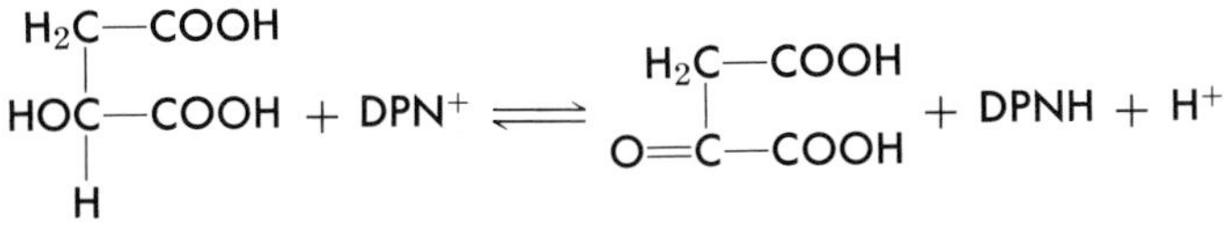

Malic acid   Oxaloacetic acid

By this reaction, the cycle is completed and the generated oxaloacetic acid made available for condensation with another mole of acetyl CoA and repetition of the process.

CITRIC ACID CYCLE—STERIC CONSIDERATIONS

Biological Asymmetry of Citric Acid. Citric acid does not have an asymmetrically substituted carbon atom and is thus devoid of optical activity. It was therefore anticipated, with reference to the reactions depicted in Fig. 16.1, that introduction of acetyl-1-^{14}C CoA into the sequence should give rise to α-ketoglutaric acid equally labeled in both carboxyl groups. However, the product actually recovered in such experiments proved to be labeled exclusively in the γ-carboxyl carbon.

$$\begin{array}{ccc}
CH_3{}^{14}COOH & & \overset{\gamma}{H_2C}\!\!-\!\!{}^{14}COOH \\
+ & & | \\
O\!\!=\!\!C\!\!-\!\!COOH & \xrightarrow[\text{1-4}]{\text{reactions}} & H_2C\beta \\
| & & | \\
H_2C\!\!-\!\!COOH & & O\!\!=\!\!C\!\!-\!\!COOH \\
& & \underset{\alpha}{}
\end{array}$$

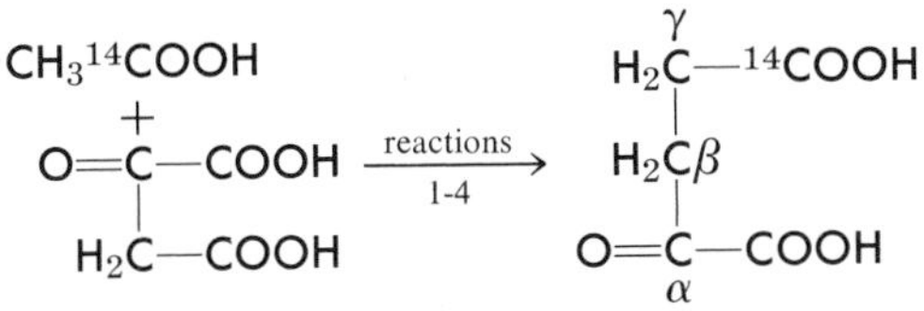

It was this observation which led Ogston to suggest that, as shown in Fig. 16.3 (page 334), the two —CH$_2$COOH groups of citric acid are not truly geometrically equivalent and that this nonequivalence becomes apparent upon three-point attachment to an enzymic surface. This problem is considered in greater detail on page 251.

Citric, cis-Aconitic, and Isocitric Acids (Reactions 2 and 3, Fig. 16.1). The reversible interconversions of these three acids, involving dehydration and hydration re-

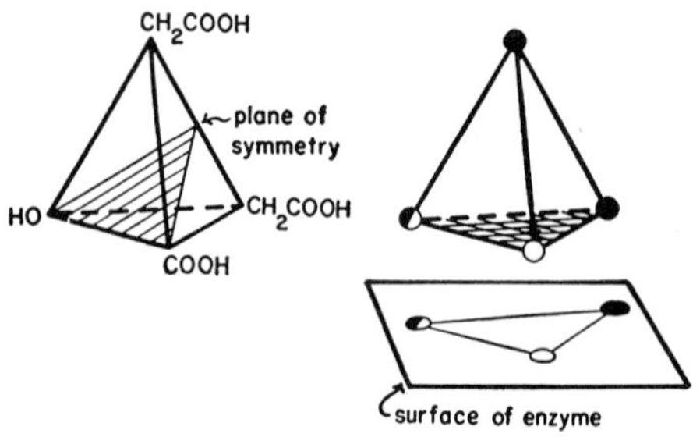

Fig. 16.3. Asymmetry apparent upon three-point attachment of citric acid to an enzymic surface. The central carbon atom of citric acid is in the center of the equilateral tetrahedron. Because of the plane of symmetry, the molecule in free solution is optically inactive. If the surface of the enzyme is asymmetric and forces attachment by —OH, —COOH, and —CH₂COOH groups in the manner shown, only one face of the tetrahedron (cross-hatched) can be accommodated on the enzymic surface. Hence, the two —CH₂COOH groups are not equivalent and only one attaches to the enzymic surface.

actions, are catalyzed by one enzyme, *aconitase*. For maximal activity, the enzyme requires reduced glutathione or cysteine, and Fe^{++}.

$$
\begin{array}{ccc}
\text{H}_2\text{C—COOH} & \text{H}_2\text{C—COOH} & \text{H}_2\text{C—COOH} \\
| & | & | \\
\text{HOC—COOH} \underset{(+\text{H}_2\text{O})}{\overset{(-\text{H}_2\text{O})}{\rightleftharpoons}} & \text{C—COOH} \underset{(-\text{H}_2\text{O})}{\overset{(+\text{H}_2\text{O})}{\rightleftharpoons}} & \text{HC—COOH} \\
| & \| & | \\
\text{H}_2\text{C—COOH} & \text{HC—COOH} & \text{HOC—COOH} \\
& & | \\
& & \text{H}
\end{array}
$$

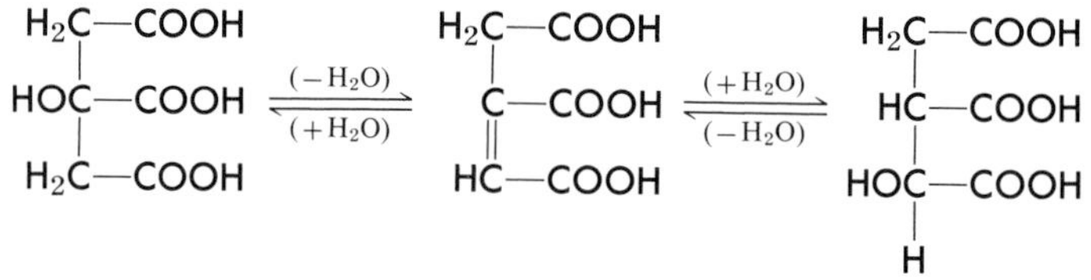

Citric acid	***cis*-Aconitic acid**	**Isocitric acid**

At equilibrium the relative abundances of the three products are citric acid, 90 per cent; *cis*-aconitic acid, 4 per cent; isocitric acid, 6 per cent. Although citric acid is favored at equilibrium, in respiring tissues the reaction sequence proceeds to the right as isocitric acid is oxidized by isocitric acid dehydrogenase in the subsequent reaction. On the basis of steric considerations outlined above, it is likely that the attachment of citric acid to aconitase occurs at three points prior to dehydration to *cis*-aconitic acid. Although the formulation shown indicates that the interconversion citrate ⇌ isocitrate occurs via aconitate, isotopic studies suggest that some direct interconversion may occur.

Respiration is inhibited by the presence of fluoroacetic acid, with resultant accumulation of citric acid. Fluoroacetic acid condenses with oxaloacetic acid to give a fluoroanalogue of citric acid which competitively inhibits citric acid utilization. The constitution of the fluorotricarboxylic acid that arises is not known, but it appears to differ from that of synthetic fluorocitric acid.

Each step in the citric acid cycle is completely stereospecific. The absolute configuration of citrate is such that, when viewed from the side opposite the middle —COOH, then moving from —OH in a counterclockwise direction, the first —CH₂COOH is that derived from acetyl CoA and the second from oxaloacetate. As we have seen, aconitase readily discriminates between the two —CH₂COOH groups of citric acid. The resulting isocitrate has the configuration

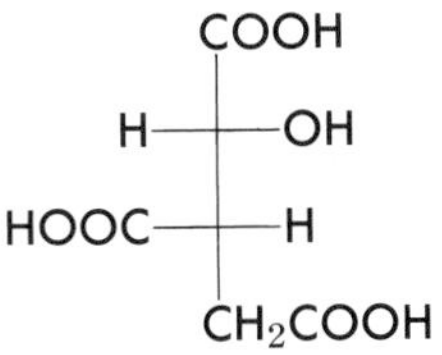

Two reactions involve hydration of a double bond, those catalyzed by aconitase and fumarase. Both involve *trans*-addition, or removal, of the elements of water. Similarly, the hydrogen atoms removed from succinate by succinic acid dehydrogenase are those in the *trans* position. Although this is a simplified summary of the cardinal features, it will suffice to show that, since the absolute structure of citrate and isocitrate are known, it is possible to specify the fate of each hydrogen atom in the entire reaction sequence.

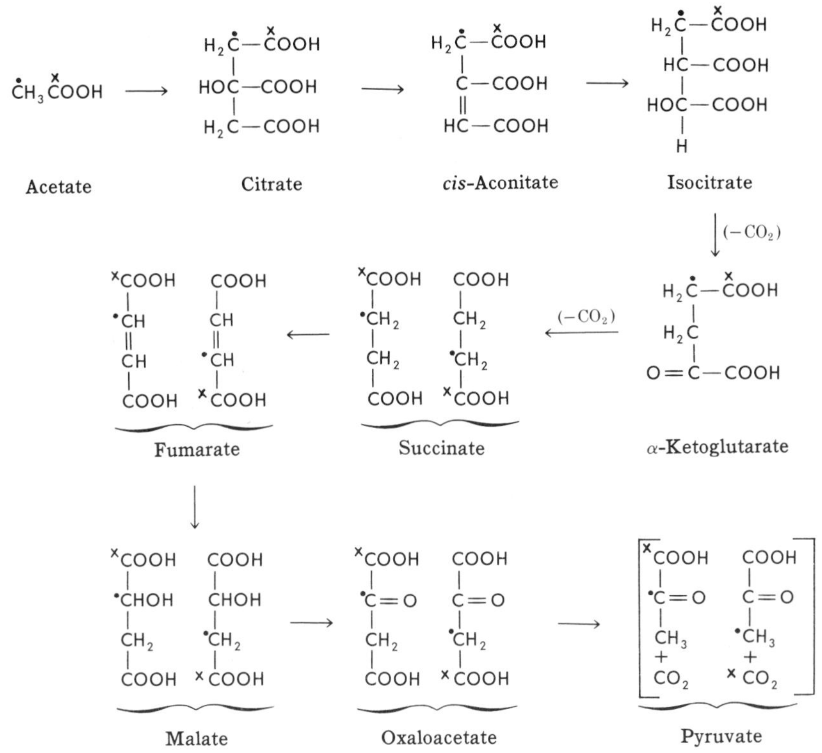

Fig. 16.4. Fate of the carbon atoms introduced as acetyl coenzyme A in a single turn of the citric acid cycle.

In Fig. 16.4 are represented the fates of the methyl (•) and the carboxyl (x) carbon atoms of acetic acid as they enter into and pass through the citric acid cycle. As a consequence of the asymmetric attachment of citrate to aconitase, no randomization of this molecule occurs about its center of symmetry. Whereas two moles of CO_2 are produced as one mole of acetyl CoA is consumed, neither of the carbon atoms lost as CO_2 actually is derived from that acetyl CoA on the first turn of the cycle. At the level of free succinic acid, randomization does occur and isotope will be symmetrically distributed about the plane of symmetry. On subsequent revolutions of the cycle, with reappearance and reutilization of oxaloacetic acid, acetyl carbon will be eliminated as CO_2.

MITOCHONDRIAL ELECTRON TRANSPORT

The enzymes of the citric acid cycle, the electron transport arrangements that deliver the electrons abstracted from intermediates of the citric cycle to oxygen, and the means by which the energy of this process is conserved by linking it to formation of ATP are all localized in mitochondria. In addition, mitochondria from various tissues contain varying amounts of other enzymes, including dehydrogenases for glutamic, pyruvic, β-hydroxybutyric, and fatty acids as well as for proline, choline, and α-glycerol phosphate, enzymes for activation and elongation of fatty acids, for synthesis of phospholipids, hippuric acid, urea, and fatty acids, and hydrolytic enzymes such as glutaminase. It is estimated that a beef heart mitochondrion has a complement of at least 50 different proteins.

Mitochondria, found in all animal and plant cells, are generally cigar-shaped bodies of the order of 0.5 by 3 μ. In the living cell, individual mitochondria may be observed to alter in shape from filaments to rods, loops, or spheres. In some tissues, mitochondria appear to be aligned in a specific manner which facilitates delivery of ATP to the energy-utilizing organelles, *e.g.*, aligned with the contractile fibers of muscle cells, in the direction of secretion in acinar pancreatic cells, or coiled about the midpiece of a spermatozoon. The number of mitochondria varies with the size and energy requirements of the cell, *e.g.*, 250 per sperm cell, 500 to 2,000 per liver cell, and 500,000 in the giant amoeba, *Chaos chaos.* In any case, mitochondria are major components of these cells; the mitochondrial protein of liver cells accounts for approximately 20 per cent of the total protein of this tissue.

The most prominent aspects of mitochondrial substructure are a limiting membrane and a system of *cristae,* closed vesicles stacked on one another like saucers. Both the limiting membrane and the membranes which comprise the cristae reveal double layers when viewed with the electron microscope. Liver mitochondria, which contain many enzymes not immediately related to the citric acid cycle or electron transport, show loosely packed cristae. In contrast, heart muscle mitochondria, which are almost entirely concerned with the citric acid cycle, fatty acid oxidation, and electron transport, exhibit extremely tightly packed cristae.

The structure of the mitochondria provides a large surface available for access to substrates. Thus, whereas a liver cell has an outer surface of about 3,000 μ^2, its mitochondria, including the foldings of the cristae, may offer a total surface of 30,000 μ^2. The membranous structure of the cristae is essential to the major function of the mitochondrion, *viz.*, conservation, as ATP, of the energy released incident to the oxidation of substrate. The cristae consist of two parallel membranes. The outer membrane separates the mitochondrial space from the cell sap. The inner membrane is studded with small hollow structures, called elementary bodies, each of which is connected to its base piece in the inner membrane by a stalk, as shown diagrammatically in Fig. 16.5. Thus within the mitochondrion there is

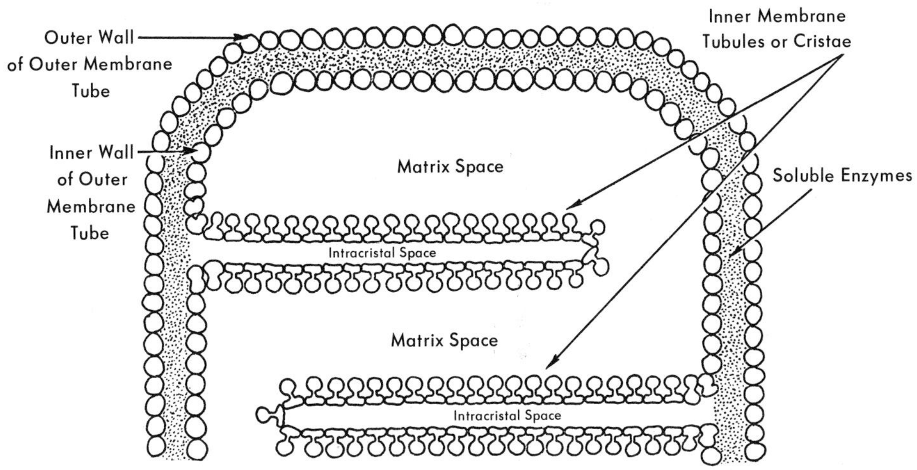

Fig. 16.5. A diagrammatic representation of the two membranes of the mitochondrion, showing particularly the relation of the cristae to the outer membrane and to the matrix space. The outer membrane is visualized as a three-dimensional tube in the lumen of which the soluble enzymes are located. (*Courtesy of Dr. David E. Green.*)

a soluble phase between inner and outer membrane separated from the soluble phase within the headpiece of the elementary body.

Relatively gentle procedures for disruption and extraction of mitochondria such as freezing and thawing, or exposure to media of low ionic strength, remove a number of enzymes. Among these are phosphatases and kinases, enzymes of the urea cycle, and cytochrome c, which is the only loosely bound cytochrome. Ultrasonic disruption of the residue destroys the double membrane structure and releases to the medium dehydrogenases, including those of the citric acid cycle except for succinic acid dehydrogenase. There then remains an insoluble, particulate material rich in lipids and containing all the cytochromes, except c, as well as dehydrogenases specific for DPNH, succinate, and, to some degree, for β-hydroxybutyrate. Particles prepared by gentler means, which retain the double membrane structure, also retain some ability to generate ATP incident to the oxidation of some substrates. These particles have been obtained either by extraction of mitochondria with digitonin or by brief sonication. Ultimate dissolution of particles is possible only by treatment with detergents. The products of such extraction will be discussed below. If the mitochondria are alternately subjected to hypotonic and isotonic media, the outer membrane ruptures and separates from the inner structure. This may also be accomplished by exposure to a phospholipase. The two types of membranes may then be separated by density gradient centrifugation since the inner membrane is more dense than the outer membrane, as it consists of less lipid per unit weight.

Chemical Composition of Mitochondria. Approximately one-third of the dry weight of mitochondria is lipid. Phospholipids containing choline and ethanolamine are generally predominant, but lesser amounts of those containing serine, inositol, and glycerol are also present. Virtually all the cardiolipin of the cell is found in the mitochondria. Heart mitochondria contain an abundance of phosphatidal ethanolamine and phosphatidal choline, whereas these are virtually absent in liver mitochondria. About two-thirds of the total fatty acids have varying degrees of unsaturation. The full significance of these lipids remains uncertain. Their structural role is apparent from the fact that the only means for further dispersal of mitochondrial particles is treatment with detergents, *e.g.*, bile salts, or with acetone, amyl alcohol, etc. A "structural protein" of high isoelectric point and molecular weight about 22,000 appears to comprise about half of the total mitochondrial protein. This protein rapidly polymerizes in solution at neutral pH, yielding a high molecular weight particle. However, it is thought that the monomeric form is built into the mitochondrial membranes. Because of its high content of basic amino acids, this protein at neutral pH is capable of binding significant quantities of cytochromes a, b, and c_1, as well as phospholipids, but does not bind cytochrome c, which is also a basic protein.

With respect to the electron transport components, mitochondrial fragments contain DPNH dehydrogenase, a high molecular weight iron-containing flavoprotein, and succinic acid dehydrogenase, which is also an iron-flavoprotein. Spectroscopic evidence indicates the presence of cytochromes b, c_1, c, a, and a_3. Of these, cytochromes b, c_1, c, and a have been isolated and examined (page 369); cytochrome a_3 may be a special form of cytochrome a. In any case, the combination of a + a_3 functions as cytochrome oxidase, the terminal member of the system. In addition there are substantial quantities of ubiquinone. Of these constituents, only cytochrome c and ubiquinone may be relatively easily extracted from the insoluble lipid-protein complex; the other components are firmly bound. The relative amounts of each of these components appear to be fixed, but they are not equimolar. Thus, it has been reported that for each molecule of DPNH dehydrogenase in beef heart mitochondria there are 1 molecule each of succinic acid dehydrogenase, cytochrome c, and cytochrome c_1, 3 molecules of cytochrome b, 6 molecules of cytochrome oxidase, and 15 molecules of ubiquinone. In addition there are perhaps two different iron-proteins (other than the two dehydrogenases which are iron-flavoproteins), but their relative amounts are unknown.

The inner and outer membranes of the mitochondrion are strikingly different. The outer membrane resembles, both in composition and in appearance as visualized by electron microscopy, the smooth membrane of the endoplasmic reticulum. The abundant phospholipids are of mixed composition, and there is also present a DPNH-cytochrome c reductase and a DPNH-cytochrome b_5 reductase, as well as glucose 6-phosphatase. All these are also present in microsomes and are not detectable in the inner membranes. In contrast, the inner membranes, which contain much less total lipid, contain all of the cardiolipin, the mitochondrial "structural protein," the enzymes of the citric acid cycle, the electron transport system, and the mechanism for oxidative phosphorylation. Thus the inner membranes, which share the general properties of bacterial cell membranes, have a protective investiture of an outer membrane which resembles the endoplasmic reticulum and the plasma membrane of mammalian cells.

The Electron Transport Chain. If mitochondria or submitochondrial particles are incubated anaerobically with intermediates of the citric acid cycle, all the above-cited components—DPN^+, flavin, nonheme iron, ubiquinone, and cytochromes b,

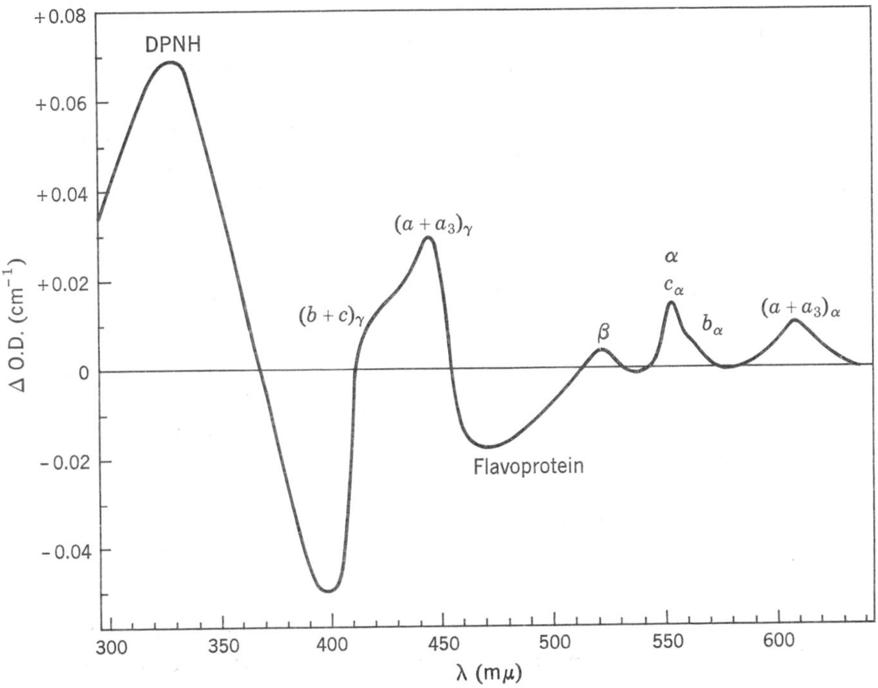

Fig. 16.6. Difference absorption spectrum between anaerobic and aerobic guinea pig liver mitochondria. The curve traces the change in optical density of a mitochondrial suspension, over the range of wave length shown, caused by complete lack of oxygen. Reduction of nonheme iron would result in varying decrease in optical density over the entire spectral range, with maximal drop at about 450 mμ. Reduction of ubiquinone would not be visible in the spectral range shown. (*Courtesy of Dr. Britton Chance and Dr. Ronald W. Estabrook.*)

c_1, c, a, and a_3—undergo reduction. Most studies of this type have relied on absorption spectrophotometry, as illustrated in Fig. 16.6. Attempts to establish the order of electron transport among these components have depended upon several stratagems. (1) Are the various components equally reduced when succinic acid or DPNH is the initial reductant? Such studies have suggested that cytochrome b may be more significant in the succinate → O_2 than in the DPNH → O_2 electron pathway. (2) Inhibitors can be utilized to interrupt electron flow in a specific manner. Thus, barbiturates, *e.g.*, amytal, prevent reduction of ubiquinone and the cytochromes, whereas antimycin A (an antibiotic from *Streptomyces griseus*) permits reduction of the flavin, nonheme iron, and ubiquinone components while blocking reduction of cytochrome c_1, c, and a_3. Rotenone inhibits DPNH dehydrogenase but not other components of the system. Cyanide, which combines with the oxidized form of cytochrome oxidase, prevents reduction of cytochromes a + a_3, but not of any other components of the system. (3) Fractionation of submitochondrial particles by detergents in Green's laboratory has been reported to yield four types of particles, which catalyze the following processes, respectively: (I) DPNH → ubiquinone; (II) succinate → ubiquinone; (III) ubiquinone → cytochrome c; (IV) cytochrome c → O_2. Each of the isolated complexes has a particle weight of 250,000 to

400,000. Complex (III) is composed of one molecule each of cytochrome c_1 and a nonheme iron protein, two molecules of cytochrome b, three or four molecules of structural protein, and 140 molecules of phospholipid. The relative amounts of each of the four complexes may vary with individual tissues; in beef heart and kidney mitochondria, complexes (IV) and (III) were found in a ratio of about 3. This arrangement of the four complexes is depicted in Fig. 16.7.

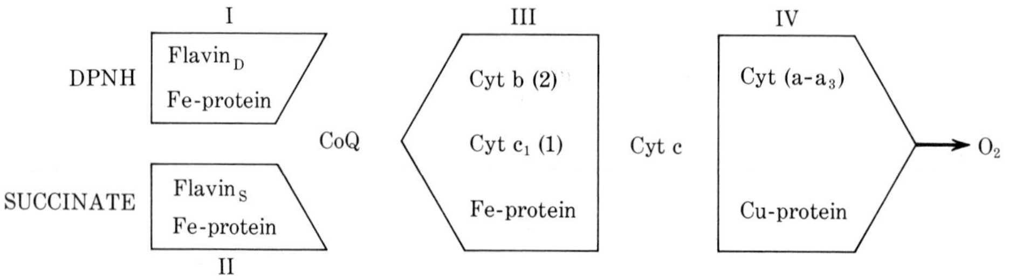

Fig. 16.7. Complexes isolated from mitochondria and participating in mitochondrial electron transport. (*From D. E. Green and R. F. Goldberger, "Molecular Insights into the Living Process," Academic Press, New York,* 1967.)

These subparticles may be recombined to reconstitute a total DPNH- or succinic oxidase system. A summary of the results of these diverse procedures permits assignment of the order of electron transport shown in Fig. 16.8. Although the details remain to be firmly established, the picture that emerges is one of a highly organized "solid-state" arrangement. Presumably, each of the electron

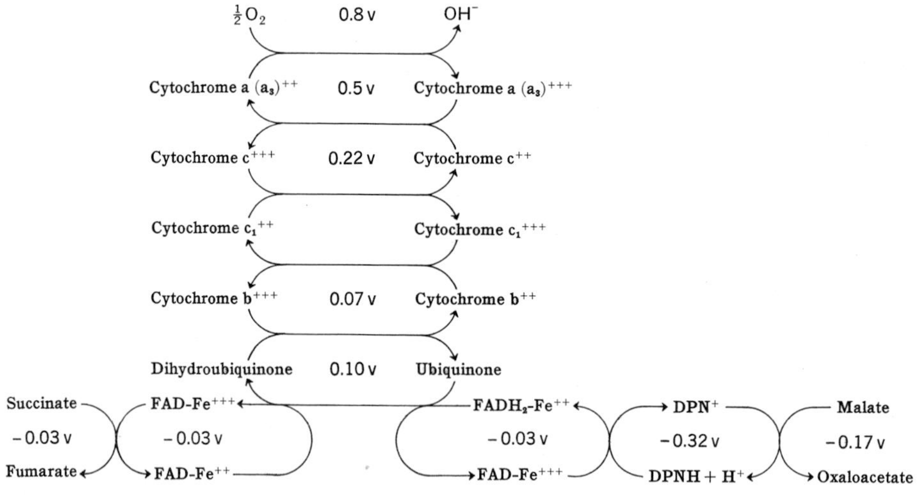

Fig. 16.8. Apparent organization and electron flow in the mitochondrial electron transport chains. Approximate values of E_0' are shown. The relative positions of cytochrome b and ubiquinone are not certain.

carriers is held rather rigidly in place and located in a manner to permit it to accept electrons from its reductant and yield these to its oxidant without need for migration from its fixed position. Ubiquinone and perhaps cytochrome c seem free to migrate in the lipid milieu, accepting electrons from the various dehydrogenase complexes. Apparently, the entire arrangement exists in a lipid, hydrophobic milieu so that only at the binding sites for the metabolic substrates, *e.g.*, pyruvate or succinate, and, perhaps, at the oxygen-binding sites, are the electron transport units exposed to an aqueous environment.

If this view of electron transport is correct, the water formed in biological oxidations arises as follows: Hydrogen atoms are transferred from the substrate to a flavoprotein, either directly or via DPN. As the nonheme iron oxidizes the reduced flavin, protons must be released to the medium.

$$\mathbf{FADH_2 + 2Fe^{+++} \longrightarrow FAD + 2Fe^{++} + 2H^+}$$

If, however, the next step is reduction of ubiquinone, protons must be recaptured.

$$\mathbf{2Fe^{++} + 2H^+ + CoQ \longrightarrow CoQH_2 + 2Fe^{+++}}$$

As the quinone is oxidized by cytochrome b, protons are again added to the medium. At the end of the chain, oxygen is reduced by cytochrome oxidase, with formation of hydroxyl ions. Combination of the latter with the previously generated protons is the immediate means of metabolic formation of water. This may be depicted as follows:

$$(a) \qquad \mathbf{CoQH_2 + 2Fe^{+++} \longrightarrow CoQ + 2H^+ + 2Fe^{++}}$$

$$(b) \qquad \mathbf{\tfrac{1}{2}O_2 + H_2O + 2Fe^{++} \longrightarrow 2OH^- + 2Fe^{+++}}$$

$$\text{Sum:} \qquad \mathbf{CoQH_2 + \tfrac{1}{2}O_2 \longrightarrow CoQ + H_2O}$$

Although it is by no means definitely established, available data suggest that for each citric acid cycle dehydrogenase complex there may be an individual transport unit to which other dehydrogenases may not have ready access. In any case, within the mitochondrion each electron transport unit must lie within easy access of a dehydrogenating complex which accepts, from its substrate, electrons to be transferred to the unit, *i.e.*, the dehydrogenases of the citric acid cycle, pyruvic acid, β-hydroxybutyric acid, choline, proline, glutamic acid, sarcosine and glycerol phosphate dehydrogenases, the fatty acyl CoA dehydrogenase and the β-hydroxy acyl CoA dehydrogenase. Of these, fatty acyl dehydrogenase has been found to have only indirect access to the transport unit; electrons must be transferred from its flavin moiety to that of an "electron-transferring flavoprotein" and thence to some component of the transport system, *e.g.*, nonheme iron, coenzyme Q, or cytochrome b.

Oxidation of Extramitochondrial Pyridine Nucleotides. Mitochondria contain a larger molar concentration of bound pyridine nucleotides than of bound cytochromes (DPN/cytochrome c = 40/1). Moreover, external (cytoplasmic) pyridine nucleotides penetrate the intact mitochondrion with great difficulty. Hence, it is intramitochondrial DPN which is reduced by the mitochondrial dehydrogenases. Accordingly, DPNH produced by reactions in the cell cytoplasm cannot, of itself, be oxidized by mitochondria. On the other hand, small molecules which do penetrate the mitochondrial barrier, can function in overcoming its impermeability to DPNH by participating in cyclic processes that may assume specific physiological roles. Such a cycle depends upon an interplay of a DPN-linked dehydrogenase which is outside the mitochondria and which utilizes DPNH for reduction of a substrate. The reduced substrate enters the mitochondria and is reoxidized by a mitochondrial enzyme that channels electrons to molecular oxygen. One such cycle, termed the α-glycerol phosphate cycle, has been demonstrated to play an important role in insect flight muscle. The cytoplasm contains an *α-glycerol phosphate dehydrogenase* which catalyzes reduction of dihydroxyacetone phosphate to α-glycerol phosphate by DPNH. The mitochondria contain another glycerol phosphate dehydrogenase which is a flavoprotein. Their coupled action accomplishes the mitochondrial oxidation of cytoplasmic DPNH.

(*a*) $\mathbf{H^+ + DPNH + dihydroxyacetone\text{-}P \longrightarrow DPN^+ + glycerol\text{-}P\ (cytoplasm)}$

(*b*) $\mathbf{Glycerol\text{-}P + FAD \longrightarrow dihydroxyacetone\text{-}P + FADH_2\ (mitochondrion)}$

Sum: $\mathbf{H^+ + DPNH + FAD \longrightarrow DPN^+ + FADH_2\ (cell)}$

This device functions with the movement of α-glycerol phosphate from cytoplasm into mitochondria, and a transfer of dihydroxyacetone phosphate in the reverse direction. However, its possible significance for mammalian cells is unknown since it has been shown that liver does not utilize a glycerol phosphate shunt for the oxidation of cytoplasmic DPNH.

A more likely shunt for hydrogen transport into mitochondria in mammalian cells appears to be one involving the mitochondrial and extramitochondrial DPN-linked malic acid dehydrogenases. This cycle involves oxaloacetate and malate as a shuttle between the soluble and mitochondrial pyridine coenzymes. The soluble DPNH is oxidized by oxaloacetate to form malate; the latter enters the mitochondria and is oxidized to oxaloacetate by the mitochondrial malate dehydrogenase and the mitochondrial-bound DPN. The oxaloacetate may then re-enter the cytoplasm for reduction by another molecule of DPNH. The mitochondrial DPNH formed would then be oxidized by the respiratory chain enzymes and this could be coupled with the formation of ATP.

Mitochondrial Metabolism of TPN. Mitochondrial content of TPN varies with the tissue of origin. Whereas approximately two-thirds of the total pyridine nucleotides of liver mitochondria is TPN, the latter constitutes only 15 to 20 per cent of the total in heart, kidney, and brain mitochondria and appears to be essentially absent from muscle mitochondria. Even in the presence of abundant oxygen and ATP, most of the TPN is in the reduced form while the DPN is largely oxidized. Mitochondria appear to be devoid of TPN-linked dehydrogenases which function in important energy-conserving oxidative pathways, and there is no direct mitochondrial path for electrons from TPNH to oxygen. TPNH oxidation can result only from retransfer to DPN or, more importantly, by such reductive, biosynthetic processes as formation of fatty acids, glutamic acid, and steroids.

Several forms of transhydrogenation have been observed. The simplest of these is catalyzed by an enzyme, *transhydrogenase,* which brings about the following reaction.

(*a*) $\mathbf{TPNH + DPN^+ \rightleftharpoons TPN^+ + DPNH}$

However, the observed activity may really reflect the operation of one or more dehydrogenases which can utilize both DPN and TPN. The consecutive reactions

$$(b)\qquad H^+ + DPNH + M \rightleftharpoons MH_2 + DPN^+$$
$$(c)\qquad MH_2 + TPN^+ \rightleftharpoons M + H^+ + TPNH$$

catalyzed by such a dehydrogenase would give the same over-all result as reaction (*a*). However, since E_0' for $DPN^+/DPNH$ and $TPN^+/TPNH$ are identical, it is clear that the process cannot account for the fact that the two systems $DPN^+/DPNH$ and $TPN^+/TPNH$ are remote from similar equilibria, the former being largely oxidized and the latter largely reduced in most living cells.

Liver and heart mitochondria conduct a transhydrogenation which may be represented as

$$(d)\qquad DPNH + TPN^+ + \text{energy} \longrightarrow DPN^+ + TPNH$$

The process is unidirectional, as shown, and irreversible. It can be observed in respiring mitochondria or, anaerobically, the energy requirement may be met by inclusion of ATP in the medium, with resultant formation of approximately 1 mole each of ADP and P_i for each mole of TPNH formed. TPNH formed in this manner is utilized for synthesis of fatty acids. This transhydrogenase process must be catalyzed by a specific enzyme, and it is of interest, therefore, that the hydrogen atom is transferred from the A side of the pyridine ring of DPNH to the B side (page 357) of TPNH.

OXIDATIVE PHOSPHORYLATION

The physiological objective of the oxidation of carbohydrates and lipids is the conservation of the free energy thus made available in physiologically utilizable form, *viz.*, as ATP. It has been demonstrated repeatedly that oxidation of intermediates of the citric acid cycle by carefully prepared mitochondia is accomplished by the formation of ATP from ADP and P_i. However, despite intensive investigation, the fundamental mechanism(s) by which this energetic transformation is accomplished remains obscure. The salient experimental observations are summarized below.

Under standardized conditions, oxidation of DPNH occurs with net formation of three molecules of ATP per molecule of DPNH or per atom of oxygen consumed; this is conventionally expressed as a *P/O ratio* of 3. Since oxidation of β-hydroxybutyric or malic acids proceeds with a similar P/O ratio, it is apparent that ATP is generated incident to the reoxidation of the DPNH formed by oxidation of the substrate and that, with the exception of the oxidation of α-ketoglutaric acid, the immediate step in which substrate is oxidized serves only as a means of providing electrons to the electron transport units.

The P/O ratio of 3 should be regarded as a minimum. Under special conditions and short experimental periods, values as high as a P/O of 6 have been occasionally obtained for oxidation of DPNH. Since the potential span, $DPNH \rightarrow O_2$, is about 1.14 volts, $\Delta F°$ for this process is about 51,300 cal. ΔF for formation of ATP under mitochondrial conditions is uncertain, but if a value of about +8000 cal. per mole is assumed, it is apparent that the free-energy release associated with the over-all process of oxidation of a mole of DPNH by O_2 could, in theory, suffice for formation of six molecules of ATP. Hence the process has a demonstrated efficiency of at least 50 per cent and may be higher.

The studies of Chance, Lehninger, Slater, and Lardy have established the approximate "sites" of energy conservation along the path of electron transport from each molecule of DPNH to oxygen. ATP formation occurs as DPNH is oxidized by DPNH dehydrogenase; a second opportunity for ATP formation is associated with the passage of electrons from ubiquinone or ferrocytochrome b to ferricytochrome c; and the third occurs in the course of oxidation of ferrocytochrome c by ferrocytochrome oxidase. Identification of these loci may be presumed to be valid, whatever the "true" value for the P/O ratio may be. Indeed, until the mechanism of oxidative phosphorylation is understood, it will not be certain that the P/O ratio is necessarily integral.

Whereas E_0' for $DPNH/DPN^+$ is -0.32 volt, that for succinate/fumarate is 0.03 volt (Table 15.1). Hence, succinate cannot be expected to reduce DPN^+ directly. Moreover, $\Delta E_0'$ between succinate/fumarate and $FADH_2/FAD$ is too small to permit ATP formation from this process. A minimum difference in potential, $\Delta E_0'$ of 0.18 volt, is required to provide the 8000 cal. per mole necessary to form ATP from ADP + P_i. Hence, the oxidation of succinic acid occasions formation of only two moles of ATP per mole of succinic acid under standard conditions; the two sites of ATP formation correspond to the last two (see above) which operate during DPNH oxidation.

Respiratory Control. It is inherent in the formulation of oxidative phosphorylation presented above that if mitochondria are "tightly coupled" so that respiration proceeds almost exclusively over phosphorylating pathways, then respiration can proceed no more rapidly than is permitted by the availability of inorganic phosphate and ADP for ATP formation. Indeed, the rate of respiration is an *inverse* function of the ratio ATP/(ADP + P_i). This has been amply demonstrated. Lardy found that a supply of ADP and P_i is essential for maximal respiration with any substrate oxidized by a mitochondrial DPN-linked dehydrogenase. Chance utilized this observation to determine, by spectrophotometric means, the effect of ADP concentration on the state of reduction of the various electron carriers and was able to ascertain the sites of energy conservation in the electron transport scheme. Some results of these studies are summarized in Table 16.1.

Table 16.1 Oxidation-Reduction Levels of Members of the Respiratory Chain in Various Metabolic Circumstances

Substrate level	O_2 level	ADP level	Respiration rate	Rate-limiting factor	Steady-state percentage reduction of				
					DPN	Flavins	Cytochromes		
							b	c	a
High	0	High	0	O_2	100	100	100	100	100
0	+	High	Slow	Substrate	0	0	0	0	0
Low	+	Low	Slow	ADP	90	21	17	7	0
High	+	Low	Slow	ADP	99	40	35	14	0
High	+	High	Fast	Respiratory chain	53	20	16	6	4

Source: Adapted from B. Chance and G. R. Williams, *Advances in Enzymol.*, **17**, 65, 1956.

The tight coupling of oxidation to phosphorylation, evident in normal mitochondria, provides a means by which the rate of oxidation of foodstuffs is regulated by the requirements of the cell for useful energy. The utilization of ATP to drive the diverse energy-requiring processes of the cell automatically increases the available supply of ADP and inorganic phosphate, which in turn become available to react in the coupling mechanism and permit respiration to proceed.

Adenylic acid kinase is a widely distributed enzyme which catalyzes the reaction

$$2ADP \rightleftharpoons ATP + AMP$$

It can serve the cell as a means of maximizing the utilization of the energy of ATP when the latter is being used more rapidly than it is synthesized. In the reverse direction, it is the first step in resynthesis of ATP from AMP. The latter is generated in a variety of synthetic processes, *e.g.*, fatty acid and amino acid activations (pages 480 and 662). The significance of AMP formation is also apparent in that it is AMP rather than ADP which serves as a positive or negative modifier controlling the rates of several enzymic processes, *e.g.*, the phosphofructokinase, diphosphofructose phosphatase, glycogen phosphorylase, and isocitric dehydrogenase reactions (Chaps. 18 and 19).

Energy Yield of the Citric Acid Cycle. The operation of the citric acid cycle includes three steps in which DPNH arises. These are isocitrate $\rightarrow$ α-ketoglutarate; α-ketoglutarate $\rightarrow$ succinyl CoA; and malate $\rightarrow$ oxaloacetate. Each of these steps provides the opportunity for formation of three moles of ATP. An additional mole of ATP is derived from succinyl CoA (page 331). The oxidation of succinate to fumarate circumvents the usual pyridine nucleotide step and yields two rather than three moles of ATP. The yields of ATP for the individual steps in the cycle are summarized in Table 16.2. It will be seen that per mole of acetyl CoA consumed, 12 moles of inorganic phosphate are utilized and 12 moles of ATP are generated. Considering ΔF for synthesis of ATP from ADP and P_i as approximately +8000 cal. per mole under physiological conditions, this represents a net energy yield, as ATP synthesized, of 96,000 cal. per mole of acetate utilized. As we have noted, this is a minimum value, and may be substantially larger.

The total oxidation of acetic acid to CO_2 and H_2O,

$$CH_3COOH + 2O_2 \longrightarrow 2CO_2 + 2H_2O$$

has a gross energy yield of 209,000 cal. per mole. Thus, (96,000/209,000) $\times$ 100, or about 45 per cent of the gross energy yield, is conserved in energy-rich phos-

Table 16.2: ENERGY YIELD OF THE CITRIC ACID CYCLE

Reaction	Coenzyme	ATP yield/mole
Isocitrate $\longrightarrow$ α-ketoglutarate + CO_2	DPN	3
α-Ketoglutarate $\longrightarrow$ succinyl CoA + CO_2	DPN	3
Succinyl CoA + ADP + P_i $\longrightarrow$ succinate + ATP	GDP	1
Succinate $\longrightarrow$ fumarate	FAD	2
Malate $\longrightarrow$ oxaloacetate	DPN	3
Total	....	12

phate compounds. The oxidation of acetyl CoA via the citric acid cycle, coupled to phosphorylation, may be represented as follows.

$$CH_3CO{-}SCoA + 2O_2 + 12ADP + 12P_i \longrightarrow 2CO_2 + CoASH + 12ATP$$

This equation disregards the water released in formation of the pyrophosphate bond of ATP and the one mole of water required for hydrolysis of the thioester.

Tissue Respiration. The oxygen consumption (microliters of gas at standard pressure and temperature) per milligram of tissue is denoted as the Q_{O_2}. As shown in Table 16.3, considerable variation is apparent among animal tissues. It is not possible to define, precisely, the factors that determine the respiratory rate of a given tissue. Since, however, the bulk of this respiration in all tissues appears to occur over the cytochrome chain and, presumably, is phosphate-linked, it may be assumed that it is the demand for energy-rich phosphate which conditions the particular respiratory rate of each tissue. Oxidations catalyzed by aerobic dehydrogenases, copper-containing oxidases, and the peroxidases are not phosphate-linked. Consequently, the rate at which these particular oxidations may proceed depends only on the usual factors that influence enzymic activity. If, then, any substance whose oxidation is catalyzed in this manner is administered to an intact animal or incubated with an isolated tissue preparation, its oxidation may be expected to occur readily, thereby increasing the rate of oxygen consumption. In contrast, administration of a substance whose oxidation is coupled obligatorily with phosphorylation, *e.g.*, feeding of excess carbohydrate, does not stimulate respiration. The superfluous calories are stored in tissues as glycogen or lipid until the demand for ATP initiates oxidation of these reserve sources of energy.

Table 16.3: RESPIRATION OF VARIOUS TISSUES AND MITOCHONDRIA

Tissue	$-Q_{O_2}$	*Tissue*	$-Q_{O_2}$
Flight muscle mitochondria	500	Duodenal mucosa	9
Heart mitochondria	80	Lung	8
Retina	31	Placenta	7
Kidney	21	Myeloid bone marrow	6
Liver fasted animal)	17	Thymus	6
Liver (fed animal)	12	Pancreas	6
Jejunal mucosa	15	Diaphragm	6
Thyroid	13	Heart	5
Testis	12	Ileal mucosa	5
Cerebral cortex	12	Lymph node	4
Hypophysis	12	Skeletal muscle	3
Spleen	12	Cornea	2
Adrenal gland	10	Skin	0.8
Erythroid bone marrow	9	Lens	0.5

Note: Q_{O_2} = microliters O_2 per milligram dry weight per hour. All values except the first two are for rat tissue slices in Ringer's phosphate + glucose.

MECHANISM OF OXIDATIVE PHOSPHORYLATION

The most widely held hypothesis to account for oxidative phosphorylation stipulates that, associated with each of several discrete oxidative steps, there must be formation of an energy-rich intermediate compound(s) that reacts with an inorganic phosphate compound, which, in turn, may react with ADP to form ATP. A tentative formulation of this general hypothesis for the synthesis of ATP is the following.

$$(a) \quad \text{Carrier a}^{\text{red}} + \text{carrier b}^{\text{ox}} + \text{X} \longrightarrow \text{carrier a}^{\text{ox}}{\sim}\text{X} + \text{carrier b}^{\text{red}}$$
$$(b) \quad \text{Carrier a}^{\text{ox}}{\sim}\text{X} + \text{Y} \longrightarrow \text{carrier a}^{\text{ox}} + \text{X}{\sim}\text{Y}$$
$$(c) \quad \text{X}{\sim}\text{Y} + \text{P}_\text{i} \longrightarrow \text{Y}{\sim}\text{P} + \text{X}$$
$$(d) \quad \text{Y}{\sim}\text{P} + \text{ADP} \longrightarrow \text{ATP} + \text{Y}$$
$$\text{Sum:} \quad \text{Carrier a}^{\text{red}} + \text{carrier b}^{\text{ox}} + \text{P}_\text{i} + \text{ADP} \longrightarrow \text{carrier a}^{\text{ox}} + \text{carrier b}^{\text{red}} + \text{ATP}$$

The "carriers" shown above may be any of the components of the electron transport chain, *viz.*, DPN, FAD, cytochromes b, c, or a, ubiquinone, or nonheme iron. No evidence clearly supports the postulate that the first activated form in this formulation is the carrier which is oxidized. With equal validity, equation (*a*) might be stated as

$$\text{Carrier a}^{\text{red}} + \text{carrier b}^{\text{ox}} + \text{X} \longrightarrow \text{carrier a}^{\text{ox}} + \text{carrier b}^{\text{red}}{\sim}\text{X}$$

and the reaction sequence continued accordingly. It must be emphasized that X and Y are hypothetical; they have not been detected or identified. The evidence which supports this formulation may be summarized as follows.

Activated Forms of Electron Carriers. Observations from diverse sources suggest the existence of "activated" forms of various electron carriers. Perhaps most convincing is the finding that respiring particles from *Alcaligenes fecalis* incubated with DPNH yield a soluble protein to which a form of DPN is bound. Addition of ADP and P_i yields ATP, and DPN^+ separates from the protein. Significantly, the latter process is reversible. There is as yet no evidence for a similar process in the mitochondria of higher forms. However, it is noteworthy that derivatives of DPN of uncertain structure have been isolated from respiring mitochondria.

Chance has shown by spectrophotometry at low temperature that mitochondria inhibited by sulfide, which binds to ferric cytochrome oxidase and thus inhibits O_2 reduction, exhibit a large band at 555 mμ attributable to an unusual form of cytochrome b. This band can be shifted to 565 mμ by addition of P_i, suggesting possible formation of an intermediate cytochrome b$\sim$P. The latter can be dissipated by imposing a requirement for energy for translocation of divalent cations (see below).

Mitochondria contain varying amounts of several types of quinones, *viz.*, ubiquinone, vitamin E, and vitamin K. It has been suggested that these might be reduced to yield quinol phosphates which would be utilized for ATP formation during reoxidation. No evidence supports this hypothesis for animal mitochondria, but it appears to be applicable in *Mycobacterium phlei* (page 374).

Thus, whereas no one of the above observations categorically supports the hypothesis of an activated carrier presented in the formulation above, the total evidence is suggestive of such a concept.

Intermediary Phosphorylated Compounds. The scheme presented would rest on firmer ground if intermediary phosphorylated compounds (Y$\sim$P) capable of reaction with ADP were identified. Despite a continuing search for such compounds, none has been detected. However, an interesting and highly suggestive model system has been described by Wang. Aerobic oxidation of ferroheme in an essentially anhydrous medium, utilizing N, N-dimethylacetamide as solvent, containing imidazole, ADP, and P_i, results in formation of ATP. In the absence of ADP, 1-phosphoimidazole accumulates; this compound is analogous to the phosphohistidyl which participates in the formation of ATP by utilizing the energy of succinyl CoA (page 331) and which had also been suggested as a participant in oxidative phosphorylation. The model is thought to involve initial formation of a ferroheme-imidazole complex which, after oxidation, spontaneously reacts with P_i. Reduction of this complex generates a phosphorylating agent as shown below.

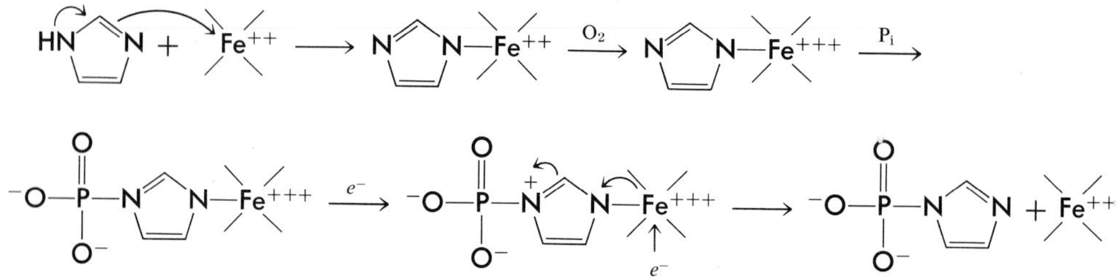

This model can readily be visualized as analogous to the events at the second two sites of phosphorylation. To account for the first site it would be necessary to postulate a phospholipid-flavinyl complex.

Inhibitors of Oxidative Phosphorylation. Oxidative phosphorylation in mitochondria can be interrupted by diverse compounds. (1) Inhibitors of electron transport, as expected, inhibit not only oxidation but associated phosphorylation. This has permitted examination of segments of the process. In the presence of antimycin A only the first coupling site in the oxidation of DPNH functions and electron flow via the cytochromes b and c is blocked. Rotenone, a powerful inhibitor of DPNH dehydrogenase, prevents phosphorylation by blocking the step between DPNH and cytochrome b. (2) Many phenols, of which 2,4-dinitrophenol is the prototype, serve as "uncouplers" of oxidative phosphorylation. In their presence, net detectable ATP formation is markedly reduced and electron transport from substrate to oxygen proceeds at the maximal velocity permitted by the enzymes of the transport system. The effectiveness of these phenols reflects their solubility in a lipid milieu and their acidic dissociation; a highly lipophilic, strong acid such as pentachlorophenol is most effective. Stimulation of any mechanism which results in markedly enhanced intramitochondrial utilization of ATP or of a precursor of a high-energy state appears to "uncouple" oxidative phosphorylation. The antibiotic valinomycin is in this category (see below). (3) In the presence of the antibiotic oligomycin, mitochondria behave as though reaction (*c*) in the above sequence (page 347) were blocked. In tightly coupled mitochondria, both phosphorylation and oxidation cease in the presence of oligomycin. (4) A fourth class of agent apparently affects the final step in the process, reaction (*d*), by preventing uptake of ADP by mitochondria. This is seen with the glycoside atractyloside, suggesting that there is a "translocating" enzyme in the mitochondrial membrane, sensitive to the glycoside and regulating the following process.

$$ATP_{in} + ADP_{out} \rightleftharpoons ADP_{in} + ATP_{out}$$

It has been suggested that oligomycin may also function as atractyloside does.

Partial Reactions of Oxidative Phosphorylation

1. Mitochondria which have not been carefully prepared or which have been exposed to dinitrophenol exhibit ATPase activity, *viz.*, they catalyze hydrolysis of ATP to ADP + P_i. Submitochondrial particles behave similarly. The fact that dinitrophenol-induced ATPase is eliminated by oligomycin suggests the existence of a reversible reaction sequence such as (*c*) and (*d*), above, with ATP hydrolysis resulting from dinitrophenol-induced instability of either X~Y or carrier ~X. This concept is further strengthened by the fact that dinitrophenol alleviates the inhibition of respiration by oligomycin.

2. Mitochondria catalyze an exchange of the oxygen of water with that of P_i:

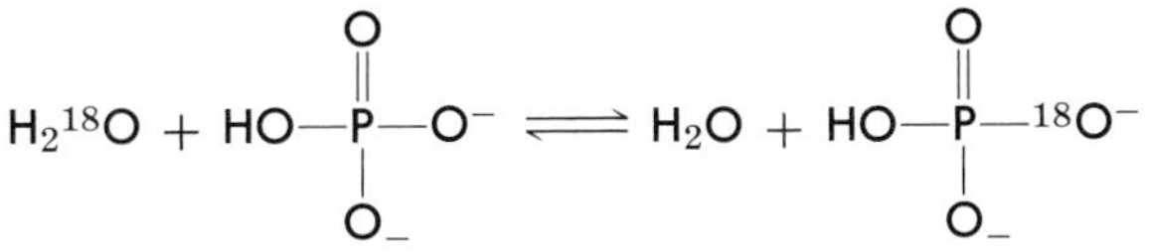

and also catalyze exchange of the oxygen of water with an oxygen on the terminal phosphate of ATP. Both processes are inhibited by oligomycin and dinitrophenol.

An adequate interpretation of this observation is not available. Were the sequence shown a satisfactory representation, then the oxygen of phosphate would be utilized in bond formation of Y~P and could be lost as YOH in the final step. Since the nature of Y is unknown, it is not clear how it exchanges oxygen with the medium. An adequate explanation of this phenomenon would do much to elucidate the mechanism involved.

3. In the absence of electron transport, mitochondria catalyze the exchange

$$ATP + {}^{32}P_i \rightleftharpoons AT^{32}P + P_i$$

This process can occur at a much greater rate than net ATP formation.

4. A similar exchange of $AD^{32}P$ with ATP is also catalyzed by mitochondria. However, whereas it would seem necessary only to invoke reaction (*d*) of the sequence, this exchange is inhibited by both oligomycin and dinitrophenol and, as in the immediately previous case, the exchange is not demonstrable if the electron transport chain is fully reduced. Hence the obvious, simplistic explanation is insufficient.

Translocation of Cations. Freshly isolated mitochondria contain an abundance of K^+, about one-third as much Mg^{++}, and lesser amounts of Na^+, Ca^{++}, Zn^{++}, Fe^{++}, and Mn^{++}. If exposed to hypotonic media or "swelling agents," these ions are leached from the mitochondria but may be reaccumulated during aerobic addition of oxidizable substrate. Indeed, K^+ may be accumulated in an amount almost 50 per cent greater than the initial concentration. This process is particularly evident in the presence of the antibiotic valinomycin, which stimulates an exchange of extramitochondrial K^+ for intramitochondrial H^+ in a ratio close to 1. The requisite energy may be supplied either by respiration or by addition of ATP. Over short intervals the ratio K^+ translocated/~P used may be as high as 7 or 8, but over longer periods this ratio approximates 3. In mitochondria pretreated so as to contain a much greater internal $[K^+]$ than that of the medium, in the presence of valinomycin the efflux of K^+ drives formation of ATP from ADP and P_i with approximately equal efficiency.

Even more impressive is the massive accumulation of Sr^{++}, Ca^{++}, or Mg^{++} by isolated mitochondria, since these ions are present in such low concentration initially. In a medium rich in Ca^{++}, for example, Ca^{++} accumulation occurs with a Ca^{++}/O of about 4 to 5. If the anion of the medium is Cl^-, Ca^{++} accumulation occurs by exchange for intramitochondrial H^+; the medium becomes acidic and the mitochondrial content, *viz.*, the intermembranous fluid, becomes alkaline. If the medium is rich in P_i, the latter follows Ca^{++} across the membrane and dense granules of a hydroxyapatite may precipitate.

The normal physiological role of this capacity of mitochondria to accumulate divalent ions is obscure. Indeed, it may never be operative at prevailing cytoplasmic concentrations of these ions. Of particular importance is the fact that, for both K^+ and Ca^{++} transport, the process may be driven either by respiration or by extramitochondrial ATP. In the former case rotenone and antimycin A inhibit cation transport; in the latter instance, oligomycin is strongly inhibitory. Most importantly, when respiration is driving cation translocation, oligomycin is *not* inhibitory.

Patently, therefore, the mitochondrion can utilize for an endergonic process the energy of an activated intermediate which is generated by electron transport and which occurs in the over-all mechanism of oxidative phosphorylation prior to the reaction in which P_i is required. Moreover, the energy required for formation of TPNH by irreversible transhydrogenation may be supplied in a similar manner. At least two other energy-requiring processes, mitochondrial contraction and reversed electron flow (see below), are also satisfied in this manner. The sum of these observations clearly suggests the existence of a useful high-energy compound formed before utilization of P_i or ADP, such as the hypothetical carrier$\sim$X or X$\sim$Y. This is substantiated by the fact that the "activated" form of cytochrome b mentioned earlier (page 347) is seen spectrophotometrically only when Ca^{++} is accumulating and by the observation that submitochondrial particles, capable of oxidative phosphorylation at the third "site," adequately transport electrons but accumulate Ca^{++} only when additional cytochrome c is provided. Although the detailed mechanism remains obscure, these findings are compatible with the concept of activated carriers.

Coupling Factors. Many attempts to analyze the process of oxidative phosphorylation have sought to fragment mitochondria into particles capable of electron transport but incapable of oxidative phosphorylation and then to attempt to restore the phosphorylative capacity with some other fraction obtained from mitochondria. These efforts have encountered some success. Racker has prepared particles capable of electron transport and limited phosphorylation. He has also extracted from a fraction, known to be the "head pieces" of the bodies fixed to the inner membrane, a soluble protein which dissociated at 0°C. into monomers of about 26,000 molecular weight and reaggregated into a decamer at room temperature. The latter exhibited ATPase activity which required Mg^{++} and was stimulated by dinitrophenol. Addition of this polymeric form to the respiring particle resulted in disappearance of the ATPase activity and significant restoration of the capacity for oxidative phosphorylation at all three sites. The mechanism of this phenomenon is unknown.

Reversed Electron Flow. A striking example of the operation of the mitochondrial system is the finding that mitochondria can catalyze the reaction

$$\text{Succinate} + \text{DPN}^+ \longrightarrow \text{fumarate} + \text{DPNH} + \text{H}^+$$

As noted earlier, from the redox potentials of these components, reaction in the reverse direction is strongly favored. However, if ATP is added under anaerobic conditions, net reduction of DPN^+ can readily be demonstrated although the stoichiometry is uncertain. The over-all process, then, may be formulated as

$$\text{Succinate} + \text{DPN}^+ + \text{ATP} \longrightarrow \text{fumarate} + \text{DPNH} + \text{H}^+ + \text{ADP} + \text{P}_i$$

Under appropriate conditions, it can be shown that neither ATP nor any compound of the class Y$\sim$P is actually required and that again an activated compound present earlier in the oxidative phosphorylation mechanism will suffice. This provides a demonstration of the equivalence of reducing power and "high-energy bonds" in the mitochondrial system. Sanadi has shown that a "coupling factor" very similar to that described above is required for reduction of DPN^+ by succinate in the presence of ATP by submitochondrial particles.

Mitochondrial Volume Changes. A large and confusing literature is concerned with factors which influence the volume of a mitochondrion. In general, three classes of volume changes may be distinguished. (1) A variety of agents and conditions injure the outer membrane either by peroxidation of the unsaturated lipids or by reaction with the proteins. Among such agents are thyroid hormones, (thyroxine, triiodothyronine), soaps, disulfide hormones (vasopressin, insulin), glutathione, Fe^{++}, Hg^{++}, and iodoacetamide. These induce swelling of large amplitude, frequently irreversible, and may be without physiological meaning. (2) The nonrespiring mitochondrion can serve as a passive semipermeable osmometer, and also exhibits volume changes of large amplitude dependent upon the tonicity of the suspending medium. Mitochondria with ample oxidizable substrate and lacking ADP also swell; the swelling may be reversed by addition of ADP or ATP. In each of the above instances, water uptake requires concurrent electron transport; respiratory inhibitors such as antimycin A or cyanide prevent water imbibition. Addition of ATP results in contraction of the swollen mitochondrion; several thousand water molecules depart from the organelle per molecule of ATP hydrolyzed. This process does not reflect water movement attendant upon expulsion of osmotically active ions but seems rather to be a physical contraction of the mitochondrion. (3) If mitochondria are induced to swell by immersion in KCl solution, contraction may be energized either by respiration or by ATP. Swelling is prevented by antimycin A but not by oligomycin, whereas contraction is prevented by oligomycin but unaffected by antimycin A. Clearly, these changes reflect the movement of cations into and out of the mitochondrion.

The action of ATP may not depend upon its hydrolysis although this is prevented by oligomycin in most circumstances. The mechanisms operative in this aspect of mitochondrial physiology and their relationship to the process of oxidative phosphorylation are of great interest but remain to be clarified.

A Proposed Chemiosmotic Basis for Oxidative Phosphorylation. It will be apparent that, in the face of a wealth of experiment and observation, the description of mitochondrial oxidative phosphorylation presented above is decidedly less than satisfactory. A series of activated intermediates is postulated, none of which has actually been observed. Nor does the postulated reaction sequence suggest an explanation for the obligatory association of oxidative phosphorylation with the double-layered membrane of complex chemical composition. An alternative approach has been proposed which rests on the following postulates: (1) the intermediates have not been found because they may not exist; (2) the first energy-rich compound in the sequence leading to ATP formation may not arise by virtue of loss or gain of an electron; (3) a membranous structure rich in lipid offers the opportunity to conduct reactions in a milieu in which the effective thermodynamic activity of water is low; (4) the over-all process may depend on reactions catalyzed by enzymes which are anisotropically arranged within the membrane so that their products can be discharged, respectively, either within or without the membrane. If the latter is poorly permeable to such products, *e.g.*, H^+ or OH^-, an electrochemical gradient may be established which can be utilized to drive an appropriate reaction, without necessity for a coupling high-energy intermediate. This concept was initially suggested by Lundegardh, was refined by Davies, and has been most fully developed by Mitchell.

If quantitative considerations and all presumed details are ignored, such a mechanism might be presumed to function as indicated in Fig. 16.9. This figure assumes the presence of a reversible ATPase so arranged in the membrane that the

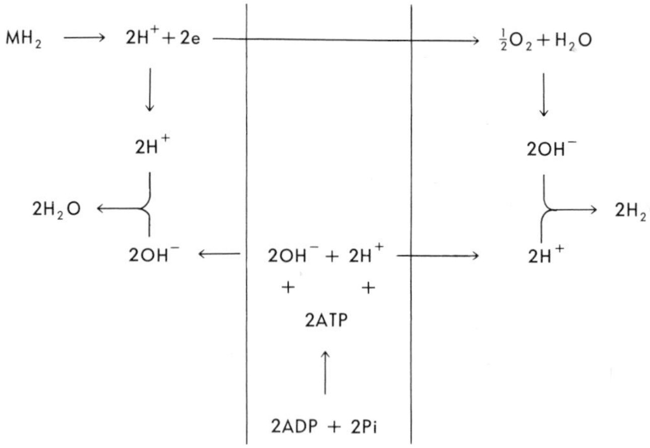

Fig. 16.9. Schematic version of a chemiosmotic mechanism for formation of ATP by a vectorially operating reversible ATPase present in the mitochondrial membrane.

protons and hydroxyl ions generated by ATP formation are discharged vectorially. If the protons are directed toward that side on which O_2 reduction occurs and the hydroxyl ions are directed toward the locus of proton formation by dehydrogenation of substrates, reformation of water would constitute the immediate driving force for ATP formation. This mechanism, which is entirely speculative, has had no direct experimental support but seems attractive because of the vigorous ion-translocating activities of the mitochondrial membrane.

REFERENCES

Books

Chance, B., ed., "Energy-linked Functions of Mitochondria," Academic Press, Inc., New York, 1963.

Florkin, M., and Stotz, E. H., "Comprehensive Biochemistry," vol. 14, Biological Oxidations, American Elsevier Publishing Company, New York, 1966.

Green, D. E., and Goldberger, R. F., "Molecular Insights into Living Processes," Academic Press, Inc., New York, 1967.

Lehninger, A. L., "The Mitochondrion: Structure and Function," W. A. Benjamin Inc., New York, 1964.

Racker, E., "Mechanisms in Bioenergetics," Academic Press, Inc., New York, 1965.

San Pietro, A., ed., "Non-heme Iron Proteins," The Antioch Press, Yellow Springs, Ohio, 1965.

Singer, T. P., ed., "Biological Oxidations," John Wiley & Sons, Inc., New York, 1966.

Slater, E. C., Kaniuga, Z., and Wojtczak, L., eds., "Biochemistry of Mitochondria," Academic Press, Inc., New York, 1967.

Umbreit, W. W., Burris, R. H., and Stauffer, J. F., "Manometric Techniques and Related Methods for the Study of Tissue Metabolism," 4th ed., Burgess Publishing Company, Minneapolis, 1964.

Review Articles

Chance, B., Techniques for Assay of the Respiratory Enzymes, in S. P. Colowick and N. O. Kaplan, eds., "Methods in Enzymology," vol. IV, pp. 273–338, Academic Press, Inc., New York, 1957.

Chance, B., The Reactivity of Hemoproteins and Cytochromes, *Biochem. J.,* **103,** 1–18, 1967.

Chance, B., and Williams, G. R., The Respiratory Chain and Oxidative Phosphorylation, *Advances in Enzymol.,* **17,** 65–134, 1956.

Ernster, L., and Lee, C. P., Biological Oxidoreductions, *Ann. Rev. Biochem.,* **33,** 729–788, 1964.

Green, D. E., and MacLennan, D. H., The Mitochondrial System of Enzymes, in D. M. Greenberg, ed., "Chemical Pathways in Metabolism," 3d ed., vol. I, pp. 47–111, Academic Press, Inc., New York, 1967.

Harris, E. J., Judah, J. D., and Ahmed, K., Ion Transport in Mitochondria, in D. R. Sanadi, ed., "Current Topics in Bioenergetics," pp. 255–278, Academic Press, Inc., New York, 1966.

Lehninger, A. L., Water Uptake and Extrusion by Mitochondria in Relation to Oxidative Phosphorylation, *Physiol. Revs.,* **42,** 467–517, 1962.

Lehninger, A. L., Carafoli, E., and Rossi, C. S., Energy-linked Ion Movements in Mitochondrial Systems, *Advances in Enzymol.,* **29,** 259–320, 1967.

Lehninger, A. L., and Wadkins, C. L., Oxidative Phosphorylation, *Ann. Rev. Biochem.,* **31,** 47–79, 1962.

Lipmann, F., Metabolic Generation and Utilization of Phosphate Bond Energy, *Advances in Enzymol.,* **1,** 99–162, 1941.

Lowenstein, J. M., The Tricarboxylic Acid Cycle, in D. M. Greenberg, ed., "Chemical Pathways of Metabolism," 3d ed., vol. I, pp. 146–270, Academic Press, Inc., New York, 1967.

Massey, V., and Veeger, C., Biological Oxidations, *Ann. Rev. Biochem.,* **32,** 579–638, 1963.

Mitchell, P., Translocations Through Natural Membranes, *Advances in Enzymol.*, **29**, 33–87, 1967.

Ochoa, S., Enzymic Mechanisms in the Citric Acid Cycle, *Advances in Enzymol.*, **15**, 183–270, 1954.

Racker, E., Mechanisms of Synthesis of Adenosine Triphosphate, *Advances in Enzymol.*, **23**, 323–399, 1961.

Sanadi, D. R., Energy Linked Reactions in Mitochondria, *Ann. Rev. Biochem.*, **34**, 21–48, 1965.

Slater, E. C., The Constitution of the Respiratory Chain in Animal Tissues, *Advances in Enzymol.*, **20**, 147–199, 1958.

Slater, E. C., Oxidative Phosphorylation, in M. Florkin and E. H. Stotz, eds., "Comprehensive Biochemistry," vol. 14, pp. 327–396, American Elsevier Publishing Company, New York, 1966.

17. Biological Oxidations. III

Oxidative Enzymes, Coenzymes, and Carriers

The preceding chapter introduced briefly the enzymes and coenzymes that participate in biological oxidations. Patently, few substances oxidized within the mammalian organism are spontaneously oxidized by oxygen at 37°C. and pH 7. Such reactions are all thermodynamically possible, *i.e.*, the E'_0 values of the substances to be oxidized are lower than that of the O_2/OH^- couple. It is the function of the enzymes that catalyze oxidations to provide a reaction pathway or mechanism which, effectively, lowers the activation energy so that the reaction can proceed.

PYRIDINE NUCLEOTIDES

The existence of diphosphopyridine nucleotide (DPN), the most abundant of the respiratory coenzymes, was noted by Harden and Young during their studies of yeast fermentation; it was isolated in the laboratory of von Euler, and its structure (page 322) was established in 1936. Triphosphopyridine nucleotide (TPN) was first described by Warburg and Christian as the coenzyme for oxidation, in erythrocytes, of glucose 6-phosphate to 6-phosphogluconic acid; its general structure was established in 1935. This nucleotide differs from DPN in that TPN contains a third molecule of phosphate esterified at C-2′ of the ribose moiety of the adenosine portion of the molecule, as shown on page 356.

Molecular models indicate that these compounds can exist in an extended form, with the adenine moiety remote from the nicotinamide, or, as depicted here, with the plane of the adenine ring lying above the plane of the nicotinamide portion of the molecule. Present evidence does not decisively indicate which form participates in enzymic reactions.

Whereas DPN is present in concentrations of 0.4 to 2.0 mg. per g. of tissue, TPN varies from 0.01 to 0.1 mg. per g., except for liver, where their concentrations are approximately equal. Most of the DPN in the living cell is present in the oxidized form; the TPN is present chiefly in the reduced state. Both nucleotides are synthesized from their components in the cells in which they occur. The biogenesis and fate of the pyridine nucleotides are presented in Chap. 27.

The abbreviated structures in Fig. 17.1 (page 356) indicate that when DPN^+ is reduced by 2-electron transfer in an enzymic reaction, the DPN^+ accepts the equivalent of a hydride ion ($H:^-$) from the oxidized substrate while a proton is liberated to the medium. The quinonoid reduced compound bears no charge on

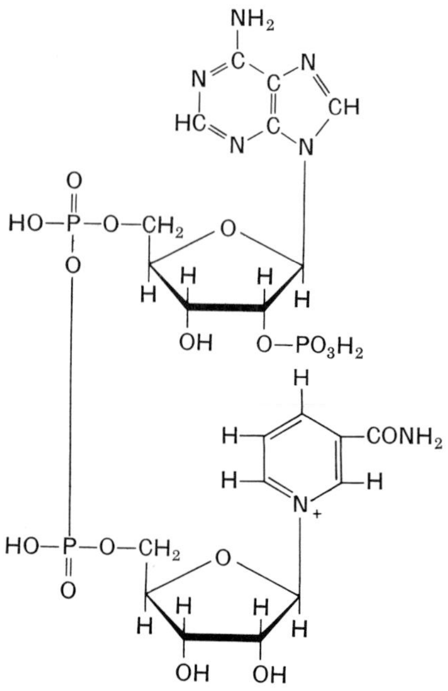

Triphosphopyridine nucleotide (TPN$^+$)

the ring nitrogen and contains one hydrogen derived from the substrate; hence the abbreviation for reduced DPN$^+$ is DPNH. Colowick demonstrated that the newly introduced hydrogen is fixed to the pyridine ring at the 4 or *para* position.

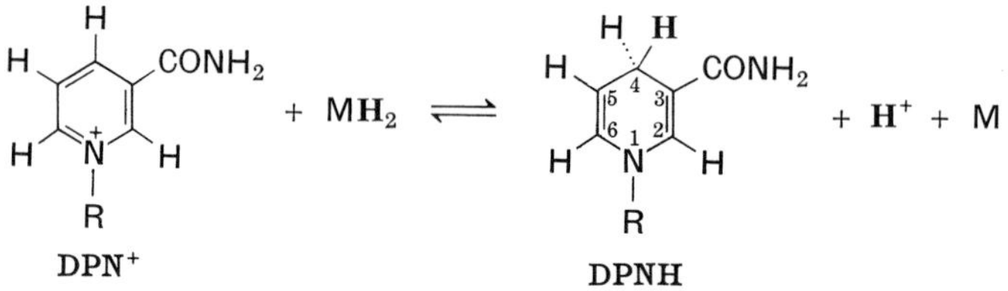

Fig. 17.1. The stereospecific reduction of diphosphopyridine nucleotide by a dehydrogenase of the A series. *R* represents the remainder of the structure (page 322).

Since the pyridine ring of DPNH is planar, the two hydrogen atoms at carbon-4 project on either side of the plane. The stereospecificity of the formation and reoxidation of DPNH was shown by the following observations of Vennesland and Westheimer. DPN^{2}H was formed nonenzymically by reduction of DNP$^+$ with dithionite in ^2H_2O, and then used, with alcohol dehydrogenase, to reduce acetaldehyde. The resulting ethanol and DPN$^+$ each contained one-half an equivalent atom of deuterium per mole. However, when DPN$^+$ was reduced by synthetic $CH_3C^2H_2OH$ in the presence of yeast alcohol dehydrogenase, DPN^{2}H containing one atom of deuterium per molecule was formed. When this DPN^{2}H was incubated with acetaldehyde and alcohol dehydrogenase, or with pyruvate and lactic acid dehydrogenase, the deuterium was quantitatively retransferred, with formation of deuteroethanol or deuterolactate, respectively. Thus, transfer of hydrogen atoms by these dehydrogenases is stereospecific with respect to the plane of the pyridine ring. It will be evident that alcohol dehydrogenase also exhibits stereospecificity with respect to the two hydrogen atoms on the hydroxyl-bearing carbon atom. Yeast alcohol dehydrogenase and heart lactic acid dehydrogenase exhibit the same stereospecificity, *i.e.*, they catalyze transfer of hydrogen to and from the same side of the pyridine ring. These are termed dehydrogenases of the A type. As shown in Table 17.1, a somewhat larger group of enzymes is presently known to exhibit stereospecific transfer to and from the opposite side of the pyridine ring. These are dehydrogenases of the B type. It has been possible to establish the absolute configuration of the two series. The hydrogen atom added by dehydrogenases of the A series projects toward the reader from the plane of the pyridine ring when that ring is oriented as in Fig. 17.1.

Table 17.1: STEREOSPECIFICITY OF PYRIDINE NUCLEOTIDE DEHYDROGENASES

Substrate	Nucleotide	Source	Stereospecificity
Isocitrate	TPN	Heart	A
Ethanol	DPN	Yeast	A
Lactate	DPN	Heart	A
Malate	DPN	Heart	A
TPN^+	DPN	Heart mitochondria	A
Farnesyl pyrophosphate	TPN	Liver	B
Glucose	DPN	Liver	B
Glucose 6-phosphate	TPN	Yeast	B
Glutamate	TPN	Muscle	B
Glutathione	TPN	Yeast	B
3α-Hydroxysteroids	DPN	Liver	B
3β-Hydroxysteroids	DPN	*Pseudomonas*	B
17β-Hydroxysteroids	DPN, TPN	Liver	B
6-Phosphogluconate	TPN	Liver	B
3-Phosphoglyceraldehyde	DPN	Yeast, muscle	B
TPN^+	DPN	*Pseudomonas*, *Escherichia coli*	B

The pyridine nucleotides participate as coenzymes in a wide variety of biological oxidations. Only a few of these known enzymes are listed in Table 17.2. No relationship is obvious between the nature of the substrate and enzyme preference for TPN or DPN. Thus, there are dehydrogenases that exhibit activity only with one of the two nucleotides, a few that catalyze reaction with either at almost equal rates, and, more frequently, dehydrogenases that exhibit catalytic activity with one coenzyme at a rate many times greater than with the other. This behavior may also vary with the species from which an enzyme is obtained. Thus, glucose 6-phosphate dehydrogenase from yeast has an absolute specificity for TPN whereas that from mammalian liver uses DPN at about 7 per cent of the rate with TPN at equivalent concentration. *In general, enzymes responsible for oxidations that supply energy to the organism utilize DPN, while those which catalyze reductive biosyntheses employ TPN.* However, it will be apparent in later discussions that there are many exceptions to this generalization.

Table 17.2: SOME DEHYDROGENASES EMPLOYING PYRIDINE NUCLEOTIDES

Substrate	Products	Coenzyme	Preparative source
Acyclic polyols	Ketoses	DPN	Rat liver, *Aerobacter*
Aldehydes	Carboxylic acids	DPN	Liver
Aspartic β-semialdehyde	β-Aspartyl phosphate	DPN	Liver
Betaine aldehyde	Betaine	DPN	Rat liver
Ethanol	Acetaldehyde	DPN	Liver, kidney, yeast
D-Glycerate	Hydroxypyruvate	DPN	Liver
α-Glycerophosphate	Dihydroxyacetone phosphate	DPN	Yeast, muscle, liver
L-Gulonate	Xylulose + CO_2	DPN	Kidney
ω-Hydroxy acids	ω-Aldehyde acids	DPN	Liver
D(−)β-Hydroxybutyrate	Acetoacetate	DPN	Liver
L(−)β-Hydroxybutyryl CoA	Acetoacetyl CoA	DPN	Liver
3β-Hydroxysteroids	3-Ketosteroids	DPN	*Pseudomonas*
17β-Hydroxysteroids	17-Ketosteroids	DPN	*Pseudomonas*
Isocitrate	α-Ketoglutarate + CO_2	DPN	Beef heart, rat liver, yeast
Lactate	Pyruvate	DPN	Muscle, other animal tissues
Malate	Oxaloacetate	DPN	Muscle, other animal tissues
Malonic semialdehyde	Malonate	DPN	*Pseudomonas*
D-Mannitol	D-Fructose	DPN	*Acetobacter*
3-Phosphoglyceraldehyde	1,3-Diphosphoglycerate	DPN	All animal tissues, yeast bacteria
Ribitol	Ribulose	DPN	Liver
Succinic semialdehyde	Succinate	DPN	Brain, bacteria
Tartronic semialdehyde	Glycerate	DPN	*Pseudomonas*
Glycerol	Dihydroxyacetone	DPN or TPN	Pig liver, rat liver, *Escherichia coli, Aerobacter aerogenes, Penicillium, Candida*
Glutamate	α-Ketoglutarate + NH_4^+	DPN or TPN	Muscle, liver, yeast
3β-Hydroxysteroids	3-Ketosteroids	DPN or TPN	Liver
Dihydrofolate	Tetrahydrofolate	TPN	Liver
D-Fructose	5-Keto-D-fructose	TPN	*Glyconobacter cirinus*
Glucose 6-phosphate	6-Phosphogluconate	TPN	Liver, erythrocytes, yeast
L-Gulonate	D-Glucuronate	TPN	Kidney
Isocitrate	α-Ketoglutarate + CO_2	TPN	Various animal tissues, yeast
Malate	Pyruvate + CO_2	TPN	Heart
Reduced glutathione (G-SH)	Glutathione (G-SS-G)	TPN	Yeast, liver
Shikimate	Dehydroshikimate	TPN	Peas, *E. coli*

The terms "prosthetic group" and "coenzyme" are rather misnomers when applied to the pyridine nucleotides. The K_m for DPN^+ and TPN^+ varies from 10^{-4} to $10^{-6}M$, whereas K_m for the "substrates" of these enzymes varies from 10^{-3} to $10^{-5}M$. In general, DPNH is bound about 10 to 100 times as firmly as DPN^+ by most dehydrogenases. The pyridine nucleotides readily dissociate from the dehydrogenases and are really cosubstrates, which serve as coenzymes only in that, physiologically, there exist *other* enzymes that catalyze reoxidation of the reduced forms, thus permitting reutilization of the nucleotides. Also, by virtue of this ease of dissociation, pyridine nucleotides, uniquely among oxidative coenzymes, may participate in dismutations, *i.e.*, reduction of one metabolite by another.

$$(a) \qquad AH_2 + DPN^+ \longrightarrow A + DPNH + H^+$$
$$(b) \qquad B + DPNH + H^+ \longrightarrow BH_2 + DPN^+$$
$$\text{Sum:} \qquad AH_2 + B \longrightarrow A + BH_2$$

Such coupling requires, therefore, that the DPN shuttle between the surfaces of the two substrate specific dehydrogenases. The major examples of this process are the participation of DPN in glycolysis (pages 398*ff.*) and the reductive formation of fatty acids, utilizing TPNH derived from oxidation of carbohydrate (page 415). An interesting possibility is offered by the coupling of a pair of dehydrogenases which are of the A and B series, respectively. For example, in glycolysis it seems likely that DPNH, reduced on the surface of triose phosphate dehydrogenase, may be used to reduce pyruvate by lactic acid dehydrogenase without leaving the surface of the former (page 402).

Formally, the reduction of DPN^+ to DPNH by MH_2 (Fig. 17.1) represents acceptance by DPN^+ of a hydride ion ($H:^-$) as a proton enters the medium. This seems a likely mechanism but has not been established unequivocally. It has been suggested that on the enzymic surface, DPN^+ and the substrate may be brought into such close proximity that a "charge transfer complex" may be formed. In such complexes, two organic nuclei approach each other and the reductant transfers an electron to the acceptor. This electron, which is generally distributed through the acceptor molecule, gives it a negative charge. The donor molecule is relatively positively charged, and the charges then serve to maintain the two molecules in an intermolecular complex.

$$A + BH_2 \longrightarrow [A^-BH_2^+] \longrightarrow AH_2 + B$$

Such complexes are highly colored and, in some instances, may be detected spectrophotometrically, but this technique has not been successful for establishing charge transfer complex formation in the reduction of DPN^+ or in the reoxidation of DPNH. Subsequent transfer of a hydrogen atom completes the reduction of the acceptor molecule.

In contrast to their oxidized forms, DPNH and TPNH both strongly absorb light at 340 mμ; this is the basis of most analytical procedures for following reactions in which these coenzymes participate. In addition, both exhibit fluorescence when activated by light at 340 mμ; the maximal emission is at about 465 mμ. The binding of DPNH or TPNH to the apoprotein of a dehydrogenase is accompanied by a shift of the absorption maximum to a lower wavelength (330 to 335 mμ) and a corresponding shift in the maximum of the emitted fluorescent light to about 440 mμ. Under these conditions, there also occurs a considerable intensification of the emitted light (an increase in quantum yield) which renders such fluorescence measurements particularly sensitive as detectors of enzyme-coenzyme complex formation.

An additional technique, *fluorescence polarization,* has provided information concerning the nature of the enzyme-coenzyme complex. When polarized light, of the proper wavelength, is used to activate small molecules, the resultant fluorescence is generally unpolarized. Light quanta can be absorbed only by those molecules appropriately oriented to the exciting wave. The subsequent fluorescence emitted also emerges parallel to the same orientation, *i.e.*, the emitted light from each molecule is polarized. However, brownian motion of small molecules is very rapid, and in the time between absorption and reemission (10^{-8} to 10^{-7} sec.), the molecules in a sample randomly redistribute their positions so that, *in sum,* the emitted light is unpolarized. In contrast, large molecules like proteins exhibit much slower brownian motion. Thus, proteins fluoresce when activated with light at 280 mμ, the absorption maximum of their tryptophan residues, and if the exciting light is plane-polarized, the emitted light is only slightly less polarized. When a dehydrogenase, *e.g.*, lactic acid dehydrogenase, to which DPNH is bound is excited by plane-polarized light at 340 mμ, the emitted fluorescent light is also strongly polarized. Hence, the coenzyme must be tightly bound to the enzyme in a rigid conformation. Such data, however, must be interpreted with caution, since they do not discriminate between binding of DPNH at the catalytic site of a dehydrogenase protein and nonspecific binding elsewhere on the molecule. On the basis of kinetic evidence, Cleland suggested the order of events shown in the following diagram for the operation of liver alcohol dehydrogenase.

$$\begin{array}{ccccccc} & \mathbf{DPN^+} & & \mathbf{RCH_2OH} & & \mathbf{RCHO} & & \mathbf{DPNH} & \\ & \downarrow & & \downarrow & & \uparrow & & \uparrow & \\ \hline \mathbf{En} & & \mathbf{En\text{-}DPN^+} & & (\mathbf{En\text{-}DPN^+\text{-}RCH_2OH} \longrightarrow \mathbf{En\text{-}DPNH\text{-}RCHO}) & & \mathbf{En\text{-}DPNH} & & \mathbf{En} \end{array}$$

The mode of coenzyme binding is not clear, but for alcohol, lactic acid, and several other dehydrogenases there is evidence that a sulfhydryl group of the protein mediates attachment to the coenzyme.

The transfer of hydrogen from substrate to pyridine nucleotide clearly denotes formation of a ternary complex of enzyme-substrate-coenzyme. Kinetic evidence also strongly suggests that transfer requires the transient existence of a ternary complex, *e.g.*, enzyme-lactate-DPN$^+$ which becomes enzyme-pyruvate-DPNH and then dissociates. In many instances there appears to be a compulsory order of binding. Thus, pyruvate cannot bind to lactic acid dehydrogenase until after formation of the enzyme-DPNH complex. This suggests that the enzymic surface, of itself, presents no pyruvate binding site and that the latter results from changes in conformation and/or charge, etc., consequent to formation of the enzyme-DPNH complex.

Several DPN-dependent dehydrogenases are zinc-proteins. Beef liver glutamic acid dehydrogenase and the alcohol dehydrogenases of yeast, horse liver, and human liver contain 4, 4, 2, and 2 gram-atoms of zinc per mole of protein, respectively. It appears likely that the metal participates in coenzyme and/or substrate binding since the three alcohol dehydrogenases bind 4, 2, and 2 moles of DPNH per mole of protein, respectively. The turnover number of the yeast enzyme is 100 times greater than that of the horse and 1,000 times greater than that of man.

The pyridine nucleotides exhibit marked specificity in structural requirements. For binding to a dehydrogenase, the adenine, carboxamide, pyrophosphate, and ribose portions are all required. Alteration in any aspect of the molecule that affects its normal spatial conformation reduces or destroys its function as a coenzyme. However, Kaplan has prepared several analogues of DPN which can substitute for DPN with various dehydrogenases. Among these are analogues in which, for the $-CONH_2$ group on the pyridine ring, has been substituted $-\underset{\underset{O}{\|}}{C}-CH_3$ or $-\underset{\underset{O}{\|}}{C}-CH_2-CH_3$ or $-\underset{\underset{S}{\|}}{C}-NH_2$, etc. Each analogue exhibits a higher K_m than does DPN itself with any particular dehydrogenase, but their relative values differ from one dehydrogenase to another. They also differ somewhat in potential from that of $DPN^+/DPNH$; hence the equilibrium position of the reaction is also altered.

The binding of substrates and coenzymes to the dehydrogenases may alter markedly their redox potentials, usually bringing them closer together. The binding sites are in such proximity and so oriented that appropriate group migration, *e.g.*, a hydride ion, is facilitated. These properties combine to lower E_a for the over-all reaction. In some instances, a group on the protein may itself participate in the reaction mechanism. One interesting observation of this type is the stereospecific transfer of tritium from ethanol to the α-carbon of a tryptophan residue of yeast alcohol dehydrogenase. (See also discussion of phosphoglyceraldehyde dehydrogenase, page 398.)

The character of the binding sites for coenzyme and substrate on the enzymic surface confers specificity to the reaction. These binding sites usually exist on a specific polymeric form of the enzyme. Thus, glutamic acid dehydrogenase (molecular weight $= 2 \times 10^6$) can be dissociated into subunits (molecular weight = 52,000) which no longer exhibit enzymic activity, but activity is evident with aggregates of molecular weight 400,000 or greater. Similarly, yeast alcohol dehydrogenase can be deaggregated into subunits that are inert enzymically. Binding of the specific coenzyme appears to maintain both the conformation and the specific polymeric nature of several dehydrogenases.

The polymeric nature of some dehydrogenases results in the presence of different but related forms of the same enzyme in animal tissues. Lactic acid dehydrogenase of mammalian tissues, molecular weight 135,000, is a tetramer. Two electrophoretically distinguishable polypeptide chains, α and β, can be equally well bound in such tetramers. Hence, it is possible to observe the presence, in different tissues, of five lactic dehydrogenases which apparently may be composed of 4α, 4β, $1\alpha + 3\beta$, $2\alpha + 2\beta$, or $3\alpha + 1\beta$ chains. These electrophoretically distinct forms of an enzyme with identical function have been termed *isozymes.* Each exhibits different K_m values for DPN and its analogues. The physiological and genetic factors that determine the mixture of the various lactic acid dehydrogenase isozymes characteristic of a given tissue or individual are not known.

In some instances, cells have more than one enzyme that can catalyze the same reaction, yet these are not isozymic forms. Thus, mammalian cells contain malic acid and isocitric acid dehydrogenases in both cytoplasm and mitochondria. The intra- and extramitochondrial forms of each differ in molecular weight, amino acid composition, and electrophoretic and immunochemical behavior.

FLAVOPROTEINS

In 1932 Warburg and Christian obtained from yeast a "yellow enzyme" capable of catalyzing the oxidation of TPNH. After the structure of riboflavin had been established, the prosthetic group of this enzyme was recognized as riboflavin 5′-phosphate (flavin mononucleotide, FMN) in 1936. In 1938, Warburg and Christian isolated the coenzyme of renal D-amino acid oxidase and demonstrated it to be flavin adenine dinucleotide, FAD (page 324). Flavoproteins serve as electron transport agents by catalyzing the consecutive reactions shown below.

(*a*) $$\mathbf{MH_2 + flavin \longrightarrow M + reduced\ flavin}$$

(*b*) $$\mathbf{Reduced\ flavin + X \longrightarrow flavin + XH_2}$$

MH_2 of reaction (*a*) may be a metabolite or the reduced form of one of the pyridine nucleotides as illustrated previously (page 359); X may be a metabolite, a pyridine nucleotide, a metal ion, a heme derivative, or O_2. For all flavoproteins, the natural acceptor, X, may be replaced in the laboratory by an appropriate reducible dye such as methylene blue, 2,6-dichlorophenol indophenol, or phenazine methosulfate. When O_2 serves as acceptor, it is reduced to peroxide.

No general mode of binding of the flavins to apoproteins has been established, but the binding is considerably stronger than that of pyridine nucleotides to their dehydrogenases. The equilibrium constant for the dissociation,

$$\mathbf{Protein\text{-}flavin \rightleftharpoons protein + flavin}$$

is generally of the order of 10^{-8} or 10^{-9}; effectively, therefore, the flavin component remains permanently attached to the enzymic protein. In at least one instance, succinic acid dehydrogenase, there is a covalent bond, of unknown nature, between FAD and a group on the protein; hence, the FAD can be removed only by proteolysis. In most other cases the flavin may be separated from the protein by treatment with ammonium sulfate in an acidic medium.

Flavoproteins exhibit characteristic absorption spectra with maxima at about 280, 380, and 450 mμ. Complete reduction is accompanied by a diminution in the absorption at 280 and 380 mμ and by complete bleaching of the band at 450 mμ, as shown in Fig. 17.2. Formation of the half-reduced form, *i.e.*, the flavin semiquinone, is evidenced by a decrease of about 50 per cent in absorbancy at 450 mμ and by the appearance of a small but broad absorption band at longer wavelengths (550 to 650 mμ). In some instances, this has been correlated with appearance of an electron spin resonance (ESR) signal at $g = 2.00$; frequently, however, the enzyme functions with a pair of flavins which may engage in spin-spin coupling so that no ESR free-radical signal is evident. Riboflavin exhibits a strong fluorescence, maximal at about 500 mμ. The fluorescence of FMN is equal to that of riboflavin, while that of FAD is only about 10 per cent as great, presumably because of internal

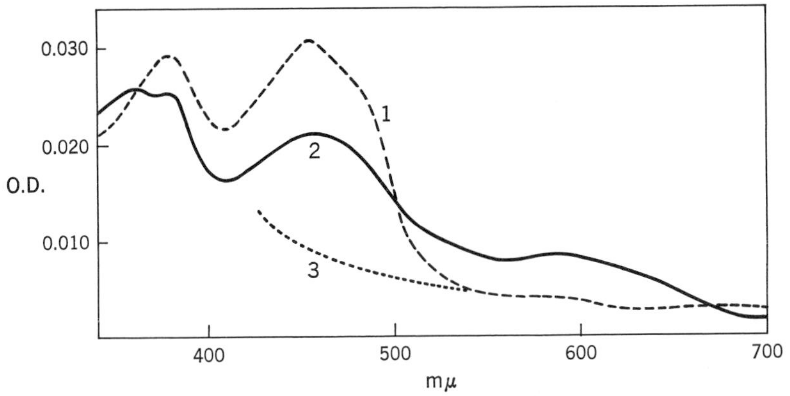

Fig. 17.2. The absorption spectrum of a flavoprotein in the fully oxidized (FAD) (1), semiquinone (FADH·) (2), and fully reduced ($FADH_2$) (3) forms. The spectrum shown is that of microsomal TPNH-cytochrome c reductase (1), reduced by its substrate (2), and by dithionite (3). (*Courtesy of Dr. H. Kamin and Dr. B. S. Masters.*)

quenching by the adenine. In general, flavoproteins show very little fluorescence. A notable exception is lipoyl dehydrogenase, whose fluorescence is almost that of free FMN, suggesting an unfolding of the dinucleotide structure. E_0' for flavoproteins varies over a wide range, the redox potential being markedly affected by the mode of binding of the coenzyme to the apoprotein.

As is evident in Table 17.3, flavoproteins vary in complexity from relatively simple proteins to enzymes that are miniature electron transport systems and may include one or more metal ions, a heme group, or an additional reducible organic group such as ubiquinone. Their molecular weights vary from 12,000 to approximately 300,000. Accordingly, there is a rich diversity of catalytic mechanisms among these enzymes.

It may be assumed that all flavoproteins catalyze removal of either a hydride ion or a pair of hydrogen atoms from the substrate. Enzymes with a single flavin are thereby fully reduced and are reoxidized to the fully oxidized form by the natural acceptor. Examples are "old yellow enzyme" and the *glucose oxidase* of *Penicillium notatum.*

$$\text{Glucose} + \text{En-FAD} \longrightarrow \text{gluconic acid} + \text{En-FADH}_2$$
$$\text{En-FADH}_2 + O_2 \longrightarrow \text{En-FAD} + H_2O_2$$

Flavoproteins such as liver D-*amino acid oxidase,* and some snake venom L-amino acid oxidases, with two flavins, are thought to have two active centers and operate with a cycle in which each flavin independently alternates between the fully oxidized and semiquinone forms, accepting only a single electron from the substrate.

$$MH_2 + \text{En-FAD} \longrightarrow \text{En-FADH}\cdot\text{-MH}\cdot$$
$$\text{En-FADH}\cdot\text{-MH}\cdot + O_2 \longrightarrow \text{En-FAD} + M + H_2O_2$$

Whereas liver L-*amino acid oxidase* contains two bound FAD molecules, that from kidney mitochondria appears to contain two bound FMN molecules per enzyme molecule. In contrast, the *TPNH-cytochrome c reductase* of microsomes shuttles between the fully reduced and the semiquinone form.

Table 17.3: SOME FLAVOPROTEINS

Substrate	Physiological electron acceptor	Other functional components	Source
D-Amino acids	O_2	...	Liver, kidney
L-Amino acids	O_2	...	Kidney, snake venoms
Diamines	O_2	...	Liver, kidney, brain, blood plasma
Glucose	O_2	...	Molds, liver
Monoamines	O_2	...	Liver, kidney, brain, blood plasma
Hydrogenase	Diverse	...	*Clostridium pasteurianum*
Acyl CoA (C_6–C_{12})	ETF*	...	Liver and heart mitochondria
Butyryl CoA	ETF*	...	Liver and heart mitochondria
Dihydrolipoic acid	DPN^+	...	Heart, liver, *Escherichia coli*
DPNH	Cytochrome b_5	...	Liver microsomes
Reduced glutathione	TPN^+	...	Liver, yeast, *E. coli*
TPNH	Cytochrome c	...	Liver microsomes
TPNH	?	...	Yeast
TPNH	O_2	...	Yeast
TPNH	?	...	Erythrocytes
Choline	Respiratory chain	Fe	Liver mitochondria
Sarcosine	ETF*	Fe(?)	Liver and kidney mitochondria
Dihydroorotic acid	DPN^+	Fe	*Zymobacterium oroticum*
DPNH	Respiratory chain	Fe	Heart mitochondria
DPNH, TPNH	Menadione	?	Heart mitochondria and cytoplasm
α-Glycerophosphate	Respiratory chain	Fe	Liver mitochondria
L-Gulono-γ-lactone	?	?	Rat liver microsomes
H_2	DPN^+	?	*Clostridium kluyverii*
Purines	O_2	Mo, Fe	Milk
Purines	O_2, DPN	Mo, Fe	Chicken liver
Succinic acid	Respiratory chain	Fe	Heart mitochondria
TPNH	NO_3^-	Mo	*Neurospora*
TPNH	Sulfite	Fe	*E. coli*
Aldehydes	O_2	Fe, Mo, ubiquinone	Liver
DPNH	NO_3^-	Heme	*E. coli, Mycobacterium tuberculosis*
DPNH	NO_3^-	Mo, Fe, heme(?)	*Pseudomonas aeruginosa*
DPNH	NO_2^-	Cu, Fe, heme(?)	*P. aeruginosa*
Formic acid	NO_3^-	Vitamin K_3, cytochrome b_1	*E. coli*
D-α-Hydroxyacids	?	Zn^{++}	Yeast
D-Lactic acid	Cytochrome c	Zn^{++}	Yeast
L-Lactic acid	Cytochrome c	Heme	Yeast
Sulfite	O_2, cytochrome c	Heme	Liver microsomes

* ETF, Electron-transferring flavoprotein (page 341).

$$H^+ + TPNH + En\text{-}2FADH\cdot \longrightarrow TPN^+ + En\text{-}2FADH_2$$
$$En\text{-}2FADH_2 + 2\ cyt\ c\text{-}Fe^{+++} \longrightarrow En\text{-}2FADH\cdot + 2\ cyt\ c\text{-}Fe^{+++} + 2H^+$$

Lipoyl dehydrogenase (page 328) is of interest in that it catalyzes hydrogen transfer between two substrates with E_0 values at least 0.3 volt below that of free FAD.

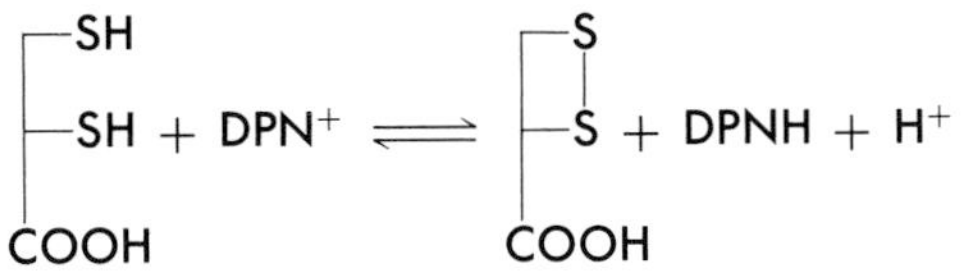

According to Massey, the two flavins of the enzyme appear to operate independently, each in conjunction with sulfhydryl groups which can be demonstrated in the presence of reducing substrate, with stabilization of the flavin semiquinone by a sulfur radical as shown in Fig. 17.3.

Metal Flavoproteins. Several flavoproteins contain firmly bound ferric iron or hexavalent molybdenum. Their mode of action appears to involve full reduction of the flavin moiety by the substrate, followed by successive 1-electron transfers to the metal component and thence to the natural acceptor. Such a simple scheme appears reasonable for the molybdenum-containing *TPNH-nitrate reductases* of *Neurospora crassa* and *E. coli*. The iron-flavoproteins contain a variable but even number of iron atoms. The absorption spectrum due to the iron is essentially identical in all cases and quite similar to that of chloroplast ferredoxin (page 368). Reduction results in characteristic bleaching of both the flavin and iron absorption

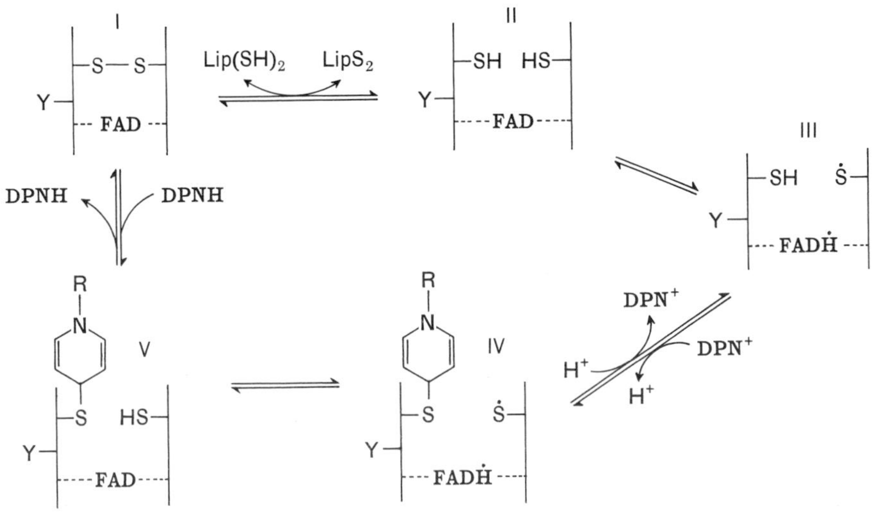

Fig. 17.3. Mechanism of action of lipoyl dehydrogenase. The oxidized enzyme is shown as I. Reduction of the enzyme disulfide by the reduced (sulfhydryl) form of lipoamide results in the disulfhydryl form (II), which transfers one electron to the enzyme-bound FAD (III). This free-radical form reacts with DPN$^+$ to form the transient intermediates IV and V. Departure of DPNH leaves the enzyme in the oxidized form I. *R* indicates the remainder of the DPN molecule (page 322).

spectra and appearance of an ESR signal at $g = 1.94$ which is due to reduced iron. The mode of iron binding to the protein is not known, but acidification results in release of sulfide in an amount stoichiometric with the iron content.

"Diaphorases." Before recognition of the possible role of ubiquinone in mitochondrial electron transport, the search for "cytochrome reductases" led to the finding of a series of flavoproteins which were termed *diaphorases* and which catalyze the reduction of dyes, *e.g.*, methylene blue, by DPNH.

$$\text{H}^+ + \text{DPNH} + \text{methylene blue} \longrightarrow \text{DPN}^+ + \text{leukomethylene blue}$$

In view of the variety of flavoproteins that catalyze reduction of metabolites by DPNH and the ease of reduction of diverse dyes by $FADH_2$ and $FMNH_2$, it will be apparent that "diaphorase" activity need not indicate a protein that normally functions between DPNH and the cytochromes. Thus, the diaphorase initially reported has proved to be the lipoyl dehydrogenase described above; diaphorase activity, therefore, is entirely artifactual.

Mitochondrial Flavoproteins. A group of flavoproteins occur as intrinsic components of the mitochondria. One of these, *DPNH dehydrogenase,* has been isolated in several forms. As a high molecular weight preparation containing a significant amount of lipid, it exhibits a ratio of FAD/Fe of 1:16. Such preparations reduce ubiquinone and ferricyanide but not cytochrome c. If further treated, the preparation loses protein, iron, and lipid, yielding what may be the elemental protein with a ratio FAD/Fe of 1:2 and exhibiting cytochrome c reductase activity. It appears likely that a nonheme iron-protein carrier is present with the dehydrogenase in the mitochondrial complex. *Succinic acid dehydrogenase* contains flavin and nonheme iron in the ratio of 1:4; this enzyme is somewhat more readily separated from the mitochondrial lipids, and also catalyzes ubiquinone reduction. Both enzymes exhibit the $g = 1.94$ ESR signal when incubated with their substrates. Neither is reoxidizable by oxygen.

In addition, there are present in mitochondria somewhat less well-characterized flavoproteins specific for oxidation of sarcosine, choline, α-glycerophosphate, and fatty acyl CoA derivatives. Apparently, none of these functions directly in the electron transport chain. Rather, they reduce another flavoprotein, the "electron-transferring protein," which, in turn, reduces the chain at the level of cytochrome b, perhaps via ubiquinone. No mitochondrial flavoprotein can directly reduce O_2.

Soluble Metalloflavoproteins. Three soluble, autoxidizable iron-flavoproteins have been studied in detail. In all three instances reduction of oxygen can be effected by the iron components, resulting in formation of oxygen radicals ($O_2^{\cdot -}$) and peroxide. The flavin may be removed from each of these enzymes, leaving iron-proteins of identical absorption spectra. The simplest of these, *dihydroorotic acid dehydrogenase,* contains 2 FAD, 2 FMN, and 4 gram-atoms of iron per mole (molecular weight = 120,000). The flavins alternate between the fully oxidized and semiquinone states. The internal transport system suggested for this enzyme is the following.

$$\text{DPNH} \rightleftharpoons \text{FAD} \rightleftharpoons \text{Fe·Fe} \rightleftharpoons \text{FMN} \rightleftharpoons \text{orotic acid}$$

The other two soluble metalloflavoproteins, *xanthine oxidase* and *aldehyde oxidase,* contain molybdenum as well as iron. It has been suggested that the molybdenum, which is the first component of these enzymes to be reduced, is involved in the initial hydroxylating attack upon the substrate. Aldehyde oxidase also contains ubiquinone, and electron transport in this enzyme has been postulated to occur as follows.

$$\text{Substrate} \longrightarrow \text{Mo}^{+6} \longrightarrow \text{FAD} \longrightarrow \text{ubiquinone} \longrightarrow \text{Fe}^{+++} \longrightarrow \text{O}_2$$

Electron spin resonance spectrometry has shown (Fig. 17.4) that the iron and molybdenum are reduced by the substrate and that the semiquinone form of one (or both) of the organic constituents participates in the reaction sequence.

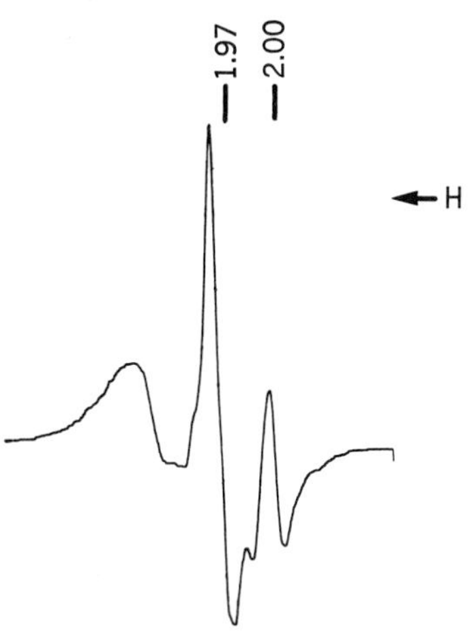

Fig. 17.4. Electron paramagnetic resonance spectrum of the reduced form of rabbit liver aldehyde oxidase. The curve traces the first derivative of the actual signal. The signals at $g = 2.00$ and 1.97 indicate the presence of an organic free radical, Mo^{+5}, and an unusual, enzyme-bound form of Fe^{++}, respectively.

Hemoflavoproteins. A few enzymes are known to possess at their catalytic sites a molecule of FAD (or FMN) and a heme group. All these enzymes are bound within the lipid structure of mitochondria or the respiratory particles of microorganisms. Best known is the L(+)-*lactic acid dehydrogenase* of yeast, which has one FMN and one iron-protoporphyrin IX per molecule. This protein is also known as *cytochrome* b_2. A cytochrome such as cytochrome c serves normally as final electron acceptor. Electron transport appears to occur in the following sequence.

$$\text{Lactate} \longrightarrow \text{FMN} \longrightarrow \text{heme-Fe}^{+++} \longrightarrow \text{cytochrome c}$$

Other hemoflavoproteins are listed in Table 17.3. Thus, "flavoproteins" vary from relatively small proteins, with a single flavin prosthetic group, to large proteins, on the surface of which is a multiple-component electron transport system.

NONHEME IRON-PROTEINS

Mention has already been made of the presence of nonheme iron in several flavoproteins. In addition, there is a group of electron carriers which are proteins containing iron in a tightly bound form and lacking an organic prosthetic group. The first of these to be recognized was the "photosynthetic pyridine nucleotide reductase" of spinach chloroplasts, now termed *chloroplast ferredoxin.* Subsequently, an iron-protein essential to nitrogen fixation in *Clostridium pasteurianum* was purified and named *ferredoxin.* Thereafter, it was rapidly shown that nonheme iron-proteins participate in all major pathways of electron transport: photosynthesis, respiration, and bacterial hydrogen and nitrogen fixation.

Iron-proteins have been isolated from a wide diversity of sources. Two major classes of these proteins may be recognized by their absorption spectra. The bacterial ferredoxins are green-brown with absorption maxima at about 280, 300, and 390 mμ. No compound of this type has yet been found in a higher organism. Chloroplast ferredoxins from algae or spinach, as well as iron-proteins from animal sources, are pink, with absorption maxima at 450 and 550 mμ. Several bacteria contain iron-proteins of both classes.

The iron content of these proteins varies from 2 to 8 gram-atoms per mole. The ligands which bind the iron to these proteins are not known. At pH 3, the iron is dislodged and, concomitantly, one mole of sulfide is released per gram-atom of iron. It seems unlikely that sulfide, per se, preexists in the iron-protein; the responsible structural arrangement remains to be identified.

The reduced form of mitochondrial and microsomal iron-proteins as well as the red iron-protein from *Azotobacter vinelandii* readily yield the $g = 1.94$ ESR signal exhibited by reduced iron-flavoproteins (page 367). Clostridial and spinach ferredoxin yield similar signals under appropriate conditions, suggesting that the mode of iron and sulfur bonding is responsible for the ESR signal of reduced iron-proteins. Since the ferredoxin of *C. pasteurianum* contains two polypeptide chains, in each of which four cysteine residues occur, separated by only two or three other residues between cysteines, it appears likely that these cysteine residues are of significance in the binding of the iron. The number of iron atoms per molecule which can be reduced by substrate has not been established. However, it is known that whereas all the iron atoms of dihydroorotic acid dehydrogenase and aldehyde oxidase are reduced by their respective substrates, the $g = 1.94$ ESR signal originates from only one of each iron pair in the former and one of each iron quartet in the latter enzyme. Table 17.4 shows that both classes of iron-proteins span a considerable range of E'_0 values. Chloroplast and bacterial ferredoxins, of almost identical E'_0, are essentially functionally interchangeable in reconstructed systems of chloroplast or clostridial enzymes. Surprisingly, *C. pasteurianum* grown on an iron-deficient medium produces a low molecular weight iron-free protein

Table 17.4: Some Electron Carrier Iron-Proteins

Name	Source	Fe/mole	E'_0, volt	Mol. wt.
Ferredoxin	*Clostridia*	5 or 7	−0.42	6,000
Ferredoxin	*Chromatia*	3	−0.49	6,000
Ferredoxin (PPNR)*	Spinach	2	−0.43	13,000
Ferredoxin	*Nostoc*	...	−0.40	
Adrenodoxin	Adrenal cortex microsomes	2	+0.15	22,000
	Heart mitochondria	2	+0.23	22,000

* PPNR, photosynthetic pyridine nucleotide reductase.

containing one FMN, termed *flavodoxin,* which can replace ferredoxin in all the electron transfer functions in which the latter is known to serve in this organism. With molecular weight of 14,600, flavodoxin is the smallest known flavoprotein. Iron-proteins which function as a storage form of iron (conalbumin, ferritin, hemosiderin), as a means of iron transport (transferrin), or as a means of oxygen transport (hemerythrin) do not contain acid-labile sulfide or exhibit the ESR signal at $g = 1.94$.

CYTOCHROMES

The discovery of the cytochromes and their general identification by Keilin have been described (page 177). It is now evident that cytochromes occur in virtually all aerobic organisms. In general, the cytochrome content of tissues parallels their respiratory activity; the highest concentrations are found in heart and other actively working muscles, such as the flight muscles of birds and insects. Lesser quantities occur in liver, kidney, brain, and nonstriated muscle; the lowest values are in skin and lung. Tumors and embryonic tissue are unusually low in cytochrome content, consistent with the fact that these tissues derive energy largely by anaerobic pathways of metabolism.

Keilin's early studies of the cytochromes demonstrated that in the presence of oxidizable substrates, and in the absence of oxygen, a series of sharp absorption bands was visible upon spectroscopic examination of a tissue suspension. Admission of oxygen resulted in disappearance of these bands, which were replaced by a group of less intense and more diffuse bands. This permitted Keilin to distinguish three cytochromes, which he designated as a, b, and c. In reduced form, each was characterized by a series of absorption maxima, summarized in Table 17.5. Of this group, Keilin demonstrated that it is cytochrome a which is autoxidizable, *i.e.*, it is reduced cytochrome a which effects the reduction of molecular oxygen. Hence, cytochrome a, which was shown to be reduced by cytochrome c, has been termed *cytochrome oxidase.* In subsequent years it has become apparent that there are numerous cytochromes. As each has been identified, it has been designated according to which of the original cytochromes it most closely resembles by these spectroscopic criteria: thus, cytochrome b_1, b_2, b_3, b_4, etc. This nomenclature should not be interpreted as indicating that their structures are *necessarily* closely related,

Table 17.5: **Properties of Mammalian Cytochromes**

Cytochrome	Absorption maxima of reduced forms			E_0', volts
	α, mμ	β, mμ	γ, mμ	
a_3	600	...	445	
a	605	517	414	+0.29
c	550	521	416	+0.25
c_1	554	523	418	+0.22
b	563	530	430	+0.07
b_5	557	527	423	+0.03

that their E_0' values are of the same magnitude, or that they serve equivalent roles in electron transport.

Addition of intermediates of the citric acid cycle to tissue suspensions, under anaerobic conditions, results in rapid reduction of the cytochromes in the order b, c_1, c, and a; admission of oxygen rapidly reoxidizes all components. However, reoxidation can be prevented by cyanide, azide, sulfide, and carbon monoxide. The respiration of most animal tissues is reduced 60 to 90 per cent by these inhibitors, presumably indicating that at least this fraction of the respiration of animal cells normally occurs over the electron chain leading to cytochrome oxidase.

Except for cytochrome b_5, all the cytochromes of animal cells are associated with the mitochondrial protein-lipid complex. Only cytochrome c is readily obtained as a homogeneous, water-soluble protein by extraction with aqueous solvents; the other cytochromes remain fixed to the insoluble particulate matter of the mitochondria. However, treatment of mitochondria or mitochondrial fragments with detergents, *e.g.*, deoxycholate or dodecylsulfate, permits isolation of each of the mitochondrial cytochromes in relatively pure form. It must be cautioned, however, that their properties in aqueous detergent solution may differ markedly from those which obtain in their native hydrophobic milieu.

The *cytochromes b* are those of lowest potential; E_0' for isolated heart cytochrome b in aqueous solution is -0.34 volt, whereas in mitochondria it behaves as if E_0 were about $+0.07$ volt. In the absence of a cationic detergent, this protein (molecular weight = 28,000) readily aggregates to a large polymer. The prosthetic group appears to be heme (iron-protoporphyrin IX, page 168). The reduced form does not autoxidize, nor does the oxidized form react with cyanide.

Cytochrome c has been obtained in crystalline form from a large variety of animal tissues. The structure of these proteins is described on page 177. At neutral pH, ferrocytochrome c does not react with CO or O_2, nor can the ferric form react with CN^-. The fully coordinated iron atom lies in a crevice, bound to an imidazole residue on one side of the plane of the heme and to some other residue on the other side and, hence, is prevented from reacting with the afore-mentioned reagents. Similar behavior is seen with cytochromes b and c_1. Thus, apparently only an electron can enter or leave these molecules; the sites of this activity are unknown. Denaturation, which results in polymerization of cytochrome c, renders it oxidizable by O_2 but not by cytochrome oxidase.

Treatment of the purified protein with the mixed lipids of mitochondria or with purified phosphatidyl ethanolamine yields a lipid-soluble complex of which only 15 per cent is cytochrome c. In this form, cytochrome c readily serves as substrate for cytochrome oxidase, and this may resemble the manner in which cytochrome c functions in mitochondria.

Cytochrome c_1 is the most recently recognized of the mitochondrial cytochromes; its spectroscopic properties closely resemble those of cytochrome c (Table 17.5). Their prosthetic groups and mode of attachment to the proteins are identical. Cytochrome c_1 is obtained as a polymer of a unit that has a molecular weight of 38,000. The ferrous form of neither the latter nor its polymer can react directly with cytochrome oxidase. Addition of a small amount of cytochrome c permits rapid oxidation, by the oxidase, of substrate amounts of cytochrome c_1. Thus, cytochromes c and c_1 react readily together, but only the former is a substrate for cytochrome oxidase. The advantage to the cell of the presence of both cytochrome c and cytochrome c_1 is not apparent.

Cytochrome oxidase, the terminal member of the cytochrome chain, is the only member capable of reducing oxygen. From the absorption spectra of yeast and heart muscle preparations treated with CO or cyanide and reduced by substrate, it had been concluded that two distinct cytochromes, a and a_3, were present at the terminus of the chain. Of these, a_3 was thought to be autoxidizable at low O_2 tension and to combine with CO and CN^- whereas cytochrome a does not. Hence, a_3 was considered to be the actual oxidase. However, purified preparations of cytochrome oxidase exhibit only the spectral properties of cytochrome a, yet are autoxidizable and react with CO and cyanide. No adequate explanation of this discrepancy is available.

Cytochrome oxidase is a polymer of subunits of molecular weight about 72,000, each of which contains one heme as well as one atom of copper, but is active only in a polymeric form complexed with mitochondrial lipid. The minimal effective polymer contains six monomers, and it appears likely that this is also the physiological form. Two of the six monomers have the spectral properties of cytochrome a_3. The prosthetic group of cytochrome oxidase, termed heme "a" and porphyrin "a," has the following structure.

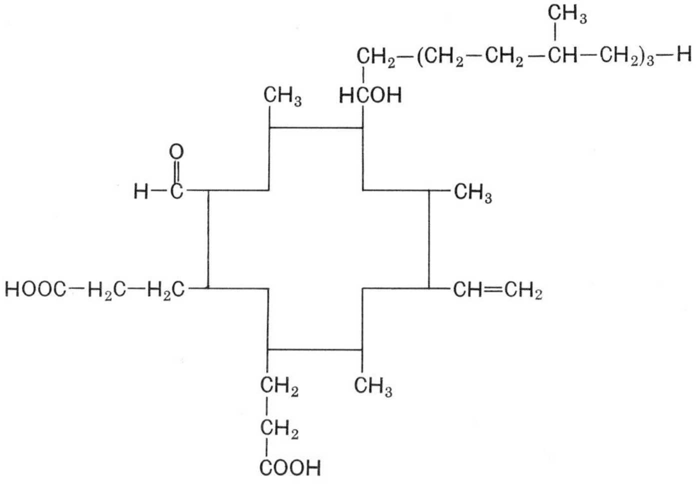

Considerable evidence indicates that cytochrome oxidase is readily autoxidizable only in the presence of cytochrome c. Thus in effect cytochrome oxidase is a cytochrome c–cytochrome a complex. The oxidized ferric form of the enzyme combines avidly with CN^- at low concentration and cannot then be reduced; this explains the high toxicity of cyanide. Reduced, *i.e.*, ferro-, cytochrome oxidase forms a stable complex with CO reminiscent of CO-hemoglobin. Absorption of a photon causes dissociation of the CO, permitting determination of the "action spectrum" of the CO-cytochrome oxidase complex even in impure mixtures. The absorption spectrum of cytochrome oxidase was initially ascertained by Warburg by this procedure. The presence of one copper atom per heme and the demonstration that the copper can be reversibly reduced in the presence of cytochrome c and reoxidized in the presence of oxygen led to the suggestion that copper atoms may be present at the immediate site of oxygen binding and reduction. Resolution of this problem may permit insight into the actual mechanism of oxygen reduction. This reaction is represented by the following equation.

$$O_2 + 2H^+ + 4e \longrightarrow 2OH^-$$

A series of observations indicate intermediate forms between fully reduced enzyme and fully reduced oxygen; none has been definitively identified. Reduction of O_2 requires delivery of four electrons. Since the process is rapid even at low oxygen tension, it seems possible that O_2 is bound at a site which affords almost simultaneous access to two Fe^{++} and two Cu^+ atoms.

MICROSOMAL ELECTRON TRANSPORT

Abbreviated electron transport chains incapable of oxidative phosphorylation occur outside the mitochondria. These chains are present in the membranous endoplasmic reticulum and may be obtained, by appropriate procedures, in particulate bodies, the "microsomes." The physiological role of this transport is only meagerly understood. The most complex system yet revealed is required for the hydroxylation of a variety of compounds, particularly the 21-hydroxylation of progesterone to deoxycorticosterone in adrenal microsomes (page 957).

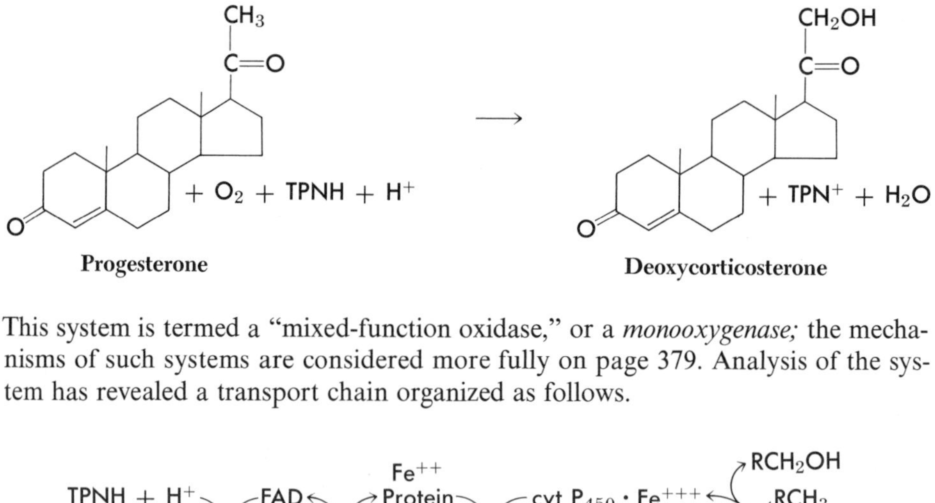

This system is termed a "mixed-function oxidase," or a *monooxygenase;* the mechanisms of such systems are considered more fully on page 379. Analysis of the system has revealed a transport chain organized as follows.

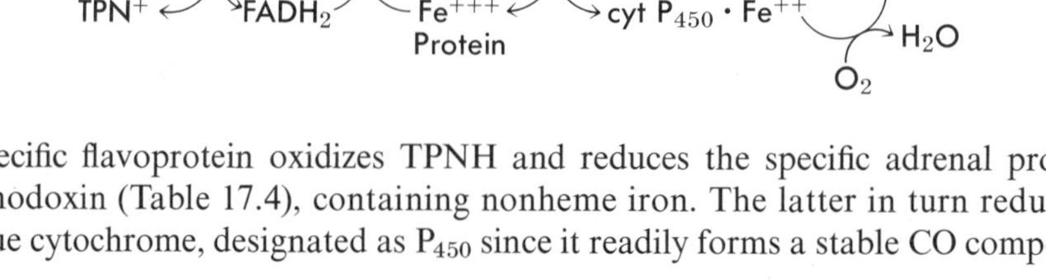

A specific flavoprotein oxidizes TPNH and reduces the specific adrenal protein, adrenodoxin (Table 17.4), containing nonheme iron. The latter in turn reduces a unique cytochrome, designated as P_{450} since it readily forms a stable CO compound with an absorption maximum at 450 mμ. The structure of this cytochrome is unknown. Although it appears to be autoxidizable, it is clear that its reoxidation by O_2 permits both formation of H_2O and generation of an oxygen radical which attacks the steroid, with resulting hydroxylation. This system is responsible for the hydroxylation of many foreign aromatic compounds introduced into the body, *e.g.*, acetanilide (page 379), and it exhibits a considerable degree of substrate specificity. Adrenal microsomes also catalyze hydroxylation of diverse steroids at various positions (page 957); it appears that a series of different enzymes may function in the terminal step in the chain.

Liver microsomes also catalyze a variety of TPNH-dependent hydroxylations, but cytochrome P_{450} has not been found in this tissue. There is present, however, a variety of abbreviated transport systems whose function is entirely unknown. Thus, liver microsomes contain a flavoprotein specific for the oxidation of DPNH by cytochrome b_5; the latter is a nonautoxidizable protein, of molecular weight 11,000 with iron-protoporphyrin IX as the prosthetic group. Cytochrome b_5 is the only cytochrome which has been resolved into its heme and apoprotein constituents and successfully reconstituted. Heme binding appears to involve an amino and an imidazole group. Reduced cytochrome b_5 is readily reoxidized by cytochrome c, but the latter is not present in microsomes. This difficulty also obscures understanding of yet a third specific microsomal flavoprotein which catalyzes reduction of cytochrome c by TPNH (page 365). None of these three microsomal flavoproteins will function in place of the others. In addition, microsomes contain substantial quantities of ubiquinone, but nothing is known of its role. It is assumed that the TPNH for these reactions is available to microsomes from the operation of the direct oxidative pathway for glucose (page 415); TPNH arises in adrenal mitochondria by the energy-dependent transhydrogenase mechanism.

BACTERIAL CYTOCHROMES AND ELECTRON TRANSPORT

The cells of higher plants, yeasts, and fungi respire in a manner similar to that of animal cells. Respiration is conducted in mitochondria which contain enzymes of the citric acid cycle, an organized electron transport system with cytochromes resembling those of animals, and an associated phosphorylating system. The efficiency of phosphorylation in these systems is not known; for example it seems likely that the "coupling site" associated with the first step in DPNH oxidation is lacking in yeast. However, in bacterial cells, organelles resembling mitochondria are not apparent. The respiratory systems of bacteria may be obtained from broken cell preparations as small lipid-rich insoluble particles of varying size which appear to be fragments of the cell membrane. These particles contain dehydrogenases and cytochromes and conduct oxidative phosphorylation. The dehydrogenases are generally similar to those of mammalian cells. However, there occurs a variety of dehydrogenases for malate: DPN-linked enzymes in the particles (*Bacillus subtilis; Pseudomonas*) and non-DPN-linked soluble dehydrogenases in cytoplasm (*S. marscens*), while some organisms (*M. lysodeikticus, A. vinelandii*) have both. On the basis of their absorption spectra, the cytochromes of bacteria may be classified as cytochromes a, b, and c, but these designations do not necessarily connote electron transport relationships similar to those in animal mitochondria. Cytochromes of the c type are most widely distributed and, in aerobes, generally resemble those of mammals. Indeed, the cytochromes c of yeast, *Neurospora,* and wheat germ have been shown to be strictly homologous with those from animal sources (page 178). Several bacteria appear to possess more than one autoxidizable cytochrome of the a class, and there is widely distributed among bacteria a heme-protein that has been designated cytochrome o. The latter, which has a prosthetic group resembling that of cytochrome c, seems to serve as the terminal oxidase in those cells in which it is present. Of singular interest is a protein in *Rhodospirillum rubrum* termed *cytochrome cc,* which has two heme moieties, is autoxidizable, and is inhibited by CO, providing a primitive arrangement suggestive of cytochrome a + a_3 of mitochondria.

Respiratory particles from *E. coli* which conduct phosphorylation with either O_2 or NO_3^- as acceptor, contain cytochromes b_1, a_1, a_2, and o. The soluble fraction obtained from these cells contains two c-type cytochromes, c_{500} and c_{552}, as well as cytochrome b_{562}. The last exhibits a potential, E_0' of 0.113 volt, somewhat higher than that of most of the b group. Particles from *A. vinelandii* contain a flavoprotein and cytochromes b_1, c_4, c_5, a_1, and a_2 in equimolar amounts as well as an eightfold molar excess of ubiquinone. Electron transport in this system is affected by antimycin A and CN^- in a manner similar to a mitochondrial system. The transport system of *Mycobacterium phlei* resembles that of higher forms and is of interest because vitamin K_9 serves as the bridge between the DPNH dehydrogenase and cytochrome b. An unidentified compound, sensitive to irradiation at 360 mμ, links succinic acid dehydrogenase to cytochrome b, which is oxidized by the usual chain containing cytochromes c_1, c, a, and a_3. Table 17.6 lists some bacterial cytochromes and certain of their characteristics.

Cytochromes are entirely absent from the cells of such obligate anaerobes as

Table 17.6: SOME BACTERIAL CYTOCHROMES

Cytochrome	Source	Absorption maxima-reduced			E_0'	Reaction with O_2
		α	β	γ		
a_1	*Escherichia coli*	590	...	...	...	+
a_2	*Azotobacter vinelandii*	652	629	460	...	+
	Pseudomonas					
b_1	*E. coli, Azotobacter*	560	530	426	0.250	−
b_{562}	*E. coli*	562	532	427	...	−
o	*E. coli*	568	...	...	...	+
	Acetobacter					
c_2	*Rhodospirillum rubrum*	550	521	416	0.32	−
c_3	*Desulfovibrio desulfuricans*	552	522	418	−0.205	+
c_4	*Azotobacter*	551	522	418	0.30	−
c_5	*Azotobacter*	555	524	418	0.32	−
c_{551}	*Pseudomonas*	551	521	416	0.286	−
c_{552}	*Chromatium*	552	523	416	0.01	+
"Cytochromoid c"	*R. rubrum*	568	...	424	−0.008	+
"Cytochromoid c"	*Chromatium*	565	...	426	−0.005	+
b_4	*Hemophilus micrococcus*	554	521	418	0.18	

clostridia and from such facultative aerobes as pneumococci and staphylococci. Oxygen consumption in the latter species reflects the activity of autoxidizable flavoproteins. Extensive studies have been conducted of the cytochromes of "oxidative anaerobes," *viz.*, the photosynthetic bacteria that live anaerobically and generate both reducing and oxidizing power by photolysis of water, and the chemosynthetic bacteria, organisms for which N_2, $SO_4^=$, NO_3^-, or Fe^{+++} serves as ultimate oxidant. All such organisms contain cytochromes which mediate reduction of the oxidant. It must be emphasized that the presence of a cytochrome c is postulated on the basis of spectroscopic evidence, *viz.*, absorption bands for the reduced compound at 550, 525, and 415 mμ. This spectrum is characteristic of the prosthetic group, whereas the redox potential appears to reflect also the mode of attachment to the protein as well as the size, conformation, and composition of the protein.

The lowest potential recorded is that of cytochrome c_3 from *Desulfovibrio desulfuricans,* with an E_0' of -0.20 volt, 0.5 volt lower than that of horse heart cytochrome c, but appropriate for functioning in the reduction of $SO_4^=$ to $SO_3^=$ ($E_0' = -0.19$ volt) by H_2 at pH 7.0 ($E_0' = -0.4$ volt). This unusual cytochrome contains two heme groups per molecule and has a molecular weight of only 12,000. Nitrate-reducing organisms, such as *Pseudomonas denitrificans,* contain cytochromes of more conventional redox potential, satisfactory for reduction of NO_3^- to NO_2^- ($E_0' = +0.5$ volt). It is not known whether ATP may be generated in the course of these reactions. Cytochromes c from photosynthetic bacteria exhibit E_0' values ranging from 0 to $+0.35$ volt. Of the many cytochromes c isolated from bacteria, only a few can serve as substrates for mammalian cytochrome oxidase, whereas cytochromes c from aerobes can be oxidized by the cytochrome oxidases of the same species. It will be recognized that only a beginning has been made in characterizing the bacterial cytochromes. Spectroscopic classification may be seriously misleading with regard to function. In an organism such as *Hemophilus parainfluenzae,* which has several cytochromes c, the cytochromes c can operate as an electron transport chain by virtue of their differences in potential, whereas the cytochrome component of *Pseudomonas aeruginosa,* apparently consisting of two cytochrome prosthetic groups, may well employ consecutive transfer from the heme of its cytochrome c-like group to that of the cytochrome a_2 group on the surface of the same protein. Despite the apparent diversity among the dozens of different bacterial cytochromes, the principles of electron transfer in mammalian mitochondria are equally evident in these more primitive organisms.

The character of oxidative phosphorylation in bacteria differs markedly from that of mammalian mitochondria. In no instance is there evidence of "respiratory control"; absence of ADP or P_i does not limit the rate of respiration. With the exception of *M. phlei,* bacterial systems are insensitive to the uncoupling action of dinitrophenol, and in *M. phlei,* dinitrophenol not only uncouples phosphorylation incident to electron transport, but also abolishes the substrate level phosphorylation associated with the oxidation of α-ketoglutaric acid. The suggestion that a derivative of *vitamin* K_1 may participate in the respiratory chain of several bacteria, notably *M. phlei,* is of particular interest. Thus, when incubated with respiratory particles of *M. phlei,* the α-chromanyl phosphate derivative of reduced vitamin K_1 reduces cytochrome c, with concomitant ATP formation.

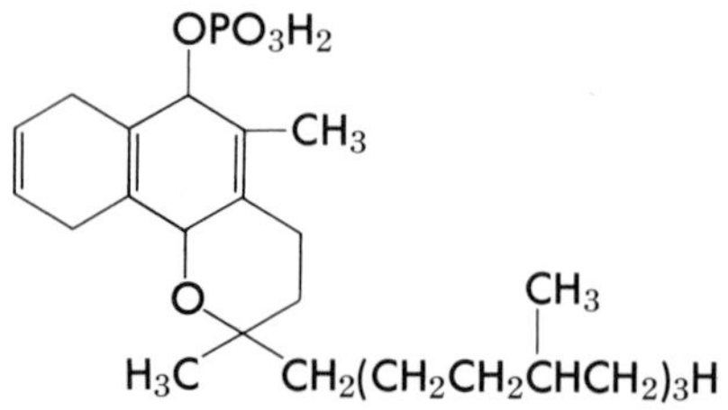

This derivative of vitamin K accumulates when *M. phlei* particles are incubated anaerobically with substrate. These findings are of interest in view of repeated reports that abnormally low P/O ratios are obtained with mitochondria from vitamin K–deficient animals, although significant quantities of vitamin K are not known to be present in animal mitochondria.

OXIDASES, OXYGENASES, HYDROXYLASES, AND PEROXIDASES

Earlier it was noted that, although oxygen is the ultimate biological oxidant, most respiration is the consequence of the activity of enzymes which catalyze reaction between substrate and a coenzyme (DPN or FAD); thereafter electrons are transported from the reduced coenzyme to oxygen via a cytochrome-containing transport system. It is this arrangement which permits conservation of energy and ATP synthesis. A large number of enzymes, however, do in fact catalyze a direct reaction between a substrate and molecular oxygen. These reactions occur in a wide variety of biosynthetic and degradative pathways, particularly of aromatic compounds and steroids. The chemical energy thus released is wasted, since there is no associated means of energy conservation. Many of these enzymes will be encountered in subsequent chapters of this textbook. Accordingly, only a general survey is provided here. The contribution of reactions of this type to the total oxygen consumption of the mammalian organism is small.

Oxidases. Two major classes of oxidases have been encountered previously. Autoxidizable flavoproteins, such as the D- and L-*amino acid oxidases,* reduce oxygen to the level of peroxide, which must be decomposed by catalase or utilized by a peroxidase (page 381); autoxidizable cytochromes serve as the termini of electron transport chains.

In principle, members of a widely distributed group of copper-containing enzymes are of the same type. It might be expected that they would function by reduction of the copper by the substrate with reoxidation by oxygen. In a few instances (*laccase, ascorbic acid oxidase*), the enzyme is blue and is bleached by substrate in the absence of oxygen. The reduced enzyme exhibits an ESR signal characteristic of Cu^+ and reacts with CO and CN^-, both of which react more readily with Cu^+ than with Cu^{++}. The most thoroughly studied copper protein is *laccase,* from the lacquer tree. This enzyme catalyzes oxidation of a large group of *o*- and *p*-dihydroxybenzenes to the corresponding quinones, coupled with the 4-electron reduction of oxygen to water. *Phenolase,* obtained from mushroom or potato, can catalyze a reaction identical with the laccase reaction but also effects oxidation of monophenols to *o*-quinones; these are further transformed to melanins (page 608).

Although the metal can transfer only one electron, the over-all process must be regarded as

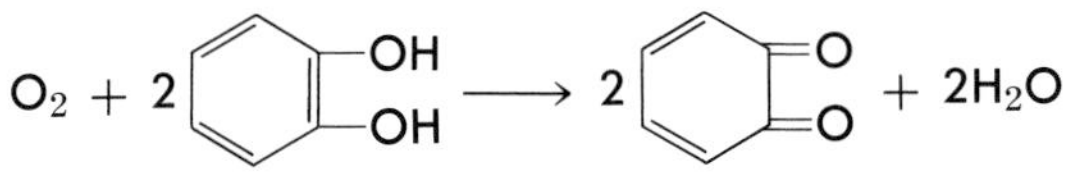

Laccase and ascorbic oxidase contain multiple copper atoms, not all of which are in the cupric state in the oxidized enzyme. These may cooperate in the reduction of O_2. Evidence has been obtained for the occurrence of a free radical of dehydroascorbate in the course of the ascorbic acid oxidase reaction. The nature of the copper-binding ligands of these enzymes has not been established. Table 17.7 shows some properties of certain copper-containing oxidases.

Table 17.7: Some Copper-containing Oxidases

Oxidase	Source	Mol. wt.	Copper atoms/molecule	Inhibited by		Cu ESR signal†
				CO	CN	
Ceruloplasmin	Blood plasma	150,000	8	±	+	+
Ascorbic acid oxidase	Squash	150,000	6	–	+	+
Laccase	Lacquer tree	120,000	4	–	+	+
Uricase	Liver	120,000	1	–	+	
Diamine oxidase	Blood plasma	225,000	4(?)	–	+	+
Dopamine hydroxylase	Adrenal	290,000	4–7	...	+	
Phenolase	Mushroom	34,000	1	+	+	±
Galactose oxidase	Molds	42,000	1	...	+	
Amine oxidase	Plants	.....	..	–	+	–

† Upon reduction by substrate and/or by chemical reagent. + indicates a signal caused by the Cu^+ state.

Source: Modified from H. R. Mahler and E. H. Cordes, "Biological Chemistry," Harper & Row, Publishers, Incorporated, New York, 1966.

Oxygenases are enzymes which utilize oxygen for oxidation of the substrate by a mechanism resulting in incorporation of both atoms of the oxygen molecule into the product. An example is the *pyrocatechase* of bacterial origin, discovered by Hayaishi, which catalyzes oxidation of catechol to *cis,cis*-muconic acid.

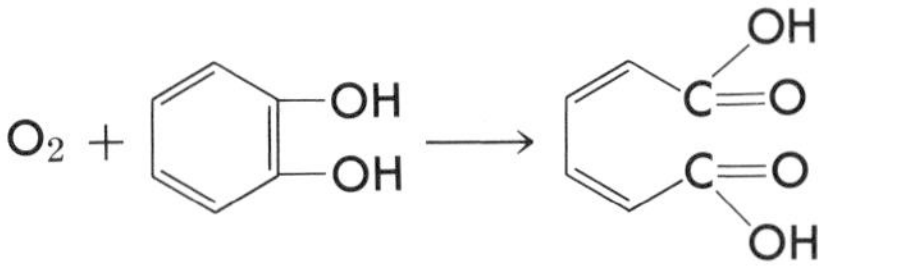

Catechol **cis,cis-Muconic acid**

Enzymes of this class catalyze a variety of processes; cleavage of an aromatic ring between two hydroxyl substituents or adjacent to one hydroxyl group, rupture of a bond joining two hydroxyl-bearing carbon atoms in an aliphatic structure, replacement of a carboxyl group by a hydroxyl group, and bridge formation across an aliphatic double bond.

Some oxygenases are listed in Table 17.8. Most of these enzymes contain

Table 17.8: SOME OXYGENASES AND THEIR DISTRIBUTION

Enzyme	Substrate	Product	Source
Cysteamine oxygenase	Cysteamine	Cysteine sulfinic acid	Rat liver
Homogentisic acid oxygenase	Homogentisic acid	Maleylacetoacetic acid	Rat liver
Hydroxyanthranilic acid oxygenase	3-Hydroxyanthranilic acid	Picolinic acid	Rat liver
Inositol oxygenase	Inositol	Glucuronic acid	Liver
Lipoxygenase	Unsaturated fatty acids	Peroxy fatty acids	Plants
Lysine oxygenase	Lysine	δ-Aminovaleramide	*Pseudomonas*
Metapyrocatechase	Catechol	α-Hydroxymuconic semialdehyde	*Pseudomonas*
Protocatechuic acid oxygenase	Protocatechuic acid	*cis,cis*-β-Carboxymuconic acid	*Neurospora*
Pyrocatechase	Catechol	*cis,cis*-Muconic acid	*Pseudomonas*
Tryptophan pyrrolase	L-Tryptophan	L-Formylkynurenine	Rat liver

ferrous iron; it is assumed that oxygen binding and reaction occur at this site on the enzyme, but the mechanism is not known. Presumably, oxygen radicals which arise by interaction of oxygen with enzyme-Fe^{++} attack the substrates of these enzymes. ESR signals with g values of 2.008 and 4.2 have been observed with lipoxygenase and metapyrocatechase; the responsible molecular species are not known with certainty.

Lipoxygenase catalyzes addition of oxygen to the double bonds of polyunsaturated fatty acids, without shortening of the chain. The conversion of β-carotene to vitamin A aldehyde (page 1049) appears to be a similar process, followed by hydrolytic bond cleavage. A unique reaction is that catalyzed by a pseudomonad enzyme, *lysine oxygenase.*

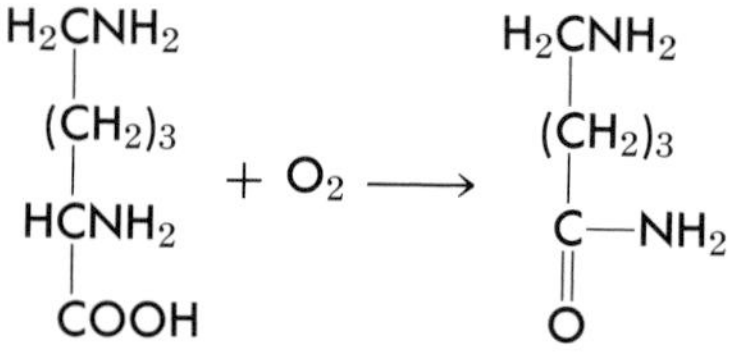

Lysine **δ-Aminovaleramide**

Hydroxylases. Many instances of hydroxylation of an organic substrate have been recognized; these are usually, but not invariably, an aromatic nucleus, *i.e.*,

$$R\text{—}H \longrightarrow R\text{—}OH$$

The newly introduced oxygen atom derives from water rather than from oxygen in at least three instances studied. These include the oxidation of purines by milk xanthine oxidase (pages 220 and 366), of N^1-methylnicotinamide by liver aldehyde oxidase, and of nicotinic acid by various bacteria. The first two enzymes are molybdoflavoproteins, and the mechanism proposed is removal of a hydride ion, which is accepted by the flavin, and replacement by a hydroxyl ion from the medium. Insertion of the hydroxyl ion is dependent on the molybdenum in the enzyme; the mechanism is not known.

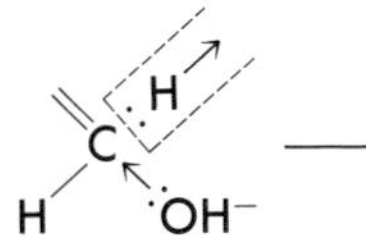

Other hydroxylations are catalyzed by copper-containing proteins, of which the best characterized is the phenolase described earlier. In this instance, the inserted oxygen derives from O_2 and a required cosubstrate, which is also a product. As shown in the equation, of each O_2 utilized, one atom is inserted into the ring and one is reduced to the level of water.

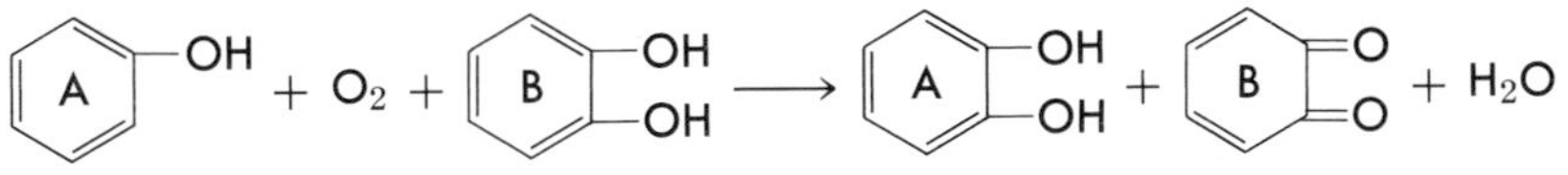

Of considerably greater importance in mammalian metabolism are the reactions catalyzed by enzymic systems present in liver and adrenal microsomes. As noted previously (page 372), these also involve incorporation of an oxygen atom from atmospheric O_2 into the substrate while the other atom is reduced to water, utilizing electrons furnished by a cosubstrate, usually a reduced pyridine nucleotide. These reactions require operation of the microsomal and adrenal mitochondrial transport systems described previously. The responsible enzymes, which are now generally termed *hydroxylases,* were originally described as "mixed-function oxidases" by Mason, who discovered this general class of reactions. Typical of reactions of this class is the hydroxylation of acetanilide by the nonspecific "aromatic hydroxylase" of rat liver microsomes.

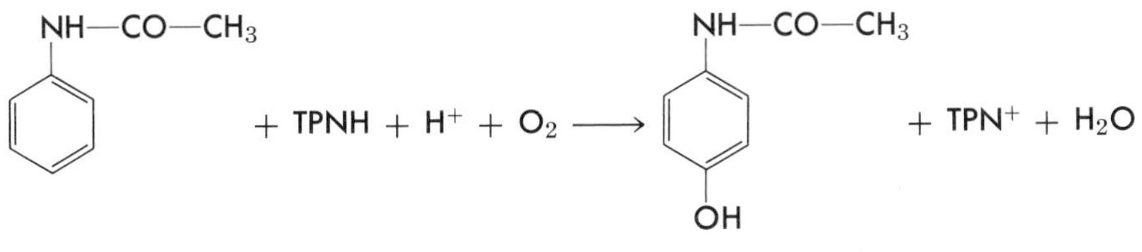

TPNH is the reducing cosubstrate for most hydroxylations, particularly in mammalian liver. Since enzymes utilizing TPNH are the terminal members of the chain, catalyzing insertion of the hydroxyl and conferring specificity only, these enzymes should be called "hydroxylases." A wide variety of aromatic compounds are hydroxylated by liver microsomes, and it is uncertain how many different hydroxylases are responsible. However, as indicated in Table 17.9, other sub-

Table 17.9: SOME HYDROXYLASES

Enzyme	Reductant	Substrate	Product	Source
Aromatic hydroxylase	TPNH	Acetanilide	*p*-Hydroxyacetanilide	Liver microsomes
Kynurenine hydroxylase	TPNH	L-Kynurenine	L-3-Hydroxykynurenine	Liver microsomes
Steroid 11-hydroxylase	TPNH	Deoxycorticosterone	Corticosterone	Adrenal mitochondria
Steroid 17-hydroxylase	TPNH	Progesterone	17 α-Hydroxyprogesterone	Adrenal microsomes
γ-Butyrobetaine hydroxylase	TPNH	γ-Butyrobetaine	Carnitine	Rat liver cytoplasm
Tyrosine hydroxylase	TPNH	Tyrosine	3,4-Dihydroxyphenylalanine	Adrenal microsomes
Cholesterol 20-hydroxylase	TPNH	Cholesterol	20-Hydroxycholesterol	Adrenal mitochondria
3-Hydroxyanthranilic acid oxidase	TPNH	3-Hydroxyanthranilic acid	Quinolinic acid	Kidney
4-Hydroxybenzoic acid oxidase	TPNH	4-Hydroxybenzoic acid	3,4-Dihydroxybenzoic acid	*Pseudomonas*
Melilotic acid oxidase	TPNH	Melilotic acid	2,3-Dihydroxyphenylpropionic acid	*Arthrobacter*
Imidazoleacetic acid hydroxylase	DPNH	Imidazoleacetic acid	Imidazoloneacetic acid	*Pseudomonas*
Kynurenic acid hydroxylase	DPNH	Kynurenic acid	Kynurenic acid 7,8-dihydrol	*Pseudomonas*
Salicylic acid oxidase	DPNH	Salicylic acid	Catechol	*Pseudomonas*
D-Camphor lactonase	DPNH	D-Camphor	D-Camphoric acid lactone	*Pseudomonas*
Steroid 21-hydroxylase	TPNH or DPNH	Progesterone	Deoxycorticosterone	Adrenal microsomes
Fatty acid hydroxylase	TPNH or DPNH	Fatty acids	ω-Hydroxy fatty acids	Liver microsomes
Fatty acyl CoA hydroxylase	?	Fatty acyl CoA	9:10-Fatty acyl CoA	Yeast
Phenolase	Ascorbic acid	3,4-Dimethylphenol	4,5-Dimethylcatechol	Mushroom
Hydroxyphenylpyruvic acid oxidase	Ascorbic acid	*p*-Hydroxyphenylpyruvic acid	Homogentisic acid	Liver
Dopamine β-hydroxylase	Ascorbic acid	3,4-Dihydroxyphenylethylamine	Norepinephrine	Adrenal
Phenylalanine hydroxylase	Dihydrobiopterin	Phenylalanine	Tyrosine	Liver

strates such as DPNH, ascorbic acid, and dihydroxyfumaric acid have been found to function in this capacity, particulary in bacterial systems. An important enzyme in mammalian metabolism, phenylalanine hydroxylase, utilizes as coelectron donor a tetrahydropteridine (page 548).

The hydroxylases of *Pseudomonas* require a flavin, usually FAD, for activity. A ternary complex, *viz.*, enzyme-flavin-substrate, is first formed; the complex is characterized by a markedly altered flavin absorption spectrum. Addition of TPNH, anaerobically, results in bleaching of the flavin absorption, with oxidation of the TPNH. Introduction of O_2 results in reoxidation of the flavin and dissociation of the hydroxylated substrate. Of interest is the fact that, in the absence of substrate, TPNH can also reduce the flavin, but for this process K_m^{TPNH} is 400 times that of the normal process.

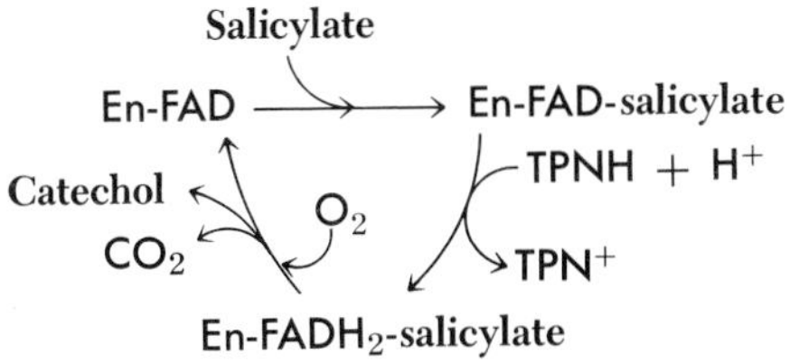

The *lactic acid oxidative decarboxylase* of mycobacteria, which utilizes FMN as the prosthetic group, is also a "mixed-function oxidase." The reaction catalyzed is the following.

$$CH_3CHOHCOOH + O_2 \longrightarrow CH_3COOH + CO_2 + H_2O$$

In this reaction an atom of oxygen is introduced into the acetic acid formed. Simultaneously, a pair of electrons is withdrawn from the *same substrate* and transferred via the flavin to the second oxygen atom, which is reduced to water.

Catalases and Peroxidases. The decomposition of peroxide by animal tissue is catalyzed by *catalase.*

$$2H_2O_2 \longrightarrow 2H_2O + O_2$$

Catalase activity is present in nearly all animal cells and organs; liver, erythrocytes, and kidney are rich sources. This activity is also present in all plant materials studied and in all microorganisms other than obligate anaerobes. In each instance, catalase is believed to prevent accumulation of noxious peroxide under circumstances otherwise favoring such an event.

Peroxidases catalyze the following reaction.

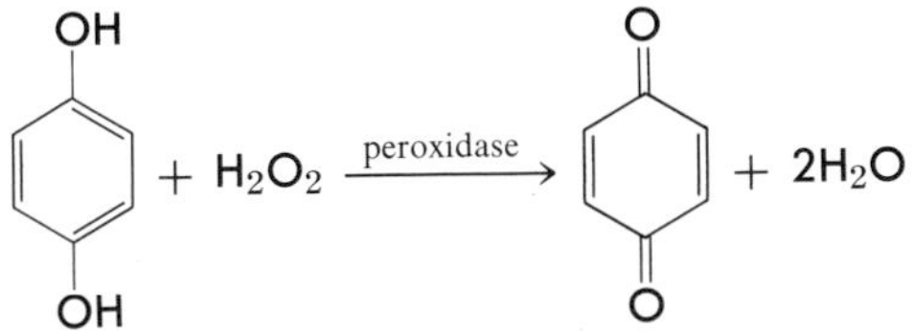

Although peroxidases are relatively rare in animal tissues, liver and kidney exhibit weak peroxidase activity, and a peroxidase has been isolated from milk; leukocytes contain a "verdoperoxidase" which is responsible for the peroxidase activity of pus.

All higher plants, on the other hand, are rich in peroxidase activity. Horseradish and turnip have been extensively employed in investigations of the peroxidases. Some peroxidases exhibit specificity for a substrate other than phenols. Thus, erythrocytes contain a peroxidase specific for oxidation of reduced glutathione; yeast and placenta contain peroxidases for ferrocytochrome c, and some bacteria possess peroxidases that oxidize DPNH. Advantage is taken of the peroxidase activity of all hemoproteins in procedures for detection of "occult" blood in feces, urine, etc. Minute amounts of blood, in the presence of peroxide, catalyze the oxidation of benzidine, gum guaiac, and other substances to colored products.

If the catalatic and peroxidatic reactions are written as

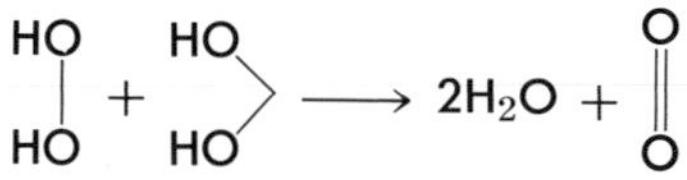

$$\begin{array}{c} \text{HO} \\ | \\ \text{HO} \end{array} + \begin{array}{c} \text{HO} \\ \diagdown \\ \text{HO} \diagup \end{array} \longrightarrow 2H_2O + \begin{array}{c} O \\ \| \\ O \end{array} \qquad \text{Catalase}$$

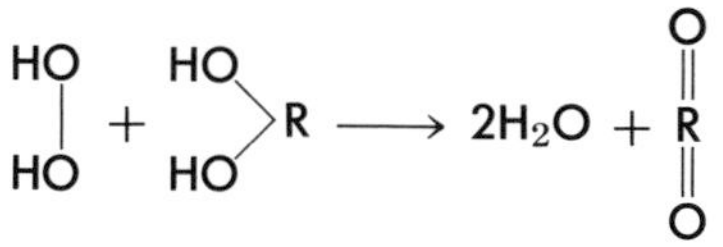

$$\begin{array}{c} \text{HO} \\ | \\ \text{HO} \end{array} + \begin{array}{c} \text{HO} \\ \diagdown \\ \text{HO} \diagup \end{array} R \longrightarrow 2H_2O + \begin{array}{c} O \\ \| \\ R \\ \| \\ O \end{array} \qquad \text{Peroxidase}$$

the analogy between the two reactions becomes apparent. The catalatic splitting of hydrogen peroxide to water and oxygen is a special case of a peroxidatic reaction in which hydrogen peroxide serves both as substrate and as acceptor. This analogy becomes more real on noting that at high concentrations of low molecular weight alcohols or formaldehyde and low peroxide concentration, catalase also exhibits peroxidatic activity. Both enzymes can utilize as substrates organic hydroperoxides with short aliphatic substituents. Since, physiologically, there may exist high concentrations of other acceptors and very low concentration of peroxide, it is conceivable that catalase serves almost exclusively as a peroxidase in animal tissues. In turn, this casts doubt upon the existence of an independent peroxidase in animal tissues. Theorell has proposed that both classes of enzymes be given the common name *hydroperoxidases*, indicating that their common substrate is hydrogen peroxide.

Horseradish peroxidase, with a molecular weight of about 44,000 and one heme group per molecule, reacts with peroxide to give consecutive complexes. Complex I exhibits an absorption maximum at 405 mμ and complex II at 425 mμ. The reaction sequence proposed by Chance is as follows.

(a) $\qquad\qquad$ En-H$_2$O + H$_2$O$_2$ $\longrightarrow$ En-H$_2$O$_2$ + H$_2$O
$\qquad\qquad\qquad\qquad\qquad$ (complex I)

(b) $\qquad\qquad$ En-H$_2$O$_2$ + MH$_2$ $\longrightarrow$ complex II + MH·

(c) $\qquad\qquad$ Complex II + MH· $\longrightarrow$ En·H$_2$O + M

It is uncertain whether complex I is with H$_2$O$_2$ or an oxidized form of the enzyme containing ferryl (Fe^{++}O) iron. In any case this is converted to complex II and back to the original enzyme in consecutive one-electron reactions, in which the substrate is oxidized to a free radical and then fully oxidized. The formation of substrate free radicals has been shown by electron spin resonance spectroscopy.

Beef liver catalase of molecular weight 248,000 has four heme groups per enzyme molecule. It may be dissociated into smaller subunits, but they lack enzymic activity. The turnover rate for catalase is one of the highest known for enzymes; one molecule of catalase can decompose 44,000 molecules of hydrogen peroxide per second. Catalase reacts with H_2O_2 to form a relatively stable enzyme-substrate complex (page 231) of uncertain structure. It is in this form only that catalase may react with the specific inhibitor 3-amino-1,2,4-triazole.

OTHER ELECTRON CARRIERS

Ubiquinone. Crane and his colleagues demonstrated the presence in mitochondria of a group of related quinones that were reduced when mitochondria were incubated anaerobically with various substrates. Accordingly, they were designated as coenzyme Q (for quinone) before their structure was established by Folkers and coworkers. The type of structure has been given previously (page 324), and their postulated role in electron transport has been presented (page 340).

Glutathione, γ-glutamylcysteinylglycine (GSH), is a ubiquitous component of animal tissues. Since free cysteine is present in only trivial quantities, glutathione is the most abundant sulfhydryl compound in cells and appears to function in maintaining many enzymes in their active conformation. Spontaneous oxidation of these enzymes may lead to disulfide formation. Consecutive disulfide exchange reactions with glutathione can serve to restore the active sulfhydryl forms, thus:

$$2\text{Enz—SH} + O_2 \longrightarrow \text{Enz—S—S—Enz}$$
$$\text{Enz—S—S—Enz} + \text{GSH} \longrightarrow \text{EnzSH} + \text{Enz—S—S—G}$$
$$\text{Enz—S—S—G} + \text{GSH} \longrightarrow \text{EnzSH} + \text{G—S—S—G}$$

The oxidized glutathione can then be reduced by the widely distributed flavoprotein *glutathione reductase* which utilizes TPNH.

$$\text{TPNH} + H^+ + \text{GSSG} \longrightarrow \text{TPN}^+ + 2\text{GSH}$$

Glutathione functions as the specific coenzyme for *glyoxalase* activity. Methylglyoxal undergoes an intramolecular oxidation-reduction to lactate in liver and muscle, although the metabolic origin of methylglyoxal is unknown. Lohmann demonstrated that glutathione is necessary for this dismutation; Quastel postulated formation of a glutathione-methylglyoxal complex which was later isolated by Racker, who also found that the consecutive action of two enzymes is required for glyoxalase activity.

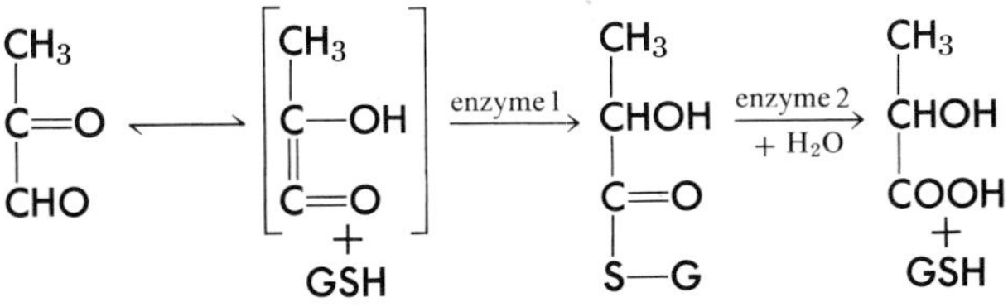

Ascorbic Acid. Ascorbic acid is reversibly oxidized by many oxidants and tissues, particularly those of plants, to dehydroascorbic acid. This reaction is catalyzed by *ascorbic acid oxidase,* a copper protein that resembles laccase in its action and reduces oxygen to water. Ascorbic acid has frequently been used as a reductant in laboratory procedures. It is required for the hydroxylation of proline (page 545); other possible physiological functions are as yet unknown.

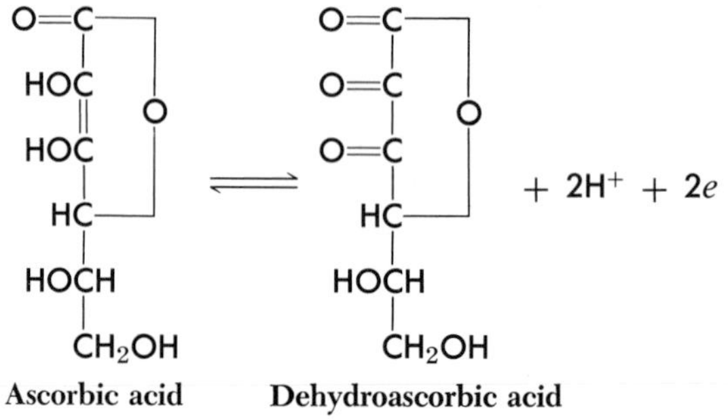

REFERENCES

Books

Boyer, P. D., Lardy, H., and Myrbäck, K., eds., "Oxidation-Reduction Enzymes," vols. 7 and 8, "The Enzymes," Academic Press, Inc., New York, 1963.

Chance, B., Estabrook, R. W., and Yonetani, T., eds., "Hemes and Hemoproteins," Academic Press, Inc., New York, 1966.

Falk, J. E., Lemberg, R., and Morton, R. E., eds., "Haematin Enzymes," Pergamon Press, New York, 1961.

Hayaishi, O., ed., "The Oxygenases," Academic Press, Inc., New York, 1962.

Keilin, D., "The History of Cell Respiration and Cytochromes," Cambridge University Press, New York, 1966.

King, T. E., Mason, H. S., and Morrison, M., eds., "Oxidases and Related Redox Systems," 2 vols., John Wiley & Sons, Inc., New York, 1965.

Kosower, E. M., "Molecular Biochemistry," McGraw-Hill Book Company, New York, 1962.

Lardy, H. A., "Respiratory Enzymes," Burgess Publishing Company, Minneapolis, 1949.

San Pietro, A., ed., "Non-heme Iron Proteins: Role in Energy Conversion," The Antioch Press, Yellow Springs, Ohio, 1965.

Singer, T. P., "Biological Oxidations," John Wiley & Sons, Inc., New York, 1967.

Slater, E. C., ed., "Flavoproteins," American Elsevier Publishing Company, New York, 1966.

Wagner, A. F., and Folkers, K., "Vitamins and Coenzymes," Interscience Publishers, New York, 1964.

Wolstenholme, G. E. W., and O'Connor, C. M., eds., "Quinones in Electron Transport," Little, Brown and Company, Boston, 1961.

Review Articles

Beinert, H., and Palmer, G., Contributions of EPR Spectroscopy to Knowledge of Oxidative Enzymes, *Advances in Enzymol.*, **27**, 105–198, 1965.

Ernster, L., and Lee, C. P., Biological Oxidoreductions, *Ann. Rev. Biochem.*, **33**, 729–788, 1964.

Hatefi, Y., Coenzyme Q (Ubiquinone), *Advances in Enzymol.*, **25**, 275–328, 1963.

Hayaishi, O., Crystalline Oxygenases of Pseudomonads, *Bacteriol. Revs.*, **30**, 720–731, 1966.

Kaplan, N. O., The Pyridine Coenzymes, in P. D. Boyer, H. Lardy, and K. Myrbäck, eds., "The Enzymes," vol. III, pp. 105–170, Academic Press, Inc., New York, 1960.

King, T. E., Reconstitution of the Respiratory Chain, *Advances in Enzymol.,* **28,** 156–236, 1966.

Lemberg, R., Cytochromes of Group A and Their Prosthetic Groups, *Advances in Enzymol.,* **23,** 265–322, 1961.

Margoliash, E., and Schejter, A., Cytochrome c, *Advances in Protein Chem.,* **21,** 113–286, 1966.

Mason, H. S., Oxidases, *Ann. Rev. Biochem.,* **34,** 595–634, 1965.

Mason, H. S., Mechanisms of Oxygen Metabolism, *Advances in Enzymol.,* **19,** 79–233, 1957.

Massey, V., and Veeger, C., Biological Oxidations, *Ann. Rev. Biochem.,* **32,** 579–638, 1963.

Shifrin, S., and Kaplan, N. O., Coenzyme Binding, *Advances in Enzymol.,* **22,** 337–415, 1960.

Theorell, H., Kinetics and Equilibria in the Liver Alcohol Dehydrogenase System, *Advances in Enzymol.,* **20,** 31–49, 1958.

Yagi, K., Mechanism of Enzyme Action—An Approach through the Study of Slow Reactions, *Advances in Enzymol.,* **27,** 1–36, 1965.

18. Carbohydrate Metabolism. I

Digestion and Absorption. Glycolysis. Anaplerosis. Reversal of Glycolysis. The Phosphogluconate Oxidative Pathway.

Many processes essential to life are *endergonic*. In animals, the energy demands of these processes must be met by chemical energy derived from the diet. Ultimately, however, this energy is derived from the sun; as will be discussed subsequently, chlorophyll-containing plants absorb solar electromagnetic energy and utilize it for synthesis of organic compounds. These compounds, in turn, serve both as energy sources for the diverse activities of plant and animal cells and as the starting materials for all other biosynthetic processes.

For mammals, energy is provided by a relatively few specific carbohydrates. As much as 60 per cent of the total food ingested may be carbohydrate, and most of this is in the form of starches, amylose and amylopectin; glycogen may be present in small amount. Other polysaccharides in the diet are of little nutritional significance in most animal species. The disaccharide sucrose which is widely distributed in the vegetable world is added in variable amounts in the preparation and seasoning of foods; the disaccharide lactose is of special importance in infant nutrition. Monosaccharides are present only in small quantities in the adult diet.

DIGESTION OF DIETARY CARBOHYDRATES

The mucosa of the gastrointestinal tract acts as a barrier against the entry into the body proper of large molecules, which, if absorbed, are not well utilized. *Digestion* is the sum of the enzymic hydrolyses of large molecules—polysaccharides, proteins, lipids, nucleic acids—to smaller components which can be absorbed and then metabolized.

Salivary Digestion of Polysaccharides. When saliva is incubated with starch and the mixture tested at intervals with iodine, the color test, initially blue, changes successively to purple, then to red-brown, and finally disappears as the *salivary α-amylase* (page 48) disrupts the starch molecules. The role of saliva in digestion of starch in the intact animal is uncertain because of variable duration of contact of enzyme and substrate. The mixing of the bolus of food with the acidic gastric juice undoubtedly terminates the action of salivary amylase, which is inactivated at low pH values.

Pancreatic Amylase. Amylolytic enzymes are not present in gastric juice. The only effects on starch during its passage through the stomach are those of possible residual amylase activity and, perhaps, some hydrolysis catalyzed by H^+ ions. The major locus of starch and glycogen digestion is the small intestine, and the most important enzyme involved in this process is *pancreatic amylase.*

Pancreatic amylase (α-1,4-glucan 4-glucanohydrolase) is indistinguishable from salivary amylase: each exhibits an absolute requirement for Cl^- ion, is stabilized by Ca^{++}, and has an optimal pH at about 7.1. This optimum may be approached in the small intestine as a consequence of the mixing of acidic gastric chyme with alkaline pancreatic and biliary secretions. The action of α-amylase shown in Fig. 3.1 (page 48) results in a mixture of maltose, isomaltose, and glucose. The α-amylases can effect digestion of the intact starch granule, and do not require preliminary rupture of the granule, *e.g.*, by cooking. Hence undigested starch granules are infrequent in feces of normal individuals on normal diets but do occur abundantly in the feces when pancreatic amylase is not entering the intestinal lumen at a normal rate.

Digestible and Nondigestible Polysaccharides. The gastrointestinal tract is deficient in enzymes capable of attacking polysaccharides other than those constructed of α-1,4 linkages. Thus the cellobiose bond of cellulose is not affected by any known mammalian enzyme. Bacterial *cellulases* can effect hydrolysis of cellulose, and in certain mammals, especially those having a rumen or large cecum, bacterial digestion of cellulose in the gastrointestinal tract makes a considerable contribution to the nutritional economy. In man the nutritional significance of dietary cellulose is negligible, and undigested vegetable fibers are demonstrable in the feces. Similarly, the vegetable pentosans are not affected by the enzymes of the mammalian gastrointestinal tract. However, some pentosans and other polysaccharides are hydrolyzed and partially degraded by bacteria in the large intestine, with formation of CO_2, alcohols, and organic acids. These acids stimulate peristalsis, while the unchanged cellulose serves as bulk or roughage, so that these plant polysaccharides are occasionally employed as mild cathartics.

Digestion of Oligosaccharides. Digestion of dietary disaccharides and the disaccharides formed by α-amylase action is completed in the small intestine. This activity is evident in the distal duodenum, is maximal in the jejunum, and continues through the proximal ileum. Disaccharide hydrolysis, however, occurs not in the intestinal lumen but in the mucosal cells. The enzymes catalyzing this hydrolysis are not well characterized. Extracts of mucosa, fractionated by gel filtration, have yielded five distinct α-specific oligosaccharases and one or two β-specific oligosaccharases. All five of the former catalyze the hydrolysis of maltose. The enzymes designated maltases I and II affect no other substrate. Maltases III and IV also catalyze the hydrolysis of sucrose (glucose and fructose in α-1, β-2 linkage). Maltase V catalyzes hydrolysis of isomaltose (glucose and glucose in α-1,6 linkage) and palatinose (glucose and fructose in α-1,6 linkage). Maltases IV and V account for about two-thirds of the total maltase activity. In addition, the β-specific enzyme(s) catalyzes hydrolysis of lactose (galactose and glucose in β-1,4 linkage) although the latter is found only in milk in the human dietary. The relative activity of the intestine in hydrolyzing these substrates is maltose 100, sucrose 30, isomaltose 30, lactose 12, palatinose 9, cellobiose 2.5.

Each of the above enzymes is present in the brush border of the mucosal epithelium in amounts sufficient to meet the demands of normal adult food consumption. However, infants fed solely human milk may ingest about 35 g. of lactose daily, which exceeds their lactase capacity. This problem disappears as the infant grows and is avoided when cow's milk, which contains much less lactose, is fed.

Genetic lack of one or more oligosaccharases has been observed in some infants and children. Intolerance to maltose itself does not occur. Children who lack sucrase activity also usually lack isomaltase activity; the basis for this is unclear, since these activities are due to seemingly independent enzymes. Lactase deficiency is both more frequent and more serious than that of the other oligosaccharidases since lactose is the major carbohydrate of infant nutrition.

INTESTINAL ABSORPTION OF CARBOHYDRATES

In the normal gastrointestinal tract, carbohydrate is converted to monosaccharides prior to transport across the intestinal mucosa. Intestinal absorption of some undigested disaccharides may occur. Sucrose enters the blood when fed in large amounts; it is treated as a foreign substance, appearing unchanged in the urine, and consequently is devoid of nutritional significance. Glucose, fructose, and galactose are efficiently transported across the intestinal barrier into the portal blood.

The intestinal absorption of simple sugars does not occur by simple diffusion only. The rapid absorption of certain hexoses is not critically dependent on the concentration gradient, as would be expected for simple diffusion. Hexoses, which diffuse physically at similar rates, are absorbed from the intestinal lumen at widely differing rates. Pentoses, lower in molecular weight, hence diffusing more rapidly, have been found to be absorbed more slowly than several hexoses. In order of decreasing rate of absorption, the data show galactose > glucose > fructose > mannose > xylose > arabinose.

Fructose, mannose, and 2-deoxyglucose appear to move across the intestine by "facilitated transport": *i.e.*, they cross more rapidly than expected by simple diffusion but cannot be concentrated against a concentration gradient. In contrast, galactose, glucose, 3-O-methylglucose, 1-deoxyglucose, 6-deoxyglucose, and 6-deoxygalactose can be concentrated against a gradient by an "active transport" mechanism. This transport system can be saturated and exhibits Michaelis-Menten kinetics. Crane has delineated the minimal structural features of sugars that are subjected to such active transport as:

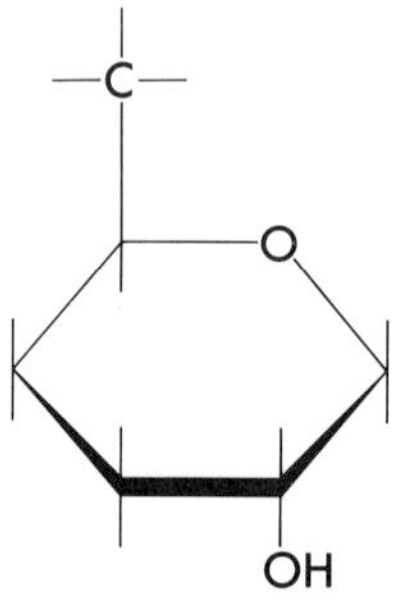

Only the hydroxyl group at position 2 of these sugars appears to be essential to the transport mechanism, which requires simultaneous movement of Na^+ in the same direction, and is also localized in the brush border of mucosal cells. The nature and mechanism of action of this "pump" are uncertain. Crane has proposed the mediation of a carrier protein, within the brush border, which must bind Na^+ in order also to bind glucose. Increasing Na^+ concentration decreases K_m for glucose for active transport, whereas K^+ and Li^+ are inhibitory. Active transport processes are considered further on pages 780*ff.*

GLUCOSE METABOLISM

Although the carbon atoms of glucose ultimately appear in CO_2, other hexoses, pentoses, lipids, amino acids, purines, pyrimidines, etc., glucose itself is subject to only one major fate in mammalian cells—phosphorylation to glucose 6-phosphate.

Phosphorylation of Glucose and Other Hexoses. *Hexokinases,* which catalyze phosphorylation of hexoses, occur universally in living cells. The hexokinases of yeast, brain, and liver are relatively nonspecific, catalyzing phosphorylation of glucose, fructose, and mannose to give the corresponding 6-phosphates.

$$\text{Glucose} + \text{ATP} \xrightarrow{Mg^{++}} \text{glucose 6-phosphate} + \text{ADP}$$

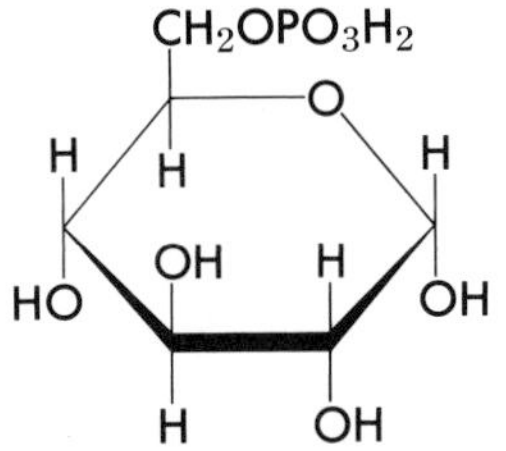

α-D-Glucose 6-phosphate

The nonspecific hexokinases exhibit K_m values for glucose of 10^{-5} to $10^{-2}M$ and have relatively high turnover numbers. In addition, mammalian tissues also contain hexose-specific kinases. The *glucokinases* catalyze phosphorylation of glucose, mannose, and 2-deoxyglucose; these exhibit higher values for K_m, *e.g.*, $10^{-2}M$. Only glucokinase is evident in the newborn rat liver; the hexokinase appears during the first 2 weeks after birth. Hexokinase is subject to allosteric inhibition by its own reaction product, glucose 6-phosphate. This may account for the fact that the maximal rate at which liver cells can metabolize glucose is limited to the rate at which it is converted to glucose 6-phosphate. However, it is unclear whether this reflects the consequence of this product inhibition or an impediment to entry of glucose into the cell.

The fructokinases of intestine, muscle, and liver and the galactokinases of liver, yeast, and *Escherichia coli* catalyze phosphorylation in the 1-position, with formation of fructose 1-phosphate and galactose 1-phosphate, respectively.

For all kinases, the actual substrate appears to be the Mg^{++}-chelate of ATP; the oxygen atom of the phosphate ester bond in the product derives from the alcohol function of the other substrate.

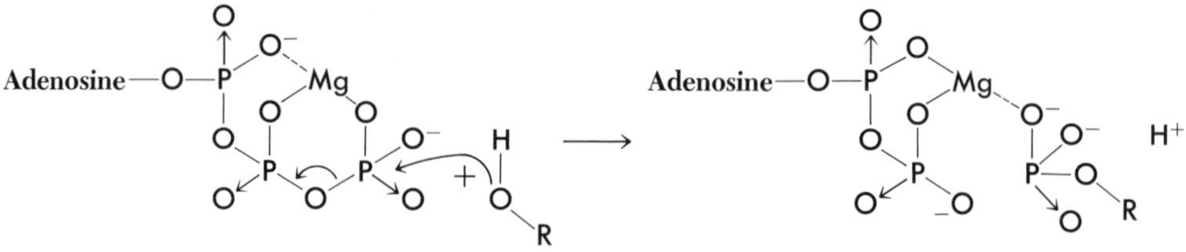

The over-all reaction is decidedly exergonic, $\Delta F^\circ = -5000$ cal., because of the relatively "low-energy" character of the hexose 6-phosphate, the removal of the proton, and the smaller stability constant of the Mg^{++}-ADP complex as compared to the Mg^{++}-ATP complex. Glucose 6-phosphate shares with other phosphate esters a limited ability to enter cells. Whereas glucose is capable of crossing cell membranes, at least in limited degree, glucose 6-phosphate is an intracellular substance. Hence, it has been suggested that the hexokinase reaction serves as a mechanism for the "capture" of glucose, locking otherwise diffusible glucose in the intracellular compartment.

Alternate Fates of Glucose. A few reactions of glucose other than phosphorylation are known in mammalian metabolism. Hepatic *glucose dehydrogenase* catalyzes the following reaction.

$$\text{Glucose} + \text{DPN}^+ \longrightarrow \text{gluconic acid} + \text{DPNH} + \text{H}^+$$

Further metabolism of gluconic acid in mammals entails phosphorylation catalyzed by a *gluconokinase;* the product, 6-phosphogluconic acid, enters the direct oxidative pathway of glucose metabolism (page 415). Direct oxidation of glucose to the level of gluconic acid occurs in various microorganisms; the enzyme catalyzing this oxidation in *Penicillium notatum* is the antibiotic *notatin,* a flavoprotein. The availability of this enzyme in quantity has permitted its use in analytical procedures for the estimation of glucose and in qualitative tests for the presence of glucose, particularly in urine. These tests depend on detection of the peroxide formed by aerobic reoxidation of the reduced, enzyme-bound flavin.

Another fate of glucose is reduction to sorbitol, catalyzed by *aldose reductase,* and oxidation of the sorbitol, catalyzed by *ketose reductase,* to yield fructose.

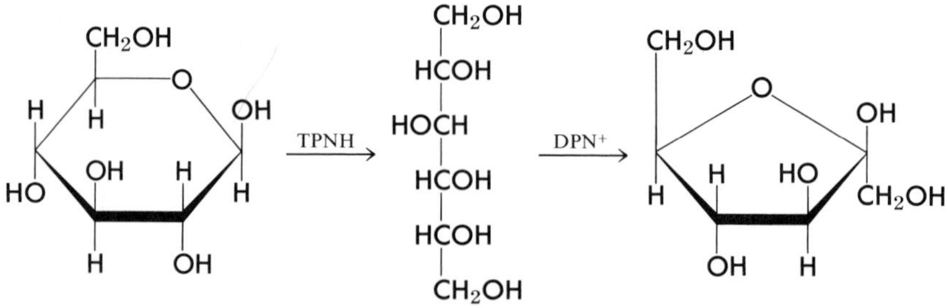

β-D-Glucose Sorbitol β-D-Fructose

This reaction sequence provides the fructose that is present in seminal plasma.

FATES OF GLUCOSE 6-PHOSPHATE

Effectively, carbohydrate metabolism may be regarded as commencing with glucose 6-phosphate, rather than with glucose itself, since it is from this ester that the important "pathways" of carbohydrate metabolism stem. These pathways are summarized in outline form in Fig. 18.1, which indicates the four prime metabolic fates possible for glucose 6-phosphate.

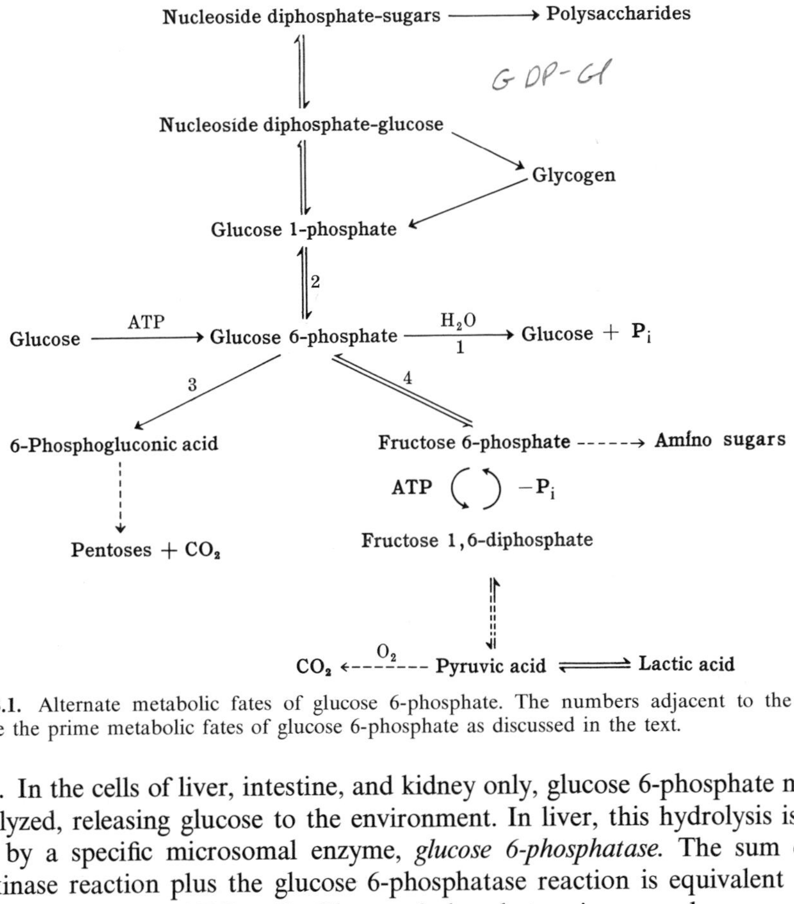

Fig. 18.1. Alternate metabolic fates of glucose 6-phosphate. The numbers adjacent to the arrows indicate the prime metabolic fates of glucose 6-phosphate as discussed in the text.

1. In the cells of liver, intestine, and kidney only, glucose 6-phosphate may be hydrolyzed, releasing glucose to the environment. In liver, this hydrolysis is catalyzed by a specific microsomal enzyme, *glucose 6-phosphatase.* The sum of the hexokinase reaction plus the glucose 6-phosphatase reaction is equivalent to the hydrolysis of ATP to ADP + P_i. Glucose 6-phosphatase is a complex enzyme, the activity of which is dependent on the lipids of its normal milieu. Moreover, the same enzyme also serves as an *inorganic pyrophosphatase,*

$$PP_i \xrightarrow[Mg^{++}]{H_2O} 2P_i$$

and as an unusual type of *kinase* capable of catalyzing the reaction

$$PP_i + \text{glucose} \longrightarrow \text{glucose 6-phosphate} + P_i$$

The latter reaction is relatively nonspecific for the sugar component. Although the activity of this enzyme would make the potential of this process for glucose

6-phosphate formation as great as that of the hepatic glucokinase, the limited supply of pyrophosphate would appear to make this route of hexose phosphate formation of relatively little physiological importance.

2. Conversion of glucose 6-phosphate to glucose 1-phosphate is the initial step in the synthesis of various nucleoside (adenosine, guanosine, uridine, thymidine) diphosphate esters of glucose. The latter are utilized for synthesis of the diphosphate esters of such sugars as galactose, galactosamine, glucuronic acid, iduronic acid, rhamnose, and fucose. These nucleoside diphosphate esters are then used for synthesis of a wide variety of polysaccharides (page 426).

3. Oxidation of glucose 6-phosphate at C-1 yields 6-phosphogluconic acid. This is the initial step in the formation of pentose and of a special oxidative pathway by which glucose may be consumed and TPNH generated for use in synthetic processes (page 415).

4. Perhaps the most prominent fate of glucose 6-phosphate is its conversion to fructose 6-phosphate, which is then phosphorylated to fructose 1,6-diphosphate. This is the initial event in *glycolysis,* the process by which some of the energy of the glucose molecule is made available to the cell as ATP and which is preparatory to the complete oxidation of glucose via the tricarboxylic acid cycle.

GLYCOLYSIS

The predominant metabolic fate of glucose in mammalian cells is entry into those reaction sequences by means of which the potential chemical energy of glucose is employed for synthesis of ATP. The maximum yield of ATP is obtained by complete oxidation of glucose to CO_2. In this process glucose is converted to pyruvic acid in the cell cytoplasm, and the pyruvic acid is then oxidized in mitochondria to CO_2 via acetyl CoA and the tricarboxylic acid cycle, with associated oxidative phosphorylation (Chap. 16). In addition, however, virtually all cells can obtain a more limited amount of energy from glucose without a requirement for molecular oxygen. In mammalian tissues this process, *glycolysis,* is accomplished by a series of reactions in which a net synthesis of two molecules of ATP occurs as each glucose molecule is converted to lactic acid.

$$\textbf{Glucose} + \textbf{2ADP} + \textbf{2P}_i \longrightarrow \textbf{2 lactic acid} + \textbf{2ATP}$$

The intermediates in this process are shown in Fig. 18.2. Table 18.1 gives the names and some of the characteristics of the enzymes participating in glycolysis. Glycolysis provides a means for rapidly obtaining ATP in a relatively anaerobic organ, such as a muscle that is suddenly required to contract vigorously, or in microbial cells that normally live in a relatively anaerobic environment. Thus, bacteria, such as those responsible for souring of milk or fermentation of cabbage to sauerkraut, also derive their energy by transformation of glucose to lactic acid. However, lactic acid is not universally the end product of anaerobic glucose metabolism. Other microorganisms anaerobically ferment glucose by pathways that also result in ATP formation but yield products such as ethanol, CO_2, acetoin, propionic acid, etc.

Study of the mechanism of alcoholic fermentation by brewer's yeast began early in the nineteenth century, attracting the attention of Lavoisier, Gay-Lussac,

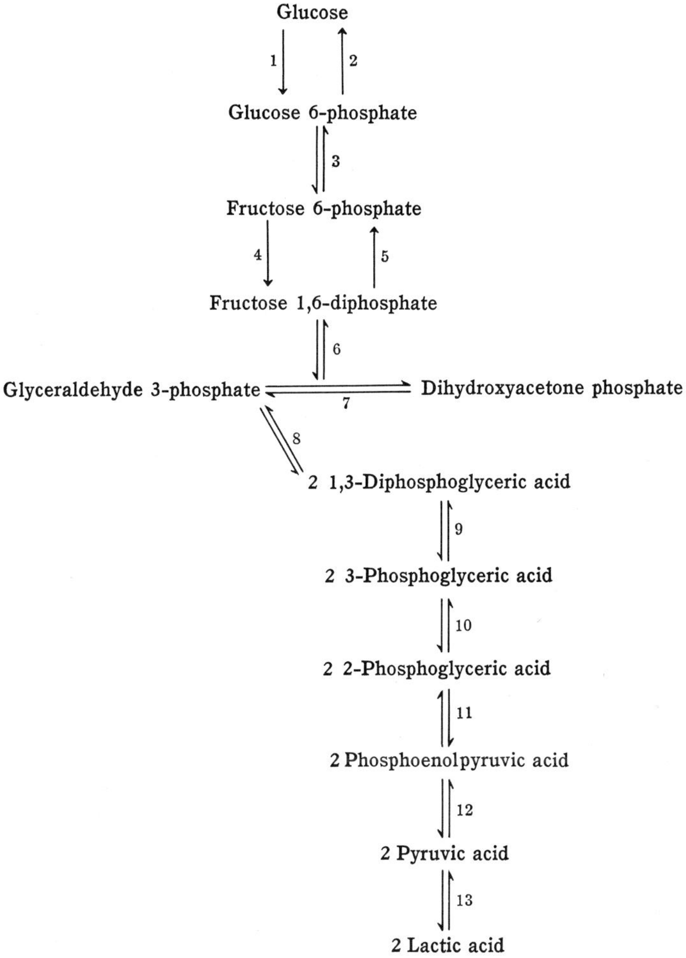

Fig. 18.2. The reactions of glycolysis. Numbers adjacent to the arrows refer to the participating enzymes as listed in Table 18.1.

Berzelius, Liebig, and Pasteur. The accidental observation, in 1890, by the brothers Buchner, that a cell-free extract of yeast could catalyze alcoholic fermentation removed the last traces of "vitalism" in biological thinking and set the stage for the beginnings of modern biochemistry in the subsequent studies of Harden and Young. The latter investigators observed the following properties of alcoholic fermentation by Buchner extracts of brewer's yeast.

1. Inorganic phosphate was essential to fermentation, which ceased when the supply of phosphate was exhausted.

2. As fermentation proceeded, a hexose diphosphate accumulated. The overall process could be summarized by the following equation.

$$\textbf{2 Glucose} + \textbf{2P}_i \longrightarrow \textbf{1 hexose diphosphate} + \textbf{2 ethanol} + \textbf{2CO}_2$$

3. When arsenate was substituted for phosphate, no hexose diphosphate

Table 18.1: ENZYMES OF GLYCOLYSIS

Enzyme*	Mol. wt.	Coenzyme or activators	Inhibitors	Equilibrium constant (approx.) at pH 7.4
1. Hexokinase, glucokinase	96,000	Mg^{++}	...	$\frac{\text{Glucose 6-phosphate} \times \text{ADP}}{\text{Glucose} \times \text{ATP}} = 6{,}300$
2. Glucose 6-phosphatase	...	Mg^{++}	Citrate	Hydrolysis strongly favored
3. Phosphoglucose isomerase	...	...	...	$\frac{\text{Glucose 6-phosphate}}{\text{Fructose 6-phosphate}} = 2.3$
4. Phosphofructokinase	360,000	Mg^{++}, ADP, AMP, P_i	ATP, citrate	
5. Diphosphofructose phosphatase	...	Mg^{++}, ATP	F^-, AMP	Hydrolysis strongly favored
6. Aldolase	150,000	None for muscle; yeast enzyme reactivated by Zn^{++}, Co^{++}, Fe^{++}, Cu^{++}	None for muscle; yeast enzyme inactivated by PP_i, cysteine	$\frac{\text{Fructose 1,6-diphosphate}}{\text{Glyceraldehyde 3-phosphate} \times \text{dihydroxyacetone phosphate}} = 10^4$
7. Phosphotriose isomerase	...	...	...	$\frac{\text{Dihydroxyacetone phosphate}}{\text{Glyceraldehyde 3-phosphate}} = 25$
8. Phosphoglyceraldehyde dehydrogenase	140,000	DPN^+	Iodoacetate	$\frac{\text{Glyceraldehyde 3-phosphate} \times \text{DPN}^+ \times \text{P}_i}{\text{1,3-Diphosphoglycerate} \times \text{DPNH}} = 1$
9. Phosphoglyceric acid kinase	...	Mg^{++}	...	$\frac{\text{3-Phosphoglycerate} \times \text{ATP}}{\text{1,3-Diphosphoglycerate} \times \text{ADP}} = 3{,}000$
10. Phosphoglyceromutase	60,000	Mg^{++}; 2,3-diphosphoglycerate	...	$\frac{\text{3-Phosphoglycerate}}{\text{2-Phosphoglycerate}} = 10$
11. Enolase	87,000	Mg^{++}, Mn^{++}	F^-, PP_i, Ca^{++}	$\frac{\text{2-Phosphoglycerate}}{\text{Phosphoenolpyruvate}} = 0.22$
12. Pyruvic acid kinase	230,000	Mg^{++}, K^+, fructose 1,6-diphosphate	Ca^{++}	$\frac{\text{Pyruvate} \times \text{ATP}}{\text{Phosphoenolpyruvate} \times \text{ADP}} = 2{,}000$
13. Lactic acid dehydrogenase	140,000	DPN^+	Oxamate	$\frac{\text{Lactate} \times \text{DPN}^+}{\text{Pyruvate} \times \text{DPNH}} = 3 \times 10^5$

* The numbers in this column correspond to the reaction numbers of Fig. 18.2.

accumulated and fermentation continued until all the glucose was converted to ethanol and CO_2.

4. The extract could be separated into a heat-labile protein fraction and a dialyzable fraction; the latter contained, as essential components, Mg^{++}, P_i, and an organic substance that was called *cozymase* (DPN).

After the studies of A. V. Hill indicated that the conversion of glycogen to lactic acid is closely related to the process of muscular contraction, Meyerhof prepared soluble extracts of muscle that catalyzed glycolysis and later demonstrated that, except for the final steps, glycolysis and alcoholic fermentation are essentially similar. These processes and their component enzymes have been intensively studied by Meyerhof, Embden, Parnas, Neuberg, and Cori, who have provided explanation of the phenomena observed by Harden and Young.

The Fructose Phosphates. Glucose 6-phosphate is converted to fructose 6-phosphate in a readily reversible reaction catalyzed by *phosphoglucose isomerase;* at equilibrium the ratio of aldose to ketose is 7:3. This isomerization of the 6-phosphates of glucose and fructose is reminiscent of the alkali-catalyzed isomerization of glucose into fructose and mannose (page 28), and it appears to involve an enzyme-bound enediol intermediate (page 268), formation of which requires the open-chain form of the hexose phosphate.

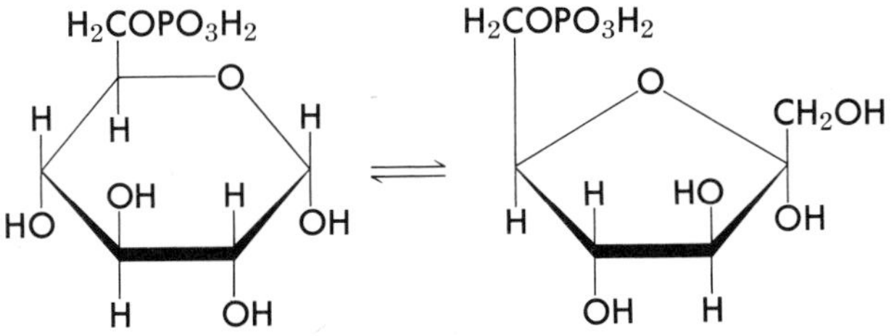

Glucose 6-phosphate

Fructose 6-phosphate

The Phosphofructokinase Reaction. The reaction which is, in effect, the committed step of the glycolytic sequence is the phosphorylation of fructose 6-phosphate with formation of fructose 1,6-diphosphate. This is the hexose diphosphate that accumulated in the Harden-Young experiments.

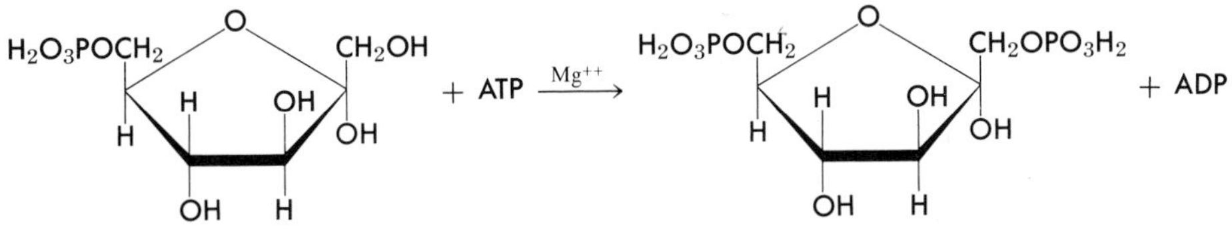

Fructose 6-phosphate

Fructose 1,6-diphosphate

Phosphofructokinases have been obtained from animal tissues, yeast, and bacteria; all show generally similar properties. The enzyme has a molecular weight of about 360,000 and tends to aggregate into various polymeric forms. The activity of the enzyme is influenced by several modifiers; these operate in such fashion that reaction may be expected to proceed when the cell requires a source of ATP and to be markedly inhibited when the adenine nucleotides of the cell exist largely as ATP. In the absence of modifiers and at low ATP concentration, a plot of enzymic activity versus fructose 6-phosphate concentration is virtually hyperbolic. At higher ATP concentrations the plot becomes sigmoid, as shown in Fig. 18.3.

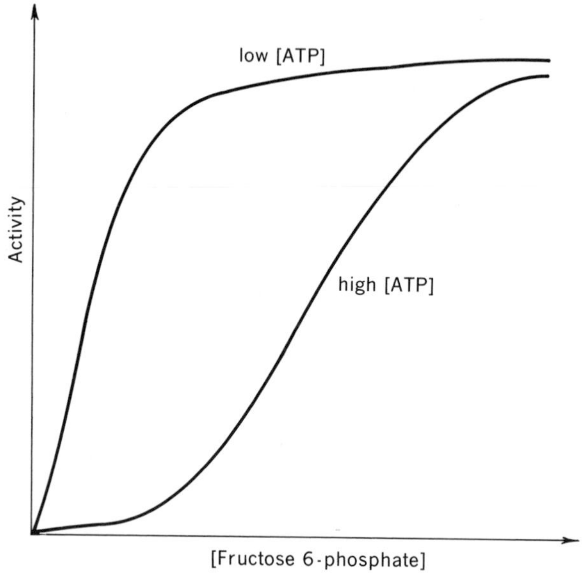

Fig. 18.3. Influence of the concentration of ATP on the kinetics of the reaction catalyzed by phosphofructokinase.

In this instance, "cooperativity" of the substrate-binding sites is the consequence of binding at modifier or allosteric sites of the negative modifier, ATP, which is also a substrate. In the presence of ATP, citrate accentuates this effect of ATP, *viz.*, still higher concentrations of substrate are required if reaction is to proceed. At any concentration of ATP, its effects are lessened by a series of positive modifiers including AMP, ADP, and P_i, AMP being most effective. It is uncertain whether positive modifiers simply compete with ATP for its modifier sites or whether there exists an independent set of sites for the positive modifiers. The binding of these modifiers shows the same cooperativity as that of the substrate, *i.e.*, binding of the second modifier molecule occurs more readily than does that of the first, etc.

Hydrolysis of Fructose 1,6-Diphosphate. Since formation of fructose diphosphate occurs with a free-energy change of about −4500 cal., this process is effectively irreversible. Yet, if glucose is to be formed under conditions which result in effective reversal of the glycolytic sequence, fructose 6-phosphate must be formed from fructose 1,6-diphosphate. This is accomplished by the hydrolytic enzyme, *diphosphofructose phosphatase.*

$$\textbf{Fructose 1,6-diphosphate} \xrightarrow[H_2O]{Mg^{++}} \textbf{Fructose 6-phosphate} + \mathbf{P_i}$$

It will be evident, however, that the concurrent presence in the cell of both this enzyme and phosphofructokinase could lead to a useless and wasteful cycle in which phosphorylation and hydrolysis are endlessly repeated. This is avoided by virtue of the special properties of the phosphatase that are, in general, opposite to those of the kinase, *viz.*, hydrolysis occurs only in the presence of relatively high concentrations of ATP whereas AMP serves as an effective negative modifier, binding of which results in a sigmoidal velocity plot. The AMP-binding site of the enzyme from *Candida utilis* has been clearly differentiated from the substrate site. Treatment of this enzyme with dinitrofluorobenzene results initially in reaction with a sulfhydryl group and then with lysyl amino groups and tyrosyl phenolic groups. If AMP is present during the reaction with dinitrofluorobenzene, the lysyl residues are unaffected but enzymic activity is abolished. If substrate is present instead, the tyrosyl groups are unaffected but the allosteric effects of AMP are abolished.

The Aldolase Reaction. The enzyme *aldolase* catalyzes the reversible cleavage of fructose 1,6-diphosphate between C-3 and C-4 to yield dihydroxyacetone phosphate and the phosphate ester of the isomeric aldotriose, glyceraldehyde. Equilibrium strongly favors the reverse reaction, *i.e.*, formation of fructose diphosphate from the two triosephosphates (Table 18.1).

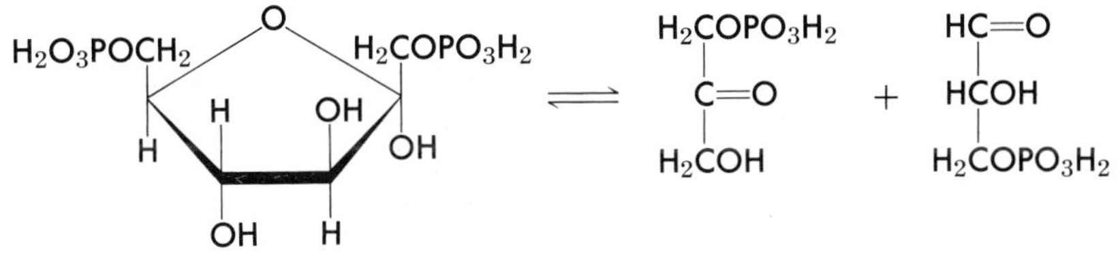

Fructose 1,6-diphosphate   Dihydroxyacetone phosphate   D-Glyceraldehyde 3-phosphate

Crystalline aldolase, obtained from rabbit muscle, is a tetramer of identical polypeptide chains; only the polymeric form exhibits enzymic activity. This enzyme can catalyze an aldol condensation between dihydroxyacetone phosphate and a wide variety of aldehydes, including 3-phosphoglyceraldehyde. Among the other reactions demonstrated in vitro have been the following.

Dihydroxyacetone phosphate + D-glyceraldehyde ⇌ D-fructose 1-phosphate
Dihydroxyacetone phosphate + L-glyceraldehyde ⇌ L-sorbose 1-phosphate
Dihydroxyacetone phosphate + acetaldehyde ⇌ methyltetrose 1-phosphate

In each instance the hydroxyl groups on the carbon atoms of the newly formed bond lie *trans* to each other. In contrast, the enzyme exhibits absolute specificity for the dihydroxyacetone phosphate component. When the enzyme is incubated with the latter substrate, in the absence of an aldehyde, one of the two hydrogen atoms on the carbon atom which is not esterified to phosphate becomes labilized and can be shown to exchange with protons in the medium. If this mixture is reduced with borohydride, the dihydroxyacetone phosphate becomes firmly bound to the amino group of a lysine residue on the enzyme, indicating the formation of a ketimine. This is thought to labilize the specific hydrogen (shown in boldface) on the adjacent carbon, followed by a possible hydride ion shift, forming a carbanion which then attacks the aldehydic carbon of the other substrate.

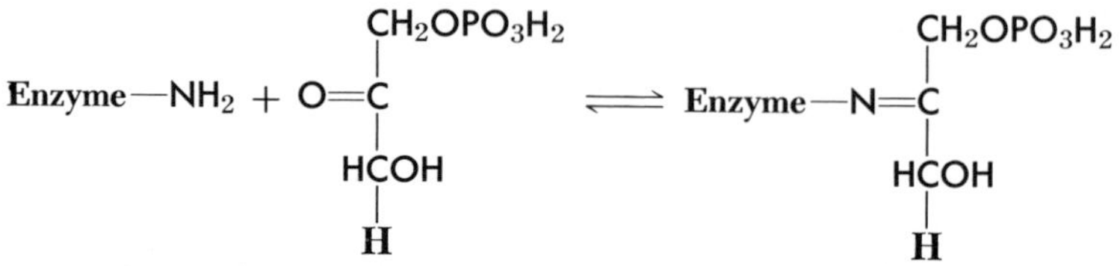

Muscle aldolase exhibits no requirement for metal ions or other cofactors, whereas the yeast enzyme is activated by Fe^{++}, Co^{++}, or Zn^{++} and is inactivated by metal-binding reagents. This suggests that the metal ion serves as a Lewis acid, bonding to the ketonic oxygen and thus facilitating reaction in a manner analogous to the ε-amino group of the lysine residue of the muscle enzyme.

Liver aldolase has properties generally similar to those of the muscle enzyme, yet is clearly a distinct enzyme. Whereas the muscle enzyme cleaves fructose diphosphate fifty times as rapidly as fructose 1-phosphate, these rates are about equal with the liver enzyme. A few differences in the amino acid sequence in the region of the lysine residue that reacts with the substrate indicate the genetic individuality of the enzymes from these two sources.

Triose Isomerase. Dihydroxyacetone phosphate and D-glyceraldehyde 3-phosphate bear the same structural relationship to each other as do fructose and glucose 6-phosphates, and, as with the hexose phosphates, the triose phosphates are interconvertible in reaction catalyzed by *triose phosphate isomerase.*

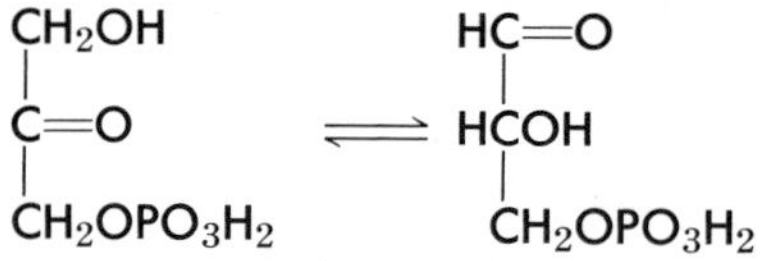

Dihydroxyacetone phosphate — **D-Glyceraldehyde 3-phosphate**

Oxidation of 3-Phosphoglyceraldehyde. The major fate of the triose phosphates stems from glyceraldehyde 3-phosphate, which is oxidized by *phosphoglyceraldehyde dehydrogenase* to the level of a carboxylic acid. P_i is required, and the actual product is the mixed acid anhydride, 1,3-diphosphoglyceric acid.

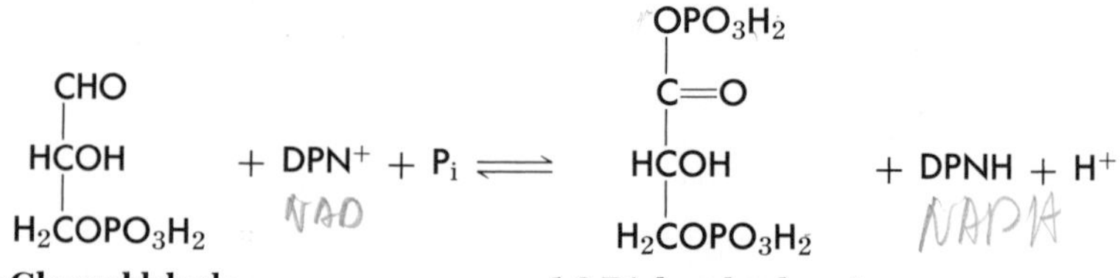

D-Glyceraldehyde 3-phosphate — **1,3-Diphosphoglyceric acid**

Phosphoglyceraldehyde dehydrogenase (triose phosphate dehydrogenase) has been obtained in crystalline form from rabbit muscle and yeast. It contains sulfhydryl groups that are essential to enzymic activity. Iodoacetate reacts with these sulfhydryl groups, inactivating the enzyme; the inhibition of glycolysis by this poison is attributed to this effect. Four moles of DPN^+ per mole of enzyme may be removed by adsorption on charcoal; reactivation of the enzyme is achieved by addition of DPN^+.

Several lines of evidence suggest that an addition compound is formed, the enzyme sulfhydryl, in effect, reducing the DPN. The resultant structure of the oxidized enzyme-coenzyme complex may be represented as

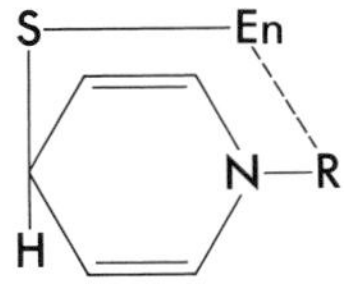

Reaction with the substrate results in an acylthioester of the enzyme to which is bound reduced coenzyme.

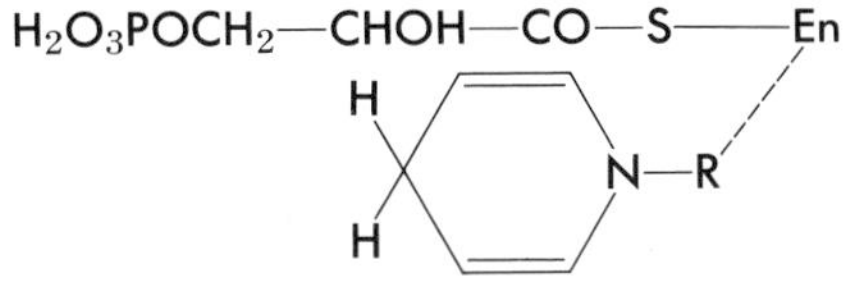

The next step appears to be replacement of the reduced coenzyme by its oxidized form. This may occur by simple exchange with DPN^+ in the medium. Even in the absence of free DPN^+, the enzyme-bound DPNH may be oxidized by interaction with pyruvate and lactic acid dehydrogenase (see below). Once this has occurred, the thioester may undergo attack by a nucleophilic agent. Physiologically, this agent is phosphate and results in formation of 1,3-diphosphoglycerate and the sulfhydryl form of the enzyme to which is bound DPN^+. If arsenate rather than phosphate is present, the product is assumed to be a highly unstable compound of the following type,

$$H_2O_3PO{-}CH_2{-}CHOH{-}CO{-}O{-}AsO_2H_2$$

1-Arseno-3-phosphoglyceric acid

which rapidly and spontaneously hydrolyzes to arsenate and 3-phosphoglycerate. Oxidized DPN^+-containing enzyme readily effects the arsenolysis of 1,3-diphosphoglycerate. In the absence of DPN^+, the diphosphoglycerate acylates the sulfhydryl in the usual manner but the acyl group is then transferred to an adjacent lysyl ϵ-amino group of the enzyme which is, thereby, inactivated.

Phosphoglyceric Acid Kinase. 1,3-Diphosphoglyceric acid is an acid anhydride, energy for the formation of which comes from the oxidation of the aldehyde. A portion of the energy generally released when an aldehyde is oxidized to a carboxylic acid is, in this case, retained in the form of chemical bond energy without passing through the energy mode of heat. Indeed, $\Delta F°$ for the simple hydrolysis of 1,3-diphosphoglycerate is approximately $-14,800$ cal. per mole, adequate to permit transfer of phosphate from the 1-position of 1,3-diphosphoglycerate to ADP. The reaction catalyzed by *3-phosphoglyceric acid kinase* is as follows.

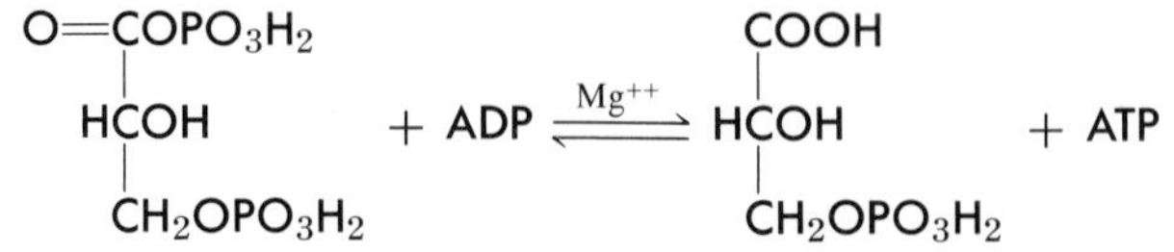

$$
\begin{array}{ccc}
\text{O}=\text{COPO}_3\text{H}_2 & & \text{COOH} \\
| & & | \\
\text{H}\overset{|}{\text{C}}\text{OH} \quad + \text{ADP} \xrightarrow{\text{Mg}^{++}} & \text{H}\overset{|}{\text{C}}\text{OH} \quad + \text{ATP} \\
| & & | \\
\text{CH}_2\text{OPO}_3\text{H}_2 & & \text{CH}_2\text{OPO}_3\text{H}_2 \\
\text{1,3-Diphosphoglyceric} & & \text{3-Phosphoglyceric} \\
\text{acid} & & \text{acid}
\end{array}
$$

The sum of the reactions catalyzed by phosphoglyceraldehyde dehydrogenase and phosphoglyceric acid kinase, respectively, is the following:

Glyceraldehyde 3-phosphate $+ P_i + DPN^+ + ADP \rightleftharpoons$
3-phosphoglyceric acid $+$ DPNH $+$ ATP $+ H^+$

The equilibrium position of the reaction lies far to the right. Indeed, it is the free-energy change in this reaction that pulls the otherwise unfavorable aldolase and triose isomerase reactions. With oxidation of triose phosphate to phosphoglyceric acid, an equivalent amount of ATP is generated. This is the best-known example of the coupling of the exergonic oxidation of a metabolite with the generation of useful chemical energy in the form of the pyrophosphate bond of ATP. Since each glucose molecule yields two triose fragments, each of which is converted to 3-phosphoglyceric acid in this manner, two molecules of ATP are thus generated per molecule of glucose.

Phosphoglyceromutase. The 3-phosphoglyceric acid that arises from the above reaction sequence is transformed into 2-phosphoglyceric acid by the action of *phosphoglyceromutase.*

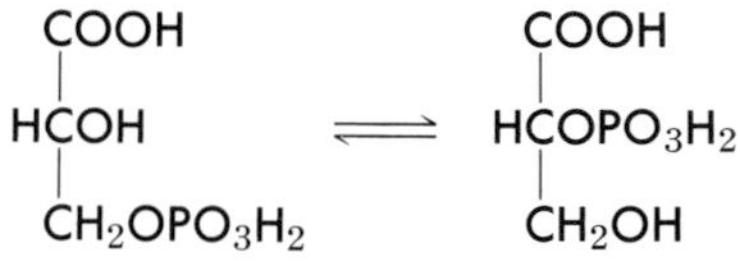

$$
\begin{array}{ccc}
\text{COOH} & & \text{COOH} \\
| & & | \\
\text{H}\overset{|}{\text{C}}\text{OH} & \rightleftharpoons & \text{H}\overset{|}{\text{C}}\text{OPO}_3\text{H}_2 \\
| & & | \\
\text{CH}_2\text{OPO}_3\text{H}_2 & & \text{CH}_2\text{OH} \\
\text{3-Phosphoglyceric} & & \text{2-Phosphoglyceric} \\
\text{acid} & & \text{acid}
\end{array}
$$

The reaction mechanism involves intermediate participation of 2,3-diphosphoglyceric acid. The rapid equilibrium of ^{32}P among all the participants and analogy to the more thoroughly studied phosphoglucomutase reaction (page 424) suggest the following mechanism.

$$
\begin{array}{ccccc}
\text{Enzyme—phosphate} & & \text{Enzyme} & & \text{Enzyme—phosphate} \\
+ & \rightleftharpoons & + & \rightleftharpoons & + \\
\text{3-phosphoglycerate} & & \text{2,3-diphosphoglycerate} & & \text{2-phosphoglycerate}
\end{array}
$$

Another type of phosphoglyceromutase also occurs in animal tissues, similar to that in seeds, and does not appear to require mediation by 2,3-diphosphoglyceric acid.

The Enolase and Pyruvic Acid Kinase Reactions. 2-Phosphoglyceric acid undergoes dehydration in the presence of *enolase* to yield the phosphoric acid ester of the enol of pyruvic acid.

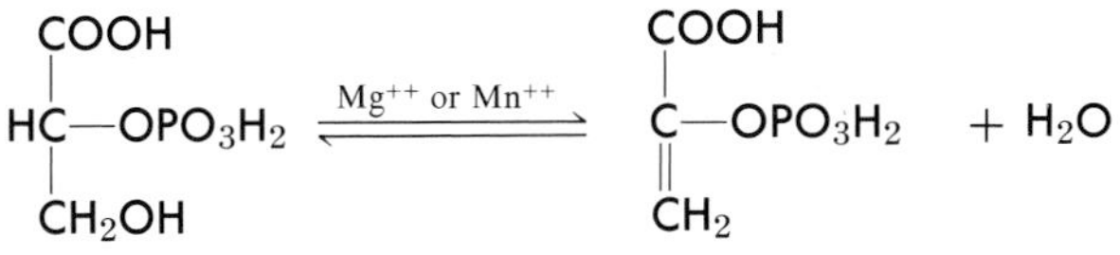

2-Phosphoglyceric acid **Phosphoenolpyruvic acid**

Even in the absence of the substrate, the cofactor, Mg^{++} or Mn^{++}, is tightly bound to the enzyme, which then undergoes a marked change in conformation. The metal then participates in binding of substrate to the enzyme and, presumably, in the subsequent electronic rearrangements. The addition of fluoride to an actively glycolyzing system abolishes the enolase reaction and results in accumulation of phosphoglyceric acids, presumably because of formation of magnesium fluorophosphate that is bound to the enzyme.

The equilibrium position of the enolase reaction is about 0.2. Hence there is almost no associated change in free energy. However, whereas the phosphate of 2-phosphoglyceric acid is esterified to a secondary alcohol, the product of the enolase reaction is the phosphate ester of the enol tautomer of pyruvic acid. At pH 7, the keto form is strongly favored and the free-energy change associated with hydrolysis of the phosphate ester of this enol is unusually high; $\Delta F°$ is approximately $-12{,}000$ cal. per mole, sufficient to permit transfer of the phosphoric acid residue to adenosine diphosphate.

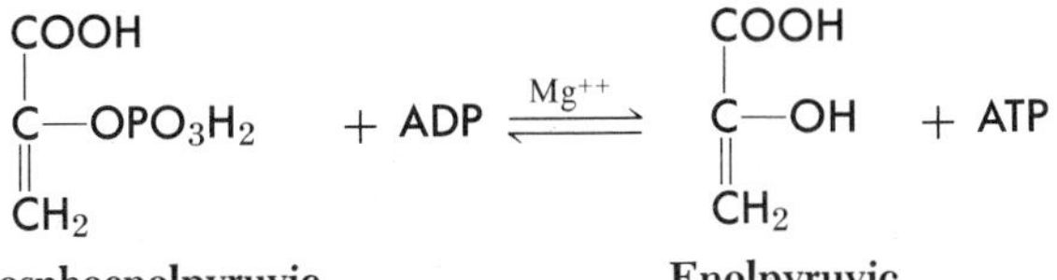

Phosphoenolpyruvic acid **Enolpyruvic acid**

The reaction is catalyzed by *pyruvic acid kinase,* a tetramer of identical subunits of molecular weight 57,000 with two substrate-binding sites. This enzyme has an absolute requirement for Mg^{++}, and like enolase, binds Mg^{++} and is inhibited by Ca^{++}. In addition, this enzyme functions only in the presence of a relatively high concentration of K^+; at least one K^+ must be bound per molecule. This requirement is most evident for the back reaction, formation of phosphoenolpyruvic acid from ATP and pyruvic acid. The turnover number in this direction is only 12, whereas for the forward reaction it is about 5,000. The metabolic consequence of this phenomenon is considered later (page 409).

In yeast, pyruvate kinase represents the second major control point in glycolysis.

The velocity versus substrate plot is sigmoidal but is much less so in the presence of fructose diphosphate, which can accelerate the forward reaction as much as fifty-fold. This "positive feed forward" is antagonized by ATP. Similar phenomena have not been observed with the mammalian enzyme.

Formation of Lactic Acid. Under aerobic conditions, the further metabolism of pyruvic acid proceeds by oxidative decarboxylation, with formation of acetyl coenzyme A, which then enters the citric acid cycle (page 329). Anaerobically, however, pyruvic acid does not accumulate. It will be recalled that the conversion of 3-phosphoglyceraldehyde to 3-phosphoglyceric acid was achieved by the reduction of DPN^+ to DPNH. Aerobically, this DPNH, like the pyruvic acid, would be oxidized by the mitochondrial electron transport system (page 339). As compared with the carbohydrate supply, DPN is present in only limited amount, and, were the DPN^+, reduced by oxidation of 3-phosphoglyceraldehyde, not reoxidized, anaerobic glycolysis would cease when all the DPN^+ was reduced to DPNH. This is prevented by the action of *lactic acid dehydrogenase* which catalyzes the following reaction.

$$\begin{array}{c} COOH \\ | \\ C{=}O \\ | \\ CH_3 \end{array} + DPNH + H^+ \rightleftharpoons \begin{array}{c} COOH \\ | \\ HCOH \\ | \\ CH_3 \end{array} + DPN^+$$

This coupled relationship between these two reactions of the glycolytic sequence may be summarized as follows.

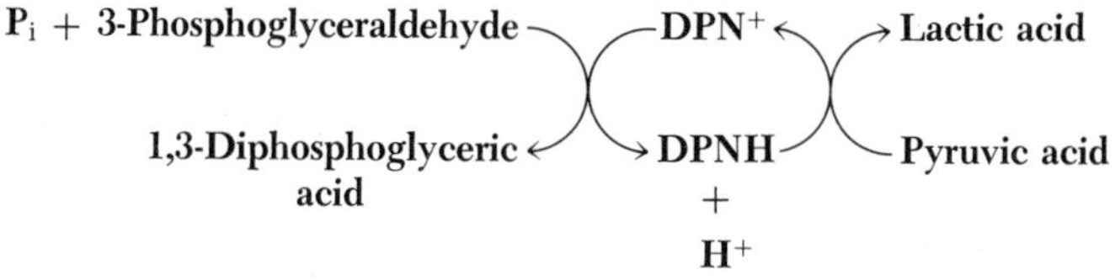

The equilibrium position of the lactic acid dehydrogenase reaction, like that of all reactions involving an alcohol and DPN, strongly favors formation of lactic acid rather than its oxidation. Hence, glycolysis proceeds with net accumulation of two molecules of lactic acid per mole of glucose. Lactic acid is a blind alley in metabolism; once formed there is no means for its further utilization other than reversal of the lactic acid dehydrogenase reaction and re-formation of pyruvic acid under aerobic conditions. In contrast to the phosphorylated intermediates of glycolysis, lactic and pyruvic acids are not locked into the cells in which they are formed. Hence, in vivo, lactic acid diffuses from actively glycolyzing muscle cells and is removed by the circulation (page 444).

Lactic acid dehydrogenase is a tetramer of four units of molecular weight 35,000 each. In animals two types of subunits are electrophoretically distinguishable. These are designated M and H, for skeletal muscle and heart respectively, since the lactic dehydrogenase of the former is mainly M_4 while that of heart is largely H_4. However, all possible hybrids, M_3H_1, M_2H_2, and M_1H_3 have been found in various tissues. The properties of H_4, which is readily inhibited by pyruvate, make it especially useful to a highly aerobic organ which removes lactate from the circulation and oxidizes it to pyruvate, the oxidation of which is completed in the mitochondria. In contrast, M_4 is not inhibited by pyruvate and hence is useful to an organ that engages in large bursts of anaerobic glycolysis.

Alcoholic Fermentation. The sequence of reactions resulting in formation of pyruvic acid in the anaerobic metabolism of glucose by many microorganisms, notably yeast, is identical with that in animal tissues. In contrast to the reversible formation of lactic acid, described above, in yeast pyruvic acid is irreversibly decarboxylated to acetaldehyde by *pyruvic acid decarboxylase,* which is not found in animal tissues.

$$CH_3COCOOH \xrightarrow{Mg^{++}} CH_3CHO + CO_2$$

This enzyme, molecular weight 175,000, binds four molecules of thiamine pyrophosphate, the essential cofactor for this reaction. As in pyruvic acid oxidation (page 328), the initial step is formation of hydroxyethyl thiamine pyrophosphate. In this instance the acetaldehyde moiety is not transferred and the complex decomposes to free acetaldehyde with re-formation of the coenzyme. The acetaldehyde thus formed is reduced by DPNH in a reaction catalyzed by *alcohol dehydrogenase.*

$$CH_3CHO + DPNH + H^+ \rightleftharpoons CH_3CH_2OH + DPN^+$$

Thus, in yeast, acetaldehyde replaces pyruvic acid as the oxidant for the DPNH that arises in the oxidation of 3-phosphoglyceraldehyde.

A variation in this mechanism was devised by Neuberg for commercial production of glycerol. If yeast fermentation occurs in the presence of sodium bisulfite, the latter combines with acetaldehyde to form an addition compound which is unavailable for reaction with DPNH. Under these conditions, the DPNH is then employed by *glycerol phosphate dehydrogenase* to reduce dihydroxyacetone phosphate.

$$\begin{array}{l} CH_2OH \\ | \\ C{=}O \\ | \\ CH_2OPO_3H_2 \end{array} + DPNH + H^+ \rightleftharpoons \begin{array}{l} CH_2OH \\ | \\ CHOH \\ | \\ CH_2OPO_3H_2 \end{array} + DPN^+$$

Dihydroxyacetone Phosphate   **α-Glycerophosphoric acid**

This is the normal source of glycerol for lipid biosynthesis (page 496). In anaerobic yeast, under these circumstances, the glycerol phosphate is rapidly hydrolyzed, by a phosphatase, to glycerol and P_i.

GLYCOLYSIS—GENERAL CONSIDERATIONS

No oxygen is consumed in the over-all process, glucose to lactate. Although two of the individual steps are oxidoreductions—*viz.*, the oxidation of glyceraldehyde 3-phosphate to 1,3-diphosphoglyceric acid (reaction 8, Fig. 18.2) and the reduction of pyruvic acid to lactic acid (reaction 13)—the participation of DPN in both these reactions results in no net oxidation or reduction.

The net yield of usable energy derived from glycolysis may now be computed. One mole of ATP is consumed at each of the two kinase steps (reactions 1 and 4) in the phosphorylation of glucose and of fructose 6-phosphate. Two moles of high-energy phosphate are delivered to ADP to regenerate ATP by reaction 9, as 1,3-diphosphoglyceric acid goes to 3-phosphoglyceric acid, and two more moles of ATP are generated by reaction 12, as phosphopyruvic acid surrenders its phosphate to ADP. Thus, per mole of glucose glycolyzed, two moles of ATP are invested initially, and four moles are ultimately generated, for a net gain of two moles of ATP.

From these relationships the over-all efficiency of glycolysis may be estimated. Some 56,000 cal. per mole are produced when the degradation of glucose proceeds only as far as lactic acid under standard conditions. Available information does not permit calculation of the actual free-energy change, ΔF, under physiological conditions. The generation of two moles of ATP from ADP and P_i consumes about 16,000 cal. under physiological conditions, from which it will be seen that $16{,}000/56{,}000 \times 100$ or about 30 per cent of the potentially available energy may be stored as chemical energy in ATP incident to the glycolysis of one mole of glucose.

It is useful, at this point, to reconsider the findings of Harden and Young (page 393). Clearly, the phosphate requirement for alcoholic fermentation in a yeast extract reflects the need for P_i in the 3-phosphoglyceraldehyde dehydrogenase reaction, the only reaction in the sequence in which P_i actually participates. In an intact cell, also, the presence of free orthophosphate must be essential for glycolysis to proceed. In such cells, as in these extracts, the amount of adenine nucleotide is small compared to the glucose available. When the ADP present is converted to ATP, the reaction sequence would halt at the stages of 1,3-diphosphoglyceric acid and phosphoenolpyruvic acid. In the living cell, the ATP is utilized in diverse endergonic processes, *e.g.*, protein synthesis or muscle contraction, thereby regenerating both ADP and P_i. In a yeast extract, however, these phenomena do not occur. The only means of utilizing ATP is the pair of kinases present in such extracts, that, together, make fructose 1,6-diphosphate from glucose. The over-all fermentation process in such extracts, therefore, may be represented as the sum of two partial systems.

(*a*)	**Glucose + $2P_i$ + 2ADP $\longrightarrow$ 2 ethanol + $2CO_2$ + 2ATP**
(*b*)	**Glucose + 2ATP $\longrightarrow$ fructose 1,6-diphosphate + 2ADP**
Sum:	**2 Glucose + $2P_i$ $\longrightarrow$ 2 ethanol + $2CO_2$ + fructose 1,6-diphosphate**

The accumulated fructose diphosphate is not metabolized further since the equilibrium of the aldolase reaction markedly favors formation of fructose diphosphate rather than the mixture of triose phosphates. The reasons for the requirement for DPN, Mg^{++}, and P_i, therefore, are apparent. It remains only to explain the arsenate effect. As already noted (page 399), 3-phosphoglyceraldehyde dehydrogenase can transfer the 3-phosphoglyceryl residue to arsenate, forming a product that spontaneously decomposes. Under these circumstances, the need for phosphate is obviated, but no ATP can be formed. Hence, the only ATP formed is that from phosphoenolpyruvic acid. The two molecules of ATP thus formed balance those required in the kinase steps; therefore, fermentation or glycolysis can proceed, in the absence of phosphate, until the glucose supply is exhausted and without accumulation of fructose diphosphate. It will be evident that such glycolysis is entirely pointless in the living cell since, although it permits glucose conversion to lactic acid—or ethanol plus CO_2—it fails to provide ATP. Clear comprehension of this circumstance affords understanding of the glycolytic machinery.

The intrinsic rate controls inherent in the glycolytic systems are imperfectly understood. Clearly the over-all rate must be determined by the availability of substrate, utilization of ATP, and the concentrations of the various enzymes. In yeast, the maximal activities of phosphofructokinase, pyruvate kinase, and aldolase are definitely limiting, being only 1 to 10 per cent of the available activities of enolase, triose phosphate dehydrogenase, and phosphoglycerokinase. In consequence some reactions operate at close to equilibrium conditions whereas those catalyzed by hexokinase, phosphofructokinase, triose phosphate dehydrogenase, and pyruvate kinase are quite remote from their own equilibria in the steady state of the cell. Indeed, the cell does not appear to attain a "steady state"with a flux through a system in which the actual concentration of each intermediate remains constant. For example, a recording of the concentration of DPNH in living yeast appears as shown in Fig. 18.4. This oscillatory behavior, which persists indefinitely, is largely the consequence of the continuing change in the velocity of the phosphofructokinase reaction as its substrate, product, and modifiers change in concentration. The pyruvate kinase reaction also contributes significantly to this pattern, and many of the other enzymes continually change in their activity, albeit less dramatically. Accordingly, plots of the concentrations of the glycolytic intermediates would show these oscillating also, some in phase with DPNH and others, necessarily, out of phase.

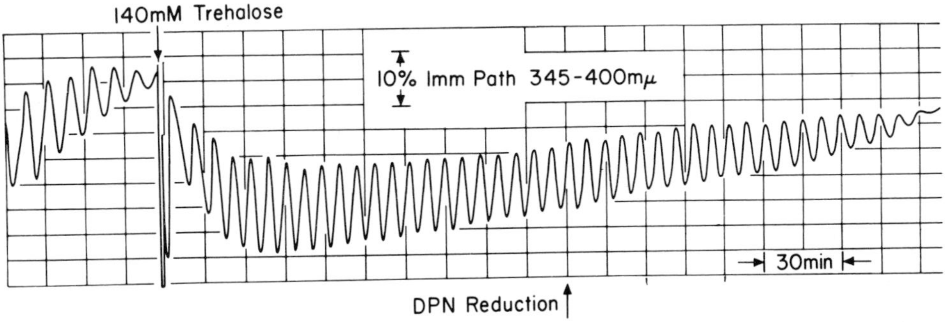

Fig. 18.4. Oscillatory behavior of DPN^+/ DPNH in a cell-free extract of yeast. As the supply of glucose neared exhaustion, trehalose, which is fermentable after hydrolysis, was introduced and the concentration of DPNH was monitored by following its absorbance. (*From K. Pye and B. Chance, Proc. Natl. Acad. Sci. U.S.*, **55**, 888, 1966.)

AEROBIC FATE OF PYRUVIC ACID

Like glucose 6-phosphate, pyruvic acid occupies a central position in metabolism, since it participates in several metabolic reactions, as shown in Fig. 18.5. Pyruvic acid may undergo reversible reduction to lactic acid (page 402), it can be converted back to glucose (page 410), used for formation of oxaloacetic or malic acids (page 408), or transaminated to form alanine, a process that is also reversible (page 541). However, the major fate of pyruvic acid in most mammalian cells is oxidation to CO_2 and acetyl coenzyme A, as described on page 327 and briefly reconsidered below.

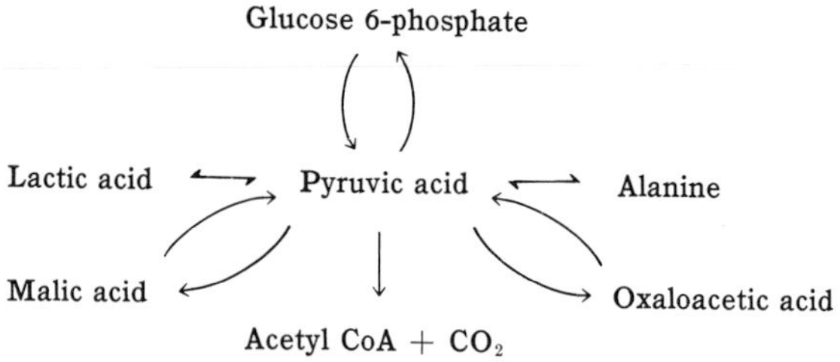

Fig. 18.5. Fates of pyruvic acid in mammals. There are several independent mechanisms for formation of oxaloacetic acid from pyruvic acid (pages 408*ff*).

Oxidative Decarboxylation of Pyruvic Acid. When the oxygen supply is not limiting, pyruvic acid is oxidatively decarboxylated to acetyl CoA (page 327) and CO_2.

$$\underset{\text{Pyruvic acid}}{CH_3COCOOH} + \underset{\text{Coenzyme A}}{CoA{-}SH} + DPN^+ \longrightarrow \underset{\text{Acetyl CoA}}{CH_3CO{-}SCoA} + CO_2 + DPNH + H^+$$

This process is essentially irreversible. Acetyl CoA may be used in a variety of reactions, such as acetylation of choline or aromatic amines, or the biogenesis of acetoacetate, long-chain fatty acids, or steroids. However, as noted previously, the major fate of acetyl CoA is condensation with oxaloacetate to form citrate, the initial step in oxidation via the citric acid cycle (Chap. 16).

Acetyl CoA is an energy-rich compound and might be employed for the generation of ATP, a reaction catalyzed by *acetate thiokinase,* the "acetate-activating enzyme" (page 326).

$$\text{Acetyl CoA} + \text{AMP} + PP_i \rightleftharpoons \text{acetate} + \text{CoA} + \text{ATP}$$

However, this potentiality is probably rarely realized physiologically, and the significance of the reaction shown derives from its reversal, providing a means for synthesis of acetyl CoA from acetate.

Under aerobic circumstances, therefore, events in the cytoplasm may be summarized as follows.

$$\text{Glucose} + \text{ADP} + 2P_i + 2DPN^+ \longrightarrow 2\text{ pyruvic acid} + 2ATP + 2DPNH + 2H^+$$

The pyruvic acid then enters the mitochondria. Oxidation of the DPNH that arises when pyruvic acid is transformed into acetyl CoA and CO_2 provides 6 moles of ATP per glucose equivalent. Complete oxidation of each mole of acetyl CoA yields

12 moles of ATP or 24 moles of ATP per glucose equivalent. Remaining for consideration is the DPNH that arose in the cytoplasm by oxidation of 3-phosphoglyceraldehyde. As indicated previously (page 341), it is not yet definitely established how extramitochondrial DPNH may be oxidized by mitochondria. However, this is theoretically possible; the 2 moles of external DPNH would provide 6 moles of ATP. As summarized in Table 18.2, this calculation permits a theoretical total yield of 38ATP per mole of glucose oxidized.

$$C_6H_{12}O_6 + 6O_2 + 38ADP + 38P_i \longrightarrow 6CO_2 + 6H_2O + 38ATP$$

Assuming that ΔF for formation of ATP under physiological circumstances is about +8000 cal. per mole, a total of about 304,000 cal. of the 686,000 cal. potentially available is actually conserved in this process. The "efficiency," therefore, is of the order of 45 per cent.

Table 18.2: Total ATP Production from Glucose in a Respiring System

Reaction sequence	ATP yield	Page
Glucose $\longrightarrow$ fructose 1,6-diphosphate	−2	395
2 Triose phosphate $\longrightarrow$ 2 3-phosphoglyceric acid	+2	400
$2DPN^+ \longrightarrow 2DPNH \longrightarrow 2DPN^+$	+6	343
2 Phosphoenolpyruvic acid $\longrightarrow$ 2 pyruvic acid	+2	401
2 Pyruvic acid $\longrightarrow$ 2 acetyl CoA + $2CO_2$		
$2DPN^+ \longrightarrow 2DPNH \longrightarrow 2DPN^+$	+6	343
2 Acetyl CoA $\longrightarrow 4CO_2$	+24	346
Net: $C_6H_{12}O_6 + 6O_2 \longrightarrow 6CO_2 + 6H_2O$	+38	

ANAPLEROSIS

Sources of Oxaloacetic Acid. According to the scheme shown in Fig. 16.1 (page 325), for each turn of the citric acid cycle, an equivalent of oxaloacetic acid is regenerated to initiate the succeeding turn of the cycle. However, for several of the intermediates of the cycle, particularly oxaloacetic acid, α-ketoglutaric acid, and succinyl CoA, there are metabolic fates other than the citric acid cycle. Among these is the decarboxylation of oxaloacetic acid to pyruvic acid, which may occur both spontaneously and by enzymic catalysis.

$$\underset{\text{Oxaloacetic acid}}{HOOCCOCH_2COOH} \longrightarrow \underset{\text{Pyruvic acid}}{CH_3COCOOH} + CO_2$$

Loss of oxaloacetic acid in this manner would, inevitably, decrease the rate at which the cycle could operate unless the loss were offset by a renewal of the supply. A readily available source is the amino acid, aspartic acid, that gives rise to oxaloacetic acid by transamination (page 541). Similarly, glutamic acid (page 541) yields α-ketoglutaric acid, which may then enter the citric acid cycle. However, in many types of growing cells and in normal liver, the converse circumstances may obtain, with both oxaloacetic and α-ketoglutaric acids being removed by transamination to form their respective amino acids. These, in turn, are the metabolites from which other amino acids can be made, particularly in plant and bacterial cells. It is essential, therefore, that other means be available to renew the supply of the keto acids. H. Kornberg has termed this process *anaplerosis*.

Several reactions are known for replenishing the supply of oxaloacetic acid by CO_2 fixation with pyruvic acid. This phenomenon was first noted by Wood and Werkman in heterotropic, nonphotosynthetic bacteria and has been well established for animal tissues. Ochoa and his collaborators demonstrated the presence of *malic enzyme,* which catalyzes the following reaction.

$$\underset{\text{Pyruvic acid}}{CH_3COCOOH} + CO_2 + TPNH + H^+ \xrightleftharpoons{Mn^{++}} \underset{\text{L-Malic acid}}{HOOCCH_2CHOHCOOH} + TPN^+$$

The mechanism of this reaction is analogous to that catalyzed by isocitric acid dehydrogenase, considered earlier (page 329). Malic acid is a normal intermediate of the citric acid cycle and is oxidized to oxaloacetic acid by DPN^+ in the presence of *malic acid dehydrogenase,* an enzyme distinct from the malic enzyme considered above. However, the high K_m for CO_2 and a free-energy change $\Delta F° = +500$ cal. for the formation of malate by malic enzyme suggest that this reaction is probably more significant as a source of TPNH for reductive biosynthesis than as an anaplerotic arrangement.

A second CO_2-fixing reaction, found by Utter, is catalyzed by *phosphoenolpyruvic acid carboxykinase.*

$$\text{Phosphoenolpyruvic acid} + CO_2 + \text{inosine diphosphate} \xrightleftharpoons{Mg^{++}} \text{oxaloacetic acid} + \text{inosine triphosphate}$$

The enzyme, which is widely distributed, has been purified from liver and muscle. The liver enzyme can utilize GDP as well as IDP; the enzyme from some sources can utilize ADP as well. The presence of acetyl CoA significantly activates this enzyme; this effect is most apparent for the reverse reaction, formation of phosphoenolpyruvate. Although this enzyme can, indeed, fix CO_2 into oxaloacetate, the relatively high K_m for CO_2 and a standard free-energy change of $+4000$ cal. again make this a path of dubious significance for CO_2 fixation. As will be seen below, the chief significance of this enzyme lies in catalysis of the reverse reaction in the course of gluconeogenesis.

The principal enzyme in animal tissues and yeast that promotes formation of oxaloacetate from pyruvate is *pyruvic acid carboxylase,* discovered by Utter and Keech. This mitochondrial enzyme catalyzes the reaction

$$\text{Pyruvic acid} + CO_2 + ATP \xrightarrow{Mg^{++}} \text{oxaloacetic acid} + ADP + P_i$$

The enzyme, a polymer of molecular weight 650,000 which deaggregates in the cold, contains four molecules of biotin, each in amide linkage to the ϵ-amino group of a lysine residue. The enzyme from animal sources has an absolute requirement for acetyl CoA. The latter does not participate in the above reaction but serves as a positive modifier essential for the first partial reaction in the catalytic cycle.

$$\text{En-biotin} + CO_2 + \text{ATP} \rightleftharpoons \text{en-biotin-}CO_2 + \text{ADP} + P_i \quad (1)$$

The structure of the biotin-CO_2 compound is presented on page 487. The second step, reaction with pyruvate, can proceed in the absence of acetyl CoA.

$$\text{En-biotin-}CO_2 + \text{pyruvate} \rightleftharpoons \text{en-biotin} + \text{oxaloacetate} \quad (2)$$

The low $K_m^{CO_2}$, the over-all equilibrium ($\Delta F^\circ = -5000$ cal.), and the allosteric effect of acetyl CoA, which should occur when cellular demand for oxaloacetate is maximal, all suggest that this is the primary mechanism for anaplerotic formation of oxaloacetate. This conclusion is further supported by numerous observations of impaired CO_2 fixation into oxaloacetate in biotin-deficient animals.

An enzyme of more limited distribution has also been found to catalyze oxaloacetate formation in some plants. The reaction is

$$\text{Phosphoenolpyruvate} + CO_2 \longrightarrow \text{oxaloacetate} + P_i$$

This is an efficient system at low CO_2 tension, and reaction occurs with a free-energy change of about -6500 cal. per mole.

REVERSAL OF GLYCOLYSIS

The reversibility of glycolysis was first suggested by A. V. Hill, who noted that isolated muscle, contracting anaerobically, converted glycogen to lactic acid and that after oxygen was introduced, the lactic acid disappeared, about one-fifth being oxidized to CO_2 while the remainder was reconverted to glycogen. Under these circumstances, or in liver receiving blood lactic acid which was formed in skeletal muscle, the adequate supply of O_2 would assure synthesis of ATP by mitochondrial oxidative phosphorylation of pyruvate and oxidation of lactate to pyruvate. Were the pyruvate kinase reaction readily reversible, the pyruvate could be converted to phosphoenolpyruvate and the latter converted to hexose by reversal of the described reactions of glycolysis. However, as indicated previously (page 401), reversal of the pyruvate kinase reaction occurs slowly. Accordingly, some alternate path from pyruvate to phosphoenolpyruvate is required if lactate or pyruvate from other metabolic sources is to be converted to hexose.

Such a pathway is made possible by enzymes encountered in the discussion of anaplerosis above. Direction is given to the process by the ATP and acetyl CoA which arise in the mitochondrial metabolism of some of the pyruvate. As indicated earlier, acetyl CoA is required for action of the mitochondrial pyruvate carboxylase, which catalyzes formation of oxaloacetate. The latter is reduced, by mitochondrial malic dehydrogenase, to malate, which diffuses into the cytoplasm, where it is reoxidized to oxaloacetate by the cytoplasmic malic dehydrogenase (page 342). Phosphoenolpyruvate carboxykinase, activated by acetyl CoA, then utilizes the energy of ITP or GTP as it catalyzes formation of phosphoenolpyruvate. This arrangement is shown in Fig. 18.6 (page 410), and the sites of the operative controls in these systems are indicated in Fig. 18.7 (page 410).

Phosphoenolpyruvate may then be converted to fructose diphosphate by reversal of reactions 11, 10, 9, 8, 7, and 6 of Fig. 18.2. DPNH required for reduction of 1,3-diphosphoglycerate is provided by concurrent oxidation of lactate to pyruvate.

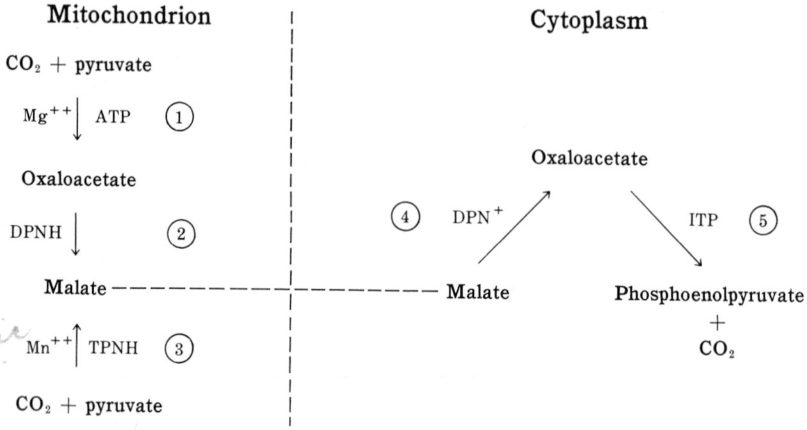

Fig. 18.6. Path of formation of phosphoenolpyruvate from pyruvate. The participating enzymes are ① pyruvate carboxylase, ② malic acid dehydrogenase, ③ malic enzyme, ④ malic acid dehydrogenase, ⑤ phosphoenolpyruvate carboxykinase.

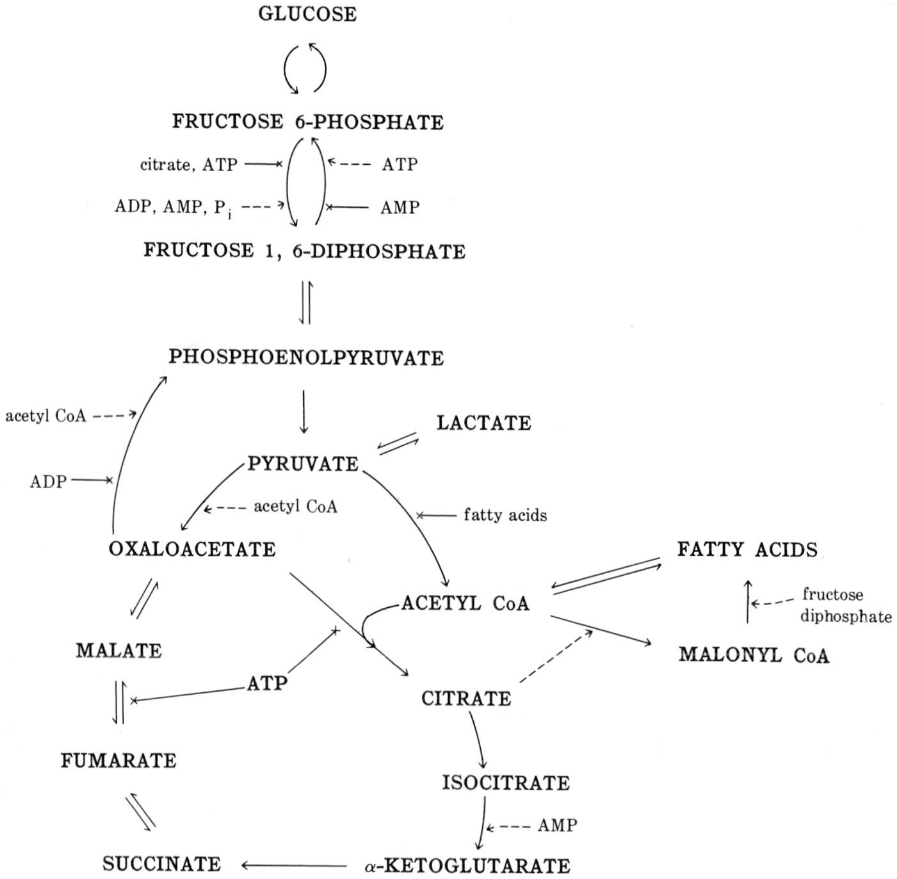

Fig. 18.7. Self-regulation of the major energy-yielding metabolic pathways. ----→ = Positive modifier; ——× = negative modifier.

The prevailing low concentration of ADP and AMP and the high concentrations of ATP and citrate would inhibit phosphofructokinase and activate diphosphofructose phosphatase (page 396), thus facilitating formation of glucose 6-phosphate, which can then be hydrolyzed by glucose 6-phosphatase.

From the description above it is evident that glucose formation from lactate is not a "reversal of glycolysis." Pyruvate kinase, phosphofructokinase, and glucokinase are not utilized, whereas pyruvate carboxylase, phosphoenolpyruvate carboxykinase, the two malic dehydrogenases, and two different phosphatases are required. While glycolysis may be summarized as

$$\text{Glucose} + 2\text{ADP} + 2\text{P}_i \longrightarrow 2\text{ lactate} + 2\text{ATP} \qquad (1)$$

the reverse process occurs as

$$2\text{ Lactate} + 6\text{ATP} \longrightarrow \text{glucose} + 6\text{ADP} + 6\text{P}_i \qquad (2)$$

For each lactate, a molecule of ATP is required in the reactions catalyzed by pyruvate carboxylase, phosphoenolpyruvate carboxykinase (actually as ITP or GTP), and phosphoglycerate kinase. The difference between the two processes, then, is the energy equivalent to four high-energy bonds, and it is this difference that drives the formation of glucose from lactate. $\Delta F°$ for the process summarized in (1) above is $-30{,}000$ cal., whereas $\Delta F°$ for the process in which energy is not conserved as ATP,

$$\text{Glucose} \longrightarrow 2\text{ lactate} + 2\text{H}^+$$

at pH 7 is about $-48{,}000$ cal. per mole. Since the energy of hydrolysis of only six ATP molecules is utilized to reverse this process, as shown in (2), the reversal would appear to be quite efficient.

INHIBITION OF GLYCOLYSIS BY OXYGEN—THE PASTEUR EFFECT

It was recognized by Pasteur, and later by Meyerhof and Warburg, that glycolysis is linked to respiration, *i.e.*, oxygen consumption. The observation repeatedly made with a wide variety of tissue preparations and microorganisms is that the rate of glycolysis, as measured by the rate of disappearance of glucose or appearance of lactic acid, is lower under aerobic conditions. The admission of oxygen to a glycolyzing system will often decrease the rate of glycolysis, or, in terms frequently employed, anaerobic glycolysis is more rapid than aerobic glycolysis. This inhibition of glycolysis by oxygen is called the *Pasteur effect.* The benefits that accrue to the organism from the operation of the Pasteur effect are clear. The process of anaerobic glycolysis (glucose → 2 lactic acid) releases only about 8 per cent of the energy that would be obtained by complete oxidation of glucose. If oxygen is available and lactic acid is oxidized to CO_2 and H_2O, the energy yield per molecule of glucose is considerably increased; hence, the energy needs of the cell can be met by consumption of considerably less glucose.

The Pasteur effect is the resultant of a variety of chemical consequences of respiration and the availability of oxygen. Several of the factors that participate in the Pasteur effect will be discussed briefly.

1. The decrease in observed glycolysis may be, in part, apparent rather than real. In many cells, in the presence of oxygen, a portion of the lactic acid formed by glycolysis is oxidized to CO_2 and H_2O, and some of the energy released is utilized to resynthesize glucose or glycogen from the remainder of the lactic acid. The net effect of this resynthesis is to decrease the net consumption of glucose as well as the accumulation of lactic acid, whereas the *rate* of glycolysis may not have diminished to quite the same extent.

2. Inorganic orthophosphate is required for both phosphorolysis of glycogen (page 438) and oxidation of 3-phosphoglyceraldehyde; the latter reaction and the pyruvic acid kinase reaction also require a supply of ADP. Orthophosphate and ADP are also consumed in the oxidation of lactate and pyruvate via the tricarboxylic acid cycle, while generating ATP. In this sense, glycolysis competes with respiration for available inorganic phosphate and ADP. When respiration is slow or absent, ample phosphate and ADP are provided for rapid glycolysis, but insofar as the oxidative reactions transform phosphate and ADP into ATP, they diminish the supply available for glycolysis. This situation can contribute to the Pasteur effect. This is supported by the observation that dinitrophenol, which "uncouples" the formation of ATP from oxidative reactions (page 348) and hence spares inorganic phosphate and ADP, enhances aerobic glycolysis in various systems.

3. As we have seen, introduction of O_2 to previously anaerobic cells would result in increased oxidation of pyruvate to acetyl CoA. The latter would stimulate the anaplerotic mechanisms, making oxaloacetate more readily available to condense with acetyl CoA to form citrate that would serve as fuel for oxidative phosphorylation, with enhanced formation of ATP from ADP, AMP, and P_i. Increased concentrations of both ATP and citrate inhibit phosphofructokinase (page 396), thereby diminishing the flow of hexose into the glycolytic sequence. This effect may well be the major basis for the Pasteur effect.

What might be described as the reciprocal of the Pasteur effect is the demonstration that high concentrations of glucose will inhibit cellular respiration, studied in isolated systems. This is known as the *Crabtree effect.*

SOME ALTERNATE PATHWAYS OF CARBOHYDRATE METABOLISM

The Glyoxylic Acid Pathway. An additional pathway for generation of dicarboxylic acids functions in plants and microorganisms but is not known to be operative in animal tissues. This pathway is made possible by two specific enzymes. *Isocitric acid lyase* catalyzes a retrograde aldol condensation.

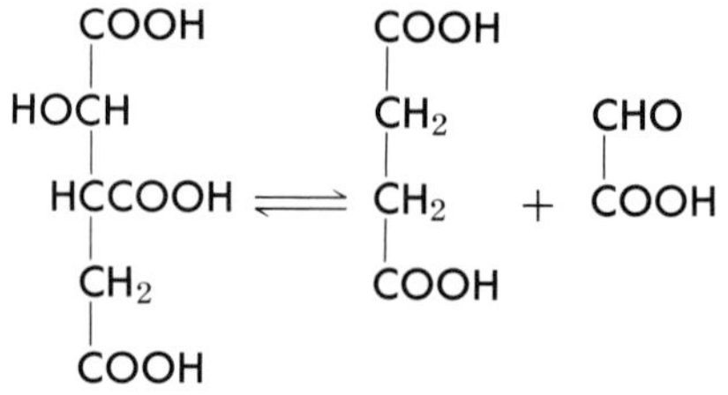

Isocitric acid — Succinic acid — Glyoxylic acid

In *E. coli,* phosphoenolpyruvate serves both as a repressor of the formation of this enzyme and as an allosteric inhibitor of existing enzyme. *Malic acid synthetase* catalyzes the condensation of acetyl CoA with glyoxylic acid to form malic acid, a reaction analogous to the formation of citric acid from oxaloacetic acid (page 329).

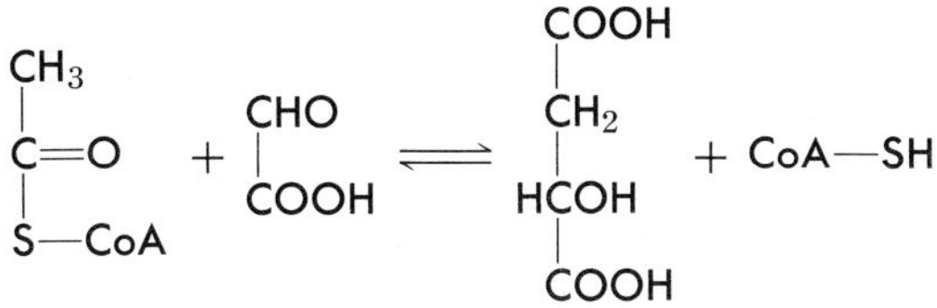

In conjunction with the previously described enzymes of the citric acid cycle, these enzymes permit the net formation of succinic acid from acetyl CoA, as shown in Fig. 18.8. The over-all reaction is as follows.

$$\textbf{2 Acetyl CoA} + \textbf{DPN}^+ \longrightarrow \textbf{succinic acid} + \textbf{2CoA} + \textbf{DPNH} + \textbf{H}^+$$

The succinic acid formed can be converted by reactions of the citric acid cycle to oxaloacetic acid, which can then react with acetyl CoA. Alternatively, the oxaloacetic acid may be converted to phosphoenolpyruvic acid and thence to glucose. In this manner the carbon of acetyl CoA, derived from the oxidation of fatty acids (Chap. 21), can be utilized for *net* formation of carbohydrate. This transformation, not possible in animals, can occur in plants and various microorganisms by virtue of the operation of isocitrate lyase and malate synthetase, which are lacking in animal tissues. The principal control of the rate of this process appears to be exercised by the concentration of phosphoenolpyruvate, which is a negative modifier of isocitrate lyase.

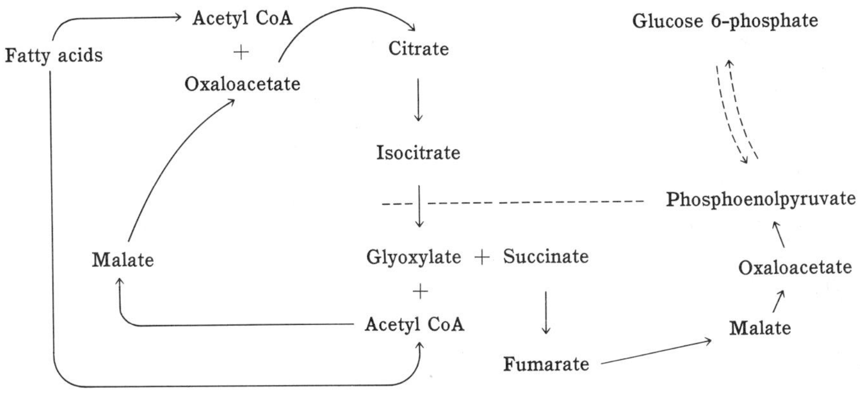

Fig. 18.8. The malate-isocitrate cycle. Oxidation of fatty acids can provide acetyl CoA, which enters the cycle at two points. Each turn of the cycle results in generation of one molecule of succinic acid. The latter can be oxidized to oxaloacetic acid by reactions of the citric acid cycle (Fig. 16.1), the oxaloacetic acid is decarboxylated and phosphorylated to phosphoenolpyruvic acid (page 410), and the latter is converted to glucose 6-phosphate by reversal of the glycolytic sequence (Fig. 18.2). Thus, four molecules of acetyl CoA are required to make two molecules of succinic acid, which yield one molecule of hexose plus two molecules of CO_2. ———— indicates influence of phosphoenolpyruvate on functioning of isocitrate lyase.

Acetyl Phosphate Metabolism in Microorganisms. Most microorganisms, like animal cells, oxidize pyruvate to acetyl CoA, which enters the citric acid cycle. An alternate pathway is the *phosphoroclastic* cleavage in *E. coli.*

$$\underset{\text{Pyruvic acid}}{CH_3COCOOH} + P_i \longrightarrow \underset{\text{Acetyl phosphate}}{CH_3COOPO_3H_2} + \underset{\text{Formic acid}}{HCOOH}$$

Acetyl phosphate may be converted to acetyl CoA by *phosphotransacetylase.*

$$\text{Acetyl phosphate} + \text{CoA} \rightleftharpoons \text{acetyl CoA} + P_i$$

In some microorganisms, the phosphate of acetyl phosphate can be used, in place of ATP, for the phosphorylation of various hexoses.

Propionic Acid Biosynthesis in Microorganisms. It has long been evident that animals can metabolize propionic acid. For example, propionic acid administered to the fasted animal soon appears, almost quantitatively, as an increment in liver glycogen. If given to a diabetic animal, propionic acid appears quantitatively in the urine as one-half mole equivalent of glucose. The pathway for this conversion remained obscure until Swick and Wood elucidated the reverse process, *viz.*, formation of propionic acid from pyruvate, in propionibacteria.

In these bacteria, the initial event in propionic acid synthesis is an unusual transfer of CO_2 from methylmalonyl CoA to pyruvic acid, catalyzed by a biotin-containing enzyme, *methylmalonyl-oxaloacetic acid transcarboxylase.*

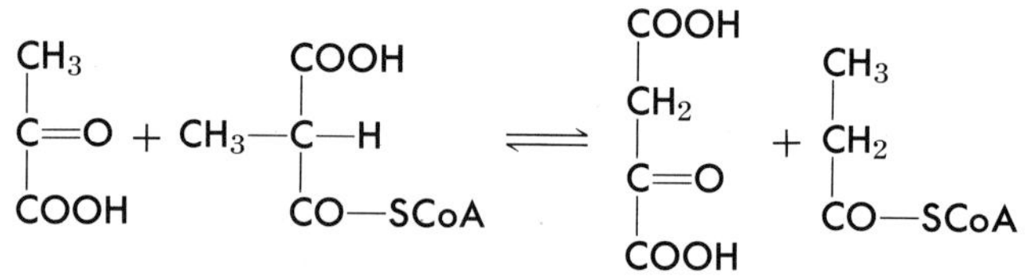

Pyruvic acid — Methylmalonyl CoA — Oxaloacetic acid — Propionyl CoA

The oxaloacetate formed is reduced to succinic acid by reversal of the usual citric acid cycle reactions, via malic and fumaric acids. A *transthioesterase* catalyzes transfer of the CoA moiety from propionyl CoA to succinic acid, forming free propionic acid and succinyl CoA.

The final step in this process is the regeneration of methylmalonyl CoA from succinyl CoA. This is catalyzed by *methylmalonyl CoA mutase,* that utilizes, as coenzyme, *dimethylbenzimidazole cobamide* (Chap. 49), a derivative of vitamin B_{12}. The mechanism of this reaction is discussed elsewhere (page 484). This reaction sequence may be summarized as follows.

(*a*) Pyruvic acid + methylmalonyl CoA $\rightleftharpoons$ oxaloacetic acid + propionyl CoA
(*b*) Oxaloacetic acid + 4(H) $\rightleftharpoons$ succinic acid + H_2O
(*c*) Succinic acid + propionyl CoA $\rightleftharpoons$ succinyl CoA + propionic acid
(*d*) Succinyl CoA $\rightleftharpoons$ methylmalonyl CoA

Sum: Pyruvic acid + 4(H) $\longrightarrow$ propionic acid + H_2O

The metabolism of propionic acid by animal tissues, essentially by a reversal of this pathway, is discussed on page 483.

Fates of Acetaldehyde. It was indicated earlier that acetaldehyde, formed by decarboxylation of pyruvate, is reduced to ethanol in the course of anaerobic fermentation. In certain aerobic fermentations, as in vinegar production, acetic acid is the terminal product. This may arise by hydrolysis of acetyl CoA or acetyl phosphate or by oxidation of acetaldehyde. Various enzymes are known that catalyze the latter reaction, including phosphoglyceraldehyde dehydrogenase (page 398). In mammals, this oxidation is catalyzed by *hepatic aldehyde oxidase,* by *xanthine oxidase,* and by a DPN-linked *aldehyde dehydrogenase.*

THE PHOSPHOGLUCONATE OXIDATIVE PATHWAY

The glycolytic pathway is one of two major reaction sequences by which glucose is metabolized in animal tissues. The second prominent pathway has been called the phosphogluconate oxidative pathway, or the "hexose monophosphate shunt." Warburg, Dickens, Lipmann, S. S. Cohen, Horecker, and Racker have contributed to its elucidation.

The phosphogluconate oxidative pathway of glucose metabolism encompasses a series of reactions that include a number of novel transformations and compounds as well as several of the intermediates and enzymes encountered in glycolysis. The enzymes of both series of reactions are found in the cytoplasm of those cells in which they function. The phosphogluconate pathway affords a means for the total combustion of glucose independent of the citric acid cycle, it is the important generator of TPNH necessary for synthesis of fatty acids and the functioning of various hydroxylases (page 378), and it serves as a source of D-ribose as well as of 4-carbon and 7-carbon sugars. Commencing with glucose 6-phosphate, there is no further requirement for ATP. In contrast to glycolysis and the citric acid cycle, the operation of this sequence cannot be visualized as a consecutive set of transformations leading in direct fashion from glucose 6-phosphate to six molecules of CO_2. To understand the phosphogluconate pathway it is desirable first to consider the participating enzymes, their substrates, and their reaction products.

Glucose 6-Phosphate Dehydrogenase. The mammalian enzyme and that from most but not all other sources catalyzes the oxidation of glucose 6-phosphate by TPN^+.

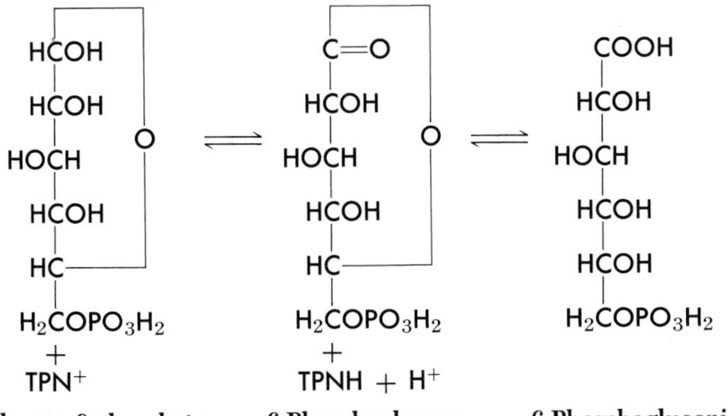

Glucose 6-phosphate | 6-Phosphoglucono-δ-lactone | 6-Phosphogluconic acid

The formation of the δ-lactone in the first step is a freely reversible reaction. Although the lactone is extremely unstable and hydrolyzes spontaneously, this process is accelerated by a specific *lactonase*. Equilibrium for the total process lies far to the right; this reaction is used to generate TPNH in experiments with purified enzymic systems.

6-Phosphogluconic Acid Dehydrogenase. Further oxidation of 6-phosphogluconic acid is catalyzed by a TPN^+- and Mn^{++}-dependent enzyme that is a β-hydroxyacid oxidative decarboxylase and is similar, therefore, to malic enzyme and isocitric acid dehydrogenase.

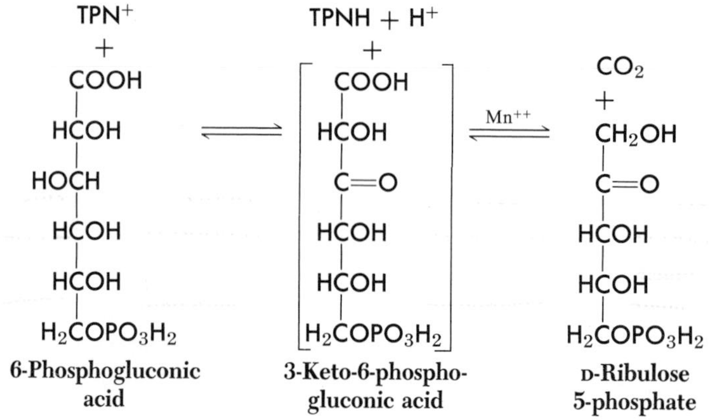

The intermediary β-keto acid has not been isolated and is presumed to decarboxylate as it is formed on the enzymic surface.

Pentose Interconversions. Specific enzymes for transformation of ribulose 5-phosphate into isomeric pentoses are an integral aspect of the operation of the phosphogluconate oxidative pathway.

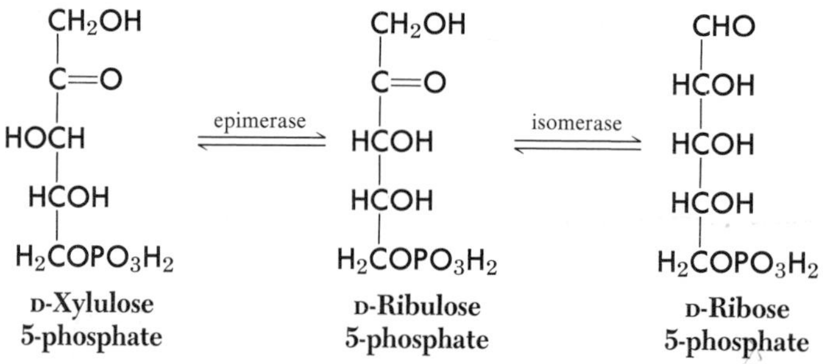

Phosphopentose epimerase catalyzes epimerization about carbon-3; the mechanism is presumed to involve transitory formation of the 2:3 enediol. *Phosphopentose isomerase* catalyzes interconversion of the ketopentose and aldopentose forms, a reaction that is formally similar to the action of hexose phosphate isomerase (page 395) and triose phosphate isomerase (page 398).

Transketolase. D-Xylulose 5-phosphate serves as a source of "active glycolaldehyde" in the *transketolase* reaction. This enzyme, which utilizes thiamine pyrophosphate as coenzyme and requires Mg^{++}, effects transfer of a 2-carbon unit from a

2-keto sugar to C-1 of various aldoses. The donor ketose must possess the L configuration at carbon-3, and the ketose product has the same configuration. Among the more important reactions catalyzed by transketolase is that in which the 7-carbon sugar sedoheptulose is formed. The "active glycolaldehyde" moiety is α,β-dihydroxyethyl thiamine pyrophosphate, a compound similar to the "active acetaldehyde" formed during oxidative decarboxylation of pyruvate (page 327), and the reaction mechanism is equivalent. No free glycolaldehyde appears during the transketolase reaction, which is readily reversible.

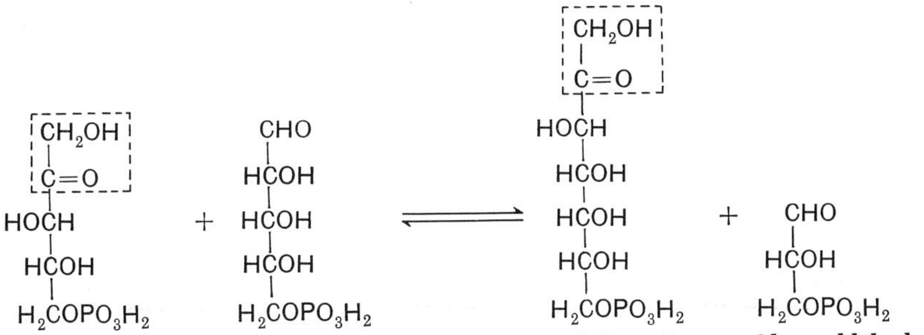

D-Xylulose 5-phosphate + **D-Ribose 5-phosphate** ⇌ **D-Sedoheptulose 7-phosphate** + **D-Glyceraldehyde 3-phosphate**

Transaldolase. Another enzyme in this series is *transaldolase,* which catalyzes transfer of carbon atoms 1, 2, and 3 of a ketose phosphate to C-1 of an aldose phosphate.

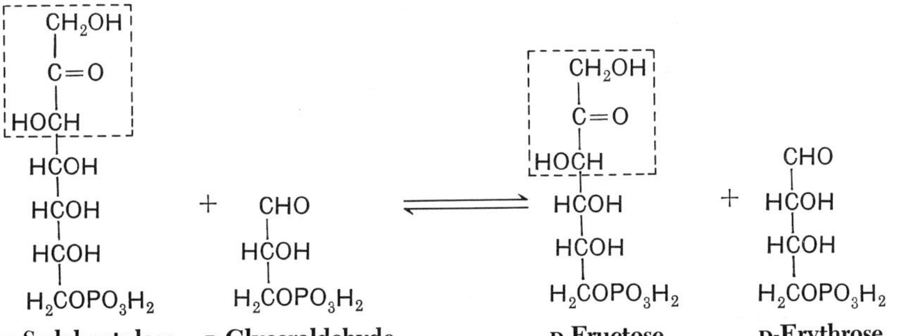

D-Sedoheptulose 7-phosphate + **D-Glyceraldehyde 3-phosphate** ⇌ **D-Fructose 6-phosphate** + **D-Erythrose 4-phosphate**

Although the reaction appears similar to the aldolase reaction (page 397), free dihydroxyacetone or its phosphate ester cannot serve as substrates, nor do they appear in the course of the reaction; the enzyme catalyzes only transfers in the manner shown above. Horecker and his colleagues demonstrated that an intermediate Schiff base is formed between the carbonyl of the transferred dihydroxyacetone moiety and the ϵ-amino group of a lysine residue of the enzyme. A similar mode of substrate binding is involved in the action of aldolase (page 263) and the reaction is thought to occur in much the same manner except for the inability to initiate reaction with free dihydroxyacetone.

Erythrose 4-phosphate formed by transaldolase has several known metabolic fates, one of which is to accept a 2-carbon unit in a transketolase-catalyzed reaction to form fructose 6-phosphate. Erythrose 4-phosphate also participates in phenylalanine synthesis (page 574).

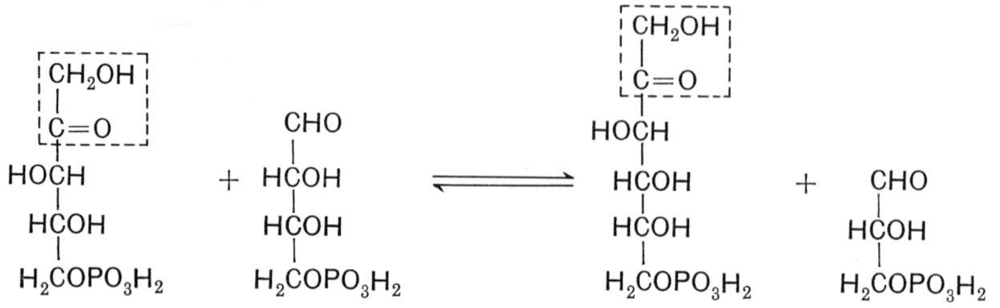

D-Xylulose 5-phosphate + D-Erythrose 4-phosphate ⇌ D-Fructose 6-phosphate + D-Glyceraldehyde 3-phosphate

Cyclic Nature of the Phosphogluconate Oxidative Pathway. The discussion above has presented all the enzymes and intermediates unique to this pathway. For its operation, however, four of the enzymes concerned in glycolysis are also required. These are triose phosphate isomerase (page 398) to catalyze interconversion of phosphoglyceraldehyde and dihydroxyacetone phosphate, aldolase to catalyze formation of fructose 1,6-diphosphate from the two triose phosphates, diphosphofructose phosphatase to hydrolyze fructose 1,6-diphosphate to fructose 6-phosphate, and hexose phosphate isomerase to convert fructose 6-phosphate to glucose 6-phosphate. These will be recognized as the enzymes involved in the normal formation of hexose from triose or pyruvate by reversal of the glycolytic sequence.

With the foregoing reactions in mind, it is possible to reconstruct a system into which hexose continually enters and from which CO_2 emerges as the sole carbon compound. This may be visualized by consideration of the set of balanced equations shown in Table 18.3, which, in sum, describe the conversion of 6 moles of hexose phosphate to 5 moles of hexose phosphate and 6 moles of CO_2. Note that the only reaction in which CO_2 is evolved is the oxidation of 6-phosphogluconic acid.

At first glance this description of the reaction sequence appears quite complex, but it may be more readily understood from the following considerations. The operation of this pathway as an *oxidizing* mechanism is accomplished entirely by reactions (*a*) and (*b*), in which TPN^+ is reduced and CO_2 is evolved. Oxidation of 6 moles of hexose by reactions (*a*) and (*b*) results in delivery of 12 pairs of electrons to TPN^+, the requisite amount for total oxidation of 1 mole of glucose to 6 moles of CO_2. As a consequence of reaction (*b*), there remain 6 moles of ribulose 5-phosphate. If these are employed for nucleotide formation, etc., they need only undergo isomerization to ribose 5-phosphate. However, if the carbon of the 6 moles of pentose were rearranged to form 5 moles of hexose, the net achievement would be the complete oxidation of one of the original 6 moles of hexose. This transformation of the carbon of the pentoses to hexose is initiated in reactions (*c*) and (*d*), with a balance of 2 moles of tetrose and 2 moles of pentose. Reactions (*e*) employ the latter to yield two additional moles of hexose and two of triose. Reactions (*f*) to (*h*) convert the two trioses to one hexose. In each series the hexose formed is fructose 6-phosphate. Finally, isomerization of the latter yields 5 moles of glucose 6-phosphate, which may now reenter the sequence at (*a*), etc.

Table 18.3: **Reactions of the Phosphogluconate Oxidative Pathway**

Step	Enzyme	Reaction	Carbon balance
a	Glucose 6-phosphate dehydrogenase	6 Glucose 6-phosphate + $6TPN^+ \longrightarrow$ 6 6-phosphogluconate + 6TPNH + $6H^+$	$6(6) \longrightarrow 6(6)$
b	Phosphogluconic acid dehydrogenase	6 6-Phosphogluconate + $6TPN^+ \longrightarrow$ 6 ribulose 5-phosphate + 6TPNH + $6H^+$ + $6CO_2$	$6(6) \longrightarrow 6(5) + 6(1)$
c	Pentose epimerase	2 Ribulose 5-phosphate $\longrightarrow$ 2 xylulose 5-phosphate	$2(5) \longrightarrow 2(5)$
	Pentose isomerase	2 Ribulose 5-phosphate $\longrightarrow$ 2 ribose 5-phosphate	$2(5) \longrightarrow 2(5)$
	Transketolase	2 Xylulose 5-phosphate + 2 ribose 5-phosphate $\longrightarrow$ 2 sedoheptulose 7-phosphate + 2 glyceraldehyde 3-phosphate	$2(5) + 2(5) \longrightarrow 2(7) + 2(3)$
d	Transaldolase	2 Sedoheptulose 7-phosphate + 2 glyceraldehyde 3-phosphate $\longrightarrow$ 2 erythrose 4-phosphate + 2 fructose 6-phosphate	$2(7) + 2(3) \longrightarrow 2(4) + 2(6)$
e	Pentose epimerase	2 Ribulose 5-phosphate $\longrightarrow$ 2 xylulose 5-phosphate	$2(5) \longrightarrow 2(5)$
	Transketolase	2 Xylulose 5-phosphate + 2 erythrose 4-phosphate $\longrightarrow$ 2 glyceraldehyde 3-phosphate + 2 fructose 6-phosphate	$2(5) + 2(4) \longrightarrow 2(3) + 2(6)$
f	Triose isomerase	Glyceraldehyde 3-phosphate $\longrightarrow$ dihydroxyacetone phosphate	$(3) \longrightarrow (3)$
g	Aldolase	Dihydroxyacetone phosphate + glyceraldehyde 3-phosphate $\longrightarrow$ fructose 1,6-diphosphate	$(3) + (3) \longrightarrow (6)$
h	Phosphatase	Fructose 1,6-diphosphate $\longrightarrow$ fructose 6-phosphate + P_i	$(6) \longrightarrow (6)$
i	Hexose isomerase	5 Fructose 6-phosphate $\longrightarrow$ 5 glucose 6-phosphate	$5(6) \longrightarrow 5(6)$
Net:		6 Glucose 6-phosphate + $12TPN^+ \longrightarrow$ 5 glucose 6-phosphate + $6CO_2$ + 12TPNH + $12H^+$ + P_i	$6(6) \longrightarrow 5(6) + 6(1)$

Experimental efforts to estimate the relative amounts of glucose catabolized via the glycolytic and phosphogluconate pathways have been based on the fact that in the phosphogluconate path, the CO_2 first formed must all be derived from C-1 of glucose, whereas in the concerted operation of glycolysis and the citric acid cycle, the initial rate of evolution of $^{14}CO_2$ from glucose 1-^{14}C and glucose 6-^{14}C should be indistinguishable. In mammalian striated muscle there appears to be no direct oxidation via phosphogluconate, and catabolism proceeds entirely via glycolysis and the citric acid cycle. In liver, an appreciable fraction, perhaps 30 per cent or more, of the CO_2 arising from glucose stems from operation of the phosphogluconate oxidative path. In mammary gland, testis, adipose tissue, leukocytes, and adrenal cortex an even larger proportion of glucose catabolism occurs via phosphogluconate oxidation. The special advantages that accrue to the organism by virtue of the operation of the direct pathway, other than pentose formation for nucleotide synthesis, appear to derive from the fact that no additional ATP is required, there is no dependence upon the availability of the 4-carbon dicarboxylic acids of the citric acid cycle, and, probably most significant, this pathway employs TPN^+ as the exclusive electron acceptor. As indicated elsewhere (page 342), it is not clear whether ATP can be efficiently formed by mitochondrial oxidation of TPNH produced in the cytoplasm, but to the extent that this does occur it (page 343) would permit generation of 3 moles of ATP per TPNH formed, or 36 moles for the total combustion of glucose. This value compares favorably with the yield of ATP from glycolysis and the citric acid cycle. However, it appears likely that most of the TPNH generated by the phosphogluconate pathway is employed as a reducing agent, particularly in the synthesis of fatty acids and steroids. This is consonant with the distribution of the enzymes concerned with these processes. Liver, mammary gland, testis, and adrenal cortex are active sites of fatty acid and/or steroid synthesis, whereas these processes are not prominent in the metabolism of striated muscle, a tissue in which phosphogluconate oxidation is not known to occur. In those cells in which phosphogluconate oxidation is possible, it appears that the rate-limiting factor is the availability of TPN^+, thus indicating a coupling of glucose oxidation with the requirements of TPNH-utilizing reactions.

Anaerobic Origin of Ribose. As described above, the oxidation of glucose 6-phosphate affords a means for ribose formation. However, any cell equipped with transketolase, transaldolase, the enzymes for interconversion of pentoses, as well as the normal enzymes of glycolysis, can also generate ribose 5-phosphate by a nonoxidative pathway, as shown by the following equations.

(*a*) **Fructose 6-phosphate + glyceraldehyde 3-phosphate** $\xrightarrow[\text{ketolase}]{\text{trans-}}$

erythrose 4-phosphate + xylulose 5-phosphate

(*b*) **Fructose 6-phosphate + erythrose 4-phosphate** $\xrightarrow[\text{aldolase}]{\text{trans-}}$

glyceraldehyde 3-phosphate + sedoheptulose 7-phosphate

(*c*) **Sedoheptulose 7-phosphate + glyceraldehyde 3-phosphate** $\xrightarrow[\text{ketolase}]{\text{trans-}}$

ribose 5-phosphate + xylulose 5-phosphate

(*d*) 2 Xylulose 5-phosphate ⟶ 2 ribulose 5-phosphate ⟶ 2 ribose 5-phosphate

Sum: 2 Fructose 6-phosphate + glyceraldehyde 3-phosphate ⟶ 3 ribose 5-phosphate

The relative contributions of the oxidative and nonoxidative pathways for net pentose formation in animal tissues are uncertain.

Formation of the D-deoxyribose component of the nucleotide units of DNA occurs by transformation of the corresponding ribonucleotides; this is presented in Chap. 27.

REFERENCES

See list following Chap. 19.

19. Carbohydrate Metabolism. II

Hexose Interconversions. Polysaccharides. Glycogen Metabolism and The Control of Carbohydrate Metabolism

Living forms utilize a wide variety of polysaccharides for the storage of energy and as structural components of cell membranes and walls. The monosaccharide units present in these polysaccharides include a diversity of aldoses, ketoses, amino sugars, deoxy sugars, and uronic acids in various combinations. It is the purpose of this chapter to describe the synthesis of these monosaccharides, their utilization for polysaccharide synthesis, the formation and degradation of storage polysaccharides in mammals, mechanisms for the control of carbohydrate metabolism, and their role in the maintenance of the concentration of blood glucose.

HEXOSE INTERCONVERSIONS

The great variety of hexoses is derived from glucose 6-phosphate by application of a surprisingly limited group of general reactions. The pathways for synthesis of some specific hexoses can be traced in Fig. 19.1; all proceed from glucose 6-phosphate, with no requirement for fragmentation or extension of the carbon chain, which remains intact in each pathway.

Aldose-Ketose Transformations. Isomerases. The interconversion of glucose 6- and fructose 6-phosphates, catalyzed by *phosphoglucose isomerase,* was described under glycolysis (page 395). Such isomerases, including *phosphomannose isomerase,* are thought to catalyze formation of an intermediate enediol. In this reaction, one of the two hydrogen atoms at C-1 of the ketose exchanges with the protons of water. Phosphoglucose isomerase and phosphomannose isomerase are both stereospecific, but only for the two different hydrogen atoms indicated, respectively, as H˙ and H* in the following structures.

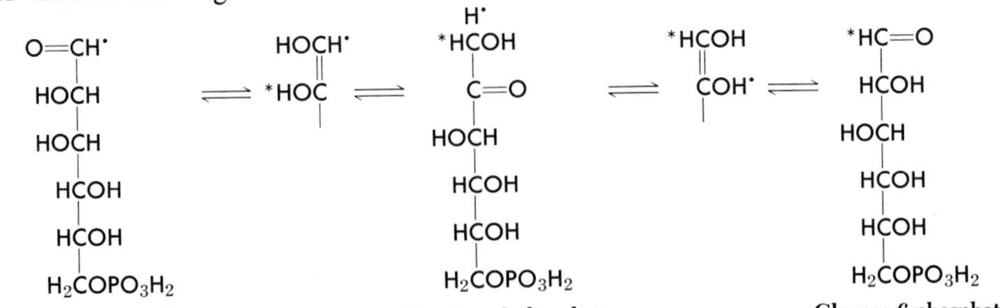

Mannose 6-phosphate Fructose 6-phosphate Glucose 6-phosphate

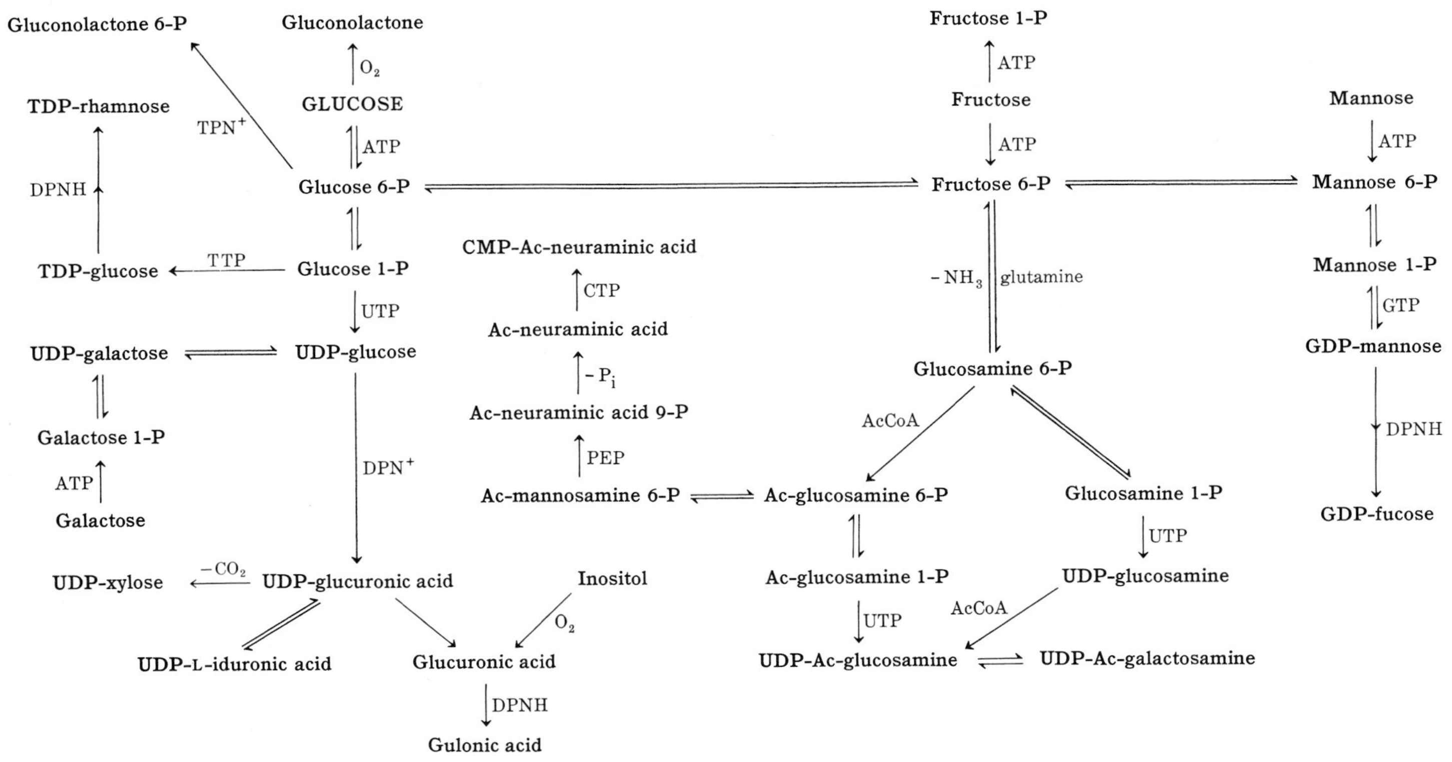

Fig. 19.1. Interconversions of some hexoses in mammalian metabolism. *Ac* in an amino sugar = N-acetyl; *PEP* = phosphoenolpyruvate.

Mutases. Mutases catalyze the apparent migration of a phosphate group from one to another hydroxyl group of the same molecule. An example is the conversion of 3- to 2-phosphoglyceric acid in glycolysis (page 400). The most studied mutase is *phosphoglucomutase,* which has been obtained in pure form from mammalian and fish muscle, yeast, and higher plants, each of which contains two distinct forms of this enzyme, as well as from several bacteria. Both forms from all sources have the same molecular weight (60,000), are single polypeptide chains, and have the same pentapeptide at the active site. The enzyme catalyzes the following reaction.

$$\text{Glucose 6-phosphate} \rightleftharpoons \text{glucose 1-phosphate}$$

At equilibrium, the 6- and 1-phosphates are present in the ratio 94:6. This is in accord with the fact that $\Delta F°$ for hydrolysis of the 1-phosphate is greater than that for the 6-phosphate. The 1-phosphate is also more labile in acid solution.

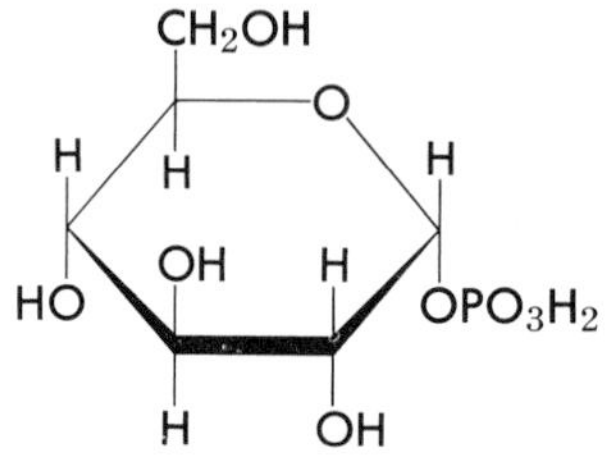

α-D-Glucose 1-phosphate

This reaction proceeds only in the presence of *glucose 1,6-diphosphate* and Mg^{++}. After addition of the ^{32}P-labeled form of any one of the three esters, equilibration of isotope occurs rapidly among all three compounds. Incubation of the enzyme with ^{32}P-labeled glucose 1,6-diphosphate resulted in labeled enzyme from which -Thr-Ala-Ser^{32}P-His-Asp- was obtained by enzymic hydrolysis. These observations suggest that the reaction may be represented as follows.

$$\text{Enzyme-OH} + \text{glucose 1,6-diphosphate} \rightleftharpoons \text{enzyme-O-}PO_3H_2 + \text{glucose 1- or 6-phosphate}$$

By repeated operation of this reaction, either monoester may be converted to the other, *viz.*, a molecule of glucose 6-phosphate reacts with the enzyme-phosphate and is esterified at the 1 position. When the diphosphate reacts with the nonesterified form of the enzyme, it surrenders the 6-phosphate group and glucose 1-phosphate remains.

A continuing source of glucose 1,6-diphosphate is supplied by the action of *phosphoglucokinase,* which catalyzes the following reaction.

$$\text{Glucose 1-phosphate} + \text{ATP} \rightleftharpoons \text{glucose 1,6-diphosphate} + \text{ADP}$$

Phosphoglucomutase can catalyze mutase reactions, albeit more slowly, with several other hexose phosphates. The initial event, with mannose 1-phosphate, is represented as follows.

$$\text{Glucose 1,6-diphosphate} + \text{mannose 1-phosphate} \rightleftharpoons \text{mannose 1,6-diphosphate} + \text{glucose 6-phosphate}$$

Sugar Esters of Nucleoside Diphosphates. Pyrophosphorylases. As shown in Table 19.1, many pathways of carbohydrate metabolism involve, as intermediates, esters formed between the C-1 of a sugar and the terminal phosphate of a nucleoside diphosphate. Virtually all glycosides and polysaccharides are synthesized from these compounds. Formation of such compounds is catalyzed by specific *pyrophosphorylases,* as shown for the synthesis of uridine diphosphate glucose (UDP-glucose).

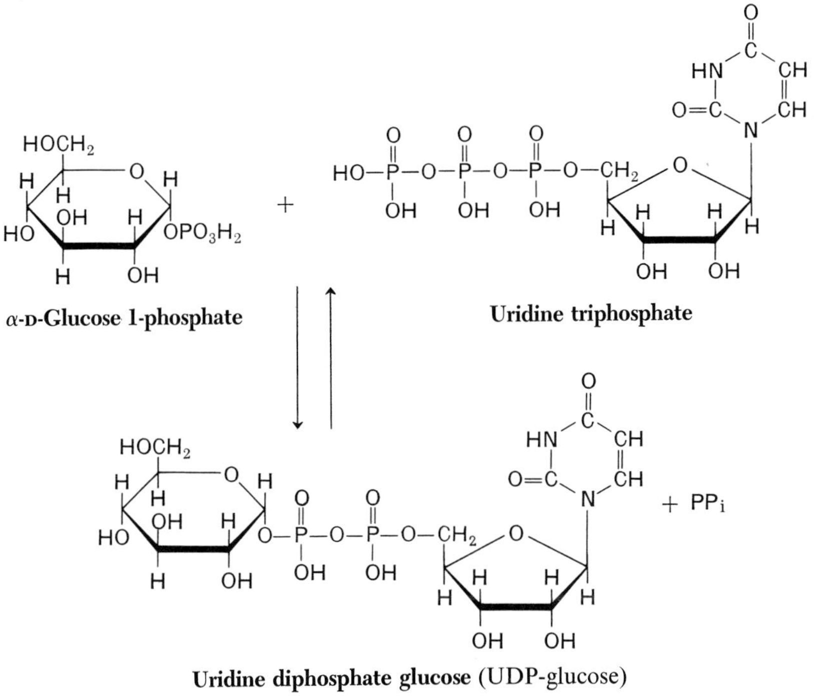

The inorganic pyrophosphate is derived from the terminal pyrophosphate of the uridine triphosphate. Since a pyrophosphate bond is broken and replaced by a new pyrophosphate linkage, the reaction is freely reversible. Physiologically, however, hydrolysis of the inorganic pyrophosphate by *pyrophosphatase* (page 218) effectively results in irreversible synthesis of UDP-glucose.

A remarkable variety of nucleoside diphosphate sugar compounds has been isolated from cells and tissues at all phylogenetic levels; some of these are listed in Table 19.1. The five major purines and pyrimidines commonly found in RNA and DNA are utilized in this manner. In all but one known instance (see below), synthesis of the nucleoside diphosphate sugar, from a simpler sugar derivative, occurs by a pyrophosphorylase-catalyzed reaction.

Nucleoside triphosphate + sugar 1-phosphate $\rightleftharpoons$ nucleoside diphosphate sugar + PP$_i$

The one exception to the general pyrophosphorylase reaction shown above is the reaction between N-acetylneuraminic acid and cytidine triphosphate to form PP$_i$ and N-acetylneuraminic acid-1′-cytidine monophosphate (page 426). In this case a nucleoside *mono*phosphate ester of the 1′-hydroxyl is formed. In the discussions to follow, the abbreviations for the nucleoside diphosphates shown in Table 19.1 will generally be employed. The pyrophosphorylase reaction is not limited to sugar phosphates. CDP-glycerol and CDP-ribitol are made similarly

Table 19.1: SOME NUCLEOSIDE DIPHOSPHATE SUGARS

Uridine diphosphate esters (UDP-X*):
Glucose,† galactose,† glucosamine,† mannosamine, N-acetylglucosamine,† N-acetylgalactosamine,† muramic acid, glucuronic acid,† iduronic acid,† galacturonic acid, xylose,† arabinose, rhamnose
Adenosine diphosphate ester (ADP-X):
Glucose
Guanosine diphosphate esters (GDP-X):
Glucose,† galactose,† mannose,† fucose,† rhamnose, colitose
Cytidine diphosphate ester (CDP-X):
Glucose, tyvelose, ascarylose, paratose, abequose
Cytidine monophosphate ester (CMP-X)‡:
N-Acetylneuraminic acid†
Deoxythymidine diphosphate esters (dTDP-X):
Glucose, galactose, mannose, glucosamine, N-acetylglucosamine, N-acetylgalactosamine, rhamnose

* X in each instance can be any one of the sugars listed in the group below each specific heading.
† Known to occur in animal tissues.
‡ A nucleoside monophosphate sugar.

from glycerol and ribitol phosphates in those bacteria in which teichoic acid is a cell wall constituent (Chap. 41). Many nucleoside diphosphate sugars are not formed directly from the sugar or sugar phosphate but can arise metabolically only by transformation of an existing nucleoside diphosphate sugar. Several examples will be found in Table 19.1.

Epimerases. Epimerization is a relatively common reaction in carbohydrate metabolism. Perhaps the best-known example is epimerization at C-4 of UDP-glucose to yield UDP-galactose, catalyzed in liver by *uridine diphosphate glucose epimerase.*

$$\text{Uridine diphosphate glucose} \rightleftharpoons \text{uridine diphosphate galactose}$$

This enzyme and all other such epimerases exhibit an absolute requirement for DPN^+. Since (1) glucose labeled with tritium at C-4 yields galactose similarly labeled and (2) an increase in fluorescence is observed during the reaction, it is thought that the corresponding 4-keto sugar and DPNH must be enzyme-bound intermediates in this reaction.

A special case of epimerization is the mutarotation (page 19) of glucose, catalyzed by *mutarotase.* This enzyme is widely distributed in animal tissues, but was first encountered in extracts of *Penicillium notatum,* in which it exists in association with glucose oxidase. The latter is specific for oxidation of β-D-glucose. The presence of mutarotase permits oxidation of the total glucose in solution more rapidly than would be possible by the relatively slow process of spontaneous mutarotation.

An epimerase widely distributed in plants catalyzes the formation of UDP-arabinose from UDP-xylose. The kidney epimerase which catalyzes interconversion of N-acetylglucosamine 6-phosphate and N-acetylmannosamine 6-phosphate appears to require the presence of ATP as a positive allosteric modifier.

"Transferases." *Phosphogalactose uridyl transferase* of liver catalyzes the following reaction.

UDP-glucose + galactose 1-phosphate $\rightleftharpoons$ UDP-galactose + glucose 1-phosphate

In liver, galactose is phosphorylated at C-1 by ATP in the presence of a *galacto-kinase.* The galactokinase of fetal liver exhibits both a K_m and V_{max} about five times those of the enzyme of adult liver. The product, galactose 1-phosphate, is used to make UDP-galactose from UDP-glucose by the above transferase reaction. The resultant UDP-galactose is converted to UDP-glucose, thus making galactose generally available for hepatic carbohydrate metabolism (this pathway is not shown in Table 19.1). Hereditary lack of phosphogalactose uridyl transferase results in *galactosemia,* a disorder of infant life characterized by inability to metabolize the galactose derived from the lactose of milk.

An alternate pathway to uridine diphosphate galactose is provided by *uridine diphosphate galactose pyrophosphorylase,* which catalyzes synthesis of UDP-galactose from UTP and galactose 1-phosphate. This enzyme is present only at a low level in fetal and infant liver but increases in amount in later years. Hence, galactosemic infants who survive develop later the capacity to metabolize galactose.

Reduction: Formation of Deoxyhexoses. Many deoxyhexoses and dideoxyhexoses have been described. In each case synthesis proceeds by a common pair of reactions. In the first step the nucleoside diphosphate sugar undergoes a dehydration catalyzed by a DPN$^+$-requiring enzyme to yield the corresponding 4-keto-6-deoxy sugar derivative. The role of DPN$^+$ is not clear. The 4-keto-6-deoxynucleoside diphosphate is then reduced by a TPNH-utilizing enzyme to form the nucleoside diphosphate 6-deoxy sugar. The synthesis of guanosine diphosphate fucose is shown as an example.

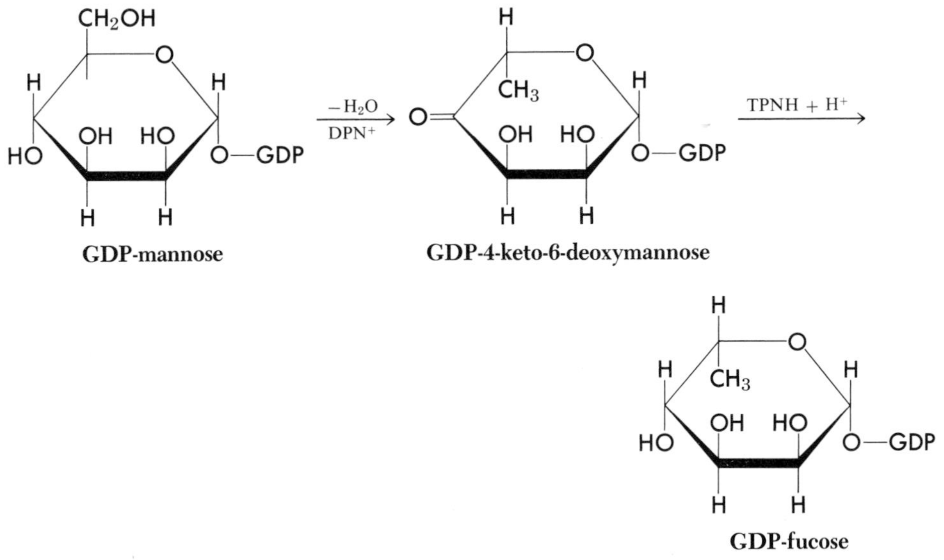

A similar reaction sequence leads to TDP-rhamnose. A second type of reduction, utilizing TPNH, occurs in diverse bacteria to yield derivatives of dideoxy sugars which are employed in cell wall biosynthesis (Chap. 41), *e.g.*, GDP-3,6-dideoxy-L-galactose (colitose), CDP-3,6-dideoxy-D-mannose (tyvelose), CDP-3,6-dideoxy-L-mannose (ascarylose), CDP-3,6-dideoxy-D-glucose (paratose).

Oxidation; Formation of Uronic Acids. Glucuronic acid is formed by the oxidation of uridine diphosphate glucose.

Uridine diphosphate glucose + $2DPN^+$ ⟶ uridine diphosphate glucuronic acid + 2DPNH + $2H^+$

One enzyme catalyzes the two consecutive oxidations required for formation of a carboxyl group from a primary alcohol; the presumed intermediate, in which C-6 should be at the aldehyde level of oxidation, has not been detected. Epimerization of UDP-D-glucuronic acid to UDP-L-iduronic acid (Table 19.1) is the source of the latter uronic acid for polysaccharide synthesis.

Decarboxylation. The only known decarboxylation of a nucleoside diphosphate sugar is the conversion of UDP-glucuronic acid to UDP-xylose. The latter is probably used for formation of the xylosides of seryl hydroxyl groups in mucoprotein synthesis (page 879). Other decarboxylations of hexoses occur as a result of β-hydroxy dehydrogenation of the free sugar or sugar phosphates (page 416).

Formation of Amino Sugars. Most amino sugars are formed by transfer of the amide group of glutamine to 6-phosphate esters of ketoses. For example:

Fructose 6-phosphate + glutamine ⟶ glucosamine 6-phosphate + glutamic acid

Presumably, it is the enediol form of fructose 6-phosphate that is the intermediate, although it has not been identified in this reaction. N-Acetylamino sugars are generally present in polysaccharides rather than the simple amino sugars. Acetylation is catalyzed by enzymes specific for each amino sugar, utilizing acetyl coenzyme A as the acylating agent; the amino sugar usually participates as a nucleoside diphosphate ester.

Aldol Condensation; Formation of Sialic Acids. Sialic acids (page 34) are constituents of a variety of mammalian glycoproteins and bacterial cell wall structures. The biosynthesis of N-acetylneuraminic acid in mammals begins with the formation of N-acetylmannosamine from N-acetylglucosamine by epimerization. The former is then converted to the 6-phosphate, utilizing ATP. In bacteria, the amino hexose 6-phosphates are epimerized. N-Acetylmannosamine 6-phosphate then condenses in an aldol-type reaction with phosphoenolpyruvic acid to yield N-acetylneuraminic acid 9-phosphate. The 9-phosphate group must be removed by hydrolysis prior to the reaction with cytidine triphosphate, which leads to formation of CMP-N-acetylneuraminic acid, as described earlier (page 426).

N-Acetylglucosamine ⇌ **N-acetylmannosamine** (mammals)

↓ ATP

(bacteria) **N-Acetylglucosamine 6-phosphate** ⇌ **N-acetylmannosamine 6-phosphate**

N-Acetylmannosamine 6-phosphate + phosphoenolpyruvic acid $\xrightarrow{-P_i}$

N-Acetylneuraminic acid 9-phosphate $\xrightarrow{-P_i}$ **N-acetylneuraminic acid + CTP** ⟶

CMP-N-acetylneuraminic acid + PP_i

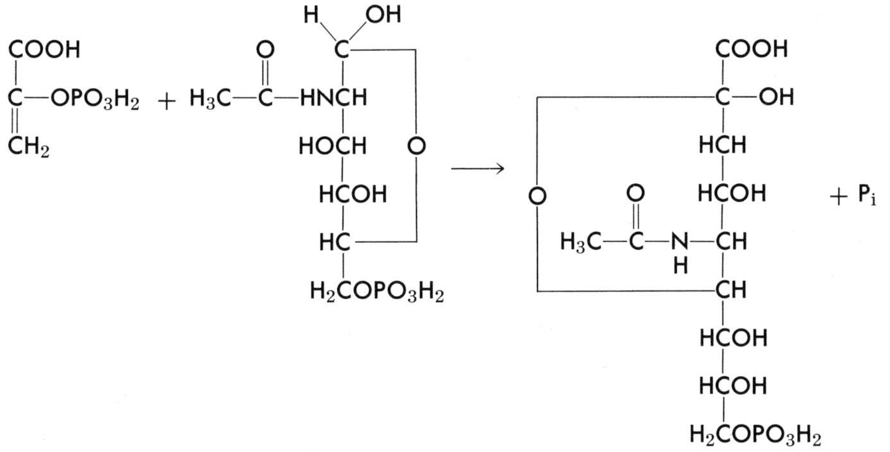

Phosphoenol-pyruvic acid **N-Acetylmannosamine 6-phosphate** **N-Acetylneuraminic acid 9-phosphate**

Metabolism of Fructose. Ingested fructose is phosphorylated in the liver by a *fructokinase* which specifically directs phosphorylation at the C-1 position of this ketose. No mutase is known which can catalyze conversion of fructose 1-phosphate to fructose 6-phosphate, nor can phosphofructokinase (page 395) effect synthesis of fructose diphosphate from fructose 1-phosphate. The only pathway available to the latter is made possible by a specific aldolase which catalyzes the reaction.

$$\textbf{Fructose 1-phosphate} \rightleftharpoons \textbf{dihydroxyacetone phosphate + glyceraldehyde}$$

The further metabolism of glyceraldehyde requires that it be reduced by DPNH to glycerol, which is then phosphorylated by glycerol kinase, using ATP, and reoxidized by DPN^+ to dihydroxyacetone phosphate. The latter then enters the usual glycolytic series.

Individuals who lack fructokinase excrete the major portion of ingested fructose in the urine. "Fructose intolerance" is a more serious illness, occasioned by genetic lack of the aldolase specific for fructose 1-phosphate, which accumulates after fructose ingestion and inhibits diverse enzyme systems.

Control of Hexose Interconversions. In the main, synthesis of other hexoses from glucose is directed toward production of compounds required for polysaccharide synthesis. As in other areas of metabolism, many, if not all, such biosynthetic pathways are self-regulating. The activated form of the hexose which is utilized in polymer formation serves as an allosteric inhibitor of an early step in its own biosynthesis. The enzymic details are as yet obscure, but Table 19.2 lists some examples of this type of metabolic control in bacteria. The specificity of these processes is illustrated by the fact that some strains of *Aerobacter* utilize mannose and others fucose in their extracellular polysaccharides; the action of GDP-mannose pyrophosphorylase is inhibited by GDP-mannose in the former and by GDP-fucose in the latter, but not vice versa.

Metabolism of Glucuronic Acid. The formation of UDP-glucuronic acid was described previously. It is not known how free glucuronic acid is made available in

Table 19.2: EXAMPLES OF METABOLIC CONTROL BY FEEDBACK INHIBITION OF HEXOSE BIOSYNTHESIS IN BACTERIA

Inhibitor	Reaction inhibited	Source
TDP-rhamnose	TTP + glucose 1-P ⟶ TDP-glucose	*Escherichia coli* *Pseudomonas*
CDP-paratose	CTP + glucose 1-P ⟶ CDP-glucose	*Salmonella paratyphi*
GDP-fucose	GTP + mannose 1-P ⟶ GDP-mannose	*E. coli*
CMP-Ac-neuraminic acid*	UTP + Ac-glucosamine 1-P ⟶ UDP-Ac-glucosamine	*E. coli*
CMP-Ac-neuraminic acid	Ac-glucosamine 6-P ⟶ Ac-mannosamine 6-P	*E. coli*
UDP-Ac-glucosamine	Fructose 6-P + glutamine ⟶ glucosamine 6-P	*E. coli*
UDP-xylose	UDP-glucose ⟶ UDP-glucuronic acid	*E. coli*

* Ac = N-acetyl.

the cell. Whatever its origin, it can then be reduced to L-gulonic acid. In plants and in animals other than primates and guinea pigs, the L-gulonic acid thus formed is employed for synthesis of ascorbic acid (Chap. 49).

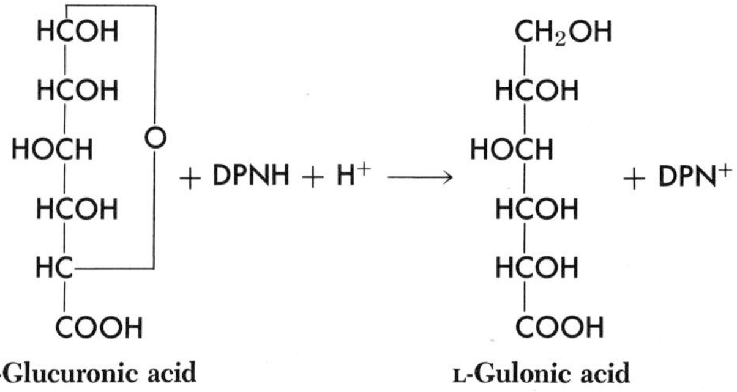

In all mammals, gulonic acid may be oxidized to L-xylulose. However, it is D-xylulose 5-phosphate which participates in the reactions of the phosphogluconate oxidative pathway described earlier (Chap. 18). As shown below, L-xylulose may be converted to D-xylulose by reduction to xylitol and reoxidation.

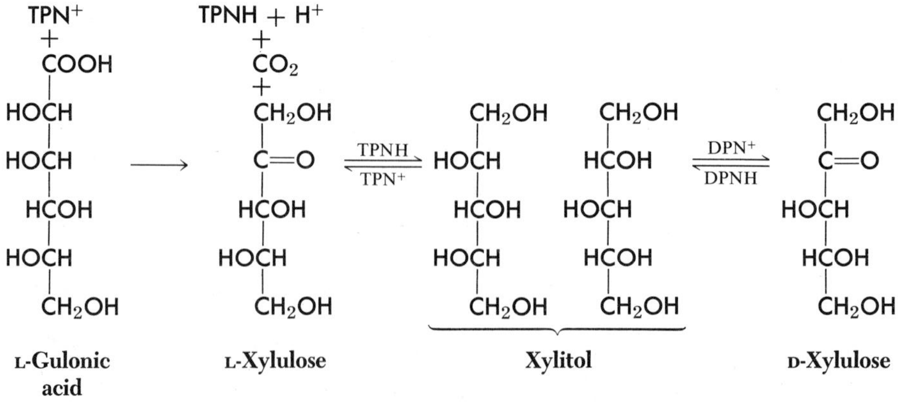

Presumably, D-xylulose may be phosphorylated by ATP and an appropriate kinase to D-xylulose 5-phosphate and then enter reactions of the phosphogluconate oxidative pathway. In consequence, this represents yet another "shunt" pathway for oxidation of glucose, bypassing the reactions of anaerobic glycolysis and the citric acid cycle. Operation of this system yields an equal mixture of TPNH and DPNH. The extent to which these reactions proceed has not been evaluated. The CO_2 arising in this process represents carbon-6 of glucose in contrast to the phosphogluconate oxidative pathway in which CO_2 arises from what had been carbon-1 of glucose.

An abnormality known as *idiopathic pentosuria* is probably attributable to the hereditary absence of the enzyme that reduces L-xylulose. As a consequence, large amounts of L-xylulose are found in the urine. In such individuals the feeding of D-glucuronic acid results in massive excretion of L-xylulose.

Polyols. Although many polyols are metabolized in bacteria, such compounds have a restricted role in mammalian metabolism. In seminal vesicles, reduction of glucose by TPNH to sorbitol with reoxidation of the latter results in formation of fructose (page 390), which is present in high concentration in seminal plasma. Reduction of other sugars in a similar manner must occur, although to a limited degree, in many tissues. Thus dulcitol, the product expected from reduction of galactose, has been found in significant quantity in the lens of galactose-fed rats.

The cyclic *myo*-inositol is found abundantly in plants as its hexaphosphate ester (page 32), and the presence of phosphatidyl inositol in brain (Chap. 37) indicates the importance of inositol in mammals. Except for the demonstration that inositol can be made from glucose in germ-free animals, little is known about the origin of this compound. In yeast, inositol synthesis is initiated with glucose 6-phosphate and requires participation of DPN. Its degradative metabolism appears to begin by oxidation to glucuronic acid, catalyzed by *inositol oxygenase* (page 378).

BIOSYNTHESIS OF GLYCOSIDES

Glycoside synthesis occurs in all living cells since these compounds are universally utilized as a form of energy storage, as constituents of cell membranes, and, among plants, in the formation of cell walls. The hydrolysis of a simple glycoside, such as maltose, proceeds with a free-energy change, $\Delta F° = -4000$ cal. per mole. Hence, formation of the glycosidic bond can occur only when the requisite energy is provided. Three general mechanisms appear to account for all known instances of glycoside synthesis. These mechanisms may first be illustrated by consideration of the synthesis of disaccharides.

Synthesis of Disaccharides. The enzyme *sucrose phosphorylase,* isolated from *Pseudomonas saccharophilia,* which had been grown in a sucrose-containing medium, was initially observed to catalyze the following reaction.

$$\textbf{Sucrose} + \textbf{P}_i \rightleftharpoons \alpha\textbf{-D-glucose 1-phosphate} + \textbf{D-fructose}$$

In the organism, formation of the monosaccharides makes them available for the metabolism of the cell. Indeed this appears generally to be true; although phosphorylases catalyze the following reversible process,

$$\textbf{Sugar-phosphate} + \textbf{sugar} \rightleftharpoons \textbf{glycoside} + \textbf{P}_i$$

in cells in which phosphorylases are operative, it is the phosphorolytic cleavage of the preexisting glycoside which is metabolically significant. However, the reverse process, disaccharide formation, is of particular interest in the present context. The glycosidic bond between the two anomeric carbon atoms in sucrose is unusual, and ΔF° for sucrose hydrolysis is -6600 cal. per mole, whereas that for glucose 1-phosphate is -4800 cal. per mole. Hence, equilibrium favors phosphorolysis of sucrose as shown above. If, however, the orthophosphate is removed from the solution, as by precipitation, sucrose formation from glucose 1-phosphate and fructose is readily demonstrated. Since glucose 1-phosphate arises in the phosphoglucomutase reaction (page 424) from glucose 6-phosphate and the latter is formed from glucose and ATP in the hexokinase reaction (page 389), the energy for formation of the glycosidic bond derives ultimately from the energy of ATP.

A second mechanism for sucrose synthesis has been observed with extracts of various plants.

(*a*) **UDP-glucose + fructose 6-phosphate $\rightleftharpoons$ sucrose 6′-phosphate + UDP**

(*b*) **Sucrose 6′-phosphate $\xrightarrow{H_2O}$ sucrose + P_i**

Since ΔF° for hydrolysis of UDP-glucose is about -7500 cal. per mole, formation of the glycosidic bond of sucrose is favored.

In every instance known, nucleotide sugars are the immediate precursors for glycoside biosynthesis. For most glycosides, formation from a nucleotide sugar precursor would proceed with a favorable free-energy change of about -3500 cal. per mole. In the case of sucrose synthesis, the formation of sucrose phosphate by reaction (*a*) is favored by only -1000 cal. but subsequent hydrolysis of the 6′-phosphate is essentially irreversible and thus assures sucrose formation.

Many instances of glycoside formation from nucleotide sugar precursors have been demonstrated; *e.g.*, lactose formation in mammary glands occurs as follows.

UDP-galactose + glucose $\longrightarrow$ lactose + UDP

This reaction is catalyzed by the concerted action of a pair of proteins. One of these, which exhibits a marked specificity for binding UDP-galactose, had formerly been known as the "lactalbumin" of milk and is structurally homologous to lysozyme (page 269). The regulatory role of this protein, α-lactalbumin, has been discussed (page 246).

The third mechanism for glycosidic bond formation is transglycosylation, a process that also occurs with sucrose phosphorylase. In addition to the reaction involving glucose 1-phosphate shown above, this enzyme catalyzes reactions of the following type (page 267).

Sucrose + L-sorbose $\rightleftharpoons$ D-glucosido-L-sorboside + fructose

Sucrose + L-arabinose $\rightleftharpoons$ D-glucosido-L-arabinoside + fructose

The enzyme appears to cleave the glycosidic bond of sucrose, forming free fructose and an enzyme-glucose compound that can react with P_i to form glucose phosphate, with fructose to form sucrose, or with various other monosaccharides to form disaccharides. The enzyme-glucose compound has been isolated but its structure has not been established. Thus the energy of the glucosidic bond of sucrose is utilized for formation of each of the other glucosides. As will be seen subsequently, transglycosylation is rarely employed, physiologically, for oligosaccharide synthesis, but is widely utilized in polysaccharide formation.

Synthesis of Other Glycosides. In many glycosides, the glucosidic bond links a mono- or disaccharide to a phenol, alcohol, or amine. Most of these are formed enzymically in a reaction between a nucleoside diphosphate sugar and the aglycone.

In mammals, a variety of phenols and alicyclic alcohols are "conjugated" with glucuronic acid and the resultant *glucosiduronide* (glucosiduronic acid) is excreted in the urine. This conjugation is catalyzed by at least two groups of microsomal enzymes of liver, with varying but broad specificity for the aglycone moiety. One group of these enzymes apparently catalyzes O-glucosiduronide formation, the other N-glucosiduronide synthesis.

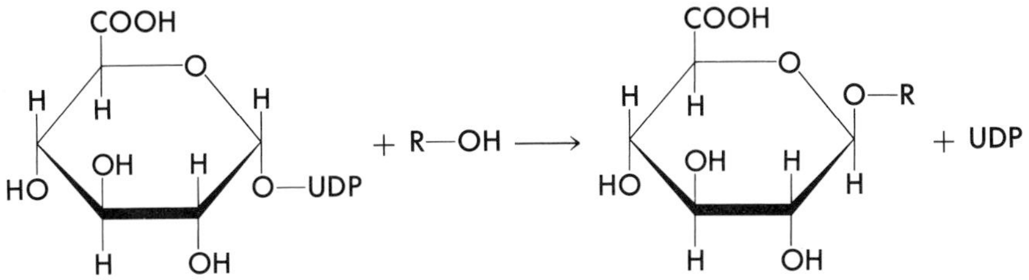

UDP-glucuronic acid A glucosiduronide

Plants do not form glucosiduronides but contain many examples of phenolic glucosides. Thus, wheat germ utilizes uridine diphosphate glucose to form β-D-glucosyl hydroquinone (arbutin). The latter can serve as substrate for a second enzyme which couples a second glucose molecule, from uridine diphosphate glucose, to the first, to form a gentiobioside [O-β-D-glucosyl(1 $\rightarrow$ 6)-O-α-D-glycosyl]. In other instances it is rhamnose (page 33) which is coupled to the aglycone, using either uridine or thymidine diphosphate rhamnose. The latter has also been found to be the rhamnose donor in the formation of the rhamnosyl derivatives of hydroxy fatty acids in some bacteria. A similar reaction, using uridine diphosphate galactose, results in formation of galactose-containing cerebrosides (Chap. 22).

Synthesis of Oligosaccharides. Specific trisaccharides are synthesized both by transglycosylation and from nucleotide sugars. Thus the *dextran sucrase* of *L. mesenteroides* (see below) catalyzes transfer of α-D-glucose from sucrose to lactulose (O-β-D-galactose 1 $\rightarrow$ 4 fructose).

$$\text{Sucrose} + \text{lactulose} \longrightarrow \text{galactosylsucrose} + \text{fructose}$$

Examples of transfer from nucleotide sugars include the following:

$$\text{GDP-fucose} + \text{lactose} \xrightarrow[\text{gland}]{\text{mammary}} \text{fucosyl } 1 \rightarrow 2 \text{ lactose} + \text{GDP}$$

$$\text{CMP-sialic acid} + \text{lactose} \xrightarrow[\text{gland}]{\text{mammary}} \text{sialyllactose} + \text{CMP}$$

$$\text{UDP-galactose} + \text{sucrose} \xrightarrow{\text{seeds}} \text{raffinose} + \text{UDP}$$

POLYSACCHARIDE BIOSYNTHESIS

Polysaccharide formation is accomplished by the same types of reaction evident in the formation of disaccharides and glycosides, *i.e.*, from nucleoside diphosphate sugars or from preexisting glycosidic structures.

Transglycosylation. Many bacteria and some plants can effect the synthesis of a linear polysaccharide by utilizing a disaccharide. Only the monosaccharide unit that contributes its anomeric carbon to the glycosidic bond of the disaccharide can be utilized for polysaccharide synthesis. Some *transglycosylases* and their bacterial origin are indicated in the reactions below.

$$n\textbf{Maltose} \underset{\textit{Escherichia coli}}{\overset{\text{amylomaltase}}{\rightleftharpoons}} \textbf{amylose} + n\textbf{glucose}$$

$$n\textbf{Sucrose} \underset{\textit{Neisseria perflava}}{\overset{\text{amylosucrase}}{\rightleftharpoons}} \textbf{amylose} + n\textbf{fructose}$$

$$n\textbf{Sucrose} \underset{\textit{Leuconostoc mesenteroides}}{\overset{\text{dextransucrase}}{\rightleftharpoons}} \textbf{dextran} + n\textbf{fructose}$$

$$n\textbf{Sucrose} \underset{\textit{Bacillus megatherium}}{\overset{\text{levansucrase}}{\rightleftharpoons}} \textbf{levan} + n\textbf{glucose}$$

Dextran is a linear polymer in which the anomeric carbon-1 of each glucose unit is in glycosidic linkage with the primary alcoholic hydroxyl at carbon-6 of the adjacent residue. Thus the repeating unit is isomaltose (page 46) rather than maltose. *Levan* is similarly built of fructose residues in 2,6-fructosidic linkage. Since the energy of the glycosidic bond of the disaccharides is conserved in the polysac-

Table 19.3: POLYSACCHARIDE SYNTHESIS FROM SOME NUCLEOSIDE DIPHOSPHATE SUGARS

Precursor	Product	Biological source*
	HOMOPOLYSACCHARIDES	
UDP-glucose	β-1,3 Glucan (callose)	Bean extracts
GDP-glucose	β-1,4 Glucan (cellulose)	Mung bean extracts
ADP-glucose	α-1,4 Glucan (starch amylose)	Wheat germ
ADP-glucose	α-1,4 Glucan (glycogen amylose)	Diverse bacteria
UDP-glucose	α-1,4 Glucan (glycogen amylose)	Liver
UDP-xylose	β-1,4 Xylan	Plants
CMP-N-acetylneuraminic acid†	Colominic acid	*Neurospora crassa*
UDP-N-acetylglucosamine	Chitin	Insects
	HETEROPOLYSACCHARIDES	
UDP-glucuronic acid + UDP-N-acetylglucosamine	Hyaluronic acid	Rous sarcoma, streptococci
UDP-glucose + UDP-glucuronic acid	Capsular polysaccharide	Type III pneumococci

* The biological source is that which has been experimentally employed.
† A nucleoside monophosphate sugar.

charides, these organisms need not furnish energy for polysaccharide synthesis if the appropriate disaccharide is available as precursor.

Synthesis of Homopolysaccharides from Nucleoside Diphosphate Sugars. In plants as well as animals, most polysaccharides are synthesized directly from hexose and pentose units rather than from disaccharides. In all instances presently known, the immediate precursor of the final polysaccharide is a nucleoside mono- or diphosphate sugar. Some of the better-known examples are listed in Table 19.3. The enzymes that catalyze formation of cellulose, amylose, and chitin appear to require the presence of either a trace of the polysaccharide or of a low molecular weight dextrin derived from the polysaccharide synthesized. These dextrins serve as "primers," in that additional hexose units are added to the priming dextrin, as acceptors, thus extending the polysaccharide chain. The synthesis of glycogen and that of starch are considered in more detail below.

All the monomer units listed in Table 19.3 are nucleoside diphosphate derivatives of either a pentose or a hexose, with the exception of CMP-N-acetylneuraminic acid, which functions in the synthesis of colominic acid, which occurs in certain strains of *E. coli,* and in mammalian glycoprotein synthesis.

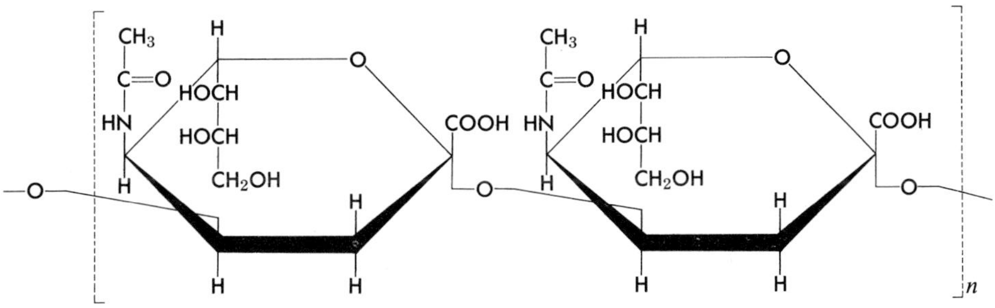

Repeating unit of colominic acid

Synthesis of Heteropolysaccharides. Little is known of the mechanisms involved in formation of heteropolysaccharides except that they proceed from nucleoside diphosphate derivatives of appropriate monosaccharides. It is surmised that the substrate utilized by the final "synthetase" is probably the nucleoside diphosphate derivative of the oligosaccharide repeating unit. This would make possible the synthesis of hyaluronic acid and other mucopolysaccharides of connective tissue (Chap. 38) as well as the simpler bacterial polysaccharides such as the antigenically specific polysaccharide of type III pneumococci (Table 19.3). Alternatively, the enzyme catalyzing synthesis might have two different binding sites, or synthesis could involve two enzymes of selective specificity.

It should be recognized that neither hetero- nor homopolysaccharides actually terminate in a reducing aldose or ketose unit. In each carefully studied instance, the potential reducing group of the chain is linked to some other molecular species. Some instances of this principle are listed in Table 19.4. The significance of such structures is further discussed in Chaps. 38 and 41.

Table 19.4: TERMINAL UNITS OF SOME POLYSACCHARIDES

Polysaccharide	*End group of chain*
Dextran	Fructose
Levan	Glucose
Cellulose	Lipid (unidentified)
Hemicelluloses	Phenols
Schardinger dextrin	Cyclic polysaccharide
Blood group substances	Glycoside of threonyl; ester of aspartic acid of protein
Hyaluronic acid	Lipid (unidentified)
Chondroitin sulfates, heparin	Glycoside of threonyl or seryl of protein
Keratosulfate	Glycoside of threonyl or seryl of protein
Glycoproteins	O-Glycosides of serine, threonine of protein
	N-Glycosides of asparagine, glutamic acid of protein
Bacterial lipopolysaccharide	Phosphoethanolamine

GLYCOGEN METABOLISM AND THE CONTROL OF CARBOHYDRATE METABOLISM

The metabolism of glycogen is subject to controls which permit regulation of both its synthesis and its degradation. These processes are accomplished by independent enzyme systems which catalyze the following over-all reactions.

$$\textbf{UDP-glucose} \xrightarrow[\text{transglycosylase}]{\text{transferase}} \textbf{glycogen} \xrightarrow[\substack{\text{glucan transferase,}\\ \text{glucosidase}}]{\text{phosphorylase}} \textbf{glucose 1-phosphate} + \textbf{glucose}$$

Uridine diphosphate glucose-glycogen glucosyl transferase (referred to as either "transferase" or "synthetase") and *glycogen phosphorylase* both exist in two different forms. Each form of both enzymes is capable of catalyzing its specific reaction, but under different conditions. Interconversion of the two forms of each of the enzymes is enzymic. Regulation of the synthetic and degradative processes is effected both by allosteric control of the primary enzymes and of the enzymes which catalyze the interconversions of the two forms of both the synthetase and the phosphorylase.

Glycogen Synthesis. Synthesis of the amylose chain of glycogen is catalyzed by *glycogen synthetase,* an enzyme discovered by Leloir that catalyzes the following reaction.

$$n\textbf{UDP-glucose} \rightleftharpoons \textbf{(glucose)}_n + n\textbf{UDP}$$

The free-energy change in this reaction is about −3200 cal. per mole of glucose equivalent; equilibrium, therefore, favors glycogen synthesis by a factor of about 250. Although UDP-glucose is undoubtedly the normal substrate, synthesis with ADP- and TDP-glucose proceeds at about 50 and 5 per cent, respectively, of the rate observed with UDP-glucose. The enzyme cannot form glycogen from UDP-glucose alone. Rather, the glucose moiety of the latter is transferred to an acceptor polyglucose chain. The preferred acceptor is glycogen itself. However, amylose, amylopectin, or a glucose oligosaccharide no smaller than the four-membered maltotetraose can also serve as acceptors.

Glycogen synthetase catalyzes formation only of α-1,4 bonds. The product formed with maltotetraose as acceptor, therefore, is amylose, the linear α-1,4 polymer of D-glucose. The enzyme has a marked affinity for glycogen and, in liver, is firmly bound to glycogen particles. Glycogen synthetase is stabilized against heat and alkaline denaturation by glucose 6-phosphate. The enzyme occurs in a phosphorylated and in a dephosphorylated form. The dephosphoenzyme, which is enzymically active, can be phosphorylated by the specific *glycogen synthetase kinase,* using ATP, to form the phosphoenzyme. The latter is less active but is markedly stimulated by glucose 6-phosphate, which increases the V_{max} of rabbit muscle phosphoenzyme about fiftyfold while slightly reducing K_m for UDP-glucose. Indeed, dog muscle phosphoenzyme appears to be totally inactive in the absence of glucose 6-phosphate. The phosphoenzyme is strongly inhibited by free UDP which could accumulate when the supply of both glucose and ATP is limited. In view of the effects of glucose 6-phosphate, the phospho and dephospho forms of the enzyme are denoted as D (dependent) and I (independent), respectively. Thus, the properties of the enzyme assure glycogen formation under appropriate circumstances.

Glycogen synthetase kinase can also exist in active and inactive forms. The inactive form is activated by low concentrations of 3′,5′-AMP (cyclic adenylic acid).

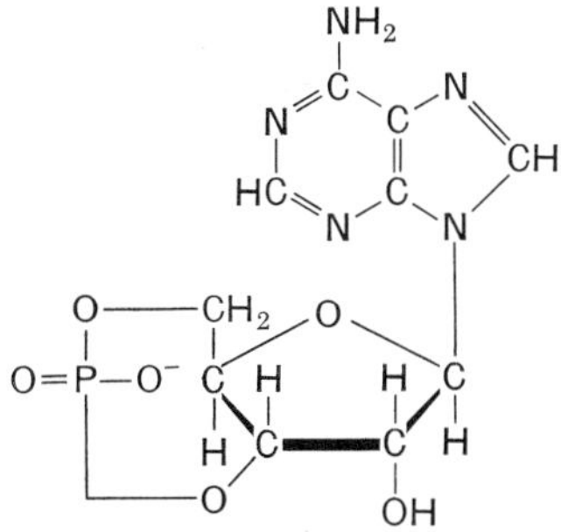

Adenosine 3′,5′-phosphate (cyclic adenylic acid)

Cyclic adenylic acid can, in turn, be formed from ATP by an enzyme system that can be activated by epinephrine. Consequently, during sudden stress the action of epinephrine is to effect the transformation of I to D glycogen synthetase so that glycogen formation may occur only if the glucose 6-phosphate concentration is sufficiently high both to assure formation of glucose 1-phosphate and to activate the synthetase. In muscle, the kinase may also be activated by a "protein factor" in the presence of an elevated [Ca^{++}]. As described on page 852, intramuscular [Ca^{++}] increases during contraction. *Glycogen synthetase phosphatase,* which catalyzes hydrolysis of the phosphoenzyme and hence is responsible for the conversion of synthetase D to synthetase I, is inhibited by glycogen itself. Accordingly, when heart or skeletal muscle glycogen is depleted, form I is dominant; as the glycogen concentration increases, the relative fraction of synthetase in the D form increases.

As noted, glycogen synthetase catalyzes formation only of α-1,6 bonds. However, glycogen has a highly branched structure, the branches resulting from the presence of α-1,6 bonds at a frequency of about every 8 to 12 glucose units. This structure is the result of the activity of a "glycogen branching enzyme," *amylo*

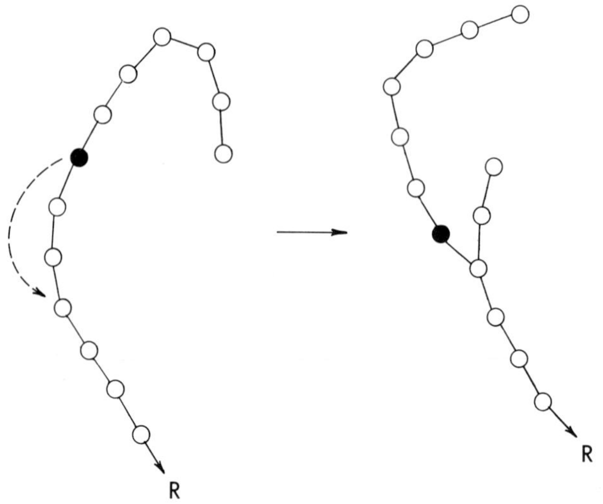

Fig. 19.2. The action of amylo-(1,4 → 1,6)-transglucosylase. *R* represents the main body of the glycogen molecule: each circle represents a glucose unit. The darkened circle represents the glucose molecule whose aldehyde carbon is transferred from 1,4 to 1,6 linkage.

(1,4 → 1,6)-transglucosylase (Fig. 19.2). This enzyme has been demonstrated in liver, muscle, and brain. It is a transglucosylase which removes terminal fragments of 6 or 7 glucose units from the main chain, or from the ends of major branches of the glycogen chain, at α-1,4 linkage, and transfers them to the same, or another, glycogen molecule but in an α-1,6 linkage, thereby creating new branches. It is the "specificity" of this "branching enzyme" which determines the interbranch distance along the polysaccharide chain.

Glycogen Phosphorolysis. In contrast to the hydrolysis of glycogen and starch in the gastrointestinal tract, within cells glycogen is degraded to α-glucose 1-phosphate by the action of *glycogen phosphorylase,* discovered by C. F. and G. T. Cori.

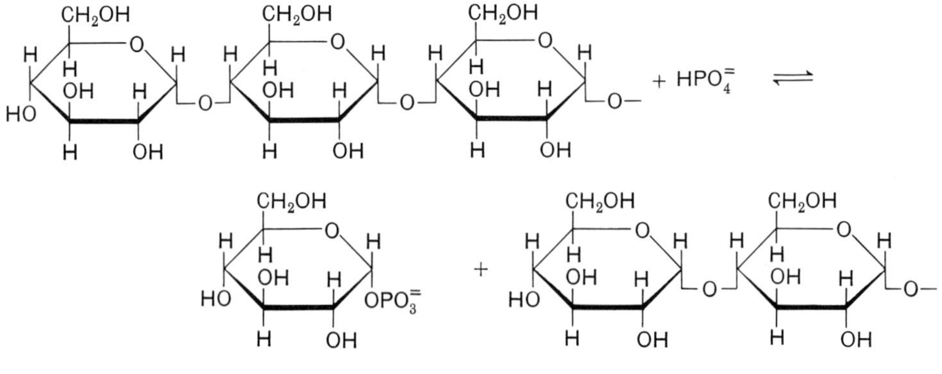

The equilibrium constant for this reaction is given by

$$K = \frac{[(C_6H_{10}O_5)_{n+1}][HPO_4{}^{=}]}{[(C_6H_{10}O_5)_n][\text{glucose 1-phosphate}]}$$

However, since each molecule of polysaccharide is both a reactant and a product, K is determined by the ratio of orthophosphate to glucose 1-phosphate. Since glucose 1-phosphate is a stronger acid than $HPO_4^=$, the value of this ratio is pH-dependent, being close to 3 at pH 7. The ready reversibility of the reaction is consonant with a value of K close to unity. The small $\Delta F°$ for the reaction is in accord with the fact that the $\Delta F°$ for hydrolysis of glucose 1-phosphate (−4800 cal. per mole) differs only slightly from that of a maltosidic bond (−4200 cal. per mole).

Phosphorylase attacks the glycogen molecule from the terminus of each chain, releasing successive glucose residues, marked ⊙ (Fig. 19.3), as glucose 1-phosphate until about four residues remain on each branch before the branch point. Activity then ceases. This results in a limit dextrin similar to that which remains after treatment of glycogen with β-amylase (page 50). A second enzyme, α-1,4 → α-1,4 glucan transferase, then transfers the trisaccharide attached to the glucose in 1,6

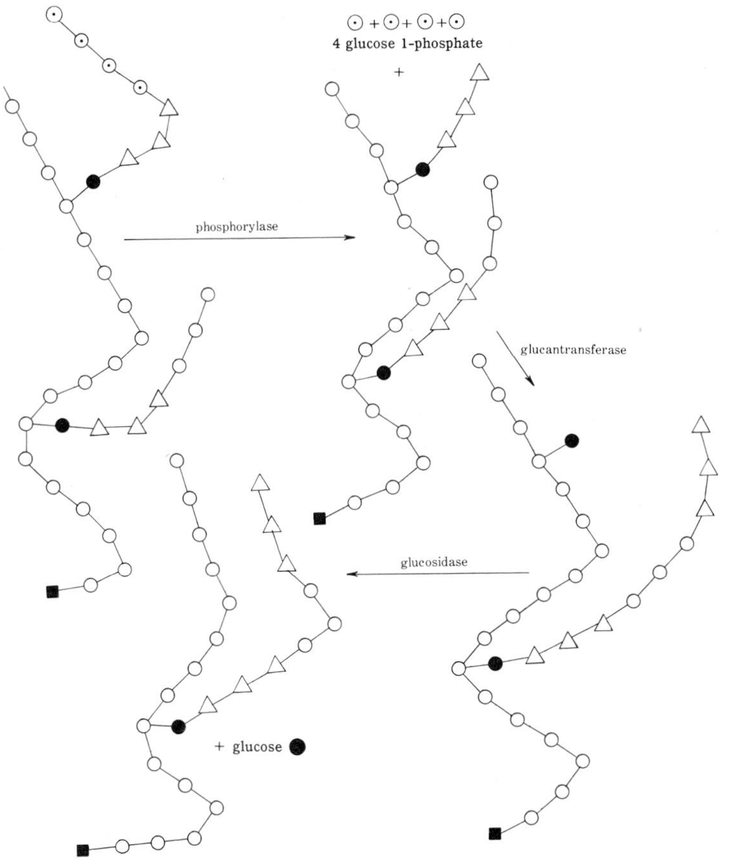

Fig. 19.3. Intracellular degradation of glycogen. Although phosphorylase can operate on all available branches of glycogen to form a limit dextrin, this diagram follows the fate of a single branch. ■ = terminal glucose with potential reducing group; ● = glucose at branch point linked 1 → 6 to the main chain; ⊙ = glucose residues removed as glucose 1-phosphate by phosphorylase; △ = three glucose units linked to glucose at the branch point.

linkage at the branch point to a different chain, extending it correspondingly and leaving a single glucose residue at the branch point. Hydrolytic removal of the glucose present in 1,6 linkage at the branch point by *amylo-(1,4 → 1,6)-transglucosylase* then permits phosphorolysis to continue until the next branch point is reached, etc.

Glycogen Phosphorylase. Muscle glycogen *phosphorylase a,* molecular weight 500,000, consists of four apparently identical polypeptide chains, each of which contains a serine residue whose hydroxyl group is esterified to phosphate, and a lysine residue, the amino group of which is present as the Schiff base (—N=CH—) of pyridoxal phosphate (page 542). The function of the pyridoxal phosphate is not understood; it is very tightly bound but may be removed by treatment with cysteine, resulting in inactivation. The serine residue bearing the phosphate ester is present in a peptide sequence which is identical with that containing the phosphorylated serine residue of glycogen synthetase, *viz.,*

```
-Glu-Ile-Ser-Val-Arg-
          |
          P
```

It remains to be established whether these enzymes derive from a common ancestral form.

Muscle also contains *phosphorylase phosphatase,* which catalyzes removal of the phosphate groups that are esterified to the serine residues of phosphorylase. This results in deaggregation of the enzyme to a dimeric form called *phosphorylase b.* The latter is enzymically inactive but, in the presence of AMP which serves as a positive allosteric modifier, it can function in the glycogen phosphorylase reaction. Conversion of phosphorylase b to phosphorylase a is accomplished by *phosphorylase kinase.*

$$\text{2 Phosphorylase b} + \text{4ATP} \longrightarrow \text{phosphorylase a} + \text{4ADP}$$

Phosphorylase kinase is stimulated by the presence of glycogen and has little activity in its absence. In turn, phosphorylase kinase also exists in an active phosphorylated and an inactive dephospho form. Phosphorylation of the inactive enzyme by ATP occurs in the presence of Mg^{++} and yet another kinase (*phosphorylase kinase kinase*) which requires activation by low levels ($10^{-8}M$) of 3′5′-cyclic adenylic acid. As in the case of glycogen synthetase, activation of *phosphorylase b kinase* can also be effected by an unidentified "protein factor" which is only active at elevated [Ca^{++}].

The fact that the systems for interconverting the two forms of glycogen synthetase and phosphorylase are subject to similar controlling factors make possible the delicately balanced regulation of muscle glycogen metabolism, some of the salient features of which are shown in Table 19.5. Resting muscle readily converts glucose to glycogen and, since [ATP] is high while [AMP] is low, glycogen phosphorolysis is minimal. Contraction, by utilizing ATP and generating AMP (page 855), reverses these relationships. Available glucose 6-phosphate enters the glycolytic pathway, leading to ATP synthesis, and there is a demand for production, from glycogen, of additional glucose 6-phosphate for this function. The increase in [AMP], decrease in [ATP], and decrease in [glucose 6-phosphate] permit operation of phosphorylase b. Epinephrine induces formation from ATP of 3′5′-cyclic AMP, which activates phosphorylase kinase kinase. This, in turn, activates

Table 19.5: THE PRIMARY ENZYMES OF GLYCOGEN METABOLISM

Resting muscle	*Contracting muscle*
Glycogen synthetase I:	Glycogen synthetase D:
Very active	Less active
Independent of glucose 6-P	Dependent on glucose 6-P
	Inhibited by UDP
Glycogen phosphorylase b:	Glycogen phosphorylase a:
Dependent on AMP	Independent of AMP
Inhibited by ATP	Insensitive to ATP
Inhibited by glucose 6-P	Insensitive to glucose 6-P
Less active	Very active

phosphorylase kinase, which is stimulated by the still abundant glycogen and, hence, converts phosphorylase b to a, thereby permitting maximal glycogen phosphorolysis.

Concurrently, the diminished [glucose 6-phosphate] has minimized the activity of existing synthetase D, and form I is converted to D by the 3′5′-cyclic AMP–activated synthetase kinase. When contraction ceases, glucose from the blood is phosphorylated, [glucose 6-phosphate] increases, and synthetase D activity is initiated. The diminished [glycogen] releases transferase phosphatase, and as conversion of synthetase D to I occurs, glycogen synthesis accelerates. Phosphorylase phosphatase converts phosphorylase a to b and, as [ATP] is restored and [AMP] declines, glycogen phosphorolysis declines markedly, permitting restoration of the glycogen stores. Figure 19.4 (page 442) depicts some of the enzymic mechanisms operative in the control of glycogen metabolism.

The control of glycogen phosphorylase in liver appears to be essentially similar to that in muscle. However, the liver and muscle enzymes are under independent genetic control (see Chap. 30) and are immunochemically different. The dephosphorylated form of liver phosphorylase is also enzymically inactive, but both active and inactive forms are identical in molecular weight, which is approximately that of muscle phosphorylase b. Inactive liver dephosphophosphorylase, which has lost two moles of phosphate per mole of enzyme, is not activated by AMP. Reactivation by a kinase and the role of 3′5′-cyclic AMP are identical with those for the enzyme in muscle. However, reactivation of liver phosphorylase is stimulated not only by epinephrine, as in the case for muscle, but also by glucagon (page 444), which is without effect on the muscle system.

Starch Formation. Starch synthesis in plants and glycogen synthesis in bacteria are analogous to that of glycogen in animal cells. The *amylose synthetase* is firmly bound to starch granules. Instead of UDP-glucose, ADP-glucose is preferentially utilized as substrate. This compound is made from ATP and glucose 1-phosphate, catalyzed by a specific pyrophosphorylase. Again, the action of the synthetase results in an amylose structure. Control of starch formation is exercised not by regulation of the amylose synthetase but rather by allosteric activation of *ADP-glucose pyrophosphorylase.* This enzyme, in plants, is stimulated as much as fiftyfold by the 3-phosphoglyceric acid and about tenfold by the fructose diphosphate which arise in photosynthesis (page 455). The branched chains of amylopectin reflect the activity of a "branching enzyme" or *transglycosylase,* analogous to that operative in glycogen synthesis.

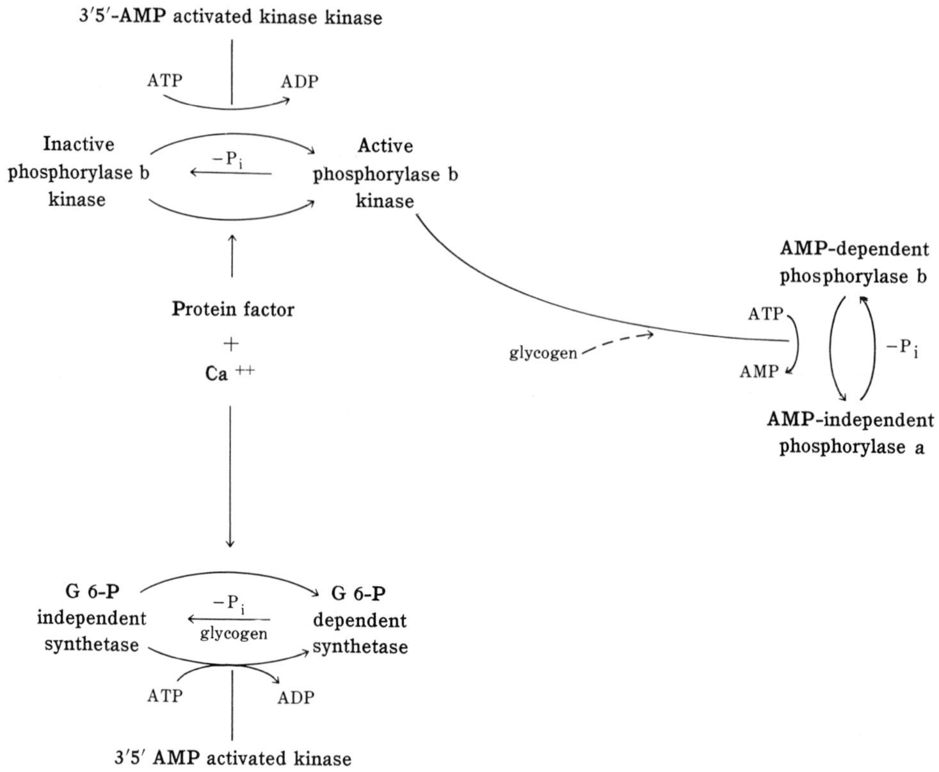

Fig. 19.4. Activation and deactivation of glycogen synthetase and glycogen phosphorylase.

PHYSIOLOGICAL ROLE OF GLYCOGEN

The large stores of chemical energy in adequately nourished tissues have little effect on osmotic pressure, since they are either water-insoluble lipids or sparingly soluble polysaccharides of very high molecular weight. The storage of large amounts of polysaccharides is characteristic of plants, where, with the exception of the seeds, lipid storage is usually scanty. Among higher animals, lipids account for the bulk of energy storage, whereas polysaccharide storage occurs only in the form of relatively small amounts of glycogen. Although glycogen occurs in most tissues, including the cells of adipose tissue, the glycogen of liver and of skeletal muscle has attracted most attention.

Glycogenesis and Glycogenolysis. Formation of glycogen is termed *glycogenesis;* nutrients that enhance the glycogen content of tissues are called glycogenic substances. These include not only the several hexoses previously discussed but also a variety of other compounds, including the glycogenic amino acids (Chap. 26), glycerol, intermediates in glycolysis, and sugar alcohols such as sorbitol and inositol. These and many other substances, when administered to fasting animals, result in some increase in the quantity of liver glycogen. In each instance, the substance in question must first be converted via glucose 6- and glucose 1-phosphate to UDP-glucose.

Glycogenolysis (glycogen breakdown) is hydrolytic in the intestine, but only glycogen phosphorolysis is significant within cells. The product, glucose 1-phosphate, is converted to glucose 6-phosphate by the phosphoglucomutase reaction and thus enters the main pathways of carbohydrate metabolism. Liver contains an α-amylase and oligosaccharases, but their activity is insignificant. Glucose formation from glucose 6-phosphate, with release to the circulating blood, is possible only in intestine, liver, and kidney, which contain a microsomal *glucose 6-phosphatase.*

$$\text{Glucose 6-phosphate} \xrightarrow{H_2O} \text{glucose} + P_i$$

The glucose 6-phosphatase activity of liver is enhanced in diabetes, during starvation, and after administration of adrenal cortical steroids (Chap. 45). This appears to reflect a change in the physical structure of the microsomes rather than a direct effect on the enzyme. Figure 19.5 indicates the over-all pathways of glycogenesis and glycogenolysis.

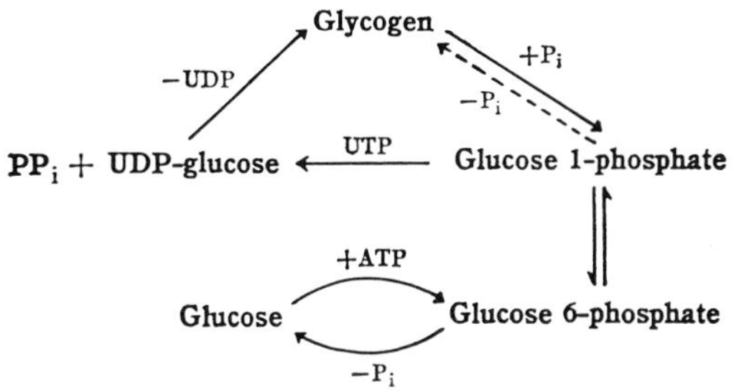

Fig. 19.5. Over-all pathways of glycogenesis and glycogenolysis. The reaction indicated by the dashed arrow, demonstrable in vitro, is not thought to be physiologically significant.

Tissue Glycogen. Glycogen of all tissues is polydisperse. As ordinarily obtained, glycogen has a range of molecular weight that may vary from 2.5×10^5 to 10^7. In addition, in various tissues the amount of glycogen that may be readily extracted, *e.g.*, with cold trichloroacetic acid solution, is always less than the total obtained by alkaline hydrolysis of the tissue. The less readily extractable fraction is present as insoluble granules. However, the behavior of administered glucose-^{14}C indicates that even this glycogen fraction is metabolically active and undergoing constant turnover.

Glycogen of Liver. In various mammalian species, between 2 and 8 per cent of the wet weight of liver is glycogen. Even in the normal animal with constant access to food, although the quantity of liver glycogen remains fairly constant, it is continuously being both formed and degraded. The liver of an animal fasted for 24 hr. is virtually depleted of glycogen. Refeeding with glycogenic materials, *e.g.*, glucose, results in prompt reaccumulation of liver glycogen.

A common procedure for determination of the effectiveness of other compounds as precursors of glucose is to fast animals until virtually no liver glycogen remains, and then administer the test substance. Any material which increases the amount of liver glycogen under these conditions is presumed also to yield glucose 6-phosphate; hence, the terms "glycogenic" and "glucogenic" are frequently synonymous.

Factors Affecting Liver Glycogen. The amount of liver glycogen depends both on the quantity of food consumed and on the composition of the diet. Thus, animals maintained on carbohydrate-poor diets generally have less glycogen than those on high-carbohydrate diets. Exercise reduces the quantity of glycogen in the liver. Experimental poisoning with phlorhizin, which lowers the renal threshold for glucose, thereby causing glucosuria, also leads to diminution of liver glycogen.

As indicated above, liver glycogen is under endocrine regulation. Administration of epinephrine (Chap. 45) or glucagon (Chap. 46) to a well-nourished animal results in prompt disappearance of much of the liver glycogen. Both hormones activate the system for 3′5′-cyclic adenylate formation and hence effect the conversion of phosphorylase b to a, thereby making available glucose 6-phosphate, which is hydrolyzed and added to the blood, with resultant *hyperglycemia.* This may be regarded as a protective mechanism in time of stress, which makes available to the musculature an abundance of nutrient when it may be required.

Insulin deficiency is accompanied by diminution in liver glycogen, which occurs for a variety of reasons which are not well understood. Among them are lack of insulin stimulus to the glycogen synthetase system, reduced acetyl CoA to stimulate the enzymes necessary for phosphoenolpyruvate synthesis in glucogenesis, reduced glucokinase activity, and accelerated glucose 6-phosphatase activity. Administration of excessive insulin might be expected to increase the quantity of liver glycogen, but this does not necessarily occur. Muscle, not liver, glycogen usually increases following insulin injection. The adenohypophysis (Chap. 47) produces one or more substances that tend to increase liver glycogen. The adrenal cortical hormones (Chap. 45), by augmenting the supply of glucose from noncarbohydrate precursors, favor accumulation of liver glycogen. Excessive administration of thyroid hormone (Chap. 43) results in a mobilization and disappearance of liver glycogen.

Muscle Glycogen. Although the normal concentration of glycogen in mammalian skeletal muscle, 0.5 to 1 per cent, is lower than that in liver, because of the large mass of muscle most of the total glycogen of the body is in this tissue. In contrast to liver glycogen, muscle glycogen is not readily depleted by fasting, even over prolonged periods. Convulsions, however, result in a dramatic decrease in muscle glycogen. Insulin administration generally enhances muscle glycogen concentration, but if the dosage is adequate to provoke hypoglycemic convulsions, the muscle glycogen content may fall to very low levels.

Epinephrine administered to a fasted animal activates the formation of cyclic adenylate in muscle as well as liver, thereby secondarily enhancing phosphorylase activity. However, glucose 6-phosphate cannot be hydrolyzed to glucose in muscle, since the requisite phosphatase is lacking. Further metabolism of glucose phosphate, *viz.*, glycolysis, ensues, and the first readily diffusible products are pyruvic and lactic acids. Rapid glycogenesis from these substances occurs in the liver. Thus, the sequence of events that follows epinephrine administration to the fasting animal is (1) a fall in muscle glycogen, (2) a rise in blood lactate, and (3) a rise in liver glycogen. Cardiac muscle, in contrast to voluntary muscle, is insensitive to insulin and the quantity of heart glycogen may vary in the opposite direction from that in other muscles.

Hereditary Disorders of Glycogen Metabolism. Mutations that lead to impairment of the catalytic activity of enzymes which function in obligatory metabolic pathways (glycolysis, citric acid cycle, protein synthesis, etc.) are necessarily lethal. Mutations may be detected in pathways which are useful but not absolutely vital. Such an alteration may cause disease but is not necessarily incompatible with continued life. The metabolism of glycogen affords an excellent example of this concept; hereditary disorders based on lack of all the enzymes of glycogen metabolism except glycogen synthetase itself have now been observed. These are summarized in Table 19.6. All are characterized by accumulation of tissue glycogen. Lack of liver glucose 6-phosphatase prevents response to the demand for addition of glucose to the blood. Lack of debranching or branching enzyme results in accumulation of glycogens with abnormal branch length. The independent failure to make liver and muscle phosphorylases indicates that these enzymes are under separate genetic control.

BLOOD GLUCOSE AND REGULATION OF GLUCOSE METABOLISM

The continual utilization of glucose requires its delivery to all tissues by the blood. The normal concentration of glucose in human blood, 8 to 12 hr. after a meal, is 70 to 90 mg. per 100 ml. Slightly higher values are obtained immediately postprandially; prolonged fasting results in little or no decline in blood glucose concentration for several days.

The dependence of various tissues on circulating blood glucose varies widely. The central nervous system is perhaps most critically dependent, since glucose is the major energy source that crosses the blood-brain barrier (Chap. 37) at a rate sufficient to sustain normal function. If the blood glucose concentration falls abruptly, the earliest symptoms observed are referable to the central nervous system. Many tissues, such as muscle, can derive a considerable portion of their chemical energy from other nutrients, such as ketone bodies (page 503), and hence are not so critically dependent on a sustained blood glucose concentration. The myocardium effectively removes fatty acids and lactic acid from the blood and utilizes them as sources of energy. This capacity of heart muscle renders it relatively insensitive to fluctuations in the level of blood glucose and confers upon the heart a great degree of adaptability to variations in composition of the nutrient medium.

From the study of the normal fluctuations and of the mechanisms operating to offset these fluctuations has grown some understanding of the diseases that result in abnormal variations in the level of blood glucose. The relative constancy of the normal blood glucose concentration, despite various disturbing factors, is an example of "homeostatic" regulation. The term *homeostasis* was coined by Cannon to describe the reaction of the body to stimuli which, by altering the concentration of some constituent in the body, initiate a series of events that tend to restore this concentration to normal. Homeostatic mechanisms are the physiological counterparts of the "inverse feedback" mechanisms of engineering.

Sources and Fate of Blood Glucose. Absorption of the products of carbohydrate digestion represents a large though variable contribution of glucose to the blood. This is a discontinuous process since most of the digestible carbohydrate is absorbed

Table 19.6: **Hereditary Disorders of Glycogen Metabolism**

Type	Enzymic defect	Glycogen structure	Organ	Eponymic name	Suggested clinical name
1	Glucose 6-phosphatase	Normal	Liver, kidney, intestine(?)	von Gierke's disease	Glucose 6-phosphatase–deficiency hepatorenal glycogenosis
2	Amylo-α-1,4-glucosidase	Normal	Generalized	Pompe's disease	α1,4 Glucosidase–deficiency generalized glycogenosis
3	α-1,4 $\longrightarrow$ 1,4-glucan transferase	Abnormal; outer chains missing or very short	Liver, heart, muscle, leukocytes	Forbes' disease	Debrancher deficiency limit dextrinosis
4	Amylo-(1,4 $\longrightarrow$ 1,6)-transglucosylase	Abnormal; very long inner and outer unbranched chains	Liver, probably other organs	Andersen's disease	Brancher deficiency amylopectinosis
5	Muscle glycogen phosphorylase	Normal	Skeletal muscle	McArdle-Schmid-Pearson disease	Myophosphorylase deficiency glycogenosis
6	Liver glycogen phosphorylase	Normal	Liver, leukocytes	Hers' disease	Hepatophosphorylase deficiency glycogenosis

Source: Adapted from J. B. Stanbury, J. B. Wyngaarden, and D. S. Fredrickson, "The Metabolic Basis of Inherited Disease," 2d ed., McGraw-Hill Book Company, New York, 1966.

within a few hours after ingestion of a high-carbohydrate meal. The continuing source of blood glucose is the hydrolysis in liver, kidney, and intestine of glucose 6-phosphate, which derives either from glycogenolysis or from other potential precursors. The formation of glucose from all precursors other than glycogen is described as *glucogenesis.* The term *gluconeogenesis* is occasionally used to designate glucose formation from noncarbohydrate precursors. Any compound that can be converted into one of the intermediates of glycolysis is potentially glucogenic, *e.g.*, many of the amino acids (Chap. 26), glycerol, and a variety of quantitatively less important substances.

The diverse fates of blood glucose are summarized in Fig. 19.6. In the normal animal, no more than a trace of glucose is lost in the urine, regardless of diet. Under various abnormal circumstances, however, urinary loss of glucose may approximate the sum of all the glucose ingested plus such additional glucose as may arise from glucogenic materials of the diet. The mechanisms which contribute to the abnormal appearance of glucose in the urine, *glucosuria,* are described in Chap. 35.

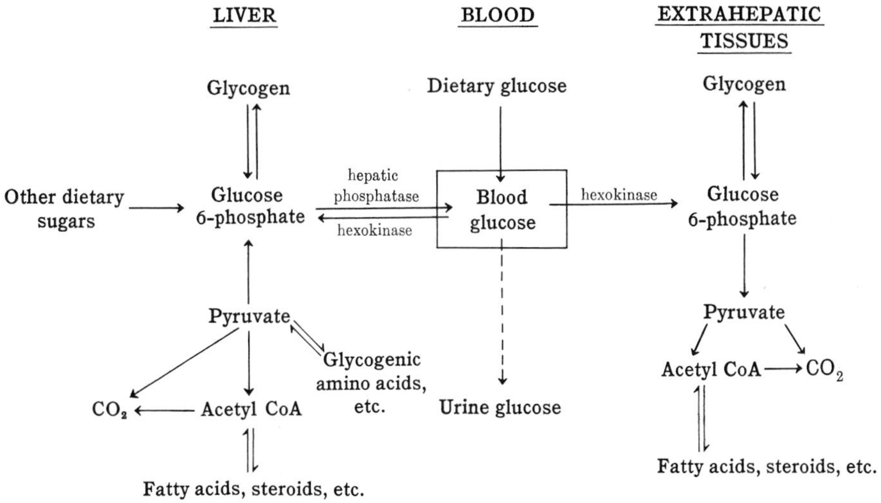

Fig. 19.6. Sources and fates of blood glucose.

Factors Influencing Blood Glucose Concentration. Figure 19.6 relates certain of the major sources and fates of blood glucose. The concentration of glucose in the blood is the resultant of the relative rates of glucose production from glycogen, amino acids, and other sources, glucose absorption from the intestinal tract, glucose utilization, and glucose loss in the urine. The rate of glucose removal from the blood is governed in part by the concentration of glucose in the extracellular compartment; the higher the concentration of glucose, the more rapid is its assimilation by tissues such as muscle and liver. Contraction of muscle is accompanied by enhanced movement of glucose into muscle cells, even at constant glucose concentration. Presumably this is because of increased activity of the mechanism which is responsible for normally facilitated glucose transport (page 388). Sustaining blood glucose at hyperglycemic levels favors formation in the body of products derived from glucose, particularly of glycogen and fatty acids in liver and adipose tissue.

Administration of insulin is followed promptly by a decline in blood glucose concentration and an increase in formation of products derived from glucose. Insulin augments the rate of transfer of glucose from the extra- to the intracellular compartment (Chap. 46). In effect K_m for glucose transport decreases as V_{max} increases. Synthesis of glycogen by muscle and of fatty acids by liver and adipose tissue is enhanced, and a rise in the fasting respiratory quotient (page 293) toward unity indicates that more of the expired CO_2 is derived from glucose. Conversely, in *diabetes mellitus,* in the pancreatectomized animal, or in the animal poisoned with alloxan (page 182), which preferentially damages pancreatic β cells, glucose underutilization is evident. Blood glucose tends to rise to glucosuric levels and glycogen production decreases. Fatty acids are synthesized at subnormal rates in liver and in adipose tissue. Administration of glucose fails to evoke the rise in blood lactic acid seen in the nondiabetic animal, and little or no rise in respiratory quotient follows glucose injection. Even in the diabetic animal, however, glucose utilization continues to occur, especially in brain and myocardium, and, furthermore, the rate of utilization is dependent on the blood glucose level. In this sense the hyperglycemia of diabetes may be regarded as serving a useful function, favoring glucose assimilation in an organism in which, because of lack of insulin, glucose entry into cells would otherwise be seriously restricted. Nevertheless, the muscles and liver of the diabetic are incapable of deriving normal nutrition from their hyperglycemic environment; this situation has been aptly described as "starvation in the midst of plenty."

Many of the consequences of diabetes resemble the effects of starvation, or more specifically of carbohydrate deprivation. The most striking difference is that the diabetic individual is hyperglycemic whereas the fasted subject may have slightly subnormal concentrations of blood glucose.

Hyperinsulinism may arise spontaneously because of hyperplasia or neoplasia of the β cells of the islets of Langerhans, or may result from injection of insulin. The responsiveness of various organs to insulin differs widely. Whereas assimilation of blood glucose by muscle cells is stimulated in hyperinsulinism, this is not the case in brain. The central nervous system is thus placed in an unfavorable position in its competition with other tissues for available glucose, and as the blood glucose level falls, disturbance of the central nervous system and convulsions may ensue.

Whereas insulin reduces the blood glucose concentration, at least four groups of hormones tend to elevate the blood glucose level. The adenohypophysis secretes substances that are, at least superficially, antagonistic to the action of insulin. Extirpation of the hypophysis of a pancreatectomized animal largely corrects its defective metabolism (Chap. 47). The hypophysectomized animal is extremely sensitive to injected insulin, and this sensitivity can be abolished by repeated injection of certain adenohypophyseal hormones. Prolonged administration of these hormones, notably somatotropin (growth hormone) and adrenocorticotropin (Chap. 47), may produce a continuing hyperglycemia and, ultimately, injure the pancreatic β cells. The injury, which may lead to permanent diabetes, is attributable to the hyperglycemia, per se, since similar injury can be produced by prolonged and excessive administration of glucose.

Glucagon from the pancreas and epinephrine from the adrenal medulla both increase the blood sugar level by release of glucose from the liver. Epinephrine rapidly elicits a hyperglycemia sometimes sufficient to exceed the renal threshold. Certain adrenal cortical steroids (Chap. 45) can also promote hyperglycemia but their effects are slower and more prolonged. Conversely, hypoglycemia occurs in the adrenalectomized animal or individual with Addison's disease if food is withheld. Hyperthyroid animals exhibit a mild diabetes with almost complete absence of liver glycogen.

The detailed mechanisms by which these hormones affect glucose metabolism are poorly understood. In general, however, it is clear that elevation of blood glucose level may result from either overproduction or underutilization. Underutilization is characteristic of the animal which is either deficient in insulin or excessively secreting hypophyseal hormones. The major defect appears to be the failure of glucose to penetrate the barrier of muscle and other cell membranes. Secondarily, this situation then sets in motion events which lead to enhanced gluconeogenesis. This is possible because of an increase in the activities of pyruvate carboxylase, phosphoenolpyruvate carboxykinase, and fructose diphosphatase, the group of enzymes, other than those of glycolysis, required for glucose synthesis from pyruvate (page 409). The increase in activity results from an absolute increase in the amounts of these enzymes which are themselves synthesized under these conditions. This synthesis requires the presence of adrenal cortical hormones and does not occur in adrenalectomized animals. Indeed, adrenal steroids administered to normal animals elicit similar synthesis of these enzymes. At the same time, the adrenal cortical hormones alter liver microsomes to enhance their glucose 6-phosphatase activity so that the glucose 6-phosphate formed from pyruvate is readily hydrolyzed to glucose. Moreover, by inhibiting protein synthesis and enhancing transaminase activity in the liver (Chap. 45), the adrenal hormones increase the supply of amino acids available for metabolism over those routes which lead to pyruvate formation, thus providing the raw material for gluconeogenesis. Finally, the liver which is releasing glucose to the circulation ceases to synthesize fatty acids from acetyl CoA (Chap. 21), partly because the decreased concentration of fructose diphosphate no longer stimulates formation of malonyl CoA (page 487), and instead utilizes fatty acids as the prime oxidative substrate. These fatty acids come both from liver lipids and from adipose tissue where lipolysis is enhanced in the absence of insulin and in the presence of both adrenal steroids and adrenocorticotropic hormone (Chap. 47). In turn, this causes an increased production of the acetyl CoA which is imperative to the operation of pyruvate carboxylase while providing the DPNH necessary for reduction of 1,3-diphosphoglyceric acid to 3-phosphoglyceraldehyde and the ATP required for the various steps in glucogenesis from pyruvate (page 409). Note that some source of DPNH is required since the starting material in question is pyruvate rather than lactate.

Insulin not only facilitates entry of glucose into cells but also, in some manner, exerts a suppressive influence on the synthesis of pyruvate carboxylase, phosphoenolpyruvate carboxykinase, and fructose diphosphatase (see above). Liver and adipose cells exposed to insulin also show increased amounts of glucokinase, enhancement of glycogen synthetase activity, increased operation of the phosphogluconate pathway which provides TPNH for fatty acid synthesis (page 485), and increased synthesis of fatty acids from either glucose or pyruvate. Insulin also promotes protein synthesis from amino acids (page 975), thus reducing the supply of gluconeogenic precursors.

Epinephrine and glucagon influence the blood glucose concentration largely by their glycogenolytic action, as described earlier (page 444). The mode of action of thyroid hormone in increasing blood glucose levels is not known.

Glucose Tolerance. The capacity of the animal to dispose of administered glucose is referred to as the *glucose tolerance.* When glucose is administered, either by mouth or by vein, its concentration in blood rises rapidly. In the oral glucose tolerance test, with the usual dose of 1 g. of glucose per kilogram of body weight, the blood glucose concentration will rise from a fasting level of about 90 mg. to a maximum of as much as 140 mg. per 100 ml. in about 1 hr. At this time, in normal man, the rate of entry of glucose into the blood will have decreased while the rate of removal by the several tissues will have increased so that the concentration in the blood begins to fall. Under these conditions, the increased rate of removal of glucose from the blood is due to the following: (1) the rate of glucose entry into cells is a direct function of the concentration of glucose in extracellular fluid; and (2) elevation in blood glucose level stimulates the normal pancreas to discharge insulin into the blood at an increased rate (Chap. 46). Glycogenesis, especially in muscle, is enhanced, glycolysis increases, as is evidenced by a rise in the level of blood lactic acid, the respiratory quotient rises toward unity, indicating a greater carbohydrate oxidation, and the blood glucose concentration falls rapidly. Generally by the end of the second hour, the blood glucose concentration returns to approximately normal and often continues to fall below the initial level, probably as a result of the continuing effect of the increased insulin secretion (Fig. 19.7). As the stimulus of hyperglycemia to the islands of Langerhans declines, insulin secretion returns to lower values.

In the course of an oral glucose tolerance test in a normal subject, the concentration of blood glucose never exceeds the renal threshold and no glucosuria occurs. In the diabetic individual, deficient in insulin, the fasting blood glucose concentra-

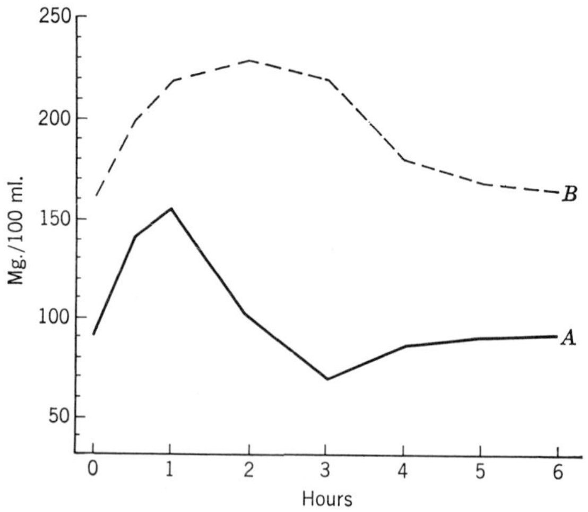

Fig. 19.7. Typical glucose tolerance curves. *A* = normal subject; *B* = diabetic subject.

tion is elevated. After oral administration of glucose, its blood level will rise even higher, often exceeding the renal threshold and provoking glucosuria. The insulin response will be deficient or lacking and, consequently, the decline in blood glucose concentration will be slow. An individual showing this type of response is said to have a decreased glucose tolerance or an elevated glucose tolerance curve (Fig. 19.7).

In individuals suffering from hyperinsulinism, the initial concentration of glucose in the blood is often lower than normal. Shortly after the blood glucose level commences to rise, following glucose ingestion, insulin is secreted. Blood glucose concentration will then start to fall and, because of the excessive insulin response, will often continue to fall to levels as low as 40 mg. per 100 ml., occasionally eliciting central nervous system symptoms such as hypoglycemic convulsions. Thus, these individuals exhibit an abnormally high glucose tolerance. Various other deviations from the normal glucose tolerance pattern have been noted. In sprue, because of a defect in absorption of glucose from the intestinal lumen, a flat curve may be obtained after oral ingestion of glucose although a normal pattern is seen after parenteral glucose administration.

REFERENCES

Books

Balazs, E. A., and Jeanloz, R. W., eds., "The Amino Sugars," vols. IIA, 1965, and IIB, 1966 Academic Press, Inc., New York.

Clark, F., and Grant, J. K., eds., "The Biochemistry of Mucopolysaccharides of Connective Tissue," Cambridge University Press, New York, 1961.

Eisenstein, A. B., ed., "The Biochemical Aspects of Hormone Action," Little, Brown and Company, Boston, 1964.

Greenberg, D. M., ed., "Metabolic Pathways," vol. 1, Academic Press, Inc., New York, 1967.

Hollman, S., and Touster, O., "Nonglycolytic Pathways of Metabolism of Glucose," Academic Press, Inc., New York, 1964.

Horecker, B. L., "Pentose Metabolism in Bacteria," John Wiley & Sons, Inc., New York, 1962.

Stacey, M., and Barker, S. A., "Carbohydrates of Living Tissues," D. Van Nostrand Company, Inc., Princeton, N.J., 1962.

Stanbury, J. B., Wyngaarden, J. B., and Fredrickson, D. S., "The Metabolic Basis of Inherited Disease," McGraw-Hill Book Company, 2nd ed., New York, 1966.

Review Articles

Archibald, A. R., and Baddiley, J., The Teichoic Acids, *Advances in Carbohydrate Chem.,* **21,** 323–375, 1966.

Ashwell, G., Carbohydrate Metabolism, *Ann. Rev. Biochem.,* **33,** 101–138, 1964.

Bernfeld, P., Enzymes of Starch Degradation and Synthesis, *Advances in Enzymol.,* **12,** 379–428, 1951.

Cohn, M., Phosphorylases, in P. D. Boyer, H. A. Lardy, and K. Myrbäck, eds., "The Enzymes," 2d ed., vol. 5, pp. 179–206, Academic Press, Inc., New York, 1961.

Ginsburg, V., Sugar Nucleotides and the Synthesis of Carbohydrates, *Advances in Enzymol.,* **26,** 35–88, 1964.

Glaser, L., Biosynthesis of Deoxysugars, *Physiol. Revs.,* **43,** 215–242, 1963.

Gunsalus, I. C., Horecker, B. L., and Wood, W. A., Pathways of Carbohydrate Metabolism in Microorganisms, *Bacteriol. Revs.,* **19,** 79–128, 1955.

Hales, C. N., Some Actions of Hormones in the Regulation of Glucose Metabolism, in P. N. Campbell and G. D. Greville, eds., "Essays in Biochemistry," vol. 3, pp. 73–104, Academic Press, Inc., New York, 1967.

Hassid, W. Z., Biosynthesis of Polysaccharides from Nucleoside Diphosphate Sugars, in D. J. Bell and J. K. Grant, eds., "The Structure and Biosynthesis of Macromolecules," pp. 63–79, Cambridge University Press, New York, 1962.

Hers, H. G., Glycogen Storage Diseases, *Advances in Metabolic Disorders,* **1,** 2–45, 1964.

Horecker, B. L., Interdependent Pathways of Carbohydrate Metabolism, *Harvey Lectures,* **57,** 35–61, 1961–1962.

Horecker, B. L., Alternate Pathways of Carbohydrate Metabolism and Their Physiological Significance, *J. Chem. Education,* **42,** 244–253, 1965.

Horecker, B. L., Biosynthesis of Bacterial Polysaccharides, *Ann. Rev. Microbiol.,* **20,** 253–290, 1966.

Imsande, J., and Handler, P., Pyrophosphorylases, in P. D. Boyer, H. A. Lardy, and K. Myrbäck, eds., "The Enzymes," 2d ed., vol. 5, pp. 281–304, Academic Press, Inc., New York, 1961.

Jonsen, J., and Laland, S., Bacterial Nucleosides and Nucleotides, *Advances in Carbohydrate Chem.,* **17,** 201–234, 1962.

Krebs, E. G., and Fischer, E. H., Molecular Properties and Transformations of Glycogen Phosphorylase in Animal Tissues, *Advances in Enzymol.,* **24,** 263–290, 1962.

Landau, B. R., Adrenal Steroids and Carbohydrate Metabolism, *Vitamins and Hormones,* **23,** 2–60, 1965.

Lardy, H. A., Gluconeogenesis: Pathways and Hormonal Regulation, *Harvey Lectures,* **60,** 261–278, 1964–1965.

Lowenstein, J. M., The Tricarboxylic Acid Cycle, in D. M. Greenberg, ed., "Metabolic Pathways," 3d ed., vol. 1, pp. 146–270, Academic Press, Inc., New York, 1967.

Manners, D. J., Enzymic Synthesis and Degradation of Starch and Glycogen, *Advances in Carbohydrate Chem.,* **17,** 371–430, 1962.

Martin, H. H., Biochemistry of Bacterial Cell Walls, *Ann. Rev. Biochem.,* **35,** 457–484, 1966.

Neufeld, E. F., and Ginsburg, V., Carbohydrate Metabolism, *Ann. Rev. Biochem.,* **34,** 297–312, 1965.

Neufeld, E. F., and Hassid, W. Z., Biosynthesis of Saccharides from Sugar Nucleotides, *Advances in Carbohydrate Chem.,* **18,** 309–356, 1963.

Randle, P. J., "Fuel and Power in Control of Carbohydrate Metabolism in Mammalian Muscle," *Soc. Expt. Biol. Symposia,* **18,** 129–155, 1965.

Sharon, N., Polysaccharides, *Ann. Rev. Biochem,* **35,** 485–520, 1966.

Sols, A., Carbohydrate Metabolism, *Ann. Rev. Biochem.,* **30,** 213, 1961.

Sutherland, E. W., Jr., The Biological Role of Adenosine-3′5′-phosphate, *Harvey Lectures,* **57,** 17–33, 1961–1962.

Wood, H. G., and Utter, M. F., The Role of CO_2 Fixation in Metabolism, in P. N. Campbell and G. D. Grenville, eds., "Essays in Biochemistry," vol. 1, pp. 1–28, Academic Press, Inc., New York, 1965.

Wood, W. A., Carbohydrate Metabolism, *Ann. Rev. Biochem.,* **35,** 485–520, 1966.

20. Carbohydrate Metabolism. III

Photosynthesis

The prime source of energy in the biosphere is the light absorbed by chlorophyll-containing cells. This energy is utilized to fix CO_2 into carbohydrate.

$$6CO_2 + 6H_2O \xrightarrow{\text{light}} C_6H_{12}O_6 + 6O_2$$

This process, the thermodynamically improbable reduction of CO_2 by H_2O, is, in effect, the reverse of the oxidation of glucose. The minimal free energy required must equal that which can be derived from glucose oxidation, *viz.*, +686,000 cal. per mole. This photosynthesis occurs in organized subcellular bodies, the *chloroplasts,* the structure of which will be considered below. If a suspension of chloroplasts is illuminated in the absence of CO_2 and placed in the dark, and CO_2 is then admitted, fixation of CO_2 into carbohydrate proceeds for a brief but significant time. It is apparent, therefore, that the process of CO_2 fixation, *per se,* is not, strictly speaking, light-dependent. Understanding of the actual photosynthetic events whereby electromagnetic energy is converted into chemical energy that may be used for endergonic biosyntheses will be facilitated if we first consider the chemical events by which, even in the dark, CO_2 fixation can be accomplished.

CO_2 Fixation: the "Dark Reaction." The fate of the carbon of CO_2 during its incorporation into organic compounds was ascertained by introduction of $^{14}CO_2$ for brief periods into suspensions of photosynthesizing algae and identification of the products formed by paper chromatography and radioautography. After only 5 sec., appreciable fixation of ^{14}C occurred; the earliest compound to become radioactive was 3-phosphoglyceric acid, labeled predominantly in the carboxyl carbon.

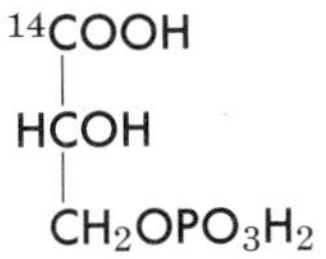

3-Phosphoglyceric acid 1-^{14}C

However, the primary acceptor of CO_2 is not a 2-carbon compound but, rather, ribulose 1,5-diphosphate. The reaction is catalyzed by *diphosphoribulose carboxylase.*

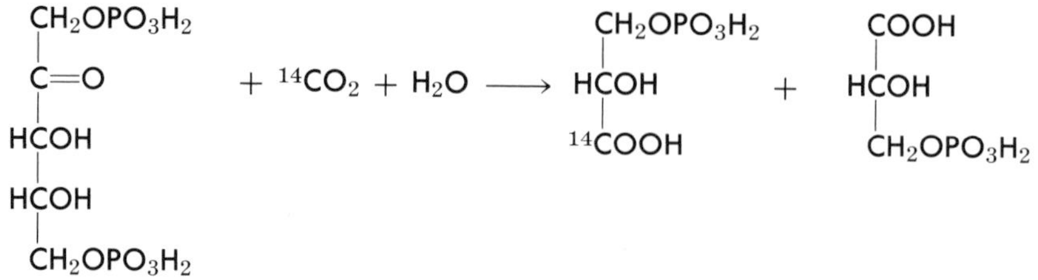

Ribulose 1,5-diphosphate 2 3-Phosphoglyceric acid

The enzyme obtained from spinach has a molecular weight of 550,000, a K_m for HCO_3^- of $0.022M$, and a turnover number (number of moles of substrate altered per unit of time) of only 1,300 per min. at V_{max}. The high, and physiologically unattainable, K_m and low turnover may be compensated for by the abundance of this enzyme, which accounts for more than 15 per cent of total spinach protein.

The carboxylation enzyme occurs in chloroplasts together with *phosphoribulokinase,* which catalyzes the following reaction.

$$\text{Ribulose 5-phosphate} + \text{ATP} \longrightarrow \text{ribulose 1,5-diphosphate} + \text{ADP}$$

All other enzymes required to complete the synthesis of hexose from 3-phosphoglyceric acid have been described in the discussion of glycolysis. However, were all the 3-phosphoglyceric acid converted to hexose by reversal of glycolysis, no ribulose diphosphate would be available to serve as acceptor for CO_2 in subsequent fixation reactions. The problem, therefore, is the converse of that considered in the operation of the phosphogluconate pathway, *viz.*, to provide a regenerative system by means of which hexose may be accumulated and ribulose 1,5-diphosphate recovered for the carboxylation reaction. This is again achieved by concerted action of the enzymes of glycolysis and those of the phosphogluconate pathway. These reactions are summarized in the balanced equations of Table 20.1.

In reaction (*b*), the two moles of 3-phosphoglycerate formed by carboxylation of ribulose 1,5-diphosphate are reduced to 3-phosphoglyceraldehyde. Of the 12 moles of triose phosphate thus formed, 10 are employed in reactions (*c*) and (*d*) to make 5 moles of fructose 6-phosphate. One of these 5 moles will represent the net gain of photosynthesis. By reactions (*e*) to (*g*), 4 moles of fructose 6-phosphate and the 2 moles of 3-phosphoglyceraldehyde still remaining yield 6 moles of xylulose 5-phosphate, which are then epimerized to ribulose 5-phosphate (*h*) and phosphorylated (*i*). The resultant 6 moles of ribulose 1,5-diphosphate may then participate in the next cycle.

The summary equation in Table 20.1 reveals the manner in which energy must be provided to this system in order to effect fixation of CO_2 into carbohydrate. Synthesis of 1 mole of hexose phosphate requires 12 moles of DPNH and 18 moles of ATP, *viz.*, fixation of one CO_2 requires 2DPNH and 3ATP. Although DPNH is the reductant for CO_2 fixation in some photosynthetic bacteria, chloroplasts from plants and algae contain a TPNH-linked 3-phosphoglyceraldehyde dehydrogenase.

The validity of this complex network of reactions, thought to occur in chloroplasts, has been shown by reconstruction of the entire process using purified enzymes. However, several observations suggest that this may not be the *sole*

Table 20.1: HEXOSE ACCUMULATION AND PENTOSE REGENERATION IN PHOTOSYNTHESIS

Step	Enzyme	Reaction	Carbon balance
a	Carboxylation enzyme	6 Ribulose 1,5-diphosphate + $6CO_2$ $\longrightarrow$ 12 3-phosphoglyceric acid	6(5) + 6(1) $\longrightarrow$ 12(3)
b	Phosphoglyceric acid kinase	12 3-Phosphoglyceric acid + 12ATP $\longrightarrow$ 12 1,3-diphosphoglyceric acid + 12ADP	12(3) $\longrightarrow$ 12(3)
	Phosphoglyceraldehyde dehydrogenase	12 1,3-Diphosphoglyceric acid + 12DPNH + $12H^+$ $\longrightarrow$ 12 3-phosphoglyceraldehyde + $12DPN^+$ + $12P_i$	12(3) $\longrightarrow$ 12(3)
c	Triose isomerase	5 3-Phosphoglyceraldehyde $\longrightarrow$ 5 dihydroxyacetone phosphate	5(3) $\longrightarrow$ 5(3)
	Aldolase	5 3-Phosphoglyceraldehyde + 5 dihydroxyacetone phosphate $\longrightarrow$ 5 fructose 1,6-diphosphate	5(3) + 5(3) $\longrightarrow$ 5(6)
d	Phosphatase	5 Fructose 1,6-diphosphate $\longrightarrow$ 5 fructose 6-phosphate + $5P_i$	5(6) $\longrightarrow$ 5(6)
e	Transketolase	2 Fructose 6-phosphate + 2 3-phosphoglyceraldehyde $\longrightarrow$ 2 xylulose 5-phosphate + 2 erythrose 4-phosphate	2(6) + 2(3) $\longrightarrow$ 2(5) + 2(4)
f	Transaldolase	2 Fructose 6-phosphate + 2 erythrose 4-phosphate $\longrightarrow$ 2 sedoheptulose 7-phosphate + 2 3-phosphoglyceraldehyde	2(6) + 2(4) $\longrightarrow$ 2(7) + 2(3)
g	Transketolase	2 Sedoheptulose 7-phosphate + 2 3-phosphoglyceraldehyde $\longrightarrow$ 4 xylulose 5-phosphate	2(7) + 2(3) $\longrightarrow$ 4(5)
h	Epimerase	6 Xylulose 5-phosphate $\longrightarrow$ 6 ribulose 5-phosphate	6(5) $\longrightarrow$ 6(5)
i	Phosphoribulokinase	6 Ribulose 5-phosphate + 6ATP $\longrightarrow$ 6 ribulose 1,5-diphosphate + 6ADP	6(5) $\longrightarrow$ 6(5)
Net: 6 Ribulose 1,5-diphosphate + $6CO_2$ + 18ATP + 12DPNH + $12H^+$ $\longrightarrow$ 6 ribulose 1,5-diphosphate + 1 fructose 6-phosphate + $17P_i$ + 18 ADP + $12DPN^+$			6(5) + 6(1) $\longrightarrow$ 6(5) + 1(6)

mechanism for carbohydrate synthesis from CO_2. Among these observations are: (1) The K_m for CO_2 for diphosphoribulose carboxylase is considerably greater than physiologically attainable CO_2 tension. (2) A reconstructed system of pure enzymes in appropriate relative amounts functions considerably more slowly than does the normal process. (3) The scheme presented predicts that hexose accumulated in the presence of $^{14}CO_2$ should be equally labeled in the 3- and 4-positions; asymmetric labeling is actually found when chloroplasts are used. (4) The scheme fails to explain the relatively rapid appearance of glycolic acid in photosynthesizing cells. (5) Since aldolase may be absent from the blue-green algae and several species of photosynthetic bacteria, formation of hexose by condensation of two molecules of triose should not be possible in these organisms. (6) In a few experimental systems, the first detectable ^{14}C-labeled compound has been alanine, suggesting that the initial event may be formation of pyruvate from CO_2 and a C_2 compound presumed to be acetyl CoA. The concomitant appearance of label in glutamic acid suggests similar formation of α-ketoglutaric acid from CO_2 and succinyl CoA. The reaction is driven by photosynthetic reduction of ferredoxin (see below), which serves as the immediate reductant in the fixation process. Details of this process and its distribution among photosynthetic organisms remain to be established. A tentative scheme for the pathway by which CO_2 fixed by these mechanisms may be accumulated as hexose is shown in Fig. 20.1.

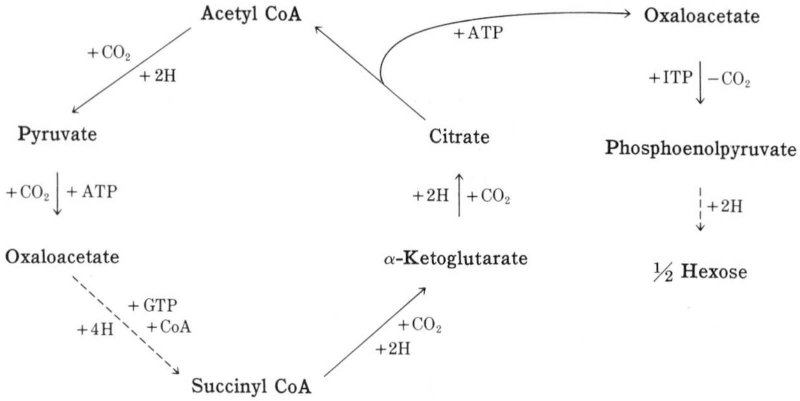

Fig. 20.1. Tentative scheme of an alternate pathway for photosynthetic accumulation of hexose.

Molecular Absorption of Light Quanta. The fundamental event in photosynthesis is the absorption of light by chlorophyll. Chlorophyll a is a magnesium-porphyrin derivative in which one pyrrole ring is partially reduced. The presence of a fifth isocyclic ring should be noted. The structure of chlorophyll a was determined largely by the work of Willstätter and by H. Fischer and coworkers. In view of the possibilities of resonance, other representations with different positions of the double bonds are possible. Both acid side chains are esterified, one as a methyl ester and the other as a phytyl ester. The structure of phytol has been presented (page 75). The non-ionic magnesium atom is held by two covalent and two coordinate linkages. In chlorophyll b, the methyl group at position 3 (in the ring at upper right) is replaced by a formyl group, —CHO.

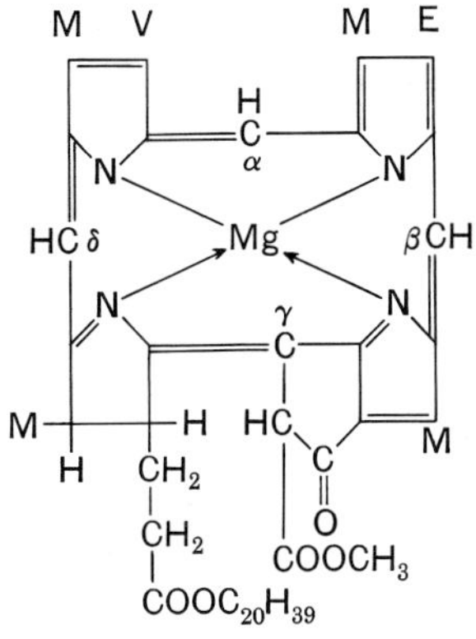

Chlorophyll a

Treatment of chlorophyll with weak acid readily removes the magnesium atom to give *pheophytin.* Saponification with alkali gives rise to *chlorophyllides,* salts of the liberated carboxyl groups. Hydrolysis of chlorophyll by the enzyme *chlorophyllase* removes the phytyl group.

Drastic degradation of the chlorophylls gives pyrroetioporphyrin.

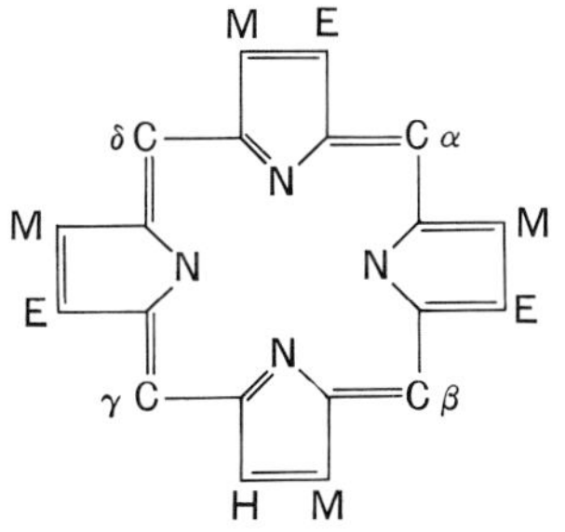

Pyrroetioporphyrin

This compound differs from etioporphyrin in containing a hydrogen atom in place of an ethyl group at position 6. The arrangement of the other groupings is identical with that in etioporphyrin III derived from protoporphyrin, indicating the close biological relationship of all these compounds.

Within the chloroplast, chlorophyll is present as a lipoprotein complex. The complex can be disrupted by nonpolar solvents which extract the lipids and carotenoid pigments. In higher plants, chlorophylls a and b are generally present in a ratio of about 3:1. In the native complex, the absorption maxima of chlorophyll are displaced about 20 mμ toward the red as compared with the absorption spectrum of pure chlorophyll in a nonpolar solvent.

The energy of a photon (one quantum of electromagnetic radiation) is given by

$$E = hc/\lambda$$

where h is Planck's constant (6.6×10^{-27} erg per sec.), c is the velocity of light (3×10^{10} cm per sec.), and λ is the wavelength in centimeters. By converting ergs to electron-volts and expressing λ in Angstrom units,

$$E = 12{,}350/\lambda \qquad \text{electron-volts}$$

For red light at 675 mμ, one quantum corresponds to 1.84 electron-volts; a mole of such quanta (one einstein) corresponds to 42,400 cal.

A molecule can absorb light quanta only of specific wavelengths. The absorbed energy is then part of the molecule, which is said to be "activated." The absorbed photon must correspond to the energy required to perturb an electron, raising it to an orbital more remote from the atomic nucleus than in the ground state. Chlorophyll has a continuing system of alternating single and double bonds. In structures of this type, the electrons in the bonds formed by overlap of carbon *p* orbitals (*e.g.*, six per benzene ring) must be assigned to the entire molecule rather than to specific interatomic bonds and, when light is absorbed, it is an electron in this "π" electron system that is in a higher energetic state. When a photon is absorbed, the molecule can exist in a transient excited state, the "singlet," with a half-life of about 10^{-9} sec. This can return to the ground state by loss of energy as heat, or by emission of a light quantum (fluorescence). Alternately, it can assume a somewhat more stable but still excited condition by transition to the "triplet" state, which has a half-life of about 10^{-5} sec. and in which not only is an electron dislocated and in a more energetic orbit but its spin is reversed.

The Chloroplast. The photosynthetic apparatus of plant and some algal cells is lodged within especially organized structures, the chloroplasts, bodies of varied shape, 3 to 10 μ long by 0.5 to 2 μ in diameter. Within these are found 10 to 100 somewhat cylindrical structures, the *grana,* which are collections of a multilayered system of lamellae. Isolated preparations of lamellae, supplemented with the more soluble proteins of the chloroplast, can conduct the photosynthetic fixation of CO_2 into hexose. The lamellae consist of stacked membranous sheets, the surfaces of which reveal a regular, almost crystalline pattern of packed units, each approximately $100 \times 150 \times 185$ Å.; and termed *quantasomes.* It is not clear whether these are discrete morphological and functional units or the minimum-sized aggregate of all the components required for photosynthesis. The composition of an average quantasome unit is shown in Table 20.2. The lipid mixture of chloroplasts and the organization of the membrane are similar to those of mitochondria. A prominent constituent of the lipids is the sulfolipids (page 68). The structures of some quinones found in chloroplasts are indicated in Fig. 20.2.

Table 20.2: Approximate Composition of an Average Spinach Quantasome

Component	*Molecules per quantasome*	*Component*	*Molecules per quantasome*
Chlorophyll a	160	Phospholipids (lecithin, phosphatidyl ethanolamine, phosphatidyl inositol, phosphatidyl glycerol)	116
Chlorophyll b	65	Sulfolipids	48
Carotenoids	48	Galactosylglycerides	500
Quinones		Cytochrome b_6	1
Plastoquinone A	16	Cytochrome f	1
Plastoquinone B	8	Plastocyanin	5
Plastoquinone C	4	Ferredoxin	5
α-Tocopherol	10		
α-Tocopherylquinone	4		
Vitamin K_2	4		

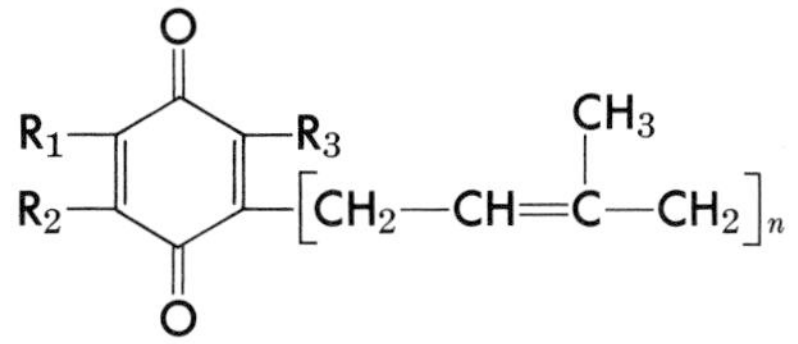

	R_1	R_2	R_3	n
Ubiquinones	$CH_3O—$	$CH_3O—$	CH_3	6–10
Plastoquinones	$CH_3—$	$CH_3—$	H	6–10
Vitamins K	H	H	CH_3	4–9

Fig. 20.2. Structures of some quinones found in chloroplasts.

The relative amounts of the nonprotein constituents of chloroplasts vary from species to species. In addition to the components shown in Table 20.2, algal chloroplasts contain conjugated proteins, *phycocyanins* and *phycoerythrins,* to which are attached tetrapyrroles called *phycobilins.*

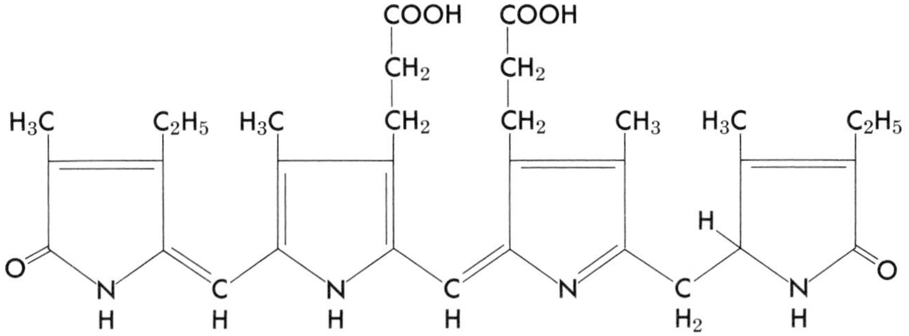

Prosthetic group of a phycocyanin

Blue-green algae, the most primitive of the algae, and photosynthetic bacteria do not contain formed chloroplasts. However, lamellae are clearly evident in the blue-green algae, distributed throughout the cytoplasm, rather than packed into discrete bodies. In photosynthetic bacteria, the photosynthetic apparatus is fixed to the cell membrane in small vesicular bodies. Disruption of the cell permits their isolation as particles termed *chromatophores,* with general lipid composition resembling that of chloroplasts. The chlorophyll is *bacteriochlorophyll,* which differs from chlorophyll a in that the double bond at $C_3—C_4$ is reduced to a single bond and the vinyl substituent at C_2 is replaced by an acetyl residue; the major absorption bands of bacteriochlorophyll are displaced far in the red, at 800 and 850 mμ. The amount of chlorophyll relative to heme proteins is reduced by a factor of almost 10 from that of chloroplasts.

Light Absorption by Chloroplasts. A quantasome is a unit within which the energy of light, at any wavelength in the visual spectrum, absorbed anywhere in the unit, can be transmitted to the specific chlorophyll molecule that actually participates in the conversion of light to chemical energy. The manner in which

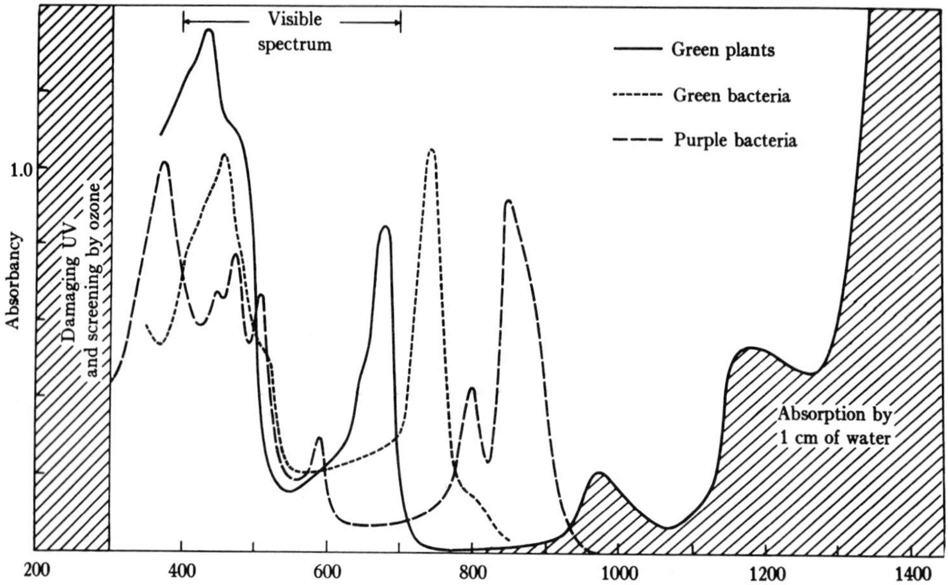

Fig. 20.3. Absorption spectra of three kinds of photosynthetic organisms, showing the regions of the spectrum that can be used for photosynthesis. The regions of intense absorption are the absorption maxima of the respective forms of chlorophyll. Carotenoid absorption is most prominent in the region 500 to 650 mμ. (*From R. K. Clayton, "Molecular Physics in Photosynthesis," Blaisdell Publishing Co., New York,* 1966.)

the overlapping absorption spectra of chloroplast pigments combine to assure utilization of light at all wavelengths in the visual spectrum is shown in Fig. 20.3. In higher plants most of the collected light is initially absorbed by the chlorophylls. Chlorophyll a has an absorption maximum at about 675 mμ; that of chlorophyll b lies at 650 mμ. Utilization of light of shorter wavelengths is possible because of the absorption spectra of accessory pigments. Thus, both the carotenoids and the chlorophylls absorb in the region 400 to 550 mμ; the phycobilins of algal chloroplasts absorb in the region 500 to 650 mμ. Bacterial chromatophores also contain accessory pigments which absorb in the latter region while their major pigment, bacteriochlorophyll, which has a small absorption band at 590 mμ, exhibits major peaks in the near infrared at 800 and 850 to 890 mμ.

The excitation energy of a photon, absorbed anywhere in the unit, spontaneously migrates, within 10^{-9} sec., to that pigment molecule in the unit with an absorption band at the longest wavelength. If that molecule readily enters some metastable state, *e.g.*, by a singlet-triplet conversion or by formation of a charge transfer complex with an adjacent molecule, it can serve as a trap for excitation energy received anywhere in the unit. Energy transfer from molecule to molecule within the unit is thought to occur by *induced resonance.* Energy is transferred as though, after absorbing a photon, each molecule fluoresces and thus passes its energy to its neighbor, although no light, as such, is actually emitted or reabsorbed. Thus, each photosynthetic unit serves as a light-gathering apparatus which, even in dim light, can collect light energy and deliver it to that chlorophyll molecule which will participate in the photochemical event.

Requirements for Photosynthesis. As developed previously, the actual enzymic process by which CO_2 is fixed into carbohydrate requires a supply of TPNH and ATP. The electromagnetic energy absorbed by the photochemical system must, therefore, be transformed into a suitable mixture of reducing potential and high-energy phosphate. In the green plant, as is evident from the over-all equation descriptive of photosynthesis, oxygen is also evolved during this process. In effect, therefore, the energy of the absorbed light must be used to reduce pyridine nucleotide, using water as the electron donor, while trapping some of the energy as ATP. These relationships may be summarized in the following equation.

$$2H_2O + 2TPN^+ + xADP + xP_i \xrightarrow{nh\nu} O_2 + 2TPNH + 2H^+ + xATP \qquad (1)$$

This is a specific case of the more general statement:

$$H_2A + TPN^+ + yADP + yP_i \xrightarrow{mh\nu} A + TPNH + H^+ + yATP \qquad (2)$$

For photosynthetic bacteria H_2A may be inorganic compounds such as H_2S or an organic compound such as succinic acid. The total external energy requirement in these instances is obviously less than that for the reversal of the oxidation of TPNH by O_2.

The Photochemical Event. Chlorophyll is universally present in cells with photosynthetic capability; in some manner, then, the chlorophyll must serve to effect a photochemical separation of oxidizing and reducing power. Such a mechanism may be schematically represented as

$$\begin{array}{c}A^0\\ \hline \text{Chl}\\ \hline B^0\end{array} \xrightarrow{h\nu} \begin{array}{c}A^0\\ \hline \text{Chl}^*\\ \hline B^0\end{array} \longrightarrow \begin{array}{c}A^-\\ \hline \text{Chl}\\ \hline B^+\end{array} \qquad (3)$$

where *A* is a potential electron acceptor and *B* is a potential electron donor situated on either side of a chlorophyll (*Chl*) molecule; *Chl** is the photoactivated form of *Chl*, A^- is the reduced form of *A*, and B^+ is the oxidized form of *B*. The simplest mechanism by which this could be accomplished is ejection of an electron from the photoactivated chlorophyll, forming a chlorophyll free radical (Chl^+). The electron is accepted by *A* as the chlorophyll radical regains an electron from *B* on its other side.

$$\begin{array}{c}A^0\\ \hline \text{Chl}^0\\ \hline B^0\end{array} \xrightarrow{h\nu} \begin{array}{c}A^0\\ \hline \text{Chl}^*\\ \hline B^0\end{array} \longrightarrow \begin{array}{c}A^-\\ \hline \text{Chl}^+\\ \hline B^0\end{array} \longrightarrow \begin{array}{c}A^-\\ \hline \text{Chl}^0\\ \hline B^+\end{array} \qquad (4)$$

Presumably no wasteful subsequent direct reaction occurs between A^- and B^+ because of the barrier of the chlorophyll, which is no longer activated and has returned to the ground state. The participation of a chlorophyll free radical is supported by observation of a characteristic electron paramagnetic resonance signal. The manner in which the photochemical event makes possible the fixation of CO_2 with evolution of O_2 in higher plants will be more readily understood by first considering the simpler case of photosynthetic bacteria. In what follows, it is assumed that equation (4) represents a satisfactory model of the operating mechanism.

Bacterial Photosynthesis. In all photosynthesizing systems, following the sepa-

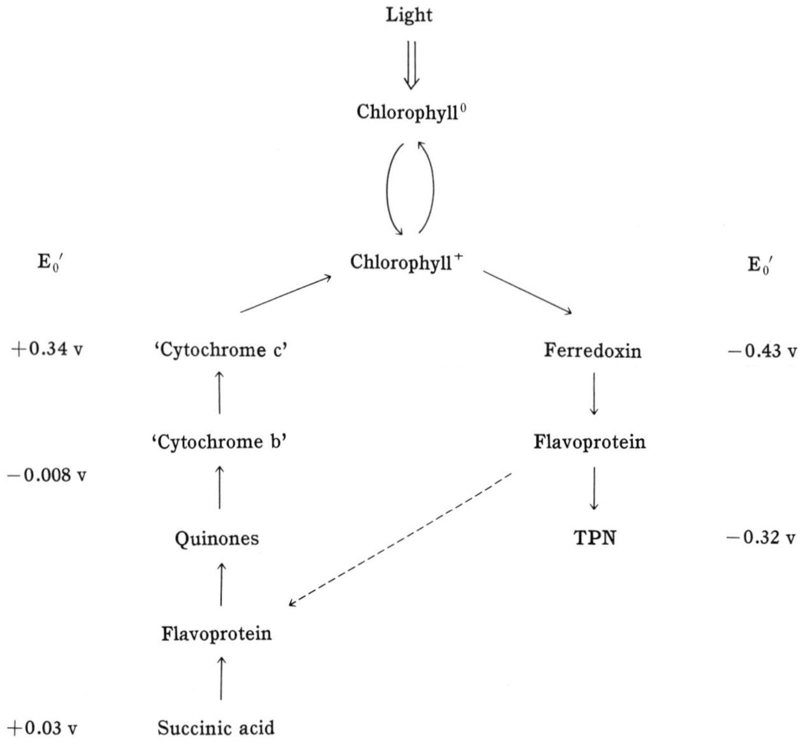

Fig. 20.4. Path of electron flow in an illuminated bacterial chromatophore.

ration of oxidizing and reducing power, as shown in equation (4), a flow of electrons is initiated along two transport systems, one of which accepts the electron delivered to *A* while the other replaces it. In a photosynthetic bacterium such as *Rhodospirillum rubrum,* electron flow occurs as shown in Fig. 20.4. In this scheme, ferredoxin serves as *A*, the immediate electron acceptor, and a cytochrome c–like protein as *B*, the immediate electron donor to the activated chlorophyll. As a net process, electrons from succinate reduce TPN^+. This electron flow is associated with the generation of at least one molecule of ATP per pair of electrons ($P_i/2e = 1$). The mechanism of this phosphorylation is considered below. The cytochrome c–like protein, in this instance, is an acidic protein with a molecular weight of about 12,000. The b-type cytochrome, molecular weight = 28,000, has two nonidentical heme groups per molecule. Several such bifunctional heme proteins, called "cytochromoids" (page 374), have been found in photosynthetic bacteria.

Thus the energy of absorbed photons drives the reaction

$$\textbf{ADP} + \textbf{P}_i + \textbf{succinate} + \textbf{TPN}^+ \longrightarrow \textbf{fumarate} + \textbf{TPNH} + \textbf{H}^+ + \textbf{ATP}$$

whereas, spontaneously, reaction would proceed in the opposite direction. Accordingly, this is an instance of photosynthesis as represented by equation (2) above. The minimum free energy required for this over-all reaction is +26,000 cal. per mole. This may be compared with 64,000 cal., the energy available in two quanta at 890 mμ, one of the absorption maxima of bacteriochlorophyll.

One other property of the system is noteworthy: a shunt, shown as a dotted line in Fig. 20.4, by which electrons can flow from the flavoprotein which oxidizes ferredoxin to the quinones which reduce the cytochromes. In the absence of oxidizable substrate, light-induced electron flow occurs along a circular path, thereby making possible ATP formation in the usual manner. This arrangement, *cyclic photophosphorylation,* employs light energy to drive ATP synthesis without accumulation of reducing power; hence it cannot effect reduction of CO_2 to hexose. It may represent the most primitive form of photosynthesis, useful to an organism in a milieu rich in organic compounds and requiring only ATP. This system functions in living bacterial cells since, in addition to the ATP and TPNH required for carbohydrate synthesis, this mechanism provides ATP to drive many other endergonic processes of the cell.

Photosynthesis in Higher Plants. In higher plant forms photochemical events meet all the requirements of photosynthesis, *i.e.*, generation of both ATP and reduced pyridine nucleotide, in the absence of an exogenous reducing substrate, with concomitant evolution of oxygen. The over-all process, then, is described by equation (1) above. The photochemical events, therefore, must provide an oxidant with an E_0' greater than that of O_2, +0.8 volt, so that it can accomplish the oxidation of H_2O, and a reductant of sufficiently low potential to effect the reduction of pyridine nucleotide. The minimum over-all difference in potential, $\Delta E_0'$, therefore, is greater than 1.2 volts, equivalent to +64,000 cal. per mole of TPNH formed. Since, as described below, two moles of ATP are also formed during this process, the net energy requirement is at least 78,000 cal. per mole TPNH formed. This may be compared with the energy equivalent of 2 einsteins of light at 675 mμ, 84,000 cal. Since a process of this efficiency is unlikely, more than two photons are required per TPN which is reduced; hence the mechanism must be more complex than that in the bacteria.

An early dissection of the process of photosynthesis in higher plants was provided by the demonstration of the "Hill reaction." Illuminated chloroplasts can reduce oxidants such as quinones and ferricyanide with evolution of O_2 and without reduction of most of the redox active constituents of the chloroplast. Indeed, this was among the earliest indications that CO_2 reduction is not intrinsic to the mechanism of oxygen evolution. The over-all reaction is the following:

$$2H_2O + 2\text{ quinone} \xrightarrow{\text{light}} O_2 + 2\text{ hydroquinone}$$

Algae with hydrogenase activity can conduct what is, in effect, the other half of the entire mechanism, *viz.*, they can photochemically reduce pyridine nucleotides and make ATP without concomitant O_2 evolution.

$$ADP + P_i + H_2 + TPN^+ \xrightarrow{\text{light}} TPNH + H^+ + ATP$$

Many observations suggest that chloroplast photosynthesis is accomplished by two photochemical processes operating in series, as shown in Fig. 20.5. Each half is analogous to the bacterial system discussed above. System II must operate at high potential since water serves as the reductant in place of the succinate considered earlier and O_2 rather than fumarate is formed. The other product is a weak

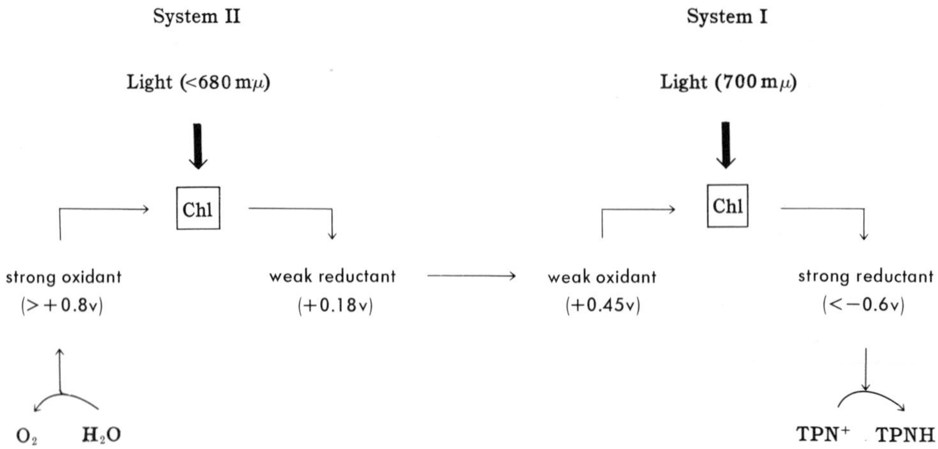

Fig. 20.5. General model for the cooperative functioning of two photochemical reaction centers in plant chloroplasts. Arrows represent the direction of electron flow.

reductant that then serves as reducing substrate for system I which operates at lower potentials and, hence, can generate TPNH.

Photosynthesis Reaction Centers. According to the scheme of Fig. 20.5, each quantasome should contain two distinct photochemical centers. This concept was first suggested by the fact that light at wavelengths longer than 700 mμ is ineffective in driving photosynthesis, whereas a mixture of light at both 689 mμ and shorter wavelengths is more effective than either alone; this is true even when the two are applied alternately.

Within each quantasome there appears to be *one* molecule of the 300 molecules of chlorophyll a which is not chemically distinguishable by ordinary extraction procedures but which, in its native milieu, is bleached by light at 700 mμ (its absorption maximum is at 683 mμ) and is then evident as a free radical by its electron spin resonance signal. Oxidation of the chloroplast by ferricyanide results in similar bleaching of only one chlorophyll a per quantasome, with an apparent E_0' of about +0.47 volt. This pigment is usually designated as P_{700}. Further, in very intense light the bulk of chlorophyll a is converted to pheophytin (page 457), but P_{700} is unaffected, indicating its unique character. The analogous reaction center in bacterial chromatophores is designated as P_{870}. It functions as does P_{700} and is unique in that it is the only molecule of bacteriochlorophyll in the chromatophore to resist oxidation by powerful oxidants such as chloroiridate. The special properties of P_{700} and P_{870} result not from a difference in structure from other molecules of chlorophyll a or bacteriochlorophyll respectively, but from the fact that, presumably, they lie in immediate proximity to the donor and acceptor molecules of the electron transport system.

However, it has not been possible similarly to identify a second reactive center in quantasomes. An early indication that two such centers exist was the observation that irradiation at 700 mμ resulted in oxidation of cytochrome f while the shorter wavelengths then caused its reduction. Additional evidence was provided by the effects of the herbicide, chlorophenyl dimethyl urea (CMU). Even when present in a concentration of only one molecule per quantasome, CMU abolishes O_2 evolution by chloroplasts, but has no effect on the function of P_{700} in the reactive center of system I in the above scheme. Thus, CMU does not affect the ability of light at 680 mμ to cause oxidation of cytochrome f but prevents reduction of cytochromes b_6 and f by blue light. Current evidence suggests that system II can bring electrons to the level of +0.18 volt, but its immediate acceptor, the weak reductant of the over-all scheme, is unknown. The detailed organization of the system as presently understood is shown in Fig. 20.6 (page 466).

The detailed arrangements of the photosynthetic electron transport system are not certain; however, the scheme shown in Fig. 20.6 is compatible with current information. The position assigned to cytochrome f is based on its E_0' of +0.36 volt, and on the fact that when chloroplasts are illuminated at 77°K., cytochrome f is immediately oxidized. Since thermal reactions are excluded at this temperature, cytochrome f must make physical contact with the photochemical center. A blue copper-protein, plastocyanin, with an E_0' of +0.4 volt, appears to lie between cytochrome f and P_{700}. Indeed, in certain preparations of spinach chloroplasts, removal of cytochrome f results in only moderately impaired reduction of TPN, whereas removal of plastocyanin results in total abolition of electron transport. It is noteworthy, however, that plastocyanin is not present in bacterial chromatophores.

Of the rich variety of quinones, only plastoquinones A and C appear to be essential to normal function; removal of the other quinones seems not to be deleterious. Both A and C appear to be required in amounts approximately stoichiometric with cytochrome f, and some evidence supports the relative positions assigned to the two quinones in Fig. 20.6. Thus, the abundant quinones of chloroplasts are as puzzling as are those of mitochondria. A red protein called *rubimedin,* with unidentified prosthetic group, has been shown to be necessary for electron transport between the two photochemical centers, but its role is unclear. The position assigned to ferredoxin in this scheme rests on its low potential, rather than on categorical demonstration that it is the immediate acceptor from P_{700}. Since illumination of P_{700} can reduce benzyl viologen, $E_0' = -0.70$ volt, system I is capable of bringing electrons to a potential even lower than that of ferredoxin.

The CMU-sensitive reaction center for system II, which is responsible for oxygen evolution, requires the presence of Mn^{++} and Cl^-. It accomplishes the oxidation of OH^- to O_2, which requires removal of four electrons per O_2. However, the mechanism of action and physical representation of this center are entirely obscure.

The Quantum Yield of Photosynthesis. Much attention has been devoted to the "quantum yield" of photosynthesis, *i.e.*, the number of light quanta required for evolution of one molecule of O_2. For some years, Warburg and his colleagues supported a value of 4 while other laboratories obtained values of 6 to 10. The operation of the scheme summarized above requires at least four photons at each of the reaction centers to achieve evolution of one O_2 with formation of 2 TPNH, or a total of 8 quanta. Assuming the average energy of the absorbed photons to be 40,000 cal. per einstein, a total of 320,000 cal. would be used per O_2 formed. Accordingly, absorption of $6 \times 320{,}000 = 1{,}920{,}000$ cal. results in formation of 12 TPNH + 6 O_2 + 24 ATP. Minimally, this process must require $(12 \times 64{,}000) +$

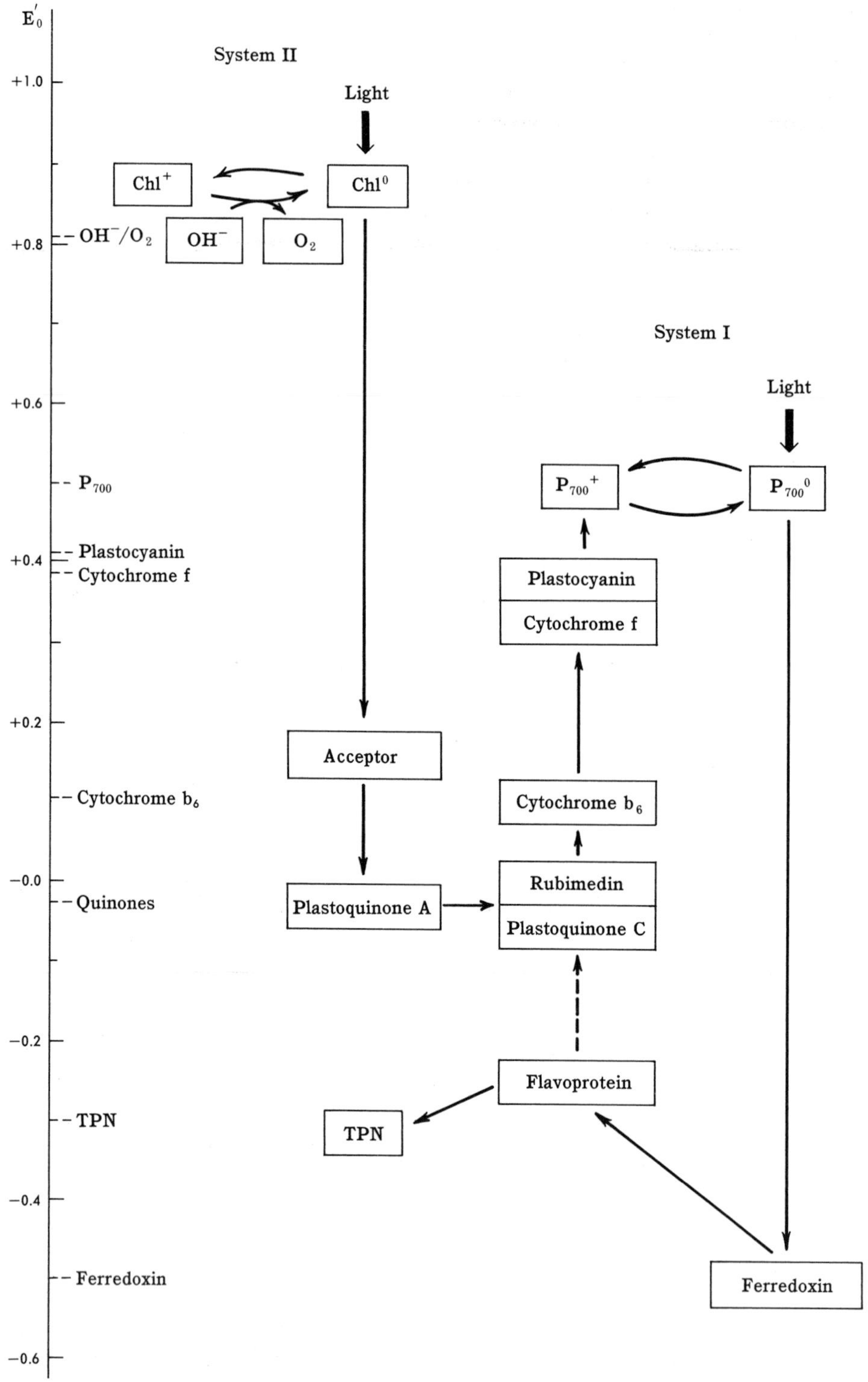

Fig. 20.6. Tentative scheme for details of electron transport in a plant photosynthetic unit. Arrows represent the direction of electron flow.

$(24 \times 8{,}000) = 960{,}000$ cal. Thus the process has an over-all efficiency of about 50 per cent and provides the cell with one hexose, 6 O_2, and a surplus of six moles of ATP to be used for other purposes.

Photophosphorylation. Electron flow through the quantasome transport chain, as in mitochondria, results in generation of ATP. Coupling of the two processes is not so tight as in mitochondria. Nevertheless, uncoupling agents (NH_4^+, dinitrophenol, hypotonicity) all effect a substantial increase in the rate of photosynthetic O_2 formation, so that the normal chloroplast does seem to be "coupled", albeit loosely. The sites of energy conservation along the chloroplast electron transfer chain have not been clearly identified. If the $P_i/2e$ value is 2, as most evidence suggests, then the logical sites appear to be in the segments quinone $\rightarrow$ cytochrome b_6 and cytochrome $b_6 \rightarrow P_{700}$.

The demands of the plant cell for ATP considerably exceed those for reduced pyridine nucleotides. As in photosynthetic bacteria, this requirement can be satisfied by cyclic photophosphorylation. The electron shunt would again appear to be from the level of the flavoprotein which serves as ferredoxin-TPN^+ reductase to some component of the ascending limb in Fig. 20.6.

The mechanism by which photon-driven electron transport achieves ATP formation appears to be generally analogous to that in mitochondria. Photophosphorylation occurs in a system constructed of an inner and outer membrane. Again, as in oxidative phosphorylation, photophosphorylation appears to be possible only when the double membrane structure is intact. Oligomycin and phlorhizin serve as inhibitors, rather than uncouplers, just as they do in mitochondria.

If chloroplasts are illuminated in the absence of ADP and P_i, and the latter are added in the dark, ATP is formed, suggesting the accumulation of an energy form that may be the equivalent of the hypothetical $X{\sim}Y$ of oxidative phosphorylation (page 347).

When illuminated, or incubated with ATP, chloroplasts contract; K^+ and Mg^{++} enter the medium, largely in exchange for H^+, and water leaves the chloroplast, adjusting the osmotic pressure. As in functioning mitochondria, illuminated chloroplasts can accept large quantities of Ca^{++}, if this ion is added to a synthetic medium. The entry of H^+ is even more impressive than with mitochondria; maximally, a gradient of three pH units across the grana membranes may be established by illumination. A gradient of this magnitude seems compatible with the "chemiosmotic" hypothesis for ATP synthesis proposed for mitochondria (page 352). Indeed, brief dark immersion of chloroplasts in a medium at pH 4.5 containing an organic anion such as succinate or barbiturate, followed by a rapid return to pH 8, effects in a few seconds a synthesis of as many as 100 molecules of ATP per quantasome. This process is prevented by the various uncouplers of photophosphorylation, indicating that the immediate mechanism of this ATP formation is probably identical with that resulting from electron flow.

One further property of this system is of interest. Not all absorbed photons result in electron flow. At low-incident light levels, about 5 per cent of the absorbed energy slowly reappears as fluorescent light, at the wavelength characteristic of chlorophyll a (700 mμ). This fluorescence continues for as long as 30 to 60 min. after illumination. Since fluorescent decay of any photoactivated state should be complete within seconds, it is clear that some other form of energy is reactivating chlorophyll which then fluoresces. Mutant forms of *Rhodopseudomonas spheroides*, that lack P_{870}, do not show this delayed fluorescence; hence the energy conversion must pass through the photochemical center. If the chloroplasts are first extracted to remove ferredoxin, the fluorescence intensity increases markedly. Deficiency of ADP + P_i results in a doubling of fluorescence. If chloroplasts are immersed in the dark in a medium of pH 4 but lacking ADP + P_i, then an unusually large fluorescence is apparent. These observations underscore the equivalence of light, electron flow, ATP, and a proton gradient as energy forms in this remarkable energy-transducing apparatus.

The description here presented has been developed only in the past few years; many details remain to be discovered, and some of those described may be incorrect. However, the general concepts now appear to be relatively secure.

REFERENCES

Books

Clayton, R. K., "Molecular Physics in Photosynthesis," Blaisdell Publishing Company, New York, 1965.

Dienes, M., ed., "Energy Conversion by the Photosynthetic Apparatus," Brookhaven National Laboratory, Upton, N.Y., 1967.

Gest, H., San Pietro, A., and Vernon, L. P., "Bacterial Photosynthesis," The Antioch Press, Yellow Springs, Ohio, 1963.

Goodwin, T. W., ed., "Biochemistry of Chloroplasts," Academic Press, Inc., New York, 1966.

Kamen, M. D., "Primary Processes in Photosynthesis," Academic Press, Inc., New York, 1963.

Kirk, J. T. O., and Tilney-Bassett, R. A. E., "The Plastids," W. H. Freeman and Co., San Francisco, 1967.

Kok, B., and Jagendorf, A. T., eds., "Photosynthetic Mechanisms of Green Plants," National Academy of Sciences, Washington, 1963.

San Pietro, A., Greer, F. A., and Army, T. J., eds., "Harvesting the Sun," Academic Press, Inc., New York, 1967.

Vernon, L. P., and Seely, G. R., eds., "The Chlorophylls," Academic Press, Inc., New York, 1966.

Review Articles

Arnon, D. I., Cell-free Photosynthesis and the Energy Conversion Process, in W. D. McElroy and B. Glass, eds., "Light and Life," pp. 489–569, The Johns Hopkins Press, Baltimore, 1961.

Arnon, D. I., Tsijimoto, H. Y., and McSwain, B. D., Photosynthetic Phosphorylation and Electron Transport, *Nature,* **207,** 1367–1372, 1965.

Bassham, J. A., Photosynthesis: Energetics and Related Topics, *Advances in Enzymol.,* **25,** 39–117, 1963.

Buchanan, B. B., The Chemistry and Function of Ferredoxins, in *Structure and Bonding,* **1,** 109–148, 1966.

Crane, F. L., and Low, H., Quinones in Energy-coupling Systems, *Physiol. Revs.,* **46,** 662–695, 1966.

Duysens, L. N. M., Photosynthesis, *Progr. in Biophysics and Molecular Biology,* **14,** 1–104, 1964.

Good, N., Izawa, S., and Hind, G., Uncoupling and Energy Transfer Inhibition in Photophosphorylation, in R. Sanadi, ed., "Current Topics in Bioenergetics," vol. 1, pp. 76–112, Academic Press, Inc., New York, 1966.

Hill, R., The Biochemist's Green Mansions: The Photosynthetic Electron Transport Chain in Plants, in P. N. Campbell and G. D. Greville, eds., "Essays in Biochemistry," vol. 1, pp. 121–152, Academic Press, Inc., New York, 1965.

Jagendorf, A., Photosynthesis, *Survey of Biol. Progr.*, **4**, 183–344, 1962.

Jagendorf, A., and Uribe, E., Photophosphorylation and the Chemiosmotic Hypothesis, *Brookhaven Symp. Biol.* **19**, 215–245, 1967.

Kok, B., Photosynthesis, in J. Bonner and J. E. Varner, eds., "Plant Biochemistry," pp. 904–960, Academic Press, Inc., New York, 1965.

Kok, B., and Cheniae, G. M., Kinetics and Intermediates of the Oxygen Evolution Step in Photosynthesis, in R. Sanadi, ed., "Current Topics in Bioenergetics," vol. 1, pp. 2–48, Academic Press, Inc., New York, 1966.

San Pietro, A., and Black, C. C., Enzymology of Energy Conversion in Photosynthesis, *Ann. Rev. Plant Physiol.,* **16,** 155–174, 1965.

Seliger, H. H., and McElroy, W. D., Photosynthesis, in "Light, Physical and Biological Action," pp. 233–255, Academic Press, Inc., New York, 1965.

Smith, J. H. C., and French, C. S., The Major and Accessory Pigments in Photosynthesis, *Ann. Rev. Plant Physiol.,* **14,** 181–224, 1963.

Van Niel, C. B., The Bacterial Photosyntheses and Their Importance for the General Problem of Photosynthesis, *Advances in Enzymol.,* **1,** 263–328, 1941.

Vernon, L. P., and Avron, M., Photosynthesis, *Ann. Rev. Biochem.,* **34,** 269–270, 1965.

21. Lipid Metabolism. I

Digestion and Absorption. Blood Lipids and Lipemia. Body Lipids. Oxidation of Fatty Acids. Synthesis of Fatty Acids. Fatty Acid Interconversions. Synthesis of Triglycerides. Regulation of Lipid Metabolism

In this and the following chapter will be described the metabolism of the triglycerides, phospholipids, sphingolipids, and the sterols. The metabolism of the "lipid-soluble" vitamins, A, D, E, and K is presented in Chap. 50.

The mammal requires, for optimal growth and maintenance, small amounts of the lipid-soluble vitamins and certain unsaturated fatty acids. With these exceptions, lipid is apparently not essential in the diet; lipids can be synthesized at a rate adequate for normal growth and health. The importance of lipid in the diet should, however, not be underestimated. Lipids are the most concentrated source of energy to the organism, yielding, per gram, over twice as many calories as do carbohydrates or proteins (page 292).

DIGESTION OF DIETARY LIPID

The bulk of dietary lipid is triglyceride. In the gastrointestinal tract a portion is hydrolyzed by *lipases* to fatty acids and glycerol.

$$\begin{array}{l} H_2COOCR \\ \quad | \\ HCOOCR \\ \quad | \\ H_2COOCR \end{array} + 3H_2O \rightleftharpoons 3RCOOH + \begin{array}{l} H_2COH \\ \quad | \\ HCOH \\ \quad | \\ H_2COH \end{array}$$

Incomplete hydrolysis yields a mixture of mono- and diglycerides in addition to the final products of the process.

Gastric Digestion of Lipid. The optimal action of gastric lipase is near neutrality, and at the low pH values in the stomach, it is essentially inactive; its significance is therefore uncertain, although some fatty acids are liberated in the stomach. It has been suggested that gastric lipase may be more important in the infant since the gastric pH is much higher in infancy and since the lipid of milk occurs in a highly emulsified state, a condition favorable to attack by a water-soluble enzyme.

Intestinal Digestion of Lipid. The major site of lipid digestion is the small intestine. In the duodenum the bolus of food encounters the *bile* and the *pancreatic juice.*

The Role of Bile. The function of bile is to promote emulsification and solubilization of lipids, and this function is due to the salts of the bile acids (page 81). Bile secretion is considered in Chap. 34. Since lipolysis involves water and water-soluble lipases, and since lipids are essentially insoluble in water, hydrolysis occurs only at the interface between the lipid droplet and the aqueous phase. The rate of reaction is in part determined by the area of this interface, and the higher the degree of emulsification, the smaller the individual lipid droplet and the larger the total available area will be. As a first approximation, per unit volume of lipid, the area of the interface will vary inversely as the radius of the average droplet and directly as the cube root of the number of droplets. Bile and pancreatic juice are somewhat alkaline and serve in part to neutralize the acidic gastric chyme. In the approximately neutral environment of the duodenal lumen the bile acids, largely taurocholic and glycocholic acids, exist as anions, and serve as detergents or emulsifying agents (page 62). In the presence of these detergents the churning effect of peristalsis results in a progressively finer and finer state of distribution of the dietary lipid in the continuous aqueous phase, facilitating lipolysis.

Lipid emulsification by the bile salts is apparently not essential for digestion, in that the lipid residue appearing in the feces when bile is totally excluded from the gastrointestinal tract is largely hydrolyzed and consists chiefly of salts of fatty acids. The intolerance for dietary lipid in patients with biliary obstruction suggests that lipid digestion is retarded in such individuals. However, the soaps as well as the monoglycerides which result from the partial hydrolysis of fats also act as detergents and thus supplement the function of the bile salts.

Pancreatic Lipase. The flow of pancreatic juice, like the flow of bile, is regulated hormonally after the introduction of gastric chyme into the duodenum. A precursor of lipase in the pancreatic juice becomes active in the intestinal lumen. The mechanism of activation of pancreatic lipase is not clear, but it has been suggested that a cofactor is needed for its activity. The degree of unsaturation (0 to 2 double bonds) and chain length (C_{12} to C_{18}) have no significant effect on the rate of hydrolysis by this enzyme. Ca^{++} has an accelerating effect on the enzyme, mainly because it forms insoluble soaps with liberated fatty acids. This prevents their inhibitory action on the enzyme and also retards resynthesis of glyceride, effectively shifting the reaction in the direction of hydrolysis.

Hydrolysis occurs predominantly at the α or α' positions, producing an α,β-diglyceride. This is then hydrolyzed to a monoglyceride, predominantly the β form (80 per cent). Indeed, since acyl migration occurs, it is not certain that pancreatic lipase can hydrolyze the β-monoglyceride. The main course of the reactions is pictured (top of page 472), where R, R′, and R″ are different fatty acid chains.

Complete hydrolysis is not a prerequisite to absorption. Fatty acid esters which are resistant to hydrolysis, *e.g.*, methyl elaidate, are absorbed and deposited in the body lipids. Apparently, such fat as is not absorbed is rather completely hydrolyzed, since in feces the fatty acids are present almost entirely as soaps.

A group of *esterases,* other than the above lipase, is also present in the pancre-

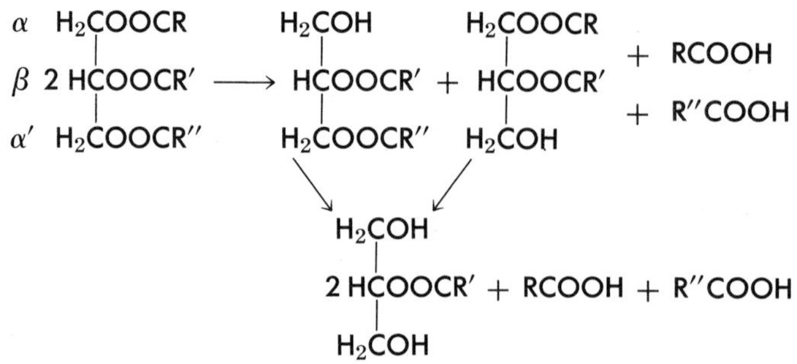

$$\begin{array}{l} \alpha \;\; H_2COOCR \\ \beta \;\; 2\,HCOOCR' \\ \alpha' \;\; H_2COOCR'' \end{array} \longrightarrow \begin{array}{l} H_2COH \\ HCOOCR' \\ H_2COOCR'' \end{array} + \begin{array}{l} H_2COOCR \\ HCOOCR' \\ H_2COH \end{array} \;\; \begin{array}{l} + \;\; RCOOH \\ \\ + \;\; R''COOH \end{array}$$

$$\begin{array}{c} H_2COH \\ 2\,HCOOCR' + RCOOH + R''COOH \\ H_2COH \end{array}$$

atic juice. These esterases catalyze hydrolysis preferentially of short-chain fatty acids, *e.g.*, tributyrin, and of other fatty acid esters, notably cholesterol esters.

A pancreatic *phospholipase A* liberates lysolecithin (page 70) from lecithin. Lysolecithin is a good detergent and aids in emulsification of the dietary lipid. Since some lecithin is present in bile, this fluid contributes a precursor of lysolecithin in addition to supplying other detergents. The presence of a phospholipase A in the intestinal mucosa has also been indicated.

INTESTINAL ABSORPTION OF LIPIDS

After ingestion of a fatty meal, the small intestine contains free fatty acids, as their soaps, together with a mixture of mono-, di-, and triglycerides well emulsified by the bile salts and the soaps themselves. A major portion of this mixture is absorbed across the wall of the small intestine. Such glycerol as is liberated is water-soluble and, together with other water-soluble compounds, including short-chain fatty acids, leaves the intestine by the portal route. The fatty acids, on the other hand, are delivered to the organism predominantly via the intestinal lymph, where they appear as triglycerides.

The demonstration of the chylous absorption of lipids was due to Munk, who in 1891 studied a patient with a lymph fistula draining at the thigh. Munk fed a variety of fats to his patient and noted that over 60 per cent of the ingested fat could generally be recovered from the discharge of the fistulous tract. Shortly after feeding of a fatty meal, the lymph, clear during fasting, became milky owing to the appearance of minute fat droplets, which subsequently were termed *chylomicra* (see below). The preponderant lipid of the lymph was triglyceride, and even when the fatty acids were fed as esters of other alcohols, it was mainly as esters of glycerol that they were recovered from the chyle.

These observations and others led to the idea that esters were completely hydrolyzed in the small intestine and that fatty acids crossed the mucosal barrier and entered the terminal lymphatics of the intestinal villi, reesterified to glycerol.

There are two independent concepts in this picture of lipid absorption: (1) total hydrolysis is a prerequisite for absorption; (2) products of lipid digestion enter the circulation exclusively by the lymphatic route. Neither of these concepts has proved to be entirely correct. Studies with isotopically labeled triglycerides indicated that approximately 40 per cent of fed triglycerides are hydrolyzed to glycerol and fatty acids, 3 to 10 per cent are absorbed as triglyceride, and the remainder is partially hydrolyzed, mainly to the β-monoglycerides.

Long-chain fatty acids (more than 14 carbon atoms), whether fed as triglycerides or as free fatty acids, appear in the chyle almost quantitatively as regenerated triglycerides in *chylomicra.* These are particles approximately 1 μ in average diameter which are lipoproteins, containing chiefly lipid and a small amount of protein. Thus during absorption, *resynthesis* of triglycerides occurs, particularly from β-monoglycerides, and is accompanied by protein synthesis required for formation of the lipoproteins, notably the β-lipoproteins (page 719) of the chylomicra. The mechanism of triglyceride synthesis is considered later (page 496) and of protein synthesis in Chap. 29.

The absorbed lipids enter the blood via the thoracic duct and accessory channels, chiefly at the angle of the left jugular and subclavian veins. Certain exceptions may be noted. In general, lipids which are liquid at body temperature are efficiently digested and absorbed. Lipids which melt significantly above body temperature are poorly digested and absorbed, and fatty acids which are solids above body temperature will not be well absorbed unless mixed with lower-melting lipids.

Fatty acids of chain length less than 10 carbon atoms are absorbed predominantly in nonesterified form by the portal route and consequently are presented directly to the liver. Since of all the common dietary lipids only those of milk are rich in fatty acids of shorter chain length, this fact is of particular interest in infant nutrition.

Factors Affecting Absorption of Lipids. Not all the factors responsible for the partition of fatty acids of different chain length to the lymph and the blood are known. However, differences in water solubility, protein interaction, micelle formation, and enzymic specificity with respect to triglyceride resynthesis (see below), in combination with permeability influences, affect the different routes of absorption. Certain factors, however, affect the passage through the intestinal mucosa, and perhaps the most important relates to the presence of detergents in the intestinal lumen.

When bile is totally excluded from the intestinal tract as a result of severe liver dysfunction, extrahepatic biliary obstruction, or biliary fistula, lipid absorption is markedly impeded. As a result, the total lipid content of acholic feces is elevated with an abundance of salts of fatty acids. The presence of these soaps, chiefly insoluble calcium salts, together with the absence of bile pigment, results in characteristic "clay-colored" stools. Not only is absorption of fatty acids impeded, but other lipid-soluble substances are also poorly absorbed. Most striking are the signs of vitamin K deficiency which are observed in biliary obstruction; these may be promptly relieved by oral administration of bile salts or by parenteral administration of vitamin K (Chap. 50). This is an important consideration in relation to surgery performed on patients with biliary disease, since vitamin K is required to ensure the normal rate of clotting of blood (Chap. 31).

The role of bile salts in fat absorption is probably associated with their detergent properties and with their formation of micelles with monoglycerides and soaps, together with small amounts of other lipid-soluble material. These micelles appear to be the major form in which the products of fat digestion are absorbed. Note that, although associated with lipid during passage across the mucosal barrier, the bile acids do not enter the lymphatic circulation. Rather, they are confined to an enterohepatic circulation, entering the portal blood, from which they are removed by the liver and reinjected with the bile into the duodenum. From this circuit relatively little bile acid is lost, little appears in the peripheral blood, and about 200 mg. appears per day in the feces.

A second factor which influences the absorption of lipids from the intestinal lumen is the metabolic activity of the intestinal mucosa per se. Enzymic systems are present in the cells of the intestinal mucosa which can convert free fatty acids and mono- and diglycerides to the triglycerides. One of these systems is similar to that in liver and is described later (page 496).

Some selectivity is exhibited by the intestinal mucosa in regard to the absorption of sterols, particularly those of plant origin. Of the principal dietary sterols, only cholesterol crosses the intestinal wall readily and is absorbed via the chylous route. Other steroids present in the diet in low concentration, *e.g.*, vitamin D and certain steroid hormones when given orally, are absorbed readily from the intestine.

The Lipids of the Feces—Steatorrhea. Lipids are present in the feces in part because of failure of quantitative absorption of dietary lipids, in part by virtue of excretion of lipids into the intestinal lumen. Lipids, notably steroids, also are present in the bile, in which they enter the intestine. In addition, direct excretion takes place across the intestinal barrier. Lipids excreted either via the bile or by the intestinal mucosa may appear in the feces, as will products formed from these lipids by intestinal bacteria. This last factor is of significance in sterol metabolism and will be considered later (page 522).

Excessive lipid in the stools is referred to as *steatorrhea.* Three important types of disturbance are recognized as resulting in steatorrhea: exclusion of bile, exclusion of pancreatic juice, and defect of the intestinal mucosa. Steatorrhea due to biliary insufficiency is, as indicated above, usually recognizable by the presence in the stools of excessive amounts of digested but unabsorbed lipid, chiefly in the form of soaps. The characteristic lack of bile pigment in the feces facilitates classification. Such steatorrhea may result from obstruction to the biliary passages caused by stone or neoplasm, biliary fistula, or severe diffuse liver disease.

The steatorrhea which results from deficiency of pancreatic juice is characterized by the presence of excessive amounts of undigested triglyceride. The pathological lesion may be *pancreatic fibrosis;* the same picture may be seen after total pancreatectomy. If the quantity of dietary fat is kept low, steatorrhea may not be observed.

Chronic disease of the pancreas must on occasion be differentiated from certain diseases of the intestinal tract proper, such as celiac disease in children or sprue in the adult. In these conditions, emulsification and digestion of fat are apparently normal, since the flow of bile and of pancreatic juice is not impaired. The steatorrhea, often of impressive proportions, is attributed to failure of the active absorptive processes in the intestinal mucosa. The lipid of the feces is predominantly in the form of soaps. In these conditions not only lipids but a variety of other nutrients fail to cross the intestinal barrier at a normal rate.

BLOOD LIPIDS AND LIPEMIA

Normal blood plasma in the postabsorptive state in man contains some 500 mg. of total lipid per 100 ml., of which about one-quarter is triglyceride. There are 180 mg. or more of cholesterol, of which some two-thirds is esterified with fatty acids and one-third is present as free sterol. Phospholipids comprise about 160 mg. per 100 ml.; the concentration of choline-containing phospholipids is higher than that of the ethanolamine phospholipids. (See also Table 31.1, page 707).

In the postabsorptive state, the mesenteric and thoracic duct lymph is a clear, watery fluid. Shortly after introduction of a fatty meal into the duodenum, the lymphatic channels, previously visualized with difficulty, become distended with a milky fluid rich in triglycerides. Most of the fat of the chyle is present as chylomicra. The discharge of this chyle into the venous blood results in a rapid rise in the lipid content of the plasma, occasionally sufficient to result in a milky opalescence. The increase in amount of blood lipid is termed *lipemia,* and specifically that which transiently follows ingestion of fat is called *absorptive lipemia.*

A number of lipoprotein fractions can be separated by repeatedly centrifuging plasma at high speeds after appropriately increasing the plasma density by addition of salts or ^2H_2O. These lipoproteins may be characterized by their flotation constants, S_f, which are analogous to the sedimentation constants of ordinary proteins (page 130), and by the densities at which they separate. Some of the properties of the lipoprotein fractions are given in Table 21.1 (also see Table 31.5, page 711). It is seen that the fractions of lowest density (high S_f) are richest in triglycerides and poorest in protein. The high water solubility of these macromolecules suggests that they may be large spheres containing lipid which is partially covered by a thin film

Table 21.1: LIPOPROTEINS OF HUMAN PLASMA

	Fraction				
	Chylomicra	Very low density	Low density	High density	Very high density
Density	<1.006	1.006–1.019	1.019–1.063	1.063–1.21	>1.21
S_f	>400	12–400	0–12		
Diameter, Å	5,000–10,000	300–700	200–250	100–150	100
Electrophoretic fraction	α_2	β_1	β_1	α_1	α_1
Amount, mg./100 ml. plasma	100–250	130–200	210–400	50–130	290–400
Approximate percentage composition					
Protein	2	9	21	33	57
Phospholipid	7	18	22	29	21
Cholesterol					
Free	2	7	8	7	3
Ester	6	15	38	23	14
Triglyceride	83	50	10	8	5
Fatty acids	...	1	1		

SOURCE: Adapted from J. L. Oncley, in F. Homburger and P. Bernfeld, eds., "The Lipoproteins: Methods and Clinical Significance," S. Karger, New York, 1958.

of hydrophilic protein. Thus the water-soluble plasma proteins play a major role in the transport of lipids. Even in the chylomicra, the small amount of protein apparently aids in stabilizing the droplets. There is probably no *free* lipid in blood plasma, *i.e.*, no lipid which is not associated with protein.

Lipoprotein Lipase. Intravenous injection of heparin markedly accelerates the elimination of turbidity of lipemic plasma in vivo. *Clearing factors* (*lipoprotein lipases*) have been found in various tissues, notably heart, lung, and adipose tissue. The lipase in blood is distinct from those which function in the cells of adipose tissue; it appears in plasma during lipemia, and its activity is enhanced by heparin, other acidic polysaccharides, or various inorganic macromolecules. Incubation of clearing factor with a bacterial *heparinase* inactivates the lipase, suggesting that heparin or a similar mucopolysaccharide is a component of the enzyme. The latter is active only in the presence of an added cation such as Ca^{++}, Mn^{++}, Mg^{++}.

Plasma lipoprotein lipase catalyzes hydrolysis of triglycerides present in chylomicra and in lipoproteins but only when they are bound to protein. Moreover, the presence of an acceptor of liberated fatty acids, such as serum albumin, is essential. Thus albumin plays a major role in the transport of unesterified or free fatty acids, which are present in a concentration of 8 to 30 mg. per 100 ml. of plasma (Table 31.1, page 707). These fatty acids have a high metabolic turnover rate. The lipoprotein lipases of the tissues are important in the mobilization of fatty acids from the lipid depots (page 500).

Lipemia. Unexplained "idiopathic" lipemia is occasionally observed. Of greater interest are the lipemias frequently seen in diabetic acidosis and glycogen storage disease. In some patients with the nephrotic syndrome, the normal utilization of lipoproteins with S_f 0 to 20 appears to be impaired. Individuals surviving a myocardial infarction have elevated levels of lower-density lipoproteins, particularly of the S_f 12 to 20 class. Such patients also may have hypercholesterolemia. However, marked alterations may occur in the normal relative concentrations of lipoproteins, *e.g.*, the ratio of α- and β-lipoproteins, without significant changes in the plasma total cholesterol values. Therefore, determinations of the distribution or relative concentrations of the lipoprotein fractions or of the lipids in these fractions are more informative than estimations of the total concentration of a specific lipid in plasma.

Males between the ages of twenty and forty show a higher plasma content of low-density lipoproteins and a lower content of high-density lipoproteins than do women in the same age group. Older males and females have higher concentrations of low-density lipoproteins than young males and females. Aspects of the postulated significance of this information to atherosclerosis will be discussed in Chap. 48.

THE LIPIDS OF THE BODY

In the normal mammal at least 10 per cent or more of the body weight is lipid, the bulk of which is *triglyceride*. This lipid is distributed in all organs as well as in certain *depots* of specialized connective tissue, the adipose tissue, in which a large fraction of the cytoplasm of the cells appears to be replaced by droplets of lipid.

Functions of Body Lipid. Body lipid is a reservoir of potential chemical energy. There is, in the normal mammal, a far greater quantity of mobilizable lipid than of mobilizable carbohydrate. This lipid yields over twice as many calories per gram as does carbohydrate (page 292) and, in addition, is stored in a relatively water-free state in the tissues, in contrast to carbohydrate, which is heavily hydrated. In most normally nourished animals the lipid depots represent by far the largest reservoir of energy, available in times of restricted nutrition for the operation of the numerous endergonic processes necessary for maintenance of life.

Much of the lipid of mammals is located subcutaneously. This protects the more thermosensitive tissues against excessive heat loss to the environment. This function, of particular importance in homoiotherms, is best exemplified in marine mammals, whose environment is both colder than body temperature and a far better thermal conductor than air. In whales a thick and continuous layer of subcutaneous adipose tissue, blubber, reduces heat losses to the environment. The subcutaneous lipid depots also insulate against mechanical trauma. A dramatic example is the depot of highly specialized lipid, spermaceti (page 75), at the cephalic extremity of the sperm whale which permits it to deliver blows of great force with its head.

The paucity of depot lipid during the intrauterine life of the mammalian fetus is of interest in relation to the functions of depot lipid during adult life. The fetus derives its nutrition across the placenta from the maternal circulation and does so continuously, in contrast to the adult, who eats intermittently. The fetus therefore can sustain itself without long-term energy reservoirs. It resides in a thermoregulated environment and needs little additional insulation. Furthermore, the fetus is well protected by amniotic fluid and maternal tissues against mechanical blows and thus requires none of the cushioning which the adult derives from its depot lipid. It is only shortly prior to term that the fetus acquires its depot lipid.

Characteristics of Depot Lipid. The depot lipid consists chiefly of triglyceride. Considerable variation is evident in different species, but within any species the composition of depot lipid is fairly uniform. More than 99 per cent of the lipid of human adipose tissue is triglyceride; the composition of the various adipose depots of man is similar, regardless of anatomical location. In general, depot lipid is richer in saturated fatty acids than is liver lipid (page 67). Lipid in the depots is in the liquid state; it appears that the lipid which is deposited subcutaneously is as saturated as is compatible with the liquid state. The more nearly saturated a sample of lipid, the larger the energy yield from oxidation. Thus, it would seem that mammals deposit under their skins that type of lipid richest in chemical potential energy and still liquid at the ambient temperature.

Although the composition of depot lipid within a species is reasonably uniform, variations may be induced by extremes of temperature or of diet. The effect of diet on the composition of depot lipid reflects the origins of depot fatty acids, *viz.*, from dietary fatty acids as well as from carbohydrate (see below). Prolonged feeding to human subjects of high corn oil-containing diets (40 per cent of calories as corn oil) resulted, beginning at 20 weeks on the diet, in significant increases in the unsaturated fatty acid content of the adipose tissue. At 160 weeks the adipose composition resembled corn oil more than it did normal adipose tissue. Earlier influences of dietary lipid on the composition of depot lipid are seen if an animal is fasted to deplete the major portion of the depots and then refed a diet rich in lipid of different physical properties from its native lipid. Thus, if hogs or rats are fed a diet rich in a highly unsaturated oil like peanut oil, the depots will contain a lipid resembling peanut oil in its high degree of unsaturation. Repletion of depot lipid by the feeding of an essentially lipid-free, high-carbohydrate diet leads to a depot lipid notably poor in unsaturated fatty acids.

Some mammals possess fat depots containing high concentrations of glycogen and a brown pigment. Such brown adipose tissue, as in the interscapular depots of the rat, is also active metabolically in a manner resembling the usual white adipose tissue and may have a different function, such as heat production. The most striking feature of brown fat cells is the presence of many lipid vacuoles surrounded by mitochondria; the cytochromes of the mitochondria probably account for the reddish-brown color of brown adipose tissue.

Metabolic Aspects of Body Lipid. When an animal is excessively nourished, the quantity of body lipid increases and, conversely, during periods of prolonged fasting the amount of body lipid decreases. It is possible, however, to adjust food intake so that the quantity of body lipid is constant over a long period of time, and indeed in most adult animals there seems to be some degree of regulation of appetite such that the lipid content of the body does not change rapidly.

Application of the technique of isotopic tracers to studies of metabolism (Chap. 13) has established the concept of continuous turnover (synthesis, degradation, and replacement of most body constituents), even at constant body composition (page 290). The lipids of the depots are continuously being mobilized, new lipid is continuously being deposited, and the constancy of the quantity of depot lipid is the result of a relatively precise adjustment of the rates of these two processes. This continuing turnover has been indicated above in the influence of dietary lipid on body lipid composition, even in a normal individual in caloric balance. Determination of the half-life of lipids has revealed that in the steady state the half-life of depot lipid in the mouse is about 5 days, in the rat about 8 days. This means that in the rat almost 10 per cent of the fatty acids in the depot lipid is replaced daily by new fatty acid. In the liver of the rat, the fatty acids have a half-life of about 2 days; in the brain, 10 to 15 days.

Adipose tissue exhibits two major metabolic features: (1) the assimilation of carbohydrate and lipids and their intermediates for fat synthesis and storage, and (2) the mobilization of lipid as free fatty acids, and to a more limited extent, as glycerol. Both these aspects are profoundly influenced by hormones. The fatty acids of adipose tissue, present as triglyceride, derive from both dietary fatty acids and carbohydrate. The pathway of fatty acid synthesis from intermediates of carbohydrate metabolism is considered below (pages 484 and 497 *ff.*), as are triglyceride formation (page 496) and the mobilization of lipids from depots (page 500). Comments here will be directed to the assimilation of lipid by adipose tissue.

The assimilative and storage activities of adipose tissue are reflected in its capacity to take up both triglycerides and free fatty acids. Glucose feeding enhances the incorporation of intravenously injected chylomicra into rat adipose tissue. The triglycerides may be taken up without prior hydrolysis, although maximal uptake by adipose tissue apparently involves triglyceride hydrolysis and reesterification. Metabolism of carbohydrate promotes esterification. Both the glycolytic and the phosphogluconate oxidative pathways operate in adipose tissue, with the latter being relatively more active than the former. The glycerol portion of the stored triglyceride is derived from glucose at the triose phosphate step in glycolysis (page 398), presumably from dihydroxyacetone phosphate via α-glycerophosphate. Adipose tissue in vitro is unable to assimilate glycerol itself because of a lack of glycerol kinase.

Many of the hormones influence the above-described processes in adipose tissue. Thus, insulin augments glucose uptake and lactate and glycerol production, as well as assimilation of triglyceride by adipose tissue (Chap. 46). There is increased lipogenesis from carbohydrate. Epinephrine acts in a manner opposite to that of insulin (Chap. 45). Other hormonal effects on adipose tissue will be described in Part Five.

OXIDATION OF FATTY ACIDS

The action of the lipases results in the hydrolysis of neutral fats to glycerol and fatty acids. Glycerol, derived from fats or phospholipids, is glycogenic, entering the glycolytic pathway via formation of α-glycerophosphate by the action of ATP and glycerokinase (page 512).

Fatty acid oxidation is accomplished by a sequence of reactions in which the fatty acyl chain is shortened in stages, two carbon atoms at a time. The fatty acyl coenzyme A derivatives, not the free fatty acids, are involved in these reactions, and the liberated 2-carbon fragments are acetyl CoA. The enzymes catalyzing these reactions have all been isolated from mitochondria.

If we begin with a fatty acid derivative of coenzyme A, the shortening of the fatty acid chain by two carbon atoms is due to four successive reactions: (1) dehydrogenation catalyzed by a flavoprotein to yield the α,β-unsaturated derivative, (2) hydration of the double bond to form the β-hydroxy compound, (3) dehydrogenation involving DPN to yield the β-keto derivative, and (4) reaction of the β-keto acyl CoA with CoA to yield acetyl CoA and a fatty acid derivative of CoA which is shorter by two carbon atoms. Successive repetitions of this sequence of four reactions result in the complete degradation of an even-numbered carbon atom fatty acid to acetyl CoA and, starting with a fatty acid containing an odd number of carbon atoms, to successive molecules of acetyl CoA and one of propionyl CoA.

The four reactions are written to show that the acyl derivatives are linked to the thiol group of CoA.

$$\mathrm{RCH_2\overset{\beta}{C}H_2\overset{\alpha}{C}H_2COSCoA + FAD \xrightarrow[\text{dehydrogenase}]{\text{acyl}} RCH_2CH{=}CHCOSCoA + FADH_2} \quad (1)$$

$$\mathrm{RCH_2CH{=}CHCOSCoA + H_2O \underset{\text{hydrase}}{\overset{\text{enoyl}}{\rightleftharpoons}} RCH_2\overset{\beta}{C}HOH\overset{\alpha}{C}H_2COSCoA} \quad (2)$$

$$\mathrm{RCH_2CHOHCH_2COSCoA + DPN^+ \underset{\text{dehydrogenase}}{\overset{\beta\text{-hydroxyacyl}}{\rightleftharpoons}} RCH_2COCH_2COSCoA + DPNH + H^+} \quad (3)$$

$$\mathrm{RCH_2\overset{\beta}{C}O\overset{\alpha}{C}H_2COSCoA + HSCoA \overset{\text{thiolase}}{\rightleftharpoons} RCH_2COSCoA + CH_3COSCoA} \quad (4)$$

Each of the major types of reaction can now be considered further.

Activation Reactions. Degradation of fatty acids requires an initial transformation to the corresponding acyl CoA derivatives; this is accomplished in two ways.

The *thiokinases* catalyze directly the formation of the CoA derivatives according to the following reaction.

$$\text{RCOOH} + \text{CoA} + \text{ATP} \underset{}{\overset{\text{Mg}^{++}}{\rightleftharpoons}} \text{RCOCoA} + \text{AMP} + \text{PP}_i$$

Several such enzymes are known, and they are named according to the length of the carbon chain of the compound which reacts most rapidly, *e.g.*, *acetic thiokinase* (acts on C_2 and C_3 fatty acids), *octanoic thiokinase* (C_4 to C_{12} fatty acids), and *dodecanoic thiokinase* (C_{10} to C_{18} fatty acids). Present knowledge of the mechanism of such reactions has already been discussed (page 327).

An acyl thiokinase that specifically utilizes GTP, instead of ATP, has been isolated from beef liver mitochondria. This enzyme activates several fatty acids, and GDP and P_i are reaction products.

A second mechanism for synthesis of acyl CoA derivatives of fatty acids is by transfer reactions catalyzed by *thiophorases*.

$$\text{Succinyl CoA} + \text{R—COOH} \overset{\text{thiophorase}}{\rightleftharpoons} \text{succinic acid} + \text{R—COSCoA}$$

Activation of fatty acids occurs mainly by thiokinase reactions in animal tissues, such as liver, heart, and kidney, whereas in some microorganisms, reactions catalyzed by thiophorases predominate.

Acyl CoA Dehydrogenases. These enzymes, which catalyze the formation of the α,β-unsaturated fatty acyl CoA derivatives [reaction (1) above], all contain FAD. Three such enzymes have been isolated from pig liver and are named for the most sensitive substrates, *viz.*, *butyryl CoA dehydrogenase, octanoyl CoA dehydrogenase,* and *hexadecanoyl CoA dehydrogenase.* The type of reaction catalyzed by each of these enzymes is the following.

$$\text{Saturated fatty acyl CoA} + \text{FAD} \longrightarrow \alpha,\beta\text{-unsaturated fatty acyl CoA} + \text{FADH}_2$$

As indicated previously (page 341), these fatty acyl dehydrogenases have only indirect access to the electron transport chain and electrons must be transferred from the flavin moiety to that of an electron-transferring flavoprotein, ETF, which in turn directs the electrons to the cytochrome system.

Enoyl CoA Hydrase (Crotonase). Only one enzyme is presently known which catalyzes reversibly the hydration of the *trans*-unsaturated fatty acyl CoA [reaction (2) above], according to the following reaction.

$$\alpha,\beta\text{-Unsaturated fatty acyl CoA} + \text{H}_2\text{O} \rightleftharpoons \text{L-}\beta\text{-hydroxyacyl CoA}$$

Although of broad specificity, the enzyme is most active with crotonyl CoA as substrate, hence the name *crotonase,* given in analogy with fumarase (page 333). Crystalline crotonase, a sulfhydryl enzyme, does not require a cofactor.

β-Hydroxyacyl CoA Dehydrogenase. Reaction (3) above requires DPN.

$$\text{L-}\beta\text{-Hydroxyacyl CoA} + \text{DPN}^+ \rightleftharpoons \beta\text{-ketoacyl CoA} + \text{DPNH} + \text{H}^+$$

Substrates of different chain length are attacked by a single enzyme.

Thiolases. The reaction catalyzed by thiolases [reaction (4) above] involves a thiolytic cleavage by CoA with formation of acetyl CoA.

$$\mathbf{C_n\text{-}\beta\text{-Ketoacyl CoA} + CoA \rightleftharpoons C_{(n-2)}\text{-fatty acyl CoA} + acetyl\ CoA}$$

Several thiolases exist which possess different chain length specificity. They are thiol enzymes, and an acyl-S-enzyme may be an intermediate in a two-step reaction of the type shown below.

$$\mathbf{C_n\text{-}\beta\text{-Ketoacyl CoA} + HS\text{-enzyme} \rightleftharpoons C_{(n-2)}\text{-fatty acyl-S-enzyme} + acetyl\ CoA}$$
$$\mathbf{C_{(n-2)}\text{-Fatty acyl-S-enzyme} + CoA \rightleftharpoons C_{(n-2)}\text{-fatty acyl CoA} + HS\text{-enzyme}}$$

Although the over-all reaction is reversible, the equilibrium position is greatly in the direction of cleavage. The equilibrium constant is 6×10^4 for formation of 2 moles of acetyl CoA from acetoacetyl CoA.

In summary, the shortening of a fatty acyl CoA derivative by two carbon atoms is shown by the following equation.

$$RCH_2CH_2CH_2COSCoA + FAD + DPN^+ + CoA \longrightarrow RCH_2COCoA + CH_3COSCoA + FADH_2 + DPNH + H^+$$

The acetyl CoA generated by fatty acid degradation mixes with acetyl CoA arising from other biochemical reactions, including the oxidative decarboxylation of pyruvate (page 327), as well as from the degradative reactions of many amino acids (Chap. 26). The numerous fates of acetyl CoA, including its entry into the tricarboxylic acid cycle, into fatty acid and sterol synthesis, etc., are described elsewhere in this text.

In addition to the CoA derivatives, the other major products of fatty acid catabolism are the reduced coenzymes DPNH and $FADH_2$. These are ultimately oxidized by the steps outlined in Chap. 16. Energy for useful work accrues to the organism in two ways as a result of fatty acid breakdown. The oxidation of $FADH_2$ and DPNH results in formation of ATP, estimated as 5 moles of energy-rich phosphate per mole of O_2 used for production of acetyl CoA by the described degradation cycle. Oxidation of the acetyl CoA produced, via the tricarboxylic acid cycle, yields an additional 12 moles of ATP per mole of acetyl CoA oxidized (Table 16.2, page 345). Net energy yield from the oxidation of a mole of palmityl CoA is calculated in Table 21.2. Assuming ΔF of +8000 cal. per mole of ATP under physiological circumstances, this represents a conservation of about 1,050,000 cal. of chemical energy in the form of energy-rich phosphate resulting from the complete oxidation of one mole of palmityl CoA. This energy yield is about one-half of the 2,400,000 cal. released when one mole (256 g.) of palmitic acid is oxidized to CO_2 and H_2O in a bomb calorimeter.

Table 21.2: ENERGY-RICH PHOSPHATE DERIVED FROM FATTY ACID OXIDATION

Reaction	*Moles ~ P formed*
Palmityl CoA + $7O_2 \longrightarrow$ 8 acetyl	35
8 Acetyl + $16O_2 \longrightarrow 16H_2O + 16CO_2$ (8 revolutions of tricarboxylic acid cycle yielding 12 ~ P per revolution)	96
Total	131

FATTY ACID CYCLE

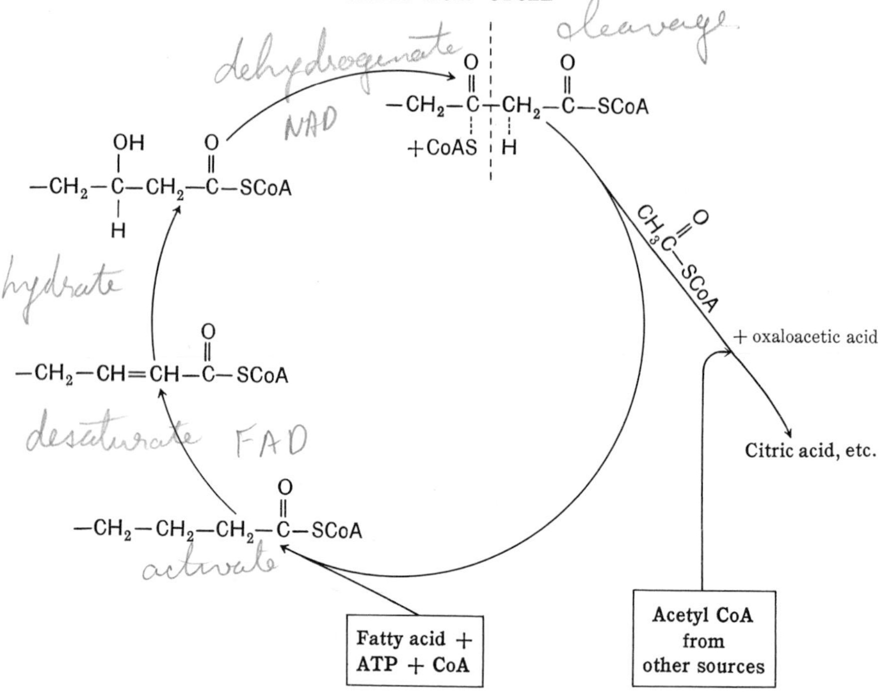

Fig. 21.1. Relationship of fatty acid degradation to the citric acid cycle.

The relationship of the cyclic reactions involving fatty acid oxidation to the tricarboxylic acid cycle is shown in Fig. 21.1.

Role of Carnitine in Fatty Acid Oxidation. Carnitine, first isolated from muscle, stimulates fatty acid oxidation in mitochondria.

$$(CH_3)_3\overset{+}{N}-\underset{\displaystyle OH}{\underset{|}{CH}}-CH_2-COO$$

Carnitine

The formation of acyl carnitines is of importance in the transport of acyl groups into and out of mitochondria. Two enzymes are involved; one catalyzes the acylation of carnitine with short-chain fatty acids, and is termed *acetyl coenzyme A–carnitine acetyl transferase.* The second enzyme catalyzes acylation of carnitine with long-chain fatty acids and has been named *palmityl coenzyme A–carnitine palmityl transferase.*

$$\textbf{Acetyl CoA + carnitine} \rightleftharpoons \textbf{acetyl carnitine + CoA}$$
$$\textbf{Palmityl CoA + carnitine} \rightleftharpoons \textbf{palmityl carnitine + CoA}$$

The above reactions are readily reversible; K is close to 1, indicating that the O-acyl bond of carnitine is a high-energy bond. The above enzymes are found in both the mitochondria and cytoplasm of cell fractions. However, they are probably located within the mitochondrial barrier, functioning as permeases (page 783).

The formation of acyl carnitine derivatives facilitates transfer of long-chain fatty acids from cytoplasm into mitochondria as acyl carnitines, where they react with CoA to form acyl CoA derivatives which are oxidized by the previously described mechanisms.

In accord with these suggested roles of carnitine in fatty acid metabolism are the findings that carnitine stimulates fatty acid oxidation by mitochondria and that the highest concentrations of carnitine, acetyl carnitine, and carnitine acyl transferases are found in tissues in which fatty acid oxidation is stimulated to the greatest degree by the addition of carnitine. Carnitine also functions in fatty acid synthesis (see below).

α-Oxidation of Fatty Acids. From the foregoing it is apparent that β-oxidation results in shortening of the fatty acid chain two carbon atoms at a time. Although this is the major oxidative fate of fatty acids, two other types of oxidation also occur, α- and ω-oxidation.

Oxidation of long-chain fatty acids to α-hydroxy acids and then to fatty acids with one carbon atom less than the original substrate has been demonstrated in brain microsomes. α-Hydroxy long-chain fatty acids are constituents of brain lipids (page 859). These hydroxy fatty acids may be converted to the α-keto acids, followed by oxidative decarboxylation, affording a pathway to the synthesis of long-chain fatty acids with an odd number of carbon atoms.

$$RCH_2—CH_2—CH_2—COOH \longrightarrow RCH_2—CH_2—CHOH—COOH \longrightarrow$$
$$RCH_2—CH_2—CO—COOH \longrightarrow RCH_2—CH_2—COOH + CO_2$$

The initial α-hydroxylation step is catalyzed by a *monooxygenase* which requires O_2, Fe^{++}, and either ascorbate or a tetrahydropteridin (page 548). Conversion of the α-hydroxy acid to an enzyme-bound α-keto acid is catalyzed by a DPN-specific dehydrogenase; the final decarboxylation involves DPN, ATP, and ascorbate.

ω-Oxidation of Fatty Acids. Fatty acids of average chain length and, to a lesser extent, long-chain fatty acids may initially undergo ω-oxidation to ω-hydroxy fatty acids which are subsequently converted to dicarboxylic acids. This series of reactions has been observed with enzymes in liver microsomes and with soluble enzyme preparations from bacteria. The initial reaction is catalyzed by a monooxygenase which requires DPNH, Fe^{++}, O_2, and a protein fraction of unknown function. Once formed, the dicarboxylic acid may be shortened from the ω end by the β-oxidation sequence described previously for fatty acid oxidation.

Propionate Metabolism. Oxidation of a fatty acid with an even number of carbon atoms results in complete degradation to acetyl CoA. Oxidation of an odd-numbered carbon atom fatty acid also yields acetyl CoA and, in addition, one equivalent of propionyl CoA. Propionic acid or propionyl CoA is also produced by the oxidation of branched aliphatic amino acids (page 605). Formation of propionyl CoA from propionic acid derived from any source is catalyzed by *acetic thiokinase* (page 480). In animal tissues the major pathway of propionyl CoA metabolism is, in part, the reverse of the mechanism whereby certain bacteria produce propionic acid (page 414). This pathway is summarized in the following three equations; each reaction will be considered in turn.

$$\text{Propionyl CoA} + \text{ATP} + \text{CO}_2 \xrightleftharpoons[]{\text{propionyl CoA carboxylase, Mg}^{++}} \text{ADP} + \text{P}_\text{i} + \begin{array}{c}\text{COOH}\\ | \\ \text{CHCH}_3 \\ | \\ \text{COSCoA}\end{array} \quad (1)$$

D-Methylmalonyl CoA

$$\text{D-Methylmalonyl CoA} \xrightleftharpoons[]{\text{methylmalonyl CoA racemase}} \text{L-methylmalonyl CoA} \quad (2)$$

$$\text{L-Methylmalonyl CoA} \xrightleftharpoons[]{\text{methylmalonyl CoA mutase}} \text{succinyl CoA} \quad (3)$$

Propionyl CoA carboxylase, as crystallized from pig heart by Ochoa and coworkers, contains one mole of bound biotin per 175,000 g. of protein of molecular weight 700,000. Biotin is covalently bound to the apoenzyme through an amide linkage to ε-amino groups of lysine residues, as in other biotin enzymes (Chap. 49). The carboxylase reaction involves formation of an enzyme-biotin-CO_2 complex similar to that described for the acetyl CoA carboxylase (page 487). *Methylmalonyl CoA racemase* by labilization of an α hydrogen atom followed by uptake of a proton from the medium, catalyzes interconversion of D- and L-methylmalonyl CoA. *Methylmalonyl CoA mutase* utilizes cobamide coenzyme (page 1040). When 2-^{14}C-methylmalonyl CoA was converted by the mutase, the label (marked * below) was found in the 3 position of succinyl CoA, indicating an intramolecular transfer of the entire thioester group, —CO—S—CoA, rather than migration of the carboxyl carbon.

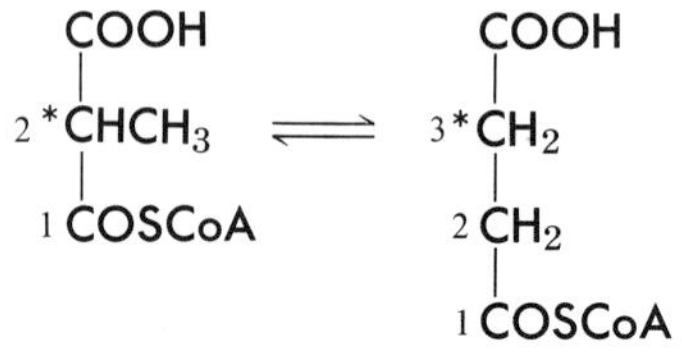

Methylmalonyl CoA Succinyl CoA

At equilibrium, formation of succinyl CoA is favored by a ratio of about 20:1 over methylmalonyl CoA. In patients with vitamin B_{12} deficiency both propionate and methylmalonate are excreted in the urine in abnormally large amounts.

The net result of the above three reactions is formation from propionyl CoA of succinyl CoA, an intermediate of the tricarboxylic acid cycle (page 331).

SYNTHESIS OF FATTY ACIDS

The chief source of fatty acids that has been considered thus far is the lipid of the diet. In addition, the mammal can synthesize the major portion of the fatty acids required for growth and maintenance. Saturated fatty acids as well as the common singly unsaturated fatty acids are readily and abundantly formed from acetyl CoA. Thus any substance capable of yielding acetyl CoA is a potential source of carbon atoms in the process of fatty acid synthesis, or *lipogenesis.*

Long before the mechanism of lipogenesis was studied, it was recognized that dietary carbohydrate could serve as a source of fatty acids. This was conclusively demonstrated by Lawes and Gilbert as early as 1860 in balance studies on fattening pigs, and these findings agree with countless observations that a high-carbohydrate intake can result in obesity. The corn-fed hog deposits more lipid subcutaneously than is contained in the diet. The demonstration that dietary protein can serve as a source of precursor for lipogenesis proved more difficult, in that it was not easy to develop obese animals on diets consisting chiefly of protein.

The occurrence of lipogenesis during the process of fattening was studied by balance experiments, in which the quantities of each class of nutrient ingested were compared with an increase in body lipid. With the isotopic technique, estimates were made of the *rate* of lipogenesis. In the animal maintained in the steady state on a high-carbohydrate, lipid-free diet, a portion of the ingested carbohydrate is used to regenerate depot lipid, which is simultaneously mobilized and catabolized. In the normal rat as much as 30 per cent of the dietary carbohydrate is utilized in lipogenic processes.

The rate of lipogenesis is not rapid under all circumstances. Synthesis of fatty acids is disturbed by a variety of conditions. If, for example, the total calories of the diet are restricted and become insufficient to maintain body weight, the rate of lipogenesis may fall to 5 per cent of its normal value. Some of the operative controls have been presented (page 478). Similar decreases in rate of lipogenesis have been observed in the thiamine-deficient animal, compatible with the role of thiamine pyrophosphate in the transformation of pyruvate into acetyl CoA (page 327).

When the pathway of fatty acid degradation was first elucidated, it was assumed that all the major steps were reversible and that fatty acid synthesis could proceed by catalysis with the same enzymes. However, there were several major difficulties with this notion. First, for isolated mitochondria or mitochondrial extracts to effect fatty acid synthesis, TPNH is required, a coenzyme unnecessary for the reactions of fatty acid oxidation. Second, fatty acid synthesis proceeds far more efficiently in nonmitochondrial (high-speed supernatant) fractions than in mitochondria. Indeed, evidence has accumulated that there are three systems for synthesis of fatty acids.

1. A cytoplasmic enzyme complex, present in many organisms, converts acetyl CoA to long-chain fatty acids when supplemented with ATP, CO_2, Mn^{++}, and TPNH. This system is referred to as the *palmitate-synthesizing system,* and catalyzes the following over-all process.

$$\text{Acetyl CoA} + 7\ \text{malonyl CoA} + 14\text{TPNH} + 14\text{H}^+ \longrightarrow$$
$$1\ \text{palmitic acid} + 7\text{CO}_2 + 8\text{CoA} + 14\text{TPN}^+ + 6\text{H}_2\text{O}$$

2. A mitochondrial enzyme system catalyzes the elongation of fatty acid chains by the successive addition of acetyl CoA units; this system utilizes both DPNH and TPNH.

$$\text{RCH}_2\text{COCoA} + \text{acetyl CoA} + \text{DPNH} + \text{TPNH} + 2\text{H}^+ \longrightarrow$$
$$\text{R(CH}_2)_3\text{COCoA} + \text{DPN}^+ + \text{TPN}^+ + \text{CoA}$$

3. A microsomal enzyme system catalyzes elongation of both saturated and unsaturated fatty acyl CoA derivatives utilizing malonyl CoA and TPNH.

$$RCH_2COCoA + \text{malonyl CoA} + 2TPNH + 2H^+ \longrightarrow R(CH_2)_3COCoA + 2TPN^+ + CoA + CO_2$$

Note that the last two systems catalyze fatty acid elongation; in contrast, the first system catalyzes *de novo* formation of fatty acids.

CYTOPLASMIC SYSTEM FOR FATTY ACID SYNTHESIS

The cytoplasmic system catalyzing the *de novo* synthesis of fatty acids is a complex containing all the enzymes required for the conversion of acetyl CoA and malonyl CoA to palmitic acid. The complex from yeast has a particle weight of 2.3×10^6; that from pigeon liver, 4.5×10^5. The enzymes of the complex are tightly bound together in mammalian tissues and in yeast, but are readily dissociable in microorganisms and in plants. The systems of *Escherichia coli,* yeast, and pigeon liver have been extensively studied; similar systems obtain in animal tissues generally.

Of major interest in the elucidation of the cytoplasmic system has been the description by Vagelos, Stumpf, Wakil, Bloch, and their collaborators of one of the proteins of the complex, termed *acyl carrier protein* (ACP), which binds acyl intermediates during the formation of long-chain fatty acids. ACP from *E. coli* is a heat-stable protein (M.W. = 10,000) that contains one free sulfhydryl group per mole, at which the acyl derivatives are bound as thioesters. The substrate-binding site of ACP has been shown to be the sulfhydryl group of a prosthetic group, 4′-phosphopantetheine (page 643). Thus ACP is a protein analogue of coenzyme A which also contains 4′-phosphopantetheine as its acyl-binding site (page 326). In ACP, the 4′-phosphopantetheine is linked to a serine residue through a phosphodiester linkage.

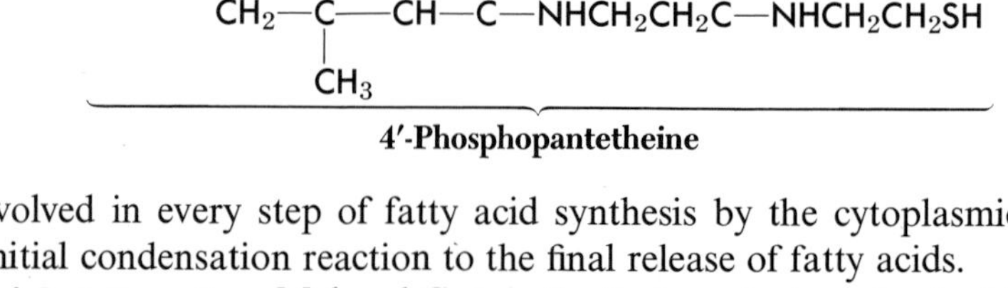

4′-Phosphopantetheine

ACP is involved in every step of fatty acid synthesis by the cytoplasmic system, from the initial condensation reaction to the final release of fatty acids.

Malonyl CoA Formation. Malonyl CoA is the first product synthesized in the sequence of reactions leading to long-chain fatty acid formation. The following reaction is catalyzed by *acetyl CoA carboxylase.*

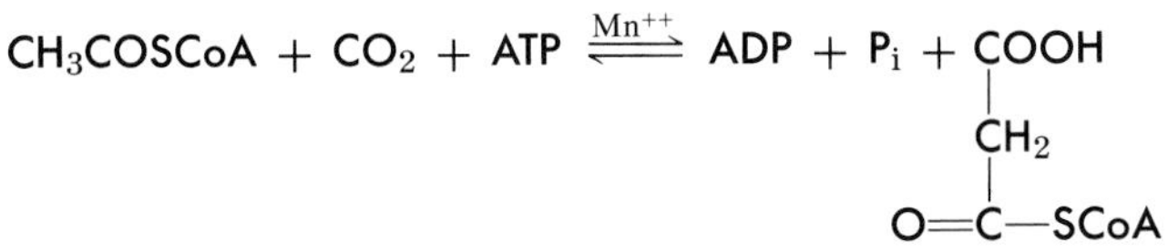

Malonyl CoA

The purified enzyme contains covalently bound biotin, which is essential for the reaction, and is specific for Mn^{++}. ATP can be replaced by UTP, but higher concentrations of the latter are required. Propionyl CoA can be carboxylated to methylmalonyl CoA (page 484) but at about one-fourth the rate of acetyl CoA carboxylation.

Malonyl CoA can be formed by reactions other than carboxylation of acetyl CoA, *viz.*, (1) by activation of malonate by a specific thiokinase in the presence of ATP and CoA, (2) by CoA transferase reactions between succinyl CoA or acetoacetyl CoA and malonate, and (3) by microbial oxidation of malonyl semialdehyde CoA to malonyl CoA. These reactions are of little significance in mammalian metabolism.

Mechanism of CO_2 Fixation. The general mechanism of action of this type of CO_2-fixation enzyme can be represented as occurring in two steps.

$$(a) \qquad \mathbf{HCO_3^- + ATP + biotin\text{-}enzyme \rightleftharpoons {}^-OOC\text{-}biotin\text{-}enzyme + ADP + P_i}$$
$$(b) \qquad \mathbf{{}^-OOC\text{-}biotin\text{-}enzyme + acetyl\ CoA \rightleftharpoons biotin\text{-}enzyme + malonyl\ CoA}$$

The structure of the carboxybiotin-enzyme intermediate was established by Lynen and coworkers.

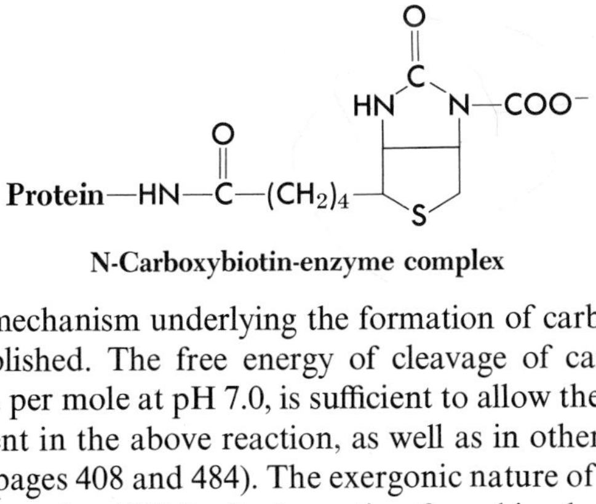

N-Carboxybiotin-enzyme complex

The chemical mechanism underlying the formation of carboxybiotin-enzyme is not definitely established. The free energy of cleavage of carboxybiotin-enzyme, $\Delta F^\circ = -4{,}700$ cal. per mole at pH 7.0, is sufficient to allow the compound to act as a carboxylating agent in the above reaction, as well as in other reactions with suitable acceptors (*cf.*, pages 408 and 484). The exergonic nature of the cleavage also explains the requirement for ATP for its formation from bicarbonate and the biotin-enzyme. Two quite different mechanisms of formation of carboxybiotin-enzyme have been proposed. They are depicted in Fig. 12.2 (page 488). Reaction (1) is a concerted reaction, whereas (2) consists of two consecutive steps, with the intermediate formation of a phosphorylated biotin-enzyme.

Control of Acetyl CoA Carboxylase Activity. The activity of acetyl CoA carboxylase is the rate-limiting step in fatty acid synthesis in mammals and is influenced by a number of metabolites. The activity of the enzyme from *E. coli* is markedly stimulated by various di- and tricarboxylic acids, notably citric and isocitric acids, as well as by α-ketoglutaric acid. Activity is also increased in the presence of long-

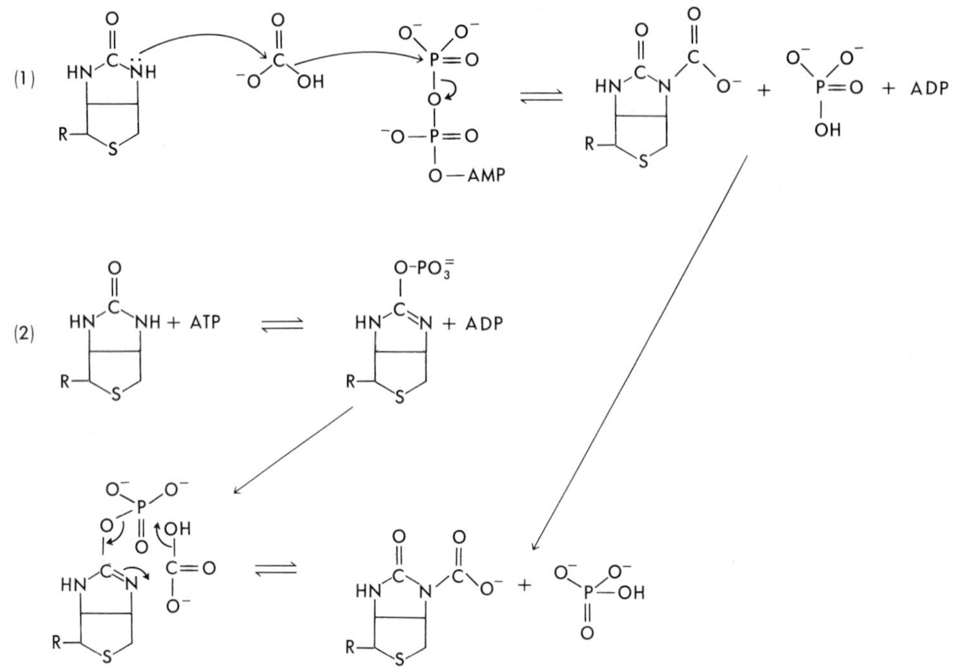

Fig. 21.2. Suggested mechanisms for the ATP-dependent carboxylation of biotin-enzyme. R = the protein portion of the biotin-containing enzyme. (*From F. Lynen, Biochem. J.,* **102**, 381, 1967.)

chain fatty acyl carnitine derivatives, *e.g.*, palmityl carnitine. In contrast, long-chain fatty acyl CoA derivatives, notably palmityl CoA, markedly inhibit the activity of acetyl CoA carboxylase. This inhibition could be relieved by addition of palmityl carnitine but not by citrate.

It has been inferred that these effects of various metabolites on acetyl CoA carboxylase activity result from conformational changes in the enzyme and may be of significance for the in vivo regulation of fatty acid synthesis. Some metabolites which stimulate enzymic activity, notably citrate and isocitrate, produce aggregation of the *E. coli* enzyme from an inactive monomer, molecular weight = 350,000, to an active aggregate of 20 units (molecular weight = 7×10^6). The inactive monomer of acetyl CoA carboxylase from liver, molecular weight = 400,000, aggregates in the presence of citrate or isocitrate to an active form consisting of 10 units (molecular weight = 4×10^6). Regulatory factors in lipid metabolism will be considered in greater detail later (pages 497 *ff.*).

Conversion of Malonyl CoA to Palmitic Acid. As indicated previously (page 485), seven moles of malonyl CoA and one mole of acetyl CoA are utilized in the overall synthesis of palmitic acid, and ACP is involved in each step of this synthetic sequence. In the following summary, ACP is written as ACP-SH to indicate its active acylation site.

$$\text{Acetyl-S-CoA} + \text{ACP-SH} \rightleftharpoons \text{acetyl-S-ACP} + \text{CoA-SH} \quad (1)$$
$$\text{Malonyl-S-CoA} + \text{ACP-SH} \rightleftharpoons \text{malonyl-S-ACP} + \text{CoA-SH} \quad (2)$$
$$\text{Malonyl-S-ACP} + \text{acetyl-S-ACP} \rightleftharpoons \text{acetoacetyl-S-ACP} + \text{ACP-SH} + CO_2 \quad (3)$$

$$\text{Acetoacetyl-S-ACP} + \text{TPNH} + \text{H}^+ \rightleftharpoons \text{D(-)}\beta\text{-hydroxybutyryl-S-ACP} + \text{TPN}^+ \quad (4)$$
$$\text{D(-)}\beta\text{-Hydroxybutyryl-S-ACP} \rightleftharpoons \text{crotonyl-S-ACP} + H_2O \quad (5)$$
$$\text{Crotonyl-S-ACP} + \text{TPNH} + \text{H}^+ \rightleftharpoons \text{butyryl-S-ACP} + \text{TPN}^+ \quad (6)$$
$$\text{Butyryl-S-ACP} + \text{malonyl-S-ACP} \rightleftharpoons \beta\text{-ketocaproyl-S-ACP} + CO_2\text{, etc.} \quad (7)$$

This series of reactions then continues as from acetoacetyl-S-ACP, reaction (4) above, with elongation of the chain by two carbon atoms in each cycle, as depicted in Fig. 21.3. Reactions (5) and (6) are analogous to those encountered previously in fatty acid oxidation (page 479). The process terminates in animal tissues with the liberation of palmitic acid by the hydrolytic action of a deacylase.

The enzymes which catalyze the above reactions have been designated as follows: (1) *acetyl transacylase;* (2) *malonyl transacylase;* (3) *β-ketoacyl-ACP* synthetase; (4) *β-ketoacyl-ACP reductase;* (5) *enoyl-ACP hydrase;* and (6) *crotonyl-ACP reductase.* Enzymes (1), (2), (3), and (5) are sulfhydryl-containing enzymes; enzyme (4) is not. Both enzymes (3) and (5) are specific for ACP derivatives. Thus acyl-ACP rather than acyl CoA is directly incorporated into long-chain fatty acids by the cytoplasmic system; the methyl group of the initial acetyl-ACP remains the terminus of the growing chain.

Propionyl CoA can replace acetyl CoA to some degree in the initial condensation reaction (1), thus providing a mechanism for biosynthesis of fatty acids containing an odd number of carbon atoms.

Points of difference between the mitochondrial fatty acid degradation system and the cytoplasmic synthetic system are noted in Table 21.3.

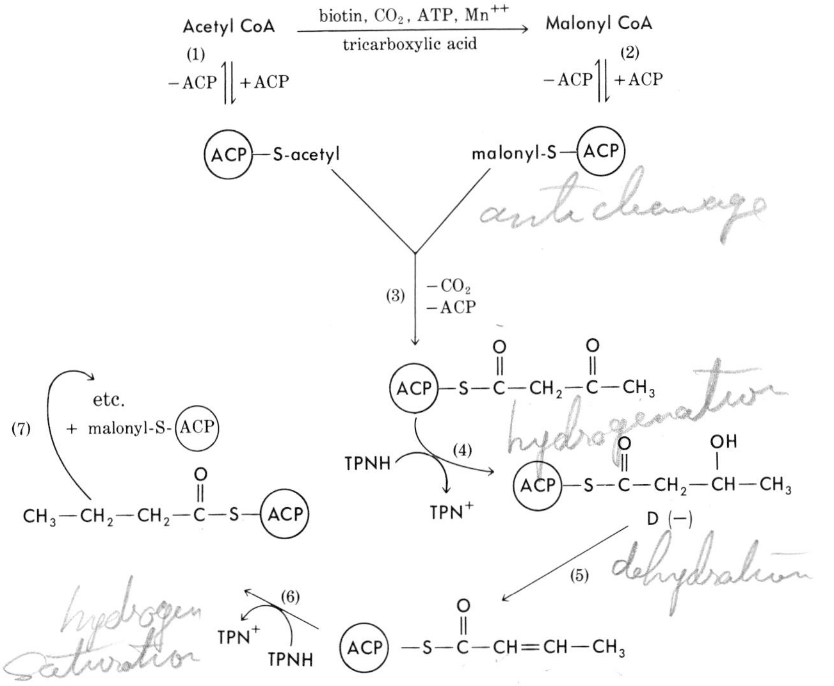

Fig. 21.3. Schematic representation of fatty acid synthesis catalyzed by cytoplasmic system. *ACP* = acyl protein carrier; the numbers in parentheses correspond to numbered steps and enzymes on the previous and this page.

Table 21.3: Comparison of Compounds Involved in Fatty Acid Metabolism

Step or component	Degradation	Synthesis
SH component.	Coenzyme A	Acyl carrier protein
Intermediate SH derivative. . .	Acetyl CoA	Malonyl-S-ACP + acetyl-S-ACP
Keto ↔ hydroxy.	DPN, L-β-hydroxybutyryl CoA	TPNH, D-β-hydroxybutyryl-ACP
Crotonyl ↔ butyryl.	FAD, electron transport system	TPNH, ACP-acyl acid

MITOCHONDRIAL SYSTEM FOR FATTY ACID SYNTHESIS

Mitochondria contain an enzyme system which catalyzes *elongation* of preformed fatty acids by successive addition of acetyl CoA units. The mitochondrial system does not require CO_2 fixation, and its products are essentially C_{18}, C_{20}, C_{22}, and C_{24} fatty acids. This pathway of fatty acid synthesis is essentially the reverse of the pathway previously described for oxidation of fatty acids.

The postulated sequence of chain elongation by four steps is as follows.

$$CH_3COSCoA + RCH_2COSCoA \underset{}{\overset{\text{thiolase}}{\rightleftharpoons}} RCH_2COCH_2COSCoA + CoA \qquad (1)$$

$$RCH_2COCH_2COSCoA + DPNH + H^+ \underset{\text{dehydrogenase}}{\overset{\beta\text{-hydroxyacyl}}{\rightleftharpoons}} RCH_2CHOHCH_2COSCoA + DPN^+ \qquad (2)$$

$$RCH_2CHOHCH_2COSCoA \underset{\text{hydrase}}{\overset{\text{enoyl}}{\rightleftharpoons}} RCH_2CH{=}CHCOSCoA + H_2O \qquad (3)$$

$$RCH_2CH{=}CHCOSCoA + TPNH + H^+ \underset{\text{reductase}}{\overset{\text{enoyl}}{\rightleftharpoons}} RCH_2CH_2CHCOSCoA + TPN^+ \qquad (4)$$

The first three steps are those which have been described as the last three reactions for fatty acid oxidation (page 479). However, reaction (4), is catalyzed by a TPNH-requiring enoyl reductase, instead of the specific FAD-requiring acyl CoA dehydrogenases, which catalyze an irreversible desaturation of fatty acids (page 480). Also, the synthetic system is active only with intermediate chain length fatty acids, C_{12}, C_{14}, C_{16}, and with unsaturated fatty acids as well. The metabolism of unsaturated compounds is considered below (page 493*ff.*)

MICROSOMAL SYSTEM FOR FATTY ACID SYNTHESIS

The microsomal system for synthesis of fatty acids apparently also provides a means of elongation of both saturated and unsaturated fatty acyl CoA derivatives, utilizing the malonyl CoA pathway rather than that with acetyl CoA. The fatty acid is first converted to acyl CoA and then reacts with malonyl CoA, followed by reduction with a TPNH-requiring enzyme. The intermediates are similar to those in the *de novo* synthetic pathway except that they are not bound to an ACP. Elongation occurs most rapidly with C_{10} to C_{16} carbon atom fatty acyl derivatives, and with C_{18} unsaturated compounds.

Sources of Reduced Nucleotides. For lipogenesis to function, there must be available sources of TPNH for the cytoplasmic and mitochondrial malonyl CoA synthetic pathway and of DPNH and TPNH for the mitochondrial and microsomal chain elongation pathways. In the cytoplasm, several sources of reduced nucleotides are available.

1. A significant source of cytoplasmic TPNH are the dehydrogenations of the phosphogluconic acid oxidative pathway. This links fatty acid synthesis to the oxidative pathway of glucose metabolism, particularly since mitochondria may not readily oxidize external TPNH.

2. A second source of cytoplasmic reduced nucleotides may derive from glycolysis.

3. A third source may arise from operation of the citric acid cycle. Citrate formed in mitochondria and transferred to the extramitochondrial compartment may be subjected to the catalytic influence of *citrate lyase* (page 500). The oxaloacetate formed could yield malate, which by reaction with malic enzyme (page 408) would yield pyruvate, TPNH, and CO_2. This sequence would again couple the DPNH arising from glycolysis to the synthesis of cytoplasmic TPNH. This is considered in greater detail later (page 499).

All the above three possible modes of providing reduced nucleotides for cytoplasmic lipogenesis reemphasize the linking of fatty acid synthesis to carbohydrate metabolism. Also, since oxidation of acetyl CoA via the citric acid cycle (3, above) depends on a source of oxaloacetate and this may arise by carboxylation of pyruvate derived from glycolysis, it is evident that fatty acid oxidation also depends on carbohydrate metabolism.

For fatty acid synthesis by the mitochondrial system, reduced nucleotides are provided by functioning of the citric acid cycle, with transhydrogenation from DPNH as the immediate source of TPNH.

Synthesis of Hydroxy Fatty Acids. Formation of α-, β-, and γ-hydroxy fatty acids as intermediates in fatty acid oxidation has been described previously (pages 479*ff.*). α-Hydroxy fatty acids, *e.g.*, cerebronic acid (page 62), are constituents of brain lipids and are synthesized by direct hydroxylation of long-chain fatty acids by an unknown mechanism. Other hydroxy fatty acids in which the hydroxyl group is more centrally located, *e.g.*, ricinoleic acid (page 62), are formed by hydration of a monoenoic acid. Extracts of *Pseudomonas* catalyzed hydration of oleic acid to 10-hydroxystearic acid. Synthesis of ricinoleyl CoA from oleyl CoA, catalyzed by soluble enzymes of the castor bean, required TPNH and molecular oxygen. Once formed, ricinoleic acid may be elongated by the soluble system (page 486), utilizing acetyl CoA.

Other Aspects of Fatty Acid Synthesis. In addition to the above-described pathways of fatty acid synthesis, other specialized pathways have been reported in bacteria and in plants. For example, in some bacteria extremely long-chain fatty acids are formed by condensation of fatty acyl CoA derivatives of various chain lengths. Thus octanoyl CoA is directly incorporated into C_{16} and C_{24} fatty acids by soluble enzymes of *Mycobacterium tuberculosis.* Fatty acids containing cyclic structures are found in bacteria, seeds, plants, and insects. A *cyclopropane synthetase,* from *Clostridium butyricum,* catalyzes the transfer of a methyl group from S-adenosylmeth-

Table 21.4: SOME ASPECTS OF PATHWAYS OF FATTY ACID SYNTHESIS

System	Cell fraction	Substrates	Cofactors	Product
De novo formation...	Soluble complex	Acetyl CoA + malonyl CoA	TPNH, ACP	$C_{16:0}$
Elongation..........	Mitochondria	Long-chain acyl CoA (sat. or unsat.) + acetyl CoA	TPNH, DPNH	$C_{n+2:y}$ acyl CoA ($y = 0$–6)
	Microsomes	Long-chain acyl CoA ($C_{10:0}$–$C_{16:0}$) + malonyl CoA	TPNH	$C_{n+2:0}$ acyl CoA
	Microsomes	Unsatd. acyl CoA ($C_{18:3} > C_{18:2} > C_{18:1}$) + malonyl CoA	TPNH	Unsatd. C_{n+2} acyl CoA
Condensation.......	Soluble complex	C_n acyl CoA (n = 8, 10, 16, 24)	DPNH, ATP, Mn^{++}	C_{2n} acyl CoA

SOURCE: Adapted from J. A. Olson, *Ann. Rev. Biochem.* **35**, 559, 1966.

ionine (page 553) to the double bond of a β- or γ-unsaturated phosphatidyl ethanolamine, forming a cyclic fatty acid. Extracts of *Pseudomonas* catalyze hydration of unsaturated fatty acids to hydroxy fatty acids.

Table 21.4 indicates some of the salient features of pathways of fatty acid synthesis.

FATTY ACID INTERCONVERSIONS

The organism obtains fatty acids from the lipid of the diet and by lipogenesis from the acetyl CoA derived from carbohydrates and certain amino acids. The composition of the mixture of fatty acids from the diet will vary considerably in the degree of unsaturation and chain length. The process of lipogenesis favors formation of saturated over unsaturated fatty acids. This is indicated both from the "hardening" effect (page 478) of a high-carbohydrate diet upon depot lipid and the observation, based upon isotopic data, of the more rapid synthesis of saturated than of singly unsaturated fatty acids.

From the mixture of fatty acids available to it, the mammalian liver produces a composite of fatty acids that is characteristic of the species. The operations involve shortening and elongation of the carbon skeleton as well as the introduction of double bonds and their elimination by reduction.

Shortening and Elongation of the Carbon Skeleton. If isotopic palmitic acid (C_{16}) is fed to rats, isotope is recovered in highest concentration in the palmitic acid of the body lipid. However, appreciable concentrations of isotope are also found in the stearic (C_{18}) and myristic (C_{14}) acids of the body. Similarly, the feeding of labeled stearic acid leads to labeling of the palmitic acid of the body lipid. These findings clearly indicate that lengthening or shortening of the carbon skeleton of saturated fatty acids occurs in the animal body and that it takes place by the gain or loss of two carbon atoms at a time. These processes find ready explanation in the series of reactions of degradation and elongation previously described.

Formation of Monoenoic Acids. When isotopic palmitic or stearic acid, both saturated, is fed, isotope can be recovered both in the saturated and in the unsaturated fatty acids of body lipid. The feeding of the C_{18}-saturated fatty acid leads to accumulation of isotope in oleic acid (C_{18}, one double bond), and feeding of the C_{16}-saturated acid yields palmitoleic acid (C_{16}, one double bond). Thus, the mammal is capable of desaturating these saturated fatty acids to give the corresponding 9,10-unsaturated derivatives. The desaturation involved in forming oleic or palmitoleic acid bears no relation to the α,β dehydrogenation that occurs in the shortening of saturated fatty acids described previously (page 480).

Introduction of a double bond may occur by aerobic or anaerobic pathways. The aerobic pathway utilizes molecular oxygen and TPNH in most organisms; in *Mycobacterium phlei,* ferrous ion and a flavin nucleotide are additional requirements. The mono-oxygenase catalyzing the aerobic reaction acts on the acyl CoA derivatives and is associated with microsomes in animal tissues, whereas a particulate fraction in other microorganisms and in plants utilizes acyl ACP derivatives as substrates. The mechanism of the aerobic desaturation reaction is not known; an oxygenated intermediate rather than a hydroxy acid appears to be involved. The anaerobic pathway for formation of a monoenoic fatty acid has been elucidated in microorganisms. The sequence consists of formation of a medium-chain β-hydroxyacyl ACP derivative, with subsequent either α,β or β,γ dehydration, followed by elongation reactions catalyzed by the fatty acid synthetase complex, of which the dehydrase, termed *β-hydroxydecanoyl thioester dehydrase,* is a component. A highly purified preparation of the dehydrase from *E. coli* has a molecular weight of 28,000 and catalyzes either α,β or β,γ dehydration, with the former predominating. The enzyme exhibits a high degree of chain length specificity, being most active with the C_{10} β-hydroxy thioester. The formation of palmitoleic acid from octanoyl ACP in *E. coli* may be depicted as follows.

$$\underset{\text{Octanoyl-ACP}}{CH_3(CH_2)_6CO{-}ACP} \ggg \underset{\beta\text{-Ketodecanoyl-ACP}}{CH_3(CH_2)_6COCH_2CO{-}ACP} \longrightarrow$$

$$\underset{\beta\text{-Hydroxydecanoyl-ACP}}{CH_3(CH_2)_6CHOHCH_2CO{-}ACP} \longrightarrow \underset{\beta,\gamma\text{-Decenoyl-ACP}}{CH_3(CH_2)_6CH{=}CHCO{-}ACP} \ggg$$

$$\underset{\text{Palmitoleyl-ACP}}{CH_3(CH_2)_6CH{=}CH(CH_2)_6CO{-}ACP} \longrightarrow \underset{\text{Palmitoleic acid}}{CH_3(CH_2)_6CH{=}CH(CH_2)_6COOH}$$

Several different acyl ACP dehydrases are present in *E. coli,* with specificity being determined by chain length. In plants, desaturation is achieved by a mechanism which differs, but the details of which are unknown.

Formation and Transformations of Polyenoic Acids. Animal tissues contain a variety

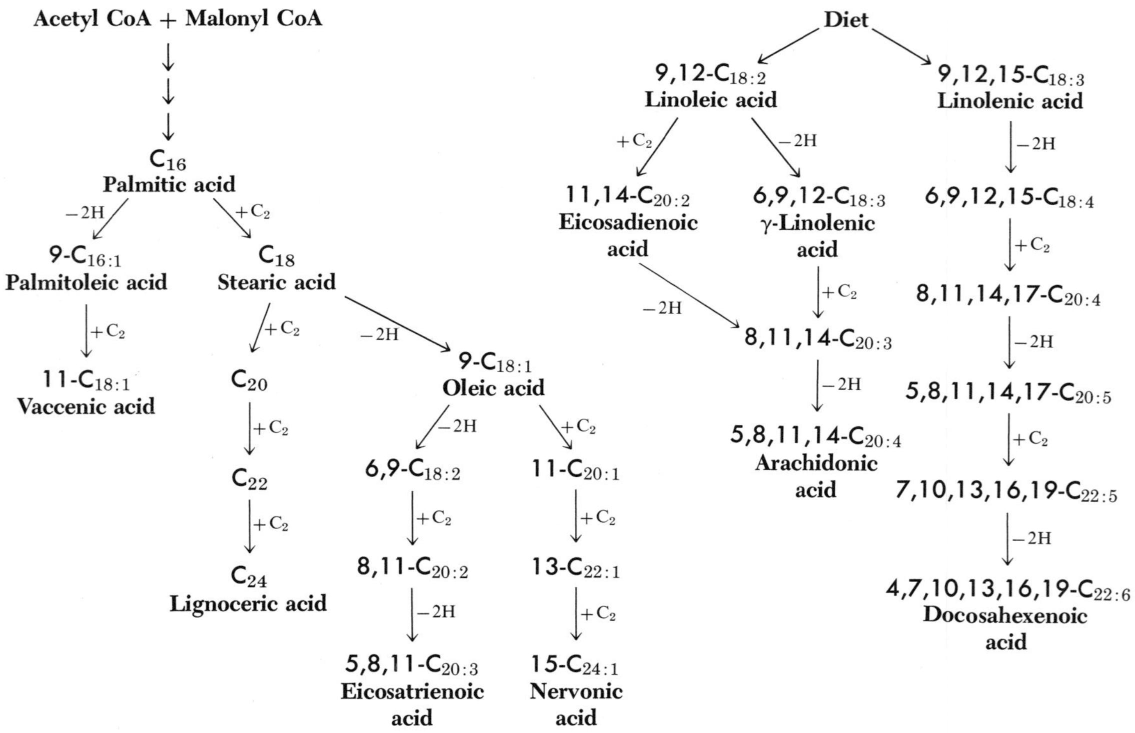

Fig. 21.4. Biosynthesis of some fatty acids in mammals. The numbers in boldface indicate the position of the double bonds; the subscripts indicate the number of carbon atoms, with the number of double bonds to the right of the colon.

of polyunsaturated fatty acids. Of these, one series can be fabricated by the animal, *de novo.* These are the fatty acids of which all the double bonds lie between the seventh carbon from the terminal methyl group and the carboxyl group. As seen from Fig. 21.4, such fatty acids may be made by alternate desaturation and chain elongation, commencing with oleic acid. However, polyunsaturated fatty acids in which one or more double bonds are situated within the terminal seven carbon atoms cannot be made *de novo.* Such polyunsaturated acids are essential in the diet and no label is found in tissue linoleic or linolenic acid (C_{18}, two and three double bonds, respectively) after administration of labeled stearic acid.

There are apparently four series of polyunsaturated acids in the mammal; two are derived from the dietary compounds, linoleic and linolenic acids, and two are synthesized from the monounsaturated acids, oleic and palmitoleic acids, these in turn being formed from the corresponding saturated acids (see above). The four series can be recognized by the distance between the terminal methyl group and the nearest double bond.

Linoleic family:	$CH_3—(CH_2)_4—CH{=}CH—$
Linolenic family:	$CH_3—CH_2—CH{=}CH—$
Palmitoleic family:	$CH_3—(CH_2)_5—CH{=}CH—$
Oleic family:	$CH_3—(CH_2)_7—CH{=}CH—$

All other polyunsaturated acids can be made from these four precursors by reaction sequences in which the chain is alternately elongated and desaturated. Elongation may occur by operation of the mitochondrial chain elongation system considered previously or by a microsomal system that utilizes malonyl CoA. Desaturation is accomplished also in the microsomal fraction by a mechanism similar to but not identical with that concerned with formation of the monounsaturated acids. Alternating double and single bonds are not formed in animal tissues. Nutritional evidence, considered in Chap. 50, suggests that pyridoxal phosphate may be essential for formation of longer-chain, polyunsaturated acids, but its role in this process is obscure. The formation of arachidonic acid is presented in more detail in Fig. 21.5, while Fig. 21.4 summarizes the operation of the various pathways which account for the diverse polyunsaturated fatty acids of animal tissues.

Desaturation and elongation reactions occur more extensively in liver than in extrahepatic tissues; fasting and diabetes are marked by an inhibition of desaturation pathways. The green alga *Chlorella vulgaris* effects a series of direct desaturations of stearic acid to give the family of *cis*-unsaturated fatty acids typical of photosynthetic tissue, oleic, linoleic and α-linoleic (9, 12, 15-octadecatrienoic) acids. The conversions involve the removal of *cis* pairs of D-hydrogens from stearic acid.

Although oxidation of polyunsaturated fatty acids occurs, these acids do not appear to be an important source of energy. Oxidation occurs in mitochondria by a modification of the previously described β-oxidation pathway, and may be depicted as follows.

$$C_{18:2(9\text{ cis, }12\text{ cis})} \xrightarrow{-3C_2} C_{12:2(3\text{ cis, }6\text{ cis})} \xrightarrow{A} C_{12:2(2\text{ trans, }6\text{ cis})} \xrightarrow{-C_2}$$

$$C_{10:1(4\text{ cis})} \xrightarrow{-C_2} C_{8:1(2\text{ cis})} \xrightarrow{B} \text{3-D-hydroxy } C_{8:0} \xrightarrow{C}$$

$$\text{3-L-hydroxy } C_{8:0} \xrightarrow{-C_2} C_{6:0} \longrightarrow 3C_2$$

$$CH_3—(CH_2)_4—CH{=}CH—CH_2—CH{=}CH—(CH_2)_7—COOH$$

Linoleic acid

$$\big\downarrow -2H$$

$$CH_3—(CH_2)_4—CH{=}CH—CH_2—CH{=}CH—CH_2—CH{=}CH—(CH_2)_4—COOH$$

γ-Linolenic acid

$$\big\downarrow +C_2$$

$$CH_3—(CH_2)_4—(CH{=}CH—CH_2)_3—(CH_2)_5—COOH$$

Homo-γ-linolenic acid

$$\big\downarrow -2H$$

$$CH_3—(CH_2)_4—(CH{=}CH—CH_2)_4—(CH_2)_2—COOH$$

Arachidonic acid

Fig. 21.5. Formation of arachidonic acid from linoleic acid.

In addition to the conventional enzymes of the β-oxidative pathway, three additional enzymes, designated *A*, *B*, and *C* on the arrows of equations on previous page, are required. *A* is a $\Delta^{3\text{-cis}}$: $\Delta^{2\text{-trans}}$ *enoyl CoA isomerase; B* a $\Delta^{2\text{-cis}}$ *enoyl CoA hydrase;* and *C* a *3-hydroxyacyl CoA epimerase.* Monoenoic and dienoic acids are oxidized at comparable rates; these reactions are stimulated by carnitine.

SYNTHESIS OF TRIGLYCERIDES

In the animal organism, free fatty acids are not found in significant quantity in the tissues or body fluids, except for blood plasma. Rather, fatty acids are present largely as esters. Synthesis of triglycerides occurs primarily in liver and in adipose tissue and in these tissues utilizes the CoA derivatives of fatty acids and α-phosphatidic acid. The latter is the precursor of both the triglycerides and the various phospholipids (Chap. 22), and is formed by the following reaction.

$$\begin{array}{cccccc} & HOCH_2 & & RCOOCH_2 & \\ & | & & | & \\ 2RCOCoA\ + & HOCH & \longrightarrow & RCOOCH & +\ 2CoA \\ & | & & | & \\ & H_2COPO_3H_2 & & H_2COPO_3H_2 & \\ \textbf{Acyl CoA} & \textbf{L-}\alpha\textbf{-Glycerophosphoric acid} & & \textbf{L-}\alpha\textbf{-Phosphatidic acid} & \end{array}$$

This reaction is absolutely specific for α-glycerophosphate and proceeds preferentially with the saturated and unsaturated C_{16} and C_{18} CoA derivatives. α-Glycerophosphate derives either from free glycerol, which is phosphorylated in the presence of glycerokinase and ATP (page 512), or from reduction of dihydroxyacetone phosphate (page 403).

Hydrolysis of α-phosphatidic acid by a *phosphatase* yields a 1,2-diglyceride, which in turn reacts with another mole of acyl CoA to form a neutral triglyceride.

$$\text{L-}\alpha\text{-Phosphatidic acid} \xrightarrow[+H_2O]{\text{phosphatase}} \text{D-1,2-diglyceride} + P_i \xrightarrow{+\text{ acyl CoA}} \text{triglyceride}$$

Note that no change in configuration is involved in conversion of L-phosphatidic acid to the D-1,2-diglyceride. As indicated previously (page 69), phosphatidic acid is designated as L because of its relationship to L-glyceryl phosphate, whereas the 1,2-diglyceride is designated as D because of its stereochemical relationship to D-glyceraldehyde.

Intestinal mucosa synthesizes triglycerides from free fatty acids and mono- and diglycerides (page 473). The path of synthesis of triglycerides from free fatty acids, and of 1,2-diglycerides, is undoubtedly the same as that given above. However, the evidence that monoglycerides can be incorporated into the triglycerides that appear in the chyle led to the discovery of a reaction that appears to be unique to intestinal mucosa. A microsomal system from rat and rabbit intestine catalyzes the following reaction.

$$\text{Monoglyceride} + \text{fatty acyl CoA} \longrightarrow \text{diglyceride} + \text{CoA}$$

Liver microsomes fail to catalyze this reaction under the same conditions.

In *Clostridium butyricum* and in *E. coli*, lysophosphatidic and phosphatidic acids can be synthesized by acylation of glycerol 3-phosphate by acyl-ACP derivatives. Indeed, the latter may be better utilized for these syntheses than are the corresponding acyl CoA derivatives.

REGULATION OF LIPID METABOLISM

The regulation of lipid metabolism may be effected or localized at a variety of metabolic transformations. Thus regulatory phenomena may be in evidence in reactions of fatty acid synthesis and oxidation, including interrelationships with carbohydrate metabolism. A second group of regulatory aspects are seen in diverse fates of the fatty acids derived from acetyl CoA and appear to be unique for particular tissues or organs. Thus, in adipose tissue, fatty acids are stored as triglycerides; in liver, fatty acids undergo rapid metabolic transformations, including formation of triglycerides, phospholipids, and cholesterol esters, as well as oxidation, whereas in brain, fatty acids are utilized for synthesis of complex lipids such as glycosphingolipids and sphingomyelins.

The factors that regulate the qualitative and quantitative synthesis and disposition of fatty acids and their metabolic products range from coarse, *e.g.*, diet, to fine, namely, an allosteric effect of a metabolite or a hormone on a rate-limiting step of a sequence of reactions of lipid metabolism.

Integration of Normal Fatty Acid and Carbohydrate Metabolism. The modifying controls operative in glycolysis, the citric acid cycle, and fatty acid synthesis combine to assure that surplus carbohydrate is directed into the formation of glycogen and fatty acids. Sites for large glycogen stores are lacking in mammals, and excess carbohydrate is channeled into fatty acids. The latter, in the absence of excessive energy demands, are stored as triglycerides, chiefly in adipose tissue.

Studies of over-all carbon balances when ^{14}C-glucose is incubated with adipose tissue indicate that the amounts of TPNH and DPNH formed in the cytoplasm of these cells by the glucose 6-phosphate oxidative pathway and glycolysis exactly equal the amount required for fatty acid biosynthesis under these circumstances. The quantity of ATP formed under these conditions, which can be calculated from the fate of the glucose and the amount of O_2 utilized, almost exactly balances that required for concomitant fatty acid formation. When the same experiments are conducted under lipolytic conditions, *viz.*, with liberation of free fatty acids from the tissue fat to the medium, there occurs an increased turnover of the tricarboxylic acid cycle, probably because of the uncoupling of oxidative phosphorylation by the free fatty acids.

Since fatty acid biosynthesis requires TPNH, and since the TPNH formed by the direct oxidative pathway under these circumstances is insufficient to account for the fatty acids formed, a fraction of the reducing power used for fatty acid synthesis must be provided by the DPNH arising in glycolysis. Modes of providing reduced nucleotides for fatty acid synthesis have been described previously (page 491).

General Aspects of Lipogenesis from Carbohydrate. Inasmuch as carbohydrate is a major source of acetyl CoA for lipogenesis, it is of interest to examine the over-all picture of fatty acid synthesis from glucose. In a net sense, palmitic acid is synthesized by condensation of 8 molecules of acetyl CoA which arise from 4 molecules of glucose. The equations below indicate that, metabolically, this is accomplished with great efficiency.

Synthesis of Palmitic Acid:

$$\text{7 Acetyl CoA} + 7\text{ATP} + 7\text{CO}_2 \longrightarrow \text{7 malonyl CoA} + 7\text{ADP} + 7\text{P}_i \qquad (1)$$

$$\text{1 Acetyl CoA} + \text{7 malonyl CoA} + 14\text{TPNH} + 14\text{H}^+ \longrightarrow \text{1 palmitic acid} + 7\text{CO}_2 + 8\text{CoA} + 14\text{TPN}^+ + 6\text{H}_2\text{O} \qquad (2)$$

Thus the requirements for synthesis of 1 molecule of palmitic acid are 8 acetyl CoA, 7ATP, and 14TPNH.

Partial Degradation of Glucose:

$$\text{4 Glucose} + 8\text{ATP} \rightarrow\rightarrow \text{4 fructose 1,6-diphosphate} + 8\text{ADP} \qquad (1)$$

$$\text{4 Fructose 1,6-diphosphate} \longrightarrow \text{8 3-phosphoglyceraldehyde} \qquad (2)$$

$$\text{8 3-Phosphoglyceraldehyde} + 8\text{DPN}^+ + 8\text{ADP} \rightarrow\rightarrow \text{8 Pyruvate} + 8\text{DPNH} + 8\text{ATP} + 8\text{H}^+ \qquad (3)$$

$$\text{8 Pyruvate} + 8\text{CoA} + 8\text{DPN}^+ \longrightarrow \text{8 acetyl CoA} + 8\text{DPNH} + 8\text{H}^+ + 8\text{CO}_2 \qquad (4)$$

$$2\text{DPNH} + 6\text{ADP} + 6\text{Pi} + \text{O}_2 + 2\text{H}^+ \longrightarrow 2\text{DPN}^+ + 6\text{ATP} + 2\text{H}_2\text{O} \qquad (5)$$

$$\text{Net:}\quad \text{4 Glucose} + 14\text{DPN}^+ + 6\text{ADP} + 6\text{Pi} + \text{O}_2 + 8\text{CoA} \longrightarrow \text{8 acetyl CoA} + 14\text{DPNH} + 14\text{H}^+ + 8\text{CO}_2 + 6\text{ATP} + 2\text{H}_2\text{O}$$

Thus the systems are in almost perfect balance. Indeed, if glycogen, rather than glucose, were the starting material, there would be a slight surplus of ATP rather than the small deficit which could be met from the metabolism of additional glucose.

Lipogenesis from carbohydrate not only is an efficient process but occurs abundantly (page 485). In contrast, the mammal is apparently unable to effect a net conversion of fatty acids into carbohydrate. This is because formation of acetyl CoA and CO_2 from pyruvate is irreversible, thus preventing direct entrance of acetyl CoA into carbohydrate precursors. Since fatty acids are degraded to acetyl CoA, the only possible route whereby the carbon atoms of fatty acids can later be located in carbohydrate is by condensation of acetyl CoA with oxaloacetate to enter the citric acid cycle. One turn of this cycle yields oxaloacetate, which contains carbon atoms initially present in the fatty acid. However, one mole of oxaloacetate is required for condensation with acetyl CoA, and only one mole of oxaloacetate is formed; hence there is no net gain of oxaloacetate. Since oxaloacetate is the potentially glucogenic product, by decarboxylation to pyruvate and reversal of glycolysis, it follows that there can be no *net increase* in available glucose as a result of fatty acid oxidation. However, plants can achieve a net synthesis of carbohydrate from stored fatty acids (page 413), a process that occurs in germinating seeds.

Inasmuch as cytoplasmic DPNH cannot directly serve in fatty acid formation, there must be a mechanism for the transfer of electrons from cytoplasmic DPNH to TPN^+. However, a cytoplasmic transhydrogenase is not known, and operation of mitochondrial transhydrogenase does not result in release of TPNH to the cytoplasm. This dilemma can be resolved by the combined operation of the following two sets of reactions, *A* and *B*. The subscripts *m* and *c* are used to designate location of the metabolite in the mitochondrial and cytoplasmic compartments, respectively.

A.

$$\text{Pyruvate}_m + \text{ATP} + CO_2 \longrightarrow \text{oxaloacetate}_m + \text{ADP} + P_i \quad (1)$$

$$\text{Oxaloacetate}_m + \text{acetyl CoA}_m \longrightarrow \text{citrate}_m + \text{CoA} \quad (2)$$

$$\text{Citrate}_m \longrightarrow \text{citrate}_c \quad (3)$$

B.

$$\text{Citrate}_c + \text{ATP} + \text{CoA} \longrightarrow \text{oxaloacetate}_c + \text{acetyl CoA}_c + \text{ADP} + P_i \quad (4)$$

$$\text{Oxaloacetate}_c + \text{DPNH} + H^+ \longrightarrow \text{malate}_c + \text{DPN}^+{}_c \quad (5)$$

$$\text{Malate}_c + \text{TPN}^+{}_c \longrightarrow \text{pyruvate}_c + \text{TPNH}_c + H^+ + CO_2 \quad (6)$$

$$\text{Net:} \quad \text{Acetyl CoA}_m + 2\text{ATP} + \text{DPNH}_c + \text{TPN}^+{}_c \longrightarrow \text{acetyl CoA}_c + 2\text{ADP} + P_i + \text{DPN}^+{}_c + \text{TPNH}_c$$

The sum of the above reactions utilizes two moles of ATP to achieve transhydrogenation from one mole of DPNH to one mole of TPN^+. The operation of this process is dependent upon the relative ease of migration of pyruvate, citrate, and malate across the mitochondrial membrane in contrast to the impermeability of the membrane to the other reactants. This scheme also accounts for the fact that whereas the oxidation of pyruvate to acetyl CoA is a mitochondrial process, in the presence of a plethora of carbohydrate acetyl CoA is made available for fatty acid biosynthesis by cytoplasmic enzymes. Finally, note that since in a steady state most of the TPN of cytoplasm is reduced whereas most of the DPN is oxidized, net transhydrogenation of DPNH to TPN can only be accomplished by the provision of energy to drive the process and ATP provides this energy.

Additional factors modulate the above reactions. Acetyl CoA has a stimulatory influence on pyruvate carboxylation [reaction (1)], and an inhibitory action on pyruvate decarboxylation, thus accelerating ultimate provision of TPNH by the above reactions and promoting fatty acid synthesis. Fasting, diabetes, and cortisol treatment also enhance pyruvate carboxylase activity. Fatty acids exert an inhibitory influence on key glycolytic enzymes in liver, notably pyruvate kinase, and could thus limit formation of cytoplasmic DPNH; this would, in turn limit potential TPNH for fatty acid synthesis. In addition, fatty acids are inhibitory to glucose 6-phosphate dehydrogenase, which catalyzes the other prime mechanism for providing TPNH. Citrate lyase [reaction (4)] is also increased in activity as a consequence of carbohydrate feeding, as well as by thyroxine. The activity of this enzyme is diminished in fasting and in diabetes.

A key enzyme in regulating the overall rates of fatty acid degradation and synthesis is isocitric acid dehydrogenase (page 329) of the citric acid cycle. The activity of this mitochondrial enzyme is subject to allosteric regulation by ATP and AMP. Under conditions when the ATP level is high, the dehydrogenase is inhibited, citrate accumulates and leaves the mitochondria, thus making citrate available in the cytoplasm for degradation to acetyl CoA and hence for lipogenesis. When the ATP level is low and AMP accumulates in the mitochondrion, the activity of the isocitric dehydrogenase is greatly enhanced since AMP is a positive effector of the enzyme. Thus the activity of the entire cycle is enhanced and little citrate leaves the mitochondrion. The allosteric regulation of the activity of the isocitric dehydrogenase would appear to play a key role in regulating the supply of ATP for cellular requirements since, as discussed earlier (page 345), the citric acid cycle is responsible for satisfying the major energy needs of the cell.

Other hormonal influences and regulatory mechanisms will be considered later in this chapter.

MOBILIZATION OF DEPOT AND LIVER LIPID

The fatty acids stored in adipose tissue as triglycerides constitute the principal reservoir of reserve substrate for oxidative metabolism. Mobilization, distribution, and oxidation of these fatty acids can occur rapidly enough to support as much as 50 per cent of the oxidative metabolism of the body. Extensive mobilization of fatty acids occurs in a variety of circumstances, *e.g.*, starvation, exposure to cold, exercise, reproduction, growth, and hibernation. The common stimulus for fatty acid mobilization in these diverse conditions may be provided by the neuroendocrine system, probably through an increased secretion of lipid-mobilizing hormones (page 501) and by a decrease in the rate of insulin secretion (see below). The major site of degradation of fatty acids to acetoacetyl CoA is the liver. Thus the transport of lipid from the depots to the liver, and vice versa, is a subject of considerable interest.

Mobilization of Depot Lipid. Even in the animal in caloric balance, a considerable fraction of the depot lipid is mobilized daily (page 478), enters the blood stream, and is delivered to the various organs. Rapid mobilization to the liver is particularly striking; here the lipid may be stored temporarily or degraded by the reactions discussed previously (pages 479*ff.*). Lipid leaving adipose cells is largely transported as unesterified fatty acids bound to plasma albumin (page 476). The release of fatty acids is accompanied by the appearance in plasma of glycerol as well, although in amounts significantly less than the equivalent of fatty acids. The fatty acids released from adipose cells derive from stored triglycerides; hydrolysis is catalyzed by a *lipase*, which is sensitive to hormones altering rates of lipolysis in adipose tissue, *e.g.*, epinephrine (Chap. 45) and insulin (Chap. 46). The mechanism of mobilization of depot lipid is unknown; factors which regulate or modify its rate are largely of two types: (1) hormonal influences and (2) substances which are toxic for hepatic cells.

Hormonal Influences on Lipid Mobilization. A number of hormones induce lipid mobilization from the depots when added to adipose tissue in vitro or when administered to experimental animals. In general, the sequence of events in vitro is (1) conversion of triglycerides to free fatty acids within the adipose tissue; (2) release of these fatty acids into the blood, a process favorably influenced by the presence of serum albumin; and (3) oxidation and, in the presence of adequate glucose, reesterification to triglyceride of a portion of the liberated fatty acids. In vivo, there occur, in addition, (4) transport of fatty acids by the circulation to diverse organs and tissues, with utilization for synthesis of triglycerides, cholesterol esters, and phospholipids, chiefly in the liver and kidney; and (5) discharge of triglycerides, cholesterol esters, and phospholipids from the liver, as lipoproteins, with resultant lipemia.

The hormones known to be active in lipid mobilization from adipose tissue, as judged by either in vitro or in vivo criteria, and occasionally by both, include epinephrine and norepinephrine (Chap. 45), adrenal steroids (Chap. 45), glucagon (Chap. 46), and the following hypophyseal hormones (Chap. 47): vasopressin, thyrotropin, adrenocorticotropin, luteotropin, somatotropin, and the adipokinetic hormones. In addition, serotonin (page 589), not generally classed as a hormone, also effects release of fatty acid from adipose tissues in vitro. These hormones stimulate lipolysis and, in general, will inhibit in vitro the action of insulin on adipose tissue, *viz.*, stimulation of triglyceride synthesis (Chap. 46).

The effects of epinephrine in vitro and in vivo on the release of fatty acids from adipose tissue are strikingly exaggerated in man when adrenal medullary tumors (pheochromocytoma, Chap. 45) result in excessive epinephrine secretion; plasma concentrations of unesterified fatty acids several hundredfold normal may be evident. The action of adrenocorticotropin on adipose tissue is a direct one, independent of the trophic influence of this hormone on the adrenal cortex (Chap. 45). Other hormones listed above all exert this type of direct action on adipose tissue. The wide diversity of humoral agents stimulating lipid release from adipose tissue provides explanation for the hyperlipemia and liver lipid deposition present in various experimental and clinical conditions (see below).

Hepatic Poisons and Lipid Mobilization. The mechanism by which a substance inducing hepatic injury can elicit mobilization of depot lipid is not understood, although the phenomenon has long been known. Chlorinated aliphatic and aromatic hydrocarbons are potent hepatoxic agents which produce lipid accumulation in the liver. Poisoning by carbon tetrachloride, because of its volatility and wide industrial use, is most often encountered in clinical medicine.

Fatty liver is frequently seen clinically following conditions other than overt poisoning. Thus in chronic infectious diseases, such as tuberculosis, and in metabolic disturbances, including starvation, the lipid content of the liver may increase markedly. The most striking cases of fatty infiltration were observed in patients with severe, untreated diabetes in the preinsulin era. As indicated below, the liver of the diabetic person is less than normally competent to synthesize fatty acids but is normally able to degrade fatty acids. The fatty liver in this disease is therefore, by exclusion, attributed to excessive migration of depot lipid to the liver, and the lipemia observed in diabetes is taken to represent in part lipid in transit from the depots. The augmenting effect of a number of hormones on lipid mobilization from depots suggests that most clinical cases of fatty liver arise from excessively rapid mobilization of depot lipid. On the other hand, liver lipid accumulation in circumstances of liver poisoning may reflect, in addition to accelerated lipid mobilization to the liver, a diminished capacity of injured hepatic cells to degrade fatty acids.

Mobilization of Liver Lipid. With the discovery of insulin, it became possible to maintain totally pancreatectomized dogs in reasonably good health. However, even when adequate insulin was supplied to control the diabetes the animals developed severe fatty livers. These fatty livers could be corrected or prevented by addition to the diet of lecithin or choline. Also, fatty livers could be produced in rats by administration of a diet poor in choline and low in protein; this fatty liver could be cured or prevented by administration of choline.

The fatty liver observed in these experimental animals arises by a different mechanism from that previously considered. Choline is a constituent of lecithin, and the capacity of the liver to generate lecithin is dependent on a supply of choline or a supply of methyl groups from S-adenosylmethionine (page 553). Certain analogues of choline, such as arsenocholine,

$$(CH_3)_3\overset{+}{N}CH_2CH_2OH \qquad (CH_3)_3\overset{+}{As}CH_2CH_2OH$$

Choline **Arsenocholine**

although foreign to nature, are incorporated into lecithin and prevent the fatty liver of choline deficiency. The term *lipotropic* substance has been assigned to compounds capable of preventing or correcting the fatty liver of choline deficiency.

These findings have suggested that, although fatty acids are delivered to the liver in a variety of forms, *e.g.*, unesterified or esterified to glycerol or to cholesterol, and are in addition, synthesized in the liver from carbohydrate or amino acid precursors, they leave the liver not only in these forms, as lipoproteins (page 475), but also as choline-containing phospholipids of lipoproteins. Indeed, the liver is the major site of synthesis of the plasma phospholipids. According to this view, when the availability of choline or methyl groups for its synthesis is restricted, the rate of lecithin synthesis decreases and consequently the rate at which fatty acids are discharged from the liver falls below normal. If other processes continue at normal rate, an accumulation of lipid in the liver results.

One finding in discord with this picture is the observation that in the hepatectomized dog, the phospholipids of the plasma are not utilized, indicating that the liver, in addition to being the main source of plasma lecithin, is also the main site of its destruction. If this is so, it is difficult to understand how, by forming lecithin, the liver effectively gets rid of fatty acids. It has been suggested that choline in the diet may enhance degradation of fatty acids, but not all the experimental evidence conforms to this contention. An alternative view is that release of triglyceride and cholesterol esters to the circulation as lipoproteins requires the synthesis of sufficient lecithin to form the stable, soluble lipoprotein complex.

Any material capable of contributing methyl groups for choline synthesis has the property of being lipotropic. Conversely, materials that deflect methyl groups from choline synthesis will, under appropriate conditions, enhance dietary fatty liver. Guanidoacetic acid (page 582) and nicotinamide (page 1026) are examples of antilipotropic substances that are irreversibly methylated in the body.

In addition to choline and methyl donors which permit choline synthesis, other substances have been reported to reduce the quantity of liver lipid in experimental animals. Exclusion of inositol from the diet may lead to a slight increase in liver lipids even in the presence of adequate choline; under these circumstances inositol becomes lipotropic. The mode of action of inositol in this regard is not known, but some phospholipids contain inositol and their formation may depend on a supply of exogenous inositol much as the synthesis of lecithin depends on a dietary supply of methyl groups or choline.

THE KETONE BODIES AND KETOSIS

Fatty acid degradation and synthesis normally proceed without significant accumulation of intermediates. Under some circumstances certain products accumulate in the blood which are traditionally but inaccurately termed *ketone bodies.* These are acetoacetic acid, β-hydroxybutyric acid, and acetone. All these products stem from acetoacetyl CoA, a normal intermediate in the oxidation of fatty acids (page 481). Moreover, it is readily formed by the reversal of the *thiolase* reaction.

$$\text{2 Acetyl CoA} \rightleftharpoons \text{acetoacetyl CoA} + \text{CoA}$$

A major fate of acetoacetyl CoA in liver is conversion to β-hydroxy-β-methylglutaryl CoA, an important intermediate in the biogenesis of cholesterol and steroids (page 517) and in the degradation of leucine (page 605).

$$\underset{\text{Acetoacetyl CoA}}{CH_3COCH_2COCoA} + \underset{\text{Acetyl CoA}}{CH_3COCoA} + H_2O \longrightarrow \underset{\beta\text{-Hydroxy-}\beta\text{-methylglutaryl CoA}}{HOOCCH_2\overset{\overset{OH}{|}}{\underset{\underset{CH_3}{|}}{C}}CH_2COCoA} + CoA$$

Cleavage of β-hydroxy-β-methylglutaryl CoA by a mitochondrial enzyme different from that catalyzing the above synthesis is the major route of formation of free acetoacetate in liver.

$$\beta\text{-Hydroxy-}\beta\text{-methylglutaryl CoA} \longrightarrow \text{acetoacetic acid} + \text{acetyl CoA}$$

A *thiophorase* reaction may also be effective to a limited degree; however, it is probably mainly concerned in resynthesis of acetoacetyl CoA in peripheral tissues.

$$\text{Acetoacetyl CoA} + \text{succinic acid} \xrightleftharpoons[]{\text{acetoacetyl-succinic thiophorase}} \text{succinyl CoA} + \text{acetoacetic acid}$$

The reduction of acetoacetate is effected by DPNH in the presence of the specific mitochondrial liver *β-hydroxybutyric acid dehydrogenase* and yields D-β-hydroxybutyric acid. In contrast, reduction of acetoacetyl CoA catalyzed by β-hydroxyacyl dehydrogenase (page 480) yields L-β-hydroxybutyryl CoA.

$$\underset{\textbf{Acetoacetic acid}}{CH_3COCH_2COOH} + DPNH + H^+ \xrightleftharpoons{\text{dehydrogenase}} \underset{\textbf{D-}\beta\textbf{-Hydroxybutyric acid}}{CH_3CHOHCH_2COOH} + DPN^+$$

Although this reaction is reversible, wide variations in the amount and in the ratio of the two acids in the circulation are encountered: in situations where there is an abundance of liver glycogen, the formation of β-hydroxybutyrate is favored; when liver glycogen is relatively low, acetoacetate predominates.

The release of acetoacetate by liver is a continuing, normal process. The total ketone body concentration in blood, expressed as β-hydroxybutyrate, is normally below 3 mg. per 100 ml., and the average total daily excretion in the urine is approximately 20 mg. This is because of efficient mechanisms for removal of acetoacetic acid by peripheral tissues, especially muscle, which can derive a sizable fraction of its total energy requirement from this nutrient. In order to be utilized acetoacetic acid must first be reconverted into its CoA derivative by transfer of a CoA residue from succinyl CoA by the action of a specific thiophorase (page 503), a pathway that appears to be minimal or absent in the liver. The acetoacetyl CoA thus formed may then be cleaved by thiolase (page 481), yielding two molecules of acetyl CoA, which then enter the citric acid cycle.

The third "ketone body" is acetone, which arises from acetoacetate by a decarboxylation catalyzed by *acetoacetate decarboxylase.* The mechanism of this reaction has been shown by Westheimer and his colleagues to involve formation of a Schiff base between enzyme and substrate followed by its decarboxylation and hydrolysis.

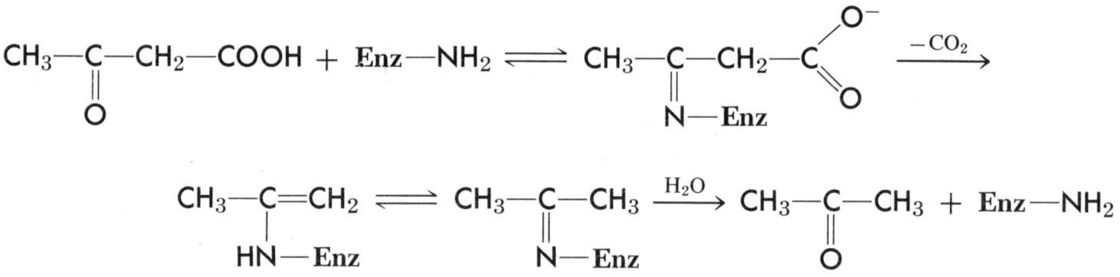

Acetone can be metabolized via two pathways; one involves cleavage to yield a 2-carbon acetyl and a 1-carbon formyl fragment, which then follow reactions characteristics of each of these groups. A second fate of acetone is conversion to propanediol, which is converted to pyruvic acid and thus contributes to all the products that may arise from pyruvic acid (pages 327*ff.* and 401*ff.*).

Interrelationships of Lipid Metabolism. The central roles of acetyl CoA and β-hydroxy-β-methylglutary CoA in lipid metabolism are depicted in Fig. 21.6. In the case of acetyl CoA, its three major fates are (1) oxidation via the citric acid cycle, (2) synthesis of fatty acids via the malonyl CoA pathway, and (3) formation of acetoacetyl CoA, thus also contributing to β-hydroxy-β-methylglutaryl CoA synthesis. The central role of this last compound is also evident. β-Hydroxy-β-methylglutaryl CoA has two major pathways of utilization, acetoacetate formation and cholesterol synthesis.

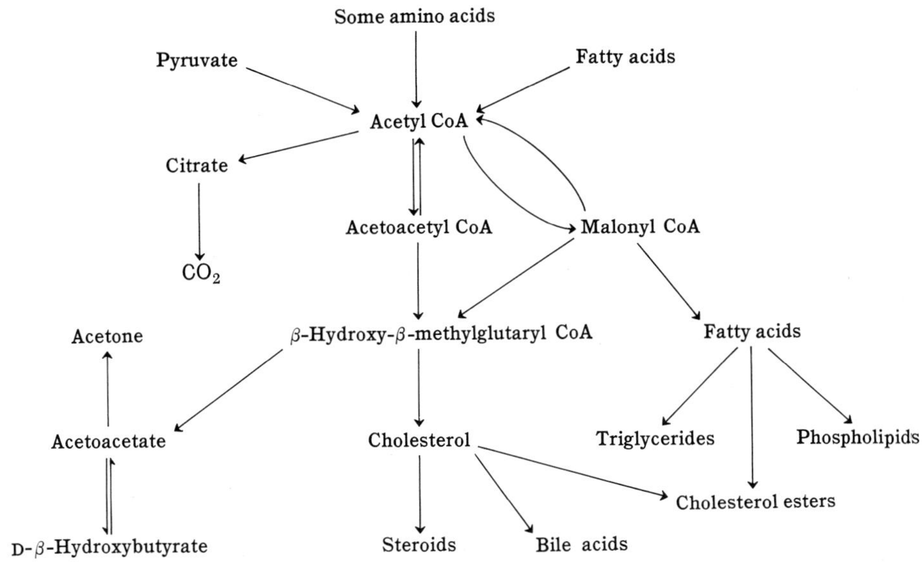

Fig. 21.6. A schematic representation of major interrelationships of the metabolism of fatty acids, cholesterol, and ketone bodies.

In circumstances of limited utilization of carbohydrate and/or excessive mobilization of fatty acids to the liver, there is a markedly diminished rate of operation of two of the three pathways for metabolizing acetyl CoA, *viz.*, the citric acid cycle and fatty acid synthesis. The former effect is apparently based on the stimulatory action of carbohydrate intermediates on citrate lyase; depressed activity of the latter enzyme could retard operation of the citric acid cycle. The limitation of fatty acid synthesis under circumstances of diminished formation of intermediates of the citric acid cycle and excess fatty acid accumulation is localized at the level of acetyl CoA carboxylase activity (pages 487 and 488). The result is a channeling of acetyl CoA into its third metabolic fate, β-hydroxy-β-methylglutaryl CoA formation and of products derived therefrom, *viz.*, cholesterol and acetoacetate. Increased formation of the latter, and of the associated β-hydroxybutyrate and acetone, elevates their concentration in the blood above normal, resulting in *ketonemia.* If the blood level exceeds the renal threshold and appreciable amounts of ketone bodies appear in the urine, *ketonuria* results. Whenever a marked degree of ketonemia and ketonuria exists, the odor of acetone is likely to be detected in the exhaled air. This triad of ketonemia, ketonuria, and acetone odor of the breath is commonly termed *ketosis.*

Causes of Ketosis. Any circumstance associated with diminished availability of the prime dietary source of energy, namely, carbohydrate, will accentuate utilization of fatty acids for this purpose. The most readily understood is *starvation.* In the absence of food, glycogen stores are rapidly depleted and survival depends largely on energy derived from depot lipid. Mobilization of this lipid is reflected in lipemia. Degradation of fatty acids in the liver proceeds more rapidly than usual, with augmented production of acetoacetyl CoA and its products. Ketosis incident to starvation is most frequently encountered clinically in gastrointestinal disturbances in infancy or pregnancy. Other circumstances in normal individuals in which excessive lipid and diminished carbohydrate are being metabolized may also lead to ketosis, *e.g.*, renal glucosuria (Chap. 35), and abrupt replacement of a normal diet by one low in carbohydrate and very rich in lipid.

In addition to the described effects of carbohydrate deprivation on lipid metabolism in the liver, it has been suggested that conditions of limited carbohydrate utilization would limit formation of α-glycerophosphate and α-phosphatidic acid (page 496) in adipose tissue, thus diminishing lipogenesis in this locus.

Clinically, by far the most important cause of ketosis is diabetes. In the diabetic individual, in contrast to the preceding situations, glucose is present in excessive amounts in the fluids of the body. The metabolic defect, *viz.*, insulin deficiency, in diabetes, however, prevents glucose utilization from operating at normal rate. From the point of view of the effect upon lipid metabolism, diabetes and starvation resemble one another. Despite the hyperglycemia in diabetes, glucose is not being catabolized at a normal rate in muscle or in liver. Excessive mobilization of depot lipid leads to lipemia and fatty liver. Generation of acetoacetyl CoA from fatty acids proceeds at an excessive rate in the liver while resynthesis of fatty acids is inhibited. If acetoacetate is formed at a rate in excess of the capacity of the extrahepatic tissues to utilize it, ketosis will develop. In diabetic individuals with severe ketosis, urinary excretion of ketone bodies may be as high as 5,000 mg. per 24 hr. and the blood concentration may reach 90 mg. per 100 ml., in contrast to normal values of $<$125 mg. and $<$3 mg., respectively.

Consequences of Ketosis. The complications encountered when more ketone bodies are formed than can be utilized relate primarily to the mode of excretion of acetoacetic and β-hydroxybutyric acids. These two substances are moderately strong acids, and even in the most acidic urine that the kidney can excrete, exist in large part as anions (Chap. 35). Cations, chiefly Na^+, are lost as these anions are excreted. Depletion of plasma and other body fluids of cations leads to acidosis (Chap. 33).

Coincident with excretion of salts of acetoacetic and β-hydroxybutyric acids (as well as of glucose, in the diabetic patient), large quantities of fluid are lost in the urine. A tendency to nausea and emesis may cause additional fluid loss, while depression of the central nervous system, leading ultimately to profound coma with areflexia, interferes with normal drinking of water. All these factors, if not corrected, complicate the acidosis by producing a state of severe dehydration.

The complications of persistent ketosis may be in large part explained by the acidosis due to loss of Na^+ and the dehydration due to uncompensated fluid loss. The dehydration and acidosis due to ketosis are distinguishable from those due to other causes by the odor of acetone on the breath and the presence of ketone bodies in the urine. If due to diabetes, the additional findings of glucosuria and hyperglycemia will almost invariably be present.

REFERENCES

See list following Chap. 22.

22. Lipid Metabolism. II

Phospholipids. Sphingolipids. Sterol Metabolism and Its Control. Disturbances of Lipid Metabolism

THE PHOSPHOLIPIDS

The phospholipids, which are ubiquitous in living organisms, fulfill a number of important biological functions. As the most polar of the lipids, with a marked solubility both in nonpolar solvents and in water, they have unique physical properties. Some of them, *e.g.*, lecithin, are dipolar ions possessing cationic and anionic groups. As such, they appear to participate uniquely in cellular structures such as mitochondria and various membranes. Together with the sphingomyelins and cerebrosides (Chap. 4), phospholipids are found in abundance in the myelin sheath of nerve (Chap. 37).

The phospholipids appear to be ideally suited for functioning as structural bridges between water-soluble proteins and nonpolar lipids, since, possessing both hydrophobic and hydrophilic ionic groups, they can interact strongly with both types of structures.

Some evidence has been presented that phospholipids play special roles in secretory processes, in ion transport, and in selective permeability. The stability of chylomicra in body fluids has been attributed in part to the electrical charges in these droplets, resulting from their content of ionic phospholipids and proteins. The possible role of the phosphatidyl ethanolamine fraction in blood clotting is discussed elsewhere (Chap. 31).

Distribution and Turnover of Phospholipids. Phospholipids are not uniformly distributed in the body lipids and are, in fact, almost totally absent from depot lipid. They do occur in the lipid of the various glandular organs, notably the liver, as well as in blood plasma, where they may comprise as much as half the total lipid. The phospholipids are abundant in certain specialized tissues, such as the myelinated portions of the nervous system, the yolks of bird eggs, and the seeds of legumes.

The turnover of various phospholipids in diverse sites in the animal has been studied with different isotopes, most frequently with ^{32}P. The biological half-lives range from less than 1 day for liver lecithin to more than 200 days for brain phosphatidyl ethanolamine. Plasma lecithin arises chiefly or exclusively in the liver and is believed to be also degraded in the liver.

FORMATION OF PHOSPHOLIPIDS

Phosphatides. The phosphatides are the major class of phospholipids (page 69) These derivatives of glycerol phosphate most frequently contain a nitrogenous base—serine, ethanolamine, and choline—which can be synthesized by the organism. A major role in phosphatide formation is played by cytidine derivatives. The most abundant phosphatide, lecithin, can be synthesized by two major pathways: (1) *de novo* synthesis, in which phosphatidyl serine serves as a precursor for other phosphatides; (2) utilization of exogenous choline, available from dietary sources or as a salvage pathway.

Phosphatidic Acid. The key substance in phosphatide synthesis is *phosphatidic acid.* Its formation from α-glycerophosphoric acid and two moles of acyl CoA has already been described (page 496). In addition, there are alternate routes for formation of phosphatidic acid from triglycerides. Over-all, these pathways may be outlined as follows.

$$\text{Triglyceride} \longrightarrow \text{1,2-diglyceride} \longrightarrow \text{phosphatidic acid}$$
$$\downarrow$$
$$\text{monoglyceride}$$
$$\downarrow$$
$$\text{phosphatidic acid} \dashrightarrow \text{phosphatides}$$

The 1,2-diglyceride is converted to phosphatidic acid by a kinase reaction.

$$\text{D-1,2-Diglyceride} + \text{ATP} \longrightarrow \text{L-α-phosphatidic acid} + \text{ADP}$$

The monoglyceride is converted by enzymes of brain tissue and intestinal mucosa to L-α-lysophosphatidic acid, which is then transformed to phosphatidic acid.

$$\text{α-(or β-) Monoglyceride} + \text{ATP} \longrightarrow \text{ADP} + \text{RCOOCH}_2$$

$$\text{L-α-phosphatidic acid} \xleftarrow{+\,\text{RCOCoA}} \begin{array}{l} \text{HOCH} \quad\quad\; \text{O} \\ \qquad\qquad\qquad\; \| \\ \text{H}_2\text{C}-\text{O}-\text{P}-\text{OH} \\ \qquad\qquad\quad\; | \\ \qquad\qquad\quad\; \text{OH} \end{array}$$

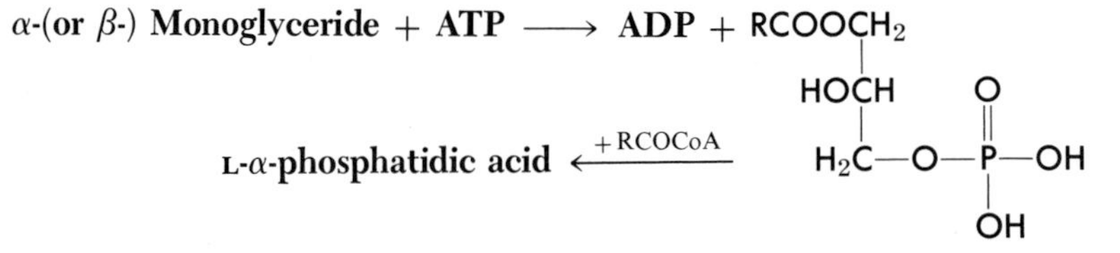

L-α-Lysophosphatidic acid

Formation of Phosphatidyl Serine. Phosphatidyl serine is formed via a cytidine intermediate, cytidine diphosphate diglyceride (CDP-diglyceride), which in turn is derived from L-α-phosphatidic acid. Formation of CDP-diglyceride has been demonstrated in mammalian liver and in bacterial sources. The reactions are the following.

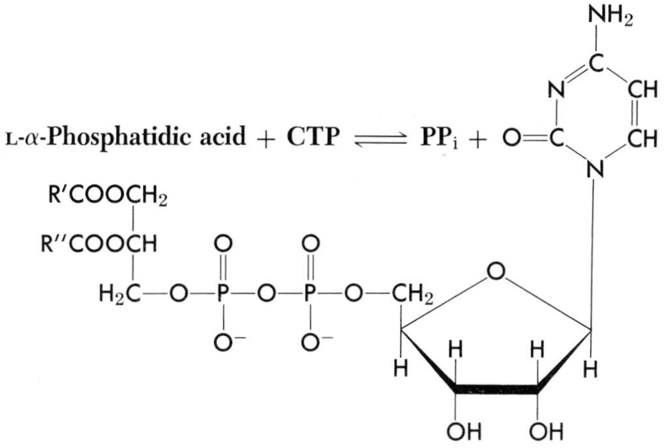

Cytidine diphosphate diglyceride

The next step results in formation of phosphatidyl serine.

$$\textbf{CDP-diglyceride} + \textsc{l}\textbf{-serine} \longrightarrow \textbf{phosphatidyl serine} + \textbf{CMP}$$

Synthesis of phosphatidyl serine by the above reaction has been reported thus far only in extracts of *Escherichia coli* but probably also occurs in mammalian tissues.

Formation of Phosphatidyl Ethanolamine. Phosphatidyl ethanolamine is formed by decarboxylation of phosphatidyl serine; the enzyme requires pyridoxal phosphate.

$$\textbf{Phosphatidyl serine} \longrightarrow \textbf{phosphatidyl ethanolamine} + CO_2$$

This decarboxylation is in accord with many earlier studies with labeled serine which showed that the ethanolamine of phosphatides is derived by decarboxylation of serine and that this occurs in the phosphatide rather than with free serine.

De Novo Formation of Lecithin. This occurs in stages by successive transfer of three methyl groups from S-adenosylmethionine (page 553) to phosphatidyl ethanolamine. This methylation occurs only in the liver and only with phosphatidyl ethanolamine as substrate. This is also the mechanism for endogenous formation of choline. In studies with ^{14}C-methyl–labeled S-adenosylmethionine, the label was incorporated in stages by successive methylation of phosphatidyl ethanolamine. Two enzymes are involved; one catalyzes initial methylation with formation of phosphatidyl monoethanolamine, and the second catalyzes introduction of the second and third methyl groups, yielding, respectively, dimethylaminoethanol- and choline-containing phosphatides. The formation of S-adenosylmethionine and its role in transmethylation are presented later (Chap. 23).

Other Routes of Phosphatide Formation. An alternate route of lecithin formation may be regarded as a salvage pathway for utilization of free choline. The key intermediate in this process is *cytidine diphosphate choline.*

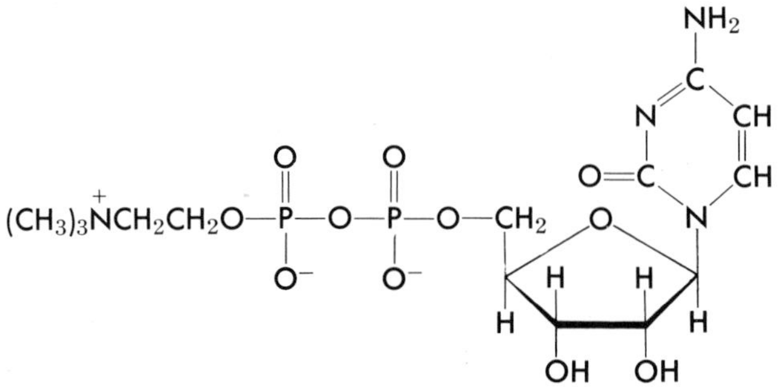

$$(CH_3)_3\overset{+}{N}CH_2CH_2O-\overset{\overset{\displaystyle O}{\|}}{\underset{\underset{\displaystyle O^-}{|}}{P}}-O-\overset{\overset{\displaystyle O}{\|}}{\underset{\underset{\displaystyle O^-}{|}}{P}}-O-CH_2$$

Cytidine diphosphate choline

The phosphorylcholine portion of the above molecule is formed by the following reaction.

$$\text{Choline + ATP} \longrightarrow {}^-O-\overset{\overset{\displaystyle O}{\|}}{\underset{\underset{\displaystyle OH}{|}}{P}}-OCH_2CH_2\overset{+}{N}(CH_3)_3 + \text{ADP}$$

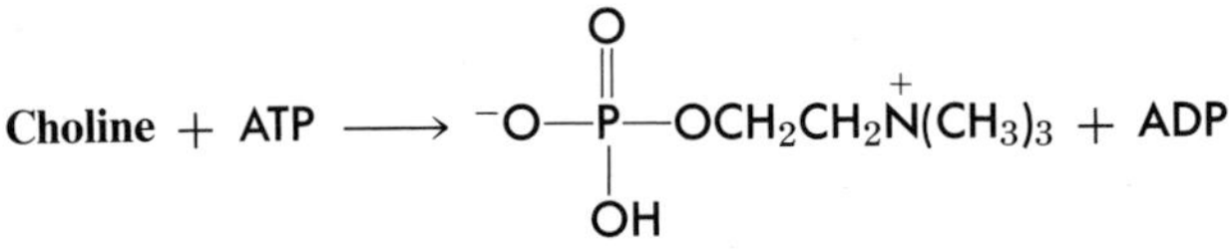

Phosphorylcholine

Formation of cytidine diphosphate choline from cytidine triphosphate and phosphorylcholine involves elimination of pyrophosphate and is thus analogous to the generation of uridine diphosphate glucose from uridine triphosphate and glucose 1-phosphate (page 425).

Cytidine triphosphate + phosphorylcholine $\rightleftharpoons$ cytidine diphosphate choline + PP$_i$

Interaction of cytidine diphosphate choline (CDP-choline) with 1,2-diglyceride yields lecithin and cytidine monophosphate.

$$\begin{array}{l} \text{R'COOCH}_2 \\ | \\ \text{R''COOCH} \\ | \\ \text{H}_2\text{COH} \end{array} \text{+ CDP-choline} \rightleftharpoons \begin{array}{l} \text{R'COOCH}_2 \\ | \\ \text{R''COOCH} \\ | \\ \text{H}_2\text{CO}-\overset{\overset{\displaystyle O}{\|}}{\underset{\underset{\displaystyle O^-}{|}}{P}}-OCH_2CH_2\overset{+}{N}(CH_3)_3 \end{array} \text{+ CMP}$$

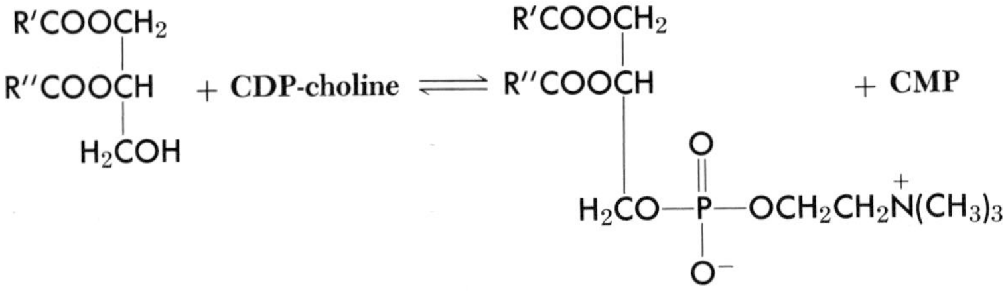

D-1,2-Diglyceride α-**Lecithin**

At the expense of ATP, cytidine triphosphate is regenerated from cytidine monophosphate.

By an analogous series of reactions, phosphorylethanolamine

$$ {}^-O-\overset{\overset{\displaystyle O}{\|}}{\underset{\underset{\displaystyle OH}{|}}{P}}-OCH_2CH_2\overset{+}{N}H_3$$

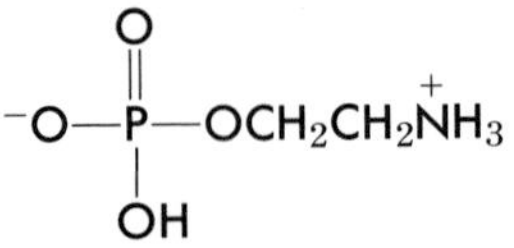

can be converted into cytidine diphosphate ethanolamine, which reacts with the diglyceride to give phosphatidyl ethanolamine.

Although phosphatidyl serine is primarily formed by the *de novo* route previously described, it can also be made in mammalian tissues by an exchange of free serine with the ethanolamine moiety of phosphatidyl ethanolamine.

$$\text{Phosphatidyl ethanolamine} + \text{L-serine} \rightleftharpoons \text{phosphatidyl serine} + \text{ethanolamine}$$

Formation of Plasmalogens. An acyl CoA derivative is probably the source of the α,β-unsaturated ether moiety of plasmalogens (page 71). An enzymic preparation from rat liver catalyzes formation of plasmalogens from CDP-choline or CDP-ethanolamine and a plasmalogenic diglyceride.

$$\text{CDP-choline} + \text{plasmalogenic diglyceride} \rightleftharpoons \text{phosphatidal choline} + \text{CMP}$$
$$\text{CDP-ethanolamine} + \text{plasmalogenic diglyceride} \rightleftharpoons \text{phosphatidal ethanolamine} + \text{CMP}$$

The formation of the plasmalogenic diglyceride may occur by reduction of the acyl group α to the alkenyl ether group. However, it is not known whether such a reduction takes place at the diglyceride stage or with the completed phospholipid.

Formation of Phosphoinositides and Phosphoglycerides. These nitrogen-free derivatives of glycerol phosphate are widely distributed in nature. Their synthesis occurs by a reaction analogous to synthesis of phosphatidyl serine.

$$\text{CDP-diglyceride} + \text{inositol} \xrightarrow{Mn^{++}} \text{phosphatidyl inositol} + \text{CMP}$$

Isotopic evidence indicates that formation of the di- and triphosphoinositides (page 72) occurs by successive phosphorylations of the monoinositide. A *phosphatidyl inositol kinase* of rat brain catalyzed these successive phosphorylations in the presence of ATP and either Mg^{++} or Mn^{++}.

The phosphoglycerides (page 71) are synthesized by a similar reaction.

(*a*) **CDP-diglyceride + glycerol 3-phosphate ⟶**

3-phosphatidyl-1′-glycerophosphate + CMP

(*b*) **3-Phosphatidyl-1′-glycerophosphate ⟶**

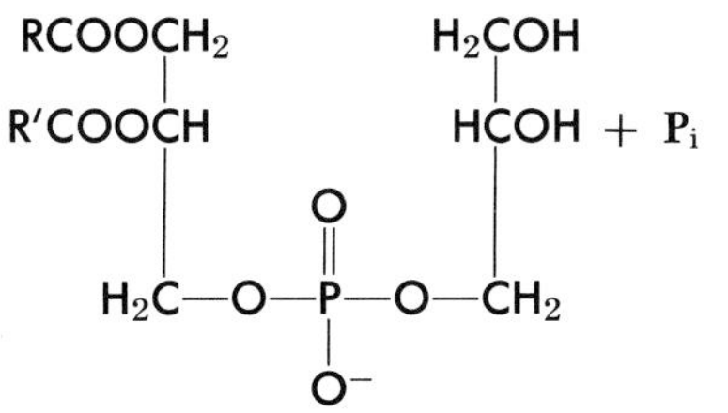

3-Phosphatidyl-1′-glycerol

The manner by which a diphosphatidylglycerol (page 71) may be formed is not known; introduction of a second molecule of phosphatidic acid might be achieved by reaction of phosphatidylglycerol with a second molecule of CDP-diglyceride.

The interrelationships of phospholipid and triglyceride synthesis are shown in Fig. 22.1.

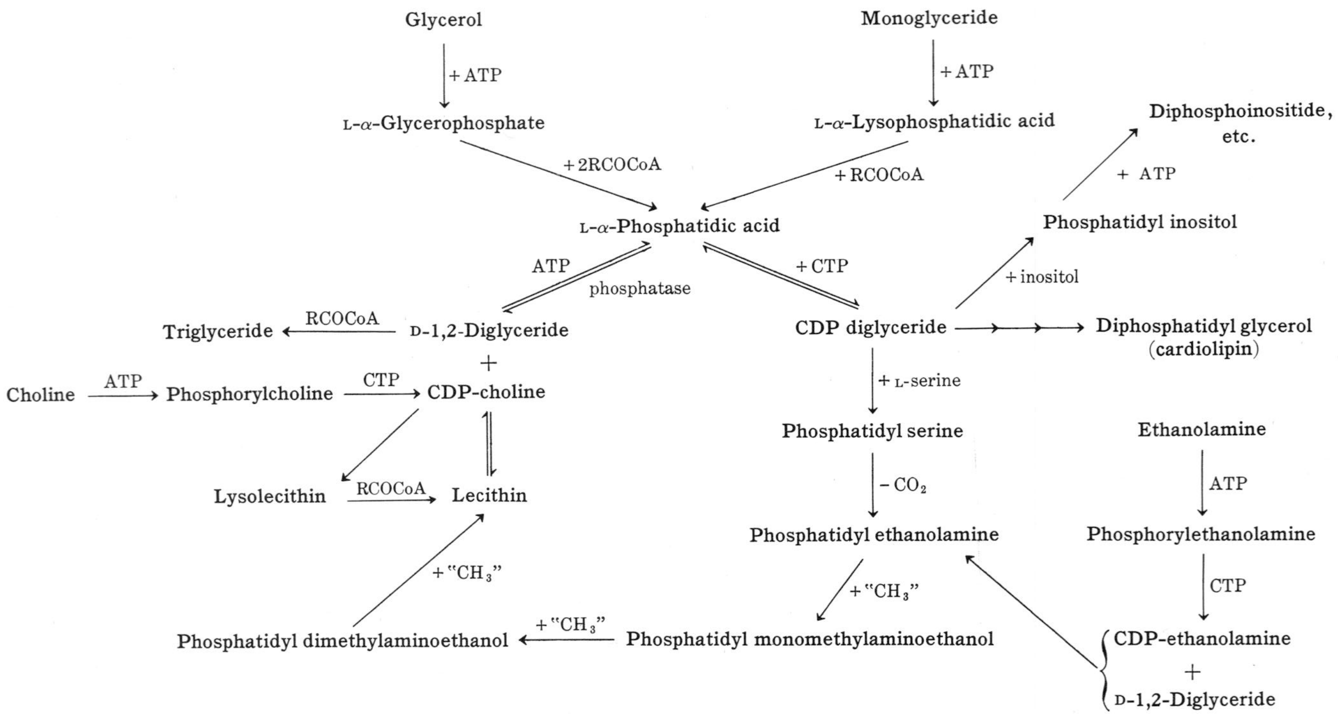

Fig. 22.1. Schematic representation of some of the interrelationships of phospholipid and triglyceride synthesis. "CH_3" designates the methyl group which is transferred from S-adenosylmethionine. It is evident from the above scheme that pathways exist by which glycerol, monoglycerides, or diglycerides can be utilized for formation of lipids or phospholipids. The key intermediate is phosphatidic acid.

ACTION OF PHOSPHOLIPASES

Several phospholipases have been described with specificity for one or more of the bonds of the phospholipids. These enzymes have been useful in the study of the structure of phospholipids as well as indicating routes of their degradation. The susceptible bonds in lecithin are indicated by the capital letters in the structure below.

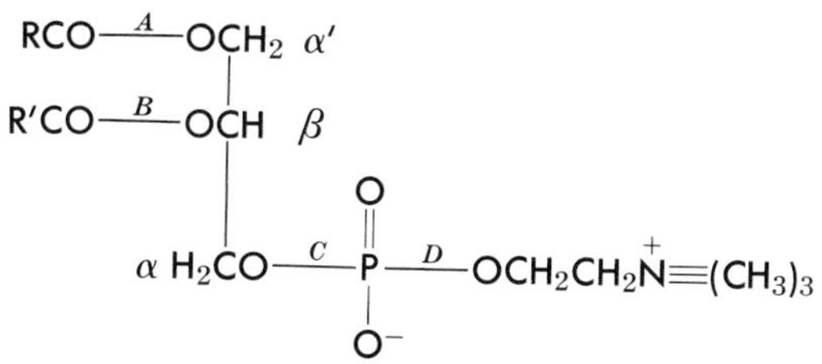

A Ca^{++}-requiring enzyme, *phospholipase A,* found in certain snake venoms and in pancreas, catalyzes hydrolysis at *B*. The product, lecithin minus one fatty acid residue, is a lysolecithin (page 70), and is a strong detergent and a potent hemolytic agent. With the aid of this enzyme, it has been shown that in some lecithins, *e.g.*, those of liver and egg, an unsaturated fatty acid is present at the β position and a saturated fatty acid at the α' position. However, lecithins from other sources possess different structures; *e.g.*, lecithin from yeast has unsaturated residues at both α' and β positions, whereas one from lung has saturated residues at both positions. Clearly, there is considerable specificity in lecithin synthesis.

Phospholipase A also cleaves bond *B* in phosphatidyl ethanolamine and in the choline- and ethanolamine-containing plasmalogens.

Lysophospholipase of pancreas and other tissues catalyzes hydrolysis of the single fatty acid ester bond, at *A*, in lysolecithin or lysophosphatidyl ethanolamine.

$$\text{Lysolecithin} + H_2O \longrightarrow \text{glycerylphosphorylcholine} + \text{fatty acid}$$

An enzyme termed *phospholipase B* catalyzes removal of the acyl groups from both the α and β positions. Also, a widely distributed mammalian enzyme has been described which catalyzes resynthesis of lecithin from lysolecithin and a fatty acyl CoA.

$$\text{Lysolecithin} + \text{RCOCoA} \longrightarrow \text{lecithin} + \text{CoA}$$

Phospholipase C catalyzes hydrolysis at position *C* (above) in a glycerylphospholipid.

$$\text{Glycerylphospholipid} + H_2O \longrightarrow \alpha,\beta\text{-diglyceride} + \text{phosphorylated nitrogenous base}$$

This enzyme also splits the ceramide-phosphate linkage in sphingomyelin (see below). The enzyme is found in the α toxin of *Clostridium welchii* and other strains of clostridia and bacilli; a similar enzymic activity is present in animal tissues (page 496).

Phospholipase D, which has been found only in plant tissues, catalyzes transphosphatidylation reactions as well as hydrolytic cleavage of the terminal diester bond (at *D*) of glycerophosphatides containing choline, ethanolamine, serine, or glycerol, with the formation of phosphatidic acid.

$$\text{Phosphatidylcholine} + H_2O \longrightarrow \text{phosphatidic acid} + \text{choline} \qquad (1)$$
$$\text{Phosphatidylcholine} + ROH \rightleftharpoons \text{phosphatidyl—OR} + \text{choline} \qquad (2)$$

Both reactions are activated by Ca^{++}. Reaction (2), a transphosphatidylation, could provide a mechanism for synthesis and turnover of phospholipids.

SPHINGOLIPIDS

Formation of Sphingosines. The long-chain, aliphatic bases, *e.g.*, sphingosine and dihydrosphingosine (page 72), which are characteristic of the sphingolipids, can be synthesized and need not be provided in the diet.

Elucidation of the sequence of reactions in sphingosine biosynthesis has been accomplished with enzymes from brain, which catalyze the following reactions.

(a) $\qquad$ **Palmityl CoA + TPNH + H$^+$ $\rightleftharpoons$ palmitic aldehyde + TPN$^+$ + CoA**

(b) $\qquad$ **Palmitic aldehyde + serine $\xrightarrow{\text{pyridoxal phosphate} \atop Mn^{++}}$ dihydrosphingosine + CO$_2$**

Carbon atoms 1 and 2 of the dihydrosphingosine and its amino group are derived from the β and γ carbon atoms of serine, whose carboxyl group is lost as CO_2. It is not clear whether reaction (b) is as depicted, since the condensation involving serine appears to occur with a reduced derivative of palmityl CoA. The remaining carbon atoms of the dihydrosphingosine derive from the palmitic aldehyde.

Dehydrogenation of dihydrosphingosine is catalyzed by a flavin enzyme.

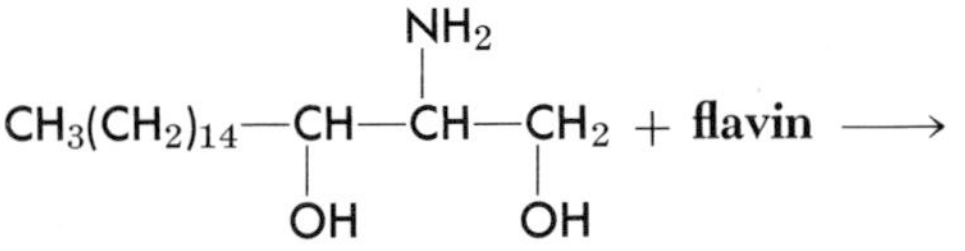

$$\begin{array}{c} NH_2 \\ | \\ CH_3(CH_2)_{14}-CH-CH-CH_2 + \textbf{flavin} \longrightarrow \\ \quad\quad | \quad\quad\; | \\ \quad\quad OH \quad\; OH \end{array}$$
Dihydrosphingosine

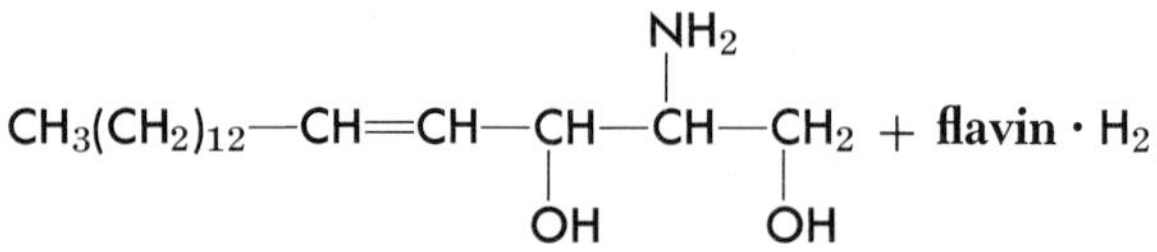

$$\begin{array}{c} NH_2 \\ | \\ CH_3(CH_2)_{12}-CH{=}CH-CH-CH-CH_2 + \textbf{flavin} \cdot H_2 \\ \quad\quad\quad\quad\quad\quad | \quad\quad\; | \\ \quad\quad\quad\quad\quad\quad OH \quad\; OH \end{array}$$
Sphingosine

A similar series of the above reactions involving the CoA derivatives of other fatty acids would yield the known C_{16}, C_{17}, C_{19}, and C_{20} sphingosines.

Formation of Ceramides. It is assumed that the ceramides (page 73) are formed by N-acylation of the sphingosines by a fatty acyl CoA.

Sphingosine + RCOCoA $\longrightarrow$ ceramide + CoA

Formation of Sphingomyelins. The synthesis of these phosphorus-containing sphingolipids results from the reaction of a ceramide with cytidine diphosphate choline.

Cytidine diphosphate choline + ceramide $\rightleftharpoons$ sphingomyelin + CMP

In addition to being N-acylated to a ceramide, sphingosine may react with UDP-galactose (page 426) to form *psychosine* (galactosylsphingosine).

Sphingosine + UDP-galactose $\longrightarrow$ psychosine + UDP

The reaction is catalyzed by a brain microsomal enzyme, *galactosyl-sphingosine transferase.* Psychosine may thus be an intermediate in cerebroside formation (see below).

Formation of Glycosphingolipids. The synthesis of these carbohydrate-containing derivatives of a ceramide involves the transfer to the ceramide of a sugar from an appropriate activated nucleotide derivative. These activated participants may be uridine diphosphate glucose or galactose, uridine diphosphate N-acetylgalactosamine, or cytidine monophosphate N-acetylneuraminic acid. The enzymes catalyzing these transfer reactions are present in brain tissue.

Cerebroside Synthesis. The gluco- and galactocerebrosides (ceramide monosaccharides) are formed by reaction of a ceramide with either uridine diphosphate glucose or uridine diphosphate galactose, respectively. Thus,

$$\textbf{Ceramide + UDP-glucose} \longrightarrow \textbf{ceramide-glucose + UDP}$$

Cerebrosides may also be formed by N-acylation of psychosine (see above).

$$\textbf{Psychosine + stearyl CoA} \longrightarrow \textbf{cerebroside + CoA}$$

Thus it appears that cerebroside formation may occur via two alternate pathways, one in which glycosylation of sphingosine is the initial reaction, followed by N-acylation, and the other in which acylation of sphingosine, to form a ceramide, precedes glycosylation.

Cerebroside sulfuric acid esters, the sulfatides (page 74), are synthesized by a reaction involving a galactocerebroside and 3′-phosphoadenosine-5′-phosphosulfate (page 563), catalyzed by *galactocerebroside sulfokinase,* a brain microsomal enzyme.

Ganglioside Synthesis. The functional significance of the gangliosides is indicated from their roles as possible cell receptor sites for viruses, as pharmacologically active substances, and as substances that accumulate in several of the lipid storage diseases (page 527). These glycosphingolipids with more than one mole of carbohydrate are formed by successive reaction of a ceramide with one or more moles of an activated carbohydrate or monosaccharide to yield either ceramide oligosaccharides or the more complex gangliosides. Some of the synthetic pathways for ganglioside formation are depicted below; NANA = N-acetylneuraminic acid; Ac = N-acetyl.

$$\textbf{Ceramide-glucose} \xrightarrow[A]{\text{UDP-galactose}} \textbf{ceramide-glucose-galactose} \xrightarrow[B]{\text{CMP-NANA, } Mg^{++}}$$

$$\textbf{ceramide-glucose-galactose-NANA} \xrightarrow[C]{\text{UDP-Ac-galactosamine, } Mn^{++}}$$

ceramide-glucose-galactose-NANA $\xrightarrow[D]{\text{UDP-galactose, } Mn^{++}}$
|
Ac-galactosamine

ceramide-glucose-galactose-NANA $\xrightarrow[E]{\text{CMP-NANA, } Mg^{++}}$
|
Ac-galactosamine-galactose

ceramide-glucose-galactose-NANA
|
Ac-galactosamine-galactose-NANA

A particulate fraction from embryonic chick brain has been shown to contain enzymes catalyzing reactions *B* and *D*, above. The enzyme for reaction *B* is a *sialyl transferase;* reaction *D* is catalyzed by a *glycolipid galactose transferase;* and the enzyme for reaction *C* can be termed a *glycolipid N-acetylgalactosamine transferase.* Each of these three enzymes is inhibited by the products of ganglioside synthesis, which bind the metal ion required for enzymic activity.

SPHINGOLIPASES

Several enzymes catalyzing hydrolysis of sphingolipids have been described recently. This list will undoubtedly grow, since these enzymes not only are of significance for structural studies but may be etiologic factors in clinical conditions in which excessive accumulation of lipid occurs (pages 527*ff.*).

Sphingomyelinase. An enzyme from human spleen catalyzes the hydrolysis of sphingomyelin or dihydrosphingomyelin with formation of ceramide and phosphorylcholine. This enzyme exhibits greater specificity than phospholipase C (see above), since it does not attack lecithin; it is also inactive toward sphingosine phosphorylcholine, indicating a specific requirement for the N-acylated sphingosine derivative. Sphingomyelinase activity is also present in liver and kidney. The activity of this enzyme may be of significance in certain lipid storage diseases (page 527).

A second sphingomyelinase, differing from that above, is also present in spleen; this is a Mg^{++}-dependent enzyme with similar substrate specificity and a higher pH optimum.

Ceramide Saccharidases. Two enzymes have been described that catalyze hydrolysis of ceramide mono- and oligosaccharides. One of these enzymes is widely distributed, since activity toward galactosylgalactosylglucosylceramide, with liberation of the terminal galactose residue, has been found in rat small intestine, brain, liver, kidney, and spleen. This enzymic activity is also present in extracts of human small intestinal tissue and is specific for glycosphingolipids. The second enzyme, also obtained from intestinal tissue, catalyzes hydrolysis of both glucosyl- and galactosylceramide, with liberation of the hexose.

STEROL METABOLISM AND ITS CONTROL

The present discussion will center about the metabolism of cholesterol, the most abundant sterol of animal tissues. Other related compounds will be mentioned only incidentally. More detailed discussion of the steroid hormones will be found in Chaps. 44 and 45.

With the exception of the sterol-like vitamins D, steroids are not essential in the diet; however, the mammal can synthesize 7-dehydrocholesterol (page 80), and it is only the requirement for radiant energy for the reaction

$$\textbf{7-Dehydrocholesterol} \xrightarrow{\text{photons}} \textbf{vitamin } \mathbf{D_3}$$

that makes vitamin D a dietary essential.

SOURCES OF BODY CHOLESTEROL

Diet. Cholesterol is present in all animal tissues, hence all carnivores ingest this sterol. Its absorption from the intestinal lumen is exclusively into the intestinal lacteals, thence to the thoracic duct. The absorption of cholesterol shows a virtually absolute dependence upon the presence of bile salts in the intestinal lumen. In the process of absorption, the major portion of the cholesterol is esterified with fatty acids and appears in lymphatic chylomicra as cholesterol esters. The absorption of sterols from the intestine has been considered previously (page 474). Two products formed from cholesterol by bacterial action in the intestine, *viz.*, β-cholestanol (page 522) and coprosterol (page 522), are poorly absorbed from the gastrointestinal tract and are found abundantly in the feces. Certain steroids, *e.g.*, β-sitosterol, have been shown to interfere with cholesterol absorption, as do a variety of other substances. Thus ferric chloride, presumably by formation of an insoluble iron salt of bile acids, as well as ion exchange resins which bind bile acids, markedly reduce absorption of intestinal cholesterol. All such agents have been studied in relation to prevention of hypercholesterolemia and its postulated role in vascular disease (page 526).

Biosynthesis of Cholesterol. All the carbon atoms of endogenous cholesterol are derived from acetyl CoA. Almost all tissues known to synthesize cholesterol, including liver, adrenal cortex, arterial wall, etc., can generate cholesterol in vitro from acetate. The enzymes involved are associated with cytoplasmic particles (microsomes) and have been obtained in water-soluble form; however, cholesterol synthesis as catalyzed by these enzymes has an absolute dependency on cofactor(s) present in the supernatant fluid obtained by sedimentation of microsomes.

The series of reactions intervening between acetate and cholesterol has been elucidated in the laboratories of Bloch, Gurin, Lynen, Popják, Rudney, Tavormina, and others. The initial series of reactions involves formation of β-hydroxy-β-methylglutaryl CoA; this compound derives principally from three moles of acetyl CoA as described previously (page 503). β-Hydroxy-β-methylglutaryl CoA is also produced in the metabolic degradation of leucine (page 524).

β-Hydroxy-β-methylglutaryl CoA is the immediate precursor of *mevalonic acid,* formed in a reaction catalyzed by a sulfhydryl-containing, TPNH-requiring enzyme, *β-hydroxy-β-methylglutaryl CoA reductase.* The reaction involves reduction of one of the carboxyl groups of hydroxymethylglutaryl CoA (Fig. 22.2) and is the committed step at which factors influencing cholesterol synthesis may exert their effects (page 524). Mevalonic acid is also formed by the condensation of acetyl CoA with acetoacetyl acyl carrier protein, an intermediate in fatty acid synthesis (page 486), forming protein-bound hydroxymethylglutarate, which may be the substrate for reduction to mevalonate. Previous consideration has been given (page 488) to the cleavage of β-hydroxy-β-methylglutaryl CoA to acetoacetate. Livers of fasting animals, which show decreased cholesterol formation, have decreased β-hydroxy-β-methylglutaryl reductase activity. This could increase relatively the cleavage reaction and explain the augmented ketogenesis of the fasting state (page 505).

Sterol biosynthesis occurs by condensation of six 5-carbon units, each derived from mevalonate. The sequence of reactions between mevalonate and squalene (the precursor of sterols) is shown in Fig. 22.2. Mevalonic acid is phosphorylated with

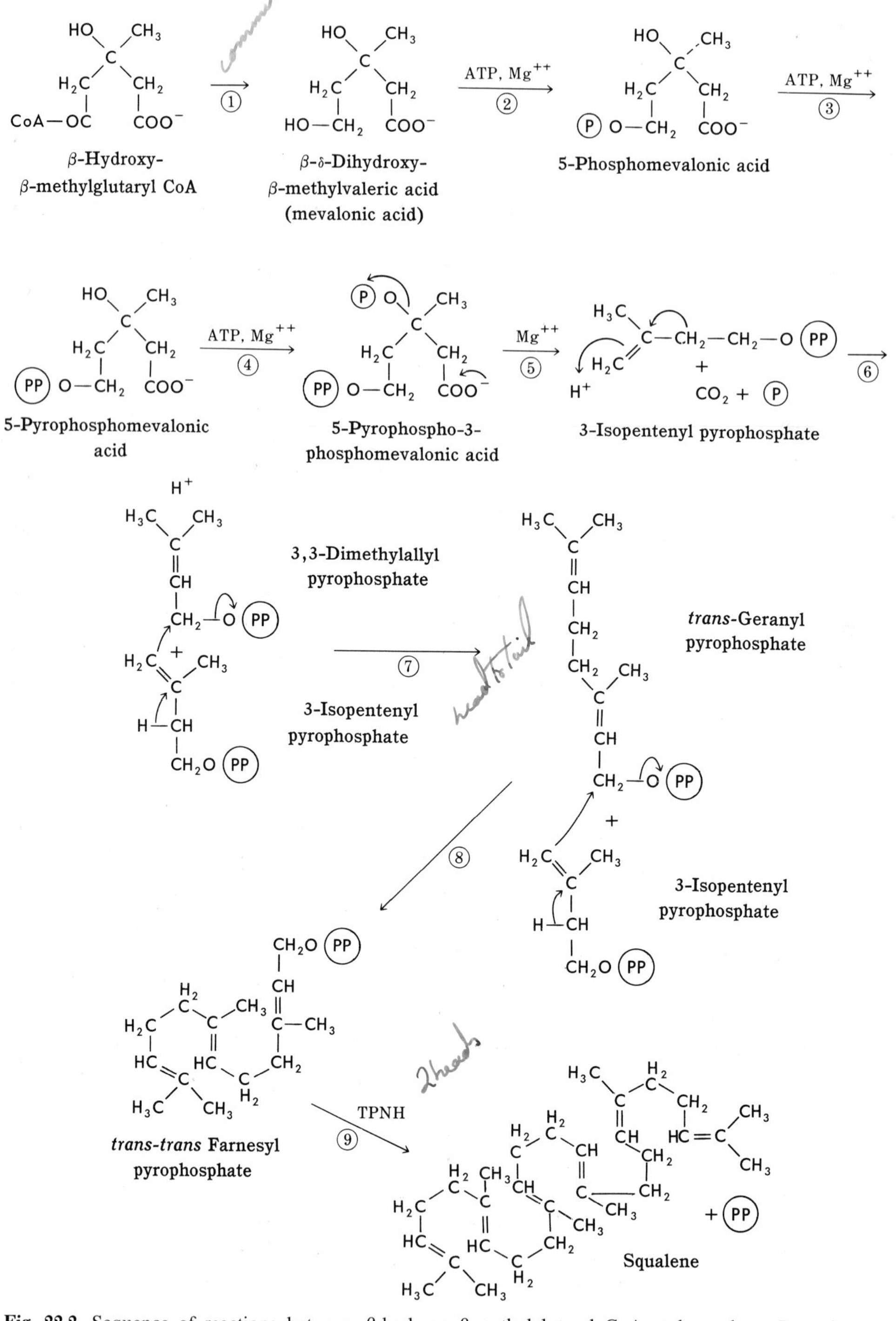

Fig. 22.2. Sequence of reactions between β-hydroxy-β-methylglutaryl CoA and squalene. P = phosphate; PP = pyrophosphate. Enzymes catalyzing the above reactions have been designated as follows: 1. β-hydroxy-β-methylglutaryl CoA reductase; 2. mevalonic kinase; 3. phosphomevalonic kinase; 4. pyrophosphophoryl mevalonic kinase; 5. isopentenyl pyrophosphate synthetase; 6. isopentenyl pyrophosphate isomerase; 7. geranyl pyrophosphate synthetase; 8. farnesyl pyrophosphate synthetase; 9. squalene synthetase.

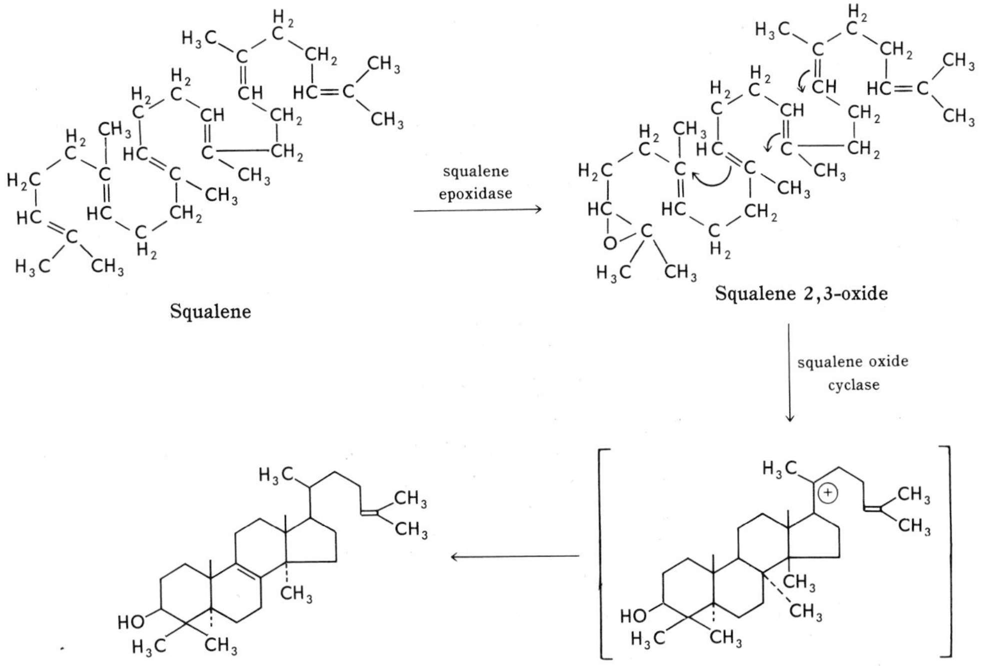

Fig. 22.3. Postulated mechanism for the conversion of squalene to lanosterol.

ATP in three successive reactions catalyzed by three distinct enzymes, with sequential formation of 5-phospho-, 5-pyrophospho-, and 5-pyrophospho-3-phosphomevalonate. The last compound is a transient intermediate which simultaneously loses the tertiary phosphate and decarboxylates to isopentenyl pyrophosphate. The latter is rapidly isomerized to dimethylallyl pyrophosphate by addition of a proton to the terminal doubly bound methylene carbon, followed by elimination of a proton from C-2. These two isomers are key substances in polyisoprenoid synthesis and are the basis for subsequent reactions involving carbon to carbon bond biosynthetic reactions in sterol formation. Dimethylallyl pyrophosphate, by virtue of its double bond and esterification with a strong acid, is an electrophilic reagent, whereas isopentenyl pyrophosphate, by virtue of its terminal doubly bound methylene carbon, is a nucleophilic reagent. The two compounds are therefore ideally suited for condensation with one another to form geranyl pyrophosphate. Repetition of this reaction between the latter and another mole of isopentenyl pyrophosphate leads to formation of farnesyl pyrophosphate. Under anaerobic conditions, with TPNH as coenzyme, reductive coupling of two moles of farnesyl pyrophosphate, an acyclic terpene, leads to formation of squalene, a symmetrical 30-carbon triterpene.

The hydrocarbon squalene is transformed into the tetracyclic steroidal configuration by the pathway depicted in Fig. 22.3. A hypothesized *squalene epoxidase* catalyzes conversion of squalene to the 2,3-oxide. The latter then undergoes an anaerobic cyclization, catalyzed by *squalene oxide cyclase,* to lanosterol. This enzymic system in liver cyclizes squalene to lanosterol exclusively, suggesting that other cyclizing systems must occur in other forms, including plants, to account for the formation of other triterpenes and sterols. The mechanism of squalene cyclization is postulated to involve an initial electrophilic attack by an oxidant, "activated molecular oxygen," of unknown identity. Concerted electron shifts following electrophilic attack lead to ring closure and the formation of a transient, intermediary carbonium ion. Lanosterol can be derived from the latter by a series of concerted hydride and methyl shifts.

The conversion of lanosterol to cholesterol (Fig. 22.4) requires the removal of three angular methyl groups, the saturation of the double bonds in the side chain and at the C,D ring juncture, and the introduction of the double bond in the 5,6 position. Removal of the methyl groups is an oxidative process. The carbon atoms eliminated appear as carbon dioxide; the mechanism of this demethylation is unknown. Formaldehyde is not an intermediate, suggesting that detachment of carbon may occur by decarboxylation, although intermediate carboxylic acids have not been isolated. The methyl group at C-14 is the first to be eliminated. Desmosterol, a postulated intermediate in cholesterol formation, has been isolated from tissues and accumulates in excessive quantities subsequent to prolonged administration of certain substances known to interfere with cholesterol biosynthesis (see below).

Alternative pathways from lanosterol to cholesterol may be postulated on the basis of the isolation of other possible intermediates (Fig. 22.4). Thus, labeled 4α-methyl-Δ^7-cholesten-3β-ol was isolated from the skin, liver, and small intestine of rats injected with ^{14}C-labeled acetate. Moreover, feeding of this sterol to rats led to its rapid conversion to cholesterol. Also, 4α-methyl-Δ^8-cholesten-3β-ol has been isolated from preputial gland tumors of mice. Finally, 24,25-dihydrolanosterol, 4α-methyl-Δ^8-cholesten-3β-ol, Δ^7-cholesten-3β-ol, and 7-dehydrocholesterol are rapidly converted to cholesterol by cell-free homogenates of liver. The order of the individual steps in the conversion of lanosterol to cholesterol is still uncertain.

Although all the carbon atoms of cholesterol may derive from acetate, in the biosynthesis of ergosterol (page 80) by yeast the additional carbon atom, a methyl group at C-24, has its origin in a methylation reaction with S-adenosylmethionine (page 553). Mevalonate has also been established as the source of the isoprenoid side chain in the biosynthesis of carotenoids (page 76) and coenzyme Q (page 324) and serves as the precursor of the rubber hydrocarbons.

Cholesterol of Body Fluids. A large portion of the cholesterol in lymph and in blood plasma is found in chylomicra. Since the dispersed state of these fat droplets is due chiefly to their content in phospholipid, it is not surprising that the ratio of phospholipid to cholesterol in the blood remains fairly constant.

Of the cholesterol in plasma, roughly two-thirds is esterified with fatty acids. The maintenance of this ratio is a function of the liver, and decreases in this value due to lowering of cholesterol ester concentration are seen in liver disease. The liver serves both as the chief source and the chief agent for disposal of plasma cholesterol, a portion of that removed from the blood appearing in the bile.

The normally high concentration of cholesterol in human bile is of consequence clinically. Though sparingly soluble in water, cholesterol readily dissolves in aqueous bile salt solutions, probably because of the formation of choleic acids,

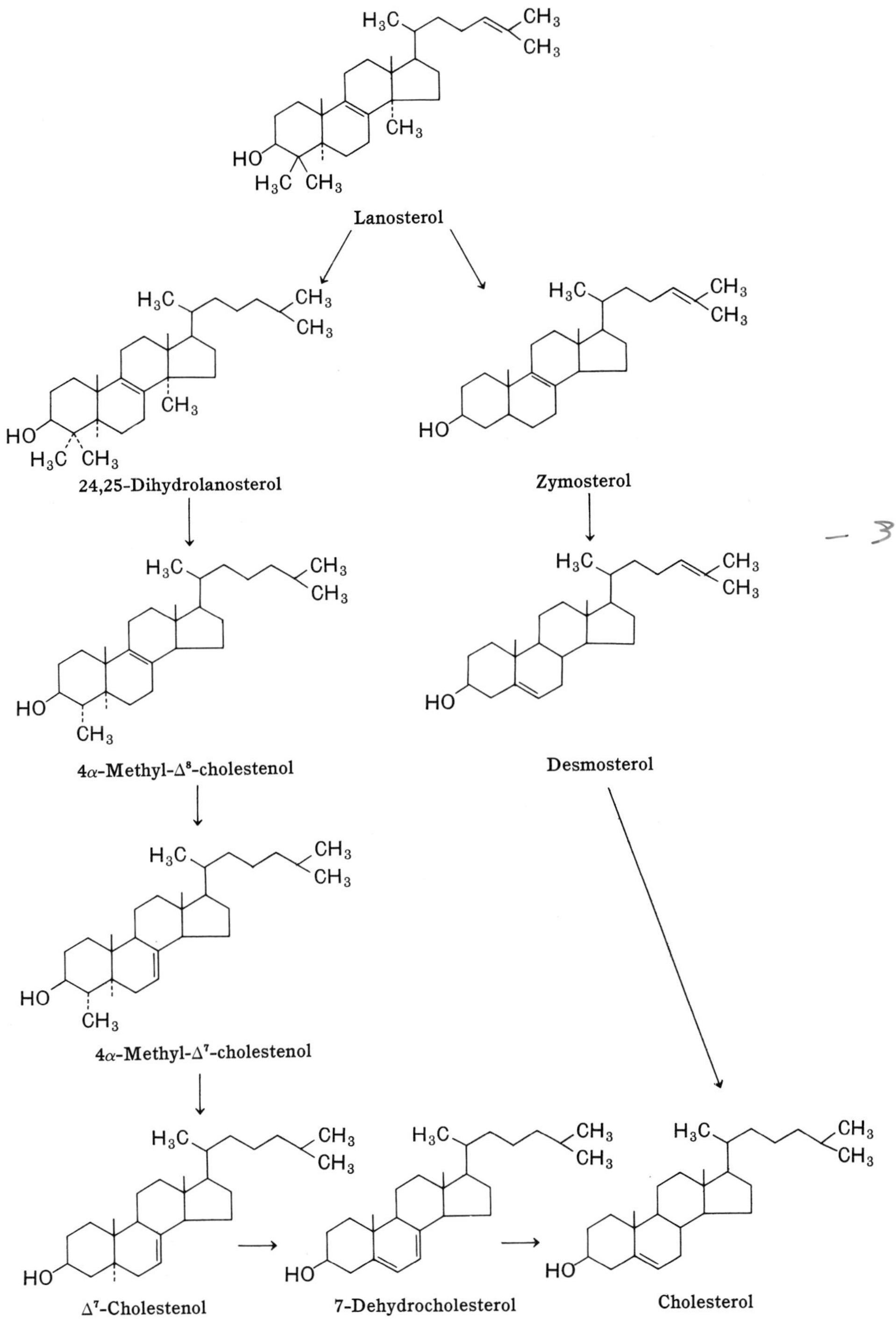

Fig. 22.4. Two theoretical pathways from lanosterol to cholesterol, based on available data. The pathway via 7-dehydrocholesterol has more supportive evidence, although both pathways, and others, may exist.

specific compounds of bile acids and sterols. In the gallbladder, both water and bile salts are reabsorbed by the action of the cholecystic mucosa, and if this process continues excessively, cholesterol crystals separate from the bile. Either biliary stasis or inflammatory disease of the gallbladder can lead to this situation. Concretions made up chiefly of cholesterol crystals are among the common *calculi* of the biliary tract, the disease being termed *cholelithiasis.* Such calculi in the gallbladder may be undetected ("silent"), but if they descend and occlude the biliary tract, particularly the common duct, a variety of clinically important events ensue.

Cholesterol enters the intestinal tract by direct excretion across the intestinal mucosa as well as via the bile. In the lumen of the gut a portion is reduced microbially to coprostanol and cholestanol via the steps indicated below and thereby is excluded from reabsorption. These two sterols, together with cholesterol, comprise the bulk of the fecal sterols. Certain of these transformations, *e.g.*, from cholestenone to cholestanol, also occur in the liver.

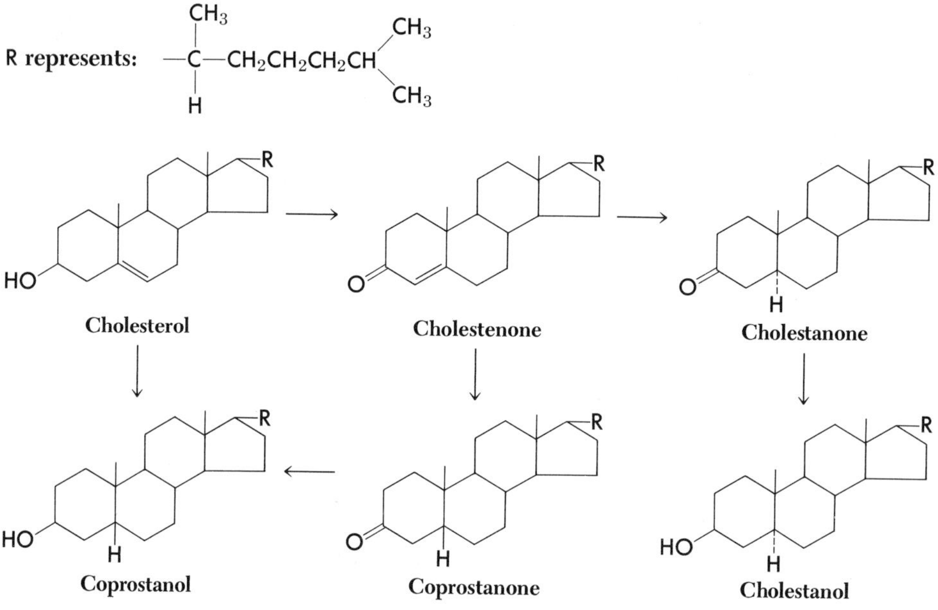

Abnormal elevations of cholesterol in the plasma are considered below.

Approximately 80 per cent of the cholesterol is transformed by liver tissue into various bile acids. In this process, hydroxylation of cholesterol is more or less completed before the degradation of the side chain is finished, although some data support the suggestion that shortening of the side chain of cholesterol may precede hydroxylation of the ring structure. The production of cholic acid from cholesterol involves successive formation of 7α-hydroxycholesterol, 3α,7α-dihydroxycoprostane, and 3α,7α,12α-trihydroxycoprostane. The last compound is then transformed into the CoA derivative of 3α,7α,12α-trihydroxycoprostanic acid; oxidation of the side chain yields cholyl CoA (Fig. 22.5). The shortening of the side chain involves initial removal of the terminal three carbon atoms. This is an oxidative sequence wherein a terminal methyl group appears as CO_2. A β-ketoacyl CoA derivative is

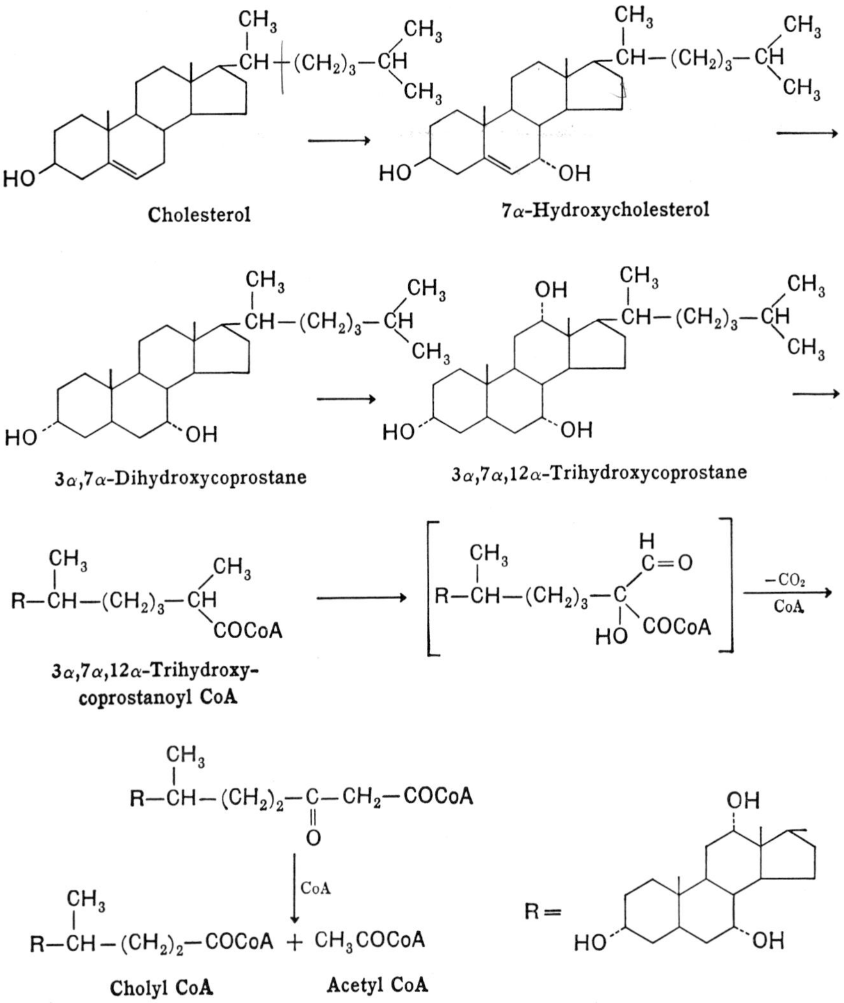

Fig. 22.5. Sequence of postulated reactions leading from cholesterol to cholyl CoA.

pictured as forming, which in a typical β-ketothiolase reaction (page 481) would yield cholyl CoA (Fig. 22.5). Conjugation to glycine or taurine occurs with the CoA derivatives of the bile acids (Chap. 34).

The bile acids, excreted in high concentration in the bile, are reabsorbed via cholecystic and intestinal mucosa, enter the portal blood, and thus return to the liver. Little bile acid escapes normally in the feces or passes the normal liver to enter the caval blood. This minor circuit of bile acids is termed the *enterohepatic circulation* and is shared by a number of metabolites in bile, *e.g.*, urobilinogen (Chap. 31).

Other Products of Cholesterol Metabolism. Another significant fate of cholesterol relates to the origin of the steroid hormones. Soluble enzymes, termed both *desmolases* and *cholesterol oxidases,* from adrenal, testis, and ovary catalyze scission of the side chain of cholesterol between C-20 and C-22, yielding pregnenolone and either isocaproic aldehyde or acid, depending on the experimental conditions. The reaction requires TPNH and molecular oxygen, and is markedly activated by either Mg^{++} or Ca^{++}. In the adrenal, 20α,22-dihydroxycholesterol has been identified as an intermediate. The sequence is depicted in Fig. 22.6. Degradation of the cholesterol side chain in organs producing steroid hormones differs from the mode of cleavage of the side chain in bile acid formation in the liver (see above). Pregnenolone formed from cholesterol serves as a precursor of steroid hormones; these transformations will be discussed in Chaps. 44 and 45. Pregnenolone exerts a feedback regulatory influence on steroidogenesis from cholesterol by inhibiting the above conversion of cholesterol to pregnenolone by desmolases.

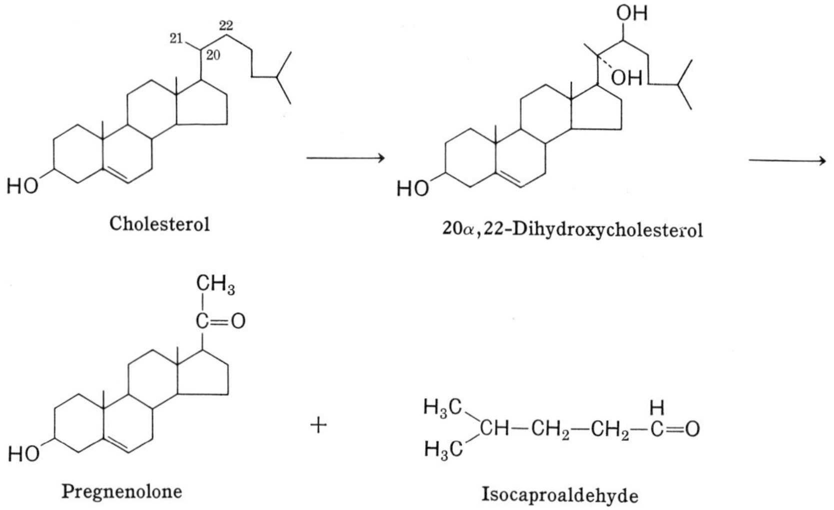

Fig. 22.6. Postulated steps in scission of the cholesterol side chain in the sequence of reactions leading from cholesterol to pregnenolone, a precursor of steroid hormones.

During intestinal absorption of cholesterol, a portion is dehydrogenated to 7-dehydrocholesterol by the intestinal mucosa. This transformation is also effected by skin and other tissues. Skin contains stored quantities of 7-dehydrocholesterol as well as having a greater content of squalene than of cholesterol.

CONTROL OF STEROL METABOLISM

Several factors play a role in the control of the metabolism of cholesterol, the chief mammalian sterol. All of these exert their effects by altering the rate of synthesis or degradation of the sterol. Two of the most important of these factors are (1) diet, including its cholesterol content; and (2) hormones.

The liver is the major site of cholesterol synthesis, although other tissues, *e.g.*, intestine, adrenals, skin, nervous tissue, aorta, and the reproductive organs, also synthesize this sterol. Except for nervous tissue, cholesterol in tissues exhibits a continuous turnover. However, relatively small, rapidly turning over pools and rather large, inert, reserve pools may exist in the same tissue, *e.g.*, the adrenal.

A major locus of control of cholesterol synthesis is at the committed step in the synthetic pathway, *viz.*, the reduction of β-hydroxy-β-methylglutaryl CoA to mevalonic acid, catalyzed by β-hydroxy-β-methylglutaryl CoA reductase (page 517). A second control site has not been localized but is in the series of reactions transforming squalene to cholesterol, possibly in the conversion of squalene to lanosterol.

Degradation of cholesterol occurs primarily in the liver and in the steroidogenic endocrine glands (adrenals and gonads). The intestinal tract is of lesser significance as a locus of cholesterol degradation.

Diet and Cholesterol Metabolism. Both fasting and cholesterol feeding markedly reduce the activity of β-hydroxy-β-methylglutaryl CoA reductase in liver. There is apparently a feedback mechanism for the control of cholesterol synthesis, since the activity of this enzyme is inversely related to the supply of dietary cholesterol and directly related to the caloric intake. Enzymic activity is elevated on refeeding, following a period of fasting. Increasing dietary carbohydrate or triglyceride augments cholesterol synthesis from acetyl CoA. Livers of diabetic animals show increased activity of the reductase; increased cholesterol synthesis is a characteristic of the diabetic state. The data suggest that although acetyl CoA is channeled toward ketone body formation in both diabetes and fasting, augmented cholesterol synthesis is evident only in diabetes.

The feeding of cholesterol, while limiting mevalonic acid formation, also has been reported to inhibit incorporation of administered mevalonate to cholesterol. A second locus of action of a feedback mechanism of dietary cholesterol has been suggested as operating at the cyclization of squalene to lanosterol.

Alterations in the level of blood cholesterol have been noted in response to changes in the degree of saturation of dietary fatty acids (Chap. 48). The more saturated the fatty acids of the diet, the higher the serum cholesterol concentration. The basis for this is not known.

The liver also plays a primary role in the degradation of cholesterol, as indicated above. The rate at which cholesterol can be converted to its metabolites, including bile acids, will influence the level of cholesterol excretion by the liver into the bile, and hence the quantity of cholesterol absorbed from the intestine. Thus this cholesterol, as well as that of dietary origin, can influence the rate of cholesterol synthesis.

Hormones and Cholesterol Metabolism. A number of hormones markedly affect the rate of cholesterol synthesis and degradation. Reference has been made previously to the augmented synthesis of cholesterol in diabetes. It is apparent that insulin lack will increase cholesterol formation, albeit indirectly as a consequence of the accelerated metabolism of fatty acids due to an inability to utilize carbohydrate as a source of energy. The rate of cholesterol synthesis is depressed by the administration of estrogens or of thyroid hormones. Cholesterol synthesis is accelerated following administration of androgens and in circumstances of inadequate secretion of the thyroid. The mechanisms underlying these hormonal effects on cholesterol metabolism are not known.

DISTURBANCES OF LIPID METABOLISM

The normal operation of lipid metabolism results in the simultaneous synthesis and deposition, mobilization, and degradation of body lipids, *i.e.*, a dynamic biological steady state. In the normal adult the quantity of body lipid may remain essentially constant for long periods of time. Disturbances of metabolism have been described that have as their basis an imbalance between the processes leading to synthesis and deposition of a lipid and those leading to its mobilization from the depots or degradation at sites of deposition.

Obesity. The sum of all the caloric requirements in the mammal, including the basal requirement plus increments for muscle work, other forms of work, and specific dynamic action, is met by the caloric yield of the diet. If the diet provides excess calories, the difference will appear in the form of stored chemical potential energy, *i.e.*, as depot lipid. Other things being equal, a surplus of 9 Cal. may be expected to result in the deposition of 1 g. of additional fat.

Although the commonest cause of obesity is overeating, other operative factors must be considered. A pathological decline in the basal metabolic rate or a change from an active to a sedentary life, if not compensated by a decreased food intake, will result in an increase in quantity of adipose tissue. Experimental damage to specific areas of the hypothalamus results in an extreme polyphagia, which, if not restricted, leads to extraordinary obesity. Poisoning with certain gold salts and with sulfanilamide has also been reported to result in obesity, and in the mouse a hereditary obesity has been discovered which is transmitted as a mendelian dominant characteristic. In the obesity which results from hypothalamic injury, as well as in the hereditary disease of mice, it appears that the defect is predominantly impaired mobilization of depot lipid rather than an excessive deposition of lipid in the depots. Whether this is true of the other forms of obesity remains to be determined. Obesity is a common finding in the history of diabetic individuals and is supposed to predispose to this and other diseases. There is no doubt that it materially decreases life expectancy in the older age groups.

The therapeutic restriction of caloric intake below the total caloric requirement necessarily leads to loss of body lipid, although, owing to transient water retention, this may not be immediately reflected in changes in body weight.

Cachexia. Failure to ingest sufficient calories may lead to the complete disappearance of gross adipose tissue from the subcutaneous and omental depots. This may occur in the course of neoplastic or chronic infectious disease, in malnutrition, or in disturbances of metabolism such as diabetes or hyperthyroidism. Specific damage to areas of the hypothalamus has been shown experimentally to produce anorexia, even in previously starved animals. *Anorexia nervosa* may often have a psychogenic component in its etiology.

Whereas the loss of body lipid incident to hyperthyroid disease is apparently attributable to excessively rapid mobilization of depot lipid, an important contribution to the cachexia of starvation, thiamine deficiency, or diabetes is a decreased capacity of the organism to synthesize fatty acids from carbohydrate precursors.

Pathological Accumulations of Cholesterol. The formation of cholesterol calculi in the biliary tract has been described (page 520). The pathological deposition of cholesterol-containing plaques in the intima of the aorta is the characteristic lesion of *atheromatosis,* and is seen in *arteriosclerosis* and *arteriolar sclerosis.* The mechanism of this deposition in man is uncertain, but concomitant changes in the composition of the blood have been noted. Although the concentration of cholesterol in plasma may not be strikingly elevated, the ratio of cholesterol to phospholipids generally is, with an increase in the lipoprotein fraction S_f 12 to 20 (page 476), which is rich in cholesterol.

Because the excessive deposition of cholesterol in vascular tissue may have undesirable consequences, efforts have been directed toward reduction of levels of blood cholesterol in the hope of lessening the degree of vascular cholesterol deposition. These attempts have utilized a variety of dietary and therapeutic procedures. Previous mention was made of the apparently favorable influence of unsaturated, compared with saturated, dietary lipids on blood cholesterol levels (page 525). The relation of dietary lipid to atherosclerosis is considered more fully in Chap. 48. Drug therapy designed to interfere with endogenous cholesterol biosynthesis has not had practical success. Certain drugs which diminish cholesterol formation lead to an accumulation of intermediates of cholesterol synthesis, including desmosterol, in the liver and in the intima of the larger blood vessels. The basis of the hypocholesterolemia of hyperthyroidism is not understood, but its occurrence has led to the search for thyroid hormone-like substances which may lower blood cholesterol without exerting the other effects of thyroidal hormones (Chap. 43). The lower incidence of atherosclerosis in the human female, as compared to that in the male, and the blood cholesterol–lowering effects of female sex hormones have similarly stimulated a search for compounds which mimic the female sex hormones in reducing blood cholesterol levels, but have minimal feminizing effects (Chap. 44).

Familial hypercholesterolemia is a genetically transmitted disorder. The genetic defect in this condition may be the failure of one of the repressor mechanisms (page 539) that function at the points of the cholesterol feedback control of cholesterol biosynthesis (page 524).

A less frequent disease in which lipid deposits often rich in cholesterol are found is *xanthomatosis*. Multiple benign fatty tumors of skin, tendon sheaths, and bone are found in this condition, associated with a lipemia and a striking hypercholesterolemia. Patients suffering from this disease have been reported to benefit by the exclusion of all animal lipid, hence of all cholesterol, from their diets.

An even rarer condition is *Schüller-Christian syndrome,* characterized by xanthomatous deposits in the flat bones of the skull, the liver, and the spleen, associated with diabetes.

Changes in the level of blood cholesterol without overt deposition are seen in certain clinical conditions. Lipemia and especially hypercholesterolemia occur in hypothyroidism (Chap. 43) as well as in the nephrotic syndrome. In neither case is the significance of the chemical change understood. Hypercholesterolemia may also be produced by increased blood levels of certain adrenal cortical steroids (Chap. 45) as well as by some hypophyseal preparations (Chap. 47).

Niemann-Pick Disease. This disease is characterized by the accumulation of unusual amounts of sphingomyelin, particularly in the spleen and liver. A total absence or a markedly lower than normal level of sphingomyelinase (page 516) activity has been reported in the spleens of some patients with Niemann-Pick disease. However, this cannot be the sole explanation for sphingomyelin accumulation in all cases of this disease, since normal enzymic activity was present in spleen in some instances.

Tay-Sachs Disease. In this condition the brain and spleen accumulate abnormally large amounts of specific gangliosides which have not been detected in normal individuals. One of these gangliosides has been identified as ceramide-glucose-galactose-(N-acetylneuraminic acid)-N-acetylgalactosamine, termed Tay-Sachs ganglioside. The second ganglioside, which is called GM-1 and accumulates in a different gangliosidosis, is ceramide-glucose-galactose-(N-acetylneuraminic acid)-N-acetylgalactosamine-galactose. It appears that either the enzymes which utilize these gangliosides as substrates for further synthetic reactions (page 515) are not exhibiting adequate activity, or there is lack of normal degradative activity.

Fabry's Disease. This disease is characterized by the abnormal accumulation of a ceramide trisaccharide, primarily in the kidney, but also in intestine and lymph nodes. Although the nature of the disturbance has not been elucidated, intestinal tissue of patients with this disease has a significantly less than normal degree of activity of ceramide trihexoside–cleaving enzyme (page 515).

Metachromatic Leukodystrophy. In this disease there is an abnormal accumulation of sulfatides, the cerebroside sulfates (page 515), in the white matter of the brain and, to a lesser degree, in the gray matter. The nature of the disturbance is not understood.

Gaucher's Disease. This disturbance of lipid metabolism is distinguished by the accumulation of glucocerebrosides in reticuloendothelial cells, notably in liver, spleen, and bone marrow. Although the biochemical basis remains to be firmly established, a marked deficiency of a specific glucocerebroside-cleaving enzyme has been detected in the spleen of some patients.

Tangier Disease. This disease has also been termed familial high-density lipoprotein deficiency, since it is characterized by almost a complete absence of plasma high-density lipoproteins as well as by the storage of excessive amounts of cholesterol esters in many tissues. There is a marked reduction in plasma cholesterol; the phospholipids are also reduced in the blood, but triglycerides are normal or elevated. The disease probably represents defective synthesis of the high-density lipoprotein.

Abetalipoproteinemia. This relatively rare disorder is characterized by an absence from the plasma of β-lipoproteins of density less than 1.063. This is associated with extensive nerve demyelinization. There is a failure of normal lipid absorption and lipid accumulates in intestinal cells; acanthocytosis is present (page 720). There is also a failure of normal chylomicra formation. The basis for these defects is not known; both a failure to synthesize the protein specific for the low-density lipoproteins and a defect in removal of dietary fat from the intestinal absorptive cells are indicated.

All the above seven disorders, as well as several of those described earlier involving excessive deposition of cholesterol, appear to be genetically transmitted.

REFERENCES

Books

Ansell, G. B., and Hawthorne, J. N., "Phospholipids," Elsevier Publishing Company, Amsterdam, distributed by American Elsevier Publishing Company, New York, 1964.

Bloch, K., ed., "Lipide Metabolism," John Wiley & Sons, Inc., New York, 1960.

Cook, R. P., ed., "Cholesterol: Its Chemistry, Biochemistry and Pathology," Academic Press, Inc., New York, 1958.

Dawson, R. M. C., and Rhodes, D. N., eds., "Metabolism and Physiological Significance of Lipids," John Wiley & Sons, Inc., New York, 1964.

Deuel, H. J., Jr., "The Lipids: Their Chemistry and Biochemistry," vol. II, 1955; vol. III, 1957, Interscience Publishers, Inc., New York.
Frazer, A. C., ed., "Biochemical Problems of Lipids," Elsevier Publishing Company, Amsterdam, distributed by American Elsevier Publishing Company, New York, 1963.
Kinsell, L. W., ed., "Adipose Tissue as an Organ," Charles C Thomas, Publisher, Springfield, Ill., 1962.
Kritchevsky, D., "Cholesterol," John Wiley & Sons, Inc., New York, 1958.
Mead, J. F., and Howton, D. R., "Radioisotope Studies of Fatty Acid Metabolism," Pergamon Press, New York, 1960.
Page, I. H., ed., "Chemistry of Lipides as Related to Atherosclerosis," Charles C Thomas, Publisher, Springfield, Ill., 1958.
Pincus, G., ed., "Hormones and Atherosclerosis," Academic Press, Inc., New York, 1959.
Popják, G., and Grant, J. K., eds., "The Control of Lipid Metabolism," Academic Press, Inc., New York, 1963.
Richards, J. H., and Hendrickson, J. B., "Biosynthesis of Steroids, Terpenes and Acetogenins," W. A. Benjamin, Inc., New York, 1964.
Rodahl, K., and Issekutz, B., eds., "Fat as a Tissue," McGraw-Hill Book Company, New York, 1964.
Schettler, G., ed., "Lipids and Lipidoses," Springer-Verlag OHG, Berlin, 1967.
Searcy, R. L., and Bergquist, L. M., "Lipoprotein Chemistry in Health and Disease," Charles C Thomas, Publisher, Springfield, Ill., 1962.
Wolf, G., ed., "Recent Research on Carnitine: Its Relation to Lipid Metabolism," The M.I.T. Press, Cambridge, Mass., 1965.

Review Articles

Ball, E. G., and Jungas, R. L., Some Effects of Hormones on the Metabolism of Adipose Tissue, *Recent Prog. Hormone Research,* **20,** 183–214, 1964.
Clayton, R. B., Biosynthesis of Sterols, Steroids, and Terpenoids, *Quart. Rev.,* **19,** 168–230, 1965.
Danielsson, H., and Tchen, T. T., Steroid Metabolism, in D. M. Greenberg, ed., "Metabolic Pathways," 3d ed., vol. 2, pp. 117–168, Academic Press, Inc., New York, 1968.
Dawson, R. M. C., The Metabolism of Phospholipids, in M. Florkin and H. S. Mason, eds., "Comparative Biochemistry," vol. IIIA, pp. 265–285, Academic Press, Inc., New York, 1962.
Dole, V. P., and Hamlin, J. T., III, Particulate Fat in Lymph and Blood, *Physiol. Revs.,* **42,** 674–701, 1962.
Eder, H., The Lipoproteins of Human Serum, *Am. J. Med.,* **23,** 269–282, 1957.
Frazer, A. C., Fat Absorption and Its Disorders, *Brit. Med. Bull.,* **14,** 212–220, 1958.
Frederickson, D. S., and Gordon, R. S., Jr., Transport of Fatty Acids, *Physiol. Revs.,* **38,** 585–630, 1958.
French, J. B., Morris, B., and Robinson, D. S., Removal of Lipids from the Blood Stream, *Brit. Med. Bull.,* **14,** 234–238, 1958.
Fritz, I. B., Carnitine and Its Role in Fatty Acid Metabolism, *Advances in Lipid Research,* **1,** 285–334, 1963.
Goodman, D. S., Cholesterol Ester Metabolism, *Physiol. Revs.,* **45,** 747–839, 1965.
Grant, J. K., Lipids: Steroid Metabolism, in M. Florkin and H. S. Mason, eds., "Comparative Biochemistry," vol. IIIA, pp. 163–203, Academic Press, Inc., New York, 1962.
Green, D. E., and Allman, D. W., Fatty Acid Oxidation, in D. M. Greenberg, ed., "Metabolic Pathways," 3d ed., vol. 2, pp. 1–37, Academic Press, Inc., New York, 1968.
Green, D. E., and Allman, D. W., Biosynthesis of Fatty Acids, in D. M. Greenberg, ed., "Metabolic Pathways," 3d ed., vol. 2, pp. 38–67, Academic Press, Inc., New York, 1968.

Hofmann, A. F., and Small, D. M., Detergent Properties of Bile Salts: Correlation with Physiological Function, *Ann. Rev. Medicine,* **18,** 333–376, 1967.

Jeanrenaud, B., Dynamic Aspects of Adipose Tissue Metabolism: A Review, *Metabolism,* **10,** 535–581, 1961.

Johnson, J. M., Recent Developments in the Mechanism of Fat Absorption, *Advances in Lipid Research,* **1,** 105–131, 1963.

Kennedy, E. P., The Metabolism and Function of Complex Lipids, *Harvey Lectures,* **57,** 143–171, 1961–1962.

Klenk, E., The Metabolism of Polyenoic Fatty Acids, *Advances in Lipid Research,* **3,** 1–23, 1965.

Krebs, H. A., The Regulation of Release of Ketone Bodies by the Liver, in G. Weber, ed., "Advances in Enzyme Reactions," vol. 4, pp. 339–354, Pergamon Press, New York, 1966.

Lindgren, F. T., and Nichols, A. V., Structure and Function of Human Serum Lipoproteins, in F. W. Putnam, ed., "The Plasma Proteins," vol. II, pp. 2–58, Academic Press, Inc., New York, 1960.

Lynen, F., The Role of Biotin-dependent Carboxylations in Biosynthetic Reactions, *Biochem. J.,* **102,** 381–400, 1967.

Mead, J. F., Lipid Metabolism. *Ann. Rev. Biochem.,* **32,** 241–268, 1963.

Olson, J. A., Lipid Metabolism, *Ann. Rev. Biochem.,* **35,** 559–598, 1966.

Olson, R. E., and Vester, J. W., Nutrition-endocrine Interrelationships in the Control of Fat Transport in Man, *Physiol. Revs.,* **40,** 677–733, 1960.

Popják, G., Some Aspects of Lipid Biochemistry, *Proc. Roy. Soc., London,* **B, 156,** 376–387, 1962.

Popják, G., and Cornforth, J. W., The Biosynthesis of Cholesterol, *Advances in Enzymol.,* **22,** 281–335, 1960.

Portman, O. W., and Stare, F. J., Dietary Regulation of Serum Cholesterol Levels, *Physiol. Revs.,* **39,** 407–442, 1959.

Rossiter, R. J., Metabolism of Phosphatides, in D. M. Greenberg, "Metabolic Pathways," 3d ed., vol. 2, pp. 69–115, Academic Press, Inc., New York, 1968.

Rudman, D., The Adipokinetic Property of Hypophyseal Peptides, *Ergeb. Physiol.,* **56,** 297–327, 1965.

Rudman, D., Hirsch, R. L., Kendall, F. E., Seidman, F., and Brown, S. J., An Adipokinetic Component of the Pituitary Gland: Purification, Physical, Chemical, and Biologic Properties, *Recent Prog. Hormone Research,* **18,** 89–123, 1962.

Scanu, A. M., Factors Affecting Lipoprotein Metabolism, *Advances in Lipid Research,* **3,** 63–138, 1965.

Senior, J. R., Intestinal Absorption of Fats, *J. Lipid Research,* **5,** 495–521, 1964.

Stumpf, P. K., and Barber, G. A., Comparative Mechanisms for Fatty Acid Oxidation, in M. Florkin and H. S. Mason, eds., "Comparative Biochemistry," vol. I, pp. 75–105, Academic Press, Inc., New York, 1960.

Van Deenen, L. L. M., and de Haas, G. H., Phosphoglycerides and Phospholipases, *Ann. Rev. Biochem.,* **35,** 157–194, 1966.

Vandenheuvel, F. A., The Origin, Metabolism, and Structure of Normal Human Serum Lipoproteins, *Can. J. Biochem. Physiol.,* **40,** 1299–1326, 1962.

Vaughn, M., The Metabolism of Adipose Tissue in Vitro, *J. Lipid Research,* **2,** 293–316, 1961.

Wenke, M., Effects of Catecholamines on Lipid Mobilization, *Advances in Lipid Research,* **4,** 69–105, 1966.

Wertheimer, E., and Shafrir, E., Influence of Hormones on Adipose Tissue as a Center of Fat Metabolism, *Recent Progr. Hormone Research,* **16,** 467–495, 1960.

23. Amino Acid Metabolism. I

Digestion of Protein. Absorption of Amino Acids. Utilization of Inorganic Nitrogen. General Aspects of Amino Acid Synthesis. Essential Amino Acids. Synthesis of Nonessential Amino Acids. Regulation of Nitrogen Metabolism.

Historically, the metabolism of amino acids has been generally viewed as a segment of protein metabolism because protein is the major source of *dietary* amino acids. However, except for information relating to the nutritive aspects of protein, *i.e.*, its role in the diet (Chap. 48), and the hydrolysis in the gastrointestinal tract of ingested protein, the prime processes of protein nitrogen metabolism, particularly in mammals, are those involving amino acids or products of their metabolism. The metabolic relation of amino acids to proteins stems from the facts that (1) amino acids are derived from protein degradation in the gut as well as in the continuing metabolism of cells, and (2) the major metabolic role of amino acids is to serve as precursors of proteins. This latter role is discussed in Chap. 29.

The present chapter considers (1) the digestion of protein and the absorption of the liberated amino acids; (2) the mechanisms available to cells for utilization of inorganic nitrogen; (3) the pathways by which mammalian cells synthesize adequate quantities of the so-called nonessential amino acids (page 540), and (4) the continuing nature of nitrogen metabolism in mammals. The following chapter presents the synthesis of those amino acids which cannot be produced by mammals and must be provided by dietary protein, but can be fabricated by plants and microorganisms. The third chapter of this group describes the contributions of the nitrogen and carbon of various amino acids to the synthesis of some important nitrogenous compounds; the fourth chapter indicates known degradative, or catabolic, pathways of mammalian amino acid metabolism.

DIGESTION OF PROTEIN

The digestion of protein in the gastrointestinal tract produces at least two significant results. The primary one is the degradation of high molecular weight protein molecules to small compounds which may be absorbed readily from the intestine. In addition, digestion of protein in mammals destroys the biological specificity, including the species differences, which makes intact proteins antigenic.

Hydrolysis of proteins in the gastrointestinal tract is accomplished by specific enzymes secreted in gastric juice and in pancreatic juice, and by enzymes of the mucosa of the small intestine. With the exception of intestinal peptidases, the proteolytic enzymes of the gastrointestinal tract are elaborated and secreted as inactive zymogens (page 252) which are converted to active enzymes in the gut. The alterations in protein structure accompanying zymogen transformation to enzyme have been described previously for the reaction chymotrypsinogen → chymotrypsin (page 252). In a broad sense, similar changes may be visualized for other zymogen conversions to active enzymes.

Gastric Digestion. Factors affecting the quantity and composition of gastric and pancreatic secretions are discussed in Chap. 34. At this point a brief consideration will be provided of protein digestion in relation to the enzymes involved, aspects of their specificity, and their effects upon dietary proteins. Previous discussion was presented of the specificity of proteolytic enzymes in relation to their use as tools in studies of protein structure (Chap. 7).

The proteinase of the gastric juice is *pepsin;* it is derived from its zymogen precursor *pepsinogen,* which is elaborated and secreted by the chief cells of the gastric mucosa (Chap. 34). The inactive zymogen (molecular weight = 40,400) is converted to the active enzyme, pepsin (molecular weight = 32,700), both by the acidity of the gastric juice and by pepsin itself; the process is therefore autocatalytic. During this conversion, 42 amino acid residues are removed from the N-terminal portion of pepsinogen as a mixture of peptides, some of which can act as pepsin inhibitors. These peptides contain 12 of the 16 basic amino acid residues of pepsinogen; their liberation results in a decrease of the isoelectric point from about pH 3.7 in pepsinogen to a value around pH 1.0 in pepsin. The latter has 40 aspartic acid and 26 glutamic acid residues; slightly more than half of these dicarboxylic acids are present in pepsin as the acid amide, leaving a marked predominance of anionic groups in the enzyme. A carboxyl group has been shown to be essential at the active site of pepsin.

The substrates for peptic activity in the stomach are either the native proteins of the diet or denatured proteins resulting from cooking of food. Pepsin rapidly initiates hydrolysis of proteins; the linkages most susceptible to this enzyme are peptide bonds involving an aromatic amino acid (phenylalanine, tryptophan, or tyrosine). In addition, pepsin acts on peptide bonds involving other amino acids, *e.g.*, leucine and acidic residues.

Although pepsin liberates free amino acids from proteins in vitro, this is a slow process. Since food remains in the stomach for a limited time, pepsin in vivo hydrolyzes dietary protein chiefly to a mixture of polypeptides. If gastric hydrochloric acid production fails to maintain gastric contents at the pH optimum of 2 to 3, necessary for peptic action, protein digestion in the stomach may be very limited. This is seen, for example, in *pernicious anemia* (Chap. 34). In *achylia gastrica,* the absence of pepsin as well as acid from gastric contents precludes protein digestion in the stomach.

Pepsin has a strong clotting action on milk. A number of plant and animal enzymes exhibit this activity, which is the initial reaction in the digestion of milk. In ruminants, clotting of milk may result from the action of a *specific* enzyme, *rennin*, which is obtained from the abomasum, or fourth stomach, of the suckling calf. The reactions involved in milk clotting are presented later (Chap. 34).

Several additional zymogens, convertible to active, distinctive proteinases, have been prepared from gastric mucosa of several species. Three of these have been designated as *pepsinogens B, C,* and *D,* respectively. A fourth gastric proteinase, termed *gastricsin,* appears to be distinct from the various pepsins.

Proteolysis in the Intestine. The material which enters the intestine comes in contact with a mixture of proteases. The pancreas secretes a slightly alkaline fluid (Chap. 34) containing several inactive zymogen precursors of proteases, *viz., trypsinogen,* three *chymotrypsinogens,* two *procarboxypeptidases,* and *proelastase.* An intestinal enzyme, termed *enterokinase,* converts trypsinogen to *trypsin,* as does trypsin itself, autocatalytically.

Trypsinogen consists of a single peptide chain, and the activation process involves scission of a single bond, liberating the hexapeptide, Val-$(Asp)_4$-Lys, from the amino-terminal end of the zymogen. Release of the peptide is accompanied by the appearance of enzymic activity and a decrease in levorotation of the protein, indicating an alteration in the conformation of the molecule. Chymotrypsinogen is converted to several active chymotrypsins; the nature of some of these transformations has been described (page 242). Procarboxypeptidases and proelastase are converted by trypsin to active carboxypeptidases and elastase, respectively.

The alkaline pancreatic juice neutralizes the acidic chyle from the stomach and provides the slightly alkaline pH optimal for the hydrolytic action of the pancreatic enzymes, each of which has characteristic specificity (page 144). *Trypsin* acts upon peptide linkages involving the carboxyl groups of arginine and lysine. Pancreas and pancreatic juice contain a trypsin inhibitor which is a polypeptide (molecular weight = 7,200); this is of unknown physiological significance. The *chymotrypsins* are most active toward peptide bonds involving phenylalanine, tyrosine, and tryptophan. Thus the action of these enzymes is additive, resulting in more complete degradation to small peptides. *Carboxypeptidase A,* a zinc-containing enzyme, rapidly liberates carboxyl-terminal amino acid residues. Its most rapid action is on residues that possess aromatic or aliphatic side chains. *Carboxypeptidase B* acts only on peptides possessing terminal arginine or lysine residues.

The mucosa of the intestine also contains enzymes that hydrolyze peptide bonds. Although these enzymes may be secreted into the intestinal juice, they function for the most part intracellularly. Hydrolysis of the smaller products of digestion may occur following their entry into the mucosal cells and during transfer across the epithelial cells. Extracts of intestinal mucosa contain a group of *aminopeptidases,* enzymes that act on polypeptides or peptide chains containing a free amino group, liberating an amino acid by scission of the peptide bond adjacent to the free amino group. Thus, *leucine aminopeptidase* has a broad specificity with respect to the N-terminal residue of the polypeptide. By successive hydrolysis of N-terminal peptide bonds, it can degrade peptides to free amino acids. This enzyme requires Mn^{++} or Mg^{++} for its action; the metal ion probably functions in a coordination complex with the enzyme and its substrate. Mucosal extracts also contain *dipeptidases,* enzymes that act specifically on certain dipeptides. An example is *glycylglycine dipeptidase,* which requires Co^{++} or Mn^{++} for its action. This enzyme does not attack the tripeptide, glycylglycylglycine. Thus the enzymic action is dependent on the presence of both amino and carboxyl groups adjacent to the sensitive peptide bond.

The successive action of the proteolytic enzymes present in the stomach and in the small intestine results in hydrolysis of most of the dietary protein to free amino acids. Although trypsin and chymotrypsin act more rapidly and more completely if preceded by the action of pepsin, these two pancreatic enzymes together can liberate amino acids from proteins. This is of some significance in individuals with a gastric resection. Indeed, while absence of gastric secretion does not seriously affect the utilization of dietary protein, exclusion of pancreatic juice from the intestine markedly impairs protein digestion. Extensive destruction of pancreatic tissue or obstruction of the pancreatic duct limits the quantity of pancreatic juice reaching the duodenum, and under these conditions significant amounts of dietary protein are not digested and appear in the feces.

ABSORPTION OF AMINO ACIDS FROM THE INTESTINE

The major products of intestinal digestion of protein are the amino acids, and these are rapidly absorbed. For example, within 15 min. after ingestion of ^{15}N-labeled yeast protein by man, significant absorption of amino acids could be demonstrated, with maximum amino acid concentration in the blood attained between 30 and 50 min. after eating. These values may be considered excessively short since they were obtained with a very small amount of fed protein and without simultaneous feeding of carbohydrate or lipid, which may delay gastric emptying. The simultaneous presence of fructose and galactose in the intestine, with the amino acids, will inhibit intestinal absorption of the latter. Nonetheless, the data reflect the speed with which ingested protein may be digested and the amino acids absorbed. Absorption of amino acids is confined chiefly to the small intestine and is an active, energy-requiring process which reflects a high degree of structural specificity. Impairment of absorption occurs with anoxia or in the presence of metabolic inhibitors or poisons. Studies with everted sacs of rat or hamster intestine and isolated segments of rat intestine have demonstrated that the L isomers of amino acids are more rapidly absorbed than the D isomers, and that, in general, the neutral and the more hydrophilic amino acids are more rapidly absorbed than are the basic and the more hydrophobic amino acids. Amino acids compete with one another for absorption sites. Thus, during the absorption of leucine there is diminished absorption of isoleucine and valine. Also, studies with mixtures of amino acids indicate that more than one transport or carrier system functions in the absorption of amino acids from the intestine.

A role for the participation of pyridoxal (vitamin B_6, Chap. 49) and Mn^{++} in amino acid concentration by cells in general has been suggested (Chap. 33). Studies in the rat reveal that variations in the endogenous vitamin B_6 economy greatly influence intestinal transport and tissue uptake of amino acids. The diminished rate of amino acid absorption from the intestine of B_6-deficient rats can be stimulated by administration of pyridoxal phosphate. Vitamin B_6, or its derivatives also participate as cofactors in a number of enzymic systems concerned with the metabolism of amino acids (see below and the following three chapters).

Present knowledge of factors influencing amino acid absorption from the intestine is based on studies in which either amino acids were administered by stomach tube to experimental animals or amino acid absorption was examined in vitro using isolated and/or everted segments of intestine. As indicated above, peptide hydrolysis in the intestine occurs, for the most part, intracellularly. It is not clear, therefore, to what extent *free* amino acids, derived from dietary protein, are presented to the intestinal mucosa for absorption.

Amino acids absorbed from the intestine enter the circulation almost wholly by way of the portal blood; very small quantities leave the intestine via the lymphatic vessels. Low molecular weight peptides may also be absorbed from the small intestine; a rise in blood peptide nitrogen occurs during protein digestion and absorption. Occasionally, native protein may also penetrate the intestinal mucosa and appear in the blood. In the young mammal the permeability of the mucosa is greater than in the adult. Also, the colostral milk (Chap. 34), secreted during the first few days after parturition, contains a protein which is a potent trypsin inhibitor. These factors, together with a low concentration of proteolytic enzymes in the digestive fluids, may lead to a degree of absorption of native proteins sufficient to cause immunological sensitization. This may be the basis for idiosyncrasies sometimes encountered toward food proteins, *e.g.*, milk proteins and egg white.

UTILIZATION OF INORGANIC NITROGEN

The prime form of inorganic nitrogen utilizable by all living cells is ammonia, which can be fixed at all phylogenetic levels, including mammals. Ammonia nitrogen derives in plants from reduction of the nitrogen of the atmosphere or of the nitrate present in the soil. Plants and many microorganisms can synthesize all the amino acids found in proteins as well as diverse other nitrogenous compounds by virtue of their capacity to utilize ammonia and to fabricate the carbon structures corresponding to each of the amino acids. Ammonia is fixed by three major reactions, the syntheses of *glutamic acid, glutamine,* and *carbamyl phosphate,* respectively. In some microorganisms alanine or aspartic acid formation may substitute for that of glutamic acid. With nitrogen fixed into these three compounds, they serve in plant and microbial cells as precursors of all the other amino acids, and participate in formation of purines, pyrimidines, and diverse other nitrogenous compounds.

Utilization of Nitrogen and Nitrate for Ammonia Formation. Nitrogen exists as N_2 in the atmosphere and, frequently, as NO_3^- in soil.

Nitrogen fixation, i.e., the utilization of N_2 with formation of NH_3, has been studied in cell-free systems of microorganisms; pyruvic acid is required as an energy source. Pyruvic acid plays a dual role, functioning as a source of ATP and as a reducing agent. Pyruvate is oxidized by a process analogous to mitochondrial oxidative phosphorylation. Ferredoxin appears to serve as the electron acceptor of lowest potential and engages in direct transfer to the *nitrogenase* system. Nitrogenase has been fractionated into two components, a molybdenum-iron protein and an iron protein. The stoichiometry of the over-all process of nitrogen fixation is as follows.

$$N_2 + 3H_2 \longrightarrow 2NH_3$$

No intermediates leave the enzymic surface so that the nitrogenase catalyzes a 6-electron reduction of N_2 analogous to the 4-electron reduction of oxygen by cytochrome oxidase (page 372) and the 6-electron reduction of sulfite to sulfide (page 564). The *nitrogenase* of *Clostridium pasteurianum* is repressible by NH_4^+.

The conversion of NO_3^- to NH_3 proceeds in two steps: (1) reduction of NO_3^- to NO_2^-, and (2) conversion of NO_2^- to NH_3. *Nitrate reductase,* which occurs in many higher plants, fungi, and microorganisms, is a pyridine nucleotide–linked molybdoflavoprotein (page 365) which catalyzes the following reaction.

$$NO_3^- + TPNH + H^+ \longrightarrow NO_2^- + TPN^+ + H_2O$$

DPNH may also be employed as the initial electron donor, depending on the source of the enzyme.

Nitrate and *hydroxylamine reductases* have been described which catalyze the second step in nitrogen assimilation, but intermediate stages from NO_2^- to NH_3, if any, are uncertain. Among the intermediates which have been considered are NO, N_2O, and NH_2OH; none of these has been found and, again, a multiple electron reduction may be catalyzed by a single enzyme without dissociation of intermediates.

Synthesis of Glutamic Acid. Glutamic acid formation occurs in a reaction catalyzed by *glutamic acid dehydrogenase,* an almost universally distributed enzyme.

$$\text{DPNH (or TPNH)} + \alpha\text{-ketoglutarate} + H^+ + NH_3 \rightleftharpoons \text{L-glutamate} + DPN^+ \text{ (or } TPN^+\text{)} + H_2O$$

The enzyme of higher plants and of most animal tissues utilizes DPN as coenzyme. Glutamic acid dehydrogenase of liver functions with either DPN or TPN. The equilibrium for this reaction strongly favors the reductive synthesis of glutamate.

Glutamic acid dehydrogenase has been obtained in crystalline form from several sources, *e.g.*, mammalian and chicken liver. The enzyme has a molecular weight of approximately 400,000 but may exist as higher aggregates. The enzyme contains zinc, which may be involved in the interaction between subunits of molecular weight of approximately 50,000.

The α-amino group of glutamic acid can be transferred to other α-keto acids, corresponding to the other normally occurring amino acids, thereby providing a mechanism for synthesis of most of the other amino acids by a process termed *transamination.* This will be considered later (page 541).

Glutamine Synthesis. Glutamine formation is catalyzed by *glutamine synthetase.*

$$\underset{\text{Glutamic acid}}{HOOC{-}CH_2{-}CH_2{-}CHNH_2{-}COOH} + ATP + NH_3 \xrightarrow{Mg^{++}} \underset{\text{Glutamine}}{H_2N{-}CO{-}CH_2{-}CH_2{-}CHNH_2{-}COOH} + ADP + P_i$$

Glutamine synthetase catalyzes a two-step process. Initially, there is a reaction involving glutamic acid and ATP with formation of an activated enzyme complex involving ADP and γ-glutamyl phosphate. In the second step, this intermediate reacts with NH_3 to form glutamine, ADP, and P_i. The enzyme in brain and liver appears to be associated with the endoplasmic reticulum.

Escherichia coli produces two distinct forms of glutamine synthetase, termed synthetase I and synthetase II. The two enzymes are identical in amino acid composition and sedimentation constant, but synthetase II contains covalently bound AMP which is absent in synthetase I. In the presence of ATP and Mg^{++}, an *adenyl transferase* catalyzes conversion of synthetase I to synthetase II. The latter is less active than the former; thus adenylation provides a mechanism of regulation of glutamine synthetase which, in *E. coli,* is also controlled by repression of enzyme synthesis and by what has been termed cumulative product feedback inhibition. The last term is used to describe the decrease in glutamine synthetase activity produced by a series of diverse products, mentioned below, which derive their nitrogen from glutamine.

The extent and rate of glutamine synthesis exceed those of all other forms of NH_3 fixation in mammals. The amide N of glutamine can be transferred to other carbon chains in diverse synthetic pathways, *e.g.,* hexosamines (page 428), carbamyl phosphate (see below), purines (page 620), histidine (page 577), tryptophan (page 573), and DPN (page 641). In several instances, the mammalian enzyme in these pathways is glutamine-specific, whereas the comparable reaction in bacteria is accomplished with ammonia. In at least two instances, the mammalian enzyme can use either glutamine or NH_3, and K_m for each is approximately the same. However, only 1 per cent of total ammonia exists as NH_3 at pH 7.4 and the remainder as NH_4^+. Thus a physiologically intolerable concentration of ammonia would be required. Glutamine, therefore, provides a means of presenting to an enzyme an unprotonated nitrogen atom at the reduction level of NH_3 and in physiologically acceptable concentration.

In addition, glutamine, synthesized from NH_3, as described above, may serve as an ammonia store. The ammonia can be released through the intervention of *glutaminase,* a widely distributed enzyme which catalyzes the following reaction.

$$\text{Glutamine} + H_2O \longrightarrow \text{glutamic acid} + NH_3$$

The enzyme of kidney is of particular interest because of its special physiological role in providing ammonium ion for conservation of cations (Chaps. 33 and 35).

Synthesis of Carbamyl Phosphate. As obtained from bacterial sources, *carbamyl phosphate synthetase* catalyzes the following reversible reaction.

$$CO_2 + NH_3 + ATP \rightleftharpoons \underset{\text{Carbamyl phosphate}}{H_2N{-}CO{-}OPO_3H_2} + ADP$$

Glutamine may function directly as the nitrogen (NH_3 from the amide group) donor in some microorganisms, replacing NH_3 in the above reaction. Carbamyl phosphate synthesis in mammalian liver is somewhat more complex and is essentially irreversible; an additional equivalent of ATP is required.

$$CO_2 + NH_3 + 2ATP + H_2O \longrightarrow H_2N{-}CO{-}OPO_3H_2 + 2ADP + P_i + H^+$$

The enzyme shows an absolute requirement for N-acetylglutamic acid.

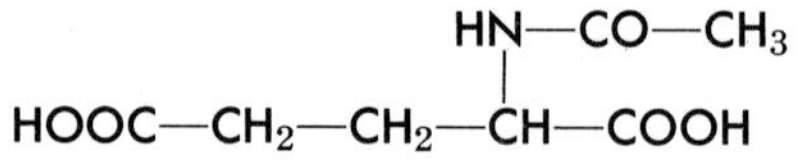

N-Acetylglutamic acid

Studies of the mechanism of the above reaction suggest that the first mole of ATP utilized activates carbonate by converting it to an enzyme-bound carboxy-phosphate. The latter complex reacts with NH_4^+ to form an intermediate enzyme-bound carbamate, with liberation of P_i. Reaction with a second mole of ATP yields carbamyl phosphate, ADP, and the enzyme. Although the precise role of acetylglutamic acid is not known, some evidence suggests that it acts as an allosteric cofactor.

Carbamyl phosphate, made by either of the above two reactions, is then available for carbamylation of amino groups, as in formation of carbamyl aspartic acid in the first step of pyrimidine synthesis (page 633) and in citrulline formation (page 545).

$$H_2N{-}CO{-}OPO_3H_2 + RNH_2 \longrightarrow H_2N{-}CO{-}NHR + P_i$$

GENERAL ASPECTS OF AMINO ACID SYNTHESIS

Before presentation of the pathways by which certain of the amino acids are synthesized, some characteristic aspects of these syntheses may be noted.

1. With NH_3 fixed into organic linkage by the three reactions described previously, amino acids can be fabricated from carbon compounds which are available as products of carbohydrate metabolism, *e.g.*, pyruvate and oxaloacetate.

2. In order for these synthetic pathways to provide a continuing supply of amino acids, each pathway must be essentially irreversible, *viz.*, it must proceed with a relatively large loss of free energy. This is accomplished by those reactions in which ATP is utilized and, effectively, split to ADP + P_i. Even more effective, in this regard, are those instances in which, over-all, ATP $\rightarrow$ AMP + PP_i, since the subsequent hydrolysis of PP_i, catalyzed by pyrophosphatase, is irreversible. In other instances, synthesis is assured by a reductive reaction, employing DPNH or TPNH, in which the equilibrium strongly favors oxidation of the reduced coenzyme.

In addition to the above aspects of synthetic pathways, the reactions involved in the synthesis of the individual amino acids are regulated by many of the metabolic control mechanisms that have been discussed previously (Chap. 13). Thus an end product of a reaction sequence often serves as a noncompetitive (regulatory) inhibitor of the enzyme which catalyzed the committed step (page 277). Some of the best examples of such inhibitors are found in pathways of amino acid synthesis (Table 23.1.), the amino acid synthesized at the final step serving as an inhibitor of the committed steps in the pathway. Another type of inhibition has been described which entails separate inhibition by end products of distinct enzymes catalyzing the same reaction. Thus two asparate kinases in *Escherichia coli,* catalyzing formation of β-aspartylphosphate, are inhibited by lysine and by threonine (Table 23.1). In contrast, in *Rhodopseudomonas capsulatas* and in other microorganisms, a single aspartokinase is inhibited weakly, if at all, by either threonine or lysine, but strongly by the two together. This has been termed *concerted feedback inhibition.*

A second mechanism of control obtains when the product of the reaction se-

Table 23.1: SOME EXAMPLES OF END-PRODUCT INHIBITION IN BIOSYNTHESIS OF AMINO ACIDS

Amino acid	Sensitive reaction	Organism or tissue	Page
Arginine	Glutamic acid ⟶ N-acetylglutamic acid	*Escherichia coli*	544
	N-Acetylglutamic acid ⟶ N-acetylglutamic-γ-semialdehyde	*Micrococcus glutamicus*	544
	N-Acetylglutamic-γ-semialdehyde ⟶ N-acetylornithine	*E. coli*	544
	Ornithine ⟶ citrulline	*E. coli*	545
Cysteine......	Serine ⟶ O-acetylserine	*Salmonella typhimurium*	547
Cystine.......	Homocysteine ⟶ cystathionine	*Rat liver*	547
Histidine	ATP + 5-phosphoribosylpyrophosphate ⟶ phosphoribosyl ATP	*S. typhimurium*	577
Isoleucine	Threonine ⟶ α-ketobutyric acid	*E. coli*	571
Leucine......	α-Ketoisovaleric acid + acetyl CoA ⟶ β-carboxy-β-hydroxyisovaleric acid	*S. typhimurium*	571
Lysine	Aspartic acid ⟶ β-aspartyl phosphate	*E. coli*	568
Proline.......	Glutamic acid ⟶ Δ^1-pyrroline-5-carboxylic acid	*E. coli*	544
Serine........	Phosphoserine ⟶ serine	*Rat liver*	549
Threonine	Aspartic acid ⟶ β-aspartyl phosphate	*E. coli*	568
	Aspartic-β-semialdehyde ⟶ homoserine	*Rhodopseudomonas capsulatus*	567
	Homoserine ⟶ 4-phosphohomoserine	*E. coli*	567
Tryptophan...	5-Phosphoshikimic acid ⟶ anthranilic acid	*E. coli*	575
Valine	Pyruvic acid ⟶ acetolactic acid	*Aerobacter aerogenes*	571

quence serves as a *repressor* (page 279) and inhibits the synthesis of the enzyme which catalyzes the reaction at the committed step. Indeed, the presence of the product may result in repression of formation of the enzymes catalyzing all the reactions at and beyond the committed step. Examples of these types of control mechanisms will be found in the material which follows. Their operation may be appreciated from the dramatic differences in the behavior of bacteria growing in media of varying composition. An organism that grows in a medium containing inorganic salts and carbohydrate does so by synthesizing all its amino acids, purines, pyrimidines, etc. If it is placed in a medium containing all 20 amino acids, these are accepted from the medium and used for protein synthesis and the cell makes a small amount of or no amino acids for itself. Similar observations may be made with addition of purines, pyrimidines, vitamins, etc., to the medium.

Living cells vary widely in their genetic capacity for synthesis of amino acids. As already noted, higher plants and many microorganisms are entirely self-sufficient in that they can synthesize all the amino acids. In contrast, for example, are the *lactobacilli,* organisms found in milk. These bacteria flourish in milk, the proteins of which provide all the amino acids. However, these cells are practically incapable of synthesizing amino acids *de novo* and remain entirely dependent upon their environment to provide amino acids for continuing metabolism and growth. Man and most other vertebrates are in an intermediary position, capable of the synthesis of a limited group of amino acids and dependent upon the environment for all others.

AMINO ACIDS NUTRITIONALLY ESSENTIAL FOR MAN

In the course of evolution, the animal organism lost the ability to synthesize the carbon chain of certain of the α-keto acids. Accordingly, the corresponding α-amino acids cannot be formed via transamination reactions. These amino acids, which must therefore be provided preformed in the diet either as free amino acids or as constituents of dietary proteins, have been termed *essential amino acids.* This term is at present used to designate those amino acids which cannot be synthesized by the organism at a rate adequate to meet metabolic requirements and must be supplied in the diet. The essential amino acids are listed in Table 23.2, as are those amino acids which need not be present in the diet. It should be emphasized that this classification is based upon growth studies in the rat (page 282). The proportion of essential amino acids provided by various dietary proteins is one factor influencing the biological value of proteins (Chap. 48).

Table 23.2: CLASSIFICATION OF THE AMINO ACIDS WITH RESPECT TO THEIR GROWTH EFFECT IN THE WHITE RAT

Essential	*Nonessential*
Arginine*	Alanine
Histidine	Aspartic acid
Isoleucine	Cystine
Leucine	Glutamic acid
Lysine	Glycine
Methionine	Hydroxyproline
Phenylalanine	Proline
Threonine	Serine
Tryptophan	Tyrosine
Valine	

* Arginine can be synthesized by the rat but not at a sufficiently rapid rate to meet the demands of *normal* growth.

The failure of young animals to grow on a diet deficient in one or more of the essential amino acids is a reflection of the inability to synthesize adequate quantities of protein under these experimental conditions. In response to an amino acid deficiency in the diet, the tissues do not make proteins lacking that particular amino acid; they simply make less protein.

The terms *essential* and *nonessential* relate only to dietary requirements and have no meaning with respect to the relative importance which the amino acids may have in metabolism. The amino acids which are essential in the diet are compounds with carbon skeletons which cannot readily be synthesized by the body. In a real sense, the so-called nonessential amino acids are of equal or greater significance for the economy of the organism in that they participate in diverse cellular reactions and functions and provide precursors for the synthesis of many important cellular constituents. Indeed, certain of the nonessential amino acids, *e.g.*, glutamic acid, have so many important metabolic roles that, were a mammal to lose suddenly its capacity to synthesize glutamic acid, serious disorganization of key reactions of metabolism might result since the animal could not wait until the next meal to replenish its supply, as would be possible in the case of the so-called "essential" amino acids.

SYNTHESIS OF NONESSENTIAL AMINO ACIDS IN MAMMALS

The nonessential amino acids are, by definition, those amino acids which need not be provided in the mammalian diet since they can be synthesized in adequate amounts. Although arginine is listed in Table 23.2 as an essential amino acid, its synthesis will be discussed in this chapter inasmuch as mammals have some capacity to fabricate arginine.

In discussion of the synthesis of glutamic acid (page 536), it was indicated that amino group transfer from glutamic acid provides a mechanism for synthesis of most of the other amino acids by transamination. Indeed, this is the prime mechanism for synthesis of the nonessential amino acids, as well as for achieving intramolecular exchange of amino groups among amino acids.

TRANSAMINATION

General Nature of Transamination. In 1937 Braunstein and Kirtzmann demonstrated that there are universally distributed enzymes, *transaminases,* which catalyze the following general reaction.

$$R{-}\overset{\displaystyle NH_2}{\overset{|}{C}H}{-}COOH + R'{-}\overset{\displaystyle O}{\overset{\|}{C}}{-}COOH \rightleftharpoons R{-}\overset{\displaystyle O}{\overset{\|}{C}}{-}COOH + R'{-}\overset{\displaystyle NH_2}{\overset{|}{C}H}{-}COOH$$

In animal tissues, one couple of the reactants is almost invariably the pair, glutamic/α-ketoglutaric acids. The most abundant of the transaminases catalyze transamination from glutamic to oxaloacetic and pyruvic acids, respectively.

L-Glutamic acid + oxaloacetic acid $\rightleftharpoons$ α-ketoglutaric acid + L-aspartic acid
L-Glutamic acid + pyruvic acid $\rightleftharpoons$ α-ketoglutaric acid + L-alanine

Transamination is the process of amino group transfer, and the enzymes that catalyze such reactions have been termed *transaminases, aminopherases,* and *aminotransferases.* The first term will be used in this book. Nomenclature of the transaminases is complicated by the participation of at least four substrates in the reversible reactions catalyzed by the enzymes. A simple procedure, used here, is to designate the enzymes by terms which include the amino acid substrates. Thus, the above two reactions are catalyzed by *glutamate-aspartate transaminase* and *glutamate-alanine transaminase,* respectively.

Liver contains transaminases specific for catalyzing transamination from glutamic acid to α-keto acids corresponding to each of the naturally occurring α-amino acids, except possibly glycine, threonine, and lysine. The glutamate-aspartate transaminase activity of several tissues is due to at least two distinct molecular forms (isoenzymes, page 361) which differ in cellular localization (mitochondria vs. soluble cytoplasm) and in physical-chemical properties. Isoenzymes of several other transaminases occur in animal tissues.

Transamination is not limited to α-amino/α-keto acids. Certain transaminases catalyze reversible transfer of an amino group from glutamic acid to specific aldehydes, with formation of the corresponding amines. Also, transaminases utilizing amino acid amides are also present in tissues (see below).

Transamination from glutamic acid provides a mechanism for synthesis of those α-amino acids in a cell which can synthesize the corresponding α-keto acids. Moreover, transamination provides a means for redistributing nitrogen. For example, in any given meal, an animal may ingest a mixture of amino acids quite different from that which is optimal to its metabolism. Thus, were the meal rich in phenylalanine and poor in aspartic acid, the following pair of reactions would provide the necessary nitrogen for aspartic acid formation.

(*a*)	**L-Phenylalanine + α-ketoglutaric acid**	$\rightleftharpoons$	**phenylpyruvic acid + L-glutamic acid**
(*b*)	**L-Glutamic acid + oxaloacetic acid**	$\rightleftharpoons$	**α-ketoglutaric acid + L-aspartic acid**
Sum:	**L-Phenylalanine + oxaloacetic acid**	$\rightleftharpoons$	**phenylpyruvic acid + L-aspartic acid**

Simultaneous and continuous operation of all the transaminase reactions accounts for the observation that shortly after administration to rats of any amino acid labeled with ^{15}N in the α-amino group, ^{15}N appears in all the amino acids except lysine, with maximal abundance of isotope in glutamic and aspartic acids and in the amide group of glutamine.

Mechanism of Transamination. Both pyridoxal phosphate and pyridoxamine phosphate are active as coenzymes in transamination.

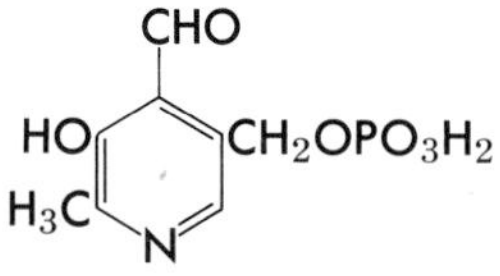
Pyridoxal phosphate

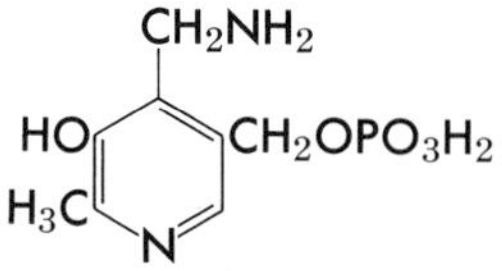
Pyridoxamine phosphate

Complete details of the mechanism of transamination are not known. Model systems have been studied in which trivalent cations, *e.g.*, Al^{+++}, plus pyridoxal phosphate catalyze transamination in vitro in the absence of enzyme, but the normal enzymic mechanism is uncertain. In a general way, the following events are indicated for a single cycle of enzymic activity. P represents the remainder of the pyridoxal phosphate nucleus bound to enzyme.

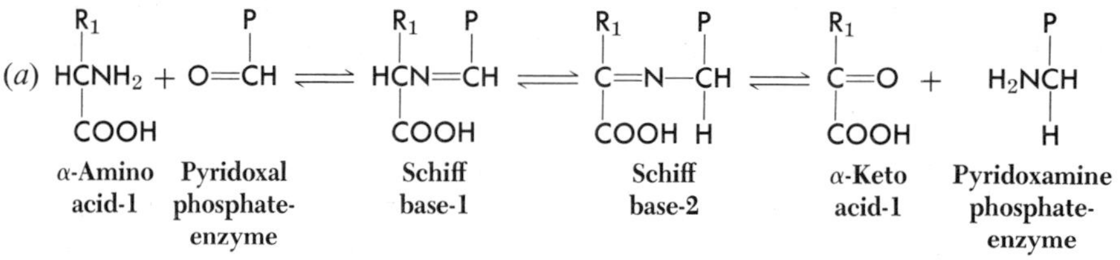

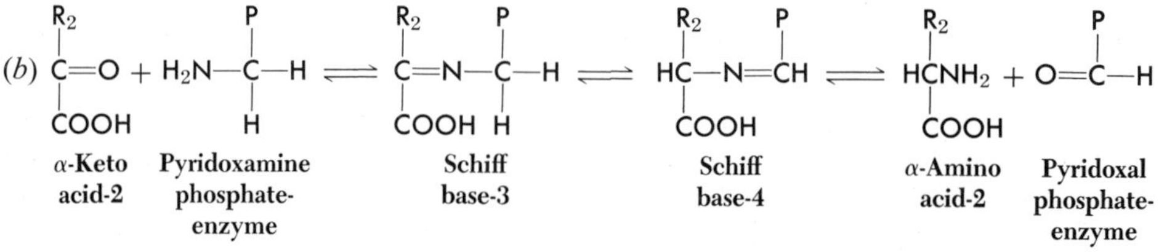

Studies with the glutamate-aspartate transaminase of pig heart suggest that this enzyme has a single binding site with which all four substrates combine in sequence. Other aspects of the transaminase mechanism were considered earlier (page 541).

Glutamine and Asparagine in Transamination. It was indicated above that amino acid amides also participate in transamination. This is the case for glutamine and asparagine, the two common amino acid amides. Thus the reaction for glutamine is a transamination, with subsequent hydrolysis of the α-keto acid-ω-amide catalyzed on the surface of the same *transaminase-deamidase.*

(*a*)
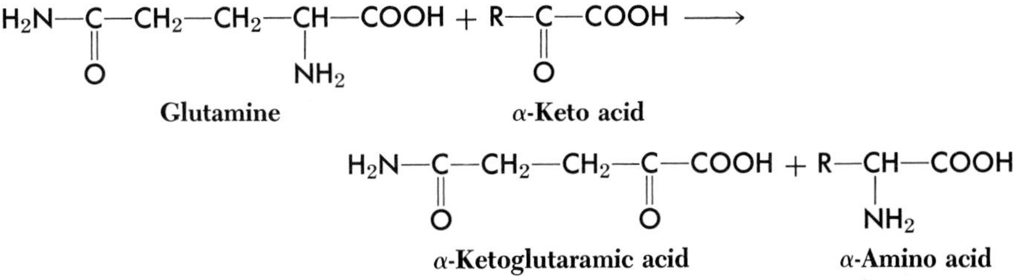

(*b*)
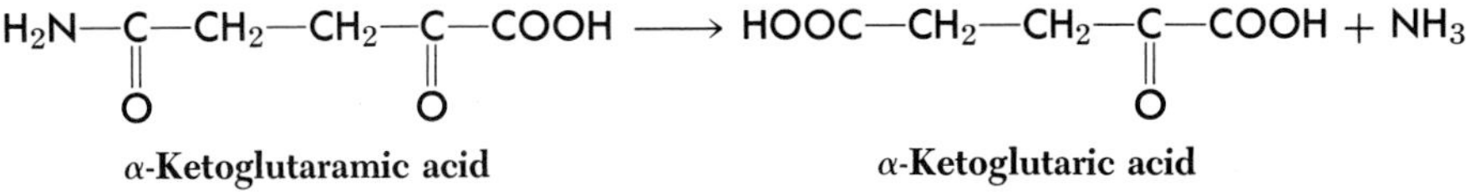

The glutamine and asparagine transaminase systems have a broad specificity; more than 30 α-keto acids are active, including the α-keto acids corresponding to many amino acids. The amide transaminase reactions, in contrast to amino acid transaminase systems, are irreversible because of rapid deamidation of the α-keto acid amides.

SYNTHESIS OF INDIVIDUAL AMINO ACIDS

Glutamic Acid, Aspartic Acid, and Alanine. Each of these amino acids is derived by transamination of an α-keto acid (α-ketoglutaric acid, oxaloacetic acid, and pyruvic acid, respectively); these α-keto acids also participate in the tricarboxylic acid cycle. The synthesis of glutamic acid has been described earlier (page 536). α-Ketoglutaric acid is the major precursor of glutamic acid, which is also formed from histidine (page 612), proline (page 595), hydroxyproline (page 595), glutamine (page 596), and, in certain microorganisms, from β-methylaspartic acid (page 1040). Aspartic acid arises primarily by transamination to oxaloacetate, and may also arise from asparagine in a hydrolytic reaction catalyzed by *asparaginase.* Alanine is synthesized chiefly from pyruvate by transamination, and is also formed by reductive amination of pyruvate in bacteria, and as a product of the decarboxylation of aspartate (page 588), and the cleavage of kynurenine (page 615).

Glutamic acid, aspartic acid, and alanine, together with glutamine, are generally the most abundant of the free amino acids in cells. Thus the concentration of glutamic acid in brain is approximately 100 to 150 mg. per g. tissue. In some microorganisms, alanine and/or aspartic acid may play the central role reserved for glutamic acid in mammalian metabolism.

Glutamine and Asparagine. The synthesis of glutamine described earlier (page

536) provides glutamine for transamination reactions (see above) as well as a reserve pool of both ammonia and glutamic acid. Asparagine is present in most proteins and occurs in abundance as the free amino acid in higher plants. Asparagine synthesis appears to occur by a mechanism similar to that of glutamine, *i.e.*, from aspartic acid, NH_3, and ATP. However, a possible alternative route obtains in which the amide-N of glutamine is transferred to aspartic acid.

Proline, Hydroxyproline, and Ornithine. Glutamic acid serves as the precursor for synthesis of proline, hydroxyproline, and ornithine. The last is not present in proteins but is the precursor for synthesis of arginine and, accordingly, participates in the urea cycle (page 560). In mammals, the initial step in the conversion of glutamate to proline and ornithine is formation of glutamic acid semialdehyde; transamination results in formation of ornithine (Fig. 23.1). Ring closure yields Δ^1-pyrroline 5-carboxylic acid, which, upon reduction by DPNH, gives proline.

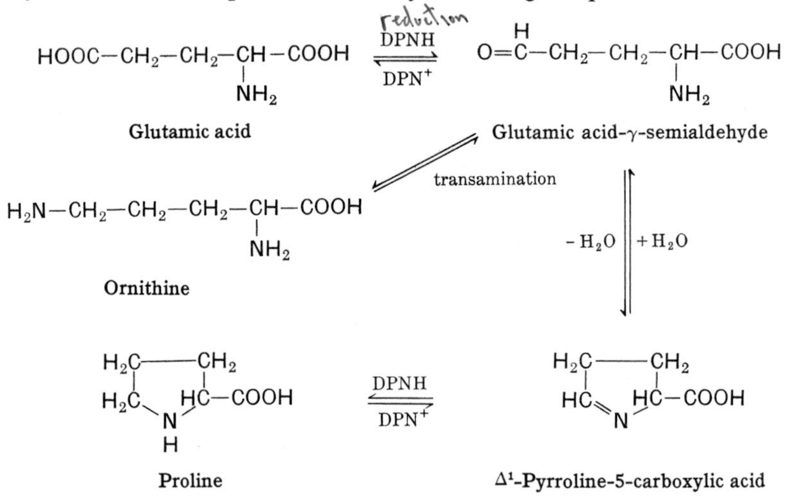

Fig. 23.1. Interrelationships in the synthesis of glutamic acid, ornithine, and proline.

The pathway for ornithine synthesis depicted in Fig. 23.1 appears to be a minor one in certain microorganisms; an alternate pathway involves N-acetylglutamic acid (Fig. 23.2). Three of the participating enzymes have been found in *E. coli* extracts: an *acetylase* that catalyzes formation of N-acetylglutamate from glutamic acid and acetyl CoA, a *transaminase* that results in formation of α-N-acetylornithine, and *acetylornithinase* that catalyzes hydrolysis of α-N-acetylornithine to ornithine. Arginine inhibits the acetylation of glutamic acid and also represses synthesis of the acetylase as well as the other enzymes involved in the pathway.

In other microorganisms, as well as in spinach leaves, yet another pathway for ornithine synthesis exists. N-Acetylglutamic acid is formed from glutamic acid and α-N-acetylornithine in a reaction catalyzed by *ornithine acetyltransferase.* Reaction with ATP yields N-acetyl-γ-glutamylphosphate, which on reduction with a TPNH-requiring enzyme gives N-acetylglutamic-γ-semialdehyde; transamination of the latter with glutamate yields α-N-acetylornithine, which is converted to ornithine by transfer of the acetyl group to glutamic acid under the influence of the ornithine

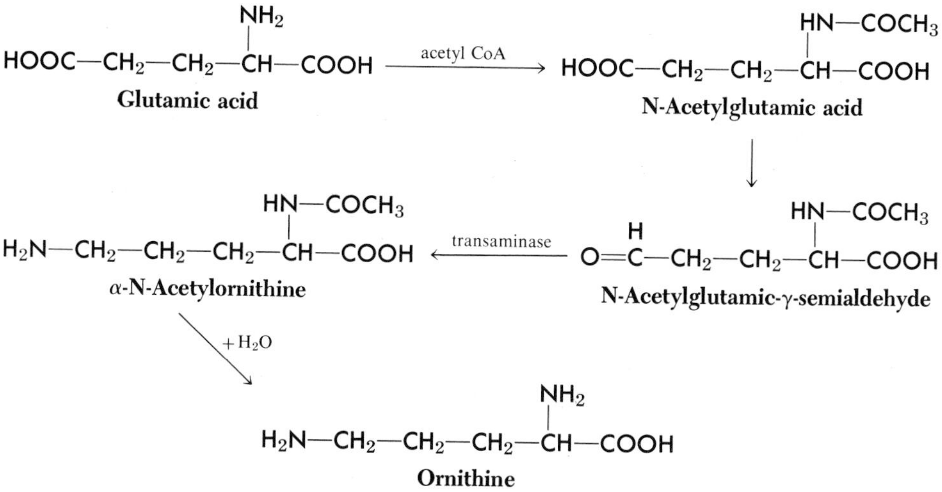

Fig. 23.2. N-Acetylglutamic acid pathway for synthesis of ornithine in some microorganisms.

acetyltransferase. This sequence of reactions has been termed an N-acetylornithine cycle.

The distributions of hydroxyproline is limited to collagen (page 871) and a few collagen-like proteins. Hydroxyproline is formed from proline by a hydroxylation reaction catalyzed by *proline hydroxylase* which utilizes molecular oxygen; Fe^{++}, ascorbic acid, and α-ketoglutaric acid are cofactors in the reaction. Proline already incorporated into a newly synthesized ribosomal polypeptide (Chap. 38) is the substrate for hydroxylation. The enzyme is specific for those proline residues in the second position after glycine, the position in which the hydroxyproline of collagen is found (Chap. 38).

As will be seen later (page 595), the relationships among proline, hydroxyproline, and arginine are, in part, reversible.

Arginine. Arginine synthesis commences with the ornithine derived from glutamic acid as described above. The first step is citrulline synthesis in a reaction catalyzed by *ornithine transcarbamylase.*

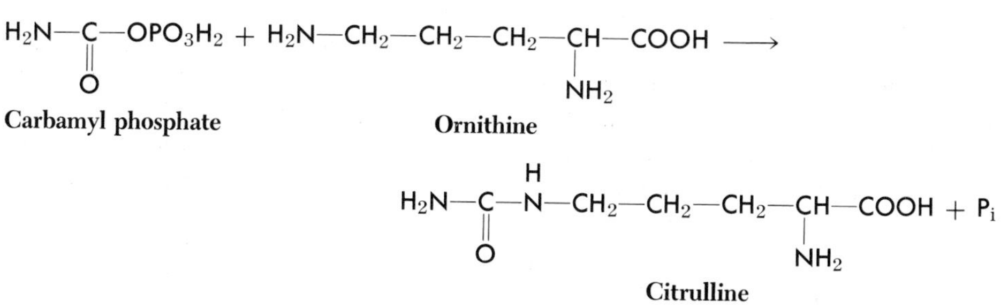

The next reaction, catalyzed by *argininosuccinic acid synthetase,* requires ATP and Mg^{++} and involves a condensation of citrulline and aspartic acid.

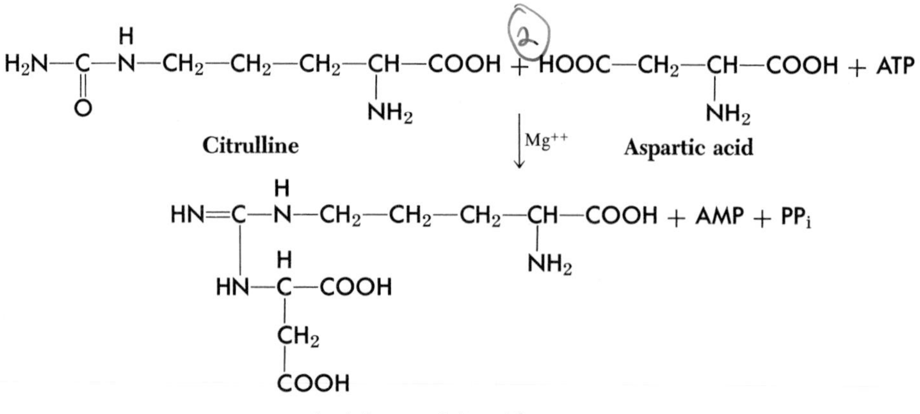

$$H_2N-\underset{\underset{O}{\|}}{C}-\underset{\underset{|}{H}}{N}-CH_2-CH_2-CH_2-\underset{\underset{NH_2}{|}}{CH}-COOH + HOOC-CH_2-\underset{\underset{NH_2}{|}}{CH}-COOH + ATP$$

Citrulline   ↓ Mg^{++} Aspartic acid

$$HN=\underset{\underset{\underset{\underset{\underset{COOH}{|}}{CH_2}}{|}}{\underset{HN-\underset{\underset{}{|}}{C}-COOH}{|}}}{C}-N-CH_2-CH_2-CH_2-\underset{\underset{NH_2}{|}}{CH}-COOH + AMP + PP_i$$

Argininosuccinic acid

Argininosuccinic acid synthetase has been partially purified from mammalian liver. Exchange studies with citrulline labeled in the ureido group with ^{18}O showed a transfer of isotope to AMP, suggesting that an enzyme-bound adenyl-citrulline intermediate is involved. The detailed mechanism remains to be elucidated.

Argininosuccinic acid is cleaved by an enzyme called *argininosuccinase* to yield arginine and fumaric acid.

Argininosuccinic acid ⇌

$$H_2N-\underset{\underset{\underset{H}{|}}{\overset{\overset{|}{N}}{\|}}}{C}-N-CH_2-CH_2-CH_2-\underset{\underset{NH_2}{|}}{CH}-COOH + \quad \underset{HOOCCH}{\overset{HCCOOH}{\|}}$$

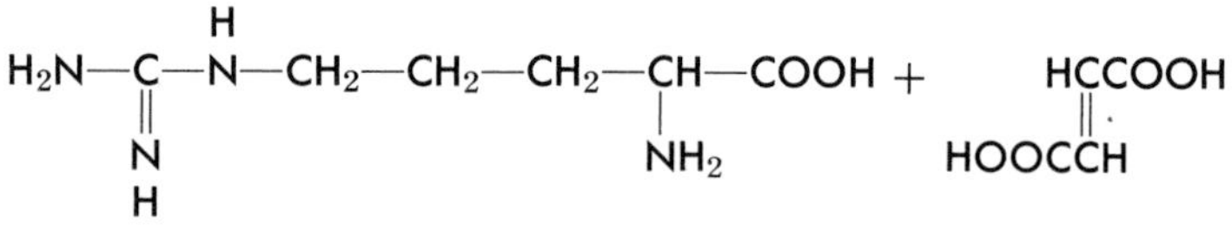

Arginine   Fumaric acid

This reaction is neither hydrolytic nor oxidative, and is readily reversible. The enzyme is widely distributed in nature; it is present in kidney, liver, and brain and has been obtained in crystalline form from bovine liver.

Since ornithine receives its δ-amino group by transamination from glutamic acid (page 544) and since aspartic acid is also formed by transamination from glutamate, two of the three N atoms of the guanido group of arginine derive from glutamic acid. The third has its origin in carbamyl phosphate; this N may also derive from glutamate via glutamine (page 536). The fumaric acid formed in the argininosuccinase reaction can be hydrated to malic acid and reoxidized to oxaloacetic acid in the citric acid cycle. The oxaloacetate can then acquire a new amino group, by transamination, with aspartate formation and repetition of the sequence.

Cysteine. Cysteine, a nonessential amino acid, is made in mammals from the essential amino acid, methionine. If sufficient methionine is fed, there is no dietary requirement for cysteine (or cystine). The reactions involved in cysteine biosynthesis are the following.

1. Demethylation of methionine to homocysteine (homologue of cysteine).

$$CH_3-S-CH_2-CH_2-\underset{\underset{NH_2}{|}}{CH}-COOH \xrightarrow{-CH_3} HS-CH_2-CH_2-\underset{\underset{NH_2}{|}}{CH}-COOH$$

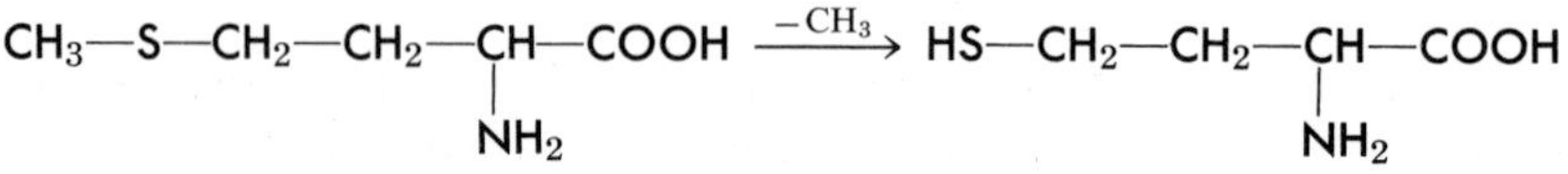

Methionine   Homocysteine

Demethylation of methionine will be considered again later in relation to trans-methylation (page 582). Since methionine biosynthesis in plants and microorganisms utilizes sulfur derived from cysteine (or cystine), the latter amino acids are the ultimate source of all amino acid sulfur.

2. Homocysteine condenses with serine in a reaction catalyzed by *cystathionine γ-synthetase*, a pyridoxal phosphate–requiring enzyme; the product is cystathionine.

$$HOOC—\underset{\underset{NH_2}{|}}{CH}—CH_2—CH_2—SH + \underset{\underset{OH}{|}}{CH_2}—\underset{\underset{NH_2}{|}}{CH}—COOH \longrightarrow$$

Homocysteine   Serine

$$HOOC—\underset{\underset{NH_2}{|}}{CH}—CH_2—S—CH_2—CH_2—\underset{\underset{NH_2}{|}}{CH}—COOH + H_2O$$

Cystathionine

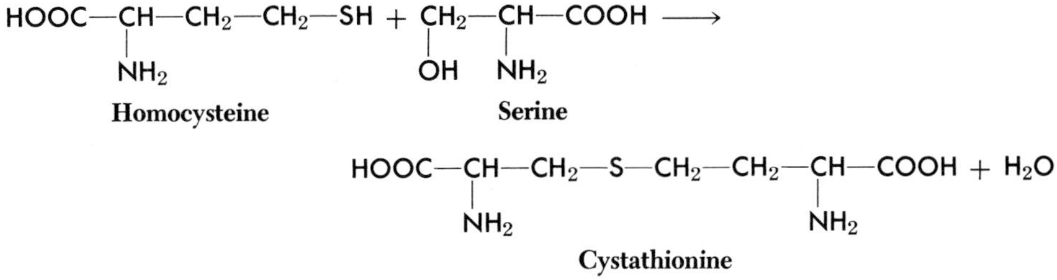

3. Cleavage of cystathionine is catalyzed by a liver enzyme, *transsulfurase (cystathionase)*, which has been obtained in crystalline form and utilizes pyridoxal phosphate. The products are cysteine, α-ketobutyric acid, and NH_3.

$$Cystathionine + H_2O \longrightarrow HOOC—\underset{\underset{NH_2}{|}}{CH}—CH_2—SH + CH_3—CH_2—\underset{\underset{O}{\|}}{C}—COOH + NH_3$$

Cysteine   α-Ketobutyric acid

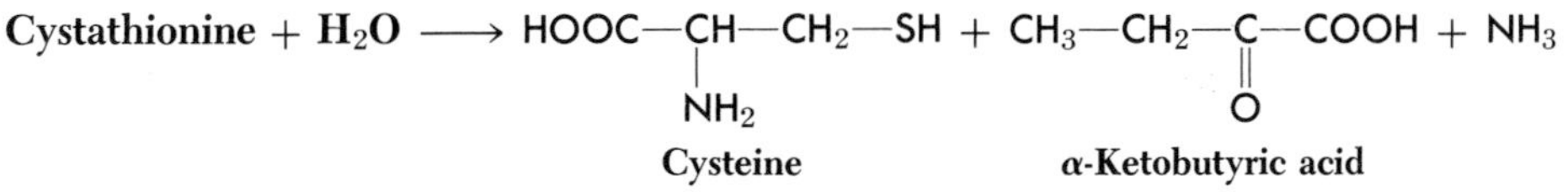

The net effect of the above reactions is an exchange of the sulfhydryl group of homocysteine with the hydroxyl group of serine in a process termed *transsulfuration.* Thus in the biosynthesis of cysteine, the carbon chain, including the amino group, arises from serine, whereas the sulfur is derived from methionine.

An alternative two-step pathway of cysteine synthesis exists in some microorganisms.

(a)
$$\underset{\underset{OH}{|}}{CH_2}—\underset{\underset{NH_2}{|}}{CH}—COOH + CH_3COCoA \longrightarrow \underset{\underset{O}{|}\underset{COCH_3}{|}}{CH_2}—\underset{\underset{NH_2}{|}}{CH}—COOH + CoA$$

Serine   Acetyl CoA   O-Acetylserine

(b)
$$\underset{\underset{O}{|}\underset{COCH_3}{|}}{CH_2}—\underset{\underset{NH_2}{|}}{CH}—COOH + H_2S \longrightarrow HS—CH_2—\underset{\underset{NH_2}{|}}{CH}—COOH + CH_3COOH + H_2O$$

O-Acetylserine   Cysteine

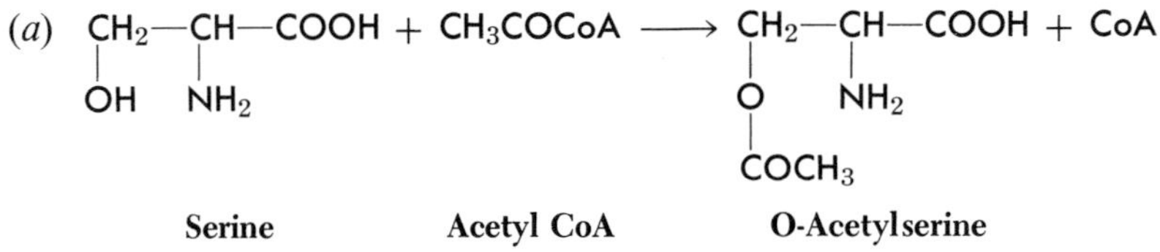

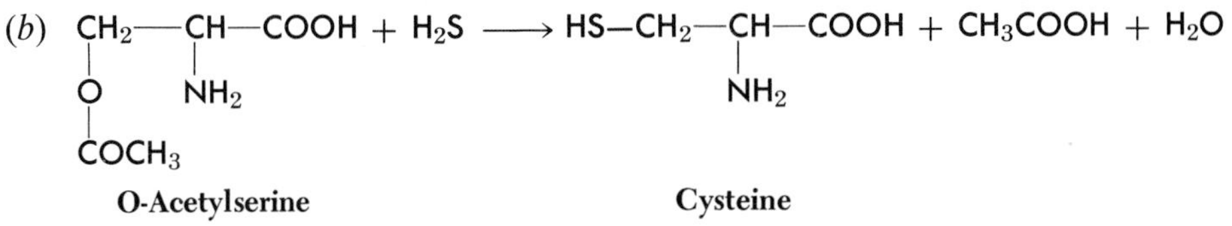

The enzyme catalyzing the first reaction is *serine transacetylase;* the second, is *O-acetylserine sulfhydrylase.* A similar pathway could explain utilization of inorganic sulfate by many microorganisms and plants for the synthesis of cysteine, since mechanisms exist for reduction of sulfate to sulfide (page 564). In yeast, *serine sulfhydrase* catalyzes formation of cysteine from serine and H_2S.

Although the dietary and metabolic equivalence of cysteine and cystine have been established, an enzymic system catalyzing their interconversion has not been described in mammalian tissue. A DPN$^+$-linked *cystine reductase* occurs in yeast and higher plants, but is different from oxidized glutathione reductase (see below). As described below, oxidized glutathione can function in a nonenzymic oxidation of cysteine to cystine.

Little or no free cystine is present in cells. The cystine of proteins is formed by oxidation of cysteine residues after their incorporation into polypeptide chains (page 657). However, in the presence of O_2 and cations such as Fe^{++} or Cu^{++}, cystine may be formed from cysteine nonenzymically. If this occurs, ready reversal can be catalyzed by *glutathione reductase.* Glutathione is γ-glutamylcysteinylglycine; its synthesis is described on page 578. Glutathione (G—SH, below) reacts nonenzymically with any disulfide, *e.g.*, cystine, to form a mixed disulfide.

$$\text{G—SH} + \text{R}'\text{—S—S—R}'' \rightleftharpoons \text{G—S—S—R}'' + \text{R}'\text{—SH}$$

A second molecule of glutathione, reacting with the mixed disulfide, yields oxidized glutathione.

$$\text{G—SH} + \text{G—S—S—R}'' \longrightarrow \text{G—S—S—G} + \text{R}''\text{—SH}$$

Glutathione reductase is a flavoprotein which catalyzes the following reaction.

$$\text{G—S—S—G} + \text{TPNH} + \text{H}^+ \rightleftharpoons \text{2G—SH} + \text{TPN}^+$$

In this manner, cystine (R′—S—S—R″ in the above reactions), which may form, is reduced back to cysteine for use by the cell.

Tyrosine. Tyrosine biosynthesis in mammals occurs by hydroxylation of phenylalanine, an essential amino acid. Much of the dietary requirement for phenylalanine is, in fact, due to the need for tyrosine. If the latter is fed, the dietary requirement for phenylalanine is reduced substantially. In this sense, tyrosine bears the same relationship to phenylalanine as cysteine does to methionine. In normal metabolism, the only known fate of phenylalanine, other than utilization for protein synthesis, is its conversion to tyrosine. The enzyme system catalyzing the hydroxylation of phenylalanine to tyrosine has been termed *phenylalanine hydroxylase* which, like other hydroxylases, catalyzes formation of hydroxylating radicals from O_2 in the presence of TPNH (page 378). The system differs from other hydroxylases in that TPNH functions as coenzyme for *dihydrobiopterin reductase,* which utilizes as substrate a pteridine occurring in liver, dihydrobiopterin.

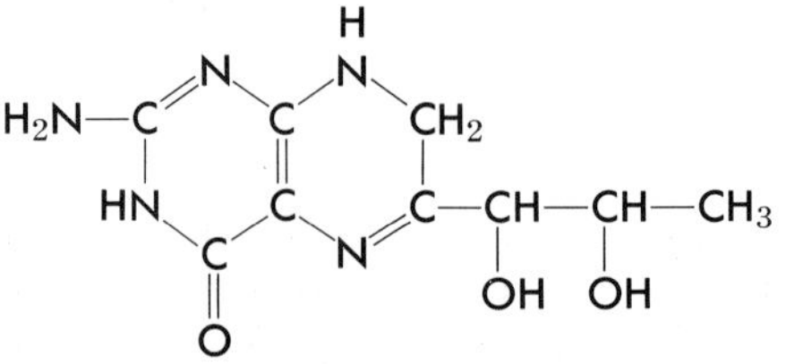

2-Amino-4-keto-6-dihydroxypropyl-7,8-dihydropteridine
(dihydrobiopterin)

The reduction product is the 4-hydroxytetrahydrobiopterin, which is a required cofactor for the hydroxylation reaction. The reactions involved may be depicted as follows.

(*a*) **Dihydrobiopterin + TPNH + H^+ $\xrightleftharpoons[\text{reductase}]{\text{dihydrobiopterin}}$ tetrahydrobiopterin + TPN^+**

(*b*) **Tetrahydrobiopterin + phenylalanine + O_2 $\longrightarrow$ dihydrobiopterin + tyrosine + H_2O**

In vitro, dihydrobiopterin reductase may be replaced by *dihydrofolate reductase* (page 550); it is not known whether the latter enzyme may function in vivo in the above hydroxylating system.

Reduced forms of biopterin or other reduced pteridine analogues are also cofactors for the hydroxylation of tyrosine (page 607) and tryptophan (page 589).

Hereditary lack of phenylalanine hydroxylase results in phenylketonuria. The recessive gene is carried by about one in every two hundred individuals. In the absence of this enzyme, minor pathways of phenylalanine metabolism, little used in normal individuals, become prominent. Transamination from phenylalanine yields phenylpyruvic acid, of which as much as 1 to 2 g. per day may be excreted. Severe mental retardation is evident early in children with phenylketonuria; restriction of their dietary intake of phenylalanine reduces the blood level of phenylalanine, abolishes excretion of phenylpyruvic acid, and prevents, in considerable degree, the mental retardation. The accumulation of phenylpyruvic acid leads also to formation and urinary excretion of phenyllactic acid, *o*-hydroxyphenylacetic acid, benzoic acid, and phenylacetic acid, the last as phenylacetylglutamine (page 580).

Serine. The carbon chain of serine derives from 3-phosphoglyceric acid formed during glycolysis. The latter may be utilized by one of two pathways which lead to either serine or its 3-phosphate ester (Fig. 23.3, page 550). Hydrolysis of 3-phosphoglyceric acid yields glyceric acid. The latter, with a DPN^+-requiring enzyme, provides hydroxypyruvic acid, which may be transaminated in a reaction catalyzed by *alanine-serine transaminase* to form serine. Alternatively, 3-phosphoglyceric acid may be oxidized to 3-phosphohydroxypyruvic acid in a DPN^+-requiring reaction catalyzed by *3-phosphoglyceric acid dehydrogenase.* Transamination with glutamic acid yields 3-phosphoserine; hydrolysis by *serine phosphatase* yields serine. The last described pathway is probably the major route of serine synthesis. 3-Phosphoglyceric acid dehydrogenase and serine phosphatase are inhibited by serine; this provides a means of regulating serine formation.

Glycine. The initial demonstration of glycine formation from serine was provided by Shemin, who administered serine, labeled with ^{15}N in the amino position and ^{13}C in the carboxyl position, together with benzoic acid, to rats and guinea pigs. The glycine excreted in the urine as benzoylglycine (hippuric acid, page 579) had a $^{15}N/^{13}C$ ratio identical with that of the administered serine, indicating formation of glycine by loss of the β-carbon of serine.

$$\underset{\text{Serine}}{\underset{|}{\overset{}{CH_2}}(OH)\text{—}\underset{|}{CH}(NH_2)\text{—}COOH} \rightleftharpoons \underset{\text{Glycine}}{\underset{|}{CH_2}(NH_2)\text{—}COOH} + \text{“}C_1\text{”}$$

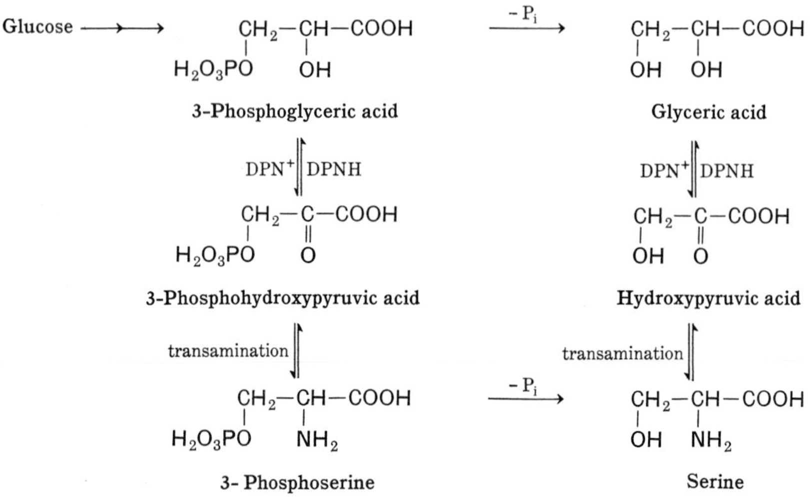

Fig. 23.3. Pathways of serine biosynthesis.

This reaction is significant not only for the synthesis of glycine but as the prime metabolic source of a group of active C_1 compounds which exist at the oxidation level of CH_3OH, HCHO, or HCOOH. Their transformations make possible the addition to diverse compounds of the groups $—CH_3$, $—CH_2OH$, and —CHO, respectively.

C_1 Compounds. The transformation of serine to glycine is catalyzed by *serine hydroxymethyl transferase* in the presence of pyridoxal phosphate and Mn^{++}. The acceptor of the C_1 unit is tetrahydrofolic acid, a reduced form of the vitamin, folic acid (page 1034). Folic acid is reduced to the corresponding dihydro compound, and then to tetrahydrofolic acid; the latter reaction is catalyzed by *dihydrofolic reductase*. The enzyme has been prepared in highly purified form from chicken liver and utilizes TPNH at neutral pH; DPNH may function at lower pH values (5.5). Dihydrofolic reductase exhibits abnormal kinetics with TPN, which both stabilizes and increases the activity of the enzyme, suggesting an allosteric effect. Although folic acid is designated as the vitamin, its tetrahydro derivative serves as the biological carrier for C_1 groups.

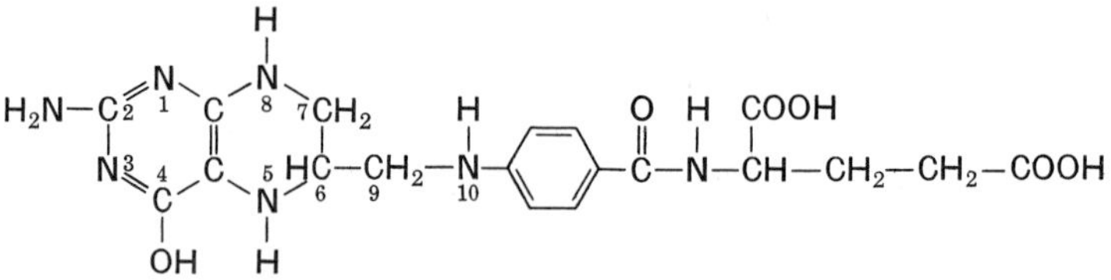

Tetrahydrofolic acid

The nitrogen atoms at the 5 and 10 positions function as the sites for subsequent reaction. In glycine formation the reaction, schematically, is the following.

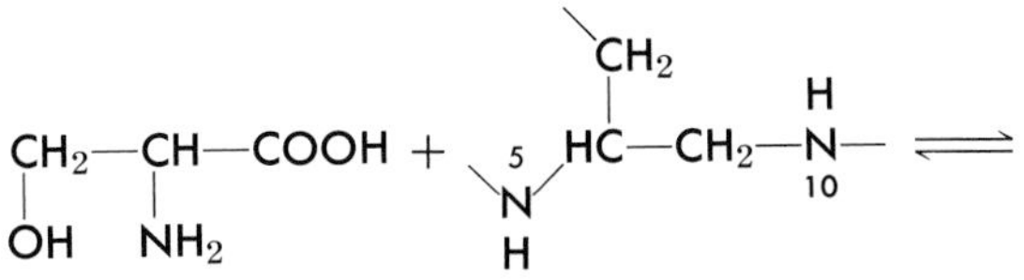

Serine N[10]-Tetrahydrofolic acid

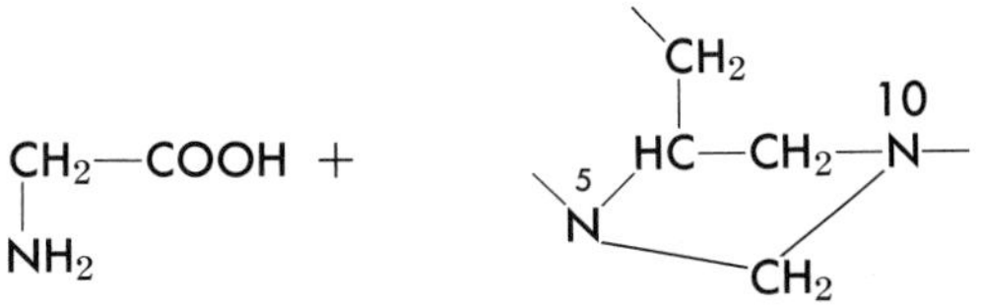

Glycine N[5],N[10]-Methylenetetrahydrofolic acid
("active formaldehyde")

Presumably, on the enzymic surface the $—NH_2$ group of serine is bound as a Schiff base to pyridoxal phosphate before transfer of the β-carbon to tetrahydrofolic acid; after the transfer, the Schiff base must hydrolyze, thereby releasing glycine. N^5,N^{10}-Methylenetetrahydrofolic acid may be oxidized by TPN^+ to yield N^5,N^{10}-methenyltetrahydrofolic acid; the reaction is catalyzed by *N^5,N^{10}-methylenetetrahydrofolate dehydrogenase.*

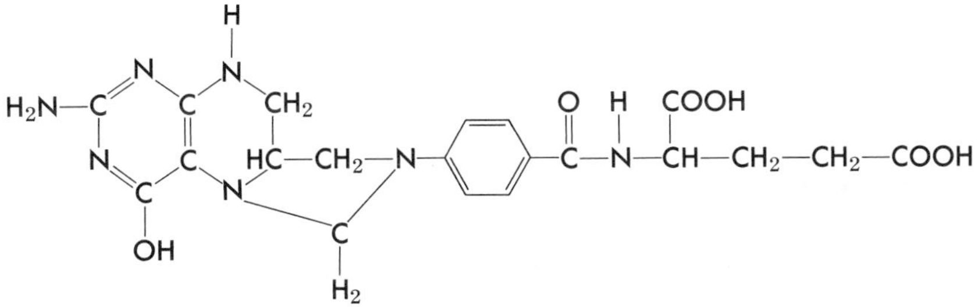

N[5],N[10]-Methylenetetrahydrofolic acid

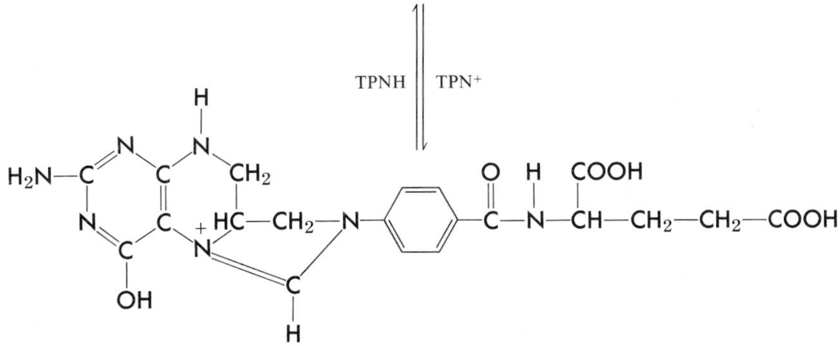

N[5],N[10]-Methenyltetrahydrofolic acid

Hydrolysis of N^5,N^{10}-methenyltetrahydrofolic acid yields N^{10}-formyltetrahydrofolic acid, the form which is utilized for formyl transfer in many synthetic pathways. This reaction is reversible and is catalyzed by N^5, N^{10}-*methenyltetrahydrofolate cyclohydrolase.*

N^{10}-Formyltetrahydrofolic acid may also be formed directly from formic acid in a reaction requiring ATP and catalyzed by *formyltetrahydrofolate synthetase.*

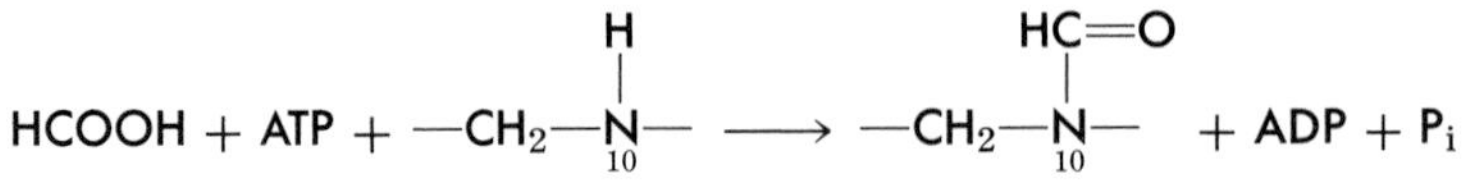

The enzyme that catalyzes this reaction has been found in pigeon liver and human erythrocytes, as well as in microorganisms. The reaction explains the appearance of the carbon of administered $H^{14}COOH$ in those compounds formed from N^{10}-formyltetrahydrofolic acid.

In some reactions, a formyl group is transferred directly from a metabolite to the N-5 position of tetrahydrofolic acid. In these instances, a second reaction requiring ATP must occur to form the N^5,N^{10}-methenyltetrahydrofolic acid, which then is converted to the N^{10} derivative or may be reduced to the N^5,N^{10}-methylene compound. In a limited number of cases, tetrahydrofolic acid may accept a formimino group, —CH=NH, at the N^5 position. This is converted by a *cyclodeaminase* to the N^5,N^{10}-methenyl compound which is utilized in the usual manner. Thus, tetrahydrofolic acid serves as a carrier for a single carbon unit at three levels of oxidation, *viz.*, $—CH_3$, $—CH_2OH$, —CHO, corresponding to methanol, formaldehyde and formic acid, respectively, as well as the formimino group, —CH=NH. These relationships are summarized in Fig. 23.4.

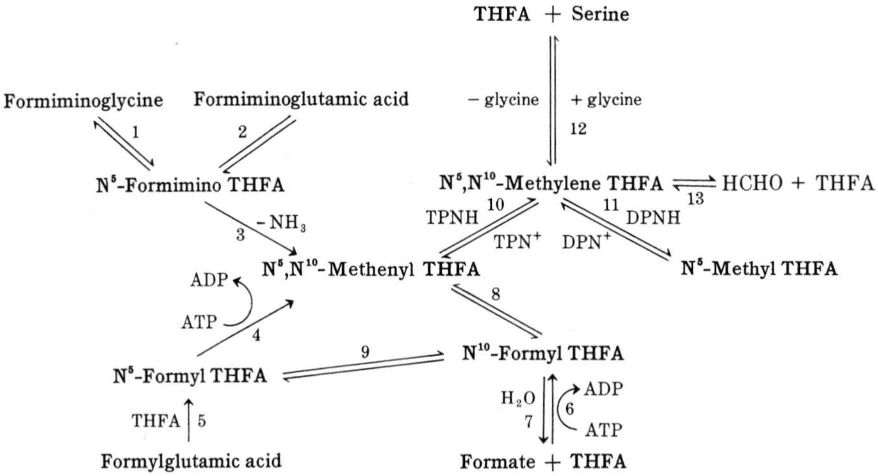

Fig. 23.4. Relationships among C_1 derivatives of tetrahydrofolic acid (THFA). Enzymes involved are 1. formiminoglycine formimino transferase; 2. formiminoglutamate formimino transferase; 3. N^5-formiminotetrahydrofolate cyclodeaminase; 4. N^5-formyltetrahydrofolate isomerase (cyclodehydrase); 5. formylglutamate formyl transferase; 6. formyltetrahydrofolate synthetase; 7. N^{10}-formyltetrahydrofolate deacylase; 8. N^5,N^{10}-methenyltetrahydrofolate cyclohydrolase; 9. N^5-formyltetrahydrofolate isomerase; 10. N^5,N^{10}-methylenetetrahydrofolate (hydroxymethyltetrahydrofolate) dehydrogenase; 11. N^5,N^{10}-methylenetetrahydrofolate reductase; 12. L-serine hydroxymethyl transferase; 13. formaldehyde-activating enzyme.

Genesis of Methyl Groups. The methylene group of N^5,N^{10}-methylenetetrahydrofolic acid is the source of the methyl group for synthesis of both methionine (page 566) and thymine (page 639). Reduction of N^5,N^{10}-methylenetetrahydrofolic acid in *E. coli* is catalyzed by a flavoprotein, *N^5,N^{10}-methylenetetrahydrofolate reductase* (reaction 11, Fig. 23.4).

$$\textbf{N}^5,\textbf{N}^{10}\textbf{-Methylenetetrahydrofolic acid} + \textbf{DPNH} + \textbf{H}^+ \underset{}{\overset{\text{FAD}}{\rightleftharpoons}} \textbf{N}^5\textbf{-Methyltetrahydrofolic acid} + \textbf{DPN}^+$$

Since N^5-methyltetrahydrofolic acid is found in liver, it is probably generated and utilized in mammalian tissues in a manner similar to that above in *E. coli.* The formation of N^5-methyltetrahydrofolic acid by the above reactions affords a pathway for *de novo* genesis of methyl groups of such diverse substances as methionine, choline, sarcosine, glycine, dimethylglycine, methanol, and acetone.

Two enzymic systems in *E. coli* catalyze the transfer of the methyl group to homocysteine to form methionine; one requires cobamide (vitamin B_{12}) and the other does not. The role of cobamide and the mechanism of this transmethylation are discussed in Chap. 49 (see Vitamin B_{12}). The methyl group of N^5-methyltetrahydrofolic acid is transferred intact.

Transmethylation. The transfer of the methyl group of methionine to appropriate acceptors, *transmethylation,* and the general metabolic significance of this reaction were first demonstrated by du Vigneaud and his colleagues. Loss of the methyl group from methionine yields homocysteine (page 546). The active form of methionine which functions in methylation reactions is S-adenosylmethionine,

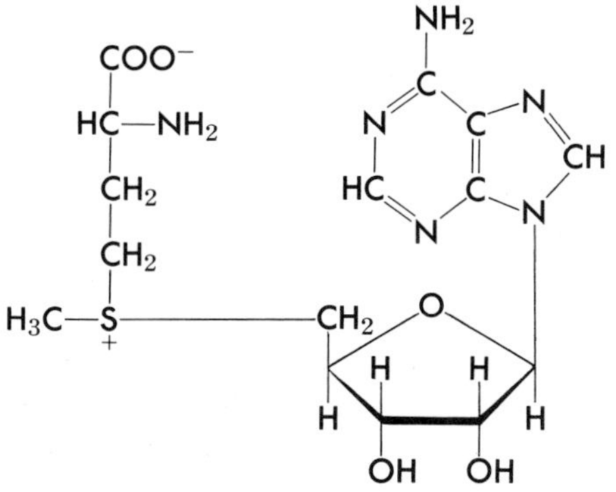

S-Adenosylmethionine

a sulfonium form of methionine, with a free energy of scission comparable to that of the pyrophosphate linkage of ATP. Synthesis of S-adenosylmethionine is catalyzed by a widely distributed enzyme which has been purified from liver.

$$\textbf{L-Methionine} + \textbf{ATP} \longrightarrow \textbf{S-adenosylmethionine} + \textbf{PP}_i + \textbf{P}_i$$

This process is unique in biological systems, being the only known instance in which, in a single reaction, the phosphates of ATP appear as a molecule of P_i plus a molecule of PP_i. The detailed enzymic mechanism is not understood. In the presence of the appropriate specific enzymes, the methyl group of S-adenosylmethionine may be transferred, for example, to guanidoacetic acid to form creatine (page 582), to phosphatidyl ethanolamine to form lecithin (page 549), or to nicotinamide to form N^1-methylnicotinamide (page 1046). In each case the other product formed from S-adenosylmethionine appears to be S-adenosyl-L-homocysteine which is cleaved to adenosine and homocysteine.

The homocysteine formed may be utilized in several ways. In plants, which, unlike animals, can also form homocysteine *de novo* (page 566), the homocysteine may be remethylated, using methyl groups formed *de novo*. In animals, another mode of re-forming methionine derives from the fact that choline, liberated from lecithin, may be oxidized in two stages to betaine; the latter then transmethylates to homocysteine with formation of methionine.

$$(a)\quad \underset{\textbf{Choline}}{(CH_3)_3\overset{+}{N}{-}CH_2{-}CH_2OH} \xrightarrow{FAD} \underset{\textbf{Betaine aldehyde}}{(CH_3)_3\overset{+}{N}{-}CH_2{-}CHO} \xrightarrow{DPN^+} \underset{\textbf{Betaine}}{(CH_3)_3\overset{+}{N}{-}CH_2{-}COO^-}$$

$$(b)\quad \underset{\textbf{Betaine}}{(CH_3)_3\overset{+}{N}{-}CH_2{-}COO^-} + \underset{\textbf{Homocysteine}}{HS{-}CH_2{-}CH_2{-}\overset{\overset{NH_2}{|}}{CH}{-}COOH} \longrightarrow \underset{\textbf{Dimethylglycine}}{(CH_3)_2N{-}CH_2{-}COOH} + \underset{\textbf{Methionine}}{H_3C{-}S{-}CH_2{-}CH_2{-}\underset{\underset{NH_2}{|}}{CH}{-}COOH}$$

If homocysteine (as homocystine) is fed to rats on a methionine-deficient diet, these processes of methylation of homocysteine provide sufficient methionine for normal growth. However, homocystine is not present in the normal diet, and, generally, homocysteine formed by transmethylation is not remethylated but is used for cysteine synthesis (page 546); the total pathway, therefore, is irreversible. Accordingly, dietary methionine is the major source of methyl groups in the animal economy. *De novo* methyl group synthesis is important for thymine synthesis (page 639), but dietary methionine supplies most of the other methyl groups in mammals.

REGULATION OF NITROGEN METABOLISM

The amino acid and nitrogen metabolism of mammals is in a dynamic steady state; amino acid synthesis and degradation proceed at all times. The absolute and relative rates of these processes, at any moment, vary with the quantity and nature of the diet and are subject to both self-regulation and modulation by several hormones. Some of these phenomena have been described previously. Some overall aspects of amino acid and nitrogen metabolism will now be presented; additional instances of regulation will be noted in subsequent chapters.

GENERAL ASPECTS OF AMINO ACID METABOLISM

The metabolism of amino acids by bacteria growing logarithmically in a rich nutrient broth is in marked contrast to that of the mammal. Bacteria accept from the medium those amino acids, and in the correct quantity, needed for maximal protein synthesis. No degradation of protein occurs and the removal of amino acids from the medium is balanced by the sum of amino acids incorporated into protein plus those utilized for synthesis of the diverse nitrogenous compounds found in such organisms. Little or no nitrogen, except for occasional hydrolytic enzymes, is returned to the medium, and amino acids not needed do not enter the cell.

No comparable situation occurs in mammalian life. At all stages in the existence of a human being, there is a continuing entry and loss of nitrogen compounds. Amino acids and lesser amounts of other nitrogenous compounds enter the body and are processed, and the metabolic products are excreted in urine and feces. During infancy and childhood, and during convalescence from a debilitating disease, the intake of nitrogen exceeds the output (*positive nitrogen balance*); the opposite situation may prevail in senescence or, relatively briefly, during starvation or certain wasting diseases (*negative nitrogen balance*). However, even in infancy or senescence the daily departure from *nitrogen equilibrium* is usually only a small fraction of the total amount of nitrogen metabolized.

Whether derived from the diet or from endogenous sources, *i.e.*, intracellular synthesis or degradation of cellular proteins, amino acids are channeled into one or more of the following pathways: (1) incorporation into protein; (2) incorporation into a small peptide; (3) utilization of the nitrogen and/or carbon for synthesis of a different amino acid; (4) utilization for synthesis of a nitrogenous compound which is not an amino acid; (5) removal of the α-amino group by transamination or oxidation, with subsequent formation of urea and oxidation of the resultant α-keto acid. The last process, (5), as well as (4), may be regarded as a disposal device for surplus amino acids. These last two pathways are essential to the animal for two general reasons: (1) Daily food consumption is related to caloric need. In satisfying that need, the animal may ingest and digest an amount of protein which provides amino acids in excess of physiological requirements. In contrast to glycogen and triglycerides, there is no equivalent storage form of protein or amino acids. Accordingly, surplus amino acids must be excreted per se or, more economically, their nitrogen must be removed and excreted and the corresponding α-keto acids oxidized for their caloric value or converted to glycogen or fatty acids for storage. (2) There is a continuing need for the biosynthesis of an array of nitrogenous compounds, *e.g.*, purines, pyrimidines, porphyrins, epinephrine, thyroxine, nicotinic acid, etc.

In some instances the nitrogen and carbon skeleton derive from a nutritionally essential amino acid; in others only the nitrogen atoms need have their origin in the dietary amino acid since the remainder of the molecule can be provided from carbohydrate precursors, as described previously for the nonessential amino acids. The capacity of the mammal to perform these conversions is apparent from the fact that a synthetic diet which provides nitrogen only in the form of the nutritionally essential amino acids, plus the small amount in the vitamins, supports maximal growth. However, in circumstances of inadequate intake of protein nitrogen, *e.g.*, starvation, ingestion of an inadequate amount of protein, or ingestion of a diet which provides adequate quantities of all but one of the essential amino acids, synthesis of nitrogenous compounds continues. For example, nicotinic acid synthesis from tryptophan would continue, albeit at a reduced level, even on a tryptophan-free diet. The tryptophan utilized would be derived from the pool of amino acids generated by the continuing hydrolysis of tissue proteins (see below). Since incomplete or imperfect proteins are not made, the amino acids remaining in the pool could not be used for resynthesis of proteins. Hence, they are catabolically degraded, and their nitrogen appears in the urine as urea. It is of interest that the extent of negative nitrogen balance is of the same magnitude, *i.e.*, about 4 to 5 g. of nitrogen per day, in human adults on a diet which provides no amino acids as it is in individuals provided a diet which lacks only one essential amino acid but is otherwise complete.

Removal of Amino Acids from the Circulation. The plasma concentration of amino acids is normally about 4 to 8 mg. of α-amino nitrogen (35 to 65 mg. of mixed amino acids) per 100 ml. (see Table 31.1, page 706). Amino acids which enter the circulation by absorption from the intestine or by intravenous administration are quickly removed and appear in all tissues and organs of the body. Thus, within 5 min., 85 to 100 per cent of a large intravenously administered quantity (5 to 10 g.) of a single amino acid may be removed by the tissues.

The liver exhibits the greatest capacity to take up circulating amino acids, with kidney also participating significantly; other tissues take up lesser quantities. Some tissues, notably brain, exhibit a selective capacity. Thus, while intravenously administered methionine, histidine, glycine, arginine, glutamine, and tyrosine rapidly appear in the brain, glutamic acid cannot readily enter this structure from the blood, and lysine, proline, and leucine do so very slowly. Rates of entry of the last three amino acids into brain are greater in young individuals than in adults.

Amino acids enter cells by an active, energy-requiring process which occurs against a concentration gradient; entry of amino acids is accompanied by accumulation of water and some Na^+. K^+ may leave the cell simultaneously, particularly when lysine or arginine enters, thus preserving electrical neutrality of the cell contents. Amino acid absorption across the intestine was considered previously (page 534).

The metabolic fate of an amino acid molecule is presumably the same whether it enters the blood from the intestine or is derived from endogenous sources. However, mixing of amino acids from these two sources may not be complete. Indeed, discrete, nonmixing "pools" of amino acids may exist within a given cell, with differing turnover rates. While this phenomenon is still of unknown physiological significance for the total organism, it is a serious variable in the interpretation of experimental data obtained with isotopically labeled amino acids, particularly in studies of protein turnover and product-precursor relationships.

Several factors influence the extent to which amino acids are distributed among the various metabolic reactions in which they may participate. For example, the growing organism or the individual convalescing from a debilitating disease uses a significant proportion of available amino acids for construction of new tissue proteins. The availability of carbohydrate and lipid influences the proportion of the total caloric requirement which must be supplied by amino acids. In addition, the pattern of the amino acid mixture supplied to the tissues, *i.e.*, the relative amounts of the various amino acids, will determine the suitability of these substances for the synthesis of a specific type of cellular protein. Finally, the influence of hormones in modifying the direction and rate of certain metabolic reactions is an important factor in amino acid and protein metabolism.

Utilization of Amino Acids for Protein Synthesis. It is abundantly clear from the increment in total body protein during growth and convalescence that dietary amino acids can be utilized for net protein synthesis. Even in the adult in nitrogen equilibrium, continual protein synthesis is required for the elaboration of digestive enzymes which may be lost in the feces, protein hormones which are made in endocrine glands and degraded elsewhere in the body, formation of plasma proteins, and replacement of the entire protein complement of those cell populations of relatively short existence, *e.g.*, erythrocytes, leukocytes, and cells of the gastrointestinal mucosa.

Demonstration that there is a continuing incorporation of dietary nitrogen into constitutive tissue proteins even in an adult animal in nitrogen equilibrium was possible by use of isotopes. In investigations by Schoenheimer and his colleagues in which amino acids labeled with ^{15}N and/or ^{2}H were given to adult rats in nitrogen balance, the labeled amino acids were invariably found in the mixed, precipitable tissue proteins. These data led to the present-day concept of protein metabolism as a dynamic process in which body proteins are continually turning over, *i.e.*, undergoing synthesis and degradation. Since the proteins remain constant in amount, the two processes must occur at equal rates.

Formation of new cells is always associated with net protein synthesis. Indeed, even in mammals, virtually all of the proteins in a given cell must be made *in situ.* The rates of turnover of proteins within existing structures vary widely, depending only in part upon the degree to which new cells are being elaborated. Protein synthesis is a constant feature of mature, nondividing cells. Even the constitutive enzymes of the normal, nondividing cells of adult liver, *e.g.*, aldolase, contain isotopically labeled amino acids, after these have been administered. Such proteins exhibit half-life times of many weeks. Less labile is the metabolism of "structural" proteins such as the myosin of muscle (Chap. 36), which exhibits a half-life of more than 6 months. This does not reflect a difference between liver and muscle with respect to all constituent proteins. Thirty minutes after injection of eight labeled amino acids into rabbits, the ratio of the specific activities of all eight was found to be the same in three different crystalline enzymes isolated from skeletal muscle. This indicates that (1) muscle cells constantly synthesize new protein, (2) the synthetic process is rapid, (3) in a given cell, all proteins are fabricated from a common pool of free amino acids, (4) this process involves *de novo* synthesis of entire protein molecules; resynthesis of proteins from partial degradation products does not occur, and (5) the individual proteins of a given cell type have different turnover rates.

Proteins which are normally found extracellularly also turn over at varying rates. In general, proteins which leave their sites of synthesis undergo replacement by new molecules relatively more rapidly than do proteins which remain as intracellular components. Thus, in man in nitrogen balance, the half-life of serum proteins is approximately 10 days. In contrast, however, the principal extracellular protein of connective tissue, collagen, exhibits almost no significant incorporation of labeled amino acids in adult animals and labeled amino acids, incorporated when the animal was younger and growing, do not disappear from this protein.

Similar studies, in which labeled amino acids were administered, have permitted estimation of the total rate of protein synthesis. In man, dog, and rat, the rates of protein synthesis, expressed as grams of nitrogen per kilogram per day, were found to be 0.6 to 1.0, 0.6, and 2.0, respectively. Thus, a 70-kg. adult man synthesizes and degrades about 400 g. of protein per day, whereas an average daily American diet provides about 100 g. of amino acids and, at any instant, the total amount of free amino acids in the body fluids is of the order of 30 g.

The fact that protein synthesis proceeds at this great rate and is exactly balanced by protein degradation is indeed remarkable. The mechanisms of protein synthesis are described in detail in Chap. 29. However, the factors which regulate the *rate* of protein synthesis and degradation are not understood. Apparently it is not the supply of amino acids which is primarily rate-limiting; net protein synthesis cannot be increased by augmenting the amino acid supply when the latter is adequate. The influence of various endocrine factors on protein synthesis will be discussed in Part Five of this book; however, the actual mechanism is not clear. In contrast to the energy- and information-requiring apparatus for protein synthesis, intracellular protein degradation occurs by simple hydrolysis catalyzed by proteinases present in the lysosomes (page 285).

REMOVAL OF AMINO GROUPS OF AMINO ACIDS

Site of α-Amino Group Removal. The liver is the major site of removal of amino groups from amino acids, although the process is a general one in all tissues studied. The importance of the liver in this metabolic process stems from its size and its receipt of amino acids absorbed from the intestine via the portal circulation as well as its enzymic capabilities. Experimental hepatectomy prevents the rapid disappearance of intravenously administered amino acids and the concomitant rise in blood urea. Moreover, when both kidneys are excised, the usual increase in blood urea, following injection of amino acids into dogs, does not occur if the liver is also absent. Blood amino acids may be elevated in patients either with severe acute liver atrophy or following surgical portocaval anastomoses with evidence of liver dysfunction. These studies established that the liver is the prime site of amino group removal and urea synthesis.

Mode of Amino Group Removal. Amino groups are removed by transamination, (pages 541*ff.*), by oxidative deamination catalyzed by L-amino acid oxidase, or by a nonoxidative deamination.

Transamination Followed by Deamination. The presence in liver of transaminases specific for the reaction between α-ketoglutaric acid and most of the individual amino acids, together with the high glutamic acid dehydrogenase activity in hepatic mitochondria, has led to the concept that the α-amino groups of most amino acids are converted to ammonia by consecutive transamination to α-ketoglutaric acid and oxidation of the glutamic acid thus formed.

(*a*) **α-Ketoglutaric acid + amino acid $\longrightarrow$ glutamic acid + α-keto acid**
(*b*) **Glutamic acid + DPN^+ + H_2O $\longrightarrow$ α-ketoglutaric acid + DPNH + H^+ + NH_3**

Sum: Amino acid + DPN^+ + H_2O $\longrightarrow$ α-keto acid + DPNH + H^+ + NH_3

Presumably, the DPNH which arises in this manner may be reoxidized by the mitochondrial electron transport system, thereby opposing the otherwise very unfavorable equilibrium of the glutamic acid dehydrogenase reaction which, of itself, markedly favors formation of glutamic acid from α-ketoglutaric acid, DPNH, and ammonia (page 536).

Oxidative Deamination of Amino Acids. Liver and kidney contain a general L-*amino acid oxidase* with low activity; flavin mononucleotide (FMN) is the prosthetic group. The general reaction is as follows.

$$(a)\quad R{-}CHNH_2{-}COOH + FMN \rightleftharpoons R{-}CO{-}COOH + NH_3 + FMNH_2$$
$$(b)\quad FMNH_2 + O_2 \longrightarrow FMN + H_2O_2$$

The peroxide formed is decomposed by *catalase* (page 381). L-Amino acid oxidases from snake venom and certain microorganisms utilize flavin adenine dinucleotide (FAD) as the prosthetic group.

L-Amino acid oxidase catalyzes oxidation of all naturally occurring L-amino acids except serine, threonine, and the dicarboxylic and dibasic amino acids. This enzyme is probably not a very significant factor in normal amino acid oxidation because of its low order of activity.

The role of the highly potent D-amino acid oxidase of liver and kidney is unclear. This cytoplasmic enzyme (page 363) also utilizes FAD as prosthetic group. It catalyzes oxidation of the unnatural D antipode of a large number of amino acids; however, D-amino acids are not known to occur in mammalian metabolism. This enzyme does provide a means of oxidizing the D-amino acids of bacterial cell walls should any of them be absorbed from the intestine. Perhaps more significant is the fact that this enzyme is identical with *glycine oxidase,* catalyzing the following reaction.

$$\underset{\textbf{Glycine}}{H_2N{-}CH_2{-}COOH} \longrightarrow \underset{\textbf{Glyoxylic acid}}{O{=}\overset{H}{\overset{|}{C}}{-}COOH} + NH_3$$

The α-carbon of glycine is a meso carbon atom (page 251); in the deamination of glycine, D-amino acid oxidase removes that hydrogen atom which bears the same relation to the amino group as does the α-hydrogen of a D-amino acid.

Liver also has *monoamine* and *diamine oxidase* (page 377) activity. These flavoproteins catalyze the aerobic oxidation of a wide variety of physiological amines to the corresponding aldehydes and NH_3. Although the amount of each individual amine is small, the total of such activity may contribute significantly to the pool of ammonia. Similarly, the several transaminases which can transfer from amino acids such as γ-aminobutyric acid contribute to the formation of glutamic acid; deamination of the latter was considered above.

Nonoxidative Deamination of Amino Acids. A group of pyridoxal phosphate-dependent *dehydrases* catalyze removal of the amino groups of serine, cysteine, homoserine, threonine, and, perhaps, homocysteine (Chap. 26); in each instance NH_3 and the corresponding α-keto acid is formed. Amino group removal may also occur with formation of the corresponding acrylic acid, as in the case of histidine (page 609). Other examples of nonoxidative removal of amino groups of amino acids will be found in Chap. 26.

Regulation of Amino Group Removal. Increased dietary intake of amino acids results in an increase in liver concentration of a variety of enzymes, specific transaminases and dehydrases, which catalyze reactions by which amino acids are degraded. Adrenal cortical hormones also achieve the same end. Hence, in the hyperadrenal cortical state, lack of adequate intake of amino acids results in a negative nitrogen balance.

The contrasting situation in man and in microorganisms with regard to the influence of amino acid intake is noteworthy. In man, ingestion of amino acids induces enzymes concerned with their disposal. In microorganisms, amino acid ingestion frequently inhibits or represses enzymes concerned with amino acid synthesis.

UREA SYNTHESIS

The nitrogen compounds of the urine of an adult in nitrogen equilibrium comprise three classes: (1) a large number of diverse compounds excreted daily in small but relatively constant amount, e.g., creatinine, uric acid, N^1-methylnicotinamide, etc.; (2) NH_3, the excretion of which is a function of the acid-base economy of the body and is generated in the kidney as a means of excreting excess protons (Chap. 35); and (3) urea, the amount of which represents the difference between the dietary intake of nitrogen and the sum of (1) + (2). Circumstances which result in positive nitrogen balance lead to a diminution in urea excretion; in those conditions in which there is excessive excretion of nitrogen, at the expense of body protein, the increment in urinary nitrogen occurs as urea. Thus, it is the formation and excretion of urea which is the "leveling device" by which nitrogen balance is maintained.

Cyclic Mechanism of Urea Formation. Urea is synthesized in *ureotelic* organisms (those which utilize urea as the major vehicle for excretion of surplus nitrogen) by adaptation of a reaction sequence that evolved in much more primitive creatures. The enzyme that, uniquely, makes urea synthesis possible is *arginase,* which catalyzes the irreversible hydrolysis of arginine to ornithine and urea.

$$H_2N-\underset{\underset{H}{\overset{\|}{N}}}{\overset{H}{\overset{|}{C}}}-N-CH_2-CH_2-CH_2-\underset{\underset{NH_2}{|}}{CH}-COOH \xrightarrow[\text{H}_2\text{O}]{\text{arginase}}$$

Arginine

$$H_2N-CH_2-CH_2-CH_2-\underset{\underset{NH_2}{|}}{CH}-COOH + H_2N-\underset{\underset{O}{\|}}{C}-NH_2$$

Ornithine   **Urea**

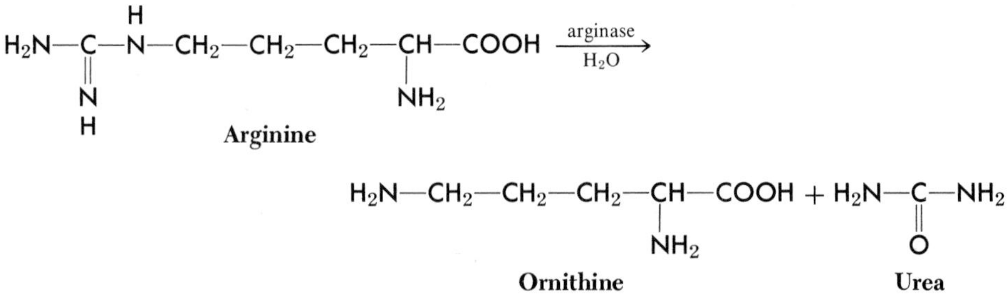

Since arginine is ubiquitous in animal cells (it is required for protein synthesis), urea can be made by any cell which possesses arginase. In mammals, the liver, which contains not only arginase but also all the other enzymes required for arginine synthesis, is the major site of urea formation. A low level of urea synthesis occurs in kidney and brain, but this seems of minor significance in the total nitrogen economy.

The catalytic role of ornithine in urea synthesis was first demonstrated by Krebs and Henseleit, who also established the position of citrulline in this process. They formulated the following cycle.

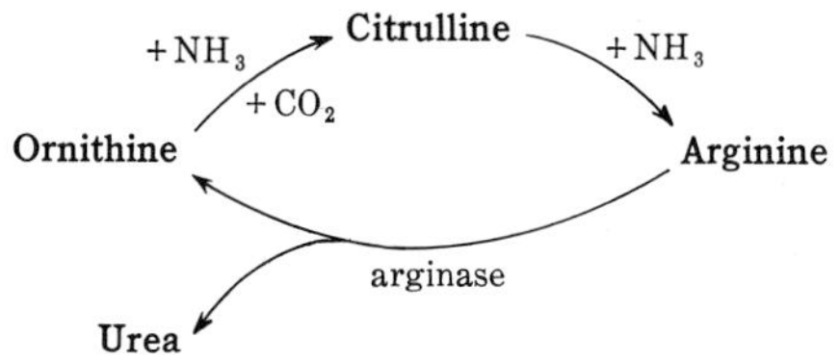

The reaction sequence: ornithine → citrulline → arginine will be recognized as the normal pathway for arginine synthesis (page 545). It is the presence of arginase in mammalian liver which converts this reaction sequence from a one-way synthetic system to a cyclic process in which the carbon chain of ornithine may be used repeatedly, with one urea molecule fabricated for each turn of the cycle.

Although the abbreviated mechanism depicted above indicates the entry of nitrogen as NH_3, it will be recalled (page 545) that citrulline formation actually utilizes carbamyl phosphate as N donor and that the second nitrogen enters as the α-amino nitrogen of aspartic acid. The energy for urea synthesis, therefore, is provided by the two molecules of ATP needed for carbamyl phosphate synthesis and the single ATP needed to make argininosuccinic acid. Since during the latter step, ATP $\rightarrow$ AMP + PP_i, and the latter is then hydrolyzed by pyrophosphatase action, $PP_i + H_2O \rightarrow 2P_i$, a total of four high-energy phosphate equivalents is required per urea molecule.

The rare hereditary disorders, *argininosuccinuria* and *citrullinuria,* are characterized by a high blood concentration and renal excretion of argininosuccinic acid and citrulline, respectively. These diseases both lead to mental retardation. Despite the excretion of these intermediates of the ornithine cycle (page 545), urea excretion is essentially normal. Although brain and kidney can make small amounts of urea, synthesis of the latter in the liver may be proceeding by an alternate, unknown pathway in such individuals. The inability to utilize argininosuccinic acid and citrulline normally is presumed to be due to a deficiency of argininosuccinase (page 546) and argininosuccinic acid synthetase (page 545), respectively.

For the liver to serve effectively as the locus of urea formation, using the α-amino nitrogen of all 20 amino acids, it must possess not only the enzymic capacity to operate the urea-forming cycle, but also the capacity to direct into the synthesis of carbamyl phosphate and aspartic acid, nitrogen from all those 20 amino acids. The role of aspartic acid in transamination and in providing one of the two nitrogen atoms for urea synthesis has been referred to earlier (page 546). Aspartic acid contributes this nitrogen atom via the synthesis of argininosuccinic acid, an intermediate in arginine formation (page 546). Aspartate formation from glutamate via transamination with oxaloacetate thus provides a mechanism for channeling amino groups from amino acids into urea. Although the other nitrogen atom of urea is depicted as arising from carbamyl phosphate via citrulline formation (page 545), the source of the ammonia for carbamyl phosphate synthesis (page 537) is somewhat uncertain. Inasmuch as nitrogen from almost all other amino acids can be transferred to glutamic acid by transamination, ammonia formation by the glutamic acid dehydrogenase reaction appears attractive. There is some evidence that glutamine participates in carbamyl phosphate synthesis. This could provide a source of the second nitrogen atom of urea; NH_3 formed from glutamic acid dehydrogenase action could be channeled into glutamine via the glutamine synthetase reaction (page 536).

Carbamyl phosphate synthetase activity and that of other enzymes of the urea cycle are under the regulatory influence of diet and of hormones. Increases in these activities are observed in rat liver consequent to a high protein intake and conditions leading to degradation of protein, *e.g.*, starvation or adrenal steroid administration.

Since urea formation is irreversible, the α-keto acids remaining after amino group removal cannot be utilized for reamination and must be degraded. Thus, in the starving animal, α-keto acids formed by amino acid group removal are oxidized for energy production (Chap. 26).

Uricotelic and Ammonotelic Organisms. Whereas mammals utilize urea for excretion of surplus nitrogen, this is not universally true among the vertebrates. Two other compounds also serve in this regard. Many species which live in the sea or in fresh water and can dispose of excreta readily and constantly excrete ammonia. Indeed this is the dominant nitrogenous excretory product among the teleost fishes; the latter also excrete urea in lesser amount, but in this case urea derives from the ultimate metabolism of purines via uric acid (page 632). The tadpole, like teleosts, excretes ammonia. During metamorphosis the enzymes of the urea cycle appear in the liver and the adult organism synthesizes and excretes urea. Thus, amphibia occupy a position between the teleosts and the mammals. Those species for which the water supply may be precarious and which commonly have a semisolid excreta, *viz.*, birds and land-dwelling reptiles, channel nitrogen metabolism into formation of uric acid. The details of this process are given in Chap. 27.

Only in ureotelic organisms is carbamyl phosphate synthesized by the irreversible synthetase reaction that requires expenditure of two molecules of ATP (page 537). Other organisms, for which it is not equally imperative that ammonia be channeled into urea via carbamyl phosphate, utilize the reversible carbamyl phosphate synthetase reaction (page 537) which can permit the nitrogen of carbamyl phosphate to return to the pool of NH_3.

REFERENCES

See list following Chap. 26.

24. Amino Acid Metabolism. II

Synthesis of Essential Amino Acids

This chapter will present the synthetic pathways, in plants and microorganisms, for those amino acids which cannot be made by mammals and hence are essential nutrients for mammals.

Methionine. The relationships of cysteine and methionine to each other are reversed in plants and animals. In plants and heterotrophic microorganisms, inorganic sulfur is fixed into organic linkage as cysteine and transferred to a different carbon chain for the synthesis of methionine. Animals, which are unable to fabricate the 4-carbon chain of homoserine, are dependent on their food for supply of methionine.

Fixation of Inorganic Sulfur. The metabolism of sulfur begins with the reduction of inorganic sulfate, the form in which it is most abundant. Reduction occurs in plants and many microorganisms, but not in most animals. The initial step is the formation from inorganic sulfate of a compound which also serves as a general agent for esterification of sulfate with alcoholic and phenolic compounds in most living forms including mammalia, *viz.*, *3′-phosphoadenosine 5′-phosphosulfate.*

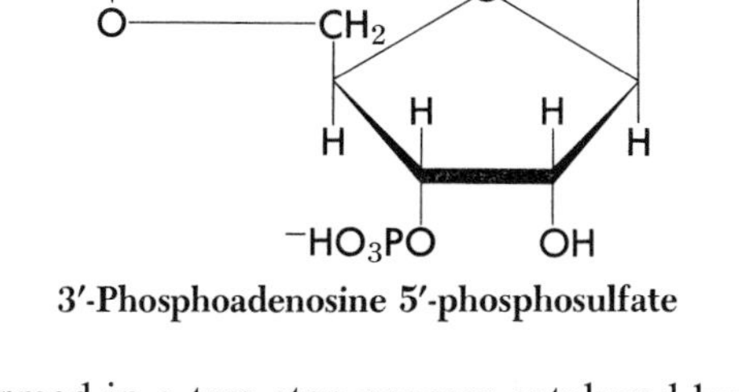

3′-Phosphoadenosine 5′-phosphosulfate

This compound is formed in a two-step process catalyzed by *ATP sulfurylase* and *adenosine 5′-phosphosulfate-3′-phosphokinase,* isolated from yeast and liver; each requires Mg^{++}.

Inorganic sulfate + ATP $\xrightarrow{Mg^{++}}$ adenosine 5′-phosphosulfate + PP_i

Adenosine 5′-phosphosulfate + ATP $\xrightarrow{Mg^{++}}$ 3′-phosphoadenosine 5′-phosphosulfate + ADP

Little is known of the reductive process other than that an enzyme, *3-phosphoadenosine 5′-phosphosulfate reductase* and TPNH are required for reduction of 3′-phosphoadenosine 5′-phosphosulfate to the level of free $SO_3^=$. In *Desulfovibrio desulfuricans,* adenosine 5′-phosphosulfate can be reduced directly to sulfite.

The *sulfite reductase* purified from *Escherichia coli* is an iron-flavoprotein which catalyzes the reduction of sulfite to sulfide or thiosulfate in the presence of TPNH. Formation of the enzyme is repressed by cysteine and cystine. Many organisms can effect the reduction of sulfite to the level of $S_2O_3^=$ (thiosulfate) and H_2S as well as elemental sulfur. The reverse processes have also been observed; various "sulfur bacteria" accomplish the aerobic oxidation of H_2S, S, $S_2O_3^=$, and $SO_3^=$ to sulfate.

Several mechanisms for fixation of sulfur, as H_2S or as sulfite, have been described. One is the reversal of the *cysteine desulfhydrase* reaction (page 598). The enzyme is present in plants and microorganisms and the reaction is reversible; pyridoxal phosphate functions as coenzyme.

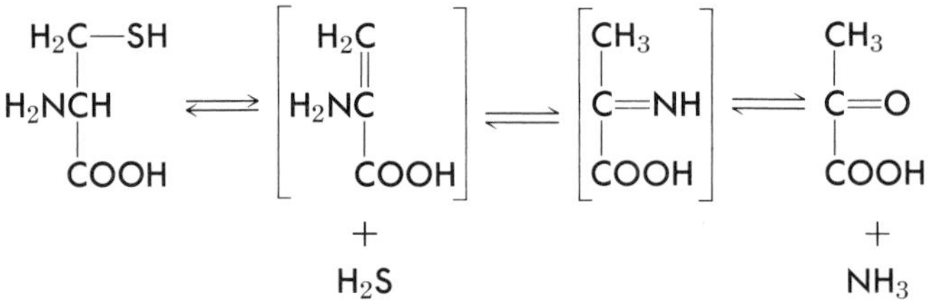

The reaction sequence to the right, as shown, is markedly favored. However, the enzyme in rat liver will catalyze the reverse reaction. Sulfide can also be utilized in a reaction catalyzed by a yeast enzyme, *serine sulfhydrase.*

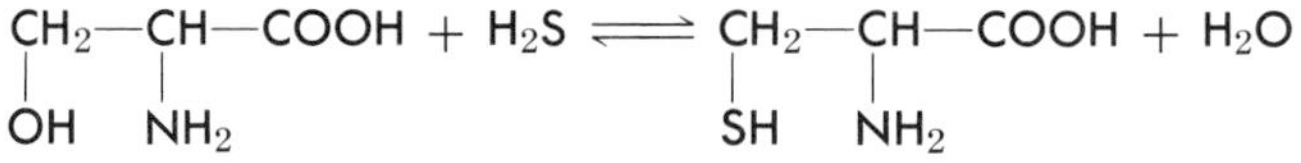

L-Serine L-Cysteine

Reduction of sulfite to thiosulfate permits reaction of the latter with serine in a reaction catalyzed by an enzyme of *Aspergillus nidulans* that requires pyridoxal phosphate, ATP, and Mg^{++}.

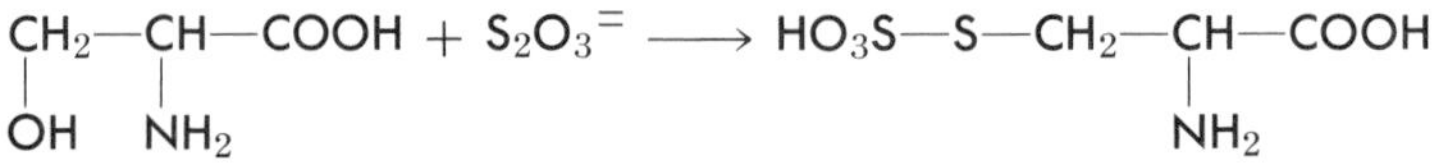

S-Sulfocysteine

S-Sulfocysteine is converted to cysteine by an unknown reaction. The pathways for sulfur fixation suggested above are summarized in Fig. 24.1.

Methionine synthesis from cysteine is accomplished by first transferring the

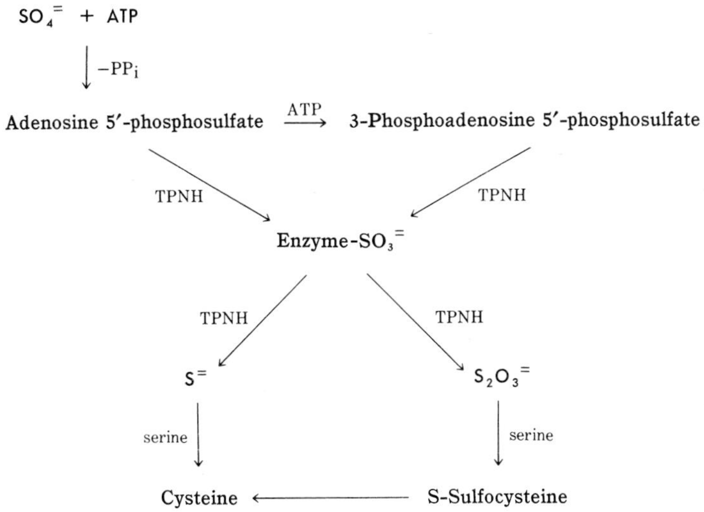

Fig. 24.1. Suggested pathways for fixation of sulfur into cysteine by plants and microorganisms.

sulfur of cysteine to a 4-carbon precursor, *homoserine,* which serves in a number of synthetic pathways and is formed as follows.

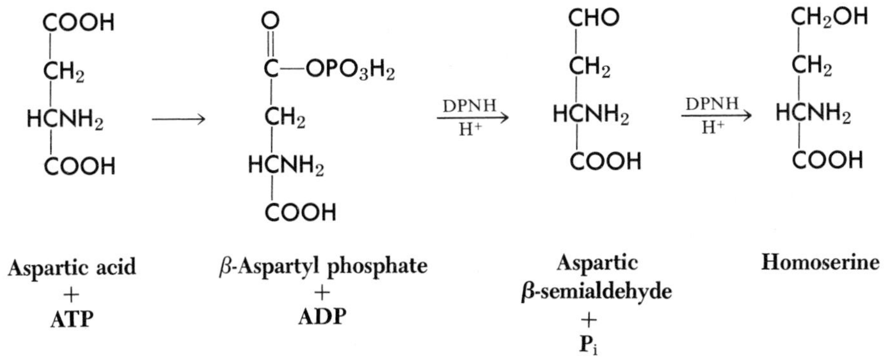

The mode of formation of aspartic β-semialdehyde resembles the reverse of the triose phosphate dehydrogenase reaction (page 398). Conversion of aspartic β-semialdehyde to homoserine is catalyzed by *homoserine dehydrogenase.* Homoserine is an intermediate not only in methionine synthesis but also of threonine (page 566), which in turn is a precursor of isoleucine (page 571). Thus homoserine dehydrogenase is one of the focal points in control of the synthesis of these amino acids; its activity is regulated through several different mechanisms. Feedback inhibition of homoserine dehydrogenase activity by threonine appears to be a common feature of these mechanisms (Table 23.1, page 539).

Homoserine is converted to O-succinylhomoserine by an acyl transfer involving succinyl CoA and catalyzed by *homoserine succinylase.* The enzyme is subject to feedback inhibition by methionine. Reaction of O-succinylhomoserine with cysteine is catalyzed by cystathionine γ-synthetase (page 547) to form cystathionine, a mixed thioether. The latter, under the catalytic influence of a *transsulfurase* (also termed cystathionine β-elimination enzyme), yields homocysteine, pyruvic acid, and NH_3. In *Neurospora,* N-acetylhomoserine may function in the synthetic pathway in lieu of O-succinylhomoserine.

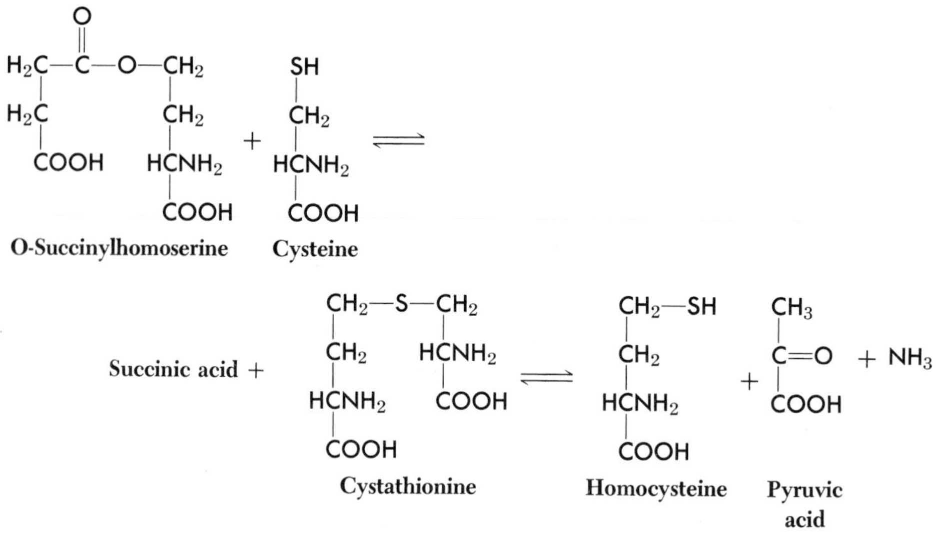

In *E. coli,* homocysteine acts as acceptor of a methyl group to complete the synthesis of methionine by either of two alternative pathways. One requires cobalamine (page 553 and Chap. 49) and will utilize N^5-methyltetrahydropteroyl mono- or triglutamate. The other pathway has no cobalamine requirement and involves transfer of the methyl group from the triglutamate form of N^5-methyltetrahydrofolic acid. *Homocysteine methyltransferases* from plant, microbial, and animal sources catalyze methyl group transfer to homocysteine from S-methyl-L-methionine, as well as from S-adenosylmethionine, but their metabolic significance is obscure.

Animals cannot synthesize homoserine from aspartic acid and, hence, are dependent on dietary provision of methionine. On the other hand, animals can form homocysteine from such ingested methionine and, by a reversal of the transsulfurase and cystathionase reactions, accomplish the synthesis of cysteine (page 547). This is reflected in the growth of rats on synthetic rations; if sufficient cysteine (cystine) is included, the dietary requirement for methionine is less than half that required in the absence of cystine. This methionine-sparing action of cystine is a result of suppression by the latter of hepatic synthesis of cystathionine γ-synthetase. This results in decreased channeling of homocysteine into formation of cystathionine and a greater fraction of homocysteine is available for methylation to methionine.

It is evident from the foregoing that bacteria use cysteine for the formation of methionine via homocysteine. Mammals, on the other hand, use methionine via homocysteine for the synthesis of cysteine (page 546).

Threonine. Threonine synthesis, like that of methionine (page 565), involves formation of homoserine from aspartic acid. Phosphorylation of homoserine by ATP yields O-phosphohomoserine in a reaction catalyzed by *homoserine kinase.*

Threonine synthetase, which requires pyridoxal phosphate, catalyzes conversion of O-phosphohomoserine to threonine. The reaction results in conversion of a compound with a γ-hydroxyl group into one with a β-hydroxyl group. This is depicted in Fig. 24.2.

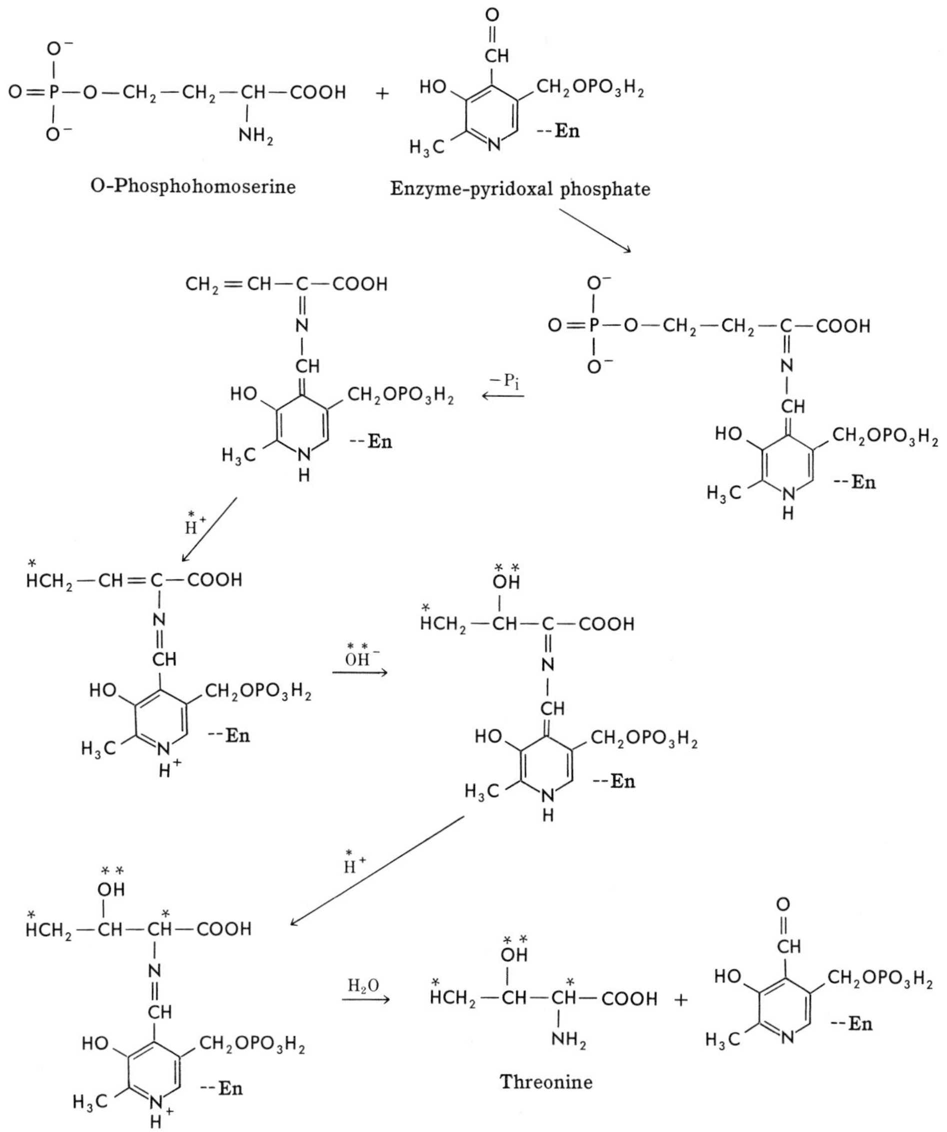

Fig. 24.2. Mechanism of reaction catalyzed by threonine synthetase. The substrate O-phosphohomoserine is bound as a Schiff base to the pyridoxal phosphate-enzyme, and the α hydrogen is labilized. Elimination of the phosphate occurs nonhydrolytically with cleavage of the C—O bond. The β-hydrogen atom adjacent to conjugated double bonds is labilized and eliminated as a proton along with the phosphate group. A proton from the medium adds to the γ position; addition of water to the α, β double bond introduces a second proton from the medium in the α position and a solvent oxygen in the β-hydroxyl position. Protons and oxygen added from the medium are designated with an asterisk.

Threonine exerts a feedback effect on homoserine synthesis in *E. coli* by inhibiting aspartyl kinase (Table 23.1) as well as by repressing synthesis of this enzyme. Multivalent inhibition of this enzyme occurs by the combined action of threonine and lysine. Under these conditions the synthesis of methionine is also blocked; addition of methionine can counteract the inhibitory effect of threonine and lysine. These are interesting examples of diverse controls of aspartyl kinase, which occupies a focal point in the synthesis of methionine (page 565) and isoleucine (page 571), as well as threonine.

Escherichia coli, liver, and kidney contain *threonine aldolase,* which utilizes pyridoxal phosphate and catalyzes the following reversible reaction.

$$\underset{\textbf{Acetaldehyde}}{CH_3{-}CHO} + \underset{\textbf{Glycine}}{\underset{NH_2}{\underset{|}{CH_2}}{-}COOH} \rightleftharpoons \underset{\textbf{Threonine}}{CH_3{-}\underset{OH}{\underset{|}{CH}}{-}\underset{NH_2}{\underset{|}{CH}}{-}COOH}$$

Since in animals acetaldehyde is available only by oxidation of ingested ethanol, it is likely that the above reaction represents a degradative process rather than a synthetic mechanism.

Lysine. Lysine synthesis occurs by two different pathways. One, termed the diaminopimelic acid pathway, is the route of lysine synthesis in bacteria, certain lower fungi, algae, and higher plants. The other, termed the aminoadipic acid pathway, occurs in other classes of lower fungi, in higher fungi, and in *Euglena.*

In the diaminopimelic acid pathway, the carbon chain of lysine is synthesized from pyruvic acid and aspartic acid. Aspartic β-semialdehyde, formed from aspartate (page 565), reacts with pyruvate in an aldol condensation catalyzed by a *condensing* enzyme, with loss of two molecules of water to form 2,3-dihydrodipicolinic acid. The latter is converted to Δ^1-piperideine-2,6-dicarboxylic acid by a TPNH-requiring *reductase;* on hydrolysis, α-amino-ϵ-ketopimelic acid is formed. In *E. coli,* succinylation with succinyl CoA, followed by transamination and deacylation, yields L,L-α,ϵ-diaminopimelic acid. The latter, by the action of a *racemase,* is converted to *meso*-diaminopimelic acid, which is decarboxylated to lysine by a pyridoxal phosphate requiring *decarboxylase* that is specific for the *meso isomer*. This synthetic pathway is shown in Fig. 24.3. In *Bacillus megaterium,* acetyl is used as the blocking group in lieu of succinyl in the diaminopimelic acid pathway.

As indicated previously, aspartic β-semialdehyde is also an intermediate in the synthesis of threonine and isoleucine. Of interest is the fact that the condensation of the semialdehyde with pyruvate, the "branching point" toward lysine synthesis, is inhibited by lysine. This feedback thus affects the first reaction in the sequence that leads only to lysine.

The aminoadipic acid pathway of lysine synthesis utilizes acetyl CoA and α-ketoglutarate as the source of the carbon chain. These two precursors condense to form homocitric acid. The latter, in reactions analogous to those of the citric acid cycle (page 325), yields *cis*-homoaconitic acid, homoisocitric acid, oxaloglutaric acid, and α-ketoadipic acid. Transamination yields α-aminoadipic acid, which, on reduction by a TPNH-requiring enzyme, forms α-aminoadipic δ-semialdehyde. Condensation of the latter with glutamate produces ϵ-N(L-glutaryl-2)-L-lysine,

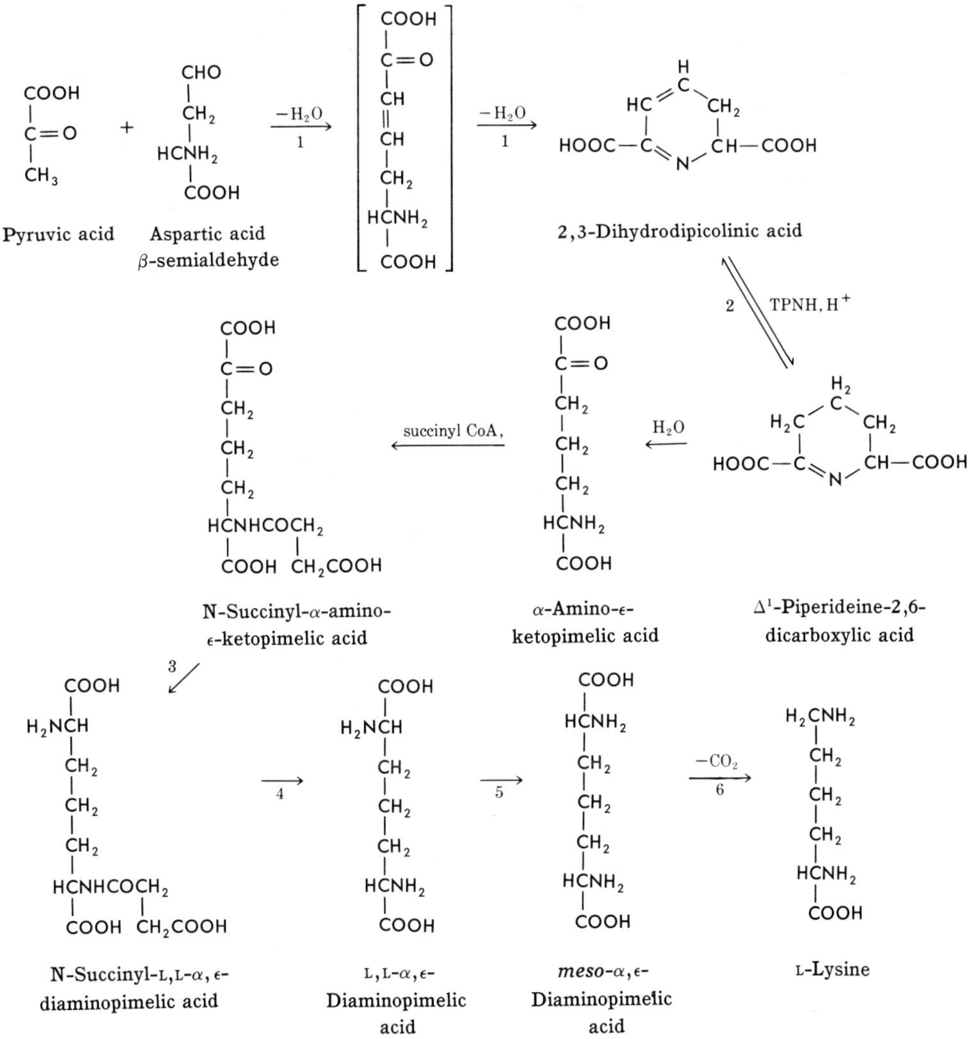

Fig. 24.3. Diaminopimelic acid pathway of lysine synthesis. Numbers at arrows indicate enzymes that have been suggested or delineated as catalysts in the pathway: 1. "condensing enzyme"; 2. 2,3-dihydropicolinic acid reductase; 3. N-succinyldiaminopimelate-glutamate transaminase; 4. N-succinyldiaminopimelate deacylase; 5. epimerase; 6. diaminopimelic acid decarboxylase.

termed saccharopine because of its recognition as an intermediate of lysine synthesis in *Saccharomyces.* Reductive and hydrolytic reactions convert saccharopine to lysine and α-ketoglutaric acid. This synthetic pathway is shown in Fig. 24.4.

Valine and Isoleucine. The synthesis of valine and isoleucine involves a reaction rare in biological systems, a pinacol rearrangement. The sequence of reactions resulting in formation of these two amino acids is similar. Indeed, four of the enzymes are common to both pathways and are probably present in the cell as a multienzyme complex bound to the particulate portion of the cell, similar, for example, to the α-keto acid dehydrogenase complex (page 331). Synthesis of valine and isoleucine has been achieved from pyruvate and α-ketobutyrate in pellet fractions obtained from *Neurospora* homogenates.

Valine. Synthesis of valine commences with condensation of the acetaldehyde moiety of a pyruvic acid molecule with a second molecule of pyruvic acid. Since thiamine pyrophosphate (ThPP) and Mg^{++} are required, it is presumed that α-hydroxyethyl thiamine pyrophosphate (page 328) serves as "active acetaldehyde"

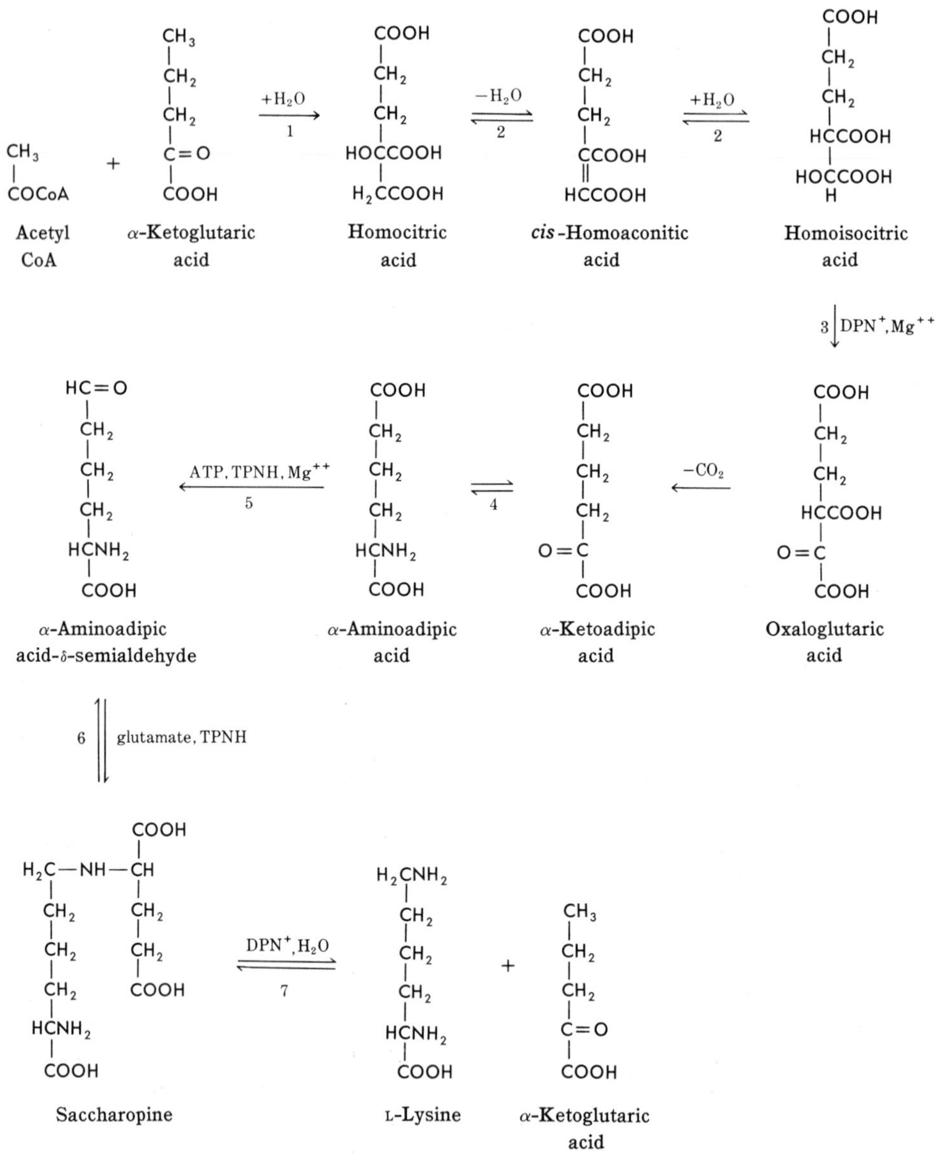

Fig. 24.4. Aminoadipic acid pathway of lysine synthesis. Numbers at arrows indicate enzymes that have been indicated or delineated as catalysts in the pathway: 1. "condensing enzyme"; 2. *cis*-homoaconitase; 3. homoisocitrate dehydrogenase; 4. α-aminoadipate-glutamate transaminase; 5. α-aminoadipic acid reductase; 6. aminoadipic semialdehyde-glutamate reductase; 7. saccharopine dehydrogenase.

in this instance. The condensation product is α-acetolactic acid. *Dihydroxy acid reductoisomerase,* requiring Mg^{++} and utilizing DPNH, catalyzes transformation to α,β-dihydroxyvaleric acid. A dehydration reaction follows, with formation of α-ketoisovaleric acid, which transaminates with glutamic acid to yield valine (Fig. 24.5). In *A. aerogenes,* which normally converts pyruvic acid to acetoin, there are two independent enzymes which catalyze the condensation to α-acetolactic acid. When grown in the presence of valine, formation of one of these is repressed, leaving enough activity to produce sufficient acetolactic acid to meet the demands for growth without wasteful formation of valine.

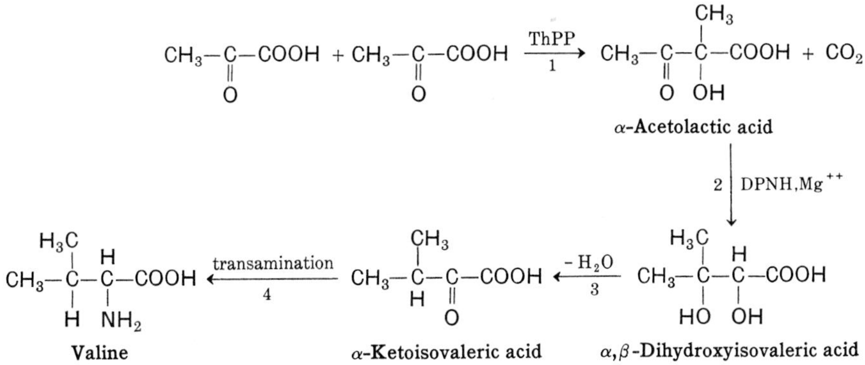

Fig. 24.5. Pathway of valine synthesis. Enzymes catalyzing reactions are indicated by numbers adjacent to arrows: 1. acetolactate synthetase; 2. dihydroxyacid reductoisomerase; 3. dihydroxyacid dehydrase; 4. branched-chain amino acid transaminase. *ThPP* = thiamine pyrophosphate.

Isoleucine. Synthesis of isoleucine occurs in a manner analogous to that of valine. The initial step is condensation of "active acetaldehyde" with α-ketobutyric acid; the latter can be derived from threonine. The *threonine deaminase* reaction, which requires pyridoxal phosphate, is completely analogous to the serine dehydrase (page 597) and cysteine desulfhydrase (page 598) reactions and produces α-ketobutyric acid in a single step. α-Ketobutyric acid may also be formed in *E. coli* from β-methylaspartic acid, which in turn can be obtained from glutamic acid in a vitamin B_{12}-requiring reaction (Chap. 49). β-Methylaspartic acid transaminates to form methyloxaloacetic acid, which on decarboxylation yields α-ketobutyric acid. The reaction sequence leading to isoleucine is shown in Fig. 24.6.

Leucine. Synthesis of leucine begins by utilizing α-ketoisovaleric acid, which is also the immediate precursor of valine. Condensation occurs with acetyl CoA, in a manner reminiscent of the formation of citric acid. Subsequent steps, leading to formation of α-ketoisocaproic acid, are analogous to the formation of α-ketoglutaric acid from citric acid. Transamination completes the reaction sequence (Fig. 24.7).

The pathways of synthesis of the branched-chain amino acids, valine, isoleucine, and leucine, are subject to a number of regulatory factors. Mention was made previously (page 568) of the factors affecting the rate of synthesis of threonine, a precursor of isoleucine. Threonine dehydrase, regulating the first step of isoleucine synthesis, is specifically inhibited by isoleucine. As indicated previously above, production of acetolactic acid in the valine synthetic pathway is inhibited by valine. In *Neurospora,* the first enzyme of the pathway of leucine synthesis, α-isopropylmal-

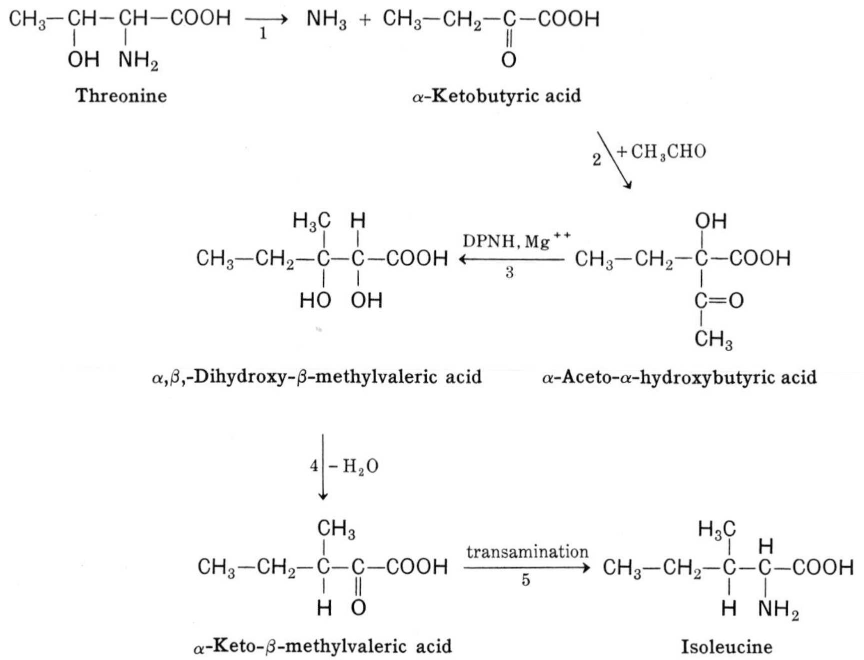

Fig. 24.6. Pathway of isoleucine synthesis. Enzymes catalyzing reactions are indicated by numbers adjacent to arrows: 1. threonine dehydrase; 2, 3, 4, and 5 are the same enzymes as 1, 2, 3, and 4, respectively, in Fig. 24.5.

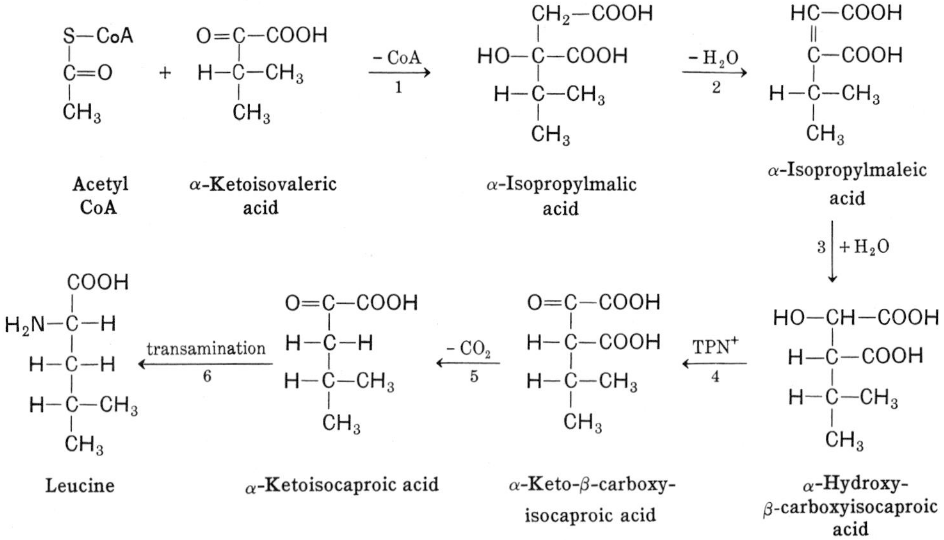

Fig. 24.7. Pathway of leucine synthesis. Enzymes catalyzing reactions are indicated by numbers adjacent to arrows: 1. α-isopropylmalate synthetase; 2. α-isopropylmalate dehydrase; 3. α-isopropylmaleate hydrase; 4. α-hydroxy-β-carboxyisocaproate dehydrogenase; 5. decarboxylase; 6. branched-chain amino acid transaminase.

ate synthetase, is repressed by leucine; the synthesis of subsequent enzymes of this pathway is induced by the product of the first reaction, and the synthesis of the inducer is controlled by feedback inhibition.

Phenylalanine, Tyrosine, and Tryptophan. These three amino acids have a common pathway of synthesis with respect to the aromatic ring. The latter is derived from the four carbon atoms of erythrose 4-phosphate; the other two carbon atoms, as well as those of the side chains of phenylalanine and tyrosine, are provided by phosphopyruvic acid. Studies with labeled precursors, and examination of the compounds which accumulate in the media of mutants blocked at various steps of the synthetic pathway, have elucidated the pathways depicted in Fig. 24.8. As indicated in the figure, chorismic acid (*Gk.*, to branch) is a branch point in the synthesis of the aromatic amino acids; conversion to anthranilic acid leads to tryptophan synthesis, while transformation to prephenic acid provides a precursor of phenylalanine and tyrosine. The latter, as indicated previously (page 548), arises in mammals by hydroxylation of phenylalanine; this reaction also occurs in certain microorganisms.

Two enzymes catalyzing formation of the first product of the above synthetic pathway, 3-deoxy-D-*arabino*-heptulosonic acid 7-phosphate, have been separated from extracts of *E. coli.* One of these is inhibited by phenylalanine and by tryptophan, the other by tyrosine, which also represses synthesis of the tyrosine-sensitive enzyme. In *Bacillus subtilis, 3-deoxy-*D*-arabino-heptulosonic acid 7-phosphate synthetase* exhibits properties of an allosteric system which is inhibited by prephenic acid and by chorismic acid, providing another feedback control mechanism. Thus this first step in the multibranched pathways for synthesis of the aromatic amino acids is subject to multiple controls.

Tryptophan. The pathway of tryptophan synthesis in higher plants is unknown. In *E. coli,* tryptophan synthesis (Fig. 24.9) begins with anthranilic acid, which is formed in the sequence of reactions leading to the other aromatic amino acids (Fig. 24.8). Two of the enzymes involved, *indoleglycerol phosphate synthetase* and *tryptophan synthetase,* have been obtained in crystalline form. The former, molecular weight 45,000, is a single polypeptide chain. Tryptophan synthetase, a pyridoxal phosphate–dependent enzyme, which catalyzes the final reactions in the sequence leading to the synthesis of tryptophan, has a molecular weight of 135,000. The reactions catalyzed by this enzyme are the following.

(*a*) **Indoleglycerol phosphate ⇌ indole + glyceraldehyde 3-phosphate**

(*b*) **Indole + L-serine** $\xrightarrow{\text{pyridoxal phosphate}}$ **L-tryptophan +** $\mathbf{H_2O}$

(*c*) **Indoleglycerol phosphate + L-serine ⇌ L-tryptophan + glyceraldehyde 3-phosphate**

It is probable that formation of tryptophan from indoleglycerol phosphate [reaction (*c*)] occurs via reactions (*a*) and (*b*) with indole as an enzyme-bound intermediate.

Studies of genetic alterations of the structure and accompanying functional characteristics of *tryptophan synthetase* have provided an important insight into genetic mechanisms (Chap. 30). Tryptophan synthetase is composed of two nonidentical and easily separated subunits, α and β (formerly termed A and B), both of which are single polypeptide chains. α is obtained as the monomer; β is obtained

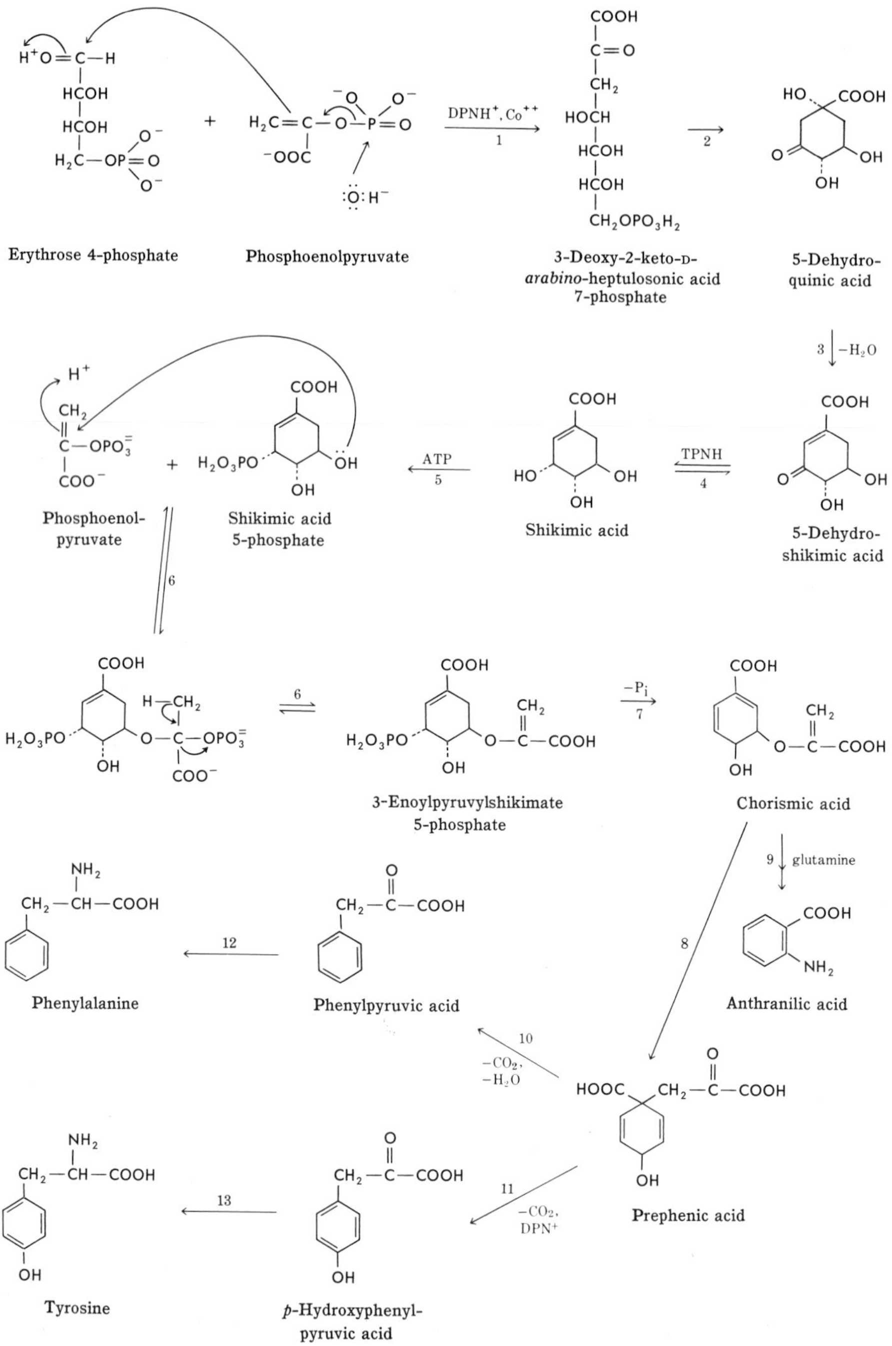

Fig. 24.8. Pathways of synthesis of phenylalanine and tyrosine, and of anthranilic acid, a precursor in tryptophan synthesis (page 575). Enzymes catalyzing reactions are indicated by numbers adjacent to arrows: 1. 3-deoxy-D-*arabino* heptulosonic acid 7-phosphate synthetase; 2. dehydroquinic acid synthetase; 3. dehydroquinic acid dehydrase; 4. shikimic acid dehydrogenase; 5. shikimate kinase; 6. 3-enolpyruvylshikimate 5-phosphate synthetase; 7. chorismic acid synthetase; 8. chorismate mutase; 9. anthranilate synthetase; 10. prephenic acid dehydratase; 11; prephenic acid dehydrogenase; 12. phenylalanine transaminase; 13. tyrosine transaminase.

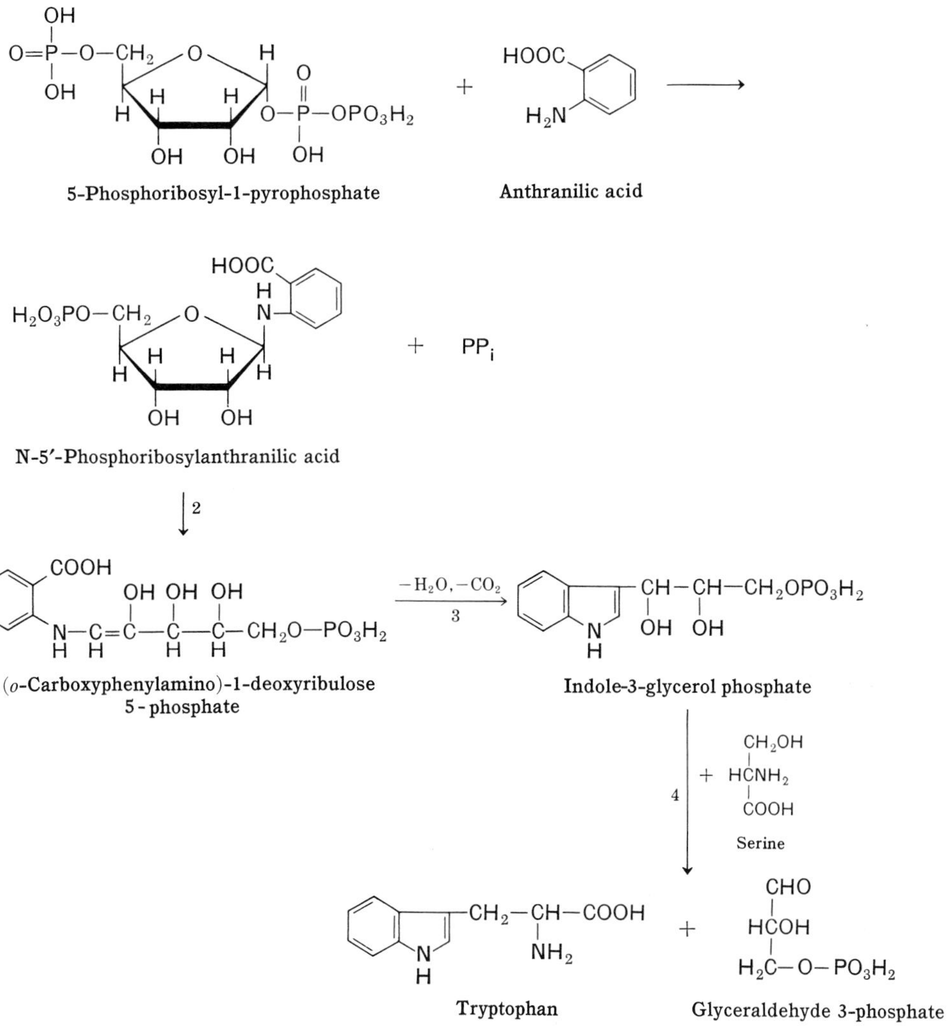

Fig. 24.9. Pathway of tryptophan synthesis in *Escherichia coli.* Enzymes catalyzing reactions are indicated by numbers adjacent to arrows: 1. anthranilate-phosphoribosylpyrophosphate transferase; 2. N-5′-phosphoribosylanthranilate isomerase; 3. indole 3-glycerol phosphate synthetase; 4. tryptophan synthetase.

as β_2. The fully associated protein is $\alpha_2\beta_2$. Reaction (*a*) can be catalyzed by α alone but only at a fraction of the rate of the associated complex; reaction (*b*) can be catalyzed by β_2 alone. Mutant α subunits will activate normal β_2 units for reaction (*b*), and most mutant β_2 units will activate normal α units for reaction (*a*). Enzyme complexes formed with normal α subunits and mutant β_2 subunits, or complexes formed with mutant α subunits and normal β_2 units, will not catalyze reaction (*c*).

The activity of anthranilate synthetase from *E. coli* (Fig. 24.8) is inhibited by tryptophan; in yeast, the amino acid both inhibits and represses the enzyme. As

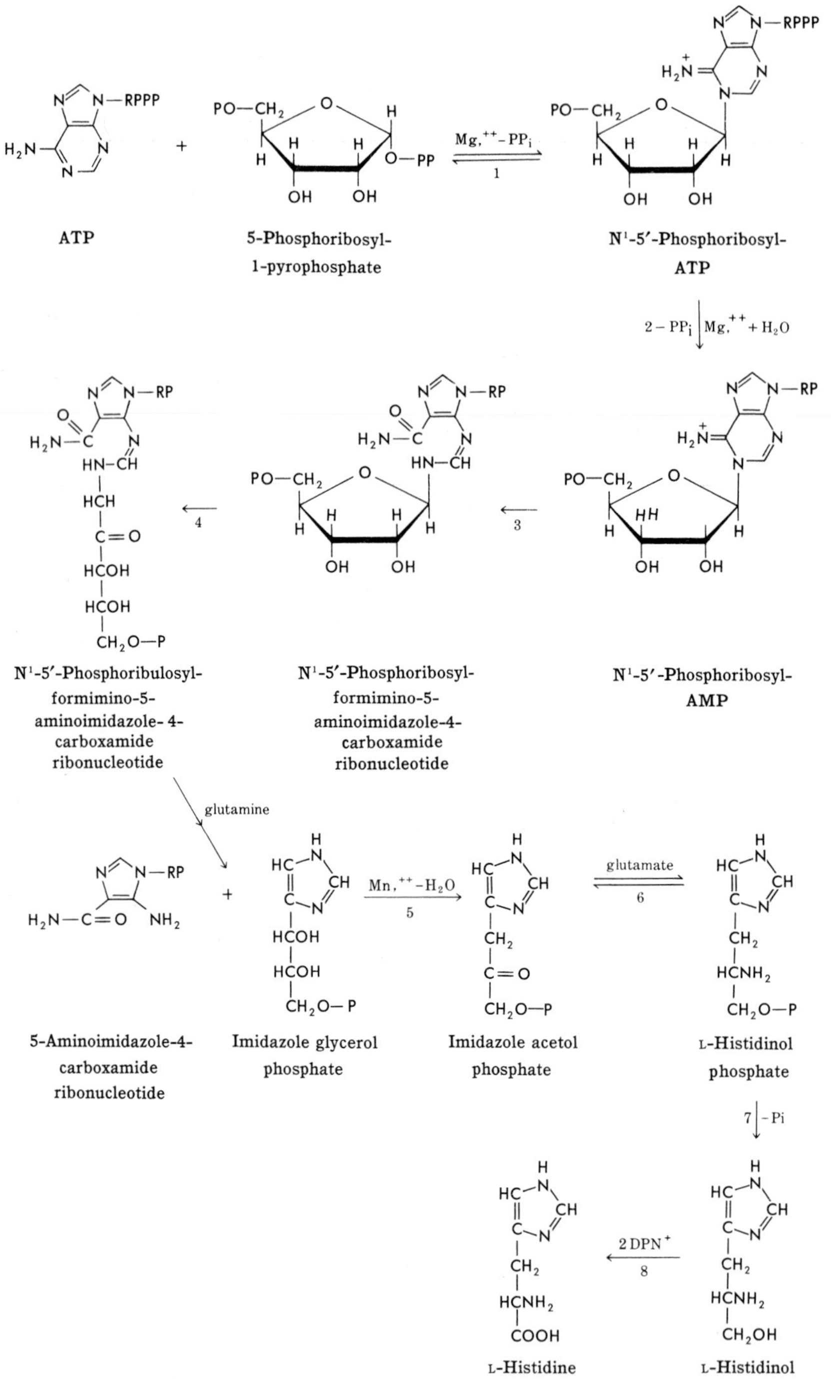

Fig. 24.10. Pathway of histidine synthesis. *RPPP* = ribose triphosphate; *RP* = ribose 5-phosphate; *PP* = pyrophosphate; *P* = orthophosphate. Enzymes catalyzing reactions are indicated by numbers adjacent to arrows: 1. phosphoribosylpyrophosphate-ATP phosphorylase; 2. pyrophosphohydrolase; 3. cyclohydrolase; 4. isomerase; 5. imidazoleglycerol phosphate dehydrogenase; 6. L-histidinol phosphate-glutamate transaminase; 7. histidinol phosphate phosphatase; 8. histidinol dehydrogenase.

mentioned previously (page 573), tryptophan also represses the enzyme in *E. coli* which catalyzes the initial step in synthesis of the aromatic amino acids.

Histidine. The pathway for synthesis of histidine, largely established by studies of mutants of *E. coli* and, particularly, *Salmonella,* has revealed a complex and remarkable series of reactions (Fig. 24.10) with a number of metabolic controls. All 10 steps (Fig. 24.10) have been identified with one of the nine genes of the histidine operon (Chap. 29). Only one intermediate, the immediate precursor of imidazoleglycerol phosphate, remains to be characterized.

The first step in histidine synthesis is inhibited specifically by histidine, which regulates the activity of *phosphoribosylpyrophosphate-ATP phosphorylase.* In the over-all synthesis of histidine, the carbon chain of ribose is affixed to the amino group of adenylic acid. The purine ring of the latter is opened, and, after provision of a nitrogen atom from glutamine, the structure breaks apart to yield imidazoleglycerol phosphate, in which the imidazole ring of histidine is fully formed and attached to a 3-carbon chain, and 5-aminoimidazole-4-carboxamide ribonucleotide, an intermediate in purine synthesis (page 623).

In the synthesis of imidazoleglycerol phosphate, the side chain and connecting two carbons of the ring derive from the five carbons of the ribose of 5-phosphoribosyl 1-pyrophosphate. The —N=C— adjoining stems from the pyrimidine portion of the fused purine nucleus. Since the carbon atom of this fragment originates, during purine synthesis, from the formyl of N^{10}-formyltetrahydrofolic acid (page 551), it derives from the β-carbon of serine or other sources of C_1 units. The final N atom is provided by the amide nitrogen of glutamine. Thus, the histidine-synthesizing system utilizes a portion of an existing purine nucleus but leaves behind a fragment (aminoimidazole carboxamide ribonucleotide), which is reconverted to purines (page 623). In the last step in the reaction sequence, a primary hydroxyl group of histidinol is oxidized by two equivalents of DPN^+ to the corresponding carboxyl group. Consecutive oxidations occur on the surface of a single enzyme without appearance in the free state of the presumed aldehydic intermediate.

REFERENCES

See list following Chap. 26.

25. Amino Acid Metabolism. III

Synthesis of Peptides and Amides. Transamidination. Transmethylation. Porphyrin Synthesis. Amino Acid Decarboxylation. Polyamine Synthesis

The prime physiological role of amino acids is in protein synthesis. However, amino acids are required by cells as starting materials for synthesis of diverse chemical compounds which are utilized for other roles in metabolism. Several such functions have already been encountered, *e.g.*, glutamic acid as a contributor of an amino group by transamination, glutamine as an ammonia donor, methionine as a source of methyl groups, and serine as a source of biologically active $—CH_2OH$ and —CHO groups. Aspartic acid also donates its amino group, as in the synthesis of the guanidine moiety of arginine, and contributes to purine synthesis (page 624). The special roles of glycine in purine synthesis and of aspartic acid in pyrimidine synthesis will be found in Chap. 27. The present chapter will present a résumé of other instances in which amino acids are employed for the synthesis of diverse nitrogenous compounds.

SYNTHESIS OF PEPTIDES AND AMIDES

All living cells can synthesize amides or peptides of amino acids. In all cases the energy required for amide bond formation is derived from ATP. However, this is utilized in a variety of ways. As described earlier (page 536), in glutamine synthesis, there is first formed an enzyme-bound acyl phosphate,

$$\textbf{ATP + RCOOH + En} \longrightarrow \textbf{En—R—CO—OPO}_3\textbf{H}_2\textbf{ + ADP}$$

which reacts with ammonia to form the amide. A similar mechanism is employed in the synthesis of the most abundant of tissue peptides, *glutathione* (γ-glutamylcysteinylglycine). This synthesis proceeds in several consecutive steps; the mechanism of the first step is unknown, but the over-all reaction is the following.

$$(a)\quad \textbf{L-Glutamic acid + L-cysteine + ATP} \xrightarrow{Mg^{++},K^{+}} \textbf{L-}\gamma\textbf{-glutamylcysteine + ADP + P}_i$$

The requirement for both Mg^{++} and K^{+} in this reaction is noteworthy. In the second reaction, an acyl phosphate is generated and then attacked by the glycine nitrogen.

(b_1) L-Glutamylcysteine + En + ATP ⟶ En-L-glutamylcysteinylphosphate + ADP

(b_2) En-L-Glutamylcysteinylphosphate + glycine ⟶ En + L-glutamylcysteinylglycine + P_i

Neither of the two enzymes [reactions (*a*) and (*b*)] is absolutely specific; the first can catalyze formation of γ-glutamylalanine or γ-glutamyl α-aminobutyric acid, and the second can catalyze formation of the various ophthalmic acids (page 899).

Carnosine. The dipeptide β-alanylhistidine, *carnosine,* originally found in muscle, is formed in a reaction, catalyzed by *carnosine synthetase,* in which the activated acyl group is the ester of AMP, analogous to the formation of acyl CoA derivatives (page 326).

β-Alanine + ATP + En ⇌ En-β-alanyl-AMP + PP_i

En-β-Alanyl-AMP + histidine ⟶ β-alanylhistidine + AMP

Carnosine synthetase also catalyzes the synthesis of anserine (β-alanyl-1-methylhistidine, page 112) when histidine is replaced by 1-methylhistidine in the above reaction, and also the corresponding 3-methylhistidine derivative. The enzyme is relatively nonspecific and catalyzes dipeptide formation either when one of a variety of amino acids is substituted for histidine or when β-alanine is replaced by other β- or ω-amino acids. For example, in brain, which has a rich supply of γ-aminobutyric acid, the enzyme catalyzes synthesis of γ-aminobutyryl histidine (*homocarnosine*) in amounts ten times that of carnosine.

Amide Synthesis. Synthesis of asparagine (page 536) and glutamine (page 536) were considered previously. Exchange of the amide nitrogen of several proteins with NH_3 and various amines is catalyzed by *transglutaminase,* which has been obtained from pig liver. The enzyme catalyzes the replacement of some amide groups of protein-bound glutamine residues.

N-Acyl Amino Acids. The formation and excretion of N-acyl derivatives of amino acids have long been known. Originally obtained from horse urine, hippuric acid (N-benzoylglycine) is present in small amount in normal human urine; synthesis occurs in the liver and can be increased greatly by administration of benzoic acid. Formation of hippuric acid and related compounds resembles that of esters in that the carboxylic acid is activated as the corresponding acyl CoA, which serves as the immediate acylating agent. Benzoyl CoA formation occurs via an intermediate, enzyme-bound acyladenylate.

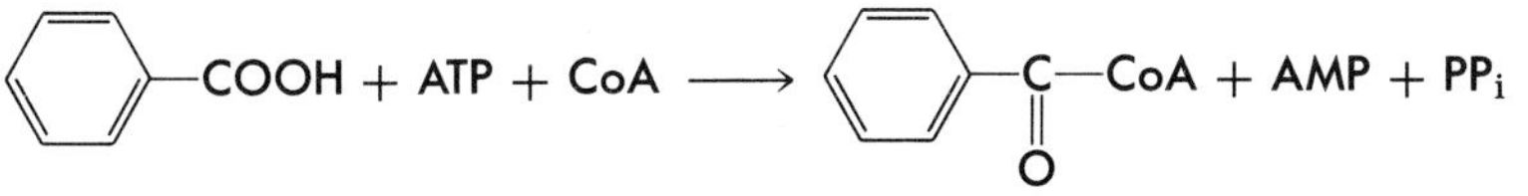

Benzoic acid Benzoyl CoA

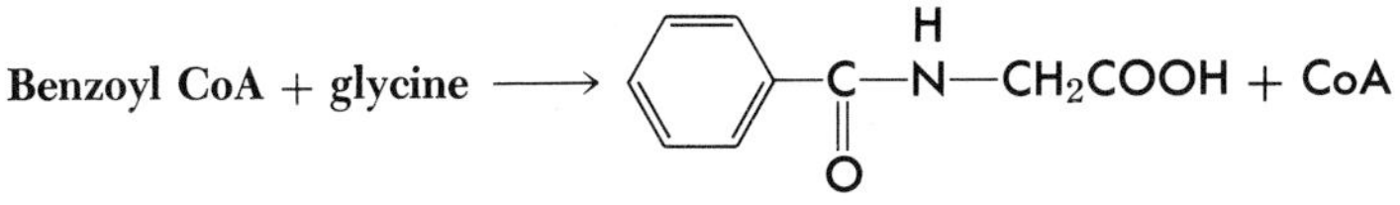

Hippuric acid

Phenylacetic acid arising in metabolism, or administered, gives rise to phenaceturic acid in the same manner.

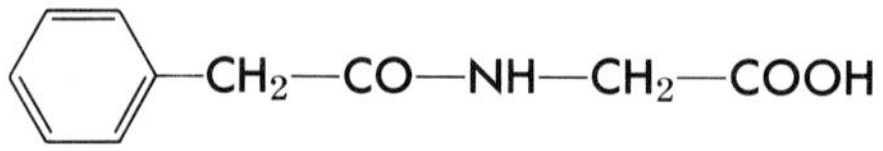

In some birds, both benzoic and phenylacetic acids are activated similarly, but then react with both the α- and δ-amino groups of ornithine to form N,N-dibenzoyl- or N,N-diphenylacetylornithine. In man, although benzoic acid is conjugated with glycine, phenylacetyl CoA reacts with the α-amino group of glutamine, yielding phenylacetylglutamine.

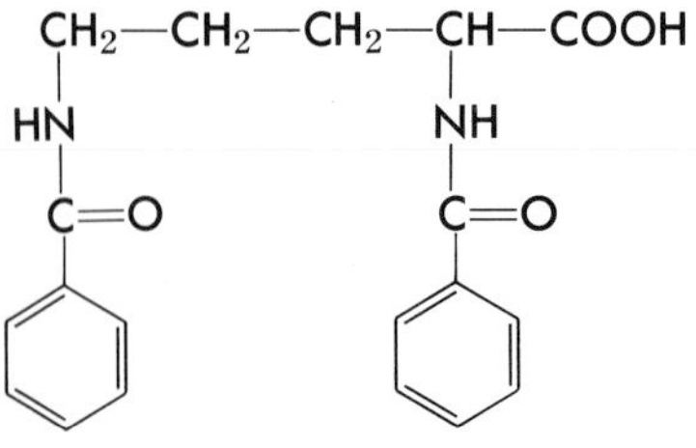

Ornithuric acid
(N,N-dibenzoylornithine)

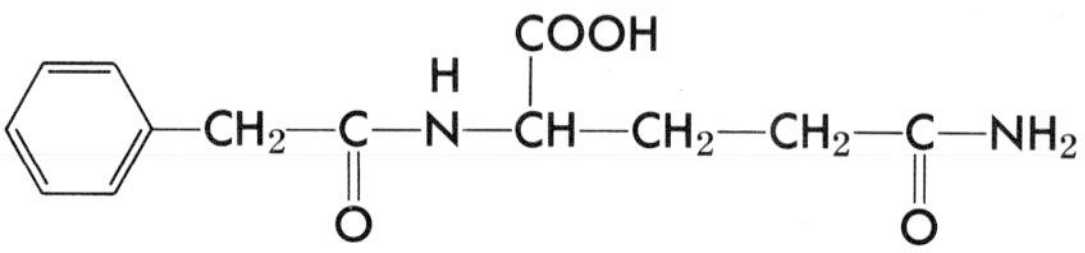

Phenylacetylglutamine

Numerous other low molecular weight aromatic acids, given to animals, form similar compounds.

Administration of excessive quantities of many α-amino acids results in excretion of the corresponding N-acetyl derivatives, probably formed by reaction with acetyl CoA. The normal physiological role of the enzymes responsible for these various syntheses is not known. A number of proteins lack a free N-terminal amino group because the N-terminal amino acid is acetylated. Available evidence indicates that acetylation occurs after synthesis of the polypeptide chain. However, N-terminal formylmethionine participates in protein synthesis (page 678).

An additional form of acylated amino acids is the group of succinylated derivatives, *e.g.*, N-succinyl-ϵ-keto-L-α-aminopimelic acid, an intermediate in lysine synthesis (page 568) and O-succinylhomoserine, an intermediate in methionine synthesis (page 565). It is presumed that these are formed by utilization of succinyl CoA (page 332).

Mercapturic Acids. Administration of halogen-substituted aliphatic or aromatic hydrocarbons, *e.g.*, bromobenzene, to animals leads to urinary excretion of mercapturic acids. In the case of bromobenzene, the product is the following.

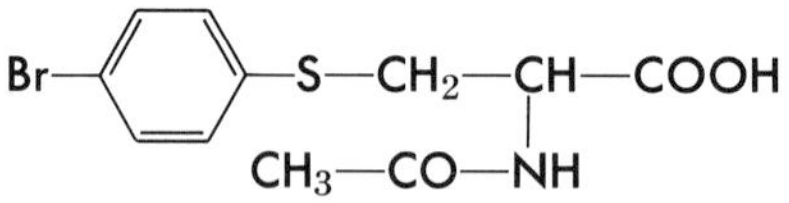

***p*-Bromophenylmercapturic acid**

Mercapturic acid formation begins in the liver with reaction of the halogenated hydrocarbon with glutathione to make the corresponding mixed thioether. The mechanism of this reaction is unknown. The glutamic acid and glycine residues are then removed by hydrolysis in the kidney, and the S-bromophenylcysteine is acetylated in the liver, utilizing acetyl CoA. Similar events make possible mercapturic acid formation from many other halohydrocarbons.

TRANSAMIDINATION

It has long been recognized that man excretes daily in the urine a quantity of creatinine, an anhydride of creatine (page 582), considerably in excess of the total creatine and creatinine ingested. Hence it followed that these compounds are synthesized in the animal organism. This process commences with the readily reversible transfer of the guanidine moiety of arginine to glycine, a process called transamidination and catalyzed by *transamidinase,* which has been purified from hog kidney.

$$H_2N-\underset{\underset{H}{\overset{\|}{N}}}{\overset{H}{\underset{}{C}}}-N-CH_2-CH_2-CH_2-\underset{\overset{|}{NH_2}}{CH}-COOH + H_2N-CH_2-COOH \rightleftharpoons$$

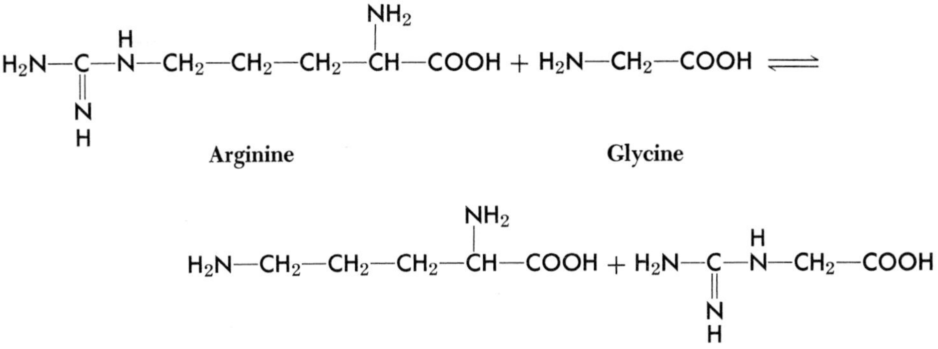

Arginine   Glycine

$$H_2N-CH_2-CH_2-CH_2-\underset{\overset{|}{NH_2}}{CH}-COOH + H_2N-\underset{\underset{H}{\overset{\|}{N}}}{\overset{H}{\underset{}{C}}}-N-CH_2-COOH$$

Ornithine   Guanidoacetic acid

Two successive reactions are involved.

(*a*)   Arginine + enzyme $\rightleftharpoons$ enzyme-amidine + ornithine
(*b*)   Enzyme-amidine + glycine $\rightleftharpoons$ guanidoacetic acid + enzyme

A stable amidine-enzyme complex was isolated after arginine, labeled with ^{14}C in the guanidine group, was incubated with transamidinase. The nature of the enzyme-amidine bond is unknown; heating the complex results in formation of urea. When glycine was added to the isolated complex, reaction (*b*) occurred. Transamidinase activity is present in human liver, pancreas, and kidney and is present in kidney but not in liver of rat, rabbit, and dog.

Guanidoacetic acid formed as described above is then methylated to form creatine (see below). This system is one of the few known instances of negative feedback regulation in mammals. Ingestion of creatine by rats markedly suppresses the level of renal transamidinase activity. Injection of creatine into the developing chicken egg also represses the level of the enzyme. That this is a physiological effect is evidenced by the fact that endogenously synthesized creatine was as effective a metabolite repressor as was exogenous creatine. The importance of this control mechanism is evident if the controlled step is bypassed by injecting guanidoacetic acid. Excessive creatine synthesis then results in a fatty liver (page 502) and impaired growth, both of which can be prevented by the administration of methionine.

Renal transamidinase can also catalyze transamidination from arginine to canaline with formation of canavanine. The last two compounds are found in the jack bean but are not known to occur in mammalian metabolism.

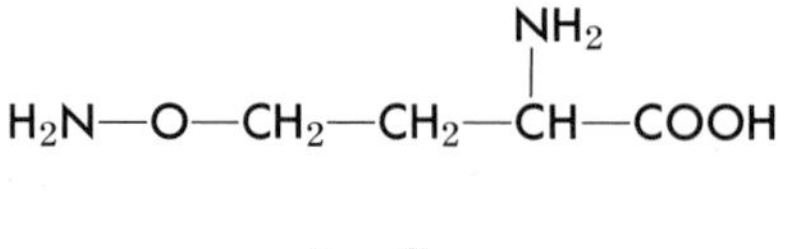

Canaline

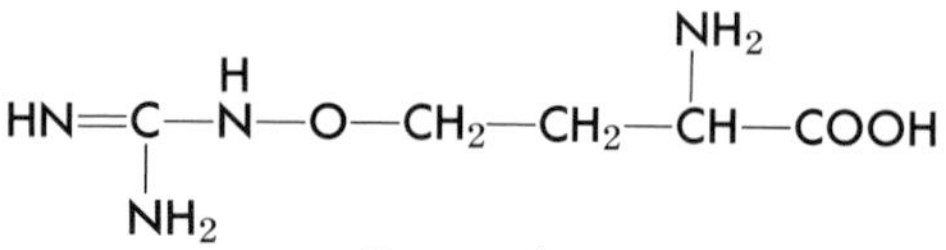

Canavanine

TRANSMETHYLATION

Transfer of a methyl group occurs in the course of many biosynthetic pathways. In most instances, the methylating agent is S-adenosylmethionine, in turn derived from methionine and ATP, as described previously (page 553). Enzymes catalyzing such methylations are designated as *methylpherases* or *methyltransferases.* Thus, creatine synthesis is completed by the action of *guanidoacetic acid methyltransferase.*

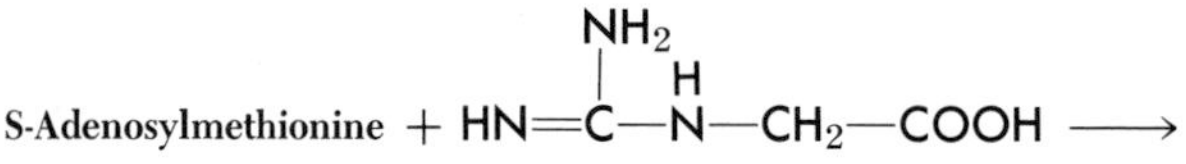

Guanidoacetic acid

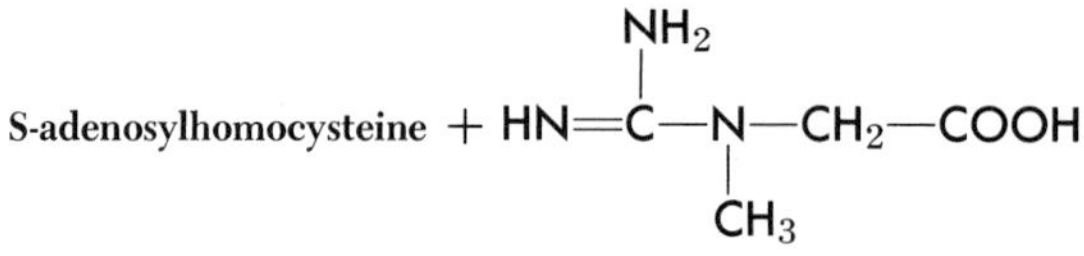

Creatine

Other methylations are cited in Table 25.1. It is evident that methyl groups can be transferred to sulfur, nitrogen, carbon, or oxygen linkage. Of interest is the requirement for three consecutive methylations to form γ-butyrobetaine.

$$H_2N{-}CH_2{-}CH_2{-}CH_2{-}COOH + 3\ \text{S-adenosylmethionine} \longrightarrow$$

γ-Aminobutyric acid

$$(CH_3)_3{-}\overset{+}{N}{-}CH_2{-}CH_2{-}CH_2{-}COOH + 3\ \text{adenosylhomocysteine}$$

γ-Butyrobetaine

Although this process has not been observed with purified enzymes, the product, γ-butyrobetaine, is hydroxylated in rat liver to form *carnitine,* which then participates in mitochondrial fatty acid transport (page 482). The requisite enzymic system requires O_2, TPNH, and Fe^{++}; hence it is a mixed-function oxidase (page 372).

$$(CH_3)_3{-}\overset{+}{N}{-}CH_2{-}CH_2{-}CH_2{-}COOH + TPNH + H^+ + O_2 \xrightarrow{Fe^{++}}$$

γ-Butyrobetaine

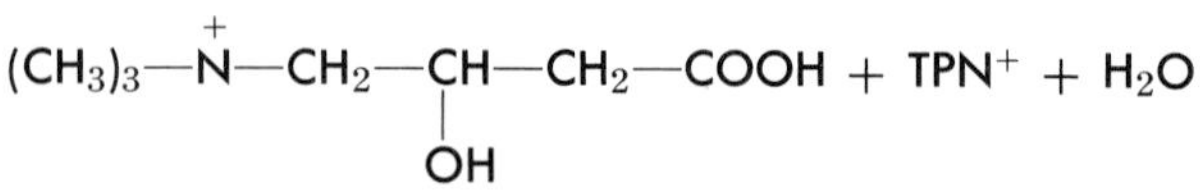

Carnitine

Table 25.1: SOME EXAMPLES OF TRANSMETHYLATION

Substrate	Product	Methylating agent	Page
Homocysteine	Methionine	N^5-Methyltetrahydrofolic acid	566
Homocysteine	Methionine	Methyl-B_{12}	566
N-Acetyl-5-hydroxytryptamine. .	Melatonin	S-Adenosylmethionine	985
γ-Aminobutyric acid	γ-Butyrobetaine	S-Adenosylmethionine	582
Carnosine.	Anserine	S-Adenosylmethionine	848
*	Ergosterol	S-Adenosylmethionine	520
*	Vitamin B_{12}	S-Adenosylmethionine	1038
Epinephrine.	Metanephrine	S-Adenosylmethionine	954
Guanidoacetic acid.	Creatine	S-Adenosylmethionine	582
Histamine.	N-Methylhistamine	S-Adenosylmethionine	588
Nicotinic acid	Trigonelline	S-Adenosylmethionine	1026
Nicotinamide.	N^1-Methylnicotinamide	S-Adenosylmethionine	1026
Norepinephrine.	Epinephrine	S-Adenosylmethionine	954
Phosphatidylethanolamine.	Lecithin	S-Adenosylmethionine	549

* The substrate which is methylated has not been identified; experimental data were obtained with yeast (ergosterol) or microorganisms (vitamin B_{12}) grown on a medium containing labeled methylating agent, followed by degradation of product for location of labeled methyl group.

A process leading to N-methylation but not involving S-adenosylmethionine has been described in a species of *Pseudomonas*. When grown on methylamine, the organism elaborated an enzyme that catalyzed the following reaction.

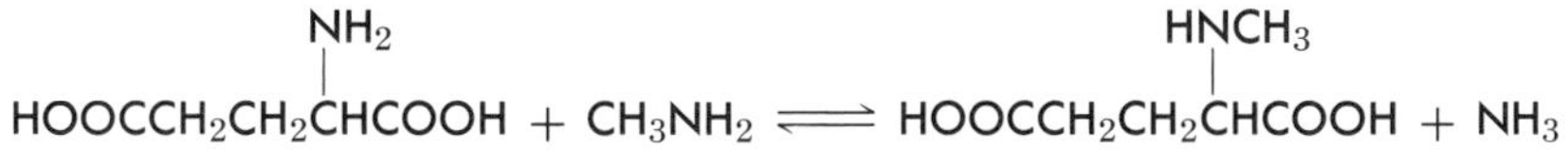

$$\underset{\text{L-Glutamic acid}}{\overset{\overset{\displaystyle NH_2}{|}}{HOOCCH_2CH_2CHCOOH}} + \underset{\text{Methylamine}}{CH_3NH_2} \rightleftharpoons \underset{\substack{\text{N-Methylglutamic}\\\text{acid}}}{\overset{\overset{\displaystyle HNCH_3}{|}}{HOOCCH_2CH_2CHCOOH}} + NH_3$$

Similar N-methylation occurred from methylamine to sarcosine, alanine, and aspartic acid.

Fate of Methyl Groups. Transmethylation from S-adenosylmethionine, a sulfonium compound, to most acceptors proceeds with a favorable change in free energy. Most compounds which serve as methyl donors in living systems also appear to be "onium" compounds. Thus, the enzymic methylation of homocysteine by betaine, with methyl group migration from a quaternary nitrogen, requires no additional energy source, such as ATP. In contrast, dimethylglycine and sarcosine cannot serve as methylating agents; their methyl groups are removed by oxidation and transfer to tetrahydrofolic acid (page 550). Similarly the thetins, which are abundant in marine algae, are effective methyl donors in such cells, whereas S-methylcysteine, found in plants such as cabbage, cannot function as a methyl donor.

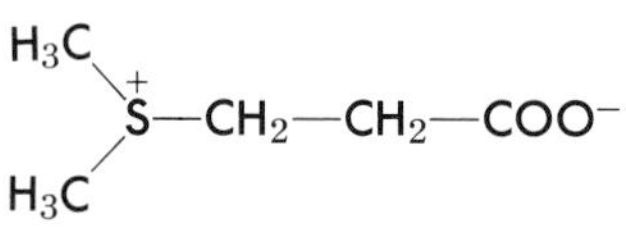

$$\underset{\textbf{Dimethylpropiothetin}}{H_3C \overset{\displaystyle +}{\underset{H_3C}{\diagdown S \diagup}}\!\!-CH_2-CH_2-COO^-} \qquad\qquad \underset{\textbf{S-Methylcysteine}}{H_3C-S-CH_2-\overset{\overset{\displaystyle NH_2}{|}}{CH}-COOH}$$

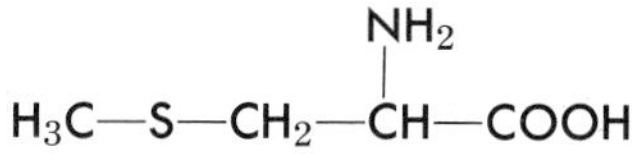

As noted earlier, choline may be oxidized to betaine (page 554) and one methyl group can then be transferred to homocysteine, when homocysteine is available. Otherwise, betaine may lose all its methyl groups, sequentially, by oxidation Each is oxidized to the level of HCHO, transferred to tetrahydrofolic acid, and thus returned to the C_1 pool in the sequence choline → betaine aldehyde → betaine → dimethylglycine → monomethylglycine (sarcosine) → glycine.

The ultimate fate of the C_1 unit in animals is not well understood. When methionine or sarcosine, labeled with ^{14}C in their methyl groups, is administered, the isotope is found in the expected positions in the various compounds already cited. However, a significant fraction of administered isotope also appears as $^{14}CO_2$, and this is the ultimate fate of this carbon, except for that lost in the urine in small amounts as creatinine, N^1-methylnicotinamide, etc. No reaction is known in animal metabolism in which there occurs direct conversion of the carbon of a C_1 derivative of tetrahydrofolic acid to carbon dioxide, yet this is the major route of C_1 metabolism. The most likely pathway by which C_1 compounds can be oxidized to CO_2 is by recombination of N^5,N^{10}-methylenetetrahydrofolic acid with glycine to form serine (page 551), which is converted to pyruvic acid under the influence of *serine dehydrase* (page 597). Oxidation of the pyruvic acid via the citric acid cycle would result in formation of CO_2 from the carbon which had once been in the C_1 pool. Some of the relationships considered above are shown in Fig. 25.1; reference should also be made to Fig. 23.4, page 552.

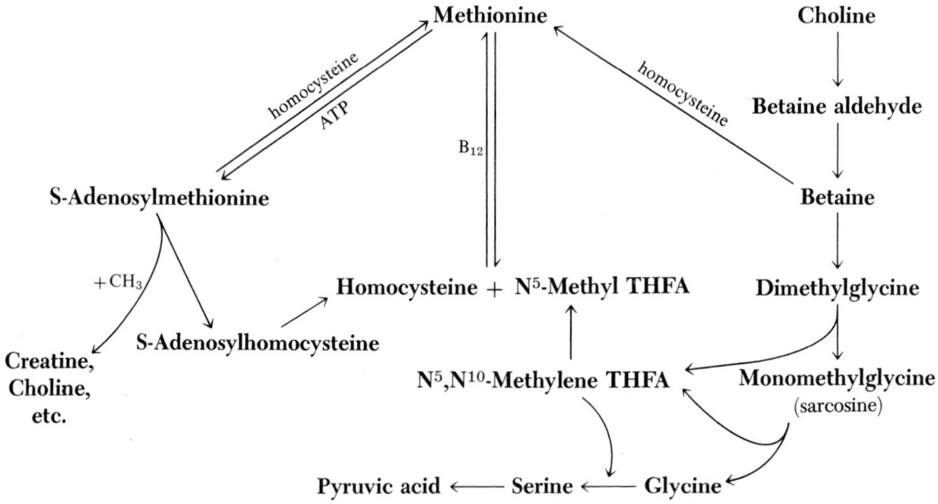

Fig. 25.1. Aspects of methyl group metabolism. THFA = tetrahydrofolic acid.

PORPHYRIN SYNTHESIS

Knowledge of the reactions by which porphyrin synthesis proceeds was originally obtained in studies in which synthesis of isotopically labeled protoporphyrin of hemoglobin was achieved either by administration of labeled glycine to experimental animals, including man, or by incubating in vitro nucleated red cells (from chicken or duck), or hemolysates of these cells, with isotopically labeled glycine.

The first step is the reaction of succinyl CoA with glycine to form α-amino-β-ketoadipic acid in a reaction catalyzed by a microsomal enzyme, *δ-aminolevulinic acid synthetase* (Fig. 25.2, reaction series *A*). The enzyme requires pyridoxal phosphate, which forms a Schiff base with the glycine. α-Amino-β-adipic acid decarboxylates on the enzyme surface to yield δ-aminolevulinic acid. Condensation of two moles of δ-aminolevulinic acid to porphobilinogen is catalyzed by *δ-aminolevulinic acid dehydrase;* the mechanism of this transformation is shown in Fig. 25.2, reaction series *B*. One molecule of δ-aminolevulinic acid forms a Schiff base with the enzyme; this is followed by a nucleophilic attack, by this intermediate anion, on the carbonyl carbon of a second molecule of δ-aminolevulinic acid. The resulting aldol loses the elements of water, and the free amino group of the second molecule of the substrate displaces the amino group of the enzyme by a "trans-Schiff" reaction, forming porphobilinogen. The reactions by which porphobilinogen is utilized for porphyrin synthesis are presented in Chapter 31. δ-Aminolevulinic acid is also the precursor of the porphyrin moiety of chlorophyll (page 457) and of the corrin ring of vitamin B_{12} (page 1038).

δ-Aminolevulinic acid dehydrase is a monomer of molecular weight 250,000 in the absence of a cation, and its kinetics have the characteristics of an allosteric enzyme. Addition of K^+, which is required for activity at low substrate concentration, promotes association of the enzyme to an equilibrium mixture of monomer, dimer, and trimer.

Since four molecules of porphobilinogen are utilized in porphyrin synthesis (page 738), all the porphyrin nitrogens are contributed by the nitrogen of glycine. Isotopic studies have established that 8 of the methylene carbon atoms of glycine are used in the synthesis of each porphyrin nucleus. Of these 8 carbon atoms, 4 become atom 2 in each pyrrole ring, *i.e.*, are in the ring adjacent to the nitrogen atom of the pyrrole nucleus, and an additional 4 are utilized for the methene bridge carbon atoms between the rings. The remaining 26 carbon atoms, not derived from glycine, have their origin in succinyl CoA.

Metabolic control of porphobilinogen synthesis, and thus of porphyrin formation, is exerted through a feedback mechanism operating at the first two steps in the pathway. Heme, as well as hemoglobin and other hemoproteins, inhibits both δ-aminolevulinic acid synthetase and δ-aminolevulinic acid dehydrase. Administration to animals of some drugs that are known to increase porphyrin excretion causes augmented synthesis of hepatic δ-aminolevulinic acid synthetase.

AMINO ACID DECARBOXYLATION

Amino acid decarboxylases are widely distributed in nature; each of these enzymes, with the exception of that catalyzing decarboxylation of histidine (page 588), requires pyridoxal phosphate as coenzyme and catalyzes removal of a carboxyl group as CO_2, generally from the α-carbon of a specific L-α-amino acid. This reaction is relatively rare in pathways of amino acid metabolism in animals but is more common in microorganisms. The reaction mechanism involves Schiff base formation on the enzymic surface between pyridoxal phosphate and the amino acid, followed by decarboxylation as a proton, from the medium, replaces the carboxyl carbon in its attachment to the α-carbon atom. Equilibrium lies far to the right, as written.

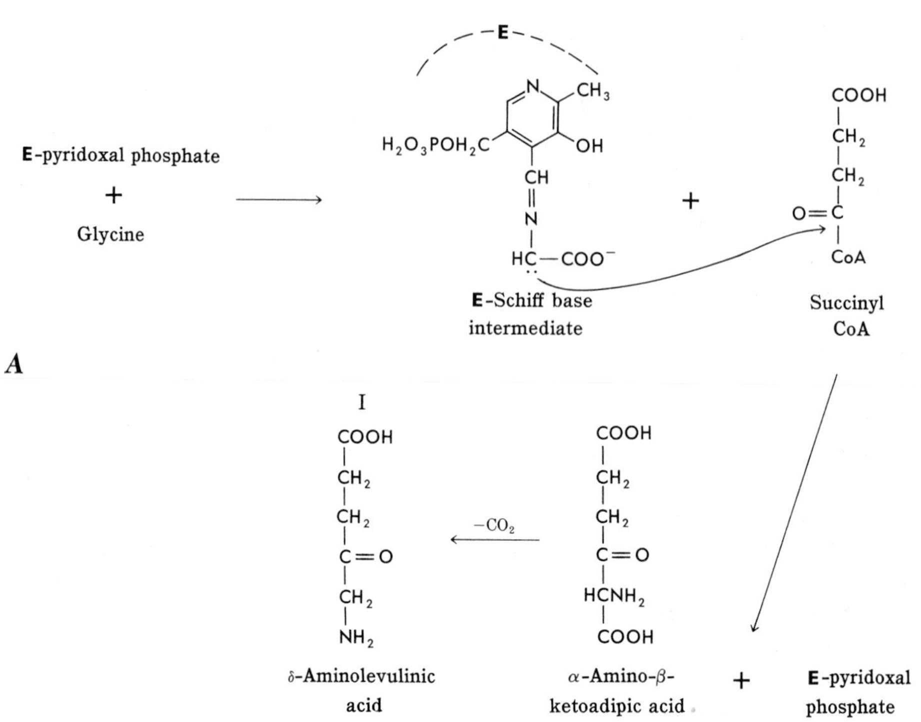

Fig. 25.2. Mechanism of synthesis of porphobilinogen. *A*, above; mechanism of formation of δ-aminolevulinic acid; E-pyridoxal phosphate represents δ-aminolevulinic acid synthetase. *B*, mechanism of condensation of two molecules of δ-aminolevulinic acid; E-NH_2 represents an amino group of δ-aminolevulinic acid dehydrase, and I and II are used to designate the two molecules of δ-aminolevulinic acid involved in the condensation reaction (see text). (*Courtesy of Dr. David Shemin.*)

$$R{-}CH_2{-}CHNH_2{-}COO^- + H^+ \longrightarrow R{-}CH_2{-}CH_2NH_2 + CO_2$$

However, it has been possible to show reversibility in at least one instance, *viz.*, *glutamic acid decarboxylase* (see below), with the demonstration that during reversal it is the proton which entered in the decarboxylation reaction which returns to the medium in the back reaction. Some metabolically significant α-amino acid decarboxylases are presented below.

γ-Aminobutyric Acid. An α-decarboxylase for L-glutamic acid is present in many bacteria, but the significance of the product, γ-aminobutyric acid, in bacterial metabolism is obscure. A highly active enzyme is present in brain (Chap. 37).

$$\underset{\textbf{Glutamic acid}}{HOOC{-}CH_2{-}CH_2{-}CHNH_2{-}COOH} \longrightarrow \underset{\textbf{γ-Aminobutyric acid}}{HOOC{-}CH_2{-}CH_2{-}CH_2NH_2} + CO_2$$

γ-Aminobutyric acid transaminates to α-ketoglutarate with formation of succinic semialdehyde. The latter is oxidized to succinic acid, which enters the tricarboxylic acid cycle.

β-Alanine. The only β-amino acid of physiological significance is *β-alanine* which, in addition to being present in many tissues and in plasma as the free amino

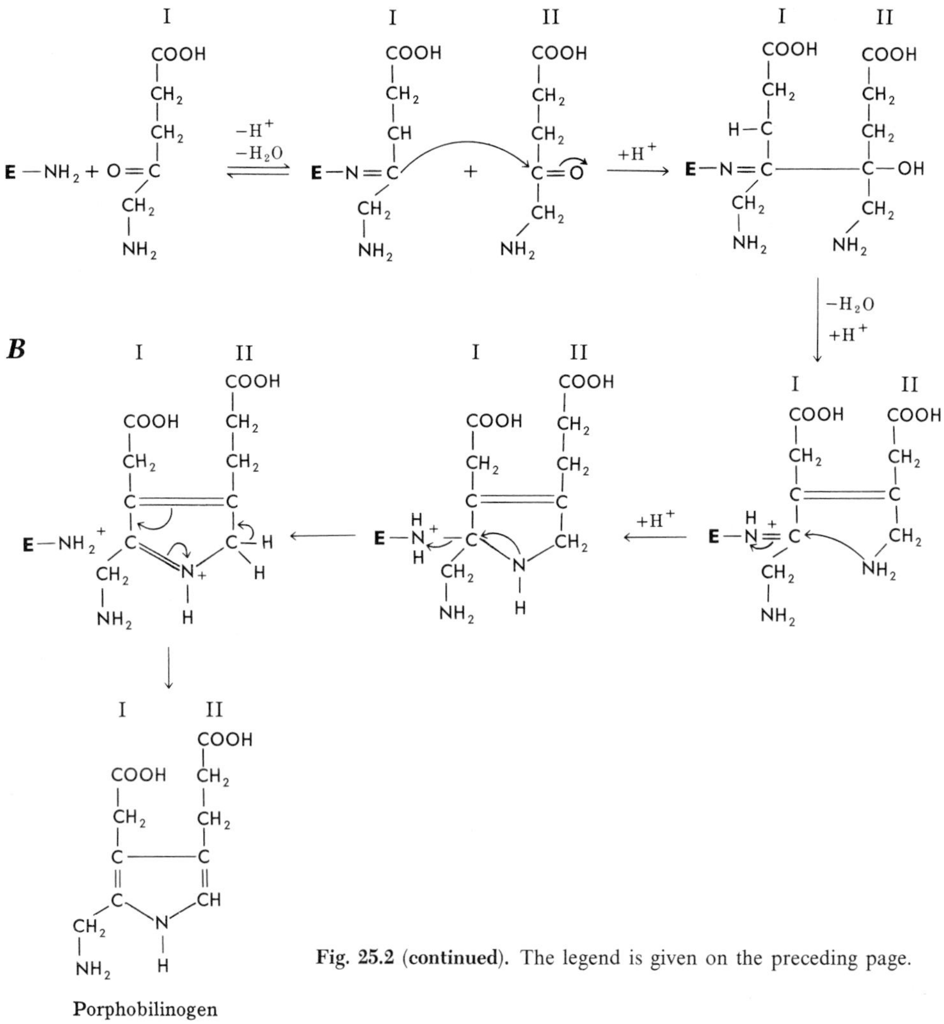

Fig. 25.2 (continued). The legend is given on the preceding page.

acid, occurs as a component of carnosine and of anserine (page 112) and as part of the pantothenic acid moiety of coenzyme A. Pantothenic acid, a vitamin, cannot be made by most animals. In bacteria, and perhaps plants, there is present a low level of *aspartic acid decarboxylase* activity which accounts for β-alanine synthesis.

$$\underset{\text{Aspartic acid}}{HOOC{-}CH_2{-}CHNH_2{-}COOH} \longrightarrow \underset{\beta\text{-Alanine}}{HOOC{-}CH_2{-}CH_2{-}NH_2} + CO_2$$

In animals, β-alanine arises from pyrimidine metabolism (page 636). Transamination from β-alanine leads to malonic semialdehyde. Oxidation of the latter to malonic acid followed by decarboxylation with formation of acetate would result in disposal via the citric acid cycle (page 325). An alternate synthetic and disposal pathway is represented by the following reaction series.

$$\underset{\text{Propionyl CoA}}{CH_3{-}CH_2{-}CO{-}SCoA} \rightleftharpoons \underset{\text{Acrylyl CoA}}{CH_2{=}CH{-}CO{-}SCoA} \rightleftharpoons \underset{\beta\text{-Hydroxypropionyl CoA}}{HO{-}CH_2{-}CH_2{-}CO{-}SCoA}$$

$$\Updownarrow$$

$$\underset{\beta\text{-Alanine}}{H_2N{-}CH_2{-}CH_2{-}COOH} \rightleftharpoons \underset{\text{Malonic semialdehyde}}{OHC{-}CH_2{-}COOH} \rightleftharpoons \underset{\beta\text{-Hydroxypropionic acid}}{HO{-}CH_2{-}CH_2{-}COOH}$$

An *aspartic acid β-decarboxylase* from microorganisms has been obtained in crystalline form and catalyzes the β-decarboxylation of aspartic acid with formation of alanine.

$$\underset{\text{Aspartic acid}}{HOOCCH_2\overset{\overset{NH_2}{|}}{C}HCOOH} \xrightarrow{-CO_2} \underset{\text{Alanine}}{CH_3\overset{\overset{NH_2}{|}}{C}HCOOH}$$

The enzyme, molecular weight 800,000, contained 16 molecules of pyridoxal phosphate per molecule of protein.

Decarboxylation of Aromatic Amino Acids. A single enzyme present in mammalian tissues can catalyze decarboxylation of histidine, tyrosine, tryptophan, phenylalanine, 3,4-dihydroxyphenylalanine, and 5-hydroxytryptophan. Thus this enzyme appears to be a general aromatic L-amino acid decarboxylase. In addition, there appear to be specific decarboxylases for the aromatic amino acids.

Histamine. The general aromatic amino acid decarboxylase has low activity toward histidine. Rather, a separate enzyme known to be present in mast cells, a major site of histamine formation, is probably the chief catalyst for histidine decarboxylation. In addition, many tissues, *e.g.*, lung, liver, muscle, and gastric mucosa have a high histamine content owing to synthesis of this amine *in situ.* Histamine is a powerful vasodilator and in excessive concentrations may cause vascular collapse. The base is liberated in traumatic shock and in localized areas of inflammation. Histamine stimulates secretion of both pepsin and acid by the stomach and is useful in studies of gastric activity. *Diamine oxidase,* a widely distributed flavoprotein (page 364), converts histamine to the corresponding aldehyde and NH_3. Some undegraded histamine is excreted in the urine as the N-acetyl and as the 1-methyl derivatives; the latter is the major metabolite of histamine in man.

Tyramine and Related Compounds. Decarboxylation of tyrosine yields tyramine; no specific physiological function is known for this compound. A specific hydroxylase, termed *tyrosinase,* catalyzes hydroxylation of tyrosine in the liver and in melanin-forming cells, to yield 3,4-dihydroxyphenylalanine (dopa). The latter is then decarboxylated by *dopa decarboxylase* to dihydroxyphenylethylamine (*o*-hydroxytyramine), an intermediate in melanin formation (Fig. 25.3). The decarboxylating enzyme is present in kidney and adrenal tissue as well as in sympathetic ganglia and nerves. In man and the rat, administration of hydroxytyramine led to excretion in the urine of homoprotocatechuic acid (3,4-dihydroxyphenylacetic acid) and 3,4-dihydroxyphenylethanol, as well as their methylated derivatives, homovanillic acid and 3-methoxy-4-hydroxyphenylethanol, respectively. The structures of these compounds are shown in Fig. 25.3. Dopa also serves as precursor in norepinephrine and

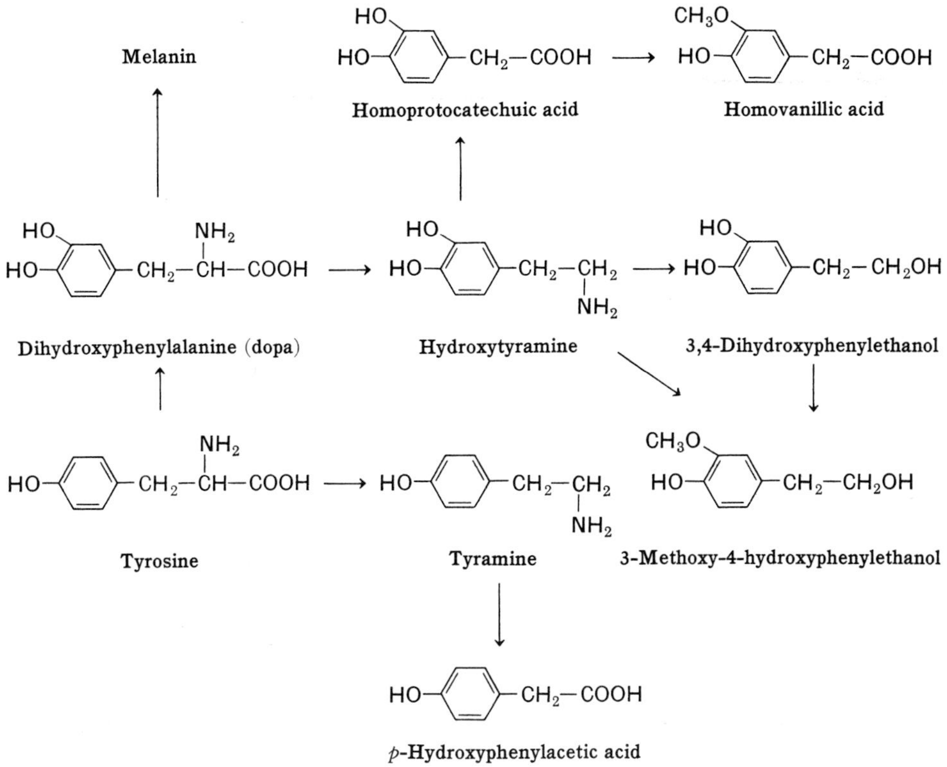

Fig. 25.3. Formation and fate of tyramine and related compounds.

epinephrine biosynthesis (Chap. 45), as well as for melanin formation.

Tryptophan Derivatives. The aromatic amino acid decarboxylase of liver has weak activity toward tryptophan. The resulting product, tryptamine, has no specific role, but when oxidized by *monoamine oxidase* yields an aldehyde which is readily oxidized by *aldehyde oxidase* to indoleacetic acid. This compound is present in small quantities in normal urine and its excretion is increased in pellagrins. However, a major route of indoleacetic acid formation, both in plants, where it serves as an *auxin* (plant growth hormone), and in animals, is by oxidative decarboxylation of indolepyruvic acid, formed from tryptophan by transamination.

Tryptophan may also be hydroxylated by phenylalanine hydroxylase (page 548) to 5-hydroxytryptophan. Decarboxylation by the enzyme *5-hydroxytryptophan decarboxylase* yields 5-hydroxytryptamine, or *serotonin,* a potent vasoconstrictor found particularly in brain, intestinal tissue, blood platelets, and mast cells. Serotonin was first isolated from blood and is also a constituent of many venoms, *e.g.*, wasp venom and toad venom. N-Methylated derivatives of serotonin, *e.g.*, bufotenin, are fairly widely distributed among amphibia and cause central nervous system damage in mammals. A possible role for serotonin as a neurohumoral agent in man has been indicated (Chap. 37). Serotonin and its major metabolic products, 5-hydroxyindoleacetic acid and 5-methoxyindoleacetic acid, are present in urine. The former can arise by a DPN-catalyzed oxidation of 5-hydroxytryplophal,

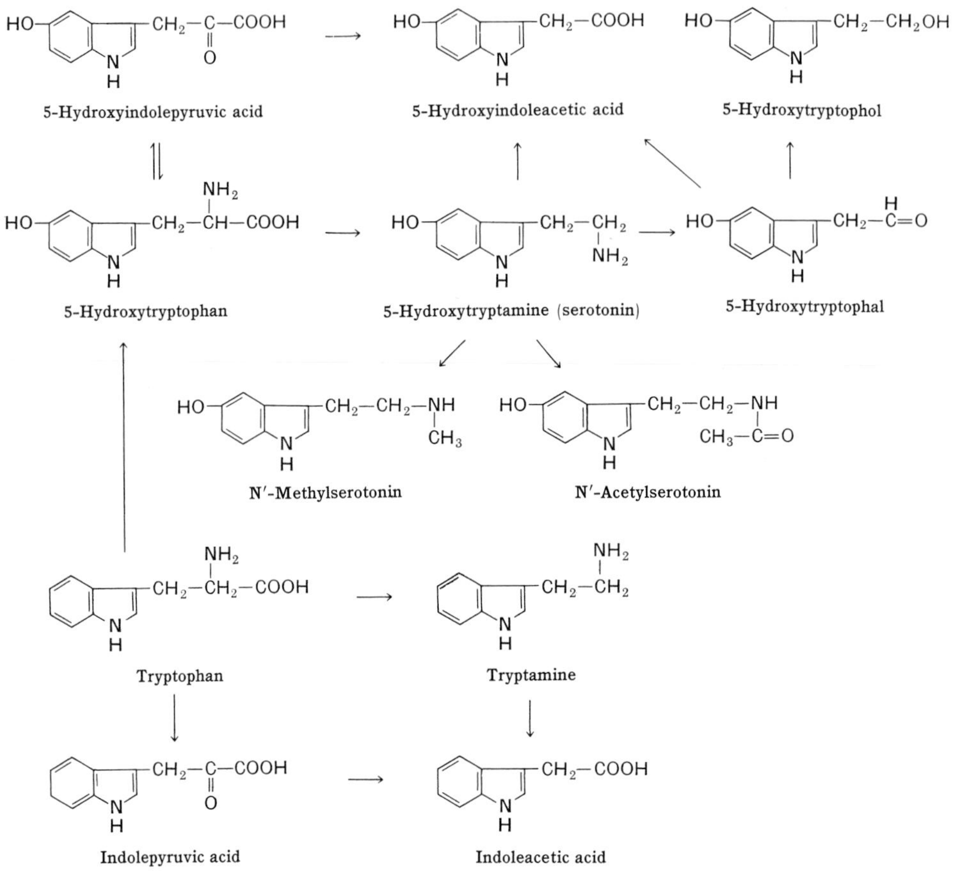

Fig. 25.4. Pathways among tryptophan and related compounds.

which arises in turn from serotonin by action of *monoamine oxidase.* Inhibition of serotonin metabolism by administration of monoamine oxidase inhibitors leads to increased formation of the N-acetyl and N-methyl derivatives of serotonin.

5-Hydroxytryptophol also is a metabolite of serotonin as a result of the action of monoamine oxidase, and is present in human urine as such, as the 5-methoxy derivative, and as the conjugates, 5-hydroxytryptophol-O-sulfate and -O-glucosiduronide. After oral administration of labeled serotonin, more than 80 per cent of the labeled products in the urine was 5-hydroxyindoleacetic acid; the total 5-hydroxytryptophol excretion represented approximately 2 per cent. 5-Methoxytryptophol is present in the pineal gland and has been shown to inhibit sexual development in the female rat.

Approximately 7 mg. of 5-hydroxyindoleacetic acid is excreted in normal urine per day. It has been estimated that 3 per cent of the dietary tryptophan is metabolized via this pathway. In patients with *malignant carcinoid,* as much as 400 mg. of 5-hydroxyindoleacetic acid is excreted daily; its excretion also rises in alkaptonuria (page 608) and in a metabolic abnormality of tryptophan metabolism, *Hartnup's disease* (page 615). Liver microsomes also catalyze hydroxylation of indole-containing compounds to the corresponding 6-hydroxy derivatives; the possible metabolic significance of this is not known. Certain of these relationships are shown in Fig. 25.4.

Numerous other instances of amino acid decarboxylation have been described. Thus, in microorganisms and in plants, arginine is decarboxylated to form agmatine. Other examples of amino acid decarboxylation will be cited elsewhere in this book. However, not all decarboxylations of amino acids occur while the amino acids are in the free state. Decarboxylation of serine occurs while serine is in ester linkage in phosphatidyl serine (page 509), with formation of phosphatidyl ethanolamine. Decarboxylation of free serine has not been described in biological systems. Somewhat analogous is the formation of coenzyme A, in which pantoic acid (page 1031) forms the pantoyl amide of cysteine, which is then phosphorylated and decarboxylated to form 4′-phosphopantetheine pantotheine (page 643). Again, no analogous decarboxylation of free cysteine to cysteamine (aminoethylmercaptan, $H_2N—CH_2—CH_2—SH$) is known to occur biologically, although products arising from cysteine metabolism are decarboxylated (page 600).

BIOSYNTHESIS OF POLYAMINES

Widely distributed among living forms are small quantities of a group of polyamines, including the following.

$H_2N—(CH_2)_4—NH_2$
1,4-Diaminobutane
(putrescine)

$H_2N—(CH_2)_5—NH_2$
1,5-Diaminopentane
(cadaverine)

$H_2N—(CH_2)_4—NH—(CH_2)_3—NH_2$
Spermidine

$$H_2N—(CH_2)_3—NH—(CH_2)_4—NH—(CH_2)_3—NH_2$$
Spermine

As implied by their names, putrescine and cadaverine have long been known, from their unpleasant odor, as a result of bacterial fermentation of protein. Cadaverine arises by decarboxylation of lysine; putrescine is formed by decarboxylation of ornithine. Putrescine is found not only in diverse bacteria but in mammalian tissues such as pancreas, lung, liver, and semen. Spermidine and spermine are much more abundant and almost universally distributed. Although synthesized by many organisms, at various phylogenetic levels, these substances are essential growth factors for a number of microorganisms, *e.g.*, *Hemophilus parainfluenzae, Aspergillus nidulans.* In general, these polyamines serve as agents which stabilize membranous structures. Thus, in their absence, the microorganisms listed above show increased permeability in hypotonic media, with loss of cell constituents. Protoplasts (bacteria without cell walls) from several species cannot withstand hypotonic media in the absence of members of this group of compounds. The swelling of mitochondria in various media is similarly prevented, and bacteriophages inactivated by chelating agents are protected by prior addition of spermine. Spermine and spermidine are present in significant amounts in ribosomes and appear to be essential to their structure and function (Chap. 29).

The synthesis of spermidine by *Escherichia coli* is accomplished in an unusual

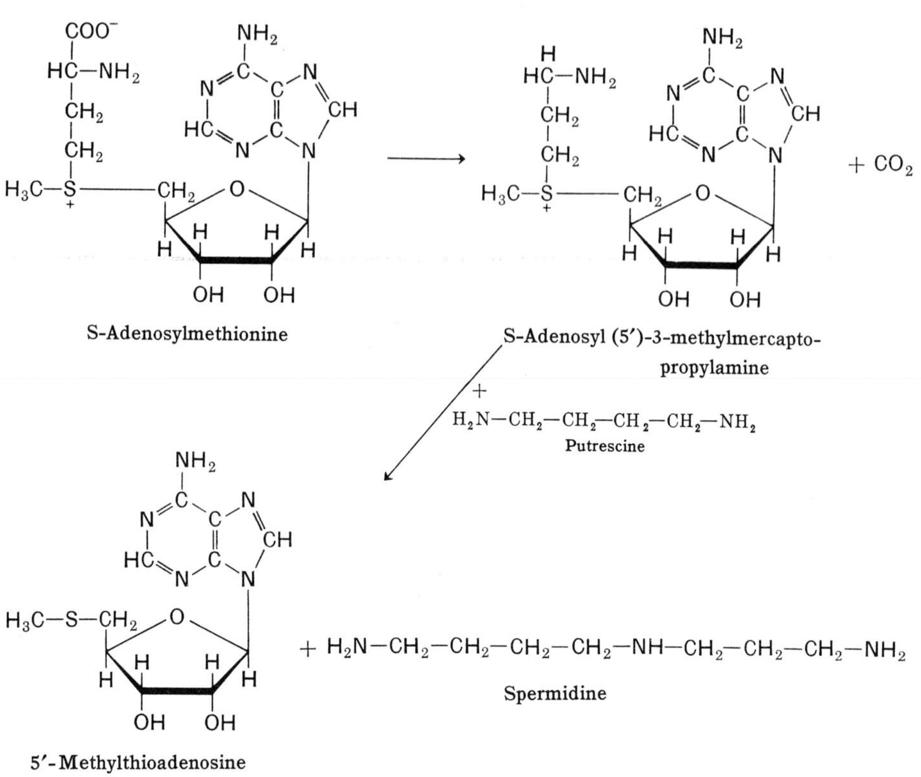

Fig. 25.5. Biosynthesis of spermidine.

reaction in which the carbon chain of methionine, including the amino group, rather than the methyl group as in transmethylations, is transferred to putrescine. The reactions are shown in Fig. 25.5. Both the decarboxylation and the transfer reaction are presently known only for this reaction sequence. Presumably, spermine is synthesized by a repetition of the decarboxylation of S-adenosylmethionine and transfer to the spermidine formed in the first transfer.

Oxidation of the polyamines, as well as certain mono- and diamines, is catalyzed by a copper-containing *oxidase* which has been obtained from beef and hog plasma in crystalline form. The molecular weights were, respectively, approximately 255,000 and 195,000, with respective copper contents of four and three atoms per molecule of protein. The enzyme catalyzed the oxidation of spermine and of spermidine.

$$\underset{\textbf{Spermine}}{H_2N(CH_2)_3NH(CH_2)_4NH(CH_2)_3NH_2} + 2O_2 + 2H_2O \longrightarrow OHC(CH_2)_2NH(CH_2)_4NH(CH_2)_2CHO + 2NH_3 + 2H_2O_2$$

$$\underset{\textbf{Spermidine}}{H_2N(CH_2)_3NH(CH_2)_4NH_2} + O_2 + H_2O \longrightarrow OHC(CH_2)_2NH(CH_2)_4NH_2 + NH_3 + H_2O_2$$

In contrast, spermidine oxidation catalyzed by an enzyme from *Serratia marcescens,* requiring FAD and an additional electron carrier, was characterized as follows.

$$H_2N(CH_2)_3NH(CH_2)_4NH_2 + O_2 \longrightarrow H_2NCH_2CH_2CH_2NH_2 + \underset{\Delta^1\text{-Pyrroline}}{\begin{array}{c} H_2C\text{——}CH_2 \\ | \quad\quad | \\ HC \quad\quad CH_2 \\ \diagdown \quad \diagup \\ N \end{array}} + H_2O$$

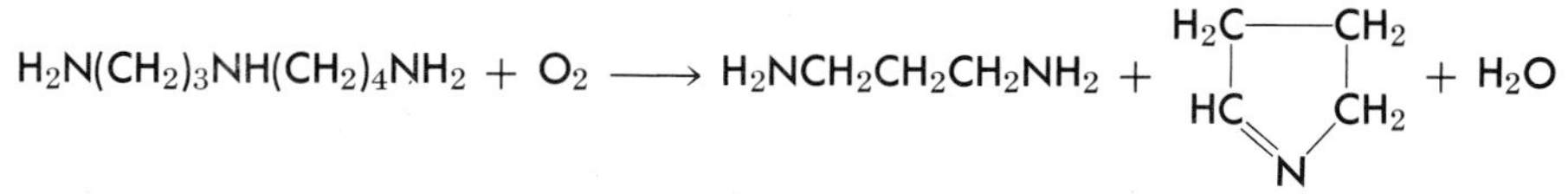

$$\underset{\text{Spermidine}}{} \qquad\qquad \underset{\text{1,3-Diaminopropane}}{}$$

REFERENCES

See list following Chap. 26.

26. Amino Acid Metabolism. IV

Glycogenic and Ketogenic Amino Acids. Metabolic Fates of Individual Amino Acids

As amino acids become available to the liver of vertebrate animals, they are utilized for protein synthesis and for some of the diverse synthetic processes considered in Chap. 25. When the liver is presented with a plethora of amino acids, *viz.*, quantities larger than those required for essential synthetic processes, this surplus cannot be stored. However, wasteful excretion of this surplus does not occur; instead, each of the amino acids is degraded by a pathway which preserves, for later use, a large fraction of the energy which would be released if the amino acids were completely oxidized. In the main, this conserving of the potential energy of the amino acids involves transformations of most of their carbon into either pyruvate or acetyl CoA. The pathways by which this is accomplished are the major focus of this chapter. However, in several instances, the degradative pathway also leads to one or more metabolites which have other, specific metabolic functions. Certain of these were presented in Chaps. 23, 24, and 25. Others will be encountered as the degradative fates of individual amino acids are considered.

Oxidative degradation of amino acids is not a significant aspect of the life of the photosynthetic plant, which makes amino acids at a rate commensurate with growth requirements. Similar considerations apply to a culture of bacteria in a medium which supports growth. However, many microorganisms can be grown in a medium in which one amino acid, or a few, constitutes the major source of carbon and from which can be fabricated all the materials essential for life. Such bacteria, notably the *Pseudomonads,* possess degradative pathways for amino acids analogous to those of the liver of vertebrates. A few examples of these pathways will be presented for comparative purposes.

GLYCOGENIC AND KETOGENIC AMINO ACIDS

The conversion of amino acids to carbohydrate precursors, and thus potentially to glucose and glycogen, or to ketone bodies, has been studied in fasted, diabetic, and phlorhizinized animals. In the last two, administration of an amino acid will increase urinary excretion of either glucose or ketone bodies. Fasted animals can be utilized to ascertain whether an administered amino acid produces an increment in liver glycogen. Studies with isotopically labeled amino acids and suspected metabolic intermediates have provided a description of the metabolic pathways for the amino acids.

In general, transformations which lead to pyruvic acid make possible net glucose formation, whereas pathways which lead to acetyl CoA or to acetoacetate result in ketone body formation. The pathways for glycogenesis (Chap. 19) and for ketogenesis (Chap. 21) have been described previously. On this basis, amino acids may, in general, be classified with respect to those which are glycogenic and those which are ketogenic (Table 26.1).

Table 26.1: GLYCOGENIC AND KETOGENIC AMINO ACIDS

Glycogenic		*Ketogenic*	*Glycogenic and ketogenic*
Alanine	Methionine	Leucine	Isoleucine
Arginine	Proline		Lysine
Aspartic acid	Serine		Phenylalanine
Cystine	Threonine		Tyrosine
Glutamic acid	Tryptophan		
Glycine	Valine		
Histidine			
Hydroxyproline			

As indicated, the initial step in the metabolism of most amino acids is removal of the α-amino group by transamination (page 558) or oxidation (page 559). Hence, it is largely the fate of the corresponding α-keto acids which is described in the following discussions.

METABOLIC FATES OF INDIVIDUAL AMINO ACIDS

Alanine; Glutamic and Aspartic Acids. Removal of the amino group of each of these three amino acids results in formation of α-keto acids which are intermediates of the tricarboxylic acid cycle. Glutamic acid may transaminate to one of a variety of keto acids or be oxidized in mitochondria by glutamic acid dehydrogenase; aspartic acid and alanine transaminate to α-ketoglutaric acid. The metabolism of the α-keto acids formed from these amino acids follow pathways previously discussed for these compounds in carbohydrate metabolism.

Ornithine, Proline, Hydroxyproline. Ornithine and proline, formed originally from glutamic acid (page 544), are reconvertible to glutamic acid. *Proline dehydrogenase* of mitochondria effects oxidation of Δ^1-pyrroline-5-carboxylic acid; this enzyme is distinct from the DPNH-linked reductase which is responsible for proline formation (page 544). Hydrolysis then yields glutamic semialdehyde, which can also be formed from ornithine by transamination. Oxidation to glutamic acid completes these sequences.

Hydroxyproline is oxidized by a DPN-linked enzyme in liver and kidney to Δ^1-pyrroline-3-hydroxy-5-carboxylic acid, which, by reactions analogous to those described for proline metabolism, leads to *erythro*-L-γ-hydroxyglutamic acid. Transamination forms α-keto-γ-hydroxyglutaric acid, which is cleaved in an aldolase type of reaction to glyoxylic acid and pyruvic acid (Fig. 26.1). The fate of glyoxylic acid will be described later (page 597).

Hydroxyproline metabolism in *Pseudomonas* follows a somewhat different pathway than that in mammals, although the nature of the reactions involved is

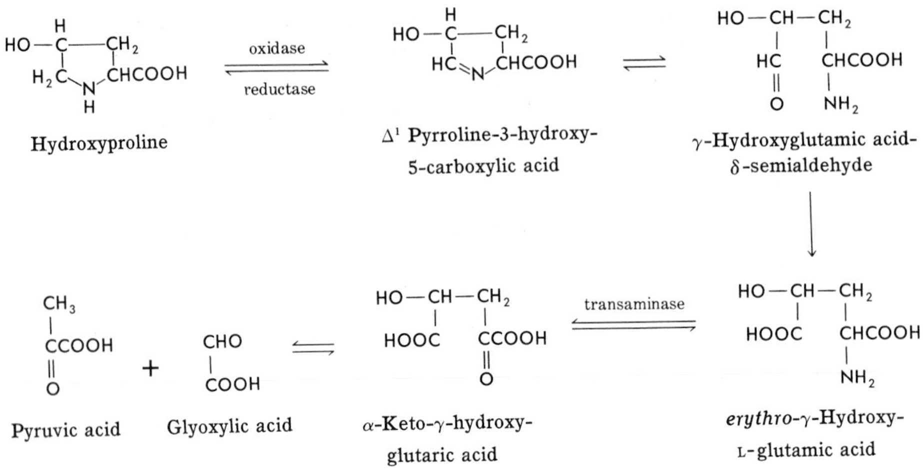

Fig. 26.1. Conversion of hydroxyproline to glyoxylic acid and pyruvic acid in mammals.

similar. Hydroxyproline is epimerized to *allo*hydroxy-D-proline, which is then oxidized to Δ¹-pyrroline-4-hydroxy-2-carboxylic acid. The latter is deaminated by a specific *deaminase* to yield α-ketoglutaric acid semialdehyde, which is oxidized by a TPN-requiring enzyme to α-ketoglutaric acid.

A single report has described a twenty- to fortyfold elevation of the concentration of free hydroxyproline in plasma and urine due to a deficiency of the enzyme mediating the oxidation of hydroxyproline to Δ¹-pyrroline-3-hydroxy-5-carboxylic acid. Mental retardation was present. Hydroxyproline excretion in the urine may be elevated in circumstances of accelerated turnover of collagen, *e.g.*, rapid bone growth in normal children or disorders affecting collagen metabolism.

Glutamine. Two pathways of glutamine metabolism have been indicated earlier. One is catalyzed by glutaminase, with conversion of glutamine to glutamic acid and ammonia (page 537). In a second pathway, glutamine transaminates with a variety of keto acids, forming the corresponding amino acids and α-ketoglutaramic acid. Hydrolysis of the latter produces α-ketoglutaric acid and ammonia (page 543). Operation of this enzyme reverses the total pathway of glutamine synthesis from α-ketoglutaric acid (page 536). The possible special role of this mode of glutamine degradation in metabolism is unclear.

Interest in glutamine metabolism is heightened by several considerations. (1) The equilibrium position of the glutamic acid dehydrogenase reaction strongly favors glutamic acid synthesis, and it is uncertain whether glutamate can serve as a prime source of ammonia. (2) α-Methylaspartic acid, a potent inhibitor of the argininosuccinate synthetase reaction, has been reported not to interfere with urea synthesis in the intact rat, indicating that a source other than aspartate, *viz.*, glutamine, may play a role in ammonia contribution to urea synthesis. (3) Renal glutaminase makes possible the conservation of cations by the organism, and the maintenance of the normal hydrogen ion concentration of body fluids. This role of glutamine and glutaminase is considered in Chap. 33. (4) Several lines of evidence have indicated that glutamine may contribute nitrogen directly to urea synthesis.

(5) Glutamine contributes to the synthesis of a variety of substances, *e.g.*, purines (Chap. 27), hexosamines (Chap. 19), and specific amino acids (Chap. 24), by transfer of its amide nitrogen.

Serine and Glycine. A major fate of serine is an α-β dehydration reaction catalyzed by a pyridoxal phosphate–requiring enzyme, *serine dehydrase;* the products are NH_3 and pyruvic acid.

$$HOCH_2-\overset{\overset{\displaystyle NH_2}{|}}{CH}-COOH \xrightarrow{-H_2O} \left[CH_2=\overset{\overset{\displaystyle NH_2}{|}}{C}-COOH \right] \longrightarrow \left[CH_3-\overset{\overset{\displaystyle NH}{\|}}{C}-COOH \right] \longrightarrow$$

Serine

$$CH_3-\overset{\overset{\displaystyle }{}}{\underset{\underset{\displaystyle O}{\|}}{C}}-COOH + NH_3$$

Pyruvic
acid

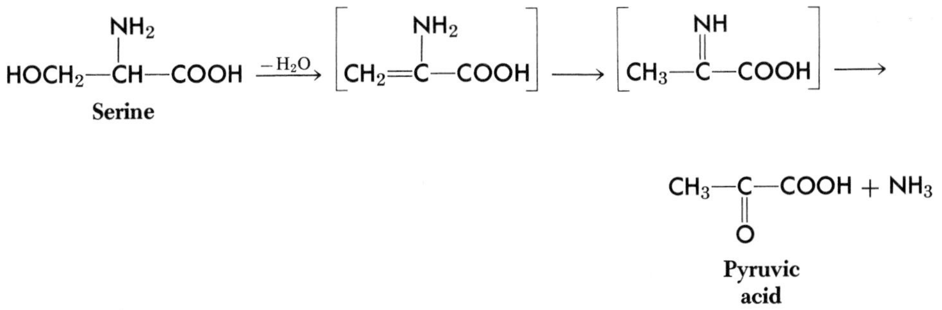

A number of other pyridoxal phosphate–requiring enzymes can catalyze an analogous deamination of serine; these include cystathionine synthetase (page 547), the β protein of tryptophan synthetase (page 573), and tryptophanase (page 616). Some degradation of serine may also occur by transamination with formation of β-hydroxypyruvic acid; the latter can give rise to glucose and glycogen (page 550). The contributions of serine to cysteine (page 547) and tryptophan synthesis (page 573) and to lipid synthesis (Chap. 22) have been considered previously.

Glycine metabolism can occur by several routes. Thus, by the action of L-*serine hydroxymethyltransferase,* glycine can accept a hydroxymethyl group from N^5,N^{10}-methylenetetrahydrofolic acid to form serine (page 551), which is directed into pyruvate metabolism as shown above. Alternatively, *glycine oxidase* (D-amino acid oxidase) catalyzes oxidation of glycine to glyoxylic acid and NH_3. Glyoxylic acid may be oxidized to formic acid and CO_2. The formic acid is then reutilized by conversion to N^{10}-formyltetrahydrofolic acid, which, among other fates, can be converted successively to N^5-methyl-, N^5,N^{10}-methenyl-, and N^5,N^{10}-methylenetetrahydrofolic acid. Reaction of the last with a second molecule of glycine yields serine (page 552). Glyoxylic acid may also be oxidized to oxalic acid and the latter excreted in the urine. These aspects of glycine metabolism are depicted in Fig. 26.2.

In primary *hyperoxaluria*, there is increased formation of oxalate from glyoxylic acid. Progressive deposition of calcium oxalate in the kidneys and other tissues leads to extensive genitourinary tract difficulty and may cause early death.

Photosynthetic plants have a *glyoxylic acid reductase* for metabolizing glyoxylate formed in the isocitric acid lyase reaction (page 412). Plants also contain a *glyoxylate transaminase,* catalyzing interconversion of glyoxylate and glycine with glutamate as an amino group donor. The metabolism of glyoxylate by the isocitratate lyase reaction and by transamination has also been described in invertebrate species.

Cysteine. The ultimate fate of cysteine in metabolism is formation of inorganic sulfate and pyruvate, which may arise via several pathways. These involve initial removal of the amino group by transamination, initial concurrent removal of both

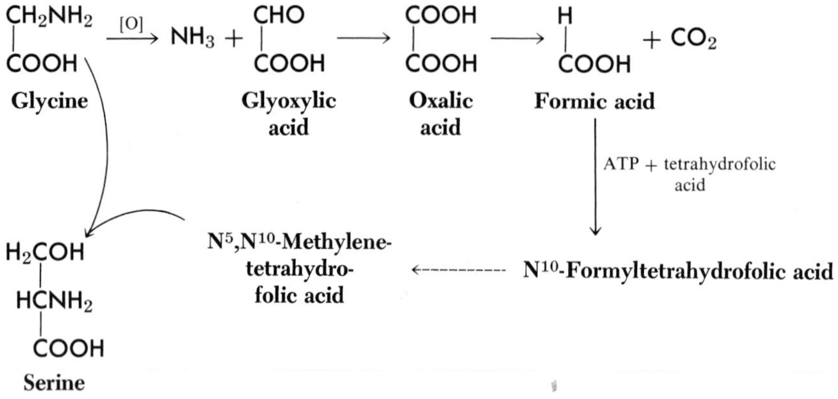

Fig. 26.2. Aspects of glycine metabolism. See also Fig. 23.4, page 552.

the sulfur and the amino group, or initial oxidation of the organically bound sulfur, respectively. These pathways are considered in turn.

Transamination of cysteine in liver and bacteria yields β-mercaptopyruvic acid, which undergoes desulfuration to form pyruvic acid. The reaction is catalyzed by *β-mercaptopyruvate transsulfurase.* The sulfur appears as H_2S because of the presence of reducing agents, *e.g.,* glutathione.

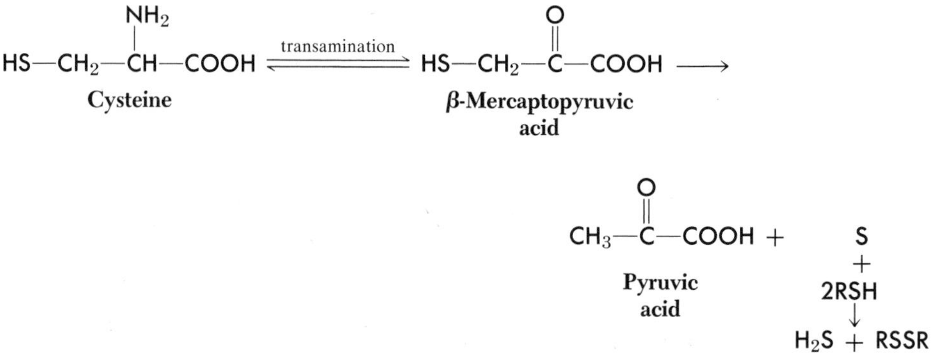

β-Mercaptopyruvic acid participates in two additional reactions which also yield pyruvic acid; the other products are, respectively, thiosulfate and thiocyanate (Fig. 26.3). Liver contains *rhodanese,* a sulfhydryl-containing enzyme which promotes formation of thiocyanate from CN^- and $S_2O_3^{=}$.

The *cysteine desulfhydrase* reaction leads to formation of pyruvic acid, H_2S, and NH_3.

$$\text{Cysteine} + H_2O \longrightarrow \text{pyruvate} + H_2S + NH_3$$

Possible reversal of this reaction has been considered previously (page 564).

A major pathway of cysteine metabolism involves oxidation of the sulfur in organic linkage. The sequence of reactions is shown on page 600.

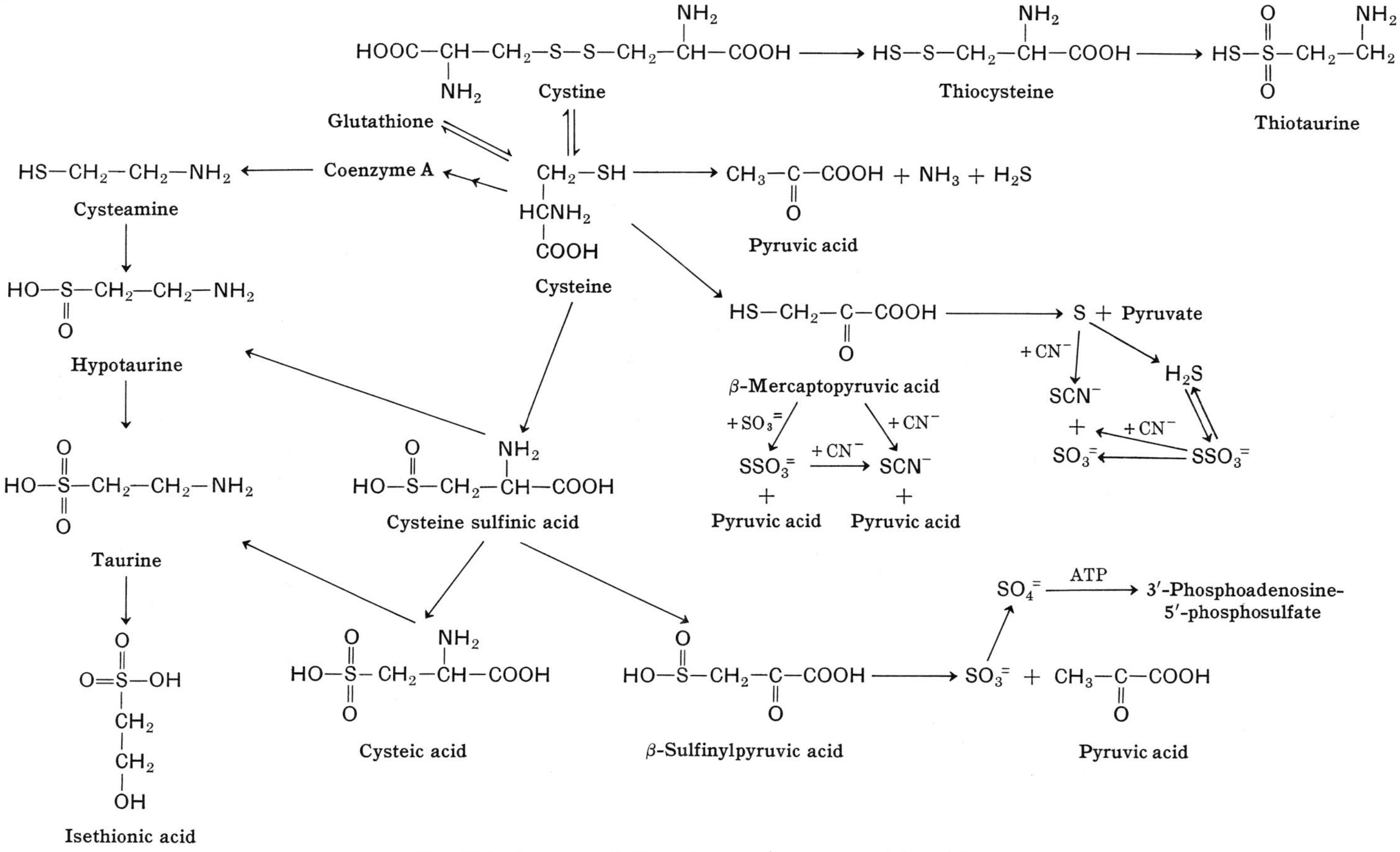

Fig. 26.3. Some metabolic pathways of cysteine and its sulfur.

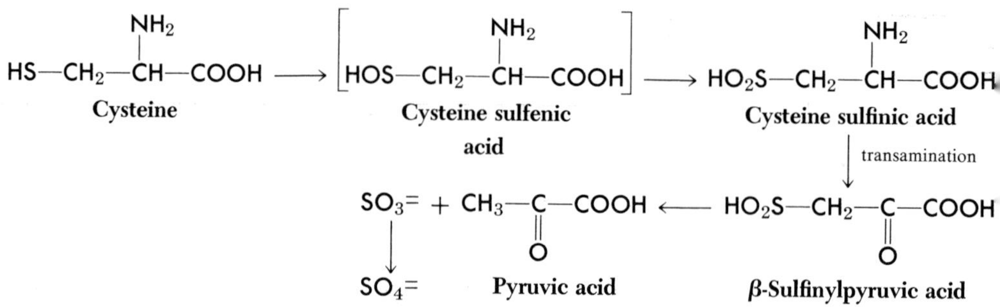

$$HS-CH_2-\underset{\underset{}{NH_2}}{CH}-COOH \longrightarrow \left[HOS-CH_2-\underset{\underset{}{NH_2}}{CH}-COOH\right] \longrightarrow HO_2S-CH_2-\underset{\underset{}{NH_2}}{CH}-COOH$$

Cysteine   Cysteine sulfenic acid   Cysteine sulfinic acid

$\downarrow$ transamination

$$SO_3^= + CH_3-\underset{\underset{O}{\|}}{C}-COOH \longleftarrow HO_2S-CH_2-\underset{\underset{O}{\|}}{C}-COOH$$

$\downarrow$

$$SO_4^= \qquad \text{Pyruvic acid} \qquad \beta\text{-Sulfinylpyruvic acid}$$

Oxidation of cysteine to cysteine sulfinic acid may occur via the hypothetical sulfenic acid and requires O_2, TPNH, and Fe^{++}.

Cysteine sulfinic acid transaminates with α-ketoglutarate; the resultant β-sulfinylpyruvic acid, in a reaction catalyzed by a specific *desulfinase,* yields pyruvic acid and sulfite. It may be presumed, by analogy with decarboxylation reactions, that it is SO_2 which is removed by the enzyme and that $SO_3^=$ is formed by hydration and ionization. The sulfite is then oxidized to sulfate by a microsomal *sulfite oxidase,* which is a hemoprotein; cytochrome c and various dyes can serve as electron acceptors. Hereditary absence of this enzyme results in mental retardation and liver disease; SO_2 may be detected in the expired air.

A direct "desulfinase" reaction has been described catalyzed by a liver preparation, with formation of alanine and sulfite. Sulfate, the ultimate form in which ingested sulfur is excreted, is also utilized for synthesis of 3'-phosphoadenosine 5'-phosphosulfate (page 563), which participates in synthesis of diverse sulfate esters.

Cysteine sulfinic acid is also utilized for the synthesis of several sulfur-containing compounds, including taurine and isethionic acid. Taurine participates in the formation of the bile acid, taurocholic acid (Chap. 34). Isethionic acid accumulates in rather large quantities in nerve tissue but its significance is unknown.

Taurine synthesis occurs in the liver via two pathways; that via hypotaurine appears to be the major one.

$$HO-\underset{\underset{O}{\|}}{S}-CH_2-\underset{\underset{}{NH_2}}{CH}-COOH \xrightarrow{-CO_2} HO-\underset{\underset{O}{\|}}{S}-CH_2-\underset{\underset{}{NH_2}}{CH_2}$$

Cysteine sulfinic acid   Hypotaurine

$\downarrow$ [O] $\downarrow$ [O]

$$HO-\underset{\underset{O}{\|}}{\overset{\overset{O}{\|}}{S}}-CH_2-\underset{\underset{}{NH_2}}{CH}-COOH \xrightarrow{-CO_2} HO-\underset{\underset{O}{\|}}{\overset{\overset{O}{\|}}{S}}-CH_2-\underset{\underset{}{NH_2}}{CH_2}$$

Cysteic acid   Taurine

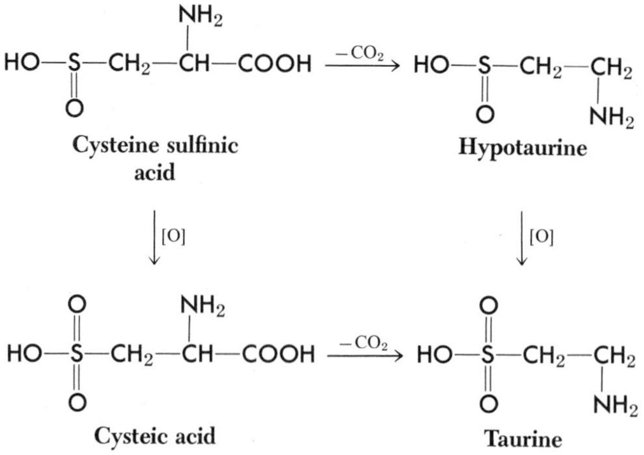

Isotopic tracer experiments reveal that isethionic acid,

$$HO-\overset{\overset{O}{\|}}{\underset{\underset{O}{\|}}{S}}-CH_2-CH_2-OH$$

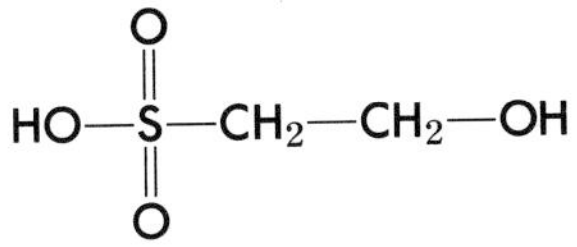

is a normal metabolite of taurine; heart tissue can effect this conversion. Taurine has been identified also as a bacterial cell wall constituent.

The thioethylamine (cysteamine) derived from cysteine and present in coenzyme A is liberated upon hydrolysis of this coenzyme. Its subsequent fate is oxidation to hypotaurine and thence to taurine. A small amount of cystine undergoes cleavage by *transsulfurase*, with formation of S-thiocysteine. Decarboxylation and oxidation of the latter result in formation of a minor quantity of thiotaurine. These pathways are summarized in Fig. 26.3.

Other pathways of cysteine metabolism are also operative in the liver but are of lesser significance. The pyridoxal phosphate–requiring *cystathionine γ-synthetase*, that catalyzes conversion of cystathionine to homoserine plus cysteine in animal metabolism, has already been considered (page 566). In the presence of cystine, this enzyme also catalyzes the reaction series shown below.

(a)
$$HOOC-\overset{\overset{NH_2}{|}}{CH}-CH_2-S-S-CH_2-\overset{\underset{NH_2}{|}}{CH}-COOH \longrightarrow$$

Cystine

$$HOOC-\overset{\overset{NH_2}{|}}{CH}-CH_2-S-SH + \text{pyruvate} + NH_3$$

Thiocysteine

(b)
$$HOOC-\overset{\overset{NH_2}{|}}{CH}-CH_2-S-SH + HS-CH_2-\overset{\overset{NH_2}{|}}{CH}-COOH \longrightarrow$$

Thiocysteine   Cysteine

$$HOOC-\overset{\overset{NH_2}{|}}{CH}-CH_2-S-S-CH_2-\overset{\underset{|}{}}{CH}-COOH + H_2S$$

Cystine

Sum: $HS-CH_2-\overset{\overset{NH_2}{|}}{CH}-COOH \xrightarrow{\text{cystine}} \text{pyruvate} + NH_3 + H_2S$

Cysteine

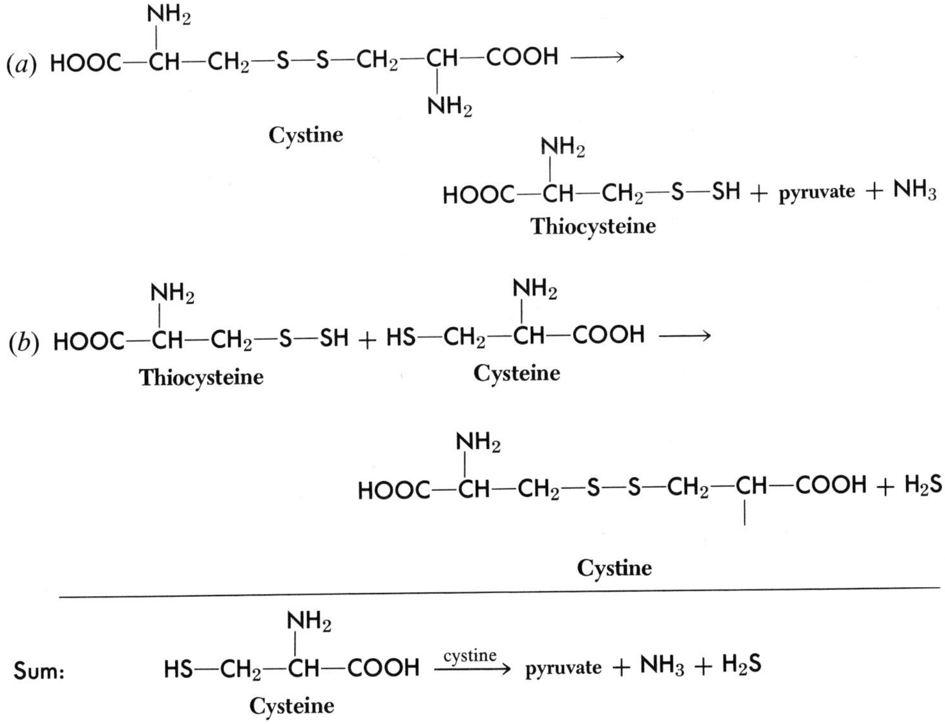

Even lesser amounts of other sulfur compounds, some of unknown origin, can be found in the urine. Among these are isobuteine, isovalthine, felinine, and S-(1,2-dicarboxyethyl)cysteine.

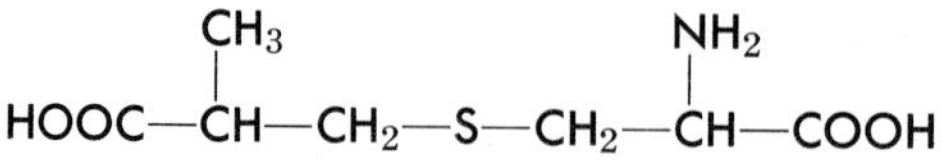

$$\underset{\text{Isobuteine}}{\text{HOOC}-\overset{\overset{\displaystyle CH_3}{|}}{CH}-CH_2-S-CH_2-\overset{\overset{\displaystyle NH_2}{|}}{CH}-COOH}$$

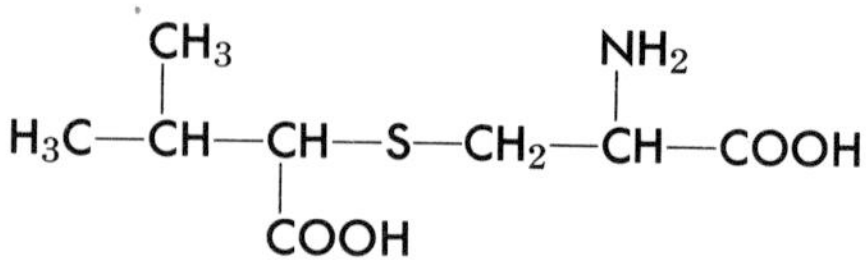

$$\underset{\text{Isovalthine}}{H_3C-\overset{\overset{\displaystyle CH_3}{|}}{CH}-\underset{\underset{\displaystyle COOH}{|}}{CH}-S-CH_2-\overset{\overset{\displaystyle NH_2}{|}}{CH}-COOH}$$

$$\underset{\text{Felinine}}{HO-CH_2-CH_2-\overset{\overset{\displaystyle CH_3}{|}}{\underset{\underset{\displaystyle CH_3}{|}}{C}}-S-CH_2-\overset{\overset{\displaystyle NH_2}{|}}{CH}-COOH}$$

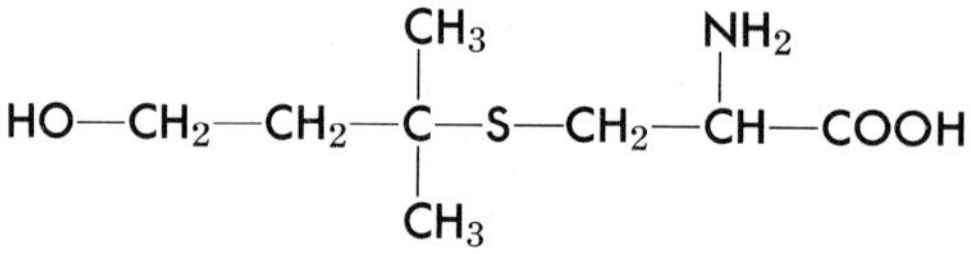

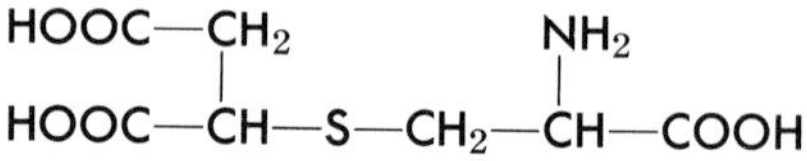

$$\underset{\text{S-(1,2-Dicarboxyethyl)cysteine}}{\overset{\displaystyle HOOC-CH_2}{HOOC-\underset{\underset{}{|}}{CH}-S-CH_2-\overset{\overset{\displaystyle NH_2}{|}}{CH}-COOH}}$$

Isovalthine was first obtained from the urine of hypercholesterolemic individuals, and its excretion in the urine of normal animals is augmented by administration of cholesterol.

Of the diverse sulfur-containing compounds described above, it is as sulfate that most of the ingested protein sulfur is excreted. Daily excretion of sulfur in the urine totals about 700 mg., of which about 75 per cent is sulfate, 5 per cent is sulfate esters (esters of phenols and oligosaccharides), and somewhat less than 20 per cent is organic sulfur in the many forms cited above.

Three hereditary disorders of cystine metabolism have been described, *viz.*, cystinuria, cystathionuria, and homocystinuria. *Cystinuria* is characterized by an abnormally high urinary excretion of cystine (page 845), as well as of several diamino acids, and is due to a defect in the renal transport system for these amino acids with consequent failure of renal tubular reabsorption. There is also a defect in the intestinal transport of these amino acids. *Cystathionuria,* with increased urinary excretion of cystathionine, is attributable to a lower than normal activity of liver transsulfurase (page 547). Available evidence suggests that in this disease the apoenzyme is present but exhibits a significantly diminished capacity to bind pyridoxal phosphate, the coenzyme. This would be a unique type of genetic alteration in a protein with loss of function. *Homocystinuria,* with excretion of homocystine in the urine, is a result of diminished activity of liver cystathionine γ-synthetase (page 547). The last two defects are associated with mental retardation and a tendency to convulsions. In the urine of six individuals with homocystinuria, an unusual homocysteine-containing compound was present, 5-aminoimidazole-4-carboxamide 5'-homocysteinyl riboside, suggesting unknown alternate pathways for the metabolism of methionine and of purines.

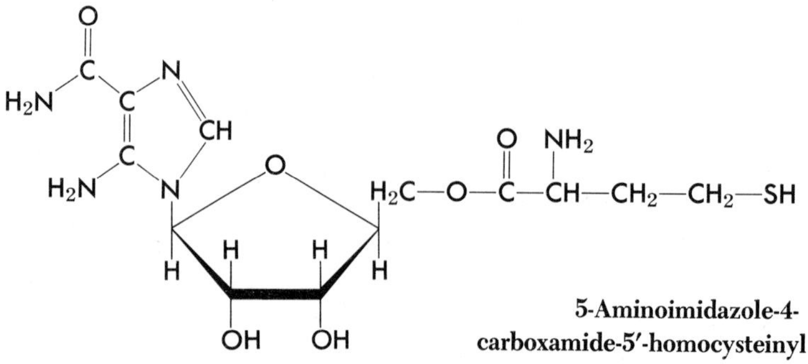

5-Aminoimidazole-4-carboxamide-5'-homocysteinyl riboside

Methionine. The major metabolic roles of methionine have already been presented; they are (1) utilization for protein synthesis (Chap. 29), (2) conversion to S-adenosylmethionine, the prime methyl group donor (page 553); and (3) conversion via the transsulfuration pathway to cystathionine, cysteine, and other derivatives. The last two of these are related; methionine is converted sequentially to S-adenosylmethionine, to S-adenosylhomocysteine, and then to homocysteine. The latter may be converted again to methionine by methylation, thus completing a cycle, or converted irreversibly to cystathionine, which is cleaved to α-ketobutyric acid and cysteine (page 547). Oxidative decarboxylation of α-ketobutyrate results in formation of propionyl CoA, the subsequent metabolism of which has already been considered (page 483). Some reactions of methionine metabolism are outlined in Fig. 26.4.

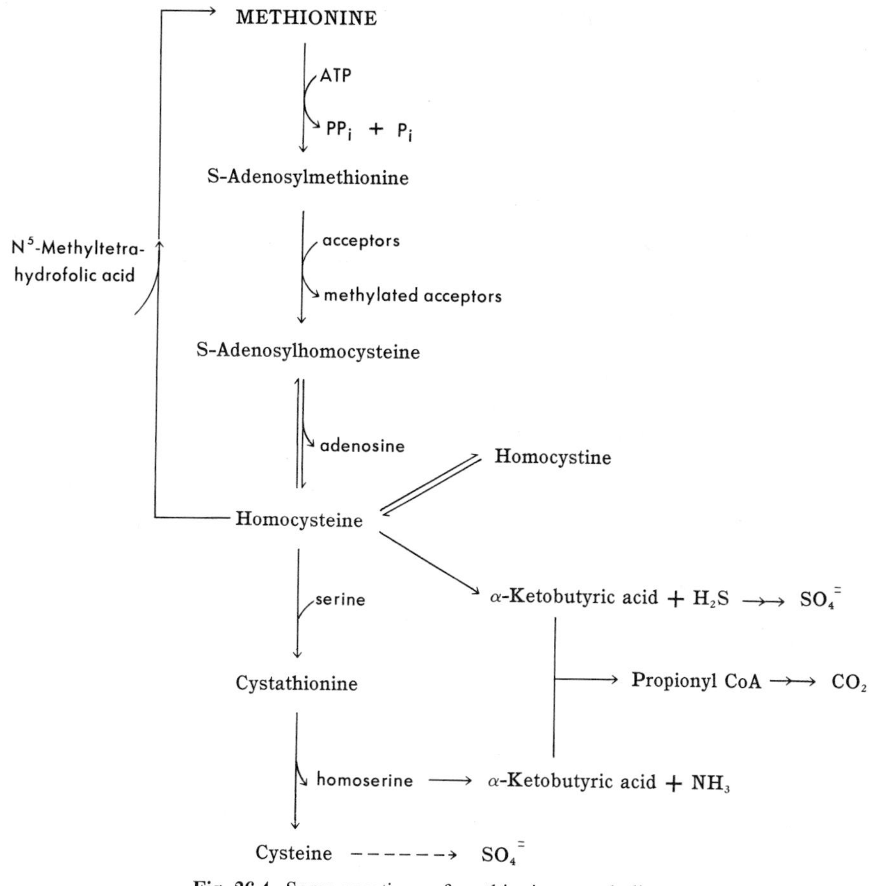

Fig. 26.4. Some reactions of methionine metabolism.

Threonine. As indicated previously, threonine does not participate in transamination (page 541). Amino group removal is catalyzed by a pyridoxal phosphate–requiring enzyme, *threonine deaminase,* to α-ketobutyrate and NH_3. The activity of this enzyme in liver increases following either threonine feeding or injection of an adrenal corticosteroid. The data suggest stimulation of enzyme synthesis; this can be prevented by administration of glucose. Threonine deaminase of *E. coli* is allosterically activated by AMP, while that from *Clostridia* is activated by ADP. In the latter organism, threonine can serve as the major source of energy for growth, yielding propionyl CoA by way of α-ketobutyrate; propionyl phosphate then generates ATP in the presence of ADP and an acyl kinase. Thus, the accumulation of ADP in the cell increases the rate of the first, and presumably the rate-limiting, reaction in the catabolic pathway for threonine, providing more energy for the synthesis of ATP.

Another pathway of threonine degradation is catalyzed by an enzyme in liver and kidney which has been termed *threonine aldolase,* a pyridoxal phosphate–requiring enzyme; the products are glycine and acetaldehyde. Acetaldehyde is oxidized to acetic acid, and the resulting acetyl CoA is utilized in the diverse manners described, as is the other product, glycine.

Some microorganisms contain *threonine dehydrogenase,* which catalyzes the following reaction.

$$\text{Threonine} + DPN^+ \longrightarrow \underset{\text{Aminoacetone}}{CH_3\overset{\overset{\displaystyle O}{\|}}{C}CH_2NH_2} + DPNH + H^+ + CO_2$$

The enzyme is activated by K^+. A similar enzymic activity has been obtained from bullfrog liver mitochondria. Aminoacetone can also be formed in microorganisms by condensation of acetyl CoA and glycine; its metabolic fate is unknown.

Threonine metabolism leading to propionyl CoA, via α-ketobutyrate, makes possible the potential appearance of three of the four carbon atoms of threonine as glucose, whereas the threonine aldolase reaction should provide two carbons to glucose, via glycine, and two carbons to ketone bodies, via acetaldehyde. However, ketone body formation *in vivo* from threonine has not been reported.

Fates of Valine, Leucine, and Isoleucine. The catabolic fates of the branched-chain,

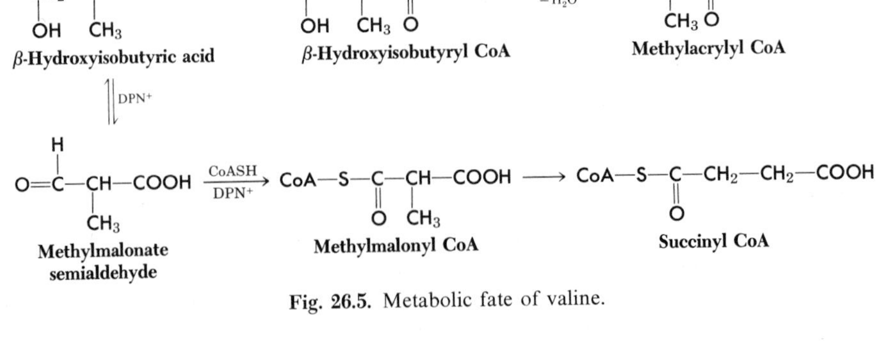

Fig. 26.5. Metabolic fate of valine.

aliphatic amino acids, valine, leucine, and isoleucine exhibit several features in common. Initial transamination yields the corresponding α-keto acids, followed by oxidative decarboxylation to the acyl CoA derivatives of one less carbon atom. Subsequent reactions, in each instance, resemble those encountered in the pathways for fatty acid oxidation.

Valine oxidation (Fig. 26.5) leads to methylmalonyl CoA, which is converted to succinyl CoA and completely oxidized, or provides three of five carbons for gluconeogenesis.

Hereditary deficiency of *α-ketoisovaleryl CoA dehydrogenase,* the enzyme catalyzing the second reaction of valine degradation, has been described, leading to isovaleric acidemia, with elevated blood and urinary levels of α-ketoisovaleric acid.

Leucine oxidation (Fig. 26.6) terminates with formation of one molecule of acetoacetic acid and one molecule of acetyl CoA, in keeping with the finding that in the diabetic animal, each mole of leucine yields 1.5 moles of acetoacetic acid. The conversion of β-methylcrotonyl CoA to β-methylglutaryl CoA is a biotin-requiring reaction with a mechanism similar to that of other biotin-CO_2 reactions (page 487). Noteworthy also is the production of β-hydroxy-β-methylglutaryl CoA in this pathway; this is also a key intermediate in steroid synthesis from acetyl CoA (page 517).

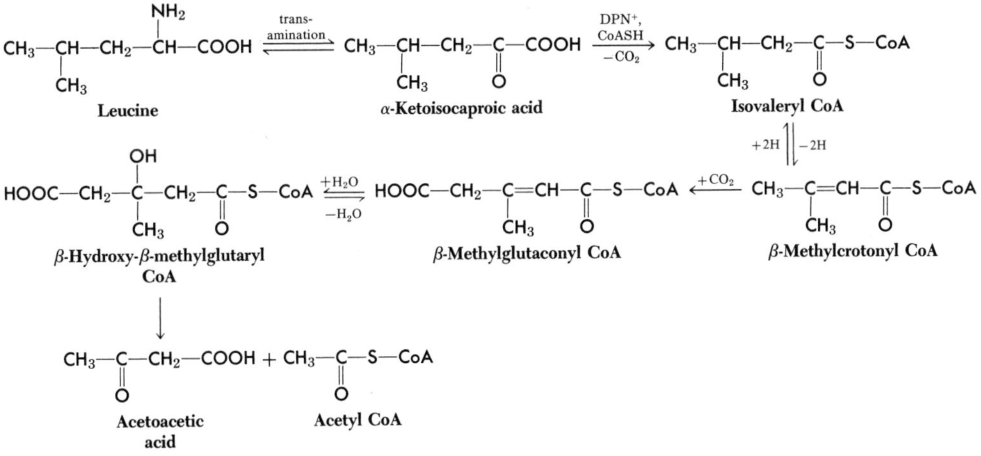

Fig. 26.6. Metabolic fate of leucine.

Isoleucine metabolism (Fig. 26.7) is concluded with formation of one molecule each of acetyl CoA and propionyl CoA, in accord with experimental evidence that this amino acid is both weakly glucogenic and weakly ketogenic.

A hereditary anomaly of the metabolism of valine, leucine, and isoleucine is described as *maple syrup urine disease,* because of the characteristic odor of the urine. A rapid deterioration is observed in the first few months of life. Some children may survive for several years but show severe mental retardation; extensive failure of myelination is evident at autopsy. The urinary odor is that of the α-keto acids corresponding to these three amino acids. Transamination is unaffected, but normal oxidative decarboxylation with acyl CoA formation does not occur. The fact that all three α-keto acids accumulate indicates (1) that a single enzyme cata-

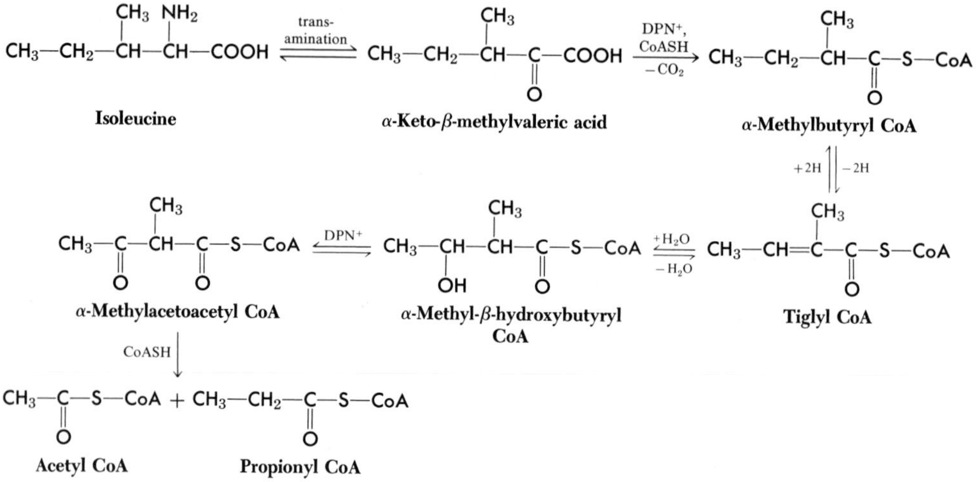

Fig. 26.7. Metabolic fate of isoleucine.

lyzes all three oxidations, and (2) that no major alternate pathway of metabolism is available.

Phenylalanine and Tyrosine. In normal individuals, almost all of the phenylalanine not utilized for protein synthesis is metabolized via tyrosine by virtue of the action of phenylalanine hydroxylase (page 548). Only in phenylketonuria (page 549) is metabolism diverted to formation of phenylpyruvic acid, which accumulates or is reduced to phenyllactic acid; a small fraction of the latter is oxidized to phenylacetic acid, which is excreted as phenylacetylglutamine. The initial event in tyrosine metabolism is transamination, catalyzed by tyrosine-glutamate transaminase, an inducible enzyme the synthesis of which increases significantly in the liver after administration of tyrosine, as well as following administration of adrenal steroids (Chap. 45), insulin or glucagon (Chap. 46). The activity of this transaminase is decreased following injection of hypophyseal somatotropin (Chap. 47). *p-Hydroxyphenylpyruvic acid oxidase,* a copper-containing protein, catalyzes oxidation of *p*-hydroxyphenylpyruvic acid to homogentisic acid. Comparison of the structures of substrate and product indicate the complexity of the reaction: hydroxylation of the ring, oxidation, decarboxylation, and migration of the side chain. The activity of this enzyme is low in fetal liver and increases slowly following birth. This may account in part for the urinary excretion of relatively large amounts of the phenyllactic acid metabolites, formed by the action of *aromatic α-keto acid reductase,* in premature infants or in normal infants who are fed tyrosine. Ascorbic acid is apparently essential for the normal activity of *p*-hydroxyphenylpyruvic acid oxidase, since ascorbic acid–deficient guinea pigs and infants excrete substantial amounts of *p*-hydroxyphenylpyruvic acid in response to an administered dose of tyrosine.

Oxidation of homogentisic acid is catalyzed by *homogentisic acid oxidase,* which requires Fe^{++} and a high concentration of a sulfhydryl compound, *e.g.,* reduced glutathione, for its action. Molecular oxygen is utilized; scission of the aromatic ring between the side chain and the adjacent hydroxyl groups occurs, with one oxygen atom becoming the 3-carbonyl oxygen of the product, maleylacetoacetate, and the other the carbonyl oxygen of the carboxyl group.

Studies of the conversion of phenylalanine to acetoacetate utilizing isotopic tracer techniques have shown that (*a*) the α-carbon atom (*) of phenylalanine becomes the carboxyl carbon of acetoacetate, (*b*) carbon atom 2 of the aromatic ring (×) is a precursor of the carbonyl carbon atom of acetoacetate, (*c*) carbon atom 1 or 3 (•) of the ring is a precursor of the terminal carbon atom of acetoacetate, and (*d*) the β-carbon atom of phenylalanine (o) becomes the α-carbon atom of acetoacetate. The evidence indicates that there is a shift of the side chain during oxidation. Pathways of phenylalanine and tyrosine metabolism are depicted in Fig. 26.8.

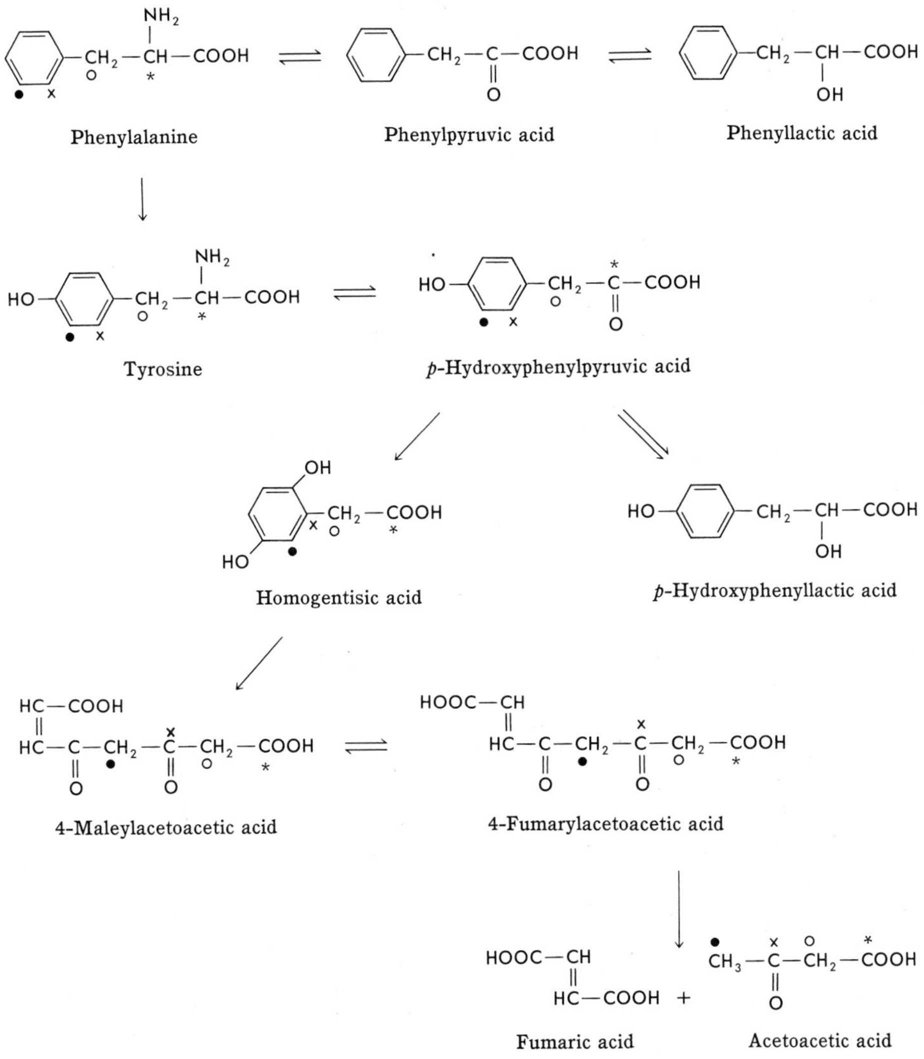

Fig. 26.8. Principal pathways of phenylalanine and tyrosine metabolism. The carbon atoms contributing to acetoacetic acid formation are indicated by the symbols (∗, o, x, •) to correlate with the text discussion.

Failure to oxidize homogentisic acid and its consequent accumulation are seen in *alkaptonuria,* a hereditary disorder. The livers of alkaptonuric individuals lack homogentisic acid oxidase but are normal with respect to all the prior and subsequent enzymes of the tyrosine catabolic pathway. The excretion of homogentisic acid is apparent since the urine, when made slightly alkaline, rapidly darkens on exposure to air due to ready oxidation of homogentisic acid to a quinone, which polymerizes to a melanin-like material. The urine is strongly reducing and gives a transitory blue color with each drop of dilute ferric chloride added. A single drop of alkalinized urine causes immediate blackening of exposed photographic paper and reduction of oxidizing agents frequently used for detection, in the urine, of glucose. In early life there is no other abnormality; in later years abnormal pigmentation of cartilage and other connective tissue (*ochronosis*) may become apparent.

Tyrosinosis appears to be another hereditary disease of tyrosine metabolism, although only one case has been reported. The metabolic block appears to be due to lack of *p*-hydroxyphenylpyruvic acid oxidase, with consequent excessive excretion of *p*-hydroxyphenylpyruvic acid.

Tyrosine as a Precursor of Melanin. In the basal layer of the epidermis are melanoblasts, in which the dark pigment, melanin, is synthesized from tyrosine by the action of copper-containing *tyrosinase.* This enzyme catalyzes the oxidation of tyrosine to dihydroxyphenylalanine (dopa) and to dopa quinone (Fig. 26.9). Tyrosinase can also catalyze the oxidation of dopa. *Hallachrome,* so named because it was isolated from a polychaete worm, *Halla parthenolapa,* is formed nonenzymically by dismutation of two moles of dopa quinone. The oxidation of indole 5,6-quinone to melanin is probably spontaneous. Melanin (Gk. *melas,* black) has not been characterized chemically, but is a polymer or group of polymers. The color of skin depends upon the distribution of melanoblasts, the melanin concentration, and perhaps its state of oxidation, since melanin can be reduced with ascorbic acid or hydrosulfite from a black to a tan form. Genetic absence of tyrosinase results in *albinism.*

Melanin is also normally found in the retina, the ciliary body, the choroid, the substantia nigra of brain, and the adrenal medulla. Melanoblasts occasionally give rise to highly malignant melanomas, which may or may not be pigmented. When stained with dopa, nonpigmented melanotic tumors rapidly darken and are therefore called *amelanotic melanomas.* Darkening of the human skin is initiated by ultraviolet irradiation of tyrosine, leading to formation of dopa. In albinos, either melanin-forming cells or tyrosinase, or both, may be entirely absent.

Lysine. Lysine does not participate in transamination. Lysine degradation in rat liver occurs primarily in mitochondria by a pathway which has features that appear to be a reversal of reactions of lysine synthesis (page 568). Saccharopine (page 570) is formed during lysine degradation, offering a direct route to α-aminoadipic-ϵ-semialdehyde and circumventing cyclic intermediates (Fig. 26.10). This semialdehyde also arises in a second pathway (Fig. 26.10) of lysine degradation initiated by oxidative deamination, catalyzed by L-amino acid oxidase of liver. The resultant keto acid cyclizes spontaneously to yield Δ^1-piperideine-2-carboxylic acid; reduction by an enzyme which utilizes either DPNH or TPNH leads to pipecolic acid. The latter is oxidized to Δ^1-piperideine-6-carboxylic acid, which, in the presence of ATP and

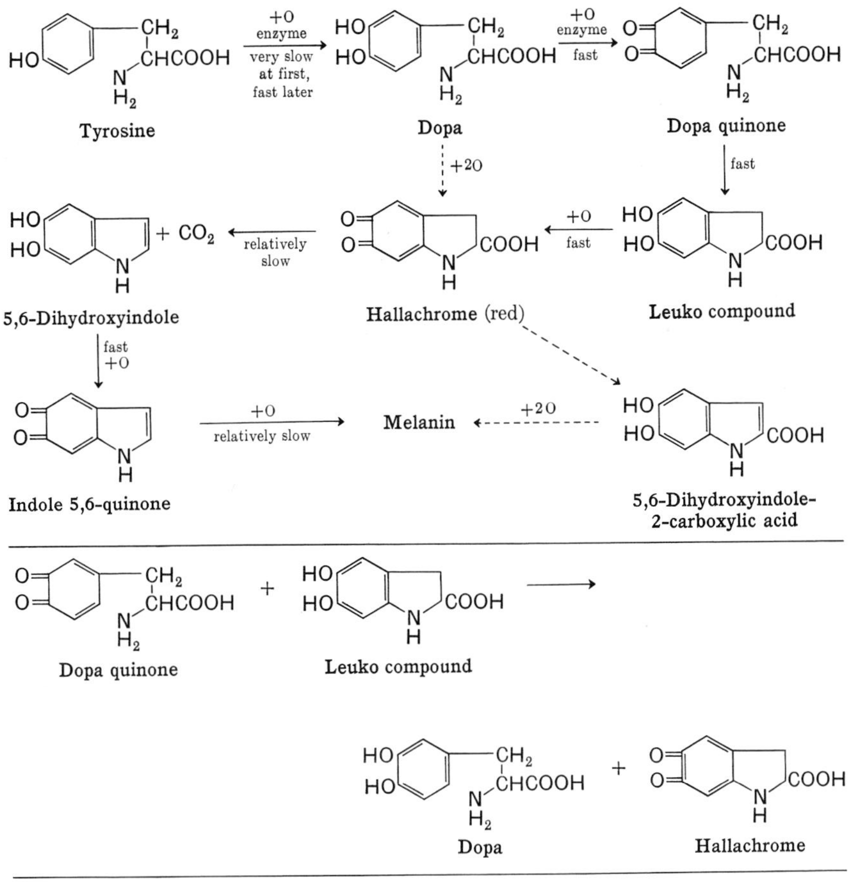

Fig. 26.9. Melanin formation from tyrosine. (*From A. B. Lerner and T. B. Fitzpatrick, Physiol. Revs.,* **30**, 91, 1950.)

Mg^{++}, can be oxidized by liver mitochondria to α-aminoadipic acid. Liver also contains an enzyme catalyzing a pyridine nucleotide–dependent oxidation of α-aminoadipic acid-ε-semialdehyde to α-aminoadipic acid. Metabolism of the latter proceeds by transamination to α-ketoglutaric acid, which, by oxidative decarboxylation, yields glutaryl CoA, which is metabolized as indicated in Fig. 26.10.

A third pathway of lysine degradation has been described in mammalian tissue as well as yeast that proceeds through acylated intermediates and would thus prevent cyclization of the keto derivative of lysine into the pipecolic acid pathway. An enzyme in liver catalyzes formation of ε-N-acetyllysine. Retention of the acyl residue occurs through steps leading to α-hydroxy-ε-acetamidocaproic acid (Fig. 26.11), which on deacylation, leads to glutarate. Alternatively, deacylation occurring at an earlier step would permit metabolism via the pipecolate pathway (Fig. 26.11).

Histidine. Although histidine can transaminate to form the corresponding imidazolepyruvic acid, the major pathway of histidine degradation is initiated by *histidase,* which catalyzes the α-β removal of a molecule of NH_3 with formation of

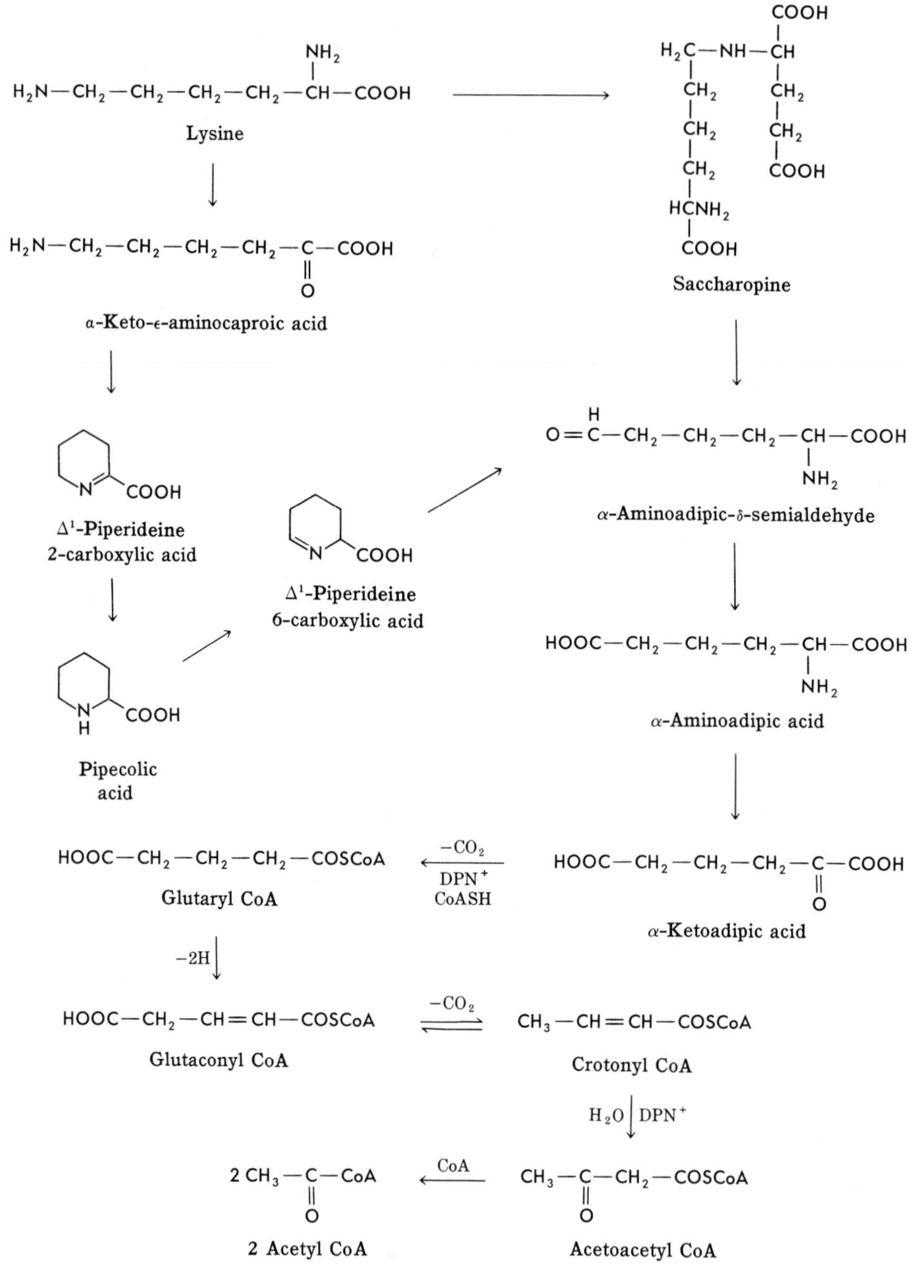

Fig. 26.10. Suggested pathways for the metabolic degradation of lysine.

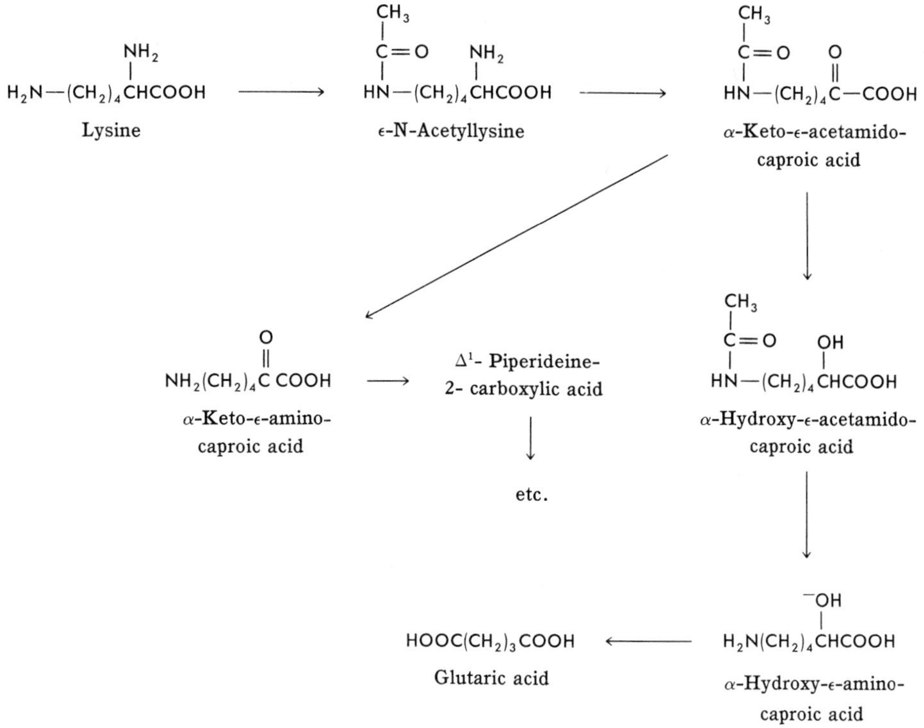

Fig. 26.11. Acylated intermediates in lysine degradation.

urocanic acid in an essentially irreversible process. The mechanism of this reaction suggests intermediate formation of an amino enzyme.

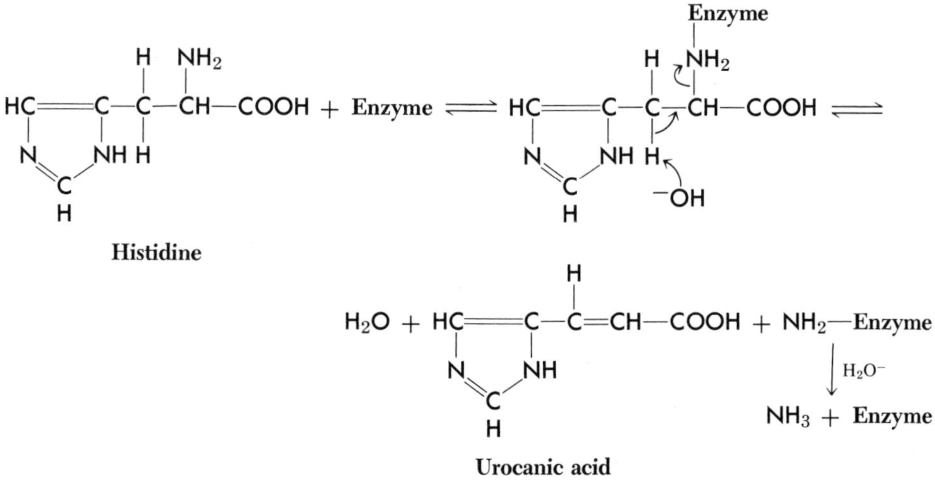

Histidase is lacking in individuals with the hereditary disorder *histidinemia,* leading to elevated blood and urine levels of histidine. A pyridoxal phosphate–requiring enzyme, *urocanase,* then catalyzes transformation of urocanic acid to imidazolone 3-

propionic acid. This reaction involves the elements of water and an internal oxidation and reduction; its mechanism remains to be elucidated. *Imidazolonepropionic acid hydrolase* effects hydrolysis of this compound to α-formiminoglutamic acid. The formimino group can then be transferred, by a specific *transferase,* to the N^5 position of tetrahydrofolic acid. The N^5-formiminotetrahydrofolic acid hydrolyzes, cyclizes to the $N^{5,10}$-methenyl derivative, and can yield the N^{10}-formyl- or the N^5,N^{10}-methylenetetrahydrofolic acid (page 552). Thus the C-2 of the histidine imidazole ring returns to the C_1 pool. Animals or human beings deficient in folic acid or in vitamin B_{12} have an excessive excretion of formiminoglutamic acid, suggesting that a major pathway for metabolism of the formimino group is conversion to a methyl group.

A small amount of imidazolone 3-propionic acid may also be oxidized to hydantoin 5-propionic acid, which is not metabolized further but is excreted in the urine. Pathways of histidine metabolism are depicted in Fig. 26.12.

In some microorganisms, formiminoglutamic acid is hydrolyzed to glutamic acid and formamide by the enzyme *formiminoglutamate hydrolase.* Studies with microorganisms have revealed control mechanisms in the pathway of histidine degradation. In *Aerobacter aerogenes,* histidase, urocanase, and formiminoglutamate hydrolase are induced by histidine and by urocanic acid and repressed by glucose and glycerol.

Ergothioneine, the betaine of 2-mercaptohistidine, is present in high concentration in human erythrocytes (20 to 30 mg. per 100 ml. whole blood). It is also found in liver and brain and in large concentration in boar semen. It is not known to serve a metabolic role in mammals, but is concentrated in the tissues after ingestion in foodstuffs of plant origin. Nothing is known of its final disposition.

Tryptophan. Tryptophan is degraded in mammals primarily by two pathways. One involves oxidation of tryptophan to 5-hydroxytryptophan followed by decarboxylation to 5-hydroxytryptamine (serotonin). This pathway was described previously, as well as that leading to indoleacetic acid (page 589). The other pathway of degradation of tryptophan involves oxidation to kynurenine, which is converted to a series of intermediates and byproducts (Fig. 26.13, page 614), all but one of which (glutaryl CoA) appear in the urine in varying amounts, and the sum of which accounts approximately for the total metabolism of tryptophan.

Oxidation of tryptophan to kynurenine is catalyzed by *tryptophan pyrrolase,* an oxygenase whose prosthetic group is heme or hematin and which utilizes molecular oxygen. Tryptophan pyrrolase is an adaptive enzyme; its activity in liver can be increased by administration of tryptophan or adrenal corticosteroids. The latter stimulate synthesis of additional enzyme, whereas tryptophan augments liver tryptophan pyrrolase activity by retarding the rate of degradation of the enzyme. The enzyme is reversibly oxygenated in the presence of tryptophan, suggesting intermediate formation of an enzyme-tryptophan-oxygen complex. Combination of tryptophan with the enzyme enhanced the reactivity of the heme toward heme-binding substances, suggesting a mechanism by which tryptophan can activate tryptophan pyrrolase. The enzyme is strongly inhibited by TPNH, which acts as an allosteric modifier and which may be considered one of the ultimate end products of tryptophan metabolism since TPNH is formed from nicotinic acid (page 662).

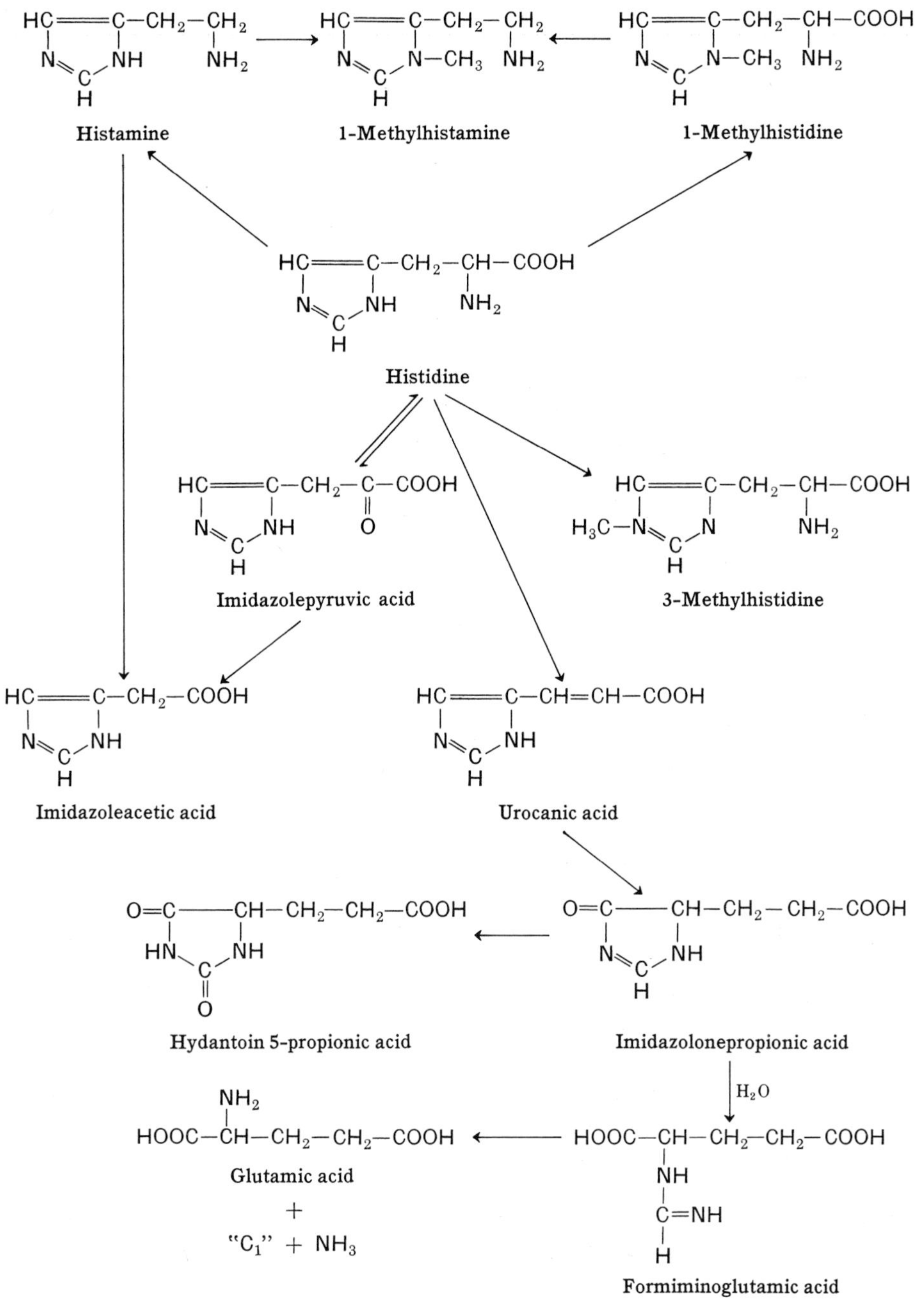

Fig. 26.12. Pathways of histidine metabolism.

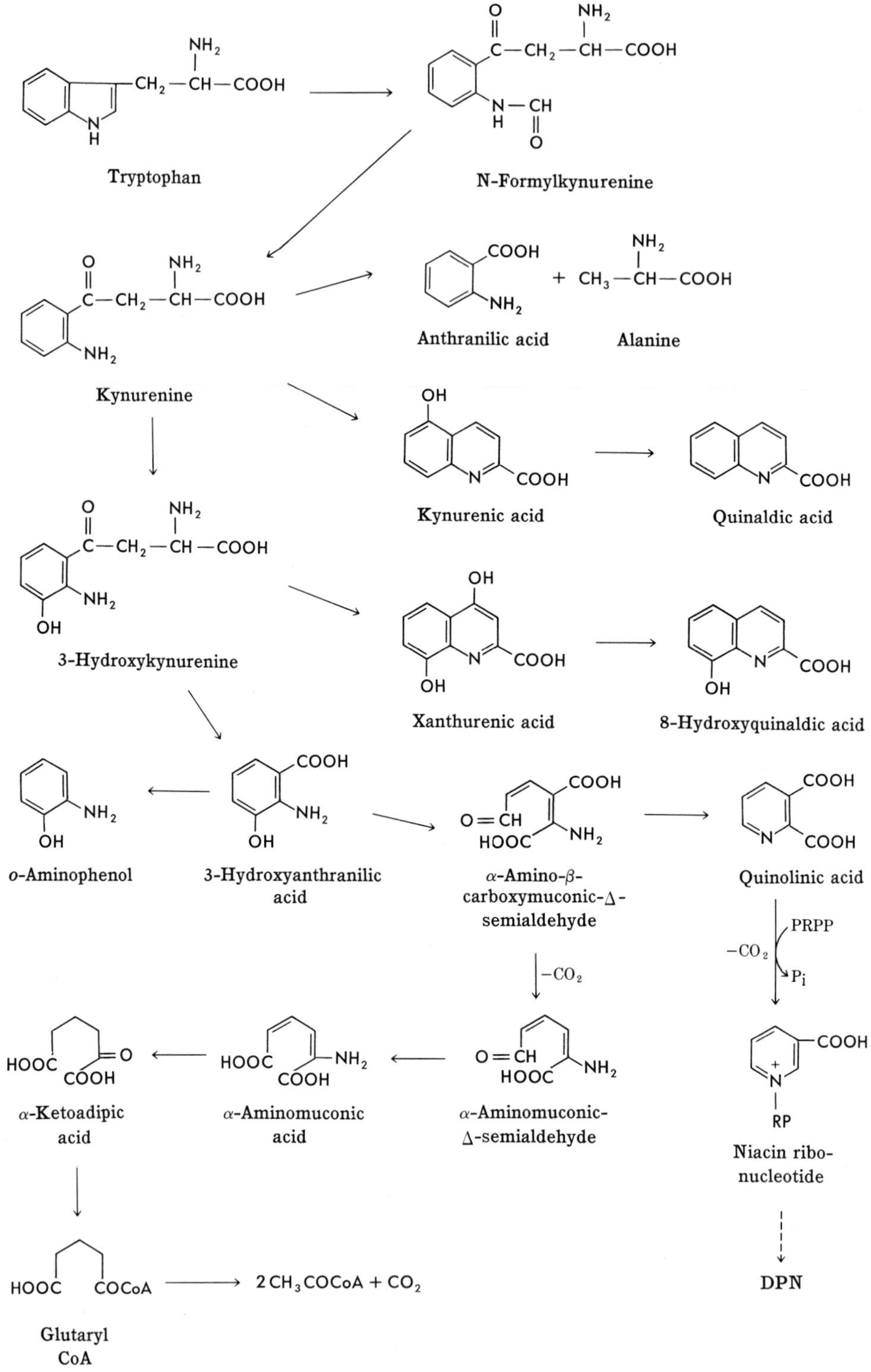

Fig. 26.13. Metabolic relationships among tryptophan and its metabolites. *PRPP* = 5-phosphoribosyl-l-pyrophosphate; *RP* = ribose 5-phosphate; *R* = ribose.

An abnormally small fraction of ingested tryptophan is oxidized to formylkynurenine owing to deficiency of tryptophan pyrrolase in *Hartnup's disease,* a hereditary disorder associated with mental retardation.

L-N-Formylkynurenine is converted to kynurenine by a liver enzyme, *kynurenine formylase.* Kynurenine is present in normal urine in trace amounts, which may be increased after tryptophan administration or in circumstances of accentuated protein catabolism.

Hydroxylation of kynurenine is catalyzed by an oxygenase, *kynurenine 3-hydroxylase,* which requires TPNH and molecular oxygen. The enzyme is specific for kynurenine, and hydroxylation occurs only at position 3. 3-Hydroxykynurenine, which in insects is also a precursor of eye pigments (ommochromes), is found in mammalian urine as the glucosiduronate, the O-sulfate, or the N-α-acetyl derivative. A pyridoxal phosphate-requiring enzyme, *kynureninase,* can catalyze cleavage of both kynurenine and 3-hydroxykynurenine to anthranilic and 3-hydroxyanthranilic acids, respectively. Alanine is the other reaction product in both instances. 3-Hydroxykynurenine is split approximately twice as rapidly as kynurenine. The latter also transaminates; the corresponding keto acid undergoes ring closure to yield kynurenic acid. This compound is dehydroxylated to quinaldic acid. Quinaldic acid accounted for approximately 30 per cent of ingested kynurenic acid in human subjects.

3-Hydroxyanthranilic acid oxidation is catalyzed by a mitochondrial enzyme, *3-hydroxyanthranilic acid oxidase,* present in liver and kidney. The enzyme is an oxygenase, requiring Fe^{++} and a sulfhydryl compound; molecular oxygen is incorporated into the product, α-amino-β-carboxymuconic-Δ-semialdehyde. The apoenzyme, free of iron, can be reactivated by addition of Fe^{++}; removal of iron is blocked by addition of substrate. Sulfhydryl, in the form of cysteine or reduced glutathione, produces conformational changes in the enzyme associated with its activation. α-Amino-β-carboxymuconic-Δ-semialdehyde is a branch point in tryptophan degradation; it may be converted to quinolinic and nicotinic acids or follow a pathway to glutaric acid and thus to acetyl CoA. These transformations are indicated in Fig. 26.13.

Formation of nicotinic acid from quinolinic acid occurs via reaction of the latter with 5-phosphoribosyl-1-pyrophosphate, with accompanying loss of CO_2, to form nicotinic acid ribonucleotide in a reaction catalyzed by *quinolinic acid transphosphoribosylase.* Nicotinic acid (niacin) ribonucleotide is directly utilized for DPN synthesis (page 641).

Pyridoxine-deficient rats and human beings excrete abnormally large amounts of kynurenine and xanthurenic acid. The latter is also excreted in abnormally large amounts in pregnancy; this may be related to elevated blood levels of estrogenic and progestational hormones, since estrogen has been reported to inhibit the conversion in vitro of kynurenine to kynurenic acid. Xanthurenic acid may suffer loss of a hydroxyl group with formation of 8-hydroxyquinaldic acid.

Indoleacetic acid (page 590) is present in the urine of mammals as indoleaceturic acid, formed by conjugation with glycine. A possible precursor is tryptamine, known to arise in the large intestine by bacterial decarboxylation of tryptophan. Indoleacetic acid and indolelactic acid are excreted in increased amounts in the urine of phenylketonuric patients (page 549). The microorganisms of the large intestine can further degrade indoleacetic acid to yield skatole, skatoxyl, indole, and indoxyl.

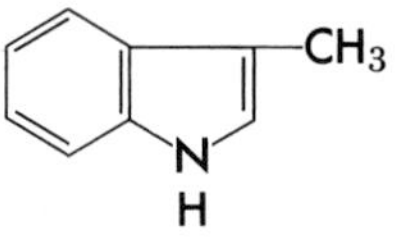

Skatole

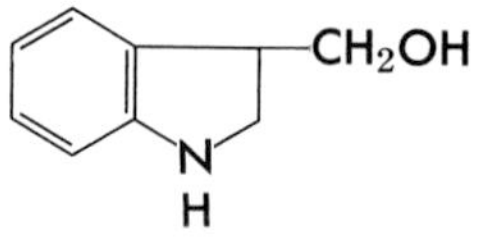

Skatoxyl

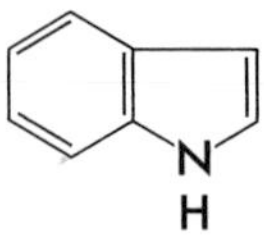

Indole

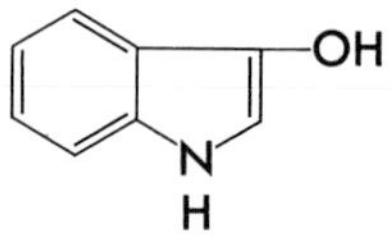

Indoxyl

Indole may also arise directly from tryptophan as a result of bacterial action. Various bacteria contain the enzyme *tryptophanase* which catalyzes cleavage of the side chain of tryptophan, with formation of indole, ammonia, and pyruvic acid. Pyridoxal phosphate is a coenzyme for the reaction. Crystalline tryptophanase of *E. coli* catalyzes several α,β-elimination and β-replacement reactions, including the following.

$$\text{Serine} \longrightarrow \text{pyruvate} + NH_3$$
$$\text{Cysteine} + H_2O \longrightarrow \text{pyruvate} + NH_3 + H_2S$$
$$\text{Serine} + \text{indole} \longrightarrow \text{tryptophan} + H_2O$$
$$\text{Cysteine} + \text{indole} \longrightarrow \text{tryptophan} + H_2S$$

Skatole and indole contribute to the unpleasant odor of feces. Small amounts of indoxyl and skatoxyl enter the circulation from the gut, are conjugated either with sulfate or glucuronic acid in the liver, and excreted in the urine as ester sulfates or as glucosiduronates. The potassium salt of indoxylsulfate is known as *indican;* the urinary concentration of indoxylsulfate has been used as a qualitative indication of the extent of bacterial activity in the large intestine.

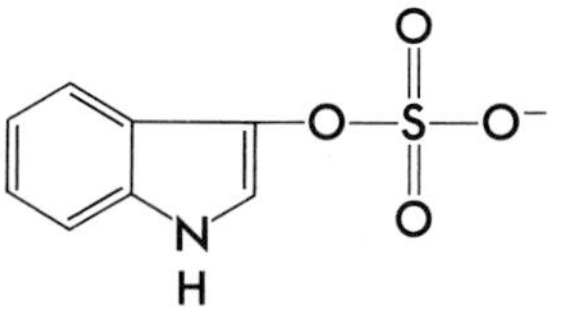

Indoxylsulfate

REFERENCES

Books

Baldwin, E., "Dynamic Aspects of Biochemistry," 3d ed., Cambridge University Press, New York, 1957.

Du Vigneaud, V., "A Trail of Research in Sulfur Chemistry and Metabolism and Related Fields," Cornell University Press, Ithaca, N.Y., 1952.

Greenberg, D. M., ed., "Metabolic Pathways," vol. II, Academic Press, Inc., New York, 1961.
Gross, F., ed., "Protein Metabolism," Springer Verlag OHG, Berlin, 1962.
Harris, H., "Human Biochemical Genetics," Cambridge University Press, New York, 1959.
McElroy, W. D., and Glass, B., eds., "Amino Acid Metabolism," Johns Hopkins Press, Baltimore, 1955.
Meister, A., "Biochemistry of the Amino Acids," 2d ed., vols. I and II, Academic Press, Inc., New York, 1965.
Munro, H. N., and Allison, J. B., eds., "Mammalian Protein Metabolism," vols. I and II, Academic Press, Inc., New York, 1964.
Nyhan, W. L., ed., "Amino Acid Metabolism and Genetic Variation," McGraw-Hill Book Company, New York, 1967.
Schoenheimer, R., "The Dynamic State of Body Constituents," Harvard University Press, Cambridge, Mass., 1942.
Shapiro, S. K., and Schlenk, F., eds., "Transmethylation and Methionine Biosynthesis," University of Chicago Press, Chicago, 1965.
Snell, E. E., Fasella, P. M., Braunstein, A., and Fanelli, R., "Chemical and Biological Aspects of Pyridoxal Catalysis," Pergamon Press, New York, 1963.
Stanbury, J. B., Wyngaarden, J. B., and Fredrickson, D. S., eds., "The Metabolic Basis of Inherited Disease," 2d ed., McGraw-Hill Book Company, New York, 1966.
Young, L., and Maw, A., "The Metabolism of Sulphur Compounds," John Wiley & Sons, Inc., New York, 1958.

Review Articles

Black, S., Biochemistry of Sulfur-containing Compounds, *Ann. Rev. Biochem.,* **32,** 399–418, 1963.
Broquist, H. P., and Trupin, J. S., Amino Acid Metabolism, *Ann. Rev. Biochem.,* **35,** 231–274, 1966.
Christensen, H. N., and Oxender, D. L., Transport of Amino Acids into and across Cells, *Am. J. Clin. Nutr.,* **8,** 131–136, 1960.
Cohen, P. P., and Brown, G. W., Jr., Ammonia Metabolism and Urea Biosynthesis, in M. Florkin and H. S. Mason, eds., "Comparative Biochemistry," vol. II, pp. 161–244, Academic Press, Inc., New York, 1961.
Cohen, P. P., and Sallach, H. J., Nitrogen Metabolism of Amino Acids, in D. M. Greenberg, ed., "Metabolic Pathways," vol. II, pp., 1–78, Academic Press, Inc., New York, 1961.
Dalgliesh, C. E., Metabolism of the Aromatic Amino Acids, *Advances in Protein Chem.,* **10,** 31–150, 1955.
Davis, B. D., Intermediates in Amino Acid Biosynthesis, *Advances in Enzymol.,* **16,** 247–312, 1955.
Davis, B. D., The Teleonomic Significance of Biosynthetic Control Mechanisms, *Cold Spring Harbor Symp. Quant. Biol.,* **26,** 1–10, 1961.
Davison, H., Physiological Role of Monamine Oxidase, *Physiol. Revs.,* **38,** 729–747, 1958.
Fasella, P., Pyridoxal Phosphate, *Ann. Rev. Biochem.,* **36,** 185–210, 1967.
Fisher, R. B., Absorption of Protein, *Brit. Med. Bull.,* **23,** 241–246, 1967.
Greenberg, D. M., Biological Methylation, *Advances in Enzymol.,* **25,** 395–432, 1963.
Guirard, B. M., and Snell, E. E., Vitamin B_6 Function in Transamination and Decarboxylation Reactions, in M. Florkin and E. H. Stotz, eds., "Comprehensive Biochemistry," vol. 15, chap. 5, Elsevier Publishing Company, Amsterdam, distributed by American Elsevier Publishing Company, New York, 1964.
Jones, M. E., Amino Acid Metabolism, *Ann. Rev. Biochem.,* **34,** 381–418, 1965.
Knox, W. E., Sir Archibald Garrod's "Inborn Errors of Metabolism," I. Cystinuria, *Am. J. Human Genet.,* **10,** 3–32, 1958.

Knox, W. E., Sir Archibald Garrod's "Inborn Errors of Metabolism," II. Alkaptonuria, *Am. J. Human Genet.*, **10,** 95–124, 1958.

La Du, B. N., and Zannoni, V., The Role of Ascorbic Acid in Tyrosine Metabolism, *Ann. N.Y. Acad. Sci.*, **92,** 175–191, 1961.

Mann, F. C., The Effects of Complete and of Partial Removal of the Liver, *Medicine,* **6,** 419–511, 1927.

Moyed, H. S., and Umbarger, H. E., Regulation of Biosynthetic Pathways, *Physiol. Revs.,* **42,** 444–466, 1962.

Quastel, J. H., Intestinal Absorption of Sugars and Amino Acids, *Am. J. Clin. Nutr.,* **8,** 137–146, 1960.

Ratner, S., Urea Synthesis and Metabolism of Arginine and Citrulline, *Advances in Enzymol.,* **15,** 319–387, 1954.

Schayer, R. W., Catabolism of Physiological Quantities of Histamine in Vivo, *Physiol. Revs.,* **39,** 116–126, 1959.

Sprinson, D. B., The Biosynthesis of Aromatic Compounds from D-Glucose, *Advances in Carbohydrate Chem.,* **15,** 235–270, 1960.

Tabor, H., and Tabor, C. W., Spermidine, Spermine, and Related Amines, *Pharmacol. Revs.,* **16,** 245–300, 1964.

Wilson, L. G., Metabolism of Sulfate, *Ann. Rev. Plant Physiol.,* **13,** 201–224, 1962.

27. Metabolism of Purines, Pyrimidines, and Nucleotides

The chemistry of the nucleic acids has been presented earlier (Chap. 9), and the importance of various nucleotides has been discussed in preceding chapters. This chapter will be concerned with the origin and metabolic fate of the nucleotides and their constituents. The biosynthesis of the nucleic acids and their role in protein synthesis are considered in the following chapters.

The purines and pyrimidines of nucleic acids are not required in the animal diet and can be synthesized in vivo. In 1874 Miescher found that salmon fast during their long migration to headwaters of rivers for spawning, while the gonads grow at the expense of muscle proteins. This gonadal hypertrophy includes synthesis of large amounts of nucleoproteins. On the other hand, many microorganisms require the presence of specific purines or pyrimidines in the culture media, and, indeed, some even appear to require, either absolutely or for optimal growth, certain nucleosides, such as thymidine. In addition, mutant strains of various synthetically competent bacteria and fungi (*Neurospora*) are unable to synthesize certain purines or pyrimidines.

DIGESTION AND ABSORPTION OF NUCLEIC ACIDS

The acidity of the gastric juice results in cleavage of the nucleoproteins found in natural dietary constituents, and digestion of the resulting proteins (histones, protamines, etc.) begins in the stomach. The nucleic acids are unaffected by gastric enzymes, and their digestion occurs mainly in the duodenum. The pancreas forms *nucleases,* and these are secreted in the pancreatic juice. Pancreatic *ribonuclease* hydrolyzes only ribonucleic acids, liberating pyrimidine mononucleotides and oligonucleotides terminating in pyrimidine nucleoside 3′-phosphate residues (page 240). *Deoxyribonuclease* acts in the presence of Mg^{++} or Mn^{++} and specifically hydrolyzes deoxyribonucleic acids to small oligonucleotides (page 190). The intestinal mucosa is also believed to form *nucleases* and *diesterases* which aid in the digestion of low molecular weight nucleic acids and oligonucleotides.

Liberated nucleotides are hydrolyzed by intestinal *phosphatases* or *nucleotidases,* yielding nucleosides and orthophosphate. Little is known concerning the individuality or specificity of such enzymes, although it is probable that many separate enzymes exist. A specific intestinal *phosphatase* cleaves adenosine 5′-phosphate but does not attack the isomeric adenosine 3′-phosphate or adenosine 2′-phosphate.

It is doubtful that nucleosides are hydrolyzed in the intestine; they are probably absorbed as such. Extracts of various tissues, *e.g.*, spleen, liver, kidney, bone marrow, etc., cleave the N-glycosidic linkage of nucleosides. The metabolism of these compounds probably occurs mainly in these tissues. The so-called *nucleosidases* have not been extensively investigated or purified, and knowledge of these enzymes is fragmentary. Present evidence indicates that there are probably specific *purine* and *pyrimidine nucleosidases.* The enzymes are of two types and may be classed as hydrolytic and phosphorolytic as shown in the following examples.

$$\textbf{Uridine} + \textbf{H}_2\textbf{O} \xrightarrow[\text{nucleosidase}]{\text{pyrimidine}} \textbf{uracil} + \textbf{ribose}$$

and

$$\textbf{Guanosine} + \textbf{phosphate} \underset{\text{phosphorylase}}{\overset{\text{nucleoside}}{\rightleftharpoons}} \textbf{guanine} + \textbf{ribose 1-phosphate}$$

The latter enzyme, *purine nucleoside phosphorylase,* is found in liver and other tissues and acts upon several purine nucleosides. The equilibrium point of the reaction suggests that this enzyme has a synthetic as well as a degradative role (see below).

PATHWAYS OF PURINE SYNTHESIS

The origin of the atoms of the purine ring was first established in the intact animal by administering suitable precursors containing labeled atoms. These studies, mainly by Buchanan and coworkers, yielded the following general picture of the origin of the purine nucleus.

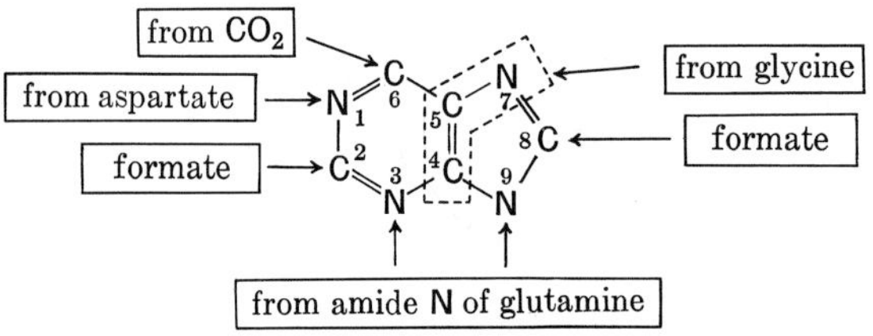

Carbon atoms 2 and 8 derive from formate or the 1-carbon unit arising from various compounds, *e.g.*, serine and glycine (page 550). Carbon atom 6 originates from carbon dioxide. Glycine contributes carbon atoms at positions 4 and 5 and the nitrogen at 7; the nitrogen atom at position 1 derives from aspartic acid, and glutamine amide nitrogen contributes the nitrogen at positions 3 and 9.

The pathways for purine biosynthesis in many species (mammals, birds, yeast, bacteria) have been studied, and the general route is essentially the same in all organisms. In essence, this process consists of a series of successive reactions by which the purine ring system on carbon 1 of ribose 5-phosphate is formed, thus leading directly to the formation of purine ribonucleotides. Neither free purines nor nucleosides appear as intermediates in this sequence. Deoxyribonucleotide synthesis is considered later.

5-Phosphoribosyl-1-pyrophosphate. This compound is a key substance in the biosynthesis of both purine and pyrimidine nucleotides. Synthesis of 5-phosphoribosyl-1-pyrophosphate occurs from ribose 5-phosphate and ATP in a reaction which is unusual in that it is catalyzed by a kinase which transfers pyrophosphate rather than phosphate.

(*a*)

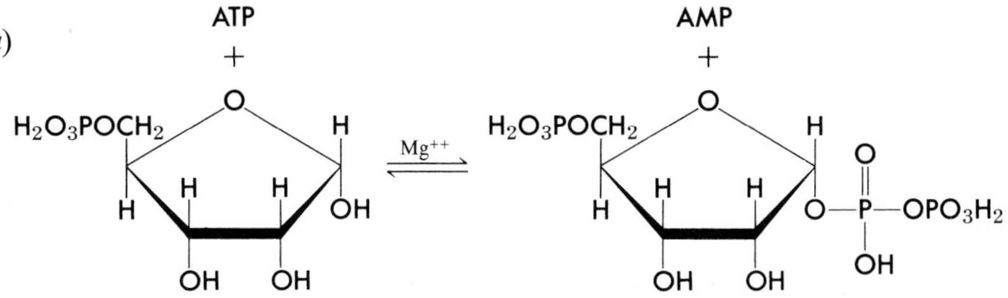

Ribose 5-phosphate **α-5-Phosphoribosyl-1-pyrophosphate**

5-Phosphoribosyl-1-amine. An acid-labile amino sugar, 5-phosphoribosyl-1-amine, is formed from glutamine and the above pyrophosphate in the presence of *glutamine phosphoribosyl pyrophosphate amidotransferase.* The enzyme from pigeon liver contains eight atoms of iron per molecule of 212,000.

(*b*)

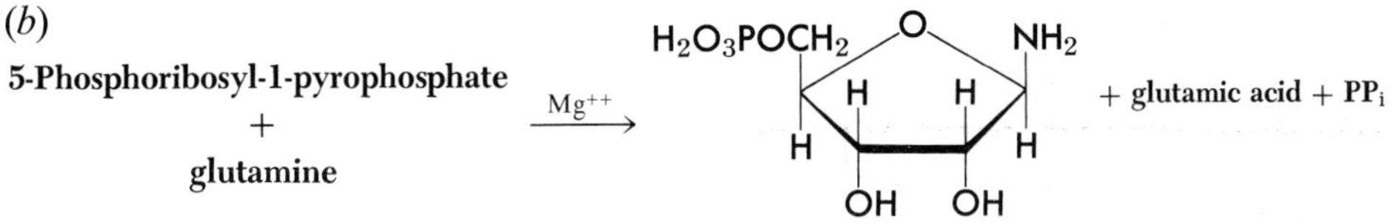

5-Phosphoribosyl-1-amine

The pyrophosphate bond in phosphoribosyl pyrophosphate is in α linkage, whereas the glycosidic bond in the purine nucleotides, as well as the configuration at C-1′ in 5-phosphoribosyl-1-amine, are of the β configuration. Thus in reaction (b) the displacement of pyrophosphate by the amino group of glutamine is accompanied by an inversion of spatial configuration at C-1′. This reaction is the "committed" metabolic step in purine biosynthesis. As will be indicated later, the rate of purine biosynthesis is subject to feedback inhibition by purine nucleotides.

Azaserine, an antibiotic isolated from a species of *Streptomyces,* inhibits utilization of glutamine for the formation of 5-phosphoribosyl-1-amine.

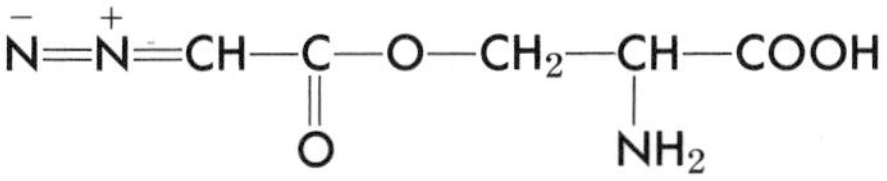

Azaserine

Azaserine inhibits the growth of certain neoplasms, because of its ability to interfere with the synthesis of purine nucleotides. Another reaction shown below, (*e*), which utilizes glutamine, is also inhibited by azaserine. Indeed, azaserine appears to inhibit each of the known enzymic reactions in which the amide N of glutamine is transferred to another carbon chain.

Glycinamide Ribonucleotide. In this next reaction, the entire structure of glycine is conjugated with 5-phosphoribosyl-1-amine. The details of the reaction remain to be elucidated. The linkage, —C(=O)—N(H)—, resembles a peptide bond.

(*c*) **Phosphoribosylamine + ATP + glycine** $\xrightarrow{Mg^{++}}$

ADP + P_i + 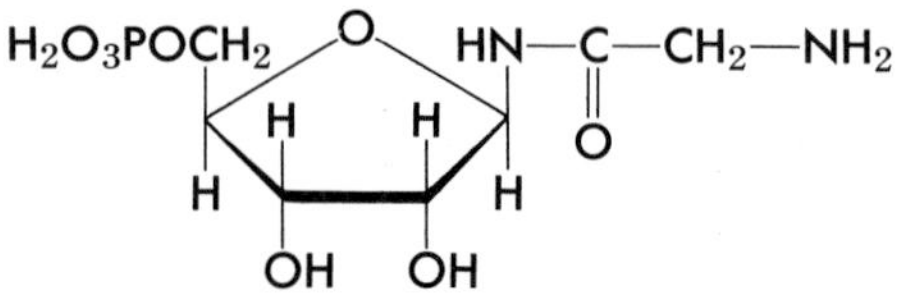

Glycinamide ribonucleotide

Formylglycinamide Ribonucleotide. Formylation of glycinamide ribonucleotide is accomplished by a transfer reaction utilizing the formyl folic acid derivative (page 551) in a reaction catalyzed by *glycinamide ribonucleotide transformylase.* Ribose-P in the formula below and in subsequent reactions represents the ribose 5′-phosphate portion of the compound.

(*d*) **Glycinamide ribonucleotide + N^5,N^{10}-methenyltetrahydrofolate** $\xrightarrow{+H_2O}$

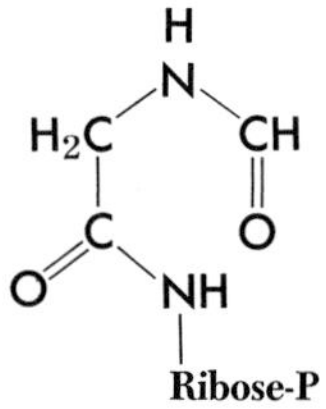

α-N-Formylglycinamide ribonucleotide + tetrahydrofolate + H^+

Formylglycinamidine Ribonucleotide. The next step involves transfer of an NH_2 group from glutamine to α-N-formylglycinamide ribonucleotide.

(*e*) 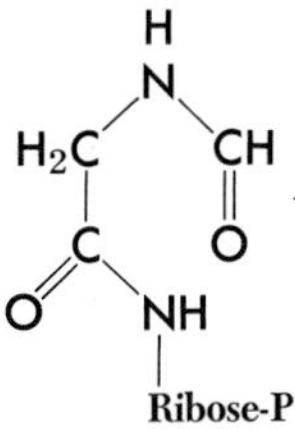

+ glutamine + ATP + H_2O $\xrightarrow{Mg^{++}}$

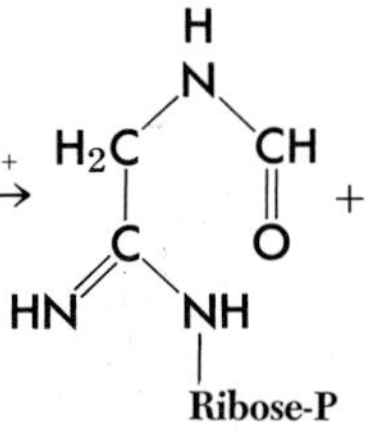

+ glutamic acid + ADP + P_i

α-N-Formylglycinamide ribonucleotide — **α-N-Formylglycinamidine ribonucleotide**

This step resembles reaction (*b*), which also involves glutamine, in being strongly inhibited by azaserine and in being irreversible. When azaserine reacts with the *amido-ligase,* which catalyzes reaction (*e*), an essential thiol group of a cysteine residue is acylated to form a thiol ether compound; after acid hydrolysis, 5-carboxymethylcysteine is recovered.

5-Aminoimidazole Ribonucleotide. The first ring closure yields an imidazole derivative.

(*f*) α-N-Formylglycinamidine ribonucleotide + ATP $\xrightarrow[K^+]{Mg^{++}}$

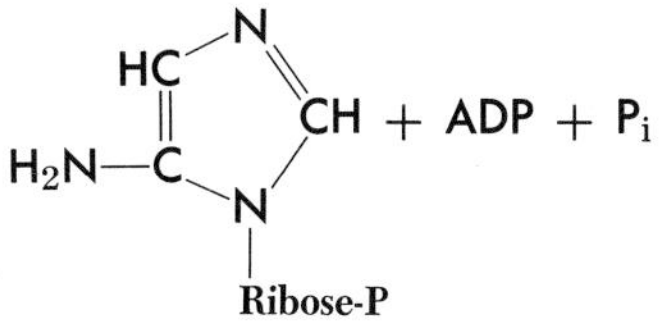

5-Aminoimidazole ribonucleotide

The enzyme catalyzing this essentially irreversible reaction also requires K^+ ions.

5-Aminoimidazole-4-carboxylic Acid Ribonucleotide. High concentrations of bicarbonate are required for the following reaction.

(*g*) 5-Aminoimidazole ribonucleotide + CO_2 $\rightleftharpoons$

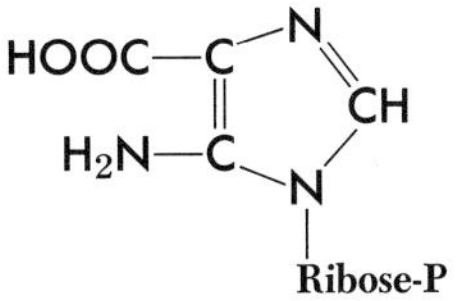

5-Aminoimidazole-4-carboxylic acid ribonucleotide

5-Aminoimidazole-4-N-succinocarboxamide Ribonucleotide. Formation of this compound is reversible.

(*h*) 5-Aminoimidazole-4-carboxylic acid ribonucleotide + ATP + aspartic acid $\overset{Mg^{++}}{\rightleftharpoons}$

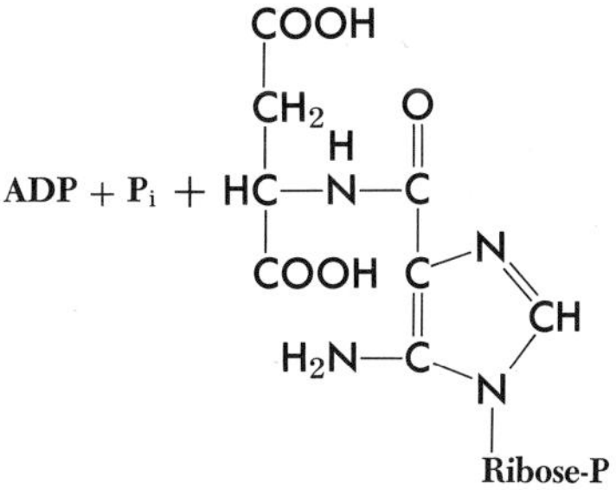

5-Aminoimidazole-4-N-succinocarboxamide ribonucleotide

5-Aminoimidazole-4-carboxamide Ribonucleotide. The enzyme catalyzing this nonhydrolytic cleavage reaction is believed to be identical with *adenylosuccinase* (page 625).

(*i*) 5-Aminoimidazole-4-N-succinocarboxamide ribonucleotide $\rightleftharpoons$

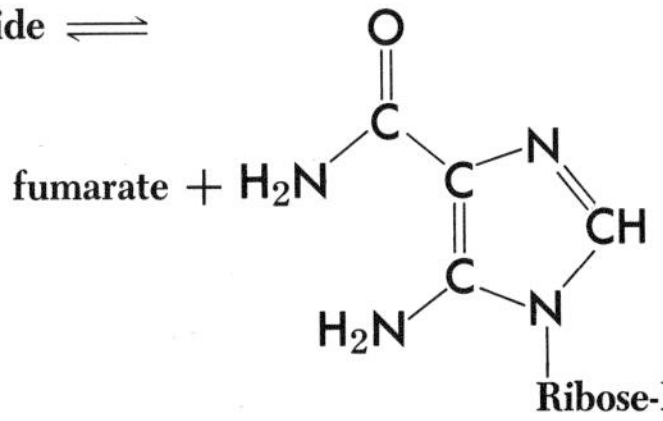

5-Aminoimidazole-4-carboxamide ribonucleotide

Aminoimidazolecarboxamide was first isolated from cultures of *Escherichia coli* which had been inhibited with sulfonamides, and later from cultures of mutant strains of other microorganisms unable to form purines. Subsequently, the ribonucleoside and ribonucleotide of this substance were isolated from sulfonamide-inhibited cultures of *E. coli*. The action of sulfonamides in preventing bacterial growth is due to inhibition of folic acid synthesis (page 241). Inasmuch as a derivative of formyl folic acid (page 552) is required for the formylation of 5-aminoimidazole-4-carboxamide ribonucleotide, as shown below, it is possible to explain the accumulation of aminoimidazolecarboxamide ribonucleotide in the presence of sulfonamides. The formation of the nucleoside and the carboxamide undoubtedly result from hydrolysis of the ribonucleotide.

5-Formamidoimidazole-4-carboxamide Ribonucleotide. The *transformylase* catalyzing this reaction requires K^+ ions.

(*j*) **5-Aminoimidazole-4-carboxamide ribonucleotide + N^{10}-formyltetrahydrofolic acid** $\xrightarrow{K^+}$

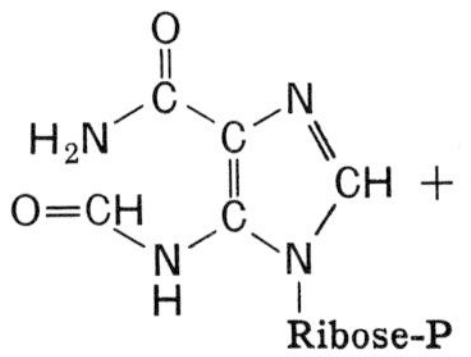

+ tetrahydrofolic acid

5-Formamidoimidazole-4-carboxamide ribonucleotide

Many metabolic antagonists of folic acid (page 644) which inhibit growth do so by inhibition of purine nucleotide biosynthesis at reaction (*j*).

Inosinic Acid. Closure of the ring by *inosinicase* yields inosinic acid (hypoxanthine ribonucleotide), the first product in the synthetic pathway with the completed purine ring structure.

Formamidoimidazolecarboxamide ribonucleotide ⇌

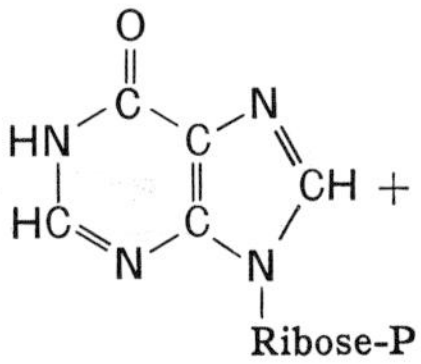

+ H_2O

Inosinic acid

Purine Nucleotide Interconversions. Formation of adenylic acid from inosinic acid occurs through the initial formation of adenylosuccinic acid, with participation of aspartic acid and guanosine triphosphate. Note that the nucleoside triphosphate derivative of one purine (guanine) is required for formation of another purine nucleotide.

Inosinic acid + GTP + L-aspartic acid $\xrightarrow{Mg^{++}}$ **GDP + P_i +**

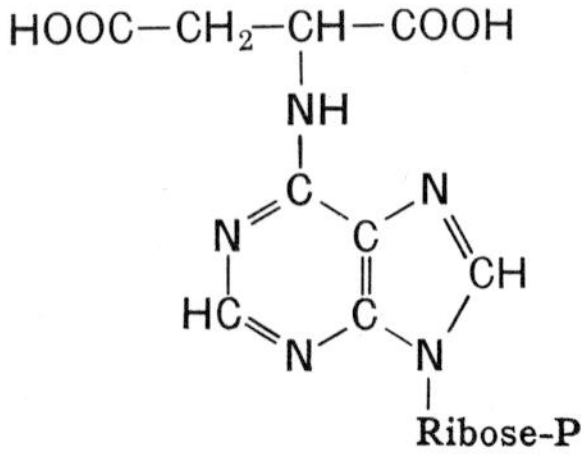

Adenylosuccinic acid

Nonhydrolytic cleavage of adenylosuccinic acid yields adenylic acid and fumaric acid.

$$\textbf{Adenylosuccinic acid} \rightleftharpoons \textbf{adenylic acid} + \textbf{fumaric acid}$$

This reaction is analogous to the cleavage of 5-aminoimidazole-4-N-succinocarboxamide ribonucleotide (reaction *i*) and is probably catalyzed by the same enzyme, *adenylosuccinase.* On extensive purification of the enzyme, the ratio of the two activities is unchanged. Moreover, mutant strains of several microorganisms which lack the ability to catalyze one reaction cannot catalyze the other.

The synthesis of guanylic acid from inosinic acid proceeds first by oxidation to xanthylic acid; this reaction requires K^+ ions and is inhibited by GMP. The next step requires glutamine and ATP.

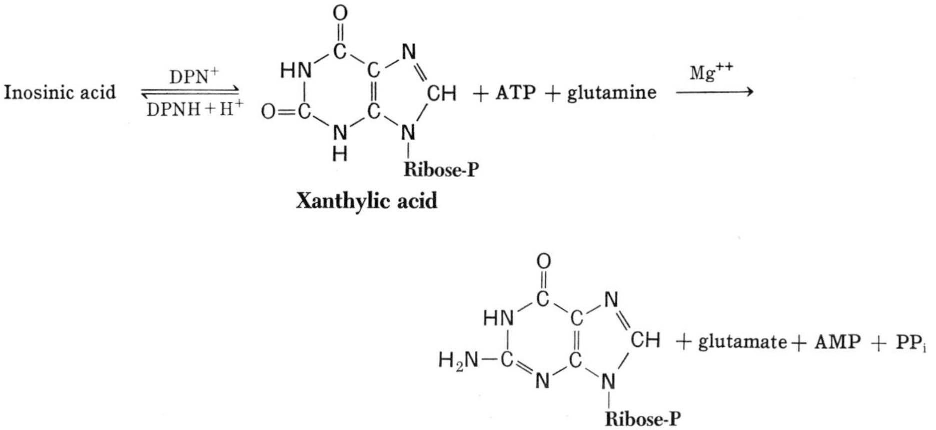

Guanylic acid

As in other reactions requiring glutamine (page 621), the formation of guanylic acid is irreversible and is inhibited by azaserine.

In summary, the synthesis of inosinic acid from elementary precursors can be regarded as the result of the sum of the somewhat artificial, composite equations (*a*) and (*b*).

(*a*) $\mathbf{2NH_4^+ + 2HCOO^- + HCO_3^- + glycine + aspartate + ribose\ 5\text{-}phosphate \longrightarrow inosinic\ acid + fumarate + 9H_2O}$

(*b*) $\mathbf{9ATP + 9H_2O \longrightarrow 8ADP + 8P_i + AMP + PP_i + 9H^+}$

The energy needed is provided from nine equivalents of ATP, including two equivalents required for glutamine formation and two equivalents for formate activation. The equations, as written, would apply only in those organisms which can activate free formate. In higher vertebrates, the one-carbon unit is derived from serine (page 551).

Other Pathways of Purine Nucleotide Formation. As described above, adenylic acid and guanylic acid are formed through the intermediate, inosinic acid. In most species, including mammals, these are undoubtedly the primary pathways for purine nucleotide formation. However, purine nucleotides can also be formed from free purines and from purine nucleosides. These routes may be regarded as salvage pathways in the tissues, permitting reutilization of purines or purine derivatives derived by breakdown of nucleic acids or nucleotides.

Free purines can react directly with 5-phosphoribosyl-1-pyrophosphate to yield 5′-nucleotides. The reversible reactions shown below are catalyzed by two distinct enzymes of liver: an *adenine phosphoribosyl transferase* and a *hypoxanthine-guanine phosphoribosyl transferase.*

$$\text{Adenine} + \text{phosphoribosylpyrophosphate} \rightleftharpoons \text{adenylic acid} + PP_i$$
$$\text{Guanine} + \text{phosphoribosylpyrophosphate} \rightleftharpoons \text{guanylic acid} + PP_i$$
$$\text{Hypoxanthine} + \text{phosphoribosylpyrophosphate} \rightleftharpoons \text{inosinic acid} + PP_i$$

Inasmuch as nucleotides are utilized for nucleic acid synthesis (Chaps. 28 and 29), these reactions can result in the incorporation of purines into nucleic acids.

Other salvage pathways involve conversion of free purines to nucleosides and nucleosides to nucleotides. Such routes may also be involved in the incorporation of purines and their respective nucleosides into nucleic acids.

Known reactions for purine nucleoside formation, catalyzed by *nucleoside phosphorylase,* are the following.

$$\text{Hypoxanthine} + \text{ribose 1-phosphate} \rightleftharpoons \text{inosine} + P_i$$
$$\text{Guanine} + \text{ribose 1-phosphate} \rightleftharpoons \text{guanosine} + P_i$$

Conversion of a nucleoside to a nucleotide occurs by the following reaction.

$$\text{Adenosine} + \text{ATP} \xrightarrow[\text{kinase}]{\text{adenosine}} \text{adenylic acid} + \text{ADP}$$

In summary, the primary pathway of formation of purine nucleotides in mammalian tissues is a *de novo* route which involves synthesis from acyclic precursors and simple compounds. There is no evidence that synthesis of free purines ever occurs. Secondary pathways may be regarded as salvage routes in which purines or nucleosides, originating from the intestinal tract or from intracellular degradative processes, are reutilized by conversion to nucleotides.

Regulation of Purine Synthesis. As already noted (page 621), the committed step in purine biosynthesis is the formation of 5-phosphoribosylamine. The enzyme, *glutamine phosphoribosyl pyrophosphate amidotransferase,* which catalyzes this reaction is inhibited by the mono-, di-, and triphosphates of guanosine and adenosine, thus retarding significantly the subsequent reactions of the entire pathway. Since mixtures of the adenine and guanine derivatives inhibit more strongly than expected from the effects of the individual compounds, separate modifier sites on the enzyme have been postulated.

Inosinic acid (IMP) is the precursor for both AMP and GMP (page 624). It has been shown that GMP inhibits formation of xanthylic acid and that AMP inhibits synthesis of adenylosuccinic acid from IMP. In effect, these feedback mechanisms serve to prevent further formation of AMP or GMP when either is present in excessive amounts, whether they arise by *de novo* synthesis or from a salvage pathway. Some of the regulatory steps in purine nucleotide synthesis are shown in Fig. 27.1. Note, however, that much of the enzymology has been obtained from studies of pigeon liver and *E. coli,* and the control mechanisms have been studied almost

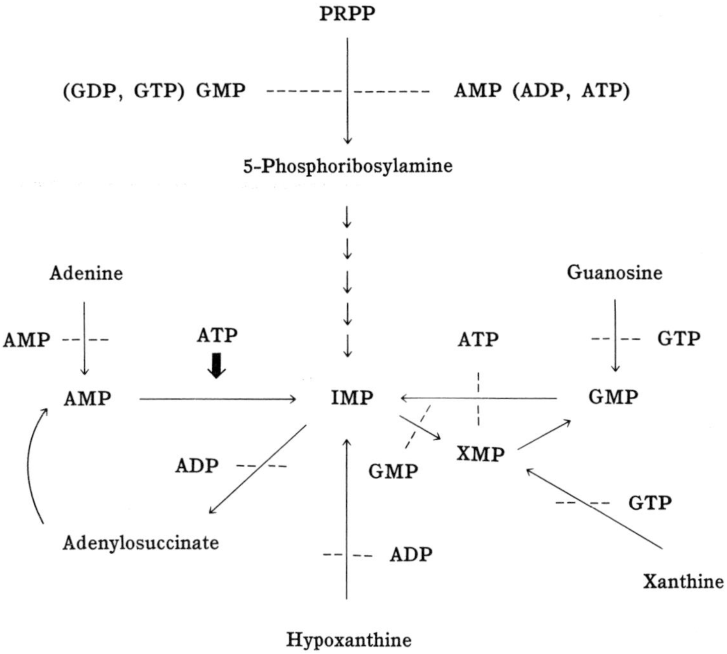

Fig. 27.1. Some interrelationships and control mechanisms in purine nucleotide biosynthesis. ⬇ indicates activation; -----indicates inhibition; **PRPP**, 5-phosphoribosyl-1-pyrophosphate; **IMP**, inosinic acid; **XMP**, xanthylic acid.

entirely with the enzymes of the microorganism. Similar control mechanisms are undoubtedly operative in mammalian tissues, but the details have not been elucidated.

CATABOLISM OF PURINES

In mammals, most of the nitrogen in the rings of administered adenine, guanine, xanthine, or hypoxanthine appears in the urine in the form of uric acid or allantoin. The purine ring is therefore not completely degraded; only small amounts of urea or ammonia are derived from this source.

Adenase and *guanase* are highly specific deaminases which act hydrolytically according to the following reactions.

$$\textbf{Adenine} + \textbf{H}_2\textbf{O} \xrightarrow{\text{adenase}} \textbf{hypoxanthine} + \textbf{NH}_3$$

$$\textbf{Guanine} + \textbf{H}_2\textbf{O} \xrightarrow{\text{guanase}} \textbf{xanthine} + \textbf{NH}_3$$

Guanase is widely distributed in liver, kidney, spleen, etc. However, adenase appears to be of more limited distribution, and deamination of adenine may occur mainly in combination in the nucleotide, adenylic acid, through the irreversible action of the enzyme, *adenylic acid deaminase*, which is present in large amounts in muscle (Chap. 36).

$$\textbf{Adenylic acid} + \textbf{H}_2\textbf{O} \xrightarrow{\text{adenylic acid deaminase}} \textbf{inosinic acid} + \textbf{NH}_3$$

Inosinic acid found in muscle reflects the high activity of the abundant adenylic acid deaminase rather than the rate of *de novo* purine synthesis. *Guanosine* and *adenosine deaminases* have also been found in animal tissues.

Xanthine oxidase (page 366), a flavin enzyme, catalyzes the oxidation of both hypoxanthine and xanthine.

$$\textbf{Hypoxanthine} + O_2 + H_2O \xrightarrow[\text{oxidase}]{\text{xanthine}} \textbf{xanthine} + H_2O_2$$

$$\textbf{Xanthine} + O_2 + H_2O \xrightarrow[\text{oxidase}]{\text{xanthine}} \textbf{uric acid} + H_2O_2$$

Uric acid formation in mammals occurs in the liver; hepatectomy results in cessation of uric acid production. In some species, uric acid is aerobically oxidized by a liver enzyme called *uricase,* a copper-protein. The over-all reaction catalyzed by this enzyme is the following.

$$\textbf{Uric acid} + 2H_2O + O_2 \longrightarrow \textbf{allantoin} + CO_2 + H_2O_2$$

The stages of purine degradation from adenine and guanine to uric acid and allantoin are summarized in Fig. 27.2.

Man and other primates do not possess uricase; hence, uric acid is the main end product of purine metabolism in these species. Other mammals which have uricase excrete allantoin as the terminal compound of purine metabolism. The Dalmatian dog, like other dogs, possesses uricase in the liver, yet excretes uric acid because of a very low renal threshold for uric acid. The pig, which is deficient in guanase, excretes guanine as well as allantoin. Indeed, guanine gout has been reported in this species; it is due to the deposition of guanine crystals in the joints and is analogous to human gout, in which monosodium urate may accumulate in cartilage (see below). Guanine is the terminal product of purine metabolism in spiders.

The utilization of free guanine or adenine for nucleotide or nucleic acid synthesis in various species probably depends on the relative activities of the corresponding deaminases and the oxidases as compared to the activities of the enzymes which can convert these purines to the nucleotides. Thus, some species can utilize adenine or guanine for nucleic acid synthesis, by employing salvage pathways (page 626), whereas others cannot do so.

A summary of the interconversions of purine derivatives (Fig. 27.3, page 630) shows that reactions may occur at the level of nucleotides, nucleosides, or with the free purines. Not all these reactions necessarily occur in every tissue or in every species, but there is evidence for each pathway shown. Of the pathways given, only two are known to be reversible by the same enzymic system. The reaction pathways include oxidative, hydrolytic, pyrophosphorolytic, and phosphorolytic reactions. The first three types of reaction may be considered to be essentially irreversible. In most instances, the biosynthetic pathways and the enzymes involved differ from the degradative reactions; this is indicated in the diagram by separate pathways of the arrows.

Uric Acid Production in Man. Uric acid production and excretion proceed at a relatively constant rate in man when the diet is free of purines. This uric acid is derived from the endogenous purine metabolism and, as with many body constituents, reflects a steady state in which the rates of purine synthesis and purine

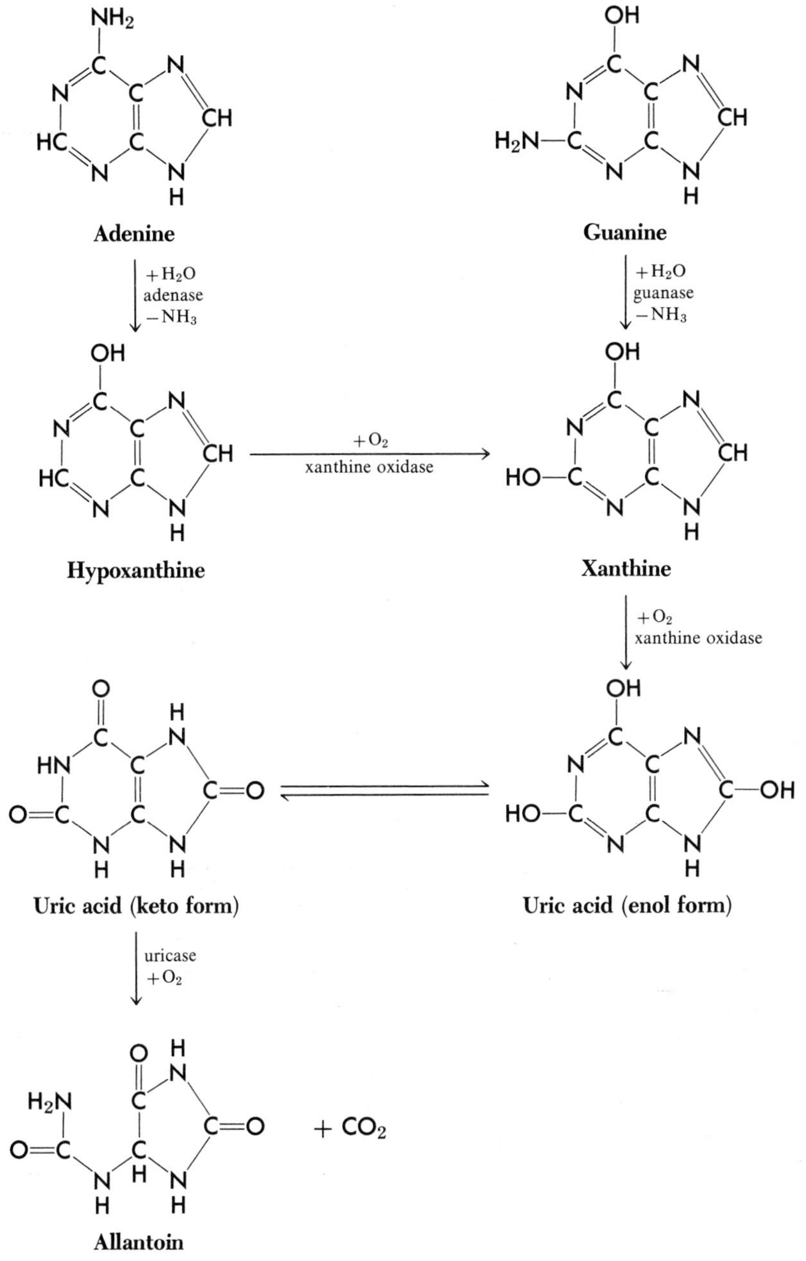

Fig. 27.2. The metabolic degradation of adenine and guanine.

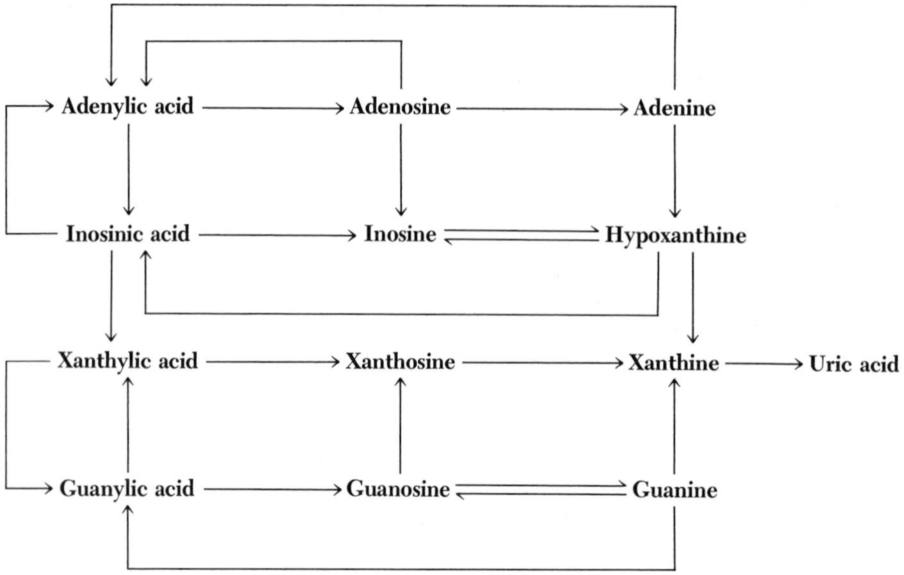

Fig. 27.3. Summary of interconversions of purine derivatives.

catabolism, measured by the excretion of uric acid, are approximately equal. Some protein foods, such as milk, cheese, and eggs, are low in purines, whereas foods rich in nucleoproteins, *e.g.*, liver and pancreas, are high in purines.

The concentration of uric acid in normal plasma is about 2 to 6 mg. per 100 ml., with an average for adults of approximately 3.5 mg. per 100 ml. for females and about 4.5 mg. per 100 ml. for males. These levels are elevated in gout, a disease in which large amounts of the sparingly soluble monosodium urate are deposited as *tophi* in cartilage. Uric acid deposits may also be found as calculi in the kidney, with resultant renal damage. A large fraction of all renal calculi consists of the sparingly soluble uric acid and its salts.

There is a large sex difference in the occurrence of gout; only about 5 per cent of the cases occur in females. In many of the cases there appears to be a familial incidence. Hyperuricemia (elevated blood uric acid) is frequently observed in the asymptomatic male relatives of gouty individuals. However, many cases of hyperuricemia may be due to other factors, *e.g.*, impaired renal function, toxemia of pregnancy, essential hypertension, and leukemia.

Studies have been made of rates of uric acid production and excretion in man. Injection of uric acid isotopically labeled with ^{15}N permits estimation of total stores of miscible uric acid from isotope dilution measurements. Stetten and coworkers have shown that normal individuals contain about 1.1 g. of uric acid, with about one-sixth of this in the plasma and about five-sixths in the extravascular water. From the rate of decline of the abundance of ^{15}N in the urinary uric acid, it was calculated that 50 to 75 per cent of the uric acid was replaced each day by newly formed uric acid. This represents a daily production of about 0.5 to 0.86 g. of uric acid, presumably arising from degradation of nucleic acids and nucleotides of the body. The production of uric acid exceeded the urinary output by 100 to

250 mg. per day. Some of the isotopic nitrogen was found in urinary urea and ammonia, indicating a partial metabolic breakdown of the purine nucleus. Similar studies in gouty individuals revealed that the miscible pool of uric acid was uniformly elevated. It appeared to include a portion of the solid-phase urate of tophi as well as urate in solution in body fluids.

The metabolic disorder of gout could represent either an increased endogenous production of uric acid or a decreased rate of elimination. Since glycine is a specific precursor of atoms 4, 5, and 7 of the purine nucleus, Stetten and his colleagues administered glycine containing ^{15}N to normal and gouty individuals and determined the abundance of ^{15}N in urinary uric acid. Most of the ^{15}N appeared in urea, as would be expected from the variety of metabolic roles of glycine. In normal individuals, after 10 days, only about 0.15 per cent had been excreted as uric acid, whereas about three times as much, 0.5 per cent, was excreted in this form by several gouty subjects. These and other experiments favor the view that in most cases of primary gout the metabolic defect is an overproduction of uric acid.

In some cases of hereditary gout, associated with excessive uric acid production, *hypoxanthine-guanine phosphoribosyl transferase* (page 626) is markedly diminished (2 to 10 per cent of normal) in the erythrocytes and leukocytes, and presumably in all cells. Thus, failure to convert guanine and hypoxanthine to the nucleotides leads ultimately to excessive production of uric acid. In other cases of gout, the overproduction of uric acid may be a consequence of imperfect operation of normal feedback control mechanisms operative in purine synthesis.

Individuals suffering from the hyperuricemia of gout and other conditions have been treated with allopurinol, an analogue of hypoxanthine. Since the compound is a specific inhibitor of xanthine oxidase, there is a gradual decrease in the levels of blood and urinary uric acid. Allopurinol is slowly oxidized to the corresponding xanthine analogue, also an inhibitor of xanthine oxidase.

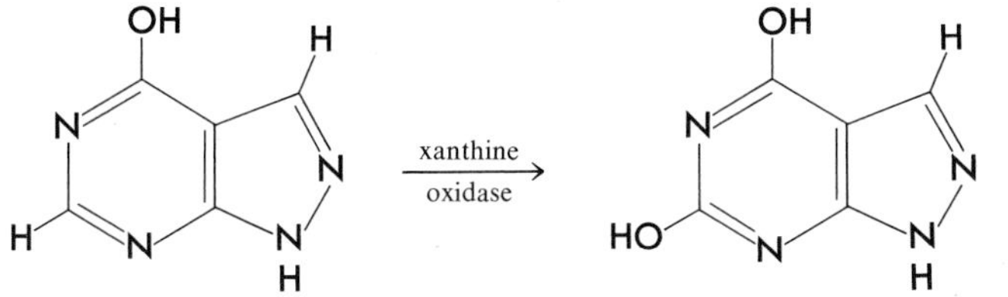

Allopurinol
[4-hydroxypyrazole (3,4-d) pyrimidine]

Alloxanthine
[4,6-dihydroxypyrazole (3,4-d) pyrimidine]

Further Degradation of Purines. In animals other than mammals, purine metabolism may proceed through further degradation reactions as a result of the action of the enzymes, *allantoinase* and *allantoicase* (see page 632). The urea formed is hydrolyzed to ammonia and CO_2 in some species, because of the presence of intestinal microorganisms which contain *urease.*

A summary of the interesting biological variation of the end products of purine metabolism is given in Table 27.1 (page 632). Birds and some reptiles, which do not synthesize urea, direct almost all amino acid nitrogen into the formation of glycine,

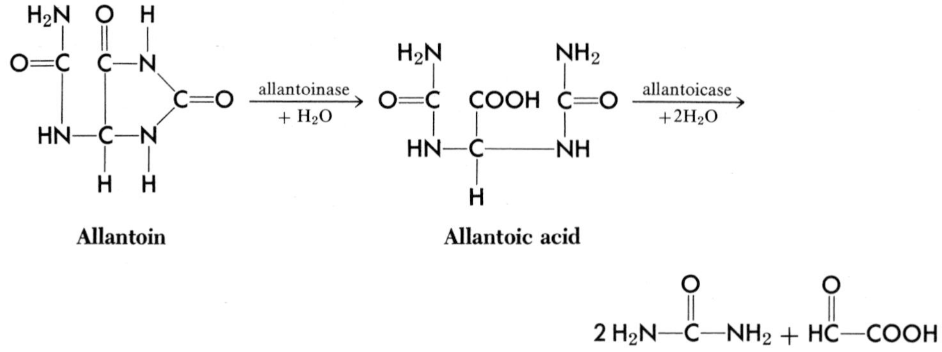

Allantoin → **Allantoic acid** → **Urea** + **Glyoxylic acid**

Table 27.1: FINAL EXCRETORY PRODUCTS OF PURINE METABOLISM

Product excreted	*Animal group*
Uric acid	Man and other primates, Dalmatian dog, birds, some reptiles (snakes and lizards)
Allantoin	Mammals other than primates, some reptiles (turtles), gastropod mollusks
Allantoic acid	Some teleost fishes
Urea	Most fishes, amphibia, fresh-water lamellibranch mollusks
Ammonia	Some marine invertebrates, crustaceans, etc.

aspartic acid, and glutamine. Total purine formation greatly exceeds actual requirements for purine nucleotides, and uric acid is the major end product of all nitrogen metabolism in these species. Those species which excrete nitrogen mainly as uric acid are called *uricotelic,* in contrast to *ureotelic* animals, which excrete nitrogen primarily as urea.

Arginase is present in the livers of vertebrates that have a ureotelic metabolism but not in the livers of animals that have a uricotelic metabolism. Thus, the end products of purine and amino acid metabolism depend on the survival of a small group of enzymes. Lack of arginase diverts amino acid nitrogen into purines in uricoteles. It may be recalled that nitrogen from glycine, glutamine, and aspartic acid is directly utilized in formation of purine nucleotides. By the reactions of transamination and glutamine formation (Chap. 23), additional amino nitrogen and ammonia can be furnished for purine formation and hence for disposal in the form of uric acid.

Because the degradation of purines is much less complete in higher animals, it is apparent that certain enzymes have been lost during animal evolution, *e.g.*, uricase, allantoinase, allantoicase, and urease.

BIOSYNTHESIS OF PYRIMIDINE NUCLEOTIDES

Administered uracil and cytosine containing ^{15}N are rapidly metabolized, and the ^{15}N is found in the excreted urea and ammonia. This is not the case for pyrimidine nucleotides. When ^{15}N-labeled nucleic acid was fed to rats, or when the mixture of mononucleotides obtained by partial hydrolysis was injected intraperitoneally, appreciable amounts of the pyrimidines of the tissue nucleic acids contained isotopic nitrogen. These results suggest that extensive breakdown to free pyrimidines did not occur, either by the oral or the intraperitoneal route, and that larger molecules, nucleosides or nucleotides, are utilized for the formation of nucleic acids. This also indicates that the intestinal digestion of nucleic acids does not proceed to the stage of free pyrimidines.

Evidence concerning the nature of pyrimidine precursors was first derived from studies with microorganisms. Orotic acid (6-carboxyuracil), first found in cow's milk, was shown to satisfy the growth requirements of "pyrimidineless" mutants of *Neurospora*. Mitchell and coworkers have isolated orotic acid from other mutant strains of *Neurospora* which are incapable of forming pyrimidines. This compound, isotopically labeled, was found by Hammarsten and collaborators to serve as a precursor of nucleic acid pyrimidines. This provided the first indication of the route by which pyrimidines are synthesized in vivo. The structures of orotic acid and uracil are shown for comparison.

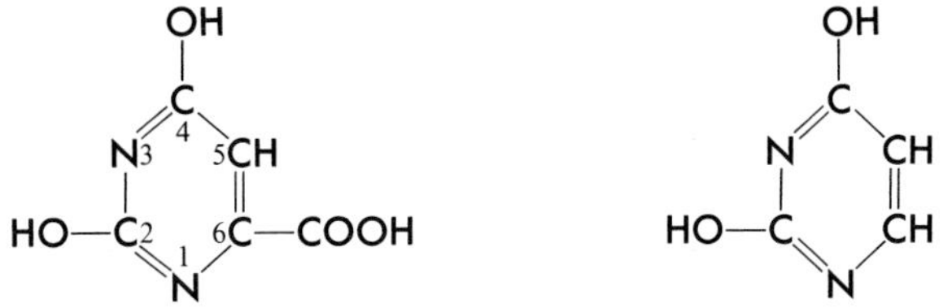

Orotic acid (6-carboxyuracil) **Uracil**

Orotic acid is a growth factor for certain microorganisms, notably *Lactobacillus bulgaricus*. For this organism, N-carbamylaspartic acid has about 10 to 20 per cent of the activity of orotic acid in stimulating growth. Carbamylaspartic acid containing ^{14}C in the ureido carbon is incorporated into the pyrimidines isolated from the nucleic acids of *L. bulgaricus;* thus the acyclic compound serves as a precursor of the pyrimidine ring.

The main enzymic pathways leading to the formation of orotic acid and of pyrimidine nucleotides have been elucidated. A key substance is carbamyl phosphate (page 537).

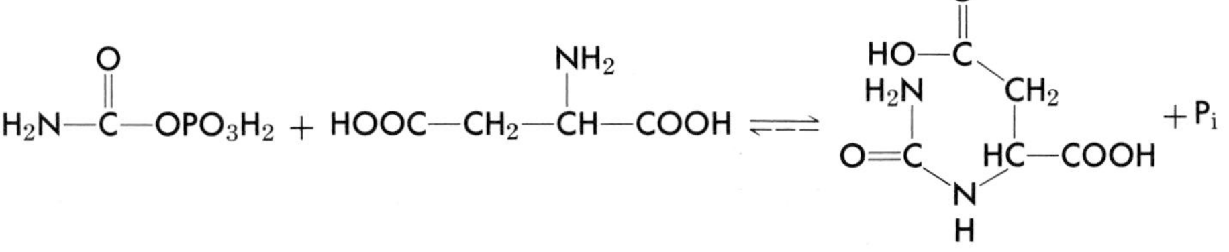

Carbamyl phosphate **L-Aspartic acid** **N-Carbamylaspartic acid**

The formation of carbamylaspartic acid, an effective precursor of pyrimidines in animals, is catalyzed by *aspartate transcarbamylase*. The equilibrium for this reaction is strongly in favor of synthesis. As obtained from *E. coli* this enzyme catalyzes the committed step in pyrimidine biosynthesis and is subject to end-product feedback inhibition by cytidine triphosphate, an ultimate product of pyrimidine biosynthesis. The mechanism of this regulation has been discussed (page 244*ff.*).

Formation of Orotic Acid. Ring closure is catalyzed by *dihydroorotase* and yields L-dihydroorotic acid. The equilibrium favors the ureido compound in the ratio 2:1.

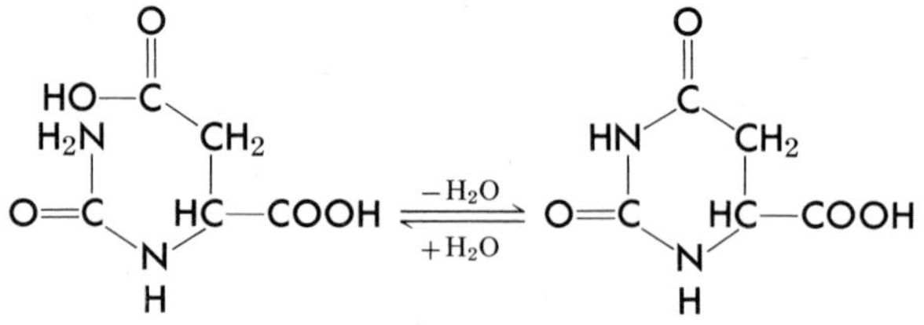

Carbamylaspartic acid **L-Dihydroorotic acid**

The dihydroorotic acid is oxidized to orotic acid by *dihydroorotic acid dehydrogenase,* an unusual enzyme which contains equal amounts of flavin mononucleotide and flavin adenine dinucleotide as well as one atom of iron per flavin (page 366); the electrons are transferred to DPN^+.

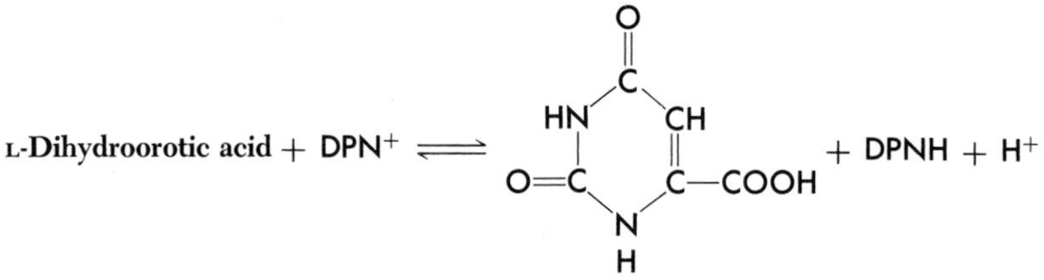

Orotic acid

Pyrimidine Nucleotide Formation. Nucleotide synthesis involves the coupling of orotic acid with 5-phosphoribosyl-1-pyrophosphate in the manner described previously for the formation of purine nucleotides (page 626). The reaction is catalyzed by *orotidine 5′-phosphate pyrophosphorylase.*

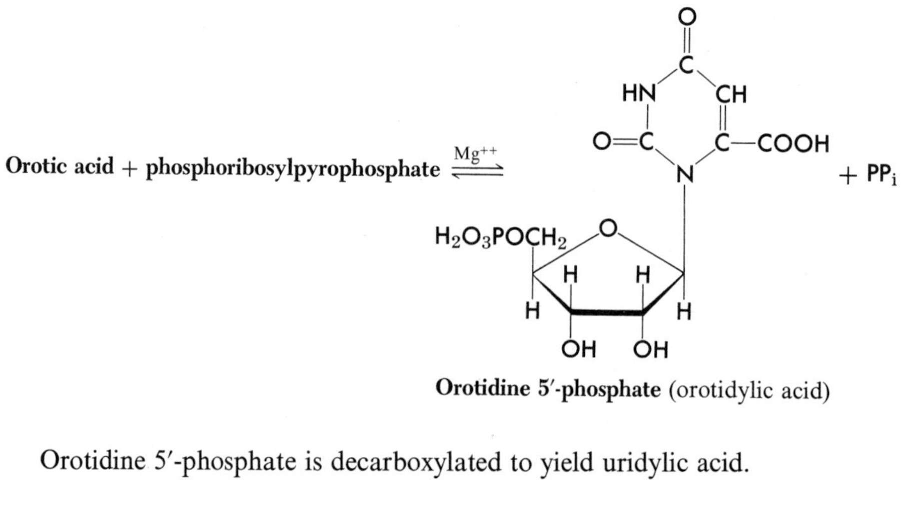

Orotidine 5′-phosphate (orotidylic acid)

Orotidine 5′-phosphate is decarboxylated to yield uridylic acid.

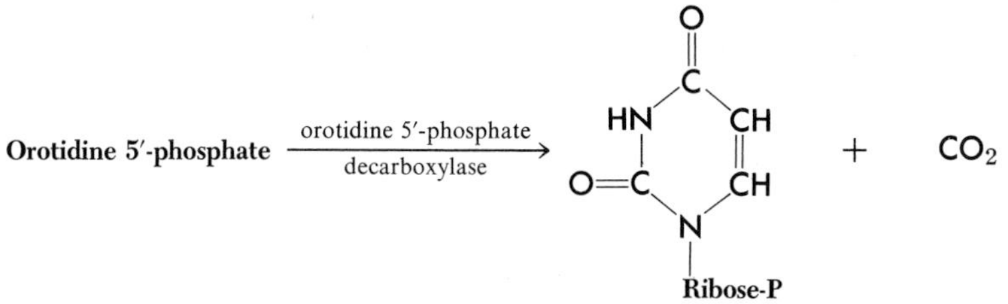

Uridine 5′-phosphate (uridylic acid)

The pathway which results in the formation of uridine triphosphate involves consecutive transfers of phosphate from ATP to uridine 5′-phosphate, with intermediate formation of uridine diphosphate.

$$\textbf{Uridine 5'-phosphate + ATP} \rightleftharpoons \textbf{uridine diphosphate + ADP}$$
$$\textbf{Uridine diphosphate + ATP} \rightleftharpoons \textbf{uridine triphosphate + ADP}$$

In the only known pathway for formation of a cytidine nucleotide, uridine triphosphate is aminated to yield the corresponding cytidine triphosphate. Ribose-PPP in structures below represents ribose triphosphate.

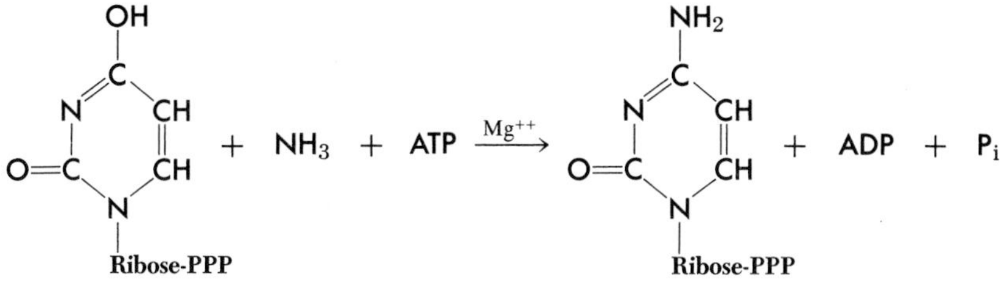

Uridine triphosphate **Cytidine triphosphate**

The stoichiometric release of one mole of inorganic phosphate from ATP suggests that a phosphorylated intermediate may be involved in the reaction. Ammonia itself is utilized in this reaction with an enzymic preparation from *E. coli;* however, in mammalian systems, indirect evidence indicates that the amino group of CTP is derived from the amide nitrogen of glutamine.

Although the above reaction is irreversible, labeled cytidine is an effective precursor of both cytosine and uracil of rat nucleic acids. Apparently, other pathways exist for interconversion of such compounds.

A hereditary disorder of pyrimidine metabolism in man, known as *orotic aciduria,* is characterized by accumulation and urinary excretion of orotic acid. The latter apparently cannot be metabolized further at a normal rate in individuals with this disorder. Administration of uracil or cytosine abolishes the excretion of orotic acid, thus providing evidence of the operation of a feedback inhibition mechanism at an early stage of pyrimidine biosynthesis in man.

Note that the pathways of formation of purine and pyrimidine nucleotides differ. In the former, all intermediates are derivatives of ribose 5-phosphate, whereas in pyrimidine nucleotide biosynthesis the pyrimidine ring is formed prior to coupling with ribose phosphate. The initial nucleotides formed in both cases, inosinic acid and orotidylic acid, are not major constituents of nucleic acids. Although purines contain a pyrimidine ring fused to an imidazole, the precursors of the two ring systems are different. 5-Phosphoribosyl-1-pyrophosphate plays a key role in formation of both types of nucleotides.

DEGRADATION OF PYRIMIDINES

The degradation of ^{14}C labeled pyrimidines, uracil and thymine, was elucidated mainly by injection of large quantities of these substances and of possible intermediates, followed by examination of urinary excretion products. Some studies have also been performed by incubation of these substances with liver slices and with individual enzymes. Figure 27.4 shows the presently known pathways for the degradation of uracil and thymine. It is believed that cytosine and methylcytosine are deaminated first to yield, respectively, uracil and thymine.

The metabolism of uracil and thymine is initiated by *reduction* reactions to give the dihydro compounds, dihydrouracil and dihydrothymine. These are then hydrolyzed by *hydropyrimidine hydrase* to the β-ureido compounds. Further hydrolysis yields the β-amino acids. The utilization and fate of β-alanine have been discussed (page 587). β-Aminoisobutyrate may yield methylmalonate; the fate of the CoA derivative of this compound is discussed on page 484. β-Aminoisobutyrate is found in the urine of some individuals in amounts up to 200 to 300 mg. per day, either because of an inherited trait or as a consequence of disease. It is excreted in increased amounts after administration of diets rich in DNA. Increased levels also occur in the urine of some patients with tumors.

FORMATION OF DEOXYRIBONUCLEOTIDES

Tissue extracts contain the 5′-mono-, di-, and triphosphates of the deoxyribonucleotides of adenine, guanine, thymine, cytosine, and methylcytosine. Evidence has also been obtained that the polyphosphates are rapidly formed from the corresponding 5′-deoxynucleotides. The questions to be considered regarding formation of the deoxynucleotides then are (1) the origin of the deoxysugar, (2) the biosynthesis of the methylated pyrimidines, thymine and methylcytosine, and (3) the formation of the 5′-polyphosphates. These problems will be considered in turn.

Although deoxyribonucleotides might be formed by reactions involving compounds analogous to those concerned in ribonucleotide synthesis, no evidence has been obtained to support this view. The demonstration by Hammarsten and coworkers that ribonucleotides are effective precursors of the deoxyribonucleotides of DNA led them to suggest the existence of pathways involving a direct conversion of ribonucleotides to deoxyribonucleotides. It was subsequently established in several laboratories that cytidine, randomly labeled with ^{14}C, is incorporated into the nucleotides of DNA, with the same relative distribution of ^{14}C in the deoxyribose and the base in thymidine as well as in deoxycytidine. In the same manner evidence has been obtained that purine deoxyribonucleotides of DNA are derived from corresponding ribonucleotides. Thus, deoxygenation of the sugar occurs at the nucleotide level.

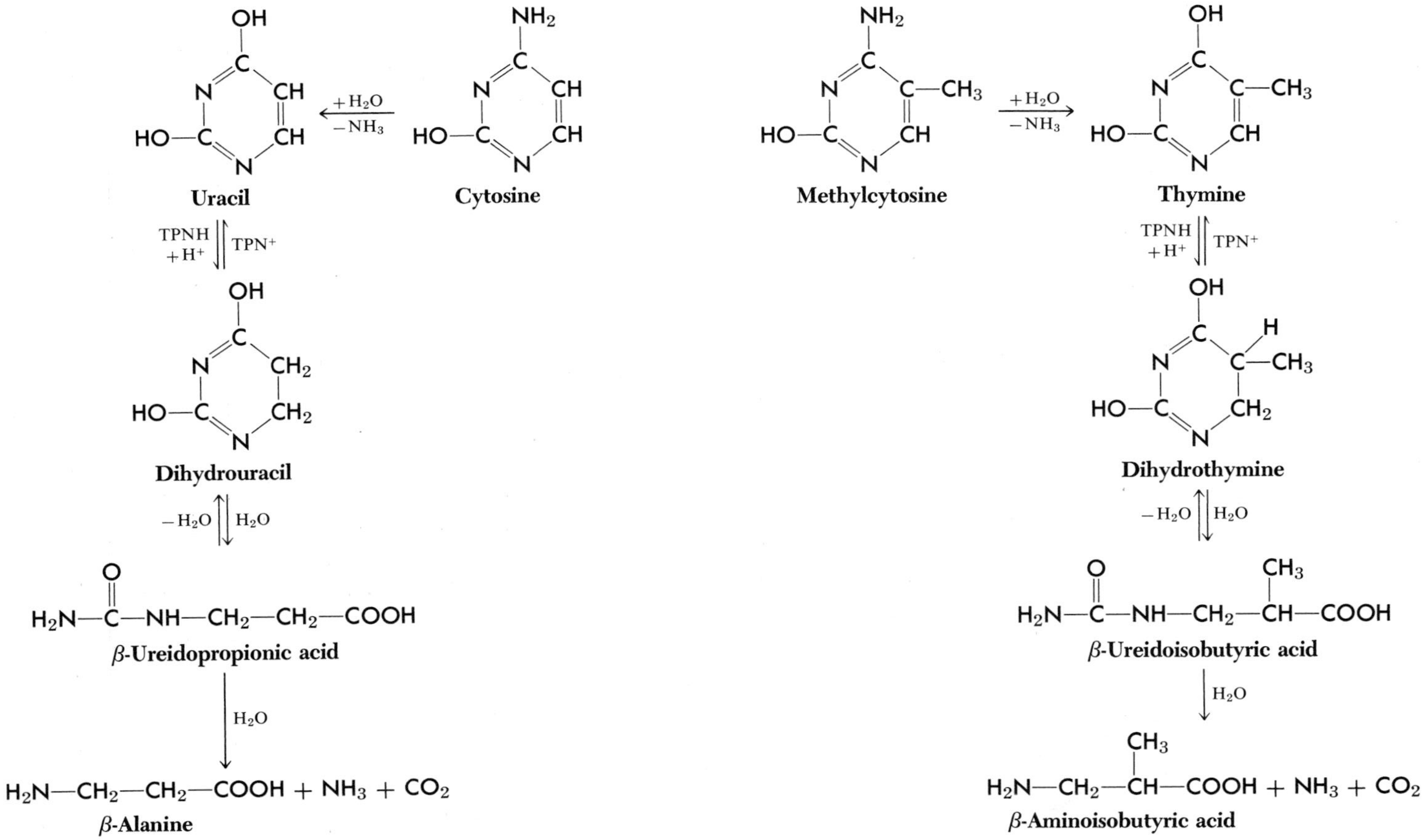

Fig. 27.4. Metabolic fate of the free pyrimidines.

The general pathway of interconversion of pyrimidine nucleotides may be summarized as follows:

$$\text{UTP} \longrightarrow \text{CTP} \longrightarrow \text{CDP} \longrightarrow \text{dCDP} \longrightarrow \text{dCMP} \longrightarrow \text{dUMP} \longrightarrow \text{dTMP}$$

$$\text{and UDP} \longrightarrow \text{dUDP} \longrightarrow \text{dUMP} \longrightarrow \text{dTMP}$$

The reduction of ribonucleotides by enzyme preparations from *Escherichia coli* has been investigated by Reichard and coworkers. The main results can be summarized as follows. (1) The substrates for reduction are the ribonucleoside diphosphates. (2) Four purified proteins are required; enzymes termed *B1* and *B2*, a sulfhydryl protein named *thioredoxin,* and *thioredoxin reductase.* (3) The specific hydrogen donor for reduction is thioredoxin in the sulfhydryl form. (4) Oxidized thioredoxin is reduced by thioredoxin reductase, the hydrogen donor being TPNH. (5) Present evidence indicates that all four nucleoside diphosphates, ADP, GDP, CDP, and UDP, are converted to the corresponding deoxy derivatives by the same reductase system. (6) The activity of the enzymic system is regulated by nucleoside triphosphates, some acting as stimulators, others as inhibitors.

The reduction of CDP illustrates the over-all nature of the process.

$$\text{Thioredoxin-S}_2 + \text{TPNH} + \text{H}^+ \xrightarrow[\text{reductase}]{\text{thioredoxin}} \text{thioredoxin-(SH)}_2 + \text{TPN}^+ \quad (1)$$

$$\text{Thioredoxin-(SH)}_2 + \text{CDP} \xrightarrow[\text{Mg}^{++}]{\text{enzymes B1 + B2}} \text{dCDP} + \text{thioredoxin-S}_2 \quad (2)$$

Reaction (1) involves thioredoxin reductase, which contains two moles of flavin adenine dinucleotide per residue of enzyme (molecular weight = 68,000). Heat-stable thioredoxin contains one mole of cystine per 108 residues and after reduction, two sulfhydryl groups. Reaction (2) is obviously complex, involving two enzyme fractions of unknown function. Mg^{++} is an obligatory requirement, and the reduction of CDP or UDP is strongly stimulated by ATP (Fig. 27.5), presumably acting as an allosteric effector. In the reduction of ADP and GDP, the reactions are stimulated by dGTP and dTTP. dATP strongly inhibits the reduction of all four ribonucleoside diphosphates. Reduction results in incorporation of a hydrogen atom at the 2′ position with no change in configuration.

The total system of effectors and inhibitors permits a balanced supply of the reduced diphosphates and hence of the triphosphates, the immediate substrates for the synthesis of DNA (Chap. 28). Although the primary pathway for formation of dUDP is via the above reactions, other pathways of interconversion exist.

A system different from that described above for reduction of ribonucleotides has been studied in various species of *Lactobacillus.* The major characteristics of this latter system are: (1) the preferred substrates are the ribonucleoside triphosphates, ATP, GTP, CTP, and UTP, (2) a single enzyme catalyzes the reduction of all four triphosphates in the presence of dihydrolipoate or the thioredoxin-thioredoxin reductase system from *E. coli,* (3) *5,6-dimethylbenzimidazole cobamide coenzyme* (vitamin B_{12} coenzyme, Chap. 49) is an obligatory requirement, (4) Mg^{++} and ATP (and dATP) stimulate the reduction of CTP and inhibit the reduction of UTP and GTP, (5) various deoxyribonucleoside triphosphates act as effectors, stimulatory and inhibitory, in a pattern which differs in detail from that of the *E. coli*

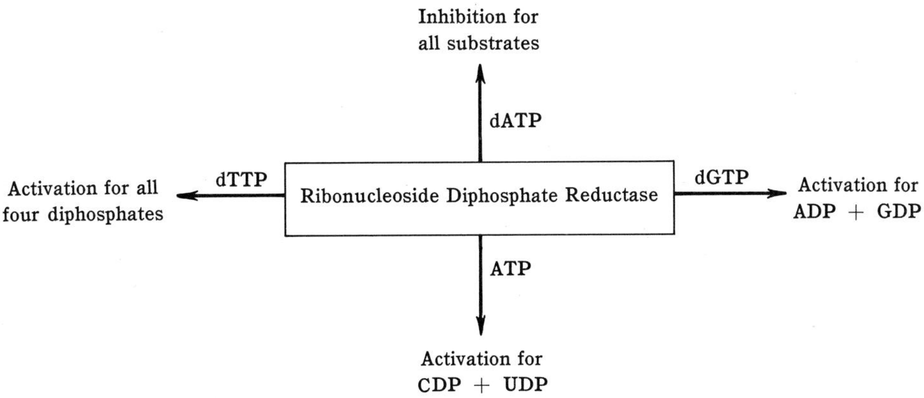

Fig. 27.5. Schematic interpretation of major allosteric effects on the ribonucleoside diphosphate reductase of *E. coli* by nucleoside triphosphates. (*Adapted from A. Larsson and P. Reichard, J. Biol. Chem.*, **241**, 2540, 1966.)

enzyme but, in principle, is similar. Thus far, the cobamide coenzyme requirement has been demonstrated only in lactobacilli.

Deamination of dCMP to yield dUMP is catalyzed by *deoxycytidylic acid aminohydrolase*. In the formulas below, d-ribose-P represents deoxyribose 5′-phosphate.

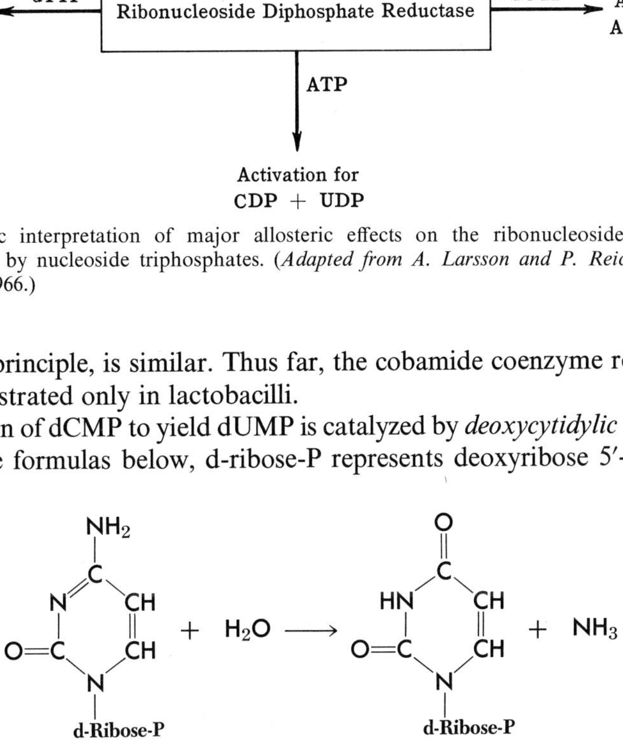

Deoxycytidine 5′-phosphate **Deoxyuridine 5′-phosphate**

This enzyme, present in liver, also catalyzes the deamination of the methyl- and hydroxymethyldeoxycytidylic acids.

5-Methyldeoxycytidylic acid + H_2O ⟶ thymidylic acid + NH_3
5-Hydroxymethyldeoxycytidylic acid + H_2O ⟶ 5-hydroxymethyldeoxyuridylic acid + NH_3

In the presence of Ca^{++}, Mn^{++}, or Mg^{++}, dCTP is an allosteric activator and dTTP is an allosteric inhibitor of the enzyme. Presumably, this is a "salvage" path, since dUMP and dTMP are mainly synthesized by the pathway described previously (page 638).

Thymidylate synthetase from *E. coli* and other sources catalyzes the following reaction.

Deoxyuridine 5′-phosphate + N^5,N^{10}-methylenetetrahydrofolate $\xrightarrow{Mg^{++}}$
thymidine 5′-phosphate + dihydrofolate

Tetrahydrofolate serves both as a carbon carrier and as a direct hydrogen donor in this complex reaction. Vitamin B_{12} coenzyme has been implicated in this reaction, but its exact role is unknown.

An analogous reaction occurs in the formation of 5-hydroxymethyldeoxycytidine 5′-phosphate in cells of *E. coli* infected with T2, T4, or T6 bacteriophages (Table 9.7, page 206), although no reduction is involved.

N^5,N^{10}-Methylenetetrahydrofolic acid + → tetrahydrofolic acid +

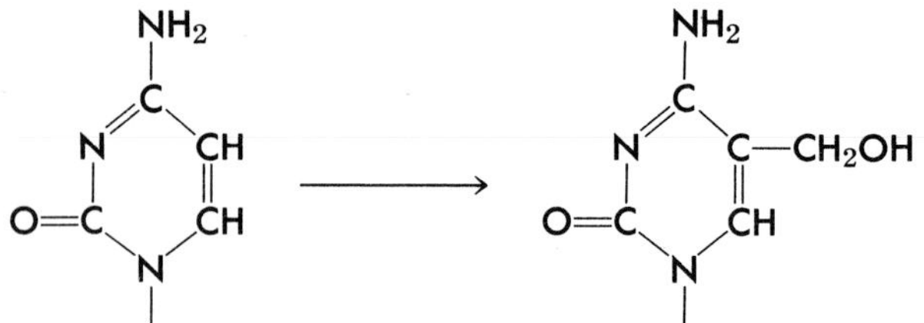

d-Ribose-P — d-Ribose-P

Deoxycytidine 5′-phosphate — 5-Hydroxymethyldeoxycytidine 5′-phosphate

Deoxyribonucleotide Polyphosphates. Formation of the deoxyribonucleotide di- and triphosphates occurs in separate steps. Enzymes from *E. coli* and liver catalyze the following transphosphorylations.

$$\text{dAMP} + \text{ATP} \xrightleftharpoons{Mg^{++}} \text{dADP} + \text{ADP}$$

$$\text{TMP} + \text{ATP} \xrightleftharpoons{Mg^{++}} \text{TDP} + \text{ADP}$$

Nucleoside diphosphokinase of calf thymus or yeast catalyzes reversibly the conversion of all ribo- and deoxyribonucleoside diphosphates to the triphosphates, *e.g.*,

$$\text{TDP} + \text{ATP} \xrightleftharpoons{Mg^{++}} \text{TTP} + \text{ADP}$$

$$\text{dADP} + \text{ATP} \xrightleftharpoons{Mg^{++}} \text{dATP} + \text{ADP}$$

$$\text{dGDP} + \text{ATP} \xrightleftharpoons{Mg^{++}} \text{dGTP} + \text{ADP}$$

REGULATION OF NUCLEOTIDE SYNTHESIS

From the foregoing it is evident that separate regulatory mechanisms are operative in controlling formation of purine nucleotides (page 626), pyrimidine nucleotides (page 633) and the deoxyribonucleotides (page 638). The ribonucleoside triphosphates are utilized in the formation of all types of RNA (Chap. 29) and the deoxyribonucleoside triphosphates are the immediate precursors for synthesis of DNA (Chap. 28). Thus balanced cellular growth and multiplication requires the synthesis of large amounts of the various nucleoside triphosphates and these must be provided in proper balance. The immediate source of energy for formation of all these compounds is ATP and regulation of its formation depends on the energy needs of the cell and is under the control of different mechanisms. However, the supply of ATP will ultimately determine the rate of formation of all ribo- and deoxyribonucleoside triphosphates since 5-phosphoribosyl-1-pyrophosphate is the starting point for formation of all nucleotides and large amounts of ATP are required for formation of this key substance. The enzyme of *E. coli* catalyzing the synthesis of phosphoribosyl pyrophosphate is strongly inhibited by ADP, thus curtailing formation of this key intermediate when the energy supply (ATP) is low. The effect of ADP as a negative effector is much greater than that of CDP or GDP. Tryptophan and histidine have little influence although 5-phosphoribosyl-1-pyrophosphate is a precursor in the synthesis of both these amino acids (Chap. 24).

FORMATION OF COENZYME NUCLEOTIDES

In addition to their incorporation into ribose nucleic acid, all the ribose-containing nucleotides found in RNA play other important metabolic roles. Thus, in various chapters of this part, metabolic reactions have been encountered which involve adenylic acid, guanylic acid, uridylic acid, cytidylic acid, and thymidylic acid or their respective 5′-polyphosphates, or other derivatives, *e.g.*, uridine diphosphoglucose and similar uridine derivatives. The biosynthesis and roles of these compounds have been considered in previous pages. Consideration has also been given to other important nucleotides which contain moieties not found in nucleic acids, *e.g.*, nicotinic acid amide, flavin, and pantothenic acid. Biosynthesis of the nucleotides containing these substances mainly involves ATP.

Flavin Nucleotides. Riboflavin, or 6,7-dimethyl-9-(1′-D-ribityl)isoalloxazine (Chap. 49), is an essential dietary constituent for mammals. As already discussed in several chapters of this Part, it functions in a mono- or dinucleotide form as the prosthetic group of a number of enzymes.

Flavin mononucleotide, riboflavin 5′-phosphate, is formed from riboflavin and ATP in a reaction catalyzed by a *flavokinase.*

$$\textbf{Riboflavin} + \textbf{ATP} \xrightarrow[\text{Mg}^{++}]{\text{flavokinase}} \textbf{flavin mononucleotide} + \textbf{ADP}$$

Flavin adenine dinucleotide is formed from the mononucleotide by a reversible reaction catalyzed by *flavin nucleotide pyrophosphorylase.*

$$\textbf{Flavin mononucleotide} + \textbf{ATP} \overset{\text{Mg}^{++}}{\rightleftharpoons} \textbf{flavin adenine dinucleotide} + \textbf{PP}_\text{i}$$

Pyridine Nucleotides. Diphosphopyridine nucleotide (DPN) (page 322) contains nicotinamide, an important dietary constituent for mammals (Chap. 49). The *de novo* pathway for formation of nicotinic acid from tryptophan (page 615) yields nicotinic acid mononucleotide directly.

In human erythrocytes, yeast, and hog liver, niacin (nicotinic acid) reacts with 5-phosphoribosyl-1-pyrophosphate to form nicotinic acid mononucleotide, which then condenses with ATP to form desamido-DPN. The latter is converted to DPN by reaction with glutamine and ATP.

(*a*) **Nicotinic acid + phosphoribosylpyrophosphate $\rightleftharpoons$ nicotinic mononucleotide + PP_i**
(*b*) **Nicotinic mononucleotide + ATP $\rightleftharpoons$ desamido-DPN + PP_i**
(*c*) **Desamido-DPN + glutamine + ATP $\rightleftharpoons$ DPN + glutamic acid + PP_i**

The *DPN synthetase* catalyzing reaction (*c*) is strongly inhibited by azaserine.

Erythrocytes and, presumably, other tissues can make nicotinamide mononucleotide from *nicotinamide* by the following reaction.

(*d*) **Nicotinamide + phosphoribosylpyrophosphate $\rightleftharpoons$ nicotinamide mononucleotide + PP_i**

Reaction (*b*) is catalyzed by *diphosphopyridine nucleotide pyrophosphorylase,* which can also catalyze the reaction:

(*e*) **Nicotinamide mononucleotide + ATP $\rightleftharpoons$ DPN + PP_i**

The responsible enzyme is in the nucleus of mammalian cells and may play a significant regulatory role in the life of the cell while also being important as a contributor to the net synthesis of the pyridine nucleotides. The enucleated mature erythrocyte contains only a trace of this enzyme.

No enzyme capable of catalyzing direct synthesis of nicotinamide from nicotinic acid has been found in plant or animal systems. However, nicotinamide can arise from nicotinic acid by consecutive operation of reactions (*a*), (*b*), and (*c*) followed by the action of *diphosphopyridine nucleotidase* (*DPNase*), which catalyzes the hydrolysis of DPN at the N-glycosidic (nucleoside) linkage between ribose and nicotinamide.

DPN + H_2O $\longrightarrow$ nicotinamide + adenosine 5′-pyrophosphoryl-5-ribose

Adenosine diphosphate ribose is hydrolyzed to adenylic acid and ribose 5-phosphate, which then follow the usual metabolic routes of these compounds.

Another enzyme catalyzing hydrolysis of dinucleotides is a *nucleotide pyrophosphatase* which cleaves various pyrophosphate linkages as follows:

DPN $\longrightarrow$ nicotinamide mononucleotide + adenylic acid (AMP)
TPN $\longrightarrow$ nicotinamide mononucleotide + adenosine 2′,5′-diphosphate
FAD $\longrightarrow$ flavin mononucleotide + adenylic acid
ATP $\longrightarrow$ ADP + P_i $\longrightarrow$ adenylic acid + P_i
Thiamine pyrophosphate $\longrightarrow$ thiamine monophosphate + P_i

The physiological role of this enzyme is unknown, but it has been useful in the elucidation of the structures of a number of coenzymes containing the pyrophosphoryl group, *e.g.*, TPN and CoA.

Triphosphopyridine nucleotide (TPN) is formed from DPN by the following enzyme-catalyzed reaction.

Diphosphopyridine nucleotide + ATP $\xrightarrow{Mg^{++}}$ triphosphopyridine nucleotide + ADP

Coenzyme A. The complete structure of coenzyme A is given on page 327. The pantothenic acid (pantoyl-β-alanine) portion of the molecule is required in the mammalian diet; its synthesis in microorganisms is described in Chap. 49.

Pantothenic acid also occurs in nature in combination with β-mercaptoethylamine (cysteamine) as pantetheine or *Lactobacillus bulgaricus* factor (LBF), so called because it is an essential nutrient for this and certain other microorganisms.

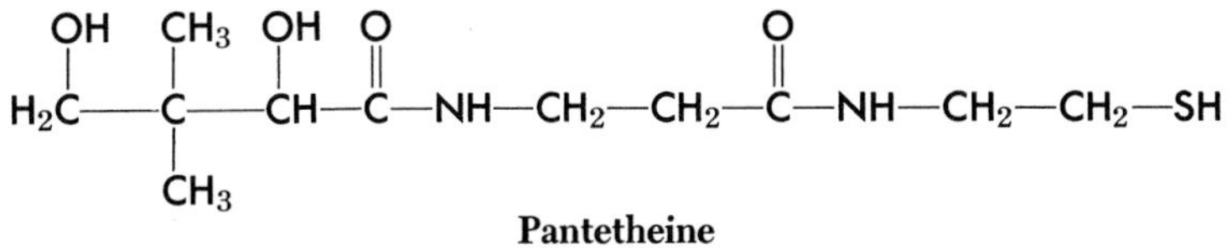

Pantetheine

Pantetheine is an intermediate in the pathway of CoA formation in mammalian liver and some microorganisms, as shown in the following reactions.

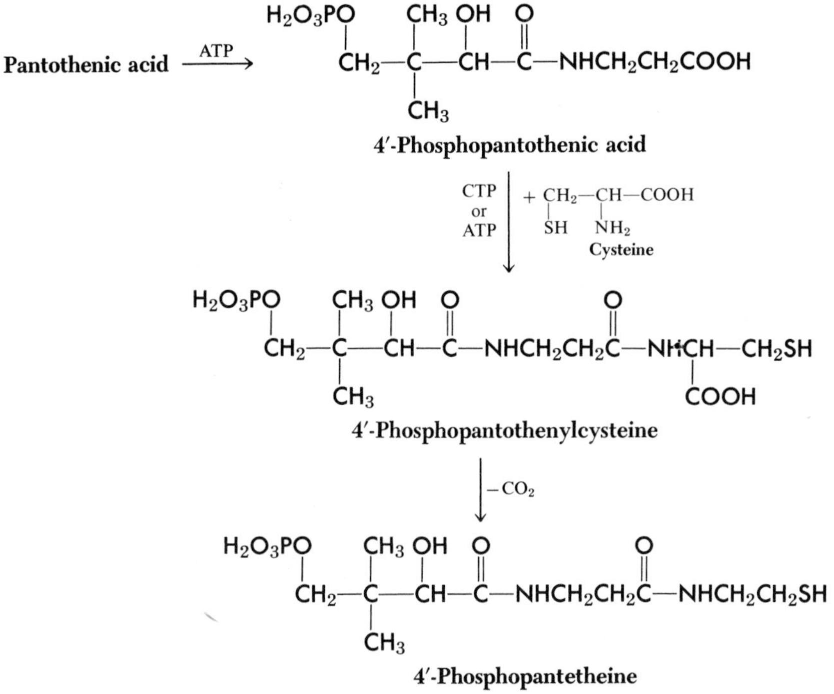

$$\text{4}'\text{-Phosphopantetheine} + \text{ATP} \xrightarrow{Mg^{++}} \text{dephospho-CoA} + PP_i$$

The 3′-phosphate of the adenosine moiety of CoA is lacking in dephospho-CoA; the latter is converted to CoA by a specific *dephospho-CoA kinase.*

$$\text{Dephospho-CoA} + \text{ATP} \xrightarrow{Mg^{++}} \text{CoA} + \text{ADP}$$

INHIBITORS OF NUCLEOTIDE SYNTHESIS

Inasmuch as rapidly dividing cells, *e.g.*, tumor cells and bacteria, have high requirements for nucleotides for formation of nucleic acids whereas adult tissues grow slowly, growth of tumors and bacteria can be inhibited by blocking nucleotide synthesis. As already noted (page 241), *sulfonamides* block formation of folic acid in organisms dependent on this process. Since folic acid is required for formylation in two steps of purine nucleotide synthesis and for thymidylate synthesis, formation of nucleotide coenzymes containing purines and nucleic acids is prevented. Similarly, since transfer of amide groups from glutamine is required in purine nucleotide synthesis, *azaserine* (page 621) and similar compounds block synthesis of purine nucleotides and hence formation of nucleic acids.

Antagonists of folic acid, *e.g., aminopterin* and *amethopterin* (Methotrexate), which inhibit tumor and bacterial growth, block the reduction of dihydrofolate to tetrahydrofolate by a specific reductase (page 550). Again, this prevents formylation at two steps in purine nucleotide synthesis and formation of thymidylate. Inhibition of bacterial growth by the above analogues of folic acid can be overcome by addition of adenine and thymine.

The purine analogue, *6-mercaptopurine,* is converted to the ribonucleotide and is a potent inhibitor of conversion of inosinic acid to both adenylic acid and xanthylic acid, a precursor of guanylic acid (page 624*ff.*). The result is prevention of nucleic acid synthesis. The inhibitor is being used as an antitumor agent in treatment of choriocarcinoma.

Some halogenated pyrimidines, *e.g., 5-fluorodeoxyuridine,* block DNA synthesis by inhibiting thymidylate synthetase (page 639). Other halogenated pyrimidines and purines act as mutagens (page 682).

REFERENCES

Books

Boyer, P. D., Lardy, H., and Myrbäck, K., eds., "The Enzymes," selected articles in vols. 2, 3, and 5, Academic Press, Inc., New York, 1960.

Chargaff, E., and Davidson, J. N., eds., "The Nucleic Acids," vol. III, Academic Press, Inc., New York, 1960.

Davidson, J. N., "The Biochemistry of the Nucleic Acids," 5th ed., John Wiley & Sons, Inc., New York, 1965.

Hutchinson, D. W., "Nucleotides and Coenzymes," John Wiley & Sons, Inc., New York, 1964.

Stanbury, J. B., Wyngaarden, J. B., and Frederickson, D. S., eds., "The Metabolic Basis of Inherited Disease," 2d ed., McGraw-Hill Book Company, New York, 1966. (See part VI, Diseases of Purine and Pyrimidine Metabolism.)

Review Articles

Buchanan, J. M., and Hartman, S. C., Enzymic Reactions in the Synthesis of Purines, *Advances in Enzymol.,* **21,** 199–261, 1959.

Glaser, L., Biosynthesis of Deoxysugars, *Physiol. Revs.,* **43,** 215–242, 1963.

Gutman, A. B., The Biological Significance of Uric Acid, *Harvey Lectures,* **60,** 35–55, 1964–1965.

Kornberg, A., Pyrophosphorylases and Phosphorylases in Biosynthetic Reactions, *Advances in Enzymol.,* **18,** 191–240, 1957.

Larsson, A., and Reichard, P., Enzymatic Reduction of Ribonucleotides, *Progress in Nucleic Acid Research and Molecular Biology,* **7,** 303–347, 1967.

Reichard, P., The Enzymic Synthesis of Pyrimidines, *Advances in Enzymol.,* **21,** 263–294, 1959.

Schmidt, G., Metabolism of Nucleic Acids, *Ann. Rev. Biochem.,* **33,** 667–728, 1964.

28. Genetic Aspects of Metabolism. I

Nature of the Gene. DNA Synthesis. Relation of DNA Sequence to Protein Sequence and Structure

Current understanding of genetic mechanisms may be regarded as the answers to a series of questions. What is the nature of the genetic material? How is the material replicated so that at each cell division, each daughter cell possesses the same complement of genetic material as the parent cell? What are the mechanisms by which this material determines the chemical, metabolic, and morphological characteristics of the individual cell or organism? In brief, all genes of higher organisms consist of DNA and its role as genetic material is to determine the nature and amount of each of the proteins synthesized, most of these being the enzymes that catalyze individual metabolic reactions.

THE GENE AND METABOLISM

Inheritance is achieved through the genes, factors which are usually carried in the chromosomes. The mode of inheritance of characteristics attributable to genes is now well known and has been studied experimentally in a large number of species which show sexual differentiation, a prerequisite for demonstrating behavior and existence of single genes in diploid organisms. Mendelian inheritance operates in man as in other species, and many characteristics caused by single gene differences have been described. Such traits as color blindness, the various blood groups, and hemophilia (Chap. 31) are familiar examples.

The general concept is that a single gene can influence only a single step in metabolism. This hypothesis arose largely from studies of anomalies in man and other species in which a metabolic reaction is blocked. An excellent example is the metabolic defect called alkaptonuria (page 608), which is inherited in man as a recessive trait. As long ago as 1908, in the book "Inborn Errors of Metabolism," Garrod concluded that homogentisic acid, excreted in the urine of alkaptonuric persons, is a normal intermediate in the oxidation of phenylalanine and tyrosine and that alkaptonuria is due to an inability to oxidize homogentisic acid. Garrod's conclusion was verified almost 50 years later by La Du, who showed an absolute lack of homogentisic acid oxidase in alkaptonuria (page 608). Studies of other metabolic anomalies of phenylalanine and tyrosine metabolism found in man, such as phenylketonuria, tyrosinosis, and albinism, revealed that these are also inherited as single genic factors. In each instance, a change in a single gene has blocked an individual metabolic reaction.

One of the first studies showing the effect of single genes in influencing chemical structures was that of R. Scott-Moncrieff and other workers on the nature of the pigments in primroses, dahlias, and other plants. The anthocyanins are blue, purple, and red glycosidic derivatives of pelargonidin which occur as oxonium salts.

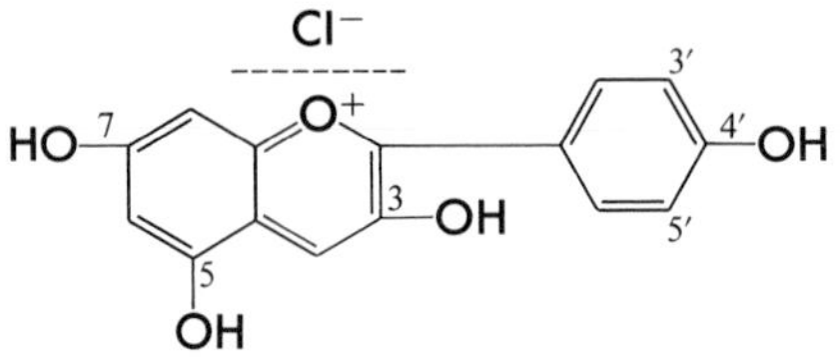

Pelargonidin

Individual genes were shown to produce the following effects: (1) oxidation to OH at 3′ (formation of cyanidin); (2) oxidation to OH at both 3′ and 5′ (formation of delphinidin); (3) methylation of the OH group at 3′; (4) methylation of the OH groups at 3′ and 5′; (5) methylation of the OH group at 7; (6) glycoside formation as 3-monosides, 3-biosides, or 3,5-diglycosides with one hexose or pentose residue, or more, being involved. Thus, *single* genes control each step in anthocyanin biosynthesis. In no case has a single genic difference been found which modifies two characteristics of the molecule simultaneously.

Genetic investigations of the bread mold, *Neurospora crassa,* initiated in 1941 by Beadle and Tatum, greatly extended the limited information concerning the effect of single genes available from other species. Since the nucleus of the sexual spore of *Neurospora* is haploid, it contains only one of each of the seven chromosomes and, correspondingly, of each gene. Each *ascus* contains eight spores which are identical in the wild type. These can be isolated and tested individually. If a mutant is crossed with the wild type, four spores will show the wild-type character and four the mutant character. The wild-type strain can be cultured on a medium containing only glucose, an inorganic source of nitrogen, salts, and biotin. Various mutant strains, produced by irradiation with ultraviolet light or x-rays, are unable to grow unless certain compounds are added to the medium, indicating loss of the ability to synthesize one or more substances.

In testing for mutants, the *Neurospora* spores can be cultured in a complete medium, *i.e.*, one containing all known growth factors, amino acids, vitamins, etc. If the strain can grow normally on the minimal medium, no mutation has occurred. If it cannot grow on this medium, it is tested further to determine whether the missing factor is a vitamin, an amino acid, etc. Tests are made first with large groups of substances, then with smaller groups, and finally with individual compounds. In this way many mutants have been found which lack the ability to synthesize one or another of the amino acids, the B group of vitamins, purines, or pyrimidines.

Let us assume that for a certain mutant strain, substance D is required for growth. Since the synthesis may proceed in a series of steps, the metabolic block may occur at any point along the chain of synthesis, beginning with a precursor A.

$$\begin{array}{ccccccc} A & \longrightarrow & B & \longrightarrow & C & \longrightarrow & D \\ & \uparrow & & \uparrow & & \uparrow & \\ & \text{Gene 1} & & \text{Gene 2} & & \text{Gene 3} & \end{array}$$

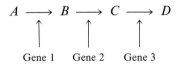

If gene 3 is altered and the block occurs between C and D, supplying D is the only way in which the strain can be grown. However, if gene 1 is altered, then addition of B or C, as well as D, will permit the growth of the mutant. In other words, alteration of several different genes will result in the requirement for D although the metabolic block may occur at any one of the different steps. Many such cases have been encountered. Although the wild type of *Neurospora* can synthesize arginine from ammonia, several mutants cannot grow unless arginine is supplied in the medium. Some "arginineless" strains can use only arginine, others can use citrulline or arginine, and still others can use ornithine, citrulline, or arginine. It is evident that the sequence of steps must be in the order $X \rightarrow$ ornithine $\rightarrow$ citrulline $\rightarrow$ arginine, exactly as it occurs in mammals (pages 545*ff.*). Similar studies greatly expedited elucidation of the pathways of biosynthesis of such amino acids as lysine, tryptophan, histidine, etc.

In every instance, it was demonstrated that a single gene controls the formation of a single enzyme. Thus, genes determine the detailed reactions of metabolism by controlling the biosynthesis of enzymes. This has frequently been stated as the one gene–one enzyme hypothesis. Studies similar to those in *Neurospora* have been made with mutant strains of various bacteria, higher plants, and animals.

THE GENE AS DNA

The above evidence led to the view that the gene controls the synthesis of enzymes but offered no information as to the chemical nature of the gene. This has been provided by a variety of studies which may be summarized as follows.

1. The experimental biologists of the late nineteenth century demonstrated, largely by excision methods, that the hereditary factors of cells are located in the nuclei. In 1902 Sutton showed that the segregation and independent assortment of the mendelian factors of inheritance can be explained on the basis that these factors are located in the chromosomes.

2. The isolation in pure form of various viruses led to the recognition that plant viruses are ribonucleoproteins and that most bacterial viruses (bacteriophages) and many animal viruses are deoxyribonucleoproteins (Chap. 9). Later, it was demonstrated that replication and the transfer of genetic information are due entirely to the nucleic acid of such viruses.

3. Mutation of genes can be produced by ultraviolet light. The action spectrum (those wavelengths of light causing this effect) which produces mutations closely corresponds to the absorption spectra of the nucleic acids. Similarly, other physical and chemical agents that are mutagenic have been shown to alter nucleic acids (page 681).

4. The various virulent strains of *Pneumococcus* have antigenically and chemically different capsular polysaccharides. Noncapsulated R cells of one type may be *transformed* into capsular organisms (S cells) of another type by the introduction of a minute amount of culture filtrate from a capsular-forming S organism. Avery, MacLeod, and McCarty isolated DNA from such filtrates and showed that minute amounts of DNA of a type III filtrate induced formation of type III polysaccharide in a type II, acapsular R strain. In a similar manner, DNA from diverse bacterial species has been used to transform variant members of the same species with respect to such characteristics as resistance to diverse antibiotics or ability to synthesize a specific compound. The DNA acts as a genetic factor in all cases since the transformed strain continues to reproduce, thereafter, as the induced type.

5. The DNA content of its set of chromosomes is an absolute constant for each species, consonant with the role of DNA as the carrier of genetic information.

All the above information led to the view that the genetic material is DNA except for certain viruses in which RNA serves this function. Therefore, DNA must possess the fundamental properties of the gene: the capacity to direct the formation of an exact replica of itself; mutability, the property of undergoing alteration without loss of reproductive function; and, finally, the ability to direct the formation of enzymes or other proteins.

SOME GENETIC METHODS

Although classical genetic studies were performed chiefly with diploid animals and plants, dramatic advances have been achieved more recently with unicellular haploid microorganisms such as various bacteria, including *Escherichia coli* and *Salmonella,* and with many bacteriophages, particularly those of the T-even types (page 205) of *E. coli.* In higher organisms, only the zygotes, eggs and sperm, are haploid and the adults are diploid. In the *Ascomycetes* (*Neurospora, Aspergillus,* various yeasts) the asexual phase of the organism is haploid. Methods have been discovered which permit exchange of genetic material in bacteria and phages and this is always followed by the vegetative haploid state of the organism. Several of these methods are described briefly below.

Tatum and Lederberg discovered that genetic recombination can occur in the K-12 strain of *E. coli.* In this organism, there are two mating types, called F^+ (male) and F^- (female). The F^+ cells possess an F-agent or sex factor which can be transferred by cellular contact to F^- cells, converting them to the F^+ state. The possession of the F^+ agent renders a cell potentially capable of transferring chromosomal material to a recipient. The chromosomal material that is transferred enters linearly in the order of the genes. By interrupting mating at various times and determining the expression of the genes in the recipient cells, the order of the genes can be mapped.

Another method of genetic exchange is provided by the process of transduction, exhibited by certain types of bacteriophages. While growing in host cells, these phages can incorporate small portions of host genetic material and subsequently transfer this material to a recipient cell upon infection. The process, originally discovered by Lederberg and Zinder, has been applied particularly to the K-12 strain of *E. coli* and to *Salmonella.*

When a bacterial cell, such as *E. coli,* is simultaneously infected by two bacteriophage particles that differ in one or more genes, a process occurs essentially similar to chromosomal recombination (crossing-over) in higher organisms. From such crosses the progeny represent the two parental types as well as the recombinant genotypes. Studies of bacteriophages T2 and T4 by this method have yielded circular maps for the single chromosomes of these viruses. Indeed, it now appears that in many bacteria and DNA viruses, a single circular chromosome is present.

Studies of genetics and genetic aspects of metabolism are particularly favorable with haploid microorganisms since a mutation is immediately expressed and since the generation times are exceedingly short. Furthermore, the growth requirements are usually simple, the number of genes is relatively small in number as compared to higher organisms, and the mutant strains are readily isolated. A particularly valuable method of isolating mutants of bacteria (*auxotrophs*) involves the use of penicillin, which kills only growing cells. When a mixture of wild type and auxotrophs is grown on a minimal medium containing penicillin, the former will grow and die whereas the latter will remain viable and resume growth when transferred to a medium supplemented with the appropriate growth factor—an amino acid, purine, pyrimidine, vitamin, etc. Reference has been made repeatedly in earlier pages of this book to the use of mutant strains of organisms which have been used for the elucidation of metabolic pathways; many of these mutant strains were isolated with the aid of the penicillin method.

REPLICATION OF DNA

As already noted (pages 193*ff.*), the Watson-Crick formulation of the structure of native DNA indicated that each DNA molecule consists of a double-stranded helix in which the two strands are bound by hydrogen bonds between amino and keto groups: adenine (A) to thymine (T), and guanine (G) to cytosine (C); hence, A = T and G = C. On the basis of this structure, it was suggested that replication might proceed by separation of the two strands and that new chains of nucleotides would be formed complementary to each of the strands. Thus, each single strand would serve as a "template" to form a new DNA strand, which is not identical but complementary, to the template. When the process is complete, two new double-stranded molecules of DNA would have been formed, one strand of each molecule having served as template, or primer, and the other strand being newly fabricated.

Consistent with this concept is an experiment by Meselson and Stahl in which *E. coli* were grown in a medium which labeled their DNA completely with ^{15}N. The bacteria were then washed carefully and, by appropriate means, permitted to undergo one synchronous cell division in a medium in which all the nitrogen was in the form of ^{14}N. During this cell division the total DNA of the culture doubled. By accurate density determinations in a concentrated solution of cesium chloride (page 198), it was shown that the DNA of this culture did not consist of two types, *i.e.*, DNA with ^{15}N and DNA with ^{14}N, but rather that all the DNA behaved as if it were composed of $^{14.5}N$. This is the result to be expected from the mechanism of replication described above. During replication each double-stranded ^{15}N-labeled DNA separated to yield two single strands. Each of the latter served as a template on which ^{14}N-labeled DNA was fabricated, yielding a new double-stranded molecule with one strand containing ^{15}N, the other containing ^{14}N. Such molecules behave physically as if they were built of $^{14.5}$N. The process is shown schematically in Fig. 28.1.

Although the Watson-Crick model provides a concept of how replication of DNA occurs, it does not illuminate the actual chemical processes involved. Knowledge of this mechanism has been provided largely by Kornberg and his collaborators.

Net synthesis of DNA can be obtained with a purified enzyme preparation, *DNA polymerase,* obtained initially from *E. coli.* In the presence of the four deoxyribonucleoside triphosphates and of primer DNA, as well as Mg^{++}, the enzyme catalyzes an over-all process which may be represented as follows.

$$\begin{matrix} \text{m TTP} \\ + \\ \text{n dGTP} \\ + \\ \text{m dATP} \\ + \\ \text{n dCTP} \end{matrix} + \text{DNA} \rightleftharpoons \text{DNA} \begin{bmatrix} \text{TMP} \\ \\ \text{dGMP} \\ \\ \text{dAMP} \\ \\ \text{dCMP} \end{bmatrix}_{2m + 2n} + 2(m + n)PP_i$$

Pyrophosphate is released in quantities equivalent to the deoxyribonucleotides utilized. The nucleoside diphosphates cannot substitute for the triphosphates. If one of the four substrates is omitted, the yield of polymer is reduced by a factor of more than 10^4. If the DNA primer is omitted, no immediate reaction takes place.

Present evidence indicates that the reaction described involves a net synthesis of a polydeoxyribonucleotide directed by the added DNA which serves as a template, the purine or pyrimidine nucleotides being added at specific loci in the growing chains by virtue of hydrogen bonding and steric fit with the complementary base on the template. As a minimum, the DNA polymerase must possess a general specificity for deoxyribonucleotide triphosphates, a binding site for a portion of a DNA strand, and an active site that catalyzes the reaction in which internucleotide linkages are formed. The physical properties of the newly synthesized DNA molecules are closely similar to those of double-stranded DNA isolated from natural sources, and have molecular weights similar to that of the primer DNA. The molecules are degraded by pancreatic deoxyribonuclease to form acid-soluble fragments. When the synthetic DNA is heated, it undergoes the characteristic changes in viscosity and in optical properties of native DNA (Chap. 9).

Heat-denatured DNA, which is single-stranded, serves as an excellent primer for formation of double-stranded DNA. Native DNA is inert unless it is pretreated in some manner to produce unfolding of the molecule. It is unknown at present how native intracellular DNA is uncoiled at the time that replication occurs intracellularly.

Evidence indicating that the added DNA serves as a primer derives from results obtained with DNA preparations which possess different base compositions (Table 28.1, page 652). The data indicate that the added primer determines the com-

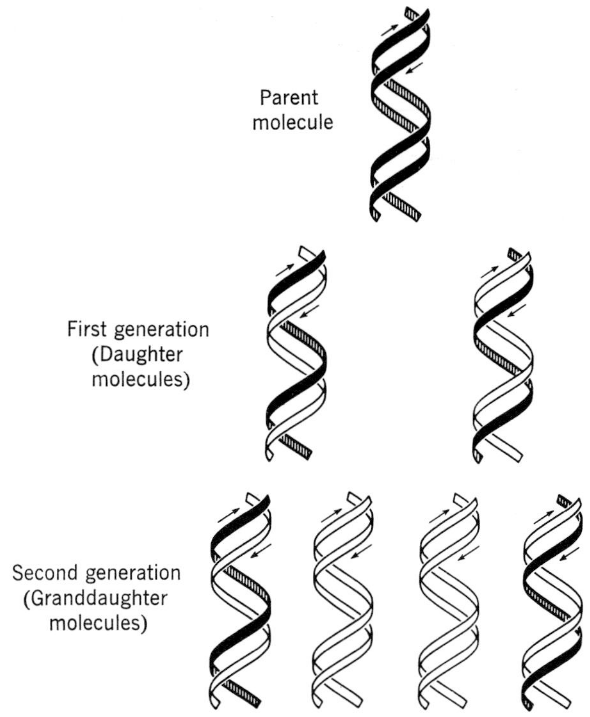

Fig. 28.1. Proposed model of DNA replication. The two strands of the parent molecule, shown in black, contain ^{15}N. In the daughter molecules, each DNA contains ^{15}N in one strand and ^{14}N in the other. In the second generation, two molecules contain ^{14}N exclusively and two molecules contain equal amounts of ^{14}N and ^{15}N. The arrows indicate the direction of the strands in the sense that phosphate diester bonds connect C-3′ of one sugar with C-5′ of the next. (*From M. Meselson and F. Stahl, Cold Spring Harbor Symp. Quant. Biol.,* **23,** 10, 1958.)

position of the enzymically synthesized DNA. Note that the use of the A-T copolymer results in synthesis of new A-T copolymer; although all four nucleoside triphosphates are present in the medium, there is no significant incorporation of nucleotides containing G or C. These results support the conclusion that base composition is replicated during the enzymic synthesis by a guiding mechanism involving hydrogen bonding of A to T and G to C (Fig. 28.2).

This conclusion would be fortified if base sequences of DNA were known. In the absence of such information, Kornberg and his colleagues resorted to an approximation called "nearest-neighbor analysis." The internucleotide bonds of enzymically produced DNA are 3′-5′, as in natural DNA. This is readily demonstrated by using ^{32}P-labeled nucleoside triphosphates in which the radioactive P is α. This P becomes the bridge between the substrate nucleotide and the nucleotide at the next position. Hydrolysis of the synthesized DNA with a mixture of a micrococcal deoxyribonuclease and splenic diesterase (page 190), yields the nucleoside 3′-phosphates quantitatively. Hence, the ^{32}P introduced at the 5′ position is transferred to the 3′ position of the neighboring nucleotide. When only one of the four added nucleoside triphosphates is labeled, *e.g.*, dATP, the ^{32}P content of each of the isolated 3′-deoxyribonucleotides is a measure of the relative fre-

Table 28.1: BASE COMPOSITION OF ENZYMICALLY SYNTHESIZED DNA

DNA	A	T	G	C	(A + G)/(T + C)	(A + T)/(G + C)
Mycobacterium phlei:						
Primer	0.65	0.66	1.35	1.34	1.01	0.49
Product	0.66	0.65	1.34	1.37	0.99	0.48
Escherichia coli:						
Primer	1.00	0.97	0.98	1.05	0.98	0.97
Product	1.04	1.00	0.97	0.98	1.01	1.02
Calf thymus:						
Primer	1.14	1.05	0.90	0.85	1.05	1.25
Product	1.12	1.08	0.85	0.85	1.02	1.29
Bacteriophage T2:						
Primer	1.31	1.32	0.67	0.70*	0.98	1.92
Product	1.33	1.29	0.69	0.70	1.02	1.90
A-T copolymer	1.99	1.93	0.05	0.05	1.03	40.00

Note: For each experiment, a different primer was used. The A-T copolymer used as primer contained equal amounts of the two bases. A, adenine; T, thymine; G, guanine; C, cytosine. The results are given as the molar ratios for each of the four bases.

* Hydroxymethylcytosine.

SOURCE: A. Kornberg, in J. M. Allen, ed., "The Molecular Control of Cellular Activity," p. 245, McGraw-Hill Book Company, New York, 1962.

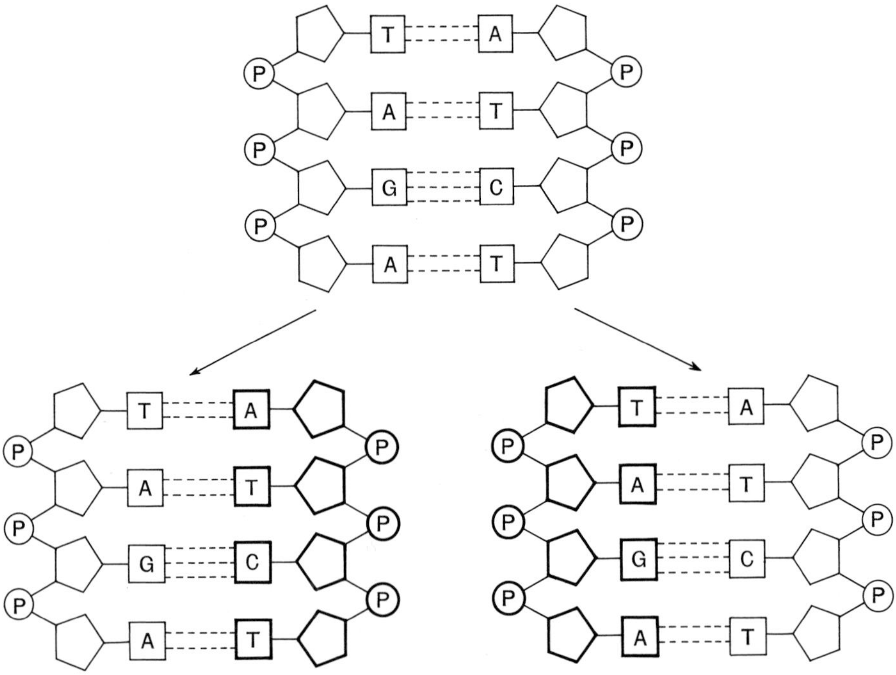

Fig. 28.2. Scheme of replication of a DNA model. Boldface chains represent the newly synthesized strands of the two daughter molecules. (*From J. Josse, A. D. Kaiser, and A. Kornberg, J. Biol. Chem.,* **236**, 864, 1961.)

quency with which the dATP reacted with each of the four available substrates during the synthesis of the DNA chains to yield the nucleotide sequences AA, AT, AG, CA. This method, when performed four times with a differently labeled substrate in each case, *i.e.*, with dATP, dGTP, dCTP, and TTP, yields the relative frequencies of all the 16 possible varieties of dinucleotide (nearest-neighbor) sequences.

The results have led to the following conclusions. All 16 possible dinucleotide (nearest-neighbor) sequences are found, and the pattern of relative frequencies is unique for each type of primer DNA used. The replication involves base pairing of A to T and of G to C. Finally, and most important, the replication produces two strands which are synthesized in opposite direction, exactly as predicted by the Watson-Crick model. The data given in Fig. 28.3 show the results obtained and those expected on the basis of this model.

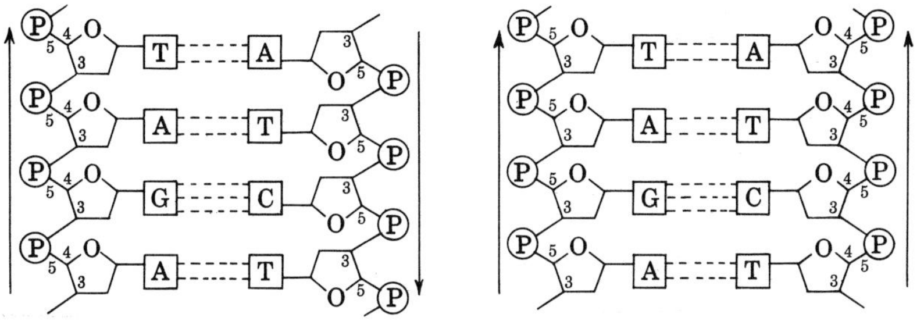

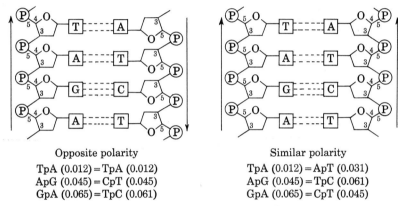

Opposite polarity	Similar polarity
TpA (0.012) = TpA (0.012)	TpA (0.012) = ApT (0.031)
ApG (0.045) = CpT (0.045)	ApG (0.045) = TpC (0.061)
GpA (0.065) = TpC (0.061)	GpA (0.065) = CpT (0.045)

Fig. 28.3. Contrast of a Watson-Crick DNA model with strands synthesized of opposite polarity with a model with strands of similar polarity. The predicted nearest-neighbor frequencies are different. Values in parentheses are sequence frequencies determined with DNA of *Mycobacterium phlei.* The strands shown are the newly synthesized ones of Fig. 28.2; for comparison, they are aligned as though they were complementary strands of the same helix. (*From J. Josse, A. D. Kaiser, and A. Kornberg, J. Biol. Chem.*, **236**, 864, 1961.)

Evidence that the polymerase furnishes a complementary copy of the DNA used as template has been provided by use of the single-stranded form of the bacterial virus ϕX174, extensively studied by Sinsheimer. The circular DNA of this virus can be freed of the outer protein shell. Intracellularly, these single-stranded forms (arbitrarily called (+) strands) are copied to make complementary (−) strands. The DNA is double stranded and both strands are in closed circular form. Separated (+) and (−) forms of the DNA are each infective to spheroplasts of *E. coli.*

Goulian, Kornberg and Sinsheimer have demonstrated that the reactions can be conducted in vitro. The procedures and results are summarized in Fig. 28.4. Isolated single-stranded DNA of ϕX174 was used with highly purified DNA polymerase, completely devoid of nuclease activity. The (+) strand was copied by the polymerase, in the presence of the deoxyribonucleoside triphosphates of adenine, cytosine, guanine and 5-bromouracil ($\overline{BU}$). (This last base will replace thymine, without loss of infectivity; see below.) This resulted in formation of a *linear,* complementary (−) strand. The ends of the linear form were joined by using an enzyme

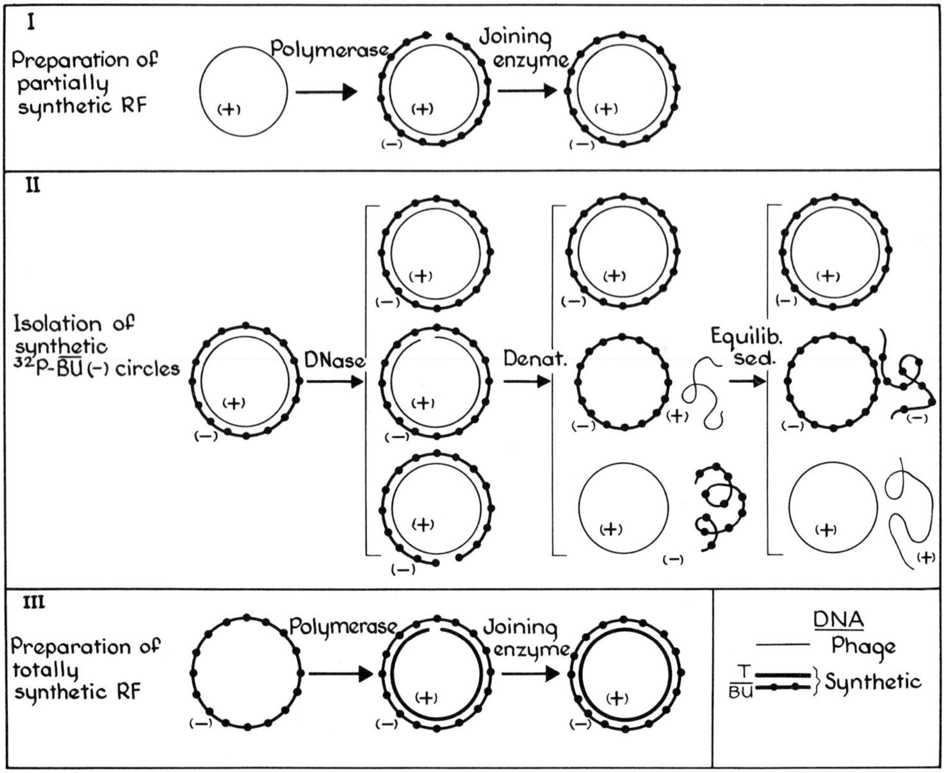

Fig. 28.4. Scheme for enzymic synthesis and isolation of ϕX174 DNA. In I the polymerase reaction occurred in the presence of the deoxyribonucleoside triphosphates of adenine, cytosine, guanine and 5-bromouracil ($\overline{BU}$). The joining enzyme (ligase) requires DPN. In II, the procedures used for isolation of intact circular forms of (+) and (−) forms are summarized. The equilibrium sedimentation was performed in a density gradient of CsCl to separate the two single-stranded forms from the remaining duplexes. Each of the single-stranded forms was then isolated by sedimentation in a sucrose gradient to separate the circles from linear forms. In III, synthetic (−) circles were used as templates for formation of duplexes in the presence of the four deoxyribonucleoside triphosphates of A, G, C and T. RF = replicative form. (*From M. Goulian, A. Kornberg and R. L. Sinsheimer, Proc. Natl. Acad. Sci., U.S.*, **58**, 2321, 1967.)

called *DNA ligase* or joining enzyme (see below) to produce the double-stranded circular form (Fig. 28.4I). The double-stranded form was treated with pancreatic deoxyribonuclease (DNase) to produce single breaks and this was then followed by heat denaturation producing some intact (+) and some intact (−) strands as well as some linear forms and residual duplex forms. The various forms of DNA molecules were then separated by density gradient centrifugation in CsCl (page 198) and in sucrose (page 131), this being possible because of the presence of the heavy bromine atoms in the (−) forms. The (+) forms had been prepared with ^{3}H labeled thymine whereas the new (−) forms contained not only bromouracil but also ^{32}P (Fig. 28.4II). These markers were important in proving the lack of contamination of the new forms. The new (−) forms were then used to prepare synthetic (+) forms from the deoxyribonucleoside triphosphates of A, C, G and T by the same methods, completing the synthesis in vitro of the double-stranded virus (Fig. 28.4III). Both the initially formed (−) strands and the final (+) strands proved to be infective to spheroplasts of *E. coli,* showing that the double replication must have been essentially free of copying errors.

Thus, complete synthesis of a biologically active DNA was accomplished with the use of only two enzymes, the polymerase serving to make complementary copies of the more than 5500 bases in the viral DNA and the joining enzyme closing the circle by linking the two terminal nucleotides together.

The DNA ligase of *E. coli* accomplishes the joining by an unusual reaction involving DPN.

$$\text{DPN} + \text{Enzyme} \rightleftharpoons \text{Enzyme-AMP} + \text{NMN}$$
$$\text{Enzyme-AMP} + \text{linear DNA} \rightleftharpoons \text{Enzyme} + \text{AMP} + \text{circular DNA}$$

$$\text{Sum:}\quad \text{DPN} + \text{linear DNA} \rightleftharpoons \text{AMP} + \text{NMN} + \text{circular DNA}$$

The pyrophosphate bond of DPN furnishes the energy for formation of a 3′-5′ phosphodiester bond in the DNA. However, a similar ligase induced in *E. coli* by T4 phage utilizes ATP in the formation of the circular form of this phage DNA. In addition to the linking of ends of a DNA chain into circular form, the ligases catalyze the repair of breaks in double-stranded DNA in which there is a break in only one strand. It is likely that a major function of such enzymes is to catalyze the repair of breaks in DNA strands after recombination of genetic material by crossing-over.

Nothing is presently understood of the phenomena that initiate DNA synthesis. The latter can begin only after the strands of the existing DNA helix begin to separate. The magnitude of this task may be evident from the structure of the DNA of *E. coli.* The "chromosome" is a single molecule of DNA in a closed circular structure. The total length of such a molecule is about 0.5 mm. or approximately 200,000 complete helical turns. Electron micrographs of replicating cells show that the two strands of DNA that constitute the double helix are gradually unwound and that a daughter strand is formed against each preexisting strand. Whatever mechanism accomplishes strand separation, once separation has occurred the formation of new, complementary strands can be accomplished by a single enzyme. The specificity of this synthesis is provided by the preexisting DNA template, base pairing being dictated by hydrogen bonding, extremely close stereospecific geometric requirements for fit into the new helical structure, and the hydrophobic and van der Waals forces which result in "stacking" of bases so that they may fit in a single, specific manner (Chap. 9). This is further indicated by the use of nucleoside triphosphates containing analogues of the naturally occurring bases. Deoxyuridine triphosphate or 5-bromodeoxyuridine triphosphate could replace TTP but not dATP, dGTP, or dCTP. Similarly, 5-methyl- and 5-bromocytidine triphosphates could specifically replace deoxycytidine triphosphate, and hypoxanthine could substitute for guanine in the nucleoside triphosphates. These findings are readily interpreted only on the basis of hydrogen bonding between polynucleotide strands involving A-T and G-C pairs. Furthermore, certain bases, *e.g.*, 2-aminopurine and 5-bromouracil, which resemble the natural ones, supplied to growing bacterial cultures not only are incorporated into DNA but prove to be mutagenic. In effect, these compounds lead to copying errors in the replication of DNA. Further evidence that mutation involves base replacement in the nucleic acid will become apparent later.

The above discussion has been concerned with DNA synthesis involving the four bases: A, T, G, and C. The *E. coli* T even bacteriophages represent a special case insofar as they contain 5-hydroxymethylcytosine (HMC) in place of cytosine (Table 9.7). When T2 DNA is used as primer in the presence of dCTP and the other triphosphates, the polymerase synthesizes a product containing C instead of HMC (Table 28.1). In T2-infected cells of *E. coli* this is prevented by an enzyme, absent in normal cells, which hydrolyzes dCTP or dCDP to dCMP.

$$\text{dCTP} + \text{H}_2\text{O} \longrightarrow \text{dCMP} + \text{PP}_\text{i}$$
$$\text{dCDP} + \text{H}_2\text{O} \longrightarrow \text{dCMP} + \text{P}_\text{i}$$

In effect, this makes more dCMP available for formation of 5-hydroxymethyl dCMP which is then converted to the triphosphate.

Kornberg and coworkers have also demonstrated that glucosylation of HMC residues occurs at the level of DNA and not with the nucleotides. These reactions are catalyzed by specific α- or *β-glucosyl transferases* present only in phage-infected cells.

$$\text{UDP-glucose} + \text{HMC-DNA} \rightleftharpoons \text{UDP} + \text{glucosyl-HMC-DNA}$$

THE NUCLEOTIDE CODE AND PROTEIN STRUCTURE

The genetic information is contained in DNA. Inasmuch as DNA consists of a linear array of bases held together in polynucleotide form, the information must be specifically conveyed by the sequences of the bases, the sugar phosphate backbones of the chains being identical in every instance. Earlier, some evidence was presented that individual genes are responsible for directing the synthesis of enzymes. Essentially similar evidence has indicated that genes direct the synthesis of all proteins, regardless of their physiological function. For example, hereditary disorders characterized by altered structure of, or failure to produce, hemoglobin, fibrinogen, or γ-globulins have been detected with increasing frequency (Chap. 30).

Genes direct not only their own replication but also the synthesis of specific proteins. The nucleic acids of plant viruses and bacteriophages may be infective without their proteins, yet newly formed virus contains specific proteins, *e.g.*, in the case of the T2 phage, cited above, the information for the synthesis of the enzymes for glucosylation of HMC and for dCTP hydrolysis is conveyed by the phage DNA. Thus the information for such specific protein synthesis is contained in the nucleic acid. Some human beings have a hemoglobin (Hb S) which, in the reduced state, is much less soluble than normal hemoglobin. Corpuscles containing Hb S undergo a change in shape known as "sickling," caused by crystallization of reduced hemoglobin. The ability to form Hb S is inherited as a single genic factor. It was demonstrated by Ingram in 1957 that Hb S differs from Hb A (normal Hb) by a single amino acid residue in the β chain (Chap. 8). Subsequently, many other abnormal hemoglobins have also been shown to differ from the normal by single amino acid replacements (Chap. 30), and similar substitutions have been found in other proteins. Furthermore, species differences among homologous proteins (see

Chap. 30) involve replacement of residues at specific loci in the peptide chains. Thus, the gene must determine the amino acid sequence of a protein; a mutation, *i.e.*, an alteration in the DNA, produces a change in the amino acid sequence.

Human hemoglobin contains α and β chains (pages 172*ff.*), each of which is under control of genes present in different chromosomes. Thus, the "one gene–one enzyme" hypothesis may be restated as one gene–one polypeptide chain, since individual enzymes and other proteins consist of only one chain or a few chains. Furthermore, the *genetic* determination of protein structure is primarily concerned with the specific kind and linear arrangement of amino acids in the polypeptide chain or chains. The "sequence concept" can now be stated, as follows: the amino acid sequence of a protein is determined by the sequence of nucleotides in a definite portion of the DNA. Some of the evidence for and implications of this hypothesis may now be examined.

Specific Protein Conformation. Inasmuch as globular proteins manifest their characteristic properties only in their native state, we may inquire whether the complex specific folding is under direct genetic control. Available evidence suggests that the amino acid sequence per se determines how the polypeptide chain folds into its native conformation. In effect, this implies that there is only one conformation of maximal stability, a view supported by observations that many highly purified proteins can be denatured, *i.e.*, assume a random form, and then under suitable experimental conditions, spontaneously regain their native properties (Chap. 7). This is seldom achieved with crude extracts of tissues in which many denatured proteins can react with one another or with the metabolites of the cell. Moreover, such extracts frequently contain proteinases which cannot readily hydrolyze native proteins but rapidly attack denatured proteins.

The reversible denaturation of hemoglobin has already been described (pages 163*ff.*). Hemoglobin lacks disulfide bonds, and the folding involves only secondary forces, *i.e.*, hydrophobic forces, hydrogen bonds, ionic interactions, etc. (pages 152*ff.*). A similar situation obtains with other proteins that lack disulfide bonds, *e.g.*, myoglobin, enolase, various amylases, cytochrome c, etc. For proteins such as ribonuclease, which contains four disulfide bridges (page 163), the situation is not entirely dissimilar. The disulfide bonds may be reduced to yield a linear polypeptide chain and, under favorable conditions, reoxidation will produce the native, active enzyme. In this case, also, folding is determined by the specific amino acid sequence which brings the correct pairs of cysteine residues into juxtaposition prior to oxidation. Thus, the *genetic* influence on conformation is exerted by determining the positions of the residues which are critical for the conformation and, therefore, for the functional properties of the protein.

Many proteins consist of two or more chains. In some, the chains are united by noncovalent forces, as in hemoglobin (page 163), aldolase (page 397), glutamic acid dehydrogenase (page 536), etc. In the last instances, dissociation and association are readily reversible by specific environmental factors in the same way that chain conformation is controlled. Thus, there are no special genetic factors for the formation of multichain proteins; their formation is a consequence of the genetic information expressed in the amino acid sequence of the constituent polypeptide chains. In a rare genetic defect of man, the synthesis of the α chain of hemoglobin is partially supressed (Chap. 31); such individuals possess Hb H, which is β_4. Clearly, not only is the β chain formed independently of the α chain, but the association of four β chains can also occur in the absence of α chains.

In a few cases, *e.g.*, insulin, chymotrypsin, and plasmin, the chains are linked by disulfide bonds. However, in these three instances the inactive precursor proteins are synthesized as a single chain. The production of the multichain active forms is the result of proteolytic cleavage leaving the residual peptide chains attached by disulfide bonds. This has been previously described for chymotrypsin (page 252) and will be described for the conversion of plasminogen to plasmin (page 734) and for the conversion of proinsulin to insulin (page 973).

Prosthetic Groups. All available evidence indicates that no direct genetic control guides the processes by which non-amino acid prosthetic groups are added to proteins. The spontaneous recombination of heme with globin to form hemoglobin (page 163) indicates that it is unnecessary to assume any special genetic control. Similarly, many enzymes that possess dissociable prosthetic groups, *e.g.*, flavins, heme, pyridoxal phosphate, DPN, TPN, metal ions, etc., are spontaneously regenerated by adding the prosthetic group to the apoenzyme (protein) under suitable conditions. A further striking example may be cited. A porphyrinless mutant of *E. coli* lacks catalase activity. If, however, ferriprotoporphyrin is added, active catalase is formed, indicating that biosynthesis of the apoenzyme has occurred without concurrent synthesis of the prosthetic group.

In the above examples, the prosthetic group is held by noncovalent bonds to the protein. When covalent bonds are formed, they may be synthesized by enzyme-catalyzed reactions, since many of the bonds may require energy for synthesis. The following examples may be cited to illustrate the problems involved: the thioether bonds linking the heme in cytochrome c (page 177), the amide bond linking biotin to the ϵ-amino group of lysine in certain enzymes (page 487), the carboxamide bond linking the polysaccharide moiety to the protein in γ-globulins (page 716), ovalbumin, and other glycoproteins, the N-acetyl of cytochrome c, ovalbumin, and other proteins, etc. In most cases, the biosynthetic processes are unknown but are controlled by genetic determination of the responsible synthetic enzymes.

Amino Acid Modification in Peptide Chains. Only 20 amino acids are generally present in proteins (page 659). Other amino acid residues are formed by subsequent modification of some residues during or after formation of the polypeptide chains. Nonphosphorylated ovalbumin is synthesized in the hen's oviduct prior to phosphorylation. Phosphorylserine (page 824), phosphorylthreonine (page 824), tyrosine-O-sulfate (page 726), hydroxyprolines (page 91), hydroxylysine (page 93), and iodinated amino acids (page 94) are formed by enzymic, energy-requiring processes after incorporation of the unsubstituted amino acid into the polypeptide chain. However, little is known of the mechanisms involved, particularly as to whether the choice of residues so affected is under specific genetic influence or is a consequence of the conformation of the proteins concerned. Covalent cross-linking of peptide chains occurs in some fibrous proteins, *e.g.*, collagen and elastin, by reactions which occur subsequent to synthesis of the polypeptide chains. Formation of the unusual amino acids, desmosine and isodesmosine, from lysine residues in elastin is an example of such a process (Chap. 38).

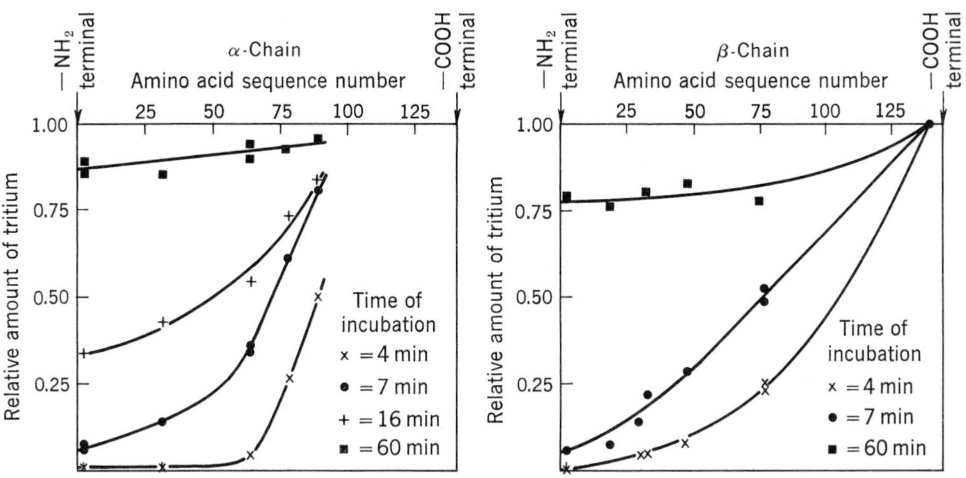

Fig. 28.5. Incorporation of labeled leucine into the α and β chains of rabbit hemoglobin. The position of each leucine residue is given at the top of each graph. (*After M. A. Naughton and H. M. Dintzis, Proc. Nat. Acad. Sci. U.S.*, **48**, 1822, 1962.)

Direction of Synthesis of Peptide Chains. Protein synthesis occurs by sequential addition of amino acids to the growing end of a peptide chain. Isolated rabbit reticulocytes continue to form hemoglobin in the presence of free amino acids. When such cells were given a 4-min. exposure to an amino acid mixture of ^{3}H-leucine with all other amino acids unlabeled, the greatest labeling occurred at the unfinished carboxyl ends of the α and β chains (Fig. 28.5). With longer exposures, progressively greater incorporation occurred at all positions, the radioactivity incorporated increasing with the distance of the particular leucine residue from the amino-terminal end. After 1 hr., the labeling approached equality at all positions. In certain bacteria and phages, mutations are known in which the growing peptide chain is terminated prematurely. The isolated peptide fragments from such mutants always include the NH_2-terminal portions of the incomplete proteins. These results clearly demonstrate that protein synthesis is initiated at the amino-terminal end of the peptide chain.

The Coding Problem. Since genetic control of protein synthesis determines only the sequence of amino acids, and since both the DNA and the peptide chain are linear polymers, they must be *colinear* in some manner—the base sequence determining the amino acid sequence. However, there are only four primary bases in DNA, *viz.*, A, G, C, and T, whereas there are 20 amino acids in proteins. Although small amounts of other bases may be present in DNA, they function as occasional substitutes for one of the usual bases. For example, methylcytosine pairs with guanine in the same manner as does cytosine. There is no evidence for any special function of the unusual bases in DNA, except in certain bacteriophages (page 205), and the coding problem can be considered in terms of the common four bases.

The 20 amino acids usually present in proteins are, in abbreviated form, Ala, Arg, Asp, Asn, CySH, Glu, Gln, Gly, His, Ile, Leu, Lys, Met, Phe, Pro, Ser, Thr, Trp, Tyr, and Val (page 108). Cystine is omitted since cysteine is incorporated as such and cystine, when present, is formed by subsequent oxidation of two cysteine residues. Glutamine and asparagine are included with their corresponding dicarboxylic acids; the incorporation into protein of each of these amino acids is under separate genetic control.

With only four bases in DNA, and 20 amino acids in proteins, the coding unit or *codon, i.e.*, the specific combination of bases controlling the incorporation of an amino acid, must be larger than one. If the codon were two bases per amino acid, this would permit only 4^2, or 16, codes. For a codon of three bases (triplet), there are 4^3, or 64, possible combinations, more than enough to code for 20 amino acids. Present evidence indicates that the codon consists of three bases (page 674*ff.*).

In order to discuss additional problems related to the code, some concepts, findings, and definitions, derived mainly from genetic studies, should be introduced. A *cistron* is that portion of the genetic material (DNA) which codes for one polypeptide chain. The *muton* is the smallest element of the cistron which can be altered by mutation; this unit comprises a single base pair in DNA.

Studies with various proteins have shown that point mutations, which are inherited in mendelian fashion, affect only a single amino acid in the polypeptide chain. This has been shown most strikingly for mutations which have altered the structure of human hemoglobin. Furthermore, mutants near one another on the genetic map (determined by crossing-over studies) produce amino acid replacements close to one another in the amino acid sequence. This has been found with the α protein of tryptophan synthetase of *E. coli* (page 690) and with the coat protein of T4 bacteriophage. Such studies provide important evidence for the colinearity of the peptide chain and the DNA of the gene (Chap. 30).

Another aspect of the coding problem should be mentioned. In an overlapping type of code, a given base forms part of the sequence of several coding units. With a triplet code and the base sequence ··· AGCTAG, the base C could be part of the coding units AGC, GCT, and CTA; in a nonoverlapping code, the units would be AGC and TAG. In an overlapping code, an alteration of C would alter three amino acids. Since known point mutations involve only a single amino acid replacement, the code is of the nonoverlapping type.

REFERENCES

See list following Chap. 30.

29. Genetic Aspects of Metabolism. II

The Gene and Protein Synthesis. Other Aspects of RNA Synthesis. The Amino Acid Code. Control of Protein Synthesis

THE GENE AND PROTEIN SYNTHESIS

This chapter is concerned primarily with the mechanisms involved in protein synthesis. Since proteins are synthesized from free amino acids, the major problems can be stated as follows. First, how is energy made available for formation of peptide bonds? Second, how are the sequences of nucleotide bases in DNA translated into the specific sequences of amino acids in proteins? Third, what are the controlling factors that determine when the genetic information is to be utilized?

Before considering these problems in detail, the main features of the over-all process of protein synthesis may be described briefly. The sites of protein synthesis are the ribosomes, particles that contain RNA and proteins. The genetic information is brought to the ribosomes by a unique type of single-stranded RNA known as "messenger" RNA, "template" RNA, or "informational" RNA; for simplicity it will be called the messenger or mRNA. This mRNA is synthesized by copying the base sequence of one strand of DNA; the process is termed *transcription.* Thus, mRNA is complementary to the DNA, in the same sense as the two strands of DNA are complementary to each other. Thus, the nucleotide sequences of mRNA determine the amino acid sequences of the proteins, a process called *translation.*

Amino acids are attached to soluble RNA molecules which, because they transfer amino acids to sites on the ribosomes, are termed transfer RNA or tRNA. The tRNA molecules are specific, and there is at least one for each of the 20 amino acids. Thus, there are at least three distinct kinds of RNA involved in protein synthesis: the low molecular weight tRNA molecules, which contain a number of unusual nucleotides (page 202); the structural RNA of the ribosomes (rRNA); and mRNA.

The mRNA is read in units of three nucleotides (triplet codon) for incorporation of a single amino acid. The tRNA possesses a triplet anticodon that is complementary and antiparallel to the codon of the mRNA. It is through such interactions that the genetic message of the mRNA is translated into a polypeptide sequence, incorporating sequentially one amino acid residue at a time beginning at the N-terminal end of the peptide chain.

Much of our present information concerning protein synthesis has come from the use of a crude cell-free system, developed mainly by Zamecnik, Hoagland, and their associates; this system permitted the study of incorporation of labeled amino acids into peptide linkage. Such cell-free preparations were obtained initially from rat liver but have since been derived from other mammalian tissues and from microorganisms and plants. The general features of preparations from all these sources have proved to be essentially similar. The system consists of the following components: ribosomes, proteins precipitable at pH 5 from the supernatant solution of a broken cell suspension, tRNA, ATP, GTP, amino acids, Mg^{++}, and soluble enzymes. The over-all requirements for protein synthesis are shown schematically in Fig. 29.1.

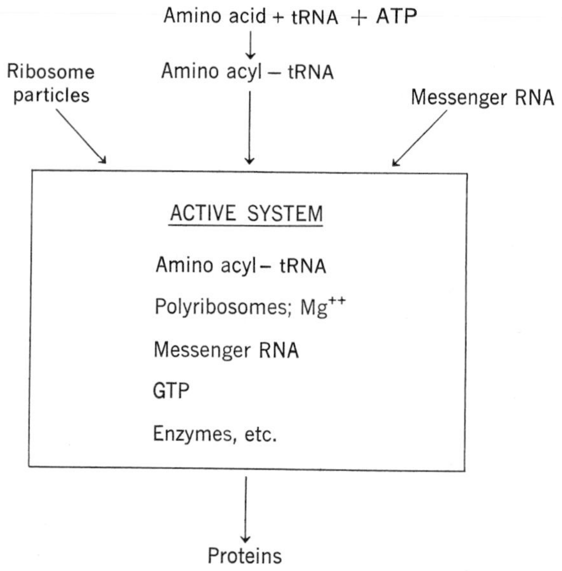

Fig. 29.1. A schematic representation of the main features of protein synthesis. Each tRNA, specific for a single amino acid, bears a sequence of three nucleotide bases complementary to a similar triplet in mRNA. Amino acids are transferred from their attachment to the tRNA to form the polypeptide sequence through the intervention of mRNA, which is attached to a polyribosome unit.

Formation of Amino Acyl-tRNA Compounds. This first step of protein synthesis, discovered by Hoagland, involves formation of an enzyme-bound amino acyl-adenylate complex [reaction (1)] in which the amino acid carboxyl forms an anhydride with the phosphate of AMP. This is followed by transfer of the amino acyl moiety to a specific tRNA [reaction (2)]; the carboxyl group of the amino acid is linked by an ester bond to a hydroxyl group of the ribose of a terminal adenylic acid residue of the tRNA. It is probably the 3′-hydroxyl group of the ribosyl group which is linked to the amino acid. Both steps are catalyzed by a single enzyme, an *amino acyl synthetase,* which is essentially specific for the amino acid involved as well as for the tRNA acceptor and requires Mg^{++}.

$$\text{Amino acid + ATP + enzyme} \rightleftharpoons \text{amino acyl-AMP-enzyme} + PP_i \quad (1)$$

$$\text{Amino acyl-AMP-enzyme + tRNA} \rightleftharpoons \text{amino acyl-tRNA + AMP + enzyme} \quad (2)$$

$$\text{Sum:} \quad \text{Amino acid + ATP + tRNA} \rightleftharpoons \text{amino acyl-tRNA + AMP} + PP_i$$

Specific amino acyl synthetases for individual amino acids have been separated, and several have been purified. Presumably, there is at least one synthetase for each amino acid in all cells. If tRNA is omitted from the reaction mixture, amino acyl adenylate can be detected in amounts equivalent to that of the enzyme.

Specificity for Amino Acids. There are two aspects of the enzymic specificity involved in formation of the amino acyl-tRNA compounds: formation of amino acyl-AMP-enzyme and transfer to tRNA. A variety of synthetic amino acyl adenylates can be utilized by *tryptophanyl-tRNA synthetase* for the reverse of reaction (1), *i.e.*, for ATP formation as induced by addition of PP$_i$ and the synthetase. However, only L-tryptophanyl adenylate is a substrate for amino acyl-tRNA formation. Similarly, *isoleucyl-tRNA synthetase* catalyzes ATP-PP$_i$ exchange with either L-isoleucine or L-valine, but the purified enzyme catalyzes formation of the tRNA compound only with L-isoleucine. That formation of the tRNA compound is specific for a *single* natural amino acid is of great importance inasmuch as these tRNAs are directly involved in transfer of the amino acid residues to their proper sites in the protein. Although they successfully distinguish among the 20 naturally occurring amino acids, the amino acyl synthetases do not have absolute specificity for their amino acid substrates, as indicated by the reported incorporation into certain proteins of such analogues as *p*-fluorophenylalanine, ethionine, selenomethionine, norleucine, and others.

At each site along its length, the template on which protein is synthesized recognizes and interacts only with a specific tRNA and not with the attached amino acyl residue. This was shown in an experiment by Lipmann and coworkers, who utilized ^{14}C-labeled cysteinyl-tRNA. This substance was oxidized to the corresponding cysteic acid derivative.

$$HS-CH_2-CH(\overset{NH_3^+}{|})-\underset{\parallel}{\overset{}{C}}(=O)-tRNA \xrightarrow[\text{acid}]{\text{performic}} \ ^-O-\underset{\parallel}{\overset{O}{S}}(=O)-CH_2-CH(\overset{NH_3^+}{|})-\underset{\parallel}{\overset{}{C}}(=O)-tRNA$$

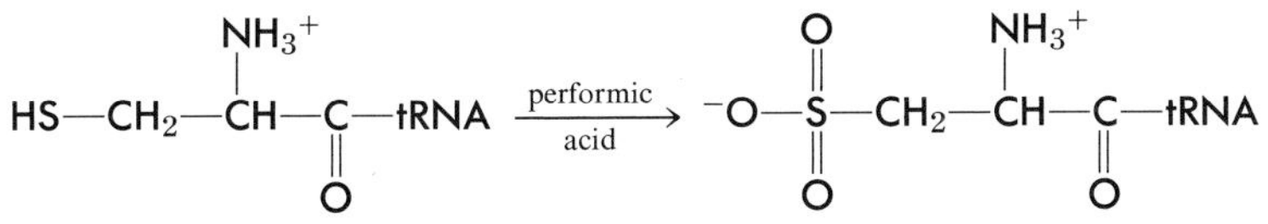

Addition of this compound to a preparation of rabbit reticulocytes capable of forming hemoglobin in vitro showed that the characteristic peptides, obtained by tryptic digestion, which ordinarily contain cysteine then contained cysteic acid. Thus, "recognition" of the amino acyl-tRNA by the protein-synthesizing system depends only on the tRNA bearing the residue, not on the attached amino acyl residue. Further evidence for this conclusion is presented later (page 676*ff.*).

Specificity of tRNA. At least one specific tRNA molecule must exist for each of the 20 amino acids. Indeed, since the code is degenerate, *i.e.*, there is more than one codon for an amino acid, there may also be more than one tRNA for a specific amino acid. Several different tRNAs have been isolated from yeast and *Escherichia coli* and the sequences of the nucleotides in several of these have been determined (page 676).

Despite the specificity for individual amino acids of the different types of tRNA, each tRNA possesses at its acceptor end an identical trinucleotide sequence:

RNA-pCpCpA. (For this method of describing polyribonucleotides, see page 189.) In many tRNAs the nonacceptor end bears a residue of guanosine 5′-phosphate. Therefore, the specificity of each tRNA must reside in the interior of the polynucleotide chain.

The simplest explanation of the specificity of amino acid transfer into the correct position in a polypeptide chain would be binding of a portion of the tRNA, such as a triplet sequence of nucleotides (anticodon), with a complementary triplet on the completely assembled system of the mRNA (codon) on the ribosomes. Such complementarity would be dependent on pairing as a result of hydrogen bonding and other forces, similar to that in DNA, except that uracil (U) in both tRNA and mRNA replaces the thymine of DNA. The triplet binding between the chains would involve the base pairs G-C and A-U. This will be considered further, below (page 674).

Inasmuch as the presumed binding of the anticodon of the tRNA with the codon of mRNA involves only a triplet, the tRNA must be folded in such a way that other triplets could not bind artifactually. This implies considerable intramolecular folding in tRNA molecules. This has been demonstrated by a large hyperchromic effect on heating, by x-ray diffraction studies, and other methods. Thus, there is considerable internal base pairing within individual tRNA molecules, involving G with C and A with U. All the tRNA molecules whose sequences are known contain many "unusual" nucleotides (page 676). Theoretical attempts to prepare folded structures of tRNA for different amino acids suggest that there may be a definite pattern in the distribution of the unusual nucleotides (page 676).

In summary, studies of the formation of amino acyl-tRNA indicate that the reaction depends on the dual specificity of the amino acyl synthetases, first, to react with a specific amino acid and, second, to couple the amino acid to a specific tRNA. The portion of the tRNA that is "recognized" by the synthetase has been tentatively assigned (page 676) and does not involve the terminal -CpCpA sequence since the latter is apparently the same for all tRNA molecules. Subsequent transfer of the amino acid from the tRNA presumably involves the specific triplet anticodon on the tRNA which binds to the codon of the mRNA on the fully assembled ribosomal system (page 676).

Ribosomes and rRNA. The first in vitro studies of amino acid incorporation into proteins showed that certain cytoplasmic particles were essential for this process, and, indeed, that the labeled amino acids became attached to these particles. In preparations from animal tissues, particularly liver, amino acid–incorporating activity is associated with *microsomes* (page 286); however, such microsomal preparations are capable of performing many kinds of metabolic reactions and can be subfractionated. Protein biosynthesis is associated with nucleoprotein particles, the *ribosomes.* In animal cells most of the ribosomes are attached to membranes representing the endoplasmic reticulum (page 284). When the microsomal fraction is treated with a detergent such as deoxycholate, lipoproteins are dissolved, and the ribosomes may be isolated by sedimentation in the ultracentrifuge. Small numbers of ribosomes have been reported to occur in the nucleus.

Ribosomes of bacterial cells are readily isolated, free of other cellular components, by repetitive differential ultracentrifugation. Ribosomes of *E. coli* consist almost entirely of ribonucleoprotein in which the rRNA content is approximately 60 to 65 per cent; ribosomes from mammalian and plant sources contain 40 to 50 per cent rRNA. Ribosomes show characteristic components in the analytical ultracentrifuge, the kinds and amounts being dependent on the Mg^{++} concentration. In 0.01 *M* Mg^{++}, ribosomes from *E. coli* possess a sedimentation constant of 70 S (S = Svedberg units, page 130), but some larger aggregates are present. In 0.001*M* Mg^{++} chiefly 70 S particles are present, with some 30 S and 50 S. In the presence of low concentration of Mg^{++}, 0.0001*M* or less, principally 30 S and 50 S particles are present. Electron micrographs as well as sedimentation studies indicate that one 30 S and one 50 S particle combine to give a 70 S particle, a complete ribosome.

These nucleoprotein particles may be dissociated by extraction with phenol or detergents to yield rRNA and protein in separate fractions. The rRNA from the 30 S particles is a molecule with 16.3 S (molecular weight = 5.6×10^5), and from the 50 S particles the rRNA consists of 23.5 S material (molecular weight = 1.1×10^6). Both types of RNA exhibit identical absorption spectra, hyperchromicity, and nucleotide composition. The composition of rRNA from *E. coli* is given in Table 29.1. Essentially similar analytical values for rRNA from other sources have also been obtained. Ribosomal proteins from different species resemble one another. It has been shown by gel electrophoresis and other methods that in *E. coli* there are at least 19 distinct proteins in the 50 S particle and at least 11 in the 30 S particle.

In addition to the 16 S and 23 S components of rRNA, there is a component of 5 S in *E. coli* containing 120 nucleotide residues. The complete sequence shows only the four usual nucleotides.

Sedimentation studies and electron micrographs of extracts of rabbit reticulocytes show that protein is synthesized on an aggregate of three to ten 70 S ribosomes. This *polyribosome,* or *polysome,* is held together by a thread of mRNA (see below). Treatment with ribonuclease destroys the thread of mRNA, but not the rRNA, liberating the individual ribosomes and terminating protein synthesis.

Table 29.1: NUCLEOTIDE COMPOSITION OF DNA AND RNA IN NORMAL AND BACTERIOPHAGE-INFECTED *Escherichia coli*

	Moles per 100 moles					$\frac{\text{Purine}}{\text{Pyrimidine}}$	$\frac{\text{A + T (or U)}}{\text{G + C}}$
	A	C	G	U or T	Minor bases		
Normal cells:							
DNA	24–25	25–26	25–26	24–25	. . . .	0.96–1.04	0.92–1.00
rRNA	25.2	21.6	31.5	21.7	. . . .	1.30	0.88
mRNA	25.1	24.1	27.1	23.7	. . . .	1.09	0.95
tRNA	20.3	28.9	32.1	15.0	3.7	1.12	0.64
Phage-infected cells:							
T2 DNA	32	17*	18	32	. . . .	0.98	1.83
mRNA	31	17	20	31	. . . .	1.06	1.68

*Hydroxymethylcytosine (see Table 9.7, page 206).

SOURCE: The data are from the work of several investigators and were compiled by F. Gros, W. Gilbert, H. H. Hiatt, G. Attardi, P. F. Spahr, and J. D. Watson in *Cold Spring Harbor Symp. Quant. Biol.,* **26,** 111, 1961.

Messenger RNA (mRNA). The abundance of RNA in cytoplasm and evidence of its important role in protein synthesis suggested that the genetic information of nuclear DNA is transmitted to an RNA which functions at the sites of protein synthesis. In 1961, Jacob and Monod postulated that control of protein formation, at least in certain microorganisms, is determined by the rate of synthesis of templates. This requires that the templates do not accumulate, in contrast to the constant presence of DNA, tRNA, and rRNA. They suggested, therefore, the transient existence of an RNA, which they called "the messenger," or mRNA. Such an RNA could represent no more than a small percentage of the total RNA of the cell, most of which can be accounted for as rRNA and tRNA. This prediction has been amply verified.

As early as 1948, S. S. Cohen observed that in cells of *E. coli* infected with bacteriophage T2, there was no significant *net* synthesis of RNA; nevertheless, about 1 to 3 per cent of the total RNA showed rapid incorporation of labeled nucleotides. Volkin and Astrachan in 1956 then demonstrated not only that this newly formed RNA is different from the bulk *E. coli* RNA but also that the nucleotide composition of the newly formed RNA is very similar to that of bacteriophage DNA, with uracil taking the place of thymine of DNA (see also below). Study of this process has indicated that the newly formed mRNA carries the genetic information for protein synthesized during phage infection. It has also been shown that in uninfected bacteria exposed for a short time to ^{32}P, RNA with base ratios similar to those of DNA is formed, with U present in place of T. These rapidly formed mRNA molecules are the genetic messengers.

Table 29.1 presents the base compositions of the nucleic acids of normal and phage-infected cells of *E. coli.* Only the mRNA molecules resemble in composition the appropriate DNA. This is particularly striking for phage-infected cells, in which the virus inhibits production of normal mRNA and utilizes host ribosomes for production of its own protein.

Direct evidence that mRNA is a complementary copy of DNA was obtained by Hall and Spiegelman. When DNA of T2 bacteriophage was heated, the helices separated and single strands were formed (page 197). When slowly cooled in the presence of mRNA of *E. coli* infected with T2 bacteriophage, hybrid DNA-RNA double-stranded helices were detected by density gradient centrifugation in CsCl. When a sample of the same mRNA was heated with genetically unrelated DNA, no evidence was obtained for hybrid formation. This suggests that base pairing occurs between one strand of DNA and the related mRNA to form a double-stranded molecule, analogous in structure to double-stranded DNA.

The mRNA molecules of *E. coli* are heterogeneous in size and have molecular weights in the range of 200,000 to 500,000 and higher. Since the polypeptide chains of most proteins range in size from approximately 10,000 molecular weight upwards, and a codon for one amino acid comprises a trinucleotide of about 350 molecular weight, mRNA of 350,000 could code for a polypeptide chain of 1,000 amino acids. As discussed later in this chapter, messenger units may include the information for the synthesis of several polypeptide chains.

5-Fluorouracil, added to *E. coli* cells, is rapidly incorporated into mRNA in place of uracil residues. This pyrimidine analogue is not incorporated into DNA, tRNA, or rRNA when the experiments are limited to short periods, *e.g.*, 1 min. at 37°C. However, the average amino acid composition of newly synthesized proteins is modified, including the production of altered enzymes. These experiments cannot be explained by postulating alterations of preexisting RNA components and are consonant with an effect on rapidly renewable molecules, such as those of mRNA.

Enzymic Synthesis of mRNA. Independently, Weiss, Hurwitz, Stevens, and their associates described a widely distributed enzyme, *DNA-dependent RNA polymerase,* which catalyzes formation of an RNA whose base composition reflects that of the DNA present in the system. Thus, this synthetic system for RNA formation fulfills the criteria for transcription of the sequence of DNA and formation of RNA.

$$\begin{matrix} \textbf{mATP} \\ + \\ \textbf{nGTP} \\ + \\ \textbf{nCTP} \\ + \\ \textbf{mUTP} \end{matrix} \xrightarrow[\text{Mg}^{++}]{\text{DNA}} \begin{matrix} \textbf{mAMP} \\ | \\ \textbf{nGMP} \\ | \\ \textbf{nCMP} \\ | \\ \textbf{mUMP} \end{matrix} + 2(\textbf{m} + \textbf{n})\textbf{PP}_i$$

The reaction is dependent on the presence of DNA; if DNA is omitted or treated first with deoxyribonuclease, no reaction occurs. Synthesis of RNA occurs only when all four ribonucleoside triphosphates are present. The synthesized RNA has normal 3′-5′ phosphodiester linkages. Most strikingly, the composition of the synthesized RNA reflects that of added DNA. This suggests that nucleotide incorporation is dependent on the ability of the bases of the ribonucleotides to pair with the bases in the primer DNA, by a mechanism similar to that for DNA replication and for the action of DNA polymerase (page 650).

Table 29.2 shows the ribonucleotide incorporation obtained when different DNA preparations are used as primers. The ratio (A + U)/(C + G) in the RNA formed is identical with the ratio (A + T)/(C + G) of the DNA added, despite the wide range of compositions of the DNA primers. Furthermore, the synthesized RNA resembles DNA in possessing a ratio of purines to pyrimidines [(A + G)/(U + C)] equal to one. Thus, these compositions correspond to those expected for messenger RNA and do not resemble the compositions of tRNA or rRNA.

Of particular interest are experiments performed with the single-stranded

Table 29.2: COMPOSITION OF RNA SYNTHESIZED IN PRESENCE OF DIFFERENT DNA PREPARATIONS

DNA added	$\frac{A+T}{C+G}$ in DNA	Nucleotide incorporation in mμmoles				$\frac{A+U}{C+G}$ observed	$\frac{A+G}{U+C}$
		AMP	UMP	GMP	CMP		
T2 phage......	1.86*	0.54	0.59	0.31	0.30	1.85	0.96
Thymus.......	1.35	3.10	3.30	2.0	2.2	1.52	0.93
Escherichia coli..	1.0	2.70	2.74	2.90	2.94	0.93	0.98
Micrococcus....	0.40	0.55	0.52	1.10	1.12	0.48	1.01

*Contains hydroxymethylcytosine instead of cytosine (Table 9.7, page 206).

SOURCE: J. Hurwitz, J. J. Furth, M. Anders, P. J. Ortiz, and J. T. August, in *Cold Spring Harbor Symp. Quant. Biol.,* **26,** 91, 1961.

DNA of the bacterial virus ϕX-174. The viral DNA has the relative base composition: A = 1, T = 1.33, G = 0.98, and C = 0.75. On the basis of complementarity, the synthesized RNA would have the expected composition U = 1, A = 1.33, C = 0.98, and G = 0.75, where the base pairing would be A → U, T → A, G → C, and C → G. This prediction was fulfilled as found by several investigators.

The in vitro experiments with *DNA-dependent RNA polymerase* presented in Table 29.2 show that both strands of DNA are copied. Such data are obtained when the DNA has been denatured (chain separation) or when the chains have been broken to produce single-stranded regions. In vivo, only one strand of the two in DNA is copied in the synthesis of mRNA; however, it has been shown that messages may be derived from discrete sections of either strand of the DNA of *E. coli.*

The data for the base compositions of mRNA of *E. coli* and of phage T2 resemble the compositions of DNA only because the average base compositions of both strands of DNA are similar in these two instances. For some organisms, the complementary strands of DNA differ widely in composition; in vivo synthesis of mRNA yields a polynucleotide whose composition is complementary to one strand and identical with the nontemplate strand, except that U replaces T.

Synthesis of mRNA by the polymerase proceeds by successive attachment from 5′ to free 3′-OH groups, as shown below.

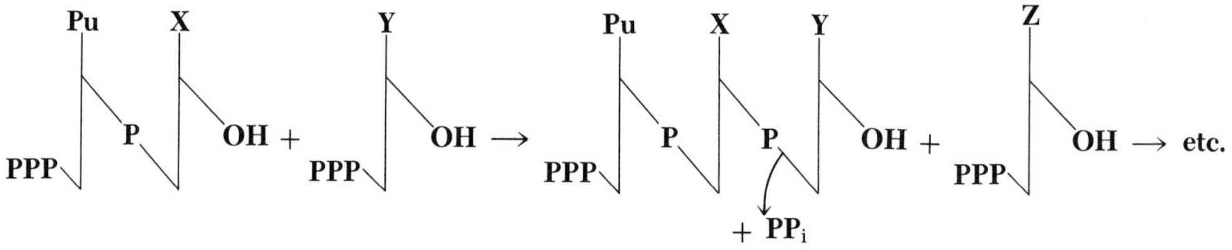

At the initiating end the 5′-nucleoside triphosphate must contain a purine; the triphosphate is subsequently hydrolyzed enzymically, leaving a 5′-phosphate and liberating pyrophosphate. In this process, the strand of DNA which serves as template and is transcribed is the one with a free 3′-OH group and the strand is *antiparallel* and complementary to the mRNA that is formed. In this respect the process resembles DNA replication. Furthermore, it has been demonstrated (page 674) that polypeptide synthesis, which is initiated at the NH_2-terminus, corresponds to the end of the mRNA bearing the 5′-phosphate group. Thus, the initiating end of the mRNA is also the initiating end for the translation of the message into a polypeptide chain.

The Transfer Enzymes. Although certain of the factors involved in protein synthesis are known, many facets of the process are still obscure. A crude *minimal* in vitro system that incorporates labeled amino acids into peptide bonds can be reconstructed as follows: (1) a mixture of the various amino acyl-tRNA compounds; (2) washed ribosomes; (3) GTP; (4) factors from the supernatant solution. The amino acyl-tRNA compounds also may be formed in the system from tRNA, amino acids, ATP, and proteins precipitated at pH 5 from supernatant solutions of broken cells (page 286). Active minimal amino acid–incorporating systems have been obtained from mammalian liver, from growing plants such as pea seedlings, from bacterial cells, from reticulocytes, and from other sources. The reticulocyte system is a particularly favorable one since the major protein formed is globin or hemoglobin, which can be isolated and studied separately.

Transfer of amino acids from their tRNA linkages to the ribosome requires enzymes from the supernatant fraction. Such *transfer enzyme* preparations are complex and have been partially resolved. The roles of the different enzymes required is still unknown. The precise role of GTP is also unknown; it is utilized in the transfer process since a mixture of GMP and GDP is formed. Available evidence suggests that one equivalent of GTP is consumed for each equivalent of amino acid transferred to the ribosome from amino acyl-tRNA.

The transfer enzymes show species specificity with respect to the source of the ribosomes. For example, with amino acyl-tRNA from *E. coli*, *E. coli* transfer enzyme was not effective with rat liver ribosomes, nor was rat liver transfer enzyme effective with *E. coli* ribosomes. However, with *E. coli* amino acyl-tRNA, transfer occurred with rat liver enzyme to rat liver ribosomes.

In summarizing the aspects of protein synthesis presented thus far, we can refer to the diagram shown in Fig. 29.2. A ribosome becomes attached to the initiating end of the mRNA. The amino acyl-tRNA responsible for chain initiation becomes attached to the 50 S component of the ribosome. The tRNA bearing the next amino acid residue binds to the 30 S unit of the ribosome; this is then transferred to the 50 S unit, forming a peptide bond and displacing the free tRNA that carried the initiating residue of the peptide chain. Amino acid residues are added sequentially, the ribosome moving along the mRNA in steps of three nucleotides for each residue added to the peptide chain. It has been postulated that the energy supplied by GTP may be utilized for the physical movement of the ribosome along the mRNA. Note that each ribosome attached to the mRNA bears a growing peptide chain; thus the mRNA functions as a template in the translation process for several polypeptide chains simultaneously. The nature of peptide chain initiation and termination, and the interactions of the tRNAs with the mRNA are considered below (amino acid code, page 673*ff.*).

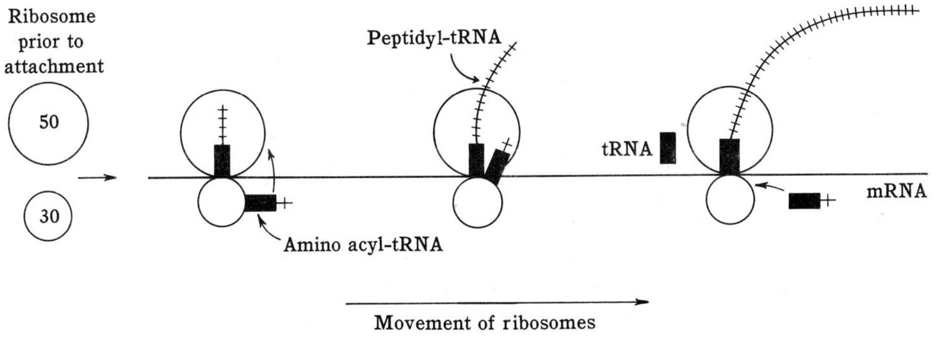

Fig. 29.2. Schematic representation of the movement of ribosomes during protein synthesis. The amino acyl-tRNA interacts with a site on the 30 S unit of the ribosome whereas the peptidyl-tRNA is attached to a site on the 50 S unit. The amino acyl-tRNA is transferred by coupling the free amino group to the carboxyl (acyl) group at the end of the peptidyl-tRNA, displacing the tRNA formerly at the end of the chain. Note that several polypeptide chains are being synthesized concurrently at different sites along the mRNA.

Inhibitors of Nucleic Acid and Protein Synthesis. As in the investigation of other complex metabolic processes, inhibitors of specific reactions have played a major role in separating the steps in the biosynthesis of nucleic acids and proteins. Some of these inhibitors are synthetic compounds; others were first isolated from the culture filtrates of various organisms as antibiotics in efforts to control infectious disorders or to inhibit the growth of cancer tissue. Substances which inhibit formation of purine and pyrimidine nucleotides block indirectly the formation of nucleic acids; these are discussed in Chap. 27.

Actinomycin D complexes with guanine residues in DNA and prevents formation of RNA by blocking the DNA-dependent RNA polymerase. Since all RNA synthesis is inhibited, protein synthesis is also completely prevented.

Inhibitors of Protein Synthesis. Formation of amino acyl-tRNA can be blocked by inhibition of the specific synthetase (page 662), *e.g.*, 5-methyltryptophan blocks formation of Trp-tRNA and hence the formation of proteins containing tryptophan.

Puromycin has a structure analogous to an amino acyl-tRNA (see below). The free amino group of the antibiotic can couple to the COOH-terminal end of the growing peptide chain. However, since puromycin contains an amide bond, —N—C(=O)— to the pentose moiety, in place of the ester of the amino acyl-tRNA, transfer to the amino group of the next incoming amino acyl-tRNA cannot occur. Hence incomplete peptide chains bearing COOH-terminal puromycin are released from the ribosomes.

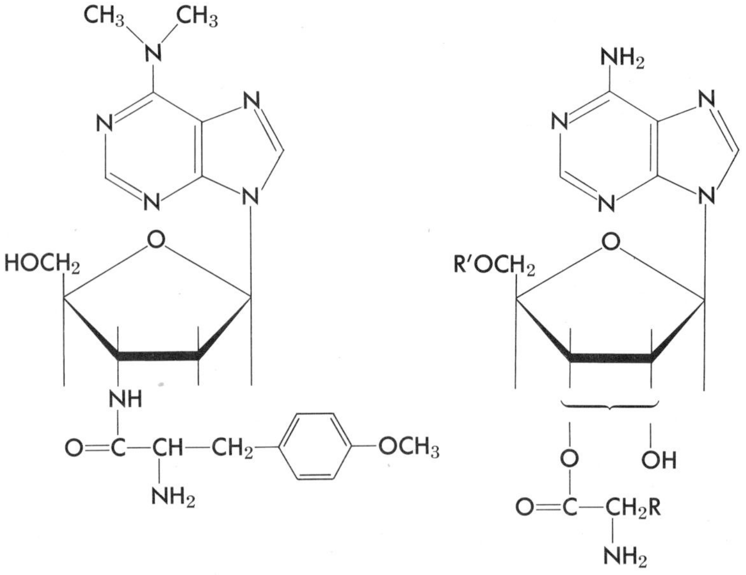

Puromycin **Amino acyl-tRNA**

The antibiotics, *chloramphenicol* and the *tetracyclines* (*Terramycin* and *Aureomycin*), inhibit protein synthesis by binding to ribosomes. Apparently these compounds prevent normal binding of mRNA to ribosomes. In any case, all these compounds are potent inhibitors of protein synthesis per se and do not prevent replication of DNA and formation of RNA.

Streptomycin and related antibiotics bind to the 50 S ribosomal subunits. This results in a decreased rate of protein synthesis and the production of faulty proteins by misreading of the codons of the mRNA (page 678).

OTHER ASPECTS OF RIBONUCLEIC ACID SYNTHESIS

The previous discussion has indicated that there are at least three known types of RNA involved in protein biosynthesis, *viz.*, tRNA, rRNA, and mRNA. Only for mRNA is there definite information concerning the specificity of its formation and the enzymic system involved.

Spiegelman and his coworkers have used the hybridization technique (page 666) to study the relationship of DNA of *E. coli* to both tRNA and rRNA. From uniformly labeled RNA containing ^{32}P or tritiated uridine, rRNA and the total tRNA were isolated. Each of these types of RNA was then heated with DNA under conditions that could break hydrogen bonds and permit mixed nucleic acid hybrid formation. Hybrid formation, detected by density gradient sedimentation in CsCl, occurred between rRNA and DNA and the total tRNA and DNA of homologous origin. When hybridization was attempted between an RNA and a DNA of genetically unrelated species, no hybrids could be detected.

These findings indicate the existence of complementary sequences in DNA and the three types of RNA, *viz.*, tRNA, rRNA, and, as previously indicated, mRNA (page 666), demonstrating that *all these types of RNA originate on DNA templates.* In essence, the role of DNA as genetic material is to provide templates for self-replication and for the synthesis of all types of RNA. Only mRNA provides coding information for specific sequences of amino acids (see section below on synthetic mRNA). The portion of DNA concerned with synthesis of rRNA and the specific tRNA molecules represents only a small part of the total DNA in *E. coli.* For tRNA this is estimated to be approximately 0.02 per cent of the DNA and for rRNA, approximately 0.15 per cent of the DNA.

It is assumed that in higher organisms, most of the RNA must be synthesized by polymerases in the nucleus since the bulk of the DNA is present in the nucleus. Certain organelles, *e.g.*, mitochondria and chloroplasts, contain some DNA. Such DNA is capable of replication and presumably carries information for formation of RNA. RNA-forming polymerases have been obtained from various microorganisms, but the mode of action and the specificity of these various enzymic preparations are still unknown.

tRNA. Although tRNA is formed on DNA templates, the synthesis of these molecules poses several problems since they contain many unusual nucleotides (page 202). Borek and Hurwitz and their coworkers have obtained evidence that methylation of various bases occurs after polynucleotide formation. Methionine is the source of the methyl groups. The thymine ribonucleotides of tRNA are formed by methylation of bound uridine nucleotides whereas the thymine of DNA is formed by methylation of free dUMP (page 639). The possible biological role of the methylation reactions is discussed below (page 676). Formation of pseudouridine may also occur at the polynucleotide level, but the mode of synthesis is unknown.

RNA Viruses. As previously noted (page 653), replication of DNA viruses, *e.g.*, the T-even series of bacteriophages, proceeds in much the same manner as in the case of DNA of cellular organisms. Also, in the case of circular viruses that possess single-stranded DNA, *e.g.*, phages ϕX-174, fl, fd, etc., a double-stranded helical structure is first formed by replication. Then one strand of the double-stranded replicative form is copied by a DNA polymerase. This seems to be analogous to the copying of one strand of DNA by the DNA-dependent RNA polymerase in vivo.

Both single- and double-stranded RNA viruses occur. Among the former type are tobacco mosaic virus (TMV), many other plant viruses, and various phages of *E. coli, e.g.*, f2, MS2, etc. (Table 9.6). Since cellular RNA molecules (tRNA, rRNA, mRNA) never serve as templates for replication, a somewhat different mechanism must exist for replication of new RNA virus strands on parental RNA. The responsible enzyme, *RNA-dependent RNA polymerase,* is formed after the viral RNA enters the cell. As in the case of both the DNA polymerase and the DNA-dependent RNA polymerase, the RNA polymerase catalyzes the formation of a complementary strand. Thus, the fundamental mechanism of template utilization for formation of new nucleic acid from the nucleoside triphosphate precursors is similar for all RNA synthesis.

Polynucleotide Phosphorylase. The first enzyme discovered which catalyzes formation of polyribonucleotides was described in 1955 by Grunberg-Manago and Ochoa. This *polynucleotide phosphorylase* was obtained initially from *Azotobacter vinelandii* and subsequently from other microorganisms; however, it is absent or present at negligible levels in tissues of higher plants and animals. The over-all reaction catalyzed is

$$n(\mathbf{XRPP}) \xrightleftharpoons{\text{Mg}^{++}} (\mathbf{XRP})_n + n\mathbf{P}_i$$

where R is ribose, P is phosphate, and X is a purine or pyrimidine base. The internucleotide linkages are 3′–5′, as in RNA. The polymerization will occur with a single nucleoside diphosphate, *e.g.*, ADP, UDP, or with a mixture of diphosphates. The action of the enzyme is in some respects analogous to that of glycogen phosphorylase (page 438). The limited biological distribution of polynucleotide phosphorylase, as well as the fact that its action is reversible, suggests that it plays a role in degradation or regulation rather than in synthesis of RNA.

The availability of the polynucleotide phosphorylase has permitted synthesis of high molecular weight polymers which have been useful as model compounds for study of nucleic acids in general. Studies of the interaction of poly A with poly U and of poly I (inosinic acid) with poly C showed complementary base pairing in the manner predicted by Watson and Crick for the two-stranded structure of DNA (page 193). Such polymers have also been used for study of the optical properties and hyperchromicity of polynucleotides and their behavior toward various enzymes. Interestingly, when mixed polynucleotides are formed, they are essentially random in composition with respect to the distribution of the various nucleotides, in contrast to the directed synthesis catalyzed by the RNA polymerase (pages 667*ff.*)

which is dependent on the presence of DNA. Polymers prepared with the aid of polynucleotide phosphorylase have proved to be of great utility as synthetic "messengers" (see below).

THE AMINO ACID CODE

The role of mRNA in protein biosynthesis and its formation under the influence of DNA have been considered (pages 666*ff.*). More direct evidence for the participation of such messengers has been obtained by use of cell-free systems and has aided in the elucidation of the code for amino acid incorporation in proteins. The fundamental observation was made by Nirenberg and Matthaei in 1961. They utilized extracts of *E. coli* freed of DNA by treatment with DNase and containing tRNA, Mg^{++}, amino acyl synthetases, transfer enzymes, and ribosomes with added GTP and amino acids. Addition of the synthetic polyribonucleotide, poly U, greatly increased incorporation of ^{14}C-phenylalanine into a product insoluble in trichloroacetic acid and indicated to be polyphenylalanine. Of 18 amino acids tested, only phenylalanine incorporation was markedly stimulated. Incorporation did not occur in the presence of inhibitors of protein synthesis, *e.g.*, chloramphenicol and puromycin, and was stopped by treatment with RNase but not DNase. Phenylalanyl-tRNA was an intermediate in the system.

From the above, it is apparent that some portion of poly U codes for phenylalanine. The codon triplet for a single residue of phenylalanine is 3U, or UUU. To determine other code compositions, homopolynucleotides and polynucleotides of mixed composition were employed independently by Nirenberg and by Ochoa and their associates. Addition of poly A to the incorporating system led to formation of polylysine, whereas poly C produced polyproline; thus, AAA and CCC are the codons for lysine and proline, respectively.

In order to illustrate the approach used with heteropolymers, some of the work with poly UG may be cited. Mixed polynucleotides were prepared with *polynucleotide phosphorylase* and the nucleoside diphosphates. Since the polymer was prepared with a reaction mixture containing 5 parts UDP to 1 of GDP, the poly UG contains the bases U and G in a ratio of 5:1. Along this linear polymer, consecutive triplets present must include UUU, UUG, UGU, GUU, UGG, GUG, GGU, GGG. These would occur in the relative frequency $UUU/U_2G = 5$ for each of the codons of the composition 2U1G and $UUU/UG_2 = 25$ for each of the codons of the composition 1U2G. With this poly UG, incorporation of cysteine, valine, leucine, glycine, and tryptophan was stimulated, in addition to phenylalanine. The experimentally determined ratios proved to be Phe/CySH = 5; Phe/Val = 5; and Phe/Leu = 8. Thus, the code compositions for cysteine, valine, and leucine were assigned the composition 2U1G. Similarly, the experimentally found ratios were Phe/Gly = 24 and Phe/Trp = 20, in accord with the composition 1U2G. In the same manner assignments of code compositions were made for almost all 20 amino acids. However, there was uncertainty of code compositions in instances in which low levels of amino acid incorporation were observed, and not all possible codes could be determined. Further, this method is limited to determining the compositions of codons and does not indicate the sequences of bases in the codons.

Another method of determining the codons and the sequences of the nucleotides within each codon was developed by Nirenberg and Leder in 1964. The procedure involves the study of specific binding of an amino acyl–tRNA to ribosomes in the presence of a specific mRNA but in the absence of protein synthesis. Indeed, a long mRNA is not required; only a trinucleotide is necessary. Presumably, recognition is achieved by specific binding of the codon on the trinucleotide and the "anticodon" on the tRNA. The interaction occurs on the 50 S subunit of the ribosome in a system containing only ribosomes, amino acyl–tRNA, 0.02M Mg^{++}, and a messenger (oligo- or polynucleotide). Poly U or $(pU)_3$ is an effective unit, *i.e.*, when present on the ribosome it specifically enhances binding of phenylalanyl-tRNA, whereas $(pU)_2$ does not, indicating that the coding ratio is 3. Similarly, $(pA)_3$ and $(pC)_3$ are effective for lysine and proline, respectively. The most effective trinucleotides are those possessing at one terminus a free 3′- (and 2′-)OH, with a 5′-phosphate at the other end, or pXpYpZ-OH, where X, Y, and Z represent purine or pyrimidine nucleosides. Thus, for phenylalanine, the codon is pUpUpU. With the complete system for protein synthesis, a synthetic polynucleotide, containing all A but with an additional C at the free 3′-OH end, or $A(pA)_n pC$, led to the synthesis of polylysine-bearing COOH-terminal asparagine. Thus, reading of the codon starts at the 5′ end for formation of the polypeptide, H_2N-Lys-Lys-----Lys-Asn-COOH. From this information and similar studies, alignment of the nucleotides in the mRNA and the synthesized polypeptide is evident.

$$\text{pApApApApApA------pApApC}^{3'}\text{-OH}$$
$$\text{H}_2\text{N-Lys-Lys------Asn-COOH}$$

Table 29.3 lists the presently known codons obtained from studies chiefly with the *E. coli* system. Some of the codons were determined from studies with polymers, others by trinucleotide binding. For several codes only indirect evidence has been obtained from observations of amino acid substitution in proteins by point mutations (Chap. 30).

Present evidence concerning the code may now be summarized and amplified as follows:

1. *The codon is a triplet.* Certain mutagens (page 682) cause the deletion or insertion of a nucleotide in the gene. Such a genetic alteration leads to formation of nonfunctional proteins because the reading of the codons is shifted from the point of nucleotide deletion (or insertion); hence the amino acid sequence is completely altered from the point of nucleotide deletion (or insertion) to the COOH-end of the peptide chain, as shown for the deletion in the arbitrary example below.

Normal reading	Phe	Lys	Pro	Phe	Gln	Lys	Gly	Asn
	UUU	AAA	CCC	~~U~~UC	CAG	AAG	GGU	AAU
After deletion	Phe	Lys	Pro	**Ser**	**Arg**	**Arg**	**Val**	

A series of three adjacent nucleotide deletions, however, would lead to the formation of a protein in which only a single amino acid is deleted, but the reading frame would remain undisturbed for the synthesis of the remainder of the pro-

Table 29.3: PRESENTLY KNOWN CODONS IN mRNA

First position	Second position U	Second position C	Second position A	Second position G	Third position
U	Phe	Ser	Tyr	Cys	U
	Phe	Ser	Tyr	Cys	C
	Leu	Ser	(CT)	(CT)	A
	Leu	Ser	(CT)	Trp	G
C	Leu	Pro	His	Arg	U
	Leu	Pro	His	Arg	C
	Leu	Pro	Gln	Arg	A
	Leu	Pro	Gln	Arg	G
A	Ile	Thr	Asn	Ser	U
	Ile	Thr	Asn	Ser	C
	Ile	Thr	Lys	Arg	A
	Met (CI)	Thr	Lys	Arg	G
G	Val	Ala	Asp	Gly	U
	Val	Ala	Asp	Gly	C
	Val	Ala	Glu	Gly	A
	Val (CI)	Ala	Glu	Gly	G

Note: The first position refers to the initial nucleotide of the triplet bearing a 5′-OH or 5′-phosphate; the third nucleotide of the triplet bears a 3′-phosphate connecting to the next triplet. (CI) refers to a chain-initiating codon for the NH_2-end of a peptide chain. (CT) refers to a chain-terminating codon for the COOH-end of a peptide chain.

tein. For three nonadjacent deletions (or insertions) of nucleotides, the amino acid sequence would be altered only between the first and last of these changes; the remaining parts of the amino acid sequence would be unchanged. Crick, Brenner, and their associates showed that in the genome of phage T4, incorporation by crossing-over in a single gene of three nucleotide deletions (or insertions) gave a functional protein whereas one or two nucleotide deletions (or insertions) did not. *This indicates that three nucleotides must be read for a single amino acid.*

The binding studies of trinucleotides with tRNA, described above, also support the view that the codon is a triplet, and these investigations have permitted determination of the sequence of nucleotides in many codons.

Khorana and coworkers have synthesized a number of copolymers of specifically determined alternating nucleotides. Thus, CUCUCUCUCUCU--- is the messenger for formation of a polypeptide in which leucine and serine alternate. If the codon consisted of two or four bases, a homopolypeptide would have been formed. Thus, these studies are in accord with a triplet code. Moreover, the results indicate CUC for leucine and UCU for serine, in agreement with trinucleotide-binding studies. Similar results, given in Table 29.4, have been obtained with other regular copolymers.

2. *Degeneracy.* For many amino acids there is more than a single codon (Table 29.3), a phenomenon called *degeneracy.* Inasmuch as the codon of the mRNA must be complementary to the anticodon of the tRNA, a corresponding multiplicity of tRNAs specific for each amino acid is implied. This is consonant with the fact that more than one tRNA has been found for many amino acids.

Table 29.4: Assignment of Codons from Regular Copolymers

Copolymer	Amino acids incorporated	Codons
CUC ¦ UCU ¦ - - -	Leu; Ser	CUC; UCU
UGU ¦ GUG ¦ - - -	Cys; Val	UGU; GUG
ACA ¦ CAC ¦ - - -	Thr; His	ACA; CAC
AGA ¦ GAG ¦ - - -	Arg; Glu	AGA; GAG

Source: H. G. Khorana, in *Federation Proc.*, **24,** 1473, 1965.

3. *Anticodons.* In view of the expected complementarity of the anticodons in tRNA to the codons in mRNA, the known structures of tRNA molecules may now be considered in terms of possible regions possessing the correct triplet sequences. Thus, for UUU, the mRNA code for phenylalanine, the expected anticodon triplet would be AAA.

In Fig. 29.3, the known sequences of Tyr-tRNA and Ala-tRNA of yeast (page 677) have been arranged to give maximal base pairing by hydrogen bonding. (Similar arrangements can be made for Ser-tRNA and Phe-tRNA.) Ala-tRNA has no unusual nucleotides in the regions in which hydrogen bonding is assumed, whereas the Tyr-tRNA has 2-MeG and ψ (pseudouridylate) (page 203) in locations that are presumed to be hydrogen-bonded. Bases that cannot hydrogen-bond, DiMeG, DiMeA, and 1-MeI, are located in areas assumed to be not hydrogen-bonded. It is implied that methylation plays a role in determining correct conformation of the tRNAs. Although these conformations are tentative, the indication that most of the dihydrouridine residues are in the same unbonded region on the left of the structures suggests a common role for these unusual nucleotides. Similarly, the single thymidine residues in each tRNA are in the right-hand loop, and ψ is only in the lower and right-hand loops. Such similar structures are possible candidates as the specific recognition sites for the activating enzymes.

The structures for these two tRNAs are arranged in the same cloverleaf form in which the triplets at the bottom are complementary and antiparallel to known codons. For Ala-tRNA, the anticodon is IGC (I being inosinate and binding as does guanylate), for Tyr-tRNA, the anticodon is GψA, for Phe-tRNA, the anticodon is OMeGAA, for Val-tRNA the anticodon is IAC, and for Ser-tRNA, the anticodon is IGA. These anticodons represent the only trinucleotides in the t-RNAs that could be complementary to known codons for these amino acids.

4. *Chain Termination.* When poly U is used as a messenger, the polyphenylalanine formed remains attached to the ribosomes but may be released by agents such as trichloroacetic acid. This suggests that normally some mechanism exists for liberating the formed polypeptide chain from the attached COOH-terminal tRNA.

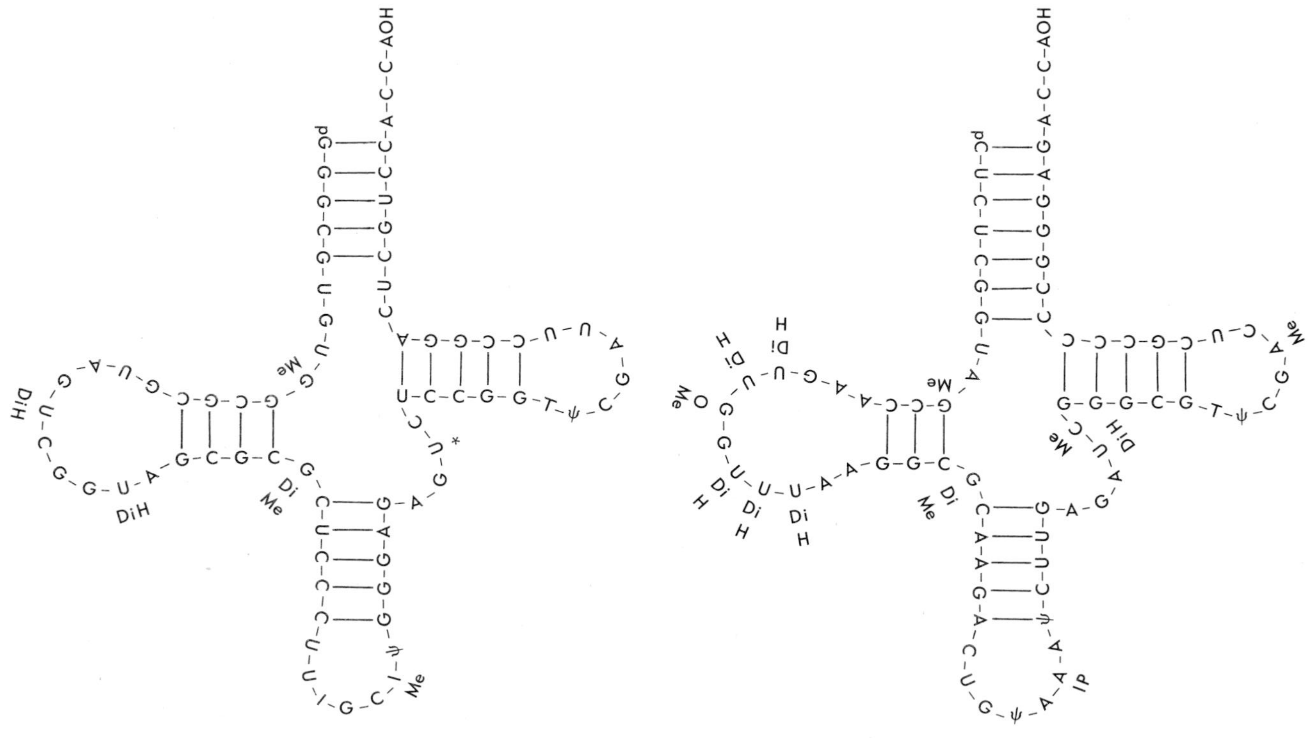

Alanyl - tRNA

Tyrosyl - tRNA

Fig. 29.3. The sequences of alanyl-tRNA and tyrosyl-tRNA arranged in similar hypothetical configuration to show the triplet anticodons at the bottom of each structure. Nucleotide bases shown opposite one another are indicated as being hydrogen bonded to each other. Abbreviations used for substituted base components in transfer RNAs are listed in Table 9.5, page 202. (*Courtesy of Dr. J. T. Madison.*)

From mutants in various phages, bacteria, and yeast that form incomplete peptide chains, it has been deduced that the codons UAA, UAG, and UGA (Table 29.3) lead to chain termination and do not code for any amino acid. The mechanism by which these three codons accomplish chain termination and polypeptide release is not understood, but it is assumed that these codons represent the normal signals for termination.

5. *Chain Initiation.* Similar to codons for chain termination, it is likely that there must be some special codon or signal for initiation of a polypeptide chain. Marcker and Sanger found with *E. coli* extracts that a Met-tRNA is rapidly and specifically N-formylated by an enzyme-catalyzed transformylation from N^{10}-tetrahydrofolate (page 552); other amino acyl-tRNA derivatives are not formylated. Subsequently, it was reported that most, if not all, of the proteins of *E. coli* at the time of their biosynthesis possess the identical N-terminal residue, namely, N-formylMet-, and that the formyl group and sometimes methionine are subsequently removed by hydrolysis, depending on the specific protein produced. The present picture of chain initiation is that there are two ribosomal binding sites: one of these is on the 30 S part of the ribosome for formylmethionyl-tRNA or for the growing chain of polypeptidyl-tRNA; the other site, on the 50 S ribosomal particle, binds to the next incoming amino acyl-tRNA to be added. According to this picture, all peptide chain initiation depends on specific binding of the N-formylmethionyl-tRNA to the 30 S particle followed by subsequent addition of amino acids from their tRNA derivatives. For binding N-formylMet-tRNA by the codon AUG, GTP, Mg^{++} and a protein fraction are required.

The only codon presently known for methionine in *E. coli* is AUG (Table 29.3), and this trinucleotide will bind both Met-tRNA and formylMet-tRNA to ribosomes. Evidence for chain initiation by the above method exists at present only for *E. coli.* Indeed, N-formylated polypeptide chains are unknown in proteins; however, a large number of proteins from various sources possess an N-acetylated terminus (page 144), but present evidence suggests that acetylation occurs after formation of the peptide chain.

The problem of chain initiation is important, since it has been shown that many messengers are of sufficient size to code for more than one polypeptide chain. Clearly, there is a mechanism for termination (discussed above) and possibly for initiation also.

The relationships of the codons and anticodons are shown schematically in Fig. 29.4. The chain-initiating codon AUG is shown at the beginning end of the mRNA. The polypeptidyl-tRNA is bound to the 50 S component of the ribosome.

6. *Reading Mistakes.* In the normal cell, reading of the code must be highly specific, with errors occurring very rarely. Insertion of the wrong amino acid at critical parts of the polypeptide chain would lead to formation of inactive enzymes or abnormal proteins. Nevertheless, translational errors of this type can be induced by agents such as the antibiotic streptomycin which, in sensitive bacteria, combines with a site on the 30 S ribosomal component. When poly U is used in vitro as a template, the most frequent error is substitution for phenylalanine (UUU) by isoleucine (AUU). This suggests that the antibiotic disturbs the position of one nucleotide, possibly by deformation of the ribosome. In cell-

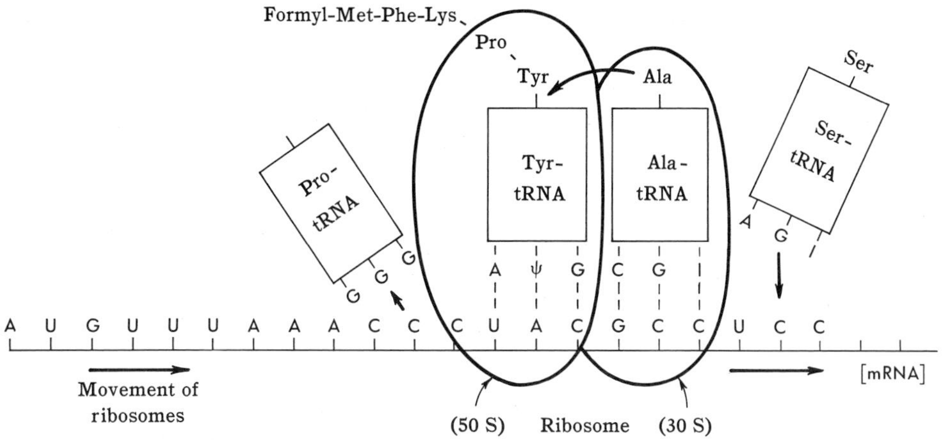

Fig. 29.4. Schematic representation of the interaction of tRNA with mRNA on the ribosome. The polypeptidyl-tRNA is bound to the 50 S component of the ribosome whereas the amino acyl-tRNA is attached to the 30 S portion of the ribosome. Each loaded tRNA is shown hydrogen-bonded by its anticodon (in *antiparallel* sequence) to the triplet codon of the mRNA.

free systems errors in translation can be induced by adding excess Mg^{++} or lowering the temperature.

7. *Suppressor Genes.* Many genetic studies have shown that harmful mutations at a site A were occasionally reversed by a second genetic change. Some of these were back mutations that simply changed an altered nucleotide sequence back to its original sequence. In some cases, however, the change occurred at a different genetic site B and yet served to suppress the effect of the harmful mutation at site A. When site B is present on a different gene, it is termed a suppressor gene. Recent work has served to elucidate the mechanism of one kind of suppressor gene.

The studies of Yanofsky and coworkers on the genetics of formation in *E. coli* of the α protein of tryptophan synthetase (page 573) have shown that an inactive protein is produced when a point mutation causes replacement of a glycine by an arginine residue at a specific locus in the peptide sequence. Introduction of an unlinked suppressor gene into this strain of *E. coli* led to formation for the most part of inactive α protein still containing Arg instead of Gly, but some of the protein was active and contained Gly in place of Arg. Study of the tRNA in such organisms has shown that the mutation in the suppressor gene has led to formation of an altered tRNA which may still possess the anticodon for Arg but which is now occasionally charged by the amino acyl synthetase with glycine rather than arginine. Thus, during formation of the α protein, in reading the messenger that bears the codon for Arg, the altered Arg-tRNA, which actually carried glycine, now binds to Arg codon and inserts glycine in the growing chain since the tRNA bears glycine. These results demonstrate, as expected from the hybridization studies (page 671), that a portion of the genome contains information for synthesis of tRNA. Further, the amino acyl synthetase must bind to a site on the tRNA entirely distinct from the anticodon triplet that is specific for the codon. This explains, at least in part, why species specificity is manifested by the amino acyl synthetase and by the transfer enzymes although the code is presumably universal (see below).

It has been demonstrated in *E. coli,* for a mutation in a phage that produces chain termination (codon UAG) in the phage protein, that a suppressor mutation can introduce tyrosine (codon UA$^{C}_{U}$), thus permitting continuation of the polypeptide chain. Isolation and sequence determinations in the two tRNAs showed that the suppressor mutation represented a change in the anticodon for Tyr-tRNA from GUA to CUA, reading in antiparallel direction as compared to the codons. This is the first instance in which a base substitution has been demonstrated directly in a nucleic acid sequence. Furthermore, since this is the only change in the nucleotide sequence, it has permitted a direct identification of the triplet anticodon site in these *E. coli* tRNAs. This triplet position corresponds exactly with the predicted positions shown in Fig. 29.3 for yeast tRNAs.

8. *Universality of Code.* Present information indicates that the code is probably universal or almost so. With cell-free extracts from many bacteria and from the tissues of various higher plants and animals, addition of poly U leads to incorporation of phenylalanine, poly C of proline, and poly A of lysine. It will be necessary to determine unequivocally the complete codons in many species before the universality of the code can be established, but the fact that these three and other codons are identical in so many organisms indicates that this is the case.

Another indication of code universality is the finding that animal viruses can be grown in bacterial cells. Thus, introduction of the DNA of vaccinia or polyoma virus in *Bacillus subtilis* leads to production of these viruses in a novel host. Since formation of new virus particles must involve utilization of much of the apparatus of the host cell, including probably the amino acyl synthetases and the tRNAs, with the messengers from the viruses, these experiments also favor the universality of the code.

More indirect but of equal cogency is the information derived from comparative studies of homologous proteins which show that large portions of the amino acid sequences have been retained even in distantly related species. To cite just one example, the cytochromes c of various species (page 695 and Chap. 8) of animals, plants, and fungi all possess approximately one-third of the residues of the sequences in common. This indicates a conservatism in evolution in which the same protein has retained essential features of its amino acid sequence and conformation for a long period of evolution (page 694*ff.*). It is difficult to visualize this degree of retention of structure if the code had changed continuously during evolution.

If the code were not essentially invariant, a change in the tRNA for cysteine, to tryptophan, for example, would lead at each site along each peptide chain to the incorporation of tryptophan instead of cysteine. Enzymes that require sulfhydryl groups for activity would be inert, disulfide bridges could not be formed, and the introduction of the bulky side chains of tryptophan would make difficult folding into normal conformation. Such a mutation leading to the wrong anticodon would obviously be lethal in haploid cells and certainly deleterious if not lethal in diploid cells. Misreading of the code in suppressor genes (discussed above) must be regarded as a relatively rare event that has survival value only when it leads to some new advantage for the cell.

Finally, it is remarkable that the mechanisms of DNA replication and the processes of RNA formation and of translation into protein are essentially the same in all organisms. Evolutionary changes with the development of enormous diversity of organisms have taken place not by major alterations of these biosynthetic processes but by the formation of new genes leading to synthesis of new enzymes or other proteins, and hence to new structures and functions. These problems will be considered in the following chapter.

Mutagenesis. Although only one strand of DNA is transcribed in formation of mRNA, it is evident that alteration of the base sequence in either strand of the DNA will lead to permanently inherited changes, since subsequent replication of DNA will perpetuate the changes. At this juncture we may consider some of the types of inherited modifications that can occur and some of the physical and chemical agents that can produce such alterations. Any alteration of the DNA of the genome will be reflected in the synthesized proteins or in the rRNA and various types of tRNA. Here we shall consider primarily the changes that affect mRNA, since these are manifested in the amino acid sequences. Examples of such changes in protein structure are discussed in Chap. 30.

Point mutations reflect a change in a single base (or base pair in the double strand) as shown by crossing-over studies in the fine structure of a single cistron. This results, of course, in a single base change in the mRNA and thus to the replacement of one amino acid by another in the sequence of a peptide chain. These are by far the most common genetic changes.

In 1927, Muller reported that x-rays produce a marked increase in mutation rate in *Drosophila.* Since then it has been shown, in all species studied, that x-rays, ultraviolet radiation, radium, and all other forms of high-energy radiation increase mutation rates, and that the rates increase with the radiation dosage. It was also shown by Muller that the spontaneous mutation rate is too high to be explained by the normal low levels of radiation prevailing in the 1920s. Thus, natural mutagenesis is presumed to be caused by intrinsic effects within the cells, and in higher organisms, within the germ cells.

Ultraviolet radiation not only produces mutations but may be lethal to cells. The effect of this radiation is chiefly due to the formation of dimers in DNA, between adjacent residues in the same chain, as shown below where R is the remainder of the nucleotide. Structure II appears to be the more probable.

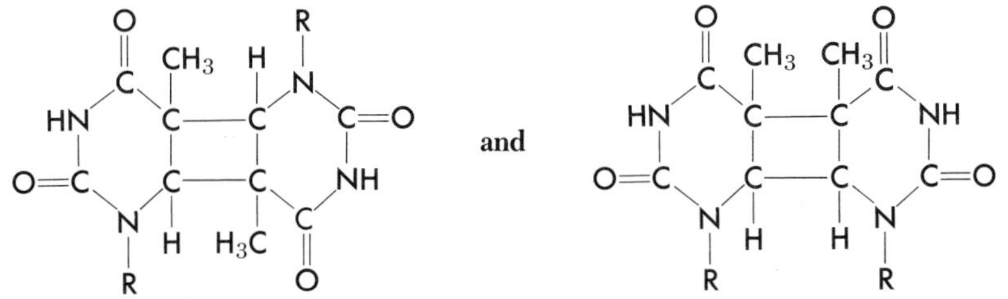

I Thymine dimers II

Normal replication in the region of the cross-linking will be prevented. In some cases, the dimer is excised and a deletion in the DNA will result. Such a defect may be repaired by enzymes which link the appropriate nucleotides, aligned by base pairing with the unaltered strand of DNA. The lethal and mutagenic effects of short wavelength ultraviolet radiation (260 mμ) may be partially reversed by exposure of the cells to longer wavelengths (330 to 450 mμ) of light. Such *photoreactivation* represents a dissociation of the dimers.

Nitrogen-mustard compounds and other strong alkylating agents that can react with amino groups are mutagenic. Particularly effective are those compounds that are bifunctional, *e.g.*, $ClCH_2CH_2N(R)CH_2CH_2Cl$, and produce cross-linking between guanine residues in the two strands of DNA. The cross-linking may prevent normal replication of DNA or may introduce copying errors.

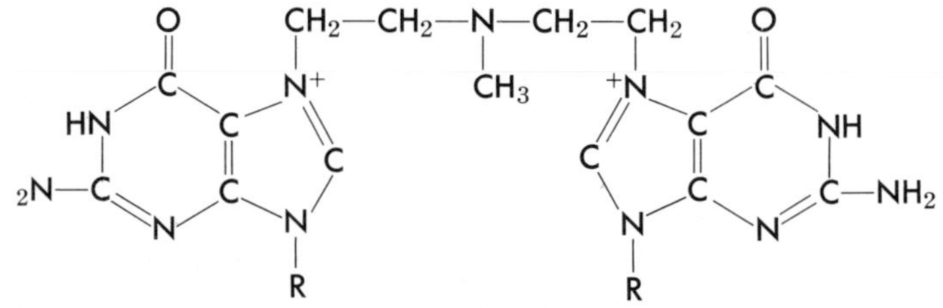

Guanine dimer cross-linked by nitrogen mustard

Mutator genes have been found in various species and lead to genetic instability, *i.e.,* a high rate of formation of apparently spontaneous mutants. Such mutator genes act on all types of genes; in *E. coli,* for example, these are mutations in genes controlling enzyme formation in the pathways for biosynthesis of amino acids, purines, pyrimidines, vitamins, and for fermentation of sugars. Such mutator genes map as discrete genes and, indeed, it has been shown in phage T4 that a mutation in the gene for DNA polymerase has the properties of a mutator gene. An alteration in this polymerase leads to copying errors that produce mutations in many genes. One mutator gene found in *E. coli* causes AT to CG *transversions* in the DNA.

Point mutations are produced by introduction of a wrong base or by copying errors. Treatment of a nucleic acid, such as the RNA of tobacco mosaic virus, with nitrous acid deaminates the bases, changing C to U and A to I (read as G). Consequent changes are produced in amino acid sequence of the TMV protein that are consonant with the predicted codon changes (page 675). Hydroxylamine (NH_2OH) deaminates C to U and is thus mutagenic. As noted earlier (pages 644 and 655) analogues of purines or pyrimidines, or the respective nucleosides or nucleotides, which become incorporated in DNA are mutagenic, since they can lead to errors in base pairing. Examples include uracil, halogenated at the 5 position with I, Br or Cl, 8-azaguanine, and 6-thioguanine.

Certain acridine dyes, *e.g.*, proflavine, which can become lodged between the bases in the strands of DNA have been demonstrated to produce nucleotide deletions (or insertions), rather than point mutations. Only when three nucleotides (or a multiple of three) are deleted (or inserted) can the messenger be translated into a polypeptide sequence which is functional; otherwise the sequence is "missense" on the carboxyl-terminal side of the deletion, since no codon will be read correctly (page 674). Deletions of amino acids are known in homologous proteins of various species and even in human hemoglobin. Although the mechanism of such "natural" deletions is unknown and can be detected only by examination of the amino acid sequences of proteins (Chap. 30), presumably they are the result of breaks and losses of nucleotides in a DNA chain. Alkylating agents may also produce chromosomal breaks by reaction with the backbone phosphates, forming labile triesters that are cleaved between the phosphate and the sugar.

Homologous chromosomes will pair during conjugation in diploid organisms. If the chromosomes are misaligned, unequal crossing over can occur; the result will be a lengthening of one chromosome (or a cistron) and a shortening of the other. One of many possible simple interpretations of this is shown below.

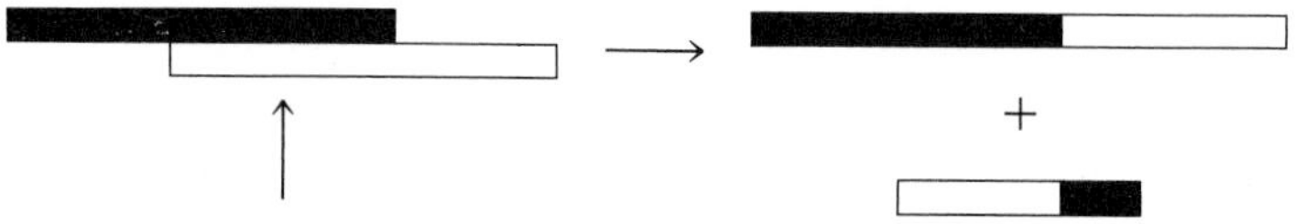

Studies of proteins have suggested that chain lengthening by this mechanism has occurred in haptoglobin (page 720), in γ-globulins (page 718), and in other proteins (Chap. 30).

A change in the carboxyl-terminating codons (Table 29.3) would lead to an extension of a polypeptide chain if the codons that follow permitted amino acid incorporation. Similarly, such extensions are also theoretically possible at the NH_2-terminus. The homologous cytochromes c of different species differ in the number of residues amino-terminal to the heme group (page 178). It is not known whether the longer chains represent a primitive type, although the fact that they occur only in plants, fungi, and invertebrates suggests that this is the case and that residues have been lost during animal evolution (see Chap. 30).

CONTROL OF PROTEIN SYNTHESIS

Some enzymes are present in cells in large amounts and others in very minute quantities. Erythrocytes contain hemoglobin and muscle cells contain myosin, yet all cells of a species or organism contain the same genetic information. Since enzyme synthesis is a direct expression of genetic information, it is of interest to ascertain which genetic and other factors regulate the intracellular concentration of various enzymes. Most of the information presently available has been derived from studies with microorganisms, but less complete studies with mammalian systems indicate that similar factors are operative.

Induced Enzyme Synthesis. Yeast cells grown on glucose do not ferment galactose, but if galactose is present, such cells will begin to ferment that sugar, after a brief lag period. The phenomenon is called *induced enzyme synthesis.* Similar induction phenomena have been observed for the formation of enzymes concerned with various aspects of metabolism in microorganisms: formation of *amylases* induced by starch, formation of *tryptophanase* in response to tryptophan, *penicillinase* formation, etc. In mammals, several enzymes have been shown to increase in activity in response to administration of their substrates. Nevertheless, not all enzymes show such changes in activity; *constitutive enzymes* remain at the same level regardless of the amount of potential inducer added to the cell culture or injected into the animal.

Many features of the induction phenomenon can be illustrated with the *β-galactosidase* of *E. coli.* Formation of the induced enzyme represents *de novo* synthesis of protein. Induction begins very rapidly, β-galactosidase synthesis at 37°C. attaining its maximal rate within 3 to 4 min. after addition of inducer. The rate is constant as long as inducer is present and returns to the uninduced rate within a few minutes after removal of the inducer. In the absence of inducer, each cell may contain 1 to 3 molecules of the enzyme; in the steady state in the presence of inducer, each cell contains about 2,000 molecules of the same enzyme.

The nature of the *inducer* for β-galactosidase has been extensively studied. Some compounds, *e.g.*, methyl β-galactoside, are excellent inducers and poor substrates; other compounds are good substrates and poor inducers. Nevertheless, the enzyme formed is the same regardless of the inducer; the latter does not confer any specificity upon the induced enzyme. Inasmuch as there is always a very small basal amount of enzyme present in uninduced cells, induction represents an increased production of an enzyme for which the cell already possesses the requisite genetic information.

Repression of Enzyme Formation. Enzyme formation can be repressed as well as induced. To illustrate the phenomenon, the pathway of arginine biosynthesis (pages 545*ff.*) may be considered.

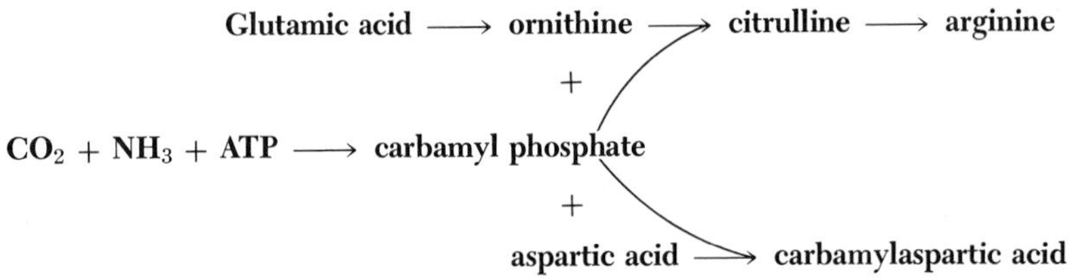

When wild-type cells of *E. coli* are grown on a minimal medium, arginine is formed and the level of *ornithine transcarbamylase* is at a constant value. When arginine is supplied from the medium, the level of this enzyme falls by dilution during subsequent growth of the culture. The presence of arginine *represses* enzyme formation. When the cells are washed and resuspended in minimal medium, they begin to form the enzyme again and the cells are said to be *"derepressed."* A most interesting feature of repression is its "feedback" aspect. Excess of arginine shuts off synthesis of all of the enzymes of the entire pathway leading to its synthesis, starting with the first committed synthetic step, as already discussed (pages 279 and 538*ff.*)

The phenomenon may also be illustrated by the biosynthesis of histidine in *Salmonella,* The last four consecutive steps leading to formation of histidine are catalyzed by (1) *imidazoleglycerol phosphate dehydrase,* (2) a *transaminase,* (3) *histidinol phosphate phosphatase,* and (4) *histidinol dehydrogenase* (page 576). When histidine is added in varying amount to wild-type cells or to various types of mutant cells, the relative activity of all four enzymes is decreased in parallel, indicating closely regulated control of their rates of formation.

As in the case of enzyme induction, enzyme formation by *derepression* involves *de novo* synthesis of the enzymes. The genes controlling the formation of each of the enzymes are always present in the organism. Indeed, in some cases the genes for a given pathway of synthesis are closely linked in the chromosome. In *Salmonella,* it has been shown by recombination studies with some 450 histidine-less mutants that the genes for all the enzymes of histidine synthesis are close together in the chromosome.

Mechanism of Induction and Repression. A general hypothesis to account for induction and repression has been proposed by Jacob and Monod. Their viewpoint may be summarized as follows. (1) Both induction and repression are primarily negative, *i.e.*, they operate by inhibiting or releasing an inhibition of protein synthesis. (2) There are two classes of genes, *structural genes,* containing the information for the structures of specific enzymes, and *regulatory genes,* primarily concerned with *rates* of protein synthesis. (3) Several linked structural genes probably represent a functional unit in that they are all active or inactive simultaneously, as in the case of the enzymes concerned in histidine biosynthesis mentioned above. (4) It is proposed that the regulatory mechanism functions at the genetic level and therefore controls the rate of synthesis of a relatively short-lived messenger (mRNA).

The regulatory gene directs the synthesis of a repressor protein which acts on an *operator* site, and the group of genes whose activity is repressed or derepressed "coordinately" is the *operon.* Among the best-studied instances are the genes for the enzymes in *E. coli* of the "lac" operon responsible for lactose metabolism; these include the genes controlling formation of *β-galactosidase, thiogalactoside transacetylase,* and a *permease* involved in transport of galactosides into the cell. The histidine operon in *Salmonella* includes 15 cistrons specifying nine enzymes, several enzymes being specified by more than one cistron.

The view that there are two types of genes is clearly established in these microorganisms. A mutation in a structural gene influences only the activity of a single enzyme. However, point mutations have been found which lead to a failure of the regulatory mechanisms and hence to a high rate of formation of all the enzymes in a metabolic pathway. Such enzymes then become *constitutive, i.e.*, their concentration is now independent of the presence of inducers or repressors. Furthermore, although mutations in structural genes follow the one gene–one enzyme hypothesis, this is not the case for regulatory genes since the levels of several enzymes may be influenced. Mutants in the operator and repressor loci have been mapped by crossing-over methods in the same manner as other mutants. The mapping relations of the genes may be shown for the lac operon in *E. coli* as follows,

where the genes are as follows: *o*, operator; *z*, β-galactosidase; *y*, permease; and *x*, transacetylase. For the z gene the NH_2-terminal side of β-galactosidase maps at the o side of the gene. Thus, mRNA and protein synthesis proceeds from the operator end through the entire operon. This has also been demonstrated for the tryptophan operon (Fig. 30.3, page 692). Note, however, that the repressor gene is not necessarily near the operon, and generally maps elsewhere on the chromosome.

In general, there is coordination of activity of all the genes in an operon, *i.e.*, the ratios of the amounts of the different enzymes in any one operon remain constant regardless of the degree of induction or repression. Indeed, all the genes of an operon are transcribed into a single polygenic strand of mRNA. *Coordinate repression* was first observed by Ames and Garry for the histidine operon and has been reported for other operons also. In the lac operon Zabin and coworkers found that the molar ratio of galactosidase to transacetylase is between 4:1 and 8:1 at various states of induction. Indeed, chain-terminating mutants (ambers) in the z gene (β-galactosidase) strongly influence the amount of transacetylase synthesized, *e.g.*, mutants near the N-terminus of galactosidase exert a major repression of acetylase synthesis, whereas mutants nearer the COOH end of the galactosidase have a lesser effect on the extent of acetylase synthesis. Thus, mRNA synthesis for all the genes in the operon is controlled unidirectionally in sequence from the operator end; this phenomenon of *polarity* is also exhibited by other operons.

The product of the regulatory gene for the lac operon has been demonstrated to be a specific protein which acts at the operator site to prevent the formation of mRNA for the entire operon. Derepression would then represent a failure to make an active repressor, as in the case of a constitutive mutant. An inducer added to an inducible strain binds with repressor, as shown for lac repressor, in order to permit the action of the operator.

Induction by derepression is not a sufficient condition, however, to ensure production of all the enzymes of an operon. The initial step in leucine synthesis in *Neurospora* is catalyzed by *isopropylmalate synthetase* (page 571); the activity of this enzyme is controlled by leucine through both repression and feedback inhibition. The remaining enzymes of the leucine pathway are induced only in the presence of the isopropylmalate synthesized by the first enzyme. Thus, a lack of leucine will not alone induce formation of the later enzymes of the pathway. There must be present enough precursor, α-ketoisovalerate (page 571), and functional first enzyme, isopropylmalate synthetase, to guarantee leucine synthesis. In this system and several others, it is now evident that three conditions must be met for induction of all the enzymes concerned in the pathway: (1) lack of end product, (2) availability of precursor for the pathway, and (3) a functional first enzyme.

Although in the original operon hypothesis Jacob and Monod suggested that control of protein synthesis is at the level of transcription, *i.e.*, by regulation of the rate of formation of mRNA, evidence has been obtained that regulation may occur at the translational level for both the histidine and valine operons. Since histidyl-tRNA is more effective than histidine in repression of the histidine operon, it has been suggested that the rate of translation of the mRNA of the operon may be the limiting factor in the rate of synthesis of the enzymes. This would imply that histidyl-tRNA represses operon function by blocking translation of the message.

Although regulatory genes and repressor mechanisms have been demonstrated only in microorganisms, it may be assumed that they are present in more complex organisms also. Differentiated cells of multicellular organisms synthesize very different levels of various enzymes, despite the identity of their genetic complements. Inductive and repressive phenomena may be involved in some of the differences in metabolic rates and in the relative importance of different pathways found in differentiated cells of the same organism.

Some indications of sequential production of proteins have been found. Certain specific regions of chromosomes (visible as puffs) synthesize RNA in temporal sequence during embryonic development and on stimulation by *ecdysone* (page 83), an insect hormone. Control of protein synthesis during embryonic development is exemplified by findings that certain noninducible enzymes are not detectable in early life but appear later. An example is hemoglobin. Fetal hemoglobin (Hb F = $\alpha_2\gamma_2$) is present in the embryo with little or no Hb A ($\alpha_2\beta_2$), indicating that β chains are not fabricated at a significant rate. In the newborn, formation of γ chains is repressed and production of β chains is accelerated.

Current understanding of the mechanism of protein synthesis has been elucidated in microorganisms. It is thought that the basic principles and participants are similar in mammals. Much less is known, however, regarding the regulation of mammalian protein synthesis. The prime regulators of metabolic phenomena in mammals are the hormones. It has not been possible to demonstrate an effect of a hormone added in vitro to a cell-free protein-synthesizing system. However, a number of hormones, including insulin, testosterone, thyroxine, growth hormone (somatotropin), and adrenal cortical steroids, when administered in vivo markedly alter the rate of protein synthesis in a specific tissue or even more generally. Some of these hormonal effects are quite rapid; within 2 hr. after injection of an adrenal cortical steroid hormone in rats, there is a significant increase in the activity and amount of liver tyrosine-glutamic acid transaminase. Effects of hormones on protein synthesis are manifest not only in the amounts of various proteins following injection of a specific hormone, but also in the synthetic activity of cell-free systems prepared from a tissue of such an animal. Evidence from the latter type of study underlies the suggestion that certain hormones, *e.g.*, thyroxine and adrenal cortical hormones, exert their influence on the rate of RNA synthesis, including an RNA with a rapid turnover, inferred to be mRNA. This is accompanied by a change in the activity of the DNA-dependent RNA polymerase and in the quantity of rough endoplasmic reticulum discernible in target cells by electron microscopy. It has been hypothesized that an effective hormone may (1) in some manner alter the degree to which the genome can function as template in the RNA synthetic system, and/or (2) alter the rate of transfer of acyl amino acids to ribosomes and their incorporation into ribosomal proteins.

REFERENCES

See list following Chap. 30.

30. Genetic Aspects of Metabolism. III

Genetic Variation of Protein Structure. Evolution of Proteins. Hereditary Disorders of Metabolism

GENETIC VARIATION OF PROTEIN STRUCTURE

Many mutations result in a change in the amino acid sequence of a protein. The first clear proof of this was provided by the study of Hb S (sickle cell hemoglobin), when Ingram demonstrated that a valine residue is substituted for a glutamic acid residue at a unique site in the β chain (page 656). Earlier, Pauling and Itano had shown that Hb A and Hb S differ not only in the solubility of the reduced forms but in electrophoretic mobility. The latter technique has been particularly valuable in detecting hemoglobin variants but fails to detect substitutions which do not result in a charge difference. Some abnormal hemoglobins for which the amino acid substitutions have been established are listed in Table 30.1. In each case, production of the variant protein is due to a point mutation representing the minimal alteration in the gene (cistron) and this is inherited in mendelian fashion, the cistrons for the α and β chains being distinct and located in different chromosomes. Most of the variant hemoglobins listed in Table 30.1 can be ascribed to mutations which represent an alteration in a single base of the DNA and are reflected in the codon of the mRNA. Because of this, these variants have been of value in verifying and predicting the base sequences in the codons. One exception is $Hb_{Freiburg}$, in which Val at β 23 is deleted. Another exception is Hb C_{Harlem}, which contains two substitutions in the β chain.

With the codons for various amino acids definitely established, it should be possible to use the amino acid substitutions of hemoglobin and other proteins to test the nature of the code compositions in various species. For example, the valine codons are GUX (Table 29.3) where X is any base. To yield the code for glutamic acid, by a single base change, as in $Hb_{Milwaukee}$ (Table 30.1), the code for glutamic acid is expected to be GAPu, where Pu is either one of the purines; hence the codon for Val must be GUPu and the change has been from U $\rightarrow$ A in the middle base of the triplet. The fact that all known amino acid substitutions in human Hb are consistent with single base changes in the codons derived from the *Esche-*

Table 30.1: SOME AMINO ACID SUBSTITUTIONS IN HUMAN HEMOGLOBIN

Type of Hb	Position	Residue in		Type of Hb	Position	Residue in	
		Hb A	Mutant			Hb A	Mutant
$J_{Toronto}$	α 5	Ala	Asp	$J_{Baltimore}$	β 16	Gly	Asp
J_{Oxford}	α 15	Gly	Asp	$G_{Coushatta}$	β 22	Glu	Ala
I	α 16	Lys	Glu	Freiburg	β 23	Val	None
$J_{Medellin}$	α 22	Gly	Asp	E	β 26	Glu	Lys
Memphis	α 23	Glu	Gln	Genova	β 28	Leu	Pro
$G_{Chinese}$	α 30	Glu	Gln	$G_{Galveston}$	β 43	Glu	Ala
$L_{Ferrara}$	α 47	Asp	Gly	K_{Ibadan}	β 46	Gly	Glu
Shimonoseki	α 54	Gln	Arg	Hikari	β 61	Lys	Asn
Mexico	α 54	Gln	Glu	$M_{Saskatoon}$	β 63	His	Tyr
Norfolk	α 57	Gly	Asp	Zürich	β 63	His	Arg
M_{Boston}	α 58	His	Tyr	$M_{Milwaukee}$	β 67	Val	Glu
$G_{Philadelphia}$	α 68	Asn	Lys	Seattle	β 70 or 76	Ala	Glu
M_{Iwate}	α 87	His	Tyr	G_{Accra}	β 79	Asp	Asn
Chesapeake	α 92	Arg	Leu	D_{Ibadan}	β 87	Thr	Lys
$J_{Capetown}$	α 92	Arg	Gln	$N_{Baltimore}$	β 95	Lys	Glu
$O_{Indonesia}$	α 116	Glu	Lys	Köln	β 98	Val	Met
Tokuchi	β 2	His	Tyr	New York	β 113	Val	Glu
C	β 6	Glu	Lys	D_{Punjab}	β 121	Glu	Gln
S	β 6	Glu	Val	O_{Arab}	β 121	Glu	Lys
$G_{San\ Jose}$	β 7	Glu	Gly	$K_{Woolwich}$	β 132	Lys	Gln
Sriraj	β 7	Glu	Lys	Hope	β 136	Gly	Asp
				Kenwood	β 143	His	Asp (or Glu)
				C_{Harlem}	β 6	Glu	Val
					β 73	Asp	Asn

SOURCE: Compiled from the work of many investigators. Positions refer to the residue from the amino-terminus of the α and β chains of Hb A (Fig. 8.1, page 173). Other abnormal hemoglobins, described under different names, have substitutions identical with one of those given above. (*See W. A. Schroeder and R. T. Jones, Fortschr. Chem. org. Naturstoffe,* **23**, 113, 1965.)

richia coli system is further indication of the universality of most, if not all, of the codons (page 680).

Inasmuch as DNA contains predominantly only the same four bases, point mutations can be expected to occur in all genetic material and will be reflected in amino acid substitutions in proteins of the same species. Relatively few of these have been reported as yet, partly because the structure of only a few proteins is known and partly because of the difficulty of isolating a pure protein from a single individual. Hemoglobin is a favorable protein for study from both viewpoints, and many single amino acid substitutions in human hemoglobins have been studied.

Tobacco mosaic virus (TMV) RNA, which can function as an mRNA (page 666), when treated with nitrous acid yields alterations in the TMV protein synthesized in the host plant by infection with the active RNA alone (page 204). The effect of nitrous acid on RNA is expected to be deamination of cytosine to uracil and of adenine to hypoxanthine, which functions in base pairing in the same manner as guanine (page 655). No effect is expected on uracil, and deamination of guanine would produce xanthine, which either would pair as guanine does or would not pair at all. Studies of this type indicate that (1) no amino acid substitution has been found which occurs in both directions; *i.e.*, where residue A replaces residue B, B does not replace A; (2) substitutions for certain amino acids have not been found, *e.g.*, a replacement for phenylalanine. Since the code for Phe is UUU, there is no mechanism by which deamination could alter this code or others containing only U or G (xanthine in place of guanine appears to be ineffective).

Yanofsky and coworkers have produced experimentally many mutants of *E. coli* that influence the activity of the α protein of tryptophan synthetase (page 573). This protein and many of its inactive variants have been isolated in pure form, and a number of amino acid substitutions have been reported.

In a single peptide obtained from the α protein, a glycine residue (I) was found to be substituted by Glu or Arg (Fig. 30.1), the enzyme containing either of these residues being totally inactive. Recombination by crossing over from the two mutant types yielded the wild-type active enzyme containing glycine. This demonstrates that the base which was altered in the code for Gly must be different in the two cases in order to yield the original code by crossing over. Other mutations from the codes for Glu or Arg have yielded additional substitutions (Fig. 30.1). A glycine residue (II) at a different locus in the protein yielded by mutations Asp and Cys. Since different amino acid substitutions were obtained at I and II, it is evident that the codons at I and II differ. The deduced codons are indicated for each substituted amino acid (Fig. 30.1).

These studies have provided important information regarding the code. The cross-over recombination, noted above, provides further evidence that the smallest mutation unit, the *muton* (page 660), is a single-base pair.

When the position of Gly I is occupied by small neutral residues, Gly, Ala, or

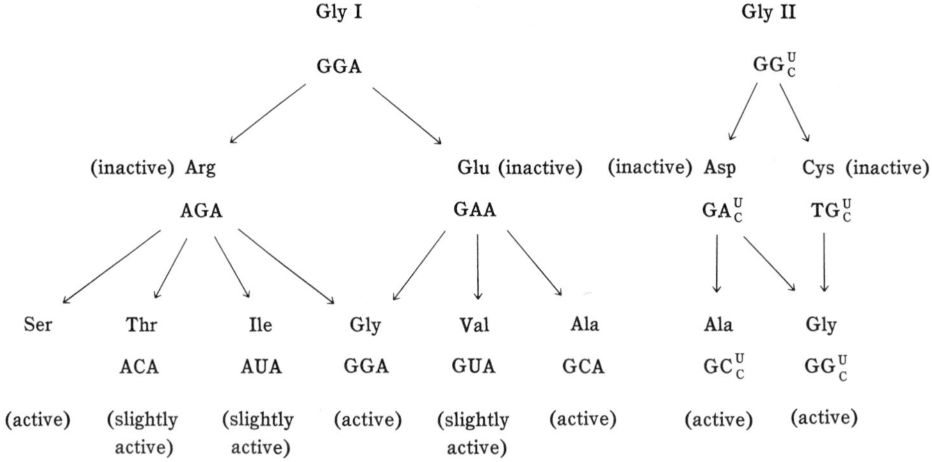

Fig. 30.1. The amino acid substitutions observed at two sites, Gly I and Gly II, are critical for synthesis of active tryptophan synthetase α protein. The associated nucleotide triplets for the mRNA were deduced from known codon relationships, all of which differ by single nucleotides. The codons for Gly I and Gly II differ since they yield different substitutions. The third base in the codon for Gly II can be either U or C. The terms "active" and "inactive" refer to the enzyme formed in these strains. (*From the studies of C. Yanofsky and coworkers.*)

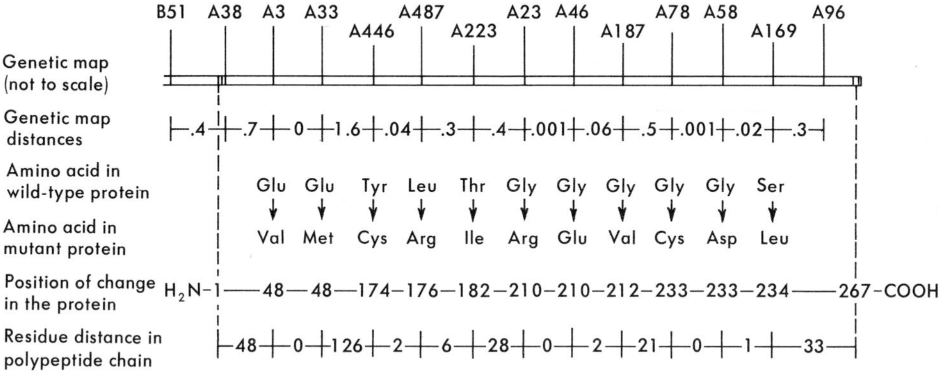

Fig. 30.2. Genetic map of the A gene and the corresponding amino acid changes in the α protein, illustrating colinearity. The positions of these changes in the amino acid sequence are also indicated. (*From C. Yanofsky, G. R. Drapeau, J. R. Guest, and B. C. Carlton, Proc. Natl. Acad. Sci. U.S.*, **57**, *296, 1967.*)

Ser, the enzyme is active. With Thr, Ile, or Val in this position, there is slight activity, but activity is lost when an ionic residue, Arg or Glu, is present.

Extensive genetic mapping studies of mutants in the gene for tryptophan synthetase α protein in conjunction with studies of the amino acid sequence of the protein have demonstrated *strict colinearity of the amino acid sequence and the genetic map.* The relationships of the mutational sites, amino acid sequence, and substitutions are shown in Fig. 30.2. All these mutant strains were detected because of the development of tryptophan-dependence for growth (inactive α protein). Evidence for colinearity has also been obtained for phage T4 "head protein" gene and for the β-galactosidase gene of *E. coli.* Thus, this fundamental concept has now been definitely established.

From the colinearity of gene and the α protein, the known sequences of nucleotides in the codons, and the direction of synthesis of mRNA (page 668) and protein (page 674), all these interrelationships may be correlated as shown in Fig. 30.3. To recapitulate briefly, mRNA is synthesized, with the 3′ strand of DNA serving as template, beginning at the 3′-OH end and in antiparallel sequence. In effect, transcription results in an mRNA nucleotide sequence that is identical with that of the 5′ DNA strand except for the presence of U instead of T in the latter. The protein is synthesized from the NH_2-terminus by transfer of one amino acid residue at a time from the respective amino acyl-tRNA bearing an anticodon that is complementary and antiparallel to the triplet codon of the mRNA. The *operator* which controls the formation of mRNA for a group of related genes (page 685) has been demonstrated to be on the NH_2-terminal side of mRNA in the case of both α protein of tryptophan synthetase and β-galactosidase of *E. coli.*

AMINO ACID SUBSTITUTION AND PROTEIN FUNCTION

Genetic variation in the structure of a protein can provide valuable information concerning the role of amino acid residues in the function of a specific enzyme or other protein. For example, certain abnormal hemoglobins were first detected

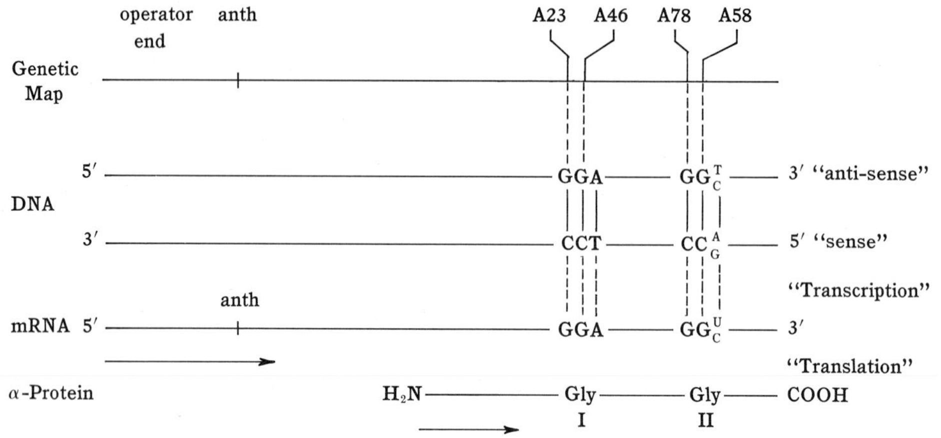

Fig. 30.3. A diagrammatic representation of the colinear relationships of the tryptophan operon, the DNA and RNA nucleotide sequences, and the amino acid sequence of the α-protein. Mapping of the mutants for Gly I and Gly II and the various recombinants has yielded the mapping relationships with regard to the amino acid sequence of the protein. The operator region is at the anthranilate synthetase (anth) end, and transcription for mRNA starts at this end, as does protein synthesis. (*From J. R. Guest and C. Yanofsky, Nature,* **210**, 799–802, 1966.)

because of the physiological disturbances which they produce. These "hereditary hemoglobinopathies" (Chap. 31) are most striking for carriers of Hb S, Hb M, and others. On the other hand, there are amino acid substitutions which appear to have no significant effect on hemoglobin function, *e.g.*, Hb $G_{\text{San José}}$, Hb I, and Hb $G_{\text{Philadelphia}}$ (Table 30.1).

The interpretation of these differences in the effects of amino acid substitutions rests on present knowledge of protein structure and function. The amino- or carboxyl-terminal ends of certain proteins may be entirely dispensable for function. Chemical treatment which alters certain amino acid side chains may produce no demonstrable effect on the function of certain proteins; *e.g.*, conversion of many or all lysine residues to homoarginine residues by reaction with O-methylisourea does not alter the enzymic activity of cytochrome c (page 176), ribonuclease, lysozyme, papain, etc. Similarly, other residues of proteins can be acylated, oxidized, or treated in other ways without significant effect on function. Nevertheless, certain groups in enzymes are absolutely critical for their functions, as in the demonstrations of "active sites," elucidated with various types of inhibitors.

Evidently, then, certain residues of proteins are essential for function; others are not. As a *minimum,* we may expect that those residues that are present at "active sites," which are essential for binding substrates, cofactors, or prosthetic groups, or which determine the folding and essential conformation of the peptide chains cannot be altered by genetic or other means without profound effects on function. The substitutions for Gly I or Gly II which result in an inactive tryptophan synthetase (page 690) are presumably due to conformational effects on the protein since a variety of nonpolar residues can substitute for Gly, but ionic residues cannot. Examination of the conformation of the myoglobin molecule (page 160) or the chains of hemoglobin (page 172) suggests that the regions at which greatest variation in amino acid sequence occur without disturbance of function are in the sections of the chains that are in the form of an α-helix. Indeed, these are the regions in which major differences in amino acid sequences occur in the myoglobins and hemoglobins of different species. It has long been known that the hemoglobins of vertebrates possess similar functional properties, yet there are substantial differences among them in amino acid composition and other properties, which nevertheless, do not affect oxygen transport.

Table 30.2 shows the variations found in the structures of the insulins of several mammals. Most of the variations occur in the internal loop of the A chain and at the carboxyl-terminal end of the B chain (Fig. 7.2, page 149). The carboxyl-terminal residue may be removed by carboxypeptidase A without loss of insulin activity. Since the variations in the A chain do not affect insulin activity, the spacing of these residues may be important but their exact chemical nature is not.

Table 30.2: AMINO ACID SUBSTITUTIONS IN INSULIN

Source	A8*	A9*	A10*	B30*
Beef	Ala	Ser	Val	Ala
Pig	Thr	Ser	Ile	Ala
Sheep	Ala	Gly	Val	Ala
Horse	Thr	Gly	Ile	Ala
Sperm whale	Thr	Ser	Ile	Ala
Sei whale	Ala	Ser	Thr	Ala
Man	Thr	Ser	Ile	Thr
Dog	Thr	Ser	Ile	Ala
Rabbit	Thr	Ser	Ile	Ser

* Positions refer to the residue number from the amino-terminus of the A or B chain (Fig. 7.2; page 149). Rat insulin resembles that of rabbit but differs in having Asp instead of Glu at A4, Lys instead of Asn at B3, and Lys or Met at B29.

The most extensive studies presently available are those on the cytochrome c of some 25 different species of animals, plants, and fungi. In those cytochromes, which possess 103 or 104 residues, only 35 residues have thus far been found to be constant (Fig. 8.4). Some of the types of substitutions have already been presented (page 178). It may be stated here in more general terms that for all proteins studied thus far, many types of "*conservative*" substitutions can occur without apparent change of functional properties of the protein or enzyme. As examples, we may cite the following: substitution of aliphatic hydrophobic residues for one another (Ile, Val, Leu, Met), substitution of polar residues for one another (Arg for Lys, Glu for Asp, Gln for Asn). Even ionic residues of opposite charge substitute for one another (Asp or Glu for Lys), presumably because such residues are mainly on the surface of the protein. Substitutions which are "*radical,*" representing different kinds of side chains, occur without deleterious effect, presumably only when the locus is not critical for function or conformation. This is discussed further below.

EVOLUTION AND PROTEIN STRUCTURE

Comparison of the proteins of various organisms reveals that, during the course of evolution, (1) those proteins which fulfill the same function at all phylogenetic levels, *e.g.*, the enzymes of glycolysis, cytochrome c, etc., synthesis of which is directed by homologous genes, have been extensively modified; (2) some proteins have disappeared; and (3) many new proteins have appeared. Since, in the immediate sense, the structure and distribution of proteins constitute the phenotypic character of a species, these three major changes represent the operation of evolution itself.

Data are at hand for only a few proteins at various phylogenetic levels. It is clear, however, that the pentapeptide obtained by tryptic hydrolysis, which includes the functionally significant serine residue of phosphoglucomutase (Table 12.2), is identical in the enzyme obtained from *E. coli,* yeast, and rabbit muscle. However, this enzyme from these diverse sources, although of approximately the same molecular weight, has many striking differences in amino acid composition. The enzymically significant sulfhydryl group of triose phosphate dehydrogenase has been found to occur in the same octadecapeptide obtained by proteolysis of the yeast and rabbit muscle enzymes, which are in other respects quite different. The many amino acid replacements which have occurred in these proteins as a result of mutagenic forces affecting the homologous DNA have not impaired their enzymic function.

Most significant, however, is the fact that no change in amino acid composition has occurred in the region of the "active site." Presumably, any substitution at that site which alters its steric fit and substrate-binding capacity or which results in loss of a residue of the group participating in the catalytic process (serine, histidine, sulfhydryl group, etc.) must destroy enzymic activity of the protein. Loss of activity might also result from replacement of a proline residue at the interruption of a helix or any other substitution that alters the conformation of the protein in a manner affecting the structure at the active site. Presumably, it is such substitutions or chain-terminating mutations (page 676) which, at some distant time, resulted in loss of specific enzymes, thus making the vertebrates dependent on their food supply for the nutritionally essential amino acids and fatty acids, as well as for the vitamins. Such are some of the mutations of tryptophan synthetase (page 690). Although the active site cannot be specified as yet, mutants created by irradiation synthesize enzymically inert proteins that are identical with the normal enzyme except for substitution of a single amino acid.

It is unknown whether vertebrates continue to synthesize similar functionless genetic derivatives of formerly active proteins which once made possible the synthesis of tryptophan, thiamine, etc. Indeed, it is not known whether the homozygotic alkaptonuric human being who possesses no active homogentistic acid oxidase, the result of a much more recent mutation, synthesizes a homologous protein which is inactive because of one or more amino acid replacements. Not all substitutions at a given site are deleterious with respect to function, as in some mutant forms of tryptophan synthetase (Fig. 30.1) or in cytochrome c (Fig. 8.4). Indeed, it is clear that there is considerable tolerance for replacement, deletion, or insertion of amino acids in many regions of various proteins. In addition to the limited information from sequence studies of proteins, examples of such tolerance are apparent from (1) the specific immunochemical behavior of similar proteins from different species; (2) the fact that many enzymes can tolerate chemical modification of various residues; (3) the proteolytic removal of a segment of a polypeptide chain without inactivation; and (4) the gross differences in amino acid composition of many enzymes and other proteins from various species.

The protein for which most information is available is cytochrome c, as described previously. Although this protein is not an enzyme, and cannot be said to have an active site in the usual sense, if the protein is to be useful physiologically, it may not be altered by mutation in a manner preventing its "fit" to cytochrome oxidase, changing its potential, or interfering with its capacity for attachment to the heme moiety. Withal, in the time since animals, plants, and fungi diverged from some common ancestral form, amino acid replacements have accumulated in the proteins of these species, without significant effect on their function in mitochondrial electron transport.

The numbers of amino acid replacements in cytochrome c of some species are compared in Table 30.3 for the 104 homologous residues. First, it is evident that the cytochromes of closely related vertebrates differ in none or only a few residues, *e.g.*, man and other primates, various ungulates, etc. Second, the greater the apparent taxonomic difference, the more the cytochromes are likely to differ. This is apparent among the vertebrates and in comparing the vertebrates with the plant (wheat), the insect, or an ascomycete (yeast, *Neurospora*). However, yeast and *Neurospora* cytochromes differ almost as much from each other as they do from those of higher organisms. This indicates either that the evolutionary pathway for these two organisms diverged a long time ago or that the greater reproductive rate of such microorganisms can lead to accumulated differences in sequence in relatively short periods. Indeed, it has been found that subtilisins (page 255) from

Table 30.3: VARIATIONS IN STRUCTURE OF CYTOCHROME C

Cytochromes compared to human cytochrome	No. of variant amino acid residues*	Cytochromes compared	No. of variant amino acid residues*
Chimpanzee	0	Horse–dog	6
Rhesus monkey	1	Horse–pig	3
Kangaroo	10	Pig–cow–sheep	0
Dog	11	Pig–whale	2
Horse	12	Horse–kangaroo	7
Chicken	13	Rabbit–pig	4
Rattlesnake	14	Rabbit–dog	5
Tuna fish	21	Dogfish–tuna	14
Dogfish	23	Dogfish–horse	16
Moth (*Samia cynthia*)	31	Moth–tuna	33
Wheat	35	Moth–yeast	48
Neurospora	43	*Neurospora*–wheat	46
Yeast	44	Yeast–*Neurospora*	39

* Most vertebrate cytochromes contain 104 residues. Comparisons are based on these residues and do not include extra residues at the NH_2-terminus in nonvertebrate cytochromes.

SOURCE: Compiled from data of several investigators. See E. L. Smith, *Harvey Lectures,* **62**, 231, 1966–1967.

two strains of *Bacillus subtilis* differ in 84 residues plus a deletion in the single chain of 275 residues, although the two enzymes do not differ significantly in other properties. Also, mutants in haploid organisms can be manifested and spread more rapidly than in diploid species. Thus far, there is little information comparable to that for cytochrome c for other proteins which can indicate such details of evolution at the molecular level.

All the cytochromes c mentioned (Table 30.3) react with yeast or mammalian cytochrome oxidase and thus do not differ functionally. Since the evolutionary divergence of animals, plants, and fungi has been estimated to have occurred approximately 2,000 million years ago, it is evident that cytochrome c is a protein of great age and has not altered in function since that time. Since the sole function of cytochrome c involves its role in mitochondrial terminal oxidation and phosphorylation, presumably a similar age must be assumed for mitochondria. The age of the earth is presently estimated at approximately 4,500 million years; hence one of the critical stages in evolution, the development of efficient aerobic metabolism, almost certainly occurred during the first half of the age of the earth. Through studies of other "old" proteins, it is hoped that further insight can be obtained regarding the evolution and development of the major cellular functions.

The increased number of proteins synthesized by higher organisms is paralleled by an increase in the amount of DNA in their genomes, and it is assumed that this reflects the occasional duplication of some or all of the genes in an individual organism independent of general mitosis. When such duplication occurs, the diploid organism possesses for a given protein four rather than two such genes, and thereafter, each pair is free to follow an independent evolutionary path. As long as one pair continues to direct synthesis of the original functional protein, the other may undergo drastic mutational change and, indeed, ultimately become responsible for synthesis of a protein which serves a different function. If the latter offers "survival value" it will be retained and the organism will flourish and perhaps come to occupy a very different ecological niche. Such a process is apparent in the case of hemoglobin.

When the amino acid sequences for the α and β chains of Hb A, the γ chain of Hb F, and the δ chain of Hb A_2 were established, the number of differences among these chains could be summarized as in Table 30.4. When the α and β chains of human Hb A were compared to those of a few other mammals, such as the horse, it was evident that there were fewer differences among α chains than among β chains. Since, moreover, the α chains are common to Hb A, Hb F, and Hb A_2, Ingram suggested that the α chain might be the oldest of these and might be homologous with the chains of lamprey oxyhemoglobin which is monomeric (like myoglobin) rather than tetrameric. Hence, from the observed number of differences, it was further suggested that succeeding gene duplications led from α chains to γ, β, and δ, in that sequence. Supporting evidence has come from a study of the hemoglobins of a single order, the primates. Living representatives of this order, according to the paleontological record and their comparative anatomy, appeared in the following sequence: tree shrews, lemurs, lorises, New World monkeys, Old World monkeys, apes, man. The α chains of hemoglobins of representa-

Table 30.4: AMINO ACID VARIATIONS IN CHAINS OF HUMAN HEMOGLOBINS

Chains compared	Number of variant amino acid residues	Time since divergence from common chain ancestor, years × 10^6
β and δ	6	44
β and γ	36	260
α and β	78	565
α and γ	83	600

SOURCE: After E. Zuckerkandl and L. Pauling, Molecular Disease, Evolution and Genic Heterogeneity, in M. Kasha and B. Pullman, eds., "Horizons in Biochemistry," Academic Press, Inc., New York, 1962.

tives of these genera show relatively few amino acid substitutions. The β chains differ increasingly from those of man as one descends to more primitive forms; most striking is the fact that the substitutions occur largely at those positions in which human β chains differ from γ chains. Indeed, the hemoglobins of adult shrews and lemurs resemble human fetal hemoglobin more closely than they resemble human Hb A, including resistance to alkaline denaturation.

Another example of gene duplications leading to the evolution of new functions is provided by the structures of trypsin, chymotrypsin A, chymotrypsin B, and elastase. The complete amino acid sequences of the first two enzymes (page 256), the partial structures of the last two, and the sequences at the disulfide bridges show considerable homology in structure, trypsin and chymotrypsin A being identical in approximately 40 per cent of their sequences. Here is an instance in which similar function, protein hydrolysis, has been retained but with considerable change in specificity with respect to the types of bonds hydrolyzed (Table 7.4 and page 250*ff.*). It is evident that for vertebrates, dependent on the presence of many preformed amino acids in the diet, there would be considerable survival value for proteinases capable of achieving maximal digestion of dietary protein.

The polypeptide hormones of the neurohypophysis (Chap. 47) provide another example of the evolution of structure and function. The structures of these hormones are given in Table 30.5. The evolution of these polypeptide hormones clearly represents single base changes in the codons, that have occurred stepwise, yielding single amino acid substitutions. The vasotocin of the frog possesses weak activity both as a vasopressin and as an oxytocin, in contrast to the powerful hormones of the mammal and other higher vertebrates. The change from vasotocin to arginine-vasopressin involves a substitution at position 3 of Ile by Phe. Similarly, from vasotocin to oxytocin, the change in position 8 is from Arg to Leu. For the evolution from a single hormone to the presence of two or more hormones in the same species, we must assume a duplication of the genetic material, as well as point mutations, in order to have independent production of two or more hormones. Similar interrelationships are found for ACTH and MSH (Chap. 47).

The structures of the hormones secretin and glucagon are sufficiently similar to indicate a common genetic origin (Fig. 30.4). This is a particularly striking example both of change of function and of alteration of control of protein synthesis. Glucagon, synthesized in the pancreas, augments blood glucose concentration by

Table 30.5: AMINO ACID SEQUENCES OF SOME NEUROHYPOPHYSEAL HORMONES

Hormone	Residue position	Species
	1* 2 3 4 5 6* 7 8 9	
Glumitocin	CyS-Tyr-Ile-Ser-Asn-CyS-Pro-**Gln**-GlyNH$_2$	Cartilaginous fishes
Isotocin	CyS-Tyr-Ile-Ser-Asn-CyS-Pro-**Ile**-GlyNH$_2$	Bony fishes
Mesotocin	CyS-Tyr-Ile-**Gln**-Asn-CyS-Pro-**Ile**-GlyNH$_2$	Amphibians
Vasotocin	CyS-Tyr-Ile-**Gln**-Asn-CyS-Pro-**Arg**-GlyNH$_2$	Bony fishes, amphibians
Oxytocin	CyS-Tyr-Ile-**Gln**-Asn-CyS-Pro-**Leu**-GlyNH$_2$	Mammals
Vasopressin	CyS-Tyr-**Phe**-**Gln**-Asn-CyS-Pro-**Arg**-GlyNH$_2$	Mammals except pig
Vasopressin	CyS-Tyr-**Phe**-**Gln**-Asn-CyS-Pro-**Lys**-GlyNH$_2$	Pig

* A disulfide bridge links residues 1 and 6 in each case. The variations in amino acid residues among these hormones are indicated in boldface type.

SOURCE: Adapted from R. Acher, *Angew. Chem.*, **5**, 798, 1966.

increasing glycogenolysis in the liver (page 977). Secretin, made in the intestinal mucosa, stimulates the flow of pancreatic juice (page 818). Presumably, gene duplication has led to the evolution of two different hormones, each under separate genetic control and each produced by specific cells.

Another instance of close structural relationship is shown by egg white lysozyme (page 269) and one of the two components of lactose synthetase (page 432) long known as α-lactalbumin of bovine milk. Of the 129 residues in lysozyme and the 123 residues in lactalbumin, at least 40 are identical in the corresponding positions and 27 represent conservative replacements. Further, the four disulfide bridges are in corresponding loci in the two proteins. On the basis of this close homology, Brew, Vanaman and Hill have proposed that the two proteins were derived by duplication of a common ancestral gene, one duplicate continuing to produce lysozyme, the other eventually giving rise by mutations to α-lactalbumin. Clearly, in this case, we have an instance of origination of new function, one enzyme catalyzing hydrolysis of a $\beta\ 1 \rightarrow 4$ glucopyranosyl linkage, the other being involved in synthesis of such a bond (page 246).

The above considerations of protein evolution provide examples of point mu-

1 5 10 15
H$_2$N-His-Ser- Asp-Gly-Thr-Phe-Thr-Ser-Glu-Leu-Ser-Arg-Leu-Arg-Asp-

——— Gln ——————— Asp-Tyr —— Lys-Tyr-Leu ——

20 25 27
Ser- Ala-Arg-Leu-Gln-Arg-Leu-Leu-Gln-Gly-Leu-Val-CONH$_2$

—— Arg —— Ala —— Asp-Phe-Val —— Trp —— Met-Asn-Thr - COOH
29

Fig. 30.4. Amino acid sequences of secretin (continuous) and glucagon showing the residues which differ in the latter hormone. Note that secretin has only 27 residues terminating in valinamide whereas glucagon has 29 residues and possesses a free α-COOH group.

tations; deletions, as in the relationship of the Hb chains (page 173) and in the NH_2-terminal parts of the cytochrome structure; and gene duplications as discussed above for the hemoglobins, pancreatic proteinases, and neurohypophyseal and other hormones. Another type of evolutionary change involves extension of sequences. Presumably, primitive proteins may have possessed relatively short polypeptide chains; extension could have occurred by unequal crossing over (page 683). This would be evident from duplications of sequence present in the same peptide chain. Such changes are difficult to detect in "contemporary" proteins since the repetitions would have become obscured in each such segment by point mutations, deletions of residues, etc. Nevertheless, such duplications are apparent in the heavy chains of γ-globulin (page 718), in ferredoxins (page 368), and in portions of the structures of two subtilisins. Extension of sequence provides yet another mechanism for the evolution of new proteins.

Ultimately, all the data on comparative structure of the proteins must be understood in terms of the changes in the physiological economy of the organism and in the total biology of the organism, *i.e.*, its adjustment with respect to both its internal and external environment. Changes in the genome of an organism can persist and spread through the species only when there is an enhanced survival value; changes resulting in an equivalent survival value can spread only if linked to an advantageous change, whereas those with diminished survival value will be eliminated in the competition for a given ecological niche.

Since 1955, when the complete structure of insulin was reported by Sanger, the progress that has been made in studying protein structure and its evolution has yielded only fragmentary information and many hints that considerable progress in our understanding of these phenomena will be forthcoming in the future. Such knowledge, supplemented with nucleic acid sequence studies now in their infancy, should continue to improve our view of the fine structure of evolution at the molecular level. Together with information from other types of biological investigations, our concepts of "man's place in nature" will acquire greater depth.

HEREDITARY DISORDERS OF METABOLISM

When Garrod, in 1908, assembled the known "inborn errors" (page 282), they were few in number. Since then many additional metabolic abnormalities have been described, and the number is growing continuously.

Earlier, attention was focused on instances in which individuals or microorganisms are unable to perform specific metabolic reactions (Chap. 28). From the relationship of the gene to protein synthesis, genetic defects represent an alteration in DNA leading either to complete failure to produce a protein or to production of a modified protein (enzyme) whose function is impaired or lacking.

For many of the metabolic defects that have been studied in human beings, the anomaly has been traced to the absence of activity of a single enzyme. Table 30.6 presents a list of some hereditary disorders in which the lacking or modified enzyme or protein has been identified. Individuals with these disorders are analogous to the mutants in *Neurospora* discussed earlier (Chap. 28), in which a single metabolic reaction is blocked. In man, most of these defects are apparent pheno-

Table 30.6: SOME HEREDITARY DISORDERS IN MAN IN WHICH THE SPECIFIC LACKING OR MODIFIED ENZYME OR PROTEIN HAS BEEN IDENTIFIED

	Affected enzyme or protein	Page reference
Acanthocytosis	β-Lipoproteins (low density)	528
Acatalasia	Catalase	
Afibrinogenemia	Fibrinogen	727
Agammaglobulinemia	γ-Globulin	718
Albinism	Tyrosinase	608
Alkaptonuria	Homogentisic acid oxidase	608
Analbuminemia	Serum albumin	723
Argininosuccinic acidemia	Argininosuccinase	561
Crigler-Najjar syndrome	Uridine diphosphate glucuronate transferase	746
Fructose intolerance	Fructose 1-phosphate aldolase	429
Fructosuria	Fructokinase	429
Galactosemia	Galactose 1-phosphate uridyl transferase	427
Glycogen storage diseases	See Table 19.6	446
Goiter (familial)	Iodotyrosine dehalogenase	925
Hartnup's disease	Tryptophan pyrrolase	615
Hemoglobinopathies	Hemoglobins	749
Hemolytic anemia	Pyruvate kinase	
Hemophilia A	Antihemophilic factor A	729
Hemophilia B	Antihemophilic factor B	729
Histidinemia	Histidase	611
Homocystinuria	Cystathionine synthetase	602
Hypophosphatasia	Alkaline phosphatase	895
Isovaleric acidemia	Isovaleryl CoA dehydrogenase	605
Maple syrup urine disease	Amino acid decarboxylase	605
Methemoglobinemia	Methemoglobin reductase	751
Orotic aciduria	Orotidine 5′-phosphate pyrophosphorylase	635
Parahemophilia	Accelerator globulin	729
Pentosuria	L-Xylulose dehydrogenase	431
Phenylketonuria	Phenylalanine hydroxylase	549
Sulfite oxidase deficiency	Sulfite oxidase	600
Wilson's disease	Ceruloplasmin	722
Xanthinuria	Xanthine oxidase	628

typically only in the homozygous state. In the heterozygote, with one normal and one mutant gene, there is usually sufficient enzyme to meet physiological needs. Clearly, the older concept of *dominance* can be explained on the basis that one normal gene of each pair can stimulate sufficient protein synthesis to serve the needs of the organism. In heterozygotes with genes for the production of one normal and one abnormal hemoglobin, *e.g.*, Hb A, Hb S, Hb M, etc., there are approximately equal amounts present of each variety. Similarly, the heterozygotic parents of galactosemic infants (page 427) possess approximately 50 per cent of the normal amount of enzyme (galactose 1-phosphate uridyl transferase).

Most of the inherited enzymic abnormalities which have been described represent cases in which a complete metabolic block of some type has occurred. The consequences of such a block may be reflected in different ways: (1) there may be a complete absence of the final product of the pathway, *e.g.*, the lack of melanin in albinism (page 608); (2) there may be an accumulation of an intermediate metabolite for which there is no significant alternate pathway, *e.g.*, certain glycogen storage disorders (page 445) in which glycogen is made but cannot be utilized, or alkaptonuria in which homogentisic acid accumulates (page 608); and (3) there may be an accumulation of products of an alternate, and ordinarily minor, pathway of metabolism, as in phenylketonuria (page 549), leading to excretion of large amounts of compounds ordinarily produced in trace amounts.

In addition to the genetic abnormalities that have been ascribed to the lack or modification of a particular enzyme or protein, there are many hereditary disorders of metabolism in which the regulatory protein has not yet been identified; some of these are listed in Table 30.7. As biochemical knowledge increases, the protein aberrations involved in these disorders will undoubtedly be identified. Furthermore, it is likely that the list will continuously increase, particularly as quantitative differences in metabolism become further defined. As in the case of the abnormal hemoglobins where the effect of the modified protein may be negligible or profound, similar findings may be expected for other altered proteins and enzymes.

Table 30.7: SOME HEREDITARY DISORDERS IN WHICH THE AFFECTED PROTEIN HAS NOT BEEN IDENTIFIED

Disorder	Biochemical manifestation	Page reference
Congenital steatorrhea	Failure to digest and/or absorb lipid	474
Cystinuria	Excretion of cystine, lysine, arginine, and ornithine	602
Cystinosis	Inability to utilize amino acids, notably cystine; aberration of amino acid transport into cells	
Fanconi's syndrome	Increased excretion of amino acids	845
Gargoylism (Hurler's syndrome)	Excessive excretion of chondroitin sulfate B	884
Gaucher's disease	Accumulation of cerebrosides in tissues	528
Niemann-Pick disease	Accumulation of sphingomyelin in tissues	527
Porphyria	Increased excretion of uroporphyrins	739
Tangier disease	Lack of plasma high density lipoproteins	528
Tay-Sachs disease	Accumulation of gangliosides in tissues	527

Chromosomal Abnormalities. In addition to the abnormalities listed in Tables 30.6 and 30.7, there are other defects which are due to the presence of an abnormal number of chromosomes. In *mongolism* one of the autosomes is present as a triploid rather than in the normal diploid condition. Individuals are also known in whom the Y chromosome is lacking or in whom excess X chromosomes are present. Since these chromosomes carry genes concerned with sexual differentiation, these individuals present syndromes associated with abnormal sexual development and usually manifest numerous other metabolic abnormalities. The hormones associated with normal sexual development are steroids (Chap. 44), and the effect of sex chromosome polysomy or deficiency suggests that some genes concerned with steroid synthesis may be present on X and Y chromosomes.

In concluding this discussion of hereditary disorders in man, some aspects of studies of human genetics warrant comment. In the early part of this century, after

Mendel's laws were rediscovered, many genetic studies were made of human traits. It soon became obvious that not all the physical, psychological, and metabolic characteristics ascribed to genetic differences are in fact inherited. The difficulties arise because of many factors. Human families are small compared with those of laboratory animals, and it is impossible to obtain accurate data for more than a few generations. Populations are highly mobile, which frequently makes it difficult or impossible to study all members of a family. Common social or environmental factors, which are largely uncontrolled, may simulate a genetic picture.

These cautionary remarks are necessary because in the study of human genetics many meaningless "pedigrees" have been compiled. In this connection, it may be noted that rickets, a disorder due to deficiency of vitamin D or to inadequate exposure of the skin to ultraviolet radiation (Chap. 50), was long ascribed to a genetic factor. Similarly, goiter, once so prevalent in parts of Switzerland and in the "goiter belt" of the United States, was believed to be inherited. It is now recognized that a similar environmental factor, deficiency of iodine, was one etiological factor operative in producing the enlarged thyroid gland. Nevertheless, it has been established that one form of goiter is inherited (Chap. 43). Clearly, it will be necessary, in other instances, as in the case of goiter, to disentangle the complex factors involved in metabolic abnormalities to determine the exact contribution of genetic, nutritional, and other environmental factors involved in each situation. Further, efforts must be devoted to learning how to correct the manifestations of the genetic disorder. The successes achieved in galactosemia (page 427) by withholding galactose from the diet of infants and in phenylketonuria by limiting the intake of phenylalanine (page 549) indicates approaches to these probems.

REFERENCES

Books

Allen, F. W., "Ribonucleoproteins and Ribonucleic Acids," American Elsevier Publishing Company, New York, 1962.

Allen, J. M., ed., "The Molecular Control of Cellular Activity," McGraw-Hill Book Company, New York, 1962.

Bryson, V., and Vogel, H. J., eds., "Evolving Genes and Proteins," Academic Press, Inc., New York, 1965.

"Cellular Regulatory Mechanisms," Cold Spring Harbor Symposia on Quantitative Biology, vol. 26, Biological Laboratory, Cold Spring Harbor, N. Y., 1961.

Dobzhansky, T., "Mankind Evolving: The Evolution of the Human Species," Yale University Press, New Haven, 1962.

Florkin, M., "A Molecular Approach to Phylogeny," American Elsevier Publishing Company, New York, 1966.

Garrod, A. E., "Inborn Errors of Metabolism," 2d ed., Henry Frowde and Hodder & Stoughton, Ltd., London, 1923.

"Genetic Code," Cold Spring Symposia on Quantitative Biology, vol. 31, Biological Laboratory, Cold Spring Harbor, N.Y., 1966.

Harris, H., "Human Biochemical Genetics," Cambridge University Press, New York, 1959.

Hayes, W., "The Genetics of Bacteria and Their Viruses," John Wiley & Sons, Inc., New York, 1964.

Ingram, V. M., "The Hemoglobins in Genetics and Evolution," Columbia University Press, New York, 1963.
Ingram, V. M., "The Biosynthesis of Macromolecules," W. A. Benjamin, Inc., New York, 1965.
Jukes, T. H., "Molecules and Evolution," Columbia University Press, New York, 1966.
Kornberg, A., "Enzymatic Synthesis of DNA," John Wiley & Sons, Inc., New York, 1962.
Lehmann, H., and Huntsman, R. G., "Man's Haemoglobins," North Holland Publishing Company, Amsterdam, 1966.
Malt, R. A., ed., "Macromolecular Synthesis and Growth," Little, Brown and Company, Boston, 1967.
McElroy, W. D., and Glass, B., eds., "The Chemical Basis of Heredity," The Johns Hopkins Press, Baltimore, 1957.
Perutz, M. F., "Proteins and Nucleic Acids," American Elsevier Publishing Company, New York, 1962.
Sager, R., and Ryan, F. J., "Cell Heredity," John Wiley & Sons, Inc., New York, 1961.
Shugar, D., ed., "Genetic Elements. Properties and Function," Academic Press, Inc., New York, 1967.
Stanbury, J. B., Wyngaarden, J. B., and Frederickson, D. S., eds., "The Metabolic Basis of Inherited Disease," 2d ed., McGraw-Hill Book Company, New York, 1966.
Taylor, J. H., ed., "Molecular Genetics," Academic Press, Inc., New York, part I, 1963; part II, 1967.
Taylor, J. H., ed., "Selected Papers on Molecular Genetics," Academic Press, Inc., New York, 1965. (A collection of 55 of the original papers which are representative of major contributions in the field.)
Vogel, H. J., Bryson, V., and Lampen, J. O., eds., "Informational Macromolecules," Academic Press, Inc., New York, 1963.
Watson, J. B., "Molecular Biology of the Gene," W. A. Benjamin, Inc., New York, 1965.
Wolstenholme, G. E. W., and O'Connor, C. M., eds., "Biochemistry of Human Genetics," Little, Brown and Company, Boston, 1959.

Review Articles

Ames, B. N., and Martin, R. G., Biochemical Aspects of Genetics: The Operon, *Ann. Rev. Biochem.,* **33,** 235–258, 1964.
Beckwith, J. R., Regulation of the Lac Operon, *Science,* **156,** 597–604, 1967.
Berg, P., Specificity in Protein Synthesis, *Ann. Rev. Biochem.,* **30,** 293–324, 1961.
Campbell, P. N., The Biosynthesis of Proteins, *Progr. in Biophysics and Molecular Biol.,* **15,** 1–38, 1965.
Clark, B. F. C., and Marcker, K. A., How Proteins Start, *Scientific American,* **218** (Jan.), 36–42, 1968.
Crick, F. H. C., The Genetic Code: III., *Scientific American,* **215,** 55–62, 1966.
Dixon, G. H., Mechanisms of Protein Evolution, in P. N. Campbell and G. D. Greville, eds., "Essays in Biochemistry," vol. 2, pp. 147–204, Academic Press, Inc., New York, 1966.
Helinski, D. R., and Yanofsky, C., Genetic Control of Protein Structure, in H. Neurath, ed., "The Proteins," 2d ed., vol. IV, pp. 1–93, Academic Press, Inc., New York, 1966.
Ingram, V. M., On the Biosynthesis of Hemoglobin, *Harvey Lectures,* **61,** 43–70, 1965–1966.
Jacob, F., and Monod, J., Genetic Regulatory Mechanisms in the Synthesis of Proteins, *J. Molecular Biol.,* **3,** 318–356, 1961.
Khorana, H. G., Polynucleotide Synthesis and the Genetic Code, *Harvey Lectures,* **62,** 79–105, 1966–1967.
Mantsavinos, R., and Zamenhof, S., The Enzymic Synthesis of Nucleic Acids, in P. Bernfeld, ed., "Biogenesis of Natural Compounds," 2d ed., pp. 537–587, The Macmillan Company, New York, 1967.

Moldave, K., Nucleic Acids and Protein Biosynthesis, *Ann. Rev. Biochem.,* **34,** 419–448, 1965.

Monod, J., Jacob, F., and Gros, F., Structural and Rate-determining Factors in the Biosynthesis of Adaptive Enzymes, in D. J. Bell and J. K. Grant, eds., "The Structure and Biosynthesis of Macromolecules," Biochemical Society Symposia, No. 21, pp. 104–132, Cambridge University Press, New York, 1962.

Nirenberg, M., Protein Synthesis and the RNA Code, *Harvey Lectures,* **59,** 155–185, 1963–1964.

Sibatani, A., Genetic Transcription or DNA-dependent RNA Synthesis, *Progr. in Biophysics and Molecular Biol.,* **16,** 17–88, 1966.

Silver, S., Molecular Genetics of Bacteria and Bacteriophages, *Progr. in Biophysics and Molecular Biol.,* **16,** 193–240, 1966.

Smith, E. L., The Evolution of Proteins, *Harvey Lectures,* **62,** 231–256, 1966–1967.

Spiegelman, S., Information Transfer from the Genome, *Federation Proc.,* **22,** 36–54, 1963.

Stent, G. S., The Operon on Its Third Anniversary, *Science,* **144,** 816–820, 1964.

Yanofsky, C., Gene Structure and Protein Structure, *Harvey Lectures,* **61,** 145–168, 1965–1966.

Also see lists following Chaps. 9 and 27.

31. Blood

Composition of Blood Plasma. Plasma Proteins. Blood Clotting. Erythrocyte and Iron Metabolism. Leukocyte Composition and Metabolism

Unicellular organisms which live in immediate contact with the external environment obtain nutrients directly from that environment and eliminate unused or unwanted materials directly into the external milieu. As organisms evolved into more complex structures, special means of communication were established which permitted a continuing integration among various tissues and organs, as well as facilitating contact with the external environment. The circulatory systems, containing the blood and the lymph, and the nervous system represent the important connecting pathways among the diverse anatomical structures of the mammalian organism. The difference in color between blood and lymph is due to the fact that the lymph, although having large numbers of leukocytes, or white cells, has very few erythrocytes, or red cells.

Both blood and lymph contain dissolved solutes as well as suspended insoluble components. The specific gravity of blood is 1.055 to 1.065, and its viscosity is approximately five to six times that of water. If blood is drawn from a vein and measures are taken to prevent clotting (page 724*ff.*), the suspended cellular elements can be separated by centrifugation. The normally clear, slightly yellow supernatant fluid is termed *blood plasma.* Should the blood be drawn and allowed to clot, there separates from the clot a clear yellowish fluid, the *blood serum.* The yellow color is due to the presence of small quantities of bilirubin, a bile pigment (page 820), and of carotenoids (page 76). The clot is composed largely of cellular elements, enmeshed in a network of fibrous strands of *fibrin* (page 727). Thus blood plasma represents blood minus its cellular elements, whereas blood serum lacks, in addition, fibrinogen, the precursor of fibrin. Lymph also clots, although somewhat more slowly than blood. The composition of lymph is discussed in Chap. 34.

COMPOSITION OF BLOOD PLASMA

The total volume of blood in the vascular system approximates 8 per cent of the body weight, about 5 or 6 liters of blood in an adult. Infants have a larger blood volume, in proportion to their body weight, than do adults; the blood volume is a function of the surface area of the body.

Dissolved in the blood plasma are solutes which comprise approximately 10 per cent of the volume. Of these solutes, proteins constitute approximately 7 per cent of the plasma, inorganic salts approximately 0.9 per cent, with the remainder of the solutes consisting of diverse organic compounds other than protein. Tables 31.1 and 31.2 indicate the concentrations of the principal nonprotein organic and inorganic components, respectively, of the blood plasma of man, with the ranges of concentration for each constituent under normal conditions. Factors influencing the concentrations of these constituents, as well as their physiological functions, are considered below as well as in other chapters of this book.

The amount of a particular blood constituent is the resultant of the rate at which addition of the component to the blood takes place, and the rate of utili-

Table 31.1: Approximate Ranges of Values for Certain of the Principal Nonprotein Organic Constituents of the Blood Plasma of Man

Constituent	*Normal range, mg./100 ml.*	*Constituent*	*Normal range, mg./100 ml.*
Nonprotein N:	25–40	Creatinine	1–2
Urea	20–30	Uric acid	2–6
Urea N	10–20	*Carbohydrates:*	
Amino acid N	4–8	Glucose	65–90
Amino acids	35–65	Fructose	6–8
Alanine	2.5–7.5	Glycogen	5–6
α-Aminobutyric acid	0.1–0.3	Polysaccharides (as hexose)	70–105
Arginine	1.2–3.0	Glucosamine (as polysaccharide)	60–105
Asparagine	0.5–1.4	Hexuronates (as glucuronic acid)	0.4–1.4
Aspartic acid	0.01–0.3	Pentose, total	2–4
Citrulline	0.5	*Organic acids:*	
Cystine	0.8–5.0	Citric acid	1.4–3.0
Glutamic acid	0.4–4.4	α-Ketoglutaric acid	0.2–1.0
Glutamine	4.5–10.0	Malic acid	0.1–0.9
Glycine	0.8–5.4	Succinic acid	0.1–0.6
Histidine	0.8–3.8	Acetoacetic acid	0.8–2.8
Isoleucine	0.7–4.2	Lactic acid	8–17
Leucine	1.0–5.2	Pyruvic acid	0.4–2.0
Lysine	1.4–5.8	*Lipids:*	
Methionine	0.2–1.0	Total lipids	385–675
1-Methylhistidine	0.1	Neutral fat	80–240
3-Methylhistidine	0.1	Cholesterol, total	130–260
Ornithine	0.6–0.8	Cholesterol, esters	90–190
Phenylalanine	0.7–4.0	Cholesterol, free	40–70
Proline	1.5–5.7	Phospholipids:	
Serine	0.3–2.0	Total	150–250
Taurine	0.2–0.8	Lecithin	100–200
Threonine	0.9–3.6	Phosphatidyl ethanolamine	0–30
Tryptophan	0.4–3.0	Plasmalogens	7–8
Tyrosine	0.8–2.5	Sphingomyelin	10–50
Valine	1.9–4.2	Total fatty acids	150–500
Bilirubin	0.2–1.4	Unesterified fatty acids	8–30
Creatine	0.2–0.9		

Table 31.2: APPROXIMATE RANGES OF VALUES FOR THE PRINCIPAL INORGANIC CONSTITUENTS OF THE BLOOD PLASMA OF MAN

Anions	*Concentration meq./liter**	*Cations*	*Concentration meq./liter**
Total	142–150	Total	142–158
Bicarbonate	24–30	Calcium	4.5–5.6
Chloride	100–110	Magnesium	1.6–2.2
Phosphate	1.6–2.7	Potassium	3.8–5.4
Sulfate	0.7–1.5	Sodium	132–150
Iodine (total)	8–15†	Iron	50–180†
Iodine (protein-bound)	6–8†	Copper	8–16†

* meq./liter, milligram equivalents per liter, or milliequivalents per liter; a meq./liter of an ion or a substance $= \frac{\text{mg./liter} \times \text{valence}}{\text{atomic or formula weight}}$.

† These concentrations are in terms of micrograms per 100 ml.

zation or removal of the substance from blood by various tissues. Although quantitative measurements of particular blood components may yield information of considerable value, the composition of the blood is but a limited indicator of the metabolic status of tissue cells. Many cellular products do not appear to a detectable degree in the circulation. In no instance is there a simple linear relationship between tissue and plasma concentrations of a given constituent.

PLASMA PROTEINS

Composition of Plasma Proteins. The total protein content of the plasma is normally in the range of 5.7 to 8.0 g. per 100 ml. Ideally, to describe the plasma proteins, a complete list should be constructed of the constituent proteins, with the amounts and function of each indicated. Such a list is not possible because information is still inadequate concerning many of the proteins present in small amounts. Although free boundary and paper electrophoresis (page 121) are convenient methods of separation and estimation of the major groups of plasma proteins, more refined procedures, *e.g.*, density gradient centrifugation (page 131), starch gel electrophoresis (page 123), and immunoelectrophoresis (page 713), have revealed a very large number of plasma proteins and fractions. Some of the members of a fraction may share only one major property in common, *e.g.*, solubility, density, molecular weight, or electrical charge at a defined pH. Some plasma proteins have been best identified or detected as a consequence of a particular biological or biocatalytic property, but have not been significantly purified. Finally, a few plasma proteins have been obtained in crystalline form and are more clearly characterizable. In the absence of a single method of delineating all plasma proteins, classification schemes have been adopted that are based largely on methods of separating these proteins. Illustrative examples will be found in the pages which follow.

One of the best methods of categorizing the proteins of plasma and their relative amounts is on the basis of their electrophoretic distribution. Figure 31.1 shows the distribution of the plasma proteins obtained in free boundary electrophoresis. The amount of each component is assessed from the area under each individual

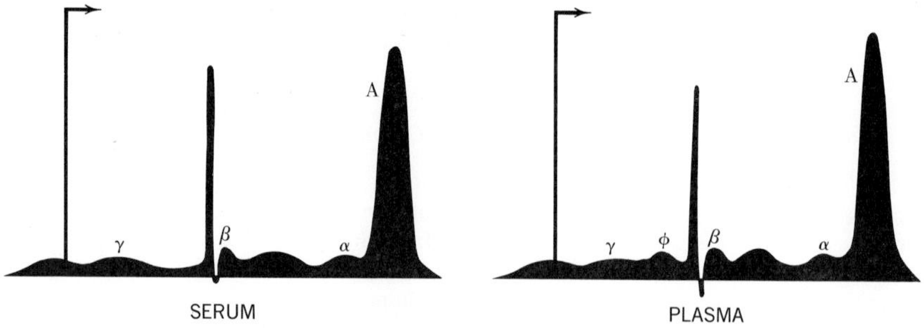

Fig. 31.1. The electrophoretic patterns obtained with (*left*) normal human serum and (*right*) plasma are shown. The main components are identified as *A* for albumin, ϕ for fibrinogen, and α, β, and γ for the different globulins. Fibrinogen is lacking in serum. Conditions of electrophoresis: sodium diethylbarbiturate (Veronal) buffer, pH 8.6; ionic strength, 0.1; potential gradient, 6 volts/cm.; time, 250 min.

boundary. Representative average values for normal human plasma are given in Table 31.3. There are only six main boundaries evident in the electrophoretic pattern obtained under the conditions of electrophoresis used (Fig. 31.1), whereas there is a much greater number of distinct proteins in plasma. Such designations as α_1-globulins, α_2-globulins, β-globulins, etc., represent groups of proteins, and each group contains several proteins which possess similar mobilities under these experimental conditions. The pH chosen for this plasma analysis was selected in order that all the proteins should have a negative charge. This minimizes the possibility of salt formation between different proteins. At the same time, the pH is sufficiently alkaline to permit migration and separation of the *major* different groups in a reasonable length of time.

Table 31.3. **Electrophoretic Distribution and Mobility of Plasma Protein Components**

(Average values obtained with large pools of normal human plasma and analyzed in sodium diethylbarbiturate (Veronal) buffer at pH 8.6. Amounts given in percentages of total protein and mobilities $\times 10^5$ in square centimeters per volt per second)

	Albumins	α_1-Globulins	α_2-Globulins	β-Globulins	Fibrinogen*	γ-Globulins
Amount....	55.2	5.3	8.7	13.4	6.5	11.0
Mobility....	5.92	4.85	3.87	2.88	2.06	1.15

* The amount of material which migrates under the "fibrinogen" boundary is much greater than the quantity of fibrinogen estimated as protein clottable by thrombin, which is about 4 per cent. The extra 2.5 per cent undoubtedly should be assigned to the γ-globulins.

Source: From S. H. Armstrong, Jr., M. J. E. Budka, and K. C. Morrison, *J. Am. Chem. Soc.*, **69,** 416, 1947.

Although precise measurements of mobility are obtained with the large electrophoretic apparatus, simpler methods are now widely used for separation and quantitation of plasma proteins. In particular, zone electrophoresis on paper or on starch blocks is generally employed for rapid, routine analysis of plasma or serum, yielding values for several of the major protein components. The use of starch gel as a supporting medium increases markedly the resolving power of zone electrophoresis. For example, the 5 conventional zones observed in paper electrophoresis can be separated into approximately 20 zones by starch gel electrophoresis. One additional component which is thus revealed has been termed pre-albumin, because its mobility, at the usual alkaline pH values used for electrophoresis, is greater than that of albumin.

The major protein constituents of the plasma are similar in all vertebrates, although the relative amounts differ considerably. The distribution of the serum proteins of several mammalian species is given in Table 31.4. The content of γ-globulins is higher in the sera of members of other species than in man and the amount of albumin is much lower in horse, cow, and pig sera.

Table 31.4: ELECTROPHORETIC DISTRIBUTION OF PROTEINS IN SOME MAMMALIAN SERA

(The data are given as percentages of the total protein and are averages of many individual determinations)

Animal	Albumins	α-Globulins	β-Globulins	γ-Globulins
Cow	41	13	8	38
Guinea pig	56	14.5	8	21.5
Horse	32	14	24	30
Pig	42	16	16	25
Rabbit	60	7	12	21
Sheep	57	11	7	25

SOURCE: From H. Svensson, *Arkiv. Kemi. Mineral Geol.*, **22A**, No. 10, 1946.

SERUM ALBUMIN

General Properties. Albumin is the smallest and most abundant of the plasma protein constituents, with a molecular weight of 69,000. Its isoelectric point, pH 4.7, is also lower than that of the other major proteins; this and its high net charge explain why the electrophoretic migration of albumin is more rapid than that of the other major plasma proteins at neutral or slightly alkaline pH values (Table 31.3). The high net charge is due to the large number of titratable groups, about 180 per mole (Table 6.1). At the pH of blood, 7.4, albumin has a net negative charge of 18. These properties aid in explaining its extremely high solubility. At pH 7.4, 40 per cent solutions of serum albumin are readily prepared.

The albumin molecule has an asymmetry corresponding to that of an ellipsoid with a diameter of 38 Å. and a length of 150 Å. It is much more symmetrical than γ-globulin or the highly elongated fibrinogen molecules. As a result, solutions of albumin have a smaller viscosity than those of fibrinogen or of the globulins since viscosity is influenced much more by shape than by molecular size. This is important since the work performed by the heart depends in large part on the viscosity of the blood. The viscosity of blood is approximately equal to that of any of the following: twice concentrated plasma, a 25 per cent solution of albumin, a 15 per cent solution of γ-globulin, or a 2 per cent solution of fibrinogen.

Osmotic Effect. The main function of albumin is its important role in osmotic regulation (Chaps. 33 and 35). Of the total osmotic effect of the plasma proteins, albumin is responsible for about 75 to 80 per cent. Although it constitutes slightly more than half the plasma proteins by weight, the effectiveness of albumin is far greater than that of the globulins because of its lower molecular weight. In addition, albumin gives a greater osmotic effect at the pH of the blood than would be expected from the ideal thermodynamic relationship where $\pi v = nRT$ (page 127). At pH 7.4 the 18 negative charges of albumin contribute strongly hydrophilic groups which cause a cluster of water molecules around each charge, thus producing a greater osmotic effect than would be given by a neutral molecule. Theoretically, the osmotic pressure should be linearly proportional to protein concentration, but albumin deviates strongly from this, again giving a greater effect than expected for its concentration in plasma. These two unusual properties contribute to the remarkable effectiveness of serum albumin as a factor in osmotic regulation.

Transport Function. Many substances which are sparingly soluble in water are readily dissolved in the presence of serum or plasma. Albumin solutions possess this property to a marked degree, and in each instance the dissolved substance is actually bound to the protein. This is the case for simple dyes as well as for a variety of substances of physiological interest, such as ions of fatty acids, naphthoquinone derivatives, bilirubin, sulfonamides, and other compounds. Albumin thus plays an important role in the transport of sparingly soluble metabolic products from one tissue to another.

Some of the properties of albumin and other important plasma proteins to be discussed are summarized in Table 31.5.

β- and γ-GLOBULINS

Interest has centered on the β- and γ-globulins of the blood because of the presence in these fractions of *antibodies* or *immune globulins,* most of which are in the γ-globulin fraction. Other β-globulins function in the transport of a variety of substances, including lipids, hormones, and inorganic ions. The properties of some of the β-globulins are given in Table 31.5 and are discussed later in this chapter.

The γ-globulin fraction of blood was originally defined as the plasma protein component which moves most slowly during electrophoresis at alkaline pH. Some of the physical and chemical properties of the γ-globulins are given in Table 31.5. γ-Globulins can be grouped into two major classes on the basis of sedimentation rates in the ultracentrifuge. Of the total number of γ-globulins, 85 to 90 per cent have a sedimentation constant of 7 S, the remainder of 19 S. The 7 S class may be divided into two subclasses which differ in carbohydrate content, amino acid sequences, and immunological properties. However, each of these categories consists of a heterogeneous group of numerous individual proteins. More sensitive immunological techniques, coupled with physical and chemical studies, have provided a more detailed classification and understanding of γ-globulins.

Some Immunological Principles. When foreign proteins, *antigens,* are injected into a suitable animal, production of specific proteins, *antibodies,* results. Although stimulation of antibody production is most commonly observed with proteins, certain polysaccharides such as the capsular components of pneumococci and other microorganisms are also antigenic. Antibodies may combine with the antigen to produce a visible precipitate; hence the term precipitin reaction. Antibodies which are

Component	Estimated amount, g./100 ml.	Sedimentation constant, Svedberg units	Molecular weight	Isoelectric point, pH	Special properties and function
Pre-albumin	0.3	4.1	61,000	....	1.3% Carbohydrate
Albumin	2.8–4.5	4.6	69,000	4.7	Osmotic regulation; transport
Globulins, total	3.0–3.5				
α_1-Globulins	0.3–0.6				
α_1-Globulin (orosomucoid)	0.075	3.1–3.5	41,000	1.8–2.7	40% Carbohydrate
α_1-Globulin (glycoprotein)	0.030	3.5	54,000	....	14% Carbohydrate
α_1-Lipoproteins*					
Density = 1.093	0.05–0.13	5.5	435,000	....	Lipid transport; 67% lipid
Density = 1.149	0.3–0.4	5.0	195,000	5.2	Lipid transport; 43% lipid
Haptoglobin, type 1–1	0.1	4.2	85,000	4.1	Binds hemoglobin; 23% carbohydrate; differing genetic types
Inter-α-globulin†	....	....	45,000	....	Thyroid hormone binding and transport
α_2-Globulins	0.4–0.9				
α_2-Globulin (glycoprotein)	....	2.6	....	3.8	16% Carbohydrate
α_2-Globulin (macroglobulin)	0.2	19.6	820,000	5.4	10% Carbohydrate
Ceruloplasmin	0.03	7.1	150,000	4.4	7% Carbohydrate; copper transport
Prothrombin‡	....	4.8	62,700	4.2	11% Carbohydrate; blood coagulation
β-Globulins	0.6–1.1				
β_1-Lipoproteins:					
Density = 0.98–1.002	0.13–0.20	....	$5–20 \times 10^6$	....	90% Lipid; lipid transport
Density = 1.03	0.20–0.25	....	3.2×10^6	....	79% Lipid; lipid transport
Lipoeuglobulin III, Density = 1.036	....	8.2	3×10^6	5.3	75% Lipid; lipid transport
β_1-Metal-binding globulin (transferrin)	0.40	5.0	85,000	5.9	5.5% Carbohydrate; iron transport
Plasminogen	....	4.7	90,000	5.6	Precursor of plasmin, a fibrinolysin (p. 733)
Antihemophilic globulin	....	....	....	6.4	
Fibrinogen	0.30	7.6	341,000	5.8	3% Carbohydrate; blood clotting (p. 726)
Cold-insoluble globulins (cryoglobulins)	....	15.0			
γ-Globulins	0.7–1.5	7.0	150,000	6.3–7.3	Antibodies
γ_1-Globulins	....	7.0	....	5.8–6.6	2.5% Carbohydrate
γ_1-Macroglobulins	0.05–0.15	19.0	1×10^6	5.1–7.7	10% Carbohydrate
γ_2-Globulins	....	7.0	150,000	7.3–8.2	3% Carbohydrate

* Other data for lipoproteins are presented in Table 21.1, page 475 † Commonly termed thyroxine-binding globulin. ‡ From bovine plasma.

SOURCE: The information in this table is largely taken from summaries by E. J. Cohn and associates, in *J. Am. Chem. Soc.*, **72**, 465, 1950; and from "The Plasma Proteins," F. W. Putnam, ed, vols. I and II, Academic Press, Inc., New York, 1960.

produced to toxins are *antitoxins*. If the antigens are cells, such as erythrocytes of another species, or bacteria, and if clumping of the cells is produced by the antibodies, the latter are *agglutinins*. If the cells are lysed, the antibodies are *lysins*. Each antigen elicits formation of different, specific antibodies.

Plasma plays an important role as the medium of transport for the antibodies in their mobilization for defense against invasion by microorganisms. Antibodies are made by cells of the reticuloendothelial and lymphoid systems. Since normal individuals are exposed during their lives to many different organisms and antigens, plasma contains a large variety of antibodies. Immunity to a virus or bacterium is associated with the presence of specific antibodies to the invading pathogen.

Artificial active immunity may be produced by injecting a nonpathogenic antigen such as killed bacteria, *e.g., Hemophilus pertussis,* the causative organism of whooping cough, or a toxoid made by treating the toxins of the diphtheria or tetanus bacilli with formaldehyde. The injected individual will develop specific antibodies which react with live organisms or untreated toxin and, thus, will possess active immunity. Temporary passive immunity is given by injection of antibodies made by immune individuals of the same species or even of another species. Antitoxins to diphtheria and tetanus toxins from horse plasma are extensively used for treatment of these diseases.

Immunoelectrophoresis. As the term indicates, immunoelectrophoresis makes possible differentiation of proteins in solution on the basis of electrophoretic and immunological properties. Electrophoretic migration is generally performed in agar gel. Immunological differentiation of the separated protein fractions is achieved by adding specific immune serum to a groove in the agar block beside the separated protein fractions. Precipitation lines or rings form within a few hours at the points of contact from diffusion of the electrophoretically separated protein fractions and of the specific immune serum in the trough. The position of these precipitation lines is determined by the electrophoretic mobility, the rate of diffusion, and the serological specificity of each of the proteins present in the solution under study. The number, shape, and intensity of the lines correspond to the incidence, nature, and extent of the precipitin reactions (Fig. 31.2).

In addition to the use of agar in rectangular dishes, an agar supporting medium in glass tubes has also been employed. After electrophoresis the agar cylinder is pushed into a tightly fitting trough previously prepared in an agar plate, and the

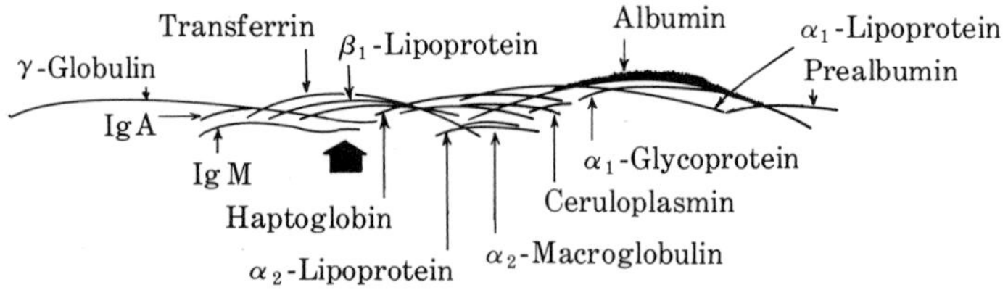

Fig. 31.2. Immunoelectrophoresis of normal human serum developed with a horse antiserum to whole normal serum. The broad vertical arrow indicates the starting point. For properties of proteins and meaning of symbols used to designate some of the proteins, see Tables 31.5, 31.6, and 31.7, pages 711 and 714.

immune serum is placed in a parallel trough to develop precipitin reactions. Similarly, another procedure combines electrophoretic separation on paper with serological testing in agar. Also, starch gel supporting medium for electrophoresis has been combined with modified precipitin gel diffusion techniques. The technique of immunoelectrophoresis has provided additional information regarding the number of serum components, the characteristics and distribution of antibodies, and the nature of γ-globulin.

IMMUNOGLOBULINS

The plasma of a normal, healthy individual who has not recently been ill with an infectious disease contains relatively small amounts of hundreds of distinct antibodies. If an antigen never previously encountered enters the body, in the course of a few days antibodies which react with the antigen appear in the blood. Repeated challenges with antigen result in increasing antibody titer of plasma. Exposure to an antigen which had previously been encountered results in much more rapid appearance, in plasma, of large quantities of antibodies. Countless natural and synthetic antigens have been observed to elicit such a response. Each initiates synthesis of antibodies which appear to react, specifically, with the administered antigen. All such antibodies are proteins. Accordingly, we may pose the following questions: How many distinct proteins serve as antibodies to a single antigen? What is the structure of an antibody? How does it react with an antigen? How does the presence, in the body, of an antigen stimulate the production of specific proteins which react with it, *viz*., specific antibodies? The presently available answers to these questions are meager and largely inadequate.

A given antigen stimulates formation of not one but several antibodies which react with the antigen. In general, the more complex the antigen, *e.g.*, a large protein, the greater the number of different antibodies produced. However, simple antigens also elicit formation of a family of antibodies. Each antibody molecule is bifunctional, *i.e.*, has two combining sites for antigen. If the antigen is also bifunctional, reaction with antibody may produce a precipitin reaction. However, the nature of the reaction cannot be specified other than to state that, since the combining site on the antibody must be provided by amino acid residues, the reaction is analogous to the formation of an enzyme-substrate complex and may involve electrostatic forces, hydrogen bonds, and hydrophobic forces. Since all antigens are relatively large, at least with a minimal size that of a tetrasaccharide or a low molecular weight protein, a large number of antibody-antigen contacts may be postulated. Nevertheless, the details of such a reaction remain to be elucidated and must await description of the three-dimensional structure of a specific antibody.

Much of the difficulty in such studies derives from the heterogeneity of the γ-globulin fraction and, until recently, lack of preparations of purified antibodies which are reasonably homogeneous. Improved procedures have resolved this difficulty in a few instances, thus permitting investigations of antibody structure.

Nomenclature and Structure of Immunoglobulins. The nomenclature of the immunoglobulins is given in Table 31.6. There are three major classes of immunoglobulins, IgG, IgA, and IgM; other, minor classes of immunoglobulins found in human

Table 31.6: NOMENCLATURE OF HUMAN IMMUNOGLOBULINS

Immunoglobulins	Subunits		Molecular formulas
	Light chains	Heavy chains	
IgG (γG)	K, λ	γ	$K_2\gamma_2$; $\lambda_2\gamma_2$
IgGa, etc.	K, λ	γa	$K_2\gamma a_2$; $\lambda_2\gamma a_2$
IgA (γA)	K, λ	α	$K_2\alpha_2$; $\lambda_2\alpha_2$
IgAa, etc.	K, λ	αa	$K_2\alpha a_2$; $\lambda_2\alpha a_2$
IgM (γM)	K, λ	μ	$(K_2\mu_2)_n$; $(\lambda_2\mu_2)_n$, where n = 5–6
IgD	K, λ	δ	$K_2\delta_2$; $\lambda_2\delta_2$
IgE	K, λ	ϵ	$K_2\epsilon_2$; $\lambda_2\epsilon_2$

plasma are designated as IgD, IgE, etc. These have been differentiated by their centrifugal and immunoelectrophoretic behavior and by the occurrence of unusual amounts of each of these classes under specific circumstances. Similar classes are found in other mammals and are probably analogous to those of the human being.

As seen from Table 31.7, most antibodies have a molecular weight of about 150,000; the macroimmunoglobulins appear to be polymers of this structure. Antibodies can be degraded by exposure to agents such as cysteine or thioethylamine to yield only two general types of subunits with molecular weight of about 23,500 and 50,000 to 70,000, respectively. Antibodies also may be degraded by treatment with proteases. Papain cleaves IgG in the presence of cysteine into three fragments, two designated Fab and the other Fc; the three represent more than 85 per cent of the molecule. Each Fab fragment has a molecular weight of 52,000 and that of Fc 48,000; the former retains full combining specificity for the antigen, indicating that the combining site is entirely contained in this fragment of the antibody. Pepsin cleaves IgG into an immunologically active fragment, termed $F(ab')_2$, with a molecular weight of about 100,000. From these findings, each major class can be distinguished from the others on the basis of subunit polypeptide structures. All immunoglobulins appear to have a structure comprised of two types of polypeptide chains which have been termed light (L) and heavy (H). A light chain may be linked by a disulfide bond to a heavy chain to give the basic monomer (LH); most classes of immunoglobulins are in the form of the dimer $(LH)_2$ (Fig. 31.3). However, light and heavy chains can interact in solution to form an immunochemically competent

Table 31.7: SOME PROPERTIES OF THE THREE MAJOR CLASSES OF IMMUNOGLOBULINS

Class or chain	Sedimentation constant	Molecular weight	Carbohydrate content, per cent	Turnover rate*	Distribution
IgG	7	150,000	2–2.5	3	Approx. 80% of total plasma immunoglobulins
IgA	7–8	160,000	8.0	10–13	Plasma; secretions of seromucous glands; milk
IgM	19	750,000	10.0	10–15	Minor constituent human plasma; major constituent shark plasma

* As per cent of total pool per day.

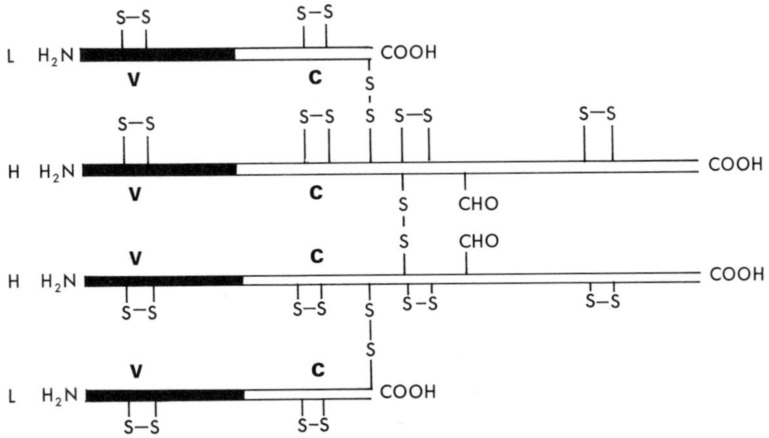

Fig. 31.3. Schematic structure of IgG. L = light chains; H = heavy chains; CHO = carbohydrate unit. The variable and constant portions of the chains, with respect to amino acid sequences, are indicated by the labels **V** and **C**, respectively. It is not known whether V is the same length in H and L chains.

dimer without formation of a covalent bond. In some immunoglobulins, *e.g.*, class IgM, the LH monomer is further polymerized to give $(LH)_{2n}$. This is indicated in Table 31.6.

Several classes of L and H chains have been described. In man, there are only two major types of L chains, K (kappa) and λ (lambda), which may be recognized serologically and by their specific amino acid sequences. All major classes of immunoglobulins may contain either K or λ chains or both; each class possesses an H chain unique for that class. In IgG, the H chain is called γ chain, in IgA, α, and in IgM, μ; the heavy chains in the minor classes of immunoglobulins, IgD and IgE, have been designated as δ (delta) and ϵ (epsilon), respectively. Although the H chains of IgG are all of the γ series, they are not identical. At least 10 similar but distinct γ chains have been noted, designated as γa, γb, γc, etc., and, presumably there are many more. Similarly there are numerous distinct H chains of the α, μ, δ, and ϵ series.

Some properties of the immunoglobulins are given in Table 31.7, and of their light and heavy chains in Table 31.8. The two light and the two heavy chains in each immunoglobulin are linked by two types of interchain disulfide bonds, one H—H disulfide and two H—L disulfide. In addition, there are 12 intrachain disulfide bonds. The carbohydrate of the molecule is present in the heavy chain (Fig. 31.3). This structure has been established for IgG; IgA probably has this structure also. IgM appears to have additional disulfide bonds, allowing five to six four-chained units of the type shown in Fig. 31.3 to be linked together.

Table 31.8: Some Properties of the Light and Heavy Chains of Immunoglobulins

Chain	Molecular weight	No. of residues	Carbohydrate content
K or λ	23,500	212–214	Not present
γ	48,000	420	All the carbohydrate of parent molecule (Table 31.7)
	50,000	420	All the carbohydrate of parent molecule (Table 31.7)
μ	70,000	600	All the carbohydrate of parent molecule (Table 31.7)

Each subunit, *i.e.*, light and heavy chains, is a single polypeptide chain. The C-terminal half of the L chains appears to be quite constant in amino acid sequence (**C**, Fig. 31.3) among all members of a major class such as IgG, as is the C-terminal three-quarters of the H chain. It is this constancy which results in the general similarities that permit their grouping as a class. In addition, the constant regions of the L and H chains of the major classes show subregions of obvious homology, approximately half of all positions being constant in all classes, indicating their common evolutionary origin.

The N-terminal half of each L chain, comprising 105 to 107 amino acids, and the N-terminal quarter of each H chain, show great variations in sequence (**V**, Fig. 31.3), and it is this variation which makes possible the large numbers of specific antibodies. These relationships are shown in Fig. 31.3. Isolated preparations of L chains do not appear to combine with antigen; H chains do react with antigen but not so tightly nor so readily as does complete antibody. Thus the H chain must make the major contribution to the antigen-combining sites.

Studies of the glycopeptides isolated from partial enzymic digestion of several types of immunoglobulins indicate that one mode of attachment of carbohydrate to the polypeptide chain is via the amide nitrogen of an asparagine residue to C-1 of N-acetylglucosamine. In addition, D-mannose, D-galactose, L-fucose, and D-N-acetylneuraminic acid residues are also components of the carbohydrate unit. There are two carbohydrate units per mole of IgG and apparently three per mole of IgM.

Mechanism of Synthesis of Immunoglobulins. Since all antibodies have the same molecular weight and general structural features, it was conceivable that specificity for antigen resided not in differences in amino acid sequence but in secondary and tertiary structure, *i.e.*, the positioning of disulfide bonds or the final conformation. This possibility was eliminated by the observation that after exposure to 4*M* guanidine in the presence of thioethylamine, removal of these reagents and reoxidation resulted in recovery of almost half of the antigen-binding capacity. Thus the three-dimensional structure, like that for ribonuclease (page 256), is the consequence of amino acid sequence. The production of specific antibodies in an animal exposed to a given antigen, therefore, represents production of immunoglobulin of the right class and with an appropriate amino acid sequence at the N-terminal region of the L and H chains.

Comparison of the amino acid sequences of a limited number of L chains reveals that at almost each position in which there occurs an amino acid replacement, it can be accounted for as the consequence of a single base change in the coding triplet (page 674). The problem then is whether, in some manner, antigen triggers production of L and H chains already encoded in the DNA but unexpressed in the absence of antigen, or whether there occurs in the immunoglobulin-producing cells an appropriate set of somatic mutations which are selected by virtue of the presence of antigen. This question has not been resolved.

Available knowledge concerning the synthesis of immunoglobulins may conveniently be considered in two stages: (1) the reactions initiated by the antigen, and (2) the synthesis of the globulin at the ribosomal level.

Role of the Antigen. Most of the available data suggests that the antigen does not play an immediate and direct role in determining the structure of the antibody, although there are opinions to the contrary. Antigen is removed from plasma by phagocytic macrophages; the latter then react with small lymphocytes in which antibody synthesis actually occurs. Each lymphocyte makes only one species of H and L chains. The immunological competence of the small lymphocyte is, in some unexplained manner, dependent upon the thymus (page 754). Apparently macrophages transfer to these cells a soluble material which is more effective than is the administered antigen. A soluble fraction from such macrophages which induced antibody synthesis in recipient lymphoid cells in tissue culture was rendered ineffective in this regard by treatment with ribonuclease, strongly suggesting an RNA-like structure for the antibody-provoking material. Such an effective RNA has also been extracted from spleen tissue of immunized mice. This suggests transfer of information from cells which have phagocytized antigen to the immunologically competent or committed lymphocyte. At this time the relationship between phagocytized antigen and the appearance of a seemingly specific form of RNA, and the subsequent role in the lymphocyte of that RNA, are not understood.

Role of the Ribosomes. The synthesis of antibody protein is accomplished by the general mechanism of protein synthesis (Chap. 29). The syntheses of the L and H chains are conducted by two different types of polyribosomes; these can be separated by density centrifugation. Polyribosomes which synthesize L chains sediment more slowly than those for H chains, although both classes of polyribosomes are of sufficient size to contain a mRNA coding for complete heavy chains. Studies of the rate of antibody synthesis suggest that the L chain is completed first and released from the polyribosomes to give a small pool of free chains which combine with partially synthesized H chains. Before H chains can be released from their polyribosomes, they must be combined with L chains. Carbohydrate is subsequently attached to the H chains. The prior requirement for combination of H with L chains for release of the former from the site of synthesis may afford a regulatory mechanism for the rate of synthesis of completed antibody. An analogous requirement obtains in hemoglobin synthesis in which the pool of α chains controls the rate of β chain release and thus of hemoglobin formation.

Variations in Globulin Production. In certain diseases there is accentuated production of immunoglobulins. These globulins are abnormal only in their quantity, rather than being new or different proteins. Indeed, in these diseases the protein produced is more homogeneous electrophoretically than is the protein of the same class in the normal individual. In *macroglobulinemia,* synthesis of large quantities of a protein of the IgM class occurs. In *multiple myeloma* the pathological globulin is of the IgA class. A globulin, first described by Bence-Jones in 1848, which appears in large amounts in the urine of patients with multiple myeloma, exhibits unusual solubility properties. The protein precipitates at 50 to 60°C., but raising the temperature to near the boiling point causes the precipitate to dissolve. Bence-Jones proteins are identical with the light chains of normal immunoglobulins and may be of either K type or λ type. Their appearance in urine results from an unbalanced synthesis of IgG (or IgA, IgM) in lymphoid cells proliferating at an abnormal rate, with L chains produced in excess of H chains. The L chains produced in excess are excreted in the urine and are identical to the chains in the circulating myeloma protein of the blood.

The amino acid sequences have been established for some type K Bence-Jones proteins. Comparisons of these from two patients revealed that each had 214 amino acid residues; the entire C terminal halves from residues 107 to 214 were identical except for substitution of valine at position 189 for a leucine residue. About 10 sequence variations occurred between residues 1 to 106, the variable portion of L chains. A λ type Bence-Jones protein contained 213 residues. As expected, K and λ differed from one another in both halves of the molecule. Similar regions of variation and constancy were disclosed from sequence studies of the L chains of type K IgM macroglobulins, which had structures similar to and, in certain regions, identical with those of type K Bence-Jones proteins.

Although complete amino acid sequences of heavy chains of immunoglobulins are not available, the data suggest that the heavy chains of IgG possess a considerable amount of amino acid sequence from the carboxyl terminus of the **C** portion (Fig. 31.3) which may be largely invariant among all antibodies of the same species. This appears to be the case for approximately 55 per cent of the total residues. Also, a considerable degree of similarity was observed in the sequences in this portion of the heavy chains of rabbit origin and the Bence-Jones proteins of human origin. However, differences in amino acid sequences have been disclosed in the H chains of IgM macroglobulins from individual patients.

Evolution of the Immunoglobulins. The information presently available shows that the K and λ chains possess many similarities in amino acid sequences. It has been postulated that these two types of light chains were probably derived from a more primitive type of L chain by gene duplication and subsequent independent evolution. Indeed, it now appears that the variable and constant halves of the L chains manifest similarities in sequence and thus may have been derived from a precursor of L chain with approximately 110 residues (Fig. 31.4). Resemblances between L and H chains are also evident, and it has been suggested that gene duplication and doubling of size occurred in the formation of H chains from L chains. These views as to the evolutionary development of immunoglobulins are supported by the fact that the L and H chains can be divided approximately into six equivalent segments, two in the L chain and four in the H chain, all six possibly being derived from a common ancestral segment. These relationships are shown in Fig. 31.4.

Antibodies in the Newborn. Newborn mammals do not appear to be able to make antibodies. However, physiological mechanisms exist for creating a temporary passive immunity in the young which permits them to resist infection until the time when the mechanism of antibody synthesis has developed. In some mammals, including man, the plasma proteins of the newborn show an electrophoretic pattern similar to that of the maternal plasma. Moreover, many specific antibodies present in maternal plasma are demonstrable at birth in the blood of the young. For these species passive immunity is achieved by transfer of antibodies from the maternal blood through the placenta to the blood of the fetus.

There is a rare, hereditary condition in which human beings lack the ability to form γ-globulins (*agammaglobulinemia,* page 700) and such proteins are lacking in plasma. These individuals are particularly susceptible to infectious agents. γ-Globulin may be administered at intervals as a therapeutic measure.

In the Ungulata—cow, horse, sheep, goat, etc.—γ-globulins are absent in the

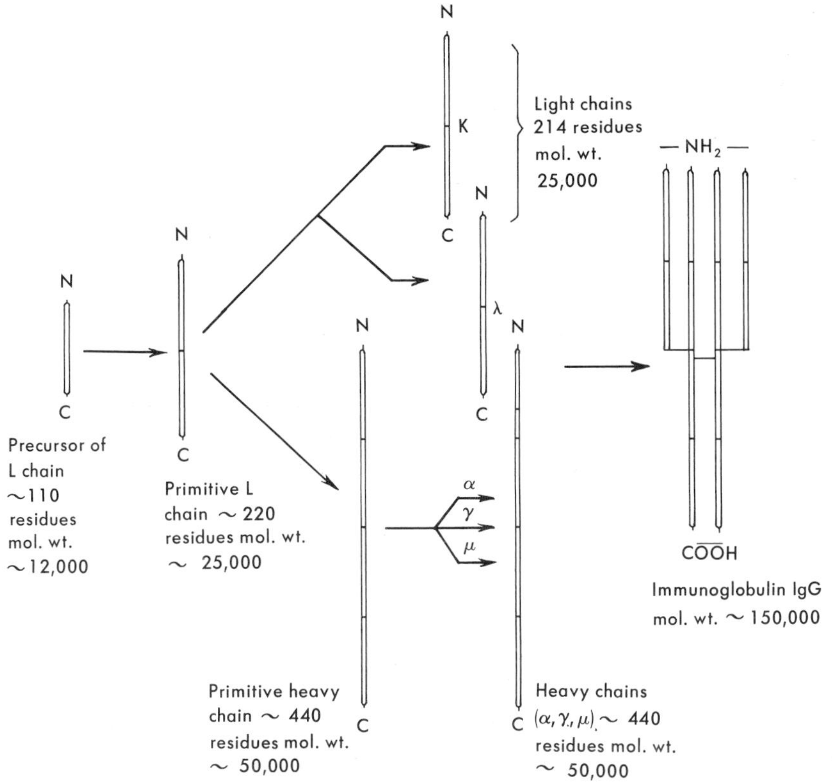

Fig. 31.4. Postulated scheme for the evolution of the immunoglobulins. N and C indicate amino and carboxyl end, respectively, of polypeptide chains. (*From R. L. Hill, R. Delaney, R. E. Fellows, Jr., and H. E. Lebovitz, Proc. Natl. Acad. Sci. U.S.*, **56**, 1762, 1966.)

serum of the newborn, and, correspondingly, no antibodies can be detected in the blood. Antibodies appear in the plasma after the young of these species have suckled and received the first milk, or colostrum. Colostrum may contain as high as 20 per cent protein, five times the amount present in milk, and the predominant fraction is immunoglobulins. In the Ungulata, the placental barrier does not permit the passage of the large antibody molecules from the maternal to the fetal circulation, whereas the newborn can absorb these antibodies from the intestine into its blood stream. Antibodies are also present in human colostrum, but they play a secondary role since the placental transmission is more important.

Newborn ungulates also possess an unusual α-globulin, *fetuin,* with a molecular weight of 40,000; it is present in large amounts in the plasma of the fetus and the young calf or pig and its content diminishes rapidly in the older animal. Fetuin is a mucoprotein containing 22 per cent carbohydrate.

THE LIPOPROTEINS

Except for the small quantity of fatty acids present as fatty acid–albumin complexes (page 476), the remainder of the plasma lipid is present as *lipoproteins.* These migrate electrophoretically with both the α- and β-globulins (Table 31.5, page 711). Lipoprotein fractions can be separated by density gradient centrifugation and are characterized by their flotation constants (page 475). The lipoproteins have been considered previously with regard to their lipid composition (Table 21.1, page 475) and their role in lipid metabolism (page 475). The lipoproteins present in serum range in molecular weight from 200,000 to 10,000,000 and contain from 4 to 95 per cent lipid. It should be stressed that the description of the classes of lipoproteins is based on empirically devised physical procedures, and that each class represents not a single molecular species but rather multiple species with similar physical properties. However, the broad chemical and structural interrelationships of lipoproteins within classes have permitted simplification of their characterization.

In some pathological conditions, the β_1-lipoprotein (low-density) is often present in increased amount. This is more readily detected by density gradient ultracentrifugal studies than by electrophoresis, since the electrophoretic β-globulin fraction is a mixture of lipoprotein and many other components. Regimens designed to lower the β- and elevate the α-lipoproteins (high-density) are based on the thesis that a lowering of the former, with its accompanying cholesterol, is a desirable prophylactic and therapeutic goal (Chap. 48).

A hereditary disorder characterized by a complete absence of plasma β-lipoproteins or by the presence of a reduced amount of low-density lipoproteins has been termed *abetalipoproteinemia* (page 528). It has also been described as *acanthocytosis,* since more than 80 per cent of the erythrocytes are *acanthocytes, i.e.*, are spherical and have numerous projecting spines and spicules (Gk. *akantha,* a thorn). Plasma lipid fractions are markedly lower than normal.

THE MUCOPROTEINS

Mucoproteins of high carbohydrate content have been detected in the plasma filtrate remaining after precipitation of the other proteins by addition of trichloroacetic acid or perchloric acid. Mucoproteins are not coagulated by heat. In Veronal buffer at pH 8.6, they migrate with the α-globulins (Table 31.5), but in acetate buffer at pH 4.0 they retain their negative charge, unlike the other plasma proteins, which possess more alkaline isoelectric points.

Orosomucoid, an acidic α_1-globulin, has been obtained in crystalline form from serum and from nephrotic urine. This protein contains about 40 per cent carbohydrate, including galactose, mannose, fucose, N-acetylglucosamine, and N-acetylneuraminic acid. Proteolytic digests of orosomucoid have yielded glycopeptides in which the β-amide of asparagine appears to form the linkage to the carbohydrate unit via N-acetylglucosamine. A small proportion of linkages to carbohydrate through the γ-carboxyl of glutamic acid may also be present.

Haptoglobins are α_2-globulins which can combine with hemoglobin to form a weak peroxidase; they comprise about one-fourth of the α_2-globulin fraction of human plasma. When hemoglobin is injected or liberated by hemolysis, it combines with haptoglobins until the capacity of the latter is exceeded. In hemolytic conditions and in acute hepatitis the plasma level of haptoglobins is diminished. A deficiency of haptoglobins, *anhaptoglobinemia,* occurs in individuals with a markedly shortened life span of erythrocytes and in pernicious anemia.

All haptoglobins appear to have a similar amount and proportion of diverse carbohydrates, including N-acetylneuraminic acid and galactose. Starch gel electrophoretic studies have shown that for each individual, the haptoglobins fall into one of three genetic groups, each of which in turn is a mixture of haptoglobins. However, all three types react equally well and identically with antibodies to each of the others. The genetic types have been termed haptoglobin 1-1, 2-2, and 2-1, respectively. Treatment of purified haptoglobin with reducing agents yields two polypeptide chains, designated as α and β. Only the α chain varies in the three phenotypes. The haptoglobin 1-1 is a single molecular species, molecular weight 80,000, and is composed of two α^1 polypeptides each of 9,000 molecular weight and one β chain, molecular weight 65,000. Each β chain has two binding sites for the α^1 unit, and each of the latter has a single site of combination with the β unit. The fusion of two α^1 units to form an α^2 unit provides a species with two sites per molecule, each capable of binding with a β unit. Thus, the haptoglobin 2-1 phenotype consists of a polymeric series of structure $(\beta 2\alpha)_M$, where $M = 2, 4, 6$, etc., while the haptoglobin 2-2 components are equivalent to $(\beta\alpha^2)_M$, with $M = 3, 5, 7$, etc.

In many metabolic disorders there is a pronounced elevation in the number of α-globulins estimated electrophoretically. This is largely because of an increase in mucoproteins. Infectious diseases, *e.g.*, pneumonia, tuberculosis, and acute rheumatic fever, and general disturbances such as cancer with metastases produce these increases, and they are frequently associated with a concurrent increase in fibrinogen and a decrease in albumin.

THE METAL-BINDING PROTEINS

A crystalline β_1-globulin, termed *transferrin,* capable of combining with iron, copper, and zinc has been isolated from plasma. This protein constitutes approximately 3 per cent of the total plasma protein and has a molecular weight of about 85,000. It contains about 5.5 per cent carbohydrate as N-acetylneuraminic acid, mannose, galactose, and N-acetylglucosamine. The protein binds two atoms of Fe^{+++} per molecule of protein but only in the presence of CO_2. The iron complex dissociates below pH 7.0. The prime function of transferrin is to transport iron. In normal individuals the circulating transferrin is one-third saturated with iron. In iron deficiency and pregnancy, there is a striking increase in the plasma concentration of transferrin. In some disease states, *e.g.*, chronic infection, liver disease, and pernicious anemia, there is a reduction in the plasma concentration of this protein. Genetic variants of transferrin are known among the human population; the common form has been designated transferrin C, and others as B, D, etc. A few cases of hereditary absence of transferrin have been described.

The interaction of copper with transferrin differs from that of iron in that at pH 7, where the protein has its maximal binding capacity for iron, its capacity to bind copper is only half maximal. On addition of iron the copper is almost entirely displaced.

A blue protein containing 0.34 per cent copper, *ceruloplasmin,* has been isolated from human plasma in crystalline form. This protein specifically binds copper and contains practically all the copper in serum; it is an α_2-globulin of molecular weight about 150,000, with 7.5 per cent carbohydrate. The 8 atoms of copper per molecule are equally distributed between Cu^+ and Cu^{++}. Dialysis against a chelating agent removes four atoms of copper, leaving a colorless protein; prolonged dialysis yields the apoprotein free of Cu. Six of the copper atoms can exchange with Cu^+ in the medium.

Ceruloplasmin can be reduced under anaerobic conditions to a colorless compound; this is completely reversible in the presence of oxygen. In the rare inherited *Wilson's disease,* which is primarily a disorder of copper metabolism, plasma ceruloplasmin is markedly reduced and copper levels increase in the liver and brain. The disorder is associated with neurological changes and liver damage. The function of ceruloplasmin is unknown, but it probably functions by reversibly binding and releasing copper at various sites in the body, thereby regulating the utilization of copper.

SYNTHESIS OF PLASMA PROTEINS

The liver is the major site of synthesis of plasma proteins. Albumin and fibrinogen formation appears to be limited to the liver; approximately 80 per cent of the total globulins also are fabricated in the liver, including the lipoproteins. The remainder of the major plasma proteins, including the γ-globulin fraction which contains most of the antibodies, are synthesized in extrahepatic tissues. Thus, γ-globulins are formed in lymphoid tissue and in the widely distributed cells of the reticuloendothelial system, notably in the spleen. Also, the specialized plasma proteins present in low concentrations, *e.g.*, protein hormones and enzymes, are produced by various tissues.

The prime role of the liver in plasma protein synthesis is reflected in the marked decrease in albumin and fibrinogen resulting from damage to hepatic tissue, seen in patients with cirrhosis, or following experimental hepatectomy. Prolonged limitation of protein intake not only diminishes the serum albumin, but depletes liver protein, with histological alteration of hepatic cells.

Studies with the isolated perfused rat liver revealed a rate of albumin synthesis of 10 to 20 mg. per hr. This would permit a daily turnover of approximately 25 per cent of the total circulating plasma protein; this value is greater than that obtained for the half-life of plasma proteins in man (page 557) but agrees with data on intact rats, which have a higher rate of metabolism than man. The rapid rate of plasma protein formation is seen in the replacement of serum albumin lost in nephritis and nephrosis. Patients may excrete 10 to 20 g. of protein daily for several months, yet in some instances no severe alteration in serum albumin concentration is evident.

The rapid turnover of plasma proteins, and the fall in serum albumin concentration caused by restricted protein intake, indicate a continuing removal of plasma proteins from the circulation and utilization by tissues. Intravenously administered albumin is efficiently used for growth in experimental animals, and for restoration of blood volume and tissue repair in man. The liver and kidney appear to play a major role in the utilization of circulating protein, as is also the case for blood amino acids. This is probably a reflection, in part, of the more rapid turnover of proteins of these organs. Incorporation of amino acids administered as serum albumin into characteristic cellular proteins has been demonstrated. Presumably, proteins, *e.g.*, serum albumin, entering cells by a mechanism involving pinocytosis, are degraded by intracellular cathepsins, with the liberated amino acids becoming available for synthesis of new protein molecules. A surprisingly large and temporary storage of intravenously administered serum albumin has been observed in the subcutaneous tissues of the rabbit.

ENZYMES OF PLASMA

The enzymes present in plasma are presumably derived from the normal dissolution of the cells of blood and other tissues. It is unlikely that plasma enzymes play specific metabolic roles in plasma, with the exception of the enzymes concerned in blood coagulation (page 724*ff.*) and removal of intravascular blood clots (page 733). The activity of certain enzymes in plasma is frequently a useful index of abnormal conditions in the tissues. Thus, *serum amylase* is elevated in cases of acute pancreatitis. The level in plasma of *acid phosphatase,* measured at pH 6, becomes very high in cases of prostatic cancer and decreases if therapy is effective. The *alkaline phosphatase* activity of blood, estimated at pH 9.0, is markedly increased in many bone diseases. In rapidly healing rickets and other conditions of rapid bone regeneration, high alkaline phosphatase levels are found; however, high alkaline phosphatase is also found in cases of hepatic obstruction.

In individuals with recent myocardial damage, the plasma level of certain enzymes, derived from cardiac tissue, is markedly increased. Estimation of *glutamic-aspartic transaminase, lactic acid dehydrogenase,* and other enzymes has proved to be of some value as diagnostic and prognostic aids for assessing cardiac damage. Elevation of the plasma level of these and certain other enzymes, *e.g.*, aldolase, also occurs in liver disease.

ABNORMAL DISTRIBUTION OF PLASMA PROTEINS

It is important to distinguish those conditions in which disturbances of the levels of the normal proteins occur and those conditions in which abnormal proteins appear in the plasma.

The normal levels of the major plasma proteins estimated electrophoretically have been given in Table 31.3. A general effect of disease, particularly when malnutrition or marked wasting occurs, is a decrease of the serum albumin level. In *nephrosis,* there is loss of albumin in the urine, and in *cirrhosis* of the liver, there is impairment of albumin synthesis. In these same circumstances, there is usually some elevation of the globulins. These changes occur in so many conditions that they are seldom useful for diagnostic purposes, but are of value in prognosis since the changes over a period of time may indicate the severity or trend of the disorder. Thus in some cases of hepatic disease, albumin may be as low as 0.3 g. per 100 ml. as compared with about 4.0 g. per 100 ml. in normal plasma. In *analbuminemia,* a hereditary disorder, synthesis of plasma albumin is impaired.

In *cirrhosis* of the liver, two- to threefold increases in γ-globulins and some increases in β-globulins may accompany the marked decrease in albumin. In many infectious disorders, increases in γ-globulins also take place, although these may be small and not specific. In *kala-azar, lymphogranuloma venereum,* and *sarcoidosis,* γ-globulin increases may be so large as to provide valuable diagnostic confirmation and useful aids in evaluating treatment. In these conditions, a marked elevation of total plasma protein may occur, with γ-globulins representing more than half the total protein. Increases in normal γ-globulins have been considered previously (page 717) in relation to Bence-Jones protein, multiple myeloma globulins, and macroglobulins.

Figure 31.5 shows the electrophoretic patterns found in some abnormal serum samples. The changes from the normal are described in the legend and indicate the usefulness of such estimations when correlated with clinical findings.

Cryoglobulins. Cryoglobulins is the name given to a group of rare serum globulins that have in common the unique property of precipitating, gelling, or even crystallizing spontaneously from a solution or serum which is cooled. Cryoglobulins occur in multiple myeloma, though relatively infrequently. They have also been found occasionally in rheumatoid arthritis. The cryoglobulins which have been studied have molecular weights similar to that of normal γ-globulin. Cryoglobulins have been found in normal sera in small amount after concentrating certain globulin fractions (Table 31.5).

C-Reactive Protein. A protein not present normally in blood may be present in the acute phase of certain infections, including rheumatic fever. Although the protein is nonspecific with respect to the inciting agent of the disease, it is detectable by a specific reaction between serum and the pneumococcal type C somatic polysaccharide in the presence of Ca^{++}. The C-reactive protein has been obtained in crystalline form; on electrophoresis it has the mobility of a β-globulin (free boundary electrophoresis) or a γ-globulin (starch zone electrophoresis). The isoelectric point is at pH 4.8, and the S value is 7.5. Since the titer of C-reactive protein in the serum is maximal in the active stage of infection and rapidly decreases during convalescence, the estimation of C-reactive protein is of considerable prognostic value.

From the foregoing, it is evident that studies of plasma protein distribution provide useful aids in clinical medicine. In certain conditions, such as multiple myeloma, nephrosis, cirrhosis, and lymphogranuloma venereum, the changes are so striking that plasma protein determinations are an essential aid in diagnosis. In many other conditions, the study of the plasma proteins is a useful adjunct in evaluating prognosis.

BLOOD CLOTTING

The major *chemical* defense against blood loss is the formation of the blood clot. Normal human blood will clot in 5 to 8 min. at 37°C. This poses the primary problems of coagulation: What factors are responsible for prevention of intravascular clotting, what changes are set in motion by removing blood from the vessels, and what substances are responsible for the clotting process?

The simplest scheme of coagulation was proposed in 1903 by Morawitz. The colorless protein mainly responsible for the coagulum is *fibrin,* which is formed from its soluble precursor, *fibrinogen.* The transformation is catalyzed by an enzyme called *thrombin.* Thrombin itself is not present in normal blood but is generated from its inactive zymogen, *prothrombin.* Conversion occurs only in the presence of Ca^{++} ions

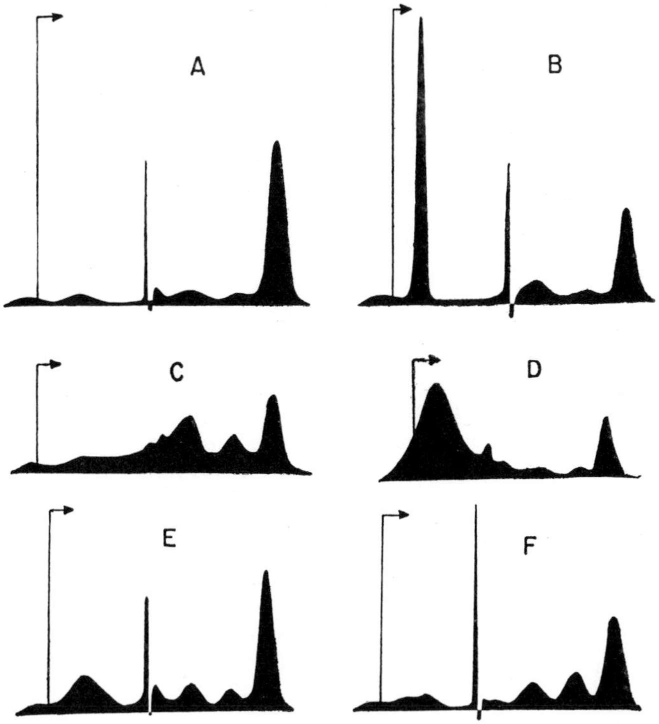

Fig. 31.5. Electrophoretic patterns of human sera. *A* is pooled serum obtained from normal individuals. *B* is from a patient with multiple myeloma; in this serum there is an abnormally large amount of γ-globulin of the class IgA. *C* is from an individual with nephrosis; the total protein is low, particularly albumin, and there are increased amounts of α_1- and α_2-globulins. *D* is the serum from a patient with cirrhosis of the liver; the albumin and total protein are very low, and the γ-globulin is elevated. *E* is from a patient with chronic rheumatoid arthritis; the γ-globulin and α-globulins are higher than found normally. *F* is from a patient with Hodgkin's disease; the α-globulins (including mucoproteins) are high, and the albumin and γ-globulin are abnormally low.

and another protein called *thromboplastin.* This minimal outline of the major changes may be formulated as two enzymic reactions.

$$\textbf{Prothrombin} \xrightarrow{Ca^{++},\ \text{thromboplastin}} \textbf{thrombin}$$

$$\textbf{Fibrinogen} \xrightarrow{\text{thrombin}} \textbf{fibrin}$$

The participation of Ca^{++} in the coagulation mechanism is easily demonstrated since clotting can be prevented by collection of blood in the presence of decalcifying agents such as oxalate, fluoride, or citrate. Use of nontoxic citrates for this purpose made possible development of large-scale preservation of whole blood and plasma for transfusions. Various pyrophosphates, metaphosphates, and chelating agents, *e.g.*, ethylenediaminetetraacetate (EDTA), which bind Ca^{++} have also been used as anticoagulants. Preservation of whole blood has been achieved by removal of Ca^{++} ions by ion exchange resins.

FIBRINOGEN AND FIBRIN

Human plasma contains about 0.3 g. of fibrinogen per 100 ml. Fibrinogen is produced in the liver. Normally, the regeneration of fibrinogen is rapid in animals depleted of this protein. Complete (*congenital afibrinogenemia*) and partial (*congenital hypofibrinogenemia*) absence of fibrinogen have been observed in a few infants. These rare diseases are characterized by a severe hemorrhagic tendency from birth and are usually fatal in early life.

Fibrinogen can be isolated by salting out in the cold by half saturation with sodium chloride or quarter saturation with ammonium sulfate. Precipitation from human plasma at 0 to $-3°C$. with 8 to 10 per cent ethanol gives a fibrinogen preparation which is about 65 per cent clottable protein; repetition of the procedure gives materials which are 90 to 95 per cent coagulable with thrombin.

The fibrinogen molecule is an elongated ellipsoid, about twenty times greater in length than in cross-sectional diameter. Some additional physical properties are given in Table 31.5 (page 711). The native fibrinogen molecule is a dimer consisting of three pairs of polypeptide chains, probably linked by disulfide bridges. The presence of carbohydrate in the molecule has some significance in the clotting mechanism (see below).

The clotting of fibrinogen is normally accomplished by thrombin, but this can also be effected by certain other enzymes, papain in particular. Trypsin does not clot fibrinogen, although it will digest fibrin clots. Certain snake venoms can cause clotting, and this is ascribed to the presence of proteolytic enzymes. Many microorganisms, notably strains of staphylococci, can also coagulate fibrinogen.

In contrast to fibrinogen, fibrin is insoluble in salt solutions. Microscopic examination of fibrin shows a fine reticulum of fibers which exhibit birefringence under polarized light. In the conversion of fibrinogen to fibrin, at least three steps may be delineated: (1) proteolysis of fibrinogen; (2) physical aggregation of fibrin monomer to form a soft clot; and (3) an enzymic process which, in the presence of Ca^{++}, results in formation of a hard clot.

Proteolysis of Fibrinogen. The action of thrombin on fibrinogen (molecular weight, 330,000) results in liberation of two peptides, termed fibrinopeptides A and B, which together have an aggregate molecular weight of approximately 9,000. Peptide A contains 18 amino acid residues, while B has 20, including tyrosine-O-sulfate. Also, approximately 18 per cent of the total sialic acid (0.6 per cent) present in fibrinogen is liberated in this conversion, as are small amounts of the other carbohydrates (mannose, galactose, and N-acetylglucosamine) present in fibrinogen. The sialic acid residues of fibrinogen can also be removed by the action of neuraminidase (page 55), with a simultaneous facilitation of clot formation. These observations suggest that the sialic acid in fibrinogen in some manner influences the action of thrombin on fibrinogen.

The protein remaining after thrombin action, *viz.*, fibrin monomer, contains four new amino-terminal glycine residues, whereas each of the liberated fibrinopeptides contains carboxyl-terminal arginine. These observations, together with studies of the action of thrombin on peptides and on esters, indicate that thrombin splits specifically arginyl bonds. Other enzymes, *e.g.*, trypsin and papain, can also catalyze hydrolysis of these peptide bonds, but in these instances, general and more widespread proteolysis of fibrin occurs. The esterase activity of thrombin has led to a method for estimation of thrombin activity of plasma, with an N-substituted arginine methyl ester as substrate.

Fibrin Monomer Aggregation; Formation of Soft Clot. The fibrin monomers resulting from the action of thrombin on fibrinogen undergo polymerization, possibly as a consequence of electrostatic attraction or of hydrogen bonding between groups unmasked by the removal of the peptides. Polymerization occurs in stages and depends on factors such as pH and ionic strength, but is independent of the presence of thrombin. Initially, end-to-end polymerization occurs with formation of primitive fibrils, followed by a side-to-side polymerization of these fibrils to form coarser fibrin strands.

The hydrolysis of fibrinogen by thrombin in the presence of a chelating agent results in the formation chiefly of a *soluble fibrin.* Clot formation, if it occurs, leads to a *soft clot,* and the fibrin present can be solubilized by buffer at pH values below 4.5 and above 9 and by 1M urea at pH 8.

Formation of Hard Clot. The formation of highly insoluble fibrin occurs in the presence of calcium and an enzyme, *fibrinase,* also designated as *fibrin stabilizing factor.* Fibrinase catalyzes a transamidation (transpeptidation) reaction that results in cross-linking γ-carboxyl groups of glutamic acid residues to ϵ-amino groups of lysine residues. The action is similar to that of *transglutaminase* (page 579). The covalent bonds between peptide chains converts the fibrin into its final, insoluble form, or hard clot. During this process, carbohydrate and ammonia are liberated. The role of carbohydrate is unknown, but clots formed from fibrinogen, previously freed of sialic acid by treatment with neuraminidase (page 55), remain soluble.

The three steps which have been described for the conversion of fibrinogen to fibrin may be summarized as follows.

(*a*) Proteolysis: $\textbf{Fibrinogen} \xrightleftharpoons{\text{thrombin}} \textbf{fibrin monomer } (f) + \textbf{fibrinopeptides A and B}$

(*b*) Polymerization: $nf \rightleftharpoons f_n$

(*c*) Clotting: $mf_n \rightleftharpoons \textbf{fibrin}$

In the above, f_n designates intermediate polymer, and n and m are variable numbers.

PROTHROMBIN AND THROMBIN

Normal plasma contains prothrombin, an enzymically inactive precursor of thrombin. Since intravenous injection of thrombin produces prompt clotting, it is evident that blood normally remains fluid in the vessels because of the absence of thrombin. However, the mechanism for preventing thrombin formation intravascularly has not been completely elucidated.

Prothrombin, prepared from human and other mammalian plasmas, is a glycoprotein containing about 4 to 5 per cent carbohydrate as hexose and glucosamine. Prothrombin has the electrophoretic mobility of an α_2-globulin and an isoelectric point at pH 4.2; the molecular weight is approximately 65,000. When dissolved in 25 per cent sodium citrate solution, prothrombin slowly dissociates into three smaller units with molecular weights of about 10,000, 15,000, and 35,000. The phenomenon is associated with the appearance of thrombin activity.

Thrombin is much less stable than prothrombin and has a similar molecular weight. "Citrate-activated thrombin" (see above) has a molecular weight of 30,000 to 35,000. Thrombin has solubility properties of a globulin; its enzymic properties and specificity were described previously (page 726*ff.*). The nature of the conversion of prothrombin to thrombin is considered below.

Plasma prothrombin deficiency (*hypoprothrombinemia*) occurs frequently in obstructive jaundice and in other liver disorders. Hepatectomy in animals has also shown the importance of the liver in the synthesis of prothrombin since blood prothrombin values decline following liver removal. Normal hepatic production of prothrombin is dependent on adequate nutritional intake of vitamin K (*Koagulation-Vitamin*) (Chap. 50). The role of this vitamin is unclear; it does not participate in the coagulation mechanism itself, and it is not a part of the prothrombin molecule.

A hemorrhagic disease of cattle is caused by Dicumarol, a derivative of coumarin, the sweet-smelling substance of clover.

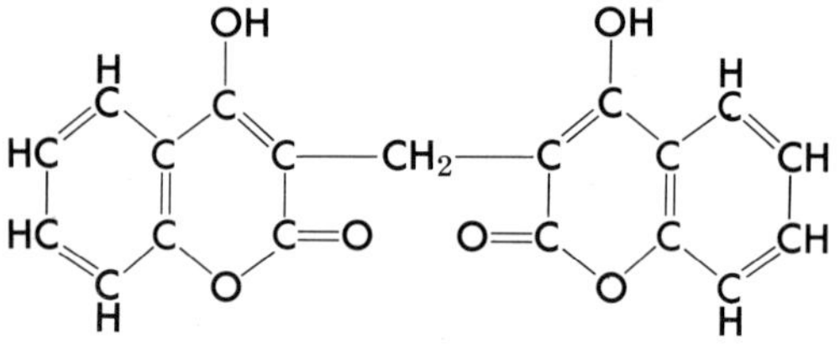

Dicumarol
[3,3′-Methylenebis-(4-hydroxycoumarin)]

Dicumarol fed to experimental animals or man causes a decrease of plasma prothrombin and is used clinically in cases of threatened thrombosis to reduce the clot-forming tendency. Dicumarol acts at the site of prothrombin formation in the liver and can be partially counteracted by feeding vitamin K. The apparent similarity in structure of the fused ring systems of vitamin K and Dicumarol suggests that a competitive metabolic effect may be involved.

THROMBOPLASTIN

The term thromboplastin was used originally to describe a substance in many tissues which, in the presence of Ca^{++}, catalyzed conversion of prothrombin to thrombin. It is now apparent that there is a group of thromboplastins, and that manifestation of the thromboplastin activity of tissue extracts requires the presence of several additional factors found in normal blood (see below). The thromboplastic activity of tissue extracts differs from that of blood. In the latter, thromboplastin activity *arises during* the clotting process, whereas this activity is evident *immediately* in certain tissue extracts, *e.g.*, of brain. This differentiation is the basis for the rapid clotting (within 12 to 15 sec.) of blood on addition of a suitable brain extract, as contrasted with the slower clotting time (5 to 10 min.) of whole blood placed in a glass tube. Lung, brain, and placenta are rich in thromboplastin activity; probably all tissues contain such material. Studies of various tissue preparations indicate that thromboplastin activity is associated both with protein (enzymic) factors and with phospholipids (see below).

Although plasma does not contain a thromboplastin, there is evidence for the existence of a prothromboplastin which can be converted to thromboplastin through the mediation of platelet factors, as well as of other factors present in normal blood.

FACTORS CONCERNED IN PROTHROMBIN ACTIVATION

In addition to Ca^{++} and thromboplastin, other factors have been implicated in the activation of prothrombin. The situation is not completely clear, partly because of the complexity of the process and partly because some of the factors have not been purified sufficiently to determine their chemical nature and relationships. Nevertheless, there is evidence for the existence of at least seven such factors. The evidence is derived mainly from the description of dyscrasias in which an individual factor is present in low concentration or missing entirely from the plasma of individuals whose blood does not manifest a normal clotting time. The independent discovery of these factors by several investigators has resulted in the use of different names for the same factor. Table 31.9 contains a list of these factors and the various synonyms used by different workers, together with the dyscrasia resulting from the deficiency of the substance.

Antihemophilic Factors. The hereditary defect known as *hemophilia* is inherited as a sex-linked recessive trait and thus occurs with significant frequency in males but is transmitted only through the female. The defect is manifested in a markedly prolonged clotting time. The hemophilic clotting system apparently contains normal amounts of fibrinogen, prothrombin, Ca^{++}, etc. Three types of hemophilia are known; the blood of one type of hemophilic added to the blood of patient with another type of hemophilia, gives a clotting time much shorter than that of either specimen alone.

In *hemophilia A,* the missing factor *antihemophilic globulin* (Factor VIII) is a labile, sparingly soluble globulin of normal plasma, usually precipitated with fibrinogen. Preparations of this globulin hasten, both in vitro and, temporarily, in vivo, the clotting of blood of individuals with hemophilia A. Thrombin formation is defective in hemophilic blood; the antihemophilic globulin is concerned with the activation of Stuart Factor (see below) which is essential for the conversion of prothrombin to thrombin.

Less is known concerning the factor deficient in individuals affected by *hemophilia B,* but some evidence suggests that this factor is concerned in thromboplastin formation. The factor is, however, distinct from antihemophilic globulin and has been termed *Christmas factor* (Factor IX), from the name of the first patient in whom the disorder was recognized; the disease is known as *Christmas disease,* and is inherited as a sex-linked recessive trait.

Bleeding symptoms are usually moderate in *hemophilia C,* and the inheritance of the trait seems to be that of an incomplete dominant. The missing substance in the trait *plasma thromboplastin antecedent* (Factor XI) is distinct from those lacking in hemophilia A or B.

Accelerator Globulin. Another factor concerned in the clotting mechanism was identified in the deficiency disease called *congenital parahemophilia.* The responsible factor involved in the transformation of prothrombin to thrombin is *accelerator* or *Ac globulin* (labile factor). The substance is present in plasma as an inactive precursor

Table 31.9: Blood Coagulation Factors

Common name	Roman numeral designation	Synonyms	Activation product	Dyscrasia or deficiency disease
Fibrinogen	I	. . .	Fibrin	Afibrinogenemia; hypofibrinogenemia
Prothrombin	II	. . .	Thrombin	
Tissue thromboplastin	III	Thrombokinase		
Calcium	IV			
Proaccelerin	V	Accelerator (Ac) globulin; labile factor	Accelerin (VI); serum Ac-globulin	Congenital parahemophilia
Proconvertin	VII	Precursor of serum prothrombin conversion accelerator (ProSPCA)	Convertin; serum prothrombin conversion accelerator (SPCA)	Result of treatment with Dicumarol; vitamin K deficiency; liver disease
Antihemophilic factor (AHF) or globulin (AHG)	VIII	Antihemophilic factor A; platelet cofactor I	Activated AHF or AHG	Classical or hemophilia A
Christmas factor	IX	Antihemophilic factor B; platelet cofactor II	Activated Christmas factor	Hemophilia B
Stuart factor	X	Prower factor	Activated Stuart factor; thrombokinase	Result of treatment with Dicumarol; liver disease; vitamin K deficiency; congenital deficiency
Plasma thromboplastin antecedent (PTA)	XI	Antihemophilic factor C	Activated PTA	Hemophilia C
Hageman factor	XII	Surface or contact factor	Activated Hageman factor	Congenital deficiency (Hageman trait)
Fibrin stabilizing factor (FSF)	XIII	Fibrinase; Laki-Lorand factor	Activated fibrin stabilizing factor (FSF′)	

proaccelerin (Factor V) which during the activation of prothrombin is transformed to the accelerator globulin (*accelerin*). Proaccelerin is thermolabile and disappears rapidly from stored plasma.

Proaccelerin may be deficient in the plasma of individuals with severe liver disease. The condition does not respond to vitamin K therapy but is corrected by administration of fresh, prothrombin-free plasma.

Proconvertin and Convertin. In contrast to proaccelerin, proconvertin is a relatively stable substance and persists in stored plasma. One hypothesis suggests that proconvertin interacts with thromboplastin in the presence of Ca^{++} to form convertin which is essential for prothrombin conversion. Another view, suggested by the name *serum prothrombin conversion accelerator,* is that the substance acts with thromboplastin to hasten prothrombin conversion.

Proconvertin deficiency occurs in patients treated with Dicumarol or in individuals with vitamin K deficiency (page 728). Vitamin K administration leads to an increase in plasma proconvertin levels.

Platelets. Suspensions or extracts of platelets (thrombocytes) accelerate coagulation of platelet-free blood, establishing the importance of platelets in blood coagulation, although their exact role is not clear. They contain small quantities of thromboplastin, and it has been suggested that rupture of platelets aids in initiation of coagulation by liberation of thromboplastin, and by release of phospholipids which activate various factors (V and VIII). The phosphatidyl ethanolamines and phosphatidyl serines contain more than half their fatty acids as arachidonic acid. The clot-promoting activity of these phospholipids is dependent on the presence of free amino groups (in the serine or ethanolamine portions) and on the presence of the unsaturated fatty acids.

Prevention of platelet dissolution may be accomplished by collection of blood with needles, tubing, and glassware which have been coated with silicones or other water-repellent polymers. Platelets remain intact on contact with these nonwettable, smooth surfaces, and coagulation is strongly retarded. In fact, if platelets are removed by high-speed centrifugation, human plasma kept in silicone containers will not coagulate for protracted periods even in the absence of calcium-binding agents. If platelet extracts or lysed platelets are added to such preparations of blood, clotting occurs rapidly.

During the dissolution of platelets, norepinephrine (Chap. 45), serotonin (page 590), and histamine (page 588) are liberated and may aid in the control of blood loss by their vasoconstrictor action.

Prolonged bleeding time may be due to *thrombocytopenia,* a deficiency of platelets. This leads to a decreased formation of active thromboplastin with deficient production of thrombin. The condition may be caused by a variety of chemical reagents including certain drugs, by ionizing radiations, by associated blood disorders, *e.g.*, certain leukemias or anemias, or by a variety of infections.

MECHANISM OF BLOOD CLOTTING

Tentative mechanisms have been proposed to account for all the various factors involved in blood clotting. One such scheme is shown in Fig. 31.6. The evidence for this "cascade" type of sequence is incomplete, but the position of each of

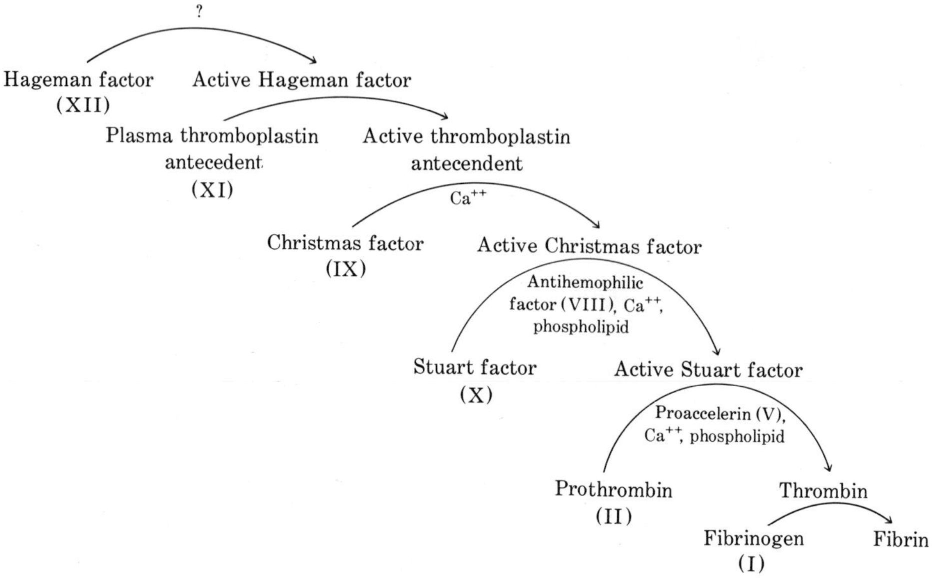

Fig. 31.6. Tentative mechanism for the initiation of blood clotting in mammalian plasma. (*Courtesy of Dr. E. W. Davie.*) The roman numerals are those in Table 31.9.

the factors is supported by some evidence. On the left side of each reaction is an inactive precursor and on the right an active factor, presumably an enzyme, except in the case of fibrin. Initiation of clotting results in the successive conversion of each inactive protein to the active enzyme by the catalytic action of each newly formed enzyme.

The first recognized event in the coagulation of shed blood is the activation of Hageman factor (XII). This is the process activated by glass or other insoluble surface-active agents. Little is known of the exact nature of the various proteins which participate in the process. Evidence that most, if not all, of them are enzymes is indicated by the requirement for Ca^{++} or other divalent cations in several of the conversions. Further, activation of Christmas factor by activated Factor XI is inhibited by diisopropylphosphofluoridate, a potent inhibitor of thrombin as well as of many other proteinases and esterases (page 254). Activated Stuart factor (X) is inhibited by soybean trypsin inhibitor. Thus, present evidence suggests that many of these enzymes may be proteolytic in action. Indeed, the activation of the inactive precursors is reminiscent of the activation of the zymogens of the proteinases, chymotrypsinogen and trypsinogen (page 252).

HEPARIN

Heparin prevents the coagulation of plasma. This substance was discovered by Howell and McLean in 1916 in crude extracts of liver. Many tissues of the body contain heparin since it specifically originates in the metachromatic granules of mast cells, which are principally found along blood vessel walls. Heparin has been isolated from lung as well as liver as the crystalline barium salt. It is an acid mucopolysaccharide of molecular weight about 17,000, which, on hydrolysis, yields glucosamine, sulfuric acid, and glucuronic acid. The probable structure of heparin has been presented previously (page 54).

Sulfation of certain polysaccharides, *e.g.*, starch, can produce products which exhibit heparin-like activity in preventing blood coagulation and also behave as clearing factors (page 476). This indicates that the repeating sulfate polymer structure in heparin is the basis of these biological activities.

Heparin acts in vivo as well as in vitro to prolong the clotting time of blood by interfering with the normal conversion of prothrombin to thrombin. However, heparin does not act alone since it has no influence on purified prothrombin. A serum protein cofactor is necessary for heparin action. Heparin inhibits activation of Christmas factor by activated Factor XI; it also inhibits the activation effect of antihemophilic factor with activated Christmas factor. These actions thus prevent the conversion of prothrombin to thrombin. Heparin also inactivates thrombin in the presence of serum. This is believed to be the basis of one of the normal mechanisms for control of the fluidity of the blood.

Thrombin causes platelets to become sticky and to adhere to each other as well as to rough surfaces, resulting in disintegration and release of thromboplastinogenase. Heparin prevents agglutination of platelets and thus aids in preventing thrombus formation.

Evidence has been obtained that plasma contains lipoproteins which act as antithrombins without addition of heparin. Such antithrombins are inactivated by extraction with ether.

Fibrin itself can adsorb thrombin. Thus clot formation immediately sets into operation a way of preventing excessive clotting by retaining the thrombin at the site of bleeding.

A summary of all the factors concerned in the coagulation mechanism still leaves us with the classical two-step reaction scheme (page 725) described by Morawitz many years ago, but with the addition of other important participants whose chemical nature and mode of action are still largely to be determined. This may be illustrated in the diagram below.

$$\text{Prothrombin} \xrightarrow{\text{blocked by inhibitors of activation}} \text{thrombin}$$

Requires: Activation of many successive factors

$$\text{Fibrinogen} \xrightarrow{\text{thrombin}} \text{fibrin (soluble)} \xrightarrow{\text{fibrinase}} \text{fibrin (insoluble)}$$

$$\text{Thrombin} + \text{antithrombins} \longrightarrow \text{inactive thrombin}$$

DISSOLUTION OF FIBRIN

Sterile blood clots usually dissolve after a few hours or days; the lysis is caused by a proteolytic enzyme. Serum becomes fibrinolytic when shaken with chloroform. The proteolytic enzyme, *plasmin* (fibrinolysin), ordinarily exists in plasma as the inactive precursor, or zymogen, *plasminogen* (profibrinolysin). Plasminogen from human blood is a protein of molecular weight approximately 90,000 and contains a single polypeptide chain. Activation of plasminogen by chloroform or other organic solvents is presumably due to separation of an inhibitor. Also effective are extracts of hemolytic streptococci or certain other bacteria which contain an enzyme termed *streptokinase* that catalyzes conversion of plasminogen to plasmin. Obviously, the above activators of plasminogen are not physiological. However, factors present in many tissues will activate plasminogen. Trypsin is an effective activator, as is *urokinase,* a proteolytic enzyme present in human blood and urine. In all instances of the activation of plasminogen to plasmin the mechanism is based on the proteolysis of the inactive precursor. A single arginyl-valyl bond of plasminogen is cleaved resulting in a two-chain molecule held together by a single disulfide bond. The transformation is reminiscent of the conversion of chymotrypsinogen to chymotrypsin (page 252) and of proinsulin to insulin (page 973). Plasmin formation from plasminogen occurs in a variety of circumstances, *e.g.*, emotional stress, during exercise, or after injection of epinephrine. Plasminogen activator also is produced in blood vessel walls, from which it is released on vascular injury.

Plasmin has a molecular weight of about 90,000 and is a true proteolytic enzyme since it acts not only on fibrin and fibrinogen but also on casein, gelatin, and other proteins. However, unlike thrombin, it does not convert fibrinogen to fibrin. Specificity studies indicate that plasmin splits arginyl linkages and thus resembles trypsin. Plasmin is also similar to trypsin in being inhibited by soybean and pancreatic trypsin inhibitors and by diisopropylphosphofluoridate (page 254).

Thus information presently available indicates that in addition to the complex, delicately balanced mechanisms for prevention of clotting, as well as for its initiation, there is also present a system for removal of intravascular clots, or thrombi. Since antiplasmins have been shown to occur in plasma, there are physiological controls for the level of proteolytic activity in plasma. The present picture may be summarized in the following outline:

$$\text{Plasminogen} \xrightarrow[\text{and other factors (?)}]{\text{tissue enzymes}} \text{plasmin} + \text{inhibitors} \rightarrow \text{inactive plasmin}$$

$$\text{Fibrin clot} \xrightarrow{\text{plasmin}} \text{soluble products}$$

THE ERYTHROCYTE AND IRON METABOLISM

The prime function of the erythrocyte, transport of oxygen and carbon dioxide (Chap. 32), is accomplished by the presence of a 34 per cent solution of hemoglobin. The erythrocyte contains the most concentrated protein solution in the body, thereby permitting the movement of about 16 g. of hemoglobin per 100 ml. of whole blood without the high viscosity which would attend the presence of an equivalent amount of protein in free solution. The mammalian adult erythrocyte contains no nucleus, possesses a relatively low respiratory metabolism, and is one of the few major cells of the body with a finite life span of established length.

DEVELOPMENT, STRUCTURE, AND COMPOSITION OF THE ERYTHROCYTE

Development of the Erythrocyte. The red cells arise from reticular cells of the bone marrow and develop in sinusoids temporarily closed to the circulation. Rapid cell multiplication occurs at the level of the pronormoblasts, which contain large nuclei with conspicuous nucleoli. Differentiation begins with the appearance of the basophilic normoblast. This cell has lost its nucleolus, the RNA content of the cytoplasm has decreased markedly, and the total protein content of the cell is at a maximum. Heme and globin syntheses are initiated at this time. At the orthochromic erythroblast stage, the nucleus is pyknotic and the hemoglobin content has been established. Mitotic figures disappear, and the nucleus then fragments, giving rise to the reticulocyte, from which the mitochondria and RNA have disappeared. Several days later the adult erythrocyte emerges containing no detectable RNA or DNA.

Structure and Composition of the Erythrocyte. The erythrocyte is a nonnucleated, biconcave disk with a diameter varying from 6 to 9 μ and a thickness of about 1 μ at the center increasing to 2 to 2.5 μ toward the periphery. The membrane of the erythrocyte is a mosaic structure approximately 60 Å. thick; its chief components are an insoluble protein, *stromatin,* a mixture of lipids including lecithin and cephalin, cerebrosides, gangliosides, and unesterified cholesterol, and a group of glycoproteins, one (or more) of which contains a sialic acid. The latter is of interest since attachment of specific viruses, *e.g.,* influenza, to erythrocytes occurs at "receptor" sites and, apparently, involves binding between virus and the sialic acid(s) of the cell membrane. Incubation of erythrocytes with *neuraminidase* (page 55), which liberates neuraminic acid (page 34), alters the receptor sites, and such cells will no longer bind virus. The erythrocyte membrane also contains the specific blood group substances (see below).

As in cells generally, the chief cation of the human erythrocyte is K^+, with lesser amounts of Na^+, Ca^{++}, and Mg^{++}, while the major anions are Cl^-, HCO_3^-, hemoglobin, inorganic phosphate, and various organic phosphates, of which 2,3-diphosphoglycerate is noteworthy (Fig. 31.7). This compound, the coenzyme of phosphoglyceromutase, is the major phosphate-containing component of the red cell. Although it could serve as a store of potentially available high-energy phosphate,

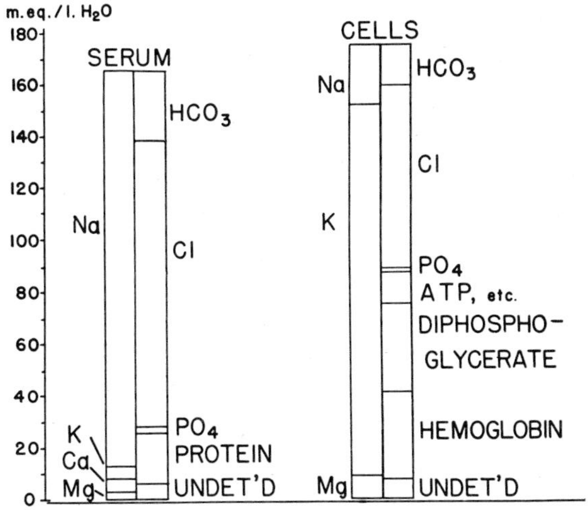

Fig. 31.7. The electrolyte structure of the oxygenated human erythrocyte. (*From G. M. Guest, Am. J. Diseases Children,* **64**, 401, 1942.)

the influence of this compound on the properties of hemoglobin (page 735) may be most significant to the erythrocyte, in contrast to the trace amounts in other cells.

The adult erythrocyte behaves as an osmometer, swelling and shrinking with decreases and increases in osmotic pressure of the bathing fluid. In a sufficiently hypotonic solution, the red cell swells, and rupture of the enclosing membrane, or *hemolysis,* occurs. Hemolysis may also be accomplished in isotonic media by various surface-active agents, *e.g.*, soaps, chloroform, and saponins (page 83). Virtually all the constituents of the human erythrocyte are in solution and diffuse into the medium upon hemolysis, leaving an insoluble residue, or *ghost,* which represents the original membrane.

The internal composition of the erythrocyte is maintained by energy-requiring mechanisms. The erythrocyte membrane shows several ATPase activities, one of which is dependent upon Na^+, K^+, and Mg^{++} and is inhibited by the presence of low concentrations of cardiac glycosides, agents which influence Na^+-K^+ transport across erythrocyte and other cell membranes. Although the glucose concentration within the human erythrocyte is generally identical with that in plasma, glucose does not enter the cell by simple diffusion but by active transport. Phosphate penetration of the erythrocyte is slow; a specific transport system has not been indicated.

Chemistry of the Blood Group Substances. Among the most interesting constituents of the red cell membrane are the "blood group substances." These mucopolysaccharides occur not only on erythrocytes but also in secretions such as saliva, gastric juice, etc. Water-soluble products showing immunochemical blood group activity have also been obtained from tissues of various species, including the human. For example, fractionation of material from ovarian cyst fluid has yielded two substances having blood group B specificity. Both contained L-fucose, D-galactose, N-acetyl-D-glucosamine, and N-acetyl-D-galactosamine, as well as 11 amino acids. Similar constituents have been described in group A substance.

All the blood group substances isolated from a number of species show striking qualitative similarities in chemical composition, despite immunological distinctions. They are composed of carbohydrate and polypeptide, and all contain the four sugars mentioned above; in addition, a sialic acid may be present. Sulfur-containing and aromatic amino acids are absent. Studies of enzymic digests of immunologically active blood group substances indicate that the polypeptide portion plays no role in immunological specificity. The group specificity is probably associated with the nature of the carbohydrate end group of the mucopolysaccharide. Thus the sole chemical difference between the sugar units of blood group A and B substances resides in the terminal sugar; in group A this is N-acetyl-D-galactosamine and in B, D-galactose. Purified preparations of blood group substances from ovarian cyst fluid and from hog gastric mucin have molecular weights ranging from approximately 260,000 to 1,800,000.

HEMOGLOBIN SYNTHESIS

Hemoglobin is the major protein of the erythrocyte. Its chemistry (pages 171*ff.*), variations in structure of its globin component (pages 688*ff.*), and the synthesis of its porphyrin precursor, porphobilinogen (pages 584*ff.*), have been considered previously. The physiological roles of hemoglobin in respiration are discussed in the following chapter.

The pathway involved in the conversion of porphobilinogen to protoporphyrin and of heme synthesis from porphobilinogen is summarized in Fig. 31.8. Enzymic preparations utilizing porphobilinogen for porphyrin synthesis have been isolated from plant, bacterial, and animal sources. A common pathway exists for the synthesis of heme and of chlorophyll leading to formation of protoporphyrin IX (page 168). Insertion of iron into the latter results in heme formation. In plants, in addition to synthesis of heme, magnesium is inserted into protoporphyrin IX to form magnesium protoporphyrin, which is converted in plastids to chlorophyll.

Protoporphyrin synthesis begins with condensation of four molecules of porphobilinogen, under the influence of a *porphobilinogen deaminase,* to a postulated intermediate, polypyrryl methane, which is converted to uroporphyrinogen I. This colorless chromogen can undergo autoxidation to form one of the urinary pigments, uroporphyrin I (page 737), or may be decarboxylated to yield coproporphyrinogen I, which in turn may be oxidized to coproporphyrin I. Alternatively, the polypyrryl methane, in the presence of an *isomerase,* may form uroporphyrinogen III, which can in a similar manner yield uroporphyrin III, coproporphyrinogen III, or coproporphyrin III. Coproporphyrinogen III may be oxidized and decarboxylated to yield protoporphyrin III (No. IX). The enzymes obtained from beef liver favor synthesis of the series III porphyrins.

Iron is incorporated at the level of protoporphyrin. No iron compounds of uroporphyrin or coproporphyrin are known to occur in nature. Extracts of nucleated, avian erythrocytes have been described which catalyze formation in vitro of either hemoglobin or myoglobin, respectively, from protoporphyrin, Fe^{++}, and either apohemoglobin or apomyoglobin. The active enzyme in such extracts has been termed *heme synthetase.* Heme formation also occurs nonenzymically in vitro from a solubilized protoporphyrin and ferrous iron. Iron may be made available in the form of transferrin (page 721) or ferritin (page 748) for incorporation into immature cells of the bone marrow.

Iron incorporation into hemoproteins, *e.g.*, the cytochromes, has been described in a variety of tissues. Apparently, all aerobic cells possess independent synthetic capability for the manufacture of hemoproteins; this activity is localized in mitochondria. Studies with rabbit bone marrow in vitro have demonstrated that the synthesis of heme and that of globin occur at approximately parallel rates, with a direct attachment of heme to the newly synthesized globin chains. However, it was possible experimentally to influence disproportionately the rates of these two synthetic processes, indicating that they are not interdependent in that stimulation or inhibition of the one need not lead to stimulation or inhibition of the other.

The rate of protoporphyrin synthesis in the developing red cell only slightly exceeds that of hemoglobin synthesis. The mature erythrocyte contains approximately 30 μg of protoporphyrin and approximately 1 mg. of coproporphyrin per 100 ml. of packed cells. In iron-deficiency anemia, red cells may contain as much as twenty times the normal content of protoporphyrin.

In vertebrates the heme of hemoglobin represents approximately 85 to 90 per cent of the total body heme; about 10 per cent is in myoglobin and less than 1 per cent in all the other hemoproteins combined, *i.e.*, cytochromes, catalase, etc.

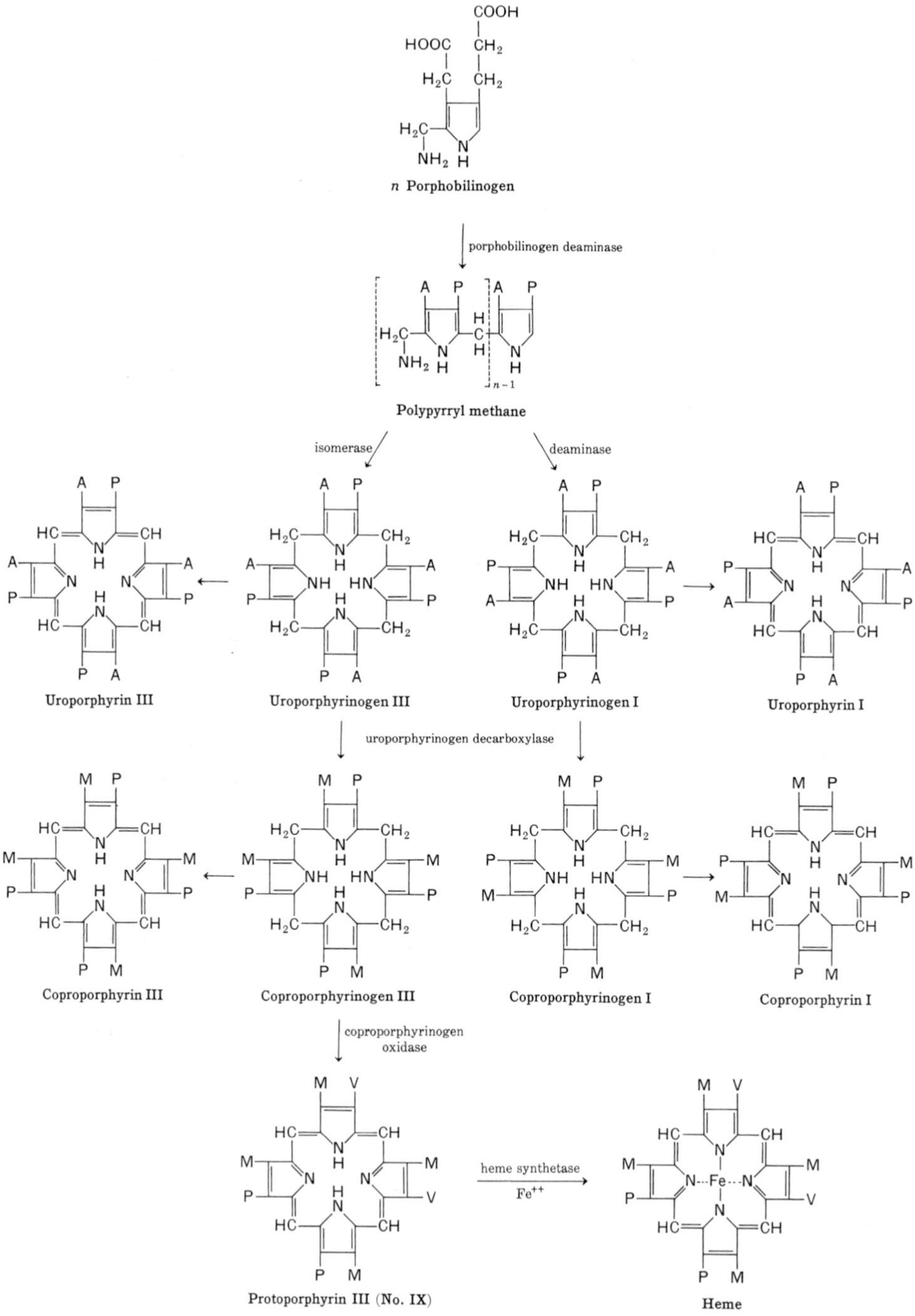

Fig. 31.8. Pathway of synthesis of heme and other porphyrins from porphobilinogen. A = acetic acid; P = propionic acid; M = methyl; V = vinyl.

Abnormalities of Heme Synthesis; Porphyrias. Some of the clinical disorders associated with or characterized by abnormalities in heme synthesis result in the appearance of unusual quantities of porphyrins in the urine and have been termed *porphyrias.* Nonspecific excretion of porphyrins in diverse disorders, *e.g.*, alcoholism, lead poisoning, and hemolytic disease, generally characterized by increased excretion of coproporphyrin, are usually termed *porphyrinurias.*

As shown in Fig. 31.8, type III porphyrins are most important in nature since their synthesis leads to protoporphyrin IX and thence to heme, or to chlorophyll. Type I porphyrins appear to be formed as by-products of heme synthesis and are not utilized by the body but are excreted in the urine (up to 300 μg per day) and in the stool (up to 600 μg per day). These colored porphyrins are oxidized forms of the colorless porphyrinogens.

Two types of porphyria, *erythropoietic* and *hepatic,* are hereditary disorders of metabolism. The former is a very rare disease in which there is excessive formation of heme precursors, notably protoporphyrin IX, in the developing red blood cells of the bone marrow. Skin photosensitivity is evident at an early age. Abnormal amounts of various heme precursors, notably uroporphyrin I and uroporphyrinogen I, are excreted in the urine. Several types of hepatic porphyria have been described. In general, in hepatic porphyrias, excessive and abnormal formation of heme precursors occurs in the liver. Urinary products present in higher than normal amounts include δ-aminolevulinic acid, porphobilinogen, uroporphyrin I, and uroporphyrinogen I. An unexplained interesting fact in hepatic porphyria is the excretion of most of the porphyrins as complexes of zinc. The urine darkens markedly on standing, because of the conversion of abnormal quantities of porphyrinogens and other heme precursors into porphyrins, porphobilins, and other unidentified pigments. In *acute* porphyria, porphyrins of type I as well as porphobilinogen are excreted in urine in unusual amounts, as much as 200 mg. per day. The porphyrin excreted in lead poisoning is coproporphyrin III; greatly increased amounts of δ-aminolevulinic acid are also present.

METABOLIC ASPECTS OF THE ERYTHROCYTE

The reticulocyte possesses functioning glycolytic and phosphogluconate oxidative pathways, the tricarboxylic acid cycle, and an intact cytochrome system and electron transfer mechanism. Moreover, the reticulocyte can synthesize hemoglobin and diverse lipids, *viz.*, cholesterol, phospholipids, and triglycerides, and can achieve the *de novo* synthesis of purine nucleotides from small precursors (page 625) as well as from 5-aminoimidazole-4-carboxamide ribonucleotide (page 623).

In contrast to the reticulocyte, the mature mammalian erythrocyte lacks mitochondria; hence the cytochrome system is absent, the tricarboxylic acid cycle is not evident although several of its enzymes are present, and synthesis of hemoglobin does not occur. There is a continuing turnover of fatty acids, indicating some lipid synthesis. Although the erythrocyte does not make cholesterol, rapid exchange occurs between its sterol and the cholesterol present in plasma lipoproteins. During maturation from the reticulocyte, the erythrocyte loses its capacity for *de novo* synthesis of purine nucleotides, but can accomplish this from 5-aminoimidazole-4-carboxamide ribonucleotide. The mature erythrocyte can also utilize preformed purines for nucleotide synthesis via the salvage pathway (page 625).

The energy of the mature erythrocyte is derived primarily from anaerobic glycolysis and the phosphogluconate oxidative pathway. Oxygen consumption of erythrocytes is low—Q_{O_2}, 0.05, consumed for the most part by the slow oxidation of hemoglobin to methemoglobin (page 172); as much as 0.5 per cent of the total hemoglobin is converted to methemoglobin in 24 hr. However, erythrocytes contain two enzymic systems which catalyze reduction of methemoglobin to hemoglobin; one utilizes DPNH, and the other, TPNH. The former coenzyme derives from glycolysis, the latter from the phosphogluconate oxidative pathway. Reduction of methemoglobin occurs only when glucose or lactate is present, and the rate of reduction in vitro is greater when methylene blue is added. Both DPNH and TPNH are oxidized by *methemoglobin reductases.* Addition of the autoxidizable dye, methylene blue, increases oxygen consumption tenfold by obviating the need for methemoglobin as the terminal oxidase in this system. A heme-containing methemoglobin reductase with a molecular weight of approximately 185,000 has been described which catalyzes reduction of methemoglobin, utilizing DPNH as a cofactor. Activity with TPNH was evident only in the presence of methylene blue. Oxygen or methylene blue may act also as terminal electron acceptors. Apparently there is no provision for generation of ATP incident to these oxidations. The role of methemoglobin reductase in familial methemoglobinemia will be considered later (page 751).

Glutathione, present in erythrocytes and absent from the plasma, plays a role in maintaining the normal integrity of the erythrocyte (page 753). The tripeptide is present largely in the reduced form and is synthesized intracellularly, showing a high rate of turnover, as reflected in a half-life of only 4 days. TPNH in the presence of *glutathione reductase* maintains glutathione in the reduced state (page 548).

Life Span of the Erythrocyte. Since mature red cells are constantly destroyed and replaced by new cells, the total red cell mass is in a dynamic state. The daily turnover, calculated from the amount of bile pigment excreted during this period, represents about 20 ml. of erythrocytes, *i.e.*, 0.85 per cent of the red cell mass. In contrast to other cellular proteins, the protein within the erythrocytes does not turn over during the life of the cell.

The life span of the red cell has been measured by various techniques. Thus, from human subjects given ^{15}N-labeled glycine, blood samples were withdrawn at intervals, and the globin and heme were analyzed for ^{15}N. The data obtained are shown in Fig. 31.9. The isotope concentration of the heme of whole blood rose rapidly to a maximum, indicating formation of red cells containing a relatively high concentration of ^{15}N during the period of isotope administration. In the time which followed cessation of ingestion of isotopic glycine, the ^{15}N concentration remained relatively constant because of maintenance in the circulation of erythrocytes of high ^{15}N concentration. These cells were not immediately removed from circulation, but new cells, unlabeled because they had been formed in the absence of labeled precursor, were slowly replacing unlabeled cells which had previously entered the circulation before the period of isotope ingestion. When the isotope concentration of the heme of the whole blood fell relatively rapidly, there must have occurred a relatively sudden replacement of cells of high ^{15}N concentration by unlabeled cells. This abrupt replacement is evidence of a finite life span for individual red cells. This has been calculated from the data of Fig. 31.9 to be 126 $\pm$ 7 days.

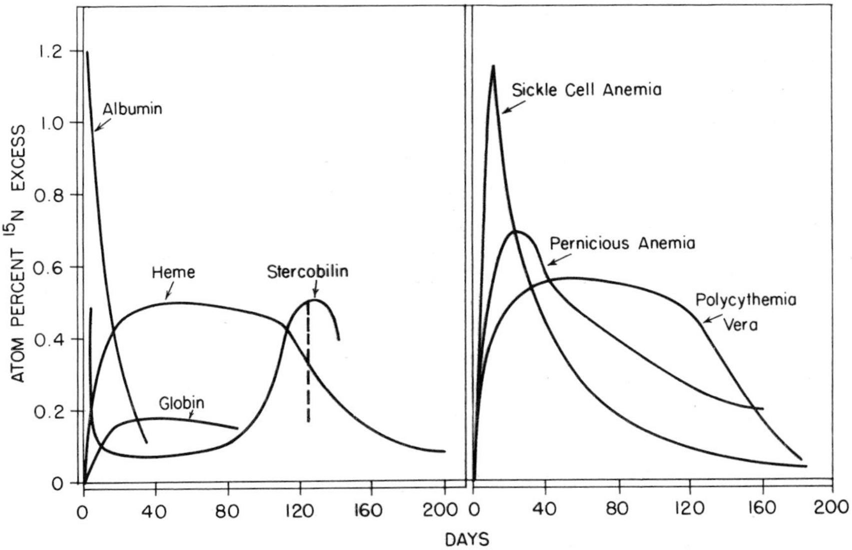

Fig. 31.9. The life span of the human erythrocyte. ^{15}N-labeled glycine was fed for 2 days in multiple doses; blood and stool samples were collected at intervals and the heme and globin of the red cells and the stercobilin of stool were analyzed for their ^{15}N content. On the left, data from a normal male. Note the life span of about 126 days, the coincidence of labeled bile-pigment formation with the death of the erythrocytes, and the early period of labeled stercobilin excretion. The behavior of serum albumin is shown to illustrate random decay. On the right, ^{15}N content of the heme of patients with sickle cell anemia, pernicious anemia, and polycythemia vera. (*Data adapted from D. Shemin and D. Rittenberg, J. Biol. Chem.*, **166**, 627, 1946; *I. M. London, R. West, D. Shemin, and D. Rittenberg, J. Biol. Chem.*, **179**, 463, 1949; **184**, 351, 1950.)

During the period in which labeled red cells were disappearing from the circulation, ^{15}N appeared in the bile pigments, thus providing direct proof of the origin of bile pigments. The early spike in the curve for bile pigment suggests rapid degradation of heme compounds not incorporated in erythrocytes, and also that a fraction of the cells developed in the marrow are prematurely destroyed, with resultant production of bile pigment.

Data showing the change in ^{15}N concentration of the serum albumin as a function of time are also presented in Fig. 31.9. The abrupt maximum and steady decline of isotope concentration are evidence of a random replacement of circulating albumin by newly formed albumin molecules of lower isotope concentration. Expressed in another way, this means that the renewal of serum albumin is a process which does not depend upon the age of the albumin molecule.

The life expectancy of the erythrocyte is not the same for all species, being approximately 107 days in the dog and 68 days in the cat and rabbit.

The above data establish that there is no turnover of the heme of the erythrocyte; the life span of hemoglobin, and of erythrocyte catalase as well, is the average life span of the erythrocyte itself. The turnover of heart muscle myoglobin and of cytochrome c of skeletal muscle is also very low. In contrast, the life span of liver cytochrome c of the rat has been found to be only about 8 days.

Fate of Erythrocytes—Bile Pigment Formation. The factors contributing to destruction of the circulating erythrocyte are not clear. However, about 120 to 130 days after emerging from the marrow, red cells are phagocytized by macrophages of the reticuloendothelial system, mainly in the spleen, liver, and bone marrow. Hemoglobin undergoes scission of the α-methene bridge to give *choleglobin,* an iron-pyrrole complex. This reaction is catalyzed by *α-methyl oxygenase;* the partially purified enzyme from liver requires TPNH, Fe^{++}, and a heat-stable activator. Removal of the globin yields *verdohemochrome;* the latter is converted to biliverdin, which apparently may still bind iron. Removal of the iron is followed by enzymic reduction of biliverdin to bilirubin. Conversion of the heme of hemoglobin to bilirubin in the body is nearly quantitative and, on the average, takes 2 to 3 hr. Hemoglobin released from erythrocytes is not reutilized as such. The liberated iron combines with plasma transferrin (page 721) and is transported to storage depots or to the bone marrow, where it is used in the synthesis of new hemoglobin. The globin is degraded and returned to the body pool of amino acids. These changes may be summarized as follows.

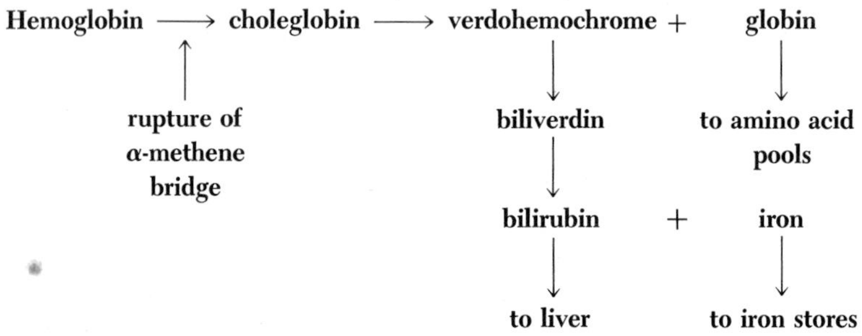

In contrast to porphyrins, which contain four pyrrole rings linked by four carbon atoms in a closed-ring system, bile pigments lack one of these carbon atoms and can be pictured as open chain compounds or as tetrapyrrole chains. The system of numbering the pyrrole rings and methene bridges in the bile pigments is derived from that used for porphyrins (Chap. 8). In the most common biological example shown below, the α-carbon has been eliminated.

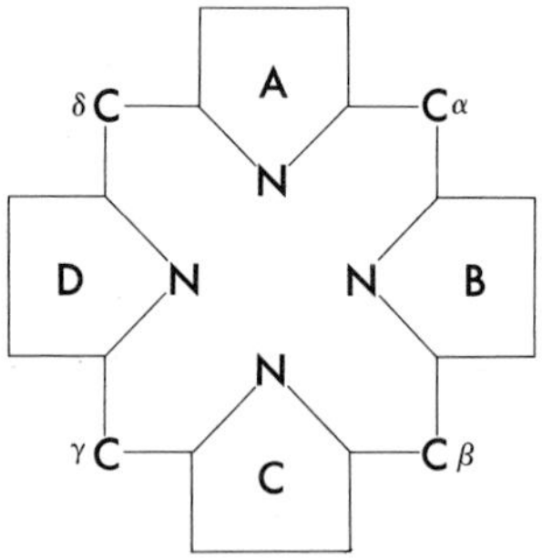

Porphyrin skeleton

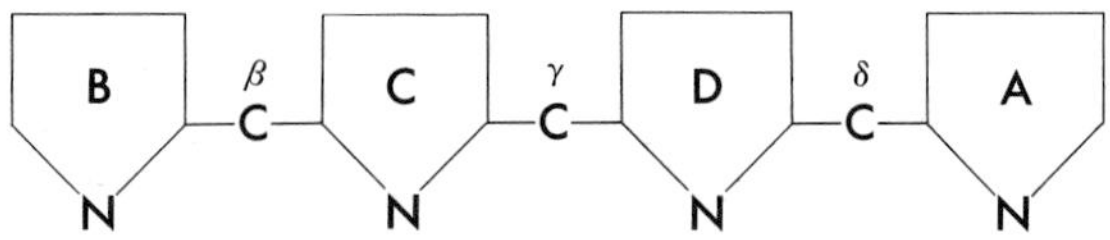

Bile pigment skeleton

Although the tetrapyrrole chain is often used for convenience in representation, the cyclic arrangement of the pyrrole nuclei, shown at the left, is a more realistic representation of the loosely planar structures of the bile pigments.

Bilirubin is transported from extrahepatic reticuloendothelial cells to the liver in combination with albumin. In the liver the protein is separated, and the bilirubin is converted into the corresponding diglucuronide by reaction with uridine diphosphoglucuronate (Chap. 19).

Bilirubin + 2 UDP-glucuronate ⟶ bilirubin diglucuronide + 2 UDP

The compound formed between bilirubin and glucuronic acid is not a glycoside (glucosiduronide), but rather an ester between each of two propionic side chains in bilirubin and the hydroxyl group at C-1 of glucuronic acid (page 433). The soluble diglucuronide passes into the bile canaliculi and thence into the bile.

In the intestine, bilirubin diglucuronide is hydrolyzed by a β-glucuronidase (page 884). As a result of the activity of the bacterial flora, bilirubin is converted to *d*-urobilinogen by reduction of the methene bridge and of one vinyl group to an ethyl group. Since resonance is no longer possible, urobilinogen is colorless. The bilinogens are all colorless until oxidized to the bilins. *d*-Urobilinogen undergoes further reduction with formation, successively, of *l*-urobilinogen, in which the remaining vinyl group is reduced to an ethyl group, and to *l*-stercobilinogen. *d*-Urobilinogen has been isolated after incubation of bilirubin with bacterial extracts and from urine shortly after withdrawal of antibiotic therapy; subsequently, the two further reduction products appeared in the urine. *l*-Stercobilinogen is present in feces; it is strongly levorotatory.

Figure 31.10 indicates the transformations described. Extensive isomerism is possible in these bile pigment derivatives. However, the naturally occurring pigments appear to be largely derivatives of protoporphyrin IX, the α-methene bridge of which is preferentially ruptured.

Urobilinogen and stercobilinogen are oxidized by air to urobilin and stercobilin, respectively, both orange-red pigments which contribute to the color of urine and feces. Approximately 1 to 2 mg. of bile pigment is excreted in the urine and as much as 250 mg. in the feces of a normal adult each day. The degradation of the erythrocyte and formation of the bile pigments are summarized in Fig. 31.11.

Van den Bergh Reaction. Bilirubin can be coupled with diazonium salts, *e.g.*, diazotized sulfanilic acid, to yield azo dyes. This reaction has been developed for the estimation of bilirubin in serum and is not given by other reduced compounds of this type, nor by biliverdin. Bilirubin diglucuronide present in serum yields immediate color upon addition of diazotized sulfanilic acid to give the "direct" van den Bergh reaction. Bilirubin bound to plasma proteins and unconjugated with glucuronic acid does not react with the reagent unless brought into solution by addition of alcohol; this is the "indirect" van den Bergh reaction. The diglucuronide is easily dissociable from the plasma proteins to which it is loosely bound, dialyzes readily, and, in consequence, appears in the urine whenever it is present in significant amounts in plasma. Bilirubin itself is strongly bound to plasma albumin, cannot be removed by dialysis, and, hence, does not appear in urine.

Jaundice. Jaundice is the accumulation of bile pigment in the plasma in amounts sufficient to impart a yellowish tint to the skin and conjunctiva. Three types of jaundice may be recognized. *Hemolytic jaundice* results from unusual destruction of red cells leading to bile pigment formation at a rate exceeding the capacity of the liver to remove the pigment from the circulation. The pigment

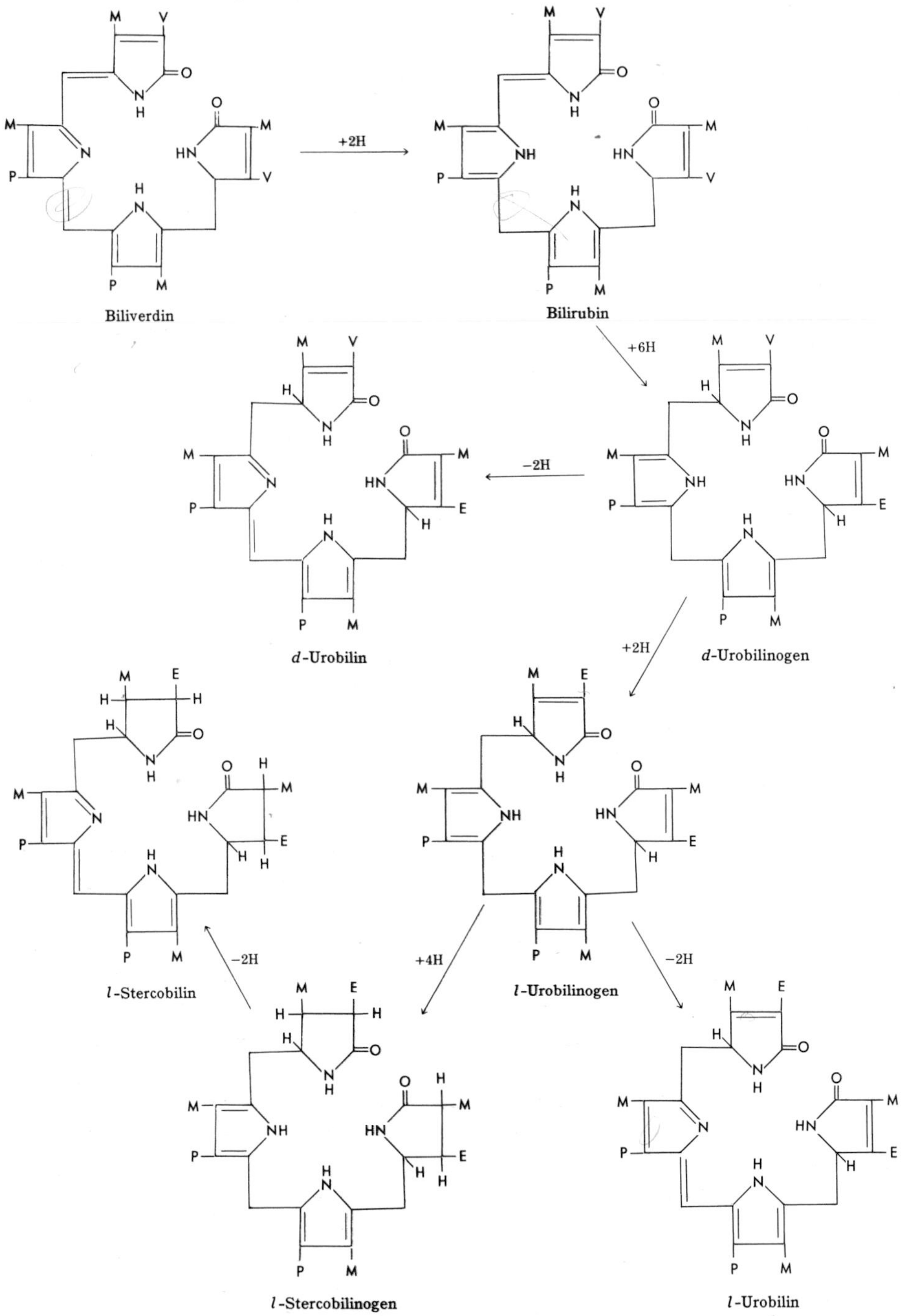

Fig. 31.10. Some products of bilirubin metabolism.

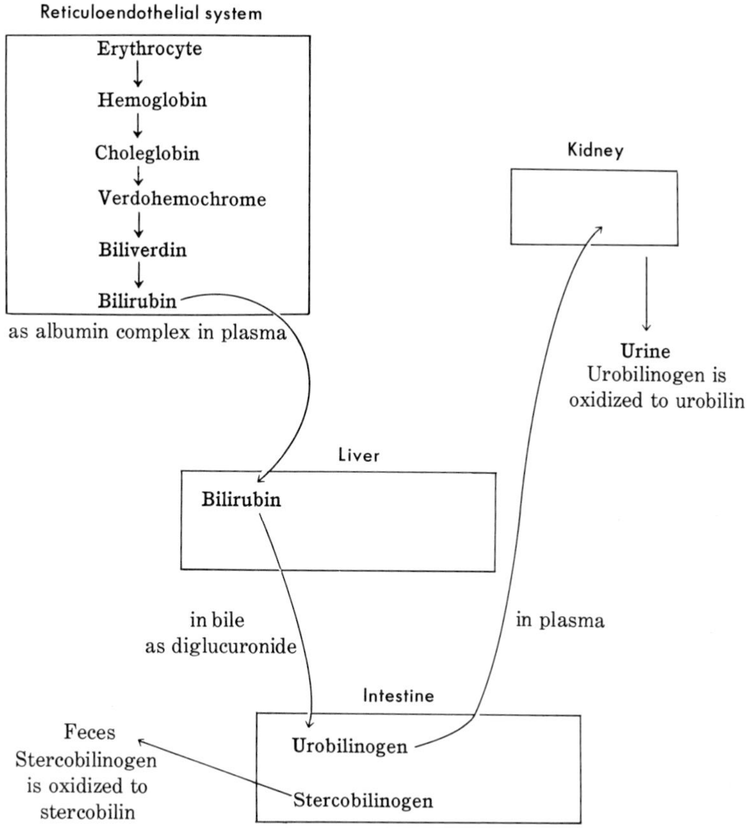

Fig. 31.11. Metabolism of the bile pigments.

which accumulates is bilirubin bound to albumin, so that the indirect van den Bergh reaction reveals the true concentration of bilirubin in hemolytic jaundice. Since bilirubin glucuronide excretion by the liver is maximal, excretion of stercobilinogen and urobilinogen is markedly elevated. In hepatic disease such as *infectious hepatitis* or *cirrhosis,* jaundice results from impaired capacity of the liver to conjugate bilirubin and secrete the diglucuronide into the bile. The indirect van den Bergh reaction reveals a high pigment concentration in these diseases. Stools may be light in color, and little urobilinogen is found in the urine. Occasionally, unusually large quantities of urobilinogen are observed in the urine of patients with hepatitis and little or no jaundice. This results from an impaired ability of the liver to reexcrete urobilinogen returning via the "enterohepatic circulation." The third type of jaundice arises from *obstruction of the biliary passages,* resulting in failure of bile to reach the lumen of the bowel. In the early stages of this disease, while liver function remains normal, the liver continues to secrete bilirubin and its glucuronide; the bile so formed is regurgitated into the circulation, and thus large quantities of bilirubin diglucuronide appear in plasma, as demonstrated by the direct van den Bergh reaction. Prolonged biliary obstruction results in liver damage so that values obtained by the indirect as well as the direct assays may be high. The stools may be clay-colored, and little or no urobilinogen can be detected in the urine, although bilirubin is excreted in the urine in large amounts.

The significance of bilirubin conjugation is particularly apparent in the *Crigler-Najjar syndrome,* a very rare and probably recessively inherited syndrome in which the liver lacks the enzyme, *uridine diphosphate glucuronate transferase,* resulting in jaundice. Apparently the capacity to conjugate bilirubin is also severely limited in the neonatal liver, and the accumulation of bilirubin (indirect pigment) in the blood of the newborn may greatly exceed that seen in adult jaundice. This frequently has deleterious consequences for the brain, where staining of the basal ganglia may occur with permanent damage (*kernicterus*). The capacity of the liver to conjugate bilirubin increases rapidly during the first few days of life.

Bile Pigments in Nature. Pigments closely related chemically to the bile pigments have also been found in some invertebrates which do not possess hemoglobin. Of interest are the pigments in the chromoproteins, phycoerythrin and phycocyanin, which function in the red and blue-green algae as light-absorbing substances together with the chlorophylls (page 459). Phycoerythrin and phycocyanin, which crystallize readily, are metal-free pigments that are at a stage of reduction intermediate between bilirubin and urobilinogen. The side chains in the β positions of the pyrrole rings are identical with those in urobilinogen, indicating the close relationship of the porphyrins, bile pigments, and related pigments in nature.

IRON METABOLISM

Dietary Requirements. The newborn infant is provided with considerably more hemoglobin than is required; both the concentration of hemoglobin in the erythrocyte and the number of red cells per unit volume are appreciably greater than in later life. This may represent a response to the relatively low oxygen tension of uterine life, as in the case of the polycythemia of individuals living at high altitudes. For some weeks after birth, red cell destruction exceeds erythropoiesis, and jaundice may be observed. However, during this period, virtually no iron appears in the excreta. Retention of iron suffices to meet the iron requirements of the infant for some months thereafter, a fortunate circumstance since milk is virtually iron-free. Subsequent to this period, the iron requirement of the child and adult reflects the needs for growth and the rate of iron loss to the environment.

As noted previously, the daily hemoglobin turnover in the adult is equivalent to approximately 20 ml. of erythrocytes or 25 mg. of iron. This is far greater than the daily increment in total body hemoglobin even during the period of maximal growth. This turnover of red cells does not lead to an extra requirement for iron since the iron released within reticuloendothelial cells from phagocytized obsolescent erythrocytes is almost entirely available for reutilization. During growth, the iron requirement for formation of hemoglobin, cytochromes, catalase, and other chromoproteins is of major significance. An extraordinary increase in total body hemoglobin is associated with puberty. The maximum iron requirement of the male thus occurs at the age of fifteen or sixteen. In the adult male, the daily iron requirement is met by replacement of relatively small losses. Although only negligible amounts of iron appear in the urine and there is no intestinal secretion of iron, there is a small daily loss through the bile.

From the menarche, the need of the female for iron is 30 to 90 per cent greater than that of the male, except for the fifteenth and sixteenth years of male life. Until the menopause, 50 per cent or more of the female's iron requirement is used in the replacement of hemoglobin lost in the menses. The average menstrual loss is about 35 ml. of blood. The replacement of this amount of blood alone requires 0.6 mg. of iron per day, in contrast to the *total* physiological iron requirement of the male adult of 0.9 mg. per day. In itself, the loss would be unimportant were it not for the fact that the average unsupplemented diet contains barely enough iron to meet the requirements. This mean catamenial requirement applies to about 60 per cent of women. In the 15 per cent with larger menstrual losses, the replacement need for iron may be almost doubled. During gestation, the requirement for iron is about 60 per cent greater than the amount lost in the menses during a similar period. An iron intake adequate to pregestational life may not be adequate to meet the demands of pregnancy. Transfer of iron to the fetus, like transfer of calcium, occurs chiefly in the last trimester of pregnancy, and iron cannot be accumulated during the earlier months of pregnancy. Therefore the mother's food must provide unusual quantities of both iron and calcium during the last 3 months of pregnancy. This rarely obtains and may lead to a hypochromic anemia. A moderate normocytic anemia, with hemoglobin values of 11 to 12 g. per 100 ml., is "physiological" during pregnancy and due to hemodilution. In circumstances of iron deficiency the tissue concentration of cytochromes may diminish before the blood level of hemoglobin, a reflection of the higher rate of turnover of the cytochromes. In summary, in adult life the iron requirement is conditioned entirely by the demand for replacement of losses, be they through the bile, placenta, uterus, or overt hemorrhage.

Although it is relatively simple to calculate the *physiological* iron requirement in terms of growth and losses, calculation of the *nutritional* iron requirement is not feasible since ingested iron is not quantitatively absorbed from the intestine. Entry of dietary iron into the body is conditioned by two factors: the chemical state of the ingested iron and the iron metabolism of the intestinal mucosa. Much of the iron ingested in natural foodstuffs is "organic iron," present in combinations that are poorly absorbed. Like calcium, iron forms numerous insoluble salts. The extent to which these are formed within the intestine will influence iron absorption. Thus, anemia due to iron deficiency readily results from incorporating large amounts of inorganic phosphate in the diet.

Many unexplained and rather striking differences in the "availability" of food iron exist. For example, iron from white flour appears to be more readily utilized than that from whole-wheat flour, while that from beef appears to be still more readily available, despite the fact that much of beef iron exists as heme compounds. For unexplained reasons, ferrous iron is more readily absorbed from the human intestine than is the ferric form. The presence of reducing agents, such as ascorbic acid, in the diet increases the availability of inorganic iron. In view of these complications, it is impossible to calculate the desired iron intake; dietary recommendations have been based on studies of iron balance and hemoglobin formation. From these it appears that the desirable dietary level of iron should be five to ten times the actual physiological requirements. Allowances of 12 mg. of iron per day for adults and 6 to 15 mg. daily for children of various age groups are liberal.

Intestinal Absorption of Iron. Studies with radioactive ^{59}Fe early demonstrated that individuals with iron-deficiency anemia absorb iron from the intestine more efficiently than do normal persons. However, when normal animals were made anemic by phlebotomy, some time elapsed before an increased efficiency of iron absorption was detected. It was postulated that normally there is a "mucosal block" to the absorption of iron, unrelated to the existence of anemia per se, and that a gastrointestinal mechanism exists which regulates passage of iron from the intestinal lumen to plasma. Granick established the relation of *ferritin* to this mucosal mechanism in the guinea pig. Ferritin is a protein containing 23 per cent of iron by weight, and isolated initially from spleen. It is composed of a protein, *apoferritin,* of molecular weight 450,000 and a ferric hydroxide-phosphate of approximate composition $[(FeOOH)_8(FeO{-}OPO_3H_2)]$. The isoelectric points and electrophoretic mobilities of ferritin and apoferritin are identical.

The intestinal mucosa of fasting guinea pigs contains only minute amounts of apoferritin. However, within 4 to 5 hr. after iron administration, there occurs a twenty- to fiftyfold increase in the amount of ferritin, implying rapid synthesis of apoferritin by the mucosal cells. Ferrous iron, entering the mucosal epithelial cell, is rapidly oxidized to ferric hydroxide, which combines with apoferritin. Within the mucosal cell, because of unknown circumstances which favor reduction of ferric to ferrous iron, breakdown of ferritin occurs, allowing absorption of additional iron from the intestine. Iron absorption may be limited by the binding capacity of the apoferritin for iron. Figure 31.12 summarizes the ferritin mechanism of iron absorption.

Iron Storage. Ferritin is of importance in yet another connection. The liver of

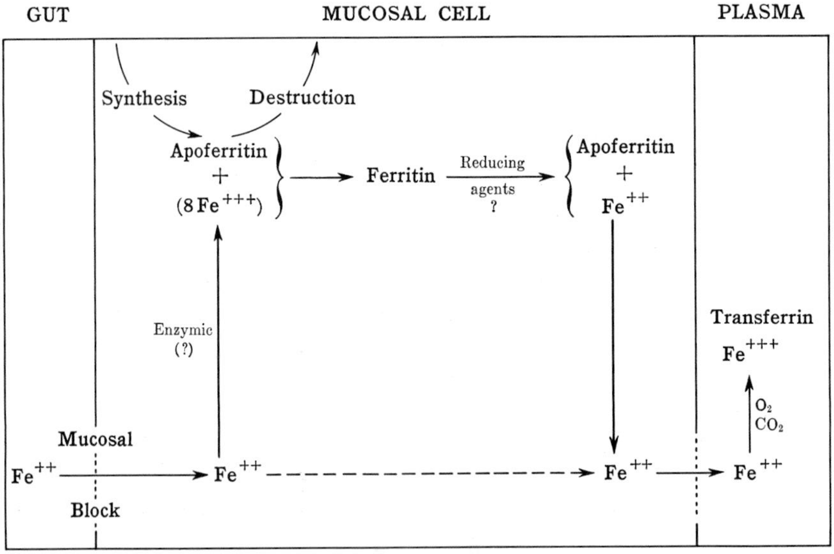

Fig. 31.12. Schematic version of the regulation of iron absorption by the mucosal cell. The amount of ferrous iron moving into the cell is regulated by the mucosal block, which is related to the level of ferrous iron in the cell and, indirectly, to the ferritin concentration. The amount of iron leaving the cell may depend upon the relative redox level of the cell, which in turn may be a function of the P_{O_2} of the blood. (*Modified from S. Granick, Physiol. Revs.,* **31,** 497, 1951.)

the adult male contains approximately 700 mg. of iron, present almost entirely as ferritin. After parenteral administration of ^{59}Fe, much of the isotope is found in the liver as ferritin. All parenterally administered iron, in excess of the ferritin storage mechanism, accumulates in the liver as *hemosiderin,* which is a normal constituent of most tissues and represents colloidal iron in the form of granules much larger than ferritin molecules. Hemosiderin granules contain up to 37 per cent of their dry weight as iron, are insoluble in water, and differ from ferritin in electrophoretic mobility. Hemosiderin granules are believed to be large aggregates of ferritin molecules with a higher content of iron. Since there is no excretory pathway for excess iron, continued administration of iron leads to hemosiderin accumulation in the liver in quantities sufficient to result in ultimate destruction of the organ. This has been observed in patients with aplastic or hemolytic anemia who have received multiple transfusions over several years.

In the absence of an excretory pathway for excess iron, the adjustment of intestinal iron absorption to need is essential to the organism. No other nutrient is known to be regulated in this manner.

Iron Transport. Another component of the iron metabolizing system is a plasma protein, *transferrin* (page 721), which facilitates iron transport and is present in a concentration of approximately 0.4 g. per 100 ml. blood (Table 31.5). At the normal level of plasma iron, 100 μg per 100 ml., transferrin is 30 per cent saturated. Transferrin exhibits a much greater affinity for iron than do the other plasma proteins. The failure of the kidney to excrete iron may result from the fact that all the plasma iron is bound to the transferrin, which is not filtrable. The homeostatic mechanisms involved in iron metabolism are summarized in Fig. 31.13.

ASPECTS OF ABNORMAL ERYTHROCYTE STRUCTURE AND FUNCTION

Aberrations in red cell structure and function represent a large area of knowledge beyond the scope of this book. However, some aspects of this topic deserve consideration because either their biochemical basis is known, or certain established biochemical alterations in erythrocytes may, in future, be shown to have significance for understanding of a particular structural or functional abnormality of the red cell.

Alterations in erythrocyte structure and function may be considered from the standpoint of the following variations from normal: (1) hemoglobin structure; (2) normal numbers of erythrocytes but with hemoglobin unavailable for transport of oxygen; (3) excessive or (4) subnormal amounts of hemoglobin accompanied by corresponding alterations in numbers of erythrocytes.

Variations in Hemoglobin Structure. In a general way, the hemoglobins exhibit three types of heterogeneity. As indicated previously (page 171), the globins are species-specific and show rather extensive differences in amino acid sequences. Within a species there is also heterogeneity based upon the normal ability to form β, γ, and δ chains, each of which may combine with α chains so that $\alpha_2\beta_2$, $\alpha_2\gamma_2$, and $\alpha_2\delta_2$ are normal to each individual. A few rare instances have been reported in which adult blood contains Hb H $= \beta_4{}^A$. In fetal life and early infancy the blood of such individuals contains $Hb_{Barts} = \gamma_4{}^F$. Since these components constitute only about 15 per cent of the total hemoglobin while the remainder is Hb F in fetal

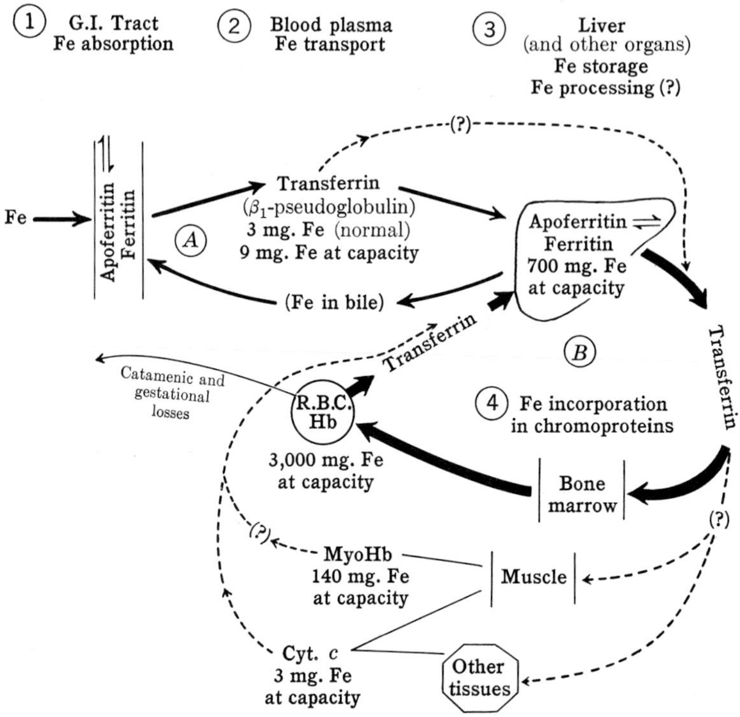

Fig. 31.13. The metabolism of iron. The cyclic movement of iron is considered in two stages, (A) the absorption, transport, storage, and excretion cycle, and (B) the storages, transport, and utilization cycle. Note the central role of the liver in both cycles. (*After D. L. Drabkin, Physiol. Revs.*, **31**, 345. 1951.)

life and Hb A thereafter, it is apparent that these individuals possess the genetic capacity to make α chains, but do not make sufficient α chains to meet their total requirements. It is noteworthy, therefore, that in the absence of α chains, a structure is formed which is a tetramer of either β or γ chains.

The third form of hemoglobin heterogeneity occurs within individuals as a consequence of a mutation which has led to an amino acid replacement in a chain (Chap. 30). This was first recognized in the case of the hemoglobin of individuals with *sickle cell anemia.* This disorder is characterized by the fact that the erythrocytes of affected individuals change from their normal shape (biconcave disk) to a sickle or crescent shape when the oxyhemoglobin of the cell is reduced to hemoglobin as a result of exposure to reduced oxygen tension. Sickling occurs in the capillaries as the hemoglobin is reduced, and about 85 per cent of the hemoglobin crystallizes from solution within the erythrocyte, thereby distorting the cell. The sickled cell exhibits markedly increased mechanical fragility and tends to hemolyze in the small diameter of the capillary, leading to reduced numbers of erythrocytes; hence the term, sickle cell anemia.

Erythrocytes which undergo sickling contain a hemoglobin distinguishable from Hb A in that it migrates differently in an electrical field. The magnitude of this altered migration rate is that expected if there were two fewer negative charges per hemoglobin molecule; this correlates well with the fact that in Hb S, a valine residue replaces the glutamic acid residue at position 6 of the β chains (Fig. 8.1, page 173).

Many other hemoglobins with amino acid replacements have since been detected. Because of their significance to the understanding of genetic mechanisms, these are summarized elsewhere (Table 30.1, page 689). Particularly noteworthy are the many different types of Hb M, in which, for example, a glutamic acid residue may replace the valine residue at position 67 or a tyrosine residue may replace the histidine residue at position 63; both replacements occur in the β chain. These changes near the heme iron result in its ready oxidation by O_2 to yield methemoglobin; hence the designation Hb M. In individuals with Hb M, no more than 30 per cent of the total hemoglobin is present as Hb M. Presumably, an individual with all his hemoglobin as Hb M could not transport O_2 in his blood at a rate sufficient to support normal functions.

Alterations in Functioning of Hemoglobin. In addition to the presence of methemoglobin in individuals with Hb M, *methemoglobinemia* may result from exposure to agents which cause oxidation of the ferrous iron of hemoglobin to the ferric state. These agents include amyl nitrite, aniline, nitrobenzene, nitrates, nitrites, etc., and methemoglobinemia has been observed clinically after administration of sulfonamides, acetanilid, phenacetin, and salicylates. A third type of methemoglobinemia, *familial methemoglobinemia* (Table 30.6, page 700), is a rare hereditary deficiency of a *DPNH-methemoglobin reductase* (page 740) in the erythrocytes of afflicted individuals. In these circumstances, as much as 25 to 40 per cent of the total hemoglobin may be present as methemoglobin, with accompanying evidence of cyanosis.

Carbon Monoxide Poisoning. In carbon monoxide poisoning, carbon monoxide displaces oxygen from oxyhemoglobin. The affinity of hemoglobin for carbon monoxide is more than 200 times as great as its affinity for oxygen. In the presence of CO the oxygen dissociation curve of oxyhemoglobin is "shifted to the left" (page 761), suggesting that the number of molecules of oxygen attached to hemoglobin is less than four (page 762), there being a partial replacement of O_2 by CO. These molecules of hemoglobin to which CO as well as O_2 is bound are more tenacious of O_2 than is the normal hemoglobin containing four molecules of oxygen, *i.e.*, $Hb(O_2)_4$. Consequently, the effects of carbon monoxide poisoning are more severe than would be expected on the basis of reaction of only a portion of the hemoglobin molecules with CO.

Poisoning by cyanide or by sulfide results from formation of complexes between these anions and the ferric iron of cytochrome oxidase. Methemoglobin but not hemoglobin forms similar complexes. Since the body contains far more hemoglobin than cytochrome, therapy for cyanide-poisoned individuals consists in accelerating methemoglobin formation, *e.g.*, by amyl nitrite administration. The resulting cyanmethemoglobin is itself nontoxic and can be slowly metabolized. The sulfmethemoglobin resulting from the combination of sulfide with the ferric iron of methemoglobin cannot be further metabolized and, once formed, remains until the cell is phagocytized.

Polycythemia. Polycythemia is the term used to describe the presence of unusually large numbers of erythrocytes in the circulation. The mechanism by which low oxygen tension stimulates erythropoiesis is unknown; its effects are seen as an adaptation to high altitudes and in the extremely high red cell content of children with congenital abnormalities of the pulmonary circulation ("blue babies"). This does not occur as a reaction of the erythropoietic tissue to hypoxia. Plasma from animals rendered anemic by various means induces polycythemia when injected into normal animals. This effect has been ascribed to the presence, in such plasma, of an agent called *erythropoietin,* a mucoprotein formed in the kidney. Purified preparations of erythropoietin, obtained from plasma and from urine, have a molecular weight of less than 15,000 and contain about 30 per cent of carbohydrate, of which about 17 per cent is present as hexosamine and 15 per cent as sialic acid. The latter is necessary for the biological activity of erythropoietin.

Administration of cobalt salts also results in polycythemia; the mechanism of this effect is unknown.

Anemia. Anemia due to a decrease in the number of erythrocytes has been mentioned previously in consideration of Hb S (see above). The presence of normal numbers of erythrocytes with lower than normal concentration of hemoglobin may also be the basis of anemia. This subject is beyond the scope of this textbook; brief reference will be made here, as well as in Part Six, to some biochemical aspects of certain of the anemias.

Anemia is more accurately described by measurements of the *total circulating red cell* mass than by the hemoglobin or red cell concentration of a blood sample since the total mass is independent of fluctuations in plasma volume. Accurate estimation of the red cell mass is possible by isotope dilution analysis. A sample of blood is incubated with glucose and ^{32}P as inorganic phosphate, or with radioactive ^{51}Cr as inorganic chromate, which is adsorbed on the surface of the red cells. The labeled blood is then injected, a second sample of blood is withdrawn after a suitable time, and the red cell mass is calculated from the extent of dilution of the administered radioactivity.

Anemias of Nutritional Origin. A group of anemias of nutritional origin are those characterized by the presence of unusually large cells in the peripheral blood (mean corpuscular volume, 95 to 160 μ^3; normal, about 87 μ^3). Since their hemoglobin concentration is normal, or greater than normal, the mean corpuscular hemoglobin content is also elevated (0.03 to 0.05 mμg; normal, about 0.03 mμg). Examples of these anemias are those due to deficiency in vitamin B_{12} (cyanocobalamine) or folic acid (Chap. 49).

Pernicious (Addison's) Anemia. The pathogenesis of pernicious anemia has long been associated with impaired gastric function. The gastric juice of persons with pernicious anemia contains no HCl and little or no pepsin. However, it is not the lack of pepsin or HCl which underlies the hemopoietic failure but rather a defect in production by the gastric mucosa of one or more mucoproteins (*intrinsic factor*), essential for the normal absorption of dietary vitamin B_{12} (*extrinsic factor*) from the intestinal tract. Purified but heterogeneous preparations of intrinsic factor from hog pyloric mucosa, with a mean molecular weight of approximately 50,000, contain one mole of vitamin B_{12} per mole of complex. The latter slowly dimerizes on incubation at 37°C. Hydrolysis of the protein yielded about 7 per cent of reducing sugars, including glucose, mannose, fucose, N-acetylglucosamine, N-acetylgalactosamine, and a sialic acid. Biological activity was rapidly destroyed by incubation with neuraminidase.

Lack of virtually any essential nutrient in the diet will contribute to anemia. Aspects of nutritional anemias will be considered in Part Six of this book.

Anemias Due to Accelerated Erythrocyte Destruction. Hemolytic anemias may occur in malaria, blackwater fever, or other infections. Hemolysis also occurs following mismatched blood transfusion and in *erythroblastosis fetalis* (hemolytic disease of the newborn). Unusual anemia of *paroxysmal hemoglobinuria* is characterized by hemolysis and hemoglobinuria in certain individuals after exposure to cold. This phenomenon appears to be due to the presence of so-called "cold" hemolysins in the blood. The blood of persons with *familial* or *congenital hemolytic jaundice* is characterized by the presence of spherical erythrocytes of less than normal diameter (spherocytes), with a markedly increased fragility to hypotonic solutions. The reasons for this appearance of the red cell are not known. The tendency to spherocyte formation is transmitted as a mendelian dominant characteristic.

Particular interest attaches to a genetically transmitted disorder termed "primaquine sensitivity." In this disturbance, erythrocytes hemolyze on administration of a wide variety of agents, including the antimalarial drug primaquine. Hemolysis seems to relate to oversensitivity to peroxides as a consequence of low concentration of reduced glutathione. In normal erythrocytes glutathione is maintained in the reduced state by glutathione reductase (page 548), which utilizes TPNH as a reductant. Primaquine-sensitive erythrocytes cannot provide TPNH at a normal rate; hence they also fail to reduce methemoglobin as it is formed. This difficulty arises consequent to any failure of the direct oxidation pathway for glucose metabolism, the prime source of TPNH. In most instances, this is due to markedly diminished activity of glucose 6-phosphate dehydrogenase. However, an essentially similar situation has been observed in the erythrocytes of individuals with genetically transmitted reduced activity of 6-phosphogluconate dehydrogenase, glutathione reductase, diphosphoglycerate mutase, phosphoenolpyruvate kinase, or the enzymes catalyzing synthesis of glutathione.

Thalassemia is a heritable disorder with increased tendency to hemolysis based on an intracorpuscular defect and characterized by large amounts of Hb $F^{\alpha_2\gamma_2}$; this is a *normal* hemoglobin appearing in *abnormal quantities.* No abnormal hemoglobin appears to be present.

In the rare disorder called *acanthocytosis* (page 528), erythrocytes are excessively susceptible to hemolysis and their life span is slightly shortened; hemolytic anemia is rare.

LEUKOCYTE COMPOSITION AND METABOLISM

The development of improved methods for separation from the blood of the two major classes of circulating leukocytes, the polymorphonuclear leukocytes and the lymphocytes, has provided adequate numbers of these cells for metabolic studies. In addition, lymphocytes are readily available either by lymphatic cannulation or by preparing cells from minced tissue which is predominately lymphocytic in cellular structure, *e.g.*, thymic tissue. Also, large numbers of polymorphonuclear leukocytes can be obtained for metabolic studies from the peritoneal cavity of suitable animals, *e.g.*, rabbits, by withdrawing peritoneal contents which have accumulated subsequent to intraperitoneal injection of an irritant such as mineral oil.

The inorganic ion content of leukocytes is unusual only in that there is present a high concentration of zinc, about twenty-five times that in erythrocytes. Unlike the erythrocytes, leukocytes possess organized systems of both respiratory and glycolytic enzymes. It is of interest that leukocytes of patients with von Gierke's disease (page 446) have a glycogen content five- or sixfold greater than normal.

The phagocytic forms of leukocytes are rich in a variety of hydrolytic enzymes, including proteinases localized in the lysosomes (page 285) of these cells. Phagocytosis is characterized by active proteolysis, although little is known of the protein metabolism of leukocytes or of the ultimate fate of the phagocytized protein. Stimulation of glycolysis is evident, and there is an increased turnover of triose phosphates, phosphatidic acid, and inositol-containing phospholipid. The stimulated turnover of specific phospholipids at the time of transfer of particles into phagocytizing cells could involve participation of acidic phosphatides in membrane functions. In addition to the above alterations, during phagocytosis a DPNH oxidase present in the granule fraction is activated.

Particular attention has centered around the role of the lymphocytes in immune phenomena. Normal lymphocytes contain a protein identical with plasma γ-globulin, and in the immunized animal antibody can be demonstrated in lymphocytes obtained from lymphoid structures. Passive immunity can be transferred to a nonimmune recipient host by transplantation of lymphocytes from a previously immunized animal. Also, heterologous tissue transplantation is more readily effected to a host thymectomized neonatally, treated with large doses of adrenal cortical steroids, or exposed to x-rays. All these procedures result in involuted lymphoid structures, with a consequent decreased production of lymphocytes and a marked lymphopenia. The role of lymphoid cells in protein synthesis and, specifically, in antibody formation has been clearly established in both in vivo and in vitro studies.

Evidence that the thymus is the site of development of the first immunologically competent cells in the newborn has indicated a possible role for humoral agents, secreted by the thymus, in immunobiological phenomena. A biologically active preparation has been obtained in partially purified form from calf thymic tissue. The product, designated as *thymosin,* stimulates lymphocytopoiesis in vivo and favorably influences immunological competence in neonatally thymectomized animals. This factor has also been demonstrated to accelerate the rate of lymphoid tissue regeneration after its involution by exposure of mice to whole-body x-radiation, and to play a role in graft vs. host reactions. Thymosin appears to be a relatively heat stable protein. This hormonal agent probably plays a role in the homeostatic control of lymphoid tissue structure and function.

Studies with cell suspensions of lymphocytes in vitro have revealed a glycolytic pathway and a significant, continuing endogenous respiration in the absence of exogenous substrate, with an R.Q. indicative of fatty acid oxidation. Lymphocytes and separated lymphocyte nuclei in vitro are capable of incorporating labeled precursors into the total proteins and nucleic acids of these structures. Addition of physiological doses ($10^{-7}M$) of a thymolytic steroid, *e.g.*, cortisol, to a lymphocyte suspension in vitro markedly inhibits this incorporation, and decreases RNA polymerase activity. Considerable attention has centered around nucleic acid metabolism of leukocytes in view of their prominent nuclei and rapid rate of cellular regeneration, both in the normal and in circumstances of exaggerated leukopoiesis. These processes exhibit a high requirement for folic acid, related to the role of this vitamin in purine biosynthesis (page 622). *Dihydrofolate reductase* required for synthesis of thymidylic acid (page 639), has been reported to be present in high amounts in leukocytes of acute leukemic and chronic myelogenous leukemic patients, but to be in very low concentration or absent in normal cells or those from individuals with chronic lymphatic leukemia. This enzyme is specifically inhibited by folic acid antagonists, *e.g.*, being affected by these substances in concentrations of $10^{-8}M$. Drugs which inhibit this enzyme, and thus folic acid utilization, have been tested as therapeutic agents in certain types of cancer (page 644).

REFERENCES

Books

Albritton, E. C., ed., "Standard Values in Blood," W. B. Saunders Company, Philadelphia, 1952.

"Antibodies," Cold Spring Harbor Symposia on Quantitative Biology, vol. 32, Biological Laboratory, Cold Spring Harbor, N.Y., 1967.

Behrendt, H., "Chemistry of Erythrocytes," Charles C Thomas, Publisher, Springfield, Ill. 1957.

Biggs, R., and MacFarlane, R. G., "Human Blood Coagulation and Its Disorders," 3d ed., Blackwell Scientific Publications, Ltd., Oxford, 1962.

Bishop, C., and Surgenor, D. M., eds., "The Red Blood Cell," Academic Press, Inc., New York, 1964.

Boutwell, T. H., and Finch, C. A., "Iron Metabolism," Little, Brown and Co., Boston, 1962.

Boyd, W. C., "Introduction to Immunochemical Specificity," Interscience Publishers, Inc., New York, 1962.

Cinader, R., ed., "Regulation of the Antibody Response," Charles C Thomas, Publisher, Springfield, Ill., 1967.

Dittmer, D. S., ed., "Blood and Other Body Fluids," Federation of American Societies for Experimental Biology and Medicine, Washington, 1961.

Elves, M. W., "The Lymphocytes," Lloyd-Luke, Ltd., London, 1966.

Good, R. A., and Gabrielsen, A. E., eds., "The Thymus in Immunobiology: Structure Function and Role in Disease," Hoeber Medical Division, Harper and Row, Publishers, Incorporated, New York, 1964.

Gray, C. H., "Bile Pigments in Health and Disease," Charles C Thomas, Publisher, Springfield, Ill., 1963.

Harris, J. W., "The Red Cell: Production, Metabolism, Destruction: Normal and Abnormal," Harvard University Press, Cambridge, Mass., 1963.

Ingram, V. M., "Hemoglobin and Its Abnormalities," Charles C Thomas, Publisher, Springfield, Ill., 1961.

Janeway, C. A., Rosen, F. S., Merler, E., and Alper, C. A., "The Gamma Globulins," Little, Brown and Company, Boston, 1967.

Kowalski, E., and Niewiarowski, S., eds., "Biochemistry of Blood Platelets," Academic Press, Inc., New York, 1967.

Landsteiner, K., "The Specificity of Serological Reactions," Harvard University Press, Cambridge, Mass., 1945.

Metcalf, D., "The Thymus," Springer-Verlag OHG, Berlin, 1966.

National Research Council, Conference on Hemoglobin, *Nat. Acad. Sci. Publ.* 557, Washington, 1958.

Putnam, F. W., ed., "The Plasma Proteins," vols. I and II, Academic Press, Inc., New York, 1960.

Schultze, H. E., and Heremans, J. F., "Molecular Biology of Human Proteins with Special Reference to Plasma Proteins," vol. I, Nature and Metabolism of Extracellular Proteins, 1966; vol. 2, Physiology and Pathology of Plasma Proteins, 1967, Elsevier Publishing Company, Amsterdam, distributed by American Elsevier Publishing Company, New York.

Seegers, W. H., "Prothrombin," Harvard University Press, Cambridge, Mass., 1962.

Seegers, W. H., ed., "Blood Clotting Enzymology," Academic Press, Inc., New York, 1967.

Sterzl, J., ed., "Molecular and Cellular Basis of Antibody Formation," Academic Press, Inc., New York, 1965.

Sunderman, F. W., and Boerner, F., "Normal Values in Clinical Medicine," W. B. Saunders Company, Philadelphia, 1949.

Williams, C. A., and Chase, M. W., eds., "Methods in Immunology and Immunochemistry," vol. I, Preparation of Antigens and Antibodies; vol. 2, Physical and Chemical Methods. Academic Press, Inc., New York, 1967.

Wolstenholme, G. E. W., and O'Connor, M., "Haemopoiesis: Cell Production and Its Regulation," J. and A. Churchill, Ltd., London, 1960.

Review Articles

A Discussion of the Chemistry and Biology of the Immunoglobulins, *Proc. Roy. Soc. London, Ser. B*, **166**, 113–243, 1966.

Berlin, N. I., Waldmann, T. A., and Weisman, S. M., Life Span of Red Blood Cells, *Physiol. Revs.*, **39**, 577–616, 1959.

Cline, M. J., Metabolism of the Circulating Leukocyte, *Physiol. Revs.*, **45**, 674–720, 1965.

Cohen, S., and Porter, R. R., Structure and Biological Activity of Immunoglobulins, *Advances in Immunol.*, **4**, 287–349, 1964.

Davie, E. W., and Ratnoff, O. D., The Proteins of Blood Coagulation, in H. Neurath, ed., "The Proteins," 2d ed., vol. II, pp. 359–443, Academic Press, Inc., New York, 1965.

Dutton, R. W., *In Vitro* Studies of Immunological Responses of Lymphoid Cells, *Advances in Immunol.*, **6**, 253–336, 1967.

Fleischman, J. B., Immunoglobulins, *Ann. Rev. Biochem.*, **35**, 835–872, 1966.

Gordon, A., Hemopoietin, *Physiol. Revs.*, **39**, 1–40, 1959.

Granick, S., Structure and Physiological Functions of Ferritin, *Physiol. Revs.*, **31**, 489–511, 1951.

Granick, S., Porphyrin Biosynthesis, Porphyria Diseases, and Induced Enzyme Synthesis in Chemical Porphyria, *Trans. N.Y. Acad. Sci.*, **25**, 53–65, 1962.

Granick, S., and Mauzerall, D., The Metabolism of Heme and Chlorophyll, in D. M. Greenberg, ed., "Metabolic Pathways," vol. II, pp. 525–616, Academic Press, Inc., New York, 1961.

Haurowitz, F., Antibody Formation, *Physiol. Revs.*, **45**, 1–47, 1965.

Heller, P., Hemoglobinopathic Dysfunction of the Red Cell, *Am. J. Med.*, **41**, 799–814, 1966.

Hoffman, J. F., The Red Cell Membrane and the Transport of Sodium and Potassium, *Am. J. Med.*, **41**, 666–698, 1966.

Karnovsky, M., Metabolic Basis of Phagocytic Activity, *Physiol. Revs.*, **42**, 143–168, 1962.

Kline, D. L., Blood Coagulation: Reactions Leading to Prothrombin Activation," *Ann. Rev. Physiol.,* **27,** 285–306, 1965.
Kunkel, H. G., Myeloma Proteins and Antibodies, *Harvey Lectures,* **59,** 219–242, 1963–1964.
Laki, K., and Gladner, J. A., Chemistry and Physiology of the Fibrinogen-Fibrin Transition, *Physiol. Revs.,* **44,** 127–160, 1964.
Lennox, E. S., and Cohn, M., Immunoglobulins, *Ann. Rev. Biochem.,* **36,** 365–406, 1967.
Link, K. P., The Anticoagulant from Spoiled Sweet Clover Hay, *Harvey Lectures,* **39,** 162–216, 1943–1944.
London, I. M., The Metabolism of the Erythrocyte, *Harvey Lectures,* **56,** 151–189, 1960–1961.
Makman, M. H., Nakagawa, S., and White, A., Studies of the Mode of Action of Adrenal Steroids on Lymphocytes, *Recent Progr. Hormone Research,* **23,** 195–227, 1967.
Marcus, A. J., The Role of Lipids in Blood Coagulation, *Advances in Lipid Research,* **4,** 1–37 1966.
Miller, J. F. A. P., and Osaba, D., Current Concepts of the Immunological Function of the Thymus, *Physiol. Revs.,* **47,** 437–520, 1967.
Moore, C. V., Iron Metabolism and Nutrition, *Harvey Lectures,* **55,** 67–101, 1959–1960.
Nisonoff, A., and Thorbecke, G. J., Immunochemistry, *Ann. Rev. Biochem.,* **33,** 355–402, 1964.
Porter, R. R., The Structure of Immunoglobulins, in P. N. Campbell and G. D. Greville, eds., "Essays in Biochemistry," **3,** 1–25, Academic Press, Inc., New York, 1967.
Prentice, C. R. M., and Ratnoff, O. D., Genetic Disorders of Blood Coagulation, *Seminars in Hematology,* **4,** 93–132, 1967.
Putnam, F. W., Structure and Function of the Plasma Proteins, in H. Neurath, ed., "The Proteins," 2d ed., vol. III, pp. 153–267, Academic Press, Inc., New York, 1965.
Ratnoff, O. D., Hereditary Disorders of Hemostasis, in J. B. Stanbury, J. B. Wyngaarden, and D. S. Frederickson, eds., "The Metabolic Basis of Inherited Disease," 2d ed., pp. 1137–1175, McGraw-Hill Book Company, New York, 1966.
Riggs, D. R., Homeostatic Regulatory Mechanisms of Hematopoiesis, *Ann. Rev. Physiol.,* **28,** 39–56, 1966.
Rimington, C., and Kennedy, G. Y., Porphyrins: Structure, Distribution, and Metabolism, in M. Florkin and H. S. Mason, eds., "Comparative Biochemistry," vol. IV, part B, pp. 557–615, Academic Press, Inc., New York, 1962.
Scheraga, H. A., and Laskowski, M., Jr., The Fibrinogen-Fibrin Conversion, *Advances in Protein Chem.,* **12,** 1–131, 1957.
Shorr, E., Intermediary Metabolism and Biological Functions of Ferritin, *Harvey Lectures,* **50,** 112–153, 1954–1955.
Singer, S. J., Structure and Function of Antigen and Antibody Proteins, in H. Neurath, ed., "The Proteins," 2d ed., vol. III, pp. 269–357, Academic Press, Inc., New York, 1965.
Watson, C. J., The Problem of Porphyria: Some Facts and Questions, *New Engl. J. Med.,* **263,** 1205–1215, 1960.
White, A., and Goldstein, A., The Thymus as an Endocrine Gland. Old Problem, New Data, *Perspectives in Biol. and Med.,* **11,** no. 3, 1968.

32. Chemistry of Respiration

Primitive organisms rely on diffusion through their environmental media to provide the oxygen needed for their metabolism and to remove the carbon dioxide produced. The active metabolism of mammalian tissues remote from the atmosphere is possible because of a mechanism which provides constant delivery of oxygen and removal of carbon dioxide. The magnitude of this task may be appreciated from the fact that a man oxidizing 3000 Cal. of mixed food per day uses about 600 liters of oxygen (27 moles) and produces about 480 liters of carbon dioxide (22 moles). Through the action of hemoglobin, oxygen is abstracted from the air, carried within a few seconds to the most distant parts of the body, and delivered to the tissues at a pressure only slightly less than that at which it existed in the atmosphere. The CO_2 produced daily by the tissues becomes H_2CO_3, an acid, in an amount equivalent to 2 liters of concentrated hydrochloric acid; yet all this acid normally pours from the tissues, through the blood, and out of the lungs with a change in the pH of blood of no more than a few hundredths of a pH unit. This chapter will describe the means by which these enormous tasks are accomplished.

THE RESPIRATORY GASES

The pressure which a gas exerts, when mixed with other gases, is the partial pressure of that gas and is denoted by the symbol P. This pressure is a function of the temperature and of the number of molecules of gas in a given volume. At constant temperature and volume, equal numbers of molecules of all ideal gases exert equal pressures, and the total pressure exerted by a gas mixture is equal to the sum of all the partial pressures in the mixture. The barometric pressure of atmospheric air is thus the sum of the partial pressures of O_2, CO_2, H_2O, N_2, etc.

The amount of any gas present in solution is proportional to the partial pressure of that gas in the total gas mixture with which the solution is in equilibrium; occasionally it is convenient to describe the concentration of a gas in solution by stating the partial pressure with which that solution might be in equilibrium. This is the *tension* of the gas in that solution and is expressed in the same units as the pressure in a gas phase, millimeters of mercury (mm. Hg). However, the actual amount of gas which will dissolve per unit volume of solvent at a given partial pressure varies with each gas. This is stated in the expression

$$C = kP$$

where C is milliliters gas per milliliter solvent, P is the partial pressure of the gas in the vapor phase in millimeters mercury, and k is the Bunsen absorption coefficient, a constant for a given gas in a given solvent at a specified temperature. The

k values for the important respiratory gases are given in Table 32.1, which also includes the effect of temperature on the solubility of these gases. The rate of diffusion of a gas through liquid (tissue in this case) varies directly with the absorption coefficient.

Table 32.1: ABSORPTION COEFFICIENTS OF RESPIRATORY GASES

Temperature, °C.	O_2	CO_2	N_2
Water:			
0	0.049	1.71	0.024
20	0.031	0.87	0.016
40	0.023	0.53	0.012
Plasma:			
38	0.024	0.510	0.012

Note: Values are ml. of gas (measured at standard conditions) which dissolve in 1 ml. of indicated solvent when the latter is equilibrated with the specified gas at 760 mm. Hg.

Inspired air mixes with the gas mixture present in the larger passages of the respiratory tract, the trachea, bronchi, and bronchioles. Some of this mixture, tidal air, is sucked into the expanding alveolar sacs, where the gases make contact with the pulmonary capillaries. From the alveolar gas mixture, O_2 diffuses across the capillary walls and into the circulating blood while CO_2 migrates in the reverse direction. On expiration, a portion of this alveolar air is forced up into the larger passages, where it mixes with the gas mixture already present, and from the tidal air a portion leaves as expired air. By proper adjustment of this tidal flow of gas, the rates of entry of O_2 and CO_2 into the alveoli equal the rates of loss, and the composition of alveolar air remains relatively constant with respect to these gases.

Although the pressure in the alveoli fluctuates rhythmically during the respiratory cycle, the mean pressure of the alveolar gas mixture is that of the atmosphere. However, alveolar air must also be saturated with water vapor evaporated from the lung surfaces. Since, at body temperature, the partial pressure of water vapor is 47 mm. Hg and is independent of the composition of the remainder of this mixture, the aqueous tension is a significant fraction of the total alveolar gas pressure and its importance must increase at diminished total pressure, *e.g.*, at high altitudes. The composition of inspired, expired, and alveolar air is shown in Table 32.2. The

Table 32.2: COMPOSITION OF THE RESPIRATORY GASES

	Inspired air		Alveolar air		Expired air	
	mm. Hg	vols. per cent	mm. Hg	vols. per cent	mm. Hg	vols. per cent
O_2	158.2	20.95	101.2	14.0	116.2	16.1
CO_2	0.3	0.04	40.0	5.6	28.5	4.5
N_2	596.5	79.0	571.8	80.0	568.3	79.2
H_2O	5.0	....	47.0	....	47.0	....
Total	760.0	99.99	760.0	99.6	760.0	99.8

composition of the alveolar gas mixture is determined by the *rate* at which alveolar air is mixed with tidal air and the latter with atmospheric air. Under normal conditions the respiratory apparatus maintains the CO_2 content of alveolar air relatively constant at 40 mm. Hg, although other components of air are not maintained with similar constancy.

Since the P_{O_2} in alveolar air is of the order of 100 mm. Hg, while that in the venous blood is about 50 mm. Hg or less, a concentration gradient exists across the capillary wall and oxygen diffuses across. As the blood rushes by, the O_2 of the alveolar gas and of arterial blood almost equilibrates and the P_{O_2} of arterial blood in man, at rest, is about 100 mm. Hg, while during vigorous exercise it may be 95 mm. Hg. This O_2 is then transported in the blood in two ways, (1) as oxygen in solution and (2) in chemical combination with the hemoglobin of erythrocytes. The limited solubility of oxygen permits transportation of only 0.3 ml. O_2 per 100 ml. of blood, and, even with a considerably increased cardiac output, this amount of oxygen does not meet metabolic requirements. However, since each gram of hemoglobin can combine with 1.34 ml. of O_2 and normal blood contains about 15 g. of hemoglobin per 100 ml., fully oxygenated blood may contain almost seventy times the amount of O_2 present in simple solution.

Thus, it will be apparent that comprehension of physiological transport of O_2 and CO_2 requires understanding of the chemistry of hemoglobin (pages 171*ff.*).

The Combination of Hemoglobin with Oxygen. The unique feature of hemoglobin is its ability to bind oxygen reversibly. Ferroporphyrin and many of its hemochromogens can also bind oxygen. However, in these instances the iron is rapidly oxidized to the ferric condition, whereas hemoglobin uniquely forms a stable oxygen complex in which the iron remains in the ferrous state. This special behavior of hemoglobin appears to be due to the fact that much of the heme of the molecule lies within a cover of hydrophobic groups of the globin, providing an environment of relatively low dielectric constant. This is suggested by the behavior of the model system previously discussed (page 175). Indeed much of the structure of hemoglobin may be required to inhibit the spontaneous oxidation of the ferrous iron by bound O_2. Hemoglobin $M_{Kankakee}$ ($\alpha_2{}^{87tyr}\beta_2$), in which the normal histidine ligand is replaced by tyrosine in the α chains, exists entirely with ferriprotoporphyrin on these chains. Although the opposing, nonbonded histidine at α^{58} or β^{63} also affords some protection, replacement by tyrosine, as in some forms of HbM, need not necessarily lead to rapid formation of methemoglobin since several invertebrate hemoglobins also contain tyrosine at this locus, yet are stable to O_2. Myoglobin and preparations of monomeric α or β chains oxidize several times more rapidly than does hemoglobin, indicating that some protection against autoxidation is provided by the tetrameric state. However, even normal hemoglobin iron is oxidized at a slow but highly significant rate, and erythrocytes require a mechanism for restoration of the Fe^{++} condition (page 740).

Factors Affecting the Combination of Hemoglobin with Oxygen. The myoglobins of both vertebrate and invertebrate muscle, like human hemoglobin, reversibly bind oxygen. Increasing P_{O_2} promotes the formation of their oxygenated forms. The relationship between oxygen tension and the formation of the oxygenated compounds is shown in Fig. 32.1. The depicted curve is a rectangular hyperbola, as expected from the mass law for the dissociation of oxymyoglobin, formulated as $MbO_2 \rightleftharpoons$

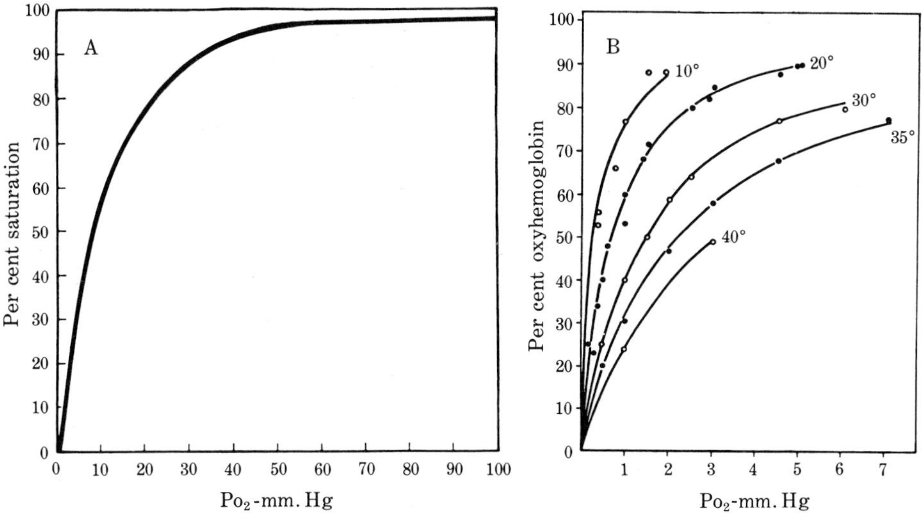

Fig. 32.1. *A,* General form of curve showing oxygenation of myoglobin as a function of oxygen tension. *B,* Effect of temperature on oxygen binding of myoglobin. (*Courtesy of Dr. E. Antonini.*)

Mb + O_2 (page 243). In contrast, the dissociation curve for the oxyhemoglobin of normal human blood and that of many other species is sigmoidal (Fig. 32.2). The sigmoidal curve was long interpreted as indicating that the presence of oxygen on one heme group of hemoglobin affects the dissociation constants of the other heme groups on the same molecule, an effect which, from the shape of the curve, must be greatest for the fourth dissociation. This behavior is described by the Hill equation,

$$Y = 100\left[\frac{(P/P_{1/2})^n}{1 + (P/P_{1/2})^n}\right]$$

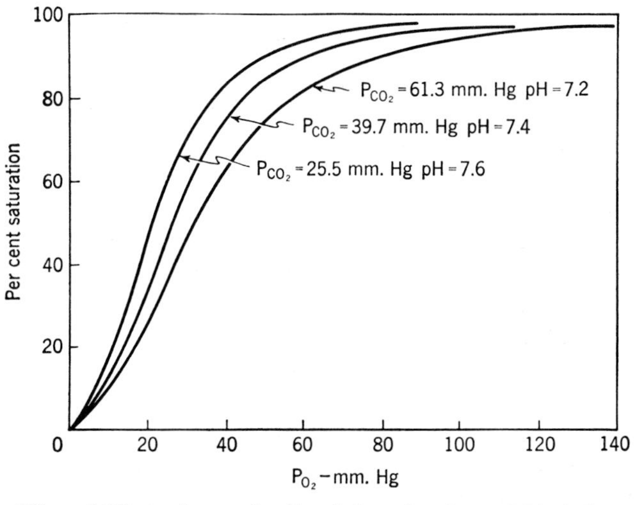

Fig. 32.2. Effect of CO_2 tension on the dissociation of oxyhemoglobin in human blood.

where Y is the per cent of hemoglobin combined with O_2 at pressure P, $P_{1/2}$ is the oxygen pressure at which 50 per cent of the hemoglobin exists as oxyhemoglobin, and n is a constant for a given species of hemoglobin. If $n = 1$, the oxygen dissociation curve is hyperbolic, as in Fig. 32.1; the greater the value for n, the more sigmoidal the curve. For normal human hemoglobin, $n = 2.9$; this constant has no physical meaning since the Hill equation is empirical, but is extremely useful for characterizing hemoglobins from different species and the effects of various treatments on the properties of any hemoglobin. Each hemoglobin molecule may be combined with 0, 2, or 4 oxygen molecules; the resulting compounds may be designated as Hb_4, Hb_4O_4, and Hb_4O_8, respectively. The sigmoidal oxygen dissociation curve is a composite of the curves shown in Fig. 32.3.

Considerable evidence indicates that the sigmoidal character of the oxygen dissociation curve reflects a significant change in the conformation of the tetrameric molecule as it binds or releases oxygen. Some of the more significant observations include: (1) Crystals of hemoglobin are observed to crack when oxygen is admitted. (2) X-ray crystallography reveals that, on deoxygenation, the β chains move, as if on a hinge, at once separating both from each other and from the α chains. (3) There

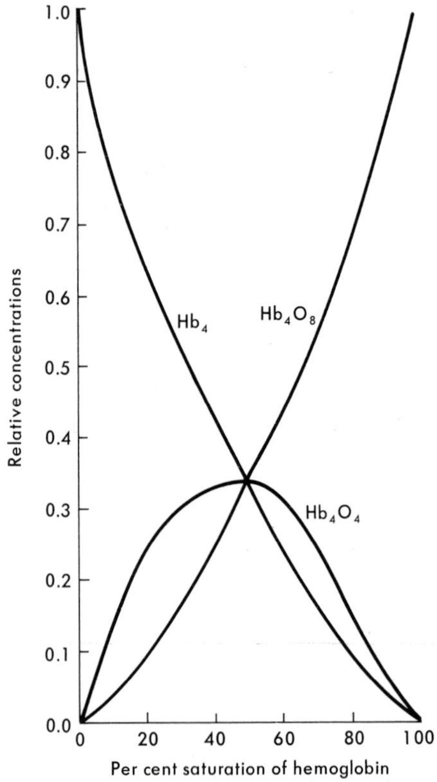

Fig. 32.3. Approximate relationships in hemoglobin-oxyhemoglobin solutions.

is a large difference between the dielectric properties of the two forms. (4) Hb H which is β_4, as well as laboratory preparations of β_4 and γ_4, show no "heme-heme" interaction, *i.e.*, $n = 1$. Hence, the binding of α to β chains is important to the phenomenon. (5) The tetrameric structure is not essential; dimeric $\alpha\beta$ shows $n = 2.6$, whereas for the monomers $n = 1$. (6) There are marked differences between Hb and HbO_2 in the rapidity of reaction with sulfhydryl reagents and the titer of carboxyl groups. (7) Carboxypeptidases A plus B remove the C-terminal 3 residues of the sequence, lysyltyrosylarginine, more rapidly from the β chains of HbO_2 than from those of Hb. The resultant hemoglobin then exhibits $n = 1$. (8) The sigmoidal curve, indicative of cooperativity or "heme-heme" interaction, is independent of the ligand and is also obtained for the binding of CO, ethyl isocyanide, and aromatic nitroso compounds. CO can be removed from Hb-CO by flash photolysis. The rate of recombination of CO with the liberated Hb is much greater than with an ordinary preparation of Hb, indicating that after loss of the CO, the tetramer briefly retains the conformation of the liganded form and that, during this period, it is more reactive with CO, and presumably with O_2, than in the normal conformation of reduced hemoglobin. (9) The latter observation indicates that although $\alpha_2\beta_2$ is in equilibrium with 2 $\alpha\beta$ and with free α and β, dissociation into dimeric or monomeric forms is unecessary to the reaction process. Indeed, $HbM_{Kankakee}$, described above, in which the α chain irons are ferric, behaves in the same manner as HbA in the CO flash photolysis system. (10) Tetrameric "Hb_4O_4" formed by oxygenation is a dimer of $\alpha\beta \cdot O_4 + \alpha\beta$. (11) Binding of O_2 to one chain of an $\alpha\beta$ dimer so greatly increases affinity for O_2 that partially oxygenated solutions of $\alpha\beta$ dimer contain only $\alpha\beta$ and $\alpha\beta \cdot O_4$ and essentially no $\alpha\beta \cdot O_2$. Similarly, as shown in Fig. 32.3, hemoglobin solutions contain no Hb_4O_2 or Hb_4O_6. Thus, the $\alpha\beta$ dimer is the functional unit. (12) The sigmoidal dissociation curve can be reconstructed utilizing the dissociation constants for the partial reactions.

The sigmoidal character of the dissociation curve of oxyhemoglobin is of great physiological significance, since, as is evident in Fig. 32.2, although the saturation of hemoglobin is affected by O_2 tension over a wide range of pressure (20 to 80 mm.), arterial hemoglobin is virtually saturated at a P_{O_2} as low as 80 mm.

Figure 32.2 indicates that the equilibrium of the hemoglobin-oxygen system may be altered by varying the P_{CO_2} in the medium surrounding the erythrocyte, a phenomenon known as the *Bohr effect*. This effect is also evident with solutions of pure hemoglobin and is attributable entirely to the change in pH effected by a change in P_{CO_2}. Oxygenation of hemoglobin results in a shift of the pK_a of some acidic group on the peptide chains from 7.71 to 6.17, *viz.*, oxyhemoglobin is a stronger acid than is reduced hemoglobin. The reversible reaction may, therefore, be schematically represented as follows:

$$HHb^+ + O_2 \rightleftharpoons HbO_2 + H^+$$

The relationship is not stoichiometric; approximately 0.65 mole of O_2 is released as one equivalent of protons is bound.

The mechanism of the Bohr effect has been the object of intensive study. Much evidence indicates that the acidic group on reduced hemoglobin is an imidazolium, and it was considered likely that this is on the heme-linked histidyl. The latter possibility has been eliminated, and the identity of the specific reactive group is presently uncertain. However, it is clear that the Bohr effect reflects the same conformational change as that responsible for the sigmoidal oxyhemoglobin dissociation curve. Supporting this conclusion are the following observations: (1) Myoglobin and preparations of monomeric α or β chains show no Bohr effect, whereas the $\alpha\beta$ dimer behaves exactly like the tetramer. (2) Hemoglobin H (β_4) shows no Bohr effect. (3) Removal of the C-terminal tripeptide abolishes the Bohr effect. (4) The magnitude of the Bohr effect, measured as change in log $P_{1/2}$ per change in pH, is determined largely by the nature of the β chains. In a general way, the smaller the vertebrate, the larger the Bohr effect observed with its hemoglobin. Accordingly, mouse and donkey hemoglobins differ significantly in this regard. When their hemoglobins were hybridized to produce $\alpha_2^{\text{mouse}}\ \beta_2^{\text{donkey}}$ and $\alpha_2^{\text{donkey}}\ \beta_2^{\text{mouse}}$, both the magnitude of the Bohr effect and the value of n for the latter combination were strikingly like that of mouse hemoglobin, while the former more closely resembled the behavior of donkey hemoglobin. (5) In any case, as shown in Fig. 32.4, the classical Bohr effect describes the consequence of only one fraction of the hemoglobin titration curve. Further addition of acid, below the physiological pH range, actually has an opposite effect. For several natural hemoglobins, this entire curve may be shifted, qualitatively, in either direction along the abscissa. Since the P_{CO_2} in pulmonary and extrapulmonary capillaries differs markedly, in order to describe the process of oxygen carriage from lungs to tissues it is necessary to construct a family of curves differing in P_{CO_2}, such as those shown in Fig. 32.2. Increased P_{CO_2} displaces the curve to the right, and at a given P_{O_2}, an increase in P_{CO_2} decreases the amount of oxyhemoglobin within the red cell.

The curves shown in Fig. 32.2 describe the behavior of human blood, *i.e.*, of hemoglobin within the erythrocyte. When purified hemoglobin is studied in dilute buffers at corresponding pH, the curves obtained are markedly shifted to the left as compared with those shown. This effect has been correlated with the anionic

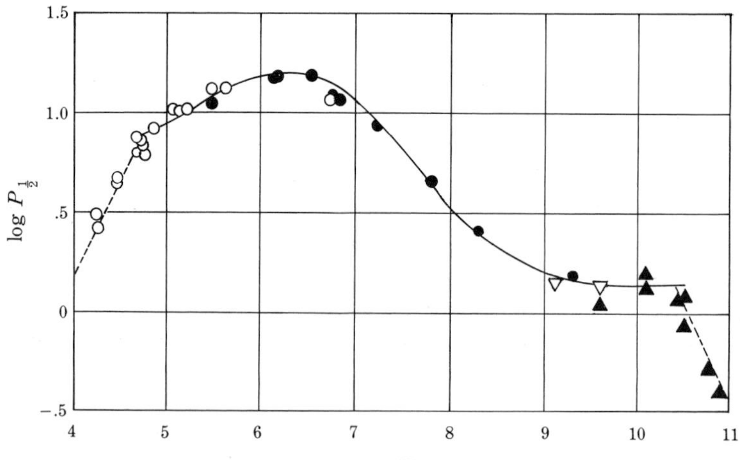

Fig. 32.4. Bohr effect in human hemoglobin. Temperature 20°C.; hemoglobin concentration 3 to 5 mg. per ml. Buffers as follows: ● = 0.2M phosphate; ○ = 0.4M acetate; ▽ = 0.05M borate; ▲ = 0.4M glycine. (*From E. Antonini, Physiol. Revs.,* **45**, 146, 1965.)

binding properties of hemoglobin. A variety of polyphosphate compounds bind to the basic groups of the globin and, presumably, subtly alter the protein conformation in such manner as to reduce the affinity for O_2. ATP, 2,3-diphosphoglycerate (page 735), and phytate (page 32) are particularly effective in this regard. In the adult human erythrocyte it is probably largely the 2,3-diphosphoglycerate which is responsible for the altered oxygen-binding properties of hemoglobin. This provides a satisfying explanation for the fact that, of all body tissues, erythrocytes contain the highest concentration of 2,3-diphosphoglycerate. Bird erythrocytes unload oxygen even more readily than do human cells; this has been shown to be due to their accumulation of phytate.

Oxygen Transport. The transport of oxygen from lungs to tissues is described by the curves shown in Fig. 32.2. In the lung, oxygen diffuses across the capillary lining in accordance with the existing gradient, then through plasma and into the erythrocytes. The P_{O_2} in erythrocytes leaving the lungs is about 100 mm. Hg, and the P_{CO_2} in the arterial blood is of the order of 40 mm. Hg. By referring to Fig. 32.2, it may be seen that the hemoglobin of arterial blood is about 96 per cent saturated.

The P_{O_2} in the interstitial fluid surrounding extrapulmonic capillaries cannot be accurately measured but is probably about 35 mm. Hg in muscle at rest, while the P_{CO_2} must be approximately 50 mm. Hg. Consequently, O_2 diffuses from red cells through plasma to interstitial fluid and then into the tissue cells, while CO_2 moves in the opposite direction. Again, despite rapid passage of blood through the capillary, equilibration is almost complete so that venous blood returning from the tissues at rest is generally found to have a P_{CO_2} of 46 mm., while the P_{O_2} is about 40 mm. Hg. Since the diffusion coefficient of CO_2 is thirty times greater than that of O_2, the pressure gradient need not be so high for the former gas. Under these circumstances, venous hemoglobin is about 64 per cent saturated with oxygen. The difference, 32 per cent of the oxygen, has been delivered to the tissues. Assuming 15 g. of hemoglobin per 100 ml. of blood, and since each gram of hemoglobin can combine with 1.34 ml. of oxygen, then

$$0.32 \times 1.34 \times 15 = 6.4 \text{ ml. of } O_2$$

has been supplied to the tissues for each 100 ml. of blood traversing the capillaries. Further inspection of the curves in Fig. 32.2 reveals that during exercise, as P_{O_2} in the tissues falls and P_{CO_2} rises, this mechanism becomes increasingly efficient for the delivery of oxygen. Delivery of the increased amounts of O_2 required during exercise is, therefore, effected by a combination of these molecular mechanisms and by acceleration of the circulation through the working muscle.

Transport of CO_2. The total CO_2 content of a blood sample may be estimated by adding mineral acid and then subjecting the sample to reduced pressure. The CO_2 content of arterial blood is about 50 ml. per 100 ml., sometimes referred to as 50 volumes per cent, while that of venous blood may be 55 to 60 volumes per cent. Thus, each 100 ml. of blood transports 5 to 10 ml. of CO_2 from tissues to lungs. Yet the difference in P_{CO_2} between arterial and venous blood at rest is only of the order of 6 mm. Hg. From the constants given in Table 32.1, it can be calculated that an increase of 6 mm. Hg in the P_{CO_2} would permit the physical solution of only an additional 0.4 volume per cent of CO_2. Further, even this increment in CO_2 content would markedly lower the pH of the venous blood, yet a much smaller change is observed. How then is the transport of CO_2 effected? To understand this process, several conditions must first be described: (1) the actual state of CO_2 in arterial and in venous blood, (2) the direct reaction between CO_2 and protein, (3) the relative behavior of hemoglobin and oxyhemoglobin as acids, and (4) the electrolyte composition of red cells and plasma.

1. Carbon dioxide in blood exists in several states. Bicarbonate, in both red cells and plasma, accounts for the major portion of all the CO_2 present. The CO_2 which diffuses across the capillary wall from the tissue space is largely in solution as CO_2 molecules since hydration to form H_2CO_3 is a slow reaction. The CO_2 generated by the various decarboxylation reactions of intermediary metabolism as molecular CO_2 diffuses from cells through interstitial fluid into the plasma largely in this form, with only a small fraction hydrated as carbonic acid. On entry into erythrocytes, hydration of CO_2 is catalyzed by *carbonic anhydrase.*

$$CO_2 + H_2O \rightleftharpoons H_2CO_3$$

The history of this enzyme is of interest. In 1928, Henriques, studying the kinetics of the hydration and dehydration of CO_2, calculated that in the absence of an intervening mechanism, the rapid passage of blood through the pulmonary capillaries would permit less than 10 per cent of the observed rate of escape of blood CO_2 into the lungs. The rate of liberation of CO_2 from serum *in vacuo* was of the order to be expected from the pH and normal diffusion, whereas escape from hemolyzed blood was many times faster. In 1933, Meldrum and Roughton obtained a preparation 100 times as active as erythrocytes and containing relatively little heme or globin, thus establishing carbonic anhydrase as a unique enzyme.

Human erythrocytes contain three isoenzymic forms of carbonic anhydrase which can be separated by electrophoresis. These have been designated as A, B, and C; B is the most abundant, while C exhibits three times the specific activity of B. Each has a molecular weight of about 30,000, with one zinc atom per molecule. Although the peptide chains are genetically homologous, the sequence of B differs markedly from that of C; there are nine substitutions in the last 20 amino acid residues at the C-terminus. Several lines of evidence suggest that the Zn^{++} atom participates in the catalytic reaction. Zn^{++} can be removed and replaced with a variety of divalent metal ions; of these only Co^{++} yields an enzymically active product. The marked affinity for anions such as Cl^- and SCN^- as well as HCO_3^- also indicates participation of the Zn^{++} in the catalytic mechanism. These enzymes catalyze not only the hydration of a carbonyl of CO_2, but also that of aldehydic carbonyls to the corresponding aldehydrols,

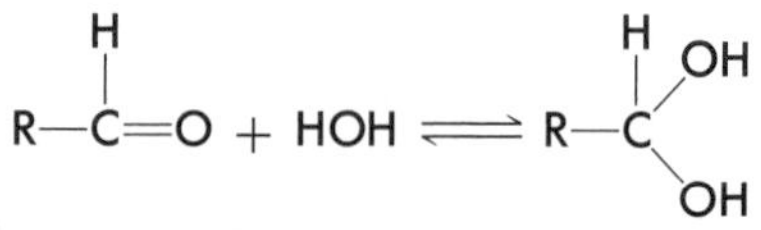

and can also function as esterases. At V_{max}, these three processes proceed at rates of approximately 1×10^6, 5×10^4, and 8×10^1 molecules substrate per molecule enzyme per minute, respectively. Sulfonamides are powerful noncompetitive inhibitors of all carbonic anhydrases; K_I for acetazolamide is about $1.5 \times 10^{-8}M$, thus permitting utilization of this drug in experiments designed to ascertain whether carbonic anhydrase activity is required for other physiological processes.

2. Carbon dioxide reacts with undissociated aliphatic amino groups to form carbamino compounds as shown in the following equation.

$$R{-}NH_2 + CO_2 \rightleftharpoons R{-}NHCOO^- + H^+$$

A fraction of the CO_2 in plasma, about 0.5 millimole (mmole) per liter, is thus bound to the plasma proteins. The trivial difference in plasma protein carbamino content between arterial and venous blood, at rest, is of little consequence in CO_2 transport from the tissues. However, the increased acidity and P_{CO_2} of venous blood during exercise permits the existence of three times as many carbamino groups in hemoglobin as in oxyhemoglobin. The heme-linked imidazole groups do not participate in this process, and the increase in hemoglobin carbamino groups appears unrelated to oxygenation per se. Arterial blood contains about 1.0 mmole of carbamino-CO_2 per liter, whereas, at rest, 1 liter of venous blood carries 1.5 to 2.0 mmoles of CO_2 as the carbamino form.

3. Like all proteins, hemoglobin is a buffer, able to react with protons or dissociate to yield protons. At the pH within corpuscles, the imidazole groups of histidine residues are chiefly responsible for the buffering action of hemoglobin. Figure 32.5 compares the titration of hemoglobin and oxyhemoglobin. These two curves

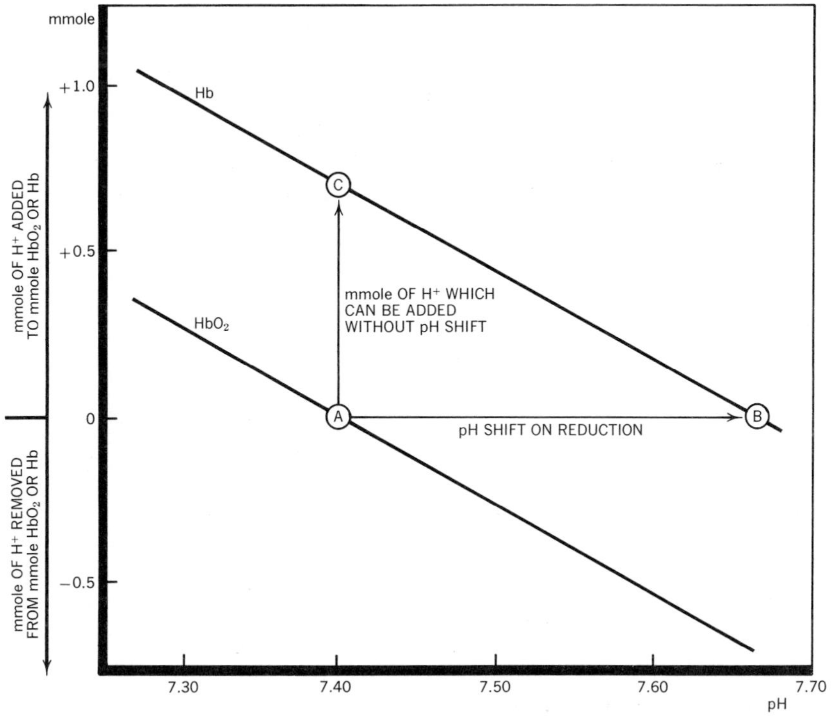

Fig. 32.5. Titration curves of hemoglobin and oxyhemoglobin; mmole denotes millimole. (*From H. W. Davenport, "The ABC of Acid-Base Chemistry," 3d ed., University of Chicago Press, Chicago,* 1950.)

are essentially parallel over the physiological range of pH. If one starts with a solution of 1 mmole of either protein at pH 7.4, the addition of 2.54 mmoles of acid or of alkali is required to change the pH of the solution by one pH unit. These curves reveal that oxyhemoglobin is a stronger acid than is reduced hemoglobin, as indicated previously (page 763). At pH 7.40, if 1 mmole of oxyhemoglobin were to be deoxygenated to hemoglobin, all other factors remaining constant, the pH would shift to (*B*) on the hemoglobin titration curve and would rise to about pH 7.67. The addition of 0.7 mmole of acid would be required to shift back along the curve to (*C*), where the pH would again be 7.40. Conversely, starting at (*C*), were 1 mmole of hemoglobin suddenly to be oxygenated, the blood pH would fall to about 7.13, and 0.7 mmole of alkali would be required to restore the pH to 7.4. These properties of hemoglobin and their physiological significance were first noted by Douglas and Haldane.

4. Both erythrocytes and plasma contain HCO_3^- and H_2CO_3. The following discussion assumes that all the CO_2 present is H_2CO_3. From the Henderson-Hasselbalch equation (page 98),

$$\text{pH} = \text{p}K + \log \frac{[\text{salt}]}{[\text{acid}]}$$

and at a normal blood pH of 7.40, since the $\text{p}K_a'$ of H_2CO_3 is 6.1, log ([salt]/[acid]) equals 1.3 and the ratio $[HCO_3^-]/[H_2CO_3]$ is 20/1. Within the physiological pH range, therefore, the main portion of the total CO_2 present in plasma and red cells exists as HCO_3^-.

A typical arterial blood sample might contain 25.5 meq. per liter of HCO_3^- in plasma and 12.7 meq. per liter of HCO_3^- within the cells. In venous blood, these values would be 26.4 meq. per liter in plasma and 13.9 meq. per liter in cells. Two factors can be cited to account for the discrepancy between the concentration of HCO_3^- within cells and plasma.

First, although red cells contain a 34 per cent protein solution, plasma is a 7.5 per cent solution of protein. Thus, if the concentrations of HCO_3^- are expressed in milliequivalents per liter of red cell and plasma *water,* instead of per liter of cells or per liter of plasma, the values are as follows: arterial blood, 27.2 meq. per liter in plasma and 19.6 meq. per liter in cells; venous blood, 28.1 meq. per liter in plasma and 21.3 meq. per liter in cells. Clearly, there remains a difference in HCO_3^- concentration between cells and plasma. This arises from the second factor, *viz.*, that within the cells, hemoglobin, which is nondiffusible, accounts for a large fraction of the total anions, while in plasma the proteins represent only a small fraction of the total anions. This results in a Gibbs-Donnan effect (page 128). Since the two solutions are in osmotic equilibrium, the total concentration of diffusible anions within cells must be smaller than the total concentration of anions in plasma. At equilibrium the ratios (r) of the concentrations of the various anions within cells (c) and plasma (p) must be constant. Thus

$$r = \frac{[HCO_3^-]_c}{[HCO_3^-]_p} = \frac{[Cl^-]_c}{[Cl^-]_p} = \text{etc.}$$

$$\frac{[HCO_3^-]_c}{[Cl^-]_c} = \frac{[HCO_3^-]_p}{[Cl^-]_p}$$

and $$\frac{[HCO_3^-]_c}{[HCO_3^-]_c + [Cl^-]_c} = \frac{[HCO_3^-]_p}{[HCO_3^-]_p + [Cl^-]_p}$$

Since $$[HCO_3^-]_p + [Cl^-]_p > [HCO_3^-]_c + [Cl^-]_c$$

it is apparent that the $[HCO_3^-]$ of plasma must exceed that within the cells, in agreement with the observed facts.

One further fact may be deduced from the Gibbs-Donnan equilibrium. Hydroxyl ions, as diffusible anions, must also be unequally distributed between red cells and plasma, whereas the product $[OH^-][H^+]$ must be identical in the two solutions.

$$[H^+]_c[OH^-]_c = [H^+]_p[OH^-]_p$$

Therefore $$\frac{[OH^-]_c}{[OH^-]_p} = \frac{[H^+]_p}{[H^+]_c} = r$$

Since r is less than 1, the concentration of hydrogen ions within the red cell is greater than the concentration in plasma and, therefore, the pH of the interior of the erythrocyte is lower than that of the surrounding plasma.

The Isohydric Shift. With the above factors in mind it becomes possible to reconstruct the events in the transport of carbon dioxide from tissues to alveolar air. When arterial blood arrives in the tissue capillaries, about 96 per cent of the hemoglobin is oxygenated. Because of the increased CO_2 tension and the decreased O_2 tension, the oxyhemoglobin dissociates; oxygen diffuses out into the interstitial fluid as CO_2 diffuses into the erythrocyte. A significant fraction of the CO_2 is immediately bound as carbamino hemoglobin. However, there still remains a large excess of CO_2 to be disposed of by other means. Under the influence of carbonic anhydrase this CO_2 is rapidly hydrated to carbonic acid, which then dissociates. Two opposing phenomena then come into play: (1) This carbonic acid would tend to lower the pH within the erythrocyte, but (2) the transformation of oxyhemoglobin to reduced hemoglobin involves a change of pK from 6.2 to 7.7, which tends to raise the pH within the erythrocyte. Consequently, protons formed in the dissociation of carbonic acid are accepted by the imidazole nitrogen of the reduced hemoglobin. The net result of these two events is to maintain the pH essentially unchanged, and K^+ ions within the erythrocyte, previously neutralized by oxyhemoglobin, are now neutralized by the newly formed HCO_3^- ions. As a result, the major portion of the CO_2 which diffused into the erythrocyte from the tissues leaves the capillary in venous blood as red cell HCO_3^-. This set of transformations is termed the *isohydric shift* and is summarized in Fig. 32.6 (page 770).

The isohydric shift entails formation of about 0.7 meq. of bicarbonate for each millimole of oxygen which dissociates from oxyhemoglobin. It will be recalled that the R.Q. for the body at rest in the fasting state is 0.82 (Chap. 14). From the values shown in Table 32.3, it appears that the quantitative operation of the isohydric shift, which is based on the difference between the acid strength of oxyhemoglobin and reduced hemoglobin, is well suited to the physiological task of removing carbon dioxide.

The Chloride Shift. Because of the isohydric shift, the ratio $[HCO_3^-]_c/[HCO_3^-]_p$

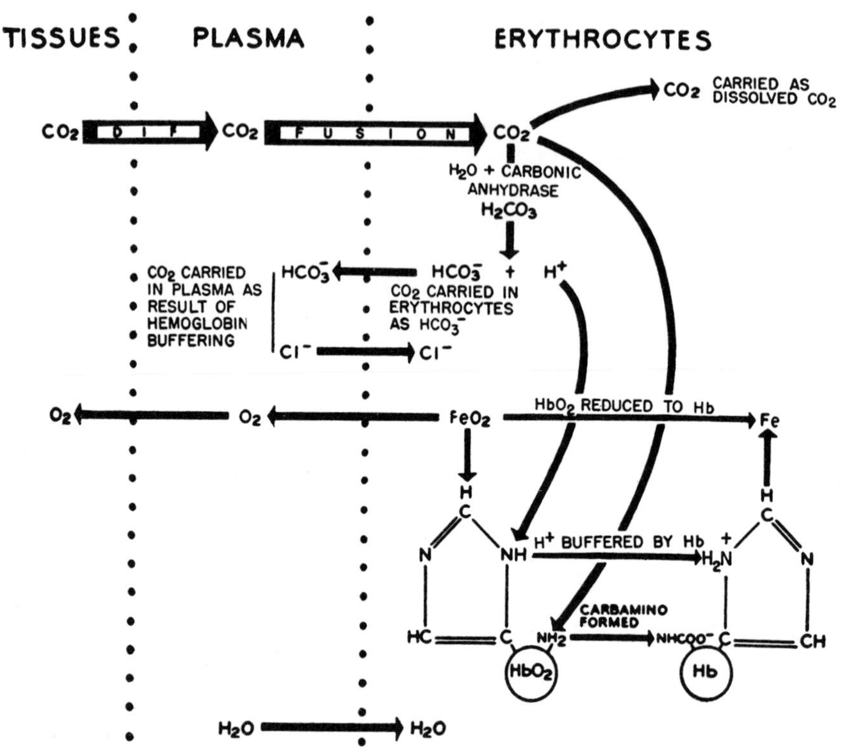

Fig. 32.6. Schematic representation of the processes occurring when carbon dioxide passes from the tissues into the erythrocytes. The imidazole group of a histidine is shown as the reactive portion of the hemoglobin molecule. (*Modified from H. W. Davenport, "The ABC of Acid-Base Chemistry," 3d ed., University of Chicago Press, Chicago,* 1950.)

is now altered, with an excess of HCO_3^- within the cells. The ratio $[HCO_3^-]_c/[Cl^-]_c$ no longer equals the ratio $[HCO_3^-]_p/[Cl^-]_p$. The escaping tendency of HCO_3^- from the cells is increased, and HCO_3^- is replaced by Cl^- from plasma until a new equilibrium is attained. The net result of this transformation is that a significant fraction of the total CO_2 which entered the erythrocyte and was hydrated and dissociated is now found in the venous plasma as HCO_3^-. Note also that while the conversion of oxyhemoglobin to reduced hemoglobin and its subsequent buffering action do not influence the osmotic pressure due to hemoglobin within the cells, since the amount of hemoglobin is unchanged, the combined result of the isohydric and chloride shifts is to increase the total number of anions and thereby increase the effective osmotic pressure within the cells. In consequence, water is redistributed between the cells and plasma so that the relative volume occupied by the erythrocytes (the hematocrit) in venous blood is appreciably higher than that in arterial blood, rising from 45 to 48 or 49 per cent by volume. A partition of the CO_2 transported by 1 liter of blood from the tissues to the lungs of a subject at rest is shown in Table 32.3. It is apparent that about 60 per cent of the total CO_2 is transported as plasma HCO_3^-, while about 32 per cent is transported as carbamino-CO_2 and HCO_3^- within erythrocytes. Directly and indirectly, therefore, hemoglobin makes possible the transport of more than 90 per cent of all the CO_2 carried by the blood.

Table 32.3: DISTRIBUTION OF TOTAL CO_2 IN ARTERIAL AND VENOUS BLOOD

	Arterial, mmole	Venous, mmole	Difference	
			mmole	ml.
Total CO_2 in 1 liter of blood	21.53	23.21	1.68	37.4
Total CO_2 in plasma of 1 liter of blood (600 ml.)	15.94	16.99	1.05	23.5
As dissolved CO_2	0.71	0.80	0.09	2.0
As bicarbonate ions	15.23	16.19	0.96	21.5
Total CO_2 in 400 ml. of erythrocytes	5.59	6.22	0.63	14.0
As dissolved CO_2	0.34	0.39	0.05	1.1
As bicarbonate ions	4.28	4.41	0.13	2.9
As carbamino-CO_2	0.97	1.42	0.45	10.0

When venous blood arrives in the pulmonary capillaries, this sequence is reversed. The lower P_{CO_2} in the alveoli results in a CO_2 concentration gradient favoring CO_2 flow from erythrocyte through plasma to the alveolar space. Simultaneously, oxygen flows from the alveoli into the erythrocyte, and, with diminished P_{CO_2} and increased P_{O_2}, the reduced hemoglobin is oxygenated. Plasma HCO_3^- moves into erythrocytes and combines with protons given up by dissociation of the newly formed oxyhemoglobin. Carbonic anhydrase catalyzes the dehydration of the carbonic acid so that CO_2 formed from the HCO_3^- of plasma can now diffuse out of the erythrocyte through the plasma and into the alveolar space. Carbon dioxide present as carbamino-CO_2 is also liberated because of the diminished CO_2 tension and the conversion of hemoglobin to oxyhemoglobin. The net result is the transport of oxygen from lungs to tissues in sufficient amount to meet metabolic requirements and delivery to the lungs of the carbon dioxide formed during metabolism without changing the acid-base pattern of the extracellular fluid or the erythrocytes.

Thus, the reversible reaction $HbO_2 + H^+ \rightleftharpoons HHb^+ + O_2$ and the conformational changes which hemoglobin undergoes as it binds and releases O_2 are of fundamental significance for the physiology of respiration. When venous blood enters the lungs, the reaction proceeds to the left since here the P_{O_2} increases, whereas the P_{CO_2}, and, consequently, the $[H^+]$ decrease and the hemoglobin is oxygenated. In capillaries outside the lungs, the increased P_{CO_2} and, therefore, increased $[H^+]$, together with the decreased P_{O_2}, favor the reaction to the right. This behavior of hemoglobin is a remarkable example of the relation of chemical structure to physiological requirements.

Fetal Respiration. For maximal oxygen transport, the loading tension at which fetal blood approaches full oxygen saturation must be in the region of the unloading tension of the maternal blood in the placenta. This is indeed the case in mammals. Fetal erythrocytes contain fetal hemoglobin, Hb $F^{\alpha_2{}^A\gamma_2{}^F}$. The γ chains differ from β chains (page 173), most notably in the unique presence of isoleucine, which is absent in Hb A, and in the presence of a single sulfhydryl group. The oxygen dissociation curve of fetal erythrocytes, compared to that of adult erythrocytes, is displaced above and to the left of the curves shown in Fig. 32.2 at any given value of CO_2 or oxygen tension. Thus at 30 mm. of O_2, 37°C., and pH 6.8, maternal blood is 33 per cent saturated whereas fetal blood is 58 per cent saturated. This increased affinity of fetal erythrocytes for O_2 is not due to the structure of Hb F. Indeed, the oxygen dissociation curve of pure Hb F does not differ significantly from that of Hb A. However, fetal erythrocytes exhibit a markedly lower concentration of 2,3-diphosphoglycerate than do adult erythrocytes. As described on page 765, this may suffice to arrange the physiologically appropriate relationships for placental oxygen transport. Accordingly, the functional advantage, if any, of the existence of fetal and adult forms of hemoglobin is presently obscure.

The presence of fetal hemoglobin is readily recognized by addition of alkali to a blood sample; hemoglobin of adult blood is rapidly converted to brown hematin while that of fetal blood remains bright red for a considerable period. The structural basis for this difference in behavior is not clear. After birth, fetal hemoglobin ordinarily disappears from the circulation and is almost entirely absent after 4 to 6 months, except in certain anemic states.

Myoglobin. The muscles of vertebrates and invertebrates contain *myoglobin,* a hemoprotein capable of reversibly binding oxygen. The structure of whale myoglobin, which is known in detail, was discussed previously (page 159). It exhibits no Bohr effect, and the oxygen dissociation curve is a rectangular hyperbola (Fig. 32.1), displaced well above and to the left of that of hemoglobin. At a venous P_{O_2} of 40 mm. Hg, at which hemoglobin is 66 per cent saturated, myoglobin is still 94 per cent saturated. At an oxygen tension of only 10 mm. Hg, hemoglobin is 10 per cent saturated whereas myoglobin is 80 per cent saturated. Related to these considerations is the fact that cytochrome oxidase can operate at V_{max} when the medium provides oxygen at P_{O_2} of about 4 to 5 mm. Hg. Thus the affinities of these three proteins for oxygen are in the order cytochrome oxidase $>$ myoglobin $>$ hemoglobin. Consequently, myoglobin can accept oxygen from hemoglobin and store it in the muscle cell for release to cytochrome oxidase when the oxygen supply becomes limiting.

In muscle at rest, oxygen probably remains fixed to myoglobin. During contraction, when the demand for oxygen is maximal and as intracellular P_{O_2} falls, oxygen dissociates from myoglobin and is available for oxidations. In man, myoglobin is present in significant quantity only in cardiac muscle and is probably of little significance in skeletal muscle. However, in diving mammals, the myoglobin content of muscles is particularly high and probably facilitates submersion for long periods. The muscle of dolphins and seals contains 3.5 and 7.7 per cent myoglobin, respectively; these relative concentrations correlate roughly with the duration of their dives. The flight muscles of birds are also rich in myoglobin. Myoglobin-like pigments have also been found in the nerves of various invertebrates; $P_{1/2}$ is of the order of 1 or 2 mm. Hg for these substances.

Carboxyhemoglobin (Carbon Monoxide Hemoglobin). Hemoglobin and the nitrogenous base derivatives of ferroprotoporphyrin bind carbon monoxide (CO) to their ferrous components in linkage analogous to that of O_2. The affinity of human hemoglobin for carbon monoxide is more than 200 times greater than its affinity for oxygen, *i.e.*, for equal formation of HbCO and HbO_2, the required partial pressure of carbon monoxide is only about one two-hundredths that of oxygen. Similarly, the affinity of CO for myoglobin is thirty-seven times that of O_2. Claude Bernard in 1858 explained the toxicity of carbon monoxide by his discovery of its combination with hemoglobin. The brilliant cherry-red color of HbCO is very distinctive and is manifested in the skin and tissues of victims of carbon monoxide poisoning.

COMPARATIVE BIOCHEMISTRY OF RESPIRATORY PROTEINS

The need for an oxygen carrier, apparent throughout the animal kingdom, has been satisfied by various means. In the most primitive animals, which are relatively small and have low metabolic rates, the carrier is enclosed in cells suspended in the coelomic fluid. With the development of a circulation there appeared oxygen carriers dissolved in the circulating plasma, and, later still, concentrated solutions of carriers were enclosed in special circulating cells, the erythrocytes. The latter represent a great advance, as they permit the presence in the circulation of large amounts of carrier without an inordinate rise in the viscosity and colloidal osmotic pressure of the circulating medium. There are a number of oxygen carriers of markedly different properties distributed throughout the animal kingdom. Almost all vertebrates have as an oxygen carrier an intracellular hemoglobin of molecular weight 66,000 with ferroprotoporphyrin III as the prosthetic group.

A great diversity is apparent among the invertebrates. In those species in which the oxygen carrier is simply dissolved in the circulating plasma, the carrier is invariably of high molecular weight, *e.g.*, 400,000 to 6,700,000. Among these are large hemoglobin-like molecules, the *erythrocruorins,* in the blood of many polychete and oligochete annelid worms and various mollusks. Details of the structures of a few of these giant molecules are available. For example, the erythrocruorin from *Limnodrilus,* molecular weight 3,000,000, consists of 108 subunits of about 28,000 molecular weight, each bearing one heme, arranged as a regular hexagonal cylinder, having the dimensions 220 Å $\times$ 160 Å.; this cylinder, in turn, consists of six lesser cylinders oriented in the same axis as the major cylinder axis. Thus, there are three layers of 36 protein subunits each in the over-all packaged molecule. Since there are four different N-terminal amino acids, it is assumed that there are four different subunit polypeptide chains in the molecule. Certain of the annelids, such as *Spirographis,* have green blood pigments, the *chlorocruorins,* of molecular weight 3,400,000, in which the porphyrin differs from protoporphyrin in that the 2-vinyl group is replaced by a formyl group. When these carriers are present in the coelomic fluid, rather than in an organized circulation, they serve to store oxygen rather than to transport it. This is evident from their low values for $P_{1/2}$, *e.g.*, that of the hemoglobin of *Ascaris* is of the order of 0.002 mm. Hg. In some instances, despite the polymeric nature of the carrier, $n = 1$, *e.g.*, that of the polychaete worm *Eupolynia;* in contrast, the polyhemoglobin of a closely related form, *Arenicola,* exhibits an $n = 6$, the highest value known. Like hemoglobin, all erythrocruorins and chlorocruorins bind one O_2 per heme.

The blood plasma of many mollusks and arthropods contains *hemocyanins,* blue pigments containing copper but no heme and with molecular weights which range from 0.5×10^6 to 10×10^6. All are polymers of much smaller monomeric units; squid hemocyanin is a decamer, molecular weight 3.75×10^6, which, on dilution or at high salt concentration, dissociates into 5 dimers and then into 10 monomers.

Each monomer contains two atoms of Cu^+ and, both in monomeric and polymeric form, binds one O_2 per two Cu^+. There is some evidence that oxygen binding and release involve oxidation and reduction of the copper, but this is uncertain. It is noteworthy that the hemocyanins of some species show a pronounced Bohr effect and values of n greater than 2.

Intracellular oxygen-carrying proteins are of relatively low molecular weight, varying from 17,500 to about 120,000; hemoglobin is the most common of these. *Hemerythrin* is present in the blood of all sipunculid worms, a few polychaete forms, and one brachiopod, *Lingula.* The most carefully studied hemerythrin, from *Sipunculus,* has a molecular weight of 105,000 and contains 16 atoms of Fe^{++}. It can be deaggregated by succinylation or treatment with N-ethylmaleimide to yield eight apparently identical subunits with two iron atoms each. It has been suggested that these are linked in part to sulfhydryl groups; treatment with dithionite elicits the $g = 1.94$ signal of many electron-carrying iron proteins (page 368). Hemerythrin binds one O_2 per pair of iron atoms; the latter may cooperate to accomplish O_2 binding in a manner analogous to the mechanism proposed for hemocyanin.

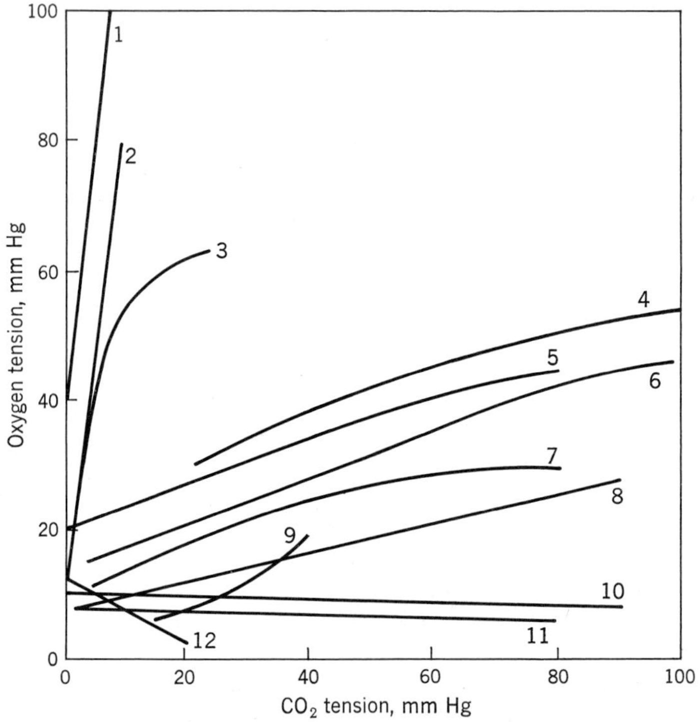

Fig. 32.7. Magnitude of the Bohr effect in the blood of various species; relationship between P_{CO_2} and $P_{1/2}$, the P_{O_2} necessary to maintain 50 per cent saturation of the oxygen-carrying protein. A sharp positive slope indicates a large Bohr effect, a flat line indicates no Bohr effect, and a negative slope indicates a negative Bohr effect. 1, squid; 2, sea robin; 3, mackerel; 4, sea lion; 5, goose; 6, dog; 7, man; 8, turtle; 9, carp; 10 *Urechis,* echiuroid worm; 11, sipunculid; 12, Busycon, conch, a marine gastropod. (*Adapted from M. Florkin, "Biochemical Evolution," Academic Press, Inc., New York,* 1949.)

Table 32.4: CHARACTERISTICS OF RESPIRATORY PIGMENTS

Pigment	Source	Location	Atoms of metal per mole	Molecular weight	Bohr effect	Dissociation curve
Myoglobin......	Mammalian muscle	Intracellular	1 Fe	17,000	None	Hyperbolic
Hemoglobin.....	*Gastrophilus* larvae	Tracheal cells	2 Fe	34,000	Positive	Hyperbolic
Hemoglobin.....	Tadpole	Intracellular	4 Fe	68,000	None	Sigmoid
Hemoglobin.....	Frog	Intracellular	4 Fe	68,000	Positive	Sigmoid
Hemoglobin.....	Man	Intracellular	4 Fe	68,000	Positive	Sigmoid
Erythrocruorin...	Mollusks	Plasma	96 Fe*	1,500,000	None	Sigmoid
Erythrocruorin...	Annelids	Plasma	192 Fe*	3,000,000	None	Sigmoid
Chlorocruorin...	Polychaete worms	Plasma	192 Fe*	3,000,000	None	Sigmoid
Hemerythrin....	*Sipunculus*	Intracellular	16 Fe	105,000	None	Hyperbolic
Hemerythrin....	*Lingula*	Intracellular	......	......	Positive	Sigmoid
Hemocyanin....	Squid	Plasma	20 Cu*	3,750,000	Positive	Sigmoid
Hemocyanin....	Snail	Plasma	200 Cu*	6,760,000	None	Sigmoid

* Approximate number.

A few vertebrates, fish of the order *Chaenichthyidae,* do not possess a respiratory pigment. The poikilothermic animals dwell in Antarctic waters, at temperatures from −2 to +2°C., and their sluggish metabolism is satisfied by the oxygen carried in solution at this low temperature.

The peculiar suitability of these diverse oxygen carriers to their task in the organisms in which they are found is best illustrated by two considerations: (1) the operation of the Bohr effect, and (2) the degree of oxygen saturation in arterial blood. The various oxygen carriers differ markedly in the extent to which a given change in P_{CO_2} affects the oxygen dissociation curve. This is illustrated in Fig. 32.7. In animals like the marine teleosts, which lead an active existence in ocean waters that are well oxygenated yet almost free of carbon dioxide, the Bohr effect is especially prominent. It is less pronounced in animals that live in fresh water or in air, whereas in those species which live in a medium poor in oxygen and rich in carbon dioxide the oxygen carrier may be absent or even exhibit an inverted Bohr effect. For each respiratory pigment it is a *sine qua non* that it should be virtually saturated with oxygen under the conditions prevailing in the arterial blood of the animal in which it functions. While this encompasses a range of P_{O_2} from 30 to 115 mm. Hg, there is no known instance in which the oxygen-carrying protein is not at least 90 per cent saturated in its native arterial blood. The characteristics of certain of the respiratory pigments mentioned above are given in Table 32.4.

It is surprising that hemoglobin, similar to that of mammalian erythrocytes, has been found in cells of plant origin, *e.g.*, in a few strains of fungi, and in some Protozoa. Of particular interest is the hemoglobin in the nodules formed by the nitrogen-fixing *Rhizobium* on the roots of Leguminosae. The pigment is a product of symbiosis since it is not formed by pure cultures of *Rhizobium* or by the plant roots in the absence of nodules. The role of hemoglobin in this biological system is not understood.

REFERENCES

Books

Barcroft, J. S., "The Respiratory Function of the Blood. II. Haemoglobin," Cambridge University Press, London, 1928.

Davenport, H. W., "The ABC of Acid-Base Chemistry: The Elements of Physiological Blood-Gas Chemistry for Medical Students and Physicians," 4th ed., University of Chicago Press, Chicago, 1958.

Haldane, J. S., and Priestly, J. G., "Respiration," Yale University Press, New Haven, 1935.

Henderson, L. J., "Blood: A Study in General Physiology," Yale University Press, New Haven, 1928.

Prosser, C. L., and Brown, F. A., Jr., "Comparative Animal Physiology," 2d ed., W. B. Saunders Company, Philadelphia, 1961.

Roughton, F. J. W., and Kendrew, J. C., eds., "Haemoglobin," Interscience Publishers, Inc., New York, 1949.

Review Articles

Antonini, E., Interrelationship between Structure and Function in Hemoglobin and Myoglobin, *Physiol. Revs.,* **45,** 123–170, 1965.

Antonini, E., Hemoglobin and its Reaction with Ligands, *Science,* **158,** 1417–1425, 1967.

Braunitzer, G., Hilse, K., Rudloff, V., and Hilschmann, N., The Hemoglobins, *Advances in Protein Chem.,* **19,** 1–73, 1964.

Drabkin, D. L., Metabolism of the Hemin Chromoproteins, *Physiol. Revs.,* **31,** 345–431, 1951.

Manwell, C., Comparative Physiology: Blood Pigments, *Ann. Rev. Physiol.,* **22,** 191–244, 1960.

Riggs, A., Functional Properties of Hemoglobins, *Physiol. Revs.,* **45,** 619–673, 1965.

Rossi Fanelli, A., Antonini, E., and Caputo, A., Hemoglobins and Myoglobin, *Advances in Protein Chem.,* **19,** 73–233, 1964.

Wang, J. H., Hemoglobin and Myoglobin, in O. Hayaishi, ed., "Oxygenases," pp. 470–516, Academic Press, Inc., New York, 1962.

Wolvekamp, H. P., The Evolution of Oxygen Transport, in R. G. MacFarland and A. H. T. Robb-Smith, eds., "Functions of the Blood," pp. 1–63, Academic Press, Inc., New York, 1961.

Wyman, J., Heme Proteins, *Advances in Protein Chem.,* **4,** 407–531, 1948.

Wyman, J., Linked Functions and Reciprocal Effects in Hemoglobin: A Second Look, *Advances in Protein Chem.,* **19,** 223–286, 1964.

33. Regulation of Electrolyte, Water, and Acid-base Balance

Fluid Compartments and Composition. Active Transport Processes. Control of Extracellular Fluid. Metabolism of Cellular Electrolytes

The ability of animals to maintain constant the composition of the *extracellular fluid,* the *milieu intérieur,* first appreciated by Claude Bernard, represents one of the most significant advances of evolution, since, with it, animals became independent of many changes in their environment. It is the purpose of this chapter to describe the nature and function of the intra- and extracellular fluids and the mechanisms which maintain their composition constant.

FLUID COMPARTMENTS OF THE BODY

In the human adult, an amount of fluid approximately equal to 50 per cent of the body weight is located within cells, while the extracellular fluid, *i.e., all* the fluid not present within cells, accounts for about 20 per cent of the body weight. Extracellular fluid may be further divided into several subcompartments, of which the largest are the interstitial fluid, which bathes most cells and represents 15 per cent of the body weight, blood plasma, the transport vehicle through which cells make contact with other cells and with the environment, amounting to about 5 per cent of the body weight, and relatively smaller volumes of cerebrospinal fluid, synovial fluid, aqueous humor, lymph, etc.

The principle of the procedures for estimation of the volume of each of the various fluid compartments is essentially the same. A material, previously found to be distributed almost exclusively within the compartment to be measured, is given intravenously in known amount. After sufficient time for mixing, a sample of plasma is obtained, the concentration of administered material measured, and, from the extent of dilution, the total volume of the particular compartment calculated after correction for the quantity excreted. *Total body water,* then, may be estimated after administration of any material which is distributed throughout the body, *i.e.,* one which passes freely through capillary endothelium, cell membranes, the blood-brain barrier, etc., followed by determination of this material in any available fluid, *e.g.,* plasma or urine. The substances which most closely meet these criteria are ^2H_2O and ^3H_2O, although total body-water measurements with these compounds also include measurement of all exchangeable hydrogen atoms in organic compounds. Other materials, particularly antipyrine, have also been employed.

Determination of *total extracellular fluid* requires a substance to which capillary walls are permeable but which fails to enter cells and, preferably, is relatively slowly excreted by the kidneys. Among the materials employed for this purpose are inulin, thiocyanate, and thiosulfate. Unfortunately, the results of these procedures are not in complete agreement; the inulin "space" appears to approximate most closely the extracellular fluid volume. Estimation of *plasma volume* requires the intravenous administration of some material which will be retained entirely within the vascular space. For this purpose, several dyes, notably Evans' blue, have been employed, as well as serum proteins labeled with ^{131}I. The volume of the *interstitial fluid,* together with all the *specialized extracellular fluids,* is calculated as the difference between the volumes of total extracellular fluid and of plasma.

COMPOSITION OF BODY FLUIDS

Figure 33.1 shows the electrolyte composition of the body fluids; the cell fluid shown is that of skeletal muscle. Although Na^+ is the chief extracellular cation, K^+ and Mg^{++} are the chief intracellular cations; Cl^- and HCO_3^- predominate outside of cells, while phosphates, sulfate, and proteins constitute the bulk of the cellular anions.

The osmotic pressure within a cell must be identical with that of the surrounding fluid, since the membranes involved are freely permeable to water and are readily ruptured by moderate pressure differentials. This is in sharp contrast to the support afforded bacterial and plant cell membranes by their rigid outer walls. The osmotic pressure considered here is that which the solution would exhibit in an osmometer with a membrane permeable solely to water, in this case about 6,000 mm. Hg. Since interstitial fluid is almost protein-free and the chief anions and cations are univalent, the height of the column in the chart indicates not only molar concentrations but also the osmolarity. In contrast, among the chief contributors to the osmotic pressure of intracellular fluid are many multivalent particles, such as Ca^{++}, Mg^{++}, protein, and phosphates. Osmotic pressure is dependent, however, solely on the total number of particles in solution, regardless of their electric charge. It follows that the concentration of electrolytes within the cells, expressed in milliequivalents per liter, is appreciably greater than that outside the cells. This may be illustrated as follows.

Imagine an extracellular fluid composed exclusively of NaCl and a cellular fluid exclusively of K^+ and protein and, further, that the protein particles bear four negative charges each. If each milliliter of extracellular fluid contained 50 Na^+ ions and 50 Cl^- ions, what then must be the composition of the cellular fluid at osmotic equilibrium? Since the extracellular fluid contains 100 particles per milliliter, the cellular fluid must also contain 100 particles per milliliter. This is possible only when the latter fluid contains 80 K^+ ions and 20 protein molecules per milliliter. Expressed in milliequivalents per liter, then, the concentration of inorganic cations within the cells is eight-fifths that of the extracellular fluid. This is an exaggeration but serves to explain the observed differences depicted on the chart. Since the protein concentration of plasma is intermediate between that of cells and extracellular fluid, plasma occupies an intermediate position with respect to its electrolyte concentration.

The composition of intracellular fluid indicated in Fig. 33.1 represents a mean for cells in general. Although this may even be a valid statement for the total content of an individual cell, it seems likely that within the cell there are areas or compartments containing unusual concentrations of one or another of the cell constituents, in part because of the variable binding capacities of different protein molecules, and, in part, reflecting specific, selective transport mechanisms.

It has been suggested that the composition of extracellular fluid resembles that of the seas during the pre-Cambrian era, when animals with closed circulations came into existence. The sea has continued to increase in salinity, while the composition of extracellular fluid has remained fixed. The present composition of sea water is shown in Fig. 33.1 for comparison.

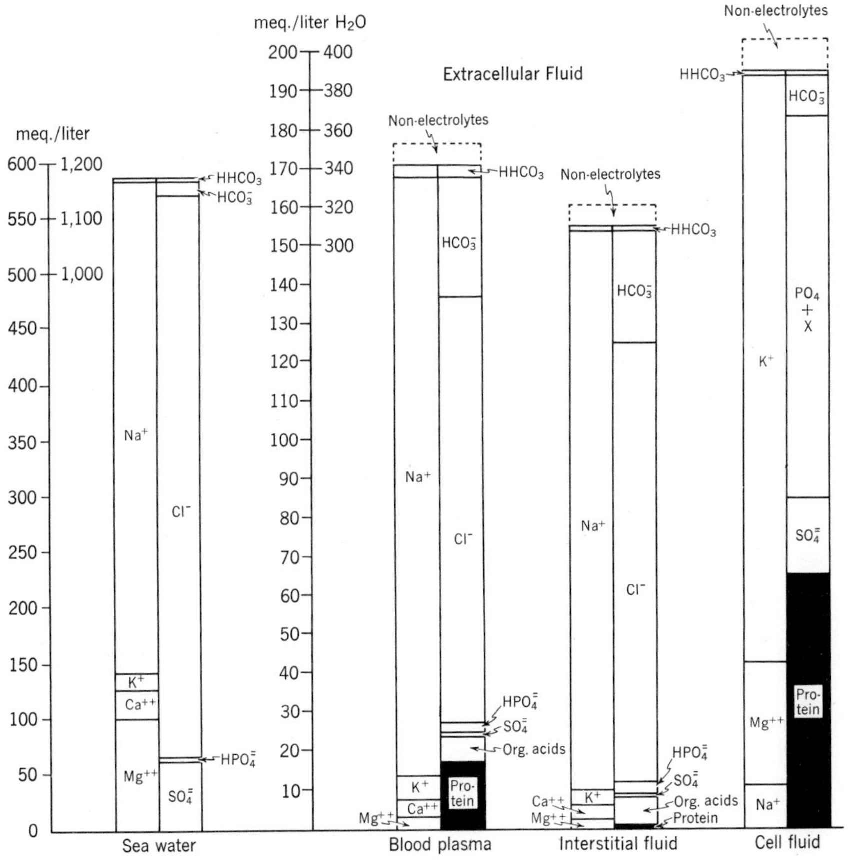

Fig. 33.1. Electrolyte composition of the body fluids. The over-all value PO_4 + X (composition of cell fluid) is obtained by subtracting the equivalents found for $SO_4^{=}$ + HCO_3^{-} + protein from the total equivalents prescribed by the cations. The composition of sea water is given for comparison. The values for cell fluid are those for skeletal muscle. (*From J. L. Gamble, "Chemical Anatomy, Physiology and Extracellular Fluid," 6th ed., Harvard University Press, Cambridge, Mass.,* 1954.)

These differences in composition of intra- and extracellular fluid are determined by the metabolism of the cell and the properties of cell membranes. Plasma may be regarded as a special subcompartment of extracellular fluid which differs only in that it contains proteins which cannot diffuse through the endothelial lining of the capillaries. Consequently, the electrolyte composition of plasma must also differ slightly from that of interstitial fluid, as formulated by the Gibbs-Donnan equilibrium.

The electrolyte composition of intracellular fluid cannot be determined with precision but must be calculated by difference. For example, an animal is first given a known quantity of sodium thiosulfate; a few minutes later samples of plasma and muscle are obtained and analyzed for thiosulfate, Na^+, K^+, and Cl^-. The thiosulfate concentration of plasma and of interstitial fluid is assumed to be identical so that the total amount of thiosulfate in the tissue sample permits calculation of its extracellular fluid volume. Knowing the latter and the concentration of Na^+, K^+, and Cl^- in plasma, one may then calculate the extracellular Na^+, K^+, and Cl^- of the tissue sample. When these are subtracted from the total Na^+, K^+, and Cl^- concentrations found in the tissue, the differences represent the intracellular ion concentrations. Obviously, there are numerous sources of error in these procedures. The technique of isotope dilution has also been employed to determine the amount and distribution of these three ions (Na^+, K^+, Cl^-). The most striking information thus obtained is that much of the total body sodium and chloride are not in the extracellular fluid. By inference, except for the quantities of these ions present in bone matrix but readily "exchangeable" with ions in plasma, this signifies intracellular concentrations of the order of 20 meq. per liter.

ACTIVE TRANSPORT PROCESSES AND THE COMPOSITION OF INTRACELLULAR FLUID

It is apparent from Fig. 33.1 that the electrolyte composition of intracellular fluid differs strikingly from that of the surrounding interstitial fluid. The cell membrane separating these fluids is an organized mosaic of protein, lipid, including phospholipid and ganglioside, and polysaccharide. Although this membrane exhibits selective permeability, its properties do not account adequately for the marked differences in composition of intra- and extracellular fluid. In general, small singly charged anions and cations pass through cell membranes, albeit with varying degrees of freedom. The membrane behaves as if it had pores such that hydrated ions less than 8 Å. in diameter pass freely, whereas the passage of larger ions is hindered. Thus, K^+, Rb^+, and Cs^+ ions can enter animal cells rather rapidly, whereas Na^+ and Li^+ ions enter with relative difficulty. Similarly, Cl^-, Br^-, and NO_3^- ions enter more readily than HCO_3^- and CH_3COO^- ions, which diffuse across the membrane relatively slowly; $SO_4^=$ ions are practically excluded from cells. These relationships were established by comparing the relative rates of entry of the appropriate radioisotopes of each of these ions into animal cell preparations at 0 and 37°C. Nevertheless, the marked disparity in composition of the fluids on either side of the cell membrane cannot be attributed solely to permeability characteristics of the membrane, which can only delay the time required for the establishment of equilibrium between the two phases with respect to each of those components which can traverse the membrane. Hence, it is necessary to postulate an "electrolyte pump" which derives its energy from the metabolic activities of the cell. The existence and operation of such a pump is indicated by many observations such as the following: human erythrocytes maintained at 0°C. lose K^+ and, in time, equilibrate with the Na^+ of the medium. If the erythrocytes are returned to 37°C., the process is reversed; Na^+ leaves the cells and K^+ reenters until the normal relationships are reestablished. However, if the cells are rewarmed in saline solution lacking glucose, redistribution of electrolytes does not occur; subsequent addition of glucose results in accumulation of K^+ and extrusion of Na^+ from the cells, and the rate at which this occurs is determined by the intracellular concentration of ATP. Essentially similar observations have been made with many animal tissues.

Considerable evidence indicates that establishment and maintenance of the normal concentrations of potassium and sodium on either side of the cell membrane result from the operation of a mechanism which, in effect, ejects Na^+ from the cell, while simultaneously effecting the reentry of K^+. This is among the most dramatic examples of the many processes which are termed "active transport," processes whereby a solute is caused to move against an electrochemical gradient, *i.e.*, from an area of relatively low to one of higher chemical potential.

The term "electrolyte pump" has been used to describe the operation of this process. The analogy to a mechanical pump is apt, in that energy is utilized to move material, in this case ions, against an opposing gradient. In all cases studied, the source of energy for this work is ATP. In erythrocytes and in fermenting yeast, which actively accumulate potassium, the energy is derived from anaerobic glycolysis. Ejection of Na^+ from muscle and nerve cells, which permits maintenance of the normal high intracellular potassium concentration, secretion of HCl by the stomach, and absorption of NaCl by the renal tubular epithelium and by the intestinal mucosa, are all processes which derive the required energy from cellular oxidative metabolism. In each instance the process is inhibited by anoxia, by cyanide (which prevents reduction of oxygen by cytochrome oxidase), and by dinitrophenol (which uncouples mitochondrial oxidative phosphorylation).

Presumably involved in this process is a protein found in erythrocytes, nerve, and many other tissues and having the apparent properties of an adenosine triphosphatase. Crude preparations of this enzyme effect the hydrolysis of ATP, in the presence of Mg^{++}, at a very low rate. Addition of either K^+ or Na^+ alone is without effect. When both the latter are present, ATP hydrolysis is markedly accelerated. The rate of hydrolysis rises with increasing $[Na^+]$; concentrations in excess of that of plasma have no further effect. Activity with $[K^+]$ passes through a maximum; excess K^+ is inhibitory. The K^+-Na^+ dependent activity is inhibited by ouabain, a cardiac glycoside which is also known to inhibit active Na^+-K^+ transport in many tissues.

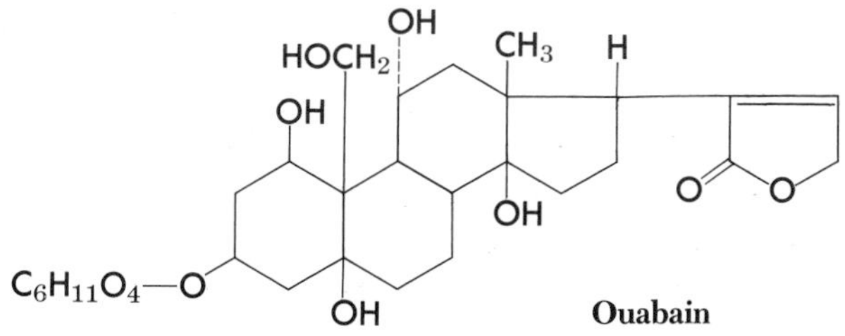

Ouabain

In the cell membrane, the protein with enzymic activity is so situated that it is affected only by Na^+ internal to the membrane and K^+ external to the membrane. This has been demonstrated with red cell "ghosts," which are leaky at low external osmotic pressures, so that their contents may be exchanged for that of their environment, and reseal at normal external osmotic pressure, permitting experimental creation of an intracellular fluid of any desired composition. Several observations suggest that this membrane protein is phosphorylated by ATP and then dephosphorylated. The phosphorylation step depends upon the presence of Na^+ and is insensitive to ouabain. Dephosphorylation requires K^+ and is inhibited by ouabain, the effect of which may be overcome by increased $[K^+]$. For each ATP hydrolyzed, 3 Na^+ are ejected and 2 K^+ are accumulated by the cell. These postulated events are summarized in Fig. 33.2. The molecular basis for these observations is obscure. The enzyme cannot be separated from the lipid matrix of the cell with retention of these properties. Accordingly, the protein is thought to exist as a lipoprotein in a lipid matrix. It must be assumed that the sodium-bound, phosphorylated protein undergoes a profound conformational change which permits it to rotate or migrate within the membrane, in turn permitting hydrolysis, release of Na^+, and binding of K^+. From this last state the protein must return to its original conformation and physical orientation so that it may release K^+, bind Na^+, etc. If this description proves to be valid, this protein is one of the "carriers" long postulated to account for active transport processes. In any case it is intimately associated with the Na^+-K^+ transport process in most animal cells and is, in some manner, also especially adapted for service in the secretory activities of such tissues as renal tubular epithelium and the salt gland of marine birds. An interesting correlation is afforded by the presence of this enzyme, in concentrations comparable to those of human erythrocytes, in the erythrocytes of a strain of sheep which maintain high erythrocyte $[K^+]$ while it is absent or present in small amount in the erythrocytes of most sheep, which have a high $[Na^+]$ and relatively low $[K^+]$. The manner of Na^+ binding to the carrier protein is unknown. A model may be provided by valinomycin and other cyclic

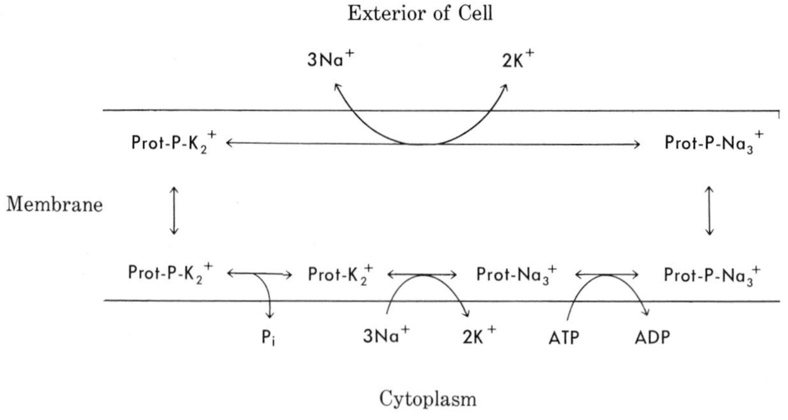

Fig. 33.2. Suggested scheme of operation of the sodium pump. Prot = the Na^+-K^+-dependent ATPase; Prot-P = a phosphorylated form of this enzyme.

oligopeptide antibiotics which trap cations in their hydrophobic interiors so that the cation moves across mitochondrial membranes with the antibiotic. Finally, Na^+-K^+ exchange is one of the dominant activities of living cells; it accounts for perhaps two-thirds of the ATP turnover of nerve and about half of the ATP utilization of resting muscle. Thus it is a major process contributing to the basal metabolic rate.

Other Active Transport Processes. Energy-dependent processes in cell membranes are responsible for maintaining the intracellular concentrations of many anions, cations, and smaller organic compounds. The concentrations of Cl^- and HCO_3^- are probably dependent only on ordinary electrochemical phenomena and are adjusted to the cation concentration. Phosphate, however, is accumulated by an unknown active process. Whereas cell membranes are poorly permeable to $SO_4^=$, there is a large sulfate gradient across almost all animal cell membranes. Active transport of sulfate is characteristic also of bacterial cells, and Pardee has crystallized from *Salmonella* cell membranes a protein, molecular weight 32,000, which spontaneously binds one sulfate per molecule. Cells in which formation of this protein is genetically repressed show a markedly reduced degree of sulfate transport activity.

As noted previously (page 685), *Escherichia coli* possess a "permease" which makes possible the accumulation of β-galactosides. From the membrane fraction of cells which had been induced by thiogalactoside, Kennedy has isolated a protein with a binding affinity for galactosides which parallels the rate at which whole cells can achieve their transport. Again, this "permease" type of transport is energy-dependent, and it is suggested that, in some manner, ATP is utilized either to dislodge the bound galactoside or to effect a conformational change which permits the permease protein to alter its position in the membrane so that the galactoside binding site is oriented inward.

Although no specific protein has yet been isolated, it is attractive to consider that an equivalent arrangement may participate in the "facilitated diffusion" of glucose into mammalian cells (page 448). Indeed, such a protein has been tentatively identified; it binds one molecule of glucose and has a histidine residue which is phosphorylated and dephosphorylated at a rate commensurate with glucose transport in muscle cells. *Escherichia coli* possess an inducible transport system for hexose phosphates and a suppressible transport protein for leucine. Glucose-, mannose-, or fructose 6-phosphate induces a transport system which is active for all three but which will not carry glucose 1-phosphate or galactose 6-phosphate, thus indicating that there must be a carrier-protein with binding characteristics highly reminiscent of those involved in the specificity of enzymic action.

Active glucose transport across the brush border of intestinal cells is linked to Na^+ movement in the same direction. Clearly, such cells require a sodium pump to eject the arriving Na^+ to keep this process operative. An increase in $[Na^+]$ decreases the K_m for glucose entry, whereas increased $[K^+]$ increases glucose K_m. Hence, although the mode of coupling is obscure, glucose transport, in this instance, is driven by the electrochemical gradients maintained by the electrolyte pump. Similar Na^+-dependent transport has been observed for amino acids across the membranes of the intestine, erythrocyte and skeletal muscle, for γ-aminobutyrate into brain mitochondria, and for choline uptake by the presynaptic termini of cholinergic nerves. Each of these transport phenomena is most readily explained by the presence of a specific protein with binding sites for Na^+ and for the organic substrate and which can rotate or migrate in the membrane only when both sites have been either filled or emptied.

Data concerning amino acid transport, *e.g.*, across the jejunum, renal tubules, diaphragm, liver, or ascites cells indicate the independent functioning of a series of amino acid–specific transport systems. All are saturatable, *viz.*, they obey Michaelis-Menten kinetics, and operate independently. Thus, increased $[H^+]$ in the medium of ascites tumor cells decreases V_{max} for taurine and glycine transport, without affecting K_m, while increasing V_{max} for glutamic acid transport. It is apparent that many active transport processes exist but that their molecular mechanisms remain to be revealed.

CONTROL OF THE BEHAVIOR OF EXTRACELLULAR FLUID

Interest in the behavior of extracellular fluid derives largely from the fact that many major disturbances of electrolyte and fluid balance originate in the extracellular fluid, and because even when the primary disturbance occurs in the behavior of intracellular fluid, it is the secondary changes in the extracellular fluid to which the physician or investigator has ready access.

Daily Requirements for Water and Electrolytes. Under usual environmental conditions, there is a daily obligatory loss of approximately 1,500 ml. of water by normal human adults. Of this, about 600 ml. is lost through the skin as insensible perspiration (page 813), 400 ml. in the expired air, and 500 ml. in the urine. Any excess of water intake over this obligatory total volume appears as an increased urine volume. To the extent to which the intake is less than this obligatory 1,500 ml., the difference must be at the expense of the total body water. Since the oxidation of glucose and lipid, in an amount sufficient to yield 2,000 cal. per day, results in formation of about 300 ml. of water, there remains an obligatory water intake of the order of 1,200 ml. per day.

In contrast, there is no equivalent obligatory loss of Na^+ or Cl^- under normal conditions. Adults on a diet devoid of Na^+ and Cl^- lose these ions in the urine for only a few days, after which the urine becomes virtually Na^+- and Cl^--free, all other circumstances remaining constant. The average diet provides 100 to 200 meq. of Na^+ and Cl^- per day, all of which, except for small amounts in sweat and feces, is excreted in the urine. In the absence of dietary K^+, urinary excretion of approximately 40 to 60 meq. of K^+ per day occurs for a few days after which urine losses diminish to about 10 meq. per day.

Disturbances of the normal relationships of extracellular fluid may be considered from four standpoints, (1) osmotic pressure, (2) volume, (3) composition, and (4) pH.

Control of Osmotic Pressure. No serious departure from the normal osmotic pressure of intracellular fluid can long be tolerated by the body; both hyper- and hypotonicity lead to irreversible and lethal changes in the central nervous system. Yet there is no mechanism for direct control of the osmotic pressure of cell contents, which are, at all times, in osmotic equilibrium with extracellular fluid. The osmotic pressure of the latter is regulated by one of the most complex homeostatic devices in the animal, which, like all homeostatic mechanisms, operates by a series of feedback devices. Adult kidneys can elaborate urine varying from 0 to 300 mmoles NaCl per liter, and the urinary salt concentration at any given time is determined by the influence of two hormones on the kidney. The antidiuretic action of vasopressin, released by the neurohypophysis (Chap. 47), enhances water reabsorption, and aldosterone (Chap. 45), from the adrenal cortex, stimulates Na^+ reabsorption. The circulating level of these hormones, in turn, is influenced by both the osmotic pressure and the $[Na^+]$ of the extracellular fluid. In consequence, the kidney discharges a dilute (hypotonic) urine when the salt concentration of plasma (which reflects extracellular fluid concentration) falls and a concentrated (hypertonic) urine when the salt concentration rises. In addition, water intake is regulated by the thirst mechanism, which is operative with even minute increases in the tonicity of extracellular fluid. The production of vasopressin and the sensation of thirst are both initiated by osmoreceptors in the hypothalamus (Chap. 47).

Control of the Volume of Extracellular Fluid. This is one of the least understood aspects of electrolyte and fluid metabolism. Of the four parameters here considered, *viz.*, osmotic pressure, volume, composition, and pH, volume is subject to greatest variation among a normal population. This makes determination of extracellular and plasma volumes of relatively little diagnostic use except in unusual instances when these determinations have been made in the same patient before onset of illness.

Plasma proteins are of prime importance in regulation of the osmotic balance between interstitial fluid and the plasma (see following chapter). Therefore, plasma volume is related usually to the amount of total circulating plasma protein, particularly albumin. Profound protein depletion results in diminution not only in serum albumin concentration but in plasma volume. However, removal of plasma or whole blood is followed by transfer of interstitial fluid to the vascular compartment with a temporary fall in serum protein concentration. Administration of concentrated albumin solution leads to a transitory increase in plasma volume.

The extracellular fluid volume is a function of the total amount of sodium available. The kidney, which responds promptly to minute changes in the concentration of many electrolytes or in pH, is relatively insensitive to alterations in the volume of this fluid. The diuresis which follows administration of isotonic NaCl solution, in contrast to water, may occur over a period of several days. However, if sodium is removed from the diet, it soon disappears from the urine and is retained with sufficient water to maintain isotonicity and, therefore, the volume of extracellular fluid. As renal mechanisms attempt to compensate for disturbance in the pH, osmotic pressure, or composition of the extracellular fluid, it is volume which is safeguarded with highest priority. However, the factors involved in the renal regulation of total extracellular fluid volume are not yet clearly defined.

Alterations in Electrolyte and Water Metabolism. If, for purposes of this discussion, changes in pH and composition are temporarily disregarded, there are six possible circumstances affecting the osmotic pressure and volume of the extracellular fluid. These are summarized in Table 33.1. Each of these may be produced readily in the laboratory and each has been observed clinically. However, many clinical situations

Table 33.1: ALTERATIONS IN VOLUME AND COMPOSITION OF BODY FLUIDS

Alteration in extracellular fluid	Volume		Plasma $[Na^+]$	Hematocrit, plasma proteins	Urinary excretion*	
	Intra-cellular	Extra-cellular			Na^+	H_2O
Hypotonic expansion	↑	↑	↓	↓	↓	↑
Isotonic expansion	...	↑	...	↓	↑	↑
Hypertonic expansion	↓	↑	↑	↓	↑	↑
Hypotonic contraction	↑	↓	↓	↑	↓	↑
Isotonic contraction	...	↓	...	↑	↓	↓
Hypertonic contraction	↓	↓	↑	↑	↑	↓

* The last two columns, showing nature of the renal response, refer to the response of the normal kidney to the stimulus of the situation summarized in the columns to the left. When the situation arises because of deranged renal function, for whatever reason, the last two columns are not applicable. ↑, increase; ↓, decrease.

SOURCE: Modified from L. G. Welt, "Clinical Disorders of Hydration and Acid-Base Equilibrium," Little, Brown & Company, Boston, 1955.

exhibit features of two or more of these alterations, which will be considered in turn.

1. *Hypotonic Expansion.* The accumulation of water without an equivalent amount of salt is occasionally encountered when copious quantities of salt-free fluids, *e.g.*, glucose solution, are given to persons with inadequate renal function. The accumulated water distributes osmotically among all the fluid compartments. The cells of the central nervous system share in this process, which may lead to convulsions ("water intoxication") and even death.

2. *Isotonic Expansion.* Accumulation of water and salt in isotonic amounts expands the extracellular fluid with no alteration in intracellular volume or composition. The fluid distributes between interstitial fluid and plasma, thereby lowering the concentration of plasma proteins and the hematocrit, and may be manifest as palpable edema of the extremities or pulmonary edema, an occasionally serious complication of parenteral fluid therapy.

3. *Hypertonic Expansion.* Accumulation or retention of sodium leads to an increase in extracellular fluid volume. If, however, this sodium is not accompanied by an equivalent amount of water, the resultant extracellular fluid is hypertonic and water transfers from cells to the extracellular compartment until osmotic equilibrium is attained. Thus, the extracellular fluid expands at the expense of the cells. This is a rare phenomenon but may be illustrated by the dramatic events occurring after ingestion of sea water, as shown in Fig. 33.3. Note that sea water contains twice as much sodium as the most concentrated urine made by the kidneys of a healthy adult. If this process continues, death may occur because of damage to the central nervous system.

4. *Hypotonic Contraction.* This results when salt is lost from the body unaccompanied by an equivalent amount of water. Several such situations are encountered in clinical practice, notably in adrenal cortical insufficiency (Chap. 45). In this instance normal renal control of sodium excretion is lost and the urine is high in salt concentration. The water which remains is distributed among all fluid compartments so that the cells expand. However, the serious aspects are those due to diminution in plasma volume, as described below.

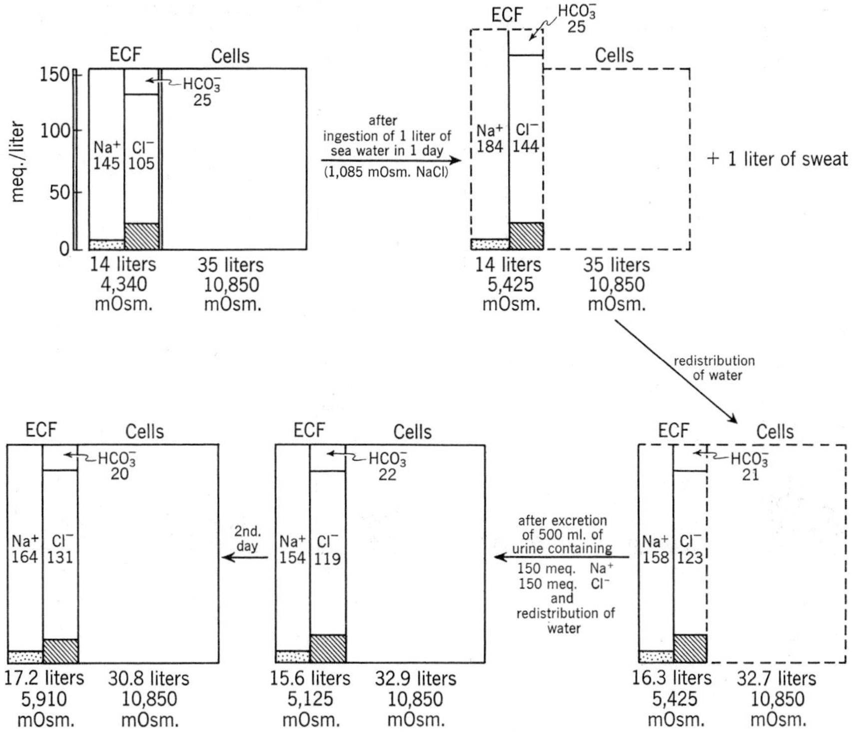

Fig. 33.3. Hypertonic expansion of extracellular fluid. The effects of sea-water ingestion. Ordinate, milliequivalents per liter. Abscissa, volume. The composition shown for extracellular fluid (ECF) is that of plasma. Only the volume and osmolar content of cellular material are presented. Broken lines denote hypothetical states, which never truly exist, as all transitions occur constantly, but which do convey the necessity for redistribution of water and the magnitude of the task confronting the kidney. The final figure (*lower left*) is based on the assumption that another liter of sea water is absorbed on the second day. Note the expansion of extracellular fluid at the expense of cellular fluid. Were no salt or water consumed, the same tendencies would be manifest except that the salt excreted in the urine would be derived from the extracellular fluid and both compartments would shrink.

5. *Isotonic Contraction.* This is the most frequently encountered of the conditions under discussion. Since there is no normal obligatory sodium loss, isotonic contraction, like hypotonic contraction, can occur only by abnormal loss of sodium from the body, most commonly in one or more of the secretions of the gastrointestinal tract. These secretions are virtually isotonic with plasma (Fig. 33.4). Moreover, as is evident in Table 33.2, the total daily production of these secretions is equal to 65 per cent of the volume of the entire extracellular fluid, and continued loss of these secretions would soon be serious. As these fluids are all isotonic, their loss does not occasion a change in intracellular volume, and the entire loss must be from the extracellular fluid, which contracts to an equivalent extent.

Interstitial fluid and plasma exist in a volume ratio of 3:1, and in isotonic contraction the fluid loss is increasingly at the expense of interstitial fluid, because of the increasing effective osmotic pressure of the plasma proteins. The clinical features of this state, frequently termed "dehydration," are due largely to the cardiovascular disturbances resulting from decreased plasma volume. Even when apparently adequate urine volumes are produced, renal insufficiency is evident by the rise in blood non-

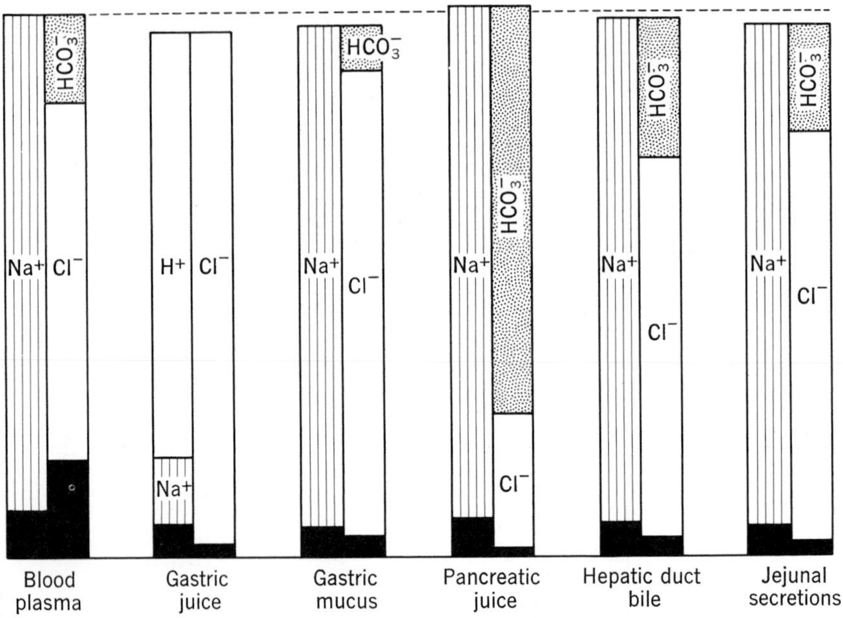

Fig. 33.4. Electrolyte composition of secretions of the gastrointestinal tract. (*From J. L. Gamble, "Chemical Anatomy, Physiology and Pathology of Extracellular Fluid," 5th ed., Harvard University Press, Cambridge, Mass.,* 1950.)

protein nitrogen concentration. The kidney responds by excreting minimal volumes of urine, but without an external supply of salt and water the extracellular fluid volume cannot be restored. The oliguria is succeeded by anuria, and finally the patient may become comatose and die of circulatory collapse. Figure 33.5 depicts changes in body fluids and electrolytes as a consequence of isotonic contraction resulting from severe diarrhea.

6. *Hypertonic Contraction.* Loss of water without an accompanying isotonic loss of sodium results in shrinkage of both the cellular and extracellular compartments. This may be expected whenever the obligatory water losses are not met, as in persons to whom no water is available, elderly debilitated patients unable to feed themselves, unattended ill persons who do not respond to the normal thirst sensation, after unusual losses of sweat uncompensated by adequate water consumption, or in persons with diabetes insipidus or mellitus who lose large amounts of water in the urine

Table 33.2: Daily Volume of Digestive Secretions of Average Adult, in Milliliters

Saliva	1,500
Gastric secretions	2,500
Bile	500
Pancreatic juice	700
Intestinal secretions	3,000
Total	8,200
Plasma	3,500
Total extracellular fluid	14,000

uncompensated by equivalent water ingestion. Since the extracellular and intracellular compartments exist in normal ratio of 2:5, the water lost is largely at the expense of the intracellular compartment, the osmotic pressure of both compartments rising in equivalent manner. Before serious impairment of function due to contraction of the plasma is manifest, changes in the central nervous system may be the dominant feature of this syndrome, as in hypertonic expansion.

In practice pure examples of these six situations are rarely encountered. Thus, although diarrhea or vomiting may give rise to isotonic contraction, the individual may fail to ingest water in sufficient quantity to meet the obligatory water losses, thus converting the situation into hypertonic contraction. The vagaries of normal existence present minor attacks on the body fluids which, if uncompensated, might lead to one of the six situations described above. The fact that the sodium concentration and volume of extracellular fluid remain so remarkably constant is evidence of the efficiency of the homeostatic mechanisms and the effectiveness of the kidney.

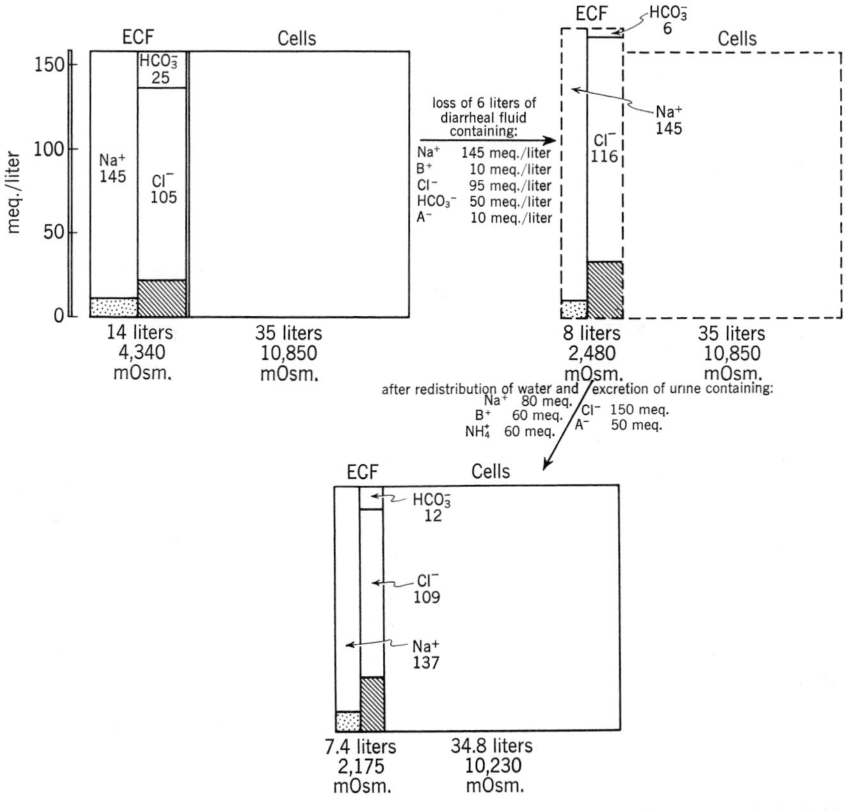

Fig. 33.5. Isotonic contraction of extracellular fluid. The effects of massive diarrhea on body fluids and electrolytes. Ordinate, milliequivalents per liter. Abscissa, volume. The composition shown for extracellular fluid (ECF) is that of plasma. The events shown occur over a 3- to 4-day period. Note the profound extracellular dehydration and relatively trivial effect on cellular volume, if water consumption is adequate. The broken lines represent a hypothetical state which never truly exists, as all transitions occur constantly. This figure does show what would happen if water redistribution and urine secretion did not occur. The final composition of extracellular fluid is conditioned by water retention and some degree of exchange of Na^+ and K^+ across cell membranes. B^+ represents the sum of cations other than Na^+ and NH_4^+; A^- represents the sum of anions other than Cl^-.

Control of the pH of Body Fluids; the Buffer Systems. Little information is available concerning the pH within cells. Data obtained by staining with intravital dyes which are also pH indicators, although subject to error, indicate that intracellular pH may vary from 4.5 in the cells of the prostate to approximately 8.5 in osteoblasts. Accurate data concerning the pH of interstitial fluid are lacking. If the interstitial $[HCO_3^-]$ is less than that of plasma, as might be expected from equilibrium considerations (page 128), and if the CO_2 tension is greater, then the mean pH of interstitial fluid may be inferred to be somewhat lower than that of venous plasma.

It is convenient to regard extracellular fluid as compounded in the following manner: Consider a solution containing mixed acids (HCl, H_2SO_4, H_3PO_4, protein, etc.) to which is added a second solution containing NaOH, KOH, etc. However, the total amount of alkali, in equivalents, exceeds that of acid. After mixing, a gas mixture containing CO_2 at 40 mm. Hg is equilibrated with the solution and maintained at the same pressure, thereby maintaining a constant $[H_2CO_3]$ in the medium. Under these circumstances an amount of HCO_3^- is generated equal to the difference between the amounts of alkali and acid in the original solutions. This concentration of HCO_3^- ion, about 25 meq. per liter in normal extracellular fluid, is a measure of the amount of alkali still available to react with additional strong acids.

The major buffer of extracellular fluid is the bicarbonate–carbonic acid system. This results from a number of factors: (1) There is considerably more bicarbonate present in extracellular fluid than any other buffer component. (2) There is a limitless supply of carbon dioxide. (3) Physiological mechanisms operate to maintain extracellular pH function by controlling either the bicarbonate or the carbon dioxide concentration of extracellular fluid. (4) The bicarbonate–carbonic acid buffer system operates in conjunction with hemoglobin, as described in Chap. 32. As in all buffered systems, pH is dependent not on absolute concentrations of buffer constituents but rather on their *ratio,* as stated in the Henderson-Hasselbalch equation. Because the $[H_2CO_3]$ is fixed only by the alveolar CO_2 tension and is unaffected by the addition of either alkali or acid, this system is considerably more efficient in maintaining pH 7.4 than are the usual buffers employed in the laboratory. This is shown by the curves in Fig. 33.6. Curve B represents the behavior of a buffer whose pK_a is 7.4. Curve A indicates the inadequacy of a buffer constructed with a nonvolatile acid of pK 6.1 in maintaining pH 7.4. Curve C demonstrates the superiority of a buffer system based on an acid which is a gas of unlimited supply and whose concentration is fixed by its partial pressure in the gas phase. Since the $[H_2CO_3]$ is fixed by the gas tension, if the gas tension is equivalent to that normally present in blood, the HCO_3^-/H_2CO_3 system is more useful at pH 7.4, with a ratio of 20, than it would be at its pK, 6.1, where the HCO_3^- would be exhausted by addition of 1.25 meq. per liter of acid, and addition of this amount of alkali would result in a rise of 0.3 pH unit.

The buffer efficiency of the bicarbonate–carbonic acid system is further enhanced by the presence of erythrocytes. This is illustrated by Fig. 33.7, which depicts the results of equilibrating two solutions with CO_2 at varying tensions. These are (1) plasma which has been separated from cells and then equilibrated (separated plasma), and (2) plasma separated carefully after equilibrating whole blood at the stated CO_2 tension (true plasma). As the CO_2 tension in separated plasma is

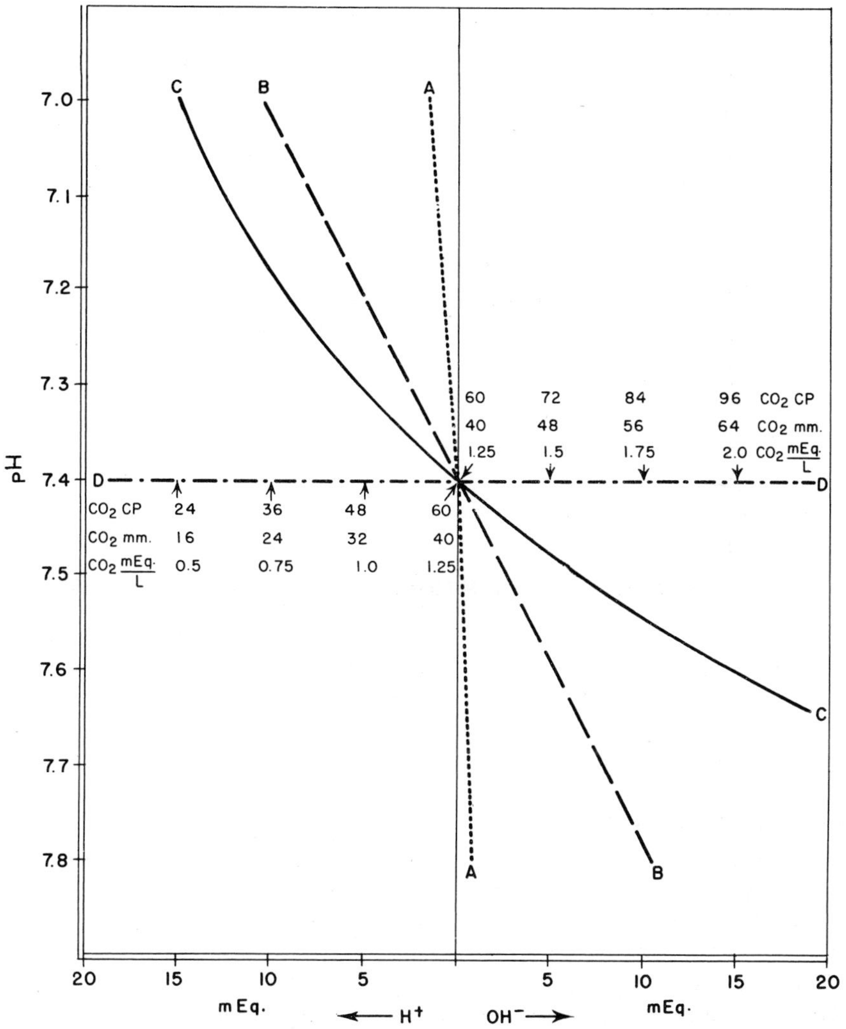

Fig. 33.6. Comparison of the HCO_3^-/H_2CO_3 buffer with the ability of other buffers to maintain pH 7.4. Each buffer is set at pH 7.4, and then 1 liter is titrated with acid or alkali as indicated. Curve *A*, a buffer system of p*K* 6.1. At pH 7.4, [A⁻] is 25 meq. per liter, and [HA] is 1.25 meq. per liter. Curve *B*, a buffer system of p*K* 7.4. At pH 7.4, [A⁻] and [HA] are 25 meq. per liter. Curve *C*, a buffer system of p*K* 6.1, one of whose components, HA, is a gas. Titration is performed in presence of an unlimited supply of gas at a partial pressure sufficient to maintain [HA] at 1.25 meq. per liter. Curve *D* is made on the assumption that HA of curve *C* is H_2CO_3 and shows the changes in P_{CO_2}, in gas phase, necessary to maintain a constant pH despite the addition of acid or alkali.

increased, CO_2 dissolves and, because of newly formed carbonic acid, the pH falls, as predicted by the Henderson-Hasselbalch equation. There is no measurable increase in bicarbonate concentration. The curve shown for "separated" plasma differs only slightly from that which would be obtained under the same experimental conditions with a solution of sodium bicarbonate in water. This difference results from the presence of other buffers in plasma, notably the proteins and phosphates, and

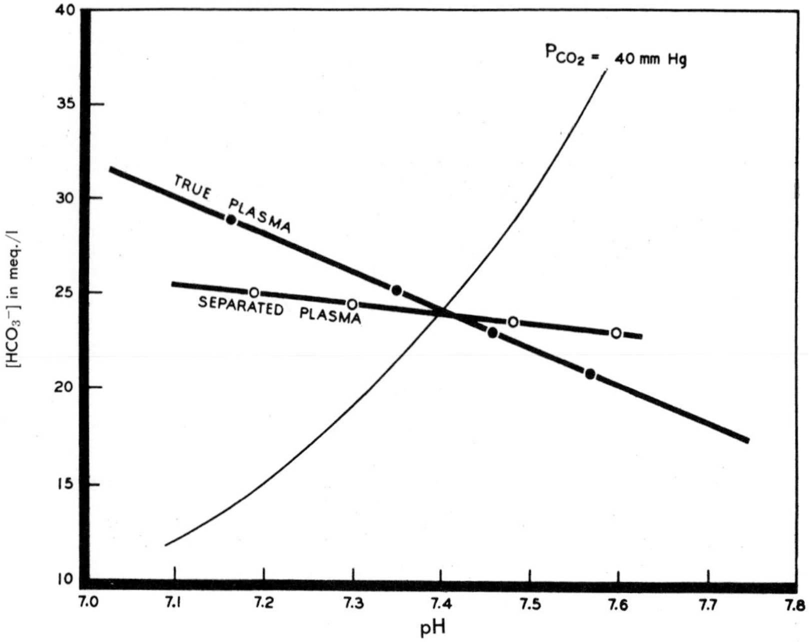

Fig. 33.7. The effect of varying CO_2 tension on the pH and $[HCO_3^-]$ of separated plasma and of true plasma (page 685). The light line is the 40-mm. CO_2 isobar, *i.e.,* at this pressure and 15 meq. per liter of HCO_3^-, pH = 7.2. Points to the left lie on isobars of increasing P_{CO_2}, those to right on isobars of lower CO_2 tension. (*After H. W. Davenport, "The ABC of Acid-Base Chemistry," 3d ed., University of Chicago Press, Chicago.* 1950.)

plasma is a somewhat better buffer than is an aqueous bicarbonate solution. The behavior of "true" plasma is in marked contrast. As the P_{CO_2} is increased, an appreciable increment in plasma $[HCO_3^-]$ occurs so that the pH does not fall as rapidly as it did in the previous instances. As the P_{CO_2} is decreased below normal, the $[HCO_3^-]$ of "true" plasma also decreases, thereby preventing the expected rise in pH.

The influence of erythrocytes on the total CO_2 content of plasma at varying CO_2 tensions was described in the preceding chapter (pages 765*ff.*). As CO_2 diffuses into cells, the H_2CO_3 reacts with hemoglobin, forming HCO_3^-, which then enters the plasma in exchange for chloride. This is not contingent upon deoxygenation of hemoglobin but is achieved more readily and with even less pH change when deoxygenation occurs simultaneously. Conversely, lowering the CO_2 tension results in a reversal of this process, with consequent diminution of plasma $[HCO_3^-]$. Although the situation does not arise under physiological conditions, only in the presence of red cells does the total CO_2 content of plasma fall to zero at a P_{CO_2} of 0 mm. This is possible because there is sufficient hemoglobin in whole blood to permit the following series of reactions to proceed to completion to the right.

$$HHb^+ + HCO_3^- \rightleftharpoons Hb^\circ + H_2CO_3 \rightleftharpoons H_2O + CO_2$$

Respiratory and Renal Regulation of the pH of Extracellular Fluid. The described combination of the properties of a buffer, one of whose components is a gas, and the automatic self-adjustments made possible by intracorpuscular hemoglobin result in the remarkably constant pH of blood plasma. In addition, the body possesses two further safeguards, the respiratory apparatus and the kidneys, which, by their control of plasma concentrations of H_2CO_3 and HCO_3^-, respectively, serve in auxiliary fashion to maintain constant the pH of extracellular fluid.

Unlike the [HCO_3^-] or the fixed anion concentration, the [H_2CO_3] is determined solely by one consideration: the partial pressure of CO_2 in the gas mixture in equilibrium with extracellular fluid, *viz.*, alveolar air. This, in turn, is dependent upon the rate at which carbon dioxide leaving pulmonary blood is diluted with atmospheric air and, hence, upon the rate and depth of respiration. These are regulated by the nervous system at the respiratory center, which appears to be sensitive to the pH and P_{CO_2} of the extracellular fluid. When the pH of extracellular fluid falls below normal because of diminished [HCO_3^-], respiration is stimulated, lowering alveolar P_{CO_2} and, hence, extracellular [H_2CO_3]. This tends to return the [HCO_3^-]/[H_2CO_3] ratio to its normal value of 20:1 and, thus, to restore pH toward 7.4. The resultant fall in plasma CO_2 tension affects the controlling nerve cells in the opposite manner, and consequently compensation would never be complete if this were the sole regulatory mechanism.

With a high plasma pH, the respiratory rate falls, alveolar P_{CO_2} and, hence, plasma [H_2CO_3] rise, and the pH moves toward 7.4. Again, perfect compensation is not attained since the increased plasma [H_2CO_3] opposes the effect of elevated pH on the controlling center. It is to be emphasized that the pH is dependent not on absolute concentrations but solely on the ratio [HCO_3^-]/[H_2CO_3].

The buffer systems of plasma can withstand the addition of 16 meq. of acid or 29 meq. of alkali per liter and still maintain pH within the range compatible with life, *viz.*, 7.0 to 7.8. With pulmonary compensation the normal pH range can be maintained despite addition of as much as 23 meq. of acid or 80 meq. of alkali per liter of plasma.

Whereas the respiratory mechanism compensates for disturbances of acid-base balance by regulating [H_2CO_3] in extracellular fluid, the kidney augments pH control by regulating [HCO_3^-]. Pulmonary compensation is extremely rapid but never complete; in contrast, renal compensation requires an extended period to be effective but may result in complete restoration of normal pH. A fall in extracellular pH due to increased alveolar CO_2 or decreased [HCO_3^-] is counteracted by two devices available to the kidney for elevating [HCO_3^-], *viz.*, excretion of acidic urine and of ammonium ions. Tendency toward alkaline extracellular fluid is counteracted by excretion of Na^+, HCO_3^-, and the dissociated forms of other weak acids.

By excreting acidic urine, the lower limit of which is approximately pH 4.6, weak acids which exist in dissociated form in plasma can be excreted in part in undissociated form. This occurs not only in pathological states but normally also, since the residual ash of the average diet is acidic. Primary phosphate is the principal acid of this ash.

$$\text{Dietary } H_2PO_4^- + \text{plasma } HCO_3^- \longrightarrow \text{plasma } HPO_4^{=} + H_2CO_3$$

The acid monovalent phosphate is transported in plasma as the divalent ion at the expense of plasma HCO_3^-. In extracellular fluid, at pH 7.4, the ratio [$HPO_4^{=}$]/[$H_2PO_4^-$] is 4/1, but in urine at pH 5.4 this ratio is 4/100. Thus, while 80 per cent

	Renal Venous Plasma	Tubular Epithelium	Glomerular Filtrate	Urine
Acidification of Urine	HCO_3^- Na^+ ←	H_2CO_3 ⇅ HCO_3^- H^+	$HPO_4^=$ Na^+ Na^+ → pH = 7.4	$H_2PO_4^-$ Na^+ pH = 4.8
Acidification of Ketotic Urine	HCO_3^- Na^+ ←	H_2CO_3 ⇅ HCO_3^- H^+	$2AcOAc^-$ Na^+ Na^+ → pH = 7.4	$HAcOAc$ Na^+ $AcOAc^-$ pH = 4.8
Ammonia Secretion	HCO_3^- Na^+ ←	H_2CO_3 ⇅ HCO_3^- H^+ NH_3	Na^+ Cl^- → pH = 7.4	Cl^- NH_4^+ pH = 4.8
Ammonia Secretion in Ketosis	$2HCO_3^-$ $2Na^+$	$2H_2CO_3$ ⇅ $2HCO_3^-$ $2H^+$ NH_3	$2AcOAc^-$ $2Na^+$ → pH = 7.4	$HAcOAc$ NH_4^+ $AcOAc^-$ pH = 4.8

Fig. 33.8. Renal compensation for acidosis. Urine formation proceeds to the right from the luminal border of tubular epithelium, and the return of electrolytes to renal venous plasma proceeds to the left. $AcOAc^-$ represents acetoacetate ion.

of the phosphate in plasma exists as $HPO_4^=$, virtually all the phosphate in acidic urine again exists as $H_2PO_4^-$, the form in which it originally entered plasma. Figure 33.8 summarizes these events. It is this $H_2PO_4^-$ which constitutes most of the acid conventionally measured as the "titratable acidity" of urine. As a result of the operation of this mechanism, the organism can cope with the acid constantly entering extracellular fluids without depleting the extracellular supply of sodium or appreciably lowering the plasma $[HCO_3^-]$. In acidosis this response of the kidney is initiated upon the lowering of extracellular pH, but if there has been a serious decrease in the plasma $[HCO_3^-]$ or increase in $[H_2CO_3]$, a considerable time is required before sufficient HCO_3^- can be regenerated by this process to restore pH to 7.4. A frequent and important illustration of this mechanism is that which occurs in the acidosis resulting from accumulation of ketone bodies (pages 503*ff.*). At pH 7.4 more than 99 per cent of acetoacetic acid exists in the dissociated form. Therefore, the following reaction occurs when this acid enters plasma,

$$CH_3{-}CO{-}CH_2{-}COOH + HCO_3^- \longrightarrow CH_3{-}CO{-}CH_2{-}COO^- + H_2CO_3$$

thereby lowering $[HCO_3^-]$ and pH. Since the pK for acetoacetic acid is 4.8, excretion of urine of pH 4.8 permits excretion of 50 per cent of the acetoacetic acid in the undissociated form. Figure 33.8 also shows these reactions. By this mechanism it is possible to generate one HCO_3^-, for return to renal venous plasma, for each two acetoacetic acid molecules formed in the liver and requiring excretion by the kidney. Renal tubular metabolism supplies the energy for this process and also provides sufficient H_2CO_3 for generation of HCO_3^- ions.

The second renal mechanism for restoring a normal extracellular pH in acidotic states is the formation and excretion of ammonium ion, a cation not present in the glomerular filtrate. Hydrolysis of glutamine is the chief source of this ammonia (page 537). This mechanism does not respond to sudden changes in extracellular pH as rapidly as does that which acidifies the urine. However, in persistent acidosis, ammonia excretion is quantitatively more significant than is acidification. Figure 33.8 schematically depicts how ammonia excretion elevates extracellular [HCO_3^-]. By this means it is possible to return to the plasma, associated with HCO_3^-, Na^+ ions which otherwise would be present in association with either the dissociated fraction of weak acids or the mineral anions of urine. Only this mechanism can compensate for acidosis occasioned by the accumulation of anions of strong acids, *e.g.*, loss of alkaline digestive secretions.

The sum of ammonium ions plus titratable acid in urine is equivalent to the Na^+ which has been returned to the extracellular fluid in association with HCO_3^-, and which otherwise would have appeared in the urine had the kidney excreted urine at pH 7.4 and been unable to make ammonia. Although in the discussion above, attention has been on the bicarbonate ion, it will be recognized that these two mechanisms entail exchange of H^+ and NH_4^+ ions, respectively, for the Na^+ of the glomerular filtrate. The mechanism of these exchanges is considered later (see below). Since a primary function of the kidney is regulation of osmotic pressure, which, in turn, is dependent upon [Na^+], loss of the amount of Na^+ represented by the titratable acid plus NH_4^+ would have forced a diminution in extracellular fluid. Conservation of Na^+ by the renal mechanisms serves to maintain both the "alkaline reserve" and the volume of plasma. Occasionally, failure of the ammonium-forming mechanism may itself be the cause of acidosis and dehydration, as in lower nephron nephrosis and Fanconi's syndrome.

One additional mechanism is available for combating acidosis but is of significance only in prolonged acidoses. This is the substitution of Ca^{++} for Na^+ in urine. The source of this calcium is the $Ca_3(PO_4)_2$ of bone, which increases in solubility with decreasing pH. As $Ca_3(PO_4)_2$ from bone enters plasma, it reacts with H_2CO_3.

$$3Ca^{++} + 2PO_4^{\equiv} + 2H_2CO_3 \longrightarrow 3Ca^{++} + 2HPO_4^{=} + 2HCO_3^-$$

The HCO_3^- ions formed are available to neutralize two molecules of an acid with a p*K* less than the p*K* of carbonic acid,

$$2HCO_3^- + 2HA \longrightarrow 2H_2CO_3 + 2A^-$$

so that the mixture in plasma may be considered to be the following.

$$3Ca^{++} + 2HPO_4^{=} + 2A^-$$

Addition of a second pair of acid molecules to the plasma and their reaction with bicarbonate gives the following.

$$2Na^+ + 2HCO_3^- + 2HA \longrightarrow 2Na^+ + 2A^- + 2H_2CO_3$$

The total mixture presented to the glomerulus is then $2Na^+$, $4A^-$, $3Ca^{++}$, and $2HPO_4^=$. If the usual acidification device is operative, $2Na^+ + 2HPO_4^= + 2H_2CO_3 \rightarrow 2Na^+$(plasma) $+ 2HCO_3^-$(plasma) $+ 2H_2PO_4^-$(urine). The over-all reaction, then, is

$$3Ca^{++} + 2HPO_4^{=} + 4HA \longrightarrow 3Ca^{++} + 2H_2PO_4^{-} + 4A^{-}$$

and one mole of tricalcium phosphate makes possible the excretion of four equivalents of acid. This constitutes an extremely effective mechanism for preventing depletion of the alkali reserve, although it may result ultimately in serious demineralization of the skeleton.

Renal compensation for circumstances which otherwise would result in a *rise* in extracellular pH is accomplished by lowering the [HCO_3^-] of extracellular fluid. This is possible only by excretion of Na^+ in association with anions other than those of the mineral acids. Such urine, therefore, is alkaline (pH 7.4 to 8.2) and contains unusual quantities of Na^+ associated with HCO_3^- and $HPO_4^{=}$. Direct excretion of Na^+ and HCO_3^- ions obviously lowers the [HCO_3^-] of extracellular fluid. Excretion of Na^+ in association with $HPO_4^{=}$ serves the same end. It will be recalled that urinary phosphate arises from the metabolism of organically bound phosphate of food, which entered extracellular fluid essentially as $H_2PO_4^-$ ions. As indicated previously, this acid radical reacts immediately in the extracellular fluid with HCO_3^-. In alkalosis, the phosphate, while never present in extracellular fluid in large concentration, is excreted with 2 Na^+ and, consequently, reduces the [HCO_3^-]. This removal of sodium from extracellular fluid is accompanied by sufficient water so that the extracellular fluid remains at normal osmotic pressure, and renal compensation for alkalosis is attended by isotonic contraction of the extracellular fluid, frequently increasing the severity of existing dehydration.

Cellular Buffering in Disturbances of Extracellular pH. Evidence indicates that cells also participate in regulation of extracellular pH. Muscle cells, renal tubular epithelium, and perhaps cells in general possess an ion exchange mechanism which mediates an exchange across the cell membrane of Na^+ for either K^+ or H^+ or both. This exchange permits the cell contents to supplement the other mechanisms which maintain extracellular pH. It is unclear whether this process is mediated by any aspect of the "electrolyte pump."

In alkalosis caused by an increase in extracellular [HCO_3^-], Na^+ enters cells in exchange for both H^+ and K^+. The protons react with extracellular HCO_3^- and the resultant CO_2 is expired. The K^+ is excreted in urine with an equivalent amount of HCO_3^-. The net result is to diminish extracellular HCO_3^- by the equivalent of the amount of Na^+ which entered cells. In acidosis, Na^+ leaves cells and both H^+ and K^+ enter. For each Na^+-H^+ exchange, a HCO_3^- ion remains in plasma, since the H^+ entering cells was derived from the dissociation of H_2CO_3. The entry of K^+ into the cell has no immediate influence on extracellular [HCO_3^-], but the diminished plasma [K^+] permits more effective acidification of the urine (page 834), thus indirectly contributing to restoration of normal extracellular pH. Simultaneously, there frequently occurs a substantial renal excretion of K^+, creating a net deficit of this cation. These events are illustrated in Fig. 33.9.

Factors Altering the pH of Extracellular Fluid. Because of the acidic nature of the ash of most foods and of the organic acids which arise in metabolic processes, acid is constantly added to extracellular fluid. As a consequence, the urine of man is usually acidic, as compared with extracellular fluid. The compensatory control exerted by the kidney prevents sodium loss, and normal extracellular fluid is remark-

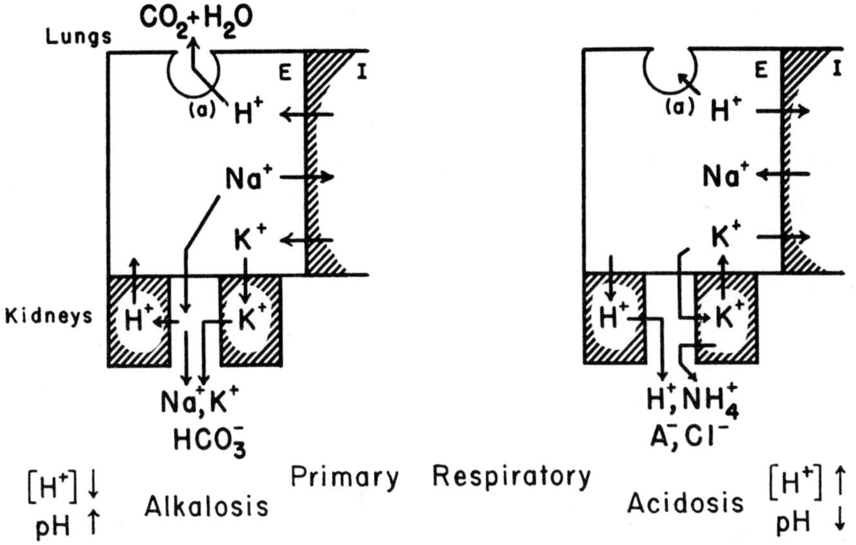

Fig. 33.9. Simplified scheme of linked transfers of cations in defense of body fluid neutrality in respiratory alkalosis and acidosis. *E* and *I* are extracellular and intracellular fluid, respectively. The primary disturbances, shown at (*a*), lead directly to respiratory alkalosis or acidosis. In alkalosis proton loss is shared by the intracellular fluid, exchanging with Na^+ from the extracellular phase. In acidosis exchanges take place in the opposite directions. (*After J. R. Elkinton and T. S. Danowski, "The Body Fluids," The Williams & Wilkins Company, Baltimore,* 1955.)

ably constant in composition, volume, and pH. Subsistence on a diet consisting largely or exclusively of fruits and vegetables results in the opposite situation, *i.e.*, addition to the extracellular fluid of an excess of alkali which is eliminated in the urine.

Alterations in [H_2CO_3]. The respiratory system has been considered in its role as compensator, but it is, on occasion, the primary malefactor. Thus, adult hysterics or children with meningitis may markedly *hyperventilate,* lowering extracellular [H_2CO_3] and, thereby, elevating pH; this is termed *respiratory alkalosis.* Since the arterial P_{CO_2} is lower than normal, operation of the hemoglobin buffer mechanism automatically decreases plasma [HCO_3^-], tending to prevent the rise in plasma pH. This cannot compensate adequately for the diminished [H_2CO_3], and hyperventilation may elevate extracellular pH to 7.65 within a few minutes. *Hypoventilation* of whatever origin (morphine poisoning, pneumonia, pulmonary edema, etc.) has the opposite effect and lowers extracellular pH. The increased P_{CO_2} results also in an increased plasma [HCO_3^-] because of the hemoglobin buffer mechanism, and individuals who are hypoventilating may immediately exhibit a low plasma pH, elevated [H_2CO_3], and elevated [HCO_3^-]; this is *respiratory acidosis.*

Compensation for either of the above circumstances of altered extracellular [H_2CO_3] is largely effected by the kidney. In the first instance, an alkaline urine is excreted, and, in the second, an acidic urine. Isotonic contraction may result from the excretion of large urine volumes but is rarely as severe as in other instances of dehydration.

The cellular exchange process also participates in these disturbances. In respiratory alkalosis, Na^+ exchanges for cellular K^+ and H^+ as described above, ameliorating the extracellular alkalosis but alkalinizing the cell contents and depleting cellular K^+, which is excreted in the urine. In respiratory acidosis, Na^+ is withdrawn from cells which are acidified by the entering protons and extracellular K^+ is diminished.

Alterations in [HCO_3^-]. More frequent and serious are those circumstances in which the alteration in pH is associated primarily with changes in [HCO_3^-]. In the simplest instances, lowering of [HCO_3^-] may be expected upon addition to the extracellular fluid of an acid stronger than carbonic acid, *e.g.*, acetoacetic acid. This is termed *metabolic acidosis* in contrast to the *respiratory acidosis* described above.

As plasma [HCO_3^-] falls in metabolic acidosis, HCO_3^- enters plasma from the red cells in exchange for Cl^-. At constant P_{CO_2} within the erythrocytes, their internal pH would be lowered and the dissociation of the hemoglobin would be repressed,

$$H_2CO_3 + Hb^\circ \rightleftharpoons HCO_3^- + HHb^+$$

thereby making more HCO_3^- available for the plasma; this would tend to restore plasma pH to normal. This is not sufficient to compensate for drastic plasma [HCO_3^-] reductions; both pulmonary and renal compensation are also required, as well as proton exchange for cellular Na^+.

Elevation in plasma [HCO_3^-] is compensated by the same mechanisms operating in reverse: the chloride shift, hypoventilation, alkaline urine, and exchange of plasma Na^+ for cellular H^+ and K^+.

The simplest means of elevating [HCO_3^-] is to administer $NaHCO_3$. This may occur in patients with peptic ulcer after overdosage with alkali or occasionally as a result of overenthusiastic use of alkali preparations for relief of gastric disorders.

Alterations Due to Fluid Loss. Those circumstances in which the effect on pH is based on unusual losses of fluid, particularly the various secretions of the gastrointestinal tract, are among the most serious and frequently encountered in clinical practice. Understanding this problem requires knowledge of the composition of these fluids and the volumes which may be involved. These have been presented in Table 33.2 and Fig. 33.4. Each of these secreted fluids is elaborated from the extracellular fluid; the chief cation is sodium, except in gastric juice. However, the anionic pattern may differ considerably from that of extracellular fluid. The effect on extracellular pH is, consequently, determined by the manner in which the fluid involved differs from extracellular fluid.

Loss of a fluid which closely resembles extracellular fluid in composition is relatively infrequent but occurs occasionally when a constant suction tube is placed in the duodenum or jejunum and also occurs in simple hemorrhage and after loss of serous exudates. In these instances, there may be no great effect on pH, but serious dehydration due to isotonic contraction may result.

The ratio of [Cl^-]/[HCO_3^-] in normal extracellular fluid is about 4. If the ratio in the lost fluid *exceeds* 4, the [Cl^-] in the remaining extracellular fluid must fall while the [HCO_3^-] rises, thereby tending to elevate pH. This may be encountered during copious loss of sweat but is most frequently seen after vomiting due to pyloric or duodenal obstruction or other causes. The result is shown in Fig. 33.10. Free acid in the vomitus is not necessary in order to develop alkalosis. All that is

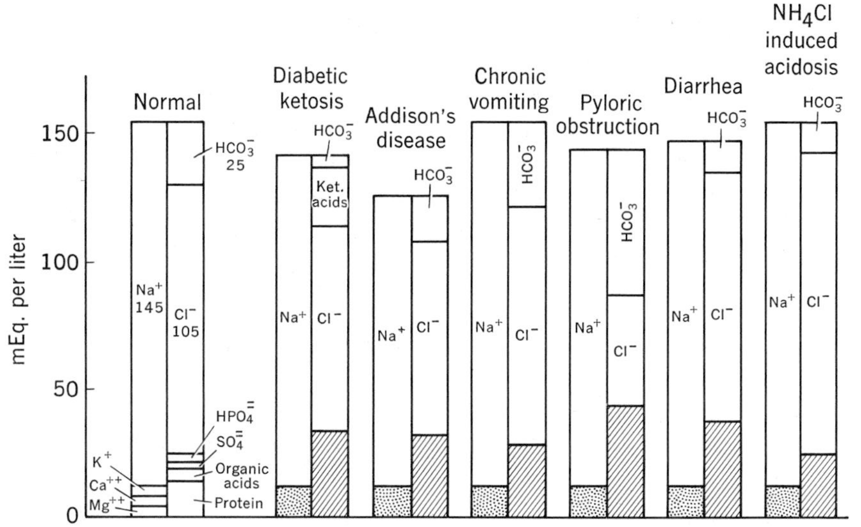

Fig. 33.10. Comparison of extracellular fluid in several pathological states. (*Modified from J. L. Gamble, "Chemical Anatomy, Physiology and Pathology of Extracellular Fluid," 5th ed., Harvard University Press, Cambridge, Mass.,* 1950.)

required is the loss of a fluid in which the $[Cl^-]/[HCO_3^-]$ ratio is greater than 4. Indeed, only very small amounts of *acidic* gastric juice are lost in prolonged vomiting; the fluid lost is largely gastric mucus, which may be contaminated with regurgitated duodenal contents. This obtains in infants who vomit and whose stomachs secrete little or no free HCl. However, loss of free acid increases the severity of the alkalosis since each mole of acid secreted results in an equivalent increase in the HCO_3^- content of extracellular fluid.

When the $[Cl^-]/[HCO_3^-]$ of the lost fluid *is less than* 4, acidosis occurs. The chief example of this type of fluid loss is diarrhea; the fluid lost is composed of mixed secretions of the pancreas and intestine as well as bile.

Alterations Due to Ketosis. It was noted earlier that acidosis may be expected whenever an acid, HA, stronger than H_2CO_3 enters the circulation at a rate greater than that at which it can be removed. This results in accumulation of the anion, A^-, and equivalent diminution in $[HCO_3^-]$ in consequence of the reaction $HA + HCO_3^- \rightleftharpoons H_2CO_3 + A^-$. This occurs in the accumulation of ketone bodies (acetoacetic and β-hydroxybutyric acids) in diabetic patients, in persons on high lipid diets, and during starvation. Ketosis usually complicates some other state and is not an isolated pathological phenomenon. Thus, in the diabetic person it complicates the dehydration already established by glucose diuresis, and renal compensation for the acidosis may further aggravate the dehydration. Ketosis also occurs readily in infants and young children who take no food and, therefore, is a frequent accompaniment of both vomiting and diarrhea in the young.

There is no comparable known situation of base production and accumulation, *i.e.*, there is no instance of accumulation of an unusual cation, such as ammonium, lithium, magnesium, etc., in quantities sufficient to disrupt the normal electrolyte balance.

PRACTICAL EVALUATION OF ACID-BASE BALANCE

Evaluation of the acid-base balance of a patient is ordinarily performed in a relatively simple manner. The minimal determinations required are urinary pH and plasma $[HCO_3^-]$. As shown in Table 33.3 these permit a decision among the four major disturbances, respiratory acidosis and alkalosis and metabolic acidosis and alkalosis, particularly if the history is known. However, at the time examination is performed the compensatory mechanisms may have restored the extracellular pH to normal, as shown by direct determination of plasma pH.

Table 33.3: EVALUATION OF ACID-BASE BALANCE

Disturbance	Urine pH	Plasma $[HCO_3^-]$, meq./liter	Plasma $[H_2CO_3]$, meq./liter
Normal	6–7	25	1.25
Respiratory acidosis	↓	↑	↑
Respiratory alkalosis	↑	↓	↓
Metabolic acidosis	↓	↓	↓
Metabolic alkalosis	↑	↑	↑

A more complete picture may be obtained if arterial blood is drawn, true plasma is separated, and the pH and total $[CO_2]$ are measured. From the nomogram in Fig. 33.11 the P_{CO_2} and $[HCO_3^-]$ can be obtained. These values are then compared with those shown in Fig. 33.12, which is interpreted as follows: The heavy line *AB* is the normal buffer line of plasma, obtained by measuring the pH and $[HCO_3^-]$ of true plasma from whole blood equilibrated at varying CO_2 tensions. *CD* is the P_{CO_2} 40-mm. isobar, and a family of isobars for varying CO_2 tensions could also be plotted but are omitted in this figure.

Point 1 of Fig. 33.12 lies on the normal buffer line to the left of the normal point and represents uncompensated respiratory acidosis. Point 2 lies on the 40-mm. isobar below the normal buffer line and represents uncompensated metabolic acidosis. Point 3 lies on the normal buffer line and to the right of the normal point and represents uncompensated respiratory alkalosis. Point 4 lies on the 40-mm. isobar and above the normal buffer line and represents uncompensated metabolic alkalosis. However, in clinical practice such data are very rare, and partial or complete compensation can be expected. Thus, point 5 lies above the normal buffer line but to the left of normal pH and must, therefore, represent partially compensated respiratory acidosis, while point 6 represents completely compensated respiratory acidosis. Point 7 lies below the normal buffer line and represents metabolic acidosis, but since it is to the right of the normal P_{CO_2} isobar, yet on the acid side of pH 7.4, it must represent partially compensated metabolic acidosis. Point 8 might denote completely compensated metabolic acidosis or respiratory alkalosis. Urine pH or other findings are necessary for decision. Point 9 represents partially compensated respiratory alkalosis, and point 10 partially compensated metabolic alkalosis.

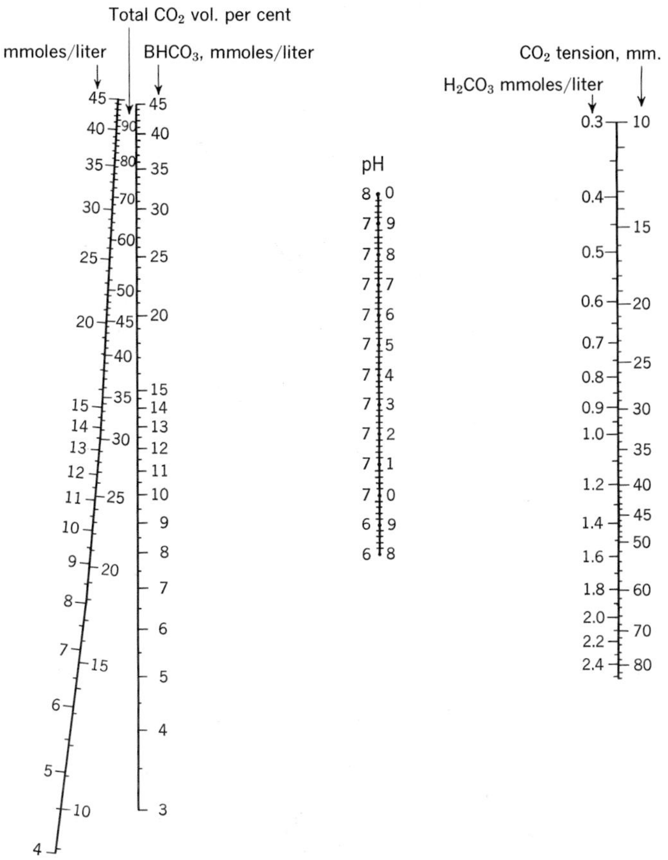

Fig. 33.11. Nomogram showing relationships between [total CO_2], [HCO_3^-], P_{CO_2}, and pH of "true" plasma. A line connecting any two of the variables will pass through the other two as formulated in the statement

$$pH = 6.10 + \log \frac{[\text{total } CO_2] - 0.0301 P_{CO_2}}{0.0301 P_{CO_2}}$$

This is derived from the Henderson-Hasselbalch equation,

$$pH = 6.10 + \log \frac{[HCO_3^-]}{[H_2CO_3]}$$

but is expressed in terms of measurements readily made in the laboratory. (*From D. D. Van Slyke and J. Sendroy, J. Biol. Chem.*, **79**, 783, 1928.)

METABOLISM OF CELLULAR ELECTROLYTES

The mean distribution of cellular electrolytes is a composite picture, obtained by analysis of a tissue, *e.g.*, muscle. Analytical data for intracellular muscle electrolytes are shown in Fig. 33.1 (page 779). No specific area within the cell is likely to have precisely this composition. Differences are to be expected in the electrolyte pattern of the cell membrane, cytoplasm, microsomes, mitochondria, nuclei, nucleoli, Golgi apparatus, etc. However, no description of the electrolyte composition of these

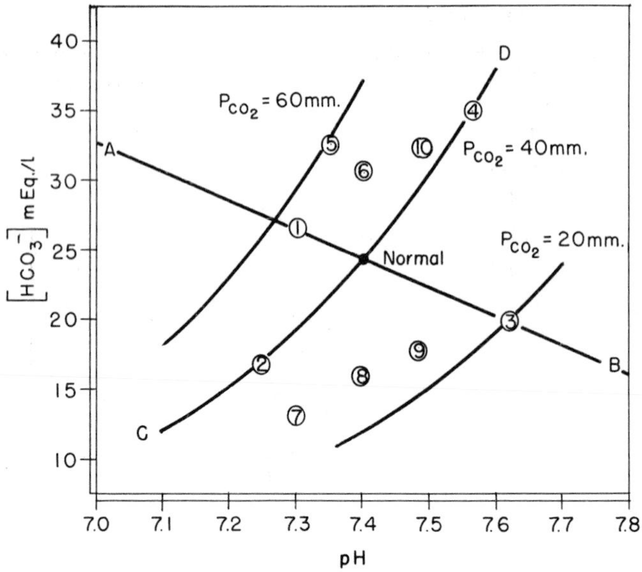

Fig. 33.12. The composition of true plasma in various disturbances of acid-base balance. Line *AB*, normal buffer line of plasma; *CD*, P_{CO_2}, 40-mm. isobar; (1) entirely uncompensated respiratory acidosis; (2) uncompensated metabolic acidosis; (3) uncompensated respiratory alkalosis; (4) uncompensated metabolic alkalosis; (5) partially compensated respiratory acidosis; (6) completely compensated respiratory acidosis; (7) partially compensated metabolic acidosis; (8) either completely compensated metabolic acidosis or respiratory alkalosis; (9) partially compensated respiratory alkalosis; (10) partially compensated metabolic alkalosis, (*After H. W. Davenport, "The ABC of Acid-Base Chemistry," 3d ed., University of Chicago Press, Chicago,* 1950.)

various cellular subdivisions is available. A significant fraction of muscle potassium is nondiffusible; presumably, this is true in all cells. At all times there is interchange between the electrolytes within and outside the cell, the rate of which may be different for each electrolyte. During the growth phase of a cell, material accumulates in relatively constant proportions. During periods of negative nitrogen balance, the cell substance is depleted, and the relative amounts of nitrogen, potassium, phosphorus, magnesium, etc., which appear in the urine in excess of the intake of these nutrients are in approximately the same proportions as those which exist within the cells.

Potassium. If data obtained with dog tissues may also be applied to man, then the mean potassium concentration is about 115 meq. per liter of cell water, while the normal serum $[K^+]$ ranges from 3.8 to 5.4 meq. per liter. The total potassium content of a 70-kg. adult is about 4,000 meq., of which only 70 meq. is in extracellular fluid.

Maintenance of normal serum $[K^+]$ is of considerable practical importance. Characteristic electrocardiographic disturbances can be correlated with serum $[K^+]$, and the symptoms of *hyperpotassemia* involve chiefly the heart. Electrocardiographic changes are easily detected at serum $[K^+]$ greater than 6 meq. per liter. At progressively higher concentrations, the alterations become more severe, and above 10 meq. per liter the heart may stop in diastole. These changes are referable solely to the extracellular accumulation of potassium and may even occur in the presence of a cellular deficit of this ion. It cannot be stated with equal certainty that the clinical picture of *hypokalemia* is due to low extracellular [K^+] since hypokalemia is usually associated with cellular deficiency of potassium as well. This situation is characterized by extreme muscular weakness, lethargy, anorexia, myocardial degenerative changes, and peripheral paralysis.

Histological and functional lesions are observed in the kidneys of potassium-deficient individuals. The convoluted tubules appear engorged and the cells develop vacuoles. Concomitantly, there is a striking diminution in concentrating ability, with a conservation of K^+ and excretion of an acidic urine containing large amounts of NH_4^+. Sodium reabsorption may even be excessive in rare instances, resulting in its accumulation with consequent edema.

Homeostatic control of the serum [K^+] is not so well regulated as that for sodium or glucose, for example. The renal mechanism involved is well designed to prevent hyperkalemia but is not equally effective in the prevention of hypokalemia. Ordinarily, 60 to 120 meq. of potassium is ingested per day. In the complete absence of dietary potassium, 30 to 60 meq. per day appears in the urine for several days and then decreases to 10 to 20 meq. per day. Values below 10 meq. per day are seen only after profound K^+ depletion. Excretion may be increased still further in renal disease, diuresis, negative nitrogen balance, acidosis, alkalosis, or adrenal cortical hyperactivity. Renal excretion effectively prevents hyperkalemia under circumstances such as increased potassium intake, tissue breakdown, or contraction of extracellular volume by dehydration.

In addition to renal excretion of potassium, other factors which may tend to decrease the serum [K^+] are limited potassium intake, dilution of extracellular fluid with potassium-free fluid, loss of potassium-containing fluids, or increased glucose uptake by cells. Loss of potassium-containing fluids is particularly prominent in emesis or gastric drainage since gastric juice may at times contain as much as 40 meq. per liter of potassium, and the intestinal digestive juices normally contain 8 to 10 meq. per liter. The fluid lost in diarrhea may contain considerably higher concentrations of potassium than these values. Consequently, potassium deficiency is more frequent clinically than is hyperkalemia. The latter obtains only in terminal states, uremia, Addison's disease, and hemoconcentration as seen in shock or following severe burns or after the injudicious administration of parenteral fluids containing potassium. Potassium deficiency, in contrast, may be expected during negative nitrogen balance, in cachexia, after the loss of digestive fluids, and as the immediate result of expansion of the extracellular fluid in the treatment of dehydration by parenteral administration of potassium-free fluids.

In several states in which there has been no primary effect on extracellular pH but a significant loss of K^+, a marked alkalosis has been observed, notably in adrenal cortical hyperactivity. The mechanism by which this alkalosis develops is not entirely certain, but it relates to the K^+-Na^+-H^+ exchange across cell membranes. Initially, K^+ leaves the cell in exchange for Na^+ and H^+, particularly the latter, thereby elevating plasma pH; the kidneys excrete a somewhat alkaline urine containing K^+, Na^+, and HCO_3^-. However, as the intracellular K^+ is depleted, serum K^+ soon falls and the kidneys excrete an acidic urine (page 834), thereby increasing the alkalosis. Therapeutic reversal of this alkalosis is possible only when the K^+ loss has been met.

Because of the widespread distribution of potassium in foods, potassium deficiency is unlikely under normal circumstances. The minimum daily requirement for potassium by man cannot be fixed but need only be sufficient to offset expected losses. The 2 to 4 g. of potassium ordinarily available in the diet per day is more than sufficient for this purpose. Experimental potassium deficiency in rats results in slow growth, thinning of hair, renal hypertrophy, necrosis of the myocardium, and death. In dogs, perhaps the most striking finding is an early ascending paralysis of the limbs.

Aspects of the metabolism of calcium and phosphate will be considered in Chap. 39 and of other inorganic ions in Chap. 48.

REFERENCES

Books

Bland, J. H., ed., "Clinical Metabolism of Body Water and Electrolytes," W. B. Saunders Company, Philadelphia, 1963.

Elkinton, J. R., and Danowski, T. S., "The Body Fluids," The Williams & Wilkins Company, Baltimore, 1955.

Gamble, L. J., "Chemical Anatomy, Physiology and Pathology of Extracellular Fluid," 6th ed., Harvard University Press, Cambridge, Mass., 1954.

Goldberger, E., "A Primer of Water, Electrolyte and Acid-Base Syndromes," Lea & Febiger, Philadelphia, 1962.

Harris, E. J., "Transport and Accumulation in Biological Systems," Butterworth and Co. (Publishers), Ltd., London, 1956.

Hoffman, J. F., ed., "The Cellular Functions of Membrane Transport," Prentice-Hall, Inc., Englewood Cliffs, N.J., 1964.

Maxwell, M. H., and Kleeman, C. R., "Clinical Disturbances of Fluid and Electrolyte Metabolism," McGraw-Hill Book Company, New York, 1962.

Murphy, Q. R., ed., "Metabolic Aspects of Transport across Cell Membranes," The University of Wisconsin Press, Madison, 1957.

Shanes, A. M., "Electrolytes in Biological Systems," American Physiological Society, Washington, 1955.

Stein, W. D., "The Movement of Molecules Across Cell Membranes," Academic Press, Inc., New York, 1967.

Welt, L. G., "Clinical Disorders of Hydration and Acid-Base Equilibrium," Little, Brown and Company, Boston, 1955.

Review Articles

Albers, R. W., Biochemical Aspects of Active Transport, *Ann. Rev. Biochem.,* **36,** 727–756, 1967.

Baker, P. F., The Sodium Pump, *Endeavour,* **25,** 166–172, 1966.

Darrow, D. C., and Hellerstein, S., Interpretation of Certain Changes in Body Water and Electrolytes, *Physiol. Revs.,* **38,** 114–138, 1958.

Heinz, E., Transport through Biological Membranes, *Ann. Rev. Physiol.,* **29,** 21–58, 1967.

Katz, A. I., and Epstein, F. H., The Physiological Role of Sodium-Potassium Activated Adenosine Triphosphatase in the Active Transport of Cations across Biological Membranes, *Israel J. Med. Sci.,* **3,** 155–165, 1967.

Koefoed-Johnson, V., and Ussing, H. H., Ion Transport, in C. L. Comar and F. Bronner, eds., "Mineral Metabolism," vol. I., pp. 169–204, Academic Press, Inc., New York, 1960.

Lipsett, M. B., Schwartz, I. L., and Thorn, N. A., Hormonal Control of Sodium, Potassium, Chloride, and Water Metabolism, in C. L., Comar and F. Bronner, eds., "Mineral Metabolism," vol. I., part B, pp. 473–550, Academic Press, Inc., New York, 1961.

Robinson, J. R., Metabolism of Intracellular Water, *Physiol. Revs.*, **40**, 112–149, 1960.

Skou, J. C., Enzymatic Basis for Active Transport of Na^+ and K^+ across Cell Membranes, *Physiol. Revs.*, **45**, 596–617, 1965.

Steinbach, H. B., Comparative Biochemistry of the Alkali Metals, in M. Florkin and H. S. Mason, eds., "Comparative Biochemistry," vol. IV, part B, pp. 677–720, Academic Press, Inc., New York, 1962.

Ussing, H. H., The Alkali Metal Ions in Isolated Systems and Tissues, in O. von Eichler and A. Farah, eds., "Handbuch der expermentellen Pharmakologie," pp. 1–195, Springer-Verlag OHG, Berlin, 1960.

Ussing, H. H., Transport of Electrolytes and Water across Epithelia, *Harvey Lectures*, **59**, 1–30, 1965.

34. Specialized Extracellular Fluids

The mechanisms available for maintaining a constant environment for the cells of the body are described in the preceding chapter. In most organs this environment is the interstitial fluid, a portion of the system of extracellular fluids which intercommunicate by way of the blood plasma. In addition to the interstitial fluid and the blood plasma, there are *lymph,* formed by filtration of tissue fluids into lymphatic capillaries, and a number of extracellular fluids serving special functions and individually elaborated by the eye, joints, skin, central nervous system, gastrointestinal tract, and mammary glands. The secretion, composition, and functions of these fluids are the subject of this chapter.

THE NATURE OF CAPILLARY EXCHANGE

The classical concept of the mechanism and dynamics of the exchange of fluid between plasma and interstitial fluid across capillaries, frequently referred to as "Starling's hypothesis," is depicted in Fig. 34.1. If plasma and a protein-free ultrafiltrate prepared therefrom (a model for interstitial fluid) are separated by a membrane permeable to all solutes except the plasma proteins, the latter exert an effective osmotic pressure of about 30 mm. Hg. This is somewhat less than the hydrostatic pressure at the arteriolar end of a capillary and somewhat more than the hydrostatic pressure at the venous end. Accordingly, a net loss of fluid through the capillary wall may be expected as blood flows through the arteriolar end of the capillary, and fluid should be regained as flow continues through the venous region of the capillary. At any area along the capillary, the pressure which determines the direction of flow through the capillary wall is given by:

$$P = (\textbf{tension}_{\text{vascular}} + \textbf{osmotic pressure}_{\text{extravascular}}) - (\textbf{tension}_{\text{extravascular}} + \textbf{osmotic pressure}_{\text{vascular}})$$

If these processes are equalized, the volume of fluid entering the capillary should equal that which leaves at the venule. This flux across the capillary is relatively slow, amounting to about 2 per cent of the plasma flow through the capillaries in most tissues.

For some time it was considered that exchange of water and solutes across the capillary is accomplished in this manner. However, although this concept of a miniature circulation about each capillary provides a mechanism by which the venous return from a tissue equals the arterial input, it does not account for the very rapid rate of exchange of water and solutes between interstitial fluid and plasma. Water, electrolytes, and small organic molecules diffuse back and forth across capillaries at rates 10 to 100 times the plasma flow. The rate varies inversely with the size of the particles, suggesting a molecular "sieving" through pores in the capillary membrane. Thus, the diffusion of Na^+ is about twice as rapid as that of glucose and ten times that of inulin. At ordinary plasma flow rates, proteins do not cross the capillaries, but when plasma flow falls the diffusion of protein becomes significant. Undoubtedly, much of this diffusion occurs through "pores" in the mucoprotein gel in the intercellular spaces. However, electron microscopic evidence indicates significant passage of large molecules by pinocytotic transport through the capillary cells. Transfer of lipid-soluble materials, *e.g.*, O_2, N_2O, ethyl ether, etc., occurs so rapidly it is presumed that they pass through lipid portions of the membrane.

Thus the exchange of materials across the capillary occurs at an enormous rate, perhaps 1,500 liters of water per minute in a 70-kg. man, *but by diffusion, in both directions, not by mass filtration and return.* This is not to deny the validity of the Starling hypothesis. The prime function of the hydrodynamic pressure at the arteriolar end of a capillary is to propel fluid along the capillary rather than to ram fluid through the capillary wall. However, since the capillary is porous, were this force unopposed, much more fluid would leave the capillary by filtration plus diffusion than would return by diffusion alone. This is largely compensated by the effective osmotic pressure of the plasma proteins, and the extent of movement of fluid is virtually identical in both directions.

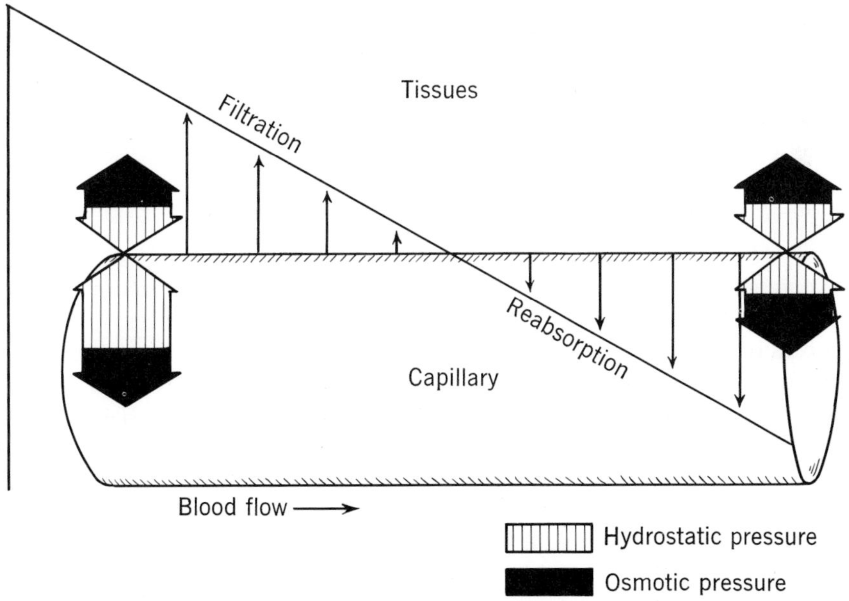

Fig. 34.1. The Starling hypothesis. At all points along the capillary, the hydrostatic pressure is opposed by the osmotic pressure of the plasma proteins. The relatively minor contributions of tissue tension and the osmotic pressure of the extravascular proteins are not considered. (*From W. A. Sodeman, "Pathologic Physiology," W. B. Saunders Company, Philadelphia,* 1950.)

Two clinically important problems arise from alteration of the normal balance between the fluids of the vascular and extravascular compartments. These are shock and edema.

Shock is circulatory failure resulting from loss of fluid from the vascular compartment by hemorrhage or an increase in capillary permeability. In persons who have been severely burned or subjected to severe traumatic injury, or who have undergone major surgery, capillaries become permeable to plasma protein, and large volumes of fluid containing albumin enter the extravascular space, thereby reducing the blood volume. The resultant diminution in blood and oxygen supply further impairs capillary permeability. Under these circumstances, the liver may release ferritin (page 748), which lowers blood pressure still further, and recovery may not occur even if fluid therapy is instituted. This is termed "irreversible shock."

Edema is the term applied to an unusual accumulation of extravascular fluid; it results from an imbalance between the transudation of fluid from the circulation and its return to the vascular system. A significant cause of edema is reduction of plasma albumin concentration which may reflect either diminished synthesis (as in malnutrition or hepatic disease) or excessive loss (as in the albuminuria of nephrosis). With hypoalbuminemia, the decreased effective osmotic pressure of the blood permits escape of fluid from the circulation. As blood volume diminishes, there is increased secretion of aldosterone, the salt-retaining hormone (Chap. 45) and a resulting increase of the effective osmotic pressure of the serum. This arouses thirst while also increasing secretion of vasopressin (Chap. 47), thus augmenting water intake and decreasing water output. The retained water and salt compose the edema fluid. "Nutritional" edema has occasionally been ascribed to hypoalbuminemia, but other significant factors must be operative as the condition has frequently been observed in children with seemingly adequate serum albumin concentrations. The *ascitic fluid,* which accumulates in the peritoneum during severe hepatic disease, is thought to result from a combination of the attendant hypoalbuminemia and portal venous hypertension.

INTERSTITIAL FLUID

Although the difference between the values for plasma volume and total extracellular fluid, determined as the ^{131}I-albumin and inulin spaces, indicates that 15 per cent of the body weight is interstitial fluid, it is not possible to obtain a direct sample for analysis. Insertion of a microneedle elicits fluid only from edematous organs, and, indeed, histological examination reveals no appreciable intercellular spaces other than the capillaries and defined lymphatic channels. It is, however, noteworthy that the fluid contained within the spaces of the cellular endoplasmic reticulum is, in part, continuous with interstitial fluid and may, in sum, account for a significant fraction of the total fluid volume estimated as "interstitial." The electrolyte structure of true interstitial fluid, therefore, can only be approximated from analyses of transudates low in protein content. As shown in Fig. 33.1 (page 779), the data indicate a slightly higher concentration of anions and lower concentration of cations than in plasma, in accord with expectations based on the Gibbs-Donnan formulation (page 128).

The intercellular cement which binds parenchymal cells together to form an organ appears to be a gel of highly polymerized hyaluronic acid (page 52). The latter is present in all organs but is most abundant in tissues of mesenchymal origin, *e.g.*, connective tissue, blood vessels, and lymphatic vessels. Materials in transit between blood and tissue cells must, therefore, diffuse through this gel, which is no obstacle to the passage of small inorganic ions, water, glucose, amino acids, etc., but does act as a barrier to large molecules such as proteins or discrete particles such as india ink, bacteria, viruses, etc. *Hyaluronidase* (page 883), derived from various sources, accelerates the subcutaneous spread of both particulate matter and solutions by depolymerizing the hyaluronic acid. Several virulent microorganisms have been found to secrete this enzyme and thus facilitate their spread in the host animal.

LYMPH

The terms lymph and interstitial fluid are frequently used interchangeably; however, since the precise nature of interstitial fluid is in doubt, it seems desirable to reserve the term lymph for the fluid which may be obtained from lymphatic ducts. The total daily lymphatic return in the normal human adult amounts to approximately 1 to 2 liters. The electrolyte composition of this fluid differs from that of plasma, as would be expected from equilibrium considerations. However, the protein content of lymph is variable, depending upon the source. Cervical lymph contains about 3 per cent protein; subcutaneous lymph, 0.25 per cent; and liver lymph, as much as 6 per cent. In each case the albumin to globulin ratio is considerably greater than that of plasma and is generally of the order of 3:1 to 5:1. There are also present in lymph sufficient fibrinogen and prothrombin to permit slow clotting.

SYNOVIAL FLUID

The electrolyte composition of synovial fluid is that of a transudate from plasma; in addition, the fluid contains mucopolysaccharide formed by cells of the synovium. Normally, the pH of synovial fluid is 7.3 to 7.4, and the specific gravity is approximately 1.010. The protein concentration is about 1 per cent, with an albumin to globulin ratio of approximately 4.0. No fibrinogen is present. The concentration of nonprotein nitrogenous substances is slightly below that of plasma; lipids are normally absent, and the glucose concentration is variable. Synovial fluid is highly viscous, varying from relative viscosities of 50 to 200, with an average of about 125, due to the presence of about 0.85 g. per 100 ml. of hyaluronic acid. Inflammatory joint disease, particularly rheumatoid arthritis, is usually accompanied by an increase in fluid volume of the joint as well as by an increased protein concentration in the fluid. The normally low protein content, and its increase when the hyaluronic acid diminishes in both concentration and average molecular weight, appear to reflect the large influence of the hyaluronic acid molecule (page 882). Electrolytes and readily diffusible substances of synovial fluid exchange with plasma, while larger particles can leave the intraarticular space only via the lymphatics. Synovial fluid, then, is an extension of interstitial fluid, and not a product of secretory activity.

SECRETION

The formation of interstitial fluid from plasma may be described in physicochemical terms based on knowledge of the diffusibility of water, the solutes of plasma, and the permeability of the capillary wall to these substances. It will be recalled that to account for the differences between the composition of intracellular and extracellular fluids, it is necessary to postulate the existence of a mechanism whereby energy, derived from metabolic processes, may be utilized to maintain the intracellular composition against an osmotic gradient. Another situation, *secretion*, may be recognized wherein cells are aligned in columnar fashion, bathed by interstitial fluid or plasma on one side and fluid of different composition on the opposite side, and in which the differences in composition of the two fluids cannot be accounted for in terms of spontaneous diffusion, osmosis, or permeability. The secretory process, operating against an osmotic, electrochemical, or hydrostatic gradient, again requires the harnessing of metabolically derived energy. It may be recognized (1) if the movement is inhibited by interruption of cellular metabolism, *e.g.*, cyanide or fluoride poisoning, (2) if, in contrast to the Donnan equilibrium, cations and anions are transported simultaneously in equivalent amounts and in the same direction, (3) if the shift takes place with nonelectrolytes, and (4) if the cells are so aligned that the transported fluid leaves by a duct and the pressure within this duct is independent of arterial pressure. The secretory activity need not involve more than one component of the secreted solution. Mammalian secretions include milk, sweat, tears, cerebrospinal fluid, aqueous humor, and the fluids of the digestive tract. Selective absorption across the intestinal mucosa, the reabsorption of water and solutes in renal tubules, and secretion into the lumen of the distal renal tubules may all be regarded from the same viewpoint. Among the more dramatic instances of secretion are the elaboration of 0.16*N* HCl by the stomach, secretion of almost pure water by sweat glands, and removal of almost all glucose and Na^+ from the urine. The fundamental mechanisms may be, in each case, an adaptation of those by means of which all cells maintain their internal composition. The mechanisms involved in these active transfers have been the subject of much investigation but remain among the major unsolved problems in biochemistry. The term "secretion" has also been generally employed to describe the behavior of the ductless endocrine glands, the activity of the liver in adding to hepatic venous blood serum albumin, prothrombin, and glucose, and the release of mucus. In these instances, although the cells "do work" in synthesizing the material, the actual transfer, cell to plasma or lumen, operates with the osmotic gradient and no work need be done to accomplish the *transfer*.

AQUEOUS HUMOR

The aqueous humor fills the anterior chamber of the eye, maintains the intraocular tension desirable for optical purposes, and nourishes the avascular cornea and lens. Its volume varies among animal species, depending on the size of the eyeball and the depth of the anterior chamber. In man, the volume is approximately 0.125 ml. The protein content is low in normal aqueous humor—about 0.025 g. per 100 ml. The albumin to globulin ratio is frequently of the same order as that in the plasma of the same subject. The concentrations of diffusible substances shown in Table 34.1 are not strikingly different from plasma. The concentration of ascorbic acid, however, is twenty times that in plasma. The components of aqueous humor enter at various points, and all depart through Schlemm's canal. Some of the fluid enters the anterior chamber by flow from the posterior chamber; some arises by diffusion from the blood vessels of the iris; and the remainder enters by the secretory activity of the ciliary body. Isotopic tracer experiments have indicated that the water exchange in the aqueous humor, per minute, is equivalent to approximately 20 per cent of the volume of the aqueous humor. The major portion of this turnover occurs by diffusion. However, only *water* and *nonelectrolytes* may be exchanged between the iris and the aqueous humor. In contrast, only about 1 per cent of the electrolyte content of the aqueous humor enters and leaves per minute, and this is entirely because of the secretory activity of the ciliary body. Increased secretion raises the intraocular pressure, giving rise to *glaucoma.* Administration of inhibitors of carbonic anhydrase, such as acetylamino-1,3,4-thiadiazole-5-sulfonamide, may relieve the elevated pressure, indicating a fundamental role for carbonic anhydrase in the secretory process.

When normal aqueous humor is removed, the anterior chamber rapidly refills with a fluid termed plasmoid aqueous humor. This fluid contains large quantities of protein, and if the paracenteses are repeated, the fluid which fills the anterior chamber becomes virtually identical with plasma.

The aqueous humor also fills the posterior chamber of the eye. However, this chamber contains in addition a gel of hyaluronic acid within a framework of collagen (page 871). Unlike the aqueous humor in the anterior chamber of the eye, that in the posterior chamber cannot be removed without causing injury to the eye, as neither the gel, originally secreted by the retina, nor the collagen can be replaced. As in the anterior chamber, there is an exchange of electrolytes and water of the posterior chamber with surrounding tissue by diffusion.

CEREBROSPINAL FLUID

The cerebrospinal fluid, contained within the subarachnoid space of the brain and spinal cord and the ventricles of the brain, originates in the choroid plexus and returns to the blood in the vessels of the lumbar region. Only a small fraction of the cells of the central nervous system actually make contact with this fluid; the remainder derive their nutrition from the blood vessels. The total volume of this fluid, about 125 ml. in a healthy adult, is renewed every 3 or 4 hr. If surgical drainage is instituted, several liters per day can be obtained. The composition of spinal fluid suggests that it is primarily a simple transudate or ultrafiltrate from plasma. Fluid taken from the lumbar region, the cisterna magna, or the ventricles is at all times in osmotic equilibrium with plasma and contains between 15 and 40 mg. of protein per 100 ml., with an albumin to globulin ratio of 4. Plasma lipids are absent.

However, the following discrepancies in composition between cerebrospinal fluid and an ultrafiltrate of plasma indicate that formation of this fluid involves secretion, presumably by the choroid plexus. While the total cation and anion composition of the fluid is in accord with the Gibbs-Donnan equilibrium, the distri-

Table 34.1: APPROXIMATE CONCENTRATIONS OF THE MAJOR ELECTROLYTES OF EXTRACELLULAR FLUIDS

Fluid	pH	Na^+	K^+	Ca^{++}	Cl^-	HCO_3^-	Protein	Other
		meq./liter of water						
Plasma	7.35–7.45	144	4.5	5.0	103	28	18 meq./liter (6.0–8.0 g./100 ml.)	Organic acids, 6 meq./liter
Edema fluid	7.4	135	3.3	3.5	105	30	<0.25 g./100 ml.	
Synovial fluid	7.3–7.4	142	4	....	117	25	1.0 g./100 ml.	
Cerebrospinal fluid	7.4	146	3.5–4.0	3.0	125	25	15–40 mg./100 ml.	
Aqueous humor	7.4	140	4.7	3.5	108	28	25 mg./100 ml.	
Tears	5.2–8.3	142	3–6	....	115	5–25	0.75 g./100 ml.	
Sweat	4.5–7.5	<85	3–6	3–5	<85	0–10	Trace	
Saliva	6.4–7.0	20–40	15–25	3–8	20–40	10–20	Variable	
Parietal gastric juice	<1.0	0	7	0	162	0	0	H^+, 155 meq./liter
Gastric mucus	7.4–7.5	145	5	....	115	30	Variable	
Mixed gastric secretions	1–2	20–60	6–7	....	145	0	Variable	H^+, 60–120 meq./liter
Pancreatic juice	7–8	148	7	6	80	80	Variable	
Jejunal fluid	7.2–7.8	142	7–10	....	105	30	Variable	
Ileal fluid	7.6–8.2	100–140	10–50	....	80	75	Variable	
Bladder bile	5.6–7.2	130	7–10	7–15	40–90	0–15	Variable	Bile salts, 50–100 meq./liter
Liver bile	7.4–8.0	145	5	5	75–110	25–50	Variable	Bile salts, 10–20 meq./liter

Note: Only the values given for plasma represent the mean of a large number of samples. In some instances a range is shown; in others a single value is quoted although the given figure is subject to appreciable variation. Since the values have been obtained in many laboratories, employing different analytical methods and sampling procedures, only the general pattern may be considered meaningful.

bution of these ions is not. Thus, the [Na^+] of cerebrospinal fluid is virtually identical with that of plasma, while the [K^+] is appreciably lower. Also, while the [Cl^-] of spinal fluid is greater than that of plasma, the [HCO_3^-] is identical in the two fluids. Data are not available concerning the free CO_2 and H_2CO_3 content of cerebrospinal fluid, but it has been assumed to be approximately equal to that of venous blood. The calcium concentration of spinal fluid appears virtually fixed and does not respond readily to changes in plasma concentration. This is particularly striking in patients with parathyroid tumors, who show markedly elevated serum calcium levels but normal spinal fluid calcium concentrations. In general, the glucose concentration of spinal fluid is lower than that of plasma but rises and falls with changes in blood glucose levels. The concentration of nonprotein nitrogenous constituents is always appreciably lower in cerebrospinal fluid than in plasma.

SWEAT

The secretion of sweat serves, through evaporation, to cool the body. When no visible perspiration is produced, the sweat glands release virtually pure water. This *insensible* perspiration may amount to 600 to 700 ml. per day. The small amount of organic and inorganic material which accumulates on the skin under these conditions is probably associated with activity of sebaceous glands rather than with that of sweat glands. In circumstances in which visible sweat is elaborated, its volume and composition vary and are determined by rate of evaporation, previous fluid intake of the individual, external temperature and humidity, and hormonal factors. Volumes as large as 14 liters per day have been recorded. Both volume and salt content of sensible perspiration are influenced by acclimatization of the individual. Persons new to an environment which is hot and humid produce copious quantities of salt-laden perspiration; [Na^+] and [Cl^-] may be as high as 75 meq. per liter. Acclimated individuals, however, produce smaller volumes with a lower salt concentration. Unreplaced loss of large volumes of perspiration may result in hypertonic contraction. Miners' or stokers' cramps result from salt loss under these circumstances and can be prevented by incorporation of small amounts of salt in drinking water. In cystic fibrosis, a congenital defect involving most or all of the glandular epithelial structures of the body, sweat and tears are characteristically rich in NaCl. This analytical difference is so striking as to be diagnostic. In hot weather, victims of this disease may succumb in a state resembling acute Addisonian crisis (Chap. 45), referable entirely to Na^+ loss, and corrected by NaCl administration.

When small volumes of visible perspiration are elaborated, its concentration of nonprotein nitrogenous materials slightly exceeds that of the plasma from which it is derived. This probably reflects evaporation of water from the elaborated sweat. However, sweat glands may possess an active mechanism for the concentration of lactic acid. The lactate concentration of the sweat of athletes far exceeds that present in plasma or urine. [K^+], [Mg^{++}], [Ca^{++}], etc., are of the order expected from those found in the plasma. Specific gravities of 1.002 to 1.005 for sweat have been reported, and the pH lies between 4.5 and 7.5.

TEARS

The fluid which normally moistens the surface of the cornea is a mixed secretion of the lacrimal glands and of the accessory sebaceous glands (the glands of Zeis and the meibomian glands). Since the surface of the cornea is exposed during waking hours, there is constant evaporation of fluid on its surface, resulting in concentration of the tear fluid. Under mild stimulus with a slow rate of tear flow, the resultant fluid appears to be hypertonic, probably because of concentration due to evaporation. When rapid tear flow is induced, the resulting solution is isotonic. In most instances this fluid has a pH of 7 to 7.4, but values from 5.2 to 8.3 have been observed; alkaline tears are shed after corneal injuries.

Diffusible nitrogenous materials and electrolytes are present in tears in concentrations similar to those of plasma. The protein concentration is generally 0.6 to 0.8 g. per 100 ml., with an albumin to globulin ratio of about 2. The presence of protein in the tears, by lowering the surface tension, enables the tears to wet epithelial surfaces. The optical properties of the eye are greatly improved by this film since microscopic irregularities in the corneal epithelium are abolished, thereby producing a perfectly smooth optical surface. Further, the film protects the eye from damage by small foreign bodies such as dust or air-borne bacteria.

An unusual component of tears is the enzyme lysozyme (pages 269), also found in nasal mucus, sputum, tissues, gastric secretions, milk, and in egg white. By catalyzing hydrolysis of the muramic acid–containing mucopeptide in the polysaccharide of the cell walls of many air-borne cocci (Chap. 41), lysozyme protects the cornea from infection.

SECRETIONS OF THE DIGESTIVE TRACT

The major portion of ingested food must be hydrolyzed into smaller components, *e.g.*, amino acids, hexoses, etc., before it can be absorbed and utilized. Many unicellular organisms, such as the fungi, secrete hydrolytic enzymes into the surrounding medium. In higher animals, digestion is initiated within the lumen of the gastrointestinal tract, supplied with digestive secretions elaborated by special glands. For the major classes of nutrients, digestion is completed within the membranes of the cells of the intestinal epithelium. In general, the glands associated with the digestive tract provide three types of secretory products, *viz.*, aqueous solutions of varying electrolyte composition and pH, enzymes, and mucus.

The organic constituents of the digestive secretions, *i.e.*, enzymes and mucoproteins, are synthesized by the secreting cells. This process may be observed histologically. In the resting cell, there is a clear vacuole within the cytoplasm. Within the vacuole there then appears a "granule," which gradually increases in size and virtually fills the vacuole. These vacuoles then migrate toward the apex of the cell and may almost fill the cell. Upon stimulation of the cell, the granules are mechanically extruded and washed down the duct. The precise chemical nature of these granules has not been defined. In organs whose secretions contain more than one enzyme, *e.g.*, the pancreas, individual cells synthesize all or most of the enzymes characteristic of this organ. On repeated stimulation of pancreatic secretion, the relative amounts of tryptic, lipolytic, and amylolytic activities remain constant in the pancreatic juice. Prolonged stimulation results in discharge of all the visible granules, but thereafter such cells are capable of continuous secretion of a fluid of constant electrolyte and enzyme content.

Saliva. Although there are numerous small glands distributed over the buccal mucosa, saliva is secreted mainly by three pairs of glands. The cells of the parotid gland are exclusively of the serous type; those of the submaxillary and sublingual glands are of mixed type. Parotid saliva is nonviscous; sublingual and submaxillary saliva are viscous because of their mucoprotein content. At low secretory rates, saliva is markedly hypotonic; the osmotic pressure rises with increasing secretory flow and may be almost isotonic at maximal rates. No hormonal stimulus is required for salivary secretion. However, epinephrine administration stimulates parotid amylase secretion; as in other instances (page 957) this effect is mediated by local formation of cyclic adenylic acid. The flow of saliva may be stimulated by local reflexes caused by mechanical factors, including presence in the mouth of foreign materials, or by conditioned reflexes, *e.g.*, sight or smell of food. This secretion moistens and thus lubricates the food mass, thereby facilitating deglutition. Human saliva contains an *α-amylase* (page 47) which catalyzes hydrolysis of polysaccharides to a mixture of oligosaccharides. Although present in saliva of many species, amylase is not secreted by the horse, dog, or cat.

The composition, pH, and volume of saliva are variable. From 1,000 to 1,500 ml. of hypotonic mixed secretions are produced daily. The $[Na^+]$ is about 20 to 40 meq. per liter; $[Cl^+]$ is subject to similar variation. Potassium is present in concentrations four to five times that of plasma. The calcium content of saliva has been reported to be from 6 to 20 mg. per 100 ml. and, at high calcium concentrations, calculi of calcium salts may form in the ducts or, in combination with organic material, may be deposited on the teeth as "tartar." The pH of saliva is generally between 6.4 and 7.0, with $[HCO_3^-]$ of 10 to 20 meq. per liter. Mucoprotein is the chief organic constituent, together with small quantities of glucose, urea, lactic acid, phenols, vitamins, and thiocyanate. Enzymes other than amylase, including a phosphatase and carbonic anhydrase, have been reported in saliva. In general, the large variations noted in composition of saliva samples may be ascribed to methods of collection, the varying stimuli employed to augment salivary flow, and the resultant variation in salivary flow.

Gastric Secretions. Secretions enter the adult human stomach from the ducts of 10 to 30 million gastric glands. Three types of cells line the gland tubules: mucous cells at the neck of the gland, the "chief" cells of the body of the gland, and the parietal, or border, cells. Parietal cells are not found in glands of the pyloric or cardiac portions of the stomach, or in the lumen of the gland. They lie between and behind the chief cells and communicate with the lumen via delicate canaliculi which pass between the chief cells. The canaliculus is itself the terminal conduit of a fine network of channels within the parietal cell. The mucous cells contain mucinogen granules and secrete a thick, viscous fluid rich in mucoprotein. The chief cells elaborate and secrete *pepsinogen* (page 532). The proteins are suspended in an essentially neutral or slightly alkaline medium in which Na^+, Cl^-, and HCO_3^- are the predominant ions. The parietal cells secrete a solution of $0.16M$ HCl and $0.007M$ KCl, with traces of other electrolytes and little or no organic material. The concentration of hydrogen ions is thus a million times greater than that of plasma. Secretion of 1 liter of such a solution, assuming plasma to be the source of H^+, K^+, and Cl^-, requires expenditure of at least 1500 cal. if the process were 100 per cent efficient.

There can be little doubt that plasma is the source of Cl^- ion. Further, as venous blood leaving the secreting stomach contains more HCO_3^- and less Cl^- than does arterial blood, the over-all process may be represented as follows.

$$NaCl + H_2CO_3 \rightleftharpoons NaHCO_{3(plasma)} + HCl_{(secreted)}$$

Since this process occurs spontaneously to the left, cellular metabolism must provide energy for its effective reversal. However, despite intensive investigation it has not been possible to ascertain the nature of the cellular mechanism. Some of the known facts include: (1) Secretion is markedly inhibited by inhibitors of carbonic anhydrase, indicating a probable role of this enzyme in the over-all process. (2) The ratio (H^+ secreted)/(O consumed) exceeds 4 by a large factor. Therefore, cellular respiration does not account for total proton production. (3) Inhibitors of cellular metabolism, *e.g.*, cyanide, iodoacetate, *p*-chloromercuribenzoate, and dinitrophenol, inhibit or abolish secretion, indicating that normal metabolic pathways provide the necessary energy. (4) A potential difference exists across the mucosa. However, these data do not permit construction of an adequate model for gastric secretion.

The parietal cells contain an unusual "ATPase," the activity of which is stimulated by Cl^- and even more by HCO_3^- but which is insensitive to Na^+ and K^+ as well as to ouabain and hence is not the ATPase of the usual electrolyte pump. The enzyme is inhibited by thiocyanate and, a concentration (100 mM) sufficient for complete inhibition completely abolishes H^+ secretion as well. Moreover, there is an excellent parallelism between the effects of dinitrophenol on the activity of this ATPase, the rate of acid secretion, and the electrical potential across the mucosa. Thus, although the secretory mechanism is uncertain, it seems likely that this ATPase participates in the process.

Secretion by the parietal cells is stimulated by histamine and by a hormone, *gastrin,* produced in the pyloric-antral mucosa of the stomach itself. As obtained from human antral mucosa, gastrin is a heptadecapeptide with the following sequence:

1 5 10 15 17
Glu-Gly-Pro-Trp-Leu-Glu-Glu-Glu-Glu-Glu-Ala-Tyr-Gly-Trp-Met-Asp-Phe-NH₂

The leucine residue in position 5 is replaced by methionyl in the hog and by valyl in sheep. The tyrosine residue at position 12 may be esterified to sulfate. The presence of pentaglutamic acid in this structure is most unusual. Especially noteworthy is the fact that all stimulatory activity for gastric acid secretion resides in the C-terminal tetrapeptide amide, which is as active as the entire molecule.

The surface epithelial cells of the stomach, the chief cells of the necks of the fundic glands, and cells of the pyloric and cardiac glands secrete mucus of complex composition. Included are mucoproteins which contain a sialic acid (page 34), blood group substances (page 736), and a relatively low molecular weight mucoprotein which serves as *intrinsic factor* (page 752). The mucoproteins produced by the surface epithelium are present as insoluble, stringy masses, while that from the glandular mucous cells is in solution, from which it may be readily precipitated by addition of acetone or alcohol.

Gastric Analysis. In clinical practice it is frequently of interest to measure the amount of acid produced in the gastric secretion. The usual practice consists of withdrawal of the residuum, followed by subcutaneous injection of histamine, sampling of gastric contents at 15-min. intervals, and titration with alkali.

Interest in these determinations arises only in instances of extreme variations. *Hypoacidity* is seldom of real consequence, but *anacidity* occurs only in pathological states, most frequently pernicious anemia and gastric carcinoma. *Hyperacidity* is considerably more frequent and is associated with chronic postprandial distress ("heartburn," "indigestion") or peptic ulcer. Peptic activity generally parallels acid secretion and is never observed in the absence of free HCl. Figure 34.2 illustrates data obtained from gastric analyses by the procedure described above.

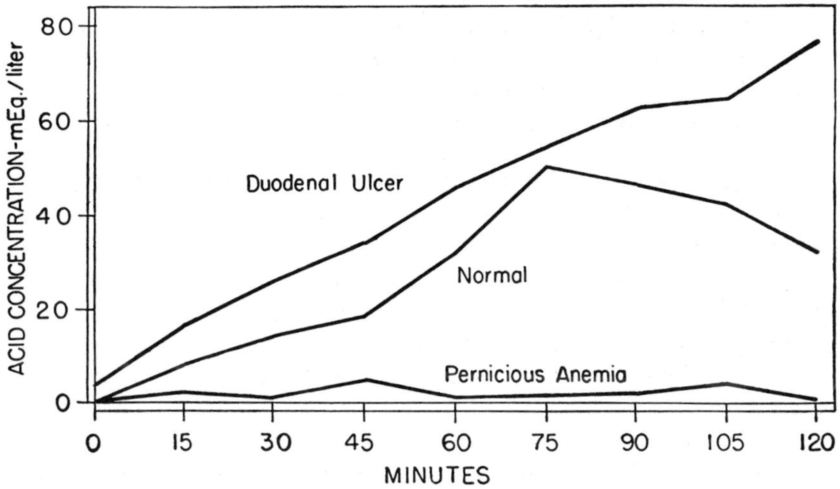

Fig. 34.2. Concentration of acid in the gastric contents of a normal subject, a patient with pernicious anemia, and a patient with a duodenal ulcer, after administration of histamine.

Gastric and duodenal ulcers result from the digestion of mucosa by pepsin-HCl. If unchecked, digestion may continue and ultimately lead to perforation with consequent peritonitis. Alternatively, erosion into a major blood vessel, such as the pancreaticoduodenal artery, may lead to serious, even fatal, hemorrhage. The mechanism of ulcer formation has not been established. This problem may be stated in a more general fashion: Why do not the stomach and other digestive organs digest themselves? This question has intrigued such illustrious investigators as John Hunter, Claude Bernard, and Ivan Pavlov, but no definitive answer has been provided. Among the more important factors are the following: (1) The digestive enzymes are isolated within the cell in secretory vacuoles, and intracellular digestion does not occur. All the major proteases of the gastrointestinal tract are stored in the cells in which they are synthesized as inactive zymogens, *viz.*, pepsinogen, trypsinogen, chymotrypsinogen, and procarboxypeptidases; hence, active proteases never truly mix with the cell contents. (2) Once in the lumen of the alimentary tract, the enzymes cannot penetrate mucosal cells because of the selective permeability of cell membranes. (3) Mucus has both acid-binding and peptic-inhibitory properties. Since mucus is steadily renewed, and only slowly digested, the enzymes are mechanically separated from tissue itself. (4) In the stomach and duodenum, the ability of mucus to protect epithelial structures is a function of pH. At unusually high $[H^+]$, the mucus is more readily digested and the epithelium exposed to the action of pepsin. (5) Resistance of the stomach to erosion is contingent upon an active blood circulation. In emotional states, such as fear or anger, the entire gastric mucosa is occasionally blanched. Ulcer formation sometimes occurs in individuals with gastric hyperacidity and whose gastric circulation responds in this manner to "psychic" stimuli. (6) Finally, hormonal secretions, *e.g.*, those of the adrenal cortex (Chap. 45), which provoke hypersecretory activity by the gastric mucosa and may inhibit mucopolysaccharide synthesis, contribute to ulcer formation. This is of concern in prolonged use of these hormones.

Pancreatic Secretion. The pancreas is a racemose gland similar to salivary glands in general structure. Wedge-shaped cells, containing secretory granules, line the alveoli. On repeated stimulation the secretory granules are discharged, leaving entirely clear cells. The daily volume of pancreatic secretion in the adult human being is 500 to 800 ml. per day. Pancreatic juice, in the dog, varies from pH 7.4 to 8.3, depending on the $[HCO_3^-]$ and $[CO_2]$, and is approximately isotonic, containing, in milliequivalents per liter, $Na^+ = 148$, $K^+ = 4$, $Ca^{++} = 6$, $Cl^- = 80$, $HCO_3^- = 80$, $HPO_4^= = 1$. In general, the sum of $[HCO_3^-] + [Cl^-]$ remains constant, the $[HCO_3^-]$ increasing at increased rates of flow. The reactions by which the pancreas produces a fluid with a HCO_3^- concentration threefold that of plasma are unknown. Studies with $NaH^{14}CO_3$ have shown that most of the HCO_3^- comes from plasma and not from the metabolism of the secreting cells. Since secretion of bicarbonate is virtually abolished by inhibitors of carbonic anhydrase, this enzyme is probably essential to the secretory process.

The pancreatic secretion contains several proteins of importance to the digestive process. These include *trypsinogen, chymotrypsinogen,* and *procarboxypeptidases,* precursors of *trypsin, chymotrypsin,* and *carboxypeptidases,* respectively, a *lipase* (*steapsin*) which is solely responsible for fat hydrolysis, an *α-amylase, deoxyribonuclease,* and *ribonuclease.* The conversion of the zymogens to active enzymes and the hydrolytic activity of each of these enzymes have been described in earlier chapters.

Secretion of pancreatic juice is under both neural and hormonal control. The presence of "secretagogues" (large polypeptides) or acid in the upper duodenum results in liberation into the circulation of a hormone, *secretin,* which markedly stimulates the flow of pancreatic juice and, to a lesser extent, of bile and intestinal juices. Since secretin is effective in the atropinized animal as well as after section of the vagus, a direct action of the hormone on the secretory cells is assumed. Secretin is a basic polypeptide whose amino acid sequence has been established (page 698). The pancreatic juice resulting from secretin stimulation is copious in volume, relatively deficient in enzymic activity, and of normal electrolyte composition. Vagal stimulation does not markedly enhance the volume of pancreatic secretion but results in a marked increase in enzymic activity. This form of response is also evoked by a second duodenal hormone, *cholecystokinin,* which, unlike secretin, stimulates secretion of enzymes by the pancreas. Cholecystokinin is a polypeptide of 34 residues which terminates in the same tetrapeptide-amide as does gastrin.

The relative amounts of the major proteolytic, lipolytic, and amylolytic enzymes remain fairly constant in the pancreas and pancreatic juices of a given species. However, the relative amounts of these enzymes secreted may be altered in response to changes in the diet, the proteases and amylase increasing with increased ingestion of protein or starch, respectively. The mechanism of this adaptation is unknown.

Intestinal Secretions. The succus entericus, the secretion of the intestinal mucosa, has been obtained for analysis by passage of suitable tubes (Miller-Abbott) in man and from loops of intestine, isolated at various levels, in experimental animals. The secreting cells are found in glands, the crypts of Lieberkühn, which are present extensively throughout the small intestine. A second type of gland, the glands of Brunner, resembling those of the stomach in appearance, is found only in the duodenum. These glands contribute a constant supply of a slightly alkaline fluid containing mucoproteins but having no enzymic activity. Four types of cells may be distinguished in the crypts; the function of each type has not been elucidated. The electrolyte composition of mixed intestinal juice is not constant but, except for a lower [HCO_3^-], resembles pancreatic juice.

Many enzymes have been demonstrated in extracts of intestine, but relatively few have been isolated in pure form and characterized. These enzymes include *enterokinase, aminopeptidases, dipeptidases, maltase, sucrase, lactase,* a *lipase, nucleases, nucleotidase, nucleosidase,* a *lecithinase,* and a *phosphatase.* Relatively little enzymic activity is demonstrable in intestinal juice, particularly from isolated loops, although the mucosa of such loops gives evidence of digestive activity. Many observations of this kind suggest that digestion is completed within the intestinal mucosa as smaller, dialyzable molecules, *e.g.*, disaccharides and di- and tripeptides, cross the villi. The stimuli to secretion by the intestinal mucosa are not so well defined as those for other digestive glands. The presence of material within the intestine leads to a constant flow of juice, but there has been no clear demonstration of the role of the abundant nerve supply.

Bile. In man, bile is continually elaborated by the polygonal cells of the liver and passes along the bile canaliculi and thence through the hepatic and cystic ducts to the gallbladder. Here it is stored and concentrated and enters the intestine through the common duct. Emptying of the gallbladder occurs continually but is accelerated when partially digested food is in the intestine. In part, this seems to be under neural control, but gallbladder contraction and emptying may be observed after complete denervation of this organ and introduction of partially hydrolyzed lipid into the duodenum. *Cholecystokinin* is released into the circulation by the duodenum and stimulates contraction of the gallbladder, with release of its contents into the duodenum.

Bile contains several compounds which are absent from all other digestive secretions, *viz.*, cholesterol, bile acids, and bile pigments. Two classes of substances may be distinguished in *hepatic* bile: (1) those which are present in concentrations differing little from those in plasma, and (2) those which may be concentrated in bile many times more than in plasma. In the first category are Na^+, K^+, Cl^-, creatinine, and cholesterol, indicating the formation of a protein-free ultrafiltrate of plasma by the polygonal cells. However, the cholesterol is synthesized by the liver and its presence in bile is not due to transfer from plasma. Representatives of the second category include bilirubin, as well as administered substances which are excreted via bile, *e.g.*, bromosulfalein (bromosulfonphthalein, BSP), *p*-aminohippurate, and penicillin. These substances are added to bile by an active secretory mechanism. Bilirubin may be concentrated as much as 1,000-fold. Since high plasma concentrations of bromosulfalein inhibit bilirubin excretion, it is thought that these two compounds compete for a single secreting mechanism. The bile acids (page 81) are made in the polygonal cells and are present in hepatic bile to the extent of 2 to 5 meq. per liter.

The capacity of the gallbladder is 50 to 60 ml. in adults; it not only is a storage sac but also concentrates bile by absorption of water and electrolytes and secretes mucoproteins. The resulting solution contains only small amounts of Cl^- and HCO_3^- and may be neutral or as acidic as pH 5.6. The $[K^+]$ appears to rise slightly during the reabsorptive process, and the final $[Ca^{++}]$ may be 15 to 30 mg. per 100 ml. The daily production of hepatic bile in normal individuals is somewhat uncertain, but biliary fistulas permit collection of 500 to 1,000 ml. per day.

The *bile acids,* synthesized in the liver, are the chief, if not the only, contribution of bile to digestion; these acids are present in bile as bile salts. In fistula bile, the concentration of bile salts may vary from 0.5 to 1.5 per cent. The role of bile salts in the emulsification, hydrolysis, and absorption of lipids has been described (page 472*ff.*). The two major components, glycocholic (cholylglycine) and taurocholic (cholyltaurine) acids (page 81), are present in a ratio of about 3:1 in human bile. Inverse ratios may be encountered in persons on very low-protein diets. Bile from carnivores contains chiefly taurocholic acid, while hog bile contains largely glycocholic acid. The daily secretion of these compounds is about 5 to 15 g. per day. Most of this is returned to the liver via the enterohepatic circulation. The daily output of bile salts falls after a few days of collection through a biliary fistula but can be reestablished by feeding cholic acid and taurine. The bile salts are formed in the liver in a manner analogous to the formation of hippuric acid (page 579).

$$\text{Cholic acid} + \text{ATP} + \text{CoA} \longrightarrow \text{cholyl CoA} + \text{AMP} + \text{PP}_i$$
$$\text{Cholyl CoA} + \text{glycine} \longrightarrow \text{cholylglycine} + \text{CoA}$$

Taurocholic acid is synthesized similarly from cholyl CoA and taurine. Presumably, analogous derivatives are similarly formed from other bile acids, *e.g.*, deoxycholic and chenodeoxycholic acids (page 81).

The bile pigments are derived from degradation of porphyrins in cells of the reticuloendothelial system, notably those of the liver (page 741). Fresh hepatic bile is golden yellow because of the bilirubin present. Bladder bile may be green, because of oxidation of bilirubin to biliverdin (page 742). On standing, all bile darkens progressively from gold to green to blue and then to brown as the pigments are oxidized. The total daily excretion of these pigments in man varies from 0.5 to 2.1 g. Coproporphyrin derived from heme (page 737) may occasionally be encountered in bile in small quantity. Bilipurpurin occurs in the bile of ruminants and appears to be derived from chlorophyll.

Bile contains three lipid constituents of limited solubility, bile salts, lecithin, and cholesterol. Bladder bile is a solution of mixed micelles of these components; hence the solubility of cholesterol is critically dependent upon the concentration of bile salts and lecithin. Unesterified cholesterol, first isolated from gallstones, is a major biliary constituent and may be present in a concentration as high as 1 per cent in bladder bile. Fatty acids also occur, as soaps, in amounts varying from 0.5 to 1.2 per cent in bladder bile, which also contains as much as 0.5 per cent of neutral fat and 0.2 per cent of phospholipids. Maintenance of this stable, supersaturated solution of cholesterol appears to be dependent on the presence of bile salts, soaps, and mucoproteins. On dialysis, bile becomes turbid, and cholesterol precipitates. Much of the cholesterol is reabsorbed in the intestine, a process entirely dependent on the presence and simultaneous reabsorption of bile salts.

The conjugated bile acids persist in the intestinal lumen until the ileum, where they are removed by an active transport process and returned to the liver for resecretion in bile. Conjugates which are hydrolyzed by amidases of intestinal bacteria are either removed by the jejunal mucosa or excreted in the stool. Lithocholic acid is formed from chenodeoxycholate by these bacteria and, although transported, cannot be readily conjugated. This acid is relatively toxic to the liver and may be important in the pathogenesis of liver damage after biliary stasis.

A number of enzymes have been found in bile, of which *alkaline phosphatase* is particularly noteworthy since a similar enzyme enters plasma from osteoblasts. In consequence, the plasma alkaline phosphatase activity may be increased either by enhanced activity of the osteoblasts or by failure of the hepatic parenchyma to secrete the enzyme, and the alkaline phosphatase activity of plasma is a useful indicator of hepatic function. Bilirubin is largely present conjugated with glucuronic acid (page 743) and with sulfate. Glucosiduronates of other cyclic alcohols are also excreted in the bile. This is a major fate of thyroxine (Chap. 43) and certain of the steroid hormones (Chaps. 44 and 45).

Gallstones are composed of normal bile components which have precipitated. Virtually all stones have an inner core of protein tinged with bile pigment. The most common stones, built of alternating layers of cholesterol and calcium-bilirubin, are about 80 per cent cholesterol. Occasionally stones are encountered which are 90 to 98 per cent cholesterol. Small calcium-bilirubin stones occur somewhat less frequently, whereas pure bilirubin or pure calcium carbonate stones are very rare in man but not uncommon in cattle. The mechanism of biliary calculus formation is not understood, but the chief contributory factors appear to be infection, biliary stasis, and perhaps the plasma concentration of cholesterol. The importance of infection is well established, and multiple cholesterol-pigment-calcium stones are generally considered of this origin. The bacteria increase the β-glucuronidase activity of bile, with resultant hydrolysis of bilirubin diglucuronide (page 743), the normal bile constituent, thereby providing the bilirubin which serves as the nucleus for stone formation.

MILK

Prior to birth, the fetus derives all its food from the mother by means of the placenta. After birth, the newborn mammal obtains its nourishment from the milk produced by the maternal mammary glands. Preparation of the mammary glands for subsequent lactation begins early in pregnancy, and secretion of milk normally begins at the end of gestation; these processes are under hormonal control (Chap. 47).

Production of milk for the newborn is a specific mammalian adaptation, and milk is unique in being an almost complete natural food from the point of view of nutrition. Its excellent nutritive quality has led to its wide use for individuals of all ages and to production of important derived foods such as cheese, butter, etc. Milk contains proteins, lipids, carbohydrates, minerals, vitamins, etc. The most significant deficiencies are the relatively low content of iron and copper and of vitamins C and D. The special nutritive properties of milk derive from the presence of several highly nutritive proteins, of lactose, of glycerides of the lower fatty acids, and of calcium and phosphorus.

Although the general composition of milk is much the same in all the Mammalia, the concentrations of certain constituents vary considerably among different species. A relation between rate of growth of the young and the protein content is readily observed by a comparison of the composition of the milk from different species. This is illustrated by some of the data of Proscher (Table 34.2) compiled some 50 years ago. The protein content of human milk is less than one-sixth that of rabbit and reindeer milk. The composition of milk varies with the time after initiation of lactation. The first milk, or *colostrum*, has unique properties, which will be discussed later.

Table 34.2: Growth Rate and Milk Composition of Different Mammals

Source	Time for doubling body weight of newborn, days	Protein content, per cent	Ash content, per cent
Man	180	1.6	0.2
Horse	60	2.0	0.4
Cow	47	3.5	0.7
Goat	19	4.3	0.8
Pig	18	5.9	0.8
Sheep	10	6.5	0.8
Dog	8	7.1	1.3
Rabbit	6	10.4	2.5

COMPOSITION OF MILK

The average composition of human and cow's milk is given in Table 34.3. The main differences are in the higher ash and protein content of cow's milk and the greater carbohydrate content of human milk. Modification of cow's milk for infant nutrition is accomplished by dilution with water to decrease the protein and ash content and addition of lactose or other carbohydrates to approximate human milk.

Table 34.3: Average Composition of Human and Bovine Milk

Constituent	Human, per cent	Bovine, per cent
Water	87.5	87
Total solids	12.5	13
Protein	1.0–1.5	3.0–4.0
Lipid	3.0–4.0	3.5–5.0
Carbohydrate	7.0–7.5	4.5–5.0
Ash	0.2	0.75

The white appearance of milk is due partly to emulsified lipid and partly to the presence of the calcium salt of casein, the main protein of milk. The occasional yellow color is caused by the pigments, carotene and xanthophyll (page 76). Fresh milk is nearly neutral; the pH is usually 6.6 to 6.8. Unsterilized milk rapidly becomes acidic because of fermentation by microorganisms.

Ash. The distribution of inorganic constituents is very similar in human and cow's milk (Table 34.4). The most noteworthy features are the high content of calcium, phosphorus, potassium, sodium, magnesium, and chlorine. Traces of other inorganic elements are also present, but, as already noted, the copper and iron content is low. These trace elements are apparently present in sufficient amounts for the needs of the infant, but a characteristic anemia develops in the growing child if milk is used as the sole food; this is due to insufficient copper and iron.

Table 34.4: PERCENTAGE DISTRIBUTION OF ASH IN MILK

Species	Ca	Mg	P	Na	K	Cl
Human	16.7	2.2	7.3	5.3	23.5	16.5
Bovine	16.8	1.7	11.6	5.3	20.7	13.6

SOURCE: From L. E. Holt, A. M. Courtney, and H. L. Fales, *Am. J. Diseases Children*, **10**, 229, 1915.

Milk is probably the ideal source of calcium and phosphorus in nutrition. These elements are essential for all cells and are needed in large quantities for the formation of bones and teeth. The secretory activity of the mammary gland achieves large differences in the concentrations of inorganic constituents in blood and milk. The molar ratios of milk to blood concentration (shown in parentheses) indicate that the sodium (0.13) and chloride (0.25) contents of milk are distinctly lower than those of plasma, whereas calcium (14), potassium (7), magnesium (4), and phosphate (7) are considerably higher.

Lactose. This disaccharide of galactose and glucose occurs primarily in milk. Since free galactose is not found in mammalian tissues or in other body fluids in significant amounts, it is evident that it is formed in the mammary gland from blood glucose. Indeed, when ^{14}C-labeled glucose is injected into goats, the label appears equally in the glucose and galactose moieties of lactose. The biosynthesis of lactose has been discussed earlier (page 432).

It has been reported that human infants fed cow's milk develop a mixed intestinal flora, whereas those that are breast-fed show a prevalence of *Lactobacillus bifidus* in the stool. Mild acid hydrolysis of human, but not of cow's, milk liberates compounds, previously nondialyzable, which act as growth factors for *L. bifidus*. The active compounds appear to be oligosaccharides containing N-acetylglucosamine and a sialic acid.

If unsterilized milk is allowed to stand, fermentation caused by *Streptococcus lactis* and related organisms produces lactic acid from lactose. After hydrolysis of the lactose, lactic acid production by these bacteria results from the reaction sequence of anaerobic glycolysis (page 402).

Lipids. The lipids of milk are chiefly triglycerides and are dispersed as very small globules. Since the fat has a lower density than the aqueous part of milk, it will slowly rise to the top to form cream, or milk can be centrifuged to accomplish this more rapidly. The fat of cow's milk contains all the saturated fatty acids with an even number of carbon atoms from butyric to stearic, with about 10 per cent of the total fat composed of glycerides of lower fatty acids. The principal fatty acids are oleic, 32 per cent; palmitic, 15 per cent; myristic, 20 per cent; stearic, 15 per cent; and lauric, 6 per cent. Small amounts of phospholipids and cholesterol are present. Human milk fat contains no fatty acids with a molecular weight lower than that of decanoic acid and differs from cow's milk fat in this respect. The quantities of most of the fatty acids are similar to those in bovine milk.

Vitamins. Milk contains all the known vitamins and is exceedingly rich in vitamin A and riboflavin. Vitamin C (ascorbic acid), vitamin D, thiamine, pantothenic acid, and niacin are present in lesser amounts. Pasteurization destroys most of the vitamin C. For young infants, additional amounts of vitamins C and D are usually supplied.

Proteins. The principal protein of bovine milk is casein, which represents about 80 per cent of the protein nitrogen. The ease of preparation of this phosphoprotein has long made it a favorite subject for investigation. Milk is centrifuged, and the cream skimmed off the top. The remaining fluid, *skim milk,* is acidified to pH 4.7, causing the casein to precipitate. The supernatant fluid is *whey,* which contains about 20 per cent of the total protein. In some types of cheese manufacture the milk is acidified by the lactic acid produced by fermentation.

Crude casein is a mixture of several related proteins of differing amino acid composition which may be distinguished electrophoretically and have been separated from one another. In order of decreasing electrophoretic mobility at alkaline pH values, these are designated as α-, β-, γ-, and κ-caseins. The γ-casein represents only about 5 per cent of the total and contains little or no phosphorus. The α- and β-caseins are rich in phosphorus. All of these are polymers in which smaller monomeric chains are linked by disulfide bridges. From tryptic digests of casein, a "phosphopeptone" has been isolated which is approximately a decapeptide; further degradation of this substance led to the isolation of phosphorylserine. As in other phosphoproteins, the phosphate is esterified to the hydroxyl group of serine. The analogous O-phosphorylthreonine has also been obtained from an acidic hydrolysate of casein. A considerable part of the phosphorus content of milk is due to casein; since casein is present mostly as calcium caseinate, milk actually contains these two important inorganic constituents largely in combination with casein. For different species, the calcium and phosphate contents appear to vary with the casein content; this approximation is to be expected from the mode of binding of these substances. In addition to binding phosphate, the caseins are conjugated to a polysaccharide of uncertain structure which contains galactose, galactosamine, and N-acetylneuraminic acid, and is present in an amount approximately 5 per cent of the weight of the protein. The mode of linkage to the protein is unknown, but galactosamine is the unit attached to an amino acid residue, while the sialic acid must be the other terminus.

As indicated above, casein is precipitated by addition of acid to the milk, thus bringing it to the isoelectric point of the casein. The abomasum (fourth stomach) of ruminants contains a protease, *rennin,* which causes clotting at pH 7. Rennin liberates, from κ-casein only, a glycopeptide which is nondialyzable but soluble in 12 per cent trichloroacetic acid. The remaining molecule, called *paracasein,* reacts with calcium to yield the insoluble curd. Rennin is not present in the human stomach; the enzyme has only weak proteolytic activity. However, other proteases (page 532) can also catalyze the conversion of casein to paracasein, and this is the initial step in casein digestion in the infant stomach. The structure of rennin is related to that of pepsin, indicating a common origin of the two proteases.

The whey proteins appear to be as numerous as those of serum, judged by the complex electrophoretic diagrams obtained with this fluid. The principal protein of bovine whey is β-lactoglobulin, which has been obtained in crystalline form and amounts to about 50 or 60 per cent of the whey protein. The heterogeneous fraction of whey proteins soluble in saturated magnesium sulfate or half-saturated ammonium sulfate is frequently designated the albumin fraction, or "lactalbumin." This is a misnomer since two proteins are present; the chief protein is β-lactoglobulin. α-Lactalbumin (page 246) has been shown to be one of the two proteins required for lactose synthesis. Other important constituents of bovine whey are the immune globulins, which carry the antibodies; these account for about 10 per cent of the whey protein. They will be discussed below under Colostrum.

Many enzymes have been found in milk, and a *lactoperoxidase* has been obtained in crystalline form. Other important enzymes are *xanthine oxidase,* a *lipase,* a *protease,* etc. Since the *alkaline phosphatase* of milk is destroyed by heat more slowly than are bacteria, the phosphatase activity of milk samples is employed as a test for efficiency of pasteurization. It is unclear whether these enzymes are significant in the physiology of milk production.

Human milk not only contains much less protein than bovine milk (Table 34.3), but the distribution of proteins is different. Casein accounts for only about 40 per cent of the proteins of human milk and the whey proteins about 60 per cent. However, the composition of caseins from human milk appears to be very similar to those of bovine origin. The other proteins of human milk have not been well characterized.

The main proteins of milk, casein and β-lactoglobulin, are unique in that they are not found in other tissues and bear no obvious relationship to any of the plasma proteins; these milk proteins are synthesized by mammary tissue from amino acids supplied by the blood. Casein and β-lactoglobulin are the most important nutritive proteins of milk. They are both preeminently suited for this function, since they are complete proteins containing all the common amino acids and are very rich in the essential ones. The relatively low sulfur content of casein, 0.78 per cent, is well balanced by β-lactoglobulin, which contains 1.6 per cent sulfur and is one of the best proteins for supporting the growth of young animals.

Milk contains a small amount of albumin, which is immunologically identical with serum albumin. The immune globulins of milk are closely related to the γ-globulins of serum, although they do not appear to be identical.

COLOSTRUM

The colostral milk, or colostrum, obtained during the first few days after parturition, differs markedly from ordinary milk in physical and biological properties. Fresh milk does not coagulate on boiling, but a surface film is formed which contains casein and calcium salts. When colostrum is boiled, a large coagulum forms. This difference in physical properties is due to the much higher protein content of colostrum and its different protein composition. Bovine milk contains about 4 per cent protein, of which 80 per cent is casein. Colostrum may contain as much as 20 per cent protein, and the predominant fraction is represented by immune globulins, which in various animals account for 40 to 55 per cent of the total protein. These globulins contain all the antibodies found in the maternal blood and are responsible for the transmission of immunity to the newborn of ungulates (page 718). The predominant change in the protein pattern which occurs in the transition from colostrum to milk is the marked decrease of the immune globulins, which occurs a few days after lactation is initiated. Correspondingly, the newborn calf can absorb these globulins from the gastrointestinal tract only during the first day or so after birth, but the passively acquired antibodies may be detected in the blood for some months. Human colostrum has only about two or three times the protein content of human milk. Although the higher concentration is largely due to the immune globulins, there is no evidence for intestinal absorption of antibodies by the suckling infant.

The lipid of bovine colostrum is usually deep yellow or orange in color, largely because of the presence of β-carotene, an important precursor of vitamin A (page 76). Colostrum contains from 50 to 100 times as much β-carotene as does ordinary milk. Larger amounts of riboflavin, niacinamide, and other vitamins are found in colostrum than in milk.

It is evident from the foregoing that colostrum serves to enhance the chances of survival of the newborn. In addition to possessing the nutritive values of milk, it also provides necessary vitamins and antibodies important to the newborn of some species. It is well to reemphasize that colostrum and milk are specific adaptations distinctive for mammals.

REFERENCES

Books

Adler, F. H., "Physiology of the Eye," The C. V. Mosby Company, St. Louis, 1950.

Babkin. B. P., "Secretory Mechanism of the Digestive Glands," Paul B. Hoeber, Inc., New York, 1944.

Beaumont, W., "Experiments and Observations of the Gastric Juice: The Physiology of Digestion," F. P. Allen, Plattsburgh, N.Y., 1933.

Cantarow, A., and Trumper, M., "Clinical Biochemistry," 6th ed., W. B. Saunders Company, Philadelphia, 1962.

Conway, E. J., "The Biochemistry of Gastric Acid Secretion," Charles C Thomas, Publisher, Springfield, Ill., 1953.

Davson, H., "Physiology of the Ocular Cerebrospinal Fluid," Little, Brown and Company, Boston, 1956.

Drinker, C. K., and Yoffey, J. M., "Lymphatics, Lymph and Lymphoid Tissue," 2d ed., Harvard University Press, Cambridge, Mass., 1956.

Gray, C. H., "Bile Pigments in Health and Disease," Charles C Thomas, Publisher, Springfield, Ill., 1961.

Review Articles

Benson, J. A., Jr., and Rampone, A. J., Gastrointestinal Absorption, *Ann. Rev. Physiol.*, **28**, 201–226, 1966.

Crane, R. K., Intestinal Absorption of Sugars, *Physiol. Revs.*, **40**, 789–825, 1960.

Czaky, T. Z., Transport through Biological Membranes, *Ann. Rev. Physiol.*, **27**, 415–450, 1965.

Davson, H., The Intra-ocular Fluids, in H. Davson, ed., "The Eye," vol. I, pp. 67–146, Academic Press, Inc., New York, 1962.

Gregory, R. A., Secretory Mechanisms of the Digestive Tract, *Ann. Rev. Physiol.*, **27**, 395–414, 1965.

Grossman, M. I., Gastrointestinal Hormones, *Physiol. Revs.*, **30**, 33–90, 1950.

Hofman, A. F., and Small, D. M., Detergent Properties of Bile Salts: Correlation with Physiological Function, *Ann. Rev. Med.*, **18**, 333–376, 1967.

Krogh, A., The Active and Passive Exchanges of Inorganic Ions through the Surfaces of Living Cells and through Living Membranes Generally, *Proc. Roy. Soc. London, Ser. B*, **133**, 140–199, 1946.

McKenzie, H. A., Milk Proteins, *Advances in Protein Chem.*, **22**, 55–234, 1967.

Macy, I. G., Kelley, H., and Sloan, R., The Composition of Milks, *Natl. Acad. Sci.-Natl. Res. Council Publ.*, **119**, 1950.

Mutt, V., and Jorpes, J. E., Contemporary Developments in the Biochemistry of the Gastrointestinal Hormones, *Recent Progr. Hormone Research*, **23**, 483–503, 1967.

Pappenheimer, J. R., Passage of Molecules through Capillary Walls, *Physiol. Revs.*, **33**, 387–423, 1953.

Pirie, A., The Vitreous Body, in H. Davson, ed., "The Eye," vol. 1, pp. 197–212, Academic Press, Inc., New York, 1962.

Schneyer, L. H., and Schneyer, C. A., eds., "Secretory Mechanisms of Salivary Glands," Academic Press, Inc., New York, 1967.

Smith, E. L., The Isolation and Properties of the Immune Proteins of Bovine Milk and Colostrum and Their Role in Immunity: A Review, *J. Dairy Sci.*, **31**, 127–138, 1948.

Smyth, D. H., ed., "Intestinal Absorption," *Brit. Med. Bull.*, **23**, 205–296, 1967.

Steele, J. M., Body Water in Man and Its Subdivisions, *Bull. N.Y. Acad. Med.*, **27**, 679–696, 1951.

35. Renal Function and the Composition of Urine

THE KIDNEY

The preceding chapter described secretions of special composition elaborated by various cells from constituents of arterial plasma. The kidney is the major secretory organ of the body, and the fluid separated from plasma by its activity is the urine. In contrast to other secretions, however, urine exhibits a remarkable range of volume and composition, and it is by virtue of its ability to alter the nature of urine with varying metabolic and environmental circumstances that the kidney aids in regulating the volume and composition of the extracellular fluid.

Each human kidney contains about 1,000,000 functional units, or nephrons. The formation of urine is the result of three processes which occur in each nephron: (1) filtration through the glomerular capillaries; (2) reabsorption of fluid and solutes in the proximal tubule, the loop of Henle, the distal tubule, and the collecting ducts; and (3) secretion into the lumen of the proximal and distal tubules.

The volume of glomerular filtrate formed by a normal 70-kg. adult is approximately 125 ml. per min. during mild water-induced diuresis. This fluid is considered to be a protein-free ultrafiltrate of plasma. Attempts have been made to estimate the maximum size of particles which can penetrate the glomerular membrane, which is built of glycoproteins and a collagen-like protein. Polypeptides and smaller proteins, including hemoglobin and myoglobin, appear readily in the urine when present in plasma, whereas serum albumin appears in urine only under unusual circumstances. These observations suggest that the basement membrane should contain pores of 20 Å. radius; however, pores have not been seen by electron microscopy. The normal glomerulus may permit passage of a significant amount of serum albumin, which is reabsorbed as it passes down the tubule. If this were of the order of 5 mg. per 100 ml. of filtrate and if none of this material were reabsorbed, a proteinuria of 9 g. per day could result. During passage through the proximal tubule, about 70 to 80 per cent of the glomerular filtrate is reabsorbed, so that about 25 to 30 ml. per min. enters the loop of Henle. This fluid is glucose-free and isosmotic and, because of the reabsorption of HCO_3^-, has a pH below that of plasma. The reabsorptive processes which occur in the proximal tubule are relatively independent of the composition and volume of the body fluids. Formation of urine is completed in the loop of Henle, the distal tubules, and the collecting ducts from which urine flows at the rate of 0.5 to 2.0 ml. per min. The cells of these structures possess facultative mechanisms for reabsorption of water, various electrolytes and nonelectrolytes, and for secretion into the urine of NH_4^+, H^+, and K^+ ions, among others. Thus, it is here that those final adjustments in the composition and volume of urine occur which serve to regulate the constancy of the *milieu intérieur.*

Clearance. This term is used to denote the removal of a substance from the blood during its passage through the kidneys and is defined as the least volume of blood or plasma which contains all of a particular substance excreted in the urine in 1 min. Clearance is, therefore, a rate with the dimensions of milliliters of plasma per minute. Thus,

$$C = \frac{U \times V}{P}$$

where U = concentration in urine, V = urine volume, milliliters per minute, P = plasma concentration, and C = clearance, milliliters per minute.

The clearance of a substance whose concentration in plasma is identical with that in the glomerular filtrate, and which is neither reabsorbed nor secreted by the tubular epithelium, is a measure of the rate of glomerular filtration. Inulin, mannitol, and, albeit less well, creatinine meet these criteria. All yield clearance values of about 125 ml. per min. per 1.73 $m.^2$ of surface area in human males, and somewhat lower values in females. It follows that any substance whose clearance is less than that of inulin, and which is not bound to plasma protein, must be reabsorbed as the glomerular filtrate flows through the tubules. The major such substances are Na^+, Cl^-, K^+, water, glucose, urea, amino acids, and uric acid. Furthermore, any substance whose clearance exceeds that of inulin must be secreted into the urine by tubular cells. NH_4^+, H^+, and N^1-methylnicotinamide behave in this manner, as do a number of other substances, notably penicillin, *p*-aminohippurate, and phenolsulfonphthalein. At low plasma concentrations, tubular secretion of *p*-aminohippurate (PAH) is so effective that it does not appear in renal venous blood. Under these circumstances, PAH clearance is a measure of effective renal plasma flow, about 650 ml. per min. per 1.73 $m.^2$. The ratio of inulin to PAH clearance is termed the *filtration fraction* and is approximately 18 per cent in normal individuals.

Transport Maximum, T_m. Another parameter of renal excretory function is the T_m, an abbreviation for *transport maximum,* the maximum ability of the kidneys either to reabsorb or to secrete a given material. For example, since essentially no glucose is excreted when the plasma glucose concentration is 100 mg. per 100 ml., the kidneys must reabsorb 125 mg. of glucose per minute in the proximal tubules. However, at an artificially elevated plasma glucose concentration of 400 mg. per 100 ml., *i.e.*, a glomerular filtration rate of 500 mg. per min., 200 mg. per min. of glucose may be excreted. The difference, 300 mg. per min., is the maximum rate at which the kidneys can reabsorb glucose and is denoted as T_m for glucose. Similarly, by raising the concentration of PAH until it appears in the renal venous blood in appreciable quantities, the maximum amount of PAH appearing in the urine per minute represents the secretory T_m for PAH.

Renal Threshold. This term denotes the plasma concentration above which a given substance appears in the urine. Thus, in man, the renal threshold for glucose varies between 125 and 160 mg. per 100 ml. plasma. This is not as precise a concept as the reabsorptive T_m described above but is more readily determined since it does not necessitate simultaneous measurements of glomerular filtration. Any statement of the threshold value for a given compound assumes a constant glomerular filtration rate; this assumption may not be valid from interval to interval. Moreover, for no substance is there a precise threshold; after the threshold concentration has been exceeded, not all the material filtered in the glomerular filtrate is necessarily excreted quantitatively. Thus, an individual with a glucose threshold of 130 mg. per 100 ml. filters and reabsorbs 163 mg. per min. at this concentration, yet at an even higher plasma concentration may exhibit a glucose reabsorptive T_m of 300 mg. per min. Table 35.1 lists the thresholds of appearance of some major plasma solutes. Evidence for a threshold of appearance (or excretion) for a diffusible substance indicates that for its reabsorption by the tubules a device exists which can be saturated.

Table 35.1: PLASMA THRESHOLDS OF APPEARANCE AND RETENTION

Substance	Threshold of appearance	Test subject	Threshold of retention	Test subject
Glucose	11.6 mmole/liter (208 mg./100 ml.)	Dog	6.84 mmole/liter (123 mg./100 ml.)	Rabbit
Glucose	7.8–11.1 mmole/liter (140–200 mg./100 ml.)	Man		
Sodium	....	......	140 meq./liter	Man, dog
Chloride	85 meq./liter	Rabbit	110 meq./liter	Dog
			100 meq./liter	Man
Bicarbonate	25 meq./liter	Dog	25 meq./liter	Man, dog
Potassium	2.8 meq./liter	Man	3 meq./liter	Man, dog
Sulfate	2–4 meq./liter	Dog	3 meq./liter	Dog
Calcium	4.25 meq./liter	Man	4.8 meq./liter	Man, dog
Phosphate	1.1–1.5 mmole/liter	Dog		

SOURCE: Adapted from A. V. Wolf, "The Urinary Function of the Kidney," Grune & Stratton, Inc., New York, 1950.

A more illuminating concept is that of the threshold of retention, the concentration at which the quantities of a given substance in plasma and urine are identical. Above this plasma concentration, urine will be more concentrated than plasma, while, below it, plasma will be more concentrated than urine. This concept describes the kidney as a regulator of the composition of the extracellular fluid more graphically than does the threshold of appearance.

RENAL EXCRETORY MECHANISMS

ELECTROLYTES

Sodium, Chloride, and Water. About 75 per cent of the Na^+, Cl^-, and water of the glomerular filtrate are reabsorbed in the proximal tubules by an active process in which Na^+ ions are selectively removed from the tubular fluid; anions move passively in accordance with the electrical gradient established by transfer of Na^+, and water moves, passively and isosmotically, with the solute. In this manner, about 25 ml. of slightly hypertonic filtrate leaves the proximal tubules per minute.

The subsequent facultative adjustment of the volume and osmolarity of the urine is made possible by (1) existence of a sodium pump in the cells of the loop of Henle and in the distal portions of the tubule and collecting duct; (2) control by the hormone, vasopressin, (Chap. 47) of the permeability to water of the distal tubules and collecting ducts; and (3) the architecture of the kidney. In the ascending limb of the hairpin-shaped loop of Henle an outwardly oriented sodium pump (chloride moves passively with the electrochemical gradient) operates while the same cells are relatively impermeable to water. Consequently, a gradient of about 200 mOsmoles (mOsm.) per liter is established between the fluid inside the ascending limb of the loop and the surrounding interstitium. This effect is multiplied as the fluid in the thin-walled, water-permeable, descending limb achieves osmotic equilibrium with the same interstitium, thereby increasing the osmolality of the fluid presented to the ascending limb. As the fluid travels through the distal tubule, it is diluted by water from the interstitium, with which it again attains osmotic equilibrium. Sodium may be removed throughout the distal tubule by an active process, the rate of which is determined by the adrenal cortical hormone, aldosterone (Chap. 45). This activity is maximal when Na^+ is to be conserved, *viz.*, when plasma $[Na^+]$ is below normal, and can result in removal of almost all Na^+ from the presumptive urine; it is minimal at elevated plasma $[Na^+]$. As the fluid proceeds down the collecting duct, it must pass once again through an area of increasing osmolarity of the surrounding tissue. The epithelium of the collecting ducts is thought to be essentially impermeable to Na^+ whereas permeability of this tissue to H_2O is regulated by the hormone, vasopressin (Chap. 47). In the absence of the latter, the duct is impermeable to H_2O; the duct fluid fails to equilibrate with the surrounding medium, and a highly dilute urine is excreted. With increasing quantities of vasopressin there is increased permeability to H_2O until, at full activity, the duct fluid is osmotically equilibrated with the hypertonic surrounding medium before entering the larger collecting passages. Vasopressin exerts similar effects on the water permeability of the toad bladder; the hormone stimulates formation of cyclic adenylic acid (page 437), which, even in the absence of hormone, can also elicit a similar increase in permeability to water. The operation of this countercurrent system is shown in Fig. 35.1 (page 832). Clearly, it is not possible to form a urine more concentrated than that of the contents of the bottom of the loop of Henle.

Potassium. The renal mechanisms involved in potassium excretion efficiently prevent potassium retention and ensure against hyperpotassemia. However, even on a potassium-free diet, normal adults may excrete 20 to 30 meq. per day. Although this could be derived from the 750 meq. filtered through the glomerulus daily, it appears likely that potassium is largely removed as fluid traverses the proximal tubules and that most urinary potassium is secreted in the distal tubules. The existence of such a secretory mechanism is suggested by the fact that K^+ clearance may exceed inulin clearance.

Whereas it seems likely that proximal reabsorption is a form of specific active transport, secretion of K^+ in the distal tubule is accomplished by exchange for Na^+. Only when Na^+ reabsorption is impaired, as in adrenal cortical insufficiency, does K^+ secretion fail and hyperpotassemia may result. Thus it appears that this exchange mechanism is one aspect of the aldosterone-controlled Na^+-reabsorptive process in

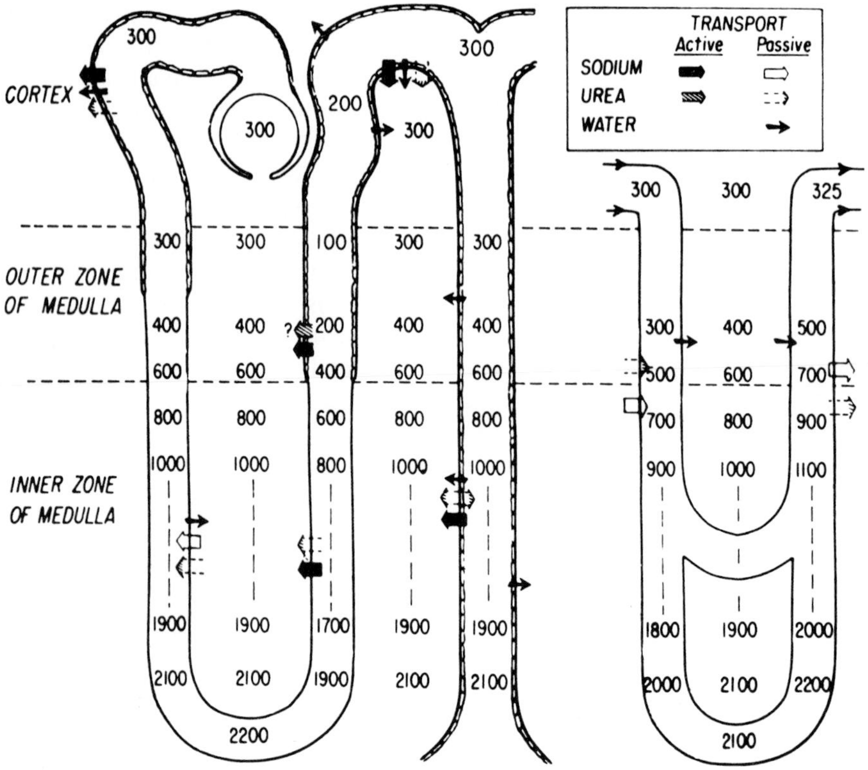

Fig. 35.1. Schematic representation of the operation of the postulated countercurrent multiplication mechanism in the formation of hypertonic urine by the kidney. The numbers represent the osmolality of the fluid in the tubule or in the surrounding interstitium.

the distal tubule (page 965). The normal operation of this mechanism enforces excretion of about 25 meq. of K^+ daily even when no potassium is ingested or at diminished plasma $[K^+]$. Exaggerated Na^+ reabsorption due to adrenal cortical hyperactivity (Chap. 45) results in augmented K^+ excretion with serious depletion of body potassium.

Acidification of Urine. In severe acidosis the $[H^+]$ of urine may be 1,000 times that of the plasma from which it is derived. This acidification begins in the proximal tubules and is completed in the distal tubules and collecting ducts. Although selective reabsorption of $HPO_4^{=}$ and dissociation of the dissolved CO_2 of the glomerular filtrate might account for acidification of urine in normal persons ingesting an average diet with an acidic ash, it cannot account for the maximum capacity of the kidney to produce acidic urine. The tubular ion exchange mechanism illustrated in Fig. 35.2, proposed by Pitts, is thought to represent the major mechanism for acidification of urine.

Essentially, the suggested mechanism includes metabolic CO_2 production, hydration to H_2CO_3 catalyzed by carbonic anhydrase, dissociation to $H^+ + HCO_3^-$, and exchange of the H^+ for Na^+ across the luminal border of the cell. Na^+ and HCO_3^- are assumed to diffuse to the opposite side of the cell, where the reverse

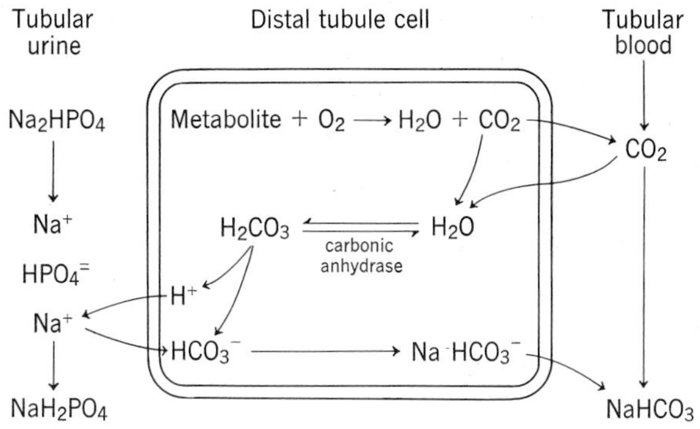

Fig. 35.2. Schematic representation of the acidification of urine by tubular cationic exchange. (*From R. F. Pitts, Am. J. Med.,* **9,** 356, 1950.)

process could operate, leading to appearance of Na^+ of the glomerular filtrate in venous blood, with HCO_3^- arising from H_2CO_3. After exchange of Na^+ and H^+ across the vascular boundary of the cell, the H^+ combines with HCO_3^-, CO_2 is produced by carbonic anhydrase catalysis, and then diffuses into the plasma. The failure of the acidification process and enhanced Na^+ excretion after administration of carbonic anhydrase inhibitors strengthen the H^+-Na^+ exchange concept and suggest that cellular CO_2 is the source of a major fraction of the secreted protons. That this is possible is suggested by the observation that the maximum rate of acidification (meq. H^+ per min.) is of the same order of magnitude as renal cellular respiration (mmoles CO_2 per min.).

Thus, both H^+ and K^+ secretion by the distal tubule are accomplished by exchange for Na^+, processes which are not unrelated. Carbonic anhydrase inhibition results not only in alkalinization of urine and diminished Na^+ reabsorption but also in a three- to fivefold increase in urine $[K^+]$. In potassium-deficiency states, with K^+ relatively unavailable for secretion, the acidification process is hyperactive; acidic urine low in $[K^+]$ is excreted, and plasma $[HCO_3^-]$ may rise to 50 or 60 meq. per liter. Conversely, when unusual quantities of K^+ are excreted, as after potassium administration, alkaline urine is produced. In respiratory acidosis, with increased plasma and cellular P_{CO_2}, urine is acidic and low in $[K^+]$, whereas in respiratory alkalosis, urine is alkaline and high in $[K^+]$. This observation suggests that there may be a competition between K^+ and H^+ for some component of the Na^+ exchange mechanism. This process may be an aspect of the general, aldosterone-regulated Na^+ absorptive process. Acidification fails in the absence of aldosterone and is accentuated when excessive aldosterone is administered or is secreted by an adrenal tumor (Chap. 45).

Bicarbonate. Ordinarily, renal excretion of bicarbonate is very low; its concentration in urine at pH 6 or below is negligible. The major portion of the bicarbonate filtered through the glomerulus is reabsorbed, largely in the proximal tubule. Although, in part, this occurs in passive fashion similar to the behavior of Cl^- in this area, it is probable that HCO_3^- reabsorption is an active process, since it exhibits a reabsorptive T_m which rises with increasing P_{CO_2} in plasma. In either case, the distal tubule is presented with fluid containing about 10 to 15 meq. per liter of HCO_3^-. Although there may be a HCO_3^--absorbing mechanism, HCO_3^- reabsorption occurs chiefly by Na^+-H^+ exchange, as described above for acidification of urine. In this case, cellular H_2CO_3 yields a proton which exchanges for Na^+; the HCO_3^- in the lumen accepts the proton, dehydrates, and the resultant CO_2 may diffuse into peritubular blood or rehydrate in the cell. In either instance, the HCO_3^- returned to blood derives from dissociation of intracellular H_2CO_3. All other circumstances remaining constant, this process operates so that, at plasma $[HCO_3^-]$ up to 24 to 28 meq. per liter, all filtered HCO_3^- returns to plasma; at higher $[HCO_3^-]$ virtually all the excess HCO_3^- remains in the urine, which is, accordingly, alkaline. In this manner, urine containing as much as 250 meq. per liter of HCO_3^- may be excreted.

The major factors which govern the behavior of the tubular epithelium with respect to the acidification of urine appear, with but a few exceptions, to be intracellular $[H^+]$ and $[K^+]$. At low P_{CO_2} (respiratory alkalosis), the ratio $KHCO_3/H_2CO_3$ in the cell is elevated, the supply of H^+ is low but $[K^+]$ is normal, and an alkaline, K^+-containing fluid is excreted. In respiratory acidosis with high P_{CO_2}, increased intracellular CO_2, and, therefore, increased $[H^+]$, an acidic urine may be expected. With an increase in plasma $[K^+]$, as after KCl administration, there is presumed to be an increased intracellular $[K^+]$ which competes favorably with cellular protons and urine is alkaline, whereas the reverse situation obtains after general depletion of body potassium. More difficult to rationalize in these terms are metabolic alkalosis and acidosis. In the former, the increased load of filtered HCO_3^- is presumed to be the dominant factor and urine pH is conditioned by the limit of 24 to 28 meq. of HCO_3^- per liter of glomerular filtrate which can be reabsorbed. This may, in part, be offset by increased P_{CO_2} due to respiratory compensation, but the latter is usually minimal and relatively ineffective. Finally, in metabolic acidosis, characterized by markedly diminished P_{CO_2}, and therefore lowered cellular $[H_2CO_3]$, the increased $[H^+]$ of the extracellular fluid is dominant and may, by exchange for cellular K^+, titrate HCO_3^- and other cellular buffers and lower cellular pH despite the low P_{CO_2}, and thus determine the pH of urine. These interrelationships are depicted in Table 35.2.

Ammonia Excretion. By measurements of the renal arteriovenous differences in glutamine concentration, it was found that, in the acidotic dog, two-thirds of the urinary ammonia was derived from the amide nitrogen of the glutamine of arterial blood. In alkalosis, virtually no glutamine was removed from blood flowing through

Table 35.2: FACTORS CONTROLLING THE pH OF URINE

	Plasma			Cells			Urine		
	$[H_2CO_3]$	$[HCO_3^-]$	$[K^+]$	$[H_2CO_3]$	$[HCO_3^-]$	$[H^+]$	$[K^+]$	$[HCO_3^-]$	$[H^+]$
Respiratory alkalosis	↓	...	↑	↓	...	↓	↑	↑	↓
Respiratory acidosis	↑	...	↓	↑	...	↑	↓	↓	↑
Potassium administration	...	...	↑	...	...	...	↑	↑	↓
Potassium depletion	...	...	↓	...	...	...	↓	↓	↑
Metabolic alkalosis	↑	↑	↓	↑	↑	↓	↑	↑	↓
Metabolic acidosis	↓	↓	↓	↓	↓	↑	↑	↓	↑

the kidney. The remaining one-third of the urinary ammonia is derived from α-amino nitrogen of amino acids. Kidney slices, in vitro, produce ammonia and α-keto acids when incubated with amino acids, by transferring the amino groups to α-ketoglutarate with subsequent oxidation of glutamate by DPN^+ (page 559). Thus, it is possible to account for all the urinary ammonia as arising from glutamine and α-amino acids present in blood traversing the kidney.

It appears likely that NH_3, rather than NH_4^+, diffuses across the tubular lining and connecting duct epithelium and is neutralized by H^+ secreted by the ion exchange process described above. This, in turn, reduces the $[H^+]$ of the urine, permitting exchange of more H^+ for Na^+, which returns to the venous blood. Moreover, if NH_3 rather than NH_4^+ is the diffusing substance, a large concentration gradient is established, thereby enhancing excretion of NH_3. Conversely, if the intraluminar fluid is alkaline, formation of NH_4^+ is depressed, and NH_3 diffusion and excretion become limited to the equilibrium concentration of NH_3, which should be low. However, the $[H^+]$ of urine is not the sole factor controlling NH_3 excretion. When acidosis ensues, several days are required before maximal ammonia excretion is obtained, despite continued formation of highly acidic urine. Similarly, when ammonia excretion was compared in dogs rendered acidotic several days in advance and dogs given NH_4Cl at the start of the experiment, and then both groups were restored to normal conditions by a slow infusion of $NaHCO_3$, the animals in the first group excreted about three times as much ammonia as those in the second group at identical urinary pH levels. This effect may, in part, be due to "adaptive" increase in renal glutaminase or amino acid transaminase activity in chronic acidosis and also appears to be mediated in some manner by the adrenal cortex, since adrenalectomized animals excrete considerably less ammonia than do normal animals after administration of NH_4Cl.

Phosphate and Calcium. Phosphate clearance is, at all times, less than inulin clearance. It has not been established which segment of the tubule is responsible for phosphate absorption. However, existence of a reabsorptive phosphate T_m, lowering of this T_m by the parathyroid hormone (Chap. 43), failure of reabsorption when serum $[K^+]$ is diminished, and the competitive behavior of arsenate, all indicate an active absorptive process. Nothing is known of the mechanisms by which phosphate excretion is enhanced in acidosis and in alkalosis.

Calcium is a threshold substance, and urine normally is virtually calcium-free. However, calcium rapidly appears in urine when the plasma concentration is elevated to only a slight degree, indicating that the tubules normally operate rather close to the capacity, or reabsorptive T_m value, for calcium.

NONELECTROLYTES

Urea. Urea is the major example of a highly diffusible substance which is neither actively reabsorbed nor secreted by the tubules. Although there is suggestive evidence that urea may be actively reabsorbed by renal tubules of humans or of herbivores during chronic ingestion of diets extremely low in protein, the behavior of urea in the nephron reflects (1) lack of any specific urea-affecting device, (2) specific mechanisms for absorption or secretion of electrolytes, and (3) osmotic equilibrium between intraluminar fluid and the renal interstitium. In the rat nephron, urea concentration increases slightly at the end of the proximal convolutions and then increases about fivefold in the loop of Henle because of the accumulation of urea in the surrounding tissue by the countercurrent mechanism. In the distal tubule, urea again tends to remain behind as water leaves. In the collecting ducts, in the region of vasopressin sensitive cells, with their high permeability to water, concentration of urea again occurs. In general, the concentration of urea in urine, like that of [Na^+], is equal to the concentration which obtains in the renal papilla.

At ordinary rates of urine flow the concentration of urea in urine is about sixty to seventy times as great as that in plasma. When urine flow is about 1 ml. per min., urea clearance is normally about 55 ml. per min. When urine excretion is 2 ml. per min. or greater, urea clearance is about 75 ml. per min., so that about 40 per cent of the urea filtered is returned to the blood, while more than 98 per cent of the water is reabsorbed. Maximal urea clearance is normally reached when the rate of urine flow is greater than 2 ml. per min.

In many renal diseases, urea clearance falls, and plasma concentration of urea rises. The fall in urea clearance merely reflects the decline in glomerular filtration. This last process is self-adjusting with respect to excretion of urea. For example, a normal individual may have a plasma urea N concentration of 10 mg. per 100 ml., a glomerular filtration of 120 ml. per min., a urea clearance of 60 ml. per min., and hence, excrete 6 mg. of urea N per minute. Thus, half the urea filtered is passively reabsorbed. In glomerulonephritis, the glomerular filtration rate may decline to as little as 60 ml. per min. Since, again, half the urea filtered is reabsorbed, then at the outset only 3 mg. of urea N is excreted per minute. In consequence, the plasma urea concentration rises. When the latter reaches 20 mg. urea N per 100 ml., at the same diminished glomerular filtration rate, 12 mg. of urea N is filtered per minute. Since half of this is reabsorbed, 6 mg. of urea N is excreted per minute, just as in the normal individual. Thus, because of the elevated plasma urea concentration, despite a striking fall in urea clearance, the nephritic person can remain in balance with respect to urea.

Creatinine. The concentration of creatinine in urine is generally thirty-five to forty times that of plasma. In man, creatinine is excreted by glomerular filtration and is partly reabsorbed by passive diffusion. However, when creatinine is injected and the plasma concentration maintained at an elevated level, creatinine clearance approaches that of inulin.

Uric Acid. The urine to plasma concentration ratio of uric acid is approximately 30. Unlike creatinine and urea, uric acid is a compound which appears in urine when the concentration in plasma exceeds values just above normal. However, it is not certain whether the excreted urate is unabsorbed or secreted. The threshold appears to be influenced by one or more of the adrenal cortical hormones, administration of which increases urinary elimination and lowers plasma concentration of uric acid.

Glucose. Glucose is reabsorbed virtually quantitatively before the glomerular filtrate reaches the loop of Henle. When, however, the glucose threshold is exceeded, as in diabetes mellitus, glucose appears in the urine. In some individuals, glucose appears in the urine sporadically at normal, or only slightly elevated, plasma concentration of glucose. This condition is termed *renal glucosuria* and is attributed to a defective mechanism for tubular glucose reabsorption. Renal glucosuria can also be induced by administration of phlorhizin (page 43), which inhibits the tubular mechanism for glucose reabsorption. With sufficient dosage of the drug, glucose clearance may almost equal inulin clearance.

TUBULAR TRANSPORT MECHANISMS

The variable composition of urine, the absence of glucose in normal urine, and direct observation of changes in composition of glomerular filtrate within the tubules have indicated the existence of facultative mechanisms for absorption of constituents of the glomerular filtrate. Direct observation of phenol red secretion by the mesonephric kidneys of frogs, and by the tubules of the metanephric kidney in tissue culture studies in the laboratories of Marshall, Richards, and Chambers, early provided evidence of tubular secretion. Clearance techniques permit ready recognition of these processes. As the plasma concentration of a compound which is actively reabsorbed is increased, the clearance remains essentially zero until the reabsorptive capacity is exceeded. Beyond that point, clearance increases with rising plasma concentration, approaching glomerular filtration rate as a limit. Clearance of a compound that is secreted by the renal tubules and not reabsorbed exceeds that possible from glomerular filtration alone, and as its concentration in plasma increases, the clearance decreases to approach glomerular filtration as a limit.

The terms "reabsorption" and "secretion" lack distinction when applied to the cell. Both processes are "secretory" and differ only in their orientation. What, then, is the nature of the cellular transport mechanisms responsible for these processes? Many observations suggest that these processes involve enzymic systems arranged to exhibit directional orientation as well as substrate specificity. Admittedly, in no instance has the renal enzyme been identified, but the parallelism between the behavior of some transport mechanisms and enzymic activity is striking.

Regardless of the details of the mechanism, energy, probably as ATP, is required to transport any substance against an electrochemical gradient. Tubular secretory mechanisms for secretion of phenol red and *p*-aminohippurate (PAH) fail in the presence of quinone, an inhibitor of dehydrogenases, of vinylacetic acid, which inhibits succinic acid oxidase, and of dinitrophenol, which prevents respiratory phosphorylation. Moreover, transport mechanisms appear to compete for available energy. Thus, whereas PAH is secreted and glucose and ascorbic acid are reabsorbed, reabsorption of both glucose and ascorbic acid is impaired when PAH is secreted at its T_m value.

There are numerous instances in which substances of similar structure compete for a common transport mechanism. Xylose clearance rises to that of inulin if the plasma glucose concentration is raised, although the total absorption of the two sugars exceeds that of either alone at the same concentration. The sets leucine-isoleucine and lysine-arginine-ornithine behave similarly, as do proline-hydroxyproline-glycine, while PAH administration diminishes secretion of penicillin. This type of competition closely resembles that of substrates competing for a single enzyme, *e.g.*, phenylethylamine and epinephrine for liver amine oxidase. An excellent example of competitive inhibition of a transport mechanism by a substance which is not itself secreted by that mechanism is the competitive inhibition of secretion of penicillin and PAH by *p*-carboxy, N,N-diisopropylsulfonamide (probenecid). The extent of secretion is dependent, at all concentrations, on the ratio of penicillin or PAH to probenecid.

A single mechanism may serve to transport material in two directions, as in the case of the H^+-Na^+ system. In this regard it is noteworthy that probenecid, which inhibits tubular secretion of PAH and penicillin, also interferes with reabsorption of uric acid and inorganic phosphate. Additional evidence of the interdependence of secretory and reabsorptive mechanisms is the fact that if the serum $[K^+]$ is lowered, as by glucose infusion, K^+ secretion is diminished, while at the same time the renal capacity for phosphate reabsorption declines.

No normal urinary constituent is secreted as efficiently by tubules as are PAH and penicillin. Indeed, only N^1-methylnicotinamide and phenylsulfate, of the normal organic urinary constituents, are known to be secreted into the tubular urine at all, and these are excreted only in milligram quantities daily. Since PAH clearance is not affected by variations of the K^+, H^+, or NH_4^+ content of urine, or by carbonic anhydrase inhibitors, the PAH-secreting device appears to be independent of mechanisms responsible for secretion of these urinary components. Although the normal substrate for the PAH-secreting mechanism is unknown, it is possible that this substrate is absorbed rather than secreted.

The only enzymes known specifically to participate in renal transport are glutaminase, which functions in secretion of ammonia, and carbonic anhydrase, which is essential for H^+-Na^+ exchange. The high mutarotase activity (page 426) of renal cortex and the fact that the deoxyglucose/glucose ratio at which mutarotase is 50 per cent inhibited also results in excretion of 50 per cent of the glucose filtered by the glomerulus have suggested that this enzyme may participate in glucose transport. However, the fact that arabinose, which is an excellent substrate for mutarotase, is not actively reabsorbed in the proximal tubule argues against this hypothesis. The high concentration of alkaline phosphatase at the luminal and vascular borders of tubular cells suggests a role for this enzyme in transport processes. Active sodium transport appears to occur along the entire length of the nephron, except in the thin, descending loop of Henle. The evidence suggests that the mechanism employed resembles closely the Na^+-K^+-H^+ exchange system of cells generally. Kidney is rich in a Na^+-K^+ requiring ATPase which, like that of erythrocyte ghosts and nerve, is sensitive to the cardiac glycosides (page 781). The relationship of K^+ secretion to renal Na^+-H^+ exchange recalls the finding that acidification of the medium during yeast fermentation is also a result of a K^+-H^+ transfer, suggesting that the process of secretion of electrolytes is basically similar in all cells. It is apparent that enzymic bases for renal tubular transport have been well established although the details remain largely unknown.

RENAL HYPERTENSION

In addition to its role in maintaining the volume and composition of extracellular fluid, the kidney is also involved in homeostatic control of arterial blood pressure. Hypertension is associated with a variety of renal disorders in man. Hypertension can be produced in dogs by clamping the renal arteries to restrict renal blood flow.

This procedure is also effective after renal denervation, indicating a humoral mechanism in the pathogenesis of this type of experimental hypertension. An enzyme, *renin,* splits a polypeptide, *hypertensin I,* from *hypertensinogen,* a serum α_2-globulin formed by the liver. Normal plasma contains a derivative of phosphatidyl serine which is a potent inhibitor of renin activity. No other known proteolytic enzyme liberates hypertensin from hypertensinogen.

Hypertensin I preparations of slightly different composition have been described, depending on the sources of the renin and the substrate used. The hypertensin I obtained by incubation of hog kidney renin with horse serum globulin is a decapeptide with the amino acid sequence Asp-Arg-Val-Tyr-Ile-His-Pro-Phe-His-Leu; this peptide exhibits no pressor activity. However, normal serum contains an enzyme which liberates the dipeptide His-Leu from the carboxyl-terminal end of the chain, yielding *hypertensin II,* the most powerful pressor agent known. All tissues, particularly intestine and kidney, exhibit peptidase activity, presumably due to leucine aminopeptidase, which rapidly destroys hypertensin II. Plasma from persons with essential hypertension contains hypertensin II in an amount sufficient to maintain an elevated blood pressure. Normal plasma is devoid of hypertensin II. Renin production and release constitute a function of the juxtaglomerular apparatus, which also functions as a pressoreceptor, thereby permitting this system to participate in the homeostatic control of arterial pressure.

Hypertensin II also acts directly on the adrenal gland to stimulate release of aldosterone (Chap. 45), resulting in Na^+ conservation as described above. Hypertensin has also been called *angiotonin.* A uniform nomenclature has been proposed which would designate this compound as *angiotensin.* In this nomenclature, the enzyme which cleaves this decapeptide (see above) is thus termed *angiotensinase* or *hypertensinase.* The precursor globulin, which is converted to angiotensin by renin, is termed *angiotensinogen.*

Experimental hypertension can also be produced by enveloping the kidneys with silk, cellophane, or acrylate resin, by subtotal nephrectomy, or by prolonged administration of salt and the adrenal cortical steroid, deoxycorticosterone. The relation of high salt intake or retention to hypertension is not understood, but dietary salt restriction has proved effective in management of human hypertensive disease. Restriction of protein intake alleviates the hypertension of partially nephrectomized rats, apparently because of failure of adrenocorticotropic hormone secretion (Chap. 47) secondary to protein deficiency. The role of adrenal cortical hormones is not clear, but adrenalectomy lowers blood pressure in a significant number of hypertensive human beings, renal hypertension cannot be induced in adrenalectomized animals, and administration of adrenocorticotropic hormone or adrenal cortical steroids produces hypertension in totally nephrectomized rats.

NATURE AND COMPOSITION OF URINE

Since the rate of formation of urine and its composition are subject to diurnal variation and to the influences of muscular activity, digestion, and even emotional phenomena, comparisons of urine specimens are generally performed by examining urines collected over a 24-hr. period.

Volume. The volume of urine voided in 24 hr. by normal adults ranges from 600 to 2,500 ml. per day. Excretion greater than this is usually indicative of disease, *e.g.*, diabetes mellitus or insipidus, nephritis, etc. Urine volume is related to fluid intake and is increased by ingestion of large volumes of fluids, by coffee or tea, which contain methyl xanthine, or by alcohol, which suppresses release of antidiuretic hormone. In the early states of kidney disease, diabetes mellitus, etc., *nocturia* is encountered. This is defined as the passage of more than 500 ml. of urine with a specific gravity below 1.018 during a 12-hr. period at night. Later in the course of kidney disease, however, as renal function becomes more severely impaired, nocturia ceases; urine volumes may become markedly diminished (*oliguria*), and, in the terminal stage of kidney disease, urine excretion may cease entirely (*anuria*). Oliguria is also seen as a result of dehydration, cardiac insufficiency, or fever.

Color. Urine is usually amber in color. The principal pigment is *urochrome,* a compound of urobilin or urobilinogen (page 743) and a peptide of unknown structure. Other pigments which may be present include uroerythrin (believed to be derived from melanin metabolism), uroporphyrins (page 739) (normally present in minute amounts), and numerous other pigments present in traces, such as riboflavin. On standing, urine usually darkens, presumably owing to oxidation of urobilinogen. An unusually dark urine is due most commonly to excretion of bilirubin. Bilirubinuria is seen in icterus of the direct van den Bergh type (page 743). This includes all instances of obstructive jaundice, as well as most intrahepatic types of icterus. A darker than normal urine may also indicate the presence of porphyrins in abnormal amounts or of homogentisic acid (page 608), which is oxidized to a black polymer when urine which is slightly alkaline stands in contact with air. "Urorosein" excretion is related to ingestion of indole compounds.

Normal Sediments. Freshly voided urine is ordinarily clear. When it has been standing, a flocculent material occasionally separates; this usually consists of a small amount of nucleoprotein or mucoprotein together with some epithelial cells from the lining of the genitourinary tract. If the urine is alkaline, a mixture of calcium phosphate and ammonium-magnesium-phosphate ("triple phosphate") may also precipitate and, occasionally, oxalates and urates, which redissolve on acidification of the urine. Uric acid may precipitate from acidic urine.

Total Solute Concentration. The combined operation of the various mechanisms already described permits elaboration of urine varying in osmolality from 50 to 1,400 mOsm. per kg., as compared with plasma at 285 mOsm. per liter. The kidneys of young children are somewhat less efficient, producing urine of from 100 to 800 mOsm. per liter. When the osmolality of urine exceeds that of plasma, the difference between their osmolalities represents the quantity of solute cleared without an equivalent loss of water. When the osmolality is less than that of plasma, one may calculate the *free water clearance.* Thus, a 24-hr. specimen of urine of exactly 2 liters at 100 mOsm. per liter represents clearance of 1.43 liters of free water during this period. In general, osmolality, and hence specific gravity, varies inversely with urine volume, the lowest osmolalities being those of persons with uncontrolled diabetes insipidus.

The pH of Urine. The pH of urine may vary between 4.8 and 8.0, but, because of the generally acidic nature of the ash of the diet, pH values between 5.5 and 6.5

are usually encountered. The acidic ash of the average diet is due to sulfuric acid, from the metabolism of the sulfur-containing amino acids, phosphoric acid from nucleic acids, phosphoproteins, and phospholipids, and intestinal absorption of anions, which, in the diet, are associated with cations that are not readily absorbable, *e.g.*, calcium or magnesium. Thus, milk yields an alkaline ash on combustion in vitro. However, in vivo much of the calcium is not absorbed in the intestine while the anions of the mineral acids of milk are absorbed; individuals restricted to a milk diet excrete urine of approximately pH 6.0. Ingestion of a diet composed largely of fruit and vegetables leads to excretion of alkaline urine. A more meaningful expression of the extent to which the kidney has secreted protons is obtained by determination of titratable acid plus ammonia; the titration is made to pH 7.4. Daily excretion by normal individuals varies from 15 to 50 meq. of titratable acid and from 30 to 75 meq. of NH_4^+; in severe acidosis these may rise to as much as 200 and 400 meq. per day, respectively, whereas in alkalosis, the urine may contain virtually no NH_3 and its pH may exceed 7.4.

Table 35.3 summarizes the composition of a 24-hr. sample of urine; the values given are averages for the American population living on an ordinary diet. In addition to the components listed in the table, urine contains small quantities of a large number of organic and inorganic materials.

Anions of Urine. Ordinarily, *chloride* is the chief anion of urine, and the amount excreted is roughly equal to that which has been ingested. On salt-poor diets, chloride may almost disappear from urine; thus, patients eating a rice diet for treatment of hypertension may excrete the equivalent of only 150 mg. sodium chloride per day. Even smaller amounts may be found in the urine of patients who have been vomiting. No limit to the maximum daily urinary chloride excretion may be stated,

Table 35.3: Composition of Average 24-hr. Urine of a Normal Adult

Component			U/P*
Sodium	2–4 g.	100–200 meq.	0.8–1.5
Potassium	1.5–2.0 g.	35–50 meq.	10–15
Magnesium	0.1–0.2 g.	8–16 meq.	
Calcium	0.1–0.3 g.	2.5–7.5 meq.	
Iron	0.2 mg.		
Ammonia	0.4–1.0 g. N	30–75 meq.	
H^+	....	4×10^{-8}–4×10^{-6} meq./liter	1–100
Uric acid	0.08–0.2 g. N	....	20
Amino acids	0.08–0.15 g. N		
Hippuric acid	0.04–0.08 g. N		
Chloride	....	100–250 meq.	0.8–2
Bicarbonate	....	0–50 meq.	0–2
Phosphate	0.7–1.6 g. P	20–50 mM.	25
Inorganic sulfate	0.6–1.8 g. S	40–120 meq.	50
Organic sulfate	0.06–0.2 g. S		
Urea	6–18 g. N	....	60
Creatinine	0.3–0.8 g. N	....	70
Peptides	0.3–0.7 g. N		

* U/P = ratio of concentration in urine to that in plasma.

but the maximum concentration which may be attained is about 340 meq. per liter.

Virtually all the *phosphorus* in urine is present as orthophosphate. The quantity excreted varies with the dietary intake. Since the amount of phosphate absorbed in the intestine seldom exceeds 70 per cent of that ingested, balance studies include estimation of fecal phosphate. Urinary excretion of phosphate may increase in acidosis, alkalosis, and primary or secondary hyperparathyroidism. Diminished phosphate excretion may be observed as the result of renal damage, in pregnancy because of the requirement of the fetus for phosphate, or in diarrhea because of failure of intestinal absorption. Patients receiving infusions of glucose or insulin also show a temporarily diminished phosphate excretion since the plasma phosphate concentration is lowered under these circumstances.

Comparatively little *inorganic sulfate* is ingested. However, about 80 per cent of the total sulfur in urine is present as inorganic sulfate, and the amount present depends upon the previous ingestion of sulfur, largely as sulfur-containing amino acids of proteins. In addition, there is present an appreciable amount of esterified sulfate as oligosaccharides (page 53) and sulfate esters of phenolic compounds, as well as a small amount of organic sulfur.

Cations of Urine. Since *sodium* and *potassium* are the major cations of the diet, they are also the major cations of human urine. Total excretion of sodium usually varies between 2.0 and 4.0 g. per day; that of potassium is about 1.5 to 2.0 g. per day. Persons on sodium-free diets or in acidosis may excrete as little as 50 mg. of sodium per day. As explained previously, however, the excretion of potassium seldom falls below 1.0 g. per day. No limit can be placed on the maximal daily output of either of these cations, but the maximal urinary concentration of sodium is approximately 300 meq. per liter, and that of potassium is about 200 meq. per liter. These concentrations can be attained only after administration of large quantities of hypertonic solutions. The daily urinary excretion of *calcium* and *magnesium* each varies between 0.1 and 0.3 g. Since the gastrointestinal tract is the major excretory pathway of these cations at normal blood levels, the amount of calcium or magnesium excreted in urine each day is not a measure of the quantities of these elements in the diet. Above threshold levels these ions are rapidly excreted in the urine. The *ammonium ion* may vary in amount from negligible quantities in alkalosis to as much as 5 g. of NH_3-nitrogen per day in severe acidoses. The amount excreted generally varies from 0.5 to 1.0 g. (35 to 70 meq.) per day.

Organic Constituents of Normal Urine. Excretion of *urea* is a direct function of total nitrogen intake and, on average diets, may vary from 12 to 36 g. per day for a 70-kg. adult. Since excretion of other nitrogenous urinary components does not vary so markedly with nitrogen intake, urea nitrogen comprises 90 per cent of the total nitrogen excretion of an individual consuming 25 g. of total dietary nitrogen, but only 60 per cent of the total urinary nitrogen of an individual eating 5 g. of total nitrogen.

On the usual American dietary, 0.7 g. of *uric acid* is excreted per day by a normal adult. Uric acid excretion rarely decreases below 0.5 to 0.6 g. even on purine-free diets but can be increased to more than 1 g. per day by ingestion of diets rich in nucleoproteins, such as glandular meats. Increased excretion of uric acid may occur in leukemia, polycythemia, hepatitis, and gout (page 630), and in response to administered aspirin, adrenal cortical steroids, or probenecid. Because of their insolubility, uric acid and its salts may precipitate in a collected urine sample or may form calculi in the lower urinary tract.

Creatinine. The amount of creatinine excreted varies, but for each individual the daily output is almost constant. This permits a simple check on the adequacy of consecutive 24-hr. urine collections. Urinary creatinine bears a direct relation to the muscle mass of the individual. This is expressed as the *creatinine coefficient,* the amount of creatinine in milligrams excreted per 24 hr. per kg. of body weight. The coefficient varies from 18 to 32 in men and from 10 to 25 in women; it is low in obese and asthenic persons and high in heavily muscled persons of average height.

Creatine excretion occurs more regularly in young children than in adults. Women may excrete more creatine and less creatinine than do men, although the creatinine excretion of women is generally as great as that of men in proportion to muscle mass. Creatine excretion rises in pregnancy and in the early post-partum period. In muscle wasting due to prolonged negative nitrogen balance, creatine excretion rises and creatinine excretion falls, with the total excretion of the two remaining roughly constant. This is seen, for example, in starvation, diabetes, hyperthyroidism, and fever. Conditions characterized primarily by muscle wasting also result in increased creatine and decreased creatinine excretion, as in various forms of muscular dystrophy. Administration of large doses of creatine leads to only small increments in urinary creatinine, the bulk of the creatine being eliminated unchanged. Administered creatinine appears quantitatively in the urine.

Hippuric acid (benzoylglycine) received its name because it was first found in equine urine. Normal excretion of hippuric acid is approximately 0.7 g. per day; ingested benzoic acid is quantitatively excreted in this form. Benzoic acid is present in natural foods, particularly fruits and berries, and is used as a preservative in various prepared foods. Since hippuric acid is formed in the liver, the rate of hippuric acid excretion after benzoic acid administration has been employed as a liver function test.

The presence of *indican* in urine is a result of bacterial action on tryptophan in the bowel, leading to formation of indole which is absorbed and oxidized in the liver to indoxyl. The latter is esterified with sulfate in the liver and excreted as the potassium salt (indican, page 616) in an amount of 5 to 25 mg. of indican per 24 hr. Increased excretion may occur in achlorhydria, because of diminished bactericidal action of the gastric juice, and in intestinal obstruction, paralytic ileus, or obstructive jaundice.

Urobilinogen is the precursor of the normal urinary pigment (page 743) and is present in small quantity in normal urine. It is detected with Ehrlich's reagent (page 107), which reacts to produce a red pigment. Normal urine gives a detectable color with this reagent in dilutions of 1:20. Undiluted urine may fail to give a visible response in biliary obstruction, while large amounts of urobilinogen are encountered in hemolytic diseases.

Glucosiduronides are normal components of urine since many compounds produced in metabolism or administered are excreted to some extent in this form (page 433), *e.g.,* chloral, menthol, phenol, morphine, aspirin, and various hormones, *e.g.,* steroids. Certain of these substances are also excreted in part as sulfate esters.

In addition to these components of urine, small quantities of other substances are normally present. These include trace elements such as copper, zinc, cobalt, fluorine, manganese, iodine, mercury, and lead. Among the organic compounds excreted in small amounts are water-soluble vitamins, peptide hormones of the hypophysis, chorionic gonadotropin in pregnancy urine, and hormones of the thyroid, gonads, and adrenal, or their metabolic products.

ABNORMAL CONSTITUENTS OF URINE

Glycosurias. The presence of an unusual amount of reducing sugar in urine is termed *glycosuria.* This is a generic term, independent of the exact carbohydrate involved. When the specific carbohydrate has been identified, the more specific description—*glucosuria, fructosuria,* etc.—is used.

Glucosuria. Freshly voided normal urine contains between 10 and 20 mg. of glucose per 100 ml. Unusual amounts of glucose in urine may be found after anesthesia or asphyxia, and in emotional states. Approximately 25 per cent of all persons with severe hyperthyroidism have glucosuria. Renal glucosuria is occasionally observed in otherwise normal individuals as well as in association with other disorders of renal tubular function. However, the most common cause of glucosuria is diabetes mellitus. The urine of diabetic patients may vary in sugar concentration from 0.5 to 12.0 per cent glucose. The mechanism of glucosuria has been considered earlier (page 836).

Pentosuria. Alimentary pentosuria occurs after eating unusual quantities of fruit or fruit juices. The pentose excreted is that which has been ingested, *e.g.*, arabinose. Of somewhat more interest is the genetic disorder *idiopathic pentosuria* (page 431). The pentose excreted is L-xylulose (page 26), which accumulates because of lack of L-xylulose dehydrogenase. There is no accompanying clinical syndrome, and this metabolic defect is apparently innocuous. Since xylulose reduces copper more rapidly than does glucose, individuals with xylulosuria invariably show positive tests for reducing sugar, thus making possible the error of mistaking pentosuria for diabetes mellitus.

Lactosuria. Moderate excretion of lactose is a frequent finding in lactating women, but lactosuria seldom occurs during pregnancy. However, glucosuria occurs during the course of about 15 per cent of all pregnancies, without accompanying hyperglycemia.

Galactosuria. Galactosuria is a consequence of galactosemia (page 427), a rare familial defect, generally detectable in very early infancy. Although no abnormality is detectable in the urine when galactose is rigorously excluded from the diet, as soon as milk, the prime dietary source of galactose, is fed, galactose may appear in the urine.

Fructosuria. Fructose rarely appears in urine. Its presence may result from a hepatic metabolic defect which may be hereditary. Fructosuria occurs less frequently than idiopathic pentosuria.

D-*Mannoheptulose,* a 7-carbon sugar, may appear in the urine of normal individuals who ingest large amounts of avocado.

Proteinuria. Normal urine contains traces of protein (including serum albumin and globulins), glycoprotein from the lining of the genitourinary tract, and mucoproteins of other origin. However, the usual clinical tests for urinary protein are negative. When proteinuria does occur, the major constituent is serum albumin, although globulins are invariably also present. The most common cause of proteinuria is renal disease, *e.g.*, acute glomerulonephritis, early chronic glomerulonephritis, the nephrotic syndrome (Fig. 35.3), and toxemia of pregnancy. In addition, albuminuria occurs in a variety of circumstances characterized by inadequate circulation to the kidney, *e.g.*, congestive heart failure, fever, anemia, liver disease, or various cardiac abnormalities. An occasional finding in otherwise normal individuals is postural or orthostatic proteinuria associated with long periods of standing or walking.

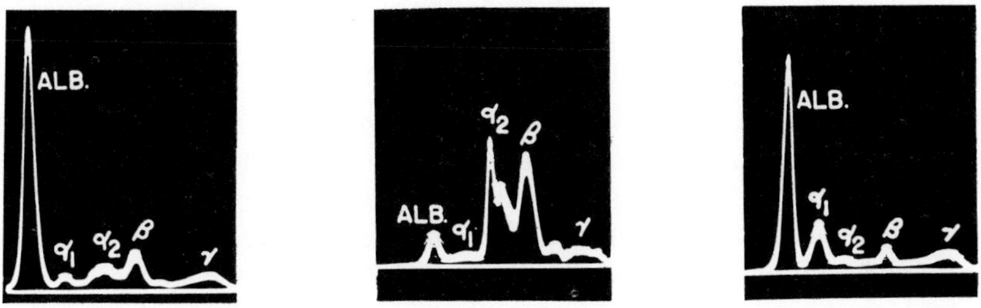

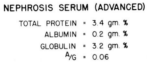

NORMAL SERUM		NORMAL RANGE	NEPHROSIS SERUM (ADVANCED)		NEPHROSIS URINE (ADVANCED)	
TOTAL PROTEIN	= 7.3 gm. %	6-8	TOTAL PROTEIN	= 3.4 gm. %	TOTAL PROTEIN	= 2.6 gm. %
ALBUMIN	= 4.4 gm. %	4.5-5.5	ALBUMIN	= 0.2 gm. %	ALBUMIN	= 1.2 gm. %
GLOBULIN	= 2.9 gm. %	1.5-3.0	GLOBULIN	= 3.2 gm. %	GLOBULIN	= 1.4 gm. %
A/G	= 1.5	1.5-2.5	A/G	= 0.06		

Fig. 35.3. Electrophoretic patterns of serum and urine from a patient with advanced nephrosis; the pattern of a normal serum is given for comparison. (*Courtesy of Dr. F. W. Putnam.*)

The urine of persons with multiple myeloma contains Bence-Jones protein, which precipitates at room temperature from acidified urine and redissolves on warming (page 717).

Other Abnormal Constituents. Other substances of metabolic origin may appear in urine when their concentration in plasma is unusually elevated. Among these are the "ketone bodies," acetoacetic acid, β-hydroxybutyric acid, and acetone, which appear during ketosis (page 503); bilirubin in hepatocellular or obstructive jaundice (page 745); urobilinogen in hemolytic disease (page 753); homogentisic acid in alkaptonuria (page 608); cystine and other amino acids, notably lysine, arginine, and ornithine, in cystinuria (page 602), and in other renal diseases, *e.g.*, Fanconi's syndrome.

Normal urine contains small amounts (up to 300 μg per day) of type I porphyrins (page 738). Excretion may increase ten- or twentyfold in liver disease and pernicious anemia. *Congenital porphyria* is a hereditary disorder characterized by an overproduction of the type I porphyrins. Uroporphyrin I and coproporphyrin I cannot be further metabolized; they are deposited in soft tissues, bones, and teeth, and as much as 100 mg. of this mixture may appear in the daily urine. *Acute porphyria* is characterized by the excretion of increased amounts of uroporphyrin III and coproporphyrin III as well as large quantities of porphobilinogen (page 737) and diverse compounds formed therefrom. Coproporphyrin III excretion is also characteristic of lead poisoning (page 737).

URINARY LITHIASIS

The low solubility of several of the normal components of urine occasionally leads to their precipitation as aggregates, or "stones." Approximately one-third of all such stones are calcium phosphate, magnesium ammonium phosphate, calcium carbonate, or a mixture of these. The formation of these stones frequently reflects chronic alkalinity of bladder and renal pelvic urine caused by infection with bacteria which hydrolyze urea, releasing ammonia. Formation of these stones will be promoted by any situation characterized by excessive excretion of calcium, *e.g.*, hyperparathyroidism (page 934), osteoporosis due to immobilization, and unusually high calcium ingestion. About half of all kidney stones are calcium oxalate, either alone or mixed with the salts of the above group. Stones of this type are frequent among individuals ingesting vegetable diets rich in such foods as spinach and rhubarb, which contain unusual amounts of oxalate. However, calcium oxalate stones are also pathognomonic of oxaluria, a hereditary disorder of glycine metabolism (page 597) in which virtually all glycine synthesized is oxidized via glyoxylic acid to oxalic acid. Less frequent are stones composed of insoluble organic compounds. Uric acid stones are common in gouty individuals (page 630). Xanthine stones are quite rare. Cystine deposits are almost invariably observed in cystinuric individuals, but are otherwise infrequent.

REFERENCES

Books

Braun-Menéndez, E., Fasciolo, J. C., Leloir, L. F., Muñoz, J. M., and Taquini, A. C., "Renal Hypertension," trans. by L. Dexter, Charles C Thomas, Publisher, Springfield, Ill., 1946.

Lewis, A. A. G., and Wolstenholme, G. E. W., eds., "The Kidney," Little, Brown and Company, Boston, 1954.

Lotspeich, W. D., "Metabolic Aspects of Renal Function," Charles C Thomas, Publisher, Springfield, Ill., 1959.

Pitts, R. F., "Physiology of the Kidney and Body Fluids," The Year Book Medical Publishers, Inc., Chicago, 1963.

Review Articles

Beyer, K. H., Functional Characteristics of Renal Transport Mechanisms, *Pharmacol. Revs.*, **2**, 227–280, 1950.

Forster, R. P., Kidney, Water and Electrolytes, *Ann. Rev. Physiol.*, **27**, 183–232, 1965.

Goldblatt, H., Renal Origin of Hypertension, *Physiol. Revs.*, **27**, 120–165, 1947.

Gottschalk, C. W., Osmotic Concentration and Dilution of the Urine, *Am. J. Med.*, **36**, 670–685, 1964.

Kirschner, L. B., Invertebrate Excretory Organs, *Ann. Rev. Physiol.*, **29**, 169–197, 1967.

Kruhffer, P., Handling of Alkali Metal Ions by the Kidney, in O. Eichler and A. Farah, eds., "Handbuch der Experimentellen Pharmakologie," pp. 233–423, Springer-Verlag OHG, Berlin, 1960.

Orloff, J., and Handler, J. S., The Cellular Mode of Action of Anti-diuretic Hormone, *Am. J. Med.*, **36**, 686–697, 1964.

Page, I. H., and Bumpus, F. M., Angiotensin, *Physiol. Revs.*, **41**, 331–390, 1961.

Pitts, R. F., Renal Production and Excretion of Ammonia, *Am. J. Med.*, **36**, 720–742, 1964.

Ulbrich, K. J., Kramer, K., and Boyland, J. W., The Countercurrent Multiplier System of the Kidney, *Progr. Cardiovascular Diseases,* **3,** 395–431, 1961.

Verney, E. B., The Antidiuretic Hormone and the Factors Which Determine Its Release, *Proc. Roy. Soc. London Ser. B,* **135,** 25–106, 1947.

Walser, M., and Mudge, G. G., Renal Excretory Mechanisms, in C. L. Comar and F. Bronner, eds., "Mineral Metabolism," vol. I, pp. 288–336, Academic Press, Inc., New York, 1960.

Weiner, I. M., and Mudge, G. H., Renal Tubular Mechanisms for Excretion of Organic Acids and Bases, *Am. J. Med.,* **36,** 743–762, 1964.

Wilson, C., ed., Physiology and Pathology of the Kidney, *Brit. Med. Bull.,* **13,** 1–70, 1957.

Windbager, E. E., and Biebisch, G., Electrophysiology of the Nephron, *Physiol. Revs.,* **45,** 214–244, 1965.

36. Muscle

The three types of muscle, striated, smooth, and cardiac, together comprise about 40 per cent of the body weight. Relatively little is known concerning possible differences in composition or metabolism among the three types of muscle. In what follows, we shall consider only those aspects of the composition and metabolism of muscle which permit it to serve as a contractile tissue.

COMPOSITION

The electrolyte composition of sarcoplasm is that shown as intracellular fluid in Fig. 33.1. Of all mammalian tissue, muscle most closely approximates this electrolyte pattern. Other than protein, the most abundant compound is glycogen, which varies from 0.5 to 1 per cent of the weight of the fresh tissue. The metabolism of muscle glycogen is considered on pages 436*ff*. Muscle contains small quantities of free amino acids, of which glutamine, glutamic acid, aspartic acid, and alanine are present in greatest quantity. This composition resembles that of most other tissues. Anserine (page 112), carnosine (page 112), and carnitine (page 582) have been isolated from skeletal muscle; the function of the first two peptides is unknown. The role of carnitine in lipid metabolism is discussed on page 482.

Muscles contain approximately 0.5 per cent of creatine, present in resting muscle largely as phosphocreatine. The biosynthesis of creatine has been described previously (page 582). Phosphocreatine is relatively unstable at the pH of sarcoplasm and is irreversibly transformed into creatinine, the anhydride of creatine (page 582). Creatinine, which serves no known metabolic function, diffuses from muscle and is excreted in the urine (page 836).

Proteins. Extraction of muscle with cold water yields a solution of proteins which consists largely of the enzymes that promote glycolysis, collectively termed "myogen." Brief extraction with alkaline 0.6*M* KCl of the residue remaining after the above procedure dissolves a protein fraction, of which *myosin* is the chief component. Szent-Györgyi observed that prolonged extraction with 0.6*M* KCl yields a myosin-containing solution of high viscosity and marked double refraction of flow, whereas brief extraction yields a solution containing almost as much protein but of much lower viscosity. In studying this phenomenon, Straub found that two different proteins are involved. These are *myosin,* the protein obtained by brief extraction with concentrated salt solution, and *actin,* which can be prepared separately by various procedures which involve preliminary removal of the mixed lipids of muscle. Both proteins have been prepared in highly purified form.

Myosin is a protein approximately 1,600 Å. in length, with a molecular weight of about 480,000. It consists of a relatively wide end section attached to a 24 Å. wide rod which is either a two- or three-stranded helical coil. These molecules readily aggregate into larger structures, some of which resemble the rod-shaped constituents of the A bands of muscle fibrils. All preparations of myosin exhibit ATPase activity, which shows two pH optima at about pH 6 and 9 and is dependent on the presence of sulfhydryl groups in the protein. Incubation of myosin with trypsin or other proteinases results in the formation of fragments called light (L-) and heavy (H-) *meromyosins,* respectively. The latter, which retain the ATPase activity of the parent molecule, consist of the head (wider) part of the molecule with a variable amount of the tail, whereas the light components derive from the tail.

Actin is a water-soluble globular protein of 60,000 molecular weight. As generally prepared, one molecule of ATP is bound to each actin molecule. If the ionic strength of a solution of this globular protein (G-actin) is increased to a value comparable to that of the sarcoplasm, actin polymerizes to a high molecular weight fibrous protein (F-actin) which is a double-stranded helix resembling the thin filament of the I bands of a myofibril (see Fig. 36.1, page 851). Simultaneously, the ATP hydrolyzes to ADP and inorganic phosphate, and one or two sulfhydryl groups, per molecule of the original G-actin, become less reactive with the usual reagents for detection of sulfhydryl groups. The inorganic phosphate is released, but the ADP remains bound to the protein. However, this dephosphorylation is not essential to polymerization. G-actin prepared containing ADP or no nucleotide also polymerizes under these conditions. Reduction of the ionic strength of the medium results in depolymerization to yield G-actin with bound ADP. Again, however, the nucleotide is unessential for the depolymerization.

From a dry powder of muscle which has been treated with nonpolar solvents to remove lipids, a small amount of fibrous protein may be obtained that differs in solubility, composition, and molecular weight from myosin and has been called *tropomyosin B.* This protein appears to be concentrated in the Z line (see Fig. 36.1, page 851). A similar protein is a prominent constituent of the "catch" muscles of mollusks. Additional lesser proteins have been described. Of these, *troponin,* may prove particularly interesting because it heightens the dependence of the ATP-induced contraction of actomyosin (see below) on Ca^{++}.

If the residue from the extraction of skeletal muscle with water is subjected to a single prolonged extraction with 0.6*M* KCl or if whole minced muscle is extracted in this manner, *actomyosin* is obtained. The latter is a complex of myosin and actin in a ratio of about 3:1. This is the material responsible for the high viscosity of certain muscle extracts, first noted by Szent-Györgyi (see above).

Actin and myosin are the materials which compose the *myofibril,* the fundamental contractile unit of muscle, and the behavior of these two proteins in the presence of ATP is the basis for contraction.

THE CONTRACTILE PROCESS

In 1939 Engelhardt observed that crude preparations of myosin containing actomyosin catalyzed the hydrolysis of ATP, the first indication of an interaction between ATP and the structural components of the contractile machinery. Szent-Györgyi observed that threads of actomyosin, prepared by squirting a solution into water, contracted while serving as an ATPase in a medium in which $[K^+]$ and $[Mg^{++}]$ resembled those of muscle. Extraction of muscle fiber bundles with glycerol provides a preparation which develops a tension comparable to contracted muscle in the presence of ATP and, at standard concentrations of other cations, is contingent upon the presence of Ca^{++}. Removal of Ca^{++} by a chelating agent then results in relaxation.

The ATPase activity of myosin and actomyosin is markedly influenced in vitro by pH, ionic strength, and the concentrations of K^+, NH_4^+, Ca^{++}, and Mg^{++}. Perhaps most significant is the fact that, whereas Ca^{++} stimulates the ATPase activities of both proteins, Mg^{++} appears to inhibit myosin ATPase but stimulates actomyosin ATPase activity. At pH 7, the phosphate portions of ATP are completely dissociated and the molecule bears four negative charges. In this form, it binds divalent cations, *e.g.*, Mg_2ATP, Ca_2ATP, etc. Since, regardless of mechanism, the hydrolysis of ATP must serve as the source of energy for muscle contraction, the mechanism of the ATPase activity of myosin has been intensively studied. On the assumption that simple hydrolysis of ATP is wasteful, the following scheme has been proposed.

$$\textbf{ATP + myosin} \xrightarrow{a} \textbf{ADP + myosin-P} \xrightarrow{b} \textbf{myosin} + \textbf{P}_i$$

Several observations, among them catalysis of exchange of $H_2{}^{18}O$ for the oxygen of P_i, support this formulation but no phosphorylated form of myosin has been definitively identified. Since $AD^{32}P$ does not exchange with ATP in the presence of myosin, such a reaction, if it occurs, cannot be reversible. Inasmuch as some group on myosin reacts with the sulfhydryl group of *p*-nitrothiophenol to form a stable addition compound only in the presence of ATP, clearly myosin and ATP do react to yield an activated form of myosin, but the nature of this activation remains unclear. The presence of myosin markedly accelerates the conversion of G- to F-actin; the mechanism of this catalysis and its relation to the contractile process, if any, are not understood.

Studies of the structure of skeletal muscle by phase contrast microscopy have contributed significantly to understanding the contractile process. Muscle fibers are built of longitudinal fibrils, about 1 μ in diameter, with alternating dark and light bands. The dense bands are birefringent and are called A (anisotropic) bands; the light bands are relatively nonbirefringent and are known as I (isotropic) bands. Within the I bands lies the dense Z line, which is continuous across the width of the fiber, holding the fibrils together and keeping the A and I bands of the many fibrils in "register." These features are schematically shown in Fig. 36.1. Electron microscopy of individual myofibrils reveals a fine structure, shown diagrammatically in Fig. 36.2. Cross sections through the denser areas of the A band show two types of filaments. The primary filaments are about 100 Å. in diameter and 200 to 300 Å. apart; around each of these are six secondary filaments about 50 Å. in diameter. These are "shared" with the surrounding six primary filaments, the fibril being built of several hundred of each type of filament. Cross sections at the H zone, the least dense portion of the A bands, show only primary filaments; in the I band

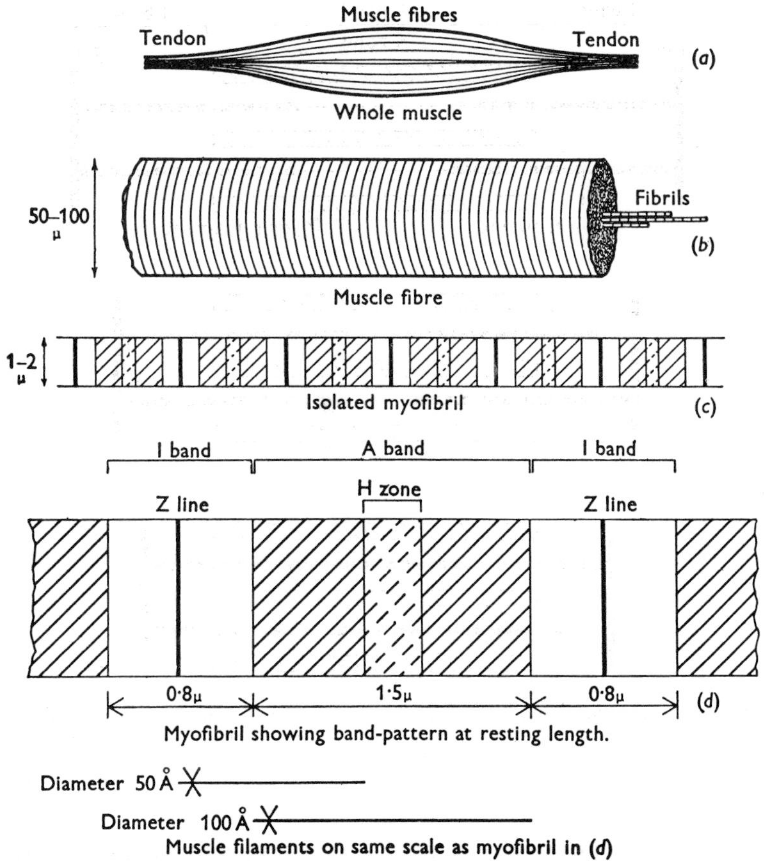

Fig. 36.1. The structure of muscle at various levels of organization; dimensions shown are for rabbit psoas muscle. (*From H. E. Huxley, Endeavour,* **15**, 177, 1956.)

only the smaller filaments are found. It has been established that the primary filaments are myosin and that the secondary filaments are F-actin.

From such studies, Hanson and Huxley suggested the following hypothesis for the contractile mechanism. In contracted fibrils, the length and diameter of the myosin filaments are unchanged, but the I bands can virtually disappear; hence the actin filaments move together into the H zone. Since the muscle can contract or relax and remain fixed at from 65 to 120 per cent of the resting length, it is suggested that there may be many points of combination between each actin and myosin filament. Thus, in vivo, there would be no single relationship between actin and myosin and the "actomyosin" isolated by prolonged extraction of muscle is, in a sense, factitious as it consists of actin and myosin in a less ordered arrangement than that which obtains in vivo while indicating the interaction of which these proteins are capable. Contraction, therefore, would be the process in which the actin and myosin filaments slide across each other. Since the surrounding fluid, in resting muscle, contains ATP in maximal concentration, myosin ATPase activity would permit the actin and myosin fibers to separate and slide across each other to

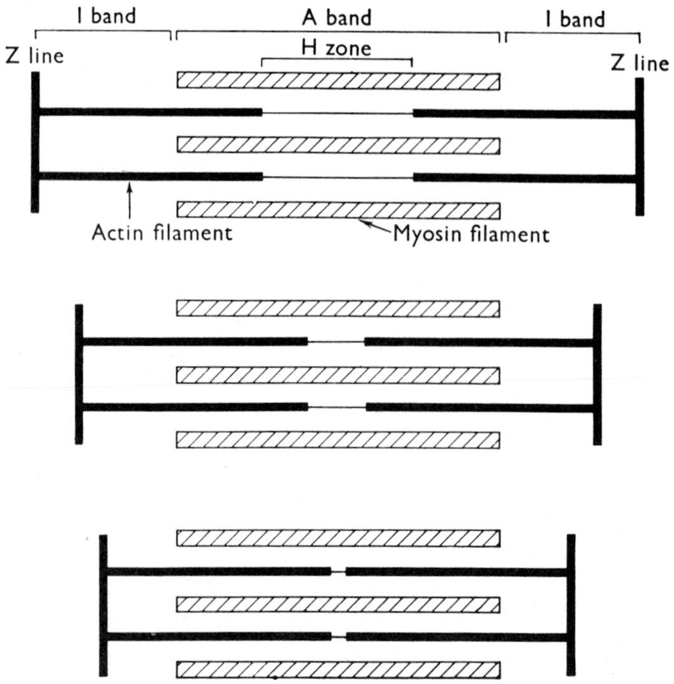

Fig. 36.2. Schematic representation of the relationships between actin and myosin filaments in extended (*above*), resting (*middle*), and partially contracted (*below*) muscle. (*Adapted from H. E. Huxley, Endeavour,* **15**, 177, 1956.)

new positions. Were this the case, initiation of contraction would be accomplished by some event which initiates the ATPase activity. This is believed to be an increase in local [Ca^{++}], the ion critical to actomyosin contraction. The molecular basis for such movement of actin and myosin filaments across each other, with subsequent bonding, is not known, nor is it understood whether the G- to F-actin conversion participates in this process.

Conduction in Muscle. Each muscle fiber is electrically polarized; in the resting state there is a potential of 90 millivolts across its membrane and the inside is negative. Upon arrival of the nervous impulse, a wave of depolarization, the action potential, travels along the membrane, the potential reversing in sign. The rapidity with which contraction is initiated suggests that the depolarizing wave must also pass into the interior of the fiber, perhaps through the transverse membrane of Z lines. The depolarization temporarily alters the permeability of the membrane to electrolytes, and it is thought that it is the resultant change in [Ca^{++}] which initiates the enzymic activity of the actin-myosin-ATP system.

Relaxation. With cessation of nervous stimulation, contraction ceases and muscle fibrils return to the resting state. When nervous stimulation ceases, the vesicular sarcotubular apparatus concentrates Ca^{++} from its surrounding fluid by an ATP-requiring active transport process. With decline in [Ca^{++}], myosin ATPase activity ceases and the resting state is reestablished.

THE SOURCE OF ENERGY FOR MUSCULAR WORK

Resting muscle, like other tissues, requires a constant supply of ATP for maintenance of the constancy of its composition and for its continuing metabolism. Muscle is unusual, however, in that large amounts of energy, as ATP, must be delivered almost instantaneously for the performance of its distinctive function. The rate of ATP formation by oxidation of carbohydrate or acetoacetate is adequate to meet the requirements of resting muscle, but not of working muscle. When working maximally, frog and mammalian muscle use approximately 10^{-4} and 10^{-3} moles of ATP per gram per minute, respectively. There is, however, only about 5×10^{-6} mole of ATP present per gram of resting muscle, an amount which cannot meet the demands of mammalian skeletal muscle for more than 0.5 sec. of intense activity. Maximal activity in frog muscle increases the rate of ATP utilization almost 1,000-fold above the basal rate; yet the maximal possible increase in oxygen consumption is no more than 100- to 200-fold. It is apparent that muscle possesses an anaerobic mechanism for rapid regeneration of ATP. In vertebrate muscles, this is the role of phosphocreatine. In the resting state, muscle has four to six times as much phosphocreatine as ATP. Muscle contains an enzyme, *creatine kinase,* which catalyzes the reversible transfer of phosphate between ADP and creatine phosphate.

$$\text{Creatine phosphate} + \text{ADP} \rightleftharpoons \text{creatine} + \text{ATP}$$

At pH 7.0, the free energy made available by hydrolysis of phosphocreatine is about 1500 cal. per mole greater than that from ATP, thus favoring formation of ATP from creatine phosphate. Phosphocreatine cannot serve as the immediate source of energy for contraction since this compound has no effect on the physical state of actomyosin, nor can it be hydrolyzed by this protein. Phosphocreatine serves, therefore, as a store of energy to be made available by way of ATP formation during contraction. Much effort has been expended in attempts to demonstrate the disappearance of ATP or creatine phosphate during a single muscle twitch, using rapid freezing techniques. The results, to date, remain unsatisfactory, but the data indicate approximately the expected decreases. The decline appears to be largely in the creatine phosphate fraction, but if the action of creatine kinase is impaired, ATP utilization is dominant.

Among invertebrates, the metabolic role of phosphocreatine is assumed by other phosphorylated guanidine compounds, most frequently phosphoarginine. Among the annelids, this role is also served by the N-phosphate derivatives of guanidoacetic acid (*Nereis diversicola*) and taurocyamine (*Arenicola*), whereas leech muscle contains N-phosphoguanidylethylserylphosphate.

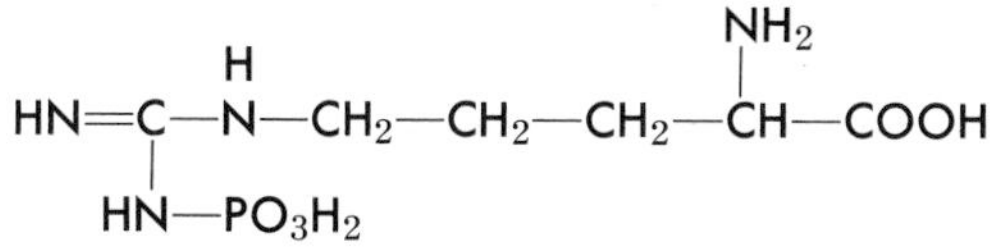

Phosphoarginine

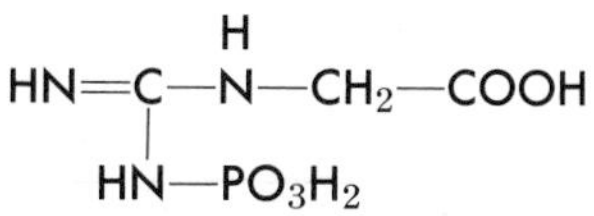

Phosphoguanidoacetic acid

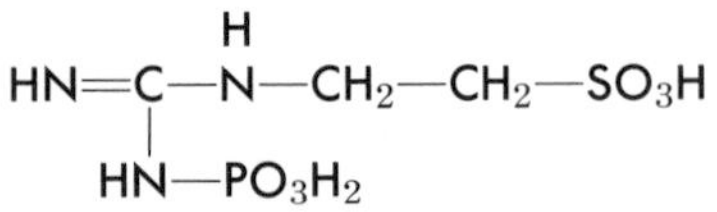

Phosphotaurocyamine

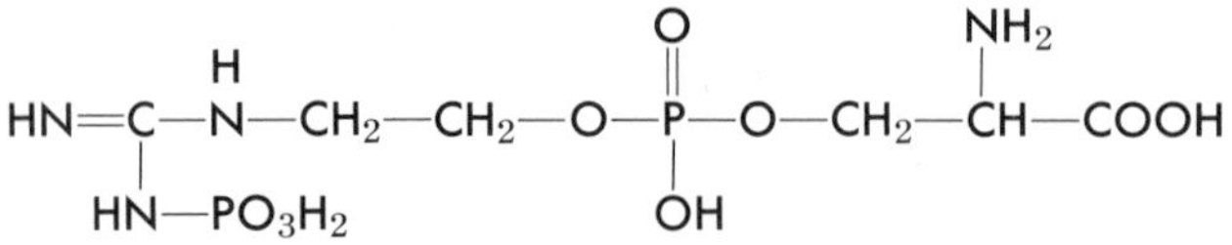

Phosphoguanidylethylserylphosphate

Despite the availability of phosphocreatine in vertebrate muscle, the total supply of high-energy phosphate available per gram of muscle cannot sustain activity for more than a few seconds, *i.e.*, about 50 twitches in isolated frog muscle preparations. Thus neither stored phosphocreatine nor the respiratory metabolism of muscle is adequate to meet the energy demands of muscle during activity which may be so intense as to render some muscles essentially anaerobic. These demands are satisfied by glycolysis of glycogen to lactic acid, which is produced by some contracting muscle even in the presence of an abundant oxygen supply. The energy liberated in this process permits net synthesis of three moles of ATP per mole of glucose equivalent, as described in Chap. 19. These quantitative relationships are illustrated by the behavior of isolated muscle preparations. If the muscle is stimulated aerobically at moderate frequency, permitting steady-state regeneration of ATP by oxidative phosphorylation, contraction continues as long as oxidizable substrate is available. If the muscle is stimulated at high frequency, tetanus persists until the glycogen and phosphocreatine have disappeared but much of the glycogen has been converted to lactic acid. Anaerobically, contraction can continue until all the glycogen and phosphocreatine have disappeared and in their place lactic acid, creatine, and inorganic phosphate accumulate. Lundsgaard demonstrated that if the preparation is stimulated anaerobically in the presence of an inhibitor of glycolysis such as iodoacetate, contraction continues until the supply of phosphocreatine and ATP is exhausted, but glycolysis stops at the triose stage and no lactic acid is formed. Figure 36.3 presents schematically the foregoing relationships with respect to the provision of energy for muscular contraction.

Whatever the nature of the actual mechanism by which ATP is used in contraction, it entails formation of ADP. Muscle possesses an enzyme, *adenylic acid kinase* (myokinase), which catalyzes the following reaction.

$$\mathbf{2ADP \rightleftharpoons ATP + AMP}$$

Studies with ^{32}P have shown that the rate of adenylic acid kinase activity is not appreciably affected by contraction. However, it seems likely that it is this reaction which generates the AMP that serves as the positive modifier for markedly accelerating the phosphofructokinase reaction (page 396), thus enhancing the rate of glycolysis. When the demand for ATP subsides, this enzyme can catalyze the reverse process, formation of ADP from AMP and ATP. No other reaction is known which can accomplish this end, *i.e.*, AMP cannot serve as a phosphate acceptor from either anaerobic glycolysis or oxidative reactions. *Adenylic acid deaminase* catalyzes hydrolysis of the adenylic acid to inosinic acid, with production of a mole of ammonia. The function of this system in muscle metabolism is not known.

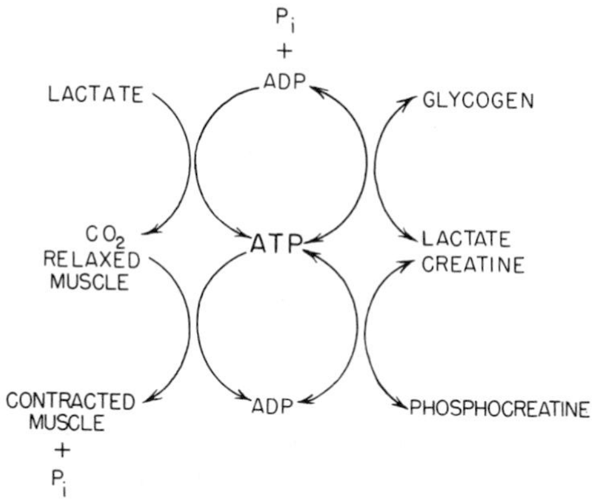

Fig. 36.3. The energy for muscular contraction. The immediate energy source is ATP, which may be regenerated from phosphocreatine and from glycolytic or aerobic phosphorylations. Resynthesis of phosphocreatine requires ATP; phosphocreatine cannot be directly employed as an energy source. In isolated muscle preparations only, resynthesis of glycogen from lactic acid requires ATP made available by complete oxidation of a fraction of the lactic acid. This is indicated in the figure. However, in vivo glycogen resynthesis in muscle is accomplished utilizing ATP gained by oxidation of either fatty acids or newly arriving blood glucose.

When isolated muscle, in vitro, is stimulated intermittently aerobically, it releases lactic acid during contraction and resynthesizes a large fraction of glycogen in the resting phase that follows. The energy for resynthesis is obtained from the simultaneous oxidation of a sufficient fraction of the lactic acid to provide the necessary ATP, but this mechanism of oxidative recovery does not obtain in vivo. In vitro, the lactic acid that diffuses into the medium during contraction reenters the muscle during recovery. In vivo, however, lactic acid diffuses into the interstitial fluid and is carried by the circulation to the liver. Here, most of the lactic acid is utilized for glycogen synthesis at the expense of ATP derived from the oxidation of about one-sixth of the lactic acid. The liver glycogen thus formed may be stored or may provide glucose to the blood. This series of reactions is illustrated in Fig. 36.4.

The ability of muscle to sustain maximal activity anaerobically leads to the accumulation of an *oxygen debt* (page 402). This may be illustrated as follows. A sprinter running at maximal speed for about 10 sec. consumes about 1 liter of oxygen, as compared with perhaps 40 ml. in a similar period of rest. Even after he has stopped running, however, the athlete will continue to breathe at a still elevated but declining rate for some time and during this period may consume an additional 4 liters of oxygen above the basal rate. Thus, the effort of running results in a total extra consumption of 5 liters of oxygen, about four-fifths of which occurs after cessation of exercise. This oxygen is utilized for the oxidation of sufficient lactic acid to convert the remaining lactic acid to glycogen and to restore the normal phos-

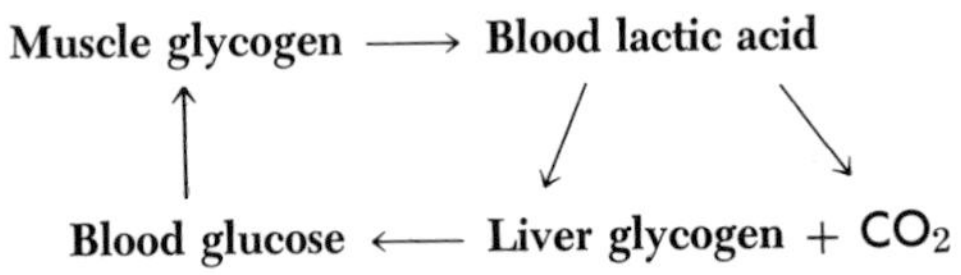

Fig. 36.4. Lactic acid, generated by working muscle in vivo, diffuses into the circulation and reaches the liver, where the lactic acid is converted to glycogen. The glycogen formed serves as a source of blood glucose, from which muscle glycogen may again be formed.

phocreatine concentration, as described above. Ultimately, therefore, all the energy for muscular work derives from the oxidation of carbohydrate. To the extent to which energy is employed for contraction in an amount greater than that which may be supplied by oxidation during the actual period of muscular exercise, an oxygen debt is incurred which must be met during the recovery period. It is noteworthy that in this sense some of the energy for muscular work is obtained through oxidations performed in the liver. The net process is such as might have occurred had the rate of exercise been sufficiently slow to permit all the demand for ATP to be met by aerobic oxidation of glucose in the muscle between contractions.

It will be seen that the operation of the mechanical aspect of this system automatically accelerates operation of the energy-yielding device. The ADP released by contraction is available both to the mitochondrial oxidative phosphorylation system and to the two energy-yielding steps in glycolysis. To the extent that ADP is converted to AMP by adenylic acid kinase, the AMP can serve as a positive effector for phosphofructokinase (page 396) and thus accelerate glycolysis. Inorganic phosphate becomes available to the mitochondria, to triose phosphate dehydrogenase, and to glycogen phosphorylase. Since the availability of ADP and orthophosphate is rate-limiting in the resting muscle, this is an example of self-regulation or "positive feedback" in a biological system.

Whereas contracting striated muscle derives its energy largely from the anaerobic transformation of glucose or glycogen to lactic acid, resting muscle derives a large portion of its energy from the oxidation of fatty acids and acetoacetate. Indeed, determinations of the arteriovenous difference in glucose concentration across resting muscle indicate only trivial consumption of glucose. Energy for the relatively slow contraction of smooth muscle similarly is derived from oxidation of fatty acids, acetoacetic acid, and, to a lesser extent, glucose. Cardiac muscle, which is rich in myoglobin, the enzymes of the tricarboxylic acid cycle, and the electron transport system, utilizes largely aerobic reactions to obtain ATP for contraction. Relatively little glucose is abstracted from the blood by cardiac muscle. Instead the latter obtains ATP by oxidation of fatty acids and, to a lesser extent, of acetoacetic acid and lactic acid. During exercise, when both skeletal and cardiac muscle metabolism is accelerated, the heart abstracts from the circulation and utilizes the lactate produced in peripheral muscles.

REFERENCES

Books

Gergely, J., ed., "Biochemistry of Muscle Contraction," Little, Brown and Company, Boston, 1964.

Mommaerts, W. H. F. M., "Muscular Contraction," Interscience Publishers, Inc., New York, 1950.

Szent-Györgyi, A., "Chemical Physiology of Contraction in Body and Heart Muscle," Academic Press, Inc., New York, 1953.

Review Articles

Bailey, K., Structure Proteins: II. Muscle, in H. Neurath and K. Bailey, eds., "The Proteins," vol. II, part B, pp. 951–1055, Academic Press, Inc., New York, 1954.

Conway, E. J., Nature and Significance of Concentration Relations of Potassium and Sodium Ions in Skeletal Muscle, *Physiol. Revs.,* **37,** 84–132, 1957.

Davies, R. E., On the Mechanism of Muscular Contraction, in P. N. Campbell and G. D. Greville, eds., "Essays in Biochemistry," vol. 1, pp. 29–56, Academic Press, Inc., New York, 1965.

Gergely, J., Contractile Proteins, *Ann. Rev. Biochem.,* **35,** 691–722, 1966.

Gergely, J., ed., The Relaxing Factor of Muscle, *Federation Proc.,* **23,** 885–940, 1967.

Huxley, H. E., The Fine Structure of Striated Muscle and Its Functional Significance, *Harvey Lectures,* **60,** 85–118, 1964–1965.

Kielley, W. W., Biochemistry of Muscle, *Ann. Rev. Biochem.,* **33,** 403–430, 1964.

Lilienthal, J. L., and Zierler, K. L., Diseases of Muscle, in R. H. S. Thompson and E. J. King, eds., "Biochemical Disorders in Human Disease," pp. 445–493, Academic Press, Inc., New York, 1957.

Mommaerts, W. H. F. M., Brady, A. J., and Abbott, B. C., Major Problems in Muscle Physiology, *Ann. Rev. Physiol.,* **23,** 529–576, 1961.

Needham, D. M., Biochemistry of Muscular Action, in G. H. Bourne, ed., "The Structure and Function of Muscle," vol. II, pp. 55–104, Academic Press, Inc., New York, 1960.

Slater, E. C., Biochemistry of Sarcosomes, in G. H. Bourne, ed., "The Structure and Function of Muscle," vol. II, pp. 105–141, Academic Press, Inc., New York, 1960.

Stracher, A., and Dreizen, P., Structure and Function of the Contractile Protein, Myosin, in D. R. Sanadi, ed., "Current Topics in Bioenergetics," vol. 1, pp. 154–202, Academic Press, Inc., New York, 1966.

Szent-Györgyi, A. G., Proteins of the Myofibril, in G. H. Bourne, ed., "The Structure and Function of Muscle," vol. II, pp. 1–54, Academic Press, Inc., New York, 1960.

Weber, A., Energized Calcium Transport and Relaxing Factors, in D. R. Sanadi, ed., "Current Topics in Bioenergetics," vol. 1, pp. 203–254, Academic Press, Inc., New York, 1966.

37. Nervous Tissue

Of the organs of the body, the brain presents one of the most complex problems of correlation of structure, composition, and function. Beginning with the work of Thudichum in the last century, it has been recognized that nervous tissue differs in several ways from other tissues. Most striking is the high content of lipid; in brain, lipids represent approximately 50 per cent of the total solids. Although studied intensively, brain composition is incompletely understood; however developments in quantitative methodology, notably application of various chromatographic techniques, have clarified some aspects of brain composition. Knowledge is still fragmentary of the manner in which the materials present in nervous tissue mediate its function. Some biochemical features of synaptic and neuromuscular conduction are well delineated, but still controversial is the picture of events occurring during axonal conduction. The biochemical basis of other brain functions, such as the origin of the nerve impulse, is unknown.

Since species differ in the timing of the chemical maturation of the nervous system, comparative discussions of the nervous system require that physiological age, not chronological age, be taken into consideration.

COMPOSITION OF NERVOUS TISSUE

As in other tissues, the lipids, proteins, polysaccharides, and nucleic acids of nervous tissue are associated with one another in a variety of linkages. Although protein-lipid complexes of nervous tissue have been isolated and characterized (see below), most information concerning the composition of the brain has been obtained by techniques designed to free each isolated material from other tissue components with which it may be associated *in situ.*

Lipids. With the exception of the neutral fats, which are not present, brain contains, in varying concentrations, representatives of all other lipid classes previously described (Chap. 4). Some of these lipids occur only in nervous tissue; others are present only in relatively limited amount elsewhere in the body. This is the case for the sphingomyelins, cerebrosides, and, to a lesser degree, the gangliosides. In other instances, lipids which are widely distributed are present in nervous tissue in uniquely high amounts. Thus, more than half of the total phosphoinositides of brain are present as triphosphoinositides, and plasmalogens comprise one-third of the total phospholipids. Cholesterol, found predominantly in other tissues and in the blood in ester form, is present in the brain of the normal adult animal only as the unesterified sterol; in the newborn and certain pathological conditions such as multiple sclerosis, a significant proportion of the sterol is as the ester. The cholesterol of peripheral nerve is also unesterified; this tissue, however, does contain triglyceride, although the latter may be in the surrounding connective tissue rather than in the nerve fiber proper.

The gangliosides of brain have been separated into two groups, containing one and two moles of sialic acid, respectively. In normal human brain, about 90 per cent of the monosialogangliosides consist of a compound with the molar ratio of ceramide to hexose to N-acetylgalactosamine to N-acetylneuraminic acid of 1:3:1:1. The hexoses are glucose and galactose. In Tay-Sachs disease (page 527) gangliosides of abnormal structures accumulate in unusually high amounts in brain.

In addition to the above lipids, brain has sulfur-containing cerebrosides (page 74). A sulfate ester of phrenosin (page 74) is abundant in white matter.

Fatty acids are widely distributed in brain as components of the compounds mentioned above. Indeed, per unit weight of brain tissue, fatty acids comprise the most abundant lipid constituent. These acids are of long chain length, C_{18} to C_{26}, and include α-hydroxy acids as well as unusually high quantities of polyenoic acids. Of a polyenoic fraction of ox brain, 43 per cent was identified as C_{22} hexaenoic acids. In the sphingomyelins from normal human brain, the proportion of stearic acid decreases with increasing age from about 80 per cent in the new born to 40 per cent in the adult. At the same time, the C_{22}-C_{26} acids increase from about 10 to 50 per cent. In some pathological states of the nervous system, the content of C_{22}-C_{26} acids is much smaller than in normal brains of the same age.

The lipids of myelin consist of cholesterol, phospholipids, and sphingolipids (largely sphingomyelins and cerebrosides) in the approximate molecular ratio of 2:2:1, respectively. Also present are phosphatidyl serine, phosphatidyl inositol, and plasmalogens. Myelin lipids are characterized by a high degree of unsaturation.

Table 37.1 presents some lipid concentrations in nervous tissue. The wide ranges of values given reflect the results of different investigators and analytical methods employed. Nonetheless, the general ranges of concentrations are indicative of the composition of nervous tissue lipids. The values in the table refer, for the most part, to mammalian tissue, although in the case of peripheral nerve some figures for frog nerve are included.

Table 37.1: CONCENTRATION OF LIPIDS IN NERVOUS TISSUE

Constituent	Gray matter	White matter	Spinal cord	Peripheral nerve
Total lipids	4.0–7.9	13.9–23.1	15.5–22.7	4.4–23.0
Total phospholipids	3.1–4.6	6.2–9.3	7.8–10.6	2.2–13.9
Cholesterol	0.6–1.4	3.6–5.4	3.9–5.9	1.1–4.8
Cerebrosides (includes all glycolipids)	0.3–1.9	4.1–7.4	3.8–6.2	1.1–4.7
Sphingomyelins	0.3–1.9	1.8–4.3	2.1–3.4	1.3–4.7

Note: All values are in grams per 100 g. fresh tissue.

SOURCE: After R. J. Rossiter, Chemical Constituents of Brain and Nerve, in K. A. C. Elliott, I. H. Page, and J. H. Quastel, eds., "Neurochemistry," 2d ed., p. 32, Charles C Thomas, Publisher, Springfield, Ill., 1962.

Proteins. The proteins of brain account for 40 per cent of its dry weight. Although tissue proteins are usually soluble in aqueous media, the relatively high proportion of lipid in nervous tissue has made difficult the mechanics of extracting proteins with such solvents. Indeed, classes of water-insoluble, nonpolar solvent-soluble proteins are now recognized as components of nervous tissue.

Proteins Soluble in Aqueous Media. Maximal extraction of protein nitrogen from brain with aqueous media is obtained at an ionic strength of about 3.0 and pH 6 to 9. The soluble proteins include nucleoproteins, ribonucleoproteins being abundant in nerve cytoplasm. Dialysis of the extract leaves a small amount of lipid-free albumins in solution, while the bulk of the extracted proteins precipitate. The latter contain approximately 20 to 25 per cent lipid, of which one-quarter is cholesterol; these proteins are therefore characterized as lipoproteins (page 118). Copper-containing, water-soluble proteins have been separated from brain tissue. The name *cerebrocuprein* has been given to one of these proteins which contains 0.3 per cent of copper.

Proteins Soluble in Nonpolar Solvents. Brain contains a class of proteins, termed *proteolipids,* lipid-protein combinations which differ from the water-soluble lipoproteins (see above) in that they are insoluble in water but freely soluble in mixtures of chloroform-methanol-water. The proteolipids differ from one another in the ratio of lipid to protein and in the nature of the lipids. Several have been obtained from the white matter of brain; proteolipid B, which has been obtained in crystalline form, contains 50 per cent each of protein and a lipid mixture consisting of equal parts of phospholipids and cerebrosides.

Insoluble Proteins; Residue Proteins. This is an operational description of the proteins remaining after thorough, successive extraction of brain tissue with polar and nonpolar solvents. Protein remaining after treatment with proteolytic enzymes, following solvent extraction, has often been included in this group; one of these is *neurokeratin.* The wide variety of procedures which have been used in its preparation indicate that neurokeratin is probably an artifact. Histochemical techniques, together with some preparative data, suggest that a neurokeratin-like substance *in situ* may represent the protein portion of some proteolipids. Amino acid analyses indicate that neurokeratin preparations differ from other keratins (page 117) with similar solubility properties.

Carbohydrates. Hexosamine determinations indicate that fresh brain tissue contains 0.15 to 0.25 per cent mucopolysaccharide; the contribution of connective tissue to this value is not clear. Chondroitin sulfate (page 52) is the chief mucopolysaccharide present. Other polysaccharides associated with proteins or glycoproteins must also be constituents of brain since there is present in small quantity a variety of hexosamines, hexuronic acids, and sialic acids. As noted above, other carbohydrates, *e.g.*, inositol, are present as components of lipids.

Electrolytes. The electrolyte composition of nerve is not remarkably different from that of intracellular fluids of other tissues except that there are not enough of the common anions, *e.g.*, phosphate, sulfate, bicarbonate, protein, etc., to balance the known cation content. The dicarboxylic amino acids, which are present in high concentration in brain tissue (see below), may partially compensate for the anionic deficit. Approximately 40 meq. of cations per liter are neutralized by the acidic phospholipids, sulfatides and sialic acids. Cations combined with these substances, however, may not be ionized, a fact which may have significance in the conduction of the nerve impulse.

METABOLISM OF BRAIN

Efforts to study the metabolism of the brain in vivo are hampered by the existence of various functional barriers to the passage of material from blood to the brain cells. The barriers have been termed the "blood-cerebrospinal fluid barrier" and the "blood-brain barrier." These terms are loosely defined and are used to refer to more than one concept. Thus, measurements of brain content of a substance shortly after its intravenous administration and, at a later time, in cerebrospinal fluid do not reflect the same aspect of the blood-brain barrier. Nor does measurement of the rate of staining of the brain by an intravenously administered dye indicate the functioning of the same blood-brain barrier as that reflected by the exchange rates of sodium between blood and brain. The usual term "blood-brain barrier" represents a complex system in which not all factors are necessarily interconnected, rather than one real barrier.

The barrier system probably consists of diverse structures utilizing several biochemical mechanisms which limit the penetration of various types of molecules, particularly those bearing a net charge, including electrolytes of extracellular fluid, and all relatively large molecules, including lipids, polysaccharides, and polypeptides. For reasons not understood, barrier permeability is occasionally altered, particularly in persons with brain tumors. This has led to methods for tumor localization. Thus iodofluoroscein does not penetrate the normal blood-brain barrier but does penetrate the blood-brain barrier at the locus of the tumor. Use of dyes containing radioactive ^{131}I permits localization of tumors by means of appropriate detecting devices.

Carbohydrate Metabolism. The energy requirement of the brain under normal conditions is met almost entirely by the oxidation of glucose. Studies with labeled glucose reveal that only a portion of the glucose obtained from the blood by the brain is oxidized directly to CO_2. Approximately one-half of the glucose carbon is incorporated into tissue constituents (see below) which are utilized later as a source of energy.

The glucose removed from the blood by the brain is utilized primarily via the glycolytic pathway. Good correlation obtains between the quantity of glucose removed and the oxygen consumption plus lactic acid produced. The phosphogluconate oxidative pathway also functions in brain; its relative significance is not established. Ultimate conversion of glucose carbon to CO_2 in brain occurs via the tricarboxylic acid cycle, as in other tissues. Brain tissue can also fix CO_2 into amino acids, presumably via oxaloacetate formation as a consequence of the presence of malic enzyme and malic acid dehydrogenase (page 408). This provides a mechanism for replenishing the supply of citric acid intermediates as a continuing source of α-ketoglutaric acid, which in turn serves as a source of glutamic acid for glutamine synthesis (see below).

The central role of the tricarboxylic acid cycle in the metabolism of nervous tissue is reflected in the neurological dysfunctions which ensue on interference with its normal operation. For example, in thiamine deficiency, with accompanying inadequate transformation of pyruvate into acetyl CoA (page 327), neurological signs of thiamine deficiency (Chap. 49) are manifest.

In *hypoglycemia,* glucose consumption by brain is reduced more than is oxygen utilization, indicating oxidation of other substrates. In brain slices and homogenates, no more than 60 per cent of the oxygen consumed is due to oxidation of carbohydrates. Also, in perfusion experiments with ^{14}C-labeled glucose during convulsions, when the total CO_2 production of the brain increased to twice the basal level, the $^{14}CO_2$ production was not similarly elevated. This suggested that the excess CO_2 produced by the brain during hypoglycemic convulsions is due to the oxidation of noncarbohydrate substrates, probably amino acids and lipid. Under these conditions there was a marked increase in turnover of proteins and lipids of brain (see below). However, in view of its availability, glucose is the principal and unique fuel of the brain.

The glycogen content of brain is approximately 0.1 per cent; metabolism of this organ thus cannot be long sustained by its carbohydrate reserves. Brain has the highest known activity of adenyl cyclase and of the specific phosphodiesterase which inactivates the cyclic nucleotide with production of 5′-AMP; the significance of this for this tissue is unknown. The low carbohydrate reserve of brain may contribute to the coma attending insulin-induced hypoglycemia. With the normal arterial blood glucose concentration of about 80 mg. per 100 ml., brain consumes 3.4 ml. of oxygen per 100 g. per min. In insulin coma, at a blood glucose level of about 8 mg. per 100 ml., oxygen consumption may be 1.9 ml. per min. At this reduced level of respiration, the supply of ATP from oxidative phosphorylation is probably inadequate for normal brain function. The large oxygen consumption of brain, about 25 per cent of the total oxygen consumption of the body at rest, accounts for the extreme sensitivity of the brain to anoxia. Coma and irreversible damage occur even after brief hypoxia.

Amino Acid and Protein Metabolism. The free amino acid concentration in the brain is significantly higher than in most tissues and is characterized by its preponderance of glutamic acid, present in 0.01*M* concentration. This amino acid, together with aspartic acid, N-acetylaspartic acid, glutamine, glutathione, and γ-aminobutyric acid, accounts for approximately 80 per cent of the nonprotein amino nitrogen content of brain. These and other nonessential amino acids are produced rapidly by brain is significantly higher than in most tissues and is characterized by its preponderance of glutamic acid, present in 0.01*M* concentration. This amino acid, together with aspartic acid, N-acetylaspartic acid, gultamine, glutathione, and γ-aminobutyric cluding liver. Under these conditions, the specific radioactivity of the free amino acid fraction of the brain, particularly glutamic and aspartic acids, was approximately ten times that of the other organs studied. The data suggest that the brain, with little reserve glycogen in relation to its high metabolic activity, may also utilize to a limited degree amino acids and proteins for energy purposes, particularly when glucose supply and utilization are inadequate. This is in agreement with the high rate of turnover of proteins in brain (see below).

γ-Aminobutyric acid arises from decarboxylation of glutamic acid (page 586) and is an inhibitor of synaptic transmission in the central nervous system. Also present in brain in higher concentrations than elsewhere in the body and of unknown significance are taurine (page 600) and cystathionine (page 547). The concentration of the latter in brain is greatest in man, less in the anthropoid apes, still less in rodents, and extremely small in invertebrates. N-Acetyl-α-L-aspartyl-L-glutamic acid has been isolated from the brain of several species, including man.

Amino acids vary in their capacity to penetrate the blood-brain barrier. Thus, tyrosine penetrates easily. Although no *net* exchange of glutamic acid, lysine, or leucine may be evident between brain and blood, these amino acids added to blood do exchange rapidly with the free amino acids in the brain. Uptake of amino acids by brain is dependent upon an active transport process (page 784). An anomalous distribution pattern of administered labeled amino acids was found in brain, suggesting compartmentalization of various free amino acids within the neurons or between neurons and glia cells.

Amino acids entering brain or arising in brain cells are rapidly incorporated into proteins by pathways of protein synthesis similar to those elsewhere (Chap. 29). Brain tissue proteins show high turnover rates, exceeded only by the albumin of plasma and the total proteins of liver. Thus, approximate half-life times of 5, 8, 10, and 14 days were found for total proteins of, respectively, the white matter, cerebral cortex, cerebellum, and spinal cord. Brain has a high RNA content (page 864), suggesting a correlation with its active protein synthesis. It has been proposed that synthesis of axonal protein, which is frequently quite remote from the cell body, is the result of the combination of axoplasmic flow and local metabolism.

Additional pathways for amino acid metabolism in brain are similar to those described previously (Chaps. 23, 25, and 26) for other tissues, including transamination operating in conjunction with intermediates of the tricarboxylic acid cycle. Of unknown metabolic significance is the evidence that brain contains all the enzymes necessary for operation of the urea cycle (page 560). Brain tissue also has transamidinase (page 581) and carnosine synthetase (page 579) activities. Thus formation of both γ-guanidobutyric acid and homocarnosine (page 579) from γ-aminobutyric acid can occur in brain.

Glutamine is synthesized by brain as in other tissues (page 536). This synthesis, which is augmented with rising blood ammonia concentrations, has led to speculation concerning the cause of coma in terminal hepatic disease. Ammonia, released either by the failing liver or by the intestinal tract and not removed by the liver, is thought to be used for glutamine synthesis by the brain; glutamine is markedly increased in amount in jugular blood under these circumstances. This may constitute a drain on the glutamic acid, and hence on the citric acid cycle intermediates of brain, limiting oxidative metabolism. Respiration declines and coma may then ensue, presumably for the same reason as the coma of hypoglycemia or hypoxia. This hypothesis, however, fails to explain why intermediates of the citric acid cycle are not replenished by CO_2 fixation (page 408). Figure 37.1 (page 864) shows some of the metabolic reactions which have been described.

Serotonin, synthesized in brain and other tissues from tryptophan (page 589),

Serotonin

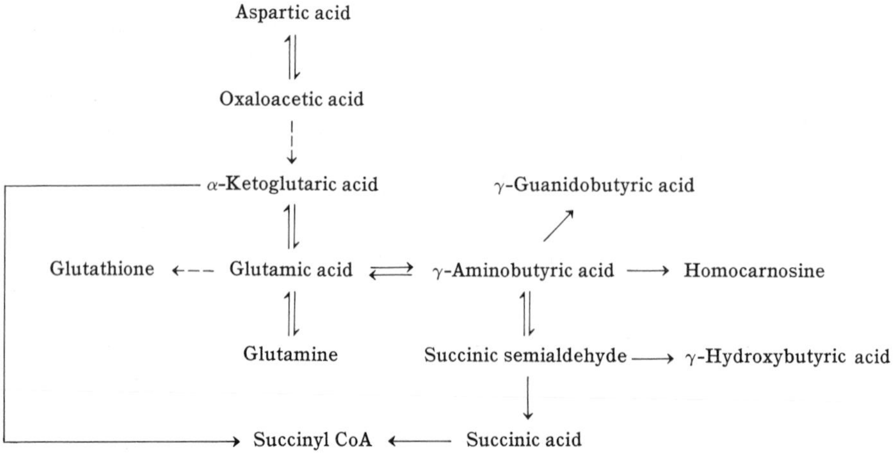

Fig. 37.1. Some aspects of amino acid metabolism in brain.

is a peripheral vasoconstrictor, stimulates uterine contraction, and may play a regulatory role in the nervous system. In the brain the compound is largely present in bound form. Certain drugs which produce marked alteration in central nervous system activity are potent antagonists of serotonin in test systems in vitro. Administration of reserpine, a drug which diminishes the activity of the central nervous system in man, augments urinary excretion of serotonin and its major metabolic product, 5-hydroxyindoleacetic acid (page 589). Monoamine oxidase inhibitors block serotonin metabolism, leading to increased formation of N-acetyl and N-methyl derivatives of serotonin (page 590).

Specialized areas of the brain are characterized by the presence of peptides with hormonal activity (Part Five). Certain of these peptides are produced in the hypothalamus and may be stored in the neurohypophysis; this is the case for vasopressin and oxytocin (Chap. 47). Others may be synthesized and function as mediators of release of adenohypophyseal hormones (Chap. 47).

Nucleic Acid Metabolism. Among the somatic cells of the organism, the large nerve cells are characterized by the highest content of ribonucleic acids and are probably among the most active producers of nucleic acids, comparable in this regard to certain actively secreting cells, *e.g.*, the exocrine cells of the pancreas. The main portion of the RNA of nerve cells seems to be in the form of liponucleoproteins, correlating in content and location with the *Nissl substance,* which is associated with the endoplasmic reticulum. The young unipolar neuroblast, two-thirds of which is nucleus, contains DNA but little RNA. As the cell proceeds through various stages of growth and development, *e.g.*, the multipolar neuroblast, the neuron, and the adult nerve cell, RNA appears and increases in quantity, with no change in the absolute amount of DNA. Concomitantly, with the increase in RNA content of the cell, the concentration of protein also rises. This sequence of events is compatible with knowledge of the interrelations of DNA, RNA, and protein synthesis (Chaps. 28 and 29).

An active nucleic acid metabolism is evident during nerve activity. Stimulation of the spinal ganglion cells of the rabbit for 5 min. produced an increase in cell size, in nucleolus size, and in the RNA concentration in the nucleolus. There was also a marked increase in RNA concentration in the cellular cytoplasm, as well as in the amount of protein in the cytoplasm. No alterations were observed in DNA concentration. When the cells were stimulated for an additional 10 min., they became smaller in volume and exhibited a markedly decreased content of cytoplasmic RNA and protein. The data suggested that a 5-min. stimulation period represented the interval of maximal activity of the cytoplasmic protein- and RNA-forming systems under these experimental conditions, and that further stimulation led to exhaustion of synthetic capacity of the cells. Essentially identical changes were observed in motor nerves of guinea pigs made to run until exhausted in a treadmill and in the cochlear ganglion of guinea pigs subjected to intense sound for several hours.

Lipid Metabolism. *Cholesterol.* Cholesterol synthesis from labeled precursors has been demonstrated in the brain of the young animal during the period of growth. Little cholesterol is synthesized in the adult brain except at sites of myelination. The pathway of cholesterol synthesis is inferred to be similar to that described previously (page 517). In young animals during the early growth period, a significant proportion of the total sterol of brain is desmosterol (page 520). Apparently not all of the brain cholesterol arises *in situ,* since labeled cholesterol injected in the young rabbit appeared in the cholesterol of the brain, as well as in other organs. This cholesterol persisted in the brain for as long as one year, in contrast to its rapid disappearance from other organs studied.

Although most of the cholesterol of adult brain is unesterified, cholesterol esters do occur in relatively high concentrations at sites of active myelination.

Phospholipids. Rapid phospholipid turnover occurs in nervous tissue via pathways similar to those of other organs (Chap. 22).

Fatty Acids. The high total fatty acid content of brain has been noted (page 859). Studies with isotopes reveal a rapid turnover of fatty acids in the brain of younger animals as compared with an appreciably slower rate in older animals. The pathways of formation and degradation of fatty acids in brain are presumed to be similar to those in other tissues (Chap. 21). N-Acetylaspartic acid, present in brain in high concentration (page 862), is thought to be involved in acetyl transport for lipogenesis in the brain, perhaps in a manner similar to that of acetylcarnitine in other tissues (page 482).

Sphingolipids. The unique lipids of brain reflect the presence of enzymic systems with high activity for the metabolism of these lipids. Much of the present knowledge, presented in Chap. 22, of the synthetic pathways for sphingosines, ceramides, glycosphingolipids, cerebrosides, and gangliosides has been obtained with enzymic preparations from brain tissue. This enzymology is also contributing to understanding of some of the lipid storage diseases (page 527). In vivo studies with labeled precursors support the in vitro data obtained with cell-free preparations. Thus, brain incorporated administered glucose, galactose, glucosamine, and serine into its sphingolipids. Both labeled galactose and sulfate were incorporated into the sulfatides of the brain of young rats.

Studies with isotopically labeled precursors indicate a high rate of synthesis in brain of inositol-containing phospholipids and gangliosides. The presence of these complex lipids in close association with proteins in proteolipids (page 860) and their high metabolic turnover have suggested that they play a role in some of the membrane structures of brain, and other tissues, and that their rapid metabolism may reflect membrane activity during function. McIlwain has suggested that gangliosides offer acidic sites in the lipid-rich membranes of cerebral tissue which function in active cation transport.

CONDUCTION AND TRANSMISSION OF THE NERVOUS IMPULSE

Conduction. Conduction in nerves is accomplished along the surface of the axon by an electrical current so weak that it cannot propagate itself unless there is some mechanism for its repeated amplification. A potential of about 75 millivolts exists between the inside of the axon and the surrounding medium. This polarization derives from the fact that the $[K^+]$ inside the cell is fifteen to thirty times that in extracellular fluid, whereas for $[Na^+]$ this relationship is reversed. During activity, a wave of negativity sweeps along the fiber and the polarity is temporarily reversed. Associated with this event, there is a marked drop in electrical impedance, Na^+ enters the fiber about 500 times more readily, and approximately 4×10^{-12} eq. per cm.^{2} per impulse enters the cell during the ascending phase of the action potential, balanced by an equivalent outward passage of K^+ during the descending phase. There follows a refractory period when no impulse may be carried while the Na^+-K^+ relationships are restored, As a depolarized point becomes negative to the adjacent area, a local current ensues and these events are repeated, thus propagating the original impulse.

What is the nature of the alteration responsible for the transient local changes in permeability of the nerve fiber membrane? Evidence has been assembled which indicates that acetylcholine is released at parasympathetic nerve endings and acts as a transmitter on the effector organ, substituting for the electric current as propagating agent. Before another impulse may be transmitted, the acetylcholine is hydrolyzed by *acetylcholine esterase.*

$$CH_3COOCH_2CH_2N(CH_3)_3 + H_2O \longrightarrow CH_3COOH + HOCH_2CH_2\overset{+}{N}(CH_3)_3$$

Crystalline acetylcholine esterase from electric tissue has an estimated molecular weight of approximately 250,000 and contains carbohydrate. The enzyme has two adjacent binding sites for the substrate: one is anionic and binds the cationic quaternary nitrogen of acetyl choline; the other is an esteratic site consisting of a serine residue and another nucleophilic group (imidazole of a histidine residue) for accepting the proton released during the reaction. The anionic site attracts the quaternary nitrogen of the substrate; in the esteratic site the electrophilic carbon of the carbonyl group forms a covalent bond with the oxygen of the serine. From the enzyme-substrate complex, the alcoholic portion, choline, is split off, leaving an acetylated esteratic site. The latter reacts rapidly with water to produce acetic acid and the regenerated enzyme. This series of events is depicted in Fig. 37.2.

Many substances which bind at the anionic site, *e.g.*, simple quaternary compounds, are reversible inhibitors of acetylcholine esterase; they block attachment of the substrate. Other inhibitors, including prostigmine and physostigmine, and

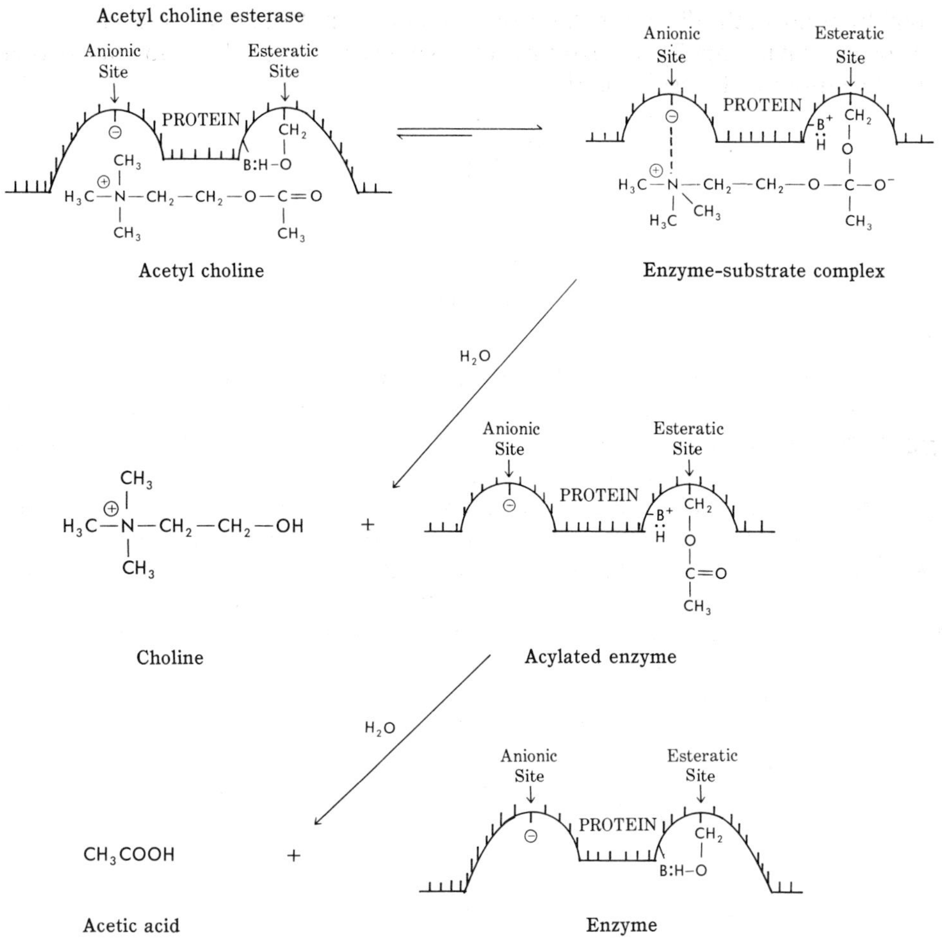

Fig. 37.2. Schematic representation of the mode of action of acetylcholine esterase. At the esteratic site, the oxygen atom of a serine residue (—CH_2—OH) is depicted in close proximity to another nucleophilic group, B, an imidazole group of a histidine residue.

related compounds that possess a carbamyl ester linkage or urethane structure, combine with the enzyme at both the anionic and the esteratic sites. Splitting off of choline leaves a carbamylated enzyme that is hydrolyzed very slowly. Consequently, this inhibition is only slowly reversible. Other inhibitors, such as diisopropylphosphofluoridate, form the stable diisopropyl phosphate ester of the serine residue at the esteratic site; the resultant ester is very stable and the enzyme is irreversibly inhibited. Various other alkyl phosphates and phosphonates react similarly; this has found application in insecticides and so-called "nerve gases."

These considerations of the mode of action of acetylcholine esterase led to formulation of the necessary criteria for a compound which might overcome such inhibition, *viz.*, a strongly cationic group at a sufficient intramolecular distance from a nucleophilic group. Several compounds were prepared from these specifications; of these, 2-pyridine aldoxime methiodide is most effective. At low concentration it rapidly removes the diisopropylphosphate group from the esteratic site and restores enzymic activity rapidly. In vivo this compound is a successful antidote for poisoning by diisopropylphosphofluoridate.

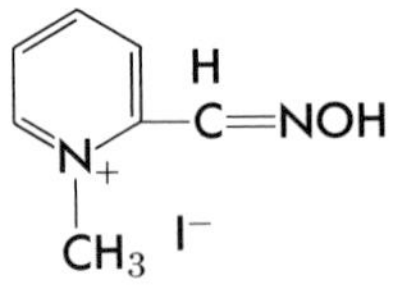

2-Pyridine aldoxime methiodide

Mechanism of Conduction. The hypothesis that acetylcholine is also involved in intracellular axonal conduction and participates in the generation of electric potentials is based on evidence that may be summarized as follows: (1) Acetylcholine and acetylcholine esterase are present in all conducting nerves. Both substrate and enzyme are localized near the surface of the axon. (2) All nerve fibers also contain *choline acetylase* which catalyzes synthesis of acetylcholine from choline and acetyl CoA. (3) Each mole of choline esterase can catalyze hydrolysis of 2×10^7 moles of substrate per minute; the known concentration of substrate and enzyme would permit complete hydrolysis in less than 100 μsec, well within the time required by the known rate of events in nervous activity. (4) Choline esterase is concentrated in the plates of the electric organs of *Torpedo* and *Electrophorus electricus.* These are the most powerful known bioelectric generators. A direct proportionality exists between enzyme concentration and voltage. (5) Neuronal conduction is abolished by inhibitors of choline esterase in *all* types of nerves. Conduction fails when enzymic activity falls to about 20 per cent of its initial level. These observations have led to the formulation by Nachmansohn of a tentative concept of the events in neuronal conduction, depicted in Fig. 37.3.

Nerve is rich in an ATPase system requiring both Na^+ and K^+ which is presumed to play a role in the cationic exchange of these cells, as in others, *e.g.*, erythrocytes (page 736).

The above picture of axonal conduction, although still incomplete and containing elements of conjecture, is in accord with a number of facts. However, there is lack of agreement with the hypothesis that acetylcholine metabolism is basic to the events underlying axonal conduction. Although acetylcholine does act on the motor endplate and on the ganglionic synapse, serious question has been raised whether it is essential for nerve conduction. A large proportion of the cells and fibers in the nervous system contain neither acetylcholine nor choline esterase. Also, these components are sequestered from the nerve membrane in the vesicular structures (see below) in the axoplasm at the nerve endings. At the terminals of such fibers, *e.g.*, the motor neurons innervating skeletal muscles, these "packets" of acetylcholine are released, not produced, by the nerve impulse. Most significant, perhaps, is the observation that axons continue to conduct impulses in the presence of drugs which strongly inhibit the synthesis of acetylcholine and after injection into the axoplasm of inhibitors of choline acetylase.

Transmission. The over-all process of synaptic and neuromuscular transmission

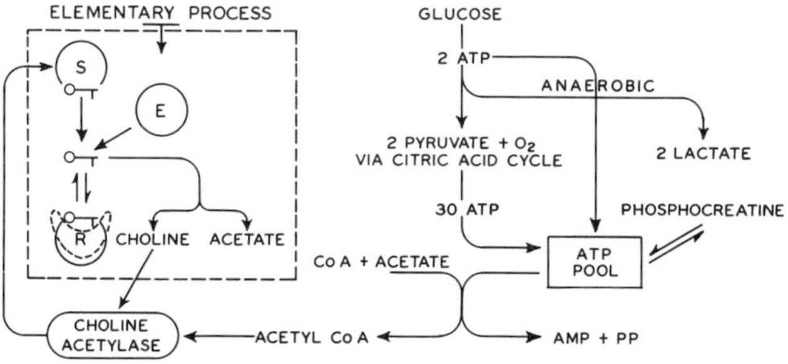

Fig. 37.3. Sequence of energy transformations associated with conduction, and integration of the acetylcholine system into the metabolic pathways of the nerve cell. The elementary process of conduction may be tentatively pictured as follows:

1. In resting condition acetylcholine ($\odot_T$) is bound, presumably to a storage protein (*S*). The membrane is polarized.

2. Acetylcholine is released by current flow (possibly H^+ movements) or any other excitatory agent. The free ester combines with the receptor (*R*), presumably a protein.

3. The receptor changes its configuration (broken line). This process increases the Na^+ permeability and permits its rapid influx. This is the trigger action by which the potential primary source of electomotive force, the ionic concentration gradient, becomes effective and by which the action current is generated.

4. The ester-receptor complex is in dynamic equilibrium with the free ester and the receptor; the free ester can be attacked by acetylcholine esterase (*E*).

5. Hydrolysis of the ester permits the receptor to return to its original shape. The permeability decreases, and the membrane is again in its original polarized condition. (*After D. Nachmansohn, Harvey Lectures,* **49,** 57, 1953–1954.)

can be broadly delineated. Acetylcholine, synthesized in the reaction catalyzed by *choline acetylase,* is stored in membrane-limited vesicles, termed *synaptic vesicles* because of their presence in large numbers at the synapse. It is not certain whether the synthesis of acetylcholine occurs in the axoplasmic fluid, with transfer to synaptic vesicles for storage, or whether the synaptic vesicles are the site of synthesis. The rupture of a vesicle liberates about 5×10^6 acetylcholine molecules and is reflected in the form of miniature end-plate potentials. The arrival of a nerve impulse at the synapse results in the coordinated rupture of many vesicles. The liberated acetylcholine diffuses across the synaptic space to the receptors located in the postsynaptic membrane. Acetylcholine esterase, in the subsynaptic web, is responsible for inactivation of the acetylcholine by hydrolysis. It is not clear to what extent the acetate and choline produced are available for resynthesis of acetylcholine.

Nerve impulses originating in the hypothalamus elicit the release of acetylcholine from the vesicles, thus regulating the liberation of neurosecretory hormones at the nerve endings (Chap. 47). The hypothalamic-neurohypophyseal system of nerve fibers thus liberates both acetylcholine and endocrine secretions, the latter being under cholinergic control.

Although acetylcholine may mediate neuronal and synaptic transmission in all nerves as well as transmit the neural stimulus to the effector organs of somatic motor fibers and parasympathetic nerves, it does not appear to be secreted by postganglionic sympathetic fibers. These secrete instead either epinephrine or norepinephrine (Chap. 45).

Some evidence suggests that γ-aminobutyrate participates in the process of transmission by inhibitory neurones. Thus both deficiency in pyridoxine and administration of isonicotinylhydrazide (Chap. 49), which decrease the level of glutamic acid decarboxylase activity in brain, result in convulsive activity which is relieved by administration of γ-aminobutyrate.

REFERENCES

Books

Caspersson, T. O., "Cell Growth and Cell Function," W. W. Norton & Company, Inc., New York, 1950.

Chagas, C., and De Carvalho, A. P., eds., "Bioelectrogenesis," Elsevier Publishing Company, Amsterdam, distributed by American Elsevier Publishing Company, New York, 1961.

Cohen, M. M., and Snider, R. S., eds. "Morphological and Biochemical Correlates of Neural Activity," Hoeber Medical Division, Harper & Row, Publishers, Incorporated, 1964.

DeRobertis, E. D. P., "Histophysiology of Synapses and Neurosecretion," The Macmillan Company, New York, 1964.

Elliott, K. A. C., Page, I. H., and Quastel, J. H., "Neurochemistry," 2d ed., Charles C Thomas, Publisher, Springfield, Ill., 1962.

Friede, R. L., "Topographic Brain Chemistry," Academic Press, Inc., New York, 1966.

Himwich, H. E., "Brain Metabolism and Cerebral Disorders," The Williams & Wilkins Company, Baltimore, 1951.

Hodgkin, A. L., "The Conduction of the Nervous Impulse," Charles C Thomas, Publisher, Springfield, Ill., 1964.

McIlwain, E., "Biochemistry of the Central Nervous System," 2d ed., J. & A. Churchill, Ltd., London, 1959.

Nachmansohn, D., "Chemical and Molecular Basis of Nerve Activity," Academic Press, Inc., New York, 1959.

Richter, D., ed., "Metabolism of the Nervous System," Pergamon Press, New York, 1957.

Van Harrefeld, A., "Brain Tissue Electrolytes," Butterworth, Inc., Washington, 1966.

Review Articles

Davison, A. N., Brain Sterol Metabolism, *Advances in Lipid Research,* **3,** 171–240, 1965.

Hawthorne, J. N., and Kemp, P., The Brain Phosphoinositides, *Advances in Lipid Research,* **2,** 127–166, 1964.

Hebb, C. O., Biochemical Evidence for Neural Function of Acetylcholine, *Physiol. Revs.,* **37,** 196–220, 1957.

Hydén, H., The Neuron, in "The Cell," J. Brachet and A. E. Mirsky, eds., vol. IV, pp. 216–323, Academic Press, Inc., New York, 1960.

Hydén, H., The Neuron and Its Glia—A Biochemical and Functional Unit, *Endeavour,* **21,** 144–155, 1962.

Martin, A. R., and Veale, J. L., The Nervous System at the Cellular Level, *Ann. Rev. Physiol.,* **29,** 401–426, 1967.

Nachmansohn, D., Chemical Control of Ion Movements across Conducting Membranes, *New Perspectives in Biol.,* **4,** 176–204, 1964.

Nachmansohn, D., Chemical Activity of the Permeability Cycle in Excitable Membranes during Electrical Activity, *Ann. N.Y. Acad. Sci.,* **137,** 877–900, 1966.

Waelsch, H., and Lajtha, A., Protein Metabolism in the Nervous System, *Physiol. Revs.,* **41,** 709–736, 1961.

38. Connective Tissue

Connective tissue is distributed throughout the body in cartilage, tendons, ligaments, the matrix of bone, the pelvis of the kidney, the ureters, and urethra; it underlies the skin, serves as binding for the blood vessels, and provides the intercellular binding substance in parenchymatous organs such as the liver and muscles. The mechanical and supportive functions of connective tissue are accomplished by extracellular, insoluble protein fibers embedded in a matrix termed the *ground substance.* The cells of connective tissue responsible for synthesis of both the insoluble fibers and the soluble matrix include not only fibroblasts, but macrophages, mast cells, and lesser numbers of other, sometimes undifferentiated, cell types.

COLLAGEN

The insoluble fiber of connective tissue is usually collagen, the most abundant protein in the body, comprising 25 to 33 per cent of the total protein and, therefore, about 6 per cent of the body weight. The amino acid composition of collagen (Table 38.1, page 872) is remarkable in that one-third of the amino acid residues are glycine; proline plus 3- and 4-hydroxyproline provide 21 per cent of the residues, and alanine another 11 per cent. Collagens are among the few proteins known to contain hydroxyprolines and δ-hydroxylysine.

In the native state, collagen fibers are of high tensile strength and swell in acidic or alkaline media. As seen with the electron microscope, collagen fibers are built of smaller fibrils which are 200 to 2,500 Å. wide and many microns long, with characteristic cross striations at 640 to 700 Å. (Fig. 38.1, page 873). Extremely thick fibers are characteristic of tendon, whereas the narrow fibrils found in cornea and the vitreous body allow minimal scatter and maximal passage of incident light. Mature collagen from most sources is essentially insoluble; however, extraction of skin of very young animals with cold salt solution or prolonged extraction of insoluble collagen with dilute acid yields a solution of the fundamental units of collagen fibrils, termed *tropocollagen,* individual molecules of which are about 15 Å. wide and 3,000 Å. long, with a molecular weight of about 300,000. These are among the most asymmetric molecules yet obtained from natural sources. Each molecule is a cable of three polypeptide chains; each chain is in a tight left-handed helix, and the triple-stranded cable, thus formed, is then twisted slightly to the right. The pitch of the helix is determined by the limited flexibility of sequences containing proline and hydroxyproline residues. The tight triple-stranded structure is possible only because of the high incidence of glycine, since in this structure, every third amino acid side chain, in each strand, is essentially within the interior of the molecule

Table 38.1: Amino Acid Composition of Collagen, Elastin, and Enamel Protein

(Residues per 1,000 total residues)

Amino acid	Human skin collagen			Elastin*		Pig embryo enamel protein
	Total	α_1	α_2	Newborn	Mature	
Glycine	330	333	337	351	352	49
Proline	128	135	120	124	124	271
3-Hydroxyproline	1	1	1	...	...	0
4-Hydroxyproline	93	91	82	22	23	0
Alanine	110	115	105	180	177	24
Aspartic acid	45	43	47	2	2	29
Glutamic acid	73	77	68	12	12	185
Serine	36	37	35	3	4	46
Threonine	18	17	19	4	5	37
Valine	24	21	33	177	174	37
Methionine	6	5	5	0	0	47
Isoleucine	10	7	14	19	20	32
Leucine	24	20	30	56	58	94
Lysine	27	30	22	4	2	11
Hydroxylysine	6	4	8	0	0	2
Histidine	5	3	10	0	0	72
Arginine	51	50	51	6	5	6
Phenylalanine	12	12	12	22	22	26
Tyrosine	3	2	5	12	12	22
Cystine/2	0	0	0	0	0	0
Desmosine and isodesmosine	...	...	...	7	11	

* Elastin preparations from aortas of chickens. Newborn, twentieth day of incubation; mature, one year old.

and only the minimal volume of glycine can fit into this structure. Accordingly, whereas there are distinct differences among the amino acid compositions of collagens from diverse sources, glycine is usually present as one-third of the total residues. The triple helix is maintained by pairs of hydrogen bonds between the parallel peptide bonds except for those which involve proline or hydroxyproline. However, the helix form is almost identical with one type of structure spontaneously assumed by synthetic single-stranded polyproline. Like the DNA helix, this structure may be "melted out"; T_m, the temperature of the helix-coil transition, rises with increasing proline and hydroxyproline content.

Heat denaturation of collagen yields the soluble preparation called "gelatin." Gelatin solutions contain varying amounts of smaller molecular species. The smallest of these, α_1 and α_2, are separable by chromatography; they are of somewhat different amino acid composition (Table 38.1) but of approximately equal molecular weight and are present in the ratio 2:1. The larger molecules, β units, are dimers of the several α chains. As shown in Fig. 38.2 (page 874), the tropocollagen molecule is built of three strands, two of α_1 and one of α_2. However, recent evidence suggests that the two α_1 strands are actually similar but distinct; if so, the cable is composed of three different α filaments, and the dimers may be of six possible types, $\beta_{1,1}$, $\beta_{1,2}$, $\beta_{1,3}$, $\beta_{2,2}$, $\beta_{2,3}$, $\beta_{3,3}$.

The β units result from formation of covalent cross-links between the primary

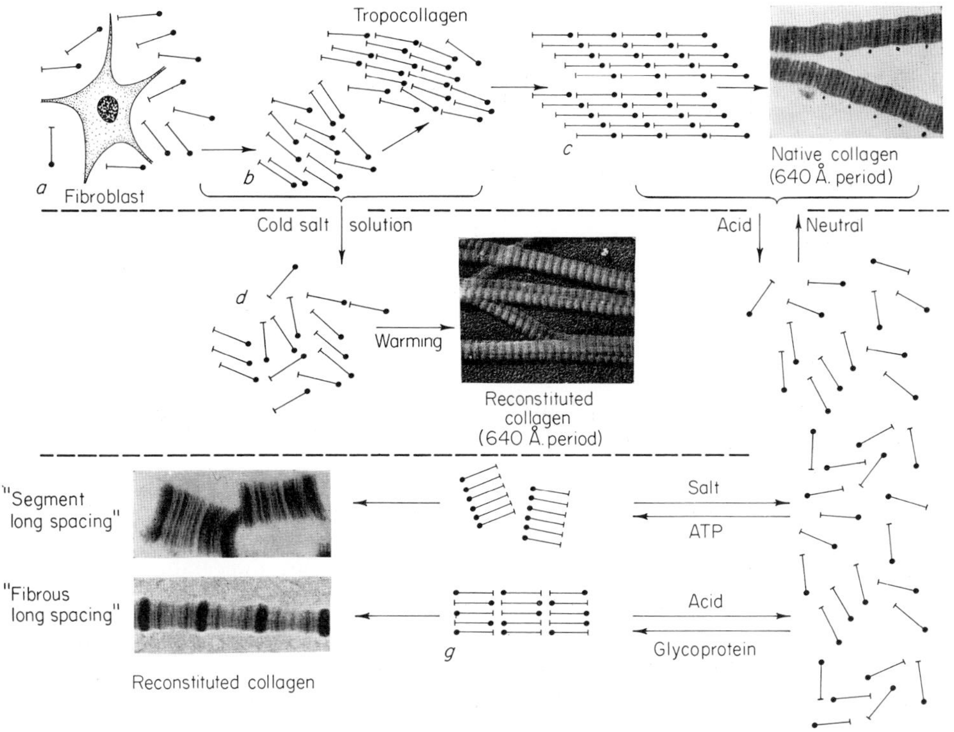

Fig. 38.1. Aggregation of tropocollagen to form collagen fibers. Tropocollagen (*b*), synthesized and released by fibroblasts (*a*), arranges itself in the surrounding medium in parallel linear arrays in which each N-terminus is displaced by about one-quarter the length of the molecule from that on either side ("quarter-staggered") (*c*). The resulting fibers show the characteristic 640–700 Å. spacing. If soluble, young collagen is dissolved, it deaggregates (*d*), and will realign normally if warmed. In the presence of ATP, reaggregation occurs with the tropocollagen units in perfect register; in the presence of protein-polysaccharides, the parallel strands may be in register but oriented randomly (*g*). The quarter-staggered array is the normal arrangement in all native collagens. (*After J. Gross, Scientific American,* **204,** (May) 120, 1961.)

strands. The nature of these bonds is not established, but several kinds of structures have been implicated. (1) Esters between aspartic and glutamic residues and hydroxyl groups of unidentified residues. (2) A bond of unknown structure which forms after oxidation of the ε-amino groups of lysine residues, close to the *N*-terminus, to α-aminoadipic acid semialdehyde. It is not clear whether this aldehyde forms Schiff bases with amino groups or interacts in aldol condensations (see below). (3) Along each chain there is a single hydroxylysine residue to which is linked, in O-glycosidic linkage, a disaccharide of glucose and galactose. It is not known whether this unit participates further in cross-linking. (4) There is evidence that collagen contains amino aldehydes corresponding to glycine, lysine, alanine, and aspartic acid. This seems possible only if these represent the C-termini of peptide chains, permitting reduction of their α-carboxyl functions. Some of the lysinal (amino aldehyde corresponding to lysine) may be deaminated to yield ε-amino, α-β dehydrohexenal (called *enosaline*), which may also participate in cross-linking.

Collagen fibrils are built by end-to-end and side-to-side joining of tropocollagen

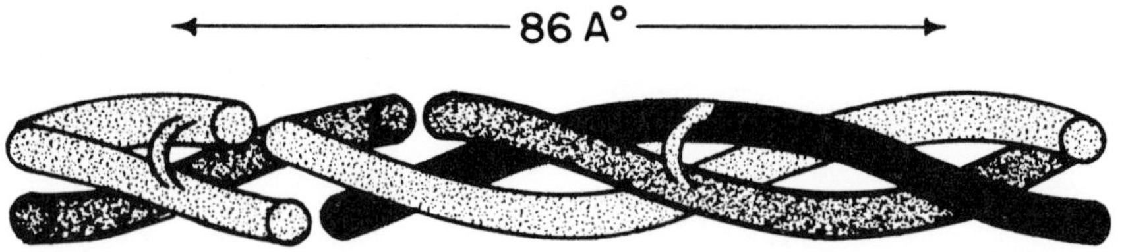

Fig. 38.2. A segment of collagen structure showing end-to-end linking of tropocollagen molecules. The tubes represent the space filled by polypeptide chains, each of which extends about 3,000 Å. The amino acid residues are 2.86 Å. apart and lie along a minor helix with a pitch of 8.6 Å. (three residues per turn). Each chain is coiled into a major helix with a pitch of 86 Å. The larger amino acid side chains extend beyond the structure shown. The U-shaped bond represents covalent cross-linking between individual strands. Each of the three chains is represented as being different.

molecules. The end-to-end "splicing" of the cables is possible because the three strands are of unequal length, as shown in Fig. 38.2. The end regions of each strand are rich in polar amino acids, *e.g.*, glutamic acid, tyrosine, and lysine, so that hydrogen and electrostatic bonds are responsible for the end-to-end alignment which results in fibrils many times longer than the tropocollagen molecule. The forces which bond adjacent triple helices to each other are uncertain, but covalent bonding is unnecessary since in dilute salt solution tropocollagen recombines to yield characteristic collagen fibrils when the solution is warmed to 40°C. (Fig. 38.1). However, covalent cross linkage to adjacent tropocollagen units occurs with time; collagen becomes increasingly difficult to dissolve and the fibrils thicken progressively as an animal ages.

Intermolecular cross-linking between adjacent triple helices to form a fibril appears to involve the same types of bonding cited above, with the aldehydes derived from lysine prominently involved. The parallel tropocollagen fibril chains must be aligned in a quarter-staggered array (Fig. 38.1) to account for the repeating structure with a spacing of about 640 Å. Formation of such cross links is prevented by thiosemicarbazide or penicillamine which disaggregate fibrils without disturbing the triple helix. Ingestion of *Lathyrus odoratus* peas has long been known to result in *lathyrism, i.e.,* in defective skeletal development with urinary excretion of unusual amounts of hydroxyproline-containing peptides. The active principle in peas is β-aminopropionitrile, $H_2N—CH_2—CH_2—CN$, which is an inhibitor of the copper-containing amine oxidase of plasma and is presumed to inhibit a similar connective tissue enzyme so that oxidation of lysyl amino groups to an aldehyde function does not occur. This may account for the failure of cross-linking between tropocollagen molecules.

Portions of the N-termini of the α chains differ from the bulk of the β chains in that they are slowly sensitive to hydrolysis by several proteases. The lysine residues responsible for intermolecular linkage are located in this area. A *collagenase* derived from the reabsorbing tail of metamorphosing tadpoles appears to cleave the α chains of tropocollagen at specific sensitive loci, resulting in fragments of a size about 200,000 and 70,000, respectively, plus some smaller peptides. Both fragments retain the triple helical conformation.

The collagen of *Ascaris* cuticle differs markedly from mammalian collagen. The amino acid composition is rather similar, but instead of a triple-stranded cable, each polypeptide chain appears to have a pair of hairpin turns so that each "α chain" is itself a triple helix, and the chain is extended by pairs of disulfide bridges between individual units.

Collagen Synthesis. Collagen formation by fibroblasts in tissue culture establishes these cells as the source of collagen. When administered, neither hydroxyproline-^{14}C nor hydroxylysine-^{14}C is incorporated into collagen, whereas administration of proline-^{14}C or lysine-^{14}C results in appearance of the respective ^{14}C-labeled hydroxy amino acids in newly formed collagen. Formation of collagen and its hydroxylated amino acids has been studied with preparations of fetal skin and induced granulomas. Appearance of hydroxyproline and hydroxylysine in the synthesized polypeptide is dependent upon the presence of ascorbic acid, Fe^{++}, and α-ketoglutarate. In their absence, there accumulates a hydroxyproline-, hydroxylysine-poor collagenous material. Available evidence suggests that ribosomal-bound procollagen is the substrate for the hydroxylation process. By using tritium-labeled proline, it was shown that only the hydrogen displaced by the hydroxyl group is exchanged with the medium during hydroxylation. Hence, the mechanism must involve hydroxylation, not consecutive dehydrogenation and hydration. Hydroxylation occurs most frequently on a proline which is one residue removed from a glycine; hence the most frequent sequences in the α chains are -Gly-Pro-Hyp- and -Gly-X-Hyp.

Once formed, the collagen of most tissues appears to be metabolically inert. If turnover occurs at all, it is extremely slow; $t_{1/2}$ is measured in years. Collagenase activity has been observed only in a few mammalian tissues, such as resorbing uterus. For experimental purposes, use has been made of the collagenase of *Clostridium histolyticum,* an enzyme which, presumably, facilitates invasion of animal tissues by this organism. Clostridial collagenase hydrolyzes bonds between glycine and proline in peptide sequences of the general structure -Pro-X-Gly-Pro-Y-; the residues at positions X and Y affect the enzymic rate but are not otherwise significant. Such sequences occur frequently in collagen. Collagenase does not attack other proteins such as casein, hemoglobin, or fibrin. It is clear that collagen can be degraded, *in situ.* Thus, remodeling of the shape of growing bone entails removal of collagen from one site while it is synthesized elsewhere. Collagen also disappears from the bone of an immobilized arm or leg. During pregnancy, growth and thickening of the uterus are accompanied by significant increases in collagen, which disappears within a few days after parturition.

ELASTIN

Elastin is the second major protein of connective tissue. Unlike collagen, it is not converted into gelatin by boiling, and, as shown in Table 38.1, its amino acid composition differs significantly from that of collagen. Collagen is the principal protein of white connective tissue; elastin predominates in yellow tissue. Thus, the collagen content of human Achilles tendon is about twenty times the elastin content, while in the yellow ligamentum nuchae the elastin content is about five times that of collagen. As the name implies, elastin is the predominant connective tissue protein of elastic structures such as the walls of the large blood vessels. Scar formation, on the other hand, is due to deposition of collagen.

Elastin is prepared by subjecting connective tissue either to mild alkaline hydrolysis or to heat; either process dissolves collagen and the mucopolysaccharides, leaving the extremely insoluble elastin. The elastin fibers thus obtained exhibit the properties of an elastomer, *e.g.*, rubber, and are quite yellow. Elastin may be solubilized by prolonged heating in 0.25*M* oxalic acid, but the resultant product has been degraded. The native fibers are built by cross-linking of repeating units of smaller compact, almost spherical molecules in essentially fibrous strands maintained by rigid cross-linking. Two forms of cross-linkage have been identified, both involving lysine. Four lysine residues are required to form the compounds *desmosine* and *isodesmosine.* Spatial considerations dictate that these must derive from no less than two independent strands. In the mode of formation postulated for these compounds (Fig. 38.3), three lysine residues must first be oxidized to the corresponding ε-aldehydes and then condense with a fourth lysyl; the mode of combination determines whether desmosine or isodesmosine is the product.

The participation of desmosine and isodesmosine in the cross-linkages of elastin is reflected in several types of evidence. The data in Table 38.1 reveal that in comparisons of the desmosine plus isodesmosine contents of elastins from aortas of

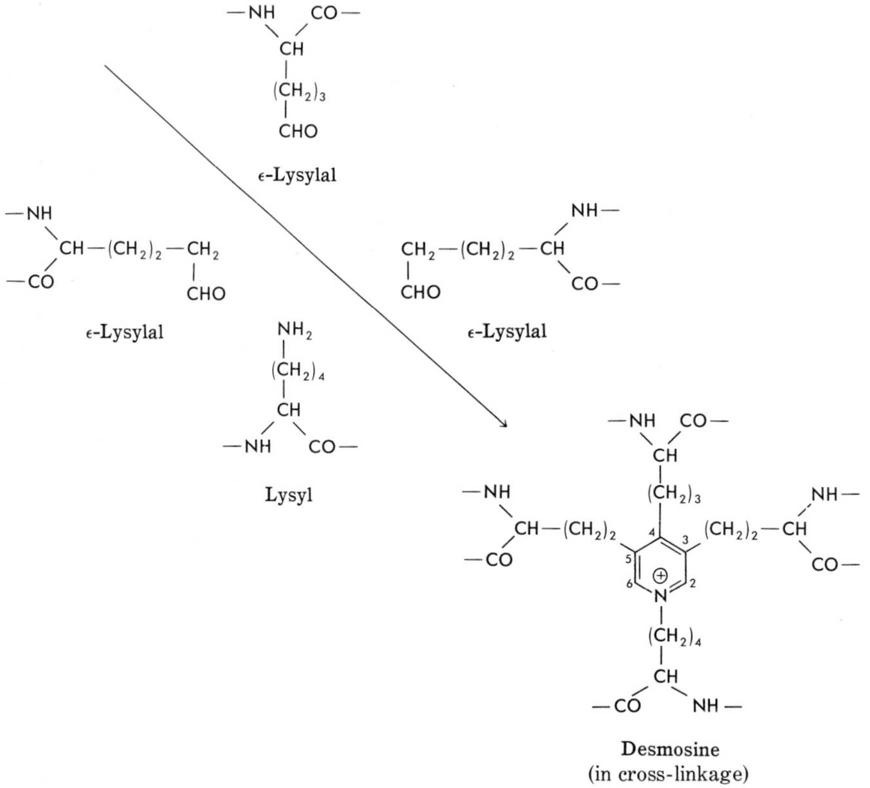

Fig. 38.3. Postulated mode of formation of desmosine from four lysine residues, three of which have previously been oxidized to the corresponding ε-aldehydes. Isodesmosine is formed similarly, with the side chains in positions 1, 2, 3, and 5 of the ring.

newborn and older chickens, the number of residues of these compounds increases with age whereas the lysine content decreases. Also, procedures which significantly diminish amine oxidase activity in growing animals, *e.g.*, pyridoxine and copper deficiencies, result in the deposition of defective elastin, particularly in the aorta, with markedly diminished tensile strength of these tissues, while the elastin exhibits a high lysine content and little or no desmosine + isodesmosine. In lathyrism (page 874) produced by β-aminopropionitrile, the latter prevents both desmosine formation and the cross-links in collagen which involve lysine aldehyde.

A second type of cross-linking in elastin is thought to involve "lysinonorleucine," the structure of which suggests that it is formed by reaction of an ϵ-amino group of one lysine residue to form a Schiff base with the aldehyde formed from oxidation of a second lysyl, with subsequent reduction of the Schiff base.

$$-OC-\underset{\underset{-NH}{|}}{CH}-(CH_2)_3-CH_2-NH-CH_2-(CH_2)_3-\underset{\underset{HN-}{|}}{CH}-CO-$$

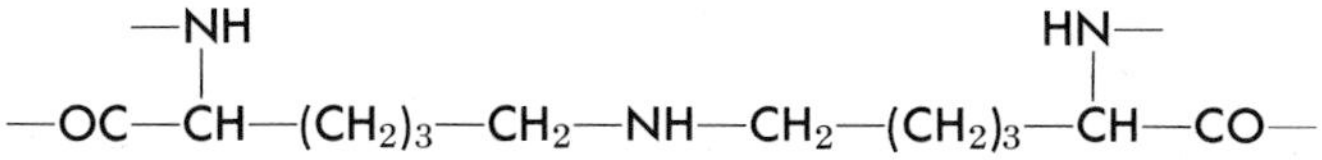

Lysinonorleucine residue

Native elastin fibers, even when stripped of mucopolysaccharide, are not digestible by trypsin or chymotrypsin but can be hydrolyzed slowly by pepsin at pH 2. However, pancreas secretes a zymogen, *proelastase* (page 533), which is activated by trypsin and by enterokinase to yield *elastase*. The latter is a protease of broad specificity, capable of hydrolyzing many proteins, but unusual in its ability to effect the hydrolysis of elastin. This proceeds after a pronounced lag phase, following which amino acids and peptides appear in solution, the fibers disappear, and a mixture of yellow, cross-linked peptides remains. Elastase is inhibited by diisopropylphosphofluoridate, and has a structure homologous with that of other pancreatic proteinases, including the amino acid sequence at the active site (page 255).

There is present an elastic protein, *resilin*, in the cuticle of insects. Resilin is present in the exoskeleton, particularly in all winged insects at the wing hinge. This structure contains 86 per cent resilin and 14 per cent chitin (page 51) plus a fibrous protein in the desert locust. Resilin is totally insoluble in the usual solvents as well as in solutions of urea or guanidine salts even at temperatures of 140°. Thus, the protein is completely in random coil form. The protein is covalently cross-linked, presumably through tyrosine residues which possess carbon to carbon bonds at positions ortho to the phenolic groups, as in the dityrosine structure below.

$$R = -CH_2-\underset{\underset{NH_2}{|}}{CH}-COOH$$

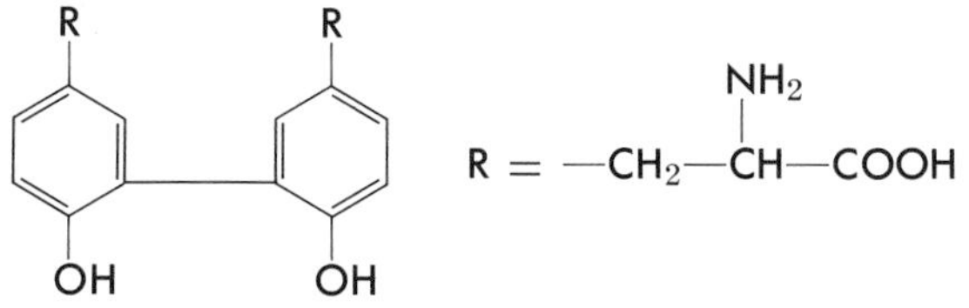

More complex structures involving tyrosine may also be present. The dityrosine structure is reminiscent of the thymine dimers in DNA formed by exposure to ultraviolet light (page 681).

Resilin resembles elastin in its rubber-like properties and in lacking secondary structure but differs in the type of cross-linking, desmosine and isodesmosine (page 876) being absent from resilin. Resilin is rich in glycine and alanine, these amino acids accounting for about half the residues, is poor in lysine and histidine, and lacks hydroxyproline and sulfur-containing residues.

MUCOPOLYSACCHARIDES

Cells of mesenchymal origin synthesize and add to their environment a variety of heteropolysaccharides. The general structures of these polymers, which were elucidated principally by Karl Meyer and his collaborators, have been presented (page 51). Table 38.2 summarizes the distribution of these polysaccharides among animal tissues. These data are not comprehensive. Of tissues thus far examined, the vitreous body alone has been found to contain only one of the seven polysaccharides (hyaluronic acid). All other sources contain mixtures of various mucopolysaccharides. The relative amounts of the components of these mixtures vary from species to species, from individual to individual, and with age. The sulfated polysaccharides vary in the extent of sulfation; a given sample may contain less sulfate than that stoichiometric with the amount expected from the structure of its characteristic repeating unit, or, as in the case of heparin and heparitin, may be excessively sulfated. Heparitin sulfate differs from heparin in that a variable fraction of the amino groups bear an N-acetyl residue instead of a sulfate (page 54). In all members of this series the 2-amino group is either acetylated or present as a sulfamic derivative; hence, it is always neutral. Some samples of keratosulfate from cornea may con-

Table 38.2: Distribution of Mucopolysaccharides

Tissue	Hyaluronic acid	Chondroitin	Chondroitin sulfate			Keratosulfate	Heparin
			A	B	C		
Skin	+	...	...	+			...
Cartilage	...	...	+	...	+	+	...
Tendon	...	...	...	+	+	...	...
Ligaments	...	...	...	+	...	...	...
Umbilical cord	+	...	...	...	+	...	...
Aqueous humor	+	...	...	...	...	...	...
Synovial fluid	+	...	...	...	...	...	...
Heart valves	+	...	...	+	...	...	...
Spinal disks	...	...	...	...	+	+	...
Bone	...	...	+	...	...	+	...
Cornea	...	+	+	...	...	+	...
Liver	...	...	...	...	...	...	+
Lung	...	...	...	...	...	...	+
Arterial wall	...	...	...	...	...	...	+
Embryonic cartilage	...	+	+	...	+	...	...
Mast cells	...	...	...	...	...	...	+

Note: + denotes positive identification of the indicated polysaccharide in the tissue shown. Absence of + need not necessarily mean complete absence of polysaccharides other than those indicated, but does signify that these are present in very small concentration, if at all.

tain galactose 6-sulfate; in addition there may be present, in unknown linkage, lesser amounts of a sialic acid and an unidentified methylpentose.

Only in the case of the hyaluronic acid of the vitreous body and synovial fluids is the polysaccharide present free in solution, unassociated with a protein. All other mucopolysaccharides are found in combination with proteins, and are termed *proteinpolysaccharides.*

The nature of these proteins, the mode of attachment of the mucopolysaccharide to the protein, and the role of the protein in the synthetic process are all subjects of intensive current investigation. The most completely characterized species is the proteinpolysaccharide which contains chondroitin sulfate A (chondroitin 4-sulfate). The minimal peptide chain, obtained from porcine costal cartilage, has a molecular weight of 10,500 (85 residues) with five or six bound polysaccharide chains. From other sources, however, proteins as large as 900,000 have been reported, binding as many as 60 polysaccharide chains. The latter vary from units of 13,500, *viz.*, about 20 repeating disaccharides, to three times as large. Thus these preparations are polydisperse and may display molecular weights of from 250,000 to 4,000,000. The heterogeneity, however, results from variation in the protein core and its binding capacity for carbohydrate chains, rather than from a huge disparity in polysaccharide chain length; this contrasts, for example, with the behavior of glycogen (page 48). Indeed, the higher values may all be somewhat artifactual and reflect isolation of polymers of a fundamental unit of five or six polysaccharide chains attached to the polypeptide of lowest molecular weight.

Each polysaccharide chain is covalently bound to the peptide backbone, in O-glycosidic linkage to a seryl hydroxyl, as follows:

—glucuronic (1 → 3) N-acetylgalactosamine 6-sulfate (1 → 4) glucuronic
(1 → 3) galactose (1 → 3) galactose (1 → 4) xylose (1 → 3) serine—

Thus at the potentially reducing end of a chain of repeating disaccharides (glucuronic acid-N-acetylgalactosamine 6-sulfate) there occurs the trisaccharide, galactosylgalactosylxylose. The carbohydrate chains are similarly bound to the protein of heparin. Threonyl as well as seryl hydroxyl groups may be so utilized in keratosulfate and submaxillary "mucoid." The nature of the polypeptide chains and the mode of polysaccharide attachment remain to be established for the other members of this group. Little is known about dermatan sulfate because this polysaccharide is always bound to a highly insoluble core, perhaps in covalent linkage to highly cross-linked, insoluble collagen.

Mucus is the secretion of mucous glands located in epithelial structures. No *specific* material is present, and mucous secretions contain a mixture of proteinpolysaccharides and *mucoproteins.* Both classes are proteins to which large amounts of carbohydrate are bound. In the former, a few relatively large carbohydrate chains are bound to each polypeptide strand; mucoproteins are polypeptide chains to which a rather large number of oligosaccharides are bound. In addition, there are undoubtedly varying amounts of the chondroitin sulfates and related materials, as well as neutral polysaccharides resembling, or identical with, some of the blood group substances (page 736). Table 38.3 lists certain properties of some proteinpolysaccharides.

Table 38.3: Properties of Some Proteinpolysaccharides

Polysaccharide	Molecular weight	No. of chains	Mol. wt. of polysaccharide chains	Charge type	Charge conformation	Constituents*	Biological source
Hyaluronic acid.	4×10^5 $2\text{–}10 \times 10^6$	1 1	4×10^5 $2\text{–}10 \times 10^6$	$—COO^-$	Equatorial	Glucuronic acid; N-Ac glucosamine	Vitreous body Synovial fluid
Chondroitin	. . .	. . .	. . .	$—COO^-$	Equatorial	Glucuronic acid; N-Ac galactosamine	Cornea
Chondroitin sulfate A	2.5×10^5 4×10^6	10–15 60	13,500	$—COO^-$ $—OSO_3^-$	Equatorial Axial	Glucuronic acid; N-Ac galactosamine 4-sulfate	Costal cartilage Ox nasal septa
Chondroitin sulfate C.	4×10^6	50	. . .	$—COO^-$ $—OSO_3^-$	Equatorial Axial	Glucuronic acid; N-Ac galactosamine 6-sulfate	Cartilage
Dermatan sulfate. . . . . (chondroitin sulfate B)	2.7×10^4	1	. . .	$—COO^-$ $—OSO_3^-$	Equatorial Axial	Iduronic acid; N-Ac galactosamine 4-sulfate	Skin
Keratosulfate.	$1\text{–}2 \times 10^4$	1	. . .	$—OSO_3^-$	Equatorial	Galactose; galactose 6-sulfate; N-Ac glucosamine 6-sulfate	Cornea
Heparitin sulfate.	. . .	. . .	. . .	$—COO^-$ $—OSO_3^-$	Equatorial Equatorial	Glucuronic acid 2-sulfate; glucosamine 6-sulfate, N-Ac or N sulfate	Aorta
Heparin.	1×10^4	. . .	. . .	$—COO^-$ $—OSO_3^-$ $—NSO_3^-$	Equatorial Equatorial . . .	Glucuronic acid 2-sulfate; glucosamine 6-O sulfate, N sulfate	Lung

Note: The data in the second and third columns are for the proteinpolysaccharide molecules, except that for hyaluronic acid, which is the only polysaccharide occurring unassociated with protein.

* N-Ac = N-Acetyl.

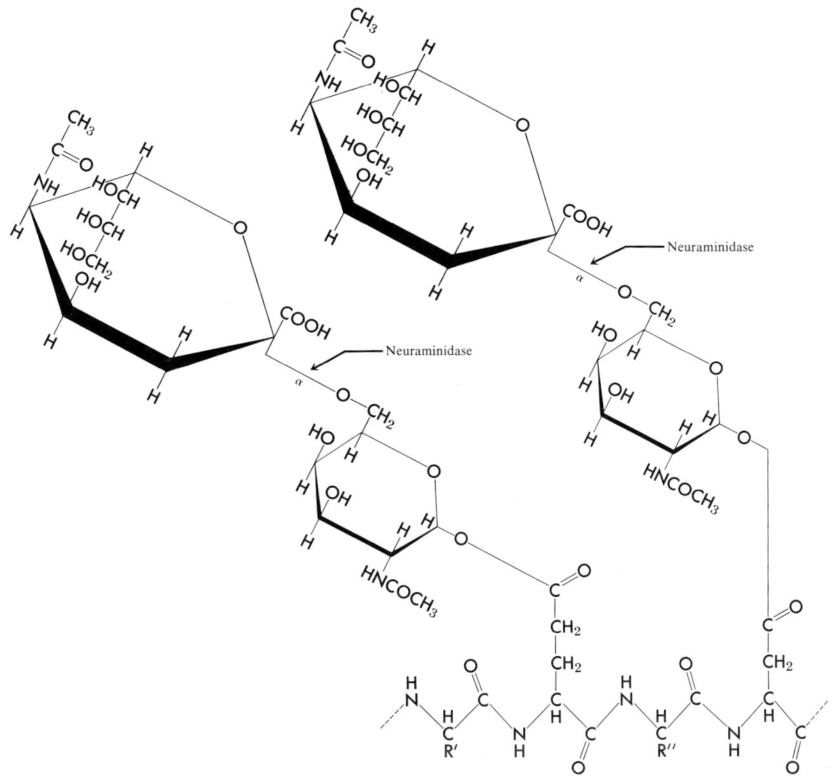

Fig. 38.4. Diagrammatic segment of bovine submaxillary gland glycoprotein, showing (1) the structure of the prosthetic group; (2) the glycosidic ester linkages joining the prosthetic groups to the β- and γ-carboxyl groups, respectively, of aspartic acid and glutamic acid residues; (3) the neuraminidase-susceptible α-ketosidic linkage within the prosthetic group.

One of the most thoroughly studied mucoproteins is that of bovine submaxillary mucin. In contrast to the proteinpolysaccharides of cartilage, in this instance about 800 disaccharide units, each of which is probably N-acetylneuraminyl $(2 \rightarrow 6)$ N-acetylgalactosamine, are attached to a single polypeptide chain, the carbohydrate accounting for about 40 per cent of the molecular weight. The type of structure is indicated in Fig. 38.4. The bonds between the disaccharides and the protein are ester linkages between the carboxyl groups of glutamic acid and aspartic acid residues and the C-1 hydroxyl group of the amino sugar moiety.

Another group of glycoproteins is of a character intermediate between the mucoproteins and the proteinpolysaccharides. This group includes the blood group substances and certain structural materials such as the insoluble major component of glomerular basement membranes. This insoluble component is a collagen-like protein, which may actually be collagen, to which is covalently bound 10 per cent of its weight in carbohydrate. The carbohydrate consists of disaccharides of glucose and galactose in glycosidic linkage to almost all of the hydroxylysine residues, and only one-tenth as many units of unknown absolute structure but containing, per unit, the following: 1 galactosamine, 5 glucosamine, 4 galactose, 3 mannose, 1 fucose, and 4 N-acetylneuraminic acid. The entire structure is extremely insoluble, but low molecular weight peptidepolysaccharides are released by collagenase treatment. This structure strikingly resembles the normal glycoprotein of human urine, molecular weight 28,000, which can bind influenza virus and serve as substrate for viral neuraminidase.

Function of Acid Polysaccharides. The most striking property of these diverse structures is that all are large polyvalent anions which attract and tightly bind cations, then termed "counterions." Even Na^+ and K^+ may be bound so effectively as to appear non-ionic. All have a tendency to aggregate; this is enhanced by polyvalent counterions such as Ca^{++}. The larger chains, particularly that of hyaluronic acid, coil up in a relatively random way and, hence, may occupy a volume largely filled with solvent water, and to which small molecules or ions have access but from which larger molecules, *e.g.*, serum albumin, may be excluded. This volume, the "*domain*" of the polysaccharide, varies with the third power of the radius of the sphere formed by the polysaccharide. For example, the end-to-end length of a hyaluronic acid molecule, molecular weight 1×10^6, is about 25,000 Å. or 2.5 μ, but it forms a sphere of effective radius of 2,000 Å.; in a 0.01 per cent solution of hyaluronic acid, the domains of these molecules would account for the total volume. In less concentrated solutions, these domains would exist as separate zones; in more concentrated solutions, the domains would collapse and interpenetrate, thus accounting for the very high viscosity of such solutions. The magnitude of these domains may be appreciated from the fact that the domain of a hyaluronic acid molecule is 75,000 times the volume occupied by three of the rigid, densely packed tropocollagen rods which, together, would have the same effective molecular weight. Hence, a solution of hyaluronate or, to much lesser degree, of the soluble protein-polysaccharides or mucoproteins with longer chains, can serve as a molecular sieve and restrict the movement of other large molecules, particularly large cations and, even at equilibrium, deny to very large molecules the volumes within the individual domains. Many of the physiological roles of the acid heteropolysaccharides are thought to reflect these properties, but there is, as yet, no explanation for the specific mixtures of such molecules found in individual tissues. Hyaluronic acid participates in maintaining the level of hydration of tissues and, in synovial fluid and the vitreous body, acts as a molecular sieve. The chondroitin sulfates are important for their ion-binding properties and function in the mineralization of the skeleton (Chap. 39). Dermatan sulfate is primarily a structural component of certain tissues and may determine the orientation of collagen fiber growth. Heparin prevents clotting in the circulation (page 733) and also activates lipoprotein lipase (page 476), although the significance of the latter process is unclear.

Metabolism of Mucopolysaccharides. Relatively little information is available concerning the synthesis of mucopolysaccharides. The origin of the uridine diphosphate esters of the monosaccharide components, *viz.*, glucuronic acid, iduronic acid, N-acetylgalactosamine, galactose, glucose, xylose, and fucose, as well as CMP-N-acetylneuraminic acid, has been described previously (Fig. 19.1, page 423). Some evidence indicates that the immediate precursors for mucopolysaccharide synthesis are in each case the appropriate derivatives of uridine diphosphate. Thus, hyaluronic acid would be formed from uridine diphosphoglucuronate and uridine diphospho-N-acetylglucosamine (page 434), in the manner already indicated for the formation of oligosaccharides and glycogen (page 433). Although it would seem that such polymers might be most readily synthesized from activated disaccharides, *e.g.*, uridine diphosphate glucuronyl N-acetylglucosamine, no evidence supports this view. Although UDP-N-acetylgalactosamine 4-sulfate has been isolated from hen oviduct and young rat cartilage, attempts to show that this compound is utilized for polysaccharide syntheses have failed. Progress in this field is hampered by the fact that it is technically difficult to prepare cell-free systems which conduct net synthesis of these macromolecules, since the enzymes are membrane-bound and inactivated by the usual manipulative procedures. Increasingly, evidence suggests that, in proteinpolysaccharide synthesis, the first several residues of each of the multiple polysaccharide chains are affixed to the polypeptide chain and that the polysaccharide chains are sulfated by 3′-phosphoadenosine 5′-phosphosulfate (page 563) as they grow. The sulfation system that is active in embryonic cartilage catalyzes formation of the C-4 sulfate ester of chondroitin present in the protein polysaccharide. However, if the latter is degraded by hydrolysis, the same enzymic system sulfates the C-6 hydroxyl groups of chondroitin.

Hyaluronidases, which effect the hydrolysis of hyaluronic acid as well as of chondroitin and chondroitin sulfates A and C, have been found in insects, snake venom, and mammalian tissues; testis is the richest source in mammals. Hyaluronidase activity has also been observed in extracts of microorganisms, spleen, ciliary body, and autolysates of skin. With the exception of the enzyme from the leech, all hyaluronidases studied appear to catalyze hydrolysis of the glucosaminidic bond between carbon-1 of the glucosamine moiety and carbon-4 of glucuronic acid. The enzyme from the leech hydrolyzes the glucuronidic bond. Like many amylases, hyaluronidase also exhibits transglycosylase activity. Hydrolysis by mammalian and leech hyaluronidase does not proceed beyond the level of a tetrasaccharide.

Although testicular and leech hyaluronidases are normal glycosidases, bacterial hyaluronidases degrade hyaluronic acid and chondroitin by cleavage of the hexosaminidic bond with formation of a 4,5 double bond in the hexuronic acid residue. Thus, bacterial hyaluronidase cleaves the tetrasaccharide resulting from mammalian hyaluronidase action into two disaccharides, hyalobiuronic acid, the repeating unit of hyaluronic acid (page 54), plus the same disaccharide with a double bond in the 4.5 position of the glucuronic acid moiety (see diagram, page 884). The water used for scission of the glycosidic bond derives from the uronic acid itself; when hydrolysis is carried out in ^{18}O-labeled H_2O, no isotope is incorporated into the product. Bacterial enzymes that hydrolyze pectin (page 51) act in a similar manner. Liver lysosomes also contain limited amounts of a hyaluronidase with an acid pH optimum analogous to that of other lysosomal enzymes. These lysosomes can "ingest" hyaluronic acid as well as chondroitin sulfates A and C, and their hyaluronidase digests these to tetrasaccharides. Ingested dermatan sulfate is not a substrate but, rather, an effective inhibitor even in low concentration, and is retained within the lysosomes.

Studies with labeled sugars, sulfate, and amino acids indicate that hyaluronic acid and the major proteinpolysaccharides are subject to constant degradation and resynthesis at varying rates which are characteristic of individual polysaccharides in

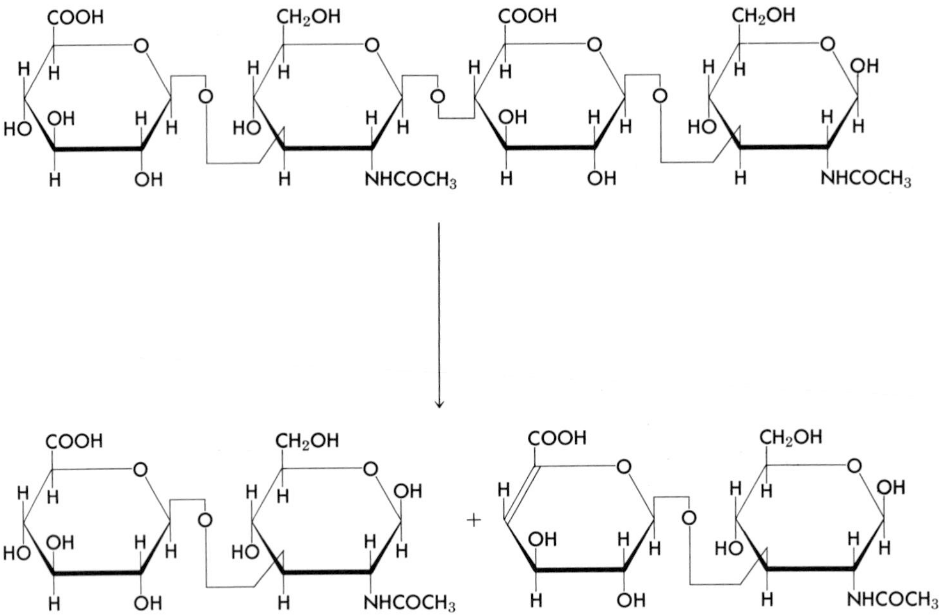

Hyalobiuronic acid

3-O-(β-D-Gluco-4,5-ene-pyranosyluronic acid)-N-acetyl-D-glucosamine

specific tissues. Only keratosulfate and the insoluble chondroitin sulfates of adult cartilage resemble collagen in being metabolically inert. This implies that these renewable materials must be released by proteolysis of their protein carriers and their polysaccharide chains degraded, probably by lysosomal hyaluronidase, to relatively low molecular weight polymers. Some of these may be hydrolyzed further by the *β-glucuronidases* of many tissues which can remove only the glucuronic moiety at the nonreducing end of these compounds. A mixture of partially sulfated oligosaccharides is then excreted in the urine.

Although liver lysosomes have a sulfatase with low activity which can slowly hydrolyze the sulfate esters of oligosaccharides, no similar enzyme is found in animal tissues generally which can catalyze removal of sulfate groups from proteinpolysaccharides. Moreover, the total daily urinary excretion of sulfate esters of low molecular weight polysaccharides is compatible with estimates of the daily total synthesis of sulfated mucopolysaccharides. In addition, urine contains significant quantities of acid heteropolysaccharides, largely chondroitin sulfates C and A, chondroitin, and heparitin sulfate, with lesser quantities of the others.

Altered mucopolysaccharide metabolism is characteristic of patients with *Marfan's* and *Hurler's diseases.* The tissues of patients with Hurler's disease contain unusually large amounts of chondroitin sulfate B and heparitin sulfate, which also appear in excessive amounts in the urine. These are the hyaluronidase-resistant polysaccharides; they accumulate in hepatic lysosomes, and secondary liver damage results. In Marfan's disease, excretion of all polysaccharides of this type is enhanced. The precise metabolic lesions in Hurler's and Marfan's diseases remain to be established.

The mucopolysaccharide composition of animal tissues varies in a rather consistent manner with aging. Thus, the keratosulfate concentration of tissues which contain this material increases throughout life, while the chondroitin sulfate content of cartilage and nucleus pulposus (intervertebral disks), as well as the hyaluronic acid of skin, decrease with aging. Administration of growth hormone (Chap. 47) at any age results in a pattern of mucopolysaccharide synthesis and composition which resembles that of the extremely young animal. Administration of testosterone (Chap. 44) appears, specifically, to increase markedly the rate of hyaluronic acid synthesis in such loci as heart valves, skin, the comb of the rooster, and the sex skin of the monkey. The hyaluronic acid of synovial fluid of rheumatic and arthritic joints is present in greater than normal amount but appears to be largely depolymerized. Administration of certain adrenal cortical steroids (Chap. 45) results in rapid repolymerization of existing hyaluronic acid while abruptly inhibiting further *de novo* synthesis. Mucopolysaccharides of the skin of the alloxan-diabetic rat exhibited a turnover rate approximately one-third that found in normal animals; this diminished rate could be restored toward normal by administration of insulin. Since in diabetes mellitus there is a significantly greater than normal susceptibility to infection, with retarded wound healing and accelerated vascular degeneration, these characteristics may reflect, in part, the decreased ability to synthesize acid mucopolysaccharides when the insulin supply is inadequate. It will be apparent from this brief survey that much remains to be learned concerning the metabolism and biological roles of these complex polysaccharides.

REFERENCES

Books

Brimacombe, J. S., and Webber, J. M., "Mucopolysaccharides," American Elsevier Publishing Company, New York, 1964.

Clark, F., and Grant, J. K., eds., "Biochemistry of Mucopolysaccharides of Connective Tissue," Cambridge University Press, New York, 1961.

Fitton Jackson, S., Harkness, R. D., Partridge, S. M., and Tristram, G. R., eds., "Structure and Function of Connective and Skeletal Tissue," Butterworth & Co. (Publishers), Ltd., London, 1965.

Gustavson, K. H., "Chemistry and Reactivity of Collagen," Academic Press, Inc., New York, 1956.

Jeanloz, R. W., and Balazs, E. A., eds., "The Amino Sugars," vols. IA, IB, IIA, IIB, Academic Press, Inc., New York, 1965–1966.

McKusick, V. A., "Heritable Disorders of Connective Tissue," 2d ed., The C. V. Mosby Company, St. Louis, 1960.

New York Heart Association, "Connective Tissue: Intercellular Macromolecules," a symposium, Little, Brown and Company, Boston, 1964.

Quintarelli, G., ed., "The Chemical Physiology of Mucopolysaccharides," Little, Brown and Company, Boston, 1967.

Springer, G. F., ed., "Polysaccharides in Biology," Transactions of the Fourth Conference of the Josiah Macy, Jr. Foundation, New York, 1961.

Wolstenholme, G. E. W., and O'Connor, M., eds., "Chemistry and Biology of Mucopolysaccharides," Little, Brown and Company, Boston, 1958.

Review Articles

Dorfman, A., and Schiller, S., Effects of Hormones on the Metabolism of Acid Mucopolysaccharides of Connective Tissue, *Recent Progr. Hormone Research,* **14,** 427–456, 1958.

Harkness, R. D., Biological Functions of Collagen, *Biol. Revs. Cambridge Phil. Soc.,* **36,** 399–463, 1961.

Harrington, W. F., and von Hippel, P. H., The Structure of Collagen and Gelatin, *Advances in Protein Chem.,* **16,** 1–138, 1961.

Hodge, A. J., and Schmitt, F. O., The Tropocollagen Macromolecule and Its Properties of Ordered Interaction, in M. V. Edds, ed., "Macromolecular Complexes," pp 19–52, The Ronald Press Company, New York, 1961.

Jakowska, S., ed., "Mucous Secretions," *Ann. N.Y. Acad. Sci.,* **106,** 157–809, 1963.

Jeanloz, R. W., Mucopolysaccharides (Acidic Glycosaminoglycans), in M. Florkin and E. H. Stotz, eds., "Comprehensive Biochemistry," vol. 5. pp. 262–296, American Elsevier Publishing Company, New York, 1963.

Kent, P. W., Structure and Function of Glycoproteins, in P. N. Campbell and G. D. Greville, eds., "Essays in Biochemistry," vol. 3, pp. 105–152, Academic Press, Inc., New York, 1967.

Mandl, I., Collagenases and Elastases, *Advances in Enzymol.,* **23,** 163–264, 1961.

Meyer, K., Chemistry of the Mesodermal Ground Substances, *Harvey Lectures,* **51,** 88–112, 1955.

Meyer, K., and Rapport, M. M., Hyaluronidases, *Advances in Enzymol.,* **13,** 199–236, 1952.

Milch, R. A., Matrix Properties of the Aging Arterial Wall, *Monographs in the Surgical Sciences,* **2,** 261–340, 1965.

Muir, H., Structure of Mucopolysaccharides, *Intern. Rev. Connective Tissue Research,* **2,** 101–154, 1964.

Partridge, S. M., Elastin, *Advances in Protein Chem.,* **17,** 227–302, 1962.

Schiller, S., Connective and Supporting Tissues: Mucopolysaccharides of Connective Tissue, *Ann. Rev. Physiol,* **28,** 137–158, 1965.

Seifter, S., and Gallop, P. M., The Structure Proteins, in H. Neurath, ed., "The Proteins," 2d ed., vol. IV, pp. 153–458, Academic Press, Inc., New York, 1966.

Veis, A., Anesey, J., and Mussell, S., A Limiting Microfibril Model of the Three-Dimensional Arrangement Within Collagen Fibers, *Nature,* **215,** 931–935, 1967.

39. Bone; Calcium and Phosphate Metabolism

CALCIUM AND PHOSPHATE METABOLISM

Although more than 99 per cent of the calcium of the body is present in the skeleton, the remaining 1 per cent serves a number of important functions unrelated to bone structure. The concentration of calcium in intracellular fluid is approximately 20 mg. per 100 g. of tissue. Its presence is essential for the activity of a number of enzymic systems, including those responsible for the contractile properties of muscle and for the transmission of the nerve impulse. The concentration of calcium is also critical in extracellular fluid, particularly for the response of muscle to neural stimuli and for functioning of the blood-clotting mechanism (Chap. 31).

Intestinal Absorption of Calcium. The major calcium salt ingested is calcium phosphate, since it is in this form that calcium is present in materials which serve as food. Calcium also occurs in nature as the carbonate, tartrate, and oxalate and, together with magnesium, as the highly insoluble mixed salt of phytic acid (the hexaphosphoric acid ester of inositol) (page 33), present in cereals.

From a nutritional standpoint, the intestinal absorption of calcium presents a major problem, largely because of the insolubility of most calcium salts. Within the body, moreover, the insolubility of calcium salts may lead to calcification of atheromatous blood vessels or calculus formation in the gallbladder or in the pelvis or tubules of the kidney. In increasing order of solubility, the three forms of calcium phosphate are $Ca_3(PO_4)_2$, $CaHPO_4$, and $Ca(H_2PO_4)_2$. At the pH prevailing within the stomach, the calcium phosphates readily dissolve; at the pH of the duodenum, the chief calcium salts are $CaHPO_4$ and $Ca(H_2PO_4)_2$.

Calcium passage across the duodenum is minimal and occurs largely by diffusion; this process seems to be inhibited, in part, in vitamin D–deficient animals. Active transport of Ca^{++} occurs across the ileal mucosa; the system is capable of operating against a fivefold concentration gradient and requires energy, presumably ATP. This process also is diminished in vitamin D deficiency and is enhanced by administration of vitamin D. This effect requires about 24 hr to become established and can be prevented by simultaneous administration of actinomycin D (page 670) or puromycin (page 670). Shortly after administration of vitamin D, incorporation of tritium-labeled orotidine into RNA in the intestinal epithelium increases about threefold. Hence, it appears that vitamin D, in some manner, derepresses synthesis of a mRNA which serves as template for formation of a protein, perhaps a Ca^{++} carrier, which is essential to active Ca^{++} transport across the ileal mucosa. Perhaps related is the capacity of mitochondria from many tissues to concentrate relatively large quantities of Ca^{++} from the medium provided that there is a source of ATP and an oxidizable substrate.

Calcium absorption from the ileum can also be enhanced by the presence of sugars and of basic amino acids. The sugar effect is relatively nonspecific. Generally, virtually all mono- and disaccharides have been absorbed before the digested food mixture arrives at the ileum. However, lactose, uniquely, may escape digestion during this passage, and its presence markedly enhances calcium absorption. The only source of lactose is milk, which also provides large quantities of calcium for the growth of the young. Much of the phosphate of milk is esterified to the hydroxyl groups of the serine residues of casein (page 824). Calcium absorption may precede the complete digestion of casein so that the concentration of inorganic phosphate liberated by digestion of casein is relatively low until the calcium has been absorbed.

The presence of substantial amounts of lysine or arginine also enhances ileal calcium absorption, but under most physiological circumstances this is of little consequence. Of these factors, therefore, only vitamin D appears essential for normal absorption of calcium. A low level of *phytase,* which catalyzes the hydrolysis of phytate (see above), is present in ileal mucosa. All phytate which escapes hydrolysis renders an equivalent amount of calcium unavailable, as do fatty acids. In steatorrhea due to biliary obstruction, sprue, or regional ileitis, insoluble calcium soaps are formed which are excreted in the stool.

Regulation of Plasma Calcium Concentration. The calcium of normal human plasma varies from 9 to 11 mg. per 100 ml. (4.5 to 5.6 meq. per liter) and exists in two major forms. The concentration of ionic calcium, the form in which calcium exerts its physiological effects, is 5 to 6 mg. per 100 ml. The second form is calcium which is not diffusible through semipermeable membranes through which ionized calcium can pass. This fraction consists largely of calcium bound to plasma proteins, particularly albumin. The amount of this fraction is a function of the total protein concentration; plasma low in protein also exhibits a low total calcium concentration. The extent of calcium binding by protein increases with increasing pH. Ionic calcium can be determined by a biological assay using the frog or turtle heart, the activity of which is proportional to the ionized calcium concentration of the medium. For clinical purposes, estimation of the ionized calcium is possible from a knowledge of the total calcium and protein concentrations, using the nomogram shown in Fig. 39.1.

Maintenance of normal neuromuscular irritability is critically dependent upon the concentration of ionic calcium as one of the factors in the ratio:

$$\frac{[K^+] + [Na^+]}{[Ca^{++}] + [Mg^{++}] + [H^+]}$$

This relationship is intended only to convey the direction of the change in irritability resulting from alterations in the concentrations of these ions. A significant decrease in concentration of ionic calcium results in tetany, while an increase may sufficiently impair function as to lead to respiratory or cardiac failure.

The plasma calcium concentration is homeostatically maintained. The mechanism includes a storage compartment, the skeleton, which may be drawn upon or

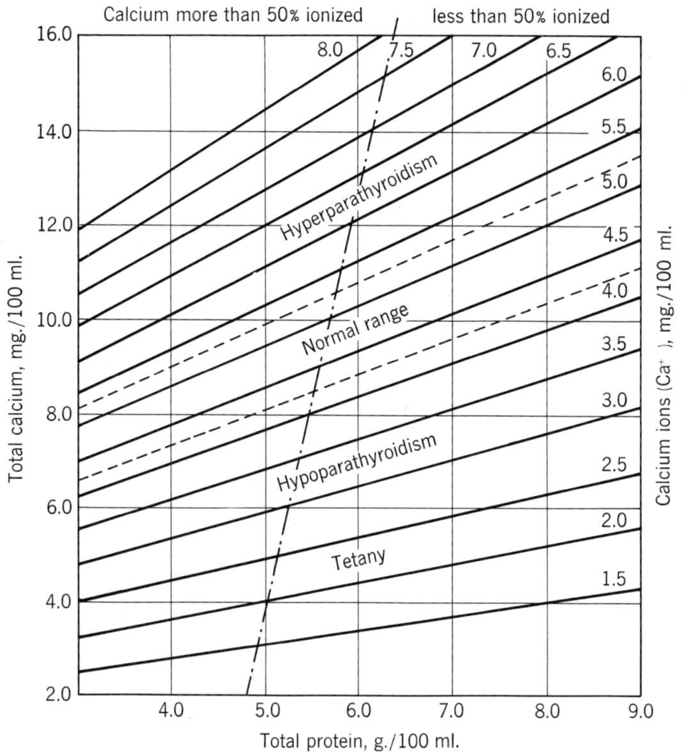

Fig. 39.1. Nomogram showing the relations between total serum calcium, protein, and ionized calcium concentrations. (*From F. C. McLean and A. B. Hastings, Am. J. Med. Sci.*, **189**, 601, 1935.)

into which excess may be deposited; the kidneys; excretion through the bile and intestine; and two hormones, parathormone and calcitonin (Chap. 43), the secretion of which is sensitive to the plasma concentration of Ca^{++}.

Serum phosphate and Ca^{++} concentrations normally exhibit a reciprocal relationship; when the phosphate concentration is diminished, $[Ca^{++}]$ is elevated, and vice versa, as would be expected were serum the soluble phase of a saturated solution. However, in hyperparathyroidism the concentrations of both ions may be elevated, whereas in juvenile rickets both may be diminished. Thus, a simple solubility product relationship does not always obtain, and the concentrations of these ions are under cellular control.

The normal major pathway of calcium excretion is the intestinal tract. Individuals on a calcium-free diet continue to excrete calcium in the feces. The calcium is a constituent of various digestive secretions, particularly the bile; the amount excreted depends upon the plasma calcium concentration. Normally, the kidney excretes little calcium. On the other hand, chronic hypercalcemia may be accompanied by sufficient calciuria to cause formation of renal calculi. In normal individuals 99 per cent of the Ca^{++} filtered through the glomeruli is reabsorbed, even at artificially elevated plasma concentrations. However, the fraction reabsorbed decreases in several pathological conditions in which bone mineral resorption is active. Bone serves as the store of calcium in the operation of a homeostatic mechanism. Under circumstances which might otherwise lead to hypocalcemia, calcium is withdrawn from the skeleton. Conversely, hypercalcemia may be prevented by deposition of calcium in bone. The means by which this is effected will be considered below.

Phosphate Metabolism. Phosphate is ubiquitous and abundant in biological materials; nutritional phosphate deficiency is not possible when food is ingested in amounts sufficient to meet the requirements for calories and protein. Most of the phosphate ingested is orthophosphate or organic phosphate which yields orthophosphate in the digestive tract. No appreciable phosphate absorption occurs in the stomach, but absorption occurs throughout the small intestine. Only the turnover of ATP is known to occur at a rate commensurate with the rate of phosphate transport, but there is no proof of its participation in this process.

The inorganic phosphate in plasma is largely orthophosphate, with $HPO_4^=$ and $H_2PO_4^-$ present in a ratio of approximately 4:1. All the plasma phosphate is diffusible and filtrable through the glomerulus. Inorganic pyrophosphate is present at a concentration of 10^{-5} to $10^{-6}M$. There also exist in plasma small quantities of hexose phosphates, triose phosphates, etc. The concentration of orthophosphate in normal plasma is 4 to 5 mg. of phosphorus per 100 ml. in children, and 3.5 to 4 mg. of phosphorus per 100 ml. in adults. This concentration is also regulated by a homeostatic mechanism. As in the case of calcium, the skeleton stores phosphate, which is withdrawn when the serum phosphate concentration falls or is deposited when the serum phosphate concentration rises. The chief excretory route is the kidney; the renal excretion of phosphate has been described earlier (page 835).

BONE

Composition of Bone. There are several different types of bone, *e.g.*, cortical bone, cancellous bone, etc. Most information concerning the nature of bone and its formation derives from studies of the long bones. When bone is allowed to stand in a dilute acidic solution, the mineral portion dissolves, leaving a flexible, tough, and almost translucent organic residue retaining the shape of the intact bone. The mineral fraction of bone is composed largely of calcium phosphate, in addition to carbonate, fluoride, hydroxide, and citrate. Most of the magnesium, about one-quarter of the sodium, and a smaller fraction of the potassium of the body are also present in bone. Bone crystals belong to the group of hydroxyapatites, of the approximate composition $Ca_{10}(PO_4)_6(OH)_2$. The crystals are platelets, or rods, about 8 to 15 Å. thick, 20 to 40 Å. wide, and 200 to 400 Å. long, with a density of 3.0. It seems likely that in bone, divalent cations other than Ca^{++} can replace Ca^{++} in the hydroxyapatite crystal lattice, whereas anions other than phosphate and hydroxyl may be adsorbed on the vast areas of surface offered by the minute crystals, or dissolved in the hydration shell about the crystal lattice.

Inorganic material comprises only about one-fourth the volume of bone, the remainder being the organic matrix. Because of the difference in density between the organic and inorganic phases, insoluble minerals comprise half the bone weight. The organic matrix consists largely of collagen; only very small quantities of mucoprotein are present in mature dense bone. Collagen fibers in the bone matrix do not differ from those present in tendon, hyaline cartilage, fibrocartilage, or other fibrous tissues. The inorganic crystal structure imparts to bone an elastic modulus similar to that of concrete.

Structure and Formation of Bone. As discussed previously (page 875), cells of mesenchymal origin (*viz.*, fibroblasts, osteoblasts) introduce into the medium about them fibrils of collagen in a milieu containing mucoprotein and mucopolysaccharides. Although this combination of collagen and mucopolysaccharide is ubiquitous in the animal body, mineralization normally occurs only in those areas destined to become bone. The mineral components must be withdrawn from the available fluid medium. It follows that this fluid phase is already supersaturated and that crystal formation is induced by "nucleation," *i.e.*, provision of a surface on which crystal lattice formation can proceed readily. Although, in chemical practice, "seeding" of supersaturated solutions to induce crystal formation is usually performed by addition of crystals of the same material, seeding for nucleation purposes can also be accomplished, and frequently more successfully, by addition of crystals of some other chemical species, *e.g.*, seeding of clouds with AgI to induce ice formation. The normal, triple-stranded quarter-staggered collagen (Fig. 38.1, page 873) serves as the nucleating agent for bone formation. Electron microscopy and low-angle x-ray diffraction studies indicate that formation of the bone mineral crystal lattice is initiated within the collagen fibers. Indeed, collagen prepared from skin or tendon and solubilized and reprecipitated under conditions which lead to formation of normal fibrils (page 874), and then placed in a mineral medium identical in composition with normal blood plasma, will initiate such nucleation and be mineralized with normal hydroxyapatite crystals.

Some evidence suggests that nucleation of collagen requires prior binding of phosphate to appropriate but unidentified amino acid residues. Dinitrophenylation of ϵ-amino groups of lysine residues prevents mineralization. The alignment of neighboring strands of the triple helix of collagen is critical. In native collagen, adjacent collagen fibrils are aligned in the fiber in a manner so that repeating groups in the tropocollagen molecules of which they are composed are not "in register" but are quarter-staggered (Fig. 39.2). Under certain conditions, starting with dissolved tropocollagen, one can prepare collagen fibers in which all the tropocollagen units are "in register" (Fig. 39.2), but such fibers fail to mineralize under the same conditions. During bone development, the crystals grow until they completely fill and surround the collagen and in turn serve as nucleating agents for deposition of hydroxyapatite in spaces between collagen fibers (Fig. 39.3).

Since collagen in connective tissue other than bone does not normally calcify in the same supersaturated medium, factors other than collagen and the mineral composition of the medium must be involved. Bone formation occurs only in the vicinity of osteoblasts, with mineral deposition advancing in cartilage which consists of collagen in a mucopolysaccharide matrix. The mucoproteins act as "plasticizers" for the collagen network, decreasing the tensile strength and increasing the swelling volume of the latter, presumably by creating interchain cross-links which are large loops. At the zone of calcification the protein polysaccharide is degraded, probably by hydrolysis of the protein backbone by lysosomal proteinases of bone cells. As crystal growth proceeds, the mineral displaces not only the domain of the muco-

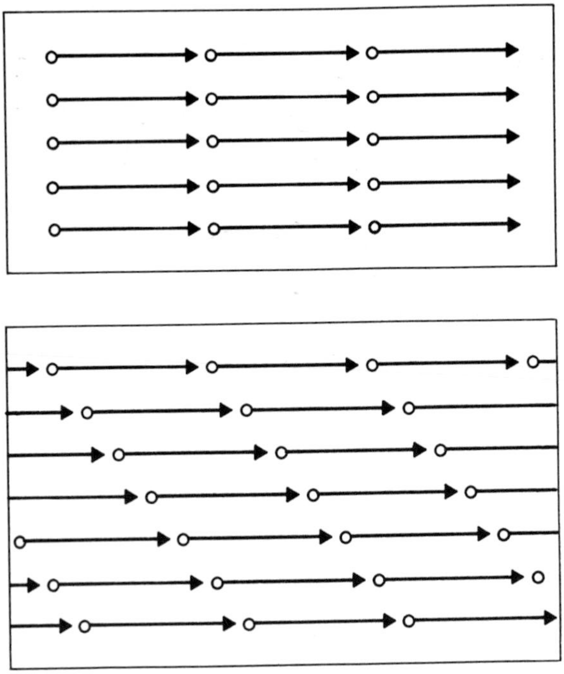

Fig. 39.2. Schematic diagram of alignment of tropocollagen molecules in collagen fibers; ▶ and ○ represent carboxyl- and amino-termini of the three strands of each tropocollagen unit. The "segment long spacing" collagen (upper figure) was prepared from tropocollagen in the presence of ATP and acetic acid. The lower figure shows the quarter-staggered alignment of tropocollagen molecules in a normal collagen fibril; the head of each tropocollagen is displaced about one-quarter of its length from that on each side.

proteins, but also the water itself. Dense, fully mineralized bone is essentially water-free; collagen accounts for 20 per cent of the weight and 40 per cent of the volume of such tissue, while the mineral constitutes the remainder. However, such a structure is interrupted by the Haversian canals lined with cells and by the vascular supply. The continuing presence of proteinpolysaccharides may explain the failure of collagen to mineralize in skin, tendon, or artery. In addition, there appear to be inhibitors of crystallization in plasma. Indeed, inorganic pyrophosphate, at concentrations of the order of $5 \times 10^{-6} M$, prevents hydroxyapatite crystal growth from solutions which are supersaturated with respect to Ca^{++} and P_i. Osteoblasts are unusually rich in alkaline phosphatase, which might serve to achieve a local increase in phosphate concentration, but the relationship of this enzyme to the calcification process is not clear.

The formation of bone as a consequence of the structure and organization of its organic matrix with subsequent nucleation of crystallization from a supersaturated medium appears to be but one instance of a general principle. As indicated in Table 39.1, mineralized tissues appear, generally, to be formed in this manner.

Bone is not a static depot of mineral but is in a dynamic state, with simultaneous osteoblastic and osteoclastic activity maintaining the constancy of bone composition. Administered $^{32}P_i$ soon appears even in the dense portions of the shaft of the large

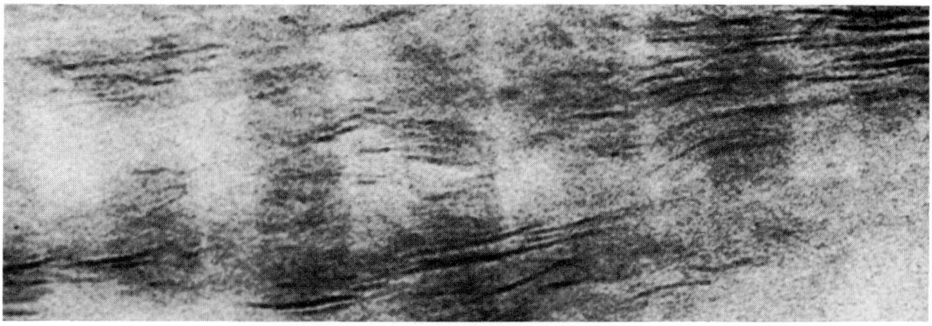

Fig. 39.3. An initial stage in bone formation. A longitudinal section through the metatarsal bone of a 12-day-old chick, magnified 200,000×, shows that nucleation and early growth of hydroxyapatite crystals is largely confined to zones occupying about 60 per cent of the axial repeat of the collagen fibrils. (*Courtesy of Dr. M. J. Glimcher and Dr. A. J. Hodge.*)

bones, as do radioisotopes of calcium and strontium. The rapidity with which these cations appear in bone is probably related to the vast surface of bone crystal exposed to extracellular fluid. Bone contains a high concentration of anions other than phosphate, *e.g.*, carbonate. Citrate may be present in an amount equal to about 1 per cent of the dry weight of bone mineral, and largely in association with Na^+. Bone appears to be a labile reservoir of Na^+ which may be drawn upon in acidosis and in which Na^+ accumulates in alkalosis or excessive Na^+ intake. These ions are adsorbed to the crystal surfaces, rather than being present as integral parts of the crystalline structure. Heavy metals which may enter the body are incorporated into the growing hydroxyapatite crystal lattice; notable examples are lead, radium, uranium, and the heavy elements derived from uranium by fission, *e.g.*, strontium.

Other Factors Influencing Bone Metabolism. Deficiency of vitamin D leads to rickets (rachitis) in the young. This vitamin promotes Ca^{++} absorption from the intestine and the manifestations of vitamin D deficiency derive largely from the resultant lack of Ca^{++}. In addition, however, a direct effect of vitamin D on bone is evident from

Table 39.1: EXAMPLES OF BIOLOGICALLY MINERALIZED TISSUES

Species	Tissue	Crystalline phase	Mineral form	Major organic matrix
Plants	Cell walls	$CaCO_3$	Calcite	Cellulose, pectins, lignins
Radiolaria	Exoskeleton	$SrSO_4$	Celestite	(?)
Diatoms	Exoskeleton	Silica	(?)	Pectins
Mollusks	Exoskeleton	$CaCO_3$	Calcite, aragonite	Protein (conchiolin)
Arthropods	Exoskeleton	$CaCO_3$	Calcite	Chitin, proteins
Vertebrates	Endoskeleton:			
	Bone	$Ca_{10}(PO_4)_6(OH)_2$	Hydroxyapatite	Collagen
	Cartilage	$Ca_{10}(PO_4)_6(OH)_2$	Hydroxyapatite	Collagen
	Tooth:			
	Dentine	$Ca_{10}(PO_4)_6(OH)_2$	Hydroxyapatite	Collagen
	Cementum	$Ca_{10}(PO_4)_6(OH)_2$	Hydroxyapatite	Collagen
	Enamel	$Ca_{10}(PO_4)_6(OH)_2$	Hydroxyapatite	Protein

SOURCE: After M. J. Glimcher, *Rev. Mod. Phys.*, **31,** 359, 1959.

the failure of calcification of rat femur, in vitro, in rachitic serum which has been fortified with calcium and phosphate.

Vitamin D toxicity, both in experimental animals and in patients who have received massive doses of vitamin D, is characterized by excessive resorption of bone, with a resultant rise in serum calcium and phosphate concentrations. Although the exact locus of vitamin D action in this regard is not clear, this effect, like that on intestinal absorption of Ca^{++}, is prevented by actinomycin D and by puromycin. The elevated serum calcium and phosphate levels lead to unusually large quantities of calcium and phosphate in the urine and formation of renal calculi. Indeed, renal failure has often been the presenting sign in patients suffering from vitamin D intoxication.

Vitamin A also affects bone development. In the young animal deprived of vitamin A, growth of the skeleton is impaired before that of the soft tissue (Chap. 50). The spinal cord may thus continue to grow after growth of the spinal column has been arrested, giving rise to compression of nerve roots as they emerge from the cord; this leads to impaired function of those parts supplied by the affected nerves. Young rats given excessive but not lethal quantities of vitamin A develop multiple fractures of the long bones, and bone deformities have been observed in children receiving excessive quantities of this vitamin. These phenomena appear to reflect depolymerization and hydrolysis of the chondroitin sulfate component of cartilage (page 878).

Ascorbic acid is also essential to the development of a normal skeleton. In ascorbic acid deficiency, the mesenchymal cells fail to elaborate normal collagen (see preceding chapter), leading to impaired calcification. Skeletal growth ceases in all deficiency states, including caloric restriction. However, only in deficiencies of calcium, phosphorus, and vitamins A, D, and C are there characteristic bone lesions other than those seen in inanition. The local chemical factors affecting mineralization are incompletely understood. Bone cells respire and glycolyze, continually producing lactic acid. Administration of parathyroid hormone increases the rate of lactic acid production, which may result in local demineralization by lowering pH. As bone crystals dissolve, citrate is released and appears in plasma. Estrogens inhibit lactic acid production, in accord with the increased bone density produced by prolonged estrogen administration (see below).

Role of the Parathyroid Hormone. The parathyroid hormone (parathormone) is important in the regulation of the metabolism of calcium and phosphate. The results of parathyroid insufficiency and of overadministration of parathyroid hormone preparations are presented in Chap. 43. The parathyroid hormone inhibits renal tubular reabsorption of inorganic phosphate, resulting in a lowering of the plasma concentration of phosphate. After administration of parathyroid extract, depolymerization of the mucoprotein aggregates occurs in the less dense areas of bone, followed by disappearance of the crystal line and matrix structures. Simultaneously, acidic mucoproteins appear in the plasma. These observations suggest that the metabolic effects of the parathyroid hormone may, in part, be mediated via the regulatory influence of the osteocytes on the structure of the ground substance. Parathyroid hormone stimulates osteoclastic activity, indeed, in the presence of this hormone, ascites tumor cells release an unidentified material which causes the dissolution of dense bone. The parathyroid gland is sensitive to serum [Ca^{++}] and secretes its active principle in response to a lowering of the concentration of circulating calcium (Chap. 43).

Role of Calcitonin. A second hormone, calcitonin (Chap. 43) is also involved in the regulation of the metabolism of calcium and phosphate. The effects of this hormone on blood [Ca^{++}] are in the opposite direction from those of parathormone. Calcitonin accelerates removal of calcium and of phosphate from the blood to bone and the rate of calcium deposition, and inhibits calcium resorption in bone. The secretion of calcitonin varies directly with the [Ca^{++}] of the serum (Chap. 43). However, the action of this hormone is not mediated by the kidney, gastrointestinal tract, hypophysis, or parathyroid gland. It appears that parathormone and calcitonin constitute a homeostatic mechanism for the regulation of the metabolism of calcium and phosphate.

Derangements of Bone Metabolism. Deviations from normal bone metabolism are characterized by either excessive or inadequate bone formation. Excessive bone formation is relatively uncommon but may be produced experimentally by estrogen administration (Chap. 44). Abnormally thick and dense bones and cerebral calcification may be seen in *chronic hypoparathyroidism* and accompanying chronic tetany. Overgrowth of bones occurs in *hypertrophic osteoarthritis,* a relatively common chronic disease occurring mainly after the age of forty years.

Abnormalities of bone metabolism leading to loss of bone may arise either from failure to mineralize the bone matrix, *osteomalacia,* or because of inadequate matrix formation, *osteoporosis.* For example, osteomalacia is seen in juvenile rickets, in which bone matrix formation proceeds normally but there is inadequate calcification of the matrix. Continuing growth of uncalcified cartilage results in bone deformity. Osteoblastic activity is greater than normal, as revealed by increased plasma alkaline phosphatase activity. The deficient supply of dietary calcium leads to diminished serum [Ca^{++}] and causes release of parathyroid hormone, which through its effect on bone and on the kidney maintains a normal serum [Ca^{++}] and a diminished serum phosphate concentration. Bone deformities are not prominent in osteomalacia of adults since skeletal growth is complete, but diminished bone density and spontaneous fractures occur frequently.

Osteoporosis is most commonly observed after the menopause, presumably reflecting decreased estrogenic activity (Chap. 44). Ascorbic acid deficiency and malnutrition also result in osteoporosis because of impaired osteoblastic formation of collagen. Hyperparathyroidism stimulates osteoclastic activity; the effect is not uniform throughout the skeleton, however, but is apparent as "punched-out" areas —*osteitis fibrosa cystica.*

Osteomalacia and fragile bones result from prolonged immobilization and disease. Immobilization of an extremity results in prompt negative calcium balance of that member; prolonged bed rest without exercise has the same influence on the entire skeleton. The underlying mechanism is not understood, but this constitutes one of the most serious problems confronting astronauts who may undertake prolonged confinement within a chamber of limited size.

Hypophosphatasia is a relatively rare, chronic familial disease of uncertain etiology, primarily affecting children, who exhibit deficient bone formation with histological changes resembling rickets. The disease is associated with a low level of alkaline phosphatase activity in the serum and tissues, and may occur in siblings; serum alkaline phosphatase activity is often low in one, or both, of the parents, without evidence of bone disease. One form of such "refractory rickets" has been shown to be transmitted as a sex-linked, dominant trait.

Calcification in Soft Tissues. Aberrant calcification may occur in a wide variety of tissues. *Myositis ossificans* is a striking example of metastatic ossification. In skin, tendon, ligaments, and heart valves, this calcification is again associated with collagen. In the intima of large vessels, such as the aorta, it is apparently associated with elastin. Calcification of the pancreas may be related to a local increase in alkalinity due to the secretory activity of acinar tissue; calculus formation at the tooth-gum margin seems to involve participation of mucoproteins.

TEETH

The tooth is built of three layers of calcified tissue. The pulp cavity, containing blood vessels and nerves, is covered with the dentine, the major calcified tissue. Where the tooth is exposed, the dentine is covered by enamel, while the submerged roots are covered with cementum. The cementum closely resembles cortical bone in composition. Dentine is hard and dense, being almost 75 per cent mineral; the enamel is more dense and harder and is almost 98 per cent mineral. The organic matrix of dentine and cementum is much like that of bone. The protein in embryonic enamel is a fibrous protein characterized by a very high proline content and the presence of hydroxylysine (Table 38.1). This is the protein which is initially laid down and calcified. None of this protein remains in fully formed enamel, which, instead, contains a very small amount of low molecular weight peptides which are almost devoid of proline and hydroxyproline. The hydroxyapatite crystals of enamel are much larger than are those of dentine, cementum, or bone. Analysis reveals less Mg^{++}, $CO_3^{=}$, Na^{+}, etc., in enamel than in dentine. Hevesy observed incorporation and disappearance of administered radioactive inorganic phosphate in teeth. Turnover of phosphate in dentine is about one-sixth as rapid as in long bones, but about fifteen to twenty times as rapid as in enamel. This relatively slow turnover of tooth minerals is consonant with their stability in potentially decalcifying conditions *e.g.*, pregnancy and vitamin D deficiency.

Fluoride and Caries. Knowledge of the relationship between fluoride and dental health stems from epidemiologic investigations commencing in 1908 of the cause of *mottled enamel.* Affected teeth are characterized by dull chalky patches distributed irregularly over the surface; they are pitted and corroded and occasionally stained yellow to dark brown. Calcification is deficient, and cement substance may be lacking. Mottled teeth occur only in persons who have drunk water containing in excess of 1.5 mg. of fluoride per liter during the years of tooth development. On the other hand, the occurrence of caries is minimal in those communities whose drinking water naturally provides 0.9 mg. of fluoride per liter. The fluoridation of other communal water supplies to bring the fluoride content to 1.0 mg. per liter has resulted in a significant decrease in the incidence of dental caries. Once the teeth have been fully formed, fluoride is without effect on dental caries. The mode of action of fluoride is not understood. Since the quantity of fluoride used is much too small to inhibit bacterial metabolism, the fluoride must, in some manner, increase the ability of teeth to withstand the usual cariogenic influences.

REFERENCES

Books

Bergsma, D., ed., "Structural Organization of the Skeleton," National Foundation, New York, 1966.

Bourne, G. H., "The Biochemistry and Physiology of Bone," Academic Press, Inc., New York, 1956.

Fitton Jackson, S., Harkness, R. D., Partridge, S. M., Tristram, G. R., eds., "Structure and Function of Connective and Skeletal Tissue," Butterworth & Co. (Publishers), Ltd., London, 1965.

Irving, J. T., "Calcium Metabolism," John Wiley & Sons, Inc., New York, 1957.

McLean, F. C., and Urist, M. R., "Bone: An Introduction to the Physiology of Skeletal Tissue," 2d edition, University of Chicago Press, Chicago, 1961.

Miles, A. E. W., ed., "Structural and Chemical Organization of Teeth," vols. I and II, Academic Press, Inc., New York, 1967.

Neuman, W. F., and Neuman, M. W., "The Chemical Dynamics of Bone Mineral," University of Chicago Press, Chicago, 1958.

Sognnaes, R. F., ed., "Calcification in Biological Systems," American Association for the Advancement of Science, Washington, 1960.

Sognnaes, R. F., ed., "Mechanisms of Hard Tissue Destruction," American Association for the Advancement of Science, Washington, 1963.

Wolstenholme, G. E. W., and O'Connor, C. M., eds., "Bone Structure and Metabolism," Little, Brown and Company, Boston, 1965.

Review Articles

Arnaud, C. D., Jr., Tennenhouse, A. M., and Rasmussen, H., Parathyroid Hormone, *Ann. Rev. Physiol.,* **29,** 349–372, 1967.

Bauer, G. C. H., Carlsson, A., and Lindquist, B., Metabolism and Homeostatic Function of Bone, in C. L. Comar and F. Bronner, eds., "Mineral Metabolism," vol. I, part B, pp. 609–677, Academic Press, Inc., New York, 1961.

Follis, R. H., Jr., A Survey of Bone Disease, *Am. J. Med.,* **22,** 469–484, 1957.

Glimcher, M. J., Molecular Biology of Mineralized Tissues with Particular Reference to Bone, *Rev. Mod. Phys.,* **31,** 359–393, 1959.

Nicolaysen, R., Eeg-Larsen, N., and Malm, O. J., Physiology of Calcium Metabolism, *Physiol. Revs.,* **33,** 424–444, 1953.

Urist, M. R., The Bone-Body Fluid Continuum: Calcium and Phosphorus in the Skeleton and Blood of Extinct and Living Vertebrates, *Perspectives in Biol. & Med.,* **6,** 75–115, 1962–1963.

Whipple, H. E., ed., "Comparative Biology of Calcified Tissue," *Ann. N.Y. Acad. Sci.,* **109,** 1–410, 1963.

40. The Eye

Light entering the eye passes through the tears on the conjunctiva and then through the cornea, aqueous humor, lens, and posterior chamber before impinging on the light-perceiving apparatus of the retina. These structures will be considered briefly, followed by a discussion of the photochemistry of vision.

STRUCTURE, COMPOSITION, AND METABOLISM

Both the *sclera* and the *conjunctiva* are of mesenchymal origin and are built largely of collagen fibers and mucoprotein. The sclera also contains chondroitin sulfates A, B, and C. The sclera is opaque because its collagen fibers are interwoven and crosslaminated, whereas in the transparent conjunctiva all fibers lie in parallel.

The *cornea* is a transparent structure composed of five separate layers: an outer epithelium and an inner endothelium, on the inside of each of which is a membrane enclosing the substantia propria. The corneal epithelium is easily isolated from the avascular stroma with its loosely attached single cell layer of endothelium. Descemet's membrane, which lies between the endothelium and the stromal connective tissue component, consists mainly of collagen. The stroma also is collagenous, but, here, the fibrils are embedded in ground substance. The mucopolysaccharides in this matrix have been well characterized but not the proteins. Keratosulfate (page 53) is the principal corneal polysaccharide. The corneal fibrils are distinguished both by their regular arrangement and by their narrow (embryonal) cross section. Corneal vascularization occurs in numerous nutritional deficiencies.

The transparency of the cornea is dependent upon its hydration. Both the aqueous humor (page 810) on the one side and the tears (page 814) on the other are slightly hypertonic, but it has been assumed that this degree of hypertonicity is sufficient to draw water at all times from the cornea and maintain it in a relatively dehydrated state. An inward-directed active transport of Na^+ in the cornea has been described, in addition to ready passage of water.

Perhaps the outstanding feature of corneal metabolism is the predominance of the phosphogluconate oxidative pathway, accounting for approximately 50 per cent of the glucose utilized. The three lactic acid dehydrogenase isoenzymes isolated from this tissue exhibit somewhat higher activity with DPNH than with TPNH. The high level of pyridine nucleotides and the high O_2 consumption ($Q_{O_2} = 6$) indicate that the cellular layers of the cornea are metabolically active.

The transparent, pale lens, which yellows with aging, is of ectodermal origin and is composed of macroscopic cells termed lens fibers; within these cells lie the gel structures. The lens is surrounded by a collagenous membrane, the lens capsule; the latter contains a collagen that is rich in hydroxylysine and hexose present in part as a glycopeptide, glucosyl-galactosyl-hydroxylysine. The interior of the lens, which contains 25 per cent protein, is a thick gel. Several proteins have been identified, α-, β-, and γ-crystallin which are soluble, and an insoluble albuminoid. The three crystallins differ in amino acid composition, but the subunits of each are about 20,000 molecular weight. The α- and β-crystallins comprise approximately 80 per cent of the total soluble protein and occur mainly as large aggregates of varying size up to 2×10^6. γ-Crystallin does not aggregate and seems to be a distinctive protein of unknown function. α-Crystallin appears to be a soluble precursor of the insoluble albuminoid. The latter is concentrated in the lens nucleus and increases with aging; the soluble proteins predominate in the outer layers. The lens is maintained in its relatively dehydrated state by means of active transport systems. The chief intracellular cation is potassium, and the extracellular cation is sodium. The lens has a low rate of metabolism, characterized primarily by glycolysis, with functioning also of the phosphogluconate oxidative pathway. Although all enzymes of the citric acid cycle are present, lens has little activity represented by this pathway.

Lenticular carbohydrate metabolism is markedly impaired in diabetes, with a marked increase in glucose and fructose content of the lens, and diminished ATP formation and amino acid incorporation into lens protein.

The outer regions of the lens contain ATP and virtually no phosphocreatine, while the nucleus contains phosphocreatine and relatively little ATP. Thus, in the central region, relatively remote from oxygen and glucose which are available only by diffusion from the aqueous humor, the lens accumulates a storage form of energy-rich phosphate which is not required in the outer layers. The usefulness of this energy is not completely clear, but one possibility is indicated below.

If kept anaerobically and aseptically, lens proteins undergo autolysis, catalyzed by two proteases which are active over a broad pH range; an active aminopeptidase is also present. The lens synthesizes protein by conventional reactions (Chap. 29), the greatest synthesis occurring in the outer peripheral layers. The lens is one of the richest sources of glutathione, the cortex containing as much as 600 mg. per 100 g. of tissue. In studies with radioactive glycine, the glutathione of the lens has been found to have a half-life of about 30 hr. Continual glutathione synthesis (page 578) would require an appreciable fraction of the ATP generated by glucose oxidation. Another reducing agent, ascorbic acid, is present in lens in a concentration of approximately 30 mg. per 100 g. tissue, or almost twenty times the plasma concentration.

Several analogues of glutathione are found in lesser amounts in the lens. In each of these tripeptides, the cysteine residue has been replaced. In ophthalmic acid, cysteine is replaced by α-amino-n-butyric acid; in norophthalmic acid, by alanine; in S-sulfoglutathione, by S-sulfocysteine; and in yet another tripeptide, by S-(α,β-dicarboxyethyl)-cysteine. The significance of these additional peptides is not known. Ophthalmic acid is a potent inhibitor of glyoxalase (page 383).

The chief expression of a deranged lens metabolism is *cataract,* or opacity, resulting from alteration of the proteins with formation of compact fibrous aggregates. In early stages of development of senile cataract, increased hydration of the lens occurs; during late stages, there is expression of fluid from the lens. During development of senile cataract, $[Na^+]$ of the lens increases while $[K^+]$ decreases. At a later time the $[Ca^{++}]$ of the lens also increases, and lens $[Ca^{++}]$ has been used as a basis for classification of the stages of cataract development. The lens then begins to swell; protein loss occurs, possibly because of increase in membrane permeability and/or disruption of synthetic processes. Amino acids diffuse from the lens, which decreases in size. The insoluble albuminoids impart increased rigidity to the lens nucleus. Similar changes have been reported for cataracts associated with diabetes in man.

Cataract is induced by a variety of states characterized by marked elevation of the plasma concentration of a monosaccharide; this occurs most readily in young animals. In diabetes mellitus, fructose and sorbitol accumulate in the lens; glucose metabolism is inhibited, lens [ATP] declines as does lens protein synthesis, but the reason for cataract is unclear. In galactosemia, dulcitol accumulates in the lens in sufficient quantity to increase osmolarity, diluting the lens proteins which then precipitate. Accumulation of galactose 1-phosphate inhibits glycolysis in these lenses. The opacity produced by xylose has been reported to be reversible, in contrast to that resulting from galactose feeding or diabetes, and was associated with restoration to normal of the phosphogluconate oxidative pathway, which is depressed in the opaque lens. Parathyroidectomy has been reported to decrease glucose utilization and lactate production in the lenses of young rats. These processes were restored to normal by addition of parathyroid hormone to lens tissue in vitro. Clouding of the lens is often seen early in clinical hypoparathyroidism (Chap. 43). The lens also becomes opaque in various nutritional deficiencies and after poisoning with several drugs.

The interior of the eye chamber is coated with a dark (light-absorbing) substance that prevents internal reflections which would obscure the image. This material is the melanin of the pigment layer between the nerve tissue of the retina and the vascular choroid tunic. The melanin also determines the color of the iris. Brown eyes result from the presence of melanin; there is no blue pigment in blue eyes, which appear blue because of absorption of longer wavelengths as light is reflected back from the iris stroma.

The *retina* is the photosensitive portion of the eye. Here the energy of light quanta is detected and impulses are sent via the optic nerve to the cerebral cortex. The retina has the highest known rate of oxygen consumption of any tissue in the body, per unit of weight, with an active phosphogluconate oxidative pathway, a high rate of anaerobic and aerobic glycolysis, and an R.Q. of 1. Lactate is the major endogenous substrate contributing to retinal respiration. All of the major pathways of carbohydrate metabolism may play roles in retinal function. Citric acid cycle activity increases markedly during maturation of the photoreceptors, localized in outer segments of the rods. Apparently, the maintenance of visual cell function is strikingly dependent on glycolysis, as evidenced by the selective toxic effect on the retina resulting from intravenous administration of iodoacetate.

The neural part of the retina is essentially a network of synapses and is, therefore, rich in acetylcholine, acetylcholine esterase, and choline acetylase.

The retina, in its nine neural layers, is quite transparent. The underlying pigment epithelium and the scattered pigment in the choroid form a nearly black absorbing layer. That light which is reflected is predominantly red as a result of the rich choroidal blood supply. Many animals possess a mirror-like structure behind the retina, called the *tapetum lucidum*. In dim light, that light which is not absorbed as it passes through the retina is reflected and passed a second time through the photosensitive cells; this affords increased sensitivity of vision in dim light. The tapetum lucidum is responsible for the glowing characteristic of the eyes of cats and dogs at night, as they reflect light. The mirror quality of the tapetum lucidum of carnivores consists of an ordered arrangement of crystals of a complex of zinc and cysteine. Many fish, amphibia, and reptiles possess a tapetum lucidum in which crystalline guanine is deposited in the iris as a reflector. The guanine crystals are rather large, with one surface oriented at 45° to the visual cells.

THE PHOTOCHEMISTRY OF VISION

Light impinging on the retina is absorbed and becomes transformed into another form of energy, presumably chemical. The substances absorbing light are, by definition, pigments, and the initial phenomena of vision are concerned with retinal pigments which absorb the light. It has been assumed that at least three chemical reactions occur initially in visual stimulation: (1) a photochemical reaction in which a pigment absorbs light and is in some manner thereby altered; (2) a second chemical process, independent of light, in which the primary photoproducts of (1) somehow initiate a nerve impulse; and (3) a chemical process, also independent of light, in which the pigment is regenerated from the products of (1) or from other substances. Once the nerve impulse is initiated, the process presumably resembles other sensory mechanisms. Largely through the efforts of Hecht and associates these ideas were given kinetic formulation, and it was shown that much of the quantitative, physiological data of human and animal vision can be explained on the basis of the above assumptions. That these postulated reactions have a chemical basis has been demonstrated by investigations of the light-sensitive pigments *rhodopsin* and *iodopsin.*

The human retina normally contains two types of receptor cells, rods and cones. Animals which have vision only in bright light ("day vision"), such as the pigeon, have only cones, and animals, like the owl, which possess only "night" or "dim vision" have only rods. Thus, at the periphery of the retina in the human eye there are only rods, associated primarily with seeing at low intensities of light. Animals having only rods do not have color vision since color perception is associated with the presence of cones. Most vertebrates possess both rods and cones; only occasional species have but one of these receptors.

ROD VISION

Rhodopsin. The rods contain *rhodopsin,* or *visual purple,* a thermolabile protein which is insoluble in water but soluble in aqueous solutions of detergents like bile salts, digitonin, saponin, and sodium oleate. The molecular weight is approximately 40,000. The absorption spectrum of rhodopsin shows the typical protein absorption near 275 mμ and, more important, a broad absorption in the visible region with a maximum at 500 mμ and a small secondary peak in the ultraviolet near 350

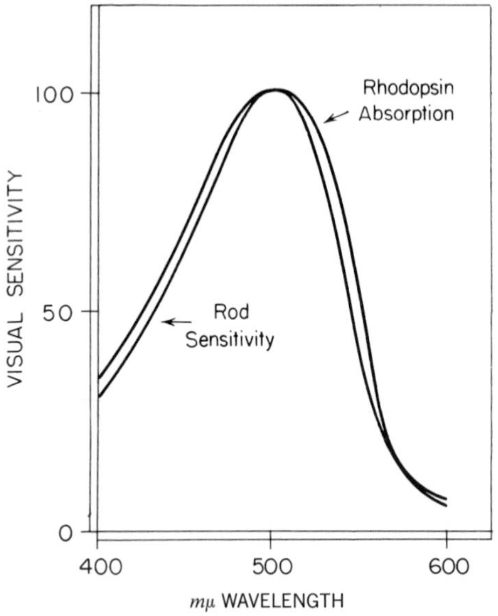

Fig. 40.1. Comparison of sensitivity of rod vision and the absorption of rhodopsin at various wavelengths. The sensitivity curve has been corrected for the effects of transmission of light through the ocular media. (*From S. Hecht, S. Shlaer, and M. H. Pirenne, J. Gen. Physiol.,* **25,** 819, 1942.)

mμ. Estimates of human rod vision in dim light show a sensitivity curve with a maximum at about 500 mμ, gradually diminishing on either side of this wavelength. The fact that the sensitivity curve for human rod vision coincides with the absorption curve for rhodopsin in the visible region (Fig. 40.1) provides strong evidence that rhodopsin is the photosensitive material of rod vision.

Upon bleaching, rhodopsin dissociates to yield the protein, *opsin,* and a carotenoid, *retinal* (formerly retinene or vitamin A_1-aldehyde). If bleached retinas are allowed to stand at room temperature, the retinal is reduced to *retinol* (vitamin A_1), the corresponding alcohol.

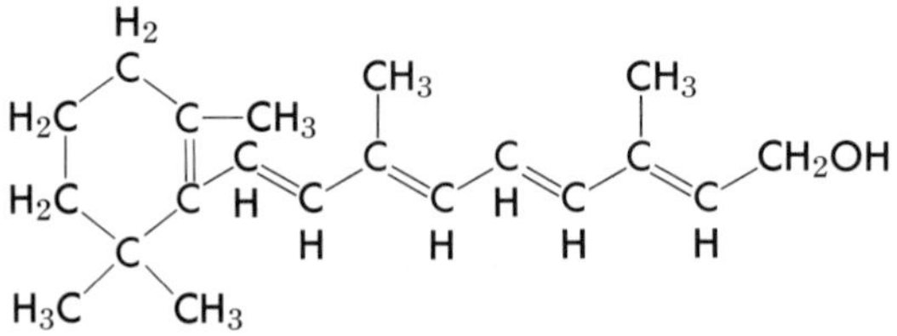

Vitamin A_1
(retinol)

Retinal and retinol contain five double bonds and can exist in various configurations. Synthetic retinol (vitamin A_1) and that in the liver of mammals is in the all-*trans* form as is the retinal liberated by bleaching rhodopsin. All-*trans*-retinal cannot condense with opsin; rhodopsin is formed by combination of opsin with 11-*cis*-retinal. These structures are shown with the hydrogen atoms omitted.

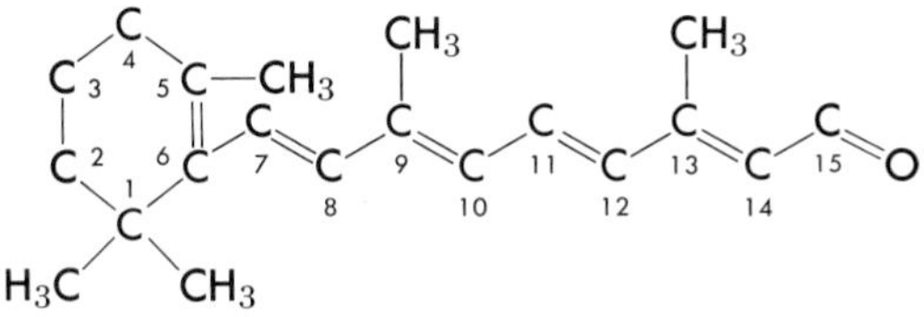

All-trans-Retinal

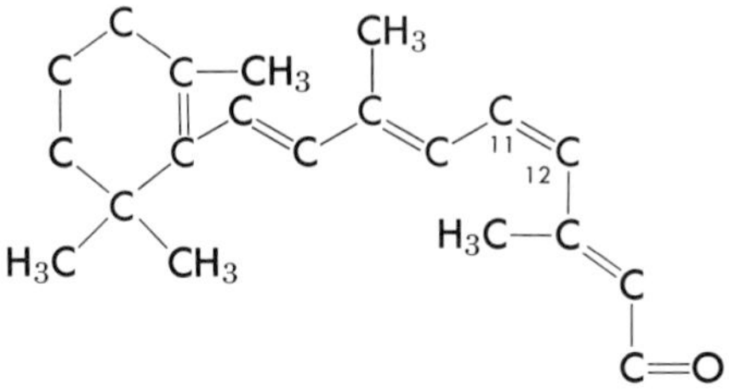

Δ11-cis-Retinal

Absorption of light by rhodopsin causes isomerization of the chromophore from the 11-*cis* to the all-*trans* configuration. This is *the* photochemical event which makes vision possible. A series of dark or thermal reactions then follows, all of them occurring spontaneously at body temperature; indeed, most of them can occur at very low temperatures, as indicated in Fig. 40.2. The approximate absorption maxima for the various intermediates are also given.

It is assumed that after the initial photoisomerization, progressive changes occur in the interaction of the opsin with the all-*trans* retinal. The two forms of metarhodopsin (deep orange and pale yellow) are in a tautomeric equilibrium

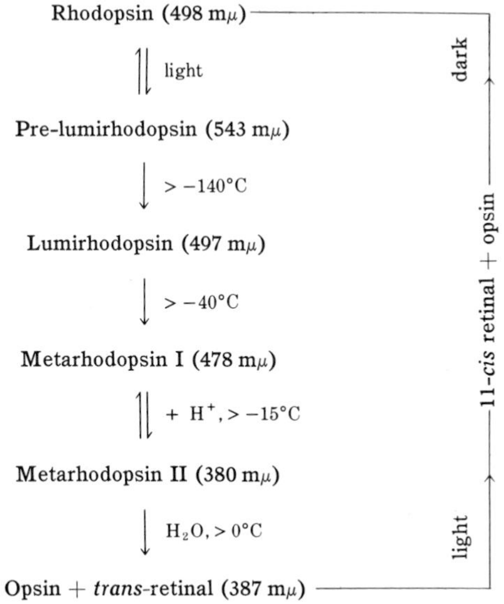

Fig. 40.2. Stages in the bleaching of rhodopsin. (*From R. G. Matthews, R. Hubbard, P. K. Brown, and G. Wald, J. Gen. Physiol.,* **47**, 215, 1963–1964. *Reprinted by permission of the Rockefeller University Press.*)

involving a H^+ ion and manifesting a large shift in the absorption maximum from 478 to 380 mμ. In the final step, metarhodopsin II is hydrolyzed to all-*trans* retinal and opsin, the hydrolysis involving cleavage of a Schiff base between the aldehyde and an ε-amino group of a lysine residue. Transformation of rhodopsin to retinal and opsin is an exergonic process, the energy for which is supplied by the absorbed photon. After pre-lumirhodopsin is formed by the isomerization of one double bond, all subsequent steps proceed spontaneously.

When rhodopsin is treated in the dark with sodium borohydride, reduction does not occur. The action of light in the presence of the reducing agent results in formation of a stable linkage, retinyl—CH_2—NH—opsin. Thus, a conformational change of the protein permits the reduction. Indeed, comparisons of bleached and unbleached rhodopsin show changes in optical rotatory dispersion, circular dichroism, and absorption spectra consistent with changes in protein conformation. Further, opsin is more labile to denaturation by acid, alkali, and heat than is rhodopsin, and bleaching exposes two sulfhydryl groups and one acid-binding group with a $pK' = 6.6$, presumably the imidazole of a histidine residue.

Of great interest are the events involved in generating the visual impulse. Although the exact step is still unknown, it must precede the relatively slow hydrolysis of metarhodopsin II, since the electrical discharge following light stimulation of the retina is more rapid. It has been suggested that the changes in charge distribution involved in the conformational changes of the protein may be responsible for the nerve excitation. Present evidence indicates that one quantum is required for bleaching each chromophoric group of rhodopsin. Although each rod of the human eye contains at least 10^7 molecules of rhodopsin, Hecht has shown that a flash of light is visible when only one rhodopsin molecule in each of five to seven rods has been affected!

Regeneration of rhodopsin is an exergonic process occurring spontaneously by reaction of opsin with 11-*cis*-retinal. However, the ultimate product of bleaching is all-*trans*-retinal, which is reduced by an alcohol dehydrogenase of the retina to retinol. The reaction is reversible with the participation of DPN or TPN.

$$\underset{\textbf{Retinol}}{C_{19}H_{27}\text{—}CH_2\text{—}OH} + DPN^+ \underset{\text{reductase}}{\overset{\text{retinal}}{\rightleftharpoons}} \underset{\textbf{Retinal}}{C_{19}H_{27}\text{—}CH{=}O} + DPNH + H^+$$

The dehydrogenase of retina (*retinal reductase*) appears to be similar to the alcohol dehydrogenase of liver, which will catalyze the above reaction in vitro. In the isolated system, the equilibrium is strongly in favor of alcohol formation.

Although all-*trans*-retinal and retinol can be reisomerized to the 11-*cis* forms by irradiation, particularly by light of short wavelengths, the reactions are relatively slow. In the eye, the reaction is catalyzed by *retinal isomerase.* However, the major site of conversion to the *cis* isomers of both the alcohol and the aldehyde may be the liver where the process must be enzyme-catalyzed. These interrelationships are shown in Fig. 40.3.

In the retina of the intact animal, under constant stimulation, a steady state exists wherein the rate of bleaching and the rate of regeneration are equal. In the dark, regeneration of rhodopsin proceeds to a maximum. Night blindness

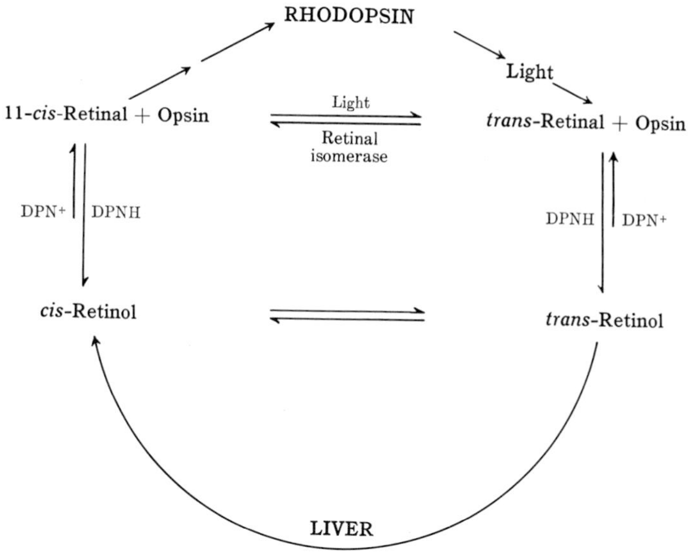

Fig. 40.3. Transformations of the carotenoids in the visual cycle.

(*nyctalopia*) in rats is produced by a diet deficient in vitamin A. The retinas of deficient rats contain less rhodopsin than do normal retinas, and the rate of rhodopsin regeneration is much slower than normal.

Dietary status with respect to vitamin A can be evaluated by measuring the *visual threshold,* the minimal light intensity required to evoke a visual sensation. Dark adaptation is the change in threshold which results from a period in the dark. In man, maximal dark adaptation of the rods requires about 25 min. In experimental studies with human subjects given vitamin A-deficient diets, the visual threshold increases, *i.e.*, higher light intensities are required to evoke the sensation of light. In clinical vitamin A deficiency, the threshold after complete dark adaptation may actually be 100 or more times higher than normal; this elevation of threshold is termed *night blindness.*

It had been known since ancient times that night blindness which develops in human beings is alleviated by feeding liver or liver extracts. It was not until the period 1920 to 1925 that the curative factor was shown to be a dietary essential, vitamin A. Night blindness commonly occurs whenever famine conditions prevail, and the condition may be widespread in parts of the world where vitamin A intake is inadequate.

Porphyropsin. The retinas of mammals, birds, amphibia, marine fish, and invertebrates contain rhodopsins with an absorption maximum near 500 mμ, whereas retinas of all true fresh-water fish have a photosensitive pigment with a maximum absorption near 522 mμ; this pigment is *porphyropsin.* The retina of salmon, which develop in fresh water and migrate to the sea, contains both pigments, with porphyropsin predominant. The retina of the adult frog contains rhodopsin, whereas the bullfrog tadpole has porphyropsin. This change in visual system occurs during metamorphosis of the tadpole to the adult form.

Like rhodopsin, the porphyropsin system undergoes cyclic changes on bleaching and regeneration in which retinol$_2$ and retinal$_2$ replace the analogues of the rhodopsin system. Porphyropsin is a conjugated protein containing the carotenoid, retinal$_2$, the absorption maximum of which is about 22 mμ more toward the red than that of retinal$_1$. A comparable displacement is found in the absorption maxima of the corresponding alcohols. Vitamin A$_2$ differs from A$_1$ by having one additional conjugated double bond in the ring.

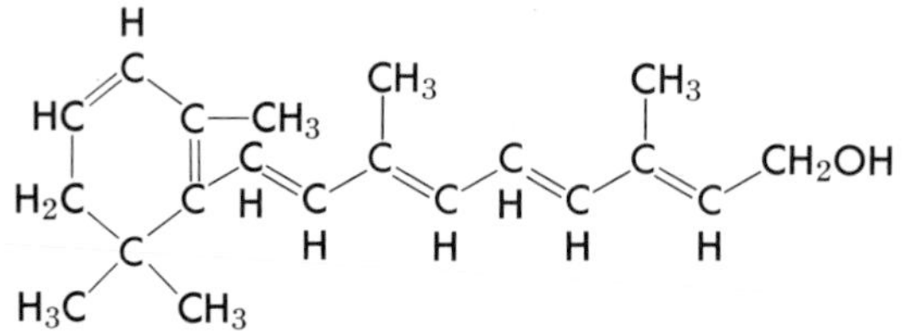

Vitamin A$_2$ (retinol$_2$)

In the case of porphyropsin, as well as all other visual pigments studied (see iodopsin below), it is the 11-*cis*-retinal which is a part of the photosensitive protein. Only the retinals derived from vitamins A$_1$ and A$_2$ are found in all visual pigments; it is the opsins which differ in the various species.

CONE VISION

The pigments responsible for cone vision are of great interest inasmuch as the cones are the dominant sensory elements in the human retina and are responsible for color vision. To account for the data of color vision, it has been assumed that there are probably at least three closely related pigments in cones; this is the Young-Helmholtz trichromatic theory. The spectral sensitivity of human cone vision is greatest at 555 mμ, which indicates that the photosensitive pigments of the cones differ from the rhodopsin found in rods.

Because of the high visual threshold of cone vision, 50 to 100 times higher than that of the rods, it might be anticipated that the cone pigments are present in small amounts. Digitonin extracts of chicken retinas, which are composed chiefly of cones, contain not only rhodopsin but also a substance with an absorption spectrum which corresponds well with the spectrum of visual sensitivity of the fovea. This pigment, *iodopsin,* has an absorption spectrum with a maximum at 555 mμ, contains retinal, but has an opsin which differs from that in the rods. Transformation of iodopsin occurs by stages similar to those described above for rhodopsin. Iodopsin contains 11-*cis*-retinal, and the final products are an opsin and the all-*trans* isomer.

Measurements have been made of the spectra in single human cones before and after bleaching. These data reveal that there are three different pigments, each present in individual cones, although some cones may contain mixtures. The approximate absorption maxima for the three pigments, after correction for absorption by the xanthophyll present in the lens and the macula, are 430 mμ (blue), 540 mμ (green), and 575 mμ (red). These three pigments are thus in accord with the expectations from the trichromatic theory. In isolated human retinas, the red- and green-sensitive pigments are regenerated after bleaching by incubation with 11-*cis*-

retinal in the dark. It is likely that the blue pigment also contains the same chromophore. Thus, it would appear that different opsins are combined with the same retinal in all systems. None of these pigments has yet been chemically separated and identified.

In each of the three types of hereditary color blindness, one of the three pigments is lacking or diminished in amount. Thus, each type of color-blind eye reveals the operation of only two of the three pigments involved in color vision. Since presumably the chromophore is the same for each pigment, the synthesis of the specific opsin is affected by mutation—the red and green opsins being controlled by genes on the X-chromosome and the gene of the blue opsin on an autosome.

REFERENCES

Books

Adler, F. H., "Physiology of the Eye," The C. V. Mosby Company, St. Louis, 1950.

Davson, H., ed., "The Eye," vols. I, II, III, and IV, Academic Press, Inc., New York, 1962.

DeReuck, A. V. S., and Knight, J., eds., "Color Vision," Little, Brown and Company, Boston, 1965.

Pirie, A., and Van Heyningen, R., "Biochemistry of the Eye," Charles C Thomas, Publisher, Springfield, Ill., 1956.

Rushton, W. A. H., "Visual Pigments in Man," Charles C Thomas, Publisher, Springfield, Ill., 1962.

Smelser, G., ed., "The Structure of the Eye," Academic Press, Inc., New York, 1961.

Wolken, J. J., "Biophysics and Biochemistry of the Retinal Photoreceptors," Charles C Thomas, Publisher, Springfield, Ill., 1966.

Review Articles

Hecht, S., Rods, Cones and the Chemical Basis of Vision, *Physiol. Revs.*, **17**, 239–290, 1937.

Hubbard, R., Bownds, D., and Yoshizawa, T., The Chemistry of Visual Photoreception, *Cold Spring Harbor Symp. Quant. Biol.*, **30**, 301–315, 1965.

Hubbard, R., and Kropf, A., Molecular Isomers in Vision, *Scientific American*, **216**, (June) 64–76, 1967.

Langham, M. E., Aqueous Humor and Control of Intra-ocular Pressure, *Physiol. Revs.*, **38**, 215–242, 1958.

Lerman, S., Metabolic Pathways in Experimental Sugar and Radiation Cataracts, *Physiol. Revs.*, **45**, 98–122, 1965.

Pirie, A., The Biochemistry of the Eye Related to Its Optical Properties, *Endeavour*, **17**, 171–189, 1958.

Wald, G., The Distribution and Evolution of Visual Systems, in M. Florkin and H. S. Mason, eds., "Comparative Biochemistry," vol. I, pp. 311–345, Academic Press, Inc., New York, 1960.

Wald, G., The Molecular Organization of Visual Systems, in W. D. McElroy and B. Glass, eds., "Light and Life," pp. 724–749, Johns Hopkins Press, Baltimore, 1961.

Wald, G., and Brown, P. K., Human Color Vision and Color Blindness, *Cold Spring Harbor Symp. Quant. Biol.*, **30**, 345–361, 1965.

41. Cell Walls of Plants and Microorganisms

The most thoroughly studied membrane of mammalian cells is that of the erythrocyte; present knowledge of this structure has been presented in Chap. 31. The separation and characterization of membranes of mammalian cells have been described previously (pages 285*ff.*). Considerably more information is available regarding the structure and synthesis of the membranous structures of plants and microorganisms. This chapter briefly summarizes this information.

All cells are bounded by complex membranes, mosaics of lipids and proteins. The components of these membranes and their manner of arrangement permit diffusion, in both directions, of some molecular species, facilitate diffusion of others, actively transport yet others, and completely prevent passage of some. In varying degree, depending on the species and cell type, enzymes localized in the membrane participate in metabolic processes such as electron transport, phosphorylation, lipid synthesis, etc. The sum of these activities determines, in large measure, the internal composition of the cell, particularly its osmolarity and osmotic pressure. Animal cells which are bathed in a solution isotonic with the cell contents nevertheless derive their characteristic sizes and shapes from whatever factors determine the dimensions of the surrounding membranes. The sizes, shapes, and cohesiveness of animal tissues and organs are, in considerable measure, determined by the arrangement of the connective tissue components which bind, surround, and interlace them. The tensile strength and resistance to compression as well as the over-all dimensions of the limbs, trunk, and head of the chordates are determined by the mineralized connective tissue which serves as the endoskeleton (Chap. 39). Insects, arthropods generally, and crustacea rely similarly on an exoskeleton which is a densely packed arrangement of fibers of chitin, a polymer of N-acetyl D-glucosamine (page 51) and in some instances of resilin (page 877).

CELL WALLS

Plant Cell Walls. The physical problems of plant cells are more severe than are those of animal cells. Plant cells may live in media of extreme hypotonicity, *viz.*, a freshwater pond, or hypertonicity, *viz.*, the sea. The mechanisms of fluid and electrolyte transport in higher plants are dependent upon the development of turgor in the cells of the conducting tissues. In the absence of a mineral skeleton, the material surrounding the cells of land plants must be capable of bearing great weight. These problems have been resolved by the secretion, from the plant cell membrane, of cellulose (page 45). The cellulose is organized in densely packed fibrils in an essentially crystalline array. Surrounding these fibers, which enmesh and encase the cell, is an amorphous matrix of other polysaccharides. These are complex materials, species specific, and of enormous diversity. The most common of these, the "*hemicelluloses,*" which bear no chemical relationship to cellulose, are xylans (page 51), linear polymers of D-xylose in β-1 $\rightarrow$ 4 linkage, bearing side chains of 4-O-methyl glucuronic acid and/or arabinose. Associated with the hemicellulose in the matrix, but not attached to it, are the *pectins* (page 51), polymers of galacturonic acid methyl ester. Individual species may also secrete into this matrix D-galactans, L-arabinans, and diverse heteroglycans. In the matrix, attached to the cellulose fibrils is a significant amount, perhaps 5 per cent of the dry weight, of a protein called *extensin,* unusual in that, like collagen it is rich in hydroxyproline which constitutes one-fifth of the amino acid residues. As in collagen synthesis, hydroxylation of proline occurs after the latter has been bound in the final peptide linkage. The formation and orientation of extensin are thought to determine the growth and direction of the polysaccharide fibers. The structural strength of this casing, in some instances, permits it to withstand an osmotic force as great as 20,000 kg. per cm^2.

The cell walls of algae, fungi, molds, and yeast need not be weight-bearing, but again must be capable of withstanding large osmotic differentials. These organisms also present a great diversity of wall materials which are homo- and heteropolysaccharides. All fungi and some yeasts employ chitin in their wall structures; yeasts also are encased in polymannans which are highly branched. Algae walls also contain large quantities of polymannuronic acids (alginic acid). Knowledge of the structure of many of these materials is incomplete, and little is known of their synthesis, but it is clear that they determine the size, shape, and physical properties of such cells.

Murein. The mechanical problem of encasing bacteria is even more complex. Again, the cell must be able to withstand severe osmotic shock, but the wall must also be capable of growing with the cell, rupturing at an appropriate moment in cell division, and rapidly resealing thereafter. This cell casing has been called a *sacculus* and is, in effect, a single "bag-shaped macromolecule," *murein,* shown in Figs. 41.1 and 41.2, which is present universally among the bacteria, with minor modifications. The repeating peptide consists of L-alanine, D-isoglutamine (the α- rather than the γ-carboxyl is an amide), L-lysine, and D-alanine. Thus, the peptide structure is continued through the γ-carboxyl of the glutamic acid residue. In many organisms, such as *Escherichia coli,* diaminopimelic acid (page 569) replaces the lysine and, depending on the species, may be in the L,L; D,D; or D,L configuration. Less commonly, this position is occupied by L-hydroxylysine (*Staphylococcus aureus*), D-ornithine (*Corynebacteria*), L-ornithine (*Micrococcus radiodurans*), or 2,4-diaminobutyric acid (*Corynebacterium tritici*). In such instances, the structure bears an additional negative charge. As indicated in Fig. 41.2, the polysaccharide exists as a linear chain, and neighboring chains, in three dimensions, can be bridged by links to the peptide moiety. In some organisms (*e.g., E. coli*) cross-links are formed directly between the carboxyl of the D-alanine residue and the amino group of the diamino acid constituent. More commonly, however, these two residues are bridged by a peptide;

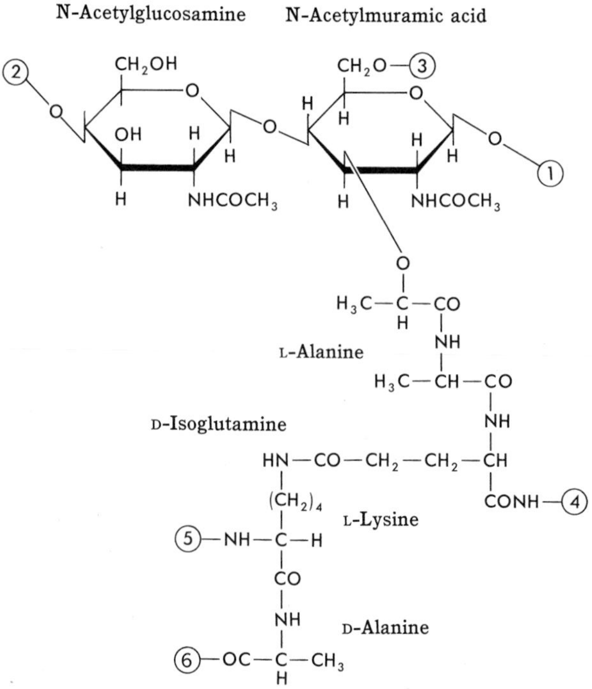

Fig. 41.1. Repeating unit of a murein (peptidoglycan). The components of the unit are indicated in the figure. In intact murein, the open bonds continue as follows: 1 and 2 continue the polysaccharide chain and lead to the next N-acetylglucosamine and N-acetylmuramic acid units, respectively. 3 represents an opportunity for linkage to another polysaccharide, in glycosidic linkage, or to a teichoic acid in phosphodiester linkage. 4 is usually filled by a hydrogen atom (isoglutamine), but occasionally this carboxyl group exists in amide linkage to glycine or another amino acid. 5 can represent an amide bond to the carboxyl group of the D-alanine in another unit 6, or two units from neighboring polysaccharide chains can be bridged by a peptide, such as pentaglycine, which leads from bond 5 on one unit to bond 6 on the next.

the larger the peptide, the more open the resulting network. This is a pentaglycine unit in *Micrococcus lysodeikticus* and *S. aureus,* as indicated in Fig. 41.2; trialanylthreonine functions in this role in *Micrococcus roseum,* while there are polyserine bridges in *Staphylococcus epidermis.*

Elucidation of these structures has been made possible by utilization of enzymes, largely from bacterial sources, each of which is specific for one of the bonds in the murein structure, designated by the encircled letters in Fig. 41.2. Such enzymes are "lysins," and bacteria exposed to them may lose their cell walls and lyse in hypotonic media.

Synthesis of Murein. Murein synthesis, as elucidated by Strominger and his coworkers, may be regarded as proceeding in three stages: synthesis of the precursor units, synthesis of the linear peptidoglycan strands, and cross-linking. The process commences with formation of UDP-N-acetylmuramic acid from UDP-N-acetylglucosamine by condensation with phosphoenolpyruvate to give an enolic ether which is converted to the lactyl ether. In consecutive steps that require ATP and, respectively, L-alanine, D-glutamic acid, and L-lysine, a UDP-N-acetylmuramyl tripeptide is constructed. D-Alanyl-D-alanine is formed separately by the pyridoxal-dependent

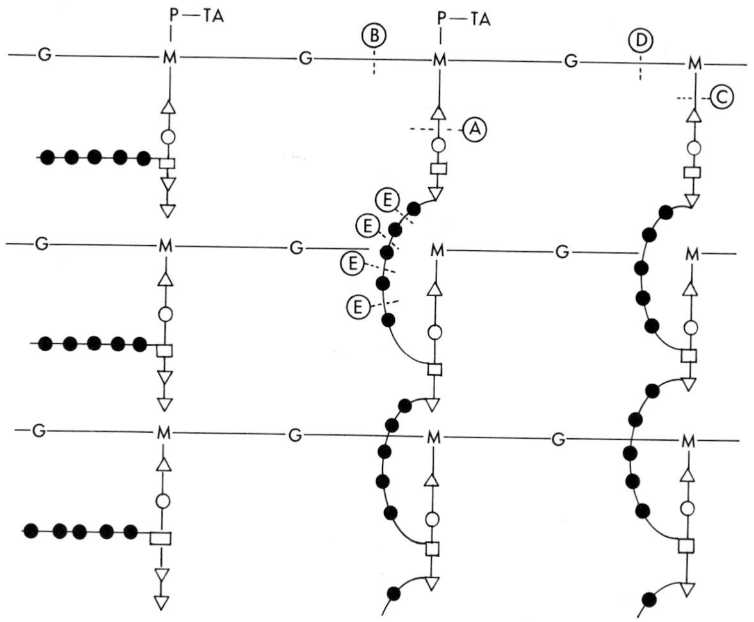

Fig. 41.2. Polymeric murein as the basic structure of a cell wall. G = N-acetylglucosamine; M = N-acetylmuramic acid; $P - Ta$ = phosphodiester to teichoic acid, △ = L-alanine, ○ = D-isoglutamine, □ = L-lysine, ▽ = D-alanine, ● = glycine. The pentaglycine chains to the left are in amide linkage to lysine but have not yet been connected to the carboxyl of D-alanine. The completed pentapeptide bridges are all shown connecting parallel polysaccharide chains in the plane of the paper. However, they can equally readily extend to equivalent murein sheets above or below the plane of the paper and thus thicken and rigidify the wall. The encircled letters A through E represent the sites of action of specific hydrolytic enzymes obtained from bacterial sources.

enzymic racemization of L-alanine and a condensation which utilizes ATP. This process is highly sensitive to the antibiotic *cycloserine.* A final ATP-dependent step results in formation of a UDP-N-acetylmuramylpentapeptide. The latter, therefore, has two D-alanyl residues at its free C-terminus, whereas only one is present in the murein structure. All these steps are catalyzed by membrane-bound enzymes.

(*a*) **UDP-N-acetylglucosamine + phosphoenolpyruvate ⟶**
UDP-N-acetylmuramic acid + P_i

(*b*) **UDP-N-acetylmuramic acid + L-alanine + ATP ⟶**
UDP-N-acetylmuramyl-L-alanine + ADP + P_i

(*c*) **UDP-N-acetylmuramyl-L-alanine + D-glutamic acid + ATP ⟶**
UDP-N-acetylmuramyl-L-alanyl-D-glutamic acid + ADP + P_i

(*d*) **UDP-N-acetylmuramyl-L-alanyl-D-glutamic acid + L-lysine + ATP ⟶**
UDP-N-acetylmuramyl-L-alanyl-D-isoglutamyl-ε-N-L-lysine + ADP + P_i

(*e*) **UDP-N-acetylmuramyl-L-alanyl-D-isoglutamyl-ε-N-L-lysine + D-alanyl-D-alanine + ATP ⟶ UDP-N-acetylmuramyl-L-alanyl-D-isoglutamyl-ε-N-L-lysyl-D-alanyl-D-alanine + ADP +** P_i

In the next stage the N-acetylmuramylpentapeptide is transferred to a membrane-bound carrier which is a C_{55} terpene, a linear sequence of 11 isoprenoid units, each of which has one double bond and terminates in an alcohol which, in the absence of the carbohydrate, is esterified to phosphate. Such structures have been named *bactoprenols.*

$$\begin{array}{c}\quad\;\; CH_3 \qquad\qquad\qquad\qquad\qquad\qquad\qquad CH_3 \\ H(H_2C{-}\overset{|}{C}H{=}CH{-}CH_2)_{10}{-}CH_2{-}\overset{|}{C}H{=}CH{-}CH_2O{-}PO_3H_2\end{array}$$

A bactoprenol phosphate

Bactoprenol phosphate + UDP-N-acetylmuramylpentapeptide ⟶ bactoprenol pyrophosphate-N-acetylmuramylpentapeptide + UMP

The carbohydrate is linked at C-1 through a pyrophosphate bridge to the lipid-soluble carrier. In this form it reacts next with UDP-N-acetylglucosamine to form the disaccharide pentapeptide, still linked to the carrier. In an ATP-dependent reaction, the α-carboxyl group of the D-glutamic acid residue is then either amidated or fixed into amide linkage with glycine.

The entire unit is thus completed and is then transferred from its membrane-bound site on the carrier to the free growing end, *i.e.*, the oxygen at C-4 of an N-acetylglucosamine residue already in place in the growing murein structure. To grow the bridging unit, as in *M. lysodeikticus,* five glycine residues are then transferred, seriatim, from glycyl t-RNA, first to the α-amino group of the lysine residue and then to the amino group of the preceding glycine. When other amino acids are utilized for the bridging structure, they are transferred from their corresponding specific t-RNA's. The final step is a transpeptidation, the terminal amino group of the pentaglycine chain displacing the C-terminal D-alanine, which is released from the structure. It is this step which is particularly sensitive to penicillin in those organisms whose growth is inhibited by this antibiotic. Thus, even after transfer of the disaccharide pentapeptide from the bactoprenol carrier, a series of reactions must occur; presumably each of these is catalyzed by an appropriate enzyme in or from the membrane.

In addition to the murein sacculus, the exterior of all cells is coated with other macromolecular structures. None of these provides the mechanical strength offered by the murein. Some are of sufficient size and afford cross-linking opportunities which permit them to thicken the basic wall structure; several are known to be covalently attached to the peptidoglycan. Many bacteria secrete "slimes" of capsular polysaccharides which vary from the relatively simple hyaluronic acid (page 51) to highly branched structures containing several monosaccharide components of various types. All these may be presumed to affect the transport of nutrients and to afford some protection against noxious agents, including potentially invasive viruses. Spore-forming organisms produce coats of tough polymers not readily digestible by enzymes likely to be encountered in their environment. Particularly noteworthy are the polymers of D-glutamic acid, linked through the γ-carboxyl, which constitute the spore coats of several of the *Bacilli.* Two major classes of secondary investiture are sufficiently widespread and significant to warrant discussion.

Teichoic Acids. The surfaces of all Gram-positive organisms contain teichoic acids, polymers of glycerol or ribitol in phosphodiester linkage. Associated with the membrane there is invariably a glycerol-type teichoic acid. In the space between membrane and murein there is frequently a teichoic acid, which may contain either

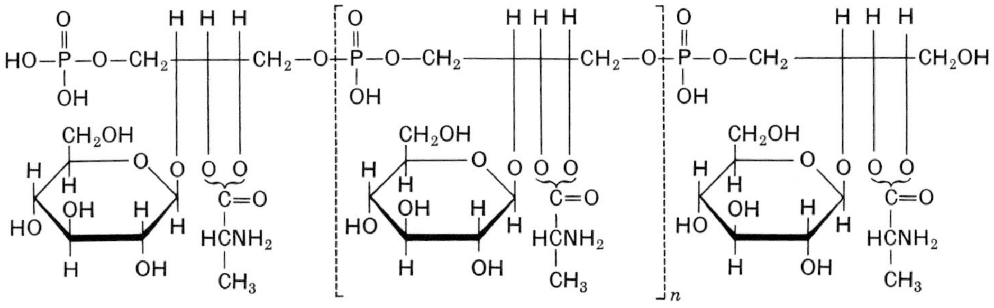

Bacillus subtilis teichoic acid (wall)

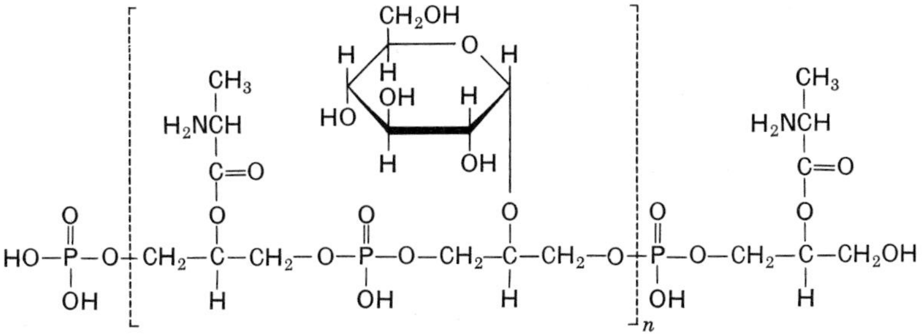

Lactobacillus arabinosus teichoic acid (intracellular)

Fig. 41.3. Proposed structures for two teichoic acids. The positions of the D-alanine residues in *Bacillus subtilis* teichoic acid are not established, but are at either C-2 or C-3 of the ribitol residues, as indicated.

glycerol or ribitol, as shown in Fig. 41.3. The available hydroxyl in the 2-position of glycerol is most frequently esterified in membrane teichoic acids to alanine (*e.g.*, *Lactobacillus arabinosus, Streptococcus lactis, S. aureus*), but it is also in β-glycosidic linkage to N-acetylglucosamine (*S. aureus*), gentiobiose (*S. aureus*), and kojitriose (*Streptococci,* group D). Glycerol-containing teichoic acids may, in addition to the above sugars, contain N-acetylgalactosamine (*S. lactis*). *Streptococcus lactis* also contains a teichoic acid in which the polymer is constructed of phosphodiester bridges between glycerol and the 1- and 4-positions of N-acetylglucosamine; the polymer structure may be represented as

-P-sugar-P-glycerol-P-sugar-P-glycerol-, etc.

When ribitol is present, the sugar substituent appears always to be linked to the 2-position of this alditol while the D-alanine can be on either of the other available positions. In a few instances, uronic substituents are present; this group is designated as *teichuronic* acids. As indicated previously, when the teichoic acid is bound to the murein, it may exist in phosphodiester linkage to the C-6 of the N-acetylmuramic acid.

The teichoic acids are synthesized by enzymes localized in the bacterial membrane. After reduction of ribulose 5-phosphate by DPNH to yield D-ribitol 5-phosphate (L-ribitol 1-phosphate), a pyrophosphorylase catalyzes formation of CDP-ribitol. CDP-glycerol is formed similarly. Polymerases then catalyze polymer formation. The sugar substituents are inserted using the appropriate UDP-derivatives.

Lipopolysaccharides. Gram-negative organisms that elaborate capsules and "slime" which cover the murein also produce complex lipopolysaccharides. The latter are of interest because they serve as the characteristic somatic O antigens of the noncapsulated organisms of this class, some of which are highly toxic to animals. Again, a great diversity of structures is found among these compounds, involving a wide variety of monosaccharides. Particular attention has been directed to the lipopolysaccharides of the *Enterobacteriaceae,* the most complex of all polysaccharides encountered to date, and of somewhat uncertain structure. This polysaccharide may be divided into three distinct regions. The outermost portion consists of long side chains of a repeating unit which is as follows.

$$\left(\begin{array}{l}\textbf{abequose}\\ \quad\,|\\ \textbf{—mannose—rhamnose—galactose}\end{array}\right)_n$$

In turn each long, branched side chain is attached to a single unit which is

$$\left(\begin{array}{r}\textbf{N-Ac-glucosamine-glucose-galactose-glucose—}\\ |\quad\;\;\\ \textbf{galactose}\end{array}\right)$$

in which the galactose-glucose linkage is α-1,3 in the longer chain and α-1,6 in the branch. The latter unit is attached to a "backbone" repeating unit which consists of two heptose molecules and an *octulosonic* acid.

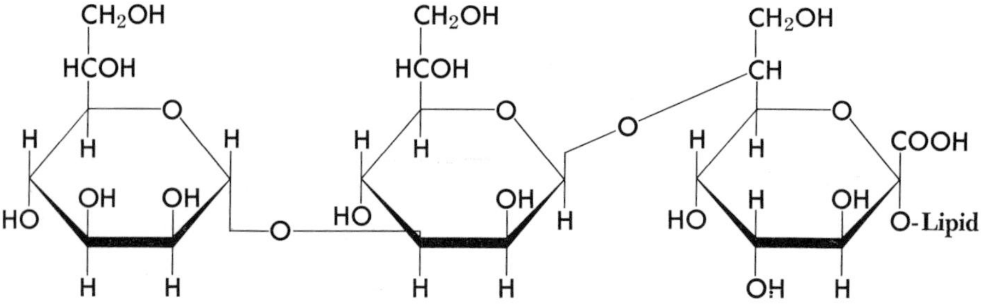

The "lipid" fraction itself includes glucosamine, phosphate, acetyl residues, and β-hydroxymyristic acid in unknown combination. The last acid is thought to be in amide linkage to the glucosamine.

The metabolic origins of all components of the side chains and core pentasaccharide unit have been presented earlier (Chap. 19). The ketooctulosonic acid arises by condensation of arabinose and phosphoenolpyruvic acid, in a manner analogous to the formation of N-acetylneuraminic acid (page 428). The synthesis of the core pentasaccharide has been demonstrated with preparations from *E. coli.* Commencing with UDP-glucose, each of the other members is attached, under the influence of a specific transferase, utilizing their UDP-derivatives. The site of attachment to the backbone of heptose and octulosonate is not known.

The synthesis of the outer, antigenic side chain occurs in a manner which parallels that of the repeating unit of the murein structure. In the initial step, UDP-galactose reacts with a terpenol phosphate which appears to be identical with bactoprenol. The other three sugars are then added, sequentially, while the growing oligosaccharide unit remains fixed to the bactoprenol.

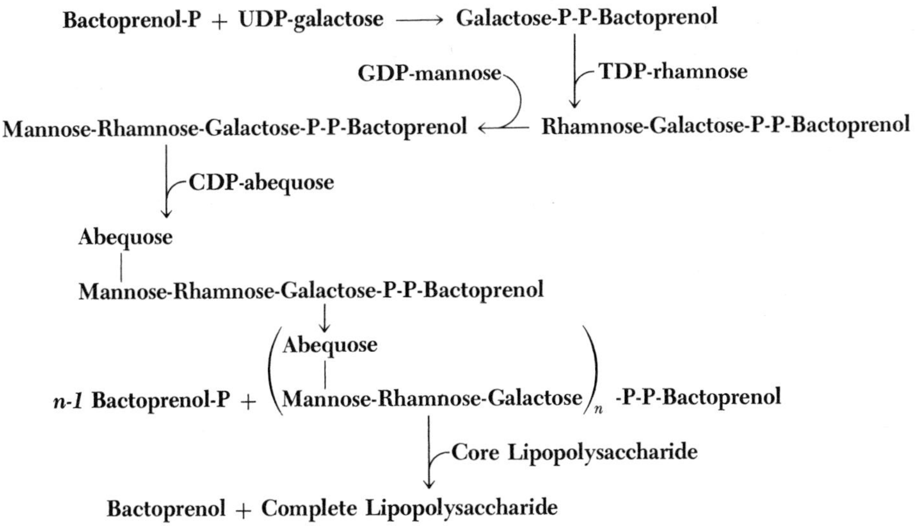

The role of bactoprenol in these synthetic processes is reminiscent of that of the acyl carrier protein in fatty acid synthesis and of tRNA in protein synthesis. Indeed, in further analogy to these processes, the completed oligosaccharide unit is fixed to the polysaccharide chain at the 'head' or potentially reducing end, thus,

$$\text{Bactoprenol}_1\text{—}\bullet + \text{Bactoprenol}_2\text{—O-O-O-O-O-O} \longrightarrow$$
$$\text{Bactoprenol}_2 + \text{Bactoprenol}_1\text{—}\bullet\text{-O-O-O-O-O-O}$$

where each circle represents a complete oligosaccharide repeating unit. In view of the role of bactoprenol in murein synthesis, it seems likely that the peptidoglycan grows in a similar manner, in contrast to the mode of growth of homopolysaccharides such as glycogen or cellulose. It remains to be established whether this process is also a model for the growth of other large heteropolysaccharides.

REFERENCES

Books

Salton, M. R. J., "The Bacterial Cell Wall," Elsevier Publishing Company, Amsterdam, distributed by American Elsevier Publishing Company, New York, 1964.

Vogel, H. J., Lampen, J. O., and Bryson, V., eds., "Organizational Biosynthesis," Academic Press, Inc., New York, 1967.

Review Articles

Archibald, A. P., and Baddiley, J., The Teichoic Acids, *Advances in Carbohydrate Chem.,* **21,** 323–376, 1966.

Horecker, B. L., Biosynthesis of Bacterial Polysaccharides, *Ann. Rev. Microbiol.,* **20,** 253–290, 1966.

Lamport, D. T. A., The Protein Component of Primary Cell Walls, *Advances in Botanical Research,* **2,** 151–218, 1965.

Luderitz, O., Staub, A. M., and Westphal, O., Immunochemistry of the *O* and *R* Antigens of *Salmonella* and Related *Enterobacteriaceae, Bacteriol. Revs.,* **30,** 192–255, 1966.

Martin, H. H., Biochemistry of Bacterial Cell Walls, *Ann. Rev. Biochem.,* **35,** 457–484, 1966.

Osborn, M. J., and Weiner, I. M., Mechanisms of Biosynthesis of the Lipopolysaccharide of *Salmonella, Federation Proc.,* **26,** 70–77, 1967.

Roelfsen, P. A., Ultrastructure of the Wall in Growing Cells and Its Relation to the Direction of Growth, *Advances in Botanical Research,* **2,** 69–147, 1965.

Salton, M. J. R., Chemistry and Function of Amino Sugars and Derivatives, *Ann. Rev. Biochem.,* **34,** 143–174, 1965.

Strominger, J. L., Izaki, K., Matsuhashi, M., and Tipper, D. J., Peptidoglycan Transpeptidase and D-Alanine Carboxypeptidase: Penicillin-sensitive Reactions, *Federation Proc.,* **26,** 9–22, 1967.

Weidel, W., and Pelzer, H., Bagshaped Macromolecules—a New Outlook on Bacterial Cell Walls, *Advances in Enzymol.,* **26,** 193–232, 1964.

42. General Considerations of the Endocrine Glands

The roles of enzymes in metabolism have been presented throughout the preceding chapters. The *hormones* also influence the velocity of cellular transformations and represent a group of regulators which have developed later in evolution. In contrast to the enzymes, which are catalysts, hormones do not *initiate* reactions but can influence the *rate* at which they proceed. However, many physiological processes may continue, although at a slower or faster rate, in the complete absence of one or more hormones. Hormones also serve as integrating influences in more complex organisms. All hormones appear to have a short physiological half-life. Therefore, for hormones to function as regulators in the maintenance of the normal state, and to achieve their regulation in feed-back systems, hormones must be continually synthesized and secreted, must exert their effects rapidly, and, in turn, be rapidly inactivated by metabolic transformations. Therefore the synthesis, secretion, effects and mode of action, and metabolic fate of hormones are topics of significance for understanding of regulatory and control phenomena of metabolism.

The term hormone was first used by Bayliss and Starling in 1902 and is derived from a Greek root meaning "to excite" or "to arouse." These investigators found that intravenous injection of an acidic extract of the duodenal mucosa into dogs prepared with a pancreatic fistula and a denervated pancreas produced a marked flow of pancreatic juice. Thus, the physiological effect could have been mediated only by way of the blood. From this developed the concept of hormones as substances produced by specific glands, secreted directly into the blood, and transported to various organs and tissues where they exert their effects.

Hormones function at low concentrations, generally less than $10^{-8}M$, are produced at a variable rate, and exert a regulatory function in response to environmental or other variations. Arbitrarily, the term hormone is limited to the products of recognized endocrine organs and does not include other agents which may fit the above description. Actually, a number of tissues synthesize, and add to the blood, substances that induce responses in structures other than those in which they are produced. Thus, the hypothalamus secretes several polypeptides with distinct simulating action on the adenohypophysis (Chap. 47). The vasopressor peptide (page 839) derived from the action of renin secreted by the kidney has cardiovascular effects. Indeed, carbon dioxide, which stimulates the respiratory center (page 793), could satisfy the above definition of a hormone. Nevertheless, the restriction to specific

Table 42.1: MAJOR ENDOCRINE GLANDS IN VERTEBRATES

Endocrine gland and hormone	Principal site of action	Principal phenomena affected
Thyroid:		
Thyroxine and triiodothyronine......	General	Metabolic rate and oxygen consumption of tissues
Calcitonin	Skeleton	Metabolism of calcium and phosphorus
Parathyroids:		
Parathormone.....................	Skeleton, kidney, gastrointestinal tract	Metabolism of calcium and phosphorus
Calcitonin	Skeleton	Metabolism of calcium and phosphorus
Testis:		
Testosterone......................	Accessory sex organs	Maturation and normal function
	General	Development of secondary sex characteristics
Seminal vesicle		
Prostaglandins*	Smooth muscle; adipose tissue	Blood pressure; smooth muscle; lipid metabolism
Ovary:		
Estrone and estradiol...............	Accessory sex organs	Maturation and normal cyclic function
	Mammary glands	Development of duct system
	General	Development of secondary sex characteristics
Corpus luteum:		
Progesterone......................	Uterus	Preparation for ovum implantation; maintenance of pregnancy
	Mammary glands	Development of alveolar system
Relaxin...........................	Symphysis pubis	Muscle tone
Placenta:		
Estrogens.........................	Same as ovarian hormones	Same as ovarian hormones
Progesterone......................	Same as corpus luteum hormone	Same as corpus luteum hormone
Gonadotropin.....................	Same as adenohypophyseal gonadotropins (LH and FSH)	Similar to but not identical with adenohypophyseal hormones
Relaxin...........................	Same as corpus luteum hormone	Same as corpus luteum hormone
Adrenal medulla:		
Epinephrine.......................	Heart muscle; smooth muscle; arterioles	Pulse rate and blood pressure; contraction of most smooth muscle
	Liver and muscle	Glycogenolysis
	Adipose tissue	Release of lipid
Norepinephrine....................	Arterioles	Contraction, increased peripheral resistance
	Adipose tissue	Release of lipid
Adrenal cortex:		
Adrenal cortical steroids............	General	
Aldosterone.....................	....	Metabolism of electrolytes and water

*Present in other tissues also (page 943).

Table 42.1: MAJOR ENDOCRINE GLANDS IN VERTEBRATES (*Continued*)

Endocrine gland and hormone	Principal site of action	Principal phenomena affected
Adrenal cortex (Cont.):		
Corticosterone; Cortisol	. . . .	Metabolism of proteins, carbohydrates, and lipids; maintenance of circulatory and vascular homeostasis; inflammation, immunity and resistance to infection; hypersensitivity
Pancreas:		
Insulin	General	Utilization of carbohydrate; stimulation of protein synthesis
	Adipose tissue	Lipogenesis
Glucagon	Liver	Glycogenolysis
	Adipose tissue	Release of lipid
Adenohypophysis:		
Luteotropin (prolactin)	Mammary gland	Proliferation and initiation of milk secretion
	Corpus luteum	Final development and functional activity, *i.e.*, secretion of progesterone
Adrenocorticotropin (ACTH)*	Adrenal cortex	Formation and/or secretion of adrenal cortical steroids
	Adipose tissue	Release of lipid
Thyrotropin (TSH)	Thyroid	Formation and secretion of thyroid hormone
	Adipose tissue	Release of lipid
Somatotropin (growth hormone)*	General	Growth of bone and muscle; anabolic effect on calcium, phosphorus, and nitrogen metabolism; metabolism of carbohydrate and lipid; elevation of muscle and cardiac glycogen
Luteinizing or interstitial cell-stimulating hormone (LH or ICSH)	Ovary	Luteinization; secretion of progesterone (see FSH)
	Testis	Development of interstitial tissue; secretion of androgen
Follicle-stimulating hormone (FSH)	Ovary	Development of follicles; with LH, secretion of estrogen and ovulation
	Testis	Development of seminiferous tubules; spermatogenesis
Neurohypophysis:		
Oxytocin	Smooth muscle, particularly uterine	Contraction, parturition
	Mammary gland, postpartum	Ejection of milk

* Present in placenta also.

Table 42.1: Major Endocrine Glands in Vertebrates (*Continued*)

Endocrine gland and hormone	Principal site of action	Principal phenomena affected
Neurohypophysis (*Cont.*):		
Vasopressin (antidiuretic hormone) . . .	Arterioles Kidney tubules	Blood pressure Water reabsorption
Pars intermedia:		
Melanocyte-stimulating hormone (MSH).	Melanophore cells	Pigment dispersal, with darkening of skin
Pineal:		
Melatonin; 5-Hydroxytryptophol.	Melanophore cells Gonads	Pigment aggregation, with lightening of skin color Gonadal function
*Thymus**:		
Thymosin.	Lymphoid tissue	Stimulation of lymphocytopoiesis
Alimentary tract†:		
Gastrin.	Stomach	Secretion of acid
Secretin.	Pancreas	Secretion of pancreatic juice
Pancreozymin‡	Pancreas	Secretion of digestive enzymes
Cholecystokinin‡.	Gallbladder	Contraction and emptying

*See pages 754*ff*; endocrine status indicated but not established.
†See pages 814*ff*.
‡Pancreozymin and cholecystokinin are the same polypeptide (see page 818).

organs and their products is useful for classification of certain chemical and biological information.

The hormone-producing organs are the glands of internal secretion, or *endocrine* glands. *Endocrinology* is the study of the structure and function of the endocrine glands and their secretory products, the hormones, including the consequences of excessive or deficient production of these hormones. The biochemist seeks to establish for each hormone its chemical structure, mechanism of secretion, metabolic effects and fates, and, ultimately, mechanism of action at cellular and enzymic levels. This last has been achieved at a cellular level for several hormones, and specific enzymic reactions have been delineated as the site of action of others. However, the precise mechanism, at a molecular level, by which hormone-enzyme interaction may alter the rate of an enzyme-catalyzed reaction has not yet been elucidated for a single hormone. It is also recognized that hormonal influence on many enzymic reactions may consist of regulating enzyme synthesis and/or substrate supply rather than interaction of hormone with the catalyst.

Not only does each endocrine gland affect a variety of processes, but these glands also affect the functioning of one another. Therefore, there exists an integrative functioning of the endocrine system of organs, reflected under normal circumstances in a high degree of hormonal balance. Derangements of this balance, either experimental or clinical, give rise to a variety of metabolic aberrations, the study of which has contributed to our understanding of health and disease.

Table 42.1 lists some information regarding the major endocrine glands. The effects listed in the table are the results observed in an intact animal, generally following systemic hormone administration. In view of the interrelationships among metabolic pathways, many hormonal actions, while of significance for the economy of the organism, are often secondary consequences of a primary, as yet unrecognized event, at cell membranes, and within cells or subcellular organelles.

REFERENCES

Books

Antoniades, H. N., ed., "Hormones in Human Plasma," Little, Brown and Company, Inc., Boston, 1960.

Butt, W. R., "Hormone Chemistry," D. Van Nostrand Company, Ltd., London, 1967.

Cannon, W. B., "The Wisdom of the Body," 2d ed., W. W. Norton & Company, Inc., New York, 1939.

Dorfman, R. I., ed., "Methods in Hormone Research," vols. I–IV, Academic Press, Inc., New York, 1962–1966.

von Euler, U. S., and Heller, H., eds., "Comparative Endocrinology," vols. I and II, Academic Press, Inc., New York, 1963.

Gray, C. H., and Bacharach, A. L., eds., 2d ed., "Hormones in Blood," 2 vols., Academic Press, Inc., New York, 1967.

Martini, L., Fraschini, F., and Motta, M., "Hormonal Steroids," Excerpta Medica Foundation, Amsterdam, 1967.

Martini, L., and Pecili, A., "Hormonal Steroids," vols. I and II, Academic Press Inc., New York, 1964, 1965.

Pincus, G., ed., "Recent Progress in Hormone Research," vols. I–, Academic Press, Inc., New York, 1945–current. An annual volume presenting reviews of current research in endocrinology.

Pincus, G., Thimann, K. V., and Astwood, E. B., eds., "The Hormones: Physiology, Chemistry and Applications," vols. I–V, Academic Press, Inc., New York, 1948–1964.

Scharrer, E., and Scharrer, B., "Neuroendocrinology," Columbia University Press, New York, 1963.

Tepperman, J., "Metabolic and Endocrine Physiology," 2d ed., Year Book Medical Publishers, Inc., Chicago, 1967.

Turner, C. D., "General Endocrinology," 3d ed., W. B. Saunders Company, Philadelphia, 1960.

Williams, R. H., ed., "Textbook of Endocrinology," 4th ed., W. B. Saunders Company, Philadelphia, 1967.

43. The Thyroid and Parathyroids

Thyroxine and Triiodothyronine. Calcitonin. Parathormone

The thyroid gland secretes thyroxine and triiodothyronine, two iodine-containing compounds that influence the rate of general metabolic activity. In addition, the thyroid produces a second type of hormone, calcitonin, also termed thyrocalcitonin, a polypeptide which participates in the regulation of calcium metabolism. The parathyroid hormone, secreted by the parathyroid glands, is a second prime factor affecting calcium metabolism.

THE THYROID. ORGANIC IODINE-CONTAINING COMPOUNDS

The iodine of the thyroid gland is present chiefly in thyroglobulin, a protein of the "colloid." Thyroglobulin has a molecular weight of 660,000 with a sedimentation constant of 19 S, an iodine content that varies from 0.5 to 1.0 per cent, and approximately 10 per cent carbohydrate as galactose, mannose, glucosamine, fucose, and a sialic acid. There are two types of polysaccharide units. The bond between the polysaccharide units and the peptide portion of thyroglobulin is probably a glycosylamine between N-acetylglucosamine and the nitrogen atom of an asparagine residue.

Thyroglobulin consists of four approximately equal subunits, some of which are held together by several disulfide bonds, and others by noncovalent bonds. Alterations in ionic strength and pH of solutions of thyroglobulin cause reversible dissociation and association of aggregates, with sedimentation constants varying over the range 6 to 27 S.

Hydrolysis of thyroglobulin yields those amino acids found commonly as constituents of proteins, and, in addition, iodinated derivatives of three amino acids, L-histidine, L-tyrosine, and L-thyronine. The 4-monoiodo-L-histidine present accounts for less than 3 per cent of the total iodine of thyroglobulin, is devoid of thyroid-like activity, and is a by-product of thyroid hormone biosynthesis. The mono- and diiodotyrosines are also without biological activity. Only the iodothyronines exert the characteristic physiological effects of thyroid preparations and are the iodinated products normally secreted by the gland. These compounds are present chiefly as constituents of the thyroglobulin; 1 per cent or less is present in the free state in the thyroid. Four iodinated thyronines (Table 43.1) have been identified in the thyroid; the most important physiologically are thyroxine and

Table 43.1: Iodine-containing Organic Compounds of the Thyroid Gland

Compound	Date of discovery in thyroid	Per cent of total iodine in human thyroid gland	Biological activity relative to thyroxine
3,5-Diiodotyrosine.......	1911	25–42	0
Thyroxine..............	1915	35–40	100
3,5,3′-Triiodothyronine...	1952	5–8	500–1,000
3,3′,5′-Triiodothyronine...	1954	<1	5
3,3′-Diiodothyronine.....	1955	<1	15–75
3-Monoiodotyrosine......	....	17–28	0

3,5,3′-triiodothyronine (page 94). The L-isomers of these compounds are the natural and biologically active forms.

BIOSYNTHESIS AND SECRETION OF IODINE-CONTAINING THYROID HORMONES

There is a continuous turnover of total iodine of the thyroid gland as a consequence of uptake of iodide from the blood and synthesis and secretion of thyroid hormones. This turnover has three aspects: (1) iodide entry into the gland, (2) conversion of iodide to iodine-containing residues of thyroid proteins, and (3) proteolysis, and secretion of thyroid hormones.

Iodide Accumulation by the Thyroid. The thyroid gland exhibits a remarkable capacity to collect injected or ingested iodide, which is accumulated rapidly in the colloid of the thyroid follicles. This occurs against a concentration gradient by an "active transport" mechanism (page 780). The gland usually concentrates iodide to at least twenty-five times its concentration in plasma, rapidly converting the iodide to organically bound iodine. Normally the thyroid contains approximately 10 μg exchangeable or free iodide, compared with 7,500 μg organically bound iodine.

The iodide-concentrating mechanism has been referred to as an "iodide pump" or "iodide trap." Its action is dependent upon a concomitant K^+ influx and Na^+ efflux, is stimulated by adenohypophyseal thyrotropic hormone (Chap. 47), and is inhibited by certain inorganic ions, *e.g.*, thiocyanate and perchlorate (see page 930). Other glands, *e.g.*, the salivary and those of the gastric mucosa, also concentrate iodide. Thus, saliva, gastric juice, and milk have iodide concentrations fourteen to forty-eight times that of plasma, but the glands producing these fluids cannot store appreciable quantities of iodine, do not form thyroxine or triiodothyronine, and do not respond to thyrotropic hormone.

Formation of Organically Bound Iodine Compounds. The reactions required for the synthesis of thyroxine in thyroglobulin involve at least three steps: (1) Iodination of tyrosine residues to monoiodotyrosine, (2) further iodination of monoiodotyrosine to diiodotyrosine, and (3) coupling of two residues of diiodotyrosine to form thyroxine and a three-carbon compound. The minimal requirements for thyroxine synthesis are tyrosine residues that can react readily with iodine and are sufficiently close to one another for coupling to occur. Alterations in the conformation of thyroglobulin, produced in response to pH, salts, etc., markedly affect the sequential conversion of tyrosine to thyroxine residues by iodination and may be of significance for the control of the in vivo synthesis of thyroidal hormones.

Protein synthesis and iodination of thyroglobulin are distinct and sequential processes. This explains the variable iodine content and iodoamino acid distribution found in thyroglobulin preparations. Thus, there are no specific tRNAs or codons for the iodoamino acids in thyroid proteins. Iodination of tyrosine residues of thyroglobulin in vitro is catalyzed by an *iodinase* and is dependent upon the prior oxidation of iodide to "active iodine," possibly iodinium ion, I^+. This reaction in the gland is catalyzed by an *iodide peroxidase* in the presence of H_2O_2. The latter is formed by autoxidation of flavoproteins (page 363). In *congenital* goitrous thyroid disorders (page 927), absence of the peroxidase results in iodide accumulation and failure to synthesize thyroid hormones. Certain "antithyroid" agents (page 930) inhibit the action of the thyroidal peroxidase.

A number of peroxidases, *e.g.*, chloroperoxidase, lactoperoxidase, myeloperoxidase, and horseradish peroxidase catalyze the iodination of thyroglobulin in the presence of added or enzymically generated H_2O_2. A similar reaction occurs with a purified thyroid peroxidase; the latter appears to be a heme protein with an average molecular weight of 64,000. This enzyme not only catalyzed iodination of thyroglobulin but resulted in significant formation of diiodotyrosine and thyroxine, as revealed by pronase digestion of the iodinated protein. Formation of 3′,3,5-triiodothyronine was not observed. The data suggest that a peroxidase enzyme may be involved in the coupling reaction, as well as in the iodination reaction, in the thyroid gland. Inasmuch as a single enzyme may catalyze oxidation of iodide and transfer to the tyrosine residues of thyroglobulin, it appears unnecessary to postulate a requirement for a separate "iodinase."

The mechanism of coupling of two diiodotyrosine residues to form a thyroxine residue, or of one monoiodo- and two diiodotyrosine residues to form a triiodothyronine residue is not known. A free-radical intermediate has been postulated. The bond between the side chain of the ring of one iodinated tyrosine residue is split, and the ring transferred to a second diiodotyrosine residue. The other product of the reaction is a three-carbon amino acid. It is apparent that the folded, compact structure of the globular, native form of thyroglobulin serves to modulate the distribution of iodine between monoiodotyrosine and diiodotyrosine and to catalyze the coupling reaction.

The data in Fig. 43.1 illustrate the rate of conversion of inorganic iodide into diiodotyrosine and thyroxine.

In vitro iodination of proteins, *e.g.*, casein and serum albumin, in slightly alkaline solution results in products which, when administered, exert a thyroxine-like effect on metabolism. Thyroxine has been isolated from hydrolysates of these iodinated proteins. This suggests that, physiologically, the coupling of iodinated tyrosine residues to form thyroidal hormones occurs uniquely in the thyroid because of the effective uptake of iodide by the gland. In addition, there is present in the thyroid the enzymic mechanism for producing iodinated thyroidal hormone.

Proteolysis of Thyroglobulin and Secretion of Thyroxine and Triiodothyronine. Thyroglobulin is the main constituent of the thyroid colloid, largely filling the lumen of the acinus, and does not enter the capillaries unless the walls are disrupted. Thyroglobulin has no hormonal properties; these appear only after release of the iodothyronines. Thyroglobulin thus provides a reserve of potentially active hormones that cannot leave the cell until proteolysis occurs.

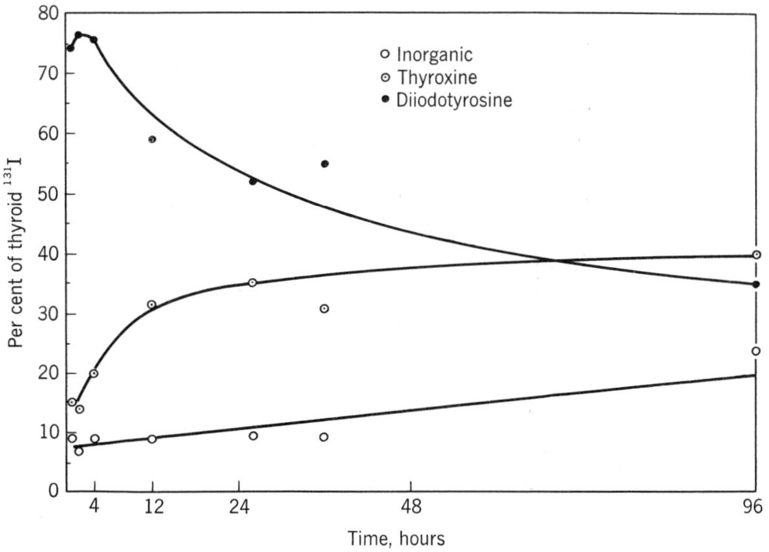

Fig. 43.1. Distribution of thyroid radioactive iodine among inorganic, diiodotyrosine, and thyroxine fractions in the normal rat following administration of ^{131}I, as iodide. (*After M. E. Morton, I. Perlman, E. M. Anderson, and I. L. Chaikoff, Endocrinology,* **30**, 495, 1942.)

Normally, proteolysis of thyroglobulin probably occurs in the lumen or the epithelial cells under the influence of intracellular proteinases; the liberated hormones are secreted into the capillaries which form a rich plexus at the base of the epithelial cells. Thyroglobulin synthesis and proteolysis occur continuously. The role of adenohypophyseal thyrotropic hormone in liberation and secretion of thyroid hormones is considered in Chap. 47.

The thyroid gland contains a microsomal *dehalogenase* that requires TPNH and catalyzes deiodination of mono- and diiodotyrosines. The liberated halogen is then available for reutilization. The enzyme is without action on the iodinated thyronines, which are secreted. In hereditary familial goiter (page 700), lack of this enzyme results in excessive loss of the iodinated tyrosines from the thyroid. In the absence of adequate functioning of this normal iodine-conserving mechanism, thyroid-deficiency symptoms (see below) may ensue.

Transport of Thyroxine and Triiodothyronine in the Blood. Thyroxine is transported in the plasma bound to three carrier proteins. The principal carrier is a specific glycoprotein with a molecular weight of 45,000 and an electrophoretic mobility between that of α_1- and α_2- globulin; hence the terms "thyroxine-binding globulin" and "inter-α-globulin." A second protein, termed prealbumin (page 709), may carry four times as much thyroxine as does the globulin, at pH 8.6, but less at physiological pH. In addition, a small amount of thyroxine is carried by albumin.

Triiodothyronine is also transported in association with the thyroxine-binding globulin, but is less tightly bound. The equilibrium between free and bound iodinated thyronines could influence the amount of hormone reaching the tissues, and this is a possible explanation for the observed greater biological activity of triiodothyronine.

The normal serum concentration of free thyroxine is approximately 4.7 μg per

100 ml, or about 1/1,000 the concentration of the bound hormone. The hormone is readily extractable from its protein carrier with butanol, but it precipitates with plasma proteins on addition of reagents that precipitate these proteins. This protein-bound iodine, designated as PBI, is a useful measure of circulating thyroid hormone; in the normal individual the PBI values range from 4 to 8 μg per ml of serum.

Thyroxine represents practically all of the total blood iodine in normal individuals; the concentration of triiodothyronine is very much less but not accurately known. 3,3′-Diiodothyronine may be present in small concentration, and occasionally monoiodotyrosine is detectable. The remainder of the total plasma iodine, usually 10 to 20 per cent, is inorganic iodide. The latter may enter the erythrocyte, which does not contain organically bound iodine.

FUNCTIONS OF THYROXINE AND TRIIODOTHYRONINE

Thyroxine and triiodothyronine accelerate cellular reactions in practically every organ and tissue of the body; this is reflected in an increase in basal metabolic rate. The ability of these hormones to accelerate growth and oxygen consumption of the whole organism is accompanied by increased activity of most enzymic systems that have been studied. For example, injection of these thyroid hormones accelerates glucose oxidation in vivo and increases the activities of liver glucose 6-phosphatase, glucose 6-phosphate dehydrogenase, and TPNH-cytochrome c reductase measured in vitro in broken cell preparations. Indeed, more than 100 enzymic systems have been reported to be altered in activity by thyroxine administration. However, few if any of these effects are likely to reflect a direct action of the hormone on the enzymes.

Induction of *de novo* synthesis of mitochondrial α-glycerophosphate dehydrogenase occurs, shortly after thyroxine administration, in all organs known to respond to the hormone by an increased oxygen consumption. This stimulation of the *de novo* synthesis of an enzyme is reflected in the ability of the iodinated thyroid hormones, either in vivo or in vitro, to stimulate protein synthesis by liver mitochondria and microsomes. In the in vitro studies, the hormonal effect has been delineated as accelerating the transfer of amino acyl-tRNA to microsomal protein. However, in view of the widespread stimulatory action of thyroxine or triiodothyronine on protein and nucleic acid synthesis, it is not presently possible to designate specific macromolecules whose synthesis may be the primary basis of hormone action.

Thyroxine added to mitochondria in vitro in concentrations of 10^{-4} to $10^{-3}M$ uncouples a portion of the phosphorylations occuring during oxidation of a specific substrate, *e.g.*, β-hydroxybutyrate. Concomitantly, water enters the mitochondria, which swell, suggesting an effect of thyroxine on the mitochondrial membrane. Uncoupling of oxidative phosphorylation channels oxidative energy into heat rather than into the synthesis of high energy phosphate-containing compounds. A similar uncoupling effect on mitochondrial oxidative phosphorylation can be obtained with dinitrophenol, which resembles thyroxine in calorigenic action. However, other compounds, *e.g.*, stilbestrol (page 949) and antimycin A, an antibiotic, also uncouple oxidative phosphorylation in vitro but are not calorigenic in the organism. Thus, the effect of thyroxine in uncoupling oxidative phosphorylation, seen in vitro with high, nonphysiological concentrations of hormone, is of unknown significance for interpretation of the physiological effects of thyroxine. Also, the increase in basal metabolic rate of the hyperthyroid organism cannot be explained in terms of any single metabolic process.

THYROID HYPO— AND HYPERFUNCTION

Thyroid *hypofunction,* both experimental and clinical, is reflected in marked slowing of body processes, *e.g.*, the basal metabolic rate and body temperature are below normal. If hypothyroidism is present at birth, *infantile myxedema,* or *cretinism,* results. In adults, the term *myxedema* has been used to describe the hypothyroid state.

The profound influence of the thyroid on growth is seen in the failure of metamorphosis of thyroidectomized tadpoles. Conversely, metamorphosis is accelerated when normal tadpoles are placed in a medium containing as little as $1 \times 10^{-6}M$ of thyroxine. The hormone is responsible for the induction during metamorphosis of the synthesis of liver enzymes essential for urea formation (page 560).

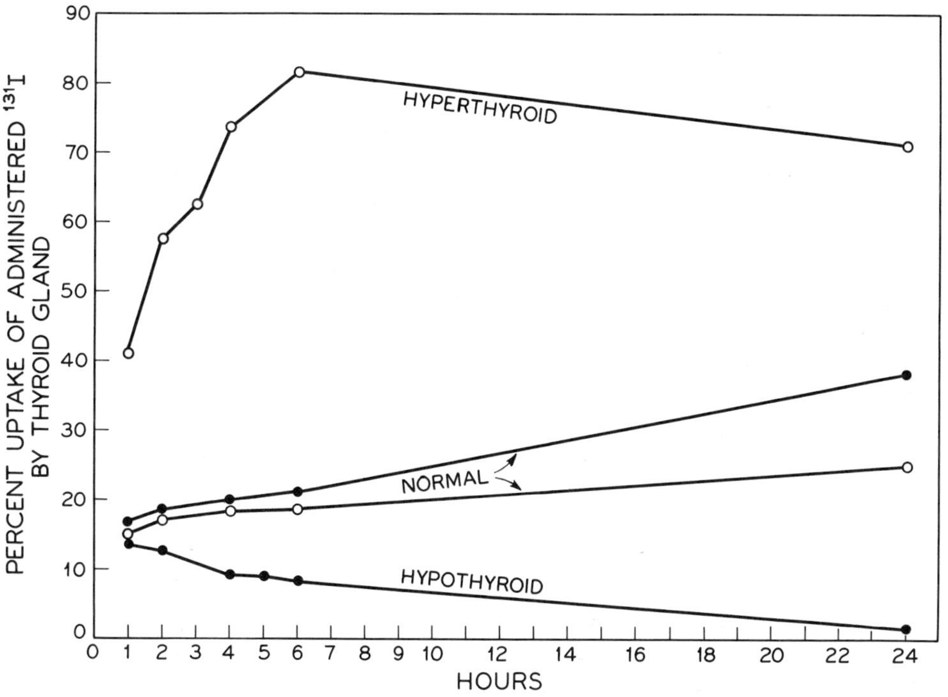

Fig. 43.2. The uptake of administered ^{131}I by the thyroids of two normal individuals and by the glands of a hypo- and of a hyperthyroid subject. ^{131}I given as iodide at time zero. (*Courtesy of Dr. L. J. Soffer.*)

The hypofunctioning thyroid gland has a lowered capacity to accumulate administered iodide (Fig. 43.2); this is of practical diagnostic value. Practical use is also made diagnostically of the binding of iodinated thyroid hormones by specific plasma proteins (page 925) by measurements of the plasma protein sites that are not occupied by hormone. Triiodothyronine, labeled with ^{131}I, is added to a sample of blood or serum. A second binding material, such as erythrocytes or an ion exchange resin, is then added; this absorbs all of the labeled hormone that cannot be bound to the plasma proteins. The amount of added hormone not bound reflects indirectly the degree to which the plasma protein binding sites were saturated initially.

Circulating iodinated thyroid hormone values in hypothyroidism are 1 to 2 μg of protein-bound iodine per 100 ml. of blood, approximately one-fourth the normal values. Blood cholesterol values are often elevated in hypothyroidism.

Thyroid *hyperfunction* is characterized by an abnormally elevated rate of most metabolic phenomena. This clinical syndrome, generally termed *Graves' disease,* has also been termed *exophthalmic goiter,* because thyroid enlargement and exophthalmos are frequently present. The basal metabolic rate may be 30 to 60 per cent above normal. Hypocholesterolemia, hyperglycemia, glucosuria, reduced glucose tolerance, and a negative nitrogen balance may obtain. Protein-bound iodine may rise to 15 to 20 μg per 100 ml. of blood, and the rate of iodine uptake by the thyroid is increased (Fig. 43.2).

METABOLIC FATE OF THYROXINE AND TRIIODOTHYRONINE

Thyroxine and triiodothyronine undergo deamination, deiodination, and conjugation. Deiodination occurs largely but not solely in extrahepatic tissues, and the other two reactions are prominent in liver, although not limited to this organ.

Hepatic Metabolism. The liver participates in the metabolism of thyroxine and triiodothyronine in two ways: (1) it removes the amino group and deiodinates the hormones, and (2) it regulates blood levels of hormone by enterohepatic circulation. As much as 40 per cent of an injected dose of thyroxine is found in the liver within 1 min. after administration. A small portion appears in the bile as iodide and as the administered hormone.

Transamination of the hormones in the liver results in formation of the corresponding pyruvic acid analogues. The latter are less active biologically. Transamination also occurs in kidney, which in addition contains a specific deaminase that requires oxygen, is not DPN-dependent, and catalyzes oxidative deamination of L-3,5,3′-triiodothyronine followed by decarboxylation with formation of triiodothyroacetic acid.

The liver also conjugates iodothyronines, with formation of glucosiduronides by linkage with the phenolic hydroxyl group of the thyronines. These conjugates are excreted in the bile. Because of the hydrolytic activity of intestinal enzymes and transmucosal transport of the liberated hormones, excretion of conjugated forms via the bile is not a significant route for their net disposal. The mechanism responsible for intestinal transport of the iodinated thyronines is similar to that for active transport of amino acids (page 784) but may in addition involve re-formation of the glucosiduronides in the intestinal mucosa. Decarboxylation of the thyronines by bacteria may occur in the intestine.

Triiodothyronine may also be transformed in part in the liver as well as in the thyroid and adrenals to the O-sulfate derivative. This sulfate ester may be the peripheral "storage" form of the hormone, since after its administration to rats it is less rapidly degraded than is triiodothyronine, and although the sulfate ester is found in plasma, it is not present in urine.

Enzymic O-methylation of iodophenols, utilizing S-adenosylmethionine, occurs in liver. Tetraiodothyroacetic acid is a good substrate; however, thyroxine and triiodothyronine are not methylated.

Deiodination of the thyroidal hormones and their corresponding thyropyruvic acids in liver is catalyzed by *thyroxine dehalogenase* (see below). Only trace amounts of the deiodinated thyronines are excreted in the urine. When thyronine was incubated with rat liver and kidney slices the products formed were tyrosine, 3′-hydroxythyronine, 3,4-dihydroxyphenylalanine, thyroacetic acid, and *p*-hydroxyphenylpyruvic acid. These findings suggest three mechanisms of thyronine metabolism: rupture of the diphenylether bridge, *o*-hydroxylation of the rings, and degradation of the alanine side chain via general pathways for amino acids.

Thus, the liver can metabolize the iodinated thyroid hormones by deiodination, removal of amino groups, and disruption of the thyronine nucleus. Also, thyroidal compounds, as well as certain of the metabolites, can be conjugated with glucuronic acid and channeled into an enterohepatic circulation, or, in specific instances, conjugated with sulfate or O-methylated.

Transformations in Peripheral Tissues. Deiodination appears to be the prime transformation of thyroxine and triiodothyronine occurring in skeletal muscle, kidney, liver, and heart; the reaction is catalyzed by *thyroxine dehalogenase (iodothyronine deiodinase).* The latter does not attack iodinated tyrosines, in contrast to the thyroid or liver *dehalogenase.* Thyroxine dehalogenase of muscle is activated by flavins and ferrous ions and attacks thyroxine at approximately three to four times the rate with triiodothyronine. Iodide is the major product from the organically bound iodine. The enzyme is distinct from thyroid dehalogenase (page 925), which is without action on iodinated thyronines. The biochemical significance of thyroxine dehalogenase is not clear. Deiodination appears to be a minor pathway for inactivating thyroid hormone, as compared with the active conjugating pathways. Suggestive evidence links the enzymic deiodination of thyroid hormones by skeletal muscle with their calorigenic action, although the exact nature of this relationship is not clear.

Excretion of Thyroidal Iodine. Only about 1 per cent of the total iodine in urine is present as iodothyronines, including iodothyropyruvic and acetic acids; the remainder is largely inorganic iodide. The utilization of L-thyroxine in man is relatively slow, with a half-life of 7 to 12 days reported with therapeutic doses. In all cases the iodine atoms in positions 3′, or 3′,5′, of the aromatic structure are more labile than those in 3, or 3,5, since the former are excreted as iodide at a distinctly higher rate. The kidneys clear approximately 33 ml. of plasma of iodide per minute.

Relative Biological Activity of Thyronines. A large number of substituted thyronines have been tested for relative biological activities. In man, administration of the acetic acid analogues results in a thyroid-like effect that is more rapid and of shorter duration than that following triiodothyronine, which in turn acts more rapidly than thyroxine. Reference has been made (page 925) to the influence of the degree of protein binding, and the rapidity of transfer into tissues, on the expression of biological activity of thyronines.

ANTITHYROID AGENTS

The term antithyroid agent is employed to designate any substance inhibiting thyroid function. These agents include the following: (1) Those which prevent release of hormone from the thyroid by a feedback mechanism (page 538). The outstanding example is thyroxine, which affects secretion and activity of hypophyseal thyrotropic hormone; this will be considered in Chap. 47. (2) Agents which retard synthesis of hormone; this group includes thiocyanate, and certain other anions, which inhibit iodide uptake by the thyroid, *viz.*, iodide, and synthetic thiocarbamides and sulfonamides, which interfere with the iodination reaction (page 924). (3) Substances that inhibit utilization of thyroid hormones, *viz.*, structural analogues. In addition, ^{131}I administration in high doses and X-irradiation of the thyroid will depress thyroid activity by destruction of thyroid tissue.

Iodide. The marked localization and role of administered iodide in the thyroid gland have been discussed. It therefore seems paradoxical that hyperactivity of the thyroid, as seen in toxic goiter, can be diminished by large doses of iodide. Elevation of plasma iodide concentration above approximately 30 μg per 100 ml prevents further iodide uptake by the thyroid. In vitro, iodide ion inhibits iodination of tyrosine by iodine, and also reduces incorporation of amino acids into thyroidal ribosomal protein. Apparently, in simple goiter *iodine* is required for hormone synthesis, whereas in toxic goiter *iodide* inhibits iodine utilization and protein synthesis by the thyroid.

Thiocyanate. Thiocyanate and certain other anions, notably perchlorate and nitrate, exert a thyroid-inhibiting effect by interfering with iodide uptake by the gland (page 923). Thiocyanate is widely distributed in nature, occuring in blood, saliva, urine, and in many plants, such as members of the genus *Brassica* (cabbages). Isothiocyanates, such as mustard oil, organic nitriles, and the widely distributed cyanogenetic glycosides may be transformed to thiocyanates in mammalian metabolism. Consumption of foods containing these substances could render iodine-deficient a diet otherwise adequate in iodine and explains the goiter endemic in certain regions of the world.

Thiocarbamides and Sulfonamides. A number of organic compounds may also be classed as antithyroid substances. Aside from structural analogues of thyroxine (see below), the antithyroid substances can be divided roughly into two main groups, the most active of which have in common a thiocarbamide grouping, as for example thiourea and thiouracil.

Thiourea

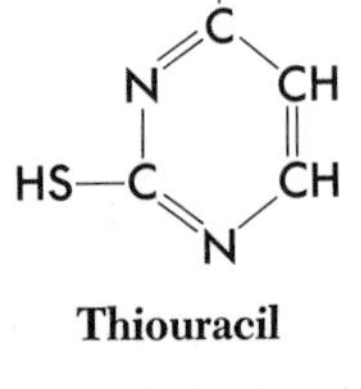

Thiouracil

A second group has an aminobenzene grouping, the best known of which are the sulfonamides; the most active of such compounds are 4,4′-diaminodiphenylmethane and 4,4′-diaminobenzil.

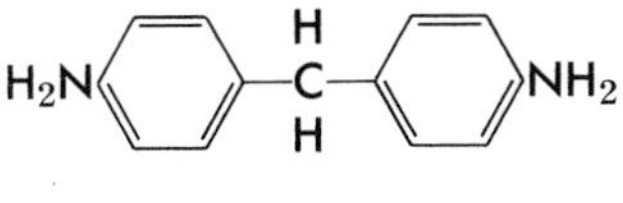

4,4′-Diaminodiphenylmethane

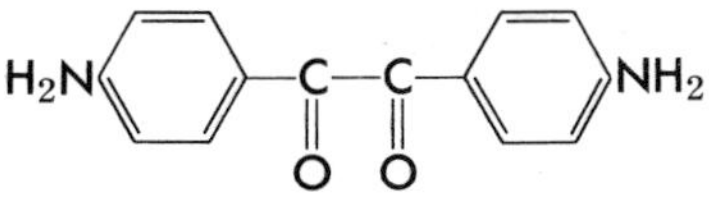

4,4′-Diaminobenzil

The thiocarbamides and sulfonamides probably act to inhibit synthesis of thyroid hormone, as does the iodide ion, by forming molecular compounds with elemental iodine in the gland, thus preventing iodination and subsequent hormone formation (Fig. 43.3). Some of the very active aromatic antithyroid substances act by forming stable substitution compounds with the gland iodine.

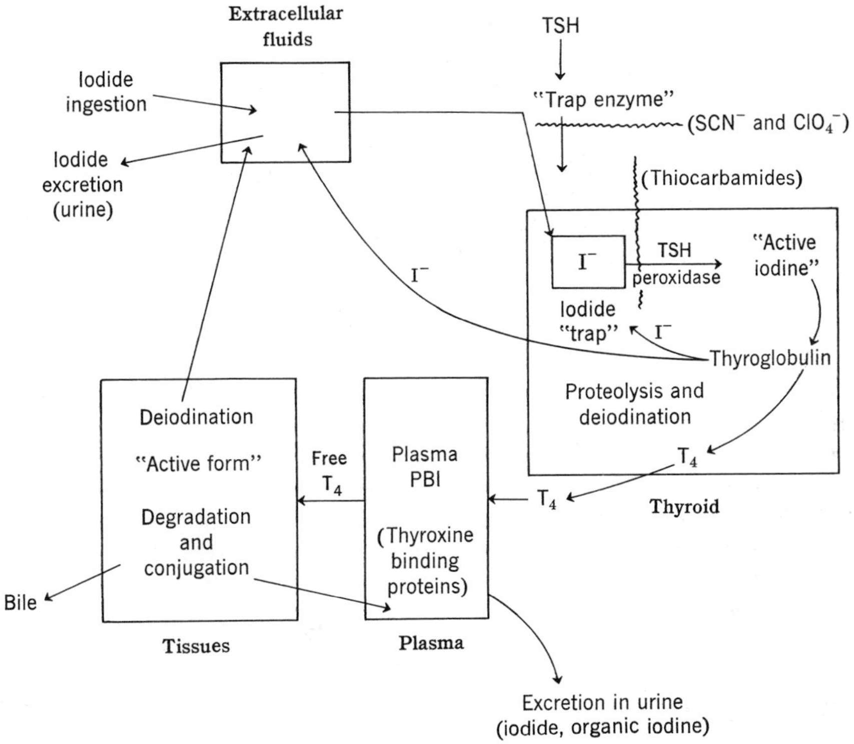

Fig. 43.3. Schematic representation of aspects of the thyroid and its iodinated hormones; T_4 designates thyroxine.

Thyronine Analogues. A number of synthetic thyronine analogues of the thyroid hormone have antithyroid or thyroid hormone-inhibiting properties. Thus, 2′,6′-diiodothyronine,

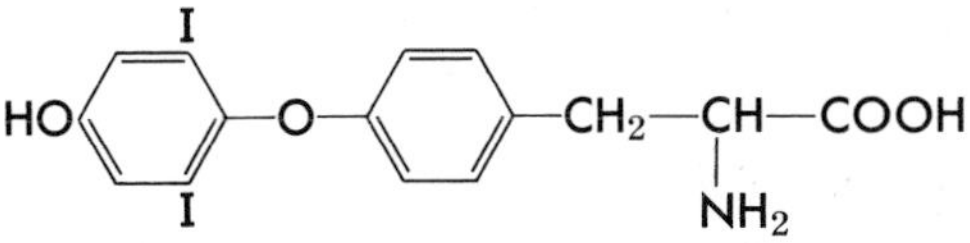

was able to antagonize thyroxine in a dose 150 times that of thyroxine and showed inhibitory action against injected thyroglobulin. Also, as in the case of compounds with thyroxine-like activity, the entire side chain is not essential for thyroxine-inhibiting action.

CALCITONIN

Calcitonin was first extracted and purified from parathyroid tissue in 1963 by Copp, who gave the active principle its name. Subsequently, calcitonin activity was demonstrated in extracts of the thyroid gland by several investigators. Indeed, the concentration of this hormone in the thyroid was significantly greater than in the parathyroids, and the active principle was renamed thyrocalcitonin. Recently, Copp has demonstrated that the production of this hormone by both the thyroid and parathyroids is a consequence of the presence in both of a common cell type which secretes calcitonin. In mammals, the number of these cells is greater in the thyroid than in the parathyroid, providing an explanation for the higher concentration of calcitonin in thyroid tissue. In lower vertebrates, calcitonin is produced primarily by the ultimobranchial body, which is embryologically related to the thyroid and parathyroid glands.

Calcitonin has been obtained in purified form from hog thyroid extracts. Activity is associated with a polypeptide of molecular weight 4,500; iodine is not present in calcitonin preparations.

Administration of calcitonin results in an acute hypocalcemia and hypophosphatemia. The effect on blood calcium levels is in the opposite direction from that of the parathyroid hormone (see below), indicating that one of the functions of calcitonin is to counteract the effects of hypercalcemia.

Calcitonin accelerates the transfer of calcium from the blood to bone; the hormone does not increase calcium loss from the body, nor is its action mediated by the kidney, gastrointestinal tract, hypophysis, or parathyroid gland. Calcitonin exerts a direct effect on bone, accelerating the rate of calcium deposition and inhibiting calcium resorption. Thus the net actions of calcitonin result in calcium retention, largely by the skeletal system.

Secretion of calcitonin by the thyroid gland, as well as by the parathyroids apparently varies with the concentration of serum calcium, being increased in hypercalcemia, thereby restoring circulating calcium levels to normal values.

THE PARATHYROIDS. PARATHORMONE

Parathormone, a second hormone secreted by the parathyroid glands, and calcitonin together comprise a hormonal mechanism for homeostatic regulation of calcium and phosphate metabolism. Products with parathormone hormonal activity have a marked calcium-mobilizing action and promote urinary excretion of phosphate.

CHEMISTRY OF PARATHORMONE

Pure parathormone, a polypeptide of molecular weight approximately 9,000, is a single peptide chain containing 83 amino acid residues and with an amino acid composition characterized by the absence of cysteine. Degradation of the hormone by dilute acid led to the identification of the region of the molecule most important for biological activity and immunolochemical properties. The portion of the hormone responsible for biological activity represents a sequence of approximately 35 amino acid residues that include the amino terminal region. Removal of two amino acid residues and a portion of the serine at the third position of the amino terminal terminus of parathormone with leucine aminopeptidase caused a loss of biological activity but retention of the immunological activity. The data suggest that a substantial portion of the sequence of the amino terminal 35 amino acids of parathormone is required for biological activity, but a considerably smaller fraction is requisite for immunological activity.

PARATHORMONE FUNCTIONS

The principal sites of regulatory influence of parathormone are the kidney, the skeleton, and the gastrointestinal tract; the hormone can influence directly and independently any one of the three.

Renal Actions. Hypersecretion of the parathyroids, or injection of active parathormone preparations induces a marked phosphaturia (page 835). This is accompanied by elevation of the plasma calcium, which is drawn from the bones, the chief reservoir of calcium ion. Conversely, hyposecretion of the parathyroids, or experimental extirpation of the glands, results in diminished phosphate excretion in the urine, with elevation of phosphate concentration in the blood, accompanied by depression of blood calcium values.

A direct influence of the parathyroids on the renal tubule was demonstrated by infusion of purified hormone into one renal artery of the dog. This led to unilateral phosphaturia without alterations in filtration rate or renal plasma flow. Of the two mechanisms which have been suggested to account for this phosphaturia, *viz.*, (1) an increase in secretion of phosphate by the distal tubule, and (2) a decrease in proximal tubular reabsorption of phosphate, considerably more experimental evidence has been accumulated in support of the first.

Administration of parathormone produces a striking decreased rate of respiration and phosphorylation in kidney mitochondria. Kidney tissue from hormone-injected animals shows an increased rate of oxidation of glucose, an increased accumulation of Ca^{++} and P_i, and an accelerated rate of incorporation of radioactive inorganic phosphate into acid-soluble organic and nucleic acid fractions. Parathormone injection markedly increases the formation and excretion of cyclic AMP by the kidney. The physiological significance of this is not yet clear.

Actions on Bone. Parathormone acts directly on bone to augment osteoclastic activity. This is evident as an increased release of Ca^{++} from bone, accompanied by loss of mucopolysaccharide from bone matrix. After prolonged administration of the hormone, serum total glycoprotein and sialic acid levels increase, as does the collagenolytic activity of bone. The calcium-mobilizing action of injected parathormone is inhibited by simultaneous administration of agents that block protein synthesis, *e.g.*, puromycin, suggesting an interdependence of the two processes. However, the true mechanism of action of parathormone remains to be defined.

Bone and kidney slices exhibited augmented glycolytic and succinic acid dehydrogenase activities when these tissues were taken from rabbits given a single injection of parathormone. The latter also augmented accumulation of citrate and lactate by bone cells. It has been postulated that elevation of acid concentration in bone by the hormone could favor, by lowering pH, transfer of calcium from bone to blood.

Parathormone injection inhibited isocitrate oxidation by kidney mitochondria as well as by bone tissue. This effect has been ascribed to an action of the hormone on the pyridine nucleotide cofactor of isocitric acid dehydrogenase. The concentration of TPN was significantly lower than normal in calvarium halves of mouse embryos cultivated in vitro in the presence of the hormone.

Thus the parathyroid secretions, parathormone and calcitonin, which is also produced by the thyroid, constitute a homeostatic mechanism for the regulation of the concentration of calcium in body fluids. The hypercalcemic effects of parathormone are counteracted by the compensatory hypocalcemic influences of calcitonin. Each hormone exerts its primary regulatory actions on calcium metabolism by a direct action on bone, altering the rate of calcium resorption (parathormone) or of calcium deposition (calcitonin).

Action on Gastrointestinal Tract. Parathormone augments calcium and phosphate absorption from the gastrointestinal tract by stimulating uptake of these ions by cells of intestinal mucosal villi, with subsequent release to the blood. The action on calcium absorption is dependent on adequate nutritional status with respect to vitamin D (Chap. 50).

REGULATION OF SECRETION OF PARATHORMONE

Secretion of parathormone by the parathyroid glands is regulated by the concentration of ionized serum calcium, varies inversely with this concentration, and rapidly responds to changes in concentration of this ion. Thus, if ionized serum $[Ca^{++}]$ is lowered by injection of fluoride, oxalate, or a calcium chelating agent, secretion of parathormone promptly rises. Conversely, elevation of serum $[Ca^{++}]$ results in diminished hormone secretion. Thus, the blood $[Ca^{++}]$ provides a basis for a "feedback" mechanism for regulation of parathormone secretion.

The half-life of circulating parathormone in man is relatively short, of the order of 20 to 30 min. This is generally the case for other polypeptide hormones of similar size, *e.g.,* insulin (Chap. 46) and adrenocorticotropic hormone (Chap. 47).

HYPO- AND HYPERPARATHYROIDISM

Hypoparathyroidism, which occurs relatively rarely in man, is due either to operative procedures on the thyroid or parathyroid glands or, even less frequently, to disease of the parathyroids. Hypocalcemia, convulsions, tetany, and death ensue in the absence of therapy, reflecting the role of calcium in maintaining normal neuromuscular irritability (page 888).

Hyperparathyroidism may result either from parathyroid hyperfunction due to a tumor or hyperplasia of the parathyroids or prolonged administration of parathyroid extracts in experimental animals. In man, severe primary clinical hyperparathyroidism leads to *osteitis fibrosa cystica,* characterized by elevation of serum $[Ca^{++}]$, diminished phosphate concentration, and markedly increased renal excretion of calcium. The augmented calcium excretion frequently causes formation of urinary calculi, with secondary impairment of renal function. When this ensues, excretion of phosphate and calcium declines, serum phosphate level rises, and calcium returns to normal levels. However, resorption of bone mineral continues during this period, and calcium and phosphate which do not appear in the urine may be excreted in the stool. Also, with excessive loss of calcium, disseminated cystic decalcification of bone results, with markedly increased susceptibility to fractures. The alterations in bone metabolism are accompanied by an increase in serum alkaline phosphatase.

Hyperparathyroidism, and its attendant symptoms, may also be encountered in man with chronic renal insufficiency. The retention of phosphate as a consequence of renal disease depresses the serum calcium level, stimulating parathormone production (see above), and may lead to parathyroid hyperplasia.

REFERENCES

Books

Gaillard, P. J., Talmadge, R. V., and Budy, A. M., eds., "The Parathyroid Glands: Ultrastructure, Secretion, and Function," University of Chicago Press, Chicago, 1965.

Pitt-Rivers, R., and Trotter, W. R., eds., "The Thyroid Gland," vols. I and II, Butterworth, Inc., Washington, 1964.

Werner, S. C., ed., "The Thyroid," 2d ed., Harper & Row, Publishers, Incorporated, New York, 1962.

Review Articles

Arnaud, C. D., Jr., Tenenhouse, A. M., and Rasmussen, H., Parathyroid Hormone, *Ann. Rev. Physiol.,* **29,** 349–372, 1967.

Aurbach, G. D., and Potts, J. T., Jr., The Parathyroids, *Advances in Metabolic Disorders,* **1,** 45–93, 1964.

Aurbach, G. D., and Potts, J. T., Jr., Parathyroid Hormone, *Am. J. Med.,* **42,** 1–8, 1967.

Barker, S. B., New Ideas of Thyroid Function, *The Physiologist,* **6,** 94–114, 1963.

Copp, D. H., Parathyroids, Calcitonin, and Control of Plasma Calcium, *Recent Progr. Hormone Research,* **20,** 59–88, 1964.

Edelhoch, H., "The Structure of Thyroglobulin and Its Role in Iodination," *Recent Progr. Hormone Research,* **21,** 1–31, 1965.

Frieden, E., Thyroid Hormones and the Biochemistry of Amphibian Metamorphosis, *Recent Progr. Hormone Research,* **23,** 139–194, 1967.

Greep, R. O., and Kenny, A. D., Physiology and Chemistry of the Parathyroids, in G. Pincus and K. V. Thimann, eds., "The Hormones: Physiology, Chemistry and Applications," vol. III, pp. 153–174, Academic Press, Inc., New York, 1955.

Greer, M. A., The Natural Occurrence of Goitrogenic Agents, *Recent Progr. Hormone Research,* **18,** 187–219, 1962.

Hoch, F. L., Biochemical Actions of Thyroid Hormones, *Physiol. Revs.,* **42,** 605–613, 1962.

Ingbar, S. H., and Freinkel, N., Regulation of Peripheral Metabolism of the Thyroid Hormones, *Recent Progr. Hormone Research,* **16,** 353–403, 1960.

Lardy, H. A., and Maley, G. F., Metabolic Effects of Thyroid Hormones in Vitro, *Recent Progr. Hormone Research,* **10,** 129–155, 1954.

Myant, N. B., ed., Thyroid Gland, *Brit. Med. Bull.,* **16,** 89–169, 1960.

Potts, J. T., Jr., Aurbach, G. D., and Sherwood, L. M., Parathyroid Hormone: Chemical Properties and Structural Requirements for Biological and Immunological Activity, *Recent Progr. Hormone Research,* **22,** 101–151, 1966.

Rasmussen, H., Parathyroid Hormone, Nature and Mechanism of Action, *Am. J. Med.,* **30,** 112–128, 1961.

Rasmussen, H., and Craig, L. C., The Parathyroid Polypeptides, *Recent Progr. Hormone Research,* **18,** 269–295, 1962.

Rawson, R. W., ed., Modern Concepts of Thyroid Physiology, *Ann. N.Y. Acad. Sci.,* **86,** 311–675, 1960.

Robbins, J., and Rall, J. E., Proteins Associated with the Thyroid Hormone, *Physiol. Revs.,* **40,** 415–489, 1960.

Symposium, Regulation of Thyroidal Function, *Federation Proc.,* **21,** 623–641, 1962.

Tata, J. R., Intracellular and Extracellular Mechanisms for the Utilization and Action of Thyroid Hormones, *Recent Progr. Hormone Research,* **18,** 187–268, 1962.

Wolff, J., Transport of Iodide and Other Anions in the Thyroid Gland, *Physiol. Revs.,* **44,** 45–90, 1964.

44. The Gonads

Androgens. Estrogens

Androgens (Gk. *andros,* male) and estrogens (Gk. *oistros,* a gadfly, hence: sting, frenzy) are generic terms for the hormones secreted chiefly by the testis and ovary, respectively, and responsible for the development of secondary sex characteristics. However, estrogens have been isolated from testis and androgens from ovarian tissue. The common embryological development of the testes and ovaries may account for the production of both male and female sex hormones by both sexes. These hormones are also found in other tissues of the body, *e.g.*, the adrenals and the placenta.

THE TESTES—MALE SEX HORMONES

Chemistry of Male Sex Hormones. The principal hormone of the testes is testosterone, produced by the interstitial cells. Two additional androgens have been isolated from testes, 15α-hydroxytestosterone and 6β-hydroxytestosterone.

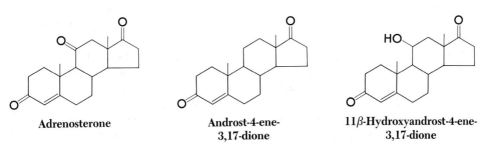

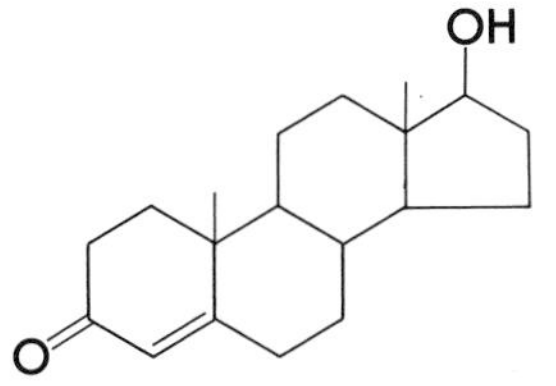

Testosterone

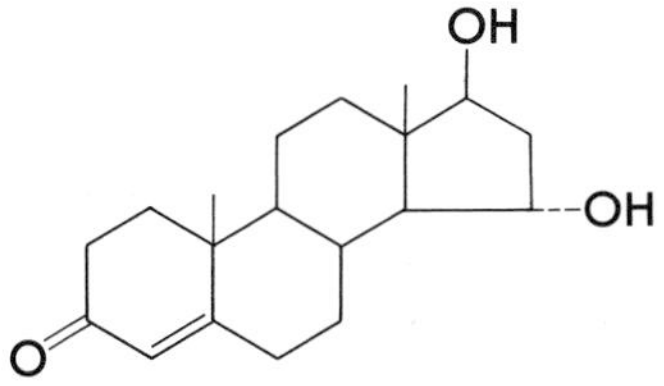

15α-Hydroxytestosterone

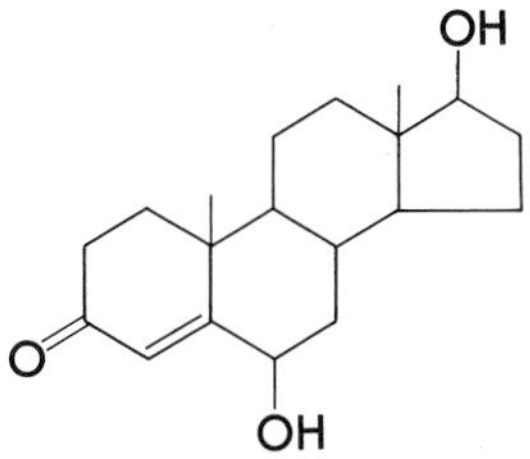

6β-Hydroxytestosterone

Compounds with androgenic activity are also synthesized in the adrenal gland. These include the following:

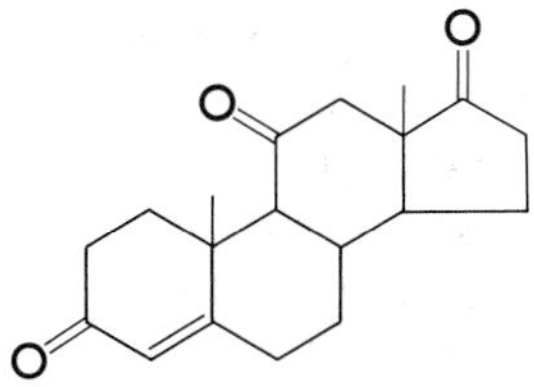

Adrenosterone

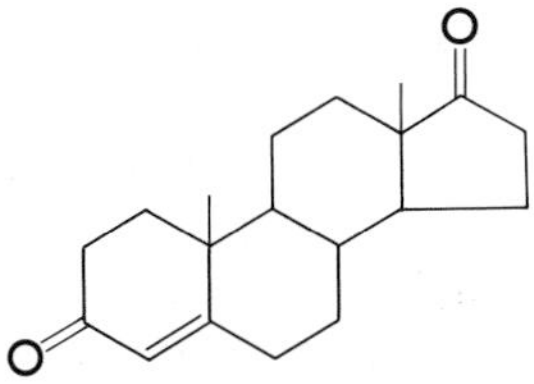

Androst-4-ene-3,17-dione

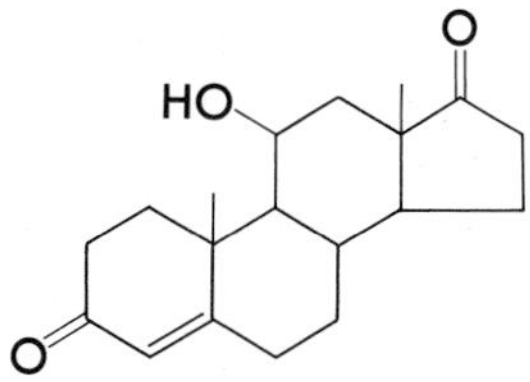

11β-Hydroxyandrost-4-ene-3,17-dione

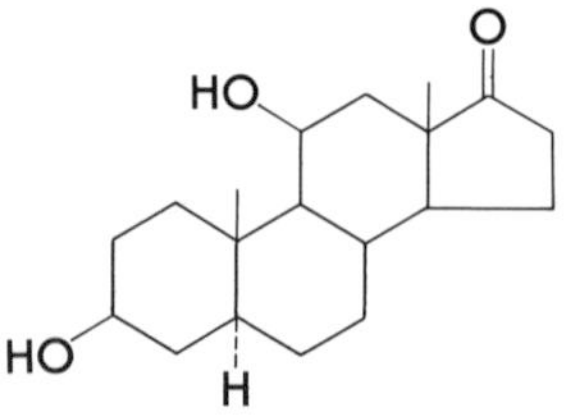

Androstane-3β,11β-diol-17-one
(11β-hydroxyepiandrosterone)

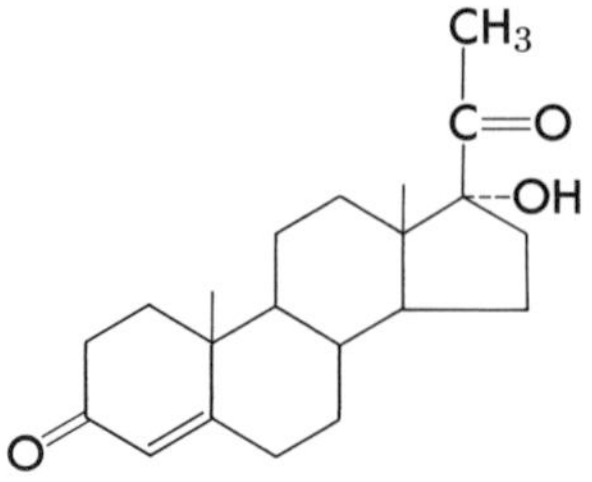

17α-Hydroxyprogesterone

The terms androgens and androgenic compounds are used to designate chemical relationships, *i.e.*, C_{19} steroids, not to imply that these substances are necessarily androgenically active steroids, since not all exhibit biological activity. On the other hand, not all steroids with androgenic activity are C_{19} compounds, *e.g.*, 17α-hydroxyprogesterone.

A number of metabolites with male sex hormone activity are present in the urine; these will be considered below.

Biogenesis of Androgens. The biogenesis of the steroid nucleus from acetate was discussed previously (page 517). Androgen biosynthesis can occur from acetate and from cholesterol. The rate limiting step in the synthesis of testosterone in the testis is the 20 α-hydroxylation of cholesterol (page 524); this reaction is influenced by hypophyseal gonadotropins (page 995). Figure 44.1, based upon in vivo and in vitro studies, illustrates the indicated pathways of testosterone synthesis in the testis as well as in other endocrine glands. In addition, androgens may be formed in peripheral tissues as a result of removal of the two-carbon side chain of circulating adrenal steroids (Chap. 45).

Functions of Androgens. Castration is followed, in the immature of all species, by a failure of development of accessory sex characteristics. In the human male, castration prior to puberty retards ossification of the epiphyses of the long bones, with consequent enlargement of the stature. The lower limbs become disproportionately long; there is increased adiposity, with the distribution of lipid resembling that in the female. The larynx is not prominent, as in the mature male, and the voice remains high-pitched. Although hair may be plentiful on the head, it fails to grow on the face and body. The penis remains infantile, and sexual feelings fail to develop. Muscular strength may be significantly diminished. If castration is performed after puberty, the changes are generally those described above for the prepubertal individual, but the effect is much less intensive.

Administration of androgenic compounds prevents the alterations produced by castration and will to a significant degree reverse the changes in the castrate animal, restoring normal conditions. Metabolic effects of androgens on the sex organs and tissues include increased fructose production by seminal vesicles and utilization of this sugar by seminal plasma, with concomitant enhancement of the activity of both aldose reductase and ketose reductase (page 390).

Androgens exert a striking anabolic influence on nitrogen and calcium metabolism, increasing tissue growth in the younger animal. Testosterone has been used

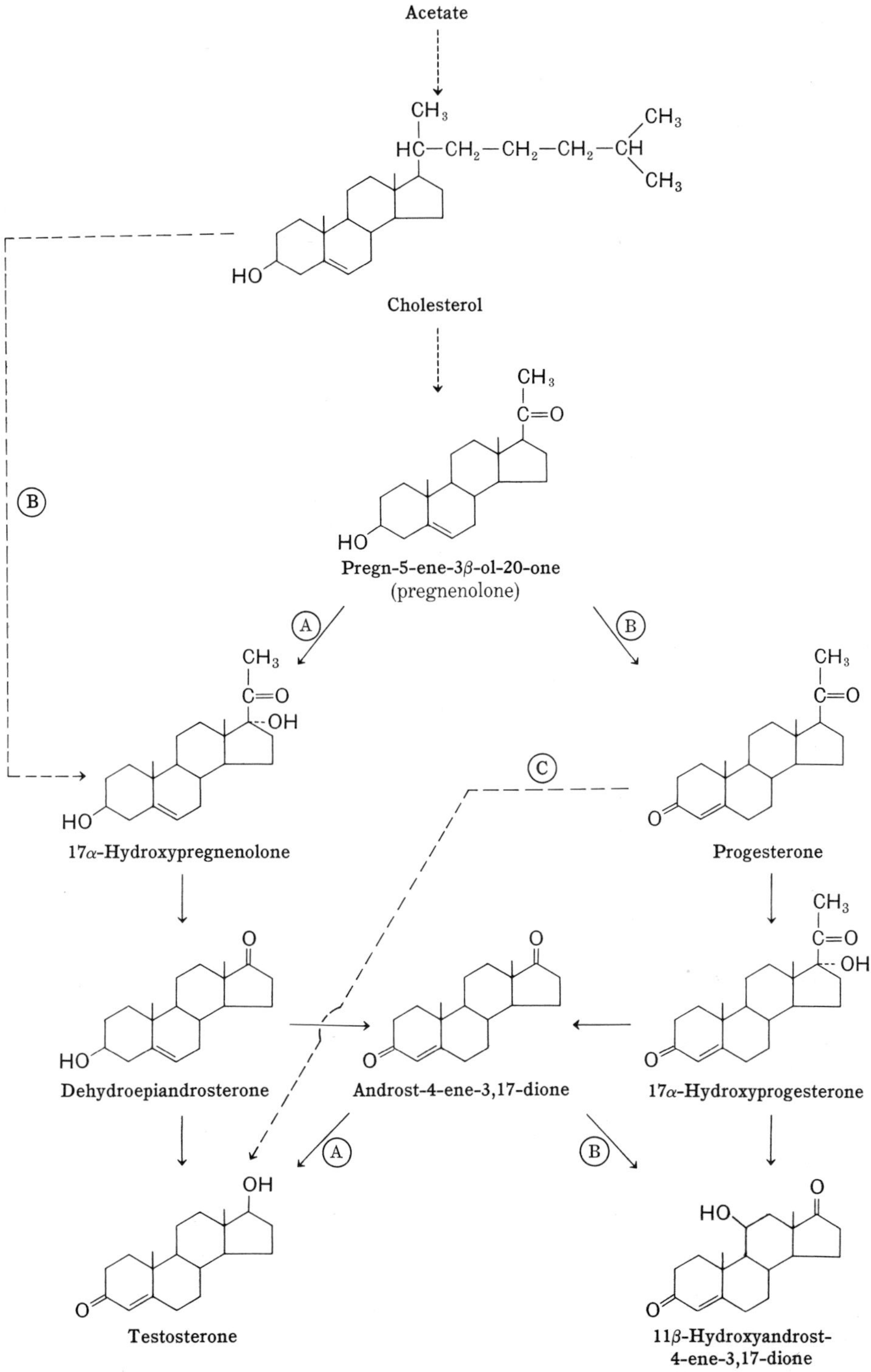

Fig. 44.1. Pathways for androgen biosynthesis; (A), in testis, ovary, adrenal gland, and placenta; (B), in adrenal gland; (C) in testis and in abnormal ovaries and adrenals via testosterone acetate as an intermediate and not involving androst-4-ene-3,17-dione.

for enhancing growth of children prior to puberty. Following testosterone administration to the rat, protein synthesis in prostatic tissue accelerates. This may be correlated with an increased number of ribosomes rich in mRNA, with a consequently augmented protein-synthesizing capacity. This may be an important mechanism of testosterone action in other organs that show an acute response to the hormone.

A biological antagonism obtains between the androgens and the estrogens. Biological responses to male sex hormones are inhibited by simultaneous administration of female sex hormones, and vice versa. Caponizing young cockerels by use of female sex hormone is an example of this antagonism, as are the resort to ovariectomy or testosterone administration for carcinoma of the breast, and the use of estrogens in the treatment of carcinoma of the prostate, frequently supplemented by orchidectomy.

Metabolism of Androgens. The following four major metabolites of testosterone are found in urine.

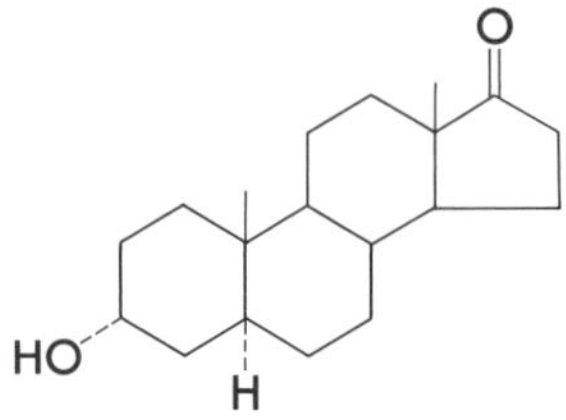

Androsterone

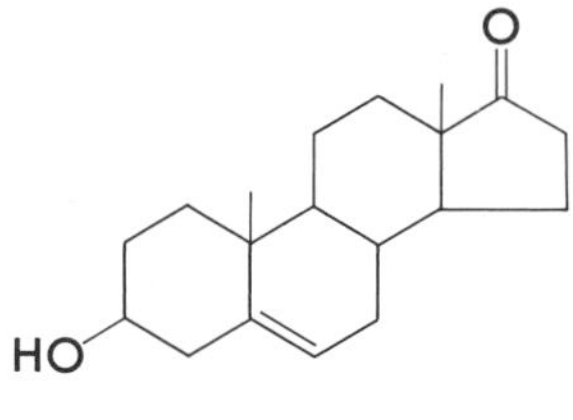

Dehydroepiandrosterone

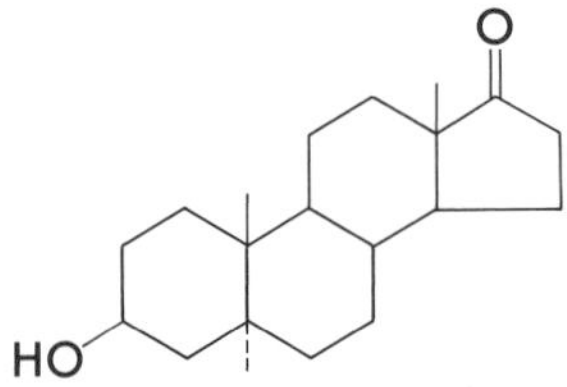

Epiandrosterone

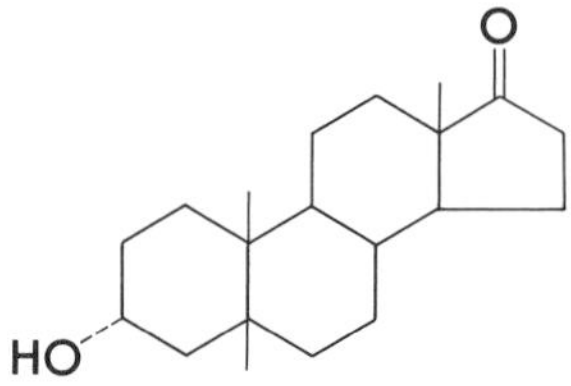

Etiocholane-3α-ol-17-one

Etiocholanolone and androsterone are quantitatively the *most important urinary androgen metabolites.* Etiocholanolone has a striking pyrogenic effect when administered to human subjects, and "etiocholanolone fever" of endogenous origin has been described clinically. Other pregnane metabolites share this pyrogenic effect.

Hydroxylation of the steroid nucleus at positions 11 and 18 also occurs in androgen metabolism; the 11- and 18-hydroxy derivatives corresponding to androsterone and etiocholanolone have been isolated from human urine. Other C-11 and C-18 hydroxylated steroids will be encountered later in discussions of estrogens and adrenal cortical steroids.

The liver is a major site of metabolic transformations of androgens. In some species (*e.g.*, the rat), the bile, as well as the urine, is an important pathway for excretion of androgen metabolites; this is somewhat less significant in man. In the liver, four major types of reaction occur in androgen metabolism; the first three indicated are catalyzed by DPN- or TPN-requiring enzymic systems: (1) a reversible conversion of testosterone to androst-4-ene-3,17-dione; (2) reduction of the 4,5 double bond of ring A by a group of enzymes designated as Δ^4-*reductases;* (3) a reversible interconversion of 3-hydroxy and 3-keto derivatives, catalyzed by a group of enzymes termed *3α- and 3β-hydroxysteroid dehydrogenases;* and (4) formation of glucosiduronide and sulfate conjugates of androgen metabolites. These conjugates are excreted in bile and urine. Thus the androgens and their metabolites are present in urine in a conjugated, water-soluble, biologically inactive form. Treatment of urine with acid and heat, or with specific hydrolytic enzymes, *i.e.*, β-glucuronidase and sulfatase, liberates the biologically active androgens, which can then be extracted with nonpolar solvents.

Other metabolic transformations of androgens occur in liver, as well as in other tissues, notably the adrenal, testes, and prostate. Hydroxylation of the steroid molecule is a particularly prominent reaction. Formation of 2β-, 6β-, 11β-, and 16α-hydroxylated derivatives of C_{19} steroids has been described in studies in which testosterone or androst-4-ene-3,17-dione has been incubated in vitro with tissue preparations. Several 16α-hydroxy C_{19} steroids have been isolated from human urine. In addition, liver and testicular tissue can effect dehydration of ring D, with formation of Δ^{16} derivatives.

Androgenic compounds other than those discussed above have been isolated from human urine in a variety of pathological conditions. The relative androgenic potency of certain compounds is shown in Table 44.1, which reveals that testosterone is the most potent of the naturally occurring biologically active androgens.

Table 44.1: RELATIVE ANDROGENIC ACTIVITY OF SOME STEROIDS ISOLATED FROM NATURAL SOURCES

Steroid		Source	Approximate amount in μg. equal to one I.U.*
Common name	Chemical name		
Testosterone	Androst-4-ene-17α-ol-3-one	Bull and stallion testes; spermatic vein blood	15
Androsterone	Androstane-3α-ol-17-one	Human, pregnant cow, and bull urine	100
Dehydroepiandrosterone	Androst-5-ene-3β-ol-17-one	Human, pregnant cow, and bull urine	300
Epiandrosterone	Androstane-3β-ol-17-one	Human and pregnant mare urine	700
Etiocholanolone	Etiocholane-3α-ol-17-one	Human urine	Inactive at 1,200
Adrenosterone	Androst-4-ene-3,11,17-trione	Adrenal cortex	500
	Androst-4-ene-3,17-dione	Adrenal cortex; spermatic vein blood	100
17α-Hydroxyprogesterone	Pregn-4-ene-17α-ol-3,20-dione	Adrenal cortex	500

* The international unit (I.U.) of androgens is established as the androgenic activity of 0.1 mg. of androsterone. The relative androgenic activities in this table are approximations on the basis of the capon's comb test and would not necessarily be similar to relative activities derived from mammalian bioassays.

SOURCE: After R. I. Dorfman, Biochemistry of Androgens, in G. Pincus and K. V. Thimann, eds., "The Hormones," vol. I, chap. 12, pp. 467–548, Academic Press, Inc., New York, 1948.

Since each of the various urinary metabolites of testosterone has a ketone group at C-17, these substances are referred to as 17-ketosteroids, and their concentration in the urine is a useful index of endogenous production of androgenic hormones. The commonly employed methods for determination of 17-ketosteroids are based on their reaction with *m*-dinitrobenzene in alkaline solution to produce a characteristic purple color.

The daily 17-ketosteroid excretion in the normal adult female (4 to 17 mg.) is approximately two-thirds that of the male (6 to 28 mg.). Prior to puberty, 17-ketosteroid excretion is about one-third that of the adult. This difference is due to the 17-ketosteroid precursors contributed by the gonads; the remainder of the 17-ketosteroids of the urine are derived from C_{19} and C_{21} steroids of the adrenal cortex. Those 17-ketosteroids derived from the adrenal cortical hormones generally have an oxygen function (hydroxyl or keto group) at C_{11}, and are termed 11-oxy-17-ketosteroids. The relative concentrations of 11-deoxy- and 11-oxy-17-ketosteroids in urine have afforded an approximate index of the relative secretory activities of the testes and the adrenal cortex, respectively.

Values reported for 17-ketosteroids in human plasma are, in μg per 100 ml., testosterone, 0.5 (men) and 0.1 (women); dehydroepiandrosterone, 90; androsterone, 60; etiocholanolone, 40; and 11β-hydroxyetiocholanolone, 15.

Androgen Analogues. Testosterone administered by mouth exhibits approximately one-sixth the potency of injected testosterone because of partial inactivation of orally administered testosterone by the liver. A synthetic androgen, methyltestosterone, has the highest known androgenic potency when administered orally, and consequently has found wide clinical use.

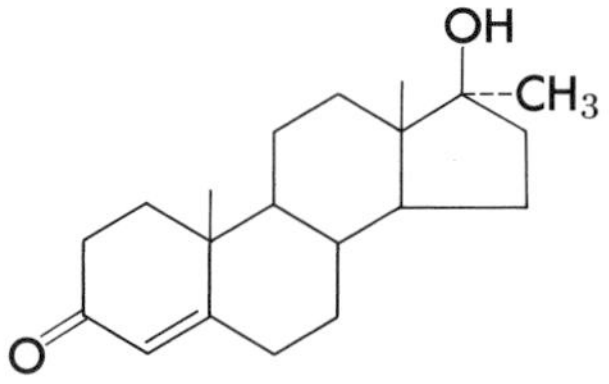

Methyltestosterone

Because of the anabolic effects of male sex hormones, as well as the therapeutic action of testosterone in carcinoma of the breast, extensive search has been made for structural analogues of androgens with the view of possible dissociation of desirable anabolic and therapeutic effects from the less desirable androgenic or masculinizing activity. Steroids without an angular methyl group in position 10 of the steroid nucleus (thus the methyl group which is C-19 is lacking) and termed *norsteroids* have proved to be of particular interest. For example, 19-nortestosterone and 17α-ethyl-19-nortestosterone,

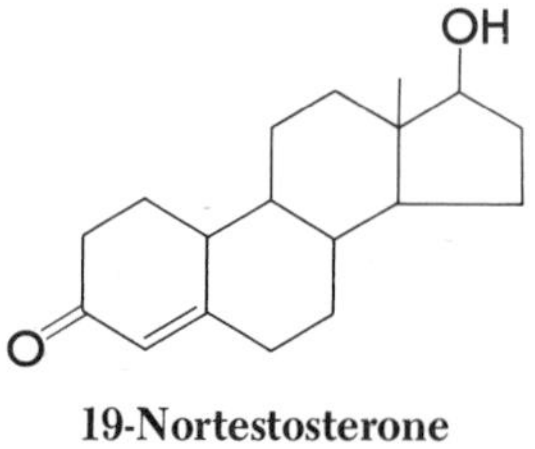

19-Nortestosterone

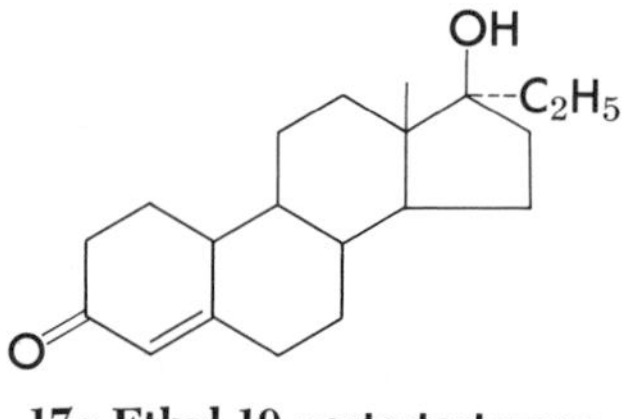

17α-Ethyl-19-nortestosterone

show on bioassay an anabolic to androgenic relative potency of about 20, whereas for testosterone propionate the ratio of these activities is approximately one. These two steroids also have a nitrogen-retaining activity in the human being, and favorably influence calcium balance.

THE PROSTAGLANDINS

The prostaglandins comprise a group of structurally related compounds whose biological effects include a lowering of blood pressure and stimulation of a variety of smooth-muscle organs. The prostaglandins, which are synthesized in the prostrate, were found initially in extracts of human seminal plasma and of the vesicular gland of sheep. Subsequently, it was shown that the prostaglandins are also present in a variety of tissues and species, *e.g.*, lung of several species, including man, calf thymus, bovine brain, and sheep iris. In addition, stimulation of structures such as nerves and the adrenal gland leads to release of prostaglandins.

Chemistry. Some 14 different prostaglandins have been isolated and characterized; all are biologically active. The prostaglandins are C_{20} fatty acids containing a five-membered ring; the parent fatty acid has been given the trivial name, prostanoic acid. The prostaglandins differ from one another in the number and position of double bonds and hydroxyl group substituents. The structure of prostaglandin E_1 is shown below.

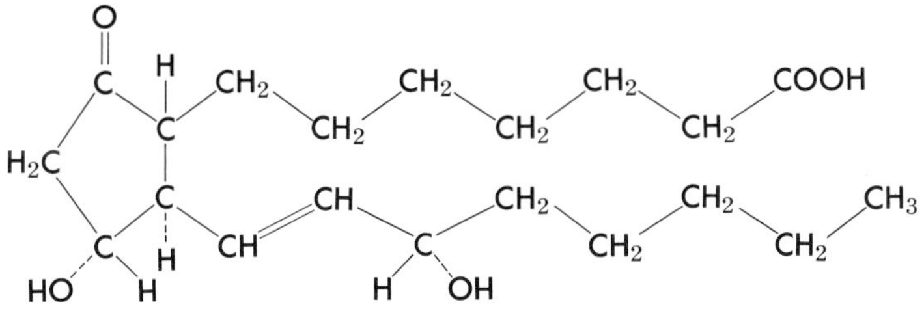

Prostaglandin E_1
(**11α,15-α-dihydroxy-9-keto-13 *trans*-prostenoic acid**)

The other prostaglandins contain one or two additional double bonds (at C-5 or at C-5 and C-17) and an additional hydroxyl group (at C-9) in place of the keto group. The double bond at C-13 has the *trans* configuration, whereas additional double bonds, when present, have the *cis* configuration. All the hydroxyl group substituents are of the α configuration.

Biosynthesis. The essential fatty acids (page 1010) serve as precursors of the prostaglandins. An enzymic system in homogenates of the vesicular gland of sheep converts arachidonic acid to prostaglandin E_2(11α, 15α-dihydroxy-9-keto-prosta-5, 13-dienoic acid). Also, homo-γ-linolenic acid (page 494) is converted into prostaglandin E_1. Molecular oxygen is utilized for formation of both the C-11 and C-15 hydroxyl groups as well as the keto group at C-9. These observations are of interest as indicating a possible approach to studies of the mechanism of action of the essential fatty acids and their possible metabolic functions.

Metabolism. Administration to the rat of prostaglandin E_1 labeled with tritium at C-5 and C-6 led to excretion of 50 per cent of labeled material in the urine and 10 per cent in the feces within 20 hr., indicating rapid excretion by way of the kidney and bile. Radioautography revealed marked concentration in liver and kidney. The identified excretory metabolites are products formed either by β oxidation, with removal of two carbon atoms from the carboxyl portion of the molecule or by reduction of the C-13 double bond and oxidation of the secondary alcohol groups at C-9 and C-15 to ketones. Enzymes catalyzing these transformations have been demonstrated in lung tissue of several species, including man.

Biological Effects. In addition to lowering blood pressure and stimulating smooth muscle, the prostaglandins inhibit release of glycerol and fatty acids from adipose tissue, as well as blocking the acceleration of this release caused by epinephrine (page 501) and glucagon (page 978). The prostaglandins have also been found to counteract the permeability response of the toad bladder to vasopressin (page 983) and to decrease nervous system excitability.

THE OVARY—FEMALE SEX HORMONES

Chemistry of Estrogens. Four compounds with estrogenic activity have been isolated from ovarian tissue and from human urine. All of these are C_{18} steroids, as compared with the C_{19} androgens. In the naturally occurring estrogens, there is no angular methyl group at position 10, and ring A is aromatic.

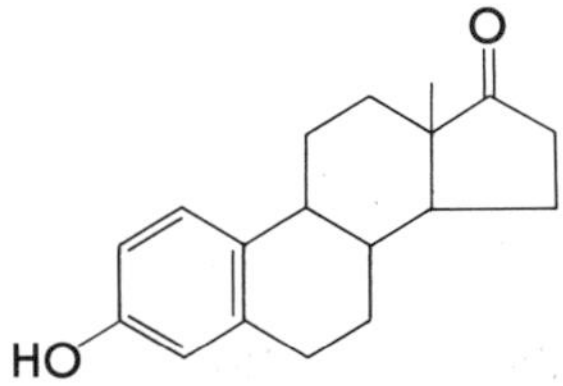

Estrone

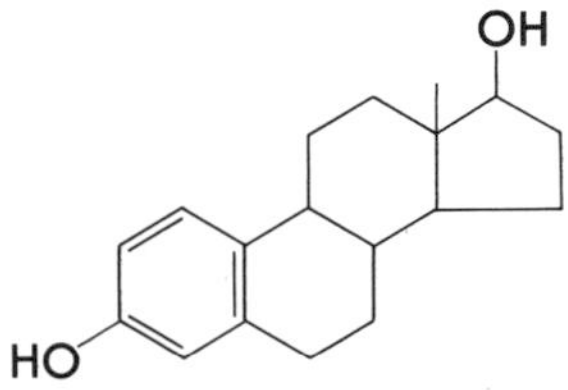

β-Estradiol

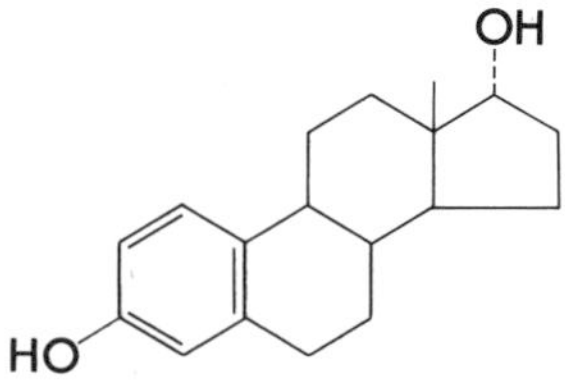

α-Estradiol

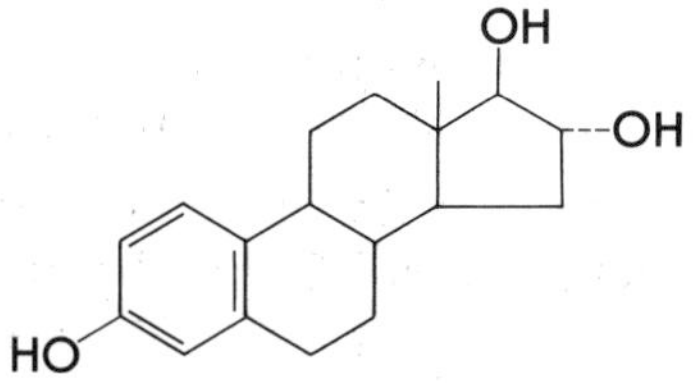

Estriol

β-Estradiol is the hormone normally secreted by the ovarian follicle. The α form, as well as estrone and estriol, are largely products of β-estradiol metabolism (see page 948). The comparative biological potencies of these compounds are shown in Table 44.2.

Although the ovaries and placenta are the chief sources of estrogenic hormones in the human being, the estrogenic content of horse *testis* is higher than that of any other endocrine organ. Indeed, stallion urine is the richest known source of β-estradiol. The physiological significance of these observations is not clear. Two estrogenic substances, obtained from pregnant mare urine, are apparently peculiar for this species. These are equilin and equilenin, with, respectively, approximately one-third and one-tenth the potency of estrone.

Table 44.2: COMPARATIVE BIOLOGICAL POTENCY OF THE ESTROGENS

Compound	Potency					
	Effective dose levels for vaginal response in rats		μg per		Relative activity	
	Subcut., μg	Oral, μg	Rat unit	Mouse unit	Spayed-rat method	Immature mouse uterine-weight method
Estrone.	0.7	20–30	1.0	1.0	100	100
β-Estradiol. .	0.3–0.4	20–30	0.08–0.125	0.05	1,000	1,000
α-Estradiol. .	. . . . . .	. . . . .	3.2–12.5	1.25	10	7.5
Estriol	. . . . . .	. . . . .	. . . .	. . . .	20	40

SOURCE: After W. H. Pearlman, The Chemistry and Metabolism of the Estrogens, in G. Pincus and K. V. Thimann, eds., "The Hormones," vol. I, chap. 10, pp. 351–405, Academic Press, Inc., New York, 1948.

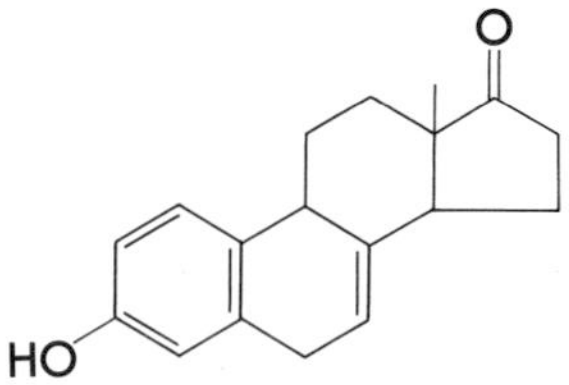

Equilin

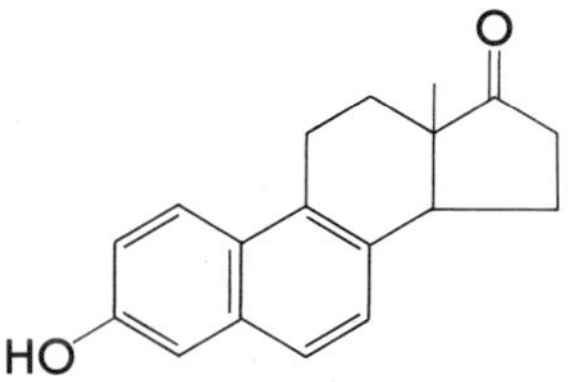

Equilenin

Biogenesis of Estrogens. The ovary is the principal site of estrogen production in the nonpregnant female. Surprisingly, the immediate precursor of the female sex hormones is the male sex hormone, testosterone. Formation of estrogens from testosterone has been observed in the presence of slices, broken cell preparations, or cell fractions of human ovarian, placental, testicular, or adrenal tissue, as well as of stallion testicular tissue. The probable pathways of estrogen biosynthesis are outlined in Fig. 44.2 and are based upon in vivo and in vitro studies with labeled compounds. The reactions leading from acetate to androst-4-ene-3,17-dione and testosterone have been indicated previously (Fig. 44.1). Transformation of 19-hydroxytestosterone to 19-oxotestosterone requires TPN, and the aromatization reaction utilizes TPNH and molecular oxygen. The final products are a phenolic C_{18} steroid and formaldehyde or formic acid.

Testosterone conversion to estrogen by other peripheral tissues is indicated by the urinary excretion of estrogen-^{14}C after administration of testosterone-^{14}C to an adrenalectomized-ovariectomized woman.

FUNCTIONS OF ESTROGENS

The Ovary and Sex Cycles in the Female. The ovaries are intimately concerned with two types of cyclic phenomena during which structural, functional, and

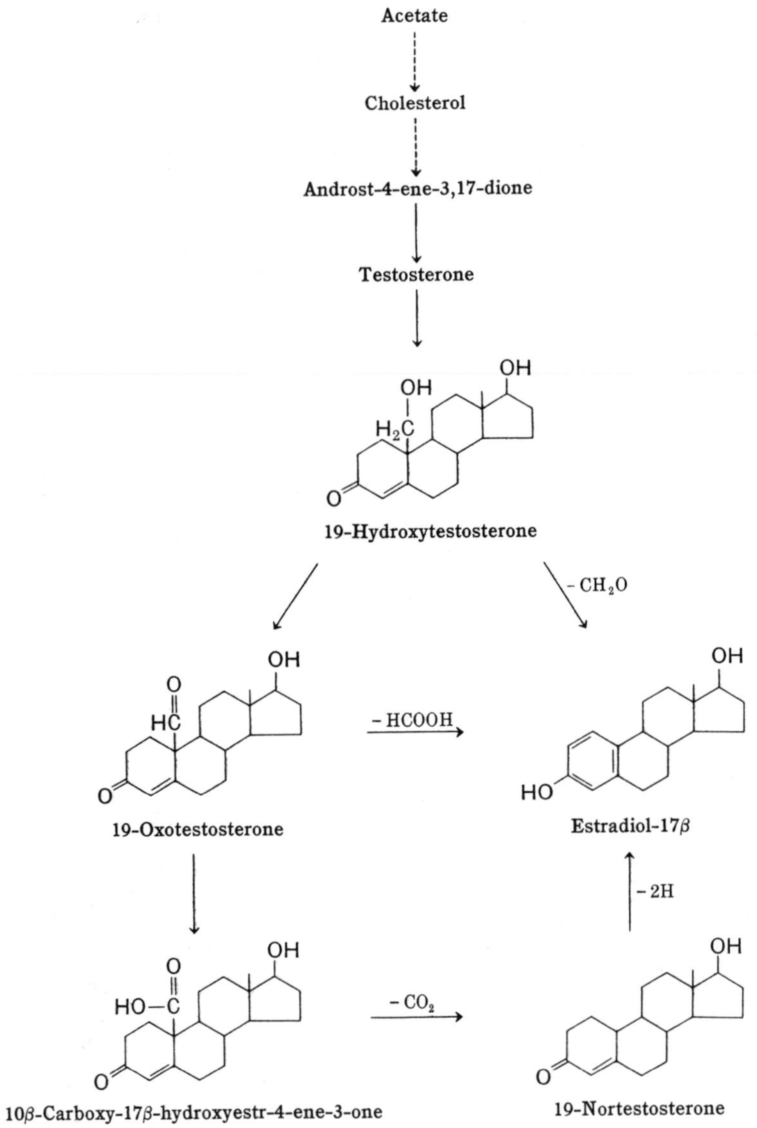

Fig. 44.2. Pathways of estrogen biosynthesis.

chemical alterations occur: (1) the estrous, or menstrual, cycles, occurring during reproductive life; and (2) the reproductive cycle.

The regulatory influence of the ovary in these cycles is the basis for the profound alterations seen following ovariectomy. If castration occurs prior to puberty, the female cycles never appear and the genital tissues remain infantile. Postpubertal castration results in cessation of the menstrual cycle, and atrophy of the uterus, vaginal mucosa, and fallopian tubes. The secondary sex characteristics disappear at varying rates after ovariectomy. In prepubertal castrates, pubic and axillary hair is scant, the typical pelvic enlargement does not take place, and lipid deposition, distinctive of mammary growth, fails to occur. Postpubertal ovariectomy causes some involution of the mammary glands, a slow change in distribution of body hair, and gradual appearance of osteoporosis. The last is seen only after prolonged deprivation of estrogenic hormones.

During the ovarian cycle, estrogens induce (1) proliferation of the vaginal epithelium, (2) augmented secretion of mucus by the cervical glands, and (3) endometrial proliferation (see also page 950). Estrogen secretion is responsible for the secondary sex characteristics of the female, stimulating growth of axillary and pubic hair, maturation of the skin, alterations in body contour, and closure of the long-bone epiphyses. Estrogens stimulate proliferation of the mammary gland during pregnancy.

Actions of Estrogens. In view of the profound physiological effects of estrogens, their wide biochemical influences are not unexpected and make difficult the discernment of primary estrogen actions from secondary effects. Estrogens have marked actions on organic metabolism, chiefly of protein, nucleic acids, and lipids, and on inorganic metabolism, notably of calcium and phosphorus. In addition, estrogens affect skin and related structures. The antagonistic action of estrogens to androgenic influences has been noted previously (page 940).

The widespread actions of estrogens on proliferative processes indicate the general anabolic influence of these steroids. This is marked in uterine and mammary gland tissue and reflected in acute stimulation of RNA and protein synthesis.

Protein and Nucleic Acid Metabolism. Administered estradiol is rapidly localized in uterine tissue; Jensen and his colleagues have reported purification from this tissue of a lipoprotein that specifically binds estradiol either injected or perfused through isolated uterine tissue. A similar estradiol-binding protein has been fractionated from rat and human mammary tumor tissue.

Within 1 hr. after estrogen administration in the rat, there is detected a pronounced increase in uterine tissue of a rapidly synthesized RNA fraction and a concomitant increase in activity of RNA polymerase. These effects can be blocked by administration of puromycin. The data suggest a mechanism of estrogen action in uterine tissue involving continuing synthesis of protein, leading to augmented RNA polymerase activity and increased synthesis of RNA with a short half-life, presumably messenger RNA.

An example of the regulatory influence of estrogens on protein synthesis is vitellin formation during development of the chick embryo. Vitellin is not produced in the absence of ovaries, but its synthesis can be initiated by estrogen injection.

Lipid Metabolism. The early effects of estrogen administration on lipid metabolism are equally dramatic, though occurring slightly later than the increase in protein and nucleic acid synthesis. In uterine tissue, estrogens produce a marked increase in phospholipid turnover. Estrogens are also lipotropic in mammals, *i.e.*, they prevent liver lipid accumulation when administered to animals on a diet deficient in a source of methyl groups (page 502).

The influence of estrogens on lipoprotein metabolism has drawn attention because of the higher incidence of coronary disease in men than in women. Estrogen administration in man decreases the level of circulating blood lipids, particularly in individuals with hyperlipemia.

Calcium and Phosphorus Metabolism. Prolonged estrogen administration results in elevation of serum calcium and phosphate levels, with calcification and hyperossification of the long bones so extensive that the marrow cavity may disappear and anemia ensue. There is an accompanying loss of calcium from the pelvic bones, which become cystic. That estrogens regulate normal bone metabolism is suggested from the bone decalcification seen in postmenopausal osteoporosis.

Miscellaneous Actions. The increased secretory activity of the skin sebaceous glands, induced by testosterone, is diminished by estrogen injection. The skin and associated structures are also affected by direct local application of estrogenic substances. This is seen in the growth of the mammary gland and in the inhibition of hair growth in rats and dogs produced by topical application of estrogens.

Metabolic Fate of Estrogens. Approximate values for estrone and β-estradiol in plasma, in μg per 100 ml., are in the female 0.03 for each steroid, and, in the male, 0.02 for estrone and 0.003 for estradiol. The liver is a major site of metabolic transformations of estrogens; interconversion occurs of estradiol-17β, estrone, and estriol. Estrone and estriol are major products of estrogen metabolism. Studies with ^{14}C-labeled estrogens have revealed a wide variety of transformations: introduction of hydroxyl groups, reversible oxidation and reduction of hydroxyl and carbonyl groups, respectively, and methylation. Quantitatively, hydroxylation at positions 2 and 16 of the estrogens appears to be a most prominent metabolic transformation. The hydroxylation mechanism is that considered previously (page 378); the specific enzymic systems (usually microsomal) utilize TPNH and molecular oxygen. However, enzymic systems hydroxylating at position 6 have been reported to utilize DPNH. Some interrelationships of estrogen metabolism are depicted in Fig. 44.3.

The interconversion of estradiol and estrone occurs in tissues other than the liver, including the placenta. In the latter, two 17β-estradiol dehydrogenase systems have been described, one DPN- and the other TPN-dependent. In addition, placenta has an estradiol-dependent transhydrogenase activity (page 342). This enzymic system is activated by estradiol; during the course of the reaction, estradiol undergoes cyclic oxidation and reduction with concomitant hydrogen transfer between the pyridine nucleotides.

The status of thyroid functioning in man markedly influences the rates of hydroxylation of steroid hormones and of the conversion of hydroxy substituents to keto groups. In hyperthyroidism the formation of 2-hydroxyestrone is increased to levels at which it becomes the major estrogen metabolite; in myxedema the amount of this steroid formed is diminished. Also, with elevated thyroidal activity, estriol formation is decreased. It appears that hydroxylation at C-2 and at C-16 are competitive reactions for the substrate, estradiol.

Estrogens and their metabolites are excreted primarily as conjugates of glucuronic and sulfuric acids, synthesized in the liver by previously described mechanisms (pages 433 and 563). The two principal conjugates are estriol glucosiduronate, with the glucosidic linkage at C-16, and estrone sulfate, with esterification at C-3. Sulfation at C-17, as well as formation of the 3,17-disulfate of estradiol, have also been described. These conjugates have been isolated from pregnancy plasma and urine, and from human placenta. They are estrogenically active when given orally; this suggests that in the conjugated form estrogens are less readily degraded by the liver.

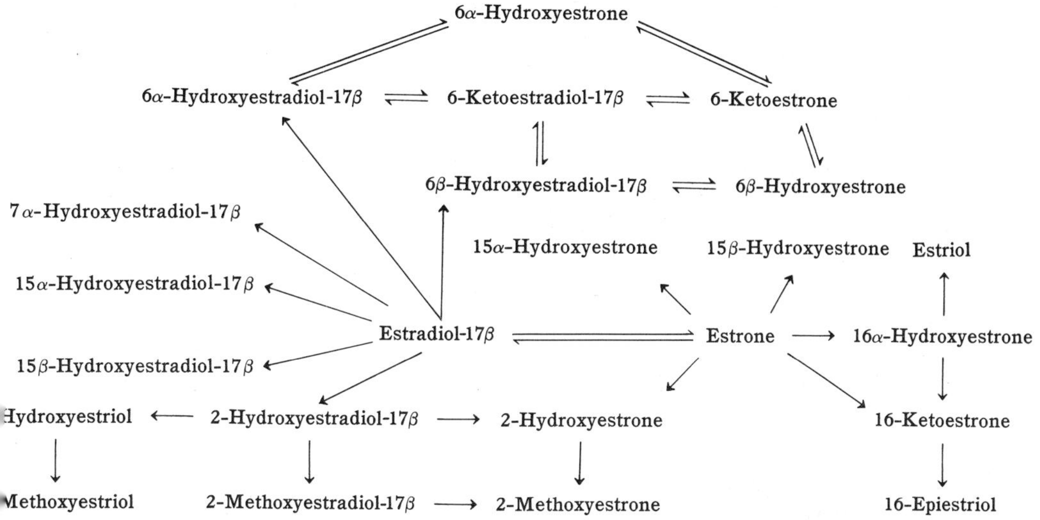

Fig. 44.3. Some interrelationships of estrogen metabolism.

A third mode of estrogen conjugation in rabbit liver involves UDP-N-acetylglucosamine. This has been reported for α-estradiol at C-17, and was dependent upon prior glucosiduronide formation at C-3. The resulting urinary conjugate was estradiol-3-glucuronoside-17α-N-acetylglucosaminide. β-Estradiol was not a substrate for the transfer of N-acetylglucosamine.

The prominent role of the liver in estrogen metabolism is reflected in a variety of experimental and clinical conditions. Nutritional factors, notably inanition, decrease the capacity of the liver to metabolize estrogens. Under these circumstances, as well as in hepatic cirrhosis, blood levels of estrogens may be persistently elevated, leading to testicular atrophy. A significant route of excretion of estrogens is the bile, and undue accumulation of these hormones may occur in circumstances of a decreased flow of bile or a diminished excretion of estrogen in the bile.

Synthetic Estrogens. Naturally occurring estrogens are most active when administered parenterally (Table 44.2). The most potent orally active estrogen is the synthetic compound, ethynylestradiol, which has approximately ten times the activity of estrone when each is administered by mouth. The high oral activity of ethynylestradiol is apparently due to its stability in the gastrointestinal tract and the liver. Stilbestrol, 4,4′-dihydroxy-α,β-diethylstilbene, is a synthetic product with approximately three to five times the estrogenic potency of estrone and is useful therapeutically because of its high potency when given orally.

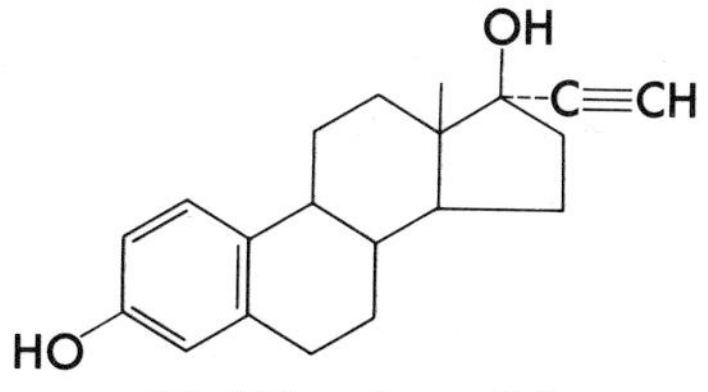

17α-Ethynylestradiol

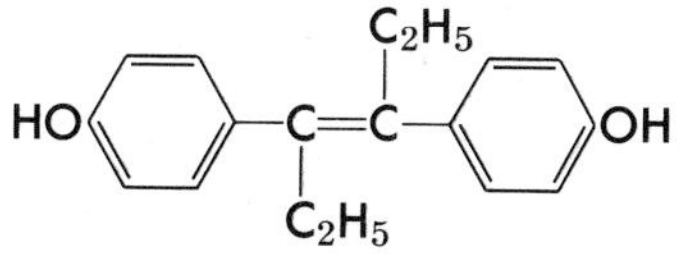

Stilbestrol

PROGESTERONE—HORMONE OF THE CORPUS LUTEUM

Chemistry and Biogenesis of Progesterone. Progesterone is produced not only by the corpus luteum, but by the placenta and adrenals as well. The structure and biosynthesis of progesterone have been given in Fig. 44.1.

Effects of Progesterone. Progesterone is secreted by the corpus luteum during the latter half of the menstrual cycle. The hormone acts upon the endometrium, previously prepared by estrogen, inducing mucus secretory activity indispensable for implantation of the ovum. If pregnancy ensues, continued secretion of progesterone is essential for completion of term. Progesterone also contributes to growth of the breasts and is thought to maintain the uterus quiescent during pregnancy. Progesterone exerts an antiovulatory effect when given during days 5 to 25 of the normal menstrual cycle; this is the basis for the use of certain synthetic progestins (page 951) as oral contraceptive agents. Progesterone in very large doses will promote retention of salt and water, its action thus resembling that of certain adrenal cortical hormones (page 965).

Metabolism of Progesterone. The position of progesterone in the conversion of acetate and cholesterol to C_{19} steroids has been indicated previously in Fig. 44.1. This explains the formation of 17α-hydroxyprogesterone, androst-4-ene-3,17-dione, and testosterone on incubation of progesterone with testicular, ovarian, adrenal, or placental tissues. These three products, therefore, can be considered as derived from progesterone metabolism. In addition, progesterone may give rise to estrone and estradiol, via testosterone (Fig. 44.2). Formation of these estrogens has been demonstrated in studies in which progesterone has been incubated in vitro with human ovarian tissue.

More significant quantitatively as a metabolite of progesterone is pregnane-3α, 20α-diol, a reduction product formed from progesterone chiefly in the liver, where it is coupled with glucuronic acid to form the chief urinary metabolite of progesterone, pregnanediol glucosiduronate.

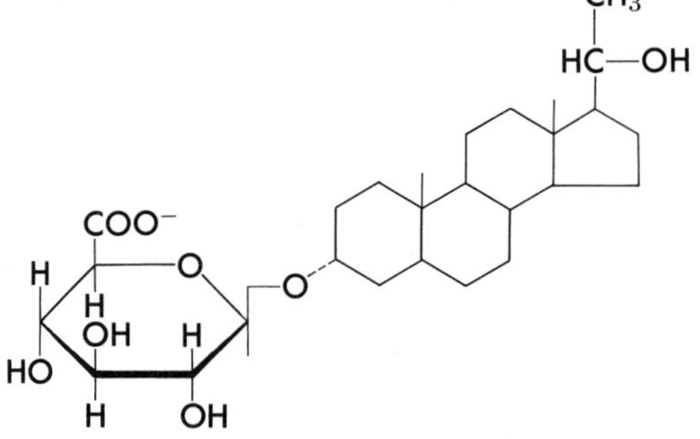

Pregnane-3α,20α-diol glucosiduronate

Reduction of the C-20 ketone group of progesterone to the 20α- and β-hydroxy isomers also occurs extrahepatically (see below).

Although pregnanediol is not the sole metabolite of progesterone and is not derived entirely from progesterone, pregnanediol excretion in the urine is a convenient semiquantitative index of progesterone elaboration and metabolism. Pregnanediol determinations are of added significance if correlated with physiological status. In women with normal menstrual cycles the corpus luteum is the major source of progesterone; during pregnancy, progesterone arises chiefly in the placenta, especially in the later periods of gestation.

Two additional metabolites of progesterone have been isolated from human ripe follicles, corpora lutea, and placenta. These are pregn-4-ene-3-one-20α-ol, with approximately one-third to one-half the progestational activity of progesterone, and the corresponding 20β-ol isomer, with one-tenth to one-fifth the activity of progesterone. Some samples of human milk may contain the 20β-ol isomer of urinary pregnanediol; this compound specifically inhibits in vitro the enzymic formation of bilirubin diglucuronide. There appears to be correlation between the occurrence of this steroid in human milk and the appearance of hyperbilirubinemia in some breast-fed infants.

Synthetic Progestins. Progesterone has high activity only when given parenterally; oral doses several hundred times the parenteral dose are required to elicit responses characteristic of progesterone. However, a number of synthetic steroids are active as progestins when given orally. The two compounds shown below have a progestational activity varying from equal to that of parenterally administered progesterone, as in the case of 17α-ethynyltestosterone, to ten times as active as progesterone, *e.g.*, 17α-ethynyl-19-nortestosterone.

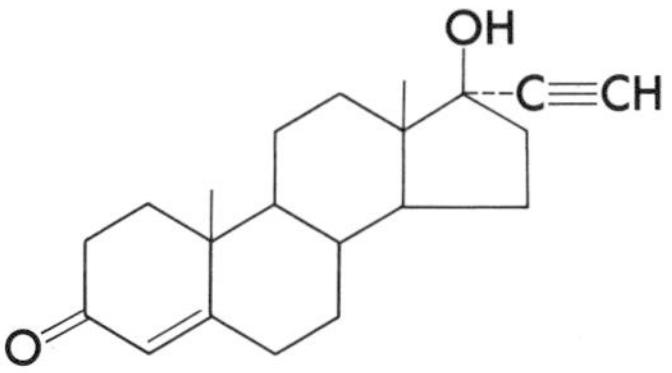

17α-Ethynyltestosterone

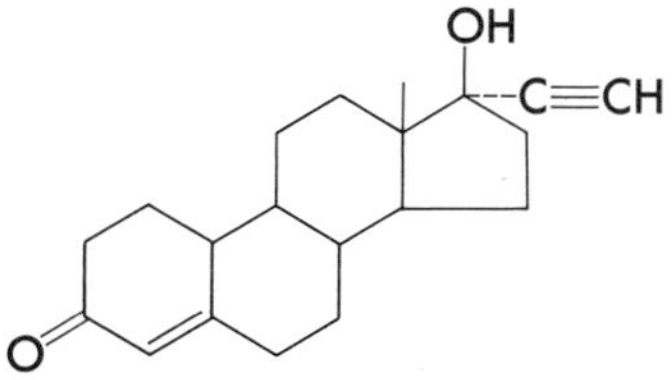

17α-Ethynyl-19-nortestosterone

RELAXIN

The corpus luteum apparently produces a second hormone in addition to progesterone. Extracts that have the capacity to relax the symphysis pubis of the guinea pig and of the mouse can be prepared from the corpora lutea of the sow, from the blood of pregnant females of a wide variety of species, including pregnant women, as well as from placenta. The active principle has been named *relaxin.* The material is active only when injected into an animal which is in normal or artificially induced estrus. Activity is associated with a basic polypeptide of molecular weight approximately 9,000.

Progesterone and estrogens also produce relaxation of the symphysis pubis of the guinea pig. However, the steroid hormones bring about relaxation only after prolonged treatment, whereas the effect with relaxin is seen in a few hours.

REFERENCES

Books

Dorfman, R. I. and Shipley, R. A., "The Androgens: Biochemistry, Physiology, and Clinical Significance," John Wiley & Sons, Inc., New York, 1956.

Dorfman, R. I., and Ungar, F., "Metabolism of Steroid Hormones," Academic Press, Inc., New York, 1965.

Young, W. C., ed., "Sex and Internal Secretions," 3d ed., vols. I and II, The Williams & Wilkins Company, Baltimore, 1961.

Zuckerman, S., "The Ovary," vol. II, Academic Press, Inc., New York, 1962.

Review Articles

Bergström, S., The Prostaglandins, *Recent Progr. Hormone Research,* **22,** 153–175, 1966.

Bergström, S., Prostaglandins: Members of a New Hormonal System, *Science,* 157, 382–391, 1967.

Bergström, S., and Samuelsson, B., eds., "The Prostaglandins," Proc. II Nobel Symp., Almqvist and Wiksells, Uppsala, 1967.

Eder, H., The Effects of Hormones on Human Serum Lipoproteins, *Recent Progr. Hormone Research,* **14,** 405–425, 1958.

Gallagher, T. F., Fukushima, D. K., Noguchi, S., Fishman, J., Bradlow, H. L., Cassouto, J., Zumoff, B., and Hellman, L., Recent Studies in Steroid Metabolism in Man, *Recent Progr. Hormone Research,* **22,** 283–303, 1966.

Hall, K., Relaxin, *J. Reprod. Fertility,* **1,** 368–384, 1960.

Jensen, E. V., On the Mechanism of Estrogen Action, *Perspectives Biol. Med.,* **6,** 47–60, 1962.

Jensen, E. V., and Jacobsen, H. I., Basic Guides to the Mechanism of Estrogen Action, *Recent Progr. Hormone Research,* **18,** 387–414, 1962.

Solomon, S., Bird, C. E., Ling, W., Iwamiya, M., and Young, P. C. M., Formation and Metabolism of Steroids in the Fetus and Placenta, *Recent Progr. Hormone Research,* **23,** 297–347, 1967.

Talalay, P., Enzymatic Mechanisms in Steroid Metabolism, *Physiol. Revs.,* **37,** 362–389, 1957.

Talalay, P., Enzymatic Mechanisms in Steroid Biochemistry, *Ann. Rev. Biochem.,* **34,** 347–380, 1965.

Williams-Ashman, H. G., Liao, S., Jancock, R. L., Jurkowitz, L., and Silverman, D. A., Testicular Hormones and the Synthesis of Riobonucleic Acids and Proteins in the Prostate Gland, *Recent Progr. Hormone Research,* **20,** 247–301, 1964.

45. The Adrenals

Epinephrine and Norepinephrine. Adrenal Cortical Hormones

Each adrenal gland is a composite of two structures, the *medulla,* of neural origin, and the *cortex,* derived from mesodermal glandular tissue. The hormonal products and their effects are distinct for each of these two portions of the adrenal.

THE ADRENAL MEDULLA

Hormones of the Adrenal Medulla. Two compounds with hormonal activity have been isolated from adrenal medullary extracts, epinephrine and norepinephrine. The natural D form of each is approximately fifteen to twenty times as active as the L isomer. These hormones are produced in chromaffin tissue generally.

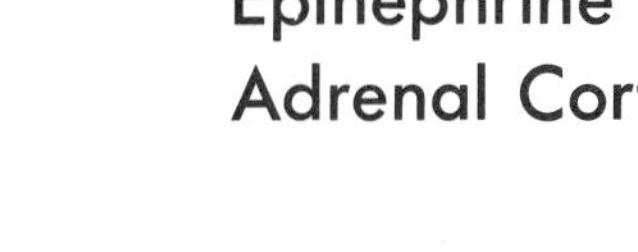

Epinephrine

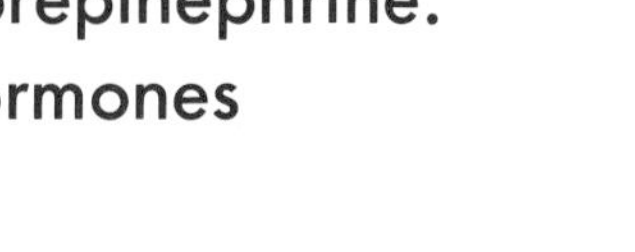

Norepinephrine

The ratios of epinephrine to norepinephrine in extracts of the medulla vary considerably with the species studied. Norepinephrine represents approximately 10 to 30, 50, and 90 to 100 per cent of total hormone of the adrenal medulla in man, cat, and whale, respectively. Commercial preparations of "epinephrine" may contain 10 to 20 per cent of norepinephrine.

Norepinephrine and epinephrine are stored in separate cells of the medulla and are released by different stimuli. Insulin releases epinephrine preferentially, while stimulation of various areas of the hypothalamus yields varying proportions of the two hormones. The release of these catecholamines from the adrenal medulla under the influence of acetylcholine is dependent upon the presence of Ca^{++}, which enters chromaffin cells readily in the presence of acetylcholine.

Biogenesis of Epinephrine and Norepinephrine. The major pathway for the formation of the medullary hormones is depicted in Fig. 45.1 (page 954). Hydroxylation of phenylalanine to form tyrosine occurs in the liver (page 548); it has not been described in the adrenal. Individuals with hereditary phenylketonuria (page 549), due to a deficiency of liver phenylalanine hydroxylase (page 549), have low plasma epinephrine concentrations.

Oxidation of tyrosine to 3,4-dihydroxyphenylalanine (dopa) in adrenals is catalyzed by *tyrosine hydroxylase,* which resembles phenylalanine hydroxylase and utilizes the same cofactors. An *aromatic L-amino acid decarboxylase,* requiring

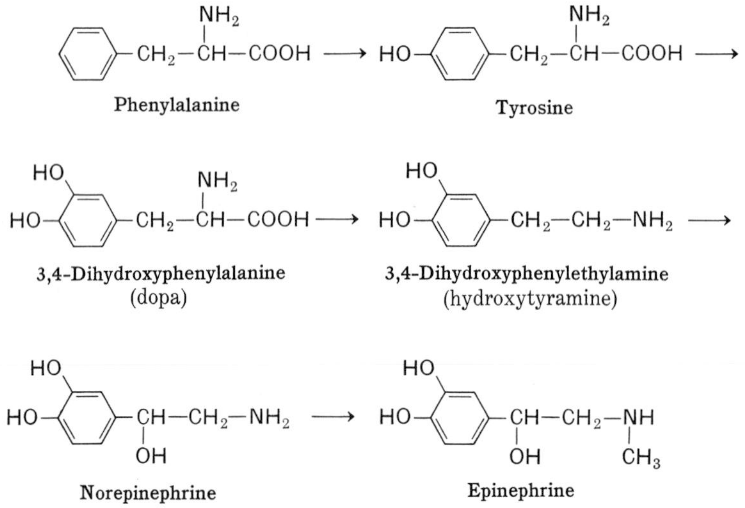

Fig. 45.1. Major pathway for biosynthesis of norepinephrine and epinephrine.

pyridoxal phosphate and present in mammalian kidney, liver, and adrenal medullary tissue, catalyzes decarboxylation of 3,4-dihydroxyphenylalanine to 3,4-dihydroxyphenylethylamine (hydroxytyramine). The latter is converted to norepinephrine under the catalytic influence of *3,4-dihydroxyphenylethylamine β-hydroxylase,* a copper-requiring mixed function oxidase that catalyzes direct addition of oxygen to the β-carbon atom. The enzyme also catalyzes β-hydroxylation of other amines structurally related to hydroxytyramine. Methylation of norepinephrine to epinephrine is catalyzed by *phenylethanolamine-N-methyl transferase,* utilizing S-adenosylmethionine as the methylating agent (page 553). The activity of this last enzyme is regulated by the pituitary–adrenal cortical system (Chap. 47); adrenal cortical steroids depress this enzymic activity. Thus the hormones secreted by the adrenal cortex may exert some of their physiological effects as a result of their action on the adrenal medulla to produce alterations in the enzymic methylation of norepinephrine to epinephrine.

In the synthesis of the medullary hormones, hydroxylation of tyrosine to 3,4-dihydroxyphenylalanine is the rate-limiting step.

Metabolism of Epinephrine and Norepinephrine. Although the medulla of man normally contains about three to ten times as much epinephrine as norepinephrine, the mean plasma concentration of epinephrine is approximately 0.06 μg per liter while the value for norepinephrine averages 0.3 μg per liter. In urine the values per 24 hr. are 10 to 15 μg for epinephrine and 30 to 50 μg for norepinephrine.

Both hormones are rapidly metabolized by three mechanisms, O-methylation utilizing S-adenosylmethionine and catalyzed by *catechol O-methyl transferase,* oxidative deamination, catalyzed by *monamine oxidase,* and conjugation. The liver is the major site of these transformations; O-methylation is the major metabolic pathway for epinephrine. The principal metabolites of epinephrine in urine are 3-methoxy-4-hydroxymandelic acid, 3-methoxyepinephrine (metanephrine), and 3-methoxy-4-hydroxyphenylglycol. The corresponding catechols, 3,4-dihydroxymandelic acid, and norepinephrine and epinephrine, are minor excretion products. Some of these metabolites are excreted in the urine and bile both as the free compounds and as sulfates or glucosiduronides; conjugation occurs at the 4-hydroxyl group of the catechols. A minor degree of N-acetylation may also occur. The major metabolic pathways are indicated in Fig. 45.2 (page 956).

The paramount importance of O-methylation in the metabolism of epinephrine is seen from an experiment in which an intravenous infusion of ^{3}H-epinephrine was given a human subject. More than 80 per cent of the radioactivity was recovered in the first 48-hr. urine as O-methylated products. This radioactivity was distributed as follows (approximate values in per cent): epinephrine, 6; metanephrine, free and conjugated, 40; 3-methoxy-4-hydroxymandelic acid, 41; 3-methoxy-4-hydroxyphenylglycol sulfate, 7; and 3,4-dihydroxymandelic acid, 2.

In other species, other metabolic reactions have been described, including N-demethylation of epinephrine in the rat and dehydroxylation of norepinephrine, both of the side chain and of the ring, in the guinea pig.

An enzymic system in the salivary gland, differing from pyrocatechase (page 377) converted epinephrine to adrenochrome in vitro. This is probably not a significant metabolic fate of epinephrine in vivo.

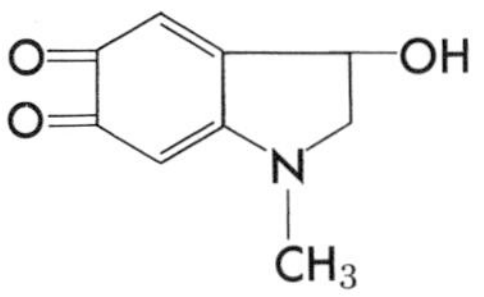

Adrenochrome

Effects of Epinephrine and Norepinephrine. The hormones of the adrenal medulla elicit a wide variety of effects, seen in Table 45.1, which also indicates certain differences in response produced by epinephrine and norepinephrine.

Cardiovascular and Muscular Effects. Epinephrine administered intravenously

Table 45.1: Comparison of the Effects of Intravenous Infusion of Epinephrine and Norepinephrine in Man

Index	Epinephrine*	Norepinephrine
Heart rate	+	−
Cardiac output	+++	0, −
Systolic blood pressure	+++	+++
Diastolic blood pressure	+, 0, −	++
Total peripheral resistance	−	++
Oxygen consumption	++	0, +
Blood glucose	+++	0, +
Blood lactate	+++	0, +
Blood nonesterified fatty acids	+++	+++
Central nervous system action	+	0
Eosinopenic response	+	0

* + = increase; 0 = no change; − = decrease.

Source: After M. Goldenberg, *Am. J. Med.,* **10,** 627, 1951.

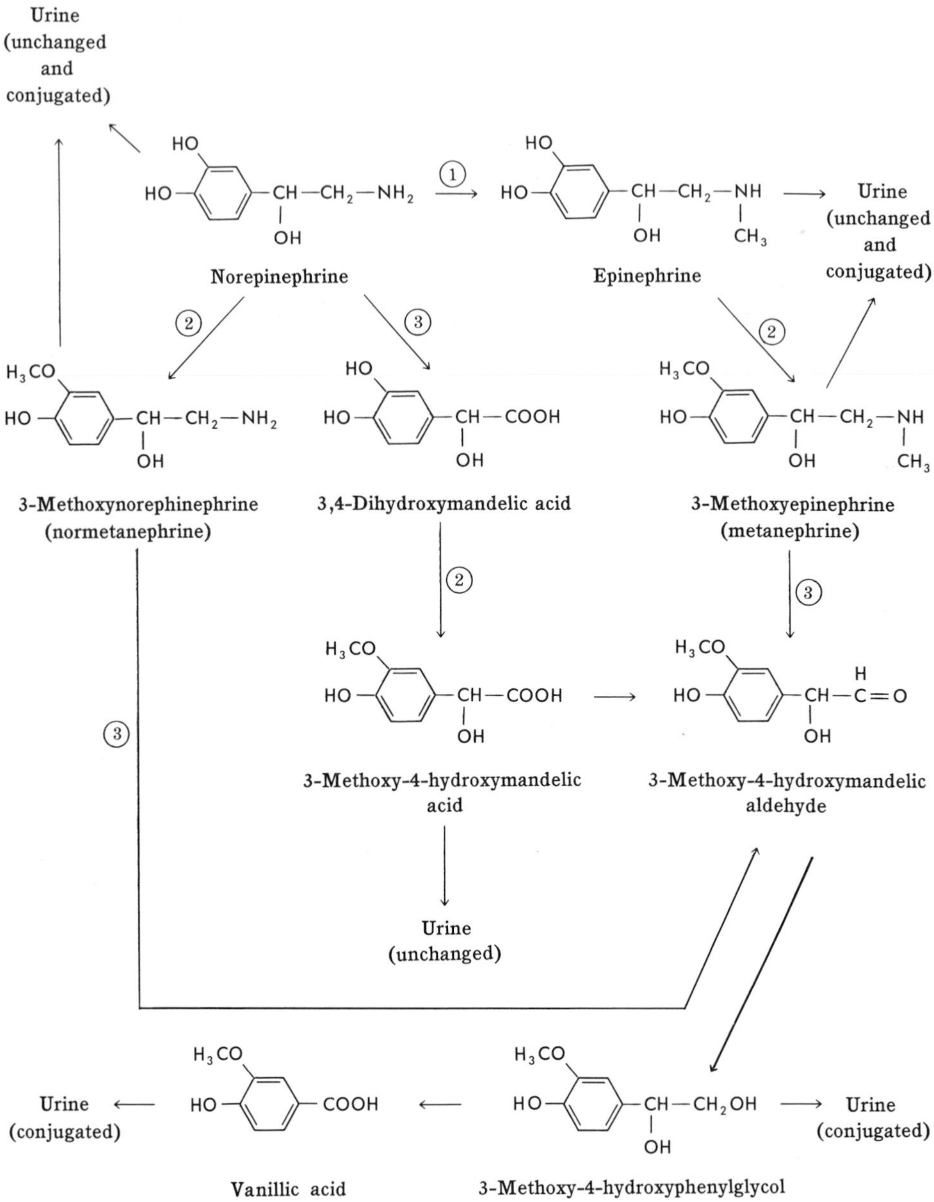

Fig. 45.2. Major pathways of metabolism of norepinephrine and epinephrine. Enzymes catalyzing individual steps are indicated by numbers at arrows: ① phenylethylamine-N-methyl transferase; ② catechol O-methyl transferase; ③ monamine oxidase. As indicated in the text (page 955), conjugates are also excreted via the bile.

produces a marked rise in blood pressure due to arteriolar vasoconstriction, and an increase in heart and pulse rate and in cardiac output. In moderate doses, epinephrine causes dilatation, rather than constriction, of the vessels of the skeletal muscles, and the coronary and visceral vessels, thus increasing blood flow in these regions.

Epinephrine produces variable effects on smooth muscle; it relaxes gastrointestinal tract musculature and causes contraction of the pyloric and ileocecal sphincters. There is a marked dilator effect on bronchial musculature.

Norepinephrine exerts much weaker inhibitory influence on smooth muscle than does epinephrine, does not relax bronchiolar musculature, and augments both systolic and diastolic blood pressure owing to increased total peripheral resistance, with little effect on cardiac output (Table 45.1).

The marked pressor effects of epinephrine and norepinephrine, as well as of intermediates in the formation of these hormones, *e.g.*, hydroxytyramine (page 954), has led to the practical utilization of α-methyldihydroxyphenylalanine as an inhibitor of aromatic L-amino acid decarboxylase (page 588) as a means of limiting the formation of pressor amines.

Effects on Carbohydrate Metabolism. The effects of epinephrine on carbohydrate and lipid metabolism have been considered previously (Chaps. 19 and 21, respectively). Epinephrine promotes glycogenolysis in muscle and liver (page 444), resulting in elevation of the blood glucose level and increased lactic acid formation in muscle. These effects are accompanied by an increased oxygen consumption, approximately 20 to 40 per cent in man, with an even greater increase in carbon dioxide production, raising the respiratory quotient. In hepatectomized animals, epinephrine exerts no effect on the falling blood sugar level. In contrast, norepinephrine has relatively little effect on carbohydrate metabolism and on oxygen consumption (Table 45.1).

Epinephrine increases synthesis of cyclic 3′,5′-AMP by the adenyl cyclase system in a variety of tissues including liver, heart, skeletal muscle, and adipose tissue. The adenyl cyclase system has been shown to be localized in the cell membrane in rat liver and adipose cells, and avian erythrocytes. The effects of epinephrine on glycogen metabolism (page 444) are mediated via adenyl cyclase. The activity of this enzyme in vitro is increased by epinephrine or norepinephrine in concentrations of 10^{-7} to $10^{-6}M$.

Effects on Lipid Metabolism. The catechol amines have a profound lipid-mobilizing activity, increasing the blood level of nonesterified fatty acids by augmenting the release of free fatty acids from adipose tissue, as a consequence of stimulation of lipolysis. There is a concomitant increase in oxygen consumption. In *pheochromocytoma* (see below), with hyperfunctioning of the adrenal medulla, blood levels of nonesterified fatty acids may be as high as several hundred times normal. Epinephrine administration to normal animals also elevates serum cholesterol and phospholipid levels and stimulates phospholipid turnover in cardiac tissue.

The effects of epinephrine and norepinephrine on blood pressure and on the release of free fatty acids from adipose tissue are counteracted significantly by the prostaglandins (page 944).

Adrenal Medullary Hyperfunction. No experimental or clinical condition corresponding to hypofunctioning of the adrenal medulla has been described. Hyperfunctioning of this structure in man is due to chromaffin tissue tumors, termed *pheochromocytomas,* and is reflected in one or several of the following features: (1) paroxysmal hypertension; (2) persistent hypertension, resembling essential or malignant hypertension; (3) a combination of hypertension, elevation of the basal metabolic rate, and glucosuria; (4) a persistent elevation of the basal metabolic rate, or hyperglycemia coexistent with intermittent hypertension. Plasma levels of norepinephrine and epinephrine may rise to more than 500 times the normal (page 954), plasma nonesterified fatty acids are elevated (page 501), and there is a marked increase in urinary levels of norepinephrine, epinephrine, and 3-methoxy-4-hydroxymandelic acid. When the tumor is extramedullary, norepinephrine may represent as much as 90 per cent of the total hormone of the tissue. When the tumor is medullary, it yields increased quantities of both hormones.

THE ADRENAL CORTEX

Disease attributable to pathological changes in the adrenal glands was first described in 1855 and is known after its discoverer as Addison's disease. It is characterized by loss of appetite, gastrointestinal disturbances (vomiting and diarrhea), rapid loss of weight, weakness and prostration, a low degree of resistance to infection, anemia, and an abnormal and characteristic pigmentation which is most pronounced in regions where normal pigmentation is greatest, *i.e.*, buccal cavity, nipples, etc. The palms of the hands and the soles of the feet remain pale. The fatal consequences of Addison's disease, as of extirpation of the adrenal glands, result solely from the lack of adequate functional *cortical* tissue.

Chemistry of Adrenal Cortical Hormones. Approximately 30 crystalline steroids have been obtained from extracts of adrenal glands. With the exception of estrone (page 944), with 18 carbon atoms, and cholesterol, containing 27 carbon atoms, all these steroids have either 19 or 21 carbon atoms. Some of the C_{19} steroids which have androgenic activity have been considered in the previous chapter (page 937). One of the steroids isolated from adrenal tissue is progesterone, the hormone of the corpus luteum (page 950); this is another example of a hormone being formed by two endocrine glands.

Of the steroids isolated from the adrenal, three are normally secreted by the adrenal cortex; their structures are given below and on the following page.

In aldosterone the customary angular methyl group at C-13 in steroids is replaced by an aldehyde group. Solutions of aldosterone contain an equilibrium mixture of the aldehyde and the hemiacetal; equilibrium favors the latter compound.

Placenta contains adrenal cortical activity equivalent to 1 or 2 per cent that of adrenal tissue, per gram of tissue. Hence the placenta can augment the supply of adrenal cortical hormones.

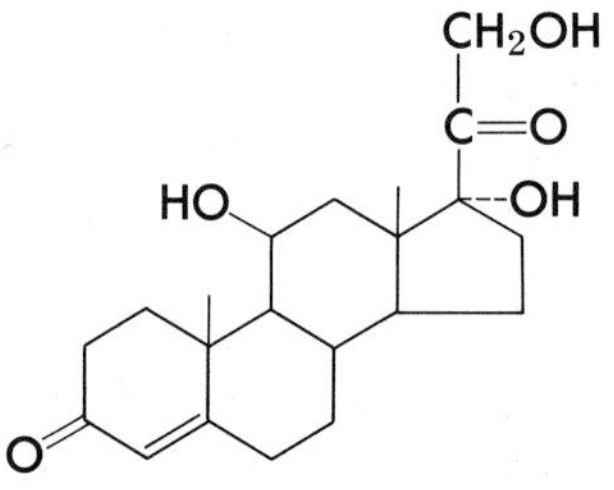

17α-Hydroxycorticosterone
(cortisol; hydrocortisone)

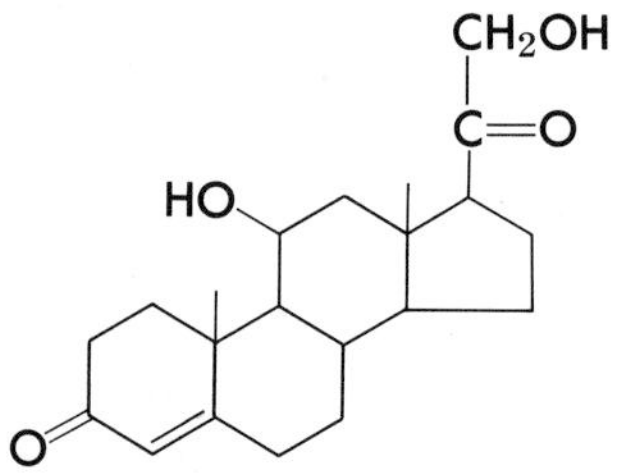

Corticosterone

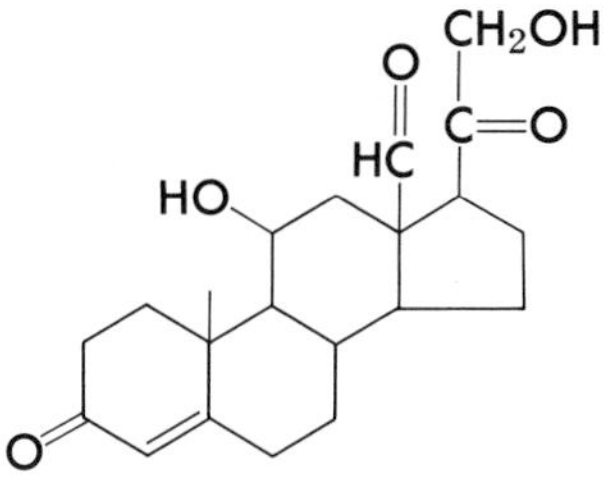

Aldosterone

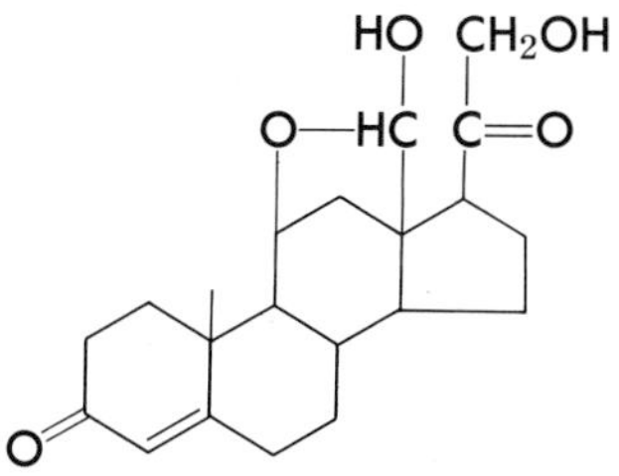

Aldosterone
(hemiacetal structure)

Biosynthesis of Adrenal Cortical Steroids. In man, the adrenal cortex normally secretes approximately 10 to 30 mg. of cortisol, 2 to 4 mg. of corticosterone, and 300 to 400 μg of aldosterone per 24 hr. The rat adrenal apparently secretes corticosterone almost exclusively. The most prominent of the androgenic steroids secreted by the human adrenal is 11β-hydroxyandrost-4-ene-3-17-dione (page 937).

Adrenal steroids are synthesized in vivo from either acetate or from cholesterol arising from the former. The cholesterol content of the adrenal is the highest found in tissues, with the exception of nervous tissue. Adrenal cholesterol is almost wholly in the esterified form; these esters have a high content of polyunsaturated, long-chain fatty acids, including an unusual fatty acid which represents the addition of a C_2 unit to the carboxyl-terminal end of arachidonic acid and has been termed adrenic acid (7, 10, 13, 16-docosatetraenoic acid). Stimulation of the adrenal cortex causes a marked decrease in cholesterol concentration at a time when maximal release of adrenal cortical steroids is occurring, indicating the role of cholesterol as precursor of the steroids. The adrenal also contains the body's highest concentration of ascorbic acid, 400 to 500 mg. per 100 g. fresh tissue.

The biosynthetic pathways for adrenal cortical steroids are indicated in Fig. 45.3 (page 960). The steps between cholesterol and pregnenolone were discussed previously (page 524), as has the pathway for synthesis of C_{19} steroids in the adrenal (page 939). Dehydroepiandrosterone, synthesized in the adrenal, is both stored and secreted by the gland as the 3-sulfate ester. Scission of the side chain of 20,22-dihydroxycholesterol, with formation of pregnenolone, is a TPNH-requiring reaction, catalyzed by a *desmolase;* this reaction is stimulated by cyclic AMP. The 3β-hydroxy dehydrogenase utilizes DPN and is present in the microsomal fraction of adrenal broken cell preparations.

The adrenal steroid hydroxylating enzyme systems, each specific for an individual position of the steroid nucleus, require TPNH and utilize molecular oxygen, as is the case for hydroxylations of other ring structures (page 378). Stimulation of hydroxylation by cyclic AMP is related to the role of this nucleotide in providing TPNH from the phosphogluconate oxidative pathway, via activation of adrenal phosphorylase.

The 11β-hydroxylating enzyme complex is mitochondrial and includes adrenodoxin, a nonheme iron protein (Table 17.4, page 368). The 18-hydroxylating system is also mitochondrial, whereas the enzyme catalyzing 21-hydroxylation is microsomal. Of the steroid-producing endocrine organs, the adrenal is the major site of 11-hydroxylation.

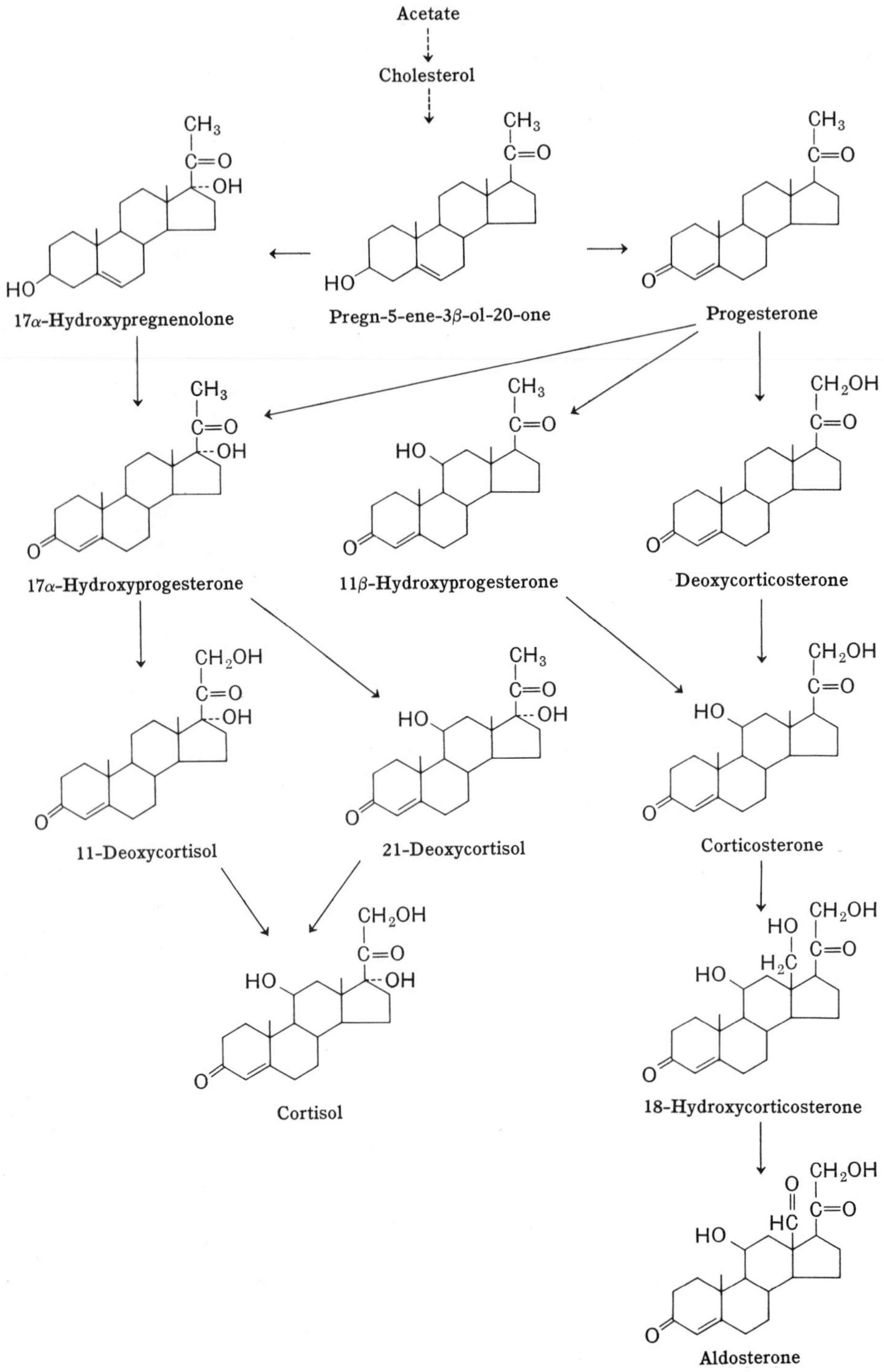

Fig. 45.3. Biosynthesis of some C_{21} adrenal cortical steroids.

Other hydroxylation reactions of C_{21} steroids have been described, including 1β-, 2α-, 2β-, 6β-, 7α-, 11α-, 15α-, and 15β-hydroxylations. 6β-Hydroxycortisol is present in human urine, and large amounts of 6β-hydroxycorticosterone were produced on incubation of hyperplastic adrenal tissue and an adrenal adenoma, both removed from a patient with primary aldosteronism (page 968). 6β-Hydroxylation of other steroids has been described, particularly when there is a deficiency in 11β-hydroxylase. This enzyme is lacking, or 11-hydroxylation is blocked, in many of the adrenal carcinomas which have been studied.

Regulation of Adrenal Cortical Secretion. The secretory activity of the adrenal cortex is regulated by hypophyseal adrenocorticotropic hormone (ACTH), particularly with regard to the production of corticosterone and cortisol (Chap. 47). However, secretion of aldosterone by the adrenal cortex, although influenced by ACTH, is significantly less dependent on the hypophysis and is markedly affected by other factors. These will be considered briefly here because of their importance in subsequent discussion of the effects of adrenal cortical steroids.

The $[Na^+]$ and $[K^+]$ in adrenal arterial blood, particularly the former, are important regulators of aldosterone secretion; decreased $[Na^+]$ is the most significant factor stimulating aldosterone release. There is, therefore, a direct, local effect of ions on the adrenal. No obligatory increase in cortisol or corticosterone secretion is evident with augmented aldosterone output. Secretion of aldosterone is also augmented by constriction of the thoracic vena cava or the hepatic portal vein, apparently because of production of angiotensin II (page 839). This influence of angiotensin II on aldosterone secretion is independent of the former's pressor action.

The adrenal becomes more responsive to ACTH under circumstances of Na^+ deficiency. Hypophysectomy does not affect aldosterone secretion evoked by Na^+ depletion but does reduce output of this steroid elicited by noxious stimuli (page 967).

Adrenal venous blood in man contains approximately 240 μg of cortisol and 80 μg of corticosterone per 100 ml. of blood. Total adrenal steroids in *peripheral blood* vary from 5 to 40 μg per 100 ml. There is a pronounced diurnal variation both in blood and in urine of these steroids; urinary concentration is maximal in early morning, slowly declines during the day, and is minimal during the night.

Transport of Adrenal Cortical Steroids. Cortisol represents approximately four-fifths of the total 17-hydroxysteroids in blood. The hormone is bound to a specific α_1-globulin which has been termed *transcortin* or *corticosteroid-binding protein* which has a single cortisol-binding site per molecule of protein (molecular weight = 52,000); this protein also binds corticosterone. In contrast, aldosterone binds weakly with transcortin and chiefly to albumin, which loosely binds cortisol but only at blood levels of the latter steroid that result in saturation of binding sites of transcortin (30 to 40 μg cortisol per 100 ml. plasma). Transcortin is produced in the liver; its concentration is lowered in liver disease and in nephrosis and elevated by thyroid hormone and by estrogen, as seen during the later stages of pregnancy. This action of estrogen is mediated by hypophyseal thyrotropic hormone (Chap. 47).

ROLE OF THE ADRENAL CORTEX

The adrenal steroids influence a wide variety of biochemical and physiological phenomena, some of which are probably a consequence of secondary effects of a primary role, and others which may be even more indirect results of initial actions. These may be grouped, arbitrarily into (1) alterations in carbohydrate, protein, and lipid metabolism, (2) effects on electrolyte and water metabolism, including an influence on circulatory homeostasis and on neuromuscular irritability, (3) hematological effects, (4) secretory action, (5) effects on inflammatory and allergic phenomena, and (6) effects on resistance to noxious stimuli.

Administration of each of the three major hormones of the adrenal cortex, *viz.*, cortisol, corticosterone, and aldosterone, produces similar effects (Table 45.2). Cortisol and aldosterone are at two extremes in terms of the responses they elicit, with corticosterone intermediate in its effects. Cortisol influences all the activities listed above, with, however, a very much weaker effect on electrolyte and water metabolism. In contrast, aldosterone exerts its prime action on electrolyte and water metabolism, but in the other areas listed has, at most, less than one-third of the activity of cortisol. Corticosterone spans in its effects all the activities listed, but with significantly less effectiveness in any one than is exhibited by cortisol or aldosterone in their respective areas of greatest potency.

The relative activities, in animal bioassays, of certain C_{21} steroids are shown in Table 45.2. That the hormones of the adrenal cortex influence a wide variety of phenomena is further indicated in Table 45.3, which lists the diverse changes seen in adrenal cortical insufficiency.

Carbohydrate Metabolism. Cortisol administration results within a few hours in increased glucose release from the liver, increased glycogenesis and augmented gluconeogenesis from amino acids, leading to glycogen deposition in liver, and a decreased peripheral utilization of glucose.

The augmented glucose release by liver in response to adrenal steroid injection is probably due to increased activity of hepatic glucose 6-phosphatase. There is an

Table 45.2: ACTIVITY RELATIVE TO CORTISONE OF NATURALLY OCCURRING CORTICAL STEROIDS IN ADRENALECTOMIZED RATS

Steroid	Life maintenance	Glycogen deposition	Sodium retention*	Muscle-work test†	Growth test	Cold test	Anti-inflammatory activity*
11-Dehydro-17-hydroxycorticosterone (cortisone)	100	100	100	100	100	100	100
Corticostosterone	75	54	255	46	108	9	3
11-Dehydrocorticosterone	58	48	.....	32	...	33	0
17-Hydroxycorticosterone (cortisol)	100	155	150	160	219	...	1,250
Deoxycorticosterone	400	0	3,000	5	...	8	0
Aldosterone	...	30	60,000	...	...	...	0

* Adrenalectomized mice.

† Adrenalectomized-nephrectomized rats.

Table 45.3: ALTERATIONS PRODUCED BY ADRENAL CORTICAL INSUFFICIENCY

Nature of change observed	*Basis for change*
Alterations in blood:	
Decrease in serum sodium and increase in serum potassium; decrease in serum chloride	Disturbed kidney function reflected in failure of tubular reabsorption of Na^+ and Cl^-
Hemoconcentration	Loss of water through kidney, accompanying electrolyte loss
Acidosis	Failure of renal Na^+-K^+-H^+ exchange mechanism
Hypoglycemia with fasting	Reduced level of liver glycogen in fasting; diminished gluconeogenesis and increased carbohydrate utilization
Increase in blood urea	Renal failure and circulatory impairment
Lymphocytosis	Lack of lymphocytolytic action of 11-oxyadrenal cortical steroids, with accompanying proliferation lymphoid tissue
Eosinophilia	Unknown
Anemia	Lack of stimulation of erythropoietic tissue and of production of intrinsic factor; gastric hyposecretion
Alterations in protein metabolism:	
Diminished urinary nitrogen in fasting	Absence of antianabolic influence of adrenal steroids
Alterations in carbohydrate metabolism:	
Reduced levels of liver glycogen in fasting	Diminished glucogenesis and gluconeogenesis
Markedly increased insulin sensitivity	Diminished glucogenesis and gluconeogenesis
Impairment of carbohydrate absorption from the gastrointestinal tract	Unknown; possibly related to disturbances in potassium metabolism*
Alterations in lipid metabolism:	
Impairment of lipid mobilization from depots.	Unknown
Generalized alterations:	
Intensification of inflammatory response and of hypersensitivity reactions	Increased influx of polymorphonuclear leukocytes and lymphocytes from blood into area of inflammation, accompanied by an augmented production of tissue substances contributing to reactions of inflammation and hypersensitivity, including destruction of fibroblasts.
Increased growth of lymphoid tissue	Lack of adrenal cortical steroid lymphocytolytic action on lymphocytes
Decreased metabolic rate	Unknown
Hypotension	Decreased extracellular volume; decreased cardiac output
Excessive pigmentation	Unknown; possibly due to intrinsic melanocyte-stimulating activity of ACTH (page 991)
Anorexia, weight loss; cessation of growth	Unknown
Muscular weakness and sensitivity to stress	Partly due to loss of potassium from cells; impaired glycogenesis.

* Passage of carbohydrate into cells is accompanied by movement of potassium in the intracellular direction. In adrenal cortical insufficiency, potassium tends to move out of cells.

increase in activities of specific liver transaminases and in pyruvate carboxylase (see below), and an increased glycogen synthetase activity, all contributing to elevation of liver glycogen stores. The early increase in pyruvate carboxylase activity suggests this enzymic step as an important locus of cortisol-augmented glycogenesis. The inhibition of amino acid incorporation into proteins in nonhepatic tissue following steroid injection also makes available additional precursors for gluconeogenesis in the liver.

Cortisol in vivo or in vitro diminishes glucose utilization by peripheral tissues, including muscle, adipose, and lymphoid tissues. Apparently this action occurs both at a transport site and an as yet unknown intracellular locus. There is also diminished mucopolysaccharide synthesis in connective tissue (page 882).

As a result of these influences of adrenal steroids on carbohydrate metabolism, there occurs, in the fasted or fed normal subject given cortisol, an elevated concentration of liver glycogen and of blood glucose. Prolonged or excessive administration of cortisol can produce a diabetic type of glucose tolerance curve (page 450) and glucosuria, and can result in permanent diabetes mellitus because of degeneration and exhaustion of pancreatic islet cells. Conversely, after adrenalectomy there is an increased sensitivity to insulin, probably related to decreased gluconeogenesis and glucose supplied by the liver. Hence, the requirement for insulin is diminished, as is the severity of any preexisting diabetes. Rapidly fatal hypoglycemia develops on fasting the hypoadrenal cortical individual, because of enhanced insulin sensitivity. In the absence of adequate cortical steroids, muscle glycogen concentration also fails to be maintained; this failure is reflected in a diminished work performance of the muscle. Work performance of adrenalectomized rats has been used for bioassay of active cortical hormones.

Protein and Nucleic Acid Metabolism. Cortisol stimulates protein synthesis in liver but markedly inhibits this process in muscle and other tissues. In those tissues in which there is inhibition of protein synthesis, there is also reduced amino acid transport. Since protein degradation in these tissues is continuing, amino acids leave these structures, resulting in wasting of soft tissues and osteoporosis. The loss of protein nitrogen from tissues consequent to elevated blood levels of cortisol is due to the action of the hormone at the level of the free amino acid pools in these cells. Under these circumstances, although protein synthesis in the liver is enhanced, it is insufficient to offset the exit of amino acids from other tissues, as reflected in an elevated plasma amino acid concentration. Oxidation of amino acids with concomitant urea synthesis is indicated by an over-all negative nitrogen balance.

The action of cortisol in augmenting liver protein synthesis is direct, since it occurs in the isolated, perfused organ when the steroid is added to the perfusion fluid. Studies with broken cell preparations demonstrated that the hormone stimulated incorporation of amino acids into ribosomal protein. A similar direct action on peripheral tissues is seen in studies with lymphoid cells. Addition of cortisol in vitro, in physiological concentrations, to thymocytes inhibited amino acid transport and incorporation into proteins of these cells.

Cortisol administration stimulates RNA synthesis in liver; in peripheral tissues, notably lymphoid and muscle, the hormone causes a decreased rate of RNA synthesis. The mechanisms of these actions of cortisol are considered below.

Lipid Metabolism. Adrenal cortical steroid administration to intact animals may increase peripheral lipogenesis, possibly because of an associated increase in insulin release. The latter hormone promotes lipogenesis (page 479). However, addition of cortisol in vitro to the epididymal fat pad of the rat causes release of free fatty acids and augments the activity of epinephrine in this system. Lipid mobilization from the depots is strikingly evident in some species, notably the rabbit, following steroid administration. This lipid-mobilizing effect could contribute to the marked ketogenic action of the hormone in adrenalectomized, depancreatized animals, as well as in human subjects with Addison's disease and concomitant diabetes.

The lipid-mobilizing action of adrenal cortical steroids is probably a reflection of an inhibition of glucose utilization. As is the case for protein synthesis (see above), the hormones influence lipid metabolism in the liver in an opposite direction from that in peripheral tissues, producing in liver an acute increase in rate of triglyceride synthesis. This is evident prior to elevation of blood fatty acid levels.

Electrolyte and Water Metabolism. The metabolism of electrolytes and water has been presented in Chap. 33. The adrenal steroids have a regulatory influence in this metabolism, particularly in relation to the concentration of sodium and potassium in extracellular fluids. Both aldosterone and deoxycorticosterone cause increased reabsorption of Na^+, Cl^-, and HCO_3^- in the distal tubules of the kidney, and by the sweat glands, salivary glands, and gastrointestinal mucosa. Thus, excessive concentrations of either of these steroids produces elevation of extracellular [Na^+] and expansion of extracellular volume and [HCO_3^-], with a decline in serum [K^+] and [Cl^-]. Sodium retention results in an exchange of intracellular K^+ for extracellular Na^+, with excretion of the K^+. Mobilization of Na^+ from connective tissue into extracellular compartments occurs. The increase in extracellular volume is accompanied by elevation of blood pressure. Lowering of serum [K^+] by cortical steroids leads to toxic effects on cardiac muscle, with characteristic electrocardiogram changes; elevation of extracellular [Na^+] causes decreased excitability of brain tissue.

Other electrolytes, *e.g.*, Ca^{++}, may appear in the urine in greater than normal concentrations following adrenal steroid administration. The augmented loss of Ca^{++} is due to the retardation of protein synthesis (see above); osteoporosis results from the decreased bone formation. Ca^{++} which otherwise would be deposited in osteoid is excreted in the urine.

In contrast to the above effects of excessive quantities of adrenal steroids, an inadequate supply of these compounds, *i.e.*, hypoadrenal corticalism, results in failure of normal tubular reabsorption of Na^+, with consequent increased excretion of Na^+, Cl^-, and water, leading to diminution of plasma volume. The consequences are those described in Chap. 33 for hypotonic contraction. Na^+ also moves into the tissues. Serum and tissue [K^+] rise because of a diminished excretion of this cation by the kidney. Reduction in extracellular volume, hemoconcentration, and increased viscosity of the blood contribute to a decreased cardiac output and hypotension.

Metabolic acidosis develops in untreated adrenal cortical insufficiency. This acidosis has a dual origin: renal and extrarenal. In the kidney, secretion of both H^+ and NH_4^+ by the renal tubule is impaired. Under conditions of acid loading, the renal response is inadequate and acidification of the urine suboptimal. With elevation of serum [K^+], there occurs a shift of bicarbonate into cells, passage of H^+ from cells into the extracellular fluid, and consequent acidosis.

The loss of extracellular sodium in adrenal cortical deficiency is reflected in undesirable effects in other tissues. For example, lowering of extracellular sodium in brain tissue is accompanied by a marked increase in excitability of this tissue. Alterations in the electroencephalogram have been described in Addison's disease in man. The hyperkalemia exerts deleterious effects on cardiac muscle, reflected in changes in the electrocardiogram; indeed, this state frequently terminates in cardiac arrest with the heart in diastole.

Certain of the alterations in serum electrolytes in the hypo- and hyperadrenal cortical states are illustrated in Fig. 45.4.

As a result of adrenal cortical insufficiency, with its consequent loss of fluid through the kidney and decrease in plasma volume, there is a decline in blood pressure with a reduced circulation through the kidneys, and renal failure. Because of the latter, the blood urea rises, as does the blood [Ca^{++}], [P_i], and [K^+]. The excretion of ingested water is also retarded, and an excessive intake of water may result in "water intoxication" (page 786).

The important role of [Na^+] and [Cl^-] in hypoadrenal cortical states is seen from the partial amelioration of adrenal cortical insufficiency by administration of NaCl. Recognition of the deleterious rise in serum [K^+], following lowering of serum [Na^+], has led to the demonstration that many of the disturbances of adrenal cortical insufficiency can be alleviated by a high sodium and low potassium intake. This regimen will rectify the mineral imbalance in adrenal cortical insufficiency and may prevent rapid deterioration and death.

Hematological Effects. Administration of the 11-oxygenated steroids other than aldosterone causes a decrease in the numbers of blood lymphocytes and eosinophils,

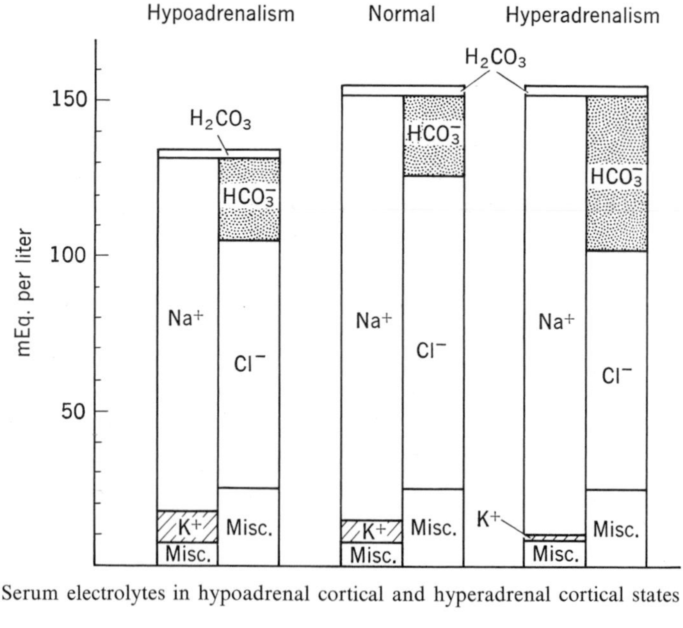

Fig. 45.4. Serum electrolytes in hypoadrenal cortical and hyperadrenal cortical states in man.

and stimulates erythropoiesis. The lymphopenia is due to the action of these steroids on lymphoid tissue, which undergoes involution because of a decrease in numbers of lymphocytes consequent to their disintegration.

The earliest biochemical alteration demonstrable in lymphoid tissue in response to exposure to steroid is a decrease in RNA polymerase activity and glucose utilization. The effects of adrenal steroids on the role of lymphoid tissue in antibody formation and immune phenomena have been noted previously (page 754). Lymphoid tissue hypertrophy and lymphocytosis may result as a consequence of adrenal cortical hypofunction.

The erythropoietic action of adrenal cortical steroids is a result of a stimulatory action of these hormones on bone marrow. The anemia resulting from adrenal cortical hypofunctioning was noted in the original description of Addison's disease.

Secretory Action. The 11-oxygenated adrenal cortical steroids augment secretion of hydrochloric acid and pepsinogen by the gastric mucosa, and trypsinogen by the pancreas. These effects may be the basis of the precipitation of ulcerative lesions of the gastrointestinal tract during prolonged steroid therapy. The pernicious anemia of Addison's disease may be due to gastric hyposecretion (see page 752).

Effects on Inflammatory and Allergic Phenomena. Cortisol and certain synthetic steroids (page 970) are anti-inflammatory substances, *i.e.,* they prevent the appearance of inflammatory responses, whether such responses are due to physical, chemical, or bacterial stimuli. These steroid effects can be demonstrated either locally, at the site of inflammation, or by their systemic administration. The steroids inhibit the influx of polymorphonuclear leukocytes from the blood into the local area of inflammation, and inhibit the localized destruction of fibroblasts, otherwise seen as a consequence of an inflammatory reaction.

Certain steroids, *e.g.*, cortisol, are markedly effective in counteracting some manifestations of the hypersensitive state, *e.g.*, anaphylactic shock. These activities of adrenal cortical steroids have led to their therapeutic use in the clinical conditions which are diseases of mesenchymal tissue, in hypersensitivity, and in acute inflammatory and allergic diseases of the eye and skin.

Adrenal Cortex and Resistance to Noxious Stimuli. The adrenal cortical-insufficient subject does not readily cope with challenges to homeostasis by noxious stimuli. For example, in amount considerably below those tolerated normally, hemorrhage, physical trauma, infectious agents, sensitizing antigens, or noxious chemicals may produce fatal consequences in adrenalectomized animals. As indicated previously (page 964), there is marked sensitivity to insulin, as well as to other hormones, *e.g.*, thyroxine. Such diverse types of noxious stimuli result in a markedly increased requirement for cortical hormones. This is evidenced by the increased protection against deleterious stimuli which is afforded adrenalectomized animals by treatment with adrenal cortical steroids.

Continued maintenance of adrenalectomized animals or human beings is possible with adrenal cortical steroids. This has led to the demonstration that partial or total adrenalectomy or hypophysectomy has an ameliorative result in some cases of malignant hypertension and in carcinoma of the breast and of the prostate.

Mechanism of Action of Adrenal Steroids. The wide variety of effects resulting from injection of adrenal steroids raises the question of which responses are primary actions, and which are indirect, or secondary effects. Although it cannot be concluded that a single primary mechanism of adrenal cortical hormone action accounts for all the phenomena produced by a steroid, it seems reasonable to ascertain whether the earliest biochemical event following hormone administration could serve as the basis for the mechanism underlying its many effects.

As described above, the primary effect of aldosterone is on Na^+ and H_2O transport, while that of the cortisol-like steroids is on organic metabolism. However, the mechanisms underlying these diverse effects may have a similar, initial basis, *viz.*, stimulation of the synthesis of a specific protein in the steroid-responsive cells. Edelman and Leaf and their colleagues have demonstrated that aldosterone combines with the nuclei of specific target cells, *e.g.*, kidney and toad bladder, inducing *de novo* synthesis of an unidentified enzyme that plays a facilitating role in Na^+ transport.

Cortisol and other steroids altering organic metabolism affect a DNA-dependent RNA synthetic mechanism, resulting in alterations in the rates of specific protein syntheses. An acute increase in RNA polymerase activity of target cells, *e.g.*, liver, as well as of other specific liver enzymes, *e.g.*, tyrosine-glutamate transaminase and pyruvate carboxylase, follows cortisol administration. The data indicate that steroid action in liver may be localized either at the level of DNA transcription by an RNA polymerase system or at an aminoacyl-tRNA translation step in protein synthesis.

At the time that protein and nucleic acid synthesis in liver are augmented in response to cortisol, these processes are decreased in lymphoid tissue (see above). An explanation of the mechanism by which a steroid may alter similar processes simultaneously in opposite directions in different tissues is not available.

Adrenal Cortical Hyperfunction in the Human Being. Hyperfunctioning of the human adrenal cortex results from cortical cell tumors in the adrenal gland or of extraadrenal locus. The consequences depend on the nature of the predominant steroid secreted by the tumor. If *cortisol* is the major compound produced, *Cushing's syndrome* results, and the alterations seen are those previously described for excessive concentrations of cortical steroids. If the steroid produced by the tumor is chiefly *aldosterone (primary aldosteronism),* there is marked Na^+ and water retention, with tendency to edema and hypertension, leading to heart failure, and extreme weakness due to low serum $[K^+]$.

In adrenal cortical hyperplasia with hypersecretion of *androgenic* steroids, the manifestations vary according to the age and sex of the patient. Adult women become masculine in appearance; hence the term *adrenal virilism.* The voice deepens, menstruation ceases, the breasts atrophy, and hair may grow on the face, chest, and limbs. In the adult male, there are overgrowth of hair, enlargement of the penis, and increased sexual desire.

When these tumors occur in young children, puberty appears prematurely. A male child of three to five years may show the sexual development of an adult, with enlargement of the penis, hair on the chest, pubis, and face, and precocious sexual desire. Unusual muscular development may be seen, and growth is rapid. In young girls, the breasts hypertrophy, pubic hair appears, the uterus develops prematurely, the clitoris is hypertrophied, and menstruation may occur.

Metabolic Fate of Adrenal Cortical Steroids. The major metabolites of adrenal

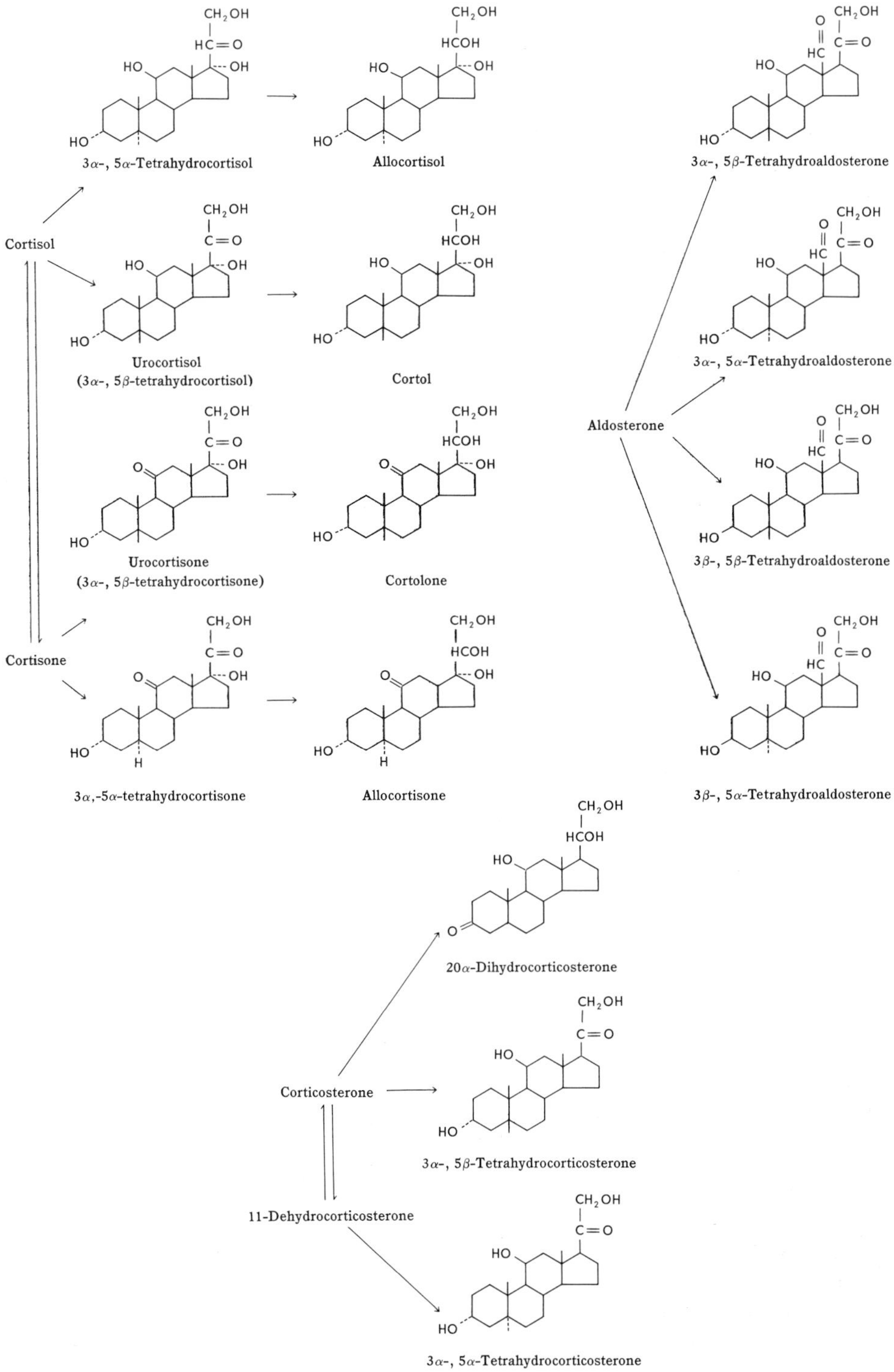

Fig. 45.5. Some metabolic products derived from adrenal cortical hormones. The chief metabolites, quantitatively, are indicated in the text.

cortical steroids result from reduction and conjugation reactions. Conjugates are primarily glucosiduronides at C-21, although some sulfate esters are also formed. The metabolic products are devoid of biological activity. The chief urinary metabolites arise in the liver by reduction reactions to form the corresponding tetra- and hexahydro compounds. Reversible oxidation-reduction also occurs at C-11 in a variety of tissues; this transformation is markedly affected by the level of circulating thyroid hormones, with oxidation favored in the hyperthyroid state. Some of the principal metabolites are indicated in Fig. 45.5.

After intravenous administration of ^{14}C-labeled cortisol in man, 93 per cent of the hormone was eliminated within 48 hr. without scission of the steroid nucleus, 70 per cent of which appeared as reduced urinary products and 20 per cent in the feces. Urocortisone, urocortisol, and the tetrahydroallopregnane (5α) derivative of cortisol were the major urinary metabolites, approximating almost half of the total products. That portion of metabolites, both free and conjugated, excreted in the bile may be reabsorbed from the intestine in an enterohepatic circulation. As in the case of other steroid hormones (page 949), liver disease impairs metabolism of the active adrenal steroids, and in such a circumstance their level may be markedly elevated in the blood.

Approximately 5 to 10 per cent of the secreted C_{21} adrenal steroids contribute normally to the total urinary 17-ketosteroids (page 942), yielding 11-oxy-17-ketosteroids, in contrast to the 11-deoxy-17-ketosteroids formed from the C_{19} androgens.

Synthetic Adrenal Cortical Steroids. A large number of synthetic steroids have been prepared in efforts to provide compounds with significantly greater activity than the natural hormones. It has also been the goal to seek substances with single actions rather than the multiplicity of responses seen with the normally secreted hormones. Certain of these synthetic compounds are many times more potent than cortisol, and some separation of biological effects has been obtained (Table 45.4).

Table 45.4: RELATIVE ACTIVITY OF SOME NATURAL AND SYNTHETIC STEROIDS TO THAT OF CORTISONE

Steroid	Sodium-retaining activity	Glycogenic activity	Anti-inflammatory activity
Cortisone	1	1	1
Cortisol	1.5	1.5	1.25
Deoxycorticosterone	30–50	0	0
Aldosterone	300–600	0	0
9α-Fluorocortisol	200–400	10–15	10–15
Δ^1-Cortisol	Minimal	3–5	4
9α-Fluoro-Δ^1-cortisol	200	30–40	10
2-Methylcortisol	100–200	5	3
2-Methyl-9α-fluorocortisol	1,000–2,000	10	10
6α-Methyl-Δ^1-cortisol	Minimal	4–6	3–5
16α-Methyl-9α-fluoro-Δ-cortisol	Minimal	25–30	25–35

REFERENCES

Books

Clark, F., and Grant, J. K., eds., "The Biosynthesis and Secretion of Adrenocortical Steroids," Cambridge University Press, New York, 1960.

von Eichler, O., and Farah, A., eds., "The Adrenal Cortical Hormones: Their Origin, Chemistry, Physiology and Pharmacology," part I, Lange and Springer, Berlin, 1962.

Eisenstein, A. B., ed., "The Adrenal Cortex," Little, Brown and Company, Boston, 1967.

von Euler, U. S., "Noradrenaline: Chemistry, Physiology, Pharmacology, and Clinical Aspects," Charles C Thomas, Publisher, Springfield, Ill., 1956.

Krayer, O., ed., "Symposium on Catecholamines," The Williams & Wilkins Company, Baltimore, 1959. (Also published in *Pharmacol. Revs.,* **11,** 241–566, 1959.)

McKerns, K. W., ed., "Functions of the Adrenal Cortex," 2 vols., Appleton-Century-Crofts, New York, 1968.

Soffer, L. J., Dorfman, R. I., and Gabrilove, J. L., "The Human Adrenal Gland," Lea & Febiger, Philadelphia, 1961.

Wolstenholme, G. E. W., and Cameron, M. P., eds., "The Human Adrenal Cortex," Little, Brown and Company, Boston, 1955.

Review Articles

Axelrod, J., The Metabolism, Storage and Release of Catecholamines, *Recent Progr. Hormone Research,* **21,** 597–622, 1965.

Berliner, D. L., and Dougherty, T. F., Hepatic and Extrahepatic Regulation of Corticosteroids, *Pharmacol. Revs.,* **13,** 329–359, 1961.

Davis, J. O., Mechanisms Regulating the Secretion and Metabolism of Aldosterone in Experimental Secondary Hyperaldosteronism, *Recent Progr. Hormone Research,* **17,** 293–352, 1961.

Denton, D. A., Evolutionary Aspects of The Emergence of Aldosterone Secretion and Salt Appetite, *Physiol. Revs.,* **45,** 245–295, 1965.

Dorfman, A., and Schiller, S., Effects of Hormones on the Metabolism of Acid Mucopolysaccharides of Connective Tissue, *Recent Progr. Hormone Research,* **14,** 427–456, 1958.

Dorfman, R. I., Biochemistry of the Adrenocortical Hormones, in H. W. Deane, subed., "Handbuch der Experimentellen Pharmakologie Erganzungswerk, vol. XIV, part I, pp. 411–513, Springer-Verlag OHG, Berlin, 1962.

Gaddum, J. H., and Holzbauer, M., Adrenaline and Noradrenaline, *Vitamins and Hormones,* **15,** 151–203, 1957.

Ganong, W. F., Biglieri, E. G., and Mulrow, P. J., Mechanisms Regulating Adrenocortical Secretion of Aldosterone and Glucocorticoids, *Recent Progr. Hormone Research,* **22,** 381–430, 1966.

Grant, J. K., Studies on the Biogenesis of the Adrenal Steroids, *Brit. Med. Bull.,* **18,** 99–105, 1962.

Hagen, P., and Welch, A. D., The Adrenal Medulla and the Biosynthesis of Pressor Amines, *Recent Progr. Hormone Research,* **12,** 27–44, 1956.

Makman, M. H., Nakagawa, S., and White, A., Studies of the Mode of Action of Adrenal Steroids on Lymphocytes, *Recent Progr. in Hormone Research,* **23,** 195–227, 1967.

Mills, I. H., Transport and Metabolism of Steroids, *Brit. Med. Bull.,* **18,** 127–133, 1962.

Sayers, G., Adrenal Cortex and Homeostasis, *Physiol. Revs.,* **30,** 241–320, 1950.

Sharp, G. W. G., and Leaf, A., Mechanism of Action of Aldosterone, *Physiol. Revs.,* **46,** 593–633, 1966.

Sharp, G. W. G., and Leaf, A., Studies on the Mode of Action of Aldosterone, *Recent Progr. Hormone Research,* **22,** 431–471, 1966.

Simpson, S. A., and Tait, J. F., Recent Progress in Methods of Isolation, Chemistry, and Physiology of Aldosterone, *Recent Progr. Hormone Research,* **11,** 183–210, 1955.

Sutherland, E. W., Jr., The Biological Role of Adenosine-3′,5′-Phosphate, *Harvey Lectures,* **57,** 17–34, 1961–1962.

Sutherland, E. W., Jr., Øye, I., and Butcher, R. W., The Action of Epinephrine and the Role of the Adenyl Cyclase System in Hormone Action, *Recent Progr. Hormone Research,* **21,** 623–646, 1965.

Yates, F. E., and Urquhart, J., Control of Plasma Concentrations of Adrenocortical Hormones, *Physiol. Revs.,* **42,** 359–443, 1962.

46. The Pancreas

Insulin. Glucagon

The pancreas produces and secretes two hormones. *Insulin* is secreted by the β type of islet cells and exerts a profound hypoglycemic effect when administered. The other hormone, *glucagon,* is produced by the α type of islet cells and exerts a hyperglycemic action. Other effects of these hormones will be considered below.

INSULIN—THE HYPOGLYCEMIC FACTOR

Chemistry of Insulin. The structure of insulin has been presented previously (page 149). The total synthesis of insulin has also been accomplished in the laboratory. The amino acid sequences of insulins of various species have been discussed (page 693). However, there is still no understanding of the relationship between the structure of the protein and its biological action. Although removal of the COOH-terminal alanine from the B chain of bovine insulin by incubation with carboxypeptidase does not alter significantly the biological activity of insulin, removal of the COOH-terminal asparagine from the A chain results in complete loss of activity. Also, digestion with trypsin leads to significant lowering of the biological potency when only one or two peptide bonds have been split. It has not been possible to prepare from insulin, by partial or complete hydrolysis, a moiety with significant hypoglycemic activity. The inactivation of insulin by digestive enzymes makes necessary parenteral injection to elicit the actions of the hormone.

Available data relating the structure of insulin to its biological action suggest that (1) the free amino and aliphatic hydroxyl groups of insulin are not essential for its activity; (2) some amide, guanidine, tyrosine phenolic groups and histidine imidazol groups can be substituted without loss of activity, although extensive iodination of phenolic groups (aromatic nuclei of tyrosine) results in marked inactivation; and (3) reduction of disulfide groups or esterification of carboxyl groups leads to loss of hormonal activity.

It has recently been established that proinsulin is the precursor of insulin. Proinsulin consists of a single polypeptide chain of approximately 73 residues with its amino portion the B chain and its carboxyl portion the A chain of insulin (page 149). Presumably, some type of specific proteolytic cleavage in a region of the chain distant from the disulfide linkages results in removal of a segment of 22 amino acid residues between the two ends of the proinsulin chain, with formation of insulin, in which the A and B chains are linked by two disulfide bonds (page 149). The conversion of proinsulin to insulin is thus similar to the activation by proteolytic cleavage of chymotrypsinogen (page 252) and plasminogen (page 734), both single chains, to chymotrypsin, comprised of three chains, and plasmin, containing two chains.

The predominance of acidic groups in insulin results in its ready combination with bases. Combination with basic proteins has been useful clinically because the insolubility of the protamine- or globin-insulin complex at tissue pH provides a relatively insoluble, slowly absorbed depot of insulin and results in a more prolonged action than that with ordinary insulin. This permits less frequent injections of insulin in clinical diabetes. Similar prolonged action is achieved with the crystalline zinc salt of protamine or globin insulin, also because of its relative insolubility. A neutral, crystalline protamine zinc insulin, described by Hagedorn in Denmark and termed NPH insulin (N, neutral; P, protamine; H, Hagedorn) and also known as isophane insulin, has been particularly useful. Lente (L., slowly) insulins, crystallized from an acetate buffer in the presence of Zn^{++}, are also long-acting, because of the relative insolubility of the crystalline product. Lente insulin resembles NPH insulin in its duration of action but has the feature of not containing foreign protein like protamine or globin. The time of onset and duration of hypoglycemic effects of various insulin preparations are indicated in Table 46.1.

Table 46.1: PROPERTIES OF VARIOUS INSULIN PREPARATIONS

Type of insulin	Time of onset of action after injection, hr.	Period of maximum action after injection, hr.	Duration of action, hr.
Regular, amorphous	1	3	6
Globin zinc insulin	2–4	8–16	16–24
Protamine zinc insulin	6–8	12–24	48–72
Isophane (NPH) insulin	2	10–20	28–30
Lente insulin	2	10–20	22–26

Effects of Insulin. The role of insulin in carbohydrate metabolism has been discussed previously (Chap. 19). The hyperglycemia and glucosuria resulting from inadequate supply of insulin are a reflection of the role of the hormone in glucose utilization by tissues.

Although the effects of insulin on carbohydrate metabolism have been the historical focus of attention, leading to the designation of the hormone as the hypoglycemic factor, insulin has effects on protein and lipid metabolism. These last two actions of insulin can be dissociated from its role in glucose utilization.

Carbohydrate Metabolism. Insulin facilitates entry of certain sugars, notably glucose, into cells. Glucose entering cells is rapidly phosphorylated; the net result is a decrease in the concentration of blood glucose and an increase in formation of intracellular hexose phosphate and of products derived from the latter, including glycogen.

The mechanism by which insulin augments transfer of glucose from extra- to intracellular compartments is not established. Increased cell membrane permeability to glucose could reflect an alteration in composition or conformation of membrane components in response to exposure to insulin. Membrane lipoproteins have been suggested as one such locus of insulin action (see also below).

Insulin also has intracellular effects on glucose utilization, independent of transport phenomena. A specific microsomal glucokinase of liver is enhanced in activity following insulin administration. Also, insulin added to rat diaphragm in vitro produced an approximately 50 per cent increase in activity of glycogen synthetase.

Protein and Nucleic Acid Metabolism. Insulin administration rapidly augments incorporation of amino acids into protein in liver and most other tissues. This effect is independent of glucose utilization; the entry of amino acids into cells is facilitated, and the incorporation of activated amino acids into ribosomal protein is enhanced. It has been postulated that insulin influences translation of messenger RNA, leading to the synthesis of a specific protein which then serves in turn to affect the rate of translation of other species of messenger RNA.

Lipid Metabolism. Insulin stimulates lipogenesis from carbohydrate in liver and extrahepatic tissues and in adipose tissue incubated in vitro following addition of insulin to the medium. The conversion of glucose carbon to fatty acids and triglycerides by adipose tissue in vitro in the presence of insulin is accompanied by inhibition of fatty acid release from the tissue, with net increase in lipid stores.

Insulin added to adipose tissue in vitro also augments phospholipid synthesis. This effect can be demonstrated in the absence of glucose in the medium.

Insulin administration stimulates transfer of Na^+, K^+, and inorganic phosphate into cells; this effect is independent of glucose utilization. Adipose cells exhibit a marked dependency upon Na^+ for their insulin-stimulated glucose metabolism. The hormone may activate an otherwise inactive carrier which requires Na^+ before accepting glucose for transport across the membrane.

Insulin Deficiency. Insulin deficiency results in *diabetes mellitus.* The consequent impairment of glucose utilization results in the demand for energy being met by a relatively greater degree of lipid and protein metabolism. There is increased gluconeogenesis from protein, with a greater excretion of nitrogen in the urine and a wasting of body tissues. Protein synthesis is depressed; ribosomes from tissues of diabetic animals show a reduced capacity to incorporate amino acids into ribosomal protein. Much of the glucose produced by gluconeogenesis is also wasted, in the absence of adequate insulin. Less pyruvate is thus available for acetyl CoA synthesis, and more of the latter is derived from lipid. The excessive mobilization of lipid from body stores leads to a lipemia and may result in a fatty liver. In addition, there is augmented production of ketone bodies, and when the production rate exceeds that of utilization, ketosis ensues. The biochemical consequences of diabetes, and their amelioration, have been discussed earlier (pages 448 and 506*ff.*).

Insulin deficiency may be due to one or more of three factors: (1) inadequate insulin production; (2) accelerated insulin destruction; and (3) insulin antagonists and inhibitors.

Inadequate Insulin Production. This is due to degeneration of pancreatic islet tissue resulting from a primary etiologic factor in the pancreas or secondarily because of continual hypersecretion of insulin by the pancreas as a consequence of a prolonged hyperglycemia. Secretion of insulin by the pancreas depends upon the level of blood glucose; as the level of blood glucose rises, insulin secretion is augmented and when the blood glucose concentration falls, insulin production by the pancreas declines. The blood glucose level may be maintained at hyperglycemic levels by increased gluconeogenesis due to excessive secretion of adenohypophyseal adrenocorticotropic hormone (page 991) or of adrenal cortical steroids (page 957) produced by an adrenal tumor. A second hypophyseal hormone, somatotropin (growth hormone, Chap. 47), also raises the level of blood glucose, and if secretion or administration of this hormone is prolonged, a secondary destruction of pancreatic β islet cells may occur. These relationships afford an explanation for the amelioration of pancreatic diabetes observed in experimental animals and in man following surgical removal of the adrenals (Chap. 45) or of the hypophysis (Chap. 47).

Insulin secretion is also markedly increased by protein ingestion, as well as by the intravenous administration of certain amino acids, notably L-arginine and, to a lesser degree, L-leucine. Arginine infusion thus provides a basis for testing insulin production by the pancreas. Release of somatotropin (Chap. 47) can also be effected by protein feeding or by administration of arginine or leucine. Thus similar stimuli, namely substrates for protein synthesis, augment the release of hormones with roles in the regulation of protein synthesis.

Accelerated Insulin Destruction. The loss of insulin activity as a result of proteolysis has been described previously (page 973). Liver and, to a lesser extent, other tissues possess an enzymic system capable of inactivating insulin. The enzyme has been termed *insulinase,* although insulin is not the only substrate, and is active both in vitro and in vivo. This aids in explaining the relatively short half-life of insulin, approximately 40 min. in man. With the purified enzyme, no hydrolysis of peptide bonds of insulin occurs.

Several enzymic systems have been described in mammalian liver and other tissues which catalyze reductive cleavage of the two interchain disulfide bonds of insulin with inactivation of the hormone, as well as disulfide exchange reactions with insulin. One of these enzymes from liver has been termed *glutathione-insulin transhydrogenase* and catalyzes reduction of the interchain disulfide bonds of insulin in the presence of reduced glutathione which functions as hydrogen donor. The reaction catalyzed by the enzyme can be coupled to that promoted by glutathione reductase (page 548), making possible oxidation of TPNH by insulin.

$$\mathrm{H^+ + TPNH + GSSG \longrightarrow TPN^+ + 2GSH}$$

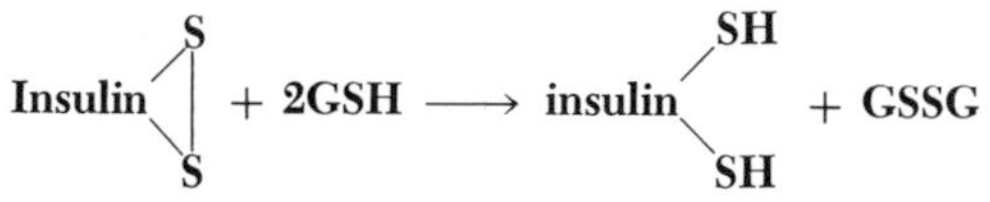

The product is a mixture of the reduced A and B chains of the hormone (page 149). The enzyme is inhibited by glucagon, the hyperglycemic factor of pancreas (see below). This could have physiological significance for maintenance of a higher level of circulating insulin in hyperglycemia. However, in view of the existence of several insulin-reducing systems in tissues, the precise significance of glutathione-insulin transhydrogenase in insulin synthesis, function, or metabolism must await further study.

Insulin Antagonists and Inhibitors. Two types of insulin antagonists which may contribute to insulin deficiency can be delineated. The insulin antibodies, produced as a result of insulin therapy, are one type. The other type is also in blood, but can be distinguished from antibodies in being present without prior administration of exogenous insulin. Either group of antagonists may function to render less effective insulin available from endogenous or exogenous sources.

In diabetic patients previously treated with insulin, a nonprecipating antibody which binds insulin is present in the β- and γ-globulin fractions of the plasma. This explains the slower rate of destruction of intravenously administered insulin in diabetic patients, and the lesser degree of response, as compared with the normal, to a given dose of insulin. The antigenic activity of insulin also explains the resistance to the bovine or porcine hormone which may develop during prolonged use. This can generally be circumvented by changing to an insulin from another species.

The second type of insulin antagonist is associated with the β_1-lipoprotein fraction (page 720) of both normal and diabetic serum. The active material reduces the effectiveness of either endogenous or exogenous insulin. Production of this insulin antagonist appears to be influenced by both the hypophysis and adrenal cortex.

Certain sulfonamide derivatives, notably substituted sulfonyl ureas, *e.g.*, 1-butyl-3-*p*-tolylsulfonylurea,

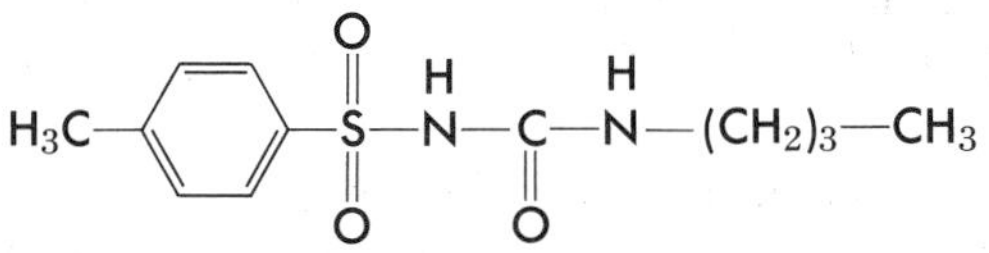

have proved useful as oral therapeutic agents in selected cases of adult diabetes. The activity of these compounds is apparently due to their ability to increase insulin release by the pancreas.

GLUCAGON—THE HYPERGLYCEMIC-GLYCOGENOLYTIC FACTOR

Administered glucagon increases blood glucose by accelerating glycogenolysis in the liver. Figure 46.1 shows the structure of bovine glucagon.

His • Ser • Gln • Gly • Thr • Phe • Thr • Ser • Asp • Tyr • Ser • Lys • Tyr • Leu • Asp •

Ser • Arg • Arg • Ala • Gln • Asp • Phe • Val • Gln • Trp • Leu • Met • Asn • Thr

Fig. 46.1. The structure of glucagon.

The structure of glucagon has been compared previously to that of secretin (Fig. 30.4, page 698), indicating a similarity in structure of the two hormones that suggests a common genetic origin.

The acceleration of glycogenolysis in the liver by glucagon is a result of the increase in activity of phosphorylase kinase (page 440). This is the basis for the hyperglycemic action of glucagon. The hormone does not affect muscle phosphorylase. The mechanism of action of the hormone is identical with that of epinephrine, *viz.*, stimulation of the formation of 3′,5′-cyclic AMP by the adenyl cyclase system. However, glucagon, in contrast to epinephrine, stimulates glycogenolysis in low, physiological concentrations, and does not cause the elevation of blood pressure characteristic of epinephrine. Consequently, glucagon has found clinical use in circumstances of either acute or persistent hypoglycemia.

Glucagon also stimulates gluconeogenesis in the liver, accompanied by increased urea production. In addition, glucagon exerts an inhibitory effect on the incorporation of amino acids into liver proteins. Daily administration of the hormone to rats produced an increased urinary excretion of nitrogen, phosphorus, and creatinine, with a loss in body weight and a decrease in liver and muscle mass.

Glucagon inhibits hepatic synthesis of fatty acids and cholesterol from acetate and stimulates ketogenesis. This is accompanied by activation of hepatic lipase, probably via the adenyl cyclase system. The net result is an increased supply of fatty acids from liver triglyceride for oxidation within the liver, increased levels of acetyl CoA and fatty acyl CoA, and augmented ketogenesis. The gluconeogenic action of glucagon may also be related to the increase in acyl CoA and acetyl CoA esters produced by the hormone, since acyl CoA esters have an inhibitory influence on pyruvate decarboxylation (page 410), and acetyl CoA directly stimulates pyruvate carboxylation (page 408).

Glucagon also stimulates glycerol and free fatty acid release from adipose tissue; this occurs when cyclic 3′,5′-AMP levels have been elevated as a result of exposure to the hormone.

Renal effects of glucagon include an increased glomerular filtration rate and renal plasma flow, thus explaining the augmented excretion of Na^+, Cl^-, K^+, phosphorus, and uric acid, reported after glucagon administration. This occurs independently of the hyperglycemic effect of glucagon.

Glucagon, like insulin, will elicit formation of nonprecipitating antibodies after prolonged injection into experimental animals.

REFERENCES

Books

Krahl, M. E., ed., "The Action of Insulin on Cells," Academic Press, Inc., New York, 1961.

Young, F. G., Broom, W. A., and Wolff, F. W., eds., "The Mechanism of Action of Insulin," Blackwell Scientific Publications, Oxford, England, 1960.

Review Articles

Ashmore, J., Cahill, G. F., Jr., and Hastings, A, B., Effect of Hormones on Alternate Pathways of Glucose Utilization in Isolated Tissues, *Recent Progr. Hormone Research,* **16,** 547–577, 1960.

Behrens, O. K., and Bromer, W. W., Glucagon, *Vitamins and Hormones,* **16,** 263–301, 1958.

de Bodo, R. C., and Altszuler, N., Insulin Hypersensitivity and Physiological Insulin Antagonists, *Physiol. Revs.,* **38,** 389–445, 1958.

Fajans, S. S., Floyd, J. C., Jr., Knopf, R. F., and Conn, J. W., Effect of Amino Acids and Proteins on Insulin Secretion in Man, *Recent Progr. in Hormone Research,* **23,** 617–662, 1967.

Foà, P. P., Galansino, G., and Pozza, G., Glucagon: A Second Pancreatic Hormone, *Recent Progr. Hormone Research,* **13,** 473–510, 1957.

Grodsky, G. M., and Forsham, P. H., Insulin and the Pancreas, *Ann. Rev. Physiol.,* **28,** 347–380, 1966.

Levine, R., ed., Symposium on Diabetes, *Am. J. Med.,* **31,** 837–930, 1961.

Levine, R., and Goldstein, M. S., On the Mechanism of Action of Insulin, *Recent Progr. Hormone Research,* **11,** 343–380, 1955.
Manchester, K. L., and Young, F. G., Insulin and Protein Metabolism, *Vitamins and Hormones,* **19,** 95–135, 1961.
Miller, L. L., Some Direct Actions of Insulin, Glucagon, and Hydrocortisone on the Isolated Perfused Rat Liver, *Recent Progr. Hormone Research,* **17,** 539–568, 1961.
Mirsky, A. I., Insulinase, Insulinase-inhibitors, and Diabetes Mellitus, *Recent Progr. Hormone Research,* **13,** 429–472, 1957.
Randle, P. J., Garland, P. B., Hales, C. N., Newsholme, E. A., Denton, R. M., and Pogson, C. I., Interactions of Metabolism and the Physiological Role of Insulin, *Recent Progr. Hormone Research,* **22,** 1–48, 1966.
Stetten, D., Jr., and Bloom, B., The Hormones of the Islets of Langerhans, in G. Pincus and K. V. Thimann, eds., "The Hormones: Physiology, Chemistry and Applications," vol. III, pp. 175–199, Academic Press, Inc., New York, 1955.
Young, F. G., On Insulin and Its Action, *Proc. Roy. Soc., Series B,* **157,** 1–26, 1962.

47. The Hypophysis

The Neurohypophysis. Regulatory Role of the Adenohypophysis. Control of Secretion of the Adenohypophysis. Thyrotropin. Adrenocorticotropin. Gonadotropins. Somatotropin. The Sexual Cycle

The hypophysis, which regulates a large portion of the endocrine activity of the organism, was initially named the *pituitary* because of the erroneous concept that it is concerned with secretion of mucus or phlegm (L. *pituita,* phlegm). The term hypophysis, from the Greek meaning undergrowth, is descriptive of the location of the gland below the brain. It consists of anterior and posterior lobes and the pars intermedia, which differ from one another embryologically, histologically, and functionally. The anterior portion, or *adenohypophysis,* is glandular and richly vascular. The posterior, or neural, portion, the *neurohypophysis,* is intimately connected with the hypothalamic areas through numerous nerve fibers and some glandular elements which comprise the hypophyseal stalk. There is little direct functional relationship between the adenohypophysis and neurohypophysis, although secretions of the latter may augment certain hormonal activities of the former. Also, each lobe has important interactions with the hypothalamus (pages 983 and 985).

The major effects of hypophysectomy appear to be due entirely to loss of adenohypophyseal functions and can be produced by removal of the adenohypophysis alone. Removal of only the neural lobe usually causes no striking dysfunction. Hypophysectomy results in the following alterations: (1) cessation of growth in young animals; in the adult, loss of body tissue, with some reversion to younger characteristics, *e.g.*, appearance of juvenile hair; (2) gonadal atrophy in either sex, with loss of secondary sex characteristics in the adult, failure of development of sex characteristics in the younger individual, and sterility; (3) atrophy of the thyroid, with metabolic alterations characteristic of hypothyroidism (page 927); and (4) atrophy of the adrenal cortex with evidence of adrenal cortical insufficiency (pages 962*ff.*). Hypophysectomy, like adrenalectomy (page 964), results in amelioration of pancreatic diabetes; this was first demonstrated by Houssay, and the hypophysectomized, depancreatized animal has been referred to as the "Houssay animal."

Uncertainty regarding the precise number of hormones secreted by the hypophysis is due to (1) evidence that this gland, as well as the hypothalamus, contains a number of polypeptides to which no known biological activity can yet be assigned, and (2) evidence that in the case of several of the hypophyseal hormones, only a portion of the molecule is essential for the hormonal functions, together with the demonstration of a similarity in portions of structures of different hormones with an accompanying overlapping of biological properties. The latter has been noted previously in connection with the evolution of structure and function of the neurohypophyseal hormones (page 697).

THE NEUROHYPOPHYSIS

The hormones of the mammalian neurohypophysis are *vasopressin,* which is responsible for pressor and antidiuretic effects, *oxytocin,* which causes smooth muscle contraction and milk ejection, and two polypeptides, each with melanocyte-stimulating activity, termed α- and *β-melanocyte-stimulating hormones* (α- and β-MSH). *Vasotocin* is produced by most nonmammalian vertebrates and shows both vasopressin- and oxytocin-like activities.

CHEMISTRY

The structures of vasotocin, vasopressin, and oxytocin have been presented in Table 30.5, page 698, together with a consideration of the species variations in the amino acid residues of these and related hormones in lower vertebrates. The naming of the vasopressins is based on the nature of the amino acid residue at position 8. The significance of this residue in the evolution of vasopressin structures was noted previously (page 697). Thus, arginine vasopressin has been isolated from human, beef, sheep, and horse hypophyseal extracts. Lysine vasopressin, with lysine substituted for arginine at position 8, occurs in the pig. On the other hand, oxytocin preparations from man, cow, and pig are identical. Oxytocin differs from the vasopressins only in residues in positions 3 and 8. Arginine vasopressin and arginine vasotocin are both present in chicken hypophyseal extracts.

Melanocyte-stimulating activity is also present in adenohypophyseal extracts. Melanocyte-stimulating material is secreted by the pars intermedia, except in those species which do not possess this structure, *e.g.*, chicken, porpoise, and whale, in which this activity is found in adenohypophyseal extracts. Overlapping biological activities of various hypophyseal principles may be related to similarities in portions of the amino acid sequences of polypeptides (see below).

The structures of the melanocyte-stimulating hormones, α- and β-MSH, are shown in Fig. 47.1. Samples of α-MSH from five species studied have the same structures; the amino-terminal serine residue is N-acetylated, and the carboxyl-terminal valine is in the form of an amide. Also, α-MSH contains 13 amino acid residues, whereas β-MSH from the same species consists of 18 amino acids, with both the amino- and carboxyl-terminal groups present as aspartic acid residues. The amino acid sequence and number of amino acid residues of β-MSH differ among species (Fig. 47.1). The amino acid sequence in positions 7 to 13 of β-MSH from monkey, horse, beef, and pig is identical with that of positions 4 to 10 of adrenocorticotropin (ACTH, page 990). The amino acid sequence of α-MSH is identical with that of the 13 N-terminal amino acid residues of ACTH, except that its N-terminus is

Hormone	Source	Amino acid sequence
α-MSH	Monkey, horse, beef, pig, sheep	*Ac-Ser·Tyr·Ser·Met·Glu·His·Phe·Arg·Trp·Gly·Lys·Pro·ValNH_2
β-MSH	Human	Ala·Glu·Lys·Lys·Asp·Glu·Gly·Pro·Tyr·Arg·Met·Glu·His·Phe·Arg·Trp·Gly·Ser Pro·Pro·Lys·Asp
	Monkey	Asp·Glu·Gly·Pro·Tyr·Arg·Met·Glu·His·Phe·Arg·Trp·Gly·Ser·Pro·Pro·Lys·Asp
	Horse	Asp·Glu·Gly·Pro·Tyr·Lys·Met·Glu·His·Phe·Arg·Trp·Gly·Ser·Pro·Arg·Lys·Asp
	Beef	Asp·Ser·Gly·Pro·Tyr·Lys·Met·Glu·His·Phe·Arg·Trp·Gly·Ser·Pro·Pro·Lys·Asp
	Pig	Asp·Glu·Gly·Pro·Tyr·Lys·Met·Glu·His·Phe·Arg·Trp·Gly·Ser·Pro·Pro·Lys·Asp

*Ac = Acetyl

Fig. 47.1. Amino acid sequences of α-MSH (five species) and β-MSH (five species). Although all β-MSH preparations, except human, contain only 18 amino acid residues, the peptide chains have been written to allow for the additional four residues in human β-MSH and displaced in order to show similarities in sequences. Amino acid substitutions among species, as compared with the human, are indicated by the underlined residues.

acetylated and the chain terminates with a valine carboxamide group. These data may account for the intrinsic melanocyte-stimulating activity of ACTH (page 991).

Since the synthesis of oxytocin was achieved by du Vigneaud and his associates, a large number of analogues of this hormone, as well as of vasopressin and MSH, have been synthesized and examined for biological activity. The data for vasopressin and oxytocin analogues permit the following conclusions regarding relationship of structure to biological activity. The cyclic structure and the chain length of nine amino acid residues are essential for biological activity. Reduction in ring size leads to the formation of inactive derivatives. Larger ring sizes, achieved by the incorporation of additional amino acid residues or methylene groups, frequently lead to structures that have inhibitory effects on the actions of the hormones. The proline residue in position 7 is essential for activity, since its replacement by other residues results in considerable decrease in or total loss of activity. The aromatic hydroxyl group in position 2 is not essential for full activity; however, the phenyl group is required for hormonal action. The amide groups in positions 4 and 5 are essential for biological activity. Amino acids with branched aliphatic side chains in position 3 are of importance for oxytocic activity. A phenylalanine in position 3 enhances vasopressor activity; in contrast, aliphatic amino acid residues are more important for the activity of oxytocin than is the phenylalanine residue for the activity of vasopressin. Amino acids with aliphatic side chains in position 8 are essential for oxytocic activity, whereas basic amino acids with aliphatic side chains are important for vasopressor activity.

These data reveal interesting relationships of structure to biological activity and could be of value in elucidation of the mechanism of action of the neurohypophyseal hormones.

BIOLOGICAL ASPECTS

The hormones of the neurohypophysis are synthesized in the supraoptic and paraventricular nuclei. The hormones, in association with a protein of molecular weight about 30,000, migrate as granules down the nerve fibers and accumulate at the nerve endings in the neurohypophysis. Secretion of the hormones is influenced by three types of stimuli: (1) action of the central nervous system, (2) the osmotic pressure of the blood, and (3) drugs. Apparently, selective release of either of the hormones may occur.

Release of vasopressin is stimulated by a variety of neurogenic stimuli, *e.g.*, pain, trauma, and emotional states. On the other hand, evidence of inhibition of vasopressin release may also be seen as a consequence of central nervous system activity, *e.g.*, excessive production of epinephrine.

With respect to the role of the blood osmotic pressure, Verney demonstrated that a 2 per cent change in osmotic pressure of the blood traversing the carotid sinus will alter the rate of vasopressin secretion. This rate diminishes or increases with hypo- or hypertonicity, respectively. Verney postulated the existence of specific osmoreceptor cells in those hypothalamic nuclei responsible for production and secretion of vasopressin. "Volume receptors," or baroreceptors, have also been postulated as regulatory factors influencing vasopressin release. This concept would explain the fact that hemorrhage is one of the most powerful known stimuli of vasopressin release. Finally, the release of vasopressin is a cholinergic-mediated response and is therefore affected by acetylcholine, anesthetics, etc.

Circulatory or Pressor Action. The pressor action of vasopressin is due to vasoconstriction in the peripheral arterioles and capillaries. There is a constriction of the coronary and pulmonary vessels, but a dilation of cerebral and renal vessels. The latter dilator effect is caused by the rise in systemic blood pressure.

Antidiuretic Action. Vasopressin exerts a marked effect on the kidneys, accelerating the rate of water reabsorption from the early part of the distal convoluted tubules to the length of the collecting tubules. The augmented water reabsorption of a fluid that is hypo-osmotic with reference to serum is accompanied by excretion of a urine which contains increased concentrations of sodium, chloride, phosphate, and total nitrogen. Glomerular filtration is apparently unaffected. Although the urine volume is less, more chloride may be excreted per unit of time. In man, as little as 0.1 μg of vasopressin will produce a maximal antidiuretic effect; the half-life of the hormone is approximately 12 min.

Vasopressin augments synthesis of cyclic AMP (page 437) in the renal tubules; cyclic AMP exerts an inhibitory effect on water transport in the isolated toad bladder. Addition of vasopressin in vitro to dog kidney preparations increased the formation of cyclic AMP, presumably by stimulating the activity of adenyl cyclase (page 957). However, the relation of this to the antidiuretic action of the hormone is not clear.

As indicated above, secretion of vasopressin is augmented in circumstances of dehydration and of increased salt intake, and is decreased when the extracellular fluid becomes hypotonic. This has been discussed in detail in Chap. 33.

The role of vasopressin is seen in man in *diabetes insipidus.* This clinical condition frequently accompanies lesions of the hypophysis or hypothalamus. It is characterized by the excretion of large quantities of urine of very low specific gravity, 1.002 to 1.006, and of low chloride content. As much as 4 to 5 liters of urine (*polyuria*) may be voided per day, and volumes several times these amounts have been reported. The patient exhibits a corresponding increase in fluid intake and a marked thirst (*polydipsia*). The condition is controlled by the parenteral administration of purified vasopressin preparations.

The relation of diabetes insipidus to disturbances in the hypothalamus is based on the neural control of neurohypophyseal function; section of the hypophyseal stalk or destruction of the supraoptic nuclei produces atrophy of the neurohypophysis and diabetes insipidus.

Oxytocic Action. The term *oxytocic* (Gk., rapid birth) is descriptive of the action of this principle in causing strong contractions of the uterus. This also occurs in the isolated uterus and is the basis of a bioassay method. A concentration of the hormone as low as 0.5 mμg per ml. will cause contraction of the isolated uterus. Oxytocin finds clinical use during and after parturition.

Oxytocin also excites the musculature of the intestine, gallbladder, ureter, and urinary bladder. The hormone also causes ejection of milk; this action is distinct from that of prolactin in stimulating milk production (page 994). The release of oxytocin in response to nipple stimulation by suckling is another example of a neuroendocrine reflex.

Certain of the biological effects of vasopressin and oxytocin are indicated in Table 47.1; their comparative potencies are given in Table 47.2.

Table 47.1: SUMMARY OF THE BIOLOGICAL EFFECTS OF VASOPRESSIN AND OF OXYTOCIN

Structure or function affected	Vasopressin	Oxytocin
Water diuresis	Inhibits	No effect
Blood pressure	Elevates	Slightly lowers
Coronary arteries	Constricts	Slightly dilates
Intestinal contractions	Stimulates	Questionable
Uterine contractions*	Stimulates	Stimulates
Ejection of milk	Slightly stimulates	Stimulates

* Response varies with species, as well as with the stage of the normal and the reproductive cycles.

Oxytocin and vasopressin added in vitro increase glucose oxidation by slices of mammary gland taken from lactating rats. This effect is abolished by puromycin, an inhibitor of protein synthesis, suggesting that protein synthesis is stimulated by the hormones and that glucose oxidation provides energy for this synthesis. Oxytocin also exerts an insulin-like effect on the utilization of glucose by adipose tissue in vitro in that it stimulates incorporation of glucose carbon into triglycerides. In contrast, vasopressin, as well as α- and β-MSH, stimulates hydrolysis of triglyceride in adipose tissue in vitro, with release of fatty acids and glycerol.

Table 47.2: POTENCY OF HIGHLY PURIFIED OXYTOCIN AND ARGININE VASOPRESSIN IN TERMS OF UNITED STATES PHARMACOPEIA STANDARD

	Oxytocic (rat uterus, without Mg^{++})	Avian depressor (fowl)	Milk-ejecting (rabbit)	Pressor (rat)	Antidiuretic (dog)
Oxytocin	500	500	500	6	3
Arginine vasopressin	25	75	100	400	400

Note: All figures are United States Pharmacopeia Units per mg.

Melanocyte-stimulating Action. α- and β-MSH cause dispersal of black pigment found in the melanophore cells of certain cold-blooded animals, thus producing a generalized blackening of the skin. Evidence that this activity is of significance in man is lacking, although injection of purified MSH in man was reported to intensify pigmentation in areas previously pigmented, but not to affect nonpigmented regions.

Melatonin, found in the pineal gland, reverses the darkening effect of MSH by stimulating aggregation, rather than dispersal, of melanin granules within melanocytes, causing lightening of skin color.

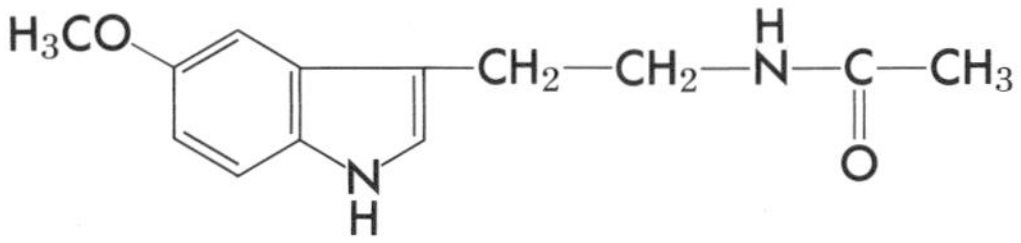

Melatonin
(N-acetyl-5-methoxytryptamine)

Melatonin also exerts a retarding influence on the estrus cycle. The compound was also obtained from hypothalamic tissue and peripheral nerve. Melatonin biosynthesis occurs by reaction of acetyl CoA with serotonin (page 589); methylation of the N-acetylserotonin formed then occurs, utilizing S-adenosylmethionine. The enzyme catalyzing this last step was found only in the pineal gland.

Melatonin administered to mice was rapidly metabolized. The major metabolic pathway involves hydroxylation at position 6 followed by conjugation primarily with sulfate and, to a small extent, with glucuronic acid.

THE ADENOHYPOPHYSIS

Regulatory Role of the Adenohypophysis. The two major regulatory systems in mammals, *viz.*, the nervous system and the endocrine glands, have multiple relationships and integrations exemplified in the regulation of secretion of at least five of the adenohypophyseal hormones, *viz.*, thyrotropin, the two gonadotropins, prolactin, and adrenocorticotropin. Prolactin is discussed with the gonadotropins because of its influence on the corpus luteum.

Control of Secretion of the Adenohypophysis. The chain of events in the control of adenohypophyseal secretion may be depicted as (1) stimulation of neuroreceptors, (2) transmission of afferent impulses to the thalamus, hypothalamus, and cortex, (3) initiation or modification of hypothalamic activity by the thalamus and cerebral cortex, and (4) release of hypothalamic neurohumoral substances which are transmitted via the hypothalamic-hypophyseal portal circulation to excite the adenohypophysis, with resulting secretion of one or several of its hormones, depending upon the nature of the initiating stimulus.

Neural stimuli reaching the hypothalamus effect the release of specific substances of molecular weight 200 to 1,200. Each is relatively specific for an individual adenohypophyseal hormone and is transmitted via the hypophyseal portal pathway. These hypothalamic compounds can, in turn, cause the release of one or more of the adenohypophyseal hormones. Hence the terms thyrotropic-releasing factor, adrenotropic hormone–releasing factor, etc. The close chemical relationships among these factors, as well as to the neurohypophyseal hormones in some instances, are seen from a degree of overlapping of their biological activities.

In addition to the above neural factors, a second set of regulatory mechanisms affects adenohypophyseal secretion. The products of the target glands, over which adenohypophyseal hormones exert regulatory influence, constitute a negative feedback regulatory mechanism in that the rate of secretion of certain adenohypophyseal hormones is inversely related to the blood concentration of the hormones produced by the target gland.

These two mechanisms of influencing the secretory activity of the adenohypophysis, *viz.*, that involving the neural pathways and that concerned with blood levels of hormonal products, are the basis for the rapid responses of adenohypophyseal secretory level to a wide variety of stimuli, ranging from environmental, *e.g.*, cold, hypoxia, trauma, noxious chemicals, etc., to psychological, *e.g.*, fear, anxiety, the presence of other animals or individuals, etc. This accounts to a degree for the numerous experimental and clinical circumstances in which a similarity in physiological effects may be seen.

THE THYROTROPIC HORMONE

Hypophysectomy in mammals results in involution of the thyroid gland, with flattening of the epithelium, and development of the previously described consequences of hypothyroidism (page 927): lowered basal metabolic rate, depressed rate of iodide uptake by the thyroid, a diminished serum concentration of protein-bound iodine, and general decreased cellular activity. Parenteral administration of hypophyseal extracts induces reparative effects in the thyroid of the hypophysectomized animal, with restoration of function and metabolic activity.

Secretion of Thyrotropic Hormone. Two mechanisms of regulation of TSH secretion exist, neural and nonneural. Neurogenic and psychogenic influences have long been recognized as affecting thyroid function and contributing to *thyrotoxicosis.* The neural influence was also shown by removal of the hypophysis from its location beneath the median eminence and transplantation to other sites; marked diminution in TSH secretion resulted. Also, appropriately placed hypothalamic lesions prevented the expected rise in TSH secretion in response to lowered levels of blood thyroxine. The latter is the basis for the nonneural mechanism of regulating TSH secretion, functioning as a feedback mechanism.

The hypothalamic thyrotropic hormone–releasing factor (TRF) has been prepared in highly purified form. Some controversy exists regarding whether this material is a simple polypeptide, as initially thought, since further purification studies have led to the isolation of a substance with which TRF activity is associated, but which does not give a ninhydrin reaction, is stable to a variety of proteolytic enzymes, and has less than one-third of the molecule as amino acids. This would suggest that one or more of the hypophyseal hypothalamic–releasing factors may not be simple polypeptides.

Chemical Properties. Highly purified preparations of thyrotropic hormone have been obtained from beef, sheep, and human hypophyses. Biological activity is associated with a glycoprotein with a molecular weight of approximately 25,000. Thyrotropin preparations have eight to nine cystine residues per mole; the disulfide groups are present as intrachain linkages. Mannose, galactose, fucose, glucosamine, and galactosamine are present, the last two as the N-acetyl derivatives; these sugars appear to be present in a single oligosaccharide unit.

Effects of Thyrotropic Hormone. The thyrotropic hormone influences the rates of the following reactions: (1) removal of iodide from blood by the thyroid; (2) conversion of iodide to thyroid hormones; and (3) release of hormonal iodine from the gland.

The hypophysectomized animal has a decreased rate of iodide uptake by the thyroid and of removal of administered iodide from plasma (Fig. 47.2). Although in the hypophysectomized animal iodide enters the gland more slowly, following entry it is rapidly converted to diiodotyrosine. *However, the rate of conversion of diiodotyrosine to thyroxine is depressed in the absence of the hypophysis* (Fig. 47.3). Hypophysectomy in the rat leads to a 50 per cent decrease in total blood hormonal iodine (PBI, page 926) within 4 days after operation. However, although thyroid size also diminishes, the quantity of stored hormones or of total iodine in the gland does not decline. Therefore, since the thyroid becomes smaller after hypophysectomy, the concentration of iodine in the gland is elevated. Thus, the lowered hormonal iodine of the plasma is not a direct stimulus to the thyroid to release its hormones, and is dependent upon TSH. The proteolysis of thyroglobulin in relation to release of thyroid hormones from the gland has been considered previously (page 924).

Administration of TSH decreases the organically bound iodine of the thyroid; there is some stimulation of iodide release also. Further, thyroid glands made hyper-

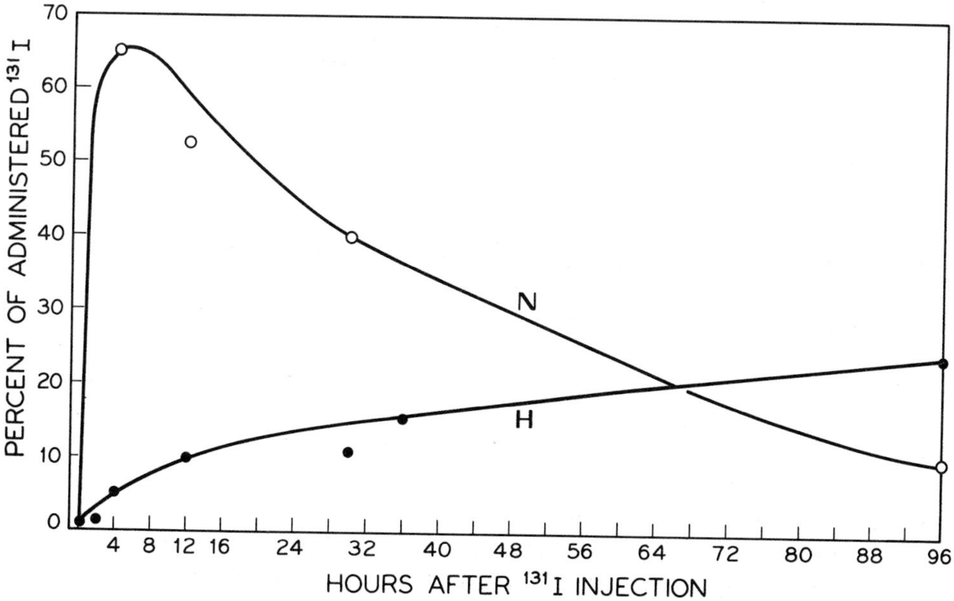

Fig. 47.2. The uptake of total radioactive iodine by the whole thyroid gland of the normal (*N*) and hypophysectomized (*H*) rat. A tracer dose of ^{131}I (as iodide) was injected intraperitoneally into each rat. (*After I. L. Chaikoff and A. Taurog, Ann. N.Y. Acad. Sci.*, **50**, 377, 1949.)

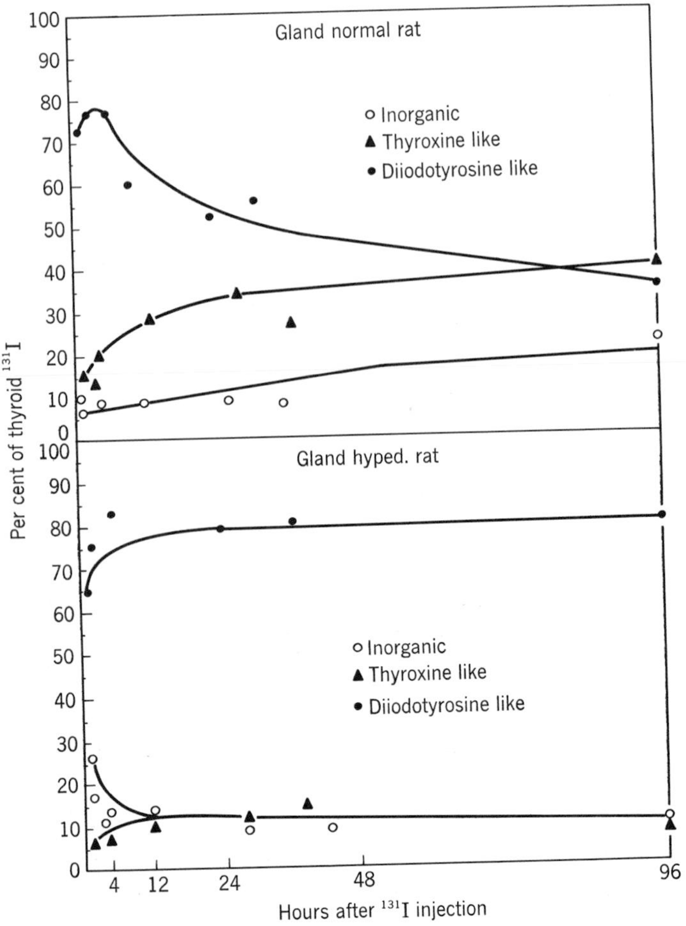

Fig. 47.3. Distribution of radioactive iodine in the thyroids of normal (*upper*) and of hypophysectomized (*lower*) rats. Each rat received intraperitoneally a tracer dose of ^{131}I as iodide. (*After I. L. Chaikoff and A. Taurog, Ann N.Y. Acad. Sci.*, **50**, 377, 1949.)

plastic with TSH administration have a greater than normal capacity to fix iodine in all fractions, with the normal distribution between diiodotyrosine and thyroxine being shifted toward thyroxine (Fig. 47.4). There is an accompanying increase in plasma hormonal iodine. The transport and metabolic fates of thyroid hormones have been discussed (pages 925 and 927).

TSH may have a direct action on orbital tissue relating to the exophthalmos of diffuse toxic goiter (page 927), although purification of TSH is accompanied by a decrease in exophthalmos-stimulating activity, and a separate hypophyseal factor has been postulated for the latter effect.

Mode of Action of TSH. The stimulatory effect of thyrotropic hormone in vivo on thyroidal activity is reflected in augmented metabolic phenomena in the gland, including TPN and TPNH formation, glucose oxidation, and phospholipid and cyclic AMP synthesis. A biologically active derivative of cyclic 3′,5′-AMP, N^6-2′-O-dibutyryl-3′,5′-AMP, which is more stable and enters cells more readily than cyclic

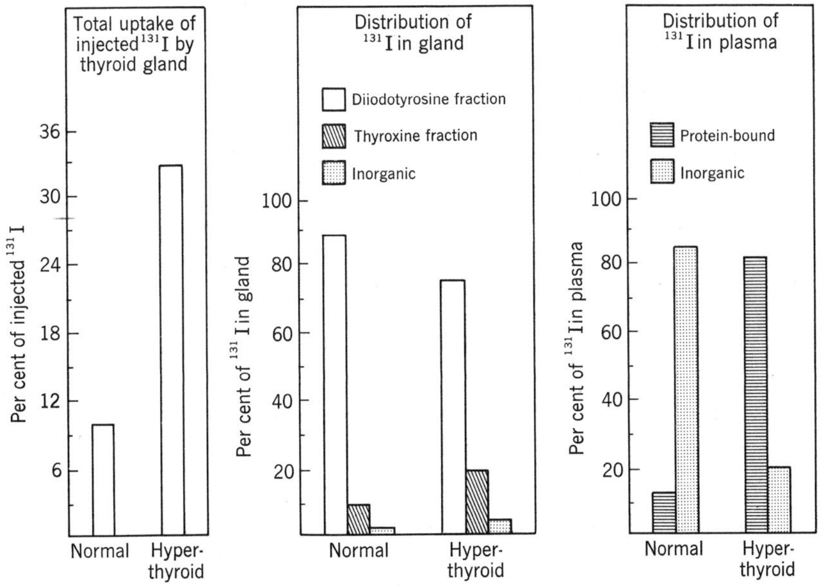

Fig. 47.4. Uptake of radioactive iodine and its distribution in the thyroid and the plasma of normal guinea pigs and of guinea pigs made hyperthyroid by injection of thyrotropic hormone. The measurements were made 16 hr. after the intraperitoneal injection of a tracer dose of ^{131}I as iodide. (*After I. L. Chaikoff and A. Taurog, Ann. N.Y. Acad. Sci.,* **50**, 377, 1949.)

3′,5′-AMP, stimulated in vitro the oxidation of glucose and the incorporation of $^{32}P_i$ into phospholipids by thyroid slices.

Thyrotropin addition to thyroid tissue in vitro stimulated oxygen consumption, glucose oxidation, amino acid entry, protein synthesis, and the phosphogluconate oxidative pathway. The hormone also increased release of free fatty acids from adipose tissue in vitro.

On the basis of morphological studies, it has been hypothesized that TSH alters the fine structure of the follicular epithelium, thereby affecting membrane permeability.

Long-acting Thyroid Stimulator. An abnormal thyroid-stimulating substance was first described in the blood of patients with hypothyroidism. This substance was distinguished from TSH by its longer duration of action and has been called the long-acting thyroid stimulator (LATS). The latter appears to be indistinguishable from 7 S γ-globulin (page 710); antisera to normal human γ-globulin will neutralize the hormonal activity of LATS. The latter, added in vitro to thyroid slices, enhances the variety of metabolic phenomena mentioned above as responding to TSH stimulation.

ADRENOCORTICOTROPIC HORMONE

The adrenocorticotropic hormone (adrenocorticotropin, ACTH, corticotropin) stimulates the secretion and growth of the adrenal cortex. The hormone will restore to normal size the atrophied adrenal cortex of the hypophysectomized animal, or, if administered postoperatively, will prevent atrophy.

Chemistry of Adrenocorticotropin. The structure of adrenocorticotropic hormone is given in Fig. 47.5. This polypeptide has been designated as α-ACTH to distinguish it from a pepsin- or acid-degraded product, β-ACTH (also termed adrenocorticotropin B).

Ser·	Tyr·	Ser·	Met·	Glu·	His·	Phe·	Arg·	Trp·	Glp·	Lys·	Pro·	Val·	Gly·	Lys·	Lys·	Arg·	Arg·	Pro·	Val·	Lys·	Val·	Tyr·	Pro·
1	2	3	4	5	6	7	8	9	10	11	12	13	14	15	16	17	18	19	20	21	22	23	24

Asp·	Ala·	Gly·	Glu·	Asp·	Gln·	Ser·	Ala·	Glu·	Ala·	Phe·	Pro·	Leu·	Glu·	Phe
25	26	27	28	29	30	31	32	33	34	35	36	37	38	39

Fig. 47.5. Amino acid sequence of human adrenocorticotropin.

The first 24 residues are identical in ACTH preparations from several species. A synthetic polypeptide containing the first 23 residues of ACTH has biological activity in vivo similar to that of the isolated hormone. Thus, residues 24 to 33 are not essential for hormonal action. However, removal of a few residues from the amino-terminal end of the molecule, by leucine aminopeptidase, destroys biological activity. The smallest synthetic polypeptide with some ACTH activity in vivo contains the amino-terminal 16 residues of adrenocorticotropin. Elongation of the peptide chain to 19 residues results in a striking increase in biological activity, indicating the importance of the basic Lys-Lys-Arg-Arg sequence for high potency.

Secretion of Adrenocorticotropin. The rate of secretion of adrenocorticotropin is also influenced by the two major mechanisms previously described (page 985). The neurohumoral mechanism provides a basis for release of ACTH as a result of a wide variety of unrelated stimuli, *e.g.*, trauma, emotional stress, drugs, chemical or bacterial toxic agents, or substances normally present in the body, *e.g.*, insulin, thyroxine, epinephrine, and vasopressin. These stimuli initiate secretion by nerve endings in the posterior hypothalamus and median eminence of corticotropin-releasing factors (CRF). The latter are transmitted via the hypophyseal portal vessels to the adenohypophysis, evoking release of ACTH. One of the factors has been designated α_2-CRF because of structural and biological similarity to α-MSH (page 982), differing chemically only in the fact that the N-substitution on serine is not an acetyl group. α_1-CRF is a larger peptide which also has some ACTH and vasopressor activity. β-CRF has 11 amino acid residues; the sequence begins with N-acetylserine, as in α-MSH, and ends with glycinamide, as in the neurohypophyseal hormones. β-CRF is significantly more activity than α-CRF as an adrenocorticotropin-releasing factor and has marked vasopressin action. Vasopressin also shows ACTH-releasing activity.

The rate of secretion of ACTH is also related inversely to the level of circulating adrenal cortical steroids. Increased rate of removal of these steroids by the tissues lowers their concentration in the blood and will evoke secretion of ACTH. On the other hand, administration of exogenous adrenal steroid will, by elevating the blood concentration of the adrenal steroids, depress ACTH secretion (page 986). This mechanism accounts for the adrenal cortical atrophy and hypofunction reported in animals and in patients treated for long periods with adrenal steroids. The locus of action of the adrenal steroids in suppressing ACTH secretion appears to be the hypothalamus, presumably by repressing release of CRF.

Biological Properties of Adrenocorticotropin. The prime role of ACTH is stimulation of the synthesis and secretion of adrenal cortical steroids. The hormone also has extra-adrenal effects which are exerted directly on specific tissues, independent of adrenal cortical mediation.

Adrenal-mediated Actions of ACTH. As a consequence of stimulation of production of adrenal steroids, ACTH injection mimics all the responses described for those hormones (pages 962*ff.*) with augmented gluconeogenesis and accompanying retardation of protein synthesis in all tissues studied, except the liver. There is increased lipid mobilization to the liver, with ketonemia, and hypercholesterolemia. Promotion of salt and water reabsorption by the kidney occurs but to a lesser degree than with aldosterone; it will be recalled that secretion of this steroid is dependent only in part on the hypophyseal principle (page 961). Lymphopenia, eosinopenia, and erythropoiesis result from ACTH administration. The hormone is effective therapeutically in the variety of clinical conditions mentioned previously which respond favorably to certain adrenal cortical steroids (page 967). Addison's disease is an exception, since in this circumstance there is a limited amount of responsive, normal adrenal cortical tissue.

Certain of the effects of ACTH administration differ from those of adrenal cortical steroid injection. ACTH releases a mixture of steroids from the adrenal, whereas injection of a single adrenal cortical steroid produces effects characteristic only of the compound injected, as well as those which follow inhibition of ACTH secretion. Consequently, prolonged ACTH administration can lead to undesirable manifestations of adrenal cortical hyperfunction (page 968), including masculinization, reflecting androgen secretion by the adrenal cortex (page 958).

Direct Effects of ACTH on Tissues. One of the direct actions of ACTH has been referred to previously, its intrinsic melanocyte-stimulating activity (page 982). This may account in part for the darkening of the skin in Addison's disease (page 958), in which the blood level of ACTH is abnormally high. One of the most striking direct effects of ACTH is its in vitro stimulation of glucose utilization and fatty acid release by adipose tissue. There is an accompanying increase in phosphorylase activity. Thus, ACTH resembles epinephrine in these actions on adipose tissue. The lipolytic effect of ACTH is due to activation of a lipase in adipose tissue. This may be an aspect of the lipid-mobilizing activity of ACTH seen in vivo (see above), since intravenous administration of ACTH to adrenalectomized rats produced a rapid increase in free fatty acids of adipose tissue and plasma. A synthetic tridecapeptide, identical in sequence with the amino-terminal portion of ACTH (page 990), also produced release of fatty acids from adipose tissue in vitro and a severalfold increase in the plasma free fatty acids of rabbits injected subcutaneously with the preparation. This so-called adipokinetic activity of these peptides and of ACTH is of interest in view of reports of a separate adipokinetic hormone in the hypophysis (page 998). ACTH also exerts a direct action on the pancreas, stimulating insulin release.

Administration of ACTH to adrenalectomized-nephrectomized rats decreased urea formation; this suggests an extra-adrenal action of the hormone on some phase of nitrogen metabolism, perhaps by facilitating amino acid transport into nonhepatic cells. ACTH administration to adrenalectomized animals or to Addisonian subjects retards the rate of disposal of both endogenous and exogenous cortisol. An inhibition of liver conjugation of injected cortisol and its metabolites has been described in adrenalectomized, ACTH-treated animals. The basis of this extra-adrenal action of ACTH is not clear. Injection of ACTH into the blood of an isolated heart-lung preparation augmented the heart rate significantly. In this regard, ACTH resembled norepinephrine and epinephrine.

Hypersecretion of Adrenocorticotropic Hormone—Pituitary Basophilism—Cushing's Disease. This clinical condition is generally attributed to hyperplasia or tumor of the basophil cells of the adenohypophysis and is characterized chiefly by an overproduction of ACTH. The main feactures of Cushing's disease are (1) obesity of the trunk (especially of the abdomen), face, and buttocks but not the limbs; purplish striae, due to distention, are present over the lower abdomen; (2) cyanosis of the face, hands, and feet, pigmentation of the skin, and excessive growth of hair; women may grow a mustache or beard; (3) demineralization of the bones; (4) hypertension; (5) loss of sexual functions; (6) hyperglycemia and glucosuria; (7) acne.

Mode of Action of ACTH. Administration of ACTH or incubation of adrenal tissue in vitro with ACTH stimulates steroidogenesis, as well as increases the adrenal tissue concentration of cyclic 3′,5′-AMP. Addition of the latter in vitro to adrenal sections augments mitochondrial steroid hydroxylation reactions. Cyclic AMP and ACTH each stimulate the transformation of cholesterol to pregnenolone in the adrenal. Increases in adrenal cyclic AMP concentrations produced by ACTH occur before increases in the rate of adrenal steroidogenesis, indicating that a primary action of the trophic hormone on the adrenal cortex is one of stimulating synthesis of cyclic AMP in the adrenal by increasing adenyl cyclase activity. Augmented levels of cyclic AMP would also provide increased amounts of TPNH for cleavage of the cholesterol side chain and for steroid hydroxylation reactions by activation of adrenal glycogen phosphorylase. However, this latter reaction is not primary to the mode of action of ACTH.

The steroidogenic response to ACTH does not occur if protein synthesis is inhibited in the adrenal, suggesting requirement for the formation of one or more proteins in the steroidogenic action of the trophic hormone. However, inhibitors of protein synthesis do not interfere with the increase in cyclic AMP concentration induced by ACTH. Thus the newly synthesized protein thought to be involved in steroidogenesis would appear to act at a site beyond adenyl cyclase.

THE GONADOTROPIC HORMONES

The hormonal influence of the hypophysis on the gonads and accessory sex organs is evident, following hypophysectomy, in the atrophy of these structures in the adult and in their failure to mature in younger individuals. This is also seen in atrophy of the gonads, amenorrhea, and impotence, in disease caused by atrophy or degeneration of the adenohypophysis.

Four gonadotropic hormones are known; two are secreted by the adenohypophysis, *viz., follicle-stimulating hormone* (FSH) and *luteinizing* or *interstitial cell–stimulating hormone* (LH or ICSH). In addition, a gonadotropin is produced by the placenta and designated as *human chorionic gonadotropin* (CG or HCG), and a gonadotropin is present in pregnant mare's serum (PMS). *Prolactin,* a separate adenohypophyseal principle, also has gonadotropic activity.

Chemistry of the Gonadotropins. Each of the hypophyseal gonadotropins has been obtained in highly purified form; some of their properties are given in Table 47.3. Four of the gonadotropins are glycoproteins whose constituents include fucose, hexoses (mannose and galactose are present in hog FSH), hexosamine, and a sialic acid. Incubation of sheep FSH or of human chorionic gonadotropin with neuraminidase liberated a sialic acid, with concomitant loss of all biological activity.

Table 47.3: SOME PROPERTIES OF GONADOTROPIC HORMONE PREPARATIONS

Hormone	Molecular weight	Isoelectric point, pH	Carbohydrate content, per cent
Luteinizing or interstitial cell–stimulating:			
Human	26,000	5.4	3.5
Sheep	30,000	7.3	16.0
Hog	100,000	7.5	5.0
Follicle-stimulating:			
Human	17,000	. . .	8.0–9.0
Sheep	33,000	4.5	8.0–9.0
Hog	29,000	5.1	7.0–8.0
Luteotropin or prolactin:			
Beef	23,500	5.7	0
Sheep	23,500	5.7	0
Human chorionic gonadotropin	30,000	3.0	28.0
Pregnant mare's serum gonadotropin	23,000	. . .	45.0

Secretion of Gonadotropins. Secretion of the hypophyseal gonadotropins is regulated by the two primary mechanisms discussed previously (page 985). Thus, a wide variety of neurogenic effects on sexual activity have been described experimentally and clinically. Release of each gonadotropin from the adenohypophysis is stimulated by specific releasing factors which are secreted in response to stimuli reaching specific loci in the hypothalamus. The luteinizing hormone-releasing factor (LRF) has not been isolated in pure form; biologically active preparations have been reported which did not give a ninhydrin reaction. Follicle-stimulating hormone–releasing factor (FSH-RF) has been partially purified; the active substance appears to have a low molecular weight ($<$1,000). Putrescine, cadaverine, and other polyamines have been reported to have FSH-RF activity.

The circulating androgens and estrogens constitute a second mechanism which influences gonadotropin secretion by acting directly on the hypothalamus to inhibit secretion of releasing factors. The androgen receptor lies in the posterior median eminence, whereas the effect of estrogens has been localized in the arcuate hypothalamic nucleus. The hypothalamic-hypophyseal mechanism, which stimulates release of FSH and LH, simultaneously inhibits release of prolactin. Conversely, liberation of prolactin by the adenohypophysis, with consequent lactogenesis, appears to result from suppression of release of other gonadotropins.

A third mechanism for regulation of gonadotropin activity appears to be of pineal origin. Melatonin and 5-hydroxytryptophol, secreted by the pineal gland, inhibit the estrus cycle (page 946). In addition, arginine vasotocin (page 981), which has also been isolated from bovine pineal glands, will inhibit the stimulatory effects of pregnant mare's serum gonadotropin.

Biological Aspects of Gonadotropins. *Follicle-stimulating hormone* produces, in the female, growth of a large number of graafian follicles with increased ovarian weight. FSH produces spermatogenesis in the testis by stimulating the epithelium of the seminiferous tubules, causing appearance of large numbers of spermatocytes in various stages of development, including mature spermatozoa. Urinary excretion of FSH is significantly elevated in castrates and following the menopause, and in malignancy of the reproductive organs.

Luteinizing hormone is concerned, in the female, with final ripening of ovarian follicles, the manifestations of heat, or estrus, and the rupture of the follicles with their transformation to corpora lutea. In the male, LH stimulates the Leydig cells, which secrete testosterone. Because of its action on the interstitial cells of both the ovaries and the testes, luteinizing hormone has also been termed interstitial cell–stimulating hormone. The effect of LH on the interstitial cells of the testes may be thought of as analogous to its action on the thecal cells of the ovary.

Prolactin was first described as an adenohypophyseal principle essential for initiation of lactation in mammals at parturition. The hormone also promotes functional activity of the corpora lutea and, thus, progesterone secretion. Prolactin functions synergistically with estrogen to promote mammary gland proliferation, in addition to its capacity to initiate secretion of milk in the hypertrophied mammary gland. In the young pigeon, prolactin produces a proliferative hypertrophy of the normally thin crop sac, providing a convenient bioassay for the hormone.

Although *chorionic gonadotropin* is of placental and not hypophyseal origin, its biological effects resemble those of the hypophyseal hormones. It appears in the urine early in pregnancy, in approximately the first week after the first missed menstrual period. This is the basis for two commonly employed tests for pregnancy. In the Aschheim-Zondek test, urine or an alcoholic precipitate of urine is injected into immature female mice or rats. Urine from a pregnant individual, containing chorionic gonadotropin, will increase ovarian weight and cause ripening of follicles and hemorrhages into some unruptured follicles. In the Friedman test, the urine is injected intravenously into female rabbits to assess its ability to produce an ovulatory response (presence of ruptured ovarian follicles).

Markedly increased amounts of adenohypophyseal-like gonadotropin are excreted in conditions other than pregnancy and give rise to a false positive Aschheim-Zondek or Friedman test. This is seen in instances of chorionepithelioma, a malignant tumor of the placental tissue, and in hydatidiform mole, a cystic degenerative disease of chorionic tissue. High titers of urinary gonadotropin are also present in the urine of males afflicted with testicular tumors composed of malignant embryonal tissue, *e.g.*, teratoma and epithelioma. Gonadotropin assays of urine are a useful diagnostic aid in these conditions.

Chorionic gonadotropin supplements the hypophysis in maintaining growth of the corpus luteum during pregnancy, although this placental principle will not prevent ovarian atrophy which follows hypophysectomy in animals. Chorionic gonadotropin also stimulates Leydig tissue and hence the male accessory organs. This has led to some success in the clinical use of this hormonal product in cryptorchidism in young males. The role of the hypophyseal gonadotropins in the sex cycle is discussed later in this chapter.

Mode of Action of Gonadotropins. The gonadotropins stimulate the synthesis of estradiol, progesterone, and testosterone by the gonads after in vivo administration or after addition to ovarian or testicular slices. Luteinizing hormone caused a rapid accumulation of cyclic AMP in slices of bovine corpus luteum. Since cyclic AMP mimicked the effect of luteinizing hormone on steroidogenesis, it has been suggested that the cyclic nucleotide is a mediator of the action of the gonadotropin. One of the sites of action of the gonadotropins in steroidogenesis is the conversion of cholesterol to pregnenolone, specifically the formation of 20 α-hydroxycholesterol (Fig. 22.6, page 524) by stimulating the 20 α-hydroxylase. This is the rate-limiting step in the cleavage of the cholesterol side chain. This action is dependent upon protein synthesis in the gonadal tissue, suggesting that regulation of protein synthesis (translational control) may be an action of the gonadotropins in directing steroidogenesis.

Prolactin added in vitro to adipose tissue stimulated glucose uptake and lipogenesis. When injected, prolactin mimics many of the actions of somatotropin (see below).

SOMATOTROPIN—GROWTH HORMONE

The adenohypophysis secretes *somatotropin,* a hormone which affects the rate of skeletal growth and gain in body weight. Hypophysectomy of younger animals results in either a greatly retarded growth rate or complete failure to grow. Growth abnormalities in man, *e.g.*, acromegaly, gigantism, and dwarfism, associated with hypophyseal dysfunction, have provided additional evidence that the secretory activity of this gland influences skeletal growth.

Chemistry of Somatotropin. Highly purified growth-promoting preparations have been obtained from hypophyses of many species, including man. Biological activity resides in a protein with a molecular weight of approximately 22,000; the complete amino acid sequence of human somatotropin has been reported. Monkey and human somatotropin exhibit similar immunological specificity and are effective biologically in all species tested. In contrast, somatotropin preparations from species other than the primate are not biologically active in human subjects. Partial chymotryptic hydrolysis of somatotropin does not destroy biological activity, suggesting that the latter is dependent upon only a portion of the protein.

Secretion of Somatotropin. A hypothalamic releasing factor for somatotropic hormone (SRF), which appears to be a polypeptide, regulates secretion of somatotropin by the adenohyphysis. Production and secretion of SRF are inversely related to the blood glucose concentration. Thus, fasting or administration of insulin results in elevated blood levels of somatotropin. Conversely, glucose administration diminishes the blood concentration of somatotropin. This indirect regulatory influence of blood glucose on somatotropin secretion, via a direct control of SRF secretion, recalls the role of blood glucose in the secretion of insulin by the pancreas (page 975), since both secretory mechanisms function in a similar manner, *i.e.*, their rate is inversely related to blood glucose concentration.

Effects of Somatotropin. The designation of a single substance as the growth-promoting hormone is inaccurate in view of the influence of other hormones, notably those of the thyroid, adrenal cortex, and pancreas, on normal growth and development. Nevertheless, repeated injection of somatotropin into young normal or hypophysectomized animals leads to acceleration of growth, including accretion of both hard and soft tissues.

The diverse effects of somatotropin are not primarily due to its influence on other endocrine glands, in contrast to the operation of other adenohypophyseal hormones. Somatotropin actions can be demonstrated in the hypophysectomized animal despite hypofunctioning of thyroid, adrenals, and gonads. However, somatotropin can act in cooperation with other hormones. For example, the anabolic effect of androgens is minimally manifest in the hypophysectomized animal, and somatotropin administration markedly enhances the nitrogen retention produced by androgens, as well as the growth of specific androgen-sensitive tissues.

In addition to its general growth-promoting and anabolic properties, somatotropin administration produces metabolic alterations which may be characterized as diabetogenic, pancreotropic, glycostatic, lipid-mobilizing and ketogenic, renal, erythropoietic, and lactopoietic.

The numerous metabolic effects of somatotropin reflect the diverse, integrated metabolic reactions contributing to growth processes. Thus, in accord with its growth-promoting action, somatotropin administration produces the following metabolic alterations:(1) Stimulation of RNA and protein synthesis in liver and peripheral tissues, reflected in nitrogen retention. This is the *anabolic effect* of somatotropin. (2) Elevated blood glucose levels, preceded by an early, acute hypoglycemia. This was considered above in relation to SRF release, and is due to the action of somatotropin on insulin release by the pancreas (*pancreatropic effect*). Continued somatotropin administration may lead to glucosuria, as well as intensify the diabetes of a diabetic subject (*diabetogenic effect*). A second pancreatropic effect of somatotropin is stimulation of glucagon secretion. This is a basis for the hyperglycemic action of somatotropin. (3) An additional action in carbohydrate metabolism is to increase muscle and cardiac glycogen (*glycostatic effect*). (4) An acute decline in plasma nonesterified fatty acids, followed by a rise in this blood fraction. Continued administration of somatotropin produces ketonemia, and ketonuria, and an increase in liver lipids, because of accelerated mobilization of depot lipid. This last effect is a result of the action of somatotropin on adipose tissue, augmenting release of fatty acids (*lipid-mobilizing effect*). (5) Increased kidney size and function, with augmented renal clearance and tubular excretion (*renotropic effect*). (6) Stimulation of reticulocytosis (*erythropoietic effect*). (7) Stimulation of milk secretion (*lactopoietic effect*). (8) Stimulation of chondrogenesis and osteogenesis, with increased synthesis of bone matrix, including chondroitin sulfate, and increased calcification. In its effects on skeletal tissues, somatotropin does not appear to act directly but functions to maintain a serum factor, termed *sulfation factor,* which is an activator of chondroitin sulfate and collagen synthesis.

The manifold metabolic effects of somatotropin pose the question of whether the hormone has a specific action on some key metabolic process which is rate-determining, an influence on cell permeability, or an action on a large number of diverse enzymic systems. Studies in vitro with cell preparations from animals which had been administered somatotropin suggest that growth hormone elicits the formation of mRNA and tRNA and stimulates incorporation of amino acids into proteins. Ribosomes from hypophysectomized animals are less active in incorporating amino acids into proteins, and the activity can be partially restored by prior treatment of the animals with somatotropin. The possible relationship of these dissected affects of the hormone on nucleic acid and protein synthesis to its other diverse metabolic actions is not presently clear.

Hypersecretion of Somatotropin in Man. In man, excessive production of somatotropin after the usual age of full skeletal growth results in *acromegaly,* and is due to an adenomatous tumor of the adenohypophysis. The characteristic features are: (1) Overgrowth of the bones of the hands, feet, and face. The feet and hands are markedly increased in size; the hands appear broadened and the fingers thickened. Bowing of the spine (kyphosis) is commonly seen. The soft tissues of the nose, lips, forehead, and scalp are thickened. There is a general overgrowth of the body hair. (2) Enlargement of the viscera (splanchnomegaly). The tongue, lungs, heart, liver, spleen, and thymus are greatly enlarged. The thyroid, parathyroid, and adrenal glands may show hypertrophy or adenomatous growths. Hyperthyroidism may be present, as well as glucosuria and hyperglycemia, suggesting a diabetes of pancreatic origin. (3) In the early stages of the disease, increased sexual activity may be evident. Later, there occur atrophy of the gonads and suppression of the sexual functions in both sexes, with impotence in men and amenorrhea in women.

If the hypophyseal adenoma occurs prior to puberty, before ossification is complete, gigantism will result. There is a general overgrowth of the skeleton, resulting in individuals of 7 or 8 ft. or more in height. The limbs are generally disproportionately long.

Hyposecretion of Somatotropin: Pituitary Dwarfism. Dwarfism, or premature arresting of skeletal development, has been considered previously in connection with hypothyroidism (cretinism, page 927). This type of dwarfing may possibly be a manifestation of a primary hypophyseal deficiency reflecting the absence of adequate secretion of the thyrotropic hormone.

Another type of arrested growth results from a deficiency of somatotropin secretion. In contrast to the cretins, these hypophyseal dwarfs do not show deformity or, as a rule, mental inferiority and frequently do not have the unattractive appearance of a cretin. The hypophyseal dwarfs are frequently immature sexually. At adult age the dwarf may be no more than 3 or 4 ft. in height. The relative proportions of the different parts of the skeleton do not deviate markedly from normal, although the head is generally large in relation to the body. Genetic dwarfism due to somatotropin deficiency has been described.

Placental Hypophyseal-like Hormones. In addition to the placental chorionic gonadotropin discussed previously, at least three additional types of hypophyseal activity have been extracted from human placenta, *viz.*, ACTH, prolactin, and somatotropin. A purified placental prolactin of molecular weight 38,000 was a dimer and reacted immunologically with antiserum against human hypophyseal somatotropin and possessed growth-promoting and lactogenic activity, as well as affecting lipid, carbohydrate, and cartilage metabolism in a manner similar to hypophyseal somatotropin. Eleven of the first 17 amino acid residues in human hypophyseal somatotropin and human placental prolactin are identical in the same positions; the remaining residues bear similarities in portions of their sequences.

Adipokinetic Hormones. Three hypophyseal peptides that stimulate fatty acid release from adipose tissue have been identified; these substances are distinct from the other recognized hypophyseal hormones, and have been termed lipotropic hormones. In their activity, the lipotropic substances resemble α- and β-MSH, and these in turn are structurally and biologically closely related to ACTH. Two of the lipotropins have been isolated from porcine tissue and designated as porcine peptides I and II. The third, isolated from sheep hypophyseal tissue, has been termed β-lipotropic peptide. The amino acid sequence of the latter has been established; it includes a sequence of seven amino acid residues common to ACTH and α- and β-MSH. β-Lipotropic peptide exhibits one-twentieth the adrenocorticotropic activity of ACTH, and melanocyte-stimulating activity equal to that of ACTH.

THE SEXUAL CYCLE

Among the most striking examples and results of endocrine interrelationships are the rhythmic sexual, or menstrual, cycles in the postpubertal female, reflecting hypophyseal-ovarian interrelationships.

The rhythmic sexual cycles initiated at puberty depend on the secretion and release of the gonadotropic hormones of the adenohypophysis. Prior to puberty, these hormones are not secreted in detectable quantities, and no significant alterations occur in either the ovaries or testes. The basis of the onset of this new activity of the hypophysis at the time of puberty is unknown. The secretion of follicle-stimulating hormone (FSH), acting synergistically with a small amount of luteinizing hormone (LH), stimulates follicle development in the ovary. As the follicle begins to ripen, growth and activity of the cells of the theca interna result in their secretion of estrogens. These estrogens induce secretion of follicular fluid by the granulosa cells and sensitize the follicle to FSH. In women, FSH stimulates maturation of a single follicle at each cycle, although occasionally two follicles may mature. The stimulated follicle continues to develop, while the other follicles that were in the process of ripening cease to grow and become, in part, atretic.

The estrogens which have been produced enter the blood stream and exert a diversified influence on the secretory activity of the adenohypophysis. FSH secretion is diminished, while secretion of LH and probably also of prolactin is augmented. LH acts upon the mature and sensitized follicle, causing its rupture, with release of the ovum (ovulation). The same hormone stimulates development of the corpus luteum in the ruptured follicle, and prolactin induces secretion by the newly formed corpus luteum. The progesterone thus produced inhibits hypophyseal secretion of LH and of prolactin; this fall in hormonal secretion coincides with menstruation. With the cessation of production of LH and prolactin, the hypophysis returns to the initial stage, *i.e.*, a new cycle begins. These events are diagrammed in Fig. 47.6.

Pregnancy. Secretion of progesterone by the corpus luteum prepares the endometrium of the uterus for implantation of the fertilized egg and for maintenance of the embryo and fetus. Progesterone also decreases the tonus of the uterus and stimulates development of the mammary gland preparatory to its secretion of milk after parturition.

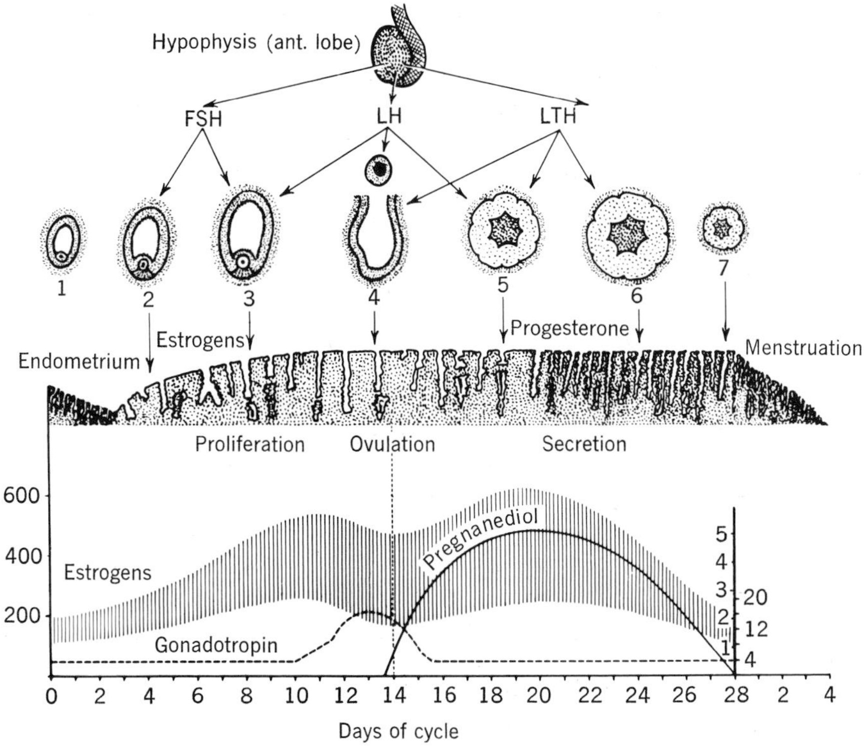

Fig. 47.6. Diagram of human female sexual cycle. The follicle-stimulating hormone (FSH) of the adenohypophysis provokes ripening of the ovarian follicle. 1, primary follicle; 2, ripening follicle; 3, ripe follicle; 4, rupture of follicle (ovulation) as a result of luteinizing-hormone (LH) action. LH stimulates corpus luteum formation (5), and prolactin (LTH) induces secretion by corpus luteum of progesterone (6). The mature corpus luteum (6) then degenerates (7) if there is no pregnancy. The endometrium proliferates under the action of estrogens in the first part of the cycle and secretes under the action of progesterone during the latter part of the cycle, which ends in menstruation. Below, urinary excretion of hormones; vertical lines, estrogens; broken line, gonadotropins; solid line, pregnanediol. On the left, the units are international units of estrogens eliminated in 24 hr.; on the right, milligrams of pregnanediol (1 to 5) and international units of gonadotropins (4 to 20) excreted in 24 hr. (*From B. A. Houssay, "Human Physiology," 2d ed., McGraw-Hill Book Company, New York,* 1954.)

If the ovum is not fertilized, it is not implanted in the uterus and the corpus luteum regresses. With implantation of the fertilized ovum, however, the corpus luteum persists in an active stage, and the normal menstrual cycle is interrupted. There is evidence that in women the corpus luteum can be removed after the fourth or fifth month of pregnancy without inducing abortion. This is related to the fact that at this time the placenta begins to produce progesterone. Urinary excretion of pregnanediol, a normal metabolite of progesterone (page 950), rises sharply at this period in pregnancy.

Production of estrogens, progesterone, and gonadotropins is augmented following ovum implantation in the uterus. The level of gonadotropin secretion is high early in gestation and forms the basis for several laboratory tests for pregnancy (page 994).

REFERENCES

Books

Berde, B., "Recent Progress in Oxytocin Research," Charles C Thomas, Publisher, Springfield, Ill., 1961.

Caldeyro-Barcia, R., and Heller, H., eds., "Oxytocin," Pergamon Press, New York, 1961.

Cole, H. H., ed., "Gonadotropins: Their Chemical and Biological Properties and Secretory Control," W. H. Freeman and Company, San Francisco, 1964.

Martini, L., and Ganong, W. F., eds., "Neuroendocrinology," vols. 1 and 2, Academic Press, Inc., New York, 1966, 1967.

Scharrer, E., and Scharrer, B., "Neuroendocrinology," Columbia University Press, New York, 1963.

Werner, S. C., ed., "Thyrotropin," Charles C Thomas, Publisher, Springfield, Ill., 1963.

Wolstenholme, G. E. W., and O'Connor, C. M., eds., "Human Pituitary Hormones," J. & A. Churchill, Ltd., London, 1960.

Young, W. C., ed., "Sex and Internal Secretions," vols. I and II, The Williams & Wilkins Company, Baltimore, 1961.

Review Articles

Daughaday, W. H., and Kipnis, D. M., The Growth-promoting and Anti-insulin Actions of Somatotropin, *Recent Progr. Hormone Research,* **22,** 49–99, 1966.

Dixon, H. B. F., Chemistry of Pituitary Hormones, in G. Pincus, K. V. Thimann, and E. B. Astwood, eds., "The Hormones: Physiology, Chemistry, and Applications," vol. V, pp. 1–68, Academic Press, Inc., New York, 1964.

Du Vigneaud, V., An Organic Chemical Approach to the Study of the Significance of the Chemical Functional Groups of Oxytocin to Its Biological Activities, *Proc. Robert A. Welch Foundation Conferences on Chem. Research,* **8,** 133–163, 1964.

Eik-Nes, K. B., Effects of Gonadotrophins on Secretion of Steroids by the Testis and Ovary, *Physiol. Revs.,* **44,** 609–630, 1964.

Engel, F. L., Extra-adrenal Actions of Adrenocorticotropin, *Vitamins and Hormones,* **19,** 189–227, 1961.

Farrell, G., Fabre, L. F., and Rauschkolb, E. W., The Neurohypophysis, *Ann. Rev. Physiol.,* **30,** 557–588, 1968.

Finkel, M. S., Human Growth Hormone, Metabolic Effects and Experimental and Therapeutic Applications, *Am. J. Med.,* **32,** 588–598, 1962.

Gaebler, O. H., Growth and Pituitary Hormones, *Newer Methods of Nutritional Biochem.,* **2,** 85–121, 1965.

Glick, S. M., Roth, J., Yalow, R. S., and Berson, S. A., The Regulation of Growth Hormone Secretion, *Recent Progr. Hormone Research,* **21,** 241–283, 1965.

Guillemin, R., The Adenohypophysis and Its Hypothalamic Control, *Ann. Rev. Physiol.,* **29,** 313–348, 1967.

Hofmann, K., and Yajima, H., Synthetic Pituitary Hormones, *Recent Progr. Hormone Research,* **18,** 41–88, 1962.

Knobil, E., and Greep, R. O., The Physiology of Growth Hormone with Particular Reference to Its Action in the Rhesus Monkey and the "Species Specificity" Problem, *Recent Progr. Hormone Research,* **15,** 1–69, 1959.

Korner, A., Growth Hormone Control of Biosynthesis of Protein and Ribonucleic Acid, *Recent Progr. Hormone Research,* **21,** 205–240, 1965.

Leaf, A., and Hays, R. M., The Effects of Neurohypophyseal Hormone on Permeability and Transport in a Living Membrane, *Recent Progr. Hormone Research,* **17,** 467–492, 1961.

Lerner, A. B., and Lee, T. H., The Melanocyte-stimulating Hormones, *Vitamins and Hormones,* **20,** 337–346, 1962.

Li, C. H., Synthesis and Biological Properties of ACTH Peptides, *Recent Progr. Hormone Research,* **18,** 1–40, 1962.

Liddle, G. W., Island, D., and Meader, C. K., Normal and Abnormal Regulation of Corticotropin Secretion in Man, *Recent Progr. Hormone Research,* **18,** 125–166, 1962.

McCann, S. M., Dhariwal, P. S. and Porter, J. C., Regulation of the Adenohypophysis, *Ann. Rev. Physiol.,* **30,** 589–640, 1968.

McKenzie, J. M., The Long-Acting Thyroid Stimulator, *Recent Progr. Hormone Research,* **23,** 1–46, 1967.

Raben, M. S., Human Growth Hormone, *Recent Progr. Hormone Research,* **15,** 71–114, 1959.

Riddle, O., Prolactin in Vertebrate Function and Organization, *J. Natl. Cancer Inst.,* **31,** 1039–1110, 1963.

Rudman, D., The Adipokinetic Property of Hypophyseal Peptides, *Ergeb. Physiol. Biol. Chem. Expt. Pharmakol.,* **56,** 297–327, 1965.

Rudman, D., Hirsch, R. L., Kendall, F. E., Seidman, F., and Brown, S. J., An Adipokinetic Component of the Pituitary Gland: Purification, Physical, Chemical and Biologic Properties, *Recent Progr. Hormone Research,* **18,** 89–123, 1962.

Savard, K., Marsh, J. M., and Rice, B. F., Gonadotropins and Ovarian Steroidogenesis, *Recent Progr. Hormone Research,* **21,** 285–365, 1965.

Sawyer, W. H., Comparative Physiology and Pharmacology of the Neurohypophysis, *Recent Progr. Hormone Research,* **17,** 437–465, 1961.

Sawyer, W. H., Neurohypophyseal Hormones, *Pharmacol. Revs.,* **13,** 225–277, 1961.

Wurtman, R. J., and Axelrod, J., The Pineal Gland, *Scientific American,* **213,** (July) 50–60, 1965.

48. The Major Nutrients

SCOPE OF THE SCIENCE OF NUTRITION

The science of nutrition may conveniently be considered in relation to the questions: What are the substances required by the animal for growth, maintenance, and reproduction, and in what quantities? What are the results of failure to meet these requirements, and what are the results of the ingestion of these substances in excess of the requirements? What is the physiological role of each of these nutrients? How does failure of this physiological function lead to the overt signs of deficiency? Which foods will enable the animal to meet these requirements, and in what amounts are these foods required?

All the major nutrients required by man have been encountered earlier in this book, in connection with the metabolic roles of these substances. What follows, therefore, is a survey of the answers presently available to the questions raised above. In the main, this discussion will be confined to "essential" nutrients, *i.e.*, those necessary for normal growth, maintenance, and reproduction and which man cannot synthesize from other constituents of the diet.

The nutritional requirements of man include water, inorganic ions, and a number of organic compounds. The requirement for water has been considered in detail (Chap. 33). The inorganic substances include the major anions and cations of the extra- and intracellular fluids and skeleton and a number of elements required in lesser amounts. Carbohydrates and lipids are required as fuels and as precursors for the synthesis of diverse substances and structures. Amino acids are needed for the synthesis of proteins and of other nitrogenous compounds. Accessory food factors called vitamins are frequently employed as structural components of coenzymes. Of the organic compounds in the body, approximately 24 have been definitely established as dietary essentials. All other substances are synthesized by the organism in the presence of an adequate supply of these essential factors. In view of the complex chemical composition of living things, the list of essential nutrients shown in Table 48.1 is perhaps more remarkable for its brevity than for its length.

ASSAY ANIMALS

Recognition of the role of nutrition in the etiology of certain diseases of man led to a search for the specific nutrients which would prevent these diseases. Among the nutrients recognized in this manner have been vitamins A, D, and B_{12}, thiamine, niacin, and ascorbic acid. These investigations required the establishment of deficiency states in experimental animals under controlled circumstances. Many animal

Table 48.1: NUTRIENTS REQUIRED BY MAN

Amino acids	Elements	Vitamins
	Established as essential	
Isoleucine	Calcium	Ascorbic acid
Leucine	Chlorine	Choline†
Lysine	Copper	Folic acid
Methionine	Iodine	Niacin‡
Phenylalanine	Iron	Pyridoxine
Threonine	Magnesium	Riboflavin
Tryptophan	Manganese	Thiamine
Valine	Phosphorus	Vitamin B_{12}
	Potassium	Vitamins A, D§, E, and K
	Sodium	
	Probably essential	
Arginine*	Fluorine	Biotin
Histidine*	Molybdenum	Pantothenic acid
	Selenium	Polyunsaturated fatty acids
	Zinc	

* Indicated to be unnecessary for maintenance of nitrogen equilibrium in adults in short-term studies but probably necessary for normal growth of children.

† Requirement met under circumstances of adequate dietary methionine.

‡ Requirement may be provided by synthesis from dietary tryptophan.

§ Requirement may be met by exposure of children to sunlight. No evidence for a requirement in adults.

species have been employed; the chicken and pigeon are used as assay animals for thiamine, the dog is used in niacin and vitamin D analyses, and the guinea pig for ascorbic acid assays. The animal most often employed in nutrition studies is the albino rat. The rat is omnivorous, has a gastrointestinal tract comparable with that of man, and in the main, exhibits nutritional requirements similar to those of man. However, it has rarely been possible to reproduce, in an experimental animal, a disease entity identical with that encountered in human deficiency states. Indeed, impaired growth of the young animal has frequently been the only criterion of deficiency. The fact that, when all nutrients but one are present in the diet in adequate amounts, the growth rate of an animal is proportional to the dietary supply of the limiting nutrient makes possible bioassays of natural foodstuffs and concentrates. During the period between the recognition of each new accessory food factor and its final identification, it has been the practice to adopt for comparative purposes a standard *unit* of activity, *e.g.*, the amount necessary to cause a specific quantitative response in a given animal. After the nutrient has been identified and is available in pure form, the data are then expressed in terms of the weight of the nutrient. However, when several structurally related and naturally occurring compounds are found to exert similar activity under the conditions of bioassay, it has been the practice to continue to express the activity of biological materials in units.

MICROORGANISMS AND NUTRITIONAL STUDIES

Microorganisms have been extremely useful in nutritional investigations. Inositol, pantothenic acid, and biotin were first recognized as required for the initial growth of certain strains of yeast. Folic acid was established as an essential factor for the growth of several strains of lactobacilli several years before its importance in mammalian nutrition was discovered. Vitamin B_{12}, the antipernicious anemia factor, was identified in one laboratory because it is essential for the growth of *Lactobacillus lactis* Dorner. Many species of bacteria, molds, yeast, and fungi have been found for which one or more of the nutrients known to be essential to man are also growth factors, thus providing a basis for rapid quantitative microbiological assays for these substances.

The bacteria which inhabit the gastrointestinal tract may supply significant amounts of vitamins to the host. This accounts for the simpler nutritional requirements of the ruminants and of those species with large ceca, *e.g.*, the horse and rabbit. Thus these animals are provided with fermentation mechanisms which produce many essential nutritional factors. To a lesser degree, this occurs also in the intestine of man, who is thereby supplied with a major fraction of his requirement for biotin and vitamin K. The role of the intestinal flora in nutrition has been established by inclusion in the diet of antibiotics. Animals raised under sterile conditions with bacteria-free intestinal tracts have also been employed in studies of this type. The inclusion of tetracycline antibiotics in the stock rations of pigs and chicks has been found to stimulate their growth, with an increased weight per pound of ingested feed. The mechanism whereby this is accomplished is not clear but probably reflects establishment of an intestinal flora which furnishes utilizable nutrients to the host animal.

NUTRITIONAL REQUIREMENTS OF MAN

Several difficulties are encountered in studies of human nutrition. The most serious of these are the impracticality of feeding human subjects rations consisting exclusively of chemically purified materials for sufficient lengths of time and the uncertainty concerning the qualitative and quantitative activities of the intestinal flora. It has been particularly difficult to establish the quantitative requirements for many nutrients known, qualitatively, to be essential. Nitrogen balance studies have yielded values for the amounts of the essential amino acids required for short periods of time for maintenance of nitrogen balance in adult human beings. However, the amino acid requirements for growth have not been established for the human species. Balance studies have also been employed to establish quantitative requirements for elements such as potassium, calcium, and iron. Perhaps the most striking result of these studies is the extreme variability found within the population. The amount of calcium required to maintain one individual in balance may be two or three times that which suffices for another. A similar situation exists with respect to iron. The quantitative requirements for the vitamins are even less securely established. The recommended dietary allowances shown in Table 48.2 represent dietary levels which, in clinical experience, are known *not* to lead to deficiency; they are not intended to represent minimal requirements but are objectives which it is presently thought desirable to seek in planning diets for normal individuals.

Table 48.2: RECOMMENDED DAILY DIETARY ALLOWANCES

	Age,† yr	Weight, lb. (kg.)	Height, in. (cm.)	Calories	Protein, g.	Calcium, g.	Iron, mg.	Thiamine, mg.	Riboflavin, mg.	Niacin equiv. mg.‡	Ascorbic acid, mg.	Vitamin A value, I.U.**	Vitamin D, I.U.
Men	18–35	154 (70)	69 (175)	2900	70	0.8	10	1.2	1.7	19	70	5,000§	
	35–55	154 (70)	69 (175)	2600	70	0.8	10	1.0	1.6	17	70	5,000	
	55–75	154 (70)	69 (175)	2200	70	0.8	10	0.9	1.3	15	70	5,000	
Women	18–35	128 (58)	64 (163)	2100	58	0.8	15	0.8	1.3	14	70	5,000	
	35–55	128 (58)	64 (163)	1900	58	0.8	15	0.8	1.2	13	70	5,000	
	55–75	128 (58)	64 (163)	1600	58	0.8	10	0.8	1.2	13	70	5,000	
	Pregnant (2d & 3d trimester)	. . .	. . .	+200	+20	+0.5	+5	+0.2	+0.3	+3	+30	+1,000	400
	Lactating	. . .	. . .	+1000	+40	+0.5	+5	+0.4	+0.6	+7	+30	+3,000	400
Infants¶	0–1	18 (8)	. . .	lb. × 52.3 ±6.8	lb. × 1.14 ±0.23	0.7	lb. × 0.45	0.4	0.6	6	30	1,500	400
Children	1–3	27 (13)	34 (87)	1300	32	0.8	8	0.5	0.8	9	40	2,000	400
	3–6	40 (18)	42 (107)	1600	40	0.8	10	0.6	1.0	11	50	2,500	400
	6–9	53 (24)	49 (124)	2100	52	0.8	12	0.8	1.3	14	60	3,500	400
Boys	9–12	72 (33)	55 (140)	2400	60	1.1	15	1.0	1.4	16	70	4,500	400
	12–15	98 (45)	61 (156)	3000	75	1.4	15	1.2	1.8	20	80	5,000	400
	15–18	134 (61)	68 (172)	3400	85	1.4	15	1.4	2.0	22	80	5,000	400
Girls	9–12	72 (33)	55 (140)	2200	55	1.1	15	0.9	1.3	15	80	4,500	400
	12–15	103 (47)	62 (158)	2500	62	1.3	15	1.0	1.5	17	80	5,000	400
	15–18	117 (53)	64 (163)	2300	58	1.3	15	0.9	1.3	15	70	5,000	400

*Allowance levels are intended to cover individual variations among most normal persons as they live in the United States under usual environmental stresses. The recommended allowances can be attained with a variety of common foods providing other nutrients for which human requirements have been less well defined. See additional discussion in chapters of this Part of allowances and of nutrients not tabulated.

† Entries on lines for age range 18 to 35 years represent the 25-yr age. All other entries represent allowances for the midpoint of the specified age periods, *i.e.*, line for children 1 to 3 is for age 2 yr (24 mo); 3 to 6 is for age 4½ yr (54 mo), etc.

‡ Niacin equivalents include dietary sources of the preformed vitamin and the precursor, tryptophan. 60 mg. tryptophan represents 1 mg. niacin.

** I.U. = international unit.

§ 1,000 I.U. from performed vitamin A and 4,000 I.U. from β-carotene.

¶ The calorie and protein allowances per pound for infants are considered to decrease progressively from birth. Allowances for calcium, thiamine, riboflavin, and niacin increase proportionately with calories to the maximum values shown.

SOURCE: Recommended by the Food and Nutrition Board, National Academy of Sciences—National Research Council, 1963. Publication 1146.

PROTEIN

Animals do not require dietary protein, per se, but rather certain of the individual amino acids derived therefrom. The concept of nutritionally essential amino acids and the manner in which they have been identified have been presented earlier (Chap. 23). The amino acids essential for growth of the rat are also those essential for adequate nutrition of the human infant. Rose established the amount of each amino acid required for the maintenance of nitrogen balance in young adults (Table 48.3). The values shown are probably high and provide an appreciable margin of safety. The fact that histidine did not appear to be necessary to maintain nitrogen balance in these studies should not be taken to mean that this amino acid is not required by the human adult, since prolonged histidine deficiency results in impaired hemoglobin production, and in eczema in infants.

Table 48.3: AMINO ACID REQUIREMENTS OF YOUNG ADULTS

Amino acid	Amount, mg./kg.	Maintenance pattern*
Arginine	0	0
Histidine	0	0
Tryptophan	7	1.0
Phenylalanine	31	4.3
Lysine	23	3.2
Threonine	14	1.9
Valine	23	3.2
Methionine	31	4.3
Leucine	31	4.3
Isoleucine	20	2.8

* Molecular ratios required relative to the tryptophan requirement.
SOURCE: From W. C. Rose, *Federation Proc.*, **8**, 546, 1949.

The values for the requirement of a single amino acid, such as those shown in Table 48.3, are not absolute but are influenced markedly by the composition of the total amino acid mixture, *e.g.*, the requirements for phenylalanine and methionine are significantly reduced by the provision of tyrosine and cystine, respectively. If only minimal amounts of essential amino acids are provided in growth studies with young rats, a striking growth stimulus results from the further provision of nonessential amino acids, *e.g.*, glutamic acid and arginine. However, addition of large quantities of such amino acids, particularly glycine, can result in serious growth depression. Optimal nutrition, therefore, requires a *balanced* amino acid mixture. Another example is afforded by experience with wheat gluten; this protein is relatively poor in lysine. If the protein is fed as 30 per cent of a rat diet, an additional 0.8 per cent of lysine is required to obtain good growth; if fed as 60 per cent of the diet, 1.3 per cent of lysine is required to achieve the same growth rate.

Biological Value of Protein. Much effort has been expended in the determination of the *biological value* of individual proteins. The term refers to the relative nutritional value of individual proteins as compared to a standard and includes a factor relating to the digestibility of a protein source as well as to its amino acid composition. Among the procedures employed are determinations of the growth rate of young rats with varying dietary levels of individual proteins, establishment of the minimal dietary level of a given protein which will permit nitrogen balance in adults of various species, and influence of a given amount of protein on the serum levels of the essential amino acids. By these procedures, proteins which lack any of the essential amino acids will have no biological value as compared with a standard protein preparation such as lactalbumin, which is known to be qualitatively complete and readily digestible. It is apparent that if the amino acid compositions of two proteins are generally similar except that *A* contains only half as much leucine as does *B*, twice as much of *A* is needed to meet the leucine requirement and the "biological value" of *B* is greater than that of *A* when the leucine supply of the diet is a limiting factor. In general, proteins of animal origin have greater biological value than do proteins of plant origin. Casein, lactalbumin, mixed muscle proteins, and mixed egg proteins are approximately equal in value. Table 48.4 shows the amount of protein from various sources which, as the sole source of protein in the diet, will permit maintenance of nitrogen equilibrium in adult men under normal conditions.

Table 48.4: ESTIMATES OF THE PROTEIN REQUIREMENTS OF ADULT MEN FOR MAINTENANCE OF NITROGEN EQUILIBRIUM

Protein source	*Required g./70 kg./day*	*Protein source*	*Required g./70 kg./day*
Beefsteak	19.2	Mixed vegetable protein 2:3, and mixed meat protein 1:3*	27.1
Whole egg	19.9	General mixed diet†	27.6
Corn germ	20.7	Potato	29.6
Haddock	21.6	Soy-white flour‡	29.8
Cottonseed flour	23.0	All-vegetable diet§	32.4
Yeast	24.0	Wheat flour	38.4
Milk	24.4	White flour	42.1
Soy flour	25.4	Whole-wheat bread	66.8
Beef	26.3		

* The amount of each food in the all-vegetable diet was decreased by one-third and replaced by meats in amount necessary to supply one-third of total nitrogen.

† A cheap complete mixed American diet in which animal proteins provided 47 per cent of total nitrogen.

‡ Thirty-six per cent soy flour protein + 64 per cent white flour protein.

§ Distribution of nitrogen: 50 per cent from white flour, 12 per cent other cereals, 13 per cent potatoes, 17 per cent other mixed vegetables, 8 per cent fruits.

SOURCE: From H. H. Mitchell, in M. Sahyun, ed., "Proteins and Amino Acids in Nutrition," chap. 2, Reinhold Publishing Corporaton, New York, 1948.

Vegetable proteins are not only generally nutritionally inferior to those of animal origin, but proteins are also present in relatively smaller concentration in plant materials. In general, plant proteins are lower in lysine, methionine, and tryptophan content, and are less readily digestible than are animal proteins. However, nutritional inadequacies of a given protein are likely to differ from those of other proteins, and there are many possible combinations of two or more inadequate proteins which are nutritionally satisfactory. In general, by providing protein from various sources, optimal benefit to the animal is assured. For example, Incaparina, a vegetable mixture prepared and distributed by the Institute of Nutrition of Central America and Panama, which contains 29 per cent whole corn, 29 per cent whole sorghum, 38 per cent cottonseed meal, 3 per cent Torula yeast, plus some $CaCO_3$ and vitamin A, affords a protein mixture of biological value only slightly less than that of cow's milk.

When inadequate proteins are combined in the diet, they must be present in the same meal. Animals do not store amino acids and can synthesize protein only when all the component amino acids are simultaneously present. For example, rats fail to grow when fed, each day, one of the essential amino acids 3 hr. after the others.

Effects of Protein Deficiency. No disease of man has yet been described which is attributable to a deficiency of a single amino acid. Indeed, relatively little is known of the possible consequences of deficiencies of single amino acids in man. The effects of such deficiencies in the rat, other than the invariable growth failure, are characterized by rapid loss of appetite and development of the disorders summarized in Table 48.5. However, if force-fed such diets in amounts equal to that consumed by well-nourished controls, a syndrome develops which includes disappearance of liver glycogen, increase in liver lipids, and atrophy of the pancreas, salivary glands, spleen, and stomach, analogous to the findings in human kwashiorkor (see below). Unexpectedly, the rate of protein synthesis in the livers of such animals is accelerated.

As the total content of a balanced amino acid mixture in the rat's diet is reduced, the first deficiency to limit health is usually that of methionine. Methionine deficiency results in fatty infiltration of the liver, progressing to cirrhosis. The liver lipids accumulate because of failure to synthesize β-lipoproteins. A similar situation is observed when the S-ethyl homologue, ethionine, is included in an otherwise adequate diet. However, this appears to be the consequence of the formation and

Table 48.5: MANIFESTATIONS OF DEFICIENCIES OF SINGLE AMINO ACIDS IN THE RAT

Amino acid	Symptoms*	
	Young rats†	Adult rats
Arginine	. . . .	Hypospermia
Cystine	Acute hepatic necrosis	
Histidine	Cataract	
Isoleucine	Anemia; hypoproteinemia	
Leucine	Hypoproteinemia	
Lysine	Anemia; sudden death	Anemia; anestrus
Methionine	Anemia; hypoproteinemia; alopecia; hemorrhagic kidneys; fatty liver and cirrhosis	Anemia; hypoproteinemia; fatty liver and cirrhosis
Threonine	Edema	
Tryptophan	Cataract; poor dentition; alopecia; gastric hyperplasia	Corneal vascularization; alopecia; testicular atrophy; fetal resorption
Valine	Locomotor dysfunction	

* All changes are reversed when the diet is supplemented with the missing component shortly following the appearance of symptoms.

† All deficiencies and any amino acid imbalance cause corneal vascularization in immature rats.

SOURCE: Modified from A. A. Albanese, *J. Clin. Nutrition,* **1,** 46, 1952.

accumulation of S-adenosylethionine (page 553) which cannot be further metabolized and, hence, makes unavailable significant amounts of hepatic adenosine. The consequent failure of protein synthesis seems then to reflect an absolute insufficiency of ATP. A similar situation occurs in rats fed large quantities of orotic acid. In both instances, reestablishment of β-lipoprotein synthesis and disappearance of liver lipids occur promptly after feeding adenosine. If choline is added to the low protein diet, fatty liver is prevented. As the dietary protein level is reduced further, an acute hepatic necrosis due to cystine deficiency appears. Addition of cystine to the ration prevents this disturbance. In order, thereafter, if casein is the protein of the diet, the amino acids which may become limiting are threonine, tryptophan, isoleucine, leucine, valine, phenylalanine, and lysine. In general, an insufficient quantity of a balanced amino acid mixture is less deleterious to the animal than an abundant supply of an unbalanced mixture.

Rats fed a low-protein, choline-supplemented diet develop pathological disturbances if the experiment is prolonged. Among the manifestations of protein deficiency are anemia, hypoalbuminemia, and edema, symptoms also commonly seen in undernourished human beings. Peptic ulcers which have been observed in rats fed diets low in protein are also unusually common in the population of certain areas of India in which no animal proteins are consumed and the vegetable proteins are of poor quality. A low-protein diet also leads to marked suppression of adenohypophyseal secretion in rats. This is reflected in a low basal metabolic rate, growth failure, diminished adrenal cortical activity, permanent anestrus, and suppression of lactation. Similar effects of protein malnutrition were noted in seriously undernourished individuals in prisoner-of-war camps, and in patients in European sanatoriums after the Second World War.

In economically undeveloped areas such as tropical America, Central and South Africa, and India, limitation of dietary protein is an important etiologic factor in a disease in children first called *kwashiorkor* in Central Africa. This name is now frequently applied to similar syndromes in other parts of the world, and is said to mean "displaced child," since it occurs in infants displaced from the breast by younger siblings. The disease is characterized by growth retardation, anemia, hypoproteinemia frequently with edema, fatty infiltration of the liver with ensuing fibrosis, and, in young Negroes, red to light brown hair. Often there is atrophy of the acinar tissue of the pancreas, with resulting diarrhea and steatorrhea. The lack of pancreatic digestive secretions makes unavailable even the small amount of dietary protein. Moreover, a renal lesion develops which leads to strikingly increased urinary excretion of free amino acids. If the diet contains much iron, the intestinal mucosa fails to restrict its absorption, and iron may accumulate in the liver. In general this disease appears in children who are fed almost exclusively on a gruel of a cereal such as plantain, taro, millet, cassava (manioc), or, most frequently, corn. In contrast to milk, which contains 5.4 g. of protein per 100 cal., these diets provide less than 2 g. of protein per 100 cal. and are deficient also in other essential dietary factors. The symptoms respond therapeutically to a high-protein diet containing considerable quantities of meat and milk products. The mortality rate in untreated children has been reported to be 30 to 90 per cent; these children are unusually susceptible to intercurrent infection and frequently succumb to acute diarrhea. It is probably significant that the incidence of primary carcinoma of the liver is considerably more frequent in individuals of these areas than in other population groups.

Kwashiorkor is undoubtedly the world's major health problem, and, as population growth outstrips the food supply of the developing nations, this problem will be exacerbated. It can be solved only by a combination of population control, improved agricultural practices, and general economic development. Nutritionally, the principal problem is the low lysine content of the above-mentioned cereals and of wheat flour. Supplementation with synthetic lysine could significantly ameliorate this situation, as could introduction of recently developed corn strains containing the genes *opaque-2* or *floury-2*. Whereas ordinary corn protein contains 2.8 g. lysine per 100 g., protein from strains positive for *opaque-2* contains 4.7 g. per 100 g. Supplementation of cereal diets with fish protein concentrate could also prevent kwashiorkor, but it is not yet certain whether such material can be produced and distributed economically and on a scale commensurate with world needs.

The negative nitrogen balance of severe protein restriction occurs, in large measure, at the expense of the liver, which may lose as much as 50 per cent of its total nitrogen. Mitochondria, microsomes, cytoplasmic enzymes, and RNA, but not DNA, are all decreased under these circumstances. Of the enzymes, liver xanthine oxidase appears to be the most labile and sensitive indicator, all activity disappearing from the rat liver after 2 weeks on a protein-free regime. Indeed, the ability to maintain xanthine oxidase activity in liver has been employed as a test for the biological value of proteins. In general the behavior of hepatic enzymes during repletion with protein after severe deficiency falls into four categories: (1) enzymes which increase in amount and then plateau (xanthine, D-amino acid oxidases); (2) enzymes which increase in activity as dietary protein increases (arginase, transaminases); (3) enzymes which decrease in amount as dietary protein increases (phosphatase); and (4) enzymes which are unaffected (cathepsins).

The chronically niacin-deficient dog illustrates the physiological priorities which exist for available protein in the realimentation of an undernourished animal. Such dogs exhibit profound anemia and hypoproteinemia, are markedly emaciated, and ingest only limited quantities of the deficient diet. Similar anemia in the protein-depleted rat appears to reflect cessation of erythropoietin formation (page 752). If niacin is given but food consumption restricted to that which the dog had previously been eating ad libitum, a marked hemopoietic response occurs with little change in body weight or serum protein concentration. If a small increase in the total food allowance is then given, the serum proteins increase in quantity although there is little or no change in body weight. Finally, if the animal is allowed to eat ad libitum, body weight is restored. These relations are shown in Fig. 48.1.

LIPID

The studies of Burr and Burr in 1929 demonstrated that the rat does not synthesize polyunsaturated fatty acids and that when these are omitted from the diet a syndrome develops characterized by a dermatitis, necrosis of the tail, sterility, and hyperemia of renal tissues with swelling of the kidneys. Subsequent studies in mice and dogs indicate that these species also have a limited ability to synthesize polyunsaturated fatty acids. Swine ingesting diets deficient in these fatty acids develop

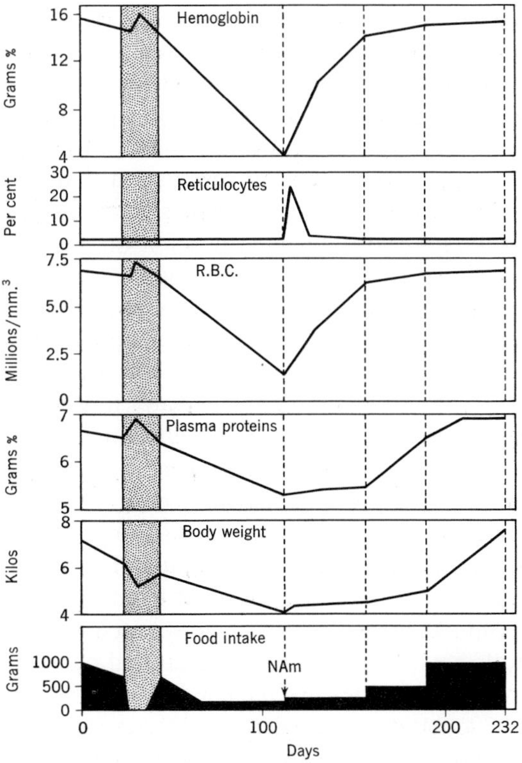

Fig. 48.1. Physiological priorities for available dietary protein. A niacin-deficient dog was carried through the blacktongue crisis with saline infusions (stippled area) (page 1027) and in 4 months of niacin deficiency became severely anemic, hypoproteinemic, and emaciated. At the arrow, nicotinamide (NAm) was given and food consumption was restricted to 250 g. per day. Note the reticulocyte response and restoration almost to normal hemoglobin concentration. When the food allowance was increased somewhat, the plasma protein concentration increased with little change in weight. When ad libitum food consumption was permitted, body weight began to increase. (*From P. Handler and W. P. Featherstone, J. Biol. Chem.* **151**, 395, 1943.)

a syndrome called parakeratosis in which erythema, seborrhea, and hyperkeratosis of the skin are manifest; a similar disorder appears to occur in zinc-deficient hogs (page 1016).

The composition of hepatic mitochondrial lipids is markedly changed by essential fatty acid deficiency in the rat. Such mitochondria exhibit diminished respiratory control, and their oxidative phosphorylation is unusually readily uncoupled by such agents as dinitrophenol and digitonin.

A deficiency syndrome in human adults resulting from lack of essential fatty acids has not been reported. Although the dermatitis of the deficient rat resembles closely the lesions of follicular hyperkeratosis in man, *viz.*, hyperplasia of surface epithelium, with well-defined layers of cutaneous fat but with the openings of the hair follicles plugged with keratin, this syndrome in man is not the result of fatty acid deficiency. However, infants fed a formula diet devoid of polyunsaturated fats developed dry, leathery, thickened skin with desquamation and oozing. The dienoic and trienoic fatty acids of their sera fell to extremely low levels. This disorder improved rapidly when trilinolein was added to the diet as 5 to 7 per cent of the total calories. When human adults are fed diets lacking in polyunsaturated fatty acids for several months, the iodine number of plasma fatty acids and the plasma concentration of polyunsaturated fatty acids decline precipitously, but no other symptoms have been observed. It seems appropriate to consider unsaturated fatty acids as essential to the human dietary, with the reservation that an absolute deficiency is unlikely in view of the wide distribution of these compounds in common foods. Although no specific function of linoleic acid is known, and although arachidonic acid, which is readily fabricated by man from linoleic acid (page 495), completely satisfies the requirement of experimental animals for polyunsaturated acids, it is convenient to designate these as a group of essential compounds rather than to regard linoleic acid as merely the precursor of arachidonic acid. Whether the essential dietary role of the polyunsaturated fatty acids relates to their serving as precursors of the prostaglandins (page 943) remains to be elucidated.

Evidence suggests that hyperlipemia and, hence, atherosclerosis may occur as a result of a *relative* deficiency in polyunsaturated fatty acids. Deposition of lipid plaques on the intima of arteries (atheroma) appears to occur most readily when the serum lipids are elevated in concentration. The material deposited has a composition similar to that of the serum lipids at the time of deposition. The concentration of plasma lipids is determined, in part, by the composition of the diet. In general, variation in the amount of dietary cholesterol, as in normal diets, is without significant influence. When greater quantities are included in experimental diets, the consequences vary with the amount of cholesterol and the composition of the diet. Large quantities (2 per cent of the diet) evoke a massive lipemia; triglyceride is mobilized from adipose tissue and deposited in the liver. Lesser quantities result in marked lipemia, including high concentrations of cholesterol itself with concomitant fall in the serum concentration of polyunsaturated fatty acids in the various lipid fractions. This is the technique used to produce experimental atherosclerosis. In contrast a diet extremely low in lipid results in prompt decline in all plasma lipid fractions, although a lipid-free, high-carbohydrate regime may elicit an increase in serum triglycerides.

When lipid is present in the diet in an amount sufficient to contribute 20 to 60 per cent of the total calories (the average American diet provides about 40 per cent of calories as lipid), the level of blood lipids appears to be influenced by the nature of the dietary lipid. When the latter is largely or exclusively saturated or monoethenoic, as when animal fats predominate in the diet, blood lipid concentrations may be as much as twice as great as when the diet is rich in polyunsaturated fatty acids. This concept was originally derived from epidemiological studies in which serum cholesterol, diet, and the incidence of atherosclerosis in the United States and Western Europe were compared with these items in economically underdeveloped areas. In the latter, vegetable oils rich in linoleic acid are the major sources of dietary lipid; the populations of these areas exhibit relatively low serum lipid concentrations, and the incidence of atherosclerotic disease in age-matched groups is stated to be decidedly lower than, for example, in the United States, England, or the Scandinavian countries. This influence of diet upon plasma lipids has been confirmed under controlled conditions in human subjects and suggests that the serum lipid concentration does not reflect the absolute level of dietary lipid but rather the *relative* contributions of saturated and unsaturated fatty acids. Safflower, corn, peanut, and cottonseed oils, in that order, appear to be most effective in maintaining a low concentration of serum lipids.

Many other factors contribute to this complex situation. As indicated previously, pyridoxine may be required for the synthesis of arachidonic acid from linoleic acid page (495). Pyridoxine deficiency in the monkey results in hyperlipemia and formation of atheromas indistinguishable from those found in human arteries. Atherosclerosis may be produced in many species by dietary means; in the rabbit and chicken atherosclerosis is induced by inclusion of cholesterol in the ration, whereas in the rat a high-lipid diet, low in unsaturated fatty acids and rich in cholesterol, is required.

Of great potential significance is recognition of the influence of eating patterns on serum lipids. Thus, rats and chicks normally eat by nibbling throughout their waking hours. If forced to eat, equicalorically, two or three meals per day, they deposit lipid in adipose tissue, the R.Q. falls, and serum lipid levels, including that of cholesterol, increase significantly. Similar studies of multiple feedings in man have not yet been reported.

Exercise significantly decreases serum cholesterol and triglyceride levels while total unsaturated fatty acids of serum appear to increase. It is perhaps significant also that diets which increase serum lipid concentrations have been reported to result in increased blood clot strength and delayed fibrinolysis. Thus, in addition, to the complex nutritional effects, genetic tendencies, emotional variability, exercise, and even anatomical relationships must be significant in the control of serum lipid concentrations. Moreover, it is not established that reduction of serum lipids will delay atherosclerosis or prevent coronary artery disease in man.

Extensive efforts have been made to determine the optimal amount of lipid in the diet. Although young animals grow successfully on a diet containing only sufficient lipid to provide minimal amounts of the essential fatty acids, it appears desirable to provide approximately 30 per cent of the total calories as lipid with a large fraction of this consisting of polyunsaturated lipids. Rats, dogs, and hogs on this lipid intake achieve a greater body weight and live slightly longer than do animals fed a minimal amount of lipid. Diets yielding more than 60 per cent of calories as lipid lead eventually to obesity in all species.

Ordinarily, triglycerides, whether hydrolyzed in the intestinal lumen or absorbed as such, are transported from the intestine in the lymphatic chyle. However, medium-chain triglycerides (C_4 to C_{10}) are readily hydrolyzed and their fatty acids enter the portal vein (page 473). Accordingly, such products may be of value in malabsorption states such as chylothorax, sprue, or ileitis.

CARBOHYDRATE

While dietary carbohydrate provides a source of energy and the substrates for many synthetic pathways, animals and man can be maintained on diets devoid of carbohydrate. However, since dietary calories may be provided most cheaply in this form, carbohydrate supplies at least 50 per cent of the calories of most human dietaries.

A carbohydrate-poor ration must provide calories in the form of lipid since protein intake is limited by the feeling of satiety provided by relatively small quantities of protein. Subjects transferred from ordinary mixed rations to a lipid-rich diet rapidly develop ketosis. After some weeks the ketosis abates and may disappear entirely. Indeed, the Eskimo normally lives on a high-lipid, low-carbohydrate ration. This problem is significant in the diabetic individual who depends on lipids for his calories, since the extent of carbohydrate utilization will depend on the amount of insulin administered. The relative tendencies of various foodstuffs to elicit or prevent ketosis are expressed as their ketogenic or antiketogenic activity. As an aid in the management of diabetes mellitus, diets are planned to contain a balance of ketogenic and antiketogenic foods.

Only rare individuals cannot utilize sucrose. Those with a genetic lack of fructokinase exhibit a benign fructosuria. However, those who lack phosphofructose aldolase cannot utilize the fructose 1-phosphate formed by fructokinase. Accumulation of fructose 1-phosphate results in a bizarre syndrome of headache and acute abdominal distress.

Lactose furnishes a significant but not the major portion of the caloric value of milk. Feeding diets high in lactose or galactose to rats results in impaired growth and cataract. The growth impairment results from a limited capacity of the rat liver to convert galactose 1-phosphate to glucose 1-phosphate. Such animals have extremely high blood sugar levels, most of which is galactose; the blood glucose concentration is markedly reduced. Similar findings are encountered in infants with "idiopathic galactosemia" (page 427). The relationship between high blood galactose concentration and cataract development is obscure (page 900). The favorable influence of dietary lactose on intestinal calcium absorption has been mentioned previously (page 888). Diets high in lactose or galactose result in demineralization of bone with resultant calciuria; the mechanism is unknown. The rare infants whose intestines are devoid of lactase grow poorly and may have severe diarrhea. Transfer to a synthetic milk containing sucrose results in prompt improvement.

MINERAL NUTRIENTS

The nutritional roles of several of the major mineral nutrients have already been considered. The elements known to be required by man, as appropriate ions, include sodium, chlorine, and potassium (Chap. 33), calcium and phosphorus (Chap. 39), iron (Chap. 31), and iodine (Chap. 43). The requirements, function, and manifestations of deficiency of these elements were described previously. In addition, man requires magnesium, manganese, and copper. The status of cobalt, zinc, fluoride, molybdenum, and selenium in human nutrition is not clearly established.

Magnesium. Magnesium is indispensable in the diet, but the daily requirement is not known. The factors which influence magnesium absorption are similar to those affecting calcium (Chap. 39). The normal serum [Mg^{++}] is 1.8 to 2.5 meq. per liter, of which about 80 per cent is ionized and diffusible and the remainder is bound to protein. The [Mg^{++}] in erythrocytes is somewhat higher—about 3.5 meq. per liter—while tissue cells generally contain about 16 meq. per liter. Most of the Mg^{++} of the body is in the skeleton. Absorbed or administered Mg^{++} is rapidly excreted in the urine. However, the major daily excretion occurs in the feces and represents unabsorbed magnesium. Ingestion of magnesium also increases excretion of calcium in both feces and urine.

Magnesium deprivation has been studied in dogs and rats fed diets which contained less than 2 parts per million of the element. Such animals gradually developed a syndrome consisting of vasodilatation, cardiac arrhythmia, hyperirritability to any external stimulus, spasticity, and tonic and clonic convulsions; death invariably followed.

Spontaneous nutritional deficiency of Mg^{++} in man is extremely unlikely. However, extensive loss of this element may occur in diarrhea and become evident if such patients are rehydrated with fluids which do not contain Mg^{++}. As the serum $[Mg^{++}]$ declines to about 1.0 meq. per liter, a syndrome resembling delirium tremens may be precipitated, *i.e.*, semicoma, tremor, carpopedal spasm, and a general tetany-like neuromuscular irritability with marked susceptibility to auditory, mechanical, and visual stimuli. Administration of Mg^{++} effects a prompt improvement.

Parenteral administration of large amounts of magnesium salts results in marked depression of both the central nervous system and peripheral neuromuscular activity. This depressive action can be antagonized by calcium; an animal anesthetized with magnesium returns to consciousness and behaves normally a few seconds after intravenous administration of sufficient calcium salt. There is no explanation for either the magnesium narcosis or the calcium antagonism. Low serum $[Mg^{++}]$ leads to tetany resembling that of hypocalcemia. Thus, while a low concentration of either of these ions in the blood leads to similar functional changes, an excess of calcium antagonizes rather than synergizes the effects of excessive magnesium.

Manganese. Manganese deficiency has been observed in the rat, chick, and pig. Male rats on diets deficient in manganese become sterile and show irreversible testicular degeneration; females on manganese-deficient diets are unable to suckle their young. In manganese-deficient pregnant sows, resorption of the fetus is common; if young are born they are undersized, weak, and ataxic. The estrus cycle of the nonpregnant animal is irregular or may cease completely. In the chick, manganese deficiency is manifested as the osteodystrophy called *perosis;* the tibial-metatarsal joint enlarges, the distal end of the tibia and the proximal end of the metatarsus are twisted, and the gastrocnemius tendon slips from its condyles. Because of the latter, this is frequently referred to as "slipped tendon disease." The chicks appear short-legged with deformed legs and spinal columns.

No precise requirement for manganese by man has been established. Manganese salts are poorly absorbed from the intestine, but after parenteral administration, manganese is concentrated in the liver and kidneys, particularly in the mitochondria, and excreted largely into the colon and bile, with only a small portion appearing in the urine. Ingestion of excessive quantities of manganese appears to interfere with absorption of iron, thus causing an anemia which is readily prevented by increasing the dietary iron.

Manganese is intimately bound to arginase of liver and activates many enzymes, *e.g.*, phosphoglucomutase, choline esterase, the oxidative β-ketodecarboxylases, certain peptidases, and muscle adenosinetriphosphatase.

Copper. Copper deficiency has been observed in infants receiving only milk; the prime manifestation is a microcytic, normochromic anemia. This is also seen in copper-deficient rats; it appears due to failure to absorb dietary iron and readily responds to parenteral administration of iron. Such animals also exhibit a diminution in hepatic cytochrome c and cytochrome oxidase as well as an increase in the iron-binding capacity of plasma. However, a macrocytic anemia, alleviated by copper administration, has been seen in infants who had been ingesting milk fortified with iron and vitamin B_{12}. In continued severe copper deficiency, rats and pigs develop a rickets-like syndrome as well as neurological disturbances. Chronic copper deficiency leads to anestrus in female rats. Earlier, they may mate but gestation terminates in fetal abortion or stillbirth. In lambs, such deficiency results in "swayback," a serious demyelinating disorder of the spinal cord. The arteries of copper-deficient pigs are fragile, owing to failure to make desmosine and isodesmosine with resultant inability to achieve an adequately cross-linked protein structure (page 876). Excessive tissue deposition of copper is seen in man in *Wilson's disease,* with hepatolenticular degeneration and characterized chemically by a deficiency of ceruloplasmin, the copper-binding globulin of normal plasma (page 721).

Known metabolic functions of copper relate to its presence in tyrosinase, uricase, and perhaps butyryl CoA dehydrogenase and cytochrome oxidase. Erythrocuprein, a copper-containing protein of unknown function, occurs in human erythrocytes.

Cobalt. Cobalt deficiency has been observed in ruminants, particularly cattle and sheep, in many areas of the world. All the effects of deficiency appear to reflect failure of synthesis by the rumen microorganisms of vitamin B_{12}, of which cobalt is a constituent (Chap. 49). There have been no reports of the production of cobalt deficiency in nonruminating animals, all of which require dietary vitamin B_{12}. The mechanism for the polycythemia induced by parenterally administered cobalt (page 752) is not understood but appears to be unrelated to the role of cobalt as a constituent of vitamin B_{12}.

Zinc. Zinc deficiency in the rat is manifested by retarded growth, alopecia, and lesions in the skin, esophagus, and cornea. Hogs fed processed peanut meal develop a syndrome called parakeratosis (page 1011), with anorexia, nausea, and vomiting. The disease is readily cured by inclusion of 0.02 per cent $ZnCO_3$ in the diet. The disease occurs only if the diet has been supplemented with calcium; a very high level of dietary calcium similarly precipitates manganese deficiency. Analyses of foodstuffs for zinc are not adequate to indicate their value as dietary sources of this element. Thus soybean, sesame, and peanut (see above) meals all contain significant quantities of zinc, yet this is unavailable when ingested. Eggs from zinc-deficient chickens produce grossly deformed, bizarre chicks. Zinc is a constituent of a number of enzymes, *e.g.*, carbonic anhydrase, alcohol and lactic acid dehydrogenases, and various peptidases. Since zinc is necessary for the activity of these enzymes, it is probably also an essential nutrient for man. However, in view of the wide distribution of zinc in foods, it is unlikely that zinc deficiency will occur in human beings eating an otherwise adequate diet.

Fluorine. The relationship of fluorine to dental caries and mottled enamel was described earlier (page 896). Since there are no other indications of a biological need for this element, classification as an essential nutrient may be largely a matter of definition. In this instance, the nutrient appears to be essential for dental health.

Molybdenum. The status of molybdenum in animal nutrition is not clear. In herbivora, ingestion of minute quantities of molybdenum results in an increased copper requirement and leads therefore, in the absence of sufficient dietary copper, to anemia as well as to lesions of bone and muscle. This effect of molybdenum is observed only when the diet provides inorganic sulfate. Molybdenum deficiency results in a diminution of intestinal and hepatic xanthine oxidase activity in the rat and may be induced readily by inclusion of tungstate in the diet. This observation led to the discovery that molybdenum is a constituent of xanthine oxidase.

Selenium. Diets in which rigorously purified casein served as the source of protein for rats resulted, after about one month, in sudden death due to hepatic necrosis. Many natural foodstuffs were found to protect against this phenomenon. Initially, this necrosis was thought to result from simultaneous deficiency of cystine (page 1008) and of tocopherol (page 1054). However, it was later demonstrated that the presence of a third material (*factor 3*) was required even when tocopherol and cystine were provided. Factor 3 has not been completely characterized but contains selenium. Indeed, selenite added to the experimental rations affords complete protection against liver necrosis. The earlier curative effects of cystine were, at least in part, attributable to traces of selenocystine, $(\text{—Se—CH}_2\text{—CHNH}_2\text{—COOH})_2$, among the naturally occurring amino acids. Some of these nutritional interrelations to vitamin E are described later (page 1055).

Selenium in somewhat larger doses is extremely toxic. Since this element will replace sulfur in cystine and methionine in many plants grown in seleniferous soils, selenium poisoning constitutes a public health and an agricultural problem in certain areas in the north central and southwestern United States.

In sum, deficiencies of trace elements appear to be rare in human nutrition. Exceptions are iodine deficiency in areas in which goiter is endemic and copper deficiency in infants on diets of unsupplemented milk.

REFERENCES

See list following Chap. 50.

49. The Water-soluble Vitamins

HISTORICAL BACKGROUND

In 1816 Magendie introduced the procedure of feeding diets of purified materials to young animals in order to observe the effects of these diets on growth. He recognized that "animals could not remain in health when fed only the staminal principles, the saccharin, oleaginous and albuminous." In 1905 Pekelharing made analogous observations on mice fed rations containing casein, egg albumin, rice flour, lard, and inorganic salts and noted that when this diet was supplemented with very small quantities of milk, the animals remained in good health. The following year, F. G. Hopkins referred to scurvy and rickets as "diseases in which for long years we have had knowledge of a dietetic factor." In 1912, Hopkins and Funk suggested the vitamin theory, *i.e.*, they postulated that specific diseases such as beriberi, scurvy, and rickets are each caused by dietary lack of specific nutritional factors.

The relationship of diet to disease had been indicated considerably earlier. Hippocrates recorded the curative effect of liver on night blindness. Cod-liver oil has been employed for treatment of rickets since the eighteenth century. In his "Treatise on Scurvy," James Lind in 1757 stated that fresh fruits and vegetables are "alone effectual to protect the body from this malady," and half a century later the British navy made routine the provision of lime juice for its seamen. The correlation between the incidence of pellagra and the ingestion of maize was recognized in Italy 160 years ago by Marzari, who stated that the disease resulted from some form of dietary inadequacy. In 1887, Takaki demonstrated that beriberi was prevented by decreasing the amount of milled rice and increasing the amounts of meat, vegetables, and milk in the ration of Japanese sailors. Ten years later, by feeding polished rice to fowls, Eijkman produced a disease similar to beriberi and noted that polyneuritis did not develop in birds fed unpolished rice and that an extract of rice polishings cured avian polyneuritis. These observations led Grijns to state that beriberi is the result of a dietary deficiency. The statements of Hopkins and Funk were extensions of this concept.

The first successful preparation of an essential food factor was a concentrate of a potent antiberiberi substance from rice polishings by Funk. Since the active factor was an amine, and necessary for life, he introduced the term "vitamine." This term has been retained as a name for accessory food factors which are neither amino acids nor inorganic elements. Since not all these substances are amines, the terminal "e" has been dropped. The earliest laboratory demonstrations of the multiplicity of such food factors were the description in 1913 by McCollum and Davis of a lipid-soluble essential food factor in butterfat and egg yolk, and the recognition 2 years later of a heat-labile water-soluble factor in wheat germ necessary for growth of young rats. These were designated "fat-soluble A" and "water-soluble B," respectively.

Although many essential factors have been added to the original list of vitamins, the distinction between lipid- and water-soluble vitamins has been retained since members of each group have certain properties in common. For example, the lipid-soluble vitamins are absorbed from the intestine with the dietary lipids, so that in steatorrhea, deficiency of lipid-soluble but not water-soluble vitamins may result. Perhaps also as a consequence of their lipid solubility, significant quantities of the lipid-soluble vitamins are stored in the liver; storage of the water-soluble vitamins is not significant. Therefore, while it is possible to administer a several weeks' supply of the lipid-soluble vitamins in a single dose, the water-soluble vitamins must be supplied more frequently. Specific metabolic functions have been demonstrated for the water-soluble vitamins which are essential portions of coenzymes, but of the lipid-soluble vitamins, a specific biochemical role has been established only for vitamin A (retinol) in the visual process (Chap. 40).

For many years vitamins were considered to be substances distinct from the major components of food, *i.e.*, carbohydrates, lipids, amino acids, minerals, and water, required in relatively minute amounts for normal nutrition, and whose absence causes specific deficiency diseases. However, niacin can be synthesized from tryptophan. Furthermore, there are substances which are presumed to be vitamins, such as pantothenic acid and biotin, but whose significance in human nutrition has never been established and for which there do not exist corresponding deficiency diseases.

In the early years of vitamin research there was confusion as to the number and nature of the water-soluble B vitamins. When only one B vitamin was known, the antiscorbutic factor was designated as vitamin C and the antirachitic factor as vitamin D. Later, the multiple nature of "vitamin B" was revealed and successively recognized factors were called vitamins B_2, B_3, B_4, etc. Most of these terms have since been discarded, either because the active factor was identified and is now known by a suitable name or because it proved to be identical with some previously recognized factor. The term "B complex" continues to be useful because in some measure these factors are found together in nature. Foodstuffs either rich or lacking in one member of the "complex" are likely to be rich or poor, respectively, in several of the others. Consequently, pure deficiencies of a single member of this group are rare in man. Moreover, the overt manifestations of deficiency of this group overlap in some degree. Thus, in the dog, individual deficiencies of niacin, riboflavin, pyridoxine, pantothenic acid, and folic acid are each characterized by a glossitis with atrophy of the lingual papillae.

THIAMINE

The structure of thiamine, which was first isolated in crystalline form by Jansen in Holland and Windaus in Germany, was established by R. R. Williams and his colleagues.

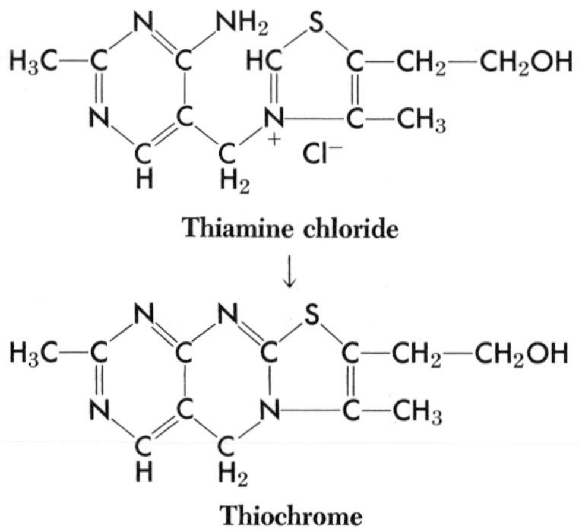

Thiamine is easily converted to *thiochrome* by mild oxidants; the blue fluorescence of thiochrome serves as the basis for measurement of thiamine concentration. Thiamine is relatively stable in acidic solutions but is rapidly inactivated by heating in neutral or alkaline solutions.

Biogenesis of Thiamine. Studies of the synthesis of thiamine by extracts of baker's yeast have revealed that the pyrimidine and thiazole moieties are formed independently. The origins of the 2-methyl-4-amino-5-hydroxymethylpyrimidine and 4-methyl-5-(β-hydroxyethyl)-thiazole shown as the starting materials in Fig. 49.1 have not been established. Thereafter, *hydroxymethylpyrimidine kinase, hydroxymethylpyrimidine phosphokinase, thiazole kinase,* and *thiamine phosphate pyrophosphorylase* catalyze the reactions indicated as *1, 2, 3,* and *4,* respectively. A phosphatase catalyzes removal of the phosphate, completing the synthesis of thiamine. Thiamine pyrophosphate, the coenzyme form of thiamine, is synthesized by direct transfer of the pyrophosphate group from ATP.

$$\textbf{Thiamine} + \textbf{ATP} \longrightarrow \textbf{thiamine pyrophosphate} + \textbf{AMP}$$

The role of thiamine pyrophosphate in the oxidative decarboxylation of α-keto acids and in the transketolase reaction has been discussed previously (pages 331 and 417).

Metabolic Fate of Thiamine. After administration of thiamine to animals, a portion may be recovered in the urine unchanged, and a part as *pyramin* (2-methyl-4-amino-5-hydroxymethylpyrimidine). The latter is presumed to arise from thiamine as the result of the activity of *thiaminase,* present in intestinal microorganisms rather than in the tissues of the host. Normal individuals, ingesting 0.5 to 1.5 mg. of thiamine daily, excrete 50 to 250 μg of the vitamin in the urine. The extent of thiaminase activity may modify the quantitative requirement for thiamine.

Thiamine Deficiency. All animals other than ruminants require a dietary supply of thiamine. In man, "dry" and "wet" beriberi have long been endemic in those areas in which polished rice is a dietary staple. The dry form is preceded by symptoms of incipient beriberi, with subsequent rapid loss of weight and muscle wasting. Marked peripheral neuritis and muscular weakness result in the patient's becoming almost helpless. Deep reflexes are lost, sensory changes may occur, and anxiety states and mental confusion are evident. The heart becomes enlarged. In wet beriberi, there is a generalized edema which may obscure the weakness and muscular wasting. Acute cardiac symptoms may develop rapidly. Since wet beriberi can respond dramatically to thiamine administration with improved cardiac function and massive diuresis, these aspects of beriberi may be specifically ascribed to thiamine deficiency. In the Western world, frank thiamine deficiency is rarely seen except in the Wernicke syndrome of chronic alcoholics; cardiac and respiratory irregularities are associated with a hemorrhagic lesion of the third and fourth ventricles of the brain, similar to lesions which have been observed in thiamine-deficient pigeons and in the brains of foxes with Chastek paralysis. A characteristic sign of thiamine deficiency in birds is head retraction (opisthotonos) which rapidly responds to administration of small quantities of thiamine.

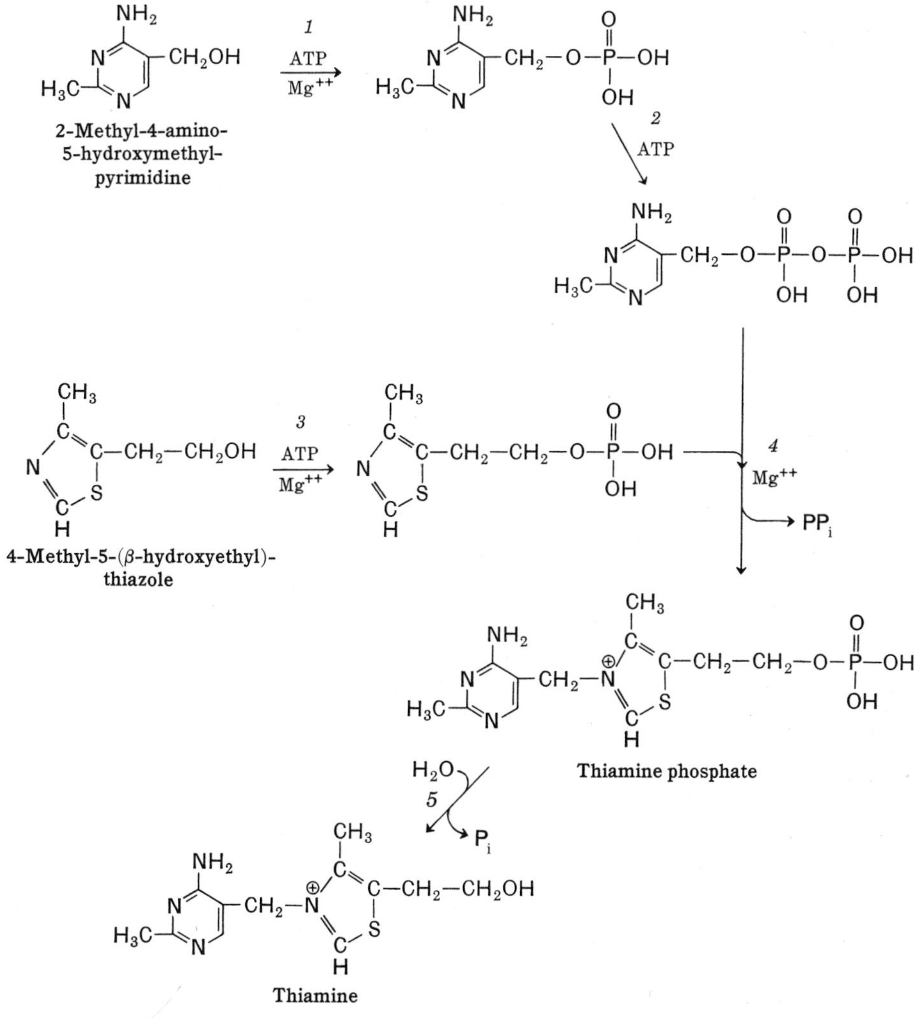

Fig. 49.1. Biosynthesis of thiamine.

The laboratory diagnosis of thiamine deficiency is unsatisfactory. A useful index appears to be the appearance of $^{14}CO_2$ when glucose 2-^{14}C is incubated with erythrocytes. CO_2 arises from glucose only by the phosphogluconate oxidative pathway in erythrocytes. Formation of $^{14}CO_2$ from glucose-1-^{14}C is unaffected in thiamine deficiency but the C-2 atom can appear as CO_2 only if the complete series of reactions is operative, including the thiamine pyrophosphate-dependent transketolase reaction (page 417). The latter has been reported to be limiting before any overt signs of thiamine deficiency are apparent.

Several thiamine analogues rapidly induce signs of thiamine deficiency. These include *pyrithiamine,* in which the pyridine group is substituted for the thiazole ring, and *oxythiamine,* in which a hydroxyl group is substituted for the free amino group on the pyrimidine ring. The effects of each of these antimetabolites are alleviated or prevented by administration of thiamine. The mechanisms by which these compounds exert their actions are unknown. Pyrithiamine inhibits synthesis in chicken erythrocytes of thiamine pyrophosphate and lowers its concentration in the brains of pyrithiamine-treated animals. Oxythiamine appears to displace thiamine in the tissues, since its administration to rats results in increased thiamine excretion.

Distribution of Thiamine. The outer layers of the seeds of plants are especially rich in thiamine. Whole-wheat bread, therefore, is an excellent source, whereas ordinary white bread is a poor source of the vitamin, since most of the thiamine is removed in the milling process. Enriching wheat flour with thiamine restores the original thiamine content, and, because of the quantities consumed, enriched bread is an important dietary source of this vitamin. Most animal tissues are useful dietary sources of thiamine; pork products are especially rich. Although the concentration of thiamine in milk is relatively low, milk is an important dietary source of this vitamin when consumed in large volumes, as in the United States. As indicated in Table 48.2, the recommended dietary allowance of thiamine is approximately 0.5 mg. per 1000 Cal. of diet. This recommendation assumes ingestion of a normal, varied diet. The thiamine requirement, however, varies with the composition of the diet. Both lipid and protein exert a thiamine-sparing action. The sparing action of lipid was thought to reflect a lesser metabolic demand for thiamine, but since the thiamine pyrophosphate content of the tissues of animals fed a high-lipid diet is considerably greater than that of animals on a high-carbohydrate diet, the lipid may in some manner protect thiamine from destruction.

Nutritional surveys indicate that for much of the American population, the thiamine intake is marginal; few adults consume more than 0.8 mg. per day, and many eat appreciably less. Thiamine intake can be augmented relatively cheaply by increased use of peas, beans, and enriched or whole-wheat bread, as well as by improved cooking practices. Prolonged cooking of peas and beans with soda results in destruction of as much as 60 per cent of the original thiamine content, and excessive cooking leaches the water-soluble thiamine from many foodstuffs.

RIBOFLAVIN

Three lines of investigation led to the isolation and identification of riboflavin: attempts to ascertain the nature of the fluorescent material present in milk whey, attempts to isolate a material from milk whey which is an essential dietary factor for the rat and initially designated as vitamin B_2, and isolation of the coenzyme of the "yellow enzyme" (page 362) from red cells.

When irradiated with ultraviolet light in alkaline solution, the yellow enzyme yielded the yellow derivative, *lumiflavin.* Since, under the same conditions, lumiflavin was also obtained from the newly isolated vitamin, the structural relationship of the vitamin and the prosthetic group of the enzyme became apparent. If irradiation is conducted in an acidic medium, *lumichrome,* which exhibits an intense blue fluorescence, is produced instead of lumiflavin. These relationships are shown below.

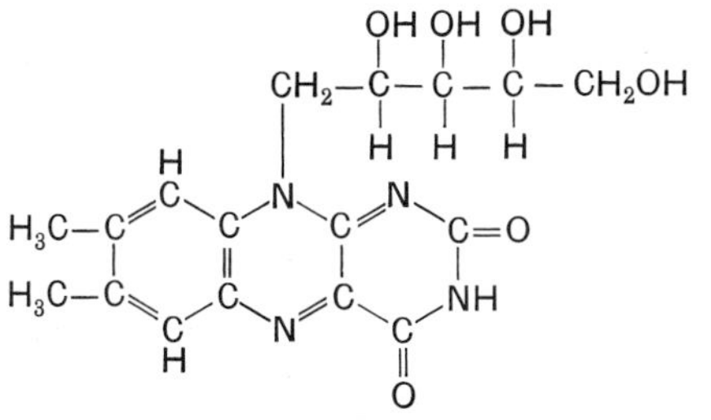

Riboflavin
(6,7-dimethyl-9-(1′-D-ribityl)-isoalloxazine)

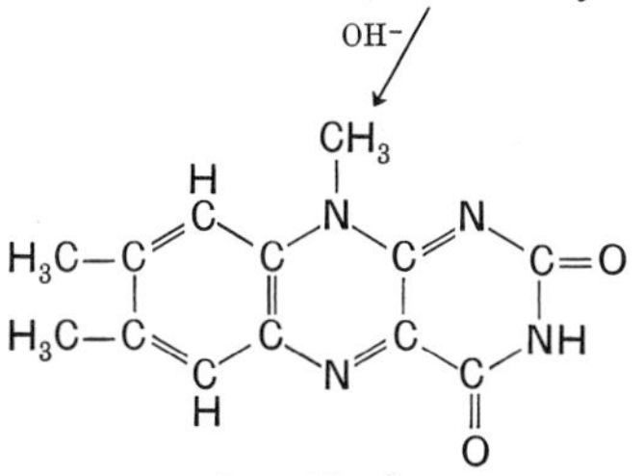

Lumiflavin

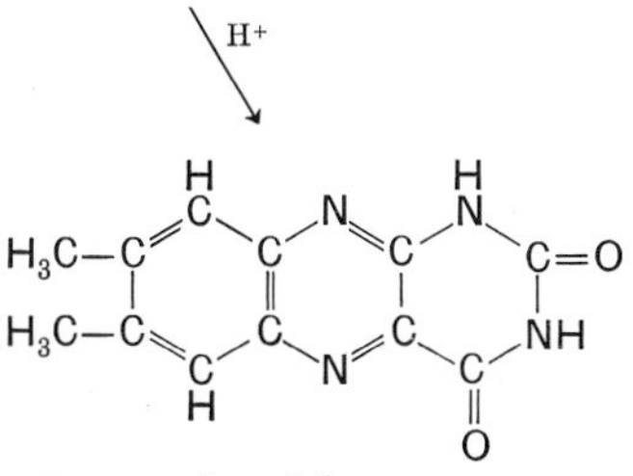

Lumichrome

Thus, the coenzyme role of riboflavin was established simultaneously with its description as an essential food factor.

Biogenesis and Metabolism of Riboflavin. Riboflavin occurs ubiquitously in biological materials. Synthesis of riboflavin is effected by all green plants and by most bacteria, yeasts, molds, and fungi but, insofar as is known, not by animals. The initial stages are identical with those of purine biosynthesis. In organisms which accumulate riboflavin, *e.g.*, the mold, *Eromothecium ashbyii,* intact purines may be utilized for riboflavin synthesis. Except for C-8 of the purine ring, all the carbon and nitrogen atoms of guanosine are found in the final product. The first established intermediate is 6,7-dimethyl-8-ribityl lumazine. Riboflavin synthesis is completed by *riboflavin synthetase,* on the surface of which are binding sites for two molecules of the substituted lumazine. In a remarkable reaction, the diazine ring of one molecule is ruptured and a 4-carbon fragment added to the second molecule, as shown by the distribution of ^{14}C indicated by the asterisks in the diagram on page 1024.

The syntheses of FMN and FAD from riboflavin have been presented previously (page 641). All biological functions of riboflavin appear to be limited to its contribution to the synthesis of these two coenzymes.

Not all bacteria can synthesize riboflavin, and it is therefore an essential

2

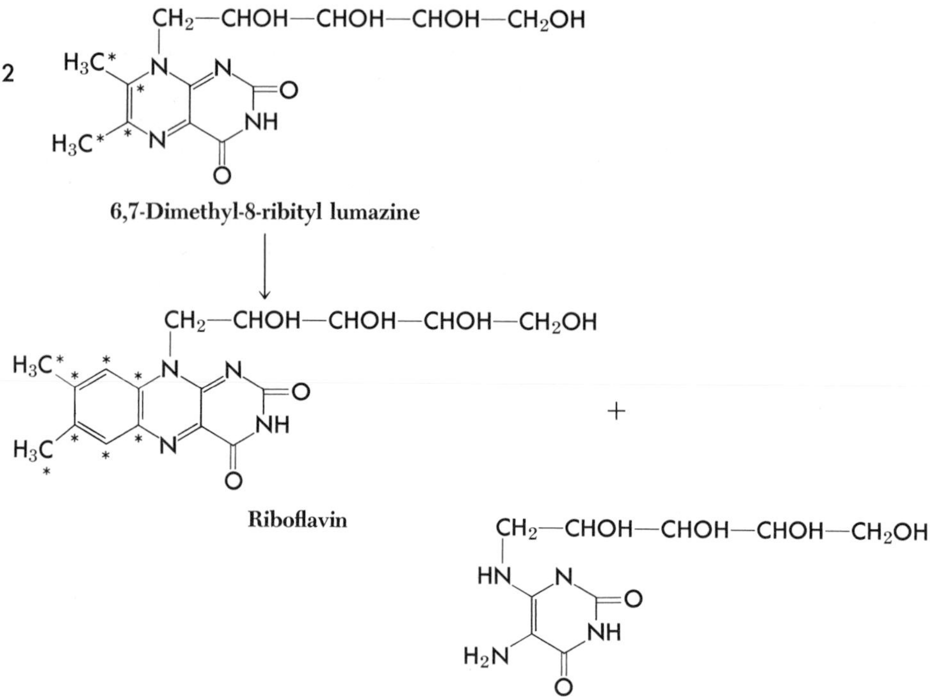

6,7-Dimethyl-8-ribityl lumazine

Riboflavin

4-Ribitylamino-5-amino-2,6-dihydroxypyrimidine

growth factor for a number of bacterial species, particularly lactobacilli. The rate of acid production by these organisms, relative to the quantity of the vitamin in the medium, is the basis of quantitative assays for riboflavin. The simplest assay entails measurement of the fluorescence of riboflavin solutions. In man, ingested riboflavin is largely excreted unchanged or as its phosphate ester, riboflavin 5′-phosphate (FMN, page 362).

Riboflavin Deficiency. Riboflavin is not known to be the prime etiologic factor in a major human disease, although patients with pellagra, beriberi, and kwashiorkor are generally also deficient in riboflavin. Uncomplicated riboflavin deficiency in man is characterized by a magenta-colored tongue, fissuring at the corners of the mouth and lips (cheilosis), a seborrheic dermatitis especially at the nasal-labial folds, and corneal vascularization. Virtually each of these signs may be reproduced in other deficiency states, particularly those due to a lack of niacin and iron, and it is therefore difficult to state to what extent riboflavin deficiency occurs in man.

Riboflavin deficiency in the rat was first observed by Goldberger and Lillie while attempting to produce pellagra in this species. The main effects seen were cessation of growth, accumulation of a dried secretion on the margin of the eyelids, and, frequently, alopecia. In more recent studies of the rat, nerve degeneration, impaired reproduction, and keratitis with cataracts have been observed. Various congenital malformations in young born to riboflavin-deficient rats have also been reported. In the dog, riboflavin deficiency is characterized by impaired growth, followed by ataxia, weakness, bradycardia, and respiratory failure, with death ensuing within 12 hr. of onset of symptoms. Riboflavin-deficient young pigs grow poorly, have coarse hair and rough skin, slightly fatty livers, ovarian degeneration, but normal hemopoiesis, whereas the deficient baboon exhibits both the skin lesions and an anemia.

Despite the fundamental role of the flavin enzymes in metabolism, riboflavin deficiency is not associated with any characteristic chemical changes useful for diagnosis. In general, FMN decreases more rapidly than does the FAD concentration, and both decrease more rapidly in liver and kidney than in heart and brain. In liver the activity of the following enzymes is lost in the following order: glycolic acid, xanthine, D-amino acid, and DPNH oxidases. Riboflavin bears some special relationship to dietary protein; on low-protein diets animals excrete increased amounts of riboflavin, resulting in a decreased concentration of the vitamin in tissues. The concentration of riboflavin in erythrocytes is perhaps the most sensitive indicator of deficiency. Normal whole blood contains 20 μg per 100 ml.

It has not been possible to establish the human requirement for this vitamin. The recommended amount was calculated from the quantity of riboflavin ingested and excreted by individuals on adequate diets. Extrapolations to man have also been made from the ratio of riboflavin to thiamine required by the rat.

There are few common foods which contain high concentrations of riboflavin; liver, yeast, and wheat germ are perhaps most noteworthy, although milk and eggs contribute a large portion of the riboflavin ingested in the usual American diet. Green leafy vegetables are also good sources of riboflavin. In most foods, riboflavin occurs as part of one of the two flavin coenzymes, and only rarely, as in retina and spleen, as the free vitamin.

Riboflavin is obtained commercially from culture media of several molds which produce the vitamin in abundance.

NICOTINIC ACID, OR NIACIN

Pellagra was recognized as early as 1735 by Don Gaspar Casal, physician to King Philip V of Spain. Although largely unknown in the United States until the twentieth century, as many as 170,000 cases were *reported* annually from 1910 to 1935 in the southeastern portion of the United States. Pellagra was established as a deficiency disease by Joseph Goldberger, who also recognized its similarity to canine blacktongue. This observation made possible the identification of nicotinic acid as the responsible factor by Elvehjem, Woolley, and their associates in 1937. Nicotinic acid, which had long been known as a product of the chemical oxidation of nicotine, was recognized as a component of TPN and DPN by Warburg and von Euler, respectively, in 1935 and 1936. Thus the metabolic role of nicotinic acid was known before its nutritional significance had been established. Niacin is the official designation for this vitamin.

Biogenesis and Metabolism of Niacin. Synthesis of niacin occurs in almost all organisms, from plants to man. The pathway for conversion of tryptophan to nicotinic acid mononucleotide in animals has been discussed (page 612). The efficiency of this conversion varies among animal species. Thus, the cat rapidly degrades all tryptophan to smaller metabolites by pathways other than those leading to quinolinic acid (page 614) and, hence, is dependent upon dietary nicotinic acid. This synthetic pathway is not operative in green plants and in many microorganisms which possess an alternate route, the details of which are unknown other than the utilization of aspartate and glycine as starting materials. The pathway leads to quinolinic acid, which is then utilized as in DPN synthesis from tryptophan.

The synthesis and metabolic fate of DPN and TPN have been presented earlier (Chap. 27). Direct synthesis of nicotinamide from nicotinic acid has not been observed in plant or animal systems. The nicotinamide structure arises in the amidation step in DPN synthesis (page 642). The glycosidic bond to the pyridine ring is cleaved in a reaction catalyzed by *diphosphopyridine nucleotidase* (DPNase), yielding nicotinamide and adenosine diphosphate ribose. The nicotinamide may then be hydrolyzed by a *nicotinamidase* and the nicotinic acid reutilized for DPN synthesis. Large doses of nicotinamide result in marked increases in hepatic DPN concentration. The subsequent decline to normal levels is prevented by "tranquilizer" drugs such as reserpine or promazine and by hypophysectomy. The mechanisms of the latter effects are unknown.

Nicotinamide undergoes irreversible methylation in the liver, with formation of N^1-methylnicotinamide (page 554), which is the major urinary metabolic product of niacin. When unusually large quantities of nicotinamide are added to the diet of rats, fatty livers and growth failure ensue, probably as a result of excessive use of methyl groups for formation of N^1-methylnicotinamide. Administration of methionine prevents the fatty liver and causes resumption of growth; choline administration will prevent the fatty liver but does not restore growth. N^1-Methylnicotinamide is oxidized by "aldehyde oxidase" (page 366) in the livers of many mammalian species, including man, with formation of the corresponding 6-pyridone (Fig. 49.2), which is then excreted. Mice also excrete the 2- and 4-pyridones.

Administration of niacin to the dog results in excretion of nicotinuric acid (nicotinoylglycine), although this does not occur in man. Birds excrete dinicotinoylornithine when fed niacin; this is analogous to ornithuric acid excretion after benzoic acid administration (page 580).

Trigonelline, present in the seeds of many plants, is devoid of antiblacktongue activity and is quantitatively excreted on administration to dogs and man. Trigonelline is formed in plants by transfer of the methyl group from S-adenosylmethionine to nicotinic acid. During germination of seeds, trigonelline is demethylated by oxidation of its methyl group, a reaction analogous to the oxidative demethylation of sarcosine (page 583). Some of the transformations occurring in the metabolism of niacin and related compounds are summarized in Fig. 49.2.

While in general, mammals can employ nicotinic acid as well as its esters and amides to satisfy their requirements for this vitamin, many bacteria show somewhat more specific requirements. Some species specifically require nicotinamide; others utilize only preformed nicotinamide ribonucleoside, mononucleotide, or DPN. *Hemophilus parainfluenzae,* which can utilize the ribonucleoside, mononucleotide, or dinucleotide, has been used for the bioassay of these compounds. In species resistant to dietary niacin deficiency, *e.g.*, the mouse and many bacteria, niacin deficiency may be induced by administration of antimetabolites such as pyridine-3-sulfonic acid, 3-acetylpyridine, or 5-thiazole carboxamide.

Niacin Deficiency. Pellagra probably reflects deficiency not only of niacin but also

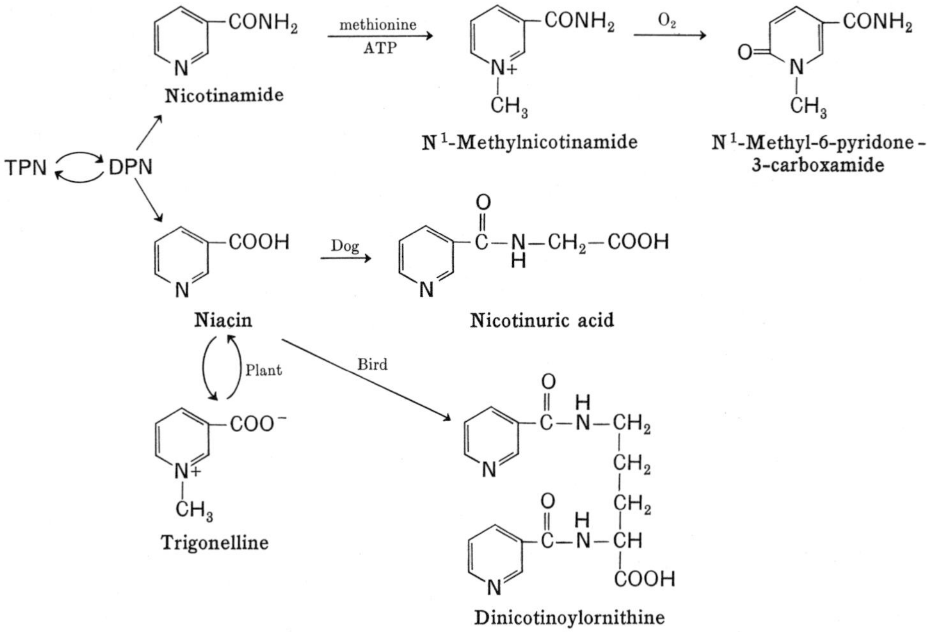

Fig. 49.2. Metabolic fates of nicotinic acid.

of several members of the B complex as well as of dietary protein. The disease is characterized by dermatitis of those areas which are exposed to sunlight, stomatitis, an atrophic, sore, magenta tongue, inability to digest food, and diarrhea. In severe cases the entire gastrointestinal tract may be hemorrhagic. There are frequently disturbances of the central nervous system, leading to dementia. The nervous system symptoms may be due in part to a concomitant thiamine deficiency.

The chief manifestations of blacktongue in dogs fed corn-containing rations are also referable to the gastrointestinal tract: stomatitis, gingivitis, thick, ropy salivation, profuse, bloody diarrhea, and severe dehydration. This syndrome may be alleviated by parenteral administration of large quantities of saline solutions. Thereafter, the dogs lose weight for several months, develop profound anemia, and die, without again showing the characteristic signs of blacktongue. If adult dogs are fed instead a synthetic ration containing no corn, typical blacktongue may not appear, but the animals slowly develop some of the deficiency symptoms described above, and die after several months. Thus canine blacktongue and presumably human pellagra are due to more than simple dietary niacin deficiency.

A possible explanation of these findings is supplied by observations with other species. Although rats on purified synthetic rations containing casein, sucrose, cottonseed oil, and salt mixture do not develop a niacin deficiency, addition to a low-protein diet of cornmeal, an unbalanced amino acid mixture, or glycine results in growth failure or weight loss, which can be alleviated by niacin. Tryptophan serves as well as niacin to support growth under these conditions; corn has long been known to be deficient in this amino acid. Thus, at least in part, the appearance of signs of niacin deficiency is related to the ingestion of an unbalanced amino acid mixture containing inadequate tryptophan. However, this explanation is not entirely satisfactory since niacin deficiency occurs in dogs on diets free of niacin but rich in casein, a tryptophan-containing protein. Addition of *free* tryptophan alleviates the niacin deficiency; therefore the fraction of tryptophan administered as the amino acid which is converted to niacin is much larger than that of dietary tryptophan present in proteins.

The only known role of niacin is formation of DPN and TPN. However, no serious impairment of oxidative reactions has been demonstrated in tissues of niacin-deficient animals, and it is not possible at present to correlate the signs and symptoms of deficiency with known metabolic functions of the coenzymes.

Niacin Requirements. In view of the facts presented above, it is difficult to assess the niacin requirement of man or of other species. Estimates must also consider the total composition of the diet, *i.e.*, the corn content, tryptophan content, etc. Without such knowledge, the recommended allowances shown in Table 48.2 provide an ample margin of safety. The extent of niacin production by the intestinal flora of man is not known.

In animal tissues, all the niacin is present as pyridine nucleotides. The vitamin is widely distributed in plant and animal sources; meat products, particularly liver, are the richest dietary sources. Milk and eggs are almost devoid of niacin; their pellagra-preventive action is probably related to their tryptophan content. Milling procedures used in production of white flour remove most of the niacin, which is replaced by addition of synthetic nicotinic acid. Cereal grains, including corn, contain unidentified substances which are converted to niacin by treatment with alkali, a fact of great nutritional significance in areas such as Mexico and Central America where corn is traditionally treated with lime before being baked into tortillas.

VITAMIN B_6

Vitamin B_6 was defined as that member of the vitamin B complex responsible for the cure of a dermatitis (acrodynia, see below) developed by young rats on a vitamin-free diet supplemented with thiamine and riboflavin. Pyridoxine was isolated from liver and yeast in 1938 and synthesized in the same year. Natural sources contain two other forms of this vitamin, *pyridoxal* and *pyridoxamine.* The three substances as a group are designated vitamin B_6; no single one is considered *the* vitamin, since all three are equally effective in animal nutrition. However, for many bacteria, particularly the lactobacilli, growth is stimulated to a much greater extent by pyridoxal and its phosphate ester than by pyridoxine. It was this observation which led to the discovery of pyridoxal and pyridoxamine by Snell.

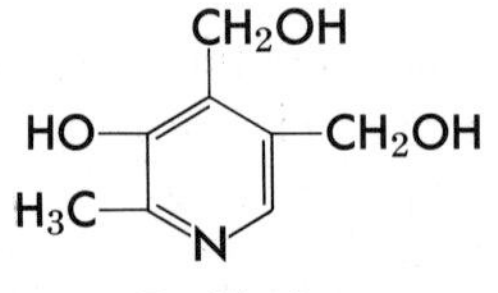

Pyridoxine

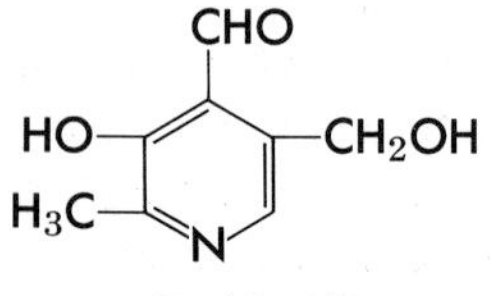

Pyridoxal

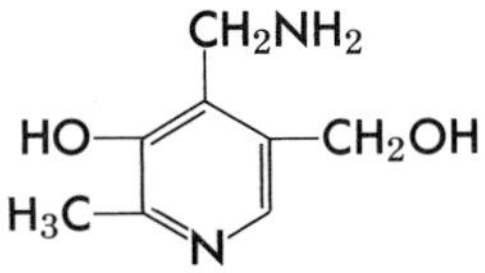

Pyridoxamine

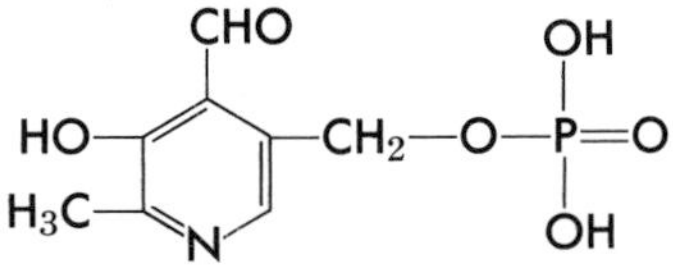

Pyridoxal phosphate

Metabolism of the B_6 Group. Pyridoxine is synthesized by as-yet-unknown reactions in green plants and many microorganisms. In liver, ingested pyridoxine is phosphorylated by a specific kinase and then oxidized to pyridoxal phosphate by a specific flavoprotein. The biological role of pyridoxal phosphate in the metabolism of amino acids (page 542) and its mechanism of action (pages 263, 542) have been discussed. Approximately 90 per cent of the pyridoxine administered to man is oxidized to 4-pyridoxic acid and excreted in this form.

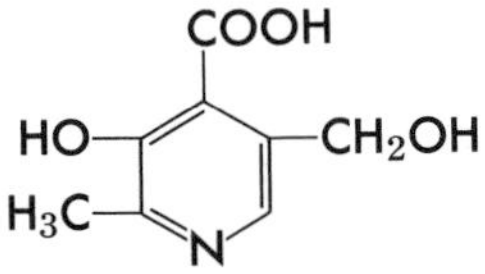

4-Pyridoxic acid

Metabolic Role of Pyridoxal. Pyridoxal phosphate plays a central role in the reactions by which a cell transforms nutrient amino acids into the mixture of amino acids and other nitrogenous compounds required for its own activities. This is most dramatically illustrated by the variation in B_6 requirements in certain bacteria. Some organisms requiring pyridoxine can grow on relatively simple media providing only a few amino acids. However, if the medium is supplemented with a mixture of amino acids, the pyridoxine requirement for maximal growth may be reduced by as much as 90 per cent. In mammals, the situation is somewhat more complex. Unlike bacteria, which absorb from their media only those amino acids which they require, animals may ingest and metabolize considerably greater quantities of amino acids than necessary for growth or nitrogen balance. Consequently, the pyridoxine requirement of animals varies directly with the protein content of the diet. In early pyridoxine deficiency, liver contains an excess of pyridoxal phosphate–requiring apoenzymes. If substantial doses of the vitamin are given, it serves as a stimulus to synthesis of additional apoenzyme; this synthesis was reported to be inhibited by puromycin but not by actinomycin (page 670).

Cells from B_6-deficient animals fail to concentrate amino acids normally, and pyridoxal accelerates amino acid concentration by ascites tumor cells in vitro. These observations led Christensen to postulate a fundamental role for pyridoxal in the active transport of amino acids across cell membranes.

Pyridoxine Deficiency. In the rat, pyridoxine deficiency is characterized by growth failure and *acrodynia,* a dermatitis on the tail, ears, mouth, and paws, accompanied by edema and scaliness of these structures. Other deficiencies lead to somewhat similar lesions, and, indeed, only the edema distinguishes acrodynia from the dermatitis of rats deficient in essential fatty acids. It is possible that this acrodynia, in part, reflects deficiency in polyunsaturated fatty acids because of failure of synthesis of arachidonic acid from linoleic acid (page 495), since pyridoxine-deficient rats accumulate only 10 per cent as much unsaturated fatty acids as their pair-fed controls. Pyridoxine deficiency in young pigs, dogs, and rats leads to microcytic, hypochromic anemia, an increase in plasma iron content, and hemosiderosis. The nervous system is also seriously affected. Rats deficient in B_6 are extremely sensitive to noise and develop epileptiform seizures; the peripheral nerves and spinal cords are demyelinated. Extensive neuropathological changes have also been found in the B_6-deficient monkey, which also develops atherosclerotic lesions.

Pyridoxine deficiency was the cause of an outbreak of convulsions in infants ingesting an inadequate commercial baby food. Similar behavior has been seen in infants with a hereditary disorder characterized by a markedly increased requirement (2 to 10 mg. per day) for pyridoxine. Their convulsions may be relieved by administration of pyridoxine or of γ-aminobutyric acid. This appears to reflect a mutant form of glutamic decarboxylase which requires an unusually high concentration of pyridoxal phosphate for activity.

No specific disease syndrome occurring spontaneously in human adults is known to be due to pyridoxine deficiency. However, deficiency signs are readily induced by administration of deoxypyridoxine and by isonicotinylhydrazide (isoniazid), a drug useful in treatment of tuberculosis.

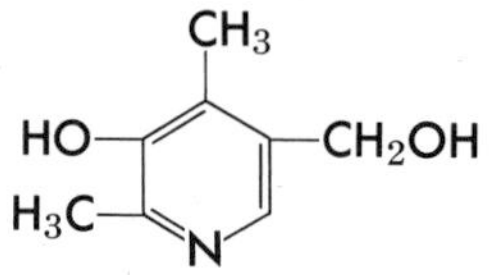

Deoxypyridoxine

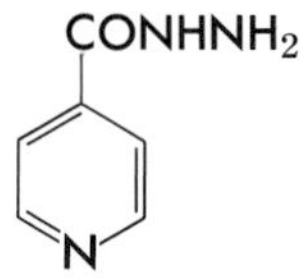

Isonicotinylhydrazide

Deoxypyridoxine is phosphorylated by pyridoxal kinase, and the resulting deoxypyridoxine phosphate competes with pyridoxal phosphate for the active site on specific apoenzymes. Isonicotinylhydrazide forms a hydrazone of pyridoxal and its phosphate ester, rendering these unavailable for enzymic reactions; the hydrazone is excreted in the urine.

Most of the signs and symptoms elicited by these compounds are also associated with other states, *e.g.*, nausea, vomiting, anorexia, seborrheic dermatitis, cheilosis, conjunctivitis, glossitis, polyneuritis, and a pellagra-like dermatitis; all these disappear upon administration of pyridoxine.

Individuals given deoxypyridoxine exhibit oxaluria, presumably due to defective glycine metabolism (page 597), a fall in the plasma concentration of tetraenoic acids compatible with the postulated role of pyridoxal phosphate in fatty acid biosynthesis (page 495), and early excretion of xanthurenic acid. The livers from pyridoxal-deficient rats showed a more pronounced fall in kynureninase activity than of the activity of the transaminase which catalyzes formation of xanthurenic acid (page 615).

The human requirement for vitamin B_6 is not known. Indeed, several studies have revealed that the urinary excretion of 4-pyridoxic acid may exceed the total known intake of vitamin B_6. It is not clear whether this reflects synthesis by intestinal bacteria or by the host. The B_6 group is widely distributed in nature, and those foods rich in other members of the B complex are excellent sources of these materials, *e.g.*, the germs of various grains and seeds, egg yolk, yeast, and meat, particularly liver and kidney.

PANTOTHENIC ACID

This factor was first recognized as a portion of the "bios" complex by R. J. Williams and his colleagues. The significance of this substance in animal nutrition was established by Jukes, Woolley, and their associates.

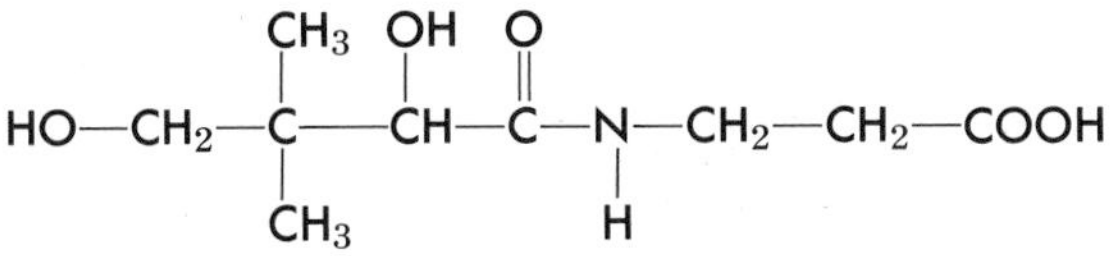

Pantothenic acid (pantoyl-β-alanine)

Metabolism of Pantothenic Acid. Pantothenic acid can be synthesized by green plants and most microorganisms but not by the rat, dog, chick, pig, monkey, mouse, and fox. The biosynthetic pathway in yeast, *Escherichia coli,* and *Neurospora* is shown in Fig. 49.3. Synthesis begins with α-ketoisovaleric acid, which is also the immediate precursor of valine. The donor of the formaldehyde is not established. β-Alanine may be formed by decarboxylation of aspartic acid or transamination of malonic semialdehyde (page 587). Oral administration of pantoic acid together with β-alanine to pantothenic acid–deficient animals is ineffective in correcting the deficiency. Since the minimum requirement for the vitamin is satisfied by the same amount of pantothenate given orally as parenterally, the amide bond must be resistant to hydrolysis in the gastrointestinal tract. The metabolic roles of pantothenic acid are its incorporation into CoA and acyl carrier protein. Nothing is known of the degradative fate of the CoA molecule. Extracts of yeast and animal tissues contain mixed disulfides of CoA and other sulfhydryl compounds, *e.g.*, glutathione or cysteine, but it is not clear whether these occur in living tissue or are artifacts of the isolation procedure.

Pantothenic Acid Deficiency. The nutritional role of this vitamin for man has not

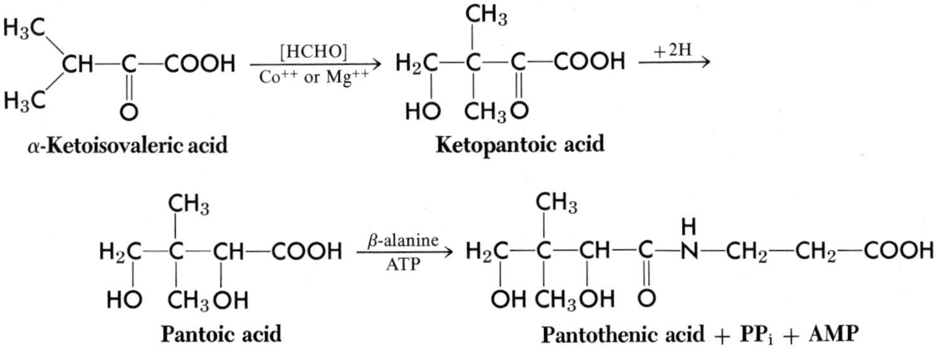

Fig. 49.3. Biosynthesis of pantothenic acid.

been determined, but pantothenic acid is an essential nutrient for all animal species which have been investigated. In the rat, pantothenic acid deficiency is characterized by retardation of growth, impaired reproduction, graying of the hair of black rats, and hemorrhagic adrenal cortical necrosis with resultant adrenal cortical hypofunction. The survival of pantothenic acid–deficient rats may be appreciably extended by administration of adequate quantities of salt or adrenal steroids. These observations are compatible with the reported synthesis of CoA from free pantothenic acid in the zona fasciculata of the adrenal cortex upon administration of ACTH to normal animals, and a 50 per cent reduction in the ACTH-stimulated secretion of corticosterone in pantothenic acid–deficient rats. Porphyrin deposits on the whiskers and under the eyes of pantothenic acid–deficient rats come from the harderian gland and are characteristic of dehydration in this species.

Shortly before death, the liver of a pantothenic acid–deficient rat may contain as little as 50 per cent of the normal CoA content and has an impaired ability to utilize pyruvate and to acetylate *p*-aminobenzoic acid. Among pantothenic acid antimetabolites, pantoyltaurine (thiopanic acid) and ω-methylpantothenic acid have been most frequently employed for production of deficiency states. Administration of the latter to man results in a syndrome consisting of postural hypotension, dizziness, tachycardia, fatigue, drowsiness, epigastric distress, anorexia, numbness and tingling of hands and feet, and hyperactive deep reflexes. However, this syndrome is not relieved by administration of pantothenic acid, although improvement is noted on a "good" high-protein diet, and these disturbances may reflect toxicity of ω-methylpantothenic acid rather than induced deficiency of pantothenic acid.

In general the distribution of pantothenic acid resembles that of the other B vitamins; yeast, liver, and eggs are among the richest sources. Meats and milk are important sources because of the concentration of the vitamin and of the quantities of these foods consumed. Most fruits and vegetables are relatively poor sources. Royal jelly, prepared by the bee colony for the nutrition of the queen bee, and fish ovaries before spawning are the richest known sources of this vitamin. The human requirement for this vitamin is not known.

BIOTIN

In 1936, Kögl and Tönnis isolated from 250 kg. of dried egg yolk 1.1 mg. of a crystalline growth factor which they named "biotin." A few years earlier a factor necessary for the growth and respiration of *Rhizobium* had been obtained and termed "coenzyme R." After the isolation of biotin, it was established that the two factors were identical. Earlier, Bateman had observed that inclusion of large amounts of raw egg white in experimental diets produced toxic symptoms in rats, and in 1926 Boas described the "egg-white injury syndrome" in rats, consisting of dermatitis, loss of hair, and muscular incoordination. She also noted that yeast, liver, and other foodstuffs contained a material which protected rats against egg-white injury. In 1940, György and du Vigneaud and their collaborators established that biotin and the anti-egg-white injury factor were identical; biotin was isolated and its structure established.

The pathway of biotin synthesis in plants and microorganisms is not known.

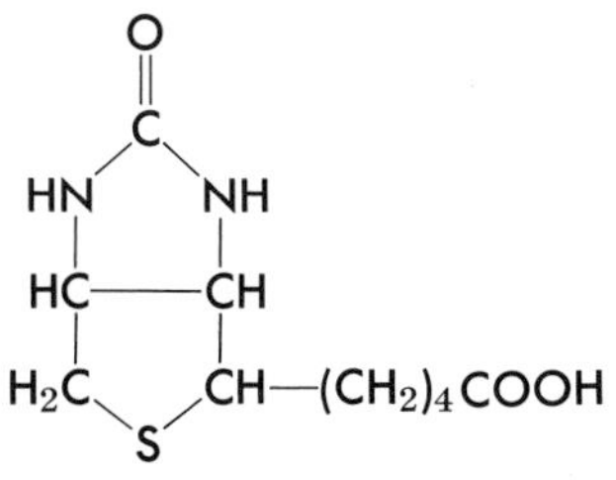

Biotin

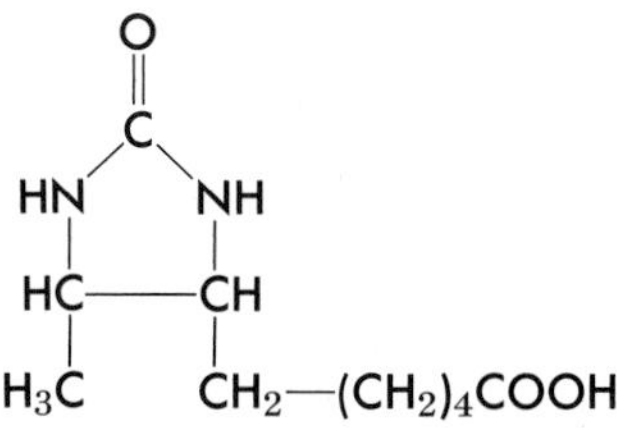

Desthiobiotin

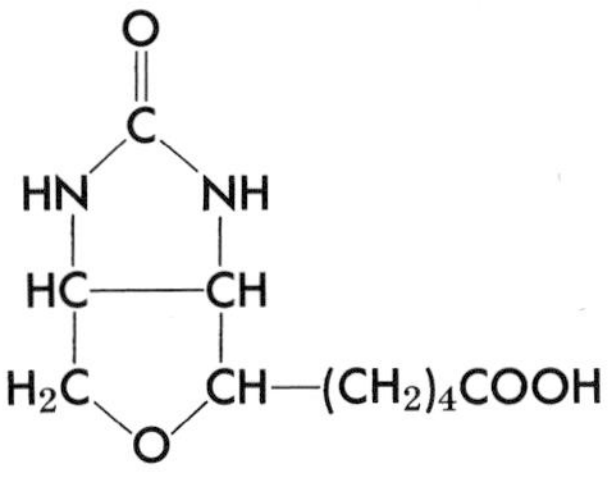

Oxybiotin

Pimelic acid can substitute for biotin in the culture medium of several microorganisms, such as *Corynebacterium diphtheriae* and *Aspergillus niger.* These and other species incorporate pimelic acid into biotin, but the total reaction sequence is obscure. *Desthiobiotin* is the immediate precursor in biotin biosynthesis; it is readily converted to biotin by bacteria. The pathway is repressed by the presence of biotin in the medium; under these circumstances desthiobiotin is not further metabolized. In those enzymes in which it functions as a coenzyme, biotin is bound covalently in amide linkage to the ϵ-amino group of a lysine residue of the apoenzyme. Proteolysis of the enzyme liberates ϵ-biotinyllysine, which has been called *biocytin.*

Biotin serves as the prosthetic group of a series of enzymes which catalyze fixation of CO_2 into organic linkage. Among these are acetyl CoA carboxylase, propionyl CoA carboxylase, methylmalonyl transcarboxylase, and pyruvate carboxylase. Consequently, the tissues of biotin-deficient animals have a reduced capacity to incorporate CO_2 into oxaloacetate, and to synthesize fatty acids. However, these tissues also show reduced activity in urea synthesis, purine synthesis, carbamylation, and tryptophan catabolism. No known biotin-enzymes account for these effects, and in contrast to the specific biotin-containing enzymes cited above, none of these activities, in normal tissue preparations, is inhibited by avidin (see below).

Biotin Deficiency. Biotin deficiency cannot be produced in most animals merely by diets deficient in this nutrient, presumably because of intestinal bacterial synthesis of this compound. However, biotin deficiency has been produced in poultry, monkeys, and calves on synthetic rations. Biotin deficiency results following sterilization of the intestinal tract, the feeding of raw egg white, and the administration of biotin antimetabolites.

Shortly after the description of the effects of feeding raw egg white, Parsons postulated that raw egg white contains a material which combines with biotin and prevents its absorption from the intestine. The active agent has been identified as a protein called *avidin,* with an estimated molecular weight of 45,000. The mode of combination of avidin and biotin has not been established, nor is the normal role of avidin understood. Denaturation of avidin abolishes its biotin-binding capacity.

Oxybiotin can substitute for biotin in the nutrition of most biotin-requiring species, including the rat and chick, and appears to be used per se, rather than being converted to biotin.

Under ordinary circumstances sufficient quantities of biotin are provided to the mammal by intestinal bacterial synthesis; the total urinary and fecal excretion of biotin by man usually exceeds the dietary intake. Biotin is widely distributed in natural products. Beef liver and yeast are among the richest sources, but most animal tissues are low in this factor. Peanuts, chocolate, and eggs contain abundant quantities. On the basis of studies of the biotin requirement of the rat, it has been suggested that the daily human metabolic requirement is approximately 10 μg but there may be no real requirement for ingested biotin.

FOLIC ACID

Existence of this nutritional factor was first suggested by Day, who found that yeast cured a nutritional cytopenia induced in monkeys by corn-containing rations of the type which produce blacktongue in the dog. Potent concentrates were obtained from spinach leaf and this led to the name "folic acid" (L., *folium*). The structure of folic acid is presented below.

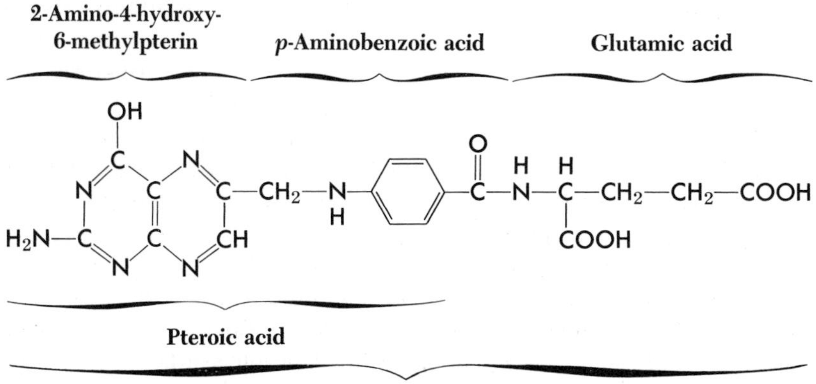

Pteroylglutamic acid (folic acid, PGA)

The molecule contains glutamic acid, *p*-aminobenzoic acid, and a substituted pterin; the combination of the pterin and *p*-aminobenzoic acid is termed *pteroic acid.* The structure shown is the pteroylglutamic acid of liver. The folic acid–active material produced by bacterial fermentation contains three glutamic acid residues combined in γ-glutamyl linkage. Many animal tissues contain pteroylheptaglutamic acid, the glutamic acid residues again being in γ-glutamyl linkage. Synthetic pteroylpolyglutamic acids, in which the glutamic acid molecules are linked in α-glutamyl bonds, are active in bacterial growth assays, while pteroyl-γ-glutamic acids are effective both in bacteria and in the treatment of macrocytic anemia in man. An enzyme in animal tissues hydrolyzes the naturally occurring pteroylpolyglutamic acid compounds to pteroylmonoglutamic acid and free glutamic acid.

Biogenesis of Folic Acid. Although the details of the biological origin of folic acid are lacking, a general picture has emerged, as summarized in Fig. 49.4. Guanosine is directly utilized for pteridine synthesis; in this process, C-8 of the purine ring is lost, whereas several of the carbons of the ribose moiety are retained. The intermediates leading to *dihydroneopterin* are speculative; the latter can be converted to a conjugated pterin, dihydropteroic acid but is not established as an essential intermediate in folic acid synthesis. Also synthesis of dihydrofolic acid from 2-amino-4-hydroxy-6-hydroxymethyldihydropteridine has been achieved with partially purified enzymes. The depicted pyrophosphate ester has not been found as an intermediate, but the synthetic compound is rapidly converted to dihydropteroic acid. Sulfonamides compete with *p*-aminobenzoic acid in the latter reaction.

Pterins are widely distributed; *xanthopterin,* present in many sources, was first isolated from butterfly wings. Insects contain a rich variety of pterins; at least five are present as eye pigments in *Drosophila.*

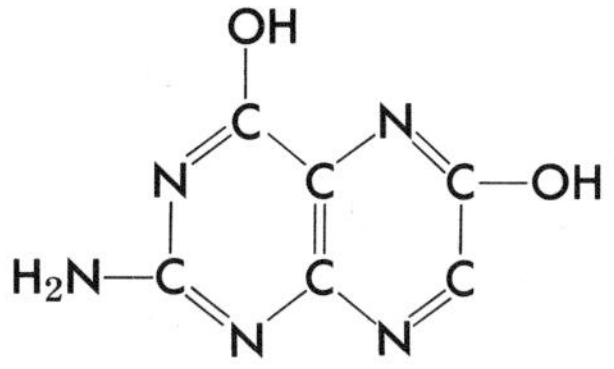

Xanthopterin

Metabolic Role of Folic Acid. The metabolic role of this vitamin has been considered earlier (page 550). As its tetrahydro derivative, it serves as a carrier of the hydroxymethyl and formyl groups. The N-5 methyl derivative of tetrahydrofolic acid is an intermediate in the *de novo* synthesis of the methyl group of methionine (page 553). Although originally synthesized as a dihydro compound and functional as the tetrahydro compound, fully oxidized folic acid is obtained when isolated because of the spontaneous, nonenzymic oxidation of several of these derivatives. As described previously (page 550), mammalian cells reduce oxidized folic acid, or the dihydro compound, to the tetrahydro derivative, utilizing TPNH in a reaction catalyzed by dihydrofolate reductase.

Because of the metabolic role of folic acid, its deficiency is reflected in failure to make the purines and thymine required for DNA synthesis. *Streptococcus faecalis,* which requires folic acid for growth, can be grown readily without this vitamin if the medium contains adenine and thymine. Cells grown in this medium possess normal amounts of DNA but no detectable folic acid, indicating that folic acid is required for purine and pyrimidine synthesis. The effects of analogues of thymine, *e.g.*, 5-bromouracil, are antagonized in bacteria both by thymine and by folic acid. The corollary is also true; growth inhibition by folic acid antimetabolites such as *aminopterin* (4-aminopteroylglutamic acid) may be alleviated in bacteria by adenine and thymine. When administered in large quantities, thymine induces a temporary hemopoietic response in patients with pernicious anemia or sprue. Cultures of both sulfonamide- and aminopterin-inhibited bacteria accumulate 5-aminoimidazole-4-carboxamide (page 624) in their media.

Folic Acid Deficiency. Folic acid deficiency is characterized by growth failure, anemia, leukopenia, or pancytopenia. Deficiency cannot always be induced in ani-

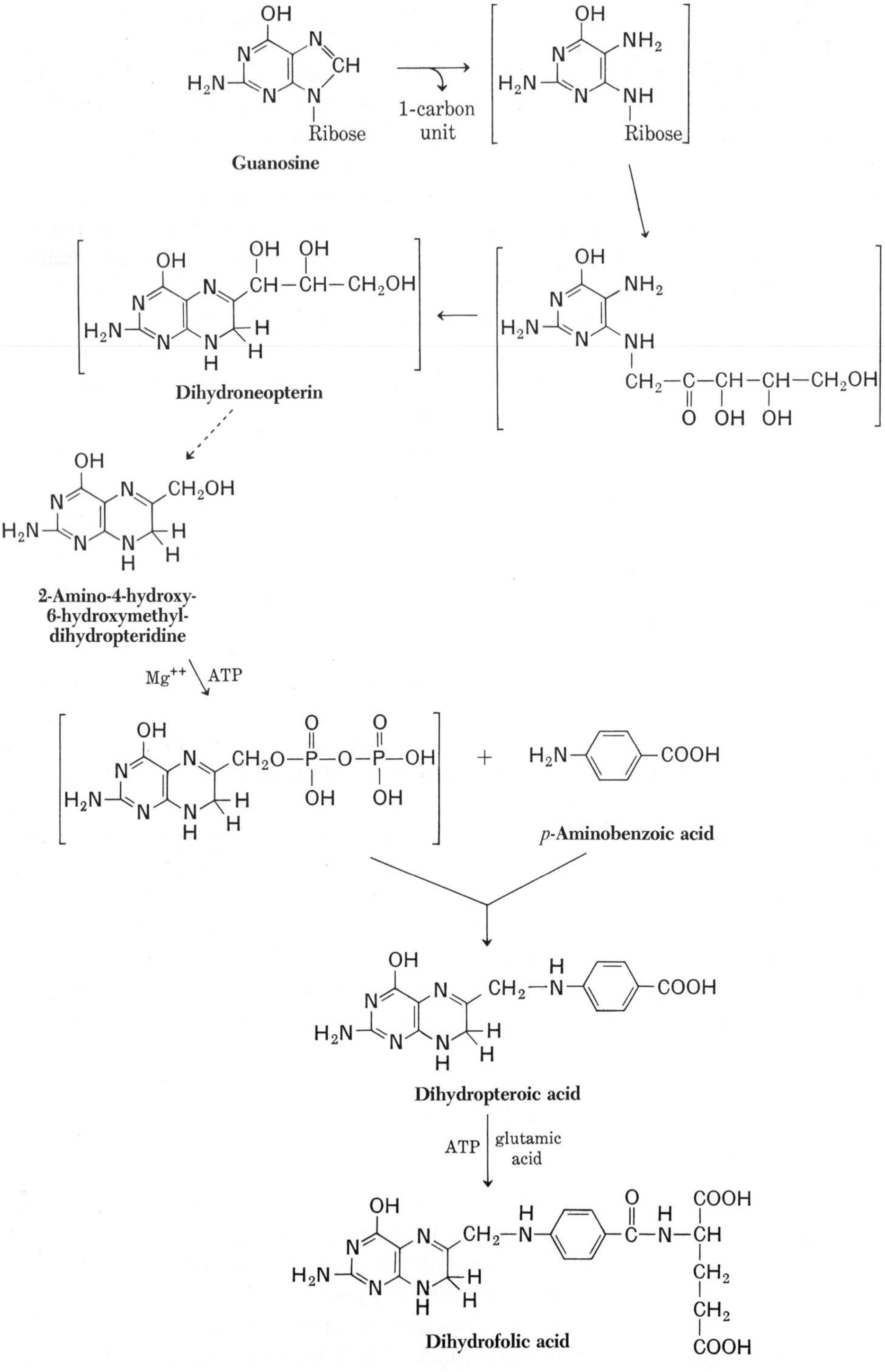

Fig. 49.4. General scheme for the biosynthesis of folic acid.

mals on folic acid–free diets. In the rat, folic acid deficiency can be more readily produced by incorporation of sulfonamides into the diet. This effect, presumably, is a result of inhibition of folic acid synthesis from *p*-aminobenzoic acid by the intestinal bacteria. In other species, deficiency symptoms have been elicited by administration of antimetabolites.

The severe anemias of folic acid–deficient animals led to trials of pteroylglutamic acid in patients with macrocytic anemias of sprue, pregnancy, and infancy. Patients with sprue generally show a striking remission of all symptoms, including the steatorrhea, with a reticulocytosis and progressive return to a normal blood picture. Essentially similar responses are frequently but not invariably obtained in infants and pregnant women with megaloblastic anemias. Thus, these three conditions appear to reflect true nutritional deficiency in folic acid. The megaloblastic anemia derives from a failure of DNA synthesis; the RNA to DNA ratio of the macrocytes is about 0.85, whereas in normal cells this ratio is 0.3. The steatorrhea of sprue is a result of an atrophic jejunum, reflecting the requirement for purines and thymine for DNA synthesis in the constantly dividing cells of this tissue.

A number of antimetabolites of folic acid have been tested for therapeutic value in human leukemia, since folic acid deficiency is characterized by a leukopenia. Aminopterin (see above) and its 9-methyl derivative inhibit the growth of almost all organisms that require folic acid; the inhibition is reversible by relatively large amounts of folic acid. Aminopterin kills rats and mice within 1 week in a concentration of 1 part per million in the diet. The animals develop a watery diarrhea, while the bone marrow rapidly becomes aplastic. Temporary remissions of acute leukemia in children and occasional remissions of lymphoid tumors in adults have been reported with use of this drug.

The status of folic acid in human nutrition is not certain. Although there is little doubt of its indispensability, the requirement for folic acid may be met, in part, by the synthetic activity of intestinal bacteria. Folic acid is widely distributed in the animal and plant world, and nutritional deficiency, therefore, should be infrequent. Although deficiency of folic acid appears to be a factor in the etiology of sprue, as well as in the development of the macrocytic anemia of pregnancy and certain macrocytic anemias of children, these events may reflect either failure to hydrolyze the natural polyglutamate forms or excessive excretion, since megaloblastic anemia observed in individuals ingesting 1 mg per day can be successfully treated with only 25 μg given parenterally.

VITAMIN B_{12}

After the demonstration by Whipple that liver promotes hemopoiesis in anemic dogs, Minot and Murphy, in 1926, showed that feeding large quantities of liver cured pernicious anemia. Later, investigators prepared from liver relatively concentrated extracts suitable for parenteral administration and therapeutically effective. The chief obstacle in subsequent studies was the difficult nature of the assay, *viz.*, the hemopoietic response of patients with pernicious anemia. In 1948, crystalline vitamin B_{12} was obtained independently by E. Lester Smith in England and by Rickes and his colleagues in the United States. The latter group employed a microbiological assay with *Lactobacillus lactis* Dorner, while Smith used the human bioassay procedure. Positive hematologic responses are obtained with doses as low as 3 μg of vitamin B_{12}. A coenzyme form of the vitamin was first isolated by Barker, who found it essential for the interconversion of glutamic and β-methylaspartic acids by extracts of species of *Clostridium* (see below).

The structure of the coenzyme form of vitamin B_{12}, *cobamide coenzyme,* is shown in Fig. 49.5. The central structure is the porphyrin-like *corrin* ring system, in which two of the pyrrole rings are linked directly rather than through a methene bridge as for the other rings and in porphyrins generally. Cobalt is in the position occupied by iron in the heme series. In the cobamide coenzyme shown, the cobalt is divalent; all but two bonds to the cobalt are of coordinate character. One of the coordinate bonds is to the nitrogen of a molecule of 5,6-dimethylbenzimidazole, shown above the plane of the corrin ring, which extends in glycosidic linkage to ribose 3-phosphate. The latter, in turn, is esterified to aminoisopropanol, the nitrogen of which is in amide linkage with the carboxyl group of a propionic acid substituted in the corrin ring. Bonded to the cobalt from below the ring is a molecule of 5′-deoxyadenosine; note that the linkage is from the cobalt to the carbon atom at 5′, the only instance of an organometallic compound in biological systems. Lesser quantities of cobamide coenzymes are found in which the 5,6-dimethylbenzimidazole is replaced by adenine, or 5-hydroxybenzimidazole, among others. The cobamide coenzymes can be isolated only under special conditions. In the presence of anions, particularly cyanide, and of light, the cobalt is oxidized to the trivalent state and the 5′-deoxyadenosine is replaced by the attacking anion. The reaction is easily followed spectrophotometrically since the cyanocobalamin exhibits an absorption peak at 360 mμ which is not evident with the coenzyme forms. Cyanocobalamin is the form in which vitamin B_{12} is ordinarily available, but the cyanide ion may be replaced by a variety of anions, *e.g.*, hydroxyl (hydroxocobalamin), nitrite (nitritocobalamin), chloride (chlorocobalamin), and sulfate (sulfatocobalamin). The derivatives all have comparable biological activities.

Neither animals nor higher plants can synthesize vitamin B_{12}. Indeed, soil microorganisms are the prime source of this vitamin; among the richest sources are activated sewage sludge (50 μg per g.), manure (0.1 μg per g.), and dried estuarine mud (3 μg per g.). Many microorganisms cannot synthesize vitamin B_{12} and require it for growth, thus providing the basis for a microbiological assay for the vitamin.

Biosynthesis of Vitamin B_{12}. *Actinomyces,* which synthesize vitamin B_{12}, make the corrin portion of the vitamin molecule from δ-aminolevulinic acid in a manner analogous to the synthesis of the porphyrin of hemoglobin (page 584). The additional methyl groups present in the vitamin are derived from S-adenosylmethionine, while the aminoisopropanol arises from threonine. The origin of 5,6-dimethylbenzimidazole ribonucleoside is unknown.

Vitamin B_{12} Deficiency. Higher plants do not concentrate vitamin B_{12} from the soil and are a relatively poor source of the vitamin as compared with animal tissues. The requirement for the vitamin is so minute that its wide distribution in foodstuffs and retention by the animal organism would seem to preclude nutritional deficiency in normal individuals. However, deficiency has been observed in individuals who abstain from all animal products, and have blood levels of 40 to 200 $\mu\mu$g of B_{12} per ml.

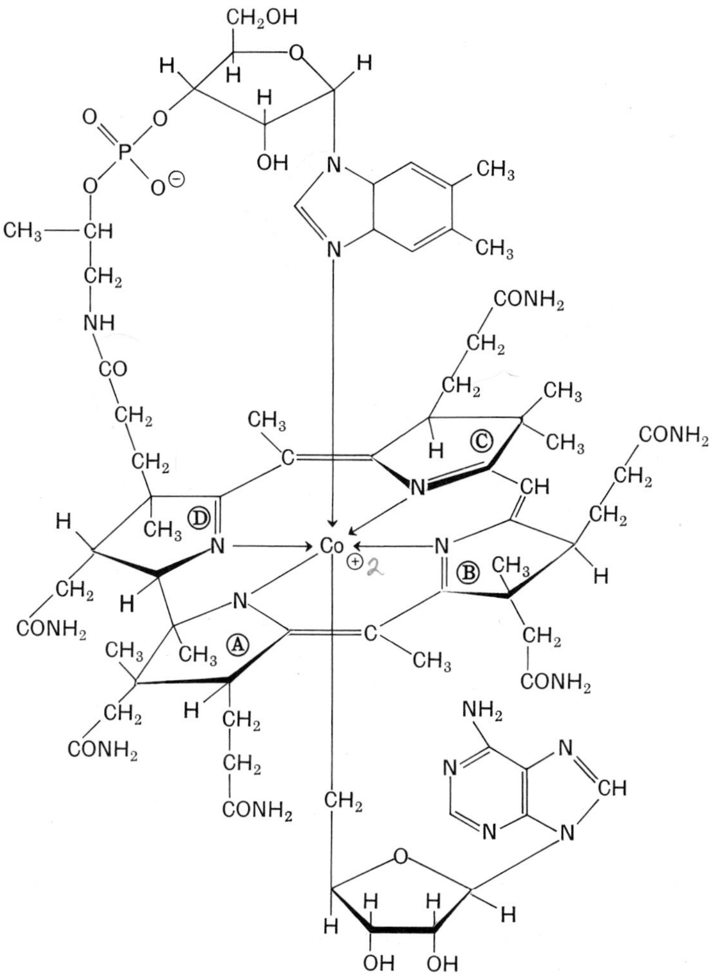

Fig. 49.5. Structure of the 5,6-dimethylbenzimidazole cobamide coenzyme.

as compared with normal values of 200 to 350 $\mu\mu$g per ml. Occasionally such individuals develop the hematological and neurological syndrome of pernicious anemia without the gastric disturbance usually responsible for this disease (see below).

Pernicious anemia results not from inadequate ingestion of vitamin B_{12}, but from defective gastric secretion (page 752). The "intrinsic factor" of gastric juice is required for the successful absorption of ingested vitamin B_{12} unless the vitamin is given in enormous excess. Indeed, the stool of patients with pernicious anemia is a rich source of the vitamin; cobalamin given orally to such individuals may be recovered quantitatively from the stool unless given with intrinsic factor. This provides the basis for a useful diagnostic test for pernicious anemia. About 0.5 μg of cyanocobalamin labeled with ^{60}Co is given orally; normal individuals excrete 5 to 30 per cent of the test dose in the stool, whereas persons with pernicious anemia excrete 75 to 95 per cent of the dose. When given with a source of intrinsic factor, normal absorption is observed. No amount of intrinsic factor can significantly increase intestinal absorption above that observed in normal persons.

Gastric juice contains a mixture of mucoproteins which bind vitamin B_{12} in varying degrees. These proteins have been fractionated and assayed for both binding capacity and ability to serve as intrinsic factor. Highly concentrated and effective preparations have been obtained; however, it is not entirely certain that a specific single mucoprotein of the gastric juice is, uniquely, intrinsic factor. There are structural differences among intrinsic factor from various species; rats respond only to rat intrinsic factor preparations, and there is evidence that human intrinsic factor is more effective in man than is that from other species. The B_{12}-intrinsic factor complex enters ileal cells; B_{12} slowly leaves in the portal blood, but the mucoprotein is either hydrolyzed or returned to the intestinal lumen. The exact role of intrinsic factor is not clear. The vitamin B_{12}-depleted tissues of persons with pernicious anemia, in relapse, take up administered vitamin B_{12} from plasma more slowly than do tissues of normal individuals; intrinsic factor preparations accelerate the uptake of vitamin B_{12} by liver slices in vitro.

Estimates of the daily requirement for cyanocobalamin are of the order of 1 μg, but there is no adequate basis for establishing this figure. The difficulty inherent in assessing this value is apparent from the fact that persons who have undergone almost total gastrectomy and no longer secrete intrinsic factor show the earliest signs of pernicious anemia only 3 to 5 years postoperatively. At sufficiently high dietary levels, *e.g.*, 200 μg per day, vitamin B_{12} can be absorbed in amounts sufficient to alleviate pernicious anemia, even without intrinsic factor.

Metabolic Role of Cobamide Coenzymes. Cobamide coenzymes participate in a series of seemingly diverse reactions. The reaction in *Clostridium tetanomorphum* studied by Barker, which led to discovery of these coenzymes, was the formation of β-methylaspartic acid from glutamic acid.

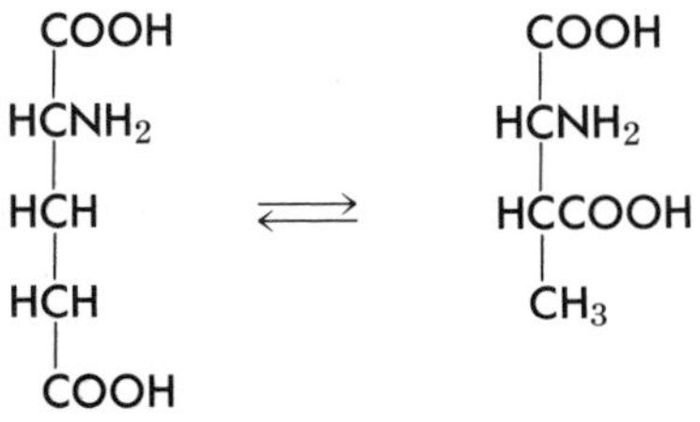

Glutamic acid **β-Methylaspartic acid**

An analogous reaction is the cobamide-catalyzed interconversion of succinyl CoA and methylmalonyl CoA (page 484). Since C-2 of methylmalonyl CoA becomes C-3 of succinyl CoA, it is evident that it is not the free —COOH carbon that migrates but rather, as indicated previously (page 484), the entire —CO—S—CoA group.

Cobamide coenzymes also participate in the dismutation of vicinal diols, *e.g.*, propane-1,2-diol, to the corresponding aldehyde.

$$CH_3CHOHCH_2OH \longrightarrow CH_3CH_2CHO$$

Propane-1,2-diol **Propionaldehyde**

Each of these reactions is of the form

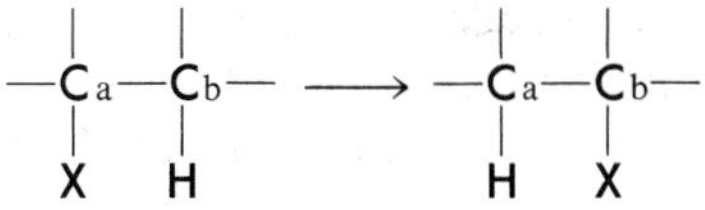

In the reactions cited above, X is, respectively, —CHNH$_2$—COOH, —CO—S—CoA, and —OH. In no instance does the migrating hydrogen exchange with protons in the medium, but it is clear that this hydrogen is transferred to and from the coenzyme. For example, if tritium-labeled propanediol and unlabeled ethylene glycol are simultaneously incubated with enzyme and coenzyme, tritium is essentially equally distributed in the resultant propionaldehyde and acetaldehyde. The deoxyribityl moiety of the deoxyadenosine probably serves as the temporary acceptor group. It is not known whether this requires rupture of the C—Co bond or whether reaction may occur as shown in Fig. 49.6 for the methylmalonyl CoA mutase reaction. Conceivably, coenzyme B$_{12}$ functions similarly in the system for reduction of ribonucleotide triphosphates to the corresponding deoxyribonucleotide triphosphates, the carbanion formed in the reaction then being more readily reduced by thioredoxin (page 638). An entirely different mechanism must be operative when coenzyme B$_{12}$ functions in the methylation of homocysteine to methionine (page 584) or of dUMP

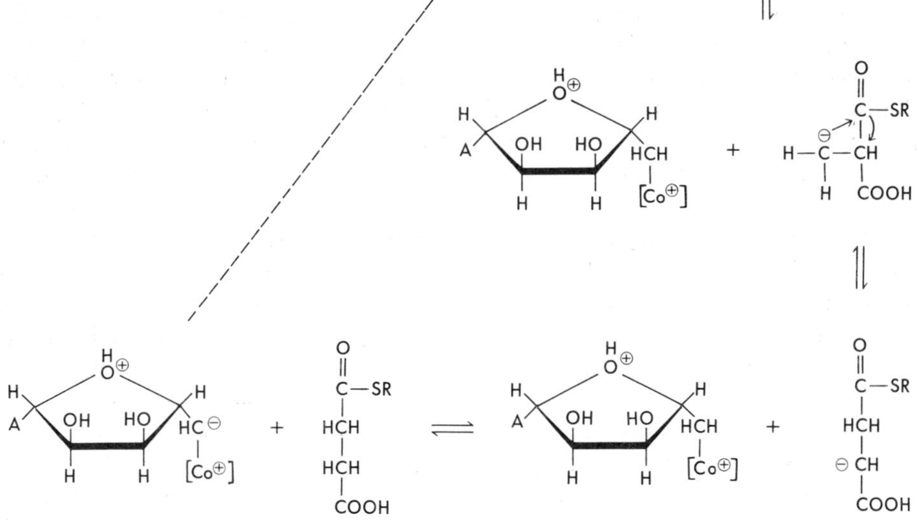

Fig. 49.6. Proposed mechanism for the methylmalonyl CoA mutase reaction. (*After H. P. C. Hogenkamp, Federation Proc.,* **25,** 1623. 1966.)

to TMP (page 639). Here the reactive form of the coenzyme is generated by a 2-electron reduction, yielding a form known as vitamin B_{12s}, in which the cobalt is monovalent. In this form it is readily alkylated by agents such as methyl iodide to yield methyl-B_{12}. The same form, generated in appropriate enzymic systems by transfer from N^5-methylfolate, is the immediate alkylating agent for homocysteine (page 553).

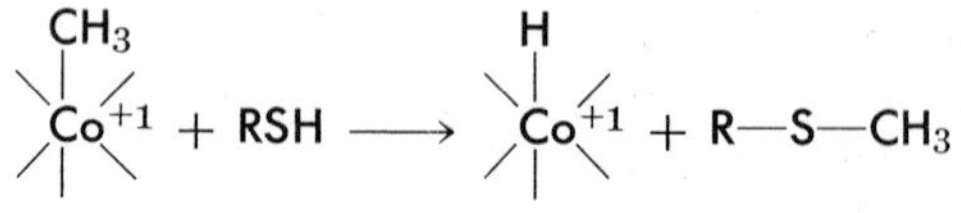

The participation of vitamin B_{12} in these reactions accounts for the ability of rats to grow on diets lacking methionine but containing homocysteine (or homocystine) and vitamin B_{12}, as well as for the limitations placed upon TMP synthesis by a deficiency of vitamin B_{12}, with resulting macrocytic anemia.

INOSITOL

Inositol may exist in one of seven optically inactive forms and as one pair of optically active isomers. Only one of these forms, *myo*-inositol (page 32), possesses biological activity.

In 1940, Woolley described alopecia in mice on inositol-deficient diets. Deficiency of this compound has also been stated to cause a "spectacled-eye" condition in rats, due to denudation about the eyes. The chick, turkey poult, pig, hamster, and guinea pig have also been reported to require inositol as an essential dietary nutrient. These observations are difficult to reconcile with the fact that when glucose-^{14}C is given to the rat, inositol-^{14}C may be isolated from its tissues, indicating an unknown synthetic pathway. However, whether or not inositol is an essential nutrient for the intact animal, studies in tissue culture have revealed that it is required for growth of fibroblasts and various strains of human cancer cells in specific experimental conditions. In yeast, inositol formation from glucose is catalyzed by two specific enzymes. A *cylase* catalyzes formation of inositol 1-phosphate from glucose 6-phosphate in the presence of DPN^+. The phosphate is removed by hydrolysis catalyzed by a specific *phosphatase.* The latter can attack the phosphoester of the axial 1-hydroxyl but has no effect on a similar ester of the equatorial 2-hydroxyl.

In special circumstances, inositol is a lipotropic agent, effectively reducing the liver lipids of rats ingesting a lipid-free, low-protein diet. However, whereas choline is effective when the low-protein diet also contains lipid, inositol is relatively ineffective in these circumstances. Although inositol-containing phospholipids are widely distributed, their relation, if any, to the lipotropic action of inositol is unexplained.

The rat can metabolize relatively large quantities of inositol; after its administration only minute amounts are recovered in the urine. Labeled inositol gives rise to labeled glucose in the urine of phlorhizinized rats. Inositol is also an antiketogenic substance. The initial step in its metabolism is oxidation by an oxygenase (page 378) with formation of D-glucuronic acid (page 429).

Inositol is widely distributed in plant and animal cells. Large quantities are found in mammalian cardiac muscle and in the skeletal muscle of sharks. It has been suggested that in the shark, inositol may have a reserve function similar to that of glycogen in other species. In plants, mono-, di-, and triphosphate esters of inositol are found in large quantities. The hexaphosphate ester, phytic acid (page 32), is found in high concentrations in grains and in the nucleated erythrocytes of birds. Dietary phytic acid is rachitogenic because formation of the insoluble salt, calcium phytate, in the gastrointestinal tract prevents normal absorption of dietary calcium (page 887). Seeds, beans and nuts contain inositol phospholipids linked glycosidically to oligosaccharides.

CHOLINE

Choline deficiency, like that of niacin, cannot be produced in animals receiving diets adequate in protein. Nevertheless, for purposes of discussion choline may be classified as a B vitamin. At least two of the disturbances in methionine-deficient animals appear specifically to relate to the need for choline.

The synthesis of lecithin in the livers of choline-deficient animals is slower than in normal animals although the rate of addition of lecithin to the plasma by the liver is approximately the same in the two groups. It is not clear why this should result in accumulation of lipid in the liver. The oxidation of long-chain fatty acids to CO_2 is impaired in liver slices from choline-deficient rats. Choline deficiency, however, is not the sole basis for fatty livers in animals on a low-protein diet, since, in the presence of limiting amounts of choline, several amino acids, notably threonine, also exert lipotropic activity. The prolonged presence of excessive lipid in the hepatic cells of animals on low-protein rations leads to necrosis and fibrosis resembling Laennec's cirrhosis. However, choline, methionine, and inositol therapy in human cirrhotic subjects has not been effective.

The pathogenesis of the hemorrhagic kidneys of choline-deficient rats is obscure. These lesions can be produced only in young rats shortly after weaning. A relation of impaired phospholipid metabolism is assumed on the basis of the observation that the lesion is produced only during that period of a few days after weaning when the rate of renal phospholipid synthesis is maximal. The hemorrhagic kidneys somewhat resemble those of "lipoid nephrosis" but are not actually identical with any known disease entity in man. Most affected rats die in uremia. Choline administration, even at the height of the disease, appears to cure many of the animals; severe hypertension and renal arteriolosclerosis may develop some months later in such animals because of heavy fibrosis of the renal capsule.

Chronically choline-deficient rats also develop anemia and hypoproteinemia, frequently with edema. These symptoms are not due to the low-protein diet used to produce choline deficiency, since the animals respond rapidly to choline administration. In poultry, deficiency in choline, as well as many other nutrients (Mn^{++}, glycine, arginine), results in perosis. Betaine is an effective substitute for choline in all species; dimethylaminoethanol is effective in place of choline in the chick but not the rat. Presumably, the demand for choline and methionine is influenced by the amount of folic acid and vitamin B_{12} also present in the diet. Choline itself is widely distributed as a constituent of lecithin. Glandular tissues are excellent dietary sources of this compound, as are nervous tissue, egg yolk, and a few vegetable oils, *e.g.*, soybean oil.

ASCORBIC ACID

In 1907 Hölst and Frölich demonstrated that guinea pigs develop scorbutic lesions on a diet of oats and bran. In 1932 King and Waugh isolated the crystalline vitamin, which had previously been termed vitamin C and has been renamed ascorbic acid.

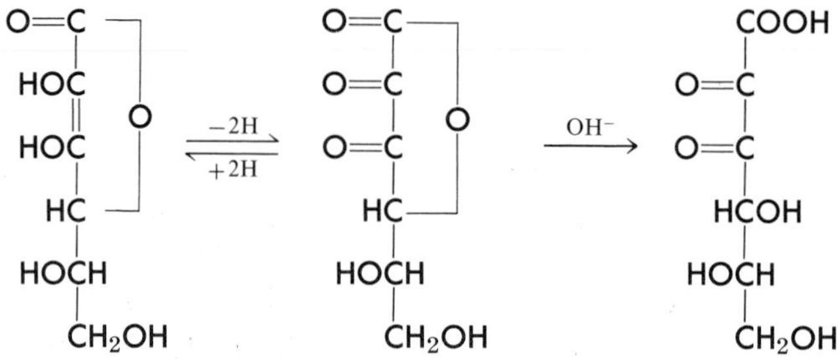

L-Ascorbic acid   L-Dehydroascorbic acid   L-Diketogulonic acid

The most prominent chemical properties of the vitamin are its acidity, due to the enolic hydrogen at C-3, and its ready oxidation to dehydroascorbic acid, catalyzed by small concentrations of metal ions. Dehydroascorbic acid is unstable in alkali, undergoing hydrolysis of the lactone ring, with formation of diketogulonic acid. Dehydroascorbic acid is readily reduced by agents such as H_2S, cysteine, and glutathione. The reducing power of ascorbic acid is the basis for most of the quantitative analytical procedures for estimation of this compound, such as titration with the dye 2,6-dichlorophenolindophenol.

Biogenesis and Metabolism of Ascorbic Acid. Ascorbic acid is a dietary essential for man, other primates, and the guinea pig but can be synthesized by all other species of animals which have been investigated. The major steps in the synthesis of ascorbic acid in animals appear to be those indicated on page 1045. Note that the two forms of L-gulonic acid represent the same structure rotated 180° in the plane of the paper. Primates and guinea pigs are thought to be unable to convert ketogulonolactone to ascorbic acid. Administered ascorbic acid-^{14}C is readily oxidized to $^{14}CO_2$ by the guinea pig, whereas in the human being, only diketogulonate-^{14}C and oxalate-^{14}C have been obtained. Oxidation to dehydroascorbic acid is catalyzed by a specific, copper-containing *ascorbic acid oxidase* present in plants but not in animals.

Metabolic Role of Ascorbic Acid. It was initially assumed that the metabolic role of ascorbic acid relates to its reversible oxidation and reduction. *No biological oxidation system has yet been described in which ascorbic acid serves as a specific coenzyme.* Nevertheless, ascorbic acid is essential to a variety of biological oxidation processes. Although the vitamin activates the oxidation of *p*-hydroxyphenylpyruvic

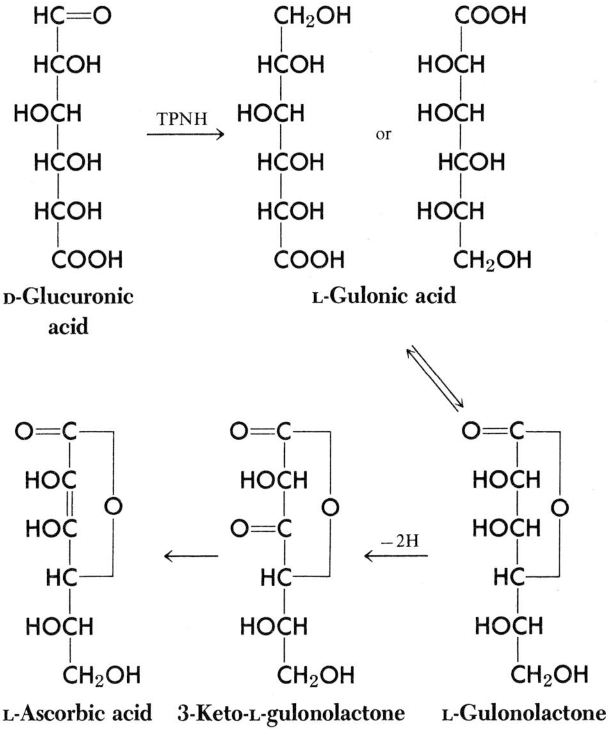

acid by rat liver homogenates (page 606), other reducing agents are equally effective, and the partially purified enzyme shows no dependence on ascorbic acid. In the presence of oxygen, solutions of ferrous ion plus ascorbate catalyze hydroxylation of a series of aromatic compounds. Thus, *p*-hydroxyphenylacetic acid yields homogentisic acid while tryptamine yields 5-hydroxytryptamine (serotonin). It is not known whether this system is operative physiologically. As indicated earlier (page 875), hydroxylation of proline and lysine in collagen synthesis is also catalyzed by a system which appears to require Fe^{++} and ascorbate.

Suggested roles for ascorbic acid in biological oxidations have been described (page 383). Ascorbic acid also facilitates removal of iron from ferritin; since the plasma iron of the anemic, scorbutic monkey is reduced to about 30 per cent of normal, this may contribute to the genesis of the anemia. The oxidized form, dehydroascorbic acid, may be enzymically reduced in animal tissues, with glutathione as the reducing agent. Ascorbic acid exists both free and in a combined form in animal tissues. In dietary deficiency of the vitamin, the free form tends to decrease in the tissues before appreciable changes occur in the concentration of bound ascorbic acid.

Ascorbic Acid Deficiency. The manifestations of ascorbic acid deficiency are largely due to functional failure in mesenchymal cells. Scurvy in adults is characterized by sore, spongy gums, loosening of the teeth, impaired capillary integrity with subcutaneous hemorrhages and edema, joint pain, anorexia, and anemia. This disease is presently rare in the Western world. However, scurvy in children is still seen occasionally as the result of poor feeding practices. Symptoms include tender, swollen joints, limitation of motion, petechial hemorrhages, inadequate tooth development, arrested skeletal development with characteristic bone disease, impaired wound healing, and anemia. Except for the anemia, impaired formation of collagen and chondroitin sulfate is the basis for all these changes. The collagen that is deposited is stated to be poor in hydroxyproline. In severe scurvy there is failure of formation of new ground substance and extensive depolymerization and solution of existing cement material, leading to breakdown of old, healed wounds. The anemia of scurvy may relate to impaired ability to utilize stored iron as well as to a secondary impairment in folic acid metabolism.

Ascorbic acid deficiency in man has been studied under controlled conditions. Enlarged keratotic hair follicles were apparent after 17 to 20 weeks; these became hemorrhagic several weeks later. The buttocks and calves were first affected, but later these lesions appeared over the entire body. After 26 weeks, hemorrhages in the gums and reduced rate of wound healing were evident. The plasma concentration of ascorbic acid per 100 ml. fell from 0.55 to less than 0.05 mg. within a few weeks, while the concentration of ascorbic acid per 100 ml. in white blood cells slowly fell from 16 to less than 1 mg. In general, scorbutic signs were apparent about 1 month after the vitamin concentration in white cells had fallen to its lowest level. Ten milligrams of ascorbic acid per day prevented these changes, while administration of 20 mg. per day to subjects showing signs of scurvy results in complete disappearance of these signs within 3 weeks. As a result of these studies, the British Medical Research Council concluded that an ascorbic acid supply of 30 mg. daily could be considered a liberal allowance. This value is less than half of that recommended by the Food and Nutrition Board of the National Research Council (Table 48.2). The higher recommended allowance, however, includes consideration not only of the appearance of overt scurvy but also of factors of well-being which cannot be readily evaluated. There is little doubt that the recommended intakes shown in Table 48.2 afford a generous margin of safety.

In man, the normal plasma ascorbic acid concentration is 0.7 to 1.2 mg. per 100 ml. However, these plasma values reflect dietary intake rather than potential deficiency in that levels below 0.5 mg. per 100 ml. are frequently encountered in persons who ingest 30 or 40 mg. of ascorbic acid daily, an amount which seems adequate for man. The concentration of ascorbic acid in the white blood cells may more closely reflect the concentration of ascorbic acid in tissues generally than does the amount in plasma. In "saturation tests," doses of about 500 mg. of ascorbic acid are administered; normal individuals excrete at least 50 per cent of the dose within 24 hr., while severely deficient subjects may excrete very little of the administered vitamin.

Fresh fruits and vegetables are important dietary sources of ascorbic acid. Cooking, canning, and other food-preparation procedures may destroy a portion of the ascorbic acid in foods. Therefore, tables showing the ascorbic acid content of various foodstuffs should be evaluated in terms of the manner of food preparation. In general, green leafy vegetables are excellent sources of ascorbic acid; new potatoes contain relatively large quantities. Because of the large quantities of potatoes consumed, these are an important source of the vitamin in the Western diet. The high concentrations in citrus fruits are well known. Some of the ascorbic acid may be lost in canning fruit juices but is largely preserved in frozen juices. In general, the ascorbic acid content of meats is relatively low, and cooking of meat destroys the vitamin.

REFERENCES

See list following Chap. 50.

50. The Lipid-soluble Vitamins

VITAMIN A

The structures of vitamin A (retinol) and of several carotenoids were presented on page 76. Vitamin A activity in mammals is exhibited by α-, β-, and γ-carotenes, by retinol and retinol$_2$ (Chap. 40), and by a few other carotenoids such as cryptoxanthin. The carotenes are useful nutritionally only insofar as they may be converted to a retinol. Symmetrical β-carotene yields, physiologically, two molecules of retinol per molecule. In α- and γ-carotenes, as well as cryptoxanthin, one of the two rings differs from that of retinol; on a weight basis these are, nutritionally, half as effective as retinol or β-carotene. Other carotenoids, such as *lycopene* or *xanthophyll,* in which neither ionone ring is identical with that of retinol, are devoid of vitamin A activity. Retinol$_2$, containing an additional double bond in the ionone ring (page 906), the form found in fresh-water fish, is as active as retinol$_1$ in promoting the growth of vitamin A–deficient animals. Generally, A_1 is found in the all-*trans* configuration but some marine crustacea contain the *neo b* (11-*cis*) form. Although pure retinol and carotene are available as standards, the vitamin A activity of foodstuffs is expressed in international units; one international unit (1 I. U.) of vitamin A is equivalent to the activity of 0.6 μg of β-carotene. The double bond systems of all carotenoids are readily oxidized by atmospheric oxygen with loss of all vitamin A activity. In natural foodstuffs protection is afforded by the presence of "antioxidants" such as vitamin E (page 1054).

Biogenesis and Metabolism of Vitamin A. The distribution of carbon atoms in retinol and the carotenes conforms to that expected from polymerization of isoprenoid units. In species in which these syntheses occur, *e.g.*, all photosynthesizing plants, the process involves the same units utilized in sterol synthesis, *i.e.*, mevalonic acid and isopentenyl pyrophosphate, but details are lacking. The rapid increase in hepatic retinol concentration after administration of β-carotene to mammals, and the apparent failure of this conversion when β-carotene was given to persons with severe liver disease, had suggested that the liver is the site of formation of retinol from carotene. Studies of this type are facilitated by the fact that, in anhydrous media, retinol reacts with $SbCl_3$ to form a vividly blue compound with an absorption maximum at 620 mμ whereas the equivalent β-carotene compound is green with an absorption maximum at 590 mμ. However, the intestinal mucosa rather than the liver is the major locus of conversion of β-carotene to retinol. Parenteral administration of β-carotene fails to alleviate deficiency symptoms in the vitamin A–depleted rat and the carotenoid is deposited, as such, in the liver. It is not clear, therefore, why individuals with liver disease have markedly impaired capacity to convert carotene to retinol.

When β-carotene, labeled with tritium on the two central carbons of the side chain connecting the ionone rings, is converted to retinol in the rat, the tritium is retained. The process is catalyzed by a iron-containing dioxygenase, followed by reduction of the resulting retinal by DPNH.

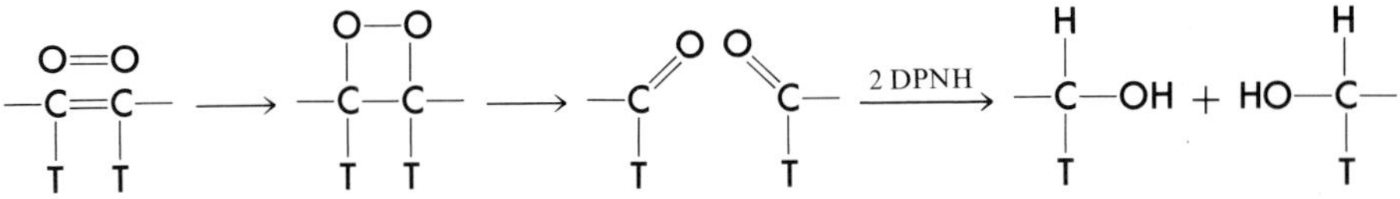

The soluble dioxygenase from rat intestinal mucosa is maximally active in the presence of detergents and is inhibited by reagents that bind iron and by sulfhydryl reagents; thus the enzyme resembles many other dioxygenases (page 377). Retinol is esterified to long chain fatty acids in the mucosa and the retinyl esters are then transported via the lymphatics in association with chylomicra to the liver.

Retinyl esters, the form present in ingested liver and fish-liver oils, are hydrolyzed in the intestine, reesterified to palmitate, transported from the intestine with chylomicra, and stored in hepatic Kupffer cells. Many tissues contain a relatively specific retinyl ester esterase. Except for the role of retinol in vision (page 901), the chemical form in which vitamin A exerts its physiological effects is not known. Retinoic acid (in which the alcohol function of retinol has been oxidized to a carboxyl) readily replaces retinol in the rat diet; this acid promotes growth of bone and soft tissues, and sperm production, but it cannot be used in the visual process and will not permit maturation of embryos. Retinoic acid is converted by the rat to an unknown form which is several times as active as the parent compound in the usual vitamin A nutritional assays. Excess retinoic acid cannot be stored and appears in the bile as the corresponding glucuronide. The ultimate metabolic fate of retinol and its derivatives is not known. These relationships may be summarized as follows:

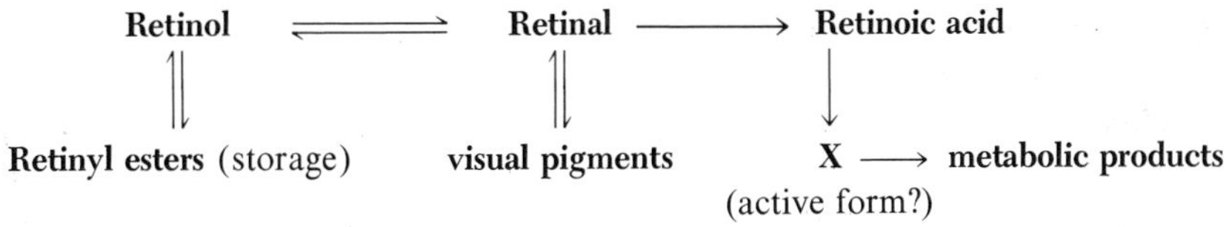

Vitamin A deficiency has a profound effect on virtually every organ. However, except for its role in the visual process (Chap. 40) and a possible role in chondroitin sulfate synthesis, referred to below, the metabolic function of vitamin A is unknown.

Vitamin A Deficiency. The sequelae of vitamin A deficiency show little species variation. Young animals invariably fail to grow; epithelial cells in all structures undergo keratinizing metaplasia. The best-known changes are in the eyes, giving rise to *xerophthalmia.* Increased susceptibility to infection of the respiratory tract is frequent. The cells of the kidney medulla may cornify and renal calculi form. The ducts of glands of all types may become blocked and the glands atrophy. Atrophy of the germinal epithelium causes sterility in the male. The vaginal smear in the female shows cornification of cells; nevertheless, normal ovulation and implantation occur, but normal young are seldom born to the vitamin A–deficient animal because of defective placentas. Rat fetuses removed by cesarean section from vitamin A–deficient females show various abnormalities. Mammary tissue of vitamin A–deficient animals exhibits an increased sensitivity to estrogens.

The earliest sign of vitamin A deficiency in man and experimental animals seems to be night blindness. There is no relation between the visual defect in night blindness and the generalized lesions of xerophthalmia. Wald was able to observe the ultimate consequences of vitamin A deficiency in the rat retina by supplying retinoic acid to protect the animal in other respects. After 5 months, the visual threshold (the luminance of a 0.02-sec. flash which evokes a perceptible electroretinogram) was increased to 2,500 times normal. In 10 months, the rats were blind. The visual cells decreased in number, and the outer segments of the rods disintegrated and then disappeared. At this time the damage could no longer be reversed by vitamin A administration.

Cessation of skeletal growth is an early manifestation of vitamin A deficiency; the bones resemble those of animals in which growth has ceased because of inanition. Failure of skeletal growth appears to reflect a defective synthesis of chondroitin sulfate which has been reported to be due to inability to make phosphoadenosine phosphosulfate (page 563), and has also been suggested to be the consequence of excessive lysosomal sulfatase activity. The early failure in the growth of endochondral bone results in lesions in the central nervous system, because of continued growth of nervous tissues after arrest of development of the bony casing. Increased cerebrospinal pressure occurs before other changes are manifest.

Hyperkeratotic papules around the hair follicles have been considered a characteristic finding in vitamin A–deficient human beings, leading in extreme cases to a "toad skin." Similar skin lesions do not appear in animals fed a diet adequate in all respects except for vitamin A. However, if the young rat is placed on a diet deficient both in vitamin A and in most of the B complex, analogous changes in the skin do occur as the innermost layers of the dermis atrophy. The skin lesions of man with vitamin A deficiency may occur only in the presence of a concomitant deficiency of the vitamins of the B complex.

Evidence of vitamin A deficiency is more difficult to produce in adult than in young animals because of two factors. First, the changes characteristic of vitamin A deficiency, particularly in the skeleton and nervous system, can be observed only in growing animals; this is also true, to some extent, with respect to epithelial changes. Second, the store of vitamin A present in the liver of normal, well-fed animals may be sufficient to meet the demands of an adult for some years. The newborn, however, does not possess a hepatic store of vitamin A. Xerophthalmia thus affects large numbers of children but relatively few adults. Indeed, tens of thousands of children, eighteen to thirty-six months old, largely in tropical countries, annually lose their vision because of this disease. Many of these also have kwashiorkor, and it may be this combination which renders these children so susceptible to infection.

In general, only impaired dark adaptation has been observed in volunteer human subjects on a vitamin A–poor diet for more than 2 years. Various reports have shown conflicting results with respect to the ease with which even impaired dark adaptation may be produced. In one study, only one subject in a group of 71 showed persistent diminution in dark adaptation during a 2-year period. In other subjects, dark adaptation impairment was variable; occasionally adaptation fell to low levels but then returned to normal, even though there had been no increase in the vitamin A allowance.

Vitamin A Toxicity. When the young rat is overdosed with vitamin A, the bones become extremely fragile, leading to numerous spontaneous fractures. The findings in children who have received overdosages of as much as 500,000 units of vitamin A per day over a considerable period are not usually so dramatic, *viz.*, tender swellings over the long bones, with limitation of motion, and definite hyperostoses. Human adults receiving 500,000 units per day show calcification of pericapsular, ligamentous, and subperiosteal structures similar to that seen in rats. At still higher dosage levels, severe headache, nosebleed, anorexia, nausea, weakness, and a dermatitis become evident within a few days. In studies of the effects of excess vitamin A on rudimentary limb buds in tissue culture, bone formation ceased, and existing cartilage was destroyed by release of lysosomal hydrolytic enzymes in the cartilage cells. Presumably similar phenomena explain the grossly malformed offspring produced by administration of massive doses of vitamin A to the pregnant rat. These manifestations of vitamin A excess, like those of iron (page 749) and of vitamin D (page 1053), are the penalty to the animal for an inability to excrete surplus quantities of these materials.

Vitamin A in Human Nutrition. Fruits and vegetables supply approximately 65 per cent of the vitamin A of the American diet. About half this vitamin intake is derived from leafy green and yellow vegetables; lettuce, spinach, chard, escarole, carrots, and sweet potatoes are particularly good sources of a mixture of carotenoids. The extent of absorption of these carotenoids by man and their conversion to retinol is not known. Steatorrhea results in fecal excretion of a large fraction of the dietary vitamin A and carotenoids. Measures which improve gastrointestinal absorption of lipids, such as prior emulsification of the lipid mixture, also improve absorption of vitamin A.

Oxidation of vitamin A and the carotenes may result in significant destruction of these compounds present in the usual diet; this can be prevented by an adequate concentration of vitamin E (see below). Fish-liver oils are the most potent natural sources of vitamin A; depending upon the species of fish and the time of year of catch, fish livers contain 2,000 to 100,000 vitamin A I.U. per gram. In contrast, human livers contain 500 to 1,000 I.U. per gram. The vitamin A content of cod-liver oil is the lowest of all fish-liver oils used as commercial sources of this vitamin; those of the shark and halibut are among the highest.

Experiments designed to establish the minimum adequate dietary level of vitamin A for adult men have been inconclusive because of the large hepatic stores of the vitamin in adults. The recommended allowances shown in Table 48.2 are liberal, and deficiency is unlikely if the vitamin A intake remains at these levels.

VITAMIN D

It had long been recognized that rickets is prevalent in areas where children are exposed to a minimum of sunlight. Experimental rickets can be produced in the rat by a diet of natural foodstuffs low in calcium. Ultraviolet irradiation of such diets prevented subsequent development of rickets; the ultraviolet-sensitive compound was identified as 7-dehydrocholesterol. It was observed that irradiation of yeast ergosterol yielded an extraordinarily potent antirachitic agent. However, whereas irradiated 7-dehydrocholesterol was as active in curing rickets in the chick as in the rat, irradiated ergosterol was much less effective in the chick than in the rat. This indicated that the active materials produced by irradiation of the two sterols were different. Irradiation of ergosterol yields a series of isomeric sterols, one of which, *calciferol,* is markedly antirachitic and is designated vitamin D_2. Further irradiation inactivates calciferol. Irradiated 7-dehydrocholesterol is designated vitamin D_3. The material for which the term vitamin D_1 was originally reserved proved to be a mixture of calciferol and other sterols, and this designation is no longer employed. The structures of these natural precursors and their biologically active photoderivatives have been given previously (page 80). There may be other sterols, active as vitamin D, distributed in nature. Thus, some fish oils contain a ketonic sterol which, in its enolic form, is about twice as active as calciferol.

Of the series of compounds obtained from irradiation of ergosterol, only calciferol (vitamin D_2) possesses antirachitic activity. However, one of the series, *tachysterol,* may be catalytically reduced to dihydrotachysterol, which is antirachitic. Irradiation of 7-dehydrocholesterol in the skin accounts for the antirachitic efficacy of sunlight and ultraviolet irradiation. Commercial vitamin D preparations are obtained by irradiation of ergosterol. By convention, one international unit of vitamin D is the activity of 0.01 ml. of average medicinal cod-liver oil; it is approximately equivalent to 0.05 μg of calciferol.

The presence of vitamin D in fish-liver oils has not been explained. If it is assumed that vitamin D can arise only through solar irradiation of sterols, then fish liver contains vitamin D only because fish ingest preformed vitamin D in their diet. The ultimate source of the vitamin in fish-liver oils would therefore appear to be the plant and animal plankton near the sea surface. However, assay of such materials has consistently given negative results, and the problem remains unsolved.

Since milk, the chief source of calcium in the diet, is widely consumed by children, this foodstuff was chosen for fortification with vitamin D. Fortification may be accomplished by irradiation with ultraviolet light, a universal practice in the preparation of evaporated milk, or by addition of vitamin D concentrates. In the former procedure, activated 7-dehydrocholesterol is produced; in the latter, calciferol is added. The vitamin D content of milk can also be increased by feeding irradiated yeast to cows; a portion of the calciferol fed appears in the milk. Since fish-liver oils are not normally included in the diet and no other foodstuffs contain quantities of vitamin D, the growing child ordinarily relies upon irradiation by sunlight to provide vitamin D. Prior to the era of routine administration of fish-liver oils or concentrates to infants, rickets was common among children of the northern latitudes, where winter is long and the sun may be obscured for several consecutive months.

Vitamin D in Metabolism. All the manifestations of vitamin D deficiency are due to defective bone formation. As indicated earlier (page 887), vitamin D mediates absorption of calcium from the intestine. Administration of vitamin D_3 stimulates formation, in cells of the intestinal mucosa, of a Ca^{++}-binding protein thought to function in intestinal Ca^{++} transport. The vitamin was shown to stimulate formation of mRNA in intestinal tissues. Actinomycin prevented formation of both mRNA and of the Ca^{++}-binding protein.

Stimulation of intestinal Ca^{++} transport, however, is not an adequate explanation for the effectiveness of the vitamin in the treatment and prevention of rickets. Rachitic cartilage fails to calcify when incubated in serum from rachitic animals even if calcium and phosphate are added, yet calcification occurs readily in serum from normal animals with identical concentrations of calcium and phosphate. Further, healing of rickets is initiated much more rapidly than can be accounted for on the basis of the additional calcium absorbed in the first several days after vitamin D therapy is initiated. At present, it can be stated only that in the presence of dietary vitamin D, calcification of the matrix of bone occurs, while in vitamin D deficiency this does not take place.

Rickets is a disease of growing bone. The last stage of bone growth, deposition of inorganic bone minerals in the newly formed matrix, fails to occur, but matrix formation continues. The provisional zone of calcification is no longer clearly demarcated but is irregular and deformed. Bowlegs, knock knees, rachitic rosary (a beaded appearance of the ribs), and "pigeon breast" are skeletal deformities seen in rachitic children.

Although severe rickets is seldom seen in the United States, mild early rickets may be prevalent. Histological evidence of rickets was found in almost 50 per cent of 700 children dying of all causes in the Johns Hopkins Hospital during the period 1940 to 1942. Since the newborn infant has virtually no reserve stores of vitamin D, vitamin D supplementation or adequate exposure to sunlight must be provided.

The ability of mammals to store vitamin D is limited. However, single doses can be administered which provide a several weeks' or months' supply of vitamin D. The fate of administered vitamin D in the mammal is not clear. Some is excreted in the bile, but the sum of urinary and fecal excretion is usually considerably less than the amount ingested. When calciferol-^{14}C is given, about 15 per cent is found in the liver, much smaller amounts appear in other tissues, and the major portion of the administered material is degraded to unidentified products.

Toxicity of Vitamin D. Excessive quantities of calciferol given either to children or to adults produce demineralization of bone, and multiple fractures may occur after only minimal trauma. Serum concentrations of both calcium and phosphate are markedly elevated, resulting in metastatic calcification of many soft tissues and formation of renal calculi. The latter may block the renal tubules sufficiently to cause a secondary hydronephrosis; frequently it is the renal disease which leads to the need for medical assistance. The basis of these effects of vitamin D overdosage is not known.

VITAMIN E

Rats grown exclusively on cow's milk are incapable of bearing young. The factor in plant oils, particularly that from wheat germ, which restores fertility in male and female rats was designated vitamin E. This vitamin inhibits both the development of rancidity and oxidative destruction of vitamin A in natural fats. Once rancidity is initiated, however, the products of the oxidation of the unsaturated fatty acids destroy vitamin E.

Vitamin E was isolated from wheat-germ oil in 1936 and given the name *tocopherol* (Gk. *tokos,* childbirth; *phero,* to bear; *ol,* an alcohol). Seven tocopherols, derivatives of the parent substance *tocol,* 2-methyl, 2-[trimethyl-tridecyl], 6-hydroxy-chromane, have been found in nature; of these, α-tocopherol has the widest distribution and greatest biological activity.

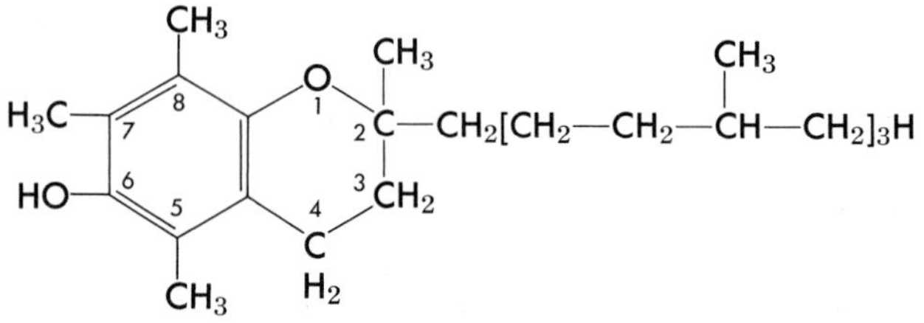

α-Tocopherol

The structures of six tocopherols are summarized below.

Tocopherol	*Substituents*
Alpha	5,7,8-Trimethyltocol
Beta	5,8-Dimethyltocol
Gamma	7,8-Dimethyltocol
Delta	8-Methyltocol
Eta	7-Methyltocol
Zeta	5,7-Dimethyltocol

Vitamin E Deficiency. The manifestations of vitamin E deficiency in laboratory animals vary considerably, and, at first observation, the diverse disturbances appear to have no common basis. The classical manifestation in the rat is infertility, but the reasons for this infertility differ in males and females. In the female rat, there is no loss in ability to produce seemingly healthy ova, nor is there a defect in the placenta or uterus. However, some time after the first week of embryonic life, fetal death occurs, and the fetuses are resorbed. This can be prevented if vitamin E is administered at any time up to the fifth or sixth day of embryonic life. In the male, the earliest observable effect of vitamin E deficiency is immotility of spermatazoa; later, there is degeneration of the germinal epithelium. However, there is no alteration in the secondary sex organs or diminution in sexual vigor, although the latter may disappear with prolonged vitamin E deficiency.

Other striking changes in the vitamin E–deficient rat include degeneration of the renal tubular epithelium, depigmentation of the incisors, and, if an unsaturated lipid like cod-liver oil is included in the diet, the appearance in lipid depots of a brown pigment. Rats fed diets deficient in vitamin E and protein develop acute hepatic necrosis, which may be prevented by cystine, methionine, tocopherol, or factor 3 (see page 1017). This hepatic necrosis occurs only if the diet contains appreciable quantities of unsaturated lipid. Another manifestation of vitamin E deficiency is hemolysis of red cells in vitro in the presence of peroxide or alloxan derivatives such as dialuric acid. Vitamin E deficiency in rats ultimately produces muscular dystrophy with progressive paralysis of the hind legs, decreased muscle creatine concentration, creatinuria, and a slight decline in creatinine excretion. Signs of vitamin A deficiency may also develop because of the destructive oxidation of vitamin A in the ration in the absence of the antioxidant properties of vitamin E.

Herbivorous animals such as the rabbit and guinea pig are more sensitive to vitamin E deficiency than is the rat even when their diets contain relatively little unsaturated lipid. In these species, muscular dystrophy develops very rapidly, and the animals may succumb after a few weeks. Similar observations have been made in calves, lambs ("white muscle disease"), and ducklings. Young chicks placed on a vitamin E–deficient feed exhibit extensive capillary damage and encephalomalacia. Eggs from vitamin E–deficient hens have a low hatchability. The embryos develop hemorrhages on the third or fourth day of incubation and are dead by the fifth day. Deficiency in the monkey results in a hemolytic anemia.

A single case of what was reported to be authentic tocopherol deficiency in an adult human being has been described. A woman with chronic xanthomatous biliary cirrhosis and impaired intestinal lipid absorption also evidenced low serum tocopherol concentration, muscular weakness, creatinuria, and red cells which were unusually fragile in the presence of dialuric acid. These signs of vitamin E deficiency disappeared after administration of α-tocopherol. Sensitivity of erythrocytes to peroxide and dialuric acid, which diminishes with tocopherol administration, has also been observed in premature infants and infants with steatorrhea.

Metabolism of Tocopherol. The property of tocopherol which appears to be related to most manifestations of deficiency is its ability to inhibit the autoxidation of unsaturated fatty acids. Methylene blue, thiodiphenylamine, and N,N,-diphenyl, *p*-phenylenediamine (DPPD), substances which participate in reversible oxidation-reduction reactions but which do not resemble tocopherol in structure, prevent many of the manifestations of vitamin E deficiency. In the chick, methylene blue prevented capillary damage and encephalomalacia; in the rat, it prevented appearance of sensitivity to the in vitro dialuric acid–hemolysis test, depigmentation of the incisors, accumulation of brown pigment in the lipid stores, and hepatic necrosis in animals on low-protein diets. Like tocopherol, methylene blue also increased hepatic storage of vitamin A in animals on diets containing marginal amounts of this vitamin. DPPD has been shown to substitute for vitamin E through three generations of rats. Many different substances whose structures bear some resemblance to that of tocopherol (coumarins, chromanes, phenols, quinones, etc.) show some vitamin E activity, and the various forms of coenzyme Q (page 324), such as hexahydro CoQ_4, are at least as active as α-tocopherol. It is not known whether the relative biological activity of these compounds correlates with their potency as antioxidants.

Liver rich in vitamin A also contains significant quantities of di-α-tocopheroquinone.

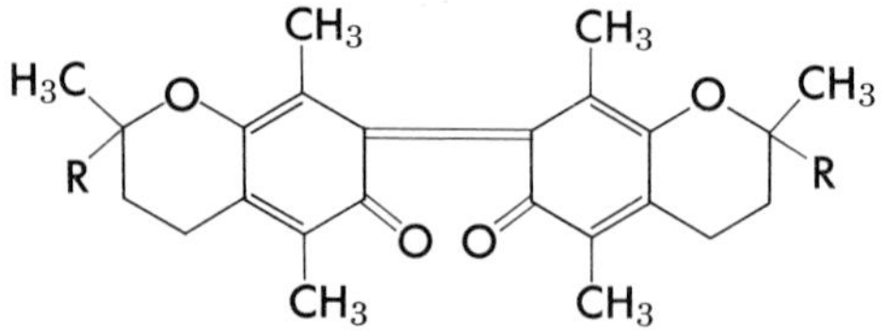

The tissues of vitamin E–deficient animals, particularly cardiac and skeletal muscle, consume oxygen more rapidly than do normal tissues. α-Tocopherol does not readily undergo reversible oxidation. Chemical oxidation yields α-tocopherylquinone and α-tocopheryloxide; the former may be reduced to α-tocopherol, but it is not known whether similar events occur in biological electron transport systems. The increased oxygen consumption of vitamin E–deficient muscle appears to relate to peroxidation of unsaturated fatty acids. In other tissues, *e.g.*, liver, this leads to disturbance of the structure of mitochondria and reduced respiration. According to one report, peroxidation of unsaturated fatty acids in the endoplasmic reticulum of muscle results in release of lysosomal hydrolases, thereby causing the muscular dystrophy. In sum, vitamin E is not utilized for biosynthesis of a coenzyme. All manifestations of vitamin E deficiency appear to be secondary to the uninhibited peroxidation of polyunsaturated fatty acids.

Both the chromane ring and the side chain of α-tocopherol are oxidized by man; the product, shown below, is excreted in the bile conjugated with two moles of glucuronic acid as the diglucosiduronate via the two hydroxyl groups.

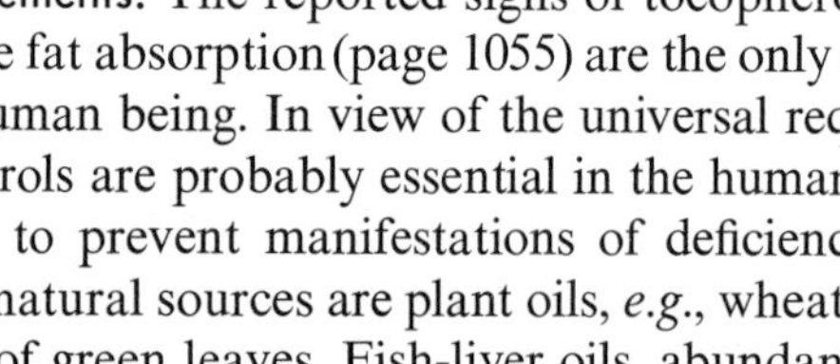

Vitamin E Requirements. The reported signs of tocopherol deficiency in a single person with defective fat absorption (page 1055) are the only evidence that vitamin E is required by the human being. In view of the universal requirement by other animal species, tocopherols are probably essential in the human diet; their widespread distribution suffices to prevent manifestations of deficiency under most circumstances. The richest natural sources are plant oils, *e.g.*, wheat germ, rice, cottonseed, as well as the lipids of green leaves. Fish-liver oils, abundant in vitamins A and D, are devoid of tocopherol.

VITAMIN K

Chicks fed synthetic rations develop a hemorrhagic condition characterized by a prolonged blood-clotting time. The preventive substance in various foodstuffs, designated vitamin K (*Koagulations* vitamin), was isolated both from alfalfa concentrates and from fish meal in 1939. The material in alfalfa was designated vitamin K_1 and that from fish meal K_2.

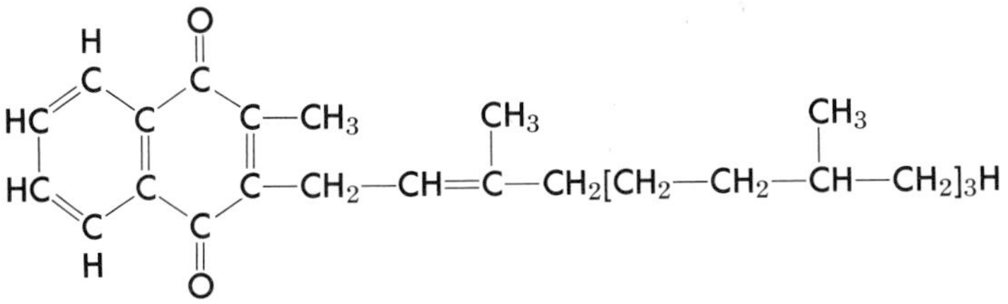

Vitamin K_1
(2-methyl-3-phytyl-1,4-naphthoquinone)

A number of vitamin K_2 analogues are known which differ only in the length of the side chain; that from fish meal has 30 carbon atoms.

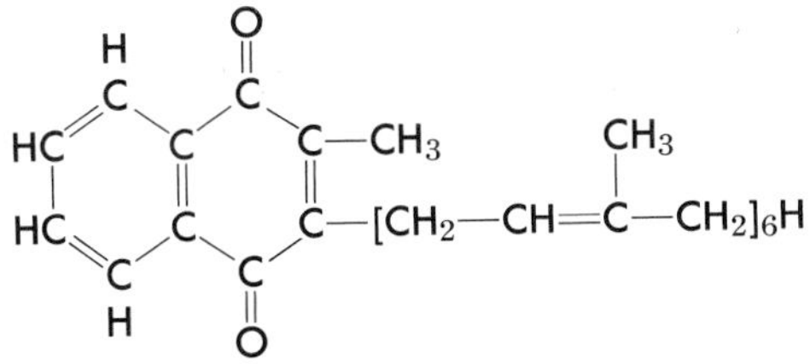

Vitamin $K_{2(30)}$
(2-methyl-3-difarnesyl-1,4-naphthoquinone)

Many related compounds also have vitamin K activity. Of these, 2-methyl-1,4-naphthoquinone (menadione) is as effective as vitamin K_1 on a molar basis.

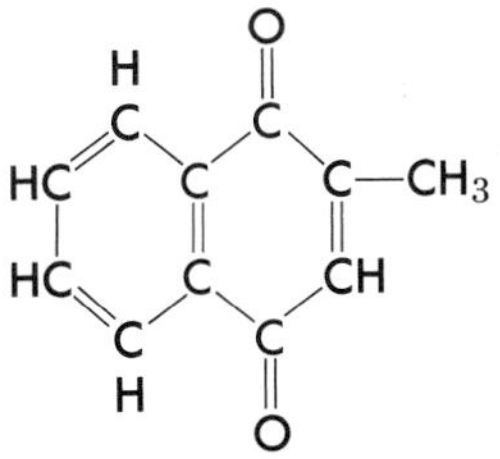

Menadione
(2-methyl-1,4,-naphthoquinone)

The reduced form of menadione, 2-methyl-1,4-naphthohydroquinone, is almost as active as the parent compound in chick assays. The nitrogen analogue, 4-imino-2-methyl-1-naphthoquinone, is three to four times as potent as vitamin K_1 and is the most effective compound yet studied.

Vitamin K Deficiency. Vitamin K deficiency resulting from dietary lack of the vitamin does not occur in usually studied rodents, not because they do not require this material, but because it is supplied by intestinal bacteria. If growth of the intestinal flora is inhibited by a sulfonamide, if germ-free animals are employed, or if care is taken to prevent coprophagy, such animals rapidly develop signs of vitamin K deficiency. In man, steatorrhea due to biliary obstruction, sprue, pancreatic failure, or other causes, with attendent diminished intestinal absorption of lipids, results in vitamin K deficiency.

The role of the vitamin in blood clotting has been discussed previously (Chap.

31). Deficiency in vitamin K is marked by diminished plasma prothrombin concentration. However, the prolonged blood-clotting time results primarily from failure of hepatic synthesis of proconvertin (page 731). Relatively minor injuries which ordinarily are innocuous may, in vitamin K–deficient animals, result in bleeding sufficient to cause shock and death. Since deficiency may exist as an indirect consequence of biliary obstruction, vitamin K is generally administered parenterally before surgery for repair of this condition.

Newborn infants may also show signs of vitamin K deficiency. This is manifested as "hemorrhagic disease of the newborn," which persists only until a bacterial flora is established in the infant's intestine. Administration of vitamin K to pregnant mothers before parturition has decreased the incidence of this disease. No manifestations of vitamin K deficiency other than impaired blood clotting have been observed. It appears that no dietary supply of this vitamin is necessary as long as intestinal absorption of lipids is adequate. The total amount required, whether provided by the diet or by the intestinal flora, is not known.

Metabolic Role of Vitamin K. The functional basis for the effect of vitamin K on hepatic synthesis of prothrombin and proconvertin (Chap. 31) is unknown. However, the wide distribution of this vitamin in plant, animal, and bacterial cells suggests a metabolic function unrelated to its special role in the blood-clotting mechanism. Uncoupling of oxidative phosphorylation has been observed in mitochondria from vitamin K–deficient chicks and after Dicumarol (page 728) administration. Dicumarol uncouples the electron transport chain both from DPNH to cytochrome c and from ferrocytochrome c to oxygen. Liver contains a flavoprotein which catalyzes reduction of menadione by DPNH. However, in view of the low concentration of vitamin K in mitochondria, the suggestion that this vitamin is required for normal oxidative phosphorylation has not gained acceptance.

The physiologically active form of vitamin K appears to be vitamin $K_{2(20)}$, found in liver after administration of menadione-^{14}C. Apparently, the side chain of vitamin K_1 is removed and a 20-carbon vitamin K_2 side chain is affixed to the ring. Presumably the enzyme catalyzing this reaction is the same as that which, in liver, also promotes addition of the side chain to the ring structure of ubiquinone (page 324). The ultimate fate of vitamin K is unknown; some menadione is excreted as a mixture of the monosulfate and diglucosiduronide of the reduced form, 2-methyl-1,4-naphthohydroquinone.

VITAMINS—SOME CONCLUDING REMARKS

It is unlikely that any new dietary requirements will be revealed by the classical techniques by which the essential food factors were discovered. There remains, however, the possibility that man and other animals require additional, as yet unknown compounds which are either provided by intestinal bacteria or, perhaps, transferred from mother to fetus in sufficient quantity to maintain adequate growth and health for one or more generations. These possibilities are illustrated by the histories of vitamin K, biotin, folic acid, and cobalamine. As yet, animals have not been grown or maintained on a diet of exclusively synthetic ingredients. The minuscule requirements for cobalamine and biotin are a reminder that as long as natural products are fed, the possible existence of undiscovered factors must be recognized.

Review of the present knowledge of the metabolic roles of the various food factors, and of the pathogenesis of the corresponding deficiency syndromes caused by lack of these factors, reveals that in virtually no instance can the deficiency syndrome be adequately explained in metabolic terms. Thus, the "three Ds" of pellagra (diarrhea, dermatitis, dementia) cannot be explained in terms of the function of the pyridine nucleotides. The mesenchymal failure of scurvy cannot yet be related to the chemical properties of ascorbic acid; the polyneuritis of beriberi cannot be explained in terms of diminished pyruvic acid dehydrogenase activity. The dermal manifestations of many deficiencies presumably reflect failure of certain of the metabolic activities in which the essential nutrients are involved, but the basis for the relationships is unknown. Indeed, the participation of vitamin A in the visual process is the best available instance of a clear relationship between the known function of a vitamin and a manifestation of the deficiency state.

Further, there is no reason to be certain that, for each of the essential food factors, all the possible manifestations of deficiency have been described. Deficiency of an essential factor may result in a profound disturbance of the function of a single organ or cell type, and the animal may die before it becomes possible to observe the effects of the deficiency in other organ systems. Nevertheless, the factor in question may be equally essential to the function of all cells. Thus, the vitamin E–deficient rabbit dies rapidly with symptoms only of muscular dystrophy and pathology of skeletal and cardiac muscle. Yet chronic vitamin E deficiency in the rat leads to disturbances in several other organs. Niacin deficiency in the dog on corn-containing rations results in blacktongue; yet this disease can be alleviated by administration of saline solution, and then a new disease syndrome develops. In each case the animal dies of failure of that essential process first affected by the dietary deficiency. To uncover other reactions influenced by the particular dietary factor, it may be necessary first to devise procedures which permit prolongation of life by circumventing in some manner the primary effects of the deficiency. An excellent illustration of the rewards of this approach is the description of the full effects of retinol deficiency in the retina of the vitamin A–deficient rat maintained on retinoic acid.

MALNUTRITION

Despite significant advances in understanding of nutrition and the fact that, from the standpoint of available information, no one need be malnourished, malnutrition is increasingly prevalent among the world's population. Nor is it entirely true that the roots of malnutrition are solely economic; in many cases ignorance is an equally important factor. Beriberi, kwashiorkor, and xerophthalmia occur on a large scale; severe rickets, pellagra, and scurvy are less frequent.

Surveys of nutritional habits of peoples in economically underdeveloped areas frequently indicate an appallingly low intake of essential nutrients. This does not, however, always result in the appearance of deficiency diseases. The explanation for this is to be found in one of the basic concepts of nutrition, the *balanced diet.*

In both man and experimental animals it is known that a balanced dietary mixture, even if restricted in *total quantity,* is preferable to a regime that is adequate in most respects but badly deficient in only *one* nutrient. Inadequate quantity may lead to delayed growth, anemia, or hypoproteinemia, but the imbalanced diet results in pellagra, beriberi, scurvy, etc. The rat fed a diet completely lacking in B vitamins survives for a surprising length of time, while a deficiency of thiamine alone may be fatal in a few weeks. Genetic differences among individuals may drastically alter the quantitative or even qualitative composition of the balanced dietary mixture desirable for optimal health. Thus, galactosemic children must eschew lactose, phenylketonuric individuals must ingest a minimum of phenylalanine, diabetics must carefully balance dietary carbohydrate and lipid. It seems likely that this list will be extended in the future.

Improved methods of laboratory diagnosis permit detection of early deficiencies of many of the essential nutrients. Application of these techniques, particularly in large-scale nutritional surveys of populations, has helped to improve the nutritional status of the Western world while revealing the enormous magnitude of the problem of undernourishment of peoples living in the equatorial belt. Indeed, in the United States it almost becomes necessary to redefine malnutrition. This term has classically implied undernutrition of one or more elements of the diet. If, instead, it implies a deviation from *balanced nutrition,* then a large segment of the American population is malnourished in that obesity is a most serious nutritional problem. It is estimated that at least 15 per cent of the population is overweight in comparison with accepted standards and that at least 6,000,000 people in the United States are pathologically overweight, *i.e.*, their weights exceed by more than 20 per cent the accepted standards for their heights and skeletal structures. If indeed hypercholesterolemia is significant in the pathogenesis of atherosclerosis, improper nutrition contributes to the leading single cause of death in the United States and Europe.

Relatively few investigations of conditions for *optimal* nutrition have been reported; even the criteria for optimal nutrition have not been established. The criterion generally employed in experimental nutrition has been the weight response of young rats. However, it is not established that ingestion of a diet which permits maximum weight gains through childhood and adolescence contributes to long and healthy adult life. Indeed, available experimental evidence suggests that diets which provide the minimal intake of essential nutrients compatible with health also promote longevity. Should this prove generally to be the case, many of the recommendations that have been given in Table 48.2 may require revision.

REFERENCES

Books

Albanese, A., ed., "Protein and Amino Acid Nutrition," Academic Press, Inc., New York, 1959.

Brock, J. F., "Recent Advances in Human Nutrition," Little, Brown and Company, Boston, 1961.

Burton, B. T., ed., "Heinz Handbook of Nutrition," 2d ed., McGraw-Hill Book Company, New York, 1965.

Ciba Foundation Study Group, "Mechanism of Action of Water-soluble Vitamins," Little, Brown and Company, Boston, 1962.
Cooper, L. F., Barber, E. M., and Mitchell, H. S., "Nutrition in Health and Disease," J. B. Lippincott Company, Philadelphia, 1960.
Crampton, E. W., and Lloyd, L. E., "Fundamentals of Nutrition," W. H. Freeman and Company, San Francisco, 1960.
Davidson, S., Meiklejohn, A. P., and Passmore, R., "Human Nutrition and Dietetics," E. & S. Livingstone Ltd., Edinburgh, 1959.
Dyke, S. F., "The Chemistry of the Vitamins," Interscience Publishers, Inc., New York, 1965.
"Endemic Goiter," World Health Organization, Publ. 44, Geneva, 1960.
Follis, R. H., Jr., "Deficiency Disease," Charles C Thomas, Publisher, Springfield, Ill., 1958.
Goldsmith, G., "Nutritional Diagnosis," Charles C Thomas, Publisher, Springfield, Ill., 1959.
Goodwin, T. W., "The Biosynthesis of Vitamins and Related Compounds," Academic Press, Inc., New York, 1963.
Griswold, R. M., "The Experimental Study of Foods," Houghton Mifflin Company, Boston, 1962.
György, P., and Pearson, W. N., eds., "The Vitamins," vols. VI and VII, Academic Press, Inc., New York, 1967.
McCance, R. A., and Widdowson, E. M., "The Composition of Foods," Her Majesty's Stationery Office, London, 1960.
Monier-Williams, G. W., "Trace Elements in Food," John Wiley & Sons, Inc., New York, 1949.
Moore, T., "Vitamin A," Elsevier Publishing Company, Amsterdam, distributed by American Elsevier Publishing Company, New York, 1957.
"Nutrition Surveys; Their Techniques and Value," *Natl. Research Council (U.S.) Bull.* 117, Washington, 1949.
"Progress in Meeting Protein Needs of Infants and Children," *Natl. Acad. Sci -Natl. Research Council,* Publ. 843, Washington, 1961.
Robinson, F. A., "The Vitamin Co-factors of Enzyme Systems," Pergamon Press, New York, 1966.
Sebrell, W. H., ed., "Control of Malnutrition in Man," American Public Health Association, New York, 1960.
Sebrell, W. H., Jr., and Harris, R. S., eds., "The Vitamins," 2d ed., vol. I, 1967; vol. II; 1968, Academic Press, Inc., New York.
Sinclair, H. M., ed., "Essential Fatty Acids," Academic Press, Inc., New York, 1958.
Smith, E. Lester, "Vitamin B_{12}," John Wiley & Sons, Inc., New York, 1965.
Watt, B. K., and Merrill, A. L., "Composition of Foods, Raw, Processed and Prepared," *U.S. Dept. Agr. Handbook* 8, Washington, 1950.
Wohl, M. G., and Goodhart, R. S., eds., "Modern Nutrition in Health and Disease," Lea & Febiger, Philadelphia, 1960.
Wolstenholme, G. E. W., and O'Connor, M., eds., "Thiamine Deficiency: Biochemical Lesions and Their Clinical Significance," J. & A. Churchill Ltd., London, 1967.

Review Articles

Best, C. H., and Lucas, C. C., Choline—Chemistry and Significance as a Dietary Factor, *Vitamins and Hormones,* **1,** 1–59, 1943.
Block, R. J., and Mitchell, H. H., Correlation of the Amino Acid Composition of Proteins with Their Nutritional Value, *Nutr. Abstr. & Revs.,* **16,** 249–278, 1946.
Braude, R., Kon, S. K., and Porter, J. W. G., Antibiotics in Nutrition, *Nutr. Abstr. & Revs.,* **23,** 473–495, 1953.

Bronte-Stewart, B., The Effect of Dietary Fats on the Blood Lipids and Their Relation to Ischaemic Heart Disease, *Brit. Med. Bull.,* **14,** 243–252, 1958.

Brown, G. M., and Reynolds, J. J., Biogenesis of the Water-soluble Vitamins, *Ann. Rev. Biochem.,* **32,** 419–462, 1963.

Burns, J. J., Ascorbic Acid, in D. M. Greenberg, ed., "Metabolic Pathways," 3d ed., pp. 394–411, Academic Press, Inc., New York, 1967.

Burr, G. O., and Barnes, R. H., Non-caloric Functions of Dietary Fats, *Physiol. Revs.,* **23,** 256–278, 1943.

Chatfield, C., and Adams, G., Proximate Composition of American Food Materials, *U.S. Dept. Agr. Circ.* 549, 1940.

Chick, H., The Causation of Pellagra, *Nutr. Abstr. & Revs.,* **20,** 523–547, 1951.

Chow, B. F., ed., Symposium on Mechanisms of Intestinal Absorption, *Am. J. Clin. Nutr.,* **12,** 161–227, 1963.

Cotzias, G. C., Manganese in Health and Disease, *Physiol. Revs.,* **38,** 503–532, 1958.

Cruickshank, E. K., Dietary Neuropathies, *Vitamins and Hormones,* **10,** 2–46, 1952.

Daniel, L. J., Inhibitors of Vitamins of the B-complex, *Nutr. Abstr. & Revs.,* **31,** 1–13, 1961.

Elvehjem, C. A., Nutritional Significance of the Intestinal Flora, *Federation Proc.,* **7,** 409–418, 1948.

Flodin, N. W., ed., Protein Nutrition, *Ann. N.Y. Acad. Sci.,* **69,** 855–1066, 1958.

Garry, R. C., and Wood, H. O., Dietary Requirements in Human Pregnancy and Lactation, *Nutr. Abstr. & Revs.,* **15,** 591–621, 1946.

Glass, G. B. J., Gastric Intrinsic Factor and Its Function in the Metabolism of Vitamin B_{12}, *Physiol. Revs.,* **43,** 529–849, 1963.

Handler, P., The Present Status of Nicotinic Acid, *Intern. Rev. Vitamin Research,* **19,** 393–451, 1948.

Hitchings, G. H., and Burchall, J. J., Inhibition of Folate Biosynthesis and Function as Basis for Chemotherapy, *Advances in Enzymol.,* **27,** 417–468, 1965.

Hogenkamp, H. P. C., Recent Developments in the Chemistry of B_{12} Coenzymes, *Federation Proc.,* **25,** 1623–1627, 1967.

Holman, W. I. M., Distribution of Vitamins within the Tissues of Common Foodstuffs, *Nutr. Abstr. & Revs.,* **26,** 277–304, 1956.

Holmes, J. O., The Requirement for Calcium during Growth, *Nutr. Abstr. & Revs.,* **14,** 597–619, 1945.

Holt, L. E., Jr., and Snyderman, S. E., Protein and Amino Acid Requirements of Infants and Children, *Nutr. Abstr. & Revs.,* **35,** 1–12, 1965.

Horwitt, M. K., Investigations of Human Requirements for B-complex Vitamins, *Natl. Research Council Bull.* 116, 1948.

Isler, O., and Wiss, O., Chemistry and Biochemistry of the K Vitamins, *Vitamins and Hormones,* **17,** 54–91, 1959.

Jaenicke, L., Vitamin and Coenzyme Function: Vitamin B_{12} and Folic Acid, *Ann. Rev. Biochem.,* **33,** 287–312, 1964.

James, A. T., and Lovelock, J. E., Essential Fatty Acids and Human Disease, *Brit. Med. Bull.,* **14,** 262–266, 1958.

Jansen, B. C. P., The Physiology of Thiamine, *Vitamins and Hormones,* **7,** 84–110, 1949.

Keys, A., The Caloric Requirement of Adult Man, *Nutr. Abstr. & Revs.,* **19,** 1–21, 1949.

Krehl, W. A., Nutritional Factors and Skin Diseases, *Vitamins and Hormones,* **20,** 121–140, 1960.

Kruse, H. D., ed., Inadequate Diets and Nutritional Deficiences in the United States, *Natl. Research Council Bull.* 109, 1943.

Leitch, I., and Hepburn, A., Pyridoxine, Metabolism and Requirement, *Nutr. Abstr. & Revs.,* **31,** 389–401, 1961.

Luhby, A. L., and Cooperman, J. M., Folic Acid Deficiency in Man and Its Interrelationships with Vitamin B_{12} Metabolism, *Advances in Metabolic Disorders,* **1,** 264–333, 1964.
Lynen, F., Role of Biotin-dependent Carboxylations in Synthetic Reactions, *Biochem. J.,* **102,** 381–400, 1967.
Mayer, J., Physiological Basis of Obesity and Leanness, *Nutr. Abstr. & Revs.,* **25,** 597–611, 871–883, 1955.
Meiklejohn, A. P., Physiology and Biochemistry of Ascorbic Acid, *Vitamins and Hormones,* **11,** 62–96, 1953.
Meites, J., and Nelson, M. M., Effects of Hormonal Imbalances on Nutritional Requirements, *Vitamins and Hormones,* **20,** 205–236, 1960.
Mistry, S. P., and Dakshinamurti, K., The Biochemistry of Biotin, *Vitamins and Hormones,* **22,** 1–56, 1965.
Morton, R. A., Ubiquinones, Ubichromenols and Related Substances, *Vitamins and Hormones,* **21,** 1–42, 1961.
Nicolaysen, R., and Eeg-Larsen, N., Biochemistry and Physiology of Vitamin D, *Vitamins and Hormones,* **11,** 29–61, 1953.
Nieman, C., and Klein Obbink, H. J., Biochemistry and Pathology of Hypervitaminosis A, *Vitamins and Hormones,* **12,** 69–101, 1954.
Pett, L. B., Vitamin Requirements of Human Beings, *Vitamins and Hormones,* **13,** 214–238, 1955.
Platt, B. S., ed., Recent Research on Vitamins, *Brit. Med. Bull.,* **12,** 1–90, 1956.
"Recommended Dietary Allowances," Publ. 3, vol. **35,** National Academy of Sciences, Washington, 1964.
Richter, C. P., Total Self-regulatory Functions in Animals and Human Beings, *Harvey Lectures,* **38,** 63–103, 1941–1942.
Rose, W. C., Amino Acid Requirements of Man, *Federation Proc.,* **8,** 546–552, 1949.
Sobel, A. E., The Absorption and Transportation of Fat-soluble Vitamins, *Vitamins and Hormones,* **10,** 47–68, 1952.
Stokstad, E. L. R., and Koch, J., Folic Acid Metabolism, *Physiol. Revs.,* **17,** 83–116, 1967.
Vasington, F. D., Reichard, S. M., and Nason A., Biochemistry of Vitamin E, *Vitamins and Hormones,* **20,** 43–88, 1960.
Vitamin B_6, Symposium on, *Vitamins and Hormones,* **22,** 321–886, 1965.
Wagner, F., Vitamin B_{12} and Related Compounds, *Ann. Rev. Biochem.,* **35,** 405–434, 1966.
Yudkin, W. H., Thiaminase, the Chastek-paralysis Factor, *Physiol. Revs.,* **29,** 389–402, 1949.

INDEX

Entries in boldface type are pages on which formula or structure of a substance is given. The term, properties, connotes physical and chemical properties and behavior. The term, composition, is used to include all constituents except enzymes. Isomeric forms of compounds, e.g., of carbohydrates and of amino acids, are entered under the most general, single term except where particular significance is indicated. Thus, glucose may include D-, L-, and α-D, etc., forms; carotene(s) includes α-, β-, and γ-forms. Acid references include the ion as well as the acid, e.g., acetoacetic acid includes acetoacetate; aspartic acid includes aspartate.